T0190312

An Introduction to Chemistry

Michael Mosher · Paul Kelter

An Introduction to Chemistry

Michael Mosher
Department of Chemistry and Biochemistry
University of Northern Colorado
Greeley, CO, USA

Paul Kelter
Longmont, CO, USA

ISBN 978-3-030-90269-8 ISBN 978-3-030-90267-4 (eBook)
https://doi.org/10.1007/978-3-030-90267-4

This Springer imprint is published by the registered company Springer Nature Switzerland AG
The registered company address is: Gewerbestrasse 11, 6330 Cham, Switzerland

Preface

To the Student

Cat Stevens got it right all those years ago, when he sang, "Oooh baby, baby it's a wild world." It is been over a half-century since this big-time singer and songwriter wrote those words, but they have never been more meaningful. As we put the finishing touches on this book, we think about the recent past and all the ways that we tried to make sense of this wild world. How did a city in the northwestern USA with an average temperature in June of 73°F, or 23°C, reach 117°F, or 47°C? What led to this ongoing worldwide fright called "climate change?" What can we do to put the climate genie back in the bottle? What about the worldwide pandemic caused by a living organism called "coronavirus"? Millions died in a little over a year. Talk about putting the genie back in the bottle. But we largely did it. We put that genie back. We used some of the best science and the best processes of engineering—working together toward a common goal of manipulating particles—atoms and molecules—so small that it takes the world's best imaging devices to see them, to save millions more lives. We understood the problem and found the solution. This is the rational study of matter—the stuff of the world and beyond—that we call "chemistry," and it is this study that is the focus of this textbook.

In many ways, our presentation of material in this book is quite similar to the way we teach this subject. We will share with you the big ideas of chemistry, and, because students often ask "Why should we know this?," we will discuss how the concepts of chemistry are applied to real-world issues, such as: energy production and storage, the launch of a rocket, the reaction of pharmaceuticals within the body, the interaction of chemicals with our climate, and the age of life itself. We will then ask questions that lead to an explanation of how these things are known.

This textbook has two primary goals. The first is to help you appreciate the depth and breadth of chemistry. The second goal is to encourage you to make connections between concepts in chemistry and our world. "How do we know?" is one of the most vital questions, no matter what your field of study. In chemistry, we often say that asking good, focused questions is vital to "thinking like a chemist." Knowing how to know and wanting to know are two essential traits of successful learning that not only are crucial to your study of chemistry but also to understanding a world beyond science.

As you read this book (and we do believe that it can be read as a wonderful story of chemistry), look for the places where questions are raised. Ask yourself why we raise each question at the point where we do. What is the key idea? Why is this useful to know? What can I now figure out that I was not able to before? What connections can I make? Does my answer make sense?

To help you learn the material, we have included many thoroughly worked examples that demonstrate not only how to find the reasonable solution but also how to connect each problem to what you have learned and sometimes draw connections to what you will learn. This is the grand challenge of learning chemistry—making connections. This is the higher-order understanding that is so important in your analysis of the issues of the day and the great social puzzles that often define our world. Unlike many subjects, chemistry is naturally cumulative. It requires you to apply previously learned information in increasingly complex ways.

Because of the connections and cumulative nature of the field, studying chemistry requires persistence—what you might call "grit"—and practice. Daily reading and problem solving are the keys. Making a serious investment with your heart as well as your head will pay off in an understanding of chemistry as well as in the enjoyment of learning. If this book enhances your desire to learn more about the world around you, then we have been successful—and so have you.

These are the things that we say to our own students. Then, together, we start learning and doing chemistry. Let us go!

To the Instructor

We are excited to present you with a different kind of textbook—one that is written from the standpoint of how an increasing proportion of chemists teach in the twenty-first century. It is our aim to complement your teaching style by giving your students a teacher's viewpoint in print. This book is about the questions we ask when we teach chemistry. It focuses on the "big ideas" of chemistry that we, as chemists, find so appealing that we choose this as our career. For example, we explore the periodic table, the energy exchanges that accompany all chemical changes, the ideas of quantum mechanics (science is about probability, not perfection), and how we can use Le Châtelier's principle to control the extent of chemical reactions. These are just some of the ideas we want students to grasp deeply when we say, "Here is what it means to think like a chemist."

A Framework of Interwoven Applications

In addition to sharing the big ideas that define science in general and chemistry in particular, we write from the belief that chemistry is vitally important to our world. In this text, we present chemistry in the context of how it is related to our everyday lives by interweaving chemistry with its impact in industry, the human body, and the environment. For example, a discussion of the energy changes that occur in chemical reactions is entwined with how these concepts are used in the petroleum and battery industries. The discussion of kinetics is applied to the fate of pesticides in the environment. We view chemistry's applications in society as generally beneficial, and we note that when the use of chemistry has led to unfortunate consequences (especially for the environment), chemistry has also been used to clean up the mess. This focus on the critical role that the ideas and practice of chemistry have in our day-to-day lives is written into the storyline of the text. We mean it when we say that this textbook is applications-based.

An Interrogative, Inquiry-Based Style

In our classrooms, we enjoy raising questions with our students, both because we like to hear their ideas and because raising questions is a key characteristic that defines the curious minds of scientists. Raising meaningful questions and finding answers is one outcome we have for students who use the text. We wrote this textbook to reflect a conversational and interrogative style, in which questions addressed to the student—often the same questions we ask in our own classes—begin and emphasize various topics. This approach involves students in the discussion, encourages them to pose questions about their world, and nurtures their curiosity about science and how it applies to society. This approach recurs throughout every chapter, as we continue to engage the student in what is most fundamental to practicing scientists: questioning the world around them.

Problem-Solving Approach

Within the text, we have expanded the use of leading questions in our worked examples. We asked ourselves—"What would we say to our students as they work through this problem? How would we prompt them to think about previous concepts?" We then added these interrogative elements and prompts to the structure of the worked examples. Throughout these examples in each chapter, we attempt to demonstrate rational thinking to guide the student to a proper solution, not just a "correct answer." This approach to problem solving encourages students to ask questions and frame them in such a way as to solve a problem effectively.

We begin each example by drawing their attention to previously covered applicable concepts and ask them to check themselves to see if they still remember them adequately. As the solution progresses, we prompt them to help make connections to critical concepts and ideas. And, once the problem is solved, we guide students in evaluating whether their answers make sense. We also give them the opportunity to test themselves and make sure they really "get it" by asking them to think about how changes in the problem might affect the outcome. Here is the basic structure of our worked examples:

- "Asking the right questions" engages the students by demonstrating the kinds of questions they should be thinking about as they begin the solution to the problem.
- Worked-out solutions guide the student step-by-step through the problem with prompts and reminders in the margins.
- "Are the answers reasonable" asks if the results meet student expectations. That is, do they make sense?
- "What if..." extends the problem by asking them to think about the impact of some change to the variables.
- Practice problems gives students additional problems, to which answers are provided at the back of the book.

These examples demonstrate how scientists think about problems and show how we work through a problem to arrive at an answer by asking questions, connecting to previously learned concepts, and making sure we really understand.

Chapter Organization

Our approach to the presentation of the material builds upon knowledge and helps to illustrate the connections between topics. For this reason, discussion of gases (Chapter 10) follows the construction of models of bonding (Chapters 8 and 9) and immediately precedes the discussion of the other phases of compounds (Chapter 11). We have taken the traditional approach to the presentation of the topics on kinetics (Chapter 13), equilibria (Chapter 14), and thermodynamics (Chapter 17). Yet, these chapters are written to be as portable as possible so that they may be moved around in a course to suit your preference. We have found that introduction of the concepts in this alternate order allows the instructor to provide a thorough discussion of equilibria from the start and aids the student in recognizing the connections between these topics. This text also includes chapters that allow exploration of nuclear chemistry (Chapter 20), coordination chemistry (Chapter 20), organic chemistry (Chapter 21), and biochemistry (Chapter 22). Many students develop strong interest in these topics that reinforces key concepts from earlier material and builds their curiosity about chemistry.

The Bottom Line

Our applications-based and interrogative approach has resulted in a book that students will likely read, reflecting how teachers actually teach. Here, then, as we say at the end of every chapter in the text, is the bottom line—in this case, a concise list of the main features of the textbook:

- Applications of chemistry are interwoven within the concepts of chemistry. Students often ask "Why do I need to know this?," so we show, at every opportunity, how chemistry is a part of our world.
- A conversational and interrogative writing style encourages students to not only read the text but to be inquisitive, both locally and globally. It involves the student in discussing and learning concepts that are important to chemistry.
- Our problem-solving approach guides students to first think about a problem, to approach the problem in a logical manner in order to arrive at a solution efficiently, and to think beyond the calculation to uncover related information about the concepts that are being explored.
- A practical student-friendly pedagogy includes writing that involves students and offers many opportunities for review (such as in the deepen your understanding through group work and the bottom line summaries). In addition, visually engaging illustrations clearly represent concepts for students and illustrate the vital connection of the world of the atom to the world in which we live.

We hope that you and your students will enjoy reading and working with this textbook as much as we enjoyed writing it.

Michael Mosher
Greeley, Colorado, USA

Paul Kelter
Longmont, Colorado, USA

Contents

About the Authors

Michael Mosher

currently holds the rank of the professor in the Department of Chemistry and Biochemistry at the University of Northern Colorado, having served as a chair in that department earlier in his career. Trained as a synthetic organic chemist, Mosher has been very interested in teaching pedagogy in the general chemistry and organic chemistry curricula. He has been particularly interested in the use of applications as tools for understanding difficult concepts in chemistry. He has received numerous awards for teaching throughout his career including the Pratt-Heins Outstanding Teaching Award at the University of Nebraska at Kearney. He has over 50 publications in chemistry and brewing science and is the author of three textbooks in those fields.

Paul Kelter

was most recently the inaugural director of the Office of Teaching and Learning as well as a professor of Science Education at North Dakota State University. An analytical chemist, Kelter, was an international leader in chemistry education, having procured substantial funding for precollege and university teacher research, development, and summer programs. He has been honored with over 20 major educational awards, including Fellow in the Education Division of the American Association for the Advancement of Science, Board of Trustees Professor at Northern Illinois University, University of Illinois Distinguished Teacher/Scholar, University of Nebraska Academy of Distinguished Teachers member, and Distinguished Teacher at the University of Wisconsin—Oshkosh.

The World of Chemistry

Contents

Supplementary Information The online version contains supplementary material available at ▶ https://doi.org/10.1007/978-3-030-90267-4_1.

M. Mosher and P. Kelter, *An Introduction to Chemistry*,
https://doi.org/10.1007/978-3-030-90267-4_1

What Will We Learn from This Chapter?

- Atoms make up all forms of matter and are too small to visually observe. Chemists have developed ways to learn about atoms and how they are arranged by studying physical and chemical properties of matter (▶ Sect. 1.2).
- The Scientific Method is a way of learning about the universe in an organized manner. It helps us to make sure the conclusions we make are scientifically valid (▶ Sect. 1.3).
- To understand science, we must know the common units that are used to make measurements: the International System (SI) (▶ Sect. 1.4).
- Dimensional Analysis can be a most useful skill when we solve scientific problems. It focuses on the units and not the numbers (▶ Sect. 1.5).
- When making measurements, we must be careful to limit ourselves to the information available and not make up information that doesn't exist. Keeping track of accuracy and precision with significant figures helps us to do that (▶ Sect. 1.6).
- The challenges of the world around us, from pollution to new medicines, provide ample opportunities for the application of chemistry and its principles (▶ Sect. 1.7).

1.1 Introduction

What goes around comes around. Crude oil appears to be plentiful, and one of its products, gasoline, is relatively cheap. It wasn't always that way. Gas prices, which rise and fall in relation to the supply of crude oil on the market, have bounced around like a ping-pong ball in response to world events and advances in oil drilling for many years. Why? It is an important question with an answer that connects science and, ultimately, chemistry, to all life on the planet.

A useful starting point compels us to travel back to 1973. The Middle East was at war over what territory belonged to which country. On one side of the crisis were Egypt and Syria, which wanted to regain land they lost in a 1967 war with the other focal point in this discussion, Israel. As with so many seemingly local conflicts, this one was much bigger, with the United States, the former Soviet Union and all of their respective allies taking sides. The stakes were enormous.

Science and Politics. Much of the world's crude oil in 1973 was continually extracted from deep in the Earth by a few countries in the Middle East, notably Iran and Saudi Arabia, among the founders of the Organization of Petroleum Exporting Countries (OPEC). This organization seeks to balance crude oil exports, supply and prices. To protest the support of Israel by the United States and other countries, OPEC reduced crude oil output.

Science and Economics. This reduction in output, an "oil embargo", sent shockwaves through the world economy. In the U.S., the price of a gallon of gasoline, which is one product of the chemical processing of crude oil, increased quickly from \$0.34 to \$0.84 (\$1.64 to \$4.04 in today's dollars). Drivers all over the world waited for hours to get to a pump, and many stations ran out of gas. In 1974, the conflict settled down and the oil embargo was lifted. But the effects were long-lasting.

Science and Public Policy. That year, the U.S. Federal government instituted a nationwide speed limit of 55 miles per hour (88 km per hour), which held until 1995. This was a response by the U.S. for energy independence. Knowing that motor vehicles get better gas mileage at 55 miles per hour than at 75 miles per hour, this seemed an easy step in reducing the demand for gasoline. Supplementing this, the U.S. Federal government introduced fuel economy requirements for cars and light trucks known as the CAFE standards (Corporate Average Fuel Economy). The CAFE standard for cars and light trucks, originally set at 18.0 miles per gallon (mpg), was changed to 37.8 mpg in 2016. In 2020, the original goal of 54.5 mpg by 2025 was adjusted to 40.0 mpg by 2026.

Science and Chemistry. Since 1973, considerable efforts have been made by scientists to develop alternative fuels and energy sources such as solar, wind, and geothermal energy that might replace the U.S. dependence on foreign supplies of crude oil. In effect, the political, economic and public policy responses to an event on the other side of the world have directed a significant amount of effort in the scientific industries.

Science is the systematic study of the universe, and we extend that via engineering to how we can use this understanding to meet our needs and wants. Science and engineering can also work together to cure the negative impact of meeting the needs and wants on a planet of nearly 8 billion people. Smog, poor air quality, and potential impacts to the climate and environment are some of the negative aspects of gasoline use. On the other hand, the positive impacts of the interaction of chemistry with crude oil are many, including the products we get from chemically processing crude oil, such as plastics, pharmaceuticals, textiles for clothing, and as we've seen, fuel for our cars.

Chemistry is one of the sciences because it systematically studies the composition, structure, and properties of matter and energy. Matter is the "stuff" of which all physical objects are made. Crude oil is an example of matter. Currently, the U.S. has estimated underground crude oil reserves of nearly 1.5 trillion gallons. Saudi Arabia has nearly 8 times as much, and Venezuela even 10% more than that. In several countries around the world, primarily the U.S., Canada, China and Argentina, companies have implemented a technique known as hydraulic fracturing, or "fracking", to get to this matter. Fracking applies a large force to break rock formations deep underground, releasing the entrapped oil and gas.

Chemistry – The systematic study of the composition, structure, and properties of the matter of our universe.

Fracking has provided the U.S. some measure of energy independence because it increases the ability to extract crude oil reserves from the ground. However, as long as the world remains a crude oil based economy, no country can be completely energy independent. Any discussion of energy independence must therefore include renewable sources of energy, including solar, wind, geothermal, and others. What is already clear is that this intersection of chemistry, humanity, and the overall health of the world's environment are intimately linked. Understanding chemistry can help us with that intersection.

This book presents a way of thinking that allows us to make scientific progress. It is an approach that is based on raising questions about how the world works and answering those questions in a systematic way. Direct answers don't always exist. Left- and right-hand turns are part of figuring out how to get where want to be. Among the most important parts of sorting it all out is communication. We share our questions and ideas as we hopefully move ahead. When we get there, we share the questions, answers, and discoveries by publishing them in scientific journals. This communication is how understanding grows among the community of scientists.

Case Study—The Community of Chemists

The community of chemists is made up of those who work in industry, government, and academia. They come from all walks of life and share their common interest in solving chemistry problems. Let's visit three such chemists.

Judith Fairclough-Baity (◘ Fig. 1.1) is an industrial chemist who worked for 19 years at one of the largest chemistry companies in the world, where she designed and made new tubing for heart–lung machines.

How did she approach the design of these tubes? Just like all chemists, she defined the key questions for which she needed answers. How flexible did the tubing need to be? How could she keep the blood from being contaminated in the process? Was the tubing compatible with living tissues? The experiments, data analyses, and conclusions that are part of this *scientific approach to knowing* led her and her coworkers to a fresh understanding—and to addi-

◘ **Fig. 1.1** Judith Fairclough-Baity at work in the laboratory. Scientists are not especially different from anyone else. They are just inquisitive people who use a different set of tools to ask questions and formulate answers

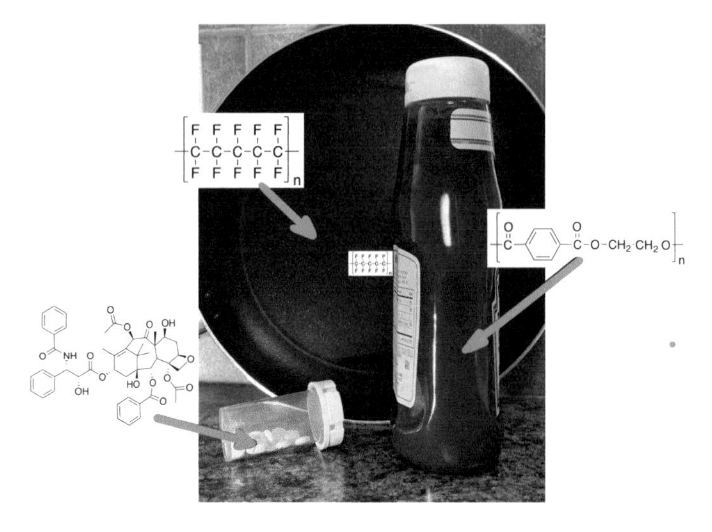

Fig. 1.2 Teflon, Taxol, and Ketchup bottles

tional questions. Eventually, after extensive experimentation and testing, she and her team arrived at a new tubing design.

Haishi Cao is professor of chemistry at the University of Nebraska at Kearney, almost in the middle of the United States. He is very involved in undergraduate education and has spent his entire career thus far working with students either in the classroom or in the laboratory. His *scientific approach to knowing* focuses on asking questions about the presence of various chemicals in different systems. With his students, he has developed procedures for determining the amount of copper, mercury, lead and fluoride using light. As an educator, in addition to asking and trying to answer questions, he is helping to prepare the next generation of scientists.

Tracy Williamson is a governmental chemist who works in the Office of Pollution Prevention and Toxics in the U.S. Environmental Protection Agency (EPA). Her role at EPA is to work closely with scientists in the U.S. to understand and implement safety procedures associated with the Toxic Substances Control Act (TSCA). This allows companies to bring new chemicals to market while following appropriate regulations for transportation, manufacturing, and sales. By asking questions about the toxic nature of compounds, she helps others understand the scientific approach to knowing.

Part of the communication between chemists and the broader community is knowing what questions to ask. This means that chemists need to have a thorough understanding of the basic principles behind what they do. By knowing the core ideas of science and having a desire to know, chemists can get to the bottom of a problem and come up with possible solutions. For example, chemists developed the Teflon in your frying pan, a very successful anticancer drug called Taxol, and the plastic used in squeezable ketchup bottles (◘ Fig. 1.2). The inquisitive approach of millions of chemists has influenced every aspect of our lives. Today, chemists continue to work together on the development of countless products, including environmentally friendly alternatives to gasoline in our cars, faster computer chips, and low-fat potato chips. Chemists solve problems, create new ones, and work to solve those.

1.2 Communication via the Language of Chemistry

What do we think of when we say the word "chemicals"? Chemists often say that **matter** is made up of chemicals. In other words, a *chemical* is just another name for the matter which makes up all physical objects in the universe. In informal discussion, the term is often used to denote something harsh or damaging, as in, "be careful of the chemicals in bleach" or, "if you accidentally ingest this chemical, seek medical help immediately!" The chemical known as sodium hypochlorite, the active ingredient in bleach, and the chemical called glyphosate, the key component of a popular herbicide, are hazardous if misused, but are helpful when used correctly. Synthetic insulin is a chemical that allows many diabetics to live normal lives, but which could be deadly if administered incorrectly. "Chemical" is a technical term for matter—not a value judgment about the safety of that type of matter.

Matter – Anything that has a mass and occupies space.

What are some other examples of matter? Nearly everything in the universe is made up of matter. This book is made up of matter. Your desk, your pencil, the coffee in your cup, and the air in your lungs are all examples of matter. Matter is anything that occupies space and has mass. Matter can take on the physical appearance of a variety of **states**.

State – The physical form of matter, typically as a solid, liquid, or gas.

What isn't made up of matter? Energy, like heat and light, is not regarded as matter because it does not occupy a specific volume of space in the conventional sense. On a more complex level, energy and matter are related, and can be interconverted, a critical idea we'll uncover later in our study of chemistry.

If chemical is another name for matter, what makes up chemicals? Chemicals, like all matter, are formed from the combination of inconceivably tiny **atoms**, the chemically indivisible particles that chemists classify by their structure into the more than 90 naturally occurring **elements** that are known in the universe. Chemists and physicists have added to this list by preparing several additional large elements that are not found in nature. You will find all the known elements, listed in the periodic table of the elements on the inside front cover of this book.

Atom – The smallest identifying unit of an element.

Element – A substance that contains only one kind of atom. All elements are listed in the periodic table of the elements.

Many of the elements in the periodic table, ◘ Fig. 1.3 are familiar because we depend on them in our day-to-day lives. For example, the element copper, made from copper atoms, forms the copper wires that carry electricity around cities and into homes. The element oxygen is part of the gas we must inhale in order to stay alive. However, oxygen gas is not composed of individual oxygen atoms. Instead, pairs of oxygen atoms are joined, or bonded, together to form oxygen molecules. A **molecule** is composed of two or more atoms bonded together.

Molecule – A particle composed of two or more atoms bonded together.

Some molecules, such as oxygen, contain identical atoms bonded together. However, many more kinds of molecules can be formed when atoms of different elements are bonded together. The molecule "H_2O" contains one atom of oxygen and two of hydrogen, and is known as a compound. A **compound** is matter that contains different elements chemically bonded together. Ethanol, found in alcoholic beverages and mixtures of some motor vehicle fuels, is a compound made up of carbon, hydrogen, and oxygen. Table sugar also has these three elements,

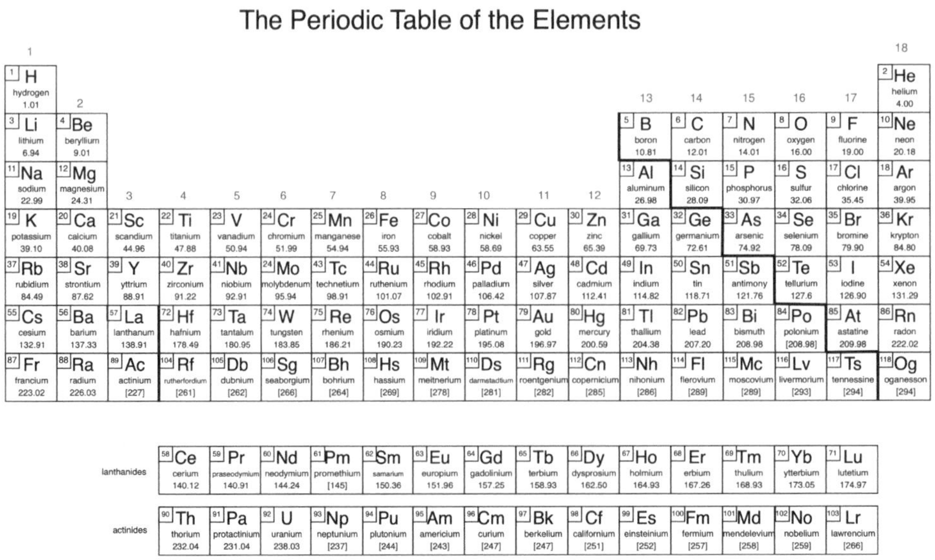

The Periodic Table of the Elements

1	2	3	4	5	6	7	8	9	10	11	12	13	14	15	16	17	18
1 H hydrogen 1.01																	2 He helium 4.00
3 Li lithium 6.94	4 Be beryllium 9.01											5 B boron 10.81	6 C carbon 12.01	7 N nitrogen 14.01	8 O oxygen 16.00	9 F fluorine 19.00	10 Ne neon 20.18
11 Na sodium 22.99	12 Mg magnesium 24.31											13 Al aluminum 26.98	14 Si silicon 28.09	15 P phosphorus 30.97	16 S sulfur 32.06	17 Cl chlorine 35.45	18 Ar argon 39.95
19 K potassium 39.10	20 Ca calcium 40.08	21 Sc scandium 44.96	22 Ti titanium 47.88	23 V vanadium 50.94	24 Cr chromium 51.99	25 Mn manganese 54.94	26 Fe iron 55.93	27 Co cobalt 58.93	28 Ni nickel 58.69	29 Cu copper 63.55	30 Zn zinc 65.39	31 Ga gallium 69.73	32 Ge germanium 72.61	33 As arsenic 74.92	34 Se selenium 78.09	35 Br bromine 79.90	36 Kr krypton 84.80
37 Rb rubidium 84.49	38 Sr strontium 87.62	39 Y yttrium 88.91	40 Zr zirconium 91.22	41 Nb niobium 92.91	24 Mo molybdenum 95.94	43 Tc technetium 98.91	44 Ru ruthenium 101.07	45 Rh rhodium 102.91	46 Pd palladium 106.42	47 Ag silver 107.87	48 Cd cadmium 112.41	49 In indium 114.82	50 Sn tin 118.71	51 Sb antimony 121.76	52 Te tellurium 127.6	53 I iodine 126.90	54 Xe xenon 131.29
55 Cs cesium 132.91	56 Ba barium 137.33	57 La lanthanum 138.91	72 Hf hafnium 178.49	73 Ta tantalum 180.95	74 W tungsten 183.85	75 Re rhenium 186.21	76 Os osmium 190.23	77 Ir iridium 192.22	78 Pt platinum 195.08	79 Au gold 196.97	80 Hg mercury 200.59	81 Tl thallium 204.38	82 Pb lead 207.20	83 Bi bismuth 208.98	84 Po polonium [208.98]	85 At astatine 209.98	86 Rn radon 222.02
87 Fr francium 223.02	88 Ra radium 226.03	89 Ac actinium [227]	104 Rf rutherfordium [261]	105 Db dubnium [262]	106 Sg seaborgium [266]	107 Bh bohrium [264]	108 Hs hassium [269]	109 Mt meitnerium [278]	110 Ds darmstadtium [281]	111 Rg roentgenium [282]	112 Cn copernicium [285]	113 Nh nihonium [286]	114 Fl flerovium [289]	115 Mc moscovium [289]	116 Lv livermorium [293]	117 Ts tennessine [294]	118 Og oganesson [294]

lanthanides	58 Ce cerium 140.12	59 Pr praseodymium 140.91	60 Nd neodymium 144.24	61 Pm promethium [145]	62 Sm samarium 150.36	63 Eu europium 151.96	64 Gd gadolinium 157.25	65 Tb terbium 158.93	66 Dy dysprosium 162.50	67 Ho holmium 164.93	68 Er erbium 167.26	69 Tm thulium 168.93	70 Yb ytterbium 173.05	71 Lu lutetium 174.97
actinides	90 Th thorium 232.04	91 Pa protactinium 231.04	92 U uranium 238.03	93 Np neptunium [237]	94 Pu plutonium [244]	95 Am americium [243]	96 Cm curium [247]	97 Bk berkelium [247]	98 Cf californium [251]	99 Es einsteinium [252]	100 Fm fermium [257]	101 Md mendelevium [258]	102 No nobelium [259]	103 Lr lawrencium [266]

Fig. 1.3 All of the chemical elements are listed in the periodic table. Each element is composed of one kind of atom. Atoms can become bonded together to form molecules of elements (such as oxygen) or molecules of compounds (such as water)

but in different proportions than in ethanol. We call the relative proportions of the elements in a compound its **composition**, so we say that the composition of ethanol is different from that of table sugar. Another term we can use is **substance**, a term often used informally, but which scientists define as a type of matter that exhibits a fixed composition.

Compound – A substance containing different elements chemically bonded together.

Composition – The relative proportions of the elements in a compound.

Substance – A type of matter that has a fixed composition.

1.2.1 Physical and Chemical Properties

Compounds on the molecular scale are far too small to see, and even when we have enough to see, their visible features are not necessarily distinctive.

Can we tell the difference between the contents of two jars based on our observations? Is one of the compounds liquid? Does one have a characteristic odor? Answering these questions gives us information about the **physical properties** of the substance. A physical property is a characteristic of a substance that can be determined without changing its composition. In other words, physical properties are descriptions of a substance, such as what it looks, smells, or tastes like. Based on these observations, we can determine which jar is more likely to contain a particular substance. We can even go further by determining the temperature at which each substance changes from solid to liquid. The melting point, the temperature at which a substance melts, is an example of a **physical change** in which the substance changes its physical properties without changing its composition. For instance, sugar changes from a solid to a liquid at 185 °C (365 °F), but when it does so, it is still sugar.

Physical change – Change in the physical state of a substance, such as a change between the solid, liquid, and gaseous states, that does not involve the formation of different chemicals.

Physical property – A characteristic of a substance that can be determined without changing its chemical composition.

What if we compared a jar of powdered starch that we use in cooking to a jar of powdered cellulose, plant fiber that is processed into paper? These are the two compounds in the jars in Fig. 1.4. They are quite different in many ways, though seem identical when we look at them. At first glance they both appear to be white, powdery

Fig. 1.4 A jar of cellulose and a jar of starch. Examining physical properties alone may not be sufficient to identify which of these jars contains cellulose and which contains starch

solids. We might ask another question to distinguish between these two substances: Do they behave in similar ways when combined with other substances? In other words, can we tell them apart by their chemical properties?

A **chemical property** is a characteristic of a substance that can be determined as it undergoes a change in its composition. For instance, the ability of gasoline to burn is one its chemical properties. A **chemical change**, the conversion of one substance into another due to a change in composition or a reorganization of atoms, often takes place during what chemists call a **chemical reaction**. For example, under the right conditions, the elements hydrogen and oxygen can combine to form water (see Fig. 1.5). This is a chemical change, because a new chemical, water, forms from the oxygen and hydrogen and is a different compound with a different arrangement of the atoms.

Chemical changes – Changes in which the reactants are changed into different chemicals, the products.

Chemical property – A characteristic of a substance that can be determined as it undergoes a change in its chemical composition.

Chemical reaction – The combination or reorganization of chemicals to produce different chemicals.

Knowing about their chemical properties can confirm the identification of the contents of each jar in Fig. 1.4. For example, under the right conditions, starch will react with the element iodine to form a deep blue/black compound, but cellulose will not. Using the physical and chemical properties of these substances, we can determine which is starch and which is cellulose. In general, knowing about the physical and chemical properties of other substances can help us distinguish between them.

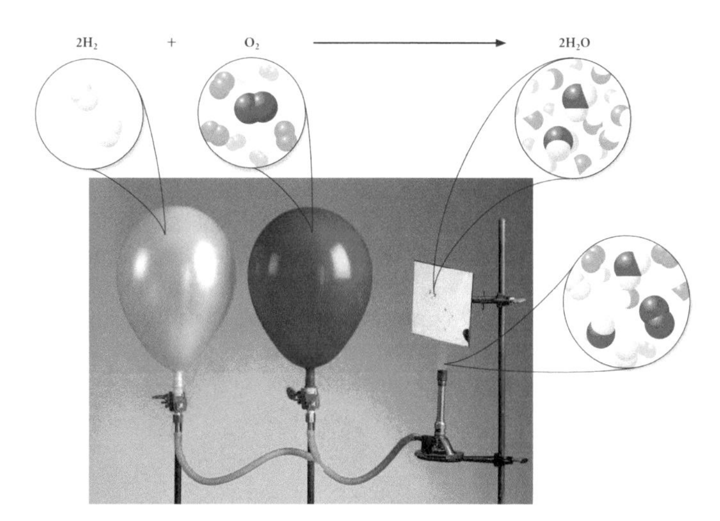

Fig. 1.5 Hydrogen and oxygen gases undergo a chemical change to make water. Note that the composition of water involves a combination of two hydrogen atoms and one oxygen atom. In this image, the reaction produces water vapor, which forms droplets on the mirror

Example 1.1—Chemical and Physical Changes

Define each of the following as a chemical or a physical change.

a. Burning automotive fuel in your car
b. Melting ice in a frying pan
c. Dissolving sugar in your tea
d. Digesting the sugar in a chocolate bar
e. Boiling water on the stove.

Asking the Right Questions

The key questions are these: What is a physical change? What is a chemical change? What is happening in each of the processes in the question?

Solution

Recall that a physical change is one in which the substance retains its chemical composition (the proportions in which its elements appear in the substance). A chemical change results in a change in the chemical composition of a substance. Therefore, as we look at each process, we think about whether changes in chemical composition occur.

a. The fuel that reacts with oxygen gas in the tank undergoes a profound change in the engine. It starts out as a liquid and is transformed into gases including carbon dioxide and water vapor. The reaction is accompanied by the release of energy. This is a chemical change.
b. Melting ice changes the physical condition of the water from solid to liquid. It is still water, so this is a physical change.
c. Dissolving sugar in your tea is a physical change. The sugar is still there; it is just mixed intimately with your tea.
d. Your body does a good job of utilizing the sugar in a chocolate bar. This is an example of a chemical change that releases energy as the sugar is converted into other substances.
e. Boiling water changes the physical condition of the water from liquid to gas. Gaseous water is 'steam'. It is still water, so this is a physical change.

Are Our Answers Reasonable?

We say that a physical change is one in which the substance retains its chemical composition, and a chemical change results in a change in the chemical composition of a substance. In parts "a" and "d" we discuss the chemical reactions that occur with each process, confirming that these are chemical processes. Although not all chemical reactions will be as familiar as these, we can use our real-world experiences with chemistry to help us know if our answers are reasonable.

What If...

What if the question asked the following: Does an energy change always accompany a chemical change? (*Ans*: *Yes*) Does an energy change always accompany a physical change? (*Ans*: *Yes*)

Practice 1.1

Indicate whether each process involves a chemical or a physical change.

a. The rusting of a copper statue
b. The yellowing of an old newspaper
c. Boiling water on the stove
d. Shaping a piece of wood into a post
e. Sharpening the blade of your lawn mower

1.2.2 Classification of Matter

We've discussed matter as anything that takes up space and possesses mass. By classifying matter into groups with similar characteristics and similar properties, we can reduce the number of questions that we need to ask to determine the chemical or physical properties of a crude oil, or any other substance.

Pure substances can be elements or compounds. For example, a sample of pure copper metal is an element, but a sample of pure water is a compound. Many of the materials we encounter are **mixtures** of substances, rather than pure substances. For example, crude oil is composed of hundreds of compounds. How do we learn about these compounds so we can eventually isolate them for our use in preparing other compounds, such as reformulating them into gasoline for cars? The first step is to classify them.

Mixture – A sample containing two or more substances.

Mixtures can be classified as either **homogeneous** or **heterogeneous** (Fig. 1.6). A homogeneous mixture, or **solution**, appears uniform all the way throughout. Both crude oil and tap water are homogeneous mixtures because one teaspoon of each mixture is the same as any other teaspoon. A heterogeneous mixture, on the other hand, contains a different composition in different places because it is not uniformly mixed.

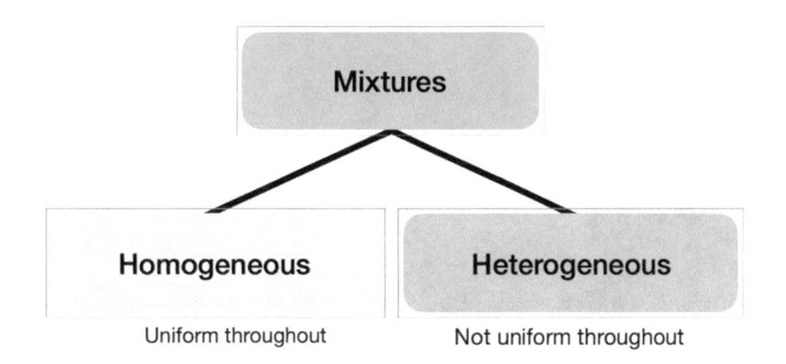

Fig. 1.6 Mixtures can be homogeneous or heterogeneous

Heterogeneous mixture – A mixture that is not uniformly mixed, so there are different proportions of the components in different parts of the mixture.

Homogeneous mixture – A mixture that is uniformly mixed so that it has the same composition throughout. Also known as a **solution**.

For example, if we consider a pot of chili, one forkful will have a different mix of ingredients than the next forkful. The mixture is heterogeneous. Grape juice is a homogeneous mixture; it is the same throughout. The same is true of a 14-karat gold ring—a homogeneous mixture of the elements gold, silver, and copper. On the other hand, cement is a heterogeneous mixture because some parts contain more sand and rock than other parts. Oil-and-vinegar salad dressing is also a heterogeneous mixture, no matter how hard you shake it. The relationships among the classifications of matter are shown in Fig. 1.7.

Mixtures of chemicals can be separated into their individual substances by methods that rely on the physical properties of the different substances. Perhaps the best example of this is in the production of safe drinking water from seawater, a problem made urgent by increasing population in areas where there is not enough fresh water. Pure water can be separated out of the homogeneous mixture called seawater by a process known as **desalination**, the removal of salts. For example, in Saudi Arabia, fresh water is so sparse that desalination of seawater has become essential to life. About 2.8 billion liters (750 million gallons) each day, or just over 50% of Saudi Arabia's usable water supply is generated via desalination. In fact, nearly 50% of the world's desalination capacity is in the Middle East and North Africa. The United States and Canada generate about 12% of the world's desalination capacity. The rest of the world accounts for the remainder. Desalination is a growing industry, and many companies are making a considerable investment in research to exploit significant improvements in the technology.

Desalination – The process that removes dissolved salts from seawater to make potable water.

One type of desalination process, known as heat desalination, uses heat to separate the mixture of seawater. This causes some of the water molecules in seawater to change their state and become gaseous water. Cooling can

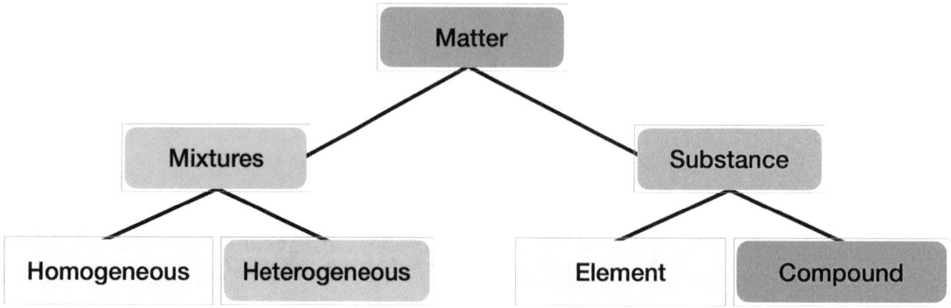

Matter
Mixtures
Substance
Homogeneous
Heterogeneous
Element
Compound

Fig. 1.7 The classification of matter

reverse the process and change the gaseous water back into liquid water. Using special equipment, the process can be operated in a way to produce large quantities of drinkable water from seawater.

To meet the needs of ever-expanding cities, many desalination facilities use the faster process known as **reverse osmosis**, in which impure water is pumped at high pressure through porous membranes. Only the water molecules are small enough to pass through the porous membrane, which leaves salts and other dissolved chemicals behind. In this way the mixture that is seawater is separated, and drinkable water is obtained.

Reverse osmosis – The process of purifying water by passing it through a semipermeable membrane.

1.3 The Scientific Method

As we have noted, chemists ask questions to discover and explore nature. This is how all scientists perform their craft. The process of science is all about asking questions and making careful observations. For example, the question “Is hydrogen gas explosive?” can be answered by preparing a little pure hydrogen, igniting it (from a safe distance!) in the presence of oxygen, and watching what happens. Although the question can be a little more complex (“How does burning gasoline impact global climate?”), the process is the same as that which we use to answer all scientific questions (Fig. 1.8).

We call this rational approach to learning and exploring the **scientific method**. The scientific method is not a rigid set of rules, because the human mind is intellectually flexible. Rather, it is an orderly way of asking questions and finding answers. Thinking about our world in this rational way ensures that we get meaningful results and steadily learn more about the topics we investigate. The method does not always provide the answer to a question immediately. Instead, many days, months, or years may be needed to arrive at an adequate answer, making the pace of progress seem slow, sometimes.

Fig. 1.8 Asking questions, experimenting, and observing are the start of the scientific questioning process

Table 1.1 The scientific method

Step 1	Formulate a question that you would like answered
Step 2	Find out what is already known about your question. This is typically done by searching the literature containing the work of other scientists
Step 3	Make observations—in other words, examine things. The observations involve the collection of both qualitative and quantitative data about the system under study. Recording the data in a systematic way is vital to this step
Step 4	Create a hypothesis that accounts for your observations
Step 5	Design and perform experiments that are carefully and deliberately set up to test your hypothesis. Specific testing is needed to ensure that the possible explanation "holds up" to scrutiny. These experiments yield results that return you to Step 3

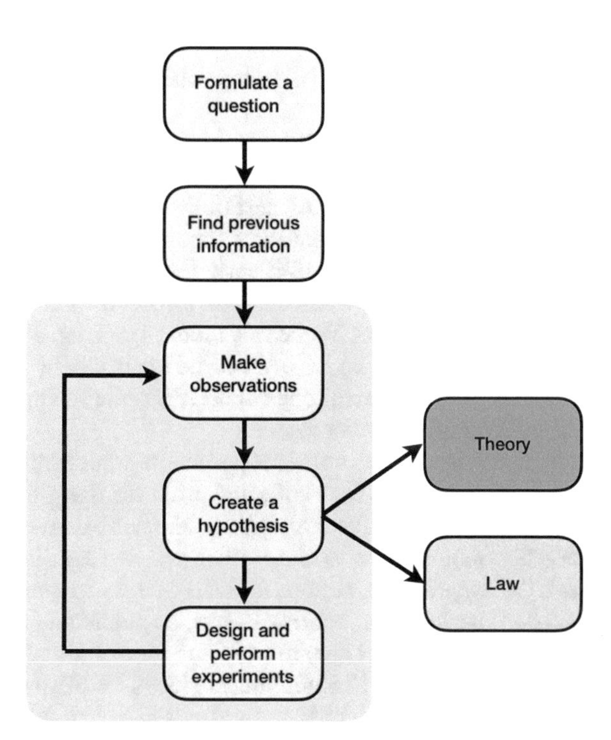

Fig. 1.9 The steps in the scientific method

Scientific method – A reliable way to find out things about nature by making use of appropriate combinations of these key activities: making observations, gathering data, proposing hypotheses, performing experiments, interpreting the results of those observations and experiments, checking to ensure that the results are repeatable, publishing the results, establishing scientific laws, and formulating theories.

The scientific method is often described as a cyclical process, illustrated stepwise in Table 1.1 and graphically in Fig. 1.9. The scientific method isn't just for scientific problems, though. You can apply this way of understanding to *any* subject or topic, whether in the classroom or in life after college.

First, we formulate a question and determine what is already known. Then, observations are made about the question and a hypothesis is created. A **hypothesis** is a testable explanation that seems to answer the question and explain what is happening. Experiments are performed to check the validity of the hypothesis. Then, we go back to Step 3 and make observations. Each iteration of these steps allows us to refine the hypothesis until it completely accounts for the observations. Many times, this refined and well-tested hypothesis provides the answer to our initial question.

Hypothesis – A tentative explanation for an observation.

Sometimes, patterns emerge in the data. When these patterns are consistently noted across a series of different systems under study, a statement called a **scientific law** can be made. The scientific law summarizes the patterns that are observed in the data. Scientific laws are succinct descriptions of the behavior of the natural world. However, a scientific law does not attempt to explain these observations.

Scientific law – Concise description of the behavior of the natural world.

Sometimes the information, observations, and repeated experimentation result in a tested explanation of the observed behavior, developed from hypotheses that have been supported by the results of experiments. This tested explanation is called a **theory**. In short, a theory is an attempt, based on scientifically-acquired evidence, to explain how or why something is happening. For example, the theory that an asteroid collided with Earth some 65 million years ago, which is based on a considerable volume of geological and chemical data, is an attempt to explain the demise of the dinosaurs. What is the key difference between a law and a theory? A law describes behavior; a theory expl*ains* behavior.

Theory – A trusted explanation of an observation, based on a hypothesis that has been tested in experiments.

A theory is a powerful statement not to be taken lightly. It is not guesswork or supposition. It is based on thoughtful and structured questions, experiments, observations, and data analysis. Some theories become trusted to such an extent that they almost acquire the status of certainties, because they are repeatedly confirmed by many different experiments over a long period of time. Other theories last for only a short time, until they are modified or even swept away by new evidence. For example, early theories on the origins of chemicals stated that chemical compounds inherent in living creatures contained a "life force" or "vital force" such that they could not be made from inanimate things such as dirt or rocks. While this theory has been abandoned because of repeatedly obtained evidence that contradicted it, some theories have yet to be scientifically refuted. One example is "quantum theory," which has led to everything from smartphones and GPS units to supercomputers and our ability to peer into the nanoworld using incredibly powerful microscopes.

As part of the scientific method, scientists perform their experiments several times. Repeating experiments time after time allows the scientist to prove that the data were collected properly, that the observations are correct within the limits of the measurements, and that the new information he or she has discovered is real. By publishing the results of their experiments, scientists also expose their work to other people's scrutiny and criticism. Other scientists then have the opportunity to check the experiments, results, and hypotheses to make sure these are valid. Communication of all types, whether via classroom teaching, writing books, or publishing journal articles, is as vital to science as performing experiments in the laboratory. There are tens of thousands of chemistry-related articles published each year worldwide; each article is intended to communicate ideas for others to consider.

The scientific method is powerful because it works. Over hundreds of years, it has yielded hypotheses, theories, and laws that describe, explain, and predict much of the behavior of the natural world. It has guided us to understand how to extract and process crude oil from the Earth, and how and why to work toward alternatives. We have been able to modify nature by removing rare metals from underground and processing them into electronic components that make smartphones, tablets, and supercomputers possible. And it has allowed us to understand much of the chemistry of life. After all, a living organism is essentially a self-contained chemical laboratory.

The scientific method can help you personally as you study and learn new subjects. ◻ Table 1.2 gives an overview of how the scientific method translates into solving problems, such as the ones in this textbook. While these

◻ Table 1.2 Solving problems using the scientific method

1. Read the problem a. Now read it again	This is the first step in solving a problem—make sure to carefully read the problem and identify what needs to be solved. Sometimes this is the most difficult part of the process!
2. Ask the Right Questions	Now that the problem has been identified, we gather the tools that we need by asking: What do I have, what do I want, and how do I get there?
3. Formulate a Solution	Apply the tools we gathered in step 2 and attempt a solution. In our Examples, we only see our approach to the solution, but in reality, we may need to make several attempts before we find the right one, we may have a reasonable approach that is different than others, but still gives the correct answer
4. Ask, "Is our Answer Reasonable?"	We need to evaluate the answer we came up with in step 3. This will help us decide whether our answer is likely to be correct, or whether a different solution may exist
5. Reinforcement—applying our understanding	High-order thinking means applying the concepts in different contexts or predicting what will happen when the ideas are applied to a different set of conditions

5 steps are not exactly the same as the ones in ◘ Table 1.1, they demonstrate a similar strategy of scientific thinking. As you apply this method of thinking and work more and more problems, you will become much better at clearly seeing the path to the solution.

Example 1.2—A Theory or a Law?

Earth's atmosphere is a mixture of gases—largely nitrogen (N_2), oxygen (O_2), and water vapor (H_2O), with much smaller amounts of argon (Ar) and a wide variety of other gases. Careful observation of these gases under controlled experimental conditions has shown that when the temperature is held constant, halving the pressure of the gas causes its volume to roughly double. As one goes up, the other goes down, as though pressure and volume were at the two ends of a seesaw. Similarly, when the pressure is doubled, the volume becomes half of the original value. We can describe this behavior by saying that at constant temperature, the volume of a gas is inversely proportional to its pressure.

Most gases deviate slightly from this generalized statement, but they all adhere to it quite closely at the temperatures found in the atmosphere, and this behavior has been checked many times with many different gases. Is this an example of a hypothesis, a theory, or a scientific law?

Asking the Right Questions

How do we differentiate among a hypothesis, a theory and a law? Which do we have in this exercise?

Solution

The statement that "at constant temperature, the volume of a gas is inversely proportional to its pressure" is a description of the behavior of a gas. It is not an explanation for the behavior being described. Therefore, this statement is an example of a scientific law rather than a theory. This particular law is actually known as Boyle's law, in honor of Robert Boyle (1627–1691), who discovered and publicized the law.

Is Our Answer Reasonable?

A scientific law describes behavior, rather than stating an explanation for the behavior, which, if overwhelming data were in place, might allow us to propose a theory. It therefore is reasonable to call this a "law."

What If...

What if the question asked us to predict the temperature of a gas as the pressure is changed? Would our resulting statement be a hypothesis, a theory, or a law? (*Ans*: *Hypothesis*)

Practice 1.2

Describe each step a scientist would follow as he or she investigated the following question using the scientific method. Carefully explain each step, describing any specific experiments and observations the scientist would make. Will the result of this application of the scientific method be a hypothesis, a theory, or a scientific law?

"There is a dark liquid in my cup. Is the liquid coffee and who put it there?"

1.4 Units and Measurement

Chemistry is a precise science that is based on our ability to properly measure the interactions of chemicals. We opened this chapter with a discussion of crude oil, and have referred to it several times, noting that its processing yields the gasoline in our cars and the plastics, pharmaceuticals, textiles and so many other products. These products are formed with chemical reactions that require specific amounts of substances to be combined for a measured amount of time at a controlled temperature. Amounts, time, and temperature are three of the most important *quantitative*, or measured, types of data that a scientist collects.

In order to communicate data and results effectively, we use units of measurement that are common among the sciences. Two major systems for measuring data are currently in use: the English system (used in Liberia, Myanmar, and the United States, where it is more formally known as the United States Customary System, or USCS) and the metric system.

1.4.1 The Système International and Its Base Units

The modification of the metric system in the 1960s gave rise to the Système International d'Unites, also known as the International System of Units (SI Units). The system defines seven **base units** used to measure seven fundamental physical quantities. ◘ Table 1.3 summarizes these units. From these seven base units, scientists can report

Table 1.3 The fundamental "base units" of the international system (SI)

Physical Quantity	Name of Unit	Symbol
Mass	Kilogram	kg
Length	Meter	m
Time	Second	s
Temperature	Kelvin	K
Amount of substance	Mole	mol
Electric current	Ampere	A
Luminous intensity	Candela	cd

measurements of all types and can effectively communicate those measurements to other scientists. During our exploration of chemistry, we will use the first five of these units extensively.

Base units – The set of seven fundamental units of the International System (SI).

1.4.2 Mass

Propylene glycol is a chemical that is a part of lipsticks, perfumes, skin care products, food, mouthwashes, e-cigarette cartridges, and baby wipes. Over 4 billion pounds of propylene glycol are produced each year from crude oil and biodiesel. Using a balance, we could measure the **mass** of a sample of this important compound. To record our scientific measurement, we would write the number from the balance and then add the base unit for our measurement. We would attach a prefix to the base unit to relate how the number differed from the original base unit by a power of 10. For example, the base unit of mass is the **kilogram (kg)**. The prefix kilo- in kilogram, one of the prefixes listed in Table 1.4, indicates that 1 kg is equal to 1000 grams (g). The kilogram is the only base unit that incorporates a prefix in its definition.

Mass – A measure of the amount of matter in a body, determined by measuring its inertia (resistance to changes in its state of motion) and expressed in the SI base unit kilogram.
kilogram (kg) – The SI base unit of mass.

$$1\ \text{kg} = 1 \times 10^3\ \text{g} = 1000\ \text{g}$$

Unfortunately, the mass of an object in kilograms is often incorrectly referred to as "weight." How do mass and weight differ? Mass is a measure of the amount of matter in an object, determined by measuring its inertia, or resistance to motion. **Weight** is a measure of the downward force exerted on a body by gravity. The mass of an object is a fundamental property of the object because it will not change, for instance, if we compare an object on Earth to the same object on the Moon. The weaker gravitational force on the Moon would cause the object to weigh less, but it would still have the same amount of mass (Fig. 1.10). A kilogram is equivalent to 2.2 lb on Earth, but only about 0.4 lb on the Moon.

Weight – A measure of the gravitational force exerted on a body.

For many measurements in the laboratory, we might only need one gram of a substance. How much is a gram? A peanut's mass is approximately 1 g. That is a pretty small amount, but 1000 peanuts have about the same mass as a kilogram. Masses that we often work with in the laboratory are much smaller than the kilogram. In fact, most masses in the laboratory will be in units of grams or milligrams. In the chemical industries, masses are often measured in tons (2000 lb) or metric tons (1000 kg, 1 Mg, or 2200 lb). The 4-billion-pound annual production of propylene glycol we noted at the beginning of this section can be expressed as 2 million tons, 1.8 million kg, or 1.8 million metric tons.

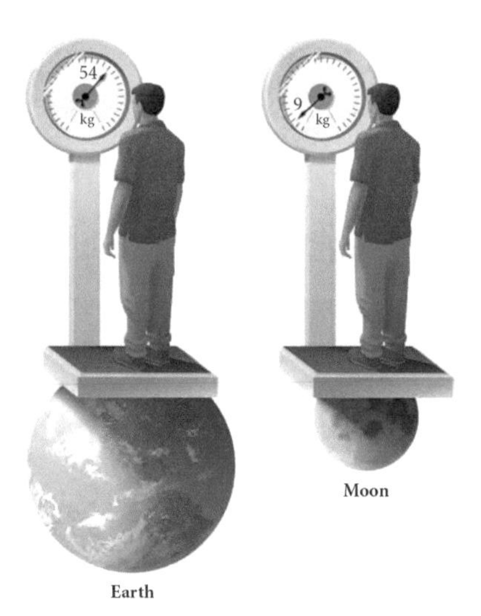

Fig. 1.10 A person's weight on the Moon is different from his or her weight on Earth. The mass of the person hasn't changed. Only the force of gravity is different

1.4.3 Length

As we seek to understand the chemical behavior of propylene glycol and the millions of other chemicals that we produce as a part of our chemical world, we often need to peer down to the molecular level—that is, the **nanoworld** of molecules, in which we talk about the tiniest fractions of length. The base unit in our own **macroworld** using the SI unit of length is the **meter (m)**. The meter is about 10% longer than a yard.

Macroworld – A term we use to describe the world that we can see that contributes to our everyday experiences.
Nanoworld – A term we use to describe the "world of the very small" at the level of individual atoms, molecules, and ions.
meter (m) – The SI base unit of length.

Often, we measure items in the laboratory in either centimeters (cm), a little more than one-third of an inch, millimeters (mm), which are one-tenth of a centimeter, or meters, which are a little more than a yard.

$$1\ \text{cm} = 10^{-2}\ \text{m} = 0.01\ \text{m}$$
$$1\ \text{mm} = 10^{-3}\ \text{m} = 0.001\ \text{m}$$
$$1\ \text{m} = 1.094\ \text{yd}$$

To measure longer distances, we often report them in kilometers (km).

$$1\ \text{km} = 10^{3}\ \text{m} = 1000\ \text{m}$$

One kilometer is equal to 1000 m, or slightly more than 0.6 mi. A runner in a "5 K" (5-km) race will run 5000 m (a little more than 3 miles). In chemical analysis, as we peer down into the nanoworld, scientists measure lengths much smaller than a meter, with sizes commonly measured in such tiny units as picometers, nanometers, and micrometers. Using the multiple from Table 1.4, there are one billion nanometers (10^9 nm) in one meter (1 m), so one nanometer is equal to 0.000000001 m.

$$1\ \text{nm} = 1 \times 10^{-9}\ \text{m} = 0.000000001\ \text{m}$$

1.4.4 Time

Chemical reactions, such as in the production of propylene glycol from biodiesel, require strict monitoring of time. The SI unit of time is the **second (s)**. The units of hour, minute, day, and year are often used in measurements of time. These are not SI units, but because they are so widely used, scientists don't often affix prefixes to

Table 1.4 The SI prefixes

Multiple	Prefix	Name
10^{24}	Y	yotta
10^{21}	Z	zetta
10^{18}	E	exa
10^{15}	P	peta
10^{12}	T	tera
10^{9}	G	giga
10^{6}	M	mega
10^{3}	k	kilo
10^{1}	da	deka
10^{-1}	d	deci
10^{-2}	c	centi
10^{-3}	m	milli
10^{-6}	μ	micro
10^{-9}	n	nano
10^{-12}	p	pico
10^{-15}	f	femto
10^{-18}	a	atto
10^{-21}	z	zepto
10^{-24}	y	yocto

the base unit 'second' to describe large time spans. For example, instead of reporting 3600 s as 3.6 ks, the scientist would be more likely to report this as 1 hour (1 h),

$$3600 \text{ s} = 3.6 \times 10^3 \text{ s} = 3.6 \text{ ks}$$

$$\Rightarrow \quad 3600 \text{ s} = 1 \text{ h}$$

Prefixes, however, are commonly used to indicate fractions of a second. For example, one millisecond (1 ms) is one-thousandth of a second.

Second (s) – The SI base unit of time.

1.4.5 Temperature

In addition to noting the time for a chemical process, such as the manufacture of propylene glycol, scientists also pay attention to the temperature of the process. For our temperature measurements, we could use the SI base unit known as the **kelvin (K)**. Notice that no "°" (degree) symbol is used with the abbreviation K. Two other temperature units are much more common, however. The English system uses the Fahrenheit scale, in which the unit of measure is the **degree Fahrenheit (°F)**. The metric system uses the Celsius scale, in which the unit of measure is the **degree Celsius (°C)**. All three systems can be used to accurately report temperatures, but they are scaled differently.

Kelvin (K) – The SI base unit of temperature.

Degree Celsius (°C) – The unit of temperature on the Celsius scale.

Degree Fahrenheit (°F) – The unit of temperature on the Fahrenheit scale.

The German physicist Gabriel Daniel Fahrenheit created the Fahrenheit scale in 1714. Fahrenheit used a mercury-based thermometer to define 0 °F as the coldest temperature he could make from a mixture of water and ammonium chloride. He called his body temperature 100 °F. Using his two reference points of cold and warm, he found the boiling point of water to be 212 °F.

The Swedish astronomer Anders Celsius developed the Celsius scale in 1742, based on dividing the difference in temperature from the freezing point of water, 0 °C, to its boiling point, 100 °C, into 100 individual units (degrees). In 1848, the British physicist Lord Kelvin (◘ Fig. 1.11) invented a temperature scale, in which one kelvin was equal in magnitude to 1 degree Celsius, although the scales have their value of zero at different points. The point we now call zero on the Kelvin scale equals –273.15 on the Celsius scale, so a temperature in Celsius can be converted into Kelvin by adding 273.15.

$$T_K = T_C + 273.15 \quad T_C = T_K - 273.15$$

We can also convert between the Celsius and Fahrenheit scales. The formula shows that the temperature of an object changes by 9 Fahrenheit degrees for every 5 Celsius degrees—a ratio of 1.8 °F/°C. The formulas required for converting between temperatures on the Celsius and Fahrenheit scales also indicate a different set point for the value of zero:

$$T_C = (T_F - 32)/1.8 \quad T_F = 1.8(T_C) + 32$$

◘ Figure 1.12 indicates the temperatures of physical changes for water in the three different units.

1.4.6 Amount of Substance

The amount of a substance is the number of particles in the sample. When designing chemical reactions, we must know how many particles of each substance is present so we can properly predict the outcome of the reaction. The SI base unit for the amount of a substance is the **mole (mol)**. One mole of a substance corresponds to 6.022×10^{23} particles, a very large number! Thus, when we work with 1 mole of atoms, we are working with 6.022×10^{23} atoms; 1 mole of molecules is 6.022×10^{23} molecules, and so on.

Mole (mol) – The SI base unit of amount of substance. One mole of entities, such as 1 mole of atoms, is approximately equal to 6.02214×10^{23} entities.

◘ **Fig. 1.11** Lord Kelvin (1824–1907) was born William Thomson in Belfast, Ireland. He was knighted in 1866 and given the title of Baron Kelvin of Largs in 1892. Even the most clever scientists don't necessarily make excellent hypotheses. Lord Kelvin wrote in 1896, "Heavier than air flying machines are impossible." And in 1900, he said "There is nothing new to be discovered in physics"

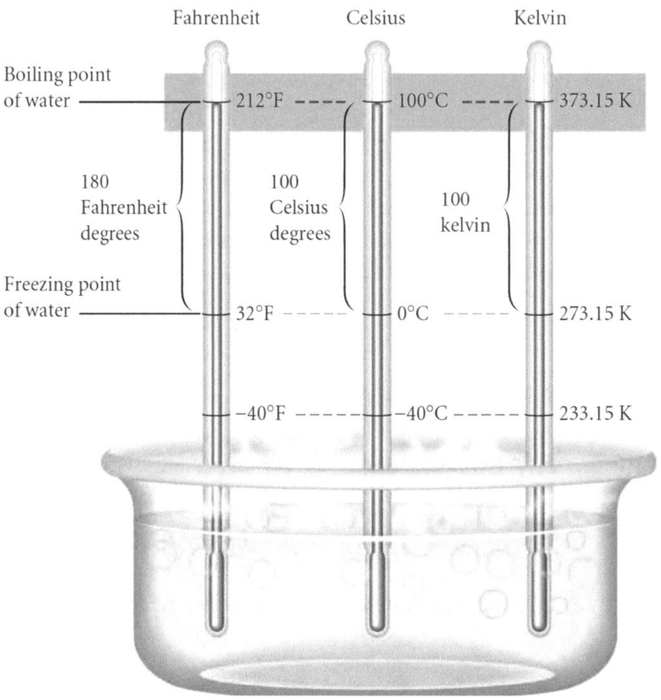

Fig. 1.12 The Fahrenheit, Celsius, and Kelvin temperature scales. Note the interesting point at –40 degrees, where the Celsius and Fahrenheit scales coincide

1.4.7 Electric Current

Some substances will transmit an electrical current and are called **conductors**. Others do not transmit a current and are therefore **insulators**. The SI base unit of electric current is the **ampere (A)**, often called the amp. This unit is used a great deal in chemistry because some chemical reactions can be used to generate an electric current (within batteries, for example). Conversely, electricity can be used to provide energy for chemical reactions.

Conductor – A substance that transmits an electrical current.

Insulator – A substance that does not transmit an electrical current.

Ampere (A) – The SI base unit of electric current.

1.4.8 Luminous Intensity

The quantity of light given off by an object, such as a light bulb, is measured using the last SI base unit, the **candela (cd)**. This unit describes the **luminous intensity**, or brightness, of something under study. For example, the Sun, which radiates the light energy that powers photosynthesis, has an estimated luminous intensity of 10^{18} candelas. Light bulbs produce a vastly smaller luminous intensity, hundreds of candelas, and small LEDs are measured in millicandelas. As you might have guessed from the name, a candle has a luminous intensity of about…1 candela!

Candela (cd) – The SI base unit of luminous intensity.

Luminous intensity – Brightness, expressed in the SI unit candela (cd).

We have spoken of incredibly large, as well as extremely small units of measure. How can we bridge the gap in sizes, so that we can communicate these differences in a consistent way? In this text, we refer to the world of individual atoms, molecules, and ions as the nanoworld, in contrast to the everyday, large-scale macroworld with which we are familiar. There is a huge change in the units that we use to report quantities as we travel between the nanoworld and the macroworld. To make the trip smoothly, we must be able to work comfortably with **scientific notation** and related mathematical operations. Scientific notation is a way of expressing a number as a base value multiplied by some number of powers of ten. For example, the number 1500 can be expressed as 1.5×1000. In scientific notation, we could express this as 1.5×10^3. You will see this method of expressing numbers used throughout this text.

Scientific notation – Presenting numbers as a base value times a number of powers of 10.

One of the most important aspects of learning about chemistry is becoming familiar with the nanoworld of chemicals and being able to visualize what happens at that scale. The fundamental particles of chemistry are atoms (◘ Fig. 1.13). These range in size from a diameter of approximately 75 pm (75×10^{-12} m $= 7.5 \times 10^{-11}$ m, using scientific notation) to over 500 pm (5.00×10^{-10} m).

How can we get some idea of what these numbers mean? Suppose we scale things up so that strontium atoms, each with a diameter of 400 pm (4.0×10^{-10} m), become the size of 2 cm pebbles. An actual pebble, then, scaled up in size to the same extent would be approximately 1000 km (about 600 miles) in diameter. That's big enough to cover all of the Great Lakes (◘ Fig. 1.14). Even with this comparison, the extraordinary difference in scale is difficult to visualize. Yet working with such comparisons should begin to give you some feeling for the scale of the nanoworld of chemistry.

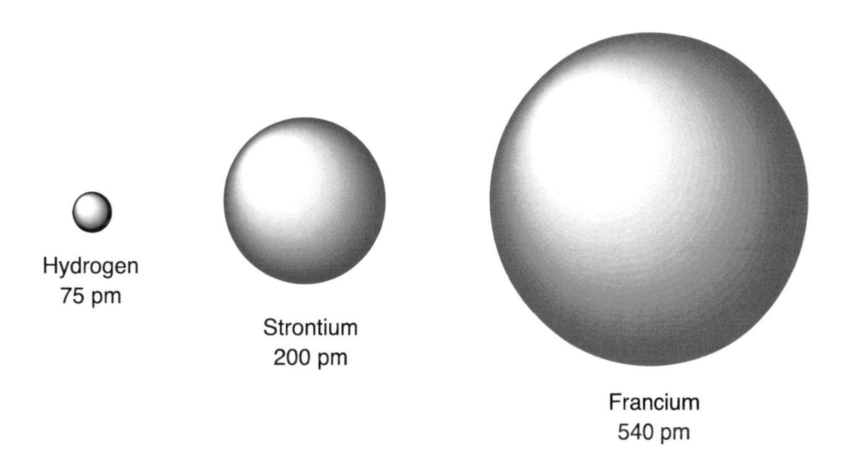

◘ Fig. 1.13 The relative size of three atoms, hydrogen (75 pm), strontium (200 pm), and francium (540 pm)

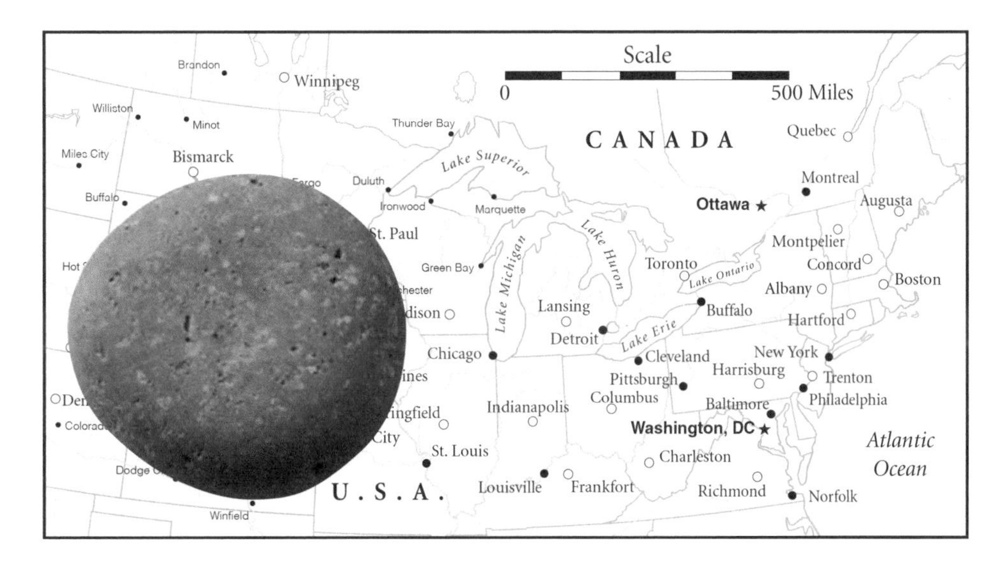

◘ Fig. 1.14 If a strontium atom were scaled up to the size of a pebble, the pebble, scaled up to the same extent, would be 1000 km wide

Example 1.3—Temperature Conversions

The chemical reactions within the human body tend to proceed optimally at a temperature of around 37 °C. Express this temperature using the Kelvin and Fahrenheit scales.

Asking the Right Questions

What is the relationship among the temperature scales? Is the Kelvin temperature larger than the Celsius temperature?

Solution

$$T_K = T_C + 273.15$$
$$T_K = 37\ °C + 273.15\ K$$
$$T_K = 310.15\ K$$
$$T_F = 1.8T_C + 32$$
$$T_F = 1.8(37) + 32\ °F = 99\ °F$$

Normal body temperature is commonly cited as equal to 98.6 °F.

Is Our Answer Reasonable?

A vital part of expanding our problem-solving ability is to ask ourselves, "Is Our Answer Reasonable?" In this exercise, we expected our Kelvin temperature to be larger than our Celsius temperature, and it was. Our answer made sense. It is a little more difficult to know if our Fahrenheit temperature made sense, because the change of 1.8 degrees Fahrenheit for each degree Celsius, in both the positive and negative directions, means that a range of answers can intuitively make sense. Repeated work with conversions such as this helps us gain a feel for what answers make sense.

What If....

What if the question asked us to covert 298 K to °F? (*Ans*: 77 °F)

Practice 1.3

Water is fed into a reaction with propylene glycol that occurs at 220 °F. Express this temperature using the Celsius and Kelvin scales.

1.4.9 Derived Units

The world's population consumes over 90 million barrels of crude oil every day. Each of those barrels has a **volume** of 42 gallons, or 0.16 cubic meters. Given this volume, the world's daily thirst for oil is about 14 billion liters. That's a lot of material to extract from the Earth every day. How much does all that oil weigh? That is, what is its mass? That's a big-picture question.

On a more personal level, how much oil does each of us consume (the so-called "*per capita*" use?) It would be helpful to know that crude oil weighs about 900 kg for each cubic meter (m^3), depending upon where the crude oil was extracted. This ratio of mass to volume is called ***density***, which for this sample is expressed as 900 kg/m^3.

Density – The mass of a substance that is present in a given volume of the substance. The SI unit for measuring density is kilograms per cubic meter (kg/m^3).

Volume and density are two of many **derived units**, formed by the combination of SI base units. Some of the more common derived units are shown in Table 1.5. We can work with these units to answer questions such as the mass of the oil and per capita consumption once we understand how we derive the derived units.

Derived units – Units formed by the combination of multiple base units.

Table 1.5 Selected derived units

Physical quantity	Derived units	Name of unit
Velocity	m s^{-1} or m/s	meters per second
Volume	m^3	cubic meters
Density	kg m^{-3} or kg/m^3	kilograms per cubic meter
Force	kg m^{-1} s^{-2} or kg/m s^2	Newton (N)
Pressure	N m^{-2} or N/m^2	Pascal (Pa)
Energy	N m	Joule (J)

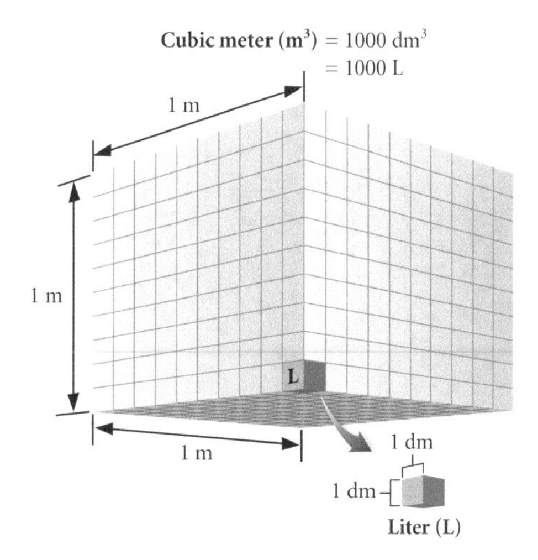

Fig. 1.15 The cubic meter, the SI unit for volume

1.4.10 Volume

Volume measures the space occupied by a sample. Geologists, economists, and news reporters, tell us about the volume of crude oil in barrels. On the other hand, in many countries, the gasoline or diesel fuel that provides energy for automobiles is measured at the "gas pump" in liters. In the U.S., we pump gas by our common unit of volume, the gallon. The actual SI base unit of volume is the cubic meter (m3) that we cited above. A cubic meter is the volume occupied by a cube of matter measuring 1 m on each side (see Fig. 1.15). We say that volume is derived from the SI unit for length, the meter.

Why do we need any other units of volume? Although the cubic meter is the SI unit for volume, this derived unit is often much too large to reflect the sample sizes that we would work with in the laboratory. A more common derived unit for volume is the cubic decimeter (dm^3), also known as the **liter (L)**. A cubic decimeter is the volume of a cube that measures 1 dm (0.1 m) on each side. This means that there are 1000 dm^3 (1000 L) in every cubic meter.

Liter (L) – A commonly used unit for volume; equal to 1 dm^3.

Another commonly used derived unit for volume is the cubic centimeter (cm^3). This unit describes a cube measuring 1 cm on each side (Fig. 1.16). In the health professions, one cubic centimeter is often abbreviated as 1 cc. There are 1000 cc in 1 L, so the cubic centimeter is also known as the **milliliter (mL)**: 1 cm^3 = 1 cc = 1 mL. Be sure that you write the abbreviation for liters as a capital L, not a lowercase l. Here is the bottom line for common volume conversions.

Milliliter (mL) – One thousandth of a liter.

$$1\ \text{L} = 1000\ \text{mL} = 1000\ \text{cm}^3 = 1000\ \text{cc} = 1\ \text{dm}^3 = 0.001\ \text{m}^3$$

1.4.11 Density

Density indicates the mass of a substance that is present in a given volume of the substance. The SI unit for measuring density is kilograms per cubic meter (kg/m^3), as we noted above when discussing the average density of crude oil. In our day-to-day world when we work with small amounts, rather than worldwide supplies, we can express density as grams per cubic centimeter (g/cm^3). Because 1 cm^3 equals 1 mL, units of density can be reported in grams per milliliter (g/mL). For example, the density of common car engine oil is 0.87 g/mL at 25 °C. As long as there are units of mass over units of volume, it is a measurement of density.

$$\text{Density} = \frac{\text{mass}}{\text{volume}}$$

Volume: 1 dm^3
"1 liter" = 1 L = 1 dm^3
1000 cm^3 = 1000 mL

Volume: 1 cm^3
1 mL

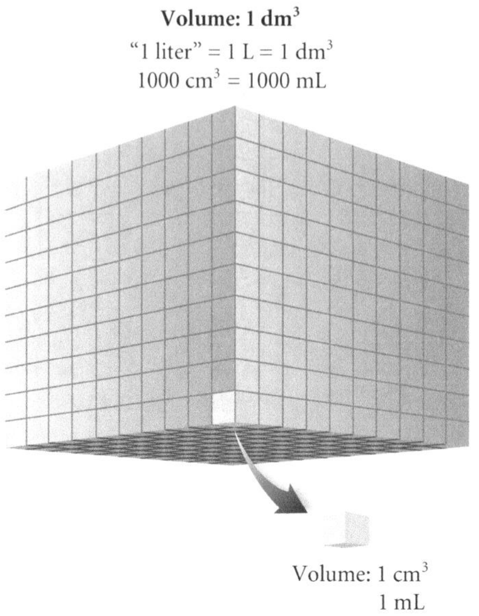

Fig. 1.16 One cubic centimeter is equal to a milliliter

Fig. 1.17 A block of silver-colored wax, a block of lead, and a block of aluminum. They look the same but are chemically quite different. We can tell which is which by measuring their densities

Figure 1.17 shows three similar blocks. However, one is silver-tinted wax, one is lead, and the third is aluminum. How can we tell the wax, lead, and aluminum apart? There are many physical properties that differ among the three blocks, such as color and hardness. Density is another. The wax has a density of 0.95 g/cm^3, the lead has a density of 11.4 g/cm^3, and the aluminum has a density of 2.70 g/cm^3. In other words, since the blocks each have the same volume, they differ greatly in their mass.

Example 1.4—Identifying a Metal via Density

A chunk of metal has a mass of 81.76 g and is added to a graduated cylinder (a type of glassware that measures volume) containing 25.0 mL of water. The total volume of the metal and water rises to 34.2 mL. What is the identity of the metal?

Metal	Density
Aluminum	2.7
Nickel	8.9
Tin	7.3
Zinc	7.1

Asking the Right Questions
In this example, we are given only enough information to tell the metals apart by their densities. The real question, then, is "What is the density of the unknown metal?"

Solution
The metal has a mass of 81.76 g. What volume does it occupy? The volume of water in the graduated cylinder increases from 25.0 to 34.2 mL when the metal chunk is added. The volume of the metal must then be 34.2 − 25.0 = 9.2 mL. We have mass and volume. We can now solve for density.

$$\text{Density} = \frac{\text{mass}}{\text{volume}}$$

$$\text{Density} = \frac{81.76\ \text{g}}{9.2\ \text{mL}} = 8.9\frac{\text{g}}{\text{mL}}$$

$$= 8.9\ \frac{\text{g}}{\text{cm}^3}$$

The metal has the same density as nickel.

Is Our Answer Reasonable?
Water has a density of approximately 1.0 g/mL. We can reasonably predict that the metal would be much more dense than water (it sinks in water). Nickel is about 9 times denser than water, so the answer seems to fit with our understanding. The answer makes sense.

What If…
What if the question asked us to determine the mass of nickel that would be needed to displace 20.0 mL of water? (*Ans*: 178 g)

Practice 1.4
How much water would be displaced if 81.76 g of aluminum, rather than nickel, were added to the graduated cylinder discussed in this exercise?

1.4.12 Extensive and Intensive Properties

Why does a log float on water? Mistakenly, we might say that the log is "lighter" than the water. Why does a pebble sink to the bottom of a lake? Again, we might mistakenly say that it is "heavier" than water. It is actually the density of an object that causes it to sink or float. The heaviness or lightness of an object is a measure of the weight of the object. This is an **extensive property**, one that is dependent on how much sample we have. Length is another extensive property. If the sample is a precious metal, such as gold or platinum, then the total cost of the metal is related to its mass (◘ Fig. 1.18), and these are both extensive properties.

Extensive property – A property of a substance that depends upon how much sample is present.

On the other hand, the price per ounce of gold is an **intensive property**, because gold will still cost the same amount of money per ounce whether you have 1 ounce (1 oz) or 1000 ounces. The density of the gold, 19.3 g/mL, is also the same no matter how much gold you have. However, the total cost of a pile of gold coins is an extensive property because it depends on how many coins you have. Large amounts cost more than small amounts. Similarly, the number

◘ **Fig. 1.18** Extensive and intensive properties of gold. The weight and cost of a chunk of gold depend on the amount (extensive properties); the color and density of the gold remain independent of the amount (intensive properties)

Table 1.6 Intensive versus extensive properties

Intensive properties	Extensive properties
Color	Mass
Density	Length
Temperature	Volume
Odor	Cost
Physical State	Energy

of barrels of crude oil on an oil tanker is an extensive property of the oil tanker. The larger the ship, the more barrels of crude oil. What about the crude oil density 900 kg/m^3? Will that change as we change the number of barrels?

Intensive property – A property of a substance that is independent of the quantity present.

Because the density does not change—it will remain at 900 kg/m^3 no matter how many barrels we have—it is an intensive property. Table 1.6 lists several intensive and extensive properties.

Example 1.5—Extensive Versus Intensive Properties

During an expedition to the top of a mountain, a camper wishes to cook some noodles for the entire group. In order to heat the noodles, the camper must boil an entire pot of water. Is the temperature at which water boils an intensive or an extensive property?

Asking the Right Questions

Initially, this problem seems confusing because we know it will take longer to bring a pot of water to a boil than to heat a cup of water to its boiling point. Let's consider the question this way: Does the amount of water make a difference in how hot it must get before it boils?

Solution

Considering the question in other ways such as that used in the Asking the Right Questions section gives the answer. The amount of water does not make a difference in how hot it must get in order to boil, so the boiling point of water is an intensive property. In other words, the pot of boiling water and the cup of boiling water have the same temperature.

Is Our Answer Reasonable?

Based on the definition of the term intensive property (a property that doesn't depend on the quantity of material), our answer does make sense.

What If...

What if the question asked whether viscosity (the resistance of a liquid to flow, or the syrupiness of a liquid) of molasses is an intensive or an extensive property? (*Ans: intensive property*)

Practice 1.5

Is the amount of time it takes to heat a cup of water to boiling an intensive property or an extensive property?

1.5 Conversions and Dimensional Analysis

The SI units typically used by scientists are often quite different from some of the units in everyday use. For instance, visitors to Canada from the United States might wish to fill their cars with gasoline ("petrol" in Canada). Initially, they might be surprised that the gasoline seems so inexpensive. The main reason for this is that the volume of Canadian petrol is reported in liters, whereas in the United States, gasoline is sold by the gallon. How can we change these different units into ones with which we are most familiar? To do the conversion, we need an expression that relates one unit to another. Such a **conversion factor** is a mathematical expression of the ratio of one unit to another. An example of a conversion factor is the relationship between pounds and kilograms. On Earth, a 2.2046-lb mass is equivalent to a 1-kg mass.

Conversion factor – A mathematical expression of the ratio of one unit to another, used to convert quantities from one system of units to another.

$$2.2046\ \text{lb} = 1\ \text{kg}$$

Because these two values are equal, we can combine them into a conversion factor that can be multiplied by a number to convert its units. The operation is identical to multiplying the original number by 1, because the numerator and denominator are equal to each other. This equality can be written as a conversion factor in two ways:

$$\frac{1\ \text{kg}}{2.2046\ \text{lb}} = 1 \qquad \frac{2.2046\ \text{lb}}{1\ \text{kg}} = 1$$

For instance, suppose we want to know the mass (in kilograms) of a book that weighs 3.5 lb. We can multiply the weight of the book by one of our conversion factors.

$$3.5\ \cancel{\text{lb}} \times \frac{1\ \text{kg}}{2.2046\ \cancel{\text{lb}}} = 1.6\ \text{lb}$$

Note that we chose to multiply the weight of the book by the conversion factor so that the unwanted units canceled. If we had multiplied by the other conversion factor, the numerical result would have been different, and the units would have been different, too! In this case, the units would not have canceled, and we would have been left with a very strange looking result:

$$3.5\ \text{lb} \times \frac{2.2046\ \text{lb}}{1\ \text{kg}} = 7.7\frac{\text{lb}^2}{\text{kg}}$$

This important method of calculation can be used to solve a great variety of problems in chemistry, and we will use this method for many of the calculations that we will do. It is known as **dimensional analysis** (or, sometimes, as the unit-conversion method, or the factor-label method) because canceling out the dimensions (or "units" or "factors") associated with each number enables us to arrive at the correct answer. The internet can be used to find some of the more common conversion factors for mass, length, volume, and time. Others may be found on the inside back cover of this book.

Dimensional analysis – A method for performing calculations by using appropriate conversion factors and allowing units (dimensions) to cancel out, leaving only the desired answer in the desired units.

We can use our conversion factors to help our American tourist in Canada. Assuming the price of gasoline in Canada is a mere 0.80 dollars per liter, what is the price in dollars per gallon?

$$\frac{\$0.80}{\text{L}} \times \frac{3.785\ \text{L}}{\text{gal}} = \frac{\$3.028}{\text{gal}}$$

When we multiply by the conversion factor, the units of liters cancels out (note the "L" in the numerator and the denominator), leaving us with units of dollars per gallon. Suddenly the price doesn't seem so low!

The three questions that are part of all unit conversion problems (and most other problems!) are:

1. What do we want to know?
2. What do we have?
3. How do we get to what we want?

We can use dimensional analysis to convert 10.25 inches into millimeters as an example of these questions. Sometimes we can't find a direct conversion from inches to millimeters. But we might find a conversion factor from inches to centimeters, and another from centimeters to millimeters.

Here is our approach:

$$\text{in} \xrightarrow{\textit{inches to centimeters}} \text{cm} \xrightarrow{\textit{centimeters to millimeters}} \text{mm}$$

Let's use this approach to construct a dimensional analysis of the units in a longer problem. We'll just have to make sure that our unwanted units cancel at each step.

$$10.25\ \cancel{\text{in}} \times \frac{2.54\ \cancel{\text{cm}}}{\cancel{\text{in}}} \times \frac{10\ \text{mm}}{1\ \cancel{\text{cm}}} = 260.4\ \text{mm}$$

To do nearly any problem, all we need to do is create our plan of attack and find the appropriate conversion factors. Let's assume we have a sample of motor oil with a density of 0.87 g/mL, but want to know the number of pounds per gallon. We can use conversion factors to solve this problem by converting one unit at a time into the units we want. Here's our plan of attack:

$$\frac{\text{g}}{\text{mL}} \xrightarrow{\textit{grams to kg}} \frac{\text{kg}}{\text{mL}} \xrightarrow{\textit{kg to pounds}} \frac{\text{lb}}{\text{mL}} \xrightarrow{\textit{milliliters to liters}} \frac{\text{lb}}{\text{L}} \xrightarrow{\textit{liters to gallons}} \frac{\text{lb}}{\text{gal}}$$

1

Starting with the initial density in grams per milliliter, we set up the problem, taking care to arrange each conversion factor so that each unwanted unit will cancel when we're done:

$$\frac{0.87\ \text{g}}{\text{mL}} \times \frac{1\ \text{kg}}{1000\ \text{g}} \times \frac{2.2046\ \text{lb}}{1\ \text{kg}} \times \frac{1000\ \text{mL}}{1\ \text{L}} \times \frac{3.785\ \text{L}}{\text{gal}} = 7.26\frac{\text{lb}}{\text{gal}}$$

The entire calculation converts the grams into pounds and the milliliters into gallons.

Example 1.6—Practice with Dimensional Analysis

Manufacturing on a global scale requires working with immense quantities of raw materials. As a close approximation of the actual amount, let's assume that 96 billion aluminum cans are produced in a given year. If each can requires 15 g of aluminum (Al), how many total kilograms of aluminum are used each day (assume 365 days in a year) to make aluminum cans?

Asking the Right Questions

As we've discussed, our key questions are:

1. What do we want to know?
2. What do we have?
3. How do we get to what we want?

Solution

We want to know: kilograms of Al/day.

From the problem we can obtain these conversion factors:

$$\frac{365\ \text{days}}{\text{year}} \quad \frac{9.6 \times 10^{10}\ \text{cans}}{\text{year}} \quad \frac{15\ \text{g Al}}{\text{can}} \quad \frac{1\ \text{kg Al}}{1000\ \text{g Al}}$$

How do we get there? If we map out our steps, there are several routes we can follow. Here is just one:

$$\frac{\text{cans}}{\text{year}} \xrightarrow{\textit{cans to grams}} \frac{\text{g}}{\text{year}} \xrightarrow{\textit{grams to kilograms}}$$

$$\frac{\text{kg}}{\text{year}} \xrightarrow{\textit{year to days}} \frac{\text{kg}}{\text{day}}$$

Using the plan, we fill in the numbers and perform the calculation:

$$\frac{9.6 \times 10^{10}\ \text{cans}}{\text{year}} \times \frac{15\ \text{g Al}}{\text{can}} \times \frac{1\ \text{kg Al}}{1000\ \text{g Al}} \times \frac{1\ \text{year}}{365\ \text{days}} = \frac{3.9 \times 10^{6}\ \text{kg Al}}{\text{day}}$$

Is Our Answer Reasonable?

Our answer is large. Should it be so? If we prepare billions of cans per year, it seems reasonable that we should use quite a bit of aluminum each day, and our answer agrees.

This doesn't mean that the math was necessarily correct, but the answer is at least reasonable. Anticipating what an answer should be, and then confirming that the answer is at least close to what we expect, is "thinking like a chemist."

What If...

What if the question asked for the mass of aluminum that is used in a week? (*Ans*: 2.7×10^{7} kg)

Practice 1.6

a. What is the mass in grams of 6.50 gallons of water that flow into a desalination plant every second? Assume the density of the water in this problem is 1.00 g/mL.
b. How many gallons of gasoline are in 58.6 L of gasoline?
c. How much would 11.2 gallons of gasoline cost in Canada if the price of gasoline in both countries is $3.34 per gallon and the exchange rate is $1.30 CN = $1.00 U.S.?

Example 1.7—Powers in Dimensional Analysis

A typical can of cola contains 355 mL of liquid. How many cubic meters would be occupied by the liquid from a 12-pack of cola?

Asking the Right Questions

Given our quantities and our goals, how do we go from milliliters to cubic meters?

Solution

We want to know how many m^3 are occupied by the cola. We've been given the number of cans and the volume of each can in mL. How do we get to what we want?

The conversion will be a little more involved than a straight conversion. We'll have to break the cm^3 term down into cm × cm × cm in order to get those units to cancel. Our plan:

$$\text{cans} \xrightarrow{\textit{cans to milliliters}} \text{mL} \xrightarrow{\textit{mL to cm}^3} \text{cm}^3 \xrightarrow{\textit{cm}^3 \textit{ to m}^3} \text{m}^3$$

$$12\ \text{cans} \times \frac{355\ \text{mL}}{\text{can}} \times \frac{1\ \text{cm}^3}{1\ \text{mL}} \times \frac{1\ \text{m}}{100\ \text{cm}} \times \frac{1\ \text{m}}{100\ \text{cm}} \times \frac{1\ \text{m}}{100\ \text{cm}} = 0.00426\ \text{m}^3$$

Note how the units cancel as we proceed with each step. Also notice how the numbers get multiplied along with the units.

Is Our Answer Reasonable?

A cubic meter is a box one meter on each side. We'd expect the volume in a dozen cans of cola to be considerably smaller than a box of these dimensions. Our answer does make sense.

What If...

What if the question asked for the volume in cubic feet? (*Ans*: 0.150 ft^3)

Practice 1.7

A house requires 1200 yd^2 of carpet. How many square meters of carpet will it require? How many square inches of carpet will it require? (Remember to square numbers when you square units.)

▶ Deepening Your Understanding Through Group Work

Recent research in college-level science education shows us that working in groups helps deepen our understanding. Why? In order to come to consensus on the best approach to solving a problem, we need to explain our ideas well enough to convince our groupmates that the ideas make sense. In doing so, we are compelled to clarify our own thinking. This is the basis of the "GROUP WORK" examples that are found in many of the chapters in this text. Here is our first group example:

Nigeria is a substantial exporter of oil. Its "Nigerian Bonny Light Oil" has an average density of 0.845 g/mL. Recently, the U.S. imported 415 million barrels (17,430 million gallons) of the oil from Nigeria, representing 8.2% of the total U.S. imports of oil. The average price of a barrel of crude oil that year was $46.03 per barrel, Knowing that there were 118 million U.S. households that year, what was the average household bill (in dollars) for the Nigerian oil, and what was the mass (in kilograms) of Nigerian oil that each household used? ◀

1.6 Uncertainty, Precision, Accuracy, and Significant Figures

Some of our discussion in this first chapter has been based on data about crude oil. We said, for example, that the worldwide average density of the liquid is 900 kg/m^3, which we can show is 0.9 g/mL by performing a dimensional analysis. We also looked at a specific sample of crude oil, Nigerian Bonny Light Oil, which has a density of 0.845 g/mL. Should we have the same confidence in the measurements? How can those who purchase and process the crude oil into our plastics, fuel for our motor vehicles, textiles, and pharmaceuticals know that these densities are correct? The confidence we have in our measurements is based on the use of **uncertainty** and **significant figures**.

Uncertainty – A measure of the lack of confidence in a measured number. An indication of how much our values may vary.

Significant figures – Those specific numbers in a measurement that have values representing actual information about the measurement.

1.6.1 Uncertainty

There are a few things in science, and in life, that we can know for certain, with no margin of error or room for doubt. For example, if we are asked how many U.S. presidents' faces are carved into Mt. Rushmore, we can be positively sure that the answer is 4. This is an **exact number**, a quantity that we count without any uncertainty. Indeed, all data that are determined by counting are exact.

Exact number – A number that can be known with absolute certainty. Exact numbers possess an infinite number of significant figures.

Measurement is different from counting, however, because every measurement is associated with a certain degree of uncertainty. There are no measurements made in which we are absolutely certain of the answer because we, and our measurement instruments, are imperfect. The uncertainty in a measurement is an indication of how much our values may vary and depends on the method used to obtain the measurement, on the quality of the measuring apparatus, and on the care taken by the person making the measurement. For example, if we are asked to measure the mass of an orange, we might obtain 245 g or 245.20 g or 245.20467 g, depending on the balance we use and on how well we know how to use the balance.

Every measured value will be associated with a particular level of uncertainty. No measurement is ever exact, no matter how careful we are.

How do scientists report the uncertainty in a measurement? There are many ways to report and keep track of uncertainty. The simplest way, which we will use in this text, is through the use of significant figures.

Even though scientists do their best to minimize the uncertainty of every measurement, there is always a level of uncertainty in their answers. Another commonly used term for uncertainty is ***error***. The two most common sources of error in measurements for properly designed experiments are:

Systematic error. We can make errors from simple clumsiness (spilling our reaction or improperly operating a piece of equipment). But it is also true that our eyes and hands and brains can manipulate labware, or read a balance, or interpret what we see only so well. This isn't limited to our natural senses; computer-controlled sensors are limited as well. Systematic error is not random. It will influence our measurement to be too high or too low, one or the other.

Random error. Measurement devices contribute random error to every measurement. No matter how well made or how carefully used, random error is unavoidable. The error will be high one time and low another time in a random distribution. There is no way to predict it.

Error – Another term for uncertainty. There are two main types of error; systematic error and random error.

1.6.2 Accuracy and Precision

As we noted in the previous section, the quality of the balance we use to measure mass is important in a measurement. More broadly, what is a high-quality measurement? We define this via the **accuracy** and **precision** of the measurements we make. To a scientist, these two terms mean completely different things. Accuracy relates how close a measurement is to its true value. Precision relates how close repeated measurements are to each other, regardless of their accuracy. A set of measurements can be accurate, precise, exhibit both of these properties, or neither. Fine instruments tend to yield measurements that are both accurate and precise.

Accuracy – The closeness of a measurement to the actual value.

Precision – The closeness of measurements to each other, irrespective of their accuracy.

Suppose four groups of students need to measure the mass of a vitamin C tablet prior to determining its vitamin C content. Let's assume that the true mass of the tablet is known to be 1.4683 g. The students all use the same type of sophisticated balance that provides four figures after the decimal point, but the first group forgets that the balance must be set to zero before they use it. These students might get these results:

Student Group 1	
Keisha	1.5674 g
Jorge	1.5673 g
Barb	1.5673 g
Myoung	1.5675 g
Average mass	**1.5674 g**
Actual mass	**1.4683 g**

Their results are very precise because the masses are in close agreement and are reported to four decimal places. They are not accurate results, either individually or when averaged, because they all contain the same systematic error, in this case because the students forgot to zero the balance before they used it. Random error is also evident in their results in that not all of the answers are the same, some are high and some are low compared to their average.

Let's assume another group of students uses the same balance and measures the mass of the tablet, but they remember to re-zero the balance before each use. They obtain these results:

Student Group 2	
Maite	1.4682 g
Maggie	1.4684 g
Winston	1.4685 g
Sabiha	1.4684 g
Average mass	**1.4684 g**
Actual mass	**1.4683 g**

The results obtained by Group 2 are both precise and accurate. Each value reported is in close agreement with the average, and all the values are accurate because they are very close to the true value for the mass. The results still contain error because the results are variable, but they are accurate and precise.

A third group of students uses a different balance to measure the tablet and remembers to zero the balance before use. Let's assume this group of students obtains these results:

Student Group 3	
Zeb	1.4673 g
Jeff	1.4691 g
Elena	1.4675 g
Alice	1.4692 g
Average mass	**1.4683 g**
Actual mass	**1.4683 g**

The results obtained vary more and are less precise than those of either Group 1 or Group 2, because they show a greater degree of uncertainty. Why did this group have a different degree of uncertainty? It could be something internal to the balance, or perhaps there was a lot of air movement around the balance that influenced the readings, or perhaps it was something else. This set of data contains more error (uncertainty) than the data from Group 2, but they yield an average value that is in agreement with the true mass.

How do we minimize error? Random error can never be eliminated; it can only be minimized. Taking the average value of a set of results subject to random error will minimize the random error. Because it is just as likely to be high as to be low, averaging many measurements will minimize the random error and can yield a very precise and accurate average result. However, systematic error (caused by, for example, a balance that is tilted) cannot be overcome by averaging, because all the results suffer from the same error. Instead, to minimize or even eliminate systematic error, we must correct whatever is causing it.

Why is it important for us to know about uncertainties in our measurements and to worry about precision and accuracy? Lives and jobs and a lot of money can be at risk if our measurements possess too much uncertainty. The dosage of drugs must be measured carefully, within known limits, to avoid doing harm instead of good. Accurate and precise measurement of the density of heart–lung machine tubing could mean the creation of a tube that doesn't crack or break apart during a long surgery. Accurately knowing the temperature at which crude oil is separated into fractions allows us to predictably control the products that we get.

1.6.3 Significant Figures

As we have just seen, it is useful to be able to assess the precision and accuracy of different values. When we combine several measurements, the result cannot be more precise than the least precise of the measurements we used to obtain the answer. This means that when we perform calculations, we cannot accept all of the numbers shown on the calculator display, because doing so would imply that we know the result to a degree of certainty that is not justified. A temperature of 325 K can be divided by 3 on a calculator, which might display an answer of

108.33333333333. How many of these digits should we write down for the answer? We don't know the temperature to 14 digits of precision. We can only report the precision that we know. In other words, some of the numbers in the calculator's display are "not significant," in the sense that they do not represent real information about the measurement. We can deal with this problem through the use of significant figures.

How do we know which values are significant? Let's follow a general chemistry student as she determines the density of a sample of oil. She uses a rough balance to obtain the mass of the oil. The balance's display says 5 g. What is the quality of that measurement? Specifically, the number means that the mass is closer to 5 g than it is to 4 g or to 6 g. This value has just one significant figure. She then uses a higher-quality balance to measure the mass. This balance says the mass of the oil is 5.125 g. This number has four significant figures. The balance's display indicates that the mass of her sample is closer to 5.125 g than it is to either 5.124 g or 5.126 g. Some uncertainty about the value still exists, but there is less uncertainty than before. The more significant figures there are in any measurement, the less uncertainty, and the more precision, there is about the measurement.

When our student writes down any value in her laboratory notebook, all of the numbers other than zero are always implied to be significant. When a zero is encountered in a measurement, determining its significance is less straightforward. All zeros that come before the nonzero digits in a number ("leading zeros") are used to locate the position of the decimal point, so they do not count as significant figures—they do not represent information about the measurement. For example, 0.005 has only one significant figure because the zeros on the left just hold the position of the decimal point.

Zeros that lie between nonzero digits, which are called 'captive zeros', are always significant. Therefore, the number 505 has three significant figures. The number 0.00505 also has only three significant figures. We can write 0.00505 in scientific notation as 5.05×10^{-3}. Scientific notation allows us to clearly see the number of significant figures in the value.

'Trailing zeros' that come at the end of a number without a decimal point are truly ambiguous—we don't know if they are significant unless we have more information about the measurement. We assume that they are not significant. But if a decimal point is in the number then all trailing zeros *are* significant. For instance, if our student reports that her oil has a mass of 5000 mg, we don't know whether she means 5×10^3 or 5.0×10^3 or 5.00×10^3. Just to be safe, we take the most careful view possible and say the value is closer to 5000 mg than it is to either 4000 mg or 6000 mg. That's a wide range of uncertainty, but we have no evidence to allow us to assume that the measurement is any more precise. The only way to resolve the ambiguity is to write a measurement in scientific notation. If the technician records the value as 5.0×10^3 mg, then the zero is significant because of the presence of the decimal point. This measurement unambiguously has two significant figures and the actual value is somewhere between 4500 and 5499. If the value is written as 5.00×10^3 mg, then the measurement has three significant figures. In this case, we would know that the actual value is between 4990 and 5010 mg. ◻ Table 1.7 summarizes the rules for determining the number of significant figures in a measured number.

◻ **Table 1.7** Rules for significant figures

Rule	Example
Nonzero digits are always significant	123 has 3 significant figures
Leading zeros are never significant	0.0123 has 3 significant figures
Captive zeros are always significant	1023 has 4 significant figures 10.203 has 5 significant figures
Trailing zeros without a decimal point are ambiguous	1230 has either 3 or 4 significant figures
Trailing zeros with a decimal point are always significant	1230.0 has 5 significant figures 0.01230 has 4 significant figures
To avoid ambiguity, use scientific notation	1.23×10^3 has 3 significant figures 1.230×10^3 has 4 significant figures 1.2300×10^3 has 5 significant figures
Counting numbers and defined values are exact (have infinite significant figures)	200 pencils is an exact number if you counted each pencil. The value of 12 inches per foot is exact because it is a defined value

Table 1.8 Calculations using significant figures (insignificant figures are in italics)

Addition/subtraction	Last significant digit in result has the same position as the highest value significant figure (underlined)	$21.3 + 1.042 = 22.3\underline{4}$ $1056 - 0.6 = 1055.4$
Multiplication/division	The result contains the same number of significant figures as the measurement with the smallest number of significant digits	$1.2 \times 4.2613 = 5.1\mathbf{1}$ $306 \times 0.023 = 7.038$ $(5.32 \times 10^6)/(3.1) = 1.7\mathit{2} \times 10^6$
Rounding rules	If the next digits are more than half-way to the next higher value, round up If the next digits are less than half-way to the next higher value, truncate If the next digit is = 5 round up or truncate so that the last digit is even	$3.2\underline{1}5350$ rounds to 3.22 $3.21\underline{5}350$ rounds to 3.215 $3.215\underline{3}50$ and $3.215\underline{4}50$ both round to 3.2154
The final answer	Only reduce to the correct number of significant figures for the final answer Keep at least 1 insignificant figure for every calculation	$(21.\underline{3} + 1.042) - (306 \times 0.023)$ $= (22.\underline{3}4) - (7.038)$ $= 15.3\mathit{02} = 15.3$

Finally, note that counting numbers (such as the number of people in a room) and certain defined conversion factors and equalities (such as 4 qt = 1 gal or 1 in = 2.54 cm) are considered to have an infinite number of significant figures. They are exact.

1.6.4 Calculations Involving Significant Figures

The mass for the sample of oil we determined earlier can be used to calculate the density of the oil. Let's assume that the volume of the oil was measured in a graduated cylinder and determined to be 6.1 mL. Dividing the mass by the volume allows us to obtain the density, which we read from the calculator display as 0.840163934 g/mL.

$$Density = \frac{mass}{volume} = \frac{5.125\ \text{g}}{6.1\ \text{mL}} = 0.840163934\ \frac{\text{g}}{\text{mL}}$$

How many of the digits displayed on the calculator are significant? The rules we use to determine this (see Table 1.8), depend on whether the number is the result of addition, subtraction, multiplication, or division. These rules ensure that our answer is no more precise than the least precise of our measurements.

Using these rules, the general chemistry student writes 0.84 g/mL in her notebook. This answer contains only two significant figures and is just as precise as the volume measurement that we obtained. If we needed to know the density to 4 significant digits (a higher level of precision), we would need to measure the density using a much more precise method. The method of measurement we choose must be determined by the precision of the results we need.

1.6.5 When to Round

If we are performing a series of calculations on a calculator, we should carry at least one of the insignificant figures through until we obtain the final answer and then do the rounding off at the end. If we cut off our insignificant digits too soon, we might lose figures that affect the accuracy of the final result. This can sometimes cause problems—for example, if you perform in one step a calculation that a textbook or multimedia package has performed in two or more steps and get an answer that differs very slightly from the solution, don't automatically assume it is wrong. The problem may lie in the different approaches taken or in rounding off or dropping the insignificant figures at different stages.

We'll emphasize this important rule about not dropping all the insignificant figures during calculations throughout this textbook. For instance, if we are working on a problem that involves many steps before the answer is obtained, we will not drop the next insignificant figure. Instead, the insignificant figures will be highlighted in color, or otherwise noted at the end of the number. This will serve as a reminder that we should carry some of the insignificant figures until the end of the calculation to avoid rounding errors at the end of the calculation.

Example 1.8—Significant Digits in Calculations
Provide the answer to this calculation. Be sure to indicate the appropriate number of significant figures.

$$\frac{288 \times 1.445}{7.9 \times 10^2} - 0.064 = ?$$

Asking the Right Questions
What is the most important concept when we work with significant figures?

Solution
The answer cannot imply more significance than the least significant quantity. The rules for significant figures help us track this accurately even when we have different mathematical functions in the problem. To start, let's complete the multiplication in the numerator. We retain at least one insignificant figure (the 2 in 416.2) in the intermediate calculation:

$$288 \times 1.445 = 416.2$$

Then we'll do the division. If this were our final computation, we would keep only two significant figures. But because we have more calculating to do, we retain a third, insignificant figure:

$$\frac{416.2}{7.9 \times 10^2} = 0.527$$

Finally, we'll subtract the two numbers. We don't round at all until all calculations are finished.

$$0.527 - 0.064 = 0.463,$$

which rounds to 0.46

Is the Answer Reasonable?
Our answer only has two significant figures. If we had rounded at the previous step, 0.527 would have become 0.53. This would have led to an incorrect answer in the final step ($0.53 - 0.064 = 0.466$), because 0.466 rounds to 0.47. Based on this understanding of significant figures, our answer seems reasonable.

What If...
What if the denominator was 799? (*Ans*: 0.457)

Practice 1.8
How many significant figures are there in each of these numbers?

a. 1.3090
b. 3450
c. 0.0020
d. 2.000
e. How many significant figures will there be in the answer to this problem?

$$\frac{2.3 + 8.91}{79.4} \times 23.324 = ?$$

1.7 Chemical Challenges for a Sustainable World

The chemist's ability to raise questions about everything around us, to design experiments, to make measurements, and to derive theories and laws via this process we call the scientific method has taught us a great deal about our world and the role of chemistry in all of its changes. The future is built by these changes, and they present many chemical challenges to us. Some of these challenges are problems caused by our use of chemistry in the past; others involve our continuing efforts to make life better, more enjoyable, and more fulfilling. As the keepers of the global commons, we can use our understanding of chemistry to work toward **sustainable development**. Some of the future challenges for chemists include those listed here.

Sustainable development – The process of meeting life's needs while permitting future generations to meet their needs.

- *The frontiers of medicine.* Chemists will continue to search for new and more effective medicines to treat cancer, AIDS, heart disease, Alzheimer's disease, rheumatoid arthritis, and many other diseases. They will also be urgently trying to create new antibiotics that enable us to "stay ahead" of the pace at which disease-causing organisms become resistant to existing antibiotics.
- *New challenges in agriculture.* Chemists have already transformed agriculture once, through their development of artificial fertilizers and pesticides. They are now heavily involved in a new and controversial transformation involving the creation of genetically modified crops. This may enable us to create new "engineered" plants with larger and more reliable yields. Nevertheless, many people are concerned about the legal, ethical, and safety issues associated with genetically modified crops.

- *The challenge of pollution.* Oil spills, smog, and pollution in general are negative aspects of our use of chemical processes—aspects that often predominate in public perceptions of chemistry. Pollution of our environment has been the price we have paid for many of the comforts and conveniences of modern life. Chemists are working hard to reduce that price by developing alternative "green" technologies and by learning how to use chemistry more effectively to clean up the problems that still arise.

- *Global climate change and ozone depletion.* A rise in levels of carbon dioxide and other greenhouse gases has been implicated in global warming. And the destruction of the protective ozone layer by chemicals called chlorofluorocarbons, or CFCs, has contributed to increases in skin cancer. Chemists will continue to research both issues to determine the causes of, and solutions to, these environmental concerns. In addition, the development of new chemical technologies will help us learn more about the hazards that caused these problems, with the goal of reducing or eliminating them.
- *Better materials.* Materials are the chemicals we use to make things, such as clothes, cars, aircraft, homes, televisions, and computers. Every material has its own set of advantages and limitations for a given application. Chemists are continually trying to enhance the advantages, such as durability, flexibility, and efficiency as electrical conductors or insulators. They are also working to overcome the limitations of some materials, such as susceptibility to corrosion, high cost, inability to recycle and reuse, and so on. That work will continue indefinitely, hopefully yielding a steady supply of new and more versatile materials to make the things we need and want.
- *New ways to supply energy.* The modern world is sustained by huge supplies of materials—such as coal, crude oil, and gas—that can provide usable energy. Unfortunately, the chemical reserves of energy are limited. Moreover, their use contributes to environmental pollution and global climate change. Chemists are learning how to get the energy we need in more sustainable and less polluting ways. Many alternative ways of supplying energy already exist, such as solar and wind power. But these ways need to be made more efficient, cost-effective, and powerful in order to find use in today's energy-hungry world.
- *Nanotechnology.* In the 1970s people began to use the term nanotechnology to describe the manipulation and machining of matter at very small scales. Some developments in this field can be described as "molecular manufacturing," involving attempts to build tiny machines, materials, and medical devices by manipulating very small assemblies of matter, even molecule by molecule or atom by atom. Traditional chemistry works with huge numbers of particles. The twenty-first century might see a new industrial revolution, in which the precise manipulation of tiny numbers of atoms, molecules, and ions becomes a routine part of our technology. Many of the first steps toward that new revolution have already been taken.
- *Understanding life.* We understand a great deal about the chemistry of life, but many processes are not fully understood. Indeed, the chemistry of what we really are—the chemistry of consciousness—is still a complete mystery to us. We have uncovered the general chemical mechanisms at the heart of all life, but there are endless intricate details still to be learned. As we learn these details, and therefore learn more about the workings of life, we will be in a much better position to fix things when the chemistry of life goes wrong and makes us ill.

These are just a few of the "hot" chemical topics that will tap the creative energy of chemists tomorrow and into the distant future. Chemists will also be involved in the operation of oil refineries, plastics factories, pharmaceutical companies, food manufacturers, cosmetics companies, makers of paints, glues, and varnish, semiconductor plants, and much, much more (see ◘ Table 1.9). The advances of science and technology depend on the continuous development and application of chemical knowledge to keep the modern world running.

Table 1.9 Some career opportunities in chemistry

Agronomist	Food technologist	Pharmacist
Anesthesiologist	Geneticist	Pharmacologist
Biochemist	Geologist	Pharmacologist sales representative
Ceramics engineer	Industrial health engineer	Physician
Consumer protection specialist	Internist	Professor
Cosmetic and perfume scientist	Medicinal chemist	Science technician
Dentist	Metallurgist	Technical writer
Dietitian	Nuclear scientist	Textiles scientist
Educator	Paint and coatings scientist	Toxicologist
Food and drug analyst	Patent examiner	Wood scientist

Possible employers of chemistry majors	
Beverage companies	Local government
Breweries	Medical laboratories
Centers for Disease Control	Medical libraries
Chemical Companies	Medical supply companies
Chemical Manufacturers	Mining companies
Department of Agriculture	National Institutes of Health
Department of Defense	Newspapers and magazines
Environmental Protection Agency	Petroleum refineries
Food Companies	Pharmaceutical companies
Food and Drug Administration	State and Federal Government
Hospitals	Technical libraries
Journals	Textile companies

The Bottom Line

- Chemists solve problems through their knowledge of how matter is constructed and how chemicals react.
- Atoms are the most fundamental particles of the chemical world. The particles of chemistry— atoms and molecules—are incredibly tiny compared to the objects we see in the everyday world.
- Molecules are composed of two or more atoms chemically bonded together. A chemical compound is any substance that contains different elements chemically bonded together.
- Matter can be divided into substances (pure materials) and mixtures, and mixtures can either be homogeneous or heterogeneous.
- Chemical changes occur when chemicals undergo reactions in which new chemical products are formed from the initial chemical reactants. Physical changes occur when chemicals undergo changes in their state.
- Scientists learn about nature, and learn how to influence it, using the scientific method. This helps ensure that their conclusions are scientifically valid.
- The International System (SI) defines the fundamental base units, and a variety of derived units, that are used to measure physical quantities. The fundamental units of the SI are: kilogram, second, meter, kelvin, mole, ampere and candela.
- Mass and weight are not the same. Mass is an amount of matter and weight is the force developed by gravity acting on a mass.
- The calculations in chemistry include the conversion of units by using factors that relate identical quantities. This can be done via a method known as dimensional analysis.
- Precision and accuracy are important to the discussion of chemistry (and science in general) because they relate the uncertainties in measurements. The number of significant digits a number contains is related to the precision of that number.
- Scientific notation is the best way to write numbers to indicate the significant figures unambiguously.
- Our understanding of chemistry can be used to solve problems in many fields, including medicine, agriculture, pollution control, and nanotechnology. In addition, we can broaden our understanding of many issues, including global warming, materials, meeting our energy needs, and life itself.

Section 1.1 Introduction

Skill Review

1. Define the term chemistry.
2. Explain why chemists need to be inquisitive.
3. In the section, we listed a number of things that chemists do. Look around the room in which you are sitting. List five additional things that you suspect are manufactured using chemistry.
4. Look on the label of ingredients for five products in your kitchen pantry. List some examples of substances that might be commercially prepared. Why did you choose these? List some substances that are added in more or less their natural form. Why did you choose these?

Chemical Applications and Practices

5. The directions for baking a particular cake include the following instruction: "Carefully add 1 cup of water to the powdered mix and stir." Use this statement to explain why measurements are important to the chemist.
6. A scientist has just identified a new drug for treating cancer. Use this statement to explain why a scientist communicating this discovery might need a chemical "shorthand."
7. Chemists often earn more than firefighters of equivalent experience, but less than physicians. What factors might account for this salary distribution?
8. On most college campuses, chemistry faculty will typically earn more than historians and English professors, but less than chemical engineers and law professors who have the same experience and productivity. What factors might account for this salary distribution?

Section 1.2 Communication via the Language of Chemistry

Skill Review

9. Define, differentiate between, and give three examples of an element and a compound.
10. What dispute might you have with the manufacturer of a cleanser that was labeled "chemical-free?"
11. How is it possible that oxygen, found in our atmosphere, can be called an element and a molecule but not a compound?
12. How is it possible that gold, found as flakes in a stream, can be called an element but not a molecule or a compound?
13. Name one characteristic that a mixture and a compound have in common.
14. Describe a critical difference between a mixture and a compound.
15. Identify each of the following chemicals as elements, molecules, or compounds. For the compounds, list the elements that are present. You may need to use the periodic table and the Internet to help you answer some of these.
    a. hydrogen gas
    b. sodium chloride (table salt)
    c. glucose
    d. neon
    e. copper sulfate
    f. titanium
16. Identify each of the following chemicals as elements, molecules, or compounds. For the compounds, list the elements that are present. You may need to use the periodic table and the Internet to help you answer some of these.
    a. nitrogen gas
    b. calcium chloride (sidewalk salt)
    c. aspirin
    d. helium gas
    e. silver metal
    f. water
17. Identify each of the following chemicals as elements, molecules, or compounds. For the compounds, list the elements that are present. You may need to use the periodic table and the Internet to help you answer some of these.
    a. sucrose (table sugar)
    b. uranium
    c. aluminum
    d. liquid nitrogen
    e. propylene glycol (used in modern antifreeze)
    f. potassium hydroxide
18. Identify each of the following chemicals as elements, molecules, or compounds. For the compounds, list the elements that are present. You may need to use the periodic table and the Internet to help you answer some of these.
    a. fluorine gas
    b. hydrazine (a rocket propellant)
    c. molten iron
    d. carbon dioxide
    e. vanadium
    f. lithium carbonate
19. Indicate whether each of these is a heterogeneous mixture or a homogeneous mixture.
    a. lake water
    b. yellow notebook paper
    c. marble
    d. soda
    e. milk
    f. dirt

20. Indicate whether each of these is a heterogeneous mixture or a homogeneous mixture.
    a. tap water
    b. apple juice
    c. beach sand
    d. paint thinner
    e. spaghetti sauce
    f. air
21. For each of these processes, indicate whether a chemical change or a physical change is taking place.
    a. molding melted chocolate into a bar
    b. heating your home with a woodstove
    c. drying your clothes in the dryer
    d. snow melting in the heat of the sun
22. For each of these processes, indicate whether a chemical change or a physical change is taking place.
    a. the yellowing of an old newspaper
    b. making hard-boiled eggs
    c. magic ink appearing on a piece of paper
    d. making dirty clothes clean in the washing machine
23. Identify each as either an element, a compound, or a mixture: (a) spaghetti sauce; (b) gold bar; (c) table salt; (d) air.
24. Identify each as either an element, a compound, or a mixture: (a) shampoo; (b) soda pop; (c) coffee; (d) charcoal.

Chemical Applications and Practices

25. Some ancient civilizations considered air an "element." Today we consider air a *mixture, compound* (select one). What is the basis of your choice?
26. Some ancient civilizations considered water an "element." Today we consider water a *mixture, compound* (select one). What is the basis of your choice?
27. Pyrotechnics are devices in which chemical reactions release energy as light, heat, and sound, and produce gases and small particles. Fireworks and flares are two examples of pyrotechnics. Chemicals are combined, often in a tube, to make a firework. Classify the changes in the substances as they are measured and combined in the firework. That is, do they remain pure substances, form mixtures, and if so, of what type?
28. A series of fireworks is set off in a controlled fireworks show. Classify the changes in the matter released from the fireworks with time. That is, as they interact with the environment, will they remain pure? Will they form mixtures? What type?
29. A "health food" store has a poster in its window that says, "Eat natural food, not chemicals." Briefly explain what is wrong with this statement, and try to summarize, in a chemically accurate way, what the writer of the poster was really trying to say.
30. A box of noodles in a "health food" store indicates that the noodles are "free of chemicals." Briefly explain what is wrong with this statement, and try to summarize, in a chemically accurate way, what the writer of that phrase was really trying to say.

Section 1.3 The Scientific Method

Skill Review

31. Suppose you sit down at your computer to draft an e-mail to a friend. However, after you have written the e-mail, it appears that it cannot be sent from your computer. Explain how you might use parts of the scientific method to help you solve this problem.
32. Suppose you wanted to determine which grade, or type, of gasoline would provide the best miles-per-gallon ratio in your car. What would be the important variables that you would control as you used the scientific method to draw your conclusion?
33. Which aspects of the scientific method are most likely to be subject to interpretation? Which, on the other hand, should be least ambiguous?
34. For what reasons, apart from the sharing of information, is it important to publish the results of scientific studies?
35. How do we distinguish a theory from a hypothesis?
36. We are familiar with how governments enact laws. How is a scientific law formed?
37. Your friend wishes to lose 20 pounds and comes to you for advice. What are the factors that go into your recommendation?
38. Your friend's physician said that he must lose 200 pounds and comes to you to compare options. How do you form your opinion on which is the best?

Chemical Applications and Practices

39. Several studies have been done to determine the effectiveness of various vitamins on specific health issues. Explain how it is possible for scientific studies on the same subject to produce conflicting results.
40. As a consumer, you encounter many claims about remedies for various human conditions from acne to balding. What aspects of the scientific method would you expect to see employed before such claims are made?
41. Assume that fish are dying in a river that runs through your town. The river begins in remote mountains, runs through many miles of farmland, and then passes through a large industrial area before entering town. What sorts of questions should be asked in an attempt to learn why the fish are dying? What investigations could be carried out to answer the questions?
42. In an attempt to treat a wart on your hand, you buy a commercially available remedy, and two weeks later your wart has disappeared. Does this prove that the remedy is effective? What steps should be performed to examine the effectiveness of the remedy in a proper scientific manner?

Section 1.4 Units and Measurement

Skill Review

43. Based on the table found in this chapter, arrange these distances in order from largest to smallest:
    1 millimeter   1 terameter   1 km   1 nm
44. Based on the table found in this chapter, arrange these masses in order from largest to smallest:
    1 microgram   1 kg   1 mg   1 picogram
45. Complete the conversions:
    a. 1.00×10^2 kg to grams
    b. 25.9 m to kilometers
    c. 25 °C to °F
    d. 3.20 mg to grams
    e. 9.11 nm to picometers
    f. 98.6 °F to °C
46. Complete the conversions:
    a. 1.50×10^2 mL to liters
    b. 8.42 g to mg
    c. 48.5 °C to °F
    d. 2.33 L to milliliters
    e. 7.00×10^2 mg to grams
    f. –20.0 °F to °C
47. Complete the conversions:
    a. 8.7 kg to milligrams
    b. 25.9 dm to meters
    c. 191 °C to °F
    d. 3.20 dL to kiloliters
    e. 9.11 s to nanoseconds
    f. 355 °F to °C
48. Complete the conversions:
    a. 3.99 mL to deciliters
    b. 8.42 Mg to kilograms
    c. –40.0 °C to °F
    d. 14.5 L to megaliters
    e. 55.5 ks to gigaseconds
    f. 75.3 °F to °C
49. Complete the conversions:
    a. 6.78 terabytes to megabytes
    b. 0.000003 L to milliliters
    c. 7.8×10^{-15} g to picograms
    d. –36.5 °F to kelvins
50. Complete the conversions:
    a. 4×10^{22} atoms to yottaatoms
    b. 37.0 μm to kilometers
    c. 46 kelvins to °C
    d. 98×10^3 s to megaseconds
51. A 2.00-L sample of nutrient agar would be able to be divided into how many equal 100.0-cm^3 media containers for a bacterial study?
52. Suppose the nutrient agar described in Problem 51 amounted to 32 containers, each holding 50.0 cm^3 of media. How many total liters of nutrient agar are in the containers?
53. In the late 1970's, "minicomputers" were 6-feet tall and typically had 8 kilobytes of memory. Current personal computers can have memories of 1 terabyte. How many times larger is the current computer memory than that in the 1970's model?
54. The North Star Horizon was a minicomputer that was introduced in late 1977. Its central processing unit (CPU) ran the various parts of the computer at a rate of 4 MHz. Current personal computers have CPUs that run at 4 GHz. How much faster are today's computers than the North Star Horizon?
55. Using scientific notation, express 327 km in:
    a. meters
    b. millimeters
    c. micrometers
    d. nanometers

56. Using scientific notation, express 499 s in:
    a. milliseconds
    b. microseconds
    c. deciseconds
    d. kiloseconds
57. Express the following numbers in scientific notation using appropriate significant figures:
    a. 1302.4 kg
    b. 0.0000450 m
    c. 844,000 s
    d. 53.05 L
58. Express the following exponentials as ordinary numbers using appropriate significant figures:
    a. 1.40×10^{2} m
    b. 8.005×10^{-2} s
    c. 0.006×10^{-6} mg
    d. 2.900×10^{9} km
59. A 62.56-g sample of mercury was added to a graduated cylinder. It had a volume of 4.60 mL. What is the density of the mercury?
60. A 22.4-g sample of a substance was added to a graduated cylinder. It caused a 18.3-mL change in the volume of water in the cylinder. What is the density of the substance?
61. A 250.0-mL sugar solution had a density of 1.37 g/mL. An additional 30.0 g of sugar was added to the solution, raising the volume by 24.6 mL. What is the density of the resulting sugar solution?
62. A chemist added 17.8 g of salt to 150.0 mL of a salt solution of unknown density. The resulting solution had a final volume of 165.9 mL and a density of 1.22 g/mL. What was the density of the original salt solution?
63. List the fundamental units that you would combine to get these derived units (you may need to look up the meaning of some of the terms).
    a. velocity
    b. acceleration
    c. volume
    d. specific heat
64. List the fundamental units that you would combine to get these derived units (you may need to look up the meaning of some of the terms).
    a. density
    b. pressure
    c. energy
    d. force

Chemical Applications and Practices

65. Which of the following rulers would provide the greatest number of significant figures in the measurement of the volume of a cardboard box?

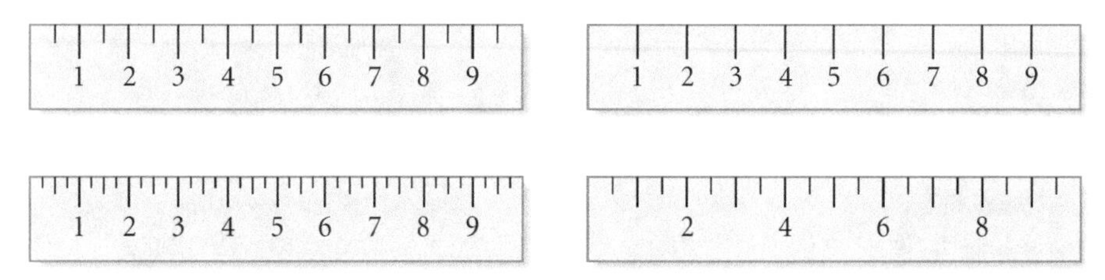

66. Which of the following beakers would provide the greatest number of significant figures for measuring 70 mL of a liquid?

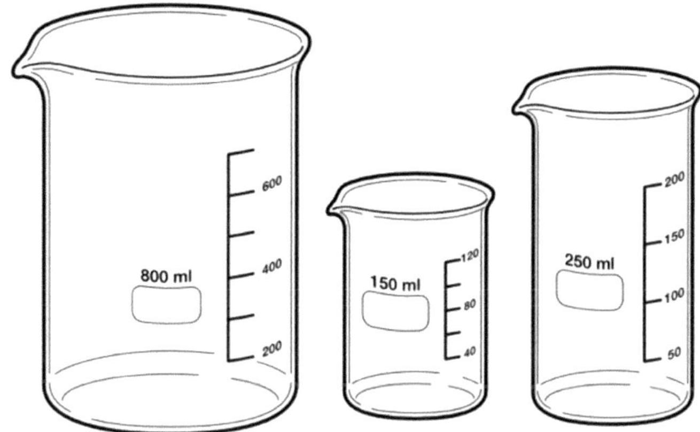

67. A fast-food restaurant wants to standardize the temperature of its coffee. It installs a coffee machine that is specified as delivering coffee at 75 °C ± 3 °C. What are the maximum and minimum temperatures of the coffee delivered by this machine in °F?

68. The temperature of the hot oil bath in a fast-food restaurant determines the quality of the french fries that are produced. The heater on the oil bath can regulate the temperature at 350 °F ± 10 °F. What are the maximum and minimum temperatures of the french fries delivered by this machine in °C?
69. Carbon atoms have an atomic radius of approximately 77 pm. If you drew a line that was 1 cm in length using a piece of charcoal, approximately how many atoms of carbon would be in this line? (For the purposes of this question, assume that the atoms are touching and that they form a straight line one atom wide.)
70. The same piece of charcoal (see Problem 69) was used to darken a box on a survey form. The box measures 5 mm on each side. How many atoms of carbon would be in this box? (For the purposes of this question, assume that the atoms are touching, that they line up into a square grid, and that they are only one layer thick on the paper.)
71. During the 1980s, several new elements were synthesized in Germany. These elements are made of unstable atoms. Half of the atoms in a sample of unstable atoms will decompose over a period of time known as the element's half-life. One of these newly discovered elements, Meitnerium (element 109), has a half-life of only 0.07 s.
    a. What is its half-life in milliseconds?
    b. What is the length of time of four of its half-lives in terms of microseconds?
72. Before filling, an empty, irregularly shaped container has a mass of 0.1956 kg. When this container is totally filled with water, the combined mass of the container and water is 305.6 g. What is the mass of water added to the container in kilograms?
73. A certain red tomato has a mass of 45.6 g. A green tomato also has a mass of 45.6 g. However, when placed in water, the green tomato floats and the red one sinks. (Try this. Depending on the degree of ripeness, it actually happens!) Which of the two is denser? Explain how two tomatoes with the same mass can have different densities.
74. Various plastics have identifiable density values. For example, you may have heard of high-density polyethylene and low-density polyethylene. If you had pieces of each that were equal in volume, which would have the greater mass?
75. Most samples of matter expand when heated. Assuming that this is the case with water at room temperature, would the density increase or decrease as the water warmed? Show the mathematical basis of your answer.
76. Ice floats on top of water. Using the information from Problem 75, explain how this could occur and show the mathematical basis of your answer.
77. In casual food preparation, one may be directed to add a pinch or a smidgeon or "just a bit" of a particular ingredient. If 10 pinches equaled 2 smidgeons and 10 "just a bits" equaled 1 smidgeon, how many "just a bits" of hot sauce should be added to a recipe that calls for 2 pinches?
78. Consider the tongue twister "Peter Piper picked a peck of pickled peppers." If two dozen pickled peppers are in a peck, and Peter Piper picked 8 pecks for his girlfriend Polly, how many peppers did Peter Piper pick?

Section 1.5 Conversions and Dimensional Analysis

Skill Review

79. Illustrate a plan of attack (in the same manner as indicated in the text) for converting units of miles per hour (mph) to meters per second.
80. Illustrate a plan of attack (in the same manner as indicated in the text) for converting units of kg m/s^2 to lb ft/s^2
81. Osmium (element 76) is one of the densest elements known. The standard density is 22.6 kg/L. What is the density in units of g/cm^3? What would be the mass of a 0.50-L sample of osmium?
82. Mercury has a density of 13.6 g/cm^3. What is the density in kg/L for mercury? Calculate the weight in pounds of 3.6 L of mercury.
83. Convert the following into the indicated units:
    a. 35 mph to kph
    b. 22.4 L/s to cm^3/s
    c. 733 mi/gal to km/L
    d. 4.184 g/°C to lb/°F
84. Convert the following into the indicated units:
    a. 12 doz/lb to gross/kg
    b. 16 lb/gal to g/mL
    c. 14.4 lb/in^2 to kg/m^2
    d. 3.04 °C/min to °F/s
85. Complete the conversions:
    a. 20.0 mpg to gpm (gallons per mile)
    b. 7.2 min/mile to km/h
    c. 1.6×10^3 ft^3 to m^3
    d. 2 °C/h to °F/s
86. Complete the conversions:
    a. 3.14×10^4 g/day to tons/year
    b. 4.8×10^{-3} g/cm^3 to µg/mm^3
    c. 38 L/s to ounces/h
    d. 4.35 dollars/L to cents/nL

87. In the vacuum of space, light travels at a speed of 186,000 miles per second. Indicate how many miles light travels through space in:
    a. 1 minute b. 1 day c. 1 year
88. The speed of sound varies greatly depending on the medium and conditions. If the speed of sound at sea level and zero degrees Celsius is approximately 1100 feet per second, indicate how far sound travels in:
    a. 1 millisecond b. 1 dekasecond c. 1 microsecond
89. Perform these conversions:
    a. 2.0 oz to pounds e. 13 ft to centimeters
    d. 96 in to meters c. 160 lb to kilograms
    b. 4.0 qt to liters f. 32 mi to kilometers
90. Perform these conversions:
    a. 3.0 kg to pounds d. 134 oz to kilograms
    b. 7400 s to days e. 75.5 cm to feet
    c. 50.34 mL to gallons f. 2.88 L to quarts
91. A typical chemistry lecture lasts for approximately 50 min. What is that time in years?... in centuries?... in decades? Report each number using a proper metric prefix that enables you not to have any zeros in your final answer.
92. Some people live to be 100 years old. What is that time in seconds?... in minutes?... in hours? Report each number using a proper metric prefix that enables you not to have any zeros in your final answer.

? Chemical Applications and Practices

93. Two cars are newly designed. The first car improves gas mileage from 8.0 to 16.0 mpg. The second improves from 45.0 to 90.0 mpg. Which car will have the greatest gas savings for a 520-mile trip as a result of these changes?
94. The semester tuition for a 16-credit load at a college is $5635, and the average class meets for a total of 15 h per credit. How much tuition is the student paying per hour of classroom instruction?
95. The mass of 6.02×10^{23} atoms of gold is approximately 197 g.
    a. What is the average mass, in grams, of just one atom of gold?
    b. Select a prefix that could more appropriately be used to report the average mass and express the mass in that unit.
96. The mass of 6.02×10^{23} atoms of carbon is approximately 12 g.
    a. What is the average mass, in grams, of just one atom of carbon?
    b. Select another prefix that could more appropriately be used to report the average mass and express the mass in that unit.
97. Suppose a football "star" signs a contract for 200 megabucks over a five-year period. How many dollars is the star paid per second? … per three-hour game?
98. Which is longer; a 100-m soccer field or a 100-yd football field? Show the mathematical proof for your answer.
99. Air is a mixture that consists mostly of oxygen and nitrogen. The density of air varies, depending on temperature and pressure. However, for this problem let's assume a reasonable value for the density of air to be 1.20 mg/cm^3. Suppose the dimensions of your dorm room are approximately 12 ft × 14 ft × 9.0 ft. What would you calculate to be the approximate mass of the air in the room?
100. Modern pewter is a mixture of tin, antimony, and copper. Formerly, pewter also contained lead. If a 1.00-lb sample of a particular pewter alloy contained, by mass, 95.0% tin and 3.4% antimony, how many grams of copper must be present in the sample?
101. Your university food service is redoing their breakfast menu. They estimate that they serve 4.0 oz of scrambled eggs per student per day, and each egg makes 1.8 oz of scrambled eggs. How much will it cost to buy enough eggs to feed 860 students if eggs cost $3.17 per dozen?
102. Your university food service is remodeling their dining room. They want to know how large to make their egg serving dish. As with problem 101, they estimate that they serve 4.0 oz of scrambled eggs per student per day, and each ounce takes up a space of 2.5 cm^3. What will be the volume of the serving dish needed to feed 860 students? To fit into the serving area, the dish must have dimensions of 0.80 m × 0.50 m. How tall must the dish be in cm?

Section 1.6 Uncertainty, Precision, Accuracy, and Significant Figures

? Skill Review

103. Some chemists are playing horseshoes. In their version of the game, each gets four chances to throw a horseshoe to hit a stake. What can you state about the accuracy and precision of the chemist that threw the horseshoes to arrive at the outcome shown here?

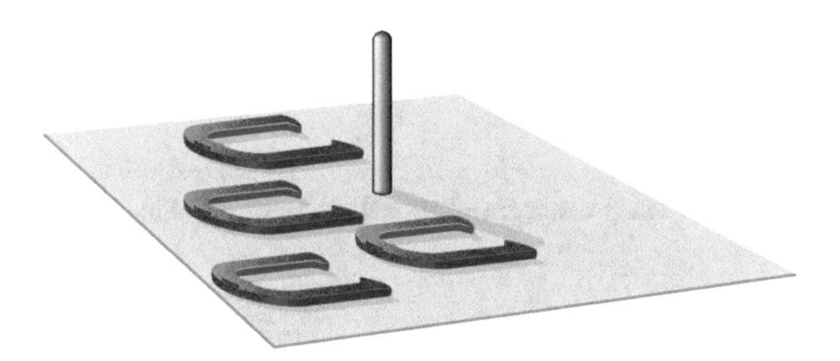

104. Two more chemists playing horseshoes each take a turn. Which of the two outcomes shown here is more precise?

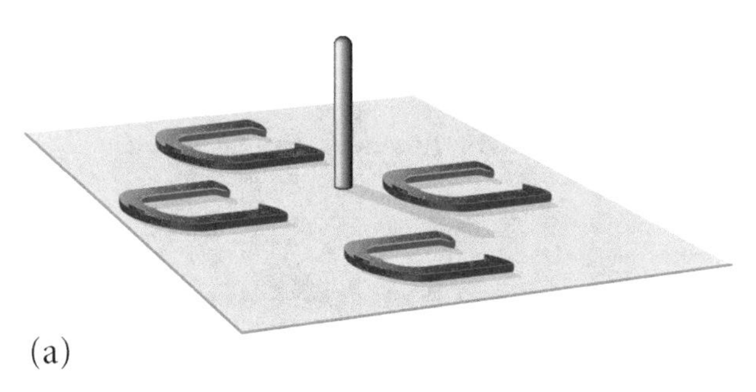

(a)

(b)

105. Three students weighed the same sample of copper shot three times. Their results were as follows:

Trial	Student 1	Student 2	Student 3
1	17.516 g	15.414 g	13.893 g
2	17.888 g	16.413 g	13.726 g
3	19.107 g	14.408 g	13.994 g

a. Calculate the average mass of the sample, as determined by each student.
b. Which set of measurements is the most precise?
c. If the true mass of the copper shot is 15.384 g, which of the students was most accurate?
d. What are the main sources of error that could have caused the differences in the values?

106. The same three students measured the volume of the same sample of water four times. Their results were as follows:

Trial	Student 1	Student 2	Student 3
1	25.55 mL	23.79 mL	25.02 mL
2	24.81 mL	24.01 mL	25.10 mL
3	23.03 mL	24.32 mL	25.07 mL
4	24.28 mL	24.19 mL	25.17 mL

a. Calculate the average volume of the sample, as determined by each student.
b. Which set of measurements is the most precise?
c. If the true volume of the water is 24.10 mL, which of the students was most accurate?
d. What are the main sources of error that could have caused the differences in the value?

1

107. Which of these values are quoted using exact numbers? How many significant figures are there in each of these numbers?
   a. 12.000000 g
   b. 3125 students
   c. 12.2 L
   d. 12 L
   e. 1 g
   f. 42 test tubes
108. Which of these values are quoted using exact numbers? How many significant figures are there in each of these numbers?
   a. 15 apples
   b. 506 people
   c. 3.2050 in
   d. 44 mi
   e. 72 °C
   f. 10 g
109. In each of these numbers, underline any zeros that are considered significant.
   a. 0.700 cm
   b. 0.101 kg
   c. 100.0 cm
   d. 100 m
   e. 0.01010 g
110. In each of these numbers, underline any zeros that are considered significant.
   a. 1.000 cm
   b. 80.2 kg
   c. 2104 cm
   d. 0.56 m
   e. 3000 g
111. How many significant figures are there in each of these values? (assume none of these are exact values)
   a. 6.07×10^{-15}
   b. 0.003840
   c. 17.00
   d. 8×10^{8}
   e. 463.8052
   f. 1406.20
   g. 0.0007
   h. 1600.0
   i. 0.0261140
   j. 1.250×10^{-3}
112. How many significant figures are there in each of these values? (assume none of these are exact values)
   a. 6.022×10^{23}
   b. 1.79×10^{-19}
   c. 3.00×10^{8}
   d. 14
   e. 0.0035020
   f. 250
   g. 13.50
   h. 101.010
   i. 12.000
   j. 550.050
113. Determine the answer for each of these problems. Report your answer using the rules for significant figures.
   a. $3.44 + 6.2$
   b. $12.57 - 3.998$
   c. $2.534 + 1.23 + 2.0500$
   d. 12.54×5.0
   e. 84×100
   f. $45.6 \div 2.4$
   g. $(754 + 0.8) \div 1.3$
   h. $(49.53 \times 1.20) + 12$
   i. $(35.865 \div 84.2) + 2.3890$
114. Determine the answer for each of the problems below. Report your answer using significant figures rules.
   a. $120 + 6.77$
   b. $453 - 0.32$
   c. $51.8 + 7.225 + 2.01$
   d. 2.54×32
   e. 36.33×0.300
   f. $140 \div 2.9375$
   g. $(135 + 3.2) \div 1.332$
   h. $(2.78 \times 1.2) + 3.96$
   i. $(14.42 - 1.023) \div 2.3$

Chemical Applications and Practices

115. Suppose your chemistry textbook fell 10.0 cm from a backpack to a desk. Then it slipped off the desk and fell another 0.91 m from the desk to the top of someone's foot. (Ouch!) Finally, it fell the remaining 40 mm to the floor. What is the total distance, with the correct number of significant digits, which the textbook has fallen?
116. A piece of cheese has a mass of 250.67 g. Four people each remove a slice of cheese from the piece. The first removes 22.5 g of cheese, the second removes 10 g, the third removes 3.557 g, and the fourth takes 80.1 g. How many grams of cheese remain? Use appropriate significant figure rules to calculate your answer.
117. Molybdenum has a melting point of over 2600 °C. It makes steel stronger, creates colors when added to molten glass, and is an integral component of some biological molecules known as enzymes. If you had a sample of the metal that had a mass of 14.56 g and a volume of 1.43 cm^3, what would you calculate as the density of molybdenum? (Remember to follow the rules for significant figures.)
118. A student obtains the density of water. A 243-mL sample of water was noted to have a mass of 235.5 g. What does the student calculate as the density of the water sample? (Remember to follow the rules for significant figures.)

Comprehensive Problems

119. What aspect of daily life do you predict will be most affected by chemistry in the generation to follow yours?
120. Name five areas of your life where chemistry plays a major role.
121. Describe two chemical changes and two physical changes that are important in growing food crops.
122. Someone remarks that the baking of a cake is a physical change. Another person says that the process is a chemical change. Who is right and why?
123. The height of horses is often measured in "hands." A "hand" is considered to be 4 inches. Furthermore, up to three-fourths, each fourth of a hand is considered to be 1 inch. Any value over three-fourths of a hand is rounded up to the next hand.
    a. What is the height of a horse, in inches, that stands 14.2 hands?
    b. How many hands is a horse that stands 1.6 m tall, from the ground to the top of the withers?
124. Write a sentence to describe what is happening when you dissolve some sugar in a cup of warm water.
125. Numbers, and their meaning, are very important to the practicing chemist.
    a. How does the meaning of an exact number differ from that of a number taken from a measurement?
    b. Beverages are often sold in six-packs. Six six-packs would have how many individual containers?
    c. Those containers hold a total of 7200 mL of the beverage. Which of the value(s) represented in this problem are exact numbers?
126. What are the two most common sources of uncertainty in measurements? Of the two, which typically produces uncertainty in a consistent direction?
127. Describe two physical and two chemical properties of water.
128. A chemist and her family go on vacation and ride the train to the mountains. They start their journey from Chicago and travel 1223 miles to Aspen, Colorado. The trip starts at 1:50 p.m. Central Time and the train arrives in Aspen at 1:53 p.m. Mountain Time the next day. How fast, in kilometers per hour, does the train average during the trip?
129. During a typical cross-country train trip, top speeds of 80.0 miles an hour are reached. What is that speed in kilometers per second?
130. A student notices that the contents of a 12 fl oz can of diet soda weigh 340 g.
    a. What is the density, in grams per milliliter, of the soda?
    b. Will this can of soda float on water or sink?
131. A dairy wishes to deliver its milk in large trucks to reduce the cost of transportation. Assuming that the delivery truck has a weight of 6500 pounds, how many gallons of milk ($d = 1.106$ g/mL) could be placed into the truck so that it could still make it over the small country bridge rated to hold a maximum of 10.0 tons?
132. Give a nanoscale interpretation of the fact that at the melting point the density of solid mercury is greater than the density of liquid mercury, and at the boiling point the density of liquid mercury is greater than the density of gaseous mercury.

Thinking Beyond the Calculation

133. In a recent study of an allegedly pre-Columbian (before Columbus) map of the North American continent, one researcher claimed that the map was a forgery because the black ink had a yellow tinge (older ink would sometimes do this) that could have been made by first laying down the yellow line and then copying a thinner black line over it. However, another researcher analyzed the difference between the boundaries of the black and yellow lines and claimed, after many measurements, that the differences were consistently so small that it could not have been done freehand.
    a. Was the second researcher using precision or accuracy to make his claim? Explain.
    b. Describe how the second researcher used the scientific method to study the map.
    c. If the black lines had an average width of 1.45 mm. What would that width be in inches? … in meters? … in yards?
134. Biological evolution is a topic that often sparks debate. Some say it is one of several theories, and others say it is a fact. The field of "chemical evolution," in which chemists study how the elements and simple compounds (such as hydrogen, methane, and ammonia) combined to form the molecules of life (including proteins and DNA) has been active for nearly a century. Look up the pioneering work of Miller and Urey using Internet or print resources. Do their experiments convince you that chemical evolution is possible? What sorts of experiments would you design to determine whether chemical evolution occurs? If so, would you call it a theory or a fact? Why?
135. While there isn't a "typical" car, plane, or ship, let's assume that the average U.S. car gets 25 miles per gallon, the average airplane gets 0.20 miles per gallon, and the average cruise ship gets 80 feet per gallon. A family of 4 is planning a trip between San Diego, California and Vancouver, British Columbia, 1400 miles away by car, 1600 miles away by cruise ship, and 1200 miles by plane. They wish to be as eco-friendly as possible. Assuming that a

plane flight would carry 150 passengers and the cruise ship has 3300 passengers on-board, which mode of transportation would be the most eco-friendly? What other factors besides mileage would go into your decision? Can you quantify any of these factors?

136. There has long been a debate about the relationship between the mathematical preparation of students and their success in first-year chemistry. The debate goes beyond standardized test scores to social issues. If you were going to study the relationship between math preparation and success in first-year chemistry, how would you approach the problem? Although you would need to consider all aspects of the scientific method in your study design, focus your answer on these concerns: What are the main questions you would ask? What experiments would you design to answer your questions? And how would you factor other, non-quantitative data into your results?

A Quest for Understanding

Contents

Supplementary Information The online version contains supplementary material available at ▶ https://doi.org/10.1007/978-3-030-90267-4_2.

M. Mosher and P. Kelter, *An Introduction to Chemistry*,
https://doi.org/10.1007/978-3-030-90267-4_2

What We Will Learn from This Chapter

- Many fundamental questions about our world can be answered by considering the smallest component of matter—the atom (▶ Sect. 2.1).
- The atom is the smallest indivisible complete unit of matter. Atoms combine and react according to the laws of Conservation of Mass and Definite Composition (▶ Sect. 2.2).
- Dalton's Atomic Theory and the Law of Combining Volumes further define how atoms can combine into compounds (▶ Sect. 2.3).
- Our modern understanding of the atom shows how electrons, protons and neutrons are arranged (▶ Sect. 2.4).
- Atoms with the same number of protons can differ in the number of neutrons. These isotopes occur in natural ratios for each of the known elements (▶ Sect. 2.5).
- The mass spectrometer can be used to determine the exact mass of an isotope. The natural abundance of each isotope can then be used to determine the average atomic mass for an element (▶ Sect. 2.6).
- The Periodic Table of Elements is a highly organized listing of the known elements. The structure of the table groups elements based on their chemical reactivity (▶ Sect. 2.7).
- When an atom gains or loses electrons, it becomes an ion. Ions combine to form ionic compounds (▶ Sect. 2.8).
- In some cases, electrons are shared between atoms to make molecules (▶ Sect. 2.9).
- A set of rules for naming compounds is known as chemical nomenclature. These rules indicate how to name binary compounds and polyatomic ions (▶ Sect. 2.10).
- We can recognize hydrates and acids by their chemical formula. They are named using nomenclature rules (▶ Sect. 2.11).

2.1 Introduction

How old is life? That's a deep question that scientists have been debating for a long time. Xenophanes of Colophon, an ancient Greek philosopher who died about 490 B.C., examined fossils of marine life to generate a hypothesis that the Earth had gone through many alternating periods of wetness and dryness in its history. Since then, scientists have been using fossils to date life on this planet. Until recently, however, almost all fossils of creatures that lived before about 2.5 billion years ago had been nearly impossible to detect because they are so tiny.

So many scientific advances have been made possible by our ability to peer ever deeper into the nanoworld, and in the early 1990s, high-powered microscopes were used to detect and identify fossils in rocks that were 3.5 billion years old. Apparently, life had existed on Earth earlier than anyone had yet imagined. In 1996 a technique was developed to examine rocks for the biosignatures of life. **Biosignatures**—traces of life left behind when a creature dies—were identified using this technique in rocks that were 3.9 billion years old. Possible traces of life have also been suggested in a 4.5 billion-year-old Martian meteorite found in Antarctica (see ◘ Fig. 2.1). Is this

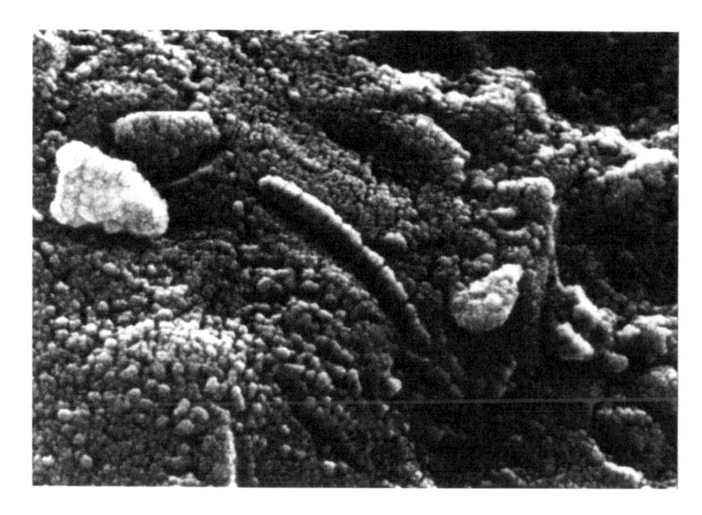

◘ **Fig. 2.1** An close-up view of ALH84001, a meteorite that is presumed to have come from the surface of Mars. In 1996, scientists discovered what appeared to be microfossils inside this meteorite. The meteorite is estimated to be about 4.5 billion years old ▶ https://web.archive.org/web/20051218192636/http://curator.jsc.nasa.gov/antmet/marsmets/alh84001/ALH84001-EM1.htm

evidence clear proof that life on Earth began over 3.5 billion years ago, perhaps seeded by microscopic life from Mars? Even more recently, Australian scientists found signatures of life in the nanoworld of "zircons," 4.4-billion-year-old crystals that contain the chemical elements carbon, oxygen, and zirconium, and are found in the semi-molten magma that flows beneath the Earth's surface. Does this mean that we know when life began? Questions are raised. Experiments are designed and explored. As we discussed in ► Chap. 1, this is the nature of vibrant science.

Biosignature – Chemical compounds, ions, or atoms that remain as evidence that life once existed in a particular location.

Why should we care about the history of life? Time is tight. Why spend it learning about the possible origins of life? If our current understanding about the origins of life is correct, it is based on the interactions of elements in the earliest stages of the Earth's history. Looking at those interactions opens the door to learning more about how primordial biology became our current biology. This may help clarify how diseases, such as the most recent SARS-CoV-2 virus, progress within us and our larger environment, and how our modern lifestyles impact the Earth. We never know where the answers may lead; this is why we raise the questions.

How can we find the biosignatures of life? All organisms, whether they are multicellular or single-celled, leave traces where they have lived. In some cases, these traces are fossils. In other cases, particularly the more ancient ones, the biosignature is simply a slight excess of one kind of a type of matter known as the atom. All organisms are made up of characteristic concentrations of specific types of atoms. By measuring larger-than-normal concentrations of these atoms, scientists can claim to have identified a biosignature of life. The biosignature that led to the discovery of life in the 3.9-billion-year-old rock is a larger-than-normal concentration of carbon-12.

What is carbon-12? Carbon-12 is a type of atom of the element carbon that contains a distinctive set of smaller particles. Although several different sets of particles can form a carbon atom, only one of these produces the carbon-12 atom. In this chapter, we'll learn about the building blocks of matter called atoms. We'll uncover the basic structure of an atom and learn how atoms are arranged in the most important organizational chart used in chemistry: the periodic table of the elements.

2.2 Early Attempts to Explain Matter

Surprisingly, it was thousands of years ago that philosophers first developed a view that the world was made of a few basic "elements," though initially these were limited to earth, air, water, and fire. The Greek philosopher Democritus (◘ Fig. 2.2) wondered how he could be sitting in one part of his house and detect that bread was baking

◘ **Fig. 2.2** An engraving of the bust of the Greek philosopher Democritus (460 B.C.–370 B.C.). His early deductions about the composition of the basic components of nature led him to believe in unseen and uncuttable particles, called *atomos* . *Source/Photographer* ► http://www.phil-fak.uni-duesseldorf.de/philo/galerie/antike/demokrit.html

elsewhere in the house. Could it be that small particles of the bread were breaking away from the loaf and traveling through the air and into his nose? Noticing that wet clothing gradually got drier and lighter led to the explanation that small, invisible pieces of water were gradually leaving the clothing. What was the smallest possible piece of this matter? In Democritus' culture, these pieces of matter were thought to be indivisible or unable to be broken down further. In their language, the particles of the basic elements were *atomos*. This is the origin of the modern word atom.

The lack of certain instruments that we routinely use today made it impossible, in Democritus' time, to perform the experiments that could either support or contradict Democritus' theories. For instance, the invention of the balance for measuring mass was one of the most significant experimental advances in the history of science. In fact, two chemical laws upon which we base so much science were discovered with the use of the balance. These laws helped to shape our current understanding of the insightful ideas put forth in Democritus' era.

2.2.1 The Law of Conservation of Mass

One chemist who faithfully followed the scientific method was Antoine Lavoisier (■ Fig. 2.3). Born to a relatively wealthy family of lawyers in France, he was destined to go to college to carry on the family tradition. At the age of 20, he graduated with a law degree, but became interested in science and measurements after attending some public lectures on chemistry. He corresponded with scientists in other countries, repeating and expanding upon their experiments. Lavoisier's careful measurements and use of the scientific method helped him arrive at better explanations of the behavior of many chemical systems. The culmination of his work was his *Traité Élémentaire de Chimie* (Elementary Treatise of Chemistry), published in 1789. Although his work was not as appreciated as it should have been during his life, we now recognize just how valuable it was in laying out the groundwork for modern chemistry.

Unfortunately, Lavoisier was born during tumultuous times in France. Due to his relative wealth, he was able to invest in a private tax-collecting company, the Ferme Générale. The execution of King Louis XVI in January 1793 following the French Revolution brought about an "Age of Terror" in France. Shortly thereafter, Lavoisier and the rest of the *Ferme Générale* shareholders were arrested and tried for treason. Due in part to Lavoisier's communication with foreign scientists, including Benjamin Franklin, he was found guilty of treason and guillotined in May 1794.

The scientific work of Lavoisier was not forgotten. One of his greatest accomplishments was showing that the mass of the **reactants** (what is consumed) in a chemical process was equal to the mass of the **products** (what is produced). In other words, he provided the first reliable experimental evidence that, matter is neither created nor destroyed in a chemical reaction. We call this tested observation the **Law of Conservation of Mass**.

Reactants – The materials that combine in a chemical reaction.

Products – The materials that are produced in a chemical reaction.

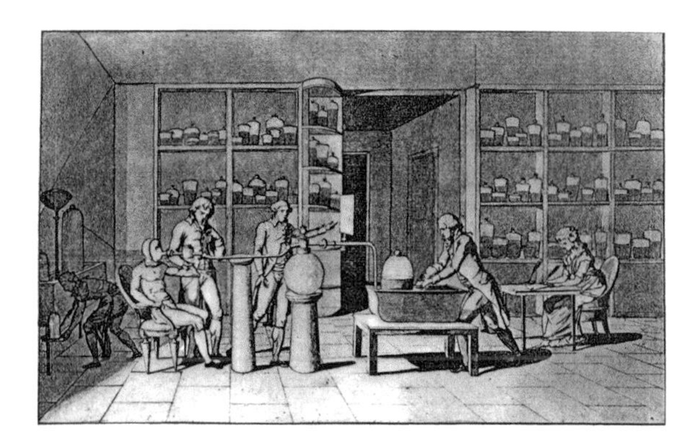

■ **Fig. 2.3** A drawing of Antoine Lavoisier's (1743–1794) famous experiment on air and human respiration. His wife, Marie-Anne Lavoisier (1758–1836) contributed to much of his work. She can be seen on the right-hand side of the illustration recording the observations that were made. Retreived from ▶ http://mattson.creighton.edu/History_Gas_Chemistry/Lavoisier.html. Larger copy from ▶ http://jn.nutrition.org/content/133/3/638/F1.large.jpg

Law of Conservation of Mass – Scientific law stating that the mass of chemicals present at the start of a chemical reaction must equal the mass of chemicals present at the end of the reaction. Although not strictly true, this law is correct at the level of accuracy of all laboratory balances.

Strictly speaking, this law is not precisely true because of the interconversion of minute amounts of mass and large amounts of energy, Albert Einstein's "mass-energy equivalence" that is the basis of the nuclear energy that has such an impact on our world. But for most reactions that we work with in the laboratory and in our everyday life, the differences in mass are so tiny that we say comfortably say that the Law of Conservation of Mass is accurate. It provides a fundamental truth about chemical reactions: The atoms with which you start are the same as the ones with which you end. When atoms participate in chemical reactions, they are not destroyed or replaced by newly made atoms but, rather, are rearranged. For example, a mixture of atoms of carbon, hydrogen, oxygen, nitrogen, and phosphorus could be arranged into long strands of DNA, the genetic code in our cells. Arranged in other ways, they are a mixture of water, carbon dioxide gas, and a fertilizer called ammonium phosphate. Arranged in still other ways, they form parts of biosignatures from billions of years ago.

2.2.2 The Law of Definite Composition

In addition to the observations and measurements made by Lavoisier, other scientists in the 1700s and 1800s were also experimenting systematically with chemical reactions. Joseph Louis Proust (◘ Fig. 2.4) was a contemporary of Lavoisier. The son of a pharmacist, he was expected to follow in his father's footsteps. For a while, he did apprentice with his father, later moving to Paris to apprentice for a renowned pharmacist. After studying there, he obtained a job at the La Salpétrière hospital and, in his spare time, experimented with chemistry, and even taught a few classes at the Royal Palace. In 1784, in front of the court of France and the King of Sweden, he ascended in a hot-air balloon, which was a notable accomplishment at the time.

This notoriety led to a position teaching chemistry in Spain in 1789. He remained there for nearly 20 years, luckily missing the violence of the French Revolution that did not spare Lavoisier. However, in 1808, Napoleon invaded Spain and Proust returned to France. With failing health and little means to make money in France, he spent most of his time experimenting with chemistry. In recognition of his accomplishments, he was elected to the Academy of Sciences in 1816, which provided a small annual salary. He returned to his hometown of Angers, studying and experimenting until his death in 1826.

His most notable discovery was what we now know as the **Law of Definite Composition** (sometimes called the Law of Constant Composition). In 1794, while he was in Spain, he published his data on a mineral that contained copper, carbon, oxygen, and hydrogen. His results indicated that every time he determined the composition of this mineral, no matter how it was prepared, where it was unearthed, or how much he started with, the masses of the elements in the mineral were always present in the same ratios. This law is true for all **chemical compounds**, irrespective of their source. We can illustrate the Law of Definite Composition using a compound containing the elements iron and sulfur, called iron(II) sulfide. If we analyze the composition of two different samples of iron(II) sulfide, we find that their mass ratios are identical.

◘ **Fig. 2.4** Joseph Proust (1754–1826) was born in Angers, France. He was a professor at different institutions in Spain. In 1794 he published his understanding of the Law of Definite Proportions. After returning to France in 1808, he spent considerable time investigating various foods and discovered the amino acid leucine, one of the building blocks of proteins

Chemical compound – A combination of elements, chemically bonded together, that has a specific identity and properties.

Law of Definite Composition – Scientific law stating that any particular chemical is always composed of its components in a fixed ratio, by mass.

Result from Analysis of Two Samples of Iron(II) Sulfide
Sample A: Compound mass = 42.0 g; Iron mass = 26.7 g; Sulfur mass = 15.3 g
Sample B: Compound mass = 100.0 g; Iron mass = 63.5 g; Sulfur mass = 36.5 g.
Mass Ratios
Sample A: 26.7 g Fe/15.3 g S = 1.74:1.00 = ratio of g Fe to g S
Sample B: 63.5 g Fe/36.5 g S = 1.74:1.00 = ratio of g Fe to g S.

We can also think of this law in the reverse fashion. Suppose we were interested in making 80.0 g of water from its elements, hydrogen and oxygen. The mass ratio of oxygen to hydrogen in water is 7.99:1.00, so we would need to chemically combine 1.00 g of hydrogen for every 7.99 g of oxygen to make the water. To make 80.0 g of water, then, we would need 8.90 g of hydrogen and 71.1 g of oxygen. Water is water; it is never made up of a different mass ratio of its components.

Preparation of Water (H_2O).
The combination of 8.90 g hydrogen and 71.1 g oxygen makes 80.0 g water.
Mass Ratio
Water: 71.1 g oxygen/8.90 g hydrogen = 7.99:1.00 g oxygen to g hydrogen.

Example 2.1—Investigating the Law of Definite Composition

Assume that you have collected two samples that you believe are pure bottled water.

a. If you separated the water into hydrogen and oxygen and obtained the following results, what would you report as the ratio of the mass of oxygen to that of hydrogen in each sample?
   Sample 1: 37.3 g of oxygen; 4.67 g of hydrogen
   Sample 2: 69.3 g of oxygen; 8.67 g of hydrogen
b. Do these data support the Law of Definite Composition?
c. If you obtained another sample of water and found it to contain 17.0 g of oxygen, how many grams of hydrogen would you expect to obtain?

Asking the Right Questions

To determine whether these two compounds are water, we need to consider the Law of Definite Composition. In other words, is the ratio of the mass of oxygen to hydrogen the same as in a sample of water (7.99:1.00)?

Solution

a. The ratios of the masses of oxygen to hydrogen in these two samples are
   Sample 1:

$$\frac{37.3 \text{ g oxygen}}{4.67 \text{ g hydrogen}} = \frac{7.99 \text{ g oxygen}}{1.00 \text{ g hydrogen}}$$

Sample 2:

$$\frac{69.3 \text{ g oxygen}}{8.67 \text{ g hydrogen}} = \frac{7.99 \text{ g oxygen}}{1.00 \text{ g hydrogen}}$$

b. Yes, the ratios are the same for different masses of the same compound, so these data support the Law of Definite Composition.
c. Because the ratio of the masses of oxygen to hydrogen is constant in the compound called water, we would expect to have a ratio of 7.99 g of oxygen to 1.00 g of hydrogen. Using words, we say, "7.99 g of oxygen is to 1.00 g of hydrogen as 17.0 g of oxygen is to how many grams of hydrogen?"

Using equations, we can write

$$\frac{7.99 \text{ g oxygen}}{1.00 \text{ g hydrogen}} = \frac{17.0 \text{ g oxygen}}{? \text{ g hydrogen}}$$

Rearranging yields

$$\frac{17.0 \text{ g oxygen} \times 1.00 \text{ g hydrogen}}{7.99 \text{ g oxygen}} = ? \text{ g hydrogen}$$

$$= 2.13 \text{ g hydrogen}$$

Is Our Answer Reasonable?

Does it make sense that our answer is 2.13 g hydrogen? Because approximately 8 g oxygen combines with 1 g of hydrogen, we would assume that 17 g oxygen (just more than twice the amount in the first example) should combine with just over 2 g hydrogen. Yes, our answer does seem reasonable.

What If…?

…we had a sample that had a ratio of 0.751 g of carbon to 1.00 g of oxygen? Using the Law of Definite Composition, how many grams of carbon would be combined with 59.3 g of oxygen? (*Ans*: 44.5 g *carbon*)

Practice 2.1

It is known that many sugars contain carbon, hydrogen and oxygen in the following ratios:

5.96 g carbon/1.00 g hydrogen
1.33 g oxygen/1.00 g carbon
7.94 g oxygen/1.00 g hydrogen.

A particular compound was analyzed and determined to have the following masses of each element:

300.275 g carbon
42.3360 g hydrogen
95.9964 g oxygen.

Is this compound a sugar? Why or why not?

The Law of Conservation of Mass and the Law of Definite Composition provide two fundamental descriptions of the behavior of chemical processes. When they were discovered, there was no satisfactory explanation for them. After all, recall that laws only describe observations of the natural world; they don't explain *why* they happen. Many scientists searched for explanations of these two laws. One of these was the English scientist John Dalton (◘ Fig. 2.5).

◘ **Fig. 2.5** John Dalton (1766–1844), a poor Quaker schoolteacher and brilliant amateur meteorologist, developed the basic points of the atomic theory. He showed early signs of brilliance and even began teaching others in his small English hometown when he was only twelve years old. *Source/Photographer* This image is available from the United States Library of Congress's Prints and Photographs division under the digital ID cph.3b12511

2.3 Dalton's Atomic Theory and Beyond

John Dalton was not the typical scientific mover and shaker of his day. While many of his contemporaries that advanced the fundamental science of chemistry were born to wealthy families with status and free time to conduct experiments, John Dalton was born in England to a working-class Quaker family. Luckily, Dalton received an excellent education—even opening his own village school to teach students of all ages when he was quite young. He excelled at mathematics and in 1781, at the age of 15, he and his brother opened a larger school that taught classes in Greek, Latin, French, and mathematics. In 1793, Dalton was appointed as a professor of mathematics and natural philosophy at New College in Manchester, England. He resigned his post in 1799 and spent the rest of his career in Manchester as a private tutor. Such a position at the time was very lucrative.

In his spare time, Dalton researched and studied various fields of science, including meteorology and chemistry. In 1794, he published research on color-blindness (an affliction he self-diagnosed and, when related to the inability to tell subtle shades of red and green apart, was long after known as Daltonism). In his later years, he was recognized by being awarded honorary degrees from Oxford and Edinburgh Universities.

His most notable accomplishments in chemistry came from his study of compounds in which two elements made more than one type of compound. This is not unusual. There are many examples of compounds that contain the same elements but in different ratios. For example, we can combine oxygen and nitrogen to make many different compounds, three of which are shown in ◘ Table 2.1. We note how each has a different mass of oxygen that combines with a given mass of nitrogen.

The results demonstrate a general law formulated by Dalton: When the same elements can produce more than one compound, the ratio of the masses of this element that combine with a fixed mass of another element is a small whole number. This is known as the **Law of Multiple Proportions**. Using the information in ◘ Table 2.1 we see that in nitric oxide, 14.0 g of nitrogen combines with 16.0 g of oxygen. A different compound of these same two elements exists in which 14.0 g of nitrogen combines with 32.0 g of oxygen. The ratio of the masses of oxygen that combine with the same mass of nitrogen (14.0 g, in this case), is 32.0 g/16.0 g, or 2:1, a small whole number. The same ratio in nitric anhydride is 40.0 g/16.0 g, or 2.5:1, or 5:2 in small whole numbers.

Law of Multiple Proportions – Scientific law stating that when the same elements can produce more than one compound, the ratio of the masses of the element that combine with a fixed mass of another element corresponds to a small whole number.

What door to understanding the structure of compounds is opened by the Law of Multiple Proportions? Let's examine ◘ Fig. 2.6. Dalton wondered whether his results could be explained by the existence, for each element, of some basic particle with a specific mass that would be the smallest part of that element. It seemed that Democritus was likely correct about the existence of atoms. Moreover, his experiments were giving results that were

◘ **Table 2.1** Comparing oxygen to nitrogen mass ratios

Common name	Fixed mass of nitrogen (g)	Mass of oxygen (g)	Ratio of mass of oxygen in Compound to mass of oxygen in nitric oxide
Nitric oxide	14	16	1:1
Nitrogen dioxide	14	32	2:1
Nitric anhydride	14	40	5:2

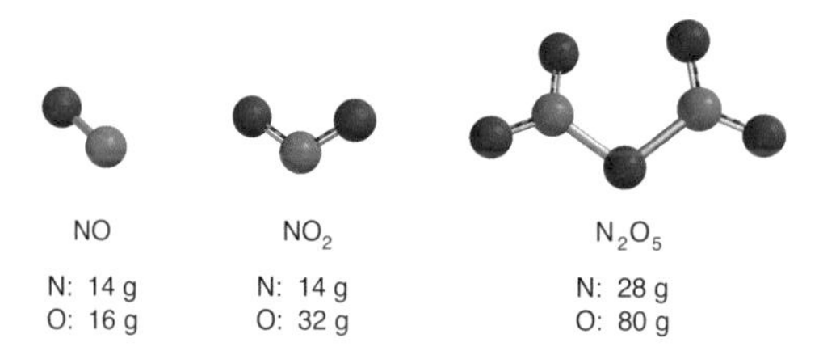

◘ **Fig. 2.6** The ratio of the mass of oxygen in N_2O_5 (nitric anhydride), and NO_2 (nitrogen dioxide) to its mass in NO (nitric oxide) is a simple whole number

Table 2.2 Dalton's atomic theory

Every substance is made of atoms
Atoms are indestructible and indivisible
Atoms of any one element are identical
Atoms of different elements differ in their masses
Chemical changes involve rearranging the attachments between atoms

entirely consistent with this idea. In 1803 he presented the results of his experiments before the Manchester Literary and Philosophical Society. His ideas, which came to be known as **Dalton's atomic theory**, are summarized in Table 2.2.

Dalton's atomic theory – The theory, developed by John Dalton, that all substances are composed of indivisible atoms.

Dalton's atomic theory was a keystone in the foundation of chemistry, but as we shall see, not every detail of this theory is correct. However, it does teach us something very important about the scientific method: Ideas do not have to be completely correct to be useful and influential. Dalton was limited by the measurements he was able to make at the turn of the nineteenth century. Our ability to easily know the mass of objects is orders of magnitude better now than in Dalton's time, and our understanding is therefore significantly better. Still, Dalton's atomic theory was a great step forward in chemistry, because it led us to the scientifically valid idea that compounds are made from little bits—atoms—combined in fixed proportions and that chemical reactions are rearrangements of these atoms.

2.3.1 The Law of Combining Volumes

One example of the refinement of ideas in the light of new information is provided by the work of the French chemist Joseph Louis Gay-Lussac (1778–1850; see Fig. 2.7). Gay-Lussac, the son of the mayor of a small town in France, was just a small child when the French Revolution began. As such, his parents kept him home instead of allowing him to go to Paris, which he did at age 16, to study for the university entrance exams. While there, he boarded with a family for whom he helped find food, which was scarce. By 1798, he was a student in the École Polytechnique (Polytechnic School) and took classes from the chemists Berthollet (the inventor of bleach) and Fourcroy (a famous lecturer of chemistry at the time).

While there, he worked with a friend, the French physicist Jean-Baptiste Biot, to construct a hot-air balloon, and rode it to a height of 7000 m (much higher than Joseph Proust had gone) in order to study the composition of the atmosphere. In 1809, he was appointed professor of chemistry at École Polytechnic. One of his most notable contributions to chemistry came about by measuring changes in volume when gases reacted to produce other gases.

Fig. 2.7 Joseph Louis Gay-Lussac (1778–1850) was the eldest of five children. His colleagues considered him a careful, elegant experimentalist. He introduced the terms "pipette" and "burette" and was the first to isolate the element boron. He is most noted for publishing Charles's law (see ▶ Chap. 10) and formulating the Law of Combining Volumes

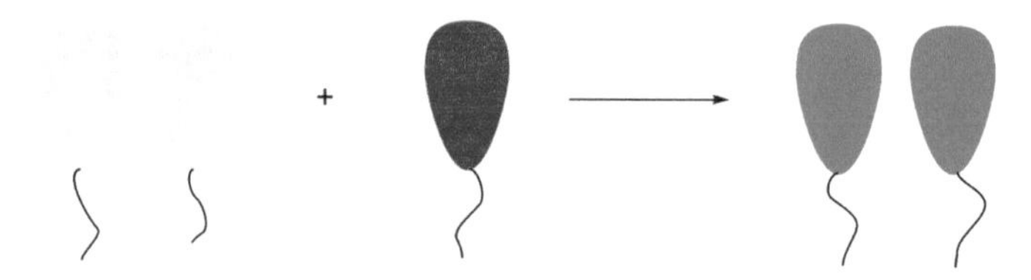

Fig. 2.8 According to the Law of Combining Volumes, when hydrogen reacts with oxygen to make water vapor, it does so such that two volumes of hydrogen react with one volume of oxygen to make two volumes of water vapor

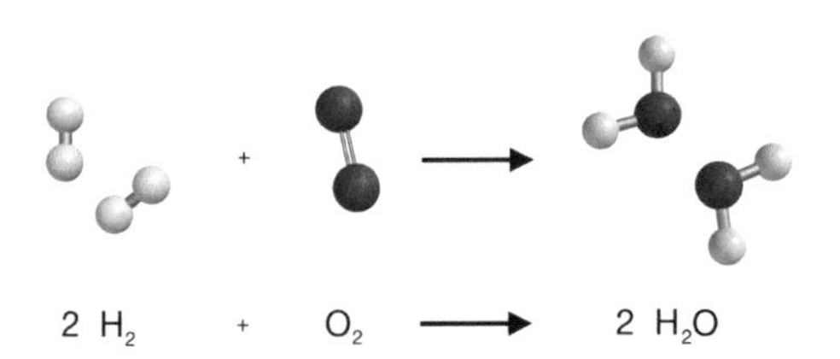

Fig. 2.9 It is because the elemental forms of hydrogen and oxygen exist as diatomic substances that the Law of Combining Volumes makes sense at the molecular level

His results are summarized in what became known as the **Law of Combining Volumes**, which states that when gases combine, they do so in small whole-number ratios, such as 1:1, 1:2, 1:3, or 3:2, provided that all the gases are at the same temperature and pressure. One example is illustrated in Fig. 2.8.

Law of Combining Volumes – Scientific law stating that when gases combine, they do so in small whole-number ratios, provided that all the gases are at the same temperature and pressure.

Dalton could not accept the results in that example because it contradicted his understanding of what would have to happen at the level of individual particles. Why did the reaction described in Fig. 2.8 cause concern to scientists like Dalton? To fit with Gay-Lussac's law, it seemed to Dalton that one oxygen atom would have to split in two in order to react with two hydrogen atoms and produce two particles of water. Dalton's atomic theory said that atoms could not be split in two, so he presumed that either Gay-Lussac's results or his reasoning must be flawed. This led to a refinement in our understanding of atoms.

Scientists theorized that perhaps the hydrogen and oxygen gases were not composed of simple indivisible atoms, but instead might be composed of two atoms stuck together. Since the **molecules** were not indivisible particles, they could be broken apart and the resulting atoms could combine with other atoms to form new molecules! This was a huge leap forward toward the modern understanding of how matter is constructed.

Molecule – A unit of matter consisting of atoms bonded together.

The apparent conflict with Dalton's atomic theory disappears if we assume that oxygen gas consists of oxygen molecules each composed of two oxygen atoms stuck together (O_2). If this is true, then hydrogen gas must also be composed of H_2 molecules and each water molecule must contain two hydrogen atoms and one oxygen atom (H_2O). One molecule of oxygen can combine with two molecules of hydrogen to create two molecules of water as shown in Fig. 2.9. Note in the figure that we can use symbols or figures to depict chemical reactions.

As our ability to observe and measure grew throughout the nineteenth and twentieth centuries, we were able to confirm these ideas about oxygen and hydrogen and build upon the ideas of Dalton, Gay-Lussac, and others. We now know that molecules can be composed of many atoms stuck together. This further illustrates how scientific knowledge is refined through experimental work and theoretical development.

▶ Here's What We Know So Far

The information gleaned by applying the scientific method to investigate the makeup of matter has provided us with a basic look at the nature of chemistry. Specifically, we know that:

- Scientific laws and theories develop over time as more information is discovered
- Matter can be neither created nor destroyed in a chemical reaction.
- Matter is composed of small indestructible particles called atoms.

- Compounds are made by combining atoms in whole-number ratios.
- The components of a compound are always present in the same ratio, by mass.
- Atoms can be combined into larger units called molecules.
- A chemical reaction rearranges atoms and molecules to form new arrangements. ◀

2.4 The Structure of the Atom

We began the chapter by discussing the history of life on our planet and how early experiments allowed us to make assertions about the nanoworld by experimenting with gases in the 19th-century macroworld. This link between the nano and macroworld continues to open doors to our understanding of atomic structure.

Archaeology, the study of past human activities provides a valuable example. This study often requires scientists to be able to assign ages to artifacts that have been unearthed. In some cases, such relics can be dated because they are buried underneath objects for which an age is known. In other cases, the archaeologists must perform laboratory experiments to estimate the age of the artifact. One discovery that helped provide a wealth of information about ancient boat construction was found buried under 6 feet of earth in Dover, England. The "Dover Boat," as it has become known, was discovered underneath the footings of an ancient city wall (◻ Fig. 2.10).

Archaeologists knew from historical records the date that the wall was constructed and, judging from its location, surmised that the boat was probably older than the wall under which it was buried. To get an accurate age of the boat, though, they needed to perform "radiocarbon dating" on small pieces of wood from the boat. What is the basis of radiocarbon dating, and what can it teach us about atomic structure?

Some of the carbon atoms in living things emit certain particles and energy (**radioactivity**) that allows the atoms to become more energetically stable. Scientists have long known that the amount of radioactive carbon

◻ **Fig. 2.10** The hull of the Bronze-Age Dover Boat is partially excavated in this figure. Note the construction of the rails running down the center of the boat. These rails were used to fasten the two halves of the boat together. Original photo by mari from Tokyo, Japan.

remains relatively constant while an organism is alive, as life's processes allow the exchange of atoms between it and its environment. For example, when a tree is alive, it is constantly exchanging molecules with the atmosphere through the actions of its leaves. (The same is true with us as we breathe, run, eat, digest, and read a chemistry textbook.) Since there is a very small, but constant, level of radioactive carbon atoms in the atmosphere, there is also a constant level of radioactive carbon atoms in the living tree. Once the tree dies, however, the radioactive carbon begins to diminish at a constant rate (speed) because no more is absorbed from the atmosphere by the tree. If we can measure the amount of radioactive carbon remaining in a piece of wood, we can estimate the year in which the tree died. The results of **radiocarbon dating** on the Dover Boat indicated that it was constructed around 1550 B.C.

Radiocarbon dating – Using measurements of the radioactivity levels of carbon-14 to determine the approximate date that an organism died.

The previous discussion implies that some carbon atoms (those that are radioactive) are different from other carbon atoms (those that are not radioactive). We see that Dalton's theory that all atoms of any one element are identical is not an adequate description. How are radioactive carbon atoms different from non-radioactive carbon atoms? What is it about the structure of the atom that gives rise to more than one type of the same element? We must head once again into the nanoworld-based structure of the atom to answer this question.

2.4.1 Electrons, Protons, and Neutrons

In the late 1800s and early 1900s, scientists were able to make measurements that changed our fundamental understanding of the structure of the atom. In the 1880s, Svante Arrhenius, a Swedish researcher working at the Academy of Sciences in Stockholm, found that some solutions contained large numbers of "electrically charged atoms." He found, for example, that when a solid substance called copper chloride was mixed into water and the positive and negative terminal of a power supply were immersed into the resulting solution, copper collected at one electrode and chlorine collected at the other. He concluded that some atoms were themselves negatively charged and others positively charged. This work provided some of the earliest evidence for the existence of what we now call **ions**—atoms or molecules that have either a positive or negative electrical charge. What caused these charges?

Ion – An atom or collection of atoms that has either a positive or negative charge.

In 1891, the English scientist G. Johnstone Stoney examined the physical properties of electricity. To represent a distinct unit of electricity, he proposed the name **electron**, from the Greek word elektron (meaning "amber"), because rubbing a rod of amber on wool gives rise to what we now call static electricity. Were electrons small, charged pieces of atoms? Did these electrons cause some ions to be negatively charged? If so, then atoms could no longer be thought of as small, indestructible or indivisible units of matter! In other words, it appeared that there was evidence showing that atoms were composed of even smaller particles.

Electron – One of the subatomic particles of which atoms are composed. Electrons carry a charge of -1.

What does the discovery of a small negatively charged particle tell us about the makeup of the atom? Many scientists noted that the discovery of the electron suggested that Dalton's atomic theory was inadequate. One such

◘ **Fig. 2.11** Joseph John ("J. J.") Thomson (1856–1940) provided the first model of the structure of the atom, known as the plum pudding model. In 1906, "in recognition of the great merits of his theoretical and experimental investigations on the conduction of electricity by gases," he was awarded the Nobel Prize in physics. *Source* GWS—The Great War: The Standard History of the All Europe Conflict (volume four) edited by H. W. Wilson and J. A. Hammerton (Amalgamated Press, London 1915) (So, the photo was taken before 1915). ► First World War.com

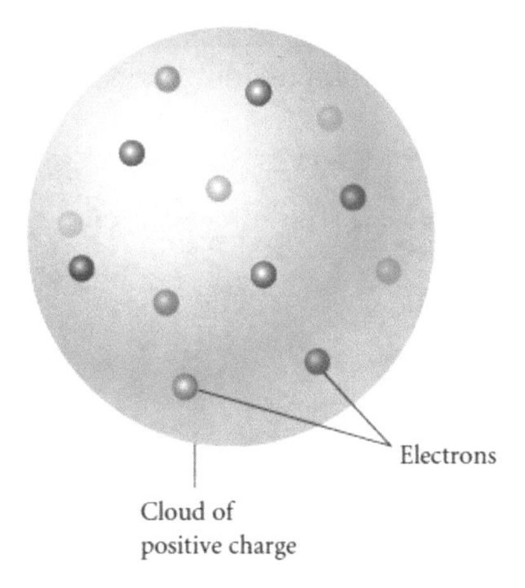

◘ **Fig. 2.12** The plum pudding model of the atom

scientist, Joseph John ("J. J.") Thomson (1856–1940; see ◘ Fig. 2.11), reasoned that if Dalton's ideas about the indestructibility of an atom were true, how could negatively charged parts of atoms (the electrons) be released from the atom? Furthermore, if negative particles were present, wouldn't positively charged pieces of atoms also exist? It makes sense that an atom with negatively charged particles inside must, in order to be electrically neutral overall, also have a balancing positive charge. Thomson envisioned such an atom. In 1904, he wrote:

» the atoms of the elements consist of a number of negatively electrified corpuscles (electrons) enclosed in a sphere of uniform positive electrification.

His model of the atom (shown in ◘ Fig. 2.12) was known as the "**plum pudding model**." In our modern times, we might think of the model as bits of strawberries mixed in yogurt, where the strawberries are the electrons in yogurt representing the positive charge of the atom. This model helped scientists start to understand what an atom might look like.

Plum pudding model – One of the initial models of the atom that incorrectly assigned electrons as particles floating in a cloud of positive charge.

Thomson based his hypothesis about the atom, containing negatively charged particles embedded in a positive material, based on data gathered with an apparatus called a "Crookes Tube", which is also called a "cathode ray tube" (◘ Fig. 2.13). This device produced invisible rays of unknown composition, which could be detected when

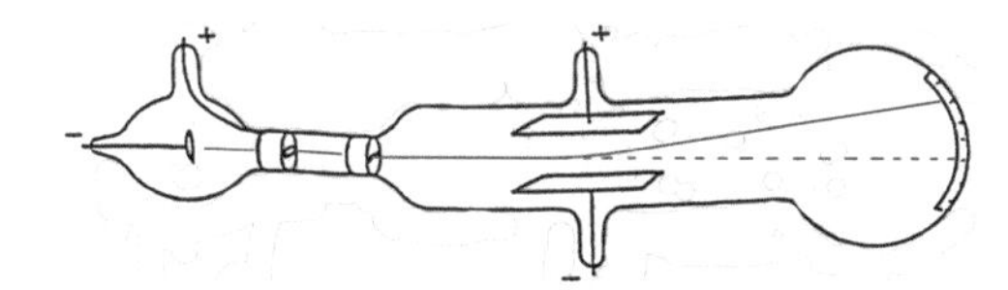

Fig. 2.13 A Crookes Tube (cathode ray tube) similar to the one that J. J. Thomson used for his experiments. The blue line represents the stream of electrons, and the dashed blue line represents the stream of positively charged particles, which were only deflected by a tiny amount. *Source* Philosophical Magazine, 44, 293 (1897)

they hit the material on the sides of the tube (much like old-fashioned "tube" televisions). Thomson used electrical and magnetic fields to deflect these rays and determined that one component of the rays seemed to behave like a negatively charged electron, and the other component was positively charged and much more massive than the electron. He judged their relative masses knowing that a field that deflected the negative component significantly would barely deflect the positive component at all.

2.4.2 Radioactivity and the Structure of the Atom

In 1909 Ernest Rutherford (1871–1937; see Fig. 2.14), a New Zealand–born physicist and former student of J. J. Thomson, and his student Ernest Marsden (1889–1970), wanted to learn more about the structure of the atom. They performed experiments in which positively charged **radiation** was directed toward a thin sheet of gold foil (Fig. 2.15). They reasoned that if Thomson's plum pudding model of the atom were correct, most of the posi-

Fig. 2.14 Ernest Rutherford (1871–1937) was born in New Zealand after his family emigrated there from Scotland in the mid-1800s. As a student of Thomson at Cambridge, he developed an instrument that could detect electromagnetic radiation. Later, while he worked at McGill University (Montréal, Canada) and at the University of Manchester (England), he investigated the actions of alpha particles. For his discoveries, he was awarded the Nobel Prize in chemistry in 1908. **Author** George Grantham Bain Collection (Library of Congress). **Permission** PD-US "No known restrictions on publication"

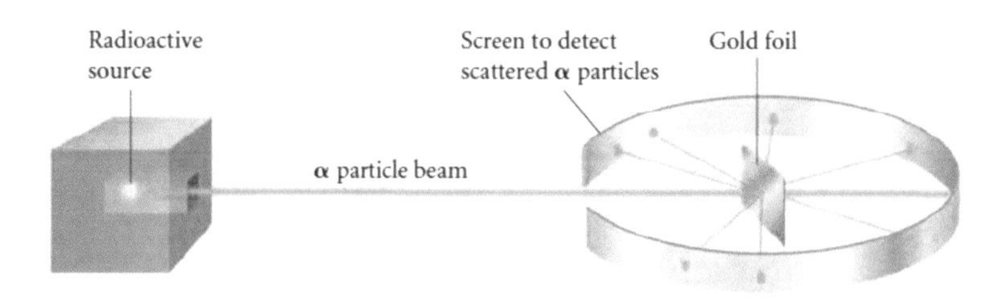

Fig. 2.15 The Rutherford gold foil experiment. Positively charged radiation (α particles) is directed toward a small piece of gold foil. Although most of the radiation passes through the foil, a noticeable percentage is scattered backward. This led Rutherford to develop a new model of the atom

tively charged radiation would be expected to pass through the foil and undergo only slight deflections as the positive charge interacted with the negatively charged electrons scattered throughout the atoms. Instead, most of the radiation went straight through the foil without deflection, and some of the radiation was scattered at very wide angles. Most amazingly—some radiation rebounded nearly straight back toward its source.

In 1911, Rutherford explained this startling result. His explanation led to a new model of the atom in which a small region of very concentrated positive charge called the **nucleus** existed within a large area of mostly empty space containing negatively charged electrons (see Fig. 2.16).

Radiation – The particles and/or energy emitted by some chemical, physical or nuclear process.

Nucleus – The center of an atom, which is very tiny compared with the overall size of the atom.

Rutherford used the term **proton** (from the Greek *protos*, meaning "first") to describe the nucleus of the hydrogen atom, and it was theorized that these protons were in the nucleus, or center, of every atom. However, his calculations of the masses of various atoms didn't agree with the experimentally determined mass. He theorized that some other kind of particle must also be in the nucleus, and he suspected that it was probably electrically neutral because the positive electric charge on the nucleus already balanced the negative charge of the electrons.

Proton – One of the subatomic particles of which atoms are composed, carrying an electrical charge of +1 and found in the nucleus.

In 1932 James Chadwick (1891–1974), an English scientist, directed electrically neutral radiation from beryllium atoms toward a block of paraffin wax. This resulted in the loss of protons from the wax. After calculating the force that would be needed to cause the ejection of the nuclei from the sample of paraffin, Chadwick realized that the radiation from the beryllium must have a mass approximately equal to that of the proton. Chadwick had discovered the **neutron**. The diagram in Fig. 2.17, reproduced from the original article in which he published his results, illustrates the process that Chadwick used.

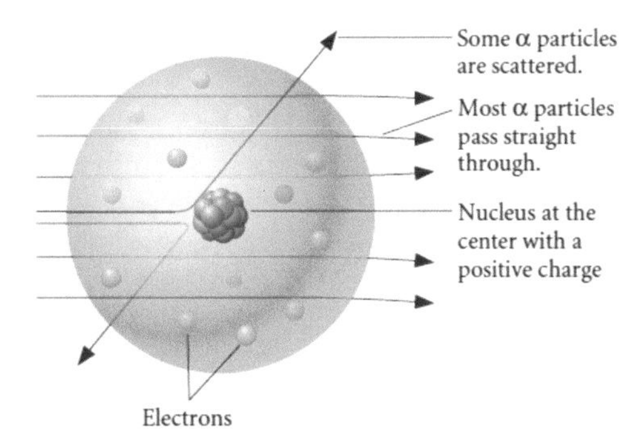

Fig. 2.16 Some of the positively charged radiation (α particles) interacts with the positive nucleus of the atom

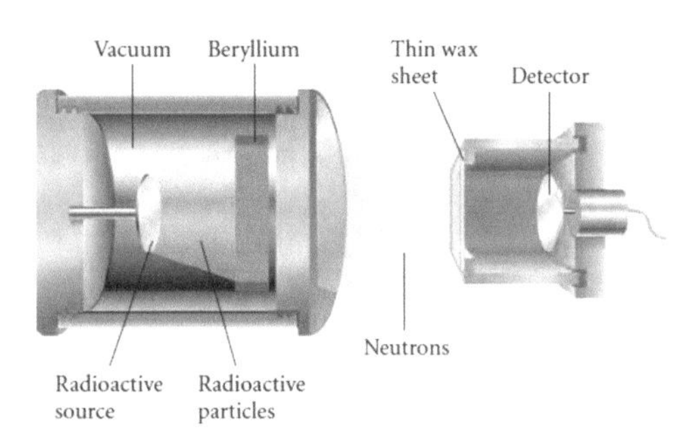

Fig. 2.17 Chadwick's experiment directed neutral radiation at a block of paraffin wax. The energy of the particles that resulted from the collision of the radiation and the wax led to the discovery of the neutron. The radiation was generated by placing radioactive polonium near a sheet of beryllium metal. The wax, shaped into a thin sheet, was placed in front of the detector

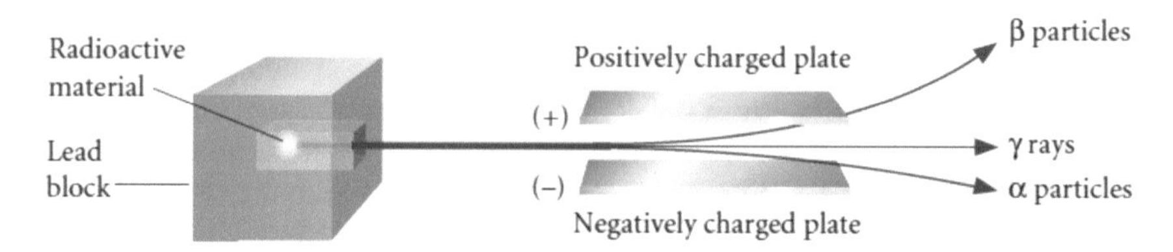

Fig. 2.18 Radiation emitted from a radioactive element placed in a lead block. Alpha particles and beta particles are deflected by an electric field. Alpha particles are positively charged and therefore are pulled toward a negatively charged plate. Beta particles are negatively charged and hence are pulled toward a positively charged plate. Gamma rays are not charged and therefore are not deflected by the electric field

Neutron – One of the subatomic particles of which atoms are composed, carrying no electrical charge and found in the nucleus.

These discoveries helped scientists draw a better picture of the structure of an atom. In addition, their work led to an understanding of the different forms of radiation. In short, some atoms are unstable and can undergo **radioactive decay** by spontaneously emitting high-energy radiation, sometimes accompanied by fast-moving particles. Collectively, this phenomenon is known as **radioactivity** a term we used at the beginning of this section to describe the energy given off by some types of carbon atoms. The three most common types of radioactivity are **gamma rays** (γ rays), **alpha particles** (α particles) and **beta particles** (β particles) (Fig. 2.18).

Radioactive decay – The process by which an unstable nucleus becomes more stable via the emission or absorption of particles and accompanying energy.

Radioactivity – The emission of radioactive particles and/or energy.

Alpha particle – A fast-moving nucleus of helium (two protons and two neutrons) emitted during the decay of a radioactive element.

Beta particle – A fast-moving electron emitted during the decay of a radioactive element. The beta particle originates from the nucleus of the decaying element.

Gamma ray – High-energy electromagnetic radiation emitted from the decay of a radioactive element.

Gamma rays are a very energetic form of electromagnetic radiation, with the same physical nature as visible light but with much higher energy. Alpha particles, the radiation used in Rutherford's experiments, are the same thing as the nucleus of a helium atom. They are composed of two protons and two neutrons and carry a total charge of +2. Beta particles are fast-moving electrons released from the nucleus of an atom, each one carrying its characteristic −1 charge. Through further work on the structure of the atom, it has been discovered that a nuclear electron is formed when a neutron splits apart into a proton and an electron.

The penetrating powers of radioactivity are put to many uses in modern life, including killing cancer cells, testing the integrity of welds in metal, and looking for hairline cracks in aircraft airframes.

▶ Here's What We Know So Far

- Atoms can be broken down into smaller parts. Those parts include the proton, neutron, and electron.
- The proton and neutron occupy the center of the atom, called the nucleus.
- The electrons occupy the space around the nucleus.
- Protons and electrons have opposite charges; protons are positively charged, and electrons are negatively charged.
- The total mass of the atom is the sum of all of the particles that make up the atom.
- Radiation is a result of the decay of atomic nuclei. Different decay products (types of radiation) exist, such as the alpha particle, the beta particle, and the gamma ray.
- Not all atomic nuclei decay naturally. Those that do decay are "radioactive". ◀

Through the work we've just described, along with that done by other scientists of the time, we now know that the nucleus of an atom contains the protons and neutrons. The proton has a tiny positive electrical charge (1.6×10^{-19} **coulombs**) and a similarly minute mass, which in the base SI unit is 1.6726×10^{-27} kg. Neutrons do not carry a charge and are a tiny bit more massive than the protons (1.6749×10^{-27} kg). The electrons, about 2000 times less massive (9.1094×10^{-31} kg) than a neutron or proton, travel within the remaining space taken up by the atom, as shown in

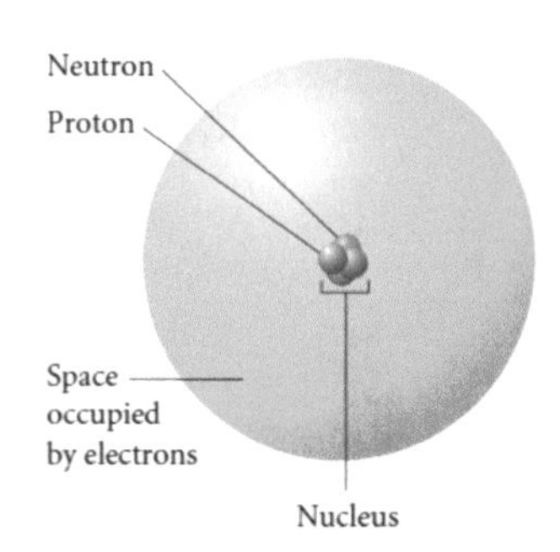

Fig. 2.19 The atom is mostly empty space. The nucleus constitutes about 1/10,000 the diameter of the atom. This drawing is not to scale. In reality, if the nucleus were as large as it is drawn on this page, the orange electron cloud would have a diameter of about 50 m!

Table 2.3 Subatomic particles

Particle	Mass (kg)	Mass (amu*)	Charge
Electron	9.1094×10^{-31} kg	5.486×10^{-4}	−1
Proton	1.6726×10^{-27} kg	1.0073	+1
Neutron	1.6749×10^{-27} kg	1.0087	0

* The atomic mass unit (amu) is an arbitrary unit used to set the masses reported in the periodic table. We will discuss it in the next section.

Fig. 2.19. The electrons also carry a negative electrical charge, equal in magnitude to, but opposite in sign from, that on the proton. To simplify counting charges, we often report the charge on the proton as +1 and that on the electron as −1 (Table 2.3).

Coulomb – The SI unit of electrical charge.

Although we now know that these subatomic particles are divisible into even smaller units, this is as far as we currently need to travel into the nanoworld to refine our model of the atom. Using this model of electrons, protons and neutrons, we can explain the individual identities of atoms and how they interact, like so many paint colors on an artist's canvas, to yield the molecules that define us and our environment. Like an artist's canvas, the mixing of the different elements might seem haphazard, only to yield a dazzling scene with so much insight, as we learn how to look at it. The picture on our canvas is the Periodic Table of the Elements, and it is the portrait that all chemists use to interpret our chemical world.

2.5 Atoms and Isotopes

The **periodic table of the elements** is a systematic arrangement of all of the known elements in the universe and could well be the single most important practical outcome in the history of chemistry. When we learn how to read it skillfully, we can assess the properties of the elements, as well as make predictions about all kinds of reactions that will occur. Because it is an essential tool for practicing chemists, as well as chemistry students, it is located on the inside front cover of this book for your easy reference. The periodic table lists the atoms of just over 90 elements that are known to occur naturally, as well as a smaller number of "heavy" elements that have been prepared in the laboratory.

Periodic table of the elements – A table presenting all the elements, in the form of horizontal periods and vertical groups; shown on the inside front cover of this book.

One of the great beauties of the periodic table is that all of the atoms of the elements in the table are constructed from the same building blocks: protons and neutrons surrounded by a cloud of electrons. Are the numbers and types of particles important in constructing an atom? Yes. Together, they define each atom and its chemical behavior. The identity of the atom is determined by the number of protons in the nucleus, which is known as the **atomic number (Z)**. For instance, all hydrogen atoms have one proton ($Z=1$), all helium atoms have two protons ($Z=2$), and all iron atoms have twenty-six protons ($Z=26$). A student identification number uniquely identifies you at your school just as the atomic number identifies a specific element in chemistry. For example, potassium has an atomic number of 19; any element with nineteen protons must be potassium.

Atomic number (Z) – The number of protons in the nucleus of an atom.

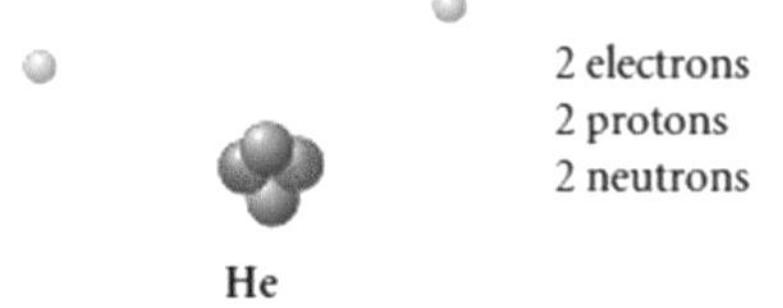

As Thomson correctly theorized, the number of electrons in an electrically neutral atom must be equal to the number of protons and, therefore, to the atomic number. However, when combining with other atoms, an atom can gain or lose electrons to form a charged entity we call an ion. Although the number of electrons that an atom contains can change in a chemical reaction, the atomic number (Z)—the number of protons it contains—always remains the same. For example, the sodium ion shown in ◘ Fig. 2.20 is still sodium, even though its electron count is no longer the same as that of a neutral sodium atom. The key question is: How many protons does the atom contain? The number of protons always identifies the element.

A positive charge indicates that there are fewer electrons than protons, meaning that the atom has lost some electrons. We typically refer to this type of ion as a **cation** (pronounced CAT-ion). Conversely, a negative charge indicates that there are more electrons than protons. This type of ion is known as an **anion** (pronounced AN-ion). If we know the charge and the atomic number of an ion, we can determine the number of electrons on the ion.

Anion – An ion with a negative charge.

Cation – An ion with a positive charge.

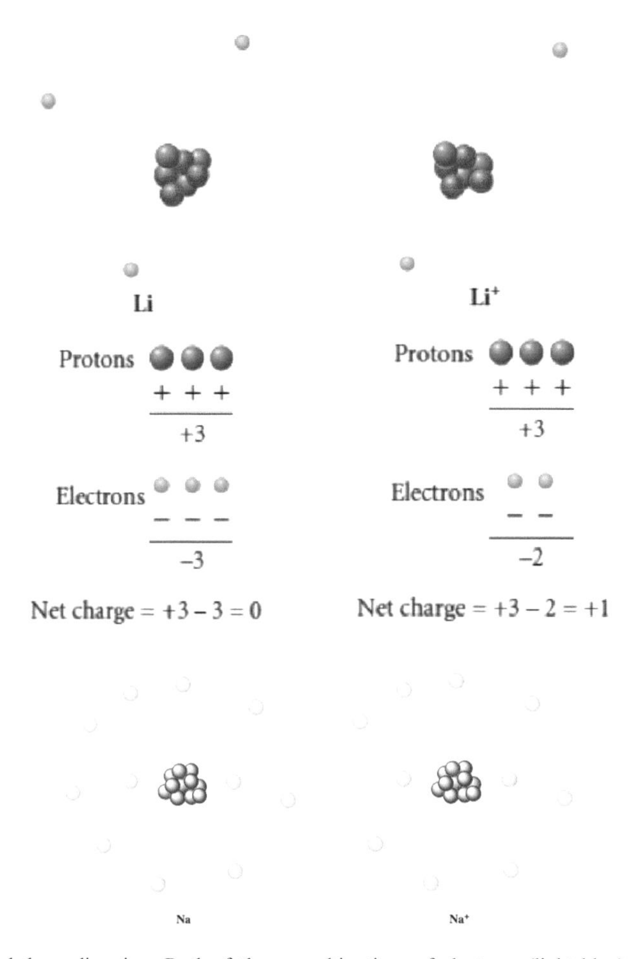

◘ Fig. 2.20 The sodium atom and the sodium ion. Both of these combinations of electrons (light blue), protons (dark blue), and neutrons (red) are sodium, despite the difference in the number of electrons

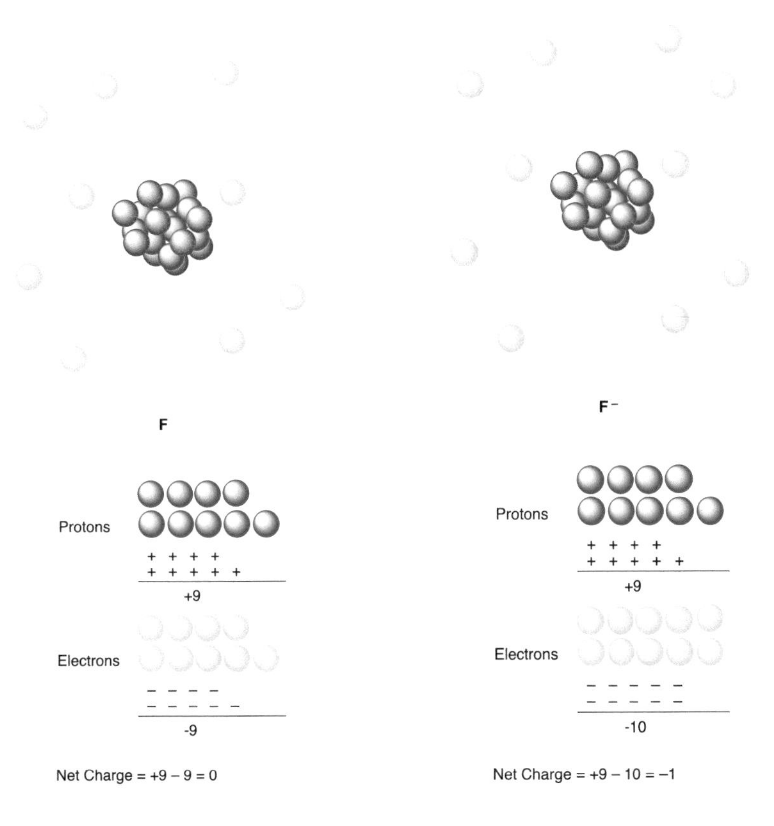

We call the sum of the number of protons and neutrons the ***mass number* (*A*)**. The number of neutrons (n) in a particular atom can be calculated by subtracting the atomic number from the mass number:

$$A - Z = n$$

Mass number (*A*) – The total number of protons and neutrons in the nucleus of an atom.

These basic components of the atoms can be written using ***nuclide notation***, as shown in Fig. 2.21. Nuclide notation lists the symbol for the element, accompanied by the atomic number and the mass number of the atom.

Nuclide notation – The shorthand notation used to represent atoms, listing the symbol for the element, accompanied by the atomic number and the mass number of the atom.

Rather than writing numbers for all of the particles that make up an atom or ion, nuclide notation provides another way to convey the quantity of subatomic particles to other scientists. This is a "chemist's shorthand", a way of expressing an important idea using symbols. We'll come across many examples of the chemist's shorthand symbolizing big ideas in this textbook and in your course.

For example, an alternative way to write the number of particles is to write the name of the element, a dash, and the mass number of the element. When we wrote "carbon-12" in the introduction to this chapter, we were indicating a neutral atom containing 6 protons, 6 neutrons, and 6 electrons.

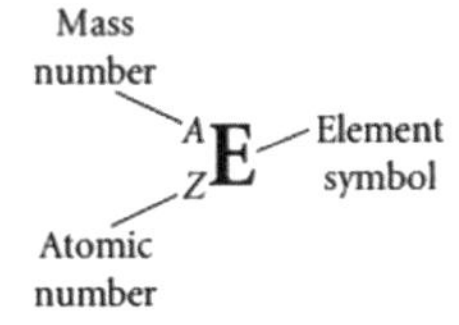

Fig. 2.21 Nuclide notation

2.5.1 Isotopes

We now have some new tools to reexamine radiocarbon dating. From our discussion we discovered that there must be at least two different types of carbon atoms: a radioactive carbon atom and a non-radioactive carbon atom. What is different about these carbon atoms?

All atoms of an element have the same number of protons, but the number of neutrons that make up the nucleus of an atom can differ. Atoms of the same element must have the same atomic number (the same number of protons); otherwise, they would be different elements. However, if they differ in the number of neutrons they contain, they are known as **isotopes**.

Isotopes – Forms of the same element that differ in the number of neutrons within the nucleus.

There are actually three naturally occurring isotopes of carbon. The most abundant (98.93%) of these isotopes is carbon-12 (6 protons and 6 neutrons). Carbon-13 (6 protons and 7 neutrons) is another, nonradioactive isotope of the element carbon; it makes up only a small amount (1.07%) of the carbon atoms in the universe. The carbon-14 isotope (6 protons and 8 neutrons) accounts for only 2×10^{-10} % of all carbon atoms and is radioactive. Carbon-14 is the isotope that can be used to determine the ages of artifacts less than about 50,000 years old.

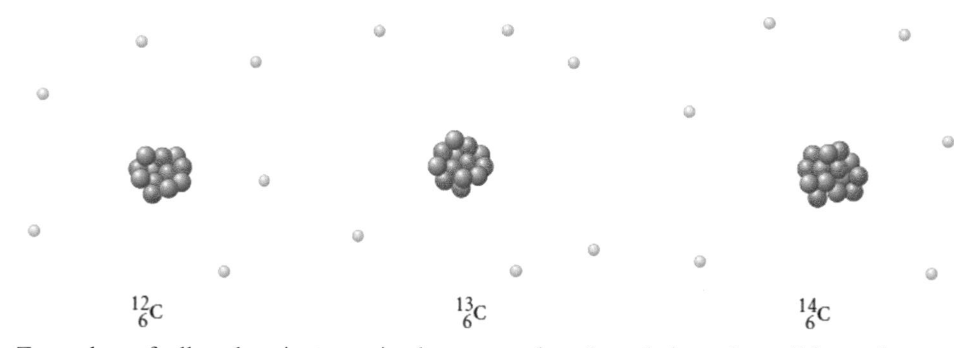

Since the Z number of all carbon isotopes is always equal to 6, and since they all have the same atomic symbol, C, we abbreviate the notation by only writing the mass number (A). Using this example of the chemist's shorthand, carbon-12 can be written as ^{12}C, and carbon-13 as ^{13}C.

Many examples of isotopes exist among the elements of the periodic table. In fact, it is rare for an element *not* to have isotopes. We list some of the isotopes of hydrogen, carbon, and oxygen in ◘ Table 2.4 in nuclide notation. We also include the relative natural abundance of each isotope. For example, 99.985% of all hydrogen atoms are ^{1}H, 0.015% are deuterium (the same single proton as hydrogen, but with one neutron), and only about 10^{-18} % of them are the radioactive isotope known as tritium (one proton and two neutrons).

Many nuclei that have either too many or too few neutrons to be energetically stable tend to give off energy and particles until they become stable. This is the reason for the radioactivity of ^{14}C. We will have much more to say about stability and nuclear particles in ▶ Chap. 20. Feel free to read ahead!

◘ **Table 2.4** Some isotopes of hydrogen, carbon and oxygen

Nuclide notation	Relative natural abundance (%)	Name of isotope	# Protons	# Neutrons	# Electrons in neutral atom
$^{1}_{1}H$	99.985	Hydrogen	1	0	1
$^{2}_{1}H$	0.015	Deuterium	1	1	1
$^{3}_{1}H$	10^{-18}	Tritium	1	2	1
$^{12}_{6}C$	98.93	Carbon-12	6	6	6
$^{13}_{6}C$	1.07	Carbon-13	6	7	6
$^{14}_{6}C$	2×10^{-10}	Carbon-14	6	8	6
$^{16}_{8}O$	99.762	Oxygen-16	8	8	8
$^{17}_{8}O$	0.038	Oxygen-17	8	9	8
$^{18}_{8}O$	0.200	Oxygen-18	8	10	8

Example 2.2—Atomic Bookkeeping

Fill in the blanks for each neutral element in the following table, and write the elemental symbols using nuclide notation.

Element	Mass Number	Protons	Neutrons	Electrons
Silicon	28	14		
Phosphorus	31		16	
Calcium			19	

Asking the Right Questions

What does the mass number indicate about a nucleus? What is true about the number of electrons in a neutral atom?

Solution

To fill in our table, we will use the relationships that define the mass number and atomic number for an element. Specifically, we recall that:

$$A - Z = n$$

A glance at the periodic table will be helpful, because the atomic number of the element indicates the symbol (and name) of the element. For example, the atomic number of silicon is 14, which is equivalent to the number of protons in atoms of the element. We use that formula to determine other things that are missing from the table. The key is to remember the relationships between mass number, protons, neutrons, and electrons.

Element	Mass number	Protons	Neutrons	Electrons	Symbol
Silicon	28	14	14	14	$^{28}_{14}Si$
Phosphorus	31	15	16	15	$^{31}_{15}P$
Calcium	39	20	19	20	$^{39}_{20}Ca$

Is Our Answer Reasonable?

A quick glance at the proton and electron columns in our table shows that they contain the same numbers (which they should, since the elements are electrically neutral). A quick addition of the proton and neutron columns should, and does, equal the numbers in the mass number column. Our answers do look reasonable.

What If...?

...an isotope of an element has exactly one-half the number of neutrons of plutonium-238 and twice the number of protons of iron? Write the nuclide notation for that isotope. (*Ans:* $^{124}_{52}Te$)

Practice 2.2

Indicate the number of protons, neutrons, and electrons in atoms of each of these elements:

$^{14}_{7}N$ $^{20}_{10}Ne$ $^{48}_{22}Ti$ Carbon-11 Lithium-7 Phosphorus-31

2.6 Atomic Mass

Plants and other organisms have a slightly greater preference for the lighter isotopes of the carbon that they use as a raw material. When a plant absorbs carbon dioxide to make cell walls, it more efficiently uses the molecules that contain the ^{12}C isotope than those that contain ^{13}C or ^{14}C. The preference isn't great at all, but the plant absorbs a slightly larger proportion of ^{12}C than of the other carbon isotopes. When the organism dies, it leaves behind some residue that shows this preference. In the millions of years that follow, the parts of the plant get compressed within layers of sand, clay and other sediment that form rock. How does the scientist in search of the origins of life determine the amount of ^{12}C versus the amount of ^{13}C in such a formation?

The distribution of isotopes in a sample of an element can be found using an instrument called a **mass spectrometer**. There are different kinds of mass spectrometers, but the simplest kind is illustrated in Fig. 2.22. In this instrument, a sample is vaporized and then converted into positive ions when a beam of fast-moving electrons knocks some of the electrons out of the atoms of the sample. The ions that are formed are accelerated by their attraction to electrically charged plates. Holes in these plates allow some of the ions to be projected through and into a magnetic field. The interaction between the charges on the ions and the magnetic field causes the ions to be deflected from their original path. The extent of this deflection depends on the masses and charges of the ions. The more massive an ion is, the less deflection it experiences, because of the greater momentum it possesses when heading out in its original straight-line path (much as a very heavy adult would not be blown so far off course by a fierce, gusty wind as a small child would).

Mass spectrometer – An instrument used to measure the masses and abundances of isotopes, molecules, and fragments of molecules.

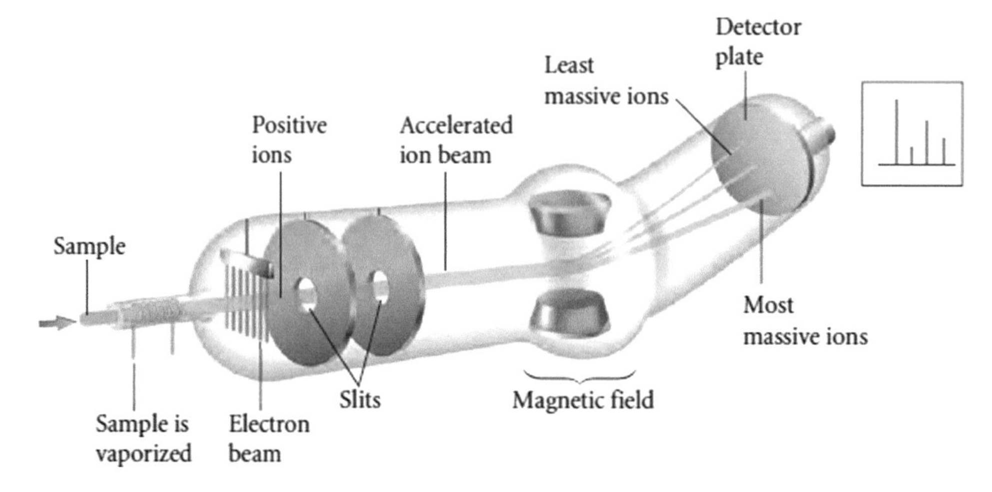

Fig. 2.22 The mass spectrometer separates ions by exploiting their deflection to different extents in a magnetic field

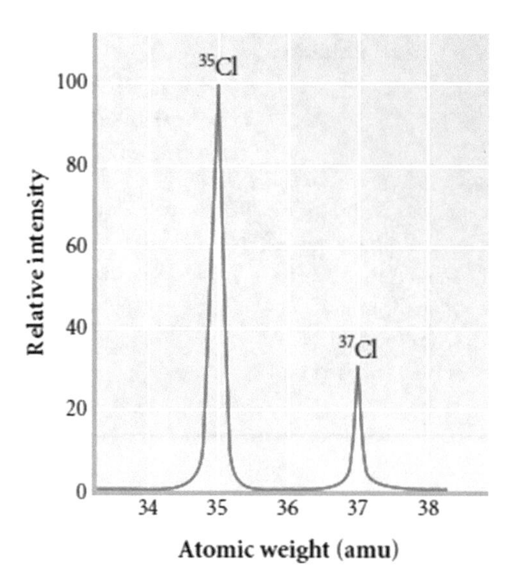

Fig. 2.23 The mass spectrum of chlorine. From this spectrum we can determine the relative abundances of ^{35}Cl (75.77%) and ^{37}Cl (24.23%)

Some atoms have two of their electrons knocked away by the beam that strikes them, converting them into cations with a +2 charge. The ions with multiple charges deflect more than otherwise identical ions with a single positive charge because of the greater interaction between their charge and the magnetic field. Within a very short distance, the original coherent beam of ions spreads out into beams of distinct ions, each beam having a different mass-to-charge ratio.

At the end of the mass spectrometer, the beams of ions are detected and the results are displayed in a **mass spectrum**, as shown in Fig. 2.23. Each peak on the spectrum indicates the relative quantity of the different ion beams that result from the original sample. Using a standard that produces an ion beam with a known mass, we can create the mass scale on the bottom of the spectrum.

Mass spectrum – The output of a mass spectrometer, in the form of a chart listing the abundance and masses of the isotopes, molecules, and fragments of molecules present.

Which element on the periodic table is the standard for all atomic masses? The most abundant isotope of carbon, ^{12}C, is defined as exactly 12 **atomic mass units (amu)**.

Atomic mass unit (amu) – The arbitrary unit of mass for elements based on the isotope carbon-12. One atom of carbon-12 is defined to have a mass of exactly 12 amu.

$$^{12}C = 12 \text{ amu}$$

An atomic mass unit (amu) is an arbitrary unit used to set the masses reported in the periodic table.

$$1 \text{ amu} = 1.66053904 \times 10^{-27} \text{ kg}$$
$$1 \text{ kg} = 6.02214086 \times 10^{26} \text{ amu}$$
$$1 \text{ g} = 6.02214086 \times 10^{23} \text{ amu}$$

Strictly speaking, the amu is not an SI unit, but its use remains sufficiently common among scientists that we will use it in this textbook.

A detailed look at the mass of individual isotopes shows that they are typically not integer values. That is, the isotope ^{63}Cu does not have a mass of exactly 63 amu. Only ^{12}C is defined with an exact mass. The isotope ^{63}Cu has a mass number of 63 but an atomic mass of 62.9296011 amu. The isotope ^{96}Mo has an atomic mass of 95.9046789 amu. Precise measurements have shown that neither a proton (roughly 1.0073 amu) nor a neutron (roughly 1.0087 amu) has a mass of exactly 1 amu. Something seems strange when you add up the exact masses of all of the protons and neutrons for a particular isotope. For example, the isotope ^{238}U, which has an atomic mass of 238.0508 amu, has an atomic mass lower than the sum of the protons and neutrons, which have a total mass of 239.9418 amu.

Mass of 92 individual protons of $^{238}\text{U} = 92 \times 1.0073 \text{ amu} = 92.6716 \text{ amu}$
Mass of 146 individual neutrons of $^{238}\text{U} = 146 \times 1.0087 \text{ amu} = 147.2702 \text{ amu}$
Total mass of individual protons and neutrons of $^{238}\text{U} = 239.9418$ amu.

Where did the rest of the mass go? When the nuclei of elements formed from the individual neutrons and protons, some of the mass—a different amount for each isotope of each element—was released as energy. That's Einstein's mass-energy equivalence we mentioned earlier.

Unfortunately, the individual masses don't translate accurately into masses that we can measure for a macroscopic quantity of an element found on a laboratory shelf. The presence of isotopes complicates the determination of mass even more. To illustrate this, let's work with a bottle of carbon powder. What mass should we use for the typical carbon atom found in this bottle, given that three different isotopes of carbon exist: ^{12}C, ^{13}C, and ^{14}C? Fortunately, in nature, the relative abundance of each isotope remains largely constant. In addition, the relative natural abundances of each of the isotopes of the elements have been determined by scientists using mass spectrometers. Knowing the percentages of each of the isotopes (and their masses), scientists have calculated the weighted average mass of each element, and that is the atomic mass listed for it in the periodic table. For our bottle of carbon powder, we would use 12.011 amu as the average atomic mass.

What is a weighted average? To better understand the concept of weighted average, we can draw an analogy to a student's grade point average (GPA). Let's look at a student who completed 10 credits of coursework in a semester. If 5 credits (50% of the total) were "A" grades (4.0, on a scale of 0.0 to 4.0), 3 credits (30% of the total) were "B" (3.0), and 2 credits (20% of the total) were "C" (2.0), then the overall GPA would be the weighted average of the individual course grades. The A grade has the greatest weight because fully half the credits were in that course. The C grade would (thankfully) have the least weight, because that course represented only 20%, or 0.20, of the whole course load. The weighted average (here, the GPA) would be:

	(4.0 × 0.50)	**+**	**(3.0 × 0.30)**	**+**	**(2.0 × 0.20)**	**=**	**3.3**
Contribution of:	A Grades		B Grades		C Grades		GPA

Does the answer make sense? Even though the student received one grade each of A, B, and C, the A was present in much greater abundance—5 credits—than the B or the C. It therefore makes sense that the GPA should be a little higher than a B average; it should fall between 3.0 and 4.0.

We can use the same strategy to calculate the atomic mass of an element. For example, mass spectrometry reveals what we have noted previously: that carbon-12 accounts for 98.93% of all carbon atoms. Carbon-13 makes up 1.07% of the carbon atoms, and carbon-14's contribution is very, very small—only 2×10^{-10} %. Remember that when we multiply by a percent, we must divide the percent by 100 before the calculation (so 98.93% becomes 0.9893). The average mass of carbon-12 is then:

The Periodic Table of the Elements

1	2	3	4	5	6	7	8	9	10	11	12	13	14	15	16	17	18
1 H hydrogen 1.01																	2 He helium 4.00
3 Li lithium 6.94	4 Be beryllium 9.01											5 B boron 10.81	6 C carbon 12.01	7 N nitrogen 14.01	8 O oxygen 16.00	9 F fluorine 19.00	10 Ne neon 20.18
11 Na sodium 22.99	12 Mg magnesium 24.31											13 Al aluminum 26.98	14 Si silicon 28.09	15 P phosphorus 30.97	16 S sulfur 32.06	17 Cl chlorine 35.45	18 Ar argon 39.95
19 K potassium 39.10	20 Ca calcium 40.08	21 Sc scandium 44.96	22 Ti titanium 47.88	23 V vanadium 50.94	24 Cr chromium 51.99	25 Mn manganese 54.94	26 Fe iron 55.93	27 Co cobalt 58.93	28 Ni nickel 58.69	29 Cu copper 63.55	30 Zn zinc 65.39	31 Ga gallium 69.73	32 Ge germanium 72.61	33 As arsenic 74.92	34 Se selenium 78.09	35 Br bromine 79.90	36 Kr krypton 84.80
37 Rb rubidium 84.49	38 Sr strontium 87.62	39 Y yttrium 88.91	40 Zr zirconium 91.22	41 Nb niobium 92.91	24 Mo molybdenum 95.94	43 Tc technetium 98.91	44 Ru ruthenium 101.07	45 Rh rhodium 102.91	46 Pd palladium 106.42	47 Ag silver 107.87	48 Cd cadmium 112.41	49 In indium 114.82	50 Sn tin 118.71	51 Sb antimony 121.76	52 Te tellurium 127.6	53 I iodine 126.90	54 Xe xenon 131.29
55 Cs cesium 132.91	56 Ba barium 137.33	57 La lanthanum 138.91	72 Hf hafnium 178.49	73 Ta tantalum 180.95	74 W tungsten 183.85	75 Re rhenium 186.21	76 Os osmium 190.23	77 Ir iridium 192.22	78 Pt platinum 195.08	79 Au gold 196.97	80 Hg mercury 200.59	81 Tl thallium 204.38	82 Pb lead 207.20	83 Bi bismuth 208.98	84 Po polonium [208.98]	85 At astatine 209.98	86 Rn radon 222.02
87 Fr francium 223.02	88 Ra radium 226.03	89 Ac actinium [227]	104 Rf rutherfordium [261]	105 Db dubnium [262]	106 Sg seaborgium [266]	107 Bh bohrium [264]	108 Hs hassium [269]	109 Mt meitnerium [278]	110 Ds darmstadtium [281]	111 Rg roentgenium [282]	112 Cn copernicium [285]	113 Nh nihonium [286]	114 Fl flerovium [289]	115 Mc moscovium [289]	116 Lv livermorium [293]	117 Ts tennessine [294]	118 Og oganesson [294]

lanthanides	58 Ce cerium 140.12	59 Pr praseodymium 140.91	60 Nd neodymium 144.24	61 Pm promethium [145]	62 Sm samarium 150.36	63 Eu europium 151.96	64 Gd gadolinium 157.25	65 Tb terbium 158.93	66 Dy dysprosium 162.50	67 Ho holmium 164.93	68 Er erbium 167.26	69 Tm thulium 168.93	70 Yb ytterbium 173.05	71 Lu lutetium 174.97
actinides	90 Th thorium 232.04	91 Pa protactinium 231.04	92 U uranium 238.03	93 Np neptunium [237]	94 Pu plutonium [244]	95 Am americium [243]	96 Cm curium [247]	97 Bk berkelium [247]	98 Cf californium [251]	99 Es einsteinium [252]	100 Fm fermium [257]	101 Md mendelevium [258]	102 No nobelium [259]	103 Lr lawrencium [266]

Fig. 2.24 The periodic table. A highly organized listing of all of the known elements

(12 amu × 0.9893)	**+ (13.00335 amu × 0.0107)**	**+ (14.00324 amu × 2 × 10⁻¹²)**	**= 12.01 amu**
Carbon-12	Carbon-13	Carbon-14	

This agrees very nicely with the mass reported in the periodic table (see Fig. 2.24). All naturally occurring samples of carbon have an atomic mass greater than 12 amu reflecting the increased mass contribution made by the small proportions of carbon-13 and carbon-14.

Example 2.3—Calculating the Atomic Mass Value

Naturally occurring chlorine is composed of two principal isotopes: chlorine-35 and chlorine-37. As the spectrum shows in Fig. 2.23, the lighter isotope is substantially more abundant than the heavier one. Use the relative abundances given in the figure, along with the mass of each isotope, to calculate the atomic mass for chlorine. Compare the mass you calculated with what is reported in the periodic table of the elements.

Asking the Right Questions

What's an isotope? How do relative abundances of isotopes relate to the average atomic mass?

Solution

Valid analogies to common situations and experiences can help us visualize difficult concepts. We can make use of the analogy to the grade point average that we discussed. It is the same calculation except the abundance of each isotope replaces the abundance of each grade.

Isotope	Mass (amu)	Abundance (%)[a]
Chlorine-35	34.968852	75.77
Chlorine-37	36.965903	24.23

$$\text{Atomic mass} = (34.968852 \times 0.7577) + (36.965903 \times 0.2423) = 35.45\ \text{amu}$$

Is Our Answer Reasonable?

The average atomic mass we calculated is between 35 and 37, as it should be (the two isotopes serve as the boundaries for a reasonable answer). Moreover, our answer is closer to the mass of chlorine-35. Since that isotope has a greater abundance, it does seem reasonable to calculate an average atomic mass closer to 35 than to 37.

What If…?

…we note that the use of the word "principal" in the statement, "Chlorine is composed of two *principal* isotopes" means that there is another naturally occurring isotope that exists, chlorine-36. This isotope is present in an abundance of about 7×10^{-11} %. What effect does that have on our calculated atomic mass of 35.45?

(*Ans: None. The abundance of chlorine-36 is far too small to affect the average atomic mass, to 4 significant figures*)

Practice 2.3

Zinc has five isotopes, which have the masses (in amu) and the abundances given below. What is the average atomic mass of zinc?

Isotope	Mass (amu)	Abundance (%)
Zinc-64	63.9296011	48.63
Zinc-66	65.9260368	27.90
Zinc-67	66.9271309	4.10
Zinc-68	67.9248476	18.75
Zinc-70	69.925325	0.62

2.7 The Periodic Table

The periodic table of the elements is the most significant document summarizing what we know about atoms. The table, shown in ◘ Fig. 2.24 is the reference chart that shows the hierarchy of atomic structure. The layout of the table summarizes some of the most crucial ideas and discoveries of chemistry. This tool will be our data source, our collaborator, and even our teacher, as we work our way through the subject of chemistry, and we will refer to it again and again as we work on solving problems. Why? As with so many things in life, it all begins with organization.

How is the Periodic Table arranged? The periodic table is arranged in **periods** (rows) and **groups** (columns). As we read from left to right along the first period, then left to right along the second and all other periods, we notice that the atomic numbers build up one proton at a time. The atoms also tend to get heavier as we move along the periods. The table seems to have strange structure in places, with unexplained spaces we must jump over as we read. The reasons will become clear as we deepen our understanding of atomic structure.

Group – One of the vertical columns in the periodic table of the elements.

Period – One of the horizontal rows in the periodic table of the elements.

The arrangement of elements in the periodic table is linked to each element's chemical properties in ways that enable us to make many useful predictions**.** One example is the heavy stair-step line on the right-hand side of the table. Why is it there? This line serves as the boundary between metal and non-metal elements. The ***metals***, located to the left of the boundary, share common characteristics. They are lustrous (shiny), malleable (can be shaped or bent), and ductile (can be stretched into wires), and they act as conductors of heat and electricity. Many of these metals are commonplace in everyday life, such as iron (Fe), the main component of steel; copper (Cu), used to make electric wiring; and silver (Ag) and gold (Au) used widely in jewelry. Many other metals are not as well known, and yet are vital to the modern lifestyle. For example, internal combustion cars would be far more polluting if not for small amounts of platinum (Pt), palladium (Pd), and Rhodium (Rh) in catalytic converters that promote the breakdown of hazardous gases into safer ones. Smartphones and laptops could not exist without the tiny amounts of tantalum (Ta) present in their microprocessors.

Metals – The largest category of elements in the periodic table. Metals exhibit such characteristic properties as ability to conduct electricity, shiny appearance, and malleability.

The **nonmetals**, located to the right of the stair-step boundary, also share some common properties that are quite different from those of the metals. Typically, they are not lustrous, break easily, and act as insulators to heat and electricity. The nonmetals are a much smaller set of elements than the metals, but life as we know it would not be possible without them. For example, the carbon (C), hydrogen (H), oxygen (O), nitrogen (N), and phosphorus (P) atoms that are combined to form DNA are all nonmetals.

The elements just on either side of the stair-step boundary between metals and nonmetals are called semimetals or **metalloids**. This group of elements exhibits some properties associated with metals and other properties associated with nonmetals. In other words, their properties are intermediate between metals and nonmetals. For example, they are lustrous but are often brittle. The semimetal silicon (Si) is called a semiconductor because its ability to conduct electricity is part way between that of a conductor and that of an insulator.

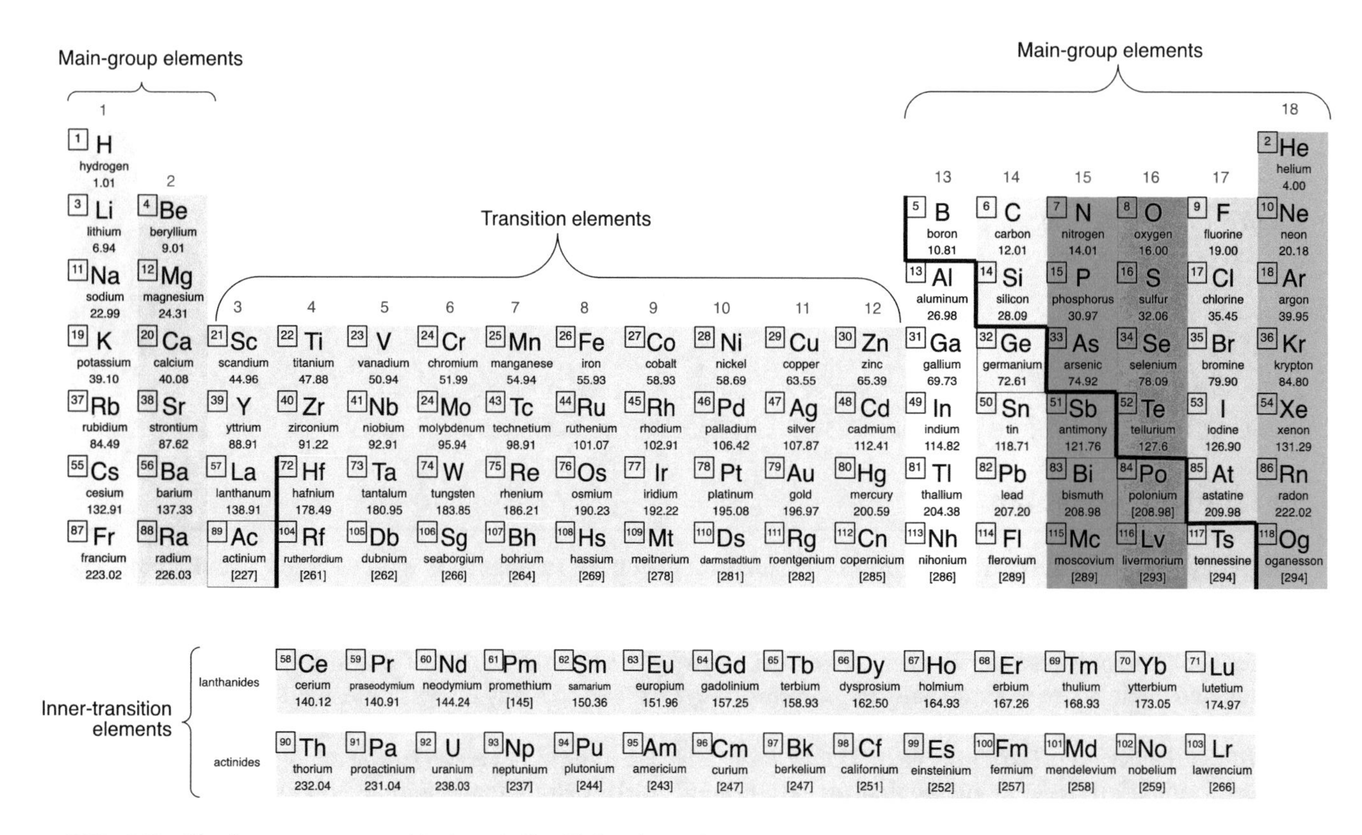

Fig. 2.25 The elements are arranged in the periodic table in columns, known as groups, that represent similar properties

Nonmetals – The elements on the right-hand side of the periodic table. Nonmetals exhibit characteristic properties that are distinct from those of the metals.

Metalloids – A small number of elements, found at the boundary between metals and nonmetals in the periodic table, that have properties intermediate between those of metals and nonmetals. Also known as semi-metals.

The elements are also located in another pattern that indicates similar properties (Fig. 2.25). Each vertical column in the periodic table contains a group of elements with similar chemical and physical properties—hence the name "group." For instance, the elements in a group often combine with other elements in a similar way, or they may have similar abilities to conduct or not conduct electricity.

All of the elements in Group VIIIA, for example, are very stable gases that are unreactive except under extreme conditions. They are known as the **noble gases**. Here, the word "noble" is used in the sense of having regal behavior (not being affected by other elements). They are also often referred to as the inert gases, although for those in the lower part of the group, this is not entirely correct. The helium used to keep dirigibles (blimps) aloft and the neon used in neon signs are two common examples of noble gases. As another example, all of the elements in Group IA are highly reactive metals known as the **alkali metals** because they form basic (or "alkaline") solutions when left in contact with water.

Alkali metals – The highly reactive metals found in Group IA of the periodic table, which form alkalis upon reaction with water.

Noble gases – The unreactive elements in the rightmost group of the periodic table (Group VIIIA). Also known as inert gases.

Looking further, we can see yet another layer of organization in the periodic table. Elements in the "A" groups of the periodic table (such as IA, IIA, VIIA, etc.) are historically known as the main-group elements, and those in the "B" groups (such as IIIB, IVB, etc.) are called transition elements or **transition metals** (Fig. 2.25). There are two rows of elements separated from the main table known as the **inner transition metals**. We will have much more to say about the historical and chemical organization of the periodic table in future chapters. At this point, however, we have the background to begin learning about how the elements combine to form more complex structures, including compounds and molecules.

Transition metals – The elements in the "B" groups between Group IIA and Group IIIA.

Inner transition metals – The two rows of elements often placed below the main table that include elements with atomic numbers 57–71 and 89–103.

2.8 Ionic Compounds

Chemical reactions consist of the rearrangement of atoms to make different substances. Although the nucleus of an atom remains unchanged in the course of a chemical reaction, the number of electrons surrounding that nucleus can change. When electrons are added to or subtracted from a neutral atom or group of atoms, a charged species known as an ion forms. For example, the neutral calcium-40 atom has 20 protons, 20 neutrons, and 20 electrons, and we can denote this using the nuclide notation $^{40}_{20}Ca$. When calcium combines with oxygen, it loses two electrons, which the oxygen atom gains. We can write the nuclide notation for the resulting calcium ion, which now has 20 protons and only 18 electrons (two more positive than negative charges) as $^{40}_{20}Ca^{2+}$.

Since the number of protons is implicit in the element symbol, we don't need to use nuclide notation for most chemical notations. In other words, calcium must have 20 protons if it is to be called calcium. We can simplify the notation of a calcium ion with only 18 electrons even further and write Ca^{2+}. Using the same simplified notation, we write the oxygen ion, which contains 8 protons, 8 neutrons and 10 electrons, as O^{2-}. Calcium now has a +2 charge and oxygen has a −2 charge. The calcium and oxygen ions form a strong bond to each other because of the opposite, attracting charges. This is known as **ionic bonding**. This strong bond means that ionic substances often have very high melting and boiling points, and the higher the charges on each ion, the more tightly they bond together, and the higher the melting and boiling points.

Ionic bonding – The result of the attractive force of oppositely charged ions that holds ions together.

Example 2.4—Ions

Fill in the complete symbols and numbers missing from this table:

Symbol	Protons	Neutrons	Electrons	Charge
$^{32}_{16}S^{2-}$				
	56	81	54	
Cl^-		20		−1

Asking the Right Questions

How is the charge of the element related to the number of electrons and protons? The charge is the difference between the number of protons (positive charge) and number of electrons (negative charge). If there are more protons than electrons, the substance will have a positive charge; more electrons than protons indicates the ion will have a negative charge. What if the number of protons and electrons are equal? You have a neutral substance. How are the number of protons and neutrons related to the mass number? Recall that the mass number is equal to the sum of the number of protons and neutrons.

Solution

Symbol	Protons	Neutrons	Electrons	Charge
$^{32}_{16}S^{2-}$	16	16	18	−2
$^{137}_{56}Ba^{2+}$	56	81	54	+2
$^{37}_{17}Cl^{-}$	17	20	18	−1

Are Our Answers Reasonable?

Just as we did in a previous example, a quick run through the table will catch any mathematical errors. Note that the sum of the electrons and charge columns should equal the values in the proton column, and the sum of the proton and neutron columns should equal the mass number for the nuclide.

What If…?

…we had an atom of radon-222, which has the symbol $^{222}_{86}Rn$? That atom can lose two protons and two neutrons in a process called "alpha-particle emission." What would be the symbol for the isotope that remained? (*Ans:* $^{218}_{84}Po$)

Practice 2.4

Fill in the missing information from this table.

Symbol	Protons	Neutrons	Electrons	Charge
$^{52}_{24}\ ^{6+}$				
	19	20	18	
	35	44		−1

Some substances, known as **ionic compounds**, are composed entirely of ions. Calcium oxide (CaO) is one example. How do we know if the compound we are examining is an ionic compound? Although the distinction is clear with respect to the presence of ions in the structure, this is often not apparent at first glance because the properly written formula for the substance doesn't include charges. However, the composition of the elements that make up the compound can provide a hint. Ionic compounds are often formed from the combination of metals and nonmetals.

Ionic compound – A compound composed of the combination of an anion and a cation.

Chemists use a **chemical formula**, containing element symbols and numbers, to represent the new combination of atoms or ions known as a chemical compound. A chemical formula is a kind of abbreviation that represents a chemical compound. It shows the symbols for the elements found in the compound and the quantity of each of those atoms or ions. The ratios of the atoms in a chemical formula arise from the Law of Multiple Proportions. For example, the formula NaCl tells us that each formula unit of this compound contains one atom of sodium and one atom of chlorine. Note that if there is only one instance of an atom, we don't write the subscript "1" – it is implied. The molecule that contains one atom of nitrogen and one of oxygen has the formula NO, rather than N_1O_1. Not listing the "1"s in the formula is part of the chemist's shorthand. The formula Sb_2S_3 tells us that two atoms of antimony (Sb) and three atoms of sulfur (S) make up the formula unit of the beautiful mineral stibnite (see Fig. 2.26) used as a lubricant in antifriction alloys. Could we have determined this formula without knowing more than the identity of the ions in the compound?

Chemical formula – A representation, in symbols, that conveys the relative proportions of atoms of the different elements in a substance.

When we are considering ionic compounds, a great clue is that the ions must combine in a ratio that makes the compound electrically neutral overall. In sodium chloride, for example, every sodium cation has a +1 charge, and every chloride anion has a –1 charge, so these ions will combine in a 1:1 ratio to form the compound. Written as individual ions, they are Na^+ and Cl^-. When we write chemical formulas, we place the positive ion first, so the formula for sodium chloride is NaCl. Note that the charges are not shown. The formula CaF_2, for the chemical calcium fluoride, indicates that for every two fluoride ions, there is one calcium ion.

How do we predict what the charge of each ion is likely to be in an ionic compound? As it so often does, the periodic table gives us insight. The **main-group elements** (the "A" groups, or groups 1–2, and 13–17 in the Periodic Table) gain or lose electrons in predictable and consistent ways when they form ionic compounds. For example, the elements in Group IIA, known as the **alkaline earth metals**, tend to lose two electrons when combining with nonmetals. This results in these atoms becoming cations with a charge of +2. The elements in Group IIIB (all metals), as well as aluminum (in Group IIIA) tend to lose three electrons when forming ionic compounds, forming cations with a charge of +3. The nonmetals of Group VA tend to gain three electrons when forming ionic compounds, to form anions with a –3 charge. The nonmetals of Group VIA, known as the **chalcogens**, tend to gain two electrons to acquire a charge of –2. The elements of Group VIIA, known as the **halogens**, tend to gain one electron when they form an ionic compound. Keep in-mind that these are just tendencies, albeit strong ones. Still, a reasonable strategy to decide how many electrons a main-group element is likely to gain or lose is to base it on the group to which the element belongs. A summary of the charges generally found on the main-group ions is shown in Table 2.5

Chalcogens – The elements found in Group VIA of the periodic table of elements.

Fig. 2.26 Stibnite is a mineral with the chemical formula Sb_2S_3. *Source* Taken by author (User:pepperedjane) on 6 January 2007 at the Carnegie Museum of Natural History

Table 2.5 Charges on typical main-group ions

Group Number	Most likely ionic charge
IA	+1
IIA	+2
IIIA	+3
IVA	+4/−4
VA	−3
VIA	−2
VIIA	−1
VIIIA	0

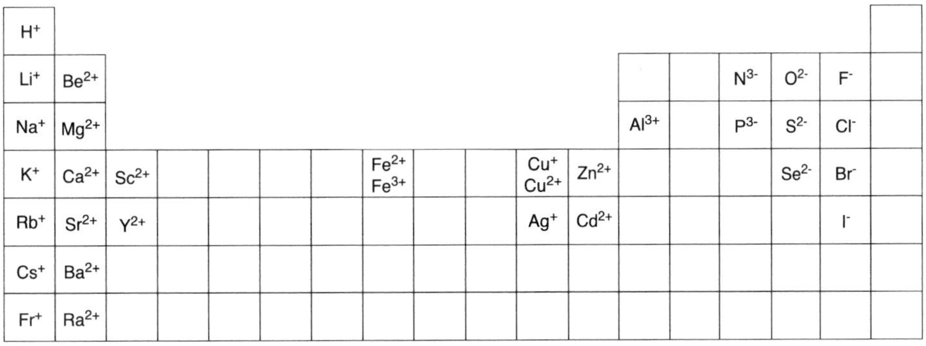

H^+																	
Li^+	Be^{2+}													N^{3-}	O^{2-}	F^-	
Na^+	Mg^{2+}											Al^{3+}		P^{3-}	S^{2-}	Cl^-	
K^+	Ca^{2+}	Sc^{2+}					Fe^{2+} Fe^{3+}			Cu^+ Cu^{2+}	Zn^{2+}				Se^{2-}	Br^-	
Rb^+	Sr^{2+}	Y^{2+}								Ag^+	Cd^{2+}					I^-	
Cs^+	Ba^{2+}																
Fr^+	Ra^{2+}																

Fig. 2.27 Some common ions and their locations in the periodic table

Alkaline earth metals – The metals found in Group IIA of the periodic table.

Halogens – The elements of Group VIIA of the periodic table.

Main-group elements – Elements in the groups of the periodic table identified with "A" (IA, IIA, etc.). Also, periodic table groups 1, 2, and 13–17.

As we step away from the main group elements and move to the transition metals of Groups IIIB through IIB (3–12), indicated by the blue-green block in Fig. 2.26, exceptions begin to appear. For example, iron commonly exists as either Fe^{2+} or Fe^{3+}, though under the right conditions, it can be Fe^{+4}, Fe^{+5}, or even Fe^{+6}. The charge of the ions in transition metals is best determined by examining the formula of the compounds they create. A list of some common ions is given in Fig. 2.27.

Example 2.5—Determining Formulas

a. The ionic compound used in some sidewalk deicers is made from calcium and chlorine. What formula would represent that compound?
b. Magnesium combines with oxygen to give a bright flash of light used in some flares and fireworks. What formula would represent the resulting compound?
c. When iron with a +3 charge combines with oxygen, it forms a red, flaky material called "rust". What formula would represent the resulting compound?

Asking the Right Questions

What are the charges on each individual element? Which element will be the cation (carry the positive charge)? Which will be the anion (carry the negative charge)? What must be true about the overall charge of the compound? Once you determine the charges on the cation and anion, you can tell how many of one will be needed to electrically balance the other to give you a neutral compound.

Solution

a. Because Ca is in Group IIA, it will form ions with a charge of +2. Chlorine is in Group VIIA and so will tend to form ions with a charge of −1. To form an electrically neutral ionic compound, there must be two chloride ions for every calcium ion. Therefore, the formula will be $CaCl_2$.

b. Magnesium is found in Group IIA, so it will form cations with a charge of +2. Oxygen is found in Group VIA and so will form anions with a charge of –2. A one-to-one combination produces an electrically neutral compound. Therefore, the formula will be MgO.
c. Iron is found in Group VIIIB and usually takes on a +2 or +3 charge. In this case, the problem states that it is a 3+ ion. Since oxygen forms anions with a charge of −2, we need to find the least common multiple to achieve a neutral compound. The least common multiple for 3 and 2 is 6. We will need two iron ions ($+3 \times 2 = +6$) to balance the charges on three oxygen ions ($-2 \times 3 = -6$) and the compound will be Fe_2O_3.

Is Our Answer Reasonable?

All of the compounds we came up with have ions that match what we expect from Fig. 2.27, and all of the compounds are neutral (the total charges on cations and anions are equal but opposite in charge), so these answers conform to the definition of an ionic compound.

What If…?

…we examine magnetite and find that it is a mixture of Fe^{2+}, Fe^{3+} and oxygen ions? We find that the formula for this compound is Fe_3O_4. What is the charge per iron ion, on average? (*Ans. The average charge is +2.67. There is one* Fe^{2+} *ion for every two* Fe^{3+} *ions*)

Practice 2.5

What is likely to be the formula of a compound resulting from the atomic combination of cesium and chlorine?

Solid ionic compounds most often exist as part of large repeating units, such as the three-dimensional crystals of NaCl that you might put on your fries or chips. The crystal is composed of equally large numbers of sodium ions and chloride ions (a 1:1 ratio). Like most ionic compounds, NaCl is a brittle solid that melts at a high temperature (801 °C). Sodium oxide (Na_2O) is an ionic compound with a melting point of 1132 °C. Ionic compounds vary in their **solubility** in water. Some, such as table salt (NaCl), are quite soluble. Others, such as strontium fluoride (SrF_2), are fairly insoluble. When soluble compounds such as NaCl are placed in water, their ions can separate and move around relatively independently. These mobile, electrically charged ions allow electricity to be conducted through the resulting solution. The ability to conduct electricity when dissolved in water is one way a substance can be identified as being ionic. Non-ionic substances do not allow the conduction of electricity when in solution.

Solubility – The ability of one substance to dissolve in another.

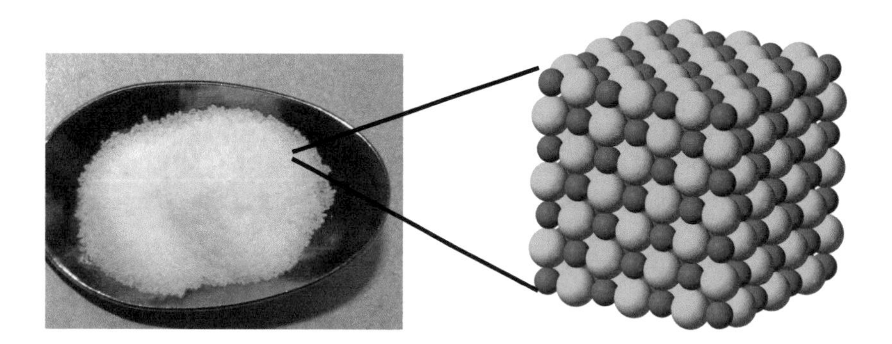

2.9 Molecules

How is carbon monoxide (CO) different from sodium chloride (NaCl)? They are both compounds that consist of two atoms in a 1:1 ratio, but sodium chloride is made of a metal and a nonmetal and has properties that we associate with ionic compounds, such as a high melting point and conducting electricity when molten or dissolved in water. Carbon monoxide, on the other hand, is made of the combination of two nonmetals, and is a gas that doesn't conduct electricity when dissolved in water. These two substances are very different in their physical properties, and this is an indication of an essential difference in chemical structure. Sodium chloride is an ionic compound. Carbon monoxide is a molecule.

Molecules are distinct units made up of two or more atoms linked together by sharing electrons between their nuclei, rather than by the transfer of electrons from one atom to another. We call bonds that are formed by sharing electrons between atoms, **covalent bonds**. Molecules such as carbon dioxide (CO_2), water (H_2O), and phosphorus pentoxide (PO_5) are composed of only nonmetals that interact via covalent bonds. They differ from ionic

Fig. 2.28 The molecular formula for water indicates that two hydrogen atoms are combined with one oxygen atom. Because water is a molecule, none of these atoms is an ion

compounds because they are not composed of ions. When we write H_2O as the **molecular formula** for water, we mean that each molecule exists independently and contains just two H atoms and one O atom (Fig. 2.28). This contrasts with the formulas for ionic compounds, which state the ratio in which ions are present in a sample. Sodium chloride, as an ionic compound, stretches into three dimensions, growing so large that we can see the crystals that we shake from a saltshaker onto our food. The number of cations and anions that define it, an extensive property, increases with size. However, the *ratio* of sodium to chloride ions will always be one-to-one. A discrete "molecule" with only one sodium cation and only one chloride anion, NaCl, just doesn't exist under normal conditions.

Covalent bonds – Bonds that atoms form by sharing electrons.

Molecular formula – A chemical formula indicating the actual number of each type of atom present in one molecule of a compound.

Molecules can range in size from diatomic molecules such as N_2 and O_2, the main components of air, to giant molecules containing many thousands, or even millions, of atoms. For example, the DNA that makes up the human genome and the protein molecules that control most of the chemistry of life are giant molecular compounds.

Even though molecules are electrically neutral like their ionic compound counterparts, they cannot be constructed using the rules for ionic compounds. Remember that these are two distinctly different types of compounds. The formulas for many ionic compounds can be predicted from the typical charges found on the ions, especially in the A, or main-group, elements. For molecules, on the other hand, experimentation is necessary to determine the molecular formulas.

One of the themes of our study of chemistry is that small differences in structure can lead to big differences in chemical behavior. An important example of this is given by two different molecules containing only oxygen atoms, one very common; the other less so. In its most common form, oxygen (O_2) is one of the **diatomic elements**, in which the normal state is in the form of molecules composed of two atoms bonded together. The other diatomic molecules are: H_2, N_2, F_2, Cl_2, Br_2 and I_2. The diatomic form of oxygen, O_2, accounts for approximately 20% of the volume of air and is the form of oxygen we need to stay alive. A different and much less common form of oxygen, O_3, contains three oxygen atoms bonded together. This form of oxygen is known as ozone and accounts for just a few molecules for every billion molecules of air in Earth's atmosphere.

Diatomic elements – Elements whose normal state is in the form of molecules composed of two atoms attached together (most notably H_2, N_2, O_2, F_2, Cl_2, Br_2, and I_2).

These two forms of oxygen molecules are **allotropes**. An allotrope is another form of a pure element with different chemical and physical properties. Examples of other elements with allotropes are sulfur (S_5, S_6, S_7, S_8) and phosphorus (P_2, P_4). Ozone and oxygen are both gases, but ozone filters out hazardous ultraviolet radiation in the upper atmosphere. Yet, when present at the earth's surface, O_3 is an irritant and poison that can cause serious damage to the tissues of the body, including the sensitive lining of our lungs, right where O_2 is needed most!

Allotropes – Forms of an element that have different chemical and physical properties, such as the allotropes O_2 and O_3. Allotropes needn't have different formulas, such as $C_{graphite}$ and $C_{diamond}$.

▶ Deepening Your Understanding Through Group Work

We now know that each element has one or more isotopes, and we can determine the ratio of those isotopes using a mass spectrometer. We also know that elements can organize into molecules of all shapes and sizes. Let's use that understanding to identify a mixture of different molecules. Assume your group has a sample of 20 mL of urine, supplied by an athlete who is being tested for prazepam, diazepam, diacetylmorphine, and cocaine. Your group's goal is to design a device, or multiple devices, that will allow you to identify each substance in the urine sample. The Internet can be a big help to find the formula, molecular mass, melting point, and boiling point of each of these drugs. Some questions to consider during your planning include: What would be the important characteristics of such a device? What might the readout of results look like? How would you confirm that the results you get truly represent the molecules you are analyzing? ◀

2.10 Naming Compounds

Communication among chemists is vital to the effective application of the scientific method we discussed in ▶ Chap. 1. This communication relies on being able to convey information in writing, over the phone, and on the Internet. Complicating matters is that over 7000 languages are spoken worldwide. How then, does a chemist in rural Colorado tell a colleague in Mexico City about a particular compound? Some compounds have "common names" that are in widespread use across a country and around the world. For instance, benzene (C_6H_6, a compound found in petroleum distillates) and taxol (the anticancer compound from ▶ Chap. 1) are known by these common names in many different countries. But this isn't universally true. As ◘ Table 2.6 illustrates for water, some compounds are so common and have been known for such a long time, that there is a different name for them in every language. For most chemical compounds, there simply isn't a name that is commonly known. Without a system to follow, it would be difficult to devise names that were meaningful to anyone else.

A simple solution to this problem was provided by the International Union of Pure and Applied Chemists (IUPAC), an organization formed in 1919 that continues to advise the scientific community on issues related to chemistry. IUPAC has a set of rules that all scientists try to follow when naming chemical compounds. Known as the rules of **chemical nomenclature**, they establish patterns to follow in naming existing compounds as well as compounds that have yet to be determined.

Chemical nomenclature – A system of rules used to assign a name to a particular substance.

2.10.1 Naming Binary Covalent Compounds

The rules for naming **binary covalent compounds** (compounds composed of only two nonmetal elements, which have atoms held together by covalent bonds) are listed in ◘ Table 2.7.

Binary covalent compounds – Compounds composed of only two nonmetal elements.

◘ **Table 2.6** The word for water in different languages.

Language	Name of H_2O	Language	Name of H_2O
English	water	German	Wasser
Spanish	agua	French	eau
Hawaiian	wai	Slovak	voda
Urdu	pani	Tagalog	tubig
Swahili	maji	Frisian	wetter
Cebuano	pagtubig	Japanese	mizu

Table 2.7 Rules for naming binary covalent compounds

Step	Rule
1	Name the first element using the exact element name
2	Name the second element by writing the stem name of the element with the suffix -ide
3	Add prefixes as shown in the list at the right, derived from Greek, to each element name to denote the subscript of the element in the formula. The prefix mono- is used only on the second element in the binary covalent compound to distinguish it from other examples containing multiple atoms. For example, CCl_4 is carbon tetrachloride
4	For readability, if the prefix ends with an 'a' or 'o', and the ion begins with an 'a' or 'o', we drop the last letter of the prefix. For example: mono- prepended to oxide would be "monoxide", not "monooxide."

Number	Prefix
½	hemi
1	mono
2	di
3	tri
4	tetra
5	penta
6	hexa
7	hepta
8	octa
9	nona
10	deca

As an example of how to use these rules, let's consider the formula for NCl_3. This compound is a molecule made of two non-metals (it has only covalent bonds) and is a binary compound (it has only two elements). As the first step using the rules in Table 2.7 we write:

nitrogen

In the second step, we write:

nitrogen *chloride*

Then, to finish the name of the molecule, we add the prefixes. Note that we do not add the prefix mono- to the *first* element name. If it is not stated, mono- is assumed:

nitrogen *tri*chloride

The following examples give you an opportunity to see these rules in practice:

N_2O_3 – dinitrogen trioxide
CO – carbon monoxide
CO_2 – carbon dioxide
SF_4 – sulfur tetrafluoride
SF_6 – sulfur hexafluoride
ClO_2 – chlorine dioxide
Cl_2O_7 – dichlorine heptoxide
P_4O_6 – tetraphosphorus hexoxide
S_2Cl_2 – disulfur dichloride.

Writing the chemical formula from the name is just the reverse of the above process: write the symbol for the first element, with a subscript number indicated by the prefix, and then do the same for the next element. For example, for the compound dinitrogen pentoxide, we would write the dinitrogen as:

$$N_2$$

And then add the pentoxide:

$$N_2O_5$$

2

Example 2.6—Naming Binary Covalent Compounds

Provide the IUPAC name or formula for these binary covalent compounds:

Name	Formula
	PCl_5
	NBr_3
Arsenic trichloride	
	XeO_4
Tetraphosphorus trisulfide	

Asking the Right Questions

What do these types of compounds have in common? These are all binary covalent compounds and the IUPAC rules, shown in ▪ Table 2.7 govern how we name them.

Solution

Name	Formula
Phosphorus pentachloride	PCl_5
Nitrogen tribromide	NBr_3
Arsenic trichloride	$AsCl_3$
Xenon tetroxide	XeO_4
Tetraphosphorus trisulfide	P_4S_3

Are Our Answers Reasonable?

Are our prefixes represented correctly? In each case, the Greek prefix agrees with the number of atoms we have. This tells us that our answers make sense.

What If…?

…we have a binary compound composed of 2 atoms of nitrogen and 5 atoms of oxygen? Name the compound. (*Ans: dinitrogen pentoxide*)

Practice 2.6

Provide either the name or the formula for each of these binary molecular compounds:

sulfur tetrafluoride	carbon tetrachloride	diphosphorus pentoxide
PCl_3	N_2O	OF_2

2.10.2 Naming Binary Ionic Compounds

The rules for naming **binary ionic compounds** (compounds typically composed of a metal and a nonmetal element, which have ions of opposite charges that attract each other) are similar to those for the binary molecular compounds, except that prefixes are never used for ionic compounds. To summarize the rules in ▪ Table 2.8 we can name a binary ionic compound by stating the name of the atom that has become a cation (positive ion) and then adding the name of the anion (negative ion). The anion derives its name from the parent atom but ends in *-ide*. Unlike binary molecular compound names, the name for a binary ionic compound does not tell us how many of each type of ion are present. How can the name tell us how many of each type of ion are in the formula unit? Remember that ionic compounds contain ions with charges that must balance to result in an electrically neutral compound. We can use these charges to determine the number of each ion.

Binary ionic compounds – Compounds typically composed of a metal and nonmetal element, which have ions that interact via electrostatic attractions.

Sodium chloride (NaCl) offers a good example. We can tell that it is likely to be an ionic compound because it is composed of a main-group metal (Na) and a nonmetal (Cl). The procedure outlined in ▪ Table 2.8 indicates that we first give the name of the metal, the element listed first in the chemical formula. Therefore, we begin with:

sodium

Then we add the name of the other element, which yields:

sodium chloride

■ **Table 2.8** Rules for naming binary ionic compounds

Step	Rule
1	Name the first element using its element name
2	Name the second element by writing the stem name of the element with the suffix *-ide*
3	For metals that can have more than one stable charge, report the charge as part of the name. Write this charge, in Roman numeral form, in parentheses immediately after the metal's name. Metals in Groups IA, IIA and IIIA, as well as aluminum (which always has a charge of +3) have only one ionic charge, and do not require Roman numerals

According to the IUPAC rules of chemical nomenclature, the name for NaCl, sodium chloride, is exactly as we expected.

Why do we need to know the charge on some metals in an ionic compound? Let's visit the metals found between Groups IIA and IIIA in the periodic table. Most of these elements, known as the transition metals, commonly exhibit more than one positively charged state. It is therefore possible for two bottles of iron chloride to contain two different ionic compounds. You might find $FeCl_2$ in one bottle and $FeCl_3$ in the other. If we considered only the first two steps in ■ Table 2.8 we would conclude that they should both be named iron chloride and we couldn't be sure which one we were referring to. However, the first bottle contains iron ions with a +2 charge. (Remember, it needs to be Fe^{2+} in order to balance the charges of the two Cl^- atoms.) The second bottle contains Fe^{3+} to balance the charges of the three Cl^- ions. To name these two compounds so we can distinguish between them, we follow Step 3 in ■ Table 2.8 and add the charge on the iron to the name of the compound. Thus, $FeCl_2$ becomes iron(II) chloride, and $FeCl_3$ becomes iron(III) chloride (■ Fig. 2.29).

The combination of titanium and oxygen to make TiO_2, a white pigment used in many consumer products, such as paint, sunscreen, and toothpaste (and even some foods!), provides another example of the use of the rules in ■ Table 2.8. This is an ionic compound because it has a transition metal (Ti) and a nonmetal (O). Following the rules outlined in ■ Table 2.8, we write the name of the metal:

titanium

Then, we add the name of the non-metal:

titanium oxide

And finally, we add the charge to the name. Knowing that one oxygen ion has a −2 charge, and that the compound has two oxygen ions, we need to make sure the titanium has a charge that can balance to make the overall charge zero. We can write the charge on the titanium in Roman numerals as IV:

titanium(IV) oxide

Writing a chemical formula from a name for an ionic compound is not much different than for a covalent compound. First, we write the symbol for the first element (the cation) and then the symbol for the second element (the anion). Then, we add subscripts so that the total positive charge EQUALS the total negative charge. Determining the charges is often the most difficult part of the process, but, as so often the case, we can use the periodic table as our guide. The process is summarized in ■ Table 2.9.

As an example of the use of ■ Table 2.9 let's write the formula for scacchite, a reddish mineral found near volcanoes, with the chemical name manganese(II) chloride. We first write the symbols for the two elements:

$$Mn_?Cl_?$$

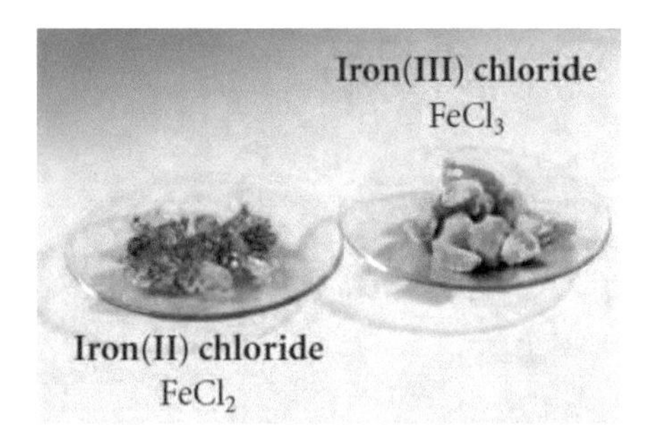

■ **Fig. 2.29** Iron(II) chloride and iron(III) chloride are distinctly different compounds. Note how the formula and the names indicate that they are different

2

Table 2.9 Rules for writing formulas from names of binary ionic compounds

Step	Rule
1	Write the symbol for the first element (the cation)
2	Write the symbol for the second element (the anion)
3	Determine the charge of the element on the left (cation) based on its location in the Periodic Table (Group IA is +1, IIA is +2 and IIIB and aluminum are +3), OR by the Roman numeral that is part of its name
4	Determine the charge of the anion by its location in the Periodic Table (Group IVA is –4, VA is –3, VIA is –2 and VIIA is –1)
5	Write subscripts for the two ions so that the total positive charge equals the total negative charge (use the least common multiple of the two ionic charges)

The Roman numeral in the name tells us that the manganese ion has a charge of +2. The location of chlorine in the periodic table (Group VIIA) indicates its charge of – 1. Therefore, two chloride ions would provide the total negative charge of −2 to balance the total manganese positive charge of +2:

$$MnCl_2$$

A yellowish-brown mineral that has been found in volcanoes in Russia is known as shcherbinaite after a prominent 20th-century Russian geochemist, V.V. Shcherbina. Its chemical composition has the name vanadium(V) oxide. We can use Table 2.9 to determine the formula for this mineral. Start with the symbols for the two elements:

$$V_?O_?$$

The Roman numeral tells us the charge on vanadium is +5. Since we know that oxygen as an anion typically has the charge of – 2, we can determine that the lowest number divisible by both a 5 and a 2 is 10. We need two vanadium ions to provide a total charge of +10, and five oxygen anions to provide a total charge of −10. We add these numbers into the formula so that the charges balance and the ionic compound has a neutral charge overall:

$$V_2O_5$$

Example 2.7—Naming Binary Ionic Compounds

Use the rules outlined in Tables 2.8 and 2.9 to supply each name or formula missing from this list.

Name	Formula
Sodium fluoride	
Calcium oxide	
	Al_2S_3
	$BaCl_2$

Asking the Right Questions

What are the steps in assigning a name? Remember that binary ionic compounds follow a set of rules that are different from the binary molecules. These rules are given in Tables 2.8 and 2.9.

Solution

Name	Formula
Sodium fluoride	NaF
Calcium oxide	CaO
Aluminum sulfide	Al_2S_3
Barium chloride	$BaCl_2$

Are Our Answers Reasonable?

One way to know is to go back and recreate the formula from the name or the name from the formula. Since we are dealing with ionic compounds, it is reasonable that our answers did not have any Greek prefixes. It is also consistent with the naming rules that the cation part in the names preceded the anion part.

What If...?

...we had a compound that was composed of two cations of an element that each contained a total of 10 electrons combined with one anion of element that contained 18 electrons? What would be the name and formula of that ionic compound? (*Ans: sodium sulfide,* Na_2S)

Practice 2.7

Supply each name or formula missing from this list.

Name	Formula
Magnesium chloride	
Lithium fluoride	
	NaBr
	Li_2O

Example 2.8—Naming Additional Compounds

Use the rules in ◘ Tables 2.8 and 2.9 to supply each name or formula missing from this list.

Formula	Name
$BaCl_2$	
$TiCl_2$	
	Copper(II) oxide
	Manganese(IV) oxide

Asking the Right Questions

How is this example different from Example 2.7? In this case, we have compounds made of transition metals that can have more than one charge. How does this impact the name? We can account for that by using a Roman numeral to represent the charge of the cation.

Solution

Formula	Name
$BaCl_2$	Barium chloride
$TiCl_2$	Titanium(II) chloride
CuO	Copper(II) oxide
MnO_2	Manganese(IV) oxide

Are Our Answers Reasonable?

Do our answers reflect the types of compounds we have? The use of a Roman numeral with the first element of a compound indicates that it is a transition metal cation in an ionic compound. That is reasonable for these compounds. The barium cation in the first compound, barium chloride, has only one charge when combining with other types of atoms, so you did not need to use a Roman numeral. The answers are, therefore, reasonable.

What If…?

…the question asked us to find the charge on the iron in Fe_2O_3?
(*Ans:* Fe^{3+})

Practice 2.8

Use the rules in ◘ Tables 2.8 and 2.9 to supply each name or formula missing from this list.

Name	Formula
$CuCl_2$	
CrO_3	
	Nickel(II) oxide
	Palladium(IV) sulfide

2.10.3 Polyatomic Ions

Many compounds are made up of more than two elements. Some of these are classified as ionic compounds, because they involve the association of ions. In some cases, the ion itself contains two or more atoms, which are covalently bonded together. The entire ion that results can carry a positive or negative charge and behave just like a monatomic ion. Ions such as this are known as **polyatomic ions** or molecular ions. ◘ Figure 2.30 shows the structures of some examples. ◘ Table 2.10 lists many formulas and names.

Polyatomic ion – An ion composed of two or more atoms covalently bonded together. Also known as molecular ions.

Polyatomic ions are everywhere. For example, the famous White Cliffs of Dover on the southern end of England are made of chalk, or what you might find sold at the store as "sidewalk chalk." Chalk has the formula $CaCO_3$ and is made up of two ions, Ca^{2+} and CO_3^{2-}. Compounds containing polyatomic ions have the same kinds of physical characteristics as binary ionic compounds (high melting point, brittle, conduct electricity when dissolved in water), because they are ionic compounds. They are a combination of a cation and an anion. When they dissolve in water, the cation and anion separate, just like other ionic compounds. If the cation or anion is a polyatomic ion, the entire polyatomic ion stays bound together and behaves in solution just as if it were a monatomic ion.

Knowing the polyatomic ions in ◘ Table 2.10 is our key step in recognizing compounds with these ions. This is the only way we will know that $CaCO_3$ does NOT consist of calcium ions, carbon ions and oxygen ions, for example. Rather, the CO_3^{2-} anion exists as a polyatomic unit called "carbonate ion" that has a charge of −2 and contains covalent bonds within the structure of the individual carbonate anions. Using our rules for binary ionic compounds (◘ Tables 2.8 and 2.9), we can arrive at the name "calcium carbonate" for the electrically neutral compound represented by $CaCO_3$.

2

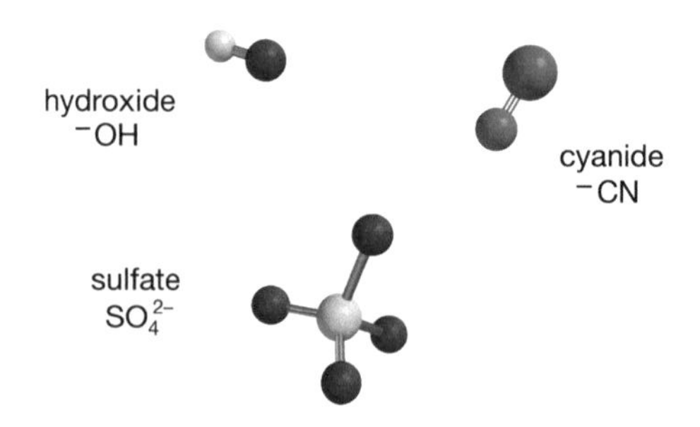

Fig. 2.30 Three polyatomic ions. These ions are made up of covalently bonded atoms. As a whole, they act as ions

Table 2.10 Selected polyatomic ions

	Formula	Name	Formula	Name
Cation with a +1 charge	NH_4^+	ammonium		
Anions with a −1 charge	ClO^-	hypochlorite	HCO_3^-	hydrogen carbonate (bicarbonate)
	ClO_2^-	chlorite	$H_2PO_4^-$	dihydrogen phosphate
	ClO_3^-	chlorate	HSO_3^-	hydrogen sulfite (bisulfite)
	ClO_4^-	perchlorate	HSO_4^-	hydrogen sulfate (bisulfate)
	NO_3^-	nitrate	MnO_4^-	permanganate
	NO_2^-	nitrite	OH^-	hydroxide
	CH_3COO^-	acetate	CN^-	cyanide
Anions with a −2 charge	HPO_4^{2-}	hydrogen phosphate	$C_2O_4^{2-}$	oxalate
	SO_4^{2-}	sulfate	CO_3^{2-}	carbonate
	SO_3^{2-}	sulfite	CrO_4^{2-}	chromate
	$S_2O_3^{2-}$	thiosulfate	$Cr_2O_7^{2-}$	dichromate
			O_2^{2-}	peroxide
Anion with a −3 charge	PO_4^{3-}	phosphate		

One common garden fertilizer has the formula $(NH_4)_3PO_4$. The ammonium cation, NH_4^+, has a +1 charge, and the phosphate anion, PO_4^{3-}, has a −3 charge. Therefore, three ammonium cations will be present for each phosphate ion. We must use parentheses in the formula to indicate that our subscript "3" applies to the entire ammonium ion, not just to the hydrogen atoms within it. If there is only one polyatomic ion present (as with the phosphate, in this case), then we do not use parentheses. What is the name this compound? Using our rules and Table 2.10, we can correctly call it ammonium phosphate. We can use the same process to determine the name from a formula containing either a polyatomic ion or a binary compound (Table 2.9), except that we must know the name, formula and charge of the polyatomic ion.

As an example, let's consider the rare lead-containing crystalline mineral known as crocoite. This mineral grows as beautiful deep red prisms. The chemical name for this compound is lead(II) chromate. What is its formula? We can use the polyatomic ion table to assist in the construction of the name. We first write the ions found from reading the name:

$$Pb_?(CrO_4)_?$$

From the name, we see that lead has a +2 charge and from Table 2.10, we see the charge on chromate is −2. Since the charges are equivalent, we'll only need one of each ion to balance the compound. The formula is:

$$PbCrO_4$$

Note that the parentheses aren't needed in the formula because only one of each polyatomic anion is used in the formula.

Example 2.9—Naming More Compounds

Provide each name or formula missing from the table.

Formula	Name
CaC_2O_4	
$Mg(NO_3)_2$	
	copper(I) sulfate
	aluminum hydroxide

Asking the Right Questions

When naming ionic compounds, our first task is to separate the compound into the cation portion and the anion portion. In every case, the cation is the first atom (or atoms) in the formula. For instance, in CaC_2O_4, the calcium is the cation. That means that the rest of the formula must be the anion, because the compound is electrically neutral. Knowing the names and charges of all the polyatomic ions is also a big help here.

Solution

Formula	Name
CaC_2O_4	calcium oxalate
$Mg(NO_3)_2$	magnesium nitrate
Cu_2SO_4	copper(I) sulfate
$Al(OH)_3$	aluminum hydroxide

Are Our Answers Reasonable?

In addition to making sure that our polyatomic ions have the correct names and formulas, we note that in each case, the combination of the anion and cation gives us a neutral compound. This suggests that our answer makes sense.

What If…?

…an ionic compound contained two sodium cations and one anion that contained 6 fluorine atoms and one silicon atom? What would be the formula of the compound?

(Ans: Na_2SiF_6)

Practice 2.9

What is the name of the compound with the formula $KMnO_4$? What is the formula for the compound called ammonium dichromate?

2.11 Naming Hydrates and Acids

Cave enthusiasts were intrigued to hear about a cave discovered near Naica, Mexico. The cave was found as workers were expanding a mineshaft at about 300 m below the surface. The natural cave, where the temperature is a balmy 57–58 °C, holds the world's largest single crystals, some of them as long as 40 m in length. The crystals are selenite, an example of a naturally occurring **hydrate** with the formula $CaSO_4 \cdot 2H_2O$. What is a hydrate?

Hydrate – Chemical compounds, often ionic in nature, which contain water molecules as part of their physical structure.

Hydrates are compounds, often ionic, which contain water molecules as part of the formula. Hydrates are not compounds that are dissolved in water; water is just not "associated" with the ionic compounds. Hydrates contain water molecules as part of the structure of the compound. In other words, the water in a hydrate is physically attached to the rest of the compound. Often, the water molecules can be easily removed by simply heating the compound. The names of these compounds help relay this fact. For example, copper(II) sulfate is often found as a hydrated compound with the formula $CuSO_4 \cdot 5H_2O$. The dot in the middle of the formula indicates that the five water molecules are included for each formula unit of copper(II) sulfate in the structure of the ionic compound. To name this class of compound we write the name of the ionic compound using the rules in ◘ Table 2.11 and then follow it with a Greek prefix and the word "hydrate" to indicate the water molecules. In this case, we write:

copper(II) sulfate pentahydrate

The chemical name for the hydrate that makes up the crystals of selenite follows the same rules of nomenclature. Calcium sulfate dihydrate ($CaSO_4 \cdot 2H_2O$) is the precursor to making gypsum plaster (also called plaster of Paris, so named because in the eighteenth century much of the construction in Europe used plaster from near Paris). When calcium sulfate dihydrate is heated, some of the water molecules are removed as steam forming calcium sulfate hemihydrate ($CaSO_4 \cdot ½H_2O$, or $CaSO_4 \cdot 0.5H_2O$). Note that not all the hydrate is eliminated entirely from this formula. While the formula hints that we have only one-half of a molecule of water in the ionic compound, this actually implies that one molecule of water exists for every two calcium sulfate formula units. When this white powder mixes with water, the water slowly re-assimilates into the crystals and reforms the dihydrate. Once reformed, the powdery gypsum plaster becomes hard, similar to the strength of the crystals of selenite.

Crystals like selenite may be hard and strong, but they can be easily destroyed by exposure to an **acid**. Acids are compounds that, when dissolved in water, create solutions with high H^+ (hydrogen cation) concentrations. Acids have a sour taste, as in vinegar (a solution of acetic acid), and can even create very corrosive solutions, such as the sulfuric acid in automotive batteries. We can identify this class of compounds because the first atom in the formula is often hydrogen. Acids containing only hydrogen and an anion, but no oxygen (such as HCl and HCN) are distinctive because they can exist as covalent compounds when in their pure form. Yet, they behave as ionic compounds when dissolved in water. Other types of acids contain oxygen atoms. In fact, the name oxygen comes from the Greek meaning "acid generator" because chemists originally thought that all acids had to have this element as part of their structure.

Acid – Chemical compounds which, when dissolved in water, produce H^+ ions.

To name acids, we follow one of two sets of rules as illustrated in ◘ Table 2.12. These rules apply if the compound has oxygen and/or if it is dissolved in water. We can discern whether the acid is dissolved in water by the

◘ **Table 2.11** Rules for naming hydrates

Step	Rule
1	Name the compound according to the Binary Ionic Naming Rules (◘ Table 2.8) using polyatomic ion names (◘ Table 2.10) as necessary
2	Use the appropriate Greek Prefix (◘ Table 2.7) to indicate how many water molecules are after the dot in the formula, followed by the word "hydrate"

◘ **Table 2.12** Rules for naming acids

Acids typically have a hydrogen atom written as the first atom in the formula	
Step	**Rule**
1	To name non-oxygen containing acids that are NOT dissolved in water, name the acid as a molecular compound
2	To name non-oxygen containing acids that ARE dissolved in water... a. Start the name with the prefix "hydro" b. Attach the name of the anion to this prefix with the suffix "-ic" c. Add the word "acid" as a second word
3	To name oxygen containing acids (whether dissolved in water or not)... a. Modify the polyatomic ion by replacing its ending... "ate" becomes "ic" "ite" becomes "ous" (examples: sulfate becomes *sulfur*ic, sulfite becomes *sulfur*ous; phosphate becomes *phosphor*ic, and phosphite becomes *phosphor*ous) b. Add the word "acid" as a second word

subscript "*aq*" written after the formula. Often, we can tell that we're dealing with an acid because the formula lists a hydrogen atom as the first atom. There are some exceptions to writing the formula with an "H" first that we will cover later in this text.

Muriatic acid is the common name of a rather hazardous commercial preparation of HCl dissolved in water that is often used to clean concrete. This preparation is made by dissolving gaseous hydrogen chloride, HCl(*g*), into water. Once dissolved, the resulting compound is called hydrochloric acid, and is represented by placing an (*aq*), for "aqueous" (meaning "in water"), after the formula, HCl(*aq*). Similarly, the deadly gas HCN(*g*), known as hydrogen cyanide, is renamed as hydrocyanic acid, HCN(*aq*), when dissolved in water. In its non-aqueous form, HF(*g*) is systematically named hydrogen fluoride. An aqueous solution of HF(*aq*) is named hydrofluoric acid and can be used to artistically etch glass.

Lakes and streams that are lined with limestone rocks have a high concentration of dissolved carbonate anions. Since the water is in direct contact with the atmosphere, there is another source of the carbonate anion: the dissolving of carbon dioxide into the water. Water that has carbon dioxide dissolved in it, such as open bodies of water and soft drinks, contains a small amount of H_2CO_3(*aq*). To name this acid, we follow the rules in Table 2.12. The ending of the anion "-ate" is changed to "-ic acid": carbonate becomes carbonic acid. Many soft drinks also contain H_3PO_4(*aq*) to add that special tartness that soda drinkers crave. This compound is known as phosphoric acid (derived from the name of the polyatomic phosphate ion, PO_4^{3-}). Note, that because these compounds contain oxygen, we do not use the prefix "hydro-" as part of the name.

Example 2.10—Acids and Hydrates

Complete the following table by supplying either the formula or the name of the indicated compound.

Formula	Name
HNO_3	
	sulfuric acid
$HClO_4$	
	hydrobromic acid
$Na_2SO_4 \cdot 10H_2O$	
$Zn(NO_3)_2 \cdot 2H_2O$	
	copper(I) sulfite trihydrate
	sodium perchlorate monohydrate

Asking the Right Questions

Are the compounds in the table molecules, ionic compounds, hydrates, or acids? Examine each of the names and/or formulas and determine their identity. Note that acids have a hydrogen atom written first in the formula or the word acid written in the name. Similarly, hydrates contain the formula for water offset by a dot, or the word hydrate in the name. Once identified, we can follow the rules in Tables 2.11 and 2.12 to determine the name or the formula.

Solution

Formula	Name
HNO_3	**nitric Acid**
H_2SO_4	sulfuric acid
$HClO_4$	**perchloric Acid**
HBr	hydrobromic acid
$Na_2SO_4 \cdot 10H_2O$	**sodium Sulfate Decahydrate**
$Zn(NO_3)_2 \cdot 2H_2O$	**zinc Nitrate Dihydrate**
$Cu_2SO_3 \cdot 3H_2O$	copper(I) sulfite trihydrate
$NaClO_4 \cdot H_2O$	sodium perchlorate monohydrate

Are Our Answers Reasonable?

Note the use of ionic compound naming to make sure we have identified the charges appropriately. In these answers, doing this makes reasonable names.

What If...?

... a compound is formed that contained the barium cation, hydroxide ion and 8 molecules of water? What are the name and formula of the compound? (*Ans: barium hydroxide octahydrate,* $Ba(OH)_2 \cdot 8H_2O$)

Practice 2.10

What is the formula or name for each of these compounds?

a. $MgSO_4 \cdot 7H_2O$
b. $Na_2CO_3 \cdot 10H_2O$
c. sodium acetate dihydrate
d. strontium hydroxide octahydrate

We've seen how a nomenclature system can allow us to communicate the names of different chemicals. This is essential because without such a system, we would need to memorize each of the names of the chemicals if we wished to talk about them. Unfortunately, this is often the case in learning the names of many of the minerals, gemstones, and rocks that exist in our world. However, the chemical makeup of these materials can be shared with others by understanding the rules for how chemical names are created. Selenite becomes calcium sulfate dihydrate; bauxite becomes aluminum oxide; and limestone becomes calcium carbonate. Knowing the formula helps us determine the name of the compound, just as knowing the name of the chemical compound can tell us the formula.

The Bottom Line

- The Law of Conservation of Mass states that the mass of the chemicals at the start of a reaction is equal to the mass of the chemicals at the end of the reaction.
- The Law of Definite Composition states that any particular chemical is always composed of its components in a fixed ratio, by mass.
- The Law of Multiple Proportions states that when the same elements can produce more than one compound, the ratio of the masses of the element that combine with a fixed mass of another element corresponds to a small whole number.
- Dalton's atomic theory states that every substance is made of atoms; atoms are indestructible; atoms of any one element are identical; atoms of different elements differ in their masses; and chemical changes involve rearranging the attachments between atoms.
- The Law of Combining Volumes states that when gases combine, they do so in small whole-number ratios, provided that all the gases are at the same temperature and pressure.
- Atoms are composed of electrons, protons, and neutrons. The modern model of the atom indicates a tiny, but dense, positively charged nucleus surrounded by a diffuse electron cloud.
- Isotopes of the same element differ in their number of neutrons. All atoms of the same element have the same atomic number (the same number of protons).
- The average atomic mass of an element (shown in the periodic table) is the weighted average of all the isotopes of that element.
- The periodic table of the elements lists all the known elements, arranged into periods and groups in a manner that reflects the chemical characteristics that particular elements share.
- We can predict whether an atom is likely to become a cation or an anion based on its location in the periodic table.
- Ions of opposite charge combine to make an electrically neutral ionic compound. Ionic compounds are generally brittle and have high melting points, and can dissolve in water by separating into cations and anions, which enables the solution to conduct electricity.
- A chemical formula makes use of the atomic symbols and subscripts, as needed, to represent the numbers of each atom or ion in a chemical compound.
- Molecules are distinct substances made up of two or more atoms linked together by sharing electrons between their nuclei.
- The purpose of chemical nomenclature is to have one name for one compound so everyone can communicate efficiently about chemicals with no confusion.
- There are different naming rules for covalent (molecular) compounds and ionic compounds.
- Some ions can be made up of many atoms and are called polyatomic ions.
- Hydrates have water molecules as part of an ionic crystal structure.
- Acids have one or more hydrogen cations (H^+) combined with an anion and have their own special naming rules.

Section 2.2 Early Attempts to Explain Matter

Skill Review

1. Exploring the unseen world often relies on indirect reasoning. Useful deductions led early philosophers and scientists to make conclusions about the possible existence of atoms. How would the formation of ice crystals on the branches of a bush support the ideas of Democritus?
2. Ancient practices, and some modern ones, involve burning incense. How would entering a room in which incense had recently been burned lead you to consider the world to be made of small, invisible particles?

3. The scientists who first used atoms to explain chemical changes assumed that the small, unseen pieces of matter could continually be rearranged into new combinations with new properties. Show how that same principle operates when the letters in the word dormitory are used to form several other words.
4. Taking the analogy in Problem 3 a bit further, explain what would have to happen in order for the exercise to illustrate the Law of Conservation of Mass.
5. In your own words, explain the Law of Conservation of Mass.
6. In your own words, explain the Law of Definite Composition.
7. Modern science is based on asking questions. In what way does a researcher determining the age of Earth ask the same basic questions posed by Democritus?
8. What connection exists between the exploration and observations of the United States space program and Democritus' questions?

Chemical Applications and Practices

9. A chemist obtains samples of many liquids containing only hydrogen and oxygen from around the world. Which of the following masses of oxygen and hydrogen would indicate that the sample was most likely water? Explain your reasoning.
   a. 2.8 g oxygen and 0.35 g hydrogen
   b. 3.69 g oxygen and 0.615 g hydrogen
   c. 8.34 g oxygen and 0.67 g hydrogen
   d. 15.0 g oxygen and 1.87 g hydrogen
10. Another chemist obtains samples of four solids containing only iron and oxygen from around the world. How many different compounds did this chemist find? Explain your reasoning.
   a. 55.8 g iron and 32.1 g oxygen
   b. 112 g iron and 48 g oxygen
   c. 14.0 g iron and 6.02 g oxygen
   d. 17.2 g iron and 7.38 g oxygen
11. Some compounds absorb water from the atmosphere in such a way that the water is chemically combined with the compound. The absorbed water is called water of hydration, and the compounds with absorbed water are known as hydrates. If 14.7 g of calcium chloride hydrate, when heated, loses 3.6 g of water, how many grams of water are lost when a 23.4-g sample is heated? To which law(s) did you refer in obtaining these results?
12. The white appearance of several magnesium-containing compounds causes an apparent resemblance among the compounds. One such compound was shown to contain 60.0% magnesium. Another sample of a magnesium-containing compound was found. What quantity of magnesium, in grams, would have to be present in a 34.6-g sample of the second compound in order for us to claim that it has the same composition as the first sample?

Section 2.3 Dalton's Atomic Theory and Beyond

Skill Review

13. Lavoisier used carefully determined masses to make discoveries about changes that occurred in chemical reactions.
   a. Which fundamental law did his work lead him to discover?
   b. Suppose Lavoisier evaporated the water from 250.0 g of seawater. If the remaining solids had a mass of 2.2 g, what would be the mass of the water that had evaporated?
14. A 200.0-g sample of a pure substance was composed of 132.9 g of copper and 67.1 g of sulfur. Another sample of the same substance, this time with a mass of 150.0 g, was brought to Joseph Proust to analyze. How many grams of copper and sulfur would you expect to find in the sample? What basic law in chemistry are you employing in order to make your determination?
15. Fill in the missing information for a compound of copper and oxygen, assuming that it is composed of a fixed ratio of copper to oxygen of 3.97:1.00.

Copper	Oxygen
17.0 g	_____ grams
_____ grams	11.5 g

16. Fill in the missing information for a compound consisting of sodium and oxygen, assuming that it is composed of a fixed ratio of sodium to oxygen of 1.44:1.00.

2

Sodium	Oxygen
6.0 g	_____ grams
_____ grams	50 g

Chemical Applications and Practices

17. Three samples containing the same number of molecules of carbon and oxygen containing compounds contain the same number of grams of carbon. If the first sample contains 1.201 g carbon and 1.60 g oxygen, what masses of oxygen are possible in the other two compounds?
18. If two samples of copper and chlorine containing compounds are compared, it is noted that the ratio of chlorine in the second sample is one and a half times the ratio of the chlorine in the first sample. What are possible formulas for these two compounds?
19. Copper and oxygen combine in more than one ratio. In one such compound it is found that 65 g of copper combine with 16 g of oxygen. If you assumed, as John Dalton might have, that the combination represented one atom of copper combining with one atom of oxygen, what would you predict as the copper-to-oxygen ratio in another compound between the two?
20. Using the first mass ratio presented in Problem 19 and your answer for the other ratio, determine how many grams of oxygen would combine with 10.0 g of copper in both compounds.
21. When the calcium is properly extracted, you can expect to obtain 40 g of calcium from 100.0 g of calcium carbonate. A 10.0-g sample of pure calcium chloride yields 6.40 g of chlorine. Which of these two compounds is a better source of calcium?
22. Three magnesium compounds have the following percentages of magnesium: Compound A: 28.6% magnesium; Compound B: 60.0% magnesium; Compound C: 41.4% magnesium. Why, according to Dalton, can there be such a great difference in the percentages of magnesium?
23. Water and hydrogen peroxide are both clear liquids that are made up of only hydrogen and oxygen. Water contains 2 g of hydrogen for every 16 g of oxygen. In hydrogen peroxide there are 2 g of hydrogen for every 32 g of oxygen. Show how these data illustrate the Law of Multiple Proportions.
24. Oxygen exists naturally in two combined forms. The diatomic oxygen that we use in breathing consists of two atoms of oxygen combined into one unit. Ozone consists of three atoms of oxygen joined into one unit. The relative masses of the oxygen and ozone units are 32 and 48. Does this agree with the Law of Multiple Proportions? Explain.

Section 2.4 The Structure of the Atom

Skill Review

25. If the discoverers of the particles that make up the atom had determined the charge on the electron to be −2, what would be the charge on the neutron and proton?
26. Describe what might have been the results of Rutherford's experiment if the charge on the electron had been reversed (that is, if it were +1 instead of −1).

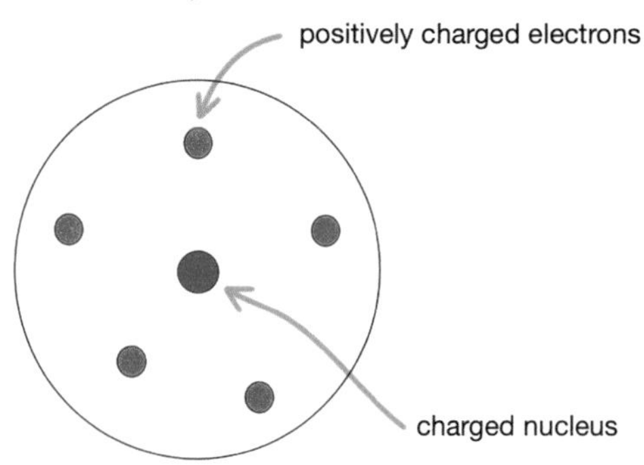

27. Draw diagrams of the models of the atom that J.J. Thomson and Rutherford, respectively, would have come up with if the electron had a +1 charge.
28. What are two objections to the explanation that positive protons compose a nucleus that is surrounded by negative electrons?

Section 2.5 Atoms and Isotopes

Skill Review

29. Explain what observations Rutherford would have seen if he had used β-particle radiation in his famous gold foil experiment.
30. Explain what observations Rutherford would have seen if he had used tin foil instead of gold foil in his experiments.
31. Fill in the information missing from the following table:

Isotope	Protons	Neutrons	Electrons	Charge
Carbon-12		6	6	0
	13	14	10	
Chlorine-35		18	–1	

32. Fill in the information missing from the following table:

Isotope	Protons	Neutrons	Electrons	Charge
Calcium-40	20		18	
	14	14		0
Sulfur-32		16	14	

33. How many electrons, protons, and neutrons are there in each of the following atoms?
    a. calcium-40, ^{40}Ca
    b. vanadium-51, ^{51}V
    c. tellurium-205, ^{205}Tl
    d. sulfur-32, ^{32}S;
    e. zinc-65, ^{65}Zn
    f. mercury-200, ^{200}Hg
34. How many electrons, protons, and neutrons are there in an atom of
    a. strontium-88, ^{88}Sr
    b. tin-118, ^{118}Sn
    c. plutonium-244, ^{244}Pu
    d. manganese-55, ^{55}Mn
    e. tungsten-183, ^{183}W
    f. indium-115, ^{115}In
35. Indicate whether each of these phrases describes the proton, neutron, or electron. (You may need to assign more than one term to each phrase.)
    a. Determines the identity of an atom.
    b. Has about the same mass as a proton.
    c. The number of particles can be changed without changing the identity of the atom.
36. Indicate whether each of these phrases refers to protons, neutrons, and/or electrons. (You may need to assign more than one term to each phrase.)
    a. Is represented by the mass number.
    b. Is represented by the atomic number.
    c. Are always present in the same quantity in an atom with 0 charge.
37. What would you calculate for the mass, in grams, of an atom that contained 1 proton and 1 electron?... 1 proton, 1 neutron, and 1 electron?
38. What would you calculate for the mass, in grams, of an atom that contained 2 protons, 2 neutrons, and 2 electrons?... 1 proton, 1 neutron, and 2 electrons?
39. What would be the charge on each of the particles listed in Problem 37?
40. What would be the charge on each of the particles listed in Problem 38?
41. Assume that an atom contained 1 proton and 1 electron. By what percentage would the mass of this atom increase if 10 electrons were added to the atom?
42. Assume that an atom contained 1 proton and 1 electron. By what percentage would the mass of this atom increase if 10 neutrons were added to the atom?

43. Which of the most abundant isotope in each pair contains the greater number of neutrons? Round the average atomic mass to the nearest whole number to obtain the mass number of the most abundant isotope. (Use the periodic table of the elements inside the front cover of your textbook.)
    a. iron or chromium
    b. tellurium or iodine
    c. cobalt or nickel
    d. helium or neon
44. Which of the most abundant isotope in each pair contains the greater number of neutrons? Round the average atomic mass to the nearest whole number to obtain the mass number of the most abundant isotope. (Use the periodic table of the elements inside the front cover of your textbook.)
    a. zinc or sodium
    b. hydrogen or sodium
    c. phosphorus or sulfur
    d. bromine or selenium
45. Arrange the following atoms in order of their quantity of protons (least number to greatest number): iron, hydrogen, calcium, fluorine, aluminum, boron.
46. Arrange these atoms in order of their quantity of protons (least number to greatest number): sodium, magnesium, copper, iodine, tin, carbon, lithium.
47. Arrange these atoms in order of their quantity of electrons (least number to greatest number): iron, hydrogen, calcium, fluorine, aluminum, boron.
48. Arrange these atoms in order of their quantity of electrons (least number to greatest number): sodium, magnesium, copper, iodine, tin, carbon, lithium.
49. Using the periodic table, determine the first element that does not have the same number of protons as neutrons. Round the average atomic mass to the nearest whole number to obtain the mass number of the most abundant isotope.
50. Using the periodic table, indicate two atoms that have the same number of protons as neutrons. Round the average atomic mass to the nearest whole number to obtain the mass number of the most abundant isotope.
51. In a neutral atom, the number of electrons must equal the number of protons. However, if an atom had two more protons than electrons, what would you report as its charge? Would this ion be called an anion or a cation?
52. If an atom had two more electrons than protons, what would you report as its charge? Would this ion be called an anion or a cation?

Chemical Applications and Practices

53. Calcium-containing ionic compounds are fairly common in our world. How many protons and electrons are found in one unit of these compounds:
    a. $CaCO_3$, limestone
    b. CaO, lime
    c. $CaCl_2$, component of sidewalk salt
    d. Ca^{2+}, the common ion of calcium
54. Give the total number of protons and electrons in an
    a. N_2 molecule
    b. Na^{-3} ion
    c. Na^{+5} ion
55. Three different isotopes of carbon are carbon-12, carbon-13, and carbon-14. How many protons, neutrons, and electrons does each isotope of carbon contain?
56. Radon-222, the radioactive gas that sometimes finds its way into our basements, is a decay product of uranium-238. How many protons, neutrons, and electrons are in these nuclides?
57. Several radioactive isotopes have medical applications. Using the nuclide representation shown here, determine the number of protons, electrons, and neutrons in an atom of each of these elements.
    Radioactive sodium is used in blood circulation studies: ${}^{24}_{11}Na$
    Radioactive iodine can be used to examine the thyroid gland: ${}^{131}_{53}I$
    Radioactive cobalt can be used to kill cancer cells: ${}^{60}_{27}Co$
    Radioactive chromium can be used to measure total blood volume: ${}^{51}_{24}Cr$
    Radioactive phosphorus is used in anti-leukemia therapy: ${}^{31}_{15}P$
58. Write nuclide notation for each of the following isotopes:
    carbon-12, carbon-13, phosphorus-32, chlorine-35, chlorine-37, iron-55.

Section 2.6 Atomic Mass

Skill Review

59. If the atomic mass of ^{12}C had been set equal to exactly 6 amu, what would be the mass of ^{16}O?... the mass of ^{1}H?

60. If the atomic mass of oxygen-16 were set equal to exactly 4 amu, what would be the mass of carbon-12?... the mass of copper-63?
61. Calculate the mass, in grams, of 1 atom of carbon-12.... of nitrogen-14.... of fluorine-19. (Assume that the mass number of these isotopes is equal to the atomic mass, rounded to the nearest whole number.)
62. Calculate the mass, in grams, of: a gross (144) of atoms of carbon-12.... nitrogen-14.... fluorine-19. (Assume that the mass number of these isotopes is equal to the atomic mass, rounded to the nearest whole number.)

Chemical Applications and Practices

63. The metal indium is used to make heat-conducting alloys. Indium typically is found with two isotopes. Based on the following information, what would you calculate as the percent relative abundance of the two isotopes? Indium-113 has a mass of 112.9043, and indium-115 has a mass of 114.9041.
64. The element neon is widely used in brightly colored electronic advertising signs. There are three common isotopes of neon with masses of 21.99, 20.99, and 19.99. One of the three isotopes makes up approximately 90% of the atoms in a neon sample. Which isotope is most likely to be responsible for the majority of neon's atomic mass reported on the periodic table?
65. The masses of three small apples are: 105.6 g, 136.1 g, and 120.5 g. What is the average mass of these apples?
66. Thirty tomatoes are placed into separate piles. In pile one, containing 12 tomatoes, the total mass is 1.08 kg. In pile two, containing just 4 tomatoes, the total mass is 0.854 kg. In pile three, containing the rest of the tomatoes, the total mass is 2.24 kg. What is the average mass of a single tomato?
67. The sulfur-containing amino acids that help us make certain proteins are critically important to our diet. Sulfur samples typically consist of four isotopes. Use the following table to determine the average atomic mass of sulfur.

Isotope	Atomic mass	Abundance (%)
Sulfur-32	31.97	95.0
Sulfur-33	32.97	0.76
Sulfur-34	33.97	4.22
Sulfur-36	35.97	0.014

Sketch a mass spectrum that illustrates the masses observed for a natural sample of sulfur.

68. Silver has many uses in our world, only one of which is as a precious metal in coinage. Use the following table to determine the average atomic mass of silver.

Isotope	Atomic mass	Abundance (%)
Silver-107	106.9051	51.84
Silver-109	108.9048	48.16

69. Bromine has two stable isotopes: ^{79}Br, and ^{81}Br, with masses of 78.9183371 amu and 80.9162906 amu, respectively. Knowing that the atomic mass of bromine is 79.904 amu, calculate the relative abundance of each isotope.
70. Chlorine has two stable isotopes that are found in nature, ^{35}Cl and ^{37}Cl, with masses of 34.96885268 amu and 36.96590259 amu, respectively. Knowing that the atomic mass of chlorine is 35.453 amu, calculate the relative abundance of each isotope.
71. Cerium has four naturally occurring isotopes. Use the following table to determine the average atomic mass of cerium.

Isotope	Atomic mass	Abundance (%)
Cerium-136	135.907	0.19
Cerium-138	137.906	0.25
Cerium-140	139.905	88.48
Cerium-142	141.909	11.08

Sketch a mass spectrum that illustrates the masses observed for a natural sample of cerium.

72. A hypothetical element was just discovered to have three isotopes. Use the following table to determine the average atomic mass of this hypothetical element.

Isotope	Atomic mass	Abundance (%)
A-100	99.754	35.25
A-102	101.688	25.75
A-103	102.599	39.00

Section 2.7 The Periodic Table

Skill Review

73. Use the descriptions given in this chapter of the organization of the periodic table to identify each of these elements.
    a. The element that is in the second column and in the second period.
    b. The transition element that is in the seventh group and in the fourth row.
    c. The third most massive noble gas.
74. Use the descriptions given in this chapter of the organization of the periodic table to identify each of these elements.
    a. The lightest alkali metal.
    b. The semimetal that is in the third group.
    c. The halogen that is a liquid at room temperature.
75. How many elements are classified as halogens?
76. Which column contains the alkali metals?…the chalcogens?
77. Identify the name of the group for each of these elements: sulfur, iodine, helium, beryllium, francium
78. Identify the name of the group for each of these elements: lithium, barium, neon, oxygen, chlorine
79. Which of these elements would you predict to be shiny?: gold, silver, lead silicon, carbon, iodine
80. Which of these elements would you predict to be brittle?: gold, silver, lead silicon, carbon, iodine

Chemical Applications and Practices

81. The next element to be discovered is likely the next element in the periodic table. To what group and period would that element belong?
82. Element 118 is a recent discovery that has been added to the periodic table. To which named group does it blong? Does that say anything about its stability?

Section 2.8 Ionic Compounds

Skill Review

83. How many electrons are there in each of the ions that make up the following compounds?
    a. NaCl
    b. KBr
    c. Na_2O
    d. CaS
84. How many electrons are there in each of the ions that make up the following compounds?
    a. $AlCl_3$
    b. $TiCl_4$
    c. WO_2
    d. LiF
85. One of the first considerations in writing ionic formulas is to determine which element is to be the cation and which is to be the anion. Decide which element in each of these pairs is more likely to become the positive ion.
    a. Ca or Br
    b. Sr or N
    c. S or Al
    d. I or Ba
86. Use the information from the previous question to determine what formula is most likely to result from the combination of each of these pairs of elements.
    a. Ca and Br
    b. Sr and N
    c. S and Al
    d. I and Ba
87. Write the formulas of the ionic compounds that form when chlorine combines individually with lithium, barium, sodium, calcium, and aluminum.
88. Write the formulas of the ionic compounds that form when oxygen combines individually with lithium, beryllium, sodium, calcium, and aluminum.
89. Ionic compounds typically dissociate into ions as they dissolve in water. For each of these compounds, predict the ions, including their charges, that will be produced when the ionic compound dissolves.
    a. $MgBr_2$
    b. KI
    c. $FeCl_3$
    d. Na_2S

90. Ionic compounds typically dissociate into ions as they dissolve in water. For each of these compounds, predict the ions, including their charges, which will be produced when the ionic compound dissolves.
    a. $CaCl_2$
    b. ZnO
    c. TiO_2
    d. $NiCl_2$.

Section 2.9 Molecules

Skill Review

91. The two major classifications of compounds are ionic and molecular. Label the descriptions given below as characteristic of an ionic compound or of a molecule.
    a. Results from a metal combining with a nonmetal.
    b. Persists as a complete entity when dissolved in water.
    c. Formula can often be determined by the identity of the atoms that combine.
92. The two major classifications of compounds are ionic and molecular. Label the descriptions given below as characteristic of an ionic compound or of a molecule.
    a. Results as atoms share their electrons.
    b. Formula represents a ratio of ions involved rather than actual numbers of atoms combined.
    c. Results from the combination of two nonmetals.
    d. Separates into individual charged particles when dissolved in water.
93. Which of these compounds are not molecular compounds?
    a. CO
    b. BrO_3
    c. K_2O
    d. PCl_3
    e. WO_2
94. Which of these compounds are molecular compounds?
    a. SO_3
    b. NiO
    c. BCl_3
    d. Na_2S
    e. SiF_4

Section 2.10 Naming Compounds

Skill Review

95. Provide the correct name for each of these binary molecular compounds.
    a. SO_2
    b. PCl_3
    c. N_2O_5
    d. CCl_4
96. Provide the correct name for each of these binary molecular compounds.
    a. NO_2
    b. N_2O
    c. PBr_5
    d. P_2O_5
97. Provide the correct name for each of these binary ionic compounds.
    a. K_2O
    b. $AlCl_3$
    c. $CaBr_2$
    d. BaS
98. Provide the correct name for each of these binary ionic compounds.
    a. CaO
    b. Al_2O_3
    c. $MgCl_2$
    d. LiF
99. Provide the correct name for each of these ionic compounds.
    a. $CaBr_2$
    b. NH_4Cl
    c. $Fe(NO_3)_3$
    d. NaCl
100. Provide the correct name for each of these ionic compounds.
    a. $KHCO_3$
    b. $Cu(NO_3)_2$
    c. $Ca(CN)_2$
    d. $MgSO_3$
101. Provide the correct formula for each of these compounds.
    a. copper(II) hydroxide
    b. chromium(III) oxide
    c. sulfur hexachloride
    d. carbon tetraiodide
    e. aluminum hydroxide
    f. magnesium sulfate
    g. sodium sulfite
    h. ammonium hydroxide
    i. boron tribromide
    j. sodium acetate

102. Provide the correct formula for each of these compounds.
   a. hydrogen chloride
   b. manganese(IV) fluoride
   c. dinitrogen tetroxide
   d. calcium perchlorate
   e. barium nitrate
   f. lead(IV) chloride
   g. potassium dichromate
   h. sodium bicarbonate
   i. lithium hydroxide
   j. titanium(IV) carbonate
103. Using the metal ion Mn^{5+}, write the correct formula for a combination with each of the following: sulfate; chloride; nitrite; carbonate; and bisulfite.
104. Using the metal ion Cr^{6+}, write the correct formula for a combination with each of the following: sulfate; chloride; nitrite; carbonate; and bisulfite.
105. Write the correct formula for phosphate compounds of the following metals: Mn^{2+}; Cu^{+}; Fe^{3+}; Mn^{5+}; and Ti^{4+}.
106. Write the correct formula for nitrate compounds of the following metals: Mn^{2+}; Cu^{+}; Fe^{3+}; Mn^{5+}; and Ti^{4+}.
107. Name these compounds.
   a. $(NH_4)_2CO_3$
   b. $Cu(HSO_3)_2$
   c. $KMnO_4$
   d. $Mg(CN)_{2\,h}$
   e. $NaHCO_3$
   f. $Ca(OH)_2$
   g. Na_3PO_4
   h. $LiClO_3$
108. Name these compounds.
   a. $Zn(CN)_2$
   b. Na_2CrO_4
   c. $Li_2Cr_2O_7$
   d. $SrSO_4$
   e. Na_2CO_3
   f. K_2HPO_4
   g. $Ba(NO_2)_2$
   h. $KClO$
109. Determine the charge on the metal cation in these compounds.
   a. $V(NO_3)_5$
   b. $TiSO_4$
   c. $W(C_2O_4)_3$
   d. $AgOH$
   e. $Ru(HCO_3)_3$
110. Determine the charge on the metal cation in these compounds.
   a. $Cr(NO_3)_3$
   b. MnO_2
   c. $Pd(C_2O_4)_2$
   d. $AuCl_3$
   e. $Co(OH)_3$
111. Write the chemical formula for the ionic compound formed between each of the following pairs of ions. Name each compound.
   a. K^{+} and SO_4^{2-}
   b. Sr^{2+} and CO_3^{2-}
   c Ca^{2+} and PO_4^{3-}
   d. Be^{2+} and NO_2^{-}
   e. Pb^{2} and CO_3^{2-}
112. Write the formula of the ionic compound produced by the combination of each of the following pairs of elements. Name each compound.
   a. potassium and iodine
   b. calcium and bromine
   c. lithium and sulfur
   d. strontium and oxygen
   e. aluminum and sulfur
   f. copper(II) and bromine

Chemical Applications and Practices

113. A lab accident caused part of the label on a bottle of a chemical to be obscured or removed. All that is remaining is "$Fe(NO_3$." Is this enough information to name the compound? What possible suggestions could you make to determine the identity of the compound?
114. The same lab accident caused part of the label on another bottle to be obscured or removed as well. The remaining part reads "NH_4C". Which of the following would you suggest as the most likely name of the compound in this bottle? Explain why you reject the others: ammonium carbonate, ammonium oxalate, ammonium chloride, ammonium copper(II).

Section 2.11 Naming Hydrates and Acids

Skill Review

115. Provide the name for each of the following hydrates:
   a. $Ba(OH)_2 \cdot 8H_2O$
   b. $HgNO_3 \cdot 2H_2O$
   c. $CuCl_2 \cdot 2H_2O$
   d. $CaSO_4 \cdot 3H_2O$
   e. $Mg(NO_2)_2 \cdot 6H_2O$
   f. $Mn(SO_4)_2 \cdot 7H_2O$

116. Provide the formula for each of the following hydrates:
   a. iron(II) fluoride tetrahydrate
   b. magnesium chromate heptahydrate
   c. copper(I) chlorate dihydrate
   d. tin(IV) chloride pentahydrate
   e. cobalt(II) bromide trihydrate
   f. lead(II) perchlorate hexahydrate
117. Provide the name for each of the following acids:
   a. HBr
   b. HF
   c. HCN
   d. HNO_3
   e. $HClO_3$
   f. H_3PO_4
118. Provide the formula for each of the following acids:
   a. acetic acid
   b. hydroiodic acid
   c. chlorous acid
   d. chromic acid
   e. sulfurous acid
   f. sulfuric acid
119. Provide the name of the acid from which each of the following polyatomic ions was obtained:
   a. HPO_4^{2-}
   b. HSO_3^-
   c. NO_2^-
   d. ClO_4^-
120. Indicate all of the ions that arise when the following acids are dissolved in water:
   a. HCl
   b. $HClO_3$
   c. $HMnO_4$
   d. H_2CO_3

Comprehensive Problems

121. What were the major objections to Dalton's atomic theory? Rewrite Dalton's atomic theory to take into account our current understanding of matter.
122. Match each chemical statement on the left-hand side of the following list with a correlating statement from Dalton's atomic theory on the right.

Chemistry	Atomic theory
a. Graphite and diamond have different properties, but both are made of only atoms of carbon	(1) Elements are made up of tiny particles called atoms
b. $2H_2 + O_2 \rightarrow 2H_2O$	(2) Atoms of a given element are identical; atoms of different elements differ in some fundamental way
c. Carbon dioxide made by fermentation has the same formula as carbon dioxide formed during a campfire	(3) Chemical reactions consist of the reorganization of the atoms involved; the atoms themselves are not changed
d. Helium has two protons in its nucleus. Lithium has three protons in its nucleus	(4) Compounds are formed when atoms combine; each compound always has the same number and types of atoms

123. When hydrogen gas combines with chlorine gas, a corrosive gas known as hydrogen chloride forms. According to Dalton, the volume ratio between the reactants and products should be 1:1:1. However, Gay-Lussac would claim a ratio of 1:1:2 (assuming that all gases are at the same temperature and pressure). Gay-Lussac's claim was proved to be valid. Which law is being used here? Explain why this explanation works better than Dalton's.
124. If the mass of a metal nut was 35.0 g and the mass of a matching threaded bolt was 7.0 g, a combined nut and bolt would have a mass of 42.0 g.
   a. What would be the mass of the threaded bolts needed to attach to 175.0 g of nuts?
   b. If someone decided to attach two bolts to a nut, what would be the resulting mass of the unit?
   c. What would be the mass of the threaded bolts needed to attach to 175.0 g of nuts under the conditions stated in part b?
   d. What is the mass ratio between bolts connected to nuts in the first unit (part a) and bolts connected to nuts in the second unit (part b)?
125. Robert Millikan determined the electrical charge for an electron, and its mass was easily calculated by using the charge-to-mass ratio. The following analogy applies roughly the same logic: Suppose you were buying a box of golf balls. What information would you need in order to know the price of each golf ball?
126. A tiny oil droplet is given an electrical charge by an electric current. The charge causes the oil droplet to be attracted to an oppositely charged metal plate above it. What two forces must be equal in order for an oil droplet to be suspended (neither rising nor falling) in the apparatus?

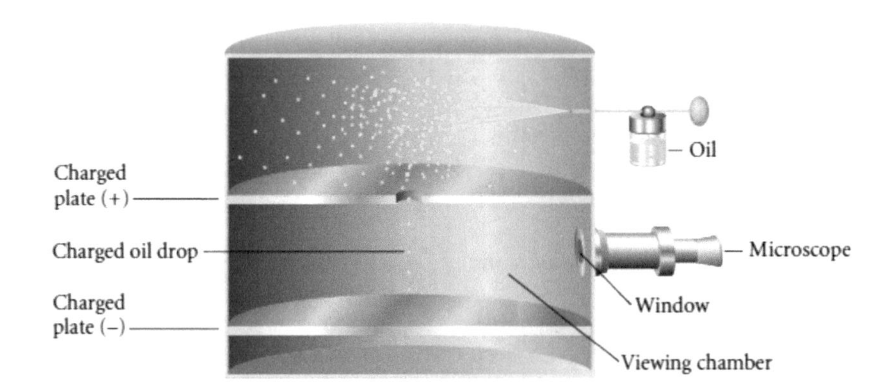

127. If the atoms in a 2.500 g sample of aluminum were lined up end to end, how long would the line of aluminum be in inches? (Assume the aluminum atoms are spheres with a radius of 143 pm.)
128. If a sample of carbon atoms contains 7.45×10^{15} atoms, how many gross of atoms are there in the sample (1 gross = 144 atoms)? If each of those atoms is 154 pm in diameter, would they stretch to 1 mile if they were lined up so that they were touching end to end?
129. During the analysis of an unknown sample of metal, a researcher places a 16.07 g chunk of the metal into a graduated cylinder containing 20.3 mL of water. The volume of the mixture was then recorded to be 22.3 mL. What is the density of the metal?
130. A new isotope of beryllium is created such that it contains 9 neutrons.
    a. How many protons and electrons would an atom of this isotope contain?
    b. What is the mass number for this new isotope?
    c. Use the information in ◘ Table 2.3 to determine the mass of the isotope in kilograms. (Assume that no mass is lost in the creation of the new isotope from the individual parts.)
131. Copper(II) oxide is a red powdery solid, while copper(III) oxide is a black sand-like substance also known as the mineral "tenorite." Which of these two ionic solids do you think will have the higher melting point, and why?
132. Solid sulfur can react with solid sodium to potentially make two different compounds.
    a. To which groups do these elements belong in the periodic table? Based on that information, which of the two elements would you expect to be shiny and malleable?
    b. If the mass of sulfur is held constant at 1.00 g, the mass of sodium in the first sample is 0.719 g. What could be the mass of the sodium in the second sample? Whose law does this represent?
    c. How many electrons, protons, and neutrons would you expect to find in a neutral atom of ^{32}S?
133. Define the term "allotrope" and do some research to find information about the possible allotropes of sulfur.
134. In the combustion of methane (CH_4), a small amount of nitrogen gas is converted to its oxides. In fact, a large number of compounds containing nitrogen and oxygen are possible from this reaction. Collectively, they are known as NO_x. Name the compounds NO, N_2O, NO_2, N_2O_2, and N_2O_4.
135. Research the mineral corundum on the Internet. What is its formula and name? If the oxygen anions were replaced with sulfur anions, what would the formula for the resulting mineral be?
136. Galena is a common mineral. Research its formula and where it is mined using the Internet. What is its chemical name? If the metal in galena had a +4 charge, what would the new formula for the resulting mineral be?

Thinking Beyond the Calculation

137. Scientists now think that much of the universe is composed of "dark matter." Investigate some scientific journals, such as Scientific American, to learn enough about dark matter to describe it. On the basis of your research, discuss the experiments that compelled scientists to draw this conclusion. Why is dark matter useful to understand?
138. In this chapter, we discussed determining the age of objects using radioactive isotopes. Scientists are constantly refining these techniques to correct for the differences between conditions long ago and those at the present time. What conditions might have changed that could require correcting? In each case, what would be the nature of the correction (that is, whether the real age is greater or less) and why?

139. Iron is often used in construction of buildings, however, unless it is coated with paint or other protectant, it can rust. This rusting results in the formation of many compounds, including iron(III) oxide.
    a. Write the formula for iron(III) oxide.
    b. Name the compound FeO. Is this an ionic or molecular compound?
    c. If iron(III) oxide were to dissolve in water, what ions would result?
    d. Assume that a sample of 100 g of iron was converted entirely to iron(III) oxide. Would the mass of the sample increase or decrease? Where did the additional mass come from (what is the source of the additional mass?)
    e. What would be the result if the iron reacted in a similar fashion with nitrogen? Name the compound.

Chemical Calculations—Introducing Quantitative Chemistry

Contents

Supplementary Information The online version contains supplementary material available at ▶ https://doi.org/10.1007/978-3-030-90267-4_3.

M. Mosher and P. Kelter, *An Introduction to Chemistry*,
https://doi.org/10.1007/978-3-030-90267-4_3

What Will We Learn from This Chapter?

- The masses of atoms and molecules can be quantified using many different units (▶ Sect. 3.2).
- Avogadro's number is the number of particles, molecules, atoms, etc. in a given amount of substance (▶ Sect. 3.3).
- The mole is a counting number used to determine the number of "things" per a unit of mass (▶ Sect. 3.3).
- The mole can be used to simplify the calculations associated with chemical equations (▶ Sect. 3.4).
- Mass percent is calculated by considering the mass of an atom or group of atoms versus the total mass of a compound (▶ Sect. 3.5).
- The formula mass for a particular compound is determined by adding the average atomic masses of the individual atoms that make up the compound (▶ Sect. 3.6).
- Chemical equations are used to relate the amount of substances that react to make products (▶ Sect. 3.7).
- The amount of product can be calculated from the chemical equation (▶ Sect. 3.8).

3.1 Introduction

A sudden, loud noise, the shock of some dramatic twist in a horror film, losing control of your car on an icy road—any one of these can cause the rapid physical responses we attribute to the release of adrenaline, including increasing heart rate, stronger heartbeat, sweating, and shifting of some blood flow from the skin to the internal organs. Adrenaline, also known as epinephrine (pronounced EP-in-EFF-rin) (◘ Fig. 3.1), is normally present in our blood at approximately 2×10^{-8} g/L and is secreted from the adrenal glands near our kidneys. In times of extreme stress, our adrenaline level increases a thousandfold to 2×10^{-5} g/L. The powerful effects of adrenaline demonstrate the significance of changing quantities of chemicals. The chemistry of the body provides many other examples. For instance, taking 600 mg of aspirin can effectively relieve pain. If you took 100 times that amount (or 60 g), the result would be fatal. Quantities in chemistry are vital.

Up to this point, we have been dealing with individual atoms and molecules. However, in the lab we typically measure quantities in liters, grams, and other "macro-sized" units. Pharmaceutical companies manufacture millions of tons of consumer health-related products each year, including about 40,000 tons of aspirin alone—truly a macroworld process. How do we relate "nanoworld" and "macroworld" quantities? Specific questions arise about our introduction to this chapter on adrenaline: What is the mass of a molecule of adrenaline? How many molecules of adrenaline exist in each liter of blood? These are the kinds of quantitative issues we explore in this chapter.

3.2 Formula Masses

In addition to being naturally present in the body, adrenaline is administered as a drug to stimulate the heart, to alleviate allergic reactions, and even to help break up fat cells during liposuction. As we might expect, control over the amounts administered is necessary to ensure safe and proper medical treatment. To manufacture, use, or detect specific quantities of adrenaline, we need information about the mass of its molecules, so we may know how many molecules are in any given amount that is administered to a patient.

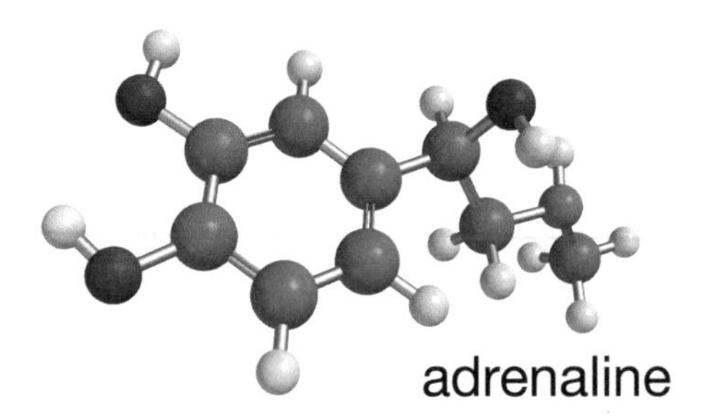

◘ **Fig. 3.1** Adrenaline (epinephrine) has the chemical formula $C_9H_{13}NO_3$. In this "ball-and-stick" representation, carbon atoms are charcoal gray, hydrogen atoms are light blue, oxygen atoms are red, and nitrogen atoms are dark blue

In ▶ Chap. 2, we looked at the method for determining the masses of atoms, and we saw that each element has a given mass (expressed in atomic mass units, amu), which is the average mass of one atom of the element. We can calculate the average **molecular mass** (or molecular weight) of a molecule, such as adrenaline, by adding together the average atomic masses of all the atoms present in the molecule. In these calculations, we tend to only keep 2 decimal places in the final answer.

Molecular mass – The mass, in atomic mass units (amu), of one molecule of a substance. It is obtained by addition of the average atomic masses of each of the atoms in the molecule.

Formula of each adrenaline molecule = $C_9H_{13}NO_3$.
In the formula we have

9 carbon atoms × 12.011 amu	=	108.099
13 hydrogen atoms × 1.008 amu	=	13.104
1 nitrogen atom × 14.0067 amu	=	14.0067
3 oxygen atoms × 15.9994 amu	=	47.9982
Average molecular mass of adrenaline	=	183.204
to 2 decimal places	=	183.20 amu

Note that we keep an extra significant figure in our calculations during the calculation for the formula mass, although we only keep 2 decimals in the final answer. The result is the molecular mass of adrenaline. **Formula mass** is a more general term than molecular mass, because it refers to compounds that do not exist in the form of individual molecules, such as ionic compounds. For example, the ionic compound sodium chloride (NaCl) has an average formula mass of 58.442 amu, found by adding the atomic mass of sodium, 22.9898 amu, to the atomic mass of chlorine, 35.453 amu.

Formula mass – The total mass of all the atoms present in the formula of an ionic compound, in atomic mass units (amu).

NaCl	
Na	22.9898 amu
Cl	35.453 amu
Total	**58.4428 amu = 58.44 amu**

The formula mass, or molecular mass, of adrenaline is the average mass of one molecule of adrenaline. Each molecule of the substance contains exactly the same number of each type of atom, but different individual molecules can have different masses, depending on which and how many isotopes of carbon, oxygen, or hydrogen they contain.

Example 3.1—Calculating Formula Masses

1. The formula for aspirin is $C_9H_8O_4$. What is the molecular mass of aspirin?

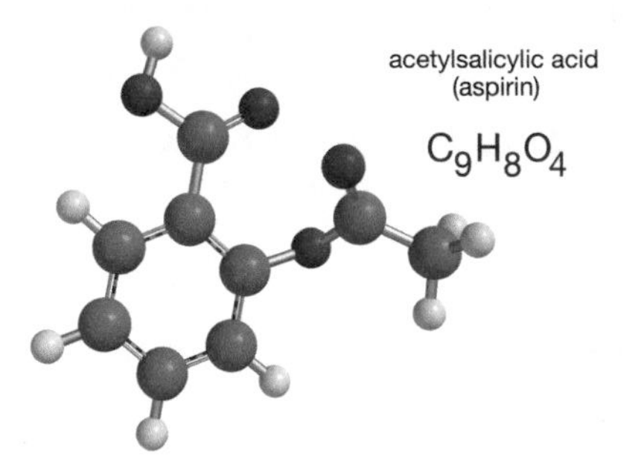

2. Many of us enjoy a good cup of coffee in the morning to help us start our day. One of the molecules that produces the aroma of coffee is C_5H_6OS (2-furylmethanethiol). Calculate the molecular mass of 2-furylmethanethiol.

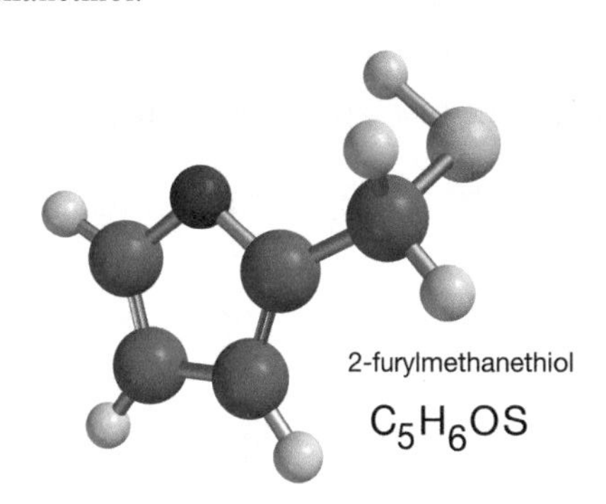

3

Asking the Right Questions

For which molecule are we calculating the molecular mass? How many of each atom exist in the molecule and what are their individual masses? The subscripts in the formula tell us the number of each atom and the periodic table contains the individual atomic masses in amu's.

Solution

1. Aspirin = $C_9H_8O_4$
   9 atoms of carbon × 12.011 amu = 108.099
   8 atoms of hydrogen × 1.008 amu = 8.064
   4 atoms of oxygen × 15.9994 amu = 63.9976
   Molecular mass = 180.1606 = 180.16 amu
2. 2-Furylmethanethiol = C_5H_6OS
   5 atoms of carbon × 12.011 amu = 60.055
   6 atoms of hydrogen × 1.008 amu = 6.048
   1 atom of oxygen × 15.9994 amu = 15.9994
   1 atom of sulfur × 32.065 amu = 32.065
   Molecular mass = 114.1674 = 114.17 amu

Are our Answers Reasonable?

Since the question asks for molecular masses, we expect the answers to be in atomic mass units, and they are, so the units are correct. Also, since we are adding these numbers together, we are limited to the least number of significant digits to the right of the decimal point. In the first case, we are limited to two digits to the right of the decimal.

What If…?

…after our daily workout, we found aspirin to be ineffective for an ankle sprain, and tried ibuprofen, $C_{13}H_{18}O_2$? Without doing the calculation, would we expect the molecule to have a greater or lesser mass than aspirin? (*Ans: Ibuprofen has a greater mass, because it has more atoms with a higher atomic mass than aspirin*) What is the mass of a molecule of ibuprofen? (*Ans*: 206.29 amu)

Practice 3.1

Calculate the formula masses of these compounds:

(a) codeine, a painkiller with the formula $C_{18}H_{21}NO_3$
(b) magnesium sulfate, found in many medical ointments, with the formula $MgSO_4$
(c) trinitrotoluene (TNT) an explosive with the formula: $C_7H_5N_3O_6$.

3.3 Counting by Weighing

When we 'do' chemistry, we generally begin by measuring our chemicals using masses. However, we need to convert these masses into some measure of the numbers of atoms, molecules, or ions in order to relate the masses to what is actually happening in the chemical nanoworld. Why can't we take a sample of adrenaline, which comes as a white powder, and use a pair of tweezers to separate out the number of individual molecules we need? Adrenaline molecules, like all molecules, are stunningly small—on the order of 1 nm across. This means that you would need to line up 100 million adrenaline molecules with your tweezers to get a sample as long as your index finger (about 10 cm)! This is why we need to establish the link between weighing things and counting them. Weighing a sample of adrenaline gives us insight into how many molecules of adrenaline we have.

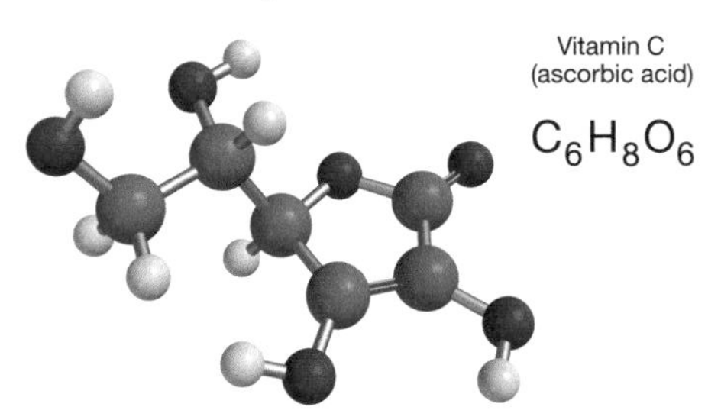

We can see how this is done on a more familiar size scale by using one tablet of vitamin C that weighs 0.75 g. If all the tablets in a jar are placed onto a balance, and if they weigh a total of 45.0 g, how many tablets are in the jar? As shown in Fig. 3.2, there are a total of 60 tablets in the jar.

$$45.0\,\text{g} \times \frac{1\ \text{tablet}}{0.75\,\text{g}} = 60\ \text{tablets}$$

We can do exactly the same thing with atoms, molecules, and ionic compounds, the only difference being that the numbers are most often extraordinarily large or small.

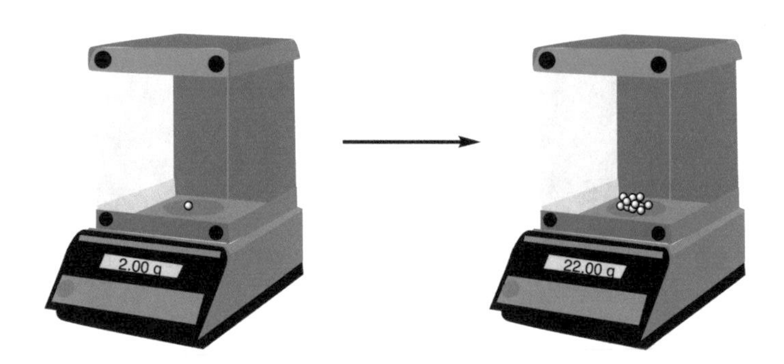

■ **Fig. 3.2** Counting tablets by weighing. If one tablet has a mass of 2.00 g, and a pile with a mass of 22.00 g should mean that there are 11 tablets in the pile

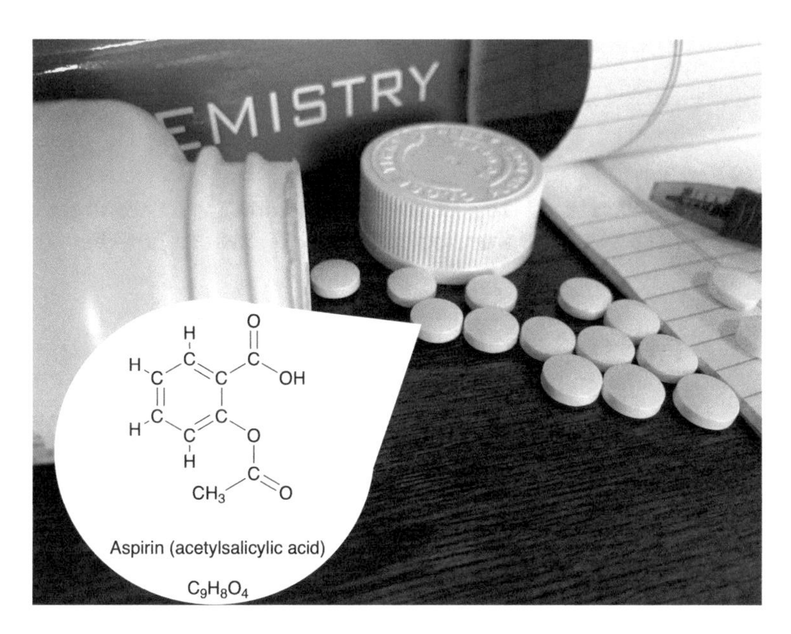

■ **Fig. 3.3** Aspirin tablets contain acetylsalicylic acid

Suppose we had just taken a pain reliever that contained aspirin, also known as acetylsalicylic acid (■ Fig. 3.3). The chemical formula for aspirin is $C_9H_8O_4$. If the tablet we used contained 500 mg (0.500 g) of aspirin, could we calculate how many molecules of pain reliever we had just ingested? Recall from the ▶ Chap. 2 that 1 amu = 1.6605×10^{-24} g (1.6605×10^{-27} kg).

The average molecular mass of aspirin, calculated in Example 3.1, is 180.16 amu.

$$\frac{180.16\ \cancel{\text{amu}}}{1\ \text{molecule}} \times \frac{1.6605 \times 10^{-24}\ \text{g}}{1\ \cancel{\text{amu}}} = 2.9915568 \times 10^{-22}\ \text{g per molecule}$$

$$(0.500\ \text{g})\left(\frac{1\ \text{molecule}}{2.9915568\ x\ 10^{-22}\ \text{g}}\right) = 1.67137 \times 10^{21}\ \textit{molecules}$$

$$= 1.67 \times 10^{21}\ \textit{molecules}$$

This quantity is approximately 300 billion times more than the current human population of the earth! The comparison serves as a reminder of just how tiny molecules must be if that many molecules have a mass of only half a gram.

Knowing that the average molecular mass of aspirin is 180.16 amu, a chemist at a pharmaceutical plant could weigh out a 180.16 g sample. How many molecules would that sample contain? To determine this, we would want to know just how many amu's are in 1 g. The mass of 1 amu = 1.6605×10^{-24} g, so we can set up a ratio and solve the equation:

$$\frac{1\ \text{amu}}{1.6605 \times 10^{-24}\ \text{g}} = \frac{x\ \text{amu}}{1\ \text{g}}$$

Fig. 3.4 The link between the amu and the gram. A single molecule of water has a mass of 18.02 amu. One mole of water has a mass of 18.02 g

$$x = 6.022 \times 10^{23} \ \frac{\text{amu}}{\text{g}}$$

That is a *huge* number! (Fig. 3.4)

How many molecules of aspirin are in 180.16 g of aspirin? We know the answer from our ratio above, but let's use the dimensional analysis method to solve the problem. First, we create a flowchart that will help us arrive at our answer.

$$\text{g aspirin} \xrightarrow{\frac{\text{amu}}{\text{grams}}} \text{amu aspirin} \xrightarrow{\frac{\text{molecules}}{\text{amu}}} \text{molecules aspirin}$$

Then we perform the calculation

$$180.16\ \cancel{\text{g aspirin}} \times \frac{1\ \cancel{\text{amu aspirin}}}{1.6605 \times 10^{-24}\ \cancel{\text{g aspirin}}} \times \frac{1\ \text{molecule aspirin}}{180.16\ \cancel{\text{amu aspirin}}}$$
$$= 6.022 \times 10^{23}\ \text{molecules aspirin}$$

This is the same huge number as the number of amu's in 1 g of a substance!

By choosing to weigh out a mass in grams that has the same numerical value as the formula mass in amu's (180.16 g and 180.16 amu, in the case of aspirin) we can ensure that the sample contains just as many molecules as there are amu's in 1 g. This gives us a very useful link between masses and numbers of molecules (or atoms or ions) that works for any chemical. Using this relationship, we can weigh out any amount of any substance and always know how many molecules there are in that sample.

Let's do the same calculation with an element, rather than a compound. The element calcium is crucial to the manufacture of healthy bones and teeth. The average atomic mass of the element calcium is 40.08 amu. How many calcium atoms are in 40.08 g of calcium?

$$40.08\ \cancel{\text{g}} \times \frac{1\ \cancel{\text{amu}}}{1.6605 \times 10^{-24}\ \cancel{\text{g}}} \times \frac{1\ \text{atom}}{40.08\ \cancel{\text{amu}}} = 6.022 \times 10^{23}\ \text{atoms}$$

The number 6.022×10^{23} appears again! This is a very important and useful number for us, because if we weigh out a mass *in grams* that is the same as the value of an element's atomic mass or a molecule's formula mass *in amu*, we will have 6.022×10^{23} atoms, molecules, or formula units.

The number 6.022×10^{23} is known as ***Avogadro's number*** (N_A) (also known as the Avogadro constant), after the Italian physicist Amadeo Avogadro (1776 – 1856; Fig. 3.5). The currently accepted value, approved by the International Union of Pure and Applied Chemists (IUPAC) in 2019, is: $6.02214076 \times 10^{23}$. However, for the calculations you will be doing, using 4 significant digits is precise enough (i.e., 6.022×10^{23}).

Avogadro's number (N_A) – The number of particles of a substance in 1 mol of that substance. The number has been defined as exactly $6.02214076 \times 10^{23}$ particles/mol.

Fig. 3.5 Lorenzo Romano Amadeo Carlo Avogadro, conte di Quaregna e di Cerreto (1776–1856), was born into a family of well-established lawyers in Italy. He, too, prepared for a legal career, obtaining his law degree in 1792 when he was only 16 years old. However, in 1800 he began studying mathematics and physics privately, completing his first research project on electricity with his brother Felice in 1803. He was hired in 1809 at the College of Vercelli, where he began his most influential work in chemistry. He eventually obtained an appointment as a professor of mathematical physics at the University of Turin, where he worked until he retired in 1850. Image in Public Domain. ▶ https://upload.wikimedia.org/wikipedia/commons/4/40/Amadeo_Avogadro.png

Example 3.2—How Many Molecules?

Codeine, with the formula $C_{18}H_{21}NO_3$, is often used as an analgesic (painkiller) for intense pain. How many molecules would be in a 0.10 g sample of codeine?

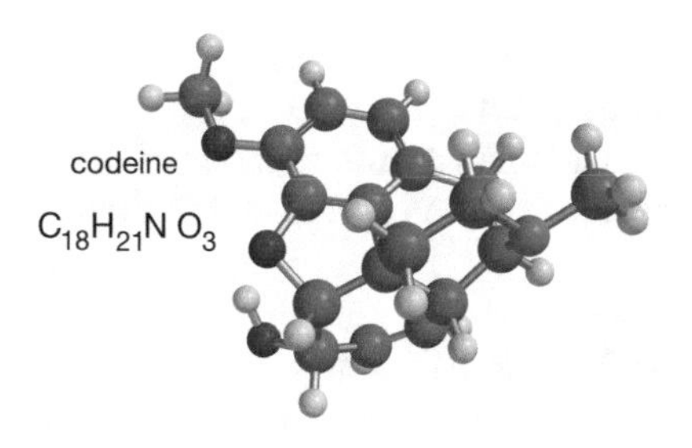

Asking the Right Questions

What is the average molecular mass of codeine? Once this mass is known, we can determine the number of molecules in that mass, using the same strategy we employed to get the number of atoms of aspirin and the number of atoms of copper.

Solution

The molecular mass of codeine, $C_{18}H_{21}NO_3$, is 299.37 amu.

$$0.10\ \text{g codeine} \times \frac{1\ \text{amu codeine}}{1.6605 \times 10^{-24}\ \text{g codeine}} \times \frac{1\ \text{molecule codeine}}{299.37\ \text{amu codeine}}$$

$$= 2.012 \times 10^{20}\ \text{molecules codeine}$$

$$= 2.0 \times 10^{20}\ \text{molecules codeine}$$

Are the Answers Reasonable?

Since we are calculating the number of tiny molecules in a mass of substance that is only a little less than a gram, we expect the result to be a very large number. Also, we see that the calculated answer is about 1000 times smaller than Avogadro's number, which is how many molecules there would be if we had 299.37 g of codeine. Therefore, this is a reasonable answer.

What If…?

…instead of codeine, we used fentanyl, $C_{22}H_{28}N_2O$, as our pain reliever. How many grams would 4.5×10^{21} molecules weigh? (*Ans*: 2.5 g)

Practice 3.2

How many atoms are there in 100.0 g of gold?

3.3.1 The Quantity We Call the Mole

An amount of any substance that contains 6.022×10^{23} "elementary things" such as molecules, atoms, or ions is given its own name in chemistry. We call that amount of things a **mole** (often abbreviated mol). Perhaps it seems odd to give a special name to what is, in effect, just a number, but this is actually a very familiar practice:

Mole (mol) – The quantity represented by 6.022×10^{23} things.

1 mole corresponds to 6.022×10^{23} things

This number is very similar to other counting numbers, such as the dozen:

1 dozen corresponds to 12 things

The mole is the basic counting unit used by chemists to indicate how many atoms, molecules, or ions they are dealing with. A chemist counts moles of particles just as a baker counts dozens of cupcakes.

To work out the mass of one mole of a chemical, we just look up its formula mass and change the units of that number from amu into grams. The resulting quantity, expressed in grams per mole $(\frac{g}{mol})$, is known as the **molar mass**—the mass of one mole of a substance.

Molar mass – The total mass of all the atoms present in the formula of a molecule, in grams per mole.

Source ► https://www.photos-public-domain.com/2010/12/17/dozen-eggs/

The recent examination of the mole has resulted in a simplification of the definition. The mole is now defined as 6.022×10^{23} particles. The original definition of the mole indicated that this was the number of carbon-12 atoms found in exactly 12 g of the isotope. A calculation using the mass of 1 amu as a conversion factor reveals how many things are in a mole:

$$12.00\ \text{g} \times \frac{1\ \text{amu}}{1.6605 \times 10^{-24}\ \text{g}} \times \frac{1\ \text{atom}}{12\ \text{amu}} = 6.022 \times 10^{23}\ \text{atoms}$$

3.4 Working with Moles

The mole is more than just a counting number. In fact, it's quite useful in calculating quantities used in reactions. Let's explore this concept more in the following section.

3.4.1 Molecules to Moles and Back Again

Figure 3.6 illustrates the quantities of 1 mol of water, 1 mol of sodium chloride ("table salt"), and 1 mol of aspirin—three different chemicals of great significance to life and medicine. Each has a different mass and occupies a different volume, but each contains the same number of formula units of the compounds concerned. The containers in the figure contain 6.022×10^{23} molecules of water, 6.022×10^{23} molecules of aspirin and 6.022×10^{23} formula units of NaCl. (Note: Because each formula unit of NaCl includes one Na^+ ion and one Cl^- ion, our sample of NaCl actually contains 6.022×10^{23} Na^+ ions and 6.022×10^{23} Cl^- ions!)

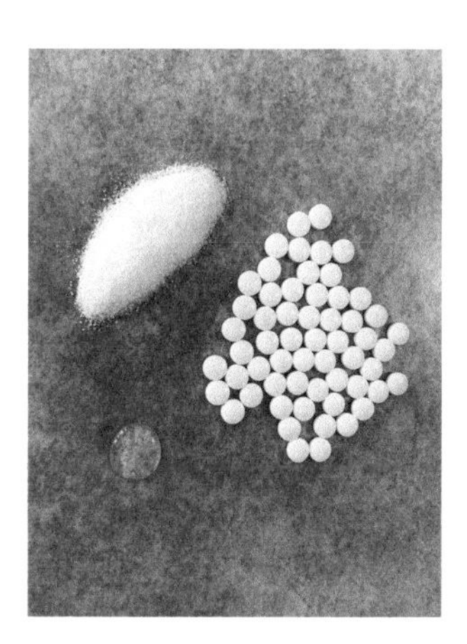

Fig. 3.6 One mole of sodium chloride ("salt") (left), one mole of aspirin (center), and one mole of water (right). 1 mol = Avogadro's number = 6.022×10^{23}

Example 3.3—Molecules and Moles

Really large numbers like 4.33×10^{24} can be quite confusing. We work with moles to make the numbers more commonplace.

a. How many moles of water are there in 4.33×10^{24} molecules of water?
b. How many moles of codeine ($C_{18}H_{21}NO_3$) are there in 4.33×10^{24} molecules of codeine?

Asking the Right Questions

What sort of answer do we expect? If a mole of water is 10^{23} water molecules, 10^{24} molecules is one power of 10 larger, so we would expect to have some number near 10. If we calculate an answer of a billion or a billionth (or even a thousandth!), we might realize that our answer would not make sense.

Solution

a. $4.33 \times 10^{24}\ \text{molecules} \times \dfrac{1\ \text{mol}}{6.022 \times 10^{23}\ \text{molecules}}$

$= 7.19\ \text{mol}$

b. Notice that the formula of the compound doesn't change the number of molecules that exist in one mole of a substance. Therefore, the answer is the same: 4.33×10^{24} molecules of any substance is equal to 7.19 mol whether we are considering aspirin, adrenaline, copper, codeine, or anything else.

Are our Answers Reasonable?

Based on our "Asking The Right Questions" discussion, we should expect an answer in the range of about 10 mol, rather than, say, a million moles or even a thousandth of a mole. Our answer of 7.19 mol therefore makes sense.

What If…?

…we had 4.33×10^{24} molecules of methane, CH_4? How many moles of methane would we have? What if we had 21.57 mol of methane? How many molecules of methane would we have? (*Ans*: 7.19 mol; 1.299×10^{25} molecules)

Practice 3.3

How many moles are there in 2.88×10^{20} formula units of sodium benzoate (C_6H_5COONa), a food preservative? How many molecules are there in 1.5 mol of sodium benzoate?

3.4.2 Grams to Moles and Back Again

How do we calculate the number of moles of a substance from its mass? Many of us sweeten our drinks by adding some "table sugar," the compound sucrose, with the formula $C_{12}H_{22}O_{11}$. If you added 10.0 g of sugar (roughly 2 heaping teaspoons) to a glass of tea, how many moles did you add? To answer this question, we must first determine the formula mass of sucrose, which tells us the mass of 1 mol and so gives us the mass-to-moles conversion factor that we need. Here is the calculation:

$$1\ \text{mol sucrose}\ C_{12}H_{22}O_{11} = 12\ \text{mol C} + 22\ \text{mol H} + 11\ \text{mol O}$$

$$12\ \cancel{\text{mol C}} \times \frac{12.011\ \text{g}}{1\ \cancel{\text{mol C}}} = 144.132\ \text{g}$$

$$22\ \cancel{\text{mol H}} \times \frac{1.008\ \text{g}}{1\ \cancel{\text{mol H}}} = 22.176\ \text{g}$$

$$11\ \cancel{\text{mol O}} \times \frac{15.9994\ \text{g}}{1\ \cancel{\text{mol O}}} = 175.9934\ \text{g}$$

These calculations tell us that the mass of 1 mol of sucrose is 342.30 g of sucrose. Mathematically, our moles-to-mass conversion factor is $\frac{1 \text{ mol sucrose}}{342.30 \text{ g sucrose}}$. Now, our sample of 10.0 g can be converted to moles:

$$10.0 \text{ g sucrose} \times \frac{1 \text{ mol sucrose}}{342.30 \text{ g sucrose}} = 0.02921 \text{ mol sucrose} = 0.0292 \text{ mol sucrose}$$

3

We can also determine how many molecules of sucrose have been added to our glass of tea. We do this using Avogadro's number, 6.022×10^{23}.

$$0.02921 \text{ mol sucrose} \times \frac{6.022 \times 10^{23} \text{ molecules sucrose}}{1 \text{ mol sucrose}}$$
$$= 1.759 \times 10^{22} \text{ molecules sucrose}$$
$$= 1.76 \times 10^{22} \text{ molecules sucrose}$$

We would have stirred this number of sugar molecules into our tea. If we write it out not in scientific notation, it emphasizes just how big this number is:

$$17,600,000,000,000,000,000,000$$

Example 3.4—Calculating Masses Corresponding to Moles

Hydrochlorothiazide (HCT) is a medicine used to lower blood pressure and deal with other illnesses related to fluid retention. Its molecular formula is $C_7H_8ClN_3O_4S_2$.

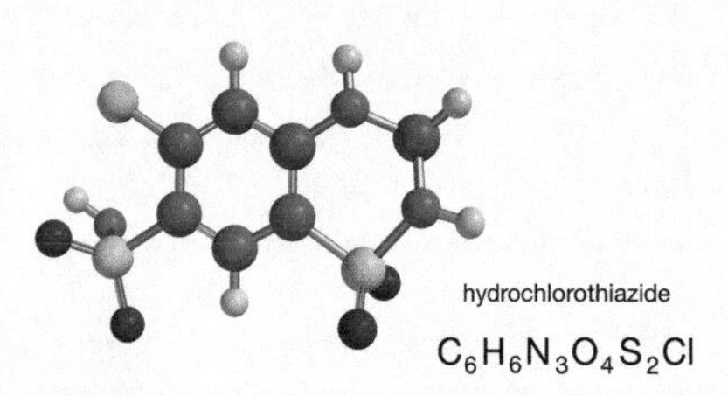

a. In a small-scale production process, a company needs 63.4 mol of the medicine. How many grams of hydrochlorothiazide is this?
b. A common individual dosage is 3.37×10^{-5} mol. How many milligrams are in the individual dose?

Asking the Right Questions

What are we calculating? In part a, we want grams. In part b, milligrams. The first represents a small-scale industrial process. The second, an individual dose. The strategy is still the same because the goal, mass, is the same. Only the prefix (grams in part a, milligrams in part b) differs. We can use the molar mass of the compound to go from moles to mass. The molar mass of the compound in grams per mole will be the same as the average molecular mass in amu.

Solution

a. $\text{HCT} \xrightarrow{\frac{\text{g HCT}}{\text{mol HCT}}} \text{g HCT}$

$$63.4 \text{ mol HCT} \times \frac{297.72 \text{ g HCT}}{1 \text{ mol HCT}} = 1.89 \times 10^4 \text{ g HCT}$$

b. $\text{mol HCT} \xrightarrow{\frac{\text{g HCT}}{\text{mol HCT}}} \text{g HCT} \xrightarrow{\frac{\text{mg}}{\text{g}}} \text{mg HCT}$

$$3.37 \times 10^{-5} \text{mol HCT} \times \frac{297.72 \text{ g HCT}}{1 \text{ mol HCT}} \times \frac{1 \text{ mg HCT}}{1 \times 10^{-3} \text{ g HCT}}$$
$$= 10.0 \text{ mg HCT.}$$

Are Our Answers Reasonable?

We expect the industrial synthesis to result in much more of the product than is in the individual dose, so this answer makes sense. We also note from other medicines we have taken that a dose of 10 to 500 mg is standard and, therefore, our answers are reasonable.

What If...?

...you needed enough 250 mg doses of HCT for a community that had 18,000 people with high blood pressure? How many moles of HCT is this? (*Ans: 15 mol HCT*)

Practice 3.4

Calculate the mass in grams of 26 mol and of 0.0025 mol of each of these compounds: table salt (NaCl), water (H_2O), and the sweetener aspartame ($C_{14}H_{18}N_2O_5$).

Example 3.5—Calculating Moles Corresponding to Masses

Methylphenidate, $C_{14}H_{19}NO_2$, is a medical drug used to stimulate the central nervous system. It is known as Ritalin, and its use in the treatment of attention deficit hyperactivity disorder (ADHD) has been the subject of controversy for years. Let's assume a company sent a 2.00×10^4 g shipment to be processed into individual 20.0 mg tablets. How many moles of Ritalin are in the shipment and in each tablet? How many molecules of Ritalin are in each tablet?

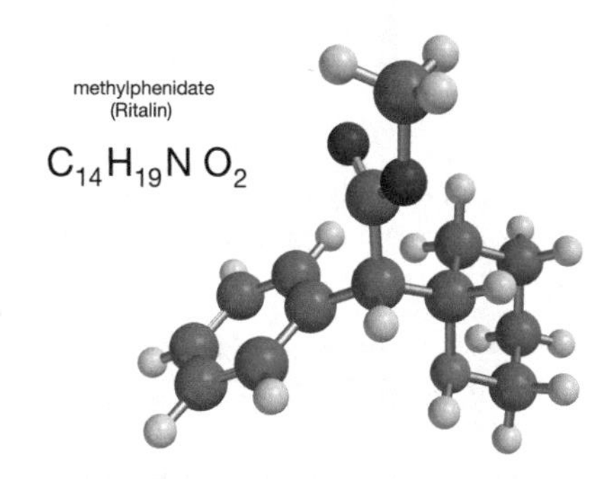

Asking the Right Questions

What do we want to know? We are asked for the number of moles of Ritalin that are in the large shipment, and in an individual 20.0 mg tablet. We are also asked for the number of molecules in each tablet. How do we proceed from the large starting sample of 2.00×10^4 g to the tablet-sized samples? We can begin by converting from grams of Ritalin in the shipment to moles. (What is the molar mass of Ritalin?)

$$\text{g Ritalin in the shipment} \xrightarrow{\frac{\text{mol Ritalin}}{\text{g Ritalin}}} \text{mol Ritalin in the shipment}$$

We can calculate the number of moles in each 20.0 mg tablet using the same general strategy, being careful to add the extra conversion factor that changes milligrams to grams before finding moles. Here's the strategy:

$$\frac{\text{mg Ritalin}}{\text{tablet}} \xrightarrow{\frac{\text{g}}{\text{mg}}} \frac{\text{g Ritalin}}{\text{tablet}} \xrightarrow{\frac{\text{mol}}{\text{g}}} \frac{\text{moles Ritalin}}{\text{tablet}}$$

Finally, we can convert from moles to molecules of Ritalin in a 20.0 mg tablet:

$$\frac{\text{moles Ritalin}}{\text{tablet}} \xrightarrow{\frac{\text{mol}}{\text{molecule}}} \frac{\text{molecules Ritalin}}{\text{tablet}}$$

Solution

We use the strategy we just outlined, along with the molar mass of Ritalin that we calculate from the formula ($C_{14}H_{19}NO_2$ is 233.312 g/mol).

$$2.00 \times 10^4 \text{ g Ritalin} \times \frac{1 \text{ mol Ritalin}}{233.312 \text{ g Ritalin}}$$

$$= 85.72 \text{ mol Ritalin}$$

$$\frac{20.0 \text{ mg Ritalin}}{\text{tablet}} \times \frac{1 \text{ g Ritalin}}{1000 \text{ mg Ritalin}} \times \frac{1 \text{ mol Ritalin}}{233.312 \text{ g Ritalin}}$$

$$= 8.572 \times 10^{-5} \frac{\text{mol Ritalin}}{\text{tablet}}$$

Now we can convert from moles to molecules of Ritalin in a 20.0 mg tablet:

$$\frac{\text{mol Ritalin}}{\text{tablet}} \xrightarrow{\frac{\text{molecules}}{\text{mol}}} \frac{\text{molecules Ritalin}}{\text{tablet}}$$

$$\frac{8.572 \times 10^{-5} \text{ mol Ritalin}}{\text{tablet}} \times \frac{6.022 \times 10^{23} \text{ molecules Ritalin}}{1 \text{ mol Ritalin}} = \frac{5.16 \times 10^{19} \text{ molecules Ritalin}}{\text{tablet}}$$

Are our Answers Reasonable?

The individual dose has considerably fewer moles than the industrial shipment, so the answers are reasonable. The number of molecules in the individual dose is a very large number, which is also reasonable, since molecules are so small.

What If…?

…a patient is directed to take two 5-mg doses of Ritalin each day. How many moles of Ritalin will the patient ingest in a week? (*Ans: 3×10^{-4} mol*)

Practice 3.5

Calculate the number of moles in 5.3 g of the compounds: lactic acid ($C_3H_6O_6$), and sulfuric acid (H_2SO_4).

As researchers have investigated the causes of heart-related health problems, one particular molecule, cholesterol, has been found to be associated with certain types of circulatory problems. Cholesterol plays a crucial role in maintaining the thin membranes that enclose every cell of your body. It also aids in the production of hormones, including estrogen and testosterone, is critical for the production of vitamin D and is also necessary for proper brain function. So again, we find that the quantity of a chemical is crucial. The chemical structure of cholesterol is shown in ◘ Fig. 3.7. Its formula is $C_{27}H_{46}O$.

According to the Centers for Disease Control, among the U.S. population, the average total cholesterol level is about 200 mg per deciliter of blood. If a blood analysis showed a cholesterol level of 1950 mg per liter of blood, how many molecules of cholesterol would be present per liter?

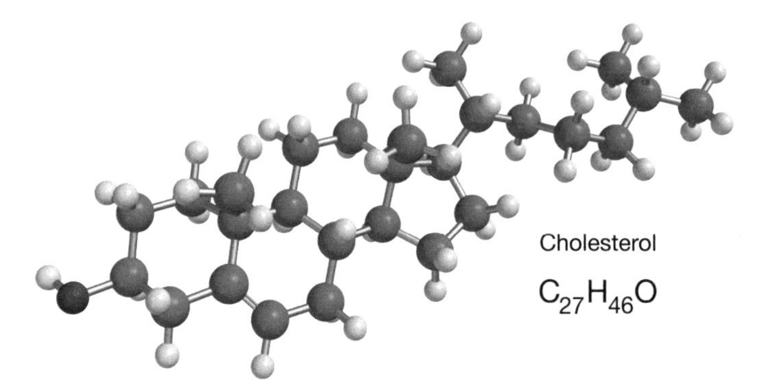

Fig. 3.7 Cholesterol drawn as a ball and stick and a space-filling model

In this case we can see that the starting measurement is in milligrams, but the question asks about the number of molecules. The strategy is to express the mass as a quantity of moles of cholesterol and then convert moles to molecules. Our dimensional analysis flowchart for this task looks like:

$$\frac{\text{mg cholesterol}}{\text{L}} \xrightarrow{\frac{\text{g}}{\text{mg}}} \frac{\text{g cholesterol}}{\text{L}} \xrightarrow{\frac{\text{mol}}{\text{g}}} \frac{\text{mol cholesterol}}{\text{L}} \xrightarrow{\frac{\text{molecules}}{\text{mol}}} \frac{\text{molecules cholesterol}}{\text{L}}$$

When we have finished our calculation, our answer should make sense. What sort of answer should we expect? Because molecules are so small, we know that there are a great many molecules in even the smallest sample of blood. We would therefore expect a very large number of molecules. Answers such as 10^{-24} molecule/L or even 10^2 molecules/L would not make sense.

Because cholesterol has a known molecular formula, $C_{27}H_{46}O$, we can determine its molar mass as 386.66 g per mole (check to make sure you agree). We will follow our flowcharted strategy to get the answer.

$$\frac{1950\ \cancel{\text{mg cholsterol}}}{1\ \text{L}} \times \frac{1\ \text{g}}{1000\ \cancel{\text{mg}}} \times \frac{1\ \cancel{\text{mol}}\ \text{cholsterol}}{386.66\ \cancel{\text{g cholsterol}}} \times \frac{6.022 \times 10^{23}\ \text{molecules}}{1\ \cancel{\text{mol}}} = \frac{3.04 \times 10^{21}\ \text{molecules cholesterol}}{\text{L}}$$

Is the answer reasonable? The quantity 1950 mg (which is 1.95 g) is less than a mole of cholesterol (386.6 g per mole), so we would have expected an answer less than Avogadro's number, but because we are counting our sample in molecules, the number should still be quite large. Our value meets these conditions, so our answer makes sense.

Example 3.6—Grams to Molecules

Sucrose ($C_{12}H_{22}O_{11}$), commonly called "table sugar," is a molecule used to sweeten our cup of coffee. If a packet of sucrose contains 1.50 g of sucrose, how many molecules is this?

Asking the Right Questions

What do we need to know and what are we given? We want the number of molecules of sucrose, and we are given a 1.50 g sample of sucrose. What is the core relationship between mass and molecules? Moles. Our strategy must include moles as the bridge between mass and molecules.

Solution

$$\text{g sucrose} \xrightarrow{\frac{\text{mol sucrose}}{\text{g sucrose}}} \text{mol sucrose} \xrightarrow{\frac{\text{molecules sucrose}}{\text{mol sucrose}}} \text{molecules sucrose}$$

$$1.50\ \text{g sucrose} \times \frac{1\ \text{mol sucrose}}{342.30\ \text{g sucrose}} \times \frac{6.022 \times 10^{23}\ \text{molecules sucrose}}{1\ \text{mol sucrose}} = 2.64 \times 10^{21}\ \text{molecules sucrose}$$

Is Our Answer Reasonable?

Do we expect to have a large number of molecules of sucrose in 1.50 g of sucrose? It is a quantity that we can see, filling part of a teaspoon when we add it to our hot tea or coffee (or latté). It is therefore reasonable to calculate a very large number of molecules in the sample of sucrose.

What If…?

…you had 5.32×10^{23} molecules of sucrose? How many teaspoons of sucrose do we have if a teaspoon can hold 4.2 g of sucrose? (*Ans: 72 teaspoons of sucrose*)

Practice 3.6

How many molecules are there in 1.50 g of water? Would you need a bathtub, a swimming pool, or a cup to hold 5.33×10^{29} molecules of water?

▶ Deepening your understanding through group work

You have solved many in-chapter and end-of-chapter examples by using dimensional analysis, and we have presented that as a reasonable problem-solving strategy because it gets us to our answer, knowing that if our units are correct, the answer is likely to be correct. Another school of thought says that dimensional analysis encourages rote problem solving without understanding what is going on in the chemical system, and that working with ratios or other methods would deepen your understanding.

In your group, consider which way of solving problems is better? Why? ◀

3.5 Percentages by Mass

Just like humans, plants undergo chemical changes that are a part of life. They have genetic material that directs the synthesis of proteins, some of which are part of cell structure and some of which make possible the chemical reactions that sustain life. Nitrogen is a vital element in the genetic material and the proteins in plants, so in order to grow, plants need plenty of nitrogen. For farms throughout the world, fertilizer supplies that nitrogen in the form of manure or industrially produced products.

Some plants need more nitrogen than others. Environmental factors, such as temperature, rainfall, and the type of soil, also affect the ability of plants to use the nitrogen in fertilizer. This means that a farmer must apply the correct amount of nitrogen for each particular situation. Among the most important factors in deciding what fertilizer to use is the percent by mass, or **mass percent**, of nitrogen in the product (◘ Fig. 3.8). Mass percent values are determined by dividing the mass of the component of interest by the total mass of the entire sample, and then multiplying the result by 100%. The component can be a compound (such as ammonium nitrate, NH_4NO_3, which is found in many liquid and solid fertilizers) or it can be an ion, atom, or element.

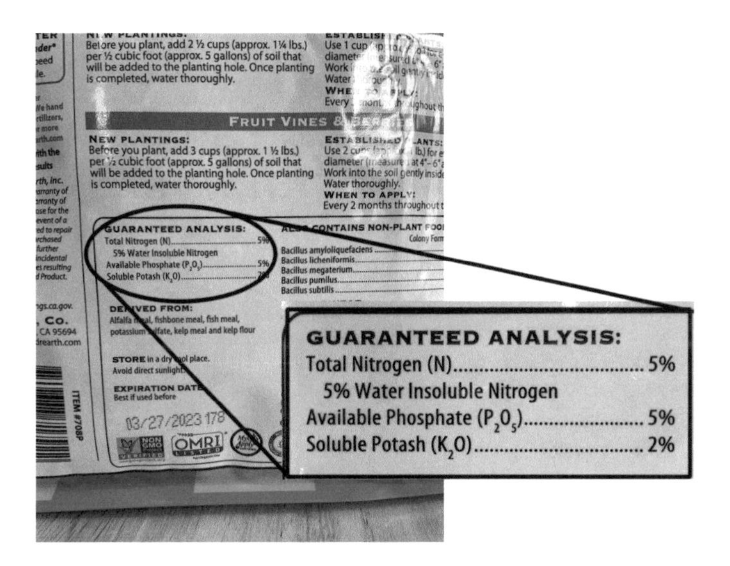

◘ **Fig. 3.8** The mass percent of nitrogen is often reported on the ingredients label of a bag of fertilizer

Mass percent – The percent of a component by mass.

$$\text{mass percent} = \frac{\text{total mass of component}}{\text{total mass of entire substance}} \times 100\%$$

3

How can mass percent be used to compare compounds? We can answer that question by solving the following problem. Which of these two compounds would supply plants with the most nitrogen per gram: ammonium nitrate, NH_4NO_3, or ammonium sulfate, $(NH_4)_2SO_4$?

We can answer the question by calculating the mass percent of nitrogen present in each compound. We need to express the mass of nitrogen in each one as a percentage of the total mass.

For NH_4NO_3

$$2\text{ mol N} \times \frac{14.0067\text{ g N}}{1\text{ mol N}} = 28.0134\text{ g N}$$

$$4\text{ mol H} \times \frac{1.008\text{ g H}}{1\text{ mol H}} = 4.032\text{ g H}$$

$$3\text{ mol O} \times \frac{15.9994\text{ g O}}{1\text{ molO}} = 47.9982\text{ g O}$$

Molar mass = 80.0436 g/mol = 80.04 g/mol (remember that 2 numbers after the decimal point on a molar mass is usually enough to answer every question.)

$$\text{mass percent nitrogen} = \frac{\text{total mass N}}{\text{total mass of } NH_4NO_3} \times 100\% = \frac{28.0134\text{ g}}{80.04\text{ g}} \times 100\% = 35.00\%\text{ N}$$

For $(NH_4)_2SO_4$

$$2\text{ mol N} \times \frac{14.0067\text{ g N}}{1\text{ mol N}} = 28.0134\text{ g N}$$

$$8\text{ mol H} \times \frac{1.008\text{ g H}}{1\text{ mol H}} = 8.064\text{ g H}$$

$$4\text{ mol O} \times \frac{15.9994\text{ g O}}{1\text{ mol O}} = 63.9976\text{ g O}$$

$$1\text{ mol S} \times \frac{32.065\text{ g S}}{1\text{ mol S}} = 32.065\text{ g S}$$

$$\text{mass percent nitrogen} = \frac{\text{total mass N}}{\text{total mass of } NH_4NO_3} \times 100\%$$
$$= \frac{28.0134\text{ g}}{132.14\text{ g}} \times 100\% = 21.20\%\, N$$

The results indicate that any chosen mass of NH_4NO_3 contains a significantly greater percentage of nitrogen (35.00%) than does the same mass of $(NH_4)_2SO_4$ (21.20%).

Example 3.7—Calculating Mass Percent

Phosphorus is another element that is vital for plant growth. Many fertilizers therefore also contain phosphorus and are sold on the basis of the percentage of phosphorus in the fertilizer. One example is calcium dihydrogen phosphate, $Ca(H_2PO_4)_2$. The fertilizer, commonly known in the agriculture industry as "triple superphosphate", is a granular substance that is made from phosphoric acid and phosphate rock and used in fertilizing grain fields and sugar cane. What is the mass percent of phosphorus in triple superphosphate?

Asking the Right Questions

What does the "mass percent" mean? It is the percentage of the entire mass that is made up of your component (phosphorus, in this example).

We can write this as a formula:

$$\text{mass percent} = \frac{\text{mass of P in the compound}}{\text{total mass of the compound}} \times 100\%$$

What is the mass, in grams, of the phosphorus? There are *2 moles of phosphorus* in each mole of the compound, therefore, its contribution is $2 \times 30.97 = 61.94$ g phosphorus per mole of triple superphosphate.

Solution

$$\text{mass percent} = \frac{61.94 \text{ g P}}{234.05 \text{ g compound}} \times 100\% = 26.46\% \text{ P}$$

Is Our Answer Reasonable?

A mass percent is the portion of the whole mass (expressed as a percent) that is made up by the phosphorus. It must therefore be smaller than the whole, or less than 100%. On that basis, the answer is reasonable. As a check of our work, we may wish to find the mass percent of all of the elements in the compound and make sure they total 100%.

What If…?

…the triple superphosphate were analyzed and found to contain 1.723% of an element? Which element would that be? (*Ans: hydrogen*)

Practice 3.7

Some mineral supplement tablets contain potassium in the form of potassium chloride (KCl). What is the mass percent of potassium in this compound?

▶ Here's what we know so far

- Formula masses are calculated by adding the average atomic masses of the individual atoms in a formula.
- Chemists count molecules in a sample by measuring its mass.
- Masses can be converted into moles by using the molar mass of a compound.
- One mole of something is 6.022×10^{23} of those things.
- The mass percent of a component in a substance is the total mass of the component divided by the total mass of the substance, and then multiplied by 100%. ◀

3.6 Finding the Formula

For many people, chocolate is close to being essential for life! Per-person consumption in the United States is estimated at over ten pounds annually. Chocolate is big business. For instance, in 2015, chocolate sales in the United States alone are estimated to be $21 billion, with the largest growth in "dark chocolate" which has a higher percentage of cocoa, since it does not contain milk. This includes not only candy bars but also powdered chocolate for drinks and baking. By 2024, worldwide sales of "premium" (expensive) chocolates will be over $33 billion. How is chocolate made? It is harvested from cacao beans from a tree aptly named Theobroma cacao. Theobroma means "food of the gods" (◘ Fig. 3.9). As much as two-thirds of the world's chocolate currently comes from the continent of Africa.

A multitude of compounds make up the flavor of chocolate. One of these is named 2,5-dimethylpyrazine. To work with this compound in chemistry, it is necessary to have its empirical and molecular formulas. These are found by experimentally determining the mass percent of the elements present in a sample of the compound. 2,5-Dimethylpyrazine, for example, contains, by mass, 66.62% carbon, 7.47% hydrogen, and 25.91% nitrogen. How can we use these results to calculate the molecular formula of 2,5-dimethylpyrazine? We can begin by calculating the ratio in which the atoms of the elements are present in the compound. To do this, we convert the mass ratio obtained from the analysis experiment into a mole ratio. The process is made simpler because the mass per-

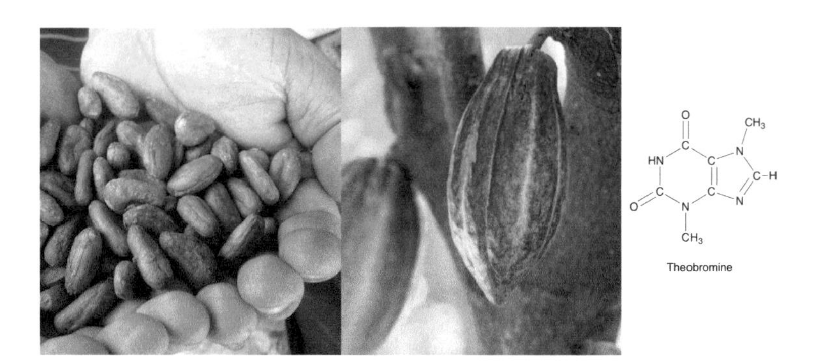

Fig. 3.9 Theobroma cacao and the food of the gods. Making chocolate begins with the seed pods of the cacao tree, whose seeds (beans) are roasted and then processed into chocolate of various types Photos from ▶ https://www.publicdomainpictures.net/pictures/220000/velka/growing-cocoa-pods-1495042625kXF.jpg and ▶ https://www.publicdomainpictures.net/pictures/40000/velka/cacoa-beans.jpg

cent data tell us how many grams of each element are present in 100 g of the compound. That is, 66.62% carbon means that every 100 g of 2,5-dimethylpyrazine is made up of 66.62 g of carbon. This is our starting point for the conversion into moles.

$$66.62\%\ \text{C} \rightarrow 66.62\ \cancel{\text{g C}} \times \frac{1\ \text{mol C}}{12.011\ \cancel{\text{g C}}} = 5.5466\ \text{mol C}$$

$$7.47\%\ \text{H} \rightarrow 7.47\ \cancel{\text{g H}} \times \frac{1\ \text{mol H}}{1.008\ \cancel{\text{g H}}} = 7.411\ \text{mol H}$$

$$25.91\%\ \text{N} \rightarrow 25.91\ \cancel{\text{g N}} \times \frac{1\ \text{mol N}}{14.0067\ \cancel{\text{g N}}} = 1.8498\ \text{mol N}$$

If we divide all these mole values by the smallest mole value (1.8498), we will obtain the simplest whole-number ratio of the atoms that make up the substance.

$$\frac{5.5466}{1.8498} = 3 \quad \frac{7.411}{1.8498} = 4 \quad \frac{1.8498}{1.8498} = 1$$

In this case, the division step yields the simplest whole-number ratio immediately. The simplest whole-number ratio of elements in a compound is called the ***empirical formula*** of the compound.

Empirical formula – The formula for a compound with the lowest whole number ratio of atoms.

We can now write the empirical formula for the compound as:

$$C_3H_4N$$

The numbers will not always work out as neatly as that, sometimes because of experimental errors and sometimes because the formula is a little more complex. For example, if for a different compound we might find that the ratio is:

1.506 mol C 3.007 mol H 1.008 mol O

they could reasonably be simplified to this ratio:

1.5 mol C 3 mol H 1 mol O

This ratio is not the simplest whole-number ratio that we need in order to write an empirical formula. However, it can be converted into the simplest whole-number ratio by multiplying all the numbers by 2. When we do this, we get:

3 mol C 6 mol H 2 mol O

This gives the empirical formula $C_3H_6O_2$. This illustrates that when we're trying to find the simplest whole-number ratio, we can multiply or divide the set of values by any number we choose, because the ratio will remain the same provided that we perform the same multiplication or division on each value. However, we always choose the number that provides the simplest whole-number ratio of atoms in the formula.

The empirical formula of 2,5-dimethylpyrazine, the chocolate flavoring molecule, is C_3H_4N. This formula tells us only the ratio in which the elements are present. It does not tell us the actual number of each type of atom in a molecule of the compound, which is the molecular formula. How do we determine the molecular formula?

To determine the molecular formula, we would need to know the average molecular mass of the compound. This could be determined experimentally via the technique of mass spectrometry that we discussed in ▶ Chap. 2. For 2,5-dimethylpyrazine, we would find a molar mass of 108.14 g. Does this match the mass of the empirical formula?

If we add the masses of the atoms in the empirical formula, we get a total of 54.072 amu (or g/mol).

$$3\ \cancel{\text{mol C}} \times \frac{12.011\ \text{g C}}{\cancel{\text{mol C}}} = 36.033\ \text{g C}$$

$$4\ \cancel{\text{mol H}} \times \frac{1.008\ \text{g H}}{\cancel{\text{mol H}}} = 4.032\ \text{g H}$$

$$1\ \cancel{\text{mol N}} \times \frac{14.0067\ \text{g N}}{\cancel{\text{mol N}}} = 14.0067\ \text{g N}$$

$$= 54.0717\ \frac{\text{g}}{\text{mol}} = 54.07\ \frac{\text{g}}{\text{mol}}$$

The average molecular mass is greater than this value, so we have to work out how many average empirical masses are contained in the average molecular mass. Because $2 \times 54.072 = 108.14$, the molecular formula for 2,5-dimethylpyrazine must be $C_6H_8N_2$, derived from $(C_3H_4N) \times 2$. There are two empirical formula units per molecule of 2,5-dimethylpyrazine. Note that we still don't have any information about the shape of the molecule or how these atoms are connected together. We will examine those questions in the chapters on bonding.

Example 3.8—Finding the Empirical and Molecular Formula

An alkyne is a special class of molecules that have a particular arrangement of atoms that make them useful in the manufacture of pharmaceuticals and polymers. One alkyne with a molar mass of 54.09 g was determined to have 11.18% hydrogen and 88.82% carbon. What is the empirical formula? What is the molecular formula?

Asking the Right Questions

What is the empirical formula? It is the *simplest* whole-number ratio of elements in the compound. What about the molecular formula? This is the *actual* whole-number ratio of elements in the compound. How do you get these ratios? Here is a strategy that will always work: assume you have 100 g of the compound and recognize that the key relationship among elements is moles to moles. This means you can convert the mass of each element into moles and compare the ratios of moles.

Solution

We first assume that we have 100 g of the alkyne. Then all the percent values just become grams (since percent is just "per-hundred"):

$$11.18\%\text{H} \rightarrow 11.18\ \text{g H} \times \frac{1\ \text{mol H}}{1.008\ \text{g}} = 11.091\ \text{mol H}$$

$$88.82\%\text{C} \rightarrow 88.82\ \text{g C} \times \frac{1\ \text{mol C}}{12.011\ \text{g}} = 7.3949\ \text{mol C}$$

We then determine the simplest whole number ratio between these two values:

$$\frac{11.09\ \text{mol H}}{7.395\ \text{mol C}} = 1.499\ \text{H} \qquad \frac{7.395\ \text{mol C}}{7.395\ \text{mol C}} = 1.000\ \text{C}$$

There are 1.499 times more H atoms than C atoms in the formula. We obtain the empirical formula for the alkyne by multiplying these by 2:

$$C_2H_3$$

The molar mass of this empirical formula is 27.04 g. However, the molar mass of 54.09 g is twice the empirical formula mass. Therefore, the molecular formula is C_4H_6.

Is Our Answer Reasonable?

Our data turned out to be simple whole-number ratios, and the molecular formula was a whole-number multiple of the empirical formula, which suggests that the answers are reasonable.

What If…?

…we have a sample of an ammonium-containing phosphate fertilizer that contains 21.21% N, 6.87% H, 23.45% P, and 48.47% O? Can you determine a likely formula for the compound? (*Ans:* $(NH_4)_2HPO_4$ *or* $N_2H_9PO_4$)

Practice 3.8

What is the empirical formula and molecular formula of a compound that contains 92.26% carbon and 7.74% hydrogen, if the molar mass of the substance is 78.11 g/mol?

3

Example 3.9—Determining the Formula of Heme

Oxygen is carried around your body bound to iron ions that are part of a "heme" chemical group, which is part of the protein we call hemoglobin. The heme portion of hemoglobin has a molar mass of 616.50 g and contains 4 nitrogen atoms and one iron atom per heme.
The percent composition of heme is:

66.4% C; 5.24% H; 9.09% Fe; 9.09% N; 10.4% O

From these data, calculate the empirical and molecular formulas for this key biochemical molecule.

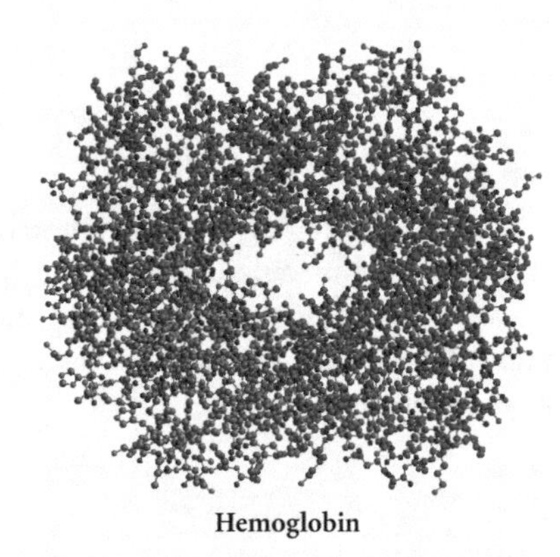

Hemoglobin

Asking the Right Questions

How is this example *similar* to Example 3.8? Our goal is to solve for the empirical and molecular formula of a compound. This suggests the same problem-solving approach as we used there. How is this example *different* than Example 3.8? The heme molecule is larger and has more atoms than the alkyne in Example 3.8. Our approach, however, is identical. The bookkeeping just gets a bit more cumbersome.

Solution

We first convert the percentages listed into grams by assuming that we have a 100.0 g sample of heme, and then convert the grams into moles by dividing each quantity by the molar mass of the appropriate element. We retain an extra (insignificant) figure throughout the intermediate calculations. We will drop all insignificant figures at the end.

$$66.4\,\text{g C} \times \frac{1\,\text{mol C}}{12.011\,\text{g C}} = 5.528\,\text{mol C}$$

$$5.24\,\text{g H} \times \frac{1\,\text{mol H}}{1.008\,\text{g H}} = 5.198\,\text{mol H}$$

$$9.09\,\text{g Fe} \times \frac{1\,\text{mol Fe}}{55.845\,\text{g Fe}} = 0.1628\,\text{mol Fe}$$

$$9.09\,\text{g N} \times \frac{1\,\text{mol N}}{14.0067\,\text{g N}} = 0.6490\,\text{mol N}$$

$$10.4\,\text{g O} \times \frac{1\,\text{mol O}}{15.9994\,\text{g O}} = 0.6500\,\text{mol O}$$

Divide each number by the smallest to determine the simplest whole number ratio:

$$\frac{5.528\,\text{mol C}}{0.1628} = 33.96\,\text{mol C} = 34\,\text{mol C}$$

$$\frac{5.198\,\text{mol H}}{0.1628} = 31.93\,\text{mol H} = 32\,\text{mol H}$$

$$\frac{0.1628\,\text{mol Fe}}{0.1628} = 1\,\text{mol Fe}$$

$$\frac{0.6490\,\text{mol N}}{0.1628} = 3.986\,\text{mol N} = 4\,\text{mol N}$$

$$\frac{0.6500\,\text{mol O}}{0.1628} = 3.993\,\text{mol O} = 4\,\text{mol O}$$

These numbers round off to this empirical formula:

$C_{34}H_{32}FeN_4O_4$

The molecular formula for heme either will be the same as the empirical formula or it will correspond to the empirical formula multiplied by some whole number. When we calculate the molar mass corresponding to the empirical formula, we find:

$$(12.011 \times 34) + (1.008 \times 32) + (55.845 \times 1) + (14.0067 \times 4) + (15.9994 \times 4) = 616.5\,\text{g per mole.}$$

Since we were told the molar mass of heme is 616.50, we can say that the molecular formula for heme is also $C_{34}H_{32}FeN_4O_4$.

Are our Answers Reasonable?

This is less obvious than in many empirical formula examples because the numbers are so large. We are helped by knowing that there are four nitrogen and one iron atom per heme. We can be reassured further by the ratio of hydrogen to nitrogen and oxygen, which is just about 8-to-1. With these data supporting the molecular formula, the answer is reasonable.

What If...?

...the molar mass of heme was 1849.5 g/mol. What would the molecular formula be for this compound? (*Ans:* $C_{102}H_{96}Fe_3N_{12}O_{12}$)

Practice 3.9

A common component of wines that have spoiled is a compound with an average molecular mass of 60.06 amu and the composition: 40.0% C; 6.70% H; 53.3% O. Calculate the empirical formula of this compound, which is called ethanoic acid (or, commonly, acetic acid).

3.7 Chemical Equations

The fundamental process of change in chemistry is the chemical reaction. For example, the final step in the manufacture of aspirin is the reaction of salicylic acid and ethanoic anhydride.

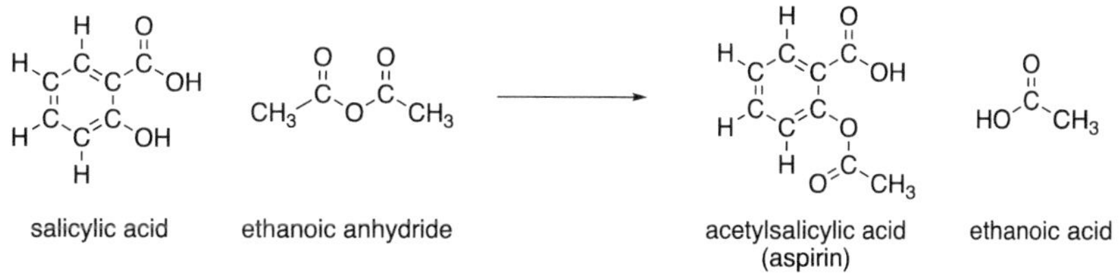

salicylic acid | ethanoic anhydride → acetylsalicylic acid (aspirin) | ethanoic acid

Salicylic acid and ethanoic anhydride (also called acetic anhydride) are the **reactants** of this reaction, which means they are the chemicals present at the start of the reaction. Aspirin and ethanoic acid (acetic acid) are the **products**, the chemicals produced as a result of the reaction. How do we know how much of our reactants we must combine to get the quantity of product we wish to make? For this or *any* other chemical reaction, the key is to understand the proportions in which the reactants combine to make the products. That information, and all the other key aspects of a reaction, can be summarized in a **chemical equation**, another example of the "chemist's shorthand."

Chemical equation – A precise quantitative description of a reaction.

Reactants – The substances located on the left-hand side of a chemical equation.

Products – The substances located on the right-hand side of a chemical equation.

A chemical equation uses chemical formulas to indicate the reactants and products of the reaction. The equation also indicates the proportions in which the chemicals involved react together and are formed. The first step in writing a chemical equation is to write the formulas or structures of all the reactants and products, with an arrow between them indicating the course of the reaction. For the synthesis of aspirin, this beginning is

$$C_7H_6O_3 + C_4H_6O_3 \rightarrow C_9H_8O_4 + C_2H_4O_2$$

From the Law of Conservation of Mass that we explored in ▶ Chap. 2, we know that atoms cannot be created or destroyed in a reaction. In other words, equations must have the same mass on each side of the arrow and the same number and types of atoms on each side of the arrow. We say that the reaction is **balanced** when this occurs. This also means that the total mass of the reactants will equal the total mass of the products (◘ Fig. 3.10).

Balanced – Appropriate numbers of each compound have been added such that the number of atoms of each element is the same in both reactants and products.

salicylic acid	ethanoic anhydride	aspirin	ethanoic acid
7 carbons	4 carbons	9 carbons	2 carbons
6 hydrogens	6 hydrogens	8 hydrogens	4 hydrogens
3 oxygens	3 oxygens	4 oxygens	2 oxygens
138 amu	102 amu	180 amu	60 amu
scales to	scales to	scales to	scales to
1 mol = 138 g	1 mol = 102 g	1 mol = 180 g	1 mol = 60 g
240 g total		240 g total	

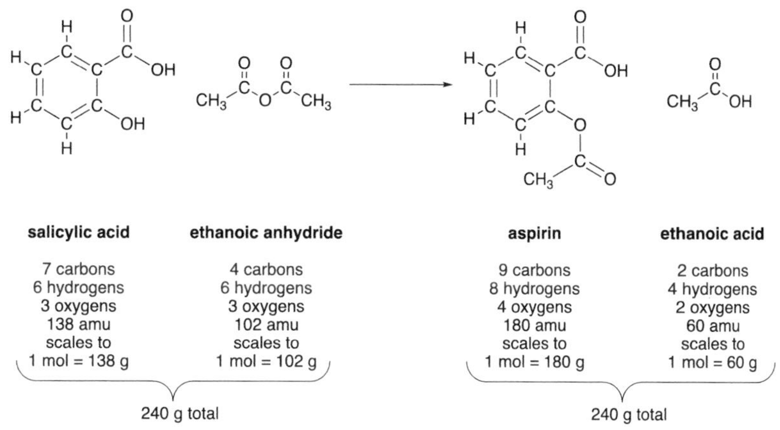

◘ **Fig. 3.10** A chemical equation has the same number and type of each atom on each side of the arrow. Notice that the mass of the products equals the mass of the reactants, because there are the same number of each type of atom on each side of the arrow. Ethanoic anhydride and ethanoic acid are commonly known as acetic anhydride and acetic acid, respectively

In this case the equation is balanced, because the number of each type of atom in the reactants is equal to the number of that type of atom in the products. The equation represents a rearrangement of the atoms, but no atom is created or destroyed; matter in the equation is conserved. Many equations, however, are not balanced initially when they are written. In order to fix this problem and balance these reactions, we have to use some reasoning to discover the true proportions in which the chemicals react and are formed.

3.7.1 Balancing Equations

The ethanoic anhydride used in the synthesis of aspirin is a very versatile compound that is used in many other chemical-manufacturing processes, such as the production of fibers, plastic materials, and pharmaceuticals. One way to make ethanoic anhydride on an industrial scale is by heating ethanoic acid in the presence of quartz or porcelain chips.

The result is a dehydration reaction in which water is removed from the reactants and becomes one of the products. To analyze the quantitative relationships among the chemicals involved in the manufacture of ethanoic anhydride, we need to write the balanced chemical equation for the process.

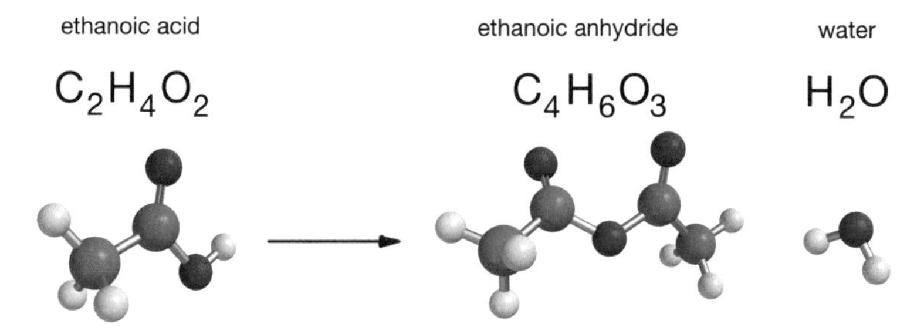

The first step in balancing the equation is to write down the formulas of the reactants and products:

$$C_2H_4O_2 \rightarrow C_4H_6O_3 + H_2O$$

But this equation is not balanced as our count of the atoms and the total molecular mass on each side of the arrow don't match (Fig. 3.11).

How do we balance a chemical equation? We'll do this by adjusting the numbers of the compounds until the numbers of atoms of each type are the same on both sides of the equation. We cannot change any of the formulas we have written; doing so would change the substances in the reaction! The only thing we can do to balance an equation is to use multiples of the correct formulas in order to find the proportions that would produce a balanced equation. For example, if we add another molecule of ethanoic acid on the left side of the reaction arrow, then we will have equal numbers of the different types of atoms on each side of the arrow:

$$C_2H_4O_2 + C_2H_4O_2 \rightarrow C_4H_6O_3 + H_2O$$

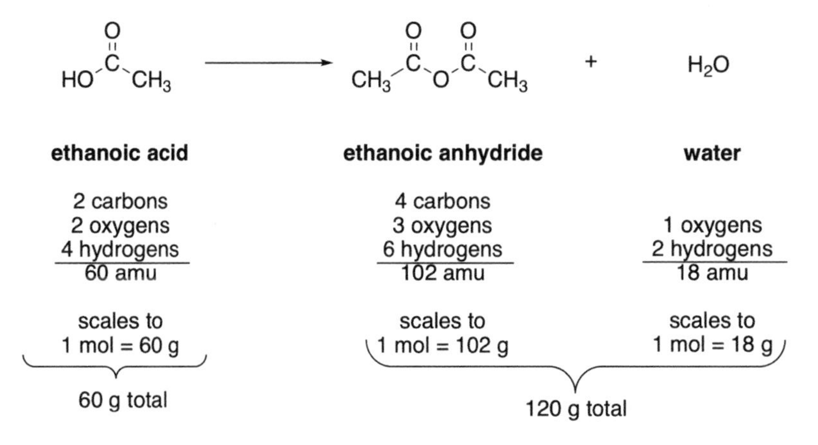

Fig. 3.11 This equation is not balanced

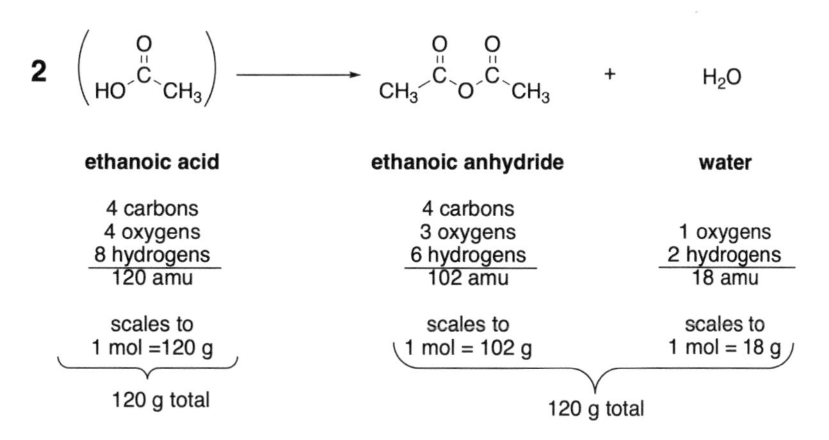

Fig. 3.12 This equation is now balanced

We indicate the desired multiples by placing numbers in front of the formulas. We call these numbers the coefficients in an equation. For example, we could change the above equation for the formation of ethanoic anhydride to read:

$$2C_2H_4O_2 \rightarrow C_4H_6O_3 + H_2O$$

Placing the **coefficient** 2 before the formula for ethanoic acid indicates that 2 molecules of ethanoic acid react to produce 1 molecule of ethanoic anhydride and 1 molecule of water. Alternatively, we can say that 2 mol of ethanoic acid react to produce 1 mol of ethanoic anhydride and 1 mol of water, since a mole is just Avogadro's number of each molecule. Check that you agree that the equation is now balanced, as shown in Fig. 3.12. You should find that each side of the equation now contains 4 carbon atoms, 8 hydrogen atoms, and 4 oxygen atoms. We have determined the proportions in which these reactants and products can be combined into a feasible chemical reaction.

Coefficient – The number placed in front of a substance in a chemical equation that reflects the specific number of those substances required to balance the equation.

Example 3.10—Balancing Equations

a. Consider the production of hydrogen via the reaction of ethane (C_2H_6) with steam. This process is completed in industry to make a mixture of carbon monoxide (CO) and hydrogen gas (H_2) known as "synthesis gas." The hydrogen can be separated from the other gases by passing it through a filter that selectively adsorbs the CO (molecules of CO stick to it). This process is known as pressure swing adsorption. What is the balanced equation for the reaction of ethane with steam?
b. Butane (C_4H_{10}) is a fuel that reacts with oxygen in the air to produce carbon dioxide and water. What is the balanced equation describing the reaction of butane and oxygen?

Asking the Right Questions

What makes an equation "balanced?" The number of atoms of each element is the same on the left (reactant) and right (product) side of the equation. After checking our work, the balanced equations should be the end result.

Solution

a. The chemical reaction is

$$C_2H_6 + H_2O \rightarrow CO + H_2$$

In this reaction, the hydrogen atoms appear in three compounds, while all other atoms are in just one compound on each side. Therefore, we will put hydrogen last in our list. Let's begin with carbon, then oxygen, and then hydrogen.

Balancing carbon: There are 2 C's on the left and one on the right, so we multiply the CO by 2.

$$C_2H_6 + H_2O \rightarrow 2CO + H_2$$

Balancing oxygen: There is 1 O on the left and 2 O's on the right, so we must multiply H_2O by 2.

$$C_2H_6 + 2H_2O \rightarrow 2CO + H_2$$

Balancing hydrogen: There are now 10 H's on the left (6+4) and 2 on the right, so we must multiply H_2 by 5 to get 10 H's on the right.

$$C_2H_6 + 2H_2O \rightarrow 2CO + 5H_2$$

Checking the totals again of all atoms:

Atom	Left side	Right side
C	2	2
O	2	2
H	10	10

This chemical equation is now balanced.

b. The reaction described is

$$C_4H_{10} + O_2 \rightarrow CO_2 + H_2O$$

Balancing carbon: There are 4 C's on the left side and 1 C on the right. We multiply CO_2 by 4 to balance carbon.

$$C_4H_{10} + O_2 \rightarrow 4CO_2 + H_2O$$

Balancing hydrogen: There are 10 H's on the left and 2 on the right, so we multiply H_2O by 5.

$$C_4H_{10} + O_2 \rightarrow 4CO_2 + 5H_2O$$

Balancing oxygen: There are now 2 O's on the left and 13 on the right, so we must multiply O_2 by 13/2 to get 13 O's on the left.

$$C_4H_{10} + \frac{13}{2}O_2 \rightarrow 4CO_2 + 5H_2O$$

Although the equation is now mathematically balanced, it is common practice to clear the fraction. In this case, we multiply all quantities by 2, giving the final equation.

$$2C_4H_{10} + 13O_2 \rightarrow 8CO_2 + 10H_2O$$

After balancing, we have:

Atom	Left side	Right side
C	8	8
H	20	20
O	26	26

Is Our Answer Reasonable?

Our check says that for each equation, the number of each atom on the left (reactant) side is the same as on the right (product) side. In addition, we have not changed any formulas. We have only changed the coefficients in front of the molecules. The answers are reasonable.

What If...?

...we had 4 molecules of butane (C_4H_{10}) instead of 2? How would that change the balanced equation? (*Ans: You would need to double all the other coefficients*)

Practice 3.10

Balance this equation for the reaction that produces phosphoric acid (H_3PO_4), which can be used as a fertilizer, and calcium sulfate, forming gypsum used in housing wallboard:

$$Ca_3(PO_4)_2 + H_2SO_4 \rightarrow CaSO_4 + H_3PO_4$$

3.7.2 More Information from the Equation

As you will learn as you continue your chemistry studies, the chemical equation contains a wealth of information about a reaction. The chemical equation can also provide information about the states, or phases, of reactants and products. In ► Chap. 1, we discussed the states of matter. The **phase** is another term for the state of matter, a part of matter that is chemically of physically homogeneous. This is often indicated by using italic letters in parentheses after each formula in the equation (◘ Table 3.1).

Phase – A part of matter that is chemically and physically homogeneous.

For example, the equation we balanced in Example 3.10, could be written to include this information. With the phases added, the equation specifically says that 13 mol of oxygen gas reacts with 2 mol of butane gas to produce 8 mol of carbon dioxide gas and 10 mol of liquid water.

$$2C_4H_{10}(g) + 13O_2(g) \rightarrow 8CO_2(g) + 10H_2O(l)$$

3.7.3 The Meaning of a Chemical Equation

An equation is a precise quantitative summary of a reaction. It indicates the ratios in which the reactants react and the products form. This means that the chemical equation shows the exact number of molecules (or atoms or ions) that react and the exact number of product molecules (or atoms or ions) that are formed. For correctly bal-

Table 3.1 The abbreviations used in equations for the phase of matter

Abbreviation	Phase of matter
(*s*)	Solid
(*l*)	Liquid
(*g*)	Gas
(*aq*)	Aqueous (solutions in water)

anced equations, the total masses on both sides of the equation (taking the coefficients into account!) should be equal. For example, we can summarize this quantitative information contained in the equation for the synthesis of aspirin:

Equation	$C_7H_6O_3(s)$	+	$C_4H_6O_3(l)$	→	$C_9H_8O_4(s)$	+	$C_2H_4O_2(l)$
Coefficients	1 mol	+	1 mol	→	1 mol	+	1 mol
Molar mass	138.118 g	+	102.088 g	→	180.154 g	+	60.052 g
	Total reactant mass = 240.206 g				Total product mass = 240.206 g		

This is useful information for a chemist who wants to make aspirin. For example, the equation indicates that it would be appropriate to add 102.088 g of ethanoic anhydride for every 138.118 g of salicylic acid to prepare 180.154 g of aspirin. That corresponds to the simple mole ratio of 1 mol of ethanoic anhydride per 1 mol of salicylic acid, as indicated by the equation. In the context of a large chemical process plant, in which millions of aspirin tablets are made daily, the reactions may require megagram quantities.

3.8 Working with Chemical Equations

The formation of ethanoic anhydride and water from ethanoic (acetic) acid can be used to explore further how we put chemical equations to work, using them to calculate things we want to know. Remember that in addition to indicating what chemicals are involved in the reaction, the equation establishes the mole ratios among all the chemicals. We found that the balanced chemical equation indicates that 2 mol of ethanoic acid ($C_2H_4O_2$) produce 1 mol of ethanoic anhydride ($C_4H_6O_3$) and 1 mol of water (H_2O).

$$\begin{array}{ccccc} 2C_2H_4O_2 & \rightarrow & C_4H_6O_3 & + & H_2O \\ 2\text{ mol} & \rightarrow & 1\text{ mol} & + & 1\text{ mol} \end{array}$$

The mole ratios that indicate how compounds react and form products are known as **stoichiometric ratios** after the Greek stoicheon for "element" and metron for "measure." The stoichiometric ratios are vital when we want to answer questions about the quantity of reactant that reacts or the quantity of product that forms. This process of asking and answering mathematical questions based on balanced chemical equations is an aspect of **stoichiometry**, which is essentially the study and use of quantitative relationships in chemical processes.

For instance, the equation relating the formation of ethanoic anhydride describes not only the compounds involved in the reaction but also the stoichiometric ratios of those compounds. We can write all the mole ratios in the form of conversion factors that will allow us to convert among the numbers of moles of the chemicals concerned.

Stoichiometric ratios – The mole ratios relating how compounds react and form products.
Stoichiometry – The study and use of quantitative relationships in chemical processes.

$$\frac{2\text{ mol ethanoic acid}}{1\text{ mol ethanoic anhydride}} \qquad \frac{1\text{ mol ethanoic anhydride}}{2\text{ mol ethanoic acid}}$$

$$\frac{2\text{ mol ethanoic acid}}{1\text{ mol water}} \qquad \frac{1\text{ mol water}}{2\text{ mol ethanoic acid}}$$

$$\frac{1\text{ mol ethanoic anhydride}}{1\text{ mol water}} \qquad \frac{1\text{ mol water}}{1\text{ mol ethanoic anhydride}}$$

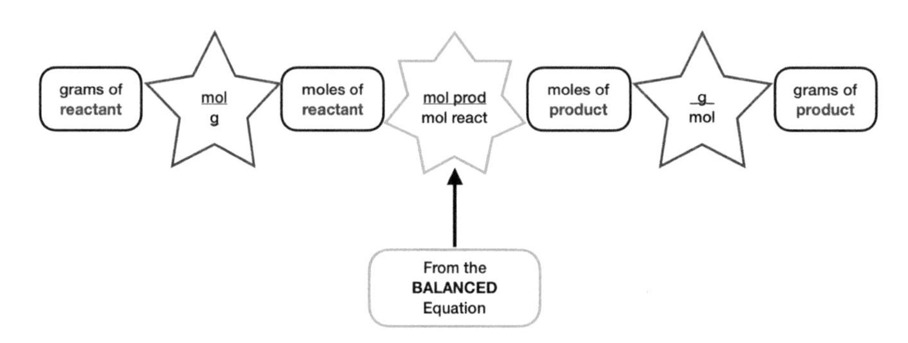

Fig. 3.13 Stoichiometry flowchart

Note that the ratios can be flipped as needed, depending on which units we are trying to cancel. Figure 3.13 shows the overall process to use these ratios, or "conversion factors" to calculate how much product we can make in a chemical reaction. Note that based on what we have, we can move to the left or the right in the flowchart in Fig. 3.13 in order to find something we want to know.

For example, let's assume we wish to know the maximum mass of ethanoic anhydride ($C_4H_6O_3$) that could be formed from 300.0 kg of ethanoic acid ($C_2H_4O_2$). The molar mass of ethanoic acid is 60.05 g and that of ethanoic anhydride is 102.1 g. We can obtain the answer using the appropriate conversion factors. We start by determining the number of grams of reactant and then follow the flowchart in Fig. 3.13 from left to right.

$$\text{kg } C_2H_4O_2 \xrightarrow{\frac{g}{kg}} \text{g } C_2H_4O_2 \xrightarrow{\frac{mol}{g}} \text{mol } C_2H_4O_2 \xrightarrow{\frac{mol\ C_4H_6O_3}{mol\ C_2H_4O_2}} \text{mol } C_4H_6O_3 \xrightarrow{\frac{g}{mol}} \text{g } C_4H_6O_3$$

$$300.0 \text{ kg } C_2H_4O_2 \times \frac{1000 \text{ g}}{1 \text{ kg}} \times \frac{1 \text{ mol } C_2H_4O_2}{60.05 \text{ g } C_2H_4O_2} \times \frac{1 \text{ mol } C_4H_6O_3}{2 \text{ mol } C_2H_4O_2} \times \frac{102.1 \text{ g } C_4H_6O_3}{1 \text{ mol } C_4H_6O_3} = 255{,}000 \text{ g } C_4H_6O_3$$

Three hundred kilograms of ethanoic acid can produce 255,000 g (which is 255.0 kg) of ethanoic anhydride by this reaction.

Example 3.11—Determining Products from Reactants

Hydrogen gas is used extensively in the food industry in a process called hydrogenation, in which the gas is added to compounds called "unsaturated fatty acids". This hardens the fatty acids of vegetable oils, making them solids at room temperature, like the fats found in butter. An example of hydrogenation is the reaction of hydrogen gas with the fatty acid called oleic acid:

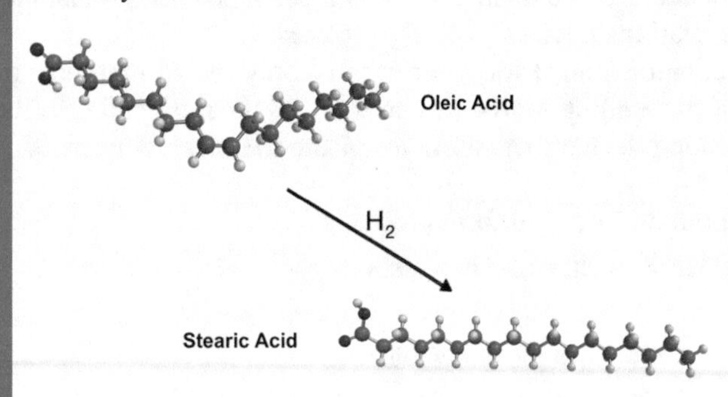

What is the maximum amount of stearic acid that can be produced from the reaction of 425 g of hydrogen gas with an excess of oleic acid?

Asking the Right Questions

First, since we are dealing with a chemical reaction, we should verify that the reaction is balanced. What is the key relationship whenever we are comparing reactants to products? *Moles-to-moles is the bridge we must always cross.* The flowchart for the strategy is:

$$\text{g } H_2 \xrightarrow{\frac{mol\ H_2}{g\ H_2}} \text{moles } H_2$$

$$\xrightarrow{\frac{mol\ C_{18}H_{36}O_2}{mol\ H_2}} \text{moles } C_{18}H_{36}O_2$$

$$\xrightarrow{\frac{g\ C_{18}H_{36}O_2}{mol\ C_{18}H_{36}O_2}} \text{g } C_{18}H_{36}O_2$$

Solution

$$425 \text{ g } H_2 \times \frac{1 \text{ mol } H_2}{2.016 \text{ g } H_2}$$

$$\times \frac{1 \text{ mol } C_{18}H_{36}O_2}{1 \text{ mol } H_2} \times \frac{284.5 \text{ g } C_{18}H_{36}O_2}{1 \text{ mol } C_{18}H_{36}O_2}$$

$$= 6.00 \times 10^4 \text{ g } C_{18}H_{36}O_2$$

Even though the mole ratio is 1-to-1, we still include it in the calculation so that the units cancel properly.

Is Our Answer Reasonable?

A molecule of stearic acid is 140 times as massive as a molecule of hydrogen, so based on the 1-to-1 mol ratio, we would expect the mass of the stearic acid product to be 140 times as large as the starting mass of hydrogen. Our answer of 60,000 g of stearic acid produced therefore makes sense.

What If…?

…we combined 425 g of oleic acid with an excess of hydrogen gas? How many grams of stearic acid can be produced? (*Ans: 428 g*)

Practice 3.11

How many grams of oleic acid are needed to produce 72.55 g of stearic acid?

Example 3.12—Determining Products from Reactants: Focus on Mole Ratios

The most common route used for the industrial synthesis of hydrogen is the reaction of water as steam with hydrocarbons—substances that contain only hydrogen and carbon, such as ethane (C_2H_6) or propane (C_3H_8). The reactants can be passed over a nickel catalyst, a substance that greatly speeds the process. The production of hydrogen from the reaction of steam with propane is:

$$C_3H_8(g) + 3H_2O(g) \xrightarrow{Ni(700-900°C)} 3CO(g) + 7H_2(g)$$

How much hydrogen can be produced from the reaction of 30.0 g of propane with excess steam?

Asking the Right Questions

What is the key mole-to-mole relationship? You want to determine the grams of hydrogen produced from propane, so the 7-to-1 mol ratio of hydrogen to propane is the key.

$$\frac{7\ \text{mol}\ H_2}{1\ \text{mol}\ C_3H_8}$$

Our flowchart that includes the key mole-to-mole bridge looks like this:

$$g\,C_3H_8 \xrightarrow{\frac{mol\,C_3H_8}{g\,C_3H_8}} mol\,C_3H_8 \xrightarrow{\frac{mol\,H_2}{mol\,C_3H_8}} mol\,H_2 \xrightarrow{\frac{g\,H_2}{mol\,H_2}} g\,H_2$$

Solution

$$30.0\ \text{g}\ C_3H_8 \times \frac{1\ \text{mol}\ C_3H_8}{44.09\ \text{g}\ C_3H_8} \times \frac{7\ \text{mol}\ H_2}{1\ \text{mol}\ C_3H_8} \times \frac{2.016\ \text{g}\ H_2}{1\ \text{mol}\ H_2}$$
$$= 9.60\ \text{g}\ H_2$$

Remember that the mole-to-mole ratio is an exact number, via the chemical equation so the resulting conversion factor does not limit your significant numbers.

Is our Answer Reasonable?

Even though hydrogen is a very light gas, the 7-to-1 mol ratio of hydrogen-to-propane means that a good deal of hydrogen will be produced from the reaction of propane with steam.

What If…?

…we needed to produce 4.00×10^4 kg hydrogen gas from methane (CH_4) using the reaction with steam to form CO and H_2? How many kilograms of methane would be required?

(*Ans: 1.06×10^5 kg*).

Practice 3.12

Using the equation in this exercise, determine how many grams of carbon monoxide can be produced from 83.8 g of water with an excess of propane.

Example 3.13—Calculations using Chemical Equations

Photosynthesis in plants generates glucose ($C_6H_{12}O_6$) and oxygen from carbon dioxide and water. How many kilograms of oxygen are produced from the photosynthesis of 330 kg of carbon dioxide?

Asking the Right Questions

In addition to seeing what the example is asking for (kilograms of oxygen), we also need to determine the equation that describes the photosynthesis reaction that generates glucose and oxygen from carbon dioxide and water. What must be true of any chemical equation? We must first balance the equation:

$$CO_2(g) + H_2O(l) \rightarrow C_6H_{12}O_6(s) + O_2(g)$$

We balance the equation as follows:

$$6CO_2(g) + 6H_2O(l) \rightarrow C_6H_{12}O_6(s) + 6O_2(g)$$

We must also develop a pathway to the answer by knowing what we are given, what we want, and how to get there:

$$kg\ CO_2 \xrightarrow{\frac{g}{kg}} g\ CO_2 \xrightarrow{\frac{mol\ CO_2}{g\ CO_2}} moles\ CO_2 \xrightarrow{\frac{mol\ O_2}{mol\ CO_2}}$$

$$moles\ O_2 \xrightarrow{\frac{g\ O_2}{mol\ O_2}} g\ O_2 \xrightarrow{\frac{kg}{g}} kg\ O_2$$

Solution

In our equation we have a 1-to-1 mol ratio of CO_2-to-O_2.

$$330\ kg\ CO_2 \times \frac{1000\ g\ CO_2}{1\ kg\ CO_2} \times \frac{1\ mol\ CO_2}{44.01\ g\ CO_2} \times \frac{1\ mol\ O_2}{1\ mol\ CO_2} \times \frac{32.00\ g\ O_2}{1\ mol\ O_2} \times \frac{1\ kg\ O_2}{1000\ g\ O_2} = 240\ kg\ O_2$$

The equation relates a 6-to-6 ratio. This is the same as 1-to-1, although we can use $\frac{6\ mol\ O_2}{6\ mol\ CO_2}$ in the equation if we wish.

Is Our Answer Reasonable?

Oxygen has a molar mass that is just a little smaller than that of carbon dioxide, so the quantity of oxygen produced should be a little lower than the starting quantity of CO_2. We did calculate a slightly smaller quantity, so the answer makes sense.

What If...?

...we started with half of the mass of CO_2 as in this example? How would that affect the number of moles of glucose generated? (*Ans*: *One-half of the mass of* CO_2 *would make one-third of the number of moles of glucose*)

Practice 3.13

What mass of CO_2 gas is released when 26 g methane (CH_4) burns in oxygen (O_2) to generate CO_2 and water (H_2O)? To solve this question, you need to use several of the techniques learned in this chapter—specifically, working out molar masses, writing balanced equations, and performing a stoichiometric calculation.

3.8.1 Limiting Reagent

We have seen that equations indicate how many moles of each reactant will react together, but chemists do not usually combine reactants in the exact proportions shown in the chemical equation. Instead, they generally use an excess amount of one or more of the reactants, perhaps the least expensive one, in order to encourage as much of the other reactants as possible to react. What limits the amount of product we can make? In this situation, the reactant that will be used up first is called the **limiting reagent**, or limiting reactant, because it is the quantity of this reagent that imposes a limit on how much product can be formed.

Limiting reagent – The reactant that is consumed first, causing the reaction to cease despite the fact that the other reactants remain "in excess." Also known as the limiting reactant.

An analogy that is helpful in understanding what we mean by a limiting reagent is the preparation of hamburgers for a frozen food company. Each hamburger must be like every other because the company's customers expect a consistent product. Each hamburger, shown in ◻ Fig. 3.14 both in parts and completed, contains *only* 1 patty, 5 pickles, 2 slices of cheese, and 1 bun.

In this analogy, we can write the preparation of the product from the "reactants" as follows:

$$1\ \text{Patty} + 5\ \text{Pickles} + 2\ \text{Cheese} + 1\ \text{Bun} \rightarrow 1\ \text{Hamburger}$$

Let's assume we have 10 patties, 50 pickles, 20 slices of cheese, and 3 buns. How many hamburgers can we make? As ◻ Fig. 3.15 shows, we can only prepare 3 complete hamburgers, because *the number of buns limits our production*. The buns are the limiting reagent in this case. How many slices of cheese are ***in excess***—that is, how many slices of cheese are left over?

In excess – The amount of a reactant that was not consumed in a reaction. The reactant in excess is not the limiting reactant.

In a manner analogous to expressing a mole ratio, we can say that according to the equation for sandwich preparation, each bun requires 2 slices of cheese, giving us a ratio:

Fig. 3.14 A hamburger and its components

Fig. 3.15 Analogy of limiting reagent to limiting ingredient in the construction of a sandwich

$$\frac{1 \text{ bun}}{2 \text{ cheese}}$$

We can also write other valid reactant "mole ratios," such as:

$$\frac{1 \text{ bun}}{5 \text{ pickles}} \quad \frac{5 \text{ pickles}}{1 \text{ patty}} \quad \frac{2 \text{ cheese}}{5 \text{ pickles}} \quad \frac{1 \text{ patty}}{1 \text{ bun}}$$

We asked how many slices of cheese are left over from our original stack of 20 slices if we use 3 buns. We could also have asked, "How many cheese slices are actually used to combine with 3 buns?" The ratio of buns to cheese is:

$$\frac{1 \text{ bun}}{2 \text{ cheese}} \text{ or } \frac{2 \text{ cheese}}{1 \text{ bun}}$$

The number of cheese slices used, then, is:

$$3 \text{ buns} \times \frac{2 \text{ cheese}}{1 \text{ bun}} = 6 \text{ cheese}$$

The number of cheese slices in excess is the number we had on the shelf (20 cheese slices) minus the number we used (6 cheese slices).

$$20 \text{ Cheese} - 6 \text{ Cheese} = 14 \text{ Slices of Cheese in Excess}$$

This process of determining how much of a reagent is used and how much is left over is the same as the process we use with chemical systems. The only difference is that we think in terms of moles rather than slices of cheese, patties, pickles, or buns. Let's consider another example, as shown in Fig. 3.16.

The formula for water is H_2O, so we need 2 model hydrogen atoms and 1 model oxygen atom per model of a water molecule. If we have available a pile of 10 oxygen atoms and a pile of 60 hydrogen atoms, we will be able to assemble only 10 water molecules before we run out of oxygen atoms. In this situation, oxygen is the limiting reagent because the number of oxygen atoms present limits to 10 the number of water molecules we can form. Hydrogen atoms, on the other hand, are "in excess," so some of them are left over once all the oxygen has been used.

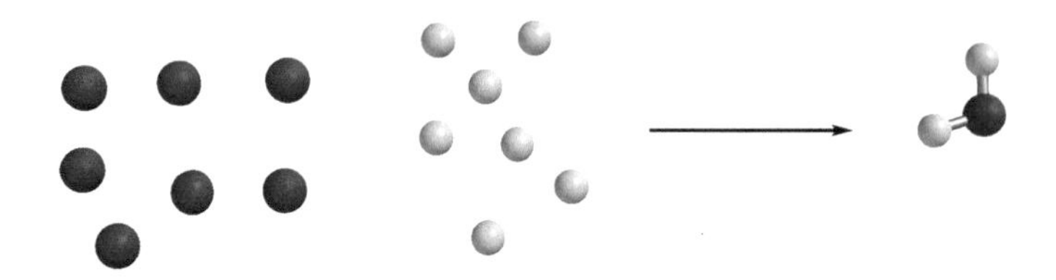

Fig. 3.16 Model atoms typically used in teaching laboratories can be used to illustrate the limiting reagent analogy

In another example, let's consider the carbon dioxide gas breathed out by the astronauts in the international space station. If it were to accumulate, it would soon rise to levels that would cause the occupants to become dizzy and confused and eventually to slip into unconsciousness. One process that has been used to remove carbon dioxide is the reaction of the gas with lithium hydroxide (LiOH) as it is drawn through special filters:

$$2LiOH\ (s) + CO_2\ (g) \rightarrow Li_2CO_3\ (s) + H_2O\ (l)$$

Will 5.0 kg of lithium hydroxide per day be sufficient to remove the carbon dioxide released by one astronaut per day (typically 1.0 kg of CO_2)?

We can answer the question by determining the limiting reagent in these circumstances. Here is the basic information we need:

$$\begin{array}{lclcl} 2LiOH\ (s) & + & CO_2\ (g) & \rightarrow & Li_2CO_3\ (s) + H_2O\ (l) \\ 2mol & + & 1\ mol & & \\ 23.95\ g/mol & & 44.01\ g/mol. & & \end{array}$$

We can calculate the amount of lithium hydroxide needed to react with the average 1.0 kg (1.0×10^3 g) of carbon dioxide released per day as follows. Remember our flowchart diagram (Fig. 3.13), and use it to construct your solution:

$$1.0 \times 10^3\ g\ CO_2 \times \frac{1\ mol\ CO_2}{44.01\ g\ CO_2} \times \frac{2\ mol\ LiOH}{1\ mol\ CO_2} \times \frac{23.95\ g\ LiOH}{1\ mol\ LiOH} = 1100\ g\ LiOH$$

If we have 5.0 kg (5.0×10^3 g) of LiOH available per day, the carbon dioxide will be the limiting reagent. The LiOH will be in excess, and there will be plenty of LiOH left over at the end of the day.

Example 3.14—Calculating Yield and Limiting Reactant

Sodium hydroxide reacts with phosphoric acid to give sodium phosphate and water. Assume that 17.80 g of NaOH is mixed with 15.40 g of H_3PO_4.

a. How many grams of Na_3PO_4 can be formed?
b. How many grams of the excess reactant remain unreacted?

Asking the Right Questions

We start by asking "What is the balanced equation?" and "Which compound limits the amount of product formed?" Then, we build our flowchart:

$$\text{g of reactant} \xrightarrow{\text{molar mass}} \text{moles of reactant} \xrightarrow{\text{mole ratio}}$$

$$\text{moles of product} \xrightarrow{\text{molar mass}} \text{g of product}$$

To determine the amount of excess reactant, we can work via moles:

$$\#\text{moles excess} = \#\text{moles original} - \#\text{moles used}$$

Then convert moles to grams.

Solution

The balanced equation for this reaction is:

$$3NaOH(aq) + H_3PO_4(aq) \rightarrow Na_3PO_4(aq) + 3H_2O(l)$$

a. Determination of limiting reagent

g Na_3PO_4 from NaOH:

$$17.80 g\ NaOH \times \frac{1\ mol\ NaOH}{40.00\ g\ NaOH} \times \frac{1\ mol\ Na_3PO_4}{3\ mol\ NaOH} \times \frac{163.94\ g\ Na_3PO_4}{1\ mol\ Na_3PO_4} = 24.318\ g\ Na_3PO_4\ \text{from NaOH}$$

g Na_3PO_4 from H_3PO_4:

$$15.40\ g\ H_3PO_4 \times \frac{1\ mol\ H_3PO_4}{97.99\ g\ H_3PO_4} \times \frac{1\ mol\ Na_3PO_4}{1\ mol\ H_3PO_4} \times \frac{163.94\ g\ Na_3PO_4}{1\ mol\ Na_3PO_4} = 25.765\ g\ Na_3PO_4\ \text{from}\ H_3PO_4$$

Therefore, NaOH is the limiting reagent and 24.32 g Na_3PO_4 are formed.

b. Determination of grams of excess reactant.

H_3PO_4 is in excess. If 24.32 g Na_3PO_4 is formed, the moles of H_3PO_4 used are given by:

$$\text{mol } H_3PO_4 \text{ used} = 24.318 \text{ g } Na_3PO_4 \times \frac{1 \text{ mol } Na_3PO_4}{163.94 \text{ g } Na_3PO_4} \times \frac{1 \text{ mol } H_3PO_4}{1 \text{ mol } Na_3PO_4}$$

$$= 0.14833 \text{ mol } H_3PO_4 \text{ used}$$

The number of moles of H_3PO_4 originally present is:

$$15.40 \text{ g } H_3PO_4 \times \frac{1 \text{ mol } H_3PO_4}{97.99 \text{ g } H_3PO_4}$$

$$= 0.15716 \text{ mol } H_3PO_4 \text{ originally present}$$

mol H_3PO_4 excess
= mol H_3PO_4 originally present – mol H_3PO_4 used
= 0.15716 mol – 0.14833 mol
= 0.00883 mol H_3PO_4 excess

$$\text{grams } H_3PO_4 \text{ excess} = 0.00883 \text{ mol } H_3PO_4 \times \frac{97.99 \text{ g } H_3PO_4}{1 \text{ mol } H_3PO_4}$$

$$= 0.87 \text{ g } H_3PO_4 \text{ excess}$$

Are Our Answers Reasonable?

One way to know if our answer makes sense is to rely on the Law of Conservation of Matter. After all is said and done, are the masses of the reactants and products the same?

We began with 17.80 g of NaOH and 15.40 g of H_3PO_4, or 33.20 total grams of reactants available. The NaOH was the limiting reactant, and there were 0.87 g of H_3PO_4 remaining after the reaction. This means that there were (33.20 g originally present – 0.87 g remaining) 32.33 g of reactants used.

What quantity of the products was formed? A total of 24.32 g of Na_3PO_4 was formed. What about grams of water? Since the mole-ratio of Na_3PO_4-to-H_2O is 1-to-3, we can determine this:

$$\text{g } Na_3PO_4 \xrightarrow{\frac{\text{mol } Na_3PO_4}{\text{g } Na_3PO_4}} \text{mol } Na_3PO_4 \xrightarrow{\frac{\text{mol } H_2O}{\text{mol } Na_3PO_4}} \text{mol } H_2O \xrightarrow{\frac{\text{g } H_2O}{\text{mol } H_2O}} \text{g } H_2O$$

$$24.32 \text{ g } Na_3PO_4 \times \frac{1 \text{ mol } Na_3PO_4}{163.94 \text{ g } Na_3PO_4} \times \frac{3 \text{ mol } H_2O}{1 \text{ mol } Na_3PO_4} \times \frac{18.02 \text{ g } H_2O}{1 \text{ mol } H_2O} = 8.020 \text{ g } H_2O \text{ formed}$$

The total mass of products formed is 8.020 g H_2O + 24.32 g Na_3PO_4 = 32.33 g of products formed. This agrees with the amount used.

What If…?

…we tripled the amount of H_3PO_4 so that the NaOH was combined with 46.20 g of H_3PO_4? How would that change the amount of product? Why? (*Ans*: *It wouldn't. The NaOH would still be the limiting reactant, so there would just be more H_3PO_4 in excess.*)

Practice 3.14

Aqueous calcium hydroxide reacts with aqueous hydrogen chloride to produce water and aqueous calcium chloride. If 12.33 g of calcium hydroxide are placed in a flask with 32.15 g of hydrogen chloride, how many grams of calcium chloride would be formed?

3.8.2 Percent Yield

Chemical reactions are rarely 100% efficient, because there are always losses due to unwanted side reactions, to some of the reactants remaining unreacted, or to some of the products being converted back into reactants once they are formed. Chemists designing a synthesis reaction can compare the actual masses obtained under various conditions with the masses predicted from the chemical equation. Doing so enables them to calculate the percent yield of a reaction and to adjust the conditions until the maximum percent yield is obtained. Often, there are compromises to be made, because it is sometimes impossible or impractical to achieve 100% yield. The maximum amount of any chemical that could be produced in a chemical reaction, which can be calculated from the equation for the reaction, is called the **theoretical yield**. The amount of product obtained experimentally is called the **actual yield**.

Theoretical yield – The maximum amount of any chemical, in grams, that could be produced in a chemical reaction. This value can be calculated from the given amounts of reactants using the balanced chemical equation.

Actual yield – The experimentally determined quantity, in grams, of product obtained in a reaction.

The **percent yield** of a reaction equals the actual yield expressed as a percentage of the theoretical yield:

Percent yield – The actual yield of a reaction divided by the theoretical yield and then multiplied by 100%.

$$\text{percent yield} = \frac{\text{actual yield}}{\text{theoretical yield}} \times 100\%$$

A non-chemical analogy is an exam score. If there were 50 points possible (the theoretical yield for a completely correct paper), and a student scored 42 (the actual yield of points), the percentage yield (the percent the student got correct) would be 42/50 × 100% = 84%. The student could have gotten 100% if the yield were complete (perfect score).

This first look at the uses of chemical equations should help us realize that chemical equations provide useful information that chemists must have if they are to do their jobs of predicting and producing desired products, whether on a relatively small scale in a laboratory or on an industrial process scale in a huge manufacturing facility.

Example 3.15—Percent Yield

In the previous exercise, we determined that 24.32 g of sodium phosphate could be prepared under the conditions given. If only 20.07 g were obtained from the reaction, what would be the percent yield of the reaction?

Asking the Right Questions

"How much did the reaction produce?" "How much could it have produced?" The percent yield is the ratio of these two values converted to a percentage.

Solution

$$\text{percent yield} = \frac{\text{actual yield}}{\text{theoretical yield}} \times 100\%$$

$$= \frac{20.07\ \text{g}}{24.32\ \text{g}} \times 100\% = 82.52\%\ \text{yield}$$

Are Our Answers Reasonable?

The actual yield is a little less than the theoretical yield, which means that the percent yield should be close to, but less than, 100%. The answer therefore makes sense.

What If...?

...we had a process in which the percent yield was known to be 48.27%? If we actually generated 113.9 g of product, what is the quantity we theoretically could have generated if the reaction had 100% yield? (*Ans: 236.0 g*)

Practice 3.15

Assuming that the yield of sodium phosphate from sodium hydroxide and phosphoric acid can never be greater than 82.52%, how many grams of sodium hydroxide would be needed to produce 100 g of sodium phosphate?

Issues and Controversies—Everyday Controversies of Quantitative Chemistry

In everyday life, we are bombarded by conflicting messages from different sides of debates about quantitative chemistry. We may not realize that quantitative chemistry is at the heart of these debates, but it is. In considering our diet, we often pose questions such as: How much food should we eat each day? How many grams of fat should be in our diet? What levels of each vitamin are too low for good health (see Fig. 3.17), and what levels are dangerously high? We know that too much alcohol in beverages is bad for us, but are small, regular amounts beneficial, or is total abstinence the most healthful choice?

These issues are complex because most chemicals have more than one effect on the body. In other words, a single chemical can participate in several different chemical reactions within the body. Some of these reactions may bring clear benefits, whereas others can pose dangers. These are all issues of quantitative chemistry, revolving around what quantities of a chemical are required to produce a certain level of benefit and what quantities produce a given level of harm. The benefits and the dangers are all caused, directly or indirectly, by the stoichiometry of the chemical reactions in which the chemicals participate.

Vitamin C is an excellent example. We know that we must consume a certain amount of this vitamin, which is normally obtained from fresh fruits and vegetables, for good health. A deficiency of vitamin C causes a variety of problems known collectively as "scurvy," which is characterized by fatigue, sore joints and muscles, hemorrhaging, and anemia. The minimum level of vitamin C needed to prevent these problems can be determined by the quantitative analysis of people's diets and by the

Nutrition Facts

Serving Size 1 cup (228g)
Servings Per Container 2

Amount Per Serving

Calories 250 Calories from Fat 110

	% Daily Value*
Total Fat 12g	**18%**
Saturated Fat 3g	**15%**
Trans Fat 3g	
Cholesterol 30mg	**10%**
Sodium 470mg	**20%**
Total Carbohydrate 31g	**10%**
Dietary Fiber 0g	**0%**
Sugars 5g	
Protein 5g	
Vitamin A	4%
Vitamin C	2%
Calcium	20%
Iron	4%

* Percent Daily Values are based on a 2,000 calorie diet. Your Daily Values may be higher or lower depending on your calorie needs.

	Calories	2,000	2,500
Total Fat	Less than	65g	80g
Sat Fat	Less than	20g	25g
Cholesterol	Less than	300mg	300mg
Sodium	Less than	2,400mg	2,400mg
Total Carbohydrate		300g	375g
Dietary Fiber		25g	30g

Fig. 3.17 The quantities of chemicals we consume are important—and what quantities are ideal is the subject of much debate

association between these diets and the appearance of symptoms of vitamin C deficiency. Even this issue, however, is linked to controversy, because different countries use different recommended levels, and nutritionists regularly debate whether the levels should be revised upward or downward. The current recommended dietary allowance (RDA) of vitamin C in the United States is 75 mg per day for adult females and 90 mg per day for adult males. Go into a drugstore, however, and you will easily find tablets available that each delivers a massive 1000-mg (1-g) dose of vitamin C. These are sold because some scientists, most notably Linus Pauling, have promoted the view that huge "megadoses" of vitamin C can protect against colds and even cancer. Others caution, however, that excessive consumption of vitamin C may cause problems such as nausea, diarrhea, iron overload—and possibly even cancer.

The debate about the appropriate amount of vitamin C to consume is mirrored by similar debates about every other vitamin and about many other components of our diet. One of the most important activities in the attempts to resolve these debates is the careful quantitative analysis of the many effects each chemical can have on the chemistry of life.

The Bottom Line

- Quantities in chemistry are crucial. Different amounts of the same chemical can have very different effects on chemical systems such as living things, environmental systems, and industrial processes.
- The formula mass (formula weight) of a chemical is the total mass of all the atoms present in its formula, in atomic mass units (amu), which is also grams per mole.

- The average molecular mass (molecular weight) of a molecule is the total mass of the molecule, in atomic mass units (amu) or in grams per mole.
- The mole is the basic counting unit of chemistry—the chemist's "dozen"—and 1 mol of any chemical contains Avogadro's number (6.022×10^{23}) of atoms, molecules or formula units.
- We use the mole to convert among atoms, molecules or formula units and grams of a substance.
- The percent, by mass, of an element in a compound is called its mass percent and is equal to:

$$\text{percent mass} = \frac{\text{mass of component}}{\text{total mass of substance}} \times 100\%$$

- The empirical formula for a compound indicates the simplest whole-number ratio in which its component atoms are present. The molecular formula indicates the actual number of each type of atom in one molecule of the compound. The molecular formula is a whole-number multiple of the empirical formula.
- A chemical equation uses chemical formulas to indicate the reactants and products of a reaction and uses numbers in front of the formulas to indicate the proportions in which the chemicals involved react together and are formed.
- A balanced chemical equation only tells you the relationships between reactants and products in a given chemical reaction. It does not tell you anything about how much of each chemical you actually have.
- Stoichiometry is the study and use of quantitative relationships in chemical processes.
- The limiting reagent in a reaction is the one that is consumed first, causing the reaction to cease despite the fact that the other reactants remain "in excess."
- The percent yield of a reaction equals the actual yield expressed as a percentage of the theoretical yield:

$$\text{percent yield} = \frac{\text{actual yield}}{\text{theoretical yield}} \times 100\%$$

Section 3.2 Formula Masses

Skill Review

1. Determine the mass, in amu, for each:
   a. Carbon monoxide, CO
   b. Silicon dioxide (the principal component in sand), SiO_2
   c. Ammonia, NH_3
   d. Sodium thiosulfate (photographer's "hypo" solution), $Na_2S_2O_3$
   e. Tristearin (a type of animal fat), $C_{57}H_{110}O_6$
2. Determine the mass, in amu, for each:
   a. Water, H_2O
   b. Sodium hydroxide, $NaOH$
   c. Fructose ("fruit sugar"), $C_6H_{12}O_6$
   d. Potassium dichromate, $K_2Cr_2O_7$
   e. Ammonium phosphate (a fertilizer ingredient), $(NH_4)_3PO_4$
3. Arrange these in order from lightest to heaviest on the basis of formula masses.
   C_6H_6 H_2O C_2H_4OH
   $CaCl_2$ CO
4. Arrange these in order from heaviest to lightest on the basis of formula masses.
   $NaBr$ KBr $KMnO_4$
   $LiOH$ Na_3PO_4

Chemical Applications and Practices

5. Saccharin ($C_7H_5NO_3S$) and aspartame ($C_{14}H_{18}N_2O_5$) have both been used as sugar substitutes.
   a. What is the mass, in amu, of one molecule of each sugar substitute?
   b. What is the mass ratio of aspartame to saccharin?
   c. What mass, in grams, of saccharin has the same number of molecules as 42.0 g of aspartame?
6. As you are reading this, you are breathing in some molecular oxygen (O_2).
   a. Express the mass of an oxygen molecule in amu.
   b. Convert that mass into grams per molecule of O_2.
   c. What is the mass in grams of 6.022×10^{23} molecules of O_2?

7. The price of gold fluctuates daily. However, if you paid $10,000 for 1 kg of gold, how much would you be paying per one atom of gold?
8. Sodium bicarbonate ($NaHCO_3$) is a useful ingredient employed in baking.
   a. What is the formula mass of $NaHCO_3$?
   b. If you paid $1.50 for 250 g of $NaHCO_3$, how much would you be paying for each formula unit?

Section 3.3 Counting by Weighing

Skill Review

9. Convert these to grams:
   a. 6.02×10^{22} atoms of silver
   b. One trillion atoms of gold
   c. 2 dozen water molecules
   d. One molecule of propane (C_3H_8)
10. Convert these to grams:
   a. 3.54×10^{21} atoms of gold
   b. A gross (12 dozen) of atoms of mercury
   c. 1.2×10^{27} molecules of glucose ($C_6H_{12}O_6$)
   d. One molecule of carbon dioxide (CO_2)
11. Indicate how many molecules (or formula units, where appropriate) there are in each of the following:
   a. 12.01 g of water
   b. 68.3 g of sodium bicarbonate ($NaHCO_3$)
   c. 100 g of methane (CH_4)
   d. 2.3 mg of glucose ($C_6H_{12}O_6$)
12. Indicate how many atoms there are in each of the following:
   a. 100 g of silver
   b. 100 g of gold
   c. 12.01 g of carbon
   d. 5.3 kg of water
13. a. One of the main waste products of animal metabolism is urea. What is the mass, in amu, of one molecule of urea (CH_4ON_2)?
   b. What is the mass in grams of 6.022×10^{23} molecules of urea?
   c. What would be the total mass in grams of 6.022×10^{22} molecules of urea?
14. Ammonium nitrate is often used as a fertilizer. Express these amounts of NH_4NO_3 in grams.
   a. one million formula units
   b. 10 mol
   c. 0.500 kg

Chemical Applications and Practices

15. Ethanol, C_2H_6O, is currently in use as both an oxygen additive and gasoline substitute due to its excellent combustion characteristics.
   a. Based on the formula, how many atoms of oxygen are there per molecule of ethanol?
   b. How many atoms of hydrogen are there in a gross (144) of ethanol molecules?
16. Adrenaline, as was noted in the chapter, causes a stronger more pronounced heartbeat. Given the structure of this compound, as indicated by the ball-and-stick model shown in Fig. 3.1, answer these questions.
   a. How many carbon atoms are there per molecule of adrenaline?
   b. How many nitrogen atoms are there per dozen adrenaline molecules?
   c. How many oxygen atoms exist in a great gross (twelve gross: 1728) of adrenaline molecules?
17. A common misconception is that since "lead is heavier than feathers", that "a pound of lead would weigh more than a pound of feathers."
   a. Assuming you have a stack of lead tokens that each weigh 33.93 g, how many lead tokens would it take to make a pound of them? (round to the nearest whole number.)
   b. Assuming each feather weighs 0.75 g, how many feathers would it take to make a pound of feathers? (round to the nearest whole number.)
   c. A pound of each should have the same mass. However, which occupies a greater volume? Which is less dense, the feather or the lead token?

18. A person was asked to guess the number of candies in a glass jar. Knowing the mass of the candies to be a total of 2.50 lbs, and the average mass of a single candy to be 25.0 g, how many candies are in the jar?
19. Acetaminophen, a non-aspirin pain reliever, has the formula $C_8H_9NO_2$. A dose of 0.500 g of acetaminophen would contain how many molecules of the pain reliever?
20. a. Methane (CH_4) is used in many laboratory burners. If a student burned 10.0 g of methane, how many molecules were consumed?
    b. Would 10.0 g of butane (C_4H_{10}) contain more or fewer molecules?

Section 3.4 Working with Moles

Skill Review

21. Which of these quantities of sodium chloride (NaCl) contains the greatest mass?
    0.100 mol 4.2×10^{23} formula units   1.60 g
22. Which of these quantities of acetaminophen ($C_8H_9NO_2$) contains the greatest mass?
    0.550 mol 9.1×10^{23} molecules   80.4 g
23. Convert these to moles:
    a. 65.0 g of CO_2
    b. 1.5×10^{22} atoms of neon
    c. 25 g of propane (C_3H_8)
    d. 4.5×10^{24} molecules of ammonia (NH_3)
24. Convert these to moles:
    a. 123 g of H_2O
    b. 9.72×10^{23} atoms of gold
    c. 25.6 mg of methane (CH_4)
    d. 3.22×10^{10} molecules of hemoglobin ($C_{2954}H_{4508}N_{780}O_{806}S_{12}Fe_4$)
25. Which of these would contain the greatest number of atoms?
    454 g of gold
    56.0 g of O_2
    245 of graphite
26. Which of these would contain the greatest number of atoms?
    100 g of tin
    17.9 mg of glucose ($C_6H_{12}O_6$)
    2 g of water
27. Convert these to the correct number of formula units:
    a. 15.0 g of $CaCl_2$
    b. 15.0 mol of $CaCl_2$
    c. 15.0 mL of a $CaCl_2$ solution that contains 0.42 mol per liter
28. Convert these to the correct number of formula units:
    a. 1.0 g of K_2CO_3
    b. 15.0 mol of K_2CO_3
    c. 7.33 kg of K_2CO_3
29. Express these quantities in grams:
    a. 1500 mol of Ne
    b. 12.5 mol of Na
    c. 0.42 mol of N_2
30. Express these quantities in grams:
    a. 0.250 mol of NaCl
    b. 0.350 mol of $(NH_4)_3PO_4$
    c. 1.5×10^{21} molecules of CO_2
31. Radioactive samples of elements are used in both the treatment and the detection of certain diseases. One such element is element 43, technetium. Nuclear medicine applications of this element include bone scans, thyroid monitoring, and brain monitoring. If a pharmacist were preparing doses to be given to patients that involved using 5.00 g of radioactive technetium, how many atoms of technetium would actually be present?

32. Two elements have been named in honor of female scientists, curium (element 96) and meitnerium (element 109). Element 109 is a relatively recent discovery. Often when synthesizing newer elements, scientists are working with extremely small samples, even only a few atoms. Suppose a certain procedure produced 100 atoms of meitnerium. How many moles is this? What would be the mass of the sample?

Chemical Applications and Practices

33. Iron is essential for the transport of oxygen in the human body and for energy production through several biochemical cycles.
    a. What is the mass, in grams, of one atom of iron?
    b. The recommended dietary allowance (RDA) for iron is approximately 15 mg. How many atoms of iron is this?
    c. Express the RDA in moles of iron.
34. How many caffeine molecules ($C_8H_{10}N_4O_2$) are found in each of these drinks?
    a. a cup of coffee with 95 mg of caffeine
    b. a cup of tea with 0.065 g of caffeine
    c. a soda with 0.038 mol of caffeine
35. Cobalt is a metal required for human health. We normally ingest small amounts of cobalt in the food we eat. Typically, you may have about 1.5 mg present in your body.
    a. How many moles is this?
    b. How many atoms of cobalt are present?
36. Male silkworms are attracted to an organic molecule, secreted by the females, that acts as an attractant for mating purposes. If the mass per mole of the compound is 238 g, and a female silkworm releases 4.20×10^{-6} g, how many molecules are available to be detected by nearby males?
37. Some elements exist in more than one molecular form. This property is called allotropy. For example, phosphorus, which is very reactive in its elemental form and is used in some explosives and some types of matches, has three different forms. The structure of white phosphorus has interlocked tetrahedrons. The simple formula of white phosphorus is P_4.
    a. The white streams of smoke from military bombings may be due to the reaction of white phosphorus with oxygen. How many grams of phosphorus would be used if an explosive device contained 10.5 mol of P_4?
    b. The most stable of the allotropic forms of sulfur is called orthorhombic sulfur (S_8). The structure resembles a puckered crown. How many moles of S_8 are equivalent to 454 g of S_8?
38. The ability of nonmetal atoms to form attachments among themselves is called catenation. Carbon exhibits this property more than any other element. Catenation is responsible for the huge variety of carbon compounds.
    a. Determine the number of moles in 100.0 g each of the following carbon compounds. (In each compound, the carbon atoms are joined to each other). C_3H_8 (propane), C_4H_{10} (butane), C_5H_{12} (pentane)
    b. Silicon, which is found directly below carbon in the periodic table of the elements, also has the ability to catenate. Determine the number of grams in 0.100 mol of each of these silicon–hydrogen compounds (called silanes): Si_2H_6 and Si_6H_{14}.
39. A vitamin that helps activate many enzymes in humans is pyridoxine (also called vitamin B_6). Its chemical formula is $C_8H_{11}NO_3$. Vitamin B_6 can be found in good supply in wheat and legumes. If a nutritionist found 0.156 g of pyridoxine in a food sample, how many moles would be reported? How many molecules would be reported? How many atoms of carbon would be contained in the vitamin sample?
40. Carbon can be found in three elemental forms: graphite, diamond, and fullerene. Because all the carbon atoms in a diamond in a ring are connected in linked tetrahedrons, a single diamond can be considered to be one large molecule (often called a macromolecule). If a diamond has a mass of 0.50 carat, how many atoms of carbon make up the ring? (1 carat = approximately 0.20 g)

Section 3.5 Percentages by Mass

Skill Review

41. Calculate the percent of carbon in each of these organic compounds:
    a. C_3H_6
    b. CH_3OH
    c. C_6H_6
    d. $C_{12}H_{22}O_{11}$
42. Calculate the percent of oxygen in each of these organic compounds:
    a. CO_2
    b. $C_6H_{12}O_6$
    c. C_3H_8O
    d. H_2CO_3

43. Arrange these sulfur-containing compounds from the greatest mass percent sulfur to the least.
    a. H_2S
    b. SO_2
    c. $Na_2S_2O_4$
    d. H_2SO_3
44. Arrange these oxygen-containing compounds from the greatest mass percent oxygen to the least.
    a. H_2O
    b. CO_2
    c. $Na_2C_2O_4$
    d. $C_6H_{12}O_2$
45. Boron forms many compounds, often with some unusual bonding properties. In each of these, determine the mass percent of boron: $B_{13}C_2$, Ti_3B_4, CaB_6.
46. Determine the mass percent of sodium in each compound: NaCl, $NaHCO_3$, NaOH.
47. Which compound in each pair has the greater percent oxygen?
    a. $FeSO_4$ or Na_2SO_4
    b. K_2FeO_4 or $KMnO_4$
    c. $Fe(NO_3)_3$ or HNO_3
48. Which compound in each pair has the greater percent carbon?
    a. CO_2 or CO
    b. NaCN or KCN
    c. C_3H_8O or $C_5H_{10}O_2$

Chemical Applications and Practices

49. The main component in kidney stones is CaC_2O_4. What is the mass percent of each element found in CaC_2O_4?
50. Hydrates are compounds that contain water molecules as part of their structure. For example, sodium carbonate decahydrate ($Na_2CO_3 \cdot 10H_2O$), washing soda, has been used as a grease remover. What percent of the compound is water? (Note: The dot between Na_2CO_3 and H_2O indicates that water is loosely attached within the crystal—in this case in a 1:10 ratio.)
51. Unsaturated fatty acids are responsible for the liquid nature of many plant oils. In this case, the term unsaturated refers to the presence of double bonds between two or more carbon atoms in the molecule. Saturated hydrocarbons, which are associated with animal fats, contain only single bonds between carbon atoms. Unsaturated compounds can be made saturated by adding hydrogen to the double-bonded areas. After such a reaction, often called hydrogenation, would the percent by mass of carbon in the saturated compound be increased or decreased over its percent by mass in the unsaturated hydrocarbon? Explain the basis for your answer.
52. The sidewalk salt that we spread on our walkways is mostly $CaCl_2$. A stable hydrate of this salt is $CaCl_2 \cdot 2H_2O$ (see Problem 50). What is the mass percent of calcium in each of these salts?

Section 3.6 Finding the Formula

Finding the Formula

Skill Review

53. Suppose an orchestra has this composition: 24 violinists, 18 brass instrumentalists, 6 cellos, and 3 percussionists. What is the "empirical formula" of this orchestra?
54. Suppose a lasagna is made from 2 boxes of noodles, 2 packages of ground beef, 2 tomatoes, and 4 packages of cheese. What is the "empirical formula" of this lasagna?
55. A compound contains only carbon and hydrogen. The mass percent of carbon in the compound is 85.7%. What is the empirical formula of the compound?
56. A compound contains only carbon, hydrogen, and oxygen. The mass percent of carbon is 52.1% and that of hydrogen is 13.1%. What is the empirical formula of the compound? If the molar mass of the compound is 92, what is the molecular formula of the compound?
57. A student forgot to simplify the following data into correct empirical formulas. What is the empirical formula for each of the following compounds?
    a. 2.33 mol C; 3.33 mol H
    b. 4.25 mol C; 8.50 mol H
    c. 2.67 mol C; 2.30 mol H; 0.66 mol O
58. A second student also forgot to simplify the following data into correct empirical formulas. What is the empirical formula for each of the following compounds?
    a. 6.24 mol C; 13.97 mol H; 0.693 mol O
    b. 6.81 mol C; 18.14 mol H
    c. 4.345 mol C; 13.02 mol H; 2.17 mol O

59. Determine the empirical formula for each of the following elemental analyses (assume the only elements present are C, H, and O.)
    a. C, 52.14%; H, 13.13%
    b. C, 85.63%; H, 14.37%
    c. C, 65.45%; H, 5.49%
60. Determine the empirical formula for each of the following elemental analyses (assume the only elements present are C, H, and O.)
    a. C, 90.51%; H, 9.49%
    b. C, 64.82%; H, 13.60%
    c. C, 38.70%; H, 9.74%

Chemical Applications and Practices

61. a. Based on this mass percent composition, determine the empirical formula of the compound: 57.1% carbon, 4.76% hydrogen, 38.1% oxygen.
    b. Which of these could possibly be the molar mass of the compound: 56, 84, or 116? (Show proof for your answer.)
62. The compound found in many small portable lighters is butane. An analysis of butane yields the following mass percent values. From these data, determine the empirical formula of this useful fuel: 82.76% C, 17.24% H.
63. Potassium argentocyanide is a toxic compound that is used in the important industrial process of silver plating. It has these mass percent values: 19.6% potassium, 54.2% silver, 12.1% carbon, and 14.1% nitrogen. What is the empirical formula of this compound?
64. In humans, ingested ethyl alcohol is enzymatically decomposed into acetaldehyde. The mass percent values of the components of acetaldehyde are 54.55% C, 9.09% H, and 36.36% O. From these data, determine the empirical formula of acetaldehyde.
65. Linoleic acid is a fatty acid that contains unsaturated carbon bonding. Generally, this trait is found in many compounds that compose vegetable oils. The molar mass of linoleic acid is 280 g. The mass percent values of the elements in linoleic acid are 77.1% carbon, 11.4% hydrogen, and 11.4% oxygen. What is the molecular formula of this helpful agricultural compound?
66. During extreme exercise, lactic acid is produced when aerobic metabolism is not available for glucose breakdown. The mass percent values of the components in lactic acid are 40.0% C, 6.72% H, and 53.3% O. The average molar mass of lactic acid is 90.1 g. What are the empirical and molecular formulas for this important biological compound?

Section 3.7 Chemical Equations

Skill Review

67. Write the proper formulas and balance the equation: Aqueous hydrochloric acid reacts with solid magnesium carbonate to form aqueous magnesium chloride, gaseous carbon dioxide, and water.
68. Balance this decomposition reaction:
    $\_\_\_ H_2O_2 \rightarrow \_\_\_ O_2 + \_\_\_ H_2O$
    How many oxygen molecules are produced for every molecule of hydrogen peroxide?
69. Provide the proper coefficients to balance each chemical equation.
    a. $\_\_\_N_2H_4 + \_\_\_N_2O_4 \rightarrow \_\_\_N_2 + \_\_\_H_2O$
    b. $\_\_\_Pb(C_2H_3O_2)_2 + \_\_\_KI \rightarrow \_\_\_PbI_2 + \_\_\_KC_2H_3O_2$
    c. $\_\_\_PCl_5 + \_\_\_H_2O \rightarrow \_\_\_H_3PO_4 + \_\_\_HCl$
    d. $\_\_\_Ba_3N_2 + \_\_\_H_2O \rightarrow \_\_\_Ba(OH)_2 + \_\_\_NH_3$
70. Provide the proper coefficients to balance each chemical equation.
    a. $\_\_\_N_2 + \_\_\_O_2 \rightarrow \_\_\_NO$
    b. $\_\_\_CH_4 + \_\_\_O_2 \rightarrow \_\_\_CO_2 + \_\_\_H_2O$
    c. $\_\_\_H_2SO_4 + \_\_\_KOH \rightarrow \_\_\_K_2SO_4 + \_\_\_H_2O$
    d. $\_\_\_CO_2 + \_\_\_H_2O \rightarrow \_\_\_C_6H_{12}O_6 + \_\_\_O_2$
71. Write the proper formulas and balance the equation: Calcium chloride reacts with sodium phosphate to produce calcium phosphate and sodium chloride.

72. a. Balance this decomposition reaction:

$$\_\_(NH_4)_2Cr_2O_7 \rightarrow \_\_Cr_2O_3 + \_\_N_2 + \_\_H_2O$$

b. How many atoms of nitrogen appear in the reactant side of the balanced equation? How many nitrogen molecules appear on the product side of the balanced equation?

73. This equation shows the proper stoichiometric ratio for all the components but one. Fill in the correct coefficient and compound to complete the equation.

$$TiCl_4 + 2H_2O \rightarrow TiO_2 + \_\_ \_\_\_\_$$

74. One method that could be used to make a chlorofluorocarbon (CFC) known as freon-12 (CCl_2F_2) is based on this reaction:

$$CCl4(g) + SbF3(g) \rightarrow CCl2F2(g) + SbCl3(s)$$

Use coefficients to balance the reaction. How many total moles of gases are involved in the balanced equation?

Chemical Applications and Practices

75. The banned insecticide DDT could be produced using this reaction:

$$\_\_C_6H_5Cl + \_\_C_2HOCl_3 \rightarrow \_\_C_{14}H_9Cl_5 + \_\_H_2O$$

a. Balance the equation.
b. How many grams of DDT ($C_{14}H_9Cl_5$) would be expected from the complete reaction of 45.0 g of C_6H_5Cl?

76. A key ingredient for many over-the-counter antacids is aluminum hydroxide, $Al(OH)_3$. Use the following chemical reaction to predict the maximum number of grams of aluminum hydroxide that could be produced from 5.20 g of $Al_2(SO_4)_3$ and unlimited NaOH.

$$Al_2(SO_4)_3 + NaOH \rightarrow Al(OH)_3 + Na_2SO_4 \text{ (not balanced)}$$

77. The "fizz" formed when some pain relief tablets are placed in water is caused by the following chemical reaction. The bubbles are produced from the release of gaseous carbon dioxide. How many grams of $NaHCO_3$ would have to be present in a tablet to react with 0.487 g of citric acid ($H_3C_6H_5O_7$)?

$$NaHCO_3(aq) + H_3C_6H_5O_7(aq) \rightarrow CO_2(g) + Na_3C_6H_5O_7(aq) + H_2O(aq)$$

78. If some sugar from grapes were to be fermented to make wine, one of the primary reactions would be

$$\_\_C_6H_{12}O_6 \rightarrow \_\_C_2H_6O + \_\_CO_2$$

a. Balance the equation.
b. How many grams of sugar would you need to make 100.0 g of ethanol (C_2H_6O)?

79. Many metals combine with halogens to produce metal halides. For example, cobalt reacts with hydrogen iodide to give cobalt(II) iodide, CoI_2.

$$Co(s) + 2HI(g) \rightarrow CoI_2(s) + H_2(g)$$

Beginning with 10.0 g cobalt, what mass of HI(*g*), in grams, is required for complete reaction? What quantity of CoI_2, in moles and in grams, is expected to be produced?

80. Potassium chlorate, $KClO_3$, is often used in the laboratory to produce oxygen gas by this reaction:

$$2KClO_3(s) \rightarrow 2KCl(s) + 3O_2(g)$$

How many grams of potassium chlorate are required to produce 2.00 g O_2?

Section 3.8 Working with Equations

Skill Review

81. Consider the following unbalanced equation:

$$\_\_C_6H_6 \rightarrow \_\_C_2H_2$$

a. Determine the molar mass for each of the compounds in the equation.
b. Assuming one mole of each compound, what is the total mass in grams of the reactant? … the product?
c. Use this data to suggest a balanced equation.

82. Consider the following unbalanced equation:

$$\_\_C_2H_2 + \_\_HBr \rightarrow \_\_C_2H_4Br_2$$

a. Determine the molar mass for each of the compounds in the equation.
b. Assuming one mole of each compound, what is the total mass in grams of the reactants? … the product?
c. Use this data to suggest a balanced equation.

83. Balance this equation:

$$\_\_Cu_2O + \_\_C \rightarrow \_\_Cu + \_\_CO$$

a. If 357.75 g Cu_2O react with 30.00 g C, how many total grams of product should be formed?
b. If 317.75 g of the product is Cu, what is the theoretical yield in grams of CO?
c. Prove that this data supports your balanced equation.

84. Balance this equation:

$$\_\_Li + \_\_H_2O \rightarrow \_\_LiOH + \_\_H_2$$

a. If 7.63 g Li react with 19.8 g water, how many total grams of product should be formed (i.e., the theoretical yield)?
b. If 26.34 g of the product is lithium hydroxide, how many grams of hydrogen should be formed?
c. Prove that this data supports your balanced equation.

85. Balance this equation:

$$\_\_H_2CO_3 \rightarrow \_\_H_2O + \_\_CO_2$$

a. How many molecules of carbon dioxide are produced for every molecule of carbonic acid that reacts?
b. How many moles of water are produced when 6.33 mol of carbonic acid decomposes?
c. If 10.0 mol of carbonic acid decompose, how many moles of water and carbon dioxide are produced?

86. Balance this equation:

$$\_\_HClO + \_\_Fe(NO_2)_2 \rightarrow \_\_Fe(ClO)_2 + \_\_HNO_2$$

a. How many molecules of nitrous acid are produced for every molecule of iron(II) nitrite that reacts?
b. How many moles of hypochlorous acid are required to react with 4 mol of iron(II) nitrite?
c. If we started the reaction with 3.225 mol iron(II) nitrite, what is the maximum number of moles of nitrous acid we could make?

87. a. Balance this equation:

$$\_\_NH_3 + \_\_O_2 \rightarrow \_\_NO + \_\_H_2O$$

b. Starting with 34.0 g of NH_3 and excess O_2, what mass of NO would be expected to be produced (i.e., the theoretical yield)?
c. How much H_2O would form?
d. If the percent yield of NO was only 35.0%, what is the maximum grams of NO that could be formed?

88. a. Balance this equation:

$$\_\_Fe_2O_3 + \_\_C \rightarrow \_\_Fe + \_\_CO$$

b. If the reaction were started with 1.00 kg of Fe_2O_3 and 0.250 kg of C, which reagent would you consider to be the limiting component?
c. Under the conditions described in part b, what would be the theoretical yield of Fe in grams?

89. a. Balance this equation:

$$\_\_Ca(OH)_2 + \_\_H_3PO_4 \rightarrow \_\_Ca_3(PO_4)_2 + \_\_H_2O$$

b. How many grams of $Ca(OH)_2$ are required to form 0.567 g of $Ca_3(PO_4)_2$?
c. How many grams of H_3PO_4 would be required for the stoichiometry described in part b?

90. This reaction takes place when some types of matches are struck:
a. Balance the equation

$$\_\_KClO_3 + \_\_P_4 \rightarrow \_\_P_4O_{10} + \_\_KCl$$

b. If 52.9 g of $KClO_3$, in the presence of excess P_4, produced 25.0 g of P_4O_{10}, what would be the calculated percent yield of the process?

91. A fast-food restaurant serves this sandwich: 2 pieces of bread, 1 meat patty, 2 pieces of cheese, and 4 pickles. If the restaurant currently had on hand the following inventory, what would be the maximum number of sandwiches that it could prepare? What ingredient would limit the sandwich production? Inventory: 200 pieces of bread, 200 meat patties, 200 cheese slices, 200 pickles.

92. A certain laboratory set-up for general chemistry requires four 250-mL beakers, six test tubes, and two graduated cylinders.
a. Write an "equation" that shows the correct combination of lab equipment to form the arrangement.
b. If a lab assistant set up 50 such arrangements, how many test tubes would be required?
c. Would 100 beakers and 100 graduated cylinders provide enough equipment for these 50 arrangements? Explain.

3

Chemical Applications and Practices

93. One method used to produce soaps includes the reaction between NaOH and animal fat. The following equation represents that process:

$$C_3H_5(C_{17}H_{35}CO_2)_3 + 3NaOH \rightarrow C_3H_5(OH)_3 + 3NaC_{17}H_{35}CO_2$$

   a. How many grams of soap, $NaC_{17}H_{35}CO_2$, would theoretically be produced by starting with 125 g of fat, $C_3H_5(C_{17}H_{35}CO_2)_3$?
   b. If the actual amount of soap produced were 31 g, what would be the percent yield of the reaction?

94. The compound $Na_2S_2O_3$, also known as "hypo", is used in developing photographs. It can be synthesized by using this chemical reaction:

$$Na_2CO_3 + 2Na_2S + 4SO_2 \rightarrow 3Na_2S_2O_3 + CO_2$$

   a. Predict the mass in grams of $Na_2S_2O_3$ that could be produced from 48.5 $_g$ of Na_2S and unlimited quantities of the other reactants.
   b. If the procedure produced only a 78.9% reaction yield, how many grams of hypo would you expect to obtain?

95. When acid reacts with a base, the typical reaction yields water and a salt product formed from the anion of the acid and the cation of the base. For example, hydrochloric acid added carefully to sodium hydroxide forms water and NaCl. Use the balanced equation shown here to calculate the maximum yield (in grams) of NaCl that can be formed when a solution containing 0.155 mol of HCl reacts with 4.55 g of NaOH.

$$HCl\,(aq) + NaOH\,(s) \rightarrow NaCl\,(aq) + H_2O\,(l)$$

96. A "magic show" demonstration often performed to illustrate volcanic action is to react vinegar with baking soda. The release of carbon dioxide gas that results produces a frothing action of bubbles. How many moles of carbon dioxide could be produced from the complete reaction of 10.0 g of baking soda ($NaHCO_3$) with excess acetic acid ($HC_2H_3O_2$) in vinegar?

$$HC_2H_3O_2\,(aq) + NaHCO_3\,(s) \rightarrow H_2O\,(l) + CO_2\,(g) + NaC_2H_3O_2\,(aq)$$

97. Different baking powders are used in cooking to achieve various results in food textures as carbon dioxide gas is released. One such baking powder contains cream of tartar ($KHC_4H_4O_6$) and sodium hydrogen carbonate ($NaHCO_3$). Typically, starch is also added to keep moisture away from these active ingredients. When water from the recipe contacts these two compounds, this reaction takes place:

$$KHC_4H_4O_6 + NaHCO_3 \rightarrow NaKC_4H_4O_6 + H_2O + CO_2$$

   What is the maximum mass in grams of $NaKC_4H_4O_6$ (also known as Rochelle salt) that could be formed if a baking powder sample contained 10.0 g each of the starting ingredients?

98. Often in dyeing cloth, a substance must be added that helps the dye to adhere to the cloth. Such a substance is known as a mordant. One such compound is aluminum sulfate octadecahydrate ($Al_2(SO_4)_3 \cdot 18H_2O$). One method to obtain this useful compound is as follows:

$$\_\_Al(OH)_3 + \_\_H_2SO_4 + \_\_H_2O \rightarrow \_\_Al_2(SO_4)_3 \cdot 18H_2O$$

   Balance the equation and determine the maximum yield (in grams) that could be obtained from reacting 25.0 g of $Al(OH)_3$ with 50.0 g of H_2SO_4 and unlimited water.

99. The following equation shows the conversion of ethanol in a wine sample into acetic acid through the action of oxygen from the air. This reaction drastically changes the taste of the wine. If 100.0 mL of wine initially contained 12.0 g of ethanol (C_2H_5OH), and after a period of time 4.00 g of acetic acid ($C_2H_4O_2$) were detected, what percent of the ethanol reacted?

$$C_2H_5OH + O_2 \rightarrow C_2H_4O_2 + H_2O$$

100. Sodium hypochlorite (NaOCl) is the active ingredient in most household bleach products. One method of preparing this important compound is as follows:

$$Cl_2(g) + 2NaOH(aq) \rightarrow NaOCl(aq) + NaCl(aq) + H_2O(l)$$

   a. Starting with 50.0 g of $Cl_2(g)$ and 50.0 g of NaOH, predict the theoretical yield (in grams) of NaOCl.
   b. If the reaction, under a given set of conditions, formed only 38.9 g of NaOCl, what would you report as the percentage yield?

101. The following reaction can be used to represent the rusting (oxidation) of iron. Annually, about 20% of the steel manufactured is used to replace steel corroded as a consequence of the oxidation of iron. Determine how many grams of oxygen and iron reacted if 454 g of Fe_2O_3 were formed. (First balance the equation.)

$$Fe(s) + O_2(g) \rightarrow Fe_2O_3(s)$$

102. You may have noticed an athletic trainer spraying an injured player with a quickly evaporating liquid during a time out. The liquid is probably chloroethane, which quickly cools down an injured area with a slight numb-

ing sensation. Chloroethane can be synthesized via the following reaction. If the reaction were begun using 85.0 g of ethene (C_2H_4) and 100.0 g of hydrogen chloride (HCl) but had only a 68.0% yield, how many grams of chloroethane (C_2H_5Cl) would be produced?

$$HCl(g) + C_2H_4(g) \rightarrow C_2H_5Cl(g)$$

103. Adding yeast cells to glucose can cause the glucose ($C_6H_{12}O_6$) to be converted to ethanol (C_2H_5OH) and CO_2. Bakers make use of this process when the CO_2 gas causes bread to rise during the baking process. Balance the following equation, and determine the number of moles of CO_2 that could be produced from 25.0 g of glucose.

$$\_\_C_6H_{12}O_6(s) \rightarrow \_\_C_2H_5OH(l) + \_\_CO_2(g)$$

104. Two students perform a chemical synthesis in their general chemistry lab. Student A obtains a 90.0% yield in the reaction. Student B obtains an 85.0% yield in the same reaction. Can you now determine which student obtained the greater mass of the product? Explain, or justify your answer.
105. Beginning with 27.5 g of aluminum and 43.6 g of hydrogen chloride, answer the following questions about this reaction:

$$2Al(s) + 6HCl(g) \rightarrow Al_2Cl_6(s) + 3H_2(g)$$

   a. Which reactant is limiting?
   b. What mass in grams of Al_2Cl_6 can be produced?
   c. What is the excess reactant, and how many grams of it remain when the reaction is completed?
106. Methyl salicylate (also known as oil of wintergreen), $C_8H_8O_3$, is produced by the reaction of salicylic acid, $C_7H_6O_3$, and methanol, CH_3OH.

$$C_7H_6O_3(s) + CH_3OH(l) \rightarrow C_8H_8O_3(s) + H_2O(l)$$

If you mix 1.00×10^2 g of each of the reactants, what is the maximum mass in grams of methyl salicylate that can be obtained?

? Comprehensive Problems

107. Review the vitamin C controversy discussed in this chapter. How is it possible that a compound can be both good and bad for your health?
108. What government agency is most concerned with setting recommended amounts of vitamins and minerals?
109. Explain why it is not totally correct to use the atomic masses given in the periodic table on the inside front cover of this book to express the mass of one molecule of any compound.
110. How big is Avogadro's number? If it were possible to place circles on notebook paper to represent atoms, how many pages would be required to complete the task? (Assume 25 lines on each side, 32 circles per line, and use of both sides of the paper.)
111. How many grams of argon would contain the same number of atoms as 10.0 g of neon?
112. Copper, silver, and gold are often referred to as the "coinage metals." If you had 454 g (approximately 1 lb) of each, which sample would contain the greater number of atoms?
113. The human hemoglobin molecule is quite large. It is known that each hemoglobin molecule contains four atoms of iron. If the iron makes up only 0.373% of the total mass of the molecule, what is the mass of 1 mol of hemoglobin?
114. When 1.00 g of one of the main components of gasoline was completely combusted (reacted with oxygen), it produced 3.05 g of CO_2 and 1.50 g of H_2O. On the basis of this information, determine the mass percent values of carbon and hydrogen in the compound and the empirical formula of the compound.
115. The main compound responsible for the characteristic aroma of garlic is allicin. From the following mass percent values, determine the empirical formula of this familiar compound: 44.4% C, 6.21% H, 39.5% S, and 9.86% O.
116. When we examine balanced equations, we find that the total of the coefficients on the left-hand side of the equation arrow does not always equal the total of the coefficients on the right-hand side. Explain why this does not violate the law of conservation of mass.
117. Examine the formula of salicylic acid in Sect. 3.7. How many molecules of salicylic acid would be required to react with 0.247 mL of ethanoic anhydride ($d = 1.080$ g/mL)?
118. A sample of ore is found to contain 14.5% aluminum oxide by mass. How many pounds of the sample would be required to make exactly 100 g of pure aluminum? (Assume the reaction to make aluminum metal and oxygen gas from aluminum oxide occurs with only a 58% yield.)
119. Zinc oxide is often used in sunscreens to help protect your skin from harmful UV rays. If a particular sunbather wishes to coat 0.75 m^2 of his exposed skin with a layer of zinc oxide that is 1.0 mm thick, how many grams of zinc oxide would be needed? (Assume the sunscreen lotion has a density of 1.10 g/mL, and that the lotion contains 18% zinc oxide by mass.)

120. Palladium(II) nitrate is a reagent that can be used to link two smaller molecules together with a covalent bond. This reagent can be made by treating metallic palladium with nitric acid.
    a. What is the balanced reaction for the preparation of the reagent?
    b. How many grams of palladium(II) nitrate can be made from 10.0 g of palladium metal and 5.6 mL of 15% (by mass) nitric acid whose density is 1.056 g/mL?

121. A researcher claims that a 1.00 g sample of gold(II) chloride contains a greater mass of chloride than does a 1.00 g sample of gold(IV) chloride.
    a. Is the researcher correct? Explain your answer.
    b. Which sample contains a greater mass of gold?
    c. If a 2.0 cm cube of gold(II) chloride was spread so that it was only 1/8 inches thick, what area would the compound occupy in in^2?
122. A sample of ethanoic acid (CH_3COOH) is made with the carbon-13 isotope such that the starred carbon is exclusively replaced with that isotope. Explain what effect, if any, that would have on the following unbalanced reaction of ethanoic acid and iodine:

$$\_\_CH_3C^*OOH + \_\_I_2 \rightarrow \_\_CH_3I + \_\_HI + \_\_C^*O_2$$

123. A liquid was burned in the presence of oxygen gas to produce carbon dioxide and water.
    a. Elemental analysis of the compound revealed it to be composed of 83.63% carbon and 16.37% hydrogen. What is the empirical formula for this compound?
    b. In another experiment, the molecular weight of the compound was determined to be 86.2 g/mol. What is the molecular formula?
    c. Balance the equation mentioned in the start of this problem.
    d. If 10.0 g of the liquid was consumed in the combustion, how many grams of carbon dioxide was produced?
124. a. Write the balanced equation for the combustion of octane (C_8H_{18}).
    b. Write the balanced equation for the combustion of ethanol (C_2H_6O).
    c. Which reaction produces more carbon dioxide per gram of fuel?
125. A gaseous compound containing only C and H was isolated. Elemental analysis provided the following data: C, 89.94%; H, 10.06%. The molecular mass of the compound was determined to be 40 g/mol.
    a. What is the empirical formula for this compound?
    b. What is the molecular formula of the compound?
    c. Write the balanced combustion equation for this compound.
    d. Using your artistic skills, provide a ball-and-stick drawing of the molecule. (hint: carbon prefers four bonds, hydrogen prefers only one bond.)

Thinking Beyond the Calculation

126. Xylene (pronounced ZI-leen) is an important organic molecule isolated from petroleum oil. It is often used as a thinner for oil-based paints.
    a. Elemental analysis of a sample of xylene shows that the mass percent of carbon is 90.51% and the mass percent of hydrogen is 9.49%. What is the empirical formula of xylene?
    b. The molar mass of xylene is 106.17 g/mol. What is the molecular formula of xylene?
    c. In the laboratory setting, xylene burns in limited oxygen environments to produce carbon monoxide and water vapor. Write the balanced equation for this reaction.
    d. If 12.5 g of xylene were combusted using the equation from part c, how many grams of oxygen gas would be required to react completely with the xylene?
    e. A vessel containing 45.8 mL of xylene (density = 0.8787 g/mL) and 31.0 g of O_2 produced carbon monoxide and water. Which reagent is the limiting reagent in this reaction? How many grams carbon monoxide are produced in the reaction?
    f. If 1.40 g of CO were actually isolated from the reaction in part e, what would be the percent yield of the reaction?

Solution Stoichiometry and Types of Reactions

Contents

Supplementary Information The online version contains supplementary material available at ▶ https://doi.org/10.1007/978-3-030-90267-4_4.

M. Mosher and P. Kelter, *An Introduction to Chemistry*,
https://doi.org/10.1007/978-3-030-90267-4_4

What Will We Learn from this Chapter?

- Why water is called "the universal solvent", and what makes it special (▶ Sect. 4.1).
- What we mean when we say that chemicals "dissolve" and "dissociate" in water, and how to distinguish between strong and weak electrolytes (▶ Sect. 4.1).
- What concentration is and what units of concentration we need to be familiar with (▶ Sect. 4.2).
- How we analyze solutions for their chemical content (▶ Sect. 4.3).
- What molecular, total and net ionic equations are and how they are useful in understanding chemical reactions (▶ Sect. 4.4).
- What general kinds of chemical reactions we should be familiar with at this point in our chemistry journey, and how we can recognize them (Sects. 4.5, 4.6 and 4.7).

4.1 Introduction

"There's gold in them thar' hills!" That statement, and others like it, conjured up images of great wealth in a worldwide gold rush that began in the late 1840s, stretching from the hills of California to Australia. Whether our excitement comes from finding a nugget in a mountain stream, buying a necklace at the store, or just watching a show about the bullion bars stored at Fort Knox, we are fascinated with gold. The use of gold is not a recent phenomenon; it has been known for a large part of recorded history. In fact, gold was first used as coinage in 700 B.C., and it continued to flourish in coin form until the early 1900s.

In our modern world, gold is prized for its rarity, beauty, and high price. In recent years, its price per troy ounce (31.1 g, or about 1.1 "avoirdupois ounces" in the units commonly used in the United States) has been as low as US $102 (1976) and as high as nearly US $1900 (2011). Gold is an excellent electrical conductor, is not very chemically reactive, and is quite malleable, making it ideal for use in electronic products, including computers. It is also highly reflective, so it is used as a shielding coating to protect astronauts from solar radiation (◘ Fig. 4.1). Even with all of our modern scientific uses, about two-thirds of the 4300 tons of gold used in 2019 went into making jewelry, with India, China and the U.S. being the top three largest consumers of gold. According to a January 2009 National Geographic article, "In all of history, only 161,000 tons of gold have been mined, barely enough to fill two Olympic-sized swimming pools. More than half of that has been extracted in the past 50 years." Yet the point of the article was less about quantities of gold than about the health hazards that gold mining has on laborers in South America and Africa.

Gold used to be found in large nuggets as relatively pure metal, and miners would dig, sort, and harvest visible chunks of gold by almost every means possible. These days, in both large- and small-scale operations, gold ore is

◘ **Fig. 4.1** Gold has many uses in today's society. It blocks harmful UV rays in an astronaut's visor and servews as corrosion-resistant connections in a computer. *Source* ▶ https://www.publicdomainpictures.net/pictures/40000/velka/--1363343892JT9.jpg

mined, pulverized into particles the size of grains of sand, and then mixed with a solution containing cyanide ion (CN^-), often by people working under hazardous conditions.

$$4Au(s) + 8NaCN(aq) + O_2(g) + 2H_2O(l) \rightarrow 4NaAu(CN)_2(aq) + 4NaOH(aq)$$

Based on the reaction equation shown above, the gold in the ore is dissolved by the cyanide solution, and the waste ore is discarded. Metallic gold is then obtained by precipitation and/or electroplating the solution. We will discuss electroplating in much more detail in the chapter on electrochemistry.

The extraction of gold using this method is known as the cyanide leaching method. It takes place in a water-based or **aqueous** solution. In fact, much of the chemistry to which we are exposed takes place in aqueous solutions. Every cell in our body is a watery sac full of reacting chemicals. Rivers, lakes, and oceans contain complex solutions, each with its own special chemistry. At home we use solutions to cook and clean.

Aqueous – Water-based; also implies that a dissolved substance interacts with a surrounding layer of water molecules.

Working with aqueous chemistry means understanding the stoichiometry of the reactions that take place in the solution. For example, although too little cyanide (CN^-) per volume of water results in an inefficient extraction of gold in the mining process, too much of it in our blood can be fatal. Long-term exposure to even low levels of cyanide can lead to health problems in those who process the gold, including central nervous system, heart, and thyroid gland damage. Chemical reactions that require us to look at measures of concentration, such as how much cyanide is in a given volume of water, deal with "solution stoichiometry," the focus of the ideas, new terms, and techniques we will explore in this chapter.

4.2 Water—A Most Versatile Solvent

When we think about the range of chemicals that must dissolve in our blood, we can see that water is an extraordinary **solvent**—a compound that typically makes up the majority of a homogeneous mixture of molecules, ions, or atoms. Dissolved in our blood, we find ions such as potassium (K^+) and chloride (Cl^-), larger molecules including glucose ($C_6H_{12}O_6$) and, to a lesser extent, cholesterol ($C_{27}H_{46}O$) and giant protein molecules composed of thousands of atoms. Water, the solvent that dissolves these things in our blood, has historically been called the universal solvent because many things dissolve in it.

Solvent – A compound that typically makes up the majority of a homogeneous mixture of molecules, ions, or atoms; dissolves the solute.

A **solution** is a homogeneous mixture (◘ Fig. 4.2). Those molecules, ions, or atoms that dissolve (are soluble) in the solvent are known as **solutes**. The solvent is typically present in the larger amount in a solution, and the solute is typically the minor component. As we work with solutions here and throughout our study of chemistry, we will

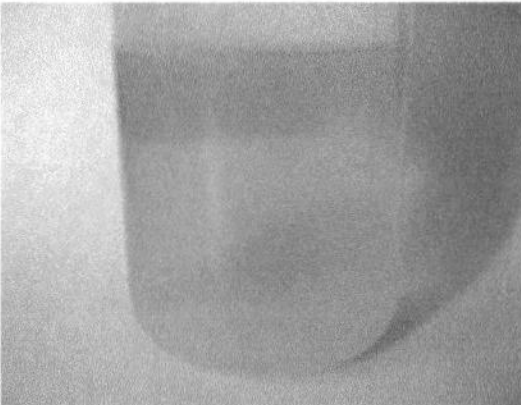

◘ **Fig. 4.2** In solution, the particles of a solute (such as the ions of copper(II) sulfate, a sample of which is shown here) form a homogeneous mixture with the particles of the solvent (water molecules, in this case). The blue color of the solute is evenly distributed. *Source* ▶ https://upload.wikimedia.org/wikipedia/commons/2/2a/Copper-Sulfate.JPG. ▶ https://upload.wikimedia.org/wikipedia/commons/2/22/CopperSulphate.JPG

be interested in the behavior and impact of the various *solutes*, such as copper(II) sulfate or sucrose, which are dissolved in water, the solvent.

Solution – A homogeneous mixture of solute and solvent.

Solutes – Molecules, ions, or atoms that are dissolved in a solvent to form a solution.

How do solutions form? Solutes dissolve in solvents because of the energetically favorable interactions between the two components of the solution. Water is a particularly versatile solvent because its structure permits it to interact with a wide variety of other chemicals. The resulting mixture is more energetically stable than if the compounds remained separate (such as layers of oil and water.) Why is this so? A closer look at the structure of water will help us answer this question.

Figure 4.3 illustrates two ways to represent water. The image on the right is known as an electrostatic potential map. This is a colorized plot of the density of electrons on the surface of the molecule. Red regions indicate locations of high electron density; blue regions indicate locations with low electron density. Where does the electron density in Fig. 4.3b appear to be greatest in water? Water is a molecule that contains an oxygen atom, which is rich in electrons. This oxygen atom, therefore, possesses a partial negative charge (δ^-) and is red in the electrostatic potential map. The hydrogen atoms attached to the oxygen atom have a lower concentration of electrons, which results in a partial positive charge (δ^+) near the blue color of the electrostatic potential map. In a neutral molecule like water, these partial charges are equal and opposite so they cancel each other and the molecule is uncharged, overall. When an ionic compound dissolves in water, it **dissociates** into individual ions that interact with the partial charges on the water molecule. The partially negative part of the water molecules interacts with the positive ions and the partially positive part of the water molecules interact with the negative ions (Fig. 4.4).

Dissociation – The process of a compound separating into its individual ions. Also called "ionization".

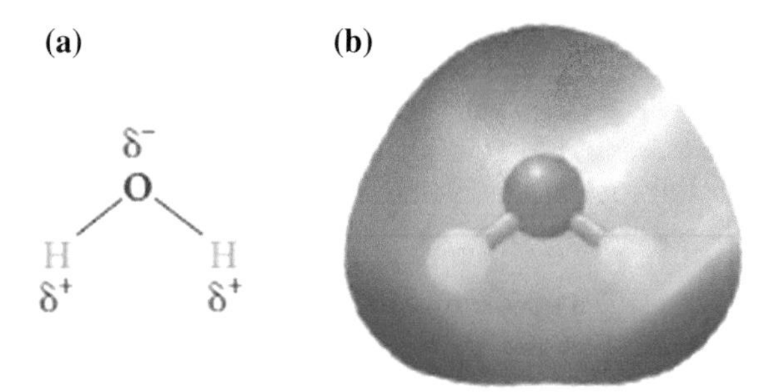

Fig. 4.3 **a** The electrons in the water molecule are distributed toward the oxygen atom. The oxygen end of water is electron-rich (δ−) and the hydrogen end of water is electron-poor (δ+). This can be represented with a colored map of electrostatic potential. **b** Red areas indicate high electron density, and blue areas represent low electron density. The colors in between (ranging from yellow to green to sky blue) indicate varying, intermediate degrees of electron density

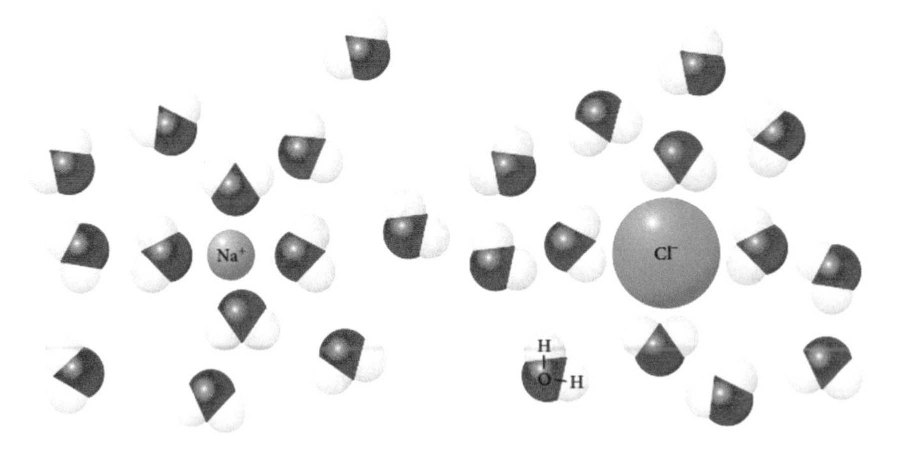

Fig. 4.4 When an ionic compound (such as NaCl) dissolves, the uneven charge distribution in water molecules facilitates the process by interacting with the positive and negative charges on the ions. Note that this is a flat drawing, but the water molecules are in a 3-dimensional arrangement around each ion, forming a "cage" of water molecules called the "hydration sphere"

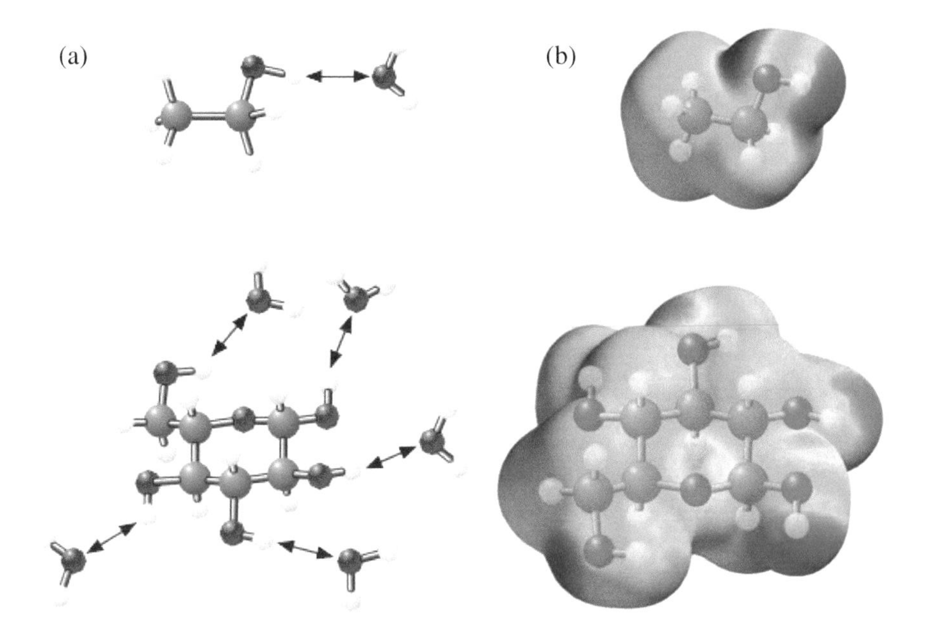

Fig. 4.5 The partial charges in **a** ethanol and **b** glucose, can interact with the opposite partial charges on water molecules and allow molecules to dissolve in water. In the electrostatic potential maps (on the right) of ethanol and glucose, the electrostatic potential is plotted on the surface of a computer-generated model of the molecules. Note how the electron density is distributed in each molecule. How do these models compare to that of water in Fig. 4.3?

As an example, let's explore the dissolution of table salt (sodium chloride, NaCl) in water. The solid salt is composed of equal numbers of sodium ions (Na^+) and chloride ions (Cl^-). Figure 4.4 shows that as the solid salt dissolves, the opposite charges of the ions and the water molecules attract. The water molecules cluster around the ions in a way that allows the ions to separate and mix with the water. The cage of water molecules around the ions is known as the **hydration sphere** of the ions, and we say that the Na^+ and Cl^- ions have become hydrated. The partial charges on the water molecules balance (and stabilize) the opposite charges on the ions they surround. The ions are not static, nor are they held in place by the water molecules. Instead, the ions separate and move about independently, always surrounded by molecules of water.

Hydration sphere – The shell of water molecules surrounding a dissolved molecule, ion, or other compound. This shell arises because of the force of attraction between the water molecules and the solute.

We can represent the hydration sphere in a chemical equation by writing the symbol (*aq*) after the hydrated molecules, ions, or atoms to specify their aqueous phase:

$$NaCl(s) \rightarrow Na^+(aq) + Cl^-(aq)$$

There are many substances that are not made up of cations and anions. Some of them dissolve in water, such as the molecule ethanol (C_2H_6O), found in alcoholic beverages, and the nutrient molecule glucose ($C_6H_{12}O_6$), shown in Fig. 4.5. Why do molecular compounds dissolve in water? These molecules possess a feature similar to water: the electrons are polarized toward specific regions of the molecule. We will examine why this occurs more closely in ▶ Chaps. 9 and 10. Note the presence of multiple locations of partial positive charge (δ^+), the blue areas, and the regions of partial negative charge (δ^-), the red areas, within the electrostatic potential maps of ethanol and glucose. These partial charges can interact with the partial charges on water molecules. The strong interaction allows the molecules to mix easily with water and dissolve as shown in Fig. 4.5. And, as is true with ions, a cage of water molecules surrounds the dissolved solute molecules.

$$C_6H_{12}O_6(s) \rightarrow C_6H_{12}O_6(aq)$$

4.2.1 Electrolytes and Nonelectrolytes

You may have encountered sports drinks sold with the claim that they help "maintain a healthy electrolyte balance" by supplying ions and water in an appropriate combination. The key ingredients quoted on the label of a

	Amount Per Serving	% Daily Value
Calories	**110**	
Total Fat	0g	0%
Saturated Fat	0g	0%
Trans Fat	0g	
Cholesterol	0mg	0%
Sodium	60mg	3%
Total Carbohydrate	27g	10%
Dietary Fiber	0g	0%
Total Sugars	1g	†
Added Sugars	0g	0%
Protein	0g	0%
Vitamin D	0mcg	0%
Calcium	57mg	4%
Iron	0mg	0%
Potassium	31mg	1%
Magnesium	31mg	7%
Manganese	2mg	87%
Chromium	25mcg	71%
Glycine	31mg	†
L-Carnosine	52mg	†
Taurine	500mg	†
Tyrosine	11mg	†
Caffeine	25mg	†

*Percent Daily Values are based on a 2,000 calorie diet.
† No % Daily Value

Ingredients: Maltodextrin, Xylitol, Natural Flavors, Taurine, Calcium Chelate, Potassium Chelate, Magnesium Chelate, Stevia Leaf Extract, Sodium Chloride, L-Carnosine, Green Tea Extract, Glycine, Tyrosine, Manganese Chelate, and Chromium Chelate.

Allergen Warning: This product is processed in a facility that also processes dairy and soy products. This product contains Xylitol.

Fig. 4.6 The key ingredients of a sports drink include sodium and potassium cations as well as anions like chloride, phosphate and citrate. *Source* ► https://www.hammernutrition.com/heed-sports-drink

leading brand of sports drink can be found in Fig. 4.6. The sodium and potassium in the drink are present as hydrated Na^+ and K^+ ions. They are accompanied by hydrated anions such as Cl^- (chloride ions) so that the entire solution has no net electrical charge. It's important to realize that no stable solution of ions can exist unless the electrical charges are balanced. Be wary of products which claim to "restore ionic balance" to your body. Your body is always in ionic balance! However, you may need to restore specific ions to your body when they are lost through perspiration or urination. Ionic compounds, such as sodium chloride and potassium chloride, which dissociate in water to release free ions, are known as **electrolytes**. The name comes from the ability of solutions containing electrolytes to conduct electricity because of the presence of mobile ions that can carry the electric current.

Electrolyte – A compound that produces ions when dissolved in water.

Electrolytes are of great medical importance, because our blood and cells must contain an appropriate mixture of electrolytes to maintain good health. The presence of electrolytes in blood and cells allows the body to absorb the right amount of water from the gut and to excrete water via the kidneys. Maintaining the optimal balance of electrolytes is especially important after strenuous exercise, in which water containing sodium ions, potassium ions, and anions like chloride, leaves the body as sweat. The ions of electrolytes also play a central role in creating the nerve impulses that allow our organs to work, our muscles to move, and our brains to think.

Substances such as sodium chloride, which dissociate completely into ions when they dissolve, are called **strong electrolytes**. Most ionic compounds that dissolve in water fall into this category of electrolyte. A few molecules,

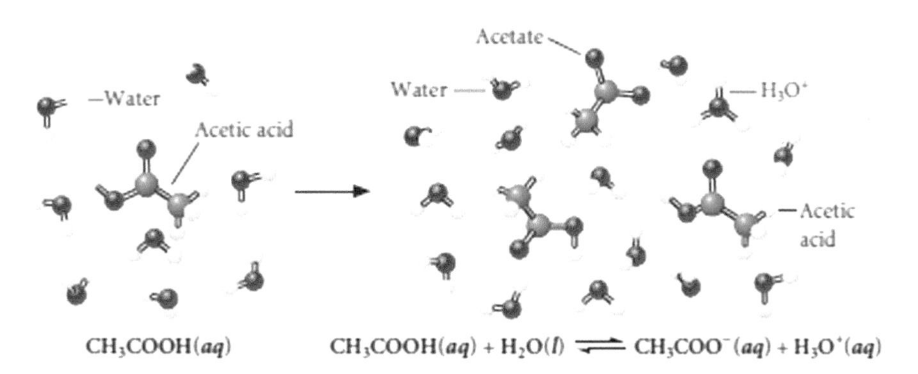

Fig. 4.7 Weak electrolytes (such as acetic acid) do not dissociate completely in water. The double arrows in the chemical equation indicate this behavior. The formula H_3O^+ is another way to represent $H^+(aq)$

such as hydrogen chloride (HCl), are also considered strong electrolytes when dissolved in water. The equation below shows that gaseous HCl first dissolves, and then completely dissociates in the water, to form hydrated ions. This illustrates the important concept that dissolving and dissociation are separate processes. A substance may completely dissolve but then only partially dissociate, or not dissociate at all.

Strong electrolyte – Any compound that completely dissociates in solution.

$$HCl(g) \rightarrow \underset{\text{dissolving (100\%)}}{HCl(aq)} \rightarrow \underset{\text{dissociating (100\%)}}{H^+(aq) + Cl^-(aq)}$$

Other electrolytes, such as acetic acid (vinegar, CH_3COOH), are called **weak electrolytes**, because even though they may dissolve completely in water, the formation of ions occurs to a much lesser extent, as shown in Fig. 4.7. Note the different type of arrow, $\rightleftharpoons$, that defines a reaction that does *not* proceed completely to products.

Weak electrolyte – Any substance that only partially dissociates in solution.

$$CH_3COOH(l) \rightarrow \underset{\text{dissolving (100\%)}}{CH_3COOH(aq)} \rightleftharpoons \underset{\text{dissociating (100\%)}}{H^+(aq) + CH_3COO^-(aq)}$$

Substances, such as glucose ($C_6H_{12}O_6$), that dissolve in water without forming ions are called **nonelectrolytes**. The nonelectrolytes dissolve and associate with water to form a hydration sphere, but they do not form ions. The equation that represents a nonelectrolyte's addition to water does not show the dissociation step (also called ionization) found in the previous two equations.

Nonelectrolyte – A compound that doesn't appreciably dissociate into ions when it dissolves.

$$C_6H_{12}O_6(s) \rightarrow \underset{\text{dissolving (100\%)}}{C_6H_{12}O_6(aq)}$$

4.2.2 Defining Electrolytes

How do we know whether a compound is a strong electrolyte, a weak electrolyte, or a nonelectrolyte? Experimentation can provide evidence as to whether a particular solution contains a strong electrolyte, a weak electrolyte, or a nonelectrolyte. One such experiment is shown in Fig. 4.8. Using a solution as the connector in a light bulb circuit, we can measure the ability of the solution to conduct electricity. The brightness of a light bulb indicates the type of electrolyte in solution. What's happening in Fig. 4.8? Because ions can "carry" an electrical charge, the solution containing the greatest number of ions will have the brightest light bulb.

We can't test chemicals for conductivity like this when they simply appear on a printed page, so how will we know if something is an electrolyte? In addition to experience, the best way to determine if something is an electrolyte is to determine if the compound is a molecule or an inorganic compound. Molecules tend to be non-electrolytes,

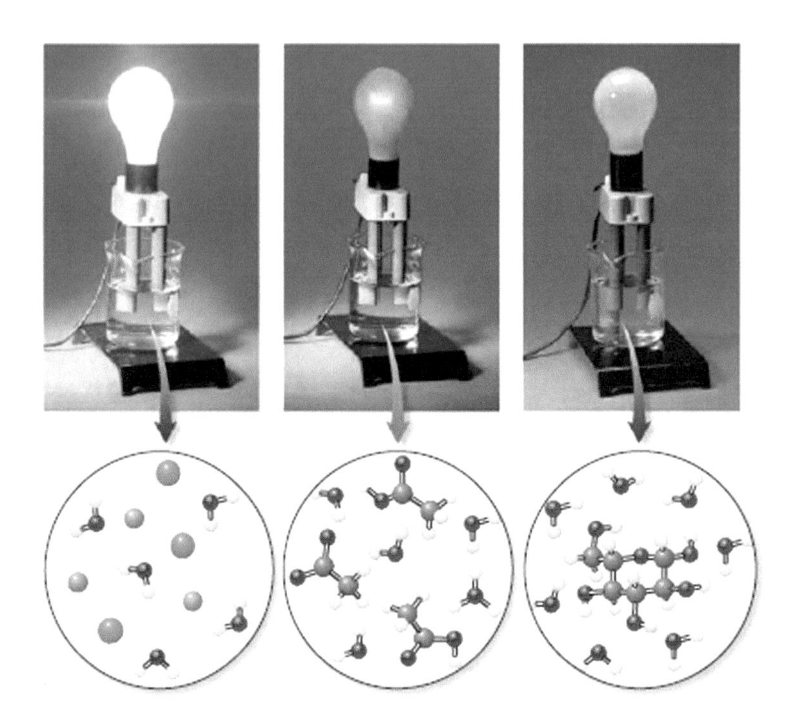

Fig. 4.8 The effect that each type of electrolyte has on the conductivity of a solution as measured by the brightness of a light bulb. From left to right are solutions of sodium chloride (NaCl), acetic acid (CH_3COOH), and glucose ($C_{12}H_{22}O_{11}$)

inorganic compounds tend to be electrolytes, and acids or bases tend to be weak electrolytes. We'll explore these in much greater detail as we progress through our studies of chemistry.

4.3 The Concentration of Solutions

Hyponatremia is a serious condition characterized by a low sodium level in the blood. The organization USA Track and Field has guidelines calling for long-distance runners to avoid drinking excessive amounts of water during long runs, because doing so could lead to excessively diluted blood and a severely reduced sodium level. The result, hyponatremia, gives rise to a high fever, nausea, and, ultimately, heat stroke. Hyponatremia occurs when the sodium concentration is less than 135 mmol (mmol $= 10^{-3}$ mol) of sodium ions per liter of blood.

The **concentration** of a solution, the amount of solute per volume of solution, is often quoted in moles of solute per liter of solution, in grams of solute per liter of solution, or in various other ways. For example, if we exert ourselves strenuously in athletic activity, you might consume sports drinks instead of water. A typical sports drink has 110 mg Na^+ per 8-oz serving, which will help keep the sodium level in your blood at an appropriate concentration. An 8-oz serving of this drink contains 110 mg sodium ions. A 32 oz drink contains 440 mg Na^+, but the concentration remains the same: 110 mg Na^+ per 8-oz serving. The concentration doesn't change as the volume of solution changes.

Concentration – An intensive property of a solution that describes the amount of solute dissolved per volume of solution or solvent.

Because chemicals react on a mole-to-mole basis, a useful concentration unit is **molarity (M)**. This unit specifically indicates the moles of solute per liter of solution, as illustrated in Fig. 4.9.

$$\text{Molarity} = \frac{\text{moles of solute}}{\text{liter of solution}} = \text{M}$$

Molarity (M) – A specific concentration term that reflects the moles of solute dissolved per liter of total solution.

For example, 600.0 mg of aspirin ($C_9H_8O_4$) in a soluble aspirin tablet might be dissolved in orange juice to form 150.0 mL (150.0 cm^3) of solution. We can calculate the molarity of the aspirin solution once we have calculated

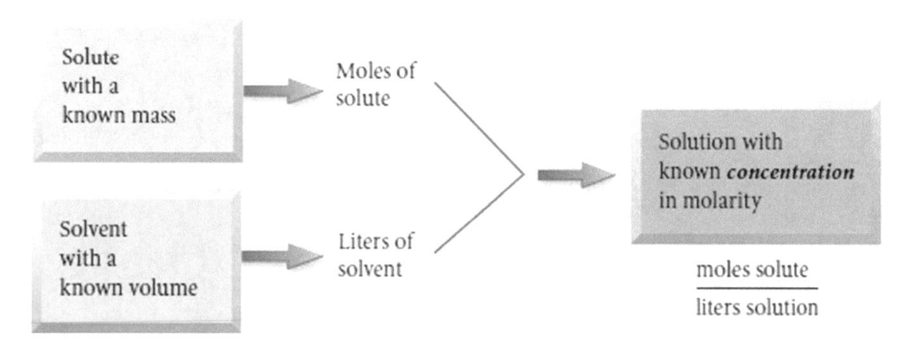

Fig. 4.9 The basic concepts of determining concentration, measured in moles per liter

the number of moles of aspirin from milligrams of aspirin and converted the volume of the juice from milliliters into liters. Let's use dimensional analysis:

$$\frac{600.0\ \cancel{\text{mg aspirin}}}{150.0\ \cancel{\text{mL solution}}} \times \frac{1\ \cancel{\text{g aspirin}}}{1000\ \cancel{\text{mg aspirin}}} \times \frac{1\ \text{mol aspirin}}{180.2\ \cancel{\text{g aspirin}}} \times \frac{1000\ \cancel{\text{mL}}}{1\ \text{L}} = \frac{0.02220\ \text{mol aspirin}}{\text{L aspirin}}$$

The molarity is 0.02220 mol of aspirin per liter of solution, or 0.02220 ***molar*** (written 0.02220 M).

Molar – The "shorthand" method of describing molarity, as in "that is a 3 molar HCl solution."

Given the molarity and known volume of a solution, the relationship: molarity $= \frac{\text{moles of solute}}{\text{liter of solution}}$ can be used to calculate the number of moles of a solute. This relationship can be used in many ways as a dimensional analysis factor. The examples and practices that follow will give you an opportunity to use this new conversion factor to solve problems.

Example 4.1—Calculating Molarity

Sodium hydroxide (NaOH) is one of the most important substances used in the chemical industry, with well over 70 billion kg manufactured annually, ranking it among the top ten chemicals produced in the world. It is used in chemical manufacturing processes, such as in making soaps and detergents. It is also used to break down the lignin that holds cellulose together in wood so that the cellulose can be made into paper. Dilute solutions of NaOH are often used in industry to verify the quality of medicines produced.

a. What is the molarity of NaOH in a solution that contains 3.48 g of sodium hydroxide in 500.0 mL of solution?
b. How many grams of NaOH are required to prepare 617 mL of 1.200 M NaOH solution?

Asking the Right Questions

How do we define "molarity?" The units include moles, rather than grams, of the solute, NaOH.

How is part "b" different than part "a?" In part "b," we are traveling in reverse, by going from molarity, a term related to moles, to grams, a mass-related term. We keep this difference in mind as we flowchart the process for each part. In both parts of the problem, we must ask ourselves: "What units do we want?" "What units do we start with?" "How do we get there?"

Solution

a. Molarity is expressed in units of $\frac{\text{moles of solute}}{\text{liter of solution}}$. We can solve for the molarity by converting the given units of $\frac{\text{grams of NaOH}}{\text{mL of solution}}$ into $\frac{\text{moles of NaOH}}{\text{liter of solution}}$ using dimensional analysis. First, we construct a flowchart such as we introduced in ▶ Chap. 1:

$$\frac{\text{grams NaOH}}{\text{mL of solution}} \xrightarrow{\frac{\text{mL}}{\text{L}}} \frac{\text{grams NaOH}}{\text{L of solution}} \xrightarrow{\frac{\text{mol}}{\text{g}}} \frac{\text{moles NaOH}}{\text{L of solution}}$$

Then we complete the calculation by matching the numbers with the appropriate units:

$$\frac{3.48\ \text{g NaOH}}{500.0\ \text{mL}} \times \frac{1000\ \text{mL}}{1\ \text{L}} \times \frac{1\ \text{mol NaOH}}{40\ \text{g NaOH}} = \frac{0.147\ \text{mol NaOH}}{\text{L}} = 0.174\ \text{M NaOH}$$

b. Given the molarity and the volume, we can solve for moles of NaOH and then change to grams of NaOH using the molar mass of the compound. Again, we'll do it all in one step by converting units.

$$\text{mL soln} \xrightarrow{\frac{\text{L}}{\text{mL}}} \text{L solution} \xrightarrow{\frac{\text{mol NaOH}}{\text{L}}} \text{mol NaOH} \xrightarrow{\frac{\text{g NaOH}}{\text{mol}}} \text{g NaOH}$$

Then we complete the calculation by matching the numbers with the appropriate units:

$$617\,\text{mL} \times \frac{1\,\text{L}}{1000\,\text{mL}} \times \frac{1.200\,\text{mol NaOH}}{1\,\text{L}} \times \frac{40.00\,\text{g NaOH}}{\text{mol NaOH}} = 29.6\,\text{g NaOH}$$

Are Our Answers Reasonable?

Part "a" asked for the molarity of a solution in which we added a little less than 0.1 mol of NaOH in a half-liter of solution. This is equal to about 0.2 mol in 1 L, or 0.2 M, fairly close to our answer.

In part "b," we want the mass of NaOH to make about two-thirds of a liter of a solution that has a concentration of a little more than 1 mol per liter, so we'd expect a mass corresponding to a bit more than two-thirds of a mole of NaOH, which would be around 27 g. Our answer of 29.6 g NaOH is therefore reasonable.

What If…?

…we had a 500.0 mL solution that contained 5.8 g of Na_2CO_3? What is the molarity of Na_2CO_3? (*Ans: 0.11 M* Na_2CO_3)

Practice 4.1

a. What is the molarity of K^+ in the typical muscle cell, which contains 6.5 g K^+ per liter of solution?
b. What is the molarity of Na^+ in a sports drink that contains 110 mg Na^+ per 8-oz serving? (1 oz = 29.6 mL)
c. How many grams of ethanol (C_2H_6O) do we need to prepare 600.0 mL of a 1.200 molar ethanol solution?

Example 4.2—Calculating Moles

a. How many moles of calcium hydroxide, $Ca(OH)_2$, are found in 250 mL of a 0.800 molar solution of this compound?
b. If we assume that $Ca(OH)_2$ is a strong electrolyte, how many moles of Ca^{2+} ions will be present in 250 mL of 0.800 M $Ca(OH)_2$? How many moles of OH^- ions will be present?

Asking the Right Questions

How is this example different from the previous one? Here, we are asked to find the number of moles in a solution that is already prepared, rather than asking for the concentration of a solution (part a of Example 4.1) or the mass of a reagent required to prepare a solution (part b). Still, the problem-solving approach is similar. We ask, "What units do we want?" "What units do we start with?" "How do we get there?" You can use a flowchart to help you stay on track.

Solution

a. Again, we use dimensional analysis to solve the problem. Note that we must convert the milliliters of solution into liters so that we can use the molarity term. Our flowchart for this calculation is constructed first:

$$\text{mL soln} \xrightarrow{\frac{\text{L}}{\text{mL}}} \text{L soln} \xrightarrow{\frac{\text{mol Ca(OH)}_2}{\text{L}}} \text{mol Ca(OH)}_2$$

Then we complete the calculation by matching the numbers with the appropriate units:

$$250\,\text{mL} \times \frac{1\,\text{L}}{1000\,\text{mL}} \times \frac{0.800\,\text{mol Ca(OH)}_2}{\text{L}} = 0.20\,\text{mol Ca(OH)}_2$$

b. We start by writing the balanced equation that describes what happens to $Ca(OH)_2$ when it is added to water.

$$\text{Ca(OH)}_2 \rightarrow \text{Ca}^{2+}(aq) + 2\text{OH}^-(aq)$$

Then we can calculate the answer by dimensional analysis. Our calculation can be accomplished using the mole ratio relating the number of moles of $Ca(OH)_2$ to the number of moles of Ca^{2+}. Our flowchart is written first:

$$\text{mL soln} \xrightarrow{\frac{\text{L}}{\text{mL}}} \text{L soln} \xrightarrow{\frac{\text{mol Ca(OH)}_2}{\text{L}}} \text{mol Ca(OH)}_2 \xrightarrow{\frac{\text{mol Ca}^{2+}}{\text{mol Ca(OH)}_2}} \text{mol Ca}^{2+}$$

Then we complete the calculation by matching the numbers with the appropriate units:

$$250\,\text{mL} \times \frac{1\,\text{L}}{1000\text{mL}} \times \frac{0.800\,\text{mol Ca(OH)}_2}{1\,\text{L}} \times \frac{1\text{mol Ca}^{2+}}{1\,\text{mol Ca(OH)}_2} = 0.20\,\text{mol Ca}^{2+}$$

For the next part, we need to use the mole ratio to indicate the relationship between $Ca(OH)_2$ and OH^-. We'll use the same flowchart for the Ca^{2+} determination, but we'll modify the last step to show the mole ratio between $Ca(OH)_2$ and OH^-. The calculation is:

$$250\ \text{mL} \times \frac{1\ \text{L}}{1000\ \text{mL}} \times \frac{0.800\ \text{mol Ca(OH)}_2}{1\ \text{L}} \times \frac{2\ \text{mol OH}^-}{1\ \text{mol Ca(OH)}_2} = 0.40\ \text{mol OH}^-$$

Are Our Answers Reasonable?

In part "a," we have only a portion (250 mL) of the original liter that contained 0.800 mol of calcium hydroxide, so our answer of 0.20 makes sense. In part "b," we understood that the ratio of hydroxide ion to calcium in the solution is 2-to-1, therefore, it makes sense that we have twice as much in the solution.

What If…?

…we had the same volume and concentration of a solution of calcium phosphate – $Ca_3(PO_4)_2$? What would the concentration of calcium ions be? What would the concentration of phosphate ions (PO_4^{3-}) be? (*Ans: 0.60 M and 0.40 M*)

Practice 4.2

How many moles of the strong electrolyte sodium phosphate (Na_3PO_4) will be in 5.00 L of a 0.77 M solution? How many moles of sodium ions will be in the solution?

Example 4.3—Calculating Volumes

What volume of a 0.150 M solution of ethanol (C_2H_6O) will contain 12.5 mol of ethanol?

Asking the Right Questions

This is another unit conversion problem that we can solve using dimensional analysis. Here, we are asked to determine the volume, rather than moles, grams or molarity, as in previous exercises. What units are we starting with? How do we change those units into the ones we want? Once again, it is not the numbers that guide us, but the UNITS.

Solution

First, we write our flowchart:

$$\text{mol ethanol} \xrightarrow{\frac{\text{L soln}}{\text{mol ethanol}}} \text{L ethanol}$$

Then we complete the calculation by matching the numbers with their proper units:

$$12.5\ \text{mol ethanol} \times \frac{1\ \text{L solution}}{0.150\ \text{mol ethanol}} = 83.3\ \text{L solution}$$

Is Our Answer Reasonable?

We start with 0.150 mol/L of ethanol, which is about one seventh of a mole for every liter. Therefore, to get 12.5 mol, we should need about 7 times that quantity of liters, or about 88 L. Our answer of 83.3 L is reasonable.

What If…?

…we started with a 0.500 M ethanol solution and wanted to have 12.5 mol of ethanol? How many liters would we need? (*Ans: 25.0 L*)

Practice 4.3

a. What volume of a 3.40 M solution of copper sulfate will contain 4.76 mol of copper sulfate?
b. What volume of a 2.25 M solution of $Ca(NO_3)_2$ will contain 5.5 mol of calcium nitrate? What volume will contain 5.5 mol of nitrate ions?

4.3.1 Parts per Million, Parts per Billion, and so on

The Environmental Protection Agency (EPA) dictates drinking water standards in the United States. These standards dictate the maximum permissible level of harmful contaminants in the water that we consume. If water in the United States has more than these maximum levels, the EPA declares it unsafe to drink. For example, arsenic, a toxic metalloid that occurs naturally, sometimes finds its way into our drinking water. In 2006, the EPA set the maximum allowed level of arsenic in drinking water as 1.3×10^{-7} M. Cities are required to test for arsenic in drinking water, so where there is excess, it is typically in rural areas. While we can use molarity to discuss the concentrations of pollutants such as arsenic, the resulting numbers are, in many cases, very small. When the concentrations get this small, we often find it easier to use an alternative measure of concentration. Specifically, we can talk about the concentration of arsenic in terms of **parts per million (ppm)**, or **parts per billion (ppb)**, or even **parts per trillion (ppt)**.

Parts per million (ppm) – One gram of solute per million grams of solution.
Parts per billion (ppm) – One gram of solute per billion grams of solution.
Parts per trillion (ppt) – One gram of solute per trillion grams of solution.

We are already familiar with the related, but larger, unit "percent." What does percent mean? In a compound that is "1% nitrogen by mass," the nitrogen contributes 1 g out of every 100 g to the total mass. In the same way, a level of one part per million (1 ppm), means the chemical contributes 1 g out of every million grams of the total mass. Similarly, one part per billion (1 ppb) corresponds to a level of 1 g out of every billion grams of the total. One part per trillion (ppt) corresponds to 1 g out of every trillion grams.

These concentration units correspond to very small levels indeed. We can get an idea of just how small from Fig. 4.10.

Let's look at these concentration units in a slightly different way. As an example, the EPA maximum level in drinking water for hexachlorobenzene (C_6Cl_6), a pesticide used on wheat in the United States before 1965, is 1 ppb. This means that the maximum allowable level in drinking water should be 1 g of C_6Cl_6 per billion grams of solution.

$$\text{ppb}\, C_6Cl_6 = \frac{\text{grams}\, C_6Cl_6}{\text{total grams of solution}} \times 10^9$$

Or, we can equivalently express this unit as:

$$\text{ppb}\, C_6Cl_6 = \frac{\text{grams}\, C_6Cl_6}{10^9\ \text{grams of solution}}$$

If we use $\frac{1\ \text{g water}}{1\ \text{mL water}}$ as the density of water and make the key assumption that *the solution is so dilute that its density is about equal to that of water*, then:

$$\text{Density of pure water and of dilute aqueous solutions} = \frac{1\ \text{g solution}}{1\ \text{mL solution}}$$

This means that we can express the concentration of hexachlorobenzene as the number of grams of C_6Cl_6 per liter of solution:

$$\frac{1\ \text{g}\ C_6Cl_6}{10^9 \text{g solution}} \times \frac{1\ \text{g solution}}{1\ \text{mL solution}} \times \frac{10^3\ \text{mL solution}}{1\ \text{L solution}} = \frac{10^{-6}\ \text{g}\ C_6Cl_6}{\text{L solution}} = \frac{1\ \mu\text{g}\ C_6Cl_6}{\text{L solution}}$$

Was our assumption about the density of the solution a good one? If the hexachlorobenzene adds one microgram of mass to an entire liter of solution, then it only changes the density from 1.00 g/mL to 1.000000001 g/mL, which is not different even to 8 significant figures! The bottom line is that *in dilute aqueous solutions*, we can express parts per billion as

$$\text{ppb solute} = \frac{\mu\text{g solute}}{\text{L solution}}$$

Because we often measure liquid in terms of volume, it is often more convenient to work with a mass/volume definition than the actual definition of parts per billion. For our hexachlorobenzene example, 1 L of solution that

(a) 1 ppm: 1 drop in a 12-gallon of water
(b) 1 ppb: one drop in a tanker truck of water
(c) 1 ppt: one drop in a 12-million gallon reservoir of water

Fig. 4.10 To visualize the idea of parts per million, per billion, and per trillion, consider that one drop of ink could be placed in these quantities of water. **a** Placing that drop of ink in a 12-gallon bucket results in 1 ppm. **b** Placing it in a tanker truck results in 1 ppb. **c** Placing it in a 12-million-gallon reservoir results in 1 ppt

Table 4.1 Common units used in expressing "parts per" concentration of a solute "X"

Unit	Mass-to-mass relationship	Mass-to-volume relationship
parts per million (ppm)	$\frac{\text{g X}}{10^6 \text{ g solution}}$	$\frac{\text{mg X}}{\text{L solution}}$
parts per billion (ppb)	$\frac{\text{g X}}{10^9 \text{ g solution}}$	$\frac{\mu\text{g X}}{\text{L solution}}$
parts per tillion (ppt)	$\frac{\text{g X}}{10^{12} \text{ g solution}}$	$\frac{\text{ng X}}{\text{L solution}}$

contains 1.0 μg hexachlorobenzene is said to have a concentration of 1.0 ppb hexachlorobenzene. We can even use this definition to convert from other concentration units to ppb. For example, what is the maximum level of arsenic, in ppb, if the molarity of a solution at this maximum level is 1.3×10^{-7} M arsenic?

Our flowchart looks like this:

$$\frac{\text{mol As}}{\text{L solution}} \xrightarrow{\frac{\text{g As}}{\text{mol As}}} \frac{\text{g As}}{\text{L solution}} \xrightarrow{\frac{1\,\mu\text{g}}{10^{-6}\,\text{g}}} \frac{\mu\text{g As}}{\text{L solution}} = \text{ppb As}$$

Then

$$\frac{1.3 \times 10^{-7}\text{mol As}}{\text{L solution}} \times \frac{74.92\,\text{g As}}{1\,\text{mol As}} \times \frac{1\,\mu\text{g As}}{\text{L solution}} = \frac{9.7\,\mu\text{g As}}{\text{L solution}} = 9.7\,\text{ppb}$$

Note the relationships among parts per million, parts per billion, and parts per trillion in Table 4.1. These relationships are ONLY valid for aqueous solutions because of the assumption that the density of the solution is 1 g/mL.

Although we more often hear about how small concentrations of compounds in water can be harmful, very small levels of some chemicals in drinking water can actually be helpful to human health. For example, the presence of tiny amounts of fluoride (F^-) in drinking water is recognized as beneficial in the prevention of tooth decay. The U.S. Department of Health and Human Services recommends a concentration of 0.7 ppm F^-. The positive effects of such small levels in preventing tooth decay have been reported as "striking" by the American Dental Association. On the other hand, high concentrations of fluoride ion can be very harmful to your health, causing bone disease and mottled teeth, which is why fluoride toothpastes have a warning on their label not to swallow the toothpaste.

Example 4.4—Converting ppm to Molarity

Many counties in the U.S. have fluoride ranges between 0.7 and 1.2 ppm. If a county has a fluoride ion (F^-) in drinking water of 1.0 ppm, what is the corresponding concentration of fluoride, in M?

Asking the Right Questions

What is the definition of "parts per million" in terms of standard units? How do we change those units into molarity (moles/liter)? Using the mass-per-volume relationship we established in Table 4.1 enables us to solve the problem with $1.0\,\text{ppm} = \frac{1.0\,\text{mg F}^-}{1.0\,\text{L solution}}$.

Solution

Remember to construct a flowchart before you start the calculation so you know how you will convert the given units (ppm) to the desired units (M).

$$\frac{1.0\text{ mg F}^-}{1.0\text{ L solution}} \times \frac{1\text{ g}}{1000\text{ mg}} \times \frac{1\text{ mol F}^-}{19.00\text{ g F}^-} = \frac{5.3 \times 10^{-5}\text{ mol F}^-}{\text{L solution}}$$

$$= 5.3 \times 10^{-5}\text{ M F}^-$$

Is Our Answer Reasonable?

We started with a very small amount of fluoride per liter of water, which should be a very small number of moles per liter. So, yes, so our answer is reasonable. We have seen that at the part per million level, concentrations of 10^{-5} to 10^{-6} M are reasonable, but will depend upon the atomic or molecular mass of the substance.

What If…?

…you wanted to make a solution of fluoride ion that was 1.5 ppm? How many moles per liter of fluoride would you need? *(Ans: 7.9×10^{-5} M)*

…you wanted to make a solution of fluoride ion that was 1.5 ppb? How many moles per liter of fluoride would you need? *(Ans: 7.9×10^{-8} M)*

Practice 4.4

Cyanide ion (CN^-) concentrations of 0.200 ppm in drinking water are considered the upper limit for safe human consumption. What is this concentration in M?

4.3.2 Dilution

How do municipal water treatment plant workers prepare water so that it contains the proper concentration of fluoride ion? They use a concentrated source of fluorine, hydrofluorosilicic acid, H_2SiF_6 ("HFSA"), which they then dilute in the drinking water. HFSA reacts with water in a fairly complex way that releases fluoride ions into the water. In Ireland, a company in the town of New Ross, about 160 km south-southwest of Dublin, receives periodic shipments of a concentrated aqueous HFSA solution from a chemical manufacturer in Spain. The concentration of this HFSA solution, shipped in 4000-gallon containers, is about 22 M HFSA. The solution is then diluted with water to 7.8 M and shipped to cities and towns all over Ireland for use in the nation's many water treatment plants. At these plants, the HFSA solution is further diluted in the water supply to a final fluoride ion concentration of 1 ppm.

How do the workers know how much to dilute? There are really two questions here. One is a chemical process question ("How do the workers prepare this solution?"), and the other is a mathematics question ("How do the workers calculate the amount of solute and solvent?"). For laboratory-scale preparation of a solution of known molarity, we use volumetric glassware (shown in ◘ Fig. 4.11), in which the etched mark on the flask indicates the listed volume to a very high precision, typically within +/−0.05% of the stated volume. We add the reagent to the flask and then dilute with solvent (water, in this case) to the mark. For industrial scale processes, much larger containers are needed, and pumps are used to move the liquids around, but the principles are similar.

We can now deal with the mathematics question. Let's assume we wish to prepare 500. mL of 7.8 M HFSA solution from a sample of 22 M HFSA. To do this, we need to calculate the volume of 22 M HFSA that can be added to the volumetric flask and diluted with water. How many moles of HFSA will be in the final solution? We are given the molarity (7.8 M) and volume (500.0 mL) of the final solution, so we can do the following calculation:

$$\frac{7.8\,\text{mol HFSA}}{\text{L HFSA soluiton}} \times 0.5000\,\text{L solution} = 3.9\,\text{mol HFSA}$$

What volume of the concentrated (22 M) HFSA solution must we dispense into the volumetric flask and dilute with water in order to give us 3.9 mol of HFSA?

$$3.9\,\text{mol HFSA} \times \frac{1\,\text{L HFSA}}{22\,\text{mol HFSA}} = 0.1773\,\text{L HFSA solution}$$

$$0.1773\,\text{L HFSA solution} = 177.3\,\text{mL HFSA solution} = 180\,\text{mL HFSA solution (to 2 significant figures)}$$

This means that we can measure 180 mL of HFSA solution and dilute to 500.0 mL in our volumetric flask and, after mixing, have a 7.8 M HFSA solution.

Note that in our calculations, we have compared the number of moles of solute in the concentrated solution that we are measuring to the number of moles of the diluted solution that we want to prepare. Mathematically, this can be written:

$$\text{Number of moles measured from the concentrated solution} = \text{Number of moles needed in the dilute solution}$$

Because the number of moles of solute does not change when we merely add more solvent, and because we can find the number of moles by multiplying the concentration of a solution times its volume we can say that

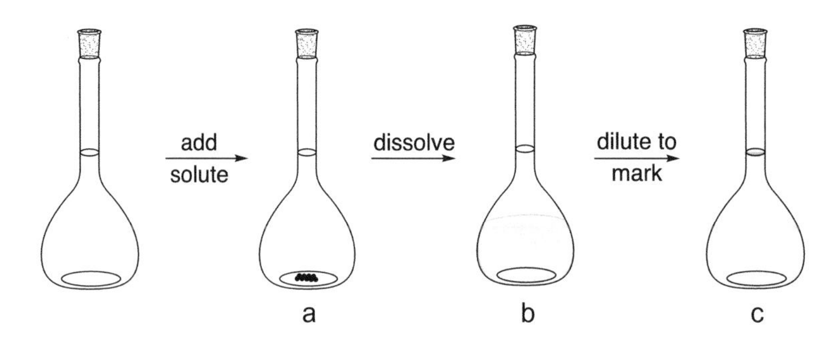

◘ **Fig. 4.11** Using a volumetric flask. **a** A solute is added to the volumetric flask and **b** dissolved in water. **c** The resulting solution is then diluted to the mark on the neck of the flask. The result is a precise and accurate volume of solution

$$C_{\text{concentrated}} \times V_{\text{concentrated}} = C_{\text{dilute}} \times V_{\text{dilute}}$$

Our goal is to find the volume of the initial solution of HFSA to dilute, so we can rearrange this relationship to solve for $V_{\text{concentrated}}$.

$$V_{\text{concentrated}} = \frac{C_{\text{dilute}} \times V_{\text{dilute}}}{C_{\text{concentrated}}}$$

Using our data for HFSA, we find that

$$V_{\text{concentrated}} = \frac{7.8\,\text{M} \times 0.5000\,\text{L}}{22\,\text{M}} = 0.177\,\text{L} = 180\,\text{mL}$$

The total volume of 22 M HFSA needed to prepare large, intercity shipments of 7.8 M HFSA is far larger, but the problem-solving strategy is the same.

Example 4.5—Practice with Dilution of Solutions

Hydrochloric acid is typically shipped as a 12 M aqueous solution. If we are to do a laboratory analysis that requires 250.0 mL of a 1.0 M solution, how many milliliters of the concentrated solution should be diluted?

Asking the Right Questions
What is the problem asking us to find? (the beginning volume of 12 M acid).
What units are we starting with (molarity), and what units do we want (milliliters)?
What is the relationship between molarity, moles and volume?

Solution
Let's begin by constructing a flowchart to help find the answer. We can start with the dilution relationship that was given above. When we are making a dilution, we aren't changing the number of moles of the substance because we are just adding a solvent (usually water):

$$\text{moles HCl from concentrated solution} = \text{moles HCl in dilute solution}$$

$$\left(\frac{\text{moles HCl}}{\text{L}} \times \text{L}\right) \text{of concentrated solution} = \left(\frac{\text{moles HCl}}{\text{L}} \times \text{L}\right) \text{of dilute solution}$$

The problem is asking us to find the volume of concentrated (12 M) solution, and we've been given everything else, so we can substitute numbers into our equation:

$$\left(\frac{12\,\text{moles HCl}}{\text{L}} \times ?\,\text{L}\right) = \left(\frac{1.0\,\text{mol HCl}}{\text{L}} \times 0.250\,\text{L}\right)$$

$$(?\,\text{L}) = \left(\frac{1.0\,\text{mole HCl}}{\text{L}} \times 0.250\,\text{L}\right) \times \frac{\text{L}}{12\,\text{moles HCl}} = 0.21\,\text{L}$$

Since the problem asked for the volume in mL, we need to convert the units:

$$0.021\,\text{L} \times \frac{1000\,\text{mL}}{\text{L}} = 21\,\text{mL}$$

Is Our Answer Reasonable?
We are diluting the hydrochloric acid by a factor of 12, so only a small amount (1/12) of the concentrated acid would need to be combined with water to make the 250.0 mL final solution. Our calculated quantity, 21 mL, is therefore a reasonable answer.

What If…?
…we started with a 36 M aqueous solution of HCl, and we wanted to make the same 250.0 mL of 1.0 M solution. How many mL of the concentrated solution would we need to dilute? (*Ans: 6.9 mL*).

Practice 4.5
Industrially prepared nitric acid (HNO_3) has a concentration of 16 molar. How many milliliters of nitric acid are required to prepare 2.00 L of 0.15 M HNO_3?

4.4 Stoichiometric Analysis of Solutions

The concepts of concentration and solution stoichiometry are used every day to examine all manner of solutions. In medicine, they are used to analyze bodily fluids for evidence of illness and to prepare pharmaceuticals in appropriate amounts. In the food industry they make it possible to measure and modify the nutritional content of foods.

For example, the amount of vitamin C in a sample of fruit juice can be determined by adding iodine and starch to the fruit juice via the following reactions:

$$\underset{\text{vitamin C colorless}}{C_6H_6O_6} + \underset{\text{brownish-red}}{I_2} \rightarrow \underset{\text{Colorless}}{C_6H_6O_6} + \underset{\text{Colorless}}{2HI}$$

$$\underset{\text{colorless}}{\text{starch}} + \underset{\text{brownish-red}}{I_2} \rightarrow \underset{\text{deep blue}}{I_2\text{-starch}}$$

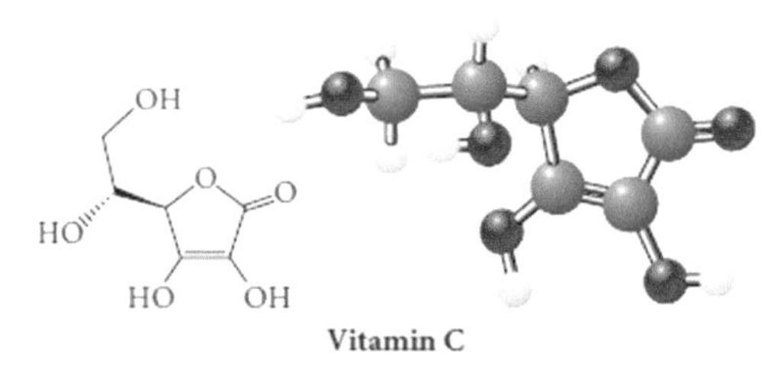

Vitamin C

► https://upload.wikimedia.org/wikipedia/commons/7/7c/Iodine-triphenylphosphine_charge-transfer_complex_in_dichloromethane.jpg

The controlled addition of a solution of known concentration to react with the substance of interest to determine its concentration is called a **titration**. In the titration of fruit juice with iodine shown in ◘ Fig. 4.12, we

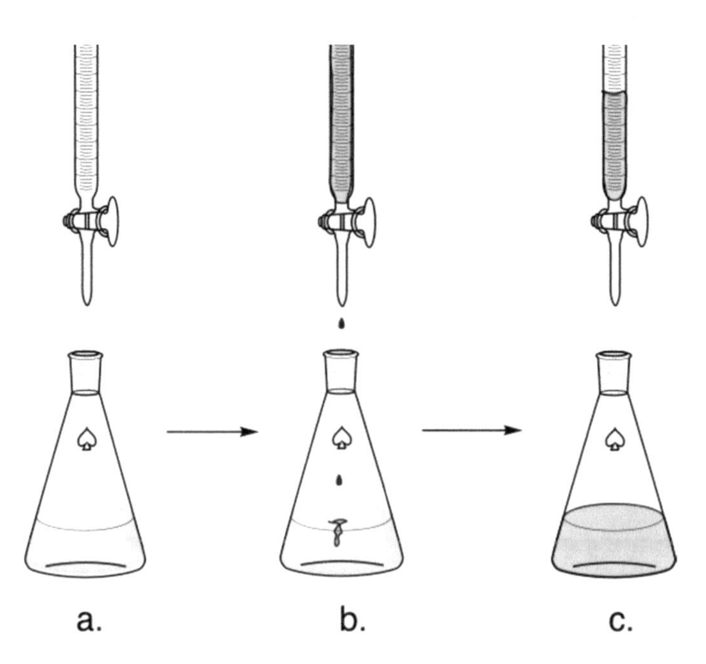

◘ **Fig. 4.12** The process of titration. **a** Fruit juice containing an unknown amount of Vitamin C is in the flask. The buret contains iodine solution. **b** The iodine is slowly added, and the flask swirled to watch for a color change. **c** When the color changes completely the titration is complete

slowly add (**titrate**) iodine solution from a **buret** (a laboratory device used to precisely and accurately add small known quantities of solution) to a sample of fruit juice. When all the vitamin C in the sample has reacted, excess iodine will begin to react with the starch and results in a deep blue color. The point at which a reactant (in this case vitamin C) has been completely consumed is known as the **equivalence point** (very close to the visual **end point**) of the titration. The end point is where the color gives us a visual cue that the reactants have been completely consumed. Even though these two points are very close together, they are not always the same and should not be used interchangeably. We can detect the end point by including a compound, in this case starch, that reacts with the excess iodine to produce a color change. Although the iodine itself is colored, it is often difficult to see this color, especially in fruit juice. However, the presence of starch results in a very obvious color change.

Buret – A laboratory device used to precisely and accurately add small known quantities of solution.

Titrate – To perform a titration.

Titration – The process of adding one reactant to an unknown amount of another until the reaction is complete; used to determine the concentration of an unknown solute.

End point – In a titration, the volume of the added reactant that causes a visual change in the color of the indicator.

Equivalence point – In a titration, the point at which all reactants have just been completely consumed.

How do we use this reaction to determine the vitamin C content of a fruit juice? We begin by preparing a solution of a known concentration of iodine. For this example, let's assume our known concentration is: 0.01052 M I_2. This is known as a **standard solution**, because we use it as a standard of known concentration against which other solutions can be tested. We then measure 50.00 mL (0.05000 L) of fruit juice and add some starch to react with excess iodine. The starch doesn't have anything to react with initially, and it will not react with the iodine until all the vitamin C has been used up. The reaction with starch will indicate that all the vitamin C is gone, so the starch is called an **indicator**. By adding our standard solution of iodine from a buret, we use up the vitamin C until a color change is observed. Typically, this process is repeated several times to minimize error.

Standard solution – A solution with a well-defined and known concentration of solute.

Indicator – A compound that changes color during a titration.

Let's assume we added an average of 16.97 mL of iodine solution before the color change was observed. How many grams of vitamin C are in the fruit juice? We can quantify this using the following strategy:

1. Write the balanced chemical equation relating the reaction of iodine with vitamin C, $C_6H_8O_6$.

$$C_6H_8O_6 + I_2 \rightarrow C_6H_6O_6 + 2\,HI$$

2. Use dimensional analysis to convert the units of 16.97 mL of iodine solution into grams of vitamin C.

$$16.97\text{ mL } I_2\text{ solution} \times \frac{1\text{ L } I_2\text{ solution}}{1000\text{ mL } I_2\text{ solution}} \times \frac{0.01052\text{ mol } I_2}{1\text{ L } I_2\text{ solution}} \times \frac{176.1\text{ g vit C}}{1\text{ mol vit C}} = 0.03144\text{ g vit C}$$

This is the number of grams of vitamin C in our 50.00 mL (0.05000 L) sample of fruit juice. We can determine the amount of vitamin C in 1 L by converting the volume to 1 L:

$$\frac{0.03144\text{ g vit C}}{50.00\text{ mL fruit juice}} \times \frac{1000\text{ mL fruit juice}}{1\text{ L fruit juice}} = \frac{0.6288\text{ g vit C}}{\text{L fruit juice}}$$

Our calculations reveal that there are 0.03144 g of vitamin C in our 50.00 mL sample of fruit juice or 0.6288 g of vitamin C per liter of fruit juice. Since orange juice is mostly water, and we assume that the density of the fruit juice is $\frac{1.0\text{ g}}{\text{mL}}$, we can calculate the concentration of vitamin C in ppm.

Recall from Table 4.1 that parts per million (ppm) of vitamin C can be expressed as $\frac{\text{mg vitamin C}}{\text{L fruit juice solution}}$. We know the vitamin C concentration in $\frac{\text{g}}{\text{L}}$, so all we need to do is convert from grams to milligrams of vitamin C.

$$\frac{0.6288\text{ g vit C}}{\text{L fruit juice}} \times \frac{1000\text{ mg vit C}}{1\text{ g vit C}} = \frac{628.8\text{ mg vit C}}{\text{L fruit juice}}$$

Remember that 1 ppm is the same as 1 mg/L, so 628.8 mg vitamin C/L fruit juice is the same as 628.8 ppm vitamin C.

How do We Know?—How to Test for Small Amounts of Water

Water is such a precious resource that countries have been known to threaten one another with war over water rights. In the United States, individual states quarrel and even sue each other over access to fresh water. For example, Nebraska and Kansas have fought for decades over who owns the water that flows from Colorado through Nebraska and into Kansas via the Republican River.

We are so used to thinking of water as essential and beneficial that it is easy to overlook the many chemical situations in which too much water can be very undesirable. What would be the effect of too much water in the Republican River? On a smaller scale, what would be the effect of too much water in a chocolate bar or potato chips? Canned cooking fats and oils certainly do not benefit from the presence of water; and there are limits to the amount of water that can be in pharmaceutical products that are ingested as tablets, gelcaps, and caplets. Therefore, it is important to be able to determine how much water is present in samples of foods, medicines, and other industrial products, even if the quantities of water involved are very small.

► https://www.flickr.com/photos/ekilby/7532577488/in/photolist-ctCqRE-ctCs8q

One chemical method used to quantify the water content of a wide range of samples, such as chewing gum, jellybeans, and peanuts, is the Karl Fischer titration, named after the scientist who devised the basic method in 1935. The Karl Fischer titration makes use of the reaction of iodine (I_2) with the water in the sample being analyzed. The reaction is performed in the presence of organic solvents such as pyridine (C_5H_5N) and methanol (CH_3OH). The precise details of the chemical reaction are quite complex, to the extent that even now, nearly 90 years after the method was first developed, the exact processes involved in the reaction are still the subject of research. We can summarize it, however, as follows:

$$(\text{Reactants}) + I_2 + H_2O \rightarrow (\text{products})$$

The chemical details vary with the actual method used. In all cases, however, the crucial quantitative fact that allows the water content to be measured is that *the iodine and water always react in a known mole ratio, which is 1:1 in the case shown above*. The end point of the titration, when we can tell that all the water has been consumed, is marked by a distinctive color change, which enables us to determine the quantity of iodine required to achieve that color change. The amount of water that must have reacted with that quantity of iodine can then be calculated.

Modern laboratory instrumentation has allowed the titration process to be automated, as shown in ■ Fig. 4.13. Instead of direct observation of a color change, the automated instruments may rely on measurement of a flow of electrons to excess iodine as soon as all the water has reacted. This flow of electrons is due to the process $I_2 + 2e^- \rightarrow 2I^-$. Another option is to use the reverse of the process shown above—namely $2I^- \rightarrow I_2 + 2e^-$—to generate the iodine needed to react with the water.

The details vary depending on the machines used, but automated Karl Fischer titrations allow accurate determination of the water content in tiny samples containing hardly any water at all. What do we mean by "hardly any" and "tiny"? The method is useful down to the level of 50 ppm water in a sample that can have a volume as small as 10 μL.

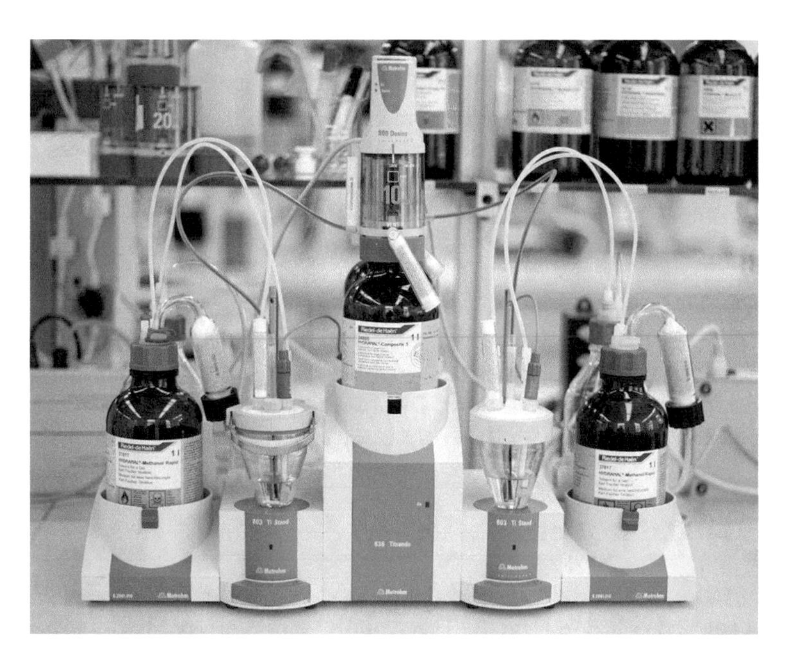

Fig. 4.13 Karl Fischer automatic titration apparatus. *Source* ▶ https://upload.wikimedia.org/wikipedia/commons/f/f1/KF_Titrator.jpg

Example 4.6—Practice with Titration

A common type of titration done by chemistry students is the reaction of a solution of sodium hydroxide (with a known concentration) with a solution of an acid (with an unknown concentration). The result can be used to determine the molarity of the acid solution. What is the molarity of a solution of 50.00 mL of HCl if 31.98 mL of 0.1253 M NaOH is required to react with it? The products of this reaction are sodium chloride and water.

Asking the Right Questions

What are we given? What do we want to know? How do we get there? We are given the volume and molarity of sodium hydroxide, along with the volume of hydrochloric acid. We are asked for the molarity of the hydrochloric acid. How can we best get there?

We have the volume of HCl (50.00 mL), but molarity is a measure of moles/volume. How can we determine the number of moles of HCl?

Solution

From the problem, we write the equation for the titration reaction of hydrochloric acid with sodium hydroxide:

$$\mathrm{HCl}(aq) + \mathrm{NaOH}(aq) \rightarrow \mathrm{H_2O}(l) + \mathrm{NaCl}(aq)$$

Our flowchart for the dimensional analysis is:

$$\text{L NaOH} \xrightarrow{\frac{\text{mol NaOH}}{\text{L soln}}} \text{mol NaOH} \xrightarrow{\frac{\text{mol HCl}}{\text{mol NaOH}}} \text{mol HCl}$$

Then we complete the calculation by dividing the number of moles of HCl by the number of liters of the solution. Our calculations reveal:

$$0.03198\,\text{L NaOH} \times \frac{0.1253\,\text{mol NaOH}}{1\,\text{L NaOH soln}} \times \frac{1\,\text{mol HCl}}{1\,\text{mol NaOH}} = 0.0040071\,\text{mol HCl}$$

$$\frac{0.0040071\,\text{mol HCl}}{0.05000\,\text{L solution}} = 0.08014\,\text{M HCl}$$

Is Our Answer Reasonable?

Since the titration used more of the HCl solution that the NaOH solution, we would expect the HCl solution to be a little more dilute. Our answer makes sense.

What If…?

…we needed 30.56 mL of HCl solution to react with the same volume and concentration of NaOH solution given above. What would the concentration of the HCl solution be? (*Ans: 0.1331 M*)

Practice 4.6

The equation for the reaction between a sodium hydroxide solution and sulfuric acid is:

$$2\mathrm{NaOH}(aq) + \mathrm{H_2SO_4}(aq) \rightarrow \mathrm{Na_2SO_4}(aq) + 2\mathrm{H_2O}(l)$$

If 22.25 mL of 0.100 M NaOH is required to react with a 25.00-mL sample of an H_2SO_4 solution, what is the molarity of the H_2SO_4 solution?

4.5 Types of Chemical Reactions

The vast number of aqueous reactions that have been identified since the beginning of chemistry would stun the ancient alchemists. Fortunately, most of these reactions fall within certain general types or categories because of key similarities among them. Three of the more important types of chemical reactions that we will learn about in this chapter are:
- Precipitation
- Acid–base
- Reduction–oxidation

In order to more fully understand the three types of processes, we will take this opportunity to build on our ability to write chemical equations in aqueous solutions. Our first step is determining what happens to the individual components when they are added to water. We'll use our knowledge of electrolytes to assist us in this process.

4.5.1 Molecular and Ionic Equations

The reaction that was the focus of Example 4.6 includes three strong electrolytes: hydrochloric acid (HCl), sodium hydroxide (NaOH), and sodium chloride (NaCl). The reaction equation, containing the individual components written as compounds, is known as the **molecular equation**. The molecular equation illustrates that the individual substances exist as hydrated compounds in the aqueous solution.

Molecular equation – A chemical equation that shows complete molecules and compounds.

The molecular equation representing the reaction of hydrochloric acid and sodium hydroxide can be written

$$\mathrm{HCl}(aq) + \mathrm{NaOH}(aq) \rightarrow \mathrm{H_2O}(l) + \mathrm{NaCl}(aq)$$

molecular equation

However, as we have already discussed, strong electrolytes dissolve in water, dissociate into their component ions, and become relatively independent within the solution. For this reason, all the individual ions can be shown separately, in what is known as a **complete ionic equation** (also known as the total ionic equation).

$$\mathrm{H^+}(aq) + \mathrm{Cl^-}(aq) + \mathrm{Na^+}(aq) + \mathrm{OH^-}(aq) \rightarrow \mathrm{Na^+}(aq) + \mathrm{Cl^-}(aq) + \mathrm{H_2O}(l)$$

complete ionic equation

Complete ionic equation – A chemical equation that indicates all of the ions present in a reaction as individual entities.

Note that only the strong electrolytes have been written as individual ions. Unlike the electrolytes, molecules such as water do not dissociate into ions to any appreciable extent. They are not written as individual ions.

What is the actual reaction that is taking place among all of the ions in solution? Complete ionic equations list all substances as they exist in the solution, irrespective of whether they are part of the reaction. For example, in the complete ionic equation above, the sodium (Na^+) and chloride (Cl^-) ions are present at both the start and the end of the reaction. They do not participate at all during the reaction; they "sit on the sidelines" like spectators at a sporting event or a concert. For this reason, they are known as **spectator ions**. The real activity during this reaction is the combination of hydrogen ions (H^+) and hydroxide ions (OH^-) to form water (H_2O). If we remove the spectator ions from the equation, we can simplify the equation to better show the combination of hydrogen ions and hydroxide ions:

$$\mathrm{H^+}(aq) + \cancel{\mathrm{Cl^-}(aq)} + \cancel{\mathrm{Na^+}(aq)} + \mathrm{OH^-}(aq) \rightarrow \cancel{\mathrm{Na^+}(aq)} + \cancel{\mathrm{Cl^-}(aq)} + \mathrm{H_2O}(l)$$

Spectator ions – Ions that do not participate in a reaction.

The result represents the overall, or *net chemical change* that occurs in this reaction and is known as the **net ionic equation**:

$$\mathrm{H^+}(aq) + \mathrm{OH^-}(aq) \rightarrow \mathrm{H_2O}(l)$$

net ionic equation

Net ionic equation – A total ionic equation written without the spectator ions.

▶ Here's what we know thus far

- *Molecular equations* show the formulas of all reactants and products but do not indicate whether any of the compounds really exist as ions in solution.
- *Total ionic equations* show all of the dissolved ions present in an equation individually, along with all other reactants and products.
- *Net ionic equations* eliminate the spectator ions and show only those dissolved ions, and other reactants and products, which actually participate in or result from the reaction concerned. ◀

Example 4.7—The Three Equations

Aqueous hydrochloric acid and aqueous sodium carbonate react to form aqueous sodium chloride, water, and gaseous carbon dioxide. Write the molecular, total ionic, and net ionic equations for this reaction.

Asking the Right Questions

How do we differentiate among the three types of reactions? Which substances are strong electrolytes?

Solution

The molecular equation is:

$$2HCl(aq) + Na_2CO_3(aq)$$
$$\rightarrow 2NaCl(aq) + H_2O(l) + CO_2(g)$$

In the complete ionic equation, all aqueous ionic compounds are written as separated ions in solution:

$$2H^+(aq) + 2Cl^-(aq) + 2Na^+(aq)$$
$$+ CO_3^{2-}(aq)2Na^+(aq) + 2Cl^-(aq)$$
$$+ H_2O(l) + CO_2(g)$$

The net ionic equation is completed by removing the spectator ions:

$$2H^+(aq) + CO_3^{2-}(aq) \rightarrow H_2O(l) + CO_2(g)$$

Are Our Answers Reasonable?

Hydrochloric acid and sodium carbonate are strong electrolytes that react to form the strong electrolyte sodium chloride, along with water and carbon dioxide. Therefore, we are reasonable in suggesting that the sodium and chloride ions will be spectator ions that do not appear in the net ionic equation.

What If…?

…we reacted potassium iodide with lead nitrate to form potassium nitrate and lead(II) iodide? Lead(II) iodide is a non-electrolyte and forms a bright yellow solid that is difficult to dissolve. Write the three equations for this reaction.

$$Ans: (Pb(NO_3)_2(aq) + 2KI(aq)$$
$$\rightarrow PbI_2(s) + 2KNO_3(aq)$$
$$Pb^{2+}(aq) + 2NO_3^-(aq) + 2K^+(aq) + 2I^-(aq)$$
$$\rightarrow PbI_2(s) + 2K^+(aq) + 2NO_3^-(aq)$$
$$Pb^{2+}(aq) + 2I^-(aq) \rightarrow PbI_2(s))$$

Practice 4.7

Identify the molecular, total ionic, and net ionic equations that describe the aqueous reaction of potassium oxalate ($K_2C_2O_4$) and nitric acid (HNO_3) to form potassium nitrate (KNO_3) and oxalic acid ($H_2C_2O_4$). Assume that oxalic acid is the only nonelectrolyte in the reaction.

4.6 Precipitation Reactions

An important challenge for chemists trying to clean up industrial wastewater is to remove the ions of "heavy metals" (with a density of 5 g/mL or greater) such as lead, mercury, and cadmium, which are toxic to humans and other life. This is particularly important in mining operations, where aqueous waste streams from the mine can contain abnormally high levels of heavy metals. One way to clean up the wastewater is to add sulfide ions (S^{2-}), because most nonalkali metal ions will combine with sulfide ions to form a solid "precipitate." The word **precipitate** is both a noun and a verb. It is used as a noun to refer to any solid material that forms within a solution, and as a verb to describe that process in action. We can say that metal sulfides form a precipitate (noun). Alternatively, we can say that metal sulfides will precipitate (verb) out of solution as soon as they form. We say that a precipitate is **insoluble** because not very much of it can dissolve into the solvent.

Precipitate – Any solid material that forms within a solution; the action describing the formation of a solid.

Insoluble – Not capable of dissolving in a solvent to an appreciable extent.

On the other hand, a relatively large amount of a **soluble** substance can dissolve in a solvent. Soluble ionic compounds are strong electrolytes, while insoluble compounds in general are weak electrolytes or nonelectrolytes. For example, about 2 kg of sucrose, table sugar, can dissolve in a liter of water at 25 °C. Sucrose is a highly soluble non-electrolyte compared to silver chloride (a very weak insoluble electrolyte), which has a solubility of about 2 mg per liter of water at 25 °C. Not all "soluble" substances are equal—they don't all dissolve to the same degree.

Soluble – The ability of a substance to significantly dissolve within a solution.

Industrial wastewater can be treated with sulfate-reducing bacteria. These bacteria convert sulfate ions into sulfide ions as part of their natural biological activity (◘ Fig. 4.14). In the presence of heavy metals, the sulfide generated by the bacteria can help to reduce the amount of soluble heavy metals by making insoluble precipitates. For example, the net ionic equation for the reaction of lead(II) ions with sulfide ions is

$$Pb^{2+}(aq) + S^{2-}(aq) \rightarrow PbS(s)$$

The ions are initially in the aqueous phase and so are dissolved in water. When they meet and combine, they form the insoluble black solid precipitate lead(II) sulfide. This is an example of a **precipitation reaction**, a reaction in which an insoluble precipitate is formed from soluble reactants. We can precipitate lead(II) sulfide in the laboratory by mixing a solution of a soluble lead(II) compound with a solution of a soluble sulfide. ◘ Figure 4.15 illustrates a similar precipitation reaction between lead(II) ions and iodide. The reaction of lead(II) nitrate and ammonium sulfide can be written using a molecular equation:

$$(NH_4)_2S(aq) + Pb(NO_3)_2(aq) \rightarrow 2NH_4NO_3(aq) + PbS(s)$$

Precipitation reaction – A reaction that forms a solid that isn't soluble in the reaction solvent.

The lead(II) sulfide is a solid that essentially doesn't dissolve or ionize, so we note this in all written forms of the equation. For example, the total ionic equation is:

$$2NH_4^+(aq) + S^{2-}(aq) + Pb^{2+}(aq) + 2NO_3^-(aq) \rightarrow 2NH_4^+(aq) + 2NO_3^-(aq) + PbS(s)$$

The net ionic equation more clearly shows this precipitation reaction. The ammonium and nitrate ions are spectator ions, so the net ionic equation reduces to:

$$Pb^{2+}(aq) + S^{2-}(aq) \rightarrow PbS(s)$$

which is simply the precipitation of lead(II) sulfide.

◘ **Fig. 4.14** The result of sulfate-reducing bacteria. Spring water in the floodplain of the Prairie Dog Town Fork of the Red River that flows through the panhandle of Texas has exposed black mud. The black sediment is formed from the interaction of dissolved metal ions and sulfide produced from sulfate-reducing bacteria

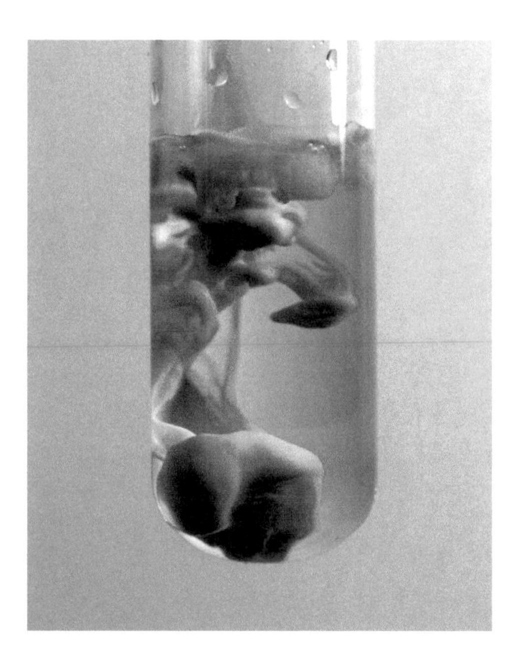

Fig. 4.15 Lead iodide precipitates from the combination of its soluble ions. *Source* ▶ https://upload.wikimedia.org/wikipedia/commons/8/80/Lead_%28II%29_iodide_precipitating_out_of_solution.JPG

Table 4.2 Solubility rules for ionic compounds

Soluble	Insoluble
Group IA and ammonium compounds are soluble	Most carbonates are insoluble except Group IA carbonates and $(NH_4)_2CO_3$
Acetate, chlorate, perchlorate, and nitrate compounds are soluble	Most phosphates are insoluble, except Group IA phosphates and $(NH_4)_3PO_4$
Most chlorides, bromides, and iodides are soluble, except those of Ag^+, Hg_2^{2+}, and Pb^{2+}	Most sulfides are insoluble except Group IA sulfides and $(NH_4)_2S$ Most sulfites are insoluble except Group IA sulfites and $(NH_4)_2SO_3$
Most sulfates are soluble except those of Ca^{2+}, Sr^{2+}, Ba^{2+}, Ag^+, Hg_2^{2+}, and Pb^{2+}	Most hydroxides are insoluble except Group IA hydroxides and $Ca(OH)_2$, $Sr(OH)_2$, and $Ba(OH)_2$

How can we predict when a precipitation reaction will occur? For most common ions, the answer has been determined experimentally and can be summarized by a set of *solubility rules*. Use of the rules outlined in Table 4.2 is helpful in determining which compounds are likely to form solids. Compounds generally follow these rules with a few exceptions, some of which are also noted in the table. We can use this information to write and balance the molecular equations for a reaction, determine whether any of the compounds form insoluble precipitates, and write the complete ionic and net ionic equations.

Example 4.8—Identifying Precipitation Reactions

Which combinations of the following aqueous solutions will produce precipitates: aluminum bromide, barium hydroxide, magnesium sulfate, and nickel(II) iodide? Write the net ionic equation for each combination that produces a precipitate. Use the solubility rules to guide you in your decision.

Asking the Right Questions

What are all the possible combinations of positive and negative ions? Which of these will produce an insoluble product? Which ions are available in the four solutions?

Solution

The four solutions in this question contain these ions: Al^{3+} and Br^-, Ba^{2+} and OH^-, Mg^{2+} and SO_4^{2-}, Ni^{2+} and I^-.

The original combinations (aluminum bromide, barium hydroxide, magnesium sulfate, and nickel(II) iodide) are all soluble because they are all aqueous solutions, so we don't need to consider them. The other possible combinations, with the "soluble" or "insoluble" verdict are shown here for each combination of the ions:

aluminum hydroxide—*insoluble*
aluminum sulfate—soluble
aluminum iodide—soluble
barium bromide—soluble
barium sulfate—*insoluble*
barium iodide—soluble
magnesium bromide—soluble
magnesium hydroxide—*insoluble*
magnesium iodide—soluble
nickel(II) bromide—soluble
nickel(II) hydroxide—*insoluble*
nickel(II) sulfate—soluble

The solubility rules indicate that we could make four insoluble precipitates by combining the various solutions in these ways:

Add aluminum bromide solution to barium hydroxide solution, to form a precipitate of aluminum hydroxide.

$$Al^{3+}(aq) + 3OH^-(aq) \rightarrow Al(OH)_3(s)$$

Add barium hydroxide solution to magnesium sulfate solution, to form a precipitate of barium sulfate, mixed with a precipitate of magnesium hydroxide.

$$Mg^{2+}(aq) + 2OH^-(aq) \rightarrow Mg(OH)_2(s)$$

$$Ba^{2+}(aq) + SO_4^{2-}(aq) \rightarrow BaSO_4(s)$$

Add nickel(II) iodide solution to barium hydroxide solution, to form a precipitate of nickel(II) hydroxide.

$$Ni^{2+}(aq) + 2OH^-(aq) \rightarrow Ni(OH)_2(s)$$

Are Our Answers Reasonable?

In this case, there's not a "common sense" test to check and see if the answers are reasonable. By relying on the rules in Table 4.2, however, we can check that our answers are reasonable.

What If...?

...we combined the following solutions? Which ones would result in the formation of a precipitate? What would the precipitate be?

a. Silver nitrate and sodium iodide.
b. Potassium sulfide and lead(II) nitrate.
c. Calcium acetate and lead(II) perchlorate.
d. Ammonium iodide and sodium chloride.

(*Ans: a. silver iodide is a solid; b. lead(II) sulfide is a solid; c. no solid forms; d. no solid forms.*)

Practice 4.8

Which combination of the following solutions will produce a precipitate:

$$AgNO_3(aq),\ NaCl(aq),\ Na_2S(aq), ZnSO_4(aq)?$$

Write the net ionic equation for each combination that produces a precipitate.

Example 4.9—The Stoichiometry of a Precipitation Reaction

Calculate the mass of the precipitate that is formed when 1.30 L of 0.0200 M $AlBr_3$ solution is added to 3.00 L of 0.0350 M NaOH solution.

Asking the Right Questions

What is the balanced molecular equation for this reaction and what is the precipitate that is formed? Which one of the reactants is the limiting reagent for the reaction? What is the maximum number of grams of the precipitate that can be formed?

Solution

The balanced molecular equation is

$$AlBr_3(aq) + 3NaOH(aq) \rightarrow Al(OH)_3(s) + 3NaBr(aq)$$

Examination of the solubility rules indicates that the aluminum hydroxide is insoluble in water.

How many grams of $Al(OH)_3(s)$ are produced if all the $AlBr_3$ is used up?

$$1.30\ L\ AlBr_3\ soln \times \frac{0.200\ mol\ AlBr_3}{1\ L\ AlBr_3\ soln} \times \frac{1\ mol\ Al(OH)_3}{1\ mol\ AlBr_3} \times \frac{78.00\ g\ Al(OH)_3}{1\ mol\ Al(OH)_3} = 2.03\ g\ Al(OH)_3$$

How many grams of $Al(OH)_3(s)$ are produced if all the NaOH is used up?

$$3.00\ L\ NaOH\ soln \times \frac{0.0350\ mol\ NaOH}{1\ L\ NaOH\ soln} \times \frac{1\ mol\ Al(OH)_3}{3\ mol\ NaOH} \times \frac{78.00\ g\ Al(OH)_3}{1\ mol\ Al(OH)_3} = 2.73\ g\ Al(OH)_3$$

Our calculations indicate that the $AlBr_3$ is the limiting reagent because it, rather than the NaOH, limits the mass of $Al(OH)_3$ formed. Mixing the two reagents together will produce 2.03 g of aluminum hydroxide as a precipitate.

Is Our Answer Reasonable?

Since the reactants are not very concentrated, we would expect a fairly small amount of product to be formed. The amount of product we calculated is only about 2 g, so this is reasonable.

What If…?

…2.80 L of 0.0200 M $AlBr_3$ solution is added to 3.00 L of 0.0350 M NaOH solution? How much aluminum hydroxide is formed? (*Ans: 2.73 g of* $Al(OH)_3$).

Practice 4.9

What mass of precipitate is produced when 0.35 L of 0.25 M sodium carbonate reacts with 0.55 L of 0.35 M barium chloride?

4.7 Acid-Base Reactions

Drainage from a gold mine is often very acidic (Fig. 4.16). In fact, the water from a gold mine can be acidic enough to seriously burn anyone who touches it. To make this water suitable for the environment, cleanup crews often add a compound that decreases the amount of acid in the waste. What is an acid? The simplest definition is that an **acid** is a substance that releases hydrogen ions (H^+) in a solution. The pain of "acid indigestion," is caused by too many hydrogen ions in the gastric fluid that fills the stomach. The chemical opposite of an acid is a **base**. A base is a substance that releases hydroxide ions (OH^-) in a solution. An aqueous base is also called an **alkali**. Acids, bases, and their solutions are explored in detail later in this text.

Acid – A compound that produces hydrogen ions (H^+) when dissolved in water.

Base – A compound that produces hydroxide ions (OH^-) when dissolved in water.

Alkali – An older term for an aqueous base. From the Arabic for "ashes from the saltwort plant".

An **acid–base reaction** is the reaction between an acid and a base. The result of many of these reactions is a solution that is neutral - neither acidic nor basic. For this reason, acid–base reactions are generally referred to as **neutralization reactions**, even when the resulting solution is not actually neutral.

Acid–Base reaction – The reaction between an acid and a base. The products are water and an ionic compound.

Neutralization reaction – An acid–base reaction.

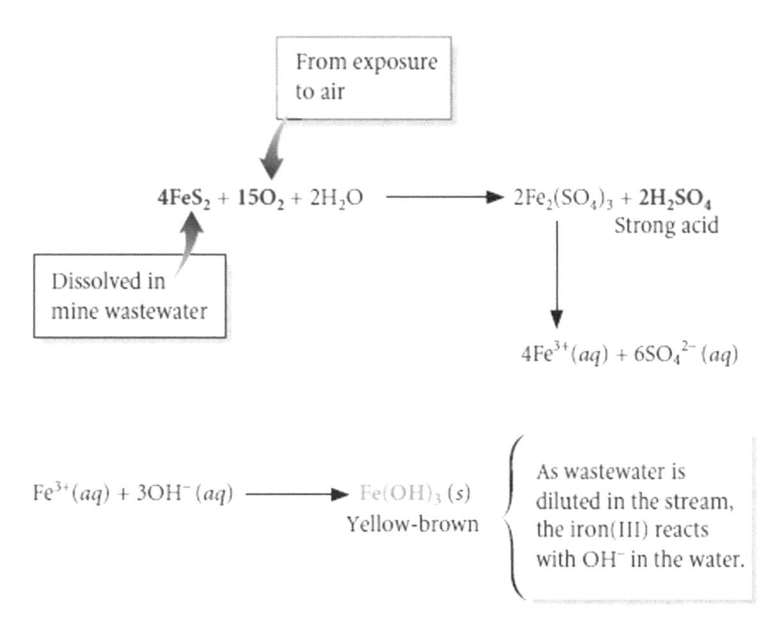

Fig. 4.16 Water that fills a gold mine dissolves small amounts of FeS_2 (iron pyrite, also known as fool's gold) that is often found in gold deposits. This compound undergoes oxidation when it reaches the air outside the mine. The result is a mixture of insoluble $Fe(OH)_3$ and sulfuric acid (H_2SO_4). Gold mine runoff typically has a rust color and extremely high acidity

4.7.1 Strong Acids and Bases

One very common acid is hydrochloric acid (HCl). In the industrial world, the enormous quantities of HCl produced (annual production of 5 billion kg, within the top 20 amounts of industrial chemicals produced) are used in the isolation and cleaning of metals and in the production of solvents and chlorinated organic compounds, as well as in acid–base reactions. HCl is a strong electrolyte. When HCl molecules dissolve in water, they completely "ionize" (dissociate completely into ions) to form hydrogen ions and chloride ions:

$$HCl(g) \rightarrow HCl(aq) \rightarrow H^+(aq) + Cl^-(aq)$$

Because HCl is acidic and ionizes completely in aqueous solution (i.e., because it is a strong electrolyte), we call it a **strong acid**. Examples of other strong acids are given in ◘ Table 4.3.

Strong acid – An acid that completely dissociates (ionizes) in solution.

A **strong base** is one that ionizes completely to produce OH^- in water. Sodium hydroxide (NaOH) is one of the more common strong bases, with a worldwide production of about 70 billion kg annually. It is often used in bathroom cleaners, in drain clog removers, and in chemical reactions that are used to prepare soaps. Because it is soluble, it is a strong electrolyte, and it dissociates completely when it is added to water:

$$NaOH(s) \rightarrow NaOH(aq) \rightarrow Na^+(aq) + OH^-(aq)$$

Strong base – A base that completely dissociates in solution.

Examples of other strong bases are given in ◘ Table 4.3.

▸ https://upload.wikimedia.org/wikipedia/commons/3/34/SodiumHydroxide.jpg

4.7.2 Weak Acids and Bases

If strong acids and bases dissociate into their ions in solution completely, what do weak acids and bases do? A **weak acid** differs from a strong acid in that it does not dissociate completely in aqueous solution. Most acids (except those in ◘ Table 4.3) are considered weak acids. For example, acetic acid is added to water to make vinegar. The reaction illustrating its acidity is shown below. Note that the reaction doesn't proceed completely to products. Rather, almost all of the acetic acid stays in solution as CH_3COOH and isn't dissociated. Because of this, weak acids are weak electrolytes.

$$CH_3COOH(l) \rightarrow CH_3COOH(aq) \rightleftharpoons CH_3COO^-(aq) + H^+(aq)$$

◘ **Table 4.3** Common Strong Acids and Bases

Strong Acids	HCl, HBr, HI, HNO_3, $HClO_4$, H_2SO_4
Strong Bases	Hydroxides of Group 1 metals and $Ca(OH)_2$, $Sr(OH)_2$, $Ba(OH)_2$

Weak acid – An acid that partially dissociates in solution.

Note from the electrostatic potential maps that the electron density in the acetate ion becomes spread out around both oxygen atoms in the molecule. Remember that blue indicates regions of positive charge and red indicates regions of negative charge.

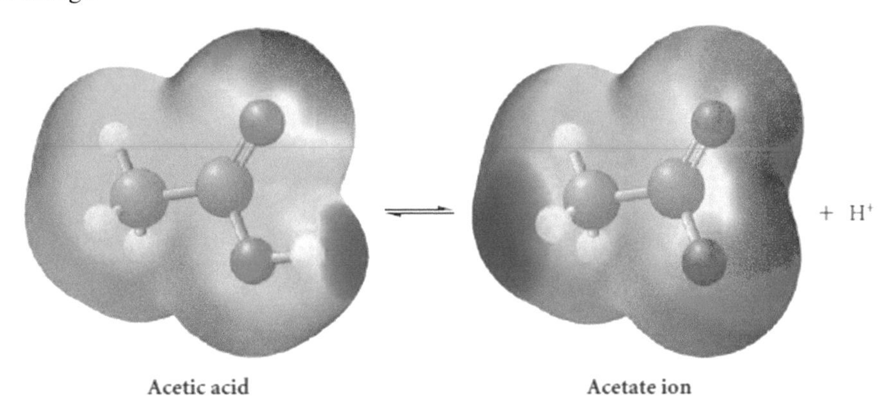

Weak bases don't react extensively to produce hydroxide ions. For example, when ammonia is added to water, only a small proportion of the ammonia reacts to produce the ammonium ion (NH_4^+) and hydroxide ion (OH^-). Most bases are weak. In a fashion similar to the weak acids, weak bases are also weak electrolytes.

$$NH_3(aq) + H_2O(l) \rightleftharpoons NH_4^+(aq) + OH^-(aq)$$

Weak base – A base that partially dissociates in solution.

4.7.3 When Acids and Bases Combine

In a reaction between a strong acid and a strong base, hydrogen ions (H^+) and hydroxide ions (OH^-) come together to form water. This is illustrated by writing the chemical equation for the reaction of HCl with NaOH. In the molecular equation, we can see that the products are an ionic compound (in this case, NaCl) and water.

$$HCl(aq) + NaOH(aq) \rightarrow NaCl(aq) + H_2O(l)$$

The total ionic equation shows exactly which ions are involved in the reaction.

$$H^+(aq) + Cl^-(aq) + Na^+(aq) + OH^-(aq) \rightarrow Na^+(aq) + Cl^-(aq) + H_2O(l)$$

The sodium and chloride ions are spectator ions, so we can simplify to the net ionic equation:

$$H^+(aq) + OH^-(aq) \rightarrow H_2O(l)$$

This is the formation of water from its ions. This is the typical outcome for strong-acid–strong-base reactions; the formation of water as the net ionic equation for the reaction indicates a neutralization reaction. In other words, the acid (H^+) and the base (OH^-) react to make a compound that is neither acidic nor basic but, rather, the neutral liquid water (H_2O).

Hydrochloric acid is also known as a **monoprotic** acid, because it produces just one mole of hydrogen ions (H^+, which is a proton) from each mole of HCl when it dissociates. This results in one mole of water, for each mole of acid, forming when the acid is completely neutralized in an acid–base reaction. Other common acids can be **diprotic** or **triprotic** and so generate *two* or *three* moles of water, respectively, for each mole of acid neutralized. Compare the molecular and net ionic equations of hydrochloric acid (HCl), sulfuric acid (H_2SO_4), and phosphoric acid (H_3PO_4), each reacting with sodium hydroxide:

Monoprotic – Can produce 1 mol of H^+ per formula unit when it dissociates.

Diprotic – Can produce 2 mol of H^+ per formula unit when it dissociates.

Triprotic – Can produce 3 mol of H^+ per formula unit when it dissociates.

$$\underset{\text{hydrochloric acid (monoprotic)}}{HCl(aq)} + \underset{\text{1 mol NaOH required}}{NaOH(aq)} \rightarrow NaCl(aq) + \underset{\text{1 mol water formed per mole acid}}{H_2O(l)}$$

$$H_2SO_4(aq) + 2NaOH(aq) \rightarrow Na_2SO_4(aq) + 2H_2O(l)$$

sulfuric acid (diprotic) — 2 mol NaOH required — 2 mol water formed per mole acid

$$H_3PO_4(aq) + 3NaOH(aq) \rightarrow Na_3PO_4(aq) + 3H_2O(l)$$

phosphoric acid (triprotic) — 3 mol NaOH required — 3 mol water formed per mole acid

Example 4.10—Stoichiometry of a Neutralization Reaction

What volume of a 0.200 M H_2SO_4 solution is needed to neutralize 25.0 mL of a 0.330 M NaOH solution?

Asking the Right Questions

What are we given, and what do we want to know? What does "neutralize" mean? What is the balanced equation for this reaction?

Solution

We should begin by writing the balanced equation for the reaction. Then we can use dimensional analysis to determine the volume of sulfuric acid solution that is needed. Remember, it's also a good idea to map out the process before you begin the calculation.

The balanced molecular equation is:

$$H_2SO_4(aq) + 2NaOH(aq) \rightarrow Na_2SO_4(aq) + 2H_2O(l)$$

Using unit conversion, we develop the entire flowchart to calculate the number of liters of sulfuric acid solution that will react with the base:

$$\text{L NaOH} \xrightarrow{\frac{\text{mol NaOH}}{\text{L soln}}} \text{mol NaOH} \xrightarrow{\frac{\text{mol } H_2SO_4}{\text{mol NaOH}}} \text{mol } H_2SO_4 \xrightarrow{\frac{\text{L } H_2SO_4 \text{ soln}}{\text{mol } H_2SO_4}} \text{L } H_2SO_4$$

$$0.0250\,\text{L NaOH} \times \frac{0.330\,\text{mol NaOH}}{1\,\text{L}} \times \frac{1\,\text{mol } H_2SO_4}{2\,\text{mol NaOH}} \times \frac{1\,\text{L } H_2SO_4\text{ soln}}{0.200\,\text{mol } H_2SO_4} = 0.0206\,\text{L } H_2SO_4$$

A total of 0.0206 L (or 20.6 mL) of 0.200 M H_2SO_4 is required. Note that the mole ratio of 2 mol of NaOH to 1 mol of H_2SO_4 is used because sulfuric acid is diprotic, having two protons that can react with sodium hydroxide.

Is Our Answer Reasonable?

Our H_2SO_4 solution is less concentrated than our NaOH solution, which suggests that we might need more than 25.0 mL of sulfuric acid to neutralize the NaOH. However, 2 mol of NaOH are neutralized for every 1 mol of H_2SO_4. Therefore, we can cut the volume we expect in half, making it less than the volume of NaOH. Our answer therefore is reasonable.

What If...?

...the sulfuric acid concentration was 0.370 M? How many mL of sulfuric acid solution would we need to neutralize 25.0 mL of a 0.330 M NaOH solution? (*Ans: 11.1 mL*).

Practice 4.10

What volume of a 0.550 M H_3PO_4 solution is needed to neutralize 50.0 mL of a 0.250 M NaOH solution?

4.8 Reduction-Oxidation Reactions

Living in an atmosphere rich in oxygen makes life possible, but it also gives us the continual challenge of preventing unwanted reactions between oxygen and chemicals on which we rely. One kind of very fast reaction with oxygen, called combustion, yields fire, and we go to great lengths to protect ourselves against unwanted fires. A slow but almost equally troublesome reaction of metals with oxygen causes corrosion, such as that which occurs when iron rusts. Rust forms when iron reacts with oxygen (in the presence of water) to form various forms of iron oxide, such as Fe_2O_3. Water is crucial for this reaction to occur at a significant rate, because the oxygen must be in solution. How do we typically reduce or eliminate the ability of iron, such as that in our patio furniture or cars, to rust? We protect the metal with a coat of paint, a plastic coating, and so on. Having other metals available to react with the oxygen can also minimize corrosion.

When chemists examine what happens in such **oxidation** reactions, they find that electrons are lost from the reactant that is being **oxidized**. Many oxidations do involve oxygen, but what is really happening is the transfer of electrons in the reaction. In forming rust, for example, elemental iron atoms lose electrons to become iron ions.

$$Fe \rightarrow Fe^{3+} + 3e^-$$

Oxidation – The process of losing electrons. Such a substance is said to be oxidized.

Oxidized – The state of having lost electrons in a redox reaction.

All processes in which chemicals lose electrons, *even if oxygen is not involved,* are known as oxidations. An oxidation cannot happen on its own, because the electrons must have somewhere to go after they are lost from the reactants that are oxidized. That is, they must be gained by some other chemical species. When rust forms, for example, the electrons from the iron atoms are transferred to molecular oxygen to form oxide ions

$$O_2 + 4e^- \rightarrow 2O^{2-}$$

We say the chemical that accepts the electrons has been **reduced** or has undergone **reduction**. A reduction is any process in which *electrons are gained by a chemical.*

Reduced – The state of having gained electrons in a redox reaction.

One useful mnemonic to remember oxidation and reduction is **OIL RIG**:
Oxidation **I**s **L**oss of electrons; **R**eduction **I**s **G**ain of electrons

Some prefer the mnemonic **LEO** says **GER**:
Loss of **E**lectrons is **O**xidation; **G**ain of **E**lectrons is **R**eduction

No matter which mnemonic you prefer, the chemical point remains the same: First, oxidation and reduction processes must occur together in **oxidation–reduction reactions** (also called **redox reactions**). And second, the chemicals concerned must react in the proportions that allow the number of electrons lost via oxidation to equal the number of electrons gained via reduction.

Oxidation–Reduction reactions – Reactions that involve the transfer of electrons from one species to another. Also known as redox reactions.

Redox reactions – Same as oxidation-reduction reactions.

The two processes that we discussed above do not exist alone. Instead, each is one-half of the complete electron-exchange process that produces iron(III) oxide (Fe_2O_3) from iron and oxygen. Still, in solving these kinds of problems, we often find it convenient to separate the overall reaction into individual oxidation and reduction half-reactions. Let's examine these **half-reactions** more closely.

Half-reaction – An equation that describes the oxidation or reduction portion of a redox reaction.

The oxidation half-reaction is: $Fe \rightarrow Fe^{3+} + 3e^-$(3 electrons are lost)

The reduction half-reaction is: $O_2 + 4e^- \rightarrow 2O^{2-}$(4 electrons are gained)

How can we obtain a complete and balanced redox reaction from the iron and oxygen half-reactions? Can we apply that concept to all redox reactions? Unfortunately, balancing a redox reaction is not as straight-forward as the trial-and-error method used to balance other reactions. This is because in *all* redox reactions we need to make sure that the number of electrons involved is the same on both sides of the equation. The number of electrons lost by the iron must be equal to the number gained by the oxygen. Charge, like mass, is conserved. *That* is the unifying concept.

To balance the redox equation, we can multiply the entire oxidation half-reaction by 4 and the entire reduction half-reaction by 3. This results in the loss of 12 electrons (the least common multiple of 4 and 3) from iron and the gain of 12 electrons by oxygen.

$$4Fe \rightarrow 4Fe^{3+} + 12e^- \text{ (oxidation)}$$

$$3O_2 + 12e^- \rightarrow 6O^{2-} \text{ (reduction)}$$

Then, the overall oxidation–reduction equation is the sum of the two half-reactions:

$$4Fe \rightarrow 4Fe^{3+} + \cancel{12e^-}$$

$$\underline{3O_2 + \cancel{12e^-} \rightarrow 6O^{2-}}$$

$$\mathbf{4Fe + 3O_2 \rightarrow 4Fe^{3+} + 6O^{2-}}$$

When the electrons are transferred in this reaction, the resulting iron ions and oxide ions combine to form Fe_2O_3. Notice that since the same electrons appear as reactants and as products, they don't appear in the overall redox equation (similar to our spectator ions in ► Sect. 4.5). The electrons just transferred from the iron to the oxygen. In this reaction, because the electrons lost by iron caused oxygen to be reduced, we say that iron is the **reducing agent**. Similarly, because oxygen caused iron to be oxidized, we say that oxygen is the **oxidizing agent**. In general:

- The chemical that loses electrons is oxidized and is called the reducing agent.
- The chemical that gains electrons is reduced and is called the oxidizing agent.
- In the oxidation half-reaction, electrons appear as products.
- In the reduction half-reaction, electrons appear as reactants.

Oxidizing agent – The species in a redox reaction that is reduced and which causes the other species to be oxidized.

Reducing agent – The species in a redox reaction that is oxidized and which causes the other species to be reduced.

4.8.1 Oxidation Numbers

In business, bookkeepers are hired to keep track of the money coming into and paid out by the company. They keep the financial books. In chemistry, our electron bookkeeping tool is called an **oxidation number**, which we assign to individual atoms on the basis of where electrons in a bond are likely to be found. For example, in the ionic compound NaCl we say that the sodium ion has an oxidation number of + 1, because it has one less electron than the sodium atom. The chloride ion has an oxidation number of – 1, because it has one more electron than the chlorine atom. In the case of ionic compounds, these are the actual charges that exist on the ions themselves.

Oxidation number – A "bookkeeping" number that reflects the charge on an ion, or the charge that an atom in a compound would have if it were charged.

We use oxidation numbers to keep track of electrons as they move among atoms, molecules, and ions in redox reactions and, in fact, in all types of reactions. The term oxidation number is often used interchangeably with the term **oxidation state**.

Oxidation state – See oxidation number.

In the iron and oxygen reaction that produces iron(III) oxide,

$$4Fe \rightarrow 4Fe^{3+} + 12e^- \text{ (oxidation)}$$

$$3O_2 + 12e^- \rightarrow 6O^{2-} \text{ (reduction)}$$

Each iron started as a neutral atom (sometimes noted with a superscript "0", Fe^0) and was oxidized, losing three electrons, to form Fe^{3+}. The oxidation number (or oxidation state) of the iron is now + 3. The oxygen molecule includes two oxygen atoms that share electrons equally—that is, neither atom exerts a strong preference for the bonding electrons—so each oxygen atom is assigned (remember, we are bookkeepers here!) an oxidation state of 0. When oxygen reacts with iron, the electrons gained produce oxide ions (O^{2-}) that have an oxidation number of – 2.

When atoms are combined to make a molecule (a covalent compound), they neither lose nor gain electrons to form ions. Instead, the electrons in molecules are shared, although not necessarily equally, between the atoms. Overall, the molecule is electrically neutral, with a net charge of zero. However, in many molecules, there is a tendency for electrons to be closer to the nuclei of some atoms than they are to others. For example, in water (H_2O), there is a marked tendency for the electrons in each of the two hydrogen-to-oxygen bonds to be found closer to the oxygen nucleus than to the hydrogen nucleus (◘ Fig. 4.3). This behavior is indicated in our electron bookkeeping system by giving oxygen an oxidation state of – 2. The electron from each hydrogen atom is drawn away from its nucleus more of the time, and we denote this by giving each hydrogen atom an oxidation number of + 1. However, keep in mind that these are not actual charges in the molecule. This merely helps us keep track of where the electrons are most likely to come from in a molecule when a redox reaction occurs. Rules that aid in our bookkeeping are listed in order of importance in ◘ Table 4.4. In cases where the rules appear to contradict each other, follow the rule that comes first in the table.

What oxidation numbers would we assign to the carbon and hydrogen atoms of methane (CH_4)? By following the rules in ◘ Table 4.4, we determine that each hydrogen atom has a + 1 oxidation number. Because the mole-

Table 4.4 Rules for assigning oxidation numbers

Oxidation number	Examples
1. An atom in an element is always 0	Na, H_2, N_2, O_2, O_3, and He
2. A monatomic ion is the same as its charge	Li^+: + 1, Ca^{2+}: + 2, Cl^-: –1, O^{2-}: –2
3. A Group IA metal in a compound: + 1 A Group IIA metal in a compound: + 2 A Group IIIB metal or Al: + 3	Sodium in NaCl: + 1 Calcium in $CaBr_2$: + 2 Aluminum in Al_2O_3: + 3
4. Fluorine in a compound: –1	Fluorine in HF: –1
5. Hydrogen in its nonmetal compounds: + 1 Hydrogen in metallic compounds: –1	In HCl, H is + 1 In NaH, H is –1
6. Oxygen in compounds: typically –2 Exception: peroxides such as H_2O_2, –1	In CO_2 and CO, O is –2 In CH_3OOCH_3, O is –1
7. Binary compounds containing metals... Group VIIA elements: typically –1 Group VIA elements: –2 Group VA elements: –3	In NaCl, Cl is –1 In FeS, S is –2 In Mg_3N_2, N is –3
8. The sum of the oxidation numbers of all atoms in the ion or molecule must equal the total charge of the ion or molecule	In HClO, H is + 1, O is –2, so Cl must be + 1 In ZnO, O is –2, so Zn must be + 2 In NH_4^+ ion, H is + 1, so N must be –3

cule is neutral overall, we assign the oxidation number – 4 to this carbon atom. The carbon atom in carbon monoxide (CO), on the other hand, is assigned a different oxidation number. According to the rules, the oxygen atom has an oxidation number of – 2. The carbon atom must, then, have an oxidation number of + 2. The higher oxidation number means that the carbon atom in carbon monoxide is more oxidized than the carbon in methane. The oxidation number of the carbon atom in each compound bears this out.

It is important that we understand the difference between the oxidation numbers of atoms in molecules and the oxidation numbers of ions in ionic compounds. In molecules, assigning oxidation states is simply an accounting procedure used to keep track of electrons; it does *not* imply that electrons have really been lost or gained by any atoms. Remember that we have just assigned the oxidation numbers on the basis of some simple rules that hypothetically assume the electrons are transferred when the molecule is made. In ionic compounds, on the other hand, the oxidation number of an ion is the real charge of the ion.

Example 4.11—Assigning Oxidation States

Assign oxidation states to all the atoms in the following elements, compounds and ions:

a. He
b. C_6H_6
c. Al_2O_3
d. Zn
e. NO_3^-
f. SF_6
g. F_2
h. CO_3^{2-}

Asking the Right Questions

What's an oxidation number? What are the rules that govern our assignment of oxidation numbers?

Solution

a. This is an element, so the oxidation number is 0.
b. This is a neutral compound. Each hydrogen atom has an oxidation number + 1 *per atom,* and because there are the same number of carbon atoms as of hydrogen atoms, *each carbon atom* must have an oxidation number of – 1.
c. This is a neutral compound. Oxygen has an oxidation number of – 2, so in this case the aluminum must have an oxidation number of + 3, because there are three oxygen atoms (total = – 6), and the oxidation numbers must sum to zero overall.
d. This is an element, so the oxidation number is 0.
e. This is an ion with a – 1 charge overall, so the oxidation number of the atoms present must sum to – 1. Oxygen has an oxidation number of – 2, so the three oxygen atoms contribute a total of – 6. Nitrogen must therefore have the oxidation number of + 5.
f. This is a neutral compound. Fluorine has an oxidation number of – 1, and there are six fluorine atoms (total = – 6), so the sulfur must have an oxidation number of + 6.
g. This is an element, so each atom has an oxidation number of 0.

h. This is an ion with a − 2 charge overall, so the oxidation number of the atoms present must sum to − 2. Oxygen has an oxidation number of − 2, so the three oxygen atoms contribute a total of − 6. The carbon, therefore, must have an oxidation number of + 4.

Are our Answers Reasonable?

There is no "common sense" check for these answers. Recognizing them as correct comes with experience and practice. However, we can at least notice that the assignment of − 2 to oxygen, + 1 to hydrogen, and − 1 to halogens like fluorine matches what we expect.

What if…?

…we considered some unusual compounds like these? What is the oxidation state of each atom?

a. CaH_2 b. H_2O_2 c. Fe_3O_4

(Ans: a. Ca = +2, H = −1; b. H = +1, O = −1; c. O = −2, Fe = +2.67; Two of the iron atoms have a charge of + 3 *and one has a charge of* + 2, *for an average of* 2.67)

Practice 4.11

Use Table 4.4 to assign the oxidation numbers of each of the atoms in the compounds below.

a. KCl
b. Fe_2O_3
c. P_4
d. CH_2Cl_2
e. Al
f. PBr_3
g. HCN.

4.8.2 Identifying Oxidation–Reduction Reactions

Redox reactions can be difficult to identify without close examination. Let's examine a redox reaction to see how it is identified as such. The reaction of hydrogen and nitrogen to make ammonia, known as the Haber process, is a redox reaction that is vital to farming. The ammonia made by this process is used to fertilize fields for the production of most crops, including corn, wheat, and sorghum (a versatile crop used to make syrup, animal feed, and even brooms!). The reaction is

$$3H_2(g) + N_2(g) \rightarrow 2NH_3(g)$$

A first look at the reaction does not reveal that any electrons are being transferred in the process, so at first we might incorrectly assume that this is not a redox reaction. However, if we examine the oxidation numbers of the individual atoms involved from Table 4.4, we can see that reduction and oxidation are taking place.

H_2 Both hydrogen atoms are assigned the oxidation number 0.
N_2 Both nitrogen atoms are assigned the oxidation number 0.
NH_3 All hydrogen atoms are assigned the oxidation number + 1; nitrogen must then be assigned the oxidation number − 3.

Overall, the hydrogen changes its oxidation state from 0 (in H_2) to + 1 (in NH_3), losing electrons in this reaction as it is oxidized. Notice that the oxidation state of hydrogen increased, which is always what happens in an oxidation. The nitrogen changes its oxidation state from 0 (in N_2) to − 3 (in NH_3), gaining electrons as it is reduced. The oxidation number of nitrogen decreased (was reduced). Because the Haber process includes reduction and oxidation, it is a redox reaction.

Redox reactions are characterized by the exchange of electrons, which makes balancing these equations difficult to do by trial and error.

Example 4.12—Identifying Redox Reactions

Is the combustion of ethane a redox reaction? Prove your answer.

$$2C_2H_6(g) + 7O_2(g) \rightarrow 4CO_2(g) + 6H_2O(g)$$

Asking the Right Questions

What should we look for that is common to all redox reactions? The single common feature is the exchange of electrons. How do we know whether there is an electron exchange? The best way is to assign oxidation numbers to each atom and see whether there are changes going from reactants to products.

Solution

Judging on the basis of the oxidation number rules in Table 4.4, we can make the following assignments:

$2C_2H_6(g)$	+	$7O_2(g)$	→	$4CO_2(g)$	+	$6H_2O(g)$
−3 +1		0		−4 +2		+1 −2

What happens to the oxidation number of each atom?
Each carbon goes from − 3 to + 4.
Each oxygen goes from 0 to − 2.
Each hydrogen does not change oxidation number.

Because oxidation numbers have changed, electrons have been exchanged. Therefore, this is a redox reaction.

Are Our Answers Reasonable?
Note that the oxygen atoms in the product have an oxidation number of -2, and the carbon atoms are $+4$. Oxygen and carbon often form stable compounds, such as water and carbon dioxide, in which they have these oxidation numbers. This tells us that our assignment of oxidation numbers is reasonable, and because oxidation numbers have changed this is a redox reaction.

What If…?
…we considered the mixture of lead(II) nitrate with sodium iodide?
(*Ans: This is not a redox reaction; the oxidation numbers do not change.*)

Practice 4.12
Is the reaction of aqueous solutions of phosphoric acid and sodium hydroxide a redox reaction?

4.9 Fresh Water—Issues of Quantitative Chemistry

Nothing is more important to us than our supplies of fresh water, which we largely draw from rivers, lakes (▫ Fig. 4.17), and underground aquifers. Managing these precious freshwater resources requires our understanding of quantitative chemistry and aqueous solutions. If fresh water is to be of use to us as a supply of drinking water or for agricultural use, we must ensure that it is clean, meaning that certain dissolved solutes must occur only in quantities that are within the acceptable limits for good health. ▫ Table 4.5 lists, for selected chemicals, the EPA's maximum permitted level, below which there is minimal risk to human health (this is called the maximum contaminant level goal, or MCLG).

▫ **Fig. 4.17** The quantitative chemistry of fresh water is a vital consideration in ensuring the safety of our water supplies and the health of the environment. *Source* ▸ https://upload.wikimedia.org/wikipedia/commons/8/87/Mineral_Creek%2C_Colorado_intensive_sampling.jpg

Table 4.5 Maximum Contaminant Level Goal (MCLG) for selected substances in safe water

Substance	MCLG (mg/L = ppm)	Potential human organ damage due to exceeding MCLG	Source of substance
As	0.010	Skin and circulatory system	Runoff from orchards and glass-manufacturing plants
Cd	0.005	Kidney	Corrosion of pipes, discharge from used batteries
CN^-	0.200	Thyroid and nerves	Discharge from gold mining, fertilizer, and plastics manufacturing
Hg	0.002	Kidney	Discharge from refineries
Dioxin ($C_{12}H_4Cl_4O_2$)	0.00000003	Reproductive system	By-product of smelting, bleaching, and pesticide manufacture

Chemists in water collection and treatment facilities check for compliance with safe water standards by quantitative analysis of the water. According to those safe-water standards, they also administer appropriate quantities of disinfecting chemicals. One such disinfectant is chlorine, which is toxic to freshwater life above about 19 parts per billion. The good news is that chlorination of drinking water has been instrumental in reducing the risk of microbial disease transmitted through a water supply.

For instance, cholera (a bacterial infection of the intestines that causes vomiting, diarrhea, and dehydration) claims tens of thousands of lives each year and is particularly invasive in countries in which the population has been uprooted as a consequence of civil war and grinding poverty. Cholera outbreaks have been reported in the past few years by the World Health Organization in several African, Asian, and Middle Eastern countries (◻ Fig. 4.18). A long-term outbreak in Haiti has been eradicated as of 2020. In those countries that disinfect their water supply with chlorine, cholera has been all but eradicated. Unfortunately, wastewater treatment effluent (runoff from the cities back into lakes, streams, rivers, etc.) has resulted in chlorine concentrations between 1 and 5 parts per million. Any effect these levels of chlorine have on the environment has yet to be determined.

Why should gold miners, such as the ones that run the mines we discussed early in this chapter, be so careful with their cyanide leaching solutions? Because the MCLG for CN^- is so low (0.200 ppm = 200 ppb), any spill of leaching solution that contains this ion in a larger concentration could spell disaster. Unfortunately, this is exactly what happened in the winter of 2000. An accidental spill in Romania resulted in approximately 22 million gallons of cyanide waste from a gold mine flowing into the nearby Tisza River. This river flows into the Danube River and through Belgrade, Yugoslavia (Now in Serbia), on its way to the Black Sea. The initial cyanide spill killed most of the life in the Tisza River and harmed much of the life in the Danube River. To compound the disaster, the Tisza River is home to 17 of Hungary's 29 protected species of fish, including the last known species of Danube sturgeon. It will be decades before the environment recovers from this accident.

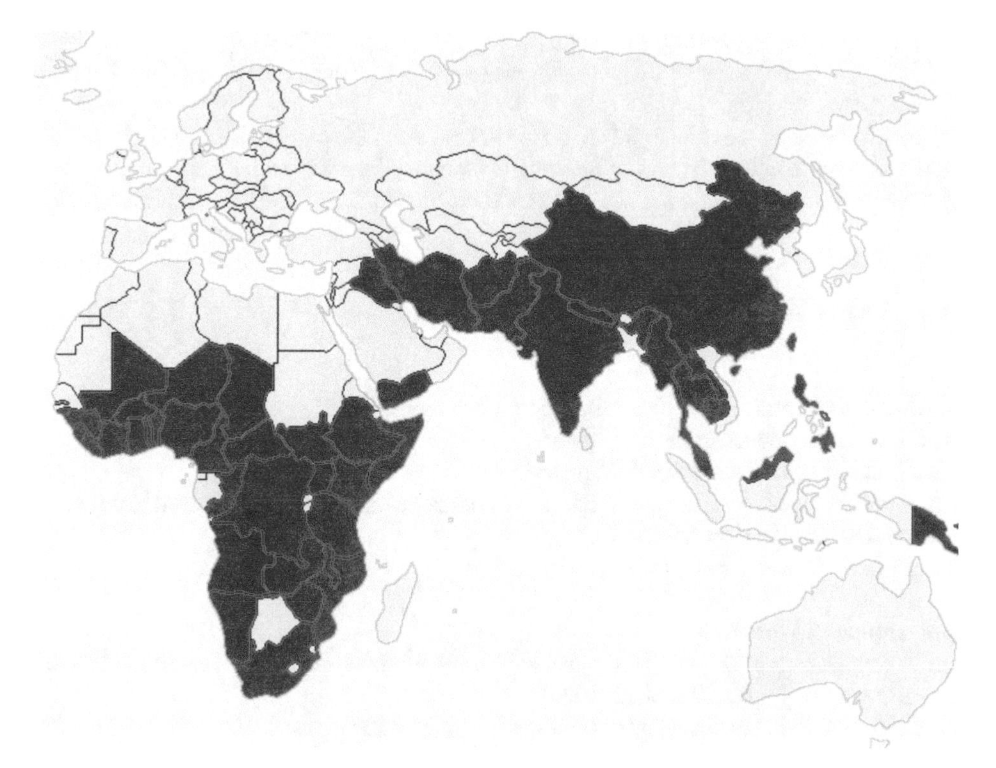

Fig. 4.18 Cholera outbreaks have been reported in many countries in Africa and Asia in the period 2010–2015

What must rivers and lakes contain in order to be healthful habitats for fresh-water life? They must contain sufficient oxygen, hold appropriate quantities of nutrients (too little or too much of these can adversely affect the water quality), and have an appropriate acid–base balance. If pollution problems are suspected in any river or lake, chemists and biologists must analyze the water and perhaps also the flesh and blood of fish, birds, and other organisms that live in or around the water. Chemical water analysis includes working with measures of concentration, including parts per million, parts per billion, and molarity, and investigating each specific type of reaction that occurs in the aqueous environment.

The Bottom Line

- Water is an extremely versatile solvent, partly thanks to its polarity, which is due to the uneven distribution of electrons in the molecule.
- When ionic compounds dissolve in water, the ions dissociate and become surrounded by water molecules—a process known as hydration.
- Strong and weak electrolytes are defined based on how much their ions dissociate in water.
- Molecular compounds dissolve in water but do not dissociate into ions and are classified as nonelectrolytes.

- The concentration of a solution can be expressed in units known as molarity, the number of moles of the solute divided by the volume of the solution in liters.
- Chemicals present at very low levels are often measured in terms of parts per million (ppm), parts per billion (ppb), or parts per trillion (ppt).
- Titration is an important chemical analysis method that can be used to determine concentrations of many different types of chemicals.
- Molecular equations show all chemicals involved in the reaction as molecular formulas.
- Complete ionic equations show each of the aqueous ions in a reaction.
- Net ionic equations show only the ions and other compounds that result in the reaction.
- Precipitation reactions involve an insoluble precipitate forming when soluble chemical species combine.
- We can judge whether an ionic compound is likely to dissolve in water using the Solubility Rules (▫ Table 4.2)
- Acid–base reactions can neutralize the acidic and basic character of each.
- Reduction–oxidation (redox) reactions are electron transfer processes in which some reactants lose electrons (are oxidized), while others gain electrons (are reduced).
- We can determine whether a reaction is a redox reaction by assigning oxidation numbers to the atoms involved (▫ Table 4.4) and seeing whether and how those oxidation numbers changed during the chemical reaction.

Focus Your Learning
Section 4.1 Water—A Most Versatile Solvent

Skill Review

1. Explain how water molecules can dissolve both cations and anions.
2. Why doesn't pure water conduct an electric current?
3. Explain what is meant by the term hydration sphere.
4. Diagram, using circles for atoms, a crystal of KCl versus the same crystal of KCl dissolved in water.
5. Explain what is meant by the term "electrolyte".
6. Is grape sugar (glucose) an electrolyte? Why or why not?

Chemical Applications and Practices

7. Earth's oceans contain tons of dissolved sodium chloride. Yet, when a ship develops an oil leak, almost none of the oil dissolves in the ocean. Explain this phenomenon.
8. When an ion dissolves, it is surrounded by a hydration sphere. If water molecules surrounded the ion so that the hydrogen portion of the water was closer to the ion, would the ion most likely be a cation or an anion?
9. Pure water does not conduct an electric current. However, aqueous solutions of some compounds conduct electricity. Explain why the presence of some solutes converts nonconducting water into a conducting solution.
10. Glycerin can be produced as a by-product in soap making. The compound dissolves so easily in water that it absorbs water from the air. This latter characteristic is why glycerin is often found in many skin lotions. As glycerin absorbs the water, the skin can be kept moist. Glycerin's structure is shown below. What aspects of glycerin's structure contribute most to its ease of dissolving in water?

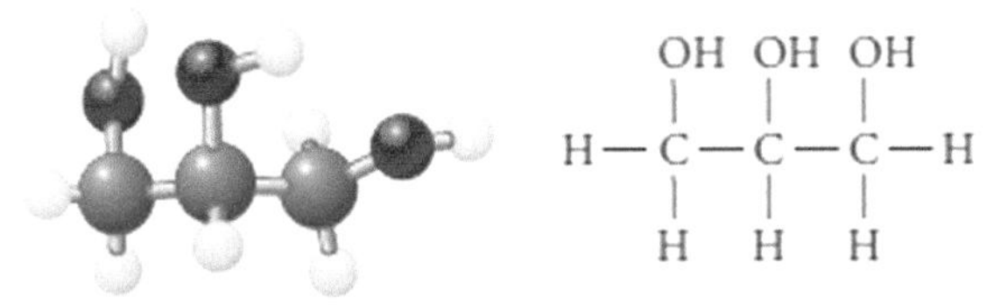

Glycerin

11. A conductivity-testing apparatus, such as the one shown in ▫ Fig. 4.8, possesses a light bulb whose brightness is related to how much current is flowing through it (and also through the solution.) A small, but measurable, amount of current must be present before the bulb becomes visibly brighter. What effect would this characteristic of the apparatus have on the classification of solutions containing strong electrolytes, containing weak electrolytes, and containing non-electrolytes?

12. When dissolved in water, which of the following would you expect to cause a conductivity tester, shown in the chapter, to produce a very bright light? (Assume that 0.50 mol of each is placed in 1.0 L of solution.)
    a. C_2H_5OH (ethanol)
    b. NaOH
    c. Na_2CO_3
    d. KCl
    e. $BaSO_4$

Section 4.2 The Concentration of Solutions

Skill Review

13. Which of the following would best represent $MgCl_2$ dissolved in water?

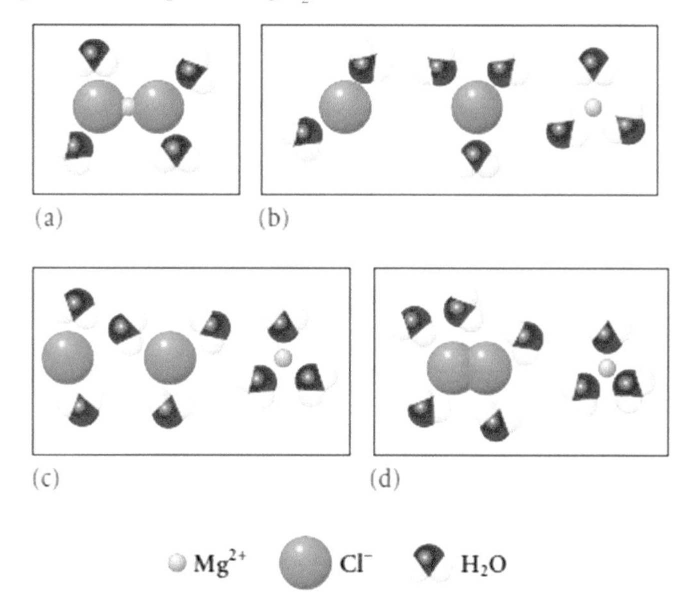

14. Which of the following would best represent H_2 dissolved in water?

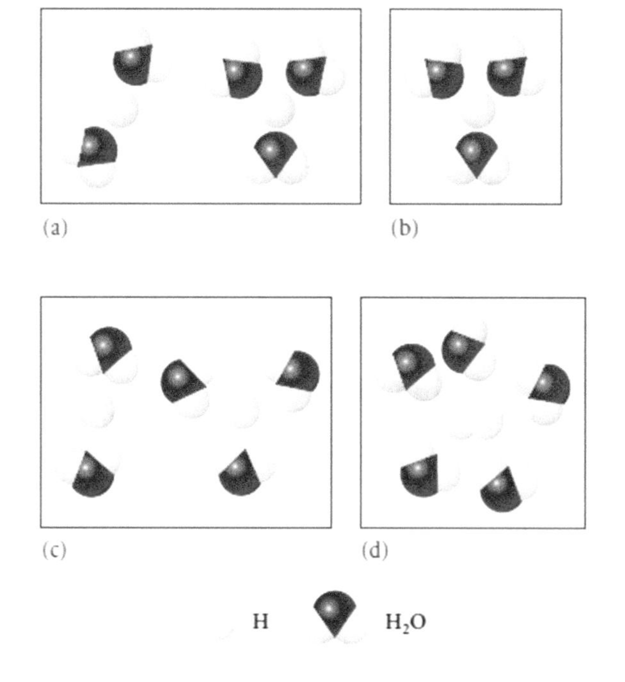

15. If 0.100 mol $MgCl_2$ is dissolved in water, how many total moles of ions are there in the resulting solution?
16. If 0.100 mol H_2 is dissolved in water, how many total moles of ions are there in the resulting solution?

17. The formula of vitamin C is $C_6H_8O_6$. Calculate the molarity of each of the following vitamin C solutions:
    a. 0.150 g of vitamin C dissolved in enough water to produce a 1.50 L solution
    b. 0.250 g of vitamin C dissolved in enough water to produce a 0.500 L solution
    c. 3.50 g of vitamin C dissolved in enough water to produce a 2.0 L solution
18. The formula of ethanol is C_2H_6O. Calculate the molarity of each of the following ethanol solutions:
    a. 0.150 g of ethanol dissolved in enough water to produce a 1.50 L solution
    b. 25 g of ethanol dissolved in enough water to produce a 0.500 L solution
    c. 100 g of ethanol dissolved in enough water to produce a 2.0 L solution

19. Determine the mass in grams of glycine ($C_2H_5NO_2$) in each of the following solutions:
    a. 100.0 mL of 0.015 *M* glycine
    b. 125.0 mL of 0.0145 *M* glycine
    c. 74.6 mL of 1.44 *M* glycine
20. Determine the volume in liters of solution needed to provide the following amounts of solute:
    a. 4.50 g ethanol (C_2H_6O) from a 2.50 *M* ethanol solution
    b. 63.7 g HCl from a 6.0 *M* HCl solution
    c. 3.0 g glucose ($C_6H_{12}O_6$) from a 0.150 *M* solution
21. Calculate the molarity of
    a. calcium ions in 1.00 L of solution containing 24.55 g of calcium chloride
    b. chloride ions in 1.00 L of solution containing 24.55 g of calcium chloride
    c. water in pure water (i.e. – water is both the solute and the solvent)
22. Calculate the molarity of
    a. HBr if 45.0 g is dissolved in 1.0 L
    b. NaOH if 32.5 g is dissolved in 500 mL
    c. chloride ions in a 5.0 L solution containing 2.00 g NaCl
23. One important source of bromine, in the form of $Br^-(aq)$, is seawater. If the molarity of Br^- in seawater is approximately 0.00081 M, how many liters of seawater would be required to obtain 1 mol of Br^-?
24. Chloride can be obtained from seawater. If the molarity of Cl^- in seawater is approximately 0.522 M, how many liters of seawater would be required to obtain 1 mol of Cl^-?
25. Determine the volume in mL of a 0.125 M solution of $MgCl_2$ needed to provide
    a. 0.10 mol $MgCl_2$
    b. 0.10 mol Mg^{2+}
    c. 2.33 mol Cl^-
26. Determine the volume in mL of a 0.0357 M solution of Na_2SO_4 needed to provide
    a. 0.10 mol Na_2SO_4
    b. 0.10 mol Na^+
    c. 4.30 mol SO_4^{2-}
27. Determine the concentration, in ppm, for each of the following solutions.
    a. 2.5 mg NaCl per 1000.0 L solution
    b. 5.25 mg Cu^{2+} ions per 500.0 mL solution
    c. 12.5 mg Cl^- ions per 300.0 mL solution
28. Determine the concentration, in ppm, for each of the following solutions.
    a. 1.20 mg KCl per 1000.0 mL solution
    b. 0.100 g Co^{2+} ions per 500.0 mL solution
    c. 1.00 micrograms CN^- ions per 450.0 mL solution
29. Perform the indicated conversions.
    a. 1.20×10^{-4} M NaCl to ppm
    b. 5.33 ppm CN^- to ppb
    c. 170 ppm NaOH to mass percent
30. Perform the indicated conversions for these aqueous solutions.
    a. 0.00250 M $CuCl_2$ to mass percent
    b. 19 ppm CN^- to M
    c. 0.0011% NaOH by mass to M

Chemical Applications and Practices

31. When growing certain bacteria, biologists must often control the acidity level of the growth medium. To do this, a variety of compounds, referred to as buffer systems, may be added in specific amounts. Which of the following has the greater molarity?
    a. 42.0 g of KH_2PO_4 dissolved in 250.0 mL of solution
    b. 21.0 g of KH_2PO_4 dissolved in 125.0 mL of solution
32. If a student seeking to prepare a 1.0 M solution of KOH added 56 g of solid KOH to exactly 1.0 L of water, would the solution be greater or less than 1.0 M? Explain the basis of your conclusion.
33. Adding antifreeze to aqueous automobile cooling systems lowers the freezing point of the resulting solution. The main ingredient in most antifreeze products is ethylene glycol ($C_2H_6O_2$). How many kilograms of ethylene glycol are dissolved in a 10.0-L solution that is 16.0 M?
34. Oxygen dissolved in water is critical for aquatic life. The level of dissolved oxygen is often used to report the quality of water in the environment. If 0.0090 g of O_2 is dissolved per liter of solution, what would be reported as (a) the molarity, and (b) the concentration in ppm, of dissolved oxygen?
35. Occasionally students enjoy a cup of coffee or tea while studying chemistry. The amount of caffeine in those beverages varies greatly. At a temperature of 65 °C, the maximum amount of caffeine ($C_8H_{10}N_4O_2$) that can be dissolved in water is 455 g/L. What would be the maximum molarity of caffeine in a 425-mL cup of 65 °C coffee?
36. The compound potassium permanganate forms intensely purple solutions that are used to react with other solutions that contain iron. How many grams of potassium permanganate ($KMnO_4$) would you need to dissolve into 0.500 L of solution if you wanted to prepare a 0.00100 M solution of $KMnO_4$?
37. The pesticide atrazine ($C_8H_{14}N_5Cl$) is slightly soluble in water. Health advisory warnings are issued if the atrazine concentration is higher than 3.0 parts per billion. How many grams of atrazine would you expect to find in 1 L of water that is 3.0 ppb in atrazine? What would be the molarity of the solution?
38. A can of non-diet soda may contain 45 g of sucrose ($C_{12}H_{22}O_{11}$) per 450 g of soda. What is the sugar concentration reported as ppm? Approximately how many grams of water would we have to add to the soda to reduce this sugar concentration to only 1 part per million?
39. A certain sports drink has a NaCl concentration of 0.20 M. If the container held 250 mL of the liquid, how many moles of Na^+ would we consume when swallowing 50.0 mL of the drink?
40. One method used to detect the presence of the toxic heavy metal barium is to precipitate the barium +2 ion as barium sulfate. If a 10.0-mL sample produced 0.565 g of barium sulfate (and assuming that no more Ba^{2+} remained dissolved) what was the original molarity of Ba^{2+} in the solution?
41. Describe how to prepare 432.0 mL of 0.7525 M potassium chromate solution starting with solid potassium chromate.
42. Describe how to prepare 2.50 L of 0.325 M calcium hydroxide solution starting with 5.00 M calcium hydroxide solution.
43. We've prepared a solution by diluting 325 mL of 0.2246 M aluminum sulfate solution with water to a final volume of 2.500 L. Calculate
    a. the number of moles of aluminum sulfate before dilution.
    b. the molarities of the aluminum sulfate, aluminum ions, and sulfate ions in the diluted solution.
44. For a solution of acetic acid (CH_3COOH) to be legally called "vinegar," it must consist of 5.00% acetic acid by mass. If vinegar is made up only of acetic acid and water, what is the molarity of acetic acid in the vinegar? The density of vinegar is 1.008 g/mL.

Section 4.3 Stoichiometric Analysis of Solutions

Skill Review

45. Assume 1:1 reaction stoichiometry in each of the following:
    a. 25.0 mL of HCl required 35.0 mL of 0.155 M NaOH to neutralize. What is the molarity of the HCl solution?
    b. 50.0 mL of $Sr(OH)_2$ required 35.0 mL of 0.0200 M sulfuric acid to neutralize. What is the molarity of the $Sr(OH)_2$ solution?
    c. 40.5 mL of nitric acid (HNO_3) required 25.0 mL of 0.35 M KOH to neutralize. What is the molarity of the HNO_3 solution?
46. Assume 1:1 reaction stoichiometry in each of the following:
    a. 33.0 mL of HCl required 17.6 mL of 2.50 M NaOH to neutralize. What is the molarity of the HCl solution?
    b. 10.25 mL of $Sr(OH)_2$ required 15.56 mL of 0.00150 M sulfuric acid to neutralize. What is the molarity of the $Sr(OH)_2$ solution?
    c. 100.0 mL of nitric acid (HNO_3) required 84.30 mL of 1.562×10^{-4} M KOH to neutralize. What is the molarity of the HNO_3 solution?

47. The label on a solution was partially obscured. It was either 0.10 M CuCl or 0.10 M $CuCl_2$. When the entire 25.0 mL of the solution was evaporated to dryness, 0.336 g of solid remained. What should the label read?
48. The label on a solution was partially obscured. It was either 0.10 M $FeCl_2$ or 0.10 M $FeCl_3$. If the solution was actually 0.10 M $FeCl_2$, how many grams of $FeCl_2$ would be obtained if 100.0 mL of the solution were evaporated to dryness?
49. First complete the following reaction using words rather than chemical symbols. Then write the molecular, total ionic, and net ionic equations that describe this aqueous reaction.
potassium hydroxide + hydrochloric acid →
50. List the spectator ions for the following aqueous reaction:
sulfuric acid + potassium hydroxide →
51. Write the molecular, total ionic, and net ionic equations for the following aqueous reaction:
sodium chloride + calcium nitrate → sodium nitrate + calcium chloride
52. Write the molecular, total ionic, and net ionic equations for the following aqueous reaction:
magnesium chloride + sodium bromide → sodium chloride + magnesium bromide

Chemical Applications and Practices

53. Using the stoichiometry found in this chapter for the vitamin C and I_2 reaction (▶ Sect. 4.3), determine which of the following two drinks contains the greater concentration of vitamin C. Which sample contains the greater amount of vitamin C?
    a. Brand A: 250.0 mL solution required 10.5 mL of 0.0855 M I_2
    b. Brand B: 300.0 mL solution required 12.0 mL of 0.0855 M I_2
54. Using the stoichiometry found in the chapter for this vitamin C and I_2 reaction (Sect. 4.3), determine which of the following two drinks contains the greater concentration of vitamin C. Which sample contains the greater amount of vitamin C?
    a. Brand A: 150.0 mL solution required 8.5 mL of 0.0650 M I_2
    b. Brand B: 100.0 mL solution required 12.5 mL of 0.0250 M I_2
55. One of the more common agents for standardizing solutions of sodium hydroxide (NaOH) is potassium acid phthalate ($KHC_8H_4O_4$), or, more accurately, potassium hydrogen phthalate, which is a monoprotic acid. The compound is often abbreviated KHP, where the P designates the phthalate ion ($C_8H_4O_4^{2-}$), not the phosphorus atom. From the following data, determine the average molarity of a NaOH solution that was titrated with KHP.
Trial A: 45.12 mL of NaOH solution neutralized 0.5467 g of KHP
Trial B: 44.89 mL of NaOH solution neutralized 0.5475 g of KHP
Trial C: 46.50 mL of NaOH solution neutralized 0.5501 g of KHP
56. One of the more common agents for standardizing solutions of potassium hydroxide (KOH) is potassium acid phthalate ($KHC_8H_4O_4$), or, more accurately, potassium hydrogen phthalate, which is a monoprotic acid. From the following data, determine the average molarity of a KOH solution that was titrated with KHP.
Trial A: 45.12 mL of KOH solution neutralized 0.2573 g of KHP
Trial B: 48.89 mL of KOH solution neutralized 0.2250 g of KHP
Trial C: 45.50 mL of KOH solution neutralized 0.2502 g of KHP
57. Hydrochloric acid has many industrial uses. The steel industry uses hydrochloric acid in a process known as "pickling steel." This is done to remove impurities on the surface of steel prior to galvanizing, in which a coating of zinc is added to the steel surface to prevent corrosion. To analyze the concentration of the "pickle liquor," a titration with sodium hydroxide may be used. Write the net ionic reaction between hydrochloric acid and sodium hydroxide, and then determine the concentration of the HCl in the pickling liquor if a 10.00 mL sample of the hydrochloric acid required 45.55 mL of a 0.9876 M solution of NaOH to completely react.
58. Nitrogen for increased yields on farms can come from a variety of sources. One common nitrogen-containing fertilizer is ammonium sulfate, $(NH_4)_2SO_4$. The reaction between ammonia (NH_3) and sulfuric acid (H_2SO_4) produces this important fertilizer.
    a. Balance the reaction.
    b. How many liters of a 1.55 M solution of sulfuric acid would be needed to completely react with 1.00 kg of ammonia?
59. One method that field geologists use to test for the presence of carbonates is to drip hydrochloric acid on a sample and note the formation of carbon dioxide gas bubbles.
    a. Balance the reaction between calcium carbonate ($CaCO_3$) and hydrochloric acid (HCl).
    b. If 5.00 mL of 0.500 M HCl completely reacted with calcium carbonate in a rock sample, how many grams of calcium carbonate did the sample contain?

4

60. Hydrogen peroxide, in very dilute concentrations, can be used as a disinfectant. The concentration of hydrogen peroxide (H_2O_2) can be determined in a titration experiment using potassium permanganate ($KMnO_4$) as shown in the following balanced equation:

$$2MnO_4^-(aq) + 5H_2O_2(aq) + 6H^+(aq) \rightarrow 5O_2(g) + 2Mn^{2+}(aq) + 8H_2O(l)$$

If a 25.0 mL sample of hydrogen peroxide required 25.2 mL of 0.353 M $KMnO_4$ solution to react with all the hydrogen peroxide, what was the molarity of the hydrogen peroxide solution?

61. If a volume of 25.45 mL HCl solution is used to completely neutralize 1.050 g Na_2CO_3 according to this equation, what is the molarity of the HCl solution?

$$Na_2CO_3(aq) + 2HCl(aq) \rightarrow 2NaCl(aq) + CO_2(g) + H_2O(l)$$

62. A student is given 0.750 g of an unknown acid, which can be one of these: oxalic acid, $H_2C_2O_4$, or citric acid, $H_3C_6H_5O_7$. To determine which acid she has, she titrates the unknown acid with 1.050 M NaOH. The acid is completely neutralized when 11.15 mL of NaOH solution are added. What is the unknown acid? Hint: oxalic acid will release two H^+ ions to react with the NaOH, while citric acid will release three H^+ ions. Write the balanced equations to help answer the question.

Section 4.4 Types of Chemical Reactions

Skill Review

63. For the following reactions, balance the molecular equation, and then write the total ionic and net ionic equations:
    Molecular: $Li_2CO_3(aq) + PbCl_2(aq) \rightarrow LiCl(aq) + PbCO_3(s)$
    Molecular: $Co(NO_3)_2(aq) + Na_3PO_4(aq) \rightarrow Co_3(PO_4)_2(s) + NaNO_3(aq)$
    Molecular: $Ba(NO_3)_2(aq) + Na_3PO_4(aq) \rightarrow Ba_3(PO_4)_2(s) + NaNO_3(aq)$
    Molecular: $NaI(aq) + Fe(NO_3)_3(aq) \rightarrow FeI_3(aq) + NaNO_3(aq)$
    Molecular: $NaOH(aq) + Fe(NO_3)_3(aq) \rightarrow Fe(OH)_3(s) + NaNO_3(aq)$
64. For the following reactions, balance the molecular equation, then write the total ionic and net ionic equations:
    Molecular: $Na_3PO_4(aq) + CaCl_2(aq) \rightarrow Ca_3(PO_4)_2(s) + NaCl(aq)$
    Molecular: $Fe(NO_3)_3(aq) + Na_3PO_4(aq) \rightarrow FePO_4(s) + NaNO_3(aq)$
    Molecular: $Ni(NO_3)_2(aq) + Na_3PO_4(aq) \rightarrow Ni_3(PO_4)_2(s) + NaNO_3(aq)$
    Molecular: $NaOH(aq) + Co(NO_3)_2(aq) \rightarrow Co(OH)_2(s) + NaNO_3(aq)$
    Molecular: $Cu(NO_3)_2(aq) + Na_3PO_4(aq) \rightarrow Cu_3(PO_4)_2(s) + NaNO_3(aq)$
65. Balance the following equations, and then write the net ionic equation for each one:
    a. $(NH_4)_2SO_4(aq) + Ba(NO_3)_2(aq) \rightarrow BaSO_4(s) + NH_4NO_3(aq)$
    b. $AgOH(s) + HCl(aq) \rightarrow AgCl(s) + H_2O(l)$
    c. $CaCO_3(s) + HCl(aq) \rightarrow CaCl_2(aq) + H_2O(l) + CO_2(g)$
    d. $CH_3CO_2H(aq) + Co(OH)_2(s) \rightarrow Co(CH_3CO_2)_2(aq) + H_2O(l)$
66. Balance the following equations, and then write the net ionic equation for each one:
    a. $Zn(s) + HNO_3(aq) \rightarrow H_2(g) + Zn(NO_3)_2(aq)$
    b. $Sr(OH)_2(s) + HCl(aq) \rightarrow SrCl_2(aq) + H_2O(l)$
    c. $HCl(aq) + CaCO_3(s) \rightarrow CaCl_2(aq) + H_2O(l) + CO_2(g)$
    d. $Na_2S(aq) + FeCl_3(aq) \rightarrow NaCl(aq) + Fe_2S_3(s)$

Section 4.5 Precipitation Reactions

Skill Review

67. According to the solubility rules (Table 4.2), which of the following salts would qualify as soluble in water?
    a. copper(II) carbonate
    b. nickel(II) sulfide
    c. $(NH_4)_2CO_3$
    d. KOH
    e. Lead(II) acetate
68. According to the solubility rules (Table 4.2), which of the following salts would qualify as soluble in water?
    a. sodium nitrate
    b. $Ba(OH)_2$
    c. magnesium sulfate
    d. $PbBr_2$
    e. AgI

69. Predict the products and write the net ionic equations for each of the following:
    a. $BaCl_2(aq) + NaNO_3(aq) \rightarrow$
    b. $Fe(NO_3)_3(aq) + (NH_4)_2SO_4(aq) \rightarrow$
    c. $CaCl_2(aq) + K_2SO_4(aq) \rightarrow$
70. Predict the products and write the net ionic equations for each of the following:
    a. $MgCl_2(aq) + KNO_3(aq) \rightarrow$
    b. $AgNO_3(aq) + NH_4Cl(aq) \rightarrow$
    c. $CaCl_2(aq) + NaOH(aq) \rightarrow$

4

71. Write out the formulas of each of the following reactants. Then predict the result of mixing the aqueous solutions for each situation. Finally, report the net ionic equation for each.
    a. copper(II) nitrate + potassium hydroxide →
    b. sodium carbonate + aluminum chloride →
    c. ammonium phosphate + zinc chloride →
72. Write out the formulas of each of the following reactants. Then predict the result of mixing the aqueous solutions for each situation. Finally, report the net ionic equation for each.
    a. barium nitrate + sodium hydroxide →
    b. ammonium carbonate + aluminum bromide →
    c. silver nitrate + zinc bromide →
73. Balance the equation for the following precipitation reaction, and then write the net ionic equation. Indicate the state of each species (*s, l, aq,* or *g*).

$$FeCl_2 + NaOH \rightarrow Fe(OH)_2 + NaCl$$

74. Predict the products of each precipitation reaction. Balance the completed equation, and then write the net ionic equation.
    a. $CoCl_2(aq) + (NH_4)_2S(aq) \rightarrow ?$
    b. $Mn(NO_3)_2(aq) + Li_3PO_4(aq) \rightarrow ?$
75. Name two soluble ionic compounds that could be used to produce each of the following insoluble salts:
    a. BaS b. $Cu(OH)_2$ c. $PbSO_4$
76. Name two soluble ionic compounds that could be used to produce each of the following insoluble salts:
    a. $AgBr$ b. $Fe(OH)_2$ c. PbS

Chemical Applications and Practices

77. A water sample is ready to be analyzed. We know the sample contains Cu^{2+}, Ba^{2+}, and Ag^{+} ions. Using Table 4.2, suggest a sequence for adding other aqueous electrolyte solutions that could separate each of these by precipitation.
78. The presence of metal ions in aqueous systems can often cause environmental complications. Suppose it was necessary to remove Cu^{2+} ions from a sample of water. Suggest at least two reagents that could be added to the aqueous system to remove the copper ions.
79. A traditional method for the analysis of dissolved ions is called gravimetric analysis. This name is derived from the process in which the ion to be analyzed is precipitated and filtered and then quantified by measuring its mass. Suppose that all of the silver ions (Ag^{+}) in a solution, present as $AgNO_3(aq)$, reacted with 25.0 mL of 0.242 M NaCl to form solid AgCl.
    a. Write the molecular and net ionic equations for the precipitate formation.
    b. Determine the grams of precipitate formed and the grams of silver present in the original solution.
80. One method of preparing the compound AgBr, used in photographic films, is to precipitate it from a solution of KBr. After first writing and balancing the molecular and net ionic equations, calculate how many grams of AgBr precipitate when 100.0 mL of 2.00 M KBr is mixed with 100.0 mL of 1.00 M $AgNO_3$.

Section 4.6 Acid–Base Reactions

Skill Review

81. Define an acid.
82. Define a base.
83. What is meant by the terms "strong" or "weak" when used to describe a base?
84. What is meant by the terms "strong" or "weak" when used to describe an acid?

85. Looking again at the reaction of iron ions in water (from Fig. 4.17), some of the iron ions will combine with water molecules, like this:

$$Fe^{3+}(aq) + 3H_2O(l) \rightarrow Fe(OH)_3(s) + 3H^+(aq)$$

In this case, are the iron ions acting like an acid or a base? Explain.

86. Ammonia is an active ingredient in many types of household cleaners. When ammonia is dissolved in water, this reaction occurs:

$$NH_3(aq) + H_2O(l) \rightleftharpoons NH_4^+(aq) + OH^-(aq)$$

Therefore, ammonia is a(n) ____________________ (acid or base?) Explain.

87. Complete and balance the following acid–base equations. Name the reactants and products. Identify the acid and the base.
    a. $CH_3CO_2H(aq) + Ca(OH)_2(s) \rightarrow$
    b. $HClO_3(aq) + NH_3(aq) \rightarrow$

88. Complete and balance the following acid–base equations. Name the reactants and products. Identify the acid and the base.
    a. $H_3PO_4(aq) + NaOH(aq) \rightarrow$
    b. $H_2C_2O_4(aq) + Mg(OH)_2(s) \rightarrow$
    ($H_2C_2O_4$ is oxalic acid, and will release two H^+ ions to react with the magnesium hydroxide.)

89. What is the net ionic reaction of the following acid–base reaction? Provide the balanced molecular equation in order to begin this problem.
    hydrogen bromide + magnesium hydroxide → water + magnesium bromide

90. What acid–base reaction would produce each of the following salts? Write the balanced chemical equation in each case.
    a. NaCl   b. K_2CO_3 c. Na_2SO_4 d. $Al(NO_3)_3$

91. a. What would be the molarity of a KOH solution if, during a titration with 0.50 M HCl, 34.5 mL of the HCl neutralized 22.4 mL of the KOH solution?
    b. How much of a 0.50 M solution of H_2SO_4 would be needed to neutralize the same amount of the KOH solution?

92. If a student mixes 25.0 mL of 0.255 M H_2SO_4 with 50.0 mL of 0.115 M KOH (assume the volumes are additive), which reagent will be in excess? What will be the concentration of the excess reagent?

Chemical Applications and Practices

93. Oxalic acid can be found in rhubarb plants. If a 0.255-g sample of purified oxalic acid required 25.7 mL of NaOH to neutralize, what would we report as the molarity of the NaOH solution? (Oxalic acid: $H_2C_2O_4$)

94. Phosphoric acid is a very versatile acid with widespread uses, from making fertilizer to soft drink ingredients.
    a. Balance the reaction between phosphoric acid (H_3PO_4) and ammonium hydroxide (NH_4OH).
    b. How many milliliters of 0.459 *M* ammonium hydroxide would be needed to neutralize 33.5 mL of 0.100 *M* phosphoric acid?

95. Carbonic acid is formed when gaseous carbon dioxide is pumped into soft drinks to establish their carbonation. Suppose you are employed at the new "ChemCola" beverage company. You must titrate the carbonic acid (H_2CO_3) in the soft drink using 0.1445 M NaOH.
    a. Write the molecular and net ionic equations.
    b. What is the molarity of carbonic acid if a 50.00 mL sample required 38.98 mL of 0.1445 M NaOH?

96. Acetic acid is the ingredient in vinegar that gives it its vinegary taste and smell. If a particular brand claims to be 5.00%, by mass, acetic acid ($HC_2H_3O_2$), how many milliliters of 0.255 M NaOH would be required to neutralize the acetic acid in a 25.0 mL vinegar sample?

97. Stomach acid (HCl) is neutralized by a variety of commercial antacids. For each of the following, determine how many grams of active antacid ingredient would be necessary to neutralize the HCl in 50.00 mL of 0.0100 M HCl (an approximation of the concentration of HCl in the stomach).
    a. $Al(OH)_3$ b. $Mg(OH)_2$ c. $CaCO_3$

98. Sulfuric acid is the acid ingredient in automobile battery acid solutions. The sulfuric acid content of such a solution can be determined through a lab analysis by reacting it in a titration with potassium hydroxide. The unbalanced reaction is

$$H_2SO_4(aq) + KOH(aq) \rightarrow K_2SO_4(aq) + H_2O(l)$$

Use the balanced reaction to fill in the missing data in the following table.

Molarity of H_2SO_4 (M)	Volume of H_2SO_4 (mL)	Molarity of KOH (M)	Volume KOH (mL)
0.25	75.0		28.6
	28.9	0.36	35.8
0.88		1.11	27.5
1.76	22.0		49.7

4

Section 4.7 Oxidation–Reduction Reactions

Skill Review

99. What does the term "oxidation" mean?
100. What does the term "reduction" mean?
101. In a compound made up only of each of the following pairs of elements, select the atom that is more likely to carry a negative or slightly negative charge.
   a. C, H
   b. O, F
   c. Na, O
   d. P, Fe
   e. Ca, O
102. In a compound made up only of each of the following pairs of elements, select the atom that is more likely to carry a negative or slightly negative charge.
   a. K, H
   b. Li, F
   c. Na, S
   d. N, Ca
   e. N, Mg
103. Assign oxidation numbers to all the elements in each of the following molecules or ions:
   a. N_2O_5
   b. PO_4^{3-}
   c. $CuCO_3$
   d. N_2
   e. H_2SO_3
104. Assign oxidation numbers to all the elements in each of the following molecules or ions:
   a. CH_4
   b. SO_4
   c. $KHCO_3$
   d. $Na_2Cr_2O_7$
   e. $KMnO_4$
105. Identify the oxidation number of each element in the following ions or compounds.
   a. IO_3^-
   b. HSO_3^-
   c. F^-
   d. MgH_2
   e. H_4AlO_4
   f. $C_2O_4^{2-}$
106. Identify the oxidation number of each element in the following ions or compounds.
   a. UO^{2+}
   b. $H_2SiO_4^-$
   c. PF_5^-
   d. N_2O_5
   e. XeO_4^{2-}
   f. $POCl_2$
107. In the following reactions, identify which species are oxidized, which are reduced and which are neither oxidized nor reduced:
   $3Fe(s) + 2O_2(g) \rightarrow Fe_3O_4(s)$
   $NaCl(s) \rightarrow Na^+(aq) + Cl^-(aq)$
   $CO(g) + H_2O(g) \rightarrow CO_2(g) + H_2(g)$
108. In the following reactions, identify which species are oxidized, which are reduced and which are neither oxidized nor reduced:
   a. $AgNO_3(aq) + NaI(aq) \rightarrow AgI(s) + NaNO_3(aq)$
   b. $C_3H_8(g) + 5O_2(g) \rightarrow 3CO_2(g) + 4H_2O(g)$
   c. $Ni(s) + Cl_2(g) \rightarrow NiCl_2(s)$
109. Which reactant is oxidized and which is reduced in the following reactions? Identify the oxidizing agent and the reducing agent.
   a. $C_3H_8(g) + 5O_2(g) \rightarrow 3CO_2(g) + 4H_2O(g)$
   b. $Si(s) + 2Br_2(l) \rightarrow SiBr_4(l)$

110. Which reactant is oxidized and which is reduced in the following reactions? Identify the oxidizing agent and the reducing agent.
  a. $FeS(s) + 3NO_3^-(aq) + 4H_3O^+(aq) \rightarrow 3NO(g) + SO_4^{2-}(aq) + Fe^{3+}(aq) + 6H_2O(l)$
  b. $3CrO_4^{2-}(aq) + 3Sn^{2+}(aq) + 24H^+(aq) \rightarrow 3Cr^{4+}(aq) + 3Sn^{4+}(aq) + 12H_2O(l)$

Chemical Applications and Practices

111. The following balanced equation depicts a reaction that can be used for the determination of iron in a steel sample. Is this a redox reaction? Prove it by showing which elements change oxidation state.

$$6Fe^{2+}(aq) + 14H^+(aq) + Cr_2O_7^{2-}(aq) \rightarrow 6Fe^{3+}(aq) + 2Cr^{3+}(aq) + 7H_2O(l)$$

112. The following reaction depicts one of the steps in obtaining the important steel alloying ingredient titanium.

$$TiCl_4 + 2Mg \rightarrow Ti + 2MgCl_2$$

  a. Identify the substance being oxidized.
  b. Identify the substance being reduced.

113. Ammonia is one of the most important industrial chemicals produced today. It is used to create a variety of products, including fertilizers and nitric acid. The Haber–Bosch process is used to create ammonia and has two major steps. First, hydrogen is produced from methane, and then hydrogen is mixed with nitrogen to produce ammonia:
  a. $CH_4(g) + H_2O(g) \rightarrow 3H_2(g) + CO(g)$
  b. $3H_2(g) + N_2(g) \rightarrow 2NH_3(g)$

  Is either of these steps an oxidation-reduction reaction? If so, identify which elements are oxidized, which are reduced and which are neither oxidized nor reduced.

114. Worldwide, the Ostwald process is used to produce over 50 million tons of nitric acid per year. In this process, ammonia is combined with oxygen to produce nitrogen dioxide, which is then used to make nitric acid. Here are the two steps of the process:
  a. $4NH_3(g) + 7O_2(g) \rightarrow 4NO_2(g) + 6H_2O(g)$
  b. $4NO_2(g) + O_2(g) + 2H_2O(g) \rightarrow 4HNO_3(aq)$

  Is either of these steps an oxidation-reduction reaction? If so, identify which species are oxidized, which are reduced and which are neither oxidized nor reduced.

115. The engine of most automobiles produces energy by the combustion of petroleum fuels. Balance the reaction below for the combustion of octane, which is a common component of gasoline:

$$C_8H_{18}(l) + O_2(g) \rightarrow CO_2(g) + H_2O(g)$$

  In this reaction, what is being oxidized, and what is being reduced?

116. Ethanol can be used as a fuel for internal combustion engines. Balance the reaction for the combustion of ethanol:

$$C_2H_5OH(l) + O_2(g) \rightarrow CO_2(g) + H_2O(g)$$

  In this reaction, what is being oxidized and what is being reduced?

Comprehensive Problems

117. Why is water called the universal solvent?
118. An interesting demonstration often used by chemistry teachers utilizes several of the principles that you have been reading about in this chapter. First, a solution of barium hydroxide is shown to conduct electricity by having the electrodes from a conductivity tester immersed in the solution and observing that a light begins to shine. Then a solution of sulfuric acid is slowly added. A white precipitate begins to form, and the light bulb dims. Eventually, the addition of the sulfuric acid causes the bulb to go out. Finally, continued addition of the sulfuric acid solution brings the bulb back on to bright light.
  a. What is the identity of the white precipitate?
  b. What is the net ionic equation for the reaction between barium hydroxide and sulfuric acid?
  c. What is the significance of the point at which the bulb goes out completely?
  d. Explain why continued addition of the sulfuric acid causes the light to come back on.
119. Some chemical reactions are best done in solution. If you had 100.0 mL of 0.230 M NaOH, how many milliliters of 0.530 M HCl would you need to have the same number of moles as found in the NaOH solution?

120. Extreme ozone pollution is described as any concentration greater than 0.28 ppm ozone, O_3. If the density of a sample of air containing that concentration of ozone were 1.30 g/L, what would be the molarity of ozone in the sample? How many molecules of ozone would be in 1 L of the air?
121. Using Table 4.2, determine the identity and formula of any and all precipitates that are likely to form when a solution containing NaOH and $(NH_4)_2CO_3$ is mixed with a solution that contains $CuNO_3$ and $BaCl_2$.
122. a. An unlabeled solution may contain either Ag^+ ions or Al^{3+} ions. Using Table 4.2, determine a suitable anion solution that, through precipitation, could be added to identify the cation present in the solution.
    b. Another unlabeled solution contains either nitrate ions or sulfate ions. Using Table 4.2, determine a solution that contains a cation that could be used in a precipitation to determine the identity of the anion in the unlabeled solution.

123. Barium-containing "milkshakes" are often used to obtain X-rays of patients suffering from intestinal problems. Barium compounds can also be toxic. Barium sulfate is insoluble in water, so it can be given to patients without concern that it would be absorbed. It is also opaque to X-rays. Write the molecular and ionic balanced equations for the formation of $BaSO_4(s)$ from $Ba(NO_3)_2(aq)$ and $Na_2SO_4(aq)$. How many grams could be obtained from mixing 125.0 mL of 0.567 M $Na_2SO_4(aq)$ and 75.0 mL of 0.786 M $Ba(NO_3)_2(aq)$?
124. The reaction of gaseous dinitrogen trioxide with water can provide aqueous nitrous acid as the product.
    a. Write the balanced chemical reaction.
    b. If 12.5 grams of dinitrogen trioxide treated with excess water produced 12.5 grams of nitrous acid, what is the percent yield of the reaction?
    c. To make the nitrous acid as described in part b, the dinitrogen trioxide was bubbled into 1.55 gallons of water. Assuming that the volume of water remains constant during the reaction, what is the concentration of nitrous acid after the reaction?
125. A chef wishes to make a very lightly sweetened tea by dissolving 1.00 g fructose ($C_6H_{12}O_6$) in 12 fluid ounces of tea.
    a. What is the concentration of fructose in the tea in molarity?
    b. What is the concentration of fructose in ppm?
    c. How many carbon atoms are there in 1.00 g of fructose?
    d. How many water molecules are in the serving of tea?
126. Nitrogen dioxide is a brown gas that is commonly found in urban pollution. It is not unusual to have tens of parts per trillion of this poisonous gas in the air over a large city like Los Angeles. Nitrogen dioxide will react with itself to form dinitrogen tetroxide.
    a. If you were to collect a 2.0 liter sample of polluted air containing 10 parts per trillion of nitrogen dioxide, how many grams of this gas would be in your sample? How many moles? How many molecules?
    b. Write a balanced equation for the conversion of nitrogen dioxide into dinitrogen tetroxide.
    c. If 3.5×10^{-9} gram of nitrogen dioxide was entirely converted into dinitrogen tetroxide, how many moles of dinitrogen tetroxide would be formed?
127. Balance these equations, and then classify each as precipitation, acid–base, or gas-forming.
    a. $Pb(NO_3)_2(aq) + NaCl(aq) \rightarrow PbCl_2(s) + NaNO_3(aq)$
    b. $Na_2CO_3(aq) + Co(NO_3)_2(aq) \rightarrow CoCO_3(s) + NaNO_3(aq)$
    c. $CaCO_3(s) + HCl(aq) \rightarrow CaCl_2(aq) + H_2O(l) + CO_2(g)$
128. Balance these equations, and then classify each as precipitation, acid–base, or gas-forming.
    a. $HNO_3(aq) + CuCO_3(s) \rightarrow Cu(NO_3)_2(aq) + H_2O(l) + CO_2(g)$
    b. $Li_3PO_4(aq) + Cu(NO_3)_2(aq) \rightarrow Cu_3(PO_4)_2(s) + LiNO_3(aq)$
    c. $Ba(OH)_2(aq) + HNO_3(aq) \rightarrow Ba(NO_3)_2(aq) + H_2O(l)$

Thinking Beyond the Calculation

129. Predict the products of the following incomplete reactions. After balancing the reaction, label each as a precipitation, acid–base, or redox reaction. For those reactions where it is possible, write the net ionic equation.
    a. $C_4H_{10}(g) + O_2(g) \rightarrow$
    b. $Ca(OH)_2(aq) + HNO_3(aq) \rightarrow$
    c. $Pb(NO_3)_2(aq) + NaCl(aq) \rightarrow$
130. Predict the products of the following incomplete reactions. After balancing the reaction, label each as a precipitation, acid–base, or redox reaction. For those reactions where it is possible, write the net ionic equation:
    a. $HCl(g) + Ca(OH)_2(aq) \rightarrow$
    b. $CH_4(g) + O_2(g) \rightarrow$
    c. $Ba(ClO_4)_2(aq) + Na_2S(aq) \rightarrow$

131. The oxalate ion ($C_2O_4^{2-}$) can be found in a variety of plants. This compound is considered toxic. Therefore, it is often important to determine the quantity in a sample. This can be done through a redox titration.

$$5C_2O_4^{2-}(aq) + 16H^+(aq) + 2MnO_4^-(aq) \rightarrow 2Mn^{2+}(aq) + 10CO_2(g) + 8H_2O(l)$$

a. Write equations that illustrate the dissolution and dissociation of oxalic acid in water.
b. Which of the species in the balanced redox reaction above contains carbon that is more oxidized, oxalate ($C_2O_4^{2-}$) or carbon dioxide?
c. Which of the species in the balanced redox reaction is more oxidized, permanganate or manganese ions?
d. How many electrons are being transferred among the reactants in the balanced redox reaction?
e. Determine which species is oxidized, and which is reduced, in the balanced redox reaction.
f. If a properly prepared plant sample required 33.5 mL of a 0.00976 M $KMnO_4$ solution to react, calculate the number of grams of oxalate ion present.

132. Aluminum is one of the most versatile metals in the world. However, it is not found in nature as a pure metal, and must be produced starting with aluminum ore, called "bauxite." Bauxite is mostly aluminum oxide, Al_2O_3, along with some impurities. In the first step of this process, bauxite is reacted with sodium hydroxide:

$$Al_2O_3(s) + 2NaOH(aq) \rightarrow 2NaAlO_2(s) + H_2O(l)$$

a. What is the oxidation state of aluminum in bauxite? What is its oxidation state in $NaAlO_2(s)$?
b. Based on the types of reactions you encountered in this chapter, how would you classify this type of reaction?
In the next step, cryolite (Na_3AlF_6) is produced:

$$NaAlO_2(aq) + 6HF(aq) + Na_2CO_3(aq) \rightarrow Na_3AlF_6(aq) + 3H_2O(l) + CO_2(g)$$

c. How would you classify this type of reaction?
d. What kind of chemical is HF? Write the chemical reaction you would expect if you dissolved HF(*g*) into water to make HF(*aq*).
In the final step of aluminum production, aluminum oxide is dissolved along with cryolite and a huge amount of electricity is passed through the mixture using carbon electrodes. The resulting reaction is:

$$2Al_2O_3(aq) + 3C(s) \rightarrow 4Al(s) + 3CO_2(g)$$

e. How would you classify this type of reaction?
f. Even though the cryolite (Na_3AlF_6) is in the mixture, it doesn't appear in the overall chemical reaction for producing aluminum metal, shown above. What do you think the purpose of the cryolite is in this process?

Energy

Contents

Supplementary Information The online version contains supplementary material available at ▶ https://doi.org/10.1007/978-3-030-90267-4_5.

M. Mosher and P. Kelter, *An Introduction to Chemistry*,
https://doi.org/10.1007/978-3-030-90267-4_5

What Will We Learn from this Chapter?

- What is energy? What are the different forms of energy and how are they transferred in a chemical reaction? (▶ Sect. 5.2)
- What is work and how is it calculated? (▶ Sect. 5.3)
- What are the units of heat and how is the joule related to the calorie? (▶ Sect. 5.4)
- Why do some objects get hotter than others when they are exposed to the same amount of heat? (▶ Sect. 5.5)
- Can we determine the amount of heat that a chemical reaction produces or absorbs? (▶ Sect. 5.6)
- Is there more than one way to calculate the heat of a reaction? (▶ Sect. 5.7)
- How do we choose the best source of energy for us as citizens? Does it matter which energy source we choose and what are the implications of our choice? (▶ Sect. 5.8)

5.1 Introduction

A rocket launch is an awe-inspiring demonstration of the ability of energy to transport large masses of material away from Earth's surface with an eye toward planetary exploration. The energy to launch the craft comes from the explosive violence of chemical reactions. When these reactions are used in a carefully controlled way, they can lift the European Space Agency's (ESA) Ariane 5 rocket (which can weigh as much as 780,000 kg) and propel the craft into orbit hundreds of miles above Earth in a very brief 10-min ride. The Ariane 5 is the current evolution of space launch vehicles produced by the ESA. It can deliver payloads weighing up to 10,200 kg into a variety of orbits hundreds or thousands of kilometers above the Earth. Between 1996 and 2020, the Ariane 5 was extremely successful, with a total of 109 launches and only two failures. It has delivered a large number of scientific and communication satellites into orbits around the Earth, and also enabled the SMART-1 (the first European Lunar Probe) to reach Earth's moon.

Two chemical reactions, indicated in ◘ Fig. 5.1, power the launch of the Ariane 5. The combustion of hydrogen gas (the combination of hydrogen and oxygen to form water) takes place in the main stage engine. At the same time, the solid rocket boosters attached to the sides host reactions involving the oxidation of aluminum by ammonium perchlorate. The major reactions are illustrated by the equations here.

$$\text{Mainstage}: 2H_2(g) + O_2(g) \rightarrow 2H_2O(g)$$
$$\text{Boosters}: 10Al(s) + 6NH_4ClO_4(s) \rightarrow 5Al_2O_3(s) + 6HCl(g) + 3N_2(g) + 9H_2O(g)$$

As the rocket sits on the launch pad, it is hard to believe that the energy that will launch it with such a spectacular display of chemical muscle is quietly present within the main fuel tank and the solid fuel of the booster rockets. Chemicals can store huge amounts of energy in this way—and then release it with very dramatic effects as soon as a chemical reaction begins.

These thoughts raise many questions about energy: What is energy? The rocket fuels used to lift rockets like the Ariane 5 contain stored energy and release it during takeoff. How is energy stored within molecules and compounds? Can we calculate how much energy is released from a molecule or compound during a rocket launch so that enough is harnessed to allow the rocket to climb to exactly the intended orbit? These questions can be answered by focusing our attention on a branch of science known as **thermodynamics**. Thermodynamics is concerned with the interconversion of different forms of energy. The specific area of thermodynamics most relevant to chemical reactions is known as **thermochemistry**, the study of energy exchanges in chemical processes.

Thermodynamics – The study of energy changes and exchanges.

Thermochemistry – The study of energy exchange in chemical processes.

◘ Fig. 5.1 The reactions that provide the energy to lift the space shuttle into orbit.

In order to properly describe energy exchanges, we must clearly state where they occur. We define the **system** as that collection of matter in which we are interested. In our discussion, the rocket and its payload make up the system. Everything else—the rest of the universe—is defined as the **surroundings**. Energy that is lost by the rocket's engines in our system is gained by the surroundings. The system and the surroundings combine to make up the **universe**. These terms are fundamental to our discussion of energy in chemistry.

System – Any set of chemicals whose energy change we are interested in.

Surroundings – Everything in contact with a chemical system.

Universe – The space consisting of both the system and the surroundings.

5.2 The Concept of Energy

The simplest commonly used definition of energy is: "the capacity to do work or produce heat." In order to understand this definition completely, we will examine the concepts of heat and work. Both of these concepts refer to ways of transferring energy.

Work is done whenever any force is used to move an object some distance. For example, suppose we use a chemical reaction to propel a heavy weight, such as the Ariane 5, upward. We are using the chemical reaction to provide the force that makes the rocket move. The chemicals that are combined in the reaction are doing work on the rocket, so they are releasing energy. Why does it take so much energy to lift the Ariane 5? You might say, "Because the rocket is very heavy," but that is only the beginning of the answer. The rocket is heavy because of the force of gravitational attraction between the particles in this system and the rest of planet Earth. To launch the rocket, we must make it move against the force of gravity, one of the three known fundamental forces of nature listed in ◘ Fig. 5.2. Since the force of gravitational attraction is so large between the Ariane 5 and Earth, a lot of energy must be supplied to lift it into orbit. The result is that a tremendous amount of work is done on the rocket.

Gravitational force – The fundamental force that causes all objects with mass to be attracted to one another.

Electromagnetic force – Also known as the electroweak force. The fundamental force responsible for the attraction between objects carrying opposite electric charges, for the repulsion between objects carrying the same electric charge, and for some transformations within subatomic particles.

Strong nuclear force – The fundamental force that holds protons and neutrons together in atomic nuclei.

Work (w) – Force acting on an object over a distance.

Work is one of the ways that energy can be transferred, but what is energy itself? We can describe **energy** as the capacity to move something against a fundamental force.

Gravitational force causes all objects with mass to be attracted to one another.

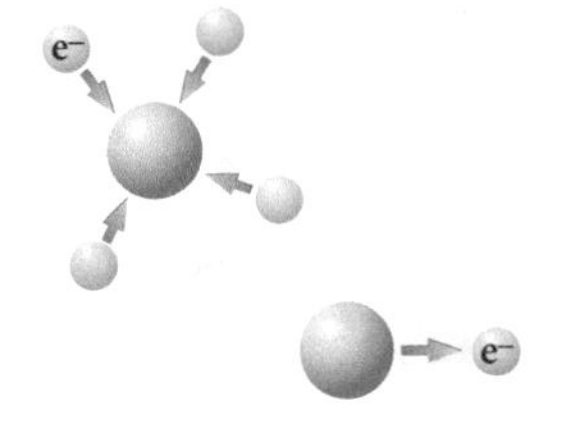

Electroweak force is responsible for the attraction between objects with opposite electric charges and the repulsion between objects with the same electric charge (when it is known as the *electric force*), the phenomena of magnetism and light, and some transformations within subatomic particles.

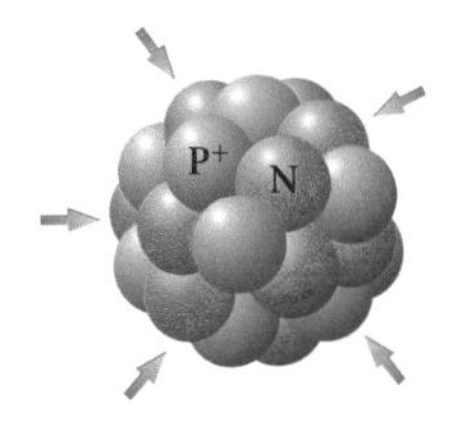

Strong nuclear force binds protons and neutrons together within atomic nuclei.

◘ **Fig. 5.2** The three known fundamental forces of nature

Fig. 5.3 Energy is required to lift an apple from the ground. The opposing force is gravity

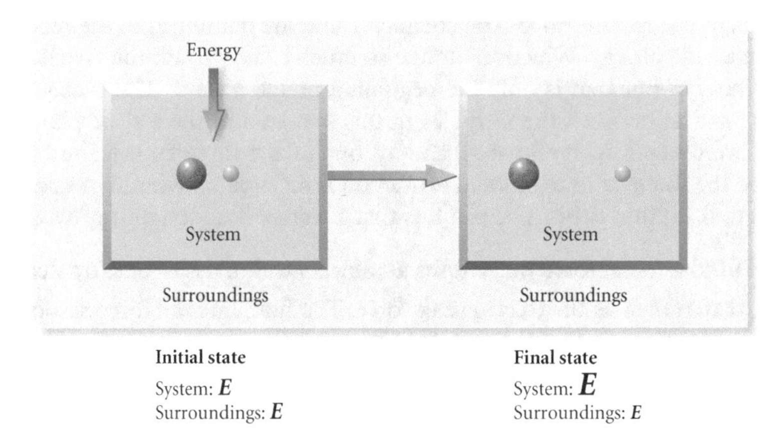

Fig. 5.4 Energy exchange between the system and the surroundings during the ionization of an atom. We must add energy to the system in order to move the electron away from the nucleus of the atom. The energy must come from the surroundings to accomplish this task

Energy – The capacity to move something against a fundamental force.

This idea can help us develop a good understanding of what energy is all about. In terms of the system and the surroundings, we can make the following statement:

- Energy is absorbed by the system from the surroundings in order to oppose a natural attraction.

This means, for example, that energy must be added to two magnets that are stuck together (a system) in order to pull them apart, because we are opposing their natural magnetic attraction. Energy must be added to an apple (a system) to lift it away from the ground because we are opposing the natural attraction due to the gravitational force (Fig. 5.3). Similarly, when we pull an electron away from a nucleus in an atom (another system), energy must be added to the system because opposite charges attract, and we are opposing that natural attraction. Figure 5.4 illustrates the energy exchange between the system and surroundings in this case. Because energy is needed to oppose a natural attraction, the reverse process allows us to draw the following opposite, perhaps counterintuitive, conclusion:

- When a natural attraction induces some kind of motion, energy is released from the system.
- Energy is released from the system to the surroundings when a natural attraction occurs.

This happens when, for example, magnets are made to stick together, an apple falls from a tree to the ground, or an electron is attracted to an atom.

5.2.1 Kinetic Energy Versus Potential Energy

Energy can be relegated to two basic forms: kinetic energy and potential energy. In addition, energy can also be propagated through space in the form of **electromagnetic radiation**, such as light, which we will discuss in detail in ▶ Chap. 6.

What is **kinetic energy**? It is the energy of movement. Everything that is moving has an amount of kinetic energy that depends on its mass and velocity, as described by the equation.

$$\textbf{Kinetic energy} = \frac{1}{2}mv^2$$

where $m = mass$ in kilograms, $v =$ velocity in meters per second.

Electromagnetic radiation – Energy that propagates through space. Examples include visible light, X-rays, and radio waves.

Kinetic energy – The energy things possess as a result of their motion.

For example, consider a 36,000-kg truck moving at the *same speed* as an 800-kg car. The truck has more kinetic energy than the car because its mass is greater. A truck moving faster than an identical truck has more kinetic energy than the slower truck because its velocity (v) is greater. The relationship of kinetic energy to mass and velocity is true for a car, a truck, a rocket—or a molecule.

If movement is associated with energy, then everything that is moving must, in appropriate circumstances, be able to do work. You can convince yourself of this by thinking about the steel ball shown in ◘ Fig. 5.5. If the ball is moving toward the slope, its energy of movement (kinetic energy) allows it to rise a certain distance up the slope. The height to which it rises depends on how fast it was originally moving. Kinetic energy is associated with the work of raising the ball, and the amount of kinetic energy that the rolling ball possesses will determine how high it can rise in defiance of the gravitational force.

The second fundamental form of energy is **potential energy**, which is the energy that an object possesses because of its position. Potential energy is stored energy, and when it is released, it can be transformed into kinetic energy. For example, if we lift an object against the force of gravity, it gains an amount of potential energy because of its position relative to the gravitational attraction to Earth's center. Recall our key idea that energy is exerted by, or transferred to, the system in order to oppose a natural attraction. The farther we pull our object away from Earth's center of gravity, the more energy is required, and the greater will be the object's stored potential energy.

Potential energy – The energy things possess as a result of their positions, such as their position in a gravitational or electromagnetic field.

As our steel ball in ◘ Fig. 5.5 rolls uphill, it begins to slow down. In fact, if we could ignore the loss of energy due to friction, the steel ball would steadily gain potential energy in an amount that is equal to the kinetic energy it loses. By the time the ball becomes stationary at the top of its climb, all the kinetic energy due to the rolling of the ball has been converted into potential energy. What happens next? The potential energy stored in the steel ball due to its relatively high altitude is released as the ball begins to run downhill. The amount of potential energy that is converted to kinetic energy, and therefore the velocity of the ball, increases. At the bottom of the hill, this potential energy has been completely converted into kinetic energy, and the ball is traveling at its maximum velocity.

This example of the interconversion between kinetic and potential energy illustrates one of the most significant fundamental laws of nature. It is known as the **Law of Conservation of Energy**.

Law of Conservation of Energy – Energy is neither created nor destroyed. It can only be transferred from place to place and transformed from one form into another.

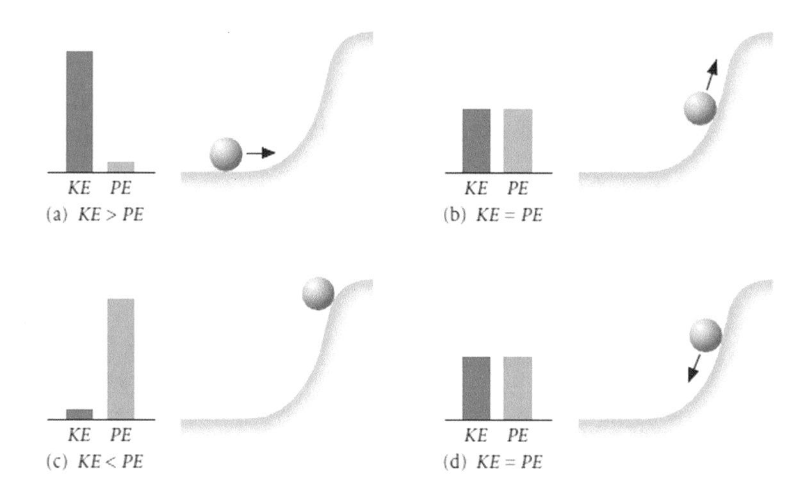

Fig. 5.5 A steel ball moving toward a hill demonstrates that energy can do work and that kinetic energy and potential energy can be interconverted. As the ball moves up the slope, its kinetic energy is converted into potential energy. When it reaches its maximum height and begins to fall back, that potential energy is converted back into kinetic energy. Work is done when the mass of the ball rises up the slope

The energy contained in molecules and compounds, such as those that power rockets, is a constantly interconverting mixture of kinetic and potential energy. The particles in the system are moving, bouncing off one another, vibrating, and rotating as the bonds between atoms stretch in and out, atoms and larger groups rotate around bonds, and entire molecules cartwheel through space. All of this motion is associated with a corresponding amount of kinetic energy. The available potential energy in chemical compounds is stored in the positions and arrangements of the electrons and nuclei that make up those compounds. For instance, as we discussed previously, it takes energy to move an electron away from an atom's nucleus, against the pull of the electric force, in just the same way as it takes energy to lift a weight up from the ground against the pull of the gravitational force.

It also requires an influx of energy to a system of atoms in order to force two or more (like-charged) electrons closer together against the repulsion due to the electric force, or to force two nuclei together against that force. The electrons and nuclei of chemicals have potential energy as a consequence of their positions in the electric force field, just as objects we lift and throw around have potential energy due to their positions in the gravitational force field.

Any set of compounds, at a given temperature and pressure, contains **chemical energy** stored as a result of the motions and positions of their atomic nuclei and electrons.

Chemical energy – Energy that is stored in a substance as a result of the motions and positions of its atomic nuclei and their electrons.

How do chemical reactions absorb or release energy? When a chemical reaction occurs in a system, new chemicals are formed with different internal motions, different electron arrangements, and a different total energy content. Any excess energy must be released to the surroundings, usually as heat and possibly as light and/or sound. Any energy acquired by the new chemicals must come from the surroundings. For example, when hydrogen and oxygen molecules rearrange to form water in the main engine of the Ariane 5, a tremendous amount of chemical energy is released. This release of energy occurs because the structure of water molecules contains much less energy than that of the oxygen and hydrogen molecules used to form the water. A small fraction of the released energy is converted into light and sound. The majority of the released energy dramatically increases the speed of motion of the particles involved, causing them to undergo a massive and explosive expansion as all of the very fast-moving particles bounce off one another. The chemical cyclone of particles within the expanding gas pounds against the inner surfaces of the engines, pushing the rocket upward, as shown in Fig. 5.6. Because this energy is released through a chemical reaction, it is often referred to as "**chemical energy,**" still a mixture of potential and kinetic energies.

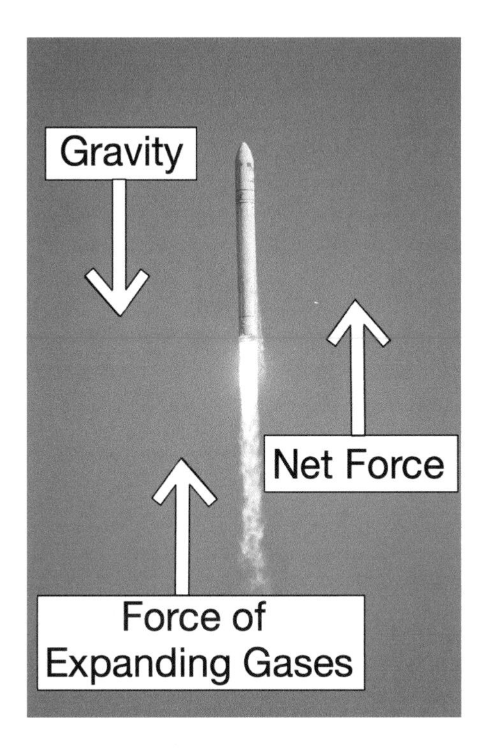

Fig. 5.6 Forces acting on a rocket engine. *Source* ▸ https://upload.wikimedia.org/wikipedia/commons/0/0d/Antares_Rocket_Test_Launch.jpg

Example 5.1—Energy in Various Forms

In our day-to-day discussions, we speak of many different kinds of energy, such as that from wind, ocean waves, or fossil fuels. Can you explain how these forms of energy (wind, waves, and fossil fuels) are related to the forms of energy discussed above?

Asking the Right Questions

Is some kind of motion associated with these forms of energy? What are the molecules in each material doing?

Solution

- Wind energy is the energy associated with the movement of air, so it is a form of kinetic energy.
- Similarly, wave energy is the energy associated with the movement of water, so it is a form of kinetic energy.
- Fossil fuel energy is the energy that we release from the fossil fuels—coal, oil and natural gas—when we burn them. The energy is stored within the chemicals but cannot be used until a chemical reaction like burning occurs. Fossil fuels contain potential energy that is released as kinetic energy when they burn.

Are Our Answers Reasonable?

In the first two examples (wind and wave energy), no chemical reaction is taking place in order for us to use the energy. Even though air and water are both collections of chemicals, we are only using the energy of their motion. Fossil fuels release kinetic energy as heat when they burn.

What If…?

…we considered energy from solar steam power ?
(*Ans: Solar energy is used as heat to boil water and produce steam. The steam is used to turn a turbine and produce electricity. This is an example of kinetic energy.*)

Practice 5.1

Explain how the "activities" of plants are related to the fundamental forms of energy. Include in your discussion the interaction of plants with the environment and their ultimate fate as food.

NanoWorld/MacroWorld—Big Effects of the Very Small: Electrons Capturing Energy in Photosynthesis

The trapping of the energy of sunlight by green plants is one of the most significant chemical processes on Earth. What makes it even more interesting is that the process is achieved by the smallest chemical change. That tiny change, powered by the absorption of energy from the Sun, is the transfer of an electron from one molecule to another. This change begins the process that provides energy to the plant and allows it to grow.

Chlorophyll *a*

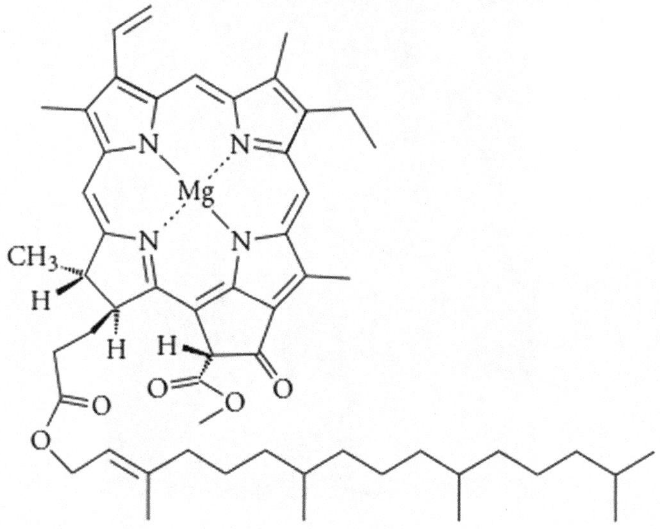

Fig. 5.7 Chlorophyll a is a complex naturally occurring molecule. The transfer of an electron from chlorophyll, powered by the energy of sunlight, is the key event of photosynthesis

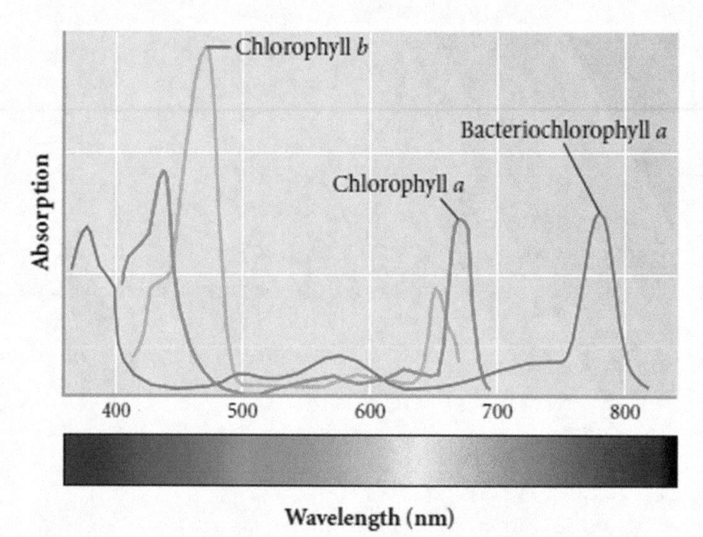

Fig. 5.8 The absorption spectra of a series of chlorophylls

Why are plants green? They contain molecules of chlorophyll, which absorb light in the red and blue parts of the visible spectrum (the range of visible colors) more than in the green parts, allowing the green light to reach our eyes. There are actually several different types of molecules known as chlorophyll. Chlorophyll *a* is shown in Fig. 5.7.

The absorption spectra of some chlorophylls, meaning the extent to which they absorb light of different wavelengths, is shown in Fig. 5.8. Because of differences in structure, every atom and molecule has its own absorption fingerprint, which we will discuss more fully in ▶ Chapter 6 (Quantum Chemistry).

When a molecule absorbs light energy, it usually raises one or more of the electrons into a higher energy configuration. When this happens to the chlorophyll in a plant, an excited higher-energy electron can be transferred to an adjacent molecule. This electron transfer requires an input of energy because the negatively charged electron moves away from a positively charged atomic nucleus within chlorophyll, against the pull of the electromagnetic force. The transferred electron carries much of this energy to the receptor molecule.

$$6\ O{=}C{=}O + 6\ H{-}O{-}H \longrightarrow C_6H_{12}O_6 + 6\ O{=}O$$

This is the tiny nanoscale event that powers photosynthesis. The transferred electron then moves among a series of other molecules involved in photosynthesis. The complex series of steps that each of the molecules goes through allows some of the energy originally absorbed from the Sun to be stored within other compounds. Ultimately, the energy from the Sun converts carbon dioxide and water into carbohydrate and oxygen—the overall energy-requiring process of photosynthesis.

The path the energy takes in completing this task is long. The chemical processes are very complex, and the involvement of huge numbers of electrons is necessary to generate significant amounts of carbohydrate and oxygen. But the crucial event begins with a single electron.

5.2.2 Heat and Reaction Profile Diagrams

How does energy flow between chemical systems and their surroundings? During a chemical change, energy is exchanged between the system and the surroundings. Much of this energy exchange is as heat. **Heat (*q*)** is the energy that is exchanged between a system and its surroundings because of a difference in temperature between the two. How does the temperature play a role? The temperature of a system is directly proportional to the kinetic energy of its particles, and the energy flow as heat between a system and its surroundings is a transfer of kinetic energy due to the collisions between particles. It is important to note that "temperature" and "heat" are not the same. Heat is a measurement of energy *transfer*, while temperature is an indicator of the average *kinetic energy content*.

Heat (*q*) – The energy that is exchanged between a system and its surroundings because of a difference in the temperature between the two.

We can portray the change in the **internal energy** content of chemicals during a reaction using **reaction profile diagrams** like those shown in ◘ Fig. 5.9. These diagrams illustrate the changes in the energy of the components of a reaction as it progresses. The horizontal axis in the diagram is known as the reaction coordinate, an arbitrary scale that denotes the progress of the reaction from start (left-hand side) to finish (right-hand side). The vertical axis represents the relative energy of the components of the system. By comparing the energy of the starting materials with the products, we can determine the direction of the flow of energy in the reaction, either from the system to the surroundings or from the surroundings to the system.

Internal energy (*U*) – The energy of a system defined as the sum of the kinetic and potential energies.

Reaction profile diagram – a graph that illustrates the energy of a reaction versus the progress of the reaction.

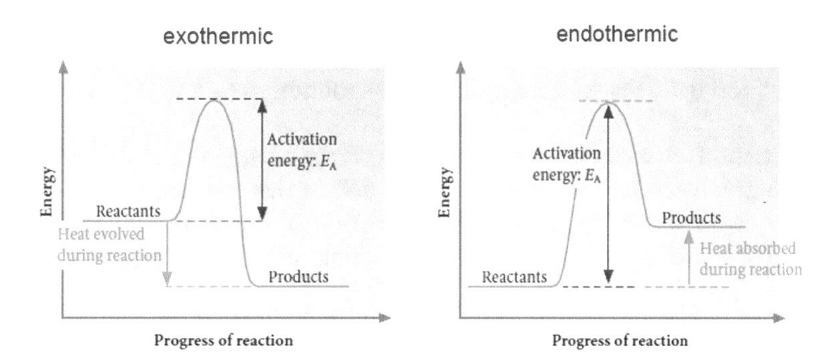

◘ **Fig. 5.9** Reaction profile diagrams. The exothermic reaction on the left involves a loss of heat. The endothermic reaction on the right requires the absorption of heat. Heat evolved is energy transferred from the system to the surroundings. Heat absorbed is energy transferred from the surroundings to the system

5

Fig. 5.10 An exothermic process. Energy is released by the burning match, which increases the temperature of the surroundings nearby, as indicated by the higher temperature near the flame. *Source* ▶ https://upload.wikimedia.org/wikipedia/commons/a/a2/License_Free_Photo_Creative_Commons_by_gnuckx_%2811644888384%29.jpg

From the diagrams in Fig. 5.9, we note that the curve of the reaction rises to a peak as the reaction proceeds. This indicates an input of energy needed to drive the reaction toward products. Virtually all reactions begin with an input of energy, known as the **activation energy**. This energy may be supplied when the overall kinetic energy of the moving particles is converted into internal potential and kinetic energy. The chemicals, after colliding, lose energy as they settle into the new chemical arrangement of the products.

Activation energy – The energy that is required to initiate a chemical reaction.

What is an exothermic reaction? A chemical reaction that releases energy as heat to the surroundings, like that shown in Fig. 5.10, is known as an **exothermic reaction**. Energy flows, in the form of heat, out of the system and into the surroundings, as illustrated by the graph on the left side of Fig. 5.9. A forest fire includes all manner of exothermic reactions, releasing torrents of energy as heat to the surroundings. Cellular respiration, a complex process, is also exothermic. Note that in an exothermic reaction, we can think of heat as a product of the reaction.

Exothermic reaction – A reaction that releases energy into the surroundings ($q < 0$).

What is an endothermic reaction? A chemical reaction that absorbs energy as heat is known as an **endothermic reaction**, as illustrated by the graph on the right side of Fig. 5.9. In this process, energy as heat flows into the system from the surroundings, as shown in the melting of ice cream (Fig. 5.11). In an endothermic reaction, such as the formation of oxygen and hydrogen from water, we can think of heat as a reactant.

Endothermic reaction – A reaction that absorbs energy from the surroundings ($q > 0$).

Many ammonium salts, such as ammonium nitrate, dissolve endothermically in water, absorbing energy as heat from the surroundings. Chemical "cold packs" that are used to cool the injured knee or elbow of an athlete use this concept.

Fig. 5.11 An endothermic process. When ice-cream melts, it absorbs energy from the surroundings, which lowers the temperature of the air nearby. *Source* ▶ https://upload.wikimedia.org/wikipedia/commons/4/4b/Ice_Cream_Dessert_%28Unsplash%29.jpg

+ Heat → +

$$2H_2O(l) + \text{heat} \rightarrow 2H_2(g) + O_2(g)$$

▶ Here's What We Know So Far

- There are three natural forces: the gravitational, electroweak, and strong nuclear forces.
- Potential energy is the energy stored within a system.
- Kinetic energy is the energy associated with motion.
- Chemical energy is a combination of potential and kinetic energy in a chemical system, due to the positions and motions of the atoms and molecules.
- The Law of Conservation of Energy states that energy can be neither created nor destroyed. Instead, energy just moves between the system and the surroundings and can be transformed from one form to another.
- Work and heat are two ways in which energy can move between the system and the surroundings.
- An exothermic process is the flow of energy as heat from the system to the surroundings.
- An endothermic process is the flow of energy as heat from the surroundings to the system.
- Changes in the energy of a reaction can be studied by examining a reaction profile diagram. ◀

5.3 A Closer Look at Work

How do we measure work? A system doesn't contain work. Mathematically, we can define work as the product of the applied force and the distance the object was moved.

$$\text{Work}\,(w) = \text{force}\,(F) \times \text{distance}\,(d)$$

To relate this equation to a chemical system, as is done by the rocket scientist, the force exerted by an expanding gas is directly related to the product of the pressure of a system and the area in which it is expanding. Pressure, as we'll discover in the chapter about the behavior of gases (▶ Chap. 11), is the force exerted per unit area by gas molecules on the walls of the container due to the physical collisions between the molecules and the walls.

$$\text{Force}\,(F) = \text{pressure}\,(P) \times \text{area}\,(A)$$

Substituting this equation into that for the calculation of the amount of work, we get:

$$\text{Work}\,(w) = P \times A \times d$$

This equation can be simplified by considering the product of the area (A) multiplied by the height (h), where height (h) is synonymous with distance (d) because they are both a measure of distance. As shown in Fig. 5.12, the base of the cylinder in the initial figure has a known area (A). Moreover, the cylinder has a known height (h), the distance from the bottom to the top of the cylinder. Therefore, the gas inside the cylinder has a known volume

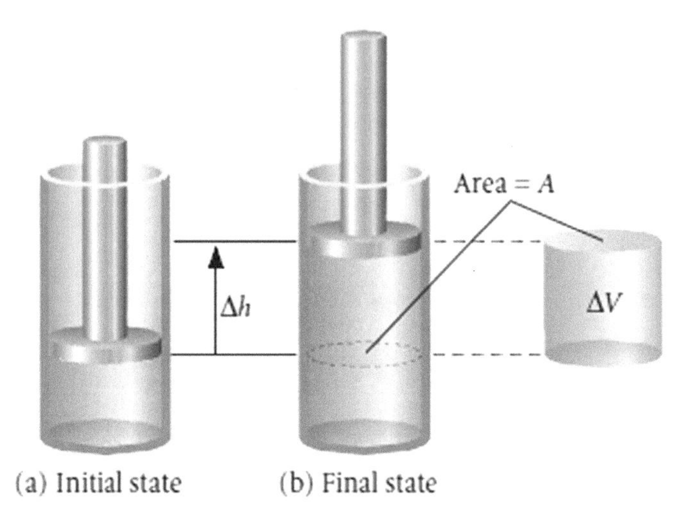

Fig. 5.12 Gas expansion does work. The initial volume of the gas increases by an amount ΔV. When the pressure remains constant, this becomes PΔV, or work

(V), because $V = A \times h$. In the final state in Fig. 5.12, the piston has been displaced by a known distance (Δh). We use the Greek letter Δ (delta) to represent the change in some quantity. Therefore, Δh means "the change in height." The resulting three-dimensional measurement is the change in the volume of the system (ΔV). Specifically, for this case, ΔV is the final volume minus the initial volume of the system: $\Delta V = V_{final} - V_{initial}$. In other words, $A \times \Delta h$ for a reaction is best represented as ΔV. Now we can restate the previous equation as:

$$\text{Work}\,(w) = \text{pressure}\,(P) \times \text{change in volume}\,(\Delta V)$$

Our final modification to the equation that defines work is an adjustment to account for the direction of energy flow. We add a negative sign to the equation to illustrate this fact:

$$w = -P\Delta V$$

This equation indicates that when the volume of a system expands (i.e., ΔV is positive), work is done by the system, and, as a result, energy is lost from the system ($w_{system} = -$). Conversely, when the volume of a system contracts (i.e., ΔV is negative), the system gains energy, and work is done on the system ($w_{system} = +$). Because pressure is recorded in atmospheres and volume in liters, the units for work (energy) are in L·atm. This can be mathematically related to the SI unit of energy known as the **joule (J),** in honor of the English physicist James Prescott Joule (1818–1889). We will derive this relationship later, but for now we can assume that 1 L·atm = 101.3 J.

Joule (J) – The SI unit of energy. In terms of base units, 1 Joule = 1 kg·m^2·s^{-2}.

Example 5.2—Work, Work, and More Work

Calculate, in L·atm and joules, the work associated with the compression of a gas from 75.0 L to 30.0 L at a constant external pressure of 6.20 atm. (1 L·atm = 101.3 J).

Asking the Right Questions

How is work related to volume change? Is the volume of the gas (the system) increasing or decreasing, and by how much?

Solution

$$\begin{aligned}\Delta V &= \text{change in volume}\\ &= V_{final} - V_{initial}\\ &= 30.0\,\text{L} - 75.0\,\text{L}\\ &= -45.0\,\text{L}\\ w &= -P\Delta V = -6.20\,\text{atm} \times (-45.0\,\text{L})\\ &= +279\,\text{L}\cdot\text{atm}\end{aligned}$$

converting this to joules, we get

$$\begin{aligned}w &= +279\,\text{L}\cdot\text{atm} \times \frac{101.3\,\text{J}}{\text{L}\cdot\text{atm}}\\ &= +2.83 \times 10^4\,\text{J} = 28.3\,\text{kJ}\end{aligned}$$

Are Our Answers Reasonable?

The sign on the value of the work term is important to us. Not only does it reflect how much the internal energy of the system changes, but it also indicates the direction in which it changes. Mathematically, the sign was determined by the direction of the change in the volume. In our example, the volume of the system is reduced, so the work done on the system is positive.

What If...?

...you considered the work done when gasoline is combusted in the piston of an engine? A piston inside of a

car's engine burns gasoline and expands the cylinder volume from 0.30 L to 0.75 L with a constant external pressure of 1.0 atm. How much work was done in L·atm and joules?
(*Ans: –0.45 L·atm; –46 J*)

Practice 5.2
Calculate, in L·atm and joules, the work associated with the expansion of a gas from 30.0 L to 300. L at a constant external pressure of 1.0 atm.

5.3.1 Internal Energy

How do we keep track of the amount of energy inside a system? The reaction of hydrogen and oxygen in the Ariane 5's main engine causes energy transfer as heat and as work. The clearest result of the work done by this system is that the rocket rises upward through Earth's atmosphere. The reaction also releases energy that heats the internal parts of the craft, along with many molecules in the ground and atmosphere that come into contact with the hot exhaust vapors from the reaction. How do we quantify just how much energy is available in our system?

The total energy of the Ariane 5, or any other rocket system, is known as the **internal energy (*U*)** of the system. The internal energy, as we noted before, is made up of the sum of the system's kinetic and potential energies. Unfortunately, we cannot calculate the absolute value for the internal energy of a system, because it is extremely difficult to account for all the individual energies that give rise to the kinetic and potential energies. However, because the *change* in the internal energy (ΔU) in a chemical process is measurable—if we can calculate the exchange of energy as heat (q) within, and the work done (w) by the system—we can determine the energy change within the system. This is the **first law of thermodynamics**.

First Law of Thermodynamics – The total change in the closed system's energy in a chemical process is equal to the heat flow (q) into the system and the work done (w) on the system.

Stating the First Law of Thermodynamics in Mathematical Terms:

$$\Delta U = q + w$$

A word of caution: We must take care with the way in which we attribute positive or negative values to q and w (see Fig. 5.13). If we are looking at things from the point of view of the system, the loss of heat from the system to the surroundings will carry a negative sign (q_{system} = "–"). Similarly, if the system does work on the

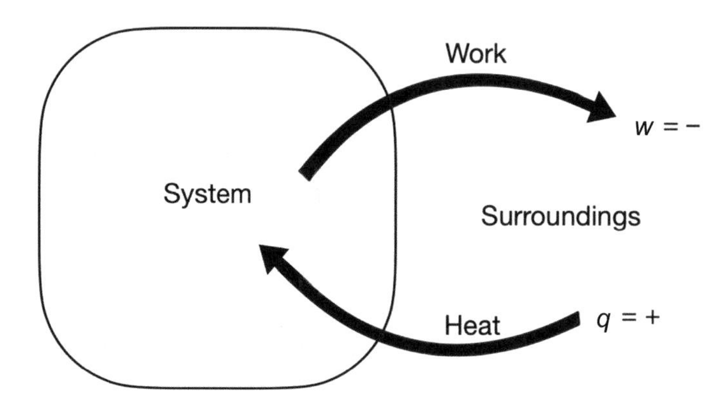

5

Fig. 5.13 From the perspective of the system, the addition of heat results in a positive value for q. If the system does work on the surroundings, the sign on w is negative

surroundings, w_{system} will have a negative sign (w_{system} = "−"). Conversely, the flow of energy as heat (q) into the system will be a positive value (q = "+"), and if work (w) is done on the system by the surroundings, w_{system} will be positive (w_{system} = "+").

Because energy can be neither created nor destroyed in a process, the total amount of energy transferred as heat and as work to or from the system is exactly equal and opposite to that transferred from or to the surroundings.

$$\Delta U_{\text{system}} = -\Delta U_{\text{surroundings}}$$

Reorganizing this equation by adding the change in internal energy of the surroundings to both sides, we get:

$$\Delta U_{\text{system}} + \Delta U_{\text{surroundings}} = 0$$

Mathematically, this is a restatement of the First Law of Thermodynamics. If we know how much energy the system loses, we automatically know that the surroundings will gain that same amount. If we know how much energy the system gains, we know that the surroundings must have lost that same amount. The net energy change in the universe (system and surroundings) is zero. Energy is conserved. The bottom line is that *the Law of Conservation of Energy is also the First Law of Thermodynamics.*

Example 5.3—The First Law of Thermodynamics

Calculate the change in the energy of a system if 51.8 J of work is done by the system with an associated heat loss of 12.3 J.

Asking the Right Questions

How are work and heat related to the internal energy of a system? Which direction are the work and energy flowing? We must pay particular attention to the sign conventions for heat and work in this problem. In this case, work is done by the system. Is this work positive or negative?

Solution

$$\Delta U = q + w$$
$$q = -12.3 \text{ J}$$
$$w = -51.8 \text{ J}$$
$$\Delta U = -12.3 \text{ J} + (-51.8 \text{ J})$$
$$= -64.1 \text{ J}$$

Are Our Answers Reasonable?

A useful system to keep in mind is you! That is, when you do work—by running, dancing, or even moving your textbooks from one class to the next—you are using energy. After the process of moving your body or your books, you have less energy than you had before, so the work has a "−" sign.

Since energy is leaving the system in the form of heat and is also leaving the system in the form of work, the total change in energy of the system must be negative. Therefore, our answer is reasonable.

What If…?

…work of 589.0 J was done on a system, and the system lost a total of 45.0 J as heat. What is the total internal energy change for the system?

(Ans: +544.0 J)

Practice 5.3

Calculate the change in the internal energy of a system if 84.7 J of work is done on the system, with an associated loss of energy as heat of 39.9 J.

5.4 The Units of Energy

In designing complex engineering equipment such as a space launch system, an aircraft, or a chemical plant, the scientist or engineer needs to know how hot the parts that are exposed to exothermic chemical reactions will become. The scientist also needs to know that the rocket engine components, the parts within an aircraft engine, or the containers that are used to hold the hot products of chemical manufacturing processes won't melt. This raises a whole series of questions, such as "How can we quantify the amount of energy being produced?" "How much heat will a chemical reaction generate?" and "How quickly will the surrounding materials heat up, and to what temperature?" Similar questions are raised in all practical investigations of energy, such as studies into the amount of energy provided to us by food or the amount of energy needed to heat a building or to operate a train, an automobile, or a jet airplane.

We often talk about the chemist's shorthand and the importance of having a common set of units with which to communicate. We can start our examination of these questions by doing a calculation of kinetic energy. Let's suppose you are riding a bicycle along a level road at 4.40 m per second (about 10 miles per hour) and that the total mass of both you and the bicycle is 85.0 kg. The kinetic energy of forward motion possessed by this system (you plus the bicycle) would be:

$$\begin{aligned}\text{Kinetic energy} &= \frac{1}{2}mv^2 \\ &= \frac{1}{2} \times 85.0\ \text{kg} \times \left(\frac{4.40\ \text{m}}{\text{s}}\right)^2 \\ &= 823\ \text{kg} \cdot \text{m}^2 \cdot \text{s}^{-2}\end{aligned}$$

This calculation yields the units of energy ($\text{kg} \cdot \text{m}^2 \cdot \text{s}^{-2}$) in terms of SI base units. However, we typically refer to this collection of units as the **joule**, and this is the unit that we will use most often when we discuss energy exchanges in chemical reactions.

$$1\ \text{joule (J)} = 1\ \text{kg} \cdot \text{m}^2 \cdot \text{s}^{-2}$$

The kinetic energy in our example is therefore 823 J.

From where did you and the bicycle get this energy? The immediate source was the series of chemical reactions in your muscles that allowed you to pedal the bike up to speed. These chemical reactions released energy that was extracted from your food. The original source of that energy is the energy of sunlight that fell on the plants that were used to make your food. In that sense, we are all solar powered!

An alternative unit used in energy exchange is related to our diet; many of us use this unit as we "count calories" to assess the amount of energy available in the food we eat. We do this by looking at the nutritional information on the food package's label, such as the one shown in Fig. 5.14. That label gives us a measure of the energy content of food in units of **Calories (Cal)**—note the capital C. In a slightly confusing convention of nomenclature, the Calorie (with a capital C) is equal to a **kilocalorie** (kcal); that is, it is 1000 times larger than the **calorie (cal)**—note the lower-case c.

$$1\ \text{Cal} = 1000\ \text{cal} = 1\ \text{kcal}$$

Calorie – Unit of energy equal to 1000 calories (1 kilocalorie) and to 4184 joules.

calorie – Unit of energy equal to 4.184 joules.

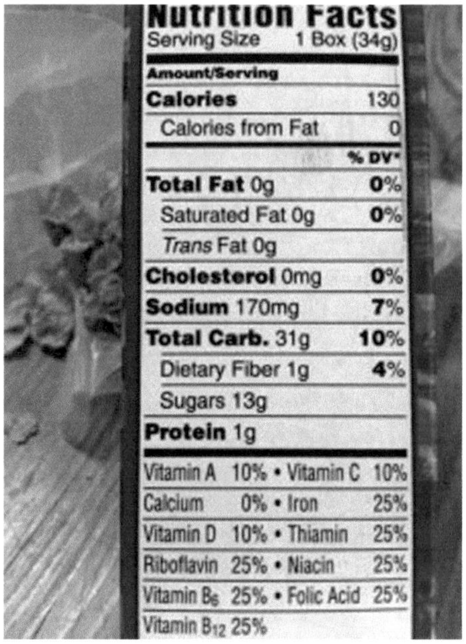

Nutrition Facts
Serving Size 1 Box (34g)

Amount/Serving	
Calories	130
Calories from Fat	0
	% DV*
Total Fat 0g	0%
Saturated Fat 0g	0%
Trans Fat 0g	
Cholesterol 0mg	0%
Sodium 170mg	7%
Total Carb. 31g	10%
Dietary Fiber 1g	4%
Sugars 13g	
Protein 1g	
Vitamin A 10% • Vitamin C	10%
Calcium 0% • Iron	25%
Vitamin D 10% • Thiamin	25%
Riboflavin 25% • Niacin	25%
Vitamin B_6 25% • Folic Acid	25%
Vitamin B_{12} 25%	

Fig. 5.14 Food labels contain information about the chemical content of the food and the amount of energy it supplies to the body. This cereal supplies 130 cal per 34-g serving

	per serving (25 ml)	%RDA* per serving
Energy	202 kJ / 48 kcal	**
Protein	0,92 g	**
Carbohydrate	10,6 g	**
Fat	0,0 g	**
Magnesium	160 mg	42%
Potassium	120 mg	6%

Fig. 5.15 Nutritional information for an energy drink. *Source* ▶ https://biotechusa.com/sites/biotechusa/documents/biotechusa_product/sf/energy-shot_eng.jpg

The values listed on most foods, even though you will often see the spelling *calories* on the label, are given in terms of the "big" Calorie, rather than the "little" calorie. In any case, these alternative units for energy have entered everyday language, in talk of "calorie-conscious lifestyles," "high-calorie food," "low-calorie snacks," and so on. How much is a calorie? One calorie (with a small c) equals the amount of energy needed to raise the temperature of one gram of water by one degree Celsius (for example, from 14.5 °C to 15.5 °C). The calorie is not an SI unit, but it can readily be converted into joules:

$$1 \text{ cal} = 4.184 \text{ J}$$

The calorie is rather a small unit to use for measuring the energy available to our bodies from the food we eat. A typical slice of bread, for example, can provide our bodies with approximately 80,000 cal of usable energy. This equals 80 kilocalories, or 80 Cal.

1 calorie (cal) = 4.184 joules
1 kilocalorie = 4184 joules = 4.184 kilojoules = 1 Calorie

In many countries, food energy is stated in kilojoules, as in the nutritional label on an energy drink that is shown in Fig. 5.15.

Example 5.4—Peanut Power!

Peanuts are a compact source of energy coming from proteins, fats, and carbohydrates found within. When these chemicals from one brand of peanuts are combined with oxygen in the cells of the body, they release 625 Calories (note the capital C) of usable energy per 1.00×10^2 g of peanuts. The following two questions refer to the food energy in a 1.00-ounce (oz.) bag of peanuts. There are 28.4 g in 1.00 oz.

a. How many kilojoules and how many joules of energy are (ideally) available to the body from the bag of peanuts?
b. A 100-watt (W) light bulb (1 W = 1 J/s) requires 100 J to run for 1 s. How many hours would the light bulb run if the energy from the peanuts were used to light the bulb?

Asking the Right Questions

What's the relationship between joules and calories? If we know the conversion factor from calories to joules (4.184 J/calorie) then we can make the conversion.

Solution

a. There are 4.184 kJ in 1 Cal and 28.4 g of peanuts per bag. We can do our calculation as follows:

$$\text{Kilojoules} = \frac{625\ \text{Cal}}{100\ \text{g}} \times \frac{4.184\ \text{J}}{1\ \text{Cal}} \times \frac{28.4\ \text{g}}{1\ \text{bag}} = \frac{743\ \text{kJ}}{1\ \text{bag}}$$

$$\text{Joules} = \frac{743\ \text{kJ}}{1\ \text{bag}} \times \frac{1000\ \text{J}}{1\ \text{kJ}} = 7.43 \times 10^5\ \text{J/bag}$$

b. We determined from part a. that we have 743,000 J of energy available from the bag of peanuts. We can solve for time in hours knowing that we need 100 J/s to power the light bulb.

$$\text{Hours} = 7.43 \times 10^5 \text{J} \times \frac{1\ \text{s}}{100\ \text{J}} \times \frac{1\ \text{h}}{3600\ \text{s}} = 2.06\ \text{h}$$

Are Our Answers Reasonable?

Remembering that a Calorie is 1000 cal, then the peanuts must contain many thousands of calories, and therefore thousands of joules. Even though one ounce of peanuts is a fairly small amount, it contains a considerable amount of chemical energy.

What If…?

…we wanted to know how long a laptop computer could be powered by a 16.0 oz jar of peanuts, if the laptop requires 24.5 watts of power.

(*Ans: 135 h. That's almost a week!*)

Practice 5.4

Apples supply us with approximately 30 Cal per 100 g, and an average apple weighs about 200 g. How many apples would you need to eat to obtain the same amount of energy as is supplied by the bag of peanuts?

5.5 Specific Heat Capacity and Heat Capacity

A scientist working with the Ariane 5's engines is often concerned with the heat generated during takeoff. The amount of heat produced by the burning rocket fuel will need to be absorbed by something if the rocket is to remain intact. The amount of insulation, and the type, will be of great importance. A chef faces similar concerns when deciding on a method by which to pick up a hot pan. Whether dealing with rocket engine or the choice of metal cookware, the general question is: How much will the temperature of an object change when it absorbs or releases a certain amount of energy as heat?

Every substance has a particular **specific heat capacity (*c*)**, often shortened to **specific heat**, which is defined as the amount of energy as heat needed to change the temperature of one gram of the substance by one degree Celsius (or one kelvin) when the pressure is constant (see ▫ Table 5.1 for some representative examples of specific heat). Substances with large values for their specific heat require more energy to change their temperature than substances with small specific heat values. For example, 1 g of water requires more than four times the energy to raise its temperature 1.0 °C than does aluminum.

Specific heat capacity (*c*) – The amount of heat needed to raise the temperature of one gram of a substance by 1 °C (or 1 kelvin) when the pressure is constant. Also known as specific heat.

The specific heat of the water in your teapot at home is $\frac{4.184\ \text{J}}{\text{g}\cdot{}^\circ\text{C}}$. This means that it will require 4.184 joules (or 1 calories) to raise the temperature of 1 gram of the water by 1 °C. Because the *change* in temperature in degrees Celsius is

Table 5.1 Specific heat capacity of selected materials

Substance	Specific heat capacity (J/g·°C)
Water	4.184
Ethanol	2.460
Aluminum	0.902
Copper	0.385
Lead	0.128
Sulfur	0.706
Iron	0.449
Silver	0.235

equal to the change in temperature in kelvins, the specific heat capacities can also be reported in units of $\frac{J}{g \cdot K}$. Given the units of specific heat capacity, we can see how the heat required to raise the temperature of any compound can be determined using the following equation:

$$q = m \times c \times \Delta T$$

where,
q = heat, in joules
m = mass, in grams
c = specific heat, in J/(g·°C) or J/(g·K)
ΔT = change in temperature, in °C or K.

This equation is useful in determining the amount of heat absorbed, or released, during a temperature change for a substance. The amount of heat needed to raise the temperature of *any* specific object by 1 °C is known as that substance's **heat capacity (*C*)**, in units of $\frac{J}{°C}$. The heat required to raise the temperature of any object by a given amount is:

$$q = C \times \Delta T$$

where,
q = heat, in joules
C = heat capacity, in J/°C or J/K
ΔT = change in temperature, in °C or K.

Heat capacity – The amount of heat needed to raise the temperature of any particular object by 1°C (or 1 K) when the pressure is constant.

Let's consider our teapot of water to help illustrate the use of these equations. Assume that we determine the heat capacity of the entire quantity of water in your teapot, when it is filled to the maximum, to be 6.27 kilojoules per °C = (6.27×10^3 J/°C). We can use dimensional analysis to work out the mass of water in the teapot, remembering that the specific heat capacity of water is 4.184 J/g·°C.

$$\text{g water} = \frac{\text{g} \cdot °\text{C}}{4.184\ \text{J}} \times \frac{6.27 \times 10^3\ \text{J}}{°\text{C}} = 1.5 \times 10^3\ \text{g}$$

We can also use this equation to determine the amount of heat required to change the temperature of a particular substance. For instance, how much energy as heat is required to raise the temperature of that teapot of water from a cold 5.00 °C to the boiling point? (The change in temperature for the water will be $\Delta T = T_{\text{final}} - T_{\text{initial}}$ = 100.0 °C – 5.00 °C = 95.0 °C.) Note that in finding our answer, we will not determine the energy needed to vaporize the water. Instead, we will calculate the heat required to arrive at a teapot full of 100.0 °C water.

The required energy as heat is $q = C \times \Delta T$. Thus,

$$\frac{6.27 \times 10^3 \text{J}}{°\text{C}} \times 95\,°\text{C} = 5.96 \times 10^5 \text{J} = 596 \text{ kJ}$$

Note how the dimensions work out to give you the proper units.

Another quantity that relates to the heat capacity of a substance is the **molar heat capacity**, which is the heat capacity of one mole of the substance, in units of $\frac{\text{J}}{\text{mol} \cdot °\text{C}}$. Knowing the specific heat of water, we can calculate the molar heat capacity of water using dimensional analysis:

$$\frac{4.184 \text{ J}}{\text{g} \cdot °\text{C}} \times \frac{18.0 \text{ g } H_2O}{\text{mol } H_2O} = \frac{75.3 \text{ J}}{\text{mol} \cdot °\text{C}}$$

Molar heat capacity – The heat capacity of one mole of a substance.

Example 5.5—How Cold is the Liquid?

If you had a glass containing 50.0 g of ethanol (also called grain alcohol) that had a temperature of 25.0 °C, and you added 20.0 g of cold water with a temperature of 10.0 °C to the ethanol, what is the final temperature of the mixture?

Asking the Right Questions

What do I assume about the final temperature of each substance? How do I know how much heat was transferred between the two substances?

Solution

The key to solving this problem is understanding three concepts: the energy lost by the ethanol will be gained by the water; even though the two substances mix, we can treat them like separate objects when calculating energy transfer as heat for both substances; and both substances will have the same final temperature.

How do we relate the energy change by each substance?

$$q \text{ lost by ethanol} = -q \text{ gained by water}$$

$$[m \times c \times (T_f - T_i)]_{\text{ethanol}} = [m \times c \times (T_f - T_i)]_{\text{water}}$$

$$\left[50.0 \text{ g} \times \frac{2.460 \text{ J}}{\text{g}\,°\text{C}} \times (T_f - 25.0\,°\text{C})\right]_{\text{ethanol}} = -\left[20.0 \text{ g} \times \frac{4.184 \text{ J}}{\text{g} \cdot °\text{C}} \times (T_f - 10.0\,°\text{C})\right]_{\text{water}}$$

Since the mixture will be a uniform temperature after mixing, T_f will be the same for both the ethanol and the water. Therefore, we can rearrange this equation using a bit of algebra:

$$123.0\, T_f - 3075 = -(83.68\, T_f - 836.8)$$
$$206.68\, T_f = 3911.8$$
$$T_f = 18.93\,°\text{C}$$

Are Our Answers Reasonable?

Since the ethanol is warmer at the beginning and the water is colder, we know that the ethanol will decrease in temperature and the water will increase in temperature. The final temperature for both is between the initial temperatures, which makes sense.

What If…?

…we mixed 50.0 g of 25.0 °C water with 20.0 g of 10.0 °C ethanol? What is the final temperature of the mixture?

(Ans: 22.1 °C)

Practice 5.5

A homebrewer has a kettle containing 40.0 L of 40 °C water, but needs to warm it up to a toasty 70 °C. What volume of 100 °C water would need to be added to accomplish this task? (Hint: water has a density of 1000 g/L).

5.5.1 Calorimetry

The thermochemist (a scientist who studies energy transfers in chemical reactions) uses the information about specific heat capacity to determine the amount of energy that is either gained or released by reactions. For example, the thermochemist could provide information about the reaction that takes place in the main engine of the Ariane 5:

$$2H_2(g) + O_2(g) \rightarrow 2H_2O(g)$$

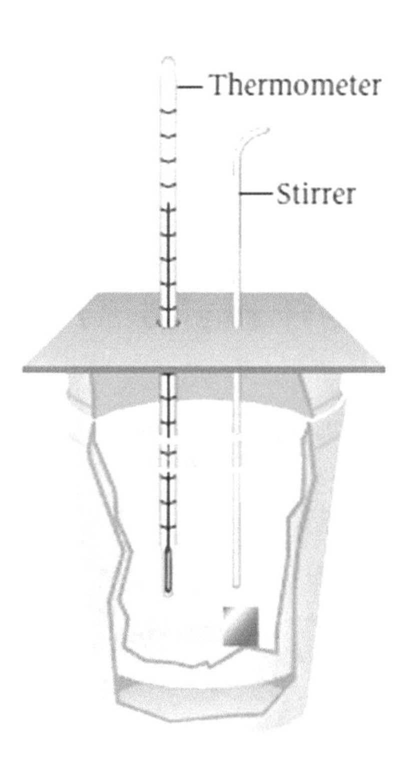

Fig. 5.16 A Styrofoam calorimeter. The inner cup of the calorimeter contains both the system and the surroundings. The outer cup provides additional insulation to keep the surroundings within the set-up

The scientist could also provide the automobile designers with information about a reaction that takes place in a gasoline engine:

$$2C_8H_{18}(l) + 25O_2(g) \rightarrow 16CO_2(g) + 18H_2O(g)$$

One of the first things that a thermochemist needs to do is to measure the amount of heat produced by a chemical reaction. How is the energy produced by a chemical reaction actually measured? By performing these reactions under carefully controlled conditions and trapping and measuring the energy as heat given off or absorbed by the system, the thermochemist can obtain information about energy transfers. This procedure is known as **calorimetry,** which means "heat measurement," and the apparatus in which it is performed is called a **calorimeter**. To illustrate the key principles, we can construct a very simple calorimeter from two Styrofoam cups as shown in Fig. 5.16. The cups act as insulation to (ideally) prevent energy exchange between the contents of the inner cup and the rest of the universe.

Calorimetry – The study of the transfer of heat in a process.

Calorimeter – An apparatus in which quantities of heat can be measured.

To understand the process, we can begin with a hot piece of iron (the system) that we add to water (the surroundings), *both within Styrofoam cups*. The water, the cups and everything outside the cups constitutes the rest of the universe. While very simple, this apparatus does a pretty good job of measuring the energy as heat transferred from the system (the piece of iron, in this case) to the surroundings (the water). Even though the water is only part of the surroundings, it is the part that is in immediate contact with our system, and it will be the most affected by the energy transfer.

Within the perfect Styrofoam calorimeter, *the energy lost by the system as heat equals the energy gained by the surroundings*. Mathematically, this is represented by:

$$q_{\text{system}} = -q_{\text{surroundings}}$$
$$m_{\text{system}} \times c_{\text{system}} \times \Delta T_{\text{system}} = -m_{\text{surround}} \times c_{\text{surround}} \times \Delta T_{\text{surround}}$$

After all of the energy as heat has been transferred, *the final temperatures of the system (iron) and that of the surroundings (water) within the calorimeter are identical.* For example, let's determine the final temperature of a sample of water when a 155 g piece of iron at 95.0 °C is added to 1.000 kg of water at 25.0 °C. We keep in mind that the energy as heat lost by the system and that gained by the surroundings are equal and opposite in magnitude; $q_{system} = -q_{surroundings}$.
Here are the data.

The Piece of Iron

$$m_{system} = 155 \text{ g of iron}$$

$$c_{system} = \frac{0.449 \text{ J}}{\text{g} \cdot {}^\circ\text{C}} \text{ (from Table 5.1)}$$

$$T_i = 95.0\ {}^\circ\text{C}$$

$$T_f = ?$$

We do not know the final temperature of the system, but we know that it will be less than 95.0 °C, because the surrounding water is initially cooler than the piece of iron. Let's call the change in temperature for the piece of iron $\Delta T_{system} = T_f - 95.0$ °C. The change in temperature will be "–" because the final temperature is less than the initial temperature.

The Water

$$m_{surround} = 1.000 \times 10^3 \text{g of water}$$

$$c_{surround} = \frac{4.184 \text{ J}}{\text{g} \cdot {}^\circ\text{C}}$$

$$T_i = 25.0\ {}^\circ\text{C}$$

$$T_f = ?$$

We do not know the final temperature of the surroundings, but we know that it will be greater than 25.0 °C, because the piece of iron (the system) is initially hotter than the water (surroundings), and will therefore transfer energy as heat into the water. Let's call the change in temperature $\Delta T_{surround} = T_f - 25.0$ °C. The change in temperature is "+" because the final temperature is greater than the initial temperature.

Finally, we know that T_f is the same for both the system and the surroundings because at that point, the energy as heat will stop flowing between them. We can now set the equations for the energy exchanges between the system and the surroundings equal to each other (but opposite in sign!) and solve:

$$\text{heat lost by piece of iron} = -(\text{heat gained by water})$$

$$m_{iron} \times c_{iron} \times \Delta T_{iron} = -m_{water} \times c_{water} \times \Delta T_{water}$$

$$155 \text{ g} \times \frac{0.449 \text{ J}}{\text{g} \cdot {}^\circ\text{C}} \times (T_f - 95.0\ {}^\circ\text{C}) = -1.000 \times 10^3 \text{ g} \times \frac{4.184 \text{ J}}{\text{g} \cdot {}^\circ\text{C}} \times (T_f - 25.0\ {}^\circ\text{C})$$

$$\frac{69.595 \text{ J}}{{}^\circ\text{C}} \times (T_f - 95.0\ {}^\circ\text{C}) = \frac{-4.184 \text{ J}}{\text{g} \cdot {}^\circ\text{C}} \times (T_f - 25.0\ {}^\circ\text{C})$$

$$69.595\ T_f - 6611.525 = -4184\ T_f + 104600$$
$$4253.595\ T_f = 111211.525$$
$$T_f = 26.145 = 26.1\ {}^\circ\text{C}$$

The final temperature of the water (and the piece of iron) will be 26.1 °C.

The preceding example only involved a temperature change, not a chemical reaction. What if the energy transfer was due to a chemical reaction instead of just a hot piece of metal? How do we measure the heat released by a chemical reaction? It turns out that we can use the very same tools, as we explore in Example 5.6.

Example 5.6—Coffee Cup Calorimetry Calculations

Hydrochloric acid reacts with sodium hydroxide in an exothermic reaction:

$$HCl(aq) + NaOH(aq) \rightarrow H_2O(l) + NaCl(aq)$$

Determine the heat, in kJ per mole of hydrochloric acid, if 15.0 mL of 1.00 M HCl is mixed with 15 mL of 1.00 M NaOH in a coffee-cup calorimeter and the temperature increases 6.95 °C. The specific heat capacity of the mixture is 4.050 J/g· °C, and has a density of 1.00 g/mL.

Asking the Right Questions

What is the system? What are the surroundings? Will any of the reactants remain after the reaction?

Solution

Our first task is to determine the limiting reagent in the reaction. So, we pick one of the products and determine the number of moles that can be made from each starting material.

For HCl:

$$15.0\ \text{mL} \times \frac{1\ \text{L}}{1000\ \text{mL}} \times \frac{1.00\ \text{mol HCl}}{1\ \text{L}} \times \frac{1\ \text{mol NaCl}}{1\ \text{mol HCl}} = 0.015\ \text{mol NaCl}$$

For NaOH:

$$15.0\ \text{mL} \times \frac{1\ \text{L}}{1000\ \text{mL}} \times \frac{1.00\ \text{mol NaOH}}{1\ \text{L}} \times \frac{1\ \text{mol NaCl}}{1\ \text{mol NaOH}} = 0.015\ \text{mol NaCl}$$

In this example, both reagents are completely consumed because both make the same number of moles of product. The "system" that we are interested in is the chemical reaction between HCl and NaOH. The "surroundings" are the resulting mixture of water and salt that exists after the reaction is complete. As before:

$$q_{\text{system}} = -q_{\text{surroundings}}$$
$$q_{\text{system}} = -(m \times c \times \Delta T)_{\text{surroundings}}$$

The mass of the system is:

$$30.0\ \text{mL} \times 1.00\ \text{g/mL} = 30.0\ \text{g}$$

$$q_{\text{system}} = -(30.0\ \text{g} \times 4.050 \frac{\text{J}}{\text{g} \cdot {}^\circ\text{C}} \times 6.95\ {}^\circ\text{C}) = -844\ \text{J}$$

Now we need to know how many moles of HCl reacted:

$$\frac{1.00\ \text{mol HCl}}{1\ \text{L}} \times \frac{1\ \text{L}}{1000\ \text{mL}} \times 15.0\ \text{mL} = 0.0150\ \text{mol HCl}$$

Therefore, the heat released per mole of HCl is:

$$\frac{-844\ \text{J}}{0.015\ \text{mol HCl}} \times \frac{1\ \text{kJ}}{1000\ \text{J}} = -56.3 \frac{\text{kJ}}{\text{mol HCl}}$$

Are Our Answers Reasonable?

The question originally stated that this reaction is exothermic, and we determined the change in heat to have a negative sign, so this is in agreement.

What If…?

…we reacted 28.6 mL of 1.00 M HCl with 28.6 mL of 1.00 M NaOH? Using information from the example problem, calculate the temperature change for this reaction.

(*Ans:* 6.95 °C)

Practice 5.6

Using information from the above example, predict the temperature increase if the same reaction is conducted, but an additional 20.0 g of water is added to the mixture. (Hint: assume that the additional water does not affect the chemical reaction or the specific heat capacity of the final mixture.)

Our Styrofoam calorimeter isn't perfect. It still allows some energy to be transferred, as either work or heat, from the system to the surroundings. In addition, the calorimeter itself can participate in the transfer of energy as heat, distorting the calculations. We can address these problems and eliminate small errors in our measurements if we arrange things so that no work is done by the system or on the system ($w=0$), and no heat is transferred to or from the inside of the calorimeter from the rest of the universe. While this is actually impossible to accomplish perfectly, we can do a much better job than we did with Styrofoam cups! If we can ensure that there is no work done, the heat for any reaction (q) will be equal to the total change in the energy of the system, because the equation $\Delta U = q + w$ will simplify to $\Delta U = q$. We can achieve this using a technique called **constant-volume calorimetry**, in which the reacting system is sealed within a steel chamber of fixed volume, called a **bomb calorimeter**, shown in ◘ Fig. 5.17. When a combustion reaction is ignited inside a bomb calorimeter, heat is transferred between the reaction and the surrounding water and calorimeter chamber. Since the bomb calorimeter cannot expand, no pressure–volume work can be done on the surroundings. We can then calculate the heat released by the reaction.

Constant-Volume calorimetry – A form of calorimetry in which the reacting system is sealed within a chamber of fixed volume, and the only way the system can release or gain energy is by the exchange of heat with the surroundings.

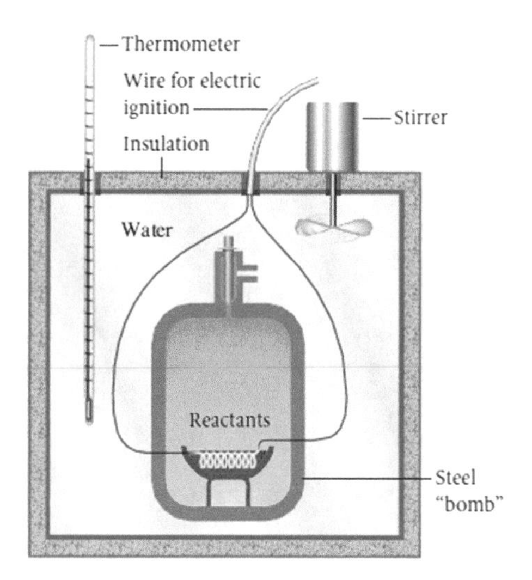

Fig. 5.17 A bomb calorimeter. The combustion reaction under study is conducted within the steel bomb submerged in the water bath. The heat given off by the reaction is measured by the change in the temperature of the water bath

Bomb calorimeter – Apparatus in which a chemical reaction occurs in a closed container, allowing the energy released or absorbed to be measured.

Example 5.7—Calculations with a Bomb Calorimeter

Each bomb calorimeter is different, but its own heat capacity can be determined experimentally using a substance that releases a known amount of energy. Once calibrated in this way, the calorimeter can be used to determine the heat output of other chemicals.

a. Glucose ($C_6H_{12}O_6$), also known as "blood sugar," is the main sugar that serves to transport chemical energy through the blood and distribute it to the body's cells. Glucose is known to release 2.80×10^3 kJ/mol when combined with excess oxygen at 298 K (25 °C). A sample of glucose weighing 5.00 g was burned with excess oxygen in a bomb calorimeter. The temperature of the calorimeter rose by 2.40 °C. Calculate the heat capacity of the calorimeter in joules per degree Celsius.
b. Propane gas has the formula C_3H_8 and can be used as a source of heat for cooking and domestic heating. For storage purposes, it is liquefied and stored in canisters, often under the name of liquefied petroleum gas (LPG). A 4.409-g sample of propane was burned with excess oxygen in the bomb calorimeter calibrated in part a. The temperature of the calorimeter increased by 6.85 °C. Calculate how much energy as heat is released per mole of propane burned under these conditions.
c. If the energy released from 1 mol of propane determined in part b were used to heat 90.0 kg (9.00×10^4 g) of water originally at 30.0 °C, what would be the final temperature of the water? The specific heat capacity of water is 4.184 J/g·°C.

Asking the Right Questions

What is the relationship between the heat released by the chemical reaction and the heat gained by the calorimeter? How does heat capacity relate to the temperature change of an object? What are the molar masses of the chemicals in the problem? For part a, how much total energy does this chemical reaction release?

Solution

a. We first need to calculate the number of moles of glucose ($C_6H_{12}O_6$) in the 5.00 g used. The molar mass of glucose is 180.0 g/mol. We calculate the number of moles used, and then use this to calculate the energy released per degree Celsius, as follows:

$$\text{moles } C_6H_{12}O_6 = 5.00 \text{ g } C_6H_{12}O_6 \times \frac{1 \text{ mol } C_6H_{12}O_6}{180.0 \text{ g } C_6H_{12}O_6} = 0.0278 \text{ mol}$$

$$\text{heat capacity of calorimeter} = \frac{2.80 \times 10^3 \text{ kJ}}{1 \text{ mol}} \times 0.0278 \text{ mol} \times \frac{1}{2.40\ °\text{C}} = \frac{32.4 \text{ kJ}}{°\text{C}}$$

b. The molar mass of propane (C_3H_8) is 44.09 g/mol. Knowing this enables us to calculate the number of moles of propane burned and, therefore, the amount of energy released. We'll need to use the heat capacity of the calorimeter to correct for the effect of the calorimeter:

$$\text{moles } C_3H_8 = 4.409 \text{ g } C_3H_8 \times \frac{1 \text{ mol } C_3H_8}{44.09 \text{ g } C_3H_8} = 0.1000 \text{ mol } C_3H_8$$

Energy change per mole C_3H_8

$$= \frac{32.4\ \text{kJ}}{°\text{C}} \times 6.85\ °\text{C} \times \frac{1}{0.1000\ \text{mol}\ C_3H_8}$$

$$= \frac{-2.22 \times 10^3 \text{kJ}\ C_3H_8}{\text{mol}\ C_3H_8}$$

We add a negative sign to the value because the energy is released from this reaction; that is, the reaction is exothermic.

c. Remember that $q = c \times m \times \Delta T$. We want to find ΔT, so we rearrange this equation:

$$\Delta T = \frac{q}{c \times m}$$

$$\Delta T = \frac{2.22 \times 10^6 \text{J}\ C_3H_8}{\frac{4.184\ \text{J}}{\text{g}\cdot°\text{C}} \times (9.00 \times 10^4 \text{g})}$$

$$= 5.90\ °\text{C}$$

The temperature of the water will rise to nearly 36 °C in a total of 90.0 L (about 24 gal) of water.

Are Our Answers Reasonable?

For part a, we find that the heat capacity of the bomb calorimeter is 32.4 kJ/°C. Since the bomb calorimeter is large and contains a lot of water, this makes sense. In part b, we find the heat of the chemical reaction to be thousands of kilojoules. Most chemical reactions release hundreds or thousands of kilojoules, so this seems to be reasonable as well. Finally, for part c, we are heating a very large amount of water, which has a relatively large specific heat capacity, so we would expect even this large amount of energy to only raise the temperature of the water by a few degrees.

What If…?

…we used the same bomb calorimeter to burn 10.0 g of octane (C_8H_{18}), and saw a temperature increase of 27.5 °C. Based on this information, calculate the heat of combustion of octane, in kJ/mol.

(Ans. 1.02×10^4 kJ/mol)

Practice 5.7

A sample of benzoic acid (C_6H_5COOH) weighing 2.442 g was reacted with excess O_2 in a bomb calorimeter. The temperature rose from 26.34 °C to 39.20 °C. The heat capacity of the calorimeter was 5.02 kJ/°C. Calculate the heat released by this reaction in units of kilojoules per mole.

▶ Here's What We Know So Far

- Changes in the internal energy of a system are determined by the sum of the energy changes as work and as heat. The signs on these terms are vital, because they define whether energy is transferred into the system or out of the system (to the surroundings).
- The specific heat capacity of a substance is the energy required to change the temperature of one gram of the substance by one degree (Celsius or Kelvin).
- The heat capacity of an object reflects the ability of the entire object to absorb heat. A large heat capacity indicates that the addition of a large amount of energy is required in order to change the temperature of the object.
- A calorimeter can be used to measure the amount of heat transfer between the system and the surroundings. We may use the equation $m_{\text{system}} \times c_{\text{system}} \times \Delta T_{system} = -m_{\text{surround}} \times c_{\text{surround}} \times \Delta T_{\text{surround}}$ to determine heat transfer. ◀

5.6 Enthalpy

In the previous section, we saw that we can measure energy changes from chemical reactions under constant-volume conditions in a bomb calorimeter. However, many of the reactions we do in the laboratory do *not* occur under constant-volume conditions. Any gases that are generated are often free to expand outward into the atmosphere. These reactions occur under constant-pressure conditions, because the release of gas or any other expansion of volume will occur until the pressure of the products of the reaction becomes equal to the atmospheric pressure.

Under constant-pressure conditions, we cannot use the simple relationship $\Delta U = q$ but instead have to work with the slightly more complex $\Delta U = q + w$ to take into account any work done by the system or done on the system. When a reaction occurs under constant-pressure conditions, the only type of work the system will be able to do on the surroundings is called "pressure–volume work" (see ▶ Sect. 5.1), such as the work done by the system when a released gas is allowed to expand. Under these conditions of constant pressure, the work done by the system is equal to $-P \times \Delta V$ (pressure multiplied by the change in volume), so the equation $\Delta U = q + w$ can be written as

$$\Delta U = q_\text{p} - P\Delta V$$

where q_p is the **heat of reaction** at constant pressure. We can rearrange that equation to get

$$q_p = \Delta U + P\Delta V$$

We introduce a new term, **enthalpy**, which is symbolized by H. Enthalpy is measured in the units of energy (joules) and is defined as the sum of the internal energy and the pressure–volume product of a system:

$$H = U + PV$$

A change in enthalpy (ΔH) can be defined as $\Delta H = \Delta U + \Delta(PV)$ and is equal to q_p.

Heat of reaction – The energy as heat released or absorbed during the course of a reaction.

Enthalpy – A thermodynamic quantity symbolized by H and defined as $H = U + PV$. The heat absorbed or released by a chemical system under constant pressure conditions.

Moreover, if the pressure of the system is constant, PV can change only as a consequence of changes in volume, so $\Delta(PV)$ is equal to $P\Delta V$. Under these constant-pressure conditions, the definition of the change in enthalpy becomes exactly the same as the definition for the heat of reaction at constant pressure, namely q_p:

$$\begin{aligned} \text{At constant pressure:} \quad & q_p = \Delta U + P\Delta V \\ \text{And} \quad & \Delta H = \Delta U + P\Delta V \\ \text{So} \quad & \Delta H = q_p \end{aligned}$$

Heats of reaction measured under constant-pressure conditions are known as changes in enthalpy. It is important to remember that this unfamiliar term really just refers to the familiar idea of energy exchange as heat for a reaction that proceeds under constant-pressure conditions.

In situations where only very small amounts of work are done by a system or on a system, the value of w is very small compared to q_p. In these situations, the easily measured heat of reaction (q_p), which also equals the enthalpy change (ΔH), is approximately equal to the total energy change of the system ΔU_{system}. This is useful when studying the chemistry of living things. For instance, most biochemical reactions occur in body fluids in which there are negligible changes in volume and, therefore, negligible contribution to the energy change of the system from work. However, if a chemical reaction produces a gas, then the work component is not necessarily negligible because the change in volume becomes large.

The most interesting and useful value to us when we are studying energy and chemistry is the total energy change (ΔU) of the chemicals in the system as they react. This is not always easy to measure directly, so one of the most significant facts about enthalpy change (ΔH) values is that they provide a readily measured approximation to the ΔU values in which we are really interested.

5.6.1 Standard Enthalpies of Reaction

Comparing changes in enthalpies for reactions is a tricky business. To be meaningful, the enthalpy changes must be measured under the same conditions. How do we compare changes in enthalpy for different chemical reactions?

Comparisons are often made with heats of reaction obtained when all of the reactants and products are in their **standard states**, as illustrated in ◘ Fig. 5.18. Then the enthalpy of the reaction becomes known as a **standard enthalpy of reaction ($\Delta_{rxn}H°$)**.

Standard state – The state of a chemical under a set of standard conditions, usually at 1 atmosphere of pressure and a concentration of exactly 1 molar for any substances in solution. Standard states are often reported at 25 °C (298 K).

Standard enthalpy of reaction ($\Delta_{rxn}H°$) – Enthalpy change of a reaction in which all of the reactants and products are in their standard states.

Standard states for thermodynamic properties are often tabulated at 25 °C. However, the definition for a substance in its standard state does not require the temperature to be 25 °C. For example, one could calculate the enthalpy of reaction at 350 °C, although to do so, one would need access to thermodynamic values tabulated at this temperature.

What is the standard state of a reactant or product? Here are the most commonly used standard states for thermodynamic work:

Fig. 5.18 Examples of some compounds in their standard states

- For a pure solid, liquid, or gas, the standard state is the state of the substance at a pressure of exactly 1 bar, which equals 100,000 Pa. Although it isn't the standard definition, many people and references still use 1 atmosphere (1 atm), which is equivalent to 101,325 Pa. The difference between 1 bar and 1 atm is not very great. (IUPAC has adopted 1 bar as the standard pressure, but 1 atm is still in widespread use.)
- For any substance in solution, the standard state is at a concentration of exactly 1 molar.

We indicate that a thermodynamic value has been determined under standard conditions by using the degree sign (°), so a standard enthalpy of reaction would be indicated as $\Delta_{rxn}H°$.

Any substance under these standard conditions is said to be in its standard state. For example, water is present all around us in three main forms: as the liquid water that runs from our taps, as the water vapor in the air, and as the solid water such as the ice in our freezers. The *standard state of water*, however, is the liquid form in which pure water exists at 1 atm of pressure. Most of the water around us is not "pure water" because it has other chemicals dissolved in it. To be in its standard state, a substance must be pure. The **reference form** of an element is the most stable form of the element at standard conditions. For the element oxygen at 1 atm and 25 °C, the reference form is O_2, rather than the less stable allotrope O_3 (ozone).

Reference form – The most stable form of the element at standard conditions.

Many different standard enthalpies of reactions have been defined. Common ones include the standard enthalpy of formation and the standard enthalpy of combustion. The **standard enthalpy of formation, $\Delta_f H°$** (also known as the **standard heat of formation**) of a substance is the enthalpy change for the formation of 1 mol of the substance in its standard state from its elements in their *reference forms*. For example, the standard enthalpy of formation of 1 mol of carbon dioxide from carbon and oxygen in their reference forms, which for carbon is graphite, can be summarized as follows:

$$C(s) + O_2(g) \rightarrow CO_2(g) \quad \Delta_f H° = -394 \text{ kJ}$$

which is the same as:

$$C(s) + O_2(g) \rightarrow CO_2(g) + 394\ kJ$$

Standard enthalpy of formation ($\Delta_f H°$) – The enthalpy change for the formation of one mole of a substance in its standard state from its elements in their reference form. Also known as the standard heat of formation.

We have to show just 1 mol of the product, because that value is embodied in the definition. This can force us to use fractional amounts of moles of some of the reactants, as in the next example, which indicates the standard enthalpy of formation of water:

$$H_2(g) + \frac{1}{2}O_2(g) \rightarrow H_2O(l) \qquad \Delta_f H° = -286 \text{ kJ}$$

We must show the oxygen in its standard state as $O_2(g)$. We cannot simply use O (because O is not the reference form of oxygen), so we must indicate half a mole of O_2. Keep in mind that when you are writing standard enthalpy equations, it is often necessary to have equations with fractional coefficients, rather than multiplying by some common factor to convert all the fractions to whole numbers.

Example 5.8—Chemistry in Space

Hydrazine (N_2H_4) is a volatile liquid that rapidly decomposes to form very hot hydrogen and nitrogen gases. This reaction is used to provide hot gases for small re-positioning thrusters on satellites orbiting the Earth.

a. Write the equation that describes the standard enthalpy of formation for this compound.
b. For hydrazine $\Delta_f H° = +95.40$ kJ/mol. Calculate the change in enthalpy accompanying the formation of 125 g of hydrazine from its elements in their reference forms.

Asking the Right Questions

What is the balanced reaction for the formation of hydrazine? How many moles of hydrazine will we make in the second reaction?

Solution

a. We need to begin with a balanced chemical equation:

$$2H_2(g) + N_2(g) \rightarrow N_2H_4(g)$$

b. Since the enthalpy of formation is given in kJ/mol, the number of moles of hydrazine must be used to calculate the enthalpy:

$$125 \text{ g } N_2H_4 \times \frac{1 \text{ mol } N_2H_4}{32.05 \text{ g } N_2H_4} = 3.90 \text{ mol } N_2H_4$$

$$3.90 \text{ mol } N_2H_4 \times \frac{95.40 \text{ kJ}}{1 \text{ mol } N_2H_4} = 372 \text{ kJ}$$

Are our Answers Reasonable?

From the information given in the question, we know that hydrazine produces heat—is exothermic—when it forms hydrogen and nitrogen. Therefore, the formation of hydrazine must be endothermic, which is what our answer indicates, so it is reasonable. Also, remember that heats of reaction are given per mole, so we have to know how many moles are reacting in order to calculate the total amount of heat that is either produced or required.

What If…?

…478 g of hydrazine decomposed. How much energy would be released?

(Ans: 1.42×10^3 kJ)

Practice 5.8

Calculate the enthalpy change for the formation of 38.0 g of liquid water. (Use the appendix to find the standard enthalpy change for the formation of water.)

Elements in their reference forms are the basic starting materials for all the reactions associated with standard enthalpies of formation. The elements themselves are not formed from any simpler chemicals, so the standard enthalpies of formation of elements in their reference forms are all equal to zero. This is to say that, using calcium as an example,

$$\text{Ca}(s) \rightarrow \text{Ca}(s) \qquad \Delta_f H^\circ = 0\ \text{kJ/mol}$$

Note that $\Delta_f H^\circ = 0$ kJ/mol does not mean that $U = 0$ kJ/mol. Also, recall that only one form of an element is its reference form. For example, $O_2(g)$ is the reference form for oxygen. When O_2 forms the allotrope ozone, O_3, in the atmosphere, $\Delta_f H^\circ$ is *not* equal to 0 kJ/mol.

5

$$3/2\,\text{O}_2(g) \rightarrow \text{O}_3(g) \qquad \Delta_f H^\circ = 143\ \text{kJ/mol}$$

The **standard enthalpy of combustion, $\Delta_c H^\circ$** (also known as the **standard heat of combustion**) of a substance is *the enthalpy change when 1 mol of the substance in its standard state is completely reacted with excess oxygen gas.* For example, consider the combustion of propane gas (C_3H_8), which we first discussed in Example 5.7.

$$\text{C}_3\text{H}_8(g) + 5\text{O}_2(g) \rightarrow 3\text{CO}_2(g) + 4\text{H}_2\text{O}(l) \qquad \Delta_c H^\circ = -2202\ \text{kJ/mol}$$

Standard enthalpy of combustion ($\Delta_c H^\circ$) – The enthalpy change when one mole of a substance in its standard state is completely burned in oxygen gas. Also known as the substance's standard heat of combustion.

When we write the equation to illustrate the standard enthalpy of combustion, we must show exactly 1 mol of the reactant, propane, being burned. The coefficients must be adjusted to fit that requirement, even as the mole *ratios* of all reactants and products necessarily remain the same.

Consider, as well, the equation for the standard enthalpy of combustion of carbon. Note that the equation is the same as the equation for the standard enthalpy of formation of carbon dioxide.

$$\text{C}(s) + \text{O}_2(g) \rightarrow \text{CO}_2(g) \qquad \Delta_c H^\circ = -394\ \text{kJ/mol} = \Delta_f H^\circ(\text{CO}_2)$$

5.6.2 Manipulating Enthalpies

When we know the value for any standard enthalpy change of a reaction, we also automatically know the value for the enthalpy change of the reverse reaction. For example, converting carbon dioxide gas back into solid carbon and oxygen gas would be the reverse of the forward reaction, the combustion of carbon to form CO_2. It should therefore be accompanied by an enthalpy change that is equal to that of the forward reaction but opposite in sign, as shown in Fig. 5.19. To help us understand what the signs on the enthalpy terms mean, we can consider this manipulation as though the enthalpy change were part of the equation. Let's examine the following equations and see how this applies to the end result. The enthalpy of combustion is negative, so it must be released from the reaction and, hence, appear on the same side as the products in this exothermic process.

In some cases, we will find it useful to multiply all of the equation coefficients by some number. Let's see what happens when we multiply the coefficients in the "combustion of carbon" equation by 2:

$$\text{C}(s) + \text{O}_2(g) \rightarrow \text{CO}_2(g) + (394\ \text{kJ/mol})$$
$$2\text{C}(s) + 2\text{O}_2(g) \rightarrow 2\text{CO}_2(g) + (2 \times 394\ \text{kJ/mol})$$

Note that although the mole ratios of the reactants and products remain the same, the enthalpy change is affected by this multiplication. Remembering that enthalpy as a product indicates a negative change in enthalpy, we note that ΔH of this reaction is now $(2 \times -394\ \text{kJ/mol}) = -788$ kJ/mol.

This ability to reverse equations and automatically know the corresponding enthalpy change is a consequence of what we know as Hess's law, which is the subject of the next section. It enables us to determine the enthalpy changes of reactions that might be very difficult to actually perform.

$$\text{C}(s) + \text{O}_2(g) \rightarrow \text{CO}_2(g) + 394\ \text{kJ/mol} \qquad \Delta_c H^\circ = -394\ \text{kJ/mol}$$

$$394\ \text{kJ/mol} + \text{CO}_2(g) \rightarrow \text{C}(s) + \text{O}_2(g) \qquad \Delta H = +394\ \text{kJ/mol}$$

Fig. 5.19 Negative enthalpy changes can be thought of as adding enthalpy to the product side of the equation. Reversing the direction of the equation changes the sign of the enthalpy change

Example 5.9—How Much Heat?

A camper wishes to warm a pot of water on a propane stove. She calculates that she'll need to produce 209 kJ to increase the temperature of her pot of water by 50 °C. How many moles of propane will be consumed to produce this much heat?

$$C_3H_8(g) + 5O_2(g) \rightarrow 3CO_2(g) + 4H_2O(l)$$
$$\Delta_c H^\circ = -2220 \text{ kJ/mol}$$

Asking the Right Questions
How much energy does she need to produce to heat the water? How many moles of propane will that require?

Solution
First, we want to know how much energy will be absorbed by the water when it is heated. We also know these relationships:

$$\text{energy released by reaction} = -(\text{energy gained by water})$$
$$\text{energy released by reaction} = -(209 \text{ kJ})$$

Accordingly, we can write this ratio:

$$\frac{1 \text{ mol } C_3H_8}{-2220 \text{ kJ}}$$

Thus:

$$-209 \text{ kJ} \times \frac{1 \text{ mol } C_3H_8}{-2220 \text{ kJ}} = 0.0949 \text{ mol } C_3H_8$$

Are Our Answers Reasonable?
Since we need about one-tenth of the energy produced by a mole of propane to heat our water, it makes sense that we needed to burn about one-tenth of a mole of propane. Of course, in real life, not all the energy produced by the propane would be transferred into the water. Some would be absorbed by the surroundings and the cooking pot. In reality, we would need to burn more propane than this amount.

What If…?
…we needed to increase the temperature of 5.00 kg of water by 50 °C? How much propane would we need to burn? Hint: you need to know the specific heat capacity of water in order to determine how much energy is required.
(*Ans: 0.471 mol*)

Practice 5.9
Calculate how many moles of the compound printed in boldface type will be needed to produce a change in enthalpy of 250 kJ for each of these reactions.

a. $\mathbf{2NH_3}(g) + 3N_2O(g) \rightarrow 4N_2(g) + 3H_2O(l)$
   $\Delta H = -1012 \text{ kJ}$
b. $2N_2O(g) \rightarrow \mathbf{O_2}(g) + 2N_2(g)$
   $\Delta H = -164 \text{ kJ}$

5.7 Hess's Law

The search for alternative fuel sources to supplement the petroleum-based fuels we currently use is a high priority in all fields of transportation, including commercial aviation, personal automobile use, and even rocketry. The search for such fuels could involve the examination of lots of new chemical reactions, some of which might never have been done before. Fortunately, one of the features of enthalpy change values works to our benefit. We can calculate the energy or enthalpy changes of reactions without having to do the actual reactions. The reason for this ease of calculation is that energy (U) and enthalpy (H) are state functions. A **state function** is a property of a system that depends only on its present state, not on how it got there. State functions are usually represented by an italic capital letter.

State function – A property of a system that depends only on its present state, not on the path by which it reached that state.

For an illustration of this idea, consider the different paths that a supply mission might take to reach the international space station, orbiting at an altitude of 240 miles above sea level. One mission may be launched directly up to a 240-mile-high orbit, ready to rendezvous with the space station. A later mission might first go into a higher orbit to release a satellite and then come back down to the 240-mile-high orbit of the space station. The eventual altitude of each mission is the same: 240 miles above sea level. In short, the final altitude is independent of the path taken. The

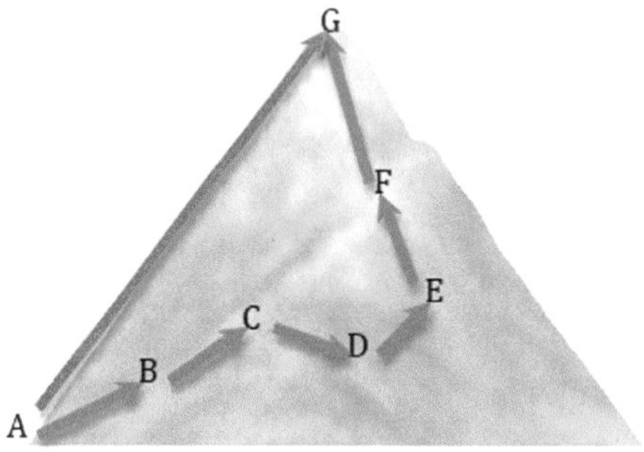

Fig. 5.20 Potential energy, like enthalpy, is a state function. If two climbers took two different paths to the summit of a mountain, their potential energy change would be the same. No matter what path you take, the change in a state function only depends on the difference between the end state and the beginning state

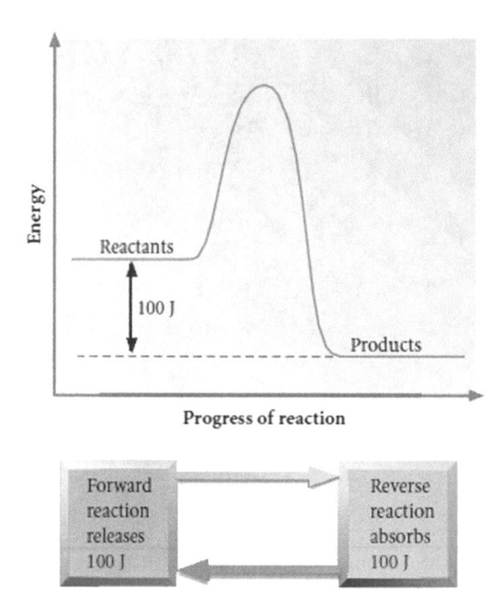

Fig. 5.21 A forward reaction has an enthalpy that is equal, but opposite in sign, to that of its reverse reaction

distances traveled by the two missions in reaching their final altitude are very different, however, because one took the direct route and the other took a more circuitous route. Altitude is a state function; distance traveled is not.

The change in the value of any state function depends *only* on the initial and final states between which it is changing. In the case of enthalpy,

$$\Delta H = H_{\text{final}} - H_{\text{initial}}$$

If a chemical reaction is carried out by two different chemical paths, the overall enthalpy *change* associated with each path will be the same although the paths might be completely different. That basic principle is known as **Hess's law** and is summarized diagrammatically in Fig. 5.20.

Hess's Law – Thermodynamic law stating that the enthalpy change of a chemical reaction is independent of the chemical path or mechanism involved in the reaction.

Enthalpy is a state function, which also justifies our earlier statement that the enthalpy change for any reaction will have the same value as the enthalpy change for the reverse reaction but will be opposite in sign. If the forward reaction has an enthalpy change of −100 J, the reverse reaction will have an enthalpy change of +100 J. This is summarized diagrammatically in Fig. 5.21.

5.7.1 Using Hess's Law

In the Salt River-Pima Maricopa Indian Community in Arizona, methane (CH_4) collected from a 10-million-ton garbage landfill is used to fuel a "green" (environmentally benign) power plant that provides electricity for 2000

Table 5.2 Sources of global methane emissions, 2020

Source	Megatons of CH_4 (1 MT = 1×10^{12} g)
Wetlands	194
Agriculture	145
Energy	134
Biomass combustion	16
Waste	68
Other	39

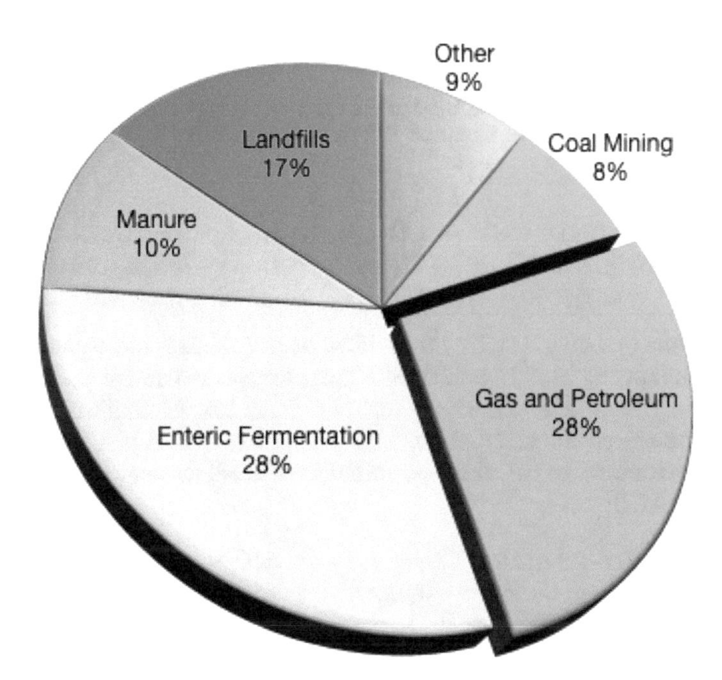

Fig. 5.22 Major sources of atmospheric methane as reported by the U.S. Environmental Protection Agency. Enteric fermentation refers to methane emissions from ruminant animals such as cattle. *Source* US EPA—public domain because it is from a US government agency

homes. Methane is a chemically very simple and convenient fuel, arising from the biological degradation of foodstuffs and other wastes in the landfill. It also makes up about 95% of the "natural gas" found above oil deposits, is burned to provide energy for domestic cooking and heating, and is a source of heat in many industries.

Table 5.2 and Fig. 5.22 list the important sources of methane emissions into the atmosphere. The data in the table suggest that it is not possible to accurately quantify the production of a gas that is emitted every day from so many sources. The numbers should be seen as rough projections of quantities that cannot be measured with great certainty. Nonetheless, harvesting some of this methane, as at the Salt River Project in Arizona, could represent an important addition to the world's energy supply.

Although the methane found in natural gas is used in a variety of different ways, one of the most common practices is to burn it to produce heat for use in industrial processes or to heat your home, in which "natural gas" is 95% methane. The combustion of methane is represented by this chemical equation:

$$CH_4(g) + 2O_2(g) \rightarrow CO_2(g) + 2H_2O(g)$$

If we use Hess's law, we can calculate the standard enthalpy change for this reaction without actually performing the reaction. To complete the calculation, we need to know the standard enthalpy changes of other reactions that will allow us to consider combusting methane indirectly. We use reference tables to obtain these standard formation reactions and their corresponding standard enthalpies. These values are

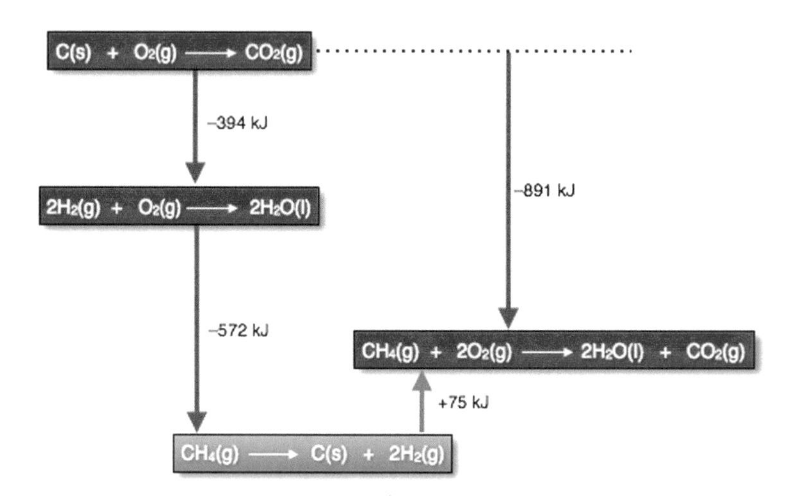

Fig. 5.23 Taking the chemical pathway described above, the standard enthalpy change for the combustion of methane is –891 kJ. No matter what path we took, or how many steps it was, we would end up with the same value for ΔH°

$$C(s) + O_2(g) \rightarrow CO_2(g) \quad \Delta_f H^\circ = -394 \text{ kJ/mol}$$
$$H_2(g) + 1/2O_2(g) \rightarrow H_2O(l) \quad \Delta_f H^\circ = -286 \text{ kJ/mol}$$
$$C(s) + 2H_2(g) \rightarrow CH_4(g) \quad \Delta_f H^\circ = -75 \text{ kJ/mol}$$

We can arrange these reactions in such a way that they allow us to calculate the enthalpy of combustion of methane. This is shown diagrammatically in Fig. 5.23. Similarly, we can rearrange the chemical reactions to mathematically arrive at the same answer. The underlying logic to Hess's law calculations is that we can multiply, and/or reverse, and then combine any chemical equations in whatever way will achieve the overall chemical change in which we are interested. For example, to calculate the enthalpy of combustion of methane, we can manipulate the three combustion equations as follows:

$$1 \times [C(s) + O_2(g) \rightarrow CO_2(g)] \quad \Delta H^\circ = -394 \text{ kJ/mol}$$
$$\mathbf{C(s) + O_2(g) \rightarrow CO_2(g)} \quad \mathbf{\Delta H^\circ = -394 \text{ kJ}}$$
$$2 \times [H_2(g) + 1/2O_2(g) \rightarrow H_2O(l)] \quad \Delta H^\circ = 2 \times (-286 \text{ kJ/mol})$$
$$\mathbf{2H_2(g) + O_2(g) \rightarrow 2H_2O(l)} \quad \mathbf{\Delta H^\circ = -572 \text{ kJ}}$$
$$-1 \times [C(s) + 2H_2(g) \rightarrow CH_4(g)] \quad \Delta H^\circ = -1 \times (-75 \text{ kJ/mol})$$
$$\mathbf{CH_4(g) \rightarrow C(s) + 2H_2(g)} \quad \mathbf{\Delta H^\circ = +75 \text{ kJ}}$$

Adding the three equations together, we calculate the overall enthalpy change:

$$C(s) + O_2(g) \rightarrow CO_2(g) \quad \Delta H^\circ = -394 \text{ kJ}$$
$$2H_2(g) + O_2(g) \rightarrow 2H_2O(l) \quad \Delta H^\circ = -572 \text{ kJ}$$
$$CH_4(g) \rightarrow C(s) + 2H_2(g) \quad \Delta H^\circ = +75 \text{ kJ}$$
$$\mathbf{C(s) + O_2(g) + 2H_2(g) + O_2(g) + CH_4(g) \rightarrow CO_2(g) + 2H_2O(l) + C(s) + 2H_2(g)} \quad \mathbf{\Delta H^\circ = -891 \text{ kJ}}$$

When we cancel the chemicals that appear in equal amounts and in the same states on both sides of the equation (O_2, C, H_2), this simplifies to the equation for the formation of methane:

$$\cancel{C(s)} + O_2(g) + \cancel{2H_2(g)} + O_2(g) + CH_4(g) \rightarrow CO_2(g) + 2H_2O(l) + \cancel{C(s)} + \cancel{2H_2(g)}$$

$$2O_2(g) + CH_4(g) \rightarrow CO_2(g) + 2H_2O(l)$$

$$\Delta_c H^\circ = -891 \text{ kJ/mol (standard enthalpy of combustion of methane)}$$

No matter which way we decide to calculate the change in enthalpy for the reaction, we should arrive at the same result. And, we don't have to actually perform the chemical reaction to be able to calculate its enthalpy change.

Example 5.10—Using Hess's Law

Given the following data, calculate the value of ΔH for the reaction shown below.

$$P_4(s) + 10Cl_2(g) \rightarrow 4PCl_5(g) \quad \Delta H = -2139\ kJ$$
$$PCl_3(g) + Cl_2(g) \rightarrow PCl_5(g) \quad \Delta H = -155\ kJ$$
$$P_4(s) + 6Cl_2(g) \rightarrow 4PCl_3(g) \quad \Delta H = ???\ kJ$$

Asking the Right Questions

What do we need to do with the given equations in order to put the molecules we want as reactants on the left side of the arrow, and the molecules we want as products on the right side of the arrow? What factors must we multiply by in order to get the desired number of moles of each reactant and product?

Solution

It is usually more productive to focus on what we want to achieve than what we want to eliminate.

$$1 \times [P_4(s) + 10Cl_2(g) \rightarrow 4PCl_5(g)]$$
$$1 \times \Delta H = -2139\ kJ$$
$$-4 \times [PCl_3(g) + Cl_2(g) \rightarrow PCl_5(g)]$$
$$-4 \times \Delta H = +620\ kJ$$

The two reactions look like this, and they add up to the desired reaction:

$$P_4(s) + 10Cl_2(g) \rightarrow 4PCl_5(g) \quad \Delta H = -2139\ kJ$$
$$4PCl_5(g) \rightarrow 4PCl_3(g) + 4Cl_2(g) \quad \Delta H = +620\ kJ$$
$$P_4(s) + 6Cl_2(g) \rightarrow 4PCl_3(g) \quad \Delta H = -1519\ kJ$$

Are Our Answers Reasonable?

Our modified equations add up to give us the desired equation, so it must be correct. Also, the resulting enthalpy is on the order of about 1000 kJ, which is a reasonable amount of energy for a chemical reaction to produce.

What If…?

…we wanted the standard enthalpy change for the formation reaction of $PCl_3(g)$ using the same equations in this example?

(Ans:−380 kJ/mol)

Practice 5.10

Given the following data, calculate the value of ΔH for the reaction shown below.

$$6Fe(s) + 4O_2(g) \rightarrow 2FeO_4(s) \quad \Delta H = -1787\ kJ$$
$$2Fe_3O_4(s) + \frac{1}{2}O_2(g) \rightarrow 3Fe_2O_3(s) \quad \Delta H = -186\ kJ$$
$$3Fe_2O_3(s) \rightarrow 6Fe(s) + \frac{9}{2}O_2(g) \quad \Delta H = ???\ kJ$$

5.7.2 Reaction Enthalpies from Enthalpies of Formation

The energy needed to maneuver the final payload stage of the Ariane 5 rocket system when it is in orbit is provided by the reaction between methylhydrazine (CH_3NHNH_2) and dinitrogen tetroxide (N_2O_4):

$$4CH_3NHNH_2(l) + 5N_2O_4(l) \rightarrow 4CO_2(g) + 9N_2(g) + 12H_2O(l)$$

One of the requirements for a chemical reaction that can be used as part of an effective propulsion system is that it has a large negative ΔH value, indicating that it releases a lot of energy. How can we determine the value of ΔH for this reaction?

One answer, based on our earlier discussion, is to perform the reaction in a calorimeter—something that we may not wish to do, especially if we are studying rocket fuel. Alternatively, we could use Hess's law to calculate the enthalpy of the reaction. However, it is often difficult to find all of the enthalpies that are needed to complete the circuitous route from reactants to products. Instead, there is another way to determine the enthalpy of the reaction. We can calculate ΔH of the reaction by means of a useful general rule about enthalpies of formation if other appropriate values of ΔH are known:

- The standard enthalpy change for a reaction can be calculated by subtracting the sum of the enthalpies of formation of the reactants from the sum of the enthalpies of formation of the products.

Remember that the enthalpy of formation of any pure element in its reference form is equal to zero kilojoules (▶ Sect. 5.4). The above statement can be expressed mathematically as follows:

$$\Delta H^\circ_{\text{reaction}} = \Sigma n_p \Delta H^\circ_{(\text{products})} - \Sigma n_r \Delta H^\circ_{(\text{reactants})}$$

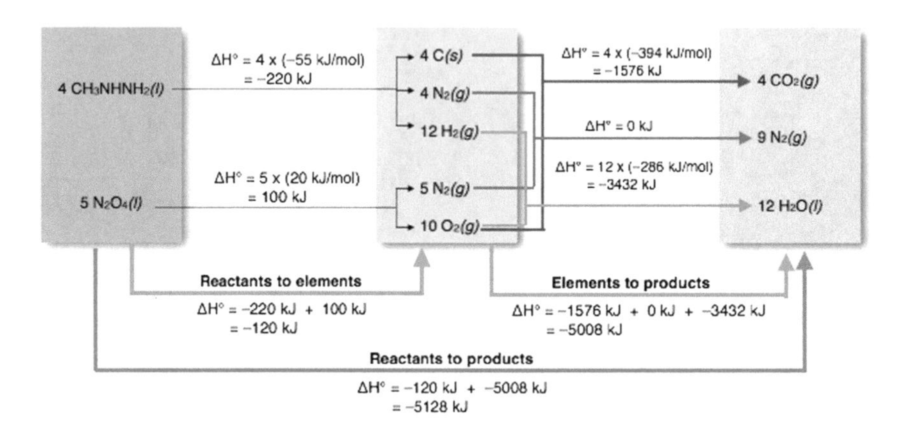

Fig. 5.24 We can consider performing a reaction by taking the indirect route, via the elements in their reference forms. Then, using Hess's law, the reactions shown here can allow us to determine the enthalpy of the reaction

where Σ (sigma) means "the sum of the following terms",
n_p = the number of moles of each product,
n_r = the number of moles of each reactant.

Since enthalpy is a state function, it doesn't matter what path we take to get from reactants to products, and the enthalpy change will be the same. One way to do this is to measure the enthalpy change from the reactants to the elements from which they are formed, and then from those elements to the products. Figure 5.24 illustrates this graphically. Mathematically, this is the same as subtracting the sum of the reactant heats of formation from the sum of the product heats of formation, as we stated above. In this example, the result of product heat of formation minus reactant heat of formation is: – 5008 kJ – 120 kJ = – 5128 kJ. Example 5.11 illustrates another example of this procedure.

Figure 5.24 outlines the reaction used to provide propulsion to maneuver the Ariane 5's payload stage in orbit. Our calculations reveal that it has a relatively large negative ΔH value equal to – 5128 kJ/mol.

Example 5.11—Thermite in Space

One reaction that may prove useful for welding platforms together in space is the thermite reaction. This is the reaction that occurs between aluminum and red iron oxide:

$$2\text{Al}(s) + \text{Fe}_2\text{O}_3(s) \rightarrow 2\text{Fe}(s) + \text{Al}_2\text{O}_3(s)$$

Calculate the standard enthalpy change of this reaction, given the following enthalpies of formation:

$$\Delta_f H^\circ(\text{Fe}_2\text{O}_3) = -824\ \text{kJ/mol};$$
$$\Delta_f H^\circ(\text{Al}_2\text{O}_3) = -1676\ \text{kJ/mol}$$

Asking the Right Questions
Why doesn't the problem give us the heats of formation for aluminum and iron? How do we find the heat change for this reaction?

Solution
Aluminum and iron are in their reference forms, and by definition, their heat of formation is zero.

To find the change in enthalpy for this reaction:

$$\Delta H^\circ_{\text{reaction}} = \Sigma n_p \Delta H^\circ_{(\text{products})} - \Sigma n_r \Delta H^\circ_{(\text{reactants})}$$

$$\Delta H^\circ_{\text{reaction}} = \left[\left(1\ \text{mol} \times -1676\frac{\text{kJ}}{\text{mol}}\right) + \left(2\ \text{mol} \times 0\frac{\text{kJ}}{\text{mol}}\right)\right] - \left[\left(1\ \text{mol} \times -824\frac{\text{kJ}}{\text{mol}}\right) + \left(2\ \text{mol} \times 0\frac{\text{kJ}}{\text{mol}}\right)\right]$$

$$= -852\ \text{kJ}$$

Are Our Answers Reasonable?
Since the question states that this reaction is used to weld metal together, it makes sense that it is an exothermic reaction so it can transfer heat to the metals being welded.

What If...?
...we considered a slightly different thermite reaction, the one between aluminum and black iron oxide:

$$8Al(s) + 3Fe_3O_4(s) \rightarrow 9Fe(s) + 4Al_2O_3(s)$$

Appendix 3 should be helpful to find the $\Delta_f H°$ for $Fe_3O_4(s)$. What would the value of $\Delta H°_{reaction}$ be?

(*Ans:−3353 kJ*)

Practice 5.11
Calculate the enthalpy change for the reaction, by using heats of formation from the Appendix.

$$4NH_3(g) + 3O_2(g) \rightarrow 2N_2(g) + 6H_2O(l)$$

Given the following reactions and $\Delta H°$ values, calculate the enthalpy change for the above reaction using Hess' Law:

$$2NH_3(g) + 3N_2O(g) \rightarrow 4N_2(g) + 3H_2O(l)$$

$$\Delta H° = -1012 \text{ kJ}$$

$$2N_2O(g) \rightarrow O_2(g) + 2\,N_2(g) \quad \Delta H° = -164 \text{ kJ}$$

5.8 Energy Choices

Humans have always had a variety of energy sources to choose from, and choices based on energy issues have always been important. In ancient times, people could keep warm by burning wood, moving south to sunnier climes, or minimizing energy losses from their bodies by wrapping themselves in animal skins and furs. Ancient people also learned to make use of wind energy to power great journeys by boat across seas and oceans. Today, the energy choices facing us are much more complex, and the possible effects of making unwise choices are much greater.

The development of modern civilization was powered largely by burning things to release heat. This heat could keep people warm and could also be used to boil water, generating the steam to power steam engines and steam turbines. The engines and turbines were used to power machinery, trains, and ships and eventually to make electricity. The greatest leaps forward in technology came with the discovery of the fossil fuels—coal, oil, and natural gas—that could be harvested from the Earth in seemingly limitless quantities and burned with few obvious disadvantages, apart from the loss of lives in the process of extracting these natural resources. Today, the burning of fossil fuels provides about 70% of the energy needed to sustain a complex industrialized country such as the United States, as shown in ▫ Fig. 5.25. However, we now know that the supplies of fossil fuels are not limit-

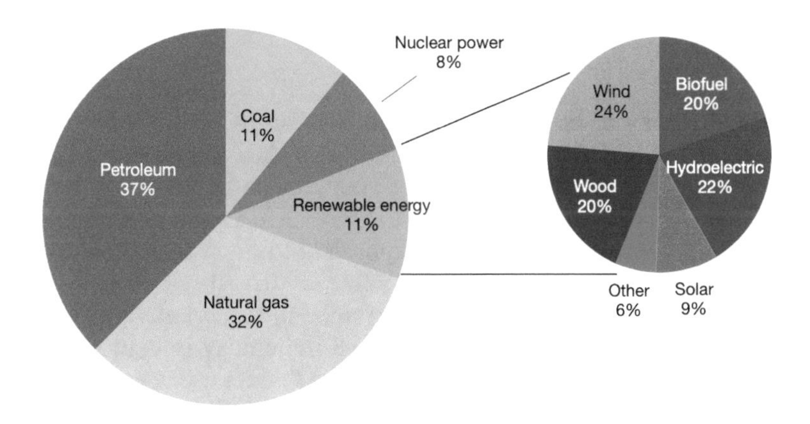

▫ **Fig. 5.25** US Energy use in 2020

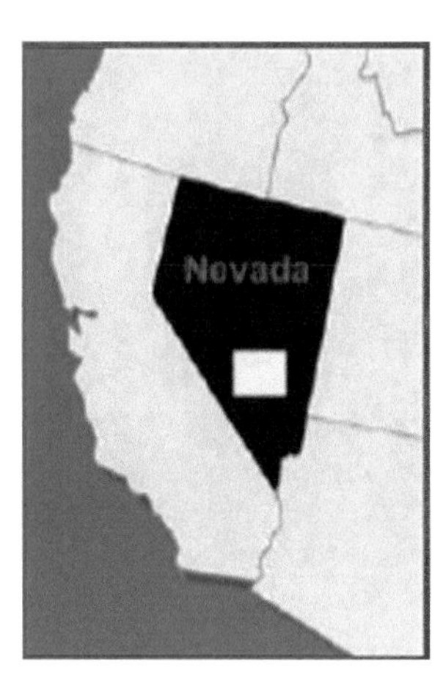

Fig. 5.26 Total area required for a solar cell power plant to meet the total U.S. annual electrical power demand is represented by the square on this map. The square keeps getting larger every year

less and that burning them to provide energy also generates problems. The most significant problem might be the demonstrated warming of the planet, caused in part by the carbon dioxide gas that is a by-product of burning these fuels.

Because fossil reserves are finite, and because the burning of these fuels can result in environmental problems, their processing, distribution and use have had a significant impact on the world's economic and political landscape. The alternative strategy of harnessing energy from nuclear processes was eagerly adopted in many countries in the latter half of the twentieth century.

One of the most attractive options to replace the use of petroleum-based fuels is to make increasing use of **renewable sources of energy**, which can be directly supplied or rapidly regenerated by natural processes. This is in contrast to nonrenewable sources, particularly the fossil fuels: coal, oil, and natural gas. Staff at the U.S. National Renewable Energy Laboratory (NREL) in Colorado have studied the feasibility of using renewable energy sources to meet all of America's energy needs. China has overtaken the United States as the largest single user of energy in the world, and as other countries modernize and raise their standard of living, they too will begin to consume more and more energy. Any plan to meet the world's needs for energy necessitates that every country makes a commitment to alternative energy sources.

Renewable sources of energy – Energy sources that can be rapidly replaced by natural processes.

According to the Renewable Energy Laboratory, the main renewable energy sources that could meet the needs of the United States include:

- **Solar cells** (also known as **photovoltaic systems**, or PVs), which convert the energy of sunlight into electricity.
- **Biomass conversion**, which releases energy from the chemical conversion— often the burning—of plants and trees that can be grown as quickly as they are consumed.
- **Hydroelectric systems**, which use the power of falling water to turn turbines that generate electricity.
- **Wind power**, which uses turbines driven by the force of the wind to generate electricity.
- **Geothermal systems**, which are drilled deep into the earth to exploit the flow of heat from the interior of the earth out toward the surface.

Each of these systems could readily make a significant contribution to power generation in the United States. They already contribute about 11% of the total use in the US, as shown in Fig. 5.25, and their contribution continues to increase yearly. One of the most interesting points made in the NREL's report is summarized in Fig. 5.26, which shows the total area that would have to be devoted to solar cells sufficient to generate all the energy required by the United States. It is a large area, equivalent to more than 11% of the area of the state of Nevada, but that is less than 25% of the area of the country that is currently paved over with roads and streets. So equipped, the United States could garner enough energy from the sun to meet all its needs.

Solar energy arrives free of charge, although providing the technology to capture, distribute, and use it entails significant costs. The energy is also very "clean" in the sense that its use does not release anywhere near the amount of pollutants released by the burning of fossil fuels. Some pollution would inevitably be associated with the manufacturing processes and other activities involved in making and maintaining the solar cells, but the overall effect of a switch to solar is likely to be a huge reduction in pollution. One of the main challenges of the future might be to develop the chemical systems needed to make full use of the free energy that floods down on the planet every day from the Sun.

When the possible contributions from biomass, hydroelectric, wind, and geothermal power are added to the equation, it all amounts to a persuasive argument for continuing to explore the use of "renewables" as an energy source for the future.

The Bottom Line

- Chemical changes are accompanied by the gain (endothermic) or release (exothermic) of energy.
- Thermodynamics is the study of energy changes and exchanges.
- Energy comes in two basic forms: kinetic energy (the energy of motion) and potential energy (positional energy).
- Energy is never created or destroyed; it is only transferred from place to place and converted from one form into another.
- All chemical reactions begin with an input of energy, the activation energy, needed to "jolt" the chemicals into reacting.
- The total change in the energy of a chemical system (U), as it undergoes a chemical reaction, is equal to the heat flow (q), known as the heat of reaction, and the work done (w): $\Delta U = q + w$.
- The work done by chemical systems is often measured as pressure-volume work: $w = -P\Delta V$.
- The SI unit of energy is the joule, which we can relate to the more familiar units of calories and Calories. 1 Calorie = 1000 calories = 4184 joules
- Each substance has a particular specific heat capacity (c), often called the specific heat, which is the amount of heat needed to change the temperature of 1 gram of the substance by 1 degree Celsius (or 1 kelvin) when the pressure is constant.
- The heat capacity of any object is the amount of heat needed to change the temperature of the entire object by 1 degree Celsius (or 1 kelvin) when the pressure is constant.
- We can measure the energy produced or absorbed by a chemical system using calorimetry, using the fundamental relationship:

$$m_{\text{system}} \times c_{\text{system}} \times \Delta T_{\text{system}} = -m_{\text{surround}} \times c_{\text{surround}} \times \Delta T_{\text{surround}}$$

- Hess's law states that the enthalpy change of a chemical reaction is independent of the chemical path or mechanism involved in the reaction.
- Enthalpy is the heat absorbed or released by a chemical system under constant pressure conditions.
- Enthalpy is a state function—it only depends on the initial and final states of the system.
- The standard enthalpy change for a reaction can be calculated by subtracting the sum of the standard enthalpies of formation of the reactants from the sum of the standard enthalpies of formation of the products of the reaction.
- Making appropriate choices about our energy sources will be a critically important part of building a successful and sustainable future for humanity.

Section 5.2—The Concept of Energy

Skill Review

1. Imagine a skateboarder at the top of a half-pipe. Describe the relative changes in potential energy and in kinetic energy as the skater goes from the top of one side of the half-pipe to the top of the other side.

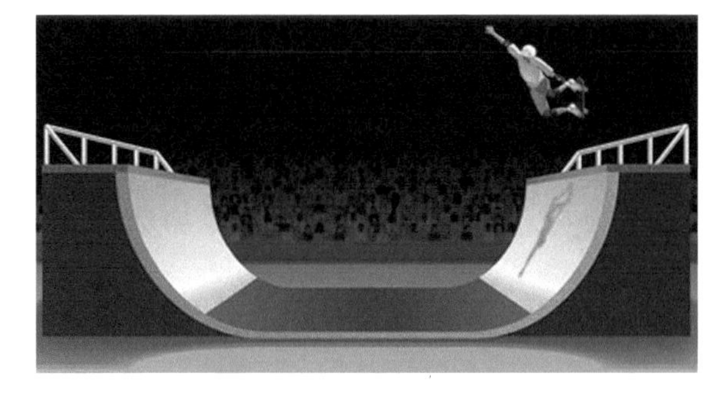

2. Describe the changes in potential and kinetic energy of a snowboarder going over a hill. If there were another uphill slope waiting at the bottom of the first hill, would the snowboarder's potential energy increase or decrease as she or he came down the first hill and immediately began to climb the second?
3. Calculate the kinetic energy, in joules, of the following:
   a. A 185-lb chemistry professor jogging at 8.0 mi/h.
   b. A 42-g hacky sack being kicked at 0.78 m/s.
   c. A molecule of CO_2 moving at 560 m/s.
4. Calculate the kinetic energy, in joules, of the following:
   a. A 2-ton truck lumbering along at 60.0 mi/h.
   b. A 200-g apple falling from a tree at 20 mi/h.
   c. A 0.25-kg ball moving at 1.75 m/s.

5. Describe the following atomic phenomena as primarily a function of the potential or kinetic energy of the system.
   a. The attractive force between an electron in a hydrogen atom and its nucleus.
   b. The vibration of two hydrogen atoms bonded in a molecule of H_2.
   c. The movement of a hydrogen molecule inside a balloon.
6. Describe the following atomic phenomena as primarily a function of the potential or kinetic energy of the system.
   a. The attraction of a hydrogen atom's electron to oxygen in water.
   b. The repulsion of a hydrogen atom's electron from the other hydrogen electron in H_2.
   c. The collision of a molecule and an atom as they react to generate a product.
7. Review the definition of heat. What is being transferred when heat moves between a system and its surroundings? How is this transfer accomplished?
8. What are the three main forms in which energy can be transferred between a system and its surroundings? Give an example of each.
9. When gasoline in an automobile engine undergoes combustion, new compounds and heat are produced. Identify the system and the surroundings in this process. Is the system gaining or losing energy in this process? What is the sign on w for this process?
10. When a cold pack is applied to a sports injury, the athlete may remark that the pack feels very cold. Identify the system and surroundings associated with the cold pack. Is the system gaining or losing energy in this process? What is the sign on q for this process?

Chemical Applications and Practices

11. Burning hydrocarbon fuels and digesting carbohydrates both involve the formation of CO_2 molecules and the release of chemical energy. Which type of compounds—hydrocarbons, carbohydrates, or carbon dioxide—would release the least amount of energy? Explain how you came to this conclusion.
12. During photosynthesis, green plants take in CO_2 and combine it with water to make carbohydrates and other compounds. This process is driven by energy from the Sun. When these food molecules are digested, energy is released. (This energy can be used to maintain healthy body temperature and to run other reactions.) Is the energy released via digestion the same as the energy that was absorbed from the Sun? Explain.
13. On a surface excursion during Apollo 14, Alan Shepard hit a few golf balls. While he originally stated that the balls flew for miles and miles, he later and more accurately estimated the distance as 200 to 400 yards. How would the amount of force needed to drive a ball 300 yards on the Moon differ from that needed to drive a ball the same distance on Earth? Base your explanation on the differences between the environments.
14. There are many forces at work in our everyday commute to work. Identify the typical forces that you'd expect to see evidence of as you drive an automobile down the street.
15. In the catabolic biochemical pathways in cells, adenosine triphosphate (ATP) is produced as part of an endothermic reaction. What is the sign for q in the reaction that represents the production of ATP? Later, the ATP can be used to help the cells do work on the surroundings. What is the sign for w in the reaction that represents that work?
16. In each of the following situations, supply the correct sign for q and w in the reaction that represents the process.
   a. In order to make soft drinks have fizz, or become carbonated, gaseous CO_2 at room temperature must be pumped, under pressure, into an aqueous solution at lowered temperatures. The system may be considered to be the resulting carbonated solution.
   b. Water placed in a microwave oven can be made to boil away into steam. Consider the water in its container to be the system.

Section 5.3—A Closer Look at Work

Skill Review

17. In a chemical reaction, the value of q was determined to be +24 joules (J). The value of work energy was determined to be + 12 J. What is the total energy change for this system? Did the surroundings gain or lose energy?
18. In a chemical reaction, the value of q was determined to be −200 joules (J). The value of work energy was determined to be + 158 J. What is the total energy change for this system? Did the surroundings gain or lose energy?
19. Calculate the total energy change for each of the following systems. Do the surroundings gain or lose energy in each process?
    a. $q = +45$ J; $w = +45$ J
    b. $q = -266$ J; $w = 1.2$ kJ
    c. $q = 23.4$ kJ; $w = -14$ kJ
    d.

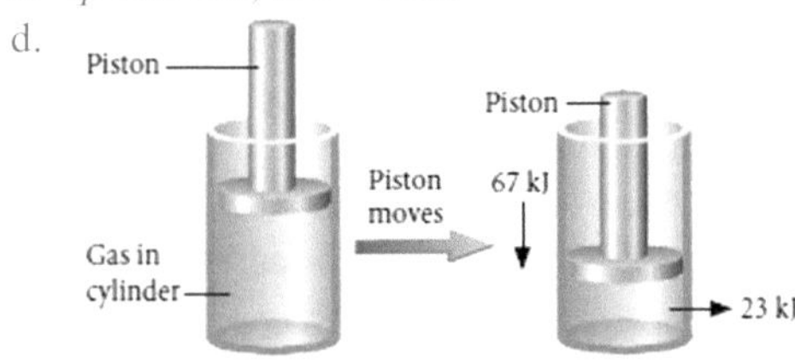

20. Calculate the total energy change for each of the following systems. Does the system gain or lose energy in each process?
    a. $q = -23$ J; $w = -37$ J
    b. $q = -88$ J; $w = +36$ J
    c. $q = +105$ J; $w = -133$ J
    d.

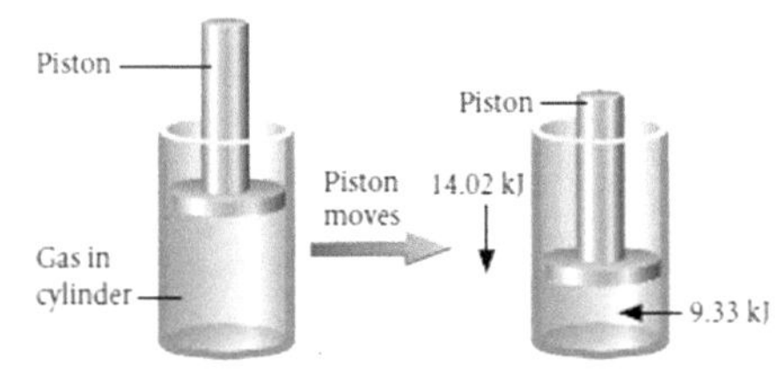

Section 5.4—The Units of Energy

Skill Review

21. What is a "joule" of energy?
22. How many joules are in one "food Calorie."
23. Suppose a 275-g apple and a 175-g orange are moving with the same kinetic energy. If the apple is moving at 15 m/s, how fast, in m/s, is the orange moving? (Maybe we can compare apples and oranges after all!)
24. Some of the best tennis players may serve a 56.9-g tennis ball at a velocity of approximately 115 mi/h. What is the kinetic energy, in joules, of the tennis ball in such a serve?
25. A molecule of N_2 moving through a room may have a velocity of approximately 420 m/s. What is the kinetic energy, in joules, of a molecule of N_2?
26. An automobile is moving at a rate of 31 m/s. At this rate it has a kinetic energy of 436 kJ. What is the mass of this automobile in kg?
27. If a person lifts a 25.0 kg weight 2.0 m high, how much energy do they use in joules? (Hint: the weight is lifted against the acceleration of gravity, which is 9.80 m/s^2.)
28. A certain backhoe machine can lift an 8,000. lb bucket of dirt to a height of 3.0 m. How much energy, in joules, is expended in this action?
29. A low-fat popcorn snack advertises that one serving provides 90 Cal. One serving is considered to be 34 g. What is the energy provided in Calories per gram? What is the energy provided in calories per gram? What is the energy provided in joules per gram?
30. A different food item advertises that one serving provides 320 Cal. One serving is considered to be 80 g. What is the energy provided in Calories per gram? What is the energy provided in calories per gram? What is the energy provided in joules per gram?

31. How many joules as heat are needed to raise 50.0 g of water from 23.0 °C to 37.0 °C?
32. How many joules as heat are required to raise 77.0 g of water from 18.0 °C to 25.0 °C?
33. The British thermal unit is used on some heating and cooling devices. The BTU is defined as the amount of energy as heat that will raise one pound of water from 58.5 °F to 59.5 °F. Express that amount of heat in joules.
34. A popular instant coffee-flavored drink provides 25 food Calories per serving. Express that quantity in calories, joules, and kilojoules.

Chemical Applications and Practices

35. One of the reactions used in some hand-warmer packets is the oxidation of iron to form rust. The formation of 1 mol of rust produces approximately 410 kJ of heat.
    a. If this heat were absorbed by 2000.0 g of water at 22.0 °C in a calorimeter, how hot could the water be made?
    b. If only 0.10 mol of rust were formed, how hot could 200.0 g of water at 22.0 °C become?
36. Ammonium nitrate can sometimes be used in the chemical cold pack applied to some sports injuries. If 10.0 g of ammonium nitrate were placed in a coffee cup calorimeter that contained 100.0 g of water and the temperature changed from 25.0 °C to 18.0 °C, what was the heat transferred in kilojoules?
37. Ethanol (C_2H_5OH) is being blended with gasoline mixtures to produce a fuel for automobiles called gasohol. If the combustion of 10.0 g of ethanol produces 268 kJ of heat, how much heat, in kilojoules, will be produced in burning 1 mol of ethanol?
38. If the combustion of 48.8 g of methane (CH_4) produces 855 kJ of heat, how much heat will be produced in burning 1 mol of methane?
39. If you are working out and lifting a 25.0 kg weight over your head (a total height of 2.0 m), and you do this 100 times, how many ounces of peanuts would you need to eat to replenish the energy that you expended during your workout? (refer to Example 5.4 and problem 29, above).
40. If the backhoe in Problem 28 burns diesel fuel (1.37×10^8 J/gal), how many gallons of diesel need to be burned to lift 1000 buckets of dirt to the full height of 3.0 m? (assume that there's no energy lost).

Section 5.5—Specific Heat Capacity and Heat Capacity

Skill Review

41. Most of the definitions that you have learned in chemistry have very specific terminology. Explain why the definition of specific heat allows you to report the value using either °C or K.
42. In order to clarify the differences among specific heat, heat capacity, and molar heat capacity, provide a brief definition of each.
43. Which of the following involves the greater amount of heat transfer?
    a. 10.0 kg of water in an automobile cooling system changes from 45.0 °C to 10.0 °C as it cools overnight.
    b. In preparation for the annual "chili cook-off," 8.0 L of water is heated from 22.0 °C to 99.0 °C.
44. If the heat capacity of a metal coffee pot filled with water were known to be 5.87 $\frac{kJ}{°C}$, what would you calculate to be the amount of heat, in kilojoules, transferred from a campfire when the temperature of the coffee pot changes from 22.0 °C to 95.0 °C?
45. Fill in the values missing from the following table.

Heat (q)	Specific Heat $\left(\frac{J}{g{\cdot}°C}\right)$	Mass (g)	ΔT (°C)
10.0 J	4.184		10.0
	0.115	10.0	5.0
15.5 J		42.5	15.0

46. Fill in the values missing from the following table.

Heat (q)	Specific Heat $\left(\frac{J}{g{\cdot}°C}\right)$	Mass (g)	ΔT (°C)
	4.184	100.0	40.0
450 J	0.315	15.0	
48.5 J		48.5	15.0

Chemical Applications and Practices

47. A student burned a cashew and held it underneath a beaker containing 51.2 g of water. A total of 0.40 g of cashew was burned, and the water was heated up by 23.0 °C. How many calories of energy did the cashew contain? How many calories per gram did the cashew contain?
48. If you wanted to determine how much energy was contained in your favorite "energy bar", you could burn a piece of it and see how much energy was released. If you burned a 1.1-g piece of your energy bar and it released enough energy to heat up 55.3 g of water by 26.0 °C, how many kilocalories does the 100.0-g bar contain?
49. Some research is being done on the production of nickel compact discs. These disks could store tremendous amounts of information and last an unusually long time. Like most metals, nickel has a low specific heat. If 35.0 g of nickel absorbs 311 J, it increases in temperature by 20.0 °C. What is the specific heat of nickel?
50. If 50.0 g of a metal alloy absorbs 471 J, it increases in temperature by 30.0 °C. What is the specific heat of this metal alloy?
51. In order to calibrate a bomb calorimeter, a 2.000-g sample of benzoic acid (C_6H_5COOH; molar mass = 122.12) was combusted. The calorimeter temperature rose by 1.978 °C. What is the heat capacity of this calorimeter? The molar energy of combustion for benzoic acid is known to be −3227 kJ.
52. A different bomb calorimeter was calibrated by combusting a 2.250-g sample of methanol (CH_3OH). The calorimeter temperature rose by 0.522 °C. What is the heat capacity of this calorimeter? The molar energy of combustion for methanol is known to be approximately −890.3 kJ.
53. a. The heat capacity of a certain bomb calorimeter was calibrated at 28.9 $\frac{\text{kJ}}{^\circ\text{C}}$. When 1.500 g of an unknown sugar was combusted in the calorimeter, the temperature rose by 2.56 °C. What was the energy of combustion of the sugar?

    b. What additional information would we need in order to report the molar heat of combustion for the sugar?
54. A student constructs a crude "coffee cup" calorimeter that contains 94.1 g of water, at 22.0 °C, in a double cup set up with a thermometer and cork cover. When an 85.8-g piece of copper at a temperature of 100.0 °C was placed in the calorimeter, the temperature was noted to equilibrate at 28.0 °C. The specific heat of copper is approximately 0.386 $\left(\frac{\text{J}}{\text{g}\cdot{}^\circ\text{C}}\right)$.

    a. Calculate the heat, in joules, gained by just the water.

    b. Determine the heat capacity, in J/°C, for the empty calorimeter.
55. The molar heat of combustion of propane is -2.2×10^3 kJ. How many grams of propane would have to be combusted to raise 1.0 kg of water (the amount you might boil to prepare a tasty macaroni and cheese dinner) from 22.0 °C to 100.0 °C? Assume that no heat is absorbed by the container or the air.
56. The specific heat of water is 4.184 J/g · °C. How many grams of water at 85.0 °C would have to be added to raise 1.00 kg of water from 25.0 °C to 50.0 °C? Assume that the container and the air absorb no heat.
57. The heat of combustion for acetylene (C_2H_2) is − 1300 kJ/mol. Methane has a heat of combustion of − 890 kJ/mol. Calculate which provides more energy per gram.
58. The heat of combustion for glucose ($C_6H_{12}O_6$) is − 2803 kJ/mol. Tristearin ($C_{57}H_{110}O_6$), a typical fat, has a heat of combustion of − 37,760 kJ/mol. Calculate which provides more energy per gram.
59. One way to test a metallic sample to see whether it is made of gold is to heat it and place it in a calorimeter to determine the specific heat of the sample. The specific heat of gold is 0.13 $\left(\frac{\text{J}}{\text{g}\cdot{}^\circ\text{C}}\right)$. What temperature change would prove that the metal was gold, if 15.0 g of the sample at 99.0 °C were placed in a calorimeter initially at 25.0 °C? The heat capacity of the calorimeter is 25.0 $\frac{\text{J}}{\text{K}}$.
60. In the chapter, a bomb calorimeter was discussed that had a heat capacity of 32.4 $\frac{\text{kJ}}{^\circ\text{C}}$. If 0.550 g of benzoic acid ($C_7H_6O_2$) were combusted in the calorimeter, what would be the expected change in temperature? (The molar energy of combustion for benzoic acid is − 3227 kJ/mol.)

Section 5.6—Enthalpy

Skill Review

61. Explain how chemical reactions that generate gases can be thought of as constant-pressure situations even though the volume of the products can be thought of as expanding?
62. When the equation $\Delta U = q_p + w$ is recast as $\Delta U = q_p - P\Delta V$, the sign between the heat and pressure–volume work changes. Explain why it is now proper to express the $P\Delta V$ with a negative sign.
63. Without using any symbols or numbers, define the term enthalpy as it is applied in chemistry.
64. What is the chief thermodynamic difference among the expressions known as "heat of reaction," "standard heat of reaction," "standard heat of formation," and "standard heat of combustion"?

65. In an exothermic reaction, does the system or the surroundings gain energy as heat?
66. What sign would characterize the value of ΔH for the reverse of an exothermic reaction?
67. What are the conditions being described when a substance is taken to be in its standard state?
68. What is the physical state of carbon dioxide when it is in standard state conditions?
69. Of the following, which would not be considered appropriate equations to represent "standard heat of formation" processes? Explain the reasons for your choices.
    a. $C(s) + \frac{1}{2}O_2(g) \rightarrow CO(g)$
    b. $N_2(g) + 3H_2(g) \rightarrow 2NH_3(g)$
    c. $CS_2(g) \rightarrow CS_2(l)$
    d. $C(g) + 4H(g) \rightarrow CH_4(g)$
    e. $4CO_2(g) + 5H_2O(l) \rightarrow C_4H_{10}(g) + 13/2\ O_2(g)$
    f.

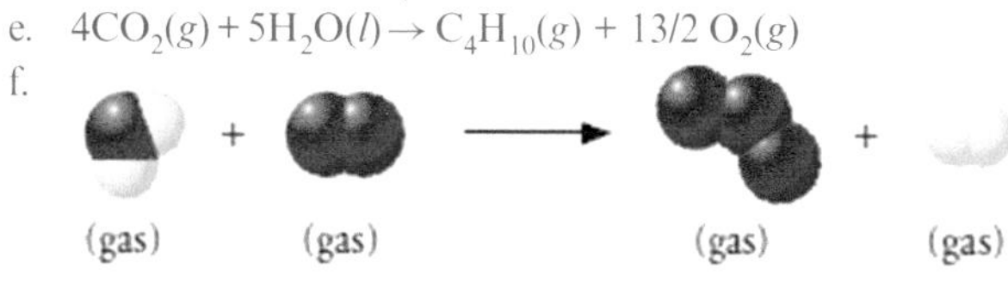

70. Of the following, which would not be considered appropriate equations to represent "standard heat of formation" processes? Explain the reasons for your choices.
    a. $C(s) + O_2(g) \rightarrow CO_2(g)$
    b. $H_2(g) + O_2(g) \rightarrow H_2O_2(g)$
    c. $\frac{1}{2}H_2(g) + O_2(g) \rightarrow H_2O(g)$
    d. $CH_4(g) + 2O_2(g) \rightarrow CO_2(g) + 2H_2O(g)$
    e. $2H(g) + O(g) \rightarrow H_2O(l)$
    f.

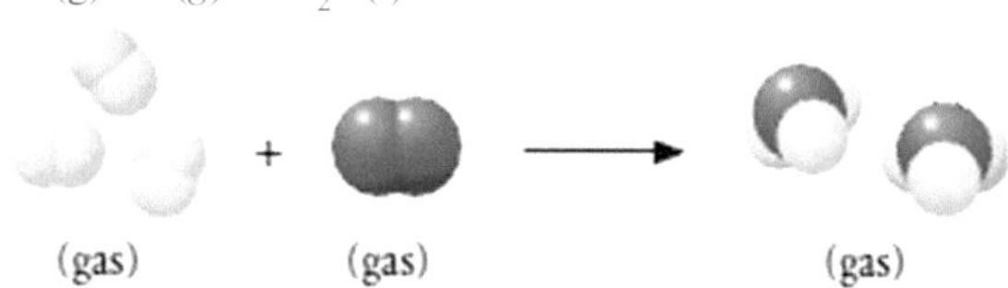

71. Write the balanced equation for the formation reaction of the following substances:
    a. $C_4H_8(g)$
    b. $CH_3COOH(l)$
    c. $NH_2OH(s)$
    d. $H_2SO_4(l)$
72. Write the balanced equation for the formation reaction of the following substances:
    a. $C_7H_{16}(l)$
    b. $C_6H_5COOH(l)$
    c. $NH_4NO_3(s)$
    d. $H_3PO_4(l)$

Chemical Applications and Practices

73. Explain why the values of H and U for reactions such as the explosion of nitroglycerin, represented here, could vary significantly.

$$4C_3H_5N_3O_9(l) \rightarrow 12CO_2(g) + 10H_2O(g) + 6N_2(g) + O_2(g)$$

74. Consider the reaction for nitroglycerin decomposition shown in Problem 73. Cite two reasons why the reverse of the reaction would not qualify as the standard heat of formation for nitroglycerin.
75. a. Write the balanced chemical equation for the standard formation of carbon monoxide.
    b. If the value for the standard heat of formation for carbon monoxide were −110.5 kJ/mol, would you consider the reaction endothermic or exothermic?
    c. What would be the ΔH value for the reverse of the reaction?
76. a. Write the balanced chemical equation for the standard formation of gaseous hydrogen peroxide (H_2O_2).
    b. If the value for the standard heat of formation for hydrogen peroxide were −136.1 kJ/mol, would you consider the reaction endothermic or exothermic?
    c. What would be the ΔH value for the reverse of the reaction?

77. The hydrocarbon fuel butane (C_4H_{10}) is used in small portable lighters. Write the balanced reaction for the standard heat of combustion for this reaction. (Recall that the two products of hydrocarbon combustion are carbon dioxide and water.)
78. Pentane (C_5H_{12}) can be blended with other hydrocarbons and additives to form gasoline. Write the balanced reaction that depicts the standard heat of formation of pentane.
79. The primary energy molecule in cells is ATP (adenosine triphosphate, $C_{10}H_{15}N_5O_{13}P_3$). Write the standard heat of formation reaction for this important biomolecule.
80. Glucose ($C_6H_{12}O_6$) is used by our bodies as an energy source. Write the standard heat of formation reaction for this important molecule.

Section 5.7—Hess's Law

Skill Review

81. What is meant by the term "state function"?
82. Since enthalpy is a state function, the change in enthalpy for the forward reaction is ______________ in magnitude, but ______________ in sign to the enthalpy for the reverse reaction.
83. The ΔH value for the following reaction is – 1012 kJ.

$$2NH_3(g) + 3N_2O(g) \rightarrow 4N_2(g) + 3H_2O(l)$$

   a. What is the value of ΔH for the reverse of the reaction?
   b. What is the value of ΔH for 1 mol of NH_3 reacting?
   c. What is the value of ΔH for 4 mol of NH_3 reacting?
84. The ΔH value for the following reaction is + 284.6 kJ.

$$3O_2(g) \rightarrow 2O_3(g)$$

   a. What is the value of ΔH for the reverse of the reaction?
   b. What is the value of ΔH for 1 mol of O_2 reacting?
   c. What is the value of ΔH for 4 mol of O_2 reacting?
85. Add the following reactions together to determine the overall reaction that occurs. What is the enthalpy change for the sum of these reactions?

$$\begin{array}{lll} \text{Reaction 1:} & C_3H_8(g) + 5O_2(g) \rightarrow 3CO_2(g) + 4H_2O(g) & \Delta H = -2043\text{kJ} \\ \text{Reaction 2:} & 3CO_2(g) \rightarrow 3C(s) + 3O_2(g) & \Delta H = +1181\text{kJ} \\ \text{Reaction 3:} & 4H_2O(g) \rightarrow 4H_2(g) + 2O_2(g) & \Delta H = +967.2\text{kJ} \\ \text{Sum:} & & \Delta H = ??? \end{array}$$

86. Add the following reactions together to determine the overall reaction that occurs. What is the enthalpy change for the sum of these reactions?

$$\begin{array}{lll} \text{Reaction 1:} & 2NO(g) + O_2(g) \rightarrow 2NO_2(g) & \Delta H = -113\text{ kJ} \\ \text{Reaction 2:} & N_2(g) + O_2(g) \rightarrow 2NO(g) & \Delta H = +183\text{ kJ} \\ \text{Reaction 3:} & 2NO_2(g) \rightarrow N_2(g) + 2O_2(g) & \Delta H = -163\text{ kJ} \\ \text{Sum:} & & \Delta H = ??? \end{array}$$

87. Determine the missing chemical reaction and its enthalpy change, given the following information:

$$\begin{array}{lll} \text{Reaction 1:} & 2H_2(g) + 2F_2(g) \rightarrow 4HF(g) & \Delta H = -1074\text{ kJ} \\ \text{Reaction 2:} & & \Delta H = ???\text{ kJ} \\ \text{Reaction 3:} & C_2H_4(g) \rightarrow 2H_2(g) + 2C(s) & \Delta H = -52.3\text{ kJ} \\ \text{Sum:} & C_2H_4(g) + 6F_2(g) \rightarrow 2CF_4(g) + 4HF(g) & \Delta H = -2486\text{ kJ} \end{array}$$

88. Determine the missing chemical reaction and its enthalpy change, given the following information:

$$\begin{array}{lll} \text{Reaction 1:} & & \Delta H = ???\text{ kJ} \\ \text{Reaction 2:} & 2MgCl_2(aq) + 2H_2O(l) \rightarrow 2MgO(s) + 4HCl(aq) & \Delta H = +282\text{ kJ} \\ \text{Reaction 3:} & 2H_2(g) + O_2(g) \rightarrow 2H_2O(l) & \Delta H = -572\text{ kJ} \\ \text{Sum:} & 2Mg(s) + O_2(g) \rightarrow 2MgO(s) & \Delta H = -1204\text{ kJ} \end{array}$$

Chemical Applications and Practices

89. The fuel used in many rural settings is propane (C_3H_8). Write the heat of formation reaction for this important fuel. Now use the combustion reaction sequence shown in the text to develop a Hess's law scheme that would allow you to calculate the ΔH of formation for propane.

90. The sugar arabinose ($C_5H_{10}O_5$) is obtained from plants with the polysaccharide gum arabic. In wheat plants, this sugar helps form important cell wall structures. The heat of formation reaction for arabinose is

$$5C(s) + 5H_2(g) + \frac{5}{2}O_2(g) \rightarrow C_5H_{10}O_5(s)$$

Without using actual kJ values, show the reactions that you could use, and the way you would use them, to obtain the arabinose formation reaction.

91. Given the following reactions and ΔH values:

$$\begin{array}{ll} B_2O_3(s) + 3H_2O(g) \rightarrow B_2H_6(g) + 3O_2(g) & \Delta H = +2035\ \text{kJ} \\ 2H_2O(l) \rightarrow 2H_2O(g) & \Delta H = +88\ \text{kJ} \\ H_2(g) + \frac{1}{2}O_2(g) \rightarrow H_2O(l) & \Delta H = -286\ \text{kJ} \\ 2B(s) + 3H_2(g) \rightarrow B_2H_6(g) & \Delta H = +36\ \text{kJ} \end{array}$$

Calculate ΔH for

$$2B(s) + \frac{3}{2}O_2(g) \rightarrow B_2O_3(s) \qquad \Delta H = ?$$

92. Use the heat of formation reaction data for water and carbon dioxide and the following reaction for the combustion of acetylene, employed in high-temperature welding applications, to find the heat of formation of acetylene (C_2H_2).

$$C_2H_2(g) + \frac{5}{2}O_2(g) \rightarrow 2CO_2(g) + H_2O(l)$$
$$\Delta_cH = -1300\ \text{kJ/mol}$$

93. a. Ethanol (C_2H_5OH) is added to gasoline to create the fuel mixture called "E10". Using heat of formation data and Hess's law, determine the heat of combustion, in kJ/mol, for ethanol.
Combustion of ethanol:

$$C_2H_5OH(l) + 3O_2(g) \rightarrow 2CO_2(g) + 3H_2O(l) \qquad \Delta_cH = ?$$

Formation of ethanol:

$$2C(graphite) + 3H_2(g) + 1/2O_2(g) \rightarrow C_2H_5OH(l) \qquad \Delta_fH = -278\ \text{kJ/mol}$$

b. Using your answer from part a, determine the heat released when 100.0 g of pure ethanol is combusted.

94. The oxidation of sulfur has many important environmental connections. Notably, acid rain is formed from sulfur oxides reacting with moisture in the air. Use the following two reactions to determine the enthalpy change when sulfur dioxide reacts with oxygen to form sulfur trioxide.

$$\begin{array}{ll} S(s) + O_2(g) \rightarrow SO_2(g) & \Delta H^\circ = -296.8\ \text{kJ} \\ 2S(s) + 3O_2(g) \rightarrow 2SO_3(g) & \Delta H^\circ = -791.4\ \text{kJ} \\ 2SO_2(g) + O_2(g) \rightarrow 2SO_3(g) & \Delta H^\circ = ? \end{array}$$

95. The octane rating on gasoline is a method of comparing fuel energy values. What is the value for the heat of combustion of octane? ($\Delta_fH^\circ(C_8H_{18}) = -249.9$ kJ/mol)

$$C_8H_{18}(l) + \frac{25}{2}O_2(g) \rightarrow 8CO_2(g) + 9H_2O(l)$$

96. A typical component of gasoline is pentane (C_5H_{12}). The standard enthalpy of combustion of pentane is − 3537 kJ/mol. Write the combustion reaction for pentane, and determine the standard enthalpy of formation for pentane.

97. Calcium metal will react in water to form calcium hydroxide, $Ca(OH)_2$. Use the data that follow and Hess's law to calculate the value of ΔH° for the reaction

$$Ca(s) + 2H_2O(l) \rightarrow Ca(OH)_2(s) + H_2(g)$$

a. $H_2(g) + \frac{1}{2}O_2(g) \rightarrow H_2O(l)$ $\Delta H° = -286$ kJ
b. $CaO(s) + H_2O(l) \rightarrow Ca(OH)_2(s)$ $\Delta H° = -64$ kJ
c. $Ca(s) + \frac{1}{2}O_2(g) \rightarrow CaO(s)$ $\Delta H° = -635$ kJ

98. Ozone is reduced by hydrogen to produce water. Use the data that follow and Hess's law to calculate the value of $\Delta H°$ for the reaction

$$3H_2(g) + O_3(g) \rightarrow 3H_2O(g)$$

a. $H_2(l) + \frac{1}{2}O_2(g) \rightarrow H_2O(l)$ $\Delta H° = -286$ kJ
b. $2H_2(g) + O_2(g) \rightarrow 2H_2O(g)$ $\Delta H° = -483.6$ kJ
c. $3O_2(g) \rightarrow 2O_3(g)$ $\Delta H° = +284.6$ kJ

Section 5.8—Energy Choices

Chemical Applications and Practices

99. Most of the world's energy consumption for power is based on fossil fuels. This source, however, is considered nonrenewable. Look back at the list of renewable energy sources and cite one advantage, in addition to this renewability, that each would have over the petroleum-based fuels used today.
100. Assuming that the fossil fuels are completely depleted at some point in the future, the world will need to depend on renewable sources of energy. Look back at the list of renewable energy sources and describe one disadvantage of using each source as the sole source of energy for a country.
101. Ethanol is increasingly popular as a more eco-friendly source of energy than gasoline, supposedly contributing less to greenhouse gas emissions. Let's take a closer look at this idea, starting with the chemical reactions for the combustion of ethanol and octane (a major component of gasoline):

$$\text{Ethanol}: 2C_2H_5OH(l) + 6O_2(g) \rightarrow 6H_2O(g) + 4CO_2(g)$$

$$\text{Octane}: 2C_8H_{18}(l) + 25O_2(g) \rightarrow 18H_2O(g) + 16CO_2(g)$$

a. Calculate the $\Delta H°$ for each of these combustion reactions. Which one produces more energy?
b. How many moles of ethanol would you need to burn to get as much energy as one mole of octane?
c. Which fuel produces more moles of greenhouse gases (water and carbon dioxide) per kJ of energy?

102. Propane ("LP gas") has been popular for many years as a more eco-friendly source of energy than gasoline, supposedly contributing less to greenhouse gas emissions. Let's take a closer look at this idea, starting with the chemical reactions for the combustion of propane and octane (a major component of gasoline):

$$\text{propane: } C_3H_8(g) + 5O_2(g) \rightarrow 4H_2O(g) + 3CO_2(g)$$

$$\text{octane: } C_8H_{18}(l) + 25O_2(g) \rightarrow 9H_2O(g) + 8CO_2(g)$$

a. Calculate the $\Delta H°$ for each of these combustion reactions. Which one produces more energy?
b. How many moles of propane would you need to burn to get as much energy as one mole of octane?
c. Which fuel produces more moles of greenhouse gases (water and carbon dioxide) per kJ of energy?

Comprehensive Problems

103. Individual atoms and molecules are so small that they have very low individual values of kinetic energy. However, given their mass and velocity, it is possible to calculate these values. What is the kinetic energy, in joules, of an oxygen molecule (O_2) in air that you are breathing if its velocity is 460 m/s? Would you expect a nitrogen molecule (N_2) moving at the same speed to have more or less kinetic energy than the oxygen molecule? Explain.
104. Distinguish between the two terms in each of the following pairs:
    a. Heat and temperature
    b. System and surroundings
    c. Exothermic and endothermic
    d. q and ΔU
105. What is the role of activation energy in a chemical process? Is it possible for the activation energy of a reaction to be greater than the overall change in energy for a chemical reaction? Explain.
106. The average temperature of a healthy human is approximately 37 °C. The average room temperature may be about 25 °C. Explain how we are able to keep our body temperature so much higher than that of our environment.

107. A 400-mL glass beaker contains 250 g of water at room temperature. As several NaOH pellets are placed in the water and begin to dissolve, you notice a warming sensation in the hand in which you are holding the beaker. Answer the following questions.
    a. If the NaOH and the water make up the system, would you consider the process endothermic or exothermic? Explain.
    b. Is the beaker part of the system or part of the surroundings?
    c. During this process, is energy flowing into or out of the system?
    d. During this process, has the kinetic energy of the water molecules been raised or lowered?
    e. What work is being done during this process?
108. A serving of Italian rice—risotto—provides 150 food Calories. What would this value be in kilojoules? Assume that this quantity of energy would be available to do the work of lifting 2.5-kg chemistry textbooks from the floor to a height of 1.5 m. How many such chemistry textbooks could, theoretically, be lifted through that height? (And later, of course, read thoroughly.)
109. Express the energy content of a 2.00-oz. candy bar that contains 247 cal in Calories per gram, in joules per gram, and in kilojoules per gram. If this energy were efficiently used to provide the kinetic energy to move a 1.5-pound chemistry book, how fast, in meters per second, could the book move?
110. One of the main reasons for eating is to obtain a supply of energy. A low-fat apple muffin may provide 170 food Calories for each 50-g muffin. How much muffin to 3 significant figures, is needed to provide 1 J?
111. a. Suppose you are heating water (225 g) in a mug that you have placed in a microwave oven so you can make a mug of hot chocolate. Calculate the amount of energy as heat that the water has absorbed as you heated it from 15.0 °C to 98.0 °C.
    b. What additional information would you need in order to determine the heat absorbed by the mug?
112. You have just removed a hot cheese pizza from the oven, and all of the ingredients are presumably at the same temperature. Without waiting for it to cool, you take a bite of the pizza. As you bite the pizza, the bread is hot on your tongue but does not burn. However, as you continue to bite, pizza sauce (mostly tomatoes and water) squeezes out and burns the roof of your mouth. Which has the higher specific heat, the bread or the sauce? Explain the basis of your answer using $q = m \times c \times \Delta T$.
113. One way to determine the heat capacity of a constant-volume calorimeter is to burn measured amounts of pure carbon in the presence of oxygen gas to form carbon dioxide. From the following experimental data, determine the heat capacity of the calorimeter. A 0.200-g sample of carbon, when completely combusted, raised the temperature of the water and the contents of the entire calorimeter from 24.0 °C to 25.5 °C. It is known that under these conditions, the heat released from the complete combustion of 1 mol of carbon is 392 kJ.
114. A student's coffee cup calorimeter, including the water it contains, has been calibrated in a manner similar to that described in Problem 51. The heat capacity was found to be 55.5 $\frac{\text{J}}{^\circ\text{C}}$. If a 65.8-g sample of an unknown metal, at 100.0 °C, was placed in the calorimeter initially at 25.0 °C, and an equilibrium temperature of 29.1 °C was reached, what is the specific heat of the metal?
115. The foods we eat provide fuel to keep us alive. Burning a 0.500-g sample of peanuts provides enough heat to raise the temperature of a calorimeter by 2.5 K. Assuming the heat capacity of the calorimeter to be 7.5 $\frac{\text{kJ}}{\text{K}}$, determine the heat of combustion for 1 g of the peanut.
116. A student performs the experiment shown graphically here. What is the specific heat of the block of metal used in the experiment? (Assume that the heat capacity of the empty calorimeter is 7.5 $\frac{\text{J}}{^\circ\text{C}}$.)

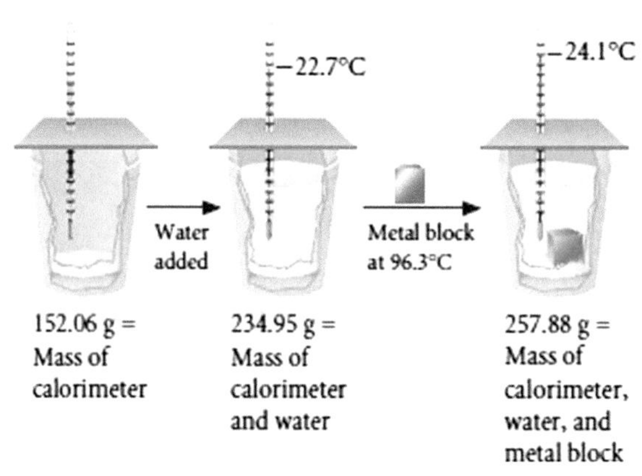

117. The reaction of the gases ethane (C_2H_6) and oxygen gives gaseous carbon dioxide and water vapor.
   a. Write the balanced chemical reaction for this combustion.
   b. What is the enthalpy change, in kilojoules, for this process?
   c. If 250.0 g of ethane are consumed in the reaction, how many grams of carbon dioxide would be produced?
   d. Why should precautions be taken to avoid performing the reaction outlined in part c in an enclosed space?
118. Under certain conditions, the reaction of chlorine gas with metallic iron can produce iron(III) chloride.
   a. Write the balanced chemical equation for this reaction.
   b. If 8.44 g of iron(III) chloride is produced, how many grams of iron were required?
   c. What is the enthalpy change, in kilojoules, required to produce 8.44 g of iron(III) chloride? ($\Delta_f H(FeCl_3) = -399$ kJ/mol)
119. Sodium hydroxide pellets can be used to unclog a drain.
   a. What is the enthalpy change when 15.0 g sodium hydroxide is added to 1.00 L of water?
   b. What is the molarity of this solution? (Assume that the volume change is negligible.)
120. A student wishes to prove the existence of silver ions in a particular solution that was supposedly made by dissolving silver nitrate in water.
   a. Explain how this could be determined using a solution of sodium sulfide and write a reaction that illustrates this method.
   b. What is the standard molar enthalpy change, in kilojoules, for the chemical reaction of a solution of silver nitrate and sodium sulfide? (Only consider the chemical reaction, not the solution process).
121. A 0.10 lb sample of sodium metal is added to 1.00 gallon of water.
   a. Assuming no change occurs in the volume of the sample, what would be the concentration of the resulting solution of sodium hydroxide?
   b. What is the enthalpy change, in kilojoules, for this process?
122. To determine the heat of combustion of ethanol, a 0.500 g sample was burned in pure oxygen in a bomb calorimeter. Write the balanced reaction for the complete combustion of ethanol in oxygen to form carbon dioxide and water. The bomb itself has a heat capacity of 570 J/°C and contained 825 g of water. The temperature of the bomb and the water rose from 25.5 °C to 29.2 °C. Calculate the enthalpy of combustion of ethanol, in kJ/mol.

? Thinking Beyond the Calculation

123. Hydrogen/oxygen fuel cells allow hydrogen and oxygen to combine to form water and energy by this reaction:

$$H_2(g) + 1/2O_2(g) \rightarrow H_2O(l)$$

   a. How many kJ per mole of hydrogen does this reaction produce?

   The combustion of octane (a primary component of gasoline) proceeds according to this reaction:

$$2C_8H_{18}(l) + 25O_2(g) \rightarrow 18H_2O(g) + 16CO_2(g)$$

   b. How many kJ per mole of octane does this reaction produce?
   c. How much energy is contained in a typical 75 L tank of gasoline, assuming it is all octane and has a density of 0.75 kg/L?
   d. Highly compressed hydrogen gas has a density of 30.0 g/L. How many liters of compressed hydrogen would you need to produce the same amount of energy contained in a 75 L tank of gasoline?
   e. Knowing that water has density of 1.00 kg/L, how many liters of water would you need to form to produce the same amount of energy that is in a 75 L tank of gasoline?
124. Phosphoric acid is used in many soft drinks to add tartness. This acid can be prepared through the following reaction:

$$P_4O_{10}(s) + 6H_2O(l) \rightarrow 4H_3PO_4(aq)$$

   a. If the value of ΔH for the reaction is − 453 kJ, what is the value of ΔH for the reverse of the reaction?
   b. How many kilojoules of energy are produced by this reaction if 10.0 g of phosphoric acid are produced?
   c. Is this reaction endothermic or exothermic?
   d. If 1.50 g of $P_4O_{10}(s)$ and 2.50 mL of water were mixed, how many grams of phosphoric acid would result?
   e. What is the enthalpy change, in kilojoules, for the process outlined in part d?
   f. If 10.0 g of P_4O_{10} were mixed with 1.00 kg of water at 25.0 °C, what would be the final temperature of the water?

Quantum Chemistry—The Strange World of Atoms

Contents

Supplementary Information The online version contains supplementary material available at ▶ https://doi.org/10.1007/978-3-030-90267-4_6.

M. Mosher and P. Kelter, *An Introduction to Chemistry*,
https://doi.org/10.1007/978-3-030-90267-4_6

What Will We Learn in This Chapter?

- What is a "quantum"? (▶ Sect. 6.1)
- How do light and matter interact? (▶ Sect. 6.2)
- What is the Bohr model of the atom? (▶ Sect. 6.3)
- How does matter exhibit both particle and wave characteristics? (▶ Sect. 6.4)
- What is the Heisenberg Uncertainty Principle and what does it mean for our understanding of the position and speed of a particle? (▶ Sect. 6.5)
- What is the Schrödinger equation and how does it explain the behavior of electrons? (▶ Sect. 6.6)
- What are the four quantum numbers for an electron in an atom? (▶ Sect. 6.7)
- How are electrons arranged within an atom? (▶ Sect. 6.8)
- Why is electron shielding important to understanding the chemical properties of atoms. (▶ Sect. 6.9)
- How can we represent the electrons in an atom and what can that tell us about the arrangement and shape of the periodic table? (▶ Sect. 6.10)

6.1 Introduction

"I need those results, STAT!" The attending physician says these words when a patient who is in desperate condition arrives in the emergency room. A sample of his blood is rushed into the hospital's clinical laboratory for analysis, and shortly, a stream of numbers that indicate the concentration of various chemicals in the patient's blood appears on the physician's computer display to use as she decides what to do to keep the patient alive. A team of clinical laboratory scientists (also known as medical technologists) processes the blood and obtains the key results. How are these analyses done? It varies depending upon what is being targeted. Tests for opiates, pain-reducing or sleep-inducing drugs derived from opium, are done differently than those for lithium, an antidepressant, in the blood. In general, the blood is first separated and then either analyzed directly or mixed with chemical reagents. Many of the analyses use instruments that measure how light interacts with the samples. Specific wavelengths of light are absorbed by different molecules in the blood causing electrons in those molecules to increase their energy. That process is detected by the instrument as a reduction in the amount of light that passes through the sample. Those processes, and many others performed in the hospital laboratory, are based on quantum chemistry. **Quantum chemistry** is the field of chemistry that models and describes the behavior of molecules and atoms at the atomic level. The physical description of their behavior is called **quantum mechanics**.

Quantum Chemistry – The study of chemistry on the atomic and molecular scale.
Quantum mechanics – The physical laws governing energy and matter at the atomic scale.

Why is it useful to know about quantum chemistry? Quantum chemistry is important because we use it to predict chemical reactivity and chemical properties. For example, we need quantum chemistry to answer questions such as, "What is the concentration of lithium in a patient's blood?"; "How is the energy of sunlight captured by plants?" and "Why do leaves and flowers have distinctive colors?" The richness in our macroworld is truly revealed only by looking deep within, at the nanoworld underneath.

The smallest complete unit of matter is a discrete unit known as an atom. You can't divide a carbon atom into a smaller unit that is still carbon. In other words, matter is quantized. A liter of water can be divided in half to give two half-liters of water. Each of those half-liters can itself be divided into halves (two quarter-liters), and so on. But the process cannot go on indefinitely because we will eventually be left with a single water molecule. We could break this molecule down into two hydrogen atoms and an oxygen atom, but the pieces would no longer be water. The H_2O molecule is the smallest unit, or **quantum**, of water. We say that matter is quantized.

Quantum – A single unit of matter or energy. The plural is quanta.

In general, a quantum is a single indivisible unit of something. The quantum of light is called a **photon**, the smallest possible unit of light energy. While many familiar things come in quanta, including eggs, cats, and chemistry books, it is the quantization of light energy into photons that plays a key part in our understanding of atomic and molecular structure, and in all the chemistry we can explain and predict as a result. Light energy is a type of **electromagnetic radiation**, so let us begin by looking at it from that point of view.

Photon – A single unit of electromagnetic radiation.

Electromagnetic radiation – A type of energy consisting of an electric field and a magnetic field, oscillating 90-degrees from each other.

6.1.1 What is Electromagnetic Radiation?

When we think of electromagnetic radiation, we might immediately think of light, that is, the wavelengths of electromagnetic radiation to which our eyes respond, but electromagnetic radiation is everywhere around us. The television signal arrives from the television station antenna via a communication satellite in the form of electromagnetic radiation, as does most of the other information we need to live our lives. Electromagnetic radiation emanates from our toaster as it heats our bagel in the morning. It originates from within the microwave to warm our soup at lunchtime and it travels among cell phones and satellites as we communicate with friends worldwide.

Electromagnetic radiation is energy that propagates in a given direction and consists of an electric field and a magnetic field. These fields are oscillating waves that are 90-degrees apart. These waves have a wavelength, λ (the Greek letter lambda), frequency, ν (the Greek letter nu), and speed (◘ Fig. 6.1). The **wavelength** of a specific packet of electromagnetic radiation is defined as *the distance from the top of one crest of a wave to the top of the next crest*. In the past, wavelengths were measured using a unit known as the **angstrom (Å),** where 1 Å $= 10^{-10}$ m. Instead of the non-SI system unit angstrom, we now use SI-based lengths such as nanometers. The **frequency** of electromagnetic radiation is defined as *the number of waves that pass a given point per second.* So, the frequency is reported in units of 1/s (s^{-1}), known as a hertz (Hz). Computer processors operate in the gigahertz (GHz) range, meaning that their internal clock cycles billions of times per second. The **amplitude** of electromagnetic radiation is the height of the oscillating wave from the crest to the trough. The amplitude of the wave increases as the number of photons of electromagnetic radiation increases.

Wavelength – A descriptor of electromagnetic radiation. Defined as the distance from the top of one crest to the top of the next crest of the electromagnetic wave.

Angstrom – A unit of distance equal to 10^{-10} m. It is symbolized by Å.

Frequency – A descriptor of electromagnetic radiation. Defined as the number of waves that pass a given point per second, in units of 1/s (s^{-1}) or hertz.

Amplitude – The distance between the highest point and the lowest point of a wave.

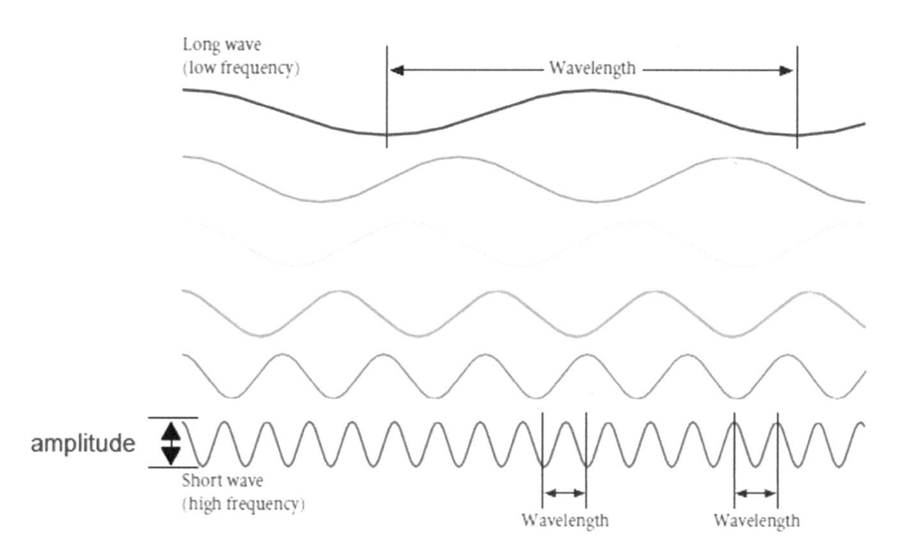

Fig. 6.1 Sine waves of similar amplitude, with wavelength and frequency illustrated

The speed of electromagnetic radiation in a vacuum is given the symbol *c* and is a fundamental physical constant equal to 299,792,458 m per second (this is usually rounded to 3.00×10^8 m·s^{-1}). We often refer to *c* as the speed of light. How fast is the speed of light? Let's consider the Sun, which is, on average, 93 million miles, or 1.5×10^{11} m, from the Earth. Sunlight takes a little more than 8 min to reach our planet. To think of this in another way, the Earth is about 25,000 miles around its equator. Light travels so fast it goes 7 times that distance each second.

6.1.2 Is There a Relationship Between Wavelength and Frequency

Is there a relationship between the wavelength, frequency, and the speed of light? The short answer is yes. They are interrelated by the equation:

$$c = \lambda\nu$$

Note the inverse relationship between frequency, ν, and wavelength, λ. As one increases, the other must decrease because the speed of light in a vacuum is a constant.

Example 6.1—Conversions Between Frequency and Wavelength

What frequency of light has a wavelength of 585 nm, which appears orange to human eyes?

Asking the Right Questions

What happens to the frequency of electromagnetic radiation as the wavelength increases? What happens to the wavelength as the frequency increases? How are these ideas shown in the formula relating frequency and wavelength?

Solution

We can use the relationship among the speed of light, wavelength, and frequency to solve for frequency:

$$\nu = \frac{c}{\lambda}$$

Before we place quantities into the formula, note that the speed of light is expressed in meters per second, so it is useful to convert the wavelength, 585 nm, to meters:

$$585\ \text{nm} \times \frac{1 \times 10^{-9}\ \text{m}}{1\ \text{nm}}$$
$$= 5.85 \times 10^{-7}\ \text{m}$$

We can now use the inverse relationship shown in the formula to solve for the frequency:

$$\frac{3.00 \times 10^8\ \text{m/s}}{5.85 \times 10^{-7}\ \text{m}} = 5.13 \times 10^{14}\ \text{s}^{-1}$$

We can also do the equivalent calculation by dimensional analysis:

$$\frac{3.00 \times 10^8\ \text{m}}{\text{s}} \times \frac{1}{5.85 \times 10^{-7}\ \text{m}}$$
$$= 5.13 \times 10^{14}\ \text{s}^{-1}$$

Is the Answer Reasonable?

In terms of the math, dividing the large exponent of the speed of light by the small exponent of the wavelength gives us a *very* large exponent of the frequency, 10^{14}. That much is reasonable. In a moment, we'll refer to ◘ Fig. 6.3 and see that the frequency of visible electromagnetic radiation is in the range of $10^{14}\ \text{s}^{-1}$, consistent with our answer.

What If…?

…we have electromagnetic radiation of frequency equal to $8.04 \times 10^{10}\ \text{s}^{-1}$? What is the wavelength of this radiation? (*Ans: 3.73×10^{-3} m*)

Practice 6.1

What is the wavelength of light that has a frequency of 7.45×10^{14} Hz? Before doing the calculation, predict: should the wavelength be shorter or longer than that in the example problem above? This prediction will help you to judge if the answer is reasonable.

6

Even though light is a wave, it isn't like a wave in the ocean; no *matter* is actually "waving" as the light travels through space. When electromagnetic radiation travels through matter, such as water, air, or a pane of glass, the electrons and nuclei of the matter tend to follow the oscillatory motion of the electromagnetic radiation, slowing the wave a little. The end result is that electromagnetic radiation travels a bit more slowly through matter than in a vacuum.

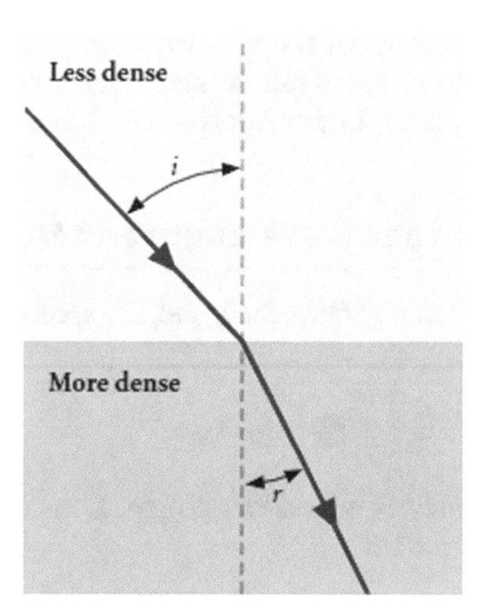

If an electromagnetic wave strikes a denser transparent material at an angle, the radiation changes direction as it enters the material. This is the effect seen in lenses made for eyeglasses, cameras, or telescopes. The angle through which a ray of light bends depends on its frequency; higher-frequency (shorter-wavelength) rays are bent more than lower-frequency ones. This is how a glass prism, shown in ◘ Fig. 6.2, is used to separate the frequencies of light to generate a spectrum, familiar to us as the colors of a rainbow.

6.1.3 Applications of Electromagnetic Radiation

The visible light spectrum of ◘ Fig. 6.2 represents only a small fraction of the entire **electromagnetic spectrum**, shown in ◘ Fig. 6.3. Our eyes do not respond to anything outside the visible range, but our technology can detect and use the invisible frequencies. For example, very long wavelengths are used for radio transmission and are called radio waves. Radio waves can also interact with atoms – the atoms absorb the radiation causing the nuclei of the atom to gain energy. As we'll uncover later, this can be used to determine the structures (organs) inside a patient. The shortest radio waves, at tenths of meters, are called microwaves and are used in cellular phone communications, radar systems, and microwave ovens.

Electromagnetic spectrum – The entire range of radiation as a function of wavelength, frequency, or energy.

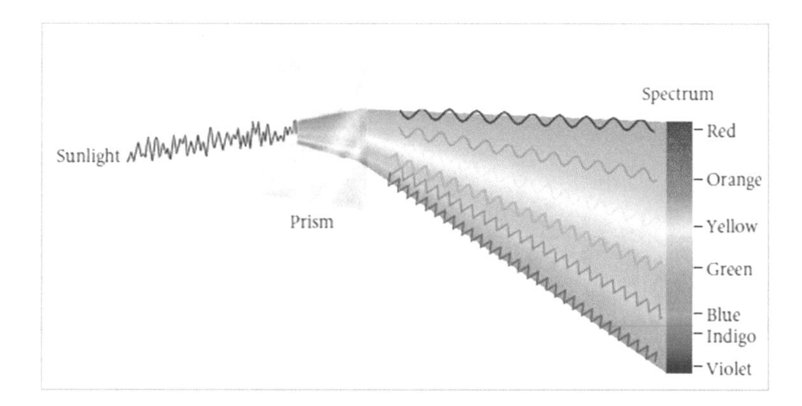

■ **Fig. 6.2** The glass prism separates out the frequencies of light from the electromagnetic spectrum by wavelength

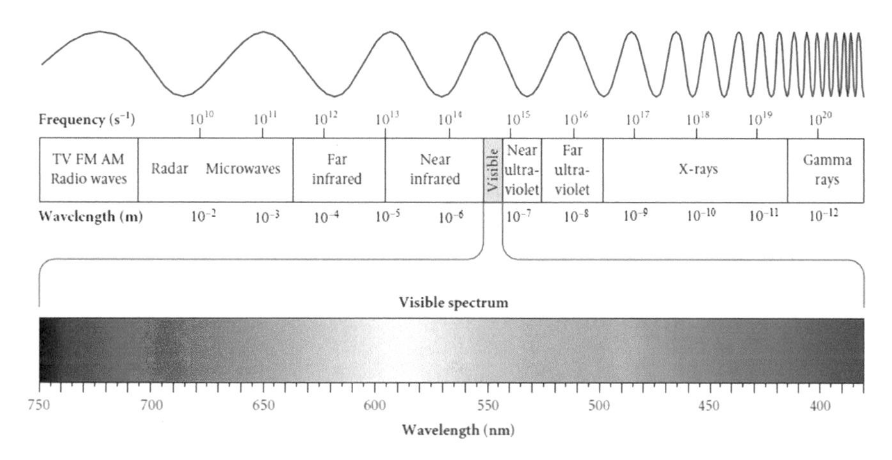

■ **Fig. 6.3** The electromagnetic spectrum as a function of frequency and wavelength

How do we use the different regions of the electromagnetic spectrum? Infrared radiation (IR), spans about 10^{-6} to 10^{-3} m in wavelength and can be detected with special "heat-sensing" scopes. IR rays are associated with the vibrations of molecules. We can feel the warmth they generate when they cause the molecules in our skin to vibrate. The clinical laboratory scientist can use IR wavelengths to determine the concentration of glucose in a patient's blood. Other scientists use IR radiation to determine the class of organic molecule with which they are working or to determine the specific brand of paint at a crime scene.

The Sun's ultraviolet (UV) radiation at approximately 10^{-7} m can cause a tan, a burn, or worse depending on the wavelength and how long we sit in the sun or under a tanning bed lamp. Many drugs and other compounds found in blood and urine also interact with UV radiation. It is this interaction that allows the clinical laboratory scientist to detect the presence and concentration of various drugs in a blood sample. The short wavelength of X-ray radiation, at about 10^{-10} m, is approximately the same size as the radius of an atom. The absorption of X-rays by dense skeletal structures results in an "X-ray" picture of the bones in our bodies. Chemists can also use this radiation to detect the spatial arrangements of atoms in solid compounds. Gamma rays have the shortest waves in the electromagnetic spectrum and are also the most energetic. They are a by-product of many nuclear reactions and are very damaging to most materials, including those that make up living things. ■ Table 6.1 gives selected commercial uses for radiation from various regions of the electromagnetic spectrum.

6.1.4 The Relationship Between Frequency and Energy

The shorter the wavelength, or the higher the frequency, of the electromagnetic radiation, the greater the energy associated with it. This relationship between wavelength, frequency, and the energy of a single photon was determined in 1900 by Albert Einstein (1879–1955) by extending the work of Max Planck (1858–1947).

$$E = h\nu = \frac{hc}{\lambda}$$

Table 6.1 Some commercial uses of electromagnetic radiation

Spectral region	Example of use
Radiowave	Television and radio
Microwave	Cooking and reheating foods, cell phones
Infrared	Heating, drying, and cameras for finding thermal "hot spots" in products
Visible	Detection of low concentrations of compounds
Ultraviolet	Germicidal action for eliminating bacteria in food, water and on surfaces
X-ray	Skeletal imaging, inspecting luggage at airports
Gamma ray	Finding defects in welds and castings, inspecting shipping containers

In this equation, E is the energy carried by each photon of the electromagnetic radiation, and h is **Planck's constant**, equal to 6.626×10^{-34} J·s. This equation enables us to calculate the energy of a single quantum of electromagnetic energy. As we'd predict, the amount of energy each photon carries is very small.

Planck's Constant – A fundamental constant equal to 6.62607×10^{-34} J·s.

Example 6.2—The Energy of Electromagnetic Radiation

Compare the amount of energy in a photon of visible radiation at 605 nm, the red light that we detect with our eyes, and that of the X-ray radiation at 2.00×10^{-10} m that we use in medical diagnosis.

Asking the Right Questions

Which photon contains greater energy, red light or X-ray? What is the quantitative relationship among energy, frequency and wavelength?

Solution

Our experience may hint that the calculation should result in a higher energy value in the X-ray. Planck's constant and the speed of light can be used to calculate the energies of each wavelength:

$$E = \frac{hc}{\lambda} = \frac{(6.626 \times 10^{-34}\ \text{J} \cdot \text{s}) \times (3.00 \times 10^{8}\ \text{m} \cdot \text{s}^{-1})}{6.05 \times 10^{-7}\ \text{m}} = 3.29 \times 10^{-19}\ \text{J for visible light}$$

$$E = \frac{hc}{\lambda} = \frac{(6.626 \times 10^{-34}\ \text{J} \cdot \text{s}) \times (3.00 \times 10^{8}\ \text{m} \cdot \text{s}^{-1})}{2.00 \times 10^{-10}\ \text{m}} = 9.94 \times 10^{-16}\ \text{J for X-rays}$$

Are the Answers Reasonable?

The energy in the X-ray is about 3000 times greater than that of the visible light. Our experience and our discussion up to this point agree that X-rays are much more powerful than visible radiation, so the answer is reasonable.

What If…?

…we had a *mole* of photons of visible radiation at 605 nm? What would be the total energy? (*Ans: 1.98×10^{-5} J*)

Practice 6.2

What is the energy associated with microwaves of wavelength 8.00 mm?

6.2 Atomic Emission, Absorption Spectroscopy and the Quantum Number

West Nile virus can be detected by the clinical laboratory scientist using an instrument that measures the amount of visible light that is absorbed when a blood sample is treated with a set of reagents. The instrument, known as a **spectrophotometer**, works by sending electromagnetic radiation into the sample. Specific wavelengths of energy are absorbed by the molecules in the sample; the instrument identifies the wavelengths that are missing from the electromagnetic radiation that passes through the sample. The clinical laboratory scientist can use this technique to determine if a patient has the virus. That information can be used by the physician to help treat the patient and hopefully minimize further damage by the virus.

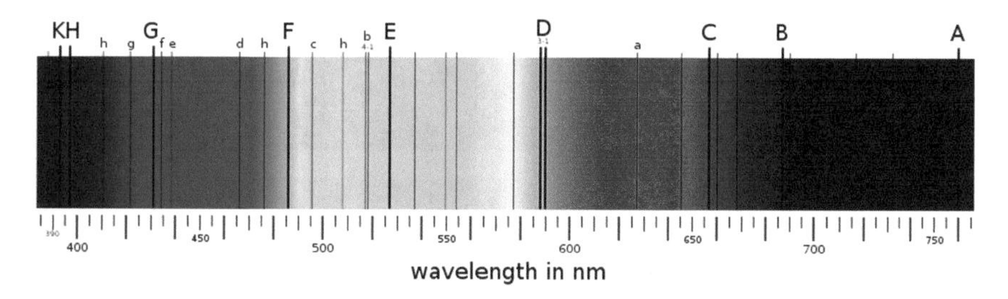

Fig. 6.4 The spectrum of electromagnetic radiation from the Sun shows that most of the wavelengths arrive at the Earth. Some of the wavelengths of radiation do not appear. These are due to the absorption by elements within the Sun. *Source* ▶ https://upload.wikimedia.org/wikipedia/commons/2/2f/Fraunhofer_lines.svg

Spectrophotometer – An instrument that can measure the intensity of electromagnetic radiation at specific wavelengths.

This technique used to detect the virus-reagent mixture is known as **spectroscopy**. Chemists can identify and quantify elements, compounds, and molecules in all types of samples—from food to water to brass to blood–with the use of this technique. Spectroscopy is *the study of how substances absorb (take in) or emit (release) electromagnetic radiation as a function of wavelength or frequency*. Specific types of spectroscopy include **emission spectroscopy** (the study of radiation *emitted* vs. wavelength) and **absorption spectroscopy** (the study of radiation *absorbed* vs. wavelength). These techniques help us determine the atomic or molecular composition of both synthetic and natural drugs, as well as the kinds of organic molecules present in interstellar space, the concentration of lead in water samples, or the amount of a substance in a sample of blood.

Spectroscopy – The measurement of how atoms and molecules interact with electromagnetic radiation as a function of the wavelength or frequency of the radiation.

Emission spectroscopy – The measurement of how atoms and molecules give off electromagnetic radiation as a function of the wavelength or frequency of the radiation.

Absorption spectroscopy – The measurement of how atoms and molecules absorb electromagnetic radiation as a function of the wavelength or frequency of the radiation.

Figure 6.4 shows the solar spectrum, the range of visible light that is emitted from the Sun. Note the discrete dark lines in the figure. These are an all-important feature in **atomic spectroscopy** because, much like each person's unique fingerprint in a crime scene analysis, each element leaves its own unique fingerprint.

Atomic spectroscopy – The study of the interaction of electromagnetic radiation with atoms.

Spectra have been obtained from the Sun since the beginning of the nineteenth century, and elemental analysis by emission spectroscopy has been done since the late 1850s. But not until the work of Johann J. Balmer in 1885 did we get a glimpse at how the wavelengths of light observed from the Sun correlated to the atomic structure of the elements.

6.2.1 Obtaining the Hydrogen Spectrum

The work of the clinical laboratory scientist relates to the early work of Balmer. Let's consider the **emission spectrum** of the hydrogen atom as we explore his work. This spectrum can be obtained using a quartz tube with a metal electrode at each end. If the tube is evacuated and then filled with hydrogen gas (H_2) and a very high voltage is applied across the electrodes, the gas in the tube glows as shown in Fig. 6.5. Why does the hydrogen tube glow? In this apparatus, electrons jump from the negative electrode to the positive electrode, colliding with and, transferring energy to, the hydrogen molecules in the tube. Enough energy is transferred to make some of the molecules dissociate into energetically excited hydrogen atoms (H*).

Emission spectrum – A plot of the intensity of radiation as a function of wavelength or frequency in an emission experiment.

$$H_2 + \text{energy} \rightarrow 2\,H*$$

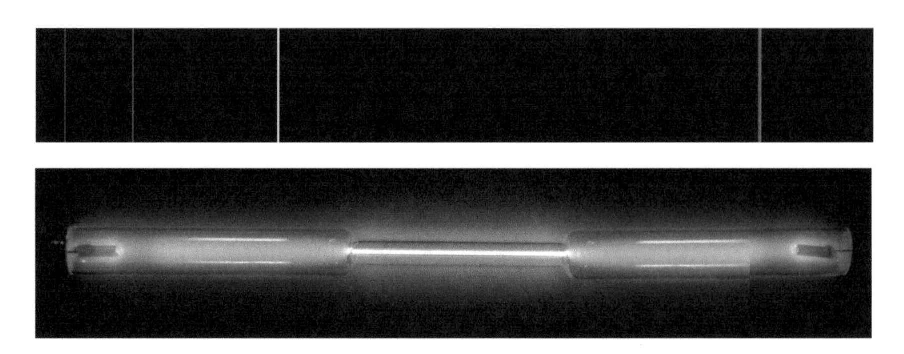

Fig. 6.5 The hydrogen lamp emits a spectrum of visible light that shows only 4 specific wavelengths. *Source* ▶ https://upload.wikimedia.org/wikipedia/commons/c/c6/Gase-in-Entladungsroehren.jpg

The excited atoms then release energy when they emit light, represented by the expression

$$H^* \rightarrow H + h\nu$$

where ν is the frequency of the light emitted by the excited hydrogen (H^*) and h is Planck's constant. If we wished, we could instead write the emission equation using the wavelength (λ) and speed of light (c), as hc/λ:

$$H^* \rightarrow H + \frac{hc}{\lambda}$$

After they lose energy by emitting radiation, the hydrogen atoms (H) eventually recombine to form H_2, the stable form of hydrogen under normal conditions.

$$H + H \rightarrow H_2$$

Each of the recombined molecules is then able to absorb more energy from the electrical discharge to dissociate into two more excited atoms, and the cycle begins again. This makes the tube glow continuously as electrical energy is converted into electromagnetic energy.

If we pass the light from the hydrogen tube through a prism, we can separate the electromagnetic radiation into the individual wavelengths that are emitted from the excited hydrogen atoms. This results in the hydrogen emission spectrum shown in Fig. 6.5. Note that unlike the Sun's emission, only specific, discrete wavelengths of light are emitted in the hydrogen emission spectrum.

6.2.2 Exploring the Hydrogen Emission Spectrum

Balmer found that all of the wavelengths emitted from the hydrogen atom in the visible portion of the spectrum could be fitted to an equation that relates the wavelength (in nanometers) to an integer (n):

$$\frac{1}{\lambda} = 1.0968 \times 10^{-2}\left[\frac{1}{2^2} - \frac{1}{n^2}\right]\text{nm}^{-1}$$

This equation is especially elegant because it introduces to us a **quantum number**, n, that can vary only in whole, positive increments. For example, the equation predicts an infinite number of wavelengths in the spectrum generated by $n=3, 4, 5 \ldots$ all the way to infinity. Balmer did not understand this entirely, but he used the equation because its results fit the data on radiation from the hydrogen atom.

Quantum number – A number used to arrive at the solution of an acceptable wave function that describes the position of an electron around an atom.

The longest wavelength of visible light the hydrogen atom emits (656.5 nm) is found when $n=3$. The next longest wavelength in the Balmer emission spectrum (486.1 nm) is at $n=4$. No light at all is emitted with a wavelength between 486.1 nm and 656.5 nm. This was a clue that helped other scientists create the modern theory of the atom.

Although Balmer's series is predominantly in the visible region of the electromagnetic spectrum, other hydrogen emission series exist that can all be fitted to a similar equation. This generalized equation was developed by the Swedish physicist Johannes Rydberg in 1888 and is known as the Rydberg formula. Notice that this formula produces the Balmer series when we set $n_f=2$. The other quantum number is variable, and we'll call it n_i.

Table 6.2 Spectral series in the hydrogen atom

Spectral series	n_f	n_i	Wavelength range
Lyman	1	2, 3, 4, …	X-ray and UV
Balmer	2	3, 4, 5, …	Visible
Ritz-Paschen	3	4, 5, 6, …	Short-wave infrared
Brackett	4	5, 6, 7, …	Long-wave infrared

The Rydberg Formula: $\frac{1}{\lambda} = 1.0968 \times 10^{-2}\left[\frac{1}{n_f^2} - \frac{1}{n_i^2}\right] \text{nm}^{-1}$

The physicist Theodore Lyman reported observing a series of wavelengths in the ultraviolet and X-ray regions. Each of those wavelengths could be calculated if n_f were held constant at 1 and if n_i were varied from 2 to 3 to 4 to 5 to ∞. In addition, there is a series with $n_f = 3$, with quantum numbers $n_i = 4, 5, 6, \dots \infty$, which begins in the infrared region; another with $n_f = 4$ and quantum numbers $n_i = 5, 6, 7, \dots \infty$; another with $n_f = 5$ and $n_i = 6, 7, 8, \dots \infty$; and so on, with n_f increasing in units of one up to infinity. Table 6.2 lists these first four series for the hydrogen atom.

It's important to remember that the equation describing the wavelengths of light found in the hydrogen emission spectrum was **empirically derived**, which means that it was derived to explain the observations from experiments rather than from theory. In other words, Balmer, Lyman, and other mathematicians and spectroscopists of the late nineteenth century made comprehensive measurements on the hydrogen emission spectrum and fitted a mathematical expression to the data.

Empirically derived – Derived from experiments and observations rather than from theory.

Example 6.3—Using the Balmer Equation

The longest wavelength for the Balmer series is found when $n = 3$:

$$\frac{1}{\lambda} = 1.0968 \times 10^{-2}\left(\frac{1}{2^2} - \frac{1}{3^2}\right) \text{nm}^{-1}$$
$$= 0.00152333 \text{ nm}^{-1}$$

$$\lambda = \frac{1}{0.00152333 \text{ nm}^{-1}}$$
$$= 656.4551 \text{ nm} = 656.46 \text{ nm}$$

What is the smallest energy of radiation for this series?

Asking the Right Questions

What is the relationship between energy and wavelength? Would we expect the longest wavelength in a series to have the lowest or highest energy?

Solution

Energy is related to wavelength by this equation, that we used in Example 6.2.

$$E = \frac{hc}{\lambda}$$

The lowest energy in the series corresponds to the longest wavelength, because energy and wavelength are *inversely* proportional. In the question, we calculated the longest wavelength to be 656.47 nm, and if we put this into SI units as 6.5647×10^{-7} m, we can convert it directly into the energy:

$$E_{\text{smallest}} = \frac{hc}{\lambda}$$
$$= \frac{(6.626 \times 10^{-34} \text{ J} \cdot \text{s})(3.00 \times 10^8 \text{ m} \cdot \text{s}^{-1})}{6.564551 \times 10^{-7} \text{m}}$$
$$= 3.03 \times 10^{-19} \text{ J}$$

Is the Answer Reasonable?

The energy of a photon of visible light has an energy of about 10^{-19} J, so the answer makes sense. In addition, in Example 6.2, the wavelength of 605 nm was a little shorter than the 656.46 nm in this exercise, and the energy was a little higher, at 3.29×10^{-19} J, compared to 3.03×10^{-19} J in this Example. This further confirms that our answer is reasonable.

What If…?

…we considered the $n = 5$ to $n = 2$ transition in the Balmer series? What would be the wavelength and energy of that transition? (*Ans: $\lambda = 434.16$ nm; $E = 4.58 \times 10^{-19}$ J*)

Practice 6.3

Calculate the wavelength of a photon corresponding to the collapse of an electron from $n = 4$ to $n = 3$; from $n = 6$ to $n = 4$; from $n = 6$ to $n = 5$.

6.2.3 Application of Emission Spectra

When scientists tried to apply the Rydberg equation to other elements, the equation didn't work. For example, ◻ Fig. 6.6 shows the visible region of the emission spectra of several elements. Note that the wavelengths of light emitted by each metal are specific to that element. In practice, however, we can use the known emission spectra for the different elements to analyze a sample for presence and concentration of those elements. The clinical laboratory scientist knows this very well and uses this information to detect and measure the concentration of lithium and potassium in a blood sample using an atomic absorption spectrophotometer.

◻ Table 6.3 gives several of the elements, the emission wavelengths typically used for detection, and the minimum concentration of the element that can be reliably detected ("detection limit"). Note that the majority of these wavelengths are in the ultraviolet region of the electromagnetic spectrum. Atomic emission, absorption, and related spectroscopy techniques are among the most important in chemistry for the analysis of elements in all

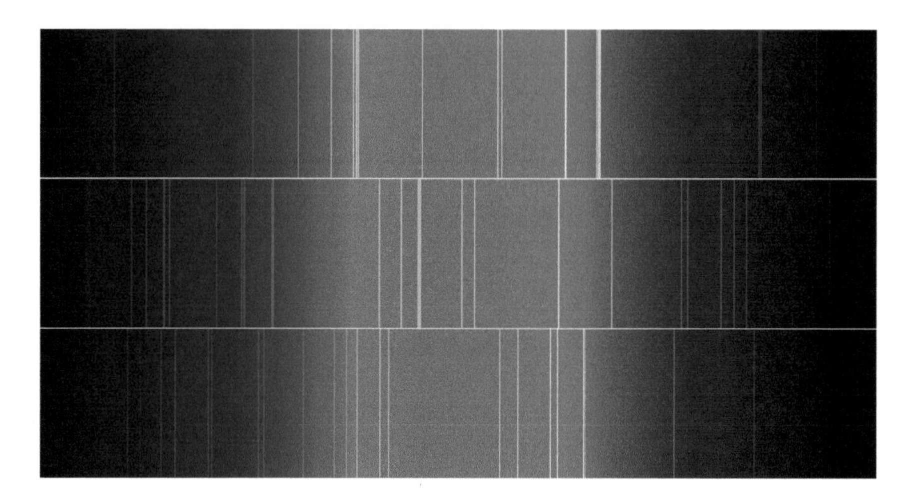

◻ **Fig. 6.6** Line spectra of carbon, selenium, and gold overlaid on the visible light spectrum. Note that each is different. While it doesn't appear that there is a pattern to the lines, scientists have determined that there is a pattern due to the locations of electrons and orbitals around each atom. *Source* ▶ https://upload.wikimedia.org/wikipedia/commons/c/cd/Carbon_spectrum_visible.png; ▶ https://upload.wikimedia.org/wikipedia/commons/6/67/Selenium_spectrum_visible.png; ▶ https://upload.wikimedia.org/wikipedia/commons/1/18/Gold_spectrum_visible.png

◻ **Table 6.3** Simultaneous determination of elements in water

Element	Wavelength (nm)	Detection limit (ppb = µg/L)
Al	396.152	1.5
As	188.980	3.5
Ba	455.403	0.04
Ca	315.887	1.5
Cd	214.439	0.3
Cr	267.716	0.5
Cu	324.754	0.3
Mo	202.032	0.8
Pb	220.353	3.0
Zn	213.857	0.3

Source Fast analysis of water samples comparing axially and radially viewed CCD simultaneous ICP-AES, varian, Inc

manner of samples from water to blood. And it is through atomic spectra that we learned core ideas about the fundamental structure of the atom, which we describe in the next section.

Emission spectra have many additional practical applications. For example, they are quite useful to the environmental chemist. By exciting atoms within a small sample of river water, the chemist can detect the presence of copper, and the intensity of the light detected by the instrument can be related to the quantity of copper in the river water sample. It is currently possible to measure nearly three dozen elements in water at the same time at the parts per billion level (ppb, micrograms, μg, of element per liter of solution) or parts per trillion level (ppt, ng/L) by simultaneously focusing on one discrete wavelength given off by each element.

6.3 The Bohr Model of Atomic Structure

The spectral lines in the emission spectra of different elements puzzled scientists initially. Sure, there was an equation to calculate the lines for the hydrogen atom, and if one knew the emission spectrum for a particular element, it could be detected in an unknown sample, but why they existed was a puzzle. The Danish physicist Niels Bohr put the puzzle's pieces together. His model of atomic structure was so important that it earned him the Nobel Prize in physics in 1922.

6.3.1 Bohr's Model

The Bohr model explained the quantized nature of atomic emissions that had been observed by Balmer and others for many years. In the model, the electron is found some distance from the nucleus (a single proton), whirling around it in a circular orbit, as shown in ◘ Fig. 6.7. The electron can be in any one of an infinite number of possible orbits around the nucleus; each orbit has its own particular distance, or radius, from the nucleus. Not just any radius will do, however. In Bohr's model, the orbits are spatially quantized. Like the layers of an onion, they are confined to specific locations, in spheres of ever-increasing radii, r, expressed in nanometers, according to the equation:

$$r = 0.052917\, n^2 \text{ nm}$$

where n can be 1, 2, 3, 4,…, ∞.

The electrons in the Bohr model orbits also have different properties. Those close to the nucleus travel much more rapidly than electrons in the outside orbits. This means that the kinetic energy of an electron in the orbits close to the nucleus is greater than the kinetic energy of those in the outer orbits. For example, some weather satellites are in a low Earth orbit of about 850 km, so they must travel at a very high speed, about 26,750 km/h, in order to maintain their orbit. On the other hand, GPS satellites, which we use to find our location on Earth when we walk, jog, bike, or drive, are 20,200 km above the Earth's surface and travel at a speed of only 13,940 km/h.

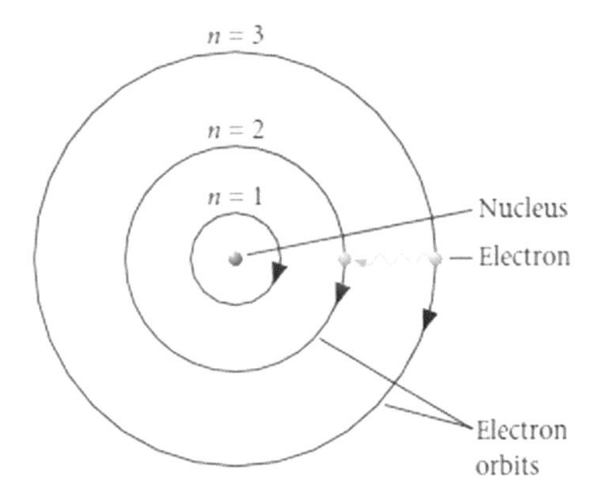

◘ **Fig. 6.7** Schematic of the Bohr model of the hydrogen atom

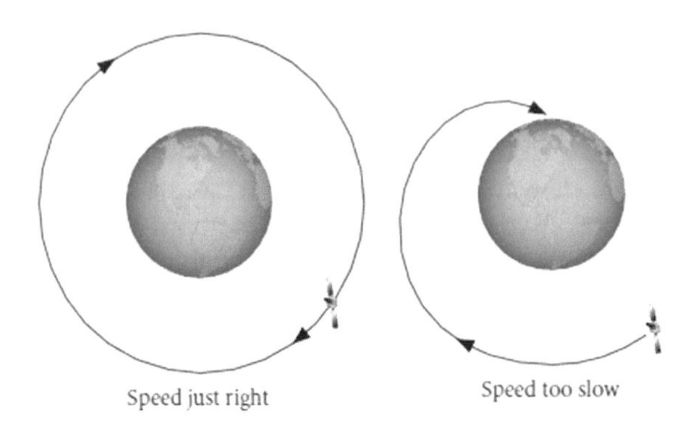

However, the satellites in higher orbits, although they are moving more slowly, have much higher potential energy because they are farther away from the center of the Earth. Overall, the total energy (potential plus kinetic) is higher for the higher orbiting satellites. In the same way, the electrons farther away from the nucleus have higher potential energy. The key idea to keep in-mind is that electrons closer to the nucleus have lower total energy than those that are farther away.

6.3.2 A Specific Energy is Associated with Each Orbit...

The allowed orbits in the Bohr atom are known as **energy levels**, because the electron in a given orbit will have a constant total energy:

Energy levels – The allowed orbits that electrons may occupy in an atom.

$$E_n = -\frac{2.1786 \times 10^{-18}\,\mathrm{J}}{n^2}$$

What is especially interesting about this equation is that n is the same quantum number we discovered in the equation for the radius. A subtle, but key point: this equation calculates the energy of a given orbit as a negative number. This is because a completely free electron (corresponding to $n=\infty$) is assigned an energy of zero. When electrons become bound within atoms, their energy drops, and because it is falling from zero, it must become negative. This means that as an electron falls from higher orbits to lower ones, its energy has increasingly larger negative values relative to zero, the value for the free electron. The lower values of the quantum number have more negative energies than the higher values. The most negative value for energy occurs when $n=1$. At this orbit ($n=1$), the electron is as strongly bound to the atom as is possible, and it is in its lowest energy state in the atom. Unless energy is supplied to the hydrogen atom, its electron will tend to be in the most strongly bound level, closest to the nucleus. As we noted previously, for the hydrogen atom, this is the smallest quantum number, $n=1$. Any atom that contains all its electrons in their lowest possible energy levels is said to be in the **ground state**, like the ground floor of a building.

Ground state – The lowest energy state of an atom.

Larger quantum numbers, $n=2, 3, 4, \ldots, \infty$, represent progressively higher atomic energy levels, just as the second, third, fourth,... floors of a building represent progressively higher gravitational energies. When electrons are moved from the ground state into these higher energy levels, we say that they occupy **excited states**. Atoms and molecules in excited states generally tend to relax back down to their ground states after a short period of time.

Excited state – Any higher energy state than the ground state.

6.3.3 Energy Changes as Electrons Move

Energy must be supplied to the hydrogen atom to promote its electron from the ground state ($n=1$) to any other state ($n>1$). Conversely, the atom must lose energy when the electron falls from any state to a lower energy state.

These **electronic transitions** can be explained very well by the quantum model of the Bohr atom. The electron must gain or lose the exact amount of energy that separates two energy levels (two orbits) in order to jump between the levels. The change in energy is expressed mathematically as

$$\Delta E = E_{\mathrm{f}} - E_{\mathrm{i}}$$

where E_{f} is the energy of the final state of the electron, E_{i} is the energy of the initial state of the electron, and ΔE is the difference in energy between these two levels.

Electronic transition – A change in atomic or molecular energy level made by an electron bound in an atom or molecule.

The change in energy made during an electronic transition can be either positive or negative. If ΔE is positive, energy is supplied to the atom to make the electron jump from a lower to a higher state. If ΔE is negative, energy is released from the atom because the electron falls from a higher to a lower energy level, as shown in Fig. 6.8.

One way to supply or release the required energy is for the atom to absorb or emit electromagnetic radiation. This is where the hydrogen emission spectrum proves useful. If we write the Rydberg equation for hydrogen emission in terms of energy instead of wavelength, we can show that the emission lines are the result of electronic transitions from the higher energy levels (or states) of the excited hydrogen atom to lower energy states. We call the quantum number of the initial energy level, n_{i} and that of the final energy level n_{f}:

$$\Delta E = -2.1786 \times 10^{-18}\,\mathrm{J}\left(\frac{1}{n_{\mathrm{f}}^2} - \frac{1}{n_{\mathrm{i}}^2}\right)$$

Note that the energy change is a discrete value. In other words, the energy is quantized. For the electron to move from one energy level to a lower energy level, it must lose this exact amount of energy. Knowing that the energy is related to the frequency indicates that $h\nu$, the electromagnetic energy, is equal to the energy lost in the transition, ΔE:

$$h\nu = \frac{hc}{\lambda} = \Delta E$$

This relationship ($\Delta E = \mathrm{h}\nu$), although somewhat similar to that proposed by Einstein, was proposed by Niels Bohr in an effort to correlate the energy emitted from a hydrogen atom during an electronic transition with the frequency of observed light. The relationship is referred to as the Bohr frequency condition.

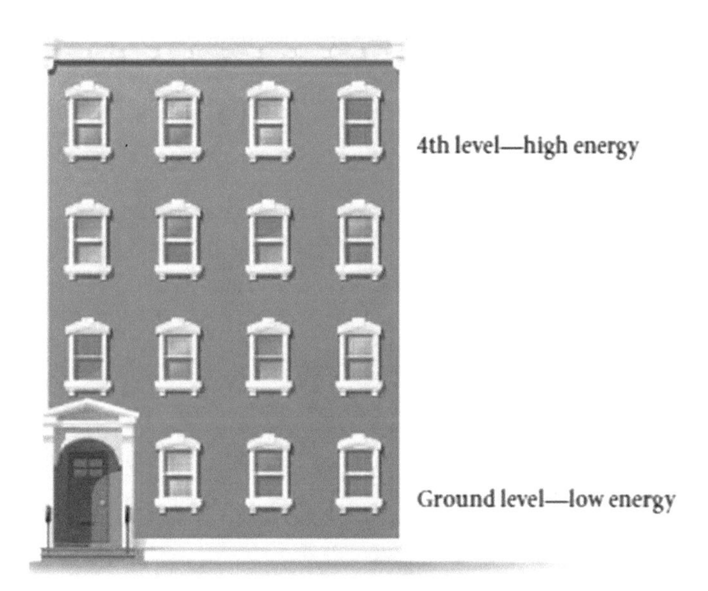

Fig. 6.8 Energy changes that occur in an atom. An increase in energy results in promotion of an electron to a higher energy level. A release of energy is observed when the electron returns to its original state.

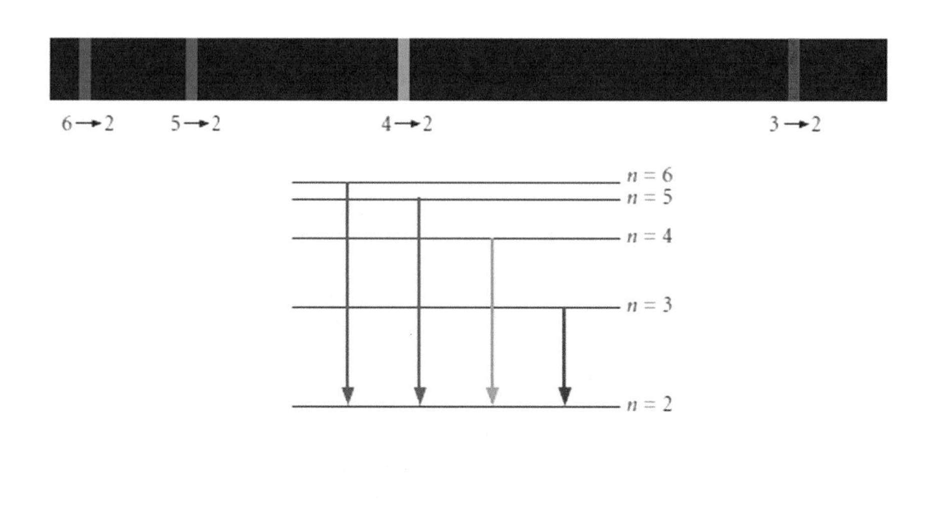

Fig. 6.9 The visible electronic transitions of the hydrogen emission spectrum. The energy of the radiation observed must exactly match the difference between energy of the excited state and that of the ground state (most energetically stable state)

If we combine the previous two equations to calculate the wavelength of radiation predicted for the hydrogen emission spectrum and convert from meters to nanometers (1 m = 10^9 nm), we get

$$\frac{1}{\lambda} = 1.0968 \times 10^7 \left[\frac{1}{n_f^2} - \frac{1}{n_i^2}\right] \text{m}^{-1} = 1.0968 \times 10^{-2} \left[\frac{1}{n_f^2} - \frac{1}{n_i^2}\right] \text{nm}^{-1}$$

This is the same equation that Rydberg, Balmer, and others found by analyzing the hydrogen emission spectrum and empirically fitting an equation to the results! When $n_f = 1$, we get the Lyman series. When $n_f = 2$, we get the Balmer series. We can continue this process until we generate all of the energies of electromagnetic radiation emitted from the hydrogen atom. These electronic transitions in the visible region (Balmer series) are depicted schematically in Fig. 6.9.

Example 6.4—Energy-to-Frequency Conversion in the Hydrogen Atom

A hydrogen atom in one of its excited states has an energy of -1.5129×10^{-20} J. What frequency of radiation is emitted when the atom relaxes down to its ground state ($n = 1$)?

Asking the Right Questions

Would we expect the energy to be lower (more negative) or higher (less negative) by the relaxation to the ground state? How do we calculate the energy change of such a transition?

Solution

An atom moving from its excited state to its ground state *releases* energy (ΔE is −), typically as light, whereas an atom that goes from the ground state to an excited state *absorbs* energy (ΔE is +).

The energy of the photon *emitted* must exactly match the energy *lost* by the hydrogen atom as the electron moves from a higher energy level to its ground state:

$$E_{\text{photon}} = -\Delta E_{\text{electron}}$$

$$h\nu = -(E_f - E_i)$$

$$\nu = \frac{-(E_f - E_i)}{h}$$

The final state energy is given by

$$E_f = -\frac{2.1786 \times 10^{-18}\ \text{J}}{n^2} = -\frac{2.1786 \times 10^{-18}\ \text{J}}{1^2} = -2.1786 \times 10^{-18}\ \text{J}$$

$$\nu = \frac{-(E_f - E_i)}{h} = \frac{-[(-2.1786 \times 10^{-18}\ \text{J}) - (-1.5129 \times 10^{-20}\ \text{J})]}{6.626 \times 10^{-34}\ \text{J} \cdot \text{s}} = \frac{-(-2.16347 \times 10^{-18}\ \text{J})}{6.626 \times 10^{-34}\ \text{J} \cdot \text{s}} = 3.265 \times 10^{15}\ \text{s}^{-1}$$

Are the Answers Reasonable?

We expected the energy to decrease as a result of the transition to the ground state, and the lower energy in the ground state confirms that. The frequency is positive, which is also reasonable.

What If…?

…we want to determine the wavelength of light emitted with this transition? What is that wavelength in nm and in what region of the electromagnetic spectrum does it belong? (*Ans: $\lambda = 91.9$ nm; in the ultraviolet region*)

Practice 6.4

What wavelength of radiation needs to be supplied to a hydrogen atom in its ground state to raise it to its excited state that has an energy of -1.5129×10^{-20} J?

6.3.4 Some Applications of Excited Atoms

We make use of the emission of light from excited atoms every day. For example, when electric current is passed through a tube containing a small amount of neon gas, it causes the neon atoms to become excited. As they return to their ground state, they emit light.

Another example of light emission from excited atoms is found in **lasers**. Lasers (the acronym stands for "light amplification by stimulated emission of radiation") provide a powerful light source from the simultaneous collapse of a large population of excited atoms to the ground state.

Laser – An acronym for "light amplification by stimulated emission of radiation".

The light emitted by each excited atom has the same phase (the peaks and troughs of the light waves line up), giving rise to a powerful emission of light. ◘ Table 6.4 sheds some light on common lasers and their uses.

6.4 Wave-Particle Duality

Funny things happen when we shrink the scale of observation down below that which we can see with our eye, even when it is aided by the most powerful optical microscope. Things that once appeared to have a single value, to occupy a single space, or have a single speed at which they moved through space no longer do any of these things. At this scale, they become smeared out over time and space so that we can discuss only the **probability** of where they might be at a given time or how fast they might be traveling when we finally catch up with them. Nothing at this level seems absolute anymore.

Probability – The likelihood of an event occurring.

Table 6.4 Lasers and their application

Lasing medium	Wavelength	Application
Diode	635 nm (red) 670 nm (deep red) 780, 800, 900, 1550 nm (Infrared)	CD players, DVDs, laser pointers, laser printers, fiber optics, dental surgery
Helium/neon (HeNe)	632.5 nm (red)	Alignment, barcode scanners, laser pointers, blood cell counting
Argon and Krypton	457.9 nm (violet) 488 nm (blue) 514 nm (green) 646 nm (red)	Forensic medicine, high-performance printing, ophthalmic surgery, holography
CO_2	1060 nm (mid-Infrared)	Industrial metal cutting, welding, surgery
HeCd	442 nm (violet) 325 nm (UV)	Testing and spectroscopy
Nd YAG Ruby	1064 nm (IR) 694 nm (red)	Materials processing, surgery, hair removal
ArK excimer	193 nm (UV)	Laser eye surgery

Because our everyday experiences do not easily prepare us for what we observe in experiments done on the atomic or molecular level, the whole idea of quantum mechanics might at first appear strange or even incomprehensible. However, the rules of quantum theory are elegant, precise, and readily understandable if we put aside our macro-scale expectations.

6.4.1 Wave-Particle Duality

We can dig deeper into the explanations that enable us to make sense of atoms if we look at some apparent contradictions between the way matter appears on the basis of everyday experience and the way it behaves at the level of individual atoms. These two views of the world are known as the classical mechanical and quantum mechanical views. In **classical mechanics** there are particles and waves, each with their own distinct characteristics. The particles do not act like waves, and the waves do not act like particles. This is the kind of behavior we see on a macroscopic level (Fig. 6.10). In quantum mechanics, we can still talk about particle and wave behavior, but the distinction between particles and waves becomes fuzzier as we shall see.

Classical mechanics – The physical description of macroscopic behavior in which particles and waves are distinct.

According to classical mechanics, particles have exact locations, and their positions can be defined precisely in space. Classically, a particle can move at any speed (below the speed of light), provided that it is supplied with the

Fig. 6.10 A physical object (like a duck) is very different from a wave. In the quantum realm the nature of objects (particles) and waves become intertwined in strange but very significant ways. *Source* ► https://www.publicdomainpictures.net/pictures/60000/velka/duck-13791602735wo.jpg

necessary energy. Also, if a particle is moving, it possesses momentum (p), which is defined as mass multiplied by velocity (v):

$$p = m \times v$$

A 100-ton train car moving at 50 km/h has considerably more momentum than a 1-ton truck moving at 50 km/h. A train traveling at 50 km/h has more momentum than an identical train that is crawling along the tracks at 10 km/h. Both mass and velocity contribute to momentum.

One of the main descriptors of a particle is mass. Because all particles can possess momentum, they must have mass. This means that we can use momentum to test whether something is behaving like a classical particle.

The Bohr model of the hydrogen atom defies the classical logic of the behavior of particles. Although electrons have momentum and therefore carry mass, they are quantized in the Bohr model in terms of their location and their energy. This quantized nature of space and energy is not the kind of behavior we expect from particles on a macroscopic level. For example, when we think of a moving racecar, we envision it as capable of possessing any velocity or mass within a continuous range. This is not true in the quantum mechanical view.

However, quantized energy states and locations have long been known in everyday life. For example, musicians who play stringed instruments quantize the energy of the vibration of the strings by shortening or lengthening the string being played. This changes the frequency of the string's vibration by setting the wavelength of the vibration, as shown in ◘ Fig. 6.11. A wave vibrating between two fixed endpoints is called a **standing wave**.

Standing wave – The constructive interference of two or more waves that results in the presence of nodes in fixed locations.

The crucial word here is *wave*. All of the cases that were known to create quantized states prior to the Bohr atom were systems that could be described by "waves," oscillating forms of energy that had a specific wavelength and frequency. Let's clarify the distinction between the particles and waves:

Particles in Nature	Waves in Nature
Particles are quantized and exist as individual particles	Waves have specific allowed energy values
The energy of a particle is NOT quantized. Particles have a range of available energies and positions	Waves occupy a range of spatial positions all at the same time (such as the positions of the vibrating violin string)
Particles have a single value of energy and position that can be measured at any one time	Waves have a continuous range of amplitudes (the magnitude of single wave units; such as the loudness of each note from the violin)
Particles carry momentum (have both mass and velocity)	Waves do not carry momentum (have no mass; the sound waves from the violin don't have a mass)

Fig. 6.11 A tuned violin string plays a note when it vibrates at a specific frequency, when a standing wave is set up between the end points of the string. A different note can be played when the player's finger is used to vary the length of the string. But only certain lengths—certain positions of the player's finger—will allow a single proper note to be played when a standing wave of a different frequency occurs. The notes are a set of quantized frequencies. *Source* ▸ https://upload.wikimedia.org/wikipedia/commons/0/04/Anders_Zorn_-_Hins_Anders_%281904%29.jpg

In the view of classical mechanics, waves are neither composed of single units nor located at exactly one place in space as are individual particles. As we can see in our picture of the violin string in Fig. 6.11, the wave exists everywhere between the bridge and the player's finger. The amplitude changes between these two sites, but the wave exists along the whole vibrating part of the violin string. It is spread out over a pretty large volume of space.

Waves can be quantized by confining them to a specific region in space, as we do when we hold the two ends of the violin string down with the bridge and our finger. In order for the amplitude of the wave to go to zero at the ends, the number of wavelengths that must fit between the two ends must be some half-integral number of the wavelength:

$$\text{Length of string} = l = n\lambda/2$$

Or

$$\lambda = 2l/n$$

In this example, the length of the string, l, sets the possible wavelengths that the vibration can have to $\lambda = 0$, $2l$, l, $(2/3)l$, $(1/2)l$, $(2/5)l$. etc. This is illustrated in Fig. 6.12, where we can see the first few allowed standing waves. Note that whereas $2l$ ($n = 1$) and l ($n = 2$) are permissible values, no wavelengths between these two values are allowed.

In our everyday macroworld, the characteristics of particles and waves are separate. But at the level of atoms (nanoworld), things we once regarded as particles can also behave like waves, and things that were traditionally thought of as waves can also behave like particles. Electrons will behave like particles and waves, and electromagnetic radiation will act like waves and particles. The distinction between particles and waves is no longer absolute at this scale. All quantum "things"—electrons, protons, photons of electromagnetic radiation—have the properties of waves and particles simultaneously. This strange behavior is called the **wave-particle duality**, which proposes that at very small dimensions, things behave like both waves and particles, regardless of how they behave on the macroscopic level.

Wave–Particle duality – The quantum mechanical theorem that states that all things have both wave and particle natures simultaneously and that both of these natures can be observed on the atomic scale.

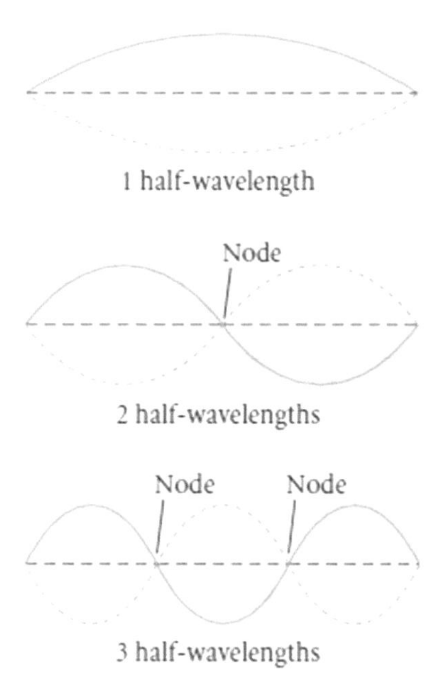

Fig. 6.12 Overtones of a violin string. In addition to the half-wavelength standing wave, the violin also produces integral multiples of the half-wavelength. These vibrations are more energetic than the half-wavelength vibration and are not deliberately played by a violinist

The earliest experiments performed on the electron investigated particle behavior by measuring the electron mass and position. However, we can show that the energy quantization observed in the hydrogen emission spectrum or predicted by the Bohr model results from the wave nature of the electron. The electron behaves as both a particle and a wave. It has mass, carries momentum, and is found as a discrete single unit—properties expected of a classical particle—but when contained in an atom or molecule it is spatially and energetically quantized—properties expected of a classical wave.

6.4.2 Can a Particle Act as a Wave?

To see why the wave nature of the electron allows its energy levels to be quantized, look at Fig. 6.13, where we have drawn a "cross section" of several possible orbits for the Bohr model. The Bohr model is three-dimensional, with spherical orbits, but that would be hard to show schematically, so we will show a circular cross section of the orbit.

Much like the standing wave vibrating on the violin string, only certain wavelengths will fit on a circle. If the radius of the circle is r, we can quantize the allowed wavelengths, λ, to be integral multiples of the circumference, which has a length of $2\pi r$:

$$\text{Circumference} = 2\pi \mathrm{r} = \mathrm{n}\lambda$$

In other words, we can fit n waves of wavelength λ around the circumference of the circle, as shown in Fig. 6.14.

Quantizing the spatial position of electrons in the Bohr hydrogen atom in three dimensions is considerably more difficult. Bohr himself never actually assigned a wavelength to his electrons. Bohr merely postulated that the momentum and radius of permissible levels were quantized because they had to be if his model were to explain the hydrogen emission spectrum. But the *reason* why space and energy are quantized for the hydrogen atom arises from the wave-like nature of the electron.

If electrons behave like waves, how do we determine their wavelengths? We can do this by using the **de Broglie formula**, proposed by Louis de Broglie in 1924:

$$\lambda = \frac{h}{p} = \frac{h}{mv}$$

The equation, which shows that the wavelength, λ, is inversely proportional to the momentum, p, through Planck's constant h, connects wave (λ) and particle (p = momentum) properties in a single formula. Again, note

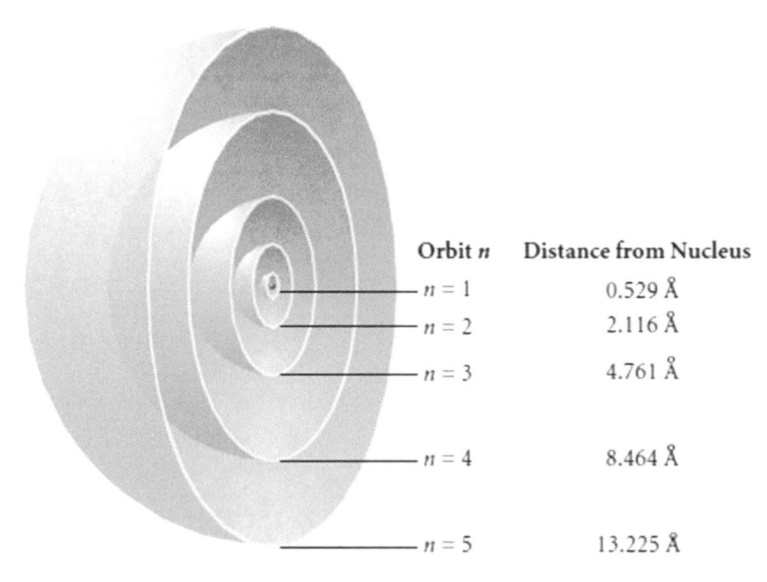

Fig. 6.13 Circular cross-sections of the atomic energy levels of the Bohr atom

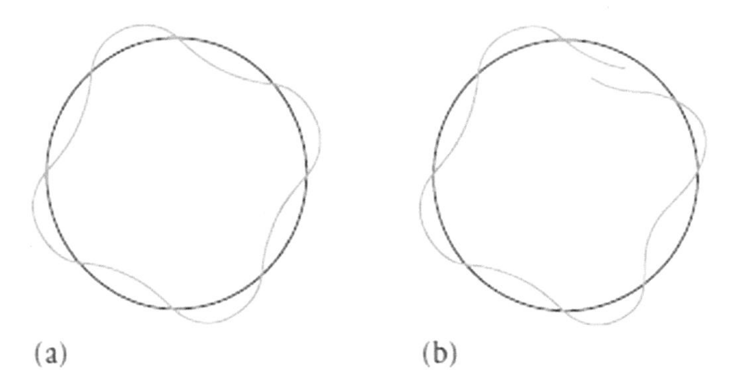

Fig. 6.14 The requirements of a wave completing a circular path. In **a**, the wavelength lines up as the circular orbit is completed. This provides a suitable orbital. In **b**, the wavelength doesn't line up and does not provide a useful orbital

that the momentum term, p, uses the velocity (v). This symbol looks similar to that for the frequency (ν), but it is vastly different. The velocity has units of m/s, whereas frequency has units of 1/s.

What is the broader purpose of the de Broglie equation? It illustrates the basic principle of wave-particle duality: Moving particles exhibit definite wave-like characteristics. The values of wavelengths calculated by this equation are most meaningful for small, fast-moving particles such as the electron. The calculation is less meaningful but still demonstrative for massive, slow-moving objects.

Example 6.5—Working with the de Broglie Equation

An electron in the ground state of the hydrogen atom has a velocity of about 2.1×10^6 m · s^{-1}. What is its de Broglie wavelength? What does the equation tell us about the electron compared to large objects, such as people?

Asking the Right Questions

What is the relationship between the wavelength and momentum (and, therefore, the mass × velocity) of an object?

Solution

The wavelength and momentum of an object are inversely related.

$$\lambda = \frac{h}{p} = \frac{h}{mv}$$

$$\lambda = \frac{h}{p} = \frac{h}{mv}$$

$$= \frac{6.626 \times 10^{-34}\ \text{J} \times \text{s}}{\left(9.109 \times 10^{-31}\ \text{kg}\right) \times \left(2.1 \times 10^{6}\ \text{m} \times \text{s}^{-1}\right)}$$

Because 1 J = 1 kg·m^{2}·s^{-2}, substituting yields

$$= \frac{6.626 \times 10^{-34}\ \text{kg} \cdot \text{m}^2 \cdot \text{s}^{-2} \cdot (\text{s})}{\left(9.109 \times 10^{-31}\ \text{kg}\right) \times \left(2.1 \times 10^{6}\ \text{m} \cdot \text{s}^{-1}\right)}$$

$$= 3.5 \times 10^{-10}\text{m} = 0.35 \text{ nm}$$

Is the Answer Reasonable?

The wavelength of the electron in the ground state of the hydrogen atom should have a wavelength within our electromagnetic spectrum, and it does. It is far, far larger than the wavelength of a person, which is also reasonable, given the inverse relationship between wavelength and momentum (mass × velocity).

What If…?

…we threw a 2.0-kg honeydew melon on to a plate at 3.0 m/s, hoping to make instant fruit salad? What would be the wavelength of the honeydew melon, and would that be consistent with the principles we discussed about the de Broglie equation? (*Ans: 1.1×10^{-34} m*)

Practice 6.5

Calculate the de Broglie wavelength for a 1200-lb car traveling 75 mi/h.

6.5 The Heisenberg Uncertainty Principle

The Bohr model regards the electron as a particle that is allowed to orbit the nucleus at only certain quantized distances. However, if it also behaves like a wave, the electron must be spread out over its entire orbit in the hydrogen atom. We can reconcile these two very different ideas by considering scale. Assume that we measure the electron's position as having a specific value of x, y, and z in our macroscopic laboratory setting. If the electron is bound in a hydrogen atom but is somehow spread out, some uncertainty will be introduced into our measurement of x, y, and z. But the uncertainty is very small, on the order of 0.1 nm or so. Our macroscopic rulers are not this precise, so it looks to us as if the electron is at a specific and absolute position in space.

If we then look at the electron on the *microscopic* level, we find that the best we can do to establish its position is to say it is somewhere in the atom, within a certain range from of the atom's center. The electron seems to behave like a particle when we are looking at it on a macroscopic scale. However, when we look at electrons on a microscopic level, they spread out like waves to fill the "container" of the atom.

6.5.1 The Limitations of Applying Quantum-Level Calculations

On a macroscopic level, we can measure both the position and the momentum of a particle as precisely as our instruments will allow. Quantum mechanics, however, proposes that, regardless of the instrument used to measure them, *there is an uncertainty in both the position and the momentum of a particle, and these uncertainties are connected.* This is called the **Heisenberg uncertainty principle** after Werner Heisenberg, who proposed it in 1925. Mathematically, we can state the Heisenberg uncertainty principle as

$$\Delta x \Delta p_x \geq \frac{h}{4\pi}$$

where Δx is the uncertainty in position, Δp_x is the uncertainty in momentum along the x-direction, and h is Planck's constant.

Heisenberg uncertainty principle – There is an ultimate uncertainty in the position and the momentum of a particle. Reducing the uncertainty in one increases the uncertainty in the other.

There is no limitation on how precisely we measure either x or p_x. Instead, the limitation is on the product of their uncertainties $\Delta x \Delta p_x$. In other words, as the uncertainty in x becomes smaller, the uncertainty in p_x gets larger. Therefore, as Δx approaches zero and we know exactly where the particle is positioned, Δp_x becomes infinite and we do not know anything about its momentum. In short, limitations are set on real-world behavior by the Heisenberg uncertainty principle, which among other things limits the maximum achievable resolution of microscopes, the ultimate size of computer chips, and, in the nanoworld, the behavior of actual atoms and molecules.

While we have focused on the electron, these principles apply to all particles, although the effect will really be *evident* only on the microscopic level. Any small particle such as the electron, proton, or neutron would be a good candidate for testing the Heisenberg uncertainty principle. Even though the same principles apply, bowling balls, elephants, and cars would not. But what about phenomena that are wave-like in our macroscopic world? What happens to classical waves on the quantum scale?

Electromagnetic radiation is wave-like on a macroscopic level. Light, in a manner similar to chemical substances, can be sub-divided only so far. Eventually, we will come to its smallest, indivisible unit. For light, the smallest unit is the photon, and, like particles, photons have been shown to carry momentum, which is a particle-like characteristic. In fact, we can calculate the momentum using the de Broglie equation. What does such a calculation tell us about the nature of light? Light, although we think of it primarily as a wave, acts like a particle too. Again, the calculation verifies the particle–wave duality.

Example 6.6—Photons and Momentum

How much momentum is carried by a single photon that emits green light and by a mole of photons that emit green light? Assume that photons of green light have wavelengths of 530 nm.

Asking the Right Questions

How should the momentum of a green photon compare to that of a mole of photons?

Solution

The mass of the photon—even a mole of photons—is so much smaller than that of a person that its momentum will be incredibly tiny compared to the person.

We can use the de Broglie equation to calculate the momentum of a single photon. Because $\lambda = \frac{h}{p}$,

$$p = \frac{h}{\lambda} = \frac{6.626 \times 10^{-34}\ \text{J} \cdot \text{s}}{5.30 \times 10^{-7}\ \text{m}}$$
$$= \frac{6.626 \times 10^{-34}\ \text{kg} \cdot \text{m}^2 \cdot \text{s}^{-2} \times (\text{s})}{5.30 \times 10^{-7}\ \text{m}}$$
$$= 1.25 \times 10^{-27}\ \text{kg} \cdot \text{m} \cdot \text{s}^{-1}$$

This is for a single photon. If we want the momentum for a mole of photons, we must multiply this value by Avogadro's number:

$$p = 1.25 \times 10^{-27}\ \text{kg} \cdot \text{m} \cdot \text{s}^{-1} \cdot \text{photon}^{-1}$$
$$\times\ (6.022 \times 10^{23}\ \text{photons} \cdot \text{mol}^{-1})$$
$$= 7.53 \times 10^{-4}\ \text{kg} \cdot \text{m} \cdot \text{s}^{-1} \cdot \text{mol}^{-1}$$

Is the Answer Reasonable?

Compare our answer for the momentum of both a single green photon and a mole of green photons to that of a 68-kg (150-lb) person ambling down the road at 2 mi/h, who has a momentum of 60 kg·m·s^{-1}. Even a mole of green photons has a much smaller momentum than our ambling person.

What If…?

…the momentum of a mole of photons was determined to be 1.13×10^{-3} $\text{kg·m·s}^{-1}\ \text{mol}^{-1}$? What color light would we see? (*Ans: λ=353 nm, violet*)

Practice 6.6

Calculate the momentum of one mole of photons with a wavelength of 200 nm, one mole of photons of 700 nm, and one mole of photons of 1000 nm. Is there a difference? Explain.

A word of caution about this type of calculation: The wavelength will have the same value, as will the frequency, whether we have one photon or a mole of photons. The wavelength and frequency of a photon are intrinsic properties—they do not change with a change in amount. The total momentum and energy carried by the electromagnetic radiation, however, are extrinsic properties because both depend on the number of photons we have. When we use the formula $E = h\nu$ we always use the energy of a single photon. Similarly, with the de Broglie equation, we always use the momentum of a single photon.

We also need to be careful of the answer that the de Broglie equation gives us for the mass of a photon. The mass that appears in the de Broglie formula is known as the **rest mass**, the mass the particle would have if it were not moving at all. Photons, however, are moving at 2.998×10^8 m/s, and slowing them down to zero speed is just not feasible. The rest mass we calculate for photons from the de Broglie formula does not make any sense for a photon at the speed of light. This is another of the strange results of looking at things at the quantum level.

Rest mass – The mass a particle would have if it were stationary.

6.6 The Mathematical Language of Quantum Chemistry

Toxicology screening of patients commonly occurs in the hospital setting. One such screening is the evaluation of blood samples for the presence of antidepressants. Many methods exist for the detection of drugs such as fluoxetine (Prozac®) and citalopram (Celexa®). One such method requires the separation of the different compounds found in blood followed by their detection using ultraviolet light. Why do only some compounds absorb ultraviolet light and allow themselves to be detected in blood? The reason lies in the quantum chemistry-based nanoworld, where the language is mathematics. Knowing a few of the more important mathematical concepts will pave the way for us to explore the current understanding of atomic structure and delve into the reasons why some molecules can absorb ultraviolet light while others cannot.

The heart of our current understanding of the atom lies within this deceptively simple-looking mathematical equation containing 3 variables:

$$\hat{\mathrm{H}}\,\psi_n = E_n\psi_n$$

This is the time-independent **Schrödinger equation**. This equation allows those who dive into such equations to calculate the exact energies, E_n, available to the electrons in an atom. To do so, they need to know about that mysterious symbol, Ψ_n, or "psi," This is known as the **wave function**. and it describes the positions and paths of the electron in its given energy level (n). Although it appears simple, the wave function, Ψ_n, is actually shorthand for a complex equation containing variables, constants, and quantum numbers. The $\hat{\mathrm{H}}$ in the Schrödinger equation is known as the **Hamiltonian operator** after William Hamilton (1805–1865), an Irish mathematician who did work that hinted at wave–particle duality. In simple terms, an operator is a mathematical term that tells us how to proceed—what operation to do—in a mathematical equation. The operator "x" tells us to multiply values. A "+" tells us to add. The Hamiltonian is one of the more complex operators that we'll encounter, however, the conclusions that result from its operation are fundamental to our proper view of atomic structure.

Schrödinger equation – The mathematical expression that relates the wave function to the energy in a quantum system.

Wave function – The mathematical equation describing a system, such as an electron, atom, or molecule, that contains all physical information that can be obtained for the system by quantum mechanics.

Hamiltonian operator – A mathematical function that is used in the Schrödinger equation.

6.6.1 The Application of Schrödinger's Equation

To illustrate the implications and utility of the Schrödinger equation, let's model electrons in an atom as waves, shown in ◻ Fig. 6.15. To keep things as simple as possible, we will allow our electron to move in only one dimension.

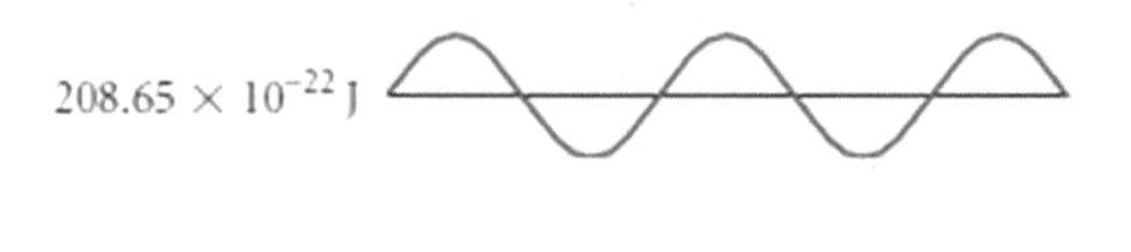

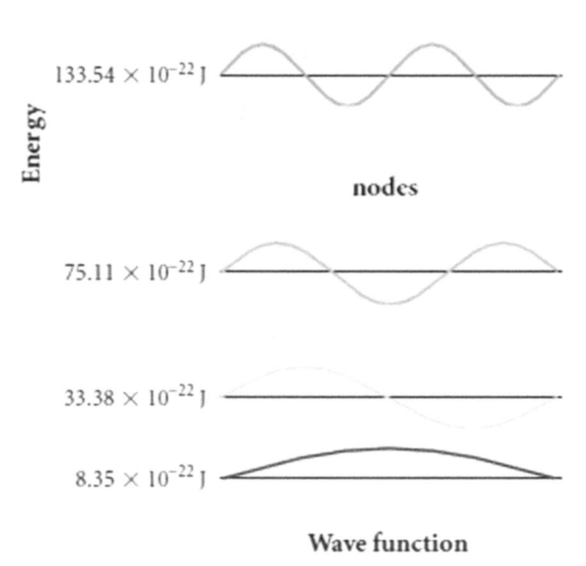

Fig. 6.15 The wave functions and energy levels for an electron confined to a one-dimensional box. The electron is defined by a wave function that has nodes at each end of the wave, indicating that it cannot be found at the wall or outside the box

In this example, our electron can move between two points within that one dimension. Think of this as a one-dimensional container or box that will confine the electron. If we use the Schrödinger equation and solve for the energy of the electron, we find the first five wave functions shown starting from the bottom of the box in Fig. 6.15. These are called standing waves, which oscillate in time just like the string on a violin, and we show them as a snapshot in time at their greatest amplitude. The first level is described by half the wavelength of a sine wave. The second level is described by a full sine wave and has a **node** in the center of the wave function—a point in the wave function where the amplitude is always zero. The next level, represented by one and a half sine wavelengths, has two nodes that divide the wave function into three segments. The wave function with $n=4$ has an additional node for a total of three, and the wave function $n=5$ has still another for a total of four.

Node – A point in the wave function where the amplitude is always zero.

Additionally, complex calculations of an electron in a one-dimensional box result in an equation illustrating the permitted energy levels for this model:

$$E_n = \frac{n^2h^2}{8mL^2}$$

where n is the quantum level ($n=1, 2, 3,...$), h is Planck's constant, m is the mass of the particle, and L is the length of the box. Although the one-dimensional model that gave rise to this equation is simple (it is based on a box containing a particle), it explains the intense colors observed in nature and in the anti-depressants we discussed at the beginning of this section. The molecule itself acts as a one-dimensional "box" because of its bonding structure. When a photon is absorbed by one of these electrons, it becomes excited and moves to a higher energy level described by our simple one-dimensional model. The absorption of the photon removes a portion of the electromagnetic spectrum that remains and results in the observed color that our eyes "see". Note that the specific amount of energy that is required in this process is related to the length of the electron path (the wave function). The energy differences between the lower energy level and the excited energy level correspond to the color we observe. The length of the chemical bond or chemical bonding structure defines the length of the box, L in the above equation.

If we calculate the energy of an electron confined to a box the size of a single covalent bond (1.5×10^{-10} m), the wavelength of the energy needed to excite the electron corresponds to ultraviolet light. If several atoms define the "box," as is seen in some types of molecules common within flower petals, then the energy is smaller (because L is bigger), and the electrons can absorb visible light. ◘ Figure 6.16 illustrates that as the length of the atoms that define the box increases, the color of the molecule changes.

6.6.2 Orbitals

If we know the probable regions around an atom where electrons reside, we could explain so many chemical and physical properties of the elements. In addition, we could determine an element's reactivity and its interactions with electromagnetic radiation, as we discussed earlier in the chapter. The Schrödinger equation gives us the ability to determine where an electron is most likely to be found within an atom. We say "most likely" because we

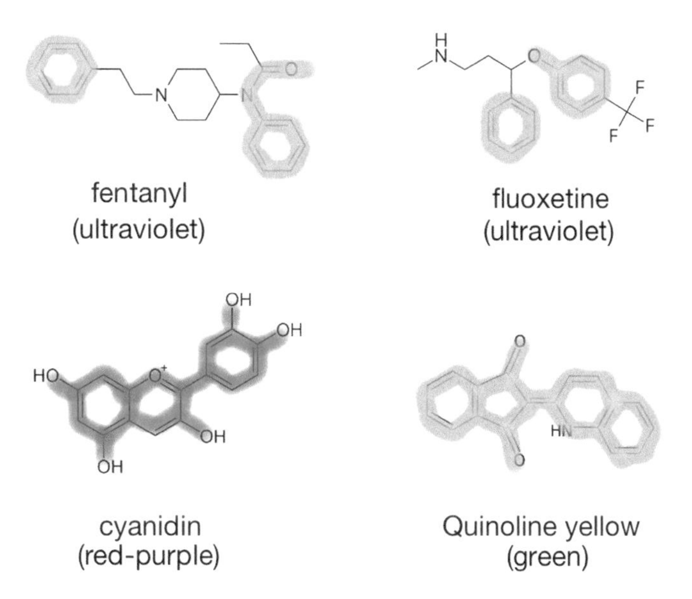

◘ **Fig. 6.16** The highlighted portions of these molecules indicates the connected bonds that can absorb electromagnetic radiation. Drugs such as fentanyl and fluoxetine have shorter areas of bonds that can absorb light and require larger energies to promote an electron. The larger energy corresponds to an absorption in the ultraviolet region (shown as yellow in the figure). As the length of the bonds increases, the absorption of light occurs within the visible region. Naturally occurring cyanidin gives leaves and flowers their color. Synthetic dyes such as Quinolone yellow, Red Dye #3, and Blue Dye #2 can be designed to give almost every color in the rainbow and are cheaper than natural dyes. The safety of synthetic dyes is an ongoing concern

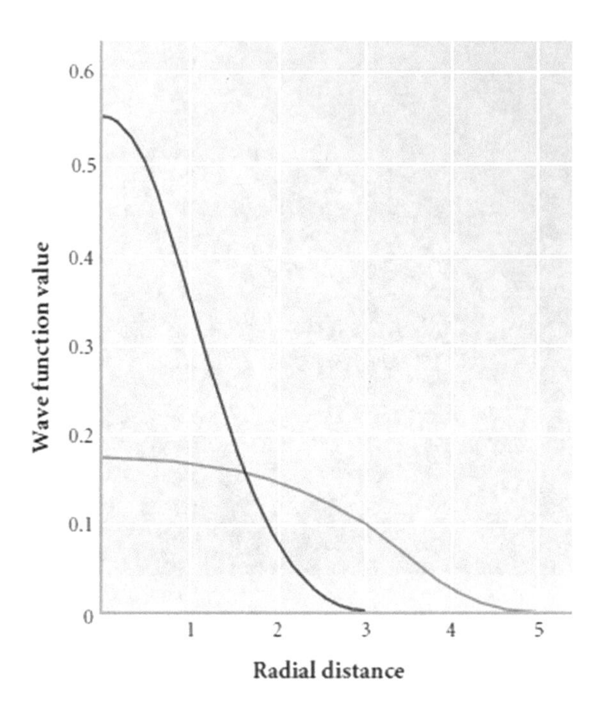

Fig. 6.17 A plot of the wave function (blue line) represented as the average distance of the electron from the nucleus. The probability function ($\Psi^*\Psi$; red line) can be used to determine the most likely shape of the orbital resulting from the wave function

can never pin the electron down exactly, as the Heisenberg uncertainty principle tells us. Instead, we can only determine the probability of finding it at each point in space. The space that the electron is allowed to occupy in a given energy level on an atom is called an **orbital**.

Orbital – The volume of space to which the electron is restricted when it is in a bound atomic or molecular energy level.

For example, if we plot the wave function of an electron around a nucleus, we get a curve similar to that shown in Fig. 6.17, where the nucleus of the atom is located at the origin of the graph. We can mathematically manipulate the wave function to create a new function depicted as $\Psi^*\Psi$, which is only slightly different from the square of the wave function (Ψ^2). The great usefulness of $\Psi^*\Psi$ is that it is proportional to the probability of finding the particle at a particular point in space. The nodes in Ψ remain at the same places in $\Psi^*\Psi$, but all values of $\Psi^*\Psi$ are positive. In short, the probability function describes the spatial distribution of the electron around an atom. We can say that the probability function ($\Psi^*\Psi$) describes the shape of an orbital.

The idea that a single particle, like an electron, doesn't exist at a single point in space, but is merely more likely to be found at certain places and less likely to be found at others, is admittedly rather odd. But it fits the ideas presented here and years of measurements that confirm these ideas. Note that there are nodes where we can guarantee that the particle will *never* be found. At these nodes, the probability function, $\Psi^*\Psi$, is equal to zero. The electron can be found on either side of the node if the probability function is positive on either side of the node, but the electron can never be found at the node. How does an electron go from one side to the other, if it is never allowed to be at the node? It does this by behaving like a wave, which has amplitude on both sides of the node: a clear demonstration of wave–particle duality.

6.7 Atomic Orbitals

Both the Hamiltonian operator and the wave functions for the hydrogen atom are known, and they allow us to develop both mathematical and visual representations of the orbitals that electrons can occupy in the three-dimensional space around the hydrogen nucleus. Specialists in quantum chemistry describe this three-dimensional system in terms of a **radial** part and an **angular** part. Because the calculations of these three-dimensional orbitals are quite complicated, we present only the results here. We will use these results to explain, over the next several chapters (and, from time to time, for the remainder of the textbook), how and why atoms and molecules interact the way they do.

Radial – Along the radius in spherical polar coordinates.

Angular – As a function of the angles that describe the orientation of the radius in spherical polar coordinates.

6.7.1 Relating Quantum Numbers to Orbitals

There are four different quantum numbers required to describe the wave function for the electron in the hydrogen atom, and they are given the symbols n, l, m_l, and m_s. These can be thought of as an "address" for an electron around an atom. The n is known as the **principal quantum number** and can be any whole number $n = 1, 2, 3, \ldots, \infty$. We are already familiar with the principal quantum number. This n is what Bohr referred to as the energy levels in his atom. Based on the possible values of n, there are an infinite number of energy levels for the hydrogen atom. The energy of each is given by

$$E_n = -\frac{2.1786 \times 10^{-18}\ \mathrm{J}}{n^2}$$

which is just what was predicted by the Bohr model. In atoms with more than one electron, any electrons with the same value for the principal quantum number (n) are said to occupy the same **principal shell**, or energy shell but only the energy levels in the hydrogen atom are given by this equation.

Principal quantum number – The quantum number, n, that describes the energy level of the orbital; n can be any whole-number value from 1 to infinity.

Principal shell – The energy level that is occupied by electrons with the same value for the principal quantum number (n).

Two of the other three quantum numbers, l and m_l, depend on the value of n. The **angular momentum quantum number**, l, can be any number from zero to $n - 1$ in whole-number steps:

$$l = 0, 1, 2, \ldots, n - 1$$

This quantum number is considered to represent the shape of the electron orbital.

Angular momentum quantum number – The quantum number, l, that describes the shape of the orbital; l can be any whole-number value from zero to $n-1$.

In the first energy level, $n = 1$, the equation used to calculate the angular momentum quantum number reveals that $l = 0$. This implies that only one shape of orbital is possible in the first energy level. In the second energy level, $n = 2$, the value of l can be 0 or 1. This means that there are two different shapes of the orbitals in the second energy level. When $n = 3$, l can be 0, 1, or 2 (three different shapes). Any electrons that share the same values for n and l are said to occupy the same **electron subshell**.

Electron subshell – The energy level occupied by electrons that share the same values for both n and l.

The m_l quantum number is the **magnetic quantum number** (also known as the orbital angular momentum quantum number). It is based on the value of l and can have any value from l to $-l$ in whole numbers.

$$m_l = -l, \ldots, -2, -1, 0, 1, 2, \ldots, l$$

This quantum number can be thought of as defining the direction in which the individual electron orbitals are pointed. For example, a specific orbital could be oriented along the x-axis or the y-axis. The value of m_l indicates this direction. If the value of m_l is zero, then the orbital has no orientation, which means that it must be a sphere.

Magnetic quantum number – The quantum number, ml, that describes the orientation of the orbital; m_l can be any whole-number value from $-l$ to $+l$. Also known as the orbital angular momentum quantum number.

n	l	m_l	Subshell notation	Number of orbitals in the subshell	Number of electrons needed to fill subshell	Maximum possible number of electrons in shell
1	0	0	1*s*	1	2	2
2	0	0	2*s*	1	2	
2	1	1, 0 –1	2*p*	3	6	8
3	0	0	3*s*	1	2	
3	1	1, 0, –1	3*p*	3	6	
3	2	2, 1, 0, –1, –2	3*d*	5	10	18
4	0	0	4*s*	1	2	
4	1	1, 0, –1	4*p*	3	6	
4	2	2, 1, 0, –1, –2	4*d*	5	10	
4	3	3, 2, 1, 0, –1, –2, –3	4*f*	7	14	32

Fig. 6.18 Allowed values for n, l, ml, and ms quantum numbers for hydrogen atomic orbitals

The number of possible values for the magnetic quantum number indicates the number of specific orbitals of a specified shape, as will be illustrated later in this chapter. For example, in the second principal shell, when $n=2$, we determined that there are two values for l: $l=0$ and $l=1$. As shown in Fig. 6.18, each value of l has a specific letter that is assigned to it. Any orbital with an l value of 0 is called an "*s*" orbital. An l value of 1 is referred to as a "*p*" orbital, and an l value of 2 is called a "*d*" orbital. Starting with an l value of 3, called an "*f*" orbital, the rest of the letters are in alphabetical order (*g*, *h*, *i*, etc.).

We need to consider each of these values to generate all the possible sets of quantum numbers. When $n=2$ and $l=0$, m_l can only be 0 (one orbital). When $n=2$ and $l=1$, m_l can be -1, 0, or $+1$. In this case, we have three specific orbitals of the same shape oriented in three different directions. Calculating all of the possible orbitals within an energy level can be a daunting task because the number of orbitals gets large fairly quickly, as shown in Fig. 6.18. But as long as we are *systematic* about what is essentially a bookkeeping task, confusion can be kept to a minimum.

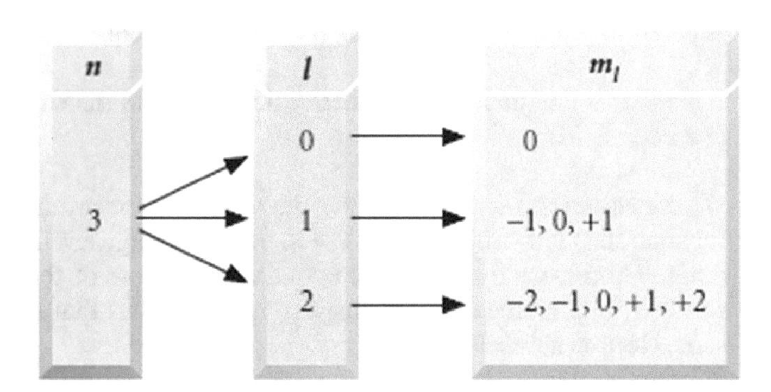

The final quantum number, m_s, is known as the **electron spin quantum number** (it is also known as the spin angular momentum quantum number). Fortunately for our bookkeeping, the only possible values for m_s are $+\frac{1}{2}$ and $-\frac{1}{2}$. The value of m_s can be either $+\frac{1}{2}$ or $-\frac{1}{2}$ for every set of *n*, *l*, and m_l allowed. Because of the restriction of only two possible values of m_s, each orbital (represented by a value of m_l) can hold a maximum of two electrons. This quantum number, indicating electron spin for two electrons in the same orbit, will be further discussed in the next section.

Electron spin quantum number (m_s) – The quantum number, m_s, that describes the spin of an electron. Also known as the spin angular momentum quantum number.

We can use shorthand to indicate the four quantum numbers that describe the position of electrons in an atom using the notation (n, l, m_l, m_s). As shown in Fig. 6.19, for $n=1$ the sets of quantum numbers for the electrons are (1, 0, 0, ½) and (1, 0, 0, –½). For $n=2$ the sets are (2, 1, 1, ½), (2, 1, 1, –½), (2, 1, 0, ½), (2, 1, 0, –½), (2, 1, –1, ½),

(n, l, m_l, m_s)	
Energy Level 1:	$(1, 0, 0, +\frac{1}{2})$; $(1, 0, 0, -\frac{1}{2})$
Energy Level 2:	$(2, 0, 0, +\frac{1}{2})$; $(2, 0, 0, -\frac{1}{2})$
	$(2, 1, 1, +\frac{1}{2})$; $(2, 1, 0, +\frac{1}{2})$; $(2, 1, -1, +\frac{1}{2})$
	$(2, 1, 1, -\frac{1}{2})$; $(2, 1, 0, -\frac{1}{2})$; $(2, 1, -1, -\frac{1}{2})$
Energy Level 3:	$(3, 0, 0, +\frac{1}{2})$; $(3, 0, 0, -\frac{1}{2})$
	$(3, 1, 1, +\frac{1}{2})$; $(3, 1, 0, +\frac{1}{2})$; $(3, 1, -1, +\frac{1}{2})$
	$(3, 1, 1, -\frac{1}{2})$; $(3, 1, 0, -\frac{1}{2})$; $(3, 1, -1, -\frac{1}{2})$
	$(3, 2, 2, +\frac{1}{2})$; $(3, 2, 1, +\frac{1}{2})$; $(3, 2, 0, +\frac{1}{2})$; $(3, 2, -1, +\frac{1}{2})$; $(3, 2, -2, +\frac{1}{2})$
	$(3, 2, 2, -\frac{1}{2})$; $(3, 2, 1, -\frac{1}{2})$; $(3, 2, 0, -\frac{1}{2})$; $(3, 2, -1, -\frac{1}{2})$; $(3, 2, -2, -\frac{1}{2})$

Fig. 6.19 Systematic bookkeeping for quantum numbers

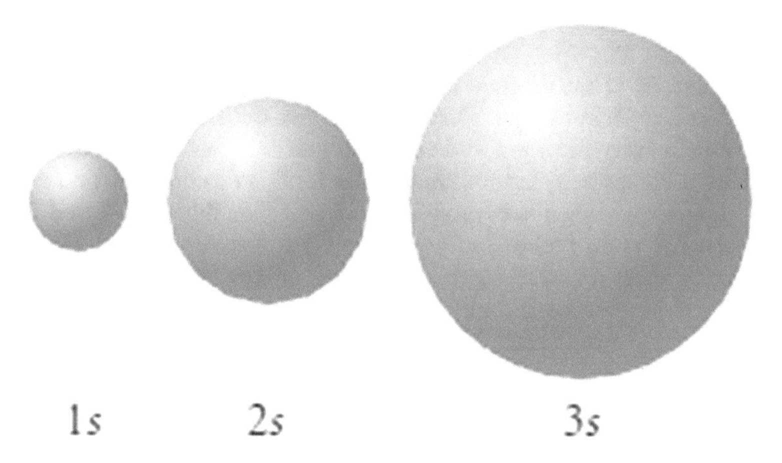

Fig. 6.20 The spatial component (with 90% probability) for the hydrogen atomic orbitals with $n = 1$, 2, and 3

(2, 1, − 1, − ½), (2, 0, 0, ½), and (2, 0, 0, − ½). If we try $n = 3$, we should get 18 sets of quantum numbers, and for $n = 4$ we should get 32. Note how this follows a general trend. The number of unique quantum number sets in each energy level is $2n^2$. Each unique quantum number set represents one electron.

How do these values help us model the electronic structure of atoms? Although the specific set of four quantum numbers represents a convenient label for a specific hydrogen atomic wave function, quantum numbers are much more than that. The principal quantum number, n, gives the energy of the system and sets the values of the l and m_l quantum numbers. Together, the n, l, and m_l tell us about the spatial distribution of the electron and define the orbital shape and size. For example, as shown in Fig. 6.20, the radial component of the orbitals with $n = 1$, 2, and 3 illustrates the difference in size of the atomic orbitals.

Orbitals that have equal energies are called **degenerate orbitals**, and the number of orbitals having the same energy is the "degeneracy". In the hydrogen atom, orbitals that have the same values of n and l have the same specific amount of energy, and these orbitals are therefore degenerate. For example, consider the orbitals within the second energy level ($n = 2$). When $n = 2$, l can be either 0 or 1. When $l = 0$, the value of m_l can be only 0; there is only one orbital, so it can't be degenerate. However, when $l = 1$, m_l can be − 1, 0, or + 1. In this case, three orbitals with the same value for l are possible. These three orbitals have the same amount of energy and are degenerate. The third energy level contains two sets of orbitals with degeneracy. One set contains three degenerate orbitals ($l = 1$, $m_l = -1, 0, +1$); the other set contains five degenerate orbitals ($l = 2$, $m_l = -2, -1, 0, +1, +2$).

Degenerate orbitals – Orbitals that have equal energies.

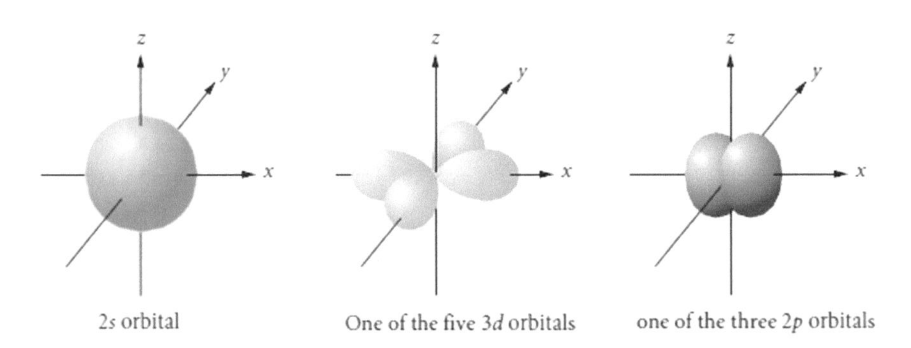

Fig. 6.21 The *s*, *p*, and *d* orbitals have very different shapes

By plotting these complex wave functions, quantum chemists have been able to model what the quantum numbers describe. For example, Fig. 6.21 shows us shapes of some of the orbitals that we have described with quantum numbers. Remember that at the atomic level, we can no longer speak in terms of absolute locations but can only give the *probability* of the quantum particle being in a given location.

6.7.2 *s* Orbitals

What can we say about the shapes of the orbitals within an atom? The orbital shown in Fig. 6.21 that has $l=0$ is radially symmetric and has a spherical shape. That is, as we look out from the nucleus, the probability distribution function looks the same in every direction and varies only with r, the distance from the nucleus. All orbitals with $l=0$ are called ***s* orbitals**. The principal quantum number is added to this as a prefix when we write the name of this specific orbital. For example, when $n=1$ and $l=0$, we write 1*s*. When $n=2$ and $l=0$, we write 2*s*, and so on. The 1*s* orbital, with $n=1$, has no nodes (locations where the electron cannot exist because the probability function for the wave function is zero) and represents the lowest energy level, or ground state, for an electron in a hydrogen atom. The 2*s* orbital has one node that is concentric around the nucleus. The 3*s* orbital has two radial nodes, the 4*s* has three, and so on, with the number of concentric nodes increasing along with the principal quantum number.

s **Orbital** – An orbital with quantum number $l=0$.

The solid spheres of Fig. 6.22 are useful in comparing relative sizes. As we expect, the 1*s* orbital is the smallest, and the size increases as n increases. This representation does not give us any help in visualizing the nodes present; the two-dimensional cross sections are more helpful in this respect. This method of representing the *s* orbitals uses a gray scale to show where the probability of finding the electron is large and where it is small. The nodes are indicated by the lightest part of the picture. In an orbital, a "radial node" is an area where the electron cannot exist, just as a node on a vibrating string is a place where the vibration itself does not occur. All *s* orbitals are spherical, with the number of radial nodes increasing from zero in the 1*s* orbital to $n-1$ in the *ns* orbital. Remember that *s* orbitals all have orbital angular momentum quantum numbers of $l=0$ and magnetic quantum numbers of $m_l=0$.

We still have to consider the Heisenberg uncertainty principle as we construct our pictures. In order to draw sensible pictures of atomic sizes, we arbitrarily assign the boundary of an atom to lie at some distance r from the center of the atom, so that when we look for the electron within that value of r from the nucleus, we find the electron a large portion (90%) of the time. We show the 1*s*, 2*s*, and 3*s* orbitals in Fig. 6.22 at the 90% probability distribution function for the electron and in several different formats, each of which is meant to emphasize some aspect of the orbital shape.

6.7.3 *p* Orbitals

For orbitals with $l=1$, which are known as ***p* orbitals**, we have a different spatial arrangement, as shown in Fig. 6.23. We use the convention of writing the principal quantum number as a prefix to the letter abbreviation for the orbital. However, because $l=1$, this allows for three possible m_l values (+1, 0, and −1). In other words, there will be three degenerate orbitals that are symmetrically equivalent. We can choose to orient the x, y, and z

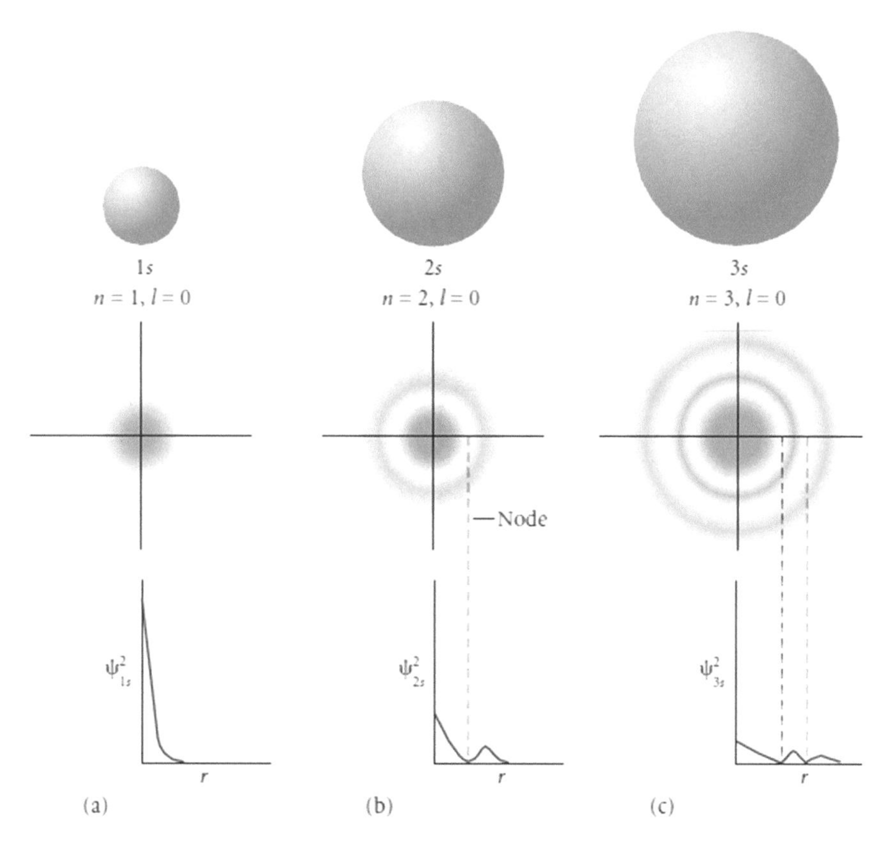

Fig. 6.22 The 1s, 2s, and 3s spatial component of the hydrogen atomic orbitals at 90% probability. Note the presence of nodes in the cutaway plots

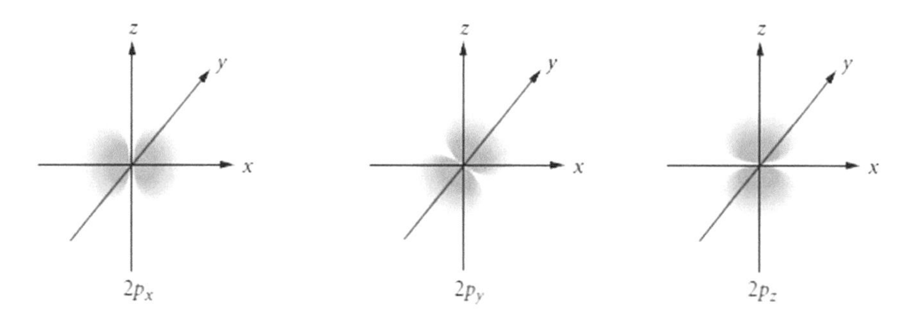

Fig. 6.23 The 2p_x, 2p_y, and 2p_z spatial component of the hydrogen atomic orbitals at 90% probability

axes with each of the three equivalent orbitals and when we do so, we call them the 2p_x, 2p_y, and 2p_z orbitals. This is shown for the lowest-energy 2p orbitals in Fig. 6.24. The 2p orbitals have no radial nodes, but each does have a planar node where the electron probability density is zero (Fig. 6.25). For the 2p_x orbital, this node is the yz plane; for the 2p_y orbital, this node is the xz plane; and for the 2p_z orbital, this node is the xy plane. We have set the size of the orbital to include 90% of the probability for finding the electron within it, just as we did with the s orbitals.

p **Orbital** – An orbital with quantum number $l = 1$.

When we increase the primary quantum number to $n = 3$, we have three 3p orbitals with the same kind of planar node structure (Fig. 6.26), but now we also have a radial node. The three 4p orbitals have the planar node plus two radial nodes, the three 5p orbitals have the planar node plus three radial nodes, and so on.

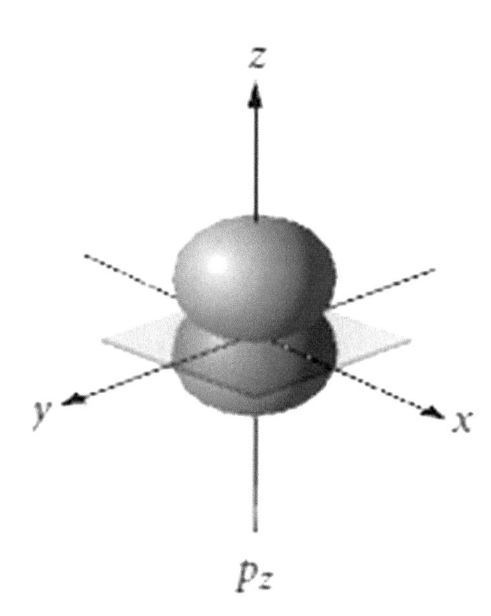

Fig. 6.24 The location of the planar node found in the $2p_z$ orbital. Although $n=2$ is used in the example, all p orbitals have a single planar node in an analogous position

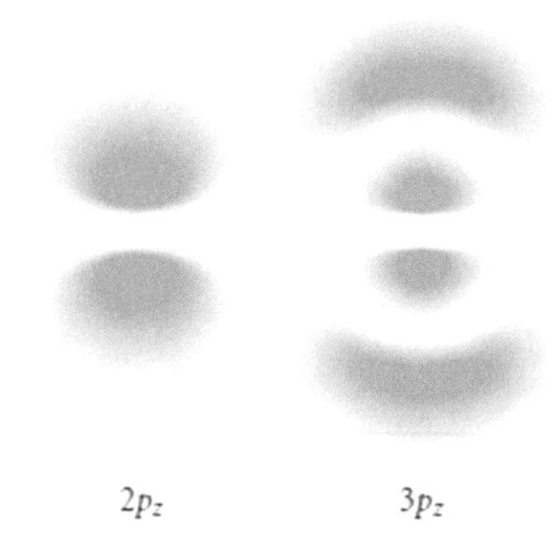

Fig. 6.25 The $2p_z$ and $3p_z$ spatial component of the hydrogen atomic orbitals at 90% probability

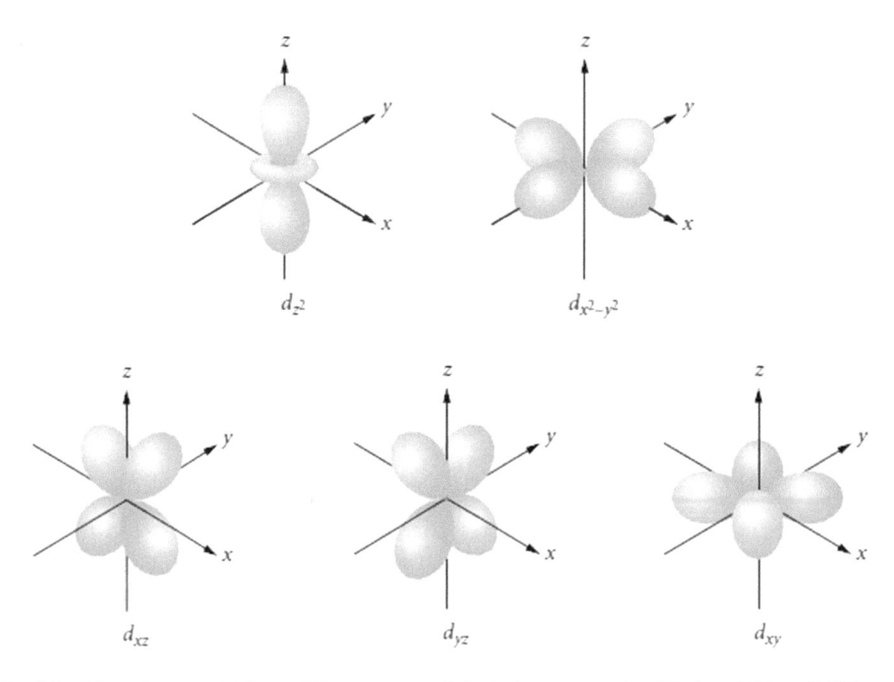

Fig. 6.26 The $3d_{xy}$, $3d_{yz}$, $3d_{xz}$, $3d_{x^2-y^2}$, and $3d_{z^2}$ spatial component of the hydrogen atomic orbitals at 90% probability

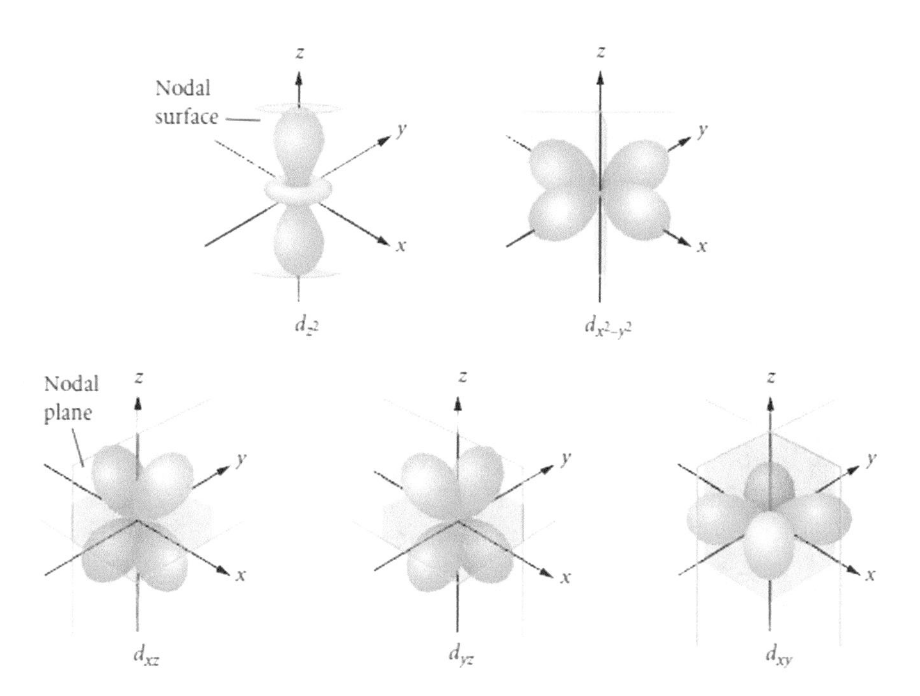

Fig. 6.27 The location of the two planar nodes found in the $3d_{xy}$, $3d_{yz}$, $3d_{xz}$, and $3d_{x^2-y^2}$ lobes and the two hyperbolic nodes orbitals of the $3d_{z^2}$ lobe. All *d* orbitals have two planar nodes in analogous positions

6.7.4 *d* Orbitals

Orbitals with $l=2$ are called ***d* orbitals** and have two nonradial nodes. The first set of quantum numbers for which l can equal 2 occurs when $n=3$. Five *d* orbitals occur, because m_l takes the values of 2, 1, 0, −1, −2 when $l=2$. For all but the $m_l=0$ *d* orbital, the nodes are planar and are at 90° to each other. These orbitals have four lobes that are labeled d_{xy}, d_{yz}, d_{xz}, and $d_{x^2-y^2}$ and are oriented as shown in Fig. 6.27. The remaining *d* orbital is labeled d_{z^2} and looks a bit different but is equivalent in energy (degenerate) to the other four *d* orbitals. The d_{z^2} orbital has two hyperbolic nodes that give the orbital a dumbbell structure with a little toroidal "donut" around the middle.

d Orbital – An orbital with quantum number $l=2$.

6.7.5 *f* Orbitals

Most of the known elements have their electrons in the *s*, *p*, and *d* orbitals of their various principal shells, but a few elements have occupied ***f* orbitals** with $l=3$ and $m_l=3, 2, 1, 0, -1, -2, -3$. These seven orbitals will be used in our description of atoms larger than barium, such as those in the lanthanide and actinide series of the periodic table. They are shown in Fig. 6.28. Note that they are multilobed and geometrically quite complicated.

f Orbital – An orbital with quantum number $l=3$.

At last, we have arrived at our current "quantum mechanical" or "wave mechanical" model of atomic structure, although we can depict it only by showing the places around the nucleus where an electron in any orbital will most probably be found. In painting a more accurate picture of what electrons are doing, we have had to acknowledge the uncertainty and strangeness—the funny things happen that we noted at the start of ▶ Sect. 6.4—down at the level of the quantum world.

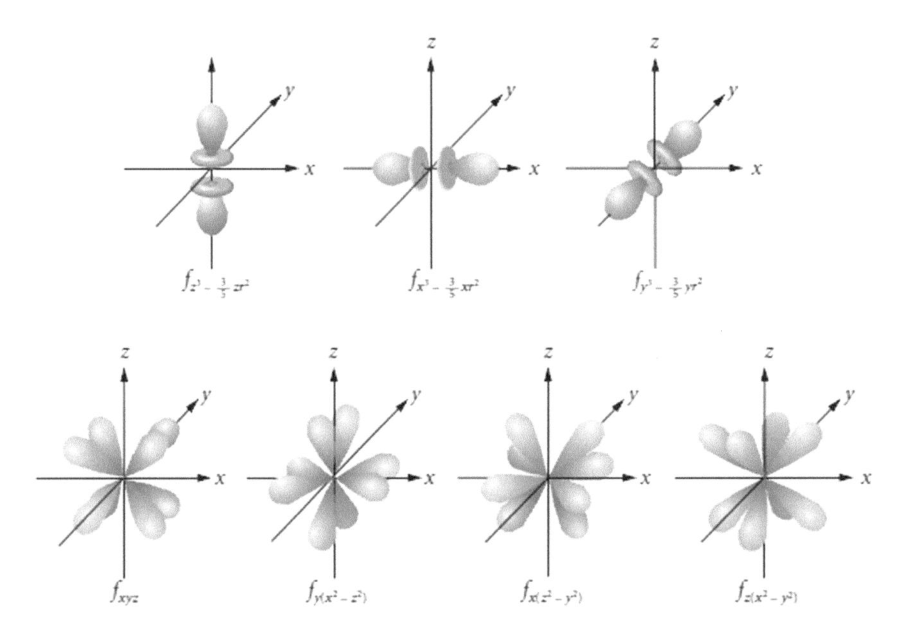

Fig. 6.28 The 4*f* spatial component of the hydrogen atomic orbitals at 90% probability

Example 6.7—Quantum Numbers

Indicate the possible values for the quantum numbers (n, l, m_l) that would correspond to the 3*d* orbital.

Asking the Right Questions

What is the sequence of allowed quantum numbers as we move from the principal (n) to the angular momentum (l) and on to the magnetic (m_l) quantum numbers?

Solution

A single 3*d* orbital would correspond to one of these sets of quantum numbers:

$n=3, l=2, m_l=-2.$
$n=3, l=2, m_l=-1.$
$n=3, l=2, m_l=0.$
$n=3, l=2, m_l=+1.$
$n=3, l=2, m_l=+2.$

Are the Answers Reasonable?

Does each set of quantum numbers follow the pattern allowed? The 3*d* set of electrons is so-called because $n=3$ and the *d* is the name given to $l=2$. The values of m_l can vary from $-l$ to l, or, here, from -2 to 2, so this follows the allowed pattern of quantum numbers.

What If…?

…we had a subshell containing 7 orbitals? What is the lowest value for the angular momentum quantum number, l, that could exist for this subshell? What is the lowest principal quantum for the subshell? Name that subshell.

(*Ans: $l=3$, or an "f" subshell. The lowest principal quantum number is $n=4$. This subshell is a 4f.*)

Practice 6.7

Indicate the orbital described by the quantum numbers, $n=2, l=1, m_l=0$.

6.8 Only Two Electrons per Orbital: Electron Spin and the Pauli Exclusion Principle

When a drug or metabolite within a patient is identified by a spectroscopic technique, the clinical laboratory scientist knows that an electron has moved from one energy level to another. Determining which electron moves requires the application of the quantum numbers to electron positions. We've talked about the other three quantum numbers (n, l, and m_l), but what about the fourth one – the electron spin quantum number (m_s)? An electron *behaves* as though it were spinning about an axis. A "spin" in one direction corresponds to an electron spin quantum number (m_s) of $+\frac{1}{2}$, and a "spin" in the opposite direction corresponds to an electron spin quantum number (m_s) of $-\frac{1}{2}$. These are the only two values of m_s available; in other words, spin is quantized into these two states, and an electron must be in one of them. The spin of an electron makes the electron behave as though it were a tiny magnet. For example, if we put hydrogen atoms into a magnetic field, electrons with $m_s=+\frac{1}{2}$, called the "up" state, will line up parallel to the field; and those with $m_s=-\frac{1}{2}$, called the "down" state, will line up antiparallel to

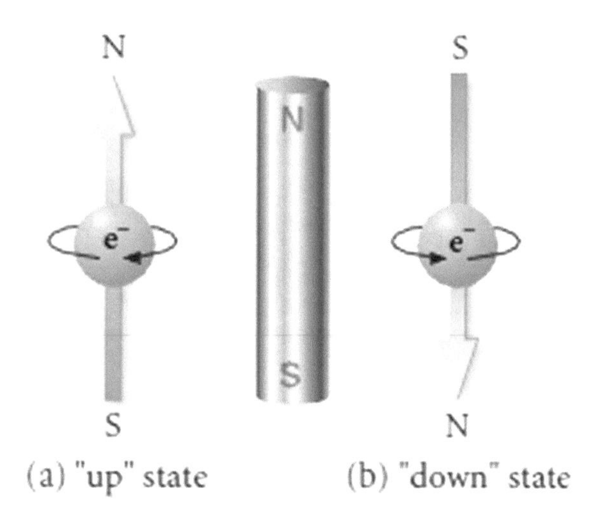

Fig. 6.29 Electrons in the spin up and spin down states

it, as shown in Fig. 6.29. We sometimes represent the up state by an arrow pointing up: ↑. Similarly, the down state is represented by an arrow pointing down: ↓.

One vital consequence of this idea of electron "spin" is that it allows two electrons to occupy the same atomic orbital. Wolfgang Pauli, in 1925, proposed a rule that allows only two electrons to occupy the same atomic orbital. According to the **Pauli exclusion principle**, no two electrons can have the same set of the four quantum numbers (n, l, m_l, and m_s) in a given atom. What does the Pauli exclusion principle tell us about the electronic structure of the atom? For two electrons to occupy the same orbital, they must have the same values for the quantum numbers n, l, and m_l. The only way that two electrons can have these quantum numbers is if their spin quantum numbers (m_s) are different. In essence, one of the electrons in the orbital is in the spin up state, and the other is in the spin down state. Because it requires four quantum numbers to describe an electron on an atom, it is not possible for more than two electrons to occupy the same orbital.

Pauli exclusion principle – In a given atom, no two electrons can have the same set of the four quantum numbers—n, l, m_l, and m_s.

6.8.1 Revisiting the Significance of Quantum Numbers

We have seen that the four quantum numbers in any electron's wave function specify everything we need to know about the electron's location in the atom. They are a bit like an electron's address in the atom. Table 6.5 summarizes the significance of all the quantum numbers.

6.9 Orbitals and Energy Levels in Multielectron Atoms

Unfortunately, the orbitals calculated in the manner that we have discussed are only absolutely correct for the orbitals of one-electron atoms and ions. We can use this approach for atomic hydrogen and its isotopes deuterium and tritium. We can also use it for He^{+}, Li^{2+}, or even Hg^{79+} if we adapt the equation slightly.

Table 6.5 The quantum numbers

n	Principal quantum number	Indicates which of the major energy levels or electron shells an orbital is in. Also, the larger the value of n, the greater the total volume of the orbital
l	Angular momentum quantum number	Indicates the shape of the orbital. The specific value of l determines whether the orbital is an *s*, *p*, *d*, or *f* subshell
m_l	Magnetic quantum number	Indicates the orientation of the orbital in 3-D space
m_s	Electron spin quantum number	Indicates the orientation of the spin of the electron

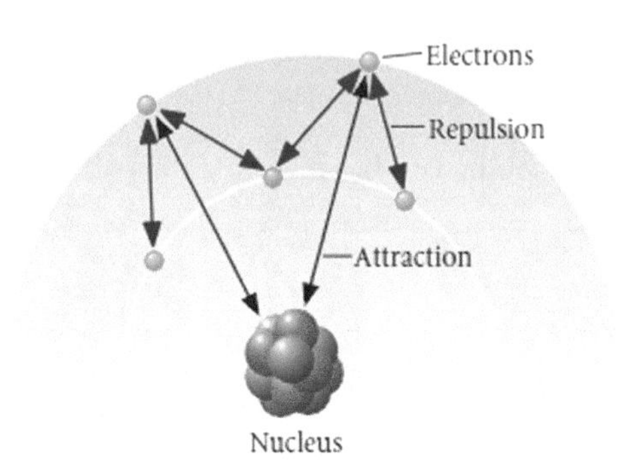

Fig. 6.30 The outer electrons are shielded from the charge of the nucleus by electrons closer to the nucleus

Problems arise when we put a second electron into an atom or ion. Because the two electrons have the same charge, they repel each other. The interaction between the two electrons makes it extremely difficult to solve for their energies and orbital wave functions exactly. Fortunately, ways exist to approximate the correct energies and electron orbitals. The approximations also work for atoms containing three or more electrons, and the results can agree very well with what we observe. How do we describe the orbitals on an atom with more than one electron? We take the hydrogen atomic orbitals, which are approximately correct because they describe the effect the nucleus has on a single electron, and we change them a little to approximate the effect of other electrons. The results are useful both because they accurately predict the physical properties of atoms, such as ionization energies and emission spectra, and because we can use the multielectron orbitals to form **molecular orbitals**, occupied by electrons in molecules. When we adapt these orbitals to account for multiple electrons, we find that they are nearly the same in almost every respect.

Molecular orbitals – Electron orbitals that are appropriate for describing bonding between atoms in a molecule.

6.9.1 The Impact of Electron Shielding in a Multielectron Atom

The largest difference between atoms is the energy associated with their individual orbitals. For the single-electron atom, the energy depends only on the principal quantum number, n. However, for the multielectron atom levels, different l values will also result in different energies.

This is because of something called **electron shielding** (also known as electron screening). As is shown in Fig. 6.30, when a second electron is present in an atom, this electron is between the first electron and the nucleus, part of the time. Because electrons are negatively charged, the positive nuclear charge experienced by the first electron is smaller than it otherwise would be, and the attractive force between that electron and the nucleus is decreased. The first electron also sometimes spends time closer to the nucleus than its partner, and when this happens, that second electron will be shielded too. For the two 1*s* electrons in helium, both are shielded the same amount, and they remain degenerate.

Electron shielding – The ability of electrons in lower energy orbitals to decrease the nuclear charge felt by electrons in higher energy orbitals.

When we focus on orbitals with different n and/or different l values, all electrons in an atom shield all other electrons, but not necessarily to the same extent. The 2*s* electrons are, on average, closer to the nucleus than the 2*p* electrons, as shown in Fig. 6.31, and the 2*s* electrons will shield the 2*p* electrons.

The nuclear charge (Z) will therefore seem to be smaller to the 2*p* electrons than to the 2*s* electrons as a result of the greater shielding. The bottom line is that the 2*p* electrons will no longer be at the same energy as the 2*s* electrons, even though they are in the same energy shell. This is shown schematically in Fig. 6.32. In this figure, electrons are represented with arrows. The $2p_x$, $2p_y$, and $2p_z$ orbitals will still be degenerate, however, because they are all at the same average distance from the nucleus and will all experience the same shielding from other electrons in the atom. This trend continues as we move down the periodic table. The *d* orbitals are shielded by the *s* and *p* orbitals and the *f* orbitals are shielded by all the others.

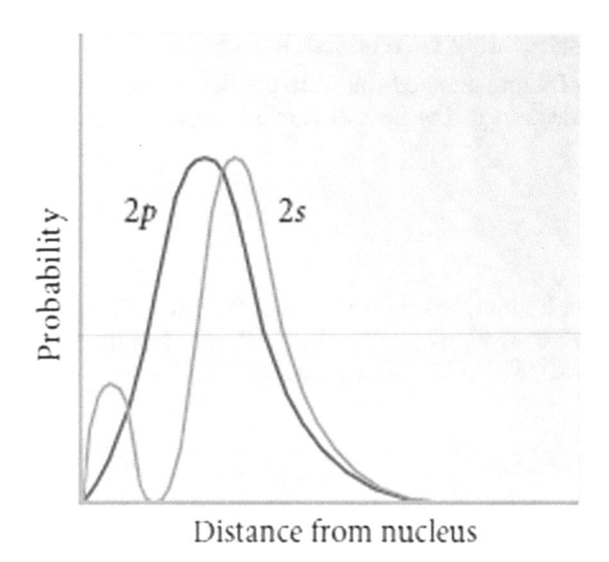

Fig. 6.31 While not easily seen in the Figure, the average electron density is closer to the nucleus in the 2*s* orbital than in the 2*p* orbital. (Remember that both the *s* and *p* orbitals have a node at the nucleus)

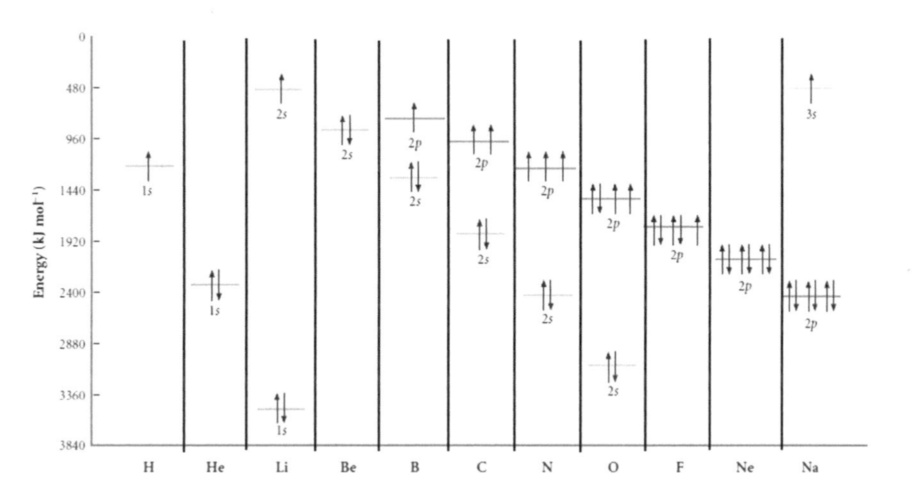

Fig. 6.32 Relative energy levels for multielectron atoms. The relative placement of the energies of the orbitals is only approximate and varies with Z, the nuclear charge, which increases to the right across the periodic table

6.10 Electron Configurations and the Aufbau Principle

Now that we have a complete set of orbitals and a general scheme for ranking their energies, we can list all the orbitals occupied by the electrons of any atom in the periodic table. The complete list of filled orbitals is called the **electron configuration** of an atom. We are often interested in the lowest energy configuration, which represents the ground state of the atom. This can be generated by adding one electron at a time to an unoccupied atomic orbital, starting with the lowest energy orbital and working our way up in energies until all electrons in the atom have been assigned an orbital. This method of filling hydrogen atomic orbitals to get multielectron configurations operates in accordance with the **Aufbau principle**, from the German word for "building up." The Aufbau principle says that as protons are added to the nucleus, electrons are successively added to orbitals of increasing energy, beginning with the lowest-energy orbitals. To convey the maximum information in the minimum space, we use a chemical notation to indicate which orbitals are occupied. The format for this notation is: $nx^{\#}$, where "n" represents the principal quantum number, "x" represents the orbital (*s, p, d, f*), and "#" represents the number of electrons in that orbital or set of orbitals. For example, in its ground state, the hydrogen atom will have its lone electron in the 1*s* orbital, this is written as: $1s^1$. Moving on to helium, both electrons can fit into the s orbital in the first energy shell, so we write the electron configuration as: $1s^2$.

Electron configuration – The complete list of filled orbitals in an atom.

Aufbau principle – The method of filling atomic orbitals to get the ground-state electron configuration of a multi-electron atom by adding each electron to the next available lowest-energy orbital; from the German phase for "building up."

$$H : 1s^1$$

$$He : 1s^2$$

The next element in the sequence, lithium, has three electrons. We have used up all the orbitals with $n = 1$ and now must start another quantum level, or shell, by placing the third electron into the $2s$ orbital. The ground-state electronic configuration for lithium is: $1s^2 2s^1$. The other elements in the same row as lithium continue to fill the second quantum shell:

Li:	$1s^2 2s^1$	N:	$1s^2 2s^2 2p^3$
Be:	$1s^2 2s^2$	O:	$1s^2 2s^2 2p^4$
B:	$1s^2 2s^2 2p^1$	F:	$1s^2 2s^2 2p^5$
C:	$1s^2 2s^2 2p^2$	Ne:	$1s^2 2s^2 2p^6$

In the above examples, note that the p orbitals fill after a total of six electrons because $2p_x$, $2p_y$, and $2p_z$ can each accommodate two electrons, one with spin up and one with spin down.

When we get to the end of the row, we will have completely filled the second energy shell. In the third row of the periodic table, the 3rd energy shell is being filled in with electrons; the 4th shell in the fourth row, etc. Now we can begin to understand how the structure of the periodic table tells us so much about the various elements, whether we are classifying elements in terms of physical and chemical properties, like Mendeleev and Meyer, or by the quantum mechanical solution to the Schrödinger equation in the order of increasing electron energy, like the quantum chemists and physicists.

Even though our compact notation for electron configuration does not indicate electron spins, there is an important detail that governs electron placement. Because of their negative charge, electrons tend to remain as far apart as possible to minimize repulsive interactions between them. When filling the $2p$ orbitals for the elements boron through neon, the electrons can stay farther apart by filling into separate $2p$ orbitals for as long as possible. The energy is minimized if the spins of the electrons are pointing in the same direction. For the first five elements in the periodic table, this doesn't present a problem when adding electrons (◘ Fig. 6.33), but when we get to carbon, there is a decision we must make about adding the 6th electron (◘ Fig. 6.34).

We could place the two $2p$ electrons in the same orbital or in different orbitals. Three unique choices are possible, as shown in ◘ Fig. 6.34, all of which represent possible electronic states for the carbon atom.

The ground state for the carbon atom is state 1, where the two $2p$ electrons occupy different orbitals with their spins aligned in the same direction. It is the lowest energy state for carbon, and it is therefore the state in which we will find a carbon atom unless we supply it with energy—by irradiating it with electromagnetic radiation, for example.

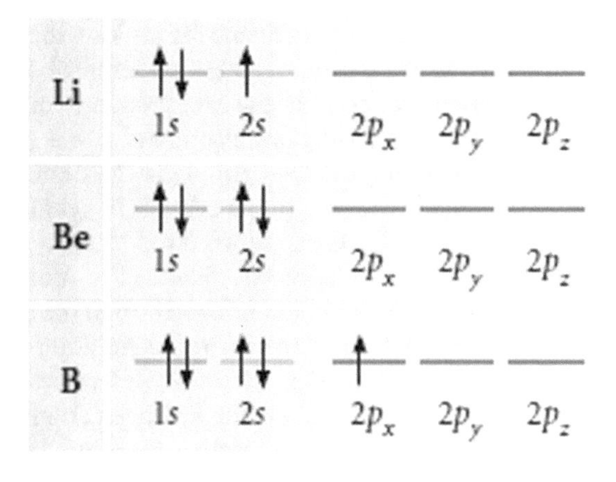

◘ **Fig. 6.33** The electron configuration of the first three elements of the second period

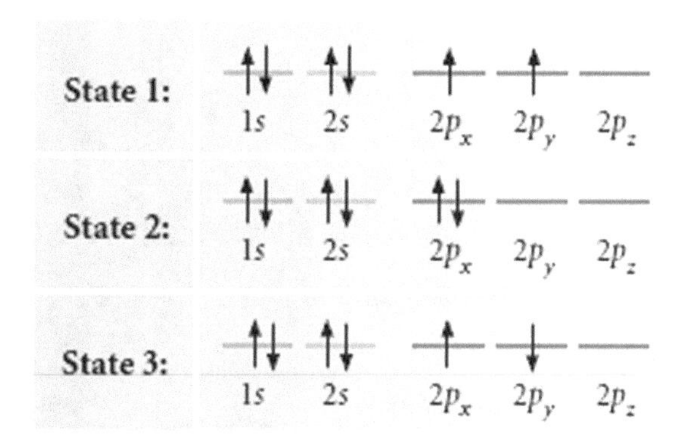

Fig. 6.34 Three possible electron configurations for carbon. State 1 is the lowest energy situation, so it represents the ground state

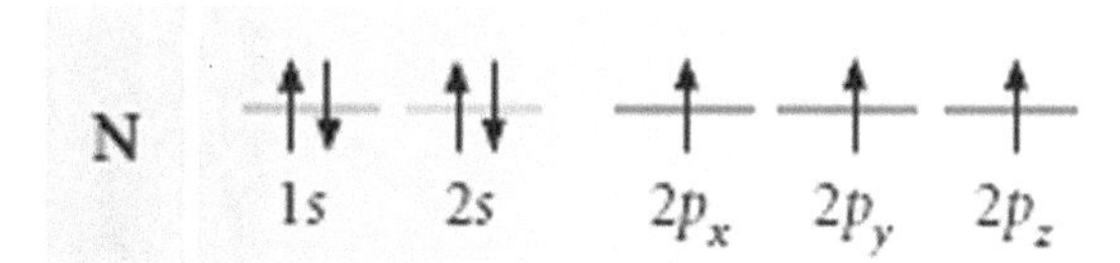

Fig. 6.35 The electron configuration for nitrogen

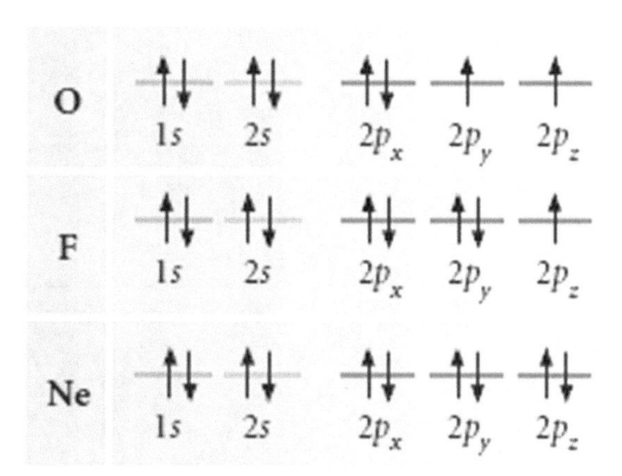

Fig. 6.36 The electron configuration for the last three elements of the second period

Overall, this behavior is summarized by what is known as **Hund's rule**. Proposed by the German physicist Friedrich Hund in 1925, this rule states that when orbitals of equal energy are available, the lowest energy configuration for an atom has the maximum number of unpaired electrons with parallel spins.

If we continue adding the electrons to the atomic orbitals, we will have the maximum number of unpaired $2p$ electrons for nitrogen, as shown in Fig. 6.35.

Hund's rule – When orbitals of equal energy are available, the lowest energy configuration for an atom has the maximum number of unpaired electrons with parallel spins.

We must again begin pairing them to continue adding electrons to the orbitals, as shown in Fig. 6.36. By the time we get to neon, the entire second shell is filled.

To proceed to sodium, we need to add the next electron to the $3s$ level. Even with our compact notation, the electron configurations are beginning to get unwieldy. However, we can say that all atoms in the 3rd row of the periodic table must have filled the $n=1$ and $n=2$ shells before electrons can be added to the $n=3$ level. This allows us to write the electron configuration in one of two ways, as given in the example for sodium shown in Fig. 6.37. The first notation shows all of the orbitals explicitly, and the second uses [Ne] to take the place of [$1s^2 2s^2 2p^6$]. We can say that the configuration of sodium is like neon, with an additional $3s^1$ electron. The electrons listed after the **core electrons** ([Ne] in this case) are called **valence electrons** and will be very important in establishing the chemical reactivity of the element.

$$1s^2 2s^2 2p^6 3s^1$$

Na

$$[Ne]3s^1$$

Fig. 6.37 The electron configuration for sodium can be written two ways

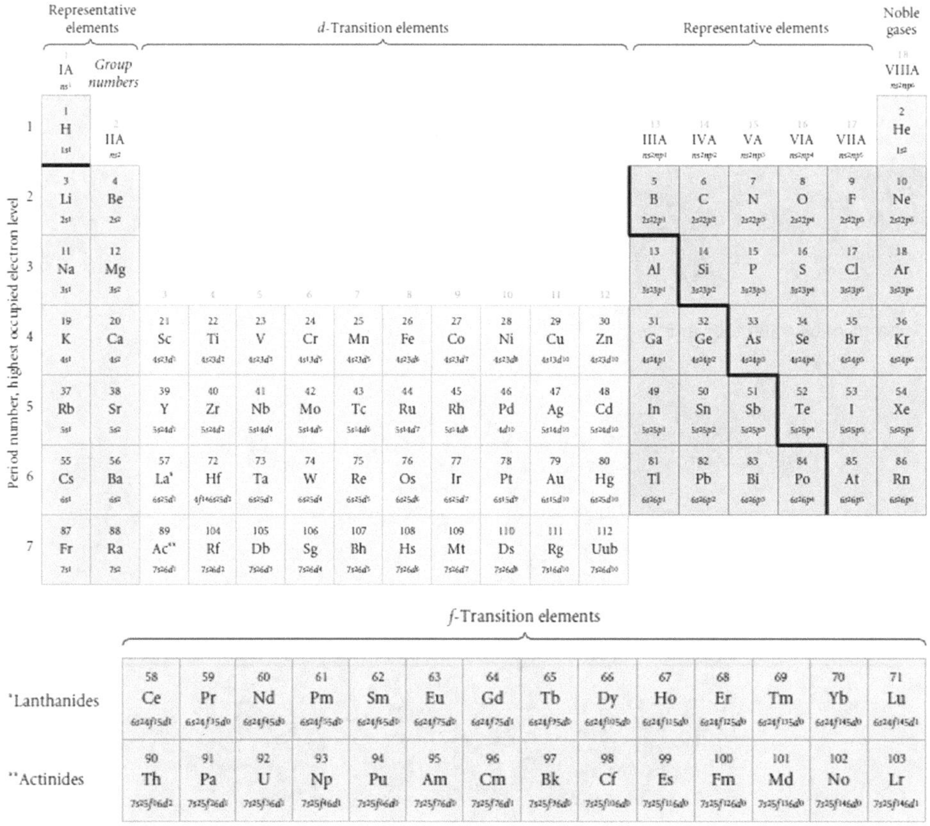

Fig. 6.38 These "partial" electron configurations show only the outermost valence electrons

Core electrons – Electrons in the configuration of an atom that are not in the highest principal energy shell.
Valence electrons – The electrons in an atom, ion, or molecule that are in the highest principal energy shell.

Similarly, the ground-state electron configuration for potassium could be written either explicitly as $1s^2 2s^2 2p^6 3s^2 3p^6 4s^1$ or as $[Ar]4s^1$. The [Ar] core notation signifies that the *electron* configuration has the same occupied orbitals as argon, plus all the others noted after it.

What happened to the 3*d* orbitals when we were constructing the electron configuration for sodium? When we noted the effect of adding more than one electron to the orbitals of the hydrogen atom, we discovered that interactions among the electrons removed the degeneracy among the electrons in the *n*-shell orbitals, so that orbitals with an increasing value of *l* had increasing energy. The 3*s* is lower in energy than the 3*p*, which, in turn, is lower than the 3*d*. The 3*d* orbital energy is *so* destabilized by the electron–electron interactions that it actually rises in energy to be just a little higher in energy than the 4*s* orbital.

Each of the elements in the first column of the periodic table have the configuration $[core]ns^1$, where [core] is the electron configuration of the noble gas directly preceding the element in the periodic table. It is this similarity in ground-state electron configuration that gives all of the elements in this column similar chemical properties, including metallic behavior, malleability, tendency to form ions with a charge of + 1, and high reactivity with water.

To form the lowest energy electron configuration, the 4*s* subshell fills before the 3*d* subshell. A similar problem is noted when we begin filling the *f* orbital subshell. The ground-state configurations for the elements are given in Fig. 6.38. The good news is that the order of filling is mostly systematic and is illustrated by the diagram shown

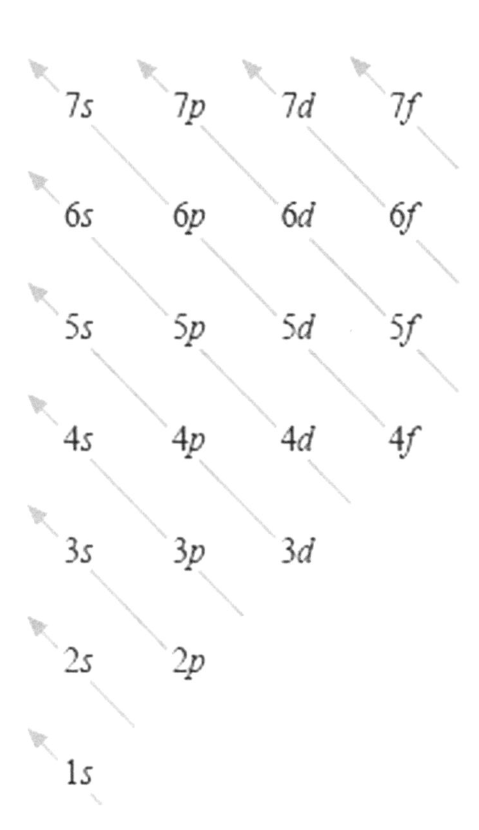

Fig. 6.39 The order in which atomic orbitals are filled. The order is determined by writing the orbitals found in each energy level in rows and then drawing diagonal lines up and to the left. The order is 1*s*, 2*s*, 2*p*, 3*s*, 3*p*, 4*s*, 3*d*, 4*p*, 5*s*, 4*d*, 5*p*, 6*s*, 4*f*, 5*d*, 6*p*, …

in Fig. 6.39. For example, we can follow the arrows in the figure and add the electrons to each subshell to obtain the ground-state electron configuration of sulfur (16 electrons):

$$S : 1s^2 2s^2 2p^6 3s^2 3p^4 = [\mathrm{Ne}]3s^2 3p^4$$

Using this information, we can obtain the electron configuration of iodine:

$$\mathrm{I} : 1s^2 2s^2 2p^6 3s^2 3p^6 4s^2 3d^{10} 4p^6 5s^2 4d^{10} 5p^5 = [\mathrm{Kr}]5s^2 4d^{10} 5p^5$$

One of the things we notice in Fig. 6.38 is that there are some exceptions to this filling order. For example, we would expect copper to have a valence shell of $4s^2 3d^9$, but in fact it is $4s^1 3d^{10}$. Exceptions to the rules for electron configuration of neutral atoms exist. And as we move down the periodic table, these exceptions become more common.

Example 6.8—Practice with Electron Configurations

Write the ground-state electron configuration for vanadium and for tellurium.

Asking the Right Questions

In what period is our element? Is its highest partially filled energy orbital an *s*, *p*, *d*, or *f* orbital?

Solution

The structure of the periodic table gives us guidance for writing the ground-state electron configuration of an element. Vanadium, atomic number 23, is in Period 4 and is the third element in the 3*d* series that begins with scandium, as shown in Fig. 6.38. It will therefore have an argon electron core with two 4*s* and three 3*d* electrons. Alternatively, we can use the filling order in Fig. 6.39 and fill the orbitals to an electron count of 23.

$$\mathrm{V}\ (23\ \text{electrons}) : [\mathrm{Ar}]4s^2 3d^3 = 1s^2 2s^2 2p^6 3s^2 3p^6 4s^2 3d^3$$

Tellurium, atomic number 52, is in Period 5 and Group VIA, in which *p* orbitals are being filled. It will have a krypton electron core, with (reading across Group V in Fig. 6.38) two 5*s* electrons, ten 4*d* electrons, and six 5*p* electrons. Alternatively, we can use Fig. 6.39, as with vanadium.

$$\mathrm{Te}\ (52\ \text{electrons}) : [\mathrm{Kr}]5s^2 4d^{10} 5p^4 = 1s^2 2s^2 2p^6 3s^2 3p^6 4s^2 3d^{10} 4p^6 5s^2 4d^{10} 5p^4$$

Are the Answers Reasonable?

We have two ways to know if our answers are reasonable. The first is to count the number of electrons in the final electron configuration, and make sure that that number is the same as the atomic number of the element. The electron configuration for vanadium has 23 electrons in it, and tellurium contains 52 electrons in its electron configuration. The more elegant way to define "reasonable" in this case is to use our understanding of the period table. Because vanadium is the third element in the 3*d* series, this transition element should have three 3*d* electrons. That's reasonable. Similarly, tellurium as the fourth element in the 5*p* main group series, it is reasonable that it has four 5*p* electrons in its outer shell.

What If…?

…we have an element that has three times as many electrons in its highest energy subshell, $l=2$, as does calcium? What is the most important fact about its chemical reactivity? (*Ans: It is chemically inert*)

Practice 6.8

Write the ground-state electron configuration for gallium and strontium.

6.10.1 Putting Quantum Chemistry into Action: The Blocks of the Periodic Table

The periodic table can be divided into four regions, or blocks, based on the filling of orbitals of the elements; see Fig. 6.40.

- *s-block*: This region includes the alkali and alkaline earth metals, in which the *s* orbitals are being filled. These Group IA and IIA elements, such as potassium and calcium, are part of the main-group elements. (Helium is grouped with the rest of the noble gases.)
- *p-block*: This region includes the remainder of the main-group elements, Groups IIIA through VIIIA, such as carbon, nitrogen, and chlorine, in which the *p* orbitals are being filled.
- *d-block*: This region includes the transition elements, in which the *d* orbitals are being filled. These include metals such as chromium, silver, and mercury.
- *f-block*: This region includes the inner-transition elements, such as uranium and plutonium, in which the *f* orbitals are being filled.

The structure of the periodic table is a very satisfying one for chemists because it gives an order to the physical properties and chemical reactivities of the elements. Although the periodic table was an outgrowth of Mendeleev's periodic law formulated in 1869, this early grouping of elements with similar properties was done empirically, just by seeing how the material behaved. By solving the Schrödinger equation, we have generated wave functions with unique sets of quantum numbers. By generating the wave functions and quantum numbers, we have created a complete set of electron orbitals, and by filling these orbitals according to the Aufbau principle, we have generated the complete periodic table—its form, the properties of its elements, everything!

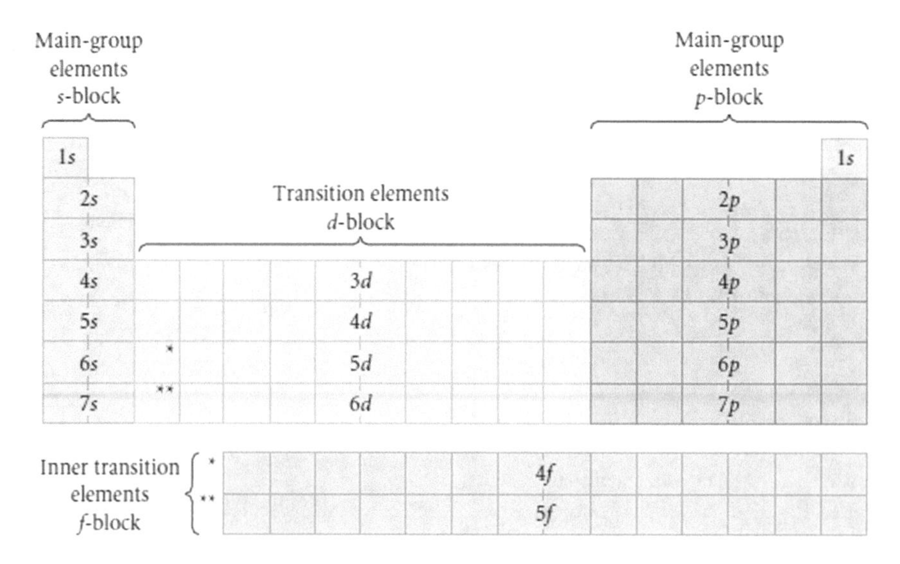

Fig. 6.40 The periodic table can be divided into regions (known as blocks) that illustrate the type of atomic orbital that is being filled

The clinical laboratory scientist and the use of spectroscopy to identify molecules in a patient's blood, urine, or other bodily fluids rely heavily on the information within the periodic table. The interaction of light with these compounds is a result of the rules for the placement of electrons around an atom.

The Bottom Line

- The smallest unit in which something can exist is a quantum. The quantum of matter is an atom. The quantum of electromagnetic radiation is a photon.
- We can interconvert among the wavelength, frequency, and energy of electromagnetic radiation.

$$E = h\nu = \frac{hc}{\lambda}$$

- Atoms and molecules absorb and emit electromagnetic radiation to gain and lose energy, but only certain wavelengths of radiation can be absorbed and emitted.
- The Rydberg equation is a general form of the equations derived from the observations of Balmer, Lyman and others and allows the calculation of the wavelength of radiation that corresponds to a specific electron transition:

$$\frac{1}{\lambda} = 1.0968 \times 10^{-2}\left[\frac{1}{n_\mathrm{f}^2} - \frac{1}{n_\mathrm{i}^2}\right]\mathrm{nm}^{-1}$$

- One of the first models of atomic structure, the Bohr model, quantized the energies and spatial locations of the electrons to explain the hydrogen emission spectrum. Although the model was not correct in other details, quantum energy and orbit locations were breakthrough concepts and are key concepts in the modern picture of the atom.
- Electronic transitions between the quantized energy levels are responsible for atomic absorption and emission spectra. The energies are given by $\Delta E = E_\mathrm{f} - E_\mathrm{i}$, where the subscripts i and f stand for initial and final states.
- At the atomic scale, all things show both wave and particle behavior. This concept is known as wave–particle duality.
- The wave and particle natures of all objects are linked in the de Broglie equation: $\lambda = h/p$.
- The Heisenberg uncertainty principle, $\Delta x \Delta p \geq h/4\pi$, places limits on how precisely we can simultaneously measure position and momentum.
- The locations of electrons on an atom can be completely described by the Schrödinger equation: $\hat{H}\Psi_n = E_n\Psi_n$.
- The probability distribution function $\Psi_n{}^*\Psi_n$ tells us where to find electrons in the atom, and the shape described by this function is called the atomic orbital.
- The principal quantum number n indicates the energy shell that an electron is in, and n, l, and m_l determine the shape and orientation of the orbital in three-dimensional space.
- Only two possible electron spin states exist: $m_s = +½$ and $m_s = -½$.
- The Pauli exclusion principle states that in a given atom, no two electrons can have the same set of the four quantum numbers—n, l, m_l, and m_s.
- Each shell of electrons shields the ones above it, reducing the attraction between the outer electrons and the nucleus.
- Electrons in multi-electron atoms occupy the available orbitals one by one, starting with the lowest energy level, according to the Aufbau principle. The complete listing of occupied orbitals in the lowest energy state of the atom is called the ground-state electron configuration.
- An electron configuration can be written out explicitly or can be written using a noble gas shell notation.
- Hund's rule for ground-state electron configurations states that when orbitals of equal energy are available, the lowest energy configuration for an atom is the one with the maximum number of unpaired electrons with parallel spins.

Section 6.1 Introducing Quantum Chemistry and Electromagnetic Radiation

Skill Review

1. At the submicroscopic level we typically discuss the probability of events and objects rather than their location and individual speeds. Describe one event or object that is best explained using probability. Is this event considered "macro" or "micro"?
2. Use the analogy of flipping a coin to explain the term probability.
3. We often speak of great changes as, "Quantum leaps." How does this term reinforce (or contradict!) our chemistry definition of "quantum?"

4. In the section narrative, we say, "Many familiar things also come in quanta, including eggs, cats and chemistry books." What things *within* your chemistry book also come in quanta?
5. As we walk through the produce section of a supermarket, we see various colors of vegetables. Let's assume that the grocer decided to organize the vegetables in order of increasing frequencies of their reflected light—a novel marketing concept. Indicate in what sequence these vegetables would appear: a red tomato, a green zucchini squash, a purple eggplant, and a yellow spaghetti squash.
6. The three colors of a traffic light are red, yellow, and green. The green light is placed at the bottom, with yellow in the center and red on top. Are these colors in order, from bottom to top, by frequency or by wavelength? Justify your answer.
7. a. What is the frequency of an X-ray with a wavelength of 1.5×10^{-2} nm?
   b. What is the energy, in joules, associated with a photon of this frequency?
   c. What would be the energy of a mole of such photons?
8. a. What is the frequency of visible light with a 400-nm wavelength?
   b. What is the energy, in joules, associated with a photon of this frequency?
   c. What would be the energy associated with a mole of these photons?
9. In what region on the electromagnetic radiation spectrum would a wavelength of 2.5×10^4 nm be placed? What would be the frequency and energy of this radiation?
10. In what region on the electromagnetic radiation spectrum would a wavelength of 5.8×10^2 nm be placed? What would be the frequency and energy of this radiation?
11. Calculate your height in nanometers, meters, and light-years. Which do you find the most convenient unit for this application?
12. Calculate the length of a 12-in ruler in nanometers, meters, and light-years. Which do you find the most convenient unit for this application?
13. In his book, *ChiRunning*, Danny Dreyer discusses harnessing the "Chi," or life force, to make your running better. The optimal running pace includes taking about 85 steps per minute, where "left–right" is defined as one step. If a runner takes 120.0 s to complete a 400.0-m lap on a track, what is her step length in meters?
14. A company manufactures potato chips, filling chip bags at the rate of 1,680 chips per second. If a bag holds 37 chips, what is the frequency of production of filled bags in seconds per bag?
15. Older cell phones operated at frequencies of 824 to 894 MHz. What range of wavelengths is this? To which part of the electromagnetic spectrum does this correspond?
16. Helium–neon (HeNe) lasers are both cheap and common. They are even sold as attachments to novelty key chains. If the helium–neon mixture lases at 0.632 μm, what is its frequency? To which part of the electromagnetic spectrum does this correspond?
17. Radio stations broadcast with frequencies given in megahertz, where $1\ \text{MHz} = 10^6\ \text{s}^{-1}$. This means that when a station advertises itself as Radio WXYZ located at 98.3 on your dial, it is broadcasting at 98.3 MHz. What wavelength, in meters, does Radio WXYZ use in its broadcast?
18. A wireless Internet connection broadcasts its signal in the 1200-MHz range. To what wavelength, in meters, does this correspond?
19. Chlorophyll, the green pigment found in plants, absorbs visible radiation best in the red region (at about 675 nm) and in the blue-violet region (at about 440 nm). What are the energies, in joules, of the photons collected by plants at these two wavelengths?
20. What is the energy, in joules, of photons emitted from a 100-MHz magnetic resonance imager?
21. How much energy, in joules, resides in 1 mol of photons whose wavelength is 440 nm?... 675 nm?
22. How much energy, in joules, resides in 2 mol of photons with a frequency of 1.5×10^{14} Hz?... in 0.75 mol?
23. It takes 495 kJ of energy to remove one mole of electrons from gaseous sodium. What is the wavelength, in nm, associated with this energy?
24. It takes 735 kJ of energy to remove one mole of electrons from gaseous magnesium. What is the wavelength, in nm, associated with this energy?

Chemical Applications and Practices

25. a. The process of photosynthesis is quite complex. However, one of the main considerations is that plant pigments absorb visible light to power reactions that convert carbon dioxide and water into food and produce oxygen. If a plant absorbs blue light that has a wavelength of 565 nm, what is the energy in joules per photon that is being absorbed?
    b. A typical ratio between photons absorbed and oxygen molecules produced is 8:1. How much energy, in joules, is required to produce one molecule of oxygen in this manner?

26. Ozone (O_3) is important in our upper atmosphere because it aids in filtering out harmful ultraviolet rays. Ultraviolet rays may be classified as UV-A, UV-B, or UV-C. The UV-B rays cause the most problems for earth-based organisms. For example, higher incidences of "jumping genes" that cause mutations may be related to exposure to UV-B.
    a. What is the energy in joules/mol of UV-B photons that have a wavelength of 312 nm?
    b. What is the energy in joules/mol of photons with a wavelength of 600 nm? To what type of radiation does this correspond?
27. UV-B radiation is responsible for "sunburn" in humans. A helpful advancement in technology is the personal UV detector. This small device uses a photoelectric response. An example is a gallium-based device to convert absorbance into an electrical signal. If the device gave a maximum reading for a photon with a wavelength of 290 nm, how much energy in joules is being absorbed? To what frequency does this correspond?
28. a. Some snakes have the ability to detect infrared radiation (IR). Are they detecting energy that is higher or lower in energy than human eyes can see?
    b. What is the source of typical IR wavelengths?
    c. A television remote control may use IR with a frequency of 1×10^{13} cycles per second. To what energy, in joules per photon, does this correspond?
29. One way to gain information about the origin and functions of stars within the universe is to study the origin and distribution of the hundreds of gamma ray sources in the sky. If one such source were producing high-energy gamma rays of 1.6×10^{-8} J/photon, what wavelength would astronomers have detected? What is the frequency of this radiation?
30. If an astronomer recorded energy from a distant star at 3.6×10^{-10} J/photon, what wavelength would he have detected? What is the frequency of this radiation?

Section 6.2 Atomic Emission, Absorption Spectroscopy and the Quantum Number

Skill Review

31. When bombarded with high-energy electrons, copper metal gives off radiation with a wavelength of 1.54 Å. What is the frequency of this radiation? To which range of the electromagnetic spectrum does this correspond?
32. Sodium arc lamps, which are used as automobile headlights and streetlights, are colored by the "sodium doublet": electromagnetic radiation produced by excited sodium atoms found at 5895 Å and 5904 Å. What color are these lights?
33. Much of the radiation striking the earth from the sun has a wavelength of approximately 500 nm. Express this wavelength in meters, angstroms, centimeters, and inches.
34. Express 280-nm ultraviolet radiation in meters, angstroms, centimeters, and inches.
35. According to the Balmer equation, the wavelength of emitted light from hydrogen can be calculated from the whole-number values of n. The difference between $n=5$ and $n=4$ is only 1. The difference between $n=2$ and $n=1$ is also only 1. Without calculating any values, explain why the two conditions don't produce the same wavelength of emitted light in hydrogen?
36. The Balmer equation can be used to calculate the wavelength of light emitted from excited hydrogen. To what initial value of n in hydrogen would an emitted wavelength of 5547 Å correspond? Explain why this value of n is never observed for hydrogen.
37. The first two wavelengths of the Balmer series of the hydrogen emission spectrum are 6562.1 Å, 4860.8 Å. What are the next three values in this series? What are the frequencies and energies, in joules, of the photons producing the emission lines?
38. What is the highest frequency of the Balmer series of the hydrogen emission spectrum? What is the lowest frequency of this series?
39. A water sample was analyzed for the presence of elements by flame emission spectroscopy. A staff member inadvertently reported the frequencies, rather than the wavelengths. He found important signals at 9.4971×10^{14} s^{-1} and 1.1206×10^{15} s^{-1}. Based on the data in Table 6.3, what elements were detected in the water sample?
40. Based on the data in Table 6.3, what are the frequencies of emission for barium and molybdenum?

Chemical Applications and Practices

41. The presence of cadmium in drinking water is undesirable because exposure to large amounts has been associated with weakening of bones and joints. The wavelength of electromagnetic radiation strongly absorbed by cadmium is 214.439 nm.

a. What is the frequency of that light?
b. In what range of the electromagnetic spectrum would you classify this frequency?
c. Which other element has an absorbance wavelength closest to (and therefore possibly difficult to distinguish from) cadmium? Consult Table 6.3 for additional information.

42. An adaptation of the Balmer equation makes it possible to calculate other emitted wavelengths from excited hydrogen. For example, if the final value for n is 3, then emitted light is in the infrared area of the spectrum. These line spectra are known as the Paschen series. Calculate the wavelength, in nm, emitted when an electron in hydrogen drops from the fourth Bohr level to the third.
43. Levels of metals in lubricating oil can be measured by Inductively Coupled Plasma spectroscopy, which allows for the analysis of dozens of metals at the same time. Assume in an analysis that aluminum was determined at BOTH 396.152 nm *and* 308.215 nm. Why might it be useful to determine concentration of a metal at two or more wavelengths?
44. An analysis for metals in blood serum was performed by Inductively Coupled Plasma spectroscopy. Which metals are likely to be present in the highest concentrations, and why?

Section 6.3 The Bohr Model of Atomic Structure

Skill Review

45. Calculate the energy, in joules, of an electron in each of these Bohr energy levels:
a. $n=1$ b. $n=3$ c. $n=5$ d. $n=7$
46. Calculate the energy, in joules, of an electron in each of these Bohr energy levels:
a. $n=2$ b. $n=4$ c. $n=6$ d. $n=8$
47. Calculate the energy, in joules, of a photon released from a hydrogen atom in each of these transitions:
a. $n=4$ to $n=1$
b. $n=3$ to $n=1$
c. $n=5$ to $n=4$
d. $n=7$ to $n=2$
48. Calculate the energy, in joules, of a photon released from a hydrogen atom in each of these transitions:
a. $n=2$ to $n=1$
b. $n=4$ to $n=2$
c. $n=4$ to $n=3$
d. $n=8$ to $n=2$
49. What region of the electromagnetic spectrum is represented by the energy in each part of problem 47?
50. What region of the electromagnetic spectrum is represented by the energy in each part of problem 48?
51. Repeat problem 47a, b, c and d for one mole of photons.
52. Repeat problem 48a, b, c and d for one mole of photons.
53. Calculate the wavelength of a photon that would cause these transitions:
a. $n=1$ to $n=2$
b. $n=3$ to $n=5$
c. $n=2$ to $n=4$
d. $n=1$ to $n=6$
54. Calculate the frequency of a photon that would cause these transitions:
a. $n=8$ to $n=10$
b. $n=3$ to $n=6$
c $n=4$ to $n=8$
d. $n=4$ to $n=5$
55. What is the shortest wavelength that can be emitted by the hydrogen atom in the Brackett spectral series? What is the longest wavelength in this series? Consult Table 6.2 for assistance with this problem.
56. According to the Bohr model of the hydrogen atom, what is the closest an electron can get to the nucleus? What is the farthest it can get from the nucleus?

Chemical Applications and Practices

57. An emission line was noted in a spectrum associated with an energy of 4.84×10^{-19} J per photon. Is this emission possible for hydrogen? (To make this problem more manageable, consider transitions within the first eight energy levels only.)
58. An emission line was noted in a spectrum associated with an energy of 7.38×10^{-19} J per photon. Is this emission possible for hydrogen? (To make this problem more manageable, consider transitions within the first eight energy levels only.)
59. To ionize a hydrogen atom is to remove its only electron. We consider a free electron to have, with respect to an electron in hydrogen, zero energy. In other words, the initial state is $n=1$ and the final state is $n=\infty$. Calculate the energy needed to remove an electron from the ground state.
60. Calculate the energy released when an electron is added to a hydrogen nucleus. Assume the transition is $n=\infty$ to $n=1$.

Section 6.4 Wave–Particle Duality

Skill Review

61. In a classical sense, compare matter and waves in terms of permissible energy values, spatial positions, and momentum.
62. How did the quantum view change the classical distinction between matter and waves? What is the quantum view called?
63. List two particle-like properties of electrons.
64. List two wave-like properties of electrons.

Chemical Applications and Practices

65. What is the momentum in kg·m·s^{-1}of an electron that is moving at 68% of the speed of light? (Use 9.1×10^{-31} kg as the mass of the electron.)
66. What is the speed in m/s of an electron that has a momentum of 9.7×10^{-23} kg·m·s^{-1}?
67. A freight train locomotive weighs about 205,000 kg and has a top speed of about 160.0 km/h. What is its momentum in kg·m·s^{-1}?
68. The average speed in m/s of an electron in the ground state of the hydrogen atom is 2.19×10^{6} m/s. What is the average momentum of an electron in this state?
69. What is the momentum in kg·m·s^{-1} of a photon with a wavelength of 540 nm?
70. What is the momentum in kg·m·s^{-1} of a photon with a frequency of 1.0×10^{14} Hz?
71. The average radius in the second Bohr orbit ($n = 2$) is 2.116×10^{-10} m. Using the Eq. $2\pi r = n\lambda$, calculate the wavelength in meters for the standing electron wave.
72. The approximate wavelength in meters for an electron in hydrogen's first energy level has been calculated to be 3.3×10^{-10} m. What is the ratio of this wavelength to the diameter of ten hydrogen atoms (7.4×10^{-11} m)? What might be some reasons why the diameter isn't closer in size to the wavelength?
73. In badminton, the object being struck by the racket is called a shuttlecock. Although the shuttlecock may be made of various materials, it mass must be close to 5.00 g. What is the wavelength in meters of a shuttlecock that is moving at 78 mi/h? Without doing calculations, explain whether the wavelength of a softball moving at the same speed would be larger or smaller.
74. If a proton, mass $= 1.67 \times 10^{-27}$ kg, were moving as fast as the electron in the ground state of hydrogen (2.1×10^{6} m/s), what would be the wavelength in meters?

Section 6.5 The Heisenberg Uncertainty Principle

Skill Review

75. In the uncertainty relationship, what do the symbols p and x represent? Explain the importance of noting that Δp and Δx multiplied must be equal to or greater than a constant.
76. Taking a photograph of a moving object—for example, a person sprinting—will cause some blurring of the actual person's position when the photograph is developed. If the entire scene were re-shot using a faster shutter speed, what information would we gain and what would we lose in the photo?
77. What is the momentum in kg·m·s^{-1} of an X-ray that has a wavelength of 6.5×10^{-10} m?
78. How does the momentum of the X-ray in Problem 77 compare to the momentum of a photon of green light that has a wavelength of 560 nm?
79. What is the momentum in kg·m·s^{-1} of a mole of X-rays with wavelength 6.5×10^{-10} m?
80. What is the momentum in kg·m·s^{-1} of a mole of photons with wavelength 560 nm?

Chemical Applications and Practices

81. Argon is used in light bulbs and as an inert atmosphere in metallurgic processes. The radius of an argon atom is 97 pm. Will the electron ever be further than 97 pm from the nucleus? Explain your answer.
82. Can an electron from an argon atom in a light bulb ever be found on the planet Jupiter? If so, *will* such an electron ever be found on Jupiter? Explain your answers.
83. Using the mass of an electron as 9.11×10^{-31} kg and velocity as 2.1×10^{6} m/s, what would you calculate as the uncertainty in the position, x, for the electron if the uncertainty in the velocity were 5.0%? What would be the answer to this if the uncertainty in the velocity were 10.0%? How does this uncertainty compare to the radius of the hydrogen atom (3.7×10^{-11} m)?

84. Which of these diagrams of an atom most agrees with the Heisenberg uncertainty principle? Explain your answer and indicate why the other two choices do not conform to the Heisenberg uncertainty principle.

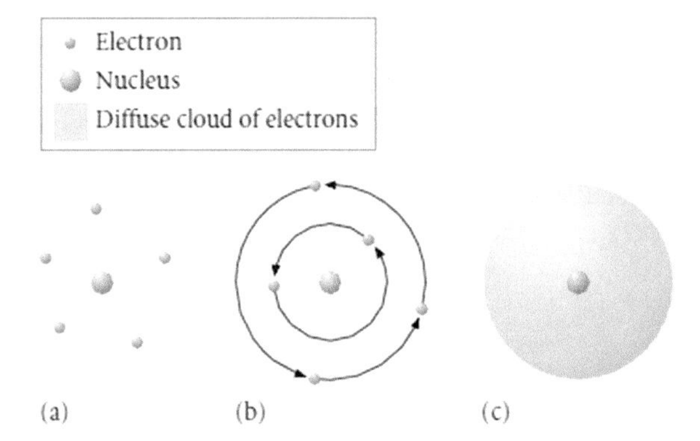

85. What is the momentum in kg·m·s^{-1} of a 190-lb chemistry professor walking through a chemistry lab at 3.0 mi/h?
86. In order for us to see the strolling professor of Problem 85, light photons would have to bounce off the professor and enter our eyes. If some of that light had a wavelength of 565 nm, what would be the momentum in kg·m·s^{-1} of the photon?

Section 6.6 The Mathematical Language of Quantum Chemistry

Skill Review

87. Acceptance of the Schrödinger equation literally knocked the electrons out of Bohr's orbit concept. However, the term orbital was maintained as a connection to the quantization of energy states proposed in Bohr's model. Explain the meaning of an electron orbital.
88. a. What is meant by the term "node" when it is applied to waves?
    b. What is the relationship among nodes, wavelength, and energy in the context of standing waves?

Chemical Applications and Practices

89. In both the Bohr model of electronic structure and the Schrödinger model, electrons appear to change energy levels by a method that does not show the electron making a gradual transition. Using the wave model, explain why a stable standing wave is not possible between two adjacent energy levels. Does this explain why there is no gradual transition between states?
90. a. What is a photon?
    b. In what way does a photon resemble particles, and in what way does it resemble light?

Section 6.7 Atomic Orbitals

Skill Review

91. The sequence in each line that follows represents values for the quantum numbers for an electron in a hydrogen atom. Select any sequence(s) that are not possible and explain the problem(s).
    a. $n = 3; l = -1; m_l = 0$
    b. $n = 2; l = +2; m_l = +1$
    c. $n = 3; l = +2; m_l = +3$
    d. $n = 1; l = +1; m_l = +1$
    e. $n = 4; l = +3; m_l = -2$
92. The sequence in each line that follows represents values for the quantum numbers for an electron in a hydrogen atom. Select any sequence(s) that are not possible and explain the problem(s).
    a. $n = 3; l = +1; m_l = 0$
    b. $n = 2; l = +2; m_l = +1$
    c. $n = 3; l = 0; m_l = -3$
    d. $n = 5; l = +4; m_l = 0$
    e. $n = 2; l = +3; m_l = -2$
93. When an electron is in the fifth energy level, how many subshells are possible? How many orbitals are possible?

94. When an electron is in the fourth energy level, how many subshells are possible? How many orbitals are possible?
95. When $l=3$, how many degenerate orbitals are possible?
96. When $l=6$, how many degenerate orbitals are possible?
97. How many sets of quantum numbers are possible when $n=5$?
98. How many sets of quantum numbers are possible when $n=3$?
99. We often use the phrase "the shape of an orbital." Indicate what restrictions apply when the phrase is used, and explain why the phrase is a bit of an abstraction.
100. Name two distinguishing features that the three *p* orbitals in the same level have in common. What property allows them to be identified separately?
101. Define and give an example of both a radial node and a planar node.
102. In which energy level would the *f* orbitals make their first appearance? (Use a quantum mechanical proof for your answer.)

Section 6.8 Only Two Electrons per Orbital: Electron Spin and the Pauli Exclusion Principle

Skill Review

103. The four quantum numbers representing an electron may be represented by: (3, 1, 0, ↑). What is the value of the fourth quantum number? What is the name given to the fourth quantum number?
104. If the four quantum numbers for an electron were (3, 2, 1, – ½) what would be the four quantum numbers for an electron in the same orbital as the first electron?

Chemical Applications and Practices

105. The magnetic properties of elements are related to the number of unpaired electrons in the atoms. Of the following, which would have unpaired electrons?

C Ca O Ne Zn

106. The magnetic properties of elements are related to the number of unpaired electrons in the atoms. Of the following, which would have unpaired electrons?

N S Na He Sc

107. If the Pauli Exclusion Principle were not used, show how the configuration of the electrons in oxygen would appear to allow no unpaired electrons.
108. Among the elements from 21 to 30, which would have the highest number of unpaired electrons?

Section 6.9 Orbitals and Energy Levels in Multielectron Atoms

Skill Review

109. Compared with a one-electron atom, name three things that will be the same in a multielectron atom's orbital descriptions, and name one critical difference.
110. How many radial and how many nonradial nodes, respectively, will be possible for each of these electron orbitals? What will be the letter designation for each of the represented orbitals?
a. $n=3$; $l=2$ b. $n=3$; $l=0$ c. $n=4$; $l=3$
111. If each orbital can hold a maximum of two electrons of opposite spin, how many electrons can each of the following subshells hold?
a. 5*f* b. 3*d* c. 4*f* d. 6*s*
112. If each orbital can hold a maximum of two electrons of opposite spin, how many electrons can each of the following subshells hold?
a. 5*f* b. 3*f* c. 4*f* d. 6*s*

Chemical Applications and Practices

113. Is it easier to remove the outermost electron from He or He^+? Explain the basis of your answer.
114. Would an electron in the 3*p* sublevel experience more nuclear pull than an electron in the 3*d* sublevel? (Assume that the electron is in the same atom and that the atom's sublevels up to 3*s* are filled.)

Section 6.10 Electron Configurations and the Aufbau Principle

Skill Review

115. a. What is the electron configuration for the ground state of the nitrogen atom?
  b. Nitrogen commonly forms a −3 ion. What is the electron configuration of the ion?
116. In the ground state of manganese, there are five electrons that would occupy the 3*d* sublevel. Show the way these electrons could be configured to follow Hund's rule and a way that would violate Hund's rule.
117. Report, for each of the following, which element is being represented.
  a. $[Ne]3s^23p^2$ b. $[Ne]3s^23p^5$ c. $[Ar]4s^1$ d. $[Kr]5s^2$
118. Report, for each of the following, which element is being represented.
  a. $[Ne]3s^2$ b. $[He]2s^22p^5$ c. $[He]2s^1$ d. $[Ar]4s^2$
119. Write the electron configuration for element 21. Explain why the correct configuration shows the 4*s* sublevel filling with electrons before the 3*d* sublevel.
120. Write out the ground-state electron configuration for element 19.

Chemical Applications and Practices

121. Which, if any, of the following contain unpaired electrons in their ground state?

K Ca Fe Zn Ne

122. Which, if any, of the following contain unpaired electrons?

K^+ Ca^{2+} Fe^{3+} Zn^{2+} Ne^+

123. Evidence indicates that copper has no unpaired electrons in the 3*d* sublevel. What ground-state electron configuration of copper would make this possible?
124. Iridium (element 77) is one of the metals that can be found in the Earth's crust not combined with other elements. It is a brittle, lustrous metal and has a melting point over 2400 °C.
  a. Judging on the basis of iridium's position in the periodic table, what other elements would it most resemble?
  b. Does iridium have any unpaired electrons?
  c. Using the Aufbau principle, determine to what sublevel the 25th electron was added to the configuration of iridium.
125. An element, "M", reacts vigorously with water to form MOH + hydrogen gas. Its outermost energy level is the same as that of another element, which is the only metal that is a liquid at room temperature. Identify the element "M", and write the shorthand notation for its electron configuration.
126. The oxide of an element, "E", has the formula EO_2, and reacts with water to form an acid, acid, HEO_3. Its outermost energy level is the same as an element that, when in acid form, is used to etch glass. Identify the element "E," and write the shorthand notation for its electron configuration.

Comprehensive Problems

127. a. The light produced in the explosion of a dramatic fireworks display reaches us before the sound of the explosion. Use another reference to determine the speed of sound. Then determine the ratio of the speed of light to the speed of sound.
  b. What is the speed of light in miles per hour?
128. When we observe the striking colors of the fireworks at a special occasion, we are observing emission spectra. Describe, chemically, what has transpired in the atoms of the elements to cause the emission of light.
129. What is the longest wavelength of the Lyman series of the hydrogen emission spectrum?
130. Explain how both the volume and the energy of an electron associated with an atom are quantized.
131. Using the equation that allows calculation of the radii of energy levels for single-electron situations in hydrogen, compare the distance between the first and second energy levels to the distance between the third and fourth energy levels and to the distance between the fifth and sixth energy levels. What trend do you notice?
132. The equations used by Bohr are valid for the hydrogen atom, but when they are applied to helium, a significant error shows up. And when they are applied to lithium and elements with higher atomic numbers, the error becomes so large that the equations offer little. However, the equations can be applied to helium and lithium ions with a fair agreement with experimental values. What helium and lithium ions would be most like the hydrogen atom?

133. When applying wave properties to electrons in atoms, we use the expression $2\pi r = n\lambda$. Explain why n can have only integer values.
134. Define the terms discrete and continuous. List five everyday items that have some property that is discrete and five that have a property that is continuous on a macroscopic scale.
135. Assign possible quantum numbers for each of these orbital pictures. Assume that each orbital is in the lowest possible principal shell.

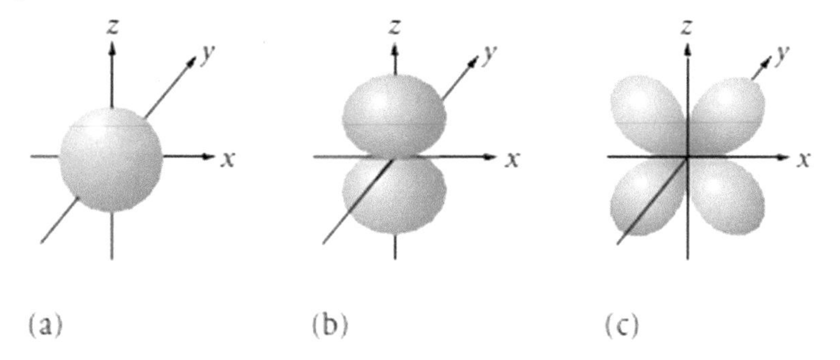

136. Compare the ground-state electron configurations of potassium and argon. Explain why it is easier to remove the outer electron of potassium than that of argon, even though potassium has more positive charge in its nucleus.
137. A 385 nm photon of light strikes a sheet of a particular metal and ejects an electron at a velocity of 6.1×10^5 m/s. What is the energy, in kJ/mol, associated with this photon?
138. A 0.30 L cup of water is placed in a microwave and irradiated with microwaves of 12.0 cm wavelength. The temperature of the water raises from 25 °C to 80 °C. How many photons are used to heat the water?
139. a. What is the frequency and energy associated with a photon of light from a ruby laser (see Table 6.4)?
    b. What mass, in ounces, would a particle exhibiting that wavelength have if it were traveling at half the speed of the photon?
140. A photon with wavelength $\lambda = 2165$ nm strikes an excited hydrogen atom.
    a. What energy, in joules, is associated with this photon? What energy, in joules, is associated with a mole of these photons?
    b. What is the frequency of the photon?
    c. Which region of the electromagnetic spectrum contains photons of this energy?
    d. If the photon is absorbed by the hydrogen atom, to what energy level is the electron promoted? To make the calculation easier, assume that the electron begins in the $n = 4$ energy level.
    e. If the hydrogen atom relaxes back to the $n = 2$ energy level from the $n = 7$ level, what would be the wavelength of the photon that was emitted? Which region of the spectrum contains this photon?
    f. What could you write as the electron configuration for the hydrogen atom described in part e, after emission of the photon?

Periodic Properties of the Elements

Contents

Supplementary Information The online version contains supplementary material available at ▶ https://doi.org/10.1007/978-3-030-90267-4_7.

M. Mosher and P. Kelter, *An Introduction to Chemistry*,
https://doi.org/10.1007/978-3-030-90267-4_7

What Will We Learn in This Chapter?

- The periodic table is a highly organized arrangement of the known elements in the world. (▶ Sect. 7.1)
- The periodic table arranges the elements into metals, nonmetals and metalloids. (▶ Sect. 7.2)
- The electronic structure, physical properties and chemical reactivity are revealed by the placement of the elements in the periodic table. (▶ Sect. 7.3)
- The number of valence electrons and their arrangement within orbitals dictates the properties of the elements in the periodic table. (▶ Sect. 7.4)
- Properties such as atomic size, electron affinity, and electronegativity are related to the electron configuration of the elements. (Sects. 7.5–7.9)
- The elements are distributed unequally throughout the Earth and atmosphere. (▶ Sect. 7.10)

7.1 Introduction

▶ https://www.publicdomainpictures.net/pictures/90000/velka/planet-earth-1401465793Ufs.jpg

We live on and within a big globe of chemicals that have interacted for well over 4 billion years to form the chemical system that we call the planet Earth. Our atmosphere provides the oxygen molecules that interact with hemoglobin in our blood and support life on this planet. The foods we eat help sustain this life, and they include molecules with common names like carbohydrates, proteins, and vitamins, as well as salts such as sodium chloride. Our quality of life is enhanced by our clothing, which is often made by combining molecules processed from crude petroleum found deep beneath the ground or seabed. The materials that we have produced for cooking, cleaning, killing, and healing, from the Stone Age to the Bronze Age to the Iron Age all the way through the Industrial Revolution of the late eighteenth and nineteenth centuries and on to the Information Age of the twenty-first century, have a common origin. The stuff of life and our way of life have been based on the interactions among the set of chemical elements listed in the periodic table.

▶ https://www.publicdomainpictures.net/pictures/150000/velka/gold-coins-1449935937Ar0.jpg

Why is oxygen (O_2), which we must breathe to stay alive, a gas, whereas the gold we can dig up from the Earth to use as jewelry is a dense solid? Why does gold last for thousands of years, unchanged, while the oxygen reacts so readily with so many other elements and compounds? Why is the oxygen that is carried around by our blood bound to the iron ions that form part of the protein hemoglobin in our red blood cells? What makes iron so well suited to this task? We need answers to these types of questions if we are to understand the natural **environment** in which we live, its effects on us, our effects on it, and how it is that we interact with it to acquire and process materials that are so much a part of our day-to-day lives.

Environment – The world around us, comprising the Earth and its atmosphere. We ourselves are part of the environment.

Our focus in this chapter is on the elements themselves and on how we use the understanding of their basic structure that we developed in the last chapter to gain insight into their chemical behavior. In the next two chapters, we will look at how and why the elements interact with each other to form the compounds that support our twenty-first-century lives. As we continue our exploration of chemistry, we keep in mind that the greatest single statement of our understanding of the behavior of the elements is the periodic table into which they have been organized. It is there that we begin our discussion.

▶ https://upload.wikimedia.org/wikipedia/commons/6/61/Chem-lab_uni-leipzig_brockhaus.jpg

7.2 The Big Picture—Building the Periodic Table

In the mid-nineteenth century, scientists initially conceived the structure of the periodic table in order to make sense of the properties and reactivities of the elements. It developed through the recording of experimental results. The quantum-based understanding of the atom, which we explored in ▶ Chap. 6, is consistent with the

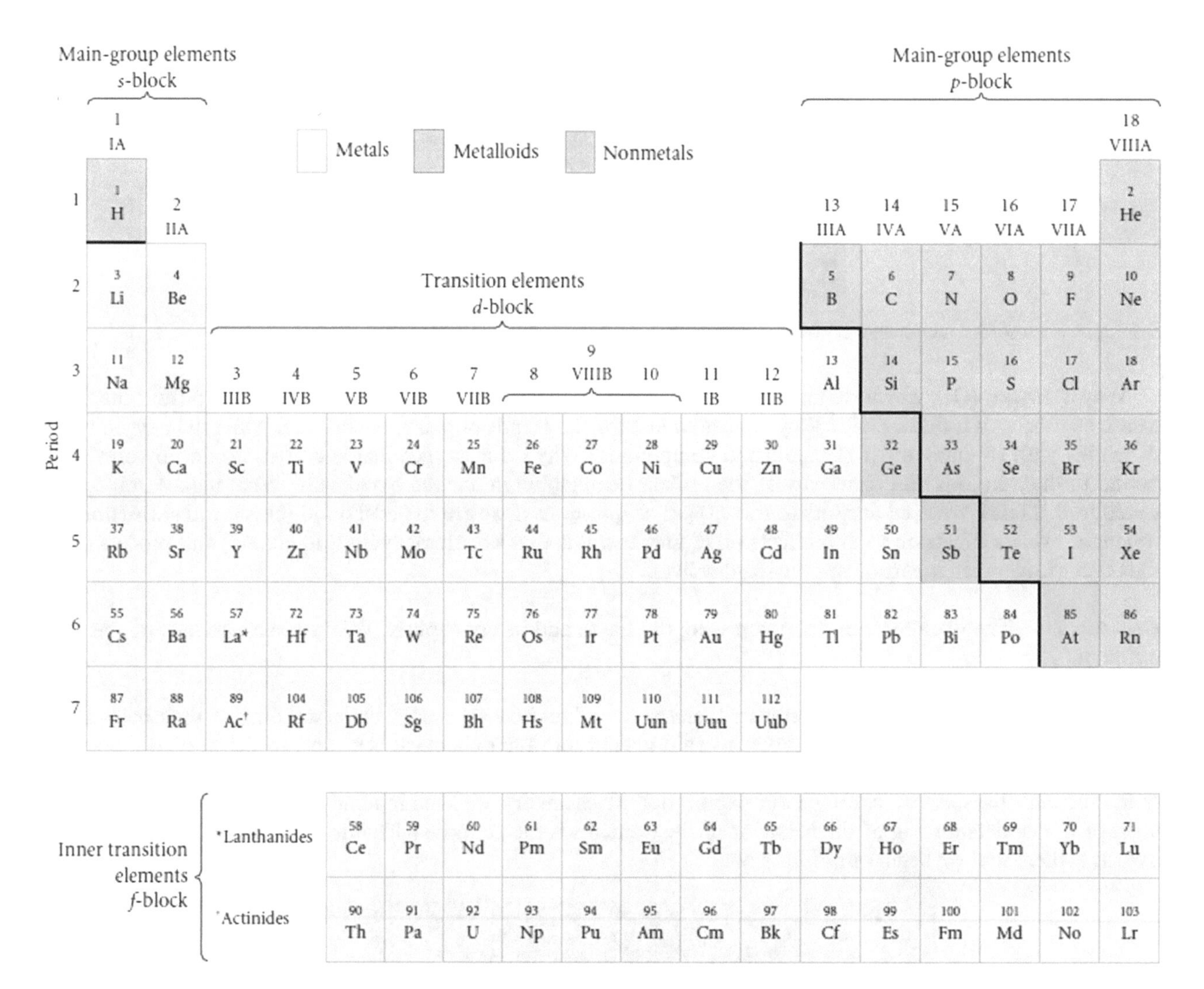

Fig. 7.1 The main ways in which we divide the periodic table into different sections. The *s*-, *p*-, *d*-, and *f* blocks contain elements with valence electrons in the same type of orbital. The horizontal rows are referred to as periods. The vertical columns are called groups. The color-coding indicates the metals, metalloids, and nonmetals. Helium is misplaced in the typical periodic table; it should be located next to hydrogen in the s block

periodic table structure derived by experimental data, and this indicates the crucial role of electron arrangement in determining an atom's reactivity. It also confirms the power of the periodic table as a classification system that reinforces its meaning and validity with each new chemical discovery.

The structure of the periodic table includes **periods**, which are horizontal rows of elements, **groups**, which are vertical columns of elements, and blocks defined in terms of which type of orbital is being filled via the Aufbau principle. This gives us the ***s*-block**, ***p*-block**, ***d*-block**, and ***f*-block**, illustrated in Fig. 7.1. Elements in the *s*-block, such as sodium, lithium, potassium, and calcium, are naturally found as positive ions. Elements in the *p*-block often form negative ions; examples include the Cl^- ions in blood and the S^{2-} ion that is a part of pyrite (FeS_2), called "fool's gold" for its gold-like appearance, and galena (PbS), mined as the main source of lead metal (Fig. 7.2). Many elements in the *d*-block can form positive ions with different charges. For example, iron can form the fairly stable Fe^{2+} and Fe^{3+} ions and the somewhat less stable Fe^{4+}, Fe^{5+}, and Fe^{6+} ions. Our further exploration of the periodic table reveals that there is some relationship between the blocks in which elements reside and their chemical and physical properties.

Periods – The horizontal rows of the Periodic Table.

Groups – The vertical columns of the Periodic Table.

s-block – The elements found in Groups 1 (IA) and 2 (IIA) where the *s* orbitals contain the outermost electrons

Fig. 7.2 Iron pyrite, FeS_2 (left), and galena PbS (right). The sulfur in these two minerals has gained electrons and exists as S^{2-} in PbS and S_2^{2-} in FeS_2. *Source* ▶ https://c.pxhere.com/photos/a5/0b/close_up_idiomorphe_crystals_iron_mineral_ore_pentagondodekaedern_pyrites_sparkle-1001290.jpg!d

p-block – The elements found in Groups 13 (IIIA) through 18 (VIIIA) where the *p* orbitals contain the outermost electrons

d-block – The elements found in Groups 3 (IIIB) through 12 (IIB) where the *d* orbitals contain the outermost electrons

f-block – The elements on the Periodic Table numbered 59–71 and 90–103 where the *f* orbitals contain the outermost electrons

Is there a reason why one horizontal period is located above or below another? The period to which an element belongs indicates how many energy levels contain electrons in the ground state of atoms of the element. Elements in Period 1 have electrons in only the first energy level (principal quantum number $n=1$). Elements in Period 2 have electrons in the first two energy levels (principal quantum numbers $n=1$ and $n=2$), and so on. The more occupied energy levels, the bigger the atom. Helium atoms, from Period 1, are tiny compared to xenon atoms, from Period 5. The tiny size of helium atoms explains why a party balloon filled with helium deflates in just a few days, because the atoms escape through little pores in the material of the balloon. Again, we see that position in the periodical table is related to what elements do.

Another key link between electron arrangement and position in the periodic table is that elements in any one **main group** (groups in the *s*-block and *p*-block) have the same number of electrons in their highest energy level. For example, every member of Group 17 (VIIA) contains seven electrons in the orbitals with the highest principal quantum number. These electrons are also known as the **valence electrons**.

Main groups – The *s*-block and *p*-block elements in the periodic table.

Valence electrons – The electrons that occupy the outer-most energy level of an atom, which interact with those of other atoms.

That is essentially why elements from each main group share some very significant chemical characteristics—why helium and neon are both unreactive gases, for example.

▶ Here's What We Know So Far

- The periodic table is organized into the *s*-block, *p*-block, *d*-block, and *f*-block. These blocks indicate the orbitals that are full or partially full in the highest energy level of the element.
- Rows in the periodic table are known as periods. The period number indicates the highest energy level in which electrons are found in the ground state of each element.
- Columns in the periodic table are known as groups. Elements in the same group have the same number of electrons in their highest energy, or valence, level.
- Elements in the same group have similar chemical and physical properties because they have similar valence electron configurations.

Example 7.1—Identifying Elements

Give the block, the period number, and the group number for each of these elements.

K, Sc, Al, C, Br

Asking the Right Questions

What is a "block?" and how does that relate to the period and group number of an element? Under what circumstances could two different elements be in the same block but in different periods and/or groups?

Solution

Potassium (K) is in Group 1 (IA) and is in the "*s*-block". By counting from the top of the column down to potassium, we see it in the 4th period.

Follow this same procedure for each of the elements in the list:

- Potassium is an *s*-block metal found in Period 4 and Group 1 (IA).
- Scandium is a *d*-block metal found in Period 4 and Group 3 (IIIB).
- Aluminum is a *p*-block metal found in Period 3 and Group 13 (IIIA).
- Carbon is a *p*-block nonmetal found in Period 2 and Group 14 (IVA).
- Bromine is a *p*-block nonmetal found in Period 4 and Group 17 (VIIA).

Are our Answers Reasonable?

We must keep in mind that the main group elements (those filling the *s* and *p* orbitals) will be in Groups 1–2 and 13–18 (IA–VIIIA). The transition elements (those filling the *d* orbitals) will not be in the A groups. This is the case in the solution, so the answers are reasonable.

What If…?

…we have an element with 7 electrons in its highest energy block, and this block is half-filled? Based on the information in Fig. 7.1, what four elements might fit this profile?

(*Ans: Eu, Gd, Am and Cm, all of which are* f^7 *in their ground state electron configurations.*)

Practice 7.1

Identify each of these elements.

a) A *p*-block element in Period 5 and Group 14 (IVA).
b) A *d*-block metal from Period 6 and Group 5 (VB).
c) An *s*-block element in Period 1 and Group 1 (IA).

7.2.1 The Historical Development of the Periodic Table

Initial attempts to organize the elements were based on the work of experimental chemists observing what the elements actually do when they react with one another. The first real progress was accomplished by the German scientist Johan Dobereiner (1780–1849) when he identified several groups of three elements (he called them **triads**) that had very similar properties. Dobereiner's triads included lithium, sodium, and potassium (now found in Group 1 (IA)), calcium, strontium, and barium (now in Group 2 (IIA)), and chlorine, bromine, and iodine (Group 17 (VIIA)).

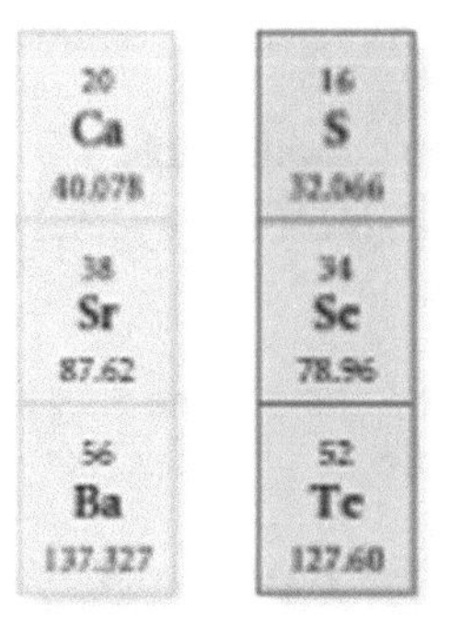

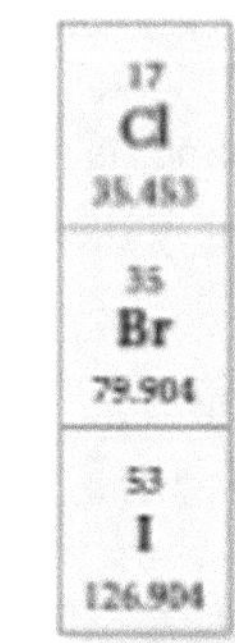

In 1864, Englishman John Newlands (1837–1898) also worked on organizing the elements into a pattern. He discovered that when the elements were arranged in order of increasing atomic mass, the properties of those elements seemed to repeat every eighth element. Newlands's system of elements, arranged in what he called **octaves** (an octave is a musical interval of eight tones) provided an early glimpse of the regular repetition of properties, or periodicity, that gives the periodic table its name. We know today that the elements have to be arranged in order of increasing atomic number (which is the number of protons in the nucleus) for the periodic nature of the periodic table to be observed.

Octave – An interval on a musical scale that is separated by eight tones.

Periodicity – The recurrence of characteristic properties as we move through a series, such as the elements of the periodic table.

Reihen	Gruppe I. — R^2O	Gruppe II. — RO	Gruppe III. — R^2O^3	Gruppe IV. RH^4 RO^2	Gruppe V. RH^3 R^2O^5	Gruppe VI. RH^2 RO^3	Gruppe VII. RH R^2O^7	Gruppe VIII. — RO^4
1	H=1							
2	Li=7	Be=9.4	B=11	C=12	N=14	O=16	F=19	
3	Na=23	Mg=24	Al=27.3	Si=28	P=31	S=32	Cl=35.5	
4	K=39	Ca=40	—=44	Ti=48	V=51	Cr=52	Mn=55	Fe=56, Co=59, Ni=59, Cu=63.
5	(Cu=63)	Zn=65	—=68	—=72	As=75	Se=78	Br=80	
6	Rb=85	Sr=87	?Yt=88	Zr=90	Nb=94	Mo=96	—=100	Ru=104, Rh=104, Pd=106, Ag=108.
7	(Ag=108)	Cd=112	In=113	Sn=118	Sb=122	Te=125	J=127	
8	Cs=133	Ba=137	?Di=138	?Ce=140	—	—	—	— — — —
9	(—)	—	—	—	—	—	—	
10	—	—	?Er=178	?La=180	Ta=182	W=184	—	Os=195, Ir=197, Pt=198, Au=199.
11	(Au=199)	Hg=200	Tl=204	Pb=207	Bi=208	—	—	
12	—	—	—	Th=231	—	U=240	—	— — — —

Fig. 7.3 Mendeleev's periodic table. The atomic mass of each of the elements is recorded in the table. Dashes in the table indicate elements that Mendeleev proposed but had not yet been discovered. The symbols at the top of the table (R^2O, RO, R^2O^3, etc.) are chemical formulas written in the style of the 1870s. *Source* ▸ https://en.wikipedia.org/wiki/File:Periodic_table_by_Mendeleev,_1871.svg

Table 7.1 Predicted (ekasilicon) and measured properties of germanium

Ekasilicon (Ek)	Germanium (Ge)
Predicted in 1871	Discovered in 1886
Atomic mass = 72	Atomic mass = 72.6
Density = 5.5 g/cm^3	Density = 5.47 g/cm^3
Density of EkO_2 = 4.7 g/cm^3	Density of GeO_2 = 4.70 g/cm^3
$EkCl_4$ will be a liquid, density = 1.9 g/cm^3	$GeCl_4$ is a liquid, density = 1.89 g/cm^3

The modern form of the table took shape through the work of German chemist Julius Lothar Meyer (1830–1895) and Russian chemist Dmitri Mendeleev (1834–1907). Both scientists were fascinated with the possibility of arranging all of the atoms into a table, and the need to publish a textbook for use by their students provided the impetus to accomplish that task. Meyer was able to find the relationship between just 28 of the 60 known elements of the time. Shortly afterwards, Mendeleev recognized the periodic relationship for all 60 of the elements. But, Mendeleev took the crucial step of predicting that certain elements must exist to occupy the gaps that were left in his table. In 1869, Mendeleev published the periodic table shown in Fig. 7.3. Within a few years of Mendeleev's predictions, three elements were discovered that matched these predictions and filled some of the gaps. For example, Mendeleev predicted the existence of ekasilicon, the element that we now know as germanium. Table 7.1 gives some of germanium's properties as predicted by Mendeleev and later as measured after the element was discovered. Two other elements (scandium and gallium) were later discovered that also matched properties predicted by Mendeleev.

Mendeleev's realization that the periodic table could be used as a predictive tool is probably the main reason why Mendeleev, rather than Meyer, is now recognized as the "father of the periodic table." What we now appreciate as inspired reasoning was, at the time, dismissed as Mendeleev's madness—a telling illustration of how the scientific method requires time for important insights to be identified among other, less fruitful, ideas. By the time of his death in 1907, the periodic table occupied a significant place in chemistry. It was even carried with Mendeleev in his funeral procession.

7.3 The First Level of Structure—Metals, Nonmetals, and Metalloids

The periodic table hangs in a highly visible place on the wall of chemistry lecture halls and laboratories all over the world. It is not there for reasons of history or adornment. It is there because it is the most important and useful work developed in the history of science for describing what we know about the chemistry of the elements. There are many versions of the table, and they vary in what information is given about each element. The basic structure of the table, however, is generally the same. A chemistry student from Russia, South Africa, Mexico, or Scotland can look at the table and communicate its ideas with any other chemistry student in the world. The language of chemistry, as illustrated by the periodic table, is universal. What does its basic structure tell us about the elements?

As we've seen before, the elements in the periodic table are arranged into three main sections—the metals, nonmetals, and metalloids (or semimetals)—as shown in Fig. 7.1. The majority of the elements in the periodic table are metals, which possess a variety of "metallic" characteristics in common.

Metals are generally:
- shiny in appearance
- solids at room temperature and pressure (except for mercury, the only metal that is a liquid at room temperature and pressure)
- good conductors of electricity
- malleable—that is, able to be hammered into various shapes
- ductile—that is, able to be stretched into wires or other objects
- likely to form positively charged ions when they react to form ionic compounds

▶ https://upload.wikimedia.org/wikipedia/commons/7/74/Stranded_lamp_wire.jpg

Metals, including iron, aluminum, and copper, supply us with some of our most widely used and versatile materials for fabrication. For example, copper is used in electric wires and in plumbing throughout our homes. Iron is the major constituent of a homogenous mixture of metals (known as an **alloy**) that makes up steel. Steel is used throughout our modern world as a structural component of cars, trains, buildings, and bridges.

Metals – A large number of elements that share certain typical characteristics, including shiny appearance, malleability, and ability to conduct electricity.

Alloy – A homogeneous mixture of metals ◻ Table 7.2

What are the industrial and commercial uses of steel? Chromium and nickel are used to create the familiar stainless steels. These steels resist corrosion and are commonly used in making tableware. Tellurium and selenium promote the machinability of steel, its ability to be easily turned and shaped into bolts and screws. Manganese makes steels that are very resistant to wear as well as to chemical reaction with water. Molybdenum is used to create hard steels for use in bearings. Careful addition of silicon creates electrical steels used in the generation and transmission

Table 7.2 Selected metals, their sources, and their uses

Metal	Major sources in nature	Uses
Al	Bauxite, clay, mica, feldspar, alumina	Food wrap, kitchenware, beverage cans
Ca	Lime (CaO), limestone ($CaCO_3$), feldspar, apatite	Manufacture of vacuum tubes, alloys, preparation of other metals
Cu	Chalcopyrite (copper iron sulfide), pure metal	Coinage metal, electric wires, plumbing pipes
Fe	Hematite (Fe_2O_3), limonite ($Fe_2O_3 \cdot 3H_2O$)	Steel
Na	Salt (NaCl), borax ($Na_2B_4O_7 \cdot 10H_2O$)	Street lamps, nuclear reactor coolant
Ni	Pentlandite and pyrrhotite (nickel–iron sulfides), garnierite (nickel-magnesium silicate)	Coinage metal, stainless steel, NiCad batteries, heating elements
Sn	Casserite (SnO_2)	Tin cans, alloys, bronze
Ti	Rutile (TiO_2), ilmenite	Aircraft parts, lightweight tank armor
W	Wolfram ochre (WO_3)	Filaments in light bulbs

of electricity. The exploration of ways in which subtle changes in the composition of steels brings about significant changes in their properties is a continuing process. That huge research and manufacturing effort all stems from the simple observation, hundreds of years ago, that letting hot charcoal mix in with molten iron improved its usefulness.

Example 7.2—Heavy Metals, Both Necessary and Toxic

Heavy metals, according to one definition, are metals with atomic masses greater than or equal to 63.546 g/mol. Some of these metals are essential for life, but many of them are toxic. Some of the heavy metals that are vital for life in low concentrations can be quite toxic in high concentrations. Heavy metals in water and soil are a major focus of environmental concern.

1. According to the definition given above, what is the lightest of the heavy metals?
2. Using the periodic table, can you identify some other heavy metals that are commonly discussed as environmental contaminants?
3. Chromium is included in some lists of heavy metals. What does that indicate about the definition of the term *heavy metals?*
4. Can you suggest why heavy metals can be particularly persistent environmental contaminants, posing problems that are difficult to correct?

Asking the Right Questions

What data about the elements are we being asked to determine? Does answering parts of this question require us to perhaps look beyond the textbook to get further information?

Solution

The question has us examine the periodic table for the atomic masses of a set of elements.

1. According to the definition supplied, copper is the lightest of the heavy metals.
2. The metallic elements heavier than copper in the periodic table include many that have received publicity as heavy metal contaminants of the environment. Examples which feature prominently in news reports on this issue are cadmium, mercury, and lead.
3. Chromium has a lower atomic mass than copper, so its inclusion in lists of heavy metals suggests that there is no universally accepted definition of the term *heavy metal*. This is in fact the case. It is a term used rather loosely to describe metallic elements of relatively high atomic mass that are also toxic.
4. Heavy metals are particularly persistent environmental contaminants because, being elements, they cannot be degraded into simpler, less toxic components. This is in contrast to toxic *compounds*, some of which are readily degraded into less harmful compounds or their component elements.

Are the Answers Reasonable?

We note the difference between heavy *metals* (elements) and *compounds*, the latter of which can be chemically degraded. However, a more precise definition of a heavy metal would be useful. In any case, our answers appear to be reasonable.

What If…?

…we were asked to propose a definition for heavy metals? How would you construct the best definition?

Practice 7.2

Which of these elements are metals? Which of these could be considered heavy metals?

Li, Si, Ni, Ce, Ge, Al, Po, Se, Rb, Cu.

NanoWorld/MacroWorld—Big Effects of the Very Small: The Diversity of Steels

Steel, shown in Fig. 7.4, is one of our most versatile construction materials, with 1.9 billion metric tons (4.2 trillion pounds) manufactured in 2019. What exactly is steel?

The most basic definition reveals that steel is a mixture of iron and small amounts of carbon, suggesting that it is a metal with a little nonmetal blended into it. Hundreds of years ago, it was discovered that the metal element iron could be changed into a tougher and more resilient material by allowing carbon from wood fires to become mixed in with it. Since then, we have discovered that adding various other elements in differing proportions can be used to vary the properties of steel in a great many ways. In fact, nowadays, we can no longer talk of steel as though it were a single thing. There are actually *over 3500 different grades of steel*, each with a particular mixture of elements added to the iron that forms the basis of them all. Looking at the differences among these grades reveals how small changes in the atoms present in the steel can have very significant effects on its properties.

There are three main types of steel, known as carbon-steel, low-alloy steel, and high-alloy steel. Carbon-steels have as little as 0.1% and sometimes more than 2% carbon added to their iron, but only very small amounts of other elements. Over 90% of the world's steel is carbon-steel, which is further categorized as high-carbon, medium-carbon, low-carbon, ultra-low carbon, and so on, depending on exactly how much carbon it contains. Alloy steels have homogeneous mixtures of metals, called alloys, added to them. The metals most commonly added to alloy steels are manganese, aluminum, copper, nickel, chromium, cobalt, molybdenum, vanadium, tungsten, titanium, niobium, zirconium, and tellurium. Steels also often contain some added nonmetals, such as silicon, selenium, nitrogen, and sulfur. The composition of several steels is shown in . Table 7.3. The numbers are reported as the mass percent of the total composition.

Fig. 7.4 Steel, such as that being poured here, has a variety of uses in the manufacture of building materials, office equipment, and automobile parts. ▶ https://upload.wikimedia.org/wikipedia/commons/9/95/Photograph_of_a_Vat_of_Molten_Pig_Iron_Being_Poured_into_a_Open_Hearth_Furnace_at_the_Jones_and_Laughlin_Steel_Company%2C_Pittsburgh%2C_Pennsy_-_NARA_-_535922.tiff

Table 7.3 The compositions of selected steels

Element	Tool steel	Basic electrical steel	Stainless steel
C	0.864	0.215	0.225
Mn	0.341	0.393	0.544
P	0.012	0.016	0.030
Si	0.185	0.211	1.00
Cu	0.088	0.211	0.226
Ni	0.230	0.248	8.76
Cr	4.38	0.017	16.7
V	1.83	0.003	0.176
Mo	4.90	0.038	0.24
W	6.28	–	–
Sn	0.029	–	–
Al	–	0.002	–
Co	–	–	0.127

Table 7.4 Selected nonmetals, their sources in nature, and some of their industrial uses

Nonmetal	Sources in nature	Industrial uses
C	Coal, graphite, diamonds	Steel manufacture, pencil "lead"
Cl	Salt (NaCl), briny water	Water purification, manufacture of dyes and explosives
N	Air	Fertilizer, gunpowder, low-temperature reactions, inert atmospheres
O	Air	Medical field, steel manufacture, combustion
P	Apatite (calcium hydroxy phosphate)	Fertilizer, chemical warfare agent, rat poison, steel manufacture
S	iron pyrite, galena, barite, pure element	Black powder, fertilizer, fireworks, rubber production

Nonmetals – possess characteristics that are generally quite different from those of metals as shown in Table 7.4. Nonmetals are generally:

- gases or solids at room temperature and pressure
- dull in appearance
- brittle—not ductile and not malleable
- poor conductors of electricity (with the exception of the forms of carbon known as graphite and fullerenes)
- likely to form negatively charged ions when they react to form ionic compounds

Despite occupying a small number of the spaces in the periodic table, the nonmetals include some of the most abundant elements found in living things, particularly the carbon, hydrogen, oxygen, nitrogen, phosphorus, and sulfur atoms from which the chemicals of life are largely made.

Nonmetals – A collection of elements that shares certain characteristics that are in contrast to those of the metals, including being gases or dull, brittle solids at room temperature.

What type of elements lie between the metals and nonmetals in the periodic table? Between the metals and nonmetals in the periodic table we find elements known as **metalloids**, or **semimetals**. Specifically, boron, silicon, germanium, arsenic, antimony, tellurium, and astatine make up the list of elements we call metalloids. These elements share some of the characteristic properties of both the metals and nonmetals, making it difficult to place them in either of these two main categories. For example, the ability of silicon and germanium to conduct electricity lies somewhere in between that of the metals and that of the nonmetals. We say that they are semiconductors, which makes them ideal for use in the computer-manufacturing industry.

Metalloids – A small number of elements that have characteristics midway between those of the metals and those of the nonmetals. Also known as semimetals.
Semimetals – A small number of elements that have characteristics midway between those of the metals and those of the nonmetals. Also known as metalloids.

The subdivision into metals, nonmetals, and metalloids is the first level of structure found within the periodic table. The next, and more fundamental, level is the subdivision into the vertical columns called groups and the horizontal rows called periods.

7.4 The Next Level of Structure—Groups in the Periodic Table

Elements within any of the vertical columns of the periodic table, which are known as groups, share some significant chemical properties. We can explain the chemical similarities of elements within each group based on each atom's electron arrangements. For instance, elements in any of the main groups (1, 2, and 13–18, traditionally known as IA through VIIIA) all have the same number of electrons in their outer energy level, which corresponds to the highest principal quantum number. The outer electrons—the valence electrons—are very significant in determining how an atom interacts with other atoms. For those elements known as transition elements, the definition of valence electrons actually includes electrons in two or three energy levels.

The International Union of Pure and Applied Chemistry (IUPAC) has made recent modifications to the periodic table. One such modification is the standardization of the numbers at the top of each of the columns of the periodic table. In this system, the columns have been numbered from 1 to 18 from left to right, although most periodic tables also retain the older group names because they convey information about electron organization. The column numbers (both the new numbers and the older group names) are shown in the periodic table on the inside front cover of this book. As we've already done, we will continue to present both the new IUPAC column numbers and the older group names in parentheses; for example, "Group 1 (1A)".

7.4.1 Group 1 (IA): Hydrogen and the Alkali Metals

The elements in Group 1 (IA), apart from hydrogen, are known as the **alkali metals** because they are highly reactive elements that combine with water to form chemicals called alkalis. *Al qili* is from the Arabic, describing the burning of the saltwort plant, which produces an ash that forms an alkaline (basic) solution. Alkali metals form alkaline solutions when they react with water. We will discuss acids and bases in ▶ Chap. 15. The alkali metals all have one valence electron in an *s* orbital, which they generally lose when reacting, to form ions with a + 1 charge. We don't find the alkali metals as free elements in the environment because they are so chemically active that their atoms have all reacted to form compounds or dissolved ions. Sodium is not naturally found in its elemental state but, rather, in the form of sodium ions (Na^+). Sodium ions are the most abundant positive ions in seawater, and they are also plentiful in the internal environment of our blood and intercellular fluids.

Alkali metals – The Group 1 (IA) elements, excluding hydrogen.

Hydrogen shares some characteristics with the alkali metals, particularly its tendency to react by forming a + 1 ion (H^+). In other respects, however, it is quite different from the alkali metals. It is not a metal; it can also form a negative ion, the "hydride" ion (H^-), when combining with other alkali metals; and two hydrogen atoms can combine by covalent bonding to form the hydrogen molecule (H_2) (◻ Table 7.5)

▶ https://upload.wikimedia.org/wikipedia/commons/2/27/Na_%28Sodium%29.jpg; ▶ https://upload.wikimedia.org/wikipedia/commons/a/a7/Calcium_metal.jpg

◻ **Table 7.5** Some commercial uses of Group 1 (IA) elements

Hydrogen	Fertilizers, plastics, pharmaceuticals, fuel, fuel cells
Lithium	Glass making, television tubes, battery electrolytes, lithium-aluminum alloys in the aerospace industry, greases
Sodium	Nuclear reactors, reagent, salt, washing soda
Potassium	Fertilizers, soaps and detergents, explosives, glass and water purification
Rubidium	Photoelectric cells
Cesium	Special glass, radiation-monitoring equipment, atomic clock
Francium	None (too rare)

7.4.2 Group 2(IIA): The Alkaline Earth Metals

The elements in Group 2 (IIA) are known as the **alkaline earth metals**. These are also reactive metals, although generally less reactive than the alkali metals. They have two valence electrons within an *s* orbital, which they generally lose when reacting, to form ions with a + 2 charge. Compounds of these metals combined with oxygen are quite common in the environment. Because these metal oxides were so readily available in the Earth, and because they form alkalis when dissolved in water, the elements in Group 2 (IIA) were originally given the name alkaline earth metals. For example, calcium oxide (CaO) is an alkaline compound used in the manufacture of cement and steel.

Alkaline earth metals – The Group 2 (IIA) elements (Table 7.6).

► https://upload.wikimedia.org/wikipedia/commons/2/2d/Rieke_magnesium.jpg; ► https://upload.wikimedia.org/wikipedia/commons/1/16/Barium_unter_Argon_Schutzgas_Atmosphäre.jpg

Table 7.6 Some commercial uses of Group 2 (IIA) elements

Beryllium	Telecommunications equipment, automatic electronics, computers, undersea communications equipment, pipe products in the oil and gas industry
Magnesium	Automotive industry, bicycles, luggage
Calcium	Metallurgic applications, lead and aluminum industries, nuclear applications, cement, soil conditioner, water treatment
Strontium	Cathode-ray tubes, automotive industry, special glass, fireworks, flares
Barium	X-ray contrast reagent, drilling fluid, oil for gas wells
Radium	Cancer treatment, luminous paint for watches and clocks

Table 7.7 Some commercial uses of the Group 13 (IIIA) elements

Boron	Glass making, soaps and detergents, nuclear reactor control rods
Aluminum	Packaging, especially soft drink containers, transportation (lightweight components in motor vehicles), window frames, aircraft parts, engines, kegs, cooking oils, indigestion tablets
Gallium	Semiconductors, microwave equipment
Indium	Display devices, low-melting-point alloys and solders, semiconductors, fire sprinkler systems
Thallium	Rat poisons and hair removers (now banned)

7.4.3 Group 13 (IIIA)

The elements in Group 13 (IIIA) are substantially less reactive metals than those in either Group 2 (IIA) or Group 1 (IA). They all have three valence electrons, two in an *s* orbital and one in a *p* orbital. When they react to form ionic compounds, they generally do so by losing the three outer electrons to form ions with a + 3 charge. Note that as the number of electrons that an atom loses to form a positive ion *increases*, the reactivity of the element *decreases*. This is our first example of a trend within the periodic table—a characteristic that varies in a logical and regular manner as we move through the table. The most significant Group 13 (IIIA) element that we gather from the environment is aluminum (63 million metric tons were produced worldwide in 2019, according to the International Aluminum Institute), found as the ore known as bauxite, which is composed of aluminum oxide (Al_2O_3) combined with varying amounts of water (Table 7.7).

▶ https://www.publicdomainpictures.net/pictures/30000/velka/metal-background-13497699070or.jpg; ▶ https://upload.wikimedia.org/wikipedia/commons/9/92/Gallium_crystals.jpg

7.4.4 Group 14 (IVA)

The elements in Group 14 (IVA), with four valence electrons, two in an *s* orbital and two in individual *p* orbitals, and they include elements of central importance to living things and to the fabricated electronic materials which make up our computer and smart phone. Carbon, at the top of the group, can be regarded as the fundamental "building block" of life, because the chemicals of life are largely based on chemical chains and rings of carbon atoms with various other atoms attached. Carbon does not generally form ions. Instead, it forms covalent bonds (we will discuss these in ▶ Chap. 8), in which electrons are shared between two atoms. Group 14 (IVA) also includes silicon, the basis of the "silicon chips" of the computer industry, and the semiconductor germanium, also used in the manufacture of computer chips. Tin and lead, at the bottom of the group, are metals and can form positive ions with either + 2 or + 4 charges (Table 7.8).

Table 7.8 Some commercial uses of the Group 14 (IVA) elements

Carbon	Commercial and military aircraft, fibers, thermoplastic matrix materials, petrochemicals, clothing, dyes, fertilizers, fuels, pharmaceuticals
Silicon	Aluminum alloys, silcones, silicon chips used in computers, semiprecious stones
Germanium	Semiconductors, transistors, catalysts in polymer production, glass for infrared devices
Tin	Coatings for other metals, bronze, soft solder, pewter, special paint used on boats to prevent barnacles
Lead	Storage batteries, cable covering, radiation shielding, pipes, pewter, pottery, additive in gasoline, lead crystal glass

► https://upload.wikimedia.org/wikipedia/commons/e/e9/SiliconCroda.jpg; ► https://upload.wikimedia.org/wikipedia/commons/6/6a/Tin-2.jpg; ► http://images-of-elements.com/tin.php

7.4.5 Group 15 (VA)

The elements in Group 15 (VA) have five valence electrons, two sharing an *s* orbital and three occupying individual *p* orbitals. When they form ions, these elements generally gain three electrons to form ions with a −3 charge. The Group 15 (VA) elements also readily participate in covalent bonding. Nitrogen, for example, makes up 80% of the volume of the Earth's atmosphere in the form of N_2 molecules. Nitrogen is also a crucial component of many covalently bonded compounds required for life, such as DNA, proteins, and many vitamins and hormones. Phosphorus is also crucial for life, being one of the atoms found in DNA, for example, and being part of the compounds found in fertilizers that are used to grow the food we consume (Table 7.9).

Table 7.9 Some commercial uses of the Group 15 (VA) elements

Nitrogen	Fertilizers, plastics, explosives, dyes
Phosphorus	Fertilizers, matches, detergents, coating to prevent corrosion
Arsenic	Pesticides, wood preservatives, semiconductors, special glass
Antimony	Flame retardants, pigments, lubricants, ammunition, used to harden other metals
Bismuth	Industrial and laboratory chemicals, pharmaceuticals, cosmetics, replacement for lead in steel alloys and for aluminum in ceramics, high-temperature superconductors, indigestion tablets

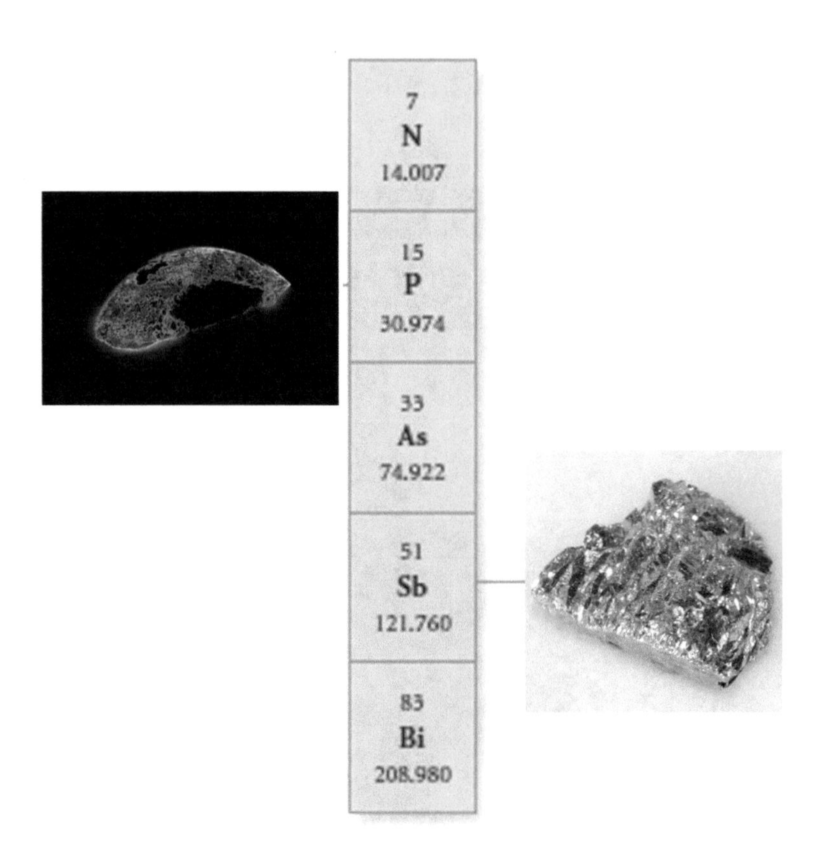

▶ https://upload.wikimedia.org/wikipedia/commons/d/da/White_phosphorus_glowing_e17.png; ▶ https://upload.wikimedia.org/wikipedia/commons/5/5c/Antimony-4.jpg; ▶ http://images-of-elements.com/

7.4.6 Group 16 (VIA): The Chalcogens

The elements in Group 16 (VIA) are known as the **chalcogens**, a name derived from the Greek *khalkos* ("copper"), because the copper ores contain several elements from this group. All of the elements in Group 16 (VIA) have six valence electrons, two in an *s* orbital and four distributed within three *p* orbitals. When they form ions, the Group 16 (VIA) elements gain two electrons to form ions with a – 2 charge.

Chalcogens – The Group 16 (VIA) elements.

In looking for Group 16 (VIA) elements of significance to ourselves and our environment, we immediately turn to oxygen. Oxygen gas, in the form of O_2 molecules, makes up about 21% of the atmosphere and is the gas we must breathe in order to stay alive. We need it because it combines with hemoglobin, as we noted in the chapter opening, and reacts with food molecules to release the energy that powers all life. We have also met, in ▶ Chap. 2, the relatively rare form of oxygen known as ozone (O_3) that is a vital part of our upper atmosphere, where the ozone layer absorbs harmful UV rays, but is a troublesome environmental pollutant at ground level (◻ Table 7.10).

◻ **Table 7.10** Some commercial uses of the Group 17 (VIIA) elements

Fluorine	Welding metals, frosting glass, insulating gas for high-power electricity transformers, additive to municipal water supplies
Chlorine	Chemical warfare, bleach, PVC plastics, water purification
Bromine	Flame retardant, pharmaceutical intermediates, swimming pool disinfectants, fuels, additives, photography
Iodine	Conductive polymers, fuel cells, dyes, photography, industrial catalysts
Astatine	Nuclear reactors

▶ https://upload.wikimedia.org/wikipedia/commons/4/44/Sulfur-sample.jpg; ▶ https://upload.wikimedia.org/wikipedia/commons/4/47/SeBlackRed.jpg

Group 16 (VIA) also contains sulfur, another element that is both vital for life and associated with harmful pollution. We need the sulfur that is a part of proteins and some vitamins and other important biochemicals, as well as fertilizers and industrial chemicals such as sulfuric acid (H_2SO_4), but oxides of sulfur such as sulfur dioxide (SO_2) are pollutants that can cause "acid rain."

▶ https://upload.wikimedia.org/wikipedia/commons/a/ac/Kawah_Ijen_-East_Java_-Indonesia_-sulphur-31July2009.jpg

7.4.7 Group 17 (VIIA): The Halogens

The elements in Group 17 (VIIA) are known as the **halogens** (from the Greek *hals* + *gen*, meaning "salt makers") because they are nonmetals that can combine with metal ions to form the chemicals we call salts. All of the elements in Group 17 (VIIA) have seven valence electrons, two in an *s* orbital and five distributed among three *p* orbitals. When they form ions, as happens when they become part of salts, they gain one electron per atom to form ions with a -1 charge. The most common example of a salt is sodium chloride (NaCl), which we know as common table salt, and we will meet many other examples throughout this book. Chloride ions (Cl^-) are abundant in seawater and in the blood and intercellular fluids of the body. Tiny amounts of fluoride ions (F^-) are important in making our teeth resistant to decay.

Halogens – The Group 17 (VIIA) elements.

See Table 7.11.

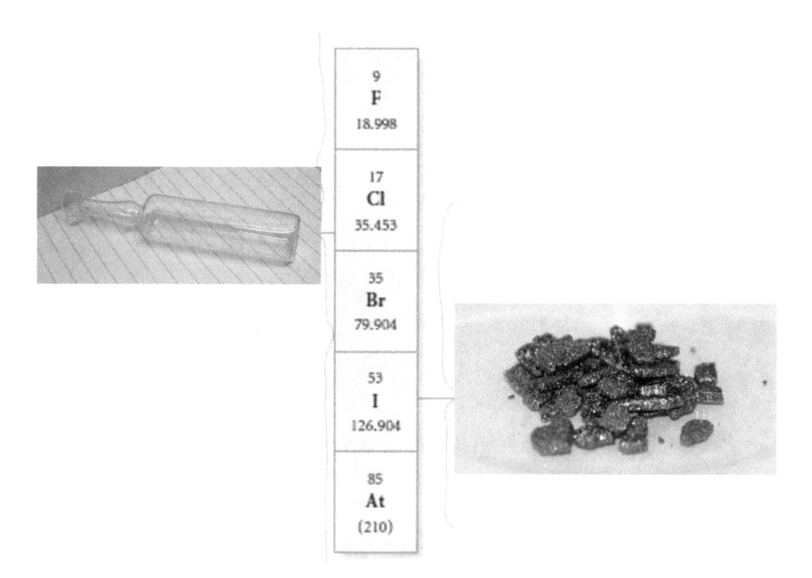

▶ https://upload.wikimedia.org/wikipedia/commons/f/f4/Chlorine_ampoule.jpg; ▶ http://woelen.homescience.net/science/index.html; ▶ https://upload.wikimedia.org/wikipedia/commons/0/0a/Sample_of_iodine.jpg

7.4.8 Group 18 (VIIIA): The Noble Gases

The elements of Group 18 (VIIIA) are all largely unreactive gases, and the group is collectively known as the **noble gases**, in the sense of aloof nobility, or **inert gases**, being distinguished from the other, much more reactive elements. The noble gases are present in very small amounts in the environment, but through their remarkable stability, they reveal to us one of the most significant secrets of chemistry. Each of the noble gases apart from helium has eight valence electrons distributed between a full *s* orbital and three full *p* orbitals. The fact that atoms of these elements do not naturally react either with themselves or with any other elements indicates that they have an exceptionally stable valence electron arrangement. This arrangement is known as a stable **octet** because of their eight valence electrons and a full energy level. Helium, with only two valence electrons, also has an exceptionally stable electron arrangement because it has a completely full energy level. This is yet another clue that tells us there is something special about an energy shell that is completely full of electrons. When we examine chemical bonding in more detail in ▶ Chap. 9, we will find that many chemical reactions can be understood in terms of the participating atoms acquiring electron arrangements that result in full energy shells, like the stable valence electron arrangements of the noble gases.

Table 7.11 Some commercial uses of the Group 16 (VIA) elements

Element	Uses
Oxygen	Steel making, metal cutting, chemical industry
Sulfur	Lead acid storage batteries, fertilizers, water treatment, petroleum refining, drying agents
Selenium	Photoreceptors, glass colorant, pigments, metallurgic and biological applications, photoelectric cells, photocopiers, semiconductors
Tellurium	Additives in steel and other metal alloys, catalyst in synthetic rubber production
Polonium	Source of alpha radiation, heat source in space vehicles

Table 7.12 Some commercial uses of the Group 18 (VIIIA) elements

Helium	Used by divers to dilute the oxygen they breathe, balloons, low-temperature research
Neon	Filling discharge tubes, ornamental lighting
Argon	Filling discharge tubes, provide an inert atmosphere for chemical reactions, light bulbs
Krypton	High-quality light bulbs
Xenon	Research purposes
Radon	Some isotopes emit alpha particles

Noble gases – The Group 18 (VIIIA) elements. Also known as inert gases.
Inert gases – The Group 18 (VIIIA) elements. Also known as noble gases.
Octet – Eight electrons in the valence shell of an atom.

Although they are very rare, we have found various good uses for the noble gases of the environment. Helium, for example, is well known as the "lighter than air" gas within party balloons, weather balloons, and airships that allows them to "float" in the air. Neon has found fame as the gas within "neon" signs. Some light bulbs use argon, krypton or xenon to prevent the tungsten filament from oxidizing. Xenon has similar uses in flash photography. Xenon and krypton are also used in the manufacture of flat panel displays for televisions and computer monitors (Table 7.12).

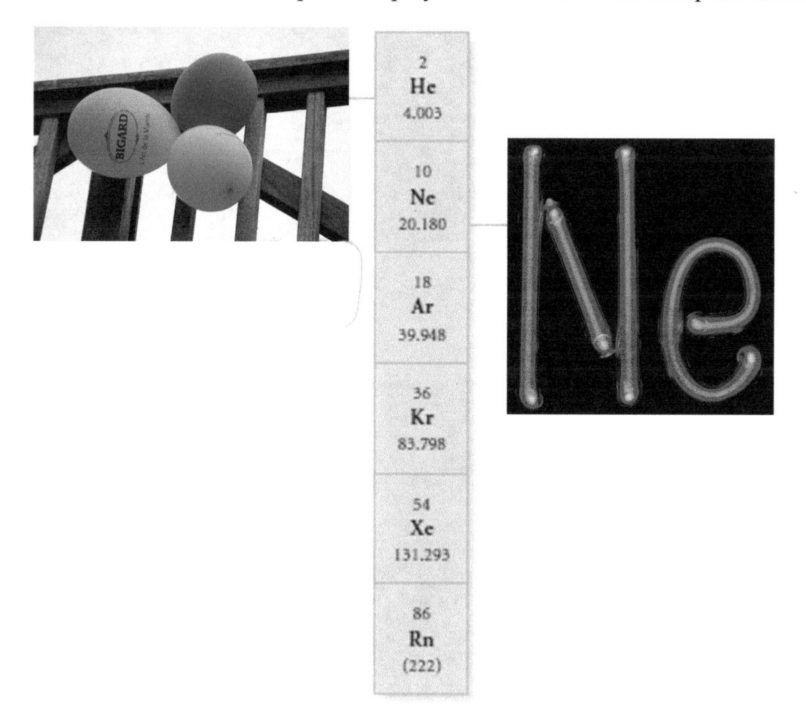

▶ https://www.publicdomainpictures.net/pictures/200000/velka/ballons-gonfles-a-lhelium--1475871773tic.jpg; ▶ https://upload.wikimedia.org/wikipedia/commons/8/88/NeTube.jpg

NanoWorld/MacroWorld—The Big Effects of the Very Small—Everblasting Gobstoppers at the Edge of Absolute Zero

Scientists at the Air Force's High Explosives Research and Development (HERD) facility at Eglin Air Force Base in Florida are using the extremely low-temperatures of liquid helium to do some amazing things with small groups of atoms and molecules.

By spraying liquid helium at very low temperature into a vacuum through a tiny nozzle, a stream of helium droplets is formed. Each droplet they create consists of between a few hundred to a few million helium atoms. These droplet clusters cool to less than 0.5 K (less than – 272 °C!) through evaporation. Scientists then use a

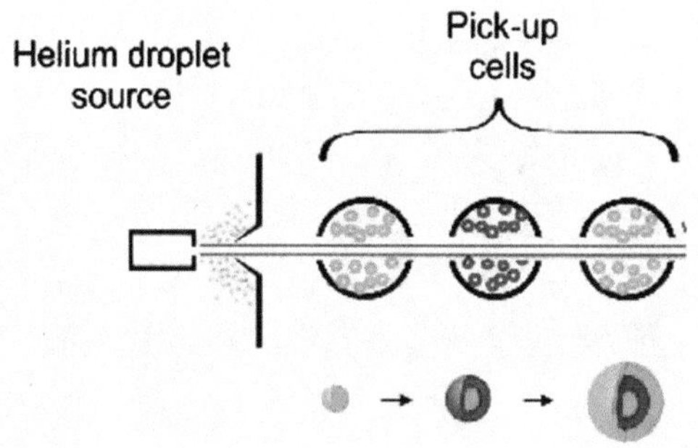

Fig. 7.5 The helium droplets are directed along a path where they collide with atoms. In this drawing, the helium droplets pick up green atoms, then red, then green atoms again

special device to shoot a stream of individual atoms into the droplet clusters, as illustrated in Fig. 7.5. The result is the placement of an individual atom inside a single helium droplet. The process can be repeated to create small clusters of atoms inside the helium droplets. Because atomic motion is so slight at these very low temperatures, the atoms don't just bounce off of each other as they likely would at room temperature. Instead a structure resembling an atomic-scale version of Willy Wonka Gobstopper candies is formed.

Being an explosives lab, one of the applications that the HERD is examining is the possibility of alternating layers of atoms and molecules that can release huge amounts of chemical energy when they react. For example, alternating atoms aluminum and molecules of iron(III) oxide would result in a laminated thermite particle (thermite was described in ▶ Sect. 6.6). At liquid helium temperatures, these chemicals will not chemically react with each other, but they might be made to react at higher temperatures, creating a potentially powerful and rapid explosion. Explosives aren't the only focus of this research. Chemical systems such as these give scientists the tools they need to study factors that govern the speed of chemical reactions (▶ Chap. 14), such as surface area and mechanisms of reaction. There is also a large and untapped area of nanochemistry that is accessible through the construction of these kinds of chemical systems. Studying nanoclusters of atoms and molecules close to absolute zero could give rise to the discovery of new and important properties of these substances.

Example 7.3—A Group Activity

How many electrons are found in the valence shell of each of these elements?

a. Be
b. As
c. I
d. In

Asking the Right Questions

How can we use our understanding of the period table and how elements are classified within it to help us find the number of valence electrons in each element?

Solution

One level of organization in the periodic table is its groups, which reflect common chemical and electron-based properties.

a) Beryllium, Be, has two valence electrons, both in the 2*s* orbital
b) Arsenic, As, has five valence electrons, two in the 4*s* orbital and three in the 4*p* orbitals
c) Iodine, I, has seven valence electrons, two in the 5*s* orbital and five in the 5*p* orbitals
d) Indium, In, has three valence electrons, two in the 5*s* orbital and one in the 5*p* orbitals

Are the Answers Reasonable?

The answers correlate with both the electron configurations and group locations of each element within the periodic table, so they are reasonable.

What If…?

…we were dealing with astatine, At, atomic number = 85? How many valence electrons are found in that element?

(*Ans: 7,* $6s^2$, $4p^5$)

Practice 7.3

Indicate which element is described by each of these phrases.

a) Two valence electrons, both in the 6*s* orbital
b) One valence electron, in the 3*s* orbital
c) Six valence electrons, two of them in the 2*s* orbital
d) Four valence electrons, and the element is a nonmetal

7.4.9 Groups 3–12 (IB–VIIIB): The Transition Elements

In the middle of the periodic table we find elements known as the **transition elements**, with the **inner transition elements** usually shown as a separate block, as in ◘ Fig. 7.1. The transition elements are arranged in short groups, historically labeled IB through VIIIB in the case of the main transition elements, and currently unlabeled for the inner transition elements.

Inner transition elements – The elements in the *f*-block of the periodic table, consisting of the lanthanides and actinides.

Transition elements – Metal elements from Groups IB through VIIIB found in the middle of the periodic table.

The elements in the first period of the inner transition elements are known as the **lanthanides**, and the second period forms the **actinides**. These inner transition elements are shown separately, primarily for the visual ease of having the periodic table fit into a convenient space. A more logical version of the periodic table is shown in ◘ Fig. 7.6, with the inner transition elements inserted where they belong in terms of their atomic numbers. The inner transition elements make up the *f*-block, and they have some specific chemical characteristics due to electrons in *f* orbitals acting as valence electrons in addition to those in the outer *s* orbital. We most often think of the inner transition elements such as uranium and plutonium in terms of their nuclear rather than their chemical properties.

Lanthanides – The first set of inner transition elements. The lanthanides include elements 58–71.
Actinides – The second set of inner transition elements. The actinides include elements 90–103.

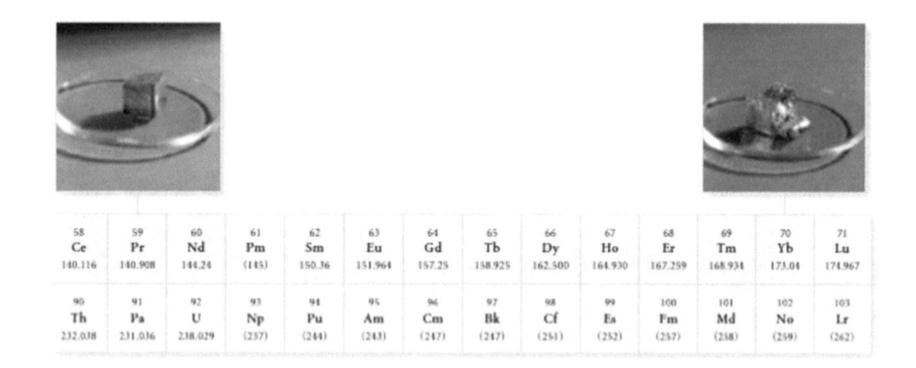

The transition elements of the *d*-block have electrons in *d* orbitals that can count as valence electrons, in addition to the electrons in the *s* orbital of their highest energy level. This situation of having valence electrons in two energy levels arises because the energies of the electrons involved are very close. It results in some interesting and important chemical characteristics, such as forming ions with various charges.

The special chemistry of the transition elements, such as iron, cobalt, copper, and zinc, makes many of them vital for life. Biochemists, for example, have become quite used to the idea that when a difficult feat of chemistry is achieved within a living thing, it is very often catalyzed by an enzyme that includes a transition metal ion at the

◘ **Fig. 7.6** Periodic table with the lanthanides and actinides where they belong in terms of electron arrangement

"active site" where the chemical reaction occurs. Transition elements are also widely used as **catalysts** in the chemical industry and, in the automotive industry, as rhodium, palladium, and platinum-based catalytic converters and advanced fuel cells in some cars.

Catalyst – A substance that greatly speeds up the rate of a reaction without being consumed by that reaction.

7.4.10 The Elements of Life

Figure 7.7 shows the periodic table with all of the elements that make up the human body highlighted in color and their relative abundances indicated, and Table 7.13 lists the relative percent abundance, by mass, of these elements within us. Note the location of these elements in the periodic table. Is there a generalization that can be made about their locations? Can you explain your generalizations?

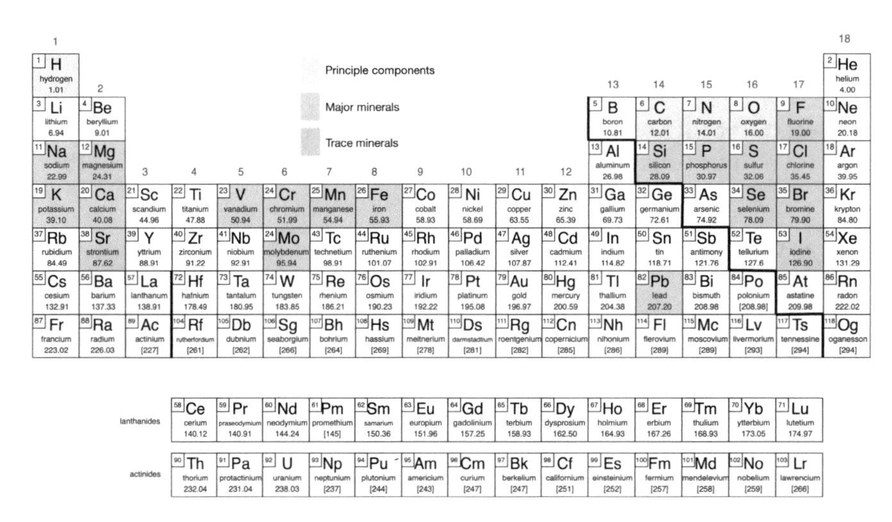

Fig. 7.7 The elements that make up you

Table 7.13 Elements of the human body

Element	Abundance (% by mass)	Element	Abundance (% m/m)
Oxygen	61.0	Carbon	23.8
Hydrogen	10.0	Nitrogen	2.6
Calcium	1.4	Phosphorus	1.1
Sulfur	0.2	Silicon	0.02
Potassium	0.02	Sodium	0.014
Chlorine	0.012	Fluorine	0.004
Magnesium	0.003	Iron	0.0004
Strontium	0.00005	Bromine	0.00003
Lead	0.00002	Manganese	Trace
Iodine	Trace	Chromium	Trace
Molybdenum	Trace	Selenium	Trace
Vanadium	Trace		

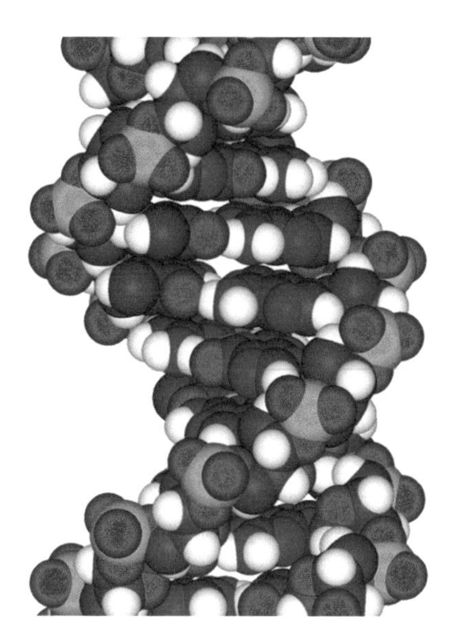

▶ https://upload.wikimedia.org/wikipedia/commons/2/2c/DNA-fragment-3D-vdW.png

The two most clear-cut generalizations are that the human body is largely constituted of elements only from the upper right-hand portion of the periodic table and that the noble gases are not part of the chemistry of life.

We can explain the absence of noble gases by looking at their chemical reactivity—or lack of it. Life is a very complex process of chemical *change* involving regulated chemical reactions. The noble gases generally do not participate in chemical change because they are so unreactive, so we would not expect to find them playing any part in the chemistry of life.

The fact that life is made primarily of the smaller elements, from a small section of the periodic table, may be linked to the fact that these elements are generally more abundant on Earth, and in the universe at large, than elements with very large atoms. In any case, nearly all of the mass of a human is made up of atoms of hydrogen (atomic number = 1), carbon (atomic number = 6), nitrogen (atomic number = 7), and oxygen (atomic number = 8). We are largely made from some of the smallest atoms of the environment. Phosphorus (atomic number = 15) and sulfur (atomic number = 16) are also important as part of proteins, DNA, and RNA.

Another feature of the elemental composition of human life is that several of the elements found within us are transition metals, although we contain these in very small amounts. The chemical versatility of transition metal ions, such as their tendency to readily form ions of differing charges, makes them useful in enzymes—biological catalysts of chemical reactions.

It is also notable that humans contain elements from every block of the periodic table except the *f*-block, which comprises the lanthanides and actinides. The elements in this block are composed of very large atoms. Many of the actinides are unstable and therefore radioactive.

▶ Here's What We Know So Far

- The elements within a group have the same number of electrons in their highest energy level.
- The elements in the periodic table are arranged into three main sections: the metals, nonmetals, and metalloids (or semimetals).
- The structure of the periodic table includes "blocks" defined in terms of which type of orbital is being filled. This gives us the *s*-block, *p*-block, *d*-block, and *f*-block elements.
- The elements that predominate in living things are carbon, hydrogen, nitrogen, and oxygen, with small concentrations of phosphorus, sulfur, and transition metals. ◀

7.5 The Concept of Periodicity

Why is the structure of the periodic table useful to know? Once we understand the trends as we traverse the groups and periods in the periodic table, we can make reasonable predictions about the chemical and physical behavior of any element in the periodic table. We will add to our understanding in subsequent chapters, even learning how to assess the likely *nuclear* behavior of an element. Once we are armed with this understanding, the periodic table serves as a most wonderful guide to the formation and interaction of substances.

The basis of the periodic table is periodicity, which will be revealed in many of the properties we explore in the remainder of this chapter. When chemists talk of "periodic properties" among the elements, they mean the way in which characteristic properties recur in a periodic manner as we move through the periodic table. For example, element number 3 in the table, lithium, is a very reactive metal that forms an alkali on reaction with water. Element 4 (beryllium), is less reactive, and as we move through elements 5 (boron), 6 (carbon), 7 (nitrogen), 8 (oxygen), 9 (fluorine), and 10 (neon), we find elements that become steadily less like lithium in chemical reactivity and physical properties. Then suddenly, with element 11 (sodium), we find another very reactive metal that forms an alkali on reaction with water. The similarities between the reactivities of lithium and sodium are so striking that it is clear that we are observing some significant repeating feature of reactivity as we move through the periodic table. If we then move on to elements 12 through 19, we find the same thing happening again. Elements 12 through 18 have properties less and less like those of sodium and lithium, and then suddenly, with element 19 (potassium), the property of being a very reactive metal that forms an alkali on reaction with water recurs. These chemical characteristics, which we call the characteristics of the alkali metals, recur in a periodic manner as we move through the periodic table.

That is the basic concept of chemical periodicity, and we could have chosen various other properties to make the same point. For example, the property of being a very unreactive gaseous element that exists as free individual atoms occurs in elements with atomic numbers 2, 10, 18, 36, 54, and 86. As with the alkali metals, we have a chemical property—the lack of reactivity, in this case—that recurs in a systematic, or periodic, manner as we move through the periodic table. The number of elements we have to pass by before a characteristic property recurs is not constant. It is 8, 8, 18, 18, 32 in the series above (see Example 7.4), but the crucial fact is that each characteristic property does recur periodically as we move through the periodic table.

Example 7.4—Explaining the Periods Between Periodicities

The periodic property of being a noble, or "inert gas" recurs as we move left to right across the periodic table. If we look at how many elements are between this recurring property, the pattern is 2, 8, 8, 18, 18, 32. Can you suggest the underlying reason for why the periodicity follows this particular pattern?

Asking the Right Questions

What is the basis for the pattern in which electrons fill orbitals? What is this pattern? How does it express itself for an inert gas?

Solution

The characteristic associated with being an inert gas is to have a stable octet of eight electrons in the atom's highest energy level, or two electrons in a completely full energy level in the case of helium. The Aufbau principle indicates that after helium, the stable octet will recur after the 2*s* and 2*p* orbitals have been filled, which means eight electrons must be added before we arrive at neon, then another eight before we arrive at argon. As we then move through Period 4, ten 3*d* electrons must be added before the 4*p* orbitals become filled, so this time we must move through 18 elements before the stable octet recurs. Similar reasoning applies to the gap between krypton and xenon. Then the gap jumps up to 32, between xenon and radon, as a consequence of the filling of *f* orbitals that must occur before the 6*p* orbitals of radon can be filled. This explanation is most clearly visualized by looking at the long version of the periodic table in ◘ Fig. 7.6 or the flower-shaped version in ◘ Fig. 7.8.

Is the Answer Reasonable?

The answer reflects the orbital filling pattern of electrons, which is the result of their filling from lowest to highest energy.

What If...?

...we were asked to present a periodic table that displayed periodicity in the clearest possible way? Would we choose the conventional periodic table, the long version, the pyramidal one, or, would it be better to design one of our own that has a different structure? Explain your choice.

Practice 7.4

Judging on the basis of the periodicity of the periodic table, indicate what would be the hypothetical atomic number of a metal that would be more reactive than Fr. How many electrons would it take to fill a noble gas in the hypothetical Period 8 of the periodic table?

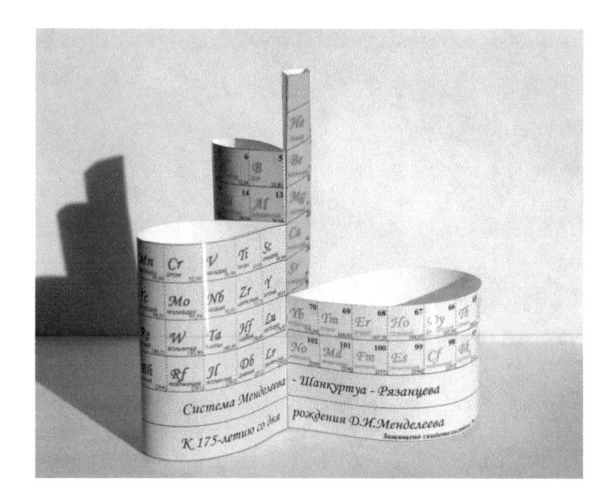

Fig. 7.8 This three-dimensional periodic table in the shape of a flower is one of many 3-D versions that are designed to illustrate the relationships between the elements. *Source* ▶ https://upload.wikimedia.org/wikipedia/commons/a/a2/Mendeleev_flower.jpg

7.6 Atomic Size

Heavy metal poisoning is unfortunately a common occurrence, particularly among children living in homes painted with lead-based paints. Once heavy metal poisoning is recognized, the patient is treated with a chelating agent that specifically grabs ions of a particular size and helps to remove them from the body. The size of the ion is extremely important in how well it binds to the chelating agent. If it is too small or too large, it will not associate with the chelating agent and be excreted from the body.

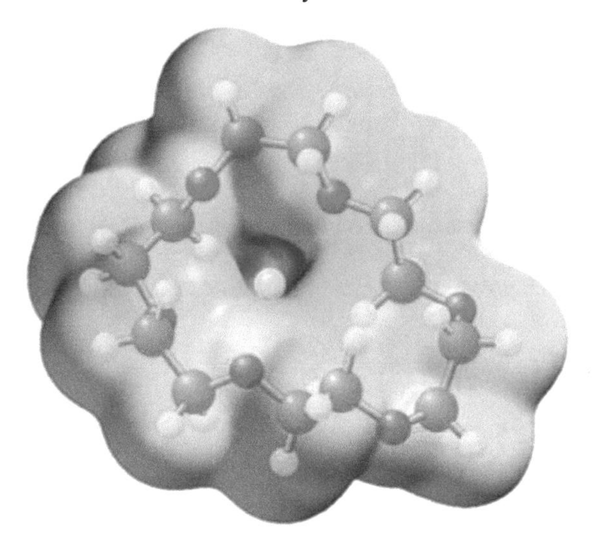

Figure 7.9 reveals the key trends and periodicities found in the size of atoms. What can you note about the size of the atoms in the periodic table? Atomic size decreases as we move from left to right along any period, and it increases as we move down any group. Figure 7.9 uses atomic radius as a measure of atomic size. The **atomic radius** is defined as half the distance between the nuclei in a molecule consisting of identical atoms. It is also known as the **covalent radius** of an atom, because it indicates the size of the atom when that atom is involved in covalent bonding. Not all elements form such molecules. Some atomic radii values must be estimated via indirect methods, such as comparing the distances between atomic nuclei when the atoms are bound within chemical compounds instead of in a molecule made of identical atoms. Values for the radii of metal atoms can also be obtained by analyzing the distance between the nuclei of the atoms within the solid structure of the metal concerned. These values are called **metallic radii**.

Atomic radius decreases →

Atomic radius increases ↓

IA	IIA	IIIA	IVA	VA	VIA	VIIA	VIIIA
H 37							He 32
Li 152	Be 113	B 88	C 77	N 70	O 66	F 64	Ne 69
Na 186	Mg 160	Al 143	Si 117	P 110	S 104	Cl 99	Ar 97
K 227	Ca 197	Ga 122	Ge 122	As 121	Se 117	Br 114	Kr 110
Rb 247	Sr 215	In 163	Sn 140	Sb 141	Te 143	I 133	Xe 130
Cs 265	Ba 217	Tl 170	Pb 175	Bi 155	Po 167	At 140	Rn 145

Fig. 7.9 Atomic radii (in picometers) for selected atoms

Atomic radius – Half the distance between the nuclei in a molecule consisting of identical atoms. Also known as the covalent radius.
Metallic radius – Half the distance between the nuclei of the atoms within the solid structure of a metal.

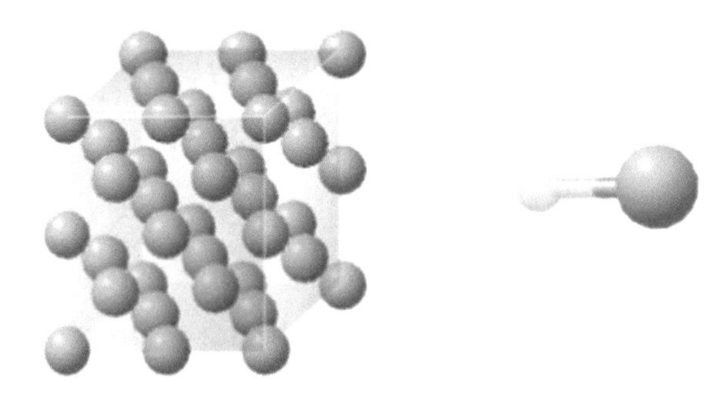

The uncertainties involved in measuring atomic radii, and the various methods that can be used, mean that the values we obtain should be regarded as approximate, and we may find slightly different values quoted in different sources. The basic trends, however, are absolutely clear: Atomic size decreases across periods and increases down groups. How can we explain these data? As we move down the group, we encounter atoms with increasing numbers of occupied electron energy levels. Each new occupied energy level, corresponding to a higher principal quantum number, includes electron orbitals of greater average radius than those of the previous level, so the atoms grow larger as we move down a group.

It might seem more difficult to explain why we find atoms of decreasing size as we move along any period, because the atoms actually contain more matter as we travel from left to right, thanks to the steady increase in protons and electrons. However, the additional electrons are being added to energy levels already occupied in previous elements of the period, and they are held by nuclei whose positive charge is steadily increasing across the period. This increased positive charge draws the electrons of the occupied energy levels closer to the nucleus as we move along a period, so the atomic radius steadily decreases.

Let's compare the two elements on either end of Period 2. Lithium has 3 protons in its nucleus and one valence electron in the second energy shell and its radius is 152 pm. At the other end of Period 2 is fluorine. Fluorine has 9 protons in its nucleus and 7 electrons in the second energy shell, and its radius is 64 pm—less than half the radius of lithium! The higher number of positively charged protons in the nucleus attracting a higher number of electrons in the same energy shell results in the electrons being pulled closer in toward the nucleus and a smaller atomic radius.

7.7 Ionization Energies

Most of the elements in the periodic table are metals. When they react with other elements to form compounds, they generally do so by losing one or more electrons to form positive ions. An input of energy is required to remove one or more electrons from an atom, and the energy involved is known as **ionization energy**.

Ionization energy – The energy needed to ionize an atom (or molecule or ion) by removing an electron from it.

The smaller the ionization energy, the more likely it is that ionization will occur. All elements, not just the metals, have ionization energy values. By examining the ionization energies of the elements, we can see *why* metals are particularly prone to form positive ions, and we can see how the differing reactivities of metals match their differing ionization energies. As we might suspect, the energy it takes to remove electrons from an atom is related to how strongly those electrons are attracted to the nucleus.

The **first ionization energy (I_1)** of an element is defined as the energy required to remove the highest energy electron from an atom of the element in the gaseous ground state. The first ionization energy is typically quoted per mole of atoms being ionized and can be symbolized in general terms, using X to represent any element, as

$$X(g) \rightarrow X^+(g) + e^-$$

First ionization energy (I_1) – The energy required to remove one electron from each atom of an element in the gaseous state.

For example, the first ionization energies for sodium and chlorine are

$$\begin{array}{ll} Na(g) \rightarrow Na^+(g) + e^- & I_1 = 495\,kJ/mol \\ Cl(g) \rightarrow Cl^+(g) + e^- & I_1 = 1255\,kJ/mol \end{array}$$

Why is the first ionization energy of sodium so much smaller than that of chlorine? Note the resulting ion's electron configuration. The sodium ion has a noble gas electron configuration. The chlorine ion does not. The first ionization energies of the main group (*s*-block and *p*-block) elements are listed in ◘ Fig. 7.10.

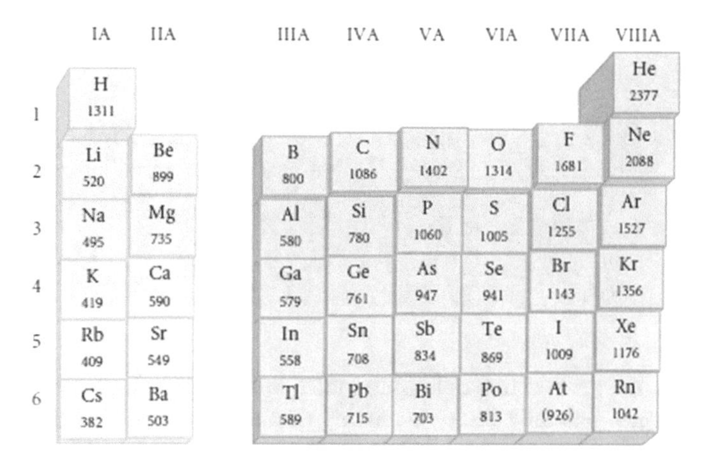

◘ **Fig. 7.10** First ionization energies for selected elements

The **second ionization energy (I_2)**, of an element is defined as the energy required to remove one electron from each singly charged + 1 ion of an element in the gaseous state:

$$X^+(g) \rightarrow X^{2+}(g) + e^-$$

The third, fourth, and subsequent ionization energies can be similarly defined, as successive electrons are removed.

Second ionization energy (I_2) – The energy required to remove one electron from each singly charged +1 ion of an element in the gaseous state.

We can see a link between ionization energy and reactivities by examining some of the alkali metals. Lithium, sodium, and potassium are all reactive metals that form alkalis on reaction with water. These metals, when they react, form ions of Li^+, Na^+, and K^+, respectively. We can get an indication of how readily the formation of these ions occurs by considering these ionization energies:

$$Li(g) \rightarrow Li^+(g) + e^- \quad I_1 = 520\,\text{kJ/mol}$$
$$Na(g) \rightarrow Na^+(g) + e^- \quad I_1 = 495\,\text{kJ/mol}$$
$$K(g) \rightarrow K^+(g) + e^- \quad I_1 = 419\,\text{kJ/mol}$$

Example 7.5—Implications of Ionization Energy Trends

The reactions of lithium, sodium, and potassium metals with water to form an alkaline solution are:

$$2Li(s) + 2H_2O(l) \rightarrow 2Li^+(aq) + 2OH^-(aq) + H_2(g)$$
$$2Na(s) + 2H_2O(l) \rightarrow 2Na^+(aq) + 2OH^-(aq) + H_2(g)$$
$$2K(s) + 2H_2O(l) \rightarrow 2K^+(aq) + 2OH^-(aq) + H_2(g)$$

Given the values listed in Fig. 7.10 for the first ionization energies of these metals, which metal will react most vigorously if added to water? If we were to put a chunk of cesium metal in water, how might it react?

Asking the Right Questions

What is ionization energy? Is it endothermic or exothermic? What is the relationship between ionization energy and the ease of a chemical reaction? What is the ionization trend as we go down the Group 1 (IA) metals? What does that trend mean for the reactivity of the metals?

Solution

Ionization energy is a measure of the energy required to remove the highest energy electron from an atom. It is an endothermic process, so when comparing elements in an otherwise equivalent reaction, the lower the ionization energy, the less energy will be required, and the more readily the reaction will occur.

The potassium metal has the lowest ionization energy of the three, so it will react most vigorously with water; sodium will react less so and lithium the least vigorously. Figure 7.11 shows the reactions of these metals with water. Note the flame that potassium makes upon reaction with water. What about cesium? Its first ionization energy of 382 kJ/mol means that it will react even more vigorously than potassium. In fact, the reaction with water is explosive. On the other hand, the first ionization energy of gold is 890.1 kJ/mol. Note its reactivity (or lack thereof) with water.

Are the Answers Reasonable?

The vigor of the reactions is consistent with the first ionization energy of each element. This relationship is especially stark when we compare cesium to gold. The answers are therefore reasonable.

What If...?

we were asked to explain the changes in first ionization energy as we move down the noble gas group? What would be your explanation?

Practice 7.5

Which of the metals in Group 2 (IIA) would you expect to react most vigorously with water? Explain your answer in terms of the ionization energies of the elements.

We have just identified one of the general trends linking ionization energy values with an element's position in the periodic table. Ionization energies generally decrease moving down any group. We can explain this trend by noting that as we move down a group, the valence electrons being removed are coming from energy levels with successively higher principal quantum numbers. Therefore, as we move down a group, the electrons being removed are farther away from the nucleus and are shielded from the nucleus by additional inner energy levels occupied with electrons.

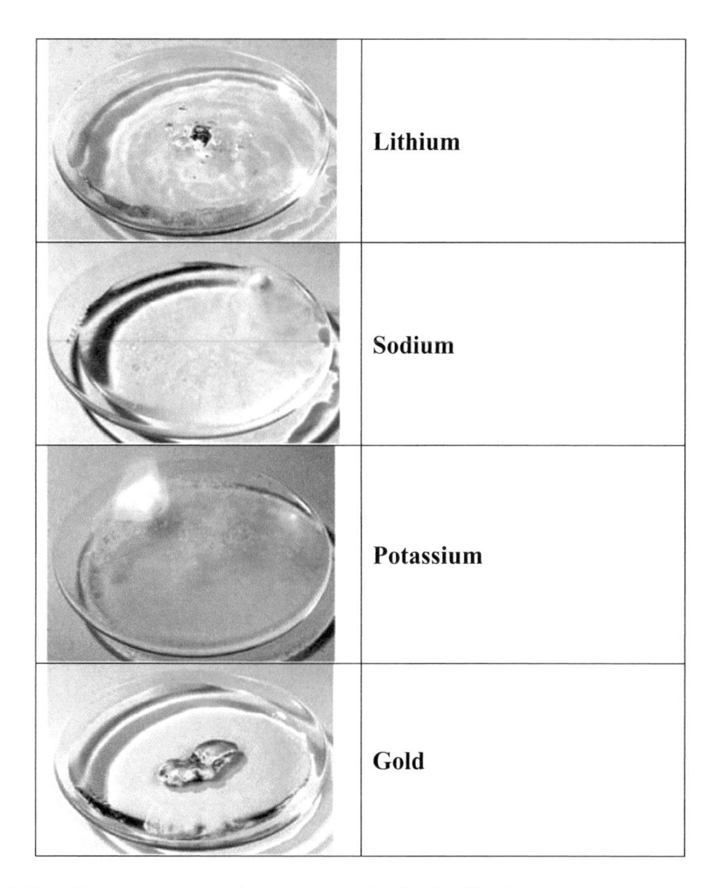

Fig. 7.11 The relative reactivities of metals can sometimes be quite clearly visualized

The other main trend in ionization energies is that they generally increase from left to right along any period. We can rationalize this as due to the extra energy needed to remove an electron from the pull of the increasing nuclear charge found as we move along a period. This is the opposite trend we noticed with atomic radii.

These trends in ionization energy can be seen in the laboratory if we choose metals with significantly different reactivities (see Fig. 7.11). The trends can also be plotted graphically, as shown in Fig. 7.12, in a way that makes the periodicity in this property apparent. As we move through Fig. 7.12, we see the ionization energies rising and falling in a periodic manner.

Example 7.6—Differences in Ionization Energies

Here are the first three ionization energies of sodium and magnesium, in kilojoules.
per mole:

	I_2	I_2	I_3
Sodium	495	4560	6920
Magnesium	735	1445	7730

There is a very large difference between the first and second ionization energies of sodium. With magnesium, however, the very large difference is seen between the second and third ionization energies. Why do the large differences in ionization energies occur where they do?

Asking the Right Questions

What does the ionization energy tell us? What is it about the electron configuration of each element that results in such a sudden increase in ionization energy?

Solution

The ionization energy tells us how much energy is required to remove the highest energy electron from the atom of an element. Sodium atoms have one outer-shell electron ([Ne]3s^1), whereas magnesium atoms have two ([Ne]3s^2). Once sodium has lost one electron, the next electron must come from a deeper electron shell, closer

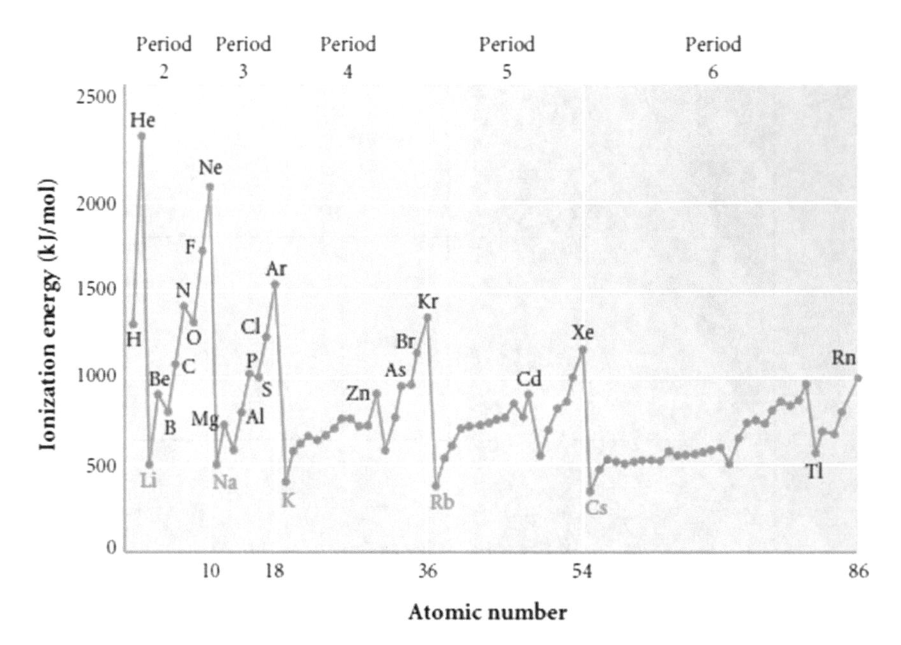

Fig. 7.12 The values of first ionization energies for the elements in the first six periods

7

to the nucleus, so a great deal more energy is required to remove that second electron. In the case of magnesium, the transition to removing electrons from an inner shell does not occur until the two outer electrons have been removed, so the big jump in ionization energy is seen between the second and third ionization energies, rather than between the first and second.

Is the Answer Reasonable?

We expect a sharp increase in ionization energy when the element has a complete shell and/or when the next electron to be removed comes from a lower energy level. With sodium, that occurs after the removal of one electron. With magnesium, that occurs after the removal of two electrons. The answer is reasonable.

What If...?

...we were given these data for the ionization energy of a representative element? In what group is this element? Explain your choice. Note that we begin with the 2nd ionization energy.

Ionization energy (kJ/mol)

2nd = 1798

3rd = 2735

4th = 4837

5th = 6043

6th = 12310

(*Ans: This element is in Group 5 (VA), because it requires a sudden, sharp increase in energy to lose its 6th electron compared to the other 5. These data are for arsenic, As.*)

Practice 7.6

Predict, and draw a graph of, the ionization energy versus electron ionized for sulfur. Compare your graph to the measured values at the WebElements site.

7.8 Electron Affinity

Our look at ionization energies concerned the energy required to generate positive ions (cations) from neutral atoms. The other side of the "ionization coin" is the generation of negative ions (anions) from neutral atoms. How does a negative ion form? It forms when an atom gains electrons, and the energy change associated with electron gain is called **electron affinity**. The electron affinity is the energy change associated with the addition of an electron to an atom in the gaseous state, and as usual, we quote the values in kilojoules per mole of atoms. For example,

$$\begin{array}{ll} F(g) + e^- \rightarrow F^-(g) & -328\ \text{kJ/mol} \\ Cl(g) + e^- \rightarrow Cl^-(g) & -349\ \text{kJ/mol} \\ Br(g) + e^- \rightarrow Br^-(g) & -325\ \text{kJ/mol} \\ I(g) + e^- \rightarrow I^-(g) & -295\ \text{kJ/mol} \\ At(g) + e^- \rightarrow At^-(g) & -270\ \text{kJ/mol} \end{array}$$

1 IA	2 IIA	13 IIIA	14 IVA	15 VA	16 VIA	17 VIIA	18 VIIIA
Li –60	Be >0	B –27	C –154	N ~0	O –141	F –328	Ne >0
Na –53					S –200	Cl –349	
K –48					Se –195	Br –325	
Rb –47					Te –190	I –295	
Cs –46					Po –183	At –270	

Fig. 7.13 Electron affinity values, in kilojoules per mole (kJ/mol), for selected elements. Electron affinity values generally decrease as we go up and to the right in the periodic table. More negative values indicate a greater affinity for electrons

Electron affinity – The energy change associated with the addition of an electron to an atom in the gaseous state.

Why is the energy of the electron affinity value negative? The addition of an electron to an atom releases energy as the electron and nucleus get closer together. Notice that the addition of an electron to the atoms of Group 17 (VIIA) returns the largest energy because the resulting anion would have the same electron configuration as a noble gas. The electron affinity values for Group 17 (VIIA) reveal one of the general trends in electron affinity and also one of the complications. The values indicate to us that adding an electron to an atom releases less energy as we move down a group. Does this trend make sense? Yes, because *added electrons are farther from the nuclei of larger atoms*, they interact less with the positive charge of the nucleus. There are many exceptions, however, such as more energy released when a chloride ion forms, in the above list, than upon formation of the fluoride anion. Why does fluorine not fit the trend? In searching for a possible explanation, we should note that fluorine is the smallest of this group of atoms, so the electrons in the outer *p* orbitals, to which the extra electron is being added, will be much closer together than in the other atoms of this group. The larger electron–electron repulsions between these closely spaced electrons, compared to those of other halogen atoms, will be associated with increased potential energy.

Figure 7.13 provides several more electron affinity values. It allows us to confirm the general trend we identified within groups, in addition to revealing some exceptions. Figure 7.13 also confirms that elements that form stable negative ions (such as those in Groups 16 (VIA) and 17 (VIIA)) have large negative electron affinity values, whereas those that form stable positive ions (Groups IA and IIA) have much lower negative electron affinity values, or even positive values.

Note, as well, that the values in Fig. 7.13 indicate that electron affinity values generally get more negative as we move from the left to the right along a period in the periodic table. Does this trend make sense? We know from our understanding of the trend associated with atomic radii that atoms get smaller as we move to the right along the period table. Because the outermost electrons are closer to the nucleus as we move across a period, the electron affinity generally tends to get more negative as we move to the right along a period.

Example 7.7—Trends

Many of the periodic properties identified in the periodic table are related to each other. Is there a relationship between electron affinity and the number of valence electrons?

Asking the Right Questions

What is electron affinity? What are the trends in electron affinity within groups and across periods?

Solution

Electron affinity is the energy change associated with the addition of an electron to an atom in the gaseous state. To see the trends, we can construct a crude periodic table with arrows pointing to the most negative electron affinity. Superimposing the number of valence electrons on this drawing will reveal any relationship.

As the number of valence electrons gets larger, the atomic radius gets smaller across a period. As the radius gets smaller, the electron affinity becomes more negative. Therefore, as the number of valence electrons increases, the electron affinity becomes more negative.

Is the Answer Reasonable?

We know that atoms of the elements on the left side of the periodic table, including sodium, potassium and calcium tend to form cations, while atoms of the elements on the right-hand side of the periodic table, such as fluorine, chlorine and oxygen tend to form anions in chemical reactions. The answer fits our experience, and is reasonable.

What If…?

…we have these electron affinity values in kJ/mol, for Group 15 (VA) elements, beginning with nitrogen?

$$-0, -72, -77, -101, -110$$

How does this relate to the electron affinities for Group 16 (VIA) and Group 17 (VIIA) elements? What does this tell us about electron affinities?

(*Ans: Although consistently negative, the electron affinity values show the opposite trend in Group 15 (VA) compared to the other groups. This shows that electron affinity is a complex measure of periodicity.*)

Practice 7.7

Is there a relationship between the trends observed in ionization energy and electron affinity? Explain.

7.9 Electronegativity

In our discussion about electron affinities, we are actually considering a rather contrived situation, because elements rarely exist in the form of free gaseous atoms. A more meaningful characteristic of elements, one that is related to their interaction with additional electrons and therefore is more firmly rooted in real chemical behavior, is their **electronegativity**. The electronegativity of an atom is a measure of the ability of the atom in a molecule to attract shared electrons to itself. Atoms in molecules share electrons to create covalent bonds. However, if the atoms sharing electrons are different, they will contain nuclei with different charges and different electron arrangements. The result is that the electrons are not shared equally. Electronegativity values have been calculated for every atom in the periodic table. However, we generally do not quote electronegativity values for the noble gases (Group 18 (VIIIA)) because, as a consequence of their exceptional stability, they do not readily form bonds.

Electronegativity – The relative ability of an atom participating in a chemical bond to attract electrons to itself.

Electronegativity values for the elements in the periodic table are one of the most powerful quantities chemists have that explain and predict the behavior of molecules and ions. Their importance cannot be overstated. In 1932, Linus Pauling defined the concept in terms of bond energies. The Pauling scale of electronegativities and the variations of it are the bases of much of our discussion about bonding in subsequent chapters. To illustrate the utility of electronegativity values, let's consider two essential vitamins that we must regularly consume: vitamin A and vitamin C. Deficiencies in these vitamins are responsible for serious health issues including night blindness (vitamin A) and anemia (vitamin C). However, our bodies require different amounts of each vitamin as part of our daily diet. Why? Vitamin A is a fat-soluble compound that our bodies retain. Vitamin C, on the other hand, is a water-soluble compound that is readily excreted from our bodies. This means that we must consume vitamin C in larger quantities than vitamin A and on a more regular basis. Therefore, in the U.S., the recommended daily allowance of vitamin C is about 70 times higher than vitamin A. As we will discuss in the next chapter, electronegativities are the basis for understanding the solubility of these vitamins.

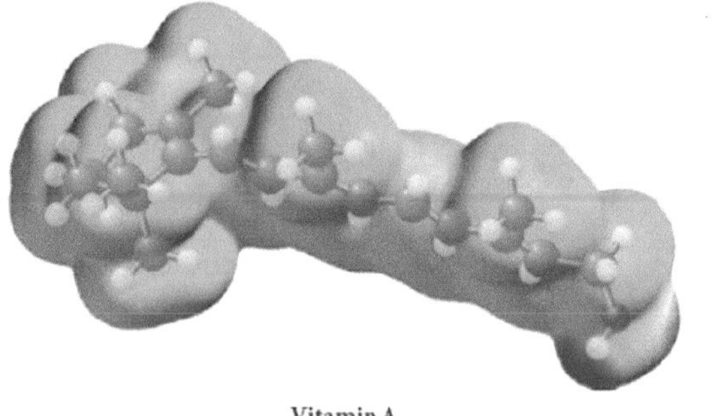

Vitamin A

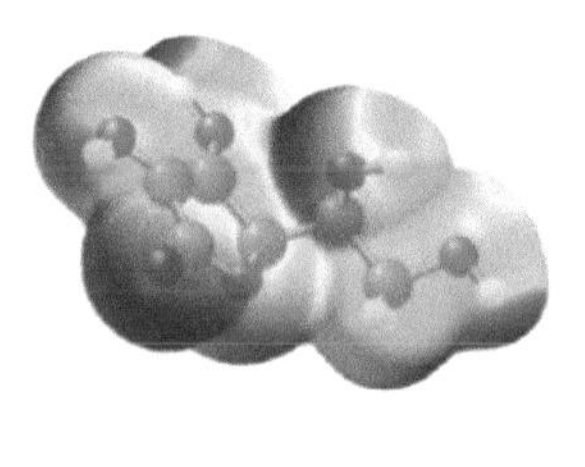

Vitamin C

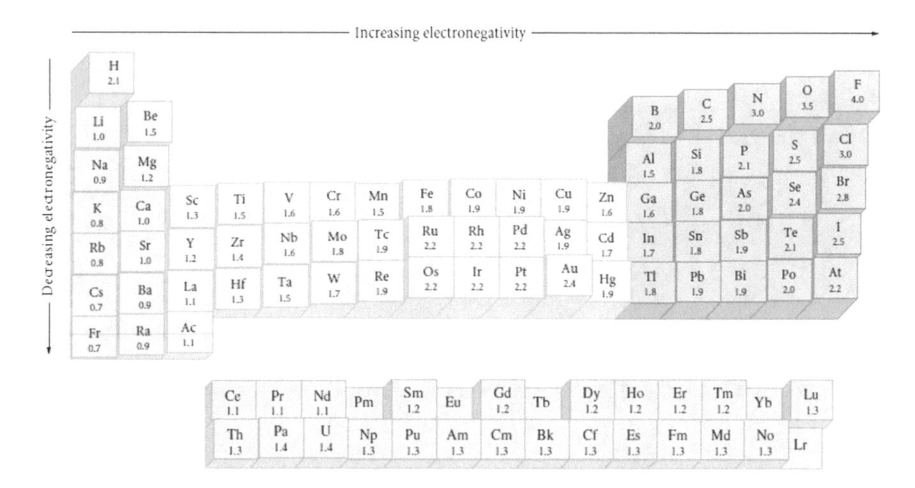

Fig. 7.14 Electronegativity values for selected elements. Electronegativity generally increases as we go up and to the right in the periodic table. Note how this trend shows the same pattern as ionization energy

Pauling derived the electronegativity values by comparing the energies required to break various chemical bonds. The values shown in Fig. 7.14 reveal two general trends. The first of these trends is that electronegativities generally increase as we move along periods all the way to Group 17 (VIIA). The second trend indicates that electronegativities generally decrease as we move down groups. This means that the lowest electronegativity values are at the bottom of Group 1 (IA) and that the element fluorine has the highest. Do these trends make sense? The increase in electronegativity along periods occurs because the atoms are progressively smaller. In addition, atoms on the right-hand side of each period possess an increased nuclear charge due to the presence of more protons in the nucleus. Both of these factors cause electrons to be attracted to the nucleus more strongly.

The decrease in electronegativity that we note as we move down groups makes sense because the atoms are growing larger. With an extra intervening shell of electrons at each step, the nuclear charge is being screened from the electrons shared in a covalent bond. The effect of the increased size and electron shielding outweighs the increase in nuclear charge as we move down a group.

The trends in electronegativity, electron affinity, and ionization energy are not without exceptions, so they are general trends rather than absolute rules. They do, however, reveal one very clear message about which atoms are most likely to attract electrons and become negative ions, and which are most likely to lose electrons and become positive ions. Atoms on the extreme left of the periodic table, especially the lower extreme left, will most readily form positive ions. Atoms on the extreme right of the periodic table, especially the upper extreme right (always remembering that Group 18 (VIIIA) is excluded), will most readily form negative ions.

Example 7.8—Trends in Electronegativity

Without examining the actual Pauling electronegativity values, predict which of these elements in each pair is more electronegative.

Li or Be N or P S or I

Asking the Right Questions

What is electronegativity? What is the trend in electronegativity values? Where is each of these elements placed on the periodic table in comparison to the trend?

Solution

Electronegativity is a measure of the ability of an atom in a molecule to attract shared electrons to itself. Electronegativity values are lowest at the bottom left-hand corner of the periodic table (cesium and francium), increasing toward the upper right-hand corner (fluorine).

Be is closer in the periodic table to fluorine. It is more electronegative.

N is only two elements away from fluorine, whereas P is three elements away. Nitrogen is more electronegative.

S is two elements away from fluorine. It is more electronegative.

Are the Answers Reasonable?

We can confirm our decisions by looking at the data in Fig. 7.14, and by noting the number of electrons in the valence shell of the atoms of each element. These data are consistent with the answers, and reasonable.

What If...?

...in the future, more than just a few atoms of one of the human-made elements, copernicum (Cn, atomic number 112) were to be synthesized. Based on its position in the periodic table, would we expect its electronegativity to be closer to that of cesium or fluorine? What element might its chemical reactions most closely resemble?

(Ans: Copernicum sits just below mercury on the periodic table, and would be expected to have chemistry similar to mercury. It has an electron configuration of $[Rn]5f^{14}6d^{10}7s^2$*, and it would likely form cations with a +2 oxidation state when in chemical reactions. Its electronegativity would likely be slightly greater than mercury given the observed trends in that group, but would be closer to 0.7 than to 4.0.)*

Practice 7.8

Which from each pair is more electronegative?

P or Na Ne or Cl N or C

7.10 Reactivity

The concept of **reactivity** is a descriptive and useful notion in chemistry. When we describe an element as "highly reactive," we mean that it readily participates in chemical reactions to form compounds. When we describe an element as "unreactive," we mean it does not readily participate in chemical reactions to form compounds. As we have already discovered, most highly reactive elements are, in their natural state, combined with other elements in the form of compounds. We will never find a hunk of soft, shiny sodium in nature. Rather, we will find it in ionic form in water or in salts on land. Unreactive elements can be found in their pure form. This has great bearing on the uses we make of elements. The metals sodium and potassium, for example, would be of little use for making automobiles or jewelry, because they react so vigorously and readily with water. Gold, on the other hand, is an ideal metal for jewelry. Because gold is so unreactive, it can be found in its free form within the earth. The scarcity of gold makes it unsuitable for the construction of automobiles, even though such automobiles might last for a very long time.

Reactivity – The ability of an element or compound to participate in chemical reactions. Reactivity is an imprecise but useful term, particularly when we are discussing relative reactivities under defined conditions.

We compromise when making automobiles, and build their metal components mostly out of steel, which is largely composed of the element iron. Iron is not totally unreactive; it corrodes into rust (iron oxide) when combined with the oxygen found in air or dissolved in water. Iron lasts long enough, however, to form the basic materials of automobiles that will last quite a few years.

Can the periodic table tell us which elements are the most reactive? The elements that react most readily generally do so by either losing or gaining electrons. The most reactive elements will be those with the greatest tendency to form positive ions (the lowest ionization energies), or with the greatest tendency to form negative ions (the largest electronegativities). These highly reactive elements are found at the extreme left of the periodic table (Groups 1 (IA) and 2 (IIA)) and on the far right (Groups 16 (VIA) and 17 (VIIA)). The least reactive elements of all are found in Group 18 (VIIIA), the noble gases; and those with intermediate and low reactivities are found between the extremes just mentioned, in the central portion of the periodic table. In Group 11 (IB), we find gold and silver, unreactive elements we use in jewelry, and also copper, which is sufficiently unreactive to be used in electric wires and the pipes of plumbing systems.

Our discussion now enables us to find more general degrees of logic in the periodic table. We now have enough understanding to give at least rudimentary answers to the questions we raised in the chapter opening about why oxygen reacts with so many elements and why gold has a very low reactivity. These questions can be answered by considering the location of these elements in the periodic table and all that this means. Another question posed in our introduction was what makes iron so well-suited to the task of carrying oxygen around the body, bound to the iron-containing protein we call hemoglobin. The iron in hemoglobin is in the form of Fe^{2+} ions, which bind *reversibly* (that is, they can continue to bind and then release, depending on conditions) to oxygen molecules (O_2). The binding has to be readily reversible, because hemoglobin must combine with oxygen where oxygen is abundant, in the blood vessels of the lungs; but must release the oxygen to the tissues of the body in which oxygen levels are much lower. A positive ion can perform the binding task by interacting with the electrons of the oxygen molecule. In order to do that in a readily reversible way, however, the ion and the atom from which it is derived must have neither too strong nor too weak an attraction for electrons. A suitable ion will be from an element of intermediate electronegativity, likely to be found around the middle of the periodic table. Iron, in the middle of the periodic table, can do the job very nicely. This is only one of several reasons why iron ions serve within hemoglobin molecules as the oxygen carriers of life, but it is surely a significant one. Other reasons include the availability of iron in the environment in which life originated and the suitability of iron to combining with the other components of hemoglobin.

One idea that we did not discuss here but will consider in the next two chapters is that judgments about the reactivity of an element cannot be made in isolation. That is, atoms react with other types of atoms, and the extent to which a reaction occurs depends on the properties of both, or even many, types of atoms. More broadly speaking, the reaction environment must be considered in deciding the nature of chemical behavior.

7.10.1 The Reactivity Series of Metals

The relative reactivities of some of the most abundant and most useful elements in the environment are very significant to us. For example, in deciding which metal is suitable for a particular industrial use, we find that the ease with which it corrodes (oxidizes in the presence of substances in the environment) will be very significant. The most common form of corrosion is reaction with oxygen, and all forms of corrosion are chemical reactions of one kind or another. The most reactive metals will generally corrode most quickly. Thinking about such issues led to the idea of listing metals in a **reactivity series** (or **activity series**), which ranks selected metals in order of reactivity (see Table 7.14).

Reactivity series – A series of elements ranked in order of reactivity. Also known as an activity series.

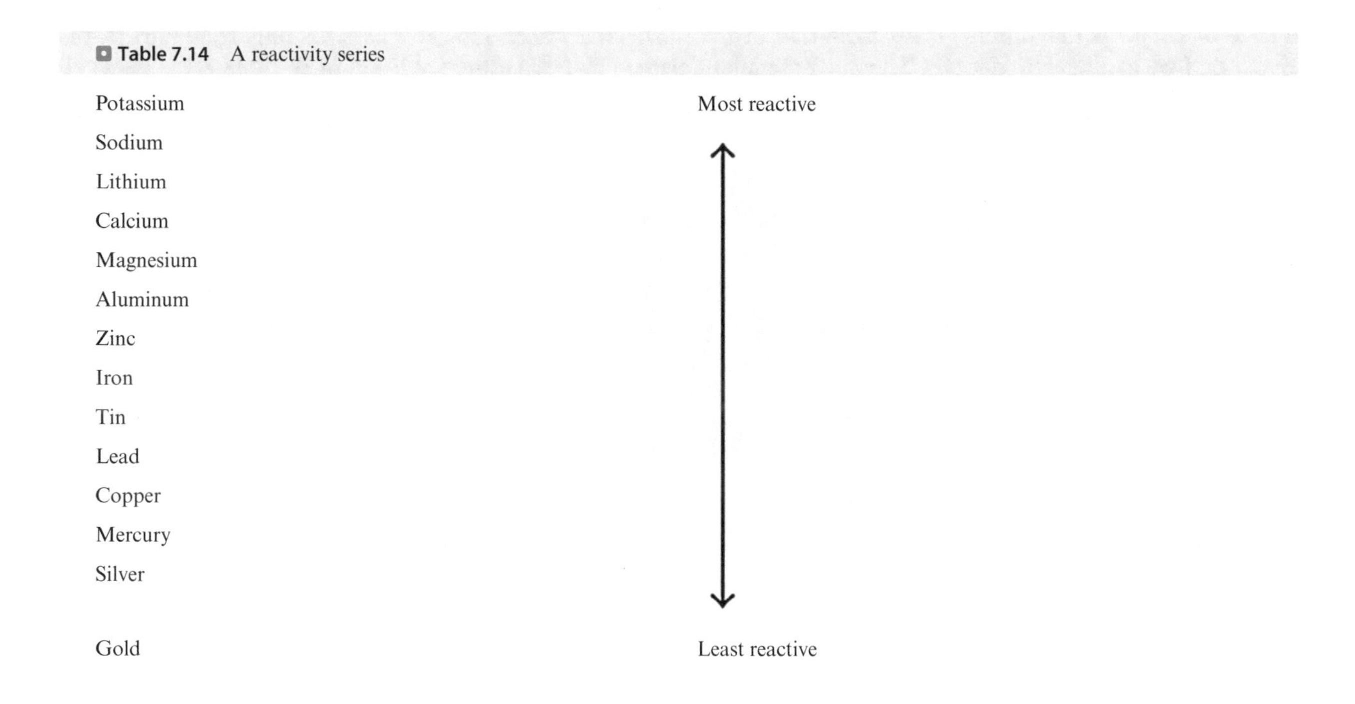

Table 7.14 A reactivity series

Potassium	Most reactive
Sodium	↑
Lithium	
Calcium	
Magnesium	
Aluminum	
Zinc	
Iron	
Tin	
Lead	
Copper	
Mercury	
Silver	↓
Gold	Least reactive

Such a list can be made by exposing metals to a range of substances, such as hydrochloric acid, water, and air, and observing how readily they react. There are some problems with this approach. For example, aluminum reacts more quickly with oxygen than does iron. The aluminum soon becomes coated, however, with a cohesive layer of aluminum oxide that protects the aluminum beneath from further corrosion. That is why aluminum cookware is quite long-lived. Iron oxide, on the other hand, does not form any protective layer but instead produces the crumbling, weak structure we know as "rust," which breaks away to reveal fresh corrosion-prone metal underneath. Nevertheless, the idea of ranking metals by reactivity has proved very useful.

An alternative, and in some ways more satisfactory, way of ranking metals by reactivity is found in the "electrochemical series" that we will discuss in ▶ Chap. 18. This is related to the tendency of each metal to give up electrons. Remember that when metals react, they generally do so by losing outer electrons to form positively charged ions. The most reactive metals are the ones that lose electrons and form ions most readily.

7.11 The Elements and the Environment

The elements of planet Earth are distributed throughout the environment in places and forms that are a consequence of their physical and chemical properties, and these physical and chemical properties can be related to the elements' positions in the periodic table. The environment in this sense comprises the entire world around us, including the Earth, its atmosphere, and ourselves.

Geologists describe the structure of the Earth in terms of five distinct regions: the core, mantle, crust, hydrosphere, and atmosphere. These are shown diagrammatically in ◘ Fig. 7.15. Although nobody has traveled to the Earth's core or has even been able to obtain samples, we know its structure from the accumulation of years of indirect evidence. There is little doubt that the core is composed largely of iron, mixed with smaller amounts of the metals nickel and cobalt and lighter elements such as carbon and sulfur. The core is largely molten, so these elements form a molten alloy (a homogeneous mixture of metals) with some nonmetals mixed within the alloy. The predominance of iron at the center of the Earth is a consequence of its high density in relation to most of the other elements that make up our planet. When the Earth formed, it went through a molten phase, during which the densest materials were drawn toward the center. The less dense materials rose outward, in just the same way as a mixture of cooking oil and water will settle into layers, the dense water beneath the less dense oil.

The mantle is another region of the Earth that has never been directly sampled, although a few different types of volcanic rocks are believed to have been derived from this region. The evidence of these rocks, combined with other indirect evidence, suggests that the mantle is composed largely of oxides of the elements iron, magnesium, silicon, calcium, and aluminum. The low-density element oxygen, which is a gas at typical atmospheric temperatures and pressures, has been retained within the Earth's mantle because of its chemical reactivity. This has caused it to react with various metals to form the much more dense compounds found in the mantle.

The Earth's crust makes up less than 1 percent of the mass of the planet and, as shown in ◘ Fig. 7.16, it has a ratio of elements that differs greatly from that of the Earth as a whole. However, it is the only solid part of the Earth that we can analyze directly. The most abundant elements of the crust are listed in ◘ Table 7.15. These el-

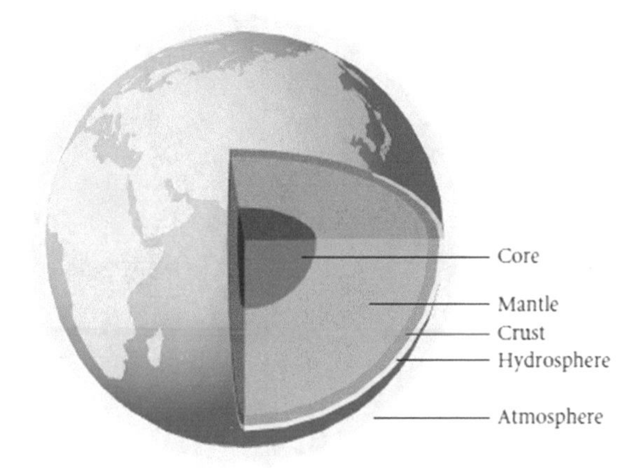

◘ **Fig. 7.15** The structure of planet Earth

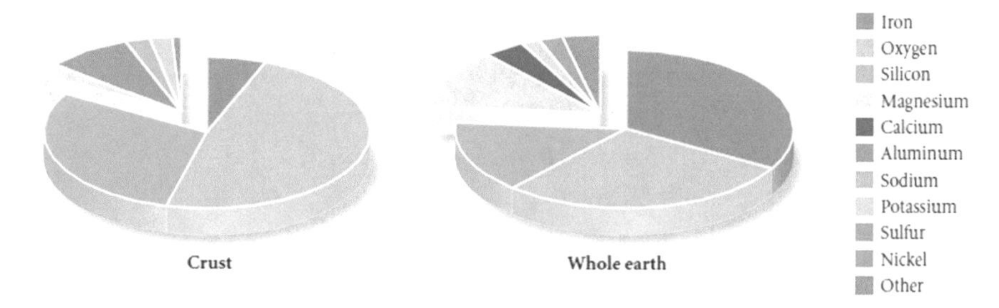

Fig. 7.16 Comparison of the elements in the Earth's crust with those in the entire Earth

Table 7.15 The abundance of elements in the earth's crust

Element	Abundance (% by mass)
Oxygen	46.60
Silicon	27.72
Aluminum	8.13
Iron	5.00
Calcium	3.63
Sodium	2.83
Potassium	2.59
Magnesium	2.09
Titanium	0.44
Hydrogen	0.14
Phosphorus	0.12
Manganese	0.10
Copper	0.0070
Gold	0.00000005

ements are largely found bound within compounds, oxides being the most predominant. Oxides of silicon are among the most common components of the crust. For example, silicon dioxide (SiO_2) is the principal component of sand and of the many types of rock from which sand is derived.

The hydrosphere is the name given to the waters of the Earth, comprising oceans, seas, lakes, rivers, and underground aquifers. In addition to the hydrogen and oxygen that make up water, the hydrosphere contains a great variety of dissolved substances that influence its properties and suitability for various uses. Seawater is unsafe for humans to drink because of its high concentration of salts, such as sodium chloride and magnesium chloride. Earth's atmosphere contains the low-density, gaseous materials that nevertheless are sufficiently dense to be retained by the gravitational pull of the planet. As shown in Table 7.16, the atmosphere is largely composed of nitrogen and oxygen, with much smaller amounts of argon, carbon dioxide, and other rare gases. Some of the most significant components of the environment, however, are too rare even to show up in most tables of the atmosphere's composition. Gases such as sulfur dioxide and nitrogen dioxide are extremely rare overall, but they cause problems over areas where they are present as environmental pollutants.

Finally, it is worth pointing out that the element hydrogen (H_2), the most abundant element in the universe, is of such low density that any hydrogen released into the atmosphere will eventually escape into space, unless it reacts with some other chemicals on the way. The only way for hydrogen to be retained on our planet is within chemical compounds, such as water (H_2O), methane (CH_4), or any one of millions of other compounds, including most of those found within living things. If chemical changes had not locked up hydrogen within such chemical compounds, we would never have evolved to ponder and study these chemical changes.

Table 7.16 Composition of dry air

Substance	Volume fraction
N_2	0.7808
O_2	0.2095
Ar	0.00934
CO_2	0.00034
Ne	1.82×10^{-5}
He	5.82×10^{-6}
CH_4	2×10^{-6}
Kr	1.1×10^{-6}
H_2	5×10^{-7}
N_2O	5×10^{-7}
Xe	8.7×10^{-8}
SO_2	$< 1 \times 10^{-6}$
O_3	$< 1 \times 10^{-7}$
NH_3, CO, NO, I_2	$< 1 \times 10^{-8}$

The Bottom Line

- The structure of the periodic table was initially developed by scientists trying to make sense of the differing reactivities of all the elements found in the natural environment.
- The structure of the periodic table includes "blocks" defined in terms of which type of orbital is being filled as we imagine filling up the available orbitals using the Aufbau principle. This gives us the *s*-block, *p*-block, *d*-block, and *f*-block.
- The horizontal "period" of the periodic table to which an element belongs indicates how many energy levels are either fully or partially occupied by electrons in atoms of that element.
- Elements in any one main group (Groups IA through VIIIA or 1, 2, and 13–18) have the same number of electrons in their highest energy level.
- The elements in the periodic table are arranged into three main sections: the metals, nonmetals, and metalloids (or semimetals).
- Metals are shiny, ductile, malleable, and good electrical conductors. They tend to form positive ions in chemical reactions.
- Nonmetals are dull, brittle, and are good electrical insulators. They tend to form negative ions in chemical reactions.
- Metalloids have properties that are mixtures of metals and nonmetals.
- The periodic table is divided into groups (columns), and elements in each group are similar in electron structure, physical properties, and chemical reactivity.
- Chemical reactions result from changing energies or locations of electrons in atoms and molecules.
- Periodicity is apparent in the way important physical and chemical characteristics recur in a periodic manner as we move through the periodic table.
- All periodic properties are related to how far the valence electrons are from the nucleus, and how strongly the nucleus can hold onto those valence electrons.
- Examples of characteristics that show clear periodicity are atomic size, ionization energy electron affinity, electronegativity, and reactivity.
- The distribution of the elements on planet Earth is a result of their physical properties and chemical reactivities.

Section 7.1 The Big Picture—Building the Periodic Table

Skill Review

1. Cite two examples wherein two elements' locations on the current periodic table would be reversed if they were placed in accordance with Mendeleev's atomic mass system of organization.
2. Some may argue that providing a list of elements in alphabetical order along with their properties would be the most useful arrangement of the known elements. Explain what advantage there is to arranging them in the periodic manner displayed in our traditional periodic table format.
3. Identify the "block" location in the periodic table for each of these elements.
   a. Sr b. Sc c. S d. Sn e. Se f. Sm g. Sb
4. Identify the "block" location in the periodic table for each of these elements.
   a. Rb b. Rn c. Ru d. Rh e. Re f. Rf g. Ra
5. Identify the "block" location in the periodic table for each of these elements.
   a. C b. Ca c. Cu d. Cr e. Cs f. Cl g. Ce
6. Identify the "block" location in the periodic table for each of these elements.
   a. He b. Be c. Fe d. Ge e. Xe f. Ne g. Te

Chemical Applications and Practices

7. The following portion of a hypothetical periodic table shows the approximate melting points of elements that are neighbors to a missing element. On the basis of the data provided, predict the melting point of the missing element.

	2610 °C	
3000 °C	?	3180 °C

8. The following portion of a hypothetical periodic table shows the densities of elements that are neighbors to a missing element. On the basis of the data provided, predict the density of the missing element.

	10.2 g/cm^3	
16.7 g/cm^3	?	21.0 g/cm^3

9. In each of these pairs, select the larger of the two elements.
   a. Mg or Ca b. P or S c. Cl or Br d. Cs or Ba
10. In each of these pairs, select the larger of the two elements.
   a. K or Ca b. O or S c. Se or Br d. Ra or Ba

Section 7.2 The First Level of Structure—Metals, Nonmetals, and Metalloids

Skill Review

11. Classify each of these as a metal, nonmetal, or metalloid.
   a. Zn b. Ga c. Ge d. Sn e. Si f. P g. Se h. As
12. Classify each of these as a metal, nonmetal, or metalloid.
   a. Li b. Ne c. Co d. Te e. C f. Bi g. I h. Sb
13. Most elements, at room temperature, are either solids or gases. However, two elements, at room temperature, are classified as liquids. What are these two elements?

14. Most of the key ingredients to alloy with iron to make various types of steel are metals. However, this is not always the case. What are two nonmetals that can be alloyed with steel?

Chemical Applications and Practices

15. a. In the production of stainless steel, small amounts of several elements are alloyed with iron. Two of the chief ingredients are nickel and chromium. Use the chart given in Table 7.3 to report the number of grams of each metal found in 100.0 g of stainless steel.
    b. The atoms of the added metals can become part of the iron structure, which changes the properties of iron. How many atoms of nickel and chromium, respectively, are present in the 100.0 g of stainless steel?
16. Assume that the other elements (besides chromium, nickel, and iron) found in stainless steel are insignificant. Based on the calculations in Problem 15, how many iron atoms are present in the sample for every atom of chromium?
17. Locate the element antimony in the periodic table. Would this element be classified as a metal, nonmetal, or metalloid? Antimony is used in alloys with lead or tin, is combined with other substances to make fireproof fabrics, is mixed with some types of glass and ceramics, and serves as the active component of some medicines. Antimony, however, does have some toxic properties. One way to obtain it is as follows:

$$Sb_2S_3 + Fe \rightarrow FeS + Sb$$

Balance this equation. The annual worldwide production of antimony is approximately 6.00×10^4 tons. How many moles of antimony is this?

18. Locate the element gold in the periodic table. Would this element be classified as a metal, nonmetal, or metalloid? Gold is used not only in jewelry but also as a wire in some computer parts and in other situations where a noncorrosive metal is needed. Gold can be solubilized in water by using a mixture of very strong acids, as follows:

$$Au(s) + HCl(aq) + NO_3^-(aq) \rightarrow [AuCl_4]^-(aq) + H_2O(l) + NO(g)$$

Balance this equation. How many grams of gold must be solubilized to produce 1 mol of $NO(g)$?

Section 7.3 The Next Level of Structure—Groups in the Periodic Table

Skill Review

19. Using the most common oxidation number for each, write out the formulas of the compounds that would be most likely to form when Li, Be, B, C, N, O, and F, respectively, combine with oxygen.
20. Write out the chemical formulas for Li, Na, K, Rb, and Cs, respectively, as if they were combined with oxygen. What accounts for the similarities in the formulas of these compounds?
21. If an element from Group 16 (VIA) were going to form an ionic compound, which type of element (metal or nonmetal) would be most likely to form the other part of the compound? Explain your answer.
22. If an element from Group 2 (IIA) were going to form an ionic compound, which type of element (metal or nonmetal) would be most likely to form the other part of the compound? Explain your answer.
23. How many unpaired electrons are found in the elements that make up Group 17 (VIIA)? How many unpaired electrons are found in the most common ions of those elements?
24. How many unpaired electrons are found in the elements that make up Group 13 (IIIA)? How many unpaired electrons are found in the most common ions of those elements?

Chemical Applications and Practices

25. The listing in Table 7.13 of elements in the human body is based on mass. Considering only the first three (oxygen, carbon, and hydrogen), indicate whether the ranking would change if the table were based on number of atoms instead of mass. Explain the basis of your answer.
26. The listing in Table 7.13 of elements in the human body includes calcium. What use does our body make of this element? In addition, calculate the quantity of calcium in a typical human in units of milligrams per kilogram of body weight.
27. Many dietary supplements contain a list of metals present. This list may include zinc, iron, cobalt, and others. Name one important role that transition metals play in human biochemistry.
28. Iron plays several roles in humans. One of its important roles involves the electron transport chain. Iron can easily change oxidation states between the +2 and +3 charges. What characteristic feature of iron's electron configuration makes this possible?

Section 7.4—The Next Level of Structure—Groups in the Periodic Table

Skill Review

29. The noble gases, except for helium, have a stable, filled octet. Explain why helium is still a noble gas even though it does not have a stable, filled octet.
30. The alkali metals have a complete electron shell of a noble gas plus one extra electron. Explain why hydrogen is often associated with this group of the periodic table.

Chemical Applications and Practices

31. Examine the full electron configuration of calcium and potassium. Explain why calcium is not as reactive as potassium.
32. The same reasoning used to answer Problem 31 applies to the relative reactivity of magnesium compared to sodium, but why is sodium not as reactive as potassium?
33. In some of the original periodic table research, tellurium was placed after iodine due to its slightly greater relative mass. However, this placed tellurium in Group 17 (VIIA) directly below bromine, chlorine, and fluorine. After examining the electron configuration of tellurium and iodine explain why the present classification is logical.
34. Nickel is placed on the periodic table immediately after cobalt. However, its average mass is less than that of cobalt. Explain why the placement is logical.

Section 7.5 Atomic Size

Skill Review

35. Place these elements in order of increasing atomic size: Kr, Br, C, Se, He.
36. Place these elements in order of increasing atomic size: Cs, P, Bi, Ba, Ne.
37. If the bond length for a C—C bond were 154 pm, what would you determine as the radius of a carbon atom?
38. If the bond length for a H—H bond were 75 pm, what would you determine as the approximate radius for an atom of hydrogen?
39. If the bond length for a C—Cl bond were 171 pm, what would you determine as the approximate radius for an atom of chlorine? (Use the answer from Problem 37 to assist you in determining the radius of a chlorine atom.)
40. If the bond length for a H—F bond were 92 pm, what would you determine as the approximate radius for an atom of fluorine? (Use the answer from Problem 38 to assist you in determining the radius of a fluorine atom.)

Chemical Applications and Practices

41. Sodium and potassium are often found as ions in sports drinks. Which element has a larger radius? Explain.
42. Helium and hydrogen have both been used in dirigibles (also known as blimps or airships). Which of these two element's atoms is larger? Explain.

Section 7.6 Ionization Energies

Skill Review

43. For this calculation, use the first ionization values of sodium and potassium found in the chapter. How many grams of potassium could be ionized using the same energy that is required to ionize 1.00 g of sodium?
44. For this calculation, use the first ionization values of calcium and potassium found in the chapter. How many grams of calcium could be ionized using the same energy that is required to ionize 5.00 g of potassium?
45. Within each of these two lists, rank the elements listed from lowest to highest first ionization energy:
    a. Al, Si, P
    b. Ne, Ar, Kr
    Review your rankings. Would you make any changes if asked to do the rankings by second ionization energy? If so describe the reasons for any changes.
46. Within each of these two lists, rank the elements listed from lowest to highest first ionization energy:
    a. Li, Be, B
    b. K, Rb, Cs
    Review your rankings. Would you make any changes if asked to do the rankings by second ionization energy? If so describe the reasons for any changes.
47. Based on the following ionization energy (I) information, predict which element(s) are most likely metals and which are nonmetals. Also, suggest the most likely periodic table group for each of these "unknown" elements. (Note that these values, all of which are in units of kilojoules per mole, may not reflect actual ionization energies of known elements.)

	I_1 (kJ/mol)	I_2 (kJ/mol)	I_3 (kJ/mol)	I_4 (kJ/mol)
Element A	740	1440	7730	8670
Element B	2100	3230	4400	5500

48. On the basis of the following ionization energy (I) information, predict which element(s) are most likely to be metals and which to be nonmetals. Also, suggest the most likely periodic table group for each of these "unknown" elements. (Note that these values, all of which are in units of kilojoules per mole, may not reflect actual ionization energies of known elements.)

	I_1 (kJ/mol)	I_2 (kJ/mol)	I_3 (kJ/mol)	I_4 (kJ/mol)
Element A	500	4600	5800	7460
Element B	1060	1900	2900	5000

Chemical Applications and Practices

49. Element 118, called Oganesson, is the largest known element to have been synthesized, though in very small amounts. To which group in the periodic table does it belong? Would its ionization energy be expected to be higher or lower than that of its next neighbor above it in the periodic table? Explain your answer.
50. Element 119 could be discovered in the near future. To which group in the periodic table would it belong? Would its ionization energy be expected to be higher or lower than that of its neighbor directly above it in the periodic table? Explain your answer.
51. The flow of potassium and sodium ions into and out of nerve cells makes it possible for signals to be sent throughout our bodies. The fact that you are reading right now is dependent on this flow of ions.
    a. What are the charges on the potassium and sodium ions, respectively?
    b. Which is the larger of the two ions, sodium or potassium?
    c. Which of the two ions, sodium or potassium, requires less energy to produce from their neutral atoms?
52. Calcium ions also have great importance in cells and move into and out of other channels in cell membranes. What would be the charge on a calcium ion? Would this ion be larger or smaller than a potassium ion? Explain.
53. Explain why the first ionization energy of sodium is lower than magnesium's first ionization energy, but the second ionization energy of sodium is higher than that of magnesium.
54. Examining the first ionization energies of the elements in Period 4 of the periodic table reveals the general tendency for an increase in ionization energy from left to right. However, there is a slight drop in ionization energy from calcium to gallium. Study the electron configuration of both of these elements and offer an explanation for this apparent deviation from the trend.
55. Many elements may be found in nature in a variety of oxidation states. However, these statements express some limitations. Explain the basis for each of these limitations.
    a. Magnesium is never found naturally as the +3 ion.
    b. Fluorine is never found naturally as the +1 ion.
    c. Hydrogen is never found naturally in the +2 oxidation state.
    d. Aluminum tends to be found naturally as the +3 ion.
56. Explain the basis for each of these limitations on the oxidation states in which certain elements appear.
    a. Arsenic can often be found with a charge of +5 but never with a charge of +6.
    b. Titanium could be found with either a +2 or a +4 charge but not with a +5 charge.
    c. Potassium is never found naturally as a neutral element.
    d. Tin can be found with a +2 or a +4 charge but not with a +5 charge.
57. The ionization energies of nitrogen are:

$$N^{+} \rightarrow N^{+} = 1402.3\ \mathrm{kJ/mol}$$
$$N^{+} \rightarrow N^{2+} = 2856.1\ \mathrm{kJ/mol}$$
$$N^{2+} \rightarrow N^{3+} = 4578\ \mathrm{kJ/mol}$$
$$N^{3+} \rightarrow N^{4+} = 7474.9\ \mathrm{kJ/mol}$$
$$N^{4+} \rightarrow N^{5+} = 9440\ \mathrm{kJ/mol}$$
$$N^{5+} \rightarrow N^{6+} = 53265.6\ \mathrm{kJ/mol}$$
$$N^{6+} \rightarrow N^{7+} = 64358.7\ \mathrm{kJ/mol}$$

Many elements have one primary oxidation number in combination with other elements. Nitrogen has many, especially when combining with oxygen. Name the following compounds of nitrogen and oxygen: N_2O, NO_2, N_2O_4, NO. What is the highest oxidation number for nitrogen in these compounds? On the basis of the ionization energies of nitrogen, can you explain why nitrogen has that highest oxidation number, and why there are so many roles in which nitrogen plays a vital part as an "element of life?"

58. The ionization energies of oxygen are:

$$O, \rightarrow O^{+} = 1313.9\ \text{kJ/mol}$$
$$O^{+} \rightarrow O^{2+} = 3388.2\ \text{kJ/mol}$$
$$O^{2+} \rightarrow O^{3+} = 5300.3\ \text{kJ/mol}$$
$$O^{3+} \rightarrow O^{4+} = 7469.1\ \text{kJ/mol}$$
$$O^{4+} \rightarrow O^{5+} = 10989.3\ \text{kJ/mol}$$
$$O^{5+} \rightarrow O^{6+} = 13326.2\ \text{kJ/mol}$$
$$O^{6+} \rightarrow O^{7+} = 71333.3\ \text{kJ/mol}$$
$$O^{7+} \rightarrow O^{8+} = 84076.3\ \text{kJ/mol}$$

When oxygen and other elements combine to form compounds, oxygen is always the anion, except when combining with fluorine. Can you explain why this is so? In the list of ionization energies for oxygen, there is a huge gap between the values for $O^{5+} \rightarrow O^{6+}$ and $O^{6+} \rightarrow O^{7+}$. Explain the gap.

Section 7.7 Electron Affinity

Skill Review

59. Which of these would have the best chance of forming a stable anion: S or Xe? Justify your choice.
60. Which of these would have the best chance of forming a stable anion: Cl or Ar?
61. Which of these has the more exothermic value for electron affinity: Cl or Si?
62. Which of these has the more exothermic value for electron affinity: O or F?
63. Arrange the following atoms on the basis of electron affinity, from least negative to most negative: Na, Mg, Al. Now arrange them on the basis of their ability to form anions. Is the order any different? If so, explain any changes you made in the ordering.
64. Arrange the following list in order from least negative electron affinity value to most negative electron affinity value: Br, Br^{-}, K.
65. Because they each have a negative charge, the electrons in an atom repel each other. Explain why the electron repulsion in a fluorine atom is a larger factor than in bromine in determining its electron affinity value.
66. Explain why ionization energy is always a positive quantity, whereas electron affinity may be positive or negative.

Chemical Applications and Practices

67. The first ionization energy of sodium is +495 kJ/mol. The electron affinity of chlorine is −349 kJ/mol. When sodium metal and chlorine gas are placed near each other, a violent reaction takes place. After a sodium atom has lost an electron and a chlorine atom has gained an electron, both are oppositely charged and have the stable electron configuration of noble gases.
    a. What is the total energy change that occurs in these two processes? Is this exothermic or endothermic?
    b. Two additional energy changes occur in this reaction: the dissociation of a chlorine atom from Cl_2 (+158.8 kJ/mol) and the energy released when the ions are combined (−1030 kJ/mol). What can you determine about the spontaneous, violent reaction between sodium and chlorine when all of the energy changes in this process are considered?
68. The first ionization energy of lithium is 5.392 eV. (1 eV = 96.485 kJ/mol). The electron affinity of fluorine is −328 kJ/mol. When lithium metal and fluorine gas are placed near each other, a reaction takes place. After the ionization of lithium and the formation of anionic fluorine, both are oppositely charged and have the stable electron configuration of noble gases.
    a. What is the total energy change that occurs in these two processes? Is this exothermic or endothermic?
    b. Two additional energy changes occur in this reaction: the dissociation of a fluorine atom from F_2 (+158.8 kJ/mol), and the energy released when the ions are combined (−1030 kJ/mol). What can you determine about the spontaneous, violent reaction between lithium and fluorine when all of the energy changes in this process are considered?

Section 7.8 Electronegativity

Skill Review

69. Arrange this list of atoms in a correct ranking from lowest electronegativity to highest electronegativity: Na, F, As, Li, S.
70. Arrange this list of atoms in a correct ranking from lowest electronegativity to highest electronegativity: K, P, O, Br, N.
71. Each of these situations depicts two atoms bonded to each other. In each case, which atom is more likely to attract electrons toward itself within the bond?
    a. N—O b. Se—S c. Br—Ge d. Cl—O
72. Each of these situations depicts two atoms bonded to each other. In each case, which atom is more likely to attract electrons toward itself within the bond?
    a. F—O b. P—S c. H—C d. C—S
73. Use Fig. 7.14 to calculate the difference in electronegativity for each of the following pairs of atoms. Also, state which atom in each pair is most likely to form positive ions. Is this consistent with your understanding based on their position in the periodic table?
    a. F, F b. Cs, O c. K, N d. S, Cl
74. Use Fig. 7.14 to calculate the difference in electronegativity for each of the following pairs of atoms. Also, state which atom in each pair is most likely to form positive ions. Is this consistent with your understanding based on their position in the periodic table?
    a. Ba, O b. P, P c. K, Br d. N, Cl

Chemical Applications and Practices

75. Using the information in the chapter, determine the change in electronegativity between Ga and Se. This change arises as the number of protons is increased by three. The change in proton number from Sc to Zn is nine. What is the change in electronegativity between Sc and Zn? Based on the change in number of protons, is this what you would expect? Why or why not?
76. In an early section we noted that the trend in electron affinity from fluorine to chlorine was a bit different than we might expect. The trend in electronegativity is, however, what we would expect; that is, it decreases from top to bottom in the group. What aspect of the definition of electronegativity helps explain the difference between the two trends?

Section 7.9 Reactivity

Skill Review

77. Arrange this list of metals in order from generally least reactive to most reactive: Na, Mg, Rb.
78. Arrange this list of nonmetals in order from generally least reactive to most reactive: S, Cl, I.
79. What is the relationship between first ionization energy and reactivity in metals and in nonmetals?
80. What is the relationship between electron affinity and reactivity in metals and in nonmetals?

Chemical Applications and Practices

81. The "coinage metals" are copper, silver, and gold.
    a. Describe the location of these elements in the periodic table.
    b. Where are they located in the activity series of metals used in the chapter?
    c. Is the location of Group 11 (IB) consistent with periodic reactivity and the activity series?
82. Aluminum metal is considered a fairly active metal. It certainly reacts with oxygen more vigorously than does iron. The reaction of both metals with oxygen produces oxides that have different characteristics. Contrast the properties of the two oxides.
83. Nonmetals such as chlorine and oxygen are both considered very reactive. However, they may also react with each other. Using any available information, decide which of the two would probably become more negative in the reaction? Explain the basis for your answer.
84. Until the 1960s the noble gases were considered chemically inert, or totally unreactive. Eventually, however, some noble gases were made to react. Among the first compounds of noble gases, fluorine was typically a reactant. Explain why fluorine was such a good candidate for this role of reactivity.

Section 7.10 The Elements and the Environment

Skill Review

85. What explanation can be given for the fact that iron is found in much greater abundance in the Earth's crust than in the mantle?
86. What substance that is common in the Earth's crust accounts for most of the silicon and oxygen (the two highest-ranking elements in the crust?)
87. a. What is the third most abundant substance, by volume, in the Earth's atmosphere?
    b. How many particles of that substance would be found in one mole of dry air?
88. a. What is the fourth most abundant substance, by volume, in the Earth's atmosphere?
    b. How many molecules of that substance would be found in one mole of dry air?

Chemical Applications and Practices

89. Table 7.15 lists the ranked abundance of elements in the Earth's crust by mass. Which element(s) in that list are most likely to be found in an uncombined state? (Uncombined, in this case, means "not combined with other elements.")
90. In Table 7.15, titanium is listed as more abundant, by mass, than hydrogen. Convert the mass to moles and number of atoms. If the ranking were to be redone by number of atoms, would hydrogen still be below titanium? Prove your answer.
91. Table 7.15 lists the abundance of elements in the Earth's crust. Check reliable Internet and other sources to find some of the principal commercial uses of these elements (either as elements or compounds), along with the prices for the products related to their uses. How do the prices of these products compare with the abundance of the elements in the Earth's crust? What factors go into determining the price of an element or resulting product?
92. Check reliable sources to determine the most important industrially used gases. How are these gases prepared for industrial use? Referring to Table 7.16, which of these gases are also among the most abundant in the Earth's crust? Is there a relationship between the most abundant gases and their ease of preparation for industrial use?

Comprehensive Problems

93. For each of these, provide a brief summary of his contribution to organizing the properties of elements into a useful arrangement.
    a. Johan Dobereiner
    b. John Newlands
    c. Dmitri Mendeleev
94. Indium (element 49) is an important element, but it is not commonly used in textbook examples. Use the Internet or another text reference to determine a chemical and a physical property that indium has in common with gallium, element 31.
95. An element with a valence shell electron configuration of s^1 is first reacted with oxygen. Then, the resulting oxide is reacted with water. From this list, predict what will be formed and explain your answer:
    a. an acid
    b. a base
    c. a precipitate
    d. a salt
    e. no reaction
96. An element with a valence shell electron configuration of s^2p^3 is first reacted with oxygen. Then, the resulting oxide is reacted with water. From this list, predict what will be formed and explain your answer:
    a. an acid
    b. a base
    c. a precipitate
    d. a salt
    e. no reaction
97. a. What is the most common form of steel used in the world?
    b. Based on composition, what is the chief difference between this most common type of steel and other types of steel?
98. Based on the characteristics listed, which type of element (metal, nonmetal, or metalloid) is being described?
    a. At room temperature the sample is a dull, brittle solid. The element is most likely to assume a negative charge in ionic compounds.
    b. At room temperature the sample is a solid that is able to conduct electricity.
99. The C–C bond is 154 pm and the H–H bond is 75 pm. Given that information, what would you predict for the bond length between a carbon and a hydrogen atom?

100. Contrast the meaning of the term valence electrons in the context of calcium and in the context of chromium.
101. a. Using the first 18 elements on the periodic table, prepare a graph with number of valence electrons on the y-axis and group number on the x-axis. Is this a periodic function? Explain why or why not.
     b. Prepare another graph of the same elements, using the most common oxidation number on the y axis and the group number on the x axis. Is this a periodic function? Explain why or why not.
102. The definition of the atomic radius of an atom is straightforward; it is one-half the distance between the nuclei of a molecule made of two identical atoms.
     a. What problems would arise if we were to define the radius as one-half the diameter of an atom?
     b. How does metallic radius differ from covalent radius?
103. Explain why, when comparing the radius of two atoms, it would be important not to base the comparison solely on which atom has the greater number of protons.
104. a. This reaction depicts an atom being ionized. To which side of the reaction (left or right) should you show the energy term for the reaction?

$$X \rightarrow X^{+} + e^{-}$$

     b. Select the correct response for each of these situations: The smaller the value for ionization energy, the (less, more) easily ionization will occur. The smaller the value for ionization energy, the (less, more) chemically reactive a metal will be.
105. Seldom do we get to describe a chemistry situation as "always" happening without exception. However, this statement is always true: Successive ionization energies in an atom are always higher than previous values. What underlying factors make this statement true?
106. Electron shielding or electron screening provides a descriptive way to explain how the attraction for an electron by the nucleus can vary in different electron arrangements. However, electron screening can have a slightly different description when applied to elements in the same row on the periodic table than when applied to elements in the same group. Contrast the type of screening, and its effectiveness, when the term is applied to a period and when it is applied to a group.
107. Arsenic and selenium are next to each other in Period 4 of the periodic table. In general the trend, from left to right on the table, is in favor of an increase in ionization energy. However, when you compare the values for As and Se, you note that the ionization drops slightly instead of increasing. What is the basis for this situation?
108. Radioactive fallout from nuclear testing can contain significant amounts of unstable strontium. This can be particularly harmful to young children, who have rapidly growing bone structures. What common charge would strontium ions have? What element is an important component in bone development? Why is exposure to radioactive strontium such a grave danger to young children?
109. a. The electron affinity of chlorine is approximately – 350 kJ/mol. Does this indicate an exothermic or an endothermic reaction?
     b. Write out the reaction, including placing the energy term in the equation, that depicts the "electron affinity reaction" for the process described in part a.
110. When summarizing the periodic trends in ionization energy, atomic radius, electronegativity, and (to some extent) electron affinity, we can typically indicate the trend with a one-directional arrow to show the trend increasing or decreasing from left to right in a row, or from top to bottom in a group. Explain why the trends for increasing reactivity start from the center of the periodic table and move outward in both directions.
111. In the chapter, we note that tool steel contains 4.38% by mass of chromium and 0.864% by mass of carbon.
     a. What is the electron configuration of chromium?
     b. If a steelmaker wishes to make 1.0 kg of tool steel, which element (carbon or chromium) would require a larger number of moles?
     c. How many atoms of chromium would be present in an 88.7 g sample of tool steel?
112. The halogens are so-named because they react with metals to make salts.
     a. How many valence electrons do each of the atoms in this group contain?
     b. Write the balanced reaction that occurs between sodium metal and iodine crystals.
113. According to Table 7.13, the human body contains the same mass percent of silicon and potassium. Which occurs in the body in a greater number of moles?
114. If the periodic table was arranged, in order, from smallest atom (based on atomic radius) to largest atom
     a. What atom would be listed first?
     b. What atom would be listed last?
     c. Would the noble gases still be aligned in a column?
     d. How many helium atoms would be needed to create a line of atoms 1.0 inches long (assuming that the atoms are just touching and placed in a straight line)?

115. Francium-210 can be made from gold-197 by bombarding the gold with oxygen-18.
   a. How many protons, neutrons, and electrons are found in one atom of francium-210?
   b. How many grams of francium-210 could be made from 2.50 g gold-197?
   c. Which has a larger atomic radius—francium-210 or gold-197? (Assume specific isotopes do not differ in their atomic radius.)

Thinking Beyond the Calculation

116. An aqueous solution of sodium bicarbonate ($NaHCO_3$) reacts with aqueous hydrochloric acid to produce carbonic acid (H_2CO_3) and sodium chloride (NaCl).
   a. Write a balanced equation illustrating this reaction.
   b. Indicate the group and period for each of the elements involved in the reaction.
   c. Indicate those elements in the reaction that can be characterized as metals.
   d. Carbonic acid decomposes in solution to produce carbon dioxide and water. Write the balanced equation for this reaction.
   e. If 10.0 mg of sodium bicarbonate is added to 275 mL of water and reacted completely with a stoichiometric amount of HCl, what concentration (in ppm) of sodium chloride will result? (Assume no volume change.)
   f. Using the information from part e, determine what the concentration (in ppm) of sodium ions will be.
   g. A pastry chef may wish to perform a reaction similar to this using sodium bicarbonate and acid. Why would the chef wish to perform this reaction?

Bonding Basics

Contents

Supplementary Information The online version contains supplementary material available at ▶ https://doi.org/10.1007/978-3-030-90267-4_8.

M. Mosher and P. Kelter, *An Introduction to Chemistry*,
https://doi.org/10.1007/978-3-030-90267-4_8

What will we learn in this chapter?

- There are three main types of chemical bonds. (► Sect. 8.2)
- The strengths of ionic bonds are related to the charges on the ions. (► Sect. 8.3)
- There are different types of covalent bonds that differ in their strength. (► Sect. 8.4)
- Lewis dot structures are one of the models we can use to draw atoms, ions and compounds. (► Sect. 8.2)
- VSEPR theory can be used to predict the 3-dimensional shapes of molecules. (► Sect. 8.5)
- Our models can help us determine if a molecule is polar or nonpolar. (► Sect. 8.6)

8.1 Introduction

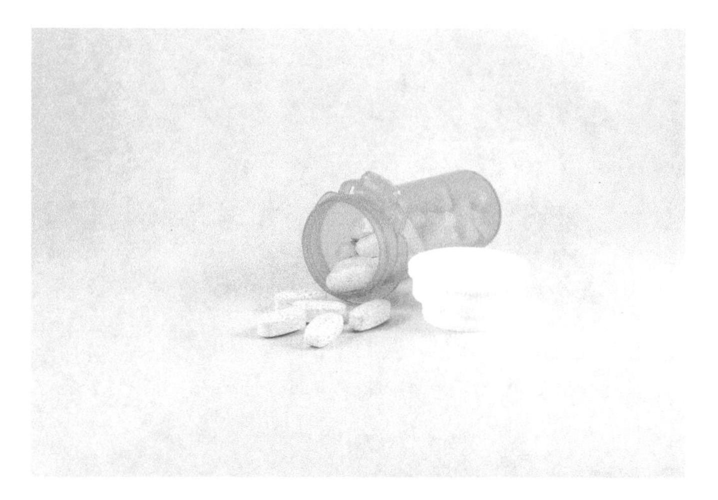

Americans spend more than $3 trillion each year on medical care, and nearly $300 billion per year on prescription and over-the-counter (OTC) medicines, about one-half of the world's total, according to the U.S. Department of Health and Human Service's Center for Medicare & Medicaid Services. Our medicine cabinets are fairly well stocked. If you and your family are typical consumers, you may have headache pills, muscle relaxers, antacids, cough syrup, and a couple of old bottles of antibiotics. A quick check of the active ingredients reveals that these bottles contain such compounds as ibuprofen, magnesium salicylate, sodium bicarbonate, and pseudoephedrine. One of the more common ingredients in a pain-relieving OTC medicine is aspirin (a molecular compound). Sodium bicarbonate (an ionic compound) is typically used as an antacid. These two compounds work in different ways to relieve common discomforts, and they are fundamentally different in their chemical makeup. They do not have the same numbers and types of atoms, although this alone isn't enough to explain why these compounds differ radically in many properties, including melting point, boiling point, solubility in water, and chemical reactivity. The main difference lies in their designation as molecular compounds or ionic compounds, and this designation is made on the basis of the way in which the atoms are bound together.

Biochemists and medicinal chemists are familiar with molecular and ionic compounds. How do the chemical bonds, the molecular or crystalline shape, and the physical properties of a drug contribute to its biological activity? In this chapter, we will answer this question and others as we explore how atoms are held together, and how this bonding determines the shapes and properties of compounds.

8.2 Modeling Bonds

Although the formula of a potential drug can be useful for determining the number and types of atoms it contains, it doesn't say much about the molecular shape. Aspirin—acetylsalicylic acid—which we discussed in ► Chap. 3, has the formula $C_9H_8O_4$. The formula alone provides no information about the shape of the aspirin molecule, or how it interacts with the body to alleviate a headache. To explore the potential interactions and build a model of aspirin, the medicinal chemist must look deeper into the atoms in the formula—into the role that the electrons play in determining how the atoms are attached. The first key idea is that the different arrangements of the electrons in a compound help determine the shape and the properties of that compound.

Acetylsalicylic acid
$C_9H_8O_4$

From careful measurements, scientists constructed a 3-D model of the aspirin molecule. The model, known as an **electron potential map (ep map)**, illustrates an uneven distribution of electron density. It also illustrates locations in the molecule that are rich in electron density (yellow and red), lack electron density (blue), and are average in electron density (green). This model provides information about the chemical and physical properties of the molecule. How does the ep map provide that information? The positions of the electrons in the molecule indicate the answer. For example, chemical reactions reposition the electrons in the reactants as products are made. Therefore, if we know where the electrons are and how they hold atoms together, we can gain insight into the potential reactions that may occur.

Electron potential map (ep map) – A 3-D model that illustrates the distribution of electron density in a molecule using colors. Red indicates more electron density and blue indicates less electron density.

8.2.1 Three Kinds of Bonds

In their search for marketable pharmaceuticals, biochemists and medicinal chemists use a wide range of **molecular models** as tools to examine the shape and properties of compounds. The advanced ep map is just one of those models. Construction of this model is fundamentally based on the understanding of the **chemical bonds**, the forces that hold atoms together, within a compound. Chemical bonds arise between atoms when some of the outermost electrons on the bonding atoms interact. How can the electrons on bonding atoms interact? In some cases, the electrons tend to congregate on one of the atoms of the bond. In other cases, the electrons are shared more or less equally between the atoms. In other words, chemical bonds describe the placement of atoms and electrons within a compound. The way the electrons are placed on bonding atoms results in three main types of chemical bonds: the covalent bond, the ionic bond, and the metallic bond.

Molecular models – Models used by chemists to explore the three-dimensional nature of molecules. Although many can be made from plastic or wood, these models can also be constructed using computers.

Chemical bonds – A sharing of electrons between two adjacent atoms. This sharing can be complete, partial, or ionic in nature.

Aspirin is an example of a compound in which the atoms are held together by covalent bonds. The **covalent bond** consists of at least one pair of electrons shared between two atoms. The positively charged nuclei on either end of the bond attract the negatively charged electrons. It is this force that holds the atoms together. If the two atoms are exactly identical, then the bond is purely covalent and the electrons are equally shared by each nucleus. If the two atoms are different, then the electrons will polarize toward the atom that is the most electronegative. Of the three types of bonds, covalent bonds are unique in that they can exist as individual units. For example, in a sample of hydrogen gas, there are individual molecules of purely covalent H_2.

Covalent bond – A sharing of electrons between two adjacent atoms. This sharing can be complete or partial.

Sodium chloride, the stuff we sprinkle on our fries, is an example of a compound with ionic bonds. An **ionic bond** results when oppositely charged ions are attracted to each other. These charges arise when the atom's electron count differs from its elemental state. In a way, an ionic bond is similar to a covalent bond where the atoms are so different that the electrons in the bond are completely polarized to just one of the two atoms. The resulting positive and negative charges attract each other strongly and hold the material together. Unlike covalent bonds, atoms held together by ionic bonds don't usually exist as individual units. In an ionic compound, the ions are surrounded by each other in a complex three-dimensional structure where oppositely charged ions are mutually attracted in all directions.

Ionic bond – A strong electrostatic attraction between ions of opposite charges.

The third type of bond, the **metallic bond**, is a special type of bond that describes how the electrons are placed when atoms from the metals are bonded together. We'll learn more about this bonding pattern in ► Chap. 11. This kind of chemical bonding is what allows metals to conduct electricity and be malleable, ductile, and shiny. Metallic bonds are like ionic bonds in that they do not exist individually in nature.

Metallic bond – A bond in which metal cations are spaced throughout a sea of mobile electrons.

8.2.2 Lewis Dot Symbols

The first step in the construction of the molecular model of a compound, such as aspirin, sodium chloride, or water, is drawing an atom itself. The "business part" of the atom is its set of valence electrons. In 1916, G. N. Lewis (◘ Fig. 8.1) developed a useful representation later called **Lewis dot symbols**. Typically, the first two dots, representing valence electrons in the *s* orbital, are placed one at a time, as a pair, to the side of the element symbol. The next three dots, representing valence electrons in the *p* orbitals, are placed individually on the other sides of the element symbol. The remaining electrons are placed alongside each of the first three dots to indicate pairs of electrons in the *p* orbitals. The Lewis dot symbols for the elements of Period 2 are shown below.

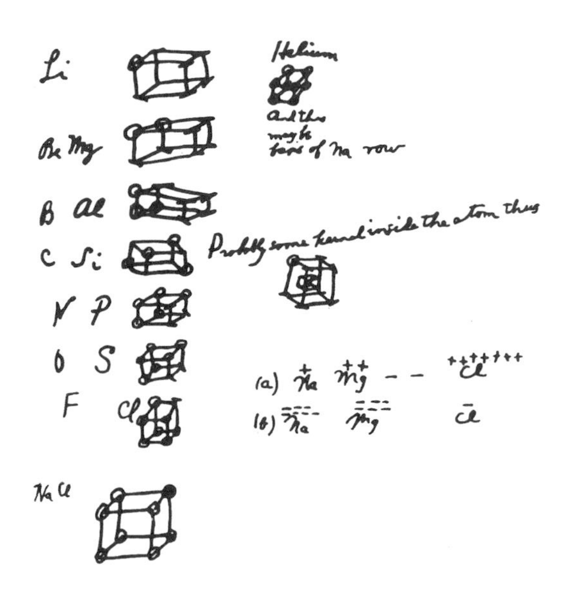

◘ **Fig. 8.1** The American chemist G. N. Lewis (1875–1946) developed a model of atomic bonding that is still used today. These notes, written by Lewis in 1902, illustrate his thinking on how electrons were arranged around an atom. The Lewis dot symbols that we use today are slightly modified from this original work. *Source*: ► https://upload.wikimedia.org/wikipedia/en/3/32/Lewis-cubic-notes.jpg

Lewis dot symbol – A drawing convention used to determine the number of valence electrons on an atom or monatomic ion.

·Li :Be :B· :C· :N· :O: :F: :Ne:

Lewis dot symbols can be used to represent ions as well. Because an ion is just an atom (or a molecule) with a different number of electrons than the uncharged species, we can draw its Lewis dot symbol. For instance, the fluoride ion is a fluorine atom plus an electron (F^-). To distinguish ions from Lewis dot symbols of atoms, we place brackets around the drawing, and we place the charge on the ion outside the brackets. Compare the Lewis dot symbols of the ions below to the atoms we discussed previously. Note in the diagram below that the nitrogen anion has three extra electrons, the oxygen anion has two extra electrons, and the fluorine anion has one extra electron. All three have eight valence electrons and the same electron configuration as the period's noble gas (Ne). This idea is important in **Lewis dot structures**, and we'll revisit it later in this chapter.

Lewis dot structure – A drawing convention used to determine the arrangement of atoms in a molecule or ion.

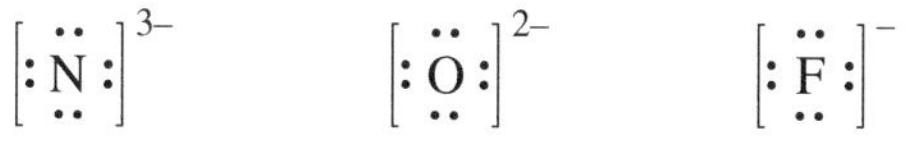

$[:N:]^{3-}$ $[:O:]^{2-}$ $[:F:]^{-}$

How might we use our drawing scheme for anions to draw cations? In a monatomic cation, electrons have been removed from the atom (or group of atoms), resulting in a positive charge. When we write the Lewis dot symbols, we indicate the number of valence electrons and the resulting positive charge. While the sodium and aluminum ions do still have core electrons, the valence electrons of the uncharged element have all been removed to form the cation, and since only the valence electrons are shown in a Lewis dot symbol, these atoms have no dots around them.

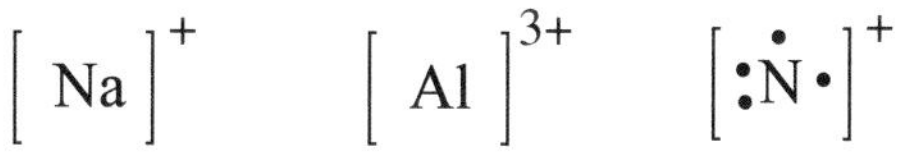

$[Na]^{+}$ $[Al]^{3+}$ $[:N\cdot]^{+}$

8.2.3 Electron Configuration of Ions

We determined the electron configuration of the elements in ▶ Chap. 6 using the Aufbau principle. Electron configurations can also be used to describe which atomic orbitals in ions contain electrons. For example, the electron configuration of the sodium ion contains one electron less than that of the sodium atom. This ion, which is necessary for proper contraction of heart tissue and electrolyte balance inside and outside of the body's cells, has an electron configuration that lacks the valence electron from the sodium atom.

$$\text{Na: } 1s^2 2s^2 2p^6 3s^1 \quad \rightarrow \quad \text{Na}^+\text{: } 1s^2 2s^2 2p^6$$

In anions such as the fluoride ion (F^-, found in toothpaste and added to most U.S. municipal water supplies to help prevent tooth decay), the electron configuration shows the addition of an electron to the valence shell:

$$\text{F: } 1s^2 2s^2 2p^5 \quad \rightarrow \quad \text{F}^-\text{: } 1s^2 2s^2 2p^6$$

We also know from our discussion of ionization energy that valence electrons are the most accessible of the electrons in an atom. The addition of an electron to an atom to make an anion occurs in the valence shell. Similarly, removing valence electrons makes a cation. How do the electron configurations of Na^+ and F^- compare with each other and with the noble gas nearest to them in atomic number on the periodic table? They have the same electron configuration as neon: $1s^2 2s^2 2p^6$. We say that the electron configurations of Na^+ and F^- are **isoelectronic** with Ne.

Isoelectronic – Having the same electron configuration.

The position of the electrons as either valence or core electrons on an atom can help us understand the structure of a molecule such as aspirin. Knowing where the electrons on the atom reside and how those electrons behave is of utmost importance in developing our three-dimensional model of a molecule.

8.2.4 Octet Rule

Why do atoms share electrons? Electrons shuffle around until each atom has its most energetically stable arrangement of electrons. As in F^- and Na^+, the most stable electron configuration of the main-group elements is isoelectronic with a noble gas. We refer to this as the **octet rule** because the stable arrangement for all of the noble gases beyond helium has eight valence electrons. For the elements H and He and the ions Li^+ and Be^{2+}, the rule is also known as the **duet rule** because of the need for only two electrons to fill the valence shell of the first-row elements. In other words, main-group (*s*-block and *p*-block) atoms typically react by changing their number of electrons in such a way as to acquire the more stable electron configuration of a noble gas. A full valence shell around an atom is a good situation for an atom, because it then has the same electron configuration as a noble gas. We'll use this rule a lot as we put together atoms to make ionic compounds and molecules.

Octet rule – The existence of eight electrons in the valence shell of an atom, creating a stable atom or ion.

Duet rule – The exception to the octet rule involving the atoms H and He. A full valence shell for the atoms H and He.

8

Example 8.1—Writing Lewis Dot Symbols

Write the electron configuration, and the Lewis dot symbol, for both P^{3-} and Al^{3+}. With which element is each isoelectronic?

Asking the Right Questions

What is the relationship between the anion (or the cation) and an atom of the neutral element? When atoms of main group elements form their most common ions, what is true about the electrons in their resulting highest energy shells?

Solution

The P^{3-} anion has three electrons more than the neutral phosphorus atom. Therefore, the electron configuration is $1s^22s^22p^63s^23p^6$. This is the same electron configuration as argon, and we therefore say that these are isoelectronic. The Lewis dot symbol shows an octet of electrons for the highest energy shell.

The Al^{3+} cation has three fewer electrons than the neutral atom. Its electron configuration is $1s^22s^22p^6$, which is isoelectronic with neon. The Lewis dot symbol also illustrates an octet of electrons.

Are Our Answers Reasonable?

One of the "Right Questions" that we posed was "what is true about the electrons in their resulting highest energy shells?" We now revisit that question, with the answer that the formation of the octet indicates a more energetically stable ion than the neutral atom. The answer of the resulting octet (lower energy) in each case is therefore reasonable.

What If…?

…we were to predict an electron configuration for the formation of a phosphorus *cation*? What would be that electron configuration? What would be the charge of the cation? With what element would it be isoelectronic?

(*Ans:* $1s^22s^22p^6$; P^{5+}; *isoelectronic with neon.*)

Practice 8.1

Write the electron configuration, write the Lewis dot symbol, and determine which atoms are isoelectronic with S^{2-}, F^-, Mg^{2+}, and Br^-.

8.3 Ionic Bonding

We noted before that sodium chloride is an ionic compound. In ancient times, it was a highly sought-after seasoning used in cooking and pickling. According to the United States Geological Survey (USGS), 283 million metric tons of sodium chloride were harvested worldwide from salt water or mined from deposits in the ground in 2019 (see Fig. 8.2). In the United States, most of the 42 million metric tons of salt that was harvested in 2019 was used to manufacture chlorine gas and sodium hydroxide. About 43% of the total is used to de-ice highways. Only 4% is used in the food industry. Even so, the average American consumes nearly than 1.3 kg of salt each year!

The human body requires only about 500 mg of sodium per day, yet the average American ingests between 2300 and 5000 mg each day. This high level of sodium consumption can contribute to hypertension, sleep apnea, and other disorders.

Sodium chloride (NaCl) is a compound held together by ionic bonds. As we saw in ► Chap. 4, sodium chloride, like all other Group IA salts, dissociates nearly completely into its ions when added to water. It is a strong

Fig. 8.2 Piles of Salt on the Salar de Uyuni, Bolivia, ready for harvest. Photo by Luca Galuzzi (▶ http://www.galuzzi.it)

Table 8.1 Important ionic compounds

Compound	Formula	Selected use
Calcium carbonate	$CaCO_3$	Limestone, chalk
Calcium chloride	$CaCl_2$	Sidewalk salt
Iron(III) oxide	Fe_2O_3	Pigment
Magnesium sulfate	$MgSO_4$	Epsom salts
Sodium bicarbonate	Na_2CO_3	Baking soda
Sodium chloride	NaCl	Table salt
Sodium fluoride	NaF	Toothpaste ingredient

electrolyte and is a good example of a typical ionic compound. Other examples of important ionic compounds are shown in Table 8.1 and Fig. 8.3. Note that some ionic compounds contain ions that have covalent bonds. For example, calcium carbonate contains calcium ions and the carbonate ions. In the carbonate ion itself, the carbon and oxygen atoms are covalently bonded to each other. We will examine these kinds of **polyatomic ions** later in this chapter. What makes sodium chloride a "typical" ionic compound? The answer lies in the nature of its bonding and properties, which we now explore.

Polyatomic ion – An ion that consists of two or more atoms covalently bonded together.

8.3.1 Description of Ionic Bonding

Sodium chloride is commercially mined or processed from brine (salt water, such as from the oceans or the Great Salt Lake in Utah). In the laboratory, we can combine sodium metal and chlorine gas in a violent reaction to make the salt. The reaction, illustrated below, requires the transfer of electrons from sodium atoms to chlorine.

$$\mathrm{Na}(s) + \frac{1}{2}\mathrm{Cl_2}(g) \rightarrow \mathrm{NaCl}(s) \quad \Delta H = -410\,\mathrm{kJ}$$

If we look at this reaction more closely, we note that one of the atoms (the metal) loses electrons to become a cation. In our reaction, the sodium atom loses an electron to become the sodium cation (Na^+).

$$\mathrm{Na}\cdot \rightarrow [\mathrm{Na}]^+ + \mathrm{e}^-$$

The other atom (the nonmetal) gains electrons to become an anion. Addition of an electron to a chlorine atom forms the chloride anion (Cl^-).

Calcium carbonate

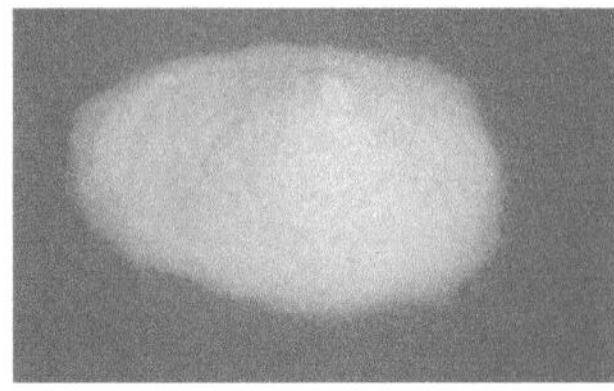

Sodium bicarbonate

Magnesium sulfate

Fig. 8.3 Some common salts

The charges on the sodium and chloride ions attract each other, causing the ions to associate with each other in an **ion pair** (Na^+Cl^-).

$$:\ddot{\underset{\cdot\cdot}{Cl}}\cdot \;+\; e^- \longrightarrow [:\ddot{\underset{\cdot\cdot}{Cl}}:]^-$$

Ion pair – Ions of opposite charges that exist in solution as an ionically bonded pair.

It doesn't stop there. Other sodium cations are also attracted to the negative charge on the chloride. Other chloride anions are attracted to the positively charged sodium. There is no directionality to these ionic attractions—they occur in all directions. The end result is a collection of alternating sodium and chloride ions arranged in a solid **crystalline lattice** (a highly ordered, three-dimensional arrangement of atoms, ions, or molecules), which is explained in detail in ▶ Chap. 11. Within a sodium chloride crystal, each of the sodium cations has six neighboring chloride anions, and each of the chloride anions has six neighboring sodium cations. The forces of attraction combine to provide a neatly packed crystal of alternating sodium and chloride ions, as shown in Fig. 8.4. So, even though we can describe sodium chloride as NaCl, because there is one sodium ion for each chloride ion, the ionic solid actually exists as a complex crystalline lattice, not as an individual ion pair.

Crystalline lattice – A highly ordered, three-dimensional arrangement of atoms, ions, or molecules into a solid.

8.3.2 Examples of Ionic Bonding

We can illustrate the formation of an ionic compound such as sodium chloride (NaCl) through the use of Lewis dot symbols. As reflected in their positions on the periodic table, sodium is a metal and chlorine is a nonmetal. Sodium has a relatively low ionization energy and loses an electron to achieve the electron configuration of a noble gas.

$$Na\cdot \rightarrow [Na]^+ + e^-$$
$$1s^2 2s^2 2p^6 3s^1 \rightarrow 1s^2 2s^2 2p^6$$

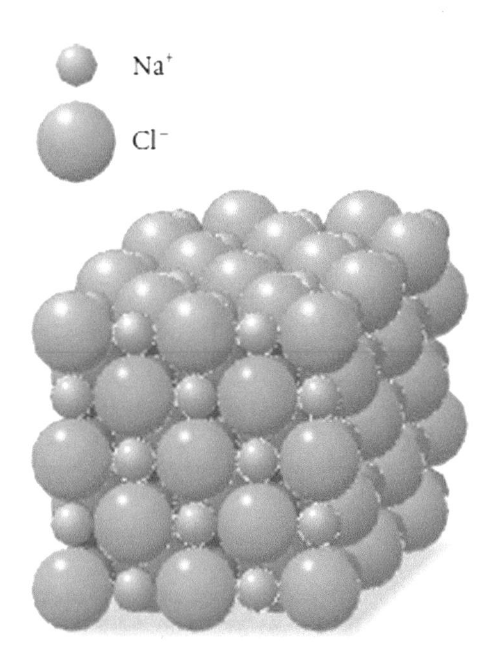

Fig. 8.4 Crystal structure of sodium chloride. In the crystal structure of NaCl, we note that each sodium ion is surrounded by six chloride ions and that each chloride ion is surrounded by six sodium ions. The resulting crystal possesses a regular close-packed pattern of ions

Chlorine, which has a very favorable electron affinity, gains an electron to achieve a noble gas electron configuration.

$$:\ddot{\underset{\cdot\cdot}{Cl}}\cdot \quad + \quad e^- \longrightarrow [:\ddot{\underset{\cdot\cdot}{Cl}}:]^-$$

$$1s^22s^22p^63s^23p^5 \rightarrow 1s^22s^22p^63s^23p^6$$

In the formation of NaCl from Na metal and Cl_2 gas, the electron from the sodium atom is added to the valence shell of the chlorine atom. We can represent this process in equation form by showing the movement of a single electron from the sodium atom to the chlorine atom with a curved half-headed arrow. The result is a sodium cation and a chloride anion that are attracted to each other. The ionic crystal that results has the formula NaCl.

$$Na\cdot \curvearrowright \; :\ddot{\underset{\cdot\cdot}{Cl}}\cdot \longrightarrow [\,Na\,]^+ + [:\ddot{\underset{\cdot\cdot}{Cl}}:]^- \longrightarrow [\,Na\,]^+[:\ddot{\underset{\cdot\cdot}{Cl}}:]^-$$

Does this process make the system more energetically stable? Energy must be added to the system to remove the electron from the sodium atom. Energy, however, is released when that electron is added to the chlorine atom, and even more energy is released when the ionic bond forms. The net result from the formation of a crystalline lattice of sodium and chloride ions is the release of 410 kJ for each mole of sodium chloride that is formed, an exothermic reaction, as shown in the balanced chemical reaction at the start of ▶ Sect. 8.3.1.

Example 8.2—Ionic Compound Formation

Calcium chloride, $CaCl_2$, is sometimes spread on our sidewalks to melt ice. Because of its ability to absorb moisture, it is used to dry gases and prevent dust buildup in mining and highway maintenance. Using Lewis dot symbols, show the transfer of electrons to form $CaCl_2$ from calcium metal and chlorine atoms.

Asking the Right Questions

What are the Lewis dot symbols for the neutral atoms of each element? Which element will form a cation and which will form the anion? What will be their charges after the electrons are transferred?

Solution

We can start by drawing the Lewis dot symbols for the calcium and chlorine atoms. Then we can show the transfer of an electron from the less electronegative calcium to the more electronegative chlorine atom (arrow).

We have a similar transfer with the other calcium electron to a different chlorine atom. We end with three ions (Ca^{2+} and two Cl^-) that have noble gas electron configurations. Note that the ions end up with either a completely full valence electron configuration (and become isoelectronic with a noble gas) or a completely empty valence shell (and become isoelectronic with a noble gas). This is the octet rule in action.

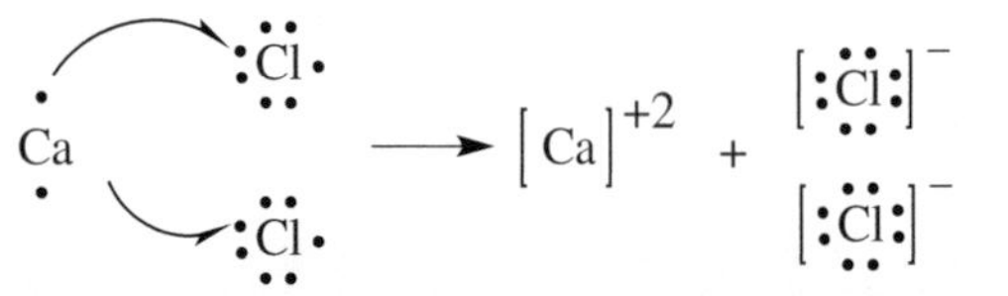

Are Our Answers Reasonable?

Does this match our expectation? Calcium is in Group IIA, with members that form cations with a +2 charge. Chlorine is in Group VIIA, and most often forms anions with a −1 charge, so this is reasonable. In addition, our final drawings show an electronically neutral compound (reasonable), and atoms that are isoelectronic with a noble gas. These are all reasonable outcomes.

What If…?

…we were working with sesquioxides, compounds that have three atoms of oxygen and two of another element? If we had the sesquioxide of aluminum, what would be the charge on aluminum?

(*Ans: Each aluminum ion must have a charge of +3*)

Practice 8.2

Use the Lewis dot structure model to show the formation of sodium oxide from atomic sodium and oxygen atoms.

Is this principle useful beyond sodium chloride? Medicinal chemists and biochemists make use of the large energy release in the formation of ionic bonds. Ethidium bromide, shown in ◘ Fig. 8.5, is a toxic compound that researchers use to stain DNA to make it easier to see under a microscope. This compound exhibits its effects by inserting itself into DNA strands, causing structural changes to the DNA molecule. The negative charges on the DNA and the positive charge of the ethidium cation help hold the compound in place. These forces of attraction explain why this compound has such a strong association with DNA. Another example of the importance of ionic bonding accounts for the source of calcium in milk. Casein, one of the major proteins found in milk, contains phosphate anions that form ionic bonds with calcium cations.

8.3.3 Ions: Does Size Matter?

Water from wells typically contains relatively high concentrations of calcium, often along with some magnesium and iron. When the water is heated, dissolved calcium bicarbonate, $Ca(HCO_3)_2$, decomposes to form rock-like deposits of calcium carbonate, $CaCO_3$, also known as **boiler scale**. The boiler scale coats the inside of the pipes that carry the water in the hot-water heater and throughout the house.

Boiler scale – The deposit of calcium carbonate (or calcium sulfate) on the inside of water pipes.

$$Ca(HCO_3)_2(aq) \rightarrow CaCO_3(s) + H_2O(l) + CO_2(g)$$

This kind of water is called "hard water." Hard water is not hazardous to our health. It is just not the best thing that could happen to our household pipes. If the buildup is bad enough, the pipes can become clogged. Many homeowners, and some cities, "soften" water by passing it through a bed of **zeolites**, which exchange the calcium ions for sodium ions that do not form insoluble salts. How do these ion-exchanging zeolites work?

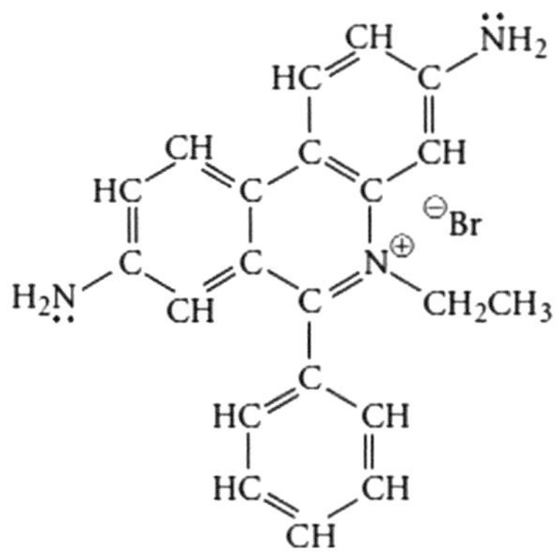

◘ **Fig. 8.5** Ethidium bromide. The stabilizing force that holds the molecule in place in DNA is the attraction between the anionic charges on the DNA and the positive charge on ethidium bromide

Zeolite – A porous solid with a well-defined structure, typically made of aluminum and silicon. Zeolites are capable of binding to ions of a specific radius based on the relative size of its pores.

► https://upload.wikimedia.org/wikipedia/commons/0/06/Limescale-in-pipe.jpg

Zeolites, like the ones shown in ◘ Fig. 8.6, are composed of aluminum and silicon oxide subunits containing Group IA and IIA cations. The resulting honeycomb arrangement of subunits results in a maze-like structure full of atom-sized holes. Ions and molecules small enough to fit through these holes can enter the zeolite, and if they are just about the same size as the holes, they get stuck inside. Energy-stabilizing interactions of the ions or molecules with the zeolite help to hold them inside. Substances that are small enough flow easily in and out, whereas ions and molecules that are too big can't enter the zeolite in the first place. Consequently, a zeolite retains only specific sizes of ions and molecules. To maintain a neutral charge, the zeolite releases ions of the same total charge as the trapped ions. For example, when a calcium (+2) ion enters a zeolite stocked with sodium ions, it displaces two sodium (+1) ions.

By carefully constructing the zeolite, researchers have been able to develop an "ionic sponge" that grabs only the ions of interest to them. In the chemical industry, this is vital for the formation of better reaction catalysts (compounds that significantly increase the rate of a chemical reaction without being consumed), the filtration of polluted air, odor control in agricultural and household settings, and the cleanup of hazardous wastes, among other applications. Experiments with the zeolite known as clinoptilolite, with a pore size of 0.4 nm, indicated that it was capable of removing radioactive cesium (^{134}Cs and ^{137}Cs) from cows affected by the 1986 Chernobyl (Ukraine) nuclear accident. This is possible because the pore size in the zeolite is large enough for the cesium to enter, but it then becomes "lost" in the internal maze of the zeolite and doesn't come back out.

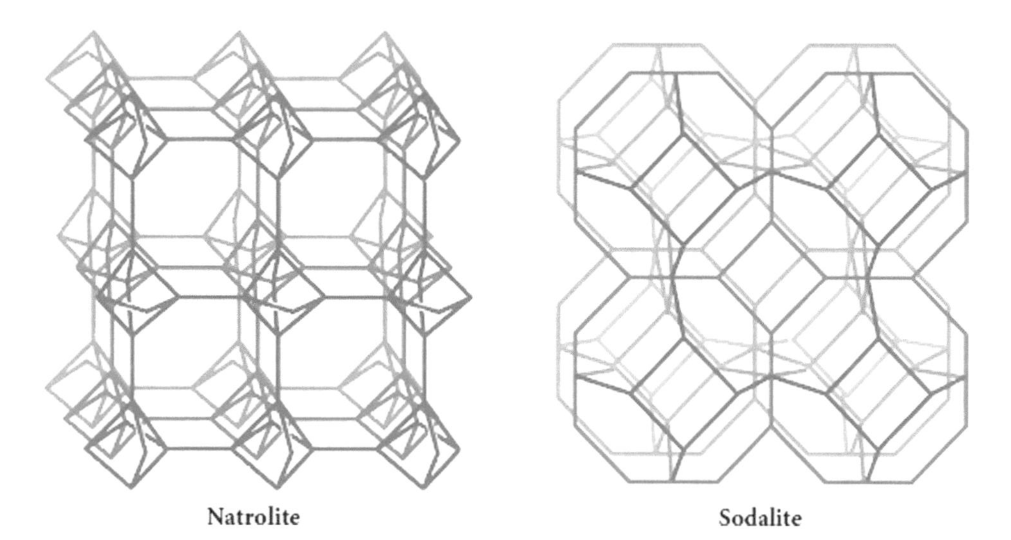

◘ **Fig. 8.6** Ions can enter, and they can sometimes leave: The structures of two natural zeolites, natrolite and sodalite, are shown here with blue lines that link the atoms of the zeolite together. The result is a network of pores that ions of the right size can enter

By far the most common use of zeolites is as an ingredient called a "builder" in laundry detergents for the removal of calcium ions from hard water. Size is a critical factor in the behavior of ionic compounds. Not only does size suggest what type of zeolite can trap a particular ion, it also plays a major role in determining the structure of the ionic crystal and the strength of the ionic bonds in that crystal.

What contributes to the size of an ion? Just as with atomic size (and the other periodic trends), it's all about the electrons. In our previous discussion of atomic size (in ▶ Chap. 7), we learned that each electron helps to balance the charge of the protons in the nucleus and shields the other electrons from the nuclear charge. When an electron is removed from an atom, the amount of shielding is reduced. There are fewer electrons left around the nucleus, so the attraction to the nucleus experienced by each remaining electron increases. Another way to say this is that the "effective nuclear charge" for each electron has increased. The total number of electron – electron repulsions decreases, so the electrons are pulled closer to the nucleus. The net effect is a reduction in the size of the electron cloud around the cation. The size of the cation is always less than the size of the atom (see ◘ Table 8.2).

The opposite happens when an electron is added to an atom. The extra electron increases the number of electron–electron repulsions and, because there are now more electrons than protons, reduces the effective nuclear charge felt by each electron. This causes the electron cloud to swell in size. The anion, then, is always larger than the atom from which it was made. These effects continue as more electrons are removed from or added to an atom. All of the ions shown in ◘ Fig. 8.7 have the same electron configuration ($1s^22s^22p^6$), but they differ in the charge of the nucleus. We can see that as the nuclear charge increases, the size of the ion decreases. In general, the cation gets smaller as more electrons are removed. Removal of the first electron from an atom greatly decreases the radius of the resulting ion. Removal of the second and all subsequent electrons from the ion has less of an effect on the radius.

As you descend within a group on the periodic table, the principal quantum number of the valence electrons increases, the distance of the electrons from the nucleus increases and the size of the ion increases. This trend is shown in ◘ Table 8.2. Notice that these trends are similar to the trends of atomic size we discussed in ▶ Chap. 7.

◘ **Table 8.2** The radius of some selected atoms and ions

Period	Group 1 (IA)	2 (IIA)	13 (IIIA)	15 (VA)	16 (VIA)	17 (VIIA)
2	Li 152	Be 113	B 88	N 70	O 66	F 64
2	Li^+ 60	Be^{2+} 31	B^{3+} 20	N^{3-} 171	O^{2-} 140	F^- 133
3	Na 186	Mg 160	Al 143	P 110	S 104	Cl 99
3	Na^+ 99	Mg^{2+} 65	Al^{3+} 50	P^{3-} 212	S^{2-} 184	Cl^- 181
4	K 227	Ca 197			Se 117	Br 114
4	K^+ 133	Ca^{2+} 99			Se^{2-} 198	Br^- 195
5	Rb 247	Sr 215			Te 143	I 133
5	Rb^+ 148	Sr^{2+} 113			Te^{2-} 221	I^- 216
6	Cs 265	Ba 217				
6	Cs^+ 169	Ba^{2+} 135				

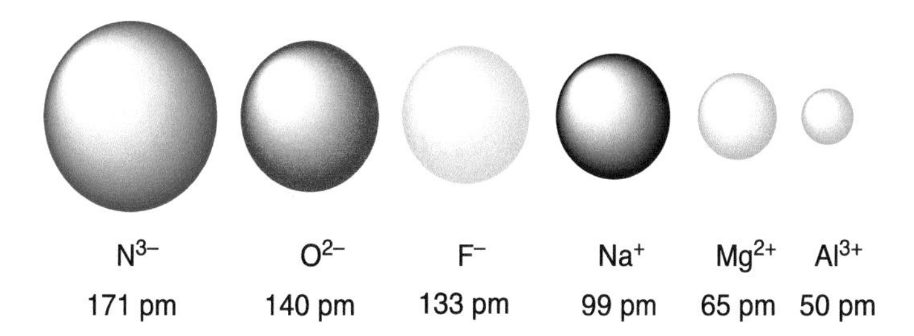

Fig. 8.7 The radii of a series of isoelectronic ions

▶ Here's What We Know So Far

- The electron configuration of an ion determines the number of electrons in its outer shell.
- We can use the electron configuration of an ion to draw the Lewis dot symbol for the ion.
- Anions, atoms with extra electrons, are larger than their corresponding atoms.
- Cations, atoms missing electrons, are smaller than their corresponding atoms.
- Ionic compounds are the combination of anions and cations to make an electrically neutral compound. The force of attraction between an anion and a cation is the ionic bond. ◀

8.3.4 Energy of the Ionic Bond

According to the American Dental Association, the addition of fluoride to toothpaste and public water systems has brought about a nationwide reduction in tooth decay. The U.S. Centers for Disease Control and Prevention suggest a "safe, effective and inexpensive" municipal waterway fluoride concentration of 0.7 parts per million. The mineral portion of teeth is hydroxyapatite, $Ca_5(PO_4)_3(OH)$. When you drink fluoridated water or brush your teeth with toothpaste containing fluoride, some of the hydroxide anions in your teeth are replaced with fluoride anions.

$$\underset{\text{Hydroxyapatite}}{Ca_5(PO_4)_3(OH)} \rightarrow \underset{\text{Fluorapatite}}{Ca_5(PO_4)_3(OH, F)}$$

The new mineral, called fluorapatite, is held together much more strongly than hydroxyapatite. What accounts for the added strength? Do the sizes of the fluoride and hydroxide ions have something to do with the strength of the mineral?

Because ionic substances exist as large crystalline lattices, the strength of the ionic bond is usually referred to as the **lattice enthalpy** of the ionic solid. The lattice enthalpy of a molecule is the amount of energy required to separate 1 mol of a solid ionic crystalline compound into its gaseous ions (see the following equation). The lattice enthalpies of some common compounds can be found in Table 8.3.

Table 8.3 Lattice enthalpies for some common ionic solids

Values are in kilojoules per mole

	F^-	Cl^-	Br^-	I^-	OH^-	O^{2-}
Li^+	1030	834	788	757	1039	2799
Na^+	923	787	747	704	887	2481
K^+	821	701	682	649	789	2238
Rb^+	785	689	660	630	766	2163
Cs^+	740	659	631	604	721	–
Mg^{2+}	2913	2326	2097	1944	2870	3795
Ca^{2+}	2609	2223	2132	1905	2506	3414
Ba^{2+}	2341	2033	1950	1831	2141	3029
Sc^{3+}	5096	4874	4711	4640	5063	13,557
Al^{3+}	5924	5376	5247	5070	5627	15,916

Lattice enthalpy – The amount of energy required to separate 1 mol of a solid ionic crystalline compound into its gaseous ions.

$$MX(s) \rightarrow M(g)^{+} + X(g)^{-}$$

Although qualitative statements can be made about the relative size of the lattice enthalpy on the basis of ionic radii, most chemists use lattice enthalpy to make some quantitative statements about the strength of an ionic bond. Unfortunately, it is quite difficult to measure lattice enthalpy accurately in an ionic crystalline solid. However, we can calculate the lattice enthalpy using Hess's law in a process known as the **Born–Haber cycle**, named after two Nobel Prize–winning German scientists (Max Born, 1882–1970, and Fritz Haber, 1868–1934).

Born–Haber cycle – A diagrammatic representation of the formation of an ionic crystalline solid using Hess's law. The cycle reveals the lattice enthalpy, which is difficult to obtain by direct measurement.

The Born–Haber cycle is a diagrammatic representation of the formation of an ionic crystalline solid. ◘ Figure 8.8 illustrates the Born–Haber cycle for the formation of sodium chloride from its elements in their standard states. Although the steps outlined in the cycle aren't necessarily the steps taken by the compounds in the reaction, the Born–Haber cycle accurately provides the enthalpy of the reaction, because enthalpy is a state function.

- At the starting point to the cycle, we sublime solid sodium metal into gaseous sodium metal. The enthalpy change ($\Delta_{sub}H$) for this process is known and is recorded on the cycle.
- Next, we dissociate the chlorine molecule into individual chlorine atoms, using the equation that relates to the bond dissociation energy for chlorine ($\Delta_{dis}H$). Because we need only one chlorine atom, only half of the enthalpy change for this process is needed.
- In the next steps, the gaseous atoms are ionized to their gaseous ions. Sodium atoms are ionized to sodium cations with an enthalpy change that corresponds to the first ionization enthalpy (*IE*). Chlorine atoms are converted into chloride anions with an enthalpy change that corresponds to the electron affinity (*EA*) for chlorine.
- In the final step, the sodium and chloride ions are brought together to make the ionic crystalline solid. This enthalpy change corresponds to the negative of the lattice enthalpy for the ionic solid ($\Delta_{lattice}H$).

When the heat of formation for the ionic solid ($\Delta_f H°$) is known by direct measurement, the direct route from the starting point to the end of the Born–Haber cycle (the lattice enthalpy, $\Delta_{lattice}H$) can be calculated by summing all of the individual enthalpy changes. Because the individual enthalpy changes are often known (heat of formation, sublimation, first ionization, bond dissociation, and electron affinity), this provides a convenient method for determining the lattice enthalpy of the ionic crystalline solid.

Lattice energy is a measurement of the force of attraction between the ions in the crystalline lattice. Mathematically, the lattice energy ($E_{lattice}$) is the amount of energy released as the solid ionic crystal is formed from its

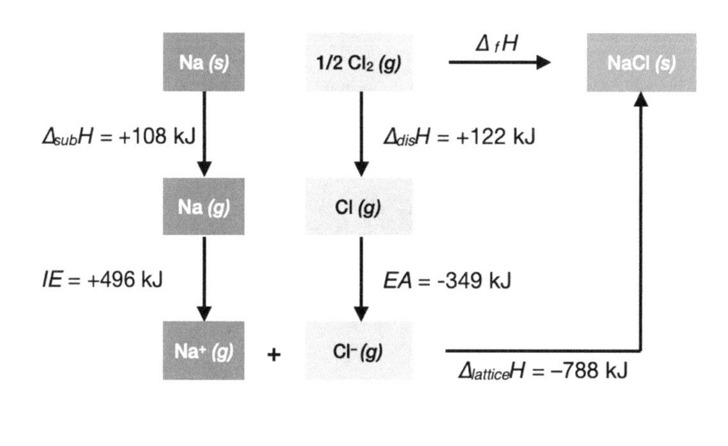

$$\Delta_f H = \Delta_{sub}H + IE + \Delta_{diss}H + EA + \Delta_{lattice}H$$

$$\Delta_f H = 108\ kJ + 496\ kJ + 122\ kJ - 349\ kJ - 788\ kJ$$
$$= -411\ kJ$$

◘ **Fig. 8.8** The Born–Haber cycle for the formation of sodium chloride gaseous ions, whereas the lattice enthalpy ($\Delta_{lattice}H$) is the amount of energy transferred as heat *at constant pressure* necessary to separate the solid ionic crystal into its gaseous ions. The good news is that these two values are just about the same, though opposite in sign; lattice energy describes a release in energy when the ionic bonds are formed ($E_{lattice}$ = "−"), whereas lattice enthalpy describes energy that is absorbed when the ionic bonds are broken ($\Delta_{lattice}H$ = "+").

Lattice energy – The amount of energy released as a solid ionic crystal is formed.

$$\Delta_{\text{lattice}}H \cong -E_{\text{lattice}}$$

Calculation of the lattice energy (not lattice enthalpy) of an ionic compound can also be accomplished using **Coulomb's law**. Coulomb's law states that the force between two particles is proportional to the product of the charges (Q) on each particle divided by the square of their distance of separation (d). The French physicist Charles Augustin Coulomb (1736–1806) proved this relationship by experiment. Because the force of attraction between two ions is related to the lattice energy, a modification of Coulomb's potential gives us the lattice energy:

$$\text{Lattice energy} = k\left(\frac{Q^+ \times Q^-}{d}\right)$$

where,
k is a proportionality constant,
Q^+ and Q^- are the charges on the ions in Coulombs, and,
d is the distance between the ions.

Coulomb's law – The force between two particles is proportional to the product of the charges (Q) on each particle divided by the square of their distance of separation (d).
Coulomb – The SI unit of charge equivalent to the SI units of 1 ampere · second. 1 Coulomb is the same as the charge on 6.24 x 10^{18} elementary protons or electrons.

The magnitude of the lattice energies or lattice enthalpies increases as the charges on the ions increase, as shown in Table 8.3. In the series $ScCl_3$, $CaCl_2$, and KCl, the charges decrease on the cation: (+3), (+2), and (+1), respectively, with a decrease in lattice enthalpies in the order $ScCl_3$ (4874 kJ/mol), $CaCl_2$ (2223 kJ/mol), KCl (701 kJ/mol). As the distance between the ions decreases, the attractions should increase and the lattice enthalpies should also increase. Evidence of this is found by comparing LiBr (788 kJ/mol), LiCl (834 kJ/mol), and LiF (1030 kJ/mol). The charges on the ions are the same in this series of compounds, but the size of the anion decreases from bromide to fluoride. Because the bromide ion (Br^-) is larger than Cl^- or F^-, LiBr has the smallest lattice enthalpy. In general, lattice enthalpies are greatest for ionic compounds that are made up of small, highly charged ions.

Let's revisit our section-opening question of why fluoride in our toothpaste is important. As we noted previously, the replacement of some hydroxy groups in the hydroxyapatite, $Ca_5(PO_4)_3(OH)$, mineral that makes up our teeth gives rise to a new mineral called fluorapatite, $Ca_5(PO_4)_3(OH,F)$. Because F^- is smaller than OH^-, the force of attraction between the Ca^{2+} and the F^- should be greater than that between Ca^{2+} and OH^- in this mineral. This is the case with nearly all binary ionic salts of fluoride compared to the hydroxide, so that the lattice enthalpy of, for example, silver fluoride (AgF), which is 953 kJ/mol, is greater than the lattice enthalpy of silver hydroxide (AgOH), 918 kJ/mol. It is therefore reasonable to consider that the lattice enthalpy of fluorapatite is greater than that of hydroxyapatite. This is one of several reasons why fluorapatite is more stable than hydroxyapatite when bathed in our acidic saliva—and more resistant to the formation of cavities. Fluorapatite formation is even more pronounced when fluoride is present in the municipal water that we drink or from dental treatments; this is related to a concept called chemical equilibrium, which we will consider in a later chapter.

Example 8.3 Predicting Lattice Enthalpies

Use the relationship of ionic sizes to predict whether calcium fluoride (found in toothpaste) or calcium chloride ("ice-melt" salt) has the greater lattice enthalpy. Also predict whether aluminum chloride or sodium chloride has the greater lattice enthalpy.

Asking the Right Questions

What is the relationship between lattice enthalpy and distance between the ions (directly related to the ionic radii) in the compound? What about lattice enthalpy and charge? In which of these do our compounds differ?

Solution

The distance between the ions is directly related to the individual ionic radii. Calcium fluoride (CaF_2) and calcium chloride ($CaCl_2$) differ only in the size of the anion bound to the calcium cation, not in the charges of the ions. From the discussion of atomic size, we noted that fluorine is smaller than chlorine. Moreover, the radius for both anions also follows this trend; fluoride is a smaller anion than chloride. Calcium fluoride, with its smaller distance between ions, should have a larger, more positive lattice enthalpy. And indeed, it does (2609 kJ/mol vs. 2223 kJ/mol).

Aluminum chloride ($AlCl_3$) and sodium chloride (NaCl) differ in the charge of the cation. The larger charge on the aluminum cation (+3) indicates that aluminum chloride should have the larger lattice enthalpy. It does (5376 kJ/mol vs. 787 kJ/mol).

Are Our Answers Reasonable?

The magnitude of the lattice enthalpy is inversely proportional to the distance between the individual ions, and directly proportional to the size of the nuclear charge on the ions. Because our answers are aligned with this fact, they make sense.

What If…?

…we had four compounds and four lattice energies? Match the compound with its lattice enthalpy (in kJ/mol).

CoI_2	$CoCl_2$	CoO	CoO
3910	2545	2691	2691

(*Ans:*

CoI_2	*$CoCl_2$*	*CoO*	*CoO*
2545	2691	2691	2691

Practice 8.3

Which has the greater lattice enthalpy, $FeCl_3$ or $FeCl_2$?

8.4 Covalent Bonding

In addition to NaCl (table salt), sucrose ($C_{12}H_{22}O_{11}$, common sugar) is another compound we may see at the dinner table (◘ Fig. 8.9). What are some physical and chemical differences between table salt and sugar? Placed side by side, they appear relatively similar to the naked eye. Both are colorless crystalline compounds. When added to water, they both dissolve. However, as we saw in ▶ Chap. 4, solutions of these two compounds have different properties. Table salt, a strong electrolyte, dissociates into sodium cations and chloride anions when dissolved in water. Sucrose, a nonelectrolyte, doesn't dissociate when it dissolves. This is why electric current passes through a salt solution and lights a bulb, while the bulb over a sugar solution remains unlit. Sucrose is an example of a compound containing only covalent bonds. The atoms that make up each sugar molecule are firmly held together; bonds between these atoms don't dissociate upon addition to water. Because of this, an aqueous solution of sucrose doesn't conduct electricity. These bonds differ from the bonds in sodium chloride, an ionic compound that dissociates in water, because the electrons in sucrose are shared.

◘ **Fig. 8.9** Common table sugar. Sugar can be crystallized in large chunks that look much like the crystals of table salt

Why is it important to know whether a bond is ionic or covalent? In short, the bonding in a substance helps us to explain its behavior. For example, **pharmacognocists** understand that whereas many ionic compounds are readily soluble in water and can be administered to patients in aqueous solutions, many covalent compounds have limited solubility in water. Drugs that are insoluble in water must be administered in other solvents, which explains why cough syrups often contain ethanol. In this section, we will examine the covalent bond up close, draw pictures of molecules that utilize this bonding scheme, and calculate the forces that hold these bonds in place.

Pharmacognocist – a person specializing in the study of naturally occurring compounds and their medicinal properties.

▶ https://upload.wikimedia.org/wikipedia/commons/thumb/2/21/Coughsyrup-promethcode.jpg/640px-Coughsyrup-promethcode.jpg

8.4.1 Description of Covalent Bonding

Within the ionic bond, one of the atoms gains electrons into its valence shell and becomes an anion while the other loses valence electrons to become a cation. In the covalent bond, the atoms share at least one pair of valence electrons to differing degrees. This sharing occurs because the nuclei of the atoms participating in the covalent bond have a similar strength of attraction for the electrons. In order to participate in a bond and at the same time become isoelectronic with a noble gas, the atoms in the covalent bond must share their electrons. Examples of compounds that contain covalent bonds include CO_2, CH_4, H_2O and H_2 (◻ Fig. 8.10).

Because of similarities in their electronegativity, the majority of covalent bonds exist between nonmetals. There are exceptions, however, such as the covalent bonds between lead and each of the four of the carbon atoms in tetraethyl lead, $Pb(C_2H_5)_4$, a compound that used to be added to gasoline to improve engine performance. Why is this an exception? The four lead–carbon covalent bonds in the compound exist as a consequence of the relatively small difference in the electronegativity between the lead and carbon atoms. Because the emissions-reducing catalytic converters required on all cars and light trucks in the United States since 1981 are destroyed by lead, and because of the harmful effects of lead on living organisms and the environment, this type of automobile fuel was banned in the United States in 1986. All of the automotive gasoline sold today in most of the world is unleaded gasoline. And as a result, the amount of lead pollution of the environment has dramatically decreased. Recent United Nations data show that this goal has largely been achieved.

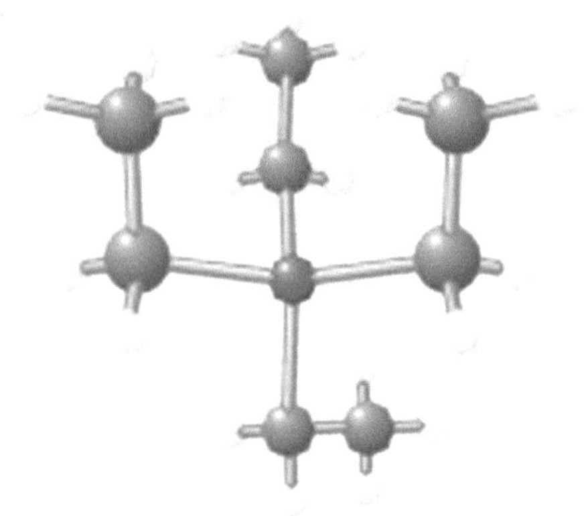

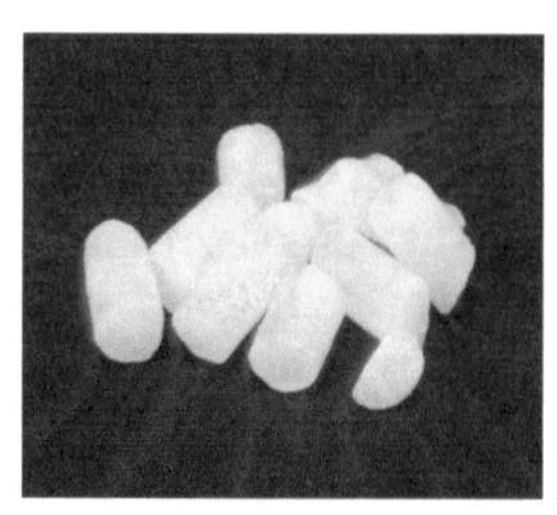

Dry ice pellets (CO_2) photo by Richard Wheeler (CC BY-SA-3.0)

Methane (CH_4) burner.

Hydrogen (H_2) used as a fuel for a car.

Glass of water (H_2O).

Fig. 8.10 How we might find CH_4, CO_2, H_2O, and H_2 in our everyday life. *Source* CO_2. ▶ https://upload.wikimedia.org/wikipedia/commons/3/36/Dry_Ice_Pellets_Subliming.jpg. ▶ https://upload.wikimedia.org/wikipedia/commons/1/11/Gas_flame.jpg. ▶ https://upload.wikimedia.org/wikipedia/commons/c/cb/Hydrogen_fueling_nozzle2.jpg

The simplest example of a compound containing a covalent bond is hydrogen gas (H_2), used in the food industry to make partially saturated oils, which contain mostly carbon–carbon and carbon–hydrogen single bonds, as well as in the manufacture of ammonia intended for agricultural use as a fertilizer. The bond between the hydrogen atoms in H_2 results from the sharing of one electron from each of the atoms. The atoms on either end of the bond have the same affinity for electrons, so the electrons in the bond are shared equally to make each hydrogen isoelectronic with a noble gas (He in this case). In other words, sharing the electrons between the two atoms effectively gives each hydrogen a $1s^2$ electron configuration. The sharing of these electrons is the defining characteristic of a covalent bond. What holds the two atoms together in a covalent bond? Each nucleus on either end of the bond exerts a force of attraction on the pair of bonding electrons. This attractive force pulls the nuclei close to form a bond. If we examine the electron cloud around the two nuclei in molecular hydrogen, we note an interesting feature of the covalent bond, which is shown in ◘ Fig. 8.11. The density of electrons is concentrated between the two atoms, but a significant amount of electron density surrounds each of the two nuclei. As shown in ◘ Fig. 8.12, the picture of electron density in HF, a compound used to etch glass, is remarkably different because the bonding electrons are not shared quite equally, although they are shared to an important extent, and the bond is still considered to be covalent.

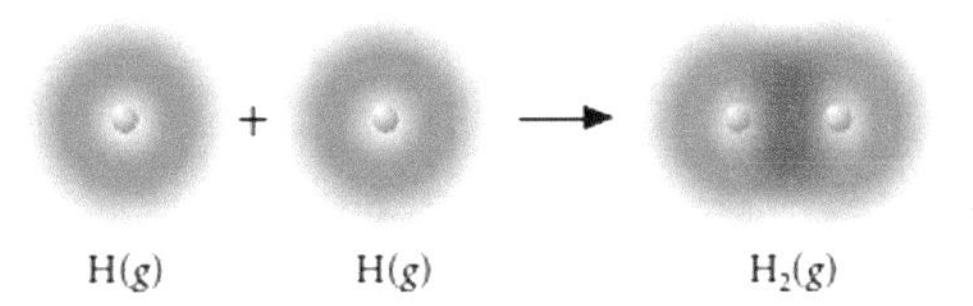

Fig. 8.11 Electrons hold the nuclei together in this model of hydrogen gas. Note that the electron density, shown in red, encircles both nuclei

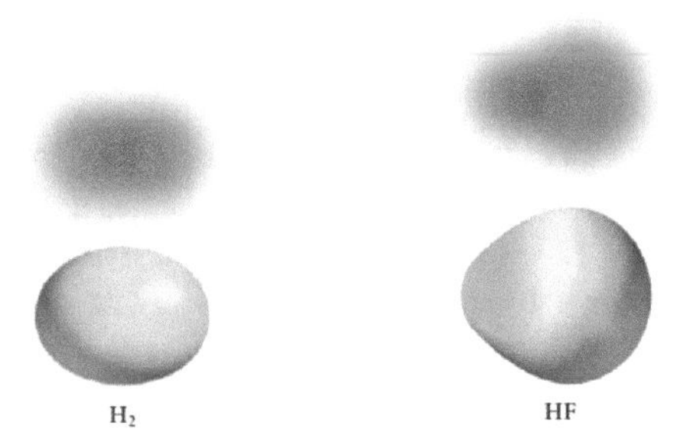

Fig. 8.12 Compare the electron density distribution in HF with that of H_2

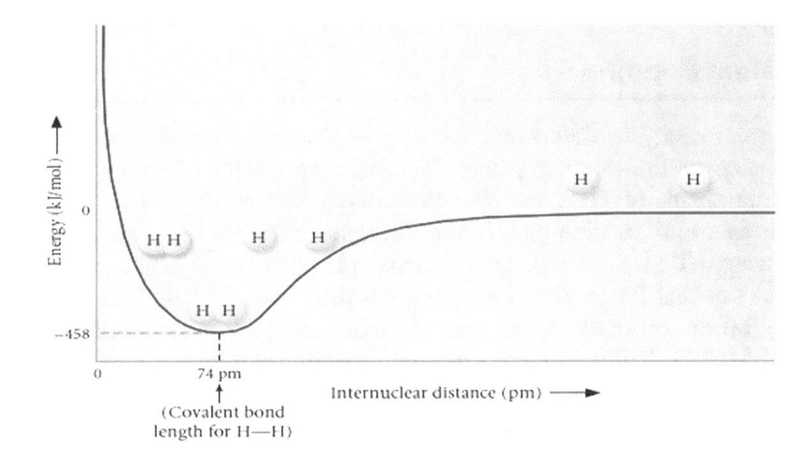

Fig. 8.13 The energy profile of a covalent bond in H_2 as a function of distance between the nuclei. Note how the attractive forces and repulsive forces balance each other at a distance of 74 pm. At this distance the interaction has the lowest energy. To break the covalent bond in H_2 and separate the atoms would require the addition of 458 kJ/mol

What keeps the nuclei from bumping into each other? While the attractive force of the nuclei for the shared electrons pulls the atoms together, a repulsive force of the positively charged nuclei pushes them apart. This repulsion keeps the nuclei from getting too close. It is the combination of attractive and repulsive forces that holds the nuclei apart at a particular average distance that we refer to as the **bond length**. A plot of the potential energy of two hydrogen atoms as a function of their distance, shown in Fig. 8.13, helps to illustrate this fact.

Bond length – The average distance between the nuclei of bonded atoms.

8.4.2 Electronegativity and the Covalent Bond

Why are electrons shared unequally in some covalent bonds such as H—F? Some atoms have a greater attraction for shared electrons than do others. The electron density in the covalent bond is pulled closer to the atom with the greater attraction for the bonding electrons. We say that some covalent bonds are **polarized** toward one of the atoms. In hydrogen fluoride, the attraction of the fluorine atom for electrons causes the bonding electrons to spend more time at the fluorine end of the molecule, resulting in a net **polarization** of the bond. The lack of electron density at the hydrogen end of the bond means that the entire charge of the nucleus isn't balanced by an equal amount of electrons on the hydrogen. The excess electron density on the fluorine end of the bond means that there is more electron charge than nuclear charge on the fluorine. So each of the atoms in the **polar covalent bond** possesses a partial charge. We usually indicate the charge separation with a lowercase Greek letter delta (δ) and a plus or minus sign:

Polarized – A molecule or bond that contains an unequal distribution of electrons.

Polarization – A molecule or bond containing electrons that are unequally distributed.

Polar covalent bond – A covalent bond in which the atoms differ in electronegativity, resulting in an unequal sharing of electrons between two adjacent atoms.

$$\overset{\delta+}{\text{H}}\text{—}\overset{\delta-}{\ddot{\underset{..}{\text{F}}}}\text{:}$$

Methods have been developed to attempt to identify the type of bond in a molecule on the basis of the electronegativity of the bonded atoms. **Electronegativity** was defined earlier as the ability of an atom in a molecule to attract shared electrons to itself (◘ Fig. 8.14). The atom with the stronger attraction for the electrons (higher electronegativity) pulls the bonding electrons closer to itself. By examining the difference in the electronegativity values of the two atoms involved in a bond, we can assess the polarization of the bonding electrons (also known as the bonding electron pair). For example, fluorine is much more electronegative than hydrogen, so the electrons spend more time on the fluorine end of the bond in HF.

Electronegativity – The ability of an atom in a molecule to attract shared electrons to itself.

8.4.3 Types of Covalent Bonding

How can we use electronegativity to determine the degree of polar character in a bond? In molecular hydrogen (H_2), the difference in the Pauling values for electronegativity (Δ) between the two atoms is zero ($\Delta = 2.1 - 2.1 = 0$). In a molecule of HCl, the electronegativity difference is less than 1 electronegativity unit ($\Delta = 0.9$). Between hydrogen and fluorine in HF, the difference is fairly large ($\Delta = 1.9$). In table salt (NaCl), the difference in electronegativity is 2.1. Molecular hydrogen is an example of a molecule with a covalent bond that is not polarized. We say that it contains a **nonpolar covalent bond** ($\Delta < 0.5$). Hydrogen chloride (HCl) and hydrogen fluoride (HF), on the other hand, contain fairly polarized covalent bonds. We say that these are **polar covalent bonds** ($0.5 < \Delta < 2.0$). If the difference in electronegativity is very large (> 2.0), the more electronegative atom's ability to attract electrons in the bond overcomes the tendency to share electrons, and we call this an ionic bond. The electronegativity difference between sodium and chlorine in table salt ($\Delta = 2.1$) is characteristic of an ionic bond. The difference in electronegativity can be useful in determining the type of bond in a compound, but the values shown here are useful guidelines only for the determination of the bonding pattern. Electronegativity differences describe a continuum, not a discrete "on-or-off," as with a light switch.

Nonpolar covalent bond – A covalent bond with a very small difference in electronegativity between the bonded atoms. The result is minimal or no charge separation in the bond.

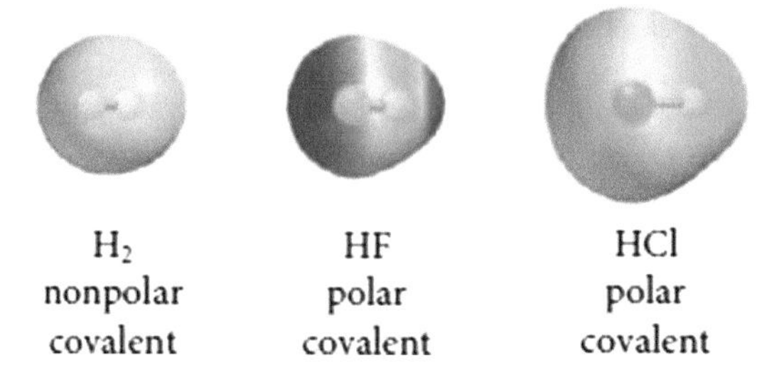

Let's revisit our example of a covalent bond between a metal and a nonmetal in tetraethyl lead. Does the electronegativity difference between the lead and carbon atoms indicate that the bond should be covalent? Examination of the table of electronegativity values indicates that the difference is only 0.6 electronegativity units. On the basis of this information, we can say that the Pb—C bonds in tetraethyl lead are polar covalent bonds and that these bonds are not as polarized as the bond in HF or HCl. We'll discuss the implications of this fact in the next section.

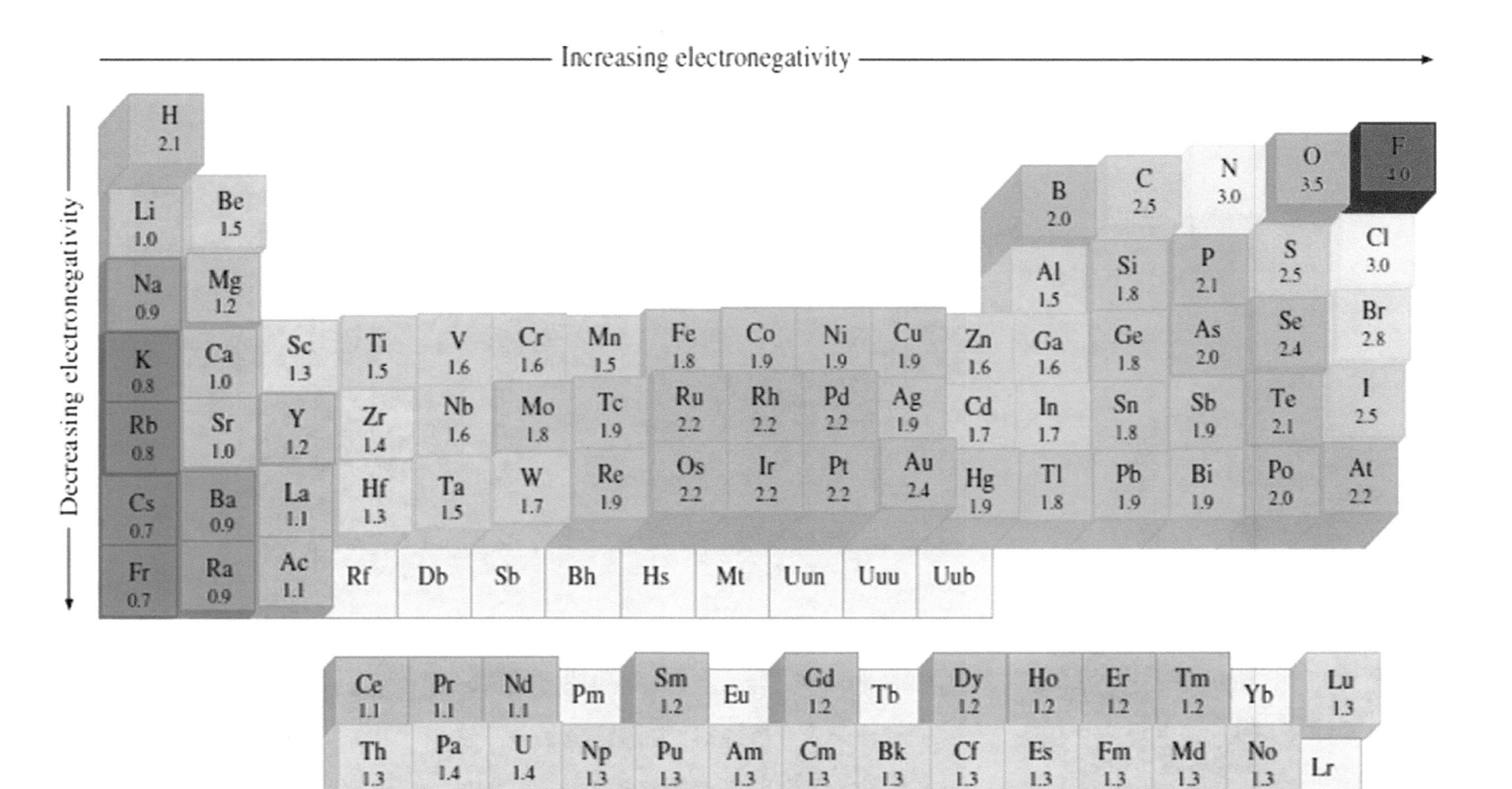

Fig. 8.14 Pauling's electronegativity values for the elements of the periodic table

Example 8.4 What Type of Bond Is It?

For each of these compounds, indicate whether the bond is an ionic bond, a polar covalent bond, or a nonpolar covalent bond.

a. KCl (a replacement for table salt in low-sodium diets)
b. H_2O (water)
c. Br_2 (used in the manufacture of brominated vegetable oils, which until recently was used to keep the citrus flavor in popular soft drinks from separating.)

Asking the Right Questions

What are the electronegativity differences (Δ) that indicate each type of bond? What is the electronegativity difference between each pair of bonded atoms or ions.

Solution

a. You'd predict an ionic bond between potassium and chloride. The two atoms differ by 2.2 electronegativity units.
b. A polar covalent bonding pattern is predicted (the difference is 1.4 electronegativity units). Hydrogen is less electronegative than oxygen, so the majority of the electron density in the bonds lies closer to the oxygen.
c. A nonpolar covalent bond is indicated here (no difference in electronegativity).

Are Our Answers Reasonable?

KCl contains two ions of very different polarity and has the physical properties that are consistent with ionic compounds, including a crystalline structure and the ability to conduct electricity when dissolved in water. The existence of water in the liquid state at normal pressure and temperature is consistent with its polar structure, and bromine is not soluble in water, suggesting that it is nonpolar. That is, the physical properties on a macroworld level are consistent with our assertions about the nature of the compound on a nanoworld level.

What If…?

…you were asked to work backward from electronegativity tables to list five binary substances that are ionic. What is that list?

(*Ans: For example, LiF, Li_2O, NaCl, BaO, BaF_2, and CsBr.*)

Practice 8.4

Predict the type of bond (covalent, polar covalent, or ionic) present in each of these binary compounds: NO, F_2, MgO.

8.4.4 Modeling a Covalent Bond—Lewis Structures

Felix Hoffman, a chemist hired by the Friedrich Bayer & Co., was charged in the 1890s with finding some better products that could be made by the company. One of the ideas that came to Hoffman was related to his father's rheumatism. At the time, this ailment and other pains were treated with large doses of salicylic acid, found in the inner bark of several types of willow trees. Unfortunately, Hoffman's father was unable to take this pain medication because its acidity irritated his stomach and throat. In 1897, Hoffman set about trying to reduce the acidity of the medicine. To do this, he needed to know how the arrangement of the atoms made the compound acidic.

Table 8.4 A set of rules for drawing lewis dot structures

1. Determine the total number of valence electrons
2. Determine the skeletal structure
   a. Hydrogen atoms are on the edges of the molecule
   b. The central atom has the lowest electronegativity. (There are many exceptions.)
   c. In oxoacids, hydrogens are usually on the oxygens
   d. Think compact and symmetric
3. Draw the Lewis dot structure
   a. Draw the bonds connecting the atoms
   b. Determine the number of electrons remaining
   c. Place the remaining electrons as lone pairs, beginning on the most electronegative atoms, until each atom has an octet
   d. All remaining lone pairs go on the central atom
   e. Assign formal charges, and redraw bonding electrons if necessary

After some experimentation, Hoffman was able to produce a compound that reduced this acidity. His product, acetylsalicylic acid ($C_9H_8O_4$) was later named aspirin by the company. Knowing how the atoms are attached, and what kind of bonds link the atoms, is important to understanding how compounds react and interact in the body. How can we illustrate the attachment of atoms in covalent molecules such as acetylsalicylic acid? One of the ways to do this is to build a model of the molecule using Lewis dot structures.

Rules describing how the atoms and electrons are placed in a Lewis dot structure enable us to draw compounds in a systematic way. These rules, found in Table 8.4, include drawing a skeletal picture of the molecule and then placing extra valence electrons in the skeleton until the model of the compound is complete. We will use these rules as we draw a Lewis dot structure of molecular hydrogen (H_2).

According to Table 8.4, the first step is to count all of the valence electrons on the atoms in the molecule. Because valence electrons are involved in bonding, knowing the total number of these electrons helps us determine the number of **bonding pairs** (electrons involved in bonding) and the number of **lone pairs** (electrons not involved in bonding).

Bonding pairs – Pairs of electrons involved in a covalent bond. Also known as bonding electron pairs.

Lone pairs – Pairs of electrons that are not involved in bonding.

In hydrogen, there are two total valence electrons (one from each atom). In step 2 (Table 8.4), we draw a skeleton for the molecule. The guidelines (steps 2a–d) for drawing a skeletal picture of a molecule are based on preferences that atoms have for particular locations in a molecule. There are only two atoms in the molecule, so we start by placing the atoms

H H

We show bonding pairs of electrons by replacing the bonding electrons with a line. We draw the line to show that the atoms are chemically bonded together in the molecule.

H—H

In step 3, we place all remaining electrons around the more electronegative atom first until the octet rule (or duet rule, for hydrogen) is satisfied. Because we do not have any remaining electrons, and because the duet rule is satisfied for both atoms, we are finished.

Hydrogen fluoride is either a fuming gas or a liquid, depending on the temperature of use (its boiling point is 19.5 °C, about 67°F), and has a host of industrial applications. For example, it is used as a raw material in the production of chlorofluorohydrocarbons (CFCs), insecticides, and fertilizers; as a catalyst in the production of pharmaceuticals; in the manufacture of semiconductors; and in etching glass (see Fig. 8.15). Let's use the rules in Table 8.4 to prepare a Lewis dot structure for HF. Step 1 requires us to count the valence electrons on all the atoms. Hydrogen has one valence electron, and fluorine has seven, so we have a total of eight valence electrons with which to work. Next, in step 2, we draw the best skeletal structure of the molecule:

H—F

We've used two electrons in our skeleton to represent the bond between the two atoms, so we have six electrons remaining (step 3). These are placed in pairs around the more electronegative atom until the octet rule (or duet rule) is satisfied. If any electrons remain, the pairs are placed around the other atoms.

H—F: (with lone pairs above, right, and below F)

Fig. 8.15 Hydrofluoric acid can etch glass like this drinking glass from the 1700s. *Source* ▶ https://upload.wikimedia.org/wikipedia/commons/b/b7/Beaker%2C_c._1780-1820%2C_etched_glass_-_Albany_Institute_of_History_and_Art_-_DSC08167.JPG

Because we used all of the electrons to fulfill the octet rule around fluorine, and because both atoms now satisfy the octet rule (or duet rule), we have completed the structure. Both fluorine and hydrogen in HF have electron configurations of a noble gas (completely filled valence shells).

Let's construct the Lewis Dot Structure for hydrogen cyanide. There are 10 valence electrons in this molecule. We start by placing carbon in the center because it is less electronegative than nitrogen (and hydrogen can never be in the center):

H–C–N

We've used up four electrons, because there are two bonding pairs. The remaining six electrons are placed first onto the more electronegative nitrogen:

$$\mathrm{H{-}C{-}\ddot{\underset{\cdot\cdot}{N}}\!:}$$

We've used up all our electrons, and hydrogen and nitrogen are satisfied, but carbon only has four electrons. We need to share two additional pairs of electrons between nitrogen and carbon to give both carbon and nitrogen a full octet. There is now a **triple bond** between carbon and nitrogen:

$$\mathrm{H{-}C{\equiv}N\!:}$$

The most common multiple bonds you will see in Lewis Dot Structures are **double bonds** (two shared pairs of electrons) as in carbon dioxide and triple bonds, as in hydrogen cyanide and N_2 (Fig. 8.16).

Triple bond – A covalent bond in which three electron pairs (six electrons) are shared between adjacent atoms.
Double bond – A covalent bond consisting of two individual bonding pairs of electrons.

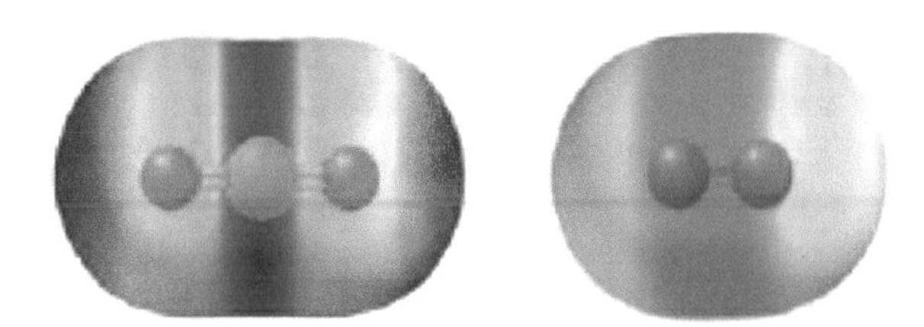

Fig. 8.16 (Left) The double bonds in carbon dioxide (CO_2). (Right) The triple bond in nitrogen (N_2)

8.4.5 Formal Charges

Methanol (also known as methyl alcohol or wood alcohol, CH_4O) is a compound being advanced as a potential substitute for gasoline. Methanol is a renewable resource obtained from the fermentation of cellulose-containing materials (such as wood). The following three structures satisfy the octet rule (or duet rule) for every atom based on the formula for methanol. Only one of them is methanol.

We can't actually see a methanol molecule, so how do we know which structure is correct? We can use the **formal charge** on the atoms in the structure to help determine the most reasonable structure for a compound.

Formal charge – The difference between the number of valence electrons on the free atom and the number of electrons assigned to that atom in the molecule.

What is a formal charge and how does it help us to select the most reasonable structure? The formal charge on an atom is the difference between the number of valence electrons on the free atom and the number of electrons assigned to the atom when it is part of a molecule. Another way to think of this is by assigning an oxidation number to each of the atoms, but the method of calculating formal charges lends us additional guidance. Oxidation numbers provide an indication of the charge an atom would have if it were actually ionic. Formal charges assume that electrons are shared equally.

Formal charge = valence electrons – # bonds – # nonbonded electrons

Mathematically, the formal charge equals the number of valence electrons on the free atom minus the number of bonds to that atom minus the number of nonbonded electrons. As a check of our math, the sum of the formal charges on all the atoms should equal the total charge of the molecule (0) or ion (+ or –). As a simple example, we can use this information to calculate the formal charge on each atom in our Lewis dot structure of HF. The fluorine atom has seven valence electrons (in Group 17, noted from the periodic table). Subtracting the number of bonds in the structure (one) and the number of nonbonded electrons (six) from this number gives a formal charge of zero for the atom. A similar calculation can be done with the hydrogen atom, which also has a formal charge of zero.

When we draw Lewis dot structures, the best structure is one that satisfies the largest number of formal charge rules. The best structure:

- has the smallest absolute magnitude for all of the formal charges
- places negative formal charges on the more electronegative atoms
- has the smallest number of nonzero formal charges

We can now consider our section-opening question: How do we decide which structure is likely to be correct? Calculating the formal charges on the central atoms in each of the possible structures of methanol enables us to choose the one on the left as the correct structure. Every atom in both structures has an octet of electrons, but only the structure on the left shows a formal charge of zero on each atom. This is one piece of evidence that the structure on the left is likely to be the most energetically stable structure of the three.

All zero formal charges
Methanol

Formal charge on C = −1
Formal charge on O = +1
Not methanol

Formal charge on C = −2
Formal charge on O = +2
Not methanol

Example 8.5 Return to Basics

Pioneers in the American West made most of their everyday items from natural sources. On the treeless plains of the Midwestern U.S. they built homes of sod, burned dried buffalo dung in the stove for heat, and made lye soap. Lye (a mixture of sodium hydroxide, NaOH, and potassium hydroxide, KOH) obtained from fireplace ashes was used to make soap from animal fat. What is the Lewis dot structure model for the hydroxide ion (OH^-)? On which atom in the hydroxide ion does the nonzero formal charge reside?

Asking the Right Questions

How do you draw the Lewis dot structure for the hydroxide ion? Then, how do you determine the formal charge of each atom? How is it related to the electronegativity of the atom?

Solution

The hydroxide ion can best be modeled by drawing the skeleton of the molecule with a single bond between the oxygen and the hydrogen. You can place the remaining electrons in pairs around the most electronegative element, giving the structure shown below.

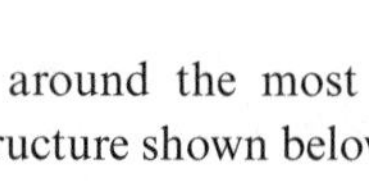

$$[:\ddot{\underset{\cdot\cdot}{O}}-H]^{\ominus}$$

Formal charge calculations indicate that the charge must reside on the oxygen (F.C. = 6 − 1 − 6 = − 1) and not on the hydrogen (F.C. = 1 − 1 − 0 = 0).

Are Our Answers Reasonable?

The Lewis dot structure contains the correct number of electrons. Oxygen is the more electronegative atom, so it is reasonable that it has the negative formal charge.

What If…?

…we were interested in carbon dioxide? What is the formal charge on the carbon atom?

(*Ans: The carbon atom has a formal charge of = 0 (4 valence electrons − 4 bonds = 0)*)

Practice 8.5

What is the formal charge on each of the atoms in OCl^-? in CH_3NH_2?

Example 8.6 Chemical Warfare and Bonding

Phosgene ($COCl_2$), a highly toxic gas, was used in World War I and several subsequent wars to kill soldiers who were hiding in places that bombs couldn't penetrate. The gas, now used in industry to make polymers, pharmaceuticals, herbicides, and other useful compounds, reacts with water and other electron-rich molecules (molecules with lone pairs of electrons). Draw the best Lewis dot structure for this molecule.

Asking the Right Questions

Which atoms are the most electronegative in the molecule? How does that affect which atom is the central atom in the structure? How do we then calculate the formal charge on each atom?

Solution

The skeletal picture of the molecule places the carbon in the center (least electronegative). The other atoms are placed symmetrically about the carbon. Of the 24 total electrons, 6 (three pairs) are used to connect the atoms. The remaining 18 electrons (nine pairs) are placed as lone pairs around the more electronegative atoms. The result is that all of the atoms, except the carbon, have a full octet. You need to share a lone pair of electrons with carbon to satisfy the octet rule on each atom. Calculation of the formal charges on each of the atoms indicates that the oxygen and the carbon atoms should share another pair of electrons.

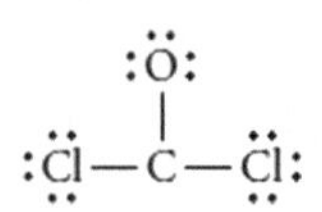

```
       ..
      :O:
       |
  ..   |   ..
 :Cl — C — Cl:
  ..       ..
```

F.C. (O) = 6 − 1 − 6 = −1
F.C. (C) = 4 − 3 − 0 = +1
F.C. (Cl) = 7 − 1 − 6 = 0

The best model of this molecule therefore shows a double bond between the carbon and the oxygen atoms and single bonds between the carbon and the chlorine atoms.

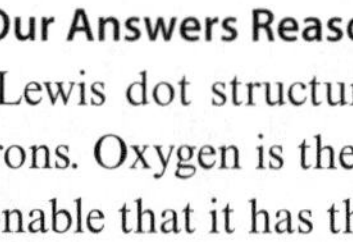

```
      :O:
       ‖
  ..   ‖   ..
 :Cl — C — Cl:
  ..       ..
```

F.C. (O) = 6 − 2 − 4 = 0
F.C. (C) = 4 − 4 − 0 = 0
F.C. (Cl) = 7 − 1 − 6 = 0

Are Our Answers Reasonable?

What would be an alternative structure, and would it be more reasonable than your current one? You could have drawn a Lewis dot structure that placed a double bond between the carbon and one of the chlorine atoms. This would have changed the formal charge on carbon to zero, but the formal charge on the chlorine atom would become + 1, and the formal charge on oxygen would remain − 1. The resulting structure would not be the best Lewis structure because of the existence of nonzero formal charges. This suggests that your structure, in which the formal charges on each of the atoms is zero, is reasonable.

What If…?

…instead of phosgene, you had carbonyl chlorofluoride, COClF, which is a decomposition product of some chlorofluorocarbon (CFC)-based refrigerants. What would be the formal charges on each atom?

(*Ans: Since the fluorine is even more electronegative than chlorine, it would behave in a similar way in the molecule. The formal charges on each atom are zero.*)

Practice 8.6

Draw the Lewis dot structures for C_2H_4 and CH_3N.

8.4.6 Resonance Structures

Carbonate ion (CO_3^{2-}) is a common polyatomic ion found in limestone, baking powder, and baking soda. Addition of acid to the carbonate ion causes the formation of carbonic acid (H_2CO_3), which decomposes rapidly into water (H_2O) and carbon dioxide (CO_2). In baking, the carbon dioxide that is released causes the bread to rise and makes its texture lighter.

Following the steps in Table 8.4 results in the structure shown below. Because of the –2 charge on the ion, we add two additional electrons to the valence electrons for a total of 24. Where did these extra electrons come from? They are probably electrons from calcium, sodium, or some other metal that donated its valence electrons. The carbon atom in our structure still needs to share electrons to satisfy the octet rule.

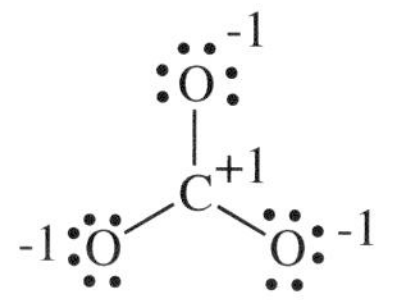

Using the formal charges on the atoms, we could rearrange our electrons to participate in a double bond with the carbon. At this point, the positive charge on the carbon atom is gone, and the octet rule is satisfied. The sum of the formal charges is equivalent to the charge on the carbonate ion. This is a good Lewis dot structure for carbonate.

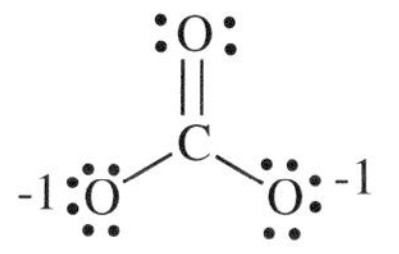

We could have shared the pair of electrons from one of the other oxygen atoms. In fact, there are two other structures that also seem to be satisfactory Lewis dot structures. When there are multiple correct Lewis dot structures possible, how do we decide which one is the most accurate model for our compound?

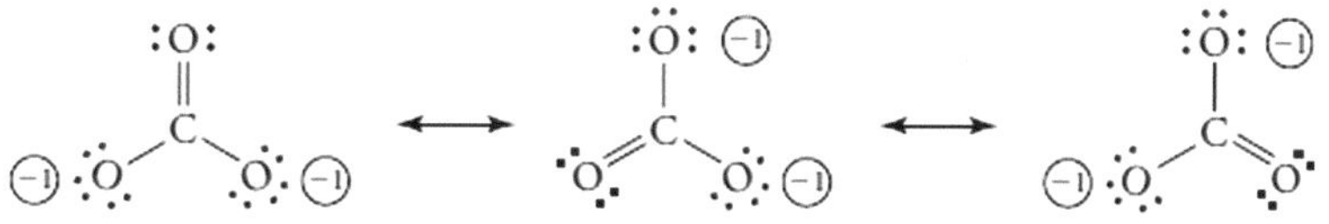

The answer is that we don't have to decide—they are all equally good representations of the carbonate ion. The three structures we've drawn are resonance structures.

A **resonance structure** is a model of a molecule in which the positions of the electrons have changed, but the positions of the atoms have remained fixed. All of the resonance structures for a molecule are correct ways to draw the Lewis dot structure. We show the relationship among these resonance structures by drawing a double-headed arrow between them. However, we shouldn't consider a single resonance structure to be a discrete entity. Because the electrons are always moving, the resonance structures must be considered together as the model of the molecule. A combination of all of the resonance structures is the best model. The resulting model is called the **resonance hybrid**. The resonance hybrid is an equal or unequal (based on experimental evidence) combination of all of the resonance structures for a molecule. For the carbonate ion, the resonance hybrid looks like this:

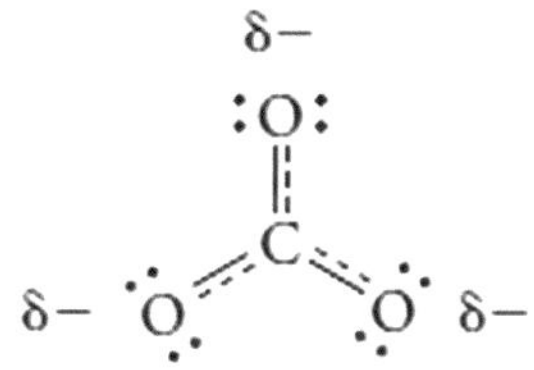

Resonance structure – A model of a molecule in which the positions of the electrons have changed, but the positions of the atoms have remained fixed.

Resonance hybrid – An equal or unequal (based on experimental evidence) combination of all of the resonance structures for a molecule.

Instead of full charges on two of the oxygen atoms, a partial charge exists on each of the oxygen atoms, and these add up to the overall −2 charge for the ion. Partial double bonds are drawn to show how the three resonance structures combined to make one resonance hybrid. In fact, we know from measurements of the bond strength in carbon dioxide that the individual bonds *are* about 33% stronger than a single bond. Our simple Lewis dot structure has accurately modeled the real molecule! This additional step of looking for resonance structures gives us the final version of our rules for drawing Lewis dot structures (Table 8.4).

We now have a sufficiently complete model to draw the Lewis dot structure for aspirin (acetylsalicylic acid): $C_9H_8O_4$.

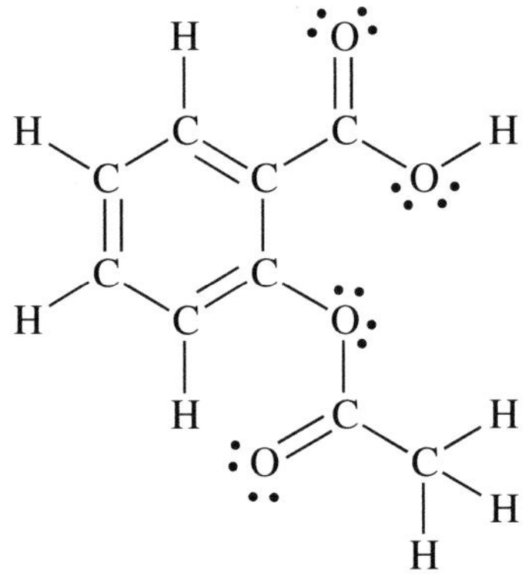

Look closely at this structure and verify for yourself that all the octets are filled and that the formal charges of all of the atoms in the structure are equal to 0.

Nanoworld-Macroworld: The Big Effects of the Very Small.—Chlorophyll, Tie-Dying and Visible Light

All molecules can absorb energy from the electromagnetic spectrum. Not all of them, however, absorb light in the visible region of the spectrum (wavelengths between ~400 nm and ~700 nm). Those that do absorb visible light appear to us as colored compounds. A series of related molecules known as the chlorophylls are probably the most important examples of this property. Why are these molecules so important? Chlorophyll-*a* is the molecule that is responsible for converting light energy to chemical energy in most plants.

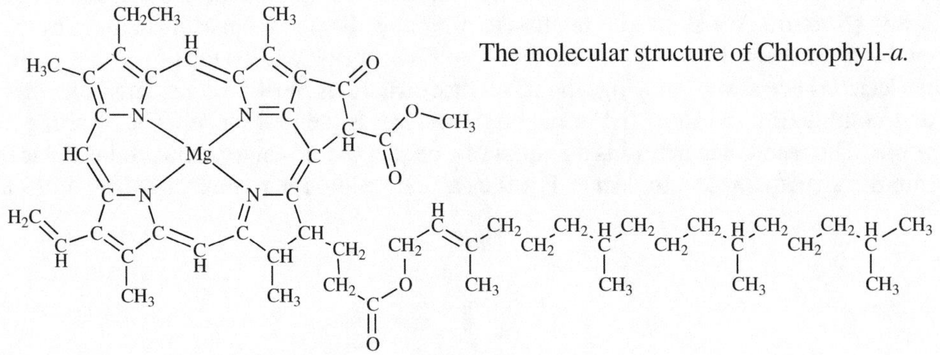

The molecular structure of Chlorophyll-*a*.

Chlorophyll-*a* absorbs light from the sun with wavelengths near 450 nm (blue) and 630 nm (red). The absorbance spectrum, a plot of all of the wavelengths of light that are absorbed versus the intensity of the absorbance, reveals the wavelengths that this molecule absorbs. Because the molecule absorbs blue and red light, the molecule appears to be green in color (the light that is not absorbed is reflected to our eyes.) By absorbing these wavelengths of light, chlorophyll-*a* is able to harness the energy of the light and pass that energy into the plant cell for use in converting carbon dioxide and water into sugars and oxygen.

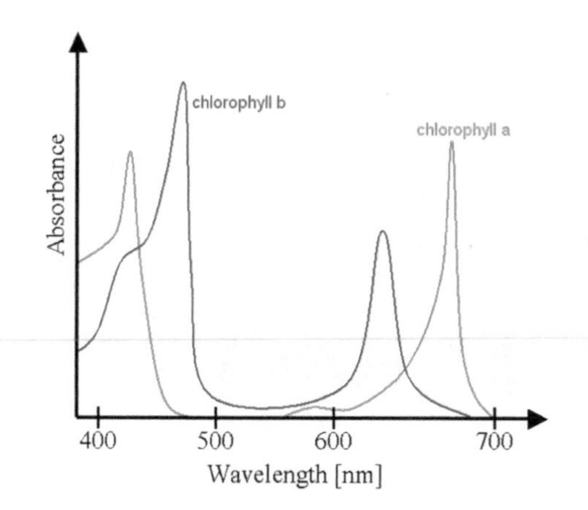

Not all molecules absorb visible light. Is there a structural similarity between molecules like chlorophyll-*a* that makes them able to do this? As we noted in earlier chapters, the absorption of energy from several regions of the electromagnetic spectrum causes electrons to move to higher energy orbitals. While the electrons around excited atoms can easily absorb light and transition from one energy level to another, the electrons in molecules that are most able to absorb light reside in the valence shell. Typically, in molecules, these electrons are found in *p* and *d* orbitals. Thus, in chlorophyll-*a*, the electrons that are found in *p* orbitals absorb light. As you'll note in the next chapter on bonding, *p* orbitals are used to make multiple bonds between atoms. As the number of multiple bonds increases in a molecule, the wavelength of the absorbed light increases. Once the number of multiple bonds becomes large enough, the absorbed light occurs in the visible region of the spectrum and the molecule appears to us as a colored compound.

Tie-dyed shirts and other colored clothes are stained with molecules that have lots of these alternating multiple-single-multiple bond networks. Indigo, for example, is one of the compounds used to give blue jeans their color. Note the number of multiple bonds in this compound. Another example, beta-carotene, is orange-red due to is extended network of multiple bonds. Cholesterol, on the other hand, lacks the ability to absorb light in the visible region because it only has one multiple bond in the structure. It appears to us as a white compound (it reflects all wavelengths in the visible region).

Indigo

cholesterol

β-carotene

Here's What We Know So Far

- The energy of an ionic bond is related to the charges on the ions and the distance separating the ions.
- Lewis dot structures can be used to draw a model of a compound. The octet rule drives our prediction of the best model for a compound. The resulting model represents the locations of electrons and atoms in a compound.
- Pairs of electrons (sometimes two or three pairs) are shared in covalent bonds.
- The atoms in a covalent bond are held at a distance that reflects the attraction for the bonding electrons and the repulsion of the adjacent nuclei.
- The Pauling electronegativity scale can be used to indicate the polarization of bonding electrons.
- We can classify covalent bonds as polar and nonpolar, depending on the electronegativity difference between the atoms in the bond.
- Formal charges help us to select the most likely structure among viable alternatives.
- Resonance structures exist in a molecule when electrons can shift positions among different atoms while the positions of the atoms remain fixed and result in multiple proper Lewis dot structures. ◀

8.4.7 Exceptions to the Octet Rule

In 1954, Robert Borkenstein, a captain in the Indiana State Police, developed a Breathalyzer like that shown in ◘ Fig. 8.17. A device similar to this was first used by police to determine alcohol levels in drunk drivers. The original Breathalyzer contained potassium dichromate and sulfuric acid that converted exhaled ethanol (CH_3CH_2OH) into acetic acid (CH_3COOH). The conversion caused the chemical reagent in the Breathalyzer to change color from yellow-orange to blue-green. The rest of the compounds in the reaction are colorless. After the test was performed, the presence of the blue-green color indicated that the suspect had been drinking. Let's examine the Lewis dot structure for sulfuric acid.

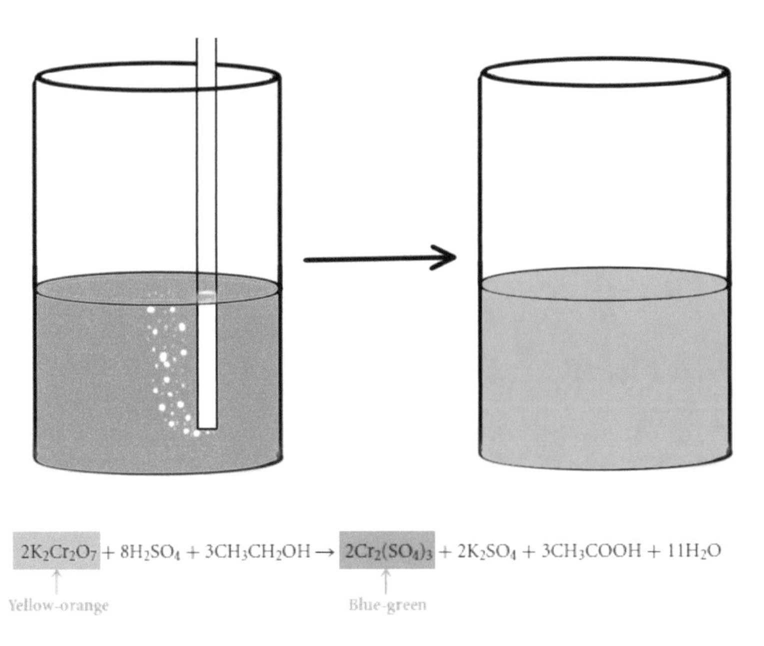

Fig. 8.17 An early breathalyzer worked by bubbling a person's breath through a solution of potassium dichromate (yellow-orange). The solution turned blue-green if that breath contained ethanol, indicating the presence of chromium(III) ions

Drawing the structure of sulfuric acid, we see that the sulfur atom has the smallest formal charge when there are six bonds from the sulfur to the adjacent atoms. Can this be correct? There is still debate about the existence of two S=O bonds in sulfuric acid (rather than single bonds to all atoms, which results in several nonzero formal charges) but the structure does seem to fit our understanding of Lewis dot structures and formal charges. Some exceptions to the octet rule must be considered if we are to accept this Lewis dot structure model.

:O:
‖
H—Ö—S—Ö—H
‖
:O:

Formal charges

F.C. (S) = 6 − 6 − 0 = 0

F.C. (O) = 6 − 2 − 4 = 0

In 1962, Neil Bartlett, then at the University of British Columbia, reported the first inert gas compound ($XePtF_6$). This was quickly followed by reports of other inert gas compounds. For example, researchers at the Argonne Laboratories prepared xenon tetrafluoride (XeF_4). In the Lewis dot structure of XeF_4, the xenon atom has twelve total electrons around it. What are the formal charges on the atoms in this model?

:F:
|
:F—Xe—F:
|
:F:

How can sulfur and xenon have more than eight electrons around them in the Lewis dot structure? Consider the orbitals that are available to valence electrons. In the valence shell of oxygen, for example, there are 2*s* and 2*p* orbitals. In the valence shell for sulfur, however, there are 3*s*, 3*p*, and 3*d* orbitals. The 3*s* and 3*p* orbitals contain electrons. A higher-energy unfilled set of orbitals (the 3*d* orbitals) can be used if needed to hold extra electrons. This means that the sulfur atom in sulfuric acid (H_2SO_4) can have eight electrons in the 3 *s* and 3*p* orbitals and place the extra four electrons in the previously empty 3*d* orbitals, which can hold a total of 10 electrons. Xenon, in XeF_4, can accommodate the extra electrons by placing them in its empty 5*d* orbitals. This is a general rule for atoms of the third row and higher of the periodic table. Atoms can have an **expanded octet**. Often this occurs when the valence electron shell includes unfilled *d* orbitals.

Expanded octet – The existence of more than eight electrons in the valence shell of an atom. Atoms of atomic number greater than 12 are capable of an expanded octet.

Boron trifluoride (BF_3) is an example of a compound in which the central atom has fewer than eight electrons around it. What is the formal charge on the boron atom in BF_3?

:F̤̈—B—F̤̈:
|
:F̤̈:

Even though the formal charges on each atom indicate that the Lewis dot structure is acceptable, the boron atom does not satisfy the octet rule. How can this be the most correct structure for the molecule? Why don't the fluoride atoms share a pair of electrons like the oxygen atoms did in the carbonate ion? To answer this, we only need to look at the formal charge on each atom. Boron, from Group IIIA, already has a formal charge of zero when it has three bonds! Therefore, the molecule is stable with only three pairs of electrons, even though its valence shell could hold a total of 8. Since it does have room for two more electrons, we might predict that BF_3 is quite reactive with molecules that are electron-rich. In fact, boron trifluoride reacts violently with molecules containing a lone pair of electrons (such as water and ammonia). The bond that forms is an example of a **coordinate covalent bond**. This bond is notable because both electrons that make the bond were donated from just one of the two atoms involved. However, after the bond is made, it is indistinguishable from a normal covalent bond.

Coordinate covalent bond – A covalent bond that results from the donation of two electrons from one of the two atoms involved in the bond. The resulting bond is indistinguishable from other covalent bonds.

:F̤̈—B(—:F̤̈:)—F̤̈: + H—N̈(—H)—H → F₃B—N(H)₃ (structure: :F̤̈: above B; :F̤̈—B—F̤̈:; B bonded to N; H—N—H; H below N)

The coordinate covalent bond is fairly common, particularly in reactions involving acids and bases. For instance, when we write the equation describing the dissolution of gaseous HCl in water, we describe the formation of a coordinate covalent bond.

$$HCl(g) + H_2O(l) \rightarrow H^+(aq) + Cl^-(aq) + H_2O(l) \rightarrow H_3O^+(aq) + Cl^-(aq)$$

The hydronium ion (H_3O^+), shown in Fig. 8.18, contains a coordinate covalent bond between the oxygen and one of the hydrogen atoms. As the strong hydrochloric acid dissolves in the water, it ionizes into H^+ and Cl^-. The hydrogen cation combines with a pair of electrons from the oxygen of water, and a covalent bond results. In the end, that bond is indistinguishable from the other O—H bonds in the hydronium ion.

Superoxide (O_2^-) is an extremely reactive anion that can cause severe damage to living tissue. To protect itself, the body has evolved an enzyme, superoxide dismutase, which quickly reacts with and destroys superoxide. What is the Lewis dot structure for superoxide, which is more accurately called the dioxygen (−1) ion? In drawing the structure, we note the presence of an odd number of total valence electrons. Because bonding electrons and lone-pair electrons are pairs of electrons, an odd number of total valence electrons in a Lewis dot structure implies that there is a lone unpaired electron (called a **radical**). The existence of a radical electron in a molecule often makes a molecule quite reactive. By placing a radical electron in a Lewis dot structure, we generate an atom that doesn't complete an octet. Calculate the formal charges in superoxide. What is the formal charge on each atom? Nevertheless, radicals like superoxide are frequently very reactive, pulling electrons from many other compounds to achieve a full valence shell.

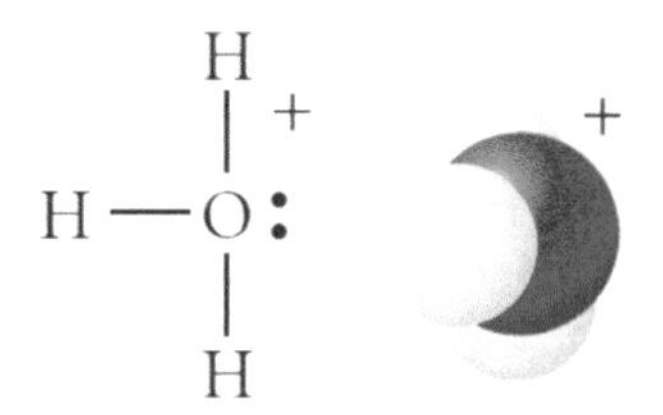

Fig. 8.18 The hydronium ion, H_3O^+, has a coordinate covalent bond between the oxygen and one of the hydrogen atoms

Radical – A molecule that contains an unpaired non-bonding electron.

8.4.8 Energy of the Covalent Bond

Angina, a sudden pain in the chest, is a symptom of a heart that doesn't have an adequate flow of oxygenated blood to work properly. Calcium channel blockers are a class of medicines used to treat this disease. These compounds fit neatly into "pockets" within proteins that control the flow of calcium ions into muscle cells. As the rate at which calcium ion passes into the heart muscle is greatly reduced, the heart relaxes, and its arteries dilate. The result is a heart that has enough oxygen to function adequately. One problem with calcium channel blockers is that the body breaks specific bonds in these drugs and makes them into biologically inactive compounds. For example, nifedipine, a common calcium channel blocker, persists in the body for only 4–8 h (◘ Fig. 8.19). Patients must continuously take the drug to prevent heart damage due to lack of oxygenated blood.

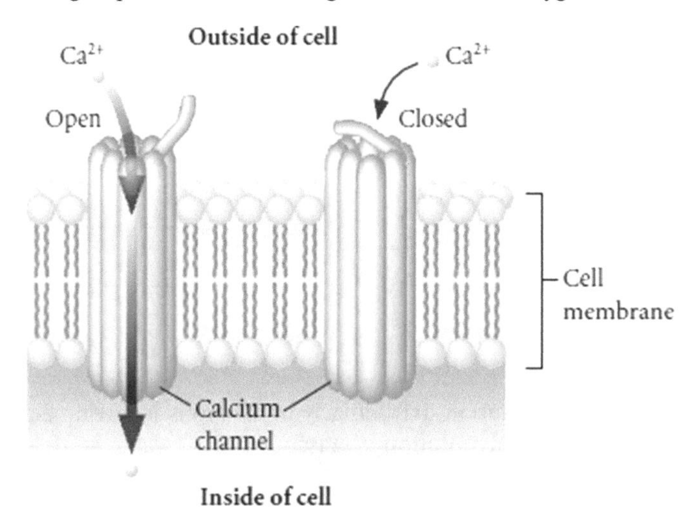

Medicinal chemists are interested in designing drugs that are similar in structure to nifedipine but resistant to metabolism. To do this, the medicinal chemist must make a compound that fits into the same pocket in the protein but contains bonds that do not break as easily as nifedipine. How do scientists know which bonds are susceptible

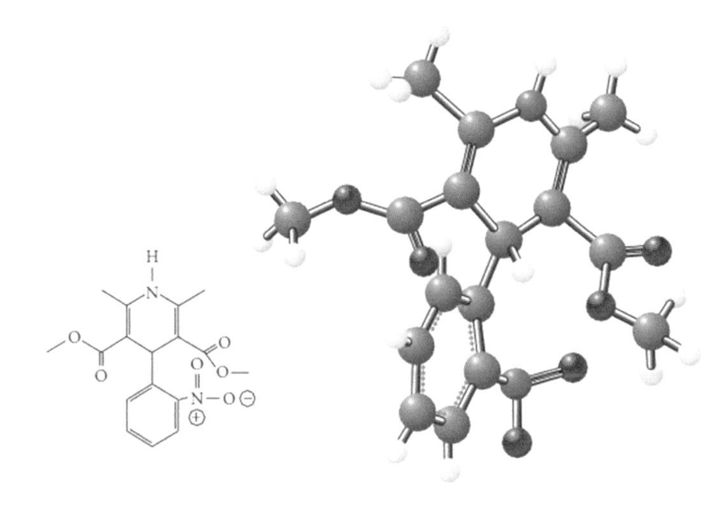

◘ Fig. 8.19 Nifedipine, a calcium channel blocker prescribed for some patients with heart trouble. Carbon atoms are black, oxygen atoms are red, nitrogen atoms are blue, and hydrogen atoms are white

to reactions within the body? Because reactions break bonds, the answer to this question is related to the strength of the different bonds in a compound.
How strong are covalent bonds? Is there a relationship between the type of bond and the strength of the bond? Is a multiple bond stronger than a single bond? To answer these questions, we must rely on our ability to estimate the amount of energy that it takes to dissociate (break) a bond. The **enthalpy of bond dissociation** ($\Delta_{dis}H$), or, simply, the **bond dissociation energy**, is the energy required to break 1 mol of bonds in a gaseous species. The equation that describes bond dissociation is shown below.

$$X - Y(g) \rightarrow X(g) + Y(g)$$

Enthalpy of bond dissociation – The enthalpy change related to breaking 1 mol of bonds in a gaseous species. Also known as the bond dissociation energy.

Bond dissociation energy – The energy required to break 1 mol of bonds in a gaseous species. Also known as the enthalpy of bond dissociation.

Bond breakage requires energy, so bond dissociation enthalpies are always endothermic and the values for bond energy will always have a positive sign. Table 8.5 lists the values for several important bonds. From the values in the table, we note that single bonds require less energy to break than their corresponding double or triple bonds. Compare the C—C single bond ($\Delta_{dis}H = 347$ kJ/mol), a C=C double bond ($\Delta_{dis}H = 614$ kJ/mol), and a C≡C triple bond ($\Delta_{dis}H = 839$ kJ/mol). Also note from the table that the length of the covalent bond shrinks as we go from a single to a double to a triple bond for a given element or pair of elements, confirming the fact that multiple bonds hold atoms more tightly and closer together. For diatomic molecules, the bond dissociation energy can be measured directly in the laboratory. The process is a little different for atoms that do not form diatomic molecules. Other atoms in a larger molecule can influence how electrons are distributed in a bond and affect the energy it takes to break the bond. The data in Table 8.5, then, are average bond dissociation energies.

Table 8.5 Average bond dssoiation energies (Δ_{dis}H) and bond lengths for some common covalent bonds

Bond	Energy (kJ/mol)	Length (pm)	Bond	Energy (kJ/mol)	Length (pm)	Bond	Energy (kJ/mol)	Length (pm)
H–H	432	75	N–H	391	101	F–F	154	142
H–F	565	92	N–F	272	136	F–Cl	253	165
H–Cl	427	127	N–Cl	200	175	F–Br	237	176
H–Br	363	141	N–Br	243	189	F–I	273	191
H–I	295	161	N–N	160	145	Cl–Cl	239	199
C–H	413	109	N=N	418	125	Cl–Br	218	214
C–F	485	135	N≡N	941	110	Cl–I	208	232
C–Cl	339	177	O–H	467	96	Br–Br	193	228
C–Br	276	194	O–F	190	142	Br–I	175	247
C–I	240	214	O–Cl	203	172	I–I	149	267
C–C	347	154	O–I	234	206	S–H	363	134
C=C	614	134	O–O	146	148	S–F	327	156
C≡C	839	120	O=O	495	121	S–Cl	253	207
C–N	305	143	C–O	358	143	S–Br	218	216
C=N	615	138	C=O	745 [a]	120	S–S	225	205
C≡N	891	116	C≡O	1072	113			

[a] C=O bond energy in CO_2 is 799 kJ/mol

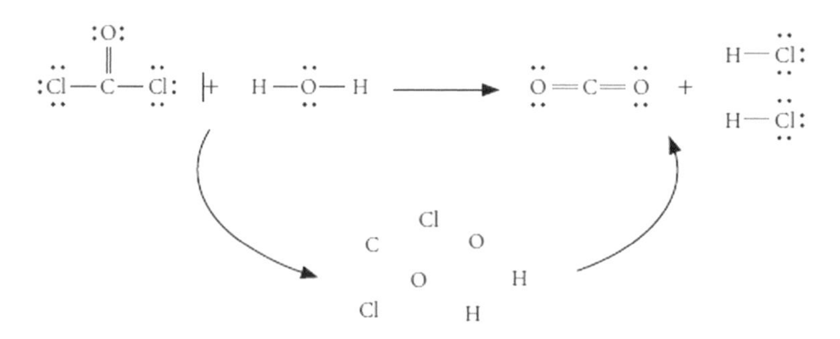

Fig. 8.20 Reaction of phosgene with water. Energy is added to the system to break all of the bonds in the reactants. Energy is released when the bonds in the products are formed

The overall enthalpy change of a reaction ($\Delta_{rxn}H$) can be estimated using bond dissociation energies. Because the enthalpy change of a reaction is a state function (▶ Chap. 5), it is the sum of all of the energy as heat added to the system minus the sum of all of the energy as heat removed from the system at constant pressure. The enthalpy change of a reaction can also be estimated by measuring the average enthalpies of bond breakage and bond formation. Breaking a bond requires an addition of energy to the system. Forming a bond results in a release of energy from a system. If we add up all of the enthalpies for the bonds that are broken and subtract all the enthalpies for the bonds that are formed, we can arrive at the enthalpy change of the reaction.

$$\Delta_{rxn}H = \Sigma(\Delta_{rxn}H)_{\text{breaking bonds}} - \Sigma(\Delta_{rxn}H)_{\text{making bonds}}$$

To use this equation, we have to know which bonds are broken and which are formed in a reaction. This is where Lewis dot structure models are very useful. By examining the Lewis dot structures of the products and reactants, we can determine which bonds are broken and which are formed. Then we can tally up the energy required to break all of the bonds in the reactants, and all of the energy released when the bonds in the products form (bond formation is *always* exothermic). The difference between the two is the enthalpy change for the reaction. This process does have limitations. First, because the tabulated data are for average bond enthalpies, the overall answer we get is an approximation of the reaction enthalpy. Second, it is important to note that chemical reactions do not actually occur by breaking all molecules into individual atoms and then combining the atoms into different configurations. In other words, this isn't how a reaction occurs.

The reaction of water and phosgene illustrates how we use this handy tool to determine approximate bond enthalpy changes during reactions:

$$H_2O(l) + COCl_2(g) \rightarrow CO_2(g) + 2HCl(aq)$$

We first draw the Lewis dot structure for each compound in the reaction and then note which bonds must be broken to make atoms, and then which bonds must form to make products. In this reaction we break one C=O bond, two C—Cl bonds, and two O—H bonds, and form two C=O bonds and two H—Cl bonds (Fig. 8.20).

The enthalpy change of the reaction is calculated below on the basis of this information. The enthalpy change for the reaction is negative; heat is released.

$$\begin{aligned}
1 \text{ mol C=O bonds} \times 745 \text{ kJ/mol} &= +745 \text{ kJ} \\
2 \text{ mol C–Cl bonds} \times 339 \text{ kJ/mol} &= +678 \text{ kJ} \\
2 \text{ mol O–H bonds} \times 467 \text{ kJ/mol} &= \underline{+934 \text{ kJ}} \\
\text{Total bonds broken (energy absorbed)} &= +2357 \text{ kJ}
\end{aligned}$$

$$\begin{aligned}
2 \text{ mol H–Cl bonds} \times (-427 \text{ kJ/mol}) &= -854 \text{ kJ} \\
2 \text{ mol C=O bonds} \times (-799 \text{ kJ/mol}) &= \underline{-1598 \text{ kJ}} \\
\text{Total bonds formed (energy released)} &= -2452 \text{ kJ}
\end{aligned}$$

$$\begin{aligned}
\text{Total energy absorbed} &= +2357 \text{ kJ} \\
\text{Total energy released} &= \underline{-2452 \text{ kJ}}
\end{aligned}$$

$$\text{Net energy change } (\Delta_{rxn}H) = \quad -95 \text{ kJ}$$

The difference between the energy absorbed and the energy released is the net energy change ($\Delta H = -95$ kJ).

Example 8.7 Calculating the Enthalpy of Combustion
Calculation of the enthalpy of combustion can be useful in determining whether a compound might make a good fuel. Calculate the enthalpy of combustion for methane (found in natural gas) using bond dissociation enthalpies.

$$CH_4(g) + 2O_2(g) \rightarrow CO_2(g) + 2H_2O(g)$$

Asking the Right Questions
What are the Lewis dot structures of all of the compounds in the reaction? Based on these structures, What bond breakage and formation occurs? What sort of energy changes do we expect?

Solution
We know that the combustion of methane is highly exothermic, supplying heat to countless homes as natural gas is consumed in heaters. You may therefore expect the net energy change to be negative, indicating a release of energy; an exothermic reaction.

$$\begin{aligned}
4 \text{ mol C–H bonds} \times 413 \text{ kJ/mol} &= +1652 \text{ kJ} \\
2 \text{ mol } O_2 \text{ bonds} \times 495 \text{ kJ/mol} &= \underline{+990 \text{ kJ}} \\
\text{Total bonds broken} &= +2642 \text{ kJ} \\
2 \text{ mol C=O bonds} \times (-799 \text{ kJ/mol}) &= -1598 \text{ kJ} \\
4 \text{ mol O–H bonds} \times (-467 \text{ kJ/mol}) &= \underline{-1868 \text{ kJ}} \\
\text{Total bonds formed} &= -3466 \text{ kJ} \\
\text{Total energy absorbed} &= +2642 \text{ kJ} \\
\text{Total energy released} &= -3466 \text{ kJ} \\
\text{Net energy change } (\Delta H) &= -824 \text{ kJ}
\end{aligned}$$

Are Our Answers Reasonable?
The reaction is exothermic. Heat is given off during the combustion of methane, so our answer makes sense.

What If…?
…you had propane, C_3H_8? What is the enthalpy of combustion for this gas, using bond dissociation enthalpies?
(*Ans:* -2057 *kJ/mol* C_3H_8)

Practice 8.7
Calculate the enthalpy change for the reaction of methanol and hydrogen bromide.

$$CH_3OH(aq) + HBr(aq) \rightarrow CH_3Br(aq) + H_2O(l)$$

▶ Here's What We Know So Far

- Atoms can exceed the octet rule if they have available *d* orbitals in their valence shell.
- Boron compounds are stable when B has just 6 electrons (but these compounds are very reactive).
- The enthalpy of a reaction can be determined from the energy absorbed during bond breakage and released during bond formation.
- We can combine the net changes in energy from bond-breaking and bond formation because bond enthalpy is a state function.
- Between two particular atoms, breaking multiple bonds requires more energy than breaking single bonds. ◀

8.5 VSEPR—A Better Model

Knowing the strength of bonds in molecules such as nifedipine is useful for determining information about the biological processes that break down the molecule and render it useless as a pharmaceutical. However, the three-dimensional structure of molecules is even more important in determining the biological activity of the compound.

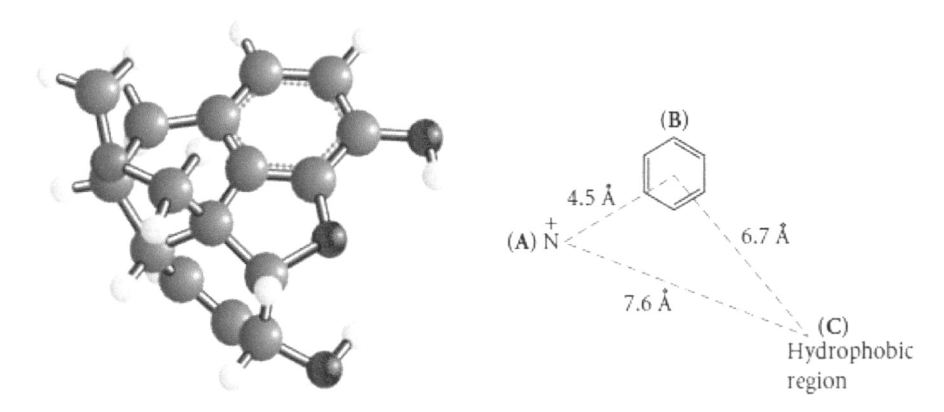

Fig. 8.21 The structure of morphine and the opioid receptor binding requirements

For instance, the opium poppy has long been used for its ability to alleviate pain. The ancient Sumerians in 3400 B.C. enjoyed its use so much that they referred to it as the "joy plant." It wasn't until 1827 that the active ingredient, morphine, was isolated from the poppy and produced commercially. The analgesic effects of morphine are one of the reasons why it continues to be used today to treat postoperative pain. However, morphine has serious side effects, which include respiratory depression, constipation, muscle rigidity, physical dependence, and a high potential for abuse. Medicinal chemists and the biochemists have been working together to try to prepare compounds that have beneficial properties similar to those of morphine but lack its harmful side effects. To develop a compound with a similar mode of action, they must determine not only the shape of the morphine molecule but also the shape of the active site on the receptor (the biological molecule to which morphine binds in the human body). Recent efforts have determined that the shape of the active site must accommodate a structure with particular dimensions, as shown in Fig. 8.21. Because there are so many different probable locations for the active site, computers were used to complete most of the work mapping the shape of the active site on the receptor.

Lewis dot structures are not very helpful in determining the shape of the morphine molecule or the active site in a case like this. What are the limitations of the Lewis dot structure models? Does a Lewis dot structure tell us anything about the three-dimensional shape of a molecule? The Lewis dot structure might lead us to believe that all molecules are flat, planar structures. But people, gerbils, roses, and guitars all occupy three dimensions, and all of these things are composed of molecules, so it makes sense that molecules occupy three-dimensional space. Even so, the idea of a "three-dimensional molecule" was fervently debated when it was first introduced. Many chemists could not believe that molecules would be anything other than flat.

Issues and Controversies—Flat molecules and ethics in science

In the 1800s, Jacobus Henricus van't Hoff (Dutch physical chemist, 1852–1911) and Achille Le Bel (French chemist, 1847–1930) independently came to the conclusion that molecules must exist as three-dimensional structures. Despite the fact that established thought on the structure of molecules asserted that they were flat, these researchers advanced their theory for public and professional scrutiny. Using careful theoretical considerations of well-studied molecules (such as tartaric acid), van't Hoff concluded that molecules had to occupy a three-dimensional structure. It was a very compelling argument for some. For others, it was heresy.

Adolph Wilhelm Hermann Kolbe (1818–1884), one of the prominent German chemists of the time, vehemently discounted the theory of three-dimensional molecules. His flat models seemed to provide the best explanation for the properties that he observed. He took such an aggressive stance on this issue that he attempted to discredit van't Hoff and his colleagues. Without experimenting to determine whether the theory was correct, he published letters in prominent chemistry journals just to tarnish the new theory. In one of his published articles, Kolbe wrote as follows:

> I have recently published an article giving as one of the reasons for the contemporary decline of chemical research in Germany, the lack of well-rounded as well as thorough chemical education. Many of our chemistry professors labor with this problem to the great disadvantage of our science. As a consequence of this, there is an overgrowth of the weed of the seemingly learned and ingenious but in reality, trivial and stu-

pefying natural philosophy. This natural philosophy, which had been put aside by exact science, is at present being dragged out by pseudoscientists from the den which harbors such failings of the human mind and is dressed up in modern fashion like a freshly rouged prostitute whom one tries to smuggle into good society where she does not belong.

A J. H. van't Hoff of the Veterinary School in Utrecht has, as it seems, no taste for exact chemical investigation. He has thought it more convenient to mount Pegasus (apparently on loan from the Veterinary School) and to proclaim in his "La chimie dans l'espace" from the top of the chemical Mount Parnassus which he ascended in his daring flight, how the atoms appeared to him to have grouped themselves throughout the Universe.

—*Journal für praktische Chemie*, 1877

This kind of public ridicule of fellow scientists does not help to advance the cause of science and research. The responsibility of scientists to maintain objectivity in the methodical practice of science carries over into the public realm as well. To practice science with less than complete objectivity results in professional ridicule and increased public skepticism.

Tens of thousands of chemistry-related articles are published worldwide each year. Highly regarded journals use the system of prepublication "peer review," in which knowledgeable scientists evaluate the validity of the research recounted by the authors of articles. While not perfect, it is the best system we know of to keep the scientific process as honest as possible. The best indication that chemistry is overwhelmingly done by ethical people is that there are so many advances in our discipline every year, and these advances are based on communicating meaningful results of chemical research.

The orientation of atoms in a molecule plays a major role in determining its properties. For instance, the Lewis dot structure of water (H_2O) can be drawn in two different ways. In one, it is a linear molecule; in the other, it is bent. Which is the correct way to draw water? If water were a linear molecule, it would have completely different properties from those that we observe. The consequences of this slight change would have drastic effects in the real world. An ocean of linear water molecules would dissolve oxygen quite well and probably kill most of the life in the sea. Even the beading of water as it runs down a window would not be possible if water were a linear molecule. Water is a bent molecule. Can we prove this with a model—one that will pass the test of properly predicting the shapes of molecules with which we are unfamiliar? As we continue this discussion, please keep in mind that models are, by their nature, simplifications that help us predict chemical behavior. Because they often simplify subtle chemistry, they do not always properly predict what happens. Yet we use models because the best ones give us insight and the best models have very good predictive power.

8.5.1 The VSEPR Model

Ronald J. Gillespie (1924–2021) and Sir Ronald Nyholm (1917–1971) introduced the **valence shell electron-pair repulsion model** (**VSEPR**; often pronounced "VES–per") in 1957 to facilitate the construction of models of three-dimensional molecular structures. The key assumption in the VSEPR model is that bonding pairs and lone pairs of electrons move away from each other and orient themselves in three-dimensional space to give minimum repulsions (lowest energy configurations). In other words, the preferred orientation is one in which the electron groups have maximum separation in three dimensions. To assist in visualizing this process, imagine pairs of electrons as balloons tied together (◻ Fig. 8.22). Two balloons tied together orient in such a way that they are opposite each other. Three balloons tied together orient themselves in a triangular shape, and four balloons push themselves to occupy the corners of a tetrahedron. What happens if you distort the shapes by pushing the balloons together and then letting go? The balloons push away again to return, ideally, to their lowest-energy (i.e.—most stable) shape.

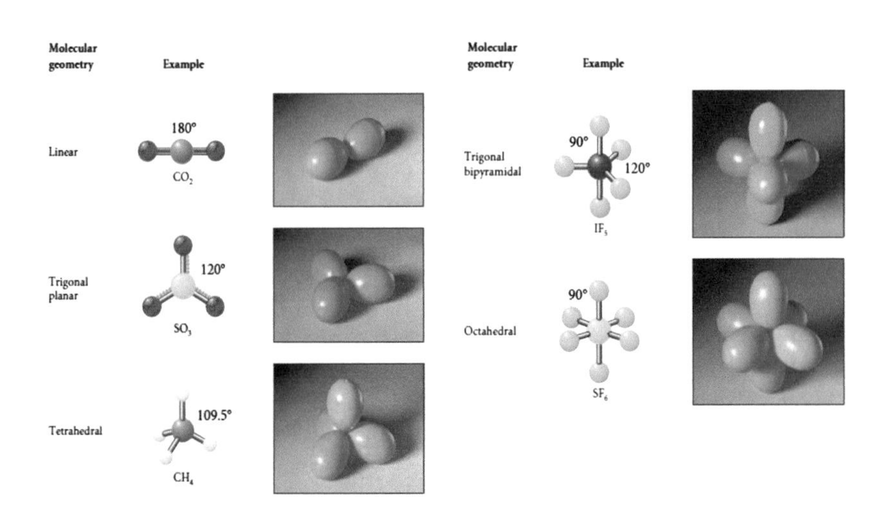

Fig. 8.22 Using balloons as an analogy for VSEPR

Valence shell electron-pair repulsion model (VSEPR) – A model that bases the three-dimensional shapes of molecules on the number of electron groups attached to a central atom.

VSEPR models determine the shape of a molecule on the basis of the number of electron groups around the central atom. The resulting model of the molecule represents the positions of the electron groups in three dimensions (the **electron-group geometry**). Electron groups include lone pairs and bonding electrons; a single bond, a double bond, or a triple bond counts as a single electron group. However, when we determine the experimental shape of a molecule, we actually depict the positions of the different nuclei (the **molecular geometry**). The actual shape of the molecule can be different from the electron-group geometry, but that doesn't mean that the molecular geometry is unattainable from the electron-group geometry. In fact, we can use the VSEPR model's electron-group geometry to determine the molecular geometry if we consider the geometry of the atoms to contain "invisible" lone pairs. The result is the molecular geometry. In other words, the molecular geometry is what the molecule would "look like" if we could see everything except the lone pairs of electrons. Let's consider some examples of the VSEPR model to illustrate and clarify this process. Table 8.6 shows the shapes of each of the electron-group geometries that we will discuss.

Electron-group geometry – The positions of the groups of electrons (lone pairs and bonding pairs) around a central atom in three dimensions.

Molecular geometry – The geometry of the atoms in a molecule or ion. The molecular geometry comes from the electron-group geometry but considers the lone pairs "invisible."

8.5.2 Examples of Three-Dimensional Structures Using the VSEPR Model

Beryllium chloride ($BeCl_2$) is a toxic white solid primarily used to make beryllium metal. The Lewis dot structure (and formal charge analysis) of this compound indicates to us that the beryllium atom is surrounded by two bonding pairs of electrons. By analogy, we can consider the central atom in $BeCl_2$ by tying two balloons together. The balloons push apart to occupy opposite orientations, so VSEPR predicts the electron-group geometry to be linear with a Cl—Be—Cl bond angle of 180°. The molecular geometry is also linear because there are no lone-pair electrons around the central beryllium atom.

Table 8.6 Electron and molecular geometries according to VSEPR

Number of Electron Groups	Electron group geometry	Number of lone pairs	Molecular geometry
2	Linear	0	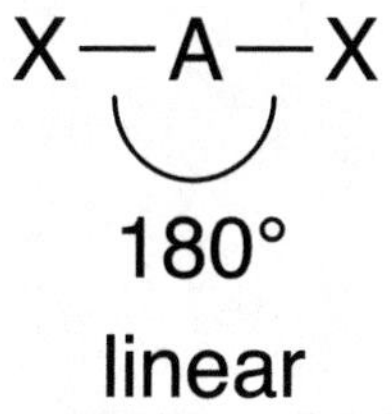 X—A—X 180° linear
3	Trigonal planar	0	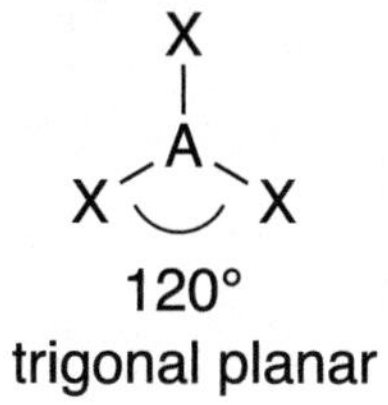 120° trigonal planar
3	Trigonal planar	1	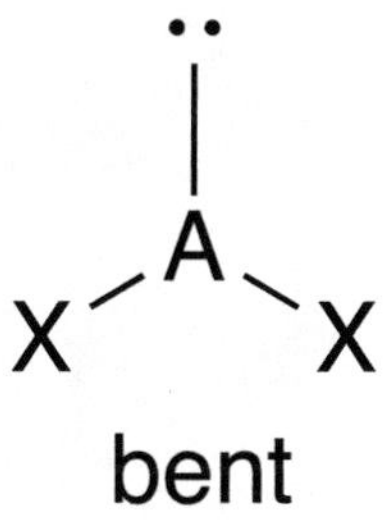 bent
4	Tetrahedral	0	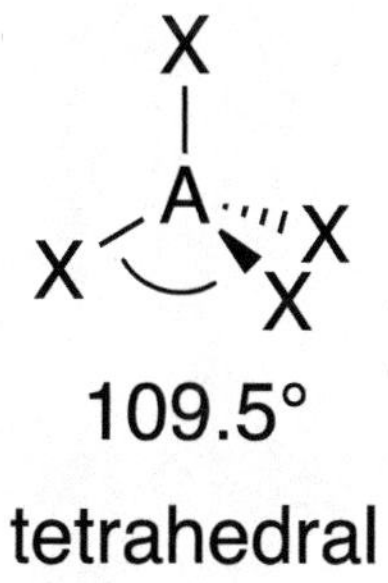 109.5° tetrahedral
4	Tetrahedral	1	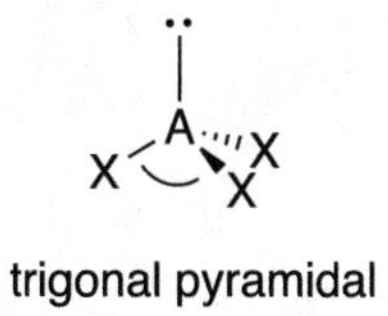 trigonal pyramidal
4	Tetrahedral	2	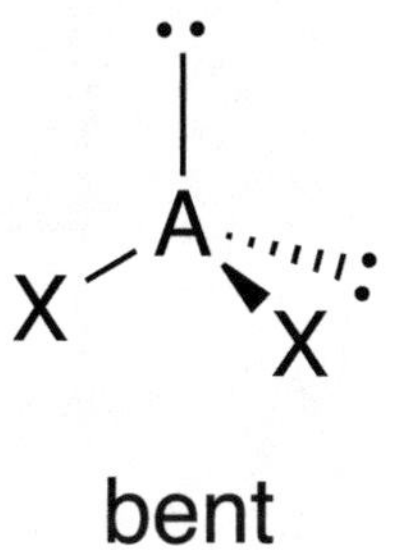 bent
5	Trigonal bipyramidal	0	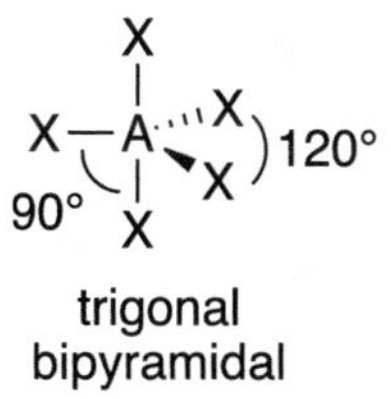 90° 120° trigonal bipyramidal

(continued)

Table 8.6 (continued)

Number of Electron Groups	Electron group geometry	Number of lone pairs	Molecular geometry
5	Trigonal bipyramidal	1	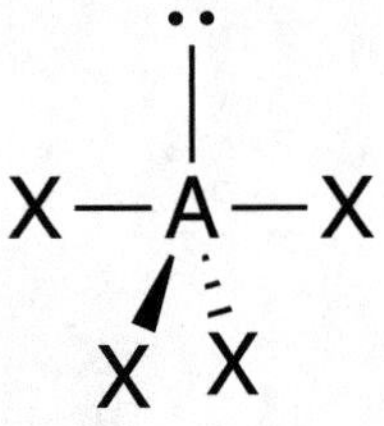 see-saw
5	Trigonal bipyramidal	2	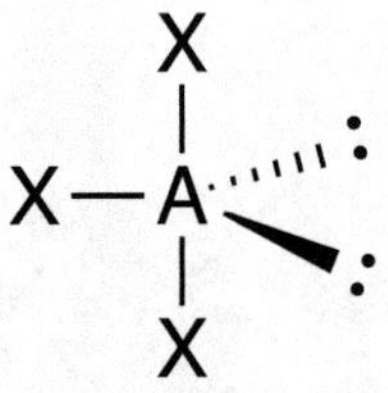 T-shaped
5	Trigonal bipyramidal	3	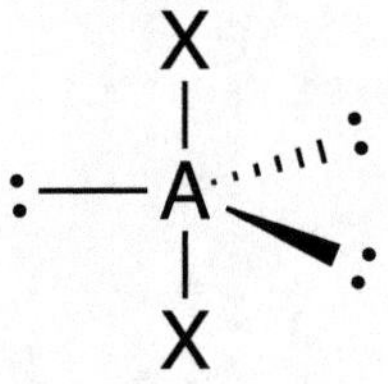 linear
6	Octahedral	0	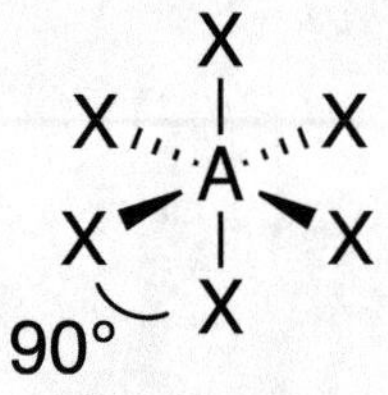 90° octahedral
6	Octahedral	1	 square pyramidal
6	Octahedral	2	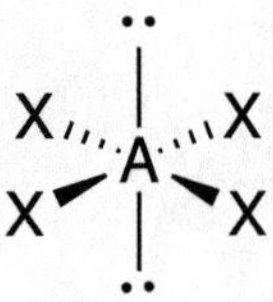 square planar

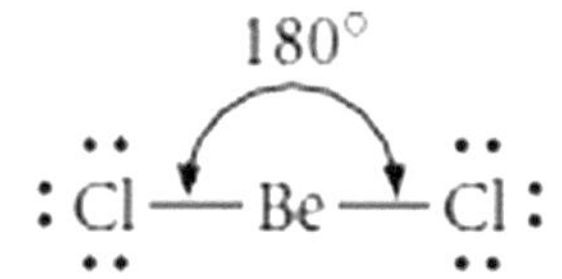

180°

:Cl—Be—Cl:

This type of geometry is also predicted for carbon dioxide (CO_2). The Lewis dot structure of CO_2 indicates that there are only two electron groups in the form of double bonds that extend from the central carbon atom. Carbon dioxide is a linear molecule.

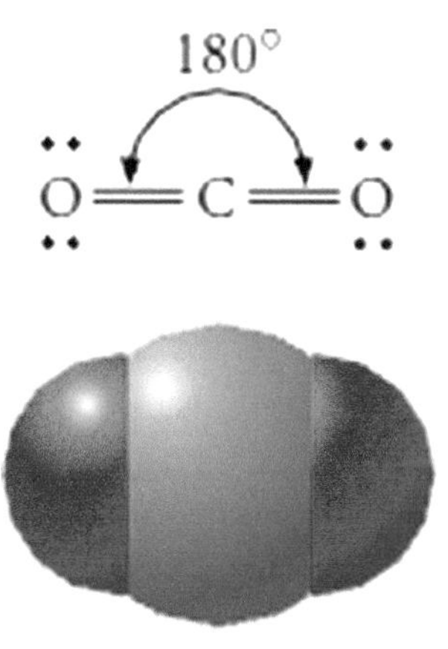

Boron trifluoride (BF_3), a compound used as a catalyst in a number of organic chemical reactions, such as polymerization (combining individual molecules into very large chains), and isomerization (changing the arrangement of atoms within a molecule), is shown below as a Lewis dot structure. This model shows that three electron pairs (the three bonds) radiate from the central boron atom. The VSEPR model dictates that the three electron pairs should occupy the corners of a triangle. Therefore, the electron-group geometry has trigonal planar geometry. The molecular geometry is also trigonal planar because there are no lone pairs on the boron. Bond angles for F—B—F are 120°.

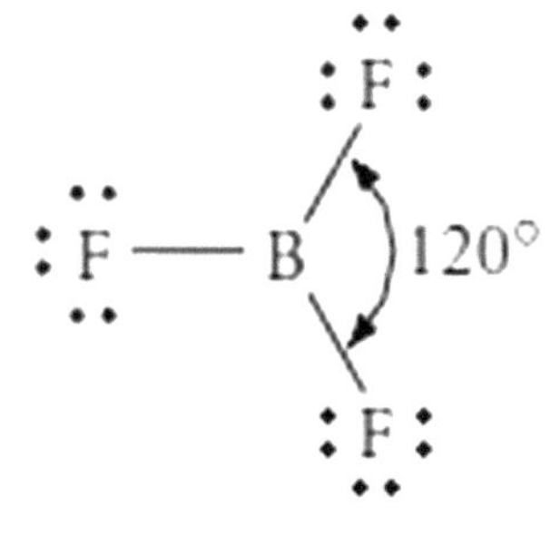

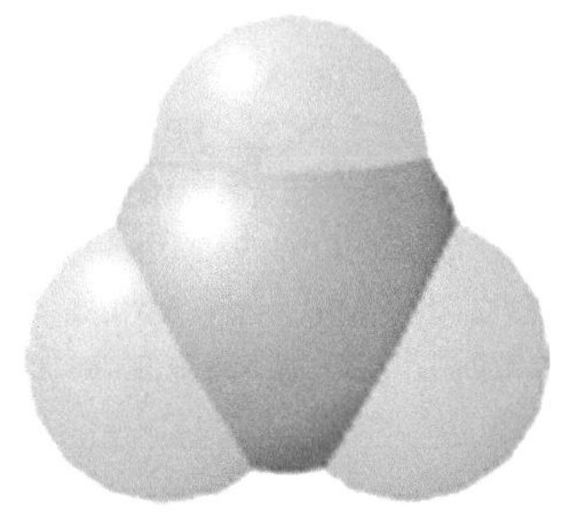

Methane (CH_4) is a molecule containing four electron pairs around the central carbon atom. As a consequence, VSEPR predicts that the bonding pairs of electrons should point to the vertices of a tetrahedron. Similarly, because of the lack of lone-pair electrons on the carbon, the molecular geometry is tetrahedral with H—C—H bond angles of 109.5°. We use a dashed wedge to illustrate that the bond points behind the plane of the paper; the filled wedge protrudes in front of the paper.

Ammonia (NH_3) is a compound used extensively in farming as a fertilizer and in livestock feed. From the Lewis dot structure, we know that the molecule possesses three **bonding electron pairs** and one lone pair. The VSEPR model allows us to predict a tetrahedral electron-group geometry. Because one of the electron groups is a lone pair, the *molecular geometry* of ammonia will be trigonal pyramidal. Similarly, the Lewis dot structure model of water indicates that there are two bonding pairs and two lone pairs on the central oxygen atom. The VSEPR model dictates that the electron groups should be arranged in a tetrahedral shape. However, the molecular geometry, determined by considering that the lone pairs are invisible, is bent, or angular. We'll discuss the subtleties of bond angles later in this section.

Bonding electron pairs – Pairs of electrons involved in a covalent bond. Also known as bonding pairs.

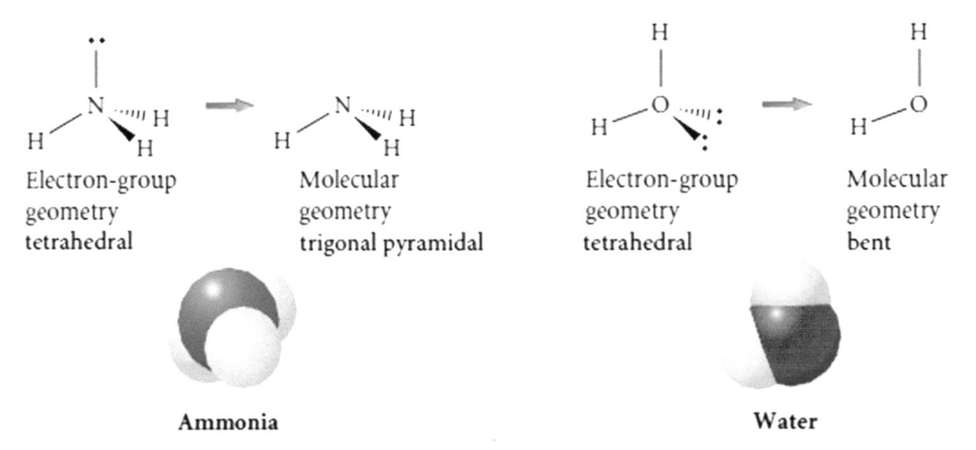

For molecules that contain more than eight valence electrons on an atom, there may be five or six electron groups about the central atom. Let's look at the structure of PCl_5. This compound is used as a catalyst in the manufacture of acetylcellulose (the plastic film on which motion pictures were printed before they went to a digital format). The Lewis dot structure of phosphorus pentachloride shows five electron groups attached to the central atom. VSEPR indicates that the molecule has a trigonal bipyramidal electron-group geometry, and the molecular geometry is also trigonal bipyramidal. This model has some interesting features. In particular, there are two distinct positions for chlorine atoms within the trigonal bipyramid. Two locations, with bond angles of 90° from one chlorine to the next nearest neighbor, occupy the **axial positions** (up and down), and three locations occupy the **equatorial positions** (outward) with bond angles of 120° between them. Molecules that contain six electron groups occupy the octahedral electron configuration. Each of the positions in the octahedron is the same. All atoms have 90° bond angles to the next closest neighbor. An example of this shape is found in SF_6.

Axial position – The position of a group when it is aligned along the *z*-axis of a molecule.
Equatorial position – The position of a portion of a molecule when it is arranged along the *x*-axis or *y*-axis of the molecule.

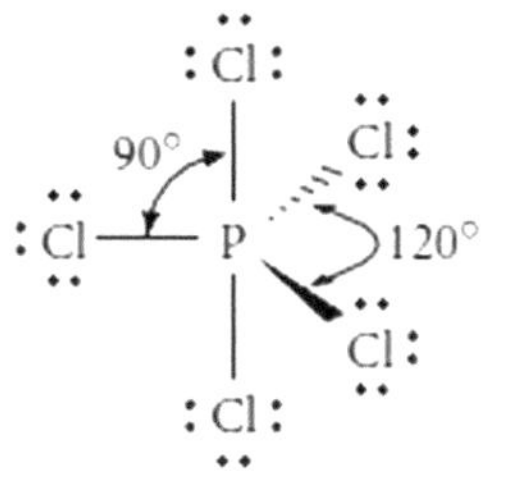

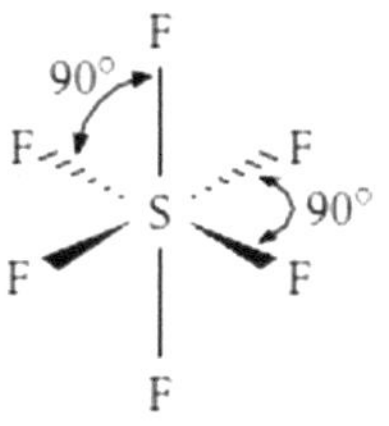

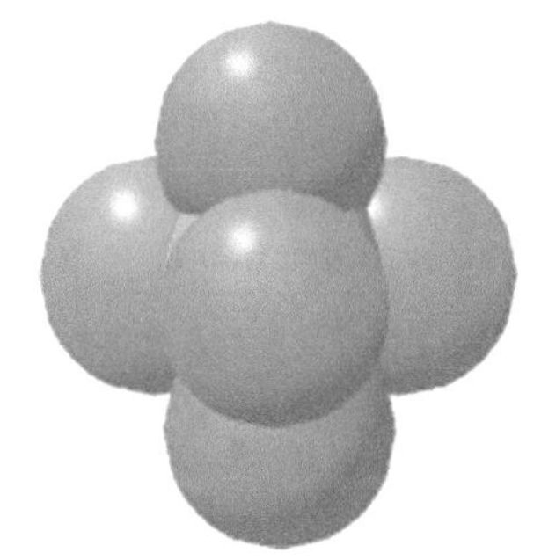

Example 8.8 Ozone and VSEPR

Ozone is an important molecule in our atmosphere because it protects us from the harsh ultraviolet rays of the sun. What is the shape of ozone (O_3)?

Asking the Right Questions

What is the relationship between the Lewis dot structure and the VSEPR structure? What does the VSEPR model add to our understanding?

Solution

The Lewis dot structure model shows a single bond and a double bond with a lone pair on the central oxygen atom. Each of these is an electron group.

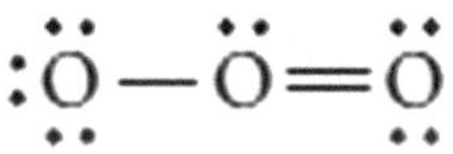

With three electron groups, ozone has a trigonal planar electron-group geometry. The VSEPR model indicates that the molecular geometry is bent.

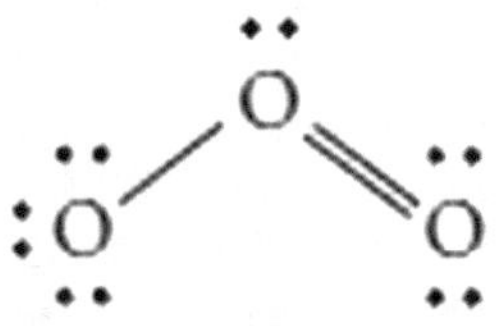

Ozone
O_3

Is Our Answer Reasonable?

The three electron groups around the central oxygen atom are consistent with the trigonal planar geometry, so the answer of a bent geometry for ozone is reasonable. Remember that "geometry" refers only to how the atoms look in the molecule, so "bent" is the appropriate shape for ozone.

What If…?

…you had a chlorate ion, ClO_3^-? What is its molecular geometry?

(*Ans: trigonal pyramidal*)

Practice 8.8

Indicate the VSEPR model shapes (electron-group and molecular) and bonding angles for HNO_3, CCl_4, and NH_3.

8.5.3 Advanced Thoughts on the VSEPR Model

VSEPR theory illustrates how electron pairs repel each other because they occupy space and have similar charge. Does the degree of repulsion depend on the type of electron pair that we're examining? To put it another way, is there a difference in the size and shape of different electron pairs? It turns out that lone pairs of electrons are big compared to bonding pairs of electrons. Gillespie illustrated this fact by noting the repulsions in terms of the size of the electron pair. The result is that the different types of electron pairs can be ordered in terms of the three-dimensional space they require:

$$\text{Lone pairs} > \text{triple bonds} > \text{double bonds} > \text{single bonds}$$

Note that the space required in the VSEPR model is not the same as the bond lengths, in which a triple bond between two carbon atoms is shorter than a double bond, which, in turn, is shorter than a single bond between the atoms.

Let's consider the physical shape of a molecule of ammonia. According to the VSEPR rules, ammonia (NH_3) has tetrahedral electron geometry. The lone pair on the nitrogen, however, is much more repulsive than the bonding pairs of electrons. In response to the repulsions and three-dimensional space requirements, the bonding pairs are pushed closer together. The result is that the H—N—H bond angles in ammonia, at 107°, are smaller than those of the true tetrahedron. The same reasoning holds for the experimentally measured bond angle in water (104.5°). The angle is severely pushed by the presence of two lone pairs on the central oxygen atom.

Gillespie noted that multiple bonds require more space than single bonds. Let's examine the structure of formaldehyde (CH_2O) as an example. The Lewis dot structure shown in ◘ Fig. 8.23 predicts three electron groups (one double bond and two single bonds) surrounding the central carbon atom. VSEPR correctly predicts trigonal planar electron-group and molecular geometry. Experimentally, though, the bond angles in formaldehyde are not 120°. Because the double bond requires more space than the single bond, we predict the H—C—H bond angle to be less than 120°. Experimental evidence suggests that the actual angle is 116°. Similarly, the H—C—O bond angle should be larger than 120°. Experimentally, it has been measured at 122°.

Using this information, we can determine the molecular geometry for sulfur tetrafluoride (SF_4) with 34 valence electrons. The molecule is typically used in the laboratory to make other fluorine-containing compounds. The VSEPR model predicts the trigonal bipyramidal electron-group geometry shown in ◘ Fig. 8.24.

However, because one of the electron groups is a lone pair, the molecular geometry is not a trigonal bipyramid. Which one of the positions, axial or equatorial, should the lone pair of electrons occupy? Because the lone pair requires more space than the bonding pairs of electrons when we are using the VSEPR model, the lone pair will occupy the position with the fewest repulsions (the most room for the lone pair). In the trigonal bipyramid,

H 122° 116° C=O H 122°

Formaldehyde

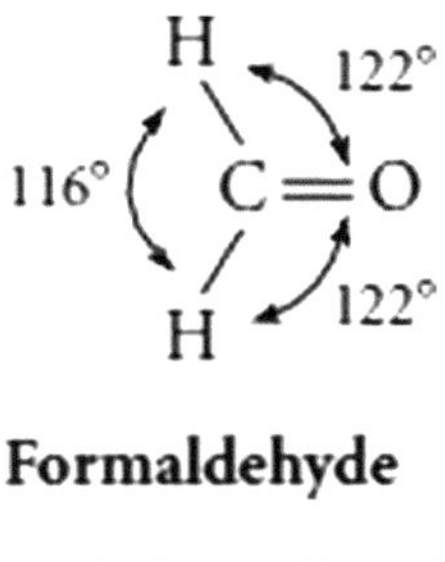

◘ **Fig. 8.23** Multiple bonds affect bond angles in formaldehyde (CH_2O). Shown here are the experimentally determined bond angles. Note that they deviate slightly from the predicted 120° angles

Fig. 8.24 Sulfur tetrafluoride is a molecule with see-saw geometry

the most unhindered site is one of the equatorial positions. Actually, lone pairs in trigonal bypyramids will always occupy the equatorial positions. The large lone pairs then push against the other electron pairs and distort them from their ideal bond angles. The result for SF_4 is referred to as a see-saw structure. The same logic can be applied to the construction of ClF_3 (T-shaped molecular geometry), BrF_5 (square pyramidal molecular geometry), and XeF_4 (square planar molecular geometry).

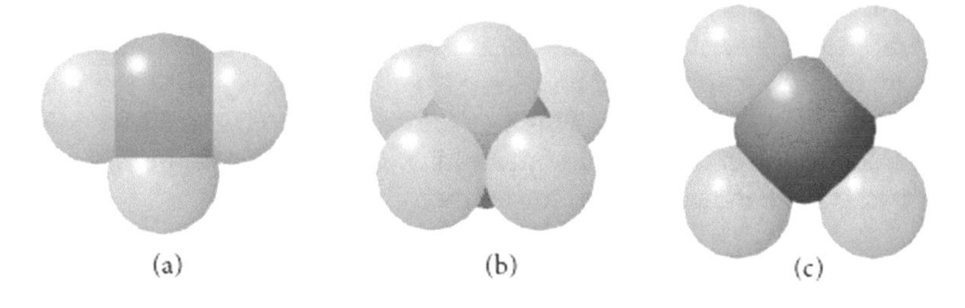

Example 8.9 A Closer Look at Ozone

When we apply the VSEPR model to ozone, our initial picture is that it is a bent molecule with an O—O—O bond angle of 120°. Given that different electron groups have different degrees of repulsions, what do you predict to be the actual bond angle in ozone (greater than or less than 120°)?

Asking the Right Questions

What are these "different electron groups," and what is the impact of each on the overall structure?

Solution

The lone pair on the central oxygen occupies more space than the bonding pairs, so the bond angle becomes smaller between the adjacent oxygen atoms. The experimentally measured angle (O—O—O) is 116.8°, compared to 120° for the theoretical bond angles in a molecule with three single bonding pairs.

Is Our Answer Reasonable?

The experimentally observed bond angle is consistent with the VSEPR model, in which the lone pair on the central oxygen atom is repulsive and requires more space, than bonding pairs. The actual bond angles in a molecule will either be the same as, or less than, the angles in the pure geometric shape. Our answer therefore is reasonable.

What If…?

…we had a molecule of oxygen difluoride, OF_2? What is its molecular geometry? What about its bond angles?

(*Ans: Bent, with a bond angle of about 103°, compared to a bond angle of about 109° for other bent molecules that have two bonding and two lone pairs electron groups on the central atom.*)

Practice 8.9

Which molecule has larger bond angles, SO_2 or H_2O?

Let's use what we now know to examine the three-dimensional structure of morphine. The formula reveals little: $C_{17}H_{21}NO_3$. As we noted at the start of this section, the Lewis dot structure, just like the formula, fails to illustrate the three-dimensional structure of morphine. The structure contains trigonal planar and tetrahedral carbon atoms arranged to make a scaffold that holds the OH groups pointed at one end of the molecule. The nitrogen resides at the opposite end of the molecule as shown in Fig. 8.25. The scaffold holds these groups at specific distances, allowing morphine to fit into the opioid receptor quite well. We can get a handle on the structure of large molecules such as morphine by analyzing the three-dimensional structure at each carbon atom. By examining each atom in the structure and determining the geometry around it, we can build the overall shape of a molecule.

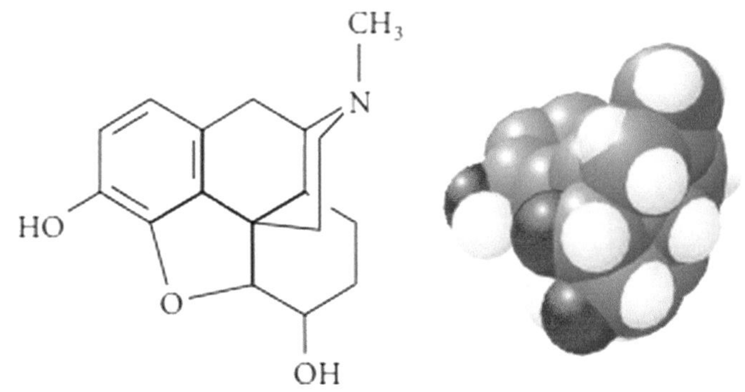

Fig. 8.25 The VSEPR model allows us to understand and interpret the three-dimensional structure of morphine, as described in the text. Note that the line drawing image on the left has omitted the carbons and hydrogens and lone pairs so that the skeleton of the structure is easier to see

Medicinal chemists need to be able to determine the shapes of the molecules they construct in order to assess the fit with an appropriate receptor in the body. This application stems from the simple rule that "structure follows function." That is, the three-dimensional arrangement of atoms within a molecule is just as important to the function (and chemical properties) of a molecule as are the identities of the atoms that make up the molecule. If the geometry of one of the bonds in morphine were not aligned just right for the opioid receptor, morphine wouldn't interact well as an analgesic agent. Knowing this information has empowered chemists to create alternative medicines with variations of the effectiveness of morphine, such as oxycodone and hydromorphone. Note the structural similarities between these molecules and morphine.

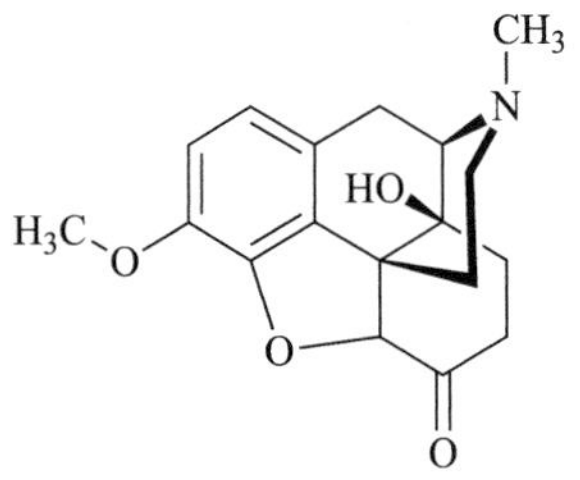

oxycodone
(Oxycontin)

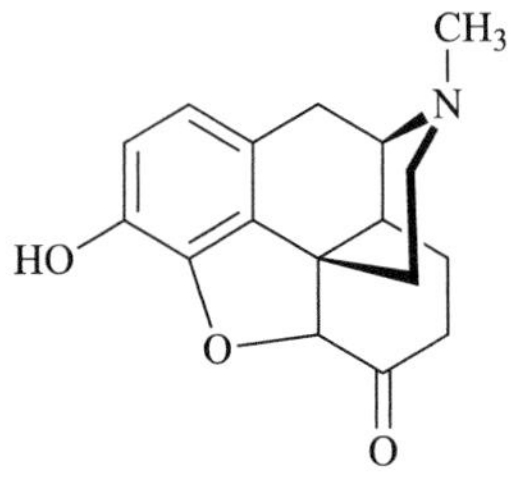

hydromorphone
(DIlaudid)

8.6 Properties of Ionic and Molecular Compounds

The cells within our body are held together by membranes that not only define the limits of the cell but also regulate the passage of materials into and out of the cell. Because most targets that interact with drugs are located inside the cell, the effectiveness of a drug is related to its ability to cross the membrane. A cross section of the cell membrane, shown in in Fig. 8.26, exposes three distinct regions with which a drug must be able to have favorable interactions in order to pass across the membrane. The first and last regions of the cell membrane will interact well only with compounds that have an overall polarization of electrons toward one end of the molecule. The middle portion of the membrane interacts well with molecules that lack this overall polarization, and this difference makes the cell membrane difficult to cross. Any new drug, then, must be carefully designed if its target lies inside the cell.

8.6.1 Bond Dipoles

The polarization of electrons in a bond is commonly referred to as a **bond dipole**. A difference in the electronegativity between the atoms on either end of the bond can be used to illustrate this polarization. In order to incorporate this information into the structure of a compound, we typically write delta plus (δ+) and delta minus

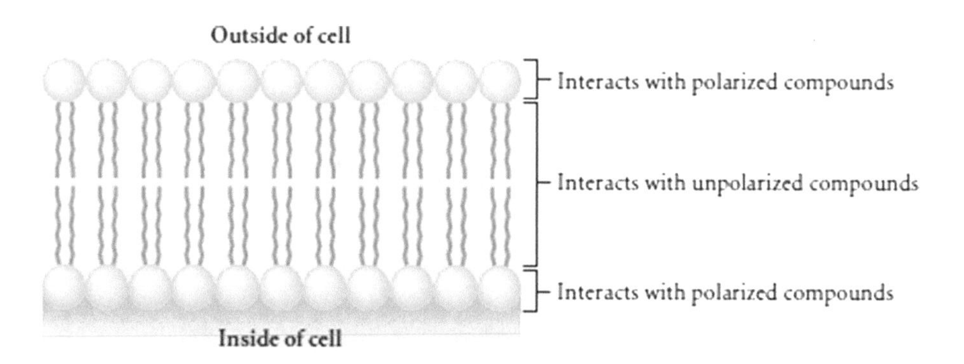

Fig. 8.26 Cross section of a cell membrane. The molecules that make up a membrane have a head and tail structure. The heads align so that they are oriented to the interior or exterior of the cell. That leaves the tails to interact in the center of the membrane

(δ −) on the atoms that make up the polar bond. Alternatively, drawing an arrow over the bond pointing to the more electronegative atom allows us to illustrate the bond dipole. For example, hydrogen fluoride (HF) contains a strongly polarized bond and can be drawn with the bond dipole to illustrate this. Note that the arrow has a plus sign built into the tail end.

Bond dipole – The polarization of electrons in a bond that results in a separation of partial charges in the bond.

H—F:

We can also draw an electrostatic potential map of HF that shows the dipole in a more visually appealing and detailed way.

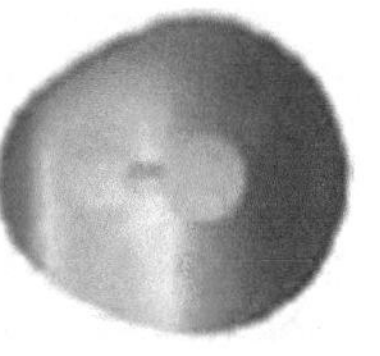

8.6.2 Dipole Moment

So far, we have handled polarization fairly qualitatively, but there is a more quantitative way to indicate polarization. The net polarization of electrons in a molecule is known as the **dipole moment.** For a molecule, this is a function of the orientation and the magnitude of each of the individual bond dipoles. Mathematically, the dipole moment (μ) is defined as.

$$\mu = Q \times d$$

Dipole moment – The polarization of electrons in a molecule that results in a net unequal distribution of charges throughout the molecule.

in which the charge at either end of the dipole (Q) times the distance between the charges (d) yields the dipole moment in Coulomb meters (C·m). In honor of Peter Debye (1884–1966), who won the 1936 Nobel Prize in chemistry for his work on molecular structures, we usually refer to dipole moments in units of **debye (D)**:

$$3.34 \times 10^{-30}\text{C} \cdot \text{m} = 1\text{ D (debye)}$$

Debye – A unit used in the measurement of dipole moment. 3.34×10^{-30} C·m = 1 D (debye)

Bond dipoles can be observed experimentally in binary molecules as the dipole moment. When the molecule is placed in an electric field, the unequal distribution of charge in the molecule (δ + on one end of the bond and δ – on the other end) causes the molecule to orient itself in such a way as to maximize electrostatic attractions. In this way, the molecule behaves similarly to a small bar magnet. The resulting orientation can be measured, and

Table 8.7 Dipole moments of some common binary compounds

Compound	Dipole moment (D)	Compound	Dipole moment (D)
H_2	0.00	ClF	0.88
HF	1.91	BrF	1.29
HCl	1.07	NO	0.16
HBr	0.79	CO	0.13
HI	0.38	O_2	0.00

Fig. 8.27 Carbon dioxide contains two bond dipoles, but overall the molecule lacks a dipole moment

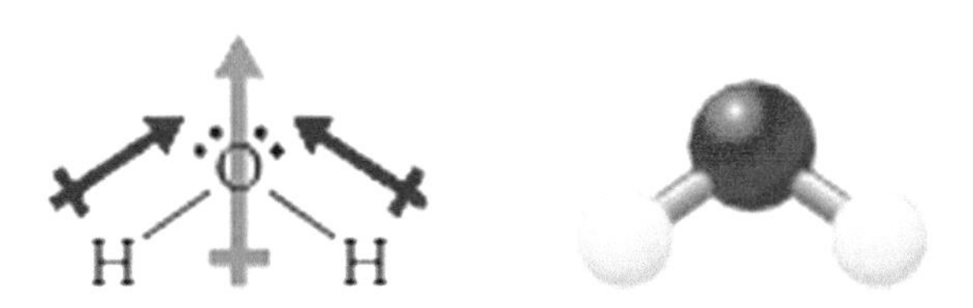

Fig. 8.28 The bond dipole in water. Note the large dipole moment (blue arrow) that arises from the sum of the two bond dipoles (red arrows)

the degree of polarization calculated. The trends observed in dipole moments of some common binary molecules (Table 8.7) seem to follow our expectations for the polarization of the bonds. As we would predict, hydrogen fluoride has the greatest dipole moment of all the hydrogen halides (HF, HCl, HBr, HI) because the electronegativity difference in this series is greatest for HF.

For molecules with more than two atoms, however, an individual bond dipole is harder to measure. Instead, the dipole moment that is measured is related to the sum of all of the bond dipoles in the molecule. For example, hydrogen cyanide (HCN) contains two bond dipoles. Both bond dipoles point in the same direction,

$$\mathrm{H{-}C{\equiv}N{:}}$$

giving rise to an overall polarization of the electrons in the molecule, and a molecule that aligns itself in an electric field with a dipole moment of about 3.0 D. Carbon dioxide also has two bond dipoles, shown in Fig. 8.27. However, they point opposite to each other. The molecule lacks the ability to align itself in an electric field, and the dipole moment of CO_2 is 0.00 D.

The dipole moment is a measure of the **polarity** of a molecule. In some cases, the bond dipoles cancel each other out, as in CO_2. The result is a **nonpolar molecule**, even if the individual bonds are polar. In other cases, the bond dipoles don't cancel out, as in HCN. The net result is a **polar molecule**. At the start of Sect. 8.4, we noted that the shape of a molecule is very important in determining its properties. In particular, we mentioned that some of water's important properties result from its shape. Is water a polar or nonpolar molecule? Water has a tetrahedral electron-group geometry and, because of the existence of two lone pairs on the oxygen atom, has a bent molecular geometry. The bond dipoles of water point toward the oxygen, but because the molecule is bent, the net dipole moment is not zero. In fact, water has a dipole moment (1.85 D) and is a polar molecule (Fig. 8.28).

Polarity – The location of a compound on a scale from polar to nonpolar. The dipole moment is a measure of the polarity of a compound.

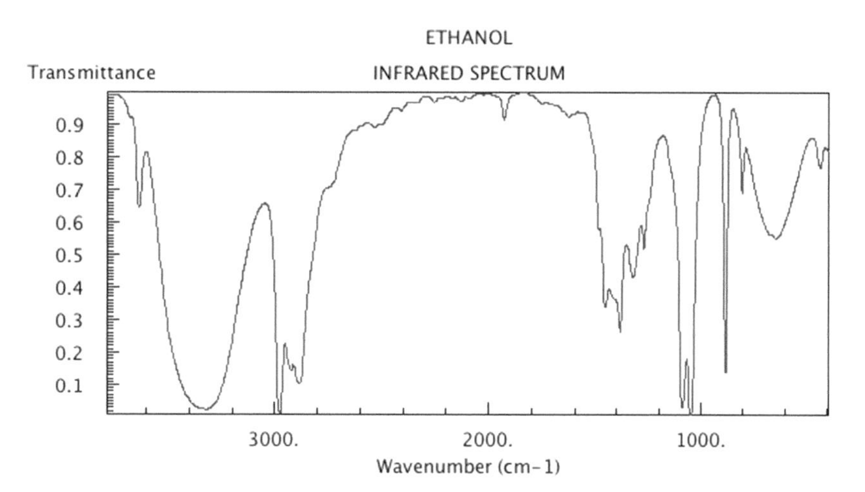

Fig. 8.29 An IR spectrum of ethanol. The frequencies of the vibrations give us information on specific types and arrangement of the atoms in the molecule. *Source* ▶ https://upload.wikimedia.org/wikipedia/commons/4/42/EtOH_Spectra.png

Nonpolar molecule – A molecule in which there is no net charge separation and no net dipole moment.
Polar molecule – A molecule in which there exists a charge separation resulting in a net dipole moment.

The dipole moment is useful in other ways as well. For example, an instrument known as an infrared (IR) spectrophotometer, shown in Fig. 8.29, utilizes dipole moments. The IR spectrophotometer records the energy associated with the vibrations of a molecule, as long as the vibration produces a change in the molecule's dipole moment. The resulting information can be used to classify compounds in terms of the types of bonds they contain, and it also reveals the identity of compounds through a comparison with known samples. IR spectroscopy has been applied to determining the quality of motor oil, finding the amount of carbon monoxide in automobile exhaust, identifying different polyester fibers from crime scenes, and examining soils and rocks on Mars via NASA's Mars Rover.

Example 8.10 Does $CHCl_3$ Have a Dipole Moment?

Does $CHCl_3$ (chloroform, a colorless liquid formerly used as the major component in cough syrups) have a dipole moment? Give the Lewis dot structure for chloroform, and show bond dipoles and the dipole moment for the molecule, if any exists.

Asking the Right Questions

What does the dipole moment measure? What does it tell you about a molecule? What are the steps you need to take to determine if a molecule has a dipole moment?

Solution

A dipole moment is a measure of the polarity of a molecule, which is related to distribution of charge within the molecule. Representing the molecule in three dimensions gives you a sense of how the charge will be distributed. Therefore, drawing the Lewis dot structure of chloroform, followed by the VSEPR-based structure, will help you to analyze the polarity.

The Lewis dot structure model of $CHCl_3$ indicates that the central carbon atom contains four electron pairs. Because they are all bonding pairs of electrons, the VSEPR model tells you that chloroform has a tetrahedral geometry. By placing individual bond dipoles on the molecule, we note that there is a net dipole moment to the molecule. Actual measurements indicate a dipole moment of 1.04 D. You would therefore predict that chloroform is a polar molecule.

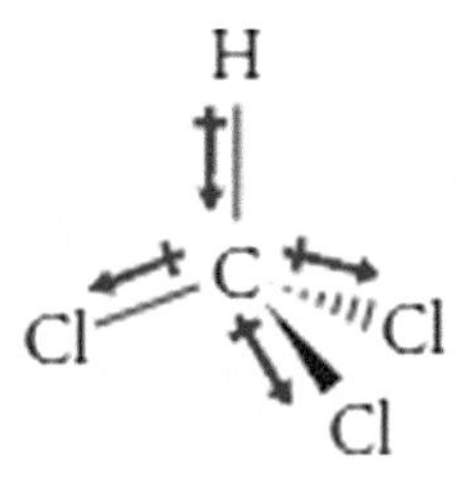

Is Our Answer Reasonable?

The dipole moment is a measure of polarity in a molecule. You can look at polarity from the reverse viewpoint, reasoning, "If the molecule is non-polar, either all the bonds have to be nonpolar, or they have to pull in directions that, as friends pulling at a central flag in a tug of war, the flag would remain in place." In this case, one friend (the C—H bond) is a little weaker than the others, so, using our imperfect analogy. the flag will move slightly. Chloroform is polar, and the answer is reasonable.

What If…?

…we took a another look at ozone, O_3? Would you predict that it has a dipole moment? Why?

(Ans: Yes, a small one, because the bonding pairs are not symmetrical around the central atom.)

Practice 8.10

Predict the bond dipoles and determine if there will be a dipole moment for N_2 and NH_3.

8.6.3 Polar Versus Nonpolar

As we mentioned before, we can use the dipole moment of a molecule to illustrate its overall polarity. In a polar molecule, a net dipole moment exists. In a nonpolar molecule, no net dipole moment exists. However, it isn't really that simple. Molecules reside on a scale of polarities from the purely nonpolar to the highly polar. Water, with its two polar covalent bonds and resulting dipole moment, is a polar molecule. However, it isn't the most polar molecule. Why do we need to know about polarity? The polarity of a molecule can be a useful predictor of the solubility of the molecule in a particular solvent. In other words, "like dissolves like." Polar molecules generally dissolve in polar solvents. Nonpolar molecules generally dissolve in nonpolar solvents. And polar molecules typically do not dissolve in nonpolar solvents. However, like all rules, this one has exceptions.

One exception is that not all molecules of the same polarity dissolve in each other. For example, water and chloroform ($HCCl_3$) do not mix together, in spite of the fact that they are both polar. In short, solubility is affected by more factors than relative molecular polarities. The structure of a molecule is very much a part of the equation, but the rules of solubility are best determined by considering the forces of interaction between dissolving molecules.

One of the major components in cell membranes is sphingomyelin (◻ Fig. 8.30). This molecule contains large bond dipoles at one end and very small bond dipoles at the other. Because the molecule is so long, one end of the molecule is polar and the other end is nonpolar. A cell membrane is made up of two layers of sphingomyelins

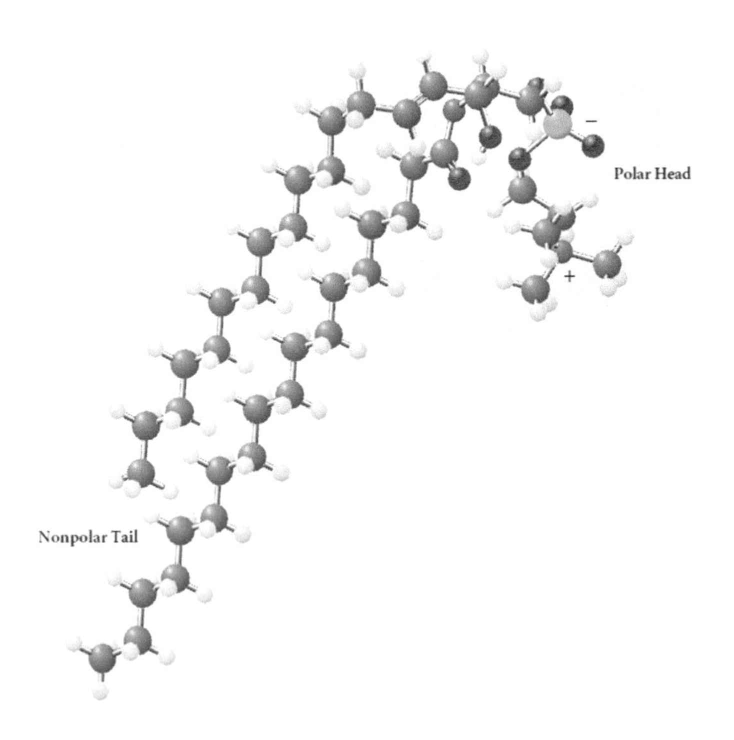

◻ **Fig. 8.30** Sphingomyelin, one of the many molecules that make up the structure of the cell membrane. Note that the tail of the molecule is nonpolar and the head is polar

with the nonpolar ends pointing toward each other. A medicinal chemist interested in developing a new drug must consider the implications of this fact. As we discussed earlier in this section, a new drug that must penetrate a cell to cause a biological response must cross the outer polar region of the cell membrane, pass through the large nonpolar region of the membrane, and then cross the inner polar region of the membrane. In other words, a drug that targets a location inside a cell must be able to dissolve in polar and nonpolar solvents. It must have some polar and some nonpolar character.

Now we have the basic understanding to revisit the biological activity of aspirin, which we introduced at the beginning of this chapter. When you are injured, one of your body's reactions is to produce chemicals called prostaglandins. There are several different types of prostaglandin molecules, and they have various effects such as inflammation, increasing pain signals sent to your brain, and other responses that help your body fight infection and heal itself. But sometimes these responses can make you uncomfortable and you'd like to minimize some of the pain and inflammation. That's where aspirin comes to your rescue. Because of aspirin's shape, it can fit into a particular location on the enzyme that creates prostaglandins, called cyclooxygenase. Low electron density (dark blue shading) is visible around the lower right-hand hydrogen in aspirin in the ep map below. This indicates that the hydrogen can chemically react with the enzyme, permanently disabling it. This means that once aspirin has done its work, your body must go to the trouble of making another enzyme to replace the one that was disabled, which is not a simple or fast process. Understanding how and why aspirin functions the way it does has enabled scientists to create other chemicals that act in a similar fashion, such as acetaminophen (Tylenol®), ibuprofen (Motrin®), and others.

Acetylsalicylic acid
$C_9H_8O_4$

Just as the construction engineer must model a building to take into account all expected mishaps, we must try to address all of the observed properties when we construct models of molecules. The models discussed in this chapter address the shape, dipole moment, and polarity of a compound. However, because they are models, they have limitations. We'll explore other models in the next chapter that attempt to address those limitations.

The Bottom Line

- Models are an important tool that chemists use to help them determine the properties of molecules.
- Better models lead to better understanding of real molecules and the ability to create new molecules with specific properties.
- Estimation of the properties of a molecule on the basis of the structure of that molecule is only as good as the model of that molecule.
- Bonding can range across the full spectrum between equal sharing and complete transfer of electrons.
- The three main types of bonds are ionic, polar covalent, and metallic.
- An anion is always bigger than the atom from which it is derived. A cation is always smaller than the atom from which it is derived.
- The lattice enthalpy is an important expression of the energetic stability of a salt.
- We can use bond enthalpy calculations to determine the approximate energy change involved in a reaction.
- Lewis dot structures are useful in constructing a simple model showing the location of atoms within a molecule.
- Resonance hybrids offer an overall picture of a molecule. Individual resonance structures do not adequately describe a molecule.

- VSEPR theory describes the shapes of molecules better than Lewis dot structures. The model drawn using VSEPR provides a three-dimensional picture of the molecule.
- The polarity of a molecule is related to the overall forces of the individual bond dipoles in the molecule and to the molecule's three-dimensional shape.

Section 8.1 Modeling Bonds

Skill Review

1. Use Lewis electron dot structures to answer these questions about aluminum and nitrogen.
   a. Which neutral atom, Al or N, has the greater number of unpaired valence electrons?
   b. Which neutral atom, Al or N, has the greater number of valence electrons?
   c. Which neutral atom, Al or N, is more likely to gain electrons to form an octet?
2. Use Lewis electron dot structures to answer these questions about carbon and sulfur.
   a. Which neutral atom, C or S, has the greater number of unpaired valence electrons?
   b. Which neutral atom, C or S, has the greater number of valence electrons?
   c. Which neutral atom, C or S, is more likely to gain electrons to form an octet?

3. Of the following, which, if any, would not have the same configuration as a hydrogen atom with two electrons?
   a. C^{2+} b. B^{3+} c. N^{3-} d. H^{+}
4. Of the following, which, if any, would not have the same configuration as a fluorine atom with eight valence electrons?
   a. Ne b. Na^{+} c. Br^{-} d. S^{2-}
5. To which group does each of these ions belong?
   a. Ion of an element with a − 2 charge and a full octet
   b. Ion of an element with a + 2 charge and a full octet
   c. Ion of an element with a − 3 charge and a full octet
6. To which group does each of these ions belong?
   a. Ion of an element with a + 1 charge and a full octet
   b. Ion of an element with a − 1 charge and a full octet
   c. Atom of an element with no charge and a full octet
7. The charges have been omitted from these Lewis structure diagrams of ions. From your knowledge of the ground state for atoms, determine the number of extra electrons that each structure is showing, and provide the proper charge for each ion.
   a. [:S:] b. [:P:] c. [:Cl:]
8. The charges have been omitted from these Lewis structures for atoms. Determine the number of extra electrons that each structure is showing, and provide the proper charge for each ion.
   a. [:O:] b. [:Br:] c. [:N:]
9. Write Lewis electron dot diagrams for each of these ions.
   a. Se^{2-} b. I^{-} c. Sr^{2+} d. Sc^{3+} e. Si^{2+}
10. Write Lewis electron dot diagrams for each of these ions.
    a. S^{2-} b. Br^{-} c. Ti^{2+} d. Ti^{4+} e. B^{3+}
11. Which of these ion(s) would have the same electron configuration as the noble gas neon?

    Cl^{-} Na^{+} F^{-} C^{2+} Al^{3+}

12. Which of these ion(s) would have the same electron configuration as fluoride (F^{-})?

    Ne^{-} O^{2-} N C^{2+} H^{+}

13. Which of these, if any, are isoelectronic?

    Ca^{2+} Sc^{+} S Ar Cl^{-}

14. Which of these, if any, are isoelectronic?

    Mg^{2+} Na Be^{2+} Ar F^{-}

15. Which neutral atoms could be represented by the Lewis dot symbol shown below

:Ẋ·

16. Which cations with a + 1 charge could be represented by the Lewis dot symbol shown below?

·X·

Chemical Applications and Practices

17. Lithium, sodium, and potassium are all very reactive alkali metals.
    a. Diagram the Lewis electron dot structure for each neutral atom.
    b. Predict the formula of the oxide compound of each.
18. Calcium, magnesium, and barium are members of the alkaline earth metals.
    a. Diagram the Lewis electron dot structure for each neutral atom.
    b. Predict the formula of the oxide compound of each.

Section 8.2 Ionic Bonding

Skill Review

19. Consider each statement below and determine whether it is true or false for compounds with ionic bonding.
    a. They are typically composed of a nonmetal and a metal.
    b. Electrons are shared among atoms in the compound.
    c. The metal typically becomes a cation.
    d. Like charges repel each other.
20. Consider each statement below and determine whether it is true or false for compounds with ionic bonding.
    a. They require only one metal and one nonmetal
    b. The metal is always written first in the formula.
    c. The nonmetal must have a − 1 charge
    d. The metal and nonmetal must be in the same period.
21. Using the octet rule, explain why aluminum loses three electrons in both aluminum oxide and aluminum chloride, yet the ratio of aluminum to the nonmetal is 1:3 in aluminum chloride and 2:3 in aluminum oxide.
22. Using the octet rule, explain why magnesium loses two electrons in both magnesium oxide and magnesium chloride, yet the ratio of magnesium to the nonmetal is 1:2 in magnesium chloride and 1:1 in magnesium oxide.
23. Within each of the series presented, rank the atoms or ions from smallest to largest radius.
    a. I^{5+} I I^{-}
    b. S^{6+} S^{4+} S^{2-}
    c. C^{+} C C^{-}
    d. Fe Fe^{3+} Fe^{2+}
24. Within each of the series presented, rank the atoms or ions from smallest to largest radius.
    a. Cu^{2+} Cu^{+} Cu
    b. Cr^{6+} Cr^{4+} Cr
    c. N^{-} N^{3-} N^{+}
    d. Pd Pd^{4+} Pd^{2+}
25. Assign each of the spheres shown here to each of the following isoelectronic atoms and ions (F^{-}, O^{2-}, Ne):

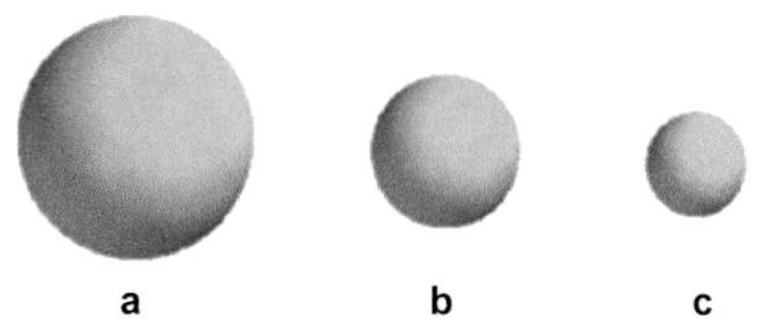

26. Assign each of the spheres shown here to each of the following isoelectronic atoms and ions (K^{+}, Ar, Ca^{2+}):

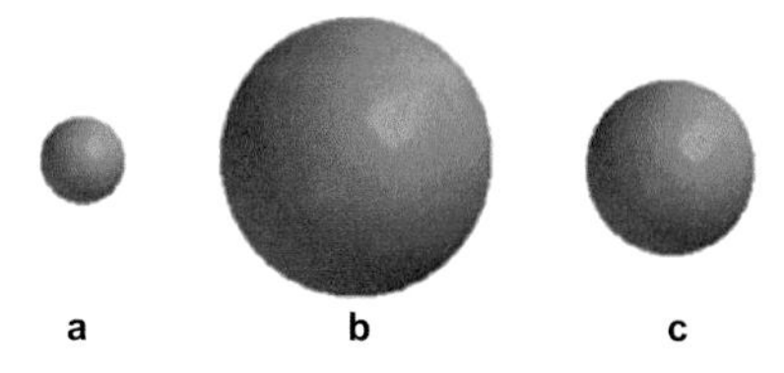

27. Using Coulomb's law, arrange these alkali halides from strongest to weakest attraction of anion and cation: KCl, NaCl, CsCl, LiCl, RbCl.
28 Using Coulomb's law, arrange these alkali halides from strongest to weakest attraction of anion and cation: LiCl, LiF, LiBr, LiI.

Chemical Applications and Practices

29. Potassium chloride (KCl) has been used as a sodium substitute in some commercial salt (NaCl) products. Show the formation of this important ionic compound using the Lewis dot symbol model.
30. When some metals oxidize, the product formed may be detrimental to the strength of the metal structure. This is the case when iron corrodes. However, aluminum forms an oxide that resists further reaction and adheres to the metal, forming a protective covering. Show the formation of this ionic compound using the Lewis dot symbol model.
31. When placed in water, the oxide compounds of many metals form alkaline, or basic, solutions. This takes place when the metal oxide reacts with water to produce hydroxide ions. Diagram the structure of the hydroxide ion that forms when lime is placed in water.
32. When placed in water, the oxide compounds of many metals form alkaline, or basic, solutions. This takes place when the metal oxide reacts with water to produce hydroxide ions. Diagram the Lewis dot structure of sodium oxide, and diagram the structure of the compound that forms when sodium oxide reacts with water.
33. If Mn^{5+} and Mn^{4+} were in a solution together, which would fit into a smaller zeolite hole?
34. Which of these copper ions would fit into a smaller zeolite hole, Cu^{+} or Cu^{2+}?
35. The melting point of KBr is 734 °C. The melting point of CsBr is 640 °C. Which of the two compounds do you predict to have the higher lattice enthalpy? Explain your answer in terms of Coulomb's law.
36. Of NaCl, KCl, and $MgCl_2$, which would you predict to have the highest lattice enthalpy? Explain your answer in terms of Coulomb's law.

8

Section 8.3 Covalent Bonding

Skill Review

37. The terms electronegativity and electron affinity are both used to explain some bonding concepts. Compare and contrast these two important terms.
38. The expression "covalent bond" is used in three different contexts within the chapter. Explain the differences among nonpolar covalent bonds, polar covalent bonds, and coordinate covalent bonds. Provide an example of a compound that illustrates each of these three types of covalent bonds.
39. Draw the outline of a periodic table. Using arrows to indicate a direction of increase, show the general periodic trend for electronegativity within a group and within a period.
40. Draw the outline of a periodic table. Using arrows to indicate a direction of increase, show the general periodic trend for atomic size within a group and within a period.
41. Diagram the Lewis structure for each of these neutral compounds. Identify any that contain an odd number of (radical) electrons

H_2O	NO	CO	NO_2
HCl	PCl_2	NBr_3	

42. Diagram the Lewis structure for each of these neutral compounds. Identify any that contain an odd number of (radical) electrons.

CS_2	N_2O	PCl_3	SO_2
H_2O_2	SeH_2	SF_4	

43. Diagram the Lewis structure for each of these ions. Identify any that contain an odd number of (radical) electrons.

OH^-	NO_2^-	Br^-	PO_4^{3-}
SO_3^{2-}	CO_3^{2-}	BrO_4^-	

44. Diagram the Lewis structure for each of these ions. Identify any that contain an odd number of (radical) electrons.

SH^- NO_3^- F^- IO_4^-
BO_3^{3-} CN^- CrO_4^{2-}

45. Diagram the Lewis structure for each of these compounds. Identify any that contain an odd number of (radical) electrons.

C_2H_6 C_3H_6 C_2H_4 C_3H_8 C_4H_{10}

46. Diagram the Lewis structure for each of these compounds. Identify any that contain an odd number of (radical) electrons.

C_2H_2 C_3H_4 C_3H_8O C_2H_6O CH_2Cl_2

47. It is possible to diagram three resonance structures for the nitrate (NO_3^-) ion. Show, by giving formal charges, that each contributes equally to the overall hybrid structure that depicts bonds of equal strength between the three oxygen atoms and one nitrogen atom.
48. It is possible to diagram many resonance structures for the phosphate (PO_4^{3-}) ion. Show, by giving formal charges, that each contributes equally to the overall hybrid structure that depicts bonds of equal strength between the four oxygen atoms and one phosphorus atom.
49. Under the proper conditions, carbon atoms can bond to form closed ring structures when the carbon atoms are bonded to each other. These are called cyclic compounds. Three four-carbon ring compounds (C_4H_4, C_4H_6, and C_4H_8) are known to exist. Diagram the Lewis dot structure of each and predict which, if any, may have a resonance structure. Diagram the resonance structures of any possible results.
50. Three straight-chain four-carbon compounds that are not cyclic (C_4H_{10}, C_4H_6, and C_4H_8) are known to exist. Diagram the Lewis dot structure of each and predict which, if any, may have a resonance structure. Diagram the resonance structures of any possible results.
51. The compounds N_2O, N_2, and N_2H_4 all contain nitrogen to nitrogen bonds. Diagram the Lewis electron structure of each. Then match these bond energies for the nitrogen-to-nitrogen bond in each. (946 kJ/mol, 160 kJ/mol, 418 kJ/mol)
52. In which structure would you expect to find the shorter carbon-to-oxygen bond, carbon monoxide or carbon dioxide? Explain the basis of your choice. In which would you expect to find the higher value for bond energy? Explain the basis of your choice.
53. Each of these compounds contains both ionic and covalent bonding. Using their Lewis dot structures, indicate where each type of bonding occurs.
    a. Na_3PO_4 b. $CaCO_3$ c. $Fe(NO_3)_2$
54. Each of these compounds contains both ionic and covalent bonding. Using their Lewis dot structures, indicate where each type of bonding occurs.
    a. NaOH b. $NaHCO_3$ c. K_2CrO_4
55. This list gives bonds that could form in compounds. Arrange them in order from least polar bond to most polar bond.

Ca–N H–C Ca–H
C–C N–O

56. This list gives bonds that could form in compounds. Arrange them in order from least polar bond to most polar bond.

H–F O–P Li–F
S–F N–S

57. a. Calculate the formal charges on the carbon atoms in ethane (C_2H_6).
    b. If one of the hydrogen atoms is replaced with an OH group, the compound becomes ethanol. What changes, if any, in the formal charge of the carbon atoms take place when this conversion is made?
    c. What is the formal charge on the oxygen atom in ethanol?
    d. If the hydrogen atom were then removed from the oxygen atom, with no other atom taking its place, what would you calculate as the formal charge for oxygen?
58. Diagram the Lewis dot structure of the sulfate ion (SO_4^{2-}) where sulfur has only 8 electrons. Then, draw another structure where it has more than 8 electrons. According to the reasoning about formal charges suggested in the chapter, which of the two structures is more likely to represent the actual structure of this important ion?

59. Identify the most polar bond in the following structure. (carbon = grey; oxygen = red; hydrogen = white)

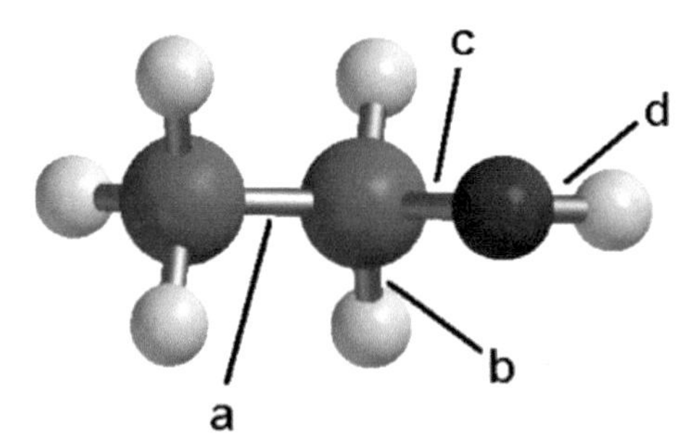

60. Identify the most polar bond in the following structure. (carbon = grey; chlorine = brown; hydrogen = white; sulfur = teal)

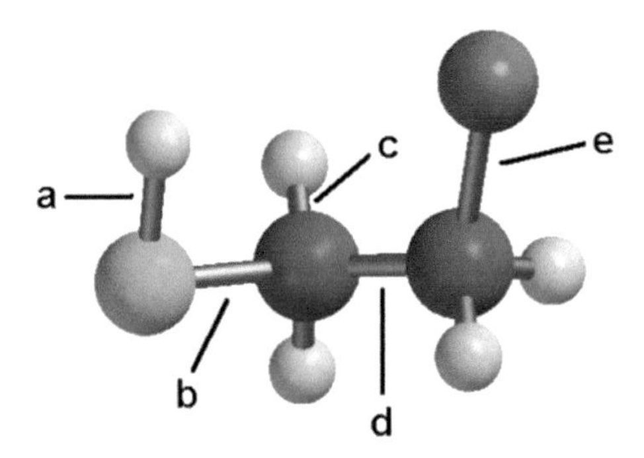

Chemical Applications and Practices

61. Ammonium chloride (NH_4Cl) is used in the manufacture of some dry cell batteries. Show the structure of this important ionic battery component using the Lewis dot symbol model.
62. One of the major manufactured compounds containing iodine is hydrogen iodide. It is used to produce other compounds that contain the iodide anion. Diagram the Lewis dot structure of this compound.
63. Diagram the Lewis structure for the air pollutant nitrogen dioxide (NO_2). Determine the formal charges on each atom in the structure. Circle and label any electrons that would be considered bonding, lone-pair, or radical.
64. Many states are blending ethanol with gasoline to produce E10 to use in automobiles. This mixture allows us to extend our dwindling gasoline supplies and improve the octane rating of the gasoline. Diagram the Lewis structure for this renewable energy extender (C_2H_5OH). Determine the formal charges on each atom in the structure. Circle and label any electrons that would be considered bonding, lone-pair, or radical.
65. The cyanide ion (CN^-) plays a critical role in reacting metals from ores in the process of producing pure metals. Diagram the Lewis structure of this important ion, and determine the formal charges on both the carbon and nitrogen atoms. When the ion combines with H^+, hydrocyanic acid is formed. On the basis of your calculations of formal charges, would it be more likely for the H^+ to attach to the carbon side or the nitrogen side of the ion? Explain your choice.
66. Hypochlorous acid is the acid that can be used to make the salt sodium hypochlorite that is found in most commercial bleach preparations. Hypochlorous acid consists of one atom each of hydrogen, chlorine, and oxygen. Diagram three different arrangements of the atoms in the molecule and use formal charge considerations to predict which is most likely.
67. Sulfurous acid (H_2SO_3) is one of the molecules that contribute to acid rain. The molecule has three oxygen atoms attached to the central sulfur atom. Diagram the Lewis structure for the acid and predict the formal charge on the sulfur atom. Is the formal charge the same as the oxidation number? Explain any differences.
68. Dimethyl sulfoxide is a solvent used in some veterinary applications. The molecule consists of a central sulfur atom bonded to an oxygen atom and two carbon atoms. Each carbon atom has three bonded hydrogen atoms attached. Each atom follows the Lewis electron dot rules. Diagram the Lewis structure and determine the formal charge on the sulfur atom.
69. The concentrations of both nitrite ion (NO_2^-) and nitrate ion (NO_3^-) must be monitored in well water. Both can be quite harmful to humans. Diagram the Lewis dot structures of both (the nitrogen atom is in the center of both, and there are no oxygen-to-oxygen bonds), and assign a formal charge to the nitrogen atom in each.

70. Phosphoric acid (H_3PO_4) is used in the production of phosphate fertilizers and is often found in soft drinks (check the labels). In the molecule, phosphorus is at the center, and the hydrogen atoms are attached to the oxygen atoms. Phosphorous acid (H_3PO_3) differs slightly in that one of the hydrogen atoms is attached to the central phosphorus atom. Diagram the Lewis electron dot structure of both and compare the formal charges on the phosphorus atoms.
71. Of the approximately 20 naturally occurring amino acids, glycine, the principal component in silk, has the simplest structure. Show the Lewis dot structure of glycine. (It is composed of two atoms of carbon, one of which is bonded to two hydrogen atoms, and a nitrogen, which is also bonded to two hydrogen atoms. The other carbon atom is bonded to two oxygen atoms, one of which is bonded to an atom of hydrogen. Examine the structure. Is there a possible resonance structure to draw for the one you have shown? If so, diagram the other resonance form.
72. Oxalic acid ($C_2O_4H_2$) is an organic acid found in rhubarb and spinach leaves. In the structure, the two carbons are connected by a single bond. Each carbon is connected to two oxygens (one with a double bond and one with a single bond). The hydrogens are on opposite sides of the structure. Diagram the structure and determine whether a possible resonance structure exists for the compound. If so, draw all of the possible resonance structures.
73. The three simplest two-carbon hydrocarbon compounds are the heating fuel ethane (C_2H_6), the plant hormone ethene (C_2H_4), and the welding gas ethyne (C_2H_2). Diagram the Lewis structure of each and predict which would have the highest bond energy. Which would have the longest carbon-to-carbon bond?
74. Absorbed light can be sufficient to break chemical bonds. Compare the bonding of Cl_2 and O_2. In which would the light absorbed have to be greater in energy? Use Planck's constant to calculate the frequency of the energy needed to break apart the molecule you selected. What would the energy need to be in order to break 1 mol of those bonds?
75. Butane (C_4H_{10}) is the fuel typically used in small, hand-held lighters to provide heat for other reactions through combustion. Use bond energies to determine the heat of the reaction between butane and oxygen to produce water and carbon dioxide. When another determination for the heat of this reaction was made, using a calorimeter rather than tabular values, the answer was slightly different. Explain why the two values may not agree, even though they were obtained for the same combustion reaction.
76. One way to produce ethanol (CH_3CH_2OH) is by the reaction of ethene (C_2H_4) and water. If the heat of this reaction equals −37 kJ/mol, what would you calculate as the bond energy for the carbon double bond in ethene? How does this compare with the tabular value for the C=C?
77. Using hydrogen gas and oxygen gas, calculate the heat of reaction both for water and for the bleaching agent hydrogen peroxide (H_2O_2). Judging on the basis of your calculation, which of the two should be more stable? Explain the basis of your choice.
78. Small portable tanks often contain propane (C_3H_8) as the fuel to provide heat for camping trips. Use bond energies and the balanced chemical equation for the combustion of propane to determine the heat of the reaction between oxygen and propane to produce carbon dioxide and water. On a per-gram basis, which fuel provides more heat, propane or butane (C_4H_{10})?

Section 8.4 VSEPR Theory—A Better Model

Skill Review

79. Identify the molecular geometry in each of the following diagrams:

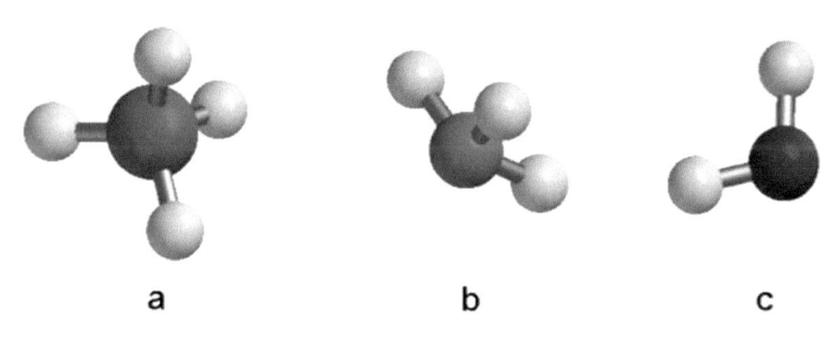

80. Identify the molecular geometry in each of the following diagrams:

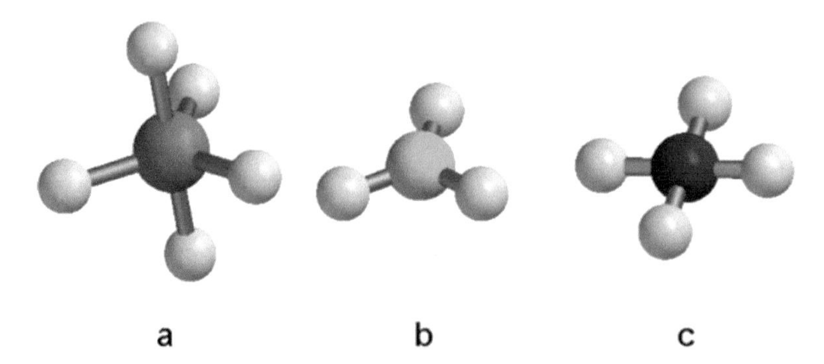

81. Identify the electron group geometry in each of the diagrams shown in problem 79.
82. Identify the electron group geometry in each of the diagrams shown in problem 80.
83. For those compounds listed in Problem 41 that contain more than two atoms, determine the electron group and molecular geometry of the central atoms.
84. For those compounds listed in Problem 42 that contain more than two atoms, determine the electron group and molecular geometry of the central atoms.
85. For those compounds listed in Problem 43 that contain more than two atoms, determine the geometry of the central atoms.
86. For those compounds listed in Problem 44 that contain more than two atoms, determine the geometry of the central atoms.
87. For those compounds listed in Problem 45, determine the geometry of each carbon atom.
88. For those compounds listed in Problem 46, determine the geometry of each carbon atom.
89. Use Lewis dot structures and the VSEPR theory to explain the bond angles in ClO_2.
90. Use Lewis dot structures and the VSEPR theory to explain the bond angles in SF_2.
91. Use VSEPR modeling to arrange this list in order from largest to smallest Cl-to-element-to-Cl bond angle:

$BeCl_2$ $AlCl_3$ CCl_4 $XeCl_4$ NCl_3

92. Use VSEPR modeling to arrange this list in order from largest to smallest F-to-element-to-F bond angle:

MgF_2 NF_3 CF_4 PF_6 IF_3

Chemical Applications and Practices

93. cis-Platin ($PtCl_2(NH_3)_2$) is a useful anticancer drug that binds to DNA. This binding causes the death of the cancerous cell. In the structure of this compound, the two chlorine atoms are adjacent.
    a. Draw a Lewis Dot Structure for each of the NH_3 molecules that are bound to the platinum.
    b. Given that the electron group geometry around the platinum is octahedral, predict the molecular geometry of this compound.
94. Ethylene oxide (C_2H_4O) is a reactive organic molecule where the two carbon atoms and the oxygen form a three-membered ring.
    a. Draw a Lewis Dot structure for this compound.
    b. Predict the molecular geometry of each atom in the ring.
    c. What would you predict for the bond angle (C—O—C) in this molecule? Does your prediction make sense? Explain your answer.
95. Carbon tetrachloride follows the "octet rule" and has tetrahedral geometry. However, another tetrachloride compound, $SeCl_4$, has a different geometry and does not follow the octet rule. Predict the shape of the second compound and diagram its Lewis dot structure.
96. Noble gases already have an octet of valence electrons. Therefore, any combination with another atom is likely to produce an exception to the octet rule. Predict the shape of the compound XeF_2 and diagram its Lewis dot structure.
97. The compound N_2H_2 has a double bond between the two nitrogen atoms. Use the VSEPR model to predict the H—N—N bond angle in the molecule. If $N_2H_2^{2+}$ had a triple bond between the two nitrogen atoms, how would that affect the same bond angle? Explain the basis of your prediction.
98. The bicarbonate ion (HCO_3^-) is essential as a buffer in your blood.
    a. Diagram the Lewis dot structure for this important ion.

b. Is there a possibility for resonance in the structure? If so, show the example(s).
c. In VSEPR modeling, what shape would this ion have?

Section 8.5 Properties of Ionic Compounds and Molecules

Skill Review

99. Examine each of the bonds depicted as dashes between atoms here. Rank them from least polar to most polar. In addition, rewrite the bond, showing an arrow in the direction of the more negative atom.

C—Cl	C—N	C—O
C—H	C—B	C—Mg

100. Examine each of the bonds depicted as dashes between atoms here. Rank the bonds from least polar to most polar. In addition, rewrite the bond, using the $\delta+$ and $\delta-$ symbols to indicate the polarity of the bond.

H—O	H—C	H—F
H—N	H—Na	H—H

101. Diagram the Lewis structures of both SF_5 and SF_6. Use the VSEPR model to predict the shape and the resulting polarity of each.
102. Diagram the Lewis structures of both PBr_5 and PBr_3. Use the VSEPR model to predict the shape and the resulting polarity of each.
103. Match the dipole moment with the molecule to which it corresponds.
    a. $\mu = 0.00$
    b. $\mu = 1.72$
    c. $\mu = 2.50$

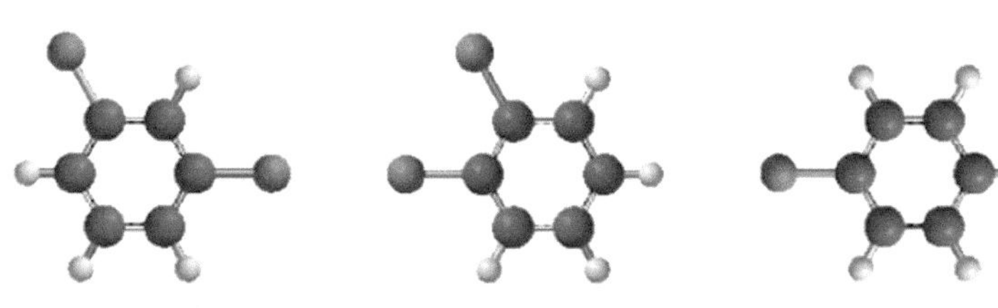

104. Match the dipole moment with the molecule to which it corresponds.
    a. $\mu = 0.82$
    b. $\mu = 1.08$ HF, HCl, HBr
    c. $\mu = 1.82$
105. Which of the following molecules would you predict to be polar?
    a. CO
    b. CO_2
    c. COS
    d. CS_2
    e. H_2CO
    f. CH_2Cl_2
106. Which of the following molecules would you predict to be polar?
    a. PBr_3
    b. PBr_5
    c. N_2O
    d. SO_2
    e. SiO_4
    f. $XeCl_2$
107. Based on the fact that polar molecules are generally soluble in polar solvents, predict which of the compounds below would be soluble in water.
    a. CO_2
    b. CO
    c. C_3H_8 (propane)
    d. C_2H_6O (ethanol)
    e. $C_2H_6O_2$ (ethylene glycol)
    f. C_6H_6 (benzene)
108. Based on the fact that polar molecules are generally soluble in polar solvents, predict which of the compounds below would be soluble in octane (C_8H_{18}).
    a. CO_2
    d. C_2H_6O (ethanol)
    c. C_4H_8 (cyclobutane)
    f. CH_2Cl_2 (dichloromethane)
    e. C_2H_2 (acetylene)
    b. O_2

Chemical Applications and Practices

109. The disinfectant hydrogen peroxide (H_2O_2) has a dipole moment greater than zero. Explain why this information supports suggesting that the molecule is not linear. If the molecule were linear, what would you predict about its dipole moment?
110. The formula $C_2H_2Cl_2$ can be drawn many different ways. Provide the Lewis dot structure that would give a molecule with the largest dipole moment. Provide the Lewis dot structure that would give a molecule that did not have a dipole moment.

Comprehensive Problems

111. A compound's dipole moment was found to be negligible ($\mu = 0.00$). Would you expect to find this compound dissolved in dichloromethane (CH_2Cl_2), ethanol (C_2H_6O), or carbon tetrachloride (CCl_4)?
112. Cyclobutane (C_4H_8) is a molecule whose carbon atoms form a four-membered ring. This molecule is very reactive compared to straight-chained butane (C_4H_{10}) that does not contain a ring.
    a. Draw the Lewis Dot structure of cyclobutane and predict the molecular geometry at each carbon center.
    b. Repeat this analysis for butane.
    c. Based on your predictions of the molecular geometry, why is cyclobutane so reactive?
113. In a previous problem, we examined cyclobutane and butane and compared their reactivities. Repeat the same analysis for cyclobutane (C_4H_8) and cyclobutene (C_4H_6). Which of these compounds would you predict to be more reactive?
114. Pentane (C_5H_{12}) is a flammable liquid used as a solvent in some chemical reactions. Each of the carbon atoms in this molecule is part of a long chain.
    a. Draw the Lewis dot structure of this compound and predict the molecular geometry at each carbon center.
    b. Do you expect this compound to be polar or non-polar?
115. These diagrams represent a hydrogen atom with different numbers of electrons. Determine which diagram represents hydride (H^-), hydrogen (H), and a proton (H^+).

116. In various compounds, nitrogen can be shown to form single, double, or even triple bonds. Diagram the Lewis dot structure of a nitrogen atom. What arrangement do you see that would allow this variety in bonding? Diagram the structure of the N_2 molecule.
117. When you inhale a breath of air, you are taking in oxygen and nitrogen molecules. Examine the Lewis dot structure of an N_2 molecule (see Problem 116). Suggest a reason why the N_2 that you inhale is unaffected and is soon exhaled as the same N_2 that you inhaled.
118. Prepare a diagram, using circles with charges inside, showing a relative size comparison between magnesium oxide (used as an abrasive) and calcium oxide (used in some plaster formulations). On the basis of your representations, explain how this could be used to predict which compound contains the larger lattice energy.
119. Lithium forms many useful salts. For example, lithium fluoride is used extensively in industry to assist in purifying aluminum. Lithium bromide has applications in the air-conditioning industry. Explain which of the two compounds has the stronger lattice energy.
120. a. What is the qualitative relationship between lattice enthalpy and melting point?
    b. On the basis of that relationship, which would you predict to have a higher melting point, MgF_2 (lattice enthalpy = 2913 kJ/mol) or NaCl (lattice enthalpy = 787 kJ/mol)? Use another reference such as the Merck Index or the Chemical Rubber Company's Handbook of Chemistry and Physics to obtain the melting points of the two compounds to check your assumptions.
121. Most ionic compounds have very high melting points. Using Table 8.2, select the ionic combination that would be likely to have the highest melting point and the one that would be likely to have the lowest melting point. Explain the basis of both selections.

122. Magnesium metal will burn so vigorously in air that it will even combine with nitrogen.
    a. What is the formula of magnesium nitride?
    b. Would the compound be ionic or covalent?
    c. Diagram the Lewis dot structure of the compound.
123. a. Explain why the bond energy listed in tables is only an approximation of the bond energy for a bond within a molecule.
    b. In which case would the carbon-to-carbon bond be expected to deviate more from the average, $H_3C—CH_3$ or $H_3C—CF_3$? Explain the basis for your reasoning.
124. The environment of an electron in a bond varies considerably depending on the type of bonding arrangement in which the electron finds itself. Electrons are attracted to the positive nucleus of an atom or ion. Using the examples of covalent, polar covalent, ionic, and metallic bonds, describe the degree of attraction of an electron to its "original" atom once it has become part of a bond with another atom.
125. Iron metal can react with oxygen gas to make iron(III) oxide.
    a. Write the balanced equation for this reaction.
    b. How many unpaired electrons are in an isolated ion of iron(III)?
    c. What is the mass percent of iron in iron(III) oxide?
126. People on low-sodium diets sometimes use a salt substitute for their meals. Suppose the salt substitute contains potassium chloride.
    a. What is the electron configuration of potassium ions?
    b. Would you expect potassium chloride to taste similar to sodium chloride? Explain.
    c. Which is larger, the potassium cation or the sodium cation?
    d. How many potassium cations are there in 1.00 g of potassium chloride?
127. Calcium carbonate is a very common mineral found in nature.
    a. Is this compound soluble in water?
    b. What is the mass percent carbon in this compound?
    c. When hydrochloric acid is added to this compound, calcium chloride and water are formed. What is the Lewis dot structure of the other product?
128. A fertilizer salesperson remarks that potassium nitrate provides more nitrogen per pound than calcium nitrate.
    a. Based on the mass percent nitrogen, is the salesperson correct?
    b. How many grams of nitrogen would be present in 1.00 lb of the potassium nitrate fertilizer?
    c. Write the Lewis dot structure for the nitrate ion.
129. The structure of nifedipine, ◘ Fig. 8.19, contains four different types of atoms.
    a. What is the molar mass of this useful drug?
    b. Draw at least one other resonance structure for this molecule.
    c. When this molecule is treated with an acid, a proton (H^+) is added to one of the nitrogen atoms in the structure. To which nitrogen does that proton bond?
130. Based on the electrostatic potential map shown for nicotine, would you expect nicotine to be soluble in water or in hexane (C_6H_{14})? Explain your answer.

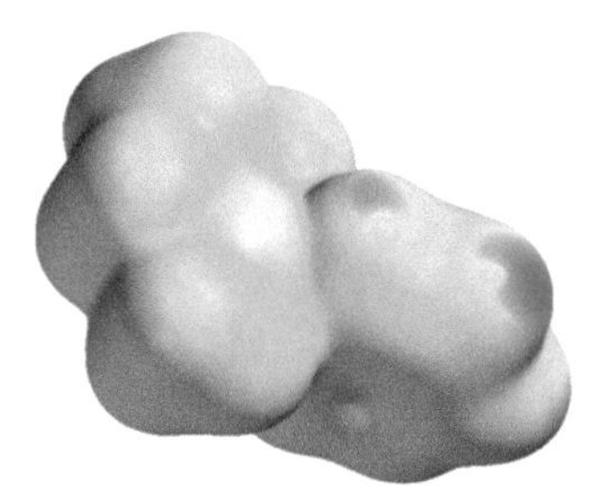

131. Based on the electrostatic potential map shown for ascorbic acid (vitamin C), would you expect ascorbic acid to be soluble in water or in hexane (C_6H_{14})? Explain your answer.

Thinking Beyond the Calculation

132. Someone proposes the use of dichloroethane ($C_2H_4Cl_2$) as a solvent to remove small wax (nonpolar hydrocarbon) buildups.
    a. Diagram a way to depict the structure that possesses an essentially zero dipole moment.
    b. Diagram another structure that would yield a dipole moment.
    c. Which of the compounds, that in part a, or that in part b, would be best suited to dissolve wax?
    d. Dichloroethane can be prepared by adding chlorine (Cl_2) to ethene (C_2H_4). Draw a Lewis dot structure for ethene.
    e. Would you predict a nonzero dipole moment for ethene?
    f. If 2.0 g of dichloroethane was prepared using the method in part d, how many grams of ethene was consumed in the reaction?

Advanced Models of Bonding

Contents

Supplementary Information The online version contains supplementary material available at ▶ https://doi.org/10.1007/978-3-030-90267-4_9.

M. Mosher and P. Kelter, *An Introduction to Chemistry*,
https://doi.org/10.1007/978-3-030-90267-4_9

What Will We Learn in This Chapter?

- Models of chemical bonding and molecular structure such as VSEPR and Valence Bond Theory allow us to discover more about the nature of molecular compounds (▶ Sect. 9.1).
- Sigma and pi bonds are key covalent bonding patterns in molecules (▶ Sect. 9.1).
- Hybridized orbitals allow us to model the structure of molecules better than VSEPR and Valence Bond Theory (▶ Sect. 9.2).
- Molecular Orbital Theory can be used to explain molecular structure and behavior that we cannot explain using simpler models (▶ Sect. 9.3)
- We can determine the bond order in a molecule and predict certain properties of molecules (▶ Sect. 9.3).
- Delocalized pi-bond systems influence the chemical and physical properties of a molecule (▶ Sect. 9.4).

9.1 Introduction

We begin our study of advanced models of bonding by peering down into a most remarkable organ, the human eye. Unlike our hands, our eyes enable us to interact with our environment without direct contact. Light reflected from other objects enters the eye through a small opening called the iris. Once there, the photons hit the retina, where they are collected, converted into electrical impulses, and sent to the brain for processing (◘ Fig. 9.1). The retina contains two types of photoreceptor cells (rods and cones) that collect the light. The number and type of rods and cones distinguish the human eye from that of other species.

Over 120 million rods occupy the human retina. These cells are extremely sensitive to light but do not distinguish color. It is these rods that allow humans some limited nighttime vision. The 7 million cones, which are much less sensitive to light than the rods, allow humans to perceive color. How are rods and cones related to molecular structure and advanced bonding models? It has to do with the way they absorb light.

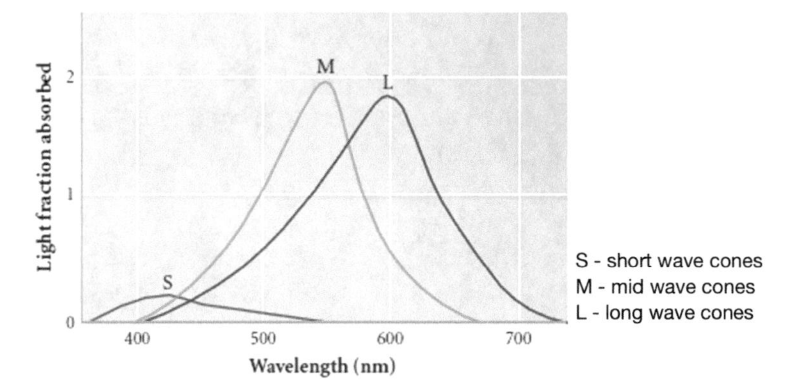

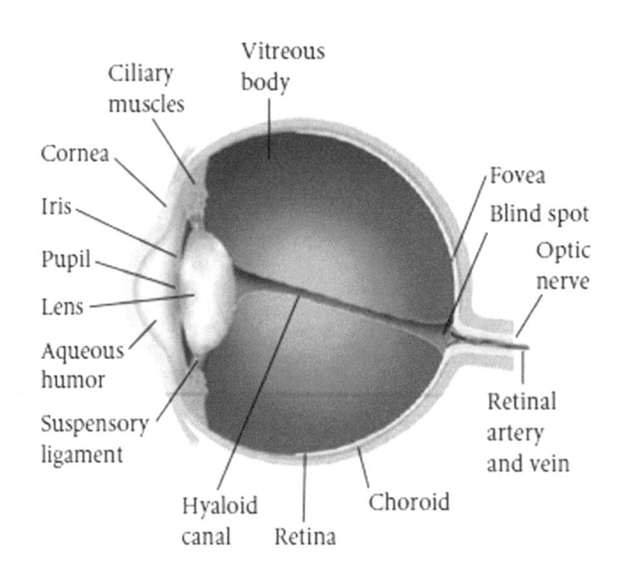

◘ **Fig. 9.1** The human eye is a complex organ. Light enters the front of the eye through the cornea and is focused by the lens on the retina at the back of the eye

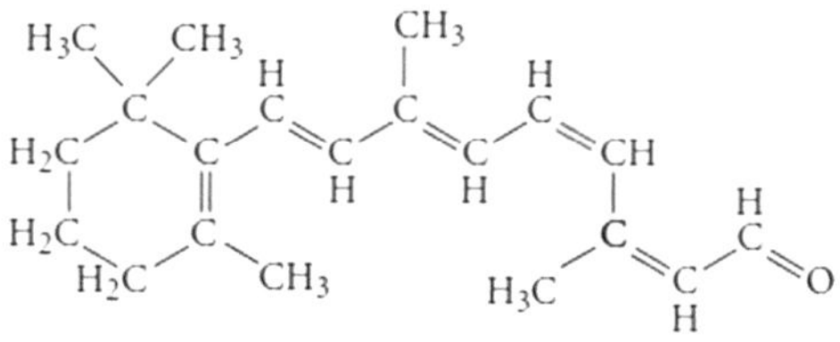

11-*cis*-Retinal

Fig. 9.2 11-*cis*-Retinal absorbs light with wavelengths near 500 nm. The absorption of light causes an electron in the double bonds to become excited, allowing the molecule to rotate about the double bond in red. The result is the formation of trans-retinal

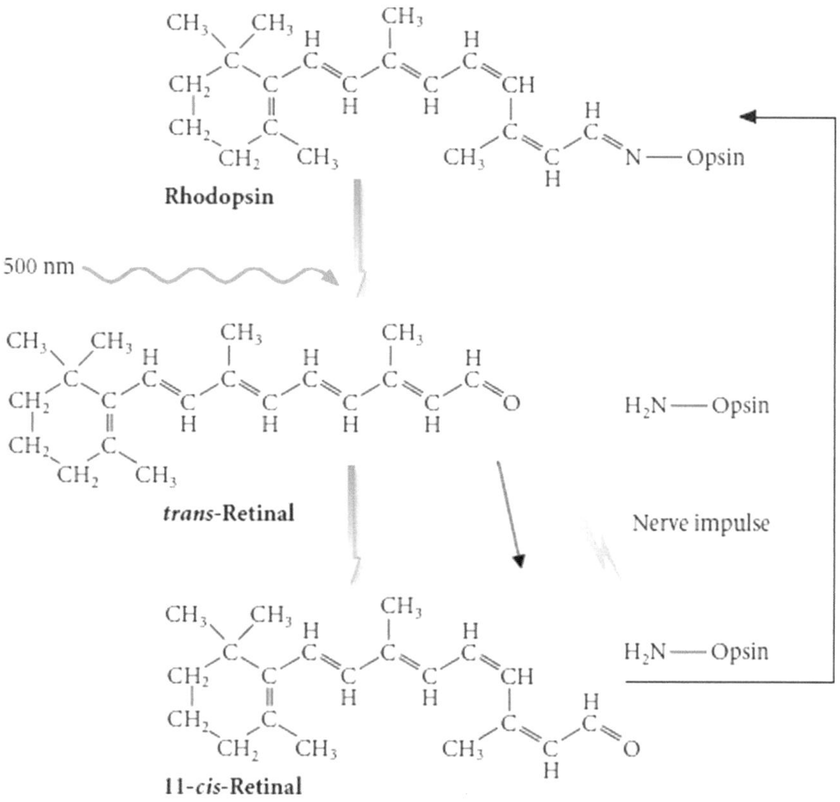

Rhodopsin
trans-Retinal
11-*cis*-Retinal

Fig. 9.3 Transformation of 11-cis-retinal to trans-retinal and back again. After the absorption of light and the reorganization of the double bond, the rhodopsin molecule undergoes a conformational change that brings about a nerve impulse. The trans-retinal is recycled by conversion back to 11-cis-retinal and reattached to an opsin molecule

Located in each rod and cone is a protein called opsin. The protein itself doesn't absorb visible light. However, when a molecule known as 11-*cis*-retinal (see Fig. 9.2) is attached to the opsin, the resulting protein, rhodopsin, becomes able to absorb wavelengths near 500 nm. A photon that strikes the rhodopsin causes a reaction that converts 11-*cis*-retinal to the more stable *trans*-retinal.

$$\text{opsin} + 11\text{–}cis\text{–retinal} \rightarrow \text{rhodopsin} \xrightarrow{500\text{ nm}} trans\text{–retinal}$$

This reaction results in a change in the position of the atoms in the retinal molecule, causing the rhodopsin to alter its shape, which assists in the generation of a nerve impulse and the release of *trans*-retinal from the protein. After the impulse, the *trans*-retinal is converted back into 11-*cis*-retinal and then attached to another opsin—ready for the next photon of light to begin the process again (see Fig. 9.3). Why is rhodopsin capable of absorbing photons? Why does 11-*cis*-retinal absorb light near 500 nm and not in some other part of the spectrum?

The answers to these questions are largely based on the structure of the retinal molecule. However, they are best understood when we examine more advanced models of bonding. Although the VSEPR model and Lewis dot structures are useful for determining the shape and connectivity of the atoms within a molecule, they do not adequately predict many of its properties. For instance, the VSEPR model of water does predict that the HO—H bond angle should be less than 109.5°. However, Lewis dot structure and VSEPR models do not properly explain why the C—C bond in ethane (CH_3—CH_3) is longer than the C—C bond in butadiene (CH_2=CH—CH=CH_2). Moreover, VSEPR doesn't explain how we can end up with equally spaced bonds. For example, in methane (CH_4), the bonding electrons occupy *s* and *p* orbitals. VSEPR can't even begin to explain how electrons in *p* orbitals (90° apart) could form 109.5° bond angles. In order to explain the reasons behind the observed bond angles, and to more accurately assess and predict the properties of molecules, we need to use better models. Our goal in this chapter is to introduce you to some of the better models currently available to chemists.

9.2 Valence Bond Theory

Just as light can be absorbed by *cis*-retinal, it can also be absorbed by the diatomic halogens. Chlorine (Cl_2) and fluorine (F_2) both absorb light according to the reactions shown in ◘ Fig. 9.4. The resulting very reactive atoms are called radicals because they contain one unpaired electron. It is this reactivity that accounts for chlorine's effects on the ozone layer and fluorine's ability to react with normally unreactive atoms such as xenon.

The Lewis dot structures of the halogens indicate that they contain nonpolar covalent bonds and single bonds between the halogen atoms. Are the single bonds in those drawings the same? They appear to be on first glance, but on the basis of the different wavelengths of light required to break the bond between these atoms (as shown in ◘ Fig. 9.4), we can reason that chlorine and fluorine must have different bond strengths. It appears that they have some differences in their bonds that the Lewis dot structure model does not identify.

Knowing the exact makeup of a bond can give us a good picture of its strength. For instance, we discussed in ▶ Chap. 8 how bond strengths can be used to arrive at a rough approximation of the enthalpy of a reaction. We also noted that bond energies were averaged to get the values we saw in the table. The fact that they were averaged implies that not all bonds between the same atoms have the same energy. We can explore this idea more deeply by studying the combustion reactions of both ethane (CH_3CH_3) and acetylene (CH≡CH). A triple bond should be stronger than a single bond, so we might expect that the combustion of acetylene would release less energy than the combustion of ethane, because it takes so much more energy to break that triple bond. Experimentally, as shown in ◘ Table 9.1, the combustion of acetylene releases only slightly less energy than the combustion of ethane (−1300 kJ/mol, compared to −1559 kJ/mol). Even though the C—C bond in ethane (376.1 kJ/mol) is much weaker than the C≡C bond of acetylene (962 kJ/mol), this difference doesn't account for the experimentally determined difference in the enthalpy of combustion between ethane and acetylene. Why are our numbers so different? Some of the missing energy is accounted for by the differences in C—H bond energy. Experimentally,

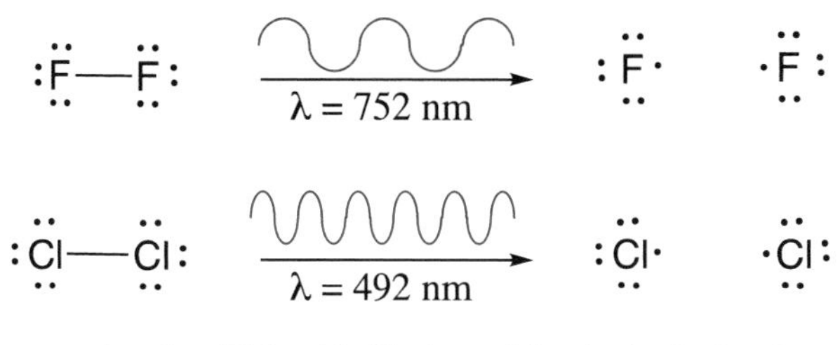

:F—F: → (λ = 752 nm) :F· ·F:

:Cl—Cl: → (λ = 492 nm) :Cl· ·Cl:

◘ **Fig. 9.4** The interaction of different wavelengths of light with chlorine and fluorine imply that these two molecules possess different bond strengths. Which bond is stronger?

◘ **Table 9.1** Calculated versus experimental $\Delta_c H$. The experimental and calculated $\Delta_c H$ values for the fuels listed do not agree. The temperature of the combustion is listed only for reference

Compound	Flame temperature (K)	$\Delta_c H$ (kJ/mol) experimental	$\Delta_c H$ (kJ/mol) calculated
Hydrogen (H_2)	2490	−242	−242
Methane (CH_4)	2285	−890	−803
Ethane (CH_3CH_3)	2338	−1559	−1429
Ethylene (CH_2=CH_2)	2643	−1411	−1324
Acetylene (HC≡CH)	2859	−1300	−1257

Fig. 9.5 Overlap of orbitals makes the bond in H_2. As two hydrogen atoms approach, their 1*s* orbitals overlap to form a bond. The overlap allows both hydrogen atoms to obtain a full valence shell

researchers have determined that the energy of the C–H bond in ethane (410 kJ/mol) is much less than that of the C–H bond in acetylene (536 kJ/mol). However, according to Lewis dot structures and to the VSEPR models, there should be no difference in energy between the C–H bonds in these two molecules. We need a better model of these compounds that takes this difference into account.

One such "better model" of a covalent bond is the valence bond model. In the 1930s, Linus Pauling, devised valence bond theory (VB theory) to address the inadequacy of the bonding models of G.N. Lewis.

9.2.1 What is a Valence Bond?

Instead of the simple sharing of a pair of electrons, Pauling's **valence bond** theory envisions a bond as the overlap of atomic orbitals on adjacent atoms. The overlapping of two half-filled orbitals supplies a pair of electrons, which are covalently shared. Let's begin by looking at the simplest chemical bond possible—the bond between two hydrogen atoms.

Valence bond theory – The theory that all covalent bonds in a molecule arise from overlap of individual valence atomic orbitals.

Molecular hydrogen's use in industry to make such products as the hydrogenated vegetable oils used in processed foods and its possible use as an alternative to gasoline makes it a meaningful compound to study. The electron configuration for a hydrogen atom is $1s^1$. When the 1*s* orbital on the hydrogen atoms overlap, a bond results. Because the electrons are shared by both orbitals, each of the hydrogen atoms possesses the $1s^2$ electron configuration, as shown in Fig. 9.5, which is a full 1*s* orbital. In this model of bonding, the more overlap there is between the two orbitals, the stronger the bond, as we'll investigate in the next section.

Example 9.1—Modifications to the Structure of HF

Hydrogen fluoride is often used by master glassworkers as they etch a design into a piece of art. Some of the world-famous Steuben art has been made via this etching technique. Using valence bond theory, describe which orbitals overlap to form a covalent bond between the hydrogen and fluorine atoms in HF.

Asking the Right Questions

How many *p* orbitals are there in the outer (valence) energy level of fluorine? Which type of orbital on fluorine might overlap with the hydrogen 1*s* to share a needed electron?

Solution

There are three orbitals in the second principal energy level of fluorine, which has a valence electron configuration of $2s^22p^5$. Overlap of the hydrogen 1*s* orbital with one of the 2*p* orbitals allows the electrons to be shared between the two atoms. The valence bond model of HF in Fig. 9.6 shows a 1*s*–2*p* orbital overlap that constitutes the covalent bond.

Is Our Answer Reasonable?

For HF, the model is consistent with a single covalent bond, so the answer is reasonable. Does the valence bond model we constructed imply anything about the properties of the bond? What is the strength of the bond? Does the *s*–*s* orbital overlap in H_2 result in a stronger bond than the *s*–*p* orbital overlap in HF? Does the overlap of the *s* orbital in the first principal energy level and the *p* orbital in the second principal energy level indicate anything about the bond? We answer these questions next.

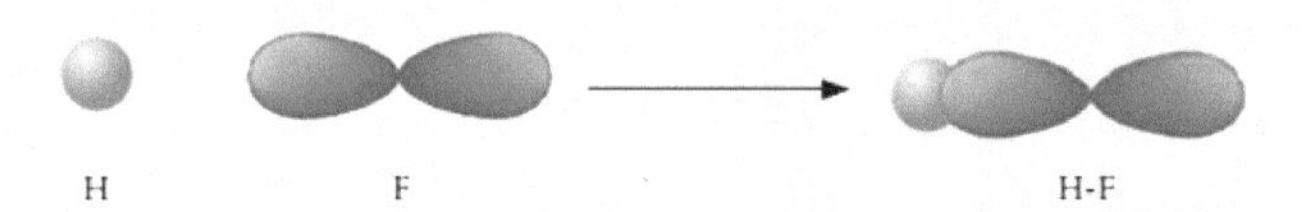

Fig. 9.6 Orbital overlap in HF

What If…?

…we note that the bond energies of H–F, H–Cl, and H–Br are, respectively, 570, 432, and 366 kJ/mol. Can you speculate about a Valence Bond Theory-based reason for this trend?

(*Ans: The amount of overlap of the hydrogen-to-halogen orbitals decreases as you go from fluorine to chlorine to bromine. This will be one focus of our next discussion.*)

Practice 9.1

Using valence bond theory, describe the orbital overlap in F_2, a very reactive molecule used to manufacture Teflon ($(C_2F_4)_n$) and other fluorinated compounds.

9.2.2 Application of Valence Bond Theory

Because the degree of electron-sharing is related to the strength of a bond, orbitals that exhibit more overlap result in stronger covalent bonds. What determines the amount of overlap? The keys are the relative energy, as shown in Fig. 9.7, and the physical size of the atomic orbitals. More specifically:

- Smaller orbitals overlap more than larger orbitals.
- Orbitals with similar sizes overlap more than orbitals with mismatched sizes.
- Orbitals with similar energies overlap more than orbitals with very different energies.

The hydrides LiH, NaH, and KH—used in the removal of oxide coatings on metals—make excellent case studies. In each of these substances, the bonds between adjacent atoms result in covalent overlap of a 1*s* orbital of the hydrogen atom and the *ns* valence orbital of the metal atom. The valence bond in LiH results from an overlap of a 2*s*–1*s* (H) orbital. The valence bond in NaH results from the overlap of a 3*s* (Na) and a 1*s* (H) orbital. Similarly, overlap in KH results from a 4*s* orbital and a 1*s* orbital. Because the size of the metal's *s* orbital is dramatically greater in potassium than in lithium, the overlap of the potassium 4*s* and hydrogen 1*s* orbitals isn't well matched (see Fig. 9.8 and Table 9.2). We would therefore predict the bond in LiH to be much stronger (better overlap) than the bond in KH.

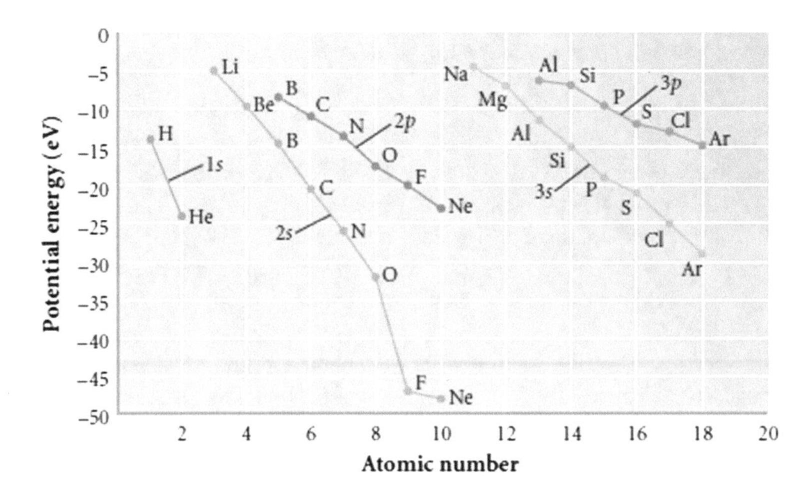

Fig. 9.7 Relative energy of the atomic orbitals. The energy level of the atomic orbitals decreases with increasing nuclear charge, and the atomic orbital becomes more stable because it has a lower potential energy

Fig. 9.8 Overlap of atomic 1*s* orbitals

Table 9.2 Bond energies resulting from orbital overlap (as the difference in size of the overlapping orbitals increases, the strength of the resulting bond decreases. Data in the table are for diatomic molecules)

Bond	$\Delta_{diss}H$ (kJ/mol)	Orbital overlap	Representation of orbital overlap
H–H	435.8	1*s*–1*s*	
H–Li	238.0	1*s*–2*s*	
H–Na	185.7	1*s*–3*s*	
H–K	174.6	1*s*–4*s*	
H–Rb	167.0	1*s*–5*s*	
H–F	569.9	1*s*–2*p*	
H–Cl	431.9	1*s*–3*p*	
H–Br	366.3	1*s*–4*p*	
H–I	298.4	1*s*–5*p*	
F–F	158.3	2*p*–2*p*	

Table 9.2 lists the bond energies for LiH, NaH, and KH as 238 kJ/mol, 185.7 kJ/mol, and 174.6 kJ/mol, respectively. So, the differing size of the overlapping *s* orbitals explains the trend in bond energy. However, there is another factor other than orbital overlap at work here, and that is the electronegativity of the two atoms that are bonded together. Note the relatively low F—F bond energy shown in Table 9.2. Although the 2*p*–2*p* orbital overlap is expected to be quite good, the electronegativity of each halogen atom competes with the orbital overlap. The electrons that participate in the bond between the two fluorine atoms are held very tightly to the individual nuclei. In addition, the lone pairs on each fluorine are very close together due to the relatively short F—F bond. Their repulsions and the electronegativity of the fluorine results in a decreased electron density between the atoms in F_2—and an unusually low bond energy.

Valence bond theory also addresses misconceptions reflected in the Lewis dot structure and VSEPR models about the lengths of bonds. There doesn't appear to be any difference in the bond length for H_2 compared to F_2 if we use only Lewis dot structures as our model. Experimentally, however, we know that the bond lengths are different. Which is longer, the bond in the hydrogen molecule or that in fluorine? The difference in bond lengths results from a difference in the orbitals that overlap to form the covalent bond. Orbitals that extend farther from the nucleus result in bonds that are longer. The key question then, is which orbital reaches farther from the nucleus, an *s* or a *p*? The end-on overlap of two *p* orbitals makes a longer bond than the overlap of two *s* orbitals, because the average electron density lies farther from the nucleus of the atom. For this reason, the bond in F_2 is longer (141.7 pm) than the bond in H_2 (74.6 pm). Overlap of two 1*s* orbitals makes a shorter bond than the overlap of two 2*s* orbitals for the same reason.

Example 9.2—Using Valence Bond Theory: Warp Drive?

According to *Star Trek* fans, dilithium crystals (Li_2) power much of the universe of the future. Judging on the basis of valence bond theory, does this compound actually exist? Does Na_2 exist? Which would have a longer bond?

Asking the Right Questions

What is the basis for a molecule to exist using Valence Bond Theory? Can this occur in dilithium?

Solution

Valence orbitals must be available to overlap so that the electrons within can interact in a covalent bond. Because each of the lithium atoms has a half-filled 2*s* orbital, we'd predict that their overlap should provide a stable dilithium molecule.

Is Our Answer Reasonable?

The answer is reasonable because it conforms to the Valence Bond theory. Dilithium does exist, but because the formation of metallic lithium is so favorable, dilithium can be observed only as a gas at high temperatures. Theoretically, disodium, Na_2, should also exist because two sodium atoms can have overlap of their 3*s* orbitals to make a bond. Because the disodium molecule is made up of bigger orbitals, we'd also predict it to have a longer bond than dilithium.

What If…?

…we were asked to draw the overlap of the orbitals in the two lithium atoms? What figure in this section best illustrates what is occurring?

(*Ans:* ◘ Figs. 9.5 and 9.8 *both represent the overlap of the s orbitals, though* ◘ Fig. 9.5 *focuses on hydrogen.*)

Practice 9.2

Which of these molecules has the longest bond: Cl_2, HCl, or F_2?

9.2.3 What's Wrong with This Model?

The valence bond theory seems to work well for a number of fairly simple molecules, but we only have to complicate matters slightly to discover a huge problem with this theory. Let's look at one of the simplest polyatomic molecules to see the nature of this problem. Methane (CH_4) is the major component of natural gas. Piped into our homes, it undergoes combustion to heat our water, cook our food, and warm our rooms. Let's build a valence bond model of methane. We start by writing the valence electron configurations of the atoms in methane. Immediately, we note that the valence electron configurations of carbon ($2s^22p^2$) and hydrogen ($1s^1$) indicate a problem. The valence shell of carbon contains a completely full 2*s* orbital, two partially filled *p* orbitals, and one completely unfilled *p* orbital. Valence bond theory indicates that we should be able to make only two bonds to the hydrogen atoms resulting from the overlap of the two partially filled *p* orbitals with the hydrogen 1*s* orbitals, as shown in ◘ Fig. 9.9 Something is wrong here.

We know that the Lewis dot structure and VSEPR models of methane correctly show four bonds. Both VSEPR and experimental data indicate that the molecule has a tetrahedral structure with four equal bonds. Therefore, each of the C—H bonds in methane must be made up of the same types of orbitals. But unless we modify the valence bond theory to account for our experimental evidence, we're sure to build a structure that will fail. This is a great example of how science develops. Theories are developed to agree with experimental evidence. Then, as more evidence is gathered, the theories are modified to fit the new evidence. If evidence arises that cannot be resolved by modifying a theory, a new theory must be found.

▶ Here's what we know so far

- Lewis dot structure and the VSEPR model do not properly explain such properties of molecules as bond lengths and bond energies. Valence bond theory (VB theory) was introduced by Linus Pauling to better explain these molecular properties.
- A valence bond is seen as the overlap of atomic orbitals on adjacent atoms.

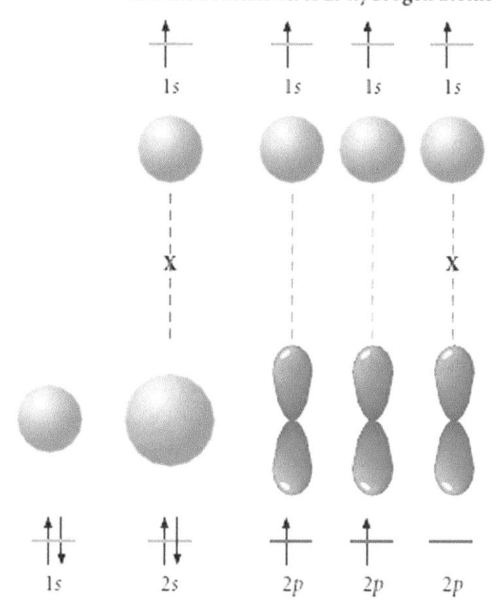

Fig. 9.9 Valence bond model of methane. Atomic orbitals that contain only one electron can overlap to form a bond. These are shown with a dotted line. The orbitals that cannot overlap to form a bond (either too many electrons or not enough electrons) have an "X" on the dotted line. The model needs some corrections because there aren't enough orbitals to make four bonds

- In VB theory:
  - smaller orbitals overlap more than larger orbitals.
  - orbitals similar in size overlap more than orbitals with mismatched sizes.
  - orbitals with similar energies overlap more than orbitals with very different energies.
- Bond length and bond energy in simple molecules can be explained by valence bond theory.
- VB theory does not properly explain bonds in polyatomic molecules, so we must modify it to make a better model. ◀

9.3 Hybridization

Linus Pauling advanced the theory of orbital **hybridization** to address the problems associated with VB theory. The result fixes the problems with the valence bond approach to model construction. In 1954, Pauling was awarded the Nobel Prize in chemistry for his years of research into the nature of the chemical bond.

Hybridization – The mathematical combination of two or more orbitals to provide new orbitals of equal energy.

9.3.1 Hybridization Defined

A hybrid is a mixture of two species, whether they are plants, paints, engines, or even orbitals. Hybrid flowers add beauty to the garden. Much as red roses and white roses can be interbred to give pink hybrids, we can mathematically combine two or more orbitals to provide new orbitals of equal energy in the process of hybridization.

▸ https://www.photos-public-domain.com/2010/11/01/red-rose-in-bloom/. ▸ https://www.photos-public-domain.com/2011/03/02/white-rose-on-rose-bush/. ▸ https://www.photos-public-domain.com/2011/08/07/pink-rose-in-full-bloom/

Recall from ▶ Chap. 6 that orbitals with the same energy are known as degenerate. What types of orbitals can be mixed together (hybridized)? The degree to which an orbital mixes with another is directly related to the difference in energy of the two orbitals. Typically, we hybridize only orbitals on the same atom that are in the same subshell, such as a hybrid made using 2*s* and 2*p* orbitals, resulting in a set of new orbitals that have the properties of all the orbitals from which they were mixed. The preparation of pink paint offers an analogy to hybridization. To make a good pink paint, we must mix red and white paint that have the same base (i.e., latex or oil). We don't get a good mixture by combining one can of latex paint and one can of oil paint.

How many orbitals do we make when we hybridize atomic orbitals? Just as if we were to mix one can of red paint and one can of white paint to get two cans of pink paint, we should expect to get two hybridized orbitals if we mix two atomic orbitals. The number of orbitals that are hybridized determines the number of new orbitals that are made. The orbitals that result from this mixing will be degenerate (have the same energy) and will have the same shape, but they will be oriented in different directions. What is the energy of the resulting hybridized orbitals? Consider our paint example again. We expect the color of the mixed paint to be the weighted average of the colors that we added. Similarly, the energy of the resulting hybridized orbitals should be the weighted average of the energies of the atomic orbitals that were mixed. Why do orbitals on an atom hybridize? An atom will produce hybridized orbitals if it is able to lower the overall energy of the molecule by doing so. For example, hybridizing orbitals on the carbon in methane (CH_4) will allow four pairs of bonding electrons to be formed and will allow those electron pairs to orient themselves as far apart as possible, providing the lowest energy configuration for the molecule. Hydrogen only has the 1*s* orbital and cannot produce hybridized orbitals, because it is able to bond with the hybridized orbitals of carbon without making any changes.

9.3.2 *sp*, *sp*^{2}, and *sp*^{3} Orbitals

To determine the hybridized orbitals on an atom, we follow a brief series of steps that convert atomic orbitals into hybridized orbitals. These steps are outlined in ◘ Table 9.3.

Let's walk through these steps in detail to show how the process works for methane (CH_4). The Lewis dot structure of methane indicates that we should have one bond from each hydrogen atom to the carbon atom.

Step 1: Write the electron configuration of the carbon atom. The electron configuration of carbon ($1s^2 2s^2 2p^2$) indicates that there is one filled 2*s* orbital and two half-filled *p* orbitals in the valence shell.

Step 2: We construct the Lewis Dot structure for the molecule by placing the carbon in the center and surround it with four hydrogen atoms. The 8 total valence electrons are placed as bonds around the carbon atom.

Step 3: In order to place four hydrogen atoms in degenerate (equal energy) bonds around the carbon atom, there must be four orbitals on carbon that will be able to interact with the four hydrogen 1s orbitals.

Step 4: We hybridize the existing valence orbitals on the carbon to make four new orbitals. Knowing that we need to mix four orbitals to make four new orbitals, we take the available 2*s* orbital and the three 2*p* orbitals and mix them to make four new degenerate orbitals, as shown in ◘ Fig. 9.10 The new orbitals are given a name in order to distinguish them from the 2*s* and 2*p* orbitals, but to show their relationship to these orbitals, we call them the $2sp^3$ orbitals. The name of the new orbitals illustrates that one 2*s* orbital and three 2*p* orbitals are mixed to make four new orbitals.

The electron configuration of our hybridized carbon atom can be written. We can write $1s^2(2sp^3)^4$ to show the hybrid orbitals. Each of the four $2sp^3$ orbitals on our carbon atom contains only one electron and is able to participate in bonding.

Step 5: The new $2sp^3$ orbitals are then overlapped with each of the hydrogen atoms to form the new $2sp^3$—1*s* bonds. The new shape of the orbitals (large at one end and small at the other) is shown in ◘ Fig. 9.10. Does the geometry of this hybridized carbon atom—that is, with four equivalent electron pairs surrounding it—agree with that obtained from the VSEPR model? In other words, does our answer make sense?

Step 6: We can now identify that there are four $2sp^3$ orbitals with a bond angle of 109° apart from each other. This means that the carbon atom is tetrahedral in its overall shape. This agrees with the result we would obtain from the VSEPR model.

Let's look at the energy changes that occurred with this hybridization. The new $2sp^3$ orbitals have energy that is in-between that of the unhybridized 2*s* and 2*p* orbitals. Energy was required to move two 2*s* electrons on carbon into the new $2sp^3$ orbitals. A small amount of energy was released when two 2*p* electrons moved to the new $2sp^3$ orbitals. The net result is that the carbon atom increased its energy in making the new hybrid orbitals. However, when the hybridized carbon combines with the four hydrogen atoms, a large amount of energy is released as

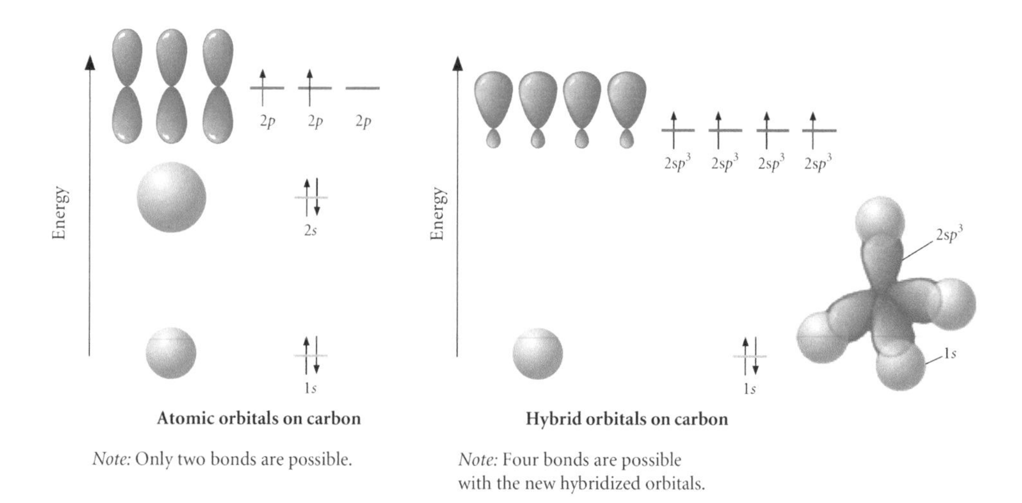

Fig. 9.10 Hybridization of the carbon atom in methane results in four hybridized orbitals that can overlap with hydrogen atoms

Table 9.3 Summary: rules for hybridizing atomic orbitals

Step 1: Determine the electron configuration	Write the electron configuration for each atom in the structure. Note the number of valence electrons
Step 2: Lewis structure	Draw the Lewis Structure for the molecule under consideration
Step 3: Orbitals needed	Count how many bonds and lone pairs of electrons are on the central atom. Each pair of electrons on the central atom needs an equivalent hybrid orbital
Step 4: Hybridize	Mix the number of valence orbitals equal to the number of hybrid orbitals needed, starting with the *s* orbital and going from there
Step 5: Form valence bonds	Overlap each hybrid orbital on the central atom with a valence orbital on one of the attached atoms
Step 6: determine geometry	Use your knowledge of molecular geometry (▶ Chap. 8) and orbitals to determine the shape of the molecule and the relative bond lengths

the new C—H bonds are formed. Although the hybridization process itself is energetically uphill, the end result (four equal bonds) is still energetically downhill (and quite favorable). In methane, the s–sp^3 overlap between each hydrogen and the central carbon atom forms the four bonds (Fig. 9.10). The outcome of the resulting valence bond model now agrees with the Lewis dot structure, the VSEPR model and, most important, with experimental evidence. Moreover, we also have an idea of the relative bond lengths and strengths in methane.

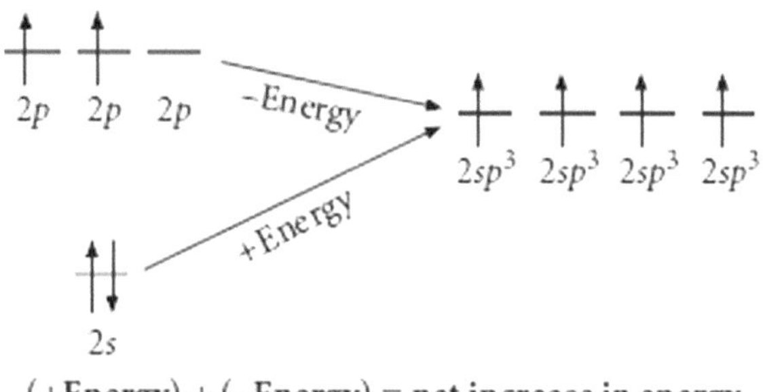

Let's apply hybridized VB theory to boron trifluoride (BF_3), a molecule that we have previously explored. This compound is particularly useful in enhancing the reactivity of molecules during the synthesis of new pharmaceutical agents. Using the ideas of hybridization and valence bond theory, what is the structure of BF_3? The electron configuration of boron is $1s^22s^22p^1$. The Lewis dot structure of boron trifluoride indicates that we should have three single bonds, so we must hybridize the orbitals on boron to make three degenerate orbitals. Following the steps in Table 9.3, we mix the 2*s* orbital and two of the 2*p* orbitals (all from boron's valence shell) to make the

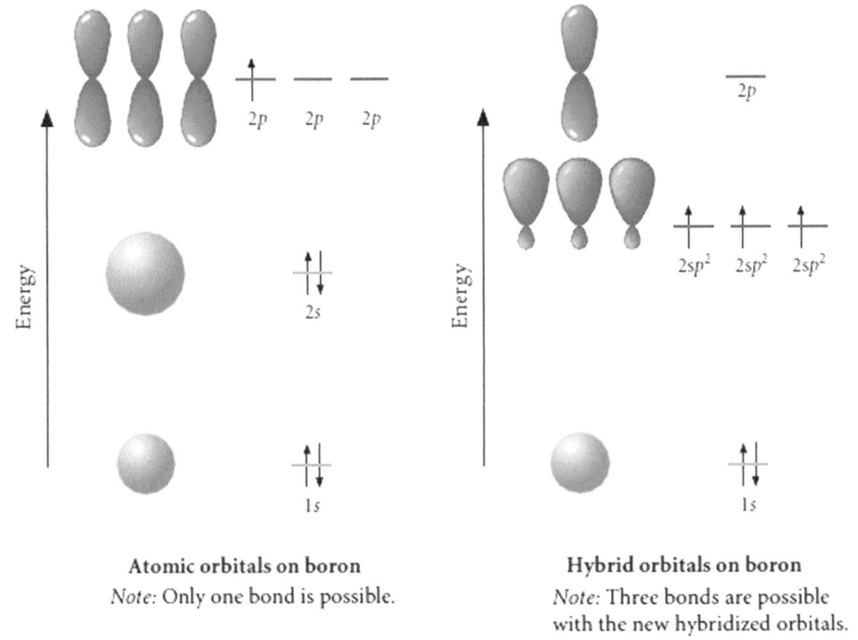

2p 2p 2p
2s
1s
Energy
Atomic orbitals on boron
Note: Only one bond is possible.
2p
2sp² 2sp² 2sp²
1s
Energy
Hybrid orbitals on boron
Note: Three bonds are possible with the new hybridized orbitals.

Fig. 9.11 Hybridization of boron to give three hybridized orbitals

new orbitals. The three resulting orbitals have the designation $2sp^2$, as shown in Fig. 9.11. Note that the superscript numbers in the name of a hybrid orbital do not indicate how many electrons are present in the orbital. We know this might be confusing, but it is the convention for this notation.

We didn't hybridize one of the 2*p* orbitals because it wasn't needed to make a bond (i.e., the empty unhybridized 2*p* orbital is still available on the boron atom). We can rewrite the electron configuration for boron trifluoride to end up with three orbitals that are available to bond with the half-filled 2*p* orbital on each unhybridized fluorine atom (p–sp^2 overlap). Now, you may ask—why didn't boron trifluoride just use the three *p* orbitals in boron's valence shell to make bonds with the fluorine atoms? This would provide pure 2*p*-2*p* orbital overlap, which ought

2p 2p 2p
2s
1s
Energy
Atomic orbitals on beryllium
Note: No bonds are possible.
2p 2p
2sp 2sp
1s
Energy
Hybrid orbitals on beryllium
Note: Two bonds are possible with hybridized orbitals.

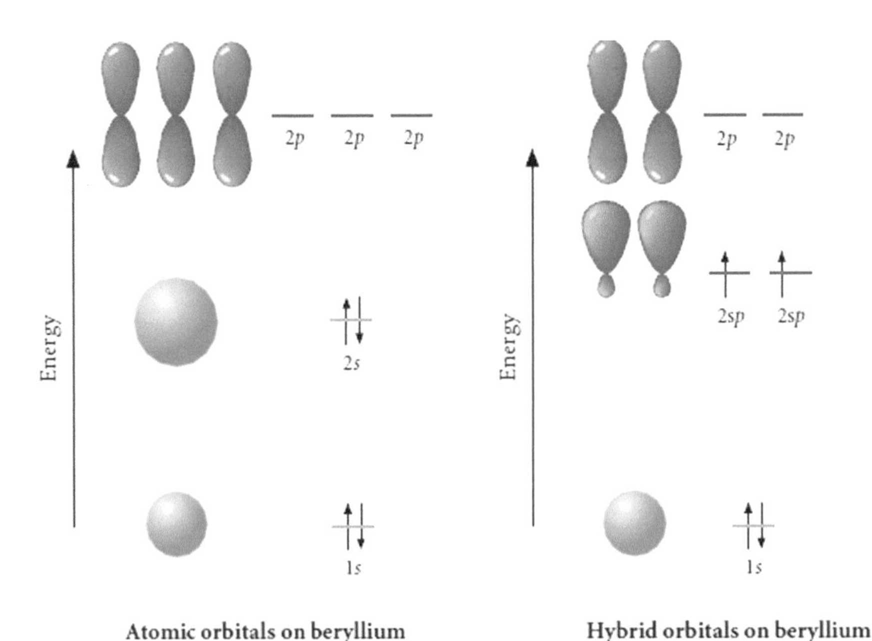

Fig. 9.12 Hybridization to make two bonds

to be good, right? The answer is because this would put the three bonding pairs of electrons only 90-degrees from each other, which would raise the overall energy of the molecule. Moreover, this is not what we experimentally observe for the three-dimensional shape of the molecule. By providing hybridized $2sp^2$ orbitals, the bonding electron pairs can be 120-degrees apart and the overall energy of the molecule is lowered.

As another example, we can build a model of beryllium hydride (BeH_2) showing covalent bonds between the beryllium ($1s^22s^2$) and hydrogen atoms. There are no electrons in beryllium's atomic $2p$ orbitals. Because we know that there are two bonds in BeH_2, we hybridize two orbitals to make two new $2sp$ orbitals, as shown in ◘ Fig. 9.12. Rewriting the electron configuration shows that we can now make bonds to the hydrogen atoms, resulting in an overlap of the $1s$ and $2sp$ orbitals. In beryllium hydride, there are two empty, unchanged $2p$ orbitals.

Example 9.3—Hybridization in Common Molecules

Describe the hybridization of the central atom in each of these molecules:

a. H_2O
b. NH_3
c. AlH_3

Asking the Right Questions

What do you need to know to determine the hybridization of the central atom? What is the relationship between the number of lone and bonded pairs around the central atom and its hybridization?

Solution

a. Lewis dot structure models for water indicate that the central oxygen has two bonds and two lone pairs. Mixing the four orbitals (one $2s$ and three $2p$) on the oxygen atom allows you to have two lone pairs of equal energy and two equal bonds to the adjacent hydrogen atoms. The oxygen atom possesses $2sp^3$ hybridized orbitals.
b. In a similar fashion, the nitrogen atom requires four electron groups (three bonds and one lone pair) in ammonia. Mixing the four orbitals in its valence shell gives you three bonds to the hydrogen atoms and one lone pair (all degenerate in energy). The nitrogen atom is $2sp^3$ hybridized.
c. Aluminum hydride has three bonds (one to each hydrogen), so you can mix the $3s$ and two of the $3p$ orbitals to make a $3sp^2$ hybridized aluminum.

Are Our Answers Reasonable?

The Lewis dot structures of water and ammonia show the same number of electron groups around their central atom. You would therefore expect them to have the same central atom hybridization, which they do, $2sp^3$. Aluminum hydride has only three electron groups around the central aluminum atom, so you would expect it to have a hybridization that requires the mixing of three orbitals, which it does via its $3sp^2$ hybridization. This answer is also reasonable.

What If…?

…we were given IF_2^+ ? What is the hybridization of the central atom, iodine?

(*Ans: $5sp^3$ hybridized*)

Practice 9.3

What is the hybridization of the central atom in each of the following substances?

a. OF_2
b. H_2S
c. NH_4^+

9.3.3 Shapes of the Hybrids

Hybrid orbitals have a shape that results from the mixing of the corresponding atomic orbitals. If we add an s orbital and a p orbital together, the result is one of the two hybrid sp orbitals. If we subtract the s orbital from the p orbital, we get the other hybrid sp orbital as shown in ◘ Fig. 9.13. Remember that mixing two orbitals gives two new orbitals. How do the angles between the resulting orbitals compare to what we expect from VSEPR rules? The sp hybridized atom has bonds with 180° angles, the sp^2 hybridized atom forms bonds with 120° angles, and the sp^3 hybridized atom contains bonding angles of 109.5° (see ◘ Fig. 9.14; see also ◘ Table 9.4). This agrees with the angles we determined from VSEPR theory.

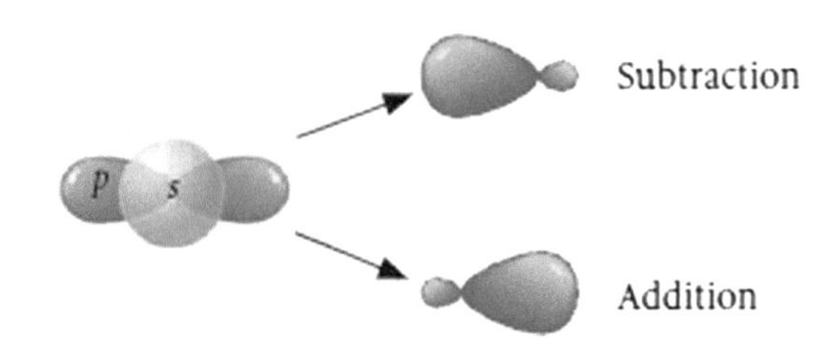

Fig. 9.13 Mixing an s and a p orbital. Addition and subtraction of the s and p orbitals gives two sp orbitals

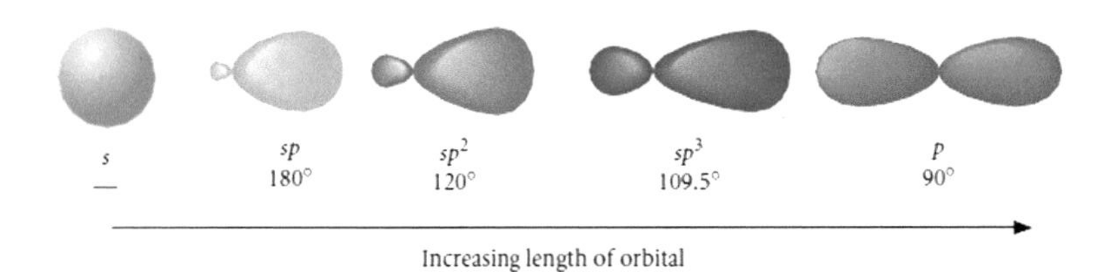

Fig. 9.14 Shapes of the hybrid orbitals

9

Table 9.4 Selected hybrid orbitals

Hybrid	Example	C—H bond length (pm)	Angle	C—C bond length (pm)
sp^3	CH_3-CH_3	109	109.5°	154
sp^2	CH_2=CH_2	108	120°	134
sp	CH≡CH	106	180°	120

Example 9.4—Shapes of the Molecules

For each of the molecules in Example 9.3, indicate the geometry around the central atom.

Asking the Right Questions

What is the relationship between the hybridization of the central atom and the molecular geometry?

Solution

a. H_2O geometry: sp^3 hybridized atoms adopt a tetrahedral geometry. Because two of the sp^3 orbitals contain lone pairs, the VSEPR model indicates that the molecule has an overall bent geometry. The bond angle should be less than 109.5° because the lone pairs repel each other more than the bonding pairs. The angle H–O–H is actually 104.5°.
b. NH_3 geometry: Again, we should have sp^3 hybridized orbitals with a tetrahedral geometry. Because one of the sp^3 orbitals contains a lone pair, ammonia has a trigonal pyramidal geometry. The bond angle (H–N–H) should be less than 109.5° because of the repulsions between the lone pair and the bonding pairs of electrons (the angle is actually 107°).
c. AlH_3 geometry: The aluminum atom is sp^2 hybridized and has a trigonal planar geometry. The bond angle is 120°.

Are Our Answers Reasonable?

The Lewis dot structure, VSEPR model-based geometry, hybridization, and observation of the actual bond angles come together to give you a consistent picture for the bonding and molecular shape of each of these molecules. This consistency resulting from several different sets of information and models means that the answers are reasonable.

What If…?

…we had H_3O^+? What is the geometry around the central oxygen atom?
(*Ans: tetrahedral*)

Practice 9.4

For each of the molecules in Practice 9.3, indicate the geometry about the central atom.

9.3.4 sp^3d and sp^3d^2 Orbitals

What if we have more than eight electrons—that is, an expanded octet—on the central atom of a molecule? Can we accommodate them with hybrid orbitals? A good example is phosphorus pentachloride (PCl_5), a reactive molecule used to make compounds containing chlorine. The Lewis dot structure model of PCl_5 indicates that there should be five bonds to the phosphorus atom. Because the combination of all of the *s* and *p* orbitals will allow only four bonds, we must include a *d* orbital in the hybridization scheme. This is why it is not a problem to have more than eight electrons in the valence shell of elements in rows 3 and higher of the periodic table—they have *d* and/or *f* orbitals available. Five orbitals on the phosphorus can be made by hybridizing the 3*s*, three 3*p*, and one of the 3*d* orbitals. The result is five degenerate $3sp^3d$ orbitals. We can mix the *d* orbital because it is in the same principal energy level and is similar in energy to the other orbitals. The resulting valence bond model of PCl_5 shows overlap of a 3*p* orbital from the chlorine atom with one of the $3sp^3d$ orbitals from the phosphorus atom (◘ Fig. 9.15). The space-filling model of the molecule shows how the orbitals fit together to give the molecule its essential structure.

Other examples are found in compounds containing xenon. The first example of a compound containing xenon, a noble gas, was made in 1962. Shortly after the discovery that xenon could make compounds with other elements, XeF_2, XeF_4, and XeF_6 were prepared. The orbital overlap required to make xenon tetrafluoride (XeF_4) shows why these compounds are possible. The Lewis dot structure of XeF_4 requires six orbitals (four for bonds and two for lone pairs). We hybridize a 5*s*, three 5*p*, and two 5*d* orbitals to make six degenerate $5sp^3d^2$ orbitals. Each bond in XeF_4 results from overlap of a 2*p* orbital from the fluorine atom with a $5sp^3d^2$ orbital from the xenon atom, with the two lone pairs on the xenon atom occupying two $5sp^3d^2$ orbitals (◘ Fig. 9.16).

9.3.5 Multiple Bonds: Sigma and Pi Bonds

Our discussion so far has centered on the use of valence bond theory to address the formation of single bonds in molecules. However, we know from ▶ Chap. 8 that atoms can participate in double and triple bonds as well. How is this possible if hybridization provides only orbitals that are oriented directly at the atom with which they are

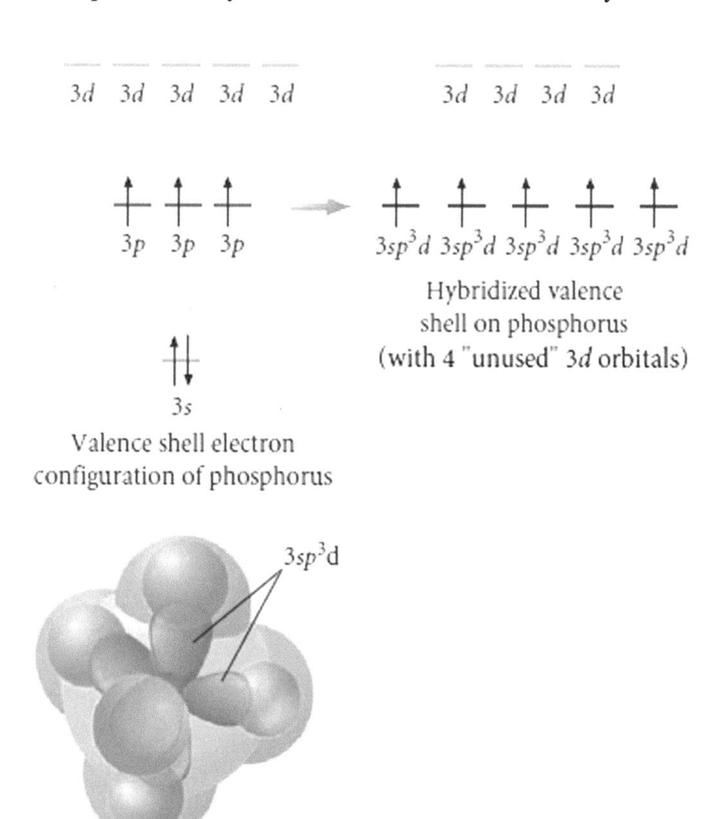

◘ **Fig. 9.15** Hybridization of the phosphorus in PCl_5. Because five bonds are needed to complete the structure, five orbitals are hybridized on the phosphorus atom to give five new hybrid orbitals

5d 5d 5d 5d 5d

5p 5p 5p

5s

Valence shell electron configuration of xenon

5d 5d 5d

$5sp^3d^2$ $5sp^3d^2$ $5sp^3d^2$ $5sp^3d^2$ $5sp^3d^2$ $5sp^3d^2$

Hybridized valence shell on xenon (with 3 "unused" 5*d* orbitals)

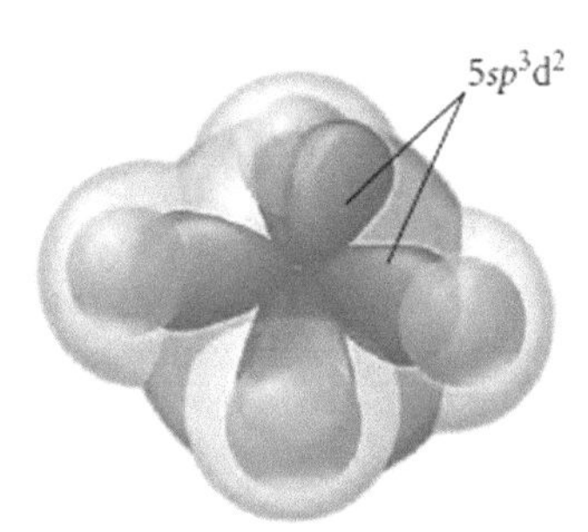

Fig. 9.16 Hybridization of xenon in XeF_4

bonded? The answer lies in the different ways in which orbitals can overlap in valence bond theory. **Sigma bonds (σ bonds)**, which make up the framework of a molecule, result from an *end-on overlap* of orbitals. Almost every "single bond" is a sigma bond. Every bond we've discussed thus far in this chapter results from end-on overlaps of orbitals, such as the four σ bonds in CH_4, each of which is an end-on overlap of an sp^3 orbital and an *s* orbital (Fig. 9.17). The resulting σ bond is a continuous cloud of electron density between the two nuclei. Remember that the orbitals are just the spaces around the nucleus where the electrons are most likely to be found. When the *s* and sp^3 orbitals overlap, they form a new "electron cloud" that is shared between the two bonding atoms.

Sigma bond (σ bond) – A covalent bond resulting from end-on overlap of orbitals. The sigma bond does not possess a nodal plane along the axis of the bond.

Pi bonds (π bonds) result from the side-to-side overlap of orbitals. This type of bond occurs primarily when unhybridized *p* orbitals on adjacent atoms overlap, as is found between atoms containing a double or triple bond. Remember that *p* orbitals have two lobes, so even though there is a portion of the *p* orbital above and below the molecule, it is just one orbital. The resulting π bond has the shape like the two halves of a hotdog bun separated

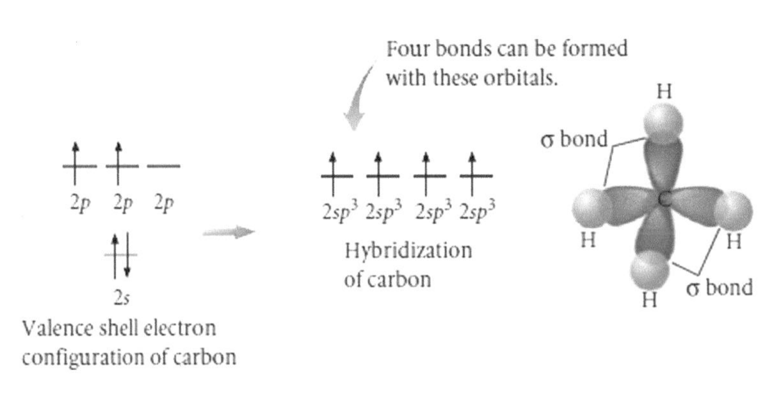

Fig. 9.17 Methane hybridization and sigma bonds

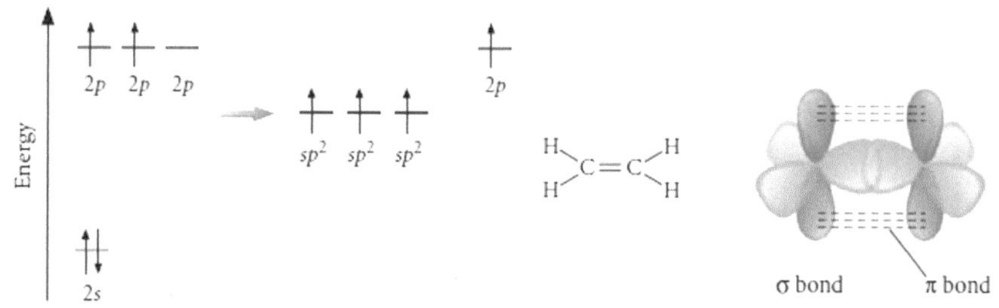

■ **Fig. 9.18** Ethylene and valence bond theory. The overlap of the orbitals between the carbons gives rise to a sigma bond and a pi bond. The sigma bond results from $2sp^2$–$2sp^2$ overlap. The pi bond results from the side-to-side overlap of the $2p$ orbitals

by empty space. That is, there is a plane that exists between the atoms that is not occupied by π electrons. This is called a "**nodal plane**" because the electron wave functions have a node between the two halves of the pi bond. Ethylene (C_2H_4) is a molecule that contains a π bond resulting from the overlap of p orbitals (■ Fig. 9.18).

Pi bond (π bond) – A covalent bond resulting from side-to-side overlap of orbitals. The pi bond possesses a single nodal plane along the axis of the bond.

Nodal planes – Flat, imaginary planes passing through bonded atoms where an orbital does not exist.

Why do we show the sp^2 and $2p$ orbitals in the diagrams in ■ Fig. 9.18 with one electron each? Shouldn't we fill the sp^2 orbitals before filling the slightly higher energy $2p$ orbital? Hund's rule would seem to suggest this, but because the energy difference between the hybridized and unhybridized orbitals of the same shell is relatively small, the electrons can be temporarily placed in higher energy $2p$ orbitals prior to bond formation. The bonds that are created end up with lower energies than they would have had in the hybridized orbitals. When there is one electron in that hybridized orbital, the bonded atom will provide the second electron used to form the bond. In addition, it is important to note that although we have discussed the hybridization process as a series of steps, no hybridization will take place without the formation of bonds. That is, in the absence of any other atoms, hybridization does not occur in an individual atom. Because the pi bond results from the side-to-side overlap of p orbitals, the overlap isn't as great as the end-on overlap in a sigma bond. The pi bond is weaker than the sigma bond.

Example 9.5—The Bonds of Acetylene

Acetylene (C_2H_2) was once used as the light in pre-electricity streetlamps. It has long been used in welding, because its combustion in oxygen produces a very hot flame. Because acetylene is a fairly small, reactive molecule, it can be used as a starting material in the preparation of other chemicals. Describe the bonding patterns in acetylene using the valence bond theory.

Asking the Right Questions

What makes a bond a sigma bond or a pi bond? What orbitals in acetylene can hybridize to form sigma bonds? What about pi bonds? What do these orbitals look like in three-dimensional space?

Solution

Sigma (σ) bonds are formed from the end-on overlap of hybrid orbitals. Review ■ Fig. 9.18 in which three σ bonds are formed from the $2sp^2$ hybrids. In addition, one electron in a $2p$ orbital of each carbon atom interacts to form a pi (π) bond, which we described as looking like two halves of a hotdog bun separated by empty space.

In acetylene, the process is the same, but the σ bonds are formed from the $2sp$ hybrids, as shown in ■ Fig. 9.19.

Each carbon atom has two electrons remaining, one in each $2p$ orbital, that can interact. As shown in the figure, these interact via two π bonds.

Is Our Answer Reasonable?

Is the bonding pattern consistent with the Lewis dot structure, and does it explain the bond angles that we observe in the molecule? The model does explain the presence of both the σ and π bonds in the system. The bond angles of 180° are also consistent with this model, so the bonding pattern is reasonable.

What If…?

…we had propyne, C_3H_4? Describe the bonding in this molecule.

(*Ans: there are 2 pi and 6 sigma bonds in the molecule.*)

Practice 9.5

Describe the valence bond theory for formaldehyde (CH_2O), a compound used in the preservation of biological tissues.

Fig. 9.19 Acetylene π bonds. There are two π bonds in acetylene that result from the overlap of two p orbitals from each carbon atom

9.3.6 Application of Hybridization Theory: Bond Lengths

Remember that our goal in coming up with scientific models of chemical bonding is to be able to match what we see in the world around us and, most important, to be able to predict the behavior of bonds and molecules that we might not have actually made yet. The hybridization of orbitals allows us to have chemical bonds with the same geometric shapes and bond angles that are explained in the VSEPR model, and Hybridization—Valence Bond theory is necessary for our models of molecules to agree with the experimentally measured bond angles.

The use of hybridized orbitals also provides information about the relative lengths of bonds in molecules. We've already discussed how *s–p* orbital overlap produces a bond that is longer than one containing *s–s* orbital overlap. This discussion can be carried further to illustrate the lengths of hybridized orbital bonds. Look back at Table 9.4 to see the relative sizes of these orbitals and some features of the resulting hybridization. From this table we see that the sp^3 hybrid orbital makes bonds that are longer than the sp^2. The sp^2 hybrid orbital is longer than the sp. Apparently, mixing more p orbitals into the hybrid produces longer hybrid orbitals. We also note that a bond made from sp^3–sp^3 orbital overlap is longer than one made from s–sp^3 overlap. The values in Table 9.4 are average values (bond lengths and bond angles) based on actual measurements from a collection of similar molecules.

Example 9.6—Bond lengths of C—C Single Bonds

The Lewis dot structures of butane (used in disposable and refillable lighters), 1–butene (a compound used to make plastic wraps), and 1,3-butadiene (the starting material used to make synthetic rubber for your car's tires) are shown in Fig. 9.20. Which has the longest C—C single bond? Which has the shortest C—C single bond?

Asking the Right Questions

What is the relationship between the hybridization of the atoms in the bond and the bond length? What is the hybridization of the atoms in each C—C bond? What can we conclude about the relative bond length based on this information?

Solution

The more p orbitals that combine to form the hybrid orbital, the longer the bond. This means that all other things being equal, an $sp^3 - sp^3$ bond is longer than the corresponding $sp^2 - sp^2$ bond. The shortest C—C bond is found in 1,3-butadiene. It results from the overlap of an sp^2 hybridized orbital with another sp^2 hybridized orbital. The longest bond is found in butane (sp^3–sp^3 orbital overlap).

Are Our Answers Reasonable?

What do the bond length data tell us? The C—C bond lengths for each type of bond are approximately: 154 pm, 150 pm, and 148 pm. This corresponds with our relative bond length assignments, so the answers are reasonable.

Hybrid overlap	Bond length (pm)	Example molecule
$sp^3 - sp^3$	154	Ethane
$sp^3 - sp^2$	150	Propene
$sp^2 - sp^2$	148	1,3-butadiene
$sp^3 - sp$	146	Propyne
sp–sp	138	1,3-butadiyne

What If...?

we want to compare the carbon-to-carbon bond lengths in ethene (C_2H_4) and ethyne (C_2H_2)? Which would have the longer bond?

(*Ans: Ethene has the longer bond, because of the overlap of sp^2 orbitals produces longer bonds[more p character] than that of sp orbitals. The measured bond lengths for the carbon-to-carbon bonds are 145 pm in ethene and 120 pm in ethyne.*)

Practice 9.6

Which has a longer N—O bond length, CH_2NOH or CH_3NHOH?

Butane | 1-Butene | 1,3-Butadiene

Fig. 9.20 Butane, 1-Butene, and 1,3-Butadiene. (The bond length in question is indicated with an arrow)

9.3.7 Advantages and Disadvantages of Hybridization

There should be some advantage gained from spending time developing the hybridized model of a molecule. In fact, there is. With this model, we are able to explain bond angles, the lengths of bonds in molecules, and the existence of pi bonds. Organic chemists put hybridization to good use. Often, they use it to describe the three-dimensional structure of a molecule to others. This approach enables organic chemists to talk with each other without drawing pictures on cocktail napkins. It also allows them to describe specific structural features of a molecule that play an important role in the chemical and physical properties of that molecule.

Unfortunately, Hybridized Valence Bond theory has some shortcomings. The relative size of many bond lengths can be accurately predicted, though some are "mystifyingly" unable to be accurately estimated. For example, why is the C—C bond shorter in 1,3-butadiene than in 1-butene? An even more refined picture of the orbitals on the molecule needs to be considered.

Example 9.7—Using Hybridization: Will the Molecule Be Solid or Liquid?

Animal fats, such as lard, tend to be solid. Vegetable oils, such as corn oil, tend to be liquid. Fats and oils are made up of compounds known as triacylglycerols. In turn, the triacylglycerols are constructed from molecules known as fatty acids. Look at the two fatty acids, stearic and oleic acids, depicted below. Describe the shapes of these molecules using valence bond theory and use this information to determine which molecule is more likely to exist as a solid at room temperature.

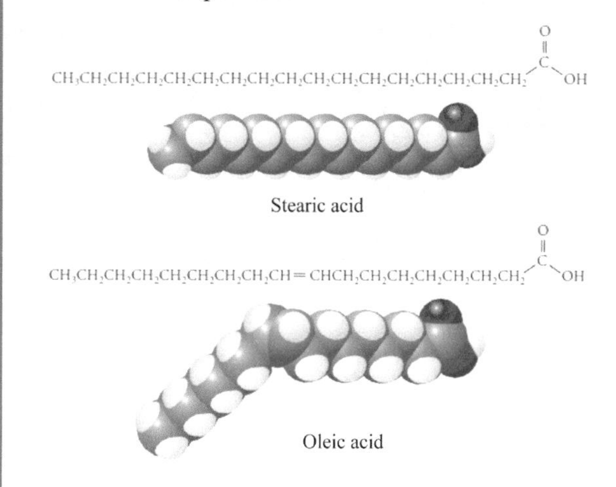

Asking the Right Questions

What makes a compound likely to be solid at room temperature? What is the relationship between the hybridization of the carbon atoms and the shape of the molecule?

Solution

If the molecules can pack easily and tightly together, they will be more likely to form a solid. If the molecules are more rigid and less able to get close together, they will be less likely to form a solid at room temperature. Stearic acid is made up of 17 carbon atoms that are sp^3 hybridized and 1 carbon atom that is sp^2 hybridized. The sp^3 hybridized carbons result in single bonds to adjacent carbons and can rotate to form a zig-zag pattern of atoms. The sp^2 hybridized carbon is more rigid in its trigonal planar configuration. Oleic acid contains mostly sp^3 hybridized carbon atoms, but it also has three sp^2 hybridized carbons. The presence of two of the sp^2 hybridized atoms in the middle of the long carbon chain causes a kink in the zig-zag structure. Stearic acid can pack tightly with other stearic acid molecules, so it is a solid at room temperature. Oleic acid, because of its bent, rigid structure doesn't stack together well with other molecules. Oleic acid is a liquid at room temperature.

Are Our Answers Reasonable?

It would seem reasonable that pencils (straight) would pack into a box in a more ordered fashion than "crazy straws" (bent). The degree of packing is related to the melting point, and so our answers are reasonable. The actual melting points can be found; oleic acid (13 °C), stearic acid (70 °C).

What If…?

…you have two fatty acids, palmitic acid, and linolenic acid? Assess if each will more likely be a solid or liquid at room temperature.

Palmitic acid: $CH_3(CH_2)_{14}COOH$.
Linolenic acid: $CH_3CH_2(CH{=}CHCH_2)_3(CH_2)_6COOH$.

(*Ans: palmitic acid is a solid (melting point* 63 °C*), and linolenic acid is a liquid (melting point* – 5 °C*) at room temperature.*)

Practice 9.7

Describe the three-dimensional shape of each of the following molecules. Solely on the basis of its shape, which do you predict will have the highest boiling point?

a. $CH_3CH_2CH_2CH_2CH_3$ (pentane).
b. $CH_3CH_2CH(CH_3)_2$ (2-methylbutane).
c. $C(CH_3)_4$ (2,2-dimethylpropane).

▶ Here's What We Know So Far

- We can mathematically combine two or more orbitals to provide new orbitals of equal energy in the process of hybridization.
- The models of many molecules agree with the experimentally measured bond angles when hybridization is taken into account.
- Although there is an energy cost to hybridization, the resulting stability of the molecule more than makes up for it.
- Hybrid orbitals can be constructed using *s*, *p* and *d* orbitals to account for atoms that contain more than eight electrons in their valence shells.
- The sigma (σ) bond does not possess a nodal plane along the axis of the bond and has a shape reminiscent of a hotdog. A pi (π) bond results from the side-to-side overlap of orbitals and is found between atoms containing a double or triple bond. It has one nodal plane along the axis of the bond, and has a shape similar to the two halves of a hotdog bun.
- Pi (π) bonds are generally weaker and more reactive than sigma (σ) bonds. ◀

9.4 Molecular Orbital Theory

Let's revisit the molecule of retinal (the light-absorbing portion of the rhodopsin protein in our eyes) that we introduced in the chapter opener. The 11-*cis*-retinal molecule contains an extended series of alternating single and double bonds known as **conjugated π bonds** (or **conjugation**), as shown in ◘ Fig. 9.21. Electrons in any bond (whether conjugated or not) can absorb a photon of light. What happens as a result of the absorption of the photon, and why is this useful to discuss at this point? These electrons become excited (more energetic) when the photon is absorbed and the electrons are promoted to a higher energy state.

Conjugated π bonds – An extended series of alternating single and double bonds.

Conjugation – The presence of conjugated π bonds—that is, a series of at least two double bonds alternating with single bonds.

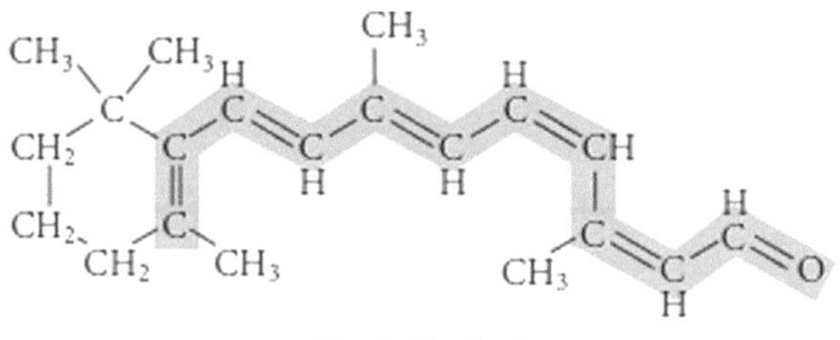

11-*cis*-Retinal

◘ **Fig. 9.21** Ethylene and valence bond theory. The overlap of the orbitals between the carbons gives rise to a sigma bond and a pi bond. The sigma bond results from $2sp^2$–$2sp^2$ overlap. The pi bond results from the side-to-side overlap of the $2p$ orbitals

The models we've already discussed say nothing about these higher energy states, so if we are to understand how they affect a molecule like 11-*cis*-retinal, which gives us the ability to see, we must examine our bonding theory once again. In this case, we will find that our very useful VB Hybridization theory is not strong enough to accommodate this expansion. We will continue to use that theory for explaining molecular shapes, etc., but to explain how higher energy states of electrons affect chemical bonds, we will need to come up with an entirely different theory. Here are the key ideas we will explore in this section:

1. Electrons can achieve higher energy states (can become excited) by the absorption of energy.
2. All electrons in a molecule can absorb photons of light with the appropriate amount of energy.
3. To become excited, core electrons (below the valence shell) require more energy (such as that of X-rays) than do valence electrons, for which the energy of UV or visible light is energetic enough.
4. Sigma-bonding electrons require more energy (UV less than 200 nm wavelengths) than π bonding electrons (visible between 200 and 700 nm) to become excited.
5. Electrons in conjugated π bonds require even less energy than unconjugated π bonds to excite an electron.
6. As the conjugated π system gets longer, the energy of the photon required to excite the electrons is reduced, moving from the blue toward the red region of the visible electromagnetic spectrum.

Chemists use this final property as they design molecules that can absorb a particular wavelength of light, which is useful for producing color film, dyes and even solar powered devices. This new theory can also be used to tell us what happens to bonds within a molecule when they absorb light, why hydrogen exists as a diatomic molecule and helium does not, and why some molecules (such as O_2) are attracted to magnetic fields and why some molecules (such as N_2) are not.

9.4.1 Molecular Orbital Theory Defined

At about the same time that Linus Pauling worked out valence bond theory, Robert S. Mulliken (1896–1986) and Erwin Schrödinger (1887–1961) began thinking about chemical bonds from the standpoint of the quantum theory that we discussed in ► Chap. 6. This theory, known as **molecular orbital (MO) theory**, treats electrons not as particles, but as waves that encompass the entire molecule. The Schrödinger equation defines the energy of each of the orbitals of an atom or molecule. Unfortunately, the Schrödinger equation cannot yet be solved exactly, except for a few very simple systems, so we usually make some approximations that allow us to arrive at a usable solution. Probably the most commonly used approximation is known as the **linear combination of atomic orbitals–molecular orbitals (LCAO–MO) theory**. In this theory, atomic orbitals are added together (both constructively and destructively) to make molecular orbitals. This is similar to hybridization theory, except instead of only combining valence orbitals on the central atom of a molecule, we will combine *all* the orbitals from all the atoms in the molecule.

Molecular orbital (MO) theory – A theory that mathematically describes the orbitals on a molecule by treating each electron as a wave instead of as a particle.
Linear combination of atomic orbitals – molecular orbitals (LCAO–MO) theory – An approximation of molecular orbital theory wherein atomic orbitals are added together (both constructively and destructively) to make molecular orbitals.

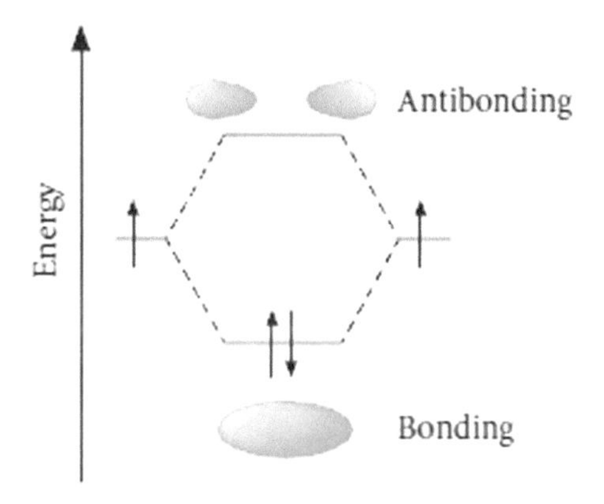

What are the key features of the molecular orbitals that are formed from atomic orbitals? We must raise several points about the formation of molecular orbitals from atomic orbitals. As shown in ◘ Table 9.5, the most

Table 9.5 Key features of MO theory

1. We restrict ourselves by insisting that only orbitals of similar energy can mix to make molecular orbitals (MOs). The 2*s* orbital, for example, can mix with a 2*s* orbital but does not mix well with the 1*s* orbital

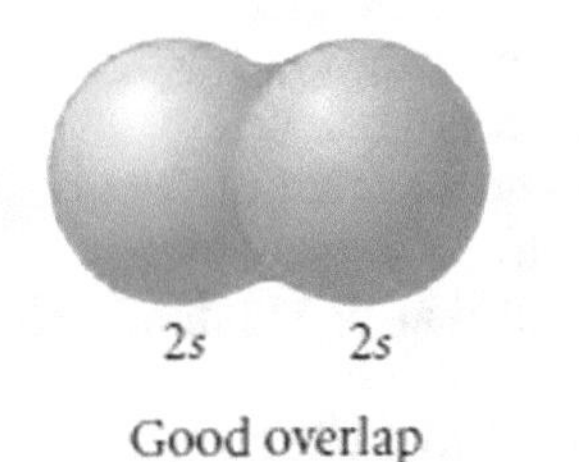

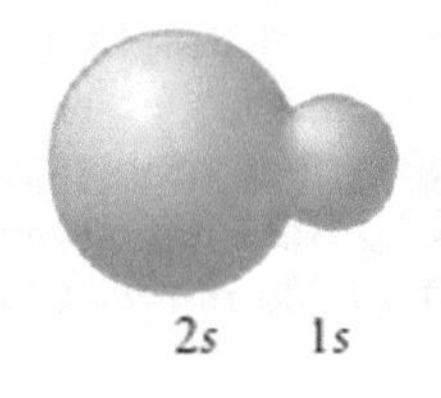

2. Only orbitals that exhibit symmetry (that is, they are similar in shape and orientation) can mix to make MOs. For example, the $2p_x$ orbital cannot mix with the $2p_z$. orbital, since they are oriented differently

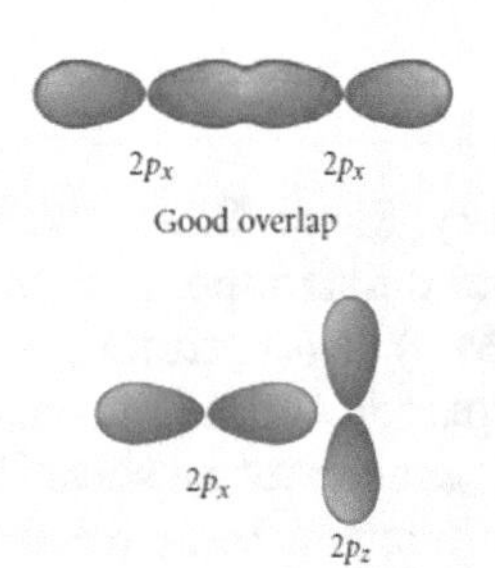

3. Mixing atomic orbitals gives the same number of MOs
- The bonding orbital is lower in energy than the atomic orbitals from which it was made
- The antibonding Orbital is higher in energy than the atomic orbitals from which it was made

4. Electrons are placed in MOs in order of increasing energy

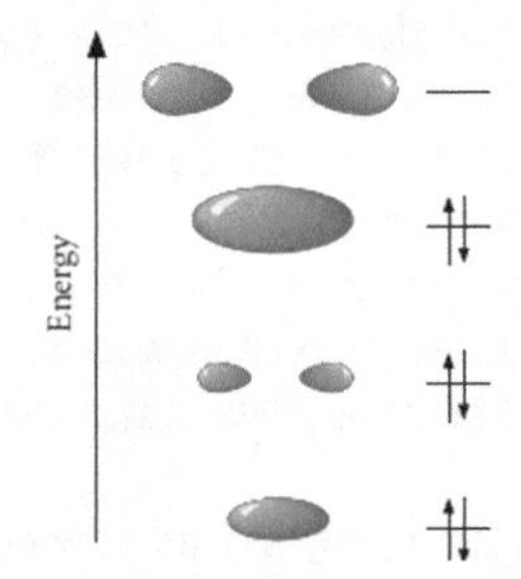

5. Each MO can hold two electrons with opposite spin
- Electrons that are placed in bonding MOs promote bonding
- Electrons that are placed in antibonding MOs favor separated atoms

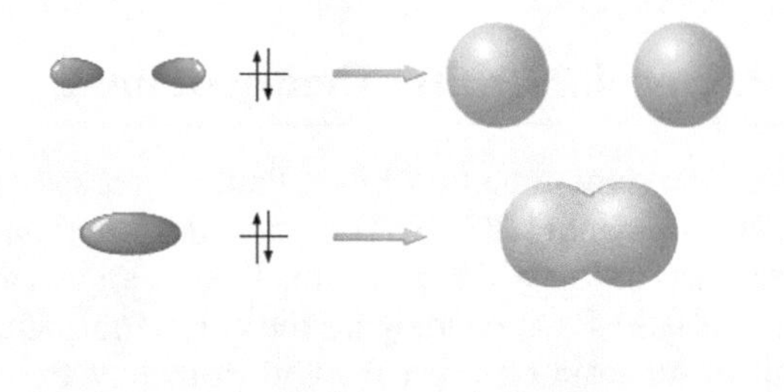

6. After mixing of the atomic orbitals, they cease to exist. Only the molecular orbitals remain

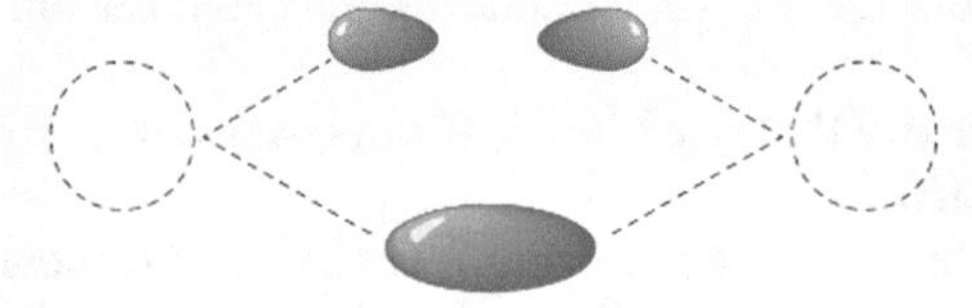

9

important point is that the combination of two orbitals gives two new orbitals, just as in the hybridization of atomic orbitals. One of the new molecular orbitals, called the **bonding orbital**, results from the addition of two overlapping atomic orbitals; the other, called the **antibonding orbital**, results from the subtraction of two overlapping atomic orbitals. The bonding orbital is lower in energy than either of the two atomic orbitals. The bonding orbital indicates that there is some electron density between adjacent nuclei (a bond exists). The antibonding orbital is higher in energy than the two atomic orbitals from which it is formed and is typically represented with an asterisk (*) to distinguish it from the bonding orbital. The antibonding orbital indicates a lack of electron density between adjacent nuclei (no bond exists). Each new molecular orbital (both bonding and antibonding) can contain two electrons. The bonding and antibonding orbitals can be thought of as two waves that can cancel each other out, but they must contain electrons to have any effect.

Bonding orbital – An orbital that indicates the presence of electron density between adjacent nuclei (a bond exists when the orbital is occupied). The bonding orbital arises from the addition of two overlapping atomic orbitals.

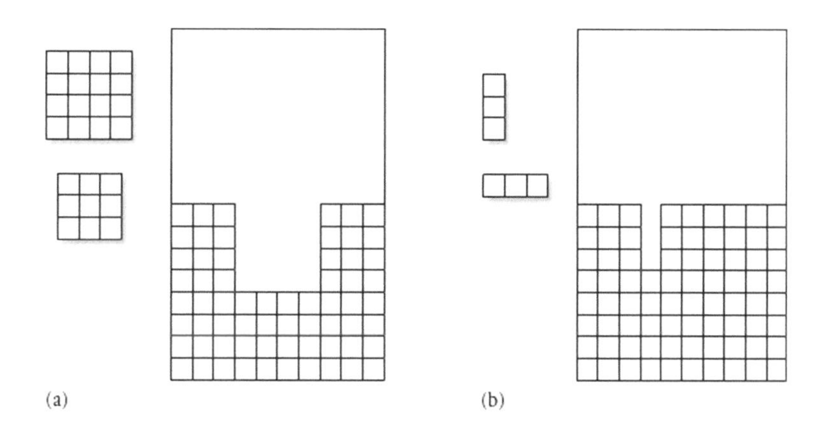

Fig. 9.22 Tetris analogy to overlapping orbitals: **a** The 4×4 square piece has the same size and shape as the hole. The 3×3 square has the same shape but the wrong size. **b** The 1×3 rectangle has the same size and shape as the hole. The horizontal 3×1 rectangle has the same size but is the wrong orientation to fit into the hole

Antibonding orbital – An orbital that indicates a lack of electron density between adjacent nuclei (no bond exists when the orbital is occupied). The antibonding orbital arises from the subtraction of two overlapping atomic orbitals.

A similarly important point in mixing atomic orbitals to make molecular orbitals is that only atomic orbitals of similar symmetry (shape and orientation) and energy provide significant overlap. This rule means that only 1*s* orbitals overlap to a great degree with 1*s* orbitals. The $2p_z$ orbital overlaps best with a $2p_z$ orbital. Their symmetry and size (i.e., energy) are similar. The 1*s* orbital doesn't overlap well with a 3*s* orbital (the size isn't a good match). The $2p_x$ orbital does not overlap well with the $2p_z$ orbital (the symmetry doesn't allow them to line up properly).

Symmetry – The property associated with orbitals that have similar size and shape.

We can think of this by imagining that overlapping orbitals are a game of Tetris. In Fig. 9.22, we can see that the best game piece to use to fill the 4×4 hole at the bottom of the screen is a 4×4 rectangle. A 3×3 rectangle wouldn't do a very good job of filling the hole, just as a 1*s* orbital and a 3*s* orbital don't overlap very well. They have the same symmetry (shape), but the size isn't right. In the second image of Fig. 9.22, we need a vertical 1×3 rectangle to fill the vertical 1×3 hole. A horizontal rectangle of the same size wouldn't work, just as a $2p_z$ orbital doesn't overlap with a $2p_x$ orbital. The symmetry (shape) of the two orbitals doesn't match, even though their size does.

9.4.2 Shapes of Orbitals and the Strengths of Bonds

What can we say about the shapes of the molecular orbitals? Molecular orbitals encompass entire molecules, and the shapes of the molecular orbitals often look similar to portions of the entire molecule. We know about their existence by calculation using some approximations of the Schrödinger equation, but how do we really know what the shape of a molecule is? X-ray crystallography is one of the methods used to take a "snapshot" of a molecule, but it relies on determining the average location of the atoms in a crystal. The scanning tunneling microscope may actually be the closest we can get to taking a picture of a single molecule.

The strength of the bond between two atoms can be represented by the **bond order**, the number of electrons in bonding orbitals minus the number of electrons in antibonding orbitals, divided by 2.

$$\text{Bond order} = \frac{\text{bonding electrons} - \text{antibonding electrons}}{2}$$

Bond order – The number of electrons in bonding orbitals minus the number of electrons in antibonding orbitals, divided by 2. The bond order indicates the degree of constructive overlap between two atoms.

A larger bond order indicates greater bond strength. The bond order typically agrees with the Lewis dot structure's expected number of bonds between adjacent atoms. However, the bond order between two atoms doesn't have to be a whole number.

To summarize, molecular orbital theory states that when the atoms of a molecule come together, atomic orbitals on each atom that are similar in symmetry and energy not only overlap but are transformed into a series of new orbitals that surround the entire molecule. These molecular orbitals have completely different shapes, sizes, and energies than the atomic orbitals from which they came. Although the atomic orbitals cease to exist after the creation of the molecular orbitals, the new molecular orbitals follow the same rules that guide us in placing electrons in the atomic orbitals.

9.4.3 MO Diagrams

Just as the civil engineer diagrams the layout of a building with a blueprint, we represent the placement of electrons in molecular orbitals with our own type of blueprints called **molecular orbital (MO) diagrams**. These diagrams graphically illustrate the energy of each of the bonding and antibonding orbitals and assist in determination of the bond order for a molecule. Graphically, the shapes of each of the molecular orbitals can be added to the diagram to get the overall picture of the molecule.

Molecular orbital (MO) diagram – A diagram used to illustrate the different molecular orbitals available on a molecule.

We can draw the blueprint—the MO diagram—for molecular hydrogen (H_2) to illustrate how this is done. We show in ◘ Fig. 9.23 that as the two hydrogen atoms approach each other, their atomic 1*s* orbitals mix to produce two new molecular orbitals. The overlap of these orbitals possesses symmetry, which means that they are equal in size and shape about the axis between the atoms (σ). One of the molecular orbitals resulting from this overlap is lower in energy (the bonding orbital, σ_{1s}, pronounced "sigma one s") and the other is higher in energy (the antibonding orbital, σ^*_{1s}, pronounced "sigma one s star"). Placement of the two electrons in H_2 follows the Pauli exclusion principle (two electrons with opposite spin per orbital) and Hund's rule (fill lowest-energy orbitals first) to fill the bonding orbital. The antibonding molecular orbital is left empty because there are no more electrons left in the molecule. The **highest-energy occupied molecular orbital (HOMO)** is the molecular orbital that contains electrons (in this case, the bonding orbital). The antibonding orbital ends up as the **lowest-energy unoccupied molecular orbital (LUMO)**. The bond order for hydrogen can be calculated as one ((2 bonding electrons−0 antibonding electrons)/2). We therefore predict that molecular hydrogen has a single bond between the two nuclei. What would happen to the bond order for this molecule if one of the electrons absorbed a photon and was promoted to the LUMO? The bond order for such a molecule would be zero ((1 bonding electron−1 antibonding electron)/2). With no net bond between the two atoms, the molecule would break into the individual hydrogen atoms. So we see that the MO theory can help us understand the impact of excited (energetic) electrons on the bonding in a molecule, which is what we need to understand how 11-*cis*-Retinal in our eyes helps us to see.

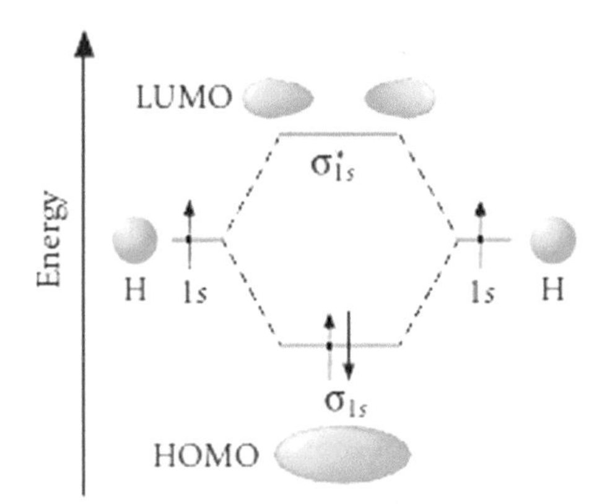

◘ **Fig. 9.23** MO diagram for H_2. This diagram graphically illustrates the energies and the resulting orbital shapes of the different molecular orbitals in the molecule

Highest-energy occupied molecular orbital (HOMO) – The most energetic molecular orbital that contains at least one electron.

Lowest-energy unoccupied molecular orbital (LUMO) – The least energetic molecular orbital that contains no electrons.

Example 9.8—MO Power

We showed in Example 9.2 that dilithium (of *Star Trek* fame) can be explained by the valence bond model. What does MO theory say about dilithium (Li_2)?

Asking the Right Questions

How do we judge using MO theory whether a molecule will exist? What is the impact of having a bonding HOMO or bonding LUMO?

Solution

We need to look at the highest-energy occupied molecular orbital (HOMO) and the lowest unoccupied molecular orbital (LUMO). If the HOMO is a bonding MO, we will therefore have a net bond between the two lithium atoms. If the HOMO is an antibonding orbital, we may not have a bond.

The MO diagram for dilithium is shown in Fig. 9.24. The HOMO is the σ_{2s} orbital and the LUMO is the σ^*_{2s} orbital. The bond order for this molecule is one ((4 – 2)/2), so MO theory suggests that Li_2 should exist as a molecule with one bond between the two lithium atoms.

Is Our Answer Reasonable?

The Li_2 molecule does exist, so to that extent, the answer is reasonable, but it quickly reacts with other lithium atoms to make lithium metal. Because of this reactivity, it is not possible to have a bottle of Li_2.

What If…?

… we considered the H_2^{2-} ion? According to MO theory, does it exist?

(Ans: No, because the σ_{1s} and the σ^*_{1s} orbital (the HOMO, in this case) are both occupied by 2 electrons.)

Practice 9.8

Use MO theory to determine whether He_2 is a stable molecule. What is the bond order in this molecule?

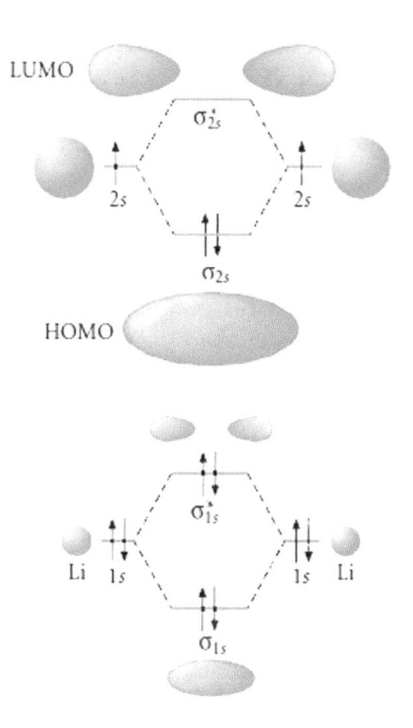

Fig. 9.24 Dilithium (Li_2) MO diagram. The MO configuration can also be written in shorthand format as $\sigma^2_{1s}\sigma^{*2}_{1s}\sigma^2_{2s}$

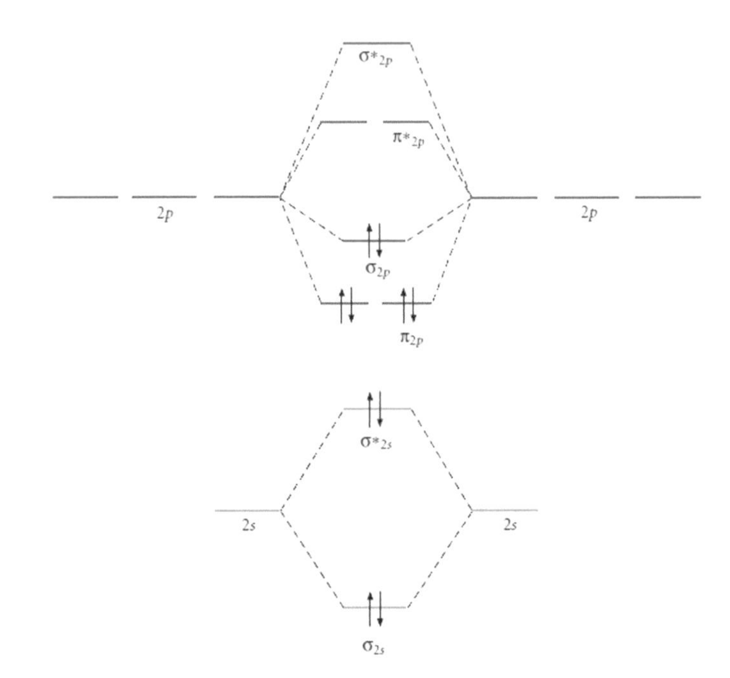

9

Fig. 9.25 MO Diagram for N_2

9.4.4 More About MO Diagrams

Lewis dot structures and valence bond theory agree that there are six bonding electrons (that is, a triple bond) between the nitrogen atoms in gaseous nitrogen (N_2). The MO diagram for the molecule should verify that the theory also gives rise to a bond order of 3. In molecules containing more than just the *s* orbitals, the MO diagram becomes much more complicated. Fig. 9.25 shows that the overlap of the $2p_y$ orbitals on each N and the $2p_z$ orbitals on each N, produces four orbitals: two π and two π*. The overlap of the $2p_x$ orbitals gives a sigma molecular orbital and a sigma-star molecular orbital. If we now place the electrons in the diagram, we see that they fill four bonding orbitals and one antibonding orbital. From the MO diagram, we can calculate a bond order of 3 for molecular nitrogen, in which the HOMO is the σ_{2p} orbital and there are two LUMOs (the degenerate π^*_{2p} orbitals).

The energy of the molecular orbitals in N_2 results in the ordering shown in Fig. 9.25. This is the same ordering that is observed in most of the diatomic molecules of the second period of the periodic table. However, a different ordering of the molecular orbitals is seen in O_2, F_2, and Ne_2, as shown in Fig. 9.26. The change in the order of the molecular orbitals arises because the difference in energy between the 2*s* and 2*p* atomic orbitals on O, F, and Ne is relatively greater than the other elements in the second row (see Fig. 9.7). Note that the two electrons in the HOMO are not paired in O_2. This allows us to explain an observed property of molecular oxygen that is unexplainable by other bonding theories. The other models of molecular oxygen correctly assign two bonds between the oxygen atoms, but oxygen possesses properties that we would not predict on the basis of those models. Oxygen, shown in Fig. 9.27, is an example of a paramagnetic molecule. The term **paramagnetism** refers to the ability of a substance to be attracted into a magnetic field. This attraction arises because of the presence of unpaired electrons within the molecule. A special case of this is **ferromagnetism** (so named because the effect is especially strong in iron), in which the paramagnetic atoms are close enough together that they reinforce their attraction to the magnetic field. The opposite of paramagnetism is **diamagnetism**. A diamagnetic molecule's electrons are paired, resulting in the molecule being repelled from a magnetic field. Nitrogen (N_2) is an example of a diamagnetic molecule because all of its electrons are paired (see Fig. 9.25).

Paramagnetism – The ability of a substance to be attracted into a magnetic field. This attraction arises because of the presence of unpaired electrons within the molecule.

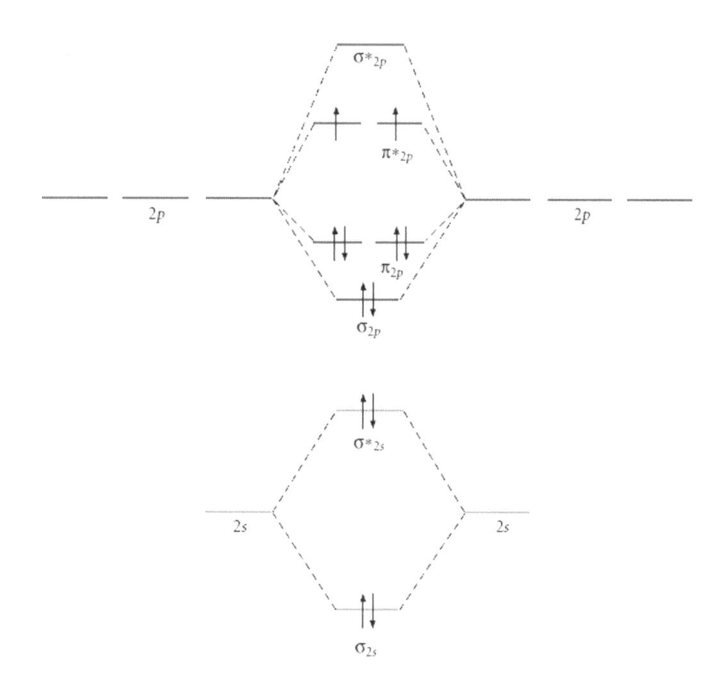

Fig. 9.26 MO diagram for O_2. *Note* that molecular oxygen has a different order for its molecular orbitals. There are also two unpaired electrons in the MO diagram. What is the bond order for O_2?

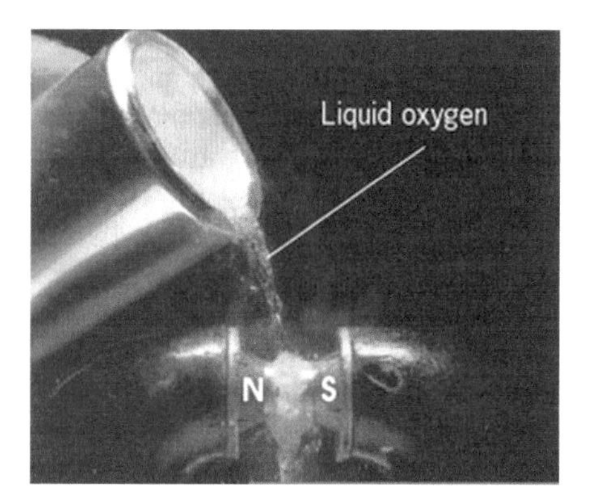

Fig. 9.27 The results of paramagnetism. Molecular oxygen is paramagnetic. Because of this, liquid oxygen is attracted to a magnetic field. *Source* ► http://www.clker.com/cliparts/8/5/d/3/15137482131641062339liquid-oxygen-magnet.hi.png

Ferromagnetism – A property of a compound that occurs when paramagnetic atoms are close enough to each other (such as in iron) that they reinforce their attraction to the magnetic field, such that the whole is, in effect, greater than the sum of its parts.

Diamagnetism – The ability of a substance to be repelled from a magnetic field. This property arises because all of the electrons in the molecule are paired.

Constructing models of molecules made with more than two atoms or molecules made from two different atoms (heteronuclear diatomic molecules) complicates the molecular orbital diagram because similar atomic orbitals on each atom have different energy levels. A further complication is added because the orbitals encompass the entire molecule and therefore must be constructed from the linear combination of the atomic orbitals on every atom in the molecule. For a 10-atom molecule, that would mean that each molecular orbital would arise from the combination of 10 different atomic orbitals. Such a diagram would be very complex indeed!

Example 9.9—MO Theory: The Power of a Photon

The halogens (F_2, Cl_2, Br_2, and so on) are extremely reactive compounds. In fact, their reaction can be initiated by a single photon of light, just like 11-*cis*-retinal in the eye. Given that the photon will promote one electron from a bonding orbital to an antibonding orbital of similar symmetry, illustrate the reaction of F_2 with that photon using MO diagrams. The MO diagram for F_2 is similar to that for O_2.

Asking the Right Questions

What is the original MO diagram for F_2? What is the bond order based on the MO diagram? What will be the effect of the promoted electron on the bond order?

Solution

The MO diagram for F_2 is shown in ◘ Fig. 9.28. Promotion of an electron from the π_{2p} orbital to an MO of the same symmetry, the σ^*_{2p} orbital, would cause the bond order for F_2 to become zero. The photon would effectively break the bond between the fluorine atoms.

Are Our Answers Reasonable?

To the extent that Molecular Orbital theory explains the observed reactivity of F_2 (as well as other halogens), the answer is reasonable.

What If…?

…we had O_2^{2-}? What would be its bond order? Would it display paramagnetism or diamagnetism?

(*Ans: Bond order = 1, and it is diamagnetic.*)

Practice 9.9

Illustrate the interaction of H_2 and O_2 with a single photon of light. After the promotion of an electron, which molecule has the larger bond order?

9

Looking back to 11-*cis*-retinal, we can produce an MO diagram that illustrates the location of the electrons in the conjugated π system. Although the diagram is very complex, the energy levels of the conjugated π system are more closely spaced as the π system increases in length. Promotion of an electron requires less and less energy as the π system lengthens, as indicated by the increasing wavelength of absorbed light in the molecules of ◘ Fig. 9.29.

This occurs because it takes less energy to promote an electron to a closer molecular orbital. We pointed out earlier that this phenomenon is used by chemists in the development of color dyes. By constructing a molecule that has a conjugated π system of just the right length, manufacturers can make molecules that absorb exactly the color needed to make dyes that appear to be any shade of color imaginable.

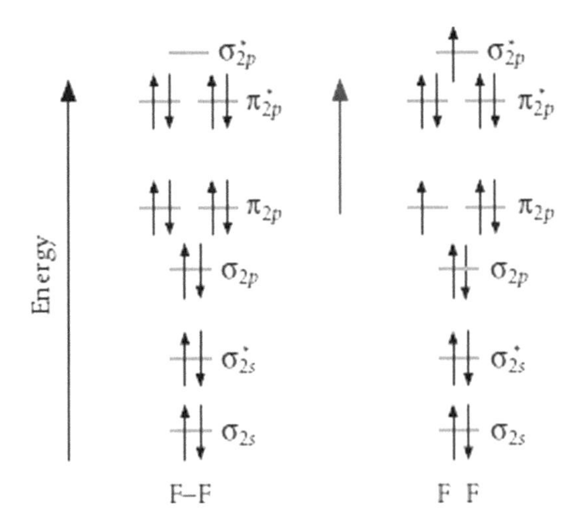

◘ **Fig. 9.28** MO diagram of F_2. The promotion of an electron from a π MO to the σ* MO reduces the bond order from 1 to 0. When this happens, there is no net bond remaining between the atoms. Addition of a photon of light to the molecule with a wavelength of 754 nm causes the molecule to break in two

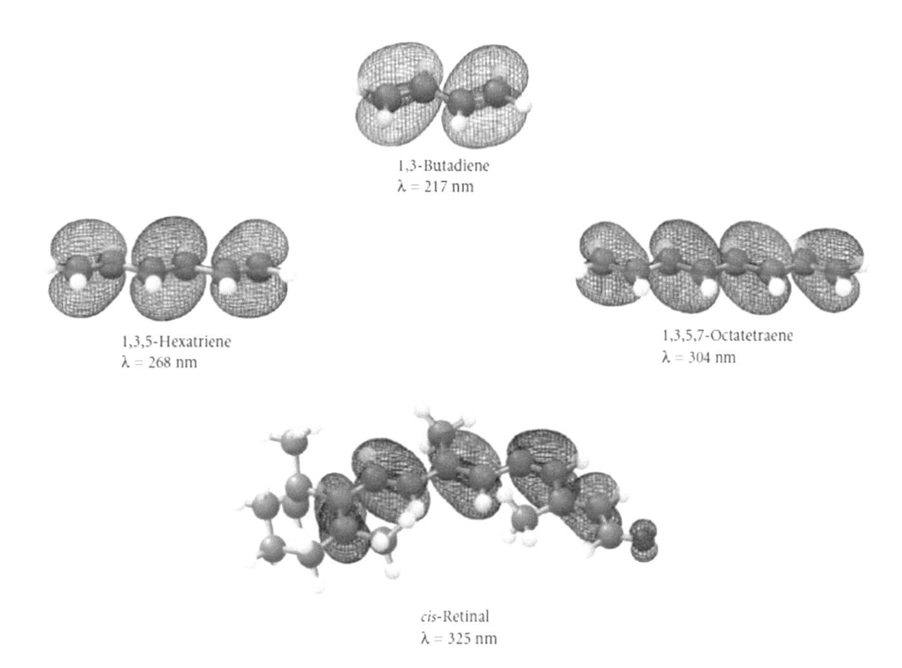

Fig. 9.29 The MOs of conjugated systems. As the length of the conjugated π system increases, the energy required to promote an electron decreases. Evidence of this is observed in the wavelength of light absorbed by the molecules. Longer π systems require less energy and a larger wavelength (energy and wavelength are inversely proportional, $E=\frac{hc}{\lambda}$)

9.5 Putting It All Together

The models of each of the bonding theories provide increasingly better correlation with the observed properties of many covalent compounds. Valence bond theory works well at showing how the electrons in sigma bonds come together to create the skeletal framework of a molecule. At the same time, the localized bonding picture described by valence bond theory conveniently determines, among other things, the lengths of bonds in a molecule. The rules of VSEPR modeling help identify the three-dimensional shape of the molecule, and Lewis dot structures rapidly reveal the connectivity of atoms in a molecule. However, molecular orbital theory works best at describing the behavior of electrons located in the bonds. An approach to drawing a molecule that uses all of these models is often followed. We can show this approach using benzene (C_6H_6).

Benzene is a carcinogenic compound that is widely used in industry as a nonpolar solvent and as a starting compound in the manufacture of other useful products such as phenol (an antiseptic compound) and nylon for ropes, fabrics and plastics. Experiments have revealed that benzene is a hexagon of six carbon atoms, each bonded to one hydrogen atom. It has also been observed experimentally that the distances between carbon atoms in the molecule are identical. In the laboratory, benzene reacts as though it has C=C bonds at each position in the molecule.

No single Lewis dot structure of benzene can be drawn to illustrate the experimentally determined bond lengths and reactivity in the molecule. Considering that benzene is best shown as a hybrid of two resonance structures, we can address the experimental observation that all C—C bonds in benzene are the same length as shown in Fig. 9.30. Valence bond theory can also be used to indicate that each carbon is sp^2 hybridized. The sigma bond network that makes up the molecule defines 120° bond angles at each flat sp^2 hybridized carbon. Bonds between carbon atoms result from the overlap of a $2sp^2$ hybrid orbital with another $2sp^2$ hybrid orbital. Each C—C bond is 140 pm in length. Bonds between the carbon and hydrogen atoms are the result of $2sp^2$–$1s$ orbital overlap. In addition, each carbon atom contains a half-filled $2p$ orbital perpendicular to the plane of the molecule.

However, the six half-filled $2p$ orbitals in benzene are best handled by molecular orbital theory as shown in Fig. 9.31. The σ bond network is separated from the π bond network in this approach. Doing so makes preparing a MO diagram a more manageable task. The six $2p$ orbitals on the carbon atoms in benzene have the same symmetry and energy, so they can be mixed to provide six new π MOs. Three bonding MOs and three antibonding MOs result (see Fig. 9.32). We place the six electrons into the new MOs, filling the three bonding MOs. The electrons in the MO diagram occupy a π bond at every carbon atom, implying that there isn't any difference in the carbons of benzene. Each C—C bond has a bond order of 1.5. The approach we have taken to draw the best model of benzene has provided us with the best (and quickest) way to represent the experimentally determined

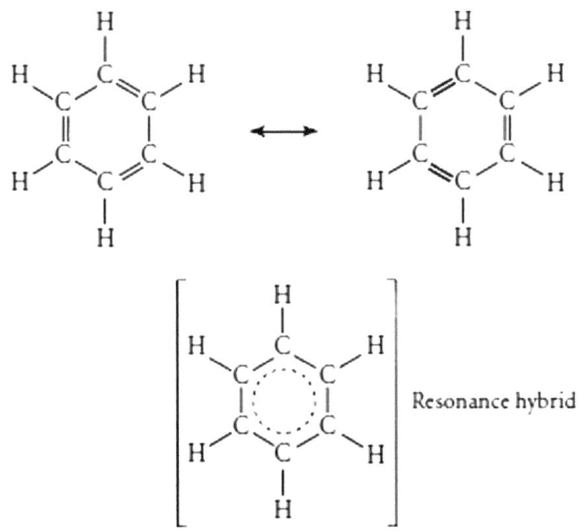

Resonance hybrid

Fig. 9.30 π bonding in benzene. Lewis dot structures and valence bond models do a good job of describing the sigma bond network in benzene

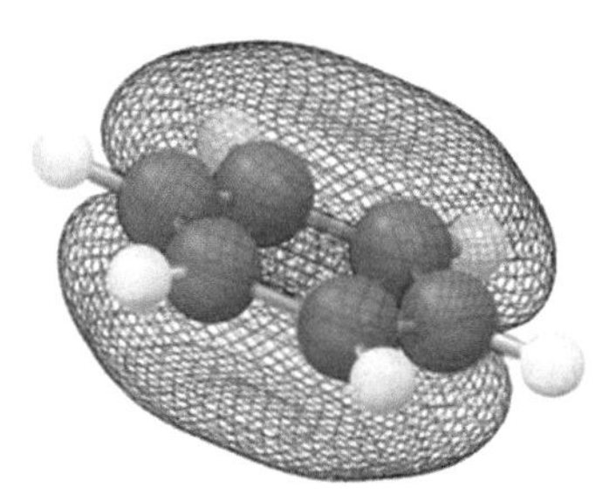

Fig. 9.31 Delocalized π bonding in benzene according to MO theory

properties of benzene (such as the C—C bond lengths and the reactivity of benzene). We no longer have to imagine different structures (arrived at from different models) of benzene to account for the observed properties.

If we more closely examine the molecular orbital model of benzene, as shown in Fig. 9.32, we find that the electrons in the π orbitals are spread out over the six atoms in the structure. This conjugated π system is said to be **delocalized**. Such delocalized systems occur any time the electron density in a molecule can be distributed between more than two atoms. Delocalization also can be seen in the carbonate ion (CO_3^{2-}) and in 11-*cis*-retinal. However, in the case of benzene (C_6H_6) and related compounds, the delocalization of the electron density imparts a particular stability to the molecule (more so than the delocalization of electrons in the carbonate ion and 11-*cis*-retinal). The stability is so important to the molecule that it is often written as shown in Fig. 9.33, with a circle to illustrate the delocalization of three π bonds in the molecule. Benzene's stability makes it a good choice for use as a reaction solvent.

Delocalized – Term used to describe a π system wherein the electron density in a molecule can be distributed among more than two atoms.

Unfortunately, this stability also makes it difficult to metabolize if accidentally ingested. One of the major problems with using benzene is that it causes cancer in humans. Table 9.6 lists certain properties of benzene and of some alternative solvents that have been used in place of it. Their properties aren't identical to those of benzene, but their substitution for benzene has done a great deal to safeguard the health and lives of industrial chemists.

Now we can finally close the chapter on the story of 11-*cis*-retinal and the science of sight. This molecule binds to a large biological molecule in the eye. It then absorbs a photon of light, an electron is promoted into a π* molecular orbital, and a π bond is broken in the molecule. Eliminating the π bond between two carbons in the molecule leaves just a σ, which allows the molecule to rotate around that bond, changing the shape of the

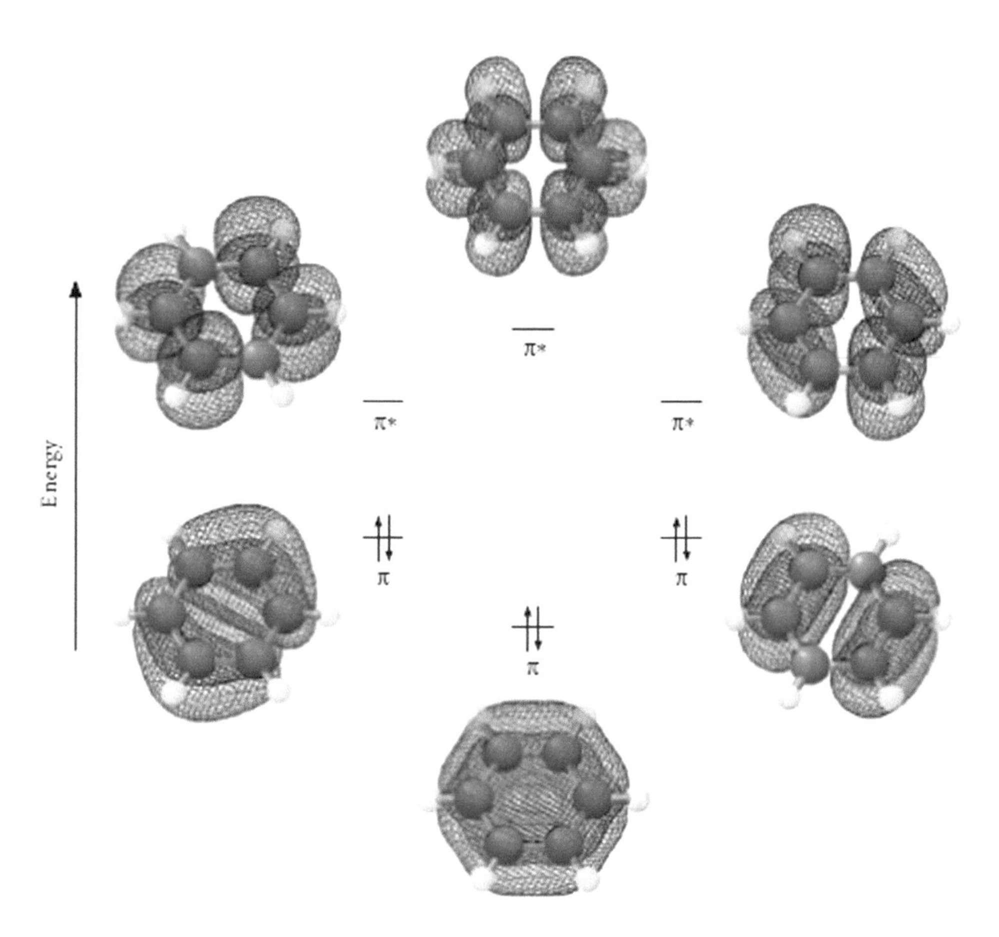

Fig. 9.32 Molecular orbital diagram for benzene

Fig. 9.33 Shorthand notation for benzene that illustrates the delocalized molecular orbitals

Table 9.6 Common solvents

Solvent	Formula	Boiling point (°C)	Dipole moment	Solubility in water
Benzene	C_6H_6	80.1	0	Very slight
Hexane	C_6H_{14}	68.7	0	Insoluble
Dichloromethane	CH_2Cl_2	40	1.60	Slight
Tetrahydrofuran	C_4H_8O	66	1.63	Miscible
Toluene	C_7H_8	111	0.36	Very slight

molecule and creating a nerve impulse. The retinal molecule is then ejected from the large biological molecule and through other reactions, rotated back to its original shape. The reformed 11-*cis*-retinal is then reattached to the large biological molecule and the process can start all over again.

The Bottom Line

- The molecular models that chemists construct increase in complexity from Lewis dot structures, to the VSEPR model, to valence bond theory, to MO theory. This increase in complexity is accompanied by an increase in satisfactory agreement with observed properties for the molecule.
- Valence bond theory, originated by Linus Pauling, defines bonds as the overlap of atomic orbitals. Sigma bonds result from end-on overlap of orbitals; pi bonds result from side-to-side overlap of orbitals.
- Hybridization of atomic orbitals gives rise to new orbitals that help explain bonding. Hybridization also results in structures consistent with the VSEPR model.
- Molecular orbital theory defines bonding with orbitals that are not confined to a single atom. Bonds result from molecular orbitals in a molecule that encompass all the atoms.
- Mixing *n* atomic orbitals gives *n* molecular orbitals. One-half of these are bonding; the other half are antibonding. Bonding MOs are lower in energy than the atomic orbitals from which they are constructed. Antibonding MOs are higher in energy than the atomic orbitals from which they are constructed.
- The Pauli exclusion principle and Hund's rule must be obeyed when placing electrons in the new MOs.
- MO diagrams can be used to identify the number of bonds (bond order) between atoms.
- Paramagnetism results from unpaired electrons in a molecule. Diamagnetism results from complete pairing of all electrons in a molecule.
- Electrons in conjugated π orbitals are said to be delocalized, because the electron density can be distributed between more than two atoms.

Section 9.2 Valence Bond Theory

Skill Building

1. a. Professional scientific journal articles, textbooks, and popular writing have all depicted chemical bonds as a typed dash between two atomic symbols. Although this does communicate that the atoms are associated with each other, cite one reason why it does not accurately describe a chemical bond.
   b. Why do chemists seek to provide better models to describe the makeup of a chemical bond?
2. The VSEPR model of chemical bonding has been very successful for some descriptions of chemical bonding, but what is the major weakness of this popular model?
3. Write out the ground-state configuration of silicon. According to the VSEPR model, how many bonds should one atom of Si be able to form? Is your answer consistent with the formula of $SiCl_4$ (a compound used in the formulation of some smoke screens)?
4. Write out the ground-state configuration of aluminum. According to the VSEPR model, how many bonds should one atom of Al be able to form? Is your answer consistent with the formula of $AlCl_3$?
5. Write the ground-state electron configuration for each of the following atoms. According to the VSEPR model, how many bonds should each atom be able to form?

   Cl Se B

6. Write the ground-state electron configuration for each of the following atoms. According to the VSEPR model, how many bonds should each atom be able to form?

   Ge Ne Mg

7. According to the text, what are the three assumptions about orbital overlap that are made in Valence Bond theory?
8. According to Valence Bond theory, which molecule should have the stronger bond: H—Cl or Cl—Cl? Why?
9. According to Valence Bond theory, carbon can only form two bonds. We know from experimental evidence that carbon actually forms four chemical bonds, which is what VSEPR predicts. Even though it fails in this respect, what does Valence Bond theory accomplish that VSEPR cannot?
10. Compare and contrast VSEPR with Valence Bond theory. What does each theory correctly predict? Where does each theory fail?

11. Draw the Lewis-dot structure of ammonia and use VSEPR theory to predict the geometry of the molecule.
    a. Based on Valence Bond theory, which orbitals on the nitrogen should be involved in bonding?
    b. If Valence Bond theory is correct, what should the geometry of ammonia be?
    c. What is required to explain the discrepancy between Valence Bond and VSEPR?
12. Draw the Lewis-dot structure of the carbonate ion and use VSEPR theory to predict the geometry of the ion.
    a. Based on Valence Bond theory, which orbitals on the carbon should be involved in bonding?
    b. If Valence Bond theory is correct, what should the geometry of the carbonate ion be?
    c. What is required to explain the discrepancy between Valence Bond and VSEPR?

Chemical Applications and Practices

13. Carbon tetrachloride (CCl_4) is seldom used as a cleaning solvent. It formerly saw widespread use, but more recently its ill effects on human health have led to its restriction. What is the ground-state designation for carbon's valence electrons? Explain why this configuration does not lend itself to forming the four equal bonds found between carbon and chlorine in this compound.
14. Tin can form compounds such as $SnCl_2$ and $SnCl_4$. What is the ground-state designation for tin's valence electrons? Explain why this configuration does not lend itself to forming the equal bonds found between tin and chlorine in one of these compounds.

Section 9.3 Hybridization

Skill Building

15. a. Write the electron configuration of hydrogen and bromine.
    b. What valence orbitals are involved in the overlap that forms the bond between H and Br?
    c. Compare the strength of this bond to the strength of the HF bond described earlier in the chapter.
16. a. Write the electron configuration of hydrogen and chlorine.
    b. What valence orbitals are involved in the overlap that forms the bond between H and Cl?
    c. Compare the strength of this bond to the strength of the HF bond described earlier in the chapter.
17. Using valence bond theory, describe the orbital overlap in Cl_2.
18. Using valence bond theory, describe the orbital overlap in F_2.
19. Using the concept of orbital overlap, predict in each pair of compounds which bond will be longer:

$$H_2 \text{ or } Cl_2 \qquad Br_2 \text{ or } Cl_2 \qquad HCl \text{ or } HBr$$

20. Using the concept of orbital overlap, predict in each pair of compounds which bond will be longer:

$$H_2 \text{ or } Li_2 \qquad F_2 \text{ or } Cl_2 \qquad LiF \text{ or } HF$$

21. Hypochlorous acid has excellent bleaching properties and must be handled with care. Using valence bond theory, show what orbitals are likely to be overlapped in HOCl. Which bond would you predict to be longer, the bond between H and O or the bond between O and Cl? Explain the basis of your prediction.
22. Look at the ground-state electron configurations of the atoms in the HOCl compound mentioned in the previous problem. Why do you think that the structure presented as HOCl more logically represents the bonding than HClO?
23. Predict the formula of potassium hydride. Which hydride would you predict to have the stronger bonds, potassium hydride or cesium hydride? Explain the basis of your prediction.
24. What would you predict as the formula of calcium hydride? Which hydride would you predict to have the stronger bonds, calcium hydride or magnesium hydride? Explain the basis of your prediction.
25. When explaining and demonstrating orbital hybridization to chemistry students, a teacher prepares "orbital omelets." This analogy involves mixing one, two, or three eggs with 100 mL of milk. How could the three omelets be used to describe three ways in which *s* and *p* orbitals can hybridize?
26. Develop an analogy similar to that in Problem 25 using an "orbital milkshake." Mix one, two, or three scoops of ice cream with one glass of milk. How could the three resulting milkshakes be used to describe three ways in which *s* and *p* orbitals can hybridize?
27. In some situations *d* orbitals may become involved in hybridization.
    a. If an atom's hybridization were designated sp^3d^2, how many orbitals would be involved?
    b. What would be the same and what would be different about these orbitals compared to hybrid orbitals made from only *s* and *p* orbitals?

28. In many sciences, including chemistry, we are sometimes forced to use the same symbols to mean different things. In each of the following examples, distinguish the meaning of the two sets of similar symbols:
    a. $2s^1 2p^1$ and $(2sp)^1$
    b. $(2sp^3)$ and $2p^3$.
29. a. Show the ground state for the valence electrons of silicon. What type of hybridization would the orbitals of silicon undergo to produce SiH_4?
    b. What would be the resulting geometric shape of a SiH_4 molecule?
30. BCl_3 and NH_3 have the same basic formula (one central atom with three attached atoms), but they differ in shape. Using the hybridization of their valence orbitals, explain the basis for their different shapes.
31. The bonding in H_2O, NH_3, and CH_4 can be explained by using sp^3 hybridization of the valence electrons in oxygen, nitrogen, and carbon, respectively. However, all three molecules have different shapes. What are the bond angles for the three molecules and what is the basis for each?
32. $SiCl_4$ and SCl_4 both contain a central atom attached to four other atoms. After examining the ground state of Si and S and bonding four chlorine atoms to each, would you predict the two molecules to have the same shape? Explain the basis for your conclusion.
33. The carbon atom in both methane (CH_4) and chloroform ($CHCl_3$) has undergone sp3 hybridization. However, the two molecules do not have exactly the same shape. Explain the cause and result of any differences.
34. The oxygen atom in both H_2O and HOF has undergone sp^3 hybridization. However, the two molecules do not have exactly the same shape. Explain the cause and result of any differences.
35. How many hybrid orbitals are formed from mixing an *s* orbital with a *p* orbital?
36. How many hybrid orbitals are formed from mixing an *s* orbital with two *p* orbitals?
37. For each hybridization listed below, indicate the number of bonds that could be formed by overlapping with the hybridized orbitals.

    sp sp^2 sp^3

38. For each type of hybridization listed below, indicate the idealized geometric shape that would be produced around a central atom having that type of hybridization.

    sp sp^2 sp^3

39. What type of hybridization would be found in the central atom of each of the following?
    a. OF_2 b. CCl_4 c. BCl_3 d. $BeCl_2$
40. What type of hybridization would be found in the central atom of each of the following?
    a. CS_2 b. H_2S c. $CSCl_2$ d. SO_2
41. Indicate what geometric shape is produced when the number of orbitals around a central atom is:
    a. 2 b. 3 c. 4 d. 5 e. 6
42. Indicate what bonding angle is produced when the number of orbitals around a central atom is:
    a. 2 b. 3 c. 4 d. 5 e. 6
43. Diagram the hybridization for the following three hydrocarbons:

    C_2H_2 C_2H_4 C_2H_6

    Energy is required to break chemical bonds. In which of the three molecules would the least amount of energy be required to break the carbon-to-carbon bond. Explain, or justify your answer.
44. Diagram the hybridization of the carbon in each of the following compounds:

    CO_2 CO CO_3^{2-}

    Energy is required to break chemical bonds. In which of the three molecules would the least amount of energy be required to break the carbon-to-oxygen bond. Explain, or justify your answer.
45. Sulfur is capable of many different oxidation states. In each of the following, determine the hybridization of sulfur and the resulting shape of the molecule or ion.
    a. SO_3 b. SO_3^{2-} c. SF_6 d. S_8
46. Nitrogen is capable of many different oxidation states. In each of the following, determine the hybridization of nitrogen and the resulting shape of the molecule or ion.
    a. NO_3^- b. NO c. NO_2 d. N_2
47. Predict the hybridization of the underlined atom and the overall shape of the following molecules.
    a. $\underline{N}H_4^+$ b. $\underline{Xe}F_4$ c. $\underline{S}F_4$ d. $\underline{N}O_2^-$

48. Predict the hybridization of the underlined atom and the overall shape of the following molecules.
    a. H<u>C</u>N b. <u>Cl</u>F_3 c. <u>Cl</u>F_5 d. <u>I</u>Cl_3
49. Predict the hybridization of each of the atoms in the following molecule.

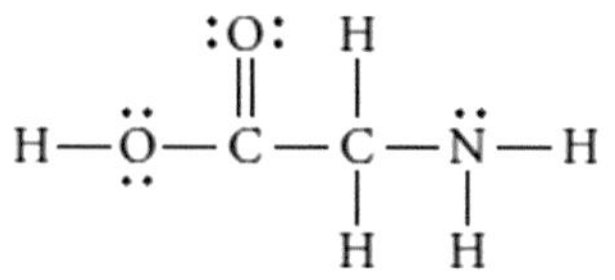

50. Predict the hybridization of each of the atoms in the following molecule.

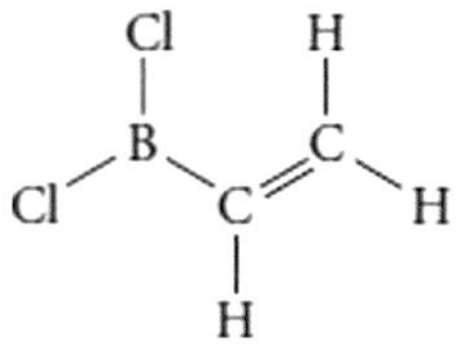

Chemical Applications and Practices

51. In the realm of science fiction, as we noted earlier, dilithium was used to propel the matter–antimatter drive of the starship Enterprise. (Some of the details are still a bit elusive.) Suppose a Klingon engineer suggests that di-potassium might also be a possible fuel. Would you argue that K_2 would have a stronger or a weaker bond than that found in Li_2? What would be the basis of your argument? (It will be best if you can cite some sound basis in valence bond theory for your argument. Klingons can be defensive when challenged.)
52. Several noble gas compounds have been synthesized under controlled conditions. Explain, using ground-state electron configurations and valence bond theory, why no diatomic molecules such as He_2, Ne_2, and the like have been made.
53. One of the first synthesized "natural" organic compounds was the nitrogen-based solid found in mammal waste: urea. From examining the formula shown here, what would you predict for the hybridization of the nitrogen atoms? What would be the bond angle from hydrogen to nitrogen to carbon?

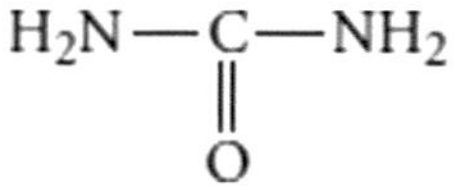

$$H_2N—C(=O)—NH_2$$

54. Three compounds composed of carbon and hydrogen can be used to illustrate the importance of hybridization and molecular shape. Diagram the Lewis dot structure for each case shown here. Then predict the hybridization in the carbon atoms and predict the bond angle from H to C to C in each structure.
    a. C_2H_2 (acetylene, used in high-temperature welding).
    b. C_2H_4 (ethylene, a plant hormone that hastens ripening).
    c. C_2H_6 (ethane, a heating fuel and an ingredient in polymers).
55. The structure of the artificial sweetener aspartame is shown below. How many pi bonds are shown in the structure? What would be the hybridization and bond angle around the carbon atom double-bonded to oxygen at the top of the molecule?

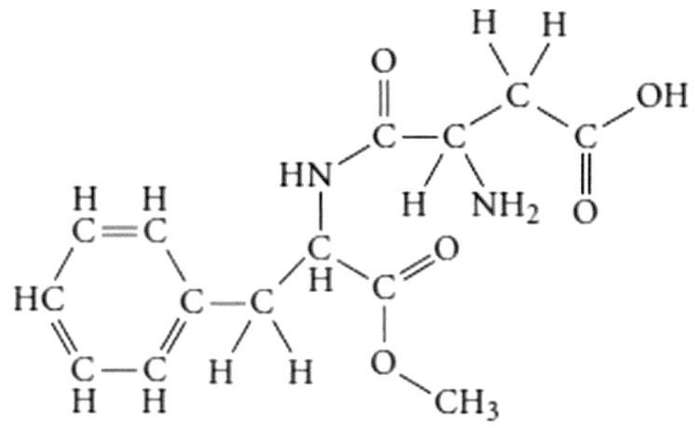

56. The male sex hormone testosterone is synthesized from cholesterol. Examine the structure shown here.

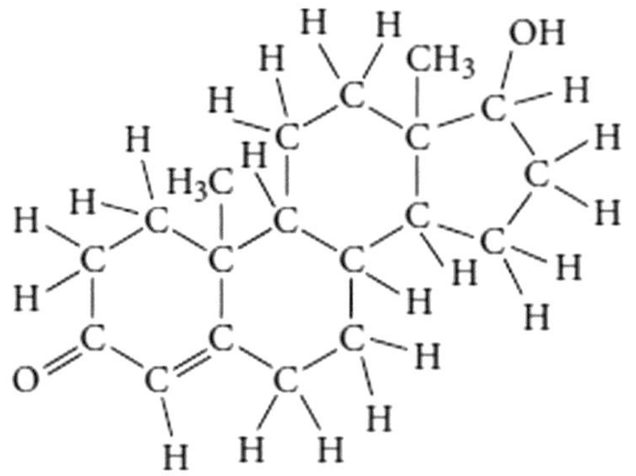

a. How many pi bonds are present in the molecule?
b. There are two carbon atoms that are bonded to oxygen. Describe the hybridization present in these two carbon atoms.
c. What are the bond angles around those two carbon atoms?

57. Antihistamines are often taken to counteract the effects of the amino acid histamine that is released in allergic reactions. Examine the structure of histamine shown below. Compare the two nitrogen atoms in the ring portion of the molecule on the basis of lone-pair electrons, bond angle, and hybridization.

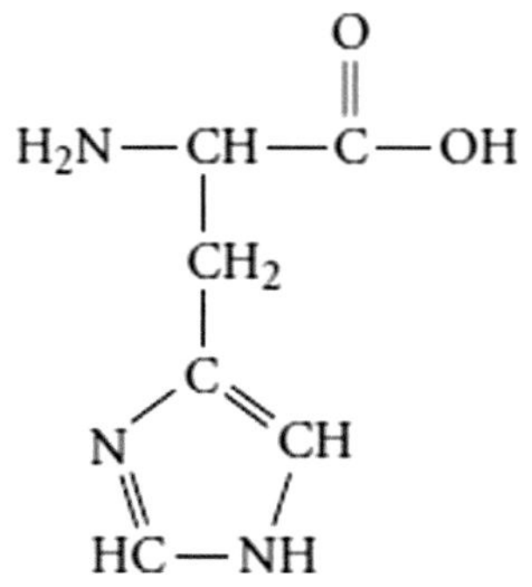

58. Vitamin B_6 (pyridoxine) is important in the metabolism of carbohydrates, proteins, and fats because it enhances the action of several enzymes. After examining the structure of this important molecule, answer the following questions.

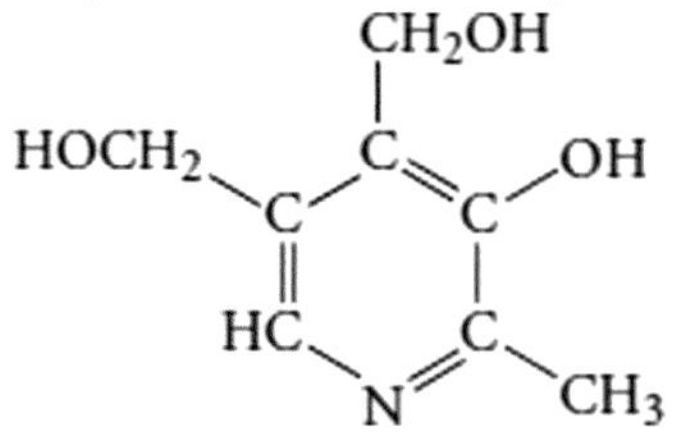

a. How many pi bonds are present in the structure?
b. Would the bond between the carbon atoms within the ring be shorter or longer than the bond between a carbon atom in the ring and one outside the ring? Explain.
c. Are the bond angles around the CH_3 group the same as or different from the bond angles around the carbon in the CH_2OH group? Explain.

59. One form of the molecule $POCl_3$ is shown below. What hybridization would be found in the phosphorus atom? What are the bond angles around the phosphorus?

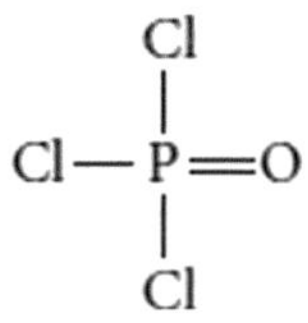

$$Cl-\overset{\overset{\Large Cl}{|}}{\underset{\underset{\Large Cl}{|}}{P}}=O$$

60. Diacetyl is one of the additives included in margarine to enhance the butter-like taste. The diagram of diacetyl shown here does not accurately depict the bond angles. What would be the correct bond angles around the two central carbon atoms? Redraw the structure to show those angles.

$$H_3C-\underset{\underset{O}{\|}}{C}-\underset{\underset{O}{\|}}{C}-CH_3$$

61. Capsaicin is one of the key ingredients in chili peppers that produce the spicy taste sensation sought by chefs. Examine its structure and answer the following questions.

a. Which carbon atoms have sp^3 hybridization?
b. How many H atoms would have to be removed to produce a double bond between the two carbon atoms that are shown farthest to the right?
c. How many pi bonds are shown in the molecule?
d. How many lone pairs of electrons are needed around the oxygen atom that is double-bonded to the carbon atom?

62. Estradiol is the active steroid used in estrogen replacement therapy.

a. Which carbon atoms have sp^2 hybridization?
b. How many H atoms would have to be removed to produce a double bond between the oxygen and carbon atoms that are shown farthest to the right?
c. How many pi bonds are shown in the molecule?
d. How many lone pairs of electrons are needed around each oxygen atom?

Section 9.4 Molecular Orbital Theory

Skill Building

63. Working from left to right, create a table in which the name of each model is correctly aligned with the scientist who devised it and a general summary of the model.

Model	Scientist	General summary
VB	G.N. Lewis	Subtract and add overlap to create π and σ bonds
MO	L. Pauling	Distribute bonding and nonbonding electron pairs
VSEPR	E. Schrödinger	Overlap and hybridization create new orbitals

64. The development of molecular orbital theory gave chemists some advantages in explaining chemical bonding.
a. What is the theoretical basis of MO theory?
b. What main advantage over VSEPR does this treatment of bonding offer?
c. What is a disadvantage of MO theory?

65. Oxygen and sulfur are both found in Group VIA on the periodic table. Consequently, they share many properties, including some bonding characteristics. However, the larger size of the *p* orbitals in sulfur prevents them from overlapping in an effective way to form pi bonds. Use orbital diagrams to show how this can be used to explain why O_2 is more stable than S_2.
66. Nitrogen and phosphorus are both found in Group VA on the periodic table. Consequently, they share many properties, including some bonding characteristics. However, the larger size of the *p* orbitals in phosphorus prevents them from overlapping in an effective way to form pi bonds. Use orbital diagrams to show how this can be used to explain why N_2 is more stable than P_2.
67. Cite two differences between bonding and antibonding orbitals.
68. Cite two similarities between bonding and antibonding orbitals.
69. Most sciences have their share of acronyms. In this chapter you have encountered several important ones related to chemical bonding. Supply the full name and meaning of each of the following:
    a. LCAO b. HOMO c. LUMO
70. In this chapter you have encountered several important symbols related to chemical bonding. Supply the full name and meaning of each of the following:
    a. π b. π^* c. σ
71. Draw the two orbitals that result from the combination of two 1*s* atomic orbitals. What makes the bonding orbital "bonding"? What makes the other one "anti-bonding"? What symbols are used for these two orbitals?
72. Draw the two orbitals that result from the "sigma" combination of two 2*p* orbitals. What makes the bonding orbital "bonding"? What makes the other one "anti-bonding"? What symbols are used for these two orbitals?
73. Draw the two orbitals that result from the "pi" combination of two 2*p* orbitals. What makes the bonding orbital "bonding"? What makes the other one "anti-bonding"? What symbols are used for these two orbitals?
74. In the oxygen molecule, why are there two pi-bonding molecular orbitals, which have the same energy?
75. Removing or adding electrons to molecules can change several properties of the bonding within the molecule. Provide complete MO diagrams for N_2^-, N_2, and N_2^+.
    a. Calculate the bond order for each.
    b. Indicate which, if any, are paramagnetic.
    c. Rank the three nitrogen species in order from shortest to longest nitrogen-to-nitrogen distance.
76. Removing or adding electrons to molecules can change several properties of the bonding within the molecule. Provide complete MO diagrams for O_2^-, O_2, and O_2^+.
    a. Calculate the bond order for each.
    b. Indicate which, if any, are paramagnetic.
    c. Rank the three oxygen species in order from shortest to longest oxygen-to-oxygen distance.
77. A diatomic homonuclear +3 ion has the following molecular orbital configuration:

$$(\sigma_{1s})^2(\sigma_{1s}^*)^2(\sigma_{2s})^2(\sigma_{2s}^*)^2(\sigma_{2p})^2(\pi_{2p})^4(\pi_{2p}^*)^3$$

    a. What is the identity of the element used to make this ion?
    b. Is the ion paramagnetic or diamagnetic?
    c. What is the bond order of the ion?
78. A diatomic homonuclear +2 ion has the following molecular orbital configuration:

$$(\sigma_{1s})^2(\sigma_{1s}^*)^2(\sigma_{2s})^2(\sigma_{2s}^*)^2(\sigma_{2p})^2(\pi_{2p})^2$$

    a. What is the identity of the element used to make this ion?
    b. Is the ion paramagnetic or diamagnetic?
    c. What is the bond order of the ion?
79. Indicate the total number of sigma and pi bonds in the following molecule.

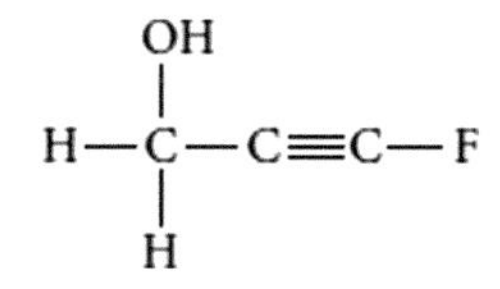

80. Indicate the total number of sigma and pi bonds in the following molecule.

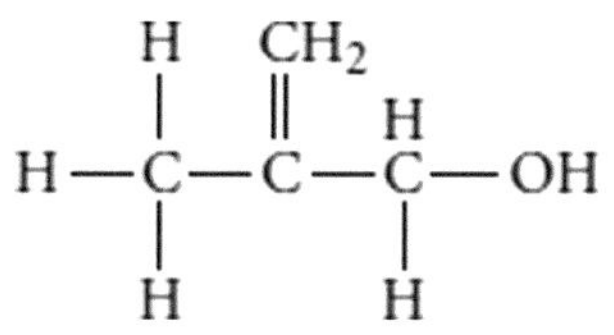

81. What is the bond order between the indicated atoms in the following molecule?

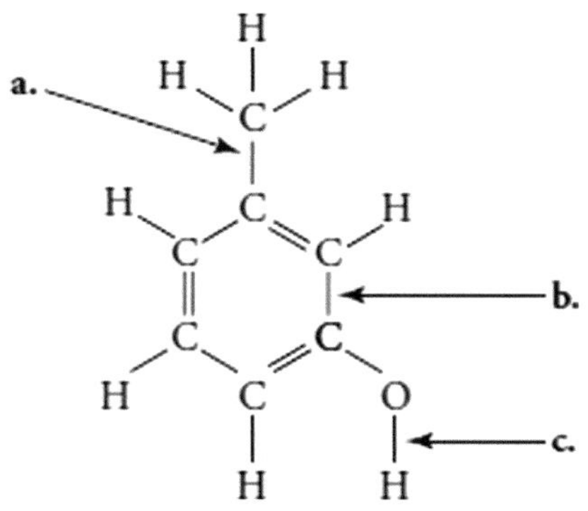

82. What is the bond order between the indicated atoms in the following molecule?

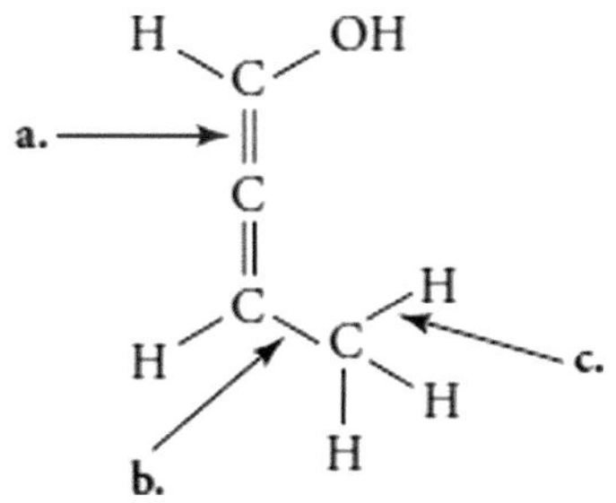

Chemical Applications and Practices.

83. Typically, phosphorus may be found in nature as P_4 (among other forms). However, P_2 is also known to exist at sufficiently lower temperatures. Using the outer valence electrons, produce the MO diagram for P_2. (You may assume the same sequence of orbitals as found in N_2.)
    a. What is the bond order of P_2?
    b. How many electrons would be found in the highest π^*?
84. Br_2 is a liquid at room temperature. Using the outer valence electrons, produce the MO diagram for Br_2. (You may assume the same sequence of orbitals as found in F_2.)
    a. What is the bond order of Br_2?
    b. How many electrons would be found in the highest π?
85. Electromagnetic casting of aluminum employs a powerful magnetic field to hold molten aluminum in a desired shape while it is cooled with a water spray. The molten aluminum is repelled by the electromagnetic field surrounding it. Based on the discussion in the text, what is likely to be true about the electron configuration of molten aluminum?
86. Assuming that the 3*s* and 3*p* valence orbitals interact in the same way as the 2*s* and 2*p* orbitals interact, what would you predict about the magnetic properties of sulfur and phosphorous?
87. a Nitric oxide (NO) is a stable, uncharged molecule that is very important in the chemistry of the lower atmosphere, particularly in the production of tropospheric ozone.
    b. Draw a Lewis-Dot structure for NO and predict the number of bonds between the nitrogen and oxygen.
    c. Using the same sequence of orbitals found in O_2 (Fig. 9.26), use MO theory to predict the bond order between N and O.
    d. Which theory predicts the more stable molecule?
    e. Will NO be paramagnetic or diamagnetic?

88. When nitric oxide (see problem 87) undergoes a chemical reaction, how many electrons is it likely to lose? Draw the molecular orbital diagram for the resulting ion and determine the bond order. What is the charge on the resulting ion? What does Molecular Orbital theory tell you about the nitric oxide ion that VSEPR Theory and Lewis structures do not?

Section 9.5 Putting It All Together

Skill Building

89. a. The term *orbital overlap* was used throughout this chapter. Describe the meaning of this important term. What exactly is being "overlapped"?
    b. Diagram the orbital overlap typical of an *s–s, s–p,* and *p–p* overlap.
90. a. The term *hybrid* was used throughout this chapter. Describe the meaning of this important term.
    b. Draw the shapes of at least three different hybrid orbitals.
91. a. Why is it necessary to draw resonant hybrids when representing the structure of ions and molecules that have delocalized electrons?
    b. Using the diagram of acetylsalicylic acid shown in Problem 101 as one example, show another example of a resonance hybrid of the molecule.
92. Using the diagram of the steroid illustrated in Problem 62 show another example of a resonance hybrid of the molecule.

Use the compound shown below to answer Problems 93–96.

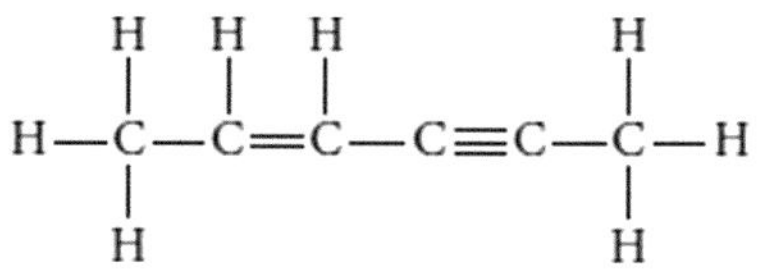

```
    H   H   H           H
    |   |   |           |
H—C—C=C—C≡C—C—H
    |                   |
    H                   H
```

93. Numbering the carbon atoms from right to left, what would be the hybridization for the second carbon?
94. Using the same numbering system, between which three atoms would you predict the bond angle to be largest?
95. How many pi bonds are in the molecule? How many sigma bonds are present?
96. Bonds between atoms can rotate if they lack a nodal plane aligned along the bond's axis. Using the same right-to-left numbering system, which carbon atom(s) would be free to rotate without affecting the movement of any other carbon atoms?

Use the compound shown below to answer Problems 97–100.

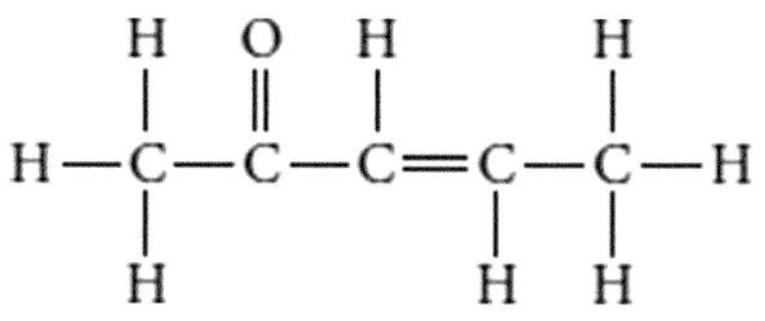

```
    H   O   H       H
    |   ‖   |       |
H—C—C—C=C—C—H
    |           |   |
    H           H   H
```

97. Numbering the carbon atoms from right to left, what would be the hybridization for the second carbon?
98. Using the same numbering system, between which three atoms would you predict the bond angle to be largest?
99. How many pi bonds are in the molecule? How many sigma bonds are present?
100. Bonds between atoms can rotate if they lack a nodal plane aligned along the bond's axis. Using the same right-to-left numbering system, which carbon atom(s) would be free to rotate without affecting the movement of any other carbon atoms?

Chemical Applications and Practices.

101. The structure for the common pain reliever acetylsalicylic acid (also known as aspirin) is shown below.

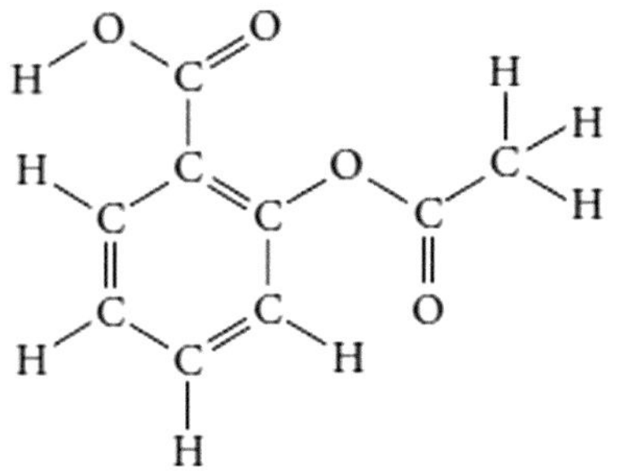

```
   O   O
H   C            H
    |            |  H
H   C   O       C
  C   C   C       H
  ‖   |   ‖
  C   C   O
H   C   H
    |
    H
```

a. Indicate which carbon atoms have associated delocalized electrons.
b. What type of hybridization do these carbon atoms have?
c. What types of orbitals are directly involved with the π bonding?

102. The versatile carbonate ion, found in seashells, in chalk, and as part of our blood buffering system, is made from three oxygen atoms bonded to a central carbon atom and two additional electrons. Diagram three resonance structures for the ion and show the resonance hybrid. What is the approximate bond order for a carbon-to-oxygen bond in the ion?
103. Naphthalene ($C_{10}H_8$) is the active ingredient in moth balls, giving them their characteristic odor. Naphthalene contains two equal-sized rings of carbon atoms that share a common side. Every carbon atom is part of a ring. Diagram the structure of this molecule and indicate the hybridization of each of the atoms.
104. Draw two resonance structures for naphthalene; see Problem 103. Then indicate the resonance hybrid for the molecule. What is the approximate bond order for a C—C bond in naphthalene?

Comprehensive Problems

105. How many unpaired electrons would be found in one atom of carbon in each of the following conditions?
 a. ground state
 b. *sp* hybridization
 c. sp^2 hybridization.
 d. sp^3 hybridization.
106. Magnesium hydride is one of the few covalent metallic hydrides. Write out the ground-state electron configuration for magnesium.
 a. What type of hybridization would be necessary to form equal bonds to the two hydrogen atoms?
 b. What orbitals would overlap to form the bonds?
 c. What shape would this molecule have?
107. Using the designated lone pairs and bonded pairs to describe electrons around a molecule's central atom, predict the shape that would be produced in each of the following:
 a. Three bonded pairs and one lone pair.
 b. Six bonded pairs and no lone pairs.
 c. Two bonded pairs and two lone pairs.
 d. Five bonded pairs and no lone pairs.
108. Nitrogen and phosphorus are both found in Group 15(VA) on the periodic table. As you would expect, they have some reactions and properties in common. For example, both nitrogen and phosphorus form trihalides such as NCl_3 and PCl_3. However, phosphorus also forms PCl_5 but nitrogen does not. Explain the reason behind this difference.
109. The nitrate ion (NO_3^-) and the nitrite ion (NO_2^-) can both be drawn showing a double bond between one oxygen atom and the central nitrogen atom. However, the bond angle is not the same in the two compounds. Which would have the smaller angle? Explain the basis for your choice.
110. The air you inhale is mostly nitrogen gas (N_2). The air you exhale is mostly the unchanged N_2 that you just inhaled. Use ground-state electron configurations and valence bond theory to show why N_2 is so stable that your biochemical processes cannot change its structure.
111. Two molecules found in petroleum oil are the straight-chain octane and the highly branched compound 2,2,3,3-tetramethylbutane. The structures of both are shown here. From what you can deduce about the three-dimensional shape of these structures, decide which one is most likely be able to associate closely with more molecules of itself and to have a higher boiling point. Explain the basis of your answer.

CH_3—CH_2—CH_2—CH_2—CH_2—CH_2—CH_2—CH_3 (octane)

$(CH_3)_3C$—$C(CH_3)_3$ (2,2,3,3- tetramethylbutane)

112. Examine the structure of oleic acid shown below (H atoms are omitted from the structure). You may have read the label on a food product that described "hydrogenated" or "partially hydrogenated" oils as an ingredient. These terms refer to the addition of hydrogen atoms to the area of the double bond(s) in an oil. How many hydrogen atoms have been left off the oleic acid structure? How many more hydrogen atoms would have to be added in order to fully hydrogenate oleic acid and remove the double bond?

C—C—C—C—C—C—C—C—C=C—C—C—C—C—C—C—C—COOH (oleic acid)

113. The compound formamide (CH_3NO) has one hydrogen atom, the nitrogen, and the oxygen attached to the carbon. In addition, the remaining two hydrogen atoms are attached to the nitrogen. Every atom satisfies the octet rule. Diagram the structure of this molecule, and explain the hybridization of both the carbon and nitrogen atoms. If you built a model of this molecule, would it lie flat on a table (planar) or have a "puckered" structure?
114. Dinitrogen monoxide, also known as nitrous oxide (N_2O), has had some use as an anesthetic and, because of some side effects, is occasionally called "laughing gas." Diagram two suitable Lewis dot structures for the compound, and state the hybridization of the nitrogen atoms in each.
115. In the introduction to this chapter, the function of 11-*cis*-retinal is described. Because of the energy required to cause the required chemical transition, it absorbs light most readily when it has a wavelength of 500 nm.
    a. What is the energy (in Joules) of one photon of 500.0 nm light?
    b. What is the energy in one mole of 500.0 nm photons?
    c. In the film, *Predator*, the alien's vision was predominantly sensitive in the infrared region of the spectrum. Assuming the alien's eyes work in a similar fashion to ours, and that the maximum sensitivity of the alien eye is at an infrared wavelength of 3000.0 nm, what is the energy (in kJ/mole) of the chemical transition in the alien's eye?
    d. Which covalent bond (see ◘ Table 8.6), if any, does this energy correspond to?
116. The ionization energies of a molecule correspond to the energy level of the orbital from which the electron is being removed. The first ionization energy is the amount needed to remove the outermost electron from a molecule. Oxygen molecules have the order of molecular orbitals shown in ◘ Fig. 9.26. Nitrogen molecules have the molecular orbital order shown in ◘ Fig. 9.25. The first ionization energy for nitrogen is 1505 kJ/mol but for the oxygen molecule it is only 1314 kJ/mol. Does this agree with what you know about the electronegativity of these elements? How can this trend be explained?
117. Compare the molecular orbital diagrams for N_2 and O_2 found in the chapter.
    a. What is the main difference between these two diagrams?
    b. What causes this difference in the energies of the molecular orbitals in the two molecules?
    c. Without regard for the bond order of the molecules, what does the order and relative energies of the molecular orbitals tell us about the inter-nuclear distances in the two molecules?
118. In metals, there are a large number of atoms packed closely together, and, correspondingly, a very large number of overlapping atomic orbitals, which form many molecular orbitals. The following trend in melting points (°C) is seen as you move from left to right across period 4 of the transition metals:

Sc	Ti	V	Cr	Mn	Fe	Co	Ni	Cu	Zn
1541	1668	1910	1907	1246	1538	1495	1455	1083	420

There is a similar trend for the other periods of the transition metals. Using what you know about Molecular Orbital theory, explain why the melting points of transition metals peak near the center of the table and then decrease.

119. Draw the Lewis structure of ethane and ethene. What does this imply about the strength of a double carbon–carbon bond vs. the strength of a carbon–carbon single bond? Look at ◘ Table 8.6 to see the enthalpies of both kinds of bonds. Does this agree with the Lewis structure prediction? Why or why not?
120. In the opening discussion of the chapter, the molecule retinal is introduced. This molecule contains a series of conjugated double bonds—double bonds that are separated by single bonds.
    a. Which is a longer bond in the molecule, the C=C bond in the ring, or the C—C bond on the opposite side of the ring?
    b. This molecule absorbs light at about 500.0 nm. What is the energy of a single photon of light with this wavelength?
    c. What is the energy of a mole of photons of this wavelength?
    d. To what color does this wavelength correspond?
121. The text of the chapter indicates that the chlorine–chlorine bond can be cleaved with 492 nm light.
    a. What is the energy of a mole of photons of this wavelength?
    b. Using Lewis dot diagrams, show how the chlorine molecule breaks apart into two chlorine atoms.
    c. What is the difference between a chlorine atom and a chloride ion?
    d. What is the hybridization of the chlorine atom in a molecule of chlorine?

9

122. Oleic acid, Example 9.7, can be treated with hydrogen gas to make stearic acid.
   a. Write the balanced molecular equation for this process.
   b. How many grams of hydrogen are needed to prepare 1.0 lb stearic acid?
   c. Stearic acid can form a film on the surface of water. In doing so, the molecule stands straight up—one end points toward the water and one end points toward the sky. Which end of stearic acid would you expect to point toward the water? Explain.

Thinking Beyond the Calculation

123. Hydrazine (N_2H_4) has many uses. For example, it is employed as a rocket fuel and as a starting material in the production of fungicides.
   a. Draw the Lewis dot structure for hydrazine.
   b. What is the VSEPR shape of the nitrogens in this molecule?
   c. What is the hybridization of each nitrogen atom?
   d. Would you expect this molecule to have a color that we can see?
   e. Hydrazine can be used as rocket fuel. The products of combustion are nitrogen gas and water vapor. Write a balanced equation showing this reaction and calculate the enthalpy of combustion for this reaction using bond energies.

The Behavior and Applications of Gases

Contents

Supplementary Information The online version contains supplementary material available at ▶ https://doi.org/10.1007/978-3-030-90267-4_10.

M. Mosher and P. Kelter, *An Introduction to Chemistry*,
https://doi.org/10.1007/978-3-030-90267-4_10

- Gases, liquids, and solids differ in the amount and strength of intermolecular forces between the individual particles (▶ Sect. 10.1).
- The pressure of a mixture of gases is the sum of the pressures of the individual gases (▶ Sect. 10.3).
- Changes in the amount, volume, temperature, and pressure of a gas are related mathematically with the combined gas laws (▶ Sect. 10.4).
- The ideal gas law can provide one of the missing parameters of a gas (amount, volume, temperature, pressure) if the other four are known (▶ Sect. 10.5).
- Kinetic-molecular theory explains why gases exert a pressure on the walls of their container (▶ Sect. 10.7).
- Effusion is the escape of gas through a pinhole in its container. The relative rate of the escape of one gas from a mixture of gases is related to the square root of the mass of the gas (▶ Sect. 10.8).

10.1 The Nature of Gases

▶ https://catalog.archives.gov/OpaAPI/media/23554288/content/stillpix/255-sts/STS128/STS128_ESC_JPG/255-STS-s128e010083.jpg

Our atmosphere is a relatively thin layer of gases surrounding our planet. It supplies all living organisms on Earth with breathable air, serves as the vehicle to deliver rain to our crops, and protects us from harmful ultraviolet rays from the Sun. Although it does contain solid particles, such as dust and smoke, and liquid droplets, such as sea spray and clouds, much of the atmosphere we live in is actually a mixture of the gases shown in ◘ Table 10.1. These gases, and the rest of the components in the atmosphere of planet Earth, are vital to our survival.

In addition to forming the atmosphere, gases are vital to our society in many other ways. Gases that are used in manufacturing, medicine, or anything else related to the economy are called **industrial gases**, shown in ◘ Table 10.2. Some, such as nitrogen and oxygen, can simply be separated from the air. Others, including sulfur dioxide, hydrogen, and chlorine, are produced by chemical reactions. No matter what uses we make of them, nearly all of these common gases behave in about the same way when present at low density. The fact that almost all gases share this characteristic enables us to work with them predictably and successfully. Why do most gases exhibit similar behavior? How can this help us understand the nature of matter? We will address these questions and others in this chapter as we learn about gases and their properties.

Industrial Gases – Gases used in industrial or commercial applications.

Table 10.1 Composition of dry air

Component	Percent by volume	Parts per million by volume
N_2	78.08	780,800
O_2	20.95	209,500
Ar	0.934	9340
CO_2	0.033	330
Ne	0.00182	18.2
He	0.000524	5.24
CH_4	0.0002	2
Kr	0.000114	1.14
H_2	0.00005	0.5
N_2O	0.00005	0.5
Xe	0.0000087	0.087
O_3	$<1 \times 10^{-5}$	<0.1
CO	$<1 \times 10^{-6}$	<0.01
NO	$<1 \times 10^{-6}$	<0.01

Table 10.2 Industrial uses of some common gases

Gas	Use	US Production (2004) (1 metric ton = 1000 kg)
Cl_2	Preparation of bleach and cleaning agents, purification of water, preparation of pesticides	12.2 million metric tons
CO_2	Refrigeration, preparation of beverages, inert atmosphere in chemical reactions and food packaging, fire extinguishers	8.1 million metric tons
H_2	Hydrogenation of oils in food and other unsaturated organic molecules, energy source in vehicles, petroleum-refining reactions, manufacturing of resins used in plastics	17.7 billion m^3
NH_3	Fertilizer and preparation of other fertilizers	10.8 million metric tons
N_2	Inert atmosphere in chemical reactions and food packaging, electronics, and metalwork; manufacture of ammonia	30.3 million metric tons
O_2	Oxidizer in chemical reactions, production of steel and acetylene, oxidizer of fuel in rockets, use in hospitals, pulp and paper industries for bleaching	25.5 million
SO_2	Production of sulfuric acid, bleaching agent in food and textile industries, control of fermentation in wines	300,000 metric tons

Many of the chemists in the eighteenth century became aware of a very interesting phenomenon as they worked with gases. They noted that dissimilar gases appeared to behave in relatively similar ways. Why is this possible? It all comes down to the common features of gases. Under ambient conditions, electrostatic interactions pull molecules together and make water a liquid and sodium chloride a solid. Gases have few such interactions, and when they occur, they are often weak and fleeting. In fact, the atoms or molecules of gases most often behave as individual units. Gases that behave as though each particle has no interactions with any other particles are called **ideal gases**; their properties are listed in Table 10.3. No gases are actually ideal, but most gases behave nearly ideally at room temperature and pressure (25 °C and 1 atmosphere), so unless otherwise noted, the quantitative relationships we will discuss assume ideal behavior. Notably, deviations from ideal behavior occur when gases are examined under conditions that encourage the interactions between molecules of a gas: high pressures and low temperatures.

Ideal Gases – Gases that behave as though each particle has no interactions with any other particle and occupies no molar volume.

Gases have an exceedingly low density compared to most liquids and solids—that is, the particles are relatively far apart, so there is very little mass in a given volume of the material. For example, at 25 °C, liquid water has a density of

Table 10.3 The ideal gas

The ideal gas is one that follows each of these rules

- The individual gas particles do not interact with each other
- The individual gas particles are assumed to have no volume
- The gas strictly obeys all of the simple gas laws
- The gas has a molar volume of 22.414 L at 1 atm of pressure and 273.15 K

1.00 g/mL, whereas dry air at sea level has a density of 0.00118 g/mL—nearly 1000 times less! Because the particles of a gas are far apart, gases can be significantly compressed in order to save space during shipping, as shown in Fig. 10.1.

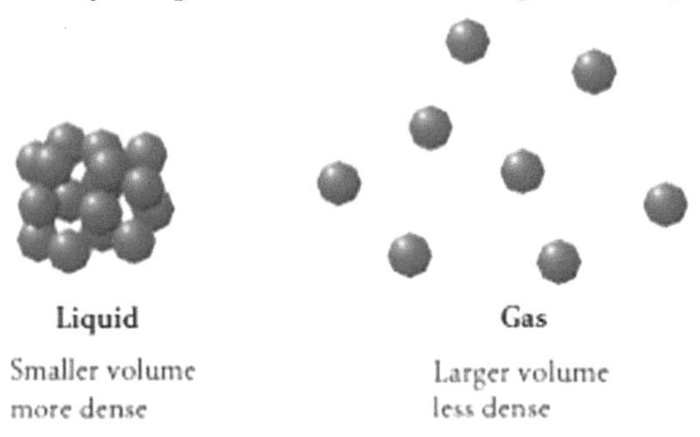

Solids and liquids are nearly incompressible and have volumes that change very little with temperature and pressure. This difference, too, is related to the nature of gases. Because gas molecules are far apart compared to molecules within a liquid or solid, the volume of a gas can be greatly affected by changes in temperature and pressure. As we will discuss later, we make use of the compressibility of gases, and the energy exchanges that accompany it, in applications from refrigeration to rocketry.

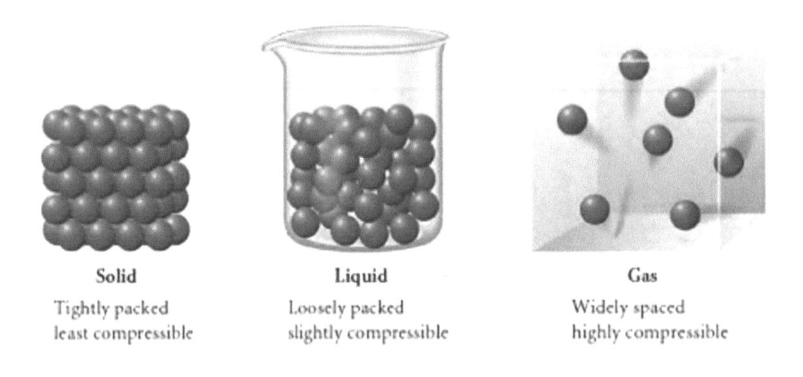

Fig. 10.1 Gases can be significantly compressed. This makes them much easier to transport and store *Source* ► https://www.publicdomainpictures.net/pictures/120000/velka/co2-cylinders-bottles.jpg

10.2 Production of Hydrogen and the Meaning of Pressure

Gas pressure is critically important when scuba diving, but what is pressure, really? Let's examine the behavior of one of the simplest and smallest gases in order to answer this question. Nearly 80 million metric tons of hydrogen gas are produced worldwide, annually. Because hydrogen gas is a very minor component of air (see ◻ Table 10.1), it must be generated via chemical processes. One method is the decomposition of water by **electrolysis**, in which an electric current passing through a solution causes a chemical reaction. The resulting hydrogen gas can be collected separately from the oxygen gas.

$$2H_2O(l) \rightarrow O_2(g) + 2H_2(g)$$

Electrolysis – A process in which an electric current passing through a solution causes a chemical reaction.

Let's assume that the process takes place at constant temperature and that the collection vessel, which has a constant volume, is empty as hydrogen begins to enter, as shown in ◻ Fig. 10.2.

What happens as the hydrogen is collected in the inverted tube? The molecules begin to collide with the walls of the container and with each other, as shown in ◻ Fig. 10.3. The **force** that each molecule applies during these collisions is equal to the mass of the particle times its acceleration:

Force – The mass of an object multiplied by its acceleration.

$$\text{Force} = \text{mass} \times \text{acceleration} = \text{kg} \times \frac{\text{m}}{\text{s}^2} = \text{kg} \cdot \text{m} \cdot \text{s}^{-2}$$

The SI unit of force is the **newton (N)**, which is equal to $1\ \text{kg} \cdot \text{m} \cdot \text{s}^{-2}$. Any moving particle will exert a force on anything it collides with. The molecules of hydrogen gas inside the collection tube are colliding with the walls of that tube (and with each other), and each collision with the wall exerts a very tiny force on it.

Newton (N) – The basic SI unit of force. $1\ \text{N} = 1\ \text{kg} \cdot \text{m} \cdot \text{s}^{-2}$.

In order to understand force (and pressure), let's take a brief detour from our discussion of gases to think about a very common force that we experience every day of our lives—gravity. Because each of us has mass and is ac-

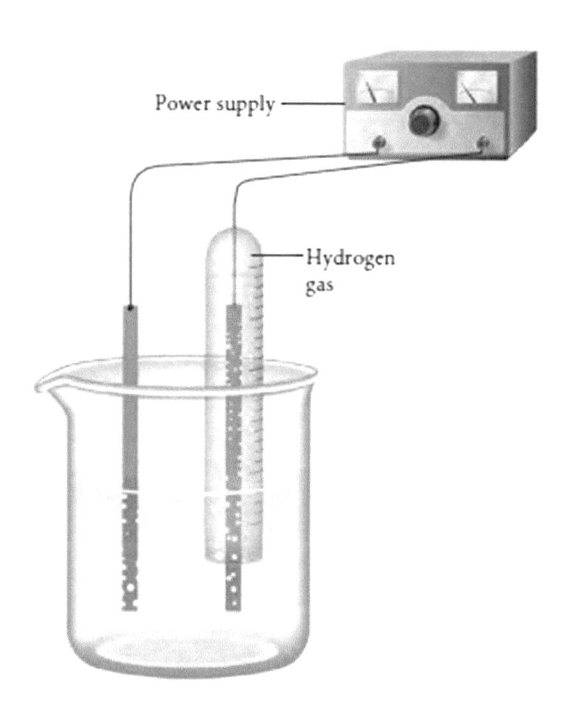

◻ **Fig. 10.2** Our hypothetical reaction set-up for the electrolysis of water. The hydrogen gas can be isolated from the reaction and sent to the collection chamber

Fig. 10.3 Molecules entering a vacant container collide with each other and with the walls of the container. The molecules have mass, acceleration, and a resulting force

celerated toward the center of the Earth by the pull of gravity $\left(\frac{9.81\ \text{m}}{\text{s}^2}\right)$, each of us exerts a force on the floor upon which we are standing. In describing our own mass × acceleration, we most often speak of force by using the familiar unit pounds, where 1 lb = 4.47 N.

Example 10.1—Calculating Force

Suppose that your mass is 55 kg and that you are on the Moon, where the acceleration due to gravity is $\frac{1.6\ \text{m}}{\text{s}^2}$, roughly one-sixth that on Earth. What force would you exert on a scale ("How much would you weigh") in newtons and in pounds? How much would you weigh on Earth, in newtons and pounds?

Asking the Right Questions

What is mass? What is force? How is the force dependent upon mass and acceleration? Once we calculate the force in newtons, how do we convert to pounds? Do we expect the force on the Moon to be the same or different than that on Earth?

Solution

Mass is a measure of the amount of substance, whereas force is the effect of acceleration, in this case as gravity, on the mass. We would therefore expect the mass to remain the same on the Moon (or on any other celestial body), whereas the force would change.

On the Moon:

$$\text{Force} = 55\ \text{kg} \times \frac{1.6\ \text{m}}{\text{s}^2} = \frac{88\ \text{kg m}}{\text{s}^2} = 88\ \text{N}$$

$$88\ \text{N} \times \frac{1\ \text{pound}}{4.47\ \text{N}} = 20\ \text{lb}$$

On Earth:

$$\text{Force} = 55\ \text{kg} \times \frac{9.8\ \text{m}}{s^2} = 539\ \text{N}$$

$$539\ N \times \frac{1\ \text{pounds}}{4.47\ N} = 121\ \text{pounds}$$

Are Our Answers Reasonable?

The mass, 55 kg, remains the same, irrespective of its location. That's reasonable. The force ("weight", in this case) is much less on the Moon because the acceleration due to gravity is much less than on Earth—this is also reasonable.

What If…?

…an astronaut who has a mass of 63.4 kg were on Mars, and weighed 52.2 lb, what would be the acceleration due to gravity?

(*Ans: 3.68 m/s²*)

Practice 10.1

A space traveler visiting the hypothetical planet X wishes to determine the gravitational pull of that planet. A 10.0 kg block of lead has been found to weigh 15 lb on planet X. What is the pull of gravity in m/s² on planet X?

The force that each molecule exerts on its container is very small, because the mass of each molecule in kilograms is very small. But when large numbers of molecules exert that force the net result is readily measurable. Scientifically, we say that the **pressure** is the amount of force applied to a given area. If we substitute the SI base units for force and area, we see that the pressure of a gas can be expressed in this way:

Pressure – The force per unit area exerted by an object on another.

$$\text{Pressure} = \frac{\text{force}}{\text{area}} = \frac{\text{N}}{\text{m}^2} = \frac{\text{kg}\cdot\text{m}\cdot\text{s}^{-2}}{\text{m}^2} = \text{kg}\cdot\text{m}^{-1}\cdot\text{s}^{-2}$$

The SI unit of pressure is the **pascal (Pa)** = 1 $\text{kg}\cdot\text{m}^{-1}\cdot\text{s}^{-2}$, named after the French mathematician Blaise Pascal (1623–1662). Measuring pressure in pascals is becoming more common, with kilopascals (kPa) often used in maritime weather reporting.

Pascal (Pa) – The basic SI unit of pressure. 1 Pa = 1 $\text{N}\cdot\text{m}^{-2}$.

What would happen to the force and pressure you exerted on the surface on which you were standing in our previous exercise if you stood on one foot instead of two? The force assuming a 150 lb person (670 N) would remain the same because the mass and the acceleration haven't changed. However, the pressure would double because the area on which you exerted your force would be cut in half (one foot rather than two). In short, you can increase pressure either by increasing force or by decreasing the area on which that force is exerted.

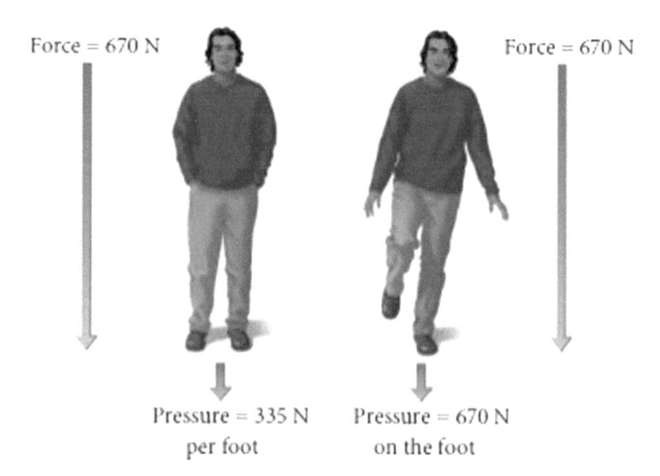

We can extend this idea to hydrogen molecules being produced by electrolysis. As we increase the number of molecules flowing into the collection vessel, more will collide with the walls in the given area. Therefore, the force per unit area—the pressure—will increase as shown in ◘ Fig. 10.4. This suggests an important relationship that we will look at a bit later: at constant volume and temperature, the pressure exerted by a gas is proportional to the number of gas molecules in the container.

We have noted that the pascal is the SI unit of pressure, but other units of pressure are often used in scientific measurements and in day-to-day conversations. For instance, we have discussed previously the **atmosphere (atm)**, which is used extensively in scientific work. One atmosphere is roughly equal to the pressure exerted by the earth's atmosphere at sea level on a typical day. Because scientific work requires a more precise standard, we can define a **standard atmosphere**, exactly 1 atm, as the pressure that supports a 760-mm column of mercury **(mm Hg)** in a barometer. Measurements have indicated that the standard atmosphere is equal to 1.01325×10^5 Pa.

Atmosphere (atm) – A unit used to measure pressure. 1 atm = 760 mm Hg.

Standard atmosphere – Exactly 1 atm, the pressure that supports 760 mm of mercury.

mm Hg – A unit used to measure pressure. 760 mm Hg represents the height of a column of mercury that can be supported by 1 atm pressure.

The barometer, a device like that shown in ◘ Fig. 10.5, can be used to measure the pressure of the atmosphere. The first barometer was made by Evangelista Torricelli in 1643, who used it to determine that the atmospheric pressure changes with changes in the weather. In recognition of his work, we say that 1 atm is equal to **760 torr**. Therefore, one standard atmosphere is equal to 760 mm Hg, to 760 torr, and to 1.01325×10^5 Pa.

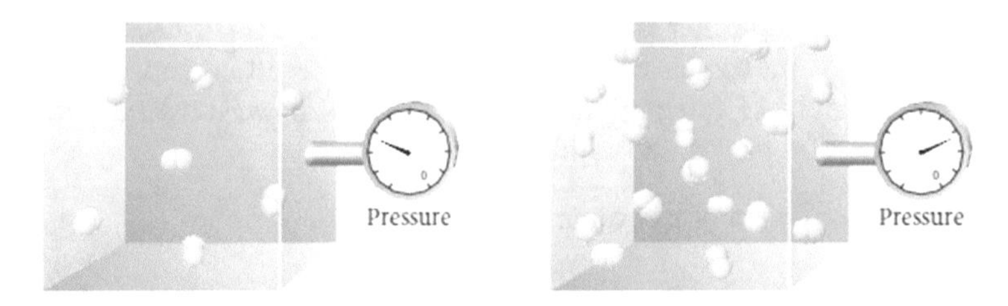

Fig. 10.4 The pressure of the hydrogen container increases as more molecules enter the vessel. The pressure is related to the force each molecule exerts on the walls of the container

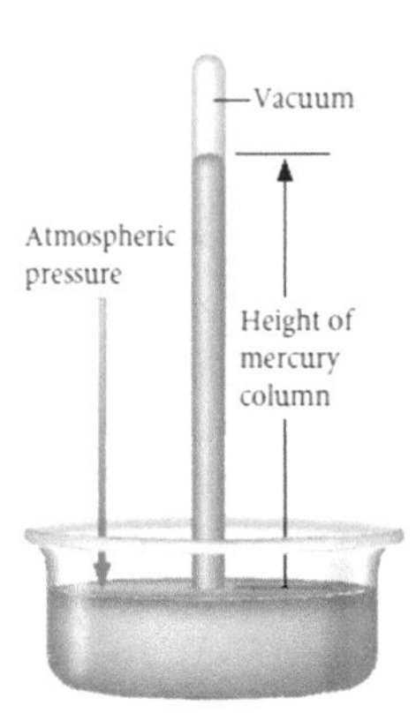

Fig. 10.5 A simple barometer. The pressure exerted by air at 1 atm supports a column of mercury 760 mm (29.9 inches) high. As the atmospheric pressure changes, so does the height of the mercury in the column

Torr – A unit used to measure the pressure of gases. 1 torr = 1 mm Hg.

Still other units are also used to describe pressure. For example, weather forecasters in some countries speak of atmospheric pressure in **bar**. A bar is equal to 1×10^5 Pa, a little less than 1 atm. In the United States, however, weather forecasters report atmospheric pressure in units of inches of mercury (which is abbreviated **in Hg**). Another unit with which we may be familiar comes into play in the United States when we inflate car or bicycle tires to their recommended pressure. That unit is known as **pounds per square inch gauge (psig)**, or, commonly, gauge pressure. This is the pressure of the gas in the tire in *excess* of the surrounding atmospheric pressure (where 1 atm = 14.7 **pounds per square inch (psi)**). This means that a tire inflated to 32.0 psig really contains air exerting a pressure of 46.7 psi (32.0 + 14.7) on the inside walls of the tire. Table 10.4 gives the relationships among different units of pressure. What they all have in common is that they are a measurement of a force exerted over an area.

Bar – A unit used to measure pressure. 1.01325 bar = 1 atmosphere (atm).

in Hg – A unit used to measure pressure. 29.921 in Hg = 1 atm.

Pounds per square inch (psi) – A unit of pressure. 14.7 psi = 1 atm.

Pounds per square inch gauge (psig) – A unit of pressure expressing the pressure of a gas in excess of the standard atmospheric pressure. Also known as gauge pressure. 14.7 psig = 2 atm.

For many years, scientists defined a **standard pressure** as 1 atm so that a standard reference was available at a common pressure. In 1982, IUPAC recommended that standard pressure be defined as equal to exactly 1 bar. However, most chemists still commonly use 1 atm as the standard pressure, and we will do so in this textbook.

Standard Pressure – Defined as 1 bar, although chemists typically use 1 atm.

Scientists also define a **standard temperature** for gases as 0 °C, or 273 K. Gases at these conditions are said to be at **standard temperature and pressure (STP).** Increasingly, scientists are beginning to use **standard ambient temperature and pressure (SATP)**, or 1 bar and 298 K, to discuss gases, although in this book we will use STP as the standard. Why use either of these standards? Each is a convenient reference point at which to compare the properties of gases. Notice that the standard temperature in STP conditions, 0 °C, is different from the 25 °C often

Table 10.4 Pressure unit conversion factors

1 standard atmosphere is equal to…

- 760 mm Hg (millimeters mercury)
- 760 torr
- 14.7 psi
- 101,325 Pa (pascals)
- 1.01325 bar
- 0.0 psig (pounds per square inch gauge)
- 29.921 in Hg (inches mercury)

used as a standard for energy measurements, as we discussed in an earlier chapter. It's always important to make sure we know what conditions are being used in a given context.

Standard Temperature – For gases, defined to be 0 °C, or 273 K.

Standard temperature and pressure (STP) – 1 atm and 273 K.

Standard ambient temperature and pressure (SATP) – 1 bar and 298 K.

Example 10.2—Conversion Among Measures of Pressure

The gas in a volleyball is measured to have a gauge pressure of 8.0 psig. What is the total pressure exerted by the gas in units of atmospheres and torr?

Asking the Right Questions

How is the gauge pressure different from the atmospheric pressure? What is the relationship among psi, atmospheres and torr?

Solution

The gauge pressure is the pressure *in excess* of the atmospheric pressure, 14.7 pounds per square inch (psi). To determine the total pressure of the gas in the volleyball, we need to add 14.7 psi to the gauge pressure.

$$P_{gas} = P_{gauge\ pressure} + P_{volleyball} = (8.0 + 14.7)\ \text{psi} = 22.7\ \text{psi}$$

We can now do our unit conversions.

$$22.7\ \text{psi} \times \frac{1\ \text{atm}}{14.7\ \text{psi}} = 1.54\ \text{atm}$$

(We retain an extra figure because we will use this value in the next calculation.)

$$1.54\ \text{atm} \times \frac{760\ \text{torr}}{1\ \text{atm}} = 1.2 \times 10^3\ \text{torr}$$

Are Our Answers Reasonable?

We expect the pressure to be greater than 1 atmosphere because of the definition of the gauge pressure. The related conversions should reflect this as well, and they do, so the answers are reasonable.

What If…?

…we were in a place where the atmospheric pressure were 725 torr and our car tire had a measured pressure of 35.0 psig? What would be the air pressure in the tire in psi, atm and torr?

(Ans: 21 psi, 1.43 atm, 1080 torr)

Practice 10.2

A weather report provides the current barometric pressure as 29.30 in Hg. What is this pressure in atm, torr, bar, and Pa?

10.3 Mixtures of Gases—Dalton's Law and Food Packaging

If gas molecules behave ideally, and don't have much interaction with each other, then what happens if we mix different kinds of gas molecules together? Using mixtures of industrial gases to prevent spoilage in food products, especially poultry, is a rapidly growing segment of the food-processing industry. Microbes, especially *Pseudomonas* bacteria, spoil the flavor and color of food and make it unhealthful to eat. Carbon dioxide gas, which in small quantities is not particularly harmful to the food or to humans, has been found to inhibit the growth of many types of microbes. Food scientists discovered this fact and have used it to develop **modified atmosphere packaging**

(MAP)—packaging the food item in an atmosphere that is different from air (recall ◘ Table 10.1). The results are impressive. For instance, meat and poultry stored in air last only a few days, but food stored under MAP can have a shelf life as long as a month, with proper refrigeration.

Modified Atmosphere Packaging (MAP) – The replacement of the air within food packaging with gases in order to prolong the shelf life of the food.

Nitrogen and argon gases are also used in meat-based MAP, and sulfur dioxide gas inhibits the growth of microbes in some beverages. Typical MAP includes mixtures containing a mole ratio of 30–35% CO_2 and 65–70% N_2 for meats, though this can vary widely. This mixture brings up an interesting and useful point. If the total pressure of the gas mixture is equal to 1.05 atm, how much does each gas contribute to the pressure?

▶ https://upload.wikimedia.org/wikipedia/commons/d/de/SelectionOfPackageMeats.jpg

John Dalton (remember his work from ▶ Chap. 2) conducted experiments that led to what we now know as **Dalton's law of partial pressures**. This law states that for a mixture of gases in a container, the total pressure is equal to the sum of the pressures that each gas would exert if it were alone. Each individual gas within a mixture contributes only a part of the total pressure, so we say that each gas exerts a **partial pressure**. Dalton's law can be summarized by the following equation:

$$P_{\text{total}} = P_1 + P_2 + \cdots + P_n$$

where

P_{total} the total pressure exerted by a mixture of gases
P_1 the partial pressure exerted by gas 1
P_2 the partial pressure exerted by gas 2
P_n the partial pressure exerted by gas n

Dalton's Law of Partial Pressures – The total pressure of a mixture of gases is the simple sum of the individual pressures of all the gaseous components.

Partial pressure – The pressure exerted by a single component of a gaseous mixture.

Even though he didn't realize it at the time, Dalton's law is a direct result of gases behaving ideally. It is precisely because the gases don't interact with each other that we can add up their individual pressures to get the total pressure inside the container.

For example, let's assume that the total pressure of the MAP gas that surrounds ground turkey in a package is 1.05 atm and that the gas is composed of 35% CO_2 and 65% N_2 by volume. Using Dalton's law of partial pressures, we see that 35% of the pressure, or 0.37 atm (0.35 × 1.05 atm) is exerted by CO_2 and 0.68 atm (0.65 × 1.05 atm) is exerted by the N_2 gas. The calculation of the partial pressures is straightforward, but Dalton's law is truly correct only for ideal gases. The actual pressures of the gases are a bit different from what we calculate because N_2 and CO_2 are not actually ideal gases; we will discuss corrections for nonideal behavior in ▶ Sect. 10.5. A key point for our current discussion is that the closer gases come to ideal behavior, the more closely they follow Dalton's law of partial pressures, and deviations occur when gases are at high pressure, are at low temperature, or are otherwise concentrated enough to exhibit intermolecular interactions with each other.

We can apply our understanding of partial pressures to the production of hydrogen gas. In addition to the electrolysis of water, another convenient way to produce small amounts of hydrogen gas is by the reaction of zinc with hydrochloric acid. The hydrogen gas produced by this reaction can be collected over water, as illustrated in ◘ Fig. 10.6.

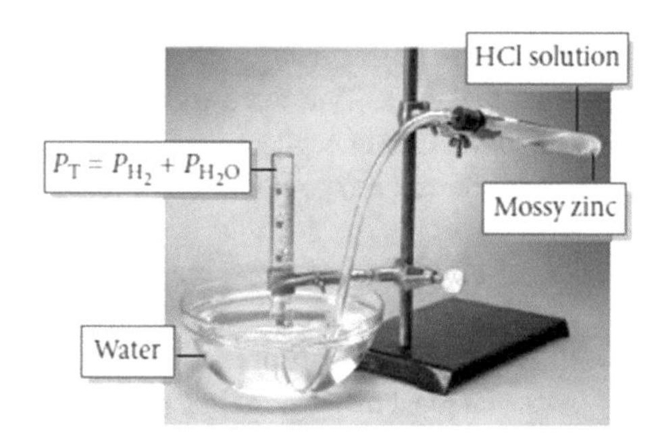

Fig. 10.6 Hydrochloric acid solution can be mixed with pieces of zinc metal to produce hydrogen gas. The gas is collected over water, so the total pressure, P_T, of the collected gas is equal to the sum of the partial pressures of the H_2 and the water vapor

$$Zn(s) + 2HCl(aq) \rightarrow ZnCl_2(aq) + H_2(g)$$

As we'll discover later, a sample of water is accompanied by a certain amount of gaseous water—water vapor—above it. When we collect a gas by bubbling it through water or by leaving it in contact with moisture, the total pressure above the liquid is the sum of the partial pressure of the water vapor and the partial pressure of the gas. At 20 °C, the vapor pressure of water is 17.5 torr. Therefore, for example, if the total pressure of the gases is 750 torr, then the partial pressure of the H_2 can be calculated as follows:

$$\begin{aligned} P_{\text{total}} &= P_{H_2O} + P_{H_2} \\ 750 \text{ torr} &= 17.5 \text{ torr} + P_{H_2} \\ 732 \text{ torr} &= P_{H_2} \end{aligned}$$

Example 10.3—Partial Pressure of Gases in the Atmosphere

A jet is cruising at 11,500 ft (3500 m) above sea level, where the atmospheric pressure outside the plane is 493 torr (0.649 atm). The plane, normally pressurized to about 650 torr (0.85 atm), suddenly has a loss of pressure until the cabin pressure equals the pressure outside. What is the partial pressure of oxygen gas (see Table 10.1) when the pressure of the gas in the plane is lowered? Most people unaccustomed to low-oxygen environments will lapse into unconsciousness, and eventually die, if the partial pressure of oxygen falls below 30 torr. With that figure in mind, if the pressure in the plane isn't quickly restored, can the passengers survive this accident?

Asking the Right Questions

What is the relationship between the partial pressure of oxygen and the total pressure? How do we compute the partial pressure of oxygen? Will the partial pressure of oxygen be high enough to support life?

Solution

The law of partial pressures relates the partial pressure of oxygen gas in air to the total pressure. According to Table 10.1, Oxygen gas makes up about 21% of air. Therefore, we multiply the air pressure at 3500 m above sea level by 0.21 to obtain the partial pressure due to O_2.

$$\begin{aligned} P_{O_2} &= P_{\text{air}} \times 0.21 \\ &= 493 \text{ torr} \times 0.21 = 1.0 \times 10^2 \text{torr} \end{aligned}$$

The partial pressure of O_2 at this altitude, about 100 torr, is less than the sea-level O_2 pressure of about 160 torr and the plane's normal O_2 pressure of about 140 torr, but it is still easily sufficient for survival.

Are Our Answers Reasonable?

Via the calculations, the partial pressure of oxygen is about 20% of the total pressure, so that is reasonable. The partial pressure of oxygen also decreased along with the total cabin pressure, so that is also reasonable. Our assertion that the oxygen pressure of 100 torr will allow the passengers to survive the accident because this is above the minimum survival level of 30 torr is also reasonable.

What If…?

…the partial pressure of nitrogen in an air sample on the plane were measured to be 384 torr? Could the passengers survive?

(*Ans: Yes, because the partial pressure of oxygen would be 103 torr*, well in excess of the *30 torr needed for survival.*)

Practice 10.3

What is the partial pressure of nitrogen gas under the same outside atmospheric pressure of 493 torr?

▶ **Here's what we know so far**

- Gases exert a force on the system that contains them. That force, exerted over a given area, is known as pressure.
- A barometer is used to measure the pressure of the atmosphere.
- STP is the abbreviation for standard pressure and temperature. For chemists, this means exactly 1 atm pressure and 0 °C
- The total pressure of a system is the sum of all of the individual pressures of the component gases. This is Dalton's law of partial pressures.
- An ideal gas is a hypothetical gas with no intermolecular forces of attraction, where the particles of the gas have no volume. No gases are ideal, but they approach ideal behavior at low pressure and high temperature. ◀

10.4 The Gas Laws—Relating the Behavior of Gases to Key Properties

Atmospheric scientists at the South Pole routinely measure the ozone (O_3) content in our atmosphere as a function of altitude. To complete this task, they employ a helium-filled balloon, such as that shown in ◘ Fig. 10.7, to carry sophisticated instruments 35 km up into the atmosphere. The balloon rises because the density of the helium gas it contains is much less than that of the air outside. As the balloon rises, the pressure and temperature of the surrounding atmosphere change, and the balloon expands. Eventually, the balloon expands so much that it bursts and falls back to Earth, allowing the scientists to recover the scientific instruments. However, in order to have a successful flight, the balloon must be filled with just the right amount of helium. Moreover, the scientists need to know how the helium gas will be affected by changes in temperature and pressure as the balloon rises through the atmosphere, so that the balloon can carry the instruments up to where they are needed. Too much gas and the balloon will burst before it reaches 35 km; too little gas and the balloon may never burst. To make their decisions, atmospheric scientists rely on some of the most fundamental properties of gases, which investigators have been aware of for hundreds of years. These long-established laws describing the behavior of ideal gases—the "gas laws"—help us to study modern concerns about gases, such as the level and fate of ozone in our atmosphere.

10.4.1 Avogadro's Law

What is the relationship between the amount of a gas and its volume? Let's explore this potential connection by taking an empty Mylar balloon and attaching it to a cylinder containing helium gas like that shown in ◘ Fig. 10.8. The gas, which is stored inside the cylinder at a pressure of 14 MPa, or about 140 atm, can be dispensed safely into the balloon with a pressure regulator. The temperature of the room is 25 °C (298 K), and the pressure is 1.0 atm.

As we begin to fill the balloon, it expands because of the gas molecules exerting force on the inside of the balloon. The Mylar balloon has relatively little tension until it is nearly full, so the balloon skin doesn't significantly

◘ **Fig. 10.7** A helium-filled balloon is used to carry sophisticated instruments into the atmosphere to monitor the levels of atmospheric ozone (O_3). The balloon rises because the helium gas it contains is much less dense than the air outside. *Source* ▶ https://upload.wikimedia.org/wikipedia/commons/7/7a/Radiosonde-wx-balloon.jpg

Fig. 10.8 A Mylar balloon can be filled with gas from a cylinder

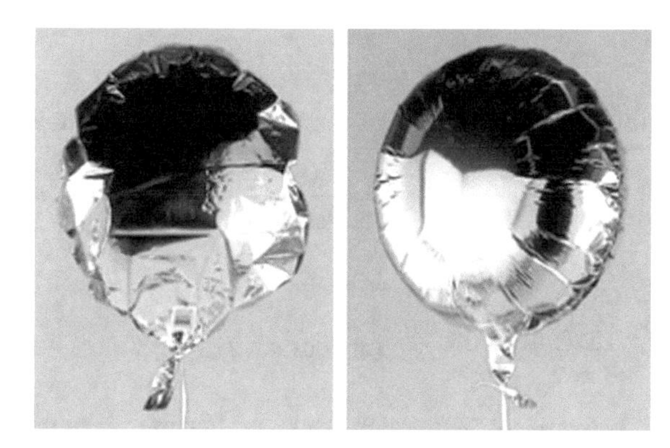

Fig. 10.9 At a volume of 1.2 L, the balloon on the left holds 0.050 mol of He. If we double the number of moles of the gas to 0.10, the volume of the balloon (shown on the right) also doubles, to 2.4 L

affect the pressure of the gas within. We find that at a volume of 1.2 L, the balloon holds 0.050 mol of He. If we double the number of moles of the gas to 0.10, the volume of the balloon also doubles, to 2.4 L, as shown in Fig. 10.9.

Now let's take another balloon and hook it up to a cylinder containing oxygen gas, also at a pressure of 140 atm. If we add 0.050 mol of O_2 we find, as with He, that the volume of the balloon is 1.2 L, also shown in Fig. 10.10. Double the number of moles, and the volume doubles to 2.4 L, just as in the hydrogen-filled balloon. This linear relationship holds irrespective of the nature of the gas, assuming ideal behavior. **Avogadro's law**, named after its discoverer Amadeo Avogadro (1776–1856), states that equal numbers of molecules are contained in equal volumes of all dilute gases under the same conditions. An implication of this law, confirmed by Fig. 10.10, is that the volume of a gas is directly proportional to the amount of gas expressed in moles, at constant temperature and pressure.

Avogadro's Law – Equal amounts of gases occupy the same volume at constant temperature and pressure.

We represent Avogadro's law as follows:

$$V = kn$$

which can be rewritten as

$$\frac{V}{n} = k \quad \text{(at constant } T, P\text{)}$$

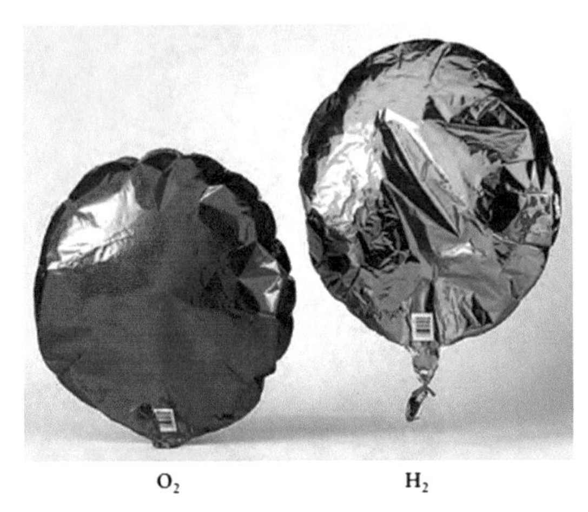

Fig. 10.10 Mylar balloons of O_2 and a balloon of He, containing 0.050 mol of gas, each has the same volume and the same pressure

where
V volume of a gas
n amount of the gas expressed in moles
k a constant

Because the constant *k* is the same value for a given temperature and pressure, we can write

$$\frac{V_{\text{initial}}}{n_{\text{initial}}} = k = \frac{V_{\text{final}}}{n_{\text{final}}}$$

or

$$\frac{V_{\text{initial}}}{n_{\text{initial}}} = \frac{V_{\text{final}}}{n_{\text{final}}} \quad \text{(at constant } T, P)$$

The conclusion seems to make sense, because our experience tells us that the more gas we add to a balloon, the larger it gets, until its skin resists.

Example 10.4—Avogadro's Law

Much of the chlorine gas produced industrially is manufactured from the electrolysis of aqueous Group 1 (IA) chlorides, such as sodium chloride. The reaction is

$$\text{NaCl}(aq) + \text{H}_2\text{O}(l) \rightarrow \text{NaOH}(aq) + \frac{1}{2}\text{H}_2(g) + \frac{1}{2}\text{Cl}_2(g)$$

Sodium hydroxide is formed along with hydrogen gas at one electrode, and chlorine gas is formed at the other electrode. If after such a reaction, 22.4 L each of hydrogen and chlorine gases, equaling 1.0 mol of each gas, is collected at STP, and then reacted to form hydrogen chloride gas, what would be the final volume of HCl gas at STP?

$$\text{H}_2(g) + \text{Cl}_2(g) \rightarrow 2\text{HCl}(g)$$

Asking the Right Questions
What is the relationship between the change in the quantity of a gas and the change in volume the gas occupies? What is the specific change in this example?

Solution
According to Avogadro's law, if 1.0 mol of each gas occupies 22.4 L at STP, then the 2.0 mol of reactants have a total volume of 44.8 L. Because 2.0 mol of gas product is produced from 2.0 mol of reactants, the product would occupy 44.8 L at STP. The numerical value would change only if the number of moles of gaseous products differed from the number of moles of reactants.

Is Our Answer Reasonable?
The number of moles of the HCl product is double the number of moles of each reactant, so we would correctly expect the volume of gas to double. The answer of 44.8 L of HCl therefore is reasonable.

What If…?
…we had the same reaction, resulting in 376 L of HCl? What is the minimum volume of hydrogen gas to produce that quantity of HCl?

(Ans: 188 L of H_2 gas)

Practice 10.4
If a 12.8-L sample of He gas contains 6.4 mol of He, how many moles would there be in a 1.5-L sample of He at the same temperature and pressure?

10.4.2 Boyle's Law

Two long-time friends, one of whom teaches at a U.S. college located at sea level (air pressure = 1.00 atm) and the other at a university in Mexico City (7340 ft above sea level, air pressure = 0.764 atm), prepare to show their students the link between moles of hydrogen and oxygen gas and the volume they occupy. Each of them adds 0.050 mol of hydrogen at 25 °C to a balloon. The U.S. professor notes that his balloon has a volume of 1.2 L. His colleague in Mexico City measures the volume of his balloon as 1.6 L. Except for the pressure of the gas in each balloon, which equals the atmospheric pressure in each city, the conditions are identical. The Mexico City balloon contains gas at lower pressure and, as a result, larger volume than the balloon at sea level in the United States. Take this to extremes and the increase in volume can cause a balloon to burst, as we described at the beginning of this section.

This inverse relationship between pressure and volume of a given amount of gas at constant temperature has been understood since 1662, when Robert Boyle (1627–1691; Fig. 10.11) demonstrated what is now known as **Boyle's law**.

Boyle's Law – The volume of a fixed amount of gas is inversely proportional to its pressure at constant temperature.

This law can be summarized as follows:

$$PV = k' \quad \text{(at constant } n, T)$$

where

P = pressure of the gas
V = the volume occupied by the gas
k = a constant (different from the constant in Avogadro's law)

Using the same logic as we did for Avogadro's law, we can rewrite Boyle's law to relate the pressure and volume of the same gas under two different conditions. For a change in either volume or pressure, we can say,

$$P_{\text{initial}}V_{\text{initial}} = P_{\text{final}}V_{\text{final}} \quad \text{(at constant } n, T)$$

Boyle investigated the relationship between pressure and volume using the apparatus shown in Fig. 10.12. Mercury was added to the open end of a U-shaped tube so that air was trapped between the mercury and the closed end. The height of the trapped air was indicative of the volume, V, of the gas. During his studies, Boyle noted that the difference in the heights of the mercury on both sides of the U-tube, plus the height of mercury at

Fig. 10.11 Robert Boyle (1627–1691) was the youngest of fourteen children born to a wealthy Irish family. He was interested in advancing his understanding of the world at a very early age. He kept very accurate observations about his experiments and was one of the first to build and use a vacuum pump. *Source* ▶ https://upload.wikimedia.org/wikipedia/commons/d/d8/PSM_V42_D450_Robert_Boyle.jpg

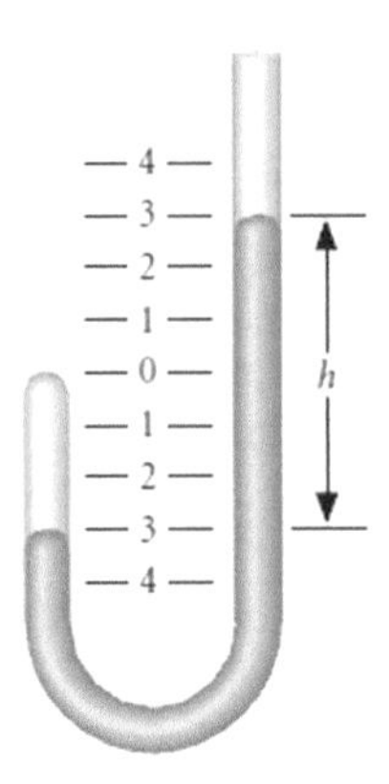

Fig. 10.12 Robert Boyle investigated the relationship between pressure and volume using the apparatus shown. Mercury was added to the open end of a U-shaped tube so that air was trapped between the mercury and the closed end. The height (h) of the trapped air was indicative of the volume, V, of the gas. The difference between the heights of the mercury on the two sides of the U-tube, plus the height of mercury at atmospheric pressure, 29.18 in. (741 mm), was indicative of the pressure, P

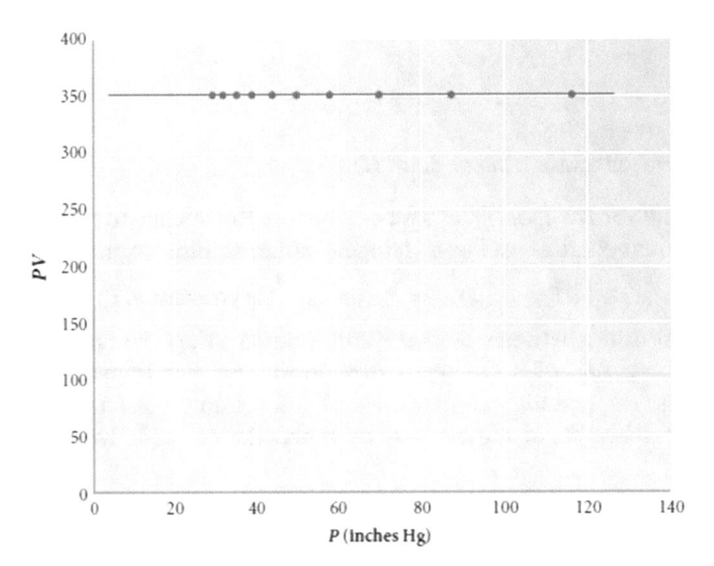

Fig. 10.13 A plot of the pressure–volume relationship using Boyle's actual data for air. The fact that the product $P \times V$ is fairly constant at these pressures testifies to the care with which Boyle did these experiments. The pressure was measured as the height of a mercury column in inches, which is proportional to the pressure of the system. The volume was measured as the height of a column of air in inches, which is directly proportional to the volume of the system

atmospheric pressure, 29 1/8 inches (740 mm) was indicative of the pressure, P, of the gas trapped in the tube. A plot of the pressure–volume relationship using Boyle's actual data for air is shown in Fig. 10.13. The plot shows that the product $P \times V$ is fairly constant at these pressures, which are not too different from normal.

We began ▶ Sect. 10.3 by using food packaging to illustrate Dalton's law of partial pressures. Food packagers also acknowledge Boyle's law when they allow for the effect of higher altitudes (lower atmospheric pressures) on food packages. In fact, the food packagers adjust for the pressure in their packages for the specific destination of the food items. For example, Fig. 10.14 shows a bag of potato chips in Estes Park Colorado, at an altitude of 2440 m (8000 ft). The bag, originally manufactured for sale near sea level, is nearly ready to burst! This is one reason why food manufacturers express the amount of product that their packages contain in terms of weight, rather than volume.

Fig. 10.14 This bag of potato chips is a lot bigger in Estes Park, Colorado, at 2440 m, than at sea level

Example 10.5—Boyle's Law

A teacher wants to know what volume her balloon will need to have if it is to hold 0.050 mol of hydrogen at 25 °C (as with the previous balloons) at a pressure of 1.3 atm. Using data for either sea level or Mexico City (provided in the main text), what is the minimum volume her balloon must hold?

Asking the Right Questions

What do we expect to happen to the volume of the balloon as a result of increasing the pressure? Is it likely to increase or decrease? How do we describe that relationship via an equation? Framing good answers to these questions before we proceed will give us an indication of whether our answer makes sense.

Solution

The volume of the balloon decreases with increasing pressure. The Boyle's Law equation describing the relationship is:

$$P_{\text{initial}} V_{\text{initial}} = P_{\text{final}} V_{\text{final}}$$

Using the Mexico City data,

$$P_{\text{initial}} = 0.764 \text{ atm } V_{\text{initial}} = 1.6 \text{ L}$$
$$P_{\text{final}} = 1.3 \text{ atm } V_{\text{final}} = ??? \text{ L}$$
$$P_{\text{initial}} V_{\text{initial}} = P_{\text{final}} V_{\text{final}}$$
$$0.764 \text{ atm}(1.6 \text{ L}) = 1.3 \text{ atm}(V_{\text{final}})$$
$$V_{\text{final}} = 0.94 \text{ L}$$

Is Our Answer Reasonable?

The pressure of the gas in this balloon is greater than in either of the other balloons. At constant temperature and with the same amount of gas in each balloon, the only way to increase the pressure of the gas is to decrease the volume of the balloon, so that the force per unit area remains the same. The answer is reasonable.

What If…?

…the pressure were constant at 1.3 atm, but the amount of the gas were doubled? What would be the resulting volume?

(*Ans: the volume would double.*)

Practice 10.5

How would the volume of the balloon in the previous exercise change if the pressure were decreased to 0.975 atm?

10.4.3 Charles's Law

As we continue to look at the gas laws, we fill a balloon at STP with equal moles each of hydrogen and oxygen gases, so the total volume of the large Mylar balloon is, according to Avogadro's law, 2.0 L. That means we've added 1.0 L of each gas. We then take the balloon outside into a frosty winter's morning where the temperature is − 10.0 °F (−23.3 °C). We notice that the balloon seems a bit smaller, as shown in Fig. 10.15. Why is the volume of the balloon less in colder temperatures?

There are several factors to consider. One of these factors involves the temperature of the gases in the balloon. Temperature is a measure of the kinetic energy of a system—the gas particles in our discussion. When the temperature

■ **Fig. 10.15** The relationship between temperature and volume of a gas (constant quantity and pressure) can be illustrated by comparing the volume of a balloon at 0 °C, on the left, with that of the balloon on the right, at − 23.3 °C

■ **Fig. 10.16** Jacques Alexander César Charles (1746–1823), a French inventor and scientist, was the first to take a voyage in a hydrogen balloon (to a height of 550 m). He invented many scientific instruments and used them in his studies, including his confirmation of Benjamin Franklin's experiments with electricity. *Source* ▶ https://upload.wikimedia.org/wikipedia/commons/2/21/Jacques_Charles.jpg

drops, the gas particles travel at lower velocities. This means that there are fewer particle collisions with the walls of the balloon in a given time (also, each collision will occur with a lower kinetic energy and velocity, and so each collision has less impact). Because each gas particle collision with the balloon's inner wall adds to the force exerted by the gas on the balloon, a smaller number of collisions per unit time indicates that the pressure of the gas is lowered. As a result, the balloon shrinks until the pressure of the gas inside the balloon equals the pressure outside the balloon. In this process, the pressure was the same at the beginning as at the end. When the temperature is lowered, the volume of the balloon gets smaller (so that the pressure remains constant). The general result that the volume of a gas is directly proportional to the temperature (at constant pressure and number of moles) was first reported by Jacques Alexander Charles (■ Fig. 10.16) in 1787 and is known as **Charles's law**.

Charles's Law – The volume of a fixed amount of gas is directly proportional to its temperature in kelvins at a constant pressure.

In the early 1800s, Joseph Louis Gay-Lussac (■ Fig. 10.17) further studied the effect of temperature on the volume of a gas and reported that gases expand by 1/273 of their volume at 0 °C for each increase in temperature of 1 °C. This relationship is shown graphically at standard pressure assuming a constant amount of gas in ■ Fig. 10.18. If we follow the line down to where the volume would become zero, we note that the temperature would be − 273.15 °C. In fact, such an extension would not be reasonable, because any gas would deviate substantially from ideal behavior and liquefy well before that point.

In any case, does − 273.15 °C seem familiar? It might. It is the temperature we know as **absolute zero**, the lowest possible temperature. This value serves as the basis for the Kelvin temperature scale, named after William Thomson (Lord Kelvin), who created this scale in 1848. On the Kelvin scale, a reading of zero kelvin (0 K) is equal

Fig. 10.17 Joseph Louis Gay-Lussac (1778–1850), a French chemist and physicist, is known for his work explaining the behavior of gases. He and Jean-Baptiste Biot were the first to ride in a hot-air balloon in 1804 to a height of 5 km. He was one of the discoverers of boron in 1808, which was later shown to be a new element. *Source* ▶ https://upload.wikimedia.org/wikipedia/commons/2/2f/Gaylussac.jpg

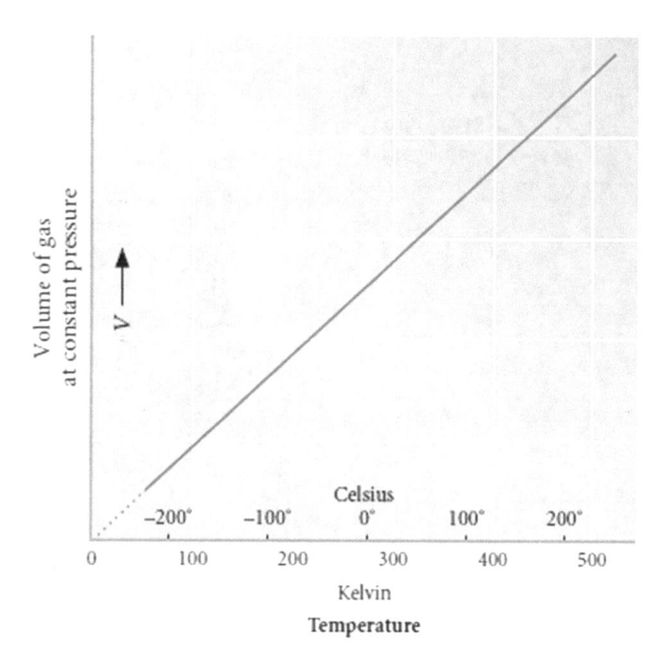

Fig. 10.18 In the early 1800s, Gay-Lussac found that ideal gases expand by 1/273 of their volume for each increase of 1 °C in temperature. If we follow the line down, we intersect the "zero volume" point at – 273 °C. In reality, such an extension would not be reasonable, because any gas would deviate substantially from ideal behavior and liquefy well before that point. This does, however, suggest that there is an absolute zero temperature.

to – 273.15 °C. At this temperature, there is no kinetic energy available, and therefore all atomic translations (movement in 3-dimensional space) stop. Is such a temperature actually achievable? Even in the far reaches of deep space, residual heat from the Big Bang keeps even the coldest objects at about 3 K. In the laboratory, however, scientists have been able to cool atoms down to only a few billionths of a degree above absolute zero.

Absolute Zero – The temperature obtained by extrapolation of a plot of gas volume versus temperature to the "zero volume" point. The lowest temperature possible, which is 0 K, or –273.15 °C.

Charles's law illustrates the relationship between volume and temperature. These two properties of a gas are directly proportional, so we can write

$$V = k''T \qquad \text{(at constant } n, P)$$

which can be rearranged to

$$\frac{V}{T} = k'' \quad \text{(at constant } n, P)$$

where
V = volume occupied by an ideal gas
T = temperature of the gas in kelvins
k'' = a constant relating the two quantities (different from the prior constants)

Just as we did with the other gas laws, we can set the two ratios of volume-to-temperature equal to each other if only the conditions of the gas have changed:

$$\frac{V_{\text{initial}}}{T_{\text{initial}}} = \frac{V_{\text{final}}}{T_{\text{final}}} \quad \text{(at constant } n, P)$$

We need to keep in mind that the relationship between volume and temperature is an approximate one (because no gas is truly "ideal") and varies from substance to substance.

Example 10.6—Charles' Law

As a demonstration of Charles's law, a Mylar balloon is filled with 1.00 L of helium gas at – 12 °C and held over a flame until the temperature of the gas within the balloon reaches 150.0 °C. What is the new volume of the balloon, assuming constant pressure?

Asking the Right Questions
What do we expect to happen to the volume of the balloon as a result of warming the gas inside? Is it likely to increase or decrease? As in the previous example, how do we describe that relationship via an equation?

Solution

The volume of the balloon will increase as the temperature increases. Mathematically, the volume is directly proportional to the kelvin temperature.

$$\frac{V_{\text{initial}}}{T_{\text{initial}}} = \frac{V_{\text{final}}}{T_{\text{final}}}$$

$$\frac{1.00\ \text{L}}{261\ \text{K}} = \frac{V_{\text{final}}}{423\ \text{K}}$$

$$\frac{1.00\ \text{L} \times 423\ \text{K}}{261\ \text{K}} = V_{\text{final}}$$

$$V_{\text{final}} = 1.62\ \text{L}$$

Is Our Answer Reasonable?
We anticipated a higher volume with an increase in temperature, and that occurred. We used the kelvin temperature to solve for the final volume. What would our solution have been if we had used the Celsius temperature? Let's consider that in the "What if…" section.

What If…?
…we had kept our temperatures in Celsius degrees when we calculated the final volume? What would be the final volume, and what would it mean?

(*Ans:* V_{final}= – 13 L; *This outcome of a "negative volume" is not reasonable, and sharply illustrates why the Celsius scale cannot be used in gas law calculations.*)

Practice 10.6
Assume the volume of the Mylar balloon described in the previous exercise is heated until its volume reaches 2.00 L. What is the temperature of the gas within the balloon?

Example 10.7—Mendeleev's Demonstration of Nonideal Behavior

Based on his own experiments done around 1870, Dmitri Mendeleev noted that for real gases, Charles's law (originally known as Gay-Lussac's Law) was not entirely correct. Mendeleev's data showed that when several gases were heated (constant n, P) from 0 to 100 °C, they occupied the following multiples of their original volume:

Air	= 1.368
"Carbonic anhydride" (CO_2)	= 1.373
Hydrogen	= 1.367
Hydrogen bromide (HBr)	= 1.386

If the gases were truly ideal, what fraction of their original volume should they occupy when heated from 0 to 100 °C? Which of these gases most nearly exhibits ideal behavior?

Asking the Right Questions

By what multiple would the volume of an *ideal* gas increase when heated from 0 to 100 °C? Which of these gases is likely to be behave closest to ideally? Why?

Solution

We can calculate the increase in volume of an ideal gas when heated from 0 to 100 °C. To keep our calculations straightforward, we assume an initial volume at 0 °C (273 K) of 1.000 L. Mendeleev used one extra figure in his work, so we can do the same.

$$V_{initial} = 1.000\ \text{L} \qquad T_{initial} = 273\ \text{K}$$
$$V_{final} = ?\ \text{L} \qquad T_{final} = 373\ \text{K}$$

$$\frac{V_{initial}}{T_{initial}} = \frac{V_{final}}{T_{final}} \qquad \text{(at constant } n, P\text{)}$$

For an ideal gas,

$$\frac{1.000\ \text{L}}{273\ \text{K}} = \frac{V_{final}}{373\ \text{K}}$$

$$V_{final} = \frac{1.000\ \text{L} \times 373\ \text{K}}{273\ \text{K}} = 1.366\ \text{L}$$

Hydrogen, at 1.367 L, is the closest to ideal behavior.

Are Our Answers Reasonable?

Should hydrogen behave closest to ideally among the four gases? This is reasonable because it is the smallest molecule and takes up the least space. Hydrogen bromide is a much larger molecule, requiring space for itself in order to exert the same pressure as hydrogen in, for example, a balloon at the higher temperature. Carbon dioxide has a lower mass than hydrogen bromide, but a higher mass than hydrogen, so its deviation from ideal behavior is less than HBr, but more than H_2. All four gases deviate from ideality, the more so the greater their molar mass.

What If…?

…we continued to heat the gases from 100 to 200 °C? How would that change the volume of an ideal gas? What about the nonideal behavior of these gases?

(*Ans: The volume of an ideal gas would increase to 1.73 L. The gases in our example (and all gases, to an extent) will continue to deviate from ideal behavior, occupying a slightly higher volume than the ideal gas.*)

Practice 10.7

In addition to its having volume, give another reason why HBr gas deviates from ideal gas behavior.

10.4.4 Combined Gas Equation

We can combine Avogadro's law,

$$\frac{V_{initial}}{n_{initial}} = \frac{V_{final}}{n_{final}}$$

and Boyle's law,

$$P_{initial}V_{initial} = P_{final}V_{final}$$

and Charles's law,

$$\frac{V_{initial}}{T_{initial}} = \frac{V_{final}}{T_{final}}$$

to come up with an equation called the **combined gas equation**. Why can we do this? The fact that all of these relationships use a proportionality constant makes this process easy. Our final equation takes volume, pressure, amount, and temperature into account:

Combined Gas Equation – The combination of Avogadro's law, Boyle's law, and Charles's law.

$$\frac{P_{initial}V_{initial}}{n_{initial}T_{initial}} = \frac{P_{final}V_{final}}{n_{final}T_{final}}$$

This equation can be used to solve for the pressure, volume, amount, or temperature of a gas, as the gas changes conditions from one state to another. If one of the conditions remains constant, that condition cancels out of the equation and simplifies our calculations. Typically, we use units of atmospheres for the pressure, liters for the volume, moles for the amount, and kelvin for the temperature.

Example 10.8—Combined Gas Equation

Recall our discussion of modified atmosphere packaging (MAP) in ▶ Sect. 10.3. If a chicken is packaged within a nitrogen–carbon dioxide gas atmosphere at STP and a volume of 300.0 mL of gas, what will be the volume of the gas (assuming no leakage through the packaging) when it is stored in a grocery freezer at a temperature of – 8.0 °C and a pressure of 1.04 atm?

Asking the Right Questions

What is changing and what is remaining the same? What is the impact of each change? Would it make sense for the volume to increase or decrease as a result of the combination of these changes? Because we are combining changes, can we use the combined gas equation?

Solution

The temperature is decreasing from 273 to 265 K. Charles's law suggests that the volume would decrease as a result. The pressure is increasing from 1.00 atm to 1.04 atm, which, via Boyle's law, would also lead to a slight reduction in volume. On balance, then, the volume should decrease. We can use the combined gas equation for our calculations.

$$P_{\text{initial}} = 1.00 \text{ atm} \quad P_{\text{final}} = 1.04 \text{ atm}$$
$$T_{\text{initial}} = 273 \text{ K} \quad T_{\text{final}} = 265 \text{ K}$$
$$V_{\text{initial}} = 0.3000 \text{ L} \quad V_{\text{final}} = ?$$
$$n_{\text{initial}} = n_{\text{final}}$$

$$\frac{P_{\text{initial}} V_{\text{initial}}}{n_{\text{initial}} T_{\text{initial}}} = \frac{P_{\text{final}} V_{\text{final}}}{n_{\text{final}} T_{\text{final}}}$$

We can rearrange variables to solve for V_{final}:

$$\frac{P_{\text{initial}} V_{\text{initial}} n_{\text{final}} T_{\text{final}}}{n_{\text{initial}} T_{\text{initial}} P_{\text{final}}} = V_{\text{final}}$$

Because $n_{\text{initial}} = n_{\text{final}}$, they cancel, and we have

$$\frac{P_{initial} V_{initial} T_{final}}{T_{initial} P_{final}} = V_{\text{final}}$$

$$\frac{1.00 \text{ atm} \times 0.3000 \text{ L} \times 265 \text{ K}}{273 \text{ K} \times 1.04 \text{ atm}} = 0.280 \text{ L}$$
$$= 280 \text{ mL} = V_{\text{final}}$$

Is Our Answer Reasonable?

We expected the volume of gas to decrease, and it did. When the physical properties change in ways that yield opposite effects—as, for example, when both temperature and pressure increase—we can judge what the outcome might be before calculating by seeing which change is greater, and that change will dominate.

What If…?

…the freezer in the grocery store broke and the temperature of the gas in the chicken package increased to 27 °C? If the maximum volume of the chicken package were 310 mL, what would be the pressure?

(*Ans: 1.06 atm*)

Practice 10.8

Assuming the same initial conditions as in Example 10.8, solve for the volume when the final pressure is 0.85 atm and the final temperature is – 11.0 °C

To this point, we have discussed relationships among the pressure, volume, temperature, and amount of a gas. We have seen that we can take into account changes in any one of these and find its effect on another variable. These relationships are summarized in ◘ Table 10.5.

◘ Table 10.5 The Gas Laws

Common name	Equation	Summary
Avogadro's law	$\frac{V}{T} = k$ (constant P, T)	Equal numbers of molecules are contained in equal volumes of all dilute gases under the same conditions
Boyle's law	$PV = k'$ (constant n, T)	There is an inverse relationship between pressure and volume of a given amount of gas at constant temperature
Charles' law	$\frac{V}{T} = k''$ (constant n, P)	The volume of a gas is directly proportional to its temperature when the pressure and amount of gas are constant
Combined gas equation	$\frac{PV}{nT} = constant$	The combination of Avogadro's, Boyle's, and Charles' laws relates initial and final pressure, volume, amount, and temperature

10.5 The Ideal Gas Equation

So far we have looked at three laws describing the behavior of gases, the laws named in honor of Avogadro, Boyle, and Charles. The balloon that carries aloft the ozone-detecting equipment (as was shown in Fig. 10.7) bursts as a consequence of the combined effects of temperature and pressure on the volume of the gas within the balloon. In order to understand why these effects occur, we need to examine how these three variables are related.

The three gas laws were summarized in the previous section as follows:

$$\frac{V}{n} = k \qquad \text{(constant } P, T\text{)}$$

$$PV = k' \qquad \text{(constant } n, T\text{)}$$

$$\frac{V}{T} = k'' \qquad \text{(constant } n, P\text{)}$$

Setting each equation equal to V yields

$$V = kn \qquad \text{(constant } P, T\text{)}$$

$$V = \frac{k'}{P} \qquad \text{(constant } n, T\text{)}$$

$$V = k''T \qquad \text{(constant } n, P\text{)}$$

We can combine all the variables on the right-hand side, because they are all proportional to V via the constants k, k' and k''. To clean it up nicely, we then combine all three of these constants into a single constant called R. This gives us this equation:

$$V = \frac{RnT}{P}$$

We clear the denominator by multiplying through by P, to give the common form of the **ideal gas equation**:

$$PV = nRT$$

Ideal Gas Equation – The equation relating the pressure, volume, moles, and temperature of an ideal gas.

The name of this equation includes the term *ideal* because the equation assumes ideal behavior of the gas. In fact, we know that gases will deviate from the ideal, yet the ideal gas equation will give us meaningful approximate results unless the gases are at very high pressures and/or low temperatures.

What is the value of R? It can be determined by measuring each of the properties (P, V, T, and n) of a gas and solving for the combined gas law shown in Table 10.5. For example, at STP (T=273.15 K, P=1.000 atm), 1 mol of an ideal gas occupies 22.414 L. The value of R, then, is 0.08206 L atm · mol^{-1} K^{-1}.

$$\frac{PV}{nT} = R$$

$$= \frac{(1.000\ \text{atm})(22.414\ \text{L})}{(1.000\ \text{mol})(273.15\ \text{K})}$$

$$= 0.08206 \frac{\text{L} \cdot \text{atm}}{\text{mol} \cdot \text{K}}$$

This is a very important number. It is known as the **ideal gas constant**, and it is used quite often in chemistry. It is helpful to commit not only the number, but also the units, to memory. Why? The units of the constant help us remember which units to use for the other variables in the equation, and they also help us confirm the units of our answers.

Ideal Gas Constant – R, the constant used in the ideal gas equation. $R = 0.08206$ L · atm/mol·K.

Example 10.9—Using the Ideal Gas Equation

A sample containing 0.631 mol of a gas at 14.0 °C exerts a pressure of 1.10 atm. What volume does the gas occupy?

Asking the Right Questions

What are we given and what are we asked to find? What conditions are changing? What equation describes the relationship among the variables?

Solution

We are given the pressure, temperature, and number of moles. The ideal gas constant, R, is always at our disposal. We are asked to find the volume at constant conditions. None of the variables are changing. So, we can use the ideal gas equation to solve for volume. By placing what we know into the equation, we can solve for the quantity that we don't know.

$$n = 0.631 \text{ mol} \quad T = 282.7 \text{ K} \quad P = 1.10 \text{ atm}$$

$$PV = nRT$$

$$1.10 \text{ atm} \times V = 0.631 \text{ mol} \times \left(\frac{0.08206 \text{ L} \cdot \text{atm}}{\text{mol} \cdot \text{K}}\right) \times 287.2 \text{ K}$$

$$V = 13.5 \text{ L}$$

Is Our Answer Reasonable?

One benchmark we may use is that at STP, 1 mol of an ideal gas occupies 22.4 L. In this example, we have 0.631 mol of gas at a temperature and pressure not too different from standard conditions, so as a first estimate, we could expect our volume to be around (0.631 mol × 22.4 L/mol) = 14 L. The answer we calculated, 13.5 L, is reasonable.

What If…?

…the temperature of the system were changed from 14.0 to – 14.0 °C? What would be the new volume?

(Ans: 12.7 L)

Practice 10.9

A 2.50-L sample of an ideal gas at 25 °C exerts a pressure of 0.35 atm. How many moles of this gas are there?

Example 10.10—The Ideal Gas Equation and Compressed Gas

How much pressure must a 48.0-L steel oxygen gas cylinder tolerate if it contains 9.22 kg of O_2 at a temperature of 25.0 °C? Given the conditions, does the gas show approximately ideal behavior?

Asking the Right Questions

What are we given? Are we asked to evaluate the impact of any changes? What are we asked to find, and how do we find it? How is this example different from Example 10.9?

Solution

There are two differences between this exercise and Example 10.9. Here, we are given mass, in kilograms, instead of the number of moles. Second, the sample is much larger than laboratory-sized samples. The question asks about ideal gas behavior. The unknown here is the pressure. Gases approach ideality if the pressure is low. High pressure creates the conditions for a great deal of interaction among gas molecules and with the container walls.

We find it convenient first to convert mass into moles and then to find pressure using the ideal gas equation. We will follow this strategy.

$$n_{O_2} = 9.22 \text{ kg} \cdot O_2 \times \frac{1000 \text{ g} \cdot O_2}{1 \text{ kg} \cdot O_2} \times \frac{1 \text{ mol} \cdot O_2}{32.0 \text{ g} \cdot O_2} = 288.1 \text{ mol } O_2$$

$$PV = nRT$$

$$P \times 48.0 \text{ L} = 288.1 \text{ mol} \times \left(\frac{0.08206 \text{ L} \cdot \text{atm}}{\text{mol} \cdot \text{K}}\right) \times 298 \text{ K}$$

$$P = 146.8 \text{ atm} = 147 \text{ atm}$$

The pressure in the steel cylinder is quite high. We would therefore expect the oxygen gas within the cylinder to deviate significantly from ideal behavior.

Are Our Answers Reasonable?

This is a relatively small cylinder for so much gas. It is, therefore, reasonable to expect a very high pressure, as we calculated here. Being a steel cylinder is one indication that the pressure of the gas inside is expected to be high.

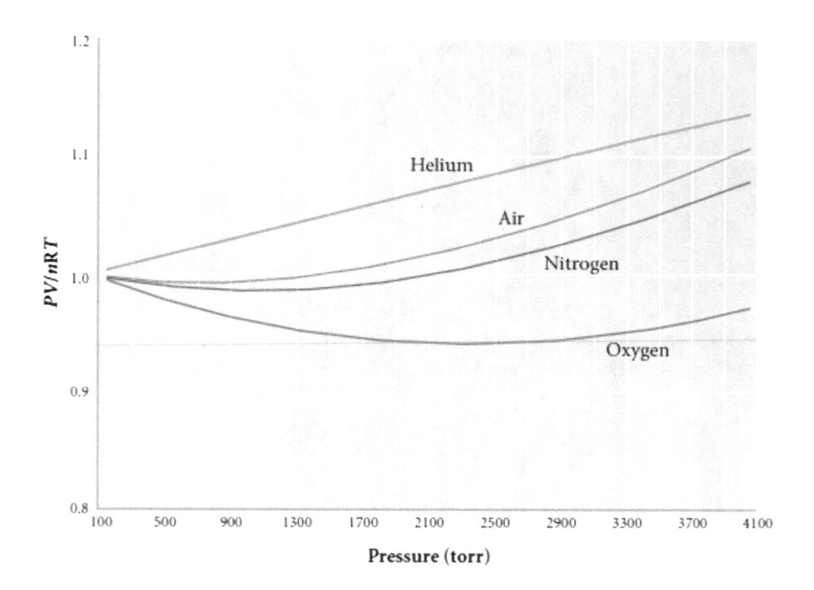

Fig. 10.19 "Ideal" is just that—a standard of perfection that no gas can ever meet. If $PV = nRT$, then $PV/nRT = 1.00$. Plotting actual values shows that all gases deviate from this equation, especially as the pressure increases. Moreover, the deviation is different for each gas

What If…?

…we had a cylinder containing the same mass of nitrogen gas instead of oxygen gas, in the same cylinder at the same temperature? What would be the pressure assuming ideal behavior?

(*Ans: The pressure would be the same, because the pressure is based on the amount of the ideal gas, rather than the nature of that gas.*)

Practice 10.10

A 0.250-mol sample of gas occupies 8.44 L at 28.7 °C. What is the pressure of the system?

We assumed ideal behavior in the previous exercise, yet we suspect that this is not likely for oxygen gas at a pressure of 147 atm. One of the themes in our discussion of gases has been the recognition that deviations, even small ones, from ideality are expected. We have chosen to neglect them in our calculations, assuming that the approximate answers we get are meaningful. However, as shown in Fig. 10.19, the ratio PV/nRT, is not constant, and the deviation becomes more severe at high pressures. Is there a way to correct for relatively high pressure or low temperatures, such as we had in Example 10.10 (compressed O_2)?

10.5.1 Volume Corrections for Real Gases

Real gas molecules occupy discrete volumes, and larger atoms and molecules take up more space than smaller ones. Therefore, the available volume of the container is not as large as if it were empty or if we assumed that the gas particles had no volume. The more moles of gas in a given space, the less free volume compared to ideal conditions, as shown in Fig. 10.20. To take this into account, we can express the available volume as

$$V_{\text{available}} = V_{\text{container}} - nb$$

in which b is a constant roughly related to the size of the molecules.

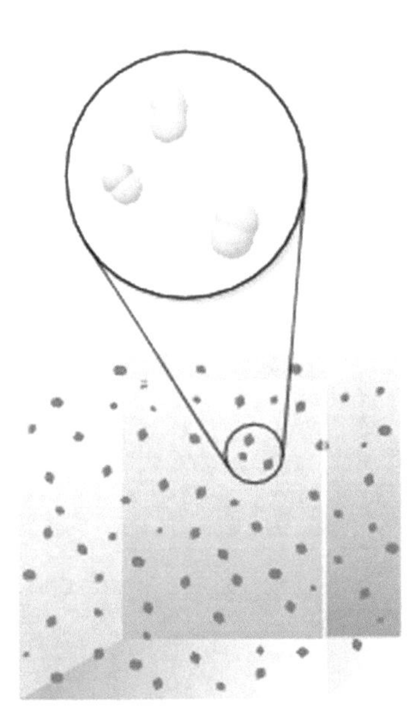

Fig. 10.20 As more and more molecules of a gas are added to a given volume, the amount of free space within the volume decreases and the gas deviates greatly from "ideal." The van der Waals equation takes this into account, as we saw in the text

10.5.2 Pressure Corrections for Real Gases

The molecules in a real gas do interact with each other. Larger atoms and molecules, and those with dipole moments, exhibit stronger interactions than smaller ones. When particles interact with each other, they do not collide as often with the walls of the container, and their force per unit area—their pressure—is lower (see Fig. 10.20. The reduction in observed pressure can be given by

$$P_{\text{actual}} = P_{\text{ideal}} - a\left[\frac{n}{V}\right]^2$$

in which a is a constant with a magnitude that is loosely indicative of the intermolecular forces that are possible in molecules. Note that the values of a in Table 10.6 are large in the relatively large molecules that have more polarizable electrons than the smaller molecules and atoms. The constant, a, is multiplied by $\left[\frac{n}{V}\right]^2$. As the volume, V, gets larger, the correction gets smaller, because the molecules are farther apart. Similarly, when the number of moles, n, is low, the correction term is reduced.

The ideal gas equation assumes the use of P_{ideal}. Therefore, in our modified model, we can substitute for P_{ideal} as follows:

$$P_{\text{ideal}} = P_{\text{actual}} + a\frac{n^2}{V^2}$$

This model of the pressure and volume deviations was put into equation form in 1873 by J. D. van der Waals. The equation he produced is

$$\left(P + a\frac{n^2}{V^2}\right)(V - nb) = nRT$$

As we noted previously, the constant b is a correction indicating that real molecules do have a finite volume and are not as far apart as we claim in ideal behavior. The more moles of gas, n, the greater will be the correction to volume, $V - nb$. The values for b are only roughly proportional to molecular size.

We can rearrange the **van der Waals equation** to solve for pressure:

$$P = \frac{nRT}{V - nb} - a\frac{n^2}{V^2}$$

Table 10.6 van der Waals constants for some gases

Gas	Formula	*a* (atm $L^2 \cdot mol^{-2}$)	*b* ($L \cdot mol^{-1}$)
Acetylene	C_2H_2	4.39	0.0514
Ammonia	NH_3	4.17	0.0371
Argon	Ar	1.35	0.0322
Carbon dioxide	CO_2	3.59	0.0427
Chlorine	Cl_2	6.49	0.0562
Ethane	C_2H_6	5.49	0.0638
Helium	He	0.034	0.0237
Hydrogen	H_2	0.244	0.0266
Methane	CH_4	2.25	0.0428
Nitrogen	N_2	1.39	0.0391
Oxygen	O_2	1.36	0.0318
Sulfur dioxide	SO_2	6.71	0.0564

van der Waals Equation – The equation that corrects the gas laws for gases that deviate from ideal behavior.

The van der Waals equation is the most often cited mathematical model for use with real gases because it works fairly well at normal pressures and temperatures. However, it doesn't fit the experimental data well at very high pressures and low temperatures. Therefore, as pressure and temperature change, different values of a and b must be chosen. Other mathematical models that fit the data even better (and are correspondingly more complex) have been derived.

Example 10.11—The van der Waals Equation

We now have the tools to better answer Example 10.10. Assuming the same volume, temperature, and number of moles of O_2 gas in the cylinder, what is a more accurate estimate of the pressure of the gas?

Asking the Right Questions

What factors contribute to the change in pressure as a result of having a real gas? What is the effect of each of the factors? How do we use the van der Waals equation to determine the corrected pressure?

Solution

We expect the actual pressure exerted by the oxygen gas to be less than the pressure exerted by an ideal gas because the gas molecules are interacting more with each other and colliding (and, therefore, interacting—the a term) less often with the walls of the container. On the other hand, the nb correction term would make the pressure greater by making the actual open volume less. The a-term correction is much more significant, so we expect the pressure to be lower than the ideal value.

$$P = \frac{nRT}{V - nb} - a\frac{n^2}{V^2}$$

$$P = \frac{288.1 \text{ mol} \cdot O_2 \times 0.08206 \text{ L} \cdot \text{atm} \cdot \text{mol}^{-1} \cdot \text{K}^{-1} \times 298 \text{ K}}{48.0 \text{ L} - \left(288.1 \text{ mol} \cdot O_2 \times 0.0318 \text{ L} \cdot \text{mol}^{-1}\right)}$$

$$- 1.36 \text{ L}^2 \cdot \text{atm} \cdot \text{mol}^{-2} \frac{(288.1 \text{ mol} \cdot O_2)^2}{(48.0 \text{ L})^2}$$

$$= 181.4 \text{ atm} - 49.0 \text{ atm} = 132 \text{ atm}$$

Is Our Answer Reasonable?

We expected a lower pressure, and the van der Waals equation confirmed this. The answer of 132 atm is therefore reasonable.

What If…?

…we had a steel tank of acetylene, C_2H_2, gas under the same conditions? What would be the pressure of the gas inside the tank? Explain why the pressure is greater or smaller than that of oxygen gas using the van der Waals terms.

(*Ans: 54.1 atm; Acetylene has a much larger a-term correction in the van der Waals model, suggesting a larger interaction with each other than oxygen molecules.*)

Practice 10.11

What is the pressure of 1.00 mol of O_2 gas in a 1.00-L balloon at 25 °C? Calculate the pressure using the ideal gas law and the van der Waals equation. Is there a difference? Then repeat the same calculations for ammonia.

10.6 Applications of the Ideal Gas Equation

Atmospheric chemists not only know the relationships that dictate how gases behave but also understand the applications of these relationships as ways to solve real-world problems. Many of their calculations—such as those that determine how high a balloon will travel—can be done using the mathematical adjustments of the van der Waals equation, but close approximations can be obtained using the ideal gas equation. Other applications of this equation are equally useful. They can be broken down into two main types: those that address physical properties and those that address chemical reactions. Examples of these applications follow.

10.6.1 Physical Change Applications

The key questions to ask when applying the ideal gas equation are the same as in any problem solving. "What do I need to find? How do I get there? What answer do I expect? Does my answer make sense?" The following examples illustrate how the ideal gas equation, density, and the molar mass of gases are related.

Example 10.12—An Application of Gas Density

The density of liquid nitrogen at – 196 °C is 0.808 g/mL. What volume of nitrogen gas at STP must be liquefied to make 20.0 L of liquid nitrogen?

Asking the Right Questions
Our target is 20.0 L of liquid nitrogen. How many moles is that? How does the density of liquid N_2 help us to calculate the number of moles? Once we have that quantity, we can recast the question to ask, "What volume of gas at STP will make that many moles of liquid nitrogen?" That's an ideal gas equation-based question.

Solution
Working backward, via the density, we can find how many moles of liquid nitrogen there are in 20.0 L of liquid nitrogen. This is the amount of gas that must be liquefied.

$$\text{g liquid N}_2 = 20.0\text{ L liquid N}_2 \times \frac{1000\text{ mL liquid N}_2}{1\text{ L liquid N}_2} \times \frac{0.808\text{ g liquid N}_2}{1\text{ mL liquid N}_2} = 16,160\text{ g N}_2$$

$$n_{\text{N}_2} = 16,160\text{ gN}_2 \times \frac{1\text{ mol N}_2}{28.0\text{ g N}_2} = 557\text{ mol N}_2$$

We use the ideal gas equation to convert from moles of N_2 to liters of the gas.

$$PV = nRT$$

$$1.00\text{ atm} \times V = 577\text{ mol} \times \left(\frac{0.08206\text{ L}\cdot\text{atm}}{\text{mol}\cdot\text{K}}\right) \times 273\text{ K}$$

$$V = 1.29 \times 10^4\text{ L N}_2$$

Is Our Answer Reasonable?
The preparation of 20 L of liquid nitrogen requires a lot of gas, because the density of liquids is often 500 times that of the corresponding gases. It is therefore reasonable that we should require over 10,000 L of N_2 gas to prepare 20 L of liquid N_2.

What If…?
…we wanted to know how much denser the liquid N_2 is than the gas? Determine the ratio of densities.

(Ans: 1.25×10^{-3} g/mL)

Practice 10.12
A gas at 16 °C and 1.75 atm has a density of 3.40 g/L. Determine the molar mass of the gas. (*Hint:* Begin by assuming 1.00 L of the gas, and then determine the number of moles that would occupy this volume.)

Dimensional analysis can lead us to a useful formula that we can use to determine the molar mass of a gas, as was requested in Practice 10.12. The symbol M is the variable used to represent molar mass, and it is also used to represent molarity. We will underline the variable for molar mass ($\underline{M}$) to help distinguish between the two. Translating the quantities back into the variables they represent, we find that the relationship between molar mass and density is

$$\underline{M} = \frac{dRT}{P}$$

$$\underline{M} = \frac{(3.40\ \text{g}\cdot\text{L}^{-1})(0.08206\ \text{L}\cdot\text{atm}\cdot\text{mol}^{-1}\cdot\text{K}^{-1})(298\ \text{K})}{1.75\ \text{atm}}$$

We see from this equation that the molar mass of an ideal gas is directly proportional to its density. This relationship approximately holds for real gases that have close to ideal behavior, but it does not hold for liquids or solids. As this equation indicates, we can determine the molar mass of a gas if we can measure its density.

10.6.2 Chemical Reactions—Automobile Air Bags

When a car crash occurs, the car stops suddenly. The occupant, however, continues to move with great velocity toward the steering wheel or dashboard. The crash often causes the release of a series of air bags within the vehicle, each designed to cushion the passengers of the car. How does the air bag reduce injuries? The deployed air bag distributes over a large area (that of the air bag) the force that the occupant would otherwise exert on the dashboard, minimizing the pressure on any particular part of the occupant's upper body. The result is an accident that causes less bodily injury than an accident in which no air bag is deployed.

▶ https://upload.wikimedia.org/wikipedia/commons/e/e7/Peugeot_306_airbags_deployed.jpg

Inside the most common style of air bag is a capsule of sodium azide (NaN_3), iron(III) oxide (Fe_2O_3), and a small detonator cap that is rigged to start the decomposition of NaN_3:

$$2NaN_3(s) \rightarrow 2Na(s) + 3N_2(g)$$

The nitrogen gas that is formed inflates the bag to its full volume (around 70 L for front air bags) within 50 ms after a crash. The sodium metal that is also generated in the reaction reacts quickly with the iron(III) oxide to form sodium oxide. This second reaction is necessary to remove dangerous sodium metal.

$$6Na(s) + Fe_2O_3(s) \rightarrow 3Na_2O(s) + 2Fe(s)$$

Example 10.13—Filling the Airbag

A technician is designing an airbag for a new model car. She wants to use 72.0 g of sodium azide (NaN_3) to inflate the bag at 22.0 °C and 1.00 atm. What is the maximum volume of the airbag under these conditions?

Asking the Right Questions

What gas fills the airbag when the sodium azide decomposes? How many moles of the gas are produced from the sodium azide? How can we convert the moles of the gas to the volume of the gas (and, therefore, the volume of the airbag)?

Solution

The airbag reaction is notable not only because the bag is filled with nitrogen gas, not air, but also because the stoichiometry of the reaction indicates that 3 mol of N_2 are formed from the decomposition of 2 mol of sodium azide, NaN_3. This must be taken into account in solving the problem.

$$\text{mol NaN}_3 = 72.0 \text{ g NaN}_3 \times \frac{1 \text{ mol NaN}_3}{65.0 \text{ g NaN}_3}$$
$$= 1.08 \text{ mol NaN}_3$$

$$\text{mol N}_2 = 1.108 \text{ mol NaN}_3 \times \frac{3 \text{ mol N}_2}{2 \text{ mol NaN}_3}$$
$$= 1.662 \text{ mol N}_2$$

$$PV = nRT$$
$$1.00 \text{ atm} \times V = 1.662 \text{ mol} \times \left(\frac{0.08206 \text{ L} \cdot \text{atm}}{\text{mol} \cdot \text{K}}\right) \times 295 \text{ K}$$
$$V = 40.2 \text{ L}$$

Is Our Answer Reasonable?

At STP conditions, 1 mol of nitrogen gas occupies 22.4 L. Here, we have a little less than 2 mol of the gas, and, at standard pressure and about room temperature, it occupies 40.2 L, a little less than twice 22.4 L. The answer is therefore reasonable.

What If…?

…we were driving a Peugeot 207 sedan that has a driver's side airbag volume of 60.0 L and a passenger's side airbag volume of 90.0 L? How many grams of NaN_3 are needed to fully inflate each airbag at 1.07 atm and 31.0 °C?

(Ans: 103 g and 155 g of NaN_3)

Practice 10.13

How many grams of sodium azide would be required if the volume of the airbag were to be 30.9 L at 31 °C and 1.00 atm?

10.6.3 Chemical Reactions—Acetylene

Acetylene (C_2H_2) is an industrial gas used in welding and in the manufacture of vinyl chloride (CH_2CHCl), acetonitrile (CH_3CN), and other industrial chemicals, including polymers such as "neoprene." The gas was first prepared inadvertently in the early 1890s by Thomas Willson, who owned a small aluminum-making company. In his quest to more effectively convert bauxite, composed mostly of aluminum oxide (Al_2O_3) to aluminum metal, he combined the ore with lime (CaO), and coal tar (mostly carbon compounds) in a furnace. A gray solid was formed that, when reacted with water, gave off a flammable gas. After receiving a sample of the solid from Willson, Francis Venable of the University of North Carolina-Chapel Hill determined that Willson had made calcium carbide (CaC_2):

$$\text{CaO}(s) + 3\text{C}(s) \rightarrow \text{CaC}_2(s) + \text{CO}(g)$$

The calcium carbide reacted with water to form acetylene (C_2H_2), which had been previously discovered by Edmund Davy in 1836.

$$\text{CaC}_2(s) + 2\text{H}_2\text{O}(l) \rightarrow \text{Ca(OH)}_2(s) + \text{C}_2\text{H}_2(g)$$

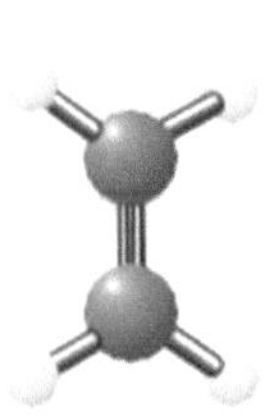

Ethylene
C_2H_4

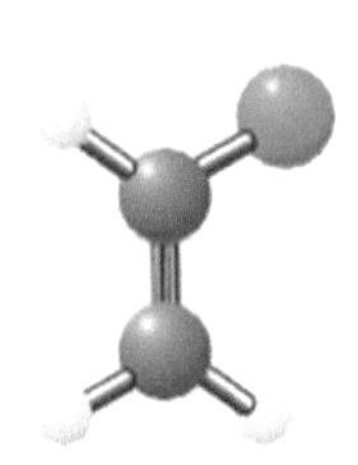

Vinyl chloride
C_2H_3Cl

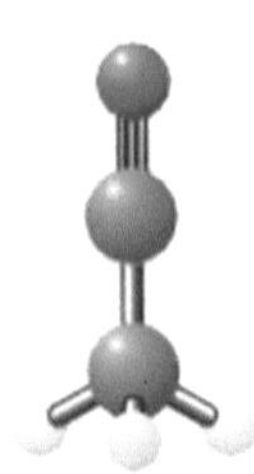

Acetonitrile
C_2H_3N

Acetylene has a great many industrial and commercial uses. Acetylene torches are commonly used to weld and cut metals. The demand for acetylene is currently met using a process different from that discovered by Willson and Davy. Acetylene can be made by pyrolysis (adding heat to convert organic solids into gases and liquids) of methane, the main component of natural gas:

$$2CH_4(g) \rightarrow C_2H_2(g) + 3H_2(g)$$

Acetylene burns with a much brighter flame than many other fuels, leading to the use of acetylene as a light source in many areas. Its use as a lamp by miners as they work underground is most notable. Acetylene also burns very hot, at about 3000 °C, when it reacts with oxygen, giving rise to its use in welding and cutting equipment.

Example 10.14—Production of Acetylene

A small chemical company is setting up the facilities to produce acetylene. Planners want to know what volume of C_2H_2 will be produced from the reaction of 4.25×10^3 g of methane if the product is to be stored at 38 °C and 1.00 atm. Assume a yield of 34%.

Asking the Right Questions

The acetylene will be produced from the pyrolysis of methane. How many moles of acetylene will be formed, assuming a 100% yield? How do we use our understanding of gas laws to convert moles of acetylene to the volume of acetylene? How do we adjust for a 34% yield?

Solution

We start by converting grams of methane to moles of methane and then to moles of acetylene, remembering the 2-to-1 mol ratio of methane to acetylene.

$$\text{mol } CH_4 = 4250 \text{ g } CH_4 \times \frac{1 \text{ mol } CH_4}{16.0 \text{ g } CH_4}$$

$$= 265.6 \text{ mol } CH_4$$

$$\text{mol } C_2H_2 = 265.6 \text{ mol } CH_4 \times \frac{1 \text{ mol } C_2H_2}{2 \text{ mol } CH_4}$$

$$= 132.8 \text{ mol } C_2H_2$$

We use the ideal gas equation to find the volume of C_2H_2 assuming 100% yield.

$$1.00 \text{ atm} \times V = 132.8 \text{ mol} \times \left(\frac{0.08206 \text{ L} \cdot \text{atm}}{\text{mol} \cdot \text{K}}\right) \times 311 \text{ K}$$

$$V = 3.39 \times 10^3 \text{ L}$$

The volume of 3.39×10^3 L assumes a 100% yield. Because the actual yield is 34%, we must multiply 3.39×10^3 L by 0.34, to get the final volume of 1.15×10^3 L of C_2H_2.

Is Our Answer Reasonable?

A thousand liters of acetylene is a lot of acetylene. However, we started with a lot of methane—425 g, so that answer seems reasonable. It is also reasonable that the 34% yield gave us a reduced amount of acetylene.

What If…?

…the small chemical company needed to produce 75,000 L of acetylene per hour to be stored at 31 °C and 0.98 atm? How many grams of methane would they need per 24-h day? Assume a reaction yield of 34%.

(Ans: 6.7×10^6 g CH_4)

Practice 10.14

How many grams of methane would be needed to produce 5.00×10^3 L of C_2H_2? Again, assume that the actual yield of the reaction is only 34% under the same storage conditions.

▶ Here's what we know so far

- Avogadro's, Boyle's, and Charles's laws can be combined into an equation that relates the initial and final states of a gas.

$$\frac{P_{\text{initial}}V_{\text{initial}}}{n_{\text{initial}}T_{\text{initial}}} = \frac{P_{\text{final}}V_{\text{final}}}{n_{\text{final}}T_{\text{final}}}$$

- The ideal gas law provides information on the current state of a gas.

$$PV = nRT$$

- The ideal gas constant (R) relates the value and the units needed to mathematically operate the ideal gas law.

$$R = 0.08206 \text{ L} \cdot \text{atm} \cdot \text{mol}^{-1} \cdot \text{K}^{-1}$$

- The molar mass of a gas can be determined using a modification of the ideal gas equation.

$$\underline{M} = \frac{dRT}{P} \blacktriangleleft$$

10.7 Kinetic-Molecular Theory

So far, our understanding of gas behavior has been based on the results of experiments. But just as we needed a model of chemical bonds to better understand and predict the structure and behavior of molecules, we need a model of gases to better understand and predict their behavior. The ideal gas laws are elegant in their simplicity and profound in their meaning. Because of their simplicity, many scientists, including Bernoulli, Clausius, Maxwell, Boltzmann, and van der Waals, sought to prove that the laws governing the behavior of ideal gases were a result of the interactions among atoms and molecules (collectively called "molecules" in the theory). The beauty of the resulting **kinetic-molecular theory** is that it enables us to arrive at the same conclusions by thinking about the nature of molecules as we do from performing well-chosen experiments. The experiments and the theory are consistent. Both are based on the following statements that model gas behavior according to the kinetic-molecular theory:

Kinetic-Molecular Theory – A theory explaining the relationship between kinetic energy and gas behavior. This theory makes possible derivation of the ideal gas laws.

1. Gases are composed of particles. The particles are negligibly small compared to their container and to the distance between each other.
2. Therefore, intermolecular attractions, which are exhibited at small distances, are assumed to be nonexistent.
3. Gases are in constant, random motion, *colliding with the walls of the container and with each other*.
4. Pressure is the force per unit area caused by the molecules colliding with the walls of the container.
5. *Because pressure is constant in a container over time*, molecular collisions are assumed to be perfectly elastic. *That is, no energy of any kind is lost upon collision.*
6. The average kinetic energy of the molecules in a system is linearly proportional to the absolute (Kelvin) temperature.

10.7.1 Using the Assumptions About Gas Behavior to Rationalize the Gas Laws

For a model to be useful, it must correctly predict actual behavior. Let's compare the predicted outcomes of the kinetic-molecular theory with our previously observed gas laws.

- Boyle's Law ($PV = \text{constant}$)

Collisions with the walls of the container give rise to pressure. What would happen if the volume of the container were increased at a constant temperature for a given number of molecules? Because temperature is a measure of the average kinetic energy of the molecules in the system, a constant temperature means that the velocity of the molecules would stay the same. The same number of molecules, traveling at the same speed, are bouncing off a larger total surface area, so the pressure on a unit of surface area would fall.

- Charles's Law ($V/T = \text{constant}$)

In this case, both the pressure and the number of molecules are constant. If the volume is lowered, how can the number of collisions per unit time (a measure of pressure) stay the same? The only way this can happen is if the velocity of the molecules becomes slower, via lowering of the temperature. As a result, the more slowly traveling molecules, which also have less kinetic energy, hit the now-closer walls at the same rate as the more rapidly moving molecules did when the walls were farther out.

- Avogadro's Law ($V/n = \text{constant}$)

How is it possible to keep the pressure and the average kinetic energy (temperature) of the molecules constant if the volume is increased? We need to add more molecules to the system—that is, increase n—so that the number of collisions per unit time in the larger system remains the same.

Example 10.15—The Importance of Assumptions

According to point 1 of the kinetic-molecular theory, "Gases are composed of molecules that are negligibly small compared to their container and to the distance between molecules." Strictly speaking, this statement is not true for real gases. How would the gas laws change if point 1 were not included?

Asking the Right Questions
What is true about the size of molecules in real gases? Why should size matter?

Solution
One answer is that if the more realistic picture of gases were taken into account, then the volume would depend on the nature of the gas. Gases composed of larger atoms and molecules take up more volume and would therefore cause more deviation from Boyle's law. The *b* term in the van der Waals model for nonideal gas behavior is one attempt to quantitatively account for this deviation.

Is Our Answer Reasonable?
To the extent that the data for the behavior of real gases is consistent with this picture, the answer is reasonable. We noted previously that the van der Waals equation is just one of several ways of correcting for the non-ideal behavior of a gas.

What If…?
…we lowered sharply the temperature of the system. According to kinetic-molecular theory, how would the behavior of the gas change?

(Ans: Lowering the temperature of the system would slow the molecules within, reducing the number of collisions per unit time with the walls of the container. The pressure of the system would be reduced, consistent with the kinetic-molecular theory.)

Practice 10.15
Explain, using the equations for kinetic energy, why two different molecules must differ in speed if the molecules have identical kinetic energy. If their kinetic energies are the same, which would be moving faster, a low-molecular-mass molecule or a high-molecular-mass molecule?

10.8 Effusion and Diffusion

In 1829, Thomas Graham (1805–1869) made measurements of the relative rates at which gases pass through small openings into a very low-pressure region, a process called **effusion** (Fig. 10.21). His results, like those that led to the other gas laws, predated the kinetic-molecular theory and confirm our conclusion regarding the relative speeds of gas particles. His measurements led to what we now know as **Graham's law of effusion**. We can state this law mathematically. For two gases of differing molar mass, $\underline{M}_1$ and $\underline{M}_2$, the ratio of their effusion rates (re_1/re_2) is equal to the square root of the inverse ratio of their molar masses ($\underline{M}_2/\underline{M}_1$).

$$\frac{re_1}{re_2} = \sqrt{\frac{\underline{M}_2}{\underline{M}_1}}$$

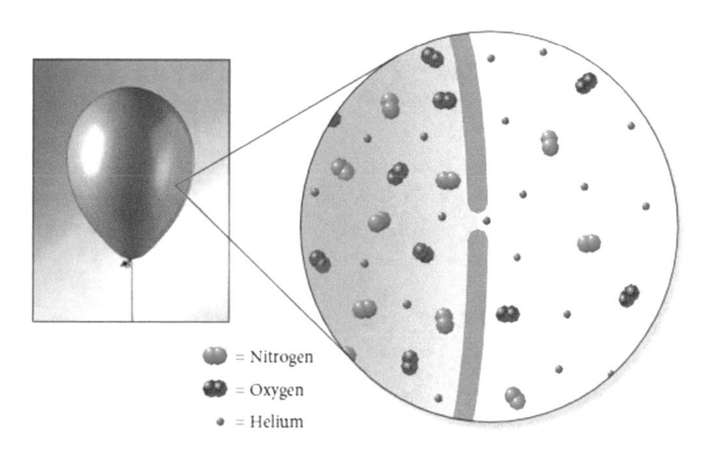

Fig. 10.21 Gases effuse through a small opening in their container. The faster molecules escape through the opening more rapidly than the slower molecules

Effusion – The process by which a gas escapes through a small hole.

Graham's law of effusion – The rates of the effusion of gases are inversely proportional to the square roots of their molar masses.

10.8.1 A Closer Look at Molecular Speeds

The kinetic energy of a molecule is equal to $\frac{1}{2}mv^2$, in which m is the mass and v is the velocity of that molecule. For a system with one type of molecule (that is, m is constant), the temperature is proportional to the square of the velocity, v^2. Molecules move faster at higher temperatures, and slower at lower ones.

For a gas containing Avogadro's number of molecules (that is, 1 mol of molecules or 6.022×10^{23} molecules), there are several additional ideas we must introduce. First, the term **velocity** means the rate of travel—the speed—in a specified direction. The molecules will be traveling in all different directions, and the net velocity (including the direction!) may well be zero meters per second, even if the individual molecules are moving very rapidly. When we talk about the rate of travel of large numbers of molecules, we are interested in their speed (how fast), not their velocity (how fast and in what direction). Also, the collisions will not be perfectly elastic, and some energy will, in fact, be transferred. There will be a distribution of kinetic energies among the many molecules and, therefore, a range of speeds that can be determined. Just as the SAT or ACT scores for students throughout the United States have a fairly wide range, the speeds of individual molecules in a sample also have a range of values. Just as there is an average SAT or ACT score, we can define the **average speed** of i molecules as we would any average value; it is the sum of the speeds divided by the number of molecules:

Average Speed – The sum of the molecular speeds of each molecule divided by the number of molecules.

$$\text{Speed}_{\text{average}} = \left(\text{speed}_1 + \text{speed}_2 + \cdots + \text{speed}_i\right)/i$$

A related value, often used in discussions about the statistics of molecular behavior, is the **root-mean-square (rms) speed**, u. This is the speed of a molecule with the average kinetic energy, and it is equal to the square root of the sum of the squares of the individual speeds divided by the number of molecules—that is:

$$u = \left[\frac{\left(\text{speed}_1^2 + \text{speed}_2^2 + \cdots + \text{speed}_i^2\right)}{i}\right]^{1/2}$$

Root-mean-square (RMS) Speed – The square root of the sum of the squares of the individual speeds of particles divided by the number of those particles.

The average score on the SAT or ACT is not necessarily the most probable; it is just the average. So it is with gases, for which there exists the most probable speed, α, of a gas. The root-mean-square speed, the average speed, and the **most probable speed** are related as 1.22 : 1.12 : 1.00.

Most Probable Speed – The most likely speed of a molecule from among a group of molecules.

The distribution of molecular speeds at different temperatures for helium and that for methane are shown in ■ Fig. 10.22. Why does methane move more slowly, on average, than helium at the same temperature? Again we revisit the factors that account for kinetic energy ($\frac{1}{2}mv^2$). If two gases are at the same temperature, their average

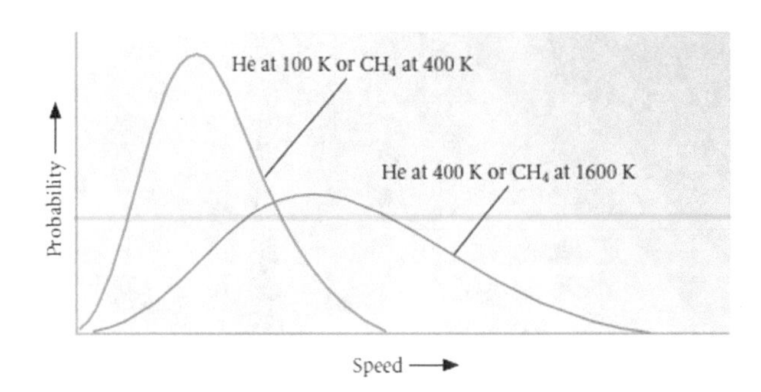

■ **Fig. 10.22** Molecular speeds are related to the mass of the gas and to the temperature, as shown here for helium and methane. Why do methane molecules move more slowly, on average, than those of helium at the same temperature?

kinetic energies will also be equal. Methane (16.0 g/mol) is heavier than helium (4.00 g/mol), so its molecules must be moving more slowly, on average.

At a given temperature (constant molecular kinetic energy = ½ mv²), heavier gas molecules move more slowly than their lighter counterparts. Although the mathematical derivation is fairly involved, it is possible, using the ideas of the kinetic-molecular theory, to state that for any sample of gas, the rms speed is related to the molar mass of the gas as follows:

$$u_{\mathrm{rms}} = \sqrt{\frac{3RT}{\underline{M}}}$$

where

T the temperature in K
$\underline{M}$ the molar mass of the gas in $\mathrm{kg \cdot mol^{-1}}$
R the ideal gas constant, $8.314\ \mathrm{J \cdot K^{-1} \cdot mol^{-1}}$
$8.314\ \mathrm{kg \cdot m^2 \cdot s^{-2} \cdot mol^{-1} \cdot K^{-1}}$

The SI units used for energy are $\mathrm{kg \cdot m^2 \cdot s^{-2}}$. This is the combined unit we know as the joule. As shown below, the units of PV are the same as well. So the ideal gas constant, R, can be written with units of $\mathrm{J \cdot K^{-1} \cdot mol^{-1}}$, units of $\mathrm{kg \cdot m^2 \cdot s^{-2} \cdot mol^{-1} \cdot K^{-1}}$, or units of $\mathrm{L \cdot atm \cdot mol^{-1} \cdot K^{-1}}$.

$$\begin{aligned} P \times V &= \left[\frac{\text{force}}{\text{area}} \times V\right] \\ &= \left[\text{mass} \times \frac{\text{acceleration}}{\text{area}} \times \text{volume}\right] \\ &= \frac{\mathrm{kg \cdot m \cdot s^{-2}}}{\mathrm{m^2}} \\ &= \mathrm{kg \cdot m^2 \cdot s^{-2}} = \mathrm{J} \end{aligned}$$

Example 10.16—The RMS Speed of Gases

Compare the rms speeds of methane and helium at 25 °C

Asking the Right Questions

Which do we expect to move faster, methane or helium? How do we calculate the speed of each? A useful analogy might be to compare the running speed of a 125-lb world-class marathon runner to that of a 600-pound sumo wrestler, where the helium is the marathon runner and the methane is the sumo wrestler.

Solution

Lighter particles move faster than heavier ones, so helium moves faster than methane.

$$\begin{aligned} u_{\mathrm{rms}} &= \sqrt{\frac{3RT}{\underline{M}}} \\ &= \sqrt{\frac{3 \times 8.314\ \mathrm{kg \cdot m^2 \cdot s^{-2} \cdot mol^{-1} \cdot K^{-1}} \times 298\mathrm{K}}{0.0160\ \mathrm{kg \cdot mol^{-1}}}} \\ &= \sqrt{4.645 \times 10^5 \mathrm{m^2 \cdot s^{-2}}} \\ &= 682\ \mathrm{m/s} \end{aligned}$$

We calculate the speed for helium:

$$\begin{aligned} u_{\mathrm{rms}} &= \sqrt{\frac{3RT}{\underline{M}}} \\ &= \sqrt{\frac{3 \times 8.314\ \mathrm{kg \cdot m^2 \cdot s^{-2} \cdot mol^{-1} \cdot K^{-1}} \times 298\mathrm{K}}{0.00400\ \mathrm{kg \cdot mol^{-1}}}} \\ &= \sqrt{1.858 \times 10^6\ \mathrm{m^2 \cdot s^{-2}}} \\ &= 1360\ \mathrm{m/s} \end{aligned}$$

Are Our Answers Reasonable?

We determined that helium is faster than methane. This is consistent with our understanding that lighter particles move faster than heavier ones. The answers are reasonable.

What If…?

…we had a sample of sulfur dioxide? How much more slowly than helium would the SO_2 molecules travel?

(Ans: The sulfur dioxide molecules would travel at ¼ the speed of the helium atoms.)

Practice 10.16

Which gas has a higher average speed at 20 °C, F_2 or CO_2?

Effusion can be used to separate mixtures of gases that would be difficult to separate by other methods. Effusion has been used to separate ^{235}U and ^{238}U from the natural mixture of uranium isotopes (0.72% ^{235}U and 99.28% ^{238}U). The separation of these two naturally occurring uranium isotopes from the uranium ore called pitchblende (UO_2) makes possible the construction of atomic fuel rods for use in nuclear power plants as well the production of atomic weapons. The fundamental process requires converting the ore into solid uranium hexafluoride (UF_6). At the processing facility, the solid is heated, and the UF_6 gas, containing the ^{235}U and ^{238}U isotopes, is fed into a series of vessels. At each stage, the lower molar mass UF_6 strikes the walls of the vessel more frequently than does the high molar mass UF_6. The vessel has semipermeable walls, so more of the lighter, ^{235}U-containing UF_6 can flow through. Hundreds of stages later, the resulting "enriched" mixture contains about 5% ^{235}U. The mixture is then solidified for use in nuclear power plants. Nuclear weapons require enrichment of the uranium mixture to at least 90% ^{235}U.

Example 10.17—Effusion

A latex balloon filled with a mixture of N_2 and He is observed. Which gas would effuse through the pores in the latex balloon more rapidly? How much more rapidly would it effuse?

Asking the Right Questions

What is the relationship between the effusion speed and the molecular mass?

Solution

The gas with the lower molecular mass, He, would effuse more rapidly. The ratio can be calculated with Graham's law of effusion:

$$\frac{re_1}{re_2} = \sqrt{\frac{M_2}{M_1}}$$

$$\frac{re_{\mathrm{He}}}{re_{\mathrm{N_2}}} = \sqrt{\frac{28.0}{4.00}}$$

$$\frac{re_{\mathrm{He}}}{re_{\mathrm{N_2}}} = 2.65$$

Therefore, the helium gas would effuse from the balloon 2.65 times faster than the nitrogen gas.

Are Our Answers Reasonable?

Helium has a much lower molecular mass than nitrogen, so it is reasonable that its particles move more rapidly than those of nitrogen.

What If...?

...a gas effused 1.66 times faster than carbon dioxide at 30 °C. Is that gas acetylene, C_2H_2, 2-methylbutane, C_5H_{12}, or methane?

(*Ans: methane*)

Practice 10.17

Which effuses more rapidly under identical conditions, He or H_2? How much more rapidly?

10.8.2 Diffusion

A related phenomenon is called **diffusion**, which involves the mixing of one gas with another or with itself. On a molecular level diffusion is far more complex, because in mixing, molecules collide with each other after moving only a very short distance—typically 5×10^{-8} m, only hundreds of times the diameter of a gas molecule itself. Therefore, mixing is a chaotic, rather slow process of the dilution of two or more gases.

Diffusion – The process by which two or more substances mix.

The food industry has a particular interest in the ability of gases to pass through small openings, such as the pores of plastic wrap that surround meats and other food items. The capacity of oxygen to pass through these pores to the food lowers its shelf life unless the packaging can maintain an inert atmosphere, as discussed in ▶ Sect. 10.3. Small condiment packages of ketchup and mustard, such as those in Fig. 10.23, are usually packaged in several layers that include plastic and foil. The foil helps prevent the exchange of air with the food, and the plastic keeps the metals in the foil away from the food, with which they might otherwise react.

Fig. 10.23 Small condiment packages of ketchup and mustard are usually packaged in several layers that include plastic and foil. The foil helps prevent the exchange of air with the food, and the plastic keeps the metals in the foil away from the food, with which they might otherwise react

▶ **Deepen Your Understanding Through Group Work**

Pneumatic retinopathy is a technique for repairing tears of the retina in the eye. In this technique, a gas bubble is injected in the eyeball, which, after laser-based attachment, allows the retina to adhere to the back of the eyeball and heal. There are four gases that are typically used: Air, sulfur hexafluoride, SF_6, perfluoroethane, C_2F_6, and perfluoropropane, C_3F_8.

Your group will need to learn enough about Pneumatic retinopathy to decide which gas to use in the case of a 56-year-old runner whose retina tore and detached just before the start of a half-marathon. He ran the race, finally undergoing surgery 3 days later, after further detachment.

What gas would work best?

What factored into your group's decision?

What were the areas of agreement and disagreement as you worked toward your decision? ◀

10.9 Industrialization: A Wonderful, yet Cautionary, Tale

The glory of science and technology is our ability to modify nature's bounty in ways that enable us to sustain a world of over 8 billion people as it may be by 2025. As consumers, we often ignore the impact that gases have on us. However, the chemical changes in our bodies, that inhaled oxygen makes possible, sustain life. The chemical change of combustion moves motorized vehicles of all shapes and sizes along our highways, on our waterways, and on a fluid bed of air in the sky. Gases can wreak havoc in the form of tornadoes, hurricanes, and natural gas explosions, or lend calm to a cool evening. They can kill when used as "nerve agents" in warfare or help heal when used as anesthetics in surgery. As we modify nature, we recognize the changes in the atmosphere that our nearly 200-year-long focus on industrialization has produced. These changes bear watching, for in the relative blink of an eye on the global time scale, we risk unraveling the protective blanket that has taken nature 2 billion years to create. Two of the greatest concerns are changes in ozone concentration and the greenhouse effect.

10.9.1 Ozone

This pale blue gas, O_3, is an allotrope of oxygen and is found in two of the four layers of the atmosphere: the stratosphere and the surface layer called the troposphere. These layers and the important gases within them are shown in Fig. 10.24. In the stratosphere, ozone absorbs much of the UV radiation that would otherwise be harmful to us. UV radiation is typically categorized into three regions based on the energy of the corresponding wavelengths. The UV-A band specifies radiation with a wavelength of 400 nm to 320 nm; UV-B indicates wavelengths of 320 nm to 290 nm; and UV-C, the most energetic, is the classification of wavelengths from 290 to 100 nm. The destruction of ozone occurs by **photodissociation**—that is, UV light of sufficient energy (UV-B) causes the molecule to dissociate, forming oxygen gas and atomic oxygen:

$$O_3(g) \overset{UV}{\rightarrow} O_2(g) + O(g)$$

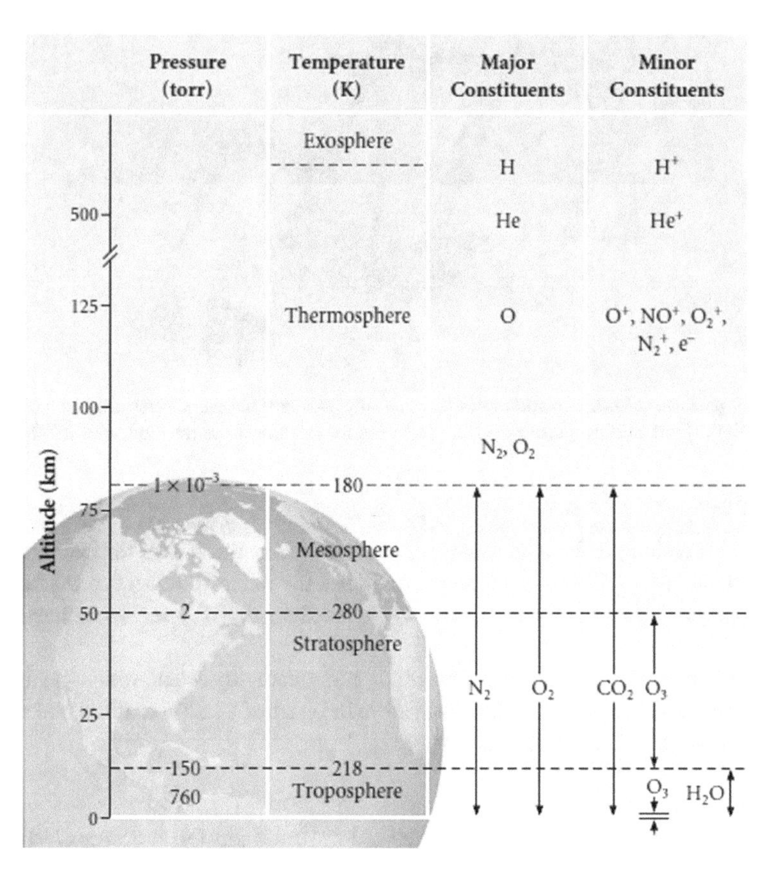

Fig. 10.24 Two of the four layers of the atmosphere contain ozone, O_3

Photodissociation – A chemical reaction where light, of sufficient energy, causes the cleavage of a bond in a molecule.

There is also a mechanism for the formation of ozone, which also contributes to the filtering of UV radiation from the sun. Shorter-wavelength UV radiation (UV-C) provides sufficient energy for the O_2 to separate into oxygen atoms:

$$O_2(g) \xrightarrow{UV} 2O(g)$$

These highly reactive individual atoms then recombine with O_2 to form ozone along with the release of heat

$$O_2(g) + O(g) \rightarrow O_3(g)$$

The resulting combination of these three equations gives rise to the Chapman mechanism, a useful description of the chemistry involving oxygen and ozone in the middle stratosphere.

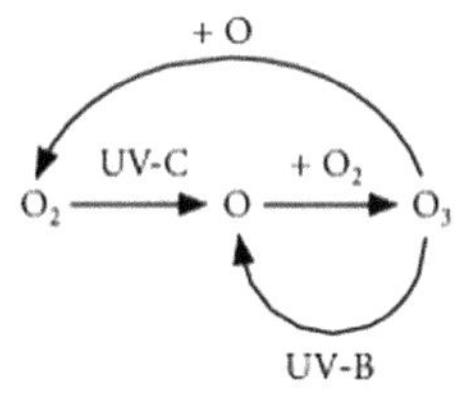

The natural balance between ozone photodissociation and ozone formation keeps most harmful UV radiation away from the Earth's surface. Within the last 50 years, however, activities related to industrialization have led to a significant reduction in the stratospheric ozone layer over many parts of the planet, as shown in Fig. 10.25. Most notable is the discovery of a "hole" in the ozone layer over Antarctica.

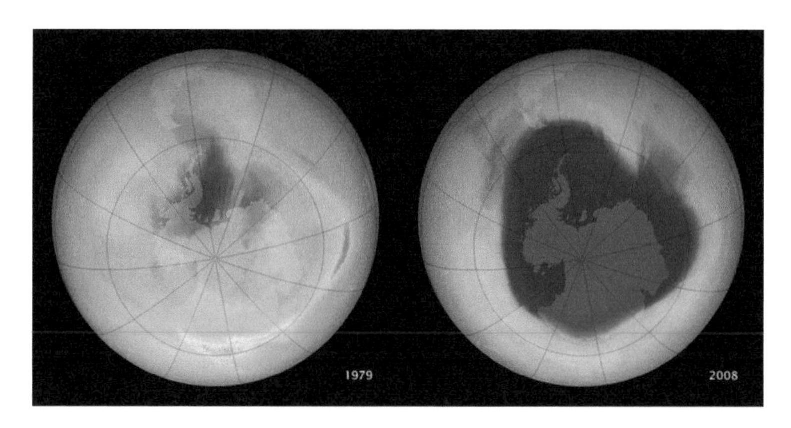

◻ **Fig. 10.25** Here are two ozone hole measurements from 1979 and 2008. Blue indicates lower ozone levels, yellow and red show higher ozone levels. NASA now uses 5 new instruments aboard the "Suomi National Polar-orbiting Partnership" (NPP) to track worldwide ozone. The ozone hole was getting worse, but has declined recently as a result global warming, as well as the United Nations 1987 Montreal Protocol international agreement on phasing out the use of chlorofluorocarbons (CFCs). Images from NASA. *Source* NASA; ▶ https://earthobservatory.nasa.gov/images/38835/antarctic-ozone-hole-1979-to-2008

◻ **Table 10.7** Common chlorofluorocarbons and halons

Formula	Common symbol	Major uses
CCl_3F	CFC-11	Polymer foams, refrigeration, air conditioning
CCl_2F_2	CFC-12	Polymer foams, refrigeration, air conditioning, aerosols, food-freezing solvents
CCl_2FCClF_2	CFC-113	Solvent
$CBrClF_2$	Halon-1211	Portable fire extinguishers
$C_2Br_2F_4$	Halon-2402	Fire extinguishers

Sherwood Rowland and Mario Molina suggested in 1974 that **chlorofluorocarbons (CFCs)**, used in refrigeration and in the formation of polymer foams, were the culprits in the destruction of the ozone layer. ◻ Table 10.7 lists common CFCs and related compounds, called **halons**, along with their uses.

Chlorofluorocarbons (CFCs) – Compounds containing only carbon, chlorine, and fluorine. CFCs are typically used as solvents and refrigerants and are known to be harmful to the ozone layer.

Halons – Compounds containing carbon and halogens. Halons are typically used in fire extinguishing applications but many are considered harmful to the ozone layer.

CFCs were initially used commonly as refrigerants. This class of compounds is chemically inert, nonflammable, inexpensive, and quite stable at the Earth's surface. These properties make them very suitable for use in homes and automobiles. However, as they are released into the atmosphere, they travel upward to the stratosphere without reacting with other compounds. Then, in the stratosphere, they are bombarded with high-energy UV radiation and do react. This produces highly reactive chlorine atoms that react with ozone as well as with individual oxygen atoms that come from ozone.

$$\begin{aligned} Cl(g) + O_3(g) &\rightarrow ClO(g) + O_2(g) \\ ClO(g) + O(g) &\rightarrow Cl(g) + O_2(g) \\ \hline \text{Overall reaction: } O_3(g) + O(g) &\rightarrow 2O_2(g) \end{aligned}$$

NanoWorld/MacroWorld The Big Effects of the Very Small: Accumulation of Ozone at the Earth's Surface

As we monitor the decrease in stratospheric ozone with growing concern, we also note the localized *increase* in surface ozone, because this gas is an irritant to the eyes, nose, and throat and damages cells in plants. Where does tropospheric ozone come from? We typically note its presence in the metropolitan areas of industrialized nations. More specifically, the reaction mechanism for the *accumulation* of ozone at the surface involves the release of nitric oxide from automobiles without catalytic converters. This is due to the unwanted combustion of nitrogen gas (present in the air) inside the combustion chambers of automobiles:

$$N_2(g) + O_2(g) \rightarrow 2NO(g)$$

Unfortunately, even automobiles with catalytic converters produce small amounts of NO. The released NO reacts with oxygen to make nitrogen dioxide:

$$2NO(g) + O_2(g) \rightarrow 2NO_2(g)$$

► https://upload.wikimedia.org/wikipedia/commons/2/24/Smog_in_the_skies_of_Delhi%2C_India.jpg

Then, when the Sun shines on a city filled with NO_2, the molecule decomposes to produce highly reactive oxygen atoms:

$$NO_2(g) \xrightarrow{\text{sunlight}} NO(g) + O(g) \quad \text{(reaction A)}$$

The atomic oxygen combines with oxygen gas in the same way that it does in the stratosphere to produce ozone:

$$O_2(g) + O(g) \rightarrow O_3(g) \quad \text{(reaction B)}$$

Then ozone is consumed in this reaction:

$$NO(g) + O_3(g) \rightarrow NO_2(g) + O_2(g) \quad \text{(reaction C)}$$

The resulting combination of reactions A, B, and C that produce and consume ozone at ground level is known as a null cycle. No net reaction occurs in a null cycle. Instead, the reactions are governed specifically by the relative concentrations of the individual reactants and products.

Especially important is the presence of organic (carbon-based) pollutants and heat, such as are present on a hot summer day. These pollutants create highly reactive organic compounds that react faster with NO than does O_3. In the presence of these organic pollutants, the null cycle is modified and excess O_3 remains! What is the effect of this "nanoworld" process on the "macroworld"? This is the impact of heat and pollution in urban areas. Ozone accumulates at the Earth's surface, especially in large cities that have lots of motor vehicle traffic. The good news is that improvements are clear. ◘ Figure 10.26a shows the change in ozone levels in one U.S. state, North Carolina, from 2000 to 2017. The national trend is shown in ◘ Fig. 10.26b.

Rowland and Molina shared the 1995 Nobel Prize in chemistry for their groundbreaking research and predictions about the fate of the ozone layer. Since that time, substitute compounds called HCFCs have been developed, including CHF_2Cl and CF_3CFH_2. These react in the lower atmosphere to release HCl and HF, never making it to the stratosphere. The loss of ozone in the stratosphere, combined with the generation of unhealthful ozone levels in large cities, is a continued cause for concern and a focus of research. The very good news is that as a result of worldwide action to reduce CFC emissions, we may have turned the corner. Since measurements were started in 1979, the largest ozone hole was measured in 2006, but it has been getting smaller since then. Many decades of continued diligence will be required before the damage to the ozone layer can be completely reversed, yet we have made an important start.

10.9.2 The Greenhouse Effect and Global Climate Change

The **greenhouse effect** is caused by the accumulation, in the atmosphere, of gases that permit light to enter but prevent some energy as heat from exiting, much like a plant greenhouse. Methane and CFCs are examples of greenhouse gases, although they are not present in sufficient concentration to be of real concern at this time.

Greenhouse Effect – The re-radiation of energy as heat from gases in the atmosphere back toward Earth. An increase in the greenhouse effect leads to an increase in global temperatures.

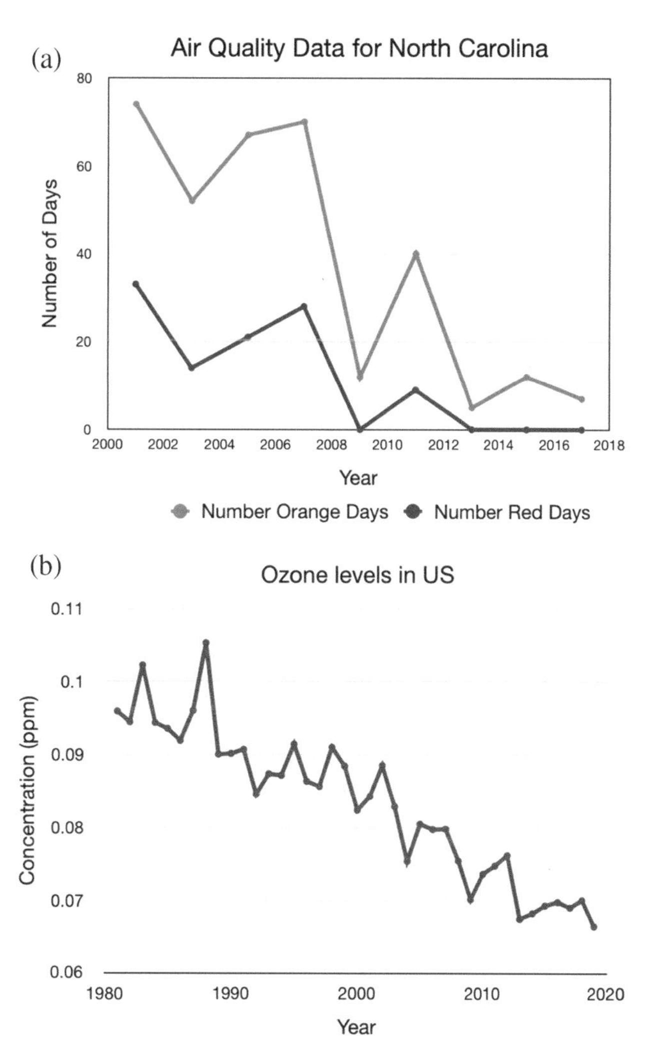

Fig. 10.26 **a** Air quality in central North Carolina from 2001 to 2017. **b** US nationwide ozone levels are decreasing with time

Carbon dioxide is entering the atmosphere in much greater amounts than are being used up, and this makes CO_2 the most important greenhouse gas.

The **carbon cycle** has historically kept the carbon in the atmosphere, seas, and land in balance over the long term. Human destruction of forests has radically curtailed the number of plants that use up CO_2 in photosynthesis:

$$6CO_2(g) + 6H_2O(g) \rightarrow C_6H_{12}O_6(s) + 6O_2(g)$$

Carbon Cycle – The natural process that describes how carbon atoms are moved among the land, sea, and atmosphere.

Burning carbon-based fuels, especially coal, oil, and natural gas, has added CO_2 to the environment, and the annual release of carbon into the atmosphere continues to rise (see Fig. 10.27). There is overwhelming agreement among scientists that the 100-year increase in the Earth's average surface temperature, shown in Fig. 10.28, is at least partially caused by the greenhouse effect due to human-produced carbon dioxide emissions. This increase in surface temperatures, as well as an increase in sea temperatures, is producing changes to the global climate. With a system as large and complex as the climate, the exact nature of these changes and their impact on local weather, growing seasons, and environmental conditions is difficult to predict, but it is certain that our climate is undergoing significant changes.

It is often said that the problems caused by the use of chemistry can also be fixed by the use of chemistry. Some good signs are on the horizon. As shown in Fig. 10.29, the fraction of global energy consumption derived from the burning of coal and oil is declining.

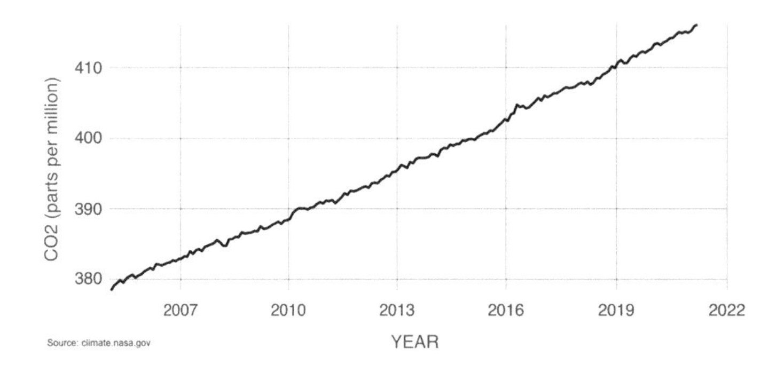

Fig. 10.27 Carbon dioxide levels have been steadily rising due to many different factors, including our insatiable thirst for fossil fuels. Source of data from NASA. *Source* ► https://climate.nasa.gov/vital-signs/carbon-dioxide/

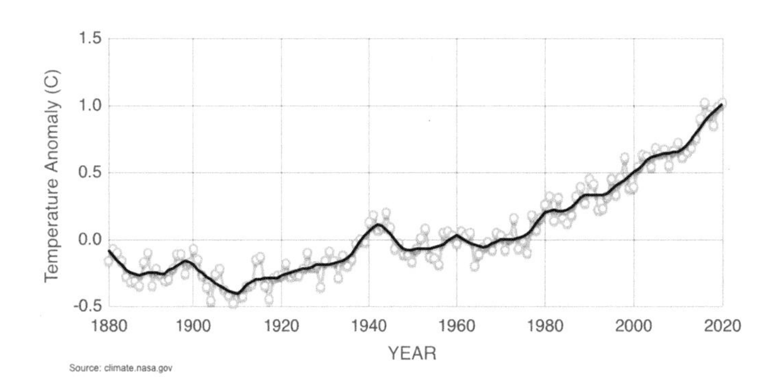

Fig. 10.28 Global land-ocean temperature index shows a steady increase in global temperatures. *Source* ► https://climate.nasa.gov/vital-signs/global-temperature/

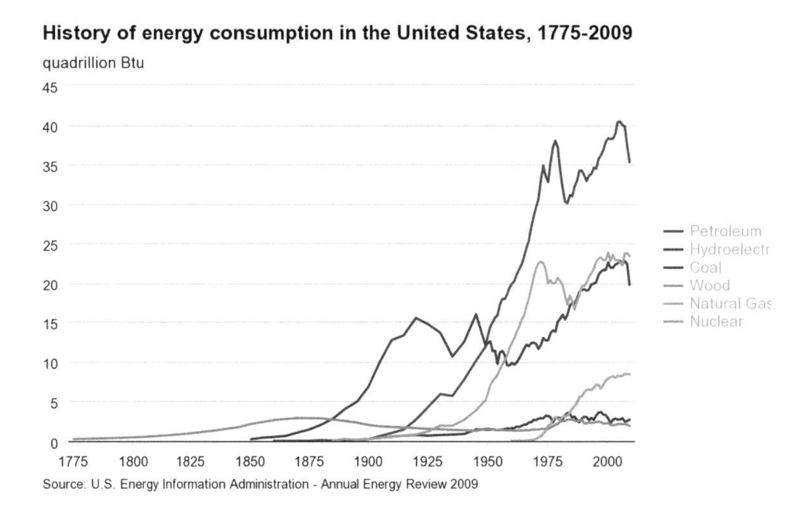

Fig. 10.29 Advances in chemistry have changed the way in which energy is produced and consumed

The interaction among gases in the atmosphere is so complex that we cannot confidently predict the long-term atmospheric effects of industrialization. The best models indicate that as we produce gases that meet the needs and the wants of the more than 8 billion people on Earth, we must be mindful of the possible impact of our current consumption. We are, after all, the keepers of the global commons.

The Bottom Line

- Industrial gases are used in manufacturing, medicine, and other industries.
- Gases that behave as though each particle has no interactions with any other are called ideal gases.
- High pressure, low temperature, and intermolecular forces of attraction between gas molecules cause gases to deviate from ideal behavior.
- Pressure is a measure of force per unit area. Pressure is the result of collisions of gas molecules with the walls of a container.
- In current usage, the standard atmosphere (1 atm) is different from standard pressure (1 bar, or about 0.987 atm).
- Dalton's law of partial pressures is concerned with the contribution of each gas to the pressure of the entire gas mixture.
- Avogadro's law deals with the relationship between the volume and number of moles of ideal gases, at constant temperature and pressure.
- Boyle's law expresses the relationship between volume and pressure of an ideal gas, at constant number of moles and temperature.
- Charles's law expresses the relationship between the volume and temperature of an ideal gas, at constant number of moles and pressure.
- The ideal gas equation combines the gas laws to interrelate the pressure, volume, amount, and temperature of an ideal gas.
- The ideal gas equation can be applied to find the molar mass and the density of an ideal gas.
- Several scientists have made mathematical models to account for ideal gas behavior. J. D. van der Waals's model is the most commonly used because it is relatively simple and takes into account corrections for pressure and volume.
- The molar mass and the density of an ideal gas are directly proportional.
- The kinetic-molecular theory shows how it is possible to start from some elementary constructs about gas behavior and derive the gas laws, therefore showing the consistency between theory and experiment.
- Graham's law of effusion shows the inverse relationship between the speed of a gas and its molar mass.
- The effects of industrialization on the atmosphere are currently being debated. Ozone levels and the greenhouse effect are two issues of greatest social concern.

Section 10.1 The Nature of Gases

Skill Review

1. Define the intermolecular forces typically experienced by molecules in the gaseous state.
2. Explain why we never see signs on delivery trucks that read "Danger—Compressed Liquid."
3. What conditions tend to favor ideal behavior in gases?
4. What are some characteristics common in nonideal gases?

Section 10.2 Production of Hydrogen and the Meaning of Pressure

Skill Review

5. If your chemistry textbook had a mass of 0.89 kg, how much force, in newtons, would the textbook exert on a desk in your classroom?
6. If your textbook (mass = 0.89 kg) measured 27 cm × 23 cm, what pressure, in pascals, would the book exert on the desk?
7. How much force, in newtons, would an 86-kg man exert on the ground?
8. A 115-lb flight attendant wears spike heels every day to work. If each heel has an area of 4.0 cm^2, what is the pressure exerted by each heel in pascals? (Assume all of the woman's weight is placed on the heels of her shoes.)
9. The pressure of a gas is measured to be 797 mm Hg. Convert this pressure into units of Pa, kPa, atm, torr, bar, and in Hg.
10. The pressure of a gas is measured to be 0.750 atm. Convert this pressure into units of Pa, kPa, torr, bar, and mm Hg.
11. An astronaut weighs 145 lb on the planet Earth, where the gravitational pull is 9.8 m/s^2. She is sent on a mission to a planet whose gravitational pull is only 0.33 m/s^2. What would a scale on this planet indicate that the astronaut weighs?
12. What is the mass in kg and weight in lb of a person who exerts a force of 375 N on a scale?

13. The air in a car tire exerts a pressure of 22.3 psig. What is the force exerted by the air if the total area inside the tire is 567 cm^2?
14. A full book bag that exerts a force of 12 N on a table takes up an area that measures 25 cm × 36 cm. Calculate the amount of pressure exerted by the bag.

Chemical Applications and Practice

15. Small inflatable sleeping mattresses make camping enjoyable for many outdoor enthusiasts. If the air pressure of a comfortable mat is measured as 17.5 psig, what is the actual pressure that the air is exerting on the inside walls of the mat? What is the pressure if reported in units of atmospheres?
16. As the altitude increases, the air pressure decreases. To prove this fact, a mountain climbing group measured the air pressure on Mount Everest at 0.67 atm. What would this pressure be if it were reported in units of kPa, bar, mm Hg, and psi?

Section 10.3 Mixtures of Gases—Dalton's Law and Food Packaging

Skill Review

17. The partial pressure of O_2 in a sample of air on a mountaintop is 115 mm Hg. The oxygen makes up 21% of the gas in the atmosphere. What is the total atmospheric pressure on the mountaintop?
18. A gaseous mixture contains 54.00% N_2, 39.00% O_2, and 7.00% CO_2 by volume. If the total pressure is 813 mm Hg at STP, what is the partial pressure of each of the gases?
19. A total of 12.4 g of N_2 and 12.4 g of O_2 are combined in a mixture that exerts a pressure of 1.23 atm. Calculate the partial pressure of N_2 and that of O_2 in the mixture.
20. A mixture of gases contains CO_2 and O_2 in a 3-to-1 ratio. If the partial pressure of CO_2 in the mixture is 2.34 atm, what is the partial pressure of O_2? What is the total pressure of the mixture?
21. A 760 mm Hg sample of gas at STP contains O_2, N_2, and Ar. If the partial pressure of O_2 is 100 mm Hg and the partial pressure of N_2 is 635 mm Hg, what is the mole fraction of each gas?
22. Calculate the mole fraction of each gas in the following mixtures at STP:
    a. 15% by volume He; 85% by volume O_2
    b. 15% by mass He; 85% by mass O_2
    c. 15.0 g He; 15.0 g O_2

Chemical Applications and Practices

23. A common way to generate small amounts of oxygen in the lab is by heating potassium chlorate in the presence of a catalyst and collecting the oxygen by bubbling it through water. If the total pressure of the collected gas was 785 mm Hg at 27 °C, what is the partial pressure of the collected oxygen? The pressure of gaseous water at 27 °C is 27 torr.
24. Producing hydrogen gas is becoming a very important technical and economic concern for food processing, and potentially as a fuel for future transportation needs. If a chemical engineer were developing a model for this purpose and collected 4.25 g of H_2 over water at 25 °C that had a total pressure of 1.15 atm, what would be the partial pressure of the H_2 collected? The pressure of gaseous water at 25 °C is 23.8 torr.
25. In a helium–neon laser, a mixture of helium and neon gas is contained in a small tube used to generate coherent light. In some applications, this type of laser is used to scan price codes at the grocery store. If the internal gas mixture is 8.85% Ne and 91.15% He by volume and the total pressure is maintained at 3.42 mm Hg, what are the partial pressures of the gas components?
26. A balloon is filled with H_2 and O_2 for a demonstration. The gases are added such that they will react completely when a flame is brought in contact with the balloon. Write the equation describing this reaction. If the total pressure of the balloon is 3.4 atm, what would the partial pressure of each gas need to be in order for the reaction to go to completion?

Section 10.4 The Gas Laws—Relating the Behavior of Gases to Key Properties

Skill Review

27. In each example, determine the missing information:
    a. $V_{initial} = 2.44$ L, $P_{initial} = 1.00$ atm; $V_{final} = 1.00$ L, $P_{final} = ?$
    b. $V_{initial} = 0.588$ L, $P_{initial} = 4.35$ atm; $V_{final} = 1.00$ L, $P_{final} = ?$
    c. $V_{initial} = 100$ mL, $P_{initial} = 563$ mm Hg; $V_{final} = 1.00$ L, $P_{final} = ?$
    d. $V_{initial} = 12.5$ L, $P_{initial} = 8.22$ atm; $V_{final} = ?$, $P_{final} = 0.44$ atm
28. In each example, determine the missing information:
    a. $V_{initial} = 2.44$ L, $T_{initial} = 200$ K; $V_{final} = 1.00$ L, $T_{final} = ?$
    b. $V_{initial} = 0.588$ L, $T_{initial} = 125$ °C; $V_{final} = 1.00$ L, $T_{final} = ?$
    c. $V_{initial} = 19.5$ L, $T_{initial} = 828$ K; $V_{final} = ?$, $T_{final} = 100$ K
    d. $V_{initial} = 22.4$ L, $T_{initial} = 32$ °C; $V_{final} = ?$, $T_{final} = 100$ °C
29. In each example, determine the missing information:
    a. $P_{initial} = 1.00$ atm, $V_{initial} = 2.44$ L, $T_{initial} = 200$ K; $P = 1.00$ atm, $V_{final} = 1.00$ L, $T_{final} = ?$
    b. $P_{initial} = 40$ atm, $V_{initial} = 123$ L, $T_{initial} = 218$ K; P = 1.00 atm, $V_{final} = 1.00$ L, $T_{final} = ?$
    c. $P_{initial} = 790$ mm Hg, $V_{initial} = 0.55$ L, $T_{initial} = -15$ °C; $P = 355$ mm Hg, $V_{final} = 1.00$ L, $T_{final} = ?$
    d. $P_{initial} = 12.5$ psi, $V_{initial} = 150.0$ mL, $T_{initial} = 12$°F; $P = ?$, $V_{final} = 6.55$ L, $T_{final} = 44$ °C
30. In each example, determine the missing information:
    a. $P_{initial} = 0.333$ atm, $V_{initial} = 0.55$ L, $T_{initial} = 273$ K; $P = ?$, $V_{final} = 0.55$ L, $T_{final} = 298$ K
    b. $P_{initial} = ?$, $V_{initial} = 2.44$ L, $T_{initial} = 200$ K; $P = 1.00$ atm, $V_{final} = 1.00$ L, $T_{final} = 25$ °C
    c. $P_{initial} = 1.00$ atm, $V_{initial} = 1.00$ L, $T_{initial} = 273$ K; $P = 6.00$ atm, $V_{final} = ?$, $T_{final} = 882$ °C
    d. $P_{initial} = 1.00$ atm, $V_{initial} = 4.07$ L, $T_{initial} = 2$ K; $P = ?$, $V_{final} = 1.00$ L, $T_{final} = 145$ K
31. At STP, a 2.5-L sample of gas contains 4.5 mol of Ne. How many moles would the sample contain if the volume were increased to 5.0 L?… if the volume were decreased to 1.0 L?
32. A balloon containing 0.50 mol of He is found to have a volume of 1.75 L. What volume would the balloon have if an additional 0.15 mol of He were added? What volume would the balloon have if 0.23 mol of He were removed from the balloon? Assume that the temperature and pressure remain constant.
33. A container holds 4.70 L of air at a pressure of 861 torr. If the gas in the container were lowered to standard pressure, what would be its volume? If the pressure of the gas were reduced to 400 torr, what would be its volume? Assume the temperature remains constant.
34. A balloon holds 2.50 L of air at a pressure of 19.5 psi. If the balloon were squeezed to a volume of 1.00 L, what would be the new pressure of the gas? If the balloon were stretched to a volume of 5.00 L, what would be the new pressure of the gas? Assume the temperature remains constant.
35. A balloon containing 1.00 L of CO_2 at 72°F is placed in a freezer at −10°F. What is the new volume of the balloon? If the balloon is placed in the oven at 250°F, what is the new volume of the balloon? Assume the pressure remains constant.
36. The volume of a fixed quantity of gas at 25 °C is changed from 0.750 L to 5.57 L. What is the resulting temperature of the gas if the pressure remains constant? If the volume of the original sample were reduced to 0.150 L, what would be the resulting temperature? Assume the pressure remains constant.
37. A 0.47-mol sample of gas at 37 °C occupies 3.20 L at 2839 mm Hg. What volume would the same gas occupy at STP?
38. Consider this experiment, wherein a gas is heated in a sealed chamber of fixed volume. What is the new pressure of the gas?

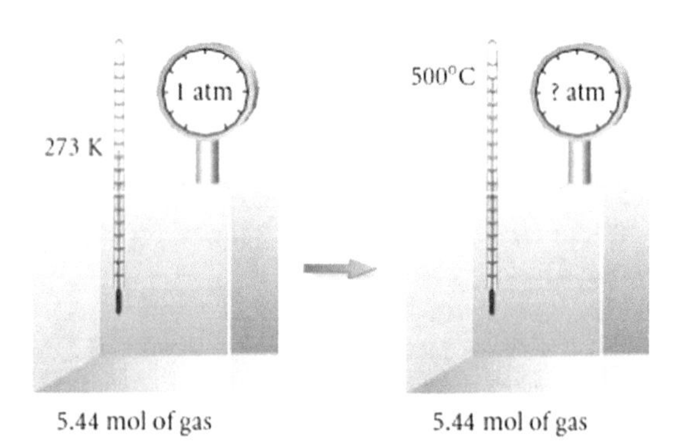

39. A balloon containing 0.15 g of H_2 at STP is pressurized to 32.0 psi. What is the new temperature of the gas inside the balloon? Assume the volume remains constant.
40. A CO gas cylinder possesses a volume of 45.0 L of gas at 2.20×10^3 psi. If the valve breaks and the gas is immediately released from the cylinder, what is the new volume of the gas? Assume the temperature remains constant and standard pressure is attained.

? Chemical Applications and Practice

41. Pressurized N_2O gas can be used to provide the "inflating" power for canned whipped cream. If such a can that is used to provide the topping for a sundae contains 0.58 g of N_2O and corresponds to 0.217 L, how many liters of N_2O will 0.33 g of N_2O fill at the same temperature and pressure?
42. Inside the cylinder of an automobile engine, a mixture of gasoline vapor and oxygen is combusted by a spark. However, the gaseous mixture is first pressurized by the action of a moving piston that decreases the volume of the gas. If the initial volume of the cylinder is approximately 485 mL and the pressure is 0.988 atm, what pressure will be required to compress the gas mixture to 50.0 mL? Assume the temperature remains constant.
43. The molecule nitrogen monoxide (NO) plays several vital roles in animal physiology. Nitrogen monoxide helps regulate blood pressure, influences blood clotting, and also influences the immune system. A researcher studying this important gas isolates a small, 0.150-mL sample at 37 °C. If this sample is cooled to room temperature (25 °C), what volume will the sample have if the pressure remains constant?
44. Because ozone (O_3) has beneficial properties at high altitudes and harmful properties at low altitudes, atmospheric scientists often study it. If a 50.0-mL sample of ozone collected at high altitude has a temperature of – 25 °C is brought back to the lab to study (where the temperature is 25 °C), what will the new volume of the gas be? Assume the pressure remains constant.

Sections 10.5 and 10.6 The Ideal Gas Equation and Its Applications

? Skill Review

45. Determine the value of the ideal gas constant, *R*, to four significant figures if 1.00 mol of gas at 1.00 atm and 273 K occupies 22.4 L. Repeat the calculation, but use each of the following as the unit for the equivalent amount of pressure.
    a. Pa b. mm Hg c. psi
46. A researcher explores the ideal gas equation. He measures the pressure of a system at different temperatures (keeping the volume and moles of the gas constant) and creates a plot of *P* versus *T*. What is the slope of the line that results if the volume is 1.0 L when 1.0 mol of gas is used in the exploration?
47. When inflated with 6.0×10^6 g of helium, a weather balloon has a pressure of 0.955 atm at 22.6 °C. What volume does the balloon occupy?
48. What would be the pressure of H_2 in a tank that has a volume of 25 L, if it contained 45 g of H_2 at 25 °C?
49. What is the volume of 44.0 g of CO_2 at STP?… at 39 °C and 0.500 atm?
50. What is the temperature of 85.0 g of N_2 if the volume is 7.49 L at 850 mm Hg?
51. How many moles of Ar would be found in a 4.33-L balloon at STP?
52. A gas is trapped in a 1.00-L flask at 27 °C and has a pressure of 0.955 atm. The mass of the gas is measured at 1.80 g. Of these, which is most likely to be the identity of the gas: NO, NO_2, or N_2O_5?
53. An unknown gas has a density of 0.600 g/L at 743 mm Hg and 66 °C. What is the molar mass of the unknown gas?
54. What is the density of CO_2 at STP?... of N_2 at STP?
55. At STP, the density of methane (CH_4) is 0.714 g/L. What is the density of methane at 25 °C and 1.15 atm?
56. If a 500.0-mL sample of air at 26.5 °C and 698 mm Hg has a mass of 0.543 g, what is the average molar mass of air?
57. A typical breath may have a volume of 450 mL. If the air you breathed were 21% oxygen gas, how many molecules of oxygen would you inhale at 37 °C and 0.922 atm?
58. If you were able to inhale 2.50 L of air at STP, how many moles of N_2 would you be breathing? Assume that air is composed of 79% nitrogen and 21% oxygen.
59. A gas at 301 K and 0.97 atm has a density of 4.48 g/L. What is the molar mass of the gas?
60. At 27.0 °C and 802 torr, a gas has a molar mass of 62.0 g/mol. What is the density of this gas? What would be the density of the gas if the pressure were changed to 1.00 atm?

61. A 5.00-g sample of a gas has a volume of 85.0 mL at 20.0 °C and 1.00 atm. What is the molar mass of this gas? Would the molar mass change if the temperature were increased to 23 °C?
62. What is the density of CO_2 at 50 °C and 0.44 atm?
63. Determine the density for each of the following gases at STP:
    a. N_2
    b. O_2
    c. CO_2
    d. SF_6
64. Determine the density for each of the following gases at 560 mm Hg and 335 K:
    a. N_2
    b. CO
    c. Br_2
    d. CH_4
65. Determine the molar mass of each of the following gases, given their density at STP. Then, predict the formula for each of the gases:
    a. $d = 1.78$ g/L
    b. $d = 0.0893$ g/L
    c. $d = 1.25$ g/L
66. Determine the molar mass of each of the following gases, given their density at 615 mm Hg and 298 K. Then, predict the formula for each of the gases:
    a. $d = 0.132$ g/L
    b. $d = 2.346$ g/L
    c. $d = 0.992$ g/L
67. Calculate the volume of oxygen gas produced in the decomposition of hydrogen peroxide based on the information given.

$$H_2O_2(aq) \rightarrow O_2(g) + H_2O(l) \qquad \text{(not balanced)}$$

    a. 2.5 g H_2O_2 decompose, O_2 gas collected at STP
    b. 175 g H_2O_2 decompose, O_2 gas collected at 585 mm Hg and 298 K
    c. 0.667 kg H_2O_2 decompose, O_2 gas collected at 300 mm Hg and 125 °C
68. Calculate the mass of calcium carbide that is needed to produce the given amounts of ethylene gas.

$$CaC_2(s) + HCl\,(aq) \rightarrow C_2H_2(g) + CaCl_2(aq) \qquad \text{(not balanced)}$$

    a. 1.0 L at STP
    b. 24.88 L at 595 mm Hg and 298 K
    c. 7.50 mL at 5.5 atm and 123 °C

? Chemical Applications and Practice

69. A student working in the lab forgets to record the temperature at which she obtained 0.675 g of $O_2(g)$. She did, however, report the pressure as 745 mm Hg and the volume as 478 mL. What would have been the temperature at those conditions?
70. Ozone (O_3) in the stratosphere helps reduce the amount of ultraviolet radiation reaching the surface of the Earth. Assuming a temperature of −25 °C and a partial pressure due to ozone of 1.2×10^{-7} atm, how many ozone molecules would be present in 1.00 L of air in the stratosphere?
71. The metabolism of glucose is the main source of energy for humans. This equation shows the overall reaction for the process:

$$C_6H_{12}O_6(s) + 6O_2(g) \rightarrow 6CO_2(g) + 6H_2O(g)$$

    Calculate the volume of CO_2 produced at 37 °C and 1.00 atm when 10.0 g glucose is oxidized.
72. Small portable lighters use compressed butane (C_4H_{10}) as their fuel. When butane is combusted, this reaction takes place:

$$2C_4H_{10}(g) + 13O_2(g) \rightarrow 8CO_2(g) + 10H_2O(g)$$

    a. At 27.0 °C and 0.888 atm, how many liters of oxygen are required for every 1.00 g of butane to react completely?
    b. If the system is compressed to 3.50 atm, keeping the temperature at 27.0 °C, how many liters of oxygen will be required to react completely with 1.00 g of butane?
73. An industrial process can be used to change ethylene into ethanol. The process uses a catalyst, high pressures (6.8 MPa), and temperatures of 3.00×10^3 K. What is the density of ethanol vapor under these conditions?

74. The familiar aroma of garlic comes mainly from the compound diallyl disulfide. Based on these measurements, determine the molar mass of this pungent compound. At 298 K, a 105-mL container holds 0.618 g of diallyl disulfide with a pressure of 0.987 atm.
75. Argon gas is being used in some incandescent light bulbs to extend the life of the tungsten filament. If a light bulb has a volume of 200.0 mL and contains 0.100 g of Ar at 25 °C, what is the pressure inside the bulb?
76. An aerosol can contains carbon dioxide as its propellant. If the can has an internal pressure of 1.3 atm at 25.0 °C, what pressure would it have if the temperature were accidentally raised to 450 °C? (Assume that the can has not yet burst, although this would probably be a very close call.)
77. The Haber process is used to produce ammonia (NH_3) for use in fertilizing corn fields in the Midwest. The reaction is

$$N_2(g) + 3H_2(g) \rightarrow 2NH_3(g)$$

   a. Which, if either, is the limiting reagent when 2.00 L of N_2 at STP is prepared to react with 2.00 L of H_2 at 1120 mm Hg and 21.0 °C?
   b. If 2.00 L of the gases, both at STP, reacted via the Haber process, how many liters of ammonia would be produced?
78. Much of the electricity in the United States is produced from the combustion of coal. However, some of the coal deposits contain FeS_2 (iron pyrite) as a contaminant. During the combustion of coal, this impurity also burns and produces harmful sulfur dioxide gas (SO_2). How many liters of SO_2 are produced from the burning of 1.00 kg of FeS_2 at 758 torr and 275 °C?

$$FeS_2(s) + 2O_2(g) \rightarrow Fe(s) + 2SO_2(g)$$

79. A gaseous sulfur compound causes the pungent aroma of rotten eggs. If 1.60 g of the gas were collected in a 1.00-L vessel at 1065 torr and 89.0 °C, what would be the molar mass of the trapped gas?
80. One common way to obtain metal samples from their impure oxide ores is to react the oxide with carbon. Generically, the equation can be written this way:

$$2MO(s) + C(s) \rightarrow 2M(s) + CO_2(g)$$

   If 5.00 g of an unknown metal oxide (MO) reacted with excess carbon and formed 0.738 L of CO_2 at 200.0 °C and 0.978 atm, what is the identity of the metal?
81. The halogen gases are caustic and toxic. Therefore, they must be handled with extreme care. Which halogen gas has a density of approximately 1.70 g/L at STP?
82. Hydrogen has an isotope known as deuterium (D) that is made up of a proton, a neutron, and an electron. Deuterium is present in measurable amounts in all hydrogen-containing substances. At STP, what are the densities of the three compounds that could exist in a hydrogen gas sample: H_2, HD, and D_2?
83. Neon lights actually do contain neon gas. But from where do we get neon? Very small amounts of Ne are present in air, less than 2.0×10^{-3} % by volume. However, when air is liquefied, the sample can carefully be separated on the basis of boiling points. This must be controlled, but because N_2, O_2, and Ar boil away in the range from 77 to 97 K, neon (with a boiling point of 27 K) can be separated. If a manufacturer collected 1.00 lb of neon at 25 °C and 789 torr, what volume would the Ne occupy?
84. Assume you are pumping air into the tires of your mountain bike before a ride. If the tire volume is 1.50 L at 28.5 °C and 6.55 atm, how many moles of air have you put inside the tire? How many molecules of air are in the tire?

Section 10.7 Kinetic-Molecular Theory

Skill Review

85. Using the tenets of the kinetic-molecular theory, explain why the absolute temperature is directly proportional to the pressure of a trapped volume of gas.
86. Does the kinetic-molecular theory assist in the explanation of Avogadro's law? If so, how?
87. Two gases known for their bleaching power are ClO_2 and Cl_2. Assume you have separate 1-L containers containing 5 g of each at 25 °C

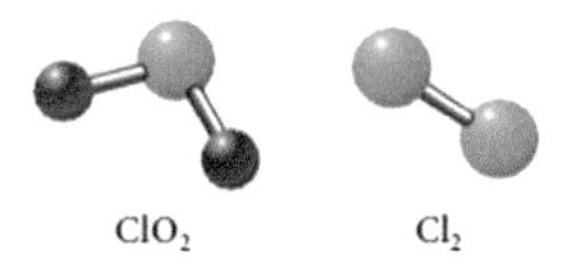

a. Compare the average kinetic energy of the molecules of each gas.
b. Compare the average speed of the molecules of each gas.
c. Which, if either, would be exerting more pressure?
d. If the samples were placed in the same container, which, if either, would have the larger partial pressure?

88. Equal masses of two gases, B_2H_6 and C_2H_2, are placed in separate containers at the same temperature.
    a. Compare the average kinetic energy of the molecules of each gas.
    b. Compare the average speed of the molecules of each gas.
    c. Which, if either, would exert a greater pressure?
    d. If the samples were placed in the same container, which, if either, would have the larger partial pressure?
89. Assuming that a single molecule of each of these gases is moving at a speed of 1100 m/s, what would you calculate as its kinetic energy?
    a. CO_2 b. NH_3 c. Ne
90. Assuming that a single molecule of each of these gases is moving at a speed of 1100 m/s, what would you calculate as its kinetic energy?
    a. H_2 b. C_2H_6 c. Ar
91. Arrange these gases in order from lowest to highest average molecular speed (assume that there are 1.00 mol of each at the same temperature and pressure).
    a. H_2 b. C_2H_6 c. Ar
92. Arrange these gases in order from lowest to highest average molecular speed (assume that there are 1.00 mol of each at the same temperature and pressure).
    a. CO_2 b. NH_3 c. Ne
93. Arrange these gases in order from highest molecular speed to lowest molecular speed. Assume each is at the same temperature.
    a. H_2 b. Ne c. O_2 d. CH_4
94. Arrange these gases in order from highest molecular speed to lowest molecular speed. Assume each is at the same temperature.
    a. N_2O b. H_2O c. COS d. UF_6
95. At 25.0 °C, what would be the root-mean-square speed of propane (C_3H_8) used in bottle gas fuel systems?
96. At extremely low temperatures, molecular motion also becomes relatively slow. What would the temperature be when the root-mean-square speed of He was 100 m/s?
97. Calculate the rms speed of CO_2 and of H_2O at 250.0 K.
98. What is the rms speed of a gas with a molar mass of 16.0 g/mol if the temperature is 45 °C?

Chemical Applications and Practice

99. A device sometimes used in teaching the gas laws consists of small marbles trapped inside a glass-walled container. The container is open at one end, where it is fitted with a moveable piston. Using each of the six points of the kinetic-molecular theory, discuss how this device compares to a gas sample in a container. In addition to the marbles obviously being larger than gas molecules, where does the analogy fit the kinetic-molecular theory and where does it not?
100. During exercise, our bodies heat up as a consequence of the increase in metabolism. Using the kinetic-molecular theory, explain how panting (breathing faster) can help reduce this increase in temperature.

Section 10.8 Effusion and Diffusion

Skill Review

101. Arrange these gases in order from lowest to highest rate of effusion.
    a. N_2 b. SO_3 c. CO_2 d. Xe
102. Arrange these gases in order from lowest to highest rate of effusion.
    a. O_2 b. Ar c. F_2 d. HF
103. A gas effuses at a rate of 0.300 m/s. A second gas has a molar mass of 2.02 and effuses 6.00 times faster than the first gas. What is the molar mass of the first gas?
104. Gas A diffuses 1.47 times as fast as gas B. If gas B has a molar mass of 54.6 g/mol, what is the molar mass of gas A?
105. If a noble gas diffuses 0.317 times as fast as does helium at the same pressure, which noble gas is it?

106. Which would take longer to effuse, 1.00 mol Xe gas or 2.00 mol CO_2 gas? Assume both are initially at the same temperature and pressure. How much faster would 1.00 mol Xe gas effuse if there were 10.0 mol CO_2 gas at the same temperature and pressure?
107. Would it be possible to separate a mixture of propane (C_3H_8) and carbon dioxide (CO_2) using Graham's law of effusion? Explain your answer.
108. Would it be possible to separate a mixture of diborane (B_2H_6) and acetylene (C_2H_2) using Graham's law of effusion? Explain your answer.

Chemical Applications and Practice

109. An interesting visual demonstration to show diffusion rates of gases is to introduce some NH_3 gas at one end of a hollow tube, while simultaneously placing some HCl at the other end. Within a few minutes, the two gases will diffuse through the air inside the tube and react to form a white NH_4Cl precipitate when they meet. If the tube was 39 cm in length, approximately where would the white precipitate form inside the tube?
110. The demonstration described in Problem 109 was repeated, this time using HF instead of HCl. Where would the precipitate form inside the tube?
111. During the Manhattan project, efforts were made to separate ^{238}U from the fissionable isotope ^{235}U. The uranium sample was converted to UF_6, which is a gas at low pressure. How much faster does $^{235}UF_6$ effuse than $^{238}UF_6$?
112. Atmospheric scientists have recently detected very low concentrations of a long-lived greenhouse gas containing sulfur and fluorine, SF_5CF_3. What is the ratio of the rates of effusion of this gas and of its suspected precursor SF_6, a material used in high-voltage insulators?

Comprehensive Problems

113. In the chapter we examine a graph of the relative speed of helium at two different temperatures. Create a plot of the rms speed of He versus temperature in Kelvin. Does this agree with your predictions about the shape of the curve?
114. If two different gases are in the same balloon, which gas will always effuse faster?
115. A particular gas has an effusion rate that is 3.16 times slower than hydrogen gas. This gas also has a dipole moment of 1.82 D. Predict the formula for this gas and explain your reasoning.
116. We mention that in the carbon cycle, carbon dioxide is consumed in photosynthesis to make sugars. What mass of sugar could be produced by photosynthesis if the only source of CO_2 was a single person breathing over the course of one day? (Assume that the person exhales 22,000 times per day and each exhale releases 0.05 L of CO_2 at STP. Also assume that no CO_2 is lost in the process.)
117. Most people can recognize the odor of hydrogen cyanide gas at about 10 ppm in air. The gas is lethal at about 50 ppm in air. Assuming that an air sample at STP containing 10 ppm HCN was composed only of nitrogen gas, what would be the partial pressure of the HCN?
118. Consider two fixed-volume containers, one containing pure O_3 and the other containing pure O_2. If both containers were exposed to UV-B light, in which container would the pressure increase? By how much would the pressure increase?
119. Examine Table 10.2 and answer these questions.
    a. Assuming SATP conditions, how many liters of ammonia gas were produced in 2004?
    b. How many liters of oxygen gas at SATP were produced?
    c. How many metric tons of hydrogen gas were produced at SATP?
    d. How many moles of N_2 gas were produced in the United States in 2004 at SATP?
120. When we exhale, we are releasing carbon dioxide, all of the nitrogen we inhaled, and unused oxygen gas. Under the same conditions of temperature and pressure, rank these three gases in order from
    a. least dense to most dense.
    b. fastest rate of diffusion to slowest rate of diffusion.
121. During exercise, we breathe faster and obtain more oxygen for metabolism. If a handball player absorbs oxygen at a rate of approximately 75 mL per kilogram of body mass per minute, how many molecules of oxygen would a player with a mass of 86.8 kg absorb during a 30.0-min match at SATP?
122. Although the ideal gas equation is sufficient for most gas law calculations, there are times when it needs to be modified. Explain the specific physical purpose behind the van der Waals constants a and b in the modified equation.
123. Hydrogen gas continues to gain attention as a possible fuel for modified cars of the near future. Compare the pressure of H_2 in a tank with a volume of 25 L that contains 45 g of H_2 at 25 °C, using calculations from the ideal gas equation and the van der Waals modification to the gas equation.

124. In a quality test, one tennis ball is filled with N_2 gas. Another tennis ball of the same volume is filled to the same pressure as the first with air. If the two tennis balls were at the same temperature, what else would also be the same?
125. If we collected some oxygen in an experiment at 298 K, how much of a change in the temperature (in kelvins), at constant pressure, would be needed to quadruple the volume? If that same sample of oxygen had its temperature lowered by 25 °C, what volume change, at constant pressure, would be expected?
126. If the CO_2 trapped above your favorite carbonated drink is exerting a pressure of 7.2 atm, yet the total pressure above the aqueous drink is 7.9 atm, what are the partial pressure and percent of water in the trapped space? (Assume that no other gas is present.)
127. The helium in a pressure tank at a carnival may be pressurized to 21 atm. If the temperature of a 27.0-L tank is 29.5 °C, how many grams of helium are in the tank? If it was your job to inflate 1.50-L balloons with the helium, how many could you inflate so that each of them would have 1.11 atm of pressure at 27.5 °C?
128. In 2004, 17.7 billion m^3 of hydrogen gas at STP was produced in the United States, as shown in ◘ Table 10.2. Most of the gas is delivered via trucks, either as a compressed gas or as a liquid stored at very low temperature. If, for example, 50% of the volume of gas produced were liquefied and stored at 20 K, and each delivery truck held 6000 gal of liquid hydrogen, how many truckloads of hydrogen would be needed in one year? (Assume the density of liquid hydrogen is 0.070 g/cm^3.)

Thinking Beyond the Calculation

129. In Problem 128 we noted that hydrogen gas can be transported via tanker trucks either as a compressed gas (at about 400 atm) or as a liquid at 20 K. If you are charged with deciding whether to ship the hydrogen as a compressed gas or as a liquid, what chemical, physical, demographic, economic, and other considerations would factor into your decision? Use the Internet and examine how some large gas production companies make these decisions.
130. Under STP conditions, 100.0 mL of an unknown hydrocarbon gas was combusted in excess oxygen. The only products of the combustion were 300.0 mL of carbon dioxide and 400.0 mL of water vapor.
    a. What are the intermolecular forces common to hydrocarbon gases?
    b. Does the behavior of these gases tend to resemble ideal gas behavior? How might these gases be expected to deviate from this behavior?
    c. How many moles of carbon dioxide are produced in the combustion?
    d. What is the formula of the hydrocarbon?
    e. If the combustion were performed at 2.50 atm and 500 °C, how many liters of water vapor would you expect to produce?

Liquids, Solids, and Intermolecular Forces

Contents

Supplementary Information The online version contains supplementary material available at ▶ https://doi.org/10.1007/978-3-030-90267-4_11.

M. Mosher and P. Kelter, *An Introduction to Chemistry*,
https://doi.org/10.1007/978-3-030-90267-4_11

What We Will Learn from This Chapter

- Many physical properties of a compound can be predicted by considering the intermolecular forces of attraction between compounds (▶ Sect. 11.2).
- London dispersion forces are the weakest of the intermolecular forces (▶ Sect. 11.3).
- Hydrogen bonding is a specific type of dipole–dipole interaction involving a hydrogen atom and atoms of oxygen, nitrogen, or fluorine (▶ Sect. 11.3).
- The phase diagram plots the temperature vs pressure for a compound and allows one to determine if a compound is solid, liquid, or gas under those conditions (▶ Sect. 11.4).
- The vapor pressure of a solvent is dependent upon the temperature (▶ Sect. 11.5).
- Solid compounds can exist as either amorphous or crystalline materials (▶ Sect. 11.6).
- Crystalline compounds arrange their components into a unit cell (▶ Sect. 11.7).

11.1 An Introduction to Water

The Tampa Bay area has a severe water shortage brought about by a fivefold increase in population since 1950. With over 3 million residents and population growth estimated at over 50,000 per year, even the abundant rainfall of Florida's coastal areas cannot keep up with the skyrocketing demand for water. Tampa Bay is not at all unique in the world, or even in the United States, in its thirst for this precious natural resource. Worldwide demand for water has increased by 600% over the past 100 years. Most of the world's increase in water consumption will be driven by population growth and rapid industrialization in Asian countries. Localized dramatic increases are also predicted in other areas with pockets of rapid population growth, such as southern Florida and the southwestern states of Nevada and Arizona. This will further stress the already limited water supplies.

What uses are there for water, and why is it so important in our lives? Domestic use for washing, cooking, and drinking generates part of the demand, but agriculture accounts for most of our consumption of water. Large amounts of water are also used for industrial production, including the generation of electricity. The patterns of water demand also reveal a great deal about the development of a country's economy.

Globally, the major use for water is agriculture. This accounts for as much as 70% of the total water use. And, as the population of the world increases, the demand for food will also increase. The result will be a dramatic in-

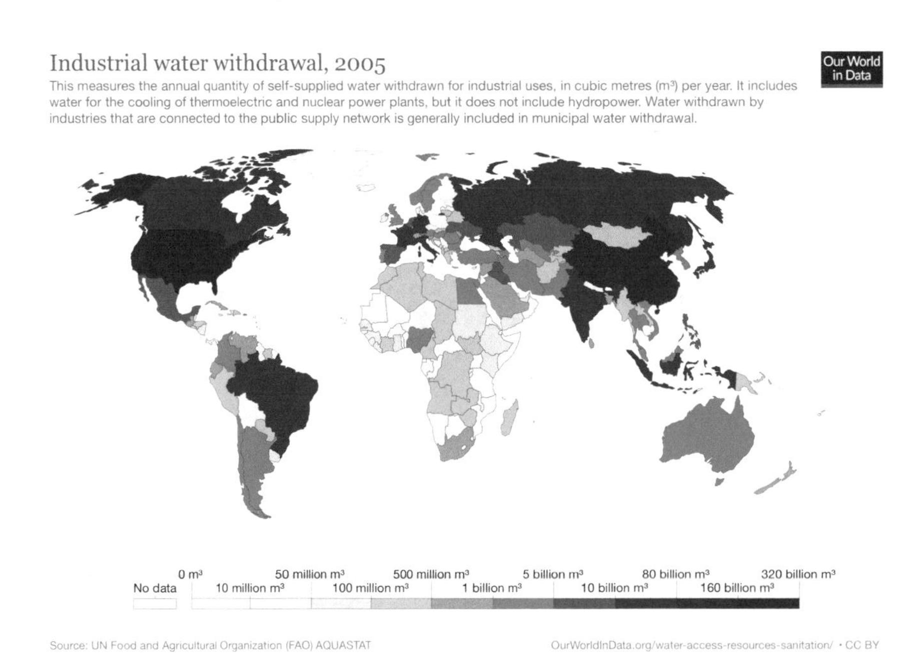

Industrial water withdrawal, 2005
This measures the annual quantity of self-supplied water withdrawn for industrial uses, in cubic metres (m³) per year. It includes water for the cooling of thermoelectric and nuclear power plants, but it does not include hydropower. Water withdrawn by industries that are connected to the public supply network is generally included in municipal water withdrawal.
Our World in Data
No data
0 m³
10 million m³
50 million m³
100 million m³
500 million m³
1 billion m³
5 billion m³
10 billion m³
80 billion m³
160 billion m³
320 billion m³
Source: UN Food and Agricultural Organization (FAO) AQUASTAT
OurWorldInData.org/water-access-resources-sanitation/ • CC BY

Fig. 11.1 *Source* ▶ https://ourworldindata.org/water-use-stress

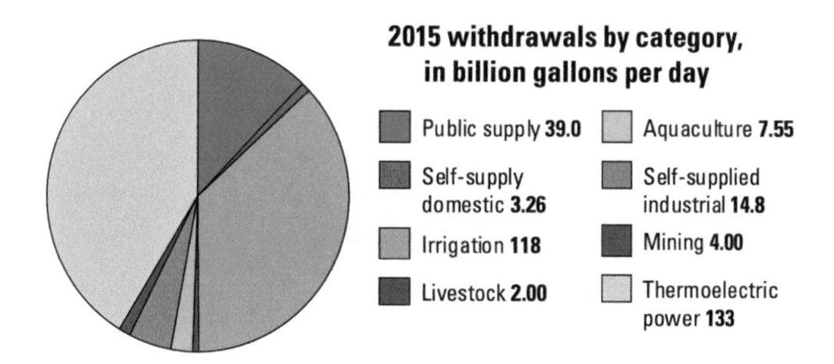

Fig. 11.2 *Source* ▸ https://pubs.usgs.gov/fs/2018/3035/fs20183035.pdf

crease in the need for water. Similarly, industry demand for water will also increase as countries become more advanced. Estimates are that by 2050, Africa will experience increases in industrial water use by as much as 800% over current levels.

Examining these data across the countries of the world, as is done in Fig. 11.1, illustrates that water is used differently in developed countries than in underdeveloped countries. For example, Fig. 11.2 shows that the United States earmarks most of its water supply for power generation and agriculture.

The study of gases and their properties in the previous chapter allowed us to consider the behavior of large numbers of molecules with very little physical interaction between them. In fact, one of the most important assumptions in the concept of the "ideal gas is that the molecules do not interact at all. This is not the case in the liquid and solid phases of matter, which are also called the "condensed phases. In liquids and solids, molecules are much closer together, and are moving much more slowly, than in the gas phase. In the condensed phases, attractions between molecules play a much bigger part in determining the macroscopic properties of the substance.

Because water is the most common liquid that we interact with on a daily basis, we will study water in this chapter as the prototypical liquid, although we will discuss other liquids as well. Water is so fundamental to life that the search for it has extended to the outer reaches of the solar system. Detecting water elsewhere in the universe could provide important information about the chemistry of the universe and, perhaps, the origin of life.

What is it about the structure and properties of water that makes it so pervasive and so vital to our world? How can these properties help us to understand water's many and varied uses—and assist us in our efforts to provide clean water for places like Tampa Bay? This chapter focuses on the nature of water. We will compare it to other liquids in order to highlight its own special character. As we might sense, water is truly unique, vital in so many ways, and it's worth knowing why.

11.2 The Structure of Liquids

When we get thirsty, we can pour ourselves a glass of water. Huge numbers of water molecules stream out of the faucet in the **liquid** state. They fall into our glass and, like the molecules of all liquids, conform to the shape of their container. We can freeze water to make it a solid, which has its own shape, or boil it, forming a **gas**, which completely fills a container no matter how much is present. To make our water hot, we pipe the water into a water heater, which may be operated using another substance vital to our way of life, **natural gas** (mostly methane, CH_4).

Liquid – A phase of a substance characterized by closely held components. The properties of a liquid include a medium density, the ability to flow, and the ability to take the shape of the container that holds it by filling from the bottom up.

Gas – The phase of a substance characterized by widely spaced components exhibiting low density, ease of flow, and the ability to occupy an enclosed space in its entirety.

Natural gas – A complex mixture of gases extracted from the Earth. The major component in many cases is methane.

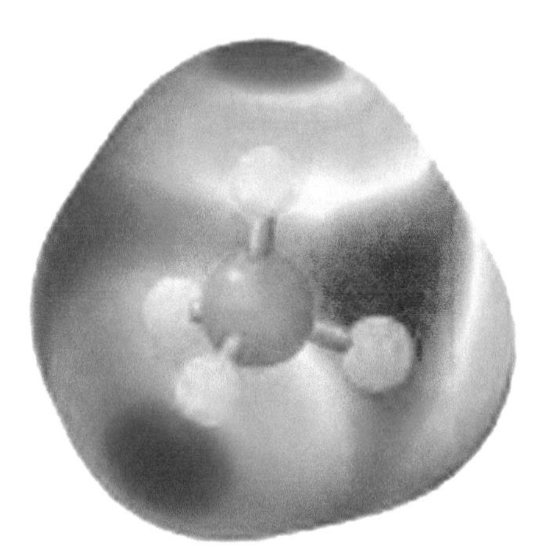

Both water and natural gas arrive at our homes via a system of underground pipes. Even though they possess similar molar masses (18 g/mol for H_2O and 16 g/mol for CH_4), water is a liquid and methane is a gas. Why are some substances liquids and some gases? The properties, chemical behavior, and day-to-day uses of these compounds are based on their structures. As we've noted in previous discussions, the structure of a molecule determines its properties.

Figure 11.3 provides several representations of a water molecule. Recall from ▶ Chap. 7 that oxygen is substantially more electronegative than hydrogen, leading to poles of charge on the individual O–H bonds in water. Using the VSEPR model from ▶ Chap. 8, we saw that water is a bent molecule, with an H–O–H bond angle of 104.5°. This led to our discovery of a net dipole moment for the polar molecule. Unlike water, methane is made up of atoms with similar electronegativity values. Additionally, the VSEPR structure of CH_4 is a perfectly balanced tetrahedron, and the molecule lacks a net dipole moment.

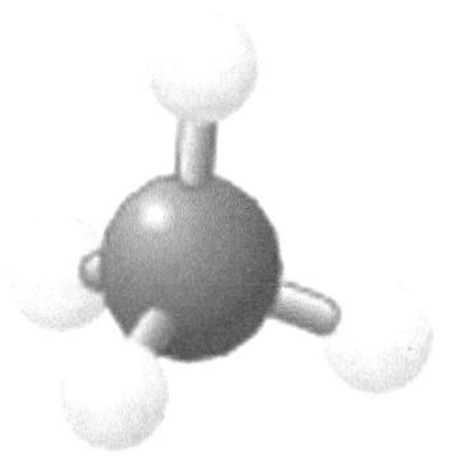

As we've seen in previous discussions, opposite charges attract. This is apparent in the structure of table salt (NaCl), in which the positive charge on the sodium ion is attracted to the negative charge on the chloride ion. But this attraction isn't limited to complete charges. Even slight distortions of electron distribution give rise to this attraction. If we recall the concept of energy as that which is needed to oppose a natural force, we can deduce that energy is released when opposite poles attract. This concept helps us to understand why water is liquid under normal conditions: The hydrogen atoms at positive poles of the dipole moment on water molecules can interact with the negative poles of oxygen atoms on other molecules. The resulting release of energy is the key to the stability of liquid water at room temperature. This interaction is an example of an **intermolecular force**. Unlike water, methane does not have sufficiently strong intermolecular forces to exist as a liquid or solid at normal conditions, so it travels to our homes as a gas.

Intermolecular force – A force of attraction between two molecules.

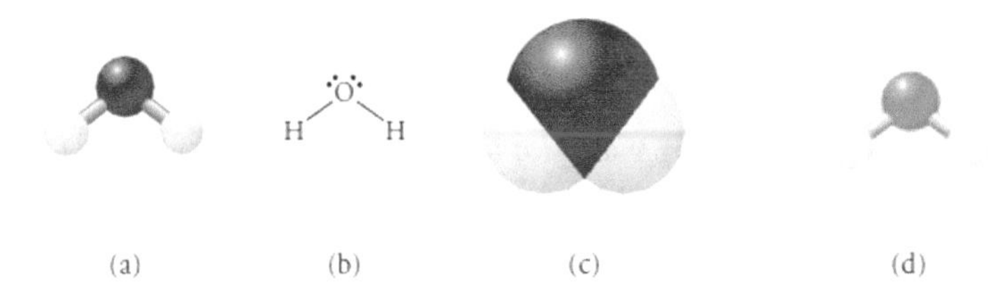

Fig. 11.3 There are several ways to represent a water molecule: **a** ball-and-stick model, **b** Lewis dot structure, **c** space-filling model, and **d** electron dot surface model

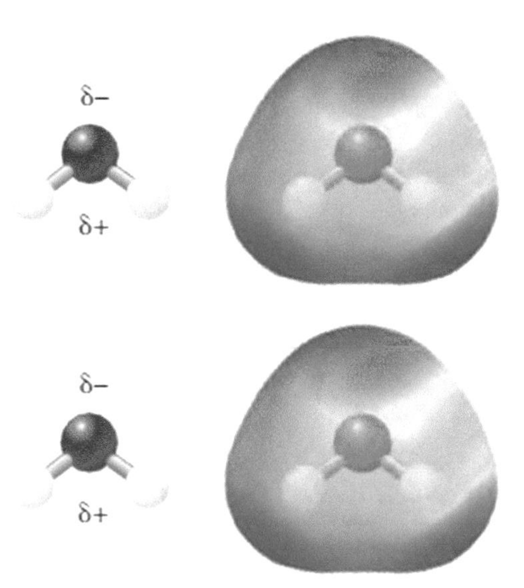

The result of the attraction between opposite poles in a sample of water is the interaction of 3-to-6 water molecules with each other at any one time, with an average of about 4.5. These molecules change partners constantly as they swirl about in the sample exchanging intermolecular forces of attraction. Although the number of interactions among the molecules of a substance does say something about its properties, the relative strength of each individual intermolecular interaction is particularly important.

How do *inter*molecular (between molecules) forces compare with the *intra*molecular (within molecules) forces that we call chemical bonds? Consider what happens when we boil water. We produce water vapor ($H_2O(g)$), but we do not produce H_2 and O_2 gas. If we did, the kitchen would be a very dangerous place, indeed! In boiling water, the individual intramolecular forces (chemical bonds) within each water molecule have not been broken. Instead, the intermolecular interactions between molecules have been disrupted. Why is this so? About 44 kJ of energy is required to convert 1 mol of liquid water to the vapor, via the breaking of intermolecular attractions. However, it takes about 940 kJ to break the O–H bonds within a mole of water. This idea can be reinforced by examining similar forces in a sample of methane. About 9 kJ of energy is needed to vaporize a mole of liquid methane, compared to 1650 kJ to break the four C–H bonds in a mole of methane! We can extend this information to make the more general statement that *intermolecular interactions are much weaker than intramolecular interactions*. Put another way, when we boil a liquid, we are merely giving the molecules enough energy to move much farther away from each other. The physical distance between individual molecules is the major distinguishing factor between the liquid phase and the gaseous phase. Less energy is required to boil a liquid than to break it into its component elements. As we've seen before, overcoming an attraction between two objects, whether it is breaking an intermolecular force or breaking a chemical bond, always requires energy.

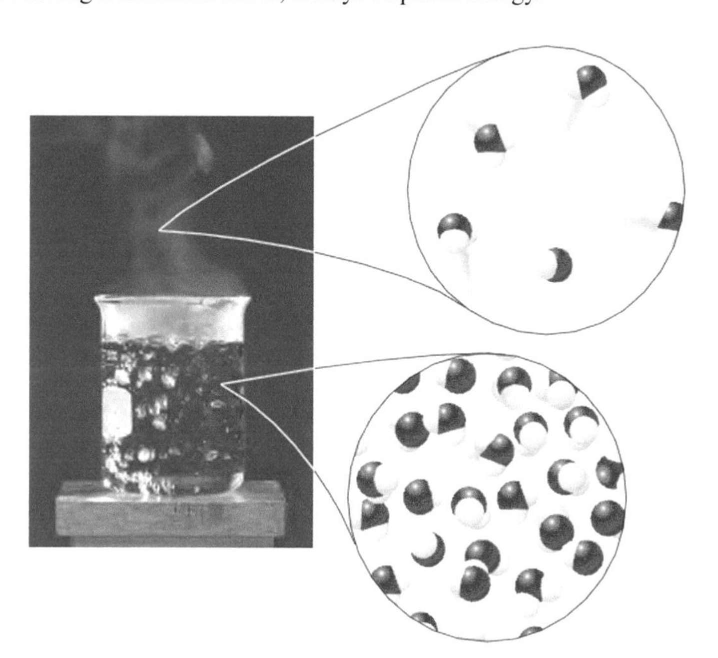

11.3 A Closer Look at Intermolecular Forces

Can we use our understanding of intermolecular forces to explain the nature of liquids other than water? In short, yes; the forces by which molecules come together as liquids are based on the intermolecular interaction of oppositely charged poles. They are collectively known as **van der Waals forces**, after the Dutch physicist Johannes Diderik van der Waals, who noted their existence in 1879 and whose correction for the behavior of real gases we discussed in ► Chap. 10. van der Waals forces can be quite weak, as reflected in the low boiling points (at 1 atm) of methane, −164 °C (109 K), and N_2, −196 °C (77 K). van der Waals forces connecting molecules, such as water, are much stronger. The stronger forces result in a much higher boiling point for water (boiling point = 100 °C at 1 atm). What are these forces and how do they vary among types of molecules?

van der Waals forces – The intermolecular forces of attraction that result in associations of adjacent substances.

11.3.1 London Dispersion Forces—Induced Dipoles

In our comparison of water and methane, we noted that the polarity of water explains why it exists as a liquid at normal temperatures. Yet the ability of nonpolar methane to liquefy at very low temperatures (boiling point = −164 °C) suggests that there is something even in nonpolar substances that can hold molecules together. The molecule octane (C_8H_{18}) is nonpolar, yet it is a liquid at room temperature. It has a boiling point of 125.7 °C, which is higher than that of our small polar water molecule. How can we reconcile these two observations?

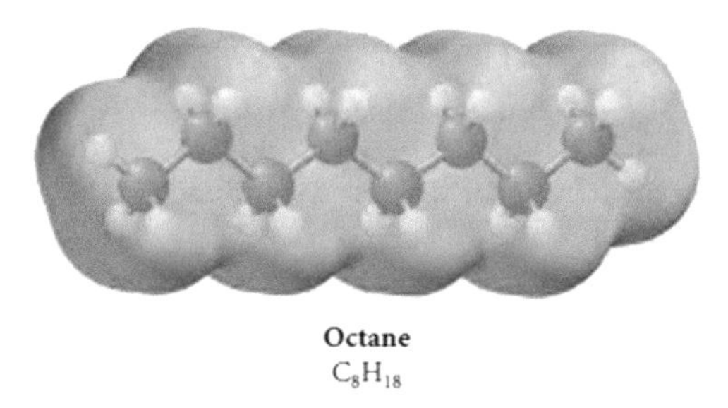

The physicist Fritz London, who defined a concept we now call **London forces**, gave the explanation to this question in 1929. The key to his concept is a property called **polarizability**, the extent to which electrons can be shifted in their location by an electric field. Polarized electrons produce an **induced dipole**, an uneven distribution of charge caused (or "induced) by the electric field. In the same way, atoms or molecules can have temporary dipoles induced by instantaneous distortions in neighboring electron positions, as shown in ◻ Fig. 11.4. The larger the number of electrons in the system, the larger the polarizability. The larger polarizability results in a larger induced dipole moment. In short, stronger attractive forces between molecules are observed when the induced dipole moment is larger.

London forces – The weakest of the intermolecular forces of attraction, characterized by the interaction of induced dipoles.

Polarizability – The extent to which electrons can be shifted in their location by an electric field.

Induced dipole – A dipole produced in a compound as a consequence of its interaction with an adjacent dipole.

Do London forces explain the differences among boiling points of nonpolar substances? ◻ Table 11.1 shows that the boiling points of substances are quite reliably related to their size, such that, for example, fluorine (18 electrons) is a gas at room temperature, whereas bromine (70 electrons) is a liquid and iodine (106 electrons) is a solid. Similarly, ethane (C_2H_6) is a gas, whereas hexane (C_6H_{14}) is a liquid and tetracosane ($C_{24}H_{50}$) is a solid.

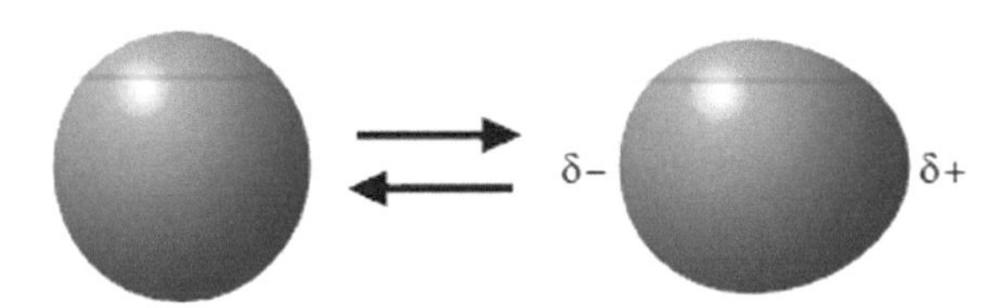

◻ **Fig. 11.4** In this representation, instantaneous induced dipoles are shown as δ+ and δ−

Table. 11.1 Boiling points of compounds at 1 atm

Compound	Name	Molar Mass (g/mol)	Boiling Point (°C)
F_2	Fluorine	38	–188.1
Cl_2	Chlorine	71	–34.6
Br_2	Bromine	160	58.8
I_2	Iodine	254	184.4
CH_4	Methane	16	–164
C_2H_6	Ethane	30	–88.6
C_4H_{10}	Butane	58	–0.5
C_6H_{14}	Hexane	86	69
C_6H_{14}	2,2-Dimethylbutane	86	50
C_6H_{14}	2,3-Dimethylbutane	86	58
$C_{10}H_{22}$	Decane	142	174.1
$C_{24}H_{50}$	Tetracosane	339	391

$CH_3—CH_2—CH_2—CH_2—CH_2—CH_3$
Hexane
bp 68°C

$CH_3—CH_2—C(CH_3)_2—CH_3$
2,2-Dimethylbutane
bp 50°C

$CH_3—CH(CH_3)—CH(CH_3)—CH_3$
2,3-Dimethylbutane
bp 58°C

Fig. 11.5 The "linear" molecule hexane has a higher boiling point than the its isomers in which some polarizable electrons are "hidden"

The structure of a molecule does have some effect on its polarizability. From Table 11.1 we see that the boiling point of hexane is 69 °C. However, 2,2-dimethylbutane, an isomer of hexane (that is, it has the same chemical formula but a different structure) shown in Fig. 11.5, boils at 50 °C, and 2,3-dimethylbutane boils at 58 °C. Why is this so? Polarizability is highly related to distance. Within a molecule of hexane, electrons are relatively close to those of other hexane molecules. On the other hand, some of the carbon atoms in the isomers of hexane are "hidden, farther away from other neighboring molecules, so the polarizability of compact molecules is not as great. Moreover, within a series of molecules of approximately the same molecular mass, those molecules that can interact more with their neighbors have a higher boiling point. Branched molecules lack some of this interaction; linear molecules have more such interaction but, because of their flexibility, do not interact perfectly; cyclic molecules have the most interaction with their neighbors and can stack together like dinner plates. This greater intermolecular interaction leads to higher boiling points.

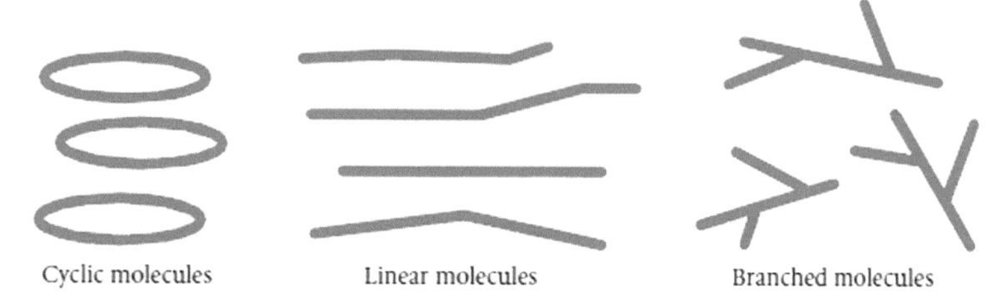

The data in Table 11.1 also show us that the number of electrons in a molecule of nonpolar substances is a major factor in their boiling points. Since more electrons means more protons in the nuclei, then we can judge the number of electrons simply by looking at the mass of the molecule. The impact of the total number of electrons (higher molar mass) on the boiling point is much greater than the effect observed in the isomers of hexanes. By comparing the boiling points of the straight-chain molecules in the table (such as ethane, −88.6 °C; butane, −0.5 °C; hexane, 69 °C; decane, 174.1 °C), we can see that as the molar mass gets larger, the boiling point increases. In fact, with a sufficiently large number of electrons, it is possible for a nonpolar molecule to have a boiling point higher than that of water. Why is it that water itself has such a high boiling point? Water must have additional intermolecular forces of attraction that the nonpolar molecules do not possess.

Example 11.1—Molar Mass, Structure, London Forces, and Boiling Point

One of the following nonpolar substances is a gas at 200 °C. All the others are liquids. Which one has the lowest boiling point at 1 atm, and why?

CH_3—CH_2—CH_2—CH_2—CH_2—CH_2—CH_2—CH_2—CH_2—CH_2—CH_2—CH_2—CH_2—CH_2—CH_3

Pentadecane
$C_{15}H_{32}$

```
        CH3      CH3       CH3
         |        |         |
  CH3—C—CH2—CH—CH2—C—CH3
         |                  |
        CH3                CH3
```

2,2,4,6,6-Pentamethylheptane
$C_{12}H_{26}$

CH_3—CH_2—CH_2—CH_2—CH_2—CH_2—CH_2—CH_2—CH_2—CH_2—CH_2—CH_3

Dodecane
$C_{12}H_{26}$

Asking the Right Questions

What are the most important criteria in determining the relative boiling points of *nonpolar* substances?

Solution

Comparing the 12-carbon compounds, we note that dodecane has a straight-chain structure, whereas 2,2,4,6,6–pentamethylheptane is highly "branched, with many of the atoms hidden from other molecules, so dodecane will have the higher boiling point of the two. Pentadecane is a longer straight-chain molecule than dodecane, so its boiling point will be the highest among the group. The actual boiling points of the compounds are (from lowest to highest):
177.8 °C – 2,2,4,6,6-Pentamethylheptane
216.3 °C – Dodecane
270.6 °C – Pentadecane.

Are Our Answers Reasonable?

The difference in structure (with the same molar mass) led to a difference of nearly 40 °C in the boiling points of 2,2,4,6,6-pentamethylheptane and (linear) dodecane. That is reasonable. The boiling point of pentadecane is higher than both of the other molecules, and that is reasonable, because it is heavier than both.

What If…?

…we have these three organic compounds: nonane (C_9H_{20}), 2,2,4,4-tetramethylpentane (C_9H_{20}), and 2,4-dimethylheptane (C_9H_{20}). The boiling points are: 134 °C, 121 °C, and 151 °C. Match the compound to its boiling point.
(*Ans: nonane= 151 °C; 2,2,4,4-tetramethylpentane= 121 °C; 2,4-dimethylheptane = 134 °C*)

Practice 11.1

Propane (C_3H_8) is a gas at room temperature, hexane (C_6H_{14}) is a liquid, and dodecane ($C_{12}H_{26}$) is a solid. Explain why these three molecules have their specific properties.

Propane C_3H_8 CH_3—CH_2—CH_3

Hexane C_6H_{14} CH_3—CH_2—CH_2—CH_2—CH_2—CH_3

Dodecane $C_{12}H_{26}$
CH_3—CH_2—CH_2—CH_2—CH_2—CH_2—CH_2—CH_2—CH_2—CH_2—CH_2—CH_3

11.3.2 Permanent Dipole–Dipole Forces

The forces that give strength to the interactions among water molecules are a result of the permanent dipoles that exist in water. Water molecules organize to maximize the energetic stability gained by attractions and to minimize the repulsions, as shown in ◘ Fig. 11.6. Dipole–dipole interactions are relatively weak compared to covalent bonds, but they can be quite strong in molecules with relatively large dipole moments. However, even permanent dipoles in a molecule may not lead to significant intermolecular attraction at normal conditions, as with the relatively small molecules HCl and H_2S, which have boiling points of only −85 °C and −61 °C. The degree of dipole–dipole interaction is dictated by the dipole moment of the molecules and the difference in electronegativity of the atoms that are bonded together. For example, there is much less of an electronegativity difference between sulfur and hydrogen, than between oxygen and hydrogen, which is why H_2S does not have interactions as strong as those

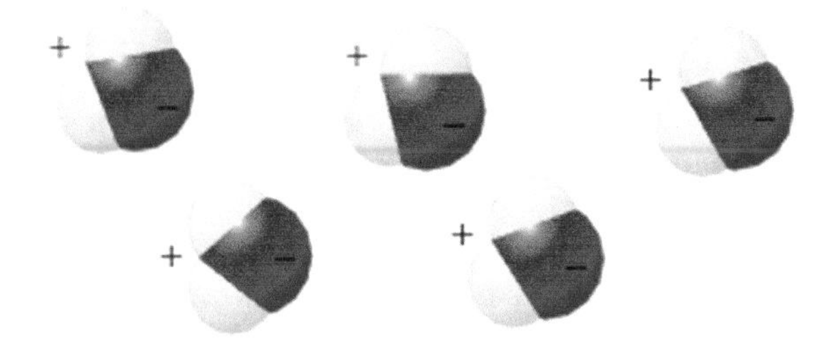

◘ **Fig. 11.6** Interaction of the oppositely charged poles of water molecules makes water a liquid at room temperature

between H_2O molecules. This is why H_2S is a gas at 1 atm and 25 °C, while H_2O is a liquid. As another example, because the electronegativity difference between hydrogen and carbon is very small (only about 0.4 on the Pauling scale), C–H bonds exhibit almost no dipole character at all. This is why the only intermolecular forces present between hydrocarbon molecules, like those in Example 11.1, are London dispersion forces (induced dipoles). Purely covalent molecules, like Cl_2 or N_2 or O_2 have zero dipole moments, so they cannot have any *permanent* dipole–dipole interactions, although they can still have induced dipole attractions between them.

11.3.3 Hydrogen Bonds

One very special type of interaction, the "hydrogen bond, is of great importance. Knowing about **hydrogen bonds**, a term first used in 1912, furthers our understanding of why water is a liquid at room temperature. This knowledge also helps us explain some very elegant and important ideas about protein and DNA structure that add to our collective insight into human biology.

Hydrogen bond – A particularly strong intermolecular force of attraction between F, O, and/or N and a hydrogen atom.

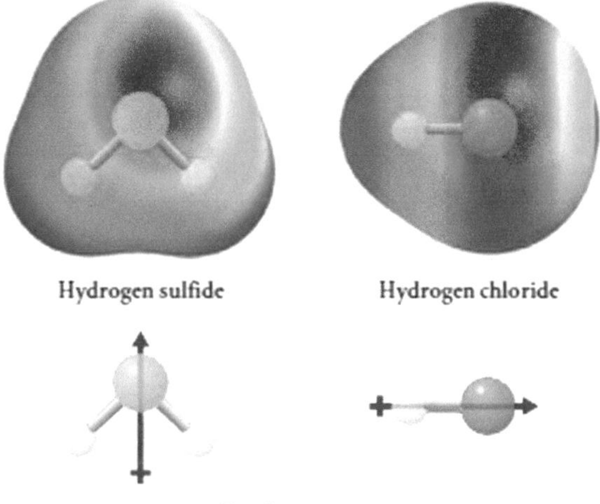

Judging on the basis of its low molar mass alone, we would never have assumed that water is a liquid at room temperature. However, water isn't unique in this regard. Other molecules, such as some of those shown in Fig. 11.7, exhibit boiling points higher than their molar masses would lead us to expect. Why are their boiling points higher than expected? Each is capable of making hydrogen bonds.

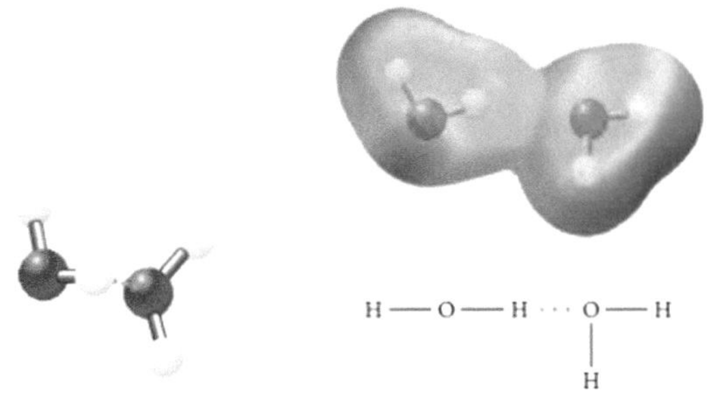

A hydrogen bond is a dipole–dipole interaction of unusual strength compared to other intermolecular forces. It is formed when a hydrogen nucleus (a proton) is shared between two highly electronegative atoms of oxygen, nitrogen, or fluorine. These electronegative atoms interact with the proton through their available lone pair (or pairs) of electrons. In essence, the electronegative atoms play tug-of-war with the proton, creating a very strong intermolecular force of attraction. We can illustrate the hydrogen bond by drawing a dotted line from one of the electronegative atoms to the hydrogen, as illustrated in Fig. 11.8.

Recent data have shown that hydrogen bonds are at least partly—as much as 10%—covalent in nature. Why is a hydrogen bond so strong? The partial positive charge on the tiny hydrogen atom gives it a relatively large charge-to-size ratio. That helps it pack a large attractive punch. This permits a strong interaction with the highly elec-

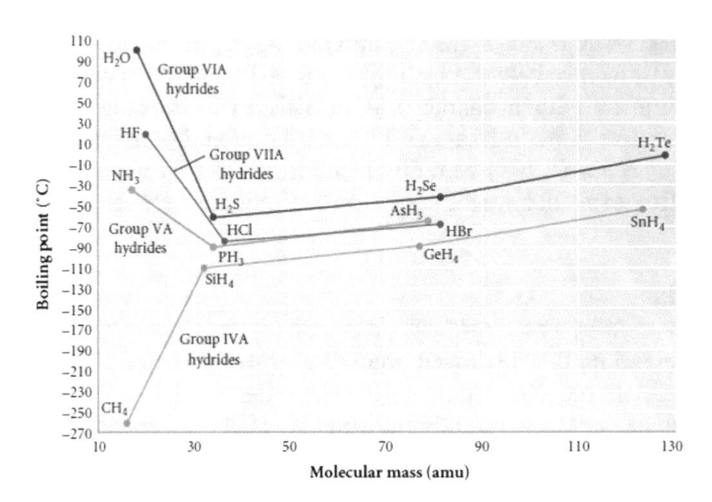

Fig. 11.7 There is a general trend of increasing boiling point with increasing numbers of electrons (and molecular mass), but hydrogen bonds keep water liquid at room temperature. In fact, hydrogen bonds are breaking and re-forming at an incredible rate

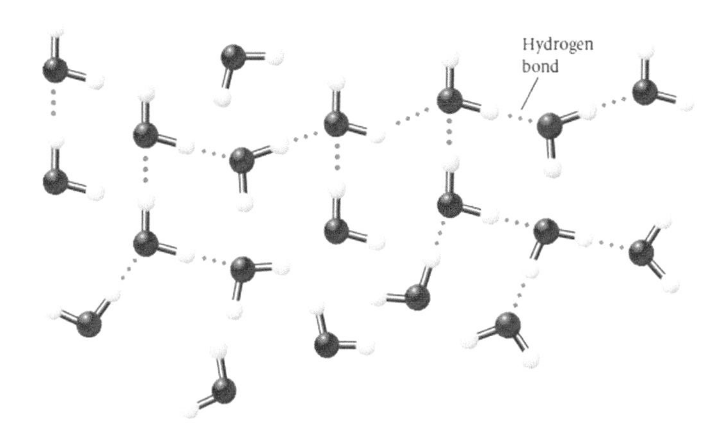

Fig. 11.8 Compounds such as water, ammonia, and hydrogen fluoride have much higher boiling points than compounds of similar size, as a consequence of hydrogen bonding

tronegative fluorine, oxygen, or nitrogen atoms. How strong are hydrogen bonds compared to covalent bonds? The O···H hydrogen bond in liquid water has a measured bond energy of about 23 kJ/mol, compared to the average O–H covalent bond energy of 470 kJ/mol, which means that this hydrogen bond is only about 5% as strong as the intramolecular O–H bond, but stronger than the 9 kJ required to vaporize a mole of methane.

What does our understanding of hydrogen bonding tell us about the properties of water? Hydrogen bonds between neighboring water molecules, as shown in Fig. 11.8, keep water in the liquid state at STP. Hydrogen bonds break and re-form billions of times per second, but on average, enough hydrogen bonds exist at any one time to keep water liquid. A flurry of activity is going on at the atomic level, even in a glass of water resting on a tabletop.

Example 11.2—Comparing the Boiling Points

Ethane (C_2H_6) is an important starting material for the industrial production of polyethylene plastics, used in items such as soft drink bottles. Ethanol (C_2H_6O), second only to water as an industrial solvent, is used in the synthesis of other compounds and in some blends of gasoline. Ethylene glycol ($C_2H_6O_2$) is the main component in conventional automobile antifreeze. Judging on the basis of their structures, arrange these compounds from lowest to highest boiling point.

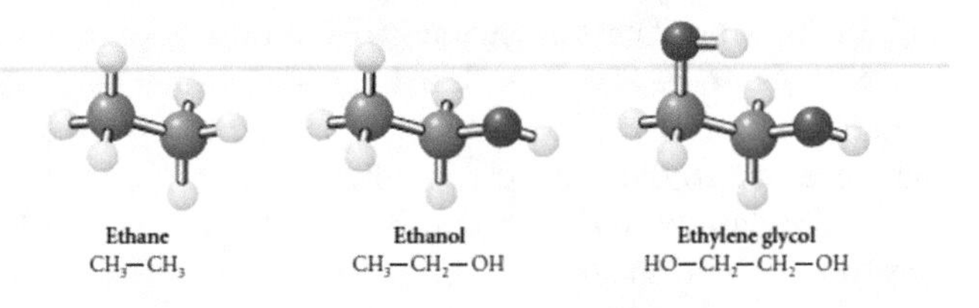

Asking the Right Questions

What structural features in these molecules can raise the boiling point? Especially noteworthy: What is the impact of hydrogen bonds?

Solution

Molecules that have –OH groups are likely candidates for hydrogen-bonding interactions. Hydrogen bonds have a substantial impact on boiling point, especially when the molecule itself is small, with many hydrogen-bonding sites.

Ethylene glycol has two polar sites at which hydrogen bonds can form, compared to one on ethanol and none on ethane. Ethane therefore has the lowest boiling point, −88.6 °C, ethanol is next at 78.5 °C, and ethylene glycol boils at 198 °C.

Are Our Answers Reasonable?

The molecules are similar except for the presence of one –OH group in ethanol and two –OH groups in ethylene glycol. The increase in boiling point that accompanies the increased opportunity for hydrogen bonding with the –OH groups is reasonable.

What If…?

…we have octane (C_8H_{18}) and heptan-1-ol ($C_7H_{16}O$) and these boiling points: 176 °C, 125 °C. Match the compound to its boiling point.

(*Ans: octane* = 125 °C; *heptan-1-ol* = 176 °C).

Practice 11.2

Recent blends of "Green" (environmentally benign) antifreeze solutions use propylene glycol ($C_3H_8O_2$) which is far less toxic than ethylene glycol. Is its boiling point higher or lower than that of ethylene glycol?

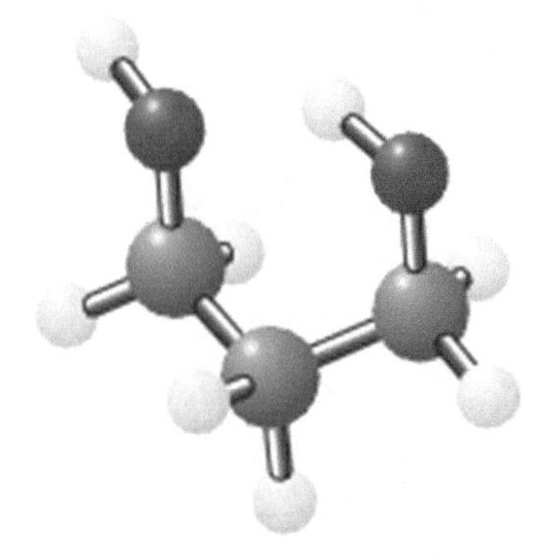

Propylene glycol

$HO-CH_2-CH_2-CH_2-OH$

▶ Here's What We Know so Far

- Whether a material is a solid, liquid or gas at 1 atm and 25 °C is completely dependent on the degree of attraction between the molecules of that material – intermolecular forces.
- Intermolecular forces (also known collectively as van der Waals forces) are the result of attractions between either induced dipoles or permanent dipoles in molecules.
- Non-polar molecules can only form temporary induced dipoles, which briefly attract other molecules. These are also called "London Dispersion forces and are individually quite weak.
- Because all molecules and atoms contain electrons, all of them have London Dispersion forces.
- Polar molecules also attract each other because of their permanent dipoles resulting in the possibility of dipole–dipole interactions.
- When hydrogen is bonded to F, O or N, the resulting dipole is quite strong, resulting in a dipole–dipole interaction referred to as "hydrogen bonding. ◀

11.4 Impact of Intermolecular Forces on the Physical Properties of Liquids: Phase Changes

Why do liquids evaporate, even when they are at room temperature, or even colder? A glass of pure water resting on a tabletop seems to be just that: at rest. But on a molecular level, the system is deliriously active. The water molecules are moving randomly and at great speed. Intermolecular hydrogen bonds are being broken and reformed at an incredible rate. Even at relatively low temperatures, water molecules on the surface of the liquid have enough energy to break free of their hydrogen bonds and go into the vapor state. With higher temperature comes increased average kinetic energy of the system. That means even greater molecular motion, more breaking of hydrogen bonds, and more molecules leaving the surface in the process we call **evaporation**. Some of the surrounding molecules return from the vapor to the liquid, attaching to the surface via hydrogen bonding, in the process of **condensation**. The corresponding escape of molecules from a solid (such as ice) to the vapor is called **sublimation**, and their return from the vapor to the solid is known as **deposition**.

Evaporation – The process of a liquid undergoing a phase transition to a gas; the opposite of condensation.
Condensation – The process of a gas undergoing a phase transition to a liquid; the opposite of evaporation.
Sublimation – The process of a solid undergoing a phase transition to a gas; the opposite of deposition.
Deposition – The process of a gas undergoing a phase transition to a solid; the opposite of sublimation.

Issues and Controversies—Using Phase Changes to Clean Water

Most campers have faced this dilemma—how do we make our water safe to drink? While most of us in the United States never worry about this problem, many people in the world have to ask this question on a daily basis. Water can be degraded with soil, waste, and even dissolved ions and molecules. The result is water that at the very least isn't pleasant to drink and in many cases even toxic. One way to accomplish purification of water is through the power of phase changes.

Water can be boiled and the vapor collected and re-condensed into liquid. The vast majority of all contaminants in this process, known as **distillation**, are removed and the water can be made safe to drink. Another phase change that can be used to make water safe is that of freezing the water to change it from a liquid to a **solid**. When water molecules align to form the crystalline structure of solid water, they push most foreign materials out of the forming ice crystal in this process called **freeze-thawing**, so that frozen water is cleared of most dissolved contaminants and is safe to consume. These processes, distillation and freeze-thawing, can be repeated many times to eliminate nearly every contaminant from the water.

Distillation – The process of boiling a liquid and condensing the vapor into liquid.

Freeze-thawing – The process of freezing and melting a substance. The solid form of the substance can be separated from impurities before melting into a liquid, thus purifying the liquid. The process also removes dissolved gasses from a liquid.

Also known to most outdoor sports enthusiasts is that rain and snow (as long as there is no visible contamination) can sometimes be safe to consume. This is because precipitation, such as rain and snow, results from the natural distillation (evaporation and condensation) of water. The result is something that is relatively free of many harmful contaminants. The old country wisdom that "there's no such thing as dirty ice is practically a scientific fact!

With appropriate equipment and education, many of the world's water issues could be mitigated by taking advantage of naturally occurring rain and snowfall. However, the major problem facing most areas in need of clean drinking water is not a lack of water, but a lack of the energy sources required to clean the water that is available, even by such simple methods as distillation or freeze-thawing. To address this issue, as well as other energy related needs, many developing nations are investing billions of dollars in renewable energy sources.

11.4.1 Vapor Pressure

If we were to let our glass of pure water sit long enough, more molecules would evaporate than would condense, and we would be left with an empty glass. Let's change the set-up by enclosing the water in a sealed flask, as shown in Fig. 11.9. How will the behavior of the system change?

Fig. 11.9 How will the behavior of the liquid and vapor change if we use a stoppered (sealed) flask?

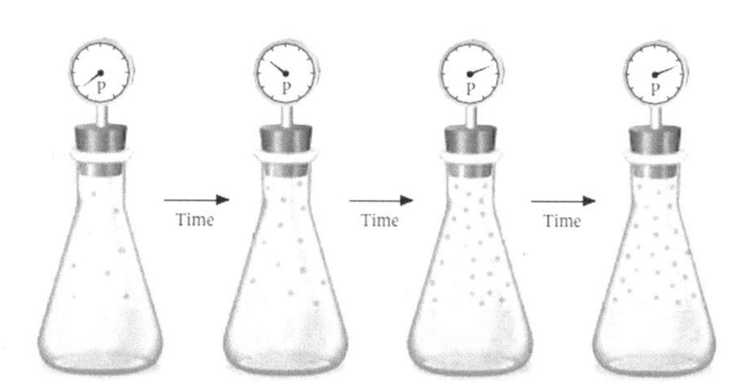

Fig. 11.10 In a sealed flask, the system will come to equilibrium, with the rate of evaporation being equal to the rate of condensation

Initially, those water molecules with sufficient energy will break free of their intermolecular hydrogen bonds and escape from the surface. The resulting vapor will collide with air molecules in the flask and with the walls of the container many times each second, exerting a force per unit area that we measure as **pressure**. Recall that a pressure of 760 torr equals 1 atm.

Pressure – The force exerted per unit area by a gas within a closed system.

In our sealed flask, as illustrated in Fig. 11.10, the more vapor that exists, the more collisions per second with the surroundings and the greater the pressure. As the concentration of vapor builds up in the flask, some of it will condense. Eventually, the air in the flask will hold as much vapor as it possibly can, and, as might occur on a sultry summer afternoon, the air will become **saturated** with water vapor. At this point, the rate of evaporation from the surface of the water will equal the rate of condensation back to the surface. When the rates of the forward (evaporation) and reverse (condensation) processes are equal, they are said to be in **equilibrium**. We indicate that both evaporation and condensation can occur by using double arrows.

Saturated – When considering a pure substance—the surrounding atmosphere contains the maximum possible amount of vapor, expressed as the vapor pressure.
Equilibrium – In a reaction or process, a condition wherein the rates of the forward and backward reactions are equal. The amounts of the reactants and products do not change, but the forward and backward reactions are still proceeding.

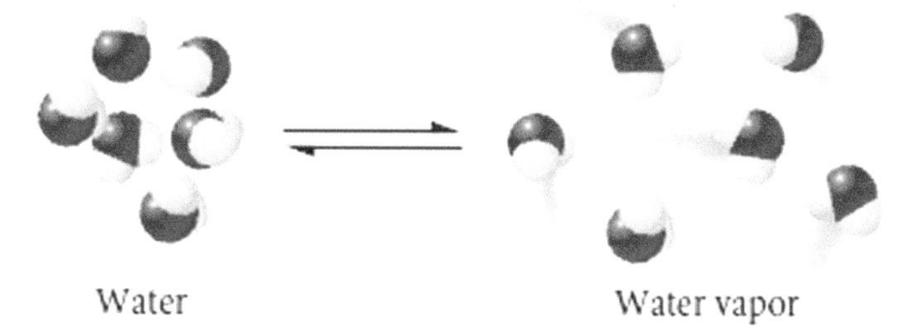

As with every process we study, bond breaking and bond forming are always vigorously occurring, even though to our eyes all seems quiet. We often indicate that the molecular-level equilibrium is dynamic, rather than static, by calling it **dynamic equilibrium**. The pressure exerted by a vapor in equilibrium with its liquid is called the **equilibrium vapor pressure**, or, commonly, the **vapor pressure**, of the liquid at a given temperature. The vapor pressure of hot water at 80 °C is 355.1 torr; that of cool water at 20 °C is 17.5 torr; this means that more water molecules are in the gas phase at relatively higher temperatures than at lower temperatures. The vapor pressure of a liquid increases with temperature. Figure 11.11 and Table 11.2 show this dependence for water. When one substance has a higher vapor pressure than another, it is said to be more **volatile**. Substances with higher volatility have higher vapor pressures.

Dynamic equilibrium – A term sometimes used in chemistry as synonymous with equilibrium to emphasize that molecular-level equilibrium is not static.

Equilibrium vapor pressure – The pressure exerted by the vapor of a liquid or solid under equilibrium conditions.

Vapor pressure – The pressure of vapor above a liquid in a closed system.

Volatile – Having a higher vapor pressure.

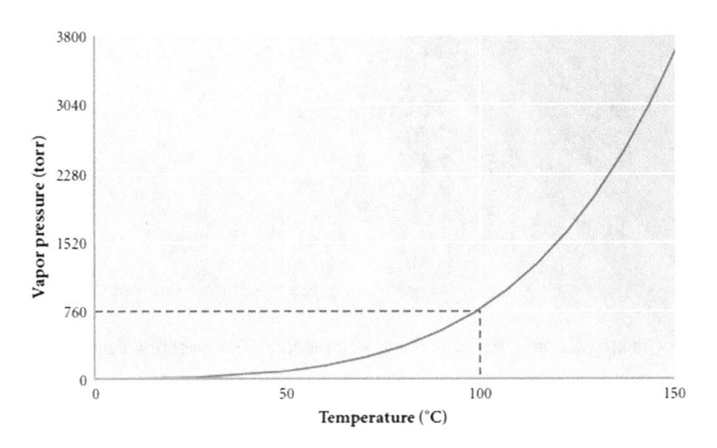

Fig. 11.11 The vapor pressure of liquids is related to temperature, as shown for water. The line on the graph indicates the boiling point of water at 1 atm or 760 torr

Table. 11.2 Vapor pressure of water at selected temperatures

Temperature (°C)	Vapor pressure (torr)
0	4.58
10	9.21
15	12.8
20	17.5
21	18.7
22	19.8
23	21.1
24	22.4
25	23.7
26	25.2
30	31.8
35	41.2
40	55.3
50	92.5
60	149.4
70	233.7
80	355.1
90	525.8

Does this make sense at the molecular level? A higher system temperature means higher average kinetic energy. Water molecules are moving faster than at lower temperature, so more molecules have enough energy to break their hydrogen bonds to neighboring molecules and evaporate. Those faster moving molecules in the vapor phase can now bounce off the sides of the container, increasing the pressure.

Example 11.3—Comparing Vapor Pressures

Which substance, water or methanol (CH_3OH), would have a lower vapor pressure at 50 °C? Explain the choice.

Asking the Right Questions

What are the structural features that relate to vapor pressure? Which of these are present in this molecule?

Solution

The key structural feature related to vapor pressure is the presence of an $-OH$ group, which can take part in hydrogen bonding. The molecule that has more extensive hydrogen bonding will have the lower vapor pressure at a given temperature. The entire structure of water encourages hydrogen bonding. Methanol can also hydrogen-bond, but it also has nearly nonpolar C—H bonds. The vapor pressure of water is 93 torr at 50 °C, compared to about 400 torr for methanol.

Is Our Answer Reasonable?
The vapor pressure of a substance is dependent upon attraction among its molecules. Hydrogen bonding sharply increases this attraction. Water is structurally well-suited for extensive hydrogen bonding, so its lower vapor pressure than methanol is reasonable.

What If…?
…we wanted a molecule that contained four carbons, and had a very high vapor pressure? What molecule might work well?
(Ans: There are several answers, with the key structural feature being polar bonds such as – OH. One example is butanetetrol, which has a vapor pressure of over 1500 torr at 20 °C, compared with 1,3-butanediol (two – OH groups), which has a vapor pressure of 8 torr at 20 °C).

Practice 11.3
Name a substance that would have a lower vapor pressure at 50 °C than either water or methanol. Explain the reasoning.

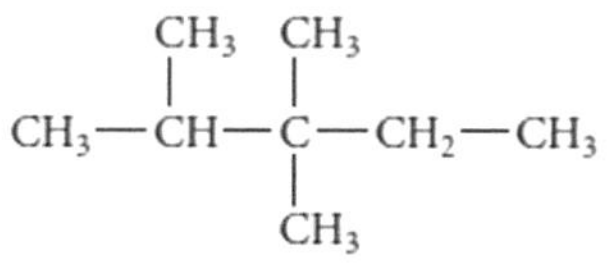

$$CH_3-CH(CH_3)-C(CH_3)_2-CH_2-CH_3$$

2,3,3-Trimethylpentane

$$CH_3-CH_2-CH_2-CH_2-CH_2-CH_2-CH_2-CH_3$$

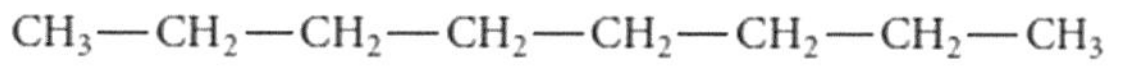

Octane

Fig. 11.12 The London forces are greater in octane than in the highly branched 2,3,3-trimethylpentane. Which would require a higher temperature to reach a vapor pressure of 400 torr?

We now know that the ability to form hydrogen bonds is one important factor leading to lower vapor pressure at a given temperature, but it is not the only factor. Numbers of electrons and structure are also important because, as we have already learned, London forces can be significant in relatively large molecules. In order to develop a vapor pressure of 400 torr, octane (C_8H_{18}) must be heated to 104 °C. The branched molecule 2,3,3–trimethylpentane, shown in Fig. 11.14, has this vapor pressure at 92.7 °C, and water reaches it at 83.0 °C (Fig. 11.12).

Example 11.4—An Implication of Vapor Pressure

Acetone (also known as propanone, C_3H_6O) is shown below. It is a polar compound with a variety of industrial uses as a solvent and in the manufacture of plastics and pharmaceuticals. It occurs naturally in plants and animals. If acetone and water were both accidentally spilled on a lab bench, which would be likely to completely evaporate first at 20 °C?

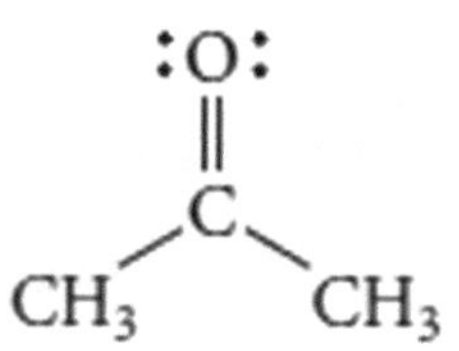

$$CH_3-C(=\ddot{O}:)-CH_3$$

Asking the Right Questions
What do we mean by "completely evaporate? What contributes to a high vapor pressure? Which molecule, water or acetone, has those key structural feature or features?

Solution
To "completely evaporate means to fully change from the liquid to the gaseous state. Even though complete evaporation is not an equilibrium process (that is, molecules are leaving the liquid's surface more rapidly than they are condensing back), we can use the vapor pressure as a guide. Regarding contribution to a high vapor pressure, a small molecule, even a polar one, that cannot form hydrogen bonds to itself will probably have a higher vapor pressure than that of water. Even without knowing the actual vapor pressure or boiling point of acetone, we can tell that it will probably evaporate faster than water.

Is Our Answer Reasonable?
Acetone at 20 °C has a vapor pressure of about 185 torr, much greater than water's 17.5 torr. Acetone (boiling point = 57 °C) will completely evaporate first. Water has a remarkably low vapor pressure for such a low molar-mass

molecule. These data are consistent with the discussion about structure and vapor pressure, and the answer is therefore reasonable.

What If…?

…we spilled decane, $C_{10}H_{22}$, and acetone on a lab bench? Which would evaporate first at 20 °C?

(Ans: Acetone would evaporate first. The vapor pressure of decane at 20 °C is about 3 torr. Even though decane is not polar, it has many more electrons than acetone and London forces will keep it from evaporating as quickly as acetone)

Practice 11.4

Compare acetone and hexane (C_6H_{12}) using the same lab bench scenario as in Example 11.4.

The relationship between temperature and vapor pressure is expressed mathematically by the **Clausius–Clapeyron equation**:

$$\ln P = \frac{-\Delta_{\text{vap}}H}{RT} + C$$

In this equation, P is the vapor pressure of the substance, $\Delta_{\text{vap}}H$ is the **enthalpy of vaporization** and C is a constant. As always, R and T are the ideal gas constant and the temperature. Therefore, this equation assumes ideal gas behavior, which means it is only approximate for any real gas. Plotting pressure vs. temperature, as we have done in Fig. 11.11, results in a curve, which is difficult to deal with when trying to predict future values or trends. Using the Clausius–Clapeyron equation allows us to produce a straight line instead of a curve by plotting ln P versus $1/RT$, where R is the gas constant (8.3145 J/mol·K) and T is the temperature in Kelvin. If we do this, the slope of the line will be $-\Delta_{\text{vap}}H$ and the y-intercept will be the constant, C. This equation and the resulting plot are quite powerful, because it allows us to predict values for which we have no actual data. For instance, we can use this equation to compare two specific combinations of vapor pressure and temperature for a particular substance. If we have the plot for a particular compound, we can determine the vapor pressure at any temperature we want.

11

Clausius-Clapeyron equation – The relationship between vapor pressure and the enthalpy of vaporization.
Enthalpy of vaporization ($\Delta_{\text{vap}}H$) – The enthalpy change associated with the phase transition from a liquid to a gas.

How do we use the Clausius–Clapeyron equation to arrive at a plot that gives us a straight line? Subtracting the equation at one temperature, T_1, (with its corresponding vapor pressure, P_1) from the equation at another temperature, T_2, with its vapor pressure, P_2, gives the following equation. Note that R and C don't change in the equations, because they are constants:

$$\ln P_2 - \ln P_1 = \left[\frac{-\Delta_{\text{vap}}H}{RT_2} + C\right] - \frac{-\Delta_{\text{vap}}H}{RT_1} + C$$

This equation condenses to:

$$\ln\frac{P_2}{P_1} = \frac{-\Delta_{\text{vap}}H}{R}\left[\frac{1}{T_2} - \frac{1}{T_1}\right]$$

Using the combined form of the equation allows us to directly compare the vapor pressure of a substance at two different temperatures, and all we need to know is the enthalpy of vaporization of the substance and the temperatures that we want to compare.

Example 11.5—Using the Clausius–Clapeyron Equation

Benzene, C_6H_6, is an important organic compound that we introduced earlier. Here are data for the vapor pressure of benzene at each temperature listed. What is the heat of vaporization?

Temperature (°C)	Pressure (torr)
0	25.3
20	75.6
30	120
40	184
70	547
90	1020

Asking the Right Questions

How can we use the temperature and pressure data to find the heat of vaporization?

Solution

According to the Clausius–Clapeyron equation,

$$\ln P = \frac{-\Delta_{vap}H}{RT} + C$$

As discussed in the text, plotting the natural log of the pressure, ln P, against $1/RT$ gives a straight line with a slope of $-\Delta_{vap}H$.

$$y = mx + b$$

$$y = \ln P;\quad m = -\Delta_{vap}H;\quad x = \frac{1}{RT};\ b = C$$

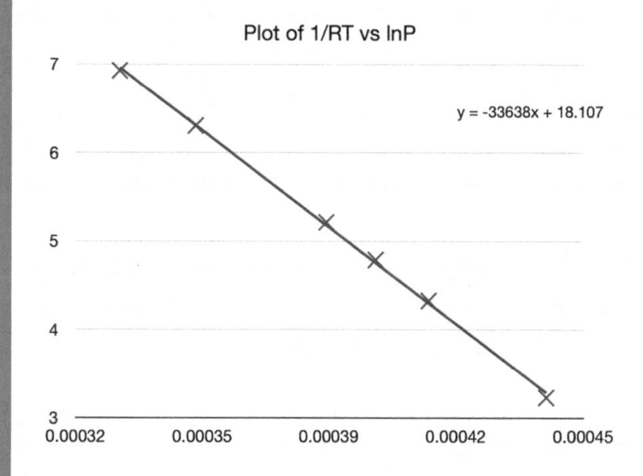

The slope of the line, $-\Delta_{vap}H = -33.6$ kJ, so the heat of vaporization for benzene is positive; $\Delta_{vap}H = 33.6$ kJ.

Is Our Answer Reasonable?

One way to decide is to go on-line and look up the data, not only for benzene, but for many organic compounds. When we do that, we note that the data are reasonable. We can also confirm that our answer is reasonable because it is a positive value and we expect that it will require energy to vaporize the liquid benzene into gas.

What If…?

…instead of plotting $1/RT$ versus ln P, we used log P? How would that affect our slope and $\Delta_{vap}H$?

(Ans: The slope would be reduced by 2.3026, (the ratio of a log to an ln of a number. We would need to multiply our slope by 2.3026 to get the $\Delta_{vap}H$*)*

Practice 11.5

Here are temperature and pressure data for tetrafluoromethane, CF_4, a low-temperature refrigerant. Determine $\Delta_{vap}H$ for CF_4.

Temp (K)	Pressure (torr)
125	19.3
140	70.2
155	191
170	425
185	824
200	1460

Example 11.6—Calculating Vapor Pressure Via the Clausius–Clapeyron Equation

The vapor pressure of propan-1-ol (also known as n-propanol) at 14.7 °C is 10.0 torr. The heat of vaporization is 47.2 kJ/mol. Calculate the vapor pressure of propan-1-ol at 52.8 °C.

Asking the Right Questions

How can the heat of vaporization help us find the vapor pressure at the higher temperature? What is the role of the Clausius–Clapeyron equation?

Solution

The Clausius–Clapeyron equation relates the vapor pressure to changes in temperature, given the heat of vaporization. Let's list what we are given.

$$\Delta_{vap}H = 47.2\ \text{kJ/mol}.$$

$$R = 8.314\ \text{J/K}\cdot\text{mol} = 0.008314\ \text{kJ/K}\cdot\text{mol}.$$

$$T_1 = 14.7\ °\text{C} = 287.9\ \text{K}.$$

$$T_2 = 52.8\ °\text{C} = 326.0\ \text{K}.$$

$$P_{vap,T_1} = 10.0\ \text{torr}.$$

$P_{vap,T_2} = x.$

Substituting,

$$\ln\left(\frac{x}{10.0}\right) = \frac{-47.2\ \text{kJ/mol}}{0.008314\text{kJ/mol}\cdot\text{K}}\left(\frac{1}{326.0\ \text{K}} - \frac{1}{287.9\ \text{K}}\right)$$

$$\ln\left(\frac{x}{10.0}\right) = -5677(-4.06\times10^{-4})$$

$$\ln\left(\frac{x}{10.0}\right) = 2.305.$$

Raising both sides to the power of e, gives:

$$\frac{x}{10.0} = e^{2.305}$$

$$e^{2.305} = 10.02$$

$$x = P_{vap,T_2} = 100.2 = 100\ \text{torr}.$$

Is Our Answer Reasonable?

The vapor pressure rose, but not dramatically. That is reasonable. Another way to see if the answer is reasonable is to find the boiling point of 1-propanol. The figure is available on reputable Internet sites as 97 °C. Our answer for pressure of 100 torr at 52.8 °C is reasonable.

What If…?

…we wanted a sample of 1-propanol at a vapor pressure of 1850 torr? What would be the temperature of the sample?

(Ans: 392 K = 119 °C)

Practice 11.6

The vapor pressure of dimethyl ether, C_2H_6O, at −18 °C is 133 torr. The heat of vaporization is 21.5 kJ/mol. Calculate the vapor pressure of dimethyl ether at 22 °C.

▶ **Here's What We Know so Far**

- The vapor pressure of a liquid increased with temperature.
- More hydrogen bonding leads to lower vapor pressure.
- All other things being equal, heavier molecules have lower vapor pressures than lighter molecules.
- Straight-chain molecules have lower vapor pressures than their branched isomers.
- When several of these factors come into play, it is hard to predict which will dominate. We then run experiments or look up information in data tables.
- The vapor pressure of a liquid increases with temperature as described by the Clausius–Clapeyron equation. ◀

11.4.2 Boiling Point

Let's return to our glass of pure water. As we heat it, the vapor pressure of the liquid increases along with the temperature. If we are at sea level on a day when the surrounding pressure is 1 atm, the liquid will start to bubble from within as the temperature approaches 100 °C. As it does so, the vapor pressure of the liquid will edge ever closer to the atmospheric pressure. When the temperature reaches 100 °C, bubbles burst forth throughout the water in a familiar phenomenon we call **boiling**. We discussed boiling earlier in the chapter, and we understand its general meaning from all the years we have been boiling water to make tea or cook vegetables. Now we are ready to look at it in more depth. Boiling is not just a surface process like evaporation, because it involves the entire liquid. We define the **boiling point** as the temperature at which the pressure of the liquid's vapor (rather than the vapor pressure, which is defined for an equilibrium process) is equal to the surrounding pressure. If that pressure is 1 **atm**, at or near which so many of life's normal activities take place, the temperature at which a liquid boils is called its **normal boiling point**.

Boiling – The process that occurs when the vapor pressure of a liquid is equal to the external vapor pressure.

Boiling point – The temperature at which the pressure of a liquid's vapor is equal to the surrounding pressure.

Normal boiling point – The temperature at which a liquid boils when the pressure is 1 atm.

A good portion of the world's population does not live at sea level, and for them, "normal is anything but. In mile-high Denver, Colorado, at 1609 m (5280 ft), the atmosphere is less dense than at sea level, so the atmospheric pressure is correspondingly lower, about 0.82 atm (620 torr). If our glass of water were heated in Denver, it would boil at about 95 °C. The difference in boiling point with pressure is even more dramatic in Mexico City, at 2240 m (7340 ft), where the atmospheric pressure is only 0.76 atm (580 torr). The boiling point of water at that altitude is only about 90 °C. ◘ Figure 11.13 shows the decrease in the boiling point of water as the altitude increases and atmospheric pressure consequently decreases. Food manufacturers take advantage of the increase in boiling point with pressure when they process foods by canning them at high pressure, allowing the food item to be heated to a relatively high temperature, typically 107 °C, for at least 3 min, killing any bacteria within. A quick way to cook soup is to use a pressure cooker, which increases the pressure within from 1 to 2 atm, raising the boiling point of

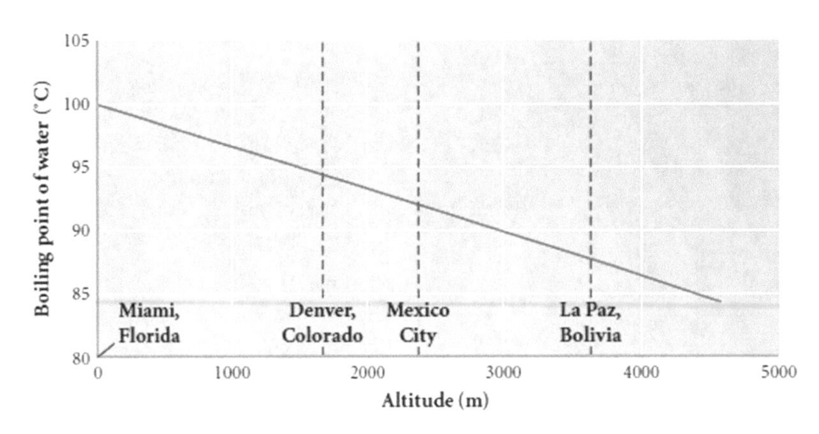

◘ **Fig. 11.13** The boiling point of water goes down as the altitude rises. This is because air pressure is lower at higher altitude (remember the definition of boiling point)

the soup to about 120 °C (250 °F). Commercial cake mixes often have two sets of instructions: one for cooking the batter at sea level (normal boiling point) and one for higher altitudes.

11.4.3 Heating Curves

We opened this chapter by looking at drinking water in the Tampa Bay area. Now let's look north, perhaps above the Arctic Circle, to a group of hikers who want to get drinking and washing water by melting some ice and then purifying it by boiling. What happens as they heat a 10.0-kg block of the ice at −10.0 °C? To fully understand the process that occurs, we must consider four changes: (1) warming the ice; (2) melting the ice; (3) heating the water, and (4) boiling the water. As we proceed, think about what happens to the energy we add, the molecular motion, and the resulting system temperature. We will assume that all of the heat goes into the ice (the system), not into the air (the surroundings).

Change 1: Warming the Ice

The temperature of our ice is −10.0 °C. As we add heat, the molecules that are fairly rigidly held in place begin to vibrate a bit more. We still have ice, but the average energy has increased, so the temperature rises. Figure 11.14 displays a **heating curve**, a plot of the temperature of a compound versus time as it is heated. The plot indicates the specific temperature ranges for solid, liquid, and gas phases.

Heating curve – A plot of the temperature of a compound versus time or energy as heat is added at constant pressure. The plot indicates the specific temperature ranges for solid, liquid, and gas phases.

The heat needed to warm 1 g of a substance enough so that its temperature rises 1.0 °C is its **specific heat** which for ice is equal to 2.05 J/g · °C. Raising the temperature of our 10.0-kg block of ice from −10.0 °C to its **melting point**, at which it changes from a solid to a liquid, requires that 205 kJ of heat, q, be added to the system.

Specific heat – The heat needed to warm 1 g of a substance so that its temperature rises 1.0 °C.

Melting point – The temperature of the phase transition as a compound changes from a solid to liquid.

$$
\begin{aligned}
q &= \text{specific heat} \times \text{mass} \times \text{change in temperature} \\
q &= \frac{2.05\ \text{J}}{\text{g}\cdot{}^\circ\text{C}} \times 10,000\ \text{g} \times \left(0.0^\circ\text{C} - \left(-10.0^\circ\text{C}\right)\right) \\
&= 205,000\ \text{J} \\
&= 205\ \text{kJ}
\end{aligned}
$$

This much heat would be supplied by, for example, burning about 4 g of methane (natural gas is typically over 90% methane) in air.

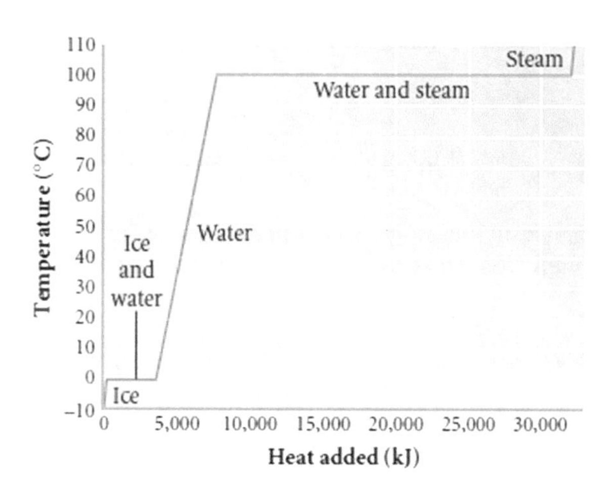

Fig. 11.14 As water is heated, it goes through changes from solid to liquid to gas. Why are some regions sloped and others flat? Why do areas of constant temperature have different lengths?

Change 2: Melting the Ice

Our ice is now at its melting point of 0.0 °C. As we add more heat, water molecules move more freely and randomly, beginning the transformation from the solid crystal to the liquid state. The temperature at this change of state, the melting point, is constant (Fig. 11.14), because the added heat is breaking hydrogen bonds, rather than increasing the average kinetic energy of the water molecules. The amount of heat needed to convert the solid to liquid at the melting point at constant pressure is called the **heat of fusion ($\Delta_{fus}H$)**, which for water is 334 J/g, or 6.01 kJ/mol. Melting is an endothermic process, which means that freezing (the opposite process) is an exothermic process. Heat is released when a liquid freezes.

Heat of fusion ($\Delta_{fus}H$) – The enthalpy change associated with the phase transition from liquid to solid.

How much heat is needed to melt 10.0 kg of ice? The calculation shown below indicates that we need to add 3340 kJ to the system.

$$
\begin{aligned}
q &= \text{mass} \times \Delta_{\text{fus}}H \\
&= 10{,}000\,\text{g}\cdot{}^{\circ}\text{C} \times \frac{334\,\text{J}}{\text{g ice}} = 3{,}340{,}000\,\text{J} = 3340\,\text{kJ}
\end{aligned}
$$

This is about the amount of heat supplied by burning about 60 g of methane in air.

Change 3: Heating the Liquid

Once the ice is completely transformed into liquid, the heat increases molecular motion, raising the temperature of the liquid (Fig. 11.14). The heat needed to raise the temperature of our 10.0-kg water sample from 0.0 to 100.0 °C can be calculated using the specific heat of water, 4.184 J/g · °C.

11

$$
\begin{aligned}
q &= \text{specific heat} \times \text{mass} \times \text{change in temperature} \\
&= \frac{4.184\,\text{J}}{\text{g}\cdot{}^{\circ}\text{C}} \times 10{,}000\,\text{g} \times \left(100.0\,{}^{\circ}\text{C} - 0.0\,{}^{\circ}\text{C}\right) \\
&= 4{,}180{,}000\,\text{J} = 4180\,\text{kJ}
\end{aligned}
$$

The total of 4180 kJ can be supplied by about 75 g of methane burning in air.

Change 4: The Boiling Point

As we continue to heat the water, it boils as the liquid is converted to vapor at constant temperature (Fig. 11.14). Analogous to the heat of fusion for melting is the **heat of vaporization ($\Delta_{vap}H$)** for converting the liquid water to vapor at the normal boiling point. The value of $\frac{2.44\,\text{kJ}}{\text{g water}}$ is over 7 times that of the heat of fusion $\left(\frac{0.334\,\text{kJ}}{\text{g ice}}\right)$. This indicates that a lot more energy is required to overcome the intermolecular forces when boiling water than when melting ice. Boiling causes all of the remaining hydrogen bonds that keep the water as a liquid to be broken. The heat needed to boil 10.0 kg of water is 24,400 kJ.

Heat of vaporization ($\Delta_{vap}H$) – The enthalpy change for the phase change from liquid to gas.

$$
\begin{aligned}
q &= \text{mass} \times \Delta_{\text{vap}}H \\
&= 10{,}000\,\text{g ice} \times \frac{2.44\,\text{kJ}}{\text{g water}} = 24{,}400\,\text{kJ}
\end{aligned}
$$

This amount of heat can be supplied by burning about 440 g of methane, which is over 3 times the total of the previous changes! The heat required from start to finish is equal to the sum of the amounts of heat required for all the changes along the way.

$$
\begin{aligned}
\text{Total heat} &= q_{\text{ice warming}} + q_{\text{ice melting}} + q_{\text{water warming}} + q_{\text{water boiling}} \\
&= 205\,\text{kJ} + 3340\,\text{kJ} + 4180\,\text{kJ} + 24{,}400\,\text{kJ} \\
&= 32{,}100\,\text{kJ}
\end{aligned}
$$

It is possible to heat the water vapor to quite a high temperature, and the amount of heat needed is related to the specific heat of water vapor, 1.84 J/g · °C. This is shown in the top right corner of Fig. 11.14.

Example 11.7—Energy Changes Upon Heating the Water

Determine the heat required to raise the temperature of a 2.50×10^2 g block of ice (equivalent to about a cup of water) in a 425 g aluminum pot from 0.0 °C to room temperature, 20.0 °C. Roughly how many grams of methane must be burned in air to supply this much heat? The heat of combustion of methane is −890 kJ/mol. Solid aluminum has a specific heat of 0.880 J/g·°C. Assume that all heat goes to the ice and the aluminum pot.

Asking the Right Questions

What processes occur as the block of ice is heated? What physical change will occur with the input of heat during each process? How do we calculate the energy as heat necessary for each process?

Solution

The input of heat will melt the ice at constant temperature (change of state) and then warm the ice from 0.0 to 20.0 °C. The aluminum pot will be warmed from 0.0 to 20.0 °C.

$$\begin{aligned}\text{Total heat} &= q_{\text{melting ice}} + q_{\text{warming liquid}} + q_{\text{warming aluminum}} \\ &= (\Delta_{\text{fus}}H \times \text{mass})_{\text{ice}} + (\text{specific heat} \times \text{mass} \times \Delta T)_{\text{water}} + (\text{specific heat} \times \text{mass} \times \Delta T)_{Al} \\ &= \{(334\,\text{J/g ice}) \times (250\,\text{g ice})\} + \{(4.184\,\text{J/g ice}\,°\text{C}) \times (250\,\text{g ice}) \times (20.0\,°\text{C} - 0\,°\text{C})\} \\ &\quad + \{(0.880\,\text{J/g Al}\,°\text{C}) \times (425\,\text{g}) \times (20.0\,°\text{C} - 0\,°\text{C})\} \\ &= 83,500\,\text{J} + 20,920\,\text{J} + 7480\,\text{J} \\ &= 111,900\,\text{J} \\ &= 112\,\text{kJ}\end{aligned}$$

$$\text{Grams of CH}_4\text{ required} = 112\text{ kJ} \times \frac{1\text{ mol CH}_4}{890\text{ kJ}} \times \frac{16.0\text{ g CH}_4}{1\text{ mol CH}_4} = 2.0\text{ g CH}_4$$

Is Our Answer Reasonable?

It is difficult to know what is reasonable in this case, because there are several processes occurring, and none of them are chemical changes. However, we note that the energy as heat generated in each step is within about an order of magnitude of each other, suggesting that the values are reasonable.

What If…?

…we did our melting in a 425 g copper pot with a specific heat of 0.0385 J/g · °C? How many grams of *propane*, C_3H_8, with a heat of combustion of 2219 kJ/mol are required to raise the temperature of the 2.50×10^2 g ice block to 20.0 °C?

(*Ans: 2.13 g of propane*).

Practice 11.7

How much energy does it take to convert 130.0 g of ice at −40.0 °C to steam at 160.0 °C?

11.5 Phase Diagrams

Much of the known universe is nowhere close to our own "normal pressure or temperature, and that affects the form water takes beyond our world. ◘ Figure 11.15 shows photographs of the moons of Jupiter taken by NASA's Galileo space probe. Why do the moons look so different from one another, from their host planet, and from anything else we've seen in the solar system? One reason apparently has to do with ice—not the type of ice we find in the refrigerator, but ice that exists under pressures of tens of thousands of atmospheres. Under these immense pressures, the molecules of ice that form large parts of the interiors of the "Jovian" moons are highly distorted. Hydrogen bonds are squeezed and shifted. Atoms are brought closer together and bond lengths shortened so that the ice becomes unusually dense. Temperature changes deep inside these moons also lead to changes in the nature of the ice. The consequences of all these changes are that the moons contract and expand, causing huge cracks and grooves. We add to our understanding of the solar system by looking at the changes in ice as temperature and pressure change.

Fig. 11.15 The moons of Jupiter as taken by NASA's Galileo space probe. *Source* ▶ https://photojournal.jpl.nasa.gov/jpeg/PIA01299.jpg

A graph showing the way the phase of a substance, or a mixture of substances, is related to temperature and pressure is called a **phase diagram**. The heating curves we discussed previously describe how water changes with temperature at constant pressure. However, as with the changes in the moons of Jupiter, a great many processes do not happen at constant pressure, and a phase diagram gives us a more comprehensive picture of these changes. In the chemical industry, phase diagrams have many uses, including determining the conditions required for manufacturing ceramics from clays. The composition of homogeneous mixtures of metals, called alloys, can also be changed on the basis of the information in phase diagrams.

Phase diagram – A plot of the phases of a compound as a function of temperature and pressure.

The phase diagram that includes the various types of ice in the core of planets is fairly complex, so let's work first with a small portion of it, that for water at more Earth-like temperatures and pressures, shown in Fig. 11.16. Note in the figure the three normal phases: solid, liquid, and gas (water vapor). There are also bold lines, called **phase boundaries**, at which two phases exist in equilibrium. We know these phase boundaries by other names. For example, any point on the phase boundary between the liquid and gas phase (line segment AB) is the "boiling point" (also the "condensation point" for the reverse reaction). Any point on the phase boundary between solid and liquid (line segment AD) is the "melting point" (or "freezing point"). How do we interpret a phase diagram? For example, how does water change state as we heat it from below 0 °C and 1 atm (the horizontal red line at "1.00" on Fig. 11.16)? Keep in mind that these are the common changes that we experience every day as we interact with water by boiling it to make tea or coffee or cause it to melt when we add it as ice to a soda.

Phase boundary – On a phase diagram, a set of specific temperatures and pressures at which a compound undergoes a phase transition.

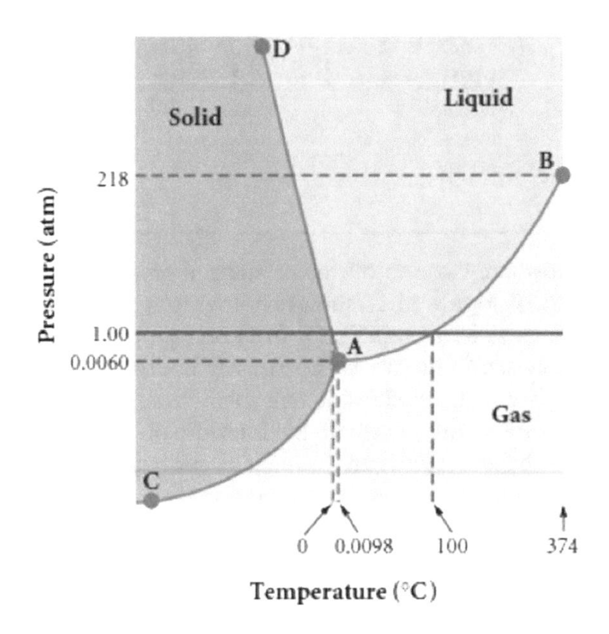

Fig. 11.16 Phase diagram of water. The red line indicates 1 atm pressure

Our water sample begins as ice. We raise the temperature to 0 °C, where it arrives at a phase boundary at which ice and liquid water exist in equilibrium. This is the normal melting (and freezing) point of water. As we continue to raise the temperature, the water remains liquid until 100 °C, the boiling point, which is at the liquid/gas phase boundary. Above 100 °C, the water will always be a gas if the pressure is kept constant at 1 atm. This is the same information we gleaned from the heating curve of water, but a phase diagram gives us so much more. We can tell, for example, that there is a **triple point** of water (Fig. 11.16 point (A) on the graph) where all three phases exist in equilibrium. For water, this occurs at 0.01 °C and 4.6 torr. The triple points of elements and compounds are useful because they serve as reference points with which temperature scales for thermometers are defined, and against which the highest-quality thermometers are calibrated. If we follow the liquid/gas phase boundary (the curved line segment AB) toward increased temperature and pressure, we get to a point above which the substance can no longer be liquefied and the gas and liquid have the same density—that is, they are indistinguishable from each other. This physical state is called a **supercritical fluid**. This point (B) is called the **critical point**, and it is associated with a **critical temperature**. The pressure at the critical temperature is the **critical pressure**.

Triple point – The temperature and pressure at which the gas, liquid, and solid phases of a substance are in equilibrium.

Supercritical fluid – This phase is seen only when the pressure and temperature of the substance are greater than the critical pressure and critical temperature. It is characterized with a density midway between that of a liquid and a gas, a viscosity similar to a gas, and the ability to completely fill the volume of a closed container.

Critical point – The temperature and pressure at which the physical properties of the liquid phase and vapor phase of a substance become indistinguishable.

Critical pressure – The highest pressure at which a liquid and gas coexist in equilibrium.

Critical temperature – The highest temperature at which a liquid and a gas coexist in equilibrium.

Example 11.8—Reading Phase Diagrams

What does a point on the line segment AD on the phase diagram of water represent? What does line AC indicate?

Asking the Right Questions

What do any lines on a phase diagram represent? What line segments separate the various phases in water?

Solution

Line segment AD represents the point of equilibrium between solid and liquid. Any point on this line is the melting (freezing) point at which the solid and liquid exist in equilibrium. Line segment AC is the point of equilibrium between solid and gas for water; sublimation and deposition occur at equilibrium anywhere along this line.

Are Our Answers Reasonable?

What happens as we move across the phase boundaries? To the left of line segment AD is solid water. To the right of the line segment is liquid. It is reasonable that the line segment is the equilibrium between these two phases. The same reasoning holds for line segment AC.

What If…?

…we are asked to interpret the CO_2 phase diagram in Fig. 11.17. Describe the changes that occur as we go from left to right from –78 to 31 °C at 60 atm.

(*Ans: CO_2 changes from solid to liquid to vapor, with only a very small region in the vapor phase*)

Practice 11.8

What is the approximate boiling point of water when the external pressure is 0.70 atm? Does this agree with our previous discussion about boiling points at high altitudes?

The phase diagram of CO_2, shown in Fig. 11.17, looks similar to that of water, with a triple point, critical temperature and pressure, and several phase boundaries. At 1 atm, CO_2 changes directly between the solid and gas phases (sublimation/deposition). Dry ice (solid CO_2) at 1 atmosphere does not melt but, rather, sublimes directly from a solid into a gas. At pressures greater than about 5 atm, CO_2 will liquefy from the solid phase. This liquid is used for dry-cleaning clothing to replace nonpolar, but environmentally hazardous, solvents such as tetrachloroethylene (C_2Cl_4). These solvents rid clothing of nonpolar grease and dirt.

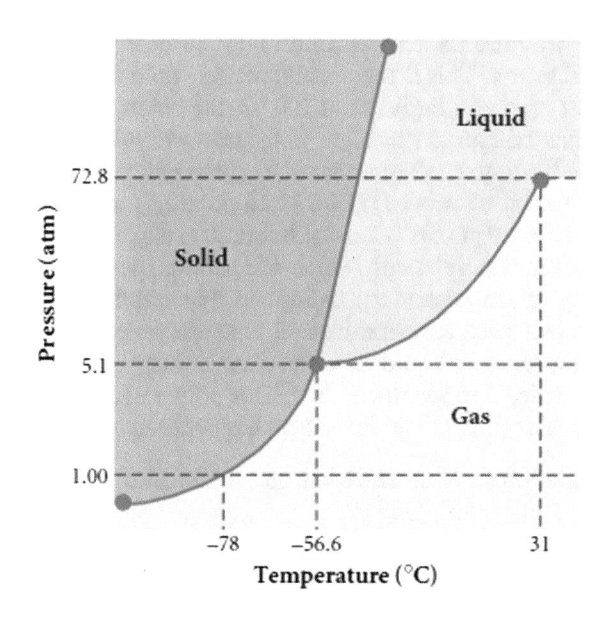

Fig. 11.17 Phase diagram of CO_2

How Do We Know—The Moons of Jupiter

The moons of the outer planets make most elegant chemical laboratories, because conditions on these worlds are so very different from those here at home. When we look at photos of Jupiter's moons, Europa and Ganymede (Fig. 11.15), we see some vexing geological features. How do we know what caused these features? We can't visit the moons, at least not directly. However, our space probes can gather various types of electromagnetic radiation—including light, ultraviolet radiation, and X-rays—that give us data from which to draw conclusions.

The probes' data seem to show that deep within the surface of these moons, there are layers of ice, each in a different phase, mixed with rocks. The phase diagram of water that we showed as Fig. 11.16 is inadequate to account for the low temperatures and massive pressures found within these moons. A more comprehensive phase diagram for water, emphasizing its solid phases, is given as Fig. 11.18. The ice that is made in our kitchen freezer or found in an ice storm—what we might call "normal ice—is technically called Ice Ih. Its structure is shown in Fig. 11.19. There are at least 17 other forms of ice that have been made in the laboratory or simulated by computer.

The phase diagram shows that as the temperature and pressure inside the moons change, different types of ice are formed. Each ice has its own density. As layers form versions of ice of different densities, they form the geological features, such as cracking and grooves, that we see on the surface of these incredible satellites. Part of the task of scientists is to develop phase diagrams that tell us about the conditions at which each type of ice exists.

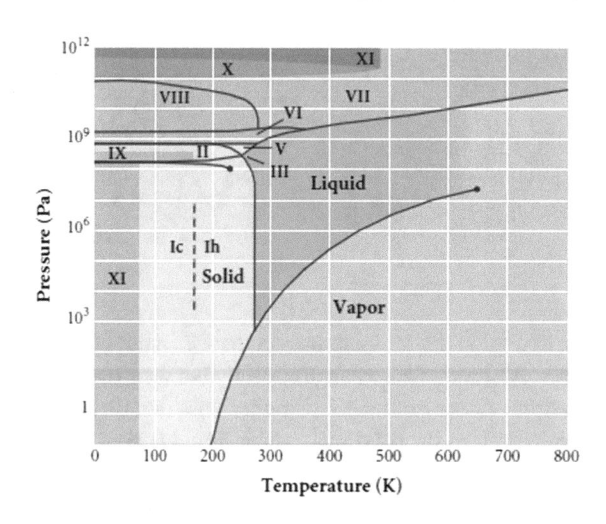

Fig. 11.18 This is an expanded phase diagram for water, taking into account the phase changes among the various solid (ice) forms. These forms of ice are thought to occur in layers beneath the surface of some of the moons of the outer planets

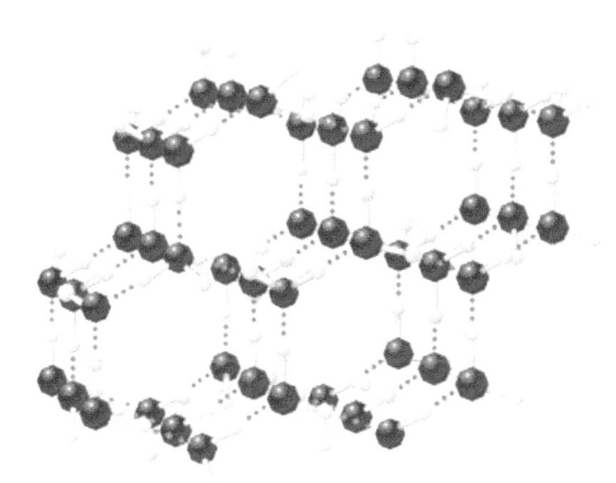

Fig. 11.19 The arrangement of water molecules within a crystal of "normal" ice (Ice Ih)

Liquid CO_2 is an excellent and relatively benign reusable alternative. Clothes are cleaned in CO_2 at 15 °C and 50 atm pressure, which is along the liquid/gas boundary in the CO_2 phase diagram. The clothes are kept close to room temperature, and about 98% of the CO_2 is recycled. The critical temperature and pressure of CO_2 are 304.2 K (31.05 °C) and 70.8 atm. Supercritical carbon dioxide, typically in the range of 310 to 400 K and 75 to 100 atm, can be used to clean industrial machinery, to replace many organic solvents in chemistry laboratories and to decaffeinate coffee beans.

11.6 Impact of Intermolecular Forces on the Physical Properties of Liquids: Other Properties

11.6.1 Viscosity

Making chocolate at home is difficult, because so much can go wrong, including unintended air bubbles, cracking, and a greasy look and feel to the final product. Having these things occur on an industrial manufacturing scale can be especially damaging and extremely costly. One of the most important properties that cause these and other inconsistencies in chocolate is called **viscosity**, which is a measure of resistance to flow. In the chocolate process industries, a viscometer is used to constantly monitor the viscosity of fully melted chocolate so that it flows into the molds and produces a final product that is consistent from batch to batch. Viscosity measurements are also important to the fuel industry, where the ability of fuels to flow is important to consider when storing, pumping and, within engines, injecting the fuel. Viscosity is reported in units of millipascal seconds (mPa·s). A highly viscous liquid, ethylene glycol ($C_2H_6O_2$), which has a viscosity of 16.1 mPa s at 25 °C and 1 atm, is combined with water to form commercial antifreeze. Pure water has a viscosity of 0.890 mPa·s at the same conditions, and the ethylene glycol/water mix has a viscosity of about 3 mPa·s. Why does the viscosity differ among liquids?

Viscosity – A measure of resistance to flow.

As with so many things related to chemistry, we gain understanding by looking at behavior at the molecular level. The molecules that constitute viscous liquids do not move well relative to each other. Why is this so? Part of the answer has to do with intermolecular forces. Stronger intermolecular forces tend to increase viscosity, but molecular size is also important. It is hard to move large molecules. As with boiling point, molecular shape is relevant because molecules must be able to move well among each other in order to flow easily. For example, water has strong intermolecular forces, but the molecules are small and can fairly easily move around each other. This gives water a relatively low viscosity. Molecules like octane (C_8H_{18}) and pentane (C_5H_{12}) have lower viscosity than water, because they have weaker intermolecular forces. For example, octane has a viscosity of 0.508 mPa·s at 25 °C and 1 atm, whereas the viscosity of pentane measures 0.224 mPa·s at these conditions.. Temperature is also a key

Fig. 11.20 A water strider (*Gerris remigis*) on the surface of a pond. Note that the water indents as it supports the weight of the water strider. *Source* Limnoporus canaliculatus; USA, TX, Travis Co.: Austin Brackenridge Field Laboratory

factor because, as we have seen, molecules with high average energy move relatively quickly and can overcome intermolecular forces, leading to low viscosity. For example, the viscosity of water changes from 0.890 mPa·s at 25 °C to 0.378 mPa·s at 75 °C. That's why chocolate, as well as oil in our car's engine, flows more smoothly at higher than at lower temperatures.

11.6.2 Surface Tension

The water strider glides from place to place along the water's surface, as shown in Fig. 11.20. Why can the insect travel on the water's surface? The key to the answer is a property of liquids called **surface tension**. This property is also responsible for the beads of water that run down our window in the rain or across the surface of our car after it has been waxed. Surface tension is a measure of the energy per unit area on the surface of a liquid. To better understand what this means, we note that hydrogen bonding in water makes it more energetically stable. Water molecules on the interior of the solution can hydrogen-bond in all directions, whereas surface molecules can form these bonds only toward the solution itself. In order to maximize the number of hydrogen bonds (and, therefore, the energetic stability of the liquid), water forms a shape with the minimum surface area, a sphere. This is why small droplets of water form beads and water striders can glide on water.

Surface tension – A measure of the energy per unit area on the surface of a liquid.

The chemical industries pay close attention to surface tension in the design of many consumer products. For example, soaps and detergents are designed to have very low surface tension in order to interact well with clothing and dishes. Surface tension is also considered in the design of pharmaceuticals and plays a large role in how tablets and gelcaps dissolve after they are swallowed.

11.6.3 Capillary Action

A **buret**, shown in Fig. 11.21, is an essential piece of glassware for measuring volume when performing laboratory analyses with liquids. Water has been added to the buret in Fig. 11.22. Why does the water seem to rise up the sides of the glass? And given this curvature, how do we read the volume of water in the buret? The answer to the first question again lies in one of our chemical themes, the impact of hydrogen bonding on the structure of water. The intermolecular forces that keep the large group of water molecules together can be called **cohesive forces** because they cause water molecules to stick to other water molecules.

Buret – A piece of laboratory glassware used to measure volume and to add discrete amounts of solution in a repetitive fashion.
Cohesive forces – The collection of intermolecular forces that hold a compound within its particular phase.

Then there is the glass in which the water sits. The glass of the buret is largely silicon dioxide (SiO_2) but with a surface containing many polar oxygen and hydrogen atoms. As is shown in Fig. 11.23, the hydrogen atoms on water interact with the oxygen atoms on the surface of the glass. The net result is that some molecules of water

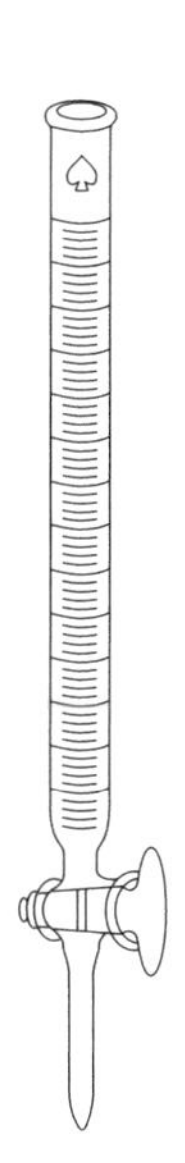

Fig. 11.21 A buret is essential labware in chemistry

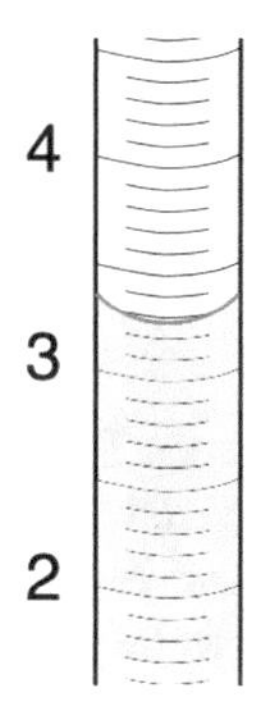

Fig. 11.22 The meniscus of water in the glass buret

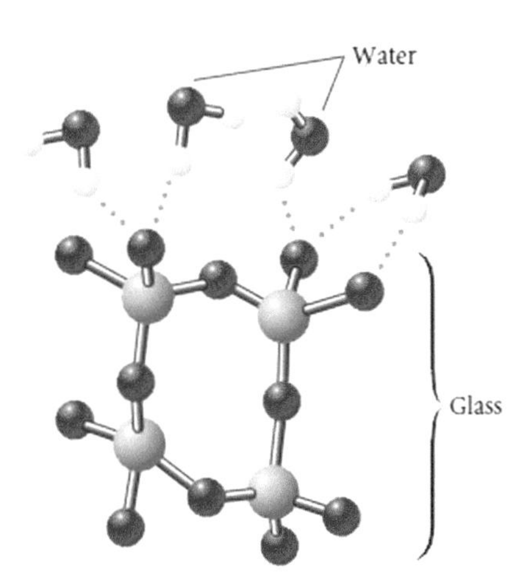

Fig. 11.23 The interaction of water with the surface of glass results in adhesion. Hydrogen bonds between the water molecules and the glass surface are illustrated

adhere to the glass, a process called **adhesion**. The effect in which water seems to crawl up the sides of a thin tube such as a buret is called **capillary action**. In this case, the adhesive forces between water and the glass are stronger than the cohesive forces within water itself, even though they are both examples of hydrogen bonding. The weight of the water prevents it from rising up even higher within the buret. Mercury in a glass tube (such as in a thermometer) shows the opposite behavior, bulging upward within the glass, in response to the dominance of cohesive over adhesive forces.

Adhesion – The interaction of a compound with the surface of another compound, such as the interaction of water with the surface of a drinking glass.

Capillary action – The effect seen when a liquid rises within a narrow tube as a consequence of adhesive forces being stronger than cohesive forces.

Adhesive forces lead to the formation of a **meniscus**. By convention, the volume of liquid delivered from the buret is read at the bottom of a concave meniscus, as shown in ◘ Fig. 11.22, so our reading of the buret is 3.45 mL. If mercury were in the buret, we would read the volume at the top of its convex meniscus.

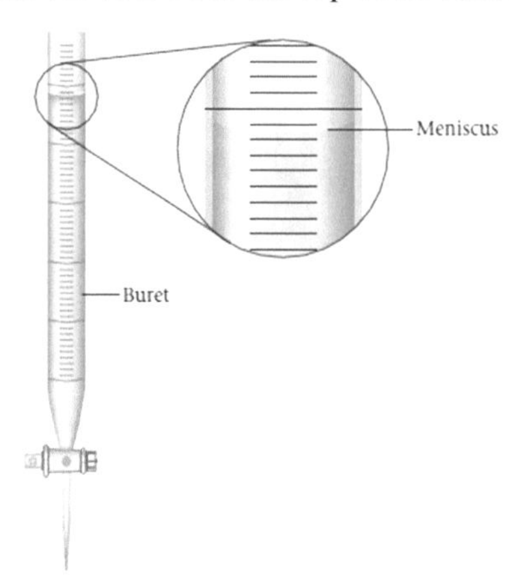

Meniscus – The concave or convex shape assumed by the surface of a liquid as it interacts with its container.

The balance between forces is not always easy to assess. For example, hexane (C_6H_{14}) shows a concave meniscus similar to that of water, though not quite so pronounced. Would we have predicted this? The surface of the glass can, presumably, induce a dipole in hexane, so we can explain the behavior. However, making predictions when several effects are important is sometimes quite challenging.

▶ Here's What We Know so Far

- A liquid fills its container from the bottom up and flows easily.
- At any phase change boundary, the different phases are in equilibrium with each other. This means they coexist, but not necessarily in equal amounts. Rather, they are changing from one phase to the other at equal rates.
- London forces are typically the weakest forces between molecules. Because every molecule has the ability to become polarized, every molecule possesses London forces. In very large molecules, these forces can be significant.
- Dipole–dipole interactions are much stronger than London forces. They exist in molecules that contain a nonzero dipole moment, and their strength is related to the strength of the dipole moment.
- Hydrogen bonds are the strongest of the van der Waals forces. They exist primarilybetween the interaction of a hydrogen atom and two atoms of F, O, and/or N.
- The strength of the intermolecular forces of attraction determines the shape of a meniscus, the surface tension of the liquid, its vapor pressure, and other physical properties associated with the liquid. ◀

11.7 The Structure of Crystals

To this point, we have considered what makes a pure substance a liquid. We've discussed intermolecular interactions, their meaning, and their implications. However, most liquids, even the most pristine water from a mountain stream, are not pure—that is, they are not homogeneous. They contain dissolved solids that, in many cases, we seek to acquire for food or manufacturing. For as long as human society has existed, the sea has played an important role in providing this food and other raw materials. As human population increases, the sea provides an increasingly important source for freshwater through evaporation, distillation or reverse osmosis (▶ Chap. 12). The majority of what must be removed from seawater to make it potable are crystalline solids, such as sodium chloride, calcium chloride and magnesium chloride, although sodium chloride is the most prevalent of these salts.

When the water is removed, these solids are left behind in a completely different form. The structure and properties of solid crystalline materials are what we will examine for the remainder of the chapter.

► https://upload.wikimedia.org/wikipedia/commons/8/8a/San_diego_bay_salt_pile.jpg

In an earlier chapter, we discussed the formation of sodium chloride from its elements. We noted the strength of the ionic bond by focusing on the lattice energy, 786 kJ/mol. To learn more about sodium chloride and other crystalline ionic compounds, however, we need to explore how the ions come together to make an ionic crystal. What can the shape of an ionic solid tell us? Knowing the locations of the ions within an ionic solid enables us to determine the density, the nature (crystalline or amorphous), and the properties of the solid (see ◻ Table 11.3).

11.7.1 Types of Solids

Why are some compounds, such as table salt or sugar, solid rather than liquid or gaseous under normal conditions? The atoms, ions, or molecules of a solid are not free to move significantly relative to each other. The materials are solids because the intermolecular forces of attraction are strong enough to hold the particles in a rigid structure. In a **crystalline solid** (◻ Fig. 11.24), the atoms, ions, or molecules are highly ordered in repeating units over long ranges. Although they have fairly rigid and fixed locations of the atoms, ions, or molecules, **amorphous solids** such as the glass in a windowpane lack the high degree of long-range order (they still have short-range order) found in a crystalline solid. Sodium chloride is an example of a crystalline solid (◻ Fig. 11.25), because it has an ordered structure with fixed locations for the sodium and chloride ions.

◻ **Table. 11.3** Different types of solids and their properties

Type of solid	Particles in solid	Forces between particles	Physical properties	Examples
Ionic	Cations and anions	Electrostatic attraction	Hard and brittle, high melting point, poor thermal and electrical conductivity	NaCl CaO $MgBr_2$
Molecular	Atoms or molecules	London dispersion, dipole–dipole, and/or hydrogen bonding	Fairly soft, low to moderate melting points, poor thermal and electrical conductivity	CH_4 $C_6H_{12}O_6$ (glucose) H_2O Kr
Network covalent	Atoms	Covalent bonds	Very hard, very high melting point, typically poor thermal and electrical conductivity	C (diamond) SiO_2 (quartz) SiC BN
Metallic	Atoms	Metallic bond	Soft to hard, low to high melting point, good thermal and electrical conductivity	Na Fe Au Al

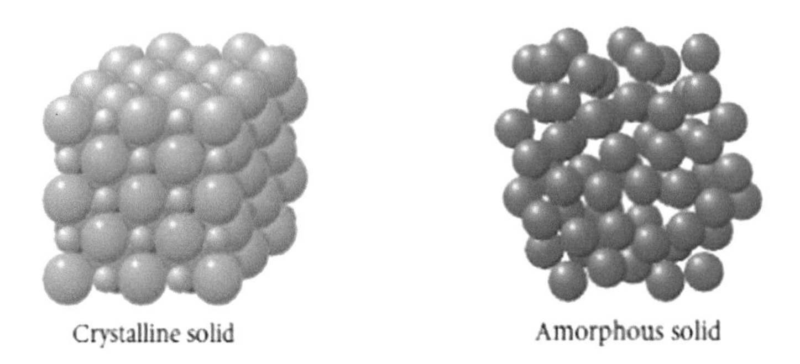

Fig. 11.24 Solids can be crystalline or amorphous

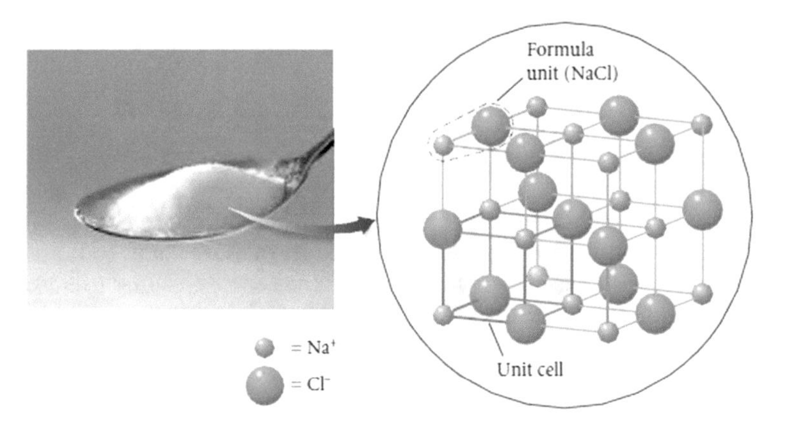

Fig. 11.25 Crystals of sodium chloride are an ionic solid

Crystalline solid – A solid made from atoms, ions, or molecules in a highly ordered long-range repeating pattern.

Amorphous solid – A solid whose atoms, ions, or molecules occupy fairly rigid and fixed locations but that lacks a high degree of order over the long term.

What does a crystal look like? Try using a magnifying glass or microscope to look at crystals of table salt or sugar. These two crystalline solids appear as solid cubes, whereas other solids take the shape of needles, plates, and a host of other faceted designs. The shape of the crystal we see is a result of the way in which the atoms and ions are arranged at the atomic level, known as the crystalline lattice. Some solids, such as table salt (NaCl), are **ionic solids**, made up of ionic compounds held together by strong electrostatic forces of attraction between adjacent cations and anions. Because these forces are very strong, the ionic solids typically have high melting points. Solutions made from ionic solids conduct electricity. We referred to them as electrolytes.

Ionic solids – Solids made up of ionic compounds held together by strong electrostatic forces of attraction between adjacent cations and anions.

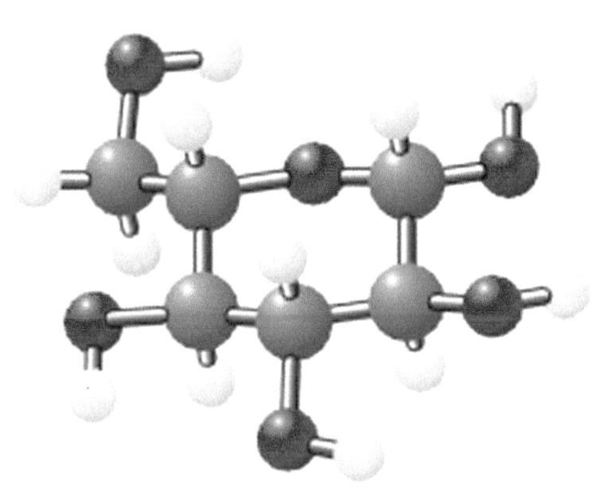

Glucose, a molecular solid
$C_6H_{12}O_6$

On the other hand, some solids, such as glucose ($C_6H_{12}O_6$), are **molecular solids**, made up of molecules held in a rigid structure. Despite the structural similarity to an ionic solid, the intermolecular forces of attraction between adjacent molecules are not as strong as the electrostatic forces found in ionic solids. Therefore, melting points of the molecular solids are not very high. In addition, because most molecules do not dissociate into ions when they are dissolved, solutions made from molecular solids typically do not conduct electricity. As we'll see in this chapter, the properties of molecular and ionic solids are different in other important ways.

Molecular solids – Solids made up of molecules that are held together by intermolecular forces.

11.7.2 Crystal Lattices

A structure known as a crystal lattice defines the shapes of crystalline solids. This **crystal lattice** marks the position of each of the atoms, ions, or molecules within the crystal. The 3-D grid comprises three pairs of parallel planes that intersect to make three-dimensional boxes, and in Fig. 11.28 we see that repeating these boxes in three dimensions gives us the lattice for a crystal. The smallest repeating unit of the crystalline lattice is the **unit cell**. In other words, the unit cell is the smallest repeating unit of atoms, ions, or molecules that would describe the entire crystal. Whenever possible, we try to arrange the corners of the lattice so that they line up with the centers of the atoms, ions, or molecules in the solid.

Crystal lattice – A highly ordered framework of atoms, molecules, or ions.

Unit cell – The repeating pattern that makes up a crystalline solid.

If the parallel sets of planes intersect at right angles to make a cube (as they do in Fig. 11.26), the lattice is known as a cubic lattice. The most basic of the cubic lattices is made up of the **simple cubic unit cell** (also known as the primitive unit cell), in which the centers of the atoms, ions, or molecules are located only on the corners of the unit cell. Polonium, a radioactive metal that shows promise, and causes great concerns, as a thermoelectric power source in space satellites, crystallizes in the simple cubic structure. The presence of an additional atom, ion, or molecule at the center of the simple cubic unit cell gives a **body-centered cubic (bcc) unit cell**. Iron metal crystallizes as a body-centered cubic structure. If an atom, ion, or molecule is located on each of the faces of the cubic lattice, the lattice is called a **face-centered cubic (fcc) unit cell**. When cooled below −182 °C, methane (the gas that heats many homes) forms crystals that have a face-centered cubic structure. Figure 11.27 illustrates each of the unit cells in the cubic lattice group.

Simple cubic unit cell – An arrangement of particles within a crystalline solid comprised of six particles occupying the corners of a cubic box.

Body-centered cubic (bcc) unit cell – A unit cell built via the addition of an atom, ion, or molecule at the center of the simple cubic unit cell.

Face-centered cubic unit cell – A unit cell built via the addition of an atom, ion, or molecule at the center of each side of the simple cubic unit cell.

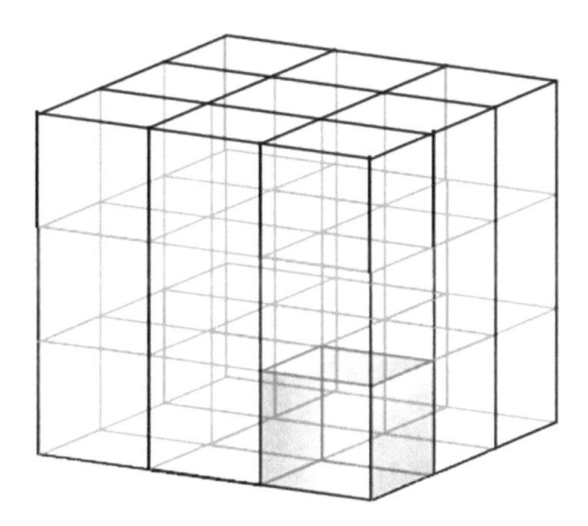

Fig. 11.26 A unit cell in a cubic lattice

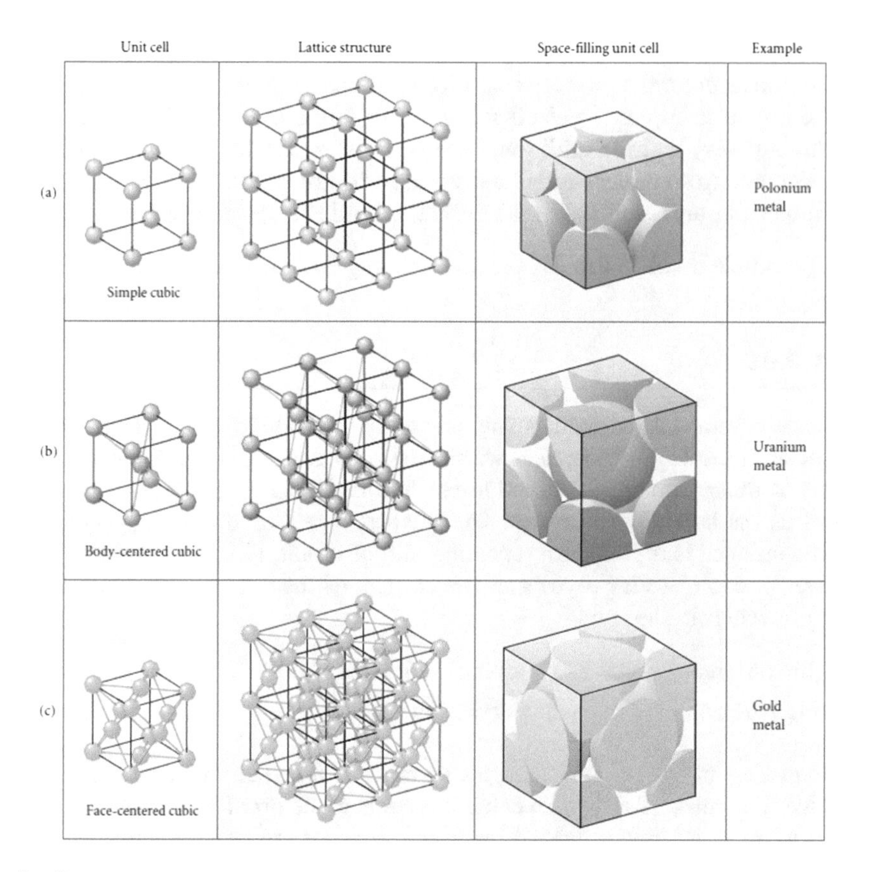

Fig. 11.27 Cubic unit cells

11.7.3 Metallic Crystals

Solid metals are made up of atoms of an element arranged in a crystalline lattice. If we imagine that the atoms are hard spherical balls, we can get a picture of how they might be arranged into a crystal. We can visualize this by placing marbles in a box. Put in just enough marbles to cover the bottom and see what happens. To utilize most efficiently all of the space available, the marbles adopt a staggered arrangement. The way they pack is called a **closest packed structure** (Fig. 11.28).

What if we want to completely fill a box with marbles? The second layer is also staggered, but it occupies the dimples in the first layer. However, there are different ways in which we can stack the third and subsequent layers of marbles. It is the placement of this third layer that determines the type of packed structure that results. If the third layer is placed in such a way that it lines up with the first layer, what results is a **hexagonal closest packed (hcp) structure** whose unit cell is a hexagonal prism (Fig. 11.29). Magnesium and zinc atoms form hexagonal closest packed solids. If the third layer is staggered in the same fashion as the second layer, so that it doesn't line up with the first layer, we obtain the cubic **closest packed (ccp) structure** (Fig. 11.30). Metals, such as gold, silver, and copper, adopt this structure and have face-centered cubic unit cells.

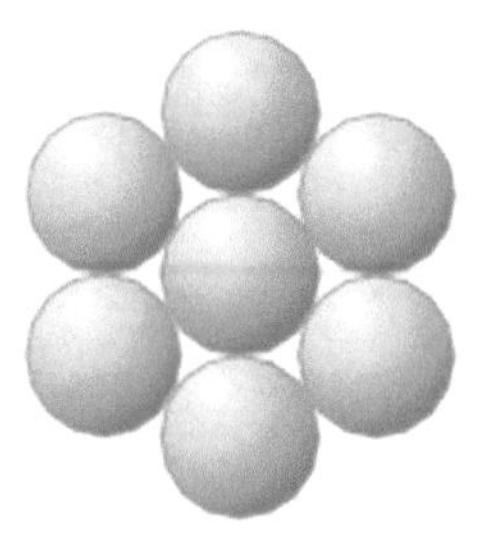

Fig. 11.28 Closest packed arrangement of spheres in two dimensions

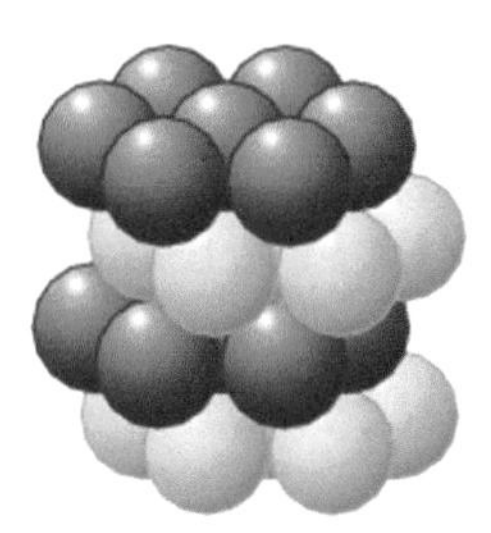

Fig. 11.29 Hexagonal closest packed structure

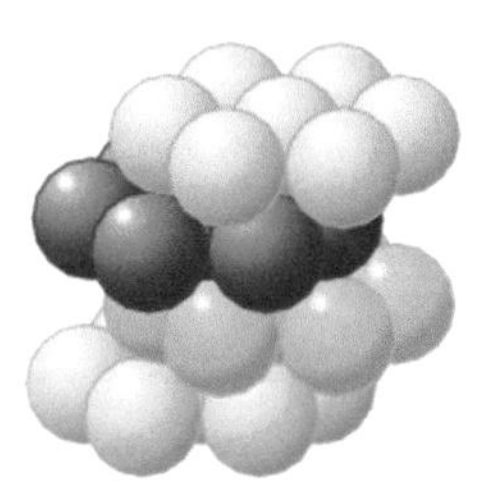

Fig. 11.30 Cubic closest packed structure

Closest packed structure – A structure wherein the atoms, molecules, or ions are arranged in the most efficient manner possible.

Cubic closest packed (ccp) structure – A closest packed structure made by staggering a second and third row of particles so that none of the rows line up.

Hexagonal closest packed (hcp) structure – A closest packed structure made by staggering a second and third row of particles in such a way that the first and third rows line up.

Example 11.9—Matching

Match each of the following metals with the figure that best represents its structure.

1. copper (fcc)
2. iron (bcc)
3. iron (hcp).

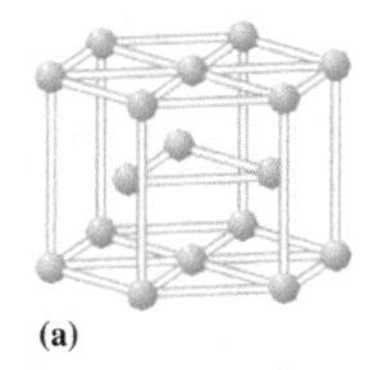

(a)

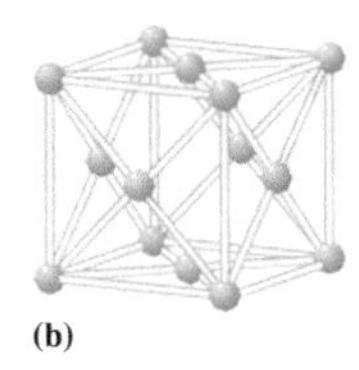

(b)

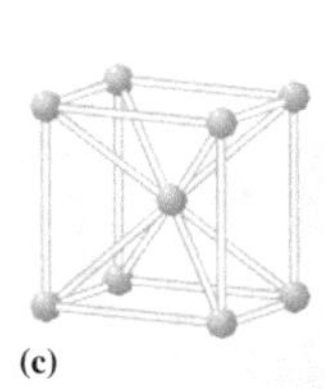

(c)

Asking the Right Questions

How are the atoms placed in each type of cell? In the face-centered cubic (fcc) unit cell, there are atoms on each of the faces of the unit cell. In the body-centered cubic (bcc) unit cell, there is an atom at the center of the unit cell. In the hexagonal closest packed (hcp) unit cell, the unit cell is a hexagonal prism, with the first and third layers lining up.

What Do I Need to Know?

Imagine that we are visiting with friends and someone asked about unit cells. (Stranger things have happened!) What would we say to the group are the key structural feature or features that would allow them to distinguish among the three types of unit cells (fcc, bcc, hcp)?

Solution

In the face-centered cubic (fcc) unit cell, there are atoms on each of the faces of the unit cell. In the body-centered cubic (bcc) unit cell, there is an atom at the center of the unit cell. In the hexagonal closest packed (hcp) unit cell, the unit cell is a hexagonal prism, with the first and third layers lining up. In the figures, copper (face-centered cubic) is (b). Iron (body-centered cubic) is (c), and iron (hexagonal closest packed) is (a).

Are Our Answers Reasonable?

The figures match our criteria for each type of unit cell, so the answers are reasonable.

What If…?

…we have a sample of lead that has an fcc unit cell? What figure would we draw to illustrate the shape?

(*Ans: Our image would look identical to* (b) *from the figure in the question*)

Practice 11.9

Provide a diagram for each of the following elements in the given unit cell.

a. manganese (simple cubic)
b. nickel (fcc)
c. tungsten (bcc).

Some metals do not form closest packed structures. For instance, the alkali metals stack together into body-centered cubic unit cells. Although we can know how a metal packs into a particular crystalline lattice, the reason why metals pack in the way they do is not well understood. Predicting the crystalline lattice structure of a metal is difficult. Why is it useful to know the structure of the unit cell? One answer has to do with the enzymes within us.

▶ **Here's What We Know so Far**

- Solids can be either amorphous or crystalline.
- Crystalline solids result from highly structured packing of the material.
- The arrangement of particles within the crystal is as a repeating unit called the unit cell.
- The simple cubic, body-centered cubic, and face-centered cubic are examples of unit cells.
- In addition, substances can also pack into closest packed structures, which include the hexagonal closest packed and the cubic closest packed structures. ◀

11.8 Crystallography and the Unit Cell

In 1997, the Nobel Prize in chemistry was awarded to Paul Boyer, John Walker, and Jens Skou "for their elucidation of the enzymatic mechanism underlying the synthesis of adenosine triphosphate (ATP). Essential to their discovery was the identification of the crystalline structure of an enzyme found in cows, bovine F1 ATP synthase (◻ Fig. 11.31). By determining the structure of this enzyme, the researchers were able to understand how ATP (a source of energy in living things) is made. This is just one example of the utility of knowing the crystal structure of a solid. Medical researchers use crystal structures to get a clear picture of the location where reactions take place, which is called the "active site of the enzyme. This information makes it possible to design drugs that will best influence the activity of the enzyme. For instance, the structure of the active site of HMG-CoA reductase (an enzyme that makes cholesterol) was used in the development of Rosuvastatin, a drug that binds to the enzyme and helps slow the formation of cholesterol. The binding of Rosuvastatin leads to lower serum cholesterol levels in patients.

The structures of these enzymes were determined by X-ray crystallography. How does this technique help us determine the structure of a crystal? We begin by examining the behavior of light. When light waves pass through a grating (a series of closely spaced narrow slits), constructive and destructive interference of the waves produces a pattern of light and dark regions. This **diffraction pattern** can be used to determine the position and widths of the slits. The most useful diffraction pattern occurs when the wavelength of the light is similar to the widths of the slits. In X-ray crystallography, a diffraction pattern is generated by using a crystal instead of a series of slits.

Diffraction pattern – A pattern of constructive and destructive interference after electromagnetic radiation passes through a solid material.

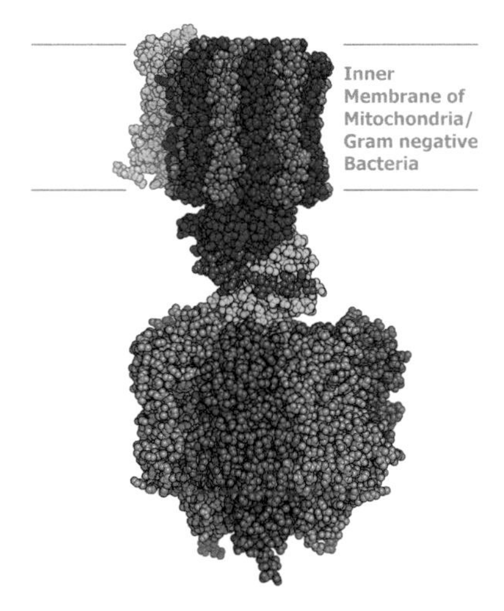

◻ **Fig. 11.31** Cartoon of the crystal structure of ATP synthase. Because there are so many atoms in this enzyme, each strand of the protein is represented as a different color. Individual atoms are not specifically defined. *Source* ▶ https://upload.wikimedia.org/wikipedia/commons/0/00/Atp_synthase.PNG

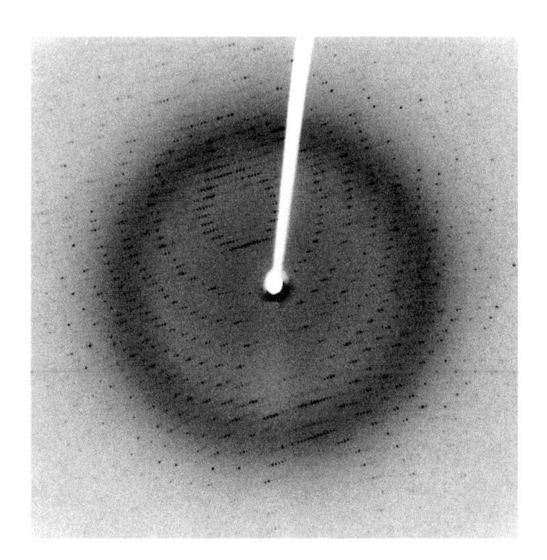

Fig. 11.32 A pattern of reflected X-rays is captured after interaction with a crystal. The location of the dots in this diffraction pattern can be used to determine the structure of a compound. *Source* ► https://upload.wikimedia.org/wikipedia/commons/7/7d/X-ray_diffraction_pattern_3clpro.jpg

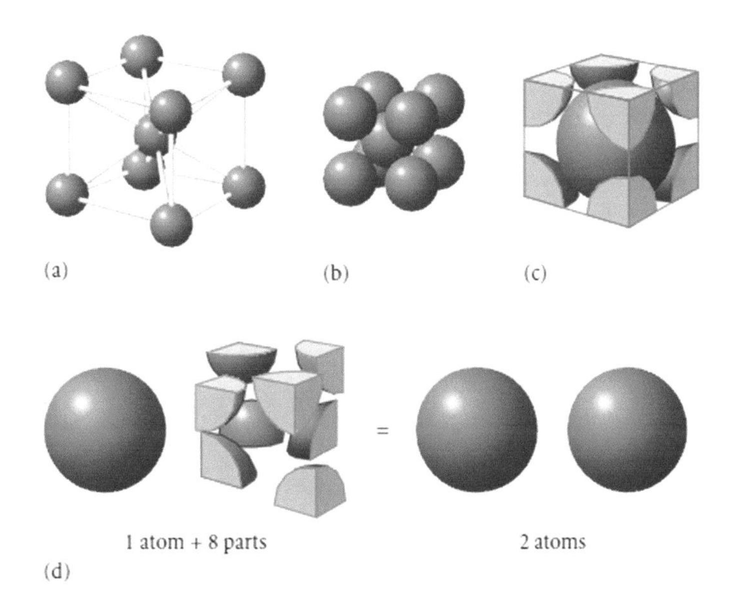

Fig. 11.33 Body-centered cubic unit cell

The unit cells contain a lot of empty space, but the spaces between the atoms or molecules can function like the slits used to create the diffraction pattern shown above. Because the spaces in the crystal are on the order of 100 pm (1 pm = 10^{-12} m), radiation of similar wavelengths is used. X-rays—electromagnetic radiation with wavelengths near 100 pm—are directed through the crystal and are diffracted as shown in Fig. 11.32. The resulting pattern of high- and low-intensity X-rays is used to determine the position and size of the atoms in the crystal. Our ability to calculate the positions of the atoms from the crystal diffraction pattern is based on the work of William Henry Bragg (1862–1942) and William Lawrence Bragg (1890–1972), a father-and-son team who shared the Nobel Prize in physics in 1915 for their work in X-ray crystallography.

Electromagnetic radiation can pass through a crystal because of the existence of empty spaces in the unit cell. How much empty space is there in a crystal? We talk about how the spheres are packed as closely together as possible, but in our unit cells we seem to have a lot of gaps. To determine the amount of empty space, we need to know how much space the atoms in a unit cell occupy and how much space the entire unit cell occupies. The difference between these two values is the empty space. How many complete atoms occupy the unit cell? To simplify the process, consider the atoms as hard spheres. Figure 11.33 shows the bcc unit cell, which contains one complete atom in the center of the unit cell. At the corners of the cell are 18 "pieces of atoms. Because there are eight corners in the unit cell, there must be the equivalent of one complete atom. The unit cell for a body-centered

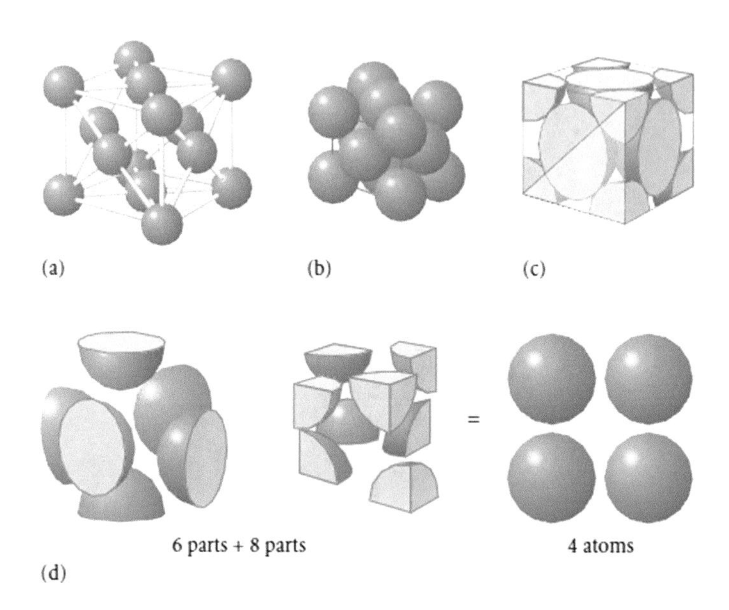

Fig. 11.34 Face-centered cubic unit cell

cubic structure contains two complete atoms: the central one and the eight 1/8's that make up the corners. We can go through a similar process for the primitive cell (one complete atom) and the face-centered cubic cell (four complete atoms).

Because we can determine how many atoms occupy a unit cell, knowing the volume of the unit cell enables us to find the amount of empty space in the cell. What is the volume of a unit cell? We can calculate this for each type of unit cell. In the face-centered cubic unit cell (Fig. 11.34), we see that the diagonal across one face of the cube is 4 times the radius of the atom ($4r$) and that the height and width of the cell are the same distance (d). To find the length d, we use the Pythagorean Theorem ($a^2+b^2=c^2$) and solve for the distance:

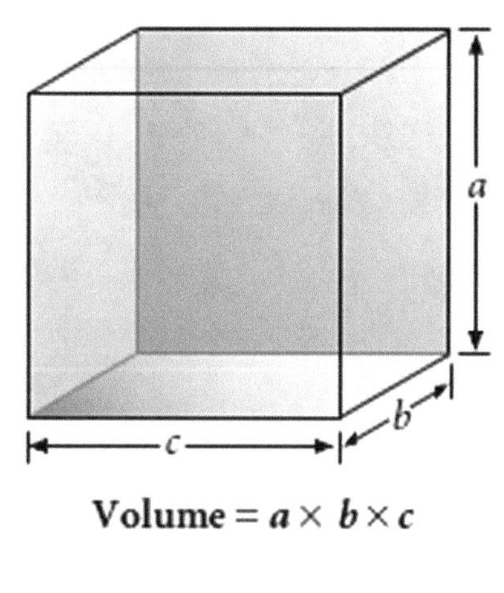

$$d^2 + d^2 = (4r)^2$$
$$2d^2 = 16r^2$$
$$d^2 = 8r^2$$
$$d = r\sqrt{8} = 2.828r$$

Therefore, the distance along one of the faces of the fcc unit cell is 2.828 times the radius of the atoms. Note that the distance along the edge of the unit cell is greater than twice the radius of the atoms. This implies, correctly, that the atoms in the corners of the unit cell do not touch the atoms in the other corners.

The volume of any parallelopiped is determined by multiplying the length times the width times the height. For the fcc unit cell, the volume would then be $(r\sqrt{8})^3$, or $22.63r^3$. To illustrate this equation, gold metal crystallizes in the cubic closest packed structure (with fcc unit cells). The volume of the unit cell for gold metal whose radius is 144 pm is

$$r = 144\,\text{pm}$$

$$\text{Volume} = (144\,\text{pm} \times \sqrt{8})^3 = 6.76 \times 10^7\,\text{pm}^3$$

Example 11.10—There's a Lot of Room in Here

Copper, silver, and gold are known as the coinage metals because of their use in making coins. Each of these metals forms a face-centered cubic solid. What percentage of a fcc unit cell is occupied by empty space?

Asking the Right Questions

How much space does a face-centered cubic unit cell occupy? Given that, how much space is occupied by atoms? How can we use this information to determine the amount of free space in the unit cell?

Solution

Because we haven't specified any particular example, let's consider a generic fcc unit cell. The volume of a fcc unit cell is $(r\sqrt{8})^3$. We know that a face-centered cubic unit cell has four complete atoms (each with a volume of $\frac{4}{3}\pi r^3$), so the total volume of the spheres is $4 \times \frac{4}{3}\pi r^3$ or $\frac{16\pi r^3}{3}$. The ratio of the two volumes will tell us the percentage of the unit cell that is taken up by atoms:

$$\%\text{ occupied} = \frac{\frac{16\pi r^3}{3}}{(r\sqrt{8})^3} = \frac{\frac{16\pi}{3}r^3}{(\sqrt{8})^3 r^3}$$

$$= \frac{\frac{16\pi}{3}}{(\sqrt{8})^3} = \frac{16.755}{22.627}$$

$$= 0.7405 \times 100\% = 74.05\%$$

The percent of occupied space in the unit cell is 74.05%, which leaves 25.95% of the unit cell vacant. That's a lot of empty space.

Are Our Answers Reasonable?

If we look at the face-centered cubic cell in Fig. 11.34c, the answer seems reasonable—that is, about a quarter of the unit cell is space.

What If...?

...we labeled each layer in the unit cell with a letter, such as "A" for the first layer. If the second layer had the same orientation as the first, we would call it "A" as well. If it had a different orientation, we would call it "B". On that basis, which of these describes the hexagonal close packing (hcp) pattern?

1. ABAABA
2. ABCABC
3. ABCCBA
4. ABABAB.

(*Ans: ABABAB*).

Practice 11.10

Polonium forms the simple cubic unit cell. How much of a primitive unit cell is empty space?

The ability to calculate the volume of the unit cell enables us to determine the density of a crystal. Mathematically, the density of a solid is the mass divided by the volume. Because we've already determined the volume of the unit cell, we need to determine its mass. How do we do this calculation? If we know how many atoms make up the unit cell and we know the mass of one atom, then we can calculate the mass of a unit cell. For example, gold crystallizes in the cubic closest packed structure (fcc unit cell). The atomic mass of gold is 197.0 g/mol, or 197 amu per atom. As we've done before, we'll keep a couple of extra figures in each calculation, rounding off only at the end.

$$\frac{197.0\,\text{g Au}}{\text{mol}} \times \frac{1\,\text{mol Au}}{6.022 \times 10^{23}\,\text{atoms}} = \frac{3.2713 \times 10^{-22}\,\text{g}}{\text{atom Au}}$$

Then

$$\frac{3.2713 \times 10^{-22}\,\text{g}}{\text{atom Au}} \times \frac{4\,\text{atoms Au}}{\text{fcc unit cell}} = 1.3085 \times 10^{-21}\,\text{g/unit cell}$$

What is the density of a block of gold? Because we can calculate the volume and the mass of any unit cell, we can determine the density of a crystalline solid if we know its unit cell. For instance, gold has a mass of 1.3085×10^{-21} g/unit cell and a unit cell volume of 6.756×10^7 pm^3. The density (mass/volume) can be calculated:

Table. 11.4 Face-centered cubic unit cell calculations

Length of side	$\left(r\sqrt{8}\right)$
Volume of unit cell	$(r\sqrt{8})^3$
Length of diagonal	$4r$
Mass of unit cell	4 × mass of one atom

$$6.756 \times 10^7\,\text{pm}^3 \times \left(\frac{1.0 \times 10^{-12}\,\text{m}}{\text{pm}}\right)^3 \times \left(\frac{100\,\text{cm}}{\text{m}}\right)^3 = 6.756 \times 10^{-23}\,\text{cm}^3$$

$$\frac{1.3085 \times 10^{-21}\,\text{g}}{6.756 \times 10^{-23}\,\text{cm}^3} = 19.4\,\text{g/cm}^3$$

The value we've calculated for gold (19.4 g/cm^3, based solely on physical measurements of the unit cell) agrees quite well with the experimental value for the density of gold (19.3 g/cm^3). Table 11.4 lists the important calculations for the face-centered cubic unit cell.

Example 11.11—He Isn't Heavy, He's Dense

Some metals are much denser than others. What makes them so dense? The arrangement of the atoms in the solid, coupled with the mass of the element, accounts for the different densities we observe. Let's calculate the density of an iron bar. Iron, whose atomic radius is 124 pm, forms bcc unit cells.

Asking the Right Questions

What data do we need to calculate the density? In particular, how do we convert the radius of the iron atom into the volume the atoms occupy in the unit cell? How many iron atoms are in a body-centered unit cell? How do we calculate the mass of iron in the unit cell with this information? How do we then calculate the density?

Solution

We can determine the density of a block of iron by dividing the mass of the appropriate number of atoms by the volume of the unit cell. The body-centered cubic unit cell has the volume $\left(r\frac{4}{\sqrt{3}}\right)^3$. For an iron body-centered cubic cell, the volume is

$$\left(124\,\text{pm} \times \frac{4}{\sqrt{3}}\right)^3 = (286.4)^3 = 2.348 \times 10^7\text{pm}^3$$

$$2.348 \times 10^7\text{pm}^3 \times \left(\frac{1\text{m}}{10^{12}\text{pm}}\right)^3 \times \left(\frac{100\text{cm}}{1\text{m}}\right)^3$$
$$= 2.348 \times 10^{-23}\text{cm}^3$$

In a bcc unit cell containing two atoms of iron, the total mass is

$$2\,\text{atoms} \times \frac{1\,\text{mol}}{6.022 \times 10^{-23}\,\text{cm}^3} \times \frac{55.847\,\text{g}}{\text{mol}}$$
$$= 1.855 \times 10^{-22}\,\text{g}$$

The density of iron is the mass of the unit cell divided by the volume of the unit cell:

$$\frac{1.855 \times 10^{-22}\,\text{g}}{2.348 \times 10^{-23}\,\text{cm}^3} = 7.90\,\text{g/cm}^3$$

The experimentally measured density of iron is 7.86 g/cm^3.

Is Our Answer Reasonable?

Our answer is very close to the experimentally determined value, so it is reasonable.

What If…?

…we were given platinum, which has an fcc unit cell, and a density of 21.45 g/cm^3? What is the radius of a platinum atom?

(*Ans:* 1.39×10^{-8} *cm = 139 pm*).

Practice 11.11

Lead is used as fishing weights because of its malleability and density. What is the density of lead (Pb)? Lead has an atomic radius of 175 pm and crystallizes in a fcc arrangement.

Other information about the crystalline solid can be determined by examining the unit cell. For example, the number of chemical species that surround a particular ion can be determined from the packing arrangement. Calculating the positions of the atoms in a unit cell based on an X-ray diffraction pattern used to be a tedious task that could take months or years to complete. Today, scientists use the X-ray diffractometer to generate X-rays and direct them at a single crystal at a variety of different angles. Powerful computers collect the data over a period of hours and store them until the diffraction pattern is complete, at which time the computers go to work deciphering the pattern into a picture of the crystal. Depending on the quality of the crystal, the location of each of the atoms in the solid can be accurately determined in a few hours.

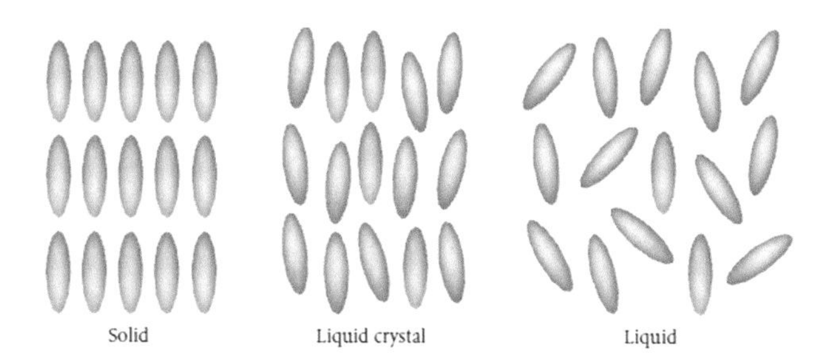

Fig. 11.35 Liquid crystals. The liquid crystal has properties that resemble the liquid and properties that resemble the solid. It is a fluid, yet highly organized phase

11.8.1 Liquid Crystals

We use liquid crystals in the displays of watches, pocket calculators, computers, and privacy windows. What is a liquid crystal and how does it work? The liquid crystal is a transitional phase between a solid and a liquid. There are some exceptions, but the molecules of the liquid crystal generally have long, rigid structures with a strong dipole moment that point along a common axis (Fig. 11.35) to align into highly ordered crystalline solids. An example of a solid that forms a liquid crystal is cholesteryl benzoate.

Cholesteryl benzoate

If the temperature of a solid with these characteristics is increased, the long, rigid structures of the molecules and the strong dipole moment continue to maintain a high degree of order within the structure. In fact, even well into the liquid state, order is maintained despite the increase in translational freedom (the ability to move in many directions). When the temperature gets hot enough, the degree of order in the structure is eliminated. In other words, in the transitional state between completely solid and completely liquid, the molecules are highly ordered and yet quite fluid (the **liquid crystal** state).

Liquid crystal – A transitional phase that exists between the solid and liquid phases for some molecules. Typically, these molecules have long, rigid structures with strong dipole moments.

The strong dipole moment is the key to the behavior of liquid crystals. We know that molecules with dipole moments align themselves within an electric field. When such a field is applied to the liquid crystal, its molecules orient themselves with the electric field. This highly ordered structure is transparent to light directed along the axis of orientation. If the electric field is removed, the translational freedom of the molecules scrambles their alignment. Visually, a liquid crystal in this state is opaque because the slight randomness scatters light. The net effect is a material that can be made transparent or opaque with the flick of a switch.

▶ Here's What We Know so Far

- The dimensions and the density of a solid can be determined if the mass and the radii of the components (be they ions, molecules, or atoms) are known.
- X-ray crystallography is a useful technique that can also determine the arrangements of atoms in the unit cell.
- The liquid crystalline state is a transitional phase that exists between the solid and liquid phase for some molecules. Typically, these molecules have long, rigid structures with strong dipole moments. ◀

11.9 Metals

To fill a cavity in a tooth, a dentist cleans out the hole and packs it with a composite material made of glass and plastic resin, or with a "silver" filling, to arrest the growth of the enamel-eating bacteria. The composite filling is popular, but because the silver filling is relatively inexpensive, there is still a need for it. Why did we put quotation marks around the word silver? The filling is actually a mixture of about 50% mercury and 50% other metals (including silver, tin, and copper). This is one of countless applications of metals in our lives. In this section, we will examine the properties of metals and learn why these elements are so useful.

Using a cast iron skillet to cook our eggs and bacon used to be the rule, not the exception. We had to have a mitten handy to pick the skillet up off the stove, or our hand would be badly burned. Metals, such as the iron in the skillet, have long been known to conduct heat and electricity. Why do metals conduct heat? One theory that explains these properties (and others) is called **band theory**. This theory addresses how adjacent metal atoms interact. For example, two lithium atoms can interact with each other by sharing their electrons. This gives rise to our postulate that dilithium can exist. A closer look at the molecular orbitals of dilithium reveals that the valence electrons come from the 2*s* orbital. A bonding orbital lies below the energy of the 2*s* orbital, and an antibonding orbital lies above the energy of the 2*s* orbital (Fig. 11.36). The valence electrons are placed in the molecular orbitals in such a way as to fill the bonding orbital and leave the antibonding orbital empty.

Band theory – A metal as a lattice of metal cations spaced throughout a sea of delocalized electrons.

What happens if three lithium atoms participate in bonding? The result of mixing three atomic valence orbitals gives rise to three valence molecular orbitals (Fig. 11.37). One of these orbitals is bonding, one is antibonding, and one is nonbonding. A **non–bonding molecular orbital** can be thought of as one that neither enhances nor diminishes the molecular bonding. The non–bonding molecular orbital arises when there is an odd number of atomic orbitals that mix in a given energy level. The result of this mixing is an odd number of MOs in the energy level (one of the orbitals is the nonbonding MO). Three lithium atoms contribute a total of three valence electrons to the valence molecular orbital diagram. Placing them in the diagram results in a full bonding MO and a half-filled nonbonding MO.

Non–bonding molecular orbital – A molecular orbital that is not involved in bonding or antibonding.

The overlap of four lithium atoms results in two bonding orbitals (completely full of electrons) and two antibonding orbitals (empty), as shown in Fig. 11.38. An equal number of bonding molecular orbitals and antibonding molecular orbitals are formed. And the bonding orbitals reside at an energy level below that of the antibonding orbitals.

If we increase the number of atoms that participate in forming molecular orbitals to eight, we find that half of the orbitals (the bonding orbitals) fill with electrons. The other half of the molecular orbitals is comprised of the empty antibonding orbitals. In addition, the orbitals within the bonding and antibonding levels begin to have noticeably different energies. This difference arises due to the requirements of the symmetry for molecular orbitals; as the size of the molecular orbital increases, the number of nodes within the molecular orbitals increases and results in a decrease in the magnitude of the molecular orbital's energy. Overall, the energy difference between the antibonding and the bonding molecular orbitals becomes smaller (bonding MOs increase in energy and antibonding MOs decrease in energy).

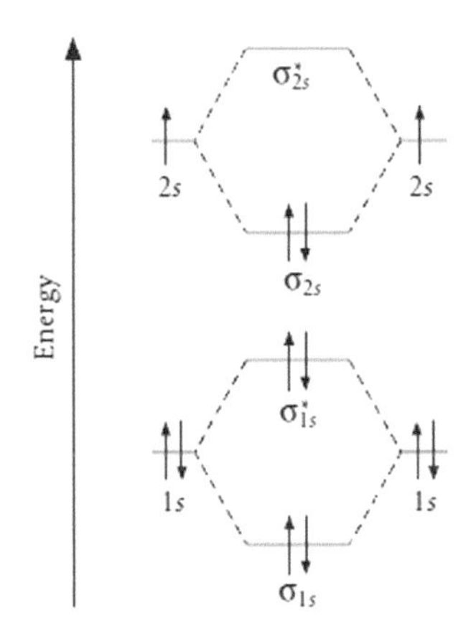

Fig. 11.36 Molecular orbital diagram of dilithium. Two bonding and two antibonding MOs make up the diagram for dilithium. Placing electrons in this MO diagram provides a full bonding MO and an empty antibonding MO

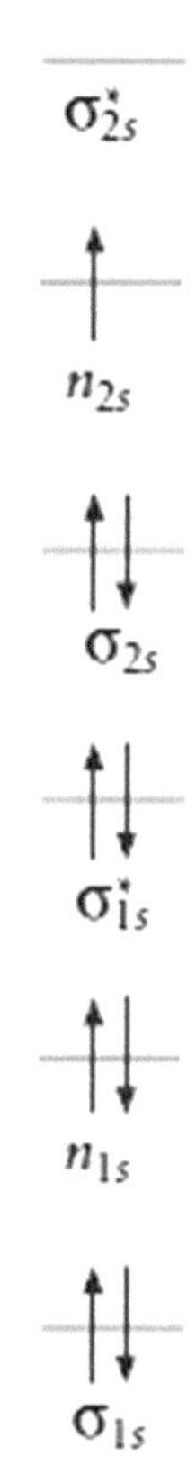

■ **Fig. 11.37** Molecular orbital diagram of trilithium. Note that the energy of the MOs derived from 2*s* orbitals is approaching the energy of the MOs derived from 1*s* orbitals

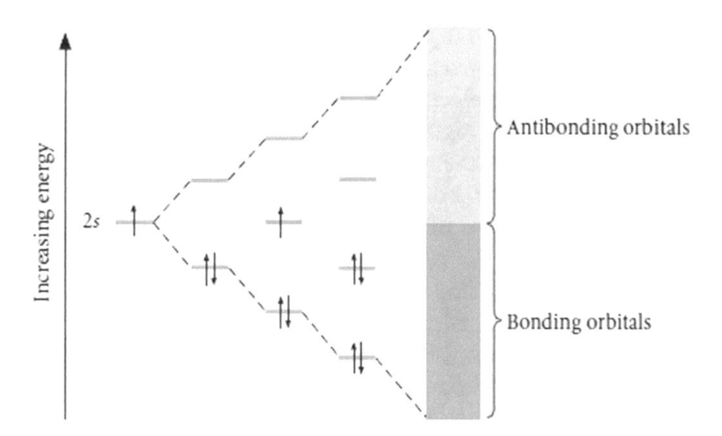

■ **Fig. 11.38** Molecular orbital diagram of lithium metal. As the number of lithium atoms participating in bonding increases, the regions of antibonding and bonding molecular orbitals become continuous

As more and more atoms combine to make a metallic crystal, the difference in energy between bonding and antibonding orbitals becomes negligible. The result is a band of molecular orbitals constructed from the overlap of *s* orbitals. In metals, the range of energies in this band is often wide enough to overlap with similarly constructed bands of *p* and *d* orbitals. Many of these bands are empty, or only partially filled, with electrons. For any metal not at absolute zero, many of the electrons occupy the unfilled portions of the higher energy bands. Because the molecular orbitals encompass the entire metallic crystal, the electrons are free to roam from one end of the metal crystal to the other.

Crystals made from nonmetal elements, however, have a slightly different arrangement. As we see from ■ Fig. 11.38, the lower energy bands are accompanied by an empty higher energy band that doesn't overlap. For instance, carbon has a band of molecular orbitals that results from the overlap of the 2*s* and 2*p* atomic orbitals. Higher in energy lies a band of molecular orbitals that result from the overlap of the 3*s* and 3*p* atomic orbitals. The lower-energy, partially filled bonding molecular orbital band is known as the **valence band**. This band contains the valence electrons. The empty higher-energy molecular orbital band is the **conduction band**. Some nonmetal elements have conduction bands that are much greater in energy than their valence band. Still others have conduction bands that overlap or nearly overlap with the valence bands.

Valence band – The collection of filled molecular orbitals on a metal.

Conduction band – The collection of empty molecular orbitals on a metal.

Band theory describes a metal as a lattice of metal cations spaced throughout a sea of delocalized electrons. How does this theory explain why metals are ductile, malleable, conductive, and shiny? Imagine, for example, deforming a metal bar with a hammer and displacing some of the metal ions. The sea of delocalized electrons can immediately adjust to the change, so the overall structure doesn't become significantly weaker as it is bent. Metals, then, are **ductile** (able to be pulled) and **malleable** (able to be pounded) to different degrees. Additionally, because electrons can easily move from one end of the molecular orbital to the other, metals can conduct electricity.

Ductile – Able to be pulled or drawn into a wire.

Malleable – Capable of being shaped.

So, to repeat our earlier question, why do metals conduct heat? Kinetic energy can be conveyed easily across a crystalline lattice that consists of positive ions within a sea of electrons; metals conduct heat easily. In other words, kinetic energy can be transferred from positive ion to positive ion with ease. This stands in contrast to the difficulty involved in transferring kinetic energy across an ionic or molecular solid, where the kinetic energy must be transferred from cation to anion to cation (or from molecule to molecule). This explains why metal ice cube trays feel colder than plastic ice cube trays. The metal trays, as thermal conductors, are better able to pull the heat from our hand. The plastic ice cube trays are made of a thermal insulator (there isn't enough thermal energy available to promote electrons into the conduction band in the plastic). Or, to say it differently, the valence electrons in plastic are tied to covalent bonds specifically joined to a given pair of atoms and are not free to move around.

Band theory also explains why metals are shiny. Because the molecular orbitals within the valence bond are so plentiful and close together, electrons can absorb and release almost any wavelength of light from the infrared to the ultraviolet. The absorption promotes the electron into an unfilled molecular orbital within the band; the release of a photon returns the electron to its original location in the band. The result is that metals reflect light, giving rise to the luster that we associate with them. White paper also reflects light but does so diffusely (in all directions). Some metals have a distinctive color. For example, gold has a yellow hue. This is due to the inefficiency of the metal when it reflects certain frequencies of absorbed light. These metals are still shiny; they just don't reflect all wavelengths of light equally.

Band theory is used in determining the color of paints, identifying the insulating value of materials in our home, and in the hospital as well, in the form of heart monitors, IV pumps, calculators, and many other instruments. What do such instruments have in common? They are all controlled by small computers, and at the heart of all computers is the element silicon. Why silicon? Its **band gap** is the key. A band gap is a small energy gap that exists between the valence band and the conduction band, as illustrated in ◘ Fig. 11.39. The existence of a band gap in silicon suggests that a certain amount of energy is needed in order to promote an electron from the valence band into the conduction band. If the gap is relatively small, as it is in silicon, the metal is a **semiconductor**. In these cases, the amount of energy needed to promote an electron into the conduction band is relatively small. (See ◘ Table 11.5 for a list of some compounds and their band gaps.) When the temperature is low, there isn't enough kinetic energy to promote many electrons into the conduction band, and the metal doesn't conduct much electricity. Increase the temperature of the semiconductor, and the conductivity increases. Silicon (band gap = 106 kJ/mol) and germanium (band gap = 64 kJ/mol) are good examples of semiconductors.

Band gap – The energy gap that exists between the valence band and the conduction band.

Semiconductor – A substance where the band gap is relatively small.

If the band gap is sufficiently large that there are barely any electrons at room temperature in the conduction band, the material is an **insulator**. Only a tiny amount of electrical conductivity can be found in an insulator.

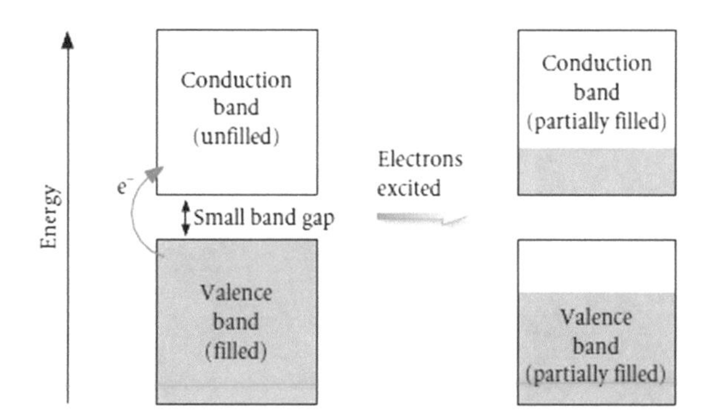

Fig. 11.39 The band gap in silicon

Table. 11.5 Band gaps of some common materials

	Material	Band gap (kJ/mol)
Insulator	Carbon (diamond)	502
(>300 kJ/mol)	Zinc sulfide	350
Semiconductor	Silicon	106
(50–300 kJ/mol)	Selenium	215
	Germanium	64
Conductor	Lithium	0
(<50 kJ/mol)	Tin (gray tin)	7.7

However, as an insulator becomes hotter, its conductivity increases. Our plastic ice cube tray is a good example of an insulator. Diamond is another good insulator (the band gap in diamond is 502 kJ/mol). In short, when the band gap is large, the material acts as an insulator. Remove the gap, and the material is a **conductor**. For example, compare the band gap in diamond to the amount of energy required to promote an electron into an unfilled orbital in a conductor such as lithium, roughly 10^{-45} kJ/mol.

Insulator – A substance that contains a very large band gap.

Conductor – A substance that has a very small or non-existent band gap.

Adding small amounts of other elements to the metal, effectively introducing imperfections in the lattice, can modify the behavior of a semiconductor. For example, if we **dope** (add an atom or other compound to) silicon (Group 14, IVA) with an element that has more valence electrons than silicon (such as phosphorus, Group 15, VA), the extra valence electrons must be placed into the conduction band (Fig. 11.40). As a consequence, the semiconductor becomes capable of conducting current. The result is called an **n-type semiconductor** because the addition of the phosphorus adds extra negative charges (electrons) to the system. However, if we dope silicon with an element containing fewer valence electrons (such as aluminum, Group 13, IIIA), there will not be enough valence electrons to fill the valence band. The result: Electrons in the valence band can be easily excited to move to an unoccupied molecular orbital. This semiconductor is electrically conductive if we apply a small electric field. We call it a **p-type semiconductor** because the addition of aluminum can be thought of as introducing positive holes in the valence band. The properties of the n- and p-type semiconductors make them quite useful for the production of small electronic devices.

Dope – To add an impurity to a pure semiconductor to alter its conductive properties.

n-type semiconductor – A semiconductor containing electrons in the conduction band.

p-type semiconductor – A semiconductor that is electrically conductive if we apply a small electric field. The semiconductor contains gaps in the valence band.

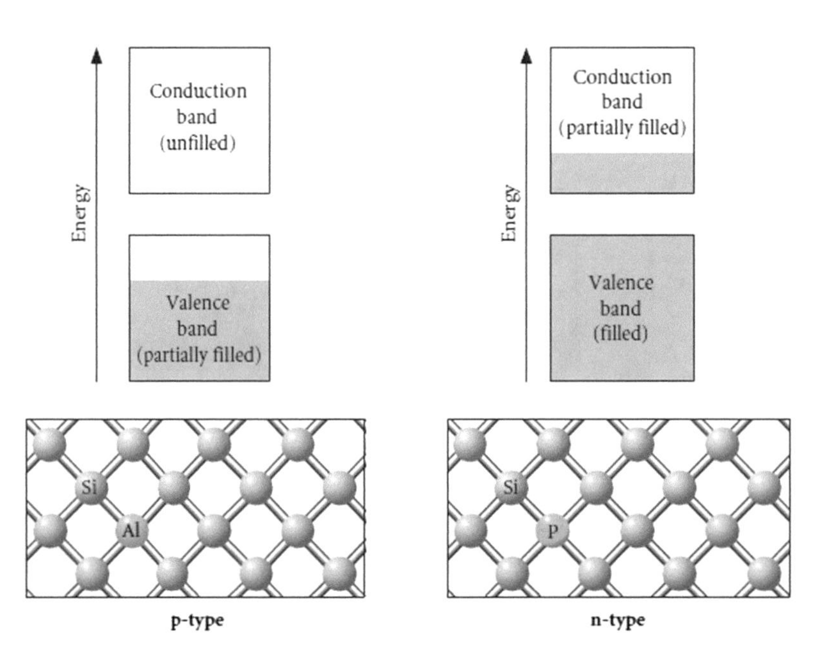

Fig. 11.40 Doping the semiconductor. An n-type semiconductor contains extra electrons added to the conduction band. A p-type semiconductor is missing some electrons from the valence band

Example 11.12—The Color of Paint

The mineral greenockite (CdS) is best known as cadmium yellow and is used as a yellow pigment in paints. Cadmium sulfide has a medium-sized band gap (4.2×10^{-19} J/photon). Judging on the basis of the information in Table 11.5, is greenockite an insulator, a semiconductor, or a conductor? What color light does cadmium sulfide absorb?

Asking the Right Questions

Calculation of the band gap will determine whether the compound is an insulator, conductor, or semiconductor. Our question should then be, what is the band gap? What wavelength of light corresponds to the energy associated with the band gap?

Solution

The band gaps in Table 11.5 are listed in units of kilojoules per mole, whereas this exercise gives the size of the band gap for greenockite in joules per photon. We can use Avogadro's number to make the units the same.

$$\frac{4.2 \times 10^{-19}\ \text{J}}{\text{photon greenockite}} \times \frac{6.022 \times 10^{23}\ \text{photons}}{\text{mol greenockite}} \times \frac{1\ \text{kJ}}{1000\ \text{J}} = \frac{250\ \text{kJ}}{\text{mol greenockite}}$$

This value places cadmium sulfide in the semiconductor region. Using the equations $E = h\nu$ and $c = \lambda\nu$, we find a frequency of 6.3×10^{14} Hz and a wavelength of 470 nm. This corresponds to the absorption of blue light. A solution of cadmium sulfide appears yellow to the eye.

Are Our Answers Reasonable?

The calculation agrees with our expectation that cadmium sulfide (aka cadmium yellow) is yellow.

What If…?

…the question asked for the band gap corresponding to a compound that absorbed light of 600 nm? What is the bad gap?

(*Ans: 3.31×10^{-19} J/photon or 199 kJ/mol*)

Practice 11.12

What is the largest wavelength of light, in nanometers, that can be absorbed by a particular semiconductor if the band gap is determined to be 0.95 eV? (1 eV/atom = 96.485 kJ/mol).

Perhaps the greatest benefit that a semiconductor offers is its ability to act as a **photovoltaic device**, in which light energy can be used to generate an electric current. Photovoltaic cells capable of over 30% efficiency have been designed. How is the sunlight converted into electricity? When silicon is doped with two elements such as

Fig. 11.41 The solar-powered car is currently being tested as a replacement for the gasoline auto that we use today. This solar-powered prototype is not too far from what we may be driving in the near future. Note the photovoltaic cells on the roof of the car. *Source* ▶ https://upload.wikimedia.org/wikipedia/commons/7/78/Solar_car3.jpg

phosphorus (Group 15, VA) and gallium (Group 13, IIIA), the semiconductor contains both electron-rich and electron-deficient areas. Light energy throughout the visible and IR range (the particular range of wavelengths depends on the chemical composition of the semiconductor) provides enough energy for electrons to "jump the band gap and enter the conduction band. This enables both negative charges (electrons) and positive holes to move so that the flow of electrons can be harvested as energy.

Photovoltaic device – A system that converts light energy into an electric current.

Photovoltaic cells have been used in many applications, from the solar-powered calculator to solar-powered water heaters, space stations, and satellites. The use of photovoltaic cells to make a solar-powered car, such as that shown in Fig. 11.41, is the next big hurdle. Because our world demands fossil fuels as a source of energy for transportation, and because of our limited supply of this natural resource, alternative-energy automobiles will one day be commonplace. Currently, auto makers and researchers have been able to develop solar-powered cars that can travel an average of 50 miles an hour relying solely on the photovoltaic cell. However, despite a precipitous drop in costs to implement the use of photovoltaic cells over the last 30 years, sales of solar-powered vehicles are still dwarfed by sales of those powered by fossil fuels and, with increasing regularity, battery-powered vehicles.

11.9.1 Alloys

Metals have properties that make them very useful for many applications. However, the limited properties of pure metals don't lend themselves well to any more than routine tasks. For example, in a medical center, surgeons have a relatively limited number of choices in materials when replacing a hip joint. They need something as strong and durable as metal, but they don't use solid copper, silver, or gold. Why don't they use a coinage metal for a hip joint? Many of the pure metals are too malleable and ductile to withstand the forces in a human body. A gold hip joint would easily bend when a patient walked with it. Other factors, such as the cost and toxicity of some of the pure metals, are also important. In fact, it's rare that we see pure metals used in the "real world". Because there are only a limited number of pure metals, and because the properties of those pure metals are often not what is required, we look to mixtures of metals to give us the properties we need.

A homogeneous mixture of two or more metals is known as an **alloy**. The elements are typically mixed together in the molten state and then allowed to cool to form the alloy. Alloys are useful because their properties can be adjusted by changing the ratio in which the elements are combined. (The Bronze Age is named after the discovery of one of the first of these metal blends.) Typically, alloys are less ductile and less malleable than pure metals. Table 11.6 lists some of the common alloys, along with a typical composition for each and several uses to which each is put. Manufacturers vary the actual composition of each type of alloy to change its properties slightly.

Alloy – A solution of two or more metals.

Two different classes of alloys exist, the **substitutional alloys** and the **interstitial alloys**, illustrated in Fig. 11.42. Substitutional alloys include metals such as brass, sterling silver, pewter, and solder. In a substitutional alloy, specific atoms within the lattice structure of a metal are replaced with other metal atoms. In the example of sterling

Table. 11.6 Common alloys and their uses

Name	Common composition (mass %)	Uses
Substitutional		
Brass	90 Cu : 10 Zn	Decorative fittings
Pewter	85 Sn : 7 Cu : 6 Bi : 2 Sb	Plates, jewelry, figurines
Medal Bronze	92–97 Cu : 1–8 Sn : 0–2 Zn	Medals, artwork
14-carat gold	58 Au : 30 Ag : 12 Cu	Jewelry
Sterling Silver	92.5 Ag : 7.5 Cu	Jewelry, kitchenware
Coinage Silver	90 Ag : 10 Cu	Silver coins
Plumber's Solder	67 Pb : 33 Sn	Pipe Solder
Interstitial		
Steel	98–99.9 Fe : 0–2 C	Building materials
Stainless steel	> 10.5 Cr : < 89.5 Fe	Eating utensils, structural materials

11

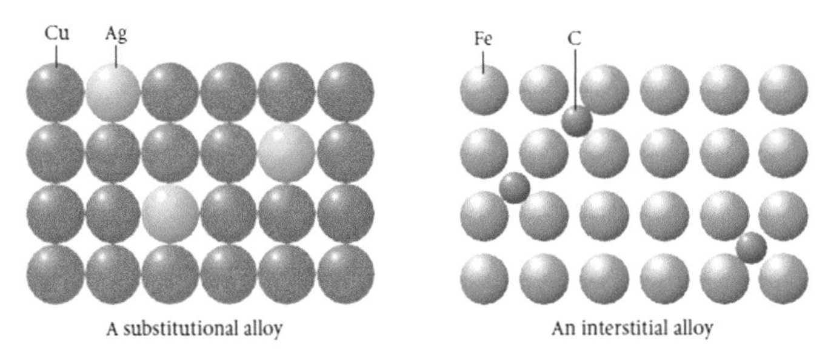

Fig. 11.42 Substitutional versus interstitial alloys

silver, an average of 7 silver atoms out of every 100 are replaced with copper atoms. Interstitial alloys are made of metals with "impurities trapped in the spaces of the lattice structure. By far the most important of these is steel. Steel, which we discussed in ► Sect. 7.2, contains carbon atoms located in the spaces between the atoms in an iron atom lattice. Once again, we see that imperfections in a metal lattice can be very useful.

Substitutional alloys – An alloy consisting of direct replacements of solvent atoms with solute atoms.

Interstitial alloys – An alloy made from the addition of solute metals placed in the existing spaces within a solvent metal.

11.9.2 Amalgams

An **amalgam** is an alloy containing the metal mercury. Dentists use an amalgam for filling teeth by mixing a powder containing silver and tin and then adding mercury. The amalgam immediately begins to form at room temperature, making a complex and very hard mixture of Ag_2Hg_3 and Sn_7Hg.

$$14Ag_3Sn + 65Hg \rightarrow 21Ag_2Hg_3 + 2Sn_7Hg$$

Other amalgams do exist. For instance, the amalgamation of gold with mercury is one way in which gold dust was extracted from powdered gravel in the California gold rush era of the mid-to-late nineteenth century. Because of the hazards of mercury poisoning, the use of mercury has been supplanted by other means of extracting gold, as we discussed earlier. Unfortunately, it continues to be used by individuals and small companies in the poorer countries of the developing world. Its use has led to mercury poisoning of the workers and the local water supplies. Other amalgams include zinc and cadmium, used in batteries to increase their lifetime.

Amalgam – A solution made from a metal dissolved in mercury.

The Bottom Line

- Water and other liquids have in common intermolecular forces called van der Waals forces that hold their molecules together. These forces are weaker than covalent or ionic bonds.
- Among these forces are dipole–dipole interactions.
- A hydrogen bond is an especially strong interaction that occurs when a hydrogen attached to an oxygen, nitrogen, or fluorine atom is close to a different atom of oxygen, nitrogen, or fluorine.
- Weaker, but collectively important intermolecular interactions in large molecules are known as London forces.
- Water is special among liquids because of intermolecular hydrogen bonds.
- The pressure exerted by the evaporation of a liquid in equilibrium with its surroundings is called its vapor pressure. The boiling point is reached when the vapor pressure is equal to the surrounding pressure.
- A heating curve describes the changes in phase and temperature that occur as a substance is heated at constant pressure.
- A phase diagram describes the changes in phase as pressure and temperature are changed.
- Intermolecular interactions can be used to explain many physical properties of water, including its viscosity, surface tension, and capillary action.
- In a crystalline solid, the atoms, ions, or molecules are highly ordered in repeating units that make up the lattice of the crystal. Amorphous solids, although their atoms, ions, or molecules inhabit rigid and fixed locations, lack the long-range order in a crystal.
- The unit cell is the basic repeating unit that, by simple translation in three dimensions, can be used to represent the entire crystalline lattice.
- The simple cubic unit cell, the body-centered cubic unit cell, and the face-centered cubic unit cell make up most of the crystalline lattices of the metallic elements.
- We can calculate the volume and density of a unit cell by means of simple geometry. In doing so, we assume the atoms are hard spheres.
- Metals often adopt a closest packed structure. These structures include the hexagonal closest packed structure and the cubic closest packed structure.
- The liquid crystalline state is a transitional phase that exists between the solid and liquid phase for some molecules. Typically, these molecules have long, rigid structures with strong dipole moments.
- Band theory describes why metals are electrical and thermal conductors, shiny, malleable, and ductile.
- The valence band and the conduction band result from the nearly infinite number of atomic orbitals that overlap to form the molecular orbitals in a metal. The band gap can be used to assess whether a compound is a conductor, a semiconductor, or an insulator.
- Alloys are mixtures of two or more metals. They include the interstitial and substitutional alloys. Amalgams are special alloys of mercury.

Section 11.1 An Introduction to Water

Skill Review

1. Describe the differences between intermolecular and intramolecular forces.
2. Describe the similarities between intermolecular and intramolecular forces.
3. Explain how intermolecular forces between molecules arise.
4. How is the strength of an intermolecular force related to the phase of matter?
5. Using electronegativity differences, rank these bonds from least to most polar:
   O–H C–H S–H N–H
6. Using electronegativity differences, rank these bonds from least to most polar:
   O–C Cl–C F–H P–O
7. In each of these statements, insert the term intermolecular or intramolecular to describe the process taking place.
   a. Water freezes when ________ attractions are forming.
   b. The production of carbon and sulfur from CS_2 indicates that ________ bonds have been broken.
   c. When dry ice sublimes, ________ bonds are broken.
   d. In general, ________ bonds are stronger than ________ bonds.
8. The terms polar and polarizable have both been used in discussions of intermolecular forces.
   a. Distinguish between these two important terms.
   b. Which of these contains the more polar bond, CO or NO?
   c. Which of these is the more polarizable atom, He or Ne?

Chemical Applications and Practices

9. In a sample of water there are two types of hydrogen-to-oxygen attractions taking place. In one attraction the average distance between the hydrogen and oxygen is 0.101 nm. In the other, the distance is 0.175 nm. Which distance represents an intermolecular attraction, and which represents an intramolecular attraction? Which of the two is the stronger attraction?
10. Carbon dioxide is generated as a result of metabolism and as a result of the combustion of carbon-based materials. This substance is a gas under "normal conditions. Explain why this is so.
11. We emphasize the theme of ▶ Sect. 11.1 by noting that at room temperature, water is a liquid and methane is a gas. State the key ideas that explain this, and relate them to the title of the section, "The Structure of Liquids: An Introduction to Molecular Forces.
12. Based on the discussion in this section, is butane, C_4H_{10}, more likely to be a liquid or gas at room temperature and 1 atmosphere pressure? Explain the answer.

Section 11.2 The Structure of Liquids

Skill Review

13. In each of these pairs of organic compounds, select the one that would have the higher boiling point. Then, using intermolecular forces, describe the basis for the selection.
    a. Pentane or 2,2-dimethylpropane
    b. Hexane or cyclohexane
    c. Pentane or hexane
    d. Pentane or water
    e. Pentane or methane

$$CH_3{-}CH_2{-}CH_2{-}CH_2{-}CH_3$$

Pentane

$$CH_3{-}CH_2{-}CH_2{-}CH_2{-}CH_2{-}CH_3$$

Hexane

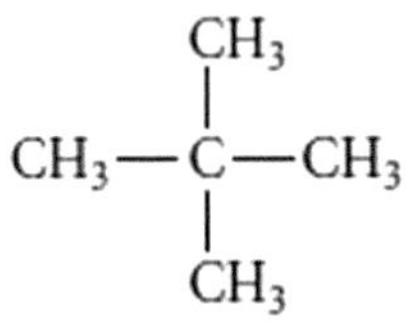

$$CH_3{-}C(CH_3)_2{-}CH_3$$

2,2-Dimethylpropane

$$(CH_2)_6 \text{ (ring)}$$

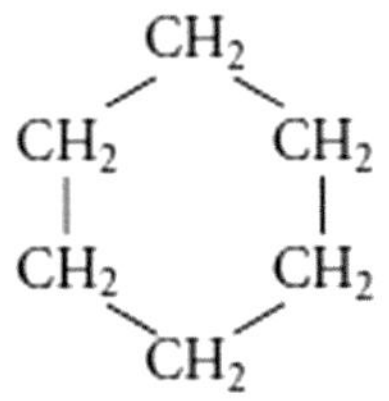

Cyclohexane

14. In each of these pairs of organic compounds, select the one that would have the higher boiling point. Then, using intermolecular forces, describe the basis for the selection.
    a. Methanol (CH_3OH) or ethanol (CH_3CH_2OH)
    b. Water or hydrogen sulfide
    c. Pentane or ethanol
    d. Hydrogen sulfide or hydrogen fluoride
    e. Ammonia (NH_3) or methanol
15. Which would have the higher melting point, CH_3OH or C_3H_8? Explain the choice and state the intermolecular forces responsible for the differences in melting points.
16. Put these in order from lowest to highest boiling point: N_2, NH_3, C_2H_6. Explain the choice and state the intermolecular forces responsible for the differences in boiling points.
17. In these pairs, compare the polarity of the bonds shown. Decide which bond in each pair is more polar and show in which direction the electron density would be shifted.
    a. B–O and B–S
    b. P–S and P–Cl
    c. O–H and O–C

18. In these pairs, compare the polarity of the bonds shown. Decide which bond in each pair is more polar and show in which direction the electron density would be shifted.
    a. C–O and C–S
    b. N–S and N–H
    c. C–H and C–Cl
19. Name the type of intermolecular forces that must be overcome in order to change each of these from the liquid state to the gas state.
    a. CH_3CH_2OH
    b. CO_2
    c. SF_6
    d. HF
20. Name the type of intermolecular forces that must be overcome in order to change each of these from the liquid state to the gas state.
    a. CH_3OH
    b. CS_2
    c. CCl_4
    d. COS
21. Select the molecule in each pair that is less polarizable.
    a. CF_4 or CCl_4
    b. H_2O or H_2Se
22. Select the molecule in each pair that is less polarizable.
    a. HI or HF
    b. CH_4 or COS
23. Which of these compounds are likely to participate in hydrogen bonding? For those that do not qualify, explain what is lacking.
    a. CH_3OH (methanol)
    b. NH_3 (ammonia)
    c. C_6H_6 (benzene)
    d. $C_2H_5OC_2H_5$ (diethyl ether)

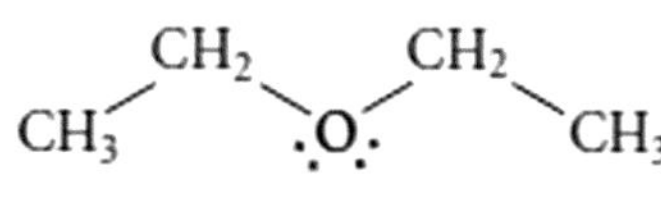

Diethyl ether

24. Which of these compounds are likely to participate in hydrogen bonding? For those that do not qualify, explain what is lacking.
    a. H_2S
    b. NCl_3
    c. B_2H_6
    d. CH_3NH_2
25. Could SO_2 be involved in hydrogen-bonding with other compounds? Explain the answer.
26. Could water hydrogen-bond with N_2? Explain the answer.

Chemical Applications and Practices

27. "Wake up and smell the coffee!" is a phrase that we may have used to catch someone's attention. Perhaps less compelling, but more accurate, would be "Wake up and smell the 2-methylfuran". This compound is one of several that are typically responsible for the aroma of coffee. Its structure is shown here.
    a. Which bond in the molecule do we think is the most polar?
    b. What intermolecular forces exist between the molecules of 2-methylfuran?

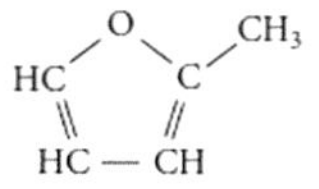

O, CH_3, HC, C, HC — CH

28. Answer these questions on the basis of the answer to Problem 27.
    a. Would we predict that this compound would have a high or a low normal boiling point?
    b. Would we predict that this compound would be a solid, a liquid, or a gas at 1 atm and 25 °C? Explain.
29. Suppose a small lab accident involved spilling the same amounts of hexane and hexanol. Which compound would we expect to evaporate first? Using intermolecular forces, explain the rationale behind the answer.

$$CH_3—CH_2—CH_2—CH_2—CH_2—CH_3$$

Hexane

$$CH_3—CH_2—CH_2—CH_2—CH_2—CH_2—OH$$

Hexanol

30. Would we predict a cup of water or a cup of pure ethanol to evaporate more rapidly? Using intermolecular forces, explain the rationale behind the answer.
31. Under the proper conditions, London forces are sufficient to liquefy the noble gases. Of Ne, Ar, and Kr, which would possess the stronger London forces? Explain the basis of the selection.
32. Which of these compounds would we predict to have the greatest intermolecular forces, F_2, Cl_2, Br_2, or I_2?
33. The Group 15 (V) elements form compounds with hydrogen as follows:

	NH_3	PH_3	AsH_3	SbH_3	BiH_3
Boiling point (°C)	−33	−88	−63	−17	16

a. Temporarily excluding NH_3, explain the general trend shown here.
b. Explain why NH_3 deviates from the general trend.

34. HF and H_2O are both compounds that exhibit hydrogen bonding. The hydrogen bonding that takes place between HF molecules is stronger than those in H_2O, yet H_2O has the higher boiling point. Explain why.

11

Section 11.4 Impact of Intermolecular Forces on the Physical Properties of Liquids: Phase Changes

? Skill Review

35. Explain why the heat of vaporization of water is so much larger than the heat of vaporization of methane.
36. Would we expect the heat of vaporization of CO to be closer to that of water or of CH_4? Explain.
37. Using this vapor pressure data for propanol, C_3H_8OH, make a graph and determine its normal boiling point.

Temperature (°C)	Vapor pressure (torr)
20.0	14.6
40.0	51.9
60.0	152.0
80.0	381.3
100.0	845.7
120.0	1697
140.0	3138

38. Using this vapor pressure data for butanol, C_4H_9OH, make a graph and determine its normal boiling point.

Temperature (°C)	Vapor pressure (torr)
20.0	4.2
40.0	14.8
60.0	59.3
80.0	163.5
100.0	389.2
120.0	823.3
140.0	1583.1

39. Use the data from problem 37 to determine the enthalpy of vaporization ($\Delta_{vap}H$) for propanol.
40. Use the data from problem 38 to determine the enthalpy of vaporization ($\Delta_{vap}H$) for butanol.
41. Using the information from problem 37 and 39, predict the vapor pressure of propanol at 0 °C.
42. Using the information from problem 38 and 40, predict the vapor pressure of butanol at 0 °C.
43. Calculate the final temperature of a system when 20.0 g of ice at 0 °C is mixed with 70.0 g of water at 15 °C.
44. How much ice would be needed to lower the temperature of 300 mL of coffee from 90.0 °C to 65.0 °C? Assume that the coffee behaves like pure water and has a density of 1.00 g/mL.

Chemical Applications and Practices

45. Water's relatively high specific heat (4.184 J/g · °C) makes it an ideal coolant for automobile engines. How much heat from an engine would be absorbed if 12.0 kg of cooling system water were heated from 25 °C to 75 °C?
46. Ethanol (C_2H_5OH) is the alcohol found in intoxicating beverages. It has a heat of vaporization of approximately 39.0 kJ/mol. How many joules would be released when 42.0 g of ethanol vapor condensed into the liquid phase?
47. Propane ($CH_3CH_2CH_3$) is a fuel typically used to operate our outdoor barbecue grills. This compound is often found inside pressurized cylinders.
    a. What type of intermolecular forces of attraction would we predict to exist in propane?
    b. Given this information, would we predict propane to exist as a gas or as a liquid at room temperature and 1.0 atm pressure?
    c. Why is propane often found in liquid form (as it is inside the barbecue grill cylinder)?
48. Chlorine exists, at typical room conditions, as a diatomic gas (Cl_2). However, under higher pressures (greater than 250 kPa), the gas can be converted to a liquid. Chlorine has many uses, chief of which is to produce bleaching agents.
    a. What type of bond exists between two chlorine atoms in Cl_2?
    b. What type of intermolecular force(s) are responsible for the gas molecules entering the liquid phase?
49. Oxygen and tellurium both form dihydrogen compounds (H_2O and H_2Te). Explain why water molecules participate in hydrogen bonding, whereas H_2Te does not.
50. a. Commercial rubbing alcohol is typically sold as a 70% solution of 2-propanol ($CH_3CHOHCH_3$). Explain why, when it is in contact with our skin, we soon feel a cooling sensation.
    b. The heat of vaporization of 2-propanol is approximately 42 kJ/mol. How much heat would be required to evaporate 5.1 g of 2-propanol?
51. Using the values given in the chapter, calculate the amount of heat needed to change a 15.0-g ice cube at – 5.0 °C to 15.0 g of steam at 100.0 °C.
52. Using the values given in the chapter, calculate the amount of heat that is liberated when a 1.0-kg bucket of steam at 100.0 °C is cooled to give a 1.0-kg bucket of water at 25.0 °C.

Section 11.5 Phase Diagrams

Skill Review

53. Use the detailed phase diagram for water (Fig. 11.18) and describe the changes that occur in pressure and in the physical state of water and ice as the pressure is lowered from 10^9 pascals to 1 Pa at 300 K.
54. Use the detailed phase diagram for water (Fig. 11.18) and describe the changes that occur in pressure and in the physical state of water and ice as the temperature is raised from 0 to 600 K at 10^6 pascals.

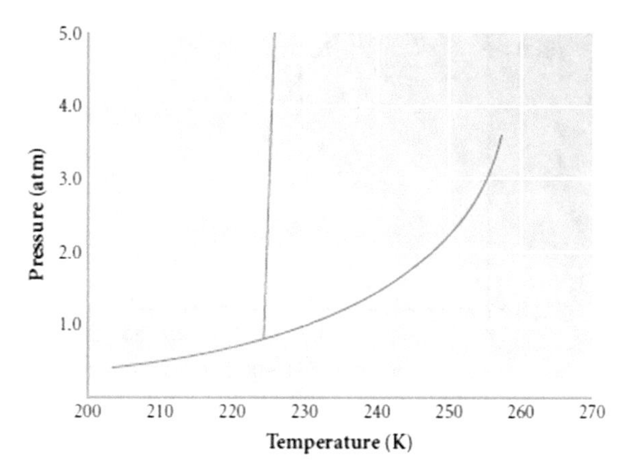

Use the phase diagram here for problems 55–62.

55. Using the phase diagram given, determine the following:
    a. The normal boiling point
    b. The triple point
    c. In what phase would we find the compound at – 25 °C and 1 atm?
56. Using the phase diagram given, determine the following:
    a. The critical pressure
    b. The critical temperature
    c. In what phase would we find the compound at –63 °C and 3 atm?
57. Using the phase diagram given, indicate what phase transition(s) occur as the compound goes from 220 K and 0.5 atm to 220 K and 3 atm.
58. Using the phase diagram given, indicate what phase transition(s) occur as the compound goes from 220 K and 1.0 atm to 240 K and 1.0 atm.
59. Using the phase diagram given, indicate what phase change will occur when the compound is heated from –45 to –53 °C at 1 atm.
60. Using the phase diagram given, indicate what phase change, if any, will occur when the pressure is adjusted from 2.0 atm to 1.0 atm at constant temperature of – 33 °C.
61. Using the phase diagram given, is it possible to liquefy a gas sample of this compound at –33 °C using increasing pressure? Explain why or why not.
62. Using the phase diagram given, is it possible to form a solid from a liquid sample when the temperature is kept constant at –53 °C? Explain the answer.

Chemical Applications and Practices

63. Ethylene (C_2H_4) is one of the most widely used compounds in the manufacture of synthetic materials. From the following data, construct a labeled phase diagram similar to the one shown for Problems 55–62. The normal boiling point is approximately –104 °C. The critical point, at 50 atm, is approximately 9.8 °C. The triple point is –169 °C and 0.0012 atm. The normal melting point is –169 °C.
64. Using the diagram constructed in Problem 63, indicate the phase in which ethylene would be found at typical room conditions.
65. Phase diagrams can measure other variables than temperature v. pressure. The figure below is an example of a binary (or two-component) phase diagram, in which the phases of two immiscible minerals changes with composition and temperature. The melting point of pure mineral "A" is 1380 °C, and pure "B" is 1470 °C. When some solid B is added to molten A, the melting point goes down (from the left side) until the lowest melting point temperature is reached, called the "eutectic point. The same thing happens as we add solid A to molten B (from the right side). What possible combinations of A and B could be prepared to have a mixture with a melting point of 1350 °C?

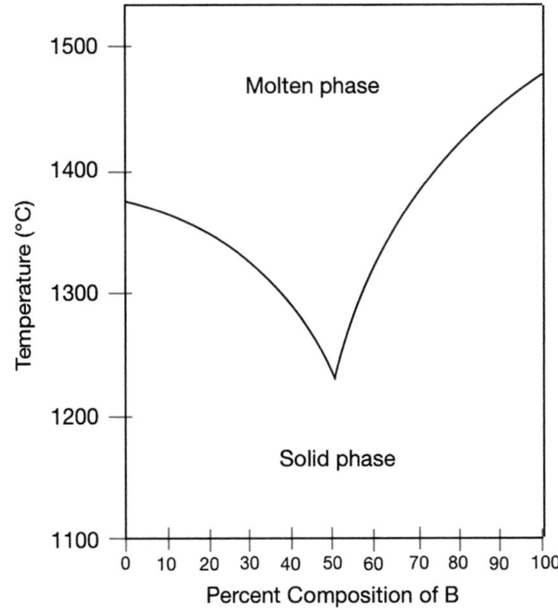

66. Using the binary phase diagram in problem 65, what would be the melting point of a mixture of minerals with a composition that is 38% A and 62% B?

Section 11.6 Impact of Intermolecular Forces on the Physical Properties of Liquids: Other Properties

Skill Review

67. Explain the differences among surface tension, viscosity, and capillary action.
68. Would we predict there to be a relationship among surface tension, viscosity, and capillary action?
69. Draw a diagram of a meniscus that would form from a buret filled with mercury.
70. Draw a diagram of a meniscus that would form from a buret filled with honey.
71. Arrange these substances in order of their increasing viscosity. Explain the order.

    honey water gasoline

72. Which of these compounds do we predict will rise highest within a capillary tube: water, hexane, or ethanol? Explain the prediction.

Chemical Applications and Practices

73. Lubricating motor oils are rated on the basis of their viscosity. The oils in these lubricants are nonpolar molecules. Explain why these hydrocarbon-based oils have a relatively high viscosity.
74. Honey has a very high viscosity. It contains (in large part) many different carbohydrates similar in molecular structure. What specific atoms would we predict to exist in a carbohydrate? Explain the reasoning.

Sections 11.7 and 11.8 The Structure of Crystals

Skill Review

75. What key feature distinguishes a crystalline solid from an amorphous solid?
76. Cite an example of a crystalline solid and of an amorphous solid.

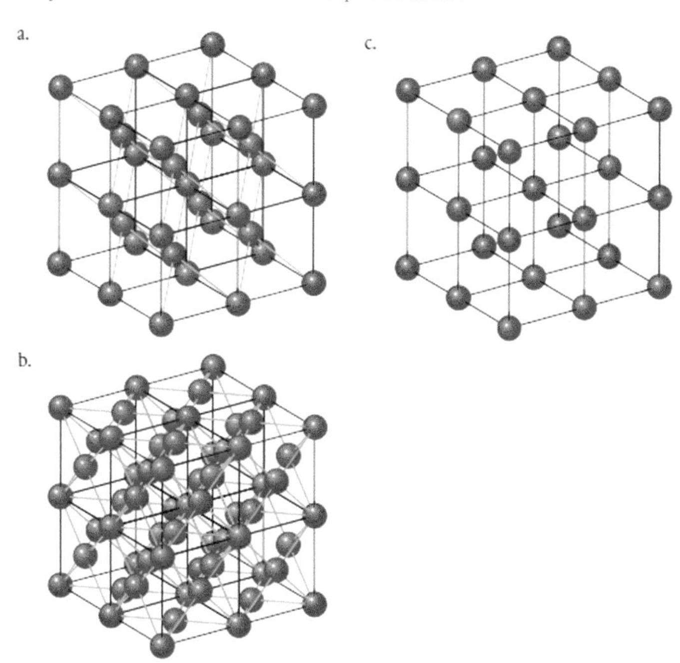

77. Which of the diagrams above represents the body-centered cubic structure of iron?

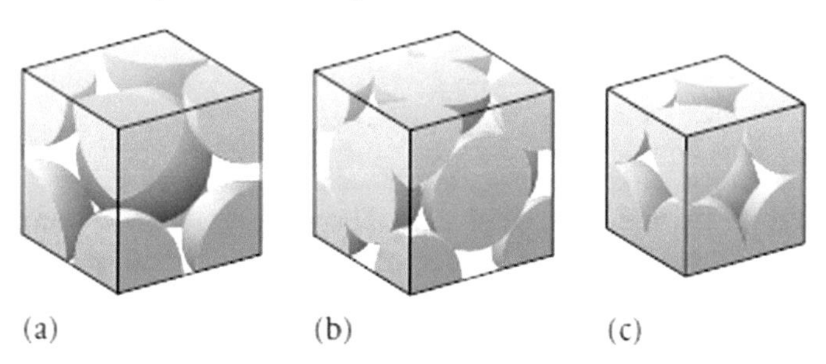

78. Which of the diagrams above represents the face-centered cubic structure of strontium?

79. How many complete atoms would we expect to find in a unit cell made from the simple cubic unit cell, but with additional atoms placed along every edge of the unit cell?
80. How many complete atoms would we expect to find in a unit cell made from the simple cubic unit cell, but with additional atoms placed only at the face of two of the sides of the unit cell?
81. Without any further information, indicate whether each of the following three solids is most likely to be molecular, ionic, or metallic: Solid A is the only one of the three that conducts electricity in the solid state. Solid B melts when placed in boiling water. Solid C dissolves in water, and the resulting solution conducts electricity.
82. As part of a display of fruit, a grocer places two types of apples (red and yellow) in a pattern. The apples are to be placed in a slanted display with a wooden frame. The first layer is made of red apples, and the second layer is placed above the first in the "dimples or gaps. Show how the third layer of red apples would be placed if the grocers, remembering their chemistry, wanted to make the display resemble hexagonal closest packing.
83. Calculate the volume of a fcc unit cell with each of the following atomic radii:
    a. 125 pm b. 210 pm c. 302 pm
84. Calculate the volume of a bcc unit cell if it is composed of atoms with each of the following radii:
    a. 200 pm b. 183 pm c. 164 pm

Chemical Applications and Practices

85. Suppose the label on a can of soup reports 220 mg of sodium per serving of soup. How many grams of NaCl will be present in one serving of the soup?
86. When passing through the small openings within crystal structures, X-rays diffract into a pattern. Say a researcher is deriving the structure of an interesting protein by using X-rays with a frequency of $1.94 \times 10^{19}\ s^{-1}$. What is the wavelength, in nanometers, of the X-rays?
87. The metal nickel has several remarkable uses. When alloyed with other metals, it can form a type of "memory metal" called nitinol, which is a nearly 1:1 mixture of nickel and titanium. Nickel crystallizes in a face-centered cubic arrangement. The density of nickel is 8.90 g/cm^3. From this information, calculate the radius of an atom of this coinage metal. (Currently the percentage of nickel in a U.S. "nickel" coin is approximately 25%.)
88. The element iridium is currently an important part of an extinction theory concerning dinosaurs. Unusual deposits of iridium indicate that a large, iridium-containing meteor may have hit our planet, resulting in a dust cloud from the impact blocking the light from the sun. Iridium crystallizes in a face-centered cubic structure. What is the density of this corrosion-resistant metal whose radius is 136 pm?
89. Titanium metal and its oxide have a wide variety of uses. Its relatively light weight and strength make it ideal in some alloys. Titanium dioxide is found in paints and some lipsticks. Pure metallic titanium crystallizes in a body-centered cubic structure. Titanium has a density of 4.5 g/cm^3. What is the radius of an atom of titanium? What percent of the unit cell is empty space?
90. Metallic chromium has an atomic radius of 125 pm. Judging on the basis of the density of chromium (7.2 g/cm^3), which of three cubic unit cells (simple, face-centered, or body- centered) would we predict for chromium?
91. One way to verify the numerical value of Avogadro's number is to use X-ray data and unit cell calculations. Although the reasoning is the same for any selection, use the following specific information about copper to make the verification. Copper crystallizes with a face-centered unit cell with an edge length of 3.6×10^{-10} m. The density of copper is approximately 8.9 g/cm^3.
92. A hypothetical metal was determined to have a fcc unit cell and a density of 0.185 mol/cm^3. What is the calculated radius of the metal?
93. Assuming that the following metals crystallize in the fcc unit cell, which would we predict to have the greater density, Co (radius = 125 pm) or Rh (radius = 135 pm)?
94. Assuming that the following metals crystallize into a bcc unit cell, which would we predict to have the greater density, V (radius = 131 pm) or Cr (radius = 125 pm)?

11

Section 11.9 Metals

Skill Review

95. Identify the metals in the following list of elements. B, Na, Al, Ca, Cr, Se, Sb, Bi
96. List two physical characteristics that would enable us to differentiate between calcium and carbon (graphite).
97. How many valence electrons are found in an atom of potassium? How many molecular orbitals are found when three atoms of potassium combine?

98. How many valence electrons are found in an atom of calcium? How many molecular orbitals are found when three calcium atoms combine?
99. The following three energies represent the band gap values for three samples. Which of the values in this list is most likely to be that of a metal?... that of a semiconductor? ... that of an insulator? 85 kJ/mol, 450 kJ/mol, 2.5 kJ/mol.
100. Which of these energies would we expect copper metal to possess as its band gap: 0.06 kJ/mol, 73 kJ/mol, or 510 kJ/mol?
101. The particular types of alloys known as amalgams use mercury as a component. Explain, using the delocalized electron bonding model, how the metal components in any alloy allow the atoms of different metals to bond.
102. How does the amalgam picture we created in Problem 101 allow for varying ratios of components, unlike the set ratio of typical compounds?

Chemical Applications and Practices

103. A grocery stocker wishes to illustrate the two types of alloys to his friends by using red and green apples to represent atoms. Explain how this would be accomplished.
104. A grocery stocker creates a stack of soup cans to display two different kinds of soups. Judging on the basis of the pattern in which they are stacked, which type of alloy do they appear to represent?

105. Sodium metal is very reactive, yet under controlled conditions its ability to absorb heat on a per-gram basis can be very useful. Some nuclear reactor designs make use of sodium instead of water as a heat transfer agent.
    a. How many valence electrons does sodium have?
    b. Using band theory, explain why thermal energy causes the electrons in sodium, like other metals, to populate antibonding orbitals.
106. Show the construction of a molecular orbital diagram from the combination of two sodium atoms, of three sodium atoms, and of four sodium atoms.
107. Semiconductors can also be used in lasers. If a laser were constructed and it produced a wavelength of 820 nm, what would be the band gap energy (in joules per photon)?
108. What would be the band gap energy (in kilojoules per mole) for a laser whose wavelength was 720 nm?
109. Doping silicon allows for the production of n-type and p-type semiconductors. Would doping silicon with indium make an n- or a p-type semiconductor? Explain the basis for the answer.
110. Silicon and selenium both have semiconductor properties. Explain why doping both with arsenic would have opposite results in terms of producing n-type or p-type semiconductors.
111. The blue color of the Hope diamond is due to small amounts of boron in the carbon lattice. Predict what electronic properties this diamond must have.
112. What properties would we expect to find in a diamond doped with small quantities of arsenic in the carbon lattice?

Comprehensive Problems

113. The compounds guanine and cytosine are two bases that make up part of the structure of our DNA. Use the structures shown to indicate how three hydrogen bonds can form between these two molecules.

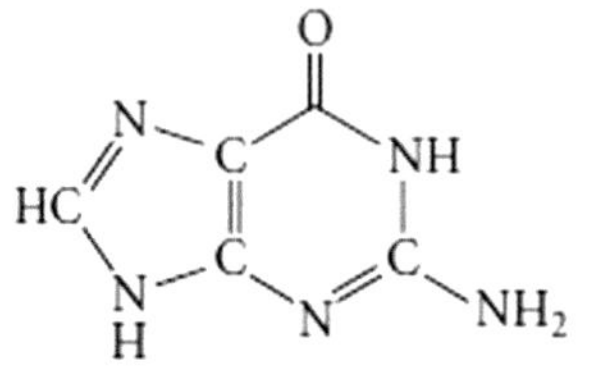

Guanine 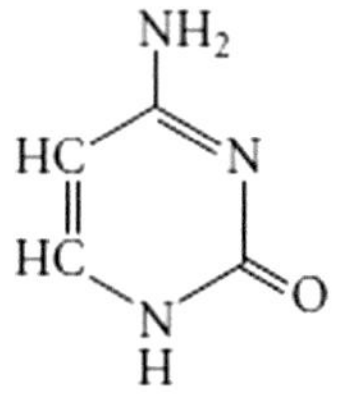   Cytosine

114. Explain why we would expect the heat of vaporization for water to be so much larger than the heat of fusion.
115. If the combustion of methane were used as the source of heat for Problem 51, how many grams of CH_4 would be needed? $\Delta_c H$ for $CH_4 = 55.5$ kJ/g.
116. Carbon tetrachloride (CCl_4) and bromoform ($CHBr_3$) both have tetrahedral structures. One has more than twice the surface tension of the other. Which has the higher surface tension, and what factor gives rise to this difference?
117. The viscosity of glycerol is over a thousand times larger than that of chloroform ($CHCl_3$). Explain what factor accounts for this large difference.
118. Suppose we are on a camping trip in the mountains of Colorado. While boiling water to cook some dried food, our friends notice, with their handy thermometer, that the water is boiling at only 90 °C. How would we explain to our perplexed friends that turning up the gas on the stove will not increase the temperature of the water?
119. The world's water consumption in 2020 was 2700 km^3/year. Assuming a density of 1.00 g/mL, how many metric tons of water will be consumed each year?
120. Hexane is a common liquid used in the chemistry laboratory as a solvent.
    a. How many carbon atoms are there in hexane?
    b. Is there a relationship between the number of carbon atoms in a straight-chain (normal) alkane and the boiling point of the alkane? If so, what is it?
    c. What intermolecular forces are present in hexane?
121. In the electron density maps shown throughout the text, the red color indicates regions of the molecule that possess partial negative charge (greater electron density). The blue color indicates regions of the molecule with partial positive charges (reduced electron density). Compare the electron density maps of water and of hydrogen-bonded water. Are there any striking differences between the two images? If so, what are those differences and what does that imply about the effects of hydrogen bonding?
122. As a beaker of water boils, bubbles develop at the bottom of the beaker and rise to the surface.
    a. What is the origin of the bubbles?
    b. How many liters of water vapor could be produced from a beaker containing 250 mL water (d = 1.00 g/mL) if the beaker is boiled to dryness? (Assume the temperature of the water vapor is 100 °C at 0.95 atm.)
    c. How much heat would be required to complete the conversion of 250 mL water at 25 °C into water vapor at 100 °C?
123. How do molecular and ionic solids, in both the molten and the dissolved state, differ in their ability to conduct an electric current?
124. When selecting molecules that make good candidates for liquid crystal displays, chemists search for (compact; long rigid) molecules with (strong; weak) dipole moments. Explain why each of your circled choices is appropriate for liquid crystals.

Thinking Beyond the Calculation

125. The fuel most commonly used in portable lighters is butane (C_4H_{10}). The normal boiling point of butane is −0.50 °C.
    a. Draw the Lewis dot structure for butane where all of the carbons are in a row.
    b. What types of intermolecular forces of attraction would we predict for this molecule?
    c. Does the low boiling point of butane make sense?
    d. Is this compound soluble in water?
    e. If 3.50 g of butane were burned in oxygen, how many moles of water would be formed? (The other product is carbon dioxide.)
    f. Considering that −0.50 °C is so much lower than typical room temperatures, why doesn't the butane in the lighter boil?
    g. Relative to butane, estimate the normal boiling point of pentane (C_5H_{12}) and propane (C_3H_8).

The Chemistry of Water: Aqueous Solutions and Their Properties

Contents

Supplementary Information The online version contains supplementary material available at (▶ https://doi.org/10.1007/978-3-030-90267-4_12).

M. Mosher and P. Kelter, *An Introduction to Chemistry*,
https://doi.org/10.1007/978-3-030-90267-4_12

What We Will Learn from This Chapter

- Water is a universal solvent because of its abundance and the fact that so many different compounds dissolve in it (▶ Sect. 12.1).
- A solution is made up of a solvent and a solute (▶ Sect. 12.2)
- The concentration of a solution can be measured using molarity, molality, parts per million, and mole percent, among others. (▶ Sect. 12.3)
- The concentration of a solute in a solvent is dependent upon the temperature of the system (▶ Sect. 12.4).
- Many properties of a solution are dependent only upon the concentration of the solute, rather than its identity (▶ Sect. 12.5).

12.1 Introduction

During those hot days of summer, our bodies cry out for something cool to drink. More and more often, we reach for a cold bottle of water to quench our thirst. Ahhh, pure, clean, natural water... Actually, "pure water" probably doesn't exist anywhere on the planet! No matter what purveyors of bottled water, water filtration systems, and many other products would have you believe, nearly every drop of water that you find on the earth has at least a small amount of something else dissolved in it. Sometimes those dissolved materials are helpful and healthful, and sometimes they are dangerous and deadly. Determining what, and just how much, is dissolved in water that is fit human consumption, as well as for agriculture and industrial applications, is one of the many goals of the water chemist. With the help of these scientists, governing agencies can determine the maximum amounts of materials that can safely be dissolved in your water. Then, it's up to the water treatment specialists to do the chemistry necessary to make sure our water is safe to use. This can be a daunting task!

In order to ensure good water quality, water treatment systems must be able to add some chemicals and remove others with the utmost attention to precision. Knowing how much of one compound to add, and how much of another chemical to remove, requires communicating and reporting with common units of measurement. We will examine these different units of measurement and how to manipulate them in this chapter. In addition, we will explore the effects that dissolved chemicals can have on the physical properties of water and how those effects can be used to our advantage.

Let's start by looking at something really close to home; your water quality. How safe is the water in your part of the world? Unfortunately, not all of us drink water that is considered safe. ◘ Table 12.1 lists the World Health Organization's assessment of the 10 countries with the worst drinking water quality, along with their key concerns. Using our knowledge of chemistry to produce clean water and air is one of the most significant ways that science can improve the lives of people around the globe.

◘ **Table 12.1** Top 10 countries with the worst drinking water quality

Rank	Country	Major concern
10	Mexico	Most of the population drinks bottled water
9	Congo	Too few have water nearby
8	Pakistan	Greatest gap in hygiene between rich and poor
7	Bhutan	Much of the water supply is contaminated
6	Ghana	Too few have access to sanitation services
5	Nepal	Much of the population defecates in the open
4	Cambodia	For some, the only safe water is delivered
3	Nigeria	Much of the population uses untreated water
2	Ethiopia	Rural areas largely lack safe drinking water
1	Uganda	Thirty minutes is the typical travel time for much of the population to reach potable water

(Data from US News & World Report, Dec 19, 2019; ▶ https://www.usnews.com/news/best-countries/slideshows/10-countries-with-the-worst-water-supply)

12.2 Water: The Universal Solvent

In ▶ Chap. 11 we raised three questions related to the projected water shortage in Tampa Bay: What is it about water that makes it unique among molecules? How can we use our understanding of the nature of water to explain its properties? And how can we use these properties to help in the search for clean water in places like Tampa Bay? We have answered the first question, focusing especially on the polar nature of water and its ability to form the special interaction called a hydrogen bond. This enabled us to answer the second question, and to explain a variety of properties of water, including its unusually high boiling point, its vapor pressure, and its phase diagram, giving us the understanding to answer much of the second question.

Implicit in the third question about the search for clean water is recognition that the water found in nature is not pure. It is filled with all manner of **solutes**, substances that dissolve in it. Water is often called the **universal solvent** because of its ability to dissolve so many chemicals to form homogeneous mixtures called **solutions**. In fact, nearly all the solutions you will use in a General Chemistry course are water-based. **Aqueous solutions** (water-based solutions) can have solutes that are gases, liquids, or solids.

Solute – The minor component in a solution.

Universal solvent – Water is often called the universal solvent because of its ability to dissolve so many chemicals to form homogeneous mixtures.

Solution – A homogeneous mixture of two or more substances.

Aqueous solution – A homogeneous mixture of a solute dissolved in the solvent water.

▶ https://www.publicdomainpictures.net/pictures/110000/velka/peaty-river-water-1.jpg

12.2.1 Why Do so Many Substances Dissolve in Water?

Water will not dissolve everything, but it does dissolve an enormous range of different materials. Why is water able to do this, and why are some materials, like most oils, not able to dissolve in water? Just as in ▶ Chap. 11, the answer lies in water's molecular structure. As an example of an aqueous solution, let's examine seawater, in which dissolved sodium chloride is one of the primary solutes. Chemically, we can understand what happens when salt dissolves in water by considering the process of dissolving as a series of discrete steps, as shown in ◼ Fig. 12.1. Although the process does not, in fact, occur in steps, we may use this procedure to enhance our understanding of the dissolution process. At each of the steps along our imaginary solution-forming pathway, think about the energy that is either required or released and ask yourself, "Is this likely to be an endothermic or an exothermic process?"

- Solute separation: In the first step, the salt must separate into sodium ions and chloride ions. The electrostatic forces (ionic bonds) that keep the ions together must be overcome. Breaking chemical bonds always requires energy, so this part of the process is endothermic ($\Delta_{solute}H = +$).
- Solvent separation: Energy is required to break hydrogen bonds holding water molecules close together in order to accommodate the incoming ions from the solute. This part of the process is also endothermic ($\Delta_{solvent}H = +$).
- Electrostatic interaction between the solvent and solute ions: Electrostatic attraction is the great stabilizer. As shown in ◼ Fig. 12.2, sodium cations are attracted to the negatively charged dipole at oxygen on each water molecule. Conversely, chloride anions are attracted to the positively charged hydrogen poles. Although there

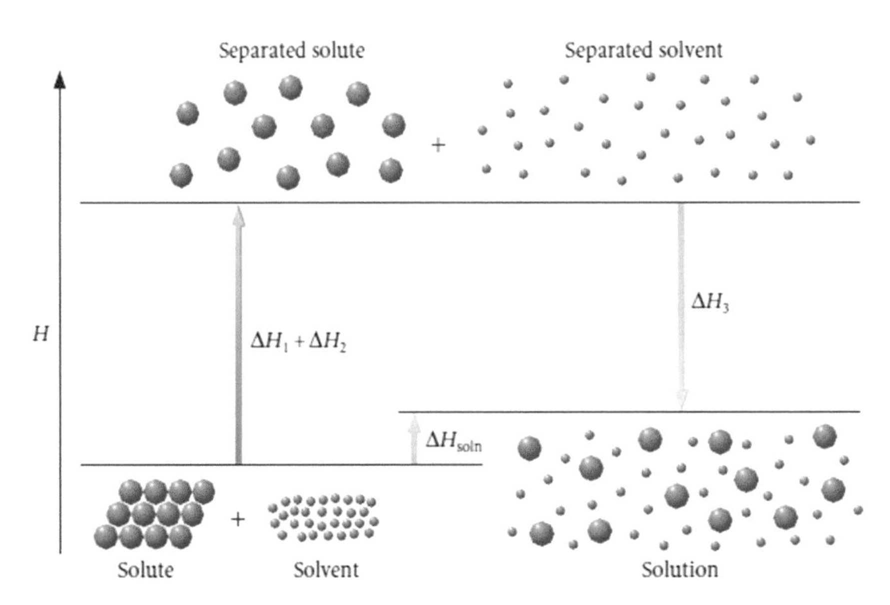

Fig. 12.1 The processes necessary, along with their energy exchanges, for the dissolution of a salt, such as NaCl, in water. The overall change in energy can be + or –, depending on the strength of the ion–dipole interaction

is constant ion movement in the solution, at any one time it is typical to have up to six water molecules surrounding each ion. This **ion–dipole interaction** contributes to the solution process in two ways. First, for the process involving sodium chloride and water, the interaction is highly exothermic ($\Delta_{solvation}H$ = "–"). In order to have favorable formation of a solution, the value of the exothermic interaction of the solute and solvent is typically larger than the endothermic processes in the first two steps. Second, dissolving salt ions in water results in dispersal of the cations and anions throughout the solution. This ion–solvent interaction is called **solvation**, and when the solvent is water, it is also known as **hydration**.

12

Ion–dipole interaction – An attraction between an ion and a compound with a dipole.

Solvation – The interaction of a solvent with its dissolved solute.

Hydration – The interaction of water (as the solvent) with dissolved ions.

- The **heat of solution ($\Delta_{sol}H$)** can be viewed as the sum of the heats of solute separation, solvent separation, and ion–dipole interaction.

$$\Delta_{sol} H = \Delta_{solute}H + \Delta_{solvent}H + \Delta_{solvation}H$$

Heat of solution (Δ_{sol}H) – The enthalpy change associated with the dissolution of a solute into a solvent (Fig. 12.2).

An overall exothermic heat of solution ($-\Delta_{sol}H$) favors solution formation but is not by itself an indicator of solubility. The additional dispersal of the ions in the is also important. For example, NaCl has a value of $\Delta_{sol}H$ = +3 kJ/mol, yet it dissolves in water. So does potassium hydroxide, KOH ($\Delta_{sol}H$ = –57.6 kJ/mol). In fact, the heat of solution of KOH is so large that adding 100 g to a liter of water will raise its temperature to nearly boiling, with $\Delta_{sol}H$ = –103 kJ for the 100 g.

As a rule of thumb, like dissolves like. Polar molecules typically dissolve many salts through ion–dipole interactions. Similarly, polar liquids mix as a consequence of the stability gained from dipole–dipole and, when possible, hydrogen-bonding interactions. Nonpolar substances, such as the industrial solvents benzene (C_6H_6) and toluene (C_7H_8) are often **miscible** (mix in all proportions) with other nonpolar substances. Part of the reason lies with London forces and part with the stability gained by dispersion as the liquids mix.

Miscible – Mixable. Two solvents that are miscible dissolve in each other completely.

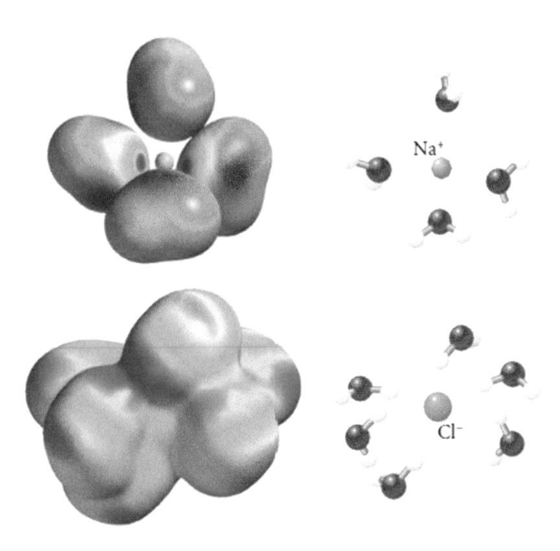

■ **Fig. 12.2** The interaction of the sodium ion with the negative end of the dipole on water and that of the chloride ion with the positive end of the dipole on water contribute to the energetic stability that permits NaCl to dissolve in water. Note in the electrostatic potential maps how the water molecule's electron potential changes as it interacts with the ions

NanoWorld/MacroWorld: The Big Effects of the Very Small—Removing Non-polar Substances from Water

Small amounts of non-polar materials, such as benzene and other carcinogens, can be dissolved in water. Sometimes even parts per trillion concentration of these chemicals can impart a strange flavor, give the water an off-color, or even be a serious health hazard. More important, wastewater containing these non-polar compounds can pose environmental hazards. Many of today's wastewater treatment facilities attempt to remove and/or reduce the amounts of these compounds in the water they treat.

As you might suspect from the idea that "like dissolves like," one of the most common methods for water treatment relies on passing the water through a material that is non-polar. This is accomplished by using a filter impregnated with carbon particles, which are non-polar and extremely insoluble in water. The non-polar carbon will attract the non-polar compounds in the water because pure carbon is completely non-polar. When molecules stick to a solid substance, the process is called **adsorption**. To ensure the maximum amount of contact with the substances in the water, the carbon particles are "activated" by various treatments. Activated carbon particles have many pores, holes, and grooves so that they have lots of surfaces to adsorb the chemicals that need to be removed from the water. The filters, referred to as "activated carbon" filters, are effective at removing even minute quantities of non-polar organic materials.

Adsorption – The attraction by intermolecular forces of molecules and atoms to a solid material.

This is how small pitcher-type water filters that you might have in your house work. Activated carbon, impregnated onto a filter inside the water purifier, holds onto organic molecules by intermolecular forces, and gives you water clean enough to drink. This process works so well that it is sometimes used as a poison antidote in the hospital. Activated carbon can be administered orally to the patient in a water-based slurry. Once ingested, the activated carbon particles are free to travel throughout the patient's body and trap the non-polar toxins so they can be removed by natural elimination processes.

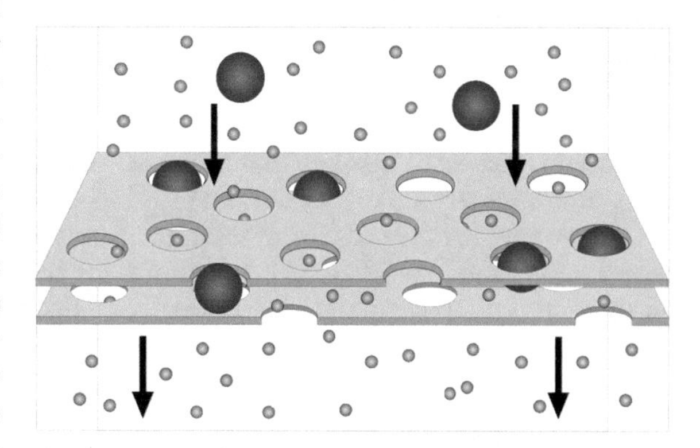

Small particles can pass through the filter, but large particles adsorb to the filter because they cannot pass through the pores. ► https://upload.wikimedia.org/wikipedia/commons/thumb/2/20/FilterDiagram.svg/800px-FilterDiagram.svg.png

Example 12.1—Predicting Solubility

Which of the following substances shown in Fig. 12.3 is (are) soluble in water: ethanol (C_2H_5OH), ethylene glycol ($C_2H_6O_2$), and cyclohexane (C_6H_{12})?

Asking the Right Questions

What determines whether a compound is soluble in water? Specifically, what attractions make solubility in water possible? Which of these molecules can have such attractions with water?

Solution

Ethanol and ethylene glycol can have electrostatic attractions, such as hydrogen bonds, with water because of the presence of the hydroxyl group, –OH, in ethanol, and two hydroxyl groups in ethylene glycol. Both are infinitely soluble in water. In fact, both can act as a solvent *or* solute with water. Cyclohexane, which lacks –OH, or any other polar group, cannot form significant electrostatic interactions with water, and is therefore not appreciably soluble in water.

Are the Answers Reasonable?

There is an idea that is especially important in organic chemistry (the chemistry of carbon-based compounds, which we will explore in ▶ Chap. 21) that "structure dictates function," and that small changes in structure can lead to important changes in function. In these molecules, the presence of the –OH makes it possible for both ethanol and ethylene glycol to mix in all proportions with water, so the answer is reasonable.

What If...?

...we are asked to change ethanol in such a way that it was not miscible (soluble in all proportions) in water, but was only slightly soluble. What might we change, and what might be the structure of our molecule?

(*Ans: We could increase the number of CH_2 groups, which do not interact with water. For example, the molecule butan-1-ol, $C_4H_{10}O$, has a solubility of 77 g/L H_2O, and 1-pentanol, $C_5H_{12}O$, has a solubility of 22 g/L.*)

Practice 12.1

Which of the following substances is (are) soluble in hexane: water, carbon tetrachloride (CCl_4), and iodine crystals (I_2)?

Ethanol

Ethylene glycol

Cyclohexane

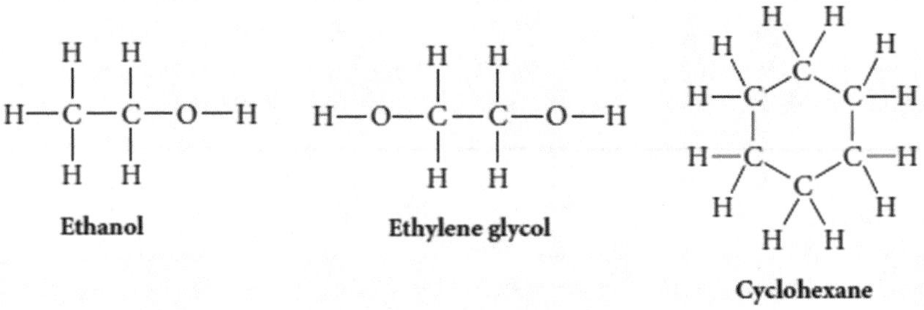

Fig. 12.3 Which of these compounds would be expected to be soluble in water?

The maximum amount of a solute that can dissolve in a solvent at a given temperature is its **solubility**, often given in grams of solute per 100 mL of solvent $\left(\frac{\text{g solute}}{\text{100 mL solvent}}\right)$. If we added any more solute, it would not dissolve. At that point, we say that the solution is **saturated** (compare this with the use of the term saturated for the vapor of the pure liquid found in ▶ Chap. 11). Sometimes the solubility of a substance in water depends on the relative size of the polar and nonpolar parts of a molecule of the substance. Table 12.2 shows that the series of alcohols gets progressively less soluble in water as the nonpolar end of each alcohol molecule becomes larger. However, even substances that we classify as "insoluble" in water can dissolve at least a little bit. One such chemical is oxygen (O_2), and that is fortunate for fish and other sea life that depend on it. Another such chemical is carbon tetrachloride, a nonpolar industrial solvent; that is not so fortunate, because even in low concentrations in water, it is considered an environmental hazard. Even crude oil, which provides the standard illustration for something that is not soluble in water ("oil and water don't mix") can dissolve very slightly in water.

Solubility – A property of a solute; the mass of a solute that can dissolve in a given volume (typically 100 mL) of solvent.

Saturated – When considering a pure substance—the surrounding atmosphere contains the maximum possible amount of vapor, expressed as the vapor pressure. When considering a solution—the solution contains the maximum concentration of dissolved solute.

Table 12.2 Solubility of some alcohol compounds in water

Compound	Name	Solubility (g/L)
CH_3OH	Methyl alcohol (methanol)	Infinite
C_2H_5OH	Ethyl alcohol (ethanol)	Infinite
C_3H_7OH	Propyl alcohol (propan-1-ol)	Infinite
C_4H_9OH	Butyl alcohol (butan-1-ol)	79
$C_5H_{11}OH$	Pentyl alcohol (pentan-1-ol)	27
$C_6H_{13}OH$	Hexyl alcohol (hexan-1-ol)	5.9
$C_7H_{15}OH$	Heptyl alcohol (heptan-1-ol)	0.9
$C_{10}H_{21}OH$	Decyl alcohol (decan-1-ol)	<0.04

12.2.2 Colloids

In the previous section, we described how materials dissolve in water. However, not everything makes a solution in water. Although a solution may possess a color due to the presence of a light-absorbing compound dissolved in the solvent, solutions, as a rule, are transparent. Sometimes, however, a solvent has materials dispersed in it that do not dissolve, and when those materials will not settle to the bottom over time, we say that the result is a **colloidal dispersion**, or **colloid**.

Colloidal dispersion – A solution of a colloid in a solvent.

Colloid – Large particles suspended throughout a solvent such that they scatter light.

Why do solutions appear transparent (even if they are colored)? The answer to this question is really just a matter of particle size. When a salt, like sodium chloride, or a molecular solid, like sucrose, dissolves in water, the "particles" that are formed are individual ions or molecules. While they may absorb light and produce colors, ions and molecules are too small to *scatter* light. The light appears to pass through the solution and the solution appears clear to us. Colloids, on the other hand, are formed when fairly large particles are dispersed evenly throughout the liquid. These particles are large enough so that light bounces off of them and is scattered in all directions. This type of interaction with light is called the **Tyndall effect**. The result is that colloids appear cloudy or opaque to us. For example, milk is mostly water, and it does have some dissolved chemicals (like calcium), but it also has a lot of insoluble materials like fats and proteins, which are present in the form of very tiny particles that are dispersed throughout the milk. They are small enough that the random movement of the water molecules will keep them from settling to the bottom of your glass, but they are much, much larger than sodium and chloride ions or sucrose molecules. Light that passes through a glass of milk is scattered and the milk appears to be opaque.

Tyndall effect – The scattering of light by a colloid.

12.3 Measures of Solution Concentration

We have said that seawater is largely an aqueous solution of dissolved sodium chloride. However, there are many other components of this solution, including a good deal of Mg^{2+}, SO_4^{2-}, and Ca^{2+}, smaller quantities of iron, phosphorus, and copper, and really small amounts of dissolved oxygen, cadmium, and even gold. What do we mean by "a good deal," "smaller quantities," and "really small amounts"? It depends on whom you ask. And "it depends" is too vague in a scientific community that requires clarity when communicating the results of measurements. We need descriptions of concentration that have consistent meaning to everyone reading the data. We also need different units of concentration for different applications.

12.3.1 Measures Based on Moles

Molarity (M)

We examined **molarity** earlier as a measure of moles of solute per liter of solution.

$$M = \frac{\text{mol solute}}{\text{L Solution}}$$

Molarity is the most commonly used concentration unit in the chemical laboratory. We can speak of the initial molarity of a solute added to a solution, as in "What is the initial molarity of the sodium chloride?" We can also discuss the actual molarity of each ion after the solute has dissolved, as in "What is the molarity of the sodium ion in the solution?"

Molarity – A concentration measure defined as moles of solute per liter of solution.

Molality (*m*)

The measure of moles of solute per kilogram of solvent is known as the **molality**.

$$\mathrm{m} = \frac{\text{mol solute}}{\text{kg solvent}}$$

The molality of a solute in solution is independent of temperature because it is based on measuring the mass of the solvent, rather than the volume of the solution. It is useful in exploring properties at a variety of temperatures, as we shall discuss in ► Sect. 12.3. And because we can measure mass accurately and precisely, molality can be determined to many significant figures, if necessary.

Molality – A concentration measure defined as moles of solute per kilogram of solvent.

Mole Fraction (χ_i) and Mole Percent

The **mole fraction** of a substance is the ratio of the number of moles of a substance present per total moles of all substances in the solution. If there are three solutes, i, j, and k, in aqueous solution, then the mole fraction of solute i is:

Mole fraction – A concentration measure defined as moles of solute per total moles of solution.

$$\chi_i = \frac{\text{mol i}}{\text{mol i} + \text{mol j} + \text{mol k} + \text{mol water}}$$

12

It is important, when we calculate the mole fraction, to include the contribution of all components in the system. Note that the denominator in the equation indicates that we add the number of moles of i, j, k, and water to obtain the total number of moles in the system. If we multiply χ_i by 100, then we have what is referred to as **mole percent**.

Mole percent – The mole fraction of a component in a solution multiplied by 100.

Example 12.2—Practice with Mole-Based Units of Concentration

Fructose is one of the three important "simple sugars," the other two being galactose and glucose. What are the values for molarity, molality, and mole fraction of 36.0 g of fructose ($C_6H_{12}O_6$) in 250.0 mL of a fruit-flavored drink? The density of the water-based drink is 1.05 g/mL and the presence of any flavorings can be neglected.

```
      CH₂OH
       |
    O=C
       |
   H—C—OH
       |
   H—C—OH
       |
   H—C—OH
       |
      CH₂OH
```

Fructose

Asking the Right Questions

How are these three measures of concentration similar and yet different? In particular, would we expect the molarity and molality of the solution to be the same? Explain.

Solution

The addition of fructose changes the density of the solution compared to pure water, so we would expect the molarity (which is based on the solution volume) and molality (based on only the solvent mass) to be different. All three measures of concentration require us to know the number of moles of fructose ($C_6H_{12}O_6$).

$$\text{mol fructose} = 36.0 \text{ g fructose} \times \frac{1 \text{ mol fructose}}{180.0 \text{ g fructose}}$$
$$= 0.200 \text{ mol fructose}$$

$$\text{Molarity of fructose} = \frac{\text{mol fructose}}{\text{L solution}} = \frac{0.200 \text{ mol fructose}}{0.2500 \text{ L solution}}$$
$$= 0.800 \text{ M}$$

The molality calculation requires that we know the mass of water. We can find this via the mass of the solution and its density.

$$\text{Mass of solution} = 250.0 \text{ mL solution} \times \frac{1.05 \text{ g solution}}{\text{mL solution}}$$
$$= 262.5 \text{ g solution}$$

(Note that we keep the extra figure for this calculation and round only the number that will be our final answer.)

$$\text{Mass of water} = \text{mass of solution} - \text{mass of fructose}$$
$$= 262.5 \text{ g} - 36.0 \text{ g} = 226.5 \text{ g water}$$

$$\text{Molality} = \frac{\text{mol fructose}}{\text{kg solvent}} = \frac{0.200 \text{ mol fructose}}{0.2265 \text{ kg solvent}}$$
$$= 0.883 \text{ m}$$

When we calculate the mole fraction, we must take into account both the moles of fructose and the moles of the solvent, water. We can covert the mass of water to moles for our mole fraction calculation.

$$\text{mol water} = 226.5 \text{ g water} \times \frac{1 \text{ mol water}}{18.0 \text{ g water}}$$
$$= 12.58 \text{ mol water}$$

$$\text{Mole fraction of fructose} = \chi_{\text{fructose}}$$
$$= \frac{\text{mol fructose}}{\text{mol fructose} + \text{mol water}}$$
$$= \frac{0.200}{0.200 + 12.58} = 0.0156$$

Are the Answers Reasonable?
The most notable concern is the difference between the molarity and the molality of the solution. Is the difference reasonable? Because the solution volume contains both water and fructose, it is a larger quantity than the mass of only water. This makes the molarity, which is divided by the larger quantity (the volume of water/fructose mixture = solution volume) smaller than the molality. That is mathematically reasonable.

What If…?
…we added 500.0 mL of water to the solution? What would be the new mole fraction of water? Assume a density of the water of 1.00 g/mL.
(*Ans:* $\chi_{fructose} = 0.00493$; $\chi_{water} = 0.995$)

Practice 12.2
What mass of potassium hydroxide (KOH) is required to prepare 600.0 mL of a 1.40 M KOH solution? What is the mole fraction of water in this solution? (Assume that the density of the solution is 1.07 g/mL.)

12.3.2 Measures Based on Mass

We discussed the two useful sets of mass-based concentration measures, weight percent and **parts per million**, **parts per billion** and **parts per trillion** earlier in this text. We present them again here for purposes of review and for application in the context of our discussion of seawater.

Parts per billion – A concentration defined as the mass of a solute per billion times that mass of solution.
Parts per million – A concentration defined as the mass of a solute per million times that mass of solution.
Parts per trillion – A concentration defined as the mass of a solute per trillion times that mass of solution.

Weight Percent (wt%)
The measure of the mass fraction of a substance in a solution, expressed as a percentage, is known as the **weight percent** or **mass percent**.

$$\text{wt\%} = \frac{\text{g substance}}{\text{g solution}} \times 100\%$$

Weight percent – A concentration unit based on the mass of the solute per total mass of solution, reported in percent. Also known as the mass percent.

Table 12.3 Average composition of major ions in seawater

Element (main form in seawater)	Parts per million (mg/L)
Chlorine (Cl^-)	19,000
Sodium (Na^+)	10,500
Magnesium (Mg^{2+})	1250
Sulfur (SO_4^{2-})	900
Calcium (Ca^{2+})	400
Potassium (K^+)	380
Bromine (Br^-)	65
Bicarbonate (HCO_3^-)	30
Strontium (Sr^{2+})	12

If we were to prepare a reference solution that has the same weight percent of sodium chloride as seawater, it would contain 29.5 g of sodium chloride per 1.00×10^3 g of solution, a weight percent of 2.95% NaCl.

$$\frac{29.5 \text{ g NaCl}}{1.00 \times 10^3 \text{ g solution}} \times 100\% = 2.95\% \text{ NaCl}$$

Another way to express this relationship is to say that every 100 g of solution contains 2.95 g of NaCl. A related unit is volume percent. As you might suspect, this is when two substances are mixed and the ratio of their volumes is used to express how much of the solute is present. For example, if 5.0 mL of ammonia is dissolved in 1000 mL of water, the resulting solution is 0.5% ammonia, by volume. Therefore, every 100 mL of the solution will contain 0.5 mL of ammonia.

12

Parts per Million, Billion, and Trillion (ppm, ppb, ppt)

In discussing the concentration of possible health hazards in water, the Environmental Protection Agency (EPA) cites its **maximum contaminant level (MCL)**, the highest acceptable level in a solution, as 10 parts per million for nitrate and 5 parts per billion for cadmium. Very low solute concentrations (sometimes called trace concentrations) are often expressed this way. The density of very dilute aqueous solutions is close enough to that of pure water, 1.0 g/mL, that we may use these volume-based conversions we derived in ▶ Sect. 12.5.2.

Maximum contaminant level (MCL) – The highest acceptable level of a contaminant in a particular solution, according to the Environmental Protection Agency.

$$\text{ppm} = \frac{1 \text{ g solute}}{10^6 \text{ g solution}} \approx \frac{1 \text{ mg solute}}{\text{L solution}}$$

$$\text{ppb} = \frac{1 \text{ g solute}}{10^9 \text{ g solution}} \approx \frac{1\ \mu\text{g solute}}{\text{L solution}}$$

$$\text{ppt} = \frac{1 \text{ g solute}}{10^{12} \text{ g solution}} \approx \frac{1 \text{ ng solute}}{\text{L solution}}$$

The average concentration, in parts per million, of the major ions that are present in seawater are listed in Table 12.3.

Example 12.3—Conversion Between Mole and Mass Concentration Units

The maximum concentration of O_2 in seawater is 2.2×10^{-4} M at 25 °C. What is this concentration in parts per million of oxygen?

Asking the Right Questions

What is the relationship between molarity and ppm? We have previously made an approximation in which we related ppm to the mass per volume of solution for very dilute aqueous solutions. What is that approximation, and can it help us decide a range for the reasonable answer?

Solution

We can solve this via dimensional analysis.

$$\frac{2.2 \times 10^{-4}\,\text{mol O}_2}{\text{L seawater}} \times \frac{32\,\text{g O}_2}{1\,\text{mol O}_2} \times \frac{1000\,\text{mg O}_2}{1\,\text{g O}_2}$$
$$= 7.0\,\text{ppm O}_2$$

Is the Answer Reasonable?

We expect the solubility of (nonpolar) oxygen to be quite low in water, and our answer confirms that. In addition, we can approximate 1 ppm O_2 as 1 mg O_2/L solution. A solution that is about 10^{-4} M has a solute (O_2, here) on the order of mg/L, another clue that the answer is reasonable.

What If…?

…instead of oxygen, we have a solution that has an aluminum ion concentration of 2.2×10^{-4} M, and was prepared by dissolving aluminum chloride in water? What is the concentration of chloride ion in the solution?
(*Ans:* 6.6×10^{-4} M)

Practice 12.3

The MCL for arsenic in drinking water is 10 ppb, according to Environmental Protection Agency guidelines. Convert this value to molarity.

▶ Here's what we know thus far…

- A solution is made up of a solute and a solvent.
- There are many ways to express the amount of a substance that is dissolved in solution. These include molarity, molality, mole fraction, percent by weight, and parts per million.
- We can convert from one concentration unit to another using the definition of the individual concentration units.

12.4 The Effect of Temperature and Pressure on Solubility

Not all solutes are infinitely soluble in a solvent. The quantity that can dissolve is also greatly affected by other factors.

12.4.1 Temperature Effects

We saw in Example 12.3 that oxygen is nearly insoluble in seawater, yet 7.0 ppm at 25 °C is still enough to allow the seas to teem with life. The concentration of oxygen in the seas and in freshwater lakes, ponds, and rivers varies as natural processes such as photosynthesis and respiration cycle oxygen into and out of the water. The other important factor that determines oxygen's solubility in water (see ◘ Table 12.4) is temperature. Note the trend, followed by all gases, that solubility of a gas decreases with temperature. ◘ Figure 12.4 shows this behavior for a few common gases. We can reason this by considering that the molecules in a hot solution are moving faster and have more kinetic energy than in a cold solution. Through collisions with the solvent molecules, gas molecules in the solution gain even more energy. With sufficient energy, the gas can escape the intermolecular attractions of the solvent molecules and return to the gas phase.

Small changes in temperature do not dramatically affect the solubility of oxygen, but a large increase in temperature can significantly lower the oxygen concentration in a waterway, and the harm done can be felt throughout the

◘ **Table 12.4** Solubility of O_2 in fresh water at selected temperatures (in air at 1 atm)

Temperature (°C)	Parts per million (mg/L)
0	14.6
5	13.1
10	11.3
15	10.1
20	9.1
25	8.3
30	7.6
35	6.9

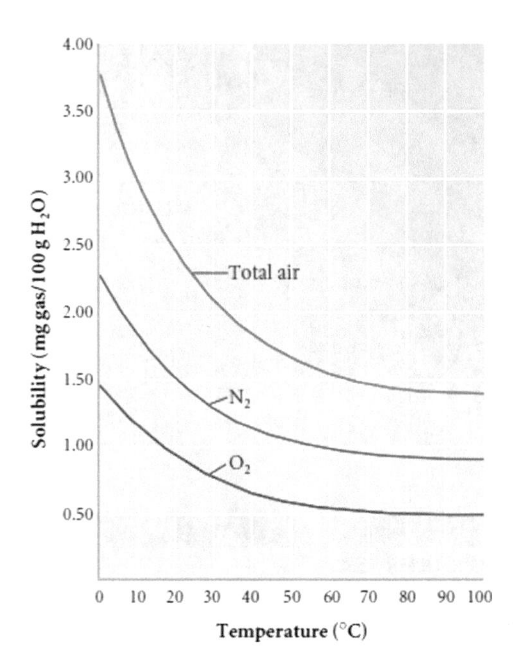

■ **Fig. 12.4** The solubility of gases decreases with temperature. This is especially important with O_2, where "thermal pollution" can have important consequences for the aquatic food chain

■ **Fig. 12.5** A cooling tower lowers the temperature of water that has been heated as part of the operation of power plants. This water is close to the temperature of the waterway from which it was taken and to which it will be returned, so the impact of thermal pollution is minimized. *Source* ▸ https://www.publicdomainpictures.net/pictures/10000/velka/IMG_1735.jpg

aquatic food chain. The artificial raising of the ambient water temperature is called **thermal pollution** and is of concern in the design of nuclear power plants, in which river water is used to cool the nuclear core of the reactor (see ▸ Chap. 20). The heated river water is passed through a **cooling tower** (■ Fig. 12.5) before it flows back to the river.

Thermal pollution – An artificial increase in the temperature of water.

Cooling tower – A large industrial apparatus used to condense gaseous vapor into liquid or to cool water that has been heated before it is discharged.

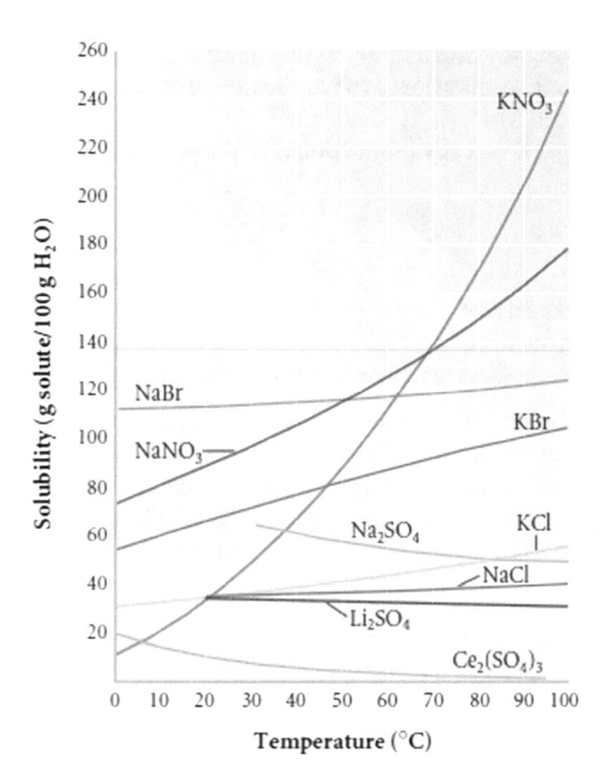

Fig. 12.6 The solubility of some salts in water as a function of temperature. Note that the solubility of the majority of these salts increases as the temperature increases. Li_2SO_4 decreases solubility as the temperature increases

In contrast to gases, the solubility of ionic solids in water generally increases with temperature, as shown in Fig. 12.6. For example, the solubility of sodium chloride in water at 0 °C is 35.7 g/100 mL, and at 100 °C it is 39.1 g/100 mL. Some solids show a much more marked increase in solubility. Potassium nitrate (KNO_3) has an aqueous solubility of 15 g/100 mL at 0 °C to 245 g/100 mL at 100 °C, a 16-fold increase. A few salts, such as cerium(III) sulfate ($Ce_2(SO_4)_3$) and sodium acetate ($NaC_3H_2O_2$), have lower solubility with increasing temperature.

12.4.2 Pressure Effects

The solubilities of solid solutes in water do not change very much with modest pressure changes. The solubility of gases in water, on the other hand, is quite sensitive to the external pressure of the particular gas that is being dissolved. This is important in the preparation of soft drinks, in which CO_2 is combined with water, sweetener, and other flavorings at pressures between about 6 and 15 atm. The "fizz" in the soda is due to the dissolved CO_2. When the bottle is opened, the CO_2 above the soda escapes from the container as the pressure of the bottle suddenly drops to atmospheric pressure. The lower pressure decreases the solubility of the CO_2 in the soda, and bubbles of CO_2 form.

► https://publicdomainpictures.net/pictures/20000/velka/a-glass-of-fizz.jpg

William Henry (1775–1836), an English chemist, noted this behavior. **Henry's law** says that, at constant temperature, the solubility of a gas is directly proportional to the pressure that the gas exerts above the solution.

Henry's law – The solubility of a gas is directly proportional to the pressure that the gas exerts above the solution.

$$P_{gas} = k_{gas}C_{gas}$$

where

P_{gas} = the pressure of the gas above the solution
C_{gas} = the concentration of the gas in the solution
k_{gas} = a constant relating the gas pressure above the solution and its concentration.

It is important to note that it is the pressure of the gas that is dissolving that is important here. If other gases are present, their pressures will NOT affect the solubility of the gas that is dissolving. In pure water at 1 atm and 0 °C, the solubility of CO_2 is 3.48 g/L. According to Henry's law, if we triple the CO_2 pressure, the solubility will roughly triple. Overall, this law holds best for non-polar gases such as N_2 and O_2 that do not significantly interact with the water. It does not hold well for a gas like HCl, which ionizes in water to form the hydrated ions H^+(aq) and Cl^-(aq).

Example 12.4—Henry's Law and CO_2 in Soda

In air, in which the pressure of CO_2 is 3.4×10^{-4} atm at 0 °C, the solubility of CO_2 in water is 1.18×10^{-3} g/L water. If the pressure of the CO_2 above the water is increased to 6.00 atm, what will be the solubility of the CO_2 in the water? Do these data lead to a conclusion consistent with our prior assertion that sodas go "whoosh" when they are opened?

Asking the Right Questions

What is the relationship between the pressure of a gas above a solution and the solubility of the gas in the solution, in words and in an equation?

Solution

The solubility of a gas is directly proportional to the pressure that gas exerts above the solution. In equation form, you can write

$$P_{gas} = k_{gas}C_{gas}$$

The CO_2 pressure is being increased by a huge factor, so you would expect the solubility to increase about proportionately. Assuming proportionality, you can eliminate the need to calculate the Henry's law constant (k_{gas}) and just solve the problem with ratios.

$$\frac{1.18\times10^{-3}\ \text{g/L}}{3.4\times10^{-4}\ \text{atm}} = \frac{x\ \text{g/L}}{6.00\ \text{atm}}$$
$$x = 21\ \text{g/L}$$

The solubility of the gas increased sharply at high external CO_2 pressure, in accordance with Henry's law. As is consistent with much of our discussion on solutions, the formulas we cite work best for very dilute solutions. When the bottle of soda is opened, the pressure of CO_2 above the liquid sharply decreases, lowering the solubility of the gas and therefore contributing to the escape of gas that you hear when you open the soda.

Are the Answers Reasonable?

The large increase in the pressure of the gas above the solution led to a proportionately large increase in the solubility of the gas in the solution. Lowering the solubility by opening the soda is, therefore, reasonable.

What If…?

…the pressure of the gas mixture is further increased, to 9.00 atm, by adding N_2? What happens to the solubility of the CO_2?

(*Ans: The solubility of CO_2 remains the same, because the its contribution to the total pressure of the gas remains the same.*)

Practice 12.4

A researcher adds 22.7 g of NaCl to 55.0 mL of water. Examine Fig. 12.6. Will all of the salt dissolve in the given amount of water at 25 °C? If not, how much water will need to be added to just dissolve the salt completely?

12.5 Colligative Properties

In April along the Oulu River in Finland, it is common to see people brave the weather (Fig. 12.7) to sit around fishing holes in the ice. The presence of the ice means that access to the center of the river is made that much easier. We take it for granted, but ice fishing is possible only because the water in freshwater rivers has a relatively

Fig. 12.7 These fishermen gather their catches on the partially frozen Oulu River in Finland in April. *Source* ▶ https://upload.wikimedia.org/wikipedia/commons/7/7a/Ice_fishing_Oulu_20090405.JPG

low concentration of dissolved solutes and has a freezing point just below 0 °C. If the Oulu River were as salty as the Red Sea or the Persian Gulf, with sodium and chloride ion concentrations equal to 40 parts per thousand, its freezing point would be about −2.5 °C, and the ability to fish on the ice would be diminsihed (or nonexistent), as anglers would have to wait for sufficient ice to form to make the trek onto the ice safe for the weight of the people and their gear. Why does the presence of the salt lower the freezing point of water? Does the amount of salt affect the freezing point, and does the nature of the salt matter? Are there any other solution properties that are affected in this way?

Let's answer the last question first. Properties of a solution that approximately depend only on the number of nonvolatile solute particles, irrespective of their nature, are called **colligative properties** (from the Latin *colligatus*, which means "collected together"). There are four useful colligative properties: vapor pressure lowering, freezing-point depression, boiling-point elevation, and osmotic pressure. Understanding vapor pressure lowering will help us answer our questions about dissolving salts in water, as well as give us insight into the other three colligative properties.

Colligative properties – Physical properties of a solution that depend only on the amount of the solute and not on its identity, including boiling-point elevation, freezing-point depression, vapor pressure lowering, and osmotic pressure.

12.5.1 Vapor Pressure Lowering

1,2,3-Propanetriol ($C_3H_8O_3$) is the systematic name for the nonvolatile substance we commonly call glycerol or glycerin. The colorless liquid is used as a lubricant and moistener, especially in cosmetics, and to reduce swelling in medical procedures, such as eye examinations. The presence of three OH groups on the molecule leads to significant hydrogen bonding, making glycerol completely soluble in water. The hydrogen bonding also leads to an increased **viscosity** for the liquid. We noted earlier that water has a vapor pressure equal to 23.8 torr at 25 °C. Glycerol has essentially no vapor pressure at room temperature. (Judging on the basis of our discussion in that section about intermolecular forces, structure, and vapor pressure, does it make sense that water should be volatile, whereas glycerol is nonvolatile?) When glycerol and water are mixed, the total vapor pressure of the resulting solution is dependent only on the vapor pressure of pure water, $P^{o}_{H_2O}$, multiplied by its mole fraction, χ_{H_2O}, in the solution.

Viscosity – A measure of resistance to flow. Can be thought of as the syrupiness of a liquid.

$$\text{Vapor pressure of the solution} = \text{Psolution} = \chi_{H_2O} P^{0}_{H_2O}$$

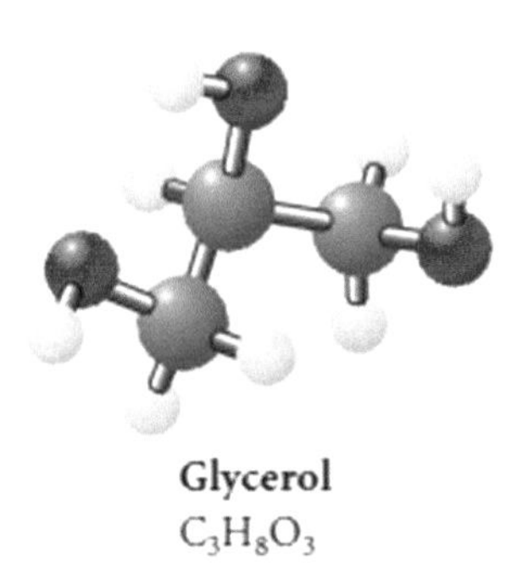

Glycerol
$C_3H_8O_3$

For example, if we add enough glycerol to water so that the mole fraction of the water is reduced to 0.900, the resulting vapor pressure of the solution will be reduced. At 25 °C, the vapor pressure of the solution would be

$$P_{\text{solution}} = \chi_{H_2O} P^{o}_{H_2O}$$
$$P_{\text{solution}} = 0.900 \times 23.8 \text{ torr} = 21.4 \text{ torr}$$

The relationship of the vapor pressure of the solution, P_{solution}, to the mole fraction, χ_{solvent}, and vapor pressure, P_{solvent}, of the volatile solvent holds true for any ideal solution containing a nonvolatile solute. It is known as **Raoult's law**, named after the French chemist Francois-Marie Raoult (1830–1901).

$$P_{\text{solution}} = \chi_{\text{solvent}} P^{o}_{\text{solvent}}$$

Raoult's law – Describes the change in vapor pressure of a solvent as solute is added to a solution.

An **ideal solution** exists when the properties of the solute and solvent are not changed by dilution. This means that other than being diluted, combining solute and solvent in an ideal solution does not release or absorb heat, and the total volume in the solution is the sum of the volumes of the solute and solvent. Only very dilute solutions approach ideal behavior, so although Raoult's law is a good first approximation, actual measurements are required to properly describe vapor pressure changes in mixtures of solutions. Figure 12.8 shows the general trend: The vapor pressure is depressed with the addition of a nonvolatile solute.

Ideal solution – A solution in which the properties of the solute and solvent are not changed by dilution. This means that other than being diluted, combining solute and solvent in an ideal solution does not release or absorb heat, and the total volume in the solution is the sum of the volumes of the solute and solvent.

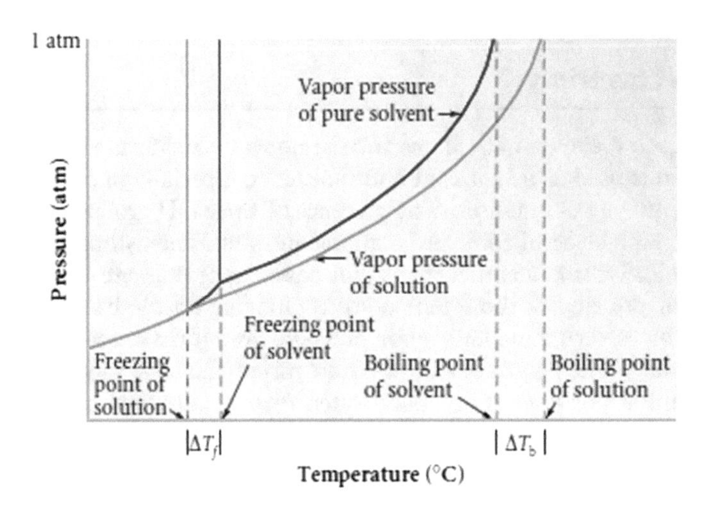

Fig. 12.8 The vapor pressure of water (red line) is lowered by the addition of a nonvolatile solute. This is described for an ideal solute by Raoult's law

Example 12.5—Vapor Pressure Lowering

Let's add 3 teaspoons (45.0 g) of sucrose, table sugar ($C_{12}H_{22}O_{11}$, molar mass = 342 g/mol), into a cup of tea containing 250.0 mL of water at 90.0 °C (density = 0.965 g/mL). What is the new vapor pressure of the solution? $P^{o}_{H_2O} = 526$ torr at 90.0 °C.

Asking the Right Questions

What is the relationship between vapor pressure and the solvent and solute concentrations? What do we need to calculate to show this relationship?

Solution

The key problem-solving hurdle is calculating the mole fraction of water in the solution. To do this, we must calculate the number of moles of each component. Remember to retain an extra, nonsignificant figure until the end of the calculations.

$$\text{mol sucrose} = 45.0\,\text{g} \times \frac{1\,\text{mol}}{342\,\text{g}}$$

$$= 0.1316\,\text{mol sucrose}$$

$$\text{g of water} = 250.0\,\text{mL}$$

$$\times \frac{0.965\,\text{g water}}{1\,\text{mL water}} \times \frac{1\,\text{mol water}}{18.02\,\text{g water}}$$

$$= 13.39\,\text{mol water}$$

$$\chi_{\text{water}} = \frac{\text{mol water}}{\text{mol glucose} + \text{mol water}}$$

$$= \frac{13.39}{0.1316 + 13.39} = 0.9903$$

$$P_{solution} = 0.9903 \times 526\,\text{torr}$$

$$= 520.9\,\text{torr} \approx 5.2 \times 10^2\,\text{torr}$$

Is the Answer Reasonable?

You expected the vapor pressure of the solution to drop as a result of adding sucrose to water, and it did. The answer is therefore reasonable.

What If…?

…we note that fructose, $C_6H_{12}O_6$, is so soluble that you can dissolve 4.3 g of fructose per g of water at 30.0 °C? What would be the vapor pressure of a saturated solution of fructose 30.0 °C, at which the vapor pressure of water is 31.8 torr?

(*Ans: The vapor pressure of the solution is* 22 torr.)

Practice 12.5

The vapor pressure of ethanol (C_2H_5OH) at 40 °C is 135.3 torr. Calculate the vapor pressure of a solution containing 26.8 g of glycerin ($C_3H_8O_3$) a nonvolatile solute, in 127.9 g of ethanol.

12.5.2 Boiling-Point Elevation

At 1 atm, the temperature of any solution must be elevated until its vapor pressure equals 1 atm in order to reach the boiling point. Concentrated solutions possess lowered vapor pressures (which we discussed earlier), but still must have a vapor pressure equal to the atmospheric pressure in order to boil. The more solute that is dissolved in a solution, the more the vapor pressure of the solution is lowered, and the higher the boiling point. The temperature of the solution must be elevated for the vapor pressure to reach that of the surroundings. For fairly dilute solutions of nonelectrolytes, the following formula approximately describes the **boiling-point elevation** due to the addition of a nonvolatile solute.

Boiling-point elevation – The increase in the boiling point due to the presence of a dissolved solute.

$$\Delta T_b = K_b m$$

where
T_b = the change in boiling point in °C
K_b = the boiling-point elevation constant, which depends on the solvent, in units of °C/m
m = the molality of the solute in moles of solute per kilogram of solvent.

The value of K_b for water is 0.512 °C/m. This means that a 2.00 molal aqueous sugar solution would have a boiling-point elevation of approximately 1.02 °C (2.00 m × 0.512 °C/m) and a boiling point of about 101 °C. We say

Table 12.5 Boiling-point elevation constants for several liquids

Solvent	K_b (°C/m)	T_b (°C)
Acetone	1.7	56.5
Benzene	2.6	80.1
Carbon tetrachloride	5.0	76.7
Ethanol (ethyl alcohol)	1.2	78.5
Methanol (methyl alcohol)	0.80	64.7
Water	0.512	100.0

"about" because the boiling-point elevation constant begins to deviate significantly from the ideal solution value at higher concentrations. Boiling-point elevation constants for several liquids are listed in Table 12.5.

Colligative properties depend on the *number*, not the *nature*, of the particles in the solution. We would expect solutions containing compounds at the same concentration to have the same elevated boiling point. However, we must consider the total number of particles that result from making the solution. For example, what would happen to the boiling point of the solution if our solute were 2.0 molal cobalt(II) chloride ($CoCl_2$) instead of sucrose? Because cobalt(II) chloride is a strong electrolyte that dissociates to form cobalt ion and chloride ion,

$$CoCl_2(s) \xrightarrow{H_2O} Co^{2+}(aq) + 2Cl^-(aq)$$

we would expect three moles of particles (ions, in this case) for every mole of $CoCl_2$ added to the solution. That is, the concentration of ions in the solution should be 6.00 molal, and the boiling point of the solution should be raised by (6.0 m × 0.512 °C/m) = 3.1 °C. This suggests that when dealing with strong electrolytes, we can modify our boiling-point elevation formula to take into account the dissociation of strong electrolytes into *i* particles.

$$\Delta T_b = i\, K_b\, m$$

12

For $CoCl_2$, $i = 3$ if the solution behaves ideally. The actual boiling-point elevation for this solution is 4.7 °C, not 3.1 °C, which tells us that at this relatively high concentration (2.0 m), the solution does not behave even close to ideally. The value *i* is known as the **van't Hoff factor**, after J. R. van't Hoff (1852–1911), a chemist from the Netherlands who suggested its use in the 1880s. Table 12.6 shows the van't Hoff factors for several electrolytes. Note that there is significant deviation from the expected values as the solute concentration increases, so results are only approximate even at relatively low concentrations.

van't Hoff factor – A factor that modifies the colligative properties on the basis of their ability to dissociate in solution.

Jacobus Henricus van't Hoff

▶ https://upload.wikimedia.org/wikipedia/commons/5/5f/Jacobus_Henricus_van_'t_Hoff_01.jpg

Table 12.6 Experimental van't Hoff factors for several electrolytes

Compound	Expected value of i	m = 0.005	m = 0.01	m = 0.05	m = 0.10	m = 0.20	m = 1.00	m = 2.00
HCl	2	1.95	1.94	1.90	1.89	1.90	2.12	2.38
NH_4Cl	2	1.95	1.92	1.88	1.85	1.82	1.79	1.80
$CuSO_4$	2	1.54	1.45	1.22	1.12	1.03	0.93	–
$CoCl_2$	3	2.80	2.75	2.64	2.62	2.66	3.40	4.58
K_2SO_4	3	2.77	2.70	2.45	2.32	2.17	–	–

Example 12.6—Boiling Point Elevation

Recipes for cooking spaghetti often call for putting a little table salt in the water before boiling it and adding the spaghetti. Does this help the spaghetti cook faster? Assume that we add 10.0 g of NaCl to 6.00 L of water and that the density of the solution is 1.00 g/mL. Also assume that this very dilute solution behaves ideally, so $i = 2$. The value of K_b for water is 0.512 °C/m.

Asking the Right Questions

Spaghetti does cook faster at *significantly* higher temperature. Does the addition of the salt increase the temperature of the water *significantly*—by at least several degrees? How do you calculate the change in temperature when the table salt is added? What is the impact of the ionization of NaCl on the chemistry, and, therefore, on the calculation?

Solution

In order to calculate the temperature change, you need the molality of the NaCl, taking into account its ionization.

$$\text{mol NaCl} = 10.0\,\text{g NaCl} \times \frac{1\,\text{mol NaCl}}{58.5\,\text{g NaCl}} = 0.171\,\text{mol NaCl}$$

$$\text{kg water} = 6.00\,\text{L water} \times \frac{1.00\,\text{kg water}}{\text{L water}} = 6.00\,\text{kg water}$$

$$\text{Molality of NaCl} = \frac{\text{mol NaCl}}{\text{kg water}} = \frac{0.171\,\text{mol NaCl}}{6.00\,\text{kg water}} = 0.0285\ \text{m NaCl}$$

The value of the van't Hoff factor is assumed to be equal to the number of particles that result from the dissolution of the NaCl; i = 2.

$$\Delta T_b = iK_b m = 2.0 \times \left(\frac{0.512\,°\text{C}}{\text{m}}\right) \times 0.0285\,\text{m}$$

$$\Delta T_b = 0.03\,°\text{C}$$

$$T_b = 100\,°\text{C} + 0.03\,°\text{C} = 100.03\,°\text{C} \approx 100\,°\text{C}$$

This insignificant change in the boiling point of water implies that the addition of salt does *not* help the spaghetti cook faster.

Is the Answer Reasonable?

There is relatively little salt being added to a large pot of water, so it is reasonable that the temperature changes only a small amount.

What If…?

…we wanted to raise the temperature of the 6.0 L of boiling water in the pot by 1.00 °C? How much NaCl would the water need to contain?
(*Ans: 340 g NaCl.*)

Practice 12.6

What is the predicted boiling point of each of these solutions? (Assume that each follows ideal behavior.)

a. 1.00 m NaCl in water
b. 0.35 m $FeCl_3$ in water
c. 1.50 m KCl in methanol

In contrast to the minimal impact of adding a dash of table salt to water when cooking spaghetti, making an aqueous solution that is 40% by volume ethylene glycol ($C_2H_6O_2$) elevates the boiling point to 105 °C (221 °F). This solution absorbs much more heat from your operating car engine than water alone would be able to absorb. However, it's not called "anti-boil." In the next section, we'll examine why it is instead called "antifreeze."

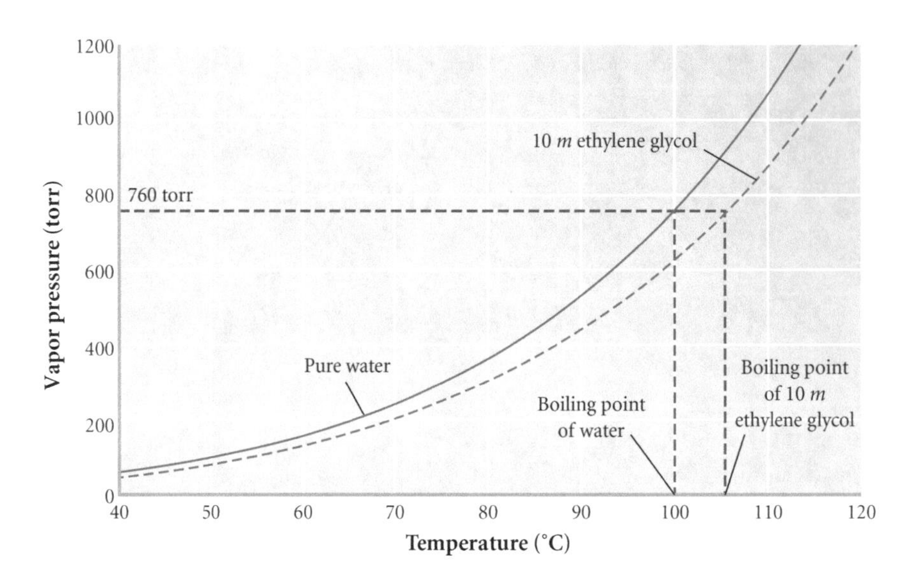

12.5.3 Freezing-Point Depression

The same antifreeze solution that raises the boiling point does double duty, because it also lowers the freezing point by about 18 °C. This **freezing-point depression** is another of the colligative properties and, like boiling point elevation, is approximately proportional to the molal concentration of the solute:

Freezing-point depression – The lowering of the melting point of a compound due to the presence of a dissolved solute.

$$T_f = i\,K_f\,m$$

where
T_f = the change in freezing point in °C
i = the van't Hoff factor
K_f = the freezing-point depression constant, which depends on the solvent, in units of °C/m
m = the molality of the solute in moles of solute per kilogram of solvent.

Raoult himself, in 1883, was the first to note that the lowering of the freezing point was the same for any nonelectrolyte solute in a given solvent. Table 12.7 lists freezing-point depression constants for several liquids. Water has a relatively low value of K_f; freezing-point depressions are much greater in other solvents. Historically, the freezing-point depression was used to determine the molar masses of substances.

Why are freezing points depressed whereas boiling points are elevated? As with all colligative properties, the key is the intermolecular forces of attraction, which affect the vapor pressure. In Fig. 12.9, the solution has a lower vapor pressure than the pure solvent. When the solvent freezes, its vapor pressure must be lowered to equal that of the solution. In order to reach that vapor pressure, the solution must be cooled below the freezing point of the pure solvent. The result is a new freezing point that is lower than the freezing point of the solvent.

Table 12.7 Freezing point depression constants for several compounds

Solvent	K_f (°C/m)	T_f (°C)
Benzene	5.1	5.5
Cyclohexane	20.0	6.5
Formic acid	2.8	8.4
Naphthalene	6.9	80.0
Phenol	7.4	43.0
Water	1.86	0

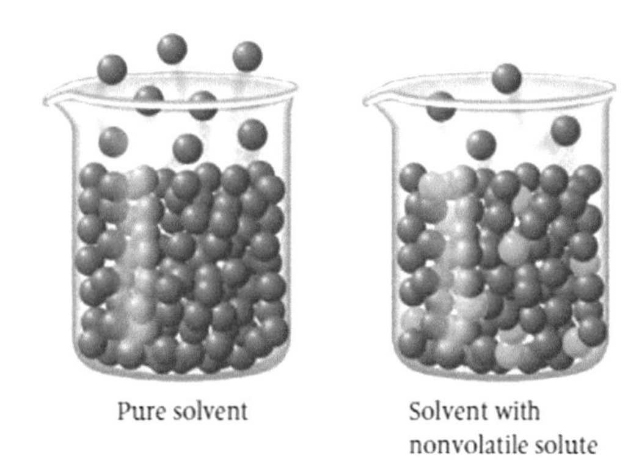

Fig. 12.9 Note the differences between the two liquids in this figure. The more concentrated solution on the right has a lower vapor pressure. We can understand this by spraying a little perfume at the front of a room. In a short time, everyone in the room will smell the perfume. Why is this so? There is a natural tendency for things to become more dilute. The perfume vapors spread out across the room as they become more dilute. The same situation occurs in vapor pressure lowering. If solvent molecules left the container on the right, the solution would become more concentrated, which would violate the natural tendency of the solution to become more dilute. Therefore, fewer molecules leave the container on the right

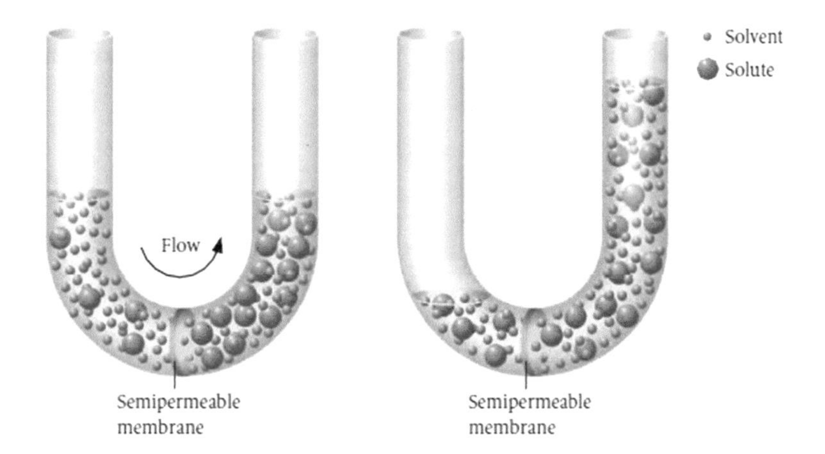

Fig. 12.10 Salt water and water are separated by a semipermeable membrane. Water molecules pass through, with more going from the purer water (on the left) to the more concentrated solution (on the right) than the other way, therefore raising the solution level. The pressure necessary to prevent this shift in solution level via osmosis is called osmotic pressure

One important consequence of freezing-point depression is the need to keep food freezers well below 0 °C. Food in a freezer contains all manner of solutes, especially various salts and sugars, so the freezing points of products such as meats and ice cream are often significantly less than 0 °C.

12.5.4 Osmosis

Another colligative property is osmosis (from the Greek *osmos*, meaning "to push"). This particular property of solutions is extremely useful in helping clean the water we consume.

To address this colligative property, let's set up the following experiment in the same manner as experiments that were done as early as the year 1748. A solution of salt water is placed in a two-chambered container separated from a sample of pure water by a **semipermeable membrane** as shown in Fig. 12.10. The membrane has very small holes in it so that small water molecules can travel through, but not larger solute particles. Why won't small ions, such as the potassium cation, travel through the semipermeable membrane? Although the holes are bigger than a potassium cation (so that the water molecules can go through), they are smaller than the entire hydrated ion of potassium. Remember that solutes dissolved in water contain a sphere of sphere of hydration—water molecules that surround the solute and help keep it dissolved. The sphere is typically much larger than a hole the size of a water molecule.

Semipermeable membrane – A thin membrane (typically made from a polymer) with pores big enough to allow solvent to pass through, but small enough to restrict the flow of dissolved solute particles.

What will happen in our experiment? According to Raoult's law, the vapor pressure of the saltwater solution will be lower than that of the pure water. In an attempt to equalize the vapor pressures, water will flow through the membrane to the solution side, diluting it and making the vapor pressures equal. In fact, water will keep flowing until the weight of the solution (the solution's "hydrostatic pressure") becomes so great that it stops the flow (see ◻ Fig. 12.10). Water can travel through the membrane in both directions, but the drive toward the most energetically stable equilibrium position makes the overall direction of flow toward dilution. This process is known as **osmosis**. The reason that the vapor pressures tend to equalize is once again related to the concept of entropy, a topic we will discuss in a later chapter.

Osmosis – The flow of solvent into a solution through a semipermeable membrane.

If we run the experiment again, but this time apply pressure to the saltwater solution in an amount that just prevents the osmosis, there will be no net change, as shown in ◻ Fig. 12.11. The pressure that must be applied on the solution side to prevent osmosis of the solvent into it is called the **osmotic pressure**, symbolized by a capital "pi", Π. In 1887, van't Hoff determined that osmotic pressure is proportional to the temperature and the molarity of the solution:

Osmotic pressure – The pressure required to prevent the flow of a solvent into a solution through a semipermeable membrane.

$$\Pi = iMRT$$

where
Π = osmotic pressure, in atmospheres
M = molarity
T = temperature in kelvins
i = the van't Hoff factor
R = the universal gas constant, 0.08206 L. atm. mol^{-1} K^{-1}

Note that this colligative property uses molarity, not molality, to express the concentration of solute in the solution. Additionally, the gas constant R is the same one we introduced when we discussed the behavior of gases.

The original measurements of osmotic pressure used animal membranes, with limited success. More reliable membranes were developed in the 1860s, incorporating a film of copper(II) ferrocyanide, $Cu_2Fe(CN)_6$, on a porous tube, and for decades, many quantitative measurements up to several hundred atmospheres were made in this

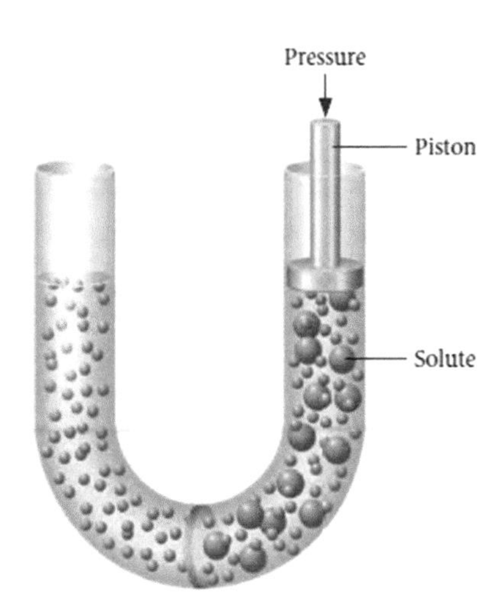

◻ **Fig. 12.11** Osmosis is prevented by a piston applying just enough pressure, called osmotic pressure, to the solution

way. As with our other colligative properties, we are mindful that the osmotic pressure equation holds well only for very dilute solutions. But with a dilute solution, one can measure the molar mass of very large polymers via the osmotic pressure experiment.

Another example of osmosis relates to red blood cells, which have a salt concentration equal to their surrounding fluid (which identifies the solutions as **isotonic**). If an intravenous solution given to a patient has a higher solute concentration than blood plasma (and is called a **hypertonic** solution), water will leave the blood cell membranes by osmosis. Hypertonic solutions are sometimes intentionally administered to patients when they have too much water in their blood plasma, such as in water intoxication, liver disease, or congestive heart failure. If the blood plasma is too concentrated, as happens in cases of severe diarrhea or excess sweating, a **hypotonic** solution (a solution that has a lower solute concentration than blood plasma) is needed so that water will flow into the cells. Many intravenous solutions given to patients in hospitals are isotonic with blood serum. Because these isotonic solutions have the same solute concentration as blood serum, they will not cause cells in the patient's body to shrink or swell. A solution **isosmotic** with blood plasma (same osmotic pressure) is 0.154 M each in Na^+ and Cl^- ions.

Hypertonic – Containing a concentration of ions greater than that to which it is judged against.

Hypotonic – Containing a concentration of ions lower than that to which it is judged against.

Isotonic – Containing the same concentration of ions.

Isosmotic – Possessing the same osmotic pressure.

Example 12.7—Osmosis and You

What is the osmotic pressure of a solution at 25 °C that is isosmotic with blood plasma, 0.154 M NaCl?

Asking the Right Questions

How is osmotic pressure related to the concentration of the solute, NaCl?

Solution

Sodium chloride dissociates into two ions, Na^+ and Cl^-, when dissolved in water. That means that the van't Hoff factor for this solution should be equal to 2.0.

$$\begin{aligned}\Pi &= iMRT \\ &= 2.0 \times 0.154\,\text{M} \\ &\quad \times 0.08206\,\text{L. atm. mol}^{-1}\,\text{K}^{-1} \\ &\quad \times 298\,\text{K} = 7.532\,\text{atm} = 7.53\,\text{atm}\end{aligned}$$

Osmosis is a fairly powerful property of solutions. If we placed this solution in an osmotic pressure apparatus with pure water, we would need 7.53 atm (110 psi) of pressure just to stop the osmosis.

Is the Answer Reasonable?

The answer tells us that a lot of pressure is needed to stop the tendency of the solvent, water, to dilute the more concentrated solution. That is consistent with the observation that osmosis is a powerful process. As we'll discover in our next discussion, reverse osmosis is another powerful—and vital—process for changing the concentration of solutions.

What If…?

…we were only able to generate 5.00 atm of pressure to stop osmosis? What would be the maximum concentration of NaCl we could have in our solution at 25 °C?
(*Ans: 0.10 M NaCl.*)

Practice 12.7

What is the osmotic pressure of each of these solutions at 298 K?

a. 0.100 M sucrose
b. 0.375 M $CaCl_2$
c. 1.250 M $(NH_4)_2SO_4$

12.5.5 Reverse Osmosis

Instead of using osmotic pressure merely to prevent the natural flow from the water to the saltwater side, let's apply additional pressure to actually push water from the saltwater solution through the semipermeable membrane, in opposition to the natural tendency toward dilution by osmosis. What will happen? The solution is losing solvent, so it will become more concentrated (saltier). Water is pushed through the membrane, so the amount of pure water on the other side of the membrane increases. We have used **reverse osmosis** to desalinate water, producing more fresh water.

Reverse osmosis – A purification process wherein pure solvent is forced (with pressure) to flow through a semipermeable membrane away from a solution.

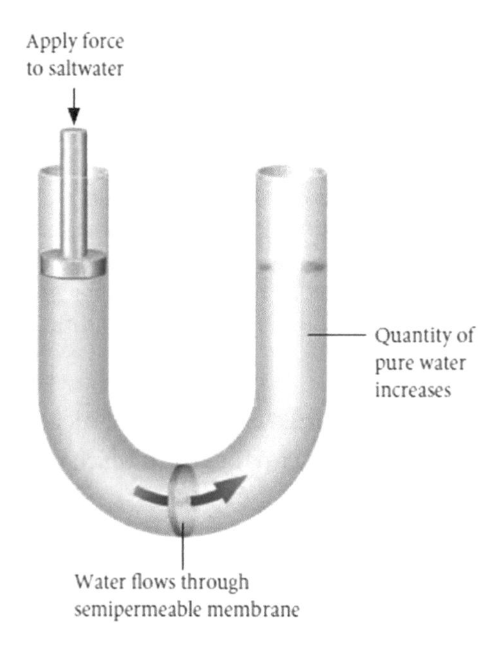

12.5.6 Back to the Future

After quite a long discussion on many aspects of water as a liquid and a solvent, we can finally answer an important question: "How can we use these properties to help in the search for clean water in places like Tampa Bay?" An important part of the answer is "by reverse osmosis." More than 26 billion gallons of freshwater are produced each day by over 20,000 desalination plants in 177 countries throughout the world. Much of this freshwater is produced by the energy-intensive processes of boiling and distilling seawater. As better and cheaper membrane materials are developed by scientists and engineers, the more energy-efficient strategy of reverse osmosis, which only requires about half the energy of distillation, will become more commonly used. The number of desalination plants that use reverse osmosis is increasing more rapidly each year. In addition, reverse osmosis has the advantage of being able to be used in the purification of wastewater and industrial sewage.

But the process doesn't have to be used on such a large scale in order to be useful. Reverse osmosis units find applications in manufacturing, food processing, and printing, as shown in Table 12.8. The use of long-lasting nylon or polyamide semipermeable membranes gives the factories, such as the one shown in Fig. 12.12, the ability to filter hundreds of thousands of gallons per day. Usable water, this precious resource, is becoming ever more scarce. It is through our understanding of its structure and its properties that we can create safe, long-term solutions that will keep fresh water flowing.

Table 12.8 Applications of reverse osmosis

Industry	Application
Cosmetics	Product preparation
Desalination	Potable water, beverage preparation
Drinking water	Mineral removal
Electronics	Water low in impurities for manufacturing
Food	Low sodium and organic chemicals in food preparation
Laboratories	Rinsing glassware
Pharmaceuticals	Pure water for large-scale production
Restaurants	Spot-free rinses, drinking water

Fig. 12.12 Reverse osmosis units can filter saltwater to produce hundreds of millions of liters of potable water per day. Such units like this one in Barcelona, Spain, are likely to play an increasingly important role in supplying growing communities with fresh water. *Source* ▶ https://upload.wikimedia.org/wikipedia/commons/d/d5/Reverse_osmosis_desalination_plant.JPG

The Bottom Line

- Water is known as the universal solvent because of its ability to form solutions with many substances.
- Solution formation can be understood in terms of specific types of energy changes.
- Solution concentration can be expressed in a variety of units, including molarity, molality, and parts per million.
- Solubility is affected by pressure and temperature in predictable ways.
- Raoult's law describes the change in vapor pressure of a solvent as solute is added to a solution.
- Colligative properties are based on the number of particles in a solution and can be understood in terms of Raoult's law and intermolecular forces.
- Colligative properties include vapor pressure lowering, boiling-point elevation, freezing-point depression, and osmosis.
- It is possible that reverse osmosis will become a vital process for supplying clean water to large cities.

Section 12.2 Water: The Universal Solvent

Skill Review

1. The solubility of NH_4Cl at 20 °C is approximately 39 g per 100 g of water. At 80 °C, 68 g per 100 g of water dissolves.
   a. Is the value for the enthalpy of solution of NH_4Cl positive or negative?
   b. Assuming a near linear relationship over the temperature range, if you saturated a solution at 40 °C, how many grams of NH_4Cl would be dissolved?
2. Organic alcohols have varying solubilities in water. Examine the following list, and explain the cause of the trend you observe.

1-Butanol	C_4H_9OH	79 g/1000 mL of water
1-Pentanol	$C_5H_{11}OH$	27 g/1000 mL of water
1-Hexanol	$C_6H_{13}OH$	5.9 g/1000 mL of water

3. Water is often considered the universal solvent. Explain this designation in your own words.
4. What properties would you associate with a solvent that was considered to be universal?
5. a. Is it possible for a compound to have a $\Delta_{soln}H=0$?
   b. Why would a solute and solvent form a solution if the value for $\Delta_{soln}H$ were 0?
6. Name the four processes that take place when a solute dissolves in a solvent, and describe each one as "endothermic" or "exothermic", as appropriate.
7. Which solute will dissolve the best in water and which would best dissolve in cyclohexane, C_6H_{12}?

   I_2 FeS KCl

8. Cyclohexane (C_6H_{12}) is often used as a nonpolar solvent. Of the following solutes, which would best dissolve in cyclohexane, and which would best dissolve in water?

NaCl	C_6H_6	CH_3OH	$C_6H_{12}O_6$	$CH_3(CH_2)_{16}COOH$
Salt	Benzene	Methanol	Dextrose	Stearic acid

9. Which of these general types of substances would you expect to be soluble in water?
   a. Sugars
   b. Hydrocarbons
   c. Metals
   d. Nonpolar molecules
   e. Polar molecules
   f. Inorganic compounds
10. Which of these pairs of liquids will be miscible?
    a. $CH_3CH_2CH_2CH_2CH_3$ and $CH_2{=}CHCH_2CH_2CH_3$
    b. C_6H_6 (benzene) and H_2O
    c. H_2O and CH_3CO_2H
    d. CH_3CH_2OH and $CH_3CH_2CH_2CH_2CH_3$
11. Of the following substances, which are soluble in water: methane (CH_4), sodium hydroxide (NaOH), and carbon dioxide (CO_2)?
12. Of the following substances, which are most likely soluble in water: COS, NH_3, and HCl?
13. Of the following substances, which is most likely soluble in benzene (C_6H_6): water, ammonia, chlorine gas, and cyclohexane (C_6H_{12})?
14. Of the following substances, which is most likely soluble in benzene (C_6H_6): methane (CH_4), sodium chloride (NaCl), acetylene (C_2H_2), and methanol (CH_4O)?

Chemical Applications and Practices

15. Two burets are filled with different liquids. Which meniscus is more likely to be representative of heptane (C_7H_{16}) and which of water?

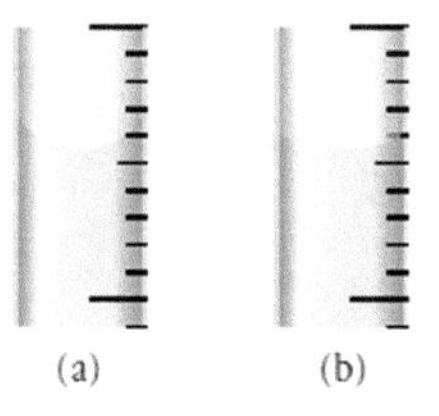

16. If you were to extract caffeine from some cola beans, you would find that you could extract much more using hot water than using cold water. Caffeine is more soluble in hot water than in cold. Explain what this fact tells you about the heat of solution for caffeine.
17. Water and ethanol are said to be miscible.
    a. Distinguish between a solute that is miscible with water and a solute that is soluble in water.
    b. What do miscible substances have in common?
18. What would you predict would occur if a saturated solution of sodium chloride were warmed? What would happen if a saturated solution of lithium sulfate were warmed?

Section 12.3 Measures of Solution Concentration

Skill Review

19. Calculate the molarity (M) of these solutions:
    a. 42.0 g of NaOH dissolved in enough water to form 0.500 L of solution
    b. 10.0 g of $C_6H_{12}O_6$ (dextrose) dissolved in enough water to form 0.250 L of solution
    c. 25.0 g of urea (NH_2CONH_2) dissolved in enough water to form 100.0 mL of solution.
20. Using the information in Problem 19 calculate the molality of each solution (assume the density of the solution is 1.0 g/mL).
21. Calculate the molality (m) of these solutions:
    a. 12.5 g of ethylene glycol antifreeze (CH_2OHCH_2OH) dissolved in 0.100 kg of water
    b. 53.0 g of sucrose ($C_{12}H_{22}O_{11}$) dissolved in 500.0 g of water
    c. 4.55 g of sodium bicarbonate ($NaHCO_3$) dissolved in 250.0 g of water.
22. Using the information in Problem 21 calculate the molarity of each solution (assume the density of each solution is 1.0 g/mL).
23. Determine the mole fraction of solute in the following:
    a. 22.7 g of benzene (C_6H_6) dissolved in 67.5 g of cyclohexane (C_6H_{12})
    b. 15.0 g of formic acid (HCOOH, found in ants) dissolved in 100.0 g of water
    c. 0.195 g of acetaldehyde (C_2H_4O, found as the product of ethanol metabolism) dissolved in 25.0 g of water.
24. Using the information in Problem 23, calculate the molality of each solution (assume the density of water is 1.0 g/mL).
25. Fill in the missing information.

Compound	Grams of compound	Grams of water	Mole fraction of solute	Molality
NH_4Cl	12.5 g	95.0 g		
KNO_3		125 g		0.155
$C_6H_{12}O_6$		250.0 g	0.115	

26. Fill in the missing information.

Compound	Grams of compound	Grams of water	Mass %	ppm
NH_4Cl	1.0 g	99.0 g		
KNO_3		125 g		125
$C_6H_{12}O_6$		300.0 g	15.8	

27. A hydrochloric acid solution was made by adding 59.85 g of HCl to 100 g water. The density of the solution was 1.20 g/mL. Calculate the concentration of the acid in molarity, molality and mole fraction.
28. What would be the mass of sodium hydroxide necessary to prepare 1.00 L of a 0.878 M aqueous solution of sodium hydroxide? What would be the mole fraction of the sodium hydroxide in the solution? Assume the density is 1.0 g/mL.
29. How would you prepare 275 mL of 0.0550 M potassium dichromate solution starting with solid potassium dichromate? Describe the process in detail.
30. Describe how you would prepare 750 mL of 0.7750 M sodium sulfate starting with solid sodium sulfate.
31. A bottle of ethanol (C_2H_5OH) is labeled "55.0% ethanol by mass; density = 0.8985 g/cm^3." Calculate the molarity, molality, and mole fraction of the ethanol in solution.
32. Glacial acetic acid has a density of 1.049 g/cm^3 and is 96.0% by weight CH_3COOH. What is the molality of this acid? What is its molarity?
33. A 2.50 L bottle of wastewater contaminated with benzene is found to contain 2.12 g of benzene. What is the solubility (in g/100 mL) of benzene in water?
34. A student records the mass of a beaker as 34.5 g, then adds some water and re-determines the mass of the beaker to be 52.7 g. A compound is then added to the water in small portions until it is no longer soluble. The beaker is then weighed again and found to be 60.4 g. What is the solubility (in g/100 mL) of the compound in water?

35. Determine the molarity of the following solutions:
    a. 2.33 g NaCl in 100.00 mL water
    b. 2.12 g benzene (C_6H_6) in 2.50 L water
    c. 73.4 g KNO_3 in 500 mL water
36. Determine the molality of the solutions in the previous question. Assume the density of the solutions is 1.050 g/mL.
37. Determine the mole fraction of the solute in the solutions in the previous question. Assume the density of water is 1.00 g/mL.
38. Determine the weight percent of the solute in the solutions in the previous question. Assume the density of water is 1.00 g/mL.
39. An aqueous solution at 25 °C, with a density of 1.061 g/mL, is found to contain 0.550 M KCl. Assume that pure water at 25 °C has a density of 1.00 g/mL. Convert this concentration to:
    a. molality
    b. mole fraction KCl
    c. mole percent KCl
    d. wt % KCl
    e. ppm KCl
40. An aqueous solution at 25 °C, with a density of 0.982 g/mL, is found to contain 0.032 m methanol (CH_4O). Assume that pure water at 25 °C has a density of 1.00 g/mL. Convert this concentration to:
    a. molarity
    b. mole fraction methanol
    c. mole percent methanol
    d. wt % methanol
    e. ppm methanol
41. An aqueous solution at 25 °C, with a density of 1.001 g/mL, is found to contain 250 ppm $CuCl_2$. Assume that pure water at 25 °C has a density of 1.00 g/mL. Convert this concentration to:
    a. molality $CuCl_2$
    b. molarity $CuCl_2$
    c. mole fraction $CuCl_2$
    d. mole percent $CuCl_2$
    e. wt % $CuCl_2$
42. An aqueous solution at 25 °C, with a density of 1.022 g/mL, is found to contain 1.4% glucose ($C_6H_{12}O_6$). Assume that pure water at 25 °C has a density of 1.00 g/mL. Convert this concentration to:
    a. molality glucose
    b. molarity glucose
    c. mole fraction glucose
    d. mole percent glucose
    e. ppm glucose
43. The MCL for some common metals as given by the Environmental Protection Agency guidelines are listed below. Convert each of these to molarity. (Assume the density of the water and of the solution is 1.00 g/mL.)
    a. Antimony, 0.006 ppm
    b. Arsenic, 0.010 ppm
    c. Barium, 2 ppm
    d. Mercury, 0.002 ppm
44. The MCL for some common compounds as given by the Environmental Protection Agency guidelines are listed below. Convert each of these to molarity. (Assume the density of the water and of the solution is 1.00 g/mL.)
    a. Carbon tetrachloride (CCl_4), 0.005 ppm
    b. Chlorobenzene (C_6H_5Cl), 0.1 ppm
    c. Xylene (C_8H_{10}), 10 ppm
    d. Vinyl chloride (C_2H_3Cl), 0.002 ppm
45. A typical cup of coffee may contain 75 mg of caffeine per 200.0 mL of water. If the density of the solution were approximately 1.09 g/mL, what would you calculate as the mass percent caffeine, the molarity, and the molality? (Hint: You'll need the formula of caffeine to answer this problem.)

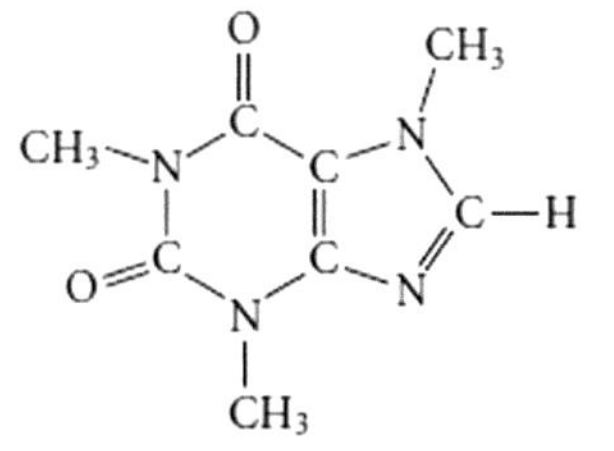

Caffeine

46. Perspiration has a slight acidity because of the presence of lactic acid ($C_3H_6O_3$). Suppose we analyze the perspiration of a chemistry student running late to class. If the density of the perspiration were 1.15 g/mL and the mass percent lactic acid were found to be 4.88%, what would you calculate as the mole fraction, molarity, and molality of the lactic acid?
47. The saline solution used in some medical procedures is a 5.00% NaCl solution. What would be the molarity of sodium ions (Na^+) in such a solution if the density were assumed to be approximately 1.02 g/mL?
48. Sodium alkylbenzene sulfonate ($C_{18}H_{29}SO_3Na$) is often used as a synthetic detergent. This ingredient helps prevent scale buildup in "hard" water applications. If a detergent solution were 0.100 M in sodium alkylbenzene sulfonate, how many grams would be dissolved per liter of solution?
49. Hard water is the name given to water containing relatively high concentrations of metal ions, such as calcium and magnesium. For instance, if a particular water sample were said to contain 130.0 ppm Ca^{2+}, it would be classified as hard water.
    a. What would be the mass percent of calcium ion in water containing 130.0 ppm Ca^{2+}?
    b. What would be the same concentration expressed as parts per billion?
50. The concentration of potassium ion in human blood cells varies over a healthy range. If the concentration of K^+ in a red blood cell were listed at 0.95 μmol/mL, what would you report as the molarity and parts per million?
51. According to the Environmental Protection Agency, if the lead concentration in more than 10% of tap water samples in a given water system exceed 0.015 ppm, the municipality must take action to reduce the lead concentration. How much lead would be contained in a 250. mL glass of water at the 0.015 ppm level?
52. According to the Environmental Protection Agency, the maximum contaminant level for benzene, C_6H_6, in drinking water, is 5 ppb. If you live in a neighborhood with a water supply that has benzene right at the legal limit, how much benzene are you drinking each day if you drink 2.0 L of tap water per day?
53. Barium salts can enter waterways as a result of metal mining operations, so cities in mining states like Colorado test for it. The health effects of ingesting high concentrations of barium (several ppm for weeks or months) include gastrointestinal problems and high blood pressure. The 2010 Water Quality Report for Denver, Colorado, included an average barium level of 37 ppb. What is the molarity of barium in the Denver drinking water?
54. Manganese is a naturally occurring mineral and is present in many water supplies. Manganese concentrations above 500 ppb are of concern, with long-term exposure above that level elevating the risk of symptoms similar to Parkinson's Disease. The 2010 Water Quality Report for Denver, Colorado, included an average manganese level of 2 ppb. What is the molarity of manganese in the Denver drinking water?
55. The City of Chicago's Comprehensive Chemical Analysis Report for 2009 lists the water hardness at its Jardine Water Purification Plant in mg/L of calcium carbonate, $CaCO_3$. The level for the north outlet of the plant is 138 mg/L. What is the mole fraction of calcium in the water? How many grams of calcium are in a 12-oz glass of water?
56. The City of Chicago's Comprehensive Chemical Analysis Report for 2009 lists an average chloride ion concentration at its Jardine Water Purification Plant of 15.7 ppm. What is the molarity of the chloride ion in the water?
57. U.S citizens average more than 22 added teaspoons of sugar (on top of natural sugars from fruits) per day. That's about 90 g. The Center for Science in the Public Interest, a healthy food and nutrition advocacy group, recommends that we eat no more than one to two added teaspoons of sugar each day. Many sodas and fruit-flavored drinks contain more than 10 teaspoons of sugar, as fructose, $C_6H_{12}O_6$, in a 12-oz can. If we consider the solution of 11 teaspoons (7.7 g per teaspoon) of fructose in a fruit-flavored drink volume of 12 oz (355 mL), determine the molarity, molality, and the mole fraction of the sugar, as fructose, in the soda. The density of the drink is 1.05 g/mL.
58. If the drink discussed in the previous problem is prepared using 11 teaspoons of sucrose (7.7 g per teaspoon), $C_{12}H_{22}O_{11}$, instead of fructose, what are the molarity, molality and mole fraction of the sucrose in the drink, which has the same density of 1.05 g/mL?

59. Potassium chloride is used in place of sodium chloride in many foods because far too many of us eat much more than the federally recommended maximum of about 2300 mg of sodium per day. The solubility of potassium chloride in water at 20 °C is 34.2 g KCl/100 g H_2O. Calculate the solubility in parts per million of potassium ion at 20 °C. In addition, calculate the molality of a saturated solution of KCl at the same temperature. Which is a more useful measure of concentration for potassium ion in water for this solution, the molality or concentration in ppm? Explain.
60. The nutrition label on a bottle of soy sauce indicates that it contains 280 mg of sodium per tablespoon. The volume of one tablespoon is 14.8 cm^3. If the solubility of sodium chloride at 20 °C is 35.0 g/100 g of water, how much sodium could one tablespoon of a saturated solution of sodium chloride contain? Assuming the soy sauce is just salt and water, what percentage of saturated is soy sauce?

Section 12.4 The Effect of Temperature and Pressure on Solubility

Skill Review

61. This represents a system at equilibrium: $O_2(aq) \rightarrow O_2(g)$
    a. In which direction will the equilibrium be shifted by an increase in pressure?
    b. In which direction will the equilibrium be shifted by an increase in temperature?
62. This a system at equilibrium: $CO_2(g) \rightarrow CO_2(aq)$
    a. In which direction will the equilibrium be shifted by an increase in pressure?
    b. In which direction will the equilibrium be shifted by an increase in temperature?
63. Are gases generally more or less soluble in water at higher temperatures? Explain why.
64. Are ionic solids generally more or less soluble in water at higher temperatures? Explain why.
65. What is Henry's Law, and what does it tell us about the ability to dissolve gases in water?
66. In the equation for Henry's Law, what does P_{gas} refer to?
67. If you were trying to dissolve chlorine gas in water, would increasing the pressure of the system by pumping air into it increase the amount of chlorine you can dissolve? Why or why not?
68. If you were trying to dissolve chlorine gas in water, would increasing the pressure of the system by pumping air into it increase the amount of chlorine you can dissolve? Why or why not?

12

Chemical Applications and Practices

69. Underwater activities require assisted breathing techniques. When using pressurized air, divers must be aware of Henry's law. Assuming that air contains 78% nitrogen, what is the concentration of N_2 in blood at 1 atm? (The Henry's law constant for $N_2(g)$ at 25 °C is 1540 atm/M.)
70. What would be the concentration of N_2 in blood if the N_2 pressure were increased to 3.0_{atm}? (The Henry's law constant for $N_2(g)$ at 25 °C is 1540 atm/M.)
71. The solubility of ammonium fluorosilicate, $(NH_4)_2SiF_6$, in water at 20 °C is 18.6 g/100 g H_2O, increasing to 40.4 g/100 mL H_2O at 60 °C. How many grams of ammonium silicate salt could you add to 500.0 g of water at 20 °C, to just reach saturation? If you then raised the temperature of the solution to 60 °C, how much more of the salt could you add before the solution becomes saturated?
72. The solubility of barium nitrate, $Ba(NO_3)_2$, in water at 20 °C is 9.02 g/100 g H_2O and increases to 27.2 g/100 g H_2O at 80 °C. What is the molarity of nitrate in 250 mL of a saturated solution of barium nitrate at 20 °C? Assume the density is 1.2 g/mL of the solution. If you raised the temperature of the solution to 80 °C, how much more barium nitrate would you need to add to saturate the solution?
73. A saturated solution of cadmium selenate, $CdSeO_4$, was prepared by adding 171 g of the salt to 250.0 g of water at 10 °C. A staff worker accidentally left the solution on a countertop, where it warmed to 30 °C. Knowing that the solubility of cadmium selenate is actually *reduced* at 30 °C to 0.231 M, how much of the salt precipitated out of the solution at the higher temperature?
74. The solubility of strontium chloride, $SrCl_2$, is 102 g/100 g H_2O at $20_{°C}$. The solubility of strontium bromide, $SrBr_2$, is 52.9 g/100 g H_2O at the same temperature, and the solubility of strontium iodide, SrI_2, is 178 g/100 g H_2O, also at the same temperature. Which of these saturated solutions has the highest halogen molarity? What is the value of the molarity? Assume the densities of all solutions are 1.0 g/mL.
75. Consider the solubility curves for nitrogen and oxygen depicted in Fig. 12.4. What does this tell you about the interactions between nitrogen and water molecules compared to those between oxygen and water molecules?
76. Carbon monoxide is about 30 times more soluble in water than carbon dioxide. Use your knowledge of intermolecular forces to explain why this might be.

77. Although most gases are less soluble in water at higher temperatures, hydrogen chloride (HCl) is not. In fact, hydrogen chloride gas is more soluble in water at higher temperatures. Explain why this is true.
78. Although most gases are less soluble in water at higher temperatures, there are some that are more soluble. Explain how this could be true for such a gas.
79. The Henry's Law constant for oxygen in water is 769 atm/M at 25 °C. Atmospheric air is 20.95% oxygen, by volume, at sea-level, where the atmospheric pressure is 1.0 atmosphere. What does Henry's Law predict the concentration of oxygen to be in a 25 °C lake that is at sea-level?
80. The Henry's Law constant for nitrogen in water is 1640 atm/M at 25 °C. Atmospheric air is 78.08% nitrogen, by volume, at sea-level, where the atmospheric pressure is 1.0 atmosphere. What does Henry's Law predict the concentration of nitrogen to be in a 25 °C lake that is at sea-level?
81. ◘ Table 12.4 shows the solubility of oxygen in water at various temperatures under 1 atmosphere of air. How would the solubility of oxygen in water compare to these values if we dissolved it under 2 atmospheres of air? Would the solubility of oxygen be significantly less, slightly less, slightly more, or significantly more? Explain your reasoning.
82. ◘ Table 12.4 shows the solubility of oxygen in water at various temperatures under 1 atmosphere of air. How would the solubility of oxygen in water compare to these values if we dissolved it under 0.5 atmosphere of air? Would the solubility of oxygen be significantly less, slightly less, slightly more, or significantly more? Explain your reasoning.

Section 12.5 Colligative Properties

Skill Review

83. Determine the melting point of the following aqueous solutions. Assume the value of the van't Hoff factor is predictable directly from the formula of the compound (i.e.; $i=3$).
    a. 0.01 m $CoCl_2$
    b. 0.05 m $CoCl_2$
    c. 1.00 m $CoCl_2$
    d. 2.00 m $CoCl_2$
84. Repeat the previous problem using the actual values for the van't Hoff factor found in ◘ Table 12.6. Compare the differences in the values.
85. Which of the following aqueous solutions would you predict to exhibit the greatest osmotic pressure at 25 °C?
    a. 0.20 M glucose ($C_6H_{12}O_6$)
    b. 0.10 M salt (NaCl)
    c. 0.15 M $Mg(NO_3)_2$
    d. 0.20 M HCl
86. Calculate the osmotic pressure of a 0.10 M $CaCl_2$ solution at 10 °C. If the temperature, in Celsius, is doubled, what is the resulting osmotic pressure of the solution?
87. Let's reverse the spaghetti question in Example 12.6. There, we considered the increase in the boiling point of the salt solution in which we were cooking spaghetti. If, instead of determining how much the boiling point of 6.0 L of water is raised by the addition of 10.0 g of NaCl, calculate how much the NaCl you would need to add to raise the boiling point of 6.0 L water by 5.0 °C. The solubility of sodium chloride in water at 100 °C is 39.2 g NaCl/100 g H_2O. On the basis of the solubility data, is it possible to raise the temperature of water by 5.0 °C by the addition of sodium chloride?
88. Using the solubility data in Problem 87, what is the maximum temperature that boiling water can be raised by saturating it with sodium chloride at 100 °C?
89. Determine the freezing point of each solution in Problem 83. You may need to use ◘ Table 12.7 to help with this problem. Assume the solutions behave ideally.
90. Determine the freezing point of the solutions in Problem 85. You may need to use ◘ Table 12.7 to help with this problem. Assume each solution behaves ideally and that the density is 1.0 g/mL for each solution.
91. In theory, what concentration of ions would you expect to find dissolved in 500.0 mL of a 0.00100 m solution of $CaCl_2$?
92. What value is expected for the van't Hoff factor in very dilute solutions of each of these?
    a. $AlCl_3$
    b. $(NH_4)_3PO_4$
    c. $Mg(OH)_2$
    d. $C_6H_{12}O_6$

93. Although several factors are necessary to explain the magnitude of the vapor pressure of a sample, some general trends can be noted. Arrange these substances in order of decreasing vapor pressure.

Water (H_2O)
Glycerol ($HOCH_2CHOHCH_2OH$)
Pentane ($CH_3CH_2CH_2CH_2CH_3$)

94. a. Complete this sentence by inserting the appropriate term in the blank: As external air pressure_______(increases, decreases) the boiling point of a liquid decreases.
    b. Explain why the term you selected is correct.
95. For each of these solutions, determine the boiling point of the solution. You may need to use Table 12.5 to help with this problem. (Assume these solutions behave ideally.)
    a. 0.75 m NaCl in water
    b. 0.040 m glucose ($C_6H_{12}O_6$) in water
    c. 0.250 M $CaCl_2$ in water (Assume the density of the solution is 1.00 g/mL.)
    d. 0.25 mol fraction naphthalene in benzene (Assume the density of the solution is 0.88 g/mL.)
96. For each of these solutions, determine the boiling point of the solution. You may need to use Table 12.5 to help with this problem. (Assume these solutions behave ideally.)
    a. 1.00 m sucrose ($C_{12}H_{22}O_{11}$) in ethanol
    b. 0.14 m Na_2SO_4 in water
    c. 0.250 M glucose ($C_6H_{12}O_6$) in water (Assume the solution's density is 1.00 g/mL.)
    d. 500.0 ppm $FeCl_3$ in methanol
97. For each of these solutions, determine the melting point of the solution. You may need to use Table 12.7 to help with this problem. (Assume these solutions behave ideally.)
    a. 0.500 m glucose ($C_6H_{12}O_6$) in water
    b. 0.055 m LiOH in water
    c. 0.125 m methanol in phenol
    d. 1.20 m benzoic acid in water
98. For each of these solutions, determine the melting point of the solution. You may need to use Table 11.7 to help with this problem. (Assume these solutions behave ideally.)
    a. 0.200 mol fraction glucose ($C_6H_{12}O_6$) in water
    b. 0.055 m $CoCl_2$ in water
    c. 0.125 m benzene in phenol
    d. 1.20 m HCl in water
99. For each of these solutions, determine the vapor pressure of the solution at STP. (Assume these solutions behave ideally.)
    a. 0.200 mol fraction glucose ($C_6H_{12}O_6$) in water
    b. 0.055 mol fraction $CoCl_2$ in water
    c. 0.125 mol fraction benzoic acid in water
    d. 1.20 m HCl in water
100. For each of these solutions, determine the osmotic pressure of the solution at STP. (Assume these solutions behave ideally.)
    a. 0.200 M $FeCl_2$ in water
    b. 0.055 m $CoCl_2$ in water (Assume the density of the solution is 1.0 g/mL.)
    c. 0.125 mol fraction glucose ($C_6H_{12}O_6$) in water (Assume the density of the solution is 1.02 g/mL.)
    d. 0.0945 mol fraction NaCl in water (Assume the density of the solution is 1.10 g/mL.)
101. Acetamide, CH_3CONH_2, has some uses in the plastics industry, is somewhat soluble in water. If you dissolve 2.20 g of acetamide in 10.0 mL of water, what is the vapor pressure of the solution at 24 °C? Assume the density of water is 1.00 g/mL.
102. At 25 °C the vapor pressure of pure water is 23.6 mm Hg and that above an aqueous glucose ($C_6H_{12}O_6$) solution is 18.9 mm Hg. Calculate the mole fraction of water and the mass in grams of glucose in the solution if the mass of water is 225 g.
103. An aqueous solution contains 10.00% glucose ($C_6H_{12}O_6$) by mass. Assume the glucose is nonionic and nonvolatile. Find the following:
    a. the freezing point of the solution
    b. the boiling point of the solution
    c. the osmotic pressure of the solution at 25 °C

104. The osmotic pressure at 20 °C is 3.76 atm for a solution prepared by dissolving 2.50 g of a non-volatile compound in enough water to give a solution volume of 100 mL. What is the empirical molar mass of the compound?
105. A 1.12% solution of potassium chloride in water freezes at −0.557 °C.
    a. Calculate the van't Hoff factor, *i*.
    b. Is this consistent with the complete dissolution of the salt in water? Explain.
106. A 0.050 m solution of copper(II) sulfate is prepared and the freezing point determined to be −0.113 °C.
    a. Calculate the van't Hoff factor, *i*.
    b. Is this consistent with the complete dissolution of the salt in water? Explain.

Chemical Applications and Practices

107. An amount of water in a closed container will have a certain vapor pressure when the temperature is held constant. Varying the shape of the container can change the surface area of the volume of water. However, this does not change the vapor pressure. Explain this observation and reconcile it with the observation that the surface area of a sample of water does change the rate of evaporation in an open container.
108. The vapor pressure of water at 25 °C is 23.76 mm Hg. What would you calculate as the new vapor pressure of a solution made by adding 50.0 g of ethylene glycol ($HOCH_2CH_2OH$, antifreeze) to 50.0 g of water? You may assume that the vapor pressure of ethylene glycol at this temperature is negligible.
109. Explain why 0.10 m solutions of nonelectrolytes have approximately the same boiling point regardless of the identity of the solute. Why doesn't this same statement apply to electrolytes?
110. Two solutions are placed in a −1.0 °C storage cooler. One solution is labeled 5.0% $C_6H_{12}O_6$ and the other 15.0% $C_6H_{12}O_6$. When they were to be retrieved the next day, the labels of both had loosened and fallen off the containers. Would you be able to identify the solutions on the basis of their freezing points? Prove your answer.
111. How many grams of antifreeze ($HOCH_2CH_2OH$) would have to be added to 5.0 kg of water in an automobile cooling system to keep it from freezing during a cold Nebraska winter with temperatures of −32 °C?
112. Freezing-point depression can be used to determine the molecular mass of a compound. Suppose that 1.00 g of an unknown molecule were added to 20.0 g of water and the freezing point of the solution determined. If the new freezing point of water were found to be −1.50 °C, what would you predict to be the molecular mass of the compound?
113. a. The outer membrane of many fruits acts as a semipermeable membrane through which osmosis can take place. Diagram the cell of a cucumber before and after it has been set in a highly concentrated salt solution.
    b. Why is it important to know the osmotic pressure of human fluids before administering any fluids to a patient?
114. One of the first stages in healing a small cut takes place when a blood clot begins to form. A key enzyme in this process is thrombin. This large enzyme has a molar mass of nearly 34,000 g/mol. What would be the osmotic pressure of a solution that contained 0.20 g of thrombin per milliliter, at 37.0 °C?
115. When preparing some solutions for analysis, chemists must be aware that some solution processes are very exothermic. Sodium hydroxide solutions are often used when analyzing acid solutions.
    a. The heat of solution for NaOH is −44.5 kJ/mole. Using three factors, describe the dissolving of solid NaOH pellets in water.
    b. What temperature change would you calculate for a solution, starting at 25.0 °C, made when 100.0 g of water is mixed with 10.0 g of NaOH? Assume the heat capacity of the solution is the same as that for water.
116. The final production of the writing paper you may be using to solve this problem involves several chemical treatments. One compound used in the process is aluminum sulfate. What would be the molarity of an industrial solution if 3250 g of $Al_2(SO_4)_3$ were dissolved in enough water to make 20.0 L of solution?
117. Sulfuric acid (H_2SO_4) solutions are used as the primary electrolyte in the lead storage batteries found in automobiles. One of the most common ways to check the charge level in such a battery is to determine the density of the solution. A sulfuric acid solution had a density of 1.58 $_g$/mL and was known to contain 35.6% by mass H_2SO_4. What is the molarity of the solution?

Comprehensive Problems

118. If humans consume 2.7×10^{15} kg of water each year, how much cadmium is the world consuming via water usage if we assume that all water systems contain the maximum contaminant limit (MCL) of 5 ppm?
119. What happens to the osmotic pressure of seawater as it is desalinated? Explain your answer.
120. Before refrigeration, many homemakers used to prepare their own jams and leave them at room temperature, possibly covered only with a cloth to keep out insects. Single celled bacteria, some of which become airborne in a busy kitchen, will not grow on the surface of the jam. Explain.
121. Which is affected more by the presence of a solute; freezing point or boiling point?

Thinking Beyond the Calculation

122. We have a can of glucose-sweetened fruit-flavored drink, and we want to determine the amount and concentration of the glucose in the drink at home, having available only a kitchen scale that weighs to an accuracy of 1.0 g, and the usual kitchen appliances (stove, microwave, fridge, etc.). Design a procedure using only what's available in the kitchen.
123. The 2010 Water Quality Report for Denver, Colorado, lists the MCLG for lead in drinking water as zero ppb. Is it possible to know for certain that there is absolutely zero lead in the drinking water sample? Should there be laws mandating a concentration of zero for any solute? Explain your answer.

Chemical Kinetics

Contents

Supplementary Information The online version contains supplementary material available at ▶ https://doi.org/10.1007/978-3-030-90267-4_13.

M. Mosher and P. Kelter, *An Introduction to Chemistry*,
https://doi.org/10.1007/978-3-030-90267-4_13

What Will We Learn from This Chapter?

- The speed of a reaction can be determined by measuring the change in the concentration of reactants or products over a given amount of time (▶ Sect. 13.2).
- A mathematical equation known as a rate law describes the rate of a reaction as a function of the concentration of reactants in the chemical equation (▶ Sect. 13.3).
- We can determine the concentration of a reactant at any point during a reaction by using the integrated rate law (▶ Sect. 13.4).
- The rate law can only be determined experimentally. Two methods to do this are the method of graphical analysis and the method of initial rates (▶ Sect. 13.5).
- The rate law can be written directly from the slow elementary step in a mechanism (▶ Sect. 13.7).
- A catalyst increases the rate of a reaction by changing the mechanism of the reaction (▶ Sect. 13.8).

13.1 Introduction

In 2019, American farmers continued an overall trend showing decades-long increases in food production, harvesting just under 14 billion bushels of corn. Production of other grains, such as wheat and rice, also generally increased compared to decades before. According to the U.S. Department of Agriculture, corn production in Iowa alone more than quadrupled, from 40 bushels per acre to 180 bushels per acre, in the nearly 85 years between 1930 and 2015. Although the introduction of the tractor and other automated farm machinery has played a large role in this increase in production, the use of **insecticides** and **herbicides** (compounds used to kill unwanted plants) has had a substantial impact. No longer do insects and weeds run rampant through cornfields and destroy crop yields. Atrazine, a herbicide, is one of the agents most commonly sprayed onto the soil from which corn crops grow in order to control weeds; approximately 80 million pounds are applied annually in the U.S., with 86% used on corn, 10% on sorghum (a grain used primarily in livestock feed and ethanol production), and 3% on sugar cane, according to the U.S. Environmental Protection Agency (EPA). When it is introduced into a farmer's field, atrazine works well to control broadleaf weeds, such as pigweed, cocklebur, velvetleaf, and certain grass weeds, without harming the corn plants. What happens to the atrazine that doesn't land on weeds? Some of it travels into the soil, where microbes and water can degrade it into by-products.

Insecticide – Any compound used to kill unwanted insects.
Herbicide – Any compound used to kill unwanted plants.

Scientists have studied this degradation process extensively. One reaction, the hydrolysis of atrazine, in which the chlorine atom on atrazine is replaced with a hydroxy (–OH) group, is of particular interest to researchers. The product, hydroxyatrazine, is rapidly metabolized by microbes living in the soil and groundwater and is viewed by the EPA as not harmful to humans. The length of time it takes atrazine to metabolize is largely determined by this initial hydrolysis reaction.

$$\text{Atrazine} + H_2O \longrightarrow \text{Hydroxyatrazine} + HCl$$

Environmental chemists study the interaction of compounds and the environment, including the chemistry of the soil, water, and air. Their work proves that **pesticides** (a pesticide is any compound used to kill unwanted organisms) do not always rapidly disappear from the environment. For example, their analyses of lakes, rivers, and streams in states that use atrazine show that it persists in the environment for quite a long time. Depending on certain environmental and biological factors (including soil depth, temperature, and the presence of microorganisms, especially fungi), the concentration of atrazine in soil can decline by half in less than a month, though about 60 days is typical. However, in natural water samples, atrazine typically degrades to one-half of its original concentration more slowly, in about 400 days. Because its degradation is relatively slow, atrazine is present in many water samples throughout the year, as shown in ◻ Fig. 13.1.

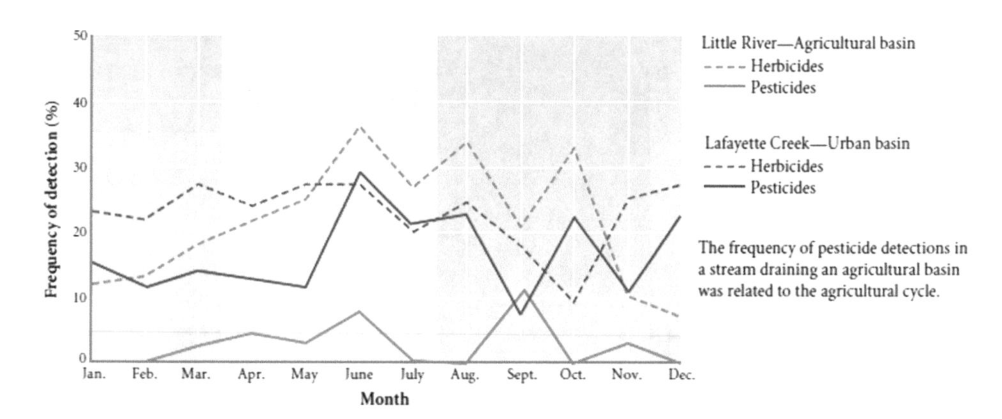

Fig. 13.1 Atrazine and other herbicides persist in the environment long after they are applied

Pesticide – Any compound used to kill unwanted organisms (whether they be animals, insects, or plants).
Environmental chemist – A scientist who studies the interactions of compounds in the environment.

Laboratory studies, however, indicate that the hydrolysis of atrazine occurs without outside assistance as an exothermic process ($\Delta H = -35$ kJ/mol). Does the value of the enthalpy change say anything about how fast this reaction occurs? The answer is a resounding NO. Values of enthalpy are only useful in determining how much heat is absorbed or released by a reaction; they do not indicate anything about the rate of the reaction. In order for us to determine how quickly the reaction occurs, we have to examine **chemical kinetics**—the study of the rates and mechanisms of chemical reactions, including the factors that influence these properties. To begin our study of chemical kinetics, we will briefly leave the farm fields and travel to the Olympic Games.

Chemical kinetics – The study of the rates and steps that occur in a chemical reaction.

13.2 Reaction Rates

The Olympic Games rely heavily on the use of accurate timekeeping. In the bobsled and luge events, such as that shown in Fig. 13.2, the accuracy of the timekeeping determines whether an athlete or team wins a gold or no medal at all. For example, at the 1998 Winter Olympic Games in Nagano, Japan, Silke Kraushaar of Germany placed first in the women's luge with a time of 50.617 s for her best run. The silver medal went to Barbara Niedernhuber, also from Germany, with a time of 50.625 s. The difference between first and second place was only a fraction of a second.

Let's examine the luge event more closely to help us introduce some new terms related to kinetics, the main topic of this chapter. The women's luge course at Nagano in 1998 was 1194 m long. Judging on the basis of her winning time, how fast did Silke Kraushaar travel? Speed is calculated by dividing the distance traveled by the change in time. Kraushaar traveled 1194 m in 50.617 s, so her **rate** of travel (speed) was 84.92 km/h (52.77 mi/h).

Rate – The "speed" of a reaction, recorded in M/s.

$$Speed = \frac{distance}{time} = \frac{1194\,\text{m}}{50.617\,\text{s}} \times \frac{1\,\text{km}}{1000\,\text{m}} \times \frac{3600\,\text{s}}{\text{h}} = \frac{84.92\,\text{km}}{\text{h}}$$

But top speeds for luge and bobsled events routinely hit 135 km/h (84 mi/h)! What we've just calculated is the **average rate** of her travel over the entire course.

Average rate – The rate of a reaction measured over a long period of time.

Chemical reactions also have rates. Whereas the speed of a bobsledder can be listed in kilometers per hour (or miles per hour), the speed of a reaction is often described in units of concentration, often molarity, per second. For instance, under certain laboratory conditions, a 0.50 M solution of the herbicide atrazine can be completely

13

Fig. 13.2 In many Olympic events, fractions of a second determine the winners. Racers in the luge event must be perfect in their form to win. Photo by Jon Wick, CC BY-2.0. *Source* ▶ https://upload.wikimedia.org/wikipedia/commons/9/98/Départ_d%27un_lugeur.jpg

hydrolyzed in 24 h. We can calculate the rate of the reaction by dividing the concentration that is consumed in the reaction by the time it took to complete the hydrolysis.

$$\textit{Average rate} = \frac{0.50\,\text{M}}{24\,\text{h}} \times \frac{1\,\text{h}}{3600\,\text{s}} = 5.8 \times 10^{-6}\,\text{M/s}$$

We can say that the average rate of this reaction over the 24 h period is 5.8×10^{-6} M/s.

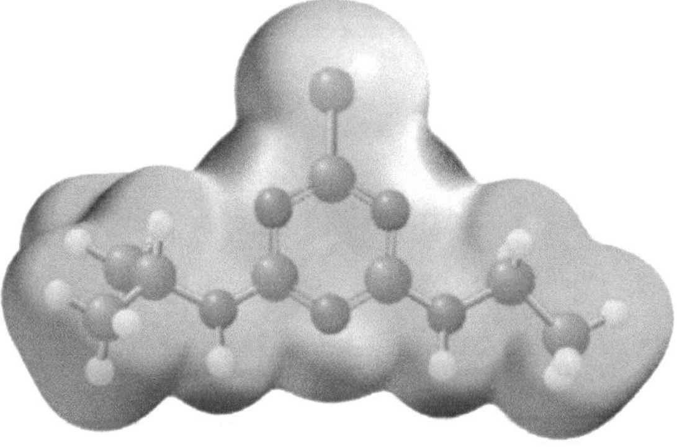

Atrazine

Determining the average rate of a reaction is one of the tasks accomplished in chemical kinetics. However, when we study kinetics, we have to be careful to note the difference between the extent of a reaction and the rate of the reaction. The extent of a reaction (which we cover in ▶ Chap. 14) is a measure of the completeness of a reaction. The rate of a reaction describes how quickly it gets to that point. For example, the combustion of methane to make the flame on our cooking stove is an essentially complete reaction (~100% of reactants are turned into products) that is also relatively fast. The oxidation of iron on a suspension bridge is also complete, but it is quite slow. Kinetics deals only with "how fast or slow and by what route." Kinetics tells us nothing about the extent of a reaction.

13.2.1 Instantaneous Rate, Initial Rate, and Average Rate

At the 2001 Grand Nationals in Chicago, Whit Bazemore in his Matco Tools Pontiac Firebird "funny car" finished the quarter-mile drag strip in 4.750 s (Fig. 13.3). Using this information (0.2500 mi in 4.750 s), we can calculate the average speed of the funny car as 0.05263 mi/s, or 189.5 mi/h. However, the speed of the funny car at the start was much less than this (0 mi/h), and the speed at the finish line was a lot faster (323.3 mi/h) than the average speed. The rate of a chemical reaction changes throughout its progress, much like the rate of travel of a funny car at a drag race. However, chemical reactions differ because they typically start with a very large rate that decreases with time. In other words, at the start of a reaction, the rate of chemical change is typically fast. As the reaction nears its end, the rate is relatively slow. Over a given time period (Δt), though, it has an average rate that describes how long it took to reach that certain point in the reaction.

$$\text{Average rate} = \frac{\Delta\text{concentration}}{\Delta t}$$

In short, just as the speed of a dragster changes during a race, the rate of the reaction changes as the reaction proceeds.

What if we consider a change in time that is negligible ($\Delta t \approx 0$)? When this happens, we are examining the **instantaneous rate** of a reaction. This is what we determine at the finish line in the funny car race (323.3 mi/h). It is also what we measure as the lights turn green at the start of the race (0 mi/h), or when the driver glances at the speedometer at any point in the race. At these times, and at any other specific time we pick, we are measuring the instantaneous rate. In a chemical reaction, the rate at the start of a reaction, when the reactant concentrations are greatest, is important. The instantaneous rate of the reaction measured at the start is referred to as the **initial rate** of reaction. Instantaneous rates can be measured if we have a plot of the reaction such as that shown in Fig. 13.4. If, on the plot of our reaction, we draw a line that is tangent to the curve (that is, a line that just touches the curve and is going in the same direction at that point), as is done in Fig. 13.5, and we measure the slope of that line (rise/run = slope), we get the instantaneous rate of the reaction at that specific time. This is a useful way to measure instantaneous rates.

Fig. 13.3 "Funny cars," with their characteristically powerful engines and oversized rear wheels, get ready to race the quarter-mile. Because of the hazards of gaining that much speed for ¼-mile, funny cars typically now race for only 1000 ft, a little under 1/5-mile. Even so, they can burn as much as 15 gallons of fuel in that short distance! Photo by Tim Felce, CC BY-SA-2.0. *Source* ▶ https://upload.wikimedia.org/wikipedia/commons/a/a9/FIA_Top_Methanol_Funny_Cars_-_Santa_Pod_2010_%284656627185%29.jpg

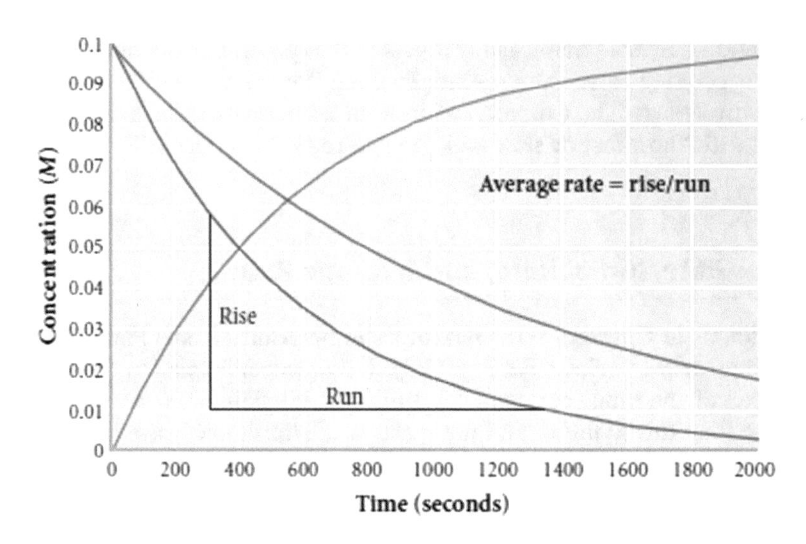

Fig. 13.4 Calculating the average rate of a reaction from a sample plot of the reactant concentration versus time. Average rates and instantaneous rates differ in the length of time used to calculate them. The average rate of a reaction is calculated by dividing the rise by the run

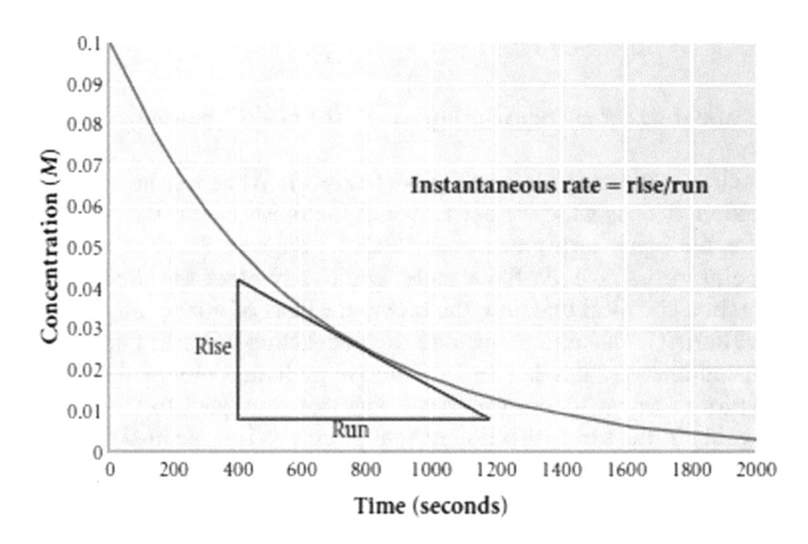

13

Fig. 13.5 Calculating the instantaneous rate of a reaction from the same sample plot used to calculate the average rate in Fig. 13.4. The slope of a tangent line can be used to find an instantaneous rate

Instantaneous rate – The rate of reaction measured at a specific instant in time. The instantaneous rate is typically measured by calculating the slope of a line that is tangent to the curve drawn by plotting [reactant] versus time.

Initial rate – The instantaneous rate of reaction when $t = 0$.

Environmental chemists often measure the rate of a reaction. For example, the environmental hydrolysis of a common herbicide called alachlor, also known as Lasso™, occurs more rapidly than the hydrolysis of atrazine. However, the rate is still slow enough that new ways to decrease the time that the herbicide spends in the environment before chemically degrading are being sought. One such way includes the reduction of alachlor to an acetanilide via electrochemical techniques.

Alachlor (Lasso) $+ H^+ + 2e^- \longrightarrow$ Acetanilide product $+ Cl^-$

The rate of this reaction can be determined in the laboratory by measuring the change in the concentration of one of the compounds over a period of time. Mathematically, the rate of the reaction is equal to the rate of the disappearance of alachlor, the rate of appearance of the acetanilide, or the rate of the appearance of Cl^-, because they are all in a 1-to-1 mol ratio. The sign placed in the rate equation is used to indicate whether the compound being measured is disappearing (minus sign) or appearing (plus sign). Remember that the Δ symbol represents a change, measured by taking the final value minus the initial value. Even though the sign of the rate for the disappearance would appear to generate a negative value for the rate, it does not, because the disappearance of the reactant is itself a negative value. The negative sign is there to make the value of the rate positive. In fact, similar to the speed of a funny car, reaction rates are always positive. Even if we measured the speed of a car moving backwards, we would still record its speed as a positive number.

$$\text{Rate} = \text{rate of disappearance of alachlor} = \frac{-\Delta[\text{alachlor}]}{\Delta t}$$

$$\text{Rate} = \text{rate of disappearance of H}^+ = \frac{-\Delta\left[\text{H}^+\right]}{\Delta t}$$

$$\text{Rate} = \text{rate of disappearance of acetanilide} = \frac{-\Delta[\text{acetanilide}]}{\Delta t}$$

$$\text{Rate} = \text{rate of appearance of Cl}^- = \frac{\Delta\left[\text{Cl}^-\right]}{\Delta t}$$

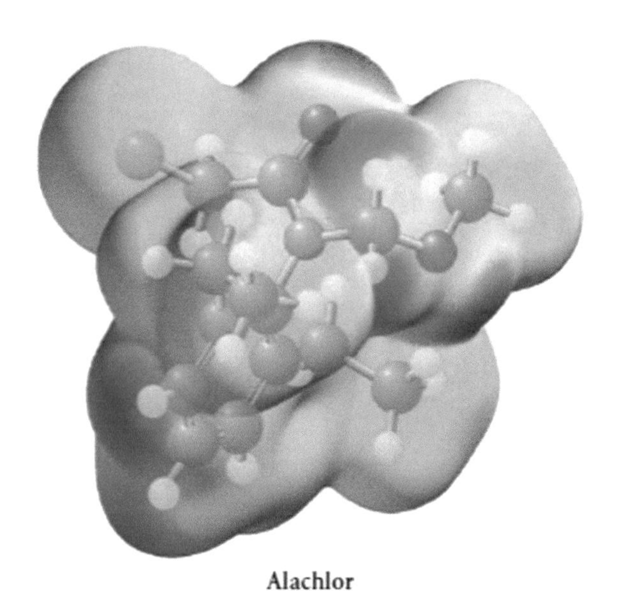

Alachlor

The rate of the decomposition of hydrogen peroxide into water and oxygen can also be written using these rules. For example, the rate of the reaction could be set equal to the rate of appearance of oxygen. In this case, the rate is the change in its concentration divided by the change in time. However, if we consider the rate of appearance of water as a measure of the rate of the reaction, we must somehow note that two molecules of water are appearing for every molecule of $O_2(g)$ that is produced.

$$2\text{H}_2\text{O}_2(\text{aq}) \rightarrow 2\text{H}_2\text{O(l)} + \text{O}_2(\text{g})$$

This is done by dividing the rate by the mole ratio of the compounds from the balanced equation. For instance, the rate of appearance of H_2O times $\frac{1\ \text{mol O}_2}{2\ \text{mol H}_2\text{O}}$ gives the rate of the reaction. Each of the rate descriptions then becomes:

$$\text{Rate} = \text{rate of disappearance of H}_2\text{O}_2(\text{g}) = \frac{-\Delta[\text{H}_2\text{O}_2]}{\Delta t} \times \frac{1}{2}$$

$$\text{Rate} = \text{rate of appearance of } O_2(g) = \frac{\Delta[O_2]}{\Delta t}$$

$$\text{Rate} = \text{rate of appearance of } H_2O(l) = \frac{\Delta[H_2O]}{\Delta t} \times \frac{1}{2}$$

Example 13.1—Rate of the Haber Process

Farmers apply ammonia to their cornfields during spring planting by injecting it directly into the soil. The ammonia that they use is made by the Haber process; the chemical reaction for this process is indicated by the equation shown below. Describe the rate of the reaction in terms of the rate of appearance or disappearance of the components of the reaction.

$$N_2(g) + 3H_2(g) \rightarrow 2NH_3(g)$$

Asking the Right Questions

What is the relationship between the rate of the reaction of each of the species and its coefficient in the reaction? That is, for each mole of nitrogen that reacts, how many moles of hydrogen react per unit time? How do we express this relationship mathematically?

Solution

The rate of the reaction can be measured by measuring the changes in concentration of each of the species. Mathematically, the rate of the reaction of N_2 is equal to one-third the rate of disappearance of H_2, because 1 mol of N_2 disappears for every 3 mol of H_2 that react. The rate of the reaction (and, therefore, of disappearance) of N_2 is also equal to one-half the rate of appearance of ammonia (NH_3), because 1 mol of N_2 disappears for every 2 mol of NH_3 that are produced. Note that the disappearance of H_2 is indicated with a minus sign and the appearance of NH_3 a plus sign.

$$\text{Rate} = -\frac{\Delta[N_2]}{\Delta t} = -\frac{1}{3}\frac{\Delta[H_2]}{\Delta t} = \frac{1}{2}\frac{\Delta[NH_3]}{\Delta t}$$

Are Our Answers Reasonable?

The best way to judge whether we've made a reasonable answer is to look at our guide, the balanced chemical equation. The answers reflect the coefficients of the equation, so the answer is reasonable.

What If…?

…two reactants, A and B, formed a product, P. A total of 0.600 mol of A reacted with 1.500 mol of B to form 0.900 mol of P in 30 min. What would be the balanced equation for the reaction and the relative rates?

$$\left(Ans\text{: } 2A + 5B \rightarrow 3P \text{ } Rate = -\frac{5}{2}\frac{\Delta[A]}{\Delta t} = -\frac{\Delta[B]}{\Delta t} = \frac{5}{3}\frac{\Delta[P]}{\Delta t}\right)$$

Practice 13.1

What is the rate of the following reaction in terms of the rates of disappearance and appearance of the components of the reaction?

$$2HI(aq) \rightarrow H_2(g) + I_2(aq)$$

We can examine this further by actually calculating the average rate of a chemical reaction. Assume that an environmental chemist begins an experiment with a 0.400 M alachlor solution. After 10 days, the concentration of alachlor is determined to be 0.350 M, and the concentration of acetanilide and that of chloride are both 0.050 M. What is the average rate of the reaction? We know the final value (0.350 M alachlor at 10 days) and the initial value (0.400 M alachlor at 0 days). Although time can be used with almost any unit, we often report rates as M/s. So, we convert the time into seconds (10 days = 864,000 s) and fill in our equations, keeping in mind that the rate measures the change in concentration per unit change in time. In this example, however, we don't know the concentration of H^+ at either time, so we can't use it to determine the rate of the reaction. Note: It would be perfectly fine to report rates with units of M/day, M/h, etc.

$$\text{Rate} = \text{disappearance of alachlor} = \frac{-(0.350\text{ M} - 0.400\text{ M})}{(864{,}000\text{ s} - 0\text{ s})} = 5.79 \times 10^{-8}\text{ M/s}$$

$$\text{Rate} = \text{appearance of acetanilide} = \frac{(0.050\text{ M} - 0.000\text{ M})}{(864{,}000\text{ s} - 0\text{ s})} = 5.79 \times 10^{-8}\text{ M/s}$$

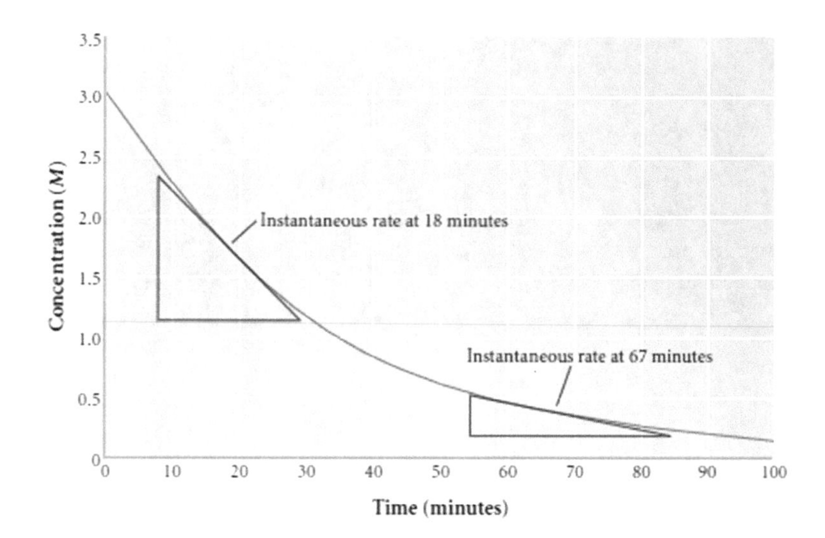

$$\text{Rate} = \text{appearance of Cl}^- = \frac{(0.050\ \text{M} - 0.000\ \text{M})}{(864{,}000\ \text{s} - 0\ \text{s})} = 5.79 \times 10^{-8}\ \text{M/s}$$

No matter which calculation we complete, we should always obtain the same rate of reaction over the same time period.

Remember that average rates, instantaneous rates, and initial rates are calculated in the same manner. The only difference is the length of time used in the calculations. Large changes in time determine the average rate. Instantaneous rates are measured when the change in time (Δt) is very small (that is, $\Delta t \rightarrow 0$). Experimentally, average rates, instantaneous rates, and initial rates can be found by measuring the color, temperature, electrochemical voltage, or some other physical property of the reactants or products that is unique within the reaction. Then, by determining the slope of a line that is tangent to the curve drawn when the concentrations of reactants or products are plotted against time, the rate can be obtained. Initial rates are measured as instantaneous rates at the start of the reaction.

Example 13.2—The Rate of a Reaction

Methoxychlor is an important insecticide used to control parasites on livestock and a variety of pests on vegetables and fruits. Its breakdown by soil microbes may proceed through the following reaction. Calculate the average rate of this reaction in M/s of methoxychlor, assuming that the concentration of CH_3OH starts at 0.000 M and after 60.0 h climbs to 0.100 M.

$2H_2O$ + Methoxychlor (H_3CO–C_6H_4–$CH(CCl_3)$–C_6H_4–OCH_3) ⟶ *bis*-Hydroxy methoxychlor (HO–C_6H_4–$CH(CCl_3)$–C_6H_4–OH) + $2CH_3OH$

Methoxychlor

bis-Hydroxy methoxychlor

Asking the Right Questions

How is the concentration of methoxychlor related to that of methanol, CH_3OH? In concept, could we use the other reactants or products to find the rate? How does the change in concentration relate to the rate?

Solution

The rate of the reaction can be determined by the rate of appearance of methanol (CH_3OH). Because methanol is forming at twice the rate of the reaction of methoxychlor, the rate of the reaction of methoxychlor is one-half the rate of appearance of methanol.

$$\text{Rate} = \frac{1}{2}\left(\frac{\Delta[\text{CH}_3\text{OH}]}{\Delta t}\right) = \frac{1}{2}\left(\frac{(0.100\ \text{M} - 0.00\ \text{M})}{(216{,}000\ \text{s} - 0.00\ \text{s})}\right)$$

$$= \frac{1}{2}\left(\frac{0.100\ \text{M}}{216{,}000\ \text{s}}\right) = 2.31 \times 10^{-7}\ \text{M/s}$$

Is Our Answer Reasonable?

To the extent that the rate equation is reasonable because it shows that the rate of formation of methanol is twice the rate of disappearance of methoxychlor, the answer is reasonable. The rate, itself, 2.31×10^{-7} M/s, won't have a lot of meaning as a specific quantity until we have had more experience calculating rates and seeing them in the framework of the reactions with which we work.

What If…?

…we measured a rate of one of the reaction species under the same conditions, and the rate was 4.62×10^{-7} M/s? Which reactant or product could this have been?

(Ans: Either water (if it was a rate of disappearance) or methanol (if it was a rate of formation).)

Practice 13.2

Calculate the average rate of the Haber process (Example 13.1), assuming that at 12.5 s the concentration of $H_2(g)$ was 0.355 M and at 83.3 s the concentration was 0.258 M.

13.3 An Introduction to Rate Laws

We pointed out in the introduction to this chapter that under certain conditions, atrazine can degrade to one-half of its original concentration in 60 days. This means that in a sample containing 5.0×10^{-7} M atrazine, the concentration reduces to 2.5×10^{-7} M after 60 days. We can calculate the rate of this reaction using the concepts we learned earlier. Can we predict the concentration of atrazine after 120 days? This question highlights a complicating factor in kinetics: The rate of a reaction typically decreases as the reaction progresses, because fewer reactant molecules exist after the reaction begins. The lower concentration of reactant reduces the likelihood that molecules will interact to make products. As a result, determining concentrations of reactants and products at a certain time during a reaction requires greater understanding of how the reaction rate changes with time. In the end, we cannot say that the concentration of atrazine should be 0 M after 120 days (see ▪ Fig. 13.6). Similarly, we cannot say that the concentration is 3.75×10^{-7} M after only 30 days.

The decomposition of hydrogen peroxide can be used to illustrate how the rate of a reaction can be related to the reaction conditions. Hydrogen peroxide, a common staple in the home medicine cabinet, is used to clean cuts and scrapes because of its ability to oxidize microbes. We observe by experiment that the rate of the reaction de-

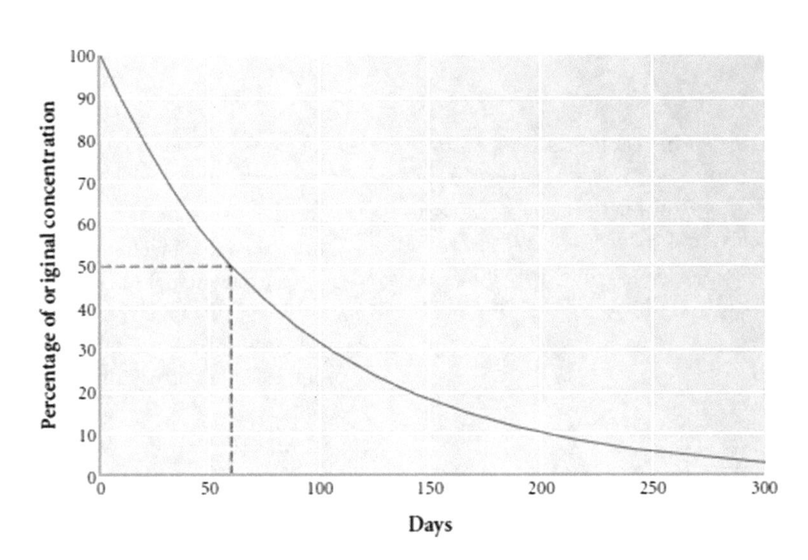

▪ **Fig. 13.6** The environmental decomposition of atrazine in groundwater. The half-life is the time it takes for a given concentration to be halved

pends on the concentration of hydrogen peroxide. As the concentration decreases, the rate of the reaction decreases. In other words, the rate is directly proportional to the concentration of hydrogen peroxide:

$$\text{Rate} \propto [H_2O_2]$$

When we examine the relationship more closely, the following mathematical equation, called a **rate law**, emerges. This rate law states that the rate of the reaction of hydrogen peroxide is equal to the product of a constant (which we call the **rate constant**) times the concentration of H_2O_2.

$$2H_2O_2(aq) \rightarrow 2H_2O(l) + O_2(g)$$

$$\text{Rate} = k[H_2O_2]$$

Rate law – An equation that indicates the molecularity of a reaction as a function of the rate of the reaction.

Rate constant – A constant that is characteristic of a reaction at a given temperature, relating to the rate of disappearance of reactants.

From the rate law, we determine that the reaction is **first order** in hydrogen peroxide. That is, the rate depends on the concentration of H_2O_2 raised to the first power. The reaction is first order overall because the rate equation for the reaction is dependent only on the concentration of H_2O_2 to the first power—that is, it is linearly related. It is important to remember that the value of the rate constant, the compounds included in the rate law, and the orders of the compounds in the equation can only be found experimentally.

First order – A reaction is first order in a particular species when the rate of the reaction depends on the concentration of that species raised to the first power.

The rate law, as we will see, can be used to determine the rate of a reaction at any reactant concentration. It will also tell us which species are the most important contributors to the rate of a reaction. The typical rate law has the following form:

$$\text{Rate} = k[A]^n[B]^m$$

where ***k*** is the rate constant,
[A] and [B] are the concentrations of reactants, and
n and ***m*** are the orders of the corresponding compounds.

Note that this method of calculating the rate depends only on the concentrations of reactants, not products. The values of n and m are measures of how dependent the rate is on the concentration of a particular reactant, and they must be experimentally determined. The reaction also can be described in terms of the overall order, which is calculated by adding n and m (and the exponents of any other reactants in the rate law). The order of a compound or the overall order of a reaction can certainly be negative or even a non-integer number. Again, it is important to remember that the order of a reactant cannot be determined just by looking at the balanced chemical equation.

The reaction of nitrogen monoxide gas with hydrogen gas is:

$$2NO(g) + 2H_2(g) \rightarrow 2H_2O(l) + N_2(g)$$

The rate law, determined by experiments in the laboratory, is:

$$\text{Rate} = k[NO]^2[H_2]$$

The reaction is said to be second order in nitrogen monoxide, because the rate depends on the concentration of nitrogen monoxide to the second power. The rate is first order with respect to hydrogen gas, because the exponent is 1, which means that there is a linear relationship between the rate and $[H_2]$. The reaction is third order (the sum of the individual orders, 2 + 1) overall. It is possible for a reaction to have fractional orders, though we will not deal with that here.

Example 13.3—The Rate Law

Environmental chemists are concerned with the damaging effects of compounds on the ozone layer. One such class of compounds is the nitrogen oxides, such as NO(g) and NO_2(g). Under certain conditions, the reaction of NO(*g*), an air pollutant released in automobile exhaust, with oxygen can produce N_2O_4(*g*). The rate law for this reaction was determined experimentally and is shown below. What is the reaction order of each of the compounds in the rate law? What is the overall order of the reaction?

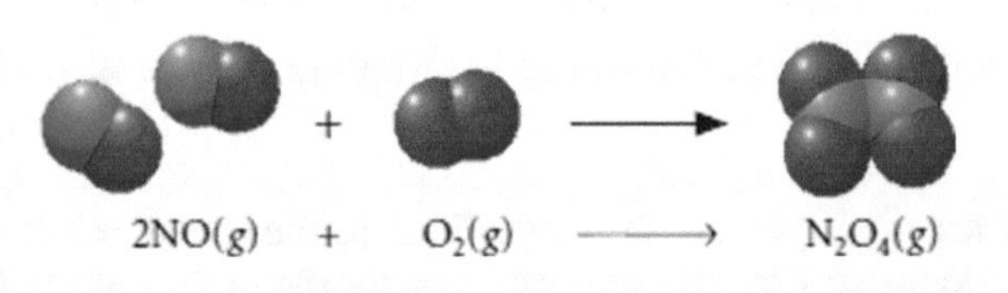

$$2NO(g) + O_2(g) \longrightarrow N_2O_4(g)$$

$$\text{Rate} = k[NO][O_2]$$

Asking the Right Questions

How is the balanced chemical equation related to the rate law? How is the experimentally determined rate law related to the order of the reaction, and each reactant in it?

Solution

The balanced chemical equation does not, by itself, guide us to the rate law. Experiments that give time vs. concentration data do that. The experimentally determined rate law says that the reaction is first order in $NO_{(g)}$ and first order in $O_{2(g)}$. Overall, then, the reaction is second order.

Are Our Answers Reasonable?

The answers reflect the rate law and each reactant within it, which is how we determine the reaction order, so the answer is reasonable.

What If…?

…we consider the reaction below, along with its rate law:

$$3ClO^- \rightarrow ClO_3^- + 2Cl^-$$

$$\text{rate} = k\left[ClO^-\right]^2$$

What is the order of the reaction?

(*Ans: second order*)

Practice 13.3

This reaction proceeds at 300 °C. What is the reaction order of the compound in the following reaction with the given rate law? What is the order of the reaction?

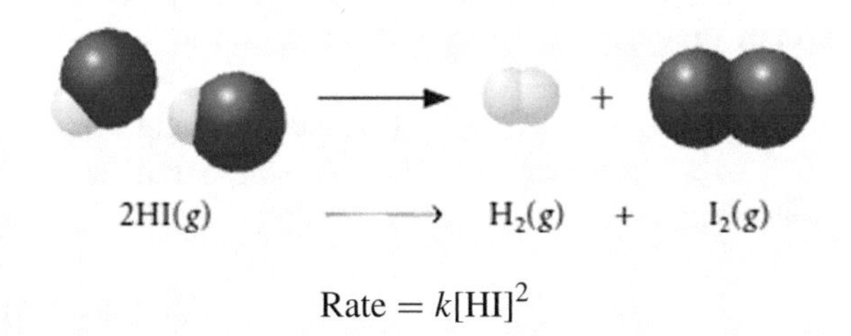

$$2HI(g) \longrightarrow H_2(g) + I_2(g)$$

$$\text{Rate} = k[HI]^2$$

13

13.3.1 Collision Theory

In Example 13.3 and Practice 13.3, the rate law is expressed as the concentration of the reactants raised to a power. This is not uncommon, but as we've mentioned before, it isn't true for reactions in general. Why do the orders of the rate law and the stoichiometric factors of a reaction often differ? Reactions are typically more complex than they appear when written on paper. To understand how this can cause the rate law to differ from what we may expect, we need to consult **collision theory**. This theory describes how the rate of a reaction is related to the number of properly oriented collisions of the molecules involved. Collision theory is heavily based on the kinetic molecular theory of gases.

Collision theory – A theory that correlates the number of properly oriented collisions with the rate of the reaction.

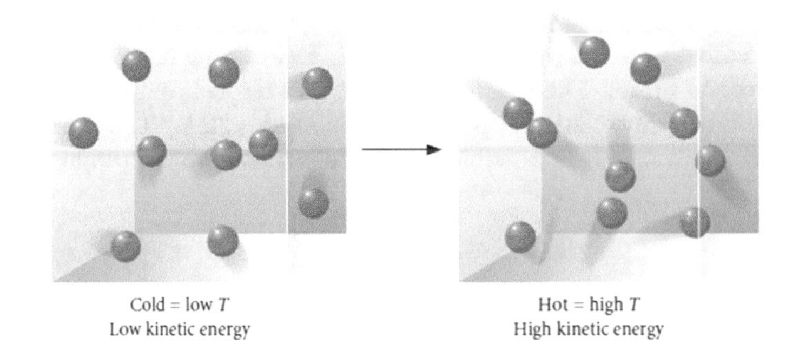

Table 13.1 Collision theory

A reaction occurs when the following conditions have been met
• Molecules collide
• Molecules have enough kinetic energy
• Molecules are properly oriented
Implications
• Larger concentrations hace faster reaction rates
• Reactions with higher temperatures have faster rates
• Rates depend on the number of properly oriented collisions
• Predicting the rate of a reaction is difficult

Kinetic molecular theory says that the thermal motion of particles (the kinetic energy) can be used to explain how a gas behaves. For instance, the pressure of a gaseous system is related to the number of collisions of the molecules with the sides of their container in a given time. If we increase the number of molecules per unit volume, the number of collisions per second also increases, assuming the temperature is constant. The pressure of the system can be increased if we raise the temperature of the gas and leave the volume of the container constant. This is because the molecules are moving faster (more kinetic energy) and engage in more collisions per second. In short, higher kinetic energy equals more collisions.

Collision theory, which is summarized in Table 13.1, requires that molecules collide in order to react. One of the more important statements in collision theory, as shown in Fig. 13.7a, say that the collisions must be energetic enough to make a product. The minimum energy required, the **activation energy (E_a)**, is specific to a particular reaction. However, an energetic collision alone is not enough to cause a reaction to occur. The collision must also occur between properly oriented molecules; see Fig. 13.7b. Because the equation we write to describe a reaction doesn't address all of these issues, rate laws are difficult to derive by simply examining the overall equation. However, what collision theory tells us is that fewer collisions will result in fewer reactions between molecules. Therefore, as the reaction progresses and there are fewer reactant molecules present, the reaction rate will slow down because there will be fewer collisions.

Activation energy (E_a) – The minimum energy of collision that reactants must have in order to successfully create the activated complex.

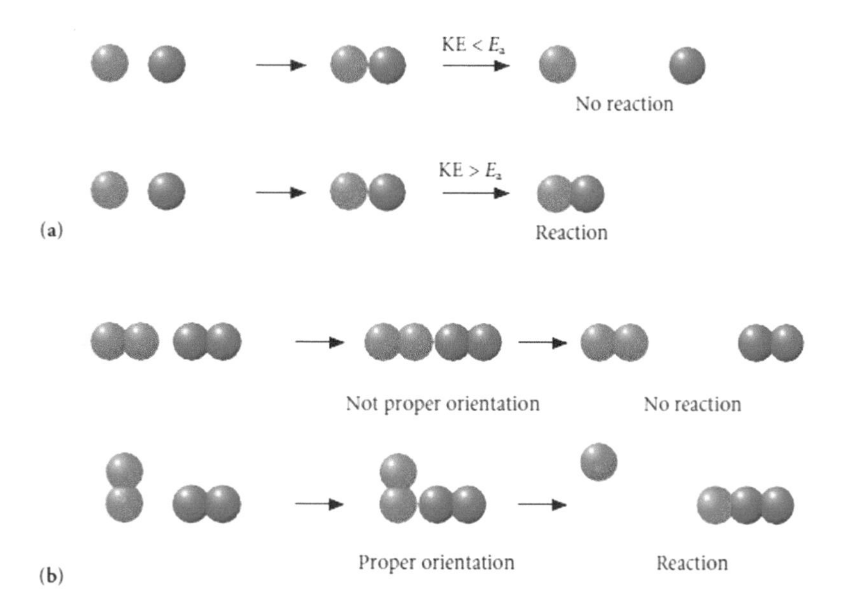

Fig. 13.7 Collision theory. **a** Collisions must be energetic enough to be considered successful. **b** Successful collisions only occur between properly oriented molecules

▶ Here's What We Know so Far

- The rate of a reaction is the change in concentration (M) per unit time. Rates are always positive numbers, often reported in M/s.
- Average rates and instantaneous rates differ only in the time measurement. Instantaneous rates have $\Delta t \approx 0$.
- The rate law (Rate $= k[A]^n[B]^m$) indicates the relationship between the rate and the concentrations of reactants.
- The rate constant is only valid for a given temperature, but the rate orders are constant under all conditions.
- The order of a reactant indicates the relationship between the reactant concentration and the rate.
- Collision theory explains why the rate of a reaction typically decreases as time passes. ◀

13.4 Changes in Time—The Integrated Rate Law

After the discovery in 1939 that DDT (1,1,1-trichloro-2,2-bis-(4′-chlorophenyl)ethane, or dichlorodiphenyltrichloroethane) can be used to control mosquito-borne malaria, its use soared. Especially important was its use to protect soldiers who were fighting in the Pacific Rim countries during World War II. Since its discovery, DDT has been sprayed to eliminate insects from cotton crops, spiders from residences, and mosquitoes from towns all across the globe (◘ Fig. 13.8). Initial testing showed that the compound wasn't very toxic to mammals. However, because the metabolism of DDT is very slow, small amounts of DDT in the environment tended to accumulate in animals (including humans) until toxic levels were present. Evidence of this caused Sweden in 1970 and the United States in 1972 to ban the use of DDT as a pesticide, although it is still used in a number of other countries, especially India and, to a lesser extent, several in Africa, to combat malaria. The hazard of DDT accumulation in the food chain versus the benefit of saving human lives by preventing malaria in the populations of, for example, East African countries is still a topic of intense debate.

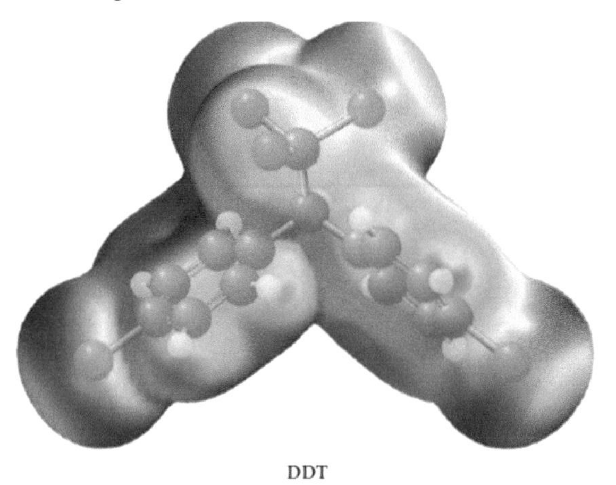

DDT

In organisms that are resistant to DDT, an enzyme known as dehydrochlorinase converts DDT into dichlorodiphenyldichloroethylene (DDE). Unfortunately, DDE is also harmful to wildlife. It can accumulate within birds and weaken their eggshells by interfering with the complex process of constructing the shells. This leads to shells that break under the normal pressures naturally associated with nesting. For this reason, several species of birds, such as the peregrine falcon, were nearly wiped out in the United States.

The rate of decomposition of DDT by soil microbes is plotted against time in ◘ Fig. 13.9. Initially the rate of decomposition is relatively high; then, as time passes, the rate begins to decline. Does this make sense according to what we know about collision theory? Yes. Reducing the number of reactant molecules reduces the number of collisions that would result in the formation of the product.

$$\text{DDT} \xrightarrow[\text{Dehydrochlorinase}]{\text{DDT}} \text{DDE} + \text{HCl}$$

Fig. 13.8 DDT and the mosquito. DDT is still one of the most cost-effective methods of controlling mosquito-borne malaria. Although many countries have banned its use because of DDT's persistence in the environment, many homes across the world are still sprayed inside and out. *Source* ▶ https://upload.wikimedia.org/wikipedia/commons/0/0e/HOUSES_IN_BEER_SHEVA_ARE_SPRAYED_WITH_DDT.jpg

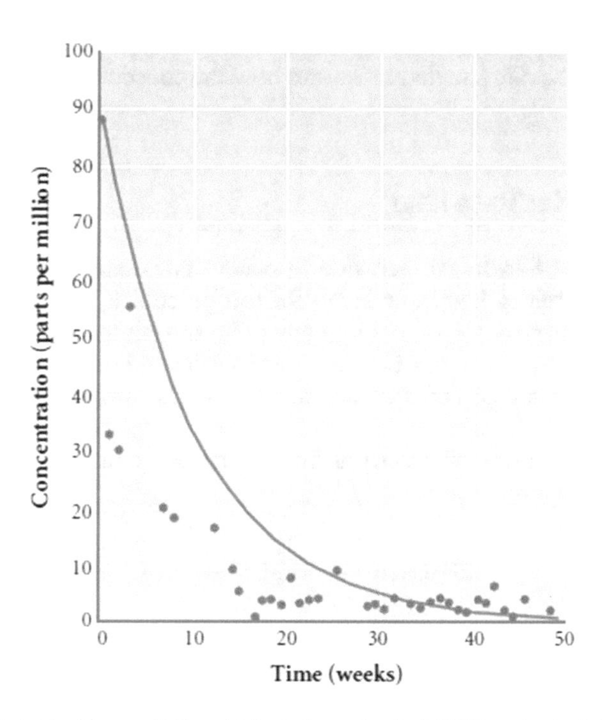

Fig. 13.9 An experiment to illustrate the bioremediation (biological recycling) of DDT-contaminated ground by soil microbes and nutrients. The decomposition of DDT occurs rapidly at first and then slows. Note that the concentration of DDT is reported in parts per million. Although the curve appears to be placed incorrectly, it is not. Computer analysis of this set of data, and additional data not shown, produced this curve

What if we're interested in determining what concentration of a chemical will exist after a certain amount of time has passed? The rate law alone won't tell us this information. For example, the Hanford Nuclear Reservation on the banks of the Columbia River just north of Richland, Washington, has produced 54 million gallons of radioactive plutonium waste over the past half-century. The waste is currently being stored in 177 underground tanks as a watery sludge. How long will it take for the radioactive waste stored at the Hanford Nuclear Reservation to decay to 1% of its current concentration? We could examine the plot shown in Fig. 13.10 to figure this

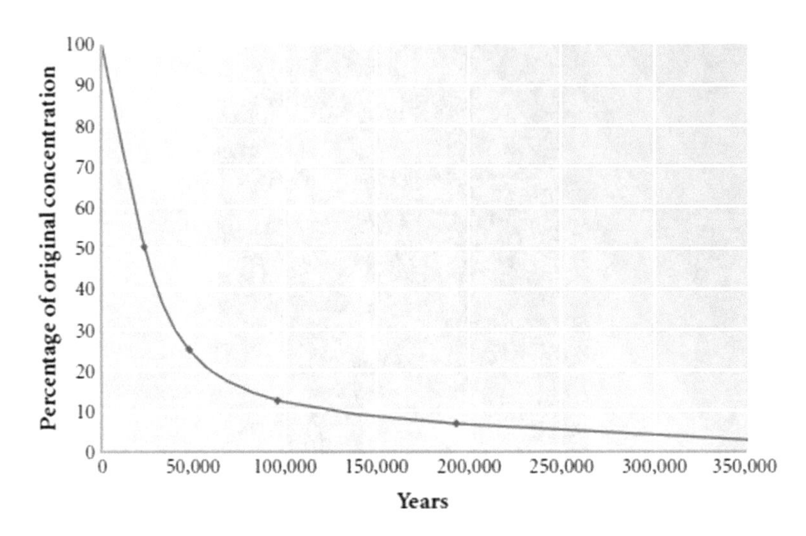

Fig. 13.10 Plot of the radioactive decay of plutonium-239

out, but we would need either to have a plot of the radioactive decay reaction or to know the rate law, the rate, and rate constant of the reaction. Can we determine the concentration at a specific time without knowing the rate or even having a plot of the reaction? Calculus comes to our rescue. With some manipulation of the rate law, we can make a more useful description of the rate of a reaction that will enable us to perform these calculations. The resulting equations are referred to as the **integrated rate laws**.

Integrated rate law – A form of the rate law that illustrates how the concentrations of reactants vary as a function of time.

13.4.1 Integrated First-Order Rate Law

13

We discussed the decomposition of hydrogen peroxide to make oxygen and water near the start of this section. Experimentally, it has been determined that the rate law for this process describes a first-order reaction. Unfortunately, this doesn't tell us the concentration of H_2O_2 at any point during the reaction, unless we know the rate of the reaction at that time and the rate constant (k). In order to calculate the concentration of H_2O_2 at any point during the reaction, we can mathematically convert the rate law into an **integrated first-order rate law**.

Integrated first-order rate law – An expression derived from a first-order rate law that illustrates how the concentrations of reactants vary as a function of time.

$$2H_2O_2(aq) \rightarrow 2H_2O(l) + O_2(g)$$

$$\text{Rate} = k[H_2O_2]$$

The integrated rate law gets its name from the mathematical process, known as integration, that we follow to generate the relationship between concentration and time. When we integrate the rate law from $t=0$ to some future time, the integrated rate law for the decomposition of hydrogen peroxide becomes:

$$\ln\left(\frac{[H_2O_2]_t}{[H_2O_2]_0}\right) = -kt$$

where:

$[H_2O_2]_0$ is the concentration of hydrogen peroxide initially,
$[H_2O_2]_t$ is the concentration of hydrogen peroxide at a particular time t, and
k is the rate constant.

In practice, we need know only three of the four values in the equation (time, rate constant, initial concentration, and concentration at time t) in order to find the fourth.

This equation is general for all first-order reactions, which we will designate using the general form:

$$A \rightarrow \text{Products}$$

where A is any single reactant. Note that the stoichiometric coefficient for A is 1. The natural logarithm (ln) of the quotient of the final reactant concentration, $[A]_t$, divided by the initial concentration $[A]_0$ is equal to the negative of the rate constant, k, times the time, t, shown in the left-hand side of the box below. We can use this equation to calculate the time required to reach a given concentration. Alternatively, we can rearrange the equation to that shown on the right-hand side of the box below if we wish to calculate the concentration at a particular time:

$$\ln\left(\frac{[A]_t}{[A]_0}\right) = -kt \quad [A]_t = [A]_0 e^{-kt}$$

We can transpose this into an equation in the form $y = mx + b$ by recognizing that

$$\ln\left(\frac{[A]_t}{[A]_0}\right) = \ln[A]_t - \ln[A]_0$$

Substituting for $\ln\left(\frac{[A]_t}{[A]_0}\right)$ yields

$$\ln[A]_t - \ln[A]_0 = -kt$$

which enables us to come up with the final straight-line (y = mx + b) form:

$$\begin{aligned} \ln[A]_t &= -kt + \ln[A]_0 \\ y &= mx + b \end{aligned}$$

where

the y axis $= \ln[A]_t$.
the x axis $= t$.
the slope $m = -k$.
the intercept $b = \ln[A]_0$.

How do we use the integrated first-order equation to find the concentration of hydrogen peroxide after 5.00 min if we have a solution in which the initial concentration, $[H_2O_2]_0$, is equal to 0.100 M? The rate constant for this reaction was determined by experiment to be $3.10 \times 10^{-3}\ s^{-1}$. Using this equation, we can calculate the amount of peroxide still remaining after 5.00 min. We must first convert our time to match the units for the rate constant (5.00 min $= 3.00 \times 10^2$ s) and then insert our known values into the integrated first-order rate law.

$$\ln[H_2O_2]_t - \ln[H_2O_2]_0 = -kt$$

$$\ln[H_2O_2]_t - \ln(0.100\ M) = -\left(3.10 \times 10^{-3} s^{-1}\right)\left(3.00 \times 10^2 s\right)$$

$$\ln[H_2O_2]_t - \ln(0.100\ \text{M}) = -0.930$$

$$\ln[H_2O_2]_t + 2.3026 = -0.930$$

$$\ln[H_2O_2]_t = -3.2326$$

$$[H_2O_2]_t = e^{-3.2326} = 0.0395\ \text{M}$$

The concentration of hydrogen peroxide remaining after 5.00 min is 0.0395 M.

Example 13.4—The Rate Law

Archaeologists near the Dead Sea in 1998 reported the discovery of a collagen sample that they believe ancient peoples used as glue. A sample of newly prepared collagen exhibits 15.2 disintegrations per minute per gram (15.2 dis/min/g) of carbon. (A disintegration is the decomposition of a radioactive nucleus such as ^{14}C.) The ^{14}C decay rate for a sample of the ancient collagen sample was found to be 5.60 disintegrations per minute per gram (5.60 dis/min/g) of carbon. What is the age of the glue? The decomposition of ^{14}C is a first-order process with a rate constant of 1.209×10^{-4} year^{-1}.

Asking the Right Questions

In solving this example, what assumption do we need to make about the activity of the fresh collagen compared to the ancient collagen? How do we use the first-order equation to solve for the age of the glue?

Solution

If we assume that fresh collagen has the same activity of ^{14}C that the ancient glue did when it was first made, we can calculate the length of time it would take a fresh sample of collagen to have the same activity of ^{14}C as the glue.

$$\ln\left(\frac{5.60\ \text{dis/min/g}}{15.2\ \text{dis/ min /g}}\right) = -(1.209 \times 10^{-4}\text{year}^{-1})t$$

$$\ln(0.3684) = -(1.209 \times 10^{-4}\text{year}^{-1})t$$

$$-0.9985 = -\left(1.209 \times 10^{-4}\text{year}^{-1}\right)t$$

$$t = 8259\ \text{year} = 8260\ \text{year}$$

Is Our Answer Reasonable?

Our answer had a positive sign – that is, we calculated an age that was +8260, rather than −8260, and that's a good sign that the answer is reasonable. Another indication of a reasonable answer is the value of 8260, rather than, for example, 6.02×10^{23} years old.

What If...?

... we were examining rocks from the moon using a radioactive nucleus, uranium-238, that decomposes via a first-order process with a rate constant of 1.55×10^{-10} year^{-1}. If a sample of moon rock has lost 43% of its initial ^{238}U-containing radioactivity, how old is the sample of moon rock?

(*Ans: 3.6 billion years old*)

Practice 13.4

Hanford's nuclear waste contains large quantities of plutonium (mostly ^{239}Pu). Researchers have determined that approximately 875 kg of solid waste plutonium are buried there. How long will it take for the mass of plutonium-239 to drop to 10% of its original value? Assume the rate constant that describes the decay of ^{239}Pu is 2.874×10^{-5} year^{-1}. How long will it take for the mass of plutonium-239 to reach half of its original concentration?

13

13.4.2 Half-Life

The persistence of pesticides and herbicides in the environment is often reported as the amount of time that it takes for half of the original concentration to decompose. In general, any reaction can be reported in this manner by calculating the amount of time it takes for the reaction to proceed to 50% completion. This value is known as the **half-life ($t_{1/2}$)** of the reaction (◘ Fig. 13.11). Perhaps we've heard this term used to express the rate of decay for a radioactive element, such as the radioactive waste stored at the Hanford Nuclear Reservation, or in accounts of the dating of ancient artifacts. The half-lives of pesticides and herbicides indicated in ◘ Table 13.2 are used to judge the safety of the compounds and to establish guidelines for the frequency of their application. How does the half-life fit in with our description of the integrated first-order rate law?

Half-life ($t_{1/2}$) – The time required for a reaction to reach 50% completion.

$$\ln\left(\frac{[\text{atrazine}]_t}{[\text{atrazine}]_0}\right) = -kt$$

Consider the first-order hydrolysis of atrazine in groundwater. The rate constant for this reaction in water has been found to be 0.001733 day^{-1}. Note that we're using a rate constant with units of day^{-1} instead of s^{-1}. This practice, using units of time that are not seconds, is acceptable as long as we make sure that our units are consistent between k and t.

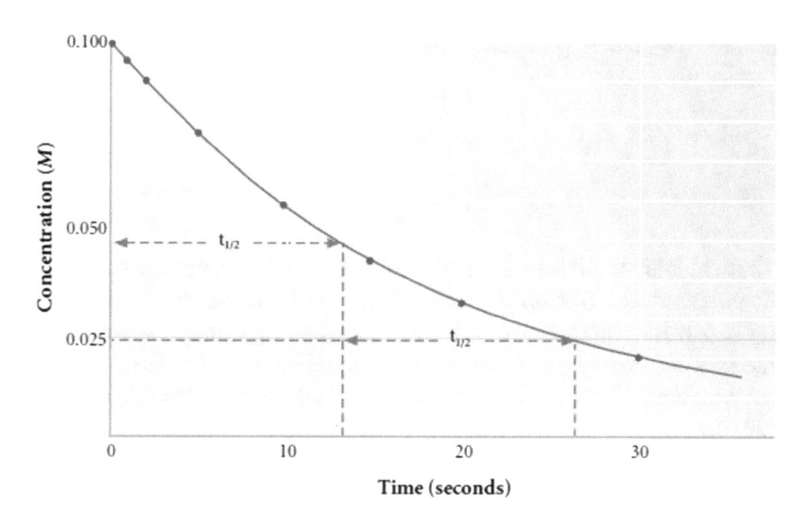

Fig. 13.11 The half-life of a reaction is the amount of time required for the reaction to reach 50% of the original concentrations. In terms of radioactive decay or the decomposition of pesticides, the passage of one half-life reduces the concentration in half. Two half-lives reduce the concentration to one-quarter of the original

Table 13.2 Solubility and half-life in soil for selected pesticides

The sorption index is the ratio of pesticide concentration bound to soil particles divided by the concentration in the aqueous phase. Pesticides with low sorption indices are more likely to be leached into groundwater supplies, because a low proportion of each binds to the soil. The half-life is reported for the pesticide in sterile soil. 2,4-D, for example, is of greater environmental concern because of its sorption index value than is DDT

Trade name/brand name	Water solubility (ppm)	Sorption index (higher = greater binding to soil)	Soil half-life (days)
2,4-D	890	20	10
Alachlor/Lasso	240	170	15
Atrazine/Aatrex	33	100	60
Dicamba/Banvel	400,000	2	14
Carbaryl/Sevin	110	300	10
Chlorsulfuron/Glean	7000	40	160
DDT	0.0055	24,000	3000
Diazinon	60	1000	40
Malathion/Cythion	145	1800	1
Metolachlor/Dual	530	200	90
Methoxychlor/Marlate	0.10	80,000	120
Pendimethalin/Prowl	<1	5000	90
Pronamide/Kerb	15	200	60
Terbacil/Sinbar	710	55	120
Terbufos/Counter	4.5	500	5
Trifluralin/Treflan	<1	8000	60

When the hydrolysis has consumed half of the original concentration of atrazine, $[\text{atrazine}]_t = ½[\text{atrazine}]_0 = 0.5[\text{atrazine}]_0$. Substituting this into the equation above, we find that

$$\ln\left(\frac{0.5[\text{atrazine}]_0}{[\text{atrazine}]_0}\right) = -kt_{1/2}$$

The concentration of atrazine ($[\text{atrazine}]_0$) cancels in the equation. Simplifying the equation further yields

$$\ln 0.5 = -kt_{1/2}$$
$$-0.693 = -kt_{1/2}$$
$$t_{1/2} = \frac{0.693}{k}$$

Our derivation reveals that the half-life ($t_{1/2}$) of a first-order reaction is, throughout the reaction, a constant that depends only on the rate constant and not on the concentration of the reactant. This means that if we know the half-life for a particular first-order reaction, we can calculate the rate constant, and vice versa. Substituting the rate constant for atrazine into this equation, we find that the half-life for the hydrolysis in water agrees with the experimental observation we noted at the start of this chapter. Half of the atrazine added to water in the environment disappears after 400 days.

$$t_{1/2} = \frac{0.693}{0.001733\ \text{day}^{-1}} = 4.00 \times 10^2\ \text{days}$$

Example 13.5—How Long Will It Take?

Diazinon crystals are sprinkled around a home's foundation to kill and repel ants from the home. How long will it take for the diazinon to decompose, assuming first-order kinetics, to 25% of its original concentration in the soil? (From Table 13.2, the half-life of diazinon is 4.0×10^1 days.)

13

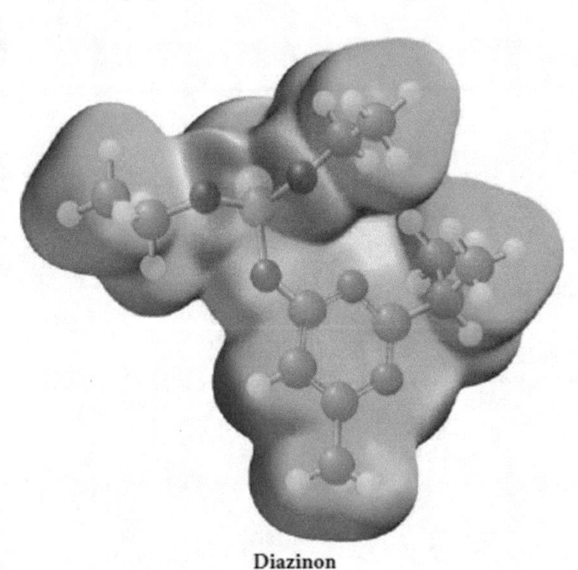

Diazinon

Asking the Right Questions

The problem doesn't state the original concentration of the diazinon in the soil. Is there a way we can work around not knowing this? How might we make use of our understanding of "%" as a ratio of the current to the initial concentration?

What Do I Need to Know?

How many half-lives pass in order for a sample that decays via first-order kinetics to lose 50% of its original concentration?

Solution

Using the half-life, we first calculate the rate constant for the reaction:

$$t_{1/2} = \frac{0.693}{k}$$

$$4.0 \times 10^1\ \text{days} = \frac{0.693}{k}$$

$$k = 0.0173\ \text{day}^{-1}$$

Note that the unit of the rate constant is reciprocal days based on the value of the half-life and we retain an extra figure in the value of k because we will use the data in the next part of the calculation.

If 25% of the diazinon remains, then:

$$ln\left(\frac{0.25[diazinon]_0}{[diazinon]_0}\right) = -(0.0173\ \text{day}^{-1})t$$
$$\ln 0.25 = -(0.0173\ \text{day}^{-1})t$$
$$-1.386 = -(0.0173\ \text{day}^{-1})t$$
$$t = 8.0 \times 10^1\ \text{days}$$

There is an alternative method by which we can calculate the answer. One half-life reduces the original concentration by 50% (50%=0.5=½). Two half-lives reduce the concentration to 25% (½ × ½ = ¼). Three reduce the concentration to 12.5% (½ × ½ × ½ = 1/8). Because we want to know the time required for the reaction to reduce the concentration of starting material to 25% of the initial concentration, we need two half-lives (4.0×10^1 days + 4 .0×10^1 days = 8.0×10^1 days). Therefore, for a first-order process, the fraction remaining, $[A]_t/[A]_0$ equals 2^{-n}, where n = number of half-lives. For example, for $n = 3$ half-lives, $[A]_t/[A]_0 = 2^{-3}$ or 0.125.

Is Our Answer Reasonable?

Two half-lives represent 80 days, the time during which 75% of the sample concentration is lost. This is consistent with our discussion about half-life. It is also reasonable computationally because both the first-order equation substitution and the half-life computation methods gave us the same answer.

What If…?

…we wanted to know if less than 1% of a sample of Dicamba remains in soil after one growing season of 128 days? What percentage actually remains, assuming a first-order kinetic decomposition? (See Table 13.2 for the soil half-life of Dicamba).

(*Ans: 0.178% remains*)

Practice 13.5

What is the length of time we would have to wait for Alachlor to decompose to 25% of its original value?... to 6.25%?... to 0.001%?

13.4.3 Other Rate Laws

Not all reactions follow first-order kinetics. Other orders do exist, even non-integer orders. For example, reactions that take place on metal surfaces typically follow zero-order kinetics. Hydrogenation of vegetable oils, the decomposition of ammonia on a tungsten wire, and the reaction of N_2O with oxygen in our car's catalytic converter are reactions on metal surfaces. They each follow zero-order kinetics.

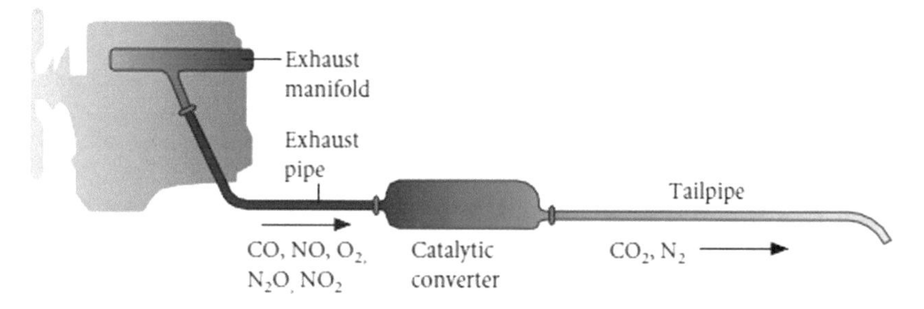

In a **zero-order reaction**, the rate law does not depend on the concentration of any of the compounds in the reaction. The rate of the zero-order reaction is constant at a given temperature. No matter what the original concentration happens to be, the reaction always proceeds at the same rate while at the temperature.

Zero-order reaction – A reaction in which the rate is not related to the concentrations of any species in the reaction.

$$\text{Rate of disappearance of A} = -\frac{\Delta[A]}{\Delta t}$$

$$\text{Rate} = k[A]^0 = k$$

The **integrated zero-order rate law**, determined by integrating the rate law over time, for a zero-order reaction is different from the integrated first-order rate law:

$$[A]_t = -kt + [A]_0$$

Substituting $0.5[A]_0$ for the concentration of A at time t, we can rearrange the equation to get the formula for the half-life of the zero-order reaction:

$$t_{1/2} = \frac{[A]_0}{2k}$$

Integrated zero-order rate law – An expression derived from a zero-order rate law that illustrates how the concentrations of reactants vary as a function of time.

The half-life of a zero-order reaction directly depends on the initial concentration of the reactant. Larger concentrations of the reactant will mean a larger half-life for the reaction. Overall, the rate of a zero-order reaction is constant but the half-life depends on the initial concentration of the reactant. For example, consider a zero-order reaction where the rate constant $k = 2.5 \times 10^{-4}$ M/s. When $[A]_0 = 1.00$ M, the half-life of the reaction is 2000s. If $[A]_0 = 0.25$ M, the half-life of the reaction is 500 s. The rate of the reaction (rate $= k$) is a constant, but the half-life changes as the concentration changes.

The **second-order rate law** can be more complicated, because there are two cases that fit the definition of a second-order reaction. In one of those cases, the reaction is second order in only one reactant:

Second-order rate law – The rate of a reaction is directly related either to the square of the concentration of a single reactant or to the product of the concentrations of two reactants.

$$\text{Rate} = k[A]^2$$

In the other case involving a second-order rate law, the reaction could be first order in two different species:

$$\text{Rate} = k[A][B]$$

In second-order reactions with only one reactant, the integrated rate law is determined using the method we have explored for the zero-order and first-order reactions. After integration and rearrangement, the **integrated second-order rate law** takes the form:

Integrated second-order rate law – An expression derived from a second-order rate law that illustrates how the concentrations of reactants vary as a function of time.

13

$$\frac{1}{[A]_t} = kt + \frac{1}{[A]_0}$$

The half-life of this type of second-order reaction can be determined by assuming that the concentration of A at some time, t, is equal to one-half the initial concentration ($[A]_t = 0.5[A]_0$). Simplifying the equation gives:

$$t_{1/2} = \frac{1}{k[A]_0}$$

The half-life for a second-order reaction (involving only one compound in the rate law) is inversely dependent on the initial concentration of the reactant. Smaller initial concentrations of the reactant will mean a longer half-life for the reaction.

If the second-order reaction contains two species in the rate law, the integrated rate law becomes much more complicated. In fact, the math gets so complicated that chemists typically manipulate the experimental conditions to reduce the number of calculations. Consider the reaction of one of the components of smog with ozone, O_3, which is also found in smog:

$$NO(g) + O_3(g) \rightarrow NO_2(g) + O_2(g)$$

$$\text{Rate} = k[NO][O_3]$$

The reaction is first order in NO, first order in O_3, and second order overall.

To calculate the concentration of a reactant, determine the rate constant, or find the time required to reach a certain concentration with this type of rate law, we conduct the reaction with a relatively small concentration of one of the reactants and a very large concentration of the other. What effect does this have on the rate law? Because one of the concentrations is very large, any change in its concentration is negligible, and we can assume that its concentration remains constant through the course of the reaction. An analogy is having someone who is rich spend \$1 in a day from a fortune of \$1 billion. The change is hardly noticeable. The fortune is essentially constant at this rate of spending. Someone who had only \$10, however, would notice the effect of spending \$1 immediately. A concentration of one component that is significantly larger than that of the other reduces the rate law to a much simpler form. For example, if we perform the reaction with 1.000 M NO and 0.001 M O_3, the concentrations of the species after the reaction has completely consumed the ozone can be determined:

$$[O_3]_f = 0.001\ \text{M} - 0.001\ \text{M} = 0.000\ \text{M}$$

$$[NO]_f = 1.000\ \text{M} - 0.001\ \text{M} = 0.999\ \text{M} \approx 1.000\ \text{M}$$

The O_3 is consumed in the reaction, but the concentration of NO is only slightly affected. We can make the assumption that the concentration of NO doesn't change. Because the final concentration of NO has essentially remained constant, it can be combined with the rate constant. And, the rate equation for the reaction can be reduced to:

$$\text{Rate} = k[NO][O_3] = k[NO]_0[O_3]$$

and, if we set $k' = k[NO]$,

$$\text{Rate} = k'[O_3]$$

where k' is the new rate constant. What is the order of our new rate law? Because of our choice of the initial concentrations, the kinetics for this reaction has taken on the form of a first-order rate law. Because of our modification, we say that this is a **pseudo-first-order rate** equation.

Pseudo-first-order rate – A modification of the second-order rate that enables one to use first-order kinetics.

We can do this manipulation with any reaction. By increasing the concentration of a particular reactant to a very large value, any order of a rate law can be simplified to a pseudo-first-order rate equation. This enables us to study reactions no matter what order they appear to be. We must remember, however, that any modification of the rate equation means that the new rate constant is not the same rate constant as in the original reaction.

▶ Here's What We Know so Far

- The integrated rate laws enable researchers to calculate one of four things about a reaction (rate, rate constant, initial concentration, or final concentration) if three of these are known.
- The half-life of a reaction indicates the time required for the reaction to reach 50% completion.
- Modifying a second-order reaction by using a large concentration of one of the reactants reduces the integrated rate law to a pseudo-first-order rate law. ◀

▶ Deepen Your Understanding Through Group Work

The United States Geological Survey has a list of dozens of pesticides that have been used for many years in the United States. It is called the "Pesticide National Synthesis Project. Use the internet to look up the compound listing at the USGS website. Then:

- Gather your team and choose one pesticide.
- Find out its structure, how it works, and its prevalence over the years, including where it is used and why.
- Discuss among yourselves why it has the use pattern with time that is does.
- What chemical, economic, and other factors go into making such a choice? ◀

13.5 Methods of Determining Rate Laws

Glyphosate herbicides are absorbed through the foliage of weeds, inhibiting a plant's ability to make new amino acids. Because these herbicides react relatively quickly with water ($t_{1/2}$ < 7 days), the rate of their biological reaction in plants is important. Weed scientists study the rates of amino acid inhibition and environmental degradation in order to suggest modifications to improve glyphosate effectiveness (◘ Fig. 13.12). The greatest benefit of the glyphosate herbicides is attributable to their rapid degradation in the environment. The rapid degradation means this class of herbicide causes less harm to the environment than other herbicides. The rates of the degradation of these compounds have been measured by experiment, and the overall rate laws have been determined from those experiments.

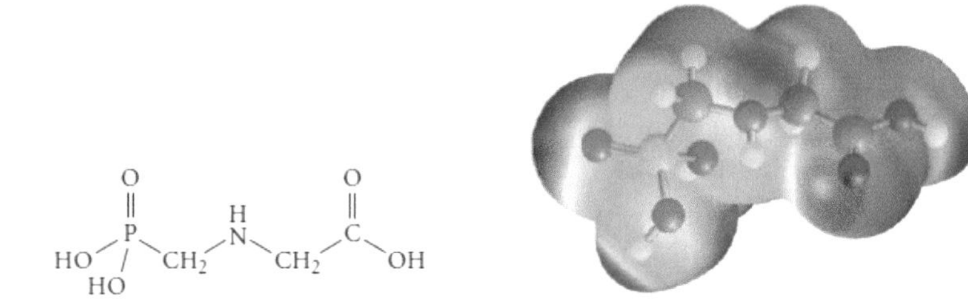

Fig. 13.12 Roundup herbicide contains glyphosate. The general herbicide is sold for use on farms and in removing unwanted weeds around the yard

We, too, can use the information from a chemical reaction to obtain the final rate law for a reaction. The first of two commonly used procedures that we will consider is the **method of initial rates**. In this approach, we measure and compare initial rates rather than comparing instantaneous rates later in the reaction. We use initial rates to determine the rate law because we can precisely measure and control the starting concentrations of the reactants and precisely identify the time for the reaction to reach a given point. In addition, limited side reactions (reactions that make products other than what we're interested in), rapid determination of the rate, and a well-defined time period make the comparison of multiple reactions experimentally straightforward.

Method of initial rates – A method of determining the rate law for a reaction where the initial rates of different trials of a reaction are compared.

We'll use as our example the reaction of nitrogen monoxide, NO(g), with oxygen to make nitrogen dioxide, NO_2(g), a compound also found in smog. We can determine the orders of the two reactants in the rate law by measuring the initial rate of this reaction as we change the concentrations of both NO(g) and O_2(g).

$$2\mathrm{NO}(g) + \mathrm{O_2}(g) \rightarrow 2\mathrm{NO_2}(g)$$

We conduct three separate reactions with different concentrations of reactants, measure the initial rate, and complete a table as shown below. Note that we have carefully chosen our starting concentrations to keep one reactant the same in two experiments while doubling the other reactant. How do we use this information to determine the orders in the rate law? We start by writing the rate law assuming all of the reactants are included in the equation:

$$\text{Rate} = k[\mathrm{NO}]^n[\mathrm{O_2}]^m$$

where n and m are the orders of the two reactants. If we compare the rate of the third reaction to the rate of the second reaction, in which [NO] is constant and [O_2] changes, the equation simplifies dramatically. To do this, we can divide the rate law for the third experiment into the rate law for the second experiment.

$$\frac{\text{rate}_2}{\text{rate}_3} = \frac{k[\mathrm{NO}]^n[\mathrm{O_2}]^m}{k[\mathrm{NO}]^n[\mathrm{O_2}]^m}$$

Experiment	[NO] (M)	[O_2] (M)	Initial rate (M/s)
1	0.0126	0.0125	1.41×10^{-2}
2	0.0252	0.0250	1.13×10^{-1}
3	0.0252	0.0125	5.64×10^{-2}

Next, we add the values from the table to the equation and simplify. Although the value of the rate constant, k, is not known, it is the same in both reactions, so it cancels out:

$$\frac{1.13 \times 10^{-1}\,\mathrm{M \cdot s^{-1}}}{5.64 \times 10^{-2}\,\mathrm{M \cdot s^{-1}}} = \frac{k(0.0252)^n(0.0250)^m}{k(0.0252)^n(0.0125)^m}$$

$$2 = \frac{(0.0250)^m}{(0.0125)^m}$$

$$2 = 2^m$$

$$m = 1$$

Why did we divide the second equation by the third? When one of the reactant concentrations is held constant ([NO], in this case) and the other concentration ($[O_2]$) is doubled, the effect on the rate is due only to the change in $[O_2]$. This effect is related to a power of 2 and reveals the order of the second reactant ($m = 1$). To continue, we divide the third rate law by the first because $[O_2]$ is constant while [NO] changes, so the effect on the rate is due only to the change in [NO]. We substitute concentrations and solve:

$$\frac{5.64 \times 10^{-2}\,\mathrm{M \cdot s^{-1}}}{1.41 \times 10^{-2}\,\mathrm{M \cdot s^{-1}}} = \frac{k(0.0252)^n(0.0125)^1}{k(0.0126)^n(0.0125)^1}$$

$$4 = \frac{(0.0252)^n}{(0.0126)^n}$$

$$4 = 2^n$$

$$n = 2$$

We are nearly ready to write the rate law for the reaction because we know the orders of the reactants. We are missing only the rate constant, which we can obtain by substituting the concentrations into the rate law. Solving for the rate constant for each set of conditions, we obtain:

$$\mathrm{Rate} = k[\mathrm{NO}]^2[\mathrm{O_2}]$$
$$1.41 \times 10^{-2}\,\mathrm{M/s} = k(0.0126\,\mathrm{M})^2(0.0125\,\mathrm{M})$$
$$k = 7.11 \times 10^3\,\mathrm{M^{-2} \cdot s^{-1}}$$

$$\mathrm{Rate} = k[\mathrm{NO}]^2[\mathrm{O_2}]$$
$$1.13 \times 10^{-1}\,\mathrm{M/s} = k(0.0252\,\mathrm{M})^2(0.0250\,\mathrm{M})$$
$$k = 7.12 \times 10^3\,\mathrm{M^{-2} \cdot s^{-1}}$$

$$\mathrm{Rate} = k[\mathrm{NO}]^2[\mathrm{O_2}]$$
$$5.64 \times 10^{-2}\,\mathrm{M/s} = k(0.0252\,\mathrm{M})^2(0.0125\,\mathrm{M})$$
$$k = 7.11 \times 10^3\,\mathrm{M^{-2} \cdot s^{-1}}$$
$$\text{Average value of } k = 7.11 \times 10^3\,\mathrm{M^{-2} \cdot s^{-1}}$$

The method of initial rates requires that we can obtain and compare the initial rates of reactions. We need at least three reactions to solve for two unknown orders. And in every case, we've used experimentally determined data to calculate the rate law for the reaction.

Note that the units for the rate constant are different depending on the order of the rate law. Specifically, if the unit of time is in seconds, the units of the rate constant are $\mathrm{M^{-(order\,-1)} \cdot s^{-1}}$.

Example 13.6—Initial Rates

The reaction of Cl_2 with NO occurs at a very rapid pace. Use the data in the table below to determine the rate law for the reaction. Then calculate the rate constant.

$$2\mathrm{NO}(g) + \mathrm{Cl_2}(g) \rightarrow 2\mathrm{NOCl}(g)$$

Experiment	[NO] (M)	$[Cl_2]$ (M)	Initial rate (M/min)
1	0.10 M	0.10 M	0.18
2	0.10 M	0.20 M	0.36
3	0.20 M	0.20 M	1.45

Asking the Right Questions

What method can we use to determine the rate law? We've discussed the method of initial rates. How can we compare rates to determine the rate law using the method?

Solution

The overall rate law, based on the reaction, is

$$\mathrm{rate} = k[\mathrm{NO}]^m[\mathrm{Cl_2}]^n$$

We can use the method of initial rates to compare the change in rate with the concentration of each substance. The order of the reaction with respect to NO is

$$\frac{\text{rate 3}}{\text{rate 2}} = \frac{1.45\ \text{M/min}}{0.36\ \text{M/min}} = \frac{k(0.20)^m(0.20)^n}{k(0.10)^m(0.20)^n}$$

$$\frac{1.45}{0.36} = \frac{(0.20)^m}{(0.10)^m}$$

$$4 = 2^m$$

$$m = 2$$

The order with respect to Cl_2 is

$$\frac{\text{rate 2}}{\text{rate 1}} = \frac{1.36\ \text{M/min}}{0.18\ \text{M/min}} = \frac{k(0.10)^m(0.20)^n}{k(0.10)^m(0.10)^n}$$

$$\frac{0.36}{0.18} = \frac{(0.20)^n}{(0.10)^n}$$

$$2 = 2^n$$

$$n = 1$$

The overall rate law is

$$\text{rate} = k[\text{NO}]^2[\text{Cl}_2]^1$$

and the value for the rate constant can then be calculated. Note that the rate is still expressed in concentration per unit time[c].

$$\text{rate} = k[\text{NO}]^2[\text{Cl}_2]^1$$

$$0.18\ \text{M/min} = k(0.10\ \text{M})^2(0.10\ \text{M})^1$$

$$k = 1.8 \times 10^2 \text{M}^{-2} \cdot \text{min}^{-1}$$

Are Our Answers Reasonable?

A good way confirm our result is to calculate a rate of reaction using the k that we computed. For example, we can calculate the rate of reaction when $[\text{NO}]=0.10\ M$ and $[\text{Cl}_2]=0.20\ M$.

$$\text{rate} = k[\text{NO}]^2[\text{Cl}_2]$$

$$\text{rate} = 1.8 \times 10^2 M^{-2} \cdot \text{min}^{-1}(0.10\ M)^2(0.20\ M)$$

$$\text{rate} = 0.36\ \text{M/ min}$$

This rate agrees with the original data in the example, so the answers are reasonable.

What If…?

…we wanted to change the initial rate for the same reaction at the same temperature and pressure? If we wanted an initial rate of 2.73 M/min with $[\text{Cl}_2]=0.20$ M, what would be [NO]?

(*Ans:* [NO] = 0.28 M).

Practice 13.6

Use the initial rates for the reaction of carbon monoxide (CO) with hemoglobin (Hb) to determine the rate law and the rate constant. What is the overall order of the reaction?

Experiment	[Hb] (M)	[CO] (M)	Initial rate (M/s)
1	2.21×10^{-6} M	1.00×10^{-6} M	0.619×10^{-6}
2	4.42×10^{-6} M	1.00×10^{-6} M	1.24×10^{-6}
3	4.42×10^{-6} M	3.00×10^{-6} M	3.71×10^{-6}

The second method for determining the rate law of a reaction has us examine only one reaction instead of a series of reactions. In this **method of graphical analysis** (also known as the method of integrated rate laws), we plot how the concentration of a reactant changes with time. Although this method does require us to measure the rate of the reaction as it proceeds over a long time period, little mathematical manipulation of the data is needed to determine the rate law. Various versions of the plot can be quickly constructed in order to establish a linear relationship between time and a measure of concentration that confirms one of our common rate laws. Figure 13.13 displays plots of zero-order, first-order, and second-order reactions. Each is a linear relationship that is derived from the integrated rate law. If our data are linear when plotted in one of these ways, it suggests that the model fits the data. Computer-based data acquisition can also help us to do statistical analyses to determine the model that best fits the data. Here are the possible outcomes that we will consider:

Method of graphical analysis – A method of determining the rate law for a reaction where the rate of a reaction is plotted versus time. Also known as the method of integrated rate laws.

- The zero-order reaction produces a linear plot when the concentration of reactant is plotted against time. The slope of the line is equal to $-k$.
- The first-order reaction produces a linear plot when the natural logarithm of the concentration of reactant is plotted against time. The slope in this case is also equal to the negative of the rate constant, $-k$.
- The second-order reaction produces a linear relationship when the reciprocal concentration is plotted against time. The slope, which is equal to the rate constant, is positive in this case.

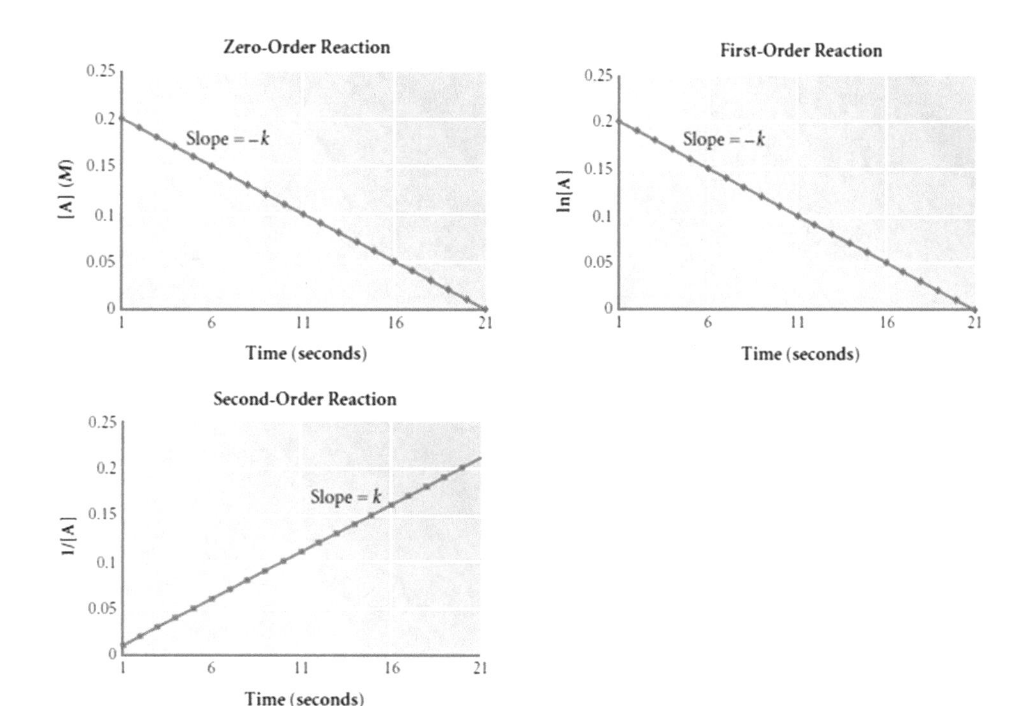

Fig. 13.13 A linear relationship exists for the data if time is plotted versus [A] for zero-order reactions, versus ln[A] for first-order reactions, and versus 1/[A] for second-order reactions. Note that the slope of the line is equal to the negative of the rate constant for zero-order and first-order plots. The slope of the second-order plot is equal to the rate constant

If the concentration of a reactant is followed as a function of time, this is the best method to use in determining the rate law. By graphing the data in three different ways, we can determine the overall order of the reaction (zero-order, first-order, or second-order).

Example 13.7—Determining the Reaction Order

The decomposition of atrazine in the presence of titanium dioxide has been studied and the rate of the reaction measured. Plot the data shown in the table, and determine whether the reaction follows zero-order, first-order, or second-order kinetics.

Time (h)	[Atrazine] (M)
0	4.65×10^{-5}
3	2.98×10^{-5}
8	1.49×10^{-5}
15	6.98×10^{-6}
22	3.67×10^{-6}

Asking the Right Questions

What are the different ways we can plot the data, and what information does each plot give us?

Solution

Using a graphical analysis software package (or three sheets of graph paper), we can plot the data for each of the three models we have discussed: zero-order, first-order, and second-order, seeing which gives us closest to a straight line. We might find it useful to add two columns to our table for ln[atrazine] and 1/[atrazine], respectively. The first graph shows the concentration is plotted versus time, the second relates the ln[atrazine] versus time, and the third illustrates 1/[atrazine] versus time. Which model best fits the data? Examination of the results leads to the conclusion that the middle plot—the one that relates first-order kinetics—is the appropriate graph. The reaction must then be first-order overall and first-order in atrazine (the reactant).

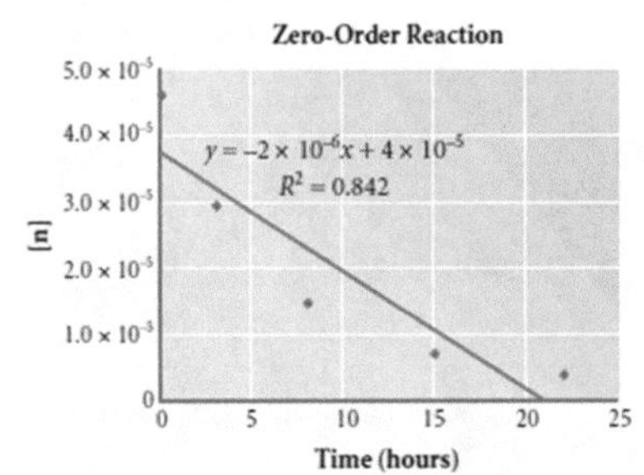

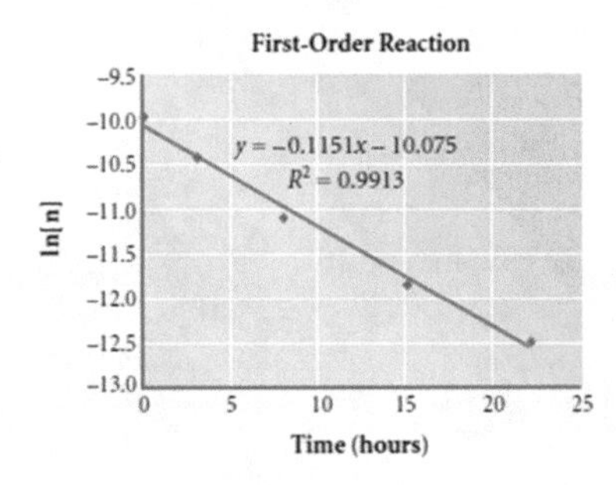

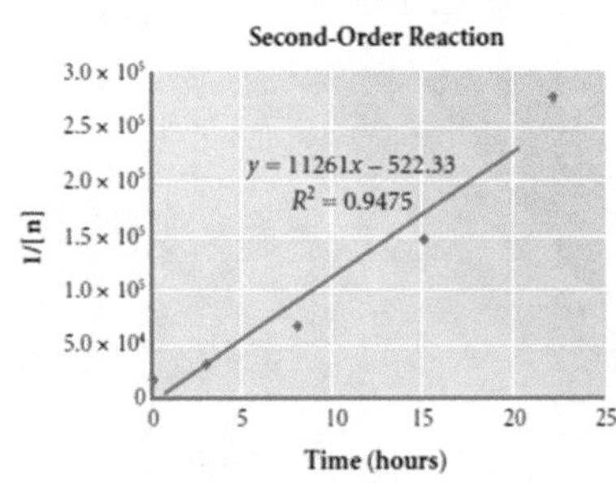

Are Our Answers Reasonable?

The data have the best fit to the line in the first-order model, so we have chosen the most reasonable answer.

What If...?

...the data did not fit any of these three models well? What might that mean, and what would we, as the head of the laboratory, do next?

(*Ans: This can either mean that the data do not fit any of these models, and another reasonable one needs to be proposed, or that there was an experimental error, and we should run the experiment a few more times to confirm the data.*)

Practice 13.7

Plot the data for the following reaction as was done in Example 13.7. Determine the overall order of this reaction. We may wish to use a graphing program to determine which graph gives the straightest line.

$$2NO_2(g) \rightarrow 2NO(g) + O_2(g)$$

Time (s)	$[NO_2]$ (M)
0	0.0100
50	0.0079
100	0.0065
150	0.0055
200	0.0048
300	0.0038
400	0.0031

13.6 Looking Back at Rate Laws

A summary of the data that we have discussed thus far is appropriate. In short, here's what we know so far:

- Reaction rates determine the speed at which a reaction progresses but do not reveal anything about the extent to which they produce a product.
- We can measure the average rate if we know the initial and final concentrations over a particular time period.
- The instantaneous rate is the rate at a given point in the reaction. It can be determined by measuring the concentrations at points when the time difference approaches zero, or it can be measured by determining the slope of a line tangent to a plot of the rate versus time.
- The initial rate is the instantaneous rate as the reaction starts.
- The rate law is experimentally determined using the method of initial rates or the method of graphical analysis.
- The rate constant is only valid at a given temperature.
- The half-life of a reaction is the time required for the concentration of a reaction to reach 50% of the initial value.
- Complex reaction orders can often be reduced to pseudo-first-order reactions by keeping the concentration of one of the reactants relatively large.

Table 13.3 Rate laws

The value of the rate is reported in the units of concentration per unit time. The value of the rate constant is different for each of the three orders (zero order = *M*/s; first-order = 1/s; and second-order = 1/M s)

Order	Rate law	Integrated rate law	Half-life	Linear plot
Zero	rate = k	$[A]_t = -kt + [A]_0$	$t_{\frac{1}{2}} = \frac{[A]_0}{2k}$	$[A]$ versus t $slope = -k$
First	rate = k [A]	$ln\left(\frac{[A]_t}{[A]_0}\right) = -k\,t$	$t_{\frac{1}{2}} = \frac{0.693}{k}$	ln[A] versus t slope = $-k$
Second	rate = k $[A]^2$	$\frac{1}{[A]_t} = k\,t + \frac{1}{[A]_0}$	$t_{\frac{1}{2}} = \frac{1}{k[A]_0}$	$\frac{1}{[A]}$ versus t slope $= +k$

Specific information for the three rate orders that we have discussed is included in Table 13.3. There are many more overall orders for reactions than those listed here.

13.7 Reaction Mechanisms

At this point, we need to do some detective work. In Example 13.2 we determined the rate of metabolism of methoxychlor, an organochlorine insecticide, using actual values for the concentrations. However, this tells us nothing about how the reaction actually happens. Upon deeper investigation, chemists have found that the reaction proceeds in two sequential steps. In the first step, methoxychlor undergoes reaction with water and cytochrome P450 (an enzyme in the liver) to make mono-hydroxymethoxychlor. In a second step, the mono-hydroxymethoxychlor reacts with another molecule of water and cytochrome P450 and is converted into the product bis-hydroxymethoxychlor.

Step 1

H_2O + Methoxychlor $\xrightarrow{\text{Cytochrome P450}}$ mono-hydroxymethoxychlor + CH_3OH

Methoxychlor

Step 2

H_2O + mono-hydroxymethoxychlor $\xrightarrow{\text{Cytochrome P450}}$ *bis*-Hydroxymethoxychlor + CH_3OH

***bis*-Hydroxy methoxychlor**

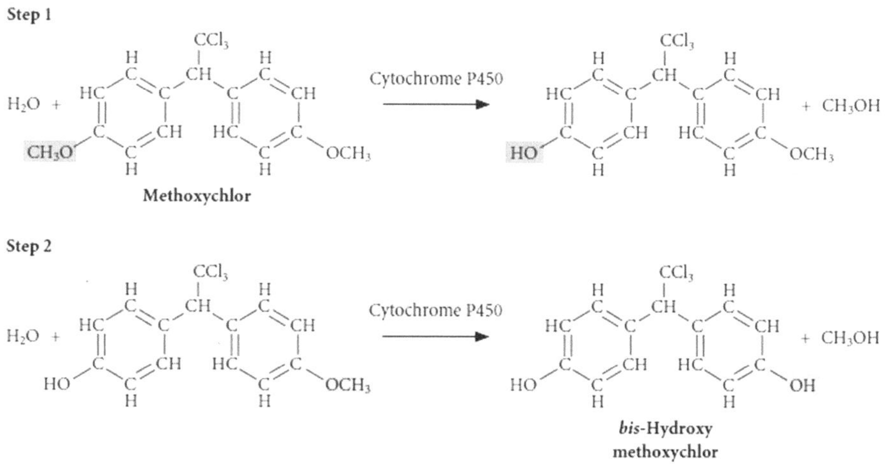

Why is it useful to know the individual steps that make up the overall reaction? Chemists often are interested in more detailed descriptions of chemical reactions, including how the reactions occur at the molecular level. We know from our previous discussions that chemical equations contain a wealth of information about a reaction. We can determine the enthalpy of the process and the stoichiometry of the reactants by examining the chemical equation. However, the overall chemical equation doesn't show how the reactants collide to become products. To address this concern, investigators study a single reaction to determine exactly how each molecule moves during the course of the process. The knowledge of exactly how a reaction proceeds is useful in predicting new reactions, determining the rates of those reactions, discovering new applications of chemistry, and learning how substances interact with humans and our environment. Their study is part of the field of mechanistic chemistry.

A **mechanism** for a reaction is the set of steps that compounds take as they proceed from reactant to product. A mechanism accounts for the experimentally determined rate of the reaction and is consistent with the overall stoichiometry of the reaction. In some cases, a mechanism invokes only one step. In others, multitude of steps exist that show how the reactants become products. In some cases, a mechanism is only one step. In others, several. However, we can't tell this by looking at the overall chemical equation.

Mechanism – The series of steps taken by the components of a reaction as they progress from reactants to products.

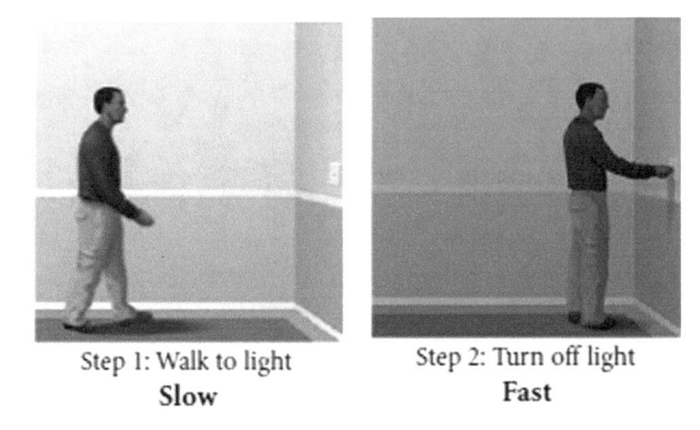

Fig. 13.14 The elementary steps completed in turning off the lights

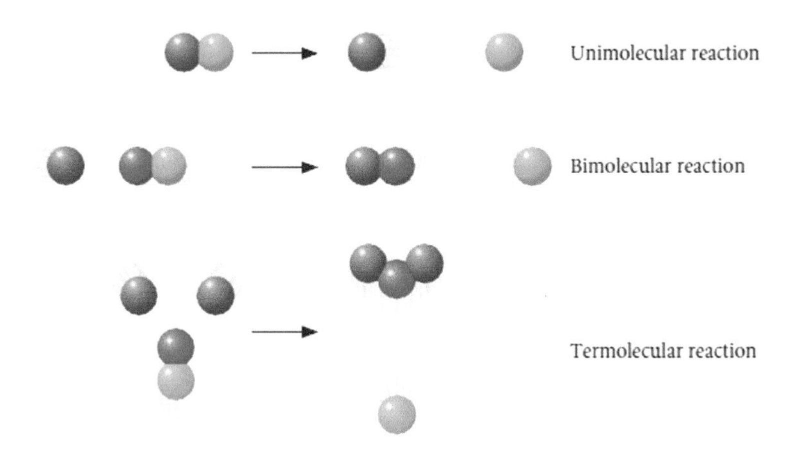

Fig. 13.15 Molecularity of reactions. The number of molecules that must collide at one time to produce the reaction determines the molecularity

Each of the single steps in the mechanism of a chemical reaction is called an **elementary step**. For example, as shown in Fig. 13.14, the mechanism of turning off the lights in a room is made up of two elementary steps: (1) walking to the light switch and (2) flipping the light switch. One of these steps, walking to the light switch, is slower. The other, flipping the switch, is so fast that its contribution to the time required to complete the overall process is negligible. Any calculations of the time required to turn off the lights, then, can be reduced to the time it takes to walk to the light switch. In other words, the time it takes to turn off the lights in the room is nearly equal to the length of time it takes to walk to the light switch. This slow elementary step in the mechanism is known as the **rate-determining step**.

Elementary step – A single step in a mechanism that indicates how reactants proceed toward products.

Rate-determining step – The slowest elementary step in a reaction sequence.

Within each elementary step, reactants come together and undergo successful collisions to make products. The number of molecules that collide in this process defines the **molecularity** of the step (Fig. 13.15). A **unimolecular reaction** involves one molecule as the only reactant. When two molecules collide, the reaction is said to be a **bimolecular reaction**. Reactions that involve the collision of three molecules simultaneously (which are called **termolecular reactions**, or trimolecular reactions) are also known, but they are rare because they require that three molecules collide at the same time and in the proper orientation. In such cases, the chance that such a collision will happen is often so small that it is prohibitive. In general, an elementary step typically has only a single bond breakage or formation.

Unimolecular reaction – A reaction in which one molecule alone is involved in the rate-determining step of the reaction.

Molecularity – A description of the number of molecules that must collide in the rate-determining step of a reaction.

Bimolecular reaction – A reaction involving the collision of two molecules in the rate-determining step.

Termolecular (trimolecular) reaction – A reaction in which three molecules collide in the rate-determining step of the reaction.

If we know the elementary steps in a reaction, we can obtain a wealth of information about the rate of the reaction. Most important, rate laws can be written directly from elementary steps. Specifically, the rate orders for the reactants in an elementary step are given by the stoichiometric coefficients of the reactants in that step. Moreover, the rates of the elementary steps in a mechanism need not be the same—indeed, in most cases they are not. A mechanism with two elementary steps typically has a fast step and a slow step, as in our example of turning on the lights. The slow step in a mechanism is the one that will determine the rate of the overall reaction.

The reaction of NO_2 and F_2 gas is a useful case study:

$$2NO_2(g) + F_2(g) \rightarrow 2NO_2F(g)$$

After careful experimentation, researchers determined a mechanism involving two elementary steps has been suggested for the reaction. This first step was determined to be slow; the second was determined to be relatively fast:

$$\begin{array}{ll} NO_2(g) + F_2(g) \rightarrow NO_2F(g) + F(g) & \text{(slow)} \\ F(g) + NO_2(g) \rightarrow NO_2F(g) & \text{(fast)} \\ \hline 2NO_2(g) + F_2(g) \rightarrow 2NO_2F(g) & \end{array}$$

Note that the sum of the elementary steps gives the balanced equation. From these steps, we can determine the rate law for the overall reaction in a way that is similar to our discussion of the length of time required to turn off the lights. The slow elementary step will determine the rate of the reaction, so we can use this rate-determining step to write the rate law. Judging on the basis of the stoichiometric coefficients of the reactants in the slow step, the overall reaction is first order in NO_2, first order in F_2, and second order overall.

$$\text{rate(for slow step)} = k[NO_2][F_2]$$

Not all mechanisms are so easy to work with. For instance, the reaction within smog of NO and O_2 to make NO_2 is much more complicated than our first glance suggests. From our discussion in from earlier in this chapter, we know the overall equation for the reaction:

$$2NO(g) + O_2(g) \rightarrow 2NO_2(g)$$

We saw that the experimentally determined rate equation for the reaction is

$$\text{rate} = k[NO]^2[O_2]$$

This implies that the reaction might be a termolecular process, which is unlikely because it would require three molecules to collide simultaneously. The mechanism of this reaction is known, and it has the elementary steps shown below.

$$\begin{array}{ll} 2NO(g) \underset{k_{-1}}{\overset{k_1}{\rightleftharpoons}} N_2O_2(g) & \text{(fast)} \\ N_2O_2(g) + O_2(g) \xrightarrow{k_2} 2NO_2(g) & \text{(slow)} \end{array}$$

From what we've just learned, we would expect the rate law for the overall reaction to be:

$$\text{rate} = k_2[N_2O_2][O_2]$$

However, that doesn't match with what we know to be the actual rate law for this reaction. The answer lies in the double arrows that we see in the fast step of the mechanism. What is the meaning of the double arrows in the fast equation? These arrows indicate that the reaction is at **equilibrium** and can proceed in both the forward direction and the reverse direction at the same time. Moreover, each of these directions has a rate constant associated with it (k_1 and k_{-1}) and, at equilibrium, the rates of the forward and backward direction are equal. The slow step in our mechanism is the reaction of N_2O_2(g) with oxygen, and we can write the rate equation of the overall reaction on the basis of this information. Note that we've included the rate constants for both reactions:

Equilibrium – A reaction proceeds both in the forward and reverse direction.

$$\begin{array}{ll} \text{rate} = k_1[NO]_2 & \text{for the fast step}(k_1 \gg k_2) \\ \text{rate} = k_2[N_2O_2][O_2] & \text{for the slow step}(k_2 \ll k_1) \end{array}$$

It follows that the rate law for the overall reaction should be:

$$\text{rate} = k_2[N_2O_2][O_2] \text{ for the overall reaction}$$

However, we know that this is not correct.

This difference arises because N_2O_2 (g) is an **intermediate** in the reaction. An intermediate is a compound that is both formed and consumed during the course of a reaction. If we examine the overall reaction, we don't see the intermediate N_2O_2 as one of the reactants. Measuring the concentration of this species, then, could be difficult.

Intermediate – A compound that is produced in a reaction and then consumed in the reaction. Intermediates are not indicated in the overall reaction equation.

We need to rewrite the rate equation so that the rate reflects only the compounds in the overall reaction. To deal with this, we will assume that the fast reaction reaches equilibrium, an assumption that greatly simplifies our determination of the overall rate law for the reaction. What does this assumption mean? Within a reaction that is at equilibrium, the rate of the forward reaction is equal to the rate of the reverse reaction.

$$k_1[NO]^2 = k_{-1}[N_2O_2]$$

Then we can rearrange our equation to solve for $[N_2O_2]$:

$$[N_2O_2] = \frac{k_1[NO]^2}{k_{-1}} = k'[NO]^2$$

where the new rate constant k' is equal to $\frac{k_1}{k_{-1}}$.

We can now substitute for the concentration of the intermediate, $[N_2O_2]$, in our original rate equation and simplify:

$$\text{rate} = k_2[N_2O_2][O_2]$$

$$[N_2O_2] = k'[NO]^2$$

13

$$\text{rate} = k_2k'[NO]^2[O_2] = k''[NO]^2[O_2]$$

where the new rate constant k'' is equal to k_2k'. This concept of equilibrium and its impact on chemical reactions is one of the most important concepts in General Chemistry. We will examine it in much more detail in later chapters.

This agrees with our experimentally determined rate equation. The mathematics used to convert our rate law containing the intermediate into the experimentally observed rate law are based on the assumption that the fast reaction is at equilibrium. This assumption requires that the rate constant for the reverse of the fast reaction be much larger than the rate constant for the slow step.

Example 13.8—Rate Laws

Write the overall reaction, identify any intermediates, and write the rate law for the following proposed mechanism for the decomposition of IBr(g) to I_2(g) and Br_2(g).

$$IBr(g) \rightarrow I(g) + Br(g) \quad \text{(slow)}$$

$$IBr(g) + Br(g) \rightarrow I(g) + Br_2(g) \quad \text{(fast)}$$

$$I(g) + I(g) \rightarrow I_2(g) \quad \text{(fast)}$$

Asking the Right Questions

How is the overall reaction related to the individual reactions? What is (are) the intermediates? The individual reactions occur at different rates. Which rate controls the overall rate of the reaction, and is therefore indicative of the rate law?

Solution

The overall reaction is the sum of the elementary steps in the mechanism.

$$2IBr(g) \rightarrow I_2(g) + Br_2(g)$$

The intermediates are produced and consumed in the reaction. They are I(g) and Br(g). Because the first step in the mechanism is the slow step, it is rate-determining. Therefore, the rate law for the reaction can be written directly from this step:

$$\text{rate} = k[IBr]$$

It is first-order in IBr(g) and first-order overall.

Are Our Answers Reasonable?
The rate law is based on the slowest reaction, and that is the case here, so that part of the answer is reasonable. The intermediates are produced and consumed in the reaction, and that is consistent with the definition of an intermediate. That they are, by their chemical nature, quite reactive, makes the answers all the more reasonable.

What If…?
…we had this mechanism for the reaction of nitrogen dioxide:

$$2NO_2(g) \rightarrow NO_3(g) + NO(g) \quad \text{(slow)}$$
$$NO_3(g) + CO(g) \rightarrow NO_2(g) + CO_2(g) \quad \text{(fast)}$$

What would be the rate law for the overall reaction?

(Ans: Rate $= k[NO_2]^2$*)*

Practice 13.8
Write the overall reaction, identify any intermediates, and write the rate law for the following proposed mechanism for the production of nitrogen dioxide (NO_2).

$$NO(g) + O_2(g) \rightleftharpoons NO_3(g) \quad \text{(fast)}$$
$$NO_3(g) + NO(g) \rightarrow 2NO_2(g) \quad \text{(slow)}$$

13.7.1 Transition State Theory

Why do some chemical reactions have faster rates than others? To understand this, we need another tool ion our chemistry tool bag: **transition state theory**. Our case study is the destruction of ozone by atomic oxygen, one of the ways in which stratospheric ozone can be depleted, through this chemical reaction:

Transition state theory – The theory that describes how the energy of activation is related to the rate of reaction.

$$O_3(g) + O(g) \rightarrow 2O_2(g)$$

By examining this reaction in the laboratory, we can determine the change in enthalpy ($\Delta H = -392$ kJ/mol) and measure the rate of the reaction. However, it is often helpful to examine a reaction by consulting a plot of the energies for the reactants, products, and any intermediates. If we make a graph of the energies as the reactants proceed along a **reaction coordinate** (the pathway describing the changes in each molecule in the reaction) to become product, we obtain a **reaction profile** like that shown in Fig. 13.16. On the basis of our earlier discussion of collision theory, we know that the reactants must have enough energy to overcome the activation energy (E_a)—assuming that the reactants collide in the proper orientation. For the reaction of atomic oxygen and ozone, the barrier is rather small ($E_{a(forward)} = 19$ kJ/mol) compared to the reverse reaction ($E_{a(reverse)} = 411$ kJ/mol). The reaction profile illustrates this.

Reaction coordinate – A measure that describes the progress of the reaction.

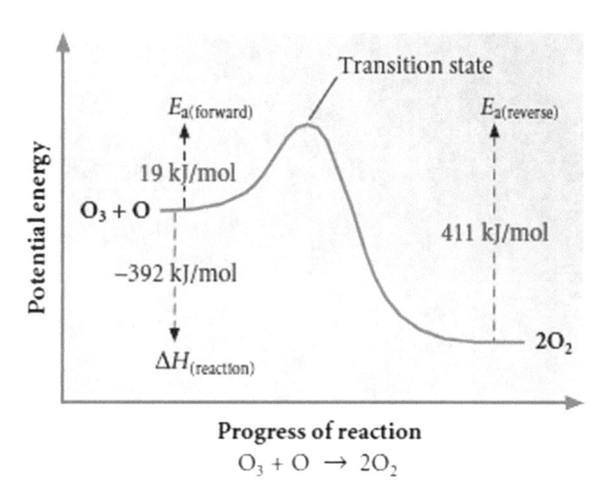

Fig. 13.16 A reaction profile diagram

Reaction profile – A plot of the progress of a reaction versus the energy of the components.

The reaction profile is used to explore transition state theory and how it applies to a reaction. This theory describes how the bonds in the reacting molecules reorganize to represent the bonds in the products. At some point on the reaction coordinate the collision occurs, and the atoms in the reactants occupy a **transition state**. The collection of atoms at the transition state, called the **activated complex**, is very energetic at this point in the reaction—more so than the reactants, products, or intermediates. The activated complex is not a separate isolatable compound. Rather, it is a snapshot of the reaction at the point in time when the molecules have collided. From the activated complex, the reaction could proceed to products, or the complex could dissociate back into the reactants.

Activated complex – The unstable collection of atoms that can break up either to form the products or to reform the reactants.

Transition state – The point along the reaction coordinate where the reactants have collided to form the activated complex.

The reaction profile shows the activation energy for the forward reaction, the activation energy for the backward reaction, and the overall change in the energy of the reaction (ΔE) in a graphical way. This change in energy (ΔE) is equal to ΔH for the reaction at constant pressure and volume. Even for cases where ΔE and ΔH are not equal (unequal volumes or pressures), the difference is usually very small. By looking at the reaction profile, we can tell whether a reaction is exothermic (Fig. 13.17) or endothermic (Fig. 13.18).

$$\Delta H_{\text{reaction}} \cong \Delta E = E_{\text{a(forward)}} - E_{\text{a(reverse)}}$$

In 1888, Svante Arrhenius (1859–1927), a Swedish chemist, studied how temperature affected the rate of a reaction. What came out of this work was a relationship between the activation energy and the rate of a reaction. The relationship that Arrhenius defined is:

$$k = Ae^{-E_a/RT}$$

13

where:
- k is the rate constant,
- A is the **frequency factor** that relates how many successfully oriented collisions occur in a particular reaction,
- E_a is the activation energy,
- T is the temperature in kelvin, and,
- R is the universal gas constant (8.314 J/mol · K).

Frequency factor – A term in the Arrhenius equation that indicates the rate of collision and the probability that colliding reactants are oriented for a successful reaction.

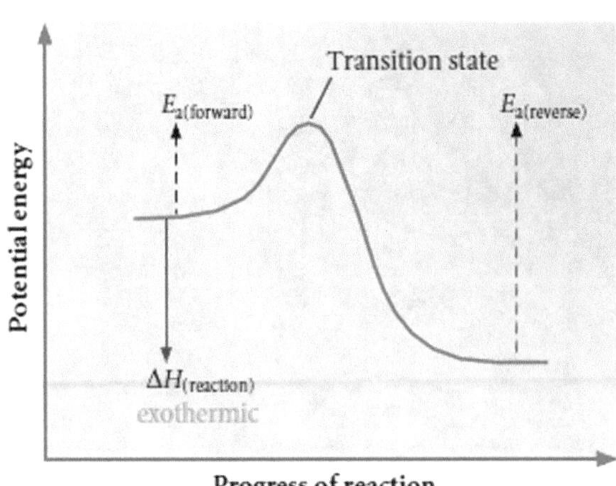

Fig. 13.17 Exothermic reaction profile. The reactants are more energetic than the products in the endothermic reaction. $E_{\text{a(reverse)}}$ is larger than $E_{\text{a(forward)}}$

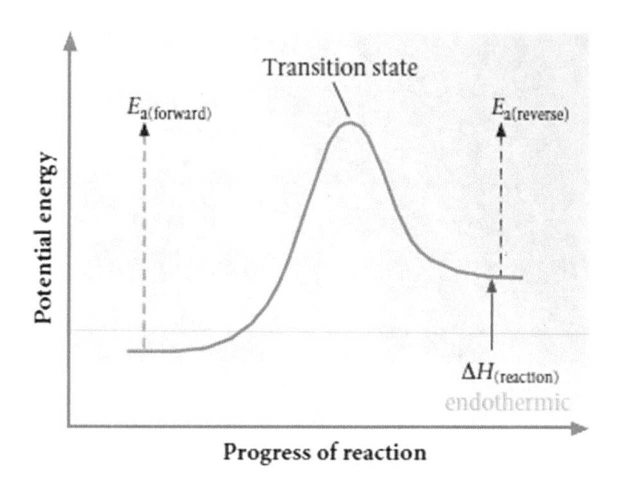

Fig. 13.18 Endothermic reaction profile. The products are more energetic than the reactants in the endothermic reaction. $E_{a(forward)}$ is larger than $E_{a(reverse)}$

This equation indicates that the rate constant of a reaction is related to the size of the activation energy, E_a. As E_a gets larger, the rate constant gets smaller and the rate of the reaction decreases.

This equation can be used to determine the rate of a reaction on the basis of the temperature of the reaction and the amount of energy the reactants require to make the activated complex. To use it, however, we must know the activation energy (E_a) and the frequency factor (A) for the reaction. Fortunately, by performing the reaction at two different temperatures, we can use this equation to calculate the activation energy without knowing the frequency factor. The equation that relates this calculation can be derived by taking the natural logarithm of the Arrhenius equation above, at two temperatures:

$$\ln\left(\frac{k_2}{k_1}\right) = \frac{E_a}{R}\left(\frac{1}{T_1} - \frac{1}{T_2}\right)$$

where k_1 and k_2 are the rate constants for the reaction obtained at two different temperatures, T_1 and T_2.

This equation provides a quick method to determine the energy of activation for a reaction, but the resulting answer can contain significant error, because its value is based on only two sets of conditions. Alternatively, we can determine the value of E_a graphically by using a series of rate constants calculated at different temperatures. To see how this is done, let's examine the original equation produced by Arrhenius:

$$k = Ae^{-E_a/RT}$$

If we take the natural log of both sides of the equation, we get

$$\ln k = \ln A - \frac{E_a}{RT}$$

Rearranging this equation as shown enables us to determine the activation energy by plotting the rate constant of a reaction at several temperatures.

$$\ln k = -\frac{E_a}{R}\left(\frac{1}{T}\right) + \ln A$$

$$y = mx \quad + \quad b$$

If we construct an "Arrhenius plot" of ln k versus $1/T$ for a series of reactions, we obtain a straight line whose slope is equal to $-E_a/R$ and whose intercept is ln A, as shown in Fig. 13.19. This method not only enables us to determine the energy of activation quite accurately but also provides a convenient way to determine the frequency factor, A.

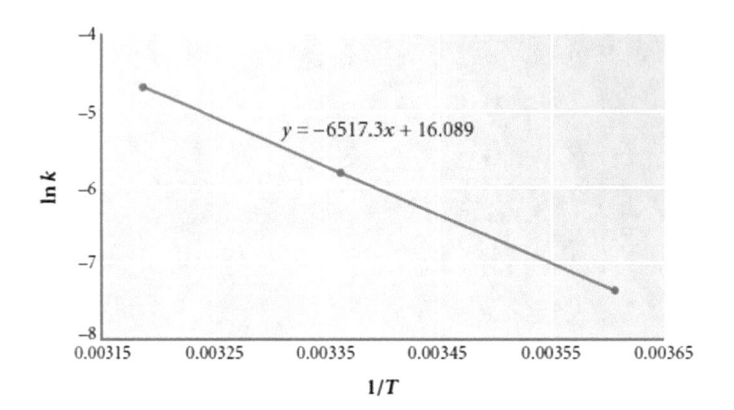

Fig. 13.19 A plot of ln k versus $1/T$ gives a straight line whose slope is $-E_a/R$. Each data point on this plot corresponds to the measurement of the temperature and rate constant in separate reactions. The slope of the line of best fit (−6517.3) is equal to $-E_a/R$. Therefore, $E_a = 54.2$ kJ/mol for this reaction. If the axes went to zero, the intercept (16.089) corresponding to ln A would be found. The frequency factor is 9,710,000

Example 13.9—Energy Barrier

The reaction of NO, a component in smog, with ozone has been extensively studied. Data for the temperature dependence are tabulated below. What is the activation energy for this reaction?

$$NO(g) + O_3(g) \rightarrow NO_2(g) + O_2(g)$$

Temperature (K)	k (1/M s)
195	1.08×10^9
298	12.0×10^9

Asking the Right Questions

How can we use the Arrhenius equation to solve for the activation energy of the reaction.

Solution

The activation energy can be calculated by substituting the values from the table into the equation:

$$\ln\left(\frac{k_2}{k_1}\right) = \frac{E_a}{R}\left(\frac{1}{T_1} - \frac{1}{T_2}\right)$$

$$\ln\left(\frac{12.0 \times 10^9 M^{-1} \cdot s^{-1}}{1.08 \times 10^9 M^{-1} \cdot s^{-1}}\right) = \frac{E_a}{8.314\ \text{J} \cdot \text{mol}^{-1} \cdot \text{K}^{-1}}\left(\frac{1}{195\ \text{K}} - \frac{1}{298\ \text{K}}\right)$$

$$2.408 = \frac{E_a}{8.314\,\text{J}\,\text{mol}^{-1} \cdot \text{K}^{-1}}(1.773 \times 10^{-3}\,\text{K}^{-1})$$

$$20.02\ \text{J} \cdot \text{mol}^{-1} = E_a\left(1.773 \times 10^{-3}\text{K}^{-1}\right)$$

$$E_a = 11292\ \text{J} \cdot \text{mol}^{-1} = 11.3\ \text{kJ} \cdot \text{mol}^{-1}$$

Is Our Answer Reasonable?

One way to determine this is to check the internet or some chemistry books in the science section of our college library for activation energy values for some reactions, so we get a sense for their range. For example, the National Institute of Standards and Technology (NIST) maintains a searchable database of 38,000 reactions.

What If…?

… we have these data:

$$k_1 = 2.00\ M^{-1} \cdot s^{-1}\ \text{at 318 K;}$$

$$k_2 = 10.0\ M^{-1} \cdot s^{-1}\ \text{at 371 K.}$$

What is the value of E_a?

(*Ans:* 30.0 kJ)

Practice 13.9

More data on the reaction of NO and O_3 are shown in this table. Calculate E_a for each pair of reactions. Are the values the same? Explain.

Temperature (K)	k (1/M s)
230	2.95×10^9
260	5.42×10^9
369	35.5×10^9

13.8 Applications of Catalysts

Environmental chemists have comprehensively explored the rate of decomposition of atrazine in aqueous solutions. Because the rate in water is so slow ($t_{1/2}=400$ days), they've spent time considering how the decomposition could be accelerated to clean up the environment more quickly. The exercise in the previous section showed that the rate of a reaction increases with an increase in temperature. That's one way to speed up a reaction. However, raising the temperature of groundwater, rivers, and lakes is not a feasible way to increase the rate of decomposition of herbicides and pesticides. Researchers have found a better way to enhance the rate of the decomposition. They have discovered that filtering atrazine-contaminated water through a container full of titanium(IV) oxide in the presence of ultraviolet light greatly increases the rate of decomposition (reducing the half-life significantly; $t_{1/2}=15$ min). What does the titanium(IV) oxide do?

The titanium(IV) oxide is used as a **catalyst** in the decomposition of atrazine. A catalyst is a compound that, when added to a reaction mixture, changes the mechanism of the reaction so that it follows a new pathway with a lower activation energy. Rather than lowering the activation energy of the existing mechanism, it creates a new set of elementary steps whose rate-determining step has a lower energy of activation than the reaction would have without the catalyst. Because E_a is lower, the new mechanism is faster, so we often say that a catalyst increases the rate of the reaction. Another useful aspect of catalysts is that they can be recovered from the reaction; catalysts are not consumed in a reaction. Because of this, they don't appear in the net chemical equation even though they do appear in the mechanism. To show that a particular reaction involves a catalyst, we typically place it over or under the arrow in the overall equation. There are two types of catalysts, homogeneous catalysts and heterogeneous catalysts. The type of catalyst, as well as the reaction profile that results, depends on how the catalyst is mixed with the reaction.

Catalyst – A substance that participates in a reaction, is not consumed, and modifies the mechanism of the reaction to provide a lower activation energy. Catalysts increase the rate of reaction.

A **homogeneous catalyst** is part of a reaction that is catalyzed in a homogeneous mixture (that is, the catalyst is intimately mixed with the reactants). For example, the destruction of ozone by chlorine radicals (a homogeneous catalyst) is illustrated by the net reaction of ozone with oxygen atoms:

Homogeneous catalyst – A specific catalyst that exists in the same physical state as the compounds in the reaction.

$$O_3 + O \xrightarrow{Cl} 2O_2$$

The mechanism of this reaction consists of two elementary steps:

Step 1: The ozone molecule reacts with elemental chlorine (the catalyst) to make a molecule of ClO and a molecule of oxygen.

Step 2: The ClO reacts with an oxygen atom (the other reactant in the net equation) to make another molecule of oxygen and regenerate elemental chlorine.

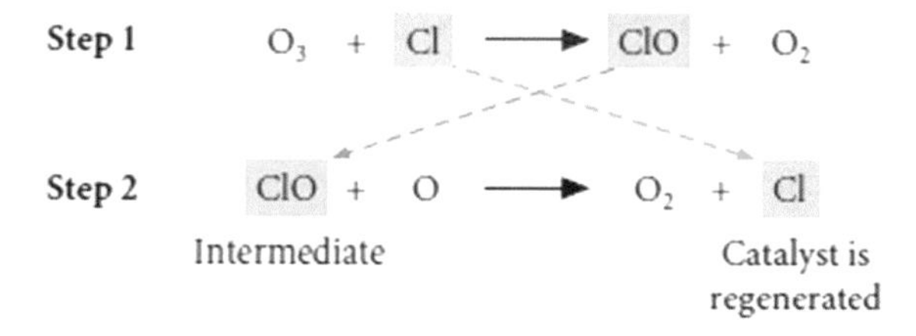

The chlorine atom appears first as a reactant and later as a product; chlorine begins and escapes the reaction without being changed and therefore is not consumed. This is exactly what catalysts do. Is ClO a catalyst too? No; ClO is an intermediate. It is formed in the course of the reaction and then consumed before products are made. The ClO never escapes the reaction. As we saw earlier, this is exactly what happens to an intermediate.

The presence of chlorine atoms causes a tremendous increase in the rate of the decomposition of ozone. How does a catalyst cause a reaction's rate to increase? As we noted earlier, the catalyst causes the reaction to follow a different mechanism with a lower activation energy. Consider the reaction profile, shown in ◻ Fig. 13.20, for the first-order decomposition of hydrogen peroxide by a homogeneous catalyst. In the non-catalyzed reaction, the reaction follows a unimolecular mechanism with a clearly defined high-energy barrier (the activation energy)

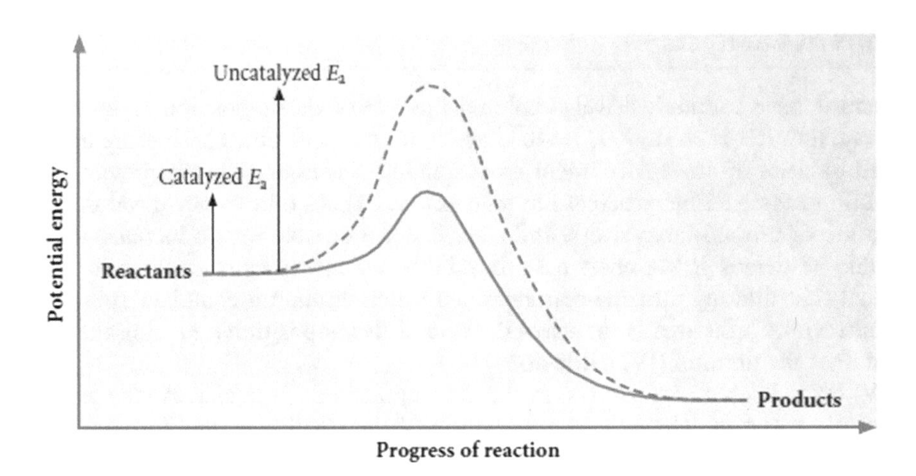

Fig. 13.20 Decomposition of H_2O_2 with and without a homogeneous catalyst. The dotted line represents the reaction profile without the catalyst; the solid line indicates the effect of a homogeneous catalyst

separating the reactants from the products. If a catalyst such as iodide is added to this reaction, the mechanism of the reaction changes. The iodide is the catalyst, and it is shown over the reaction arrow to identify it as such. What is the role of OI^- in the reaction? The reaction makes an intermediate, which is more stable than the activated complex. The solid line in the reaction profile in Fig. 13.20 illustrates the new reaction.

Step 1	$H_2O_2 + I^- \longrightarrow H_2O + OI^-$
Step 2	$OI^- + H_2O_2 \longrightarrow H_2O + O_2 + I^-$
Net reaction	$2H_2O_2 \xrightarrow{I^-} 2H_2O + O_2$

What is the outcome of adding a catalyst? Remember that our catalyst has produced a new mechanism for the reaction. If we plot the reaction profile for this new mechanism, we can see that the activation energy is lower. This means that more molecules will have the necessary energy to become activated complexes in the mechanism. Although the overall reaction enthalpy (ΔH) hasn't changed, the activation energy has been reduced dramatically. A reduction in the activation energy causes the rate of the reaction to increase.

In systems with a **heterogeneous catalyst**, the catalyst and the reactants are in *different* physical states. The catalyst is typically a metal in solid form, whereas the reactants are typically in gaseous, aqueous, or liquid form. The TiO_2 decomposition of atrazine is an example of this type of catalysis. The catalytic converter (usually made up of platinum, palladium, and/or rhodium metal) in our automobile is another. One of the reactions in the catalytic converter cuts down on smog-forming NO by reducing it to N_2. The general unbalanced reaction is:

$$NO(g) \rightarrow N_2(g)$$

Heterogeneous catalyst – A specific catalyst that exists in a different physical state than the reaction.

The heterogeneous catalyst works in a three-step process. In the first step shown in Fig. 13.21, the reactant molecules **adsorb** to the catalyst—that is, they stick to the catalyst's surface. Note that the word adsorb is different than the word **absorb**, which means being taken up or mixed into a substance. A small activation energy barrier must be crossed to achieve the surface binding of the reactant. Often, there is only a very small barrier for binding of a reactant to the surface of a catalyst. Then the reactants migrate around on the surface until they collide to make the product. A larger, yet still low, activation energy barrier exists for this step. In the final step, the products are **desorbed**, or released from the surface of the catalyst (the reverse of being adsorbed).

Adsorb – To be associated, through intermolecular forces of attraction, with a surface.

Absorb – To be associated, through intermolecular forces of attraction, by being taken up or mixing into a substance.

Desorbed – Released from a surface.

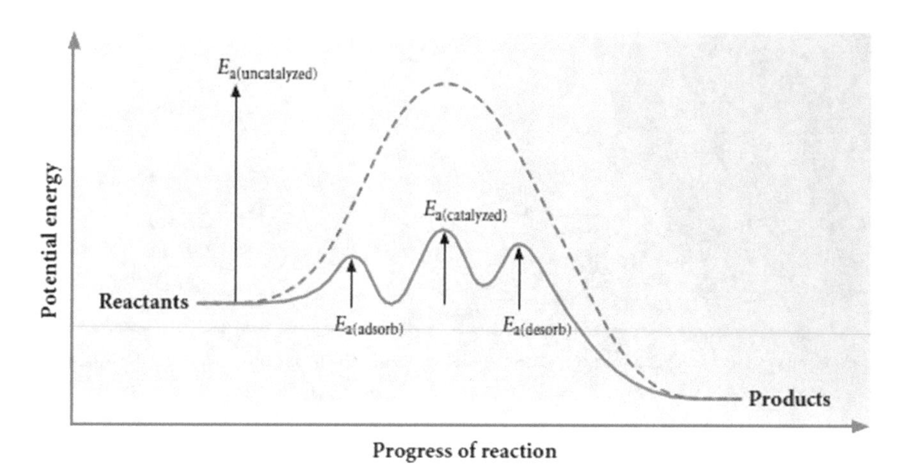

Fig. 13.21 Heterogeneous catalysis of NO

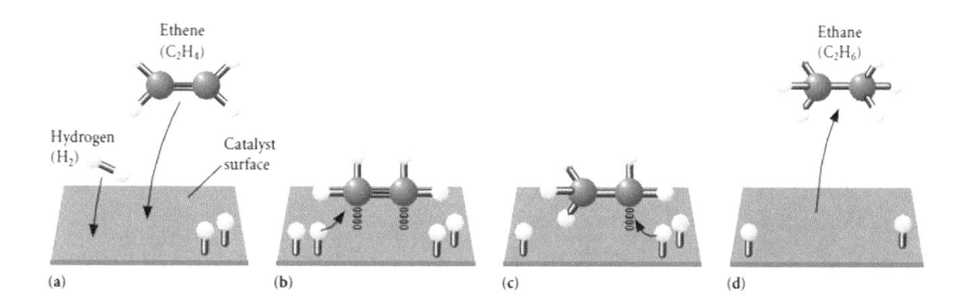

NanoWorld/MacroWorld: The Big Effects of the Very Small: Enzymes—Nature's Catalysts

The modern confectionery industry operates a booming business, helping to satisfy that sweet tooth in most of us. Fanciful desserts require sucrose as a sweetener. Fortunately, the American farmer can meet the large demand for sweeteners. A total of 32 million tons of sugar beets are grown annually in the U.S., primarily in Colorado, Montana, Nebraska, Wyoming, North Dakota and Minnesota, yielding about 8 million tons of sucrose per year. Sugar cane production, primarily in Hawaii, Louisiana, Texas, and Florida, adds another 32 million tons to the total. But even more sugar is needed to meet the demand in the U.S. One of the ways to meet our appetite for sweet things involves cornstarch and a biological molecule known as D-glucose isomerase. The product of these molecules is sweeter than sucrose alone. It is known as high-fructose corn syrup, and while its use is declining in the last few years within the U.S., it still surpasses that of sucrose in the confectionery industry.

Like other enzymes, D-glucose isomerase acts as a catalyst that speeds up a reaction. The enzymes work by binding selectively to a particular molecule, forcing it into just the right shape, and then assisting in the reaction that makes the product. They increase the value of A (the frequency factor from the Arrhenius equation) and decrease the energy of activation (E_a) at the same time. This activity arises because the backbone of the amino acid polymer weaves the enzyme into a structure similar to a catcher's mitt. Along the inside of the catcher's mitt (the active site of the enzyme) lie portions of the enzyme that are polar and portions that are nonpolar. The arrangements of the polar and nonpolar groups provide a template that exactly matches that of the reactant they bind (also known as the substrate). When the substrate binds, the enzyme bends it into a conformation similar to that of the product of the reaction. Then, when the conformation is just right, the reaction takes place. After releasing the product, the enzyme returns to its original shape, ready to accept another substrate (Fig. 13.22). The net result is an increase in the reaction rate without an increase in temperature, which is particularly useful in the food industry.

For example, lactase (an enzyme that converts lactose into glucose and galactose) is used in the dairy industry to make digestible milk products. Because many of these products spoil at temperatures warmer than those found in a refrigerator, the use of enzymes to speed the reaction without a temperature increase is quite helpful. After the lactase has been added to milk, lactose-intolerant people can drink all of the milk they want. And they owe their settled stomach to one of nature's catalysts.

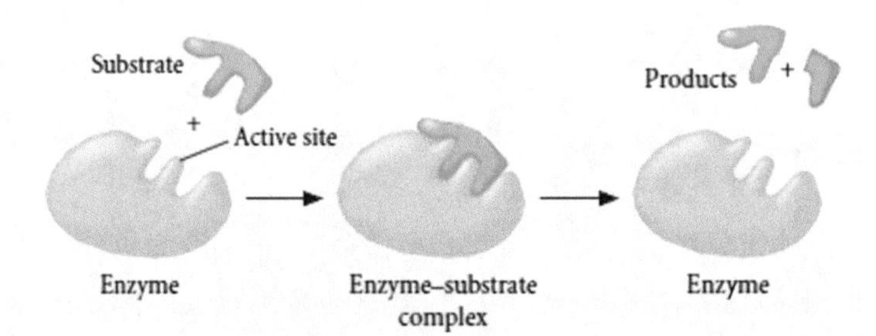

Fig. 13.22 Diagram of an enzyme-catalyzed reaction. The substrate binds to the active site on an enzyme, where the reaction is catalyzed

The principles of chemical kinetics are important not only to chemists but to anyone concerned with the rate of a process. The rate of decay at the landfill, the rate of ozone depletion, and the persistence of pesticides and herbicides in the environment can easily be determined by examining the rate laws associated with these processes. Our work in the lab can provide the rate laws for these reactions.

The Bottom Line

- Reaction rates determine the speed at which a reaction progresses but do not reveal anything about the extent to which they produce a product.
- The average rate can be measured if we know the initial and final concentrations over a particular time period.
- The instantaneous rate is the rate at a given point in the reaction. It can be determined by measuring the concentrations at points when the time difference approaches zero, or it can be measured by determining the slope of a line tangent to a plot of the rate versus time.
- The initial rate of a reaction is the instantaneous rate as the reaction starts.
- The rate law ($\text{Rate} = k[A]^n[B]^m$) indicates the relationship between the rate and the concentrations of reactants.
- The rate constant is only valid for a given temperature, but the rate orders are constant under all conditions.
- The order of a reactant indicates the relationship between the reactant concentration and the rate.
- Collision theory explains why the rate of a reaction typically decreases as time passes.
- The half-life of a reaction is the time required for the concentration of a reaction to reach 50% of the initial value.
- Complex reaction orders can often be reduced to pseudo-first-order reactions by keeping the concentration of one of the reactants large.
- The rate law is experimentally determined using the method of initial rates or the method of graphical analysis.
- Transition state theory, which is based on collision theory, describes the energy of a reaction during the course of a reaction. A plot of the reaction coordinate versus the energy can give meaningful information, such as the activation energy, the presence of any intermediates, the effect of a catalyst, and the enthalpy for the forward and reverse process.
- Catalysts speed up a reaction without being consumed in the reaction.

Section 13.2 Reaction Rates

Skill Review

1. Use the analogy of a runner in a 10-km road race to define each of these pertinent reaction kinetics terms:
   a. Extent of reaction
   b. Average rate
   c. Instantaneous rate
   d. Initial rate
2. Use the analogy of firing a bullet from a gun to define each of these kinetics terms:
   a. Extent of reaction
   b. Average rate
   c. Instantaneous rate
   d. Initial rate
3. For the following reaction in a 4.00-L container, it was found that 1.00×10^{-4} mol of C_4H_8 reacted over a time period from 10:58 A.M. to 11:15 A.M.

$$C_4H_8(g) \rightarrow 2C_2H_4(g)$$

What is the average rate, in *M*/s, for this reaction? Is it possible to have an instantaneous rate faster than the average rate? Explain.

4. The reaction described in Problem 3 was performed a second time. Into a 1.5-L flask was placed 3.25 mol of C_4H_8(g). After 56 min, 1.00 mol of C_2H_4(g) was obtained. What is the average rate, in *M*/s, for this reaction?
5. Using the same reaction as in Problem 3 ($C_4H_8 \rightarrow 2C_2H_4$), compare the rate of appearance of C_2H_4 to the rate of disappearance of C_4H_8. Write out this relationship using the style presented in the chapter that depicts the rate, as it would be expressed based on either of the components.
6. The following reaction shows PH_3 decomposing into two products.

$$4PH_3(g) + 3O_2(g) \rightarrow 4P(g) + 6H_2O(g)$$

Write the rate expression that depicts the rate of disappearance of the reactant when compared to the appearance of each product.

7. Determine the rate of reaction, in *M*/s, for each of these systems over the time period indicated. Assume that the reaction is

$$A \rightarrow 2B$$

a. $[A]_o = 0.350$ M; $[A]_t = 0.300$ M; $t_0 = 0$ s; $t = 100$ s
b. $[A]_o = 1.522$ M; $[A]_t = 0.350$ M; $t_0 = 0$ min; $t = 15$ min
c. $[A]_o = 0.050$ M; $[A]_t = 0.010$ M; $t_0 = 0$ days; $t = 399$ days
d. $[A]_o = 0.280$ M; $[A]_t = 0.140$ M; $t_0 = 35$ min; $t = 2.5$ h

8. Determine the rate of reaction, in *M*/s, for each of these systems over the time period indicated. Assume that the reaction is

$$A \rightarrow 2B$$

a. $[A]_o = 0.350$ M; $[A]_t = 0.280$ M; $t_0 = 0$ s; $t = 100$ s
b. $[A]_o = 2.50$ M; $[A]_t = 0.250$ M; $t_0 = 0$ s; $t = 15$ min
c. $[A]_o = 1.750$ M; $[A]_t = 0.010$ M; $t_0 = 0$ days; $t = 250$ days
d. $[A]_o = 0.125$ M; $[A]_t = 0.105$ M; $t_0 = 15$ s; $t = 2.5$ min

9. In the general reaction shown below, the rate of disappearance of A is 4.5×10^{-2} M/s.

$$2A + 3B \rightarrow C + 2D$$

a. What is the rate of disappearance of B?
b. What is the rate of appearance of C?
c. What is the rate of appearance of D?10. In the general reaction shown below, the rate of disappearance of A is 1.85×10^{-4} M/s.

$$A + 2B \rightarrow 2C + 3D$$

a. What is the rate of disappearance of B?
b. What is the rate of appearance of C?
c. What is the rate of appearance of D?

11. Either on a piece of graph paper or using a computer, plot the data given in the accompanying table, which are from a drag race. Then determine:
   a. The initial speed
   b. The instantaneous speed at $t = 8$ s
   c. The instantaneous speed at $t = 25$ s
   d. The average speed of the dragster
   e. In words, how does this compare to the typical plot of a reaction?

Distance (mi)	Time (s)
0	0
0.055	5
0.193	10
0.611	20
1.120	30

12. Plot the accompanying data on a piece of graph paper or using a computer. Then determine the following:
    a. The initial rate, in M/min, of the reaction
    b. The instantaneous rate, in M/min, at $t = 15$ min
    c. The instantaneous rate, in M/min, at $t = 45$ min
    d. The average rate, in M/min, of the reaction

[Pesticide]	Time (min)
0.10000 M	0
0.055294 M	15
0.030575 M	30
0.016906 M	45
0.009348 M	60

Chemical Applications and Practice

13. The common disinfectant hydrogen peroxide (H_2O_2) can decompose according to the following balanced equation:

$$2H_2O_2(aq) \rightarrow 2H_2O(l) + O_2(g)$$

    a. How much faster is the appearance of H_2O than the appearance of O_2?
    b. Write the expression that defines the rate of the reaction based on the rate of disappearance of H_2O_2 and the rates of appearance of H_2O and O_2.

14. Reducing the level of nitrogen monoxide compounds emitted in automobile exhaust is a priority of environmentally concerned citizens. One response to this has been the development of catalytic converters. The NO in the exhaust passes over a rhodium-containing converter and is changed to the components already present in clean air.

$$2NO(g) \rightarrow N_2(g) + O_2(g)$$

If the rate of appearance of N_2 were 1.5×10^{-6} mol/s, what would we calculate as the rate of disappearance of NO?

15. The relationship between ozone and humans has a rather unique feature. Whether ozone benefits humans depends largely on proximity. Ozone is considered beneficial when it occurs in the upper atmosphere, but it can cause serious respiratory problems if it is in the atmosphere layer closest to us. Ozone can be formed by the following reaction:

$$O_2(g) + O(g) \rightarrow O_3(g)$$

This graphical representation shows the decomposition of oxygen over time. Redraw the graph and sketch a line that would depict the appearance of ozone over the same time period.

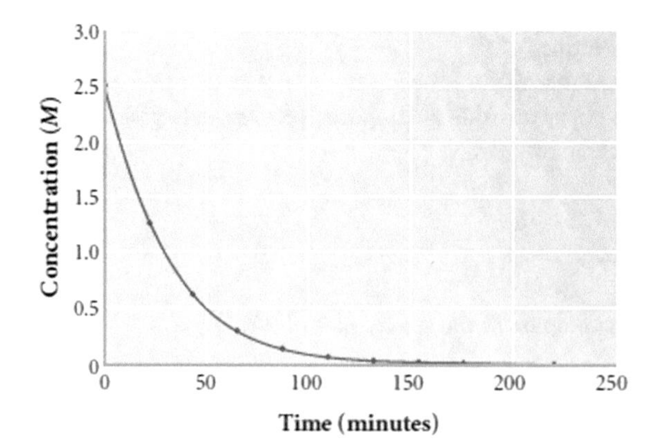

16. As we have seen, the reaction that is used to produce the agriculturally critical fertilizer ammonia combines nitrogen and hydrogen gas in the following manner:

$$3H_2(g) + N_2(g) \rightarrow 2NH_3(g)$$

a. Using the following graph of the disappearance of N_2 as a guide, sketch two additional lines that factor in the ratios of the different species for the disappearance of H_2 and appearance of NH_3, respectively.
b. If the disappearance of N_2 in a particular reaction were 1.2×10^{-3} mol/min, what would be the rate of appearance of NH_3?

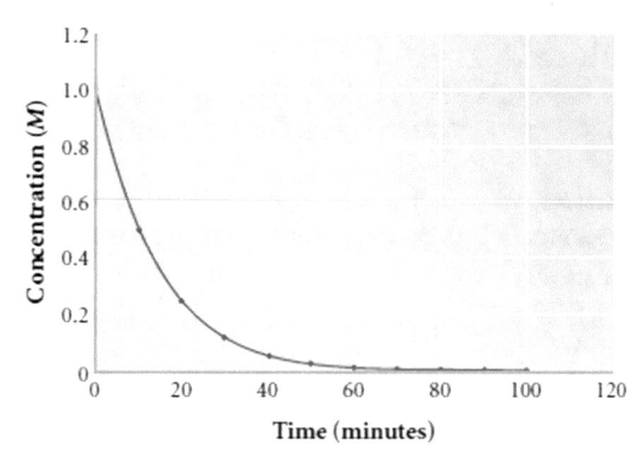

Section 13.3 An Introduction to Rate Laws

Skill Review

17. Judging on the basis of collision theory, indicate whether each of these modifications would increase the rate of a reaction. Explain the answers.
    a. increasing the temperature
    b. decreasing the temperature
    c. increasing the initial concentration of reactant
    d. diluting the reaction with more solvent
18. Judging on the basis of collision theory, indicate whether each of these modifications would have an effect on the rate of a reaction. Explain the answers.
    a. increasing the volume of the reaction
    b. decreasing the number of reactant molecules
    c. adding some product molecules to the reaction
    d. removing the product molecules as they form
19. The following equation represents the rate law for the decomposition of an important pesticide (symbolized here as Pest).
    a. What is the order of the reaction?
    b. What is the meaning of k?
    c. What effect would doubling the concentration of pesticide have on the rate of the reaction?
    d. What effect would doubling the concentration of the pesticide have on the value of k?

$$\text{Rate} = k[\text{Pest}]^2[\text{H}^+]$$

20. Report the overall order of each of these reactions:
    a. $\text{Rate} = k[\text{NO}][\text{O}_3]$
    b. $\text{Rate} = k[\text{NO}][\text{H}_2][\text{H}_2\text{O}]^{-1}$
    c. $\text{Rate} = k[\text{H}_2][\text{Cl}_2]^{1/2}$
21. Calculate the rate of the following reaction, if the rate constant is $3.95 \times 10^{-4}\ \text{s}^{-1}$ and $[\text{A}] = 0.509$ M. The reaction is first order in A and zero order in B.

$$\text{A} + \text{B} \rightarrow 2\text{C} + \text{D}$$

22. Calculate the rate constant for a reaction if, when $[\text{A}] = 0.672$ M, the rate of the reaction is 2.99×10^{-3} M/s. The reaction is second order in A.

$$\text{A} \rightarrow \text{B}$$

23. The rate of a reaction, expressed by the following equation, doubles when the concentration of A is doubled. What is the order of the reaction with respect to A?

$$2\text{A} + \text{B} \rightarrow \text{C}$$

24. The rate of a reaction, expressed by the following equation, is observed to quadruple when the concentration of B is doubled. What is the order of the reaction with respect to B?

$$2A + B \rightarrow C$$

25. If a reaction is third order with respect to a reactant, and the concentration of that reactant is increased by 50%, by what factor does the rate increase?
26. A certain reaction is second order with respect to "B". The initial rate is 2.5×10^{-3} M/s when the concentration of B is 1.0 M. What is the initial rate when the concentration of B is 2.5 M?

Chemical Applications and Practices

27. The hypochlorite ion (ClO^-) is present in commercial bleach. One aqueous reaction in which this ion participates is

$$3ClO^- \rightarrow ClO_3^- + 2Cl^-$$

Without any further information, give two reasons why we could not claim that the rate law is rate $= k[ClO^-]$.
28. If the rate law for the hypochlorite reaction from the equation in Problem 27 ($3ClO^- \rightarrow ClO_3^- + 2Cl^-$), is rate $= k[ClO^-]^2$, what would we report as the order of the reaction? If the concentration of hypochlorite were tripled, what effect would this have on the rate of the reaction (assuming no change in temperature)? The order of the reaction does not match the stoichiometry of the equation. What does this indicate?
29. The decomposition of $N_2O_4(g)$ into $NO_2(g)$ is known to be second order with respect to N_2O_4:
   a. Write a balanced equation for this reaction.
   b. Write the rate law for this reaction.
   c. When the concentration of N_2O_4 is 1.50 M, the rate of decomposition is 2.32×10^2 M/s. Calculate the value and units of the rate constant.
   d. What is the rate of this reaction when the concentration of N_2O_4 is 0.50 M?
30. If the decomposition of HI(g) into $H_2(g)$ and $I_2(g)$ is known to be second order with respect to HI:
   a. Write a balanced equation for this reaction.
   b. Write the rate law for this reaction.
   c. When the concentration of HI is 2.00 M, the rate of decomposition is 1.58×10^{-2} M/s. Calculate the value and units of the rate constant.
   d. What is the rate of this reaction when the concentration of HI is 1.50 M?
31. The following reaction is first order with respect to NO and first order with respect to O_2, and the rate constant is $5.0 \times 10^4\ M^{-1}\ s^{-1}$:

$$2NO(g) + O_2(g) \rightarrow N_2O_4(g)$$

   a. Write the rate law for this reaction.
   b. What is the rate of the reaction when [NO] = 0.125 M and $[O_2] = 0.250$ M?
   c. What is the overall reaction order?
32. The following reaction is second order with respect to NO and first order with respect to H_2, and the rate constant is $3.5 \times 10^5\ M^{-2}\ s^{-1}$:

$$2NO(g) + 2H_2(g) \rightarrow N_2 + 2H_2O(g)$$

   a. Write the rate law for this reaction.
   b. What is the rate of the reaction when $[NO] = 2.5 \times 10^{-3}$ M and $[H_2] = 1.50 \times 10^{-3}$ M?
   c. What is the overall reaction order?

Section 13.4 Changes in Time—The Integrated Rate Law

Skill Review

33. Determine the rate constant for each of these first-order reactions:
   a. decomposition of peroxyacetyl nitrate; $t_{1/2} = 1920$s
   b. decomposition of sulfuryl chloride; $t_{1/2} = 525$ min
   c. radioactive decay of ^{40}K; $t_{1/2} = 1.25 \times 10^9$ years
   d. radioactive decay of ^{14}C; $t_{1/2} = 5730$ years

34. Determine the half-life for each of these first-order reactions:
    a. radioactive decay of ^{131}I; $k = 0.08619\ day^{-1}$
    b. radioactive decay of ^{24}Na; $k = 0.0473\ h^{-1}$
    c. decomposition of DDT; $k = 2.31 \times 10^{-4}\ day^{-1}$
    d. metabolism of malathion; $k = 0.693\ day^{-1}$
35. In the following first-order reaction, the half-life was determined to be 43 min. How long would it take for the concentration of A to drop,
    a. to 50% of the original amount? b. to 25% of the original amount? c. to 10% of the original amount?

$$A \rightarrow B$$

36. The radioactive decay of ^{14}C is a first-order reaction.
    a. If a sample originally contains 1.59×10^{-5} M ^{14}C, how long will it take before the concentration is 0.795×10^{-5} M? The half-life of ^{14}C is 5730 years.
    b. How long will it take before the concentration is 1.00×10^{-6} M?
37. For each case below, calculate the concentration of A at the time indicated. The reaction is first order with a half-life of 3.95×10^2 s.
    a. $[A]_0 = 0.100$ M; $\Delta t = 50.0$ s
    b. $[A]_0 = 0.100$ M; $\Delta t = 100.0$ s
    c. $[A]_0 = 0.200$ M; $\Delta t = 50.0$ s
    d. $[A]_0 = 0.200$ M; $\Delta t = 20.0$ s
38. For each case below, calculate the time required to reach the concentration shown. The reaction is first order with a half-life of 554 s.
    a. $[A]_0 = 0.100$ M; $[A] = 0.010$ M
    b. $[A]_0 = 0.350$ M; $[A] = 0.095$ M
    c. $[A]_0 = 0.200$ M; $[A] = 0.010$ M
    d. $[A]_0 = 0.250$ M; $[A] = 0.100$ M
39. What was the initial concentration of A if $[A] = 0.0388$ M after 2.75 days? Assume that the half-life of the first-order reaction is 1.18 days.

$$A \rightarrow B$$

40. Determine the rate constant (in s^{-1}) for the first-order decomposition of 2,4-D, if the initial concentration was 2.73×10^{-3} M and the concentration after 60 days was 8.44×10^{-4} M.
41. In the following cases of first-order reactions, something is wrong with the data. Determine which value is incorrect and indicate why.
    a. $[A]_0 = 0.100$ M; $[A]_t = 0.900$ M; $t_0 = 0$ s; $t = 10$ s
    b. $[A]_0 = 0.090$ M; $[A]_t = 0.010$ M; $t_0 = 10$ s; $t = 0$ s
42. In the following cases of first-order reactions, something is wrong with the data. Determine which value is incorrect and indicate why.
    a. $[A]_0 = 0.100$ M; $[A] = 0.050$ M; $t_0 = 0$ s; $t = 10$ s; $t_{1/2} = 24$ min
    b. $[A]_0 = 0.900$ M; $[A] = 0.450$ M; $t_0 = 0$ s; $t = 10$ s; $k = 0.085\ s^{-1}$
43. What is the half-life of the decomposition of ammonia on a metal surface, a zero-order reaction, if $[NH_3] = 0.0333$ M initially and $[NH_3] = 0.0150$ M at $t = 450$ s?
44. Determine the concentration of ammonia from Problem 43 when $t = 800.0$ s.
45. What is the value of the rate constant for a zero-order reaction with $t_{1/2} = 3.55$ h? Assume the original concentration $[A]_0 = 0.100$ M.
46. What is the value of the rate constant for a second-order reaction with $t_{1/2} = 300.0$ days? Assume the original concentration $[A]_0 = 1.50$ M.
47. What is the half-life of a second-order reaction if the concentration of A drops to 10% of its original value of 2.00×10^{-3} M in 520 min?

$$2A \rightarrow B$$

48. What is the concentration of A after 3.5 h in the second-order reaction illustrated in Problem 47, if $[A]_0 = 0.100$ M and $k = 1.2 \times 10^{-4}\ M^{-1} \cdot s^{-1}$?

49. Determine the concentration of the reactant in each of these cases. Assume that the reaction is second order with $k = 0.0312\ M^{-1}\ min^{-1}$.
    a. $[A]_0 = 0.100$ M; $t = 10.0$ s
    b. $[A]_0 = 0.500$ M; $t = 10.0$ s
    c. $[A]_0 = 0.339$ M; $t = 200.0$ s
    d. $[A]_0 = 0.0050$ M; $t = 24$ days
50. Determine the concentration of the reactant in each of these cases. Assume that the reaction is second order with $k = 0.410\ M^{-1}\ h^{-1}$.
    a. $[A]_0 = 0.100$ M; $t = 10.0$ h
    b. $[A]_0 = 0.500$ M; $t = 10.0$ h
    c. $[A]_0 = 0.222$ M; $t = 1.75$ h
    d. $[A]_0 = 0.0010$ M; $t = 14$ days
51. A graphical plot of concentration of a reactant versus time during a reaction will reveal, for reactions above zero order, a changing rate. Use collision theory to explain why the rate slows over time.
52. When we examine the three integrated rate expressions mentioned in the chapter, we can easily note that all three contain the symbol k. What would be the units for k in each of the zero-order, first-order, and second-order rate expressions? (If necessary, use any time unit merely as "time.")
53. Suppose two students are discussing their recent chemistry lab experiment. The first student, Pablo, remarks that his first-order reaction has a half-life of 25 min. The other student, Peter, replies that coincidently, his second-order reaction also has a half-life of 25 min. Then both go back to repeat their experiments, but with different amounts of starting materials. Neglecting experimental error, explain any differences, or lack of, that they might find this time.
54. a. What advantage does changing a reaction to pseudo-first order kinetics give to an experimenter?
    b. Describe how to alter a kinetics experiment in order to study it in the pseudo-first-order kinetic model.
    c. Explain why the concentration of one component in the pseudo-first-order model can be made to be part of the specific rate constant of a reaction.

Chemical Applications and Practices

55. An agricultural chemist is attempting to detect the decomposition of a new herbicide. The chemist notes that after application, the compound decays from 100.0% potency to 75.0% potency over a time period of 1 week (168 h). Assume that the original concentration $[A]_0 = 1.00 \times 10^{-3}$ M.
    a. What would be the specific rate constant if the reaction were first order with respect to the herbicide?
    b. What would be the specific rate constant if the reaction were second order with respect to the herbicide?
    c. What would be the specific rate constant if the reaction were zero order with respect to the herbicide?
56. Use the values obtained in Problem 55 to determine the half-life of the herbicide, assuming:
    a. first-order kinetics
    b. second-order kinetics
    c. zero-order kinetics
57. Assume that the fermentation of glucose by yeast, to produce ethanol, is a first-order process. Under certain conditions the value of the specific rate constant is $0.00205\ h^{-1}$. If the initial concentration of glucose were 0.980 M, what would be the concentration after 244 h of fermentation?
58. The precipitation of metal ions with sulfide is often used as an identifying technique for the metal ions. The production of hydrogen sulfide to be used in the precipitation, however, can be dangerous. Consequently, H_2S can be generated by placing thioacetamide (CH_3CSNH_2) in an aqueous acid solution. If the first-order decay constant for thioacetamide under those conditions were $0.46\ min^{-1}$, what would we calculate as the time required for 0.100 M thioacetamide to reach a concentration of 0.0100 M?
59. All nuclear decay processes (alpha, beta, and gamma decay) follow first-order kinetics. The isotope americium-241 is used in many smoke detectors. It has a half-life of approximately 241 years. How long would it take the americium in a smoke detector to decay (into neptunium) from its initial radiation level to 66.6% of its original value?
60. Explain why nuclear decay processes can be considered always to follow first-order kinetics?

Section 13.5 Methods of Determining Rate Laws

Skill Review

61. Plot the following data and determine whether the reaction follows zero-order, first-order, or second-order kinetics.

Time (s)	Concentration (M)
0	0.500
5	0.274
10	0.189
20	0.116
30	0.084

62. Plot the data in Problem 12 and determine whether the reaction follows zero-order, first-order, or second-order kinetics.
63. Explain the change in the rate of the reaction in each case below.

$$\text{Rate} = k[\text{A}][\text{B}]$$

a. We double [A].
b. We double [B].
c. We double [A] and [B].

64. Explain the change in the rate of the reaction in each case below.

$$\text{Rate} = k[\text{A}]^2[\text{B}]$$

a. We double [A].
b. We double [B].
c. We triple [A].

65. Use the method of initial rates to determine the order of each component, along with the general rate law and the value of the rate constant, in the following hypothetical reaction. What is the overall order of the reaction?

$$\text{A} + \text{B} \rightarrow 2\text{C}$$

Experiment	[A] (M)	[B] (M)	Initial rate (M/s)
1	0.10	0.10	0.222
2	0.10	0.20	0.444
3	0.20	0.20	0.444

66. Use the method of initial rates to determine the order of each component, along with the general rate law and the value of the rate constant, in the following hypothetical reaction. What is the overall order of the reaction?

$$\text{A} + \text{B} \rightarrow 2\text{C}$$

Experiment	[A] (M)	[B] (M)	Initial rate (M/s)
1	0.10	0.10	0.286
2	0.10	0.20	0.143
3	0.20	0.20	0.286

67. Use the method of initial rates to determine the order of each component, along with the general rate law and the value of the rate constant, in the following hypothetical reaction. What is the overall order of the reaction?

$$\text{A} + 2\text{B} \rightarrow 2\text{C}$$

Experiment	[A] (M)	[B] (M)	Initial rate (M/s)
1	0.10	0.10	0.105
2	0.10	0.20	0.420
3	0.20	0.20	0.840

68. Use the method of initial rates to determine the order of each component, along with the general rate law and the value of the rate constant in the following hypothetical reaction. What is the overall order of the reaction?

$$2A + 2B \rightarrow 2C$$

Experiment	[A] (M)	[B] (M)	Initial rate (M/s)
1	0.050	0.10	0.074
2	0.10	0.20	0.888
3	0.050	0.20	0.222

69. Use the method of graphical analysis to determine the order for the hypothetical reaction $A \rightarrow C + D$. What is the rate constant (with appropriate units) for this reaction?

[A] (M)	Time (min)
0.432	0
0.385	1
0.291	3
0.197	5
0.103	7

70. Use the method of graphical analysis to determine the order for the hypothetical reaction $A \rightarrow B$. What is the rate constant (with appropriate units) for this reaction?

[A] (M)	Time (s)
0.100	0
0.050	62
0.025	124
0.013	186
0.0065	248

71. The following two diagrams represent two different containers. Within each container, two different compounds are placed.

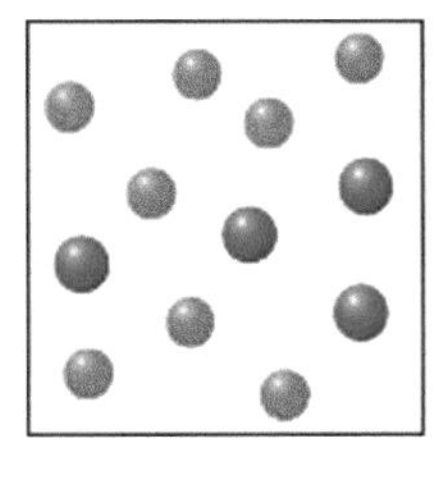
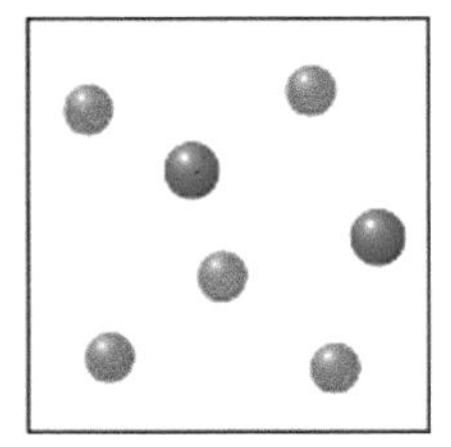

a. If the reaction about to take place is first order with respect to the large dark atoms and zero order with respect to the small light atoms, which graphical plot would yield a straight line for each component?
b. How would the overall reaction rate compare between the first and second containers?
c. What units would we assign to the rate constant for the reaction between the two components? (Please use seconds for the time unit.)

72. The following diagram represents a container with two gas components in their initial conditions. The reaction is zero order with respect to the large atoms and first order with respect to the small atoms. Prepare a similar diagram but show what would be needed to produce a reaction that would be three times faster than the first (assuming no change in temperature or any other reaction conditions).

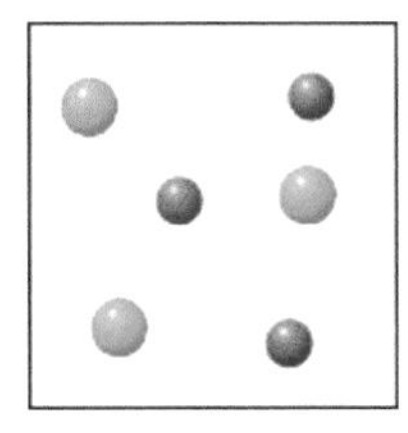

Chemical Applications and Practices

73. The reaction below was monitored by noting the concentration of chlorine that was formed as the reaction proceeded.

$$H_2O + COCl_2 \rightarrow CO_2 + 2HCl$$

Initially $[COCl_2]_{initial} = 0.100$ M, and the following data were obtained.

t (min)	[HCl] M
0.00	0.000
1.00	0.019
2.00	0.036
3.00	0.052
4.00	0.066
5.00	0.078
6.00	0.090
8.00	0.110
10.00	0.126
15.00	0.155
20.00	0.173

a. Create plots of [HCl] and $[COCl_2]$ versus *t*.
b. Determine the rate law for this reaction.
c. At what time has 50% of the $COCl_2$ reacted?
d. Plot the pH of the solution as a function of time.

74. The molecule HOF decomposes over time into HF and oxygen gas.

$$2HOF(g) \rightarrow 2HF(g) + O_2(g)$$

The initial concentration of HOF was 0.75. The concentration was determined as a function of time:

time (min)	[HOF] (M)
0	0.75
2	0.712
5	0.668
20	0.489
60	0.201

a. Create plots of [HF] and $[O_2]$ versus *t*.
b. Determine the rate law for this reaction.
c. At what time has 50% of the HOF reacted?

75. Municipal water supplies are often treated with chlorine compounds in order to take advantage of the oxidizing ability of chlorine to react with potential pollutants. In some industrial settings, other oxidizing substances are used. One example of such an application is use of the powerful oxidizing potential of iron(VI) compounds. Given the information below, what order would we report for the reaction of the iron(VI) compound? What would we report as the value for *k*?

$[Fe^{6+}]$ (M)	Time (min)
0.100	0.00
0.0682	1.00
0.0517	2.00
0.0300	5.00
0.00920	10.00

76. The gas phase hydrogenation of ethene is shown in the following reaction:

$$C_2H_2(g) + 2H_2(g) \rightarrow C_2H_6(g)$$

Experiment	$P_{C_2H_2}$ (atm)	P_{H_2} (atm)	Initial rate (atm/s)
1	0.10	0.10	11
2	0.20	0.10	22
3	0.10	0.20	22
4	0.20	0.05	11

a. If the following data were collected, using the method of initial rate method, for four experiments at a fixed temperature, what would we determine as the rate law for the reaction?
b. What would we calculate as the rate constant for the reaction?
c. If the experiment were run once again, at the same temperature, with initial pressures of 0.020 atm for ethane and 0.020 atm for hydrogen, what would we determine as the rate?

77. Suppose the following represents the kinetic study of the oxidation of a new plastic stabilizer being proposed for use in automobile upholstery.

$$\text{Stabilizer}(aq) + H^+(aq) + O_2(g) \rightarrow \text{oxidation product}$$

Experiment	Stabilizer (*aq*) (M)	$H^+(aq)$(M)	$O_2(g)$ (M)	Rate (M/s)
1	0.400	0.300	0.560	7.14×10^{-4}
2	0.100	0.500	0.200	4.55×10^{-5}
3	0.100	0.100	0.200	4.55×10^{-5}
4	0.400	0.300	0.750	1.28×10^{-3}
5	0.100	0.300	0.560	3.57×10^{-4}

a. What is the rate law for the reaction?

b. What is the rate constant for the reaction at this temperature?
c. What would we calculate as the rate of the reaction, at the same temperature, if the respective initial concentrations were 0.111, 0.200, and 1.00 for the stabilizer, hydrogen ion, and oxygen, respectively?
d. If the rate of the same reaction, at the same conditions, were determined to be 2.55×10^{-4} M/s when the initial concentration of stabilizer was 0.300 M with a hydrogen ion concentration of 0.100 M, what would the initial oxygen pressure have to be?

78. Brewed coffee left on a warming device will gradually change flavor as a consequence of several complex reactions that occur as the flavor oils decompose. Suppose a flavor chemist for Dr. Beans Coffee Company collected the following data on one such oil as it decomposed over time.

Concentration (M)	Time (min)
0.00128	0
0.00120	10.0
0.00113	20.0
0.00107	30.0
0.000781	100.0
0.000561	200.0

a. Using graphical techniques, determine the order of the decay process.
b. What is the rate constant for the decay?
c. Starting with the initial concentration, how long would it take for half the ingredient to decompose (half-life)?

79. The naturally occurring isotope of hydrogen known as tritium has two neutrons and one proton. The reaction between tritium and deuterium may provide the basis for controlled fusion reactions. The half-life of the radioactive tritium isotope is approximately 12 years. How much radioactive tritium, starting with a sample producing a beta decay count rate of 545 cpm (counts per minute), would remain after 3 years?

80. Sodium hypochlorite (NaOCl) is the active ingredient in many aqueous commercial bleaching products. If the NaOCl added to a particular sewage treatment process had a half-life of 15 days, what percentage of the original concentration would remain after 1 year? (Assume first-order kinetics.)

Section 13.6 Looking Back at Rate Laws

Skill Review

81. What mathematical effect would there be on the three graphs for first-, second-, and third-order, respectively, if before plotting the terms on each y axis we first subtracted the initial concentration, ln of the initial concentration, and 1/(initial concentration), respectively, from each reading?
82. A researcher mistakenly plots the data from a first-order reaction using log [A] instead of ln [A]. What effect would this have on the resulting graph?
83. What sign do we expect for the slope of the line in a plot of [A] versus t for a zero-order reaction?
84. What sign do we expect for the slope of the line in a plot of ln[A] versus t for a first-order reaction?
85. Which indicates a faster reaction for a first-order process: $k = 1.66 \times 10^{-2}\ s^{-1}$ or $k = 8.95 \times 10^{-2}\ s^{-1}$?
86. Which indicates a faster reaction for a first-order process: $t_{1/2} = 24$ days or $t_{1/2} = 15$ days?
87. Why does the half-life expression for a first-order reaction have a value of 0.693 in it?
88. In the expression for the half-life of a second-order reaction, why is there not a factor of ½ anywhere?
89. If we're performing an experiment that has two reactants, the concentration of both reactants will obviously change throughout the course of the reaction. How do we examine the effect of changing the concentration of just one of the reactants?
90. If an aqueous second-order reaction involves water and another compound as reactants, the reaction can be assumed to be a pseudo-first order reaction. Explain.

Chemical Applications and Practice

91. The decay of m-amsacrine, a useful anti-cancer drug, is first-order with a half-life of only 23 min in blood. Assuming that only 4 molecules of m-amsacrine were present in a patient 23 min ago, how many molecules exist in the patient currently? How many molecules will be present in the patient 23 min from now? ...in 46 min?
92. Selenium-82 has an extremely long half-life: 1.08×10^{20} years. Assuming the age of the earth to be about 4.5 billion years, and assuming that there were two atoms of selenium-82 present when the earth was formed, how many atoms of selenium-82 would be present in the earth today?
93. One method of producing yellow lead chromate pigment is via a reaction of sodium chromate. Chromium(III) ions can be made to undergo a three-electron oxidation to the chromate ion (CrO_4^{2-}) with the addition of cerium(IV) ions. The rate law was found to be

$$\text{Rate} = k\left[Cr^{3+}\right]^2\left[Ce^{4+}\right]\left[CrO_4^{2-}\right]^x$$

Suppose the concentration of each component in the rate law were doubled. If this caused the rate to quadruple, what would we calculate as the value of x?

94. An approximation that chemists may sometimes casually use is that increasing the temperature of a system by 10 degrees Celsius doubles the rate of a reaction. Suppose we wanted to increase a reaction rate by using only concentration changes. How much would a concentration have to change to cause a reaction to become ten times faster? (Assume the reaction is first order.)

Section 13.7 Reaction Mechanisms

Skill Review

95. What does the term "bimolecular reaction" mean?
96. Everything else being equal, which reaction do we think would be the slowest: a unimolecular, bimolecular or termolecular reaction? Explain.
97. Explain the relationship among a reaction mechanism, elementary steps and the rate-determining step.
98. Explain how it is possible for the proposed mechanism of a reaction to be acceptable, yet perhaps not represent the actual occurrence of events in the mechanism of that reaction.

99. We are seated at the kitchen table. Describe each of the steps involved in answering the telephone in the front room (be very detailed). Which of these steps is the rate-determining step for this process?
100. What is the rate-determining step in the process used to go to chemistry class from home?
101. Define "transition state theory".
102. How is transition state theory different from collision theory?
103. What three factors directly influence the size of the rate constant for a given chemical system.
104. A temperature increases, what does collision theory predict about the speed of a reaction? Does this agree with Arrhenius' equation?
105. What is the activation energy for a hypothetical reaction ($A \rightarrow B$) if $k = 1.74 \times 10^{-2}\ s^{-1}$ at 300 K and $k = 4.22 \times 10^{-2}\ s^{-1}$ at 400 K?
106. Calculate the rate constant of a first-order reaction at 35 °C, given that $k = 8.5 \times 10^{-4}\ s^{-1}$ at 25 °C and $E_a = 144$ kJ.

Chemical Applications and Practices

107. The aqueous reaction between hydrogen peroxide and iodide in an acid solution is represented here:

$$H_2O_2 + 3I^- + 2H^+ \rightarrow 2H_2O + I_3^-$$
$$\text{Rate} = k[H_2O_2][I^-][H^+]$$

One of the following mechanisms can be accepted for the above rate law, and one cannot. Select the acceptable mechanism and show proof for the selection as well as giving the basis for rejection of the other.

Mechanism A		Mechanism B	
$H^+ + I^- \rightarrow HI$	(fast)	$H_2O_2 + I^- \rightarrow H_2O + OI^-$	(slow)
$H_2O_2 + HI \rightarrow H_2O + HOI$	(slow)	$H^+ + I^- \rightarrow HI$	(fast)
$HOI + H^+ + I^- \rightarrow H_2O_+ I_2$	(fast)	$H^+ + OI^- + HI \rightarrow H_2O_+ I_2$	(fast)
$I^- + I_{2\rightarrow} I_3^-$	(fast)	$I_2 + I^- \rightarrow I_3^-$	(fast)

108. When one reactant produces two products, the reaction is said to be a disproportionation reaction. The following shows the disproportionation of the hypochlorite ion used to make commercial bleaching products:

$$3ClO^-(aq) \rightarrow 2Cl^-(aq) + ClO_3^-(aq)$$

On the basis of the following proposed mechanism, what would we write as the rate law for this interesting reaction?

$$ClO^- + ClO^- \rightarrow ClO_2^- + Cl^- \quad \text{(slow)}$$
$$ClO^- + ClO_2^- \rightarrow ClO_3 + Cl^- \quad \text{(fast)}$$

109. The solvent acetone has many industrial uses and is also a common ingredient in nail polish remover. It does decompose to a weak acid, carbonic acid. At 10.0 °C, the rate constant for that decomposition is approximately 6.4×10^{-5} L/mol s. At 78 °C, the rate constant for the reaction has a value of 2.03×10^{-1} L/mol · s.
    a. What is the activation energy for this reaction?
    b. What is the rate constant for this reaction at 37 °C?
110. At 78 °C, what would we calculate as the value of the Arrhenius frequency factor, A, for the decomposition of acetone (Problem 109)?

Section 13.8 Applications of Catalysts

Skill Review

111. Provide clear and concise definitions for each term:
    a. Reaction intermediate
    b. Activated complex
    c. Homogeneous catalyst
    d. Heterogeneous catalyst
112. a. Explain how a catalyst increases the rate of a chemical reaction.
    b. Explain why a catalyst does not appear as part of the stoichiometry in a balanced equation.

113. Consider the following elementary steps for the decomposition of N_2O.

$$N_2O \rightarrow N_2 + O$$
$$N_2O + O \rightarrow N_2 + O_2$$

a. What is the overall reaction?
b. Indicate any intermediates or catalysts involved in the reaction.
c. Experimental evidence reveals that the rate law for this reaction is Rate$=k[N_2O]$. Which of the steps is the rate limiting step in the mechanism?

114. The following mechanism has been proposed for the reaction of ICl(g) and H_2(g).

$$H_2 + ICl \rightarrow HI + HCl$$

$$HI + ICl \rightarrow I_2 + HCl$$

a. What is the overall reaction?
b. Indicate any intermediates or catalysts involved in the reaction.
c. The first step in the mechanism is slow compared to the second. What is the rate law for the reaction?

115. A schematic mechanism for the reaction of lactose and lactase (an enzyme) to produce glucose and galactose is shown below.

$$\text{lactose} + \text{lactase} \rightarrow (\text{lactase} - \text{lactose})$$
$$(\text{lactase} - \text{lactose}) \rightarrow (\text{lactase} - \text{glucose} - \text{galactose})$$
$$(\text{lactase} - \text{glucose} - \text{galactose}) \rightarrow \text{lactase} + \text{glucose} + \text{galactose}$$

a. What is the overall reaction?
b. Indicate any intermediates or catalysts involved in the reaction.
c. The first step in the mechanism is slow compared to the rest. What is the rate law for the reaction?

116. Each of the following represents an elementary step in a different mechanism. Classify each as unimolecular, bimolecular, or termolecular.

a. $^{40}K \rightarrow {}^{40}Ar$
b. $N_2 + Fe \rightarrow FeN_2$
c. $2NO \rightarrow N_2O_2$
d. $2NO + Cl_2 \rightarrow 2NOCl$

Chemical Applications and Practices

117. The industrial production of ammonia makes use of catalysts. One generalized mechanism may be presented as follows:

$$N_2(g) + Fe(s) \rightarrow FeN_2(s)$$

$$3H_2(g) + 3Fe(s) \rightarrow 3FeH_2(s)$$

$$FeN_2(s) + 3FeH_2(s) \rightarrow 4Fe(s) + 2NH_3(g)$$

a. What is the overall stoichiometry of the reaction?
b. What intermediates, if any, are present?
c. What catalyst is present?
d. Is the catalyst homogeneous or heterogeneous?

118. Diagram a reaction profile that illustrates the exothermic reaction whose mechanism is shown here. Next use a dotted line to show the same profile with any changes that would reflect the role of a homogeneous catalyst.

$$A(aq) + B(aq) \rightarrow C(aq) \quad (\text{slow})$$
$$C(aq) \rightarrow D(g) \quad (\text{fast})$$

❓ Comprehensive Problems

119. In the reaction shown below, the initial concentration of H_2O_2 is 0.250 M, and 8 s later the concentration is 0.223 M. What is the initial rate of this reaction expressed in M/s and M/h?

$$2H_2O_2(aq) \rightarrow 2H_2O(l) + O_2(g)$$

120. What is the half-life for the first-order decomposition of dimethyl ether at 500 °C if the rate constant for the reaction is 2.567×10^{-2} min^{-1}?

$$(CH_3)_2O \rightarrow CH_4 + H_2 + CO$$

121. A first-order reaction, A → B, has a rate of 0.0875 M/s when [A] = 0.250 M.
   a. What is the rate constant?
   b. What is the half-life for this reaction?
122. If the order of a reaction component were − 1, what would be the effect of tripling the concentration of that component?
123. If the $\Delta H°$ value for a reaction were + 125 kJ and the activation energy for the reverse of the reaction were 75 kJ, what would we calculate as the value for the activation energy for the forward reaction?
124. Consider the following reaction mechanism.

$$C_2H_6O(aq) + HCl(aq) \rightleftharpoons C_2H_7O^+(aq) + Cl^-(aq) \text{ (fast)}$$
$$C_2H_7O^+(aq) + Cl^-(aq) \rightarrow C_2H_5Cl(aq) + H_2O(l) \text{ (slow)}$$

   a. Write the overall reaction that is indicated by the mechanism.
   b. What would we predict as the rate law for the reaction?
125. Consider the following reaction mechanism.

$$C_2H_4(aq) + HCl(aq) \rightarrow C_2H_5^+(aq) + Cl^-(aq) \text{ (slow)}$$
$$C_2H_2^+(aq) + Cl^-(aq) \rightarrow C_2H_5Cl(aq) \text{ (fast)}$$

   a. Write the overall reaction that is indicated by the mechanism.
   b. What would we predict as the rate law for the reaction?

Thinking Beyond the Calculation

126. Draw a hypothetical reaction profile for the Haber process involving 1 mol of nitrogen and 3 mol of hydrogen in a 1.0 L flask.

$$N_2(g) + 3H_2(g) \rightarrow 2NH_3(g)$$

   a. Use information we have learned in other chapters to determine whether the reaction is exothermic or endothermic.
   b. Is this reaction spontaneous at room temperature? At what temperature is the reaction at equilibrium?
   c. On the reaction profile we drew, indicate what we'd expect to see if the reaction were homogeneously catalyzed.
   d. Draw what we'd expect to see if the reaction were heterogeneously catalyzed.
   e. If the catalyzed reaction were accomplished at 300 °C using the quantities outlined in the start of this question, what would be the yield, in grams, of ammonia ($NH_3(g)$)? Assume the reaction produces only a 45% yield.
   f. If 10.0 kg of nitrogen and 30.0 kg of hydrogen were combined in the catalyzed reaction, what would be the theoretical yield, in kilograms, of ammonia?

Chemical Equilibrium

Contents

Supplementary Information The online version contains supplementary material available at ▶ https://doi.org/10.1007/978-3-030-90267-4_14.

M. Mosher and P. Kelter, *An Introduction to Chemistry*,
https://doi.org/10.1007/978-3-030-90267-4_14

What We Will Learn from This Chapter

- A reaction at equilibrium is dynamic. The forward reaction and the reverse reaction continue to operate but the concentrations of the individual components remain constant (▶ Sect. 14.2).
- Chemical equilibria are very common (▶ Sect. 14.3).
- The equilibrium constant is a measure of the extent to which a reaction proceeds toward products (▶ Sect. 14.4).
- Adjusting the stoichiometry of a reaction, or adding multiple reactions together adjusts the equilibrium constant (▶ Sect. 14.5).
- The ICEA table is a good bookkeeping tool that aids in determining the concentration of a component of a reaction at equilibrium (▶ Sect. 14.6).
- The concentrations of reactants and products change in response to a stress applied to a reaction at equilibrium (▶ Sect. 14.7).

14.1 Introduction

"I'm running late." We seem to hear that expression so often in our hypercharged, have-to-do-it-now world. Running late might mean being stuck in traffic or on mass transit. There is not much we can do in a sea of cars and trucks. But sometimes we actually do run to get where we are going. As we struggle to make up time, we breathe more heavily than normal, desperately sucking air into our lungs. Our body's efforts to get more oxygen to the muscles are greatly enhanced by a protein in muscle cells called myoglobin. When we are resting, the myoglobin becomes loaded up with oxygen molecules. These oxygen molecules can be quickly released when our muscle cells undergo activity. How does the myoglobin acquire and release oxygen when we need it?

Myoglobin (Mb), shown in ◘ Fig. 14.1, is a protein with a molecular weight of about 16,900 g/mol. A key part of the protein is the **heme group**, shown in ◘ Fig. 14.2. The heme contains an Fe^{2+} ion that can form a bond with one molecule of oxygen, creating a myoglobin–oxygen complex as follows:

$$\mathrm{Mb} + \mathrm{O_2} \rightarrow \mathrm{MbO_2}$$

Myoglobin (Mb) – A biochemical compound responsible for storing and releasing oxygen in a living organism

Heme group – A compound that, when bound to hemoglobin and iron cations, is responsible for binding oxygen.

◘ **Fig. 14.1** Myoglobin (Mb) is a protein with a molecular weight of about 16,900 g/mol. A key part of the protein is the heme group, highlighted on the right, which can bind to an oxygen molecule such as that shown in red. *Source* ▶ https://upload.wikimedia.org/wikipedia/commons/6/60/Myoglobin.png

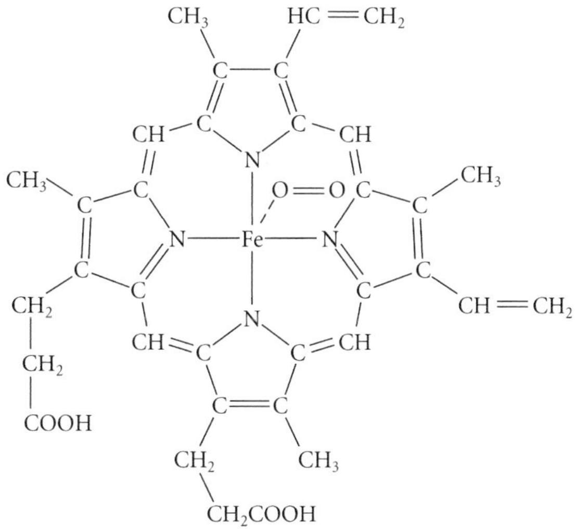

■ **Fig. 14.2** This drawing of a heme group highlights the Fe^{2+} ion, which can bind to one molecule of oxygen as shown

When our cells need the oxygen, it is released from the complex by reversing the direction of the reaction:

$$MbO_2 \rightarrow Mb + O_2$$

The free myoglobin is then available to bind more oxygen, which it can later release as needed, and continue in this reversible cycle of bind–release–bind–release. We can examine this reaction more closely to help us understand the concepts of reversible reactions and the essential details of reactions such as this, which are known as chemical equilibria.

14.2 The Concept of Chemical Equilibrium

Why does myoglobin's reversible cycle of binding and releasing oxygen take place? Just as a falling object or a pan balance reaches a stable state at the lowest achievable energy, chemical systems also proceed to a state of lowest energy, called "free energy". We'll learn later in this text that the concept of "free energy" is more complicated than our normal idea of energy. When we consider myoglobin and oxygen in living systems, we find that the two competing reactions exist. In one, myoglobin combines with oxygen. In the other, the MbO_2 complex releases oxygen. The end result is that the molecules in each reaction will never be completely converted into products, because these products are always being converted back into reactants. These reversible reactions will settle into a state in which the forward and reverse reactions occur at equal rates, where no net change in the concentrations occurs, even though both reactions continue. We describe this state as the position of dynamic chemical equilibrium, or just **equilibrium** (the plural is equilibria).

Equilibrium – A system of reversible reactions in which the forward and reverse reactions occur at equal rates, such that no net change in the concentrations occurs, even though both reactions continue.

What makes chemical equilibrium different from the equilibrium state reached by a balance? There are several things, but the main difference is that chemical systems are always in motion. Even when the concentrations of reactants and products look like they aren't changing, they aren't static. Reactants are still turning into products and products are still turning back into reactants, but since this happens simultaneously, it is difficult to perceive anything happening in the system, except with very specialized analytical equipment. Think of a teller at a bank that begins and ends the day with $20,000 in her cash drawer. During the day, she may be performing numerous transactions that add and subtract from the cash drawer. But by the end of the day, there's still $20,000 in the drawer.

We symbolize a system that can settle into equilibrium by using double arrows, to represent the fact that both the forward and reverse reactions occur. For example, the myoglobin and oxygen reaction equilibrium is written

$$Mb + O_2 \rightleftharpoons MbO_2$$

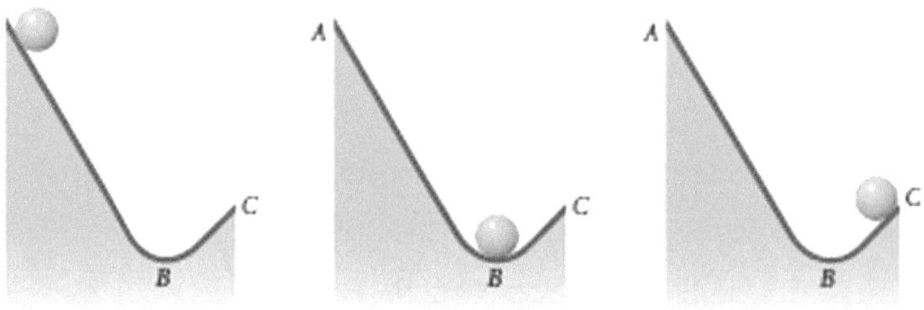

Fig. 14.3 Like a ball finding its lowest energy point in a valley, reactions find their lowest energy point. This is the equilibrium position (B). Unlike this ball, however, chemical systems are dynamic. Even at equilibrium, the forward and backward reactions are always working

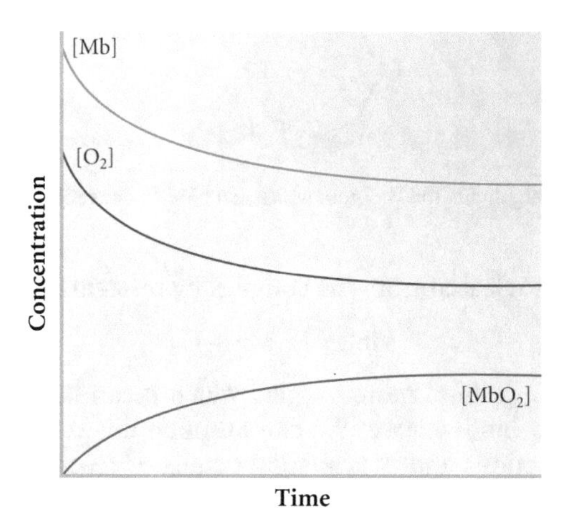

Fig. 14.4 When myoglobin and oxygen are mixed, the concentration of oxygen decreases, as does that of free myoglobin, as shown. The concentration of the MbO_2 complex increases from 0_M. As soon as the MbO_2 complex forms, however, it begins to dissociate back into Mb and O_2

As we have done previously, the molar concentration of a substance will be expressed by placing its chemical symbol in brackets, [ ], such as [Mb] or [MbO_2].

We can examine what happens to the concentrations of the reactants and products as the lowest energy point is approached (see Fig. 14.3). To make things as straightforward as possible, we will consider the reaction occurring in a beaker, rather than in the human body where all kinds of complications (such as "running late") affect the chemistry.

Let's assume that we place oxygen in our beaker so that its concentration is 3.0×10^{-4} M. We would say $[O_2]_0 = 3.0 \times 10^{-4}$ M, where the subscript 0 means the concentration at "time = 0," which is another way of saying the initial concentration. In fact, the amount of myoglobin normally present in muscle cells varies a great deal among, and within, biological species. In humans, a mid-range value is $[Mb]_0 = 2.0 \times 10^{-4}$ M. We will assume for this discussion that we have a closed system (not the human body) and that there is no MbO_2 complex initially, so $[MbO_2]_0 = 0$ M. As the reaction proceeds, oxygen binds to myoglobin. The forward reaction proceeds with a rate constant of k_1, known in this reaction as the **binding rate constant**.

$$\mathrm{Mb} + \mathrm{O_2} \xrightarrow{k_1} \mathrm{MbO_2}$$

Binding rate constant – The rate constant that indicates the association of two molecules.

The concentration of oxygen decreases, as does that of free myoglobin, as shown in Fig. 14.4. The concentration of the MbO_2 complex increases from zero as the reaction proceeds. As soon as the MbO_2 complex forms, however, it begins to dissociate back into Mb and O_2 with a rate constant of k_{-1}, known in this reaction as the **release rate constant**.

Release Rate Constant – The rate constant of the reaction in which oxygen is released from the myoglobin-oxygen reaction.

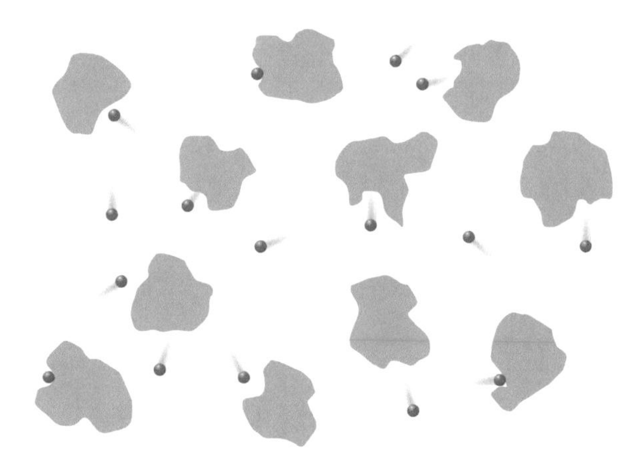

Fig. 14.5 Equilibrium is a dynamic process. The forward and reverse reactions are occurring at equal rates. The result is no net change in concentration of Mb (the blue shapes), O_2 (the red dots), or MbO_2 at equilibrium

$$MbO_2 \xrightarrow{k_{-1}} Mb + O_2$$

Both reactions proceed until the system has reached the minimum energy, which occurs when the rates of the forward and reverse reactions are equal. That statement embodies two key ways in which we define chemical equilibrium.

14.2.1 Definitions of Equilibrium

A chemical system that is at equilibrium can be defined in the following two ways:

1. The concentrations of reactants and products are constant.
2. The rates of the forward and reverse reactions are equal. This means that the rate constant expressions of the forward and reverse reactions are equal to each other.

$$\text{rate}_f = \text{rate}_r$$
$$k_1[Mb][O_2] = k_{-1}[MbO_2]$$

This can be graphically illustrated as in Fig. 14.5. The forward and reverse reactions that bind and release oxygen are still occurring. In fact, they are occurring at the same rate. Therefore, equilibrium is a dynamic process. And there is no net change in the concentration of Mb, O_2, or MbO_2 at equilibrium.

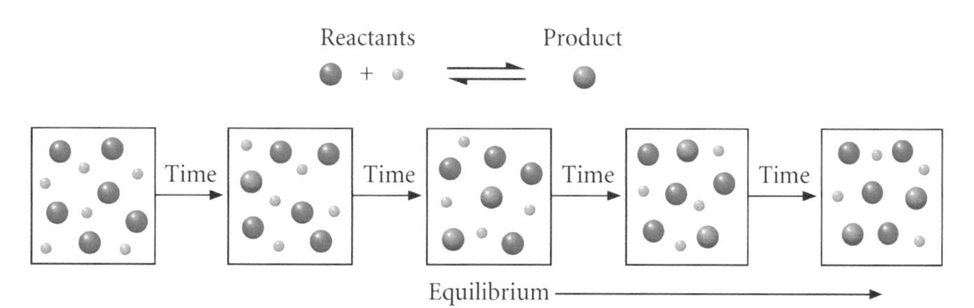

Example 14.1—Equilibrium Concentrations

Hydrogen and iodine gases can be combined to form hydrogen iodide,

$$H_2(g) + I_2(g) \rightleftharpoons 2HI(g)$$

Given these initial and final concentrations, calculate the equilibrium concentration of HI.

$[I_2]_0 = 0.037$ M $[H_2]_0 = 0.121$ M
$[HI]_0 = 0.099$ M $[I_2] = 0.022$ M
$[H_2] = 0.106$ M $[HI] = ???$ M

Asking the Right Questions

Note the 1-to-1-to-2 mol ratio for the reactants and product in the balanced chemical equation. How is this useful in our calculation? How do the changes in the concentrations of the I_2 and H_2 guide us toward the equilibrium concentration of HI?

Solution

The mole ratios of reactants to product are 1-to-1-to-2, so 1 mol each of H_2 and I_2 react to produce 2 mol of HI in a constant volume. Because the volume remains constant, we can think of the reaction stoichiometry this way: 1 M H_2 and 1 M I_2 react to produce 2 M HI. The concentrations of I_2 and H_2 both decrease by 0.015 M.

$$0.015\,\mathrm{M\,H_2\,(or\,I_2)} \times \frac{2\,\mathrm{M\,HI}}{1\,\mathrm{M\,H_2(or\,I_2)}} = 0.030\,\mathrm{M\,HI\,produced}$$

$$[\mathrm{HI}] = [\mathrm{HI}]_0 + 0.030 = 0.13\,\mathrm{M}$$

Is Our Answer Reasonable?

Perhaps the first test of "reasonable" is that the concentration of the product, [HI], should have increased, because that of each reactant decreased. The hydrogen iodide concentration did increase. In addition, the [HI] increased twice as much as $[H_2]$ and $[I_2]$, which is also reasonable, given the 1-to-1-to-2 mol ratios of reactants and product.

What If…?

…in a separate experiment, we had the same starting concentrations, but the final [HI] = 0.35 M? What are the final $[I_2]$ and $[H_2]$?

(*Ans:* $[I_2]=0.05\ M$ *and* $[H_2]=0.089\ M$)

Practice 14.1

Calculate the equilibrium concentration of HI if $[I_2]_0=0.050$ M, $[H_2]_0=0.044$ M, $[HI]_0=0.206$ M, and $[I_2]=0.041$ M.

It has been shown by experiment that the equilibrium concentrations of myoglobin, oxygen, and the MbO_2 complex are related by the ratio of the product to reactant concentrations known as the **mass-action expression** or, often, the **equilibrium expression**, of the reaction.

Mass-Action Expression – The ratio of product to reactant concentrations raised to the power of their stoichiometric coefficients. This expression relates the equilibrium concentrations to the equilibrium constant. It is also known as the equilibrium expression.

Equilibrium Expression – The ratio of product to reactant concentrations raised to the power of their stoichiometric coefficients. This expression relates the equilibrium concentrations to the equilibrium constant. It is also known as the mass-action expression.

$$K = \frac{[\mathrm{MbO_2}]}{[\mathrm{Mb}][\mathrm{O_2}]}$$

The value K is called the **equilibrium constant** for the reaction at a given temperature. We can get the equivalent result by separating the rate constants from the concentrations in the rate expressions as follows. At equilibrium, the rates of the forward and reverse reaction are equal.

Equilibrium Constant (K) – The value of the equilibrium expression when it is solved using the equilibrium concentrations of reactants and products.

$$k_1[\mathrm{Mb}][\mathrm{O_2}] = k_{-1}[\mathrm{MbO_2}]$$

Rearranging the equation to put the rate constants on one side gives

$$\frac{k_1}{k_{-1}} = \frac{[\mathrm{MbO_2}]}{[\mathrm{Mb}][\mathrm{O_2}]}$$

Therefore, the equilibrium constant, K, is the ratio of the forward rate constant to the reverse rate constant.

$$K = \frac{k_1}{k_{-1}}$$

The value for k_1 in the myoglobin–oxygen reaction at 20 °C is $1.9 \times 10^7\mathrm{M^{-1}} \cdot \mathrm{s^{-1}}$, and k_{-1} is 22 $\mathrm{M^{-1}} \cdot \mathrm{s^{-1}}$, so we can calculate the value of K:

$$K = \frac{k_1}{k_{-1}} = \frac{1.9 \times 10^7}{22} = 8.6 \times 10^5$$

We typically would express a calculation such as this by including the units. However, the equilibrium constant does not have units. This is true because the actual thermodynamic definition of an equilibrium constant is based on the ratio of a substance's concentration to a standard state concentration. Even though the reasoning is subtle, the bottom line is that the equilibrium constant is typically given without units. In the reaction of myoglobin and oxygen, K is simply 8.6×10^5.

The last two results—that there exists a mass-action expression to find the equilibrium concentrations of Mb, O_2, and MbO_2, and that there exists an equilibrium constant for this expression at a particular temperature—can be extended to any reversible reaction. For example, in a general equilibrium (reactants A and B yielding products C and D) represented with stoichiometric coefficients m, n, p, and q,

$$m\text{A} + n\text{B} \rightleftharpoons p\text{C} + q\text{D}$$

The general mass-action expression is

$$K = \frac{[\text{C}]^p[\text{D}]^q}{[\text{A}]^m[\text{B}]^n}$$

Example 14.2 explains why each equilibrium concentration is raised to the power of its coefficient in the mass-action expression.

Example 14.2—The Mass-Action Expression

The production of ammonia for use in fertilizers, via the Haber process, can be depicted as

$$\text{H}_2(g) + \text{H}_2(g) + \text{H}_2(g) + \text{N}_2(g) \rightleftharpoons \text{NH}_3(g) + \text{NH}_3(g)$$

Write the mass-action expression for the formation of ammonia.

Asking the Right Questions

What is a mass-action expression? How do we write one based on the reactants and products of an equation?

Solution

A mass-action expression is the ratio of the product to reactant concentrations. If we base the mass-action expression on the reaction exactly as it is written, we get this mass-action expression:

$$K = \frac{[\text{NH}_3][\text{NH}_3]}{[\text{H}_2][\text{H}_2][\text{H}_2][\text{N}_2]} = \frac{[\text{NH}_3]^2}{[\text{H}_2]^3[\text{N}_2]}$$

The exponential form on the right-hand side is what we would have obtained had we written the Haber process in the usual way:

$$3\text{H}_2(g) + \text{N}_2(g) \rightleftharpoons 2\text{NH}_3(g)$$

This shows why the mass-action expression includes the equilibrium concentration of each substance raised to the power of its coefficient. If the coefficient is 1, as with N_2, it is not shown as an exponent, because raising anything to the power of 1 does not change its value.

Is Our Answer Reasonable?

The mass-action expression is in the proper form of product and reactant concentration ratios, taking into account the coefficients (which are raised to a power equal to the coefficient in the balanced chemical equation), so the answer is reasonable.

What If…?

…we had this equilibrium (mass-action) expression:

$$K = \frac{[\text{H}_2\text{O}]^2[\text{CO}_2]}{[\text{CH}_4][\text{O}_2]^2}$$

What is the balanced chemical equation that results in such a equilibrium expression?

(*Ans:* $CH_4(g)+2O_2(g) \rightleftharpoons CO_2(g)+2H_2O(g)$)

Practice 14.2

One of the ways to produce hydrogen gas industrially is with the reaction of methane with high-temperature steam. The reaction can be written as

$$\text{CH}_4(g) + \text{H}_2\text{O}(g) \rightleftharpoons \text{CO}(g) + \text{H}_2(g) + \text{H}_2(g) + \text{H}_2(g)$$

Write the mass-action expression for this reaction.

Example 14.3—Practice with Mass-Action Expressions

Write the mass-action expression for each of these reactions:

a. $PCl_5(g) \rightleftharpoons PCl_3(g) + Cl_2(g)$
b. $S_8(g) \rightleftharpoons 8S(g)$
c. $Cl_2O_7(g) + 8H_2(g) \rightleftharpoons 2HCl(g) + 7H_2O(g)$

Asking the Right Questions

As with the previous exercise, the key question is, "What is a mass-action expression? How do we write one based on the reactants and products of an equation?" In particular, note the coefficients and their place in the mass-action expression.

Solution

In each case, the mass-action expression for the general reaction

$$mA + nB \rightleftharpoons pC + qD$$

is of the form in which the equilibrium concentration of each reactant or product is raised to the power of its coefficient.

$$K = \frac{[C]^p[D]^q}{[A]^m[B]^n}$$

Products ↑

$$\text{a. } K = \frac{[Cl_2][PCl_3]}{[PCl_5]}$$

↓ Reactant

Coefficient of S ↑

$$\text{b. } K = \frac{[S]^8}{[S_8]}$$

Coefficient of H_2O ↑ Coefficient of HCl ↑

$$\text{c. } K = \frac{[H_2O]^7[HCl]^2}{[Cl_2O_7][H_2]^8}$$

↓ Coefficient of H_2

Mathematically $[C]^p [D]^q = [D]^q [C]^p$, so the order in which we write these terms doesn't matter, as long as we keep each exponent with its term. For instance, the exponent p must remain with [C].

Are Our Answers Reasonable?

The answers have the coefficients of the balanced chemical equation as exponents in the mass-action expression, and that is reasonable. This means that there is a huge impact from the concentration of sulfur in equation "b" and hydrogen and water in equation "c" on the concentrations of the other reactants and products in the equations.

What If…?

…we changed our general reaction to this:

$$mA + nB + zR \rightleftharpoons pC + qD + tV$$

What would be the mass-action expression for the equation?

(*Ans: it would be* $\frac{[C]^p[D]^q[V]^t}{[A]^m[B]^n[R]^z}$)

Practice 14.3

Write the mass-action expression for each of these reactions:

a. $NO(g) + O_3(g) \rightleftharpoons NO_2(g) + O_2(g)$
b. $HCl(aq) + NH_3(aq) \rightleftharpoons NH_4Cl(aq)$
c. $C_2H_2(g) + 2H_2(g) \rightleftharpoons C_2H_6(g)$

14

Concentration-based mass-action expressions are generally suitable representations for the equilibria that occur in chemical reactions. However, in some aqueous solutions, especially where the solute concentrations are high, chemists often make a correction using activities instead of molarities, to get the most meaningful results. For example, if we prepare a solution that contains 3.0 mol of hydrochloric acid (HCl) in a liter of water, we say that the HCl concentration is 3.0 molar. We expect this strong acid to dissociate into H^+ and Cl^- ions, and we therefore assert that in the solution, the concentration of each ion, H^+ and Cl^-, is 3.0 M. However, this is not completely true. Some of the hydrogen cations do interact with the chloride anions in solution, and the ions interact with the water solvent through ion–dipole interactions. This means that the effective concentration is likely to be somewhat different from the intended concentration, especially in relatively concentrated solutions. This effective concentration of the solute is called its activity. We will generally not consider the impact of activity in our discussions, but it is important for us to know that such an idea exists and can be dealt with quantitatively.

▶ Here's What We Know so Far

- Most reactions do not go to completion. Rather, they reach a minimum energy state. We call this the point the position of chemical equilibrium.
- At chemical equilibrium, the rates of the forward and reverse reactions are equal.
- Chemical equilibrium is a dynamic, not static, condition.
- A reaction at chemical equilibrium can be described by a mass-action expression for which there is an equilibrium constant, K, that depends on temperature. ◀

14.3 Why Is Chemical Equilibrium a Useful Concept?

The reaction of myoglobin with oxygen is just one of countless examples of equilibrium processes in living systems. In fact, the chemistry of blood is filled with equilibria. The chemistry of the environment also involves equilibrium chemistry. As we will explore later in this chapter, we can control the position of equilibrium—that is, we can make it possible for reactions to proceed almost all the way toward products or to reach a point at which mostly reactants exist. For example, we can select the conditions so that the greatest possible amount of ammonia is formed from hydrogen and nitrogen in the Haber process. We can also maximize the amount of sulfur trioxide that is formed from the reaction of sulfur dioxide with oxygen as part of the industrial-scale preparation of sulfuric acid. We can attempt to reduce the concentrations of acid in lakes and streams. We can learn about the impact of chlorofluorocarbons (CFCs) on stratospheric ozone levels. We can prepare pharmaceuticals to work in the most effective ways. We can analyze for the presence of an extraordinary variety of substances, from silver to steroids. We can do these things because we understand the fundamental ideas of equilibrium. Our ability to use equilibrium concepts to control the extent of chemical processes has the following vital implications:

1. Economic implications via the trillions of dollars of manufactured products prepared and sold by the chemical industry.
2. Environmental implications for the quality of air, water, and land in the closed system that is Earth.
3. Personal implications for our health.

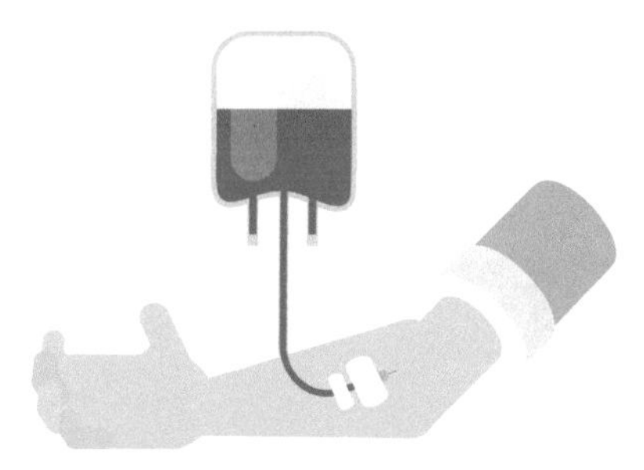

▶ https://publicdomainvectors.org/en/free-clipart/Blood-transfusion/74205.html

Table 14.1 lists some significant processes that involve chemical equilibria. Among them is an analytical technique that is used to separate mixtures and it is the subject of the boxed feature on equilibrium and chromatographic analysis.

14.4 The Meaning of the Equilibrium Constant

Sulfuric acid (H_2SO_4) is the most-produced industrial chemical in the world, with an annual production of nearly 300 billion kg. Sulfuric acid has a wide range of uses, the most important being the production of agricultural fertilizers. Its industrial and agricultural importance makes examining the manufacture of sulfuric acid an excellent way to illustrate the meaning of the equilibrium constant. Most of the sulfuric acid produced in is made via the **Contact process**, which began to be used on an industrial scale in 1880. The process is a set of steps performed on readily available materials.

Contact Process – An industrial method used to produce sulfuric acid from elemental sulfur.

Table 14.1 Chemical equilibrium processes

Category	Process	Important reaction
Industrial	Contact process	$2SO_2(g) + O_2(g) \rightleftharpoons 2SO_3(g)$
Industrial	Haber process	$3H_2(g) + N_2(g) \rightleftharpoons 2NH_3(g)$
Biological	Myoglobin uptake of oxygen	$Mb(aq) + O_2(aq) \rightleftharpoons MbO_2(aq)$
Analytical	Chromatographic analysis	$A_{\text{mobile phase}} \rightleftharpoons A_{\text{stationary phase}}$
Analytical	Metal analysis with EDTA	$Ca^{2+} + EDTA^{4-} \rightleftharpoons CaEDTA^{2-}$
Environmental	Leaching of lead into water	$PbCO_3(s) + 2H^+(aq) \rightleftharpoons Pb^{2+}(aq) + CO_2(g) + H_2O(l)$

Step 1: The process begins with the sulfur provided by minerals such as iron pyrite (FeS_2) or hydrogen sulfide gas (H_2S). The sulfur combines with oxygen from the air.

$$S(s) + O_2(g) \rightleftharpoons SO_2(g)$$

Step 2: In this step, sulfur trioxide (SO_3) is generated by the oxidation of SO_2 produced in step 1.

$$2SO_2(g) + O_2(g) \rightleftharpoons 2SO_3(g)$$

Sulfur dioxide gas is passed over a catalyst bed containing 6–10% vanadium(V) oxide (V_2O_5) at 600 °C and then at 400 °C. Temperature has a major effect on the equilibrium of this reaction, as we will see.

Step 3: Sulfur trioxide (SO_3) is then combined with concentrated sulfuric acid and water to give more sulfuric acid in a net process that can be written as

$$SO_3(g) + H_2O(l) \rightleftharpoons H_2SO_4(aq)$$

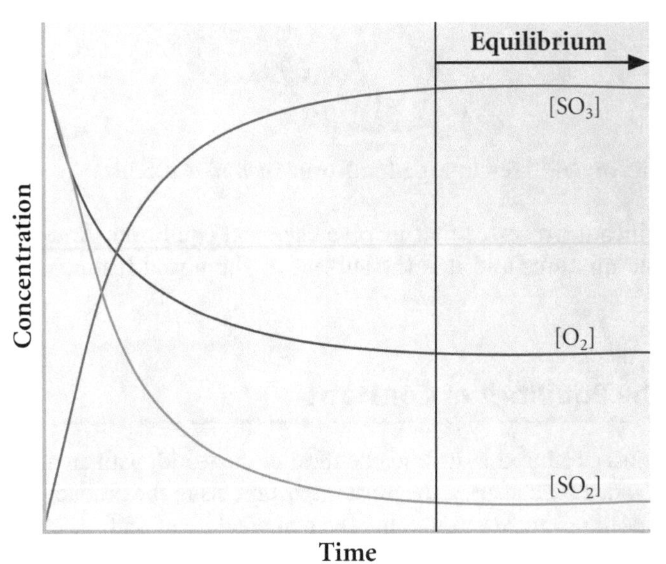

14

Let's revisit step 2 of the Contact process, the reaction of SO_2 and oxygen to make SO_3. The equilibrium constant, K, for this reaction at 27 °C is 4.0×10^{24}. What does the value of the equilibrium constant tell us about the relative amounts of reactants and products when the reaction reaches equilibrium? When the value for K is very large, the equilibrium lies essentially all the way to the products side. (When we say, "reactants and products side," we are always referring to these in terms of the forward reaction, written from left to right.) The reaction goes almost, but not fully, to completion. We can use a line chart, as shown above, to visualize the meaning of the equilibrium constant for this reaction by comparing the equilibrium position (E) to the starting position (S). We start with only reactants (R) and have no products (P). In practice, the Contact process is not carried out at 27 °C because it takes too long for this reaction to reach equilibrium at that temperature. In the chemical industry, time is money. To speed up the process the temperature is raised, which lowers the equilibrium constant for this reaction. We will examine the question of how the equilibrium constant changes with temperature. Unfortunately, for this and many other equilibria, the extent of the reaction indicated by the equilibrium constant and the kinetics of the reaction (how fast it gets there), can force us into a compromise between reaction speed and reaction yield.

Every reaction has its own set of temperature-dependent values of K (which is also known as K_c when only concentrations are used to determine its value), the equilibrium constant. Our chapter-opening reaction of myoglobin with oxygen has $K = 8.6 \times 10^5$ at 20 °C, which is fairly large. The opposite, a small equilibrium constant, is also possible. For example, an equation can be written that illustrates the formation of silver and chloride ions in a saturated solution of silver chloride from solid AgCl. Recall from ▶ Chap. 12 that a saturated solution is one in which no more solute will dissolve. The small equilibrium constant of 1.7×10^{-10} indicates that the product concentrations are quite small, $[Ag^+] = [Cl^-] = 1.3 \times 10^{-5}$ M. This implies that very little silver chloride dissolves at 25 °C before equilibrium (the saturated solution) is reached.

$$AgCl(s) \rightleftharpoons Ag^+(aq) + Cl^-(aq) \qquad K = 1.7 \times 10^{-10}$$

When the value for K is very small, the equilibrium lies nearly all the way to the reactants side. We can show this using our line chart. In this case, the equilibrium position is quite close to the silver chloride reactant, and we would say that silver chloride isn't very soluble in water.

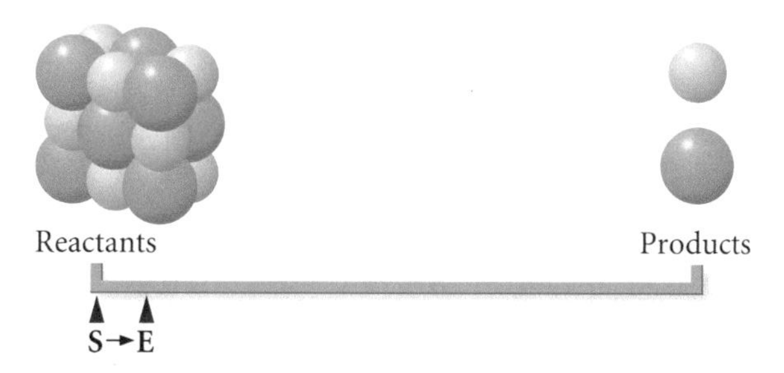

How Do We Know?—Equilibrium and Chromatographic Analysis

If we were to take a vote among chemists as to the single most important technique for finding out what we have, and how much of it there is in a sample, the winner might well be **chromatography**. This technique is used in many chemical analyses. For instance, the analysis of steroids and other banned substances in the urine of baseball players, football players, and Olympic athletes is done by separating the chemicals via chromatography. As another example, the compounds that make up gasoline can be separated and identified chromatographically. The technique has even been used to identify the gases on the planet Venus.

Chromatography – A chemical technique involving the partition of a solute between a stationary phase and a mobile phase. The technique can be used to separate or purify mixtures of solutes.

In a chromatograph, the components of a sample can be separated on the basis of how they distribute themselves between two chemical or physical phases.

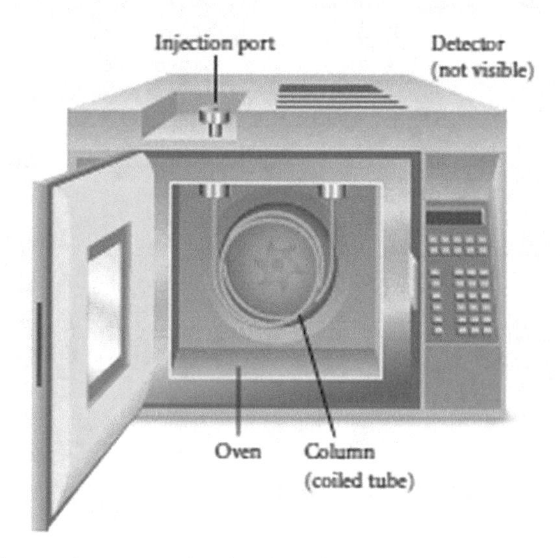

Fig. 14.6 The essential parts of a gas chromatograph. There is an inlet connected to a column into which the sample is fed. The sample is then pushed through the column by a carrier gas such as helium (in gas chromatography) or by a liquid, often an aqueous solution (in liquid chromatography). This phase moves, so itis called the mobile phase. It passes through the column containing a stationary phase, so called because it stays in place on the column

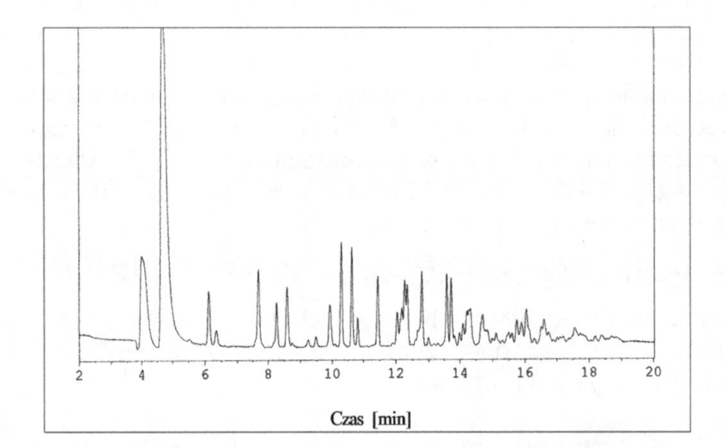

Fig. 14.7 A chromatogram obtained from analyzing the waste gases emitted from a phosphoric acid production plant. Image produced by Joanna Kośmider. CC0 1.0. *Source* ▶ https://upload.wikimedia.org/wikipedia/commons/9/91/Chromatogram_KF.png

Figure 14.6 shows the essential parts of a gas chromatograph. The sample to be analyzed is injected into the instrument and pushed by an inert gas (in **gas chromatography**) or by a liquid (in liquid chromatography) into a long tube known as a column. The gas or liquid that pushes the sample moves, so it is called the **mobile phase**. The sample and the mobile phase pass through the column packed with a **stationary phase**, so called because it stays in place on the column.

Gas Chromatography – A specific chromatography technique in which the mobile phase is a gas and the stationary phase is a solid.

Mobile Phase – In chromatography, the phase that moves.

Stationary Phase – In chromatography, the phase that does not move.

On the ride through the column, all of the components (called the analytes) of the sample interact physically or chemically with the stationary phase. Here is where our connection to equilibria comes in. The interaction of each analyte, "A," with the mobile and stationary phases can be described by the reversible reaction

$$A_{\text{mobile phase}} \rightleftharpoons A_{\text{stationary phase}}$$

The equilibrium constant (called a **distribution constant**, K_D) has the mass-action expression

$$K_D = \frac{[A_{\text{stationary phase}}]}{[A_{\text{mobile phase}}]}$$

Distribution Constant – The equilibrium constant that describes the partitioning of a solute between two immiscible phases.

If an analyte interacts considerably with the stationary phase, the concentration of the analyte in the stationary phase will be greater than its concentration in the mobile phase. Therefore, the value of K_D will be a number greater than 1. In such cases, the analytes will slow down and take more time to travel through the column. If the analytes do not interact well with the stationary phase, the value of K_D will be small, and the analytes will move quickly through the column. For a given set of conditions in the chromatograph, each analyte will have its own distribution constant (K_D) and will exit the column at a different time.

Figure 14.7 shows a chromatogram of the compounds in the waste gases emitted from phosphoric acid production. The components in a sample from an athlete or a factory can be identified by using chromatography. However, the method has even more uses. For instance, chromatography is often one of the first steps in many sophisticated analyses. Once the analytes are separated in a chromatography instrument, they can immediately be fed into other instruments, such as a mass spectrometer or infrared spectrometer, to confirm their identity. In some cases, the information from the other instruments is used to determine the identity of an unknown component in a mixture. (Fig. 14.8 shows two of the more important multistep analyses, which are commonly referred to as hyphenated techniques.) For this reason, chemical equilibrium, the basis of chromatography, is often the most important process in a multistep instrumental analysis.

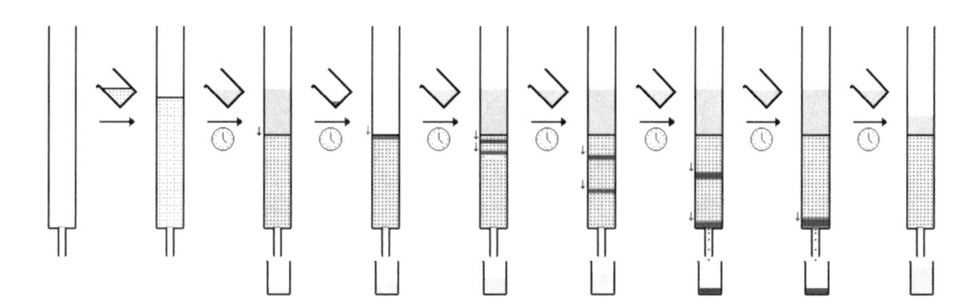

Fig. 14.8 Chromatography is based on the different interactions of the individual components with the chromatography system. A second analysis can be added to the end of the chromatography system to determine when the individual compounds elute. These are known as hyphenated techniques. They include techniques such as GC–MS (gas chromatography–mass spectrometry) and LC-NMR (liquid chromatography–nuclear magnetic resonance). These techniques are used to separate, identify, and quantitate the solutes within a solution. *Source* ▸ https://upload.wikimedia.org/wikipedia/commons/3/3f/Column_chromatography_sequence.png

Let's examine another example of an equilibrium constant. The ionization of sulfurous acid (H_2SO_3) to form hydrogen ions and bisulfite ions (HSO_3^-) has a value for K of 0.0120 at 25 °C.

$$H_2SO_3(aq) \rightleftharpoons H^+(aq) + HSO_3^-(aq) \qquad K = 0.0120$$

This is an intermediate value; that is, K is relatively close to 1. At equilibrium, there remain significant amounts of both reactants and products. When the value for K is close to 1, the equilibrium lies somewhere between the reactants and products sides. Again, our line chart helps us see the equilibrium position.

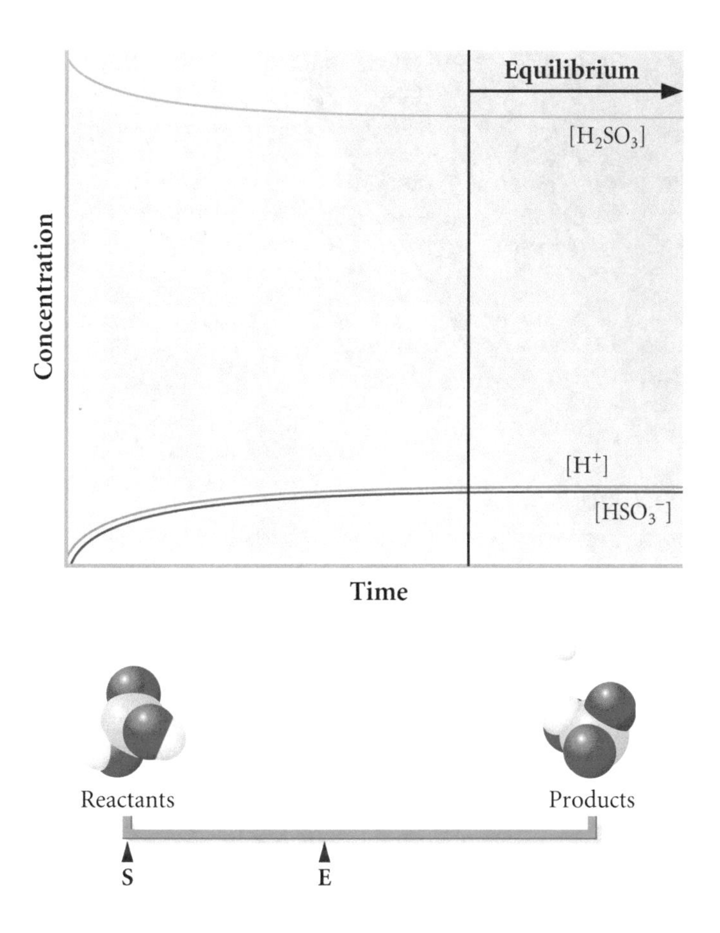

Here's What We Know so Far

- Processes that have very large values of K are those in which mostly products are present at equilibrium.
- Processes that have very small values of K have mostly reactants present at equilibrium.
- Processes with K values not too far from 1 have significant amounts of both reactants and products at equilibrium. ◀

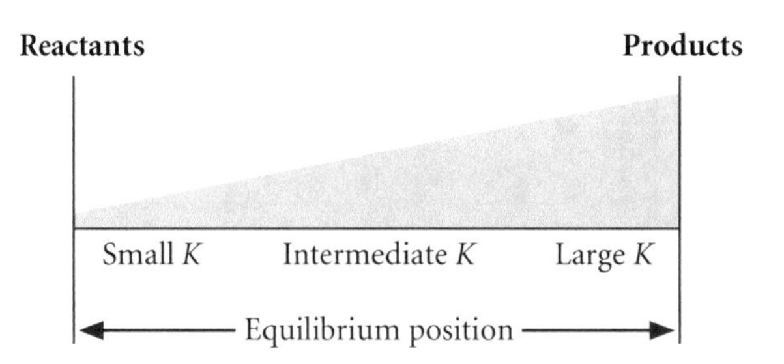

Fig. 14.9 The relative equilibrium points of typical reactions based on their equilibrium constants. High values of K favor product formation. Small values of K favor the reactants side. Intermediate values favor a middle-of-the-road position

The meaning of the equilibrium constant is shown graphically in Fig. 14.9. As we continue to build our understanding of equilibrium, we will learn that the equilibrium position for a reaction is related to the equilibrium constant as well as to the temperature, the pressure, and changes in the concentration of the reactants and products. As we explore these relationships, our goal is not only to interpret the meaning of K but also to understand how we can manipulate our reaction conditions to produce what we want (products or reactants) under the best possible conditions. In industrial-scale manufacturing, such as the production of sulfuric acid, these "best possible conditions" can include chemical, environmental, and economic factors.

Example 14.4—Interpreting *K*

One of the most important chemicals in the toolbox of analytical chemists has the chemical name ethylenediaminetetraacetic acid, or, more simply, EDTA. A chemist would like to determine the concentration of several metal ions, including Co^{2+}, Zn^{2+}, Ni^{2+}, and Ca^{2+} in a solution by titrating them with the ionic form of EDTA known as $EDTA^{4-}$, shown below. The values of K for the reaction of the metal ion with $EDTA^{4-}$ are listed next to each cation. Judging solely on the basis of these values, and assuming the equilibrium is established very quickly, would EDTA be a reasonable choice to react almost completely with each of these metal ions?

Ethylenediaminetetraacetate

(*example reaction*) $Ca^{2+}(aq) + EDTA^{4-}(aq) \rightleftharpoons CaEDTA^{4-}(aq)$

Element	Cation	K
Cobalt	Co^{2+}	2×10^{16}
Zinc	Zn^{2+}	3×10^{16}
Nickel	Ni^{2+}	4×10^{18}
Calcium	Ca^{2+}	5×10^{10}

Asking the Right Questions

How do we judge if an equilibrium constant is large enough to indicate an essentially complete reaction? Are these K values large enough?

Solution

Because each of the equilibrium constants are considerably larger than 1, $EDTA^{4-}$ will react essentially completely with these metals.

Are Our Answers Reasonable?

Looking only at the value for K (and ignoring how quickly or slowly the reactions might occur) suggests that the reactions will be essentially complete. The values for K in each case are so large that our assertion of essentially complete reactions between the EDTA and the metals is reasonable.

What If…?

…we were doing an analysis for Co^{2+} via reaction with EDTA, and we halved the initial concentration of EDTA. How would that affect the equilibrium constant?

(*Ans: It would not affect the value of the equilibrium constant.*)

Practice 14.4

Which of these reactions provide(s) mostly products? … mostly reactants?

a. $NH_4^+(aq) + H_2O(l) \rightleftharpoons NH_3(aq) + H_3O^+(aq)$
$K = 5.6 \times 10^{-10}$

b. $HF(aq) + H_2O(l) \rightleftharpoons H_3O^+(aq) + F^-(aq)$
$K = 7.2 \times 10^{-4}$

c. $HOCl(aq) + H_2O(l) \rightleftharpoons H_3O^+(aq) + OCl^-(aq)$
$K = 3.5 \times 10^{-8}$

d. $HClO_4(aq) + H_2O(l) \rightleftharpoons H_3O^+(aq) + ClO_4^-(aq)$
$K = 1.0 \times 10^{7}$

14.4.1 Homogeneous and Heterogeneous Equilibria

Let's look again at the reaction of the calcium and EDTA ions in the previous example and compare that to the dissolution of silver chloride, as well as the Haber process for the combination of H_2 and N_2 to produce ammonia (NH_3), the focus of Example 14.2.

$$Ca^{2+}(aq) + EDTA^{4-}(aq) \rightleftharpoons CaEDTA^{2-}(aq)$$

$$3\,H_2(g) + N_2(g) \rightleftharpoons 2\,NH_3(g)$$

$$AgCl(s) \rightleftharpoons Ag^{+}(aq) + Cl^{-}(aq)$$

Note the phases of each substance. In the EDTA titration of calcium ion, the reactants and products are all in the aqueous phase. When all of the phases are the same in a reaction, we say that the reaction establishes a **homogeneous equilibrium**. The formation of ammonia is another example of a homogeneous equilibrium because all of the substances are gases. On the other hand, the dissolution of silver chloride in water begins with a solid and forms ions in the aqueous phase. When there are different phases present in the reaction, we have a **heterogeneous equilibrium**.

Homogeneous Equilibrium – An equilibrium that results from reactants and products in the same phase, or physical state.

Heterogeneous Equilibrium – An equilibrium that results from reactants and products in different phases or physical states.

The mass-action expressions for homogeneous and heterogeneous equilibria differ in one vital aspect, which we can examine by focusing on the formation of a saturated solution of silver chloride. The density of solid AgCl is 5.56 g/cm^3. We can think of density as a measure of the concentration, because it is the amount of substance per unit volume. Density is independent of the quantity of substance; therefore, the concentration of pure AgCl is the same whether we have 10 g or 100 tons. Constants, such as the concentration of any pure solid or liquid, are said to be part of the equilibrium constant and not included in the mass-action expression for the equilibrium constant. This is a general result that is valid for pure solids and pure liquids and is approximately correct for solvents such as water, but only when the solutes are very dilute. This means that for the dissolution of silver chloride, our original mass-action expression,

$$K = \frac{[Ag^{+}][Cl^{-}]}{[AgCl]}$$

can be simplified by recognizing that AgCl is a solid:

$$K = \left[Ag^{+}\right]\left[Cl^{-}\right]$$

We can apply this same reasoning to derive the mass-action expression for one step in the large-scale cleanup of sulfur dioxide emitted from industrial smokestacks. In this process, SO_2 is converted to solid $CaSO_4$.

$$2\,CaO(s) + 2\,SO_2(g) + O_2(g) \rightleftharpoons 2\,CaSO_4(s)$$

$$K = \frac{1}{[SO_2]^2[O_2]}$$

14.5 Working with Equilibrium Constants

In this section, we begin to learn how we can work with reaction, mass-action expressions, and equilibrium constants. This is the next step in learning how to control experimental conditions by using equilibria to achieve desired chemical outcomes. As with our previous section, we begin by looking at the reaction of sulfur dioxide and oxygen.

Here are three ways of writing the balanced reaction of SO_2 with O_2:

$$2\ SO_2(g) + O_2(g) \rightleftharpoons 2\ SO_3(g)\ \text{(reaction A)}$$
$$4\ SO_2(g) + 2\ O_2(g) \rightleftharpoons 4\ SO_3(g)\ \text{(reaction B)}$$
$$SO_2(g) + 1^2\ O_2(g) \rightleftharpoons SO_3(g)\ \text{(reaction C)}$$

We multiplied each coefficient in reaction A by 2 to obtain reaction B. We then divided each coefficient by 4 to obtain reaction C. However, all the reactions are the same because the mole ratios within each reaction do not change. That is, the ratio of SO_2 to O_2 is 2 to 1 in each reaction. What about the mass-action expressions?

$$K \text{ for reaction A} = \frac{[SO_3]^2}{[SO_2]^2[O_2]}$$

$$K \text{ for reaction B} = \frac{[SO_3]^4}{[SO_2]^4[O_2]^2}$$

$$K \text{ for reaction C} = \frac{[SO_3]}{[SO_2]\,[O_2]^{1/2}}$$

The mass-action expressions are all different! Let's pick a set of typical Contact process equilibrium concentration values and substitute them into each expression: $[O_2] = 1.5 \times 10^{-3}$ M; $[SO_2] = 1.3 \times 10^{-14}$ M; and $[SO_3] = 1.0 \times 10^{-3}$ M. To calculate the equilibrium constant, we substitute the equilibrium concentrations of all substances into the mass-action expression and do the math indicated by the expression.

$$K_{\text{reaction A}} = \frac{[SO_3]^2}{[SO_2]^2[O_2]} = \frac{\left[1.0 \times 10^{-3}\right]^2}{\left[1.3 \times 10^{-14}\right]^2\left[1.5 \times 10^{-3}\right]} \approx 4.0 \times 10^{24}$$

$$K_{\text{reaction B}} = \frac{[SO_3]^4}{[SO_2]^4\,[O_2]^2} = \frac{\left[1.0 \times 10^{-3}\right]^4}{\left[1.3 \times 10^{-14}\right]^4\left[1.5 \times 10^{-3}\right]^2} \approx 1.6 \times 10^{49}$$

$$K_{\text{reaction C}} = \frac{[SO_3]}{[SO_2]\,[O_2]^{1/2}} = \frac{\left[1.0 \times 10^{-3}\right]}{\left[1.3 \times 10^{-14}\right]\left[1.5 \times 10^{-3}\right]^{1/2}} \approx 2.0 \times 10^{12}$$

How do the values of K differ, and what is the relationship to the change in reaction coefficients? We doubled the coefficients going from reaction A to reaction B, and $K_{\text{reaction B}} = (K_{\text{reaction A}})^2$. In going from reaction B to reaction C we divided the coefficients by 4, and $K_{\text{reaction C}} = (K_{\text{reaction B}})^{1/4}$. In general, then, when we change the coefficients of a reaction by a factor of n,

$K_{\text{new}} = (K_{\text{old}})^{n}$.

Example 14.5—Reversing the Reaction

We opened the chapter by discussing the importance of the interaction of human myoglobin with oxygen in muscle cells.

$$Mb + O_2 \rightleftharpoons MbO_2 \qquad K = 8.6 \times 10^{-7}$$

Many biochemists and molecular biologists cite the "dissociation constant" for the reaction written in reverse:

$$MbO_2 \rightleftharpoons Mb + O_2$$

Given the relationship between the change in coefficients and the change in the value of K just discussed, can we determine the equilibrium constant for the reverse (dissociation) reaction?

Asking the Right Questions

How do we change the mass-action expression when we reverse the chemical equation? What is the effect of this change on the value of the equilibrium constant?

Solution

Reversing the reaction requires us to take the inverse of the original mass-action expression. That is,

$$Mb + O_2 \rightleftharpoons MbO_2 \qquad \text{(original reaction)}$$

$$K = \frac{[MbO_2]}{[Mb][O_2]} \qquad \text{(original reaction)}$$

14

$$MbO_2 \rightleftharpoons Mb + O_2 \qquad \text{(reverse reaction)}$$

$$K = \frac{[Mb][O_2]}{[MbO_2]} \text{(inverse mass action expression)}$$

Mathematically,

$$K_{dissociation} = \left(K_{original}\right)^{-1}$$

$$K_{dissociation} = (8.6 \times 10^{-7})^{-1} = 1.2 \times 10^6$$

Is Our Answer Reasonable?

The original reaction had a very small equilibrium constant, meaning that the reaction favored the reactants side. When the reaction was reversed, the reactants of the original reaction became the products of the reversed (dissociation) reaction. The equilibrium now favors the forward reaction, and a large value of K is therefore reasonable.

What if…?

…the equilibrium constant for the reaction of myoglobin with oxygen was measured as 3.0×10^{-7}? What would the coefficients for each substance in the chemical equation have been?

(*Ans:* $\frac{1}{4}Mb + \frac{1}{4}O_2 \rightleftharpoons \frac{1}{4}MbO_2$)

Practice 14.5

Given the value of K in Example 14.5, what is the equilibrium constant for this reaction?

$$1/2MbO_2 \rightleftharpoons 1/2Mb + O_2$$

14.5.1 Combining Reactions to Describe a Process

One of the more important questions that chemists answer is "How much do I have?" In the chemical industry, this issue is often related to quality control: "Is my vitamin C tablet actually filled with vitamin C?" "Does my aspirin tablet contain the right amount of aspirin?" "Does my potato chip have too much water in it?" Chemical titration is the chemist's tool often used to answer such questions. However, two key features of a titration experiment are necessary for the titration to be effective. First, the titration reaction has to be essentially complete, meaning that the equilibrium constant, K, must be quite large. In addition, a titration reaction should be fast, even when the concentrations of the reactants are low.

In the analytical laboratory, the concentration of acetic acid (CH_3COOH, known technically as ethanoic acid) can be measured by titration with a solution of sodium hydroxide of known concentration. Can we show that the equilibrium constant, which we will call $K_{titration}$, is very large at 25 °C? The reaction is

$$CH_3COOH(aq) + OH^-(aq) \rightleftharpoons CH_3COO^-(aq) + H_2O(l)$$

The mass-action expression is

$$K_{titration} = \frac{[CH_3COO^-]}{[CH_3COOH][OH^-]}$$

We can consider this reaction as the sum of two separate equations for which we know the equilibrium constants at 25 °C. Using the known equations, we can calculate the value for $K_{titration}$.

$$CH_3COOH(aq) + H_2O(l) \rightleftharpoons CH_3COO^-(aq) + H_3O^+(aq) \qquad K_1 = 1.8 \times 10^{-5}$$

$$H_3O^+(aq) + OH^-(aq) \rightleftharpoons 2H_2O(l) \qquad K_2 = 1.0 \times 10^{14}$$

$$CH_3COOH(aq) + OH^-(aq) \rightleftharpoons CH_3COO^-(aq) + H_2O(l) \qquad K_{titration} = ?$$

The mass-action expressions for the known reactions are

$$K_1 = \frac{[CH_3COO^-][H_3O^+]}{[CH_3COOH]} \qquad K_2 = \frac{1}{[H_3O][OH^-]}$$

If we multiply the two known mass-action expressions, we get

$$\frac{[CH_3COO^-][H_3O^+]}{[CH_3COOH]} \times \frac{1}{[H_3O][OH^-]} = \frac{[CH_3COO^-][H_3O^+]}{[CH_3COOH][H_3O][OH^-]}$$

$$K_1K_2 = \frac{[CH_3COO^-]}{[CH_3COOH][OH^-]}$$

This is the mass-action expression for the titration reaction! For this process, then,

$$K_{titration} = K_1K_2 = (1.8 \times 10^{-5})(1.0 \times 10^{14}) = 1.8 \times 10^9$$

In short, when the overall reaction is the sum of other reactions, the overall equilibrium constant is the product of the equilibrium constants for these other reactions. In the case of the titration of acetic acid with sodium hydroxide, $K_{titration}$ is sufficiently large (and the reaction is experimentally determined to be quite fast). Therefore, the titration of acetic acid with sodium hydroxide should (and does!) work well.

14.5.2 Calculating Equilibrium Concentrations from *K* and Other Concentrations

We know from our previous discussion that it is possible to calculate the equilibrium constant by substituting equilibrium concentrations of reactants and products into the mass-action expression. We have one equation, the mass-action expression, and one unknown, *K*. In our day-to-day work as a chemist, biologist, medical technologist, or soil scientist, the more likely scenario is that the value of *K* is already known and we will need to calculate the equilibrium concentration of one or more of the reactants or products.

Let's use the Contact process for producing sulfuric acid to show how this is done for one unknown concentration when the other equilibrium concentrations and *K* are known. At 27 °C, $K=4.0 \times 10^{24}$ for the reaction.

$$2SO_2(g) + O_2(g) \rightleftharpoons 2\ SO_3(g)$$

If the equilibrium concentrations of SO_2 and O_2 are 6.4×10^{-12} M and 1.2×10^{-3} M, respectively, what is the equilibrium concentration of SO_3? We have one equation with one unknown, so we can solve for $[SO_3]$. We start by writing the equilibrium constant expression:

$$K = \frac{[SO_3]^2}{[SO_2]^2[O_2]}$$

Then, we substitute the numbers we know:

$$4.0 \times 10^{24} = \frac{[SO_3]^2}{[6.4 \times 10^{-12}]^2[1.2 \times 10^{-3}]}$$

And solve for what we want to know:

$$[SO_3]^2 = 4.0 \times 10^{24}(6.4 \times 10^{-12})^2(1.2 \times 10^{-3})$$

$$[SO_3] = \sqrt{0.197} = 0.443 \approx 0.44\ M$$

Does our answer make sense? We ought to check our math, and we can do so by substituting the values back into the mass-action expression to verify the value of the equilibrium constant.

$$\frac{[SO_3]^2}{[SO_2]^2[O_2]} = \frac{[0.443]^2}{[6.4 \times 10^{-12}]^2[1.2 \times 10^{-3}]} = 4.0 \times 10^{24}$$

We can make such substitutions because the reaction is at equilibrium. When this is not the case, a different way of thinking, the subject of ▶ Sect. 14.6, will prove useful.

14.5.3 Using Partial Pressures

Most of the equilibrium constants that chemists employ are derived using concentrations measured in molarity because much of the chemistry we do is in solution. However, when working with gases, we can use partial pressures in units of atmospheres. What is the relationship between the pressure of a gas and its concentration? We can use the ideal gas equation, which holds well for gases in low concentration.

$$PV = nRT$$

$$\text{Concentration} = \frac{\text{moles}}{\text{volume}} = \frac{n}{V} = \frac{P}{RT}$$

The mass-action expression for the reaction of oxygen with sulfur dioxide can be written using molarity, as usual, or by substituting the equivalent variables from the ideal gas law ($\frac{P}{RT}$) in place of the concentration.

$$2\ SO_2(g) + O_2(g) \rightleftharpoons 2\ SO_3(g)$$

$$K = \frac{[SO_3]^2}{[SO_2]^2[O_2]} = \frac{\frac{P^2_{SO_3}}{(RT)^2}}{\frac{P^2_{SO_2}}{(RT)^2} \times \frac{P_{O_2}}{(RT)}}$$

We can combine all of the RT terms and put them on the side with the pressure-based mass-action expression.

$$K = \frac{P^2_{SO_3}}{P^2_{SO_2}P_{O_2}}(RT)^1$$

where

- K = the mass-action expression solved using only molar concentrations
- P = the pressure in atmospheres
- T = the temperature in kelvins
- R = the universal gas constant.

If we set K_p equal to the pressure-based mass-action expression and rearrange the equation, we get

$$K_p = \frac{P^2_{SO_3}}{P^2_{SO_2}P_{O_2}}$$

$$K = K_p\,(RT)^1$$

We have introduced a new equilibrium constant, K_p, which is based on the mass-action expression for the partial pressures of substances in the reaction. K_p is often calculated with partial pressures, P, in units of atmospheres. As a matter of nomenclature, we use K_p when dealing with partial pressures and K (or, in some literature, K_c, where "c" stands for "concentration") when dealing only with concentrations. Note in this equation that the value of the exponent of RT indicates the difference in the number of moles of gas between reactants and products. In general, then, we can write the relationship between K and K_p in two ways:

$$K = K_p(RT)^{-\Delta n} \qquad K_p = K(RT)^{\Delta n}$$

where

- Δn = the change in the number of moles of gas (products minus reactants),
- R = 0.08206 L atm mol^{-1} K^{-1}, and,
- T = temperature in kelvins.

In the reaction of sulfur dioxide with oxygen, there is one fewer mole of gaseous products (2 SO_3) than of reactants (2 SO_2 and O_2), so the change in the number of moles of gas from reactant to product, Δn, is equal to −1.

At 673 K and 1 atm total system pressure, the value of K_p for the oxidation of SO_2 to SO_3 is 1.58×10^5, measured using the equilibrium partial pressure of each gas. We can calculate K using this information. Note that, although the values for K and K_p are slightly different due to the shift from the use of concentration to the use of pressure, they are both still large numbers in this case. This is true in general because both "versions" of the equilibrium constant tell us the same thing about the reaction: the ratio of products to reactants at equilibrium.

$$K = K_p(RT)^{-\Delta n}$$

$$K = (1.58 \times 10^5)\left\{(0.08206\text{L atm mol}^{-1}\,\text{K}^{-1})(673\text{K})\right\}^{-1} = 8.73 \times 10^6$$

Example 14.6—Converting *K* to K_p

Calculate K_p for the production of ammonia via the Haber process.

$$3H_2(g) + N_2(g) \rightleftharpoons 2NH_3(g) \qquad K = 0.060\ (\text{at } 500\ ^\circ\text{C})$$

Asking the Right Questions

Why do we use K_p? How might its use help us remember the formula for the conversion of *K* to K_p?

Solution

K_p can be used when we are dealing with partial pressures. Since the conversion of the concentration of a gas to its partial pressure requires *RT* via the ideal gas equation, that understanding can help us remember the conversion formula. In this example, we lose 2 mol of gas going from reactants to products, so:

$$\Delta n = -2$$

$$K_p = K(RT)^{\Delta n}$$

$$K_p = 0.060 \times (0.08206\ \text{L atm/mol}^{-1} \cdot \text{K}^{-1} \times 773\text{K})^{-2}$$

$$K_p = 0.060 \times (62.43)^{-2}$$

$$K_p = 1.5 \times 10^{-5}$$

Is Our Answer Reasonable?

The equilibrium constant remained less than 1, indicating that the reactants are still favored over the products, so the answer is reasonable.

What If…?

…the equilibrium constant, *K*, for the above reaction at 1200 °C was 2.50. What would be the value of K_p?
(*Ans:* = 1.71×10^{-4})

Practice 14.6

Calculate the value of K_p at 25 °C for the reaction of chlorine and carbon monoxide.

$$CO(g) + Cl_2(g) \rightleftharpoons COCl_2(g)$$

$$K = 3.7 \times 10^9 (\text{at } 25\ ^\circ\text{C})$$

14.6 Solving Equilibrium Problems—A Different Way of Thinking

The world is not at equilibrium. Its shifts are seen and felt in the massive upheavals of earthquakes, volcanoes, and hurricanes and in the smallest electronic interactions of atoms. Life itself is a process in which reversible reactions within us keep shifting their equilibrium positions, always chasing a moving target, in ways that make our survival possible. If one aspect of being human is our ability to understand our world, then one expression of this understanding is the ability to manipulate our starting reaction conditions to control the position of equilibrium. We cannot stop earthquakes, but we can reliably make ammonia, sulfuric acid, and a whole host of other chemicals. We can examine the complex interactions of carbon with the environment. We can understand the interaction of myoglobin and oxygen. We can pick the best conditions for chemical analyses. All these things are based on knowing how the conditions change from the starting point to equilibrium. How can we do this? We must first develop a different way of thinking.

This way of thinking is based on two questions: "Given the chemical and physical conditions, how do I judge what is likely to happen?" and "How can I use my understanding to change these conditions to obtain the outcome I want?" Here are some judgments we must make when we combine substances:

1. What chemical reactions will occur?
2. Which among these are important? Which are unimportant?
3. Given the initial concentrations of substances, in which direction (toward the formation of more reactants or toward the formation of more products) is the reaction likely to proceed?
4. How can we solve for the amounts of substances of interest that are present at equilibrium?
5. How can we manipulate the conditions to maximize the concentration of the desired components and minimize that of the others?

14.6.1 A Case Study

We discussed the analysis of acetic acid (CH_3COOH) previously. Here, we will look at the equilibrium established when an acetic acid solution is prepared with an initial concentration of 0.500 M. Remember the overarching question: "Given the chemical and physical conditions, how do I judge what is likely to happen?" Specifically, what are the concentrations of all the substances in my flask at equilibrium? Let's approach this by answering the other questions, in order.

1. What chemical reactions will occur?
   One reaction is the ionization of the acid as shown in reaction A, and this will supply CH_3COO^- and H^+. Another reaction that always occurs in aqueous solution is the ionization of water itself, also supplying H^+, as shown in reaction B.
   A: $CH_3COOH(aq) \rightleftharpoons CH_3COO^-(aq) + H^+(aq)$ $\quad K = 1.8 \times 10^{-5}$
   B: $H_2O(l) \rightleftharpoons H^+(aq) + OH^-(aq)$ $\quad K = 1.0 \times 10^{-14}$
2. Which among these are important? Which are unimportant?
   The "important chemistry" is that which affects the outcome of the overall process. The unimportant chemistry (in this context) is that which has no significant bearing on the outcome. The extent of the reactions and their relative importance to us are indicated by their equilibrium constants. Even though neither reaction proceeds very far, the ionization of acetic acid is far more significant than the ionization of water (we will test this assertion later.) We can make this claim because the value of K for equation A is about 10^8 times (100 million times) larger than the value of K for equation B. The mass-action expression for the ionization of acetic acid (equation A) is

$$K = \frac{[CH_3COO^-][H_3O^+]}{[CH_3COOH]} = 1.8 \times 10^{-5}$$

   In many systems, it is entirely possible that more than one reaction is important to our calculations. However, in this chapter, we will typically deal with one important reaction in each process that we discuss.
3. Given the initial concentrations of substances, in which direction is the reaction likely to proceed?
   We can predict the direction of change in a reaction by using the **reaction quotient (Q)**. This is the numerical outcome of the mass-action expression using initial concentrations, which are designated $[\,]_0$.

Reaction Quotient (Q) – The ratio of product concentrations to reactant concentrations raised to the power of their stoichiometric coefficients for a reaction that is not at equilibrium.

$$Q = \frac{[CH_3COO^-]_0[H_3O^+]_0}{[CH_3COOH]_0}$$

Q is then compared to the equilibrium constant, K, in order to determine the direction in which the reaction will proceed. Keep in mind that Q deals with initial conditions and K deals with equilibrium conditions. Calculation of the reaction quotient indicates which way the reaction will proceed in order to establish equilibrium.

- If Q is equal to K, the system is at equilibrium.
- If Q is greater than K, there is too much product present. The system will shift to the left to reach equilibrium. Mathematically, the numerator is much larger, and the denominator much smaller, than the equilibrium values.
- If Q is less than K, there is too much reactant present. The system will shift to the right to reach equilibrium.

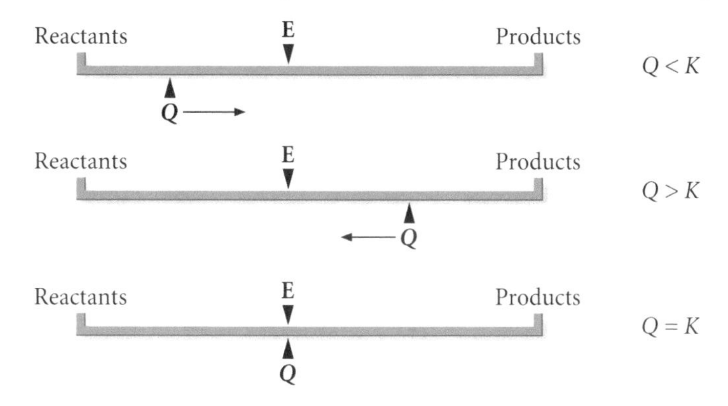

In our current example, the reaction must go toward formation of products ("to the right") because we initially have no products ($Q=0$), and this will be the case in many equilibrium problems we will encounter.

If we were to change the initial concentration of our substances, such that $[H^+]_0 = 0.0100$ M, $[CH_3COO^-]_0 = 0.200$ M and $[CH_3COOH]_0 = 0.150$ M, we could calculate Q for these initial concentrations.

$$Q = \frac{[CH_3COO^-]_0[H_3O^+]_0}{[CH_3COOH]_0} = \frac{(0.200\ M)(0.0100\ M)}{(0.150\ M)} = 0.0133$$

Here, Q is much greater than K (1.8×10^{-5}), so the reaction will shift toward the formation of reactants ("to the left"). Mathematically, this makes sense, because reducing the product concentration and increasing the reactant concentration will reduce the quotient of the mass-action expression. What does this indicate about the chemistry? If chemical change is an inevitable part of our world, it is especially useful to be able to control that chemical change in order to have a desirable impact on manufacturing, human biology, and the environment.

Example 14.7—Predicting the Direction of Equilibrium

Predict the direction in which the reaction of myoglobin and oxygen will proceed to reach equilibrium for each set of conditions.

$Mb + O_2 \rightleftharpoons MbO_2$ $\quad K = 8.6 \times 10^5$

a. $Q = 1500$

b. $Q = 8.3 \times 10^6$

c. $[Mb]_0 = 2.04 \times 10^{-4}$ M; $[O_2]_0 = 3.00 \times 10^{-6}$ M; $[MbO_2]_0 = 2.50 \times 10^{-4}$ M

d. $[Mb]_0 = 5.00 \times 10^{-5}$ M; $[O_2]_0 = 9.21 \times 10^{-7}$ M; $[MbO_2]_0 = 1.00 \times 10^{-4}$ M

e. $[Mb]_0 = 1.0 \times 10^{-4}$ M; $[O_2]_0 = 2.25 \times 10^{-6}$ M; $[MbO_2]_0 = 1.94 \times 10^{-4}$ M

Asking the Right Questions

What is the significance of the reaction quotient? What is the direction that a reaction will go when K and Q are not equal?

Solution

The reaction quotient, Q, helps us see in which direction the reaction will go to reach equilibrium. When the value for K is small, it is true that the equilibrium position will favor the reactants. However, if we started with *no* products at all, even such a reaction will move (slightly) to the right to form a little product. We can show this graphically (note the myoglobin is a large molecule; the oxygen molecule is highlighted in yellow to make it visible in this image):

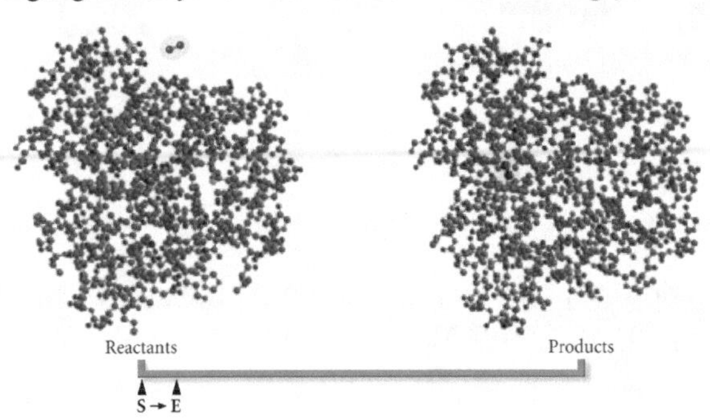

a. $Q < K$, so the reaction will form more products and shift to the right.

b. $Q > K$, so the reaction will form more reactants and shift to the left.

c. $$Q = \frac{[MbO_2]_0}{[Mb]_0[O_2]_0} = \frac{(2.50 \times 10^{-4})}{(2.04 \times 10^{-4})(3.00 \times 10^{-6})} = 4.08 \times 10^5.$$

$Q > K$, so the reaction will form more products and shift to the right.

d. $$Q = \frac{(1.00 \times 10^{-4})}{(5.00 \times 10^{-5})(9.21 \times 10^{-7})} = 2.17 \times 10^6$$

$Q > K$, so the reaction will form more reactants and shift to the left.

e. $$Q = \frac{(1.94 \times 10^{-4})}{(1.0 \times 10^{-4})(2.25 \times 10^{-6})} = 8.6 \times 10^5$$

$Q = K$, so the reaction is at equilibrium.

Are Our Answers Reasonable?

We can look at it mathematically and say that whenever Q is greater than K, it means that the numerator (reactant part) of the mass-action expression is too high, so it must be lowered while making the denominator (product part) of the mass-action expression larger. The reaction will therefore form more reactant, and shift the reaction to the left. Similar reasoning holds when Q is less than K. On this basis, the answers are reasonable.

What If...?

...the temperature of the system changed so that the equilibrium constant, $K = 5.02 \times 10^6$? How would that change the answers in parts a. – e.?

(*Ans: Only part "d" would change, because Q would now be less than K, shifting the reaction direction toward the product side.*)

Practice 14.7

Under certain specific conditions, the formation of ozone from NO_2 has an equilibrium constant $K = 120$. Predict the direction of this reaction, given the conditions in each case below.

$NO_2(g) + O_2(g) \rightleftharpoons NO(g) + O_3(g) \quad K = 120$

a. $Q = 1.5 \times 10^5$

b. $[O_3] = 1.0 M$; $[O_2] = 1.0 M$; $[NO] = 0.50 M$; $[NO_2] = 0.25 M$

c. $[O_3] = 0.0010 M$; $[O_2] = 2.55 M$; $[NO] = 1.78 \times 10^{-6} M$; $[NO_2] = 5.4 \times 10^{-3} M$

d. $[O_3] = 0.033 M$; $[O_2] = 0.019 M$; $[NO] = 9.3 M$; $[NO_2] = 0.044 M$

4. How can we solve for the amounts of substances of interest that are present at equilibrium?

The magnitude of the value of *K* determines our approach to solving this question.

14.6.2 Small Value of *K*

We have a system that contains 0.500 M CH_3COOH, which ionizes with the relatively small *K* of 1.8×10^{-5}. What will be the equilibrium concentrations of acetate ions (CH_3COO^-) and hydrogen ions (H^+)?

$$CH_3COOH(aq) \rightleftharpoons CH_3COO^-(aq) + H^+(aq) \qquad K = 1.8 \times 10^{-5}$$

The lack of products initially present, along with the small equilibrium constant, suggests that the reaction will proceed to the right, but not very far. Our view of where we start and where we end looks something like this on our equilibrium line chart:

Tracking the changes that take place as we go from the start to equilibrium requires a bit of bookkeeping. Bookkeepers keep track of a company's finances using tables that show changes in financial data in an organized, clear way. We will do the same thing for our equilibrium data by setting up a table that contains the following rows:

Initial row: The "initial" row includes the initial concentration of each species. Notice that we have inserted concentration values in this row. When solving equilibrium problems, we must always work with concentrations, not just numbers of moles.

	$CH_3COOH(aq)$	$\rightleftharpoons$	$CH_3COO^-(aq)+$	$H^+(aq)$
→Initial	0.500 M		0 M	0 M

Change row: The "change" row describes the change in concentration of each species that occurs in order to reach equilibrium. This must take into account the stoichiometric ratios among reactants and products in the reaction (all 1-to-1 in this example) and the magnitude of the equilibrium constant, which in this case is small enough to indicate that the reaction will not go very far before equilibrium is reached. Because we don't yet know the amount of acetic acid that will ionize, we call it "*x*." The concentration of acetic acid will decrease by "*x*" M, and the acetate and hydrogen ion concentrations will increase by "*x*" M. Again, the stoichiometry of the equation is included in this line of the table.

	$CH_3COOH(aq)$	$\rightleftharpoons$	$CH_3COO^-(aq)+$	$H^+(aq)$
Initial	0.500 M		0 M	0 M
→Change	$-x$		$+x$	$+x$

Equilibrium row: When the free energy change of the reaction equals zero, such that the energy itself is at a minimum, the reaction will be at equilibrium. The value for each substance in the "equilibrium" row equals the initial amount plus the change (so that 0.500+ "$-x$"$=0.500-x$). This is sufficient for us to proceed with the problem, especially via programmable calculators, which can solve the necessary equations with a few keystrokes.

	$CH_3COOH(aq)$	$\rightleftharpoons$	$CH_3COO^-(aq)+$	$H^+(aq)$
Initial	0.500 M		0 M	0 M
Change	$-x$		$+x$	$+x$
→Equilibrium	$0.500-x$		x	x

Assumptions row: Whether or not we have a programmable calculator, we can get a deeper understanding of the equilibria in solution by making key assumptions. We can simplify the problem solving by assuming that because the value for K is small, the value of "x" will also be small—that is, there is negligible ionization compared to our initial acetic acid concentration. We'll assume that it will be so small as to be unimportant compared to the initial concentration of 0.500 M. Making that assumption enables us to say that $0.500-x \approx 0.500$. However, because "x" is not unimportant compared to 0M (any quantity is infinitely large compared to zero!), we cannot neglect "x" in the other columns of the table. The "assumptions" row shows the final equilibrium amounts with any assumptions we make.

	$CH_3COOH(aq)$	$\rightleftharpoons$	$CH_3COO^-(aq)+$	$H^+(aq)$
Initial	0.500 M		0 M	0 M
Change	$-x$		$+x$	$+x$
Equilibrium	$0.500-x$		x	x
→Assumptions	0.500		x	x

The table that we have worked with is an expanded version of what is often called an ICE table, which stands for Initial, Change, Equilibrium table. Our expanded version includes the Assumptions row, so we will call this an **ICEA table**. We will often use ICEA tables in our equilibrium problem solving because they give us an organized way to assess the changes in concentration that occur in our reactions. We can now substitute the equilibrium concentrations generated in the assumptions row of our table into the mass-action expression and solve for "x."

ICEA Table – A bookkeeping approach to determining concentration of different reaction components in an equilibrium.

14

$$K = \frac{[CH_3COO^-][H_3O^+]}{[CH_3COOH]}$$

$$1.8 \times 10^{-5} = \frac{(x)(x)}{0.500}$$

$$1.8 \times 10^{-5} = \frac{x^2}{0.500}$$

$$9.0 \times 10^{-6} = x^2$$

$$3.0 \times 10^{-3} = x$$

If we return the value of "x" to our ICEA table and solve for the assumptions row, we obtain the equilibrium concentrations of all species in the reaction, assuming our assumption is justified.

$$x = [CH_3COO^-] = [H^+] = 3.0 \times 10^{-3}\,M$$

The actual equilibrium concentration of acetic acid is $0.500-3.0\times10^{-3}=0.497$ M, or 0.50 M to two significant figures. Do our results make sense? That is, are they reasonable? We can determine that our results are mathematically correct by substituting the concentration values back into the equilibrium expression to show that this results in K.

$$\frac{(3.0 \times 10^{-3})^2}{0.50} = 1.8 \times 10^{-5}$$

It is very important to always ask ourselves "was our assumption justified"? In other words, was "x" negligible compared to the original acetic acid concentration? We say that our assumption is valid if the change as a result of the assumption is less than 5% of the original concentration.

$$\frac{x}{\text{original concentration}} \times 100\% \leq 5\%$$

Although we will use this "5% rule" in our work, professional chemists sometimes require a smaller tolerance, and sometimes a larger, depending on the process with which they are working. With our data,

$$\frac{3.0 \times 10^{-3}}{0.500} \times 100\% = 0.6\%$$

and using the 5% rule, we find that our assumption of negligible ionization (0.6%, in this case) was valid.
Let's revisit our equilibrium line chart and see what 0.6% ionization means in terms of the extent of the reaction.

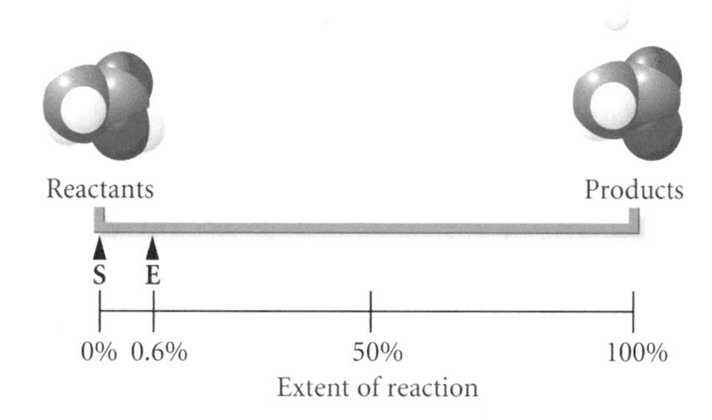

This confirms our initial thinking that with the very small value for K, the reaction would not proceed—acetic acid would not ionize—appreciably. To be sure our calculations are valid, we should also test our other assumption, that the ionization of water is unimportant in this case. This is a more complex issue that we will consider later.

Example 14.8—Concentration of Lead Ion in Saturated Lead Bromide

Determine the concentration of lead ion in a saturated solution that would result from the dissolution of solid lead(II) bromide, $PbBr_2(s)$.

$$PbBr_2(s) \rightleftharpoons Pb^{2+}(aq) + 2\,Br^-(aq)$$

$$K = 4.6 \times 10^{-6} \text{ at } 25\,°C$$

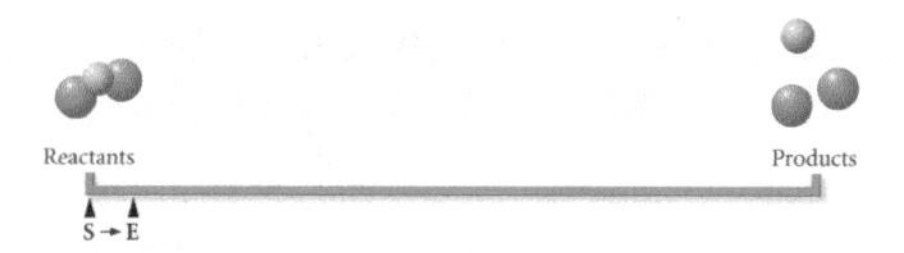

Asking the Right Questions

What is the important chemistry that occurs in solution? Because there is only $PbBr_2$ in the solution initially, in which direction will the reaction proceed? How can we solve for the amounts of substances of interest that are present at equilibrium?

Solution

The only important chemistry of interest is the dissolution of the lead bromide, which is a largely insoluble salt, judging by the small value of K. Even though the equilibrium constant is quite small, the reaction will proceed (minimally) to the products side because there were no products present initially.

Our equilibrium line chart is similar to that for the dissociation of acetic acid. What are the equilibrium concentrations of substances? We can use the same style of table as we did with the acetic acid dissociation. Our only modification arises because lead bromide is a pure solid, so it is not part of the equilibrium expression. We show this in the ICEA table below.

	$PbBr_2(s) \rightleftharpoons$	$Pb^{2+}(aq)+$	$2Br^-(aq)$
Initial	–	0 M	0 M
Change	–	$+x$	$+2x$
Equilibrium	–	x	$2x$

In this table, "x" represents the amount of $PbBr_2(s)$ that dissolves. This value is therefore equal to the solubility of the salt. Because the initial amount of each product was

0 M, the amount gained, "x" and "$2x$," is not negligible compared to 0 M; therefore, we made no assumptions and have omitted this line from the table. Note that we use "$2x$" to represent the change in $[Br^-]$ because bromide has a stoichiometric factor of 2 in the equation. This stoichiometric factor will also come into play when we write the mass-action expression.

To solve for the amounts of substances of interest that are present at equilibrium, we can substitute our concentrations into the mass-action expression for the equation:

$$K = [Pb^{2+}][Br^-]^2 4.6 \times 10^{-6}$$
$$= (x)(2x)^2 4.6 \times 10^{-6} = 4x^3 \left[Pb^{2+}\right]$$
$$= x = 1.048 \times 10^{-2}\,M$$
$$\approx 1.0 \times 10^{-2}\,M \left[Br^-\right] = 2x$$
$$= 2.096 \times 10^{-2}\,M \approx 1.0 \times 10^{-2}\,M$$

We check our answer by solving for K.

$$K = (1.048 \times 10^{-2}) \times (2.096 \times 10^{-2})^2$$
$$= 4.6 \times 10^{-6}$$

Is Our Answer Reasonable?

The concentrations for the ions that we calculated are quite low, and that makes sense, given the very low value for K.

What If…?

…instead of a divalent cation (a charge of +2) we had a compound with a trivalent cation? Calculate the $[F^-]$ in a saturated solution of ScF_3, for which $K = 5.8 \times 10^{-24}$. What is the $[Sc^{3+}]$ in the solution?

(*Ans:* $[F^-] = 2.0 \times 10^{-6}$ M, $[Sc^{3+}] = 6.8 \times 10^{-7}$ M)

Practice 14.8

What is the concentration of lead ion in a saturated solution of $PbCl_2$?

$$PbCl_2(s) \rightleftharpoons Pb^{2+}(aq) + 2\,Cl^-(aq)$$
$$K = 4.6 \times 10^{-6} \text{ at } 25\,°C$$

In Example 14.8, only one important equilibrium needed to be considered. As our understanding of equilibrium deepens, we will find that more than one reaction may well be important. For example, although it wasn't illustrated, the chemical interactions of Pb^{2+} and Br^- with water weren't considered. In both of these cases, the interactions are minimal, so the dissolution of $PbBr_2$ was the only equilibrium of importance. However, if instead of $PbBr_2$ we had attempted to calculate the concentration of lead ion from the dissolution of PbS, the reaction of S^{2-} with water would have been very important, as illustrated by the equilibria below. In this case, both equations would need to be considered.

14

$$PbS(s) \rightleftharpoons Pb^{2+}(aq) + S^{2-}(aq) \qquad K = 7 \times 10^{-29}$$
$$S^{2-}(aq) + H_2O(l) \rightleftharpoons HS^-(aq) + OH^-(aq) \qquad K = 0.083$$

Including the formation of HS^- in our calculations would modify the solubility of lead(II) sulfide. We will learn how to deal quantitatively with multiple equilibria later.

14.6.3 Large Value of *K*

The reaction of sulfur dioxide and oxygen to form sulfur trioxide at 400 °C serves as an excellent model with which to examine systems that have large equilibrium constants, in which the reaction is essentially complete.

$$2\,SO_2(g) + O_2(g) \rightleftharpoons 2\,SO_3(g) \qquad K = 8.7 \times 10^6$$

If the initial concentrations of SO_2 and O_2 are 1.0×10^{-3} M and 2.0×10^{-3} M, respectively, what is the equilibrium concentration of SO_3? Given the large equilibrium constant, we can assume that the reaction goes nearly to completion, except for a small amount that remains unreacted. An equilibrium line chart can help us visualize the extent of reaction.

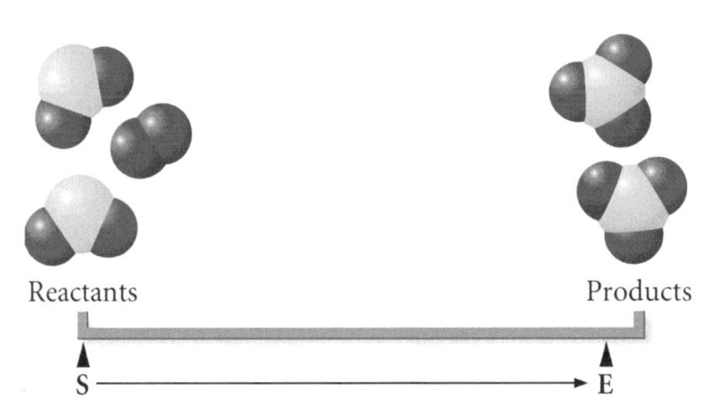

This is, in effect, a limiting reactant problem in which SO_2 is the limiting reactant. On the basis of the reaction stoichiometry and the large value of *K*, we can develop an ICEA table. We complete the first row of the table by writing the initial concentration of each species in the reaction.

	$2SO_2(g)$ +	+	$O_2(g)$	⇌	$2SO_3(g)$
→ Initial	1.0×10^{-3} M		2.0×10^{-3} M		0 M

The "change" row in our ICEA table will be a little different than before. We have said that the equilibrium constant is large, and this means the reaction will go essentially to completion. However, we must hold a tiny amount of each reactant back because the reaction will not go all the way to products.

Let's indicate the small amount that remains behind as "$2x$" for SO_2 and "x" for O_2. That is the "$+2x$" and "$+x$" that we place in the respective concentrations in the change row. Other than those small amounts that remain, all of the SO_2 that can react will do so. That means, based on the reaction stoichiometry, that 1.0×10^{-3} M SO_2 and 0.5×10^{-3} M O_2 will react (less the "$2x$" and "x" amounts of reactants that remain) to form almost 1.0×10^{-3} M SO_3 (less the "$2x$" that is unreacted). The total $[SO_2]$ that will react, then, is *all of it* except for the $2x$ that remains. This is where we get the change of $-1.0 \times 10^{-3} + 2x$. In other words, all of it forms products except for the small amount, "2x," that does not react. Because of the 2-to-1 mol ratio of SO_2 to O_2, 0.5×10^{-3} M O_2 will react with 1.0×10^{-3} M SO_2. And if 2_x mol of SO_2 remains unreacted, x mol of O_2 will be unreacted. This is why the change in concentration of O_2 is $-0.5 \times 10^{-3} + x$.

	$2SO_2(g)$	+	$O_2(g)$	⇌	$2SO_3(g)$
Initial	1.0×10^{-3} M		2.0×10^{-3} M		0 M
→Change	$-1.0 \times 10^{-3} + 2x$		$-0.5 \times 10^{-3} + x$		$1.0 \times 10^{-3} - 2x$

The equilibrium row is, again, the result of adding together the initial and change entries in each column.

	$2SO_2(g)$	+	$O_2(g)$	⇌	$2SO_3(g)$
Initial	1.0×10^{-3} M		2.0×10^{-3} M		0 M
Change	$-1.0 \times 10^{-3} + 2x$		$-0.5 \times 10^{-3} + x$		$1.0 \times 10^{-3} - 2x$
→ Equilibrium	$2x$		$1.5 \times 10^{-3} + x$		$1.0 \times 10^{-3} - 2x$

In the last row of the table, the "assumptions" row, we make the assumption that "x" is small compared to other values to which it is being added. If this is true, then $1.5 \times 10^{-3} + x \approx 1.5 \times 10^{-3}$. And, if that is true, then $1.0 \times 10^{-3} - 2x \approx 1.0 \times 10^{-3}$ must be true as well.

	$2SO_2(g)$	+	$O_2(g)$	⇌	$2SO_3(g)$
Initial	1.0×10^{-3} M		2.0×10^{-3} M		0 M
Change	$-1.0 \times 10^{-3} + 2x$		$-0.5 \times 10^{-3} + x$		$1.0 \times 10^{-3} - 2x$
Equilibrium	$2x$		$1.5 \times 10^{-3} + x$		$1.0 \times 10^{-3} - 2x$
→ Assumptions	$2x$		1.5×10^{-3}		1.0×10^{-3}

We then write the mass-action expression, plug our assumptions row values into the equation, and solve for "x":

$$K = \frac{[SO_3]^2}{[SO_2]^2[O_2]} = \frac{(1.0 \times 10^{-3})^2}{(2x)^2(1.5 \times 10^{-3})} = 8.7 \times 10^6$$

$$4x^2 = 7.66 \times 10^{-11}$$

$$x = 4.38 \times 10^{-6}\text{M}$$

$$2x = [SO_2] = 8.75 \times 10^{-6} \approx 8.8 \times 10^{-6}\text{M}$$

Does this answer make sense? Substituting back into the mass-action expression shows that our equation is mathematically valid.

$$\frac{(1.0 \times 10^{-3})^2}{(8.75 \times 10^{-6})^2(1.5 \times 10^{-3})} = 8.7 \times 10^6$$

Is our assumption valid? That is, is the value of "x" negligible compared to1.5×10^{-3} M?

$$\frac{x}{\text{original concentration}} \times 100\% = \frac{4.38 \times 10^{-6}\text{M}}{1.5 \times 10^{-3}\text{M}} \times 100\% = 0.29\%$$

This easily passes the 5% test, so our assumption that "x" is negligible compared to the initial concentrations of O_2 and SO_3 is valid.

Example 14.9—The Myoglobin–Oxygen System

Solve for the equilibrium concentrations of myoglobin, oxygen, and the MbO_2 complex, given the following initial concentrations. Assume that the myoglobin-oxygen reaction is the most important reaction in this system. In other words, we may neglect the ionization of water because its equilibrium constant is relatively very small.

$[\text{Mb}]_0 = 2.0 \times 10^{-4}$ M; $[\text{O}_2]_0 = 1.9 \times 10^{-5}$ M;

$[\text{MbO}_2]_0 = 0$ M

$\text{Mb} + \text{O}_2 \rightarrow \text{MbO}_2$

$K = 8.6 \times 10^5$

Asking the Right Questions

What does the value of K tell us about the extent of the reaction and its position at equilibrium? What information do we get from the balanced equation? Because the value of K is much greater than 1, the reaction favors product formation. Note that the initial concentration of the reactants is not the same. Why is that important? How do we deal with it in our thinking and our computations?

Solution

The value of the equilibrium constant, K, is large, so the reaction will be nearly complete at equilibrium. We have a 1-to-1 mol ratio of myoglobin to oxygen, but we don't have enough oxygen ($[\text{O}_2]_0 = 1.9 \times 10^{-5}$ M) to react with all of the myoglobin ($[\text{Mb}]_0 = 2.0 \times 10^{-4}$ M). Therefore, this is a limiting reactant problem, with oxygen as the limiting reactant—there is excess myoglobin.

We set up the ICEA table, write the mass-action expression, and solve for "x." Our assumption, that "x" is small compared to 1.9×10^{-5}, will need to be checked.

	Mb	+	O_2	$\rightleftharpoons$	MbO_2
Initial	2.0×10^{-4} M		1.9×10^{-5} M		0 M
Change	$-1.9 \times 10^{-5} + x$		$-1.9 \times 10^{-5} + x$		$1.9 \times 10^{-5} - x$
Equilibrium	$1.81 \times 10^{-4} + x$		x		$1.9 \times 10^{-5} - x$
Assumptions	1.81×10^{-4}		x		1.9×10^{-5}

$$K = \frac{[\text{MbO}_2]_o}{[\text{Mb}]_o[\text{O}_2]_o} = \frac{(1.9 \times 10^{-5})}{(1.81 \times 10^{-4})(x)} = 8.6 \times 10^5.$$

$$x = [\text{O}_2] = 1.22 \times 10^{-7}\text{M} \approx 1.2 \times 10^{-7}\text{ M}$$

$$[\text{Mb}] = 1.81 \times 10^{-4} + x \approx 1.8 \times 10^{-4}\text{ M}$$

$$[\text{MbO}_2] = 1.9 \times 10^{-5} - -x \approx 1.9 \times 10^{-5}\text{ M}$$

The value of "x" is well under 5% of our initial concentrations, so our assumption that "x" is negligible is valid. Checking our math yields:

$$\frac{(1.9 \times 10^{-5})}{(1.81 \times 10^{-4})(1.22 \times 10^{-7})} = 8.6 \times 10^5$$

Are Our Answers Reasonable?

When we solve this type of example, in which we are given initial concentrations and seek to find final concentrations, there are two powerful ways to know if our answer makes sense (and must necessarily be correct!). First, we compare the changes in the reactant and product concentrations to our expectation. Given the value of K, did most of the limiting reactant change to product? Here, it did. That's reasonable. Second, upon substituting the equilibrium data into the equilibrium expression, did we calculate the correct value for K? Then our answer is not only reasonable, it is necessarily correct.

What If…?

…instead of the starting the reaction with no MbO_2, we had an initial concentration of 0.050 M? Describe (don't calculate) how the reaction and the results would have been different than the problem we solved.

(*Ans: The reaction would have had an equilibrium point that was shifted to the left compared to the example we solved, with the equilibrium concentration of each reactant higher than that which we originally calculated.*)

Practice 14.9

Determine the equilibrium concentration of each compound in the following reaction at 25 °C, given the data indicated.

$$[C_2H_4O_2]_0 = 0\,M;\ [C_2H_3O_2^-]_0 = 0.100\,M;$$
$$[H^+]_0 = 0.0500\,M$$
$$C_2H_3O_2^-(aq) + H^+(aq) \rightleftharpoons C_2H_4O_2(aq)$$
$$K = 5.6 \times 10^{-4} \text{at } 25\,°C$$

14.6.4 Intermediate Value of *K*

Sulfurous acid dissociates in water to produce ions. The equilibrium constant is not very small, nor is it very large:

$$H_2SO_3(aq) \rightleftharpoons H^+(aq) + HSO_3^-(aq) \qquad K = 0.0120\ (\text{at } 25\,°C)$$

Other chemistry also occurs in an aqueous solution of sulfurous acid, including the ionization of water and the ionization of the HSO_3^- ion:

$$H_2O(l) \rightleftharpoons H^+(aq) + OH^-(aq) \qquad K = 1.0 \times 10^{-14} (\text{at } 25\,°C)$$
$$HSO_3^-(aq) \rightleftharpoons H^+(aq) + SO_3^{2-}(aq) \qquad K = 1.0 \times 10^{-7} (\text{at } 25\,°C).$$

The equilibrium constants of these reactions are quite small compared to the initial ionization of sulfurous acid, so we will focus only on the first equation.

Although the value of *K* is intermediate (close to 1), we will assume that the we can neglect the dissociation of H_2SO_3 (that is, we assume that the dissociation reaction does occur, but the change in concentration, "*x*," is negligible compared to the concentrations of our initial substances).

	$H_2SO_3(aq)$	$\rightleftharpoons$	$H^+(aq)$	+	$HSO_3^-(aq)$
Initial	1.50 M		0 M		0 M
Change	$-x$		$+x$		$+x$
Equilibrium	$1.50 - x$		x		x
Assumption	1.50		x		x

$$K = \frac{[H^+][HSO_3^-]}{[H_2SO_3]}$$

$$0.0120 = \frac{x^2}{1.50}$$

$$x = 0.134\,M = [H^+] = [HSO_3^-]$$

Before going on, we must test our assumption that "*x*" is negligible compared to 1.50 M, because our assumptions row in the ICEA table indicates that $1.50 - x = 1.50$. We do this by dividing "*x*" by 1.50 and multiplying by 100%. Unfortunately, "*x*" (0.134) is 9% of 1.50! Our assumption, in this case, is not valid. This suggests that our equilibrium line chart shows more than a little forward reaction.

Reactants Products

S E

To arrive at the correct answer for this problem, we must solve the problem explicitly—that is, without making the assumption that the dissociation of H_2SO_3 is unimportant compared to its original concentration of 1.50 M. In order to solve the problem, we must substitute the equilibrium row of our ICEA table into the mass-action expression:

$$0.0120 = \frac{x^2}{1.50 - x}$$

Unfortunately, the presence of both an "x" and an "x^2" in the same equation complicates the math somewhat. One useful way to solve for "x" in these situations is to arrange the values so that we have a **quadratic equation** of the general form $ax^2 + bx + c$. For our example, we multiply both sides of the equation by $(1.50 - x)$:

Quadratic equation – A mathematical equation written in the form $ax^2 + bx + c = 0$.

$$0.0120 \times (1.50 - x) = x^2$$
$$0.0180 - 0.0120x = x^2$$

Then we collect all of our terms onto one side of the equals ("=") sign by adding $0.0120 \times$ and subtracting 0.0180 from both sides of the equation. The result has set our equation equal to zero:

$$x^2 + 0.0120x - 0.0180 = 0$$
$$ax^2 + \quad bx + \quad c = 0$$

Comparing this to the general form for a quadratic equation, we obtain the values for the constants a, b, and c:

$$a = 1; \qquad b = +0.0120; \qquad c = -0.0180$$

Some programmable calculators will solve for the values of "x" in the equation by entering the constants a, b, and c. The other, perfectly satisfactory option is to employ the **quadratic formula**, the equation used to solve for "x":

$$x = \frac{-b \pm \sqrt{b^2 - 4ac}}{2a}$$

Quadratic formula – The method used to solve for x from the quadratic equation,

$$x = \frac{-b \pm \sqrt{b^2 - 4ac}}{2a}$$

Inserting our values for a, b, and c into this equation and solving, we obtain two values for x:

$$x = \frac{-0.0120 \pm \sqrt{(0.0120)^2 - 4(1)(- - 0.0180)}}{2(1)}$$

$$x = \frac{-0.0120 \pm \sqrt{0.000144 - 0.072}}{2}$$

$$x = \frac{-0.0120 \pm 0.2686}{2}$$

$$x = 0.128, - -0.140$$

The existence of two values for "x" would appear to be a problem. However, it is not. One of these values gives equilibrium concentrations that are physically impossible to obtain. Let's determine the concentrations in both cases and see why only one works:
If "x"= −0.140, then

$$[H_2SO_3] = 1.50 + 0.140 = 1.64\,M;\ [H^+] = \left[HSO_3^-\right] = -0.140\,M$$

If "x"=0.128, then

$$[H_2SO_3] = 1.50 - 0.128 = 1.37\,M; \qquad [H^+] = [HSO_3^-] = 0.128\,M$$

When "x"= −0.140, we obtain negative values for both $[H^+]$ and $[HSO_3^-]$. This is impossible, so the correct value for "x" must be 0.128. And, the correct answer to this problem must be $[H_2SO_3]$=1.37 M and $[H^+]$=$[HSO_3^-]$=0.128 M. Using these values, we need to check our answer:

$$K = \frac{(0.128)^2}{1.37} = 0.0120$$

The invalid answer from the quadratic equation will generate a negative value for a concentration. We must never assume, however, that a negative value of "x" will always produce the invalid answer. In the end, the key to doing equilibrium problems is to make assumptions when we can to simplify the math. However, we must test the assumptions that we make. When the assumptions fail, we use the equilibrium row in the ICEA table to solve for "x" explicitly.

14.6.5 In Summary

We have seen that our ability to interpret the meaning of equilibrium constants and to make assumptions that simplify our problem solving are the ideas at the core of this different way of thinking about chemistry. We raised five questions at the beginning of this section:

1. What chemical reactions will occur?
2. Which among these are important? Which are unimportant?
3. Given the initial concentration of substances, in which direction (toward the formation of more reactants or toward the formation of more products) is the reaction likely to proceed?
4. How can we solve for the amounts of substances of interest that are present at equilibrium?
5. How can we create the conditions to maximize the concentration of the desired components and minimize that of the others?

We have now answered the first four of these questions. We are now ready for question 5, which concerns the human control of chemical processes—one big payoff of our study of equilibrium.

14.7 Le Châtelier's Principle

We have discussed understanding chemical equilibrium as "a different way of thinking." **Le Châtelier's principle**, described by Henry Louis Le Châtelier (1850–1936; ◘ Fig. 14.10) in 1884, extends this to changes in a system at equilibrium. Its implications are profound. This useful principle can be summarized as follows:

Le Châtelier's Principle – If a system at equilibrium is changed, it responds by returning toward its original equilibrium position.

› If a stress is applied to a system at equilibrium, it will change in such a way as to partially undo the applied stress and restore the equilibrium.

Although the system moves back toward its original set of equilibrium conditions, it never quite makes it, so there is a net change (often substantial) in equilibrium concentrations as a result of changes in the system. For people,

Fig. 14.10 Henry Louis Le Châtelier (1850–1936), a French chemist, was a mining engineer before working as a professor. In addition to inspiring his work on thermodynamics, his interest in high temperatures, dating from his studies of mineralogy, led to the development of the oxyacetylene torch for cutting and welding steel. CC BY-SA-4.0. *Source* ▶ https://upload.wikimedia.org/wikipedia/commons/1/17/La_Sorbonne_M_le_professeur_H_Le_Chatelier%2C_membre_de_l%27Institut.jpg

Fig. 14.11 Le Châtelier's principle is somewhat analogous to this situation. The two wrestlers push against each other. As the one on the left pushes the defender out of position (applying a stress), the defender pushes back trying to restore his original position, but not fully succeeding. This funerary relief depicts two Greek wrestlers from about 500 BC. Photo by Fingalo. CC BY-SA 2.0. *Source* ▶ https://upload.wikimedia.org/wikipedia/commons/5/5f/07Athletengrab.jpg

the equivalent saying might be "Push me and I'll push back." This is shown graphically in Fig. 14.11. This principle has several implications for a system at equilibrium:

- If the concentration of a component is changed, the system will respond in such a way that the concentration returns toward (but doesn't quite make it to) its original equilibrium value.
- If the pressure of a system is changed, it will respond in such a way as to return the pressure toward (but not to) the original equilibrium value.
- If the temperature of a system is changed, it will respond by exchanging heat, such that the temperature returns toward the original equilibrium value.

Let's look at the effect on equilibrium of each of these changes—concentration, pressure, and temperature—in more detail.

14.7.1 Changes in Concentration

If the concentration of a component is changed, the system will respond in such a way that the concentration returns toward its original equilibrium value. We began this chapter by discussing the interaction of oxygen and myoglobin to form a complex that stores and transports oxygen within muscle cells. How is the control of oxygen levels in the muscle cells indicative of this statement of Le Châtelier's principle? Let's look at it mathematically, using two separate sets of conditions. We saw one set of equilibrium conditions in Example 14.9. Recall the data for the reaction:

$$\mathrm{Mb} + \mathrm{O_2} \rightleftharpoons \mathrm{MbO_2} \qquad K = 8.6 \times 10^{-7}$$

$$[\mathrm{Mb}] = 1.8 \times 10^{-4}\,\mathrm{M}; \quad [\mathrm{O_2}] = 1.2 \times 10^{-7}\,\mathrm{M}; \quad [\mathrm{MbO_2}] = 1.9 \times 10^{-5}\,\mathrm{M}$$

Let's change the conditions so that the oxygen level in the blood increases to 1.0×10^{-5} M. How will the system respond to this change? We can use our reaction quotient, Q, to determine which way the reaction will go to reestablish equilibrium.

$$Q = \frac{[\mathrm{MbO_2}]_0}{[\mathrm{Mb}]_0[\mathrm{O_2}]_0} = \frac{(1.9 \times 10^{-3})}{(1.8 \times 10^{-4})(1.0 \times 10^{-3})} = 1.1 \times 10^4$$

Because $Q < K$, the reaction will shift to the right to produce more product and reduce the concentration of the reactants. Let's solve for the concentrations of myoglobin, oxygen, and the MbO_2 complex at this new equilibrium position. The ICEA table can be set up as follows:

	Mb	+	O_2	$\rightleftharpoons$	MbO_2
Initial	1.8×10^{-4} M		1.0×10^{-5} M		1.9×10^{-5} M
Change	$-x$		$-x$		$+x$
Equilibrium	$1.8 \times 10^{-4} - x$		$1.0 \times 10^{-5} - x$		$1.9 \times 10^{-5} + x$

When we solve using the quadratic equation, we find that "x" $= 9.8 \times 10^{-6}$ M, and

$$[\mathrm{Mb}]_{\mathrm{new}} = 1.7 \times 10^{-4}\,\mathrm{M}; \quad [\mathrm{O_2}]_{\mathrm{new}} = 2.0 \times 10^{-7}\,\mathrm{M}; \quad [\mathrm{MbO_2}]_{\mathrm{new}} = 2.9 \times 10^{-5}\,\mathrm{M}$$

compared to the starting position,

$$[\mathrm{Mb}]_0 = 1.8 \times 10^{-4}\,\mathrm{M}; \quad [\mathrm{O_2}]_0 = 1.0 \times 10^{-5}\,\mathrm{M}; \quad [\mathrm{MbO_2}]_0 = 1.9 \times 10^{-5}\,\mathrm{M}$$

The reactant concentrations (myoglobin and oxygen) have decreased, and the concentration of the product (the myoglobin–oxygen complex) has increased. In other words, we added more reactant and the system shifted to the right to consume that additional reactant and make more product. In the end, the system moved to restore the equilibrium. In our human analogy, the system was pushed, and it pushed back.

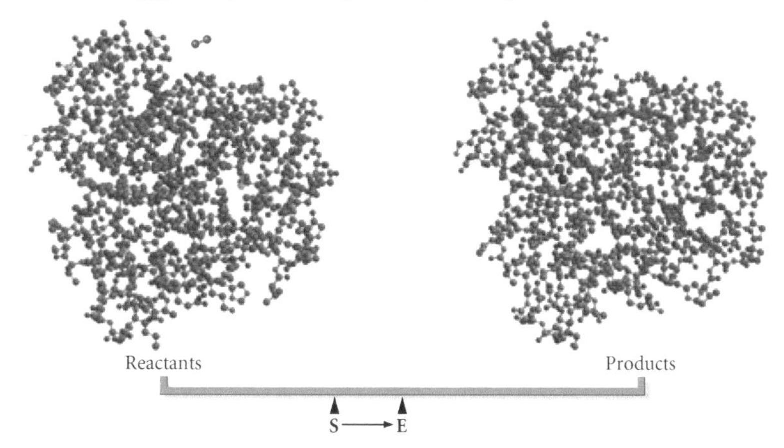

During exercise, the body uses large amounts of oxygen. Some of that oxygen is supplied by the equilibrium position shifting to the left, in response to the fall in free oxygen levels that occurs because so much oxygen is being consumed. This releases more oxygen and free myoglobin, enabling athletes to run freely, for example. When the need for oxygen is less, the free myoglobin combines with oxygen (the reaction shifts to the right) and forms more of the MbO_2 complex. The complex can act as a storage location for oxygen, making it ready for release when it is needed.

Concentration changes make possible the industrial-scale manufacturing of chemicals. As the chemicals are produced in a reaction, they are removed, forcing the reaction to produce products continually in an "attempt" to restore the original equilibrium position. This is an important part of the Contact process, in which the SO_3 that is produced from the reaction of SO_2 and O_2 is continually removed, forcing continued production of SO_3.

14.7.2 Changes in Pressure

If the pressure of a system is changed (at constant temperature and volume), by changing the amounts of reactant or product gases, it will respond in such a way as to return the pressure toward the original equilibrium value.

To illustrate the importance of this statement of Le Châtelier's principle, let's look at a reaction in which changes in pressure are meaningful—one that involves gases. Recall the reaction for the manufacture of ammonia from hydrogen and nitrogen via the Haber process:

$$3H_2(g) + N_2(g) \rightleftharpoons 2\,NH_3(g)$$

If the pressure of the system is substantially increased from 1 to 300 atm, how will it respond? The system will shift in a direction that will lower the pressure toward the original equilibrium value. Recall from our discussion of gases (▶ Chap. 10) that the pressure of a gas is approximately proportional to the number of moles of a real gas. Fewer moles means lower pressure, so the reaction used in the Haber process will shift to the right when the pressure is increased. Typically, we observe changes in volume as the reaction seeks to accommodate pressure changes. As the volume is decreased, the pressure increases, and as the volume is increased, the pressure increases.

An increase in pressure favors the side with fewer moles of gas.

A decrease in pressure favors the side with more moles of gas.

How does this help in the manufacture of ammonia? ◘ Table 14.2 shows the mole percent of ammonia present at equilibrium at various pressures for the Haber process. At each temperature, higher pressure means a shift toward ammonia. In practice, the industrial manufacture of ammonia is done between 100 and 300 atm.

14.7.3 Changes in Temperature

If the temperature of a system is changed, it will respond by exchanging energy as heat, which will result in a new ratio of products to reactants being established. Changing the temperature of a system at equilibrium is different than changing concentrations or pressures. If the temperature of a system is changed, the value of the equilibrium constant changes. As we discovered in the chapter on kinetics, the forward and reverse rates of a reaction change with temperature, and the equilibrium constant also changes with temperature.

The reaction for the manufacture of ammonia is exothermic:

$$3H_2(g) + N_2(g) \rightleftharpoons 2\,NH_3(g) \qquad \Delta H^\circ = -92.0\,\text{kJ}$$

How will the system respond if the temperature is raised from 200 to 500 °C at constant pressure? The change in enthalpy for the formation of ammonia is negative. Therefore, the reaction gives off heat. Another way to picture an exothermic reaction is to say that heat is kind of like a product of the reaction. What shift is necessary for an equilibrium to respond to the increase in heat? The reaction will proceed in the direction that absorbs heat rather than proceeding in the direction that releases more heat, which would raise the system temperature. In the formation of ammonia, equilibrium would be restored with a shift to the left resulting in the formation of more reactants. Therefore, to increase the production of ammonia, we would need to lower the temperature of the system, causing the reaction to shift to the right and produce more heat. As we can see in ◘ Table 14.2, the production of ammonia is enhanced at lower temperatures.

◘ **Table 14.2** Mole percent ammonia at different pressures and temperatures

	Pressure (atm)					
Temperature (°C)	**1**	**10**	**50**	**100**	**300**	**1000**
200	15.3	50.7	74.4	81.5	89.9	98.3
400	0.4	3.9	15.3	25.1	47.0	79.8
600	–	0.5	2.3	4.5	13.8	31.4

Table 14.3 Dependence of the equilibrium constant (K_p) on temperature

$2\,SO_2(g) + O_2(g) \rightleftharpoons 2\,SO_3(g)$	$\Delta H = -198$ kJ
Temperature (°C)	K_p
400	1.6×10^5
500	2.3×10^3
600	91
700	6.9
800	0.84
900	0.15
1000	0.034
1100	0.0096

- Exothermic reactions shift toward the left (the reactants side) when heated. This can be seen in Tables 14.2 and 14.3.
- Endothermic reactions shift toward the right (the products side) when heated.

As illustrated in Example 14.10, equilibrium constants are temperature dependent. Table 14.3 shows the change in K_p at various temperatures for the exothermic Contact process addition of oxygen to SO_2 to form SO_3. Although the data in the table show that the value of K decreases as the temperature increases, there are many cases in which K increases with increasing temperature. The direction of the change is dependent on the enthalpy of the process.

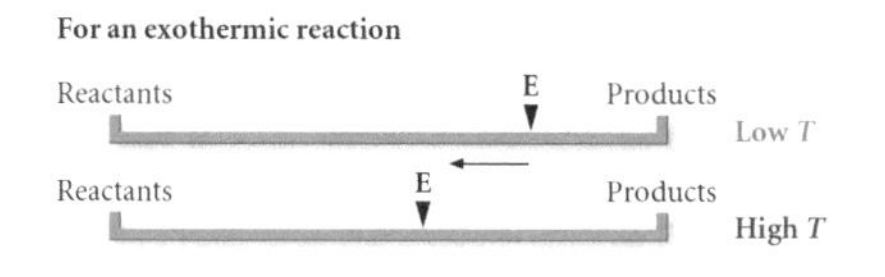

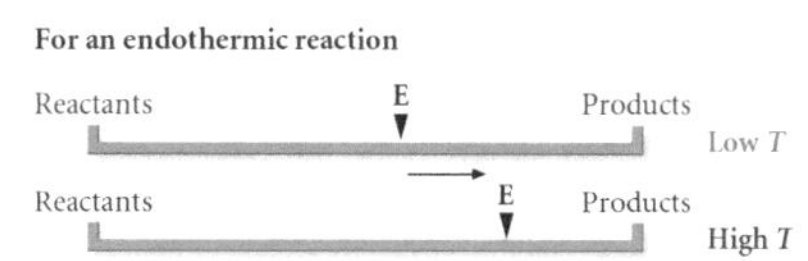

Example 14.10—Le Châtelier's Principle

The reaction for the combination of hydrogen gas and oxygen gas to give water vapor is

$$2H_2(g) + O_2(g) \rightleftharpoons 2H_2O(g) \qquad \Delta H^o = -484\ \text{kJ}$$

Predict and explain the effect of each of these changes to the system on the direction of equilibrium.

a. H_2O is removed as it is being generated.
b. H_2 is added.
c. The pressure on the system is decreased.
d. The system is cooled.

Asking the Right Questions

What type of change occurs in each case? What is the impact of each type of change on the position of equilibrium?

Solution

a. Moves to the right (removal of product forces Q to be less than K, and the system compensates by making more product).
b. Moves to the right (same as above, except Q is less than K because $[H_2]_0$ is larger than at the previous equilibrium).
c. Moves to the left because decreasing the pressure favors the side with more moles of gas.
d. Moves to the right because an exothermic reaction gives off heat. Cooling the system increases the temperature gradient between the system and the surroundings, therefore allowing a continued flow of energy as heat to the surroundings.

Are Our Answers Reasonable?

In each case, we looked at the individual change, and noted how it affected the equilibrium position. The explanation that we gave for each decision indicates that the answers make sense, especially those in which we were able to compare the impact of Q versus K via the mass-action expression.

What If…?

…we had a side reaction that removed oxygen as it was being generated? What effect would that have on the position of equilibrium?

(*Ans: It would shift it to the left.*)

Practice 14.10

Predict the effect of each of these changes to this system on the direction at equilibrium.

$$2C_2H_6O(aq) + 2CO_2(g) \rightleftharpoons C_6H_{12}O_6(aq) \quad \Delta H = -70\,\text{kJ}$$

a. Heat is added.
b. Some ethanol (C_2H_6O) is removed from the reaction.
c. A catalyst is added.
d. More glucose ($C_6H_{12}O_6$) is added to the flask.

▶ **Deepen Your Understanding Through Group Work**

On the prairies of the Midwestern US, one of the food chains starts with grasses. Moving up the food chain are insects, spiders, birds, foxes, and finally coyotes.

Does this food chain represent an equilibrium process? Pick one part of the food chain. Write an equilibrium expression for it. Describe the effect of changing the concentration of a species on the concentration of other species. What would be the effect of changing temperature on the equilibrium? ◀

14.7.4 The Importance of Catalysts

Economics play an important part in the chemical industry. A production process that reaches equilibrium more rapidly will allow more product to be manufactured. More product means the opportunity for increased sales. Catalysts—substances that substantially increase the rate of a reaction without being consumed—are therefore vital to the chemical process industries.

A catalyst does not affect the equilibrium position, but it allows the process to achieve equilibrium more rapidly than it would without a catalyst.

Fritz Haber used magnetite (Fe_3O_4) for the preparation of ammonia. Iron-related catalysts are still used for ammonia production (via the Haber process), although other metals can also be used, including osmium and mixtures of ruthenium and barium.

The nitrogen gas used in the synthesis of ammonia is obtained by distilling liquid air. The hydrogen gas for the Haber process is derived from the iron(II) oxide– and chromium(III) oxide–catalyzed reaction of carbon monoxide (derived from coal tar, a thick liquid that is a by-product of fuel production) with water. The equation illustrating the formation of hydrogen by this method is

14

$$CO(g) + H_2O(g) \overset{FeO+Cr_2O_3}{\rightleftharpoons} CO_2(g) + H_2(g) \qquad (500\,^\circ C)$$

The nickel-catalyzed reaction of methane (from natural gas) and water vapor also produces hydrogen:

$$CH_4(g) + H_2O(g) \overset{Ni\ 400\ psi}{\rightleftharpoons} CO(g) + 3\,H_2(g) \qquad (800\,^\circ C)$$

Table 14.4 Impact of external changes on the equilibrium direction

Change	System response	Change in K
Concentration	Shifts toward restoring initial concentration	None
Pressure (of the system)		
Increase	Shifts toward side with fewer moles of gas	None
Decrease	Shifts toward side with more moles of gas	None
Temperature		
Exothermic reaction	Raising temperature shifts toward reactant	Decreases
	Lowering temperature shifts toward product	Increases
Endothermic reaction	Raising temperature shifts toward product	Increases
	Lowering temperature shifts toward reactant	Decreases
Catalyst	No change in direction; reaches equilibrium more rapidly	None

This means that when the price of coal or natural gas rises, the cost of fertilizer can rise as well. This is another example of the relationship between industrial chemical processes and international economics.

The Contact process uses vanadium(V) oxide on pumice as the catalyst for the conversion of SO_2 to SO_3. Although the catalyst can last for up to 20 years, replacing it each year helps ensure efficiency in the reaction. Platinum is a better catalyst but costs much more. Platinum catalysts can also become less effective in the presence of impurities such as arsenic.

Le Châtelier's principle is one of the most important ideas in all of chemistry. Through its application, we understand and control reactions that have the most far-reaching importance, from the essential reactions in the human body to the production of billions of kilograms of material goods used in everything from fertilizers to plastics. Table 14.4 summarizes the impact of changing the concentration, pressure, or temperature, as well as the impact of the addition of a catalyst, on the direction of chemical equilibria.

We will apply the concepts we learned in this chapter, including the nature of equilibrium, our new way of thinking about problem solving, and Le Châtelier's principle, to look at equilibria. Our immediate focus will be on a most practical, and therefore, important, set of reactions—acid–base equilibria.

The Bottom Line

- Reactions can proceed reversibly toward the products or back toward the reactants.
- The point in a reaction at which there is no net change in the concentration of reactants or products is known as chemical equilibrium—or, often, simply as equilibrium.
- The energy of a reaction is at a minimum at equilibrium.
- The rates of the forward and reverse reactions are equal at equilibrium.
- The mass-action expression relates the equilibrium concentrations of reactants and products in a reaction.
- The equilibrium constant is temperature dependent.
- The size of the equilibrium constant gives us information about the extent of a reaction.
- Modifying the coefficients of a reaction modifies the value of its equilibrium constant.
- The equilibrium constant can be converted for use with partial pressures or molarities.
- The equilibrium constant for the sum of multiple chemical reactions is the mathematical product of the individual K values.
- We can use the equilibrium constant and mass action expression to calculate the equilibrium concentration of substances in a reaction.
- Solving problems relating to reaction equilibria involves asking and answering a series of systematic questions.
- We can use the reaction quotient, Q, to assess which way a reaction will proceed to reach equilibrium.
- Le Châtelier's principle concerns the impact of changing the pressure, temperature, and concentration conditions of a reaction at equilibrium.
- A catalyst does not affect the equilibrium position. It changes the reaction mechanism in such a way as to speed up the reaction.

Section 14.2 The Concept of Chemical Equilibrium

Skill Review

1. The word dynamic refers to changes. Explain how the descriptive term dynamic equilibrium can be applied to a chemical system where the concentrations of reactants and products do not change.
2. Which of the following situations can be described as being in "dynamic equilibrium"?
   a. The number of people at the ticket windows as a stadium is filling for a sporting event.
   b. The number of people sitting down at a music concert.
   c. The number of loaves of bread on the shelf of our local grocery store when the store is open for business.
   d .The number of loaves of bread on the shelf of our local grocery store when the store is closed for the holidays.
3. When describing a reacting system, a scientist may say that the reaction "does not go to completion." Use 'lowest energy' and equilibrium to explain the meaning of that phrase.
4. Complete the following sentences with the words "product" and/or "reactant", for this reaction: $A \rightleftharpoons B$
   In the figure below, compound A is represented by blue spheres and compound B is represented by red spheres. Choose the picture that matches with the following conditions:
   a. Before the reaction starts happening: ______________
   b. The system at equilibrium, if the value of K is large: ______________
   c. The system at equilibrium, if the value of K is small: ______________

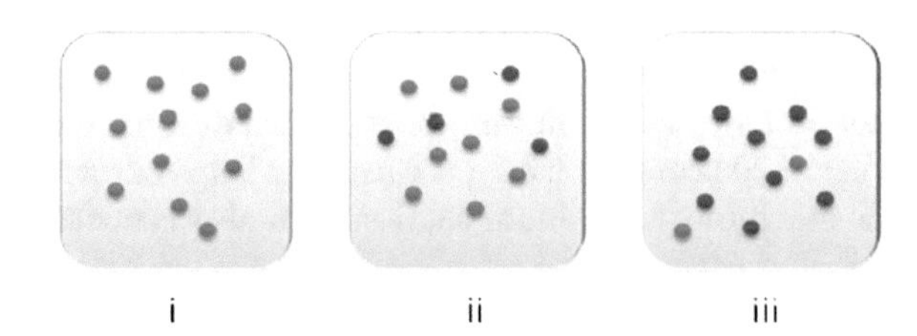

5. What is true about the forward and reverse rates of a chemical reaction when the reaction is at equilibrium? What does this imply about the concentrations of reactants and products when the reaction is at equilibrium?
6. Assume the expression "$rate_1$" represents the rate of a reaction, and "$rate_2$" represents the expression for the rate of the reverse of that reaction. Which of these statements is not true of a reaction at equilibrium, and why?
   a. $rate_1/rate_2 = 1$
   b. $rate_2/rate_1 = -1$
   c. $rate_1/rate_2 = K^2$
   d. $rate_2 = rate_1 \times K$
7. Write the mass-action expression for each of these reactions:
   a. $HCl(g) + C_2H_4(g) \rightleftharpoons C_2H_5Cl(g)$
   b. $CH_4(g) + 2O_2(g) \rightleftharpoons CO_2(g) + 2H_2O(g)$
   c. $2H_2(g) + O_2(g) \rightleftharpoons 2H_2O(l)$
8. Write the mass-action expression for each of these reactions:
   a. $CaCO_3(s) \rightleftharpoons CaO(s) + CO_2(g)$
   b. $SO_3(aq) + H_2O(l) \rightleftharpoons H_2SO_4(aq)$
   c. $4NH_3(g) + 7O_2(g) \rightleftharpoons 4NO_2(g) + 6H_2O(g)$

Chemical Applications and Practices

9. Phosphoric acid (H_3PO_4) is used in soft drinks and in producing fertilizers. As shown in the following reaction, phosphoric acid can be produced by the action of sulfuric acid on rocks that contain calcium phosphate.

$$Ca_3(PO_4)_2(s) + 3H_2SO_4(aq) \rightleftharpoons 3CaSO_4(aq) + 2H_3PO_4(aq)$$

   Describe the system at equilibrium, using each of these concepts:
   a. Reaction rates
   b. Concentration conditions
10. Batteries in cars, watches, and the like all depend on a drive from reactants to products that produces electricity. When the production of electricity stops, we typically say that the battery is "dead." A chemical way to express this is to say that the battery has attained equilibrium. Explain why this chemical statement also describes why the battery no longer produces electricity.
11. Hydrogen chloride gas (used in the production of hydrochloric acid) can be produced directly by the combination of hydrogen and chlorine gas as follows:

$$H_2(g) + Cl_2(g) \rightleftharpoons 2HCl(g)$$

   At equilibrium, $k_1[H_2][Cl_2] = k_{-1}[HCl]^2$. Write the mass-action expression (the equilibrium expression) for the reaction.
12. Chlorofluorocarbons, including Freon-12, have been used in air conditioning units. However, their use is being phased out in most countries as a consequence of their breakdown into chlorine atoms that attack our planet's protective ozone layer. Atmospheric scientists studied the following reaction to understand the breakdown process:

$$CCl_2F_2(g)\ (\text{Freon-12}) \rightleftharpoons CClF_2(g) + Cl(g)$$

   At equilibrium, $k_1[CCl_2F_2] = k_2[CClF_2][Cl]$. Write the mass-action expression, i.e. equilibrium expression for the reaction.

Section 14.3 Why Is Chemical Equilibrium a Useful Concept?

Skill Review

13. Explain the economic outcome to the petroleum industry if equilibria could not be controlled.
14. What types of interactions might exist in a chromatographic system to make a solute interact strongly with the mobile phase instead of the stationary phase?

Chemical Applications and Practices

15. The following chromatogram was developed when ink from a black felt-tip pen was drawn on a plate containing a stationary phase. The plate was then dipped in solvent, and the solvent moved up the plate by capillary action. After development of the plate, it was obvious that the original black ink had separated into the different dyes used to make the ink black.
    a. Which color dye in the ink has the largest value of K_D?
    b. Which color dye interacts least with the stationary filter paper?

16. The hypothetical gas chromatogram shown below illustrates five different compounds in a mixture. The first compound is air. The remaining compounds are labelled "A", "B", "C", and "D". From the chromatogram, which component has the lowest value for K_D? Which has the most interaction with the stationary phase?

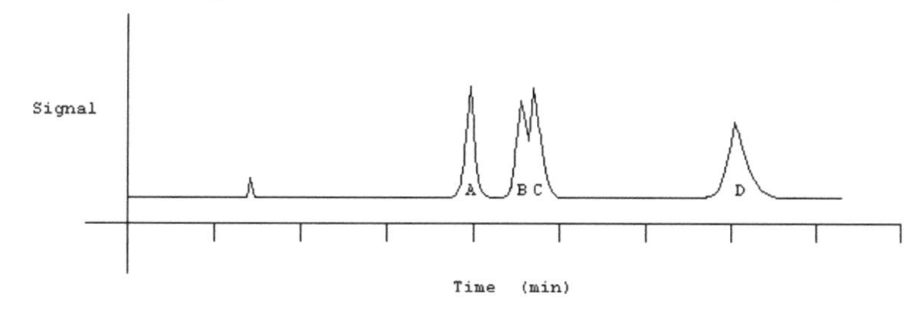

17. If a chemical reaction has a very large value of K (the equilibrium constant), what do we know about the chemical reaction when it is at equilibrium? Does the value of K tell us anything about how quickly the system will achieve equilibrium?
18. If a chemical reaction has a very small value of K (the equilibrium constant), what do we know about the chemical reaction when it is at equilibrium? What does this imply about the reverse of this chemical reaction?

Section 14.4 The Meaning of the Equilibrium Constant

Skill Review

19. For each of the following systems, an equilibrium expression is given. If any errors are present in the expressions, correct them and rewrite the equilibrium expression.
    a. $2PbO(s) + O_2(g) \rightleftharpoons 2PbO_2(s)$ $K = [PbO_2]/[PbO]^2[O_2]$
    b $H_2O(l) + SO_3(g) \rightleftharpoons H_2SO_4(aq)$ $K = 1/[SO_3]$
20. For each of the following systems, an equilibrium expression is given. If any errors are present in the expressions, correct them and rewrite the equilibrium expression.
    a. $H_2CO_3(aq) \rightleftharpoons H_2O(l) + CO_2(g)$ $K = 1/[CO_2]$
    b. $H_2O(l)+NH_3(g) \rightleftharpoons NH_4^+(aq)+OH^-(aq)$ $K = [NH_4^+][OH^-]/[NH_3][H_2O]$

21. Write the mass-action expression for each reaction.
    (a) $2\ H_2O_2(g) \rightleftharpoons 2\ H_2O(g) + O_2(g)$
    (b) $2\ NO_2(g) \rightleftharpoons N_2O_4(g)$
    (c) $C_6H_6(g) + 3\ H_2(g) \rightleftharpoons C_6H_{12}(g)$
    (d) $H_2(g) + I_2(g) \rightleftharpoons 2\ HI(g)$
22. Write the mass-action expression for each reaction.
    (a) $2\ O_3(g) \rightleftharpoons 3\ O_2(g)$
    (b) $2\ HI(g) \rightleftharpoons H_2(g) + I_2(g)$
    (c) $K_2CO_3(aq) + H_2O(aq) \rightleftharpoons 2K^+(aq) + 2\ OH^-(aq) + CO_2(g) + H_2O(g)$
    (d) $AgNO_3(s) \rightleftharpoons Ag^+(aq) + NO_3^- - (aq)$
23. Write the mass action expression (K) for each of the following reactions:
    (a) $2\ HI(g) \rightleftharpoons H_2(g) + I_2(g)$
    (b) $CaCO_3(s) \rightleftharpoons CaO(s) + CO_2(g)$
    (c) $CO_2(g) + H_2(g) \rightleftharpoons CO(g) + H_2O(g)$
    (d) $H_2(g) + I_2(g) \rightleftharpoons 2HI(g)$
24. Write the mass action expression (K) for each of the following reactions:
    (a) $CO(g) + Cl_2(g) \rightleftharpoons COCl_2(g)$
    (b) $NH_3(aq) + HCl(g) \rightleftharpoons NH_4Cl(aq)$
    (c) $2\ SO_3(g) \rightleftharpoons 2\ SO_2(g) + O_2(g)$
    (d) $2H_2O_2(l) \rightleftharpoons O_2(g) + 2H_2O(l)$
25. Using the equilibrium line chart shown below for the reaction of carbon monoxide and water:

$$CO(g) + H_2O(g) \rightleftharpoons CO_2(g) + H_2(g) \qquad K = ????$$

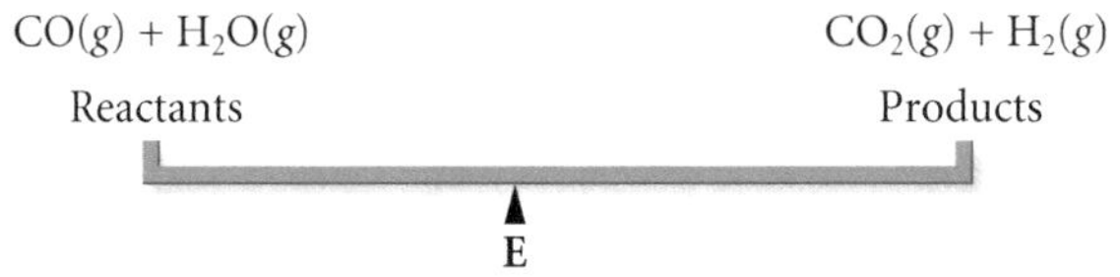

$CO(g) + H_2O(g)$ $CO_2(g) + H_2(g)$

Reactants Products

E

    a. Estimate the value of K.
    b. Based on this information, what can we say about the relative concentrations of carbon monoxide and carbon dioxide when the reaction has reached equilibrium?

26. Based on the equilibrium line charts for the following hypothetical reactions, which reaction would produce fewer moles of product? Explain why this is possible. (Assume that each reaction has the same initial conditions.)

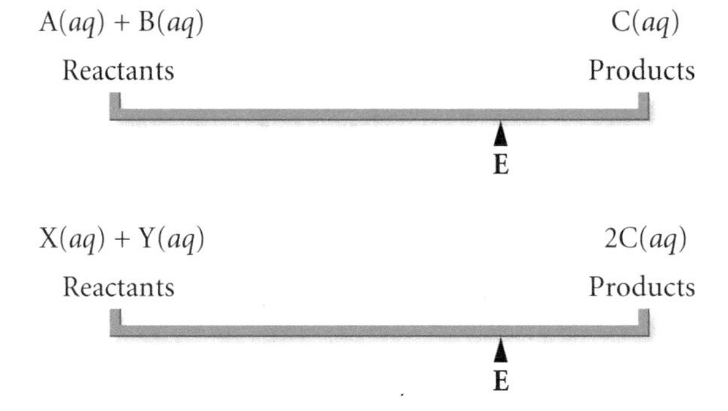

Chemical Applications and Practices

27. Some camp stoves and portable burners operate via the combustion of propane (C_3H_8). Balance the following combustion equation and write the equilibrium expression for the reaction.

$$C_3H_8(g) + O_2(g) \rightleftharpoons CO_2(g) + H_2O(g)$$

28. Ethanol (C_2H_5OH) is widely used as a fuel additive in gasoline.
    a. Balance the following combustion equation and write the equilibrium expression for this important reaction.

$$C_2H_5OH(g) + O_2(g) \rightleftharpoons CO_2(g) + H_2O(g)$$

    b. Would we expect the value for K for this reaction, at combustion temperatures, to be large or small? Explain.

29. Hydrogen peroxide, in a dilute solution, is a very commonly used disinfectant. The following balanced equation shows the decomposition reaction for gas-phase hydrogen peroxide.

$$2H_2O_2(g) \rightleftharpoons 2H_2O(g) + O_2(g)$$

Given the following initial and final concentrations, determine the equilibrium concentration of O_2.

$$[H_2O_2]_0 = 0.100; \quad [H_2O]_0 = 0.050; \quad [O_2]_0 = 0.025$$
$$[H_2O_2] = 0.050; \quad [H_2O] = 0.100; \quad [O_2] = ?$$

30. The Haber process, at 450 °C, has an equilibrium constant $K = 0.30$. What is the equilibrium concentration of hydrogen in a system containing the following equilibrium con0centrations of nitrogen and ammonia?

$$[N_2] = 0.100; \ [NH_3] = 0.100; \ [H_2] = ?$$

31. Molybdenum, element 42, has many uses, perhaps most notably in the production of a strong type of steel. The reaction shown here is involved in one of the steps in recovering molybdenum from its natural ore.

$$2MoS_2(s) + 5O_2(g) \rightleftharpoons 2MoO_3(s) + 4SO(g)$$

Write the proper mass-action expression for the equilibrium constant for this reaction.

32. The following reaction shows the dissociation of magnesium hydroxide into ions. This compound can be found in several commercial antacids.

$$Mg(OH)_2(s) \rightleftharpoons Mg^{2+}(aq) + 2OH^-(aq)$$

Write the proper mass-action expression for the equilibrium constant for this reaction.

33. At 450 °C, the Haber process has a K value of approximately 0.30. At 25 °C, the reaction between NO from airplane exhaust and ozone has a K value of approximately 3.0×10^{34}. The ionization reaction of acetic acid in vinegar, at 25 °C, is approximately 1.8×10^{-5}. Complete the following table and compare the three systems.

System	K	Reactant favored	Product favored	Mixture
Ammonia synthesis	0.30			
Ozone depletion	3.0×10^{34}			
Acid ionization	1.8×10^{-5}			

34. Aqueous solutions of hydrofluoric acid (HF) have the unique property of being able to dissolve glass. The ionization of HF in water has an equilibrium value, at 25 °C, of approximately 3.5×10^{-4}. The decomposition of gaseous dinitrogen tetroxide ($N_2O_4(g)$), rocket fuel used on the lunar landers of the Apollo missions, has a K value of 0.133. The reaction of chlorine gas (Cl_2) and carbon monoxide (CO) to form phosgene ($COCl_2$), a deadly nerve gas, has a K value of 3.7×10^9 at 25 °C. Complete the following table and compare the three systems.

System	K	Reactant favored	Product favored	Mixture
Ionization of HF	3.5×10^{-4}			
Decomposition	0.133			
Phosgene reaction	3.7×10^9			

35. Water hardness is a property of water that depends on the concentration of calcium and magnesium ions. One method used to determine these concentrations is titration with a known solution of EDTA. The goal in the titration is to produce a reaction between the dissolved metal ions and EDTA that converts essentially all the dissolved ions into compounds involving EDTA.
    a. Explain why this is a desirable outcome in the analysis of water hardness.
    b. Explain why K values on the order 10^{10} for the reactions between calcium and EDTA give us more confidence in the analysis than we would have if the K values were on the order of 10.

36. In aqueous solutions, acids dissociate to produce hydronium ions and a negative ion. If the reverse reaction is favored, the acid is considered weak. Judging on the basis of the following two acid dissociation reactions, which acid is the weaker acid?

Acetic acid (found in vinegar) $K = 1.8 \times 10^{-5}$
Benzoic acid (found in berries) $K = 6.5 \times 10^{-5}$

37. At elevated temperatures, hydrogen peroxide establishes the following equilibrium.

$$2H_2O_2(aq) \rightleftharpoons O_2(g) + 2H_2O(aq)$$

The equilibrium concentrations of the gases at 1250 K are 0.124 mol/L for H_2O_2, 0.018 mol/ L for O_2, and 0.036 mol/L for H_2O. Calculate the value of K.

38. At elevated temperature and pressure, the reaction indicated by the following equation has an equilibrium constant, K, equal to 1.211.

$$2CO_2(g) \rightleftharpoons 2CO(g) + O_2(g)$$

Calculate the equilibrium constant, K, for the reverse equation.

39. Phosgene decomposes to carbon monoxide and chlorine.

$$COCl_2(g) \rightleftharpoons CO(g) + Cl_2(g)$$

K is 0.337 at 225 °C. If we place 0.500 mol of $COCl_2$ in a 1.00-L flask, seal the flask and then heat it to 225 °C, what are the equilibrium concentrations of $COCl_2$, CO, and Cl_2? What percentage of the original $COCl_2$ decomposed at this temperature?

40. Carbon dioxide decomposes to carbon monoxide and oxygen.

$$2CO_2(g) \rightleftharpoons 2CO(g) + O_2(g)$$

K is 6.1×10^{-8} at 300 °C. If we place 0.50 mol of CO_2 in a 2.00-L flask, seal the flask and then warm it to 300 °C, what are the equilibrium concentrations of CO_2, CO, and O_2? What percentage of the original CO_2 decomposed at this temperature?

41. An equilibrium mixture of NOCl, NO, and Cl_2 at a high temperature contains the gases at the following concentrations: $[NOCl]=2.52 \times 10^{-2}$ mol/L, $[NO]=8.10 \times 10^{-3}$ mol/L, and $[Cl_2]=4.05 \times 10^{-3}$ mol/L. Calculate the equilibrium constant, K, for the reaction.

$$2\ NOCl(g) \rightleftharpoons 2nO(g) + Cl_2(g)$$

42. A mixture of CO, Cl_2, and $COCl_2$ is placed in a reaction flask. The equilibrium concentrations of each of the gases are: $[CO]=0.0102$ mol/L, $[Cl_2]=0.00609$ mol/L, $[COCl_2]=0.0294$ mol/L. Calculate the equilibrium constant, K, for the reaction.

14

$$CO(g) + Cl_2(g) \rightleftharpoons COCl_2(g)$$

Section 14.5 Working with Equilibrium Constants

Skill Review

43. Water is a product in many chemical reactions. It can be directly made from its elements:
    a. Write the equilibrium expression for the reaction.

$$2H_2(g) + O_2(g) \rightleftharpoons 2H_2O(g)$$

    b. If the equilibrium concentrations for the compounds, at a specific temperature, were as follows, what would be the numerical value for K?

$$[H_2] = 0.134\ M; \qquad [O_2] = 0.673\ M; \qquad [H_2O] = 1.00\ M$$

    c. What is the value for K for the reverse reaction?
    d. What is the value of K for the reaction of only 1 mol of hydrogen and 0.5 mol of oxygen to make 1 mol of water (that is, if we halve the coefficients in this reaction)?

44. If a solid bar of zinc is placed in a solution containing 1 M silver ions (Ag^+), a spontaneous reaction takes place producing zinc ions and silver metal with an equilibrium constant value of approximately 8×10^{52}.
    a. What would we calculate as the equilibrium constant for the reverse reaction?
    b. What would we calculate as the equilibrium constant if the stoichiometric coefficients were doubled?

45. A gaseous reaction mixture contains 0.10 atm N_2, 0.42 atm H_2, and 0.50 atm NH_3 in a 5.0-L container. $K=0.0250$ for the equilibrium system

$$N_2(g) + 3\ H_2(g) \rightleftharpoons 2\ NH_3(g)$$

(a) Is the system at equilibrium?

(b) If it is not at equilibrium, in which direction will the system move to reach equilibrium?

46. A gaseous reaction mixture contains 0.33 atm NO_2 and 0.12 $_{atm}$ N_2O_4 in a 2.0-L container. $K=310$ for the equilibrium system

$$2\ NO_2(g) \rightleftharpoons N_2O_4(g)$$

(a) Is the system at equilibrium?

(b) If it is not at equilibrium, in which direction will the system move to reach equilibrium?

47. Consider the following equilibria involving $H_2O_2(g)$ and their corresponding equilibrium constants.

$$2H_2O_2(l) \rightleftharpoons O_2(g) + 2H_2O(l) \qquad K_1$$
$$1/2O_2(g) + H_2O(l) \rightleftharpoons H_2O_2(l) \qquad K_2$$

What is the relationship between K_1 and K_2?

48. Consider the following equilibria involving NH_3(g) and their corresponding equilibrium constants.

$$N_2(g) + 3\ H_2(g) \rightleftharpoons 2\ NH_3(g) \qquad K_1$$
$$NH_3(g) \rightleftharpoons 1/2N_2(g) + 3/2\ H_2(g) \qquad K_2$$

What is the relationship between K_1 and K_2?

Chemical Applications and Practices

49. When scientists seek information about air pollution in large cities, one of the reactions they study is:

$$2NO(g) + O_2(g) \rightleftharpoons 2NO_2(g)$$

If K for the reaction, at 25 °C, is approximately 1.7×10^{12} and the specific rate constant for the reverse reaction is approximately 6.6×10^{-12}, what would we calculate as the value of the rate constant for the forward reaction? (Units have been omitted for this problem.)

50. Ethyl acetate, a common laboratory solvent, can be prepared by the following reaction of acetic acid and ethanol:

$C_2H_4O_2$	$+ C_2H_6O \rightleftharpoons$	$C_4H_8O_2 + H_2O$
acetic	ethanol	ethyl acetate

If K for the reaction, at 25 °C, is 2.2 and the specific rate constant for the forward reaction is 4.22×10^{10}, what is the value of the rate constant for the reverse reaction? (Units have been omitted for this problem.)

51. Producing industrially useful amounts of acetylene (C_2H_2) is important for more than acetylene torches. Acetylene is also used as a starting compound for many important polymers, such as vinyl chloride (used in PVC materials) and acrylonitrile.
   a. Using the equation $CH_4(g) \rightleftharpoons 1/2\ C_2H_2(g) + 3/2\ H_2(g)$, write the appropriate mass-action equilibrium expression.
   b. Rewrite the equation showing the stoichiometric ratios for the production of 1 mol of $C_2H_2(g)$. Write the appropriate mass-action equilibrium expression for this equation.
   c. If the mathematical value for the equilibrium constant for the reaction in part a were known, how would it be mathematically converted to the equilibrium constant for the reaction in part b?

52. One of the reactions used to produce formaldehyde (CH_2O) is

$$CH_3OH(g) + 1/2O_2(g) \rightleftharpoons CH_2O(g) + H_2O(g)$$

   a. Write the mass-action equilibrium expression for the reaction.
   b. Rewrite the reaction, keeping the same stoichiometric ratio, but showing the reaction utilizing 1 mol of oxygen. Write the mass-action equilibrium expression for this reaction.
   c. How are the two equilibrium expressions mathematically related?

53. The reaction that allows many biochemical reactions to take place involves the breakdown of ATP (adenosine triphosphate) into ADP (adenosine diphosphate). The equilibrium constant of this reaction, at 37 °C, is approximately 1.4×10^5. One of the early steps in the breakdown of glucose from food is the attachment of a phosphate group. The equilibrium constant for this process is approximately 4.7×10^{-3}. In living cells these two reactions are combined. What would be the equilibrium constant for the resulting combined reaction?

54. The production of tin is important because it has many practical uses, from plating iron objects to helping deliver fluoride in toothpaste as SnF_2. Combine the first two of the following reactions, and use the appropriate equilibrium constants, to obtain the equilibrium constant for the third reaction.

$$SnO_2(s) + 2CO(g) \rightleftharpoons Sn(s) + 2CO_2(g) \qquad K_1 = 14$$
$$CO(g) + H_2O(g) \rightleftharpoons CO_2(g) + H_2(g) \qquad K_2 = 1.3$$
$$SnO_2(s) + 2H_2(g) \rightleftharpoons Sn(s) + 2H_2O(g) \qquad K_3 = ?$$

55. Many compounds have more than one important use. Such is the case with the weak acid phenol (C_6H_5OH). It can be used, in dilute form, as an antiseptic and as a component in making some plastics. In water, phenol dissociates slightly as shown here:

$$C_6H_5OH(aq) \rightleftharpoons H^+(aq) + C_6H_5O^-(aq)$$

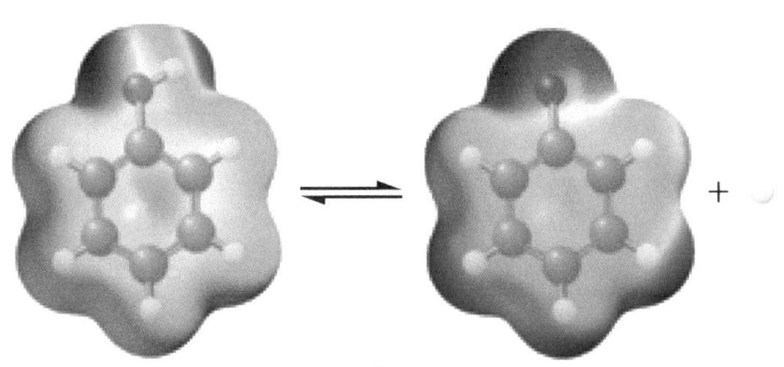

The equilibrium constant for this reaction is 1.3×10^{-10}. To analyze a solution containing phenol, we may carefully add measured amounts of aqueous NaOH. Recall that in water, the following reaction also takes place:

$$H^+ + OH^- \rightleftharpoons H_2O \qquad K = 1.0 \times 10^{14}$$

Adding these two equations produces a third that represents the reaction between sodium hydroxide and phenol. Using equilibrium constants, evaluate the feasibility of the analysis—that is, whether the reaction will proceed toward products enough for the analysis to be performed.

56. Procaine ($C_{13}H_{20}N_2O_2$) can be used to produce the anesthetic novocaine ($C_{13}H_{21}N_2O_2Cl$). Procaine is a weak base that undergoes the following reaction in aqueous solutions:

$$C_{13}H_{20}N_2O_2 + H_2O \rightleftharpoons OH^- + C_{13}H_{20}N_2O_2H^+ \qquad K = 7.1 \times 10^{-6}$$

If a solution of procaine were to be analyzed by adding carefully measured amounts of hydrochloric acid (HCl), explain how we could use the following reaction to assist in the evaluation of the viability of the analysis.

$$H^+ + OH^- \rightleftharpoons H_2O \qquad K = 1.0 \times 10^{14}$$

57. Using information from Problems 55 and 56, determine the equilibrium constant value for the reaction between phenol and procaine. Would it be a reasonable analytical strategy to employ phenol as a reactant to determine the concentration of procaine in solution?
58. Using the information from Problems 55 and 56, determine the equilibrium constant value for the reaction between HCl and NaOH. Would it be a reasonable analytical strategy to employ HCl as a reactant to determine the concentration of a NaOH solution?
59. At temperatures near 400 °C, the K_p value for the synthesis of ammonia is 2.5×10^{-4}. At 400 °C, what would be the approximate value of K for the following reaction?

$$3H_2(g) + N_2(g) \rightleftharpoons 2NH_3(g)$$

60. The reaction of nitrogen monoxide and chlorine gas has a large value for the equilibrium constant, K, at 298 K. However, it is often easier to measure the partial pressures of each component of a gaseous system than it is to measure their concentration. What is the value of K_p at this temperature?

$$2NO(g) + Cl_2(g) \rightleftharpoons 2NOCl(g) \qquad K = 6.3 \times 10^4 \text{ (at 298 K)}$$

Section 14.6 Solving Equilibrium Problems—A Different Way of Thinking

Skill Review

61. Consider the acetic acid system described earlier in the text:

$$CH_3COOH(aq) \rightleftharpoons CH_3COO^-(aq) + H^+(aq) \qquad K = 1.8 \times 10^{-5}$$

What is the equilibrium hydrogen ion concentration, $[H^+]$, given each of these initial concentrations?

a. $[CH_3COOH]_0 = 0.500$ M; $[CH_3COO^-]_0 = 0.0$ M
b. $[CH_3COOH]_0 = 0.100$ M; $[CH_3COO^-]_0 = 0.100$ M
c. $[CH_3COOH]_0 = 0.010$ M; $[CH_3COO^-]_0 = 0.0$ M

62. Consider the chemical system:

$$COCl_2(g) \rightleftharpoons CO(g) + Cl_2(g) \qquad K = 1.6 \times 10^{-5}$$

What is the equilibrium concentration of carbon monoxide, [CO], given each of these initial concentrations? Assume the initial concentration of CO is 0.00 M.
a. $[COCl_2]_0 = 0.50$ M; $[Cl_2]_0 = 0.0$ M
b. $[COCl_2]_0 = 1.5$ M; $[Cl_2]_0 = 2.00$ M
c. $[COCl_2]_0 = 2.25$ M; $[Cl_2]_0 = 1.2$ M

63. At some temperature, the reaction $H_2 + I_2 \rightleftharpoons 2HI$ has $K = 617$. Predict in which direction, forward or reverse, the reaction would proceed when:
a. $[H_2]_0 = 0.240$ M; $[I_2]_0 = 0.080$ M; $[HI]_0 = 0.20$ M
b. $[H_2]_0 = 0.030$ M; $[I_2]_0 = 0.100$ M; $[HI]_0 = 1.50$ M

64. At some temperature, the reaction $H_2 + I_2 \rightleftharpoons 2HI$ has $K = 617$. Predict in which direction, forward or reverse, the reaction would proceed when:
a. $[H_2]_0 = 0.990$ M; $[I_2]_0 = 0.280$ M; $[HI]_0 = 0.500$ M
b. $[H_2]_0 = 0.250$ M; $[I_2]_0 = 1.000$ M; $[HI]_0 = 0.500$ M

65. Using the value of $K = 8.6 \times 10^{-5}$ for the myoglobin and oxygen reaction described in the text, $Mb + O_2 \rightleftharpoons MbO_2$, indicate which of the systems described in the following table are at equilibrium. If an example is not at equilibrium, predict the direction of change (forward or reverse) that would take place to attain the equilibrium condition.

	[Mb] (M)	$[O_2]$ (M)	$[MbO_2]$ (M)
a.	3.5×10^{-4}	2.5×10^{-4}	0
b.	1.0×10^{-4}	1.0×10^{-4}	8.6×10^{-13}
c.	2.0×10^{-4}	1.5×10^{-4}	2.6×10^{-11}
d.	0	2.5×10^{-4}	1.0×10^{-10}

66. The decomposition of 2 mol of carbon dioxide into carbon monoxide and oxygen gas can occur under certain conditions. If, at a particular temperature, $K = 4.5 \times 10^{-5}$, predict the direction of change (forward or reverse) that would take place to attain the equilibrium condition. (Hint: Start by writing the balan10ced equation.)

	$[CO_2]$ (M)	[CO] (M)	$[O_2]$ (M)
a.	0.44	1.0×10^{-5}	1.0×10^{-5}
b.	1.0×10^{-4}	0.22	1.5×10^{-8}
c.	6.3×10^{-8}	3.9×10^{-2}	4.7×10^{-12}

67. Barium sulfate ($BaSO_4$) is a slightly soluble compound that has some medical applications when used to diagnose gastrointestinal problems. Because barium can be toxic, it is important to keep the concentration of barium ions at a minimum. Given the following reaction and equilibrium value, determine the maximum equilibrium concentration of $Ba^{2+}(aq)$ when in the presence of solid $BaSO_4$.

$$BaSO_4(s) \rightleftharpoons Ba^{2+}(aq) + SO_4^{2-}(aq) \qquad K = 1.1 \times 10^{-10}$$

68. Many municipal water supplies are treated with fluoride ions (F^-) in an attempt to strengthen dental enamel. If the water contains significant amounts of calcium ions, a precipitation reaction can take place. When solid calcium fluoride is present in water, the following equilibrium is established:

$$CaF_2(s) \rightleftharpoons Ca^{2+}(aq) + 2F^-(aq) \qquad K = 4.0 \times 10^{-11}$$

What would be the concentration of fluoride ion present at equilibrium?

69. One of the least soluble compounds known is antimony sulfide (Sb_2S_3). If the concentration of antimony ion in a solution in which Sb_2S_3 was present were 2.2×10^{-19}, and no other solute was present, what would we calculate as the equilibrium constant for the following reaction?

$$Sb_2S_3(s) \rightleftharpoons 2Sb^{3+}(aq) + 3S^{2-}(aq)$$

70. The equilibrium constant for a reaction is very temperature dependent. Assume 0.060 is the equilibrium constant for the Haber process at a given temperature.

$$N_2(g) + 3H_2(g) \rightleftharpoons 2NH_3(g)$$

What is the equilibrium concentration of H_2 if the equilibrium concentrations of N_2 and NH_3 are found both to be 0.0010 M?

71. Using the same value of K for the ammonia synthesis reaction in Problem 70, solve for:
   a. The equilibrium concentration of NH_3 when $[N_2]=0.0010$ M and $[H_2]=0.010$ M
   b. The equilibrium concentration of H_2 when $[NH_3]=0.020$ M and $[N_2]=0.015$ M

72. Calculate the $[NO_2]$ given the equilibrium concentrations based on the reaction
   $2NO(g)+O_2(g) \rightleftharpoons 2NO_2(g)$ $K=1.71\times10^{12}$.
   a. $[NO]=0.00020$ M; $[O_2]=0.000050$ M; $[NO_2]=?$
   b. $[NO]=0.00010$ M; $[O_2]=0.000010$ M; $[NO_2]=?$

Chemical Applications and Practices

73. Codeine ($C_{18}H_{21}NO_3$), an analgesic drug obtained by prescription, produces the following reaction when added to water.

$$C_{18}H_{21}NO_3(aq) + H_2O(l) \rightleftharpoons OH^-(aq) + C_{18}H_{21}NO_3H^+(aq) \quad K = 1.6\times10^{-6}$$

   a. What reactions are taking place in the solution?
   b. Which among these are important? Which are unimportant?
   c. If a solution had the following concentrations, in which direction would it proceed?
   $[C_{18}H_{21}NO_3]_0 = 0.10$ M; $[OH^-]_0 = 0$ M; $[C_{18}H_{21}NO_3H^+]_0 = 0$ M
   d. Once equilibrium was achieved, what would be the concentrations of the species mentioned in part c?

74. Acetylsalicylic acid ($C_9H_8O_4$, also known as aspirin) dissociates in water with an equilibrium constant, at 25 °C, of 3.0×10^{-4}.

$$C_9H_8O_4(aq) \rightleftharpoons H^+(aq) + C_9H_7O_4^-(aq)$$

   a. If the initial concentration of $C_9H_8O_4$ were 0.10 M, what would we calculate as the amount of $C_9H_8O_4$ remaining at equilibrium?
   b. What percentage of $C_9H_8O_4$ reacted?
   c. Draw an equilibrium line chart that illustrates this system.

75. One method to obtain silver metal from impure lead samples is named after the work of Samuel Parkes (1761–1825). A silver-containing lead sample is melted, and zinc is added to the molten sample. The molten zinc makes a coating on the surface. The molten silver is approximately 300 times more soluble in the molten zinc than in the impure molten lead. (The zinc–silver mixture is later removed, and pure silver is obtained by distilling away the zinc.) An equilibrium constant can be written for the concentration of silver in the lead mixture versus the amount in the zinc: $K_D=[Ag(Zn)]/[Ag(Pb)]$. If the [Ag(Zn)] was 0.0010 M and the [Ag(Pb)] was 0.000011 M, would it be wise to wait to see whether more Ag could be extracted? Or has the extraction for this system reached a maximum? Explain the answer.

76. The solubility of a solute in one solvent compared to another can be used to our advantage. The ratio of the dissolved amounts of solute to solvent, called a partition coefficient, may be used in a similar way as the distribution constant described earlier. The K_D value for a compound between a water layer and an ether layer is 0.024. (Note: This represents the amount dissolved in ether divided by the amount dissolved in water.) Suppose that the compound was ether extracted from a plant and is now 0.015 M in ether. What will be the molarity of the compound that will form, in water, when water is placed in contact with the ether?

77. At relatively high temperatures, the following reaction can be used to produce methyl alcohol:

$$CO(g) + 2H_2(g) \rightleftharpoons CH_3OH(l) \quad K = 13.5$$

   a. If the concentration of CO, at equilibrium, were found to be 0.010 M, what would be the equilibrium concentration of hydrogen gas?
   b. Draw an equilibrium line chart that illustrates this system.

78. Pyruvic acid is produced as an intermediate during the metabolism of carbohydrates in cells. In water it undergoes the following reaction:

$$CH_3COCOOH(aq) \rightleftharpoons H^+(aq) + CH_3COCOO^-(aq) \quad K = 6.6\times10^{-3}$$

a. If the equilibrium concentration of $CH_3COCOOH$ were found to be 0.0010 M, what would be the equilibrium concentration of CH_3COCOO^-?
b. Draw an equilibrium line chart that illustrates this system.

79. At high temperatures, methane (CH_4) can be reacted with steam to produce carbon monoxide and hydrogen gas, as shown in the following reaction:

$$CH_4(g) + H_2O(g) \rightleftharpoons CO(g) + 3H_2(g)$$

If the equilibrium constant is 0.25, at a specific temperature and the equilibrium concentrations of $[CH_4]=0.11$ M, $[H_2O]=0.28$ M, and $[CO]=0.75$ M, what would we calculate as the $[H_2]$?

80. The aroma of rotten eggs can be partially attributable to the foul-smelling sulfur compound H_2S. Hydrogen sulfide can dissolve in water to form an aqueous solution:

$$H_2S(g) \rightleftharpoons H_2S(aq)$$

The equilibrium constant for the process at a specific temperature is 0.0200. If the initial concentration of $H_2S(g)$ was 0.00100 M in a sealed flask, what would we determine to be the equilibrium concentrations of H_2S as a gas and in the solution?

81. One source of ethylene (C_2H_4) is the hydrogenation of acetylene (C_2H_2). The equilibrium constant for the reaction varies greatly with temperature. If the equilibrium constant for the following reaction is 4.2×10^{15}, what would be the equilibrium concentration of hydrogen gas in a system when the equilibrium concentrations of C_2H_2 and C_2H_4 were, respectively, 1.2×10^{-5} M and 0.025 M?

$$C_2H_2(g) + H_2(g) \rightleftharpoons C_2H_4(g)$$

82. Butanoic acid ($CH_3CH_2CH_2COOH$) has the aroma of spoiled butter. A better-smelling compound, an ester, can be made when butanoic acid is treated with methanol and an acid catalyst.

$$CH_3CH_2CH_2COOH + CH_3OH \rightleftharpoons CH_3CH_2CH_2COOCH_3 + H_2O$$

Under certain conditions, the equilibrium constant for the reaction is 1.5×10^{-2}. What would be the equilibrium concentration of the product ester if the initial concentrations of each reactant were 0.500 M? Note that water is not the solvent and must be included in *K*.

83. The element vanadium is unusually resistant to corrosion. Alloyed with iron (approximately 5% vanadium) it produces a useful type of steel. One reaction used to obtain vanadium has a *K* value of 14 at 298 K.

$$VO^+(aq) + 2H^+(aq) \rightleftharpoons V^{3+}(aq) + H_2O(l)$$

Starting with $[VO^+]_0=0.15$ M and $[H^+]=0.100$ M, what would we calculate as the equilibrium concentrations of $VO^+(aq)$, $H^+(aq)$, and $V^{3+}(aq)$?

84. Using the reaction in Problem 85, decide in which direction the reaction would proceed when the following concentrations were known. (Prove the answers.)
a. $[VO^+] = 0.10$ M; $[H^+] = 0.25$ M; $[V^{3+}] = 0.85$ M
b. $[VO^+] = .0025$ M; $[H^+] = 0.10$ M; $[V^{3+}] = 0.025$ M

85. Using the *K* value of 8.6×10^{-5} presented in the text for the myoglobin and oxygen reaction $Mb + O_2 \rightleftharpoons MbO_2$, what would we calculate as the [Mb], $[O_2]$, and $[MbO_2]$ when the initial concentrations were as follows: $[Mb]=0.00100$ M; $[O_2]=0$ M, and $[MbO_2]=0.000020$ M?

86. Some compounds have the same chemical formula and are known as isomers. One such compound is 4-pentan-1-ol. Under certain conditions, it can adopt either a linear or cyclic structure:

$$C_5H_8O \text{ (linear)} \rightleftharpoons C_5H_8O \text{ (cyclic)} \qquad K = 2.2 \text{(at 25 °C)}$$

What is the equilibrium concentration of each component of the mixture if the initial concentrations of each component in the mixture were 0.100 M?

Section 14.7 Le Châtelier's Principle

Skill Review

87. Newton's first law of motion can be stated like this: "A body at rest tends to stay at rest, and a body in motion tends to stay in motion, unless acted upon by an external force." How does this relate to what we now know about chemical systems in regard to equilibrium?

88. Newton's third law of motion can be stated like this: "For every action, there is an equal and opposite reaction." How does this relate to chemical equilibrium and Le Châtlier's Principle?
89. The text states that when a system is pushed away from it's equilibrium state by adding one compound to the system, it will attempt to restore concentrations, pressures, etc., but they don't quite make it back to what they were previously. What is restored to it's original value when a system moves back to equilibrium?
90. If a chemical system is at equilibrium, what is true about the forward and reverse reaction rates? If the system is pushed away from equilibrium by increasing the concentration of products, what will temporarily happen to the reverse reaction rate?
91. Consider the following reaction at 200 °C.

$$A(s) + 2B(g) \rightleftharpoons 2C(g) + 2D(l) \qquad K_p = 4.0$$

A 20.0L reaction vessel at 200 °C initially holds 10.0 g of A, 10.0 g of D, 0.200 atm of B and 0.600 atm of C. Calculate the partial pressures of B and C when equilibrium is established.
When the following changes are made, will it cause a shift to the right, the left, or will there be no effect at all?
   a. increase the amount of D
   b. decrease the concentration of C
   c. increase the concentration of B
92. For the same chemical system as the one described in problem 91, predict the effect on the partial pressure of C (will it increase or decrease?) when the following changes are made after the system has achieved equilibrium:
   a. double the mass of A
   b. double the size of the container
   c. decrease the concentration of B
93. Which of the changes to the given equilibrium will NOT result in a shift toward products?

$$N_2(g) + 3\,H_2(g) \rightleftharpoons 2\,NH_3(g) \quad \Delta H = -92.2\,\text{kJ}$$

   (a) removal of ammonia
   (b) addition of hydrogen
   (c) decreasing the temperature
   (d) increasing the temperature
   (e) decreasing the size of the container
94. Which of the changes to the given equilibrium will NOT result in a shift toward products?

$$2\,SO_3(g) \rightleftharpoons 2\,SO_2(g) + O_2(g) \quad \Delta H = 198\text{kJ}$$

   (a) removal of SO_3
   (b) removal of oxygen
   (c) decreasing the temperature
   (d) increasing the temperature
   (e) increasing the size of the container
95. Using Le Châtelier's principle, decide whether each of these changes would cause the equilibrium of the system presented to shift to the left, would cause it to shift to the right, or would have no effect on the equilibrium.

$$CH_4(g) + 2O_2(g) \rightleftharpoons CO_2(g) + 2H_2O(g)$$

   a. Removing CO_2 from the system
   b. Adding heat to the system
   c. Decreasing the volume of the container
   d. Adding $H_2O(g)$ to the system
   e. Adding inert He to the system
96. Calcium oxide (CaO) is also known as lime. It is one of the leading chemicals produced worldwide, thanks to its many uses in plant and animal foods, insecticides, paper making, and plaster products. It is produced, at high temperatures, from calcium carbonate.

$$CaCO_3(s) \rightleftharpoons CaO(s) + CO_2(g) \quad \text{(endothermic reaction)}$$

If the system were in a closed container, what effect (shift to the left, shift to the right, or no effect) would each of these changes have on the favored direction of the reaction?
   a. Removing CO_2
   b. Adding CaO

c. Raising the temperature
d. Enlarging the size of the container
e. Adding a suitable catalyst

Chemical Applications and Practices

97. The following equilibrium constant values are found for dissolving two solids in water to produce aqueous solutions of the ions shown. Note that the stoichiometry for the process is the same in both systems.

$$AgCl(s) \rightleftharpoons Ag^+(aq) + Cl^-(aq) \qquad K = 1.7 \times 10^{-10}$$
$$CuCl(s) \rightleftharpoons Cu^+(aq) + Cl^-(aq) \qquad K = 1.9 \times 10^{-7}$$

a. In which system would we find the greater number of moles of dissolved ions?
b. Would adding more solid to the other system increase the number of dissolved ions so that it might equal the total found in the choice for part a? Explain any reasoning.

98. The production of ethylene is important in the manufacture of polyethylene products. The following reaction shows how ethylene can be made from ethane by removing hydrogen from ethane. The reaction typically utilizes a catalyst (not shown in the equation).

$$CH_3CH_3(g) \rightleftharpoons CH_2CH_2(g) + H_2(g)$$

a. At 298 K, the equilibrium constant is 0.96. If a 1-L container initially contained 0.10 M CH_3CH_3, what would be the equilibrium concentration of all three species?
b. After equilibrium is reached, an additional 0.010 mol of H_2 is injected into the container without changing its volume. What would be the concentration of all three species when equilibrium was once again restored?

99. A dilute solution of $Na_2Co(H_2O)_6Cl_4$ has a faint pink color and can be used as "invisible ink." When some of the loosely held water is driven off, with gentle heating, Na_2CoCl_4 forms, with a visible change to blue. If we stored our "invisible ink" solution in a refrigerator, would it appear pink or blue? Explain the basis of the answer.

100. The recognizable aromas of many fruits are due to a group of compounds known as esters. For example, the following reaction shows the production of pentyl acetate. (Assume that water is not the solvent.)

$$\underset{\text{Pentanol}}{CH_3CH_2CH_2CH_2CH_2OH} + \underset{\substack{\text{Ethanoic acid} \\ \text{(acetic acid)}}}{CH_3COOH} \rightleftharpoons \underset{\substack{\text{Pentyl ethanoate} \\ \text{(pentyl acetate)}}}{CH_3CH_2CH_2CH_2CH_2OOCCH_3} + H_2O$$

Which of these methods would increase the amount of product formed, and why?
a. Add water so that the pentyl ethanoate would dissolve better.
b. Add magnesium sulfate so it would react with any water formed.

101. One way to produce rubbing alcohol ($CH_3CHOHCH_3$) is from propene:

$$CH_3{-}CH{=}CH_2 + H_2O \underset{\text{Catalyst}}{\rightleftharpoons} CH_3{-}\overset{\overset{\displaystyle OH}{|}}{C}H{-}CH_3$$

Explain what the role of the catalyst is, in relation to the equilibrium, in this reaction.

102. Tungsten's unusually high melting point (over 3000 °C) and its efficiency at producing light from electrical energy led to its use in light bulb filaments. It can be obtained via the following reaction:

$$WO_3(s) + 3H_2(g) \rightleftharpoons W(s) + 3H_2O(g)$$

Like most reactions in which a metal is obtained from another compound, this reaction is endothermic. Explain why pressure changes are not considered significant factors when tungsten is obtained in this manner.

103. For the following reactions, predict whether the pressure of the reactants or products increases or remains the same when the volume of the reaction vessel is increased.
   (a) $2H_2O(g) \rightleftharpoons O_2(g) + 2H_2(g)$
   (b) $C_2H_4O(g) + H_2O\ (l) \rightleftharpoons C_2H_2(g) + O_2(g)$
   (c) $Cl_2(g) + H_2(g) \rightleftharpoons 2HCl(g)$
104. For the following reactions, predict the direction the equilibria will shift when the volume of the reaction vessel is decreased.
   (a) $CaCO_3(s) \rightleftharpoons CaO(s) + CO_2(g)$
   (b) $2O_3(g) \rightleftharpoons 3O_2(g)$
   (c) $2CO_2(g) \rightleftharpoons 2\ CO(g) + O_2(g)$
105. What will happen to the sodium ion concentration if more solid sodium carbonate is added to the flask?

$$Na_2CO_3(s) \rightleftharpoons 2Na^+(aq) + CO_3^{2-}(aq)$$

106. What will happen to the chlorine gas pressure if the pressure of the entire system is increased through the addition of argon gas to the flask? Explain the answer.

$$PCl_5(g) \rightleftharpoons PCl_3(g) + Cl_2(g)$$

107. Heating limestone in a sealed vessel decomposes the compound into calcium oxide and carbon dioxide.

$$CaCO_3(s) \rightleftharpoons CaO(s) + CO_2(g)$$

Predict the effect on the equilibrium of each change listed below.
   (a) add $CaCO_3$
   (b) add CO_2
   (c) add CaO
   (d) raise the temperature
   (e) decrease the volume of the reaction
108. The decomposition of a generic compound is endothermic and produces oxygen gas.

$$2A(g) \rightleftharpoons 2B(g) + O_2(g)$$

Predict the effect of each of these changes on the position of the equilibrium.
   (a) add O_2
   (b) add A
   (c) decrease the volume of the reaction
   (d) add B
   (e) raise the temperature

109. The value of the mass action expression (K) at 25 °C for

$$COCl_2(aq) + 2\ H_2O(l) \rightleftharpoons H_2CO_3(aq) + 2\ HCl(g)$$

is 6.5×10^{13}. Explain what will happen when phosgene ($COCl_2$) is mixed with water at this temperature.

110. The value of the equilibrium constant for the following reaction at a specific temperature is 4.9×10^{-33}.

$$CaCO_3(s) \rightleftharpoons CaO(s) + CO_2(g)$$

Explain what will happen when a sample of calcium carbonate is sealed in a flask at this temperature.

Comprehensive Problems

111. Acids react by donating hydrogen ions. The strength of the acid reaction is determined by the ability the acid has to donate the H^+ to water. Symbolically, the reaction in water can be represented as follows:

$$HA \rightleftharpoons H^+ + A^-$$

The following is a list of some common weak acids followed by their water reaction equilibrium values. All are compared at the same temperature. Which acid in the list is the weakest? Which is the strongest?
Acetic acid, $HC_2H_3O_2$ (found in vinegar) $K = 1.8 \times 10^{-5}$
Formic acid, $HCHO_2$ (found in ants) $K = 1.8 \times 10^{-4}$
Benzoic acid, $HC_7H_5O_2$ (found in some berries) $K = 6.3 \times 10^{-5}$

112. Scientists have investigated hydrazine (N_2H_4) and nitrogen monoxide (NO) in their study of rocket fuels. Use the following reaction to write the mass-action expression for this hydrazine reaction.

$$N_2H_4(g) + 2NO(g) \rightleftharpoons 2N_2(g) + 2H_2O(g)$$

113. Suppose there are two synthetic routes by which a pharmaceutical company can make the same prescription drug. Method A uses expensive starting materials but has a large value for K. Method B uses inexpensive starting materials but has a small value for K. We have been asked to discuss briefly, in a planning meeting, the various considerations involved in deciding between these two methods. What would we say about the arguments for making a profit with either method?
114. Is it possible for K_p to equal K? Explain the answer.
115. Using only whole-number coefficients, write a balanced chemical equation that represents the conversion of oxygen molecules into ozone. If the equilibrium constant for that reaction were 7×10^{-58}, what would we calculate as the equilibrium constant for the reaction that produces 1 mol of ozone?
116. The electrochemical reaction powering a nickel–cadmium rechargeable battery has an equilibrium constant value, at 298 K, of approximately 1.5×10^{11}, as written below. What is the value for the reverse reaction of this process? The reaction takes place in an alkaline solution.

$$Cd(s) + NiO_2(s) \rightleftharpoons Ni(OH)_2(s) + Cd(OH)_2(s)$$

117. An owner of a coffee shop, who happens to be a former chemistry student, notices that there are some similarities between customers (that is, people in the shop and those not entering the shop) and the reactants and products at equilibrium in a chemical system. What type of customer movement would represent equilibrium for the shop? What type of equilibrium would the profit-minded owner prefer, one with a large value of K or a small value of K? Explain. If Q were less than K, would it be a good business day or a poor business day? Explain.

$$K = \frac{[\text{customers}]}{[\text{people on the street}]}$$

118. As we have noted previously, the presence of fluoride in some municipal water systems can bring about the precipitation of CaF_2 if the concentration of Ca^{2+} is very high.

$$CaF_2(s) \rightleftharpoons Ca^{2+}(aq) + 2F^-(aq) \qquad K = 4.0 \times 10^{-11}$$

   a. How many moles per liter of CaF_2 would dissolve in pure water?
   b. If the concentration of Ca^{2+} in a water sample were 0.0050 M, as might be present in hard water samples, how many moles per liter of CaF_2 would dissolve?
119. The following reaction has historical importance as it was once used to produce a useful fuel called "water gas." The process involved using steam to convert coal into carbon monoxide and hydrogen gas.

$$C(s) + H_2O(g) \rightleftharpoons CO(g) + H_2(g) \qquad K_p = 21 (\text{at } 1000\,°C)$$

If the initial partial pressure of $H_2O(g)$ were 52 atm, what would we calculate as the equilibrium partial pressures of $H_2O(g)$; $CO(g)$ and $H_2(g)$?
120. Carbonic acid, present in carbonated beverages, can participate in two reactions that both involve removing a H^+ from the molecule.

$$H_2CO_3(aq) \rightleftharpoons H^+(aq) + HCO_3^- - (aq) \qquad K = 4.3 \times 10^{-7}$$

$$HCO_3^- - (aq) \rightleftharpoons H^+(aq) + CO_3^{2-}(aq) \qquad K = 5.6 \times 10^{-11}$$

Starting with an initial concentration of H_2CO_3 of 0.10 M, what would we calculate as the equilibrium concentrations of $[H_2CO_3]$, $[HCO_3^-]$, $[H^+]$, and $[CO_3^{2-}]$? Be sure to justify any assumptions we made to solve the problem.
121. Examine the hypothetical reaction illustrated below. A snapshot of the reaction was taken at specific times during the course of the reaction. Which frame represents the first frame in which equilibrium has been reached? Explain the answer.

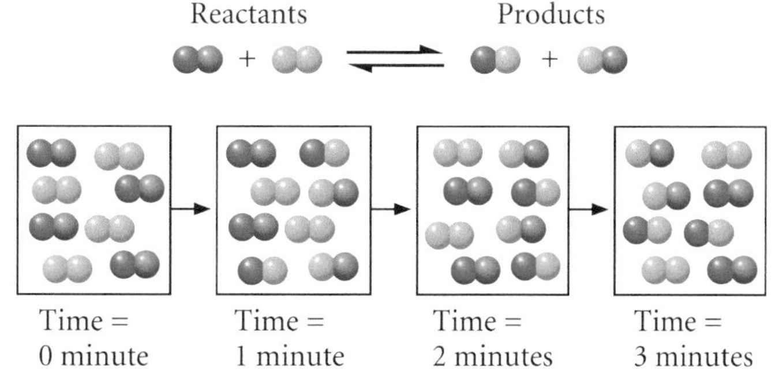

122. We began this chapter discussing the interaction of myoglobin with oxygen. Hemoglobin is a much more familiar molecule, but we intentionally chose myoglobin rather than hemoglobin for our introduction. Use the internet to examine the structure of hemoglobin. Why is myoglobin a more reasonable molecule than hemoglobin to choose for an introduction to equilibrium?
123. Look up the equilibrium constants for the solubility of lead(II) iodide and strontium oxalate in the Appendix at the back of the book.
    a. Based on these values, draw equilibrium line charts that include the starting and equilibrium positions of the reaction.
    b. Next, calculate the solubility of each.
    c. Based on the answers, do our equilibrium line charts require any revision? Explain the answer.
124. Essay question: The quadratic equation is a useful way to solve for concentrations when we have an intermediate value of an equilibrium constant. But our simple way of solving for the concentration does not pass the 5% rule. Given programmable calculators, many of which have quadratic equation solvers, is it ever worth introducing simplifying assumptions, or should we just solve all equilibrium problems using the quadratic equation?
125. We note in Example 14.4 that EDTA-metal ion complexes have very high equilibrium constants. Can we surmise why this is so based on the structure of the $EDTA^{4-}$ ion shown in the exercise?
126. At 3000 K, a 1.00 mol sample of CO_2 in a 1.00-L container is 54.8% decomposed to carbon monoxide and oxygen at equilibrium:

$$2CO_2(g) \rightleftharpoons 2CO(g) + O_2(g)$$

    What is the value of K?
127. Two gaseous substances A and B react to produce gaseous substance C.
    When reactant A decreases by molar amount x, product C increases by molar amount x. When reactant B decreases by molar amount x, product C increases by molar amount 2x. Write the balanced chemical equation for the reaction.
    An equilibrium mixture of these three substances is suddenly compressed so that the concentrations of all substances initially double. In what direction does the reaction shift as a new equilibrium is attained?

Thinking Beyond the Calculation

128. The industrial preparation of methanol, a potential gasoline substitute, is accomplished by the hydrogenation of carbon monoxide. The reaction is

$$CO(g) + 2H_2(g) \rightleftharpoons CH_3OH(g)$$

    a. If the value of K for this reaction is 35.2 under certain conditions at 373 K, what is the value of K_p?
    b. What is the value of K for the reverse reaction?
    c. If the rate of the forward reaction was determined to be 3.56×10^{-12} M/s, what is the rate of the reverse reaction at equilibrium?
    d. Under certain conditions at 0 °C, the value of K_p is 3.77×10^{-6}. If a researcher tried to make methanol under these conditions by adding 1.5 mol of CO and 3.5 mol of H_2 to a 5.0-L sealed flask, what would be the equilibrium pressure of methanol?
    e. What effect, if any, would the addition of more CO have on the equilibrium concentration of methanol at 373 K?
    f. With added catalyst and removal of the methanol as it is formed in the reaction, the reaction can quickly produce 100% yield. If 1.0 L of $H_2(g)$ at 25 °C cost \$0.10 and 1.0 L of $CO(g)$ at 25 °C cost \$0.30, what would it cost to produce 1.0 L of $CH_3OH(l)$ at 25 °C? Assume the expense associated with any experimental procedure used to make the compounds is negligible.

Acids and Bases

Contents

Supplementary Information The online version contains supplementary material available at (▶ https://doi.org/10.1007/978-3-030-90267-4_15).

M. Mosher and P. Kelter, *An Introduction to Chemistry*,
https://doi.org/10.1007/978-3-030-90267-4_15

What We Will Learn From This Chapter

- Acids and bases are common and can be found everywhere (▶ Sect. 15.1).
- Acids and bases can be defined using Arrhenius Theory, Brønsted-Lowry Theory, or Lewis Theory (▶ Sect. 15.2).
- The structure of an acid and base determines the strength of the acid or base (▶ Sect. 15.3).
- The pH scale is used to report the concentration of the hydronium ion (H_3O^+) in solution (▶ Sect. 15.4).
- The pH of a solution containing weak acids or weak bases can be calculated using an ICEA table (Sects. 15.5–15.6).
- Polyprotic acids can provide three equivalents of hydrogen cations when dissolved in aqueous solution (▶ Sect. 15.7).
- Some salts produce ions that affect the pH of aqueous solutions (▶ Sect. 15.8).
- Anhydrides react with water in aqueous solution and result in a change in the pH of the solution (▶ Sect. 15.9).

15.1 Introduction

"Home is where the heart is." Home is also where the phosphoric acid is and where other important acids are, including nitric, sulfuric, acetic, and even acetylsalicylic and ascorbic acids. Most homes also have products containing bases, among which we count ammonia and sodium hydroxide as the two most important. Given the continuing increase in world and US population (see ◘ Fig. 15.1), the resulting need for more food and the increase in sales of consumer products, we can understand why many acids and bases are among the top 20 chemicals produced in the United States, as shown in ◘ Table 15.1.

Why are acids and bases worth knowing about? As we will see in this chapter, they are a part of every household, the people in it, and the house itself. Acid–base reactions in our blood keep us alive. Amino acids combine to form

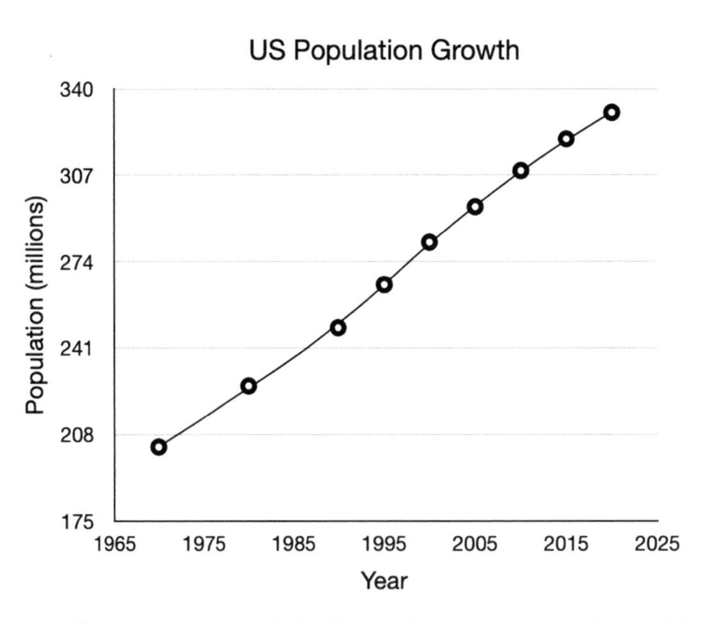

◘ **Fig. 15.1** The rate of increase in US population closely mimics that for the world. For example, in 2015, the world population was estimated at 7.3 billion people. By 2025, nearly 9 billion people are expected to populate the world

◘ **Table 15.1** Acids and bases in top chemicals produced in US Industry in 2009

Substance	Key commercial uses	Billions of kg produced
Sulfuric acid	Fertilizer	29.5
Ammonia	Fertilizers, plastics	9.29
Phosphoric acid	Fertilizers, animal feed	8.56
Sodium hydroxide	Industrial synthesis, cleaners	6.41
Nitric acid	Fertilizers, explosives	6.33

proteins that are enzymes, hormones, and structural materials in our bodies. DNA and RNA are formed from nucleotides that are made by the chemical combination of phosphoric acid, a sugar, and a base. Our food is mostly acidic. Many of us take daily supplements of vitamin C (ascorbic acid) for its health benefits. We clean our homes with products that include both acids, such as acetic acid (CH_3COOH) and sulfuric acid (H_2SO_4), and bases, such as ammonia (NH_3) and sodium hypochlorite ($NaOCl$). The walls of our house or apartment are likely to be made of **gypsum**, another name for calcium sulfate ($CaSO_4$), formed from the reaction of phosphate rock and sulfuric acid. Municipal water supplies are monitored, and (when required) chemically treated, to maintain safe acid concentrations.

Gypsum – Calcium sulfate dihydrate, $CaSO_4 \cdot 2H_2O$.

▶ https://upload.wikimedia.org/wikipedia/commons/4/40/Supermarket_full_of_goods.jpg

Acids, such as aluminum sulfate ($Al_2(SO_4)_3$) and bases, such as sodium hydroxide ($NaOH$) and sodium carbonate (Na_2CO_3), are used in the manufacture of the paper on which we may be reading these ideas. Or, we might be reading them on a computing device that depends on a computer chip to operate, and the manufacturing of computer chips requires the use of nitric (HNO_3), sulfuric (H_2SO_4), and hydrofluoric (HF) acids. Other acids and bases are used in the manufacture of plastics and polymer-based clothing, such as polyester and rayon. Fertilizer for the garden is often made with ammonia-based compounds. Gold is isolated from its ores with a base known as sodium cyanide ($NaCN$). Toothpaste contains a base known as tetrasodium pyrophasphate ($Na_4P_2O_7$). The list goes on and on. Whether the context we discuss is biological, medical, agricultural, environmental, or industrial, acids and bases will play a role (see ◘ Table 15.2).

15.2 What Are Acids and Bases?

There are several models we can use to define an acid. One of the earliest of these is the Arrhenius model, named after the Swedish chemist Svante Arrhenius (1859–1927), in which an **acid** is any species that produces hydrogen ions in aqueous solution and a **base** is any species that produces hydroxide ions in aqueous solution. For example, nitric acid can be classified as an Arrhenius acid because it produces hydrogen ions in aqueous solution.

Acid – According to the Arrhenius model, a species that produces hydrogen ions in solution. Compare the definitions of Brønsted–Lowry acid and Lewis acid.

◘ **Table 15.2** Common acids and bases and examples of their uses

Acid or base	Example of use
Acetylsalicylic acid	Aspirin
Ammonia	Household cleaners, fertilizer, manufacture of fertilizers
Lauric acid, $CH_3(CH_2)_{10}COOH$	Key ingredient in toothpaste and shampoo manufacture
Nitric acid	Key ingredient in acid precipitation
Nicotine (a base)	Tobacco products
Sodium hydroxide	Manufacture of soap, many industrial compounds

Base – According to the Arrhenius model, a species that produces hydroxide ions in solution. Compare the definitions of Brønsted-Lowry base and Lewis base.

$$HNO_3(aq) \rightleftharpoons H^+(aq) + NO_3^-(aq)$$

Similarly, sodium hydroxide is an Arrhenius base because it produces hydroxide ions in aqueous solution.

$$NaOH(aq) \rightarrow Na^+(aq) + OH^-(aq)$$

In truth, hydrogen ions (H^+) and hydroxide ions (OH^-) do not exist free in solution. Rather, they always interact with the solvent, often water. Just as metal cations, such as Na^+, are surrounded by water molecules through ion–dipole interactions, hydrogen cations (which are just protons because they have no neutrons or electrons) similarly interact with water molecules that surround them. In fact, protons interact so strongly that they become directly incorporated into the fabric of one or more water molecules and are easily passed from one water molecule to another. Experimental data show that each proton is immediately surrounded by at least four water molecules and is most likely surrounded by many more. Research by several scientific teams suggests that, when H^+ is surrounded by twenty-one (21) water molecules, a most wonderful structure, shown in ◘ Fig. 15.2, is formed! However, in room temperature aqueous solution, no single well-defined structure exists, because hydrogen bonds are being continually broken and formed. For simplicity, chemists typically refer to the hydrogen ion as either H^+ or H_3O^+ (the hydronium ion), understanding all the while that H^+ actually exists surrounded by several water molecules.

A more useful way of describing acids and bases is known as the Brønsted–Lowry model, named after the Danish scientist Johannes Nicolaus Brønsted (1879–1947, ◘ Fig. 15.3) and the English chemist Thomas Lowry (1874–1936), who proposed the definition in 1923.

In this model, a **Brønsted–Lowry acid** is defined as any species that donates a hydrogen ion (proton) to another species. This definition of an acid is applicable to all solvents in which protons can be exchanged, and, being more far-reaching than the Arrhenius model, it is the definition most often used by chemists. A **Brønsted–Lowry base** is any species that can accept a hydrogen ion (proton) from an acid. As an example of a Brønsted–Lowry acid and base, consider what happens when hydrogen chloride gas is dissolved in water.

Brønsted–Lowry acid – Any species that donates a hydrogen ion (proton) to another species.

Brønsted–Lowry base – Any species that accepts a hydrogen ion (proton) from another species.

$$\underset{\text{Hydrochloric acid}}{HCl(g)} + H_2O(l) \rightleftharpoons Cl^-(aq) + \underset{\text{Hydronium ion}}{H_3O^+(aq)}$$

◘ Figure 15.4 shows the Lewis structures of the reactants and products. In the Brønsted–Lowry model, HCl is the acid and water is the base. The chloride ion (Cl^-) is called the **conjugate base** of HCl because it is the base that

15

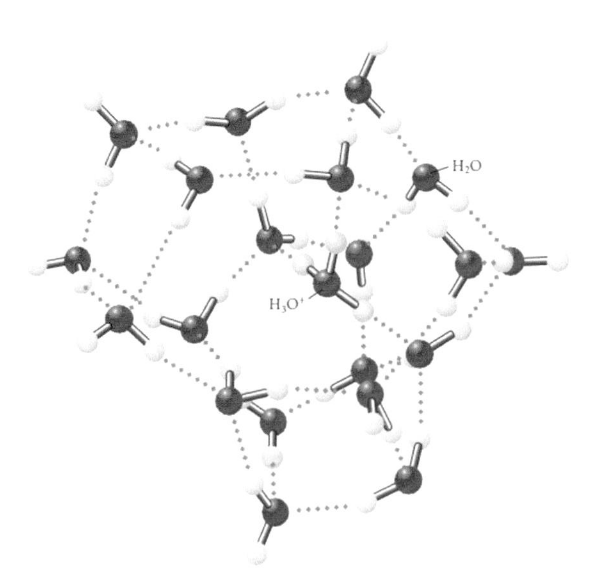

◘ **Fig. 15.2** Data show that hydrogen ions interact with several water molecules. One possible structure involves a shell of 21 water molecules surrounding a H_3O^+ ion

Fig. 15.3 Johannes Nicolaus Brønsted (1879–1947) was a Danish physical chemist. In addition to studying quantum mechanics and developing an acid–base theory, he became interested in politics during the German occupation of Denmark in World War II. He was elected to the Danish parliament in 1947 but became ill and died before he could take office. *Source* ▶ https://upload.wikimedia.org/wikipedia/commons/2/24/Johannes_Brønsted.jpg

$$HCl(aq) + H_2O(l) \rightleftharpoons H_3O^+(aq) + Cl^-(aq)$$

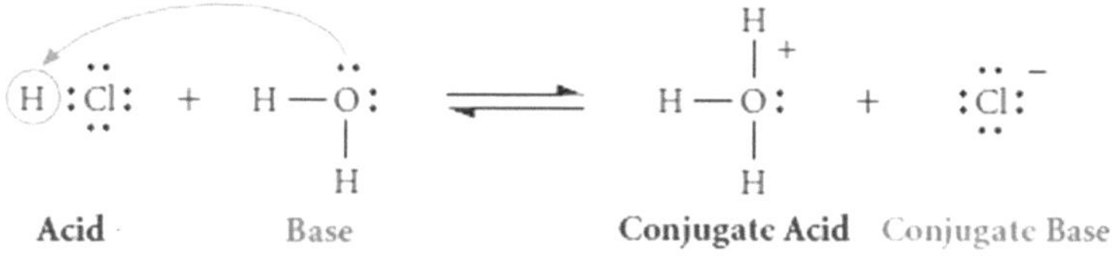

Acid Base **Conjugate Acid** Conjugate Base

Fig. 15.4 When hydrochloric acid gas is added to water, the HCl molecule ionizes and interacts with the solvent as shown here with Lewis structures

results after the HCl donates a proton to water. Similarly, H_3O^+ is the **conjugate acid** of water because it is the acid that results after the water accepts the proton from HCl.

HCl and Cl^- are a conjugate acid–base pair.

H_3O^+ and H_2O are a conjugate acid–base pair.

Conjugate acid – The acid that results after accepting a proton.
Conjugate base – The base that results after donating a proton.

Note that a conjugate base is paired with an acid (Cl^- and HCl) and a conjugate acid is paired with a base (H_3O^+ and H_2O). Also, note that the terms "acid" and "base" are used for reactants, while the terms "conjugate acid and base" are used for products.

$$\underset{[\text{Acid}]}{HCl(g)} + \overset{[\text{Base}]}{H_2O(l)} \rightleftharpoons \underset{[\text{Conjugate base}]}{Cl^-(aq)} + \overset{[\text{Conjugate acid}]}{H_3O^+(aq)}$$

Example 15.1—Acids, Bases, and Their Conjugates

Complete the equation for the reaction of formic acid (HCOOH) and water. Identify the conjugate acid–base pairs.

$$HCOOH(aq) + H_2O(l) \rightleftharpoons$$

Asking the Right Questions
What does an acid do when it encounters water? What role does water play in this process? That is, is water an acid or a base in this process? What is a conjugate, and what are the conjugate pairs in this reaction?

Solution

Formic acid donates a proton (hydrogen cation) to water, meaning that water is a base in this reaction. The products will be the formate and hydronium ions. A conjugate is the acid or base that results after the base or acid has accepted or donated a proton. When deciding on the conjugate pairs, we think about what each reactant forms when it either donates or accepts a hydrogen ion. Note the conjugate pairs: formic acid and the formate anion are one pair. Water and the hydronium ion are the other pair.

$$\underset{[acid]}{\overset{[base]}{HCOOH(aq) + H_2O(l)}} \rightleftharpoons \underset{[Conjugate\ base]}{HCOO^-(aq)} + \overset{[Conjugate\ acid]}{H_3O^+(aq)}$$

Are our answers Reasonable?

Our reaction shows that formic acid did what acids do—it donated a proton to a base (water, in this reaction), resulting in the conjugate base (formate ion) and acid (hydronium ion). The equation describes reasonable acid–base behavior.

What If…?

…a weak base, NH_3, and water reacted with each other? Indicate the complete reaction and list the conjugate pairs.

(*Ans:* $NH_3(aq) + H_2O(l) \rightleftharpoons NH_4^+(aq) + OH^-(aq)$
Base $= NH_3$ *Acid* $= H_2O$
Conjugate acid $= NH_4^+$ *Conjugate base* $= OH^-$)

Practice 15.1

Complete the equation for the reaction of ethylenediamine with water. Identify the conjugate acid–base pairs.

$$H_2NCH_2CH_2NH_2(aq) + H_2O(l) \rightleftharpoons$$

We previously mentioned that ammonia (NH_3) is an important chemical, and the annual worldwide production of more than hundreds of billions of kg would seem to confirm this. Ammonia is a Brønsted–Lowry base because it accepts a hydrogen ion from water. The unshared pair of electrons on the nitrogen can form a covalent bond with the hydrogen, forming the ammonium ion, as shown in Fig. 15.5.

Example 15.2—Ammonia as a Base

Identify the conjugate acid–base pairs in the reaction of ammonia with water. How does water behave differently in this reaction than when it reacts with HCl?

Asking the Right Questions

How is this reaction similar to that in Example 15.1? How is it different? How does the role of water in this example compare to its role in Example 15.1?

Solution

This contrasts with Example 15.1 because ammonia (NH_3) is a base. Water takes on a very different role than in the previous exercise, that of an acid. Yet the concept of the conjugates is still the same—both ammonia and water will have conjugate pairs.

$$\underset{[Base]}{NH_3(aq)} + \overset{[acid]}{H_2O(l)} \rightleftharpoons \underset{[Conjugate\ acid]}{NH_4^+(aq)} + \overset{[Conjugate\ base]}{OH^-(aq)}$$

Is our Answer Reasonable?

Remember the definition of a conjugate: the acid or base that results after the base or acid has accepted or donated a proton. In this reaction, there are still conjugates, with ammonia as the starting base, and its conjugate of the ammonium ion. The process yields a correct acid–base reaction along with chemically valid conjugate acid–base pairs, so the answer is reasonable.

We devote many words in this text to the important properties of water. Here we see another one. As the "universal solvent," water can be an acid or, as in Example 15.1, a base, depending on the solute. Any substance that can be an acid or a base is called amphiprotic, or amphoteric. We will explore this property later in the chapter.

What If…?

…aqueous hydrochloric acid reacted with the ammonia? Write the equation and identify the conjugate pairs.

(*Ans:* $NH_3(aq) + HCl(aq) \rightleftharpoons NH_4^+(aq) + Cl^-(aq)$
Acid $= HCl$ *Base* $= Cl^-$
Coniugate acid $= NH_4^+$ *Conjugate base* $= NH_3$)

Practice 15.2

Identify the conjugate pairs in the reaction of lauric acid $(CH_3(CH_2)_{10}COOH)$ with water.

Fig. 15.5 Ammonia is an Arrhenius base because it generates OH^- in water. It is also a Brønsted–Lowry base because it accepts a proton from water

G. N. Lewis (remember him from the development of Lewis structures) proposed a third model of acids and bases in which a **Lewis base** donates a previously non-bonded pair of electrons (a lone pair) to form a coordinate covalent bond. The electrons in the bond are not transferred entirely but, rather, are shared to make a covalent bond. Reduction and oxidation do not take place when a Lewis base donates a pair of electrons. All Brønsted–Lowry acids are also Lewis acids, and all Brønsted–Lowry bases are also Lewis bases. However, many metal ions, such as Al^{3+}, are **Lewis acids**, because they can accept electrons, although they do not donate protons. For example, when Al^{3+} is in aqueous solution, it combines with water as a Lewis acid:

$$Al^{3+}(aq) + 6H_2O(l) \rightleftharpoons Al(H_2O)_6^{3+}(aq)$$

Lewis acid – Accepts a previously nonbonded pair of electrons (a lone pair) to form a coordinate covalent bond.
Lewis base – Donates a previously nonbonded pair of electrons (a lone pair) to form a coordinate covalent bond.

It can then act as a Brønsted–Lowry acid:

$$Al(H_2O)_6^{3+}(aq) + H_2O(l) \rightleftharpoons \left[Al(H_2O)_5OH\right]^{2+}(aq) + H_3O^+(aq)$$

In another example, we could identify borane (BH_3) In the reaction below as a Lewis acid and ammonia as a Lewis base. This reaction and the specific bonding patterns were discussed in detail earlier in this text. Note that in some Lewis acid–base reactions, there is only one product, so there is no "conjugate" formation as with the previous examples.

$$\underset{\text{Lewis acid}}{BH_3(g)} + \underset{\text{Lewis base}}{NH_3(g)} \rightleftharpoons H_3B - NH_3(s)$$

Notice that these three models of acid–base behavior get progressively more flexible, but they don't contradict each other. The Arrhenius model is the most restrictive because it only applies to aqueous solutions. The Brønsted–Lowry model works in any solvent where protons can be donated, but when water is the solvent then all Brønsted–Lowry acids and bases are also Arrhenius acids and bases.

We mentioned in the opening of the chapter that aluminum sulfate ($Al_2(SO_4)_3$, a Lewis acid) is added to wood pulp as it is formed into paper. The compound is used as a "sizing agent," a substance that prevents ink from spreading. Unfortunately, this acid slowly breaks down the cellulose in paper, causing it to turn light brown with age. That is why many newspapers and paperback books look old even after only weeks or months. Paper meant for long-term archival purposes is "acid-free" and uses other types of sizing agents.

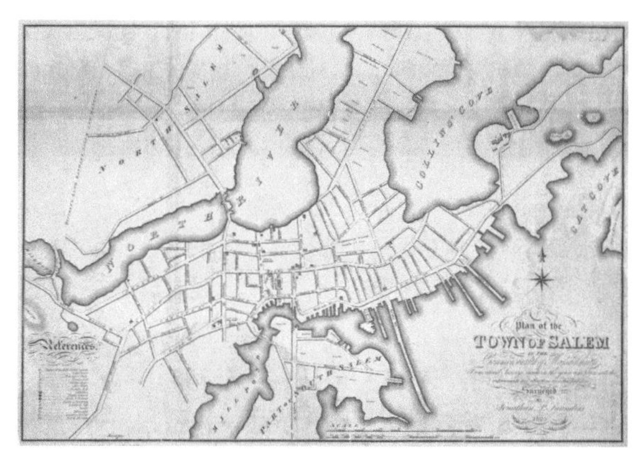

► https://upload.wikimedia.org/wikipedia/commons/6/66/1820_Salem_Massachusetts_map_bySaunders_BPL_12094.png

We will have more to say about Lewis acids in the next chapter. In our present discussion, because we are dealing primarily with water as a solvent, and because Lewis acids become Brønsted–Lowry acids in water, the Brønsted–Lowry model will be the most useful to us.

How do we recognize an acid when we see one? Even though nearly all common foods (including milk!) are acidic, we often associate acids specifically with citrus juices. Orange juice and grapefruit juice can taste sour. This is a characteristic property of acids. Bases, on the other hand, are slippery to the touch and often taste bitter.

Table 15.3 Properties of acids and bases

Acids	Bases
Taste sour	Taste bitter
Donate protons during an acid–base reaction	Accept protons during an acid–base reaction
React with some metals to produce hydrogen gas and metal ions	Form insoluble hydroxides with many metal ions

Acidic – Having a pH less than 7.00 in aqueous solution at 25 °C.

Hydrochloric acid reacts with zinc to produce hydrogen gas:

$$Zn(s) + 2HCl(aq) \rightarrow Zn^{2+}(aq) + 2Cl^{-}(aq) + H_2(g)$$

This is typical of another characteristic of Brønsted–Lowry acids: that they react with many metals to form solutions and often release hydrogen gas. In this sense, acids corrode metals. Bases often react with metal ions to produce insoluble hydroxides, such as $Fe(OH)_3$. A vital property that acids and bases share is that they modify the structure of some types of organic molecules, often found in plants, to cause color changes. In the next chapter, we will discuss how we use these color changes to help us analyze the amount of substances present in a sample. All of these things—sour or bitter, slimy, reactions with metals, color changes, and more—are properties that signal that a material may be acidic or basic (see Table 15.3).

15.3 Acid Strength

We now know what acids and bases are, and we have seen some typical acid–base reactions. We have shown that acids and bases are present throughout our world and within ourselves. However, just as people come in all shapes and sizes, acids and bases come in different strengths. These differences have a profound impact on their chemical behavior and their uses.

We can begin to understand acid (and, by extension, base) strength by recalling that nearly all of our foods are acidic. Let's turn that statement around and see whether it still makes sense. If nearly all foods are acidic, are nearly all acids suitable as foods? That is, we know that citrus fruits contain citric acid ($C_6H_8O_7$) and that ingesting it in reasonable amounts is safe. Oranges and grapefruits also contain ascorbic acid ($H_2C_6H_6O_6$), which is also known as vitamin C. Aspirin ($C_8H_8O_4$) is known chemically as acetylsalicylic acid. We eat these things. Some are necessary for good health (vitamin C); some can relieve headaches and may even help prevent heart attacks (aspirin). Yet ingesting sulfuric acid (H_2SO_4) at the concentration found in a car battery would be harmful and could be lethal. This is also true of the hydrochloric acid that is used as a 30% solution in water to clean stains from concrete.

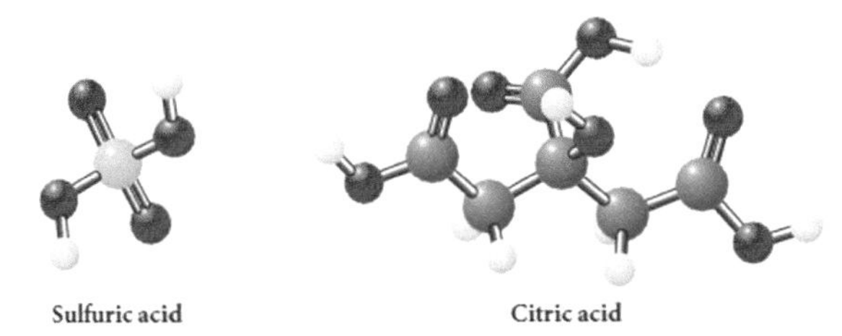

There are two issues here. The first is that sulfuric acid and hydrochloric acid are strong acids (a term defined below). Many acids found in foods are weak acids. The other point is that in batteries and concrete cleaners, the acids are relatively concentrated, whereas in an orange or other food, the acids are fairly dilute. Diluting the hydrochloric acid by a factor of 1000 with water would render it less effective for cleaning concrete, even though it would remain a strong acid. Some acids are inherently strong, and some are inherently weak. Solute concentration, whether concentrated or dilute, doesn't change the inherent strength of the acids or bases.

15.3.1 Strong and Weak Acids

In aqueous solution, a **strong acid** is one that dissociates (or, in some cases, ionizes) essentially completely. "Dissociates" means separates or, in this case, loses a hydrogen cation—a proton. Molecules in which the hydrogen atom

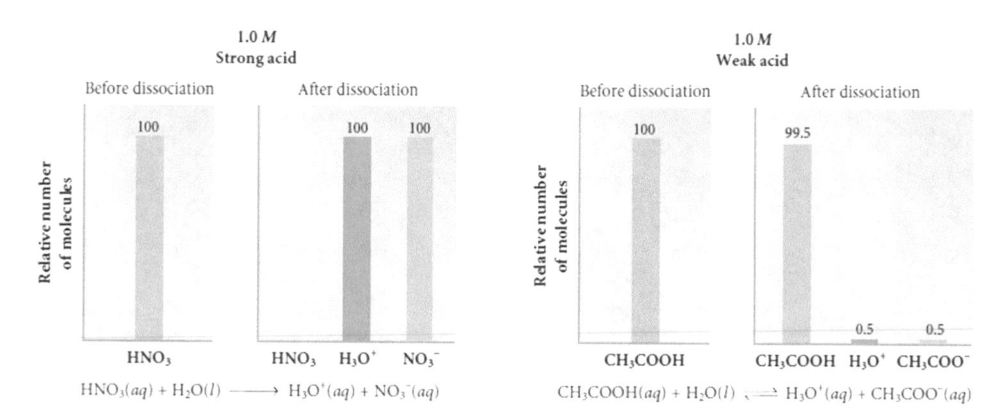

Fig. 15.6 There is a competition between acetic acid (CH_3COOH) and hydronium ion (H_3O^+) to get rid of a hydrogen ion. The hydronium ion is a stronger acid, so the acetic acid does not dissociate to a great extent. The single-headed arrow in the ionization of nitric acid indicates that the reaction goes essentially to completion

is covalently bonded (like HCl) actually do ionize when they go into solution. Whether the process is called dissociation or ionization, the result is the same: the production of H^+ in the solution. A **weak acid** only partially ionizes (or dissociates) in aqueous solution. We can reinforce these ideas by comparing a strong acid, nitric acid (HNO_3), to a weak acid, acetic acid (CH_3COOH). The ionization equations for 1 molar aqueous solutions of these two acids can be written as follows:

$$HNO_3(aq) + H_2O(l) \rightleftharpoons NO_3^-(aq) + H_3O^+(aq)$$

$$CH_3COOH(aq) + H_2O(l) \rightleftharpoons CH_3COO^-(aq) + H_3O^+(aq)$$

Strong acid – An acid that fully dissociates in water, releasing all of its acidic protons.
Weak acid – An acid that only partially dissociates in water.

The ionization of nitric acid proceeds essentially all the way to products, so it is a strong acid. The reverse reaction will not occur to any extent because nitric acid donates a proton to water (forward direction) much more effectively than the hydronium ion (H_3O^+) donates a proton to nitrate ion (reverse direction). Acetic acid only slightly dissociates (typically less than 1%, depending on its initial concentration), so it is a weak acid. This occurs because H_3O^+ can donate a proton to the acetate ion (CH_3COO^-), the conjugate base of acetic acid, regenerating the initial reactants (see Fig. 15.6.)

The equilibrium constant for the dissociation of an acid is called the **acid dissociation constant (K_a)** and has the same relationship to the extent of an acid–base reaction as any other equilibrium constant to that of its own reaction. This is a key point:

The principles of equilibrium are consistent for every chemical equation.

Acid dissociation constant (K_a) – The equilibrium constant for the dissociation of an acid.

The equilibrium constant may be called K_a for acids, K_b for bases, or K_{sp} for solids; and there are others. But their fundamental meaning as equilibrium constants, and how we use them in our problem solving, are the same. Table 15.4 lists several weak acids and their K_a values.

The mass-action expressions for the reactions shown for nitric and acetic acids with water can be written in the following "shorthand" form (neglecting the presence of water because it is the solvent, and neglecting activity effects).

$$K_a = \frac{[NO_3^-][H_3O^+]}{[HNO_3]} \, or \, \frac{[NO_3^-][H^+]}{[HNO_3]}$$

$$K_a = \frac{[CH_3COO^-][H_3O^+]}{[CH_3COOH]} \, or \, \frac{[CH_3COO^-][H^+]}{[CH_3COOH]}$$

Table 15.4 K_a values for selected weak acids

Formula	Name	K_a
HSO_4^-	Hydrogen sulfate ion	1.2×10^{-2}
$HClO_2$	Chlorous acid	1.2×10^{-2}
HF	Hydrofluoric acid	7.2×10^{-4}
HNO_2	Nitrous acid	7.0×10^{-4}
$HC_3H_5O_3$	Lactic acid	1.3×10^{-4}
$HC_2H_3O_2$ (CH_3COOH)	Acetic acid	1.8×10^{-5}
$[Al(H_2O)_6]^{3+}$	Hydrated aluminum ion	1.4×10^{-5}
HOCl	Hypochlorous acid	3.5×10^{-8}
HCN	Hydrocyanic acid	6.2×10^{-10}
NH_4^+	Ammonium ion	5.6×10^{-10}
HOC_6H_5	Phenol	1.6×10^{-10}

Nitric and other strong acids have K_a values much greater than 1, often regarded as infinity in aqueous solution. Weak acids, such as acetic acid ($K_a = 1.8 \times 10^{-5}$), have K_a values less than 1. We can use the equilibrium line c6hart that we used in ▶ Chap. 14 to see the extent of the reaction for HNO_3, the strong acid, compared to CH_3COOH, the weak acid. Remember that the "S" indicates the starting point and the "E" represents the equilibrium point.

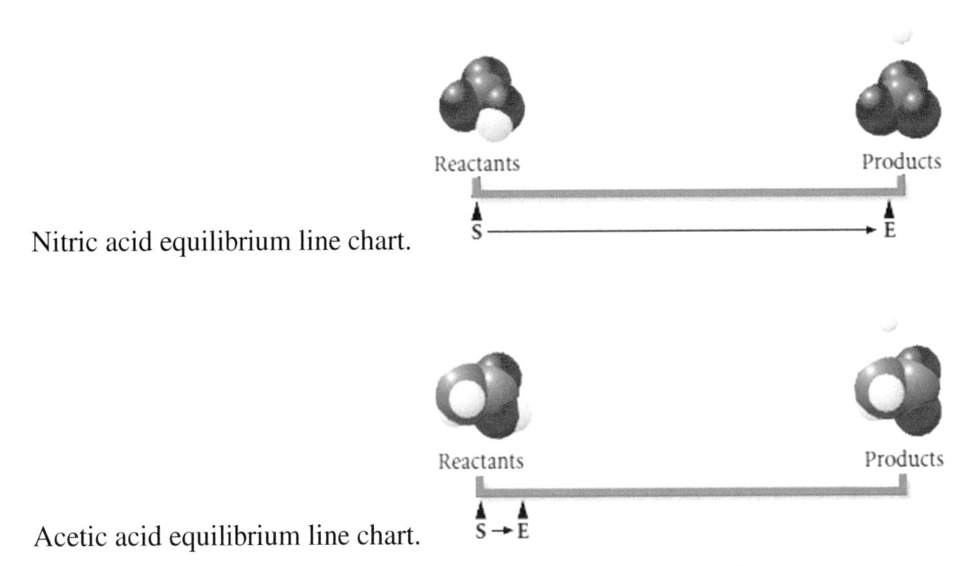

Nitric acid equilibrium line chart.

Acetic acid equilibrium line chart.

There are differences in acid strength among strong acids. For example, $HClO_4$ is inherently stronger than HCl. However, this difference is not observable in water, in which HCl, or any other strong acid, appears to be just as strong as $HClO_4$, because their strength makes their reaction with water essentially complete. It is possible to use nonaqueous solvents to show differences in strength even among strong acids.

Example 15.3—Relative Acid Strength

Arrange the following acids in order from weakest to strongest. How is our placement of them related to their K_a values?

Acid	K_a
$HClO_2$	1.2×10^{-2}
CH_3COOH	1.8×10^{-5}
H_2SO_4	>1
HCN	6.2×10^{-10}

Asking the Right Questions

What is the difference between a strong and a weak acid? How does that relate to the K_a value?

Solution

Recall that a strong acid is one that dissociates (or ionizes) completely at 1 molar concentration, whereas a weak acid does not. This is measured by the value of K_a. The larger the value of K_a, the stronger the acid. Based on our values of K_a, the order of acids from weakest to strongest is $HCN < CH_3COOH < HClO_2 < H_2SO_4$.

Is our Answer Reasonable?

Acid strength is indicated by the K_a value, and these acids are in the correct order. The first and last acids on our list have K_a values that are over 10 orders of magnitude apart! That is quite a difference in acid strength, and means the order is reasonable.

What If...?

...instead of H_2SO_4, we had lactic acid, $HC_3H_5O_3$ among our list of acids? (See Table 17.4.) Where would lactic acid fit in our order?

(*Ans. It is stronger than acetic acid and weaker than chlorous acid*).

Practice 15.3

The Appendix of this textbook has a list of many acids. On the basis of the K_a values, pick three acids that are weaker than hydrofluoric acid (HF) and three that are stronger than hypochlorous acid (HOCl).

Given the relative acid strengths of nitric acid and acetic acid, what might we conclude about the strengths of their conjugate bases, nitrate ions and acetate ions? Nitric acid is strong, so its ionization equilibrium to produce H^+ will lie all the way to the right. In other words, the nitrate ion is such a weak base that it essentially doesn't go back to form nitric acid at all. Another way of expressing this is to say that in the tug of war for protons between water (which acts as a base on the reactant side) and nitrate (which acts as a base on the product side), the water wins because water is a much stronger base than the nitrate ion. This battle, and its outcome, are shown in Fig. 15.7.

Let's consider acetic acid, which we call a weak acid because its acid ionization reaction in water doesn't go very far to the right. This means that most of the acetic acid stays as CH_3COOH when it is added to water. When water (on the reactant side) and acetate ion (on the product side) enter into the tug of war for protons, the acetate ion wins because it is a much stronger base than water. These reactions are shown in Fig. 15.8.

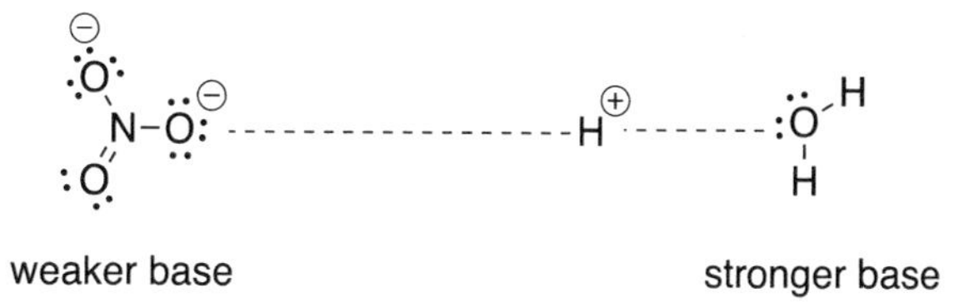

weaker base — stronger base

Fig. 15.7 There is a "tug of war" for hydrogen ions between water and the nitrate ion. The water wins because water is a much stronger base than the nitrate ion

stronger base — weaker base

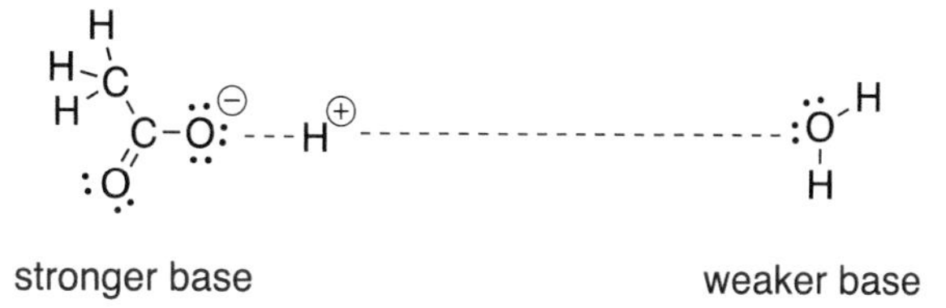

Fig. 15.8 In the reaction of acetate (CH_3COO^-) and water, the acetate wins because it is a stronger base than water

Example 15.4—Strength of the Conjugates

For each of the following acids in water—hydrofluoric acid ($HF, K_a = 7.2 \times 10^{-4}$), acetic acid ($CH_3COOH, K_a = 1.8 \times 10^{-5}$), and hypochlorous acid ($HOCl, K_a = 3.5 \times 10^{-8}$):

a. Write the mass-action expression for dissociation of each acid.
b. Place the conjugate bases of these acids in order from strongest to weakest.
c. State which of the conjugate bases are stronger than water and which are weaker than water.

Asking the Right Questions

What is the relationship between the strength of an acid and its conjugate base? Given that relationship, what is the trend in conjugates if the acids are placed from strongest to weakest? How does the behavior of water as an acid or base fit in this discussion?

Solution

a.

$$K_a = \frac{[F^-][H_3O^+]}{[HF]}$$

$$K_a = \frac{[CH_3COO^-][H_3O^+]}{[CH_3COOH]}$$

$$K_a = \frac{[OCl^-][H_3O^+]}{[HOCl]}$$

b. Based on the K_a values, the acid strengths are ordered as follows:

$$HF > CH_3COOH > HOCl$$

The relative strengths of the conjugate bases are therefore in reverse order:

$$OCl^- > CH_3COO^- > F^-$$

c. Each of these bases is stronger than water.

Are Our Answers Reasonable?

Our list has the conjugate bases showing a strength trend that is opposite to that of the acids. That is reasonable, knowing the inverse relationship between the strength of conjugates. None of the acids is strong, so all of the conjugate bases will have basic activity in which they will win the "tug of war" for protons with water. Generally, only the conjugate bases of strong acids will be weaker than water.

What If...?

...we had the acid H_3O^+? How would the conjugate base of that ion compare to that of water?

(*Ans: The conjugate base of H_3O^+ is water!*)

Practice 15.4

Using ◘ Table 15.4, pick two acids with conjugate bases that are weaker than the acetate ion (CH_3COO^-).

15.3.2 The Different Strengths of Acids

As with so many answers to chemical questions, the key to differing acid strengths lies in structure. For binary acids such as HCl or HF (shown in ◘ Fig. 15.9), where the electronegative atom is bonded directly to the hydrogen, smaller atoms have the valence electrons present in a smaller space. This higher electron density results in stronger bonds between the electronegative atom and hydrogen, which makes these acids weaker (less likely to donate the proton). That is why HF is a much weaker acid than HCl. However, if the sizes of the atoms bonded to hydrogen are about the same, the acidity increases with increasing electronegativity of the atom bonded to hydrogen, because the polarity of the bond also increases. This is why HF is a stronger acid than H_2O, which, though it is not binary, has two H–O bonds.

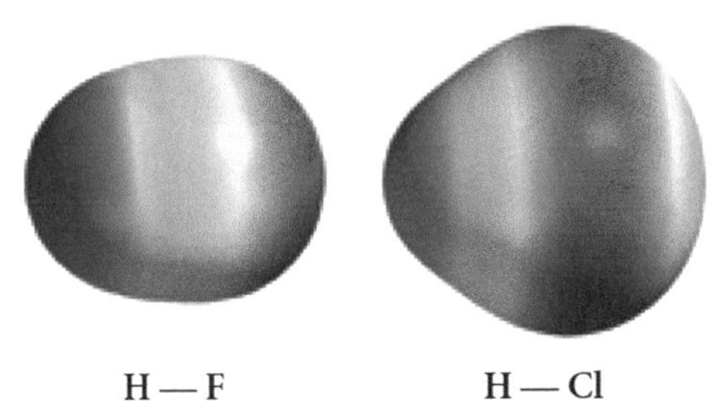

Consider a "generic" oxygen-containing compound with a central atom, A, as shown in ◘ Fig. 15.10. If A has a high electronegativity, then it will have a tendency to form a covalent bond with oxygen, which is also highly electronegative, while weakening the bond between the oxygen and hydrogen. The hydrogen can then be easily removed, which means the compound is acidic. The more electronegative the atom (A) is, the more acidic the compound becomes (see ◘ Fig. 15.11). Chlorine is more electronegative than sulfur, which, in turn, is more electronegative than phosphorus. This means that perchloric acid ($HClO_4$) is inherently stronger than sulfuric acid (H_2SO_4), which is stronger than phosphoric acid (H_3PO_4). We do not see the difference between perchloric and sulfuric acids in aqueous solution, but phosphoric acid is noticeably weaker in water than either of these other compounds.

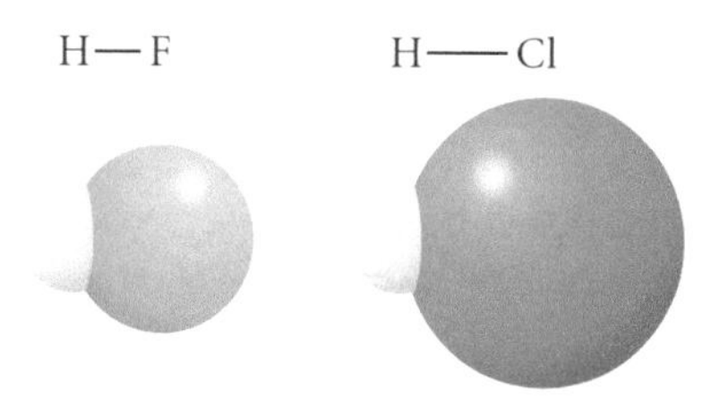

■ **Fig. 15.9** Compare the relative sizes of the atoms in HCl and HF

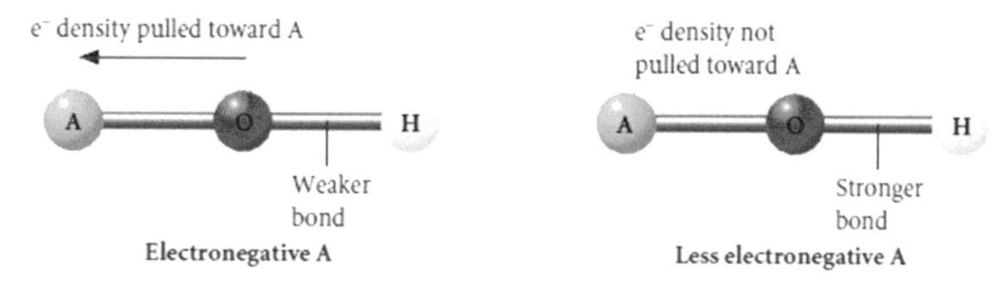

■ **Fig. 15.10** In this "generic" oxygen-containing compound the central atom, A, is bonded to an oxygen atom, which is itself bonded to a hydrogen atom. If A is highly electronegative, it will weaken the bond between oxygen and hydrogen

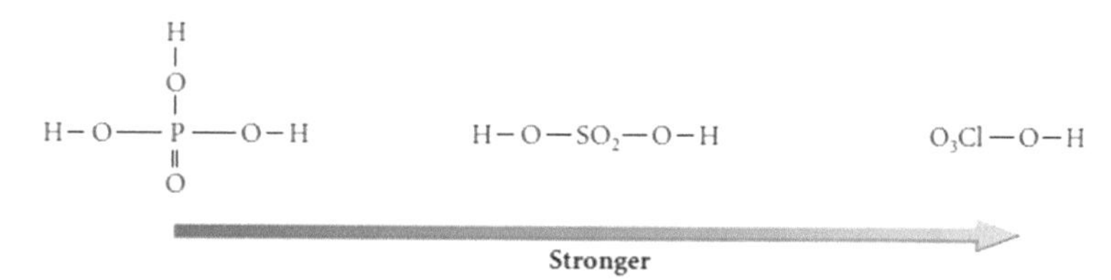

■ **Fig. 15.11** Chlorine is more electronegative than sulfur, which, in turn, is more electronegative than phosphorus. The result is that $HClO_4$ is inherently stronger than sulfuric acid (H_2SO_4), which is stronger than H_3PO_4. The listed structures are examples of resonance structures of each molecule

For the same central atom (sulfur, for example), the higher the oxidation state, the higher the attraction for electrons and the stronger the covalent bond between the sulfur and oxygen atom. This tends to weaken the O–H bond in these compounds, as described above. This is why H_2SO_4 (with sulfur in the +6 oxidation state) is a stronger acid than H_2SO_3 (where sulfur is in the +4 oxidation state). For the same reason, HNO_3 is stronger than HNO_2, and the strength of so-called chlorine "oxoacids" is $HClO_4 > HClO_3 > HClO_2 > HClO$.

This model also explains why a compound such as NaOH is basic. Let's look again at ■ Fig. 15.10, where "A" is Na. Sodium has a relatively low electronegativity and therefore will not form a covalent bond with the oxygen atom. The bond between the oxygen and hydrogen is stronger and will remain intact. The hydroxide ion is therefore readily released to the solution, making NaOH a strong base in water. The same model holds for all Group IA bases.

Example 15.5—Which Acid is Stronger?

Given the two acids HIO and HIO_3 and two values for K_a, 0.17 and 2×10^{-11}, indicate which K_a value goes with which acid.

Asking the Right Questions

How do we judge which of the two acids will be stronger? Which K_a value goes with the stronger acid?

Solution

HIO_3 has more oxygen atoms than HIO. These additional oxygen atoms will draw electrons away from the O–H bond, weakening it. The acidic hydrogen will come off more easily from HIO_3 than the one in HIO. HIO_3 is therefore the stronger acid. Its K_a value is 0.17. The K_a value for HIO is 2×10^{-11}.

Is our Answer Reasonable?

We discussed the connection between oxidation state of the central atom and its tendency to strengthen the interaction between the central atom and oxygen atoms (allowing for the release of protons—that is, acidic behavior). It therefore is reasonable that HIO_3 is the stronger acid.

What If…?

…we were asked to describe why sodium hydroxide is a base, but phosphoric acid, which has three hydroxide groups connected to its central phosphorus atom, is an acid. What are the factors that we should take into account in our discussion?

(*Ans: The factors include the electronegativity of phosphorus compared to sodium, the oxidation state of the phosphorus, and the likelihood that the remaining anion would be stable after releasing a proton.*)

Practice 15.5

Which acid is stronger, H_2S or HBr? Explain.

Will the hydrogen ion concentration always be greater in a solution of a strong acid than in a solution of a weak acid? We might envision that a solution of a strong acid generates a greater hydrogen ion concentration than a similar solution of a weak acid. However, recall our previous discussion of the terms strong, *weak*, concentrated, and *dilute*. The concentration of an acid, be it weak or strong, must be considered if we are to determine the concentration of hydrogen ions in solution. In fact, a concentrated solution of a weak acid can produce more hydrogen ions in solution than a dilute solution of a strong acid.

NanoWorld-MacroWorld: The Big Effects of the Very Small—Superacids and Superbases

We've explored acidity and basicity in this chapter with compounds that are both strong and weak. Strong acids, as we've noted, dissociate essentially to completion upon addition to water. A closer examination shows that, while the strong acids to dissociate nearly to completely solvated ions, there is still a very small, but measurable, concentration of undissociated acid in solution. This 'problem' is solved by a class of compounds that carries dissociation to its corresponding ions to the extreme. Enter the **superacid** and superbase.

Superacids – Amazingly strong acids that can be used to add hydrogen ions to organic molecules that are otherwise impervious to such reaction.

Many examples of superacids have been discovered. The rather lengthy list includes fluorosulfonic acid (FSO_3H), methanesulfonic acid (CH_3SO_3H), and trifluoromethanesulfonic acid (triflic acid, CF_3SO_3H) that are often used when performing reactions in the organic chemistry laboratory. These are acids we don't want to spill on our hand! For example, triflic acid has an acid dissociation constant that indicates a very powerful acid, $K_a = 8 \times 10^{14}$. Compared to HCl with a $K_a = 1 \times 10^7$, triflic acid is approximately ten million times more acidic! Superbases have also been discovered and used in the laboratory. A particular class of superbases, known as 'proton sponges' for their ability to bind free hydrogen ions, are particularly useful in organic chemistry reactions. As we note from their structure, these compounds exist as neutral compounds, and the binding of a proton to their structure results in a fairly stable cation. For example, the protonated version of the proton sponge known as N,N,N′,N′-tetramethylnaphthalene-1,8-diamine has a $K_a = 7.9 \times 10^{-13}$ Compared to the acidity of protonated ammonia, the ammonium ion (NH_4^+) with its $K_a = 5.8 \times 10^{-10}$ is roughly 1000 times less basic.

N,N,N',N'-tetramethylnaphthalene-1,8-diamine + H-OH → (protonated diamine)$^+$ + $^-$OH

Example 15.6—Concentrated Versus Dilute Strong Acids

Fish and other aquatic organisms are very sensitive to the acid concentration in their surroundings. Lake trout, for example, can flourish when the hydrogen ion concentration in a lake is 1×10^{-5}M, but they cannot survive if the hydrogen ion concentration becomes greater than 1×10^{-4}M. Indicate whether a 1000.0 L tank, filled to capacity, would have too great a hydrogen ion concentration for lake trout to survive if the tank contained:

a. 0.0365 g of HCl
b. 6180 g of boric acid (H_3BO_3). This much boric acid in 1000.0 L of water ionizes about 0.008%=0.008 mol of H^+ produced per 100 mol of H_3BO_3 added. *Note:* We are strictly concerned with the acid concentration here, not with the possible impact of the boric acid itself on the fish.

Asking the Right Questions

What is the maximum hydrogen ion concentration in which lake trout can survive? How much hydrogen ion is in a 1000.0 L tank at that concentration? Which of our acids supply more than this quantity of hydrogen ions to the solution (that is, which ones will kill the fish)?

Solution

In order for the lake trout to survive in the tank, the [H^+] must be less than or equal to 1×10^{-4} M. Because there are 1000.0 L of water in the tank, the number of moles of hydrogen ion supplied by the acid in each case must not be more than strong by the addition of very electronegative groups to the central atom.

$$\frac{1 \times 10^{-4}\text{mol H}^+}{1 \text{ L solution}} \times 1000.0\text{L solution} = 0.1\text{mol H}^+$$

a. $$0.0365\text{g HCl} \times \frac{1 \text{ mol HCl}}{36.5 \text{ g HCl}} \times \frac{1\text{mol H}^+}{1 \text{ mol HCl}}$$
$$= 0.00100\text{mol H}^+$$

As this result shows, the strong acid solution is so dilute that the fish can survive.

b. $$6180\, g\, H_3BO_3 \times \frac{1 mol\, H_3BO_3}{61.8 g\, H_3BO_3} \times \frac{0.008 mol\, H^+}{100 mol\, H_3BO_3}$$
$$= 8 \times 10^{-3}\text{molH}^+$$

In this case, the fish would not die from the acid level caused by the addition of this very weak acid.

Are our answers Reasonable?

Although HCl is a strong acid, there is very little of it in the tank, so the $[H^+]$ is low. Although H_3BO_3 is a weak acid, there is a lot of it in the container, so the $[H^+]$ is higher than with that from the HCl solution. This is the point we seek to make (our sense of what is "reasonable" here)—that "strong" and "concentrated" are not the same thing; the same is true for "weak" and "dilute."

What If…?

…instead of HCl, we had the same amount (0.0365 g) of acetic acid, $K_a = 1.8 \times 10^{-5}$? Would the lake trout survive?

(*Ans: Yes. Acetic acid is a weak acid, and the final* $[H^+]$ *would be far less than that from the* HCl *solution.*)

Practice 15.6

What is the hydrogen ion concentration of a solution in a 500.0 L tank filled to capacity with an aqueous solution that contains 2.38 g of the strong acid HNO_3?

As we saw in Example 15.6, just because an acid is strong does not necessarily mean that the resulting hydrogen ion concentration due to this acid will be substantial. Instead, the equilibrium hydrogen ion concentration will depend on both the strength *and* the initial concentration of the acid. The acidity of solutions, whether food, shampoos, or samples of acid rain, is properly discussed in measures of equilibrium hydrogen ion concentration.

15.4 The pH Scale

Maintenance of a public swimming pool is often fairly involved. Not only does the public agency need to make sure the pool is leak-free, but daily measurements of the water in the pool are conducted to ensure that the water meets a set of standards for health. The job of monitoring water quality at the pool often falls to the lifeguards, who measure and record the chlorine concentration and pH of the water. Why is this process done? The pH of the water is essential to maintaining a chlorination level that keeps the water safe for swimmers; too low or too high and swimming becomes unsafe.

What does pH measure? The pH is a measure of the acidity of the water. Although the concentration of acid in the pool can be expressed in terms of molarity, we can describe the acid concentration more conveniently with a mathematical operator that gets rid of the exponent. This is done via a term "p," that is the first part of the common expression **pH**. The pH can be used to describe the acidity of a waterway, as in "The pH of the water in the lake is about 4.5," or in advertisements for shampoo: "It will protect your hair; it is pH-balanced." What is pH? We can use the standard mathematical definition: Note that we use H^+ as shorthand for H_3O^+.

pH – A numerical value related to hydrogen ion concentration by the relationship $pH = -\log[H^+]$ or $pH = -\log[H_3O^+]$.

$$pH = -\log[H^+] \text{ or } pH = -\log[H_3O^+]$$

The term "p" is a mathematical operator that in practice means "take the negative of the base-10 logarithm of." The combined term "pH" is interpreted as "the negative of the base-10 logarithm of the hydrogen ion concentration." For example, if $[H^+] = 6.4 \times 10^{-2}$ M, then $pH = -\log(6.4 \times 10^{-2}\text{ M})$. To solve this on the calculator, we would do the following:

1. Enter 6.4×10^{-2} into the calculator.
2. Press the "log" button. (The calculator display will now read "– 1.1938….")
3. Press the "+/–" key, to give the final answer of 1.19, rounded to two significant figures past the decimal point.

This means that for a solution in which $[H^+] = 6.4 \times 10^{-2}$ M, the pH is 1.19. In subsequent chapters we will deal with related terms such as pCl, which is roughly equal to –log [Cl^-], and pCa, which approximately represents –log [Ca^{2+}]. The next exercise helps illustrate why "p" is a convenient operator with which to express concentrations.

The equation (pH = –log [H^+]) can also be used in reverse. For instance, if we know that the pH of a solution is 1.19, we can calculate the [H^+]. Mathematically, the conversion is

$$pH = -\log[H^+]$$

$$1.19 = -\log[H^+]$$

$$-1.19 = \log[H^+]$$

$$10^{-1.19} = [H^+]$$

$$[H^+] = 6.4 \times 10^{-2}\,M$$

So, in general, to calculate the [H^+] from the pH:

$$[H^+] = 10^{-pH}$$

We need to make a note about significant figures and logarithms. The exponent on a number written in scientific notation is not significant, as we have seen. The numbers to the left of the decimal point in a logarithm carry the same information as the exponent in scientific notation. That is, they are not considered significant figures. For instance, from our discussion above, each

$$6.4 \times 10^{-2}\,M \text{ and } pH = 1.19$$

15

possesses only two significant figures.

Example 15.7

Using the Operator "p" to Determine Useful Quantities.
Calculate the desired term.

a. pH for $[H^+] = 3.2 \times 10^{-11}$ M
b. pK_a for $K_a = 1.8 \times 10^{-5}$
c. pOH for $[OH^-] = 8.8 \times 10^{-13}$ M
d. pH for $[H^+] = 3.2 \times 10^{-10}$ M
e. $[H^+]$ for pH = 12.73
f. $[OH^-]$ for pOH = 5.08

Asking the Right Questions

What does the operator "p" mean? How do we convert between concentration and pH?

Solution

a. $pH = -\log(3.2 \times 10^{-11}\,M) = 10.49$
b. $pKa = -\log(1.8 \times 10^{-5}) = 4.74$
c. $pOH = -\log(8.8 \times 10^{-13}\,M) = 12.06$
d. $pH = -\log(3.2 \times 10^{-10}\,M) = 9.49$
e. $[H^+] = 10^{-12.73} = 1.9 \times 10^{-13}\,M$
f. $[OH^-] = 10^{-5.08} = 8.3 \times 10^{-6}\,M$

Are Our Answers Reasonable?

One way to tell if the answers are reasonable is to see if the "p" value is within one unit of the exponent digit of the concentration. For example, in part "e," the pH is 12.73, within one unit of the exponent value of 13 (from 10^{-13}). This is true for all of our answers. The other measure of "reasonable" is to make sure that the higher the p value, the lower the concentration. This is the true for each part of the example.

What If…?

…we have a solution with a sodium concentration of 2.4M? What is pNa?

(Ans: $pNa = -0.38$)

Practice 15.7

Determine each of the desired terms.

a. pH for $[H^+] = 4.61 \times 10^{-3}$ M

b. pK_a for $K_a = 2.77 \times 10^{-9}$

c. pOH for $[OH^-] = 3.22 \times 10^{-6}$ M

d. $[H^+]$ for pH = 3.92

e. $[H^+]$ for pH = 1.49

f. $[OH^-]$ for pOH = 9.93

Converting from $[H^+]$ to pH enables us to express acid concentration in a way that does not require exponents. The use of the log scale has another important implication. If we look again at parts a and d of Example 15.7, we see that the value of $[H^+]$ changes by a factor of 10 and the pH changes by one unit. This is the impact of taking the log of a number. The pH will change by one unit for each power-of-10 change in $[H^+]$ concentration. A decrease in $[H^+]$ from 1×10^{-5} to 1×10^{-6} M is indicated by an increase in pH from 5 to 6. This power-of-10 relationship is shown in Fig. 15.12, in which we begin with $[H^+] = 1 \times 10^{-8}$ M in water and increase by powers of 10 until we get to $[H^+] = 1$ M. Also note that a small value of pH corresponds to a large value for $[H^+]$, and vice versa.

Example 15.8—[H+] and pH: One Implication

In Exercise 15.6, we discussed the sensitivity of aquatic species to the hydrogen ion concentration. Let's extend this discussion. Mussels can survive in waterways with a pH of 6.8. They cannot survive at pH 5.2. What is the ratio of the hydrogen ion concentrations in the two solutions? That is, how many times greater is one than the other?

Asking the Right Questions

How does the pH change as $[H^+]$ changes? How is the ratio of the hydrogen ion concentrations different in meaning than the ratio of pH values? Which solution has the higher hydrogen ion concentration, $[H^+]$?

Solution

The pH scale is logarithmic. That is, each pH unit represents a difference in $[H^+]$ of a factor of 10. The pH values of 6.8 and 5.2 therefore have very different hydrogen ion concentrations. In order to find the ratio of the concentrations, we must convert the pH of each solution into its hydrogen ion concentration. We cannot compare the pH values, themselves, as a measure of the hydrogen ion concentration ratio. In addition, The lower the pH, the higher the $[H^+]$, so the pH 5.2 solution has a greater $[H^+]$ than the pH 6.8 solution.

$[H^+] = 10^{-pH}$, so for the waterway that has a pH = 6.80,

$$[H^+] = 10^{-6.80} = 1.6 \times 10^{-7}\ M$$

For the waterway that has a pH = 5.20,

$$[H^+] = 10^{-5.20} = 6.3 \times 10^{-6}\ M$$

The ratio of the hydrogen ion concentrations is the quotient of the two values:

$$\frac{6.3 \times 10^{-6}\ M}{1.6 \times 10^{-7}\ M} = 39$$

Are Our Answers Reasonable?

The pH values of the waterways differ by 1.6 units. A pH difference of 1 unit means a 10-to-1 $[H^+]$ ratio, and a pH difference of 2 units means a 100-to-1 (that is, 10×10) $[H^+]$ ratio. It is reasonable that a pH difference of 1.6 should be within those boundaries—and that small differences in pH have great impact on life.

What If…?

…we want to change the hydrogen ion ratio of a sample of the waterways to 250-to-1, while keeping the more acidic sample at pH = 6.80? What would be the pH of the more acidic sample?

(*Ans*: *pH* = 9.20)

Practice 15.8

If the hydrogen ion concentration of a body of water is 500 times that of another waterway, and the pH of the less acidic water is 8.84, what is the pH of the more acidic water?

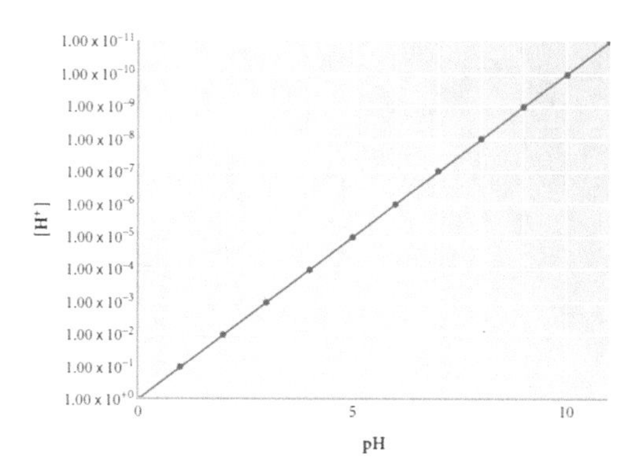

Fig. 15.12 Relationship between pH and $[H^+]$. The pH unit is a logarithm term. Therefore, a 1-unit increase represents a tenfold increase in $[H^+]$

15.4.1 Water and the pH Scale

Pure water can undergo **autoprotolysis**, proton transfer within only the solvent itself. Such a proton transfer would indicate that one of the water molecules is acting as a base and the other is acting as an acid.

$$H_2O(l) + H_2O(l) \rightleftharpoons H_3O^+(aq) + OH^-(aq) \quad K_w = 1.00 \times 10^{-14} \text{ at } 24\,°C$$

Autoprotolysis – Proton transfer between molecules of the same chemical species, as in the autoprotolysis of water.

The ability of a compound to act both as an acid and as a base (not necessarily in the same reaction) isn't unique to water. Those compounds capable of such a feat are called **amphiprotic**. They include, among many others, water, ammonia (NH_3), and, as we shall discover in ▶ Sect. 15.7, the amino acids that make up the proteins in the human body. In the autoprotolysis of water, the reaction is often simplified as

$$H_2O(l) \rightleftharpoons H^+(aq) + OH^-(aq) \quad K_w = 1.00 \times 10^{-14} \text{ at } 24\,°C$$

Amphiprotic – Having the ability to act as either an acid or a base in different circumstances.

It is understood that the H^+ and OH^- really aren't naked ions; they are solvated by the aqueous solution. This equation has the mass-action expression

$$K_w = \left[H^+\right]\left[OH^-\right] = 1.00 \times 10^{-14}$$

Taking the log of both sides gives

$$pK_{ux} = pH + pOH = 14.00$$

In pure water, the only source of $[H^+]$ and $[OH^-]$ is water itself, so the concentrations of H^+ and OH^- (formed in a 1:1 ratio) will be equal. If we call each "x," then

$$K_w = \left[H^+\right]\left[OH^-\right] = x^2 = 1.00 \times 10^{-14}$$

$$x = \left[H^+\right] = \left[OH^-\right] = 1.00 \times 10^{-7}\ M$$

This means that in pure water at 24 °C, $\left[H^+\right] = \left[OH^-\right] = 1.00 \times 10^{-7}$ M, and the solution will have pH = 7.0. We define this as **neutral pH** when the solvent is water and the system is at 24 °C. Reported to two significant digits, the pH is 7.00 at 25 °C also. This pH value also equals pOH because both $[H^+]$ and $[OH^-]$ are equal to 1.0×10^{-7} M under these conditions. The K_w value is temperature-dependent, as shown in Table 15.5.

15

Neutral pH – A pH of 7.00 in aqueous solution at 25 °C.

Unless otherwise stated, we will assume a temperature of 24 °C for the remainder of our discussion, so that $K_w = 1.00 \times 10^{-14}$. In aqueous solution, we can always determine pH, $[H^+]$, pOH, or $\left[OH^-\right]$ in a solution if any one of these factors is known.

Table 15.5 K_w at several temperatures

Temperature (°C)	K_w	pK_w
0	1.14×10^{-15}	14.94
10	2.92×10^{-15}	14.54
20	7.81×10^{-14}	14.11
24	1.00×10^{-14}	14.00
25	1.01×10^{-14}	14.00
30	1.47×10^{-14}	13.83
50	5.47×10^{-14}	13.26
60	9.71×10^{-14}	13.01

Example 15.9—Water in a Pristine Lake

Pure water has a pH = 7.0. But "pristine" rainwater (unaffected by pollutants from sources such as nitrogen oxides from automobile tailpipe emissions or sulfur oxides from industrial smokestack emissions) generally has a pH of between 5.5 and 6.0. This results from the dissolving and equilibrating of carbon dioxide from the atmosphere and the subsequent release of a hydrogen ion to the water.

$$CO_2(g) + H_2O(l) \rightleftharpoons H_2CO_3(aq) \rightleftharpoons H^+(aq) + HCO_3^-(aq)$$

Many waterways in the United States have pH values significantly lower than this range as a consequence of acid precipitation, in which the stronger acids HNO_3 and H_2SO_4 combine with the water. We discussed the implications of this in Exercises 15.6 and 15.8. If the pH of the water in a pristine lake is 5.90, determine the value of $[H^+]$, $[OH^-]$, and pOH in this water.

Asking the Right Questions

How do we convert among pH, pOH, $[H^+]$ and $[OH^-]$? What happens to $[H^+]$ and pH as $[OH^-]$ decreases and pOH increases?

Solution

There are several ways to do this problem. We'll show just one.

$$pOH = 14.00 - pH = 14.00 - 5.90 = 8.10$$

$$[H^+] = 10^{-pH} = 10^{-5.9} = 1.3 \times 10^{-6}\ M$$

$$[OH^-] = 10^{-pOH} = 10^{-8.10} = 7.9 \times 10^{-9}\ M$$

Are Our Answers Reasonable?

One calculation that suggests that the answers are reasonable is that $[H^+][OH^-] = K_W$ and $(1.3 \times 10^{-6})(7.9 \times 10^{-9}) = 1.0 \times 10^{-14}$.

What If...?

...the hydrogen ion concentration in the lake were doubled? Would the pH be halved, doubled, or neither? Explain.

(*Ans: Neither. The pH is not directly proportional to pH, Rather, it is an inverse logarithmic relationship. The pH would become 5.60.*)

Practice 15.9

What are the pH, $[H^+]$, and $[OH^-]$ values for a solution with pOH = 12.35?

The sum of the pH and pOH values is 14.00. This is the basis of the common pH scale, shown in Fig. 15.13. Note that the hydrogen ion and hydroxide ion concentrations are inversely related. Aqueous solutions with a pH less than 7.0 are said to be acidic, and those with a pH greater than 7.0 are **basic**. Also, as the solutions become more acidic or basic, their pH moves farther away from 7.0, as shown in the figure. The pH values of some common substances are also listed in Table 15.6 and illustrated graphically in Fig. 15.14. Note where food products tend to be located on the pH scale. What types of products tend to be on the basic side of the pH scale?

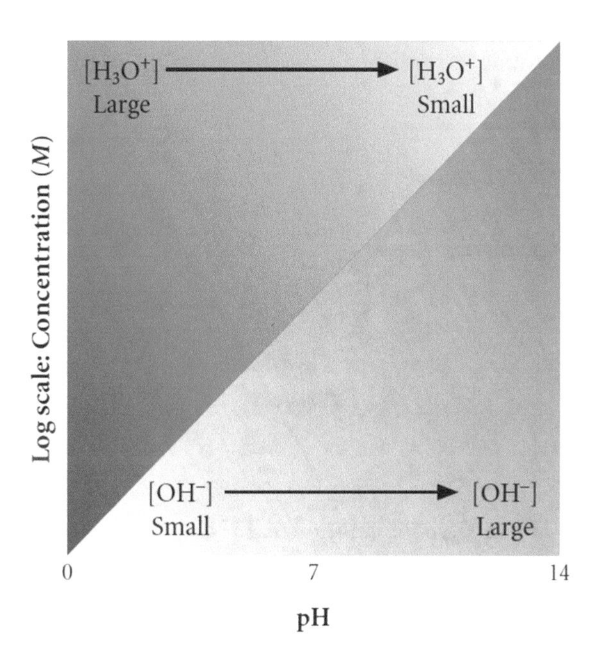

Fig. 15.13 The sum of the pH and pOH values is 14.00 in water at 24 °C. This is the basis of the common pH scale, shown here. Note the inverse relationship between the concentration of hydrogen and that of the hydroxide ion

Table 15.6 pH of some common substances

Substance	Contains this acid or base	pH
Battery acid	Sulfuric acid	1.3
Stomach acid	Hydrochloric acid	1.5–3.0
Vinegar	Acetic acid	2.5
Wine	Tartaric acid	2.8–3.8
Apples	Malic acid	2.9–3.3
Food preservative	Benzoic acid	3.1
Cheese	Lactic acid	4.8–6.4
Milk	Lactic acid	6.5–6.8
Blood	Carbonate ion and others	7.3–7.5
Baking soda	Bicarbonate ion	8.3
Detergents	Carbonate and phosphate ions	10–11
Milk of magnesium	Magnesium hydroxide	10.5
Drain cleaner	Sodium hydroxide	13+

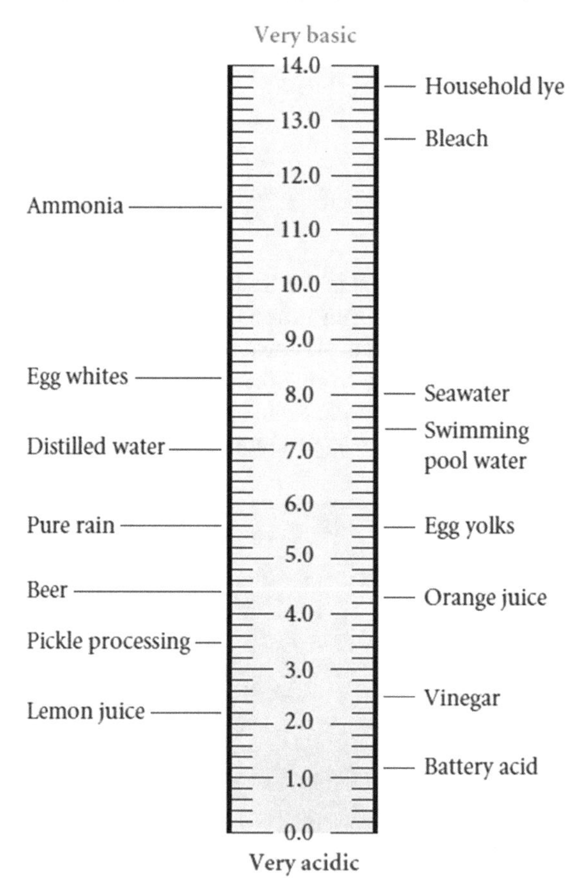

Fig. 15.14 pH of some common substances

Basic – Having a pH greater than 7.00 in aqueous solution at 25 °C.

▶ Here's What We Know So Far

- Acids and bases are substances that can donate or accept hydrogen atoms.
- Strong and weak refers to the extent to which the acid or base forms products in an acid–base reaction.

- Concentrated and dilute refers to the quantity of acid or base that is present in a solution.
- The pH scale is a frame of reference with which we can categorize solutions of acids and bases.
- Many common substances fit within this frame of reference. ◀

At its heart, chemistry is about the interactions of substances. The applications of chemistry concern the changes that result from these interactions. To understand the nature of change in acid–base chemistry, we first need to consider how to determine the changes that occur when an aqueous acidic or basic solution is prepared. This will be the focus of the remainder of this chapter. In the next chapter, we will consider how changing a prepared acidic or basic solution affects its chemical behavior.

15.5 Determining the pH of Acidic Solutions

We have learned that strong acids, such as the HCl solution used to clean the concrete patio, essentially completely dissociate in aqueous solution ($K_a \gg 1$) and that weak acids, such as the aqueous solution of acetic acid found in vinegar, dissociate only partially ($K_a < 1$). We also noted that the concentration of [H$^+$] (and therefore the pH) in solution is determined by both the strength and the initial concentration of the acid.

15.5.1 PH of Strong Acid Solutions

A **monoprotic acid** is an acid (such as HCl) that contains only one acidic hydrogen ion, or proton. H_2SO_4, by contrast, contains two acidic protons and is called a **diprotic acid**. For a strong monoprotic acid, the hydrogen ion concentration in dilute aqueous solution roughly equals the initial concentration of the acid (neglecting activity effects). This is approximately true at any concentration between about 10^{-6} M and 1 M. When the strong acid concentration is less than this, the autoprotolysis of water supplies a significant number of hydrogen ions to the solution, and the pH becomes more difficult to calculate. Even with an HCl solution of concentration 10^{-8} M or less, the pH is still just below 7.00 because of the supply of H$^+$ from the autoprotolysis of water.

Monoprotic acid – An acid that contains one acidic proton.

Diprotic acid – An acid that contains two acidic protons.

Example 15.10—Calculating the pH of a Strong Acid Solution

Our stomach contains hydrochloric acid. If we were going to prepare a solution that had a hydrogen ion concentration within the range of that in the stomach, we might work with a 3×10^{-2} M aqueous solution of HCl. What is the pH of this solution?

Asking the Right Questions

Is HCl a strong or weak acid? How does that relate to the extent of its dissociation and the pH calculation?

Solution

This strong acid would essentially completely dissociate to give $[H^+] = 3 \times 10^{-2}$ M and $[Cl^-] = 3 \times 10^{-2}$ M. This means that the [H$^+$] in solution equals the original HCl concentration.

We can calculate the pH of this solution:

$$pH = -\log[H^+] = -\log(3 \times 10^{-2}\ M) = 1.5$$

Is Our Answer Reasonable?

We have a moderately concentrated strong acid, so we would expect the pH to be in the range of, perhaps, 1 to 3. We can narrow our expected answer further by noting that [H$^+$] is between 10^{-1} and 10^{-2} M, which means that the pH will be between 1 and 2. Our answer to this problem makes sense.

What If…?

…instead of HCl, we had a 3×10^{-2} M aqueous solution of acetic acid, CH_3COOH, $K_a = 1.8 \times 10^{-5}$? If we tried to light a light bulb by using each solution as an electrolyte, what would we observe? Explain.

(*Ans: The HCl solution would allow the bulb to shine much more brightly than would the acetic acid solution. Acetic acid is a weak acid, and would not dissociate completely.*)

Practice 15.10

What is the pH of a 0.0010 M HNO_3 solution?... of a 0.0000250 M $HClO_4$ solution?

15.5.2 Le Châtelier's Principle and the Supply of Hydroxide Ion in Acidic Solutions

Consider a strong acid in an aqueous solution with a pH = 3.0. On the basis of our previous discussion, we know that the pOH is 11.0, and $[OH^-] = 1 \times 10^{-11}$ M (because 14 = pH + pOH). Where does that small amount of hydroxide ion come from? The only source is the autoprotolysis of water:

$$H_2O(l) \rightleftharpoons H^+(aq) + OH^-(aq) \quad K_W = 1.00 \times 10^{-14}$$

If the liquid were pure water, $[OH^-]$ would equal 1×10^{-7} M. However, the addition of H^+ from the strong acid imposed a stress on the aqueous equilibrium, to which it responded by lessening the extent of the dissociation of water. To put it another way, when the strong acid was added, the reaction shifted to the left to compensate. This is an example of Le Châtelier's principle, which we discussed in ▶ Sect. 14.6, and is also known as the **common-ion effect**. We added an ion *common* to (the same as) one of the products (H^+) of the autoprotolysis and HCl dissociation reactions, and the result was that the autoprotolysis reaction did not proceed to the right as much as it would have without the acid. We will look at many of the practical outcomes of Le Châtelier's principle and the common-ion effect in the next chapter. The key in this introduction to acids and bases is that the addition to an aqueous solution of any substance that produces H^+ will reduce the supply of H^+ and OH^- from the autoprotolysis of water. Therefore, in all but the very dilute acid solutions, the autoprotolysis of water as a source of H^+ is generally negligible.

Common-ion effect – The addition of an ion common to one of the species in the solution, causing the equilibrium to shift away from production of that species.

For example, let's consider the pH of a 1.0×10^{-8} M solution of HCl, a strong acid. We might predict that the HCl produces 1.0×10^{-8}M H^+ in solution, which is correct.

$$HCl(aq) \rightarrow H^+(aq) + Cl^-(aq)$$

$$1.0 \times 10^{-8}\text{M} \rightarrow 1.0 \times 10^{-8}\text{M} \quad 1.0 \times 10^{-8}\text{ M}$$

Determining the pH of this solution gives

$$pH = -\log[H^+]$$

$$pH = -\log(1.0 \times 10^{-8})$$

$$pH = 8.00$$

15

Does our answer make sense? No it doesn't. Adding HCl to water, even a very small amount, shouldn't cause the pH to increase! What have we forgotten to consider? The concentration of hydrogen ions in the solution is made up of *all* sources of $[H^+]$.

$$[H^+]_{total} = [H^+]_{HCl} + [H^+]_{water}$$

This means we should add all of the sources of hydrogen ion together and then determine the pH of the solution. What is the [H +] due to water? Recall our autoprotolysis equation:

$$H_2O(l) \rightleftharpoons H^+(aq) + OH^-(aq) \quad K_W = 1.00 \times 10^{-14}$$

On the basis of this, we might be inclined to say that $[H^+]_{water} = [OH^-]_{water} = 1.0 \times 10^{-7}$ M. However, we must keep in mind the effect of Le Châtelier's principle, in which having hydrogen ions supplied by the HCl suppresses the ionization of H_2O. Therefore, $[H^+]_{water}$ will be less than 1.0×10^{-7} M. Although we won't go into the details here, it is possible to determine that in this solution, $[H^+]_{water} = 9.5 \times 10^{-8}$ M.

$$[H^+]_{total} = [H^+]_{HCl} + [H^+]_{water}$$

$$[H^+]_{total} = (1.0 \times 10^{-8}\text{ M}) + (9.5 \times 10^{-8}\text{ M})$$

$$[H^+]_{total} = 1.05 \times 10^{-7}\text{ M}$$

$$pH = 6.98$$

This is more reasonable than our initial answer of pH = 8 (after all, an acid solution should be acidic, not basic!), and it also shows the importance of Le Châtelier's principle as well as the autoprotolysis of water.

15.5.3 pH of Weak Acid Solutions

Lactic acid ($HC_3H_5O_3$) is a weak acid that is produced in our muscle cells when we work too strenuously to maintain aerobic respiration (respiration in the presence of sufficient oxygen). Recent evidence shows that contrary to previously held assumptions, lactic acid is a normal product of metabolism and not a barrier to athletic performance. Lactic acid can be prepared commercially and is used in products ranging from biodegradable polymers to a spray to help extend the shelf life of beef strips. Its wide range of biological and commercial applications makes lactic acid a useful prototype for our discussion about the pH of weak acids.

Let's determine the pH of a 0.10 M solution of lactic acid. We considered the steps in accomplishing this task in ▶ Chap. 14. This is an equilibrium problem, and we may solve it using the same principles that we use to solve any other equilibrium problem.

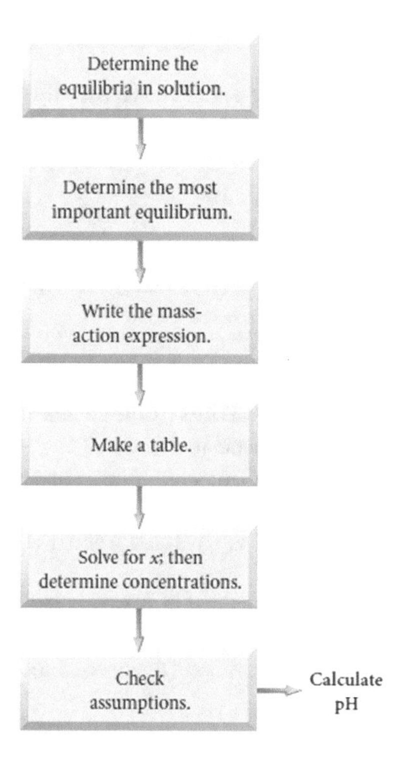

Steps in Solving for the pH of a Weak Acid in Aqueous Solution

Step 1: Determine the equilibria, and the resulting species, that are in the solution.
There are two important equilibria occurring simultaneously in this solution: the ionization of lactic acid ($HC_3H_5O_3$, written here as HL for simplicity) and the autoprotolysis of water.

$$HL(aq) \rightleftharpoons H^+(aq) + L^-(aq) \quad K_a = 1.38 \times 10^{-4}$$

$$H_2O(l) \rightleftharpoons H^+(aq) + OH^-(aq) \quad K_W = 1.00 \times 10^{-14}$$

Step 2: Determine the equilibria that are the most important contributors to $[H^+]$ in the solution.
As we discussed previously, an acid that is much stronger than water will significantly depress the ionization of water in accordance with Le Châtelier's principle. Let's assume that this holds true for a relatively concentrated solution of the weak acid *so that we can safely neglect the autoprotolysis of water as a contributor of* H^+. We will test this assumption later.
Step 3: Write the equilibrium expression for the important contributors to $[H^+]$ (which usually means just the weak acid).

$$K_a = \frac{[L^-][H^+]}{[HL]}$$

Step 4: Set up a table of the initial and equilibrium concentrations of each pertinent species.
Weak acids are called weak because the extent of their ionization is normally quite small. As a first approximation, we may assume that "x" in our table is far less than 0.100 M, or, equivalently, that $[HL] \approx [HL]_0$. We do this because it greatly simplifies our problem solving, but we will also test this assumption later. The assumption will generally work if K_a is at leas t 100 times smaller than $[HL]_0$. In the present problem, the K_a of 1.38×10^{-4} is nearly 1000 times smaller than the $[HL]_0$ value of 0.10 M, so our assumption is probably valid.

	HL	⇌	H+	+	L-
Initial	0.10 M		0 M		0 M
Change	$-x$		$+x$		$+x$
Equilibrium	$0.10 - x$		$+x$		$+x$
Assumptions	0.10		$+x$		$+x$

Step 5: Solve for the estimated concentration of each species. We use the "assumptions" line for now and we'll test those assumptions later:

$$1.38 \times 10^{-4} = \frac{x^2}{0.10}$$

$$x = \left[H^+\right] = \left[C_3H_5O_3^-\right] = 3.7 \times 10^{-3}\ M$$

$$[HC_3H_5O_3] = 0.10 - 3.7 \times 10^{-3} = 0.096\,M$$

Step 6: Check our assumption that the extent of ionization of lactic acid is negligibly small.
This is done using the "5% rule," which we introduced in ▶ Sect. 14.5. Also as discussed in that section, we can ignore the 5% simplification if we want to program our calculator (or want to write it out by hand) to solve the quadratic equation in all cases. We present the 5% rule because it shows that it is possible to use our understanding of chemistry to arrive at a solution strategy that is straightforward whether or not a programmable calculator is available.

If less than 5% of the original amount of the weak acid ionizes, this is considered "negligible." If more than 5% ionizes, then we must take this into account in our problem-solving during Step 4. We will show how this is done below. What percentage of the lactic acid has ionized (or dissociated)?

$$\%\ dissociated = \frac{[C_3H_5O_3^-]}{HC_3H_5O_3} \times 100\% = \frac{3.7 \times 10^{-3}}{0.10} \times 100\% = 3.7\%$$

We can consider the ionization to be negligible because less than 5% of the original acid reacted to form products.

We have one more assumption to check—that the contribution of water to the H^+ concentration is negligible. We see that lactic acid contributes 3.7×10^{-3} M. As we discussed previously in this chapter, unless the H^+ concentration from the acid is 10^{-6} M or less, the contribution of water is negligible. Our assumption is verified in this case, then.

Step 7: Solve for the pH of the solution.

$$pH = -\log\left[H^+\right] = -\log\left(3.7 \times 10^{-3}\right) = 2.43$$

Remember that the number of significant figures after the decimal point in the pH is equal to the number of significant figures in the pre-exponential term. Because the pre-exponential term is 3.7, we use two figures after the decimal point ("0.43"). The pH = 2.43.

Example 15.11—pH of a Weak Acid

We have mentioned aqueous hydrofluoric acid (HF) several times in this chapter, and we have found it to be a weak acid, $K_a = 7.2 \times 10^{-4}$. It is employed in the chemical industry to produce compounds in gasoline and fluorocarbons used as refrigerants. Calculate the pH in a 1.0×10^{-3} M HF solution.

Asking the Right Questions

What are the equilibria that contribute to $[H^+]$? Which of these are likely to contribute enough $[H^+]$ to be considered important? How does the mass-action expression for the weak acid help us solve the example?

Solution

Step 1: Equilibria in solution

The two important equilibria in the solution are the ionization of HF and the autoprotolysis of water.

$$HF(ag) \rightleftharpoons H^+(ag) + F^-(aq) \quad K_a = 7.2 \times 10^{-4}$$

$$H_2O(1) \rightleftharpoons H^+(ag) + OH^-(aq) \quad K_{ux} = 1.00 \times 10^{-14}$$

Step 2: Determine the equilibria that are the most important contributors to $[H^+]$ in the solution

We can assume that the HF ionization is the only important contributor to $[H^+]$ in solution because its K_a value is far greater than the K_w value of water.

Step 3: Write the equilibrium expression for the important contributors to [H+]

$$K_a = \frac{[F^-][H^+]}{[HF]}$$

Step 4: **Table of concentrations**

	HF	H^+	F^-
Initial	1.0×10^{-3} M	0 M	0 M
Change	$-x$	$+x$	$+x$
Equilibrium	$1.0 \times 10^{-3} - x$	$+x$	$+x$
Assumptions	1.0×10^{-3}	$+x$	$+x$

Although we can assume that the ionization of HF is negligibly small (that is, "x" is essentially irrelevant compared to 1.0×10^{-3} M), the assumption is hazardous because K_a, 7.2×10^{-4}, is not less than about $x = [H^+] = [F^-] = 8.5 \times 10^{-4}$ M. Rather, it is nearly equal to $[HF]_0$! Let's determine the implications of making the assumption.

Step 5: Solve

$$7.2 \times 10^{-4} = \frac{x^2}{1.0 \times 10^{-3}}$$

$$x = [H^+] = [F^-] = 8.5 \times 10^{-4} \text{ M}$$

Step 6: Check our assumption of negligible dissociation

$$\% \ dissociated = \frac{[F^-]}{[HF]} \times 100\%$$

$$= \frac{8.5 \times 10^{-4}}{1.0 \times 10^{-3}} \times 100\%$$

$$= 85 \ \%$$

Wow! This is certainly not negligible compared to 5%, so we may not claim that the ionization of HF is negligibly small. We must solve the following equation, which uses the "equilibrium" row of the ICEA table, rather than the "assumptions" row.

$$7.2 \times 10^{-4} = \frac{x^2}{1.0 \times 10^{-3} - x}$$

We can set up the quadratic equation by multiplying K_a, 7.2×10^{-4}, by $[HF]$, $1.0 \times 10^{-3} - x$, to give

$$x^2 = 7.2 \times 10^{-7} - 7.2 \times 10^{-4}x$$

Setting the equation = 0 so we may solve for x yields

$$x^2 + \left(7.2 \times 10^{-4}\right)x - 7.2 \times 10^{-7} = 0$$

Then, using the quadratic equation:

$$x = \frac{-b \pm \sqrt{b^2 - 4ac}}{2a}$$

$$a = 1 \quad b = 7.2 \times 10^{-4} \quad c = -7.2 \times 10^{-7}$$

$$x = \frac{-(7.2 \times 10^{-4}) \pm \sqrt{(7.2 \times 10^{-4})^2 - 4(1)(-7.2 \times 10^{-7})}}{2(1)}$$

$$x = \frac{-(7.2 \times 10^{-4}) \pm 1.843 \times 10^{-3}}{2}$$

$$x = [H^+] = [F^-] = 5.62 \times 10^{-4} \text{ M} \approx 5.6 \times 10^{-4} \text{ M}$$

$$[HF] = 1.0 \times 10^{-3} - \left(5.62 \times 10^{-4}\right) = 4.38 \times 10^{-4}M \approx 4.4 \times 10^{-4}M$$

Step 7: Solve for the pH of the solution

$$pH = -\log [H^+] = -\log \left(5.62 \times 10^{-4}\right) = 3.25$$

Is our Answer Reasonable?

We can check our math by substituting back into the mass-action expression.

$$K_a = \frac{(5.62 \times 10^{-4})^2}{4.38 \times 10^{-4}} = 7.2 \times 10^{-4}$$

Our answer is confirmed.

What If...?

...the solution contained 0.125 mol of HF in 250.0 mL of solution? What would be the solution pH?

(*Ans*: $pH = 1.72$)

Practice 15.11

What is the pH of a 0.250 M acetic acid solution $(K_a = 1.8 \times 10^{-5})$?

15.5.4 pH of a Mixture of Monoprotic Acids

If we ingest some vinegar, 3% acetic acid, does it significantly change the pH of our HCl-containing stomach? This question falls within the general topic of the pH of a mixture of acids that differ greatly in strength. Consider a solution that is 0.50 M HCl and 0.90 M acetic acid (CH_3COOH), $K_a = 1.8 \times 10^{-5}$. To solve for the pH of the mixture, we employ our usual approach. The relevant equilibria in the solution are

$$HCl(ag) \rightleftharpoons H^+(ag) + Cl^-(ag) \quad K_a >> 1$$

$$CH_3COOH(aq) \rightleftharpoons H^+(aq) + CH_3COO^-(aq) \quad K_a = 1.8 \times 10^{-5}$$

$$H_2O(l) \rightleftharpoons H^+(aq) + OH^-(aq) \quad K_w = 1.00 \times 10^{-14}$$

This says that there are three sources of hydrogen ion: HCl, CH_3COOH, and H_2O. We can write this in equation form:

$$[H^+]_{total} = [H^+]_{HCl} + [H^+]_{CH_3COOH} + [H^+]_{H_2O}$$

The hydrochloric acid is a substantially stronger acid than either the acetic acid or the water and is present in a concentration that is large enough so that we may make the *assumption* that neither the acetic acid nor the water ionizes significantly. In fact, the presence of hydrogen ion (the common ion) from HCl further suppresses the acetic acid and water ionization in accordance with Le Châtelier's principle.

$$[H^+]_{total} \approx [H^+]_{HCl} \approx 0.50\,M$$

We will test this assumption later.

What is the hydrogen ion contribution due to the acetic acid? We know that $[H^+]_{total} = 0.50\,M$. We also know that the $[H^+]$ contributed by the acetic acid will be essentially equal to the $[CH_3COO^-]$. The concentration of acetic acid at equilibrium will be essentially equal to its initial concentration.

$$[CH_3COOH] = 0.90\,M$$

$$[H^+]_{acetic\,acid} = [CH_3COO^-] = \text{“}x\text{”}$$

We can use the mass-action expression for acetic acid ionization to solve for $[H^+]_{acetic\ acid}$.

$$K_a = \frac{[CH_3COO^-][H^+]}{[CH_3COOH]}$$

$$1.8 \times 10^{-5} = \frac{x(0.50)}{0.90}$$

$$x = [CH_3COO^-] = [H^+]_{acetic\,acid} = 3.2 \times 10^{-5}\,M$$

This shows that acetic acid is not a significant contributor to the hydrogen ion concentration of the solution, and although we did not quantify it here, the water is even less important. Further, if we had had an aqueous solution of only 0.90 M CH_3COOH (without the HCl), the equilibrium concentration of the acetate ion, $[CH_3COO^-]$, would have been

$$K_a = \frac{[CH_3COO^-][H^+]}{[CH_3COOH]}$$

$$1.8 \times 10^{-5} = \frac{x^2}{0.90}$$

$$x = [H^+] = [CH_3COO^-] = 4.0 \times 10^{-3}\,M$$

This is over 100 times higher than the equilibrium acetate ion concentration of 3.2×10^{-5} M in the HCl solution. This confirms what we said previously, that "the presence of hydrogen ion (the common ion) from HCl further suppresses the acetic acid and water ionization in accordance with Le Châtelier's principle." The main message to keep in mind from this discussion is that when we calculate the pH of a mixture of acids, we must look for the dominant equilibrium, as well as taking into account the concentrations of the acids. This will determine the pH of the solution.

15.6 Determining the pH of Basic Solutions

We have noted in this chapter the impact of acids and bases on ourselves and our world. A useful, though perhaps unfortunate, example concerns tobacco, which, according to the World Health Organization, prematurely kills millions of people worldwide each year. Tobacco companies have used scientists' understanding of acid–base chemistry to increase by a factor of 100 the amount of nicotine that leaves burned tobacco particles and enters the gas phase to be absorbed by the lungs. The two important compounds involved in this chemistry are the nicotine itself ($C_{10}H_{14}N_2$) and ammonia (NH_3), both of which are bases (see Fig. 15.15). We will look at how ammonia can be used to increase the amount of gaseous nicotine that enters the lungs of smokers. As a first step in our analysis, we need to see how and why ammonia acts as a base.

Ammonia has many applications, including those shown in Table 15.1. It is easily in the top 10 industrial chemicals, with over 235 billion kg produced worldwide in 2019. Ammonia is quite soluble in water, acting as a weak base. Using the Arrhenius model of base behavior, it supplies hydroxide ions to the solution, and it's also a Brønsted–Lowry base and a Lewis base because it accepts a proton and donates a pair of electrons.

Ammonia is typical of many bases in that it has an unshared pair of electrons on the nitrogen atom that can accept a proton from, in this case, water molecules.

$$NH_3(aq) + H_2O(l) \rightleftharpoons NH_4^+(aq) + OH^-(aq) \quad K_b = 1.8 \times 10^{-5}$$

Ammonia is considered a weak base because the K_b value is much less than 1. On the other hand, Group 1 (IA) hydroxides, such as NaOH and KOH, completely dissociate in water ($K_b \gg 1$) to supply hydroxide ion and are considered strong bases.

$$NaOH(s) \rightarrow Na^+(aq) + OH^-(aq)$$

The solubility of Group IIA hydroxides increases from the essentially insoluble $Be(OH)_2$ to the more soluble (0.01 M) $Ca(OH)_2$ to the still more soluble (up to 0.1 M) $Ba(OH)_2$. Soluble Group 2 (IIA) hydroxides, such as $Ca(OH)_2$, are strong bases like their Group 1(IA) counterparts, but they supply two moles of hydroxide for every mole of base that dissolves.

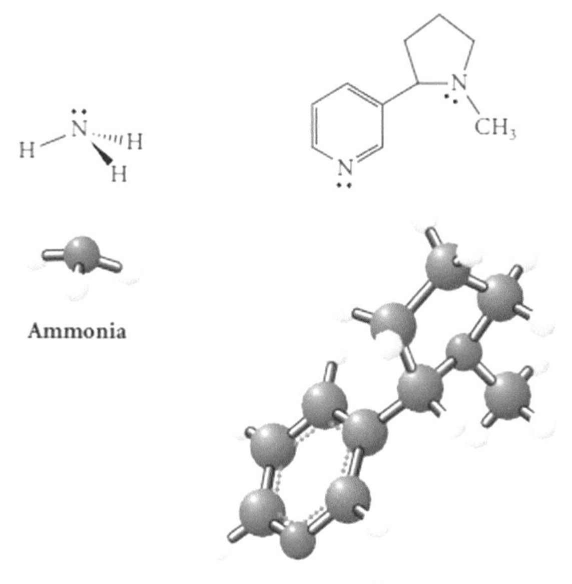

Fig. 15.15 Ammonia and nicotine are both bases

$$Ca(OH)_2(s) \rightleftharpoons Ca^{2+}(aq) + 2OH^-(aq)$$

Drawing on our understanding of equilibrium (that is, what are the right questions to ask in order to understand what happens in the solution?) and our sense of the pH scale, let's try the following two examples dealing with the pH of a strong base solution and of a weakly basic ammonia solution. Even though we are now looking at bases, rather than acids, the equilibrium process and the thinking involved are largely the same.

Example 15.12—pH of a Strong Base

Calculate the pH of a 0.60 M aqueous solution of KOH.

Asking the Right Questions

What differentiates a strong base from a weak base? Are there other reactions to consider that provide $[OH^-]$ to the solution?

Solution

As a strong base, KOH completely dissociates. This is consistent with the behavior of Group IA and soluble Group IIA hydroxides.

$$KOH(s) \rightarrow K^+(aq) + OH^-(aq) \quad K \gg 1$$

The other acid–base reaction that occurs in solution is the autoprotolysis of water, although it is unimportant as a source of OH^- because the K value is so small compared to that of the strong base.

$$H_2O(l) \rightleftharpoons H^+(aq) + OH^-(aq) \quad K_w = 1.0 \times 10^{-14}$$

Therefore,

$$[OH^-] = 0.60\ M$$

$$pOH = 0.22;\ pH = 13.78$$

Is Our Answer Reasonable?

We expect the pH to be quite high, analogous to the very low pH of a strong acid, so our answer makes sense. In addition, the pH is considerably higher than with the same concentration of ammonia, a weak base. This is shown in the next exercise.

What If…?

…the concentration of KOH were halved, so that it was 0.30 M? A student did the pH calculation twice, first claiming that the pH = 6.89 (one-half of the original value, to match the concentration change), and, upon reconsideration, calculating 14.52. What was the error in each case, why weren't the student's answers reasonable, and what is the actual pH of the 0.30 M KOH solution?

(*Ans: In the first case, the pH indicates an acid (less than 7).In addition, pH is not linearly related to concentration. In the second case, the student added, rather than subtracted, the pH change to 14, implying that the pH increased when the KOH solution was diluted. The actual pH=13.48.*)

Practice 15.12

Calculate the pH of a 1.06 M solution of NaOH.

15

Example 15.13—pH of a Weak Base

Determine the pH of a 0.600 M solution of ammonia.

Asking the Right Questions

What are the equilibria that contribute to $[OH^-]$? Which of these are likely to contribute enough $[OH^-]$ to be considered important? How does the mass-action expression for the weak acid help us solve the example?

Solution

We may proceed as with any equilibrium problem, by listing the equilibria that occur in solution.

$$NH_3(aq) + H_2O(l) \rightleftharpoons NH_4^+ + OH^-(aq) \quad K_b = 1.8 \times 10^{-5}$$

$$H_2O(l) \rightleftharpoons H^+(aq) + OH^-(aq) \quad K_w = 1.00 \times 10^{-14}$$

The autoprotolysis of water is insignificant compared to the hydrolysis of ammonia. In fact, the OH^- produced by the ammonia reaction will further depress the autoprotolysis of water (via Le Châtelier's principle), so the OH^- and H^+ due to water will be less than in neutral solution, as discussed previously. The only relevant equilibrium expression is

$$K_a = \frac{[NH_4^+][OH^-]}{[NH_3]}$$

$$NH_3 \rightleftharpoons NH_4^+ + OH^-$$

	NH_3	NH_4^+	OH^-
Initial	0.600 M	0 M	0 M
Change	$-x$	$+x$	$+x$
Equilibrium	$0.600 - x$	$+x$	$+x$
Assumptions	0.600	$+x$	$+x$

a

$$1.8 \times 10^{-5} = \frac{x^2}{0.600}$$

$$x = [NH_4^+] = [OH^-] = 3.3 \times 10^{-3}\ M$$

$$pOH = 2.48;\ pH = 11.52$$

The percent dissociation was:

$$\frac{3.3 \times 10^{-3}}{0.600} \times 100\% = 0.55\%$$

Our assumption was valid.

Is our Answer Reasonable?

The pH of this weakly basic solution is about 2 units lower than that of the equally concentrated strong base in the previous example. (Remember the distinction between "strong" and "concentrated!") That means the mathematical solution (indicated by the pH) matches the chemical solution that it is a weak base, and the answer is reasonable. Were we justified in neglecting the autoprotolysis of water as a source of hydroxide ions? The only source of hydrogen ions in this solution is the autoprotolysis of water. If pH = 11.52, then $[H^+]_{water} = 3.0 \times 10^{-12}$ M. This also equals $[OH^-]_{water}$, and this is insignificant compared to 3.3×10^{-3} M, the total $[OH^-]$. Is our answer reasonable? To the extent that our pH reflects a moderately basic solution, yes, it is reasonable.

What If...?

...we added three times the volume of a 2.0 molar NaOH solution to the ammonia solution? What would happen to the pH? Speculate: Which would be the important process(es) that contributes to $[OH^-]_{total}$?

(Ans: The pH would rise. The dissociation of NaOH to Na^+ to OH^- would be the only important process, because OH^- is so much stronger a base than is NH_3.)

Practice 15.13

Calculate the pH of a 0.500 M solution of dimethylamine, $(CH_3)_2NH$, $K_b = 5.9 \times 10^{-4}$.

One class of alkaloids, called "ephedrines," are used as dietary supplements to help in weight loss, to enhance body building, and to increase the user's energy level. In doses of more than 8 mg per serving, the compounds are considered dangerous for routine use. Another of the alkaloids, cocaine, has a severe psychological impact. It can be made basic to create the free-base form called "crack."

15.7 Issues and Controversies—Nicotine and pH Control in Cigarettes

Nicotine can exist in three forms, depending on the pH of its environment (see Fig. 15.16). Below pH 3, it is in the diprotonated form. Between about pH 3 and pH 8, most of it is in the monoprotonated form.

If the pH is above 8, most of the nicotine exists in the volatile "free-base" form, which readily evaporates at the temperature of burning tobacco. This free-base is effectively absorbed by the lung tissue. Ammonia is added to tobacco leaves to raise the pH, making available more free-base nicotine for inhalation. The Food and Drug Administration has also determined that pH levels have been manipulated in smokeless tobacco products. The products for new users are at a relatively low pH, so there is not that much free-base nicotine available. Smokeless tobacco for "experienced" users has a higher pH, which leads to a higher level of nicotine absorption. Recent research shows that the pH levels in e-cigarettes (vaping products), varies over at least 4 pH units.

Nicotine is an **alkaloid**, one of many nitrogen-containing bases found in vegetables and other plants. Table 15.7 lists the structures and uses of some alkaloids, many of which are commonly used names. They often have a substantial impact on the central nervous system and brain.

Alkaloid – Nitrogen-containing bases found in vegetables and other plants.

$pk_{a_1} = 3.12$ (a) $pk_{a_2} = 8.02$ (b) (c)

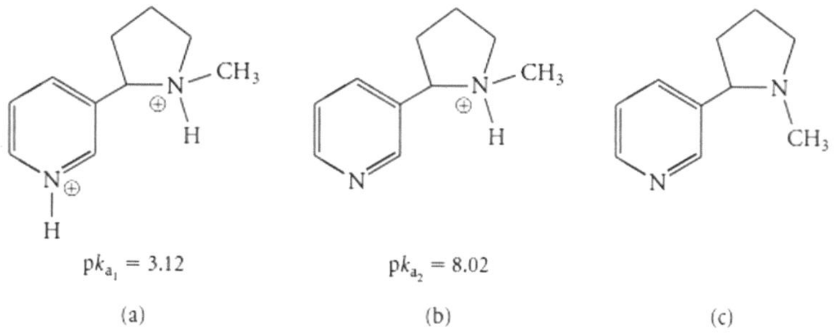

Fig. 15.16 The three forms of nicotine. **a** The form that predominates below pH 3.12. **b** The form that predominates between pH 3.12 and 8.02. **c** The form that predominates above pH 8.02. This "free-base" form is readily vaporized and absorbed into the lungs when the cigarette burns

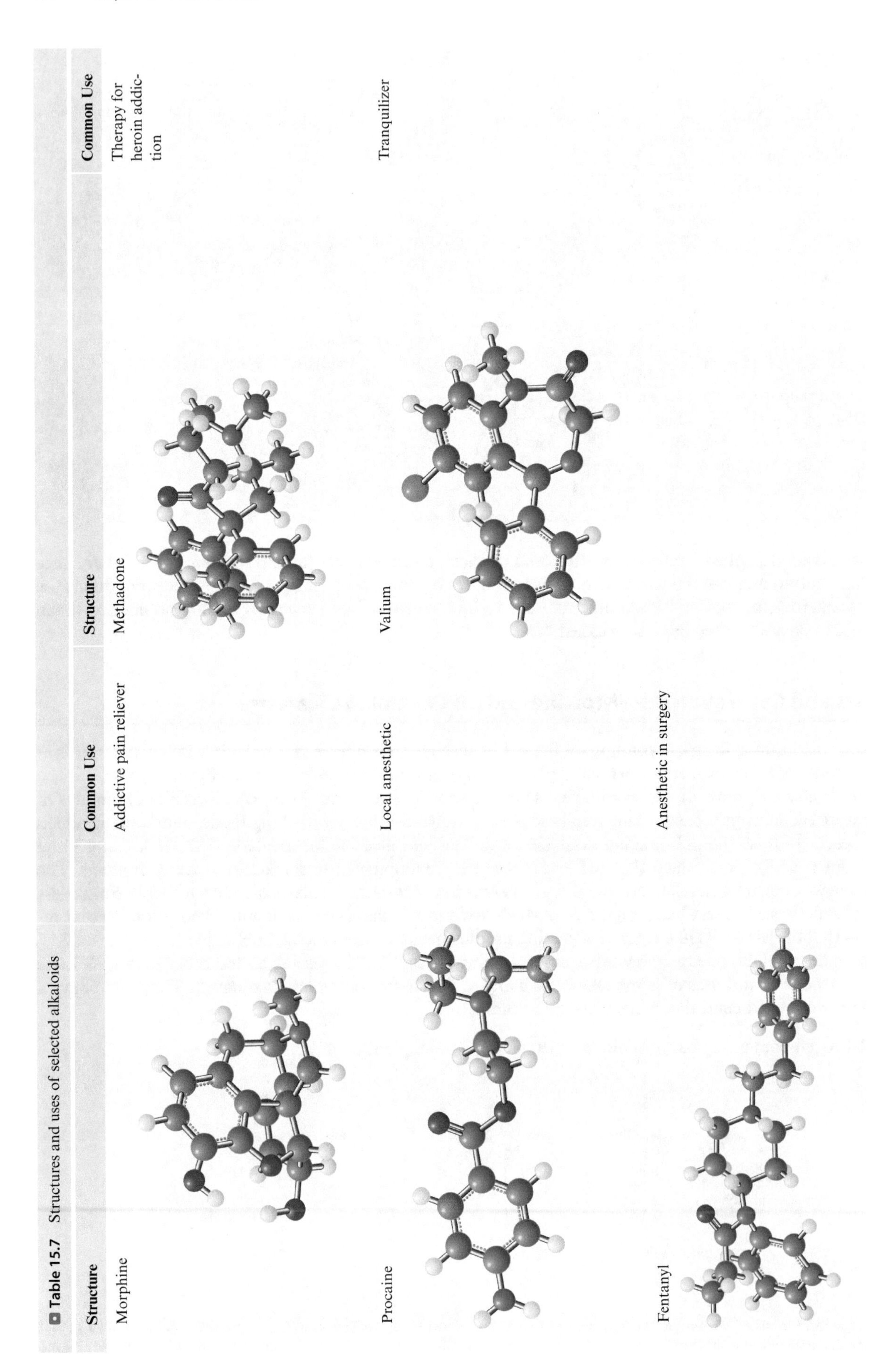

Table 15.7 Structures and uses of selected alkaloids

Structure	Common Use	Structure	Common Use
Morphine	Addictive pain reliever	Methadone	Therapy for heroin addiction
Procaine	Local anesthetic	Valium	Tranquilizer
Fentanyl	Anesthetic in surgery		

15.7 Polyprotic Acids

Our theme is impact—the impact of acids and bases on the chemical industry, on our environment, and on ourselves. Phosphoric acid (H_3PO_4) fits well with that theme. It is different from most of the acids we have mentioned thus far, because it is a **polyprotic acid**. Like all polyprotic acids, phosphoric acid contains more than one acidic hydrogen. The structure of this **triprotic acid** (three acidic hydrogen atoms) acid is shown in Fig. 15.17, with its three acidic hydrogen atoms highlighted. The acid is produced from "phosphate rock," which is largely composed of fluoroapatite ($Ca_5(PO_4)_3F$) and other compounds containing iron, calcium, silicon, aluminum, and fluorine.

Polyprotic acid – An acid that can release more than 1 mol of hydrogen ions per mole of acid.

Triprotic acid – An acid that contains three acidic protons.

Annually, around 30 million tons of marketable phosphate rock was mined in the United States, mostly from Florida and North Carolina, according to the U.S. Geological Survey. Significant amounts of phosphates are also produced in China and in the Morocco and Western Sahara regions of Africa. This ore is converted into phosphoric acid for use in our society. Table 15.8 indicates the wide variety of uses of phosphoric acid. Phosphoric acid can be used to convert fluoroapatite into a soluble fertilizer (calcium dihydrogen phosphate) and a monoprotic acid (hydrofluoric acid), as shown in the following reaction:

$$2Ca_5(PO_4)_3F + 14H_3PO_4 \rightleftharpoons 10Ca(H_2PO_4)_2 + 2HF$$

In the human body, phosphate ion is important in maintaining the pH of blood in a fairly narrow range of 7.3–7.5.

Phosphoric acid is one of two inorganic polyprotic acids that have a worldwide impact on our ability to convert what nature has given us into products that sustain and improve our quality of life. Sulfuric acid is the other.

Sulfuric acid is a diprotic acid that is prepared by the Contact process, as we discussed in ▶ Sect. 14.3. Sulfur is burned in oxygen to form sulfur dioxide, which is then converted into sulfur trioxide via a catalyst such as vanadium:

$$S(s) + O_2(g) \rightleftharpoons SO_2(g)$$

$$2SO_2(g) + O_2(g) \rightleftharpoons 2SO_3(g)$$

The sulfur trioxide is then combined with water to give sulfuric acid:

$$SO_3(g) + H_2O(l) \rightleftharpoons H_2SO_4(l)$$

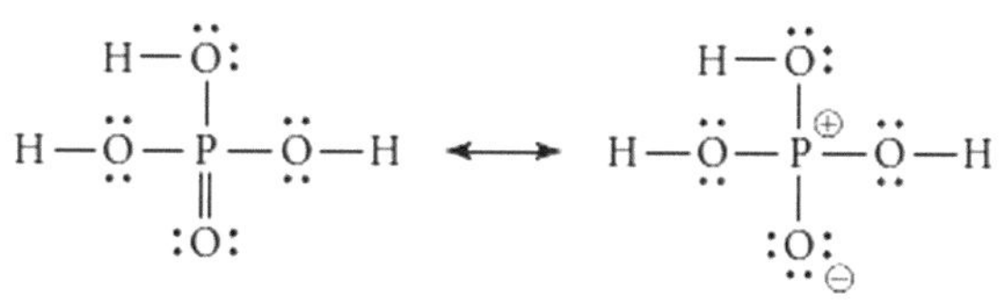

Fig. 15.17 Phosphoric acid (H_3PO_4) is triprotic acid because it contains three acidic hydrogen atoms

Table 15.8 Uses of phosphoric acid in manufacturing

Uses
Fertilizer
Dentrifrices
Soaps
Detergents
Fire control agents
Soft drinks
Incandescent light filaments
Corrosion inhibitors in metals
Organic chemicals such as ethylene and propylene

Amount Per Serving		% Daily Value
Calories	30	
Total Carbohydrate	7 g	3% †
Total Sugars	6 g	**
Includes 6 g Added Sugars		12% †
Vitamin D (as cholecalciferol)	25 mcg (1000 IU)	125%
Calcium (as tricalcium phosphate)	500 mg	38%
Phosphorus (as tricalcium phosphate)	200 mg	16%
Sodium	10 mg	< 1%

† Percent Daily Values are based on a 2,000 calorie diet.
** Daily Value not established.

Fig. 15.18 Calcium supplements often contain the tribasic phosphate ion. This supplement contains a mixture of calcium-containing minerals including $Ca_3(PO_4)_2$ and $CaCO_3$. *Source* https://www.gummyvites.com/-/media/vms/vitafusion/supplements/in/calcium-supp-fact.jpg?h=3000&w=3000&la=en&hash=E026E581E3E09B15E4C0D01FA8ACDEB4F2572867

Among the many uses for sulfuric acid is a century-old process for the conversion of phosphate rock to the fertilizer calcium dihydrogen phosphate monohydrate, $Ca(H_2PO_4)_2{\cdot}H_2O$. The reaction can be summarized as follows:

$$2Ca_5(PO_4)_3F + 7H_2SO_4 + 3H_2O \rightleftharpoons 3Ca(H_2PO_4)_2 \cdot H_2O + 7CaSO_4 + 2HF$$

Note that the product of this reaction is similar to the one produced by the treatment of phosphate rock with phosphoric acid.

Among the other uses of sulfuric acid is in the manufacture of phosphoric acid itself from fluoroapatite, with the resulting production of calcium sulfate dihydrate (gypsum) under reaction conditions that are different from those in the previous reaction.

$$2Ca_5(PO_4)_3F + 10H_2SO_4 + 20H_2O \rightleftharpoons 10CaSO_4 \cdot 2H_2O + 6H_3PO_4 + 2HF$$

The gypsum produced in this reaction is quite valuable for use as wallboard or drywall in the construction of homes. Furthermore, HF is a useful reagent in the glass industry.

In the reactions we have just shown, the phosphates and sulfates are present in a variety of forms: as polyprotic acids (H_3PO_4 and H_2SO_4); as a **monobasic salt** (it can accept one acidic hydrogen atom—$Ca(H_2PO_4)_2{\cdot}H_2O$); and as a **dibasic salt** (it can accept two acidic hydrogen atoms—$CaSO_4$). Another common compound is calcium phosphate, $Ca_3(PO_4)_2$, a **tribasic salt** (it can accept three acidic hydrogen atoms). Calcium phosphate is one of several calcium salts, including calcium carbonate, $CaCO_3$, and calcium citrate, $Ca_3(C_6H_5O_7)_2{\cdot}4H_2O$, that many people take daily as a calcium supplement (Fig. 15.18). Each of these species affects the acid concentration of solutions in predictable ways. Understanding these effects helps us to see how these acids are employed in manufacturing the products that we use.

Monobasic salt – A salt that can accept one hydrogen ion.

Dibasic salt – A salt that can accept two hydrogen ions.

Tribasic salt – A salt that can accept three acidic hydrogen atoms.

15.7.1 The pH of Polyprotic Acids

Phosphoric acid is so common that it represents an important and useful model for our introduction to the pH of polyprotic acids. The ingredient label on our can or bottle of soda will tell us that we are exposed to this acid on a daily basis. Phosphoric acid is a polyprotic acid; it contains more than one acidic hydrogen in its structure. Let's begin by calculating the pH of 4.0 M aqueous phosphoric acid (neglecting activity effects).

As with any equilibrium problem, the first step is to establish the equilibria of the species in solution that contribute hydrogen ions. Phosphoric acid can donate three hydrogen ions per molecule, the second being more difficult to donate than the first because of the negative charge on the resultant dihydrogen phosphate anion

Table 15.9 K_a values for selected polyprotic acids

Formula	Name	K_{a1}	K_{a2}	K_{a3}
H_3PO_4	Phosphoric acid	7.4×10^{-3}	6.2×10^{-8}	4.8×10^{-13}
H_3AsO_4	Arsenic acid	5.0×10^{-3}	8.0×10^{-8}	6.0×10^{-10}
H_2CO_3	Carbonic acid	4.3×10^{-7}	5.6×10^{-11}	
H_2SO_4	Sulfuric acid	>1	1.2×10^{-2}	
H_2SO_3	Sulfurous acid	1.5×10^{-2}	1.0×10^{-7}	
H_2S	Hydrosulfuric acid	1.0×10^{-7}	1.0×10^{-15}	
$H_2C_2O_4$	Oxalic acid	6.5×10^{-2}	6.1×10^{-5}	
$H_2C_6H_6O_6$	Ascorbic acid	7.9×10^{-5}	1.6×10^{-12}	

($H_2PO_4^-$). The third will be tougher still as a consequence of the greater charge on the monohydrogen phosphate anion (HPO_4^{2-}). This is reflected in the values for K_a for each step in the process, in which K_{a_1} is the equilibrium constant for the first acidic ionization, K_{a_2} refers to the second ionization, and K_{a_3} describes the third. Table 15.9 lists K_a values for several polyprotic acids. K_{a_1} values are not only larger than those of K_{a_2}, but they are often *considerably* so. Water is also a source of hydrogen ions in an aqueous solution of phosphoric acid.

$$H_3PO_4 \rightleftharpoons H_2PO_4^- + H^+ \quad K_{a_1} = 7.5 \times 10^{-3}$$

$$H_2PO_4^- \rightleftharpoons HPO_4^{2-} + H^+ \quad K_{a_2} = 6.2 \times 10^{-8}$$

$$HPO_4^{2-} \rightleftharpoons PO_4^{3-} + H^+ \quad K_{a_3} = 4.8 \times 10^{-13}$$

$$H_2O \rightleftharpoons OH^- + H^+ \quad K_W = 1.0 \times 10^{-14}$$

The total concentration of hydrogen ion will equal the sum of the contributions from all of the reactions.

$$[H^+]_{total} = [H^+]_{H_3PO_4} + [H^+]_{H_2PO_4^-} + [H^+]_{HPO_4^{2-}} + [H^+]_{H_2O}$$

Which reactions are important contributors of hydrogen ions to the solution? The K_{a_1} value is considerably larger than the others, so we may assume that it is the only important equilibrium. The K_a values tell us that the first equilibrium contributes about 100 million times more H^+ than the second equilibrium, and more than 10 trillion times more H^+ than the third equilibrium! That is,

$$[H^+]_{total} \cong [H^+]_{H_3PO_4}$$

We will eventually want to prove that the other reactions are not important contributors if we make this assumption.

We may now solve the problem, determining the pH of a 4.0 M solution of H_3PO_4, as we have other equilibrium problems. We'll remember to test any assumptions we make along the way.

$$K_{a_1} = \frac{[H_2PO_4^-][H^+]}{[H_3PO_4]}$$

	H_3PO_4	$\rightleftharpoons$	H^+	+	$H_2PO_4^-$
Initial	4.0 M		0 M		0 M
Change	$-x$		$+x$		$+x$
Equilibrium	$4.0-x$		$+x$		$+x$
Assumptions	4.0		$+x$		$+x$

$$7.4 \times 10^{-3} = \frac{x^2}{4.0}$$

$$x = \left[H_2PO_4^-\right] = \left[H^+\right] = 0.17\ M$$

$$pH = 0.76$$

$$[H_3PO_4] = [H_3PO_4]_0 - \left[H_2PO_4^-\right] = 4.0 - 0.17 = 3.83\ M \approx 3.8\ M$$

We originally made the assumption that "x" is negligible relative to 4.0 M. In this case,

$$\frac{0.17}{4.0} \times 100\% = 4.3\%$$

Our assumption is valid, based on the "5% rule".

We must now test our assumptions that the other equilibria were not important contributors of hydrogen ion to the solution. We may begin with water, knowing that

$$[H^+]_{water} = [OH^-]_{total} = \frac{K_w}{[H^+]_{total}} = \frac{1.0 \times 10^{-14}}{0.17} = 5.9 \times 10^{-14}\ M$$

Our assumption that water was a negligible source of hydrogen ion was fine (with thanks to Le Châtelier!). What about $[H^+]_{HPO_4^{2-}}$, which results from the loss of a hydrogen ion by $H_2PO_4^-$?

$$H_2PO_4^- \rightleftharpoons HPO_4^{2-} + H^+ \quad K_{a_2} = 6.2 \times 10^{-8}$$

We showed above that $\left[H_2PO_4^-\right]$ is equal to 0.17 M. What is $[H^+]$ in this equation? It is roughly equal to the hydrogen ion concentration resulting from the first dissociation, that of H_3PO_4, $\left[H^+\right]_{total} \approx [H^+]_{H_3PO_4}$, and that is also equal to 0.17 M. We can substitute these values into the equilibrium expression to determine $[H^+]_{H_2PO_4^-}$ (which equals $[HPO_4^{2-}]$):

$$K_{a_2} = 6.2 \times 10^{-8} = \frac{[HPO_4^{2-}][H^+]}{[H_2PO_4^-]} = \frac{[HPO_4^{2-}](0.17)}{(0.17)}$$

Therefore,

$$\left[HPO_4^{2-}\right] = K_{a_2} = 6.2 \times 10^{-8}\ M$$

Also, $[H^+]_{HPO_4^{2-}}$ will equal $[HPO_4^{2-}]$, because they are both produced in the same reaction.

$$[H^+]_{HPO_4^{2-}} = 6.2 \times 10^{-8}\ M$$

This confirms the assumption that only the first equilibrium, the dissociation of H_3PO_4, is an important contributor to $\left[H^+\right]_{total}$.

We can take this one step further and calculate the hydrogen ion concentration due to the dissociation of the dibasic anion HPO_4^{2-}:

$$HPO_4^{2-} \rightleftharpoons PO_4^{3-} + H^+ \quad K_{a_3} = 4.8 \times 10^{-13}$$

In this case, $[H^+]_{total}$ is still 0.17 M and $\left[HPO_4^{2-}\right] = 6.2 \times 10^{-8}$M. Substituting into the equilibrium expression for K_{a_3} yields

$$K_{a_3} = 4.8 \times 10^{-13} = \frac{[PO_4^{3-}][H^+]}{[HPO_4^{2-}]} = \frac{[E_4^{3-}](0.17)}{6.2 \times 10^{-8}}$$

$$x = \left[PO_4^{3-}\right] = \left[H^+\right]_{PO_4^{3-}} = 1.8 \times 10^{-19}\ M$$

In summary, we have accounted for all of the phosphate species and all of the hydrogen ion.

$$[\text{phosphate species}] = [H_3PO_4] + [H_2PO_4^-] + [HPO_4^{2-}] + [PO_4^{3-}]$$

$$4.0\,M = 3.83\,M + 0.17\,M + 6.2 \times 10^{-8}\,M + 1.8 \times 10^{-19}\,M$$

$$[H^+]_{total} = [H^+]_{H_3PO_4} + [H^+]_{H_2PO_4^-} + [H^+]_{HPO_4^{2-}} + [H^+]_{H_2O}$$

$$0.17\,M \approx 0.17\,M + 6.2 \times 10^{-8}\,M + 1.8 \times 10^{-19}\,M + 5.9 \times 10^{-14}\,M$$

We note that when the concentration of acid is much greater than the value of K_{a_1}, only the first dissociation of phosphoric acid itself contributes significantly to the hydrogen ion concentration. The concentration of each phosphate species changes in solution as the pH changes, with less of the acidic forms (H_3PO_4 and $H_2PO_4^-$) and more of the basic forms (HPO_4^{2-} and PO_4^{3-}) present at higher pH levels.

Example 15.14—Concentration of Species in a Polyprotic Acid Solution

Oxalic acid ($H_2C_2O_4$), which is found in beet leaves, rhubarb, and spinach, is used in the bookbinding industry, as well as in dye and ink manufacturing. In the analytical laboratory it can be used as a primary standard against which to determine the molarity of sodium hydroxide. Using the information in Table 15.9, determine the pH and $[H_2C_2O_4]$ of a 1.40 M solution of oxalic acid.

$$H_2C_2O_4(aq) \rightleftharpoons HC_2O_4^-(aq) + H^+(aq) \quad K_{a_1} = 6.5 \times 10^{-2}$$

$$HC_2O_4 - (aq) \rightleftharpoons H^+(aq) + C_2O_4^{2-}(aq) \quad K_{a_2} = 6.5 \times 10^{-5}$$

Asking the Right Questions

What are the major species in solution? What are the equilibria in which they are a part? Which of these equilibria are important? Which can we neglect?

Solution

The major species in solution are $H_2C_2O_4$ and H_2O. There are a number of equilibria that will occur in the solution. Judging on the basis of the values of K_{a_1}, K_{a_2}, and K_w, by far the most significant equilibrium will be the dissociation of $H_2C_2O_4$.

	$H_2C_2O_4 \rightleftharpoons$	$HC_2O_4^-(aq)+$	$H^+(aq)$
Initial	1.40 M	0 M	0 M
change	−x	+x	+x
equilibrium	1.40 − x	+x	+x
assumptions	1.40	+x	+x

As always, we may test our "negligible ionization" assumption ($1.40 - x \approx 1.40$). However, because K_a is *not* less than $10^{-2} \times [H_2C_2O_4]_0$, our assumption may well not be valid.

$$K_{a_1} = \frac{[HC_2O_4^-][H^+]}{[H_2C_2O_4]} = 6.5 \times 10^{-2} = \frac{x^2}{1.40}$$

$$x = [H^+] = [H_2C_2O_4] = 0.302\,M$$

Testing the 5% rule, we find that.

$$\frac{0.302}{1.40} \times 100\% = 21.5\%!$$

This confirms our prediction based on K_a and $[H_2C_2O_4]_0$. Therefore, $[H_2C_2O_4] \neq [H_2C_2O_4]_0$ but rather equals "$1.40 - x$." In other words, we can't use the assumption.

$$6.5 \times 10^{-2} = \frac{x^2}{1.40 - x}$$

We can solve using the quadratic formula. Clearing the fraction and setting equal to zero, we get

$$x^2 + 0.065(x) - 0.091 = 0$$

which gives $a = 1$, $b = 0.065$, and $c = -0.091$. Solving yields

$$x = \frac{-0.065 \pm \sqrt{(0.065)^2 - 4(1)(-0.091)}}{2(1)}$$

$$x = 0.271\,M, \text{ and } -0.336\,M$$

we eliminate the negative answer, because $[H^+]$ cannot be negative. So,

$$[H^+] = [HC_2O_4^-] = 0.271\,M$$

$$[H_2C_2O_4] = 1.40 - 0.271 = 1.13\,M$$

$$pH = -\log(0.271) = 0.57$$

Checking the math, we find that

$$K_{a_1} = \frac{(0.271)^2}{1.13} = 0.0650$$

Are our Answers Reasonable?

Although we may classify oxalic acid as "weak" because its K_{a_1} value is less than 1 (and its K_{a_2} value is even smaller), it is still stronger than many common weak acids such as acetic and citric acids. It is therefore reasonable that this relatively concentrated solution should have a low pH. The oxalic acid concentration at equilibrium is a little lower than its initial concentration, and that is reasonable, given the low K_{a_1} value for this weak acid.

What If...?

...we had an acid, "HA," with the same concentration as the oxalic acid (1.40 M), but with a K_{a1} that is larger than that of oxalic acid. Would [HA] be larger, smaller, or about the same as $[H_2C_2O_4]$?

(*Ans: [HA] would be smaller than* $[H_2C_2O_4]$*, because the larger* K_{a_1} *would mean that the acid dissociates to a greater extent.*)

Practice 15.14

Determine the pH of an aqueous 0.200 M H_3PO_4 solution.

We have seen that we can get a sense of what the pH of a solution will be by assessing the competing equilibria that occur in solution. Using this understanding, the pH of seemingly complex systems can be solved in a structured and meaningful fashion. We can extend this understanding to salts that contain an anion and cation that have acid–base behavior.

15.8 Assessing the Acid–Base Behavior of Salts in Aqueous Solution

Salts, such as NaCl, NH_4NO_3, and $NaNO_2$, are ionic compounds. When they dissociate in water they may exhibit acid–base behavior. The key questions we need to ask when assessing whether a salt will be acidic, basic, or neutral in aqueous solution are: What are the acid–base properties of the cation and anion parts of the salt? and Which is more influential, the acid strength of the cation or the base strength of the anion? The positive ion is always acidic, and the negative ion is always basic. It's just a question of relative strength. Whichever is stronger will determine whether the salt solution is acidic or basic.

Sodium nitrite ($NaNO_2$) is an important example because it is a food additive that helps retard spoilage in meat and also is used in many industrial applications, including the production of nitrogen-containing dyes as well as anticorrosion agents. Its use in the food industry has been restricted by the US Food and Drug Administration to 200 parts per million in meat, most fish and poultry that is ready for sale, because sodium nitrite was implicated in the 1970s as a possible precursor for some cancer-causing compounds. The salt essentially completely dissociates in aqueous solution to give Na^+ and NO_2^- ions. These are the main species in solution in addition to H_2O. What are the acid–base properties of each of these species? Remember the conjugate acid–base relationships:

- Strong acids and bases have extremely weak conjugates. Na^+ and other alkali and alkaline earth metal ions exhibit no important acidic properties.
- The nitrite ion (NO_2^-) is the conjugate base of the weak acid HNO_2 ($K_a = 4.6 \times 10^{-4}$). Remember that *weak* is a relative term. HNO_2 is weak compared to HNO_3, which has $K_a \gg 1$, but is far stronger than HCN ($K_a = 6.2 \times 10^{-10}$). The NO_2^- ion will act as a weak base.
- Water has relatively little acid–base effect.

Therefore, an aqueous solution of $NaNO_2$ should be slightly basic. We can now determine how basic by applying our understanding of equilibrium.

15.8.1 The Relationship of K_a to K_b

Consider an aqueous 0.500 M $NaNO_2$ solution. The process in which the nitrite ion (or any base) reacts with water to produce the conjugate acid and hydroxide ion is called **base hydrolysis**. The important equilibrium is

$$NO_2^-(aq) + H_2O(l) \rightleftharpoons HNO_2(aq) + OH^-(aq) \quad K_b = ?$$

Base hydrolysis – The process in which a base reacts with water to produce its conjugate acid and hydroxide ion.

Note that this equation describes the equilibration of a base with water. What is the value for K_b? When we look up an entry for NO_2^- in the tables, we do not find it. However, we do find a value for the K_a of nitrous acid:

$$HNO_2(ag) \rightleftharpoons H^+(ag) + NO_2^-(ag) \quad K_a = 7.0 \times 10^{-4}$$

Is it possible to relate the two equilibria to get a value for Kb of the nitrite ion hydrolysis? The short answer is yes. If we take the expression for K_b and multiply by $\frac{[H^+]}{[H^+]}$, we get

$$K_b = \frac{[HNO_2][OH^-]}{[NO_2^-]} \times \frac{[H^+]}{[H^+]} = \frac{[HNO_2][OH^-][H^+]}{[NO_2^-][H^+]}$$

Because $K_w = [H^+][OH^-]$, we can substitute K_w into the equation, which gives

$$K_a = \frac{[HNO_2]K_w}{[NO_2^-][H^+]}$$

If we write the mass-action expression for K_a for the ionization of nitrous acid, we note that $\frac{[HNO_2]}{[NO_2^-][H^+]}$ is equal to $1/K_a$. This means that the K_b expression may be rewritten as

$$K_b = \frac{K_w}{K_a}$$

or, as more often cited,

$$K_w = K_a \times K_b$$

This means that we can determine the K_a or K_b value for the conjugate of any weak acid or base, given its equilibrium constant. For the nitrite ion,

$$K_b = \frac{K_w}{K_a} = \frac{1.0 \times 10^{-14}}{7.0 \times 10^{-4}} = 1.4 \times 10^{-11}$$

We may now determine the pH of this weak base as we would any other weak base in solution. We will do this as Exercise 15.15.

Example 15.15—pH of a Salt containing a Cation with No Acidic Properties

Determine the pH of a 0.500 M $NaNO_2$ solution.

Asking the Right Questions

What determines whether a salt solution is acidic, basic, or neutral? What are the roles of the cation and anion in this determination?

Solution

We noted previously that the Na^+ cation has no acid—base properties, and the hydrolysis of the NO_2^- ion will produce OH^- ions, leading to a basic solution.

$$NO_2-(aq) + H_2O(l) \rightleftharpoons HNO_2(aq) + OH^-(aq)$$

$$K_b = \frac{[HNO_2][OH^-]}{[NO_2^-]} = 1.4 \times 10^{-11}$$

We may now proceed as with any other weak-base problem.

$$NO_2-(aq) + H_2O(l) \rightleftharpoons HNO_2(aq) + OH^-(aq)$$

	$NO_2-(aq) + H_2O(aq) \rightleftharpoons$	$HNO_2(aq)+$	$OH^-(aq)$
Initial	0.500 M	0 M	0 M
Change	−x	+x	−x
Equilibrium	0.500 − x	−x	−x
Assumptions	0.500	−x	−x

In this exercise, "x" $= [HNO_2] = [OH^-]$

$$1.4 \times 10^{-11} = \frac{x^2}{0.500}$$

$$x = [OH^-] = [HNO_2]$$

$$= 2.6 \times 10^{-6}\ M$$

This passes the 5% test (K_b is so small!).

$$pOH = -\log\left(2.6 \times 10^{-6}\right) = 5.59$$

$$pH = 14 - 5.59 = 8.41$$

Is our Answer Reasonable?

We have a weak base, and the pH is indicative of this. Therefore, our calculation is reasonable.

What If...?

...instead of $NaNO_2$, we have a 0.350 M NH_4F solution? Is this solution acidic or basic?

(Ans: It is acidic, because the NH_4^+ ion is a slightly stronger acid than the F^- ion is a base.)

Practice 15.15

Determine the pH of a 0.250 M sodium acetate (CH_3COONa) solution.

What happens when both the cation and the anion have acid–base properties? That is, what if the cation can react to supply significant amounts of hydrogen ion to the solution and the anion can supply significant amounts of hydroxide ion? In a broad sense, we can determine whether the solution will be acidic or basic from the relative strengths of the acidic and basic parts of the salt. For example, we can view ammonium cyanide (NH_4CN) as undergoing two important equilibrium reactions, because the ammonium ion is acidic and the cyanide ion is basic:

$$NH_4^+(aq) \rightleftharpoons NH_3(aq) + H^+(aq) \quad K_a = \frac{K_w}{K_{b(NH_3)}} = 5.6 \times 10^{-10}$$

$$CN^-(aq) + H_2O(l) \rightleftharpoons HCN(aq) + OH^- \quad K_b = \frac{K_w}{K_{a(HCN)}} = 1.6 \times 10^{-5}$$

Based on the equilibrium constants for the reactions, the CN^- ion is a much stronger base than the NH_4^+ ion is an acid, and we would therefore expect the solution to be basic. However, unlike the previous example using $NaNO_2$, in which only the nitrite ion had any acid–base behavior, here both the cation and the anion contribute to the pH of the solution, and we must take both equilibria into account. In the sense that we have one substance that acts as an acid and one that acts as a base, this is similar to an amphiprotic substance. The derivation for the pH of a substance that has both acid and base properties, such as this type of salt or an amphiprotic substance such as Na_2HPO_4, is fairly complex. The resulting formula, however, is simple and useful:

$$[H^+] = \left(K_{a(NH_4^+)} \times K_{a(HCN)}\right)^{1/2}$$

$$pH = 1/2\left(pK_{a(NH_4^+)} + pK_{a(HCN)}\right)$$

If the concentration of the substance is greater than 0.100 M, the pH of the aqueous salt solution is approximately concentration independent (true for amphiprotic substances as well). Let's use these equations to calculate the pH of a 0.800 M NH_4CN solution.

$$[H^+] = \left(K_{a(NH_4^+)} \times K_{a(HCN)}\right)^{1/2} = \left(5.6 \times 10^{-10} \times 6.2 \times 10^{-10}\right)^{1/2} = 5.9 \times 10^{-10}\ M \quad pH = 9.23$$

Amino acids are biologically vital compounds that have the ability to act as an acid and as a base within the same molecule. That is, one part of the molecule is acidic and a different part of the molecule is basic. Amino acids can polymerize into large units to form proteins such as hemoglobin (which transports oxygen), pepsinogen (which digests other proteins), and human growth hormone (which promotes normal growth). Amino acids are water-soluble because they carry both positive and negative charge in aqueous solution. An example of how this can happen, involving the amino acid alanine, is shown in Fig. 15.19. When dissolved in water, the carboxylic acid group loses a hydrogen ion, and the amine group gains a hydrogen ion. Why does this happen? The $-NH_2$ is a stronger base than the $-COO^-$, so the hydrogen ion will move from the COOH to the amine group. Ions that are doubly ionized in this way are called **zwitterions**. They can act in the same way as any other amphiprotic substance, donating a hydrogen ion to water or accepting one from water.

Neutral form ⇌ Zwitterion

Fig. 15.19 Amino acids are soluble in water because they often exist as zwitterions—doubly charged amino acids. As shown here with alanine, this can occur because the —NH_2 is a stronger base than the —COO^-, so the hydrogen ion will move from the COOH to the amine group

Zwitterion – A molecular ion carrying both an acid group, which generates a negative ion, and a basic group, which generates a positive ion.

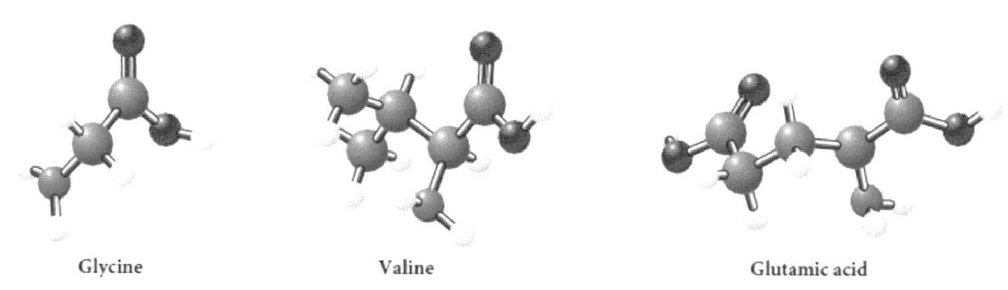

Example 15.16—pH of Zwitterions

Given the following reactions and equilibrium constants, calculate the pH of a solution of 0.200 M alanine.

$$CH_3(NH_3^+)COO^-(aq) \rightleftharpoons CH_3(NH_2)COO^-(aq) + H^+(aq) \quad K_{a_2} = 1.4 \times 10^{-10}$$

$$CH_3(NH_3^+)COO^-(aq) + H_2O(l) \rightleftharpoons CH_3(NH_3^+)COOH(aq) + OH^-(aq) \quad K_{b_2} = 2.2 \times 10^{-12}$$

Asking the Right Questions

How is this example similar to Example 15.15? Which of the equilibria is stronger? Based on the favored equilibrium, will the solution be acidic or basic? How do we calculate the pH for this system?

Solution

As in Example 15.15, we are dealing with competing equilibria. The donation of hydrogen ion by alanine is favored over its acceptance of a hydrogen ion. We therefore would expect the solution to be somewhat, though not strongly, acidic.

We must find K_{a_1} for the second equation, just as in the ammonium cyanide case described previously.

$$K_{a_1} = \frac{K_w}{K_{b_2}} = \frac{1.0 \times 10^{-14}}{2.2 \times 10^{-12}}$$

$$= 4.5 \times 10^{-3}$$

$$[H^+] = \sqrt{K_{a_1} \times K_{a_2}}$$

$$= \sqrt{1.4 \times 10^{-10} \times 4.5 \times 10^{-3}}$$

$$= 7.9 \times 10^{-7}\ M$$

$$pH = 6.10$$

Is our Answer Reasonable?

We asserted that the solution should be somewhat acidic because the acid-producing equilibrium is favored over the base-producing equilibrium. Both K values are relatively small, so we would expect the pH to be fairly close to 7, and the answer is reasonable.

What If…?

…instead of $NaNO_2$, we have a 0.350_M NH_4F solution? What is the pH of this solution?

(*Ans: 6.20*)

Practice 15.16

What is the pH of a 0.150 M alanine solution?

Table 15.10 Effect of cation and anion on the acidity of a salt

Cation	Anion	Aqueous Solution	Example
Acidic	Neutral	Acidic	NH_4NO_3
Neutral	Basic	Basic	Na_2CO_3
Neutral	Neutral	Neutral	NaCl
Acidic	Basic	Depends on the relative strength of each	NH_4CO_3

The principles that we have introduced for the evaluation of the pH of a salt can be extended from the two cases we have dealt with to a third case, in which the anion has no basic properties but the cation has acidic properties. The thinking—that is, the questions that we raise—is the same. These questions about the nature of equilibria in solution represent the unifying problem-solving theme in this chapter as well as in the previous one dealing with chemical equilibrium.

Table 15.10 qualitatively summarizes the effect of the cation and of the anion on the pH of a salt.

15.9 Anhydrides in Aqueous Solution

We started this chapter saying, "Home is where the heart is." The "home" that we discussed was not only our individual residence, but ourselves and our world. In a sense, our final application, the set of reactions that cause acid deposition from the atmosphere, is a proper place to conclude the discussion, because the Earth is our communal home.

In order to understand acid deposition, of which acid rain is one form, we need to understand the reactions of acidic and basic anhydrides (also known as acid and basic oxides) with water.

Basic anhydrides are binary compounds formed between metals with very low electronegativity and oxygen (see ▶ Sect. 15.2 for review of why these reactions would form bases). Strong bases are formed when Group IA and Group IIA anhydrides (*an* = "without" + *hydro* = "water") react with water. Metal hydroxides are often prepared this way instead of by reaction of the metal with water, which can sometimes be violent, as with the reaction of cesium with water to produce cesium hydroxide and hydrogen gas. Examples of anhydride and water reactions are

$$Li_2O(s) + H_2O(l) \rightleftharpoons 2LiOH(aq)$$

$$CaO(s) + H_2O(l) \rightleftharpoons Ca(OH)_2(aq)$$

Basic anhydrides – Binary compounds that are formed between metals with very low electronegativity and oxygen and that react vigorously with water.

Acid anhydrides are binary compounds formed between nonmetals and oxygen. Examples are SO_2, SO_3, NO_2, P_4O_{10}, and CO_2. These compounds react with water to form acids, their acid strength being related to the electronegativity of the nonmetal combined with oxygen. One example is the reaction of sulfur dioxide generated in industrial smokestacks with water:

$$SO_2(g) + H_2O(l) \rightleftharpoons H_2SO_3(aq)$$

Acid anhydrides – Binary compounds formed between nonmetals and oxygen that react with water to form acids.

The sulfurous acid generated in this reaction is not strong. However, dust in the air can catalyze the reaction between SO_2 and oxygen:

$$2SO_2(g) + O_2(g) \rightleftharpoons 2SO_3(g)$$

In a reaction analogous to the Contact process, the sulfur trioxide reacts with water vapor in the air to form sulfuric acid:

$$SO_3(g) + H_2O(l) \rightleftharpoons H_2SO_4(aq)$$

The acid that is formed can fall to Earth on a variety of surfaces, including snow, rain, and fog, and deposit on trees, lakes, and the like, and that is why we call the process **acid deposition**.

Acid deposition – The precipitation of acidic compounds from the atmosphere. This includes wet deposition as rain and dry deposition of particles. ▶ https://upload.wikimedia.org/wikipedia/commons/0/0c/Acid_rain_woods1.JPG

Nitrogen and oxygen released from the tailpipes of motorized vehicles during operation can react to form nitric oxide, which then slowly reacts with atmospheric oxygen to form nitrogen dioxide:

$$N_2(g) + O_2(g) \rightleftharpoons 2NO(g)$$

$$2NO(g) + O_2(g) \rightleftharpoons 2NO_2(g)$$

The nitrogen dioxide that is produced then reacts with water vapor to produce nitrous and nitric acids:

$$2NO_2(g) + H_2O(l) \rightleftharpoons HNO_2(aq) + HNO_3(aq)$$

The nitric acid adds to the problem of acid deposition. As can be seen in ◘ Fig. 15.20, a lot of deposition that occurs throughout the United States has an unusually low pH. However, some places (for example, in the western United States) seem to be less affected by acid deposition than other places. Why?

The atmosphere acts as a large mixing chamber, and we've considered only the acidic inputs so far. Ammonia gas from agriculture and animal-feeding operations can react with water vapor to form aqueous ammonia, increasing the pH of the precipitation:

$$NH_3(g) + H_2O(l) \rightleftharpoons NH_4^+(aq) + OH^-(aq)$$

There are more natural sources of ammonia gas in the western United States than sources of sulfur dioxide and nitrogen dioxide, so the pH of precipitation in this region is higher. The ability of some lakes to mitigate the effects of acid deposition has to do with **acid-neutralizing capacity**. This represents the theme of our next chapter: how and why acid–base and other types of reactions in aqueous solution are used in chemical analysis. This discussion will include acid–base neutralization, buffers, and titrations. Home is where the heart is. But we have shown that the home, and the people within it, are, chemically speaking, where the acids and bases also reside. Our global home, Spaceship Earth, is ever changing. The Earth, and we who occupy it, owe much of this change to the acids and bases we have discussed here. In the next chapter, we will learn how.

Acid-neutralizing capacity – The capacity of a solution, such as lake water, to neutralize acidity.

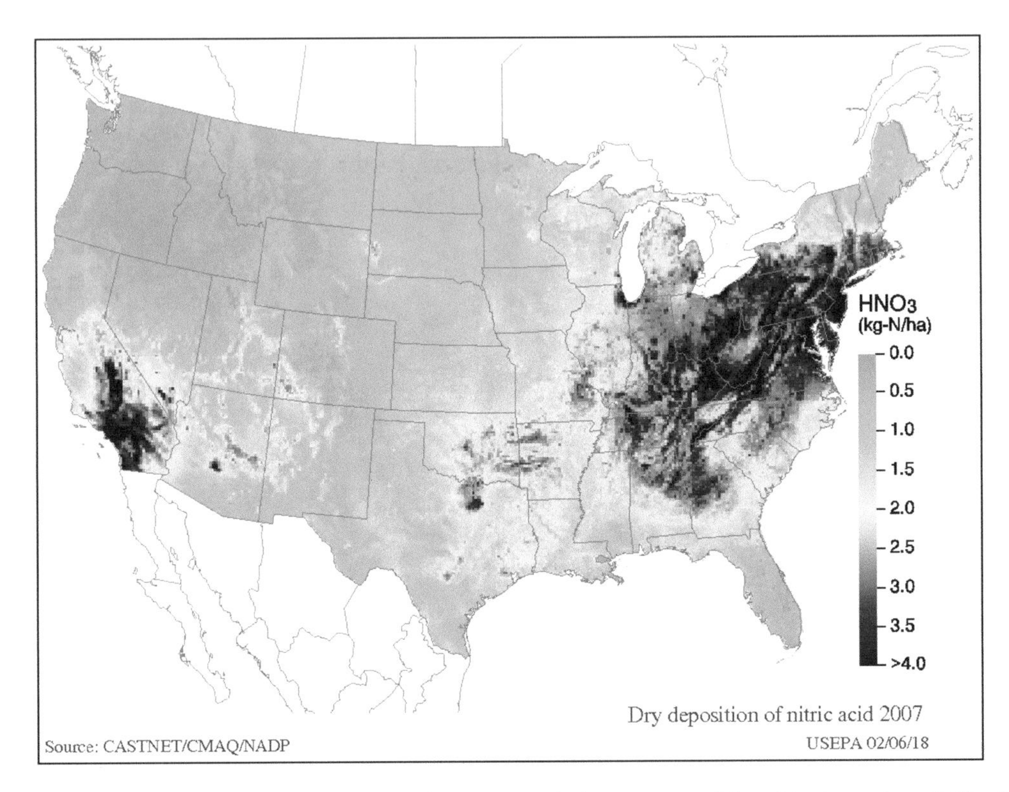

Fig. 15.20 The deposition of nitric acid (HNO_3) in the US in 2007. Note the large amount of deposition in southern California and from the Ohio River to New England. Recent analyses have shown that most of the dry deposition has been declining for the last 10 years. Image from the U.S. EPA. *Source* https://gaftp.epa.gov/castnet/tdep/images/hno3_dw/hno3_dw-2007.png. Access from the National Atmospheric Deposition Program. Since this is a US government publication, it is in the public domain. *Source* ▶ http://nadp.slh.wisc.edu/committees/tdep/tdepmaps/

The Bottom Line

- Acids and bases can be defined using three different models: Arrhenius, Bronsted-Lowry, and Lewis.
- Acids and bases have conjugates pairs whose behavior is related to that of the acid or base from which they are derived.
- Acids and bases come in different strengths.
- Acid strength and acid concentration are separate concepts, which must both be considered when determining H^+ concentration.
- The stronger the acid, the weaker its conjugate base. The stronger the base, the weaker its conjugate acid.
- We use pH as our common measure of acidity. $\mathrm{pH} = -\log\left[\mathrm{H}^+\right]$
- Numbers to the left of the decimal in a pH value are NOT significant figures.
- There is no lower or upper limit on the pH scale.
- We can interconvert among H^+, OH^-, pH, and pOH for a given acidic or basic solution.
- We can calculate the pH of strong and weak acids and bases in aqueous solutions:
  Step 1: Determine the equilibria, and the resulting species, that are in the solution.
  Step 2: Determine the equilibria that are the most important contributors to $[H^+]$ in the solution.
  Step 3: Write the equilibrium expression for the important contributors to $[H^+]$
  Step 4: Set up a table of the initial and equilibrium concentrations of each pertinent species.
  Step 5: Solve for the estimated concentration of each species.
  Step 6: Check our assumptions.
- We can solve for the pH of a polyprotic acid or base, including salts.
- Salts are just ionic compounds.
- When the concentration of a polyprotic acid is much greater than the value of K_{a_1}, only the first dissociation of the acid contributes significantly to the hydrogen ion concentration.
- K_a and K_b are related via K_w. For an acid/base conjugate pair: $K_\mathrm{a} \cdot K_\mathrm{b} = K_\mathrm{w}$
- For the solution of a salt with both acidic and basic properties: $\mathrm{pH} = 1/2\left(\mathrm{p}K_{\mathrm{a(NH_4^+)}} + \mathrm{p}K_{\mathrm{a(HCN)}}\right)$
- The reaction of an acid anhydride with water results in an acid, and the reaction of a basic anhydride with water results in a base.

Section 15.2 What Are Acids and Bases?

Skill Review

1. Explain how ammonia (NH_3) qualifies as a base in both the Brønsted–Lowry and Arrhenius acid–base models.
2. Explain how HCl and NaOH are classified in both the Brønsted–Lowry and Arrhenius acid–base models.
3. Pick a Brønsted-Lowry acid and discuss, via reaction with water, how it illustrates the differences and similarities among Brønsted-Lowry, Arrhenius and Lewis acids.
4. Pick an Arrhenius base and discuss, via reaction with water, how it illustrates the differences and similarities among Brønsted-Lowry, Arrhenius and Lewis acids.
5. Complete the equation for the reaction of these acids or bases with water.
   a. (acid) $C_6H_5COOH(aq) + H_2O(l) \rightleftharpoons$
   b. (base) $C_2H_5NH_2(aq) + H_2O(l) \rightleftharpoons$
   c. (acid) $ClCH_2COOH(aq) + H_2O(l) \rightleftharpoons$
   d. (base) $N_2H_4(aq) + H_2O(l) \rightleftharpoons$
6. Complete the equation for the reaction of these acids or bases with water.
   a. (acid) $H_3PO_3(aq) + H_2O(l) \rightleftharpoons$
   b. (base) $NH_2SO_3^-(aq) + H_2O(l) \rightleftharpoons$
   c. (base) $(CH_3CH_2)_3N(aq) + H_2O(l) \rightleftharpoons$
   d. (acid) $HIO_3(aq) + H_2O(l) \rightleftharpoons$
7. Indicate the conjugate base of each of these acids.
   a. HNO_3 b. HBr c. H_2O d. $HClO_4$
8. Indicate the conjugate base of each of these acids.
   a. HCl b. NH_4^+ c. CH_3OH d. H_2SO_4
9. Indicate the conjugate acid of each of these bases.
   a. NaOH b. NH_3 c. H_2O d. NaF
10. Indicate the conjugate acid of each of these bases.
    a. KBr b. CH_3OH c. KNO_2 d. KSH
11. Acids react with active metals to produce hydrogen gas. Complete, balance, and name each of the products for these reactions.
    a. $Al(s) + HCl(aq) \rightarrow$ ______ + ______
    b. $Ca(s) + HNO_3(aq) \rightarrow$ ______ + ______
    c. $Na(s) + HCl(aq) \rightarrow$ ______ + ______
    d. $K(s) + HNO_3(aq) \rightarrow$ ______ + ______
12. Many bases react with metals to form insoluble hydroxides. Write the formula for each of these hydroxides.
    a. aluminum hydroxide
    b. copper(II) hydroxide
    c. barium hydroxide
    d. strontium hydroxide
    e. iron(III) hydroxide
    f. calcium hydroxide
13. Being strong usually refers to the ability to carry out a task or produce an effect. (For example, a spice may have a strong flavor, and a weightlifter may display a lot of strength.) How is this term used when applied to acids? When an acid is said to be strong, what effect is being described?
14. What is the definition of a Lewis acid? Why are many metal ions capable of behaving as Lewis acids? Write out the reaction that depicts an aluminum cation behaving as a Lewis acid in an aqueous solution.
15. Balance and identify each of the species in the following re- actions as either acid, base, conjugate acid, or conjugate base.
    a. $HCl + H_2O \rightarrow Cl^- + H_3O^+$
    b. $NaOH + CH_3COOH \rightarrow CH_3COONa + H_2O$
    c. $H_2SO_4 + Mg(OH)_2 \rightarrow MgSO_4 + H_2O$
16. Balance and identify each of the species in the following reactions as either acid, base, conjugate acid, or conjugate base.
    a. $HCOOH + NH_3 \rightarrow NH_4^+ + HCOO^-$
    b. $KOH + CH_3OH \rightarrow CH_3OK + H_2O$
    c. $H_3PO_4 + Ca(OH)_2 \rightarrow Ca_3(PO_4)_2 + H_2O$

Chemical Applications and Practices

17. One common antacid used in relief of upset stomachs is $Mg(OH)_2$. It helps to neutralize excess hydrochloric acid. Balance the following representative equation, and identify the conjugate acid–base pairs.

$$Mg(OH)_2 + HCl \rightarrow MgCl_2 + H_2O$$

18. Prizes in cereal boxes used to include little submarines that, when filled with baking soda ($NaHCO_3$) and placed in a cup of water containing a little vinegar (CH_3COOH), would rise and fall as though by magic. Balance the reaction of baking soda and vinegar, and identify the conjugate base and acid pairs.

$$CH_3COOH + NaHCO_3 \rightarrow CH_3COONa + H_2CO_3$$

Section 15.3. Acid Strength

Skill Review

19. Describe the characteristics of a strong acid with regard to each of the following:
    a. The numerical value of its K_a.
    b. The ability of its conjugate base to regain a H^+.
    c. The approximate percent dissociation of a 0.1 M solution.
20. Describe the characteristics of a weak acid with regard to each of the following:
    a. The numerical value of its K_a.
    b. The ability of its conjugate base to regain a H^+.
    c. The approximate percent dissociation of a 0.1 M solution.
21. Judging on the basis of electron density and electronegativity, which acid, HBr or HI, would we expect to be stronger? Explain the choice.
22. Which conjugate base, $Br^-(aq)$ or $I^-(aq)$, would we expect to be stronger? Explain the choice.
23. Using molecular structure and electronegativity, explain which acid, $HClO_4$ or $HBrO_4$, would be the weaker.
24. Which acid, phosphoric (H_3PO_4) or phosphorous (H_3PO_3), would we expect to be stronger? Explain.
25. ■ Table 15.4 gives K_a values for several weak acids. Comparing hydrofluoric acid and hydrocyanic acid,
    a. Write the mass-action expression for the dissociation of each acid;
    b. Which has the stronger conjugate base?
    c. Which conjugate is stronger base than water?
26. Use the Appendix to find K_b values for pyridine and diethylamine. Then,
    a. Write the mass-action expression for the reaction of each with water;
    b. Which has the stronger conjugate acid?
    c. Which conjugate is stronger acid than water?
27. In separate containers we have 500. mL each of 3.50 M acetic acid, CH_3COOH, $K_a = 1.8 \times 10^{-5}$, and 0.001 M HCl. Which solution has the higher hydrogen ion concentration? Explain.
28. In separate containers we have 250. mL each of 0.002 M sodium hydroxide, NaOH, and 5.00 M dimethylamine, $(CH_3)_2NH$, $K_b = 5.9 \times 10^{-4}$. Which solution has the higher hydroxide ion concentration? Explain.

Chemical Applications and Practices

15

29. Formic acid, found in ants, and acetic acid, found in vinegar, have the K_a values 1.8×10^{-4} and 1.8×10^{-5}, respectively.
    a. Which acid has the stronger conjugate base?
    b. Which acid, if both were in 0.10 M solutions, would have the higher percent dissociation?
30. The K_a value for an acid provides information about one type of reaction of the acid, its ability to provide H^+. However, a weak acid may have other very important reactions. For example, hydrofluoric acid, $K_a = 7.2 \times 10^{-4}$, has the ability to etch glass. Phenol, $K_a = 1.6 \times 10^{-10}$, can be used as a disinfectant.
    a. Which of these two is the weaker acid?
    b. Which has the stronger conjugate base?
    c. If both were 0.10 M, which would produce the greater concentration of H^+?
31. Propanoic acid is a weak acid that can be used to prepare a type of mold retardant. What is the value for K_a of the acid if, in a solution, the following equilibrium concentrations were found: [acid] = 0.10 M, [conjugate base] and $[H^+] = 0.0011$ M?
32. At 25 °C the K_b for ammonia (NH_3) is 1.8×10^{-5}. What is the hydroxide concentration when $[NH_3] = 0.103$ M and $[NH_4^+] = 0.00205$ M?

Section 15.4 The pH Scale

Skill Review

33. Calculate the pH of each of these solutions.
    a. $[H^+] = 4.55 \times 10^{-3}$ M
    b. $[H^+] = 3.27 \times 10^{-6}$ M

c. $[H^+] = 8.11 \times 10^{-9}$ M

34. Calculate the $[H^+]$ for each of these solutions.
    a. pH = 1.50
    b. pH = 10.25
    c. pH = 5.38
    d. pH = 7.00
35. Calculate the $[OH^-]$ for each of these solutions.
    a. pOH = 11.65
    b. pH = 3.35
    c. $[H^+] = 4.47 \times 10^{-3}$ M
36. Calculate the pH for each of these solutions.
    a. $[H^+] = 8.04 \times 10^{-13}$ M
    b. pOH = 9.72
    c. $[OH^-] = 0.025$ M
37. Calculate the values missing from the following table

	$[H^+]$ (M)	pH	$[OH^-]$ (M)	pOH
a		4.42		
b	0.0056			
c			0.000078	
d				10.10

38. Calculate the values missing from the following table.

	$[H^+]$ (M)	pH	$[OH^-]$ (M)	pOH
a		12.50		
b	0.000035			
c			0.00383	
d				3.75

39. If the pH value in an aqueous sample were doubled, what effect would be detected in the hydronium ion concentration?
40. What would be the effect on the hydroxide ion concentration if the pH were doubled?
41. Two acid solutions are at pH 4.50 and 8.95, respectively. Which is more strongly acidic? What is the ratio of acid strength, stronger to weaker?
42. Two acid solutions are at pH 2.06 and 5.11, respectively. Which is more strongly acidic? What is the ratio of acid strength, stronger to weaker?
43. How many grams of the strong base NaOH are necessary to prepare 500.0 mL of a solution at pH = 13.00?
44. How many grams of the strong acid HNO_3 are necessary to prepare 750.0 mL of a solution at pH = 1.60?

Chemical Applications and Practices

45. Paper is produced from the processed fibers of trees. Part of the process of sulfate pulping uses sodium hydroxide. What would be the pH of a solution in a wood-pulping mill if it contained 10.0 g of OH^- ions for every 10.0 L of solution?
46. One method to increase oil production in areas drilled through limestone deposits is to use hydrochloric acid to increase drainage channels through the stone. If such a solution had a pH of 2.59, what would we calculate as the grams of HCl dissolved per liter of solution?
47. Two samples of rainwater are being analyzed for an environmental impact study. What is the hydronium concentration in each sample? What is the pH of each sample?
    a. 500.0 mL containing 1.55×10^{-5} mol of H^+
    b. 250.0 mL containing 7.25×10^{-6} mol of H^+

48. The pH of human blood must be maintained within a very narrow range to ensure proper health. The following blood samples were analyzed to determine their pH values. What would we calculate as the hydronium concentration in each?
   a. pH = 7.42
   b. pH = 7.38
   c. pH = 7.51
49. Assuming a negligible change in volume, how many moles of either OH^- or H^+ would have to be added to change the pH of 1.00 L of a solution from 4.35 to 5.85?
50. Formic acid has a pK_a value of 3.74. Benzoic acid has a pK_a value of 4.20. Compare the electrostatic potential maps of formic acid and benzoic acid. Which is the stronger acid? Which of the two acids would have the weaker conjugate base?

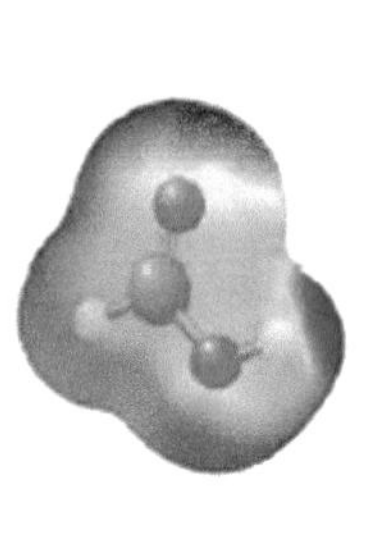

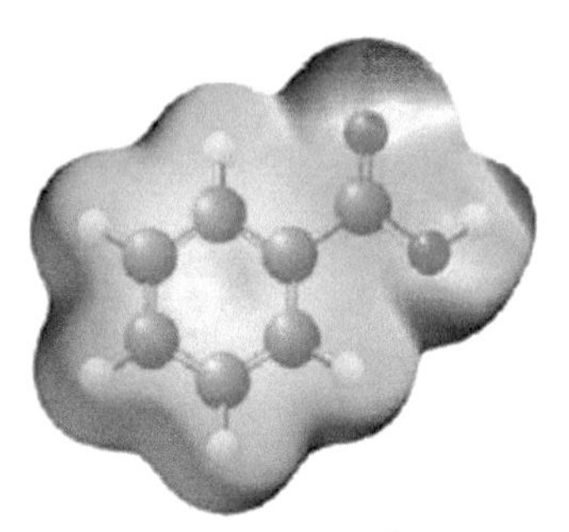

Formic acid Benzoic acid

Section 15.5 Determining the pH of Acidic Solutions

Skill Review

51. Determine the pH of each of these solutions of strong acids.
   a. 0.45 M HCl
   b. 0.045 M HCl
   c. 0.000487 M HNO_3
   d. 0.00026 M HBr
52. Determine the pH of each of these solutions of strong bases.
   a. 0.550 M NaOH
   b. 0.00089 M KOH
   c. 0.00388 M KOH
   d. 0.015 M KOH
53. Determine the pH of each of these solutions. (Use the table in the text to find the values for the appropriate K_a.)
   a. 0.45 M HOCl
   b. 0.0250 M CH_3COOH
   c. 0.18 M HF
   d. 0.0010 M HCOOH

15

54. Determine the pH of each of these solutions. (Use the table in the text to find the values for the appropriate K_a.)
   a. 0.299 M HOCl
   b. 0.18 M CH_3COOH
   c. 0.45 M lactic acid ($HC_3H_5O_3$)
   d. 0.050 M HCN
55. Determine the value of pK_a for:
   a. $K_a = 3.75 \times 10^{-5}$
   b. $K_a = 1.84 \times 10^{-2}$
   c. $K_a = 4.59 \times 10^{-8}$
56. Determine the value of K_a for:
   a. pK_a = 3.50
   b. pK_a = 4.74
   c. pK_a = 6.17
57. If the pH of a 0.015 M solution of codeine, a drug used in some pain relievers, is 10.19, what is the value of K_b for codeine ($C_{18}H_{21}NO_3$)?
58. What would be the resulting pH when a solution was made that was 0.0100 M in HCl and 0.100 M in HCN (hydrocyanic acid, a deadly poison, which has a K_a value of 6.2×10^{-10})?
59. How many mL of glacial (water-free) acetic acid, CH_3COOH, (density = 1.049 g/mL, $K_a = 1.8 \times 10^{-5}$) are needed to prepare 10.0 L of an acetic acid solution of pH = 2.04?
60. What would be the pH of a solution prepared by combining 13.6 g of crystalline benzoic acid, C_6H_5COOH, with water to a final volume of 2.00 L? K_a for benzoic acid = 6.5×10^{-5}

Chemical Applications and Practices

61. Among the growth requirements for bacteria is the proper range of aqueous hydrogen ion concentration. Suppose a microbiologist was preparing growth media to study a specific bacterium. Examine both of the following situations and determine in which case the $H^+(aq)$ would be greater than 1.0×10^{-6} M?

a. A 0.500-L solution containing 1.00 g of benzoic acid ($K_a = 6.5 \times 10^{-5}$; molar mass = 122 g)
b. 100.0 mL of 0.0001 M sulfuric acid

62. Benzoic acid is often used to prepare a preservative known as sodium benzoate. If the K_a value of benzoic acid is 6.5×10^{-5}, what is the hydrogen ion concentration when the acid concentration is 0.0040 M and the conjugate base concentration is 0.0024 M?
63. The following reaction depicts an industrial process to manufacture gaseous hydrogen fluoride.
    a. How many grams of HF can be made from 1.00 kg of fluorospar (CaF_2)?
    b. HF can then be used to prepare fluorocarbon compounds. What would be the H^+, F^-, and OH^- concentrations in a solution that was 0.25 M in HF? (K_a of HF $= 7.2 \times 10^{-4}$)
64. An aspirin tablet may contain 250 mg of acetylsalicylic acid ($pK_a = 3.522$, 180.16 g/mol). What would be the approximate pH when two tablets were dissolved in 275 mL of water?
65. A vitamin C tablet may contain 500.0 mg of ascorbic acid ($C_6H_8O_6$). What would be the pH of a solution made from dissolving one such tablet in 355 mL of solution? (K_{a_1} of ascorbic acid $= 8.0 \times 10^{-5}$) Does the 5% rule assumption apply in this example? Show proof of the answer.
66. Benzoic acid ($K_a = 6.5 \times 10^{-5}$) and propionic acid ($K_a = 1.3 \times 10^{-5}$) can both be used to produce food preservatives. A 0.10 M solution of one of the acids has $[H^+] = 0.00255$ M. Which acid was used?

Section 15.6 Determining the pH of Basic Solutions

Skill Review

67. Determine the pH of each of these solutions.
    a. 0.100 M aniline ($K_b = 3.8 \times 10^{-8}$)
    b. 0.0100 M NaOH
    c. 0.250 M ammonia
68. Determine the pH of each of these solutions.
    a. 0.0333 M methylamine ($K_b = 5.9 \times 10^{-4}$)
    b. 0.0150 M $Ca(OH)_2$
    c. 0.016 M ammonia
69. Compare the pH of a 0.75 M sodium hydroxide solution with a 0.75 M diethylamine solution, $(CH_3CH_2)_2NH$, $K_b = 7.1 \times 10^{-4}$. How much stronger (by ratio of the concentrations of hydroxide ion) is the strong base than the weak base?
70. Compare the pH of a 0.75 M acetic acid solution, CH_3COOH, $K_a = 1.8 \times 10^{-5}$, with a 0.75 M diethylamine solution, $(CH_3CH_2)_2NH$, $K_b = 7.1 \times 10^{-4}$. How much stronger an acid is the acetic acid than is the diethylamine? Only compare acid strengths, i.e., the respective $[H^+]$ ratios.

Chemical Applications and Practices

71. Pyridine is a weak base that can be used to make a product used in some mouthwash preparations. The K_b value of pyridine is approximately 1.4×10^{-9}. What would be the hydroxide ion concentration, the pOH, and the pH of a solution that was 0.0010 M pyridine?
72. The base $Ca(OH)_2$ is known as slaked lime. It is widely used in the paper industry and in steel making. When properly heated, it gives off a bright light. In the 1800s, this light was used to illuminate some theaters. Actors began appearing in the "limelight."
    a. Write out the equation representing the dissociation of lime in water.
    b. $Ca(OH)_2$ is not very soluble in water. If 0.025 mol could dissolve per liter, what would we calculate as the pH of the solution?
73. The typical "fish aroma" is due to the production of amine compounds. What would be the K_b value of ethylamine ($CH_3CH_2NH_2$) if a 500.0 mL solution that contained 1.90 g of ethylamine had a pH of 11.87?
74. Metacaine is used to anesthetize groups of fish when scientific studies on them are conducted. The active ingredient of metacaine is a base known as ethyl 3-aminobenzoate. What would be the K_b value of ethyl 3-aminobenzoate ($C_9H_{11}NO_2$) if a 750.0 mL solution containing 1.00 g had a pH of 9.30?

Section 15.7 Polyprotic Acids

Skill Review

75. Write out the three hydrogen dissociation steps for phosphorous acid (H_3PO_3). What is the oxidation number of phosphorus in H_3PO_3?
76. Write out the two hydrogen dissociation steps for carbonic acid (H_2CO_3). What is the oxidation number of carbon in H_2CO_3?
77. The respective K_a values for the three dissociation steps of phosphoric acid are 7.4×10^{-3}, 6.2×10^{-8}, and 4.8×10^{-13}. What would be the pH and HPO_4 concentration of a 1.0 $_M$ solution of H_3PO_4?
78. What are the pH and the HSO_4^- concentration of a 0.750 M solution of H_2SO_4?
79. Calculate $[H_2C_2O_4]$, $[HC_2O_4^-]$, $[C_2O_4^{-2}]$, and $[H^+]$ in a 0.20 M oxalic acid solution. Oxalic acid, $H_2C_2O_4$ has $K_{a_1} = 6.5 \times 10^{-2}$ and $K_{a_2} = 6.1 \times 10^{-5}$.
80. Calculate $[H_2C_2O_4]$, $[HC_2O_4^-]$, $[C_2O_4^{-2}]$, and $[H^+]$ in a 0.40 M oxalic acid solution. Oxalic acid, $H_2C_2O_4$ has $K_{a_1} = 6.5 \times 10^{-2}$ and $K_{a_2} = 6.1 \times 10^{-5}$.

Chemical Applications and Practices

81. Sulfurous acid (H_2SO_3) is a by-product of burning sulfur-containing coal. SO_2 is produced during the process and, when combined with water in the air, can produce H_2SO_3. Write the balanced equations that show the two-stage ionization of sulfurous acid. Identify the conjugate base produced in each stage.
82. Carbonic acid can be found in carbonated drinks. It forms when CO_2 reacts with water.
    a. Write the balanced equation that shows the formation of carbonic acid from dissolved carbon dioxide.
    b. Write out the equilibrium expressions for both ionization steps for this diprotic acid.
    c. Use Le Châtelier's principle to explain why increasing the pressure of CO_2 produces a lower pH in the solution.
83. Nicotine is dibasic because of the presence of two nitrogen atoms, each of which may accept hydrogen ions. The respective K_b values, at 25 °C, are approximately 7.0×10^{-7} and 1.1×10^{-10}. What would be the pH of a solution that was 0.045 M in nicotine.
84. Tartaric acid ($H_2C_4H_4O_6$) is a diprotic acid used in some baking preparations. For the successive hydrogen ionizations, $K_{a_1} = 9.2 \times 10^{-4}$ and $K_{a_2} = 4.3 \times 10^{-5}$.What would be the pH of a solution that was 1.0 M in tartaric acid?

Section 15.8 Assessing the Acid–Base Behavior of Salts in Aqueous Solution

15

Skill Review

85. Arrange the following 0.10 M solutions in order of decreasing pH: NH_4Cl, NaCl, $NaC_2H_3O_2$.
86. Which of these ions could produce a basic aqueous solution?

$$S^{2-}, Cl^-, NO_3^-, NO_2^-, CO_3^{2-}, OCl^-$$

87. Two acids, HX and HY, have pK_a values of 4.55 and 5.44, respectively. Which salt, NaX or NaY, will produce the more basic aqueous solution when prepared as 0.10 M?
88. Two sodium salts, symbolized as NaW and NaY, are completely dissolved to produce 0.20 M solutions. The respective pH values of the two solutions are 8.55 and 9.55. Which acid, HW or HY, is stronger? Explain.
89. What is isoelectric pH? Why is it useful information to know about amino acids?
90. If the isoelectric pH for an amino acid were above 7, what would that indicate about the relative values of its K_a and K_b?
91. The K_a value for the dissolved cation $Zn(H_2O)_6^{2+}$ is approximately 2.4×10^{-10}. What is the pH of a solution that is 0.10 M $ZnCl_2$? (Hint: What happens when $ZnCl_2$ dissolves?)
92. Will each of these salts be acidic, basic, or neutral?
    a. KI
    b. NH_4F
    c. $(NH_4)_3PO_4$
93. Determine the value of K_a if:
    a. $K_b = 4.26 \times 10^{-5}$
    b. $K_b = 8.36 \times 10^{-9}$
    c. $pK_a = 2.85$

94. Determine the value of K_b if:
   a. $K_a = 6.90 \times 10^{-3}$
   b. $K_a = 1.77 \times 10^{-12}$
   c. $pK_a = 4.74$
95. Determine the pH of a solution that is 0.050 M in HCOONa.
96. Determine the pH of a solution that is 0.136 M in KNO_2.

Chemical Applications and Practices

97. The salt ammonium chloride is used in some chemical "cold packs" to absorb heat and cool muscle wounds. Ammonium chloride can be produced from an acid–base reaction.
   a. Write out the reaction using an acid and a base that would produce ammonium chloride.
   b. Would the resulting solution of ammonium chloride be acidic, basic, or neutral? Explain.
   c. Determine the pH of a solution of ammonium chloride that is 0.136 M.
98. The salt sodium hypochlorite (NaOCl) can be used in some bleaching actions needed for disinfecting aqueous systems.
   a. Write out the balanced equation that represents the complete dissociation of the salt in water.
   b. Would this be likely to produce an acidic, basic, or neutral solution?
   c. Determine the pH of a solution that is 0.250 M NaOCl.
99. Sodium carbonate, sometimes called soda ash, is used industrially in the manufacture of glass, paper, and soaps. What would be the pH of a 0.15 M solution of sodium carbonate?
100. Sodium bicarbonate, also known as baking soda or sodium hydrogen carbonate, is produced industrially by the addition of carbon dioxide to soda ash. Sodium bicarbonate has widespread uses in antacids, paper manufacturing, and some fire extinguishers, as well as to remove some harmful gases during coal combustion. What would be the pH of a 0.15 M solution of sodium bicarbonate?
101. Sodium benzoate, a salt of benzoic acid sometimes used as a food preservative, dissolves in water to produce the benzoate ion, $C_6H_5CO_2^-$. The K_a value for benzoic acid is approximately 6.5×10^{-5}.
   a. .Write out the reaction of the benzoate ion in water and calculate the K_b value for the reaction.
   b. .What would be the pH of a 0.010 M solution of sodium benzoate?
102. Novocain is often used as a local anesthetic. The compound is actually a salt of the base procaine. Procaine has a K_b value of 7.13×10^{-6}.
   a. What would be the K_a of Novocain?
   b. What would be the pH of a 0.010 M solution of Novocain?
103. Lysine is considered an essential amino acid. Essential amino acids are those that are not synthesized by humans and must, therefore, be part of a healthful diet. Lysine can be found in beans. The formula of lysine is given below. Rewrite the formula showing lysine as a zwitterion.

$$H_2N-CH_2-CH_2-CH_2-CH_2-\underset{\displaystyle COOH}{\overset{\displaystyle H}{\overset{|}{\underset{|}{C}}}}-NH_2$$

104. Glycine has the simplest structure of the amino acids.
   a. Write out the reactions that show glycine acting as an acid and acting as a base.
   b. What would be the pH of a 0.10 M solution of glycine? (The approximate K_{a_2} and K_{b_2} values, at 25 °C, needed are 2.0×10^{-10} and 2.2×10^{-12}.)

$$H-\underset{\displaystyle H}{\overset{\displaystyle NH_2}{\overset{|}{\underset{|}{C}}}}-COOH$$

Section 15.9 Anhydrides in Aqueous Solution

Skill Review

105. Give the structure of the anhydride of sulfuric acid.
106. Indicate the structure of the substance that would be the anhydride of $Ba(OH)_2$.

Comprehensive Problems

107. The hydrogen ion donated by Brønsted–Lowry acids is typically represented in aqueous solutions as $H_3O^+(aq)$.
    a. Explain the origin of the positive charge on this ion.
    b. What other forms could the hydrogen ion take in water?
    c. Would the shape of H_3O^+ be more likely to be flat or pyramidal? Explain.
108. Acetic acid, found in vinegar, has a small equilibrium constant. Hydrochloric acid is known as a strong acid. Which of these two representations depicts acetic acid? (Note: In the boxes, HX represents a general acid structure where X^- represents the conjugate base.)

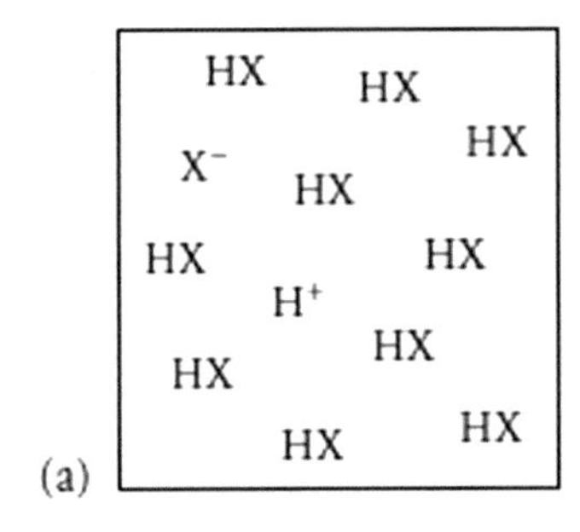

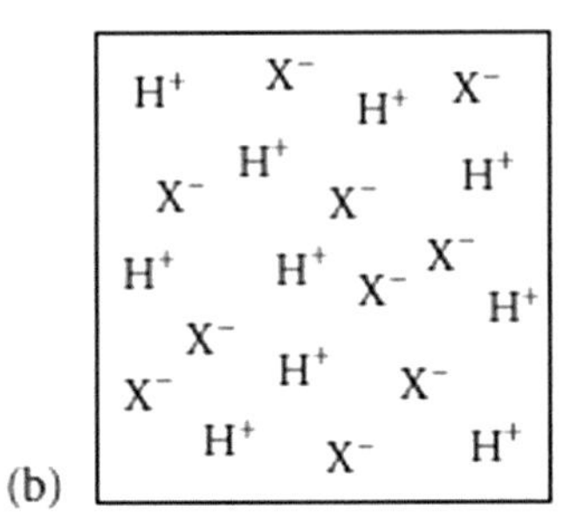

109. Is it possible for a concentrated solution of a weak acid to have the same level of hydronium ion concentration as a dilute solution of a strong acid such as HCl?
110. Water, which has hydrogen bonds to oxygen, can form both H^+ ions and OH^- ions. KOH, which also can hydrogen bond to oxygen, produces only OH^- ions in solution. Using structure and electronegativity, explain why this situation occurs.
111. When comparing the strengths of strong acids, one must use a solvent other than water. One such solvent is acetic acid. Explain why acetic acid would make a better solvent than water for comparing acid strength among strong acids.
112. At 25 °C the K_w value for water is 1.0×10^{-14}. Using that value and the autoprotolysis of water, determine what percent of water molecules actually dissociate under these conditions.
113. Explain why the second ionization constant of a diprotic acid is typically much smaller than the first.
114. Only very pure phosphoric acid is used in food products. An older term for soft drinks is phosphates. This name was applied because pure phosphoric acid was used to produce a tart taste in some beverages and to help dissolve the other ingredients.
    a. What is the pH of a solution that is known to be 5.44 M in H_3PO_4?
    b. What would be the approximate concentration of $PO_4^{3-}(aq)$ in that solution?
115. Lactic acid is produced in muscle tissue during vigorous exercise. It is also found in spoiled milk. The K_a value of lactic acid ($CH_3CHOHCOOH$), at 25 °C, is 1.3×10^{-4}. What is the $H^+(aq)$ concentration in a sample of cellular fluid that is 0.0000033 M in lactic acid and 0.0027 M in lactate ion?

Thinking Beyond the Calculation

116. A salt of sorbic acid ($HC_6H_7O_2$) has been used as mold inhibitor. The salt most commonly used in this practice is potassium sorbate ($KC_6H_7O_2$).
    a. Will a solution of potassium sorbate produce a neutral, acidic, or basic solution?
    b. Write a balanced chemical reaction that describes the processes that take place when potassium sorbate is dissolved in water.
    c. The K_b value of the sorbate ion is 5.88×10^{-10}. What is the pH of a 0.0100 M solution of potassium sorbate?
    d. If a researcher wanted to make 500.0 mL of a solution of potassium sorbate with a pH of 9.44, how many grams of this compound would need to be added to the water?
    e. A solution is prepared that contains both 0.01000 M NH_3 and 0.01000 M potassium sorbate. What is the pH of the solution? What effect, if any, does the ammonia have on the pH of the solution (compare the answers to parts c and e).

Applications of Aqueous Equilibria

Contents

Supplementary Information The online version contains supplementary material available at (▶ https://doi.org/10.1007/978-3-030-90267-4_16).

M. Mosher and P. Kelter, *An Introduction to Chemistry*,
https://doi.org/10.1007/978-3-030-90267-4_16

What We Will Learn in This Chapter

- Titrations are reactions where a reactant is added to another in a measured amount until the reaction is complete (▶ Sect. 16.1).
- Buffers are solutions that resist changes in pH (▶ Sect. 16.2).
- We can determine the concentration of components within a solution using a titration (▶ Sect. 16.3).
- A precipitate can be formed during a titration that can be helpful in determining the concentration of analytes (▶ Sect. 16.4).
- Chemical complexes are useful in titrations (▶ Sect. 16.5).

16.1 Introduction

Water, so essential to life, makes up 70% of the Earth's surface and a nearly equal proportion of our own body mass. Water in living organisms is useful as a medium in which to dissolve the compounds necessary for life, and it also acts as a barrier to keep some compounds out of our bodies. As fundamental as it is to life, there are some everyday processes in which the presence of water can be harmful. Small amounts of water in our gas tank can reduce the efficiency of our car's engine; water in our motor oil increases the rate of decomposition of the lubricating properties (which can lead to the breakdown of the interior of our engine); and water in the coolant used in the manufacture of metal parts can damage the tooling machines, resulting in imperfections in the parts.

▶ https://www.publicdomainpictures.net/pictures/260000/velka/olive-oil-1525310655O9n.jpg

One job people working as chemical technicians have is to perform tests many times each day to determine the quality of the coolant, oil, or gasoline that their company uses or sells. Indeed, one of the many measures of quality oil is that it contains only a negligible amount of water. Technicians use the analytical method of titration, which we first discussed in ▶ Sect. 4.3, to measure the concentration of water in oil and to measure the quantity of a whole host of compounds and ions in water samples. A **titration**, shown in the photograph at the beginning of this chapter, is the controlled addition of just enough solution of known concentration, called a **titrant**, to react with all of an **analyte** (the substance of interest) so that we can determine its concentration. Titrating oil to determine the amount of water present is only one of the multitudes of applications of titrations. These applications fall into several different categories, including those that:

- cause a reduction or oxidation to occur in an analyte
- result in the formation of a precipitate
- form a complex ion
- involve an acid and a base reacting together

Analyte – A solute whose concentration is to be measured by a laboratory test.

Titrant – The solution being added to a solution of an analyte during a titration.

Titration – The process used to determine the exact concentration of an analyte.

Food and pharmaceutical manufacturers use titrations for **quality control**—making sure that the product contains what it is supposed to, and in the proper amounts. Environmental chemists use titrations for analyzing trace (very small) amounts of hazardous metals and other potentially harmful substances. ◘ Table 16.1 lists some analy-

Table 16.1 Selected titration-based analyses

What the titration determines	Primary reagent
Acidity	Sodium hydroxide
Alkalinity	Hydrochloric acid
Vitamin C	Iodine
Chloride ion	Silver nitrate
Water hardness	EDTA
Dissolved oxygen	Sodium thiosulfate
Salinity	Mercuric nitrate
Water	Iodine, sulfur dioxide, primary amines

ses that are commonly accomplished using titrations. At the core of most titrations are many of the principles of aqueous equilibrium that we discussed in ► Chaps. 14 and 15, and we will put those concepts to good use here. A good starting point is buffer solutions because they are commonplace both in industrial titration analyses and, more broadly, in biochemical systems (including us!).

Quality control – The practice in industry of ensuring that the product contains what it is supposed to, and in the proper amounts.

16.2 Buffers and the Common-Ion Effect

What is water "hardness" and how is it determined? The degree of "hardness" of water, a measure of the concentration of calcium (sometimes including magnesium and iron) in household water supplies, is determined by titration. When hard water is heated, the calcium ions form rock-hard carbonate and sulfate precipitates. The resulting solids, known as **boiler scale**, build up inside pipes and restrict water flow. In addition, ice cubes made with "hard" water melt to produce a cloudy precipitate, as shown in Fig. 16.1. In addition, ice cubes made with "hard" water melt to produce a cloudy precipitate. Furthermore, very hard water has a bitter taste that many people find unappealing.

Boiler scale – A buildup of calcium and magnesium salts within pipes and water heaters. Typically composed of calcium carbonate and magnesium carbonate.

The concentration of calcium ions in water can be determined by titration with ethylenediaminetetraacetic acid (EDTA), which we first discussed in ► Chap. 14 and will consider in greater detail later in this chapter. The equation describing the titration is:

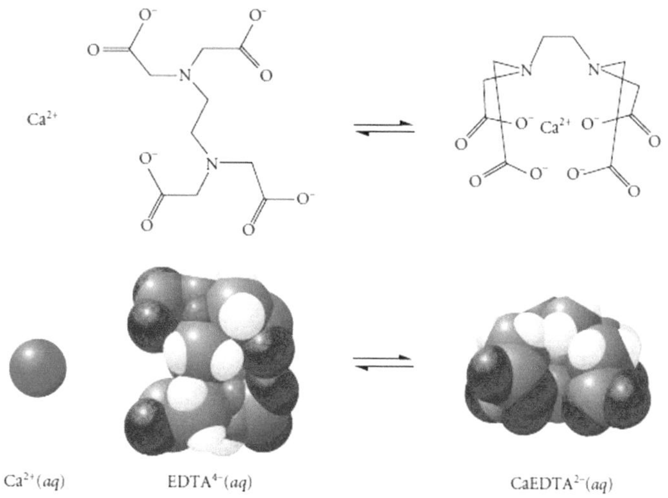

$Ca^{2+}(aq)$ EDTA$^{4-}(aq)$ CaEDTA$^{2-}(aq)$

$$Ca^{2+}(aq) + EDTA^{4-}(aq) \rightleftharpoons CaEDTA^{2-}(aq) \quad K = 5.0 \times 10^{10}$$

Fig. 16.1 Hard water can deposit solids in pipes, on faucets, and in a glass of water. *Source* ▶ https://upload.wikimedia.org/wikipedia/commons/b/bb/Hard_water_and_drop.jpg

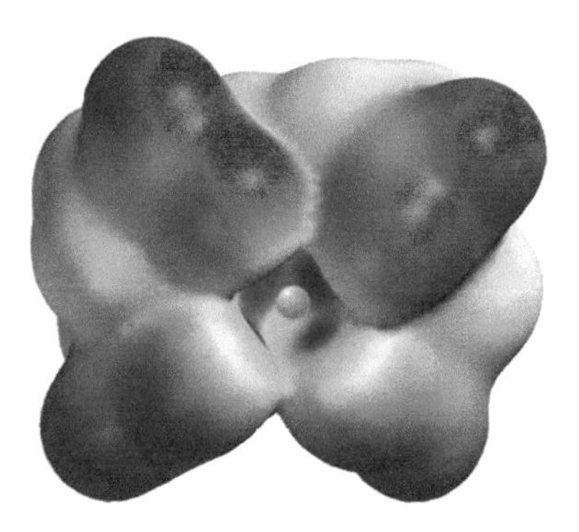

In order to ensure that the EDTA exists as the tetraanion, the entire system must remain basic during the titration. An added buffer maintains the alkaline solution. A **buffer** is a chemical system that is able to resist changes in pH. A buffer is the combination of a weak acid and its conjugate base or of a weak base and its conjugate acid. Buffers do not exist if a strong acid or strong base is paired with its conjugate. However, buffers can accommodate the addition of strong acid and base, and can also withstand dilution, without a large change in the solution pH.

Buffer – A solution containing a weak acid and its conjugate base or a weak base and its conjugate acid. Buffers resist changes in pH upon the addition of acid or base or by dilution.

One standard hard-water analysis protocol requires the addition of a buffer made from ammonia (a weak base) and ammonium chloride (a source of ammonia's conjugate acid). The required pH of the buffer for this analysis is 10. According to the protocol, this can be achieved when the initial concentrations are $[NH_3]_0 = 8.44\,M$ and $[NH_4^+]_0 = 1.27\,M$. How do these initial concentrations produce a buffer, and how does a buffer maintain a relatively constant pH?

To answer this question, we proceed as we do with any equilibrium process, by first examining the possible reactions that can take place in the aqueous solution. In doing so, we recognize that ammonium chloride (NH_4Cl) will dissociate in water to give the ammonium ion (NH_4^+) with chloride (Cl^-) as a spectator ion. The key reactants in the buffer are NH_4^+, NH_3, and H_2O. Their specific reactions are given below.

Reaction 1: Ammonia, a weak base, reacts with water.

$$NH_3(aq) + H_2O(l) \rightleftharpoons NH_4^+(aq) + OH^-(aq) \quad K_b = 1.8 \times 10^{-5}$$

Reaction 2: Ammonium ion, the conjugate acid of ammonia, reacts with water.

$$NH_4^+(aq) + H_2O(l) \rightleftharpoons NH_3(aq) + H_3O^+(aq) \quad K_a = 5.6 \times 10^{-10}$$

(Recall from ▶ Sect. 14.7 that, for the ammonium ion, $K_a = K_w/K_b$, where K_b is the dissociation constant for NH_3.)

Reaction 3: Water undergoes autoprotolysis.

$$2H_2O(l) \rightleftharpoons H_3O^+(aq) + OH^-(aq) \quad K_w = 1.0 \times 10^{-14}$$

The last reaction has a relatively small equilibrium constant, compared to the other two, so its contribution to the pH of the solution is unimportant. Judging by their respective equilibrium constants, the base, ammonia, is stronger than its conjugate acid, ammonium ion, so we expect reaction 1 to be dominant. The reaction produces hydroxide ion, so we expect the solution to be basic.

16.2.1 The Impact of Le Châtelier's Principle on the Equilibria in the Buffer

Reaction 1 produces NH_4^+. However, we already have a relatively high initial concentration of ammonium ion ($[NH_4^+]_0 = 1.27\,M$). How will this affect the extent of reaction 1? Le Châtelier's principle suggests that the presence of the ammonium ion (a "common ion," in this case) will shift the equilibrium of reaction 1 to the reactants side, as shown in ◘ Fig. 16.2. Reaction 1 will be drastically suppressed because of the common-ion effect, an outcome of Le Châtelier's principle.

What about reaction 2? It produces ammonia (NH_3). However, the initial concentration is also quite high, $[NH_3]_0 = 8.44$ M. As with reaction 1, the presence of a large concentration of ammonia will shift the reaction to the left, toward the reactants side (◘ Fig. 16.3). Reaction 2 will also be drastically suppressed because of the common-ion effect, an outcome of Le Châtelier's principle.

The impact of reactions 1 and 2 being suppressed is that we may assume that the *equilibrium* concentrations of NH_3 and NH_4^+ are approximately equal to their initial concentrations.

$$[NH_3] \approx [NH_3]_0 = 8.44\,M \quad \text{and} \quad [NH_4^+] \approx [NH_4^+]_0 = 1.27\,M$$

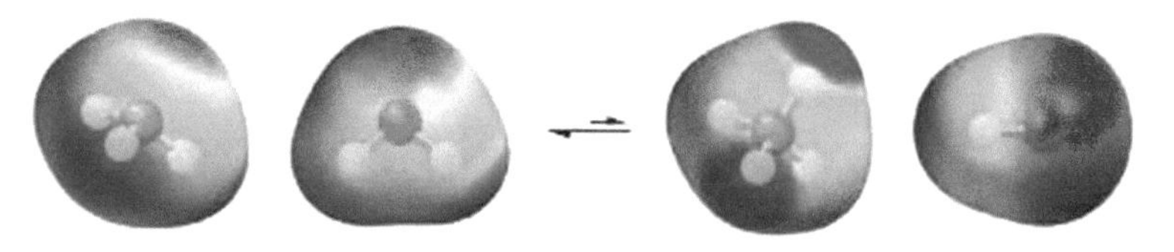

◘ **Fig. 16.2** Addition of NH_4^+ shifts the equilibrium of reaction 1 to the left

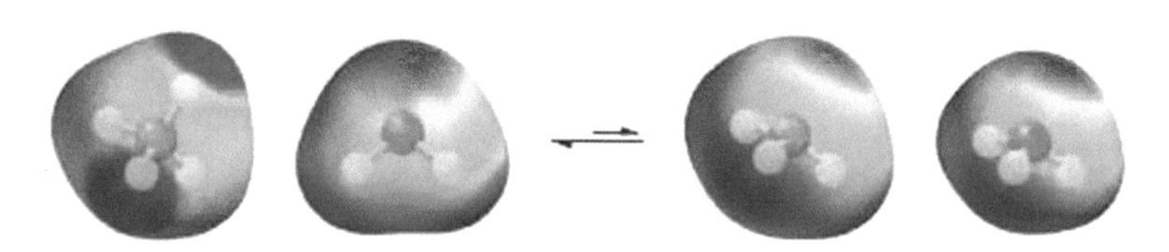

◘ **Fig. 16.3** Addition of NH_3 shifts the equilibrium of reaction 2 to the left

Because of this assumption (which we will test later), we can solve for the pH of this buffer solution using the mass-action expression derived from reaction 1, the hydrolysis of ammonia.

$$K_b = \frac{[NH_4^+][OH^-]}{[NH_3]} \quad 1.8 \times 10^{-5} = \frac{(1.27)[OH^-]}{(8.44)}$$

$$[OH^-] = 1.2 \times 10^{-4}\,M$$

$$[H^+] = \frac{K_w}{[OH^-]} = \frac{1.0 \times 10^{-14}}{1.2 \times 10^{-4}} = 8.3 \times 10^{-11}\,M$$

$$pH = 10.08$$

Alternatively, we could have used the mass-action expression derived from reaction 2 (the dissociation of ammonium) to solve for the pH of the solution, as we will discover in Example 16.1.

What about our claim that reaction 1 was suppressed by the presence of the common ion, ammonium? To show the impact of the common ion, let's calculate $[OH^-]$ in an 8.44 M NH_3 solution that has no added NH_4^+—in other words, not a buffer, just a weak base—using the principles we learned in ▶ Chaps. 14 and 15.

$$NH_3(aq) + H_2O(l) \rightleftharpoons NH_4^+(aq) + OH^-(aq) \quad K_b = 1.8 \times 10^{-5}$$

$$K_b = \frac{[NH_4^+][OH^-]}{[NH_3]} \quad 1.8 \times 10^{-5} = \frac{[NH_4^+][OH^-]}{(8.44)}$$

$$1.8 \times 10^{-5} = \frac{x^2}{(8.44)}$$

$$x = [NH_4^+] = [OH^-] = 0.012\,M$$

$$pH = 12.09$$

Our calculations reveal that the total hydroxide and ammonium ion concentrations, which are equal in this solution of weak base, are $[OH^-]=[NH_4^+]=0.012$ M. Moreover, when we compare the hydroxide ion concentration of the buffer, in which $[OH^-]=1.2\times10^{-4}$ M, to the value we just calculated for the weak base alone, we note that it is 100-fold less. The presence of the ammonium ion in the buffer has suppressed reaction 1, the hydrolysis of ammonia, by 99%! If we were to do a similar calculation with the ammonium ion, we would find that the buffer suppresses the acid dissociation of NH_4^+ by over 99.99%. Le Châtelier's principle has again proved its worth as a formidable part of the chemist's toolbox.

Our goal was to show how a mixture of these concentrations of ammonia and ammonium ion results in a buffer solution with a pH of about 10. Example 16.1 shows the implications of using a slightly different approach to achieve the same goal.

Example 16.1—Alternative Route to the pH of the Buffer

Instead of calculating the pH of the system using reaction 1, calculate the pH of the buffer using reaction 2, the acid dissociation of the ammonium ion. What does the result tell us about solving for the pH of a buffer?

Asking the Right Questions

What is true about the extent of the reaction in a buffer system compared to a system in which a weak acid is added to water? How does the equilibrium position compare to the starting position in each case? What does that mean about the equilibrium concentration of ammonium ion?

Solution

We may still assert that the extent of reaction in a buffer system is negligible; the starting position is the same as the equilibrium position, as shown by the equilibrium line chart.

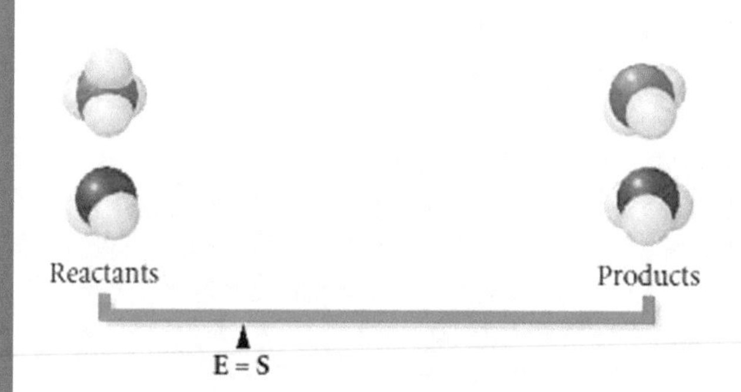

The implication is that the starting and equilibrium concentrations are essentially equal, so, as before,

$$[NH_3] \approx [NH_3]_0 = 8.44\,M$$

and

$$[NH_4^+] \approx [NH_4^+]_0 = 1.27\,M$$

Additionally, we are directed to solve the problem using reaction 2, so we can use its mass-action expression to solve for $[H^+]$ and pH.

$$NH_4^+(aq) + H_2O(l) \rightleftharpoons NH_3(aq) + H_3O^+(aq)$$

$$K_a = 5.6 \times 10^{-10}$$

$$K_a = \frac{[NH_3][H_3O^+]}{[NH_4^+]}$$

$$5.6 \times 10^{-10} = \frac{(8.44)[H_3O^+]}{1.27}$$

$$[H_3O^+] = 8.4 \times 10^{-11}\,M$$

$$pH = 10.08$$

Our answer is the same whether we use the mass-action expression for the hydrolysis of the base (ammonia) or the dissociation of its conjugate (ammonium ion). This is a useful outcome that is a result of the common-ion effect of suppressing both the acid reaction and the base reaction in the buffer. When working with buffers, we may use either the acid dissociation or the conjugate base hydrolysis mass-action expression, but we often work with the expression describing the reaction of the stronger conjugate (the ammonia hydrolysis, in this case).

Are Our Answers Reasonable?

Because the equilibrium is suppressed in the buffer, the pH will be the same whether solving using the weak base or its conjugate acid, so our answer is reasonable. In addition, we may note that the stronger conjugate is the ammonia (K_b for ammonia is considerably larger than K_a for the ammonium ion), leading to a basic pH for the buffer system.

What If…?

…the initial concentration of ammonium ion were doubled, to 2.54 M? Before doing the calculation, predict: Will the pH of the buffer be higher or lower than when $[NH_4^+]$ = 1.27 M? Then do the calculation.
(*Ans: The pH will be slightly lower. The final pH* = 9.77.)

Practice 16.1

Determine the pH of a buffer made from 1.50 M NH_3 and 3.50 M NH_4^+.

Example 16.2—Practice Calculating the Initial pH of a Buffer

We have shown that an ammonia–ammonium ion buffer can have a pH of about 10. Are all buffers basic? To help answer the question, calculate the pH of a buffer that contains 0.200 M each of acetic acid (CH_3COOH) and its conjugate base, the acetate ion (CH_3COO^-).

Asking the Right Questions

What is the relationship between the strength of the conjugates in the buffer and the acidity or basicity of the solution? Which conjugate is stronger in this buffer? How do we solve for the pH of the buffer system?

Solution

We noted in the last example that we normally use the reaction and mass-action expression for the stronger conjugate when calculating the pH of the buffer.

The relevant conjugate reactions (using the shorthand form for the acid dissociation) and equilibrium constants are

$$CH_3COOH(aq) \rightleftharpoons H^+(aq) + CH_3COO^-(aq)$$
$$K_a = 1.8 \times 10^{-5}$$

$$CH_3COO^-(aq) + H_2O(l) \rightleftharpoons CH_3COOH(aq) + OH^-(aq) \quad K_b = 5.6 \times 10^{-10}$$

We will use the mass-action for the stronger conjugate, acetic acid. As with other buffer systems, we recognize that both reactions are suppressed as we have explained using Le Châtelier's principle, so the equilibrium concentrations are about equal to the initial concentrations of the respective components.

The mass-action expression for the reaction of acetic acid is

$$K_a = \frac{[CH_3COO^-][H^+]}{[CH_3COOH]}$$

Solving for $[H^+]$ and then pH, we find that

$$[H^+] = \frac{K_a[CH_3COOH]}{[CH_3COO^-]} = \frac{(1.8 \times 10^{-5})(0.200)}{(0.200)} = 1.8 \times 10^{-5} M$$

$$pH = 4.74$$

Are our Answers Reasonable?

Acetic acid is a stronger conjugate than the acetate ion, therefore, the pH should be acidic. That's reasonable. Our answer of pH = 4.74 is consistent with this expectation and is therefore reasonable.

What If…?

…the K_a of a weak acid and the K_b of its conjugate base were equal? What would be the pH of the buffer?
(*Ans: pH* = 7.0).

Practice 16.2

Would we predict that a buffer prepared from the mixture of 0.300 M formic acid (HCOOH) and 0.400 M sodium formate (HCOONa) would be acidic or basic? Why? Prove any assertion by calculating the pH of this buffer. K_a of formic acid = 1.8×10^{-4}.

We have seen from this discussion that it is possible to determine the approximate pH of a buffer. Extending that idea a step further, it is also possible to pick a buffer that will be in the pH range we want by looking at whether the acid or the base conjugate is the stronger. There is often much more to the selection of a buffer than just pH, because we must consider factors such as whether the buffer will interact with the substances we are studying and whether the buffer's presence in the reaction system has any unintended health consequences. Once these factors are taken into account, we need to know how to prepare the buffers in order to use them.

▶ Here's What We Know So Far

- A buffer is a chemical system that is able to resist changes in pH.
- A buffer consists of the combination of a weak acid and its conjugate base, or of a weak base and its conjugate acid, in similar concentrations.
- A buffer contains a high enough concentration of each conjugate that acid–base equilibria are effectively suppressed (shifted towards reactants), which is an outcome of Le Châtelier's principle.
- Solving for the pH using either the weak base reaction or the weak acid reaction gives equivalent results. Therefore, we use either method, depending on whether we have access to the value of K_a or K_b.
- It is possible to calculate the pH of the buffer system by applying the principles of equilibrium and acid–base chemistry that we learned in the last two chapter. ◀

16.2.2 Buffer Preparation

The chemical technician working in the food industry often uses a pH meter to determine the acidity of the food that is being prepared. Sodas, for instance, often have phosphoric acid added to them to provide a tart taste, and pH control is vital (◘ Fig. 16.4).

One of the most common uses of a buffer solution is to calibrate pH meters. This ensures that the reading on the meter accurately represents the pH of a solution with which we are working. Recipes exist for the preparation of standard calibration buffer solutions. One source for these recipes is the *The Pharmacopeia of the United States of America/The National Formulary*, multi-thousand-page compendium of "standards and specifications for materials and substances that are used in the practice of the healing arts." Organized in 1884 and produced and updated by medical and pharmaceutical experts ever since, the resulting United States Pharmacopeia ("USP" for short) contains standard analysis procedures for a great many substances.

The USP recipe to prepare calibration buffers in the range of 2.2–4.0 recommends using a solution of potassium hydrogen phthalate, or "KHP" ($KHC_6H_4(COO)_2$), which dissociates in water to form K^+ and the phthalate ion, $HC_6H_4(COO)_2^-$, which we will abbreviate as HP^-. The conjugate acid of this weak base, phthalic acid ("H_2P") is generated by addition of hydrochloric acid (a source of H^+) to the HP^-.

$$HP^-(aq) + H^+(aq) \rightleftharpoons H_2P(aq) \quad K \approx 900$$

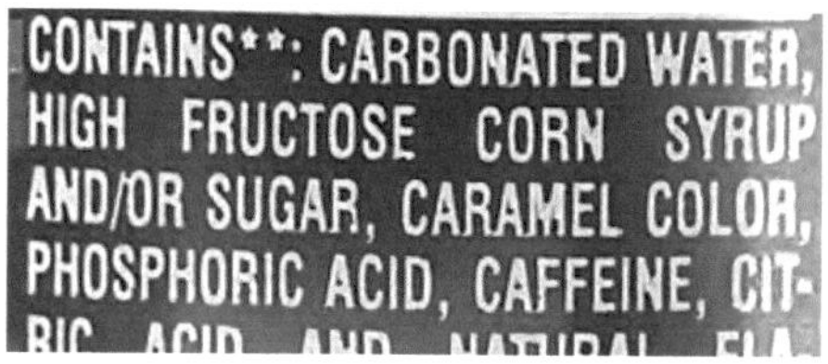

Fig. 16.4 Phosphoric acid is often found in soft drinks

The equilibrium position of the reaction of the weak base with the strong acid is so far toward the formation of products that we can say the reaction is essentially complete. That is, although we will normally write the reaction to show that it settles at some equilibrium point,

$$HP^-(aq) + H^+(aq) \rightleftharpoons H_2P(aq)$$

we may also write it to show that the reaction is essentially complete,

$$HP^-(aq) + H^+(aq) \rightarrow H_2P(aq)$$

Suppose we wish to prepare a buffer with pH = 3.0, and assume that we want the total concentration of phthalate species to be 0.0500 M. That is,

$$[H_2P] + [HP^-] = 0.0500\,M$$

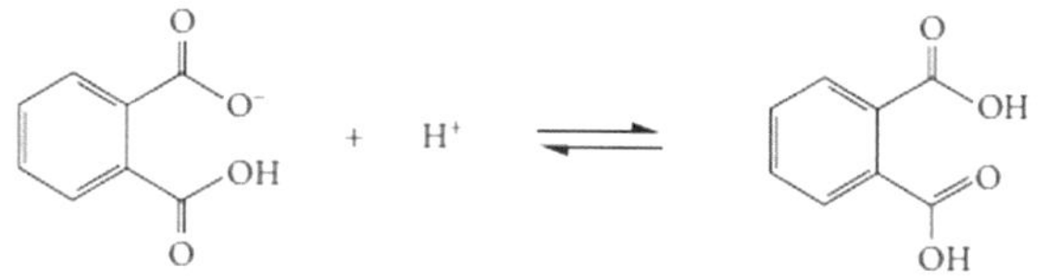

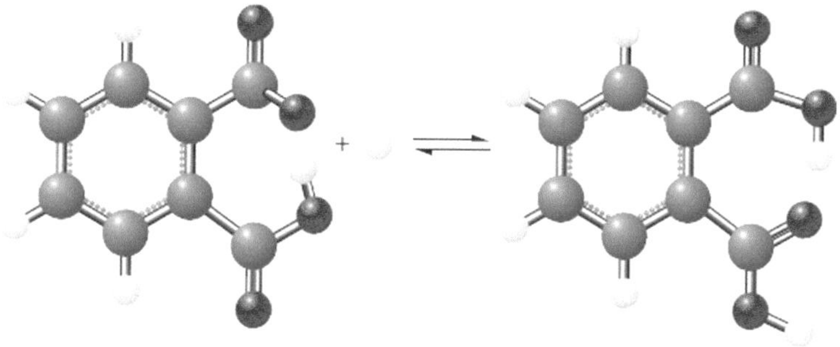

Once the proper ratio of conjugate acid to base is present, the buffer will be at pH = 3.00. How do we find the proper ratio to create the buffer solution? How do we then find the final concentrations of H_2P and HP^-?

As has been true throughout these three chapters on equilibrium (▶ Chap. 14, 15 and thischapter), the answer lies in writing down the relevant equilibrium reaction and its mass-action expression. For the phthalate buffer system, the important reaction is the dissociation of phthalic acid (H_2P) in water (prove that the K_b for HP^- is too small to compete with H_2P):

$$H_2P(aq) + H_2O(l) \rightleftharpoons HP^-(aq) + H_3O^+(aq) \quad K_{a_1} = 1.12 \times 10^{-3}$$

We can write it using our shorthand form:

$$H_2P(aq) \rightleftharpoons HP^-(aq) + H^+(aq) \qquad K_{a_1} = 1.12 \times 10^{-3}$$

The mass-action expression for the reaction is

$$K_{a_1} = \frac{[HP^-][H^+]}{[H_2P]}$$

Dividing both sides by $[H^+]$ gives us an expression for the ratio of conjugate base to acid:

$$\frac{K_{a_1}}{[H^+]} = \frac{[HP^-]}{[H_2P]}$$

Solving, with the use of the pH = 3.00, $[H^+] = 1.0 \times 10^{-3}$ M,

$$\frac{(1.12 \times 10^{-3})}{(1.00 \times 10^{-3})} = \frac{(1.12)}{(1.00)} = \frac{[HP^-]}{[H_2P]}$$

we find that the ratio of the base, HP^-, to its conjugate acid, H_2P, is 1.12 to 1. We'll call this Eq. 16.1.

$$[HP^-] = 1.12[H_2P] \quad (16.1)$$

We said initially that we had set the sum of concentrations of the two species to be a total of 0.0500 M. We'll call this Eq. 16.2.

$$[H_2P] + [HP^-] = 0.0500\,M \quad (16.2)$$

Thus, we have two equations (Eqs. 16.1 and 16.2) and two unknowns ($[H_2P]$ and $[HP^-]$). We can solve Eq. 16.2 by substituting $1.12[H_2P]$ in place of $[HP^-]$, as allowed by Eq. 16.1:

$$[H_2P] + 1.12[H_2P] = 0.0500\,M$$

$$2.12[H_2P] = 0.0500\,M$$

$$[H_2P] = 0.0236\,M$$

$$[HP^-] = 0.0500M - 0.0236\,M$$

$$[HP^-] = 0.0264\,M$$

Therefore, in order to obtain a buffer at pH = 3.00 with a total molarity of 0.0500 M, we'll have to make up a solution that is 0.0236 M in H_2P and 0.0264 M in KHP. We can check that these quantities are reasonable by substituting back into the mass-action expression and solving for $[H^+]$.

$$[H^+] = \frac{K_{a_1}[H_2P]}{[HP^-]} = \frac{(1.12 \times 10^{-3})(0.0236)}{(0.0264)}$$

$$[H^+] = 1.00 \times 10^{-3}\,M$$

16

To recap: If we started with a solution that was 0.0500 M KHP and added 0.0236 mol of HCl, we would produce a solution with a pH = 3.00.

Example 16.3—Practice with Buffer Preparation

How many grams of sodium formate (HCOONa), molar mass = 68.01 g/mol, must be dissolved in a 0.300 M solution of formic acid (HCOOH), to make 400.0 mL of a buffer solution with a pH = 4.60? Assume that the volume of the solution remains constant when we add the sodium formate.

Asking the Right Questions

We are asked to find the mass of sodium formate needed to combine with formic acid to prepare the buffer. Let's develop a stepwise approach to the problem.

Step 1: Our solution is at pH 4.60. What is the ratio of the concentrations of the acid and the conjugate base, $\frac{[HCOO^-]}{[HCOOH]}$, needed to produce a buffer with a pH = 4.60? To solve for the ratio of concentrations of acid and conjugate base needed, we can first write the important processes that occur in the solution. Two equilibria are important in this buffer system:

$$HCOOH(aq) + H_2O(l) \rightleftharpoons H_3O^+(aq) + HCOO^-(aq)$$

$$K_a = 1.8 \times 10^{-4}$$

$$HCOO^-(aq) + H_2O(l) \rightleftharpoons HCOOH(aq) + OH^-(aq)$$

$$K_b = 5.6 \times 10^{-11}$$

As we discussed in Exercise 16.1, we can use either reaction to solve for the acid to-conjugate-base ratio.

Step 2: Noting that the concentration of formic acid is 0.300 M (given in the example), can we then find the concentration of the conjugate base? Remember that the sodium salt will completely dissociate in solution, leaving the formate anion ($HCOO^-$) as the conjugate base and Na^+ as the spectator ion.
Step 3: Having calculated the concentration of the conjugate base, can we find the number of moles of the base (the formate ion, $HCOO^-$), in 400.0 mL of solution?
Step 4: Finally, can we convert from moles of formate (as the sodium formate salt) to grams of sodium formate?

Solution

Step 1: What is the ratio of the concentrations of the acid and the conjugate base, $\frac{[HCOO^-]}{[HCOOH]}$, needed to produce a buffer with a pH = 4.60?

$$K_a = \frac{[HCOO^-][H^+]}{[HCOOH]}$$

The pH of the solution is 4.60, so $[H^+] = 2.51 \times 10^{-5}$ M.

$$1.8 \times 10^{-4} = \frac{[HCOO^-](2.51 \times 10^{-5})}{[HCOOH]}$$

$$\frac{[HCOO^-]}{[HCOOH]} = \frac{1.8 \times 10^{-4}}{2.51 \times 10^{-5}} = \frac{7.17}{1.00}$$

Step 2: Noting that the concentration of formic acid is 0.300 M (given in the example), can we then find the concentration of the conjugate base?

$$\left[HCOO^-\right] = 7.17 \times [HCOOH]$$

$$\left[HCOO^-\right] = 7.17 \times 0.300\,M$$

$$\left[HCOO^-\right] = 2.15\,M$$

Alternatively, we could have combined steps 1 and 2 to find $[HCOO^-]$ directly by substituting 0.300 M into the mass-action expression:

$$1.8 \times 10^{-4} = \frac{[HCOO^-](2.51 \times 10^{-5})}{0.300}$$

$$\left[HCOO^-\right] = 2.15\,M$$

Step 3: Having calculated the concentration of the conjugate base, can we find the number of moles of the base (the formate ion, $HCOO^-$), in 400.0 mL of solution?

$$\frac{2.15\,\text{mol}\,HCOO^-}{\text{L solution}} \times 0.4000\,\text{L solution}$$
$$= 0.860\,\text{mol}\,HCOO^- = 0.860\,\text{mol}\,HCOONa$$

Step 4: Finally, we convert from moles of formate (as the sodium formate salt) to grams of sodium formate.

$$0.860\,\text{mol}\,HCOONa \times \frac{68.01\,\text{g}\,HCOONa}{1\,\text{mol}\,HCOONa}$$
$$= 58\,\text{g}\,HCOONa$$

Is our Answer Reasonable?

Is it reasonable that the buffer system should be acidic? Formic acid (HCOOH) is stronger than its conjugate, so the acid dissociation reaction will dominate, resulting in an acidic buffer solution. This is an appropriate conjugate pair to use to prepare an acidic buffer solution, so the answer is reasonable.

What If…?

…we add 10.0 mL of 0.500 M HCl to the solution? What important changes would occur in the solution?
(*Ans: the HCl would react with the formate ion to form more formic acid, lowering the pH of the buffer solution. We will discuss soon how to determine if the addition of strong acid will have a major or minor impact on the pH of a buffer.*)

Practice 16.3

How many grams of sodium formate (HCOONa) must be dissolved in a 0.150 M solution of formic acid (HCOOH) to make 500.0 mL of a buffer solution with pH = 3.95? Assume that the volume of the solution remains constant when we add the sodium formate.

Example 16.4—More Practice with Buffer Preparation

How many milliliters of 0.200 M HCl must be added to 50.0 mL of a 0.200 M solution of NH_3 in order to prepare a buffer that has a pH of 8.60?

$$NH_3 + H_2O(l) \rightleftharpoons NH_4^+ + OH^- \quad K_b = 1.8 \times 10^{-5}$$

Asking the Right Questions

We have a solution of NH_3. As in the previous example, we can ask, what is the ratio of the conjugates needed to make the buffer have the desired pH? Once we have established that ratio, how much HCl do we need to the existing amount of NH_3 in order to produce enough NH_4^+ to make that ratio?

As we think through the problem-solving strategy, we may want write down what we know about the system.

Let's outline the key questions in a stepwise fashion:

Step 1: What is the ratio of the weak base to the conjugate acid, $\frac{[NH_3]}{[NH_4^+]}$, in the pH = 8.60 buffer ($[H^+] = 2.5 \times 10^{-9}$

M)? This ratio will be the first equation we need to solve in order to find the amounts of NH_3 and NH_4^+. The base hydrolysis equation contains $[OH^-]$ rather than $[H^+]$. We can solve for $[OH^-]$ from $[H^+]$ and K_w. We will keep an extra significant figure for the intermediate calculation.

$$[OH^-] = \frac{K_w}{[H^+]} = \frac{1.00 \times 10^{-14}}{2.51 \times 10^{-9}} = 3.98 \times 10^{-6}\ M$$

We can then set up our mass-action expression to solve for $\frac{[NH_3]}{[NH_4^+]}$.

$$NH_3(aq) + H_2O(l) \rightleftharpoons NH_4^+(aq) + OH^-(aq)$$

$$K_b = 1.8 \times 10^{-5}$$

$$K_b = \frac{[NH_4^+][OH^-]}{[NH_3]}$$

$$\frac{K_b}{[OH^-]} = \frac{[NH_4^+]}{[NH_3]}$$

Inverting both sides of the equation, we get

$$\frac{[OH^-]}{K_b} = \frac{[NH_3]}{[NH_4^+]}$$

Step 2: What is the sum of the moles of NH_3 and NH_4^+ in the final solution? It must equal the initial 0.100 mol of NH_3 used to prepare the original solution because we are changing the NH_3 to NH_4^+, not getting rid of any from the reaction flask. This is the second equation we need to solve to find the amounts of NH_3 and NH_4^+.

$$\text{mol } NH_4^+ + \text{mol } NH_3 = 0.0100\ \text{mol}$$

Step 3: Finally, how much 0.200 M HCl will be needed to react with our original 0.100 mol of NH_3 to generate the proper number of moles of NH_4^+?

Solution

Step 1: Finding the ratio of $\frac{[NH_3]}{[NH_4^+]}$ in the solution.

$$\frac{[OH^-]}{K_b} = \frac{[NH_3]}{[NH_4^+]}$$

Substituting for $[OH^-]$ and K_b yields

$$\frac{3.98 \times 10^{-6}}{1.8 \times 10^{-5}} = \frac{0.221}{1.00} = \frac{[NH_3]}{[NH_4^+]}$$

Because the volume of the buffer solution is the same for NH_3 and NH_4^+, we can work with moles rather than molarity (the liters will cancel from the calculation).

$$\text{mol } NH_3 = 0.221 \times (\text{mol } NH_4^+)$$

Step 2: Finding moles of NH_4^+ and NH_3 so that the sum equals 0.0100 mol.

$$\text{mol } NH_4^+ + \text{mol } NH_3 = 0.0100\ \text{mol}$$

$$\text{mol } NH_4^+ + [0.221 \times (\text{mol } NH_4^+)] = 0.0100\ \text{mol}$$

$$1.221 \times (\text{mol } NH_4^+) = 0.0100\ \text{mol}$$

$$\text{mol } NH_4^+ = 0.00819\ \text{mol}$$

$$\text{mol } NH_3 = 0.00181\ \text{mol}$$

As a check of our work, we find that substituting these values back into the mass-action expression results in the value of K_b.

Step 3: Finding how many milliliters of 0.200 M HCl we need to add to produce 0.00819 mol of NH_4^+ from NH_3.

$$0.00819\ \text{mol HCl} \times \frac{1\ \text{L HCl solution}}{0.200\ \text{mol HCl}} \times \frac{1000\ \text{mL HCl solution}}{1\ \text{L HCl solution}} = 41\ \text{mL HCl solution}$$

Is our Answer Reasonable?

One way to decide what is a reasonable pH when we add HCl to NH_3 is to think about the pH of a solution of NH_3 without HCl. Such a solution has a pH of between 11 and 12, depending upon the NH_3 concentration. By adding HCl to the solution, we would expect the pH to drop as we produce NH_4^+, but buffers resist a change in pH, so the pH should still be weakly basic. Our answer is therefore reasonable.

What If...?

... the mole ratio of our solution was changed so instead of being $\frac{[NH_3]}{[NH_4^+]} = \frac{0.221}{1.00}$, it was $\frac{[NH_3]}{[NH_4^+]} = \frac{1}{0.221}$? What would be the pH of the buffer?

(*Ans: pH*=9.91).

Practice 16.4

How many milliliters of 0.10 M HCl must be added to 25.0 mL of a 0.2000 M solution of NH_3 to prepare a buffer that has a pH of 8.80?

In practice, even fairly dilute buffers with fairly low conjugate concentrations (not far above their equilibrium constants) will have pH values similar to those with equal ratios of higher conjugate concentrations. If the buffer is too dilute, other factors cause the pH to move toward neutrality.

16.2.3 The Henderson–Hasselbalch Equation: We Proceed, but with Caution

We have seen how the mass-action expression is used to find the approximate ratio of base to acid in a buffer. In 1902, seven years before Peter Sørenson coined the term pH, a Massachusetts physician named Lawrence Joseph Henderson, who was studying buffers in blood, published work relating $[H^+]$ to the acid and base concentrations in a buffer. For an acid, HA, and its conjugate base, A^-, Henderson noted that $[H^+] = \frac{K_a[HA]}{[A^-]}$, as we showed in Example 16.2 for the acetic acid–acetate ion buffer system. Written a slightly different way for visual clarity, this is:

$$[H^+] = K_a\frac{[HA]}{[A^-]} = K_a\frac{[\text{weak acid}]}{[\text{conjugate base}]}$$

This is just a rearrangement of the same K_a expression we have been using. Note the relationship between the acid and the base. The acid is any weak acid, and the base is the conjugate base of that weak acid. In 1916, K. A. Hasselbalch (pronounced "hassle-back"), a physiologist at the University of Copenhagen, used Sørenson's "new" pH term, taking the negative log of both sides of Henderson's equation to give:

$$\text{pH} = -\log(K_a) - \log\left(\frac{[\text{weak acid}]}{[\text{conjugate base}]}\right)$$

This can be slightly modified by using pK_a (the same as $-\log(K_a)$) and changing the sign before the log term, which inverts the concentration within the log term. This gives the final form of what is commonly known as the **Henderson–Hasselbalch equation**:

$$\text{pH} = pK_a + \log\left(\frac{[\text{conjugate base}]}{[\text{weak acid}]}\right)$$

Henderson–Hasselbalch equation – A shorthand equation used to determine the pH of a buffer solution. pH = pK_a + log(base/acid).

The equation is not necessary; we can solve any buffer problem without it, including the ammonia–ammonium ion system and the KHP buffer system we discussed earlier in this section. However, scientists in the medical and biological professions use the equation because it can be a timesaver for calculating the pH. We added the phrase "we proceed, but with caution" to the title of this subsection because, in spite of how common the use of the Henderson–Hasselbalch equation is, there are two reasons why we must be cautious with its use. First, a system must contain both an acid and its conjugate base in order for the equation to be valid. If we don't have a buffer system, using this equation will lead us to incorrect results. Second, and more troublesome, is the tendency accidentally invert the ratio of acid to base within the log term. Proceed with caution!

Example 16.5—The Henderson–Hasselbalch Equation

Use the Henderson–Hasselbalch equation to find the ratio of conjugate base to weak acid in an acetic acid–acetate buffer solution with a pH of 5.0. $K_a = 1.8 \times 10^{-5}$.

Asking the Right Questions

How is the Henderson–Hasselbalch equation related to the mass-action expression? Based on the "What do I need to know?" section for this example, is there a higher concentration of acetic acid or acetate ion in the pH = 5.0 solution?

Solution

The Henderson–Hasselbalch equation is a different form of the mass-action expression.

$$\text{pH} = pK_a + \log\left(\frac{\text{base}}{\text{acid}}\right)$$

$$5.0 = 4.74 + \log\left(\frac{[CH_3COO^-]}{[CH_3COOH]}\right)$$

$$0.26 = \log\left(\frac{[CH_3COO^-]}{[CH_3COOH]}\right)$$

$$10^{0.26} = \frac{[CH_3COO^-]}{[CH_3COOH]} = 1.8$$

This means there is 1.8 times as much acetate ion as acetic acid in a solution with a pH just above the pK_a.

Is our Answer Reasonable?

We expected there to be a higher concentration of base than acid in the solution, and this is what we calculated. The answer therefore is reasonable.

What If...?

...we were using the Henderson–Hasselbalch equation to calculate the pH of a buffer and we had inadvertently inverted the base and acid ratio (that is, used an acid-to-base ratio of 1.8)? What would have been the pH? Based on the pH value, how would we know that we inverted the concentrations?

(Ans: pH = 4.48; The pH is less than the pK_a, which indicates that we have a higher concentration of acid to base, the opposite of the case in the example.)

Practice 16.5

Determine the pH of an acetic acid–acetate buffer containing 0.500 M acetic acid and 0.250 M sodium acetate. $K_a = 1.8 \times 10^{-5}$.

The conclusion to Exercise 16.5 is important, because it says that in order for the pH of the buffer to be higher than the pK_a, there must be more conjugate base than weak acid. The opposite statement is also true (as illustrated in Practice 16.5); that is, pH values lower than the pK_a value require more conjugate acid than base. Take a quick look back at Examples 16.1 through 16.3 to see that this is so. In Example 16.2, the concentrations of the weak acid and conjugate base are equal. When this is so, the pH is equal to the pK_a, and the buffer that results has the greatest possible capacity to neutralize strong acids and bases, which is also referred to as maximum buffer capacity.

16.2.4 Buffer Capacity

The job of the chemical technician may include monitoring the emissions from a smokestack. Recent laws enacted to protect the environment have made this a priority. To reduce the emissions, businesses often make use of a **scrubber** attached to the smokestack. Scrubbers are used to remove sulfur dioxide from the smoke emitted by combustion of sulfur-rich coal. Chemically, the process can be accomplished by **flue-gas desulfurization**, in which a limestone slurry is combined with sulfur dioxide in a multistep process:

$$CaCO_3(s) + SO_2(g) \rightleftharpoons CaSO_3(s) + CO_2(g)$$

Scrubber – A device that removes harmful impurities from smokestack gases.

Flue-gas desulfurization – A process used to remove sulfur dioxide (and other sulfur oxides) from combustion smoke.

The calcium sulfite that is formed is further oxidized and hydrated to produce gypsum ($CaSO_4{\cdot}2H_2O$), which is used to make wallboards for building construction.

$$CaSO_3(s) + 1/2\,O_2(g) + 2H_2O(l) \rightleftharpoons CaSO_4 \cdot 2H_2O$$

To make this conversion of sulfur dioxide to calcium sulfate dihydrate more efficient, a mixture of formic acid (HCOOH) and sodium formate (HCOONa) is often added to the scrubber. The low pH of this buffer (Example 16.3) enhances the reaction efficiency by converting some of the sulfite in solution to bisulfite, (HSO_3^-). Doing so allows the more soluble calcium bisulfite ($CaHSO_3$) to form, as well as minimizing calcium sulfite buildup on the processing machinery. How can the formic acid–formate ion buffer system keep the system pH within a small range? The **buffer capacity** of a system is a measure of the number of moles of strong acid or strong base that can be added while keeping the pH relatively constant. IUPAC defines "relatively constant" as between +/−1 pH unit. Others have different parameters. Because this discussion serves as an introduction to acid–base titrations, we will define "relatively constant" as that point at which all of the conjugate acid (or base) is reacted.

Buffer capacity – The degree to which a buffer can "absorb" added acid or base without changing pH.

Suppose we wish to determine the buffer capacity of 100.0 mL of a solution containing 0.200 M HCOOH and 0.400 M HCOONa. We first address this problem by noting the relevant equilibria in this aqueous solution:

$$2H_2O(l) \rightleftharpoons H_3O^+(aq) + OH^-(aq) \qquad K_w = 1.0 \times 10^{-14}$$

$$HCOOH(aq) \rightleftharpoons HCOO^-(aq) + H^+(aq) \qquad K_a = 1.8 \times 10^{-4}$$

Again, the autoprotolysis of water is insignificant, based on the relative sizes of the equilibrium constants. Considering only the equilibrium between formic acid and formate ion, the pH of this system is 4.05, shown by our calculation:

$$[H^+] = \frac{K_a[HCOOH]}{[HCOO^-]} = \frac{1.8 \times 10^{-4}(0.200)}{(0.400)}$$

$$[H+] = 9.0 \times 10^{-5}\,M \qquad pH = 4.05$$

Let's look at the impact on the pH of adding strong acid and then strong base.

16.2.4.1 Change 1: Addition of a Strong Acid

What happens to the pH of our buffer if we add 10.0 mL of a 1.00 M HCl solution to 100.0 mL of buffer (total solution volume = 110.0 mL)? We can set the stage by calculating the moles of each component initially in solution.

$$\text{mol HCOOH}_{\text{initial}} = 0.1000\,\text{L HCOOH} \times \frac{0.200\,\text{mol HCOOH}}{\text{L HCOOH solution}} = 0.0200\,\text{mol HCOOH}$$

$$\text{mol HCOO}^-_{\text{initial}} = 0.1000\,\text{L HCOO}^- \times \frac{0.400\,\text{mol HCOO}^-}{\text{L HCOO}^-\ \text{solution}} = 0.0400\,\text{mol HCOO}^-$$

We continue by finding out how many moles of HCl were added.

$$\text{mol HCl}_{\text{added}} = 0.0100\,\text{L} \times \frac{1.00\,\text{mol}}{\text{L}} = 0.0100\,\text{mol HCl}$$

The addition of the strong acid supplies H^+ to the solution, and this H^+ essentially completely reacts with $HCOO^-$ to form HCOOH ($K = 5.6 \times 10^3$).[1]

How much HCOOH will be formed? We can organize our thinking using a table. At the top of our table we will write the equation that indicates the reaction of the added HCl (a source of H^+) with the conjugate base of formic acid ($HCOO^-$). We note that this is a limiting-reactant calculation in which the HCl is limiting and the $HCOO^-$ is in excess.

	$HCOO^-$ (aq)	+	H^+ (aq)	⇌	$K = 5.6 \times 10^3$
Moles initial	0.0400		–		0.0200
Moles added	–		0.0100		–
Change	−0.0100		−0.0100		+0.0100
Moles at equilibrium	0.0300		≈0		0.0300

After addition of the HCl, which threw the system out of equilibrium, it returns again to its new equilibrium position, shown by

$$HCOOH(aq) \rightleftharpoons HCOO^-(aq) + H^+(aq)$$

We still have lots of conjugate acid and base—a buffer system, for which we can find the pH:

$$[H^+] = \frac{K_a[HCOOH]}{[HCOO^-]} = \frac{1.8 \times 10^{-4}(0.300)}{(0.300)} = 1.8 \times 10^{-4}\,M \qquad pH = 3.74$$

The buffer has responded to the addition of HCl by having only a slightly lowered pH, from 4.05 to 3.74. Note that the pH = pK_a when the concentration of acid and base are equal.

---

1 The value for the equilibrium constant for the reaction of the strong acid with the formate anion ($HCOO^-$) to form HCOOH can be calculated by combining the following two equations, as we did in ▶ Sect. 15.4.

$$HCOO^-(aq) + H_2O(aq) \rightleftharpoons HCOOH(aq) + OH^- \quad K_b = K_W/K_a = 5.6 \times 10^{-11}$$

$$H^+(aq) + OH^-(aq) \rightleftharpoons H_2O(aq) \quad K = 1/K_W = 1.0 \times 10^{14}$$

$$HCOO^-(aq) + H^+(aq) \rightleftharpoons HCOOH(aq) \quad K = (K_b \times 1/K_W) = 1/K_a = 5.6 \times 10^3$$

We use this method several times in this chapter to determine the equilibrium constant for the addition of strong acids or bases to weak bases or acids in a titration.

16.2.4.2 Change 2: Addition of a Strong Base

What happens to the pH of our original solution if we add 15.0 mL of a 1.00 M NaOH solution (total volume = 115.0 mL)? We already know how much formic acid (0.0200 mol) and formate ion (0.0400 mol) we started with. How many moles of NaOH were added?

$$\text{mol NaOH}_{\text{added}} = 0.0150\,\text{L NaOH} \times \frac{1.00\,\text{mol NaOH}}{\text{L NaOH solution}} = 0.0150\,\text{mol NaOH}$$

The addition of the strong base supplies OH^- to the solution, and this OH^- essentially completely reacts with HCOOH to form $HCOO^-$ ($K = 1.8 \times 10^{10}$). Therefore, we will set up a table based on the reaction of OH^- with HCOOH.

	HCOO(aq)	+	OH^-(aq)	⇌	$HCOO^-$(aq)	$K = 1.8 \times 10^{10}$
Moles initial	0.0200				0.0400	
Moles added			0.0150			
Change	−0.0150		−0.0150		+0.0150	
Moles at equilibrium	0.0050		≈0		0.0550	

We still have both acid and conjugate base in the buffer, although the acid concentration is a little low. We can solve for the pH.

$$[H^+] = \frac{K_a[\text{HCOOH}]}{[\text{HCOO}^-]} = \frac{1.8 \times 10^{-4}(0.0050)}{(0.0550)} = 1.64 \times 10^{-5}\,\text{M} \quad \text{pH} = 4.79$$

The pH of the buffer has gone up, but the solution still has the capacity to keep the pH close to the original value of 4.05.

16.2.4.3 Change 3: Exceeding the Buffer Capacity

At what point will the buffer no longer have the capacity to keep the pH reasonably close to the original value? In other words, when will the pH begin to rise or fall sharply? How much will the pH change before stabilizing? The addition of 50.0 mL 1.00 M HCl solution to the original buffer solution will stress the system. Let's see how much, again using our table.

	$HCOO^-$aq	+	H^+aq	⇌	HCOOH(aq)	$K = 5.6 \times 10^3$
Moles initial	0.0400				0.0200	
Moles added			0.0500			
Change	−0.0400		−0.0400		+0.0400	
Moles at equilibrium	≈0		0.0100		0.0600	

How did we calculate the moles at equilibrium in this case? This is still a limiting-reactant problem, with the $HCOO^-$ instead of the H^+ limiting the amount of product formed. We have 0.0500 mol of H^+, but only 0.0400 mol of $HCOO^-$ with which to react! This means that 0.0400 mol will react to form the HCOOH product, and there will be 0.0100 mol of H^+ in excess. Note as well that since we no longer have a buffer solution (there is no conjugate base remaining), we cannot use the Henderson-Hasselbach equation.

What is the final pH of the resulting solution? At equilibrium, we have a solution that contains 0.0600 mol of HCOOH and 0.0100 mol of H^+. The formic acid is so much weaker than the hydrochloric acid (judging on the basis of their K_a values) that its presence in a similar concentration to the HCl will not affect the final hydrogen ion concentration. Only the H^+ from the HCl is important. Therefore, the pH is based solely on the $[H^+]$ due to the ionization of the strong acid. The total volume of 150.0 mL (0.1500 L) is the sum of the initial 100.0 mL of solution and the 50.0 mL HCl solution added to it.

$$[H^+] = \frac{(0.0100\,\text{mol})}{(0.1500\,\text{L})} = 0.0667\,\text{M} \qquad \text{pH} = 1.18$$

This reveals that the buffer's ability to resist a change in pH has been exceeded; the pH has changed dramatically. In the original solution, the formate ion ($HCOO^-$) could react with up to 0.0400 mol of the strong acid without having any excess H^+ to greatly lower the pH. This is its buffer capacity toward strong acid. Similarly, the formic acid could react with up to 0.0200 mol of strong base without having any excess strong base to greatly raise the pH. This is its buffer capacity toward strong base. As we saw with the addition of too much HCl, when the buffer capacity is exceeded, the pH changes quite sharply. For a monoprotic acid or base, the buffer capacity is greatest when $[HA]=[A^-]$. When this occurs, the buffer can neutralize equal amounts of either strong base or strong acid.

Example 16.6—Buffer Capacity

Let's look at the buffer capacity of an ammonia–ammonium buffer system like the one we used for our calcium–EDTA analysis at the beginning of this chapter. We will start with 200.0 mL of a solution containing 0.500 M NH_3 and 0.200 M NH_4^+. What will be the initial pH of the buffer? Will we exceed the buffer capacity by adding 75.0 mL of 2.00 M NaOH? What will be the pH after that addition?

Asking the Right Questions

Do we expect the pH of the buffer to be acidic or basic? How do we solve for the pH? What is the "buffer capacity," and how do we know when our solution has exceeded it?

Solution

Ammonia is a stronger conjugate than the ammonium ion, so the solution will be basic with a total of 0.100 mol of NH_3 and 0.0400 mol NH_4^+. The solution has more conjugate base than acid, so the pH should be higher than the pK_a for ammonium, 9.26. We can use the expression for K_a of NH_4^+ so to solve directly for $[H^+]$:

$$NH_4^+(aq) \rightleftharpoons NH_3(aq) + H^+(aq)$$

$$K_a = 5.6 \times 10^{-10}$$

$$[H^+] = \frac{K_a[NH_4^+]}{[NH_3]} = \frac{5.6 \times 10^{-10}(0.200)}{(0.500)}$$

$$[H^+] = 2.2 \times 10^{-10} \quad pH = 9.65$$

We now take the system out of equilibrium by adding 75.0 mL of 2.00 M NaOH (0.150 mol of OH^- added). The OH^- will react with the weakly acidic ammonium ion to give more ammonia. Given the amount of OH^- added, what is the limiting reactant?

	$NH_4^+(aq)$	+	$OH^-(aq)$	$\rightleftharpoons$	$NH_3(aq)$	+	$H_2O(l)$	$K = 5.6 \times 10^4$
Moles initial	0.0400				0.100			
Moles added			0.150					
Change	−0.0400		−0.0400		+0.0400			
Moles at equilibrium	≈ 0		0.110		0.140			

NH_4^+ is the limiting reactant, and we will have 0.110 mol of OH^- and 0.140 mol of NH_3 in excess. The $[OH^-]$ will determine the final pH because it is a much stronger base than NH_3. The total solution volume consists of the 200.0 mL with which we started and the 75.0 mL of strong base that we added, for a total of 275.0 mL.

$$[OH^-] = \frac{(0.0110\,\text{mol}\,OH^-)}{(0.2750\,\text{L}\,OH^-\ \text{solution})} = 0.0400\,\text{M}$$

$$pOH = 1.40 \quad pH = 12.60$$

The excess of OH^-, and the resulting sharp increase in the pH, are clear signs that we have exceeded the buffer capacity.

Are our Answers Reasonable?

The initial pH is basic, as we predicted, based on the relative strengths of the conjugates. Enough strong base was added to exceed the buffer capacity (by converting all of the ammonium ion to ammonia, and still having OH^- left over), so the pH should have risen sharply, which it did, a reasonable result.

What If…?

… after adding the strong base, we added 0.0110 mol of HCl in an extra 50 mL? What would be the pH of the resulting solution?

(*Ans: We will have used up all of the strong base with strong acid, leaving us with a solution of* 0.140 *mol of* NH_3 *in a total volume of* 325 *mL* $[NH_3]=0.431$ *M.* $pH=11.44$.)

Practice 16.6

What is the pH of the ammonia–ammonium ion buffer in this exercise after the addition of 30.0 mL of 0.100 M HCl?... after the addition of 50.0 mL of 1.50 M HCl?

Example 16.7—Keeping the pH Within Specified Limits

Our goal in buffer preparation is often to keep the pH within specific limits upon the addition of a strong acid or base. How many milliliters of 0.100 M HCl may be added to 100.0 mL of a buffer containing 0.150 M each of formic acid (HCOOH) and formate ion ($HCOO^-$) so that the pH will not change by more than 0.20?

Asking the Right Questions

Even though the example asks for mL of HCl, in chemistry, our central reaction-based relationship concerns moles. Therefore, the key question is, "How many moles of HCl can we add to the system in order to keep the pH of the system within 0.20 of the original pH?" Before doing that calculation, we need to know, "What is the original pH, and what pH is 0.20 lower?".

Solution

What are the initial pH and the final pH values? The initial amounts of formic acid and formate ion are equal:

$$\text{mol HCOOH} = \text{mol HCOO}^-$$

$$= 0.1000\,\text{L} \times \frac{(0.150\,\text{mol})}{(1.00\,\text{L})}$$

$$= 0.0150\,\text{mol of each}$$

$$[\text{H}^+] = K_a = 1.8 \times 10^{-4}\,\text{M}$$

$$\text{pH} = \text{p}K_a = 3.74$$

When we add HCl solution, the pH is not to drop below 3.54 (0.20 from the original pH). The concentration of H^+ at this pH is $[H^+]=10^{-3.54}=2.88 \times 10^{-4}$ M. We can solve for the ratio of acid to base:

$$\frac{[\text{H}^+]}{K_a} = \frac{2.88 \times 10^{-4}}{1.8 \times 10^{-4}} = \frac{[\text{HCOOH}]}{[\text{HCOO}^-]} = \frac{1.60}{1}$$

Therefore, retaining an extra figure in this intermediate calculation,

$$[\text{HCOOH}] = 1.60[\text{HCOO}^-]$$

Because the volumes are equal,

$$\text{mol HCOOH} = 1.60(\text{mol HCOO}^-)$$

The total amount of formic acid and formate ion equals 0.0300 mol (it was originally 0.01500 mol each). We can substitute for HCOOH and solve.

$$1.60(\text{HCOO}^-) + \text{HCOO}^- = 0.0300\,\text{mol}$$

$$2.60(\text{HCOO}^-) = 0.0300\,\text{mol}$$

$$\text{mol HCOO}^- = \frac{0.0300\,\text{mol}}{2.60} = 0.0115\,\text{mol}$$

$$\text{mol HCOOH} = 1.60(0.0115\,\text{mol HCOO}^-)$$
$$= 0.0184\,\text{mol}$$

Because our original amounts of HCOOH and $HCOO^-$ were 0.01500 mol each, we can add up to 0.0035 mol of HCl and still have the pH stay within 0.20 of the original pH. We can now determine the maximum HCl solution volume.

$$\text{mL HCl} = 0.0035\,\text{mol HCl} \times \frac{1.000\,\text{L}}{0.1000\,\text{mol}}$$
$$= 0.035\,\text{L} = 35\,\text{mL}$$

We see the importance of working in moles with buffer calculations. Keep in mind that this is possible only because the volumes of the conjugates cancel. This will be the case only when dealing with buffers.

Is our Answer Reasonable?

The addition of HCl lowered the pH (reasonable), and the solution pH stayed within 0.20 of the original pH. This is reasonable for a buffer.

What If…?

…the original concentrations of the formic acid and formate ion were doubled to 0.300 M in the same volume? How many milliliters of 0.100 M NaOH may be added so that the pH will not change by more than 0.20?

(*Ans: 70 mL*).

Practice 16.7

Solve the same problem using the Henderson–Hasselbalch equation instead of the mass-action expression for the buffer.

16

▶ Here's What We Know Thus Far…

- A buffer should not react with the system it is buffering.
- The pK_a of a buffer should be as close as possible to the pH we want to maintain.
- The buffering capacity of a buffer must be sufficient to accommodate the addition of a strong acid or a strong base. ◀

16.2.5 Food for Thought: Are Strong Acids and Bases Buffers?

A buffer, as we noted before, is a mixture of a weak acid and its conjugate base or of a weak base and its conjugate acid. Is it possible that a solution containing only a strong acid (such as HCl) or a strong base (such as NaOH) could be a buffer? For example, a solution of 0.10 M HCl has a pH of 1.0. The solution pH remains close

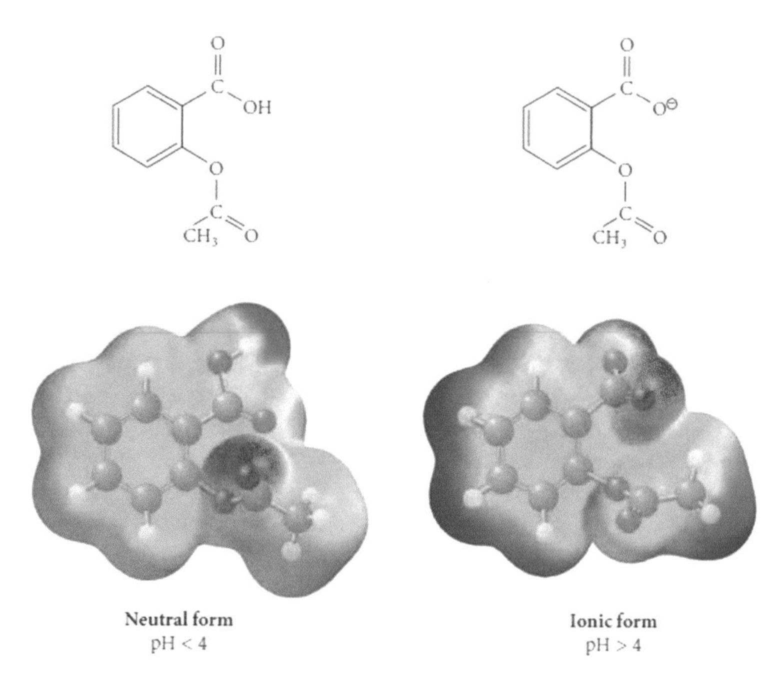

Fig. 16.5 The neutral and ionic forms of aspirin

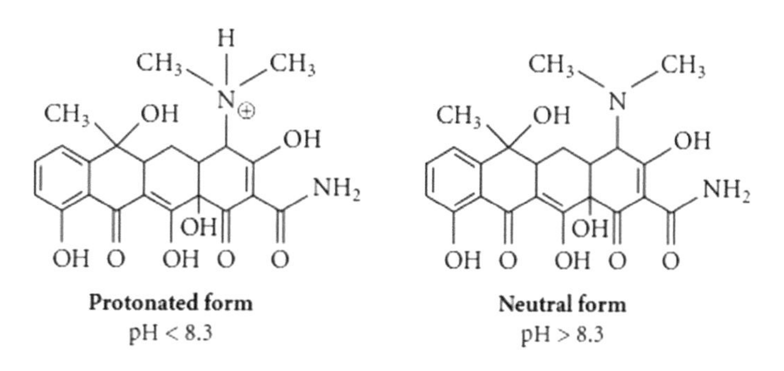

Fig. 16.6 Neutral and ionic forms of tetracycline

to 1.0 even when a little NaOH is added. Similarly, the pH of 0.10 M NaOH remains close to 13.0 even when a little strong acid is added. In that sense, they meet the criterion of keeping the pH of the solution fairly constant upon addition of strong acid or base. However, buffers should also keep their pH constant when diluted, and this is where strong acids and bases fail as buffers. Dilution of a strong acid or a strong base solution causes the pH to change by roughly 1 unit with every tenfold dilution. On the other hand, buffers made from conjugate acid–base pairs show very little change in pH with dilution. This is critical in biochemical processes, which require a fairly constant pH whatever the solution concentration.

The importance of buffers in medicine is illustrated by a class of drugs called antacids. These drugs are compounds that neutralize gastric acid and exert a buffering effect in the stomach for temporary relief of acid indigestion. Similarly, magnesium carbonate is added to certain brands of aspirin specifically to alter stomach pH. The magnesium carbonate, by neutralizing the "gastric juice," increases stomach pH. The net result is that the aspirin exists predominantly in its ionized form, as shown in Fig. 16.5, and reduces the risk of aspirin-induced stomach bleeding or ulcers. Although this is desirable in some cases, it can reduce the amount of drug absorption. Unfortunately, antacids alone may also raise the pH of the stomach above the pK_a of other drugs, resulting in their ionization and in reduced absorption through the stomach. This reduced absorption means that the drug treatment is not effective. Consequently, many prescription medications bear the warning "Do not take antacids containing (hydroxides of) aluminum, calcium, or magnesium (e.g., Mylanta, Maalox) while taking this medication."

The buffering effects of antacids and other buffered medicines can give rise to a number of undesirable drug interactions with prescription medications. Tetracycline, as another example, is an antibiotic drug that is absorbed primarily through the stomach lining in its protonated form, shown in Fig. 16.6, thanks to the normally low pH of the stomach.

NanoWorld/MacroWorld—Big Effects of the Very Small: Buffers in Biochemical Studies and Medicine

Enzymes, which catalyze nearly all the chemical reactions that occur in living organisms, are often highly sensitive to the hydrogen ion concentration of the environment in which they are found. Slight increases or decreases in pH can have a dramatic effect on an enzyme's ability to carry out its unique function. Consequently, living organisms typically have buffering systems in place to maintain a relatively constant pH. Technological advances in biochemistry and molecular biology have enabled chemists to prepare, purify, and study just about any enzyme they desire outside of its normal cellular environment, as long as they can maintain the pH at around 7. For example, members of the class of enzymes called cytochrome P450 are chemically active only between pH 6.5 and pH 8.5, with optimum activities usually occurring between pH 6.8 and pH 7.5. These enzymes are present in the human liver, where they help the body rid itself of foreign chemicals, including most pharmaceuticals.

In 1966, Norman Good and coworkers reported on the design of a dozen new weak acids for preparing buffers to use in biological research. These so-called **Good (or Good's) buffers** are now widely used, because they are fairly chemically stable in the presence of enzymes or visible light and do not interact with biological compounds. Moreover, they are easy to prepare. Several of the Good buffers and boric acid are among the most commonly used in the study of enzyme behavior. Their abbreviated names, structures, and pK_a values are listed in ◻ Table 16.2. An additional set of highly effective biological buffers were synthesized in the late 1990s and are now commercially available. Not too surprisingly, they are called "better buffers."

Good (Good's) buffers – Buffers typically used in biochemical research because they are chemically stable in the presence of enzymes or visible light and are easy to prepare.

Here's What We Know So Far

- Many reactions are pH-sensitive and require buffers to control pH.
- A buffer resists change in pH upon addition of a strong acid or a strong base, or upon dilution.
- Buffers are composed of weak conjugate acid–base pairs.
- We can solve for the pH of buffers in a straightforward way by recognizing the importance of Le Châtelier's principle and the common-ion effect.
- We can calculate the approximate ratio of conjugate acid to base in order to prepare a buffer of a known pH.
- Solving for the pH of a buffer upon addition of strong acid or base is really solving a limiting-reactant problem.
- It is possible to exceed the buffer capacity, in which case the pH will move sharply higher (with excess base) or sharply lower (with excess acid). ◀

We have seen that buffers are used to maintain the pH of all kinds of systems, ranging from municipal scrubbers to our body. Our overarching application of aqueous equilibria in this chapter is analysis of calcium in hard water via titration with EDTA. We are one step closer to completing our task. We next focus on titrations.

16.3 Acid–Base Titrations

We discussed the wide range of titrations, of which acid–base titration is one important type, in the opening section of this chapter. A small sample of the commercial, environmental and biological uses of acid–base titrations includes: analysis of the acidity of food and drink, determination of the pH of water supplies, measurement of the solubility of pharmaceuticals, determination of amino acids in blood, and determination of the acidity or basicity (called the "total acid" or "total base" number) of motor oils.

How do we perform an acid–base titration? In the lab, we typically set up an acid–base titration by monitoring the pH as shown in ◻ Fig. 16.7. This normally includes a buret to accurately measure the volume of titrant delivered, a beaker or flask, and a calibrated pH meter. Industrial laboratories often use automated titrators to increase efficiency. The typical acid–base titrations fall into one of these main categories:

- strong-acid–strong-base titrations
- strong-acid–weak-base titrations
- weak acid–strong base titrations

A fourth type, weak-acid–weak-base titrations, is typically not used because the equilibrium constant for the overall reaction is not nearly as large as with the other systems, and the indication of the end of the titration is too gradual to tell us when the titration is complete.

Table 16.2 Common biological buffers: good buffers and boric acid

Abbreviation	pKa	Structure	Ball and Stick model
MES	6.15	O(ring)N–CH_2–CH_2–S(=O)(=O)–OH	
MOPS	7.20	O(ring)N–CH_2–CH_2–CH_2–S(=O)(=O)–OH	
HEPES	7.55	$HOCH_2CH_2$–N(ring)N–CH_2–CH_2–S(=O)(=O)–OH	

(continued)

Table 16.2 (continued)

Abbreviation	pKa	Structure	Ball and Stick model
Tris	8.30	$H_2N-C(CH_2OH)_3$	
CAPS	10.40	$C_6H_{11}-NHCH_2CH_2CH_2-S(=O)_2-OH$	
Boric acid	9.24	H_3BO_3	

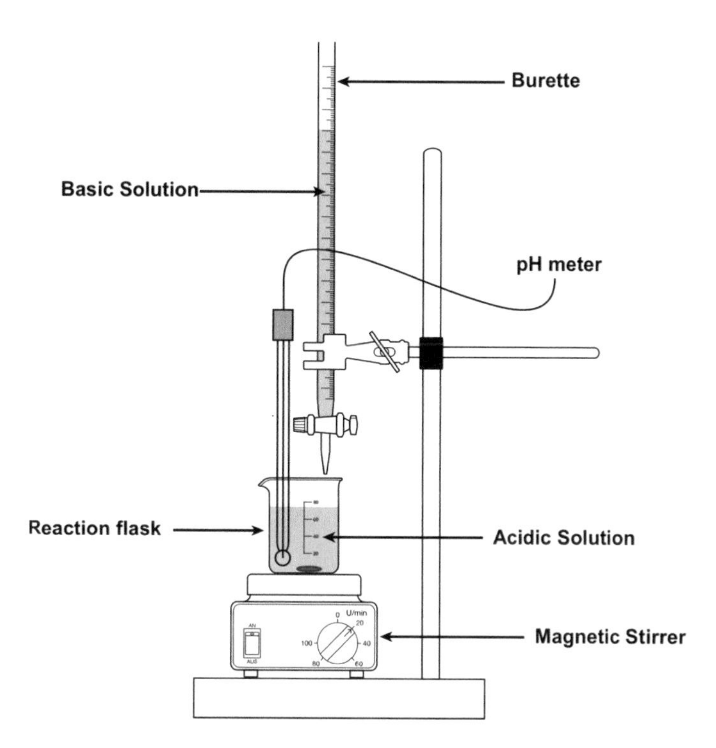

Fig. 16.7 A typical set-up for an acid–base titration monitored by a pH meter includes a buret to accurately measure the volume of titrant delivered, a beaker, and the calibrated pH meter. An automatic titrator is used when there are many titrations to be done

16.3.1 Strong-Acid–Strong-Base Titrations

The determination of HCl molarity based on titration with NaOH is a common process throughout all levels of chemistry and from the academic to the industrial laboratory setting. Let's examine the changes that take place during a strong-acid–strong-base titration by assuming that we wish to titrate 50.00 mL of 0.1000 M HCl by adding known amounts of 0.2000 M NaOH. The results of the titration are graphically shown in Fig. 16.8a–f, which illustrates the relationship between pH and volume of OH^- added. We'll start by determining the initial pH of the analyte solution.

16.3.1.1 Part 1: Initial pH

We will calculate the pH of the initial 0.1000 M solution of HCl as we would that of any other strong acid:

$$pH = -\log\left[H^+\right] = -\log(0.1000) = 1.0$$

We enter this on the graph shown in Fig. 16.8a.

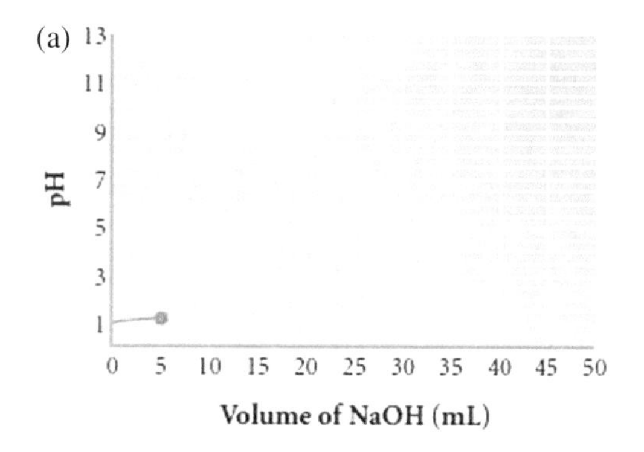

Fig. 16.8 Each plot follows the pH changes as we add 0.2000 M NaOH solution to 50.00 mL of 0.1000 M HCl solution. The initial pH is shown here, followed in turn by the pH after addition of the listed volumes

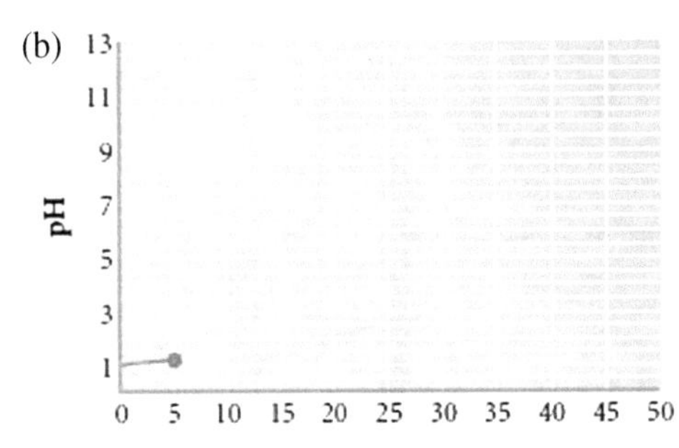

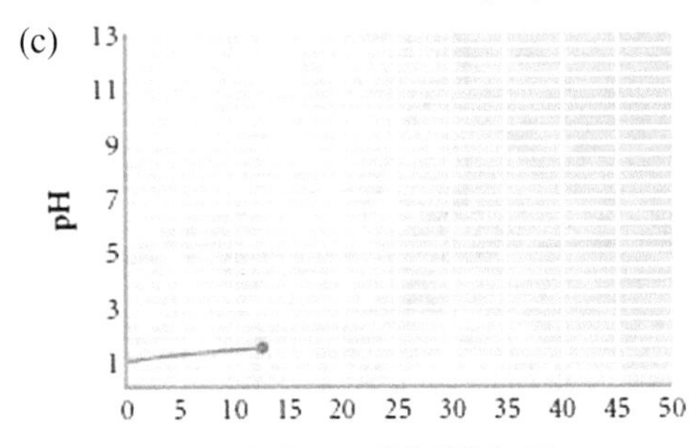

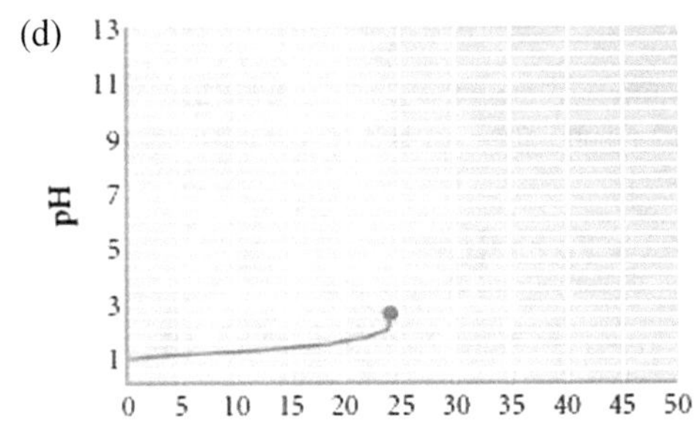

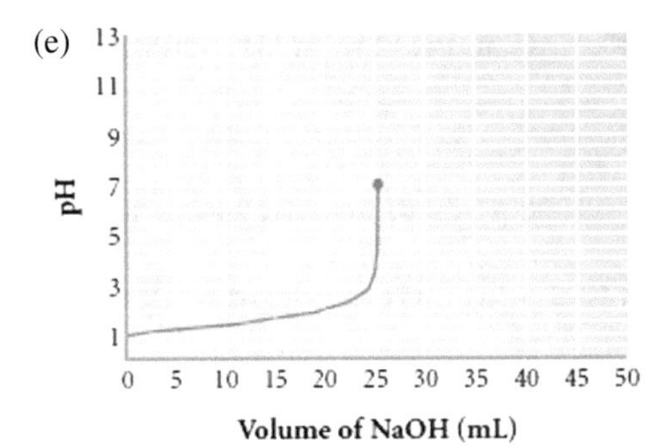

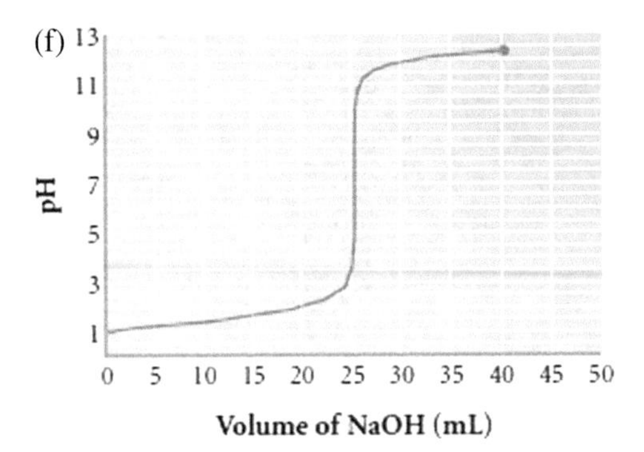

Fig. 16.8 (continued)

16.3.1.2 Part 2: Addition of 5.00 mL of NaOH Solution

What will be the reaction of the strong acid with the strong base? We can write the reaction in molecular form:

$$HCl(aq) + NaOH(aq) \rightleftharpoons H_2O(l) + NaCl(aq)$$

However, the net ionic form gives a better sense of what is going on in the solution:

$$H^+(aq) + OH^-(aq) \rightleftharpoons H_2O(l) \qquad K = 1.0 \times 10^{14}$$

The equilibrium constant is very high and the reaction is also fast, both good features to have when doing a titration. What, and how much, will be left over after addition of the NaOH titrant to the HCl solution? This is a limiting-reactant problem, and we can use the same type of table that we used when discussing buffers to help us sort it all out. Because we are adding sodium hydroxide solution, and changing the volume of the solution being titrated, we will keep track of the moles of acid and base in our tables. For example, the initial concentration of our strong acid is

$$[H^+] = 0.1000\,M = \frac{(0.1000\,mol)}{(1.000\,L)}$$

How many moles of H^+ are in the solution? Using the concentration in moles per liter and the volume in liters, we find that

$$\frac{(0.1000\,mol)}{(1.000\,L)} \times 0.05000\,L = 0.005000\,mol\,H^+$$

We can calculate the moles of strong base, OH^-, added to the solution, in the same way.

$$\frac{(0.2000\,mol)}{(1.000\,L)} \times 0.05000\,L = 0.001000\,mol\,OH^-$$

Putting our values in tabular format shows that we still have plenty of strong acid in the solution. Water is the solvent, so it will not enter into the mass-action expression, and we can ignore it in our calculations.

	$H^+(aq)$	+	$OH^-(aq)$	$\rightleftharpoons$	$H_2O(l)$	$K = 1.0 \times 10^{14}$
Moles initial	0.005000					
Moles added			0.001000			
Change	−0.001000		−0.001000			
Moles at equilibrium	0.004000		≈0			

We can calculate the pH, keeping in mind the total solution volume is now 55.00 mL (50.00 mL of HCl solution + 5.00 mL of added NaOH solution).

$$[H^+] = \frac{(0.004000\,mol)}{(0.05500\,L)} = 0.0727\,M$$

$$pH = 1.14$$

The addition of the strong acid has raised the pH, but not much, because we still have plenty of excess acid. We have entered the data onto our graph in Fig. 16.8b.

16.3.1.3 Part 3: Addition of a Total of 12.50 mL of NaOH Solution

Half of the H^+ is neutralized at this point.

	H^+aq	+	OH^-aq	$\rightleftharpoons$	H_2O	$K = 1.0 \times 10^{14}$
Moles initial	0.005000					
Moles added			0.002500			
Change	−0.002500		−0.002500			
Moles at equilibrium	0.002500		≈0			

The total volume is 62.50 mL (50.00 mL of HCl solution + 12.50 mL of NaOH solution), which results in a pH of 1.4, as shown here and in Fig. 16.8c.

$$[H^+] = \frac{(0.002500\,\text{mol})}{(0.06250\,\text{L})} = 0.0400\,\text{M}$$

$$\text{pH} = 1.40$$

16.3.1.4 Part 4: Addition of a Total of 24.00 mL of NaOH Solution

By the time we add 24.00 mL of 0.2000 M NaOH to the acid, we have neutralized nearly all of the hydronium ion, as shown in the following data table.

	H^+aq	+	OH^-aq	⇌	H_2O	$K = 1.0 \times 10^{14}$
Moles initial	0.005000					
Moles added			0.004800			
Change	−0.004800		−0.004800			
Moles at equilibrium	0.000200		≈0			

The total volume is 74.00 mL (50.00 mL of HCl solution + 24.00 mL of NaOH solution), which results in a pH of 2.57, as shown.

$$[H^+] = \frac{(0.000200\,\text{mol})}{(0.07400\,\text{L})} = 2.70 \times 10^{-3}\,\text{M}$$

$$\text{pH} = 2.57$$

As we get very close to neutralizing the acid, the pH starts to rise sharply.

16.3.1.5 Part 5: Addition of a Total of 25.00 mL of NaOH Solution

At this point, all of the strong acid has been neutralized by the strong base. This is called the **equivalence point** of the titration, the exact point at which the reactant has been neutralized by the titrant.

Equivalence point – The exact point at which the reactant in a titration has been neutralized by the titrant.

	H^+aq	+	OH^-aq	⇌	$H_2O(l)$	$K = 1.0 \times 10^{14}$
Moles initial	0.005000					
Moles added			0.005000			
Change	−0.005000		−0.005000			
Moles at equilibrium	≈0		≈0			

We say that there are "≈ 0" ("about zero") hydrogen and hydroxide ions in the solution. This really says that the amount is insignificant when compared to the original amounts of acid and base that were mixed. How much is "≈ 0" in this neutral solution?

We have 75.00 mL of water at equilibrium. The only reaction that is important in our solution at this point is the autoprotolysis of water. Finally, the equilibrium constant for this reaction has become very important. Using the equation for this reaction, and the corresponding mass-action expression, we can solve for the concentration of H^+ in the solution.

$$H_2O(l) \rightleftharpoons H^+(aq) + OH^-(aq) \qquad K_w = 1.0 \times 10^{-14}$$

$$\left[H^+\right] = \left[OH^-\right] = 1.0 \times 10^{-7}\,\text{M}$$

$$\text{pH} = 7.00$$

The answer to our question is $[H^+] = 1.0 \times 10^{-7}$ M, which is insignificant compared to the initial concentration of strong acid. However, in the absence of other sources of H^+, this is very significant, since it is the ONLY source of H^+. The solution now has a neutral pH, as shown in Fig. 16.8e. Note the sharp rise to the equivalence point, which makes it easy to identify.

A word of caution: the pH at the equivalence point will equal 7.00 only in monoprotic strong-acid–strong-base titrations. We will show why this is so immediately after Example 16.8.

16.3.1.6 Part 6: Addition of a Total of 40.00 mL of NaOH Solution

At this point, the additional strong base is just being added to pure water (neglecting the spectator ions Na^+ and Cl^-, which have no effect on the pH), so we would expect the pH to rise sharply. We are adding 15.00 mL of 0.2000 M NaOH, or 0.00300 mol, past the equivalence point, and our pH is calculated as shown for the total solution volume of 90.00 mL (50.00 mL of HCl solution + 40.00 mL of NaOH added).

$$[OH^-] = \frac{(0.003000\text{ mol})}{(0.09000\text{ L})} = 0.03333\text{ M}$$

$$pOH = 1.48$$

$$pH = 12.52$$

The solution has become quite basic. The complete strong-acid–strong-base **titration curve** is shown in Fig. 16.8f.

Titration curve – A plot of the pH of the solution versus the volume (or concentration) of titrant.

To summarize, we have shown that the addition of a strong base will only slightly increase the pH of a strong acid until very close to the equivalence point, where it will rise sharply to pH = 7. After the equivalence point, the pH will continue its sharp increase to a point that depends on the concentration of the strong-base titrant.

Example 16.8—Titrating Sodium Hydroxide with Hydrochloric Acid

Calculate and draw a graph showing the relationship between the pH and the volume of HCl for the titration of 25.00 mL of 0.2500 M NaOH with the following total volumes of 0.1250 M HCl:

a. 0 mL
b. 20.00 mL
c. 49.80 mL
d. 50.00 mL
e. 60.00 mL.

Asking the Right Questions

What reaction describes the important chemistry that occurs when HCl is added to NaOH? How much of what will remain at each stage of the titration process? How do we determine the pH of the solution at each stage?

Solution

The net ionic reaction is the same as in the addition of NaOH to HCl, except that the reactant (now OH^-) and the titrant (now H^+) have switched roles.

$$OH^-(aq) + H^+(aq) \rightleftharpoons H_2O(l) \quad K = 1.0 \times 10^{14}$$

a. 0 mL of acid added: The pH of this strong base can be found from the hydroxide ion concentration, which is equal to the initial concentration of the NaOH solution.

$$[OH^-] = 0.2500\text{ M} \quad pOH = 0.60 \quad pH = 13.40$$

b. 20.00 mL of acid added: We can use the same table setup that we used previously to calculate the moles of each after reaction. There are initially

$$\frac{(0.2500\text{ mol})}{(1\text{ L})} \times 0.02500\text{ L} = 0.006250\text{ mol OH}^-$$

in the solution. We add

$$\frac{(0.1250\text{ mol})}{(1\text{ L})} \times 0.02000\text{ L} = 0.002500\text{ mol H}^+$$

in a total solution volume of 45.00 mL (25.00 mL of base + 20.00 mL of acid) = 0.04500 L.

	$OH^-(aq)$	+	$H^+(aq)$	$\rightleftharpoons$	$H_2O(l)$	$K = 1.0 \times 10^{14}$
Moles initial	0.006250					
Moles added			0.002500			
Change	−0.002500		−0.002500			
Moles at equilibrium	0.003750		≈ 0			

$$[OH^-] = \frac{(0.003750\,mol)}{(0.04500\,L)} = 0.08333\,M$$

$$pOH = 1.08$$

$$pH = 12.92$$

c. 49.8 mL of acid added: Using the same type of calculations, we find that

	OH^- (aq)	+	H^+ (aq)	$\rightleftharpoons$	$H_2O(I)$	$K = 1.0 \times 10^{14}$
Moles initial	0.006250					
Moles added			0.006225			
Change	–0.006225		–0.006225			
Moles at equilibrium	0.000025		≈ 0			

$$[OH^-] = \frac{(0.0000250mol)}{(0.07480L)} = 3.34 \times 10^{-4}\,M$$

$$pOH = 3.48$$

$$pH = 10.52$$

d. 50.00 mL of acid added:

	OH^- (aq)	+	H^+ (aq)	$\rightleftharpoons$	$H_2O(I)$	$K = 1.0 \times 10^{14}$
Moles initial	0.006250					
Moles added			0.006250			
Change	–0.006250		–0.006250			
Moles at equilibrium	≈ 0		≈ 0			

The solution is at the equivalence point, and pH = 7.0 for a strong-base–strong-acid titration.

e. 60.00 mL of acid added: We have added 10.00 mL of excess acid, which is a total of 0.00125 mol. The total solution volume is 85.00 mL (25.00 mL of base + 60.00 mL of added acid) = 0.08500 L. We can calculate the pH from this information.

$$[H^+] = \frac{(0.001250\,mol)}{(0.08500\,L)} = 0.01471\,M$$

$$pH = 1.83$$

The titration curve shown in Fig. 16.9 has a shape similar to what we might have expected. It is fairly flat until close to the equivalence point, where it drops quite sharply and over a large pH range, so it is easily detectable.

Are our Answers Reasonable?

As we look over our answers, it is good practice to think about where we are in the titration process. We began with a strong base, and the pH was high, a reasonable outcome. Some strong acid was added, but not enough to neutralize the entire amount of strong base, so the pH remained quite basic. Only when enough acid was added to get close to the equivalence point did the pH begin to drop. When excess strong acid was present in the solution, the pH dropped sharply, a reasonable outcome.

What If…?

…the temperature of the reaction were not 24 °C, as we assume for equilibrium processes, but was, rather, 50 °C? At the equivalence point, what would be the pH and why?

(*Ans: Table 16.5 shows that* K_w *increases with temperature. At* 50 °C, $K_w = 5.47 \times 10^{-14}$ *and* $pK_w = 13.26$. *At the equivalence point,* $pH = pOH = 13.26/2 = 6.63$.)

Practice 16.8

Calculate and draw a graph of the relationship between the pH and the volume of NaOH solution for the titration of 25.00 mL of 0.2500 M HCl with these total volumes of 0.2500 M NaOH:

a. 0 mL
b. 10.00 mL
c. 20.00 mL
d. 25.00 mL
e. 30.00 mL
f. 40.00 mL

16.3.2 Acid–Base Titrations in Which One Component is Weak and One is Strong

Although the general problem-solving strategy is the same when we titrate a weak acid with a strong base (or a weak base with a strong acid) as when we perform a strong-acid–strong-base titration, there is a critical difference, the K_a (or K_b) of the analyte, which affects both the pH and the pH change at the equivalence point. Our ability to do a titration analysis depends on having a large, sharp break in the equivalence point. We can best see this by comparing the data from our HCl–NaOH titration with those from the titration of 50.00 mL of a 0.1000 M

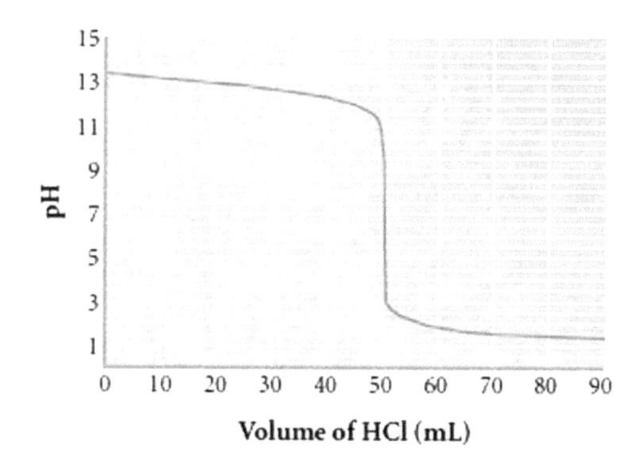

Fig. 16.9 The titration curve for the addition of 0.125 M HCl to 25.00 mL of 0.250 M NaOH

acetic acid solution (CH_3COOH, $K_a = 1.8 \times 10^{-5}$) with a 0.2000 M sodium hydroxide solution. The amounts and concentrations are the same. Only the acid has been changed, from strong to weak.

16.3.2.1 Part 1: Initial pH

We can determine pH in this weak acid as we would in any other weak acid.

$$CH_3COOH(aq) \rightleftharpoons CH_3COO^-(aq) + H^+(aq) \qquad K_a = 1.8 \times 10^{-5}$$

$$K_a = \frac{[H^+][CH_3COO^-]}{[CH_3COOH]} \qquad 1.8 \times 10^{-5} = \frac{x^2}{(0.1000)}$$

$$x = \left[CH_3COO^-\right] = \left[H^+\right] = 1.3 \times 10^{-3}\,M$$

$$pH = 2.87$$

We already see a difference in the titration curve (. Fig. 16.10a) because the weak acid's pH is nearly 2 units higher than the initial pH of the strong acid of the same concentration.

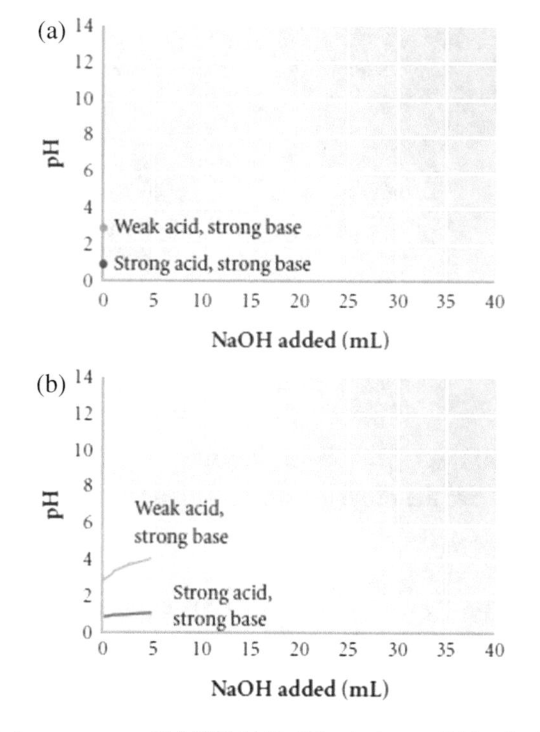

Fig. 16.10 Each plot follows the pH changes as we add 0.2000 M NaOH solution to 50.00 mL of 0.1000 M acetic acid solution. The pH values at each volume are superimposed on those from Fig. 16.9 to show the difference in the nature of the titration curve between the strong acid and the weak acid

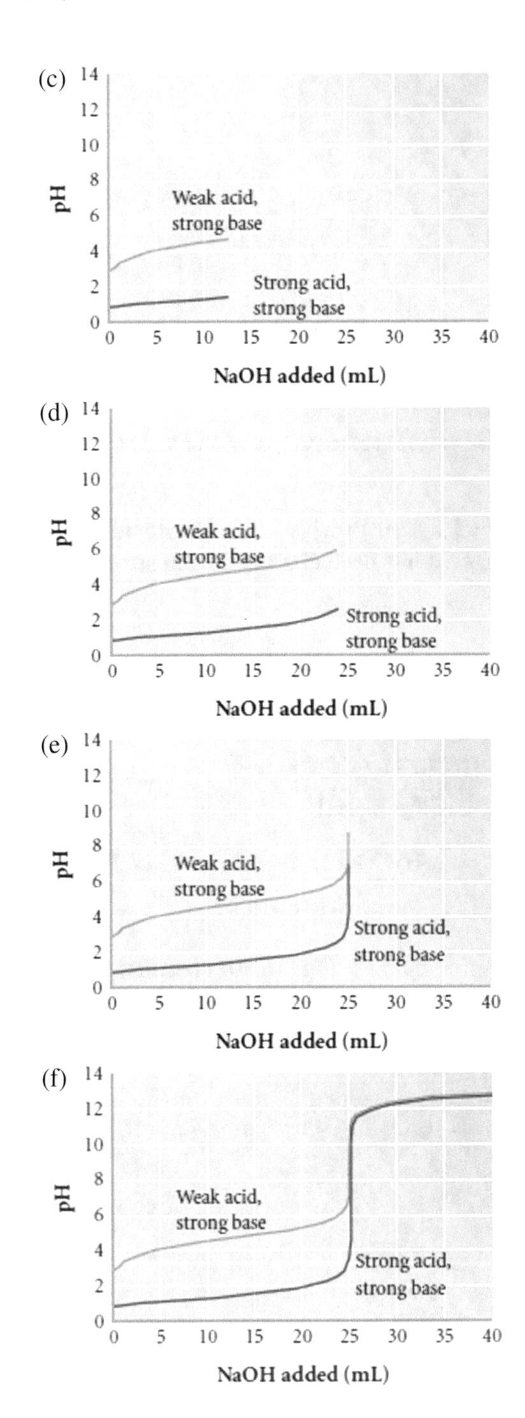

16

. Fig. 16.10 (continued)

16.3.2.2 Part 2: Addition of 5.00 mL of NaOH Solution

We will use the same thinking—and ask the same questions—to assess the chemistry here as in the strong-acid–strong-base titration. We begin with

$$\left(50.00\,\text{mL} \times \frac{0.1000\,\text{mol}}{\text{L}}\right) = 0.005000\,\text{mol acetic acid}$$

to which we add: $\left(5.000\,\text{mL} \times \frac{0.2000\,\text{mol}}{\text{L}}\right) = 0.001000\,\text{mol of OH}^-$.

Question 1: What will be the reaction of the acetic acid with the NaOH solution? The net ionic form of the reaction is useful because spectator ions are not part of the acid–base chemistry.

$$CH_3COOH(aq) + OH^-(aq) \rightleftharpoons CH_3COO^-(aq) + H_2O(l) \qquad K = 1.8 \times 10^9$$

The equilibrium constant is the same whether we describe a reaction by giving its molecular or net ionic form. For this reaction, *K* is quite high and the reaction is fast.

Question 2: What, and how much, will be left over after addition of the titrant? We can use a table to clarify the amounts involved when the system reaches its new equilibrium position.

	$CH_3COOH(aq)$	+	$OH^-(aq)$	⇌	$CH_3COO^-(aq)$	+	$H_2O(l)$
Moles initial	0.005000				≈0		
Moles added			0.001000				
Change	−0.001000		−0.001000		+0.001000		
Moles at equilibrium	0.004000		≈0		0.001000		

Because we have both a weak acid and its conjugate base, we have produced a buffer, and we can solve for the pH as with any buffer system. Note that we have substituted moles into the equation instead of molarities because the total volume of the solution will cancel out.

$$[H^+] = \frac{K_a[CH_3COOH]}{[CH_3COO^-]} = \frac{1.8 \times 10^{-5}(0.004000)}{(0.001000)} = 7.2 \times 10^{-5}\,M$$

$$pH = 4.14$$

We have entered the data point onto our graph in Fig. 16.10b. How is the curve shaping up compared to that of the strong-acid–strong-base titration? Because the solution is a buffer at this point, this part of the titration is called the **buffer region**. We expect the pH to be relatively constant within the buffer region.

Buffer region – The region of a titration indicated by the presence of a weak acid or base and its conjugate. The pH changes little within this region.

16.3.2.3 Part 3: Addition of 12.50 mL of NaOH Solution

Proceeding as we did in part b, we can generate the following table and solve for pH. Again, in our calculation of the hydrogen ion concentration, we needn't worry about the volumes of acid and base because they will cancel out during our calculation.

	$CH_3COOH(aq)$	+	$OH^-(aq)$	⇌	CH_3COO^- (aq)	+	$H_2O(l)$
Moles initial	0.005000				≈0		
Moles added			0.002500				
Change	−0.002500		−0.002500		+0.002500		
Moles at equilibrium	0.002500		≈0		0.002500		

$$[H^+] = \frac{K_a[CH_3COOH]}{[CH_3COO^-]} = \frac{1.8 \times 10^{-5}(0.002500)}{(0.002500)} = 1.8 \times 10^{-5}\,M$$

$$pH = 4.74$$

We are still in the buffer region, having neutralized precisely one-half of the acetic acid and forming an equal amount of acetate ion, the conjugate base. This is called the **titration midpoint**, at which the pH equals the pK_a of acetic acid. It is also possible to work backward, finding the pK_a of an acid from pH at the titration midpoint. As before, we have entered the data point onto our graph (see Fig. 16.10c). How is the curve shaping up compared to that of the strong-acid–strong-base titration?

Titration midpoint – The pH of the titration where the concentration of weak acid or base is equal to the concentration of its conjugate. At this point, the pH is equal to the pK_a of the analyte.

16.3.2.4 Part 4: Addition of a Total of 24.00 mL of NaOH Solution

Continuing, we generate the following table, which shows that we are still— though barely—in the buffer region. The pH is beginning to rise on the way to the equivalence point.

	$CH_3COOH(aq)$	+	$OH^-(aq)$	⇌	CH3COO⁻(aq)	+	$H_2O(l)$
Moles initial	0.005000				≈0		
Moles added			0.004800				
Change	−0.004800		−0.004800		+0.004800		
Moles at equilibrium	0.000200		≈0		0.004800		

$$[H^+] = \frac{K_a[CH_3COOH]}{[CH_3COO^-]} = \frac{1.8 \times 10^{-5}(0.000200)}{(0.004800)} = 7.5 \times 10^{-7} M$$

$$pH = 6.12$$

Continue to keep an eye on the data comparison (◘ Fig. 16.10d) between this titration and the strong-acid–strong-base titration.

16.3.2.5 Part 5: Addition of a Total of 25.00 mL of NaOH Solution

The completed table shows that we are at the equivalence point. We converted all of the acetic acid to acetate ion.

	$CH^3COOH(aq)$	+	$OH^-(aq)$	⇌	CH_3COO^- (aq)	+	$H_2O(l)$
Moles initial	0.005000				≈0		
Moles added			0.005000				
Change	−0.005000		−0.005000		+0.005000		
Moles at equilibrium	≈0		≈0		0.005000		

How do we calculate the pH of this solution? To answer that, we need to go back to our most important questions: "What is in the solution?" and "What equation describes their behavior?".

The answer to the first question is "0.005000 mol of acetate ion is present in a total of 75.00 mL of solution." This is always true: at the equivalence point, the only reactants present in the system are the conjugate base (or acid, depending on what we are titrating) and water. We can calculate the concentration of acetate ion:

$$[CH_3COO^-] = \frac{(0.005000\,\text{mol})}{(0.07500\,\text{L})} = 0.06667\,\text{M}$$

We have neutralized all of the weak acid, and we now have a solution of its conjugate base. This acetate ion will react with water to form acetic acid.

$$CH_3COO^-(aq) + H_2O(l) \rightleftharpoons CH_3COOH(aq) + OH^-(aq)$$

We can calculate the equilibrium constant for the hydrolysis of this conjugate base of acetic acid, K_b, from K_w and the acetic acid K_a.

$$K_b = \frac{K_w}{K_a} = \frac{1.0 \times 10^{-14}}{1.8 \times 10^{-5}} = 5.6 \times 10^{-10}$$

We can now solve for the pH of the weak base.

$$K_b = \frac{[OH^-][CH_3COOH]}{[CH_3COO^-]}$$

$$5.6 \times 10^{-10} = \frac{x^2}{(0.06667)}$$

$$x = [CH_3COOH] = \left[OH^-\right] = 6.1 \times 10^{-6}\,M$$

$$pOH = 5.21$$

$$pH = 8.79$$

The pH of the weak base is greater than 7 at the equivalence point. Compare this to the pH = 7 equivalence point of the strong-acid–strong-base titration. There, only water was able to affect the pH at the equivalence point. In this titration, however, we have an aqueous solution of the conjugate base when we reach the equivalence point. The pH for this titration did rise sharply near the equivalence point, making it easy to determine, as shown in ◻ Fig. 16.10e.

Part 6: Addition of a Total of 40.00 mL of NaOH Solution

We now add a 0.2000 M solution of a strong base to a very weak base. The weak base will not be an important contributor to the total hydroxide ion concentration because it is so weak. The hydroxide ion concentration will be strictly determined by the excess moles of OH^- and the total solution volume in which it is contained.

$$\text{excess mol OH}^- = 0.01500 \times \frac{(0.2000\,\text{mol})}{(1.000\,\text{L})} = 0.003000\,\text{mol}$$

$$[\text{OH}^-] = \frac{(0.003000\,\text{mol})}{(0.09000\,\text{L})} = 0.0333\,\text{M}$$

$$\text{pOH} = 1.48$$

$$\text{pH} = 12.52$$

The pH has increased sharply. Compare this data point with that from the equivalent volume of strong base added to the strong acid in the previous titration, shown in ◻ Fig. 16.10f. They are the same! In both cases, after the equivalence point, we added the same volume of the same concentration of strong base.

16.3.3 Summarizing the Key Ideas of the Weak-Acid–Strong-Base Titration Discussion

There are four key areas that define the titration curve and the information it yields.

Initial pH: For equal concentrations, this will be closer to neutral for weaker acids, further away for stronger acids.

Buffer region: This is where we have generated enough conjugate base, while still having the weak acid available, to produce a buffer. This part of the titration curve will have a gently sloping pH until quite near the equivalence point. An important point in the buffer region is the titration midpoint where we have neutralized exactly half of the weak acid ($\text{pH} = \text{p}K_a$). There is an inflection point that we do not see with strong acid-strong base titrations (compare ◻ Figs. 16.9 and 16.11 in this regard).

Equivalence point: The starting weak acid has been exactly neutralized, and the resulting solution is a weak conjugate base. The stronger the acid (the larger the K_a value), the sharper will be the break at the equivalence point. ◻ Figure 16.11 shows that when the K_a value of the weak acid is too small, the pH break at the equivalence point is too small to be of practical use.

Post-equivalence point: Excess strong base sharply raises the pH of the system.

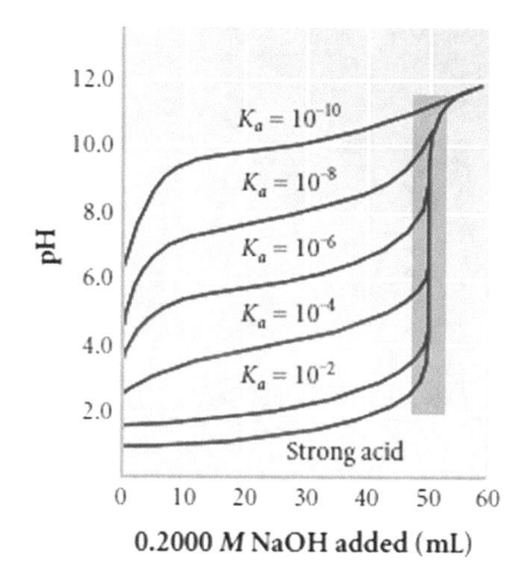

◻ **Fig. 16.11** This plot shows the relationship between K_a value for 50.00 mL of a 0.2000 M acid and the change in pH at the equivalence point when titrated with a 0.2000 M NaOH solution. Notice two items in the plot. First, the concentration of the acid, not its relative strength, determines the equivalence point. Second, the larger the K_a value, the sharper the pH break at the equivalence point

Example 16.9—Titrating a Weak Base with a Strong Acid

To further compare systems, calculate the pH values and draw the curve for the titration of 50.00 mL of 0.1000 M ammonia with 0.2000 M of hydrochloric acid using the same volumes of titrant as in our previous titrations:

a. 0 mL
b. 5.00 mL
c. 12.50 mL
d. 24.00 mL
e. 25.00 mL
f. 40.00 mL

Asking the Right Questions

How does the reaction of the weak base with a strong acid differ from the reaction of a strong base with a strong acid? What are we left with after the addition of each portion of acid? How do we solve for the pH at each point?

Solution

a. Initial pH. The important reaction is the hydrolysis of ammonia.

$$NH_3(aq) + H_2O(l)$$
$$\rightleftharpoons NH_4^+(aq) + OH^-(aq) \quad K_b = 1.8 \times 10^{-5}$$
$$K_b = \frac{[NH_4^+][OH^-]}{[NH_3]}$$
$$1.8 \times 10^{-5} = \frac{x^2}{0.1000}$$
$$x = [OH^-] = 1.34 \times 10^{-3} M$$
$$pOH = 2.87$$
$$pH = 11.13$$

b. Addition of 5.00 mL of HCl solution. The system is thrown out of equilibrium by the addition of the HCl solution, much as the acetic acid system had to compensate for the addition of NaOH. We will generate a buffer, shown in the table.

	$NH_3(aq)$	$\rightleftharpoons$	$NH_4^+(aq)$	+	$OH^-(aq)$	K_b = 1.8 10–5
Moles initial	0.005000				≈ 0	
Moles added			0.001000			
Change	–0.001000		–0.001000		+0.001000	
Moles at equilibrium	0.004000		≈ 0		0.001000	

When the system returns to equilibrium,

$$NH_3(aq) \rightleftharpoons NH_4^+(aq) + OH^-(aq) \quad K_b = 1.8 \times 10^{-5}$$
$$[OH^-] = \frac{K_b[NH_3]}{[NH_4^+]}$$
$$\frac{(1.8 \times 10^{-5})(0.004000)}{(0.001000)} = 7.2 \times 10^{-5} M$$
$$pOH = 4.14$$
$$pH = 9.86$$

c. Addition of 12.50 mL of HCl solution. Just as above, we can generate the following table and solve for pH.

	$NH_3(aq)$	+	$H^+(aq)$	$\rightleftharpoons$	$NH_4^+(aq)$	K_b = 1.8 10–5
Moles initial	0.005000				≈ 0	
Moles added			0.002500			
Change	– 0.002500		–0.002500		+0.002500	
Moles at equilibrium	0.002500		≈ 0		0.002500	

When the system returns to equilibrium,

$$NH_3(aq) \rightleftharpoons NH_4^+(aq) + OH^-(aq) \quad K_b = 1.8 \times 10^{-5}$$
$$[OH^-] = \frac{K_b[NH_3]}{[NH_4^+]}$$
$$\frac{(1.8 \times 10^{-5})(0.002500)}{(0.002500)} = 1.8 \times 10^{-5} M$$
$$pOH = pK_b = 4.74$$
$$pH = 9.26$$

d. Addition of a total of 24.00 mL of HCl solution. We are nearing the equivalence point. How do the various titration curves, shown in ▣ Fig. 16.12, compare?

	$NH_3(aq)$	+	$H^+(aq)$	⇌	$NH_4^+(aq)$	$K_b = 1.8\ 10–5$
Moles initial	0.005000				≈ 0	
Moles added			0.004800			
Change	–0.004800		–0.004800		+0.004800	
Moles at equilibrium	0.000200		≈ 0		0.004800	

When the system returns to equilibrium,

$$NH_3(aq) \rightleftharpoons NH_4^+(aq) + OH^-(aq) \qquad K_b = 1.8 \times 10^{-5}$$

$$[OH^-] = \frac{K_b[NH_3]}{[NH_4^+]}$$

$$\frac{(1.8 \times 10^{-5})(0.000200)}{(0.004800)} = 7.5 \times 10^{-7}\,M$$

$$pOH = 6.12$$

$$pH = 7.87$$

e. Addition of a total of 25.00 mL of HCl solution. We are at the equivalence point, at which we have added exactly the same number of moles of HCl as there were moles of ammonia at the start of the titration. We now have a solution that is weakly acidic as a consequence of the conversion of ammonia to the ammonium ion.

	$NH_3(aq)$	+	$H^+(aq)$	⇌	$NH_4^+(aq)$	$K_b = 1.8\ 10–5$
Moles initial	0.005000				≈ 0	
Moles added			0.005000			
Change	–0.005000		–0.005000		+0.005000	
Moles at equilibrium	≈ 0		≈ 0		0.005000	

$$[NH_4^+] = \frac{(0.005000\,\text{mol})}{(0.07500\,\text{L})} = 0.06667\ \text{M}$$

$$K_a = \frac{K_w}{K_b} = \frac{1.0 \times 10^{-14}}{1.8 \times 10^{-5}} = 5.6 \times 10^{-10}$$

We can now solve for the pH of the weak acid.

$$K_a = \frac{[H^+][NH_3]}{[NH_4^+]}$$

$$5.6 \times 10^{-10} = \frac{x^2}{0.06667}$$

$$x = [NH_3] = [H^+] = 6.1 \times 10^{-6}\,\text{M}$$

$$pH = 5.21$$

f. Addition of a total of 40.00 mL of HCl solution.

$$\text{mol excess H}^+ = 0.01500\,\text{L} \times \frac{(0.2000\,\text{mol})}{(1.000\,\text{L})}$$

$$= 0.003000\,\text{mol}$$

$$[H^+] = \frac{(0.003000\,\text{mol})}{(0.09000\,\text{L})} = 0.0333\,\text{M}$$

$$pH = 1.48$$

Are our Answers Reasonable?

The data conform to those in Figs. 16.11 and 16.12 (see below) for the titration of a weak base with a strong acid, showing the characteristic buffer region, equivalence point and post-equivalence point pH ranges. Note how the pH changes show the opposite trends to those in Fig. 16.10 a–f. These outcomes suggest that the answers are reasonable.

What If…?

…the concentration of HCl were doubled to 0.4000 M? What volume of HCl would have been required to reach the equivalence point?
(*Ans:* 12.50 mL).

Practice 16.9

Calculate the pH values and draw the curve for the titration of 25.00 mL of 0.2500 M acetic acid with 0.2500 M sodium hydroxide using the same volumes of titrant as in our previous titrations (K_a (acetic acid) = 1.8×10^{-5}):

a. 0 mL
b. 5.00 mL
c. 12.50 mL
d. 24.00 mL
e. 25.00 mL
f. 40.00 mL.

We have seen that a successful monoprotic acid–base titration has an analyte (what we are titrating) that, when titrated with a strong acid or strong base, has a large *K*. The reaction is typically fast and reproducible. Polyprotic acid and base titrations are common, and the analysis of their titration curves presents interesting challenges, although we will not deal with these here.

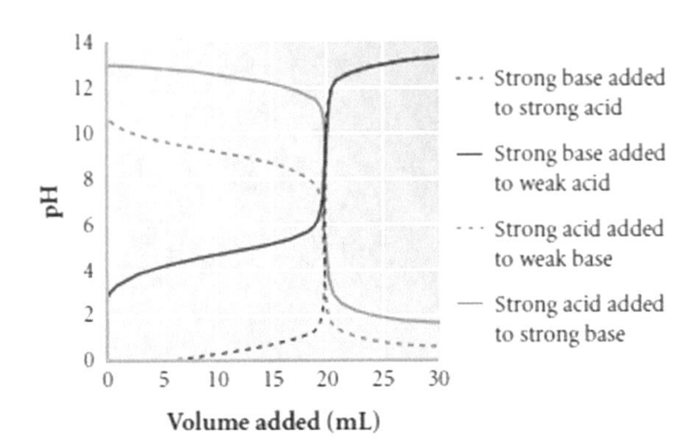

Fig. 16.12 This plot shows a comparison of the pH for strong-acid–strong-base, weak-acid–strong-base, and weak-base–strong-acid titrations using equal concentrations of acid and base

16.3.4 Indicators

How do we know when we have reached the equivalence point in an acid–base titration? The pH meter can provide very accurate readings of the pH. Making successful a measurement using a pH meter, however, relies on its having been properly maintained, calibrated, and operated. For speed and simplicity in a titration, especially when out in the field, the chemical technician often relies on a chemical pH reporter. This class of compounds, known as **acid–base indicators** or **pH indicators**, visually indicates the change in pH as we approach, reach, and pass the equivalence point. Indicators themselves are conjugate acid–base pairs of organic molecules that change color as they change between their acid and base forms. Perhaps the best known example is phenolphthalein, which changes from colorless to rose-pink as the pH changes from 8.5 to 9.5. Figure 16.13 shows that phenolphthalein actually has several structural changes (and therefore color changes) from very low to very high pH.

Acid–base indicator – A compound that changes color on the basis of the pH of the solution in which it is dissolved. The color change is often a result of structural changes due to protonation or deprotonation of acidic groups within the compound.

pH indicator – A compound that changes color on the basis of the acidity or basicity of the solution in which it is dissolved. Also known as an acid–base indicator.

Why is phenolphthalein such a popular acid–base indicator? Its color change occurs in the pH region where many acids titrated with strong bases reach their equivalence point. For example, acetic acid has an equivalence point, at which it is essentially all changed to acetate ion, at about pH 9, and the pH goes from about 6–11 within a small volume of titrant on either side of the equivalence point. The titration of a strong acid with strong base (or vice versa) also changes very rapidly in the range of pH 6 to 11. An indicator that itself changes color anywhere in this area would, generally speaking, be a suitable indicator of the equivalence point.

16

In a given titration, how do we know which acid–base indicator to choose? We want to choose an indicator with a pK_a as close as possible to the pH at the equivalence point. A distinct color change is also useful. For example, the pK_a of phenolphthalein is about 9.5. This is well within the region of the pH at the equivalence point of titration of acetic acid by sodium hydroxide (pH $\approx$ 9, depending on acetic acid concentration). As a rule of thumb, the color changes of pH indicators are visible to $+/-1$ pH unit on either side of the indicator pK_a. This means that phenolphthalein will be completely colorless below a pH of 8.5 and will then be rose-pink from pH 8.5 to about pH 11. It will turn colorless again above pH 11. A selection of common pH indicators, their color changes, and their pH ranges is given in Fig. 16.14.

During the chemical technician's analysis, only a few drops of an indicator solution are added to the analyte solution. Why so little? Because the indicator also undergoes an acid–base reaction. This means that in addition to adding the required volume to neutralize the analyte, we add just a bit more to cause the color change in the indicator. For example, if we titrate a solution of acetic acid containing phenolphthalein indicator, we might need 35.27 mL of a sodium hydroxide solution to react with the acetic acid itself. Changing the indicator's structure (and therefore its color) might require an additional 0.02 mL of the strong base. The equivalence point is at 35.27 mL, but the point at which we see the change in indicator color that tells us the titration is finished, which is called the **titration endpoint**, is at 35.29 mL. Therefore, we would use 35.29 mL as our number for calculating the concentration of analyte. This could lead to very large errors if we had a lot of extra indicator in the solution. These errors are reduced if the endpoint is as close as possible to the equivalence point.

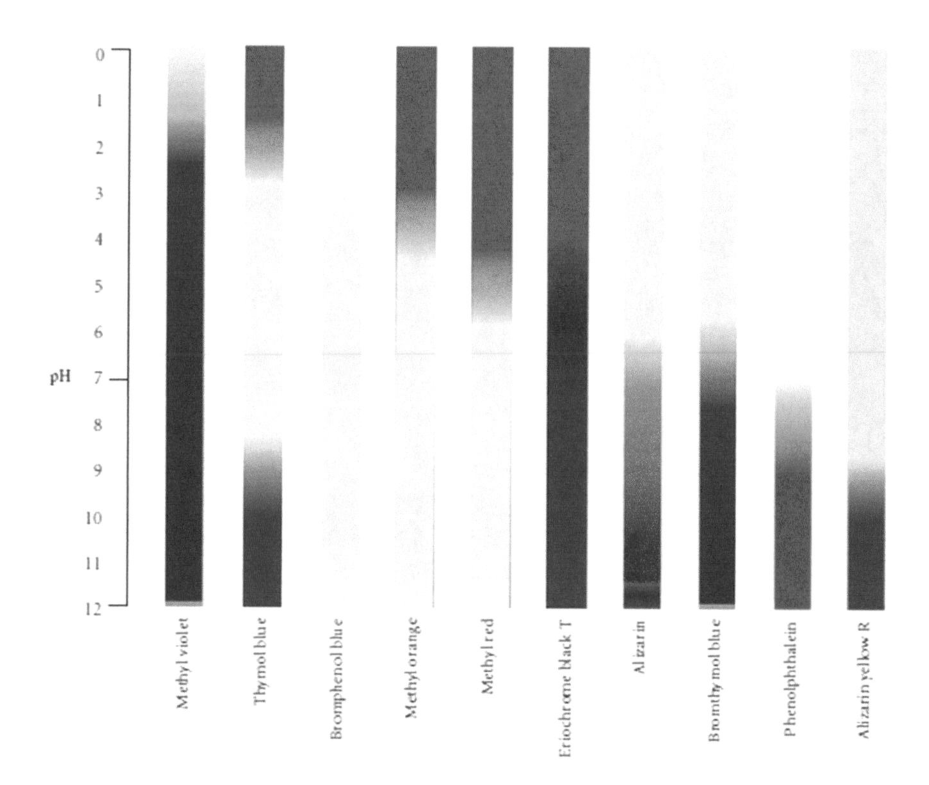

Fig. 16.14 A selection of common pH indicators, their color changes, and their pH ranges. Our key criterion in selecting the proper indicator is that its pK_a should be as close as possible to the equivalence point of the titration in which it is used. Note that most indicators exhibit a color change over less than 2 pH units

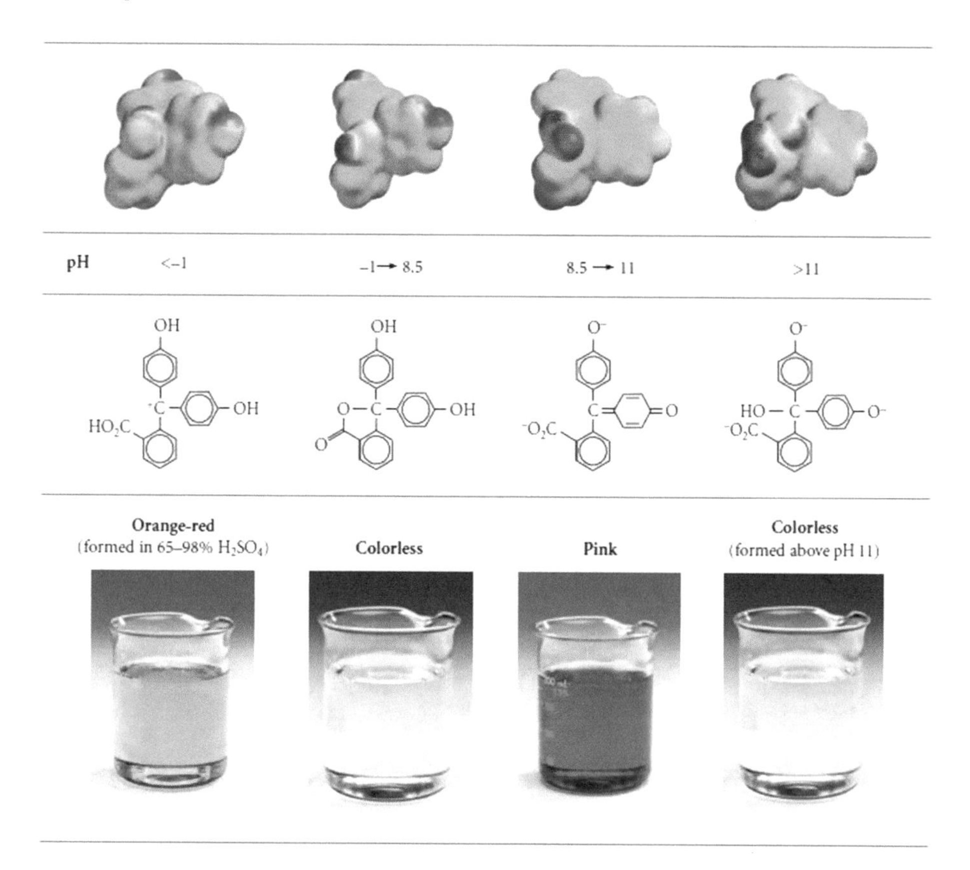

Fig. 16.13 The structure of phenolphthalein changes as we increase the pH from very low to very high

Titration endpoint – The volume and pH at which the indicator has changed color during a titration. The endpoint is not always the same as the equivalence point.

Example 16.10—Picking an Indicator

What would be a reasonable indicator for the titration of 0.10 M ammonia with 0.10 M HCl?

Asking the Right Questions

The key question is "What is the pH of the solution at the equivalence point?".

Solution

When essentially all of the ammonia is converted to ammonium ion, the pH will be around 5.2, as we saw in Exercise 16.9, part e. As shown in ◘ Fig. 16.14, several indicators change color around pH = 5.2, and methyl red appears to be a good choice.

Is our Answer Reasonable?

Based on the color change in the region of the equivalence point, the answer is reasonable.

What If…?

…we didn't have methyl red handy. What would be an alternative?

(Ans: Methyl orange would be a good choice, as its color changes in the range of pH of the equivalence point of the titration.)

Practice 16.10

What would be a reasonable indicator for the titration of 0.10 M NaOH with 0.10 M HCl?

Some natural and commercially prepared indicators are made up of several colorful organic molecules, and they change color throughout the pH range. Notable among the natural indicators are the **anthocyanins**, which are responsible for most of the different colors found in vegetables and flowers. There are over 150 naturally occurring anthocyanins in foods such as the red cabbage that we cook and serve with dinner. The juice from the red cabbage can therefore be used as an indicator. ◘ Figure 16.15 shows the wide range of colors that can be obtained by adjusting the pH of red cabbage juice. Commercially prepared solutions of mixtures of indicators can mimic these color changes, but far more intensely, so far less is required. For instance, the commercially prepared "universal indicator" is a mixture of thymol blue, methyl red, bromthymol blue, and phenolphthalein. Because each indicator gains or loses protons at a different pH, the universal indicator has color changes over a wide pH range, as shown in ◘ Fig. 16.16.

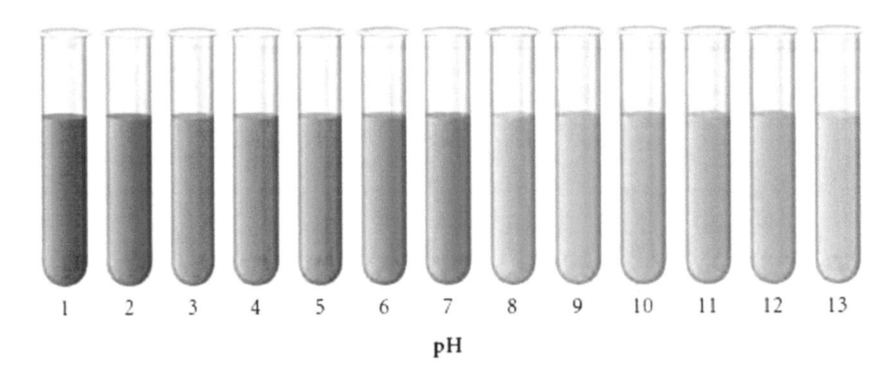

◘ **Fig. 16.15** The spectrum of colors in each sample of red cabbage juice is caused by the changes in structure of the anthocyanins as the pH moves from 1 to 13

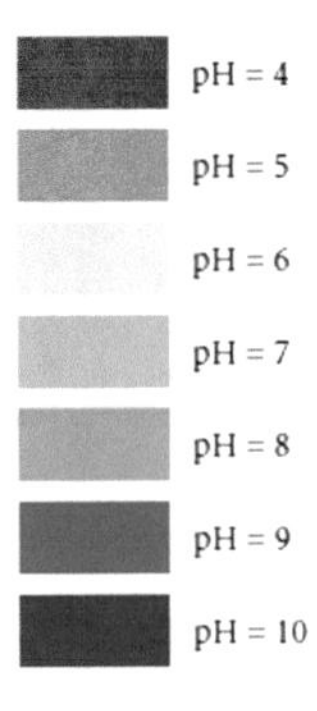

◘ **Fig. 16.16** A typical universal indicator is a mixture of several common indicators

Anthocyanins – A naturally occurring class of compounds responsible for many of the colors of plants. These compounds often act as acid–base indicators.

▶ Here's what we know so far

- Strong-acid–strong-base titrations show a relatively level pH until near the equivalence point, where the pH rises dramatically.
- Titration curves in which one component is weak and the other is strong contain four regions: the initial pH, the buffer region, the equivalence point region and the post–equivalence point region.
- The buffer region contains a point at which one-half of the analyte has been converted to its conjugate. This is called the titration midpoint, and the pH at this point is equal to the p*K* of the analyte.
- The larger the p*K* of the analyte, the sharper will be the change in pH at the equivalence point.
- We can use an indicator to "see" the equivalence point of a titration.
- We add only a few drops of an indicator to the titration solution so that the equivalence point and titration endpoint can be as close together as possible. ◀

16.4 Solubility Equilibria

The Pacific Ocean is an incredibly complex heterogeneous system. The bottom layers of this and other massive waterways are covered with a variety of soils and sediments, including **calcareous oozes**, calcium-containing detritus from dead single-celled, calcium-based sea life. One of the important compounds within the oozes is calcium carbonate ($CaCO_3$), some of which is in contact with ocean water, dissociating to form calcium and carbonate ions. The equation relating this dissociation is written so that the solid is a reactant and the dissolved ions are products:

$$CaCO_3(s) \rightleftharpoons Ca^{2+}(aq) + CO_3^{2-}(aq)$$

Calcareous oozes – The calcium-containing detritus from dead single-celled, calcium-based sea life.

The mass-action expression that can be used to determine the solubility of $CaCO_3$ is called the **solubility product** when the equation is written as shown above. The solubility product is equal to the product of the concentrations of the ions (remember that the $CaCO_3$ is a solid and is not written as part of the mass-action expression).

$$K_{sp} = \left[Ca^{2+}\right]\left[CO_3^{2-}\right]$$

Solubility product – The mass-action expression for the solubility reaction of a sparingly soluble salt.

This equilibrium constant is called the **solubility product constant (Ksp)** and has the same conceptual meaning as any other equilibrium constant along with its mass-action expression. Because the values of K_{sp} tend to be very small, the concentrations of ions are quite low. ◘ Table 16.3 lists representative K_{sp} values for some of the sparingly soluble salts.

Solubility product constant (K_{sp}) – The constant that is part of the solubility product mass-action expression. Typically, K_{sp} values are much less than 1.

The difficulty with describing solubility using a single mass-action expression is that there are so many other processes that enter into the chemistry that our typically simple mass-action expression often just won't do. The simple calculations we can perform do not always agree with what we observe in real systems. Let's take a look at the calcium carbonate system in a somewhat nonmathematical approach as we discover the factors that affect the solubility of solids.

Table 16.3 Selected K_{sp} values at 25 °C

Ionic solid	K_{sp} (at 25 °C)	Ionic solid	K_{sp} (at 25 °C)	Ionic solid	K_{sp} (at 25 °C)
Fluorides		Hg_2CrO_4	2×10^{-9}	$Co(OH)_2$	2.5×10^{-16}
BaF_2	2.4×10^{-5}	$BaCrO_4$	8.5×10^{-11}	$Ni(OH)_2$	1.6×10^{-16}
MgF_2	6.4×10^{-9}	Ag_2CrO_4	9.0×10^{-12}	$Zn(OH)_2$	4.5×10^{-17}
PbF_2	4.0×10^{-8}	$PbCrO_4$	2×10^{-16}	$Cu(OH)_2$	1.6×10^{-19}
SrF_2	7.9×10^{-10}	**Carbonates**		$Hg(OH)_2$	3×10^{-26}
CaF_2	4.0×10^{-11}	$NiCO_3$	1.4×10^{-7}	$Sn(OH)_2$	3×10^{-27}
Chlorides		$CaCO_3$	8.7×10^{-9}	$Cr(OH)_3$	6.7×10^{-31}
$PbCl_2$	1.6×10^{-3}	$BaCO_3$	1.6×10^{-9}	$Al(OH)_3$	2×10^{-32}
$AgCl$	1.6×10^{-10}	$SrCO_3$	7×10^{-10}	$Fe(OH)_3$	4×10^{-38}
Hg_2Cl_2	1.1×10^{-18}	$CuCO_3$	2.5×10^{-10}	$Co(OH)_3$	2.5×10^{-43}
Bromides		$ZnCO_3$	2×10^{-10}		
PbB_{r2}	4.6×10^{-6}	$MnCO_3$	8.8×10^{-11}	**Sulfides**	
$AgBr$	5.0×10^{-13}	$FeCO_3$	2.1×10^{-11}	MnS	2.3×10^{-13}
Hg_2Br_2	1.3×10^{-22}	Ag_2CO_3	8.1×10^{-12}	FeS	3.7×10^{-19}
Iodides		$CdCO_3$	5.2×10^{-12}	NiS	3×10^{-21}
PbI_2	1.4×10^{-4}	$PbCO_3$	1.5×10^{-15}	CoS	5×10^{-22}
AgI	1.5×10^{-16}	$MgCO_3$	1×10^{-15}	ZnS	2.5×10^{-22}
Hg_2I_2	4.5×10^{-29}	Hg_2CO_3	9.0×10^{-15}	SnS	1×10^{-26}
Sulfates		**Hydroxides**		CdS	1.0×10^{-28}
$CaSO_4$	6.1×10^{-5}	$Ba(OH)_2$	5.0×10^{-3}	PbS	7×10^{-29}
Ag_2SO_4	1.2×10^{-5}	$Sr(OH)_2$	3.2×10^{-4}	CuS	8.5×10^{-45}
$SrSO_4$	3.2×10^{-7}	$Ca(OH)_2$	1.3×10^{-6}	Ag_2S	1.6×10^{-49}
$PbSO_4$	1.3×10^{-8}	$AgOH$	2.0×10^{-8}	HgS	1.6×10^{-54}
$BaSO_4$	1.5×10^{-9}	$Mg(OH)_2$	8.9×10^{-14}	**Phosphates**	
Chromates		$Mn(OH)_2$	2×10^{-15}	Ag_3PO_4	1.8×10^{-18}
$SrCrO_4$	3.6×10^{-5}	$Cd(OH)_2$	2.5×10^{-14}	$Sr_3(PO_4)_2$	1×10^{-31}
		$Pb(OH)_2$	1.2×10^{-15}	$Ca_3(PO_4)_2$	1.3×10^{-32}
		$Fe(OH)_2$	1.8×10^{-15}	$Ba_3(PO_4)_2$	6×10^{-39}
				$Pb_3(PO_4)_2$	1×10^{-54}

16.4.1 Side Reactions that Affect Our Reaction of Interest

The solubility of many ions, when dissolved in an aqueous system, is affected by side reactions. The solubility of calcium carbonate in a large ocean-based system is no exception (Fig. 16.17).

1. Hydrolysis of carbonate ion. The carbonate ion formed from the dissolution of calcium carbonate is a base that reacts with water to form bicarbonate ion and hydroxide ion.

$$CO_3^{2-}(aq) + H_2O(l) \overset{K_{b_1}}{\rightleftharpoons} HCO_3^-(aq) + OH^-(aq)$$

   Because the carbonate ion is involved in this reaction, some of it is removed from the calcium carbonate solubility equilibrium. The net result, in accordance with Le Châtelier's principle, is that more of the calcium carbonate dissolves than we would predict.

2. The interplay between the atmosphere and the ocean water. Dissolved CO_2 from the air mixes with ocean water to form carbonic acid and, ultimately, hydrogen ion and bicarbonate ion.

$$CO_2(aq) + H_2O(l) \overset{K}{\rightleftharpoons} H_2CO_3(aq) \overset{K_{a_1}}{\rightleftharpoons} H^+(aq) + HCO_3^-(aq)$$

■ Fig. 16.17 Many processes, including the formation of bicarbonate ion and the reaction of hydrogen and hydroxide ions, affect the solubility of calcium carbonate in the ocean. These stromatolites are formations of calcium carbonate. Photo by Vincent Poirier *Source* ▶ https://upload.wikimedia.org/wikipedia/commons/9/91/20130118-HighbormeCay-Stromatolite-03.JPG

This interaction generates the bicarbonate ion, which influences the carbonate ion equilibrium shown in the hydrolysis of carbonate ion step above. The net result is to reduce the effect of the interaction of carbonate ions with water. The exact change depends on the amount of carbon dioxide dissolved in seawater.

3. The formation of water from the reaction of hydrogen and hydroxide ions. These are produced from the processes described in reactions 1 and 2.

$$H^+(aq) + OH^-(aq) \overset{1/K_w}{\rightleftharpoons} H_2O(l)$$

The effect of this reaction is an increase in the concentration of the bicarbonate ion generated from the carbon dioxide equilibrium.

As we can see, these three equilibria interact with others in the ocean as part of a remarkably complex system in which temperatures and concentrations change, making the calculation of calcium carbonate solubility in the ocean most challenging.

Issues and Controversies—Side Reactions and Activities

Often, in chemistry as in other sciences, situations are more complex than they appear. Shifting our focus from the oceans to a more controlled setting, we find that side reactions can still confound apparently simple systems. Consider the solubility of lead(II) iodide (PbI_2) in distilled water. The amount of precipitated lead(II) iodide is related to the initial concentration of iodide as shown in ■ Fig. 16.18.

Not accounting for these changes can lead to significant errors in calculating the solubility of lead iodide. We can take a simple view of the solubility of a salt such as calcium sulfate by considering only the dissociation reaction:

$$CaSO_4(s) \overset{K_{sp}}{\rightleftharpoons} Ca^{2+}(aq) + SO_4^{2-}(aq)$$

Our mass-action expression is:

$$K_{sp} = \left[Ca^{2+}\right]\left[SO_4^{2-}\right]$$

However, Meites, Pode, and Thomas wrote as early as 1966 that even in the absence of side reactions, the concentrations we would calculate would be wrong by about 60%, largely because of the tendency of the calcium and sulfate ions to stay together as individual **ion pairs**. When the small amount of calcium sulfate dissolves, most of it does not form individual ions. Rather, the ions associate intimately with each other. This is especially important in salts of highly charged (±2 or ±3) cations and anions. The bottom line is that even in the absence of side reactions, such as the addition of H^+ to SO_4^{2-} to make HSO_4^- in acidic solution, there are several factors that affect solubility at the molecular level, making many such calculations challenging. These factors include.

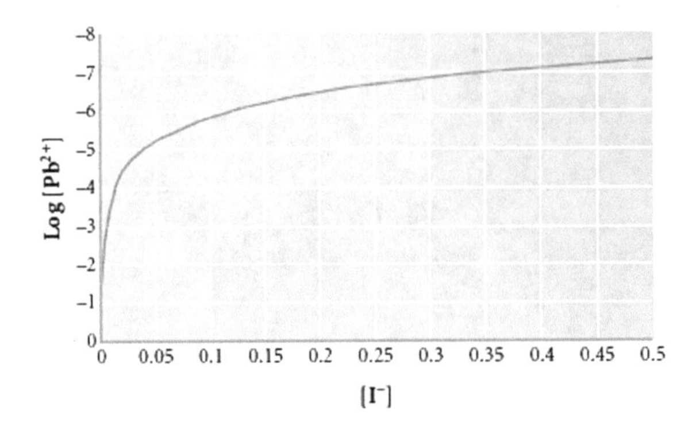

Fig. 16.18 Even a system that seems as simple as the solubility of lead(II) iodide isn't that simple. Most of the Pb^{2+} ions precipitate upon the addition of a small amount of iodide. However, a significant concentration of Pb^{2+} remains

Ion pair – Ions in solution that associate as a unit.

1. Formation of ion pairs, as mentioned in the previous paragraph,
2. Ion activities, a measure of the effective concentration of ions in solution,
3. Thermodynamic measures, including enthalpy and entropy changes in the solution process, which we explore in a later chapter.

Even though solubility systems can actually be quite complex, we will focus in the text on chemicals that conform much closer to our expectations so we can practice applying our knowledge of aqueous equilibrium systems. Even in the more complex systems like we've mentioned above, the principles of equilibrium still dictate the outcomes, but sometimes other principles affect things significantly.

16.4.2 Calculating Molar Solubility

Some simple univalent (singly charged cation and anion) systems do give reasonable answers when we do solubility calculations. In these cases, and others as well, the total number of moles of solute that dissolve per liter of solution is often called the molar solubility. For example, we can calculate the **molar solubility** of silver bromide (AgBr) by using its solubility product constant and mass-action expression.

16

$$\text{AgBr(s)} \rightleftharpoons \text{Ag}^+(\text{aq}) + \text{Br}^-(\text{aq}) \qquad K_{\text{sp}} = 5.0 \times 10^{-13}$$

$$K_{\text{sp}} = \left[\text{Ag}^+\right]\left[\text{Br}^-\right]$$

Molar solubility – The total number of moles of solute that dissolve per liter of solution.

We can set up our table, as we've done before. Although the solid AgBr will not enter into the mass-action expression, we'll include it in the K_{sp} ICE tables because the number of moles of silver bromide that will dissolve into solution and the number of moles of silver and bromide ions produced will all be equal, because there is a 1-to-1-to-1 mol ratio in the reaction. We can designate the molar solubility of AgBr as *s*, in which case the equilibrium concentrations, $[Ag^+]$ and $[Br^-]$, will also be *s*.

	AgBr(s)	⇌	$Ag^+(aq)$	+	$Br^+(aq)$
Initial	–				
Change	$-s$		$+s$		$+s$
Equilibrium	–		s		s

$$K_{sp} = [Ag^+][Br^-]$$
$$5.0 \times 10^{-13} = s^2$$
$$s = 7.1 \times 10^{-7}M = [Ag^+] = [Br^-]$$
$$s = \text{molar solubility of AgBr} = 7.1 \times 10^{-7}\text{M}$$

Qualitatively, what would we expect to happen to the solubility if we added a little sodium bromide, in which the added bromide is a common ion? According to Le Châtelier's principle, addition of an ion common to the product would push the dissociation reaction back to the left, decreasing the solubility. Caution: If we add to much bromide ion, the solubility of silver bromide could actually increase, as a consequence of the formation of soluble species such as $AgBr_2^-$.

Example 16.11—How Much Dissolves?

One method of analyzing groundwater for nitrate requires that the chloride ions in the water sample be removed first. This is typically done by adding a solution of silver ions (Ag^+) in order to precipitate the sparingly soluble silver chloride salt (AgCl). What silver ion concentration is present when silver chloride is added to water?

$$AgCl(s) \rightleftharpoons Ag^+(aq) + Cl^-(aq)$$
$$K_{sp} = 1.6 \times 10^{-10}$$

Asking the Right Questions

How is the silver ion content related to the solubility of silver chloride? How do we solve for the solubility of silver chloride?

Solution

The silver ion concentration in the solution is due to the solubility of silver chloride, so solving for the solubility will give us the silver ion concentration. Setting up the table, we get:

	AgCl(s)	⇌	Ag^+(aq)	C l − (aq)	K s p = 1.6 10^{-10}
Initial	–		0		0
Change	–*s*		+*s*		+*s*
Equilibrium	–		*s*		*s*

Then, as usual, we can find our equilibrium concentrations by solving the mass-action expression:

$$K_{sp} = [Ag^+][Cl^-]$$
$$1.6 \times 10^{-10} = (s)(s) = s^2$$
$$s = 1.3 \times 10^{-5}\,\text{M}$$

Is our Answer Reasonable?

Silver chloride is an insoluble salt, meaning very little dissolves when added to water. Our very small calculated value is consistent with this, and is therefore reasonable. When we substitute our solubility value back into the mass-action expression, we get K_{sp}, confirming our answer.

What If…?

…the solution contained enough table salt to make $[Cl^-]=0.10$ M? What would be $[Ag^+]$?
(*Ans:* $[Ag^+]=1.6\times10^{-9}$ M).

Practice 16.11

Calculate the molar solubility of barium fluoride (BaF_2), $K_{sp}=2.4\times10^{-5}$.

Example 16.12—Calculating K_{sp}

The concentration of calcium ions in a saturated solution of calcium fluoride was found to be 2.15×10^{-4} M. What is the apparent value for the solubility product constant, K_{sp}?

Asking the Right Questions

This problem is asking the same question as the previous exercise, but in the reverse direction. The problem gives us the molar solubility of calcium ions. How can we use that value, via the equilibrium expression, to determine K_{sp}? How do we write the solubility in the equilibrium expression to account for the 1-to-2 mol ratio of calcium to fluorine?

Solution

The equilibrium under consideration is

$$CaF_2(s) \rightleftharpoons Ca^{2+}(aq) + 2F^-(aq)$$

The mole ratio of Ca^{2+} to F^- is 1-to-2, so if the equilibrium concentration of Ca^{2+} is 2.15×10^{-4} M, that of F^- must be twice as large, or $[F^-]=4.30 \times 10^{-4}$ M. We can substitute these values into the mass-action expression.

$$K_{sp} = \left[Ca^{2+}\right]\left[F^-\right]^2$$

$$K_{sp} = \left(2.15 \times 10^{-4}\right)\left(4.30 \times 10^{-4}\right)^2 = 3.98 \times 10^{-11}$$

Is our Answer Reasonable?

The value for K_{sp} is very low, consistent with the nature and meaning of the solubility product. In addition, the value of K_{sp} agrees with that in Table 16.3. Our solubility product value is not only reasonable, it is correct, based on the solubility product table!

What If…?

…we added enough sodium fluoride to make the equilibrium concentration of fluorine equal to 4.30×10^{-1}? What would be the K_{sp} and $[Ca^{2+}]$?

(*Ans: We added enough sodium fluoride to make the equilibrium concentration of fluorine equal to* 4.30×10^{-1}? *What would be the* K_{sp} *and* $[Ca^{2+}]$?).

Practice 16.12

Calculate the apparent value of K_{sp} for lead bromide ($PbBr_2$) if the concentration of bromide in a saturated solution is 2.1×10^{-2} M.

16.4.3 Solubility, Precipitation, and Gravimetric Analysis

Chemical technicians can determine the concentration of substances in solution by causing them to form insoluble salt precipitates and weighing these precipitates or their related solids in a technique called **gravimetric analysis**. The quantitation of a sample on the basis of its mass is among the most powerful tools at the disposal of chemical technicians because good balances are both highly accurate and precise, and weighing a sample is fast and inexpensive. For example, the amount of chloride in a sample is routinely determined by combining the chloride ion with silver ion, forming the solid silver chloride as described in the Exercise 16.11.

$$Cl^-(aq) + Ag^+(aq) \rightleftharpoons AgCl(s)$$

Gravimetric analysis – A laboratory technique in which the concentration of substances in solution is determined by forming insoluble salt precipitates and weighing them or their related solids.

Other substances can be routinely determined by gravimetric analysis. Aluminum ion concentrations can be found via the formation of aluminum hydroxide, $Al(OH)_3$. Igniting the aluminum hydroxide (driving off water at high temperature) forms aluminum oxide (Al_2O_3), which can then be weighed. Aluminum can also be determined by reaction with 8-hydroxyquinoline (C_9H_7ON) to form $Al(C_9H_6ON)_3$ without subsequent ignition.

16

$$Al^{3+}(aq) + 3C_9H_7ON(aq) \rightleftharpoons Al(C_9H_6ON)_3(s) + 3H^+(aq)$$

The solid forms good crystals that can be weighed after drying. Sulfur can be determined by reaction of the sulfate (SO_4^{2-}) with barium ion to form barium sulfate:

$$SO_4^{2-}(aq) + Ba^{2+}(aq) \rightleftharpoons BaSO_4(s)$$

Calcium concentrations can be measured by reaction to form calcium oxalate (CaC_2O_4).

$$Ca^{2+}(aq) + C_2O_4^{2-}(aq) \rightleftharpoons CaC_2O_4(s)$$

In each of these analyses there are complicating factors, such as the presence of other elements that can react with the precipitating agents, as well as the complex nature of the precipitation process, so the procedures are a bit more involved than the simple reactions suggest. In fact, acid–base and other equilibria are nearly always a vital part of chemistry. Despite all of these concerns, a host of elements can be determined via precipitation. Understanding solubility equilibria and how we can affect them makes the analyses all the more meaningful.

Precipitation is also used in **metal recovery**, in which dissolved metals are reclaimed from processing wastes. Many metals in industrial effluents (runoff from manufacturing) are worth recovering because they are environmental hazards or they waste finite metal resources. Recycling these metals saves money and the environment. Such metals, which include copper, mercury, lead, and zinc, are typically recovered as their sulfide salts, although some metals can be precipitated as their corresponding carbonate salt.

Metal recovery – Recycling of metals from waste streams by complexation with chelating agents.

16.4.4 To Precipitate or Not to Precipitate

Often, the formation of a precipitate is not as obvious as simply mixing two solutions containing ions that form a sparingly soluble salt. Let's say our chemical technician is interested in mixing two solutions, one containing silver ions and one containing chloride ions, yielding final concentrations of $[Ag^+] = 1.0 \times 10^{-4}$ M and $[Cl^-] = 1.0 \times 10^{-4}$ M. Will mixing these solutions cause the formation of the insoluble AgCl? To answer this question, we can perform a calculation using our mass-action expression. However, because the concentrations do not reflect equilibrium conditions, we will be calculating the reaction quotient related to the solubility product. We call this Q_{sp}.

$$AgCl(s) \rightleftharpoons Cl^-(aq) + Ag^+(aq) \qquad K_{sp} = 1.6 \times 10^{-10}$$

$$Q_{sp} = [Ag^+]_0 [Cl^-]_0$$

$$Q_{sp} = \left(1.0 \times 10^{-4}M\right)\left(1.0 \times 10^{-4}M\right) = 1.0 \times 10^{-8}$$

The K_{sp} value for AgCl is 1.6×10^{-10}. This value is based on the equilibrium conditions of the sparingly soluble salt. The reaction quotient we calculated is 60 times greater than the equilibrium constant for the precipitation, so the AgCl precipitate forms. If the reaction quotient were smaller than the equilibrium value, no precipitate would form. To reiterate:

- If $Q_{sp} > K_{sp}$, a precipitate forms and continues to form until $Q_{sp} = K_{sp}$.
- If $Q_{sp} < K_{sp}$, no precipitate forms, and we could dissolve more of the solid if we wanted to.

Example 16.13—Will It Make a Solid?

A chemical technician wishes to precipitate the lead ions in 100. mL of a water sample. If the water sample contains 3.10×10^{-10} M Pb^{2+}, and 100. mL of a solution containing 7.0×10^{-4} M Cl^- is added, will a precipitate form? K_{sp} for $PbCl_2 = 1.6 \times 10^{-5}$.

Asking the Right Questions

How do we judge whether a precipitate will form? How is the mass-action expression related to that decision?

Solution

If the product of the desired concentrations raised to the appropriate power in the mass-action expression exceeds the value of K_{sp}, a precipitate will form.

The solubility equilibrium is

$$PbCl_2(s) \rightleftharpoons Pb^{2+}(aq) + 2Cl^-(aq)$$

The mass-action expression is

$$K_{sp} = \left[Pb^{2+}\right]\left[Cl^-\right]^2$$

When the two solutions are mixed, the total volume is doubled, to 200 mL. Therefore, the final concentrations of the respective ions after the two solutions are mixed become

$$\left[Pb^{2+}\right] = 1.55 \times 10^{-10} M$$

$$\left[Cl^-\right] = 3.5 \times 10^{-4} M$$

and Q_{sp} is

$$Q_{sp} = \left(1.55 \times 10^{-10}\right)\left(3.5 \times 10^{-4}\right)^2 = 1.90 \times 10^{-17}$$

No, the chloride solution will not precipitate the lead in the sample.

Is our Answer Reasonable?

The K_{sp} shows that the salt is sparingly soluble, but more soluble than many salts with the same mole ratios of cation and anion that have much lower K_{sp} values. The concentration of lead in solution is so low that it is reasonable, given the K_{sp} value, that the ions do not precipitate.

What If…?

…we wanted the solution to contain a lead concentration of 1.0×10^{-3} M? What is the maximum concentration of chloride ion that can exist in the solution for this to occur?

(Ans: $[Cl^-] = 0.063$ M).

Practice 16.13

Will a precipitate form when equal volumes of the lead solution in the exercise above and 3.5×10^{-4} M sulfide ion are mixed? K_{sp} for $PbS = 7.0 \times 10^{-29}$.

Fig. 16.19 The process of sedimentation can be used to clarify water for drinking. *Source* ▶ https://upload.wikimedia.org/wikipedia/commons/9/95/Decantação.jpg

16.4.5 Acids, Bases, and Solubility

The chemical technician who works at the water treatment center is often responsible for treating wastewater before it is returned to the environment. One such treatment is **sedimentation**, in which aluminum sulfate, $Al_2(SO_4)_3$ and calcium hydroxide, $Ca(OH)_2$, are added to help clarify and purify the wastewater (Fig. 16.19). The aluminum and hydroxide ions in the solution form a gelatinous precipitate.

$$Al^{3+}(aq) + 3OH^-(aq) \rightleftharpoons Al(OH)_3(s) \qquad K_{sp} = 2 \times 10^{-32}$$

Sedimentation – The process of removing dissolved organic matter, heavy metals, and other impurities from water.

The solid settles, carrying with it some dissolved organic material, microorganisms and other undesirable substances in a process called **coagulation**. Iron(III) hydroxide can also be used in this way.

Coagulation – The precipitation of a solid along with some dissolved organic material, microorganisms and other undesirable substances.

On the other hand, increasing the acidity of waterways can increase the concentration of undesirable metals. For example, lead can be found naturally as the insoluble sulfide, PbS. When acidic waters contact the natural lead sulfide, hydrogen ions compete via Le Châtelier's principle to form hydrogen sulfide, H_2S. This allows the lead ion to enter the waterway as Pb^{2+}. The equilibrium constant for the process is not especially high ($K \approx 10^{-7}$), but the leaching of metals, including lead, mercury, cadmium and aluminum into waterways, even at low concentrations, is of concern.

16.5 Complex-Ion Equilibria

We noted before that adding a common ion to a solution of a sparingly soluble salt affects its solubility. We said that if too much of the common ion were added, the sparingly soluble salt could dissolve instead of precipitate. This interesting phenomenon arises because of the formation of a chemical complex, which typically consists of one or more metal cations bonded to one or more Lewis bases known as **ligands** (recall that Lewis bases donate electrons). Examples of ligands include Cl^-, F^-, OH^-, CN^-, NH_3, and H_2O. The complexes can be anions, as with $AgCl_4^{3-}$, or cations, such as $Co(NH_3)_6^{3+}$. Each of these ionic complexes has one or more ions of opposite charge to balance the total charge in the solution.

Ligand – A compound that associates with a metal ion through coordinate covalent bonds.

16.5.1 Introducing the Formation Constant

Our first example of a chemical complex begins with the zinc cation, which exists in aqueous solution bonded to four water molecules, written as $Zn(H_2O)_4{}^{2+}$. In an ammonia–ammonium ion buffer, an ammonia molecule replaces a water molecule.

$$Zn(H_2O)_4^{2+}(aq) + NH_3(aq) \rightleftharpoons ZnNH_3(H_2O)_3^{2+}(aq) + H_2O(l)$$

We will next simplify the expression by assuming the presence of water, as we often do in acid–base reactions, and simplify the expression.

$$Zn^{2+}(aq) + NH_3(aq) \rightleftharpoons Zn(NH_3)^{2+}(aq) \qquad K_{f_1} = 190$$

The equilibrium constant for the formation of the zinc–ammonia complex is called its **formation constant (Kf)** or **stability constant**, and conceptually, it means the same thing as any other equilibrium constant. Le Châtelier's principle teaches us that as we add more ammonia, more NH_3 ligands will form coordinate covalent bonds with the central atom, each step having its own formation constant.

Formation constant (K_f) – The equilibrium constant describing the formation of a stable complex. Typically, K_f values are large. Also known as the stability constant.

Stability constant – See formation constant.

$$Zn(NH_3)^{2+}(aq) + NH_3(aq) \rightleftharpoons Zn(NH_3)_2^{2+}(aq) \qquad K_{f_2} = 220$$

$$Zn(NH_3)_2^{2+}(aq) + NH_3(aq) \rightleftharpoons Zn(NH_3)_3^{2+}(aq) \qquad K_{f_3} = 250$$

$$Zn(NH_3)_3^{2+}(aq) + NH_3(aq) \rightleftharpoons Zn(NH_3)_4^{2+}(aq) \qquad K_{f_4} = 110$$

◘ Figure 16.20 shows the distribution of the various zinc–ammonia complexes as the ammonia concentration is increased. Because the formation constants of the steps are so similar, several different zinc–ammonia species are typically present in solution, as shown in the figure. Going from Zn^{2+} to $Zn(NH_3)_4{}^{2+}$does not give a clear, sharp endpoint, and in general, metal-ion concentrations cannot be analyzed by titration that involves a process with multiple formation constants. Is it possible to have a titrant that will completely combine with a single reaction in a 1-to-1 mol ratio with the metal ion, in order to determine its concentration? This is where EDTA, introduced as a titrant in ► Sect. 16.1, comes in.

16.5.2 Extending the Discussion to EDTA

EDTA is a very useful compound for titrations. It is typically used in its most basic form, shown in ◘ Fig. 16.21. It has *six* pairs of electrons, one pair on each of the two nitrogen atoms and one pair on each of four oxygen atoms, which form between four and six coordinate covalent bonds to a single metal ion such as Ca^{2+}. Substances

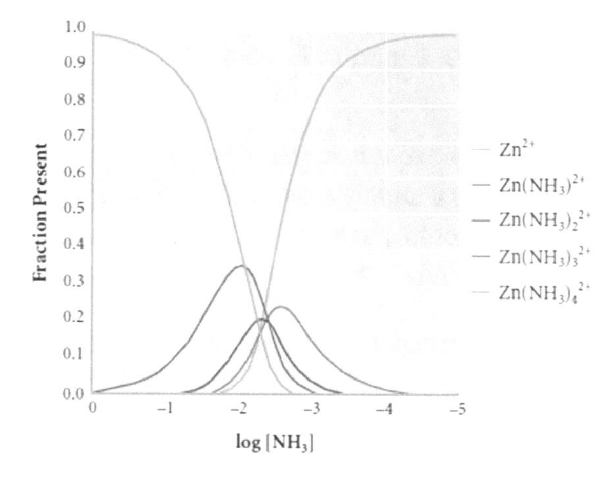

◘ **Fig. 16.20** Here are the changes that occur in the equilibrium concentrations of the various zinc–ammonia complex ions as we increase the concentration of ammonia. The x-axis displays log $[NH_3]$, so that each factor of 10 by which we change $[NH_3]$ occupies equal space in the plot

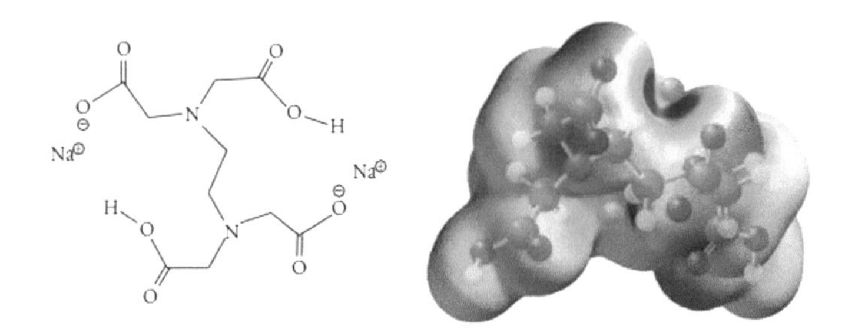

Fig. 16.21 EDTA is a most important chelating agent. Solutions of EDTA are typically prepared as the disodium salt (Na_2EDTA), with $EDTA^{2-}$ shown here as a Lewis dot structure (left) and in a free-energy-minimized configuration (right), including the two sodium ions

Table 16.4 Formation constants of some metal-EDTA complexes

Element	Cation	K_f
Silver	Ag^{+}	2.1×10^{7}
Calcium	Ca^{2+}	5.0×10^{10}
Cobalt	Co^{2+}	2.0×10^{16}
Zinc	Zn^{2+}	3.0×10^{16}
Iron(II)	Fe^{2+}	2.1×10^{14}
Nickel	Ni^{2+}	3.6×10^{18}
Bismuth	Bi^{3+}	8.0×10^{27}
Iron(III)	Fe^{3+}	1.7×10^{24}
Vanadium	V^{3+}	8.0×10^{25}

that form multiple bonds in this way are called **chelates** (the Greek word *chele* means "claw") or chelating agents because they grab the metal ion like a set of claws.

Chelates – Substances capable of associating through coordinate covalent bonds to a metal ion. Also known as chelating agents.

These substances are further characterized by the number of coordinate covalent bonds they make to the metal ion. Ammonia (NH_3) is a **monodentate ligand** (one-toothed ligand). EDTA can be a **tetradentate ligand** (four-toothed ligand) or a **hexadentate ligand** (six-toothed ligand). In ▶ Chap. 14, we saw the very high equilibrium constants (formation constants) for the reactions of EDTA with metals, shown in Table 16.4. Note in the table how the higher charge on the Fe^{3+} results in a dramatically higher formation constant with EDTA compared to that of Fe^{2+}. We now see that the great stability of such metal–chelate complexes is a result of the **polydentate** nature of the ligand. Several industrially important polydentate chelating agents are listed in Table 16.5. These compounds are so useful that well over a billion kg are used annually in a host of different products and applications, some of which are listed in Table 16.6.

16

Monodentate ligand – A ligand that makes one coordinate covalent bond to a metal ion.

Hexadentate ligand – A ligand that makes six coordinate covalent bonds to a metal ion.

Polydentate – Capable of forming several coordinate covalent bonds from a single ligand to a metal.

Tetradentate ligand – A ligand that makes four coordinate covalent bonds to a metal ion.

Complex formation with EDTA and its chemical relatives can be used to determine the concentration of many metals, such as zinc, aluminum, nickel, cobalt, iron, and, in the study of the hardness of water, calcium, and magnesium. We can write the reaction of the calcium ion with EDTA in ionic equation form:

$$Ca^{2+}(aq) + EDTA^{4-}(aq) \rightleftharpoons CaEDTA^{2-}(aq)$$

Table. 16.5 Industrially important aminopolycarboxylic acid chelating agents

Name	Abbreviation	Structure	Ball and Stick
Ethylenediaminetetraacetic acid	EDTA	$(HO{-}C({=}O){-}CH_2)_2N{-}CH_2{-}CH_2{-}N(CH_2{-}C({=}O){-}OH)_2$	
Diethylenetriaminepentaacetic acid	DTPA	$(HO{-}C({=}O){-}CH_2)_2N{-}CH_2{-}CH_2{-}N(CH_2{-}C({=}O){-}OH){-}CH_2{-}CH_2{-}N(CH_2{-}C({=}O){-}OH)_2$	
N-(Hydroxyethyl)-ethylenediamine-triacetic acid	HEDTA	$(HO{-}C({=}O){-}CH_2)_2N{-}CH_2{-}CH_2{-}N(CH_2{-}CH_2{-}OH)(CH_2{-}C({=}O){-}OH)$	

(continued)

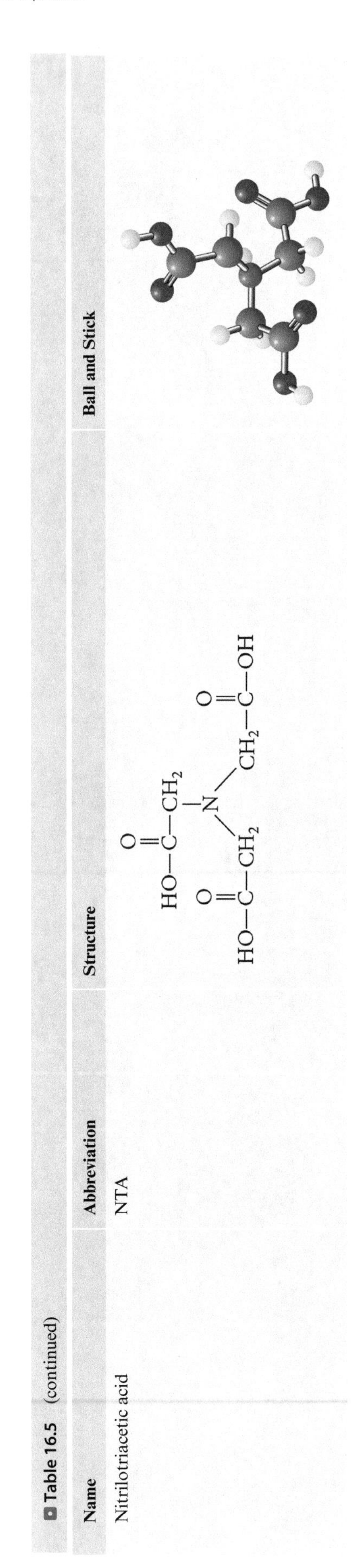

Table 16.5 (continued)

Name	Abbreviation	Structure	Ball and Stick
Nitrilotriacetic acid	NTA	$N(CH_2-C(=O)-OH)_3$	

16

Table. 16.6 Products and applications of chelating agents

Application	Benefits of using chelating agents
Foods and beverages Canned seafood products Dressings, sauces, spreads Canned beans Beverages	Protects the natural flavor, texture, color, and nutritive value of food products Improves shelf life and consumer appeal
Cleaning products Heavy-duty laundry detergents Hard-surface cleaners	Better foaming, detergency, and rinsing in hard water Helps remove metal oxides and salts from fabrics Enhances shelf life by inhibiting rancidity, clouding, and discoloration Improved consumer appeal and product value Improved germicidal action
Personal care products Creams, lotions Bar and liquid soaps Shampoos Hair preparations	Better lathering in shampoos and soaps, particularly in the presence of hard water Improves shelf life and consumer appeal Prevents softening, brown spotting, and cracking in bar soaps Improves stability of fragrances, fats, oils, and other water-soluble ingredients
Pharmaceuticals Treatment for lead poisoning Drug stabilization	EDTA is approved by the US FDA for use in treatment of heavy-metal poisoning Deactivates metal ions that interfere with drug performance
Pulp and paper Mechanical pulp bleaching Chemical pulping Reduction of paper yellowing	Higher brightness and/or lower bleaching costs Less need to overbleach to ensure specified brightness level
Water treatment Boilers Heat exchangers	Dissolves common types of scale during normal operation Improves process efficiency and reduces downtime Works over a wide range of temperatures, pH levels, and pressures
Metalworking Surface preparation Metal finishing and plating	Improved product performance in hard water Improved high-temperature performance
Textiles Preparation Scouring Bleaching	Less need to overbleach to ensure specified brightness level Dye shade stability
Agriculture Chelated micronutrients	Excellent water solubility makes metal chelants more readily utilized by plants than the inorganic forms of metals
Polymerization Styrene-butadiene polymerization PVC polymerization	Stable polymerization rates Reduced polymer buildup in reactors Better polymer stability and shelf life
Photography Developers Bleachers	Higher-quality prints and negatives Enhanced silver recovery Increased longevity of prints and negatives
Oilfield applications Drilling Production Recovery	Prevents plugging, sealing, and precipitation by deactivating metal ions

or we can give it more visual clarity by giving Lewis structures as well as space-filling models, as shown in Sect. 16.1. Because the equilibrium constant is so high, the reaction is essentially complete. This is important when designing a titration. It is also vital in another of EDTA's important uses: combining with metals in food products so that they are unavailable to participate in spoilage processes. To put it in a more formal way, the metal ions are **sequestered** (tied up) by EDTA.

Sequester – To tie up an ion or compound by chelation and make it unavailable for use.

The hard-water analysis at pH 10 allows EDTA to react with both calcium and magnesium ions in the water, which is a measure of "total hardness." At pH values above 12, magnesium ions precipitate as the hydroxide, and the analysis allows determination of the calcium ion concentration alone. Here again, as is our theme in this discussion, pH control is vital. Let's look more closely at the relationship of pH to the reaction. Why should the titration of calcium with EDTA be more complete at basic than at acidic pH values?

16.5.3 The Importance of the Conditional Formation Constant

Equilibrium represents a competition—a wrestling match between substances to acquire and release ions and molecules in their quest for energetic stability. In an aqueous solution of calcium and EDTA, the primary competitor is hydrogen ion. The acidic H_4EDTA can lose four protons in stepwise fashion to form $EDTA^{4-}$.

$$\mathrm{H_4EDTA} \rightleftharpoons \mathrm{H_3EDTA^-} + \mathrm{H^+} \quad K_{a_1} = 1.0 \times 10^{-2}$$

$$\mathrm{H_3EDTA^-} \rightleftharpoons \mathrm{H_2EDTA^{2-}} + \mathrm{H^+} \quad K_{a_2} = 2.2 \times 10^{-3}$$

$$\mathrm{H_2EDTA^{2-}} \rightleftharpoons \mathrm{HEDTA^{3-}} + \mathrm{H^+} \quad K_{a_3} = 6.9 \times 10^{-7}$$

$$\mathrm{HEDTA^{3-}} \rightleftharpoons \mathrm{EDTA^{4-}} + \mathrm{H^+} \quad K_{a_4} = 5.5 \times 10^{-11}$$

The higher the pH, the greater the fraction of $EDTA^{-4}$ in the aqueous solution as hydrogen ions are sequentially removed from the molecule. Figure 16.22 shows that as the pH goes down (the solution is made more acidic), the fraction of EDTA present as $EDTA^{4-}$ (highlighted in the figure) diminishes drastically. The tendency to form protonated EDTA reduces the stability of the calcium–EDTA complex with reactions such as this:

$$\mathrm{CaEDTA^{2-}} + \mathrm{H^+} \rightleftharpoons \mathrm{Ca^{2+}} + \mathrm{HEDTA^{3-}}$$

We show this reduction in stability via the **conditional formation constant (K′)**, which takes into account the fractions of the free (uncomplexed) metal ion and the $EDTA^{4-}$.

$$K' = K_f \, \alpha_{Ca^{2+}} \alpha_{EDTA^{4-}}$$

where $\alpha_{Ca^{2+}}$ = fraction of free Ca^{2+} (1, in this case).

$\alpha_{EDTA^{4-}}$ = fraction of EDTA present as $EDTA^{4-}$

Conditional formation constant (K′) – The formation constant that accounts for the free metalion and its associated ligand.

For example, with our buffer system at pH = 10.0, the formation constant, K_f, for the titration reaction is 5×10^{10} and $\alpha_{Ca^{2+}} = 1$. It is possible to calculate $\alpha_{EDTA^{4-}}$, but we will simply estimate it from Fig. 16.22 as being roughly 0.10.

$$K' = K_f \, \alpha_{Ca^{2+}} \alpha_{EDTA^{4-}} = 5 \times 10^{10}(1)(0.10) = 5 \times 10^9$$

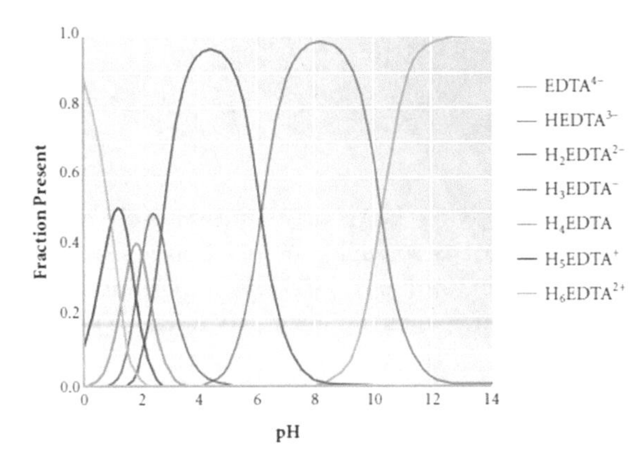

Fig. 16.22 As the pH is lowered, the fraction of $EDTA^{4-}$ sharply decreases, reducing the conditional stability constant of any metal–EDTA titration. This is one of several factors that is important to consider when selecting the best pH for this type of analysis

The conditional formation constant is still sufficiently high for the titration to be essentially complete. If, however, we calculate the conditional formation constant at pH = 3.0, at which point very little of the EDTA is present as $EDTA^{4-}$ and $\alpha_{EDTA^{4-}} = 2 \times 10^{-11}$, we get

$$K' = K_f \alpha_{Ca^{2+}} \alpha_{EDTA^{4-}} = 5 \times 10^{10}(1)(2 \times 10^{-11}) = 1$$

The conditional formation constant is far too low for the titration to be feasible.

Example 16.14—Conditional Formation Constant

With some metals, the competition between EDTA and ammonia to bond to the metal ion can become important when we are judging the feasibility of a titration. If we were titrating zinc instead of calcium in the pH = 10 ammonia buffer, we would find that the zinc would form complex ions ranging from $Zn(NH_3)^{2+}$ to $Zn(NH_3)_4^{2+}$. This would make the fraction of free Zn^{2+}, $\alpha_{Zn^{2+}}$, less than 1, lowering the conditional formation constant.

$$Zn^{2+}(aq) + EDTA^{4-}(aq) \rightleftharpoons ZnEDTA^{2-}(aq)$$

$$K_f = 3 \times 10^{16}$$

Calculate the conditional formation constant for the titration of zinc with EDTA in the pH = 10 buffer in which $\alpha_{Zn^{2+}} = 8 \times 10^{-6}$, and $\alpha_{EDTA^{4-}} = 0.10$. Is this titration still feasible in spite of the low fraction of uncomplexed zinc?

Asking the Right Questions

How is the conditional formation constant related to the formation constant? Based on this relationship, how can we solve for the conditional formation constant?

Solution

The conditional formation constant is the formation constant adjusted by multiplying by the fraction of free (uncomplexed) metal ion and free $EDTA^{4-}$. We can solve for the conditional formation constant, K_f, just as we did with the calcium–EDTA system. Here, however, the fraction of free metal ion is very low.

$$K' = K_f \alpha_{Zn^{2+}} \alpha_{EDTA^{4-}}$$
$$= 3 \times 10^{16}(8 \times 10^{-6})(0.10) = 2.4 \times 10^{10}$$

This value for K' is still quite large, and the titration works well.

Is our Answer Reasonable?

The value for K' is substantially less than that of K_f, and that is reasonable, given the small fraction of free zinc and $EDTA^{4-}$. Given the large size of K', we reasonably concluded that the titration is feasible.

What If…?

…the solution were pH = 3 ($\alpha_{EDTA^{4-}} = 2 \times 10^{-11}$)? Would the titration be feasible?

(*Ans: No, because* $K' = 4.8$).

Practice 16.14

Calculate the conditional formation constant, K_f, for the system in Exercise 16.14 at pH = 3.0.

In the titration of metals with EDTA, we have seen equilibrium principles all come together in one of the most important types of aqueous analyses at the disposal of the chemical technician. Recall Table 16.1 at the beginning of the chapter, in which we listed the calcium–EDTA titration-based hard-water analysis among several ones that are commonly done. Equilibrium principles make these processes ideal for use in many everyday venues. That EDTA titrations are so very common in industrial and academic settings is testimony to the universal utility of equilibrium theory and practice.

Example 16.15—Data for a Calcium–EDTA Analysis

Here are some data obtained from the titration of calcium ion in water with EDTA. How many milligrams of calcium are there per liter of water? The EDTA solution was prepared by combining the disodium salt Na_2H_2EDTA with water. Its molarity is 0.01944 M. About 5 mL of an ammonia–ammonium ion buffer are combined with a 50.00 mL aliquot of the water. The resulting solution is titrated with EDTA, and it requires 31.88 mL to reach the Eriochrome Black T indicator endpoint, shown by a color change from wine red to blue.

Asking the Right Questions

What is the mole ratio of calcium to EDTA in the titration? How do we account for the 50.00 mL aliquot when the example calls for the concentration in milligrams of calcium per liter?

Solution

Calcium and EDTA combine in a 1-to-1 mol ratio. We begin by finding the mass of calcium in the 50.00 mL aliquot.

$$\text{Grams of Ca in the 50.00 mL aliquot}$$
$$= 0.03188\ \text{L EDTA solution}$$
$$\times \frac{0.01944\ \text{mol EDTA}}{1\ \text{L EDTA soln}} \times \frac{1\ \text{mol Ca}}{1\ \text{mol EDTA}} \times \frac{40.08\ \text{g Ca}}{1\ \text{mol Ca}}$$
$$= 0.02484\ \text{g Ca in the 50 mL aliquot.}$$

To convert to the 1 L sample, we use dimensional analysis to scale up to the 1 L volume:

$$\frac{0.02484\ \text{g}}{50\ \text{mL}} \times \frac{1000\ \text{mL}}{1\ \text{L}} = \frac{0.497\ \text{g}}{\text{L}} \approx 0.500\ \text{g/L Ca}$$

This corresponds to fairly hard water.

Is our Answer Reasonable?

This is in the range of normal hard water samples (as opposed to calculating a concentration of, for example, 5×10^{8} ppm or 3×10^{-8} ppm), so the answer is reasonable.

What If...?

... the aliquot were 25.00 mL? How would that have changed the volume of EDTA used in the titration?

(*Ans:* 15.94 mL *of EDTA would have been used.*)

Practice 16.15

How many milliliters of the EDTA solution described in this exercise would be needed to titrate a 50.00 mL sample of water containing 123.8 ppm Ca?

The Bottom Line

- A buffer resists change in pH upon addition of a strong acid or strong base or upon dilution.
- We can solve for the pH of buffers in a straightforward way by recognizing the importance of Le Châtelier's principle and the common-ion effect.
- We can calculate the approximate ratio of conjugate acid to base in order to prepare a buffer of a known pH.
- Solving for the pH of a buffer upon addition of strong acid or base is really solving a limiting reactant problem.
- It is possible to exceed the buffer capacity, in which case the pH will go sharply higher (with excess base) or lower (with excess acid).
- A titration is a technique used to find out how much of a substance is in a solution.
- There are several types of titrations, including reduction–oxidation, precipitation, complex-formation, and acid–base titration.
- Strong-acid–strong-base titrations show a relatively level pH until near the equivalence point, where the pH dramatically changes.
- Titration curves in which one component is weak and the other is strong contain four regions: the initial pH, the buffer region, the equivalence-point region and the post–equivalence point region.
- The buffer region contains a point at which one-half of the analyte has been converted to its conjugate. This is called the titration midpoint, and the pH is equal to the pK_a of the analyte.
- An indicator is used to visually detect the equivalence point of a titration.
- Only a few drops of an indicator are added to the titration solution so that the equivalence point and endpoint can be as close together as possible.
- Solubility equilibria can often be complex, involving several side reactions and molecular-level processes that make calculations challenging.
- The effects of ion-pairing, activity, and other thermodynamic considerations add to the challenge of properly calculating the concentration of dissolved salts in aqueous solution.
- Gravimetric analysis is based on weighing the precipitate that includes the substance of interest.
- The pH of an aqueous solution can significantly affect the solubility of the substances in that solution.
- The formation constant is a measure of the extent of reaction between a Lewis base and metal ion in aqueous solution.
- The reaction of chelating agents and metal ions has a very high formation constant.
- The analysis of calcium in hard water by EDTA titration is an important application of complex-ion equilibrium.

Section 16.2 Buffers and the Common-Ion Effect

Skill Review

1. Write the equations that would describe the equilibria present in each of these solutions:
   a. 0.10 M NH_3
   b. 0.250 M $Fe(OH)_3$
   c. 0.125 M HCOONa
2. Write the equations that would describe the equilibria present in each of these solutions:
   a. 0.30 M PbS
   b. 0.150 M NH_4Cl
   c. 0.050 M CH_3COOH
3. Decide which, if any, of these pairs could be used to prepare a buffer solution.
   a. HF and NaF
   b. CH_3COOH and NH_3
   c. NH_4Cl and HF
   d. H_2SO_4 and $NaHSO_4$
   e. NH_4NO_3 and NH_3
4. Decide which, if any, of these pairs could be used to prepare a buffer solution.
   a. KBr and HBr
   b. NaOH and CH_3COOH
   c. HCOOH and HNO_3
   d. $NaNO_3$ and HNO_3
   e. CH_3COONa and HCl
5. Without using the Henderson–Hasselbalch equation, calculate the pH of a buffer made from each of these pairs. Assume that the concentrations given are those in the final mixture.
   a. 1.00 M NH_3 and 1.00M NH_4Cl
   b. 4.50 M NH_4Cl and 0.50M NH_3
   c. 2.50 M CH_3COOH and 0.75M CH_3COONa
6. Without using the Henderson–Hasselbalch equation, calculate the pH of a buffer made from each of these pairs. Assume the concentrations given are those in the final mixture.
   a. 2.33 M NH_3 and 1.00M NH_4Cl
   b. 2.50 M HCOOH and 1.50 M HCOONa
   c. 0.100 M CH_3COOH and 0.75M CH_3COONa
7. Suppose an ammonia–ammonium buffer has pH of 10.1. Indicate the effect, if any, of each of these changes:
   a. Adding NH_3
   b. Adding NH_4^+
   c. Adding Cl^-
8. Suppose an acetic acid–acetate buffer has pH of 4.74. Indicate the effect, if any, of each of these changes:
   a. Adding NH_3
   b. Adding Na^+
   c. Adding HCl
9. Suppose we were to prepare a buffer solution using acetic acid and sodium acetate. What would be the molar ratio of acid to its conjugate when the pH was adjusted to each of these values?
   a. pH = 3.74
   b. pH = 4.74
   c. pH = 5.74
10. Suppose we were to prepare a buffer solution using ammonia and ammonium chloride. What would be the molar ratio of acid to its conjugate when the pH was adjusted to each of these values?
    a. pH = 10.10
    b. pH = 9.26
    c. pH = 8.40
11. How many milliliters of 0.20 M HCl would we have to add to 100.0 mL of 0.2500 M ammonia in order to prepare a buffer that has each of these pH values?
    a. pH = 9.26
    b. pH = 10.5
    c. pH = 8.5
12. How many milliliters of 0.150 M NaOH would we have to add to 100.0 mL of 0.100 M acetic acid in order to prepare a buffer that has each of these pH values?
    a. pH = 4.26
    b. pH = 3.75
    c. pH = 5.25
13. Indicate the approximate pH of a buffer made from equal concentrations of each of these pairs. The appendix may be useful to determine K_a values.
    a. NH_3/NH_4^+
    b. CH_3COOH/CH_3COO^-
    c. $HCOOH/HCOO^-$
14. Indicate the approximate pH of a buffer made from equal concentrations of each of these pairs. The appendix may be useful to determine K_a values.
    a. $C_6H_5COOH/C_6H_5COO^-$
    b. $CH_3NH_2/CH_3NH_3^+$
    c. $H_3BO_3/H_2BO_3^-$
15. List and explain the factors that determine the pH of a buffered system.
16. List and explain the factors that determine the buffer capacity of a buffered system.

17. A buffer is prepared using chloroacetic acid ($ClCH_2COOH$, $K_a = 1.4 \times 10^{-3}$) and potassium chloroacetate ($ClCH_2COOK$).
    a. Write out the key equilibrium expressions.
    b. Calculate the pH of a solution made by diluting 1.5 g of potassium chloroacetate with 100.0 mL of 0.10 M chloroacetic acid.
18. A buffer is prepared using pyridine (C_5H_5N, $K_b = 1.7 \times 10^{-9}$) and pyridinium chloride (C_5H_5NHCl).
    a. Write out the key equilibrium expressions.
    b. Calculate the pH of a solution made by dissolving 2.50 g of pyridine and 1.25 g of pyridinium chloride into a solution with a final volume of 100.0 mL.
19. The K_b value of methylamine (CH_3NH_2) is 4.3×10^{-4}. The conjugate acid of this weak organic base is the methylammonium ion ($CH_3NH_3^+$, $K_a = 2.3 \times 10^{-11}$). Calculate the pH of a solution that is made from a solution containing 0.10 mol of methylamine and 0.20 mol of methylammonium ion, first using the K_a approach and then using the K_b approach.
20. The K_a value of benzoic acid (C_6H_5COOH) is 6.46×10^{-5}. The conjugate base of this weak organic acid is the benzoate anion ($C_6H_5COO^-$). Calculate the pH of a solution containing 0.025 mol of benzoic acid and 0.250 mol of benzoate anion, first using the K_a approach and then using the K_b approach.
21. Determine the pH of an ammonia–ammonium buffer ($K_b = 1.8 \times 10^{-5}$) with each of these concentrations:
    a. $[NH_3] = 0.10$ M; $[NH_4^+] = 0.10$ M
    b. $[NH_3] = 0.20$ M; $[NH_4^+] = 0.050$ M
    c. $[NH_3] = 1.50$ M; $[NH_4^+] = 0.10$ M
    d. $[NH_3] = 0.050$ M; $[NH_4^+] = 0.750$ M
22. Determine the pH of a phenol (C_6H_5OH)/phenoxide ($C_6H_5O^-$) buffer ($K_a = 1.28 \times 10^{-10}$) with these concentrations:
    a. $[C_6H_5OH] = 0.20$ M; $[C_6H_5O^-] = 0.050$ M
    b. $[C_6H_5OH] = 1.00$ M; $[C_6H_5O^-] = 1.00$ M
    c. $[C_6H_5OH] = 0.050$ M; $[C_6H_5O^-] = 0.10$ M
    d. $[C_6H_5OH] = 0.70$ M; $[C_6H_5O^-] = 0.45$ M
23. What is the pH of a 500.0 mL solution containing 32.08 g of benzoic acid, $C_7H_6O_2$, and 50.81 g of sodium benzoate? What is the pH after 25.0 mL of 3.00 M HCl have been added? Did the system exceed its buffer capacity after the addition of the HCl?
24. What is the pH of 250.0 mL a solution containing 15.64 g of chloroacetic acid, $ClC_2H_3O_2$, ($K_a = 1.4 \times 10^{-3}$)? What is the pH after 8.87 g of NaOH have been added (assume no change in volume)? Did the system exceed its buffer capacity after the addition of the NaOH?
25. We wish to make a buffer of pH = 4.63. Which of the following acids, used along with its conjugate base, would be the most useful?
    a. CH_3COOH ($K_a = 1.8 \times 10^{-5}$)
    b. C_6H_5OH ($K_a = 1.6 \times 10^{-10}$)
    c. $ClCH_2COOH$ ($K_a = 1.4 \times 10^{-3}$)
    d. C_6H_5COOH ($K_a = 6.5 \times 10^{-5}$)
26. We wish to make a buffer of pH 9.02. Should we use a solution made by combining sodium acetate with acetic acid, or one made by combining ammonia with ammonium chloride? Why?
27. What volume, in mL, of 0.150 M NaOH must be added to 50.0 mL of 0.100 M CH_3CO_2H to obtain a buffer with a pH of 5.00? Additional information from the chapter may be needed to solve this problem.
28. What mass of CH_3COONa must be added to 50.0 mL of 0.150 M HCl to obtain a buffer with a pH of 4.75? Additional information from the chapter may be needed to solve this problem.

Chemical Applications and Practices

29. Why does Le Châtelier's Principle help us explain the equilibria in buffer systems?
30. Explain how Le Châtelier's Principle affects the extent of the dissociation of acetic acid in a buffer that is 0.10 M acetic acid and 0.10 M acetate ion.
31. What is the pH of 200.0 mL of a 0.15 M HCl solution? What is the pH after addition of 30.0 mL of a 0.20 M NaOH solution? Is this a buffer solution? Explain the answer.
32. What is the pH of 400.0 mL of a 0.25 M NaOH solution? What is the pH after addition of 20.0 mL of a 0.50 M HCl solution? Is this a buffer solution? Explain the answer.
33. Physiologically important buffers help maintain proper pH levels within our cells. Although the actual buffer system is a complex mixture, we can focus on one particular system that involves phosphate ions. The pH of human

blood must be maintained at approximately 7.40. What would we calculate as the ratio of dihydrogen phosphate ($H_2PO_4^-$) to monohydrogen phosphate (HPO_4^{2-}) at that pH? We will need to consult the acid dissociation table for the appropriate equilibrium value.

34. Another important buffer system for humans is formed between carbonic acid and the bicarbonate ion. Calculate the molar ratio of bicarbonate ion to carbonic acid present at pH 7.40. Obtain the necessary equilibrium constant from the table of acid dissociation constants.
35. When studying bacterial growth, microbiologists must determine the optimum pH range for maximum growth. Then, during subsequent culturing, this range can be maintained through proper application of buffer chemistry. The K_a value of formic acid (HCOOH) is 1.8×10^{-4}.
    a. If this acid and its salt, sodium formate (HCOONa), were selected as the main buffer in a bacteria growth medium, what would be the resulting pH of the media when 500.0 mL of 0.20 M formic acid was mixed with 0.45 g of sodium formate?
    b. Would this buffer be better equipped to resist changes in an acidic or basic direction? Explain.
36. Referring to the same situation as presented for the microbiologist in Problem 35, calculate the volume of 2.5 M NaOH that would be needed to neutralize the 0.20 M formic acid solution to prepare a formic acid–sodium formate buffer with a pH of 3.85.
37. Dairy products such as yogurt, buttermilk, and sour cream are made with the aid of bacteria that convert lactose (milk sugar) to lactic acid. During production, a yogurt sample may attain a pH of 4.00 as a consequence of the presence of lactic acid. If a lactic acid–potassium lactate buffer was produced with the following amounts, what would be the resulting pH? Lactic acid ($CH_3CH(OH)COOH$) = 0.020 mol; potassium lactate ($CH_3CH(OH)COOK$) = 0.015 mol; in 0.500 L. The K_a value of lactic acid is 1.4×10^{-4}.
38. Assume that the bacteria mentioned in Problem 37 produced an additional 0.010 g of lactic acid in a 0.500-L sample of yogurt that already contained [lactic acid] = 0.050 M and [lactate] = 0.050 M. What would be the resulting pH?
39. Propanoic acid (CH_3CH_2COOH) is naturally produced by *Propionibacter shermanii*, a bacterium responsible for the holes in Swiss cheese. The K_a value of propanoic acid is 1.3×10^{-5}. If propanoic acid and its sodium salt were chosen to prepare a buffer system, what would be the pH at which the buffer would have equal ability to resist acidic and basic changes?
40. If 100.0 mL of a propanoic acid–propanoate buffer solution contained 0.50 mol of acid and 0.50 mol of propanoate, how many milliliters of 0.10 M NaOH would be required to exhaust the buffer capacity?

Section 16.3 Acid–Base Titrations

Skill Review

41. In each of these strong-acid–strong-base titrations, determine the volume of titrant that would result in a neutralization.
    a. 0.045 L of 0.23 M HCl titrated with 0.15 M NaOH
    b. 50.0 mL of 0.50 M NaOH titrated with 0.23 M HCl
    c. 20.0 mL of 0.20 M H_2SO_4 titrated with 0.15 M KOH
    d. 0.050 L of 0.10 M NaOH titrated with 0.23 M H_2SO_4
42. In each of these weak-acid–strong-base titrations, determine the volume of titrant that would result a neutralization.
    a. 0.055L of 0.13 M CH_3COOH titrated with 0.15 M NaOH
    b. 50.0 mL of 0.50 M HCOOH titrated with 0.23 M NaOH
    c. 25.0 mL of 0.10 M $ClCH_2COOH$ titrated with 0.45 M KOH
    d. 0.045 L of 0.83 M C_6H_5COOH titrated with 0.70 M KOH
43. Determine the pH of the following titration at each of the points indicated. A 75.0 mL solution of 0.137 M NaOH is titrated with 0.2055 M HCl.
    a. initial pH
    b. after addition of 10.0 mL of HCl
    c. after addition of 25.0 mL of HCl
    d. after addition of 50.0 mL of HCl
    e. after addition of 100.0 mL of HCl
44. Determine the pH of the following titration at each of the points indicated. A 175-mL solution of 0.060 M HCl is titrated with 0.10 M NaOH.
    a. initial pH

b. after addition of 10.0 mL of NaOH
c. after addition of 50.0 mL of NaOH
d. after addition of 105.0 mL of NaOH
e. after addition of 150.0 mL of NaOH

45. Determine the pH of the following titration at each of the points indicated. A 50.0-mL solution of 0.100 M NH_3 is titrated with 0.125 M HCl.
a. initial pH
b. after addition of 10.0 mL of HCl
c. after addition of 20.0 mL of HCl
d. after addition of 40.0 mL of HCl
e. after addition of 50.0 mL of HCl

46. Determine the pH of the following titration at each of the points indicated. A 100.0-mL solution of 0.017 M CH_3COOH ($K_a = 1.8 \times 10^{-5}$) is titrated with 0.025 M NaOH.
a. initial pH
b. after addition of 10.0 mL of NaOH
c. after addition of 34.0 mL of NaOH
d. after addition of 68.0 mL of NaOH
e. after addition of 100.0 mL of NaOH

47. Perform the necessary calculations and sketch a titration curve diagram for the following strong-acid–strong-base titration: 25.0 mL of 0.250 M KOH using 0.150 M HNO_3 as the titrant.
a. initial pH
b. after adding 2.00 mL of HNO_3
c. after adding 20.0 mL of HNO_3
d. after adding 40.0 mL of HNO_3
e. after adding 41.7 mL of HNO_3
f. after adding 43.0 mL of HNO_3
g. after adding 50.0 mL of HNO_3

48. Using 0.25 M NaOH as the titrant, calculate the pH of the resulting solution, and sketch the "pH versus volume of titrant" titration curve, for the neutralization of 50.0 mL of 0.10 M formic acid (HCOOH, $K_a = 1.8 \times 10^{-4}$) at each of these points:
a. initial pH
b. after adding 2.00 mL of NaOH
c. after adding 10.0 mL of NaOH
d. after adding 19.0 mL of NaOH
e. after adding 20.0 mL of NaOH
f. after adding 21.0 mL of NaOH
g. after adding 30.0 mL of NaOH

49. From the list of indicators provided in the chapter, select the best choice for an indicator to use in each of these titrations:
a. HCl analyte with NH_3 as the titrant
b. Propanoic acid analyte with KOH as the titrant
c. Nitric acid analyte with NaOH as the titrant

50. From the list of indicators provided in the chapter, select the best choice for an indicator to use in each of these titrations:
a. Acetic acid analyte with NaOH as the titrant
b. Ammonia analyte with HCl as the titrant
c. Phenol analyte with NaOH as the titrant, K_a (phenol) = 1.28×10^{-10}

51. Determine the color of each of the following indicators in their respective solutions.
a. phenolphthalein; pH = 2.5
b. bromthymol blue; distilled water
c. methyl orange; 0.0056 M HCl
d. methyl violet; 0.049 M NH_3

52. Determine the color of each of the following indicators in their respective solutions.
a. alizarin; 0.025 M NaOH

b. bromthymol blue; 0.15 M NH_3 and 0.15 M HCl
c. thymol blue; 0.15 M HCOOH
d. methyl red; 0.15 M acetic acid and 0.15 M acetate

53. A 25.00 mL sample of base "B" requires 22.06 mL of 0.1204 M HCl to reach the equivalence point.
    a. What is the concentration of "B"? Assume a 1:1 mol ratio for the reaction.
    b. If the K_b of the base is 1.0×10^{-9}, please select a suitable indicator for the titration.
54. A 40.00 mL sample of acid "A" requires 37.77 mL of 0.1056 M NaOH to reach the equivalence point.
    a. What is the concentration of "A"? Assume a 1:1 mol ratio for the reaction.
    b. If the K_a of "A" is 3.0×10^{-4}, please select a suitable indicator other than phenolphthalein for the titration.
55. A solution containing 0.2962 g of a weak acid required 27.24 mL of 0.1811 M NaOH to reach the equivalence point. What is the molar mass of the acid? Assume a 1:1 mol ratio for the reaction.
56. A solution containing 0.500 g of a weak acid required 49.75 mL of 0.0750 M NaOH to reach the equivalence point. What is the molar mass of the acid? Assume a 1:1 mol ratio for the reaction.

Chemical Applications and Practices

57. Which titration does this diagram of pH v. volume of titrant represent?
    a. 0.1 M sodium acetate solution with 0.20 M hydrochloric acid solution

    or

    b. 0.1 M sodium benzoate with 0.20 M hydrochloric acid solution

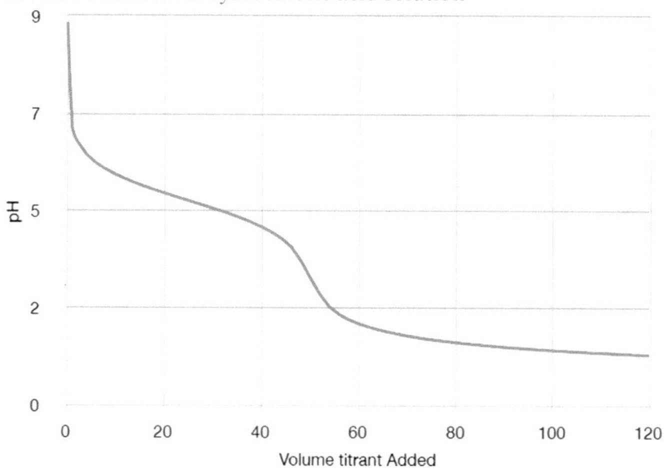

58. Which titration is represented by this diagram of pH v. volume of titrant?
    a. 0.30 M sodium hydroxide solution with 0.15 M hydrochloric acid solution

    or

    b. 0.30 M sodium hydroxide solution with 0.15 M acetic acid solution.

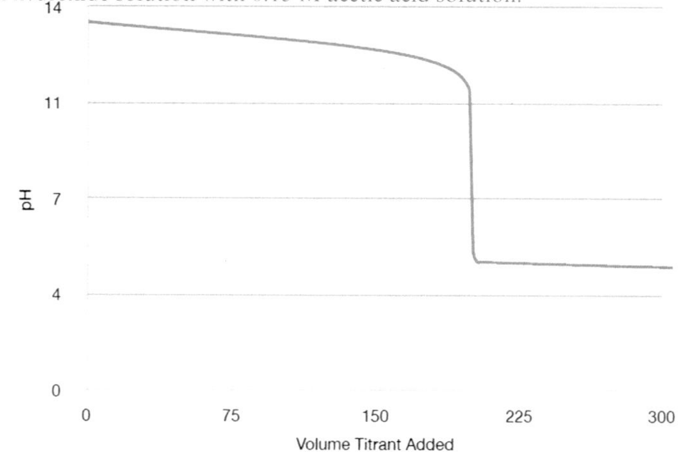

59. Vinegar is a dilute solution of acetic acid in water. A 50.00-mL vinegar sample was found to require 20.0 mL of 0.15 M NaOH in order to change the phenolphthalein indicator to pink.

a. What is the pH of the sample after the reaction?
b. What is the molarity of the vinegar sample?
c. What percent of the original vinegar solution is acetic acid (CH_3COOH, $K_a = 1.8 \times 10^{-5}$)? (Assume the density of the solution is 1.00 g/mL.)

60. A yogurt dessert was found to contain lactic acid by chemical analysis. Analysis revealed that 100.0 mL of the dessert required 22.43 mL of 0.0156 M NaOH in order to react completely with the acid.
    a. What is the pH of the sample after the reaction?
    b. What is the best indicator to use for this titration?
    c. What mass/volume percent of the dessert is lactic acid ($CH_3CH(OH)COOH$, $K_a = 1.4 \times 10^{-4}$)?
61. A chemist has isolated a potential acid–base indicator from a specific type of tealeaf.
    a. Using the following data, determine the approximate pK_a of the indicator. The extracted compound shows a bright red color when in a solution that has a pH of 7.85. At a pH of 9.85, the color has shifted totally to green.
    b. Using our estimated pK_a value, determine the ratio of the red form to the green form at a pH of 9.50.
    c. If we used this new indicator in a titration of HCl with NaOH, would the resulting endpoint be accurately indicated? Explain why or why not.
62. A vinegar solution, which contains acetic acid as the active ingredient (CH_3COOH, $K_a = 1.8 \times 10^{-5}$), was mixed with some sodium acetate. The pH was found to be 4.15, and the concentration of acetic acid was determined to be 0.0125 M at equilibrium.
    a. What is the equilibrium concentration of sodium acetate?
    b. What color is bromthymol blue at this pH?
    c. What is the pH if another 10.0 g of sodium acetate (CH_3COONa) is added to 500 mL of the solution? (Assume the volume of the solution does not change.)

Section 16.4 Solubility Equilibria

Skill Review

63. Write out the reaction that describes the dissolution of each of these sparingly soluble salts. Then write the corresponding mass-action expression for the equilibria.
    a. AgI b. Ag_2CrO_4 c. Al_2S_3 d. $Ca_3(PO_4)_2$
64. Write out the reaction that describes the dissolution of each of these sparingly soluble salts. Then write the corresponding mass-action expression for the equilibria.
    a. $PbCl_2$ b. $NiCO_3$ c. MnS d. $Zn(OH)_2$
65. Use the following data to calculate the molar solubility for each of these solids.
    a. CuS, $K_{sp} = 8.5 \times 10^{-45}$
    b. Ag_3PO_4, $K_{sp} = 1.8 \times 10^{-18}$
    c. $FeCO_3$, $K_{sp} = 2.1 \times 10^{-11}$
66. Use the following data to calculate the molar solubility for each of these solids.
    a. Ag_2S, $K_{sp} = 1.6 \times 10^{-49}$
    b. $Fe(OH)_2$, $K_{sp} = 1.8 \times 10^{-15}$
    c. MgF_2, $K_{sp} = 6.4 \times 10^{-9}$
67. Use the following data to calculate the solubility product constant (K_{sp}) for each of these solids.
    a. NiS, $s = 5.5 \times 10^{-11}$ M
    b. $PbCrO_4$, $s = 1.41 \times 10^{-8}$ M
    c. Ag_2CO_3, $s = 2.0 \times 10^{-4}$ M
68. Use the following data to calculate the solubility product constant (K_{sp}) for each of these solids.
    a. CoS, $s = 2.2 \times 10^{-11}$ M
    b. $Zn(OH)_2$, $s = 2.24 \times 10^{-6}$ M
    c. $CaSO_4$, $s = 7.81 \times 10^{-3}$ M
69. Use the table of K_{sp} values in the text to determine which of these salts has the greatest molar solubility.
    a. CaF_2 b. BaF_2 c. MgF_2
70. Use the table of K_{sp} values in the text to determine which of these salts has the greatest molar solubility.
    a. PbI_2 b. AgI c. Ag_3PO_4

71. Indicate whether a solid will form in each of these mixtures. Start by finding the concentrations of each ion in the resulting solution.
    a. 125 mL of 0.100 M $BaCl_2$ and 10.0 mL of 0.050 M Na_2SO_4
    b. 35 mL of 0.0045 M $Ag(NO_3)$ and 100.0 mL of 0.0038 M NaCl
72. Indicate whether a solid will form in each of these mixtures. Start by finding the concentrations of each ion in the resulting solution.
    a. 100.0 mL of 0.015 M K_2CO_3 and 100.0 mL of 0.0075 M $BaBr_2$
    b. 25 mL of 0.57 M $Ni(NO_3)_2$ and 100.0 mL of 0.150 M NaOH
73. Iron(III) hydroxide has a very low K_{sp} value: 1.6×10^{-39}. Explain the effect of changing the pH of an iron(III) hydroxide solution on the solubility of this salt.
74. Copper(II) carbonate has a K_{sp} value of 2.5×10^{-10}. Explain the effect of changing the pH of a copper(II) carbonate solution on the solubility of this salt.
75. Indicate whether a solid will form if 25.0 mL of 0.12 M $Ba(NO_3)_2$ are added to 50.0 mL of 0.0850 M Na_2CO_3. How would our answer change if the solution contained 25.0 mL of 0.2047 M KNO_3?
76. We add 0.12 M $Ba(NO_3)_2$ to 50.0 mL of 0.0850 M Na_2CO_3. At what volume will a precipitate just begin to form? How would our answer change if the solution contained 30.0 mL of 0.1000 M KNO_3?

Chemical Applications and Practices

77. When sources of the fluoride ion were being considered for use in fluoridated toothpastes, compounds such as calcium fluoride (CaF_2) could have been among them. This salt has a K_{sp} value of approximately 4.0×10^{-11}. Write out the mass-action K_{sp} expression, and calculate the concentration of fluoride ion in a saturated solution.
78. To maintain good health, we need to have certain amounts of several dissolved metal ions. Zinc ions are critical for the role they play with several hundred enzymes that function in digestion, immune systems, and fertility. However, some foods bind the zinc ions and prevent them from being absorbed from food.
    a. Write out the mass-action K_{sp} expression for zinc sulfate, the form of zinc in many vitamin supplements.
    b. If phytic acid, which is found in some foods, reacted with zinc ions, how would this affect the solubility of zinc sulfate?
79. Calcium ions can be mineralized to form bones and teeth. Neglecting any other considerations, which of the following solids would provide the greatest number of calcium ions in a saturated solution? Use the mass-action expression for each to calculate the value for each. Explain the impact of alternative equilibria on our answers.
    Calcium carbonate, $K_{sp} = 3.3 \times 10^{-9}$
    Calcium iodate, $K_{sp} = 7.1 \times 10^{-7}$
    Calcium phosphate, $K_{sp} = 1.2 \times 10^{-29}$
80. The cadmium +2 ion, which is considered a toxic heavy metal ion, has been a by-product of several mining operations. One method that could remove it from aqueous systems would be to tie the ions up as a cadmium hydroxide precipitate. The K_{sp} value of cadmium hydroxide is 2.5×10^{-14}.
    a. What is the molar solubility of cadmium hydroxide?
    b. Explain how the presence of dissolved Cd^{2+} might be a greater problem at acidic or basic soil pH values.

Section 16.5 Complex-Ion Equilibria

Skill Review

81. Write equilibria equations that describe the stepwise formation of each of these complex ions.
    a. $Ag(NH_3)_2^+$
    b. $Ni(NH_3)_6^{2+}$
82. Write equilibria equations that describe the stepwise formation of each of these complex ions.
    a. $CuBr_3^{2-}$
    b. $Ag(S_2O_3)_2^{3-}$
83. How many milliliters of a 0.0156 M $EDTA^{4-}$ solution would be needed to completely complex each of these metals in solution?
    a. 100.0 mL of 0.150 M Zn^{2+}
    b. 50.0 mL of 0.740 M Ca^{2+}
    c. 20.0 mL of 0.050 M Mg^{2+}

84. What would be the molar concentration of an $EDTA^{4-}$ solution in each case if it took exactly 25.00 mL to react completely with each of these solutions?
   a. 30.0 mL of 0.250 M Zn^{2+}
   b. 40.0 mL of 0.700 M Ca^{2+}
   c. 20.0 mL of 0.150 M Mg^{2+}

Chemical Applications and Practices

85. As we noted in the chapter, the fraction of species for the $EDTA^{4-}$ ion is critical when we consider EDTA–metal ion titrations. The available metal ion, uncomplexed with other ligands such as ammonia, can also be an important consideration. The formation constant for the Zn^{2+}–EDTA complex was given as 3×10^{16}. What would be the conditional formation constant if the fraction of free zinc ion were 1×10^{-4} and the fraction of $EDTA^{4-}$ species were 0.05?
86. A 25.0-mL sample of water is treated with ammonia buffer and Eriochrome Black T indicator. The sample requires 15.0 mL of 0.0185 M EDTA solution to reach the endpoint. What is the concentration of Ca^{2+} ion in moles per liter and ppm?
87. A dilute solution of hydrated Cu^{2+} ions will appear blue and without a precipitate. However, the addition of some ammonium hydroxide will cause precipitation of some light blue copper(II) hydroxide, often within a deep blue solution. Further addition of ammonium hydroxide will dissolve the precipitate and form a dark blue solution of $Cu(H_2O)_2(NH_3)_4^{2+}$.
   a. What do these observations suggest about the relative value of the formation constant for $Cu(NH_3)_4^{2+}$?
   b. Write out the stepwise formation of the copper–ammonia complex.
88. Although very toxic, cyanide compounds such as KCN and NaCN can be used to extract gold from ores because the gold will form soluble complexes with cyanide ions. The cumulative formation constant for $Au(CN)_2^-$ is approximately 1.6×10^{38}. In this extraction process, the cyanide extracting solution must be kept very alkaline to prevent the formation of HCN. Very small amounts of gold can be extracted in this manner thanks to the very high value of the formation constant.
   a. Write out the mass-action expression for the formation constant of $Au(CN)_2^-$.
   b. If the concentration of CN^- were maintained at 0.010 M in an extract and the gold complex ion concentration were found to be 8×10^{-5} M, what would we estimate as the concentration of Au^+?
89. Under proper medical supervision, one treatment for lead poisoning may involve reaction with a soluble EDTA salt. This is called chelation therapy. If the following reactions were part of a successful treatment, which complex, the calcium–EDTA ion or the lead–EDTA ion, would we predict to have the higher formation constant?

$$CaEDTA^{2-}(aq) + Pb^{2+}(aq) \rightleftharpoons PbEDTA^{2-}(aq) + Ca^{2+}(aq)$$

90. How many milliliters of a solution of 0.0010 M $EDTA^{4-}$ would be used to titrate the lead in 1000 mL of a 0.0020 M solution of $Pb(NO_3)_2$? (Assume a 1:1 reaction between Pb^{2+} and EDTA.)

16

Comprehensive Problems

91. Cite three processes in which the use of a buffer would be necessary to maintain the pH within a fixed range.
92. A buffer commonly found in biochemical labs and known as TRIS is $(CH_2OH)_3CNH_2$, $pK_b = 5.70$. If a 1.0-L solution were made with 0.15 mol of TRIS, how many grams of $TRISH^+$ chloride salt, $(CH_2OH)_3CNH_3Cl$, would have to be added to the solution to make a buffer with pH = 8.1?
93. a. A compound of highly oxidized iron may be used to break down certain environmental wastes. However, the reactions of the iron compound (K_2FeO_4) are highly pH dependent. In order to do feasibility studies of this compound for possible wastewater treatment, phosphate buffers could be used to maintain a fairly constant pH value. Obtain the K_a values of H_3PO_4, $H_2PO_4^-$, and HPO_4^{2-}. Which two would be the best combination to use if the study were to be done at pH = 8.20?
   b. What would be the ratio of the components at that pH?
94. The presence of lead ions in the environment can pose a hazard. One gravimetric method to test for the presence of lead in a sample is to precipitate the lead as lead sulfate. The K_{sp} value for lead sulfate is 1.3×10^{-8}.
   a. Write out the mass-action K_{sp} expression for this compound.
   b. If the sulfate concentration is made sufficiently high, lead ions will be almost completely precipitated from the solution. If a solution had a lead ion concentration of 0.0010 M initially, what concentration of lead would remain after precipitation if the sulfate concentration were maintained at 0.010 M?

95. a. Calcium, zinc, and cobalt all form +2 ions. However, from Table 16.4 we can see that the formation constant for calcium–EDTA is considerably less than that for the zinc- EDTA and cobalt–EDTA complexes. Explain why this contrast is logical.
    b. Explain why the formation constant for Fe^{2+}–EDTA is less than that for Fe^{3+}–EDTA.
96. By adjusting solution pH, a biologist may manipulate the charges on side chains of enzymes. A biologist studying the function of an enzyme responsible for a step in the conversion of atmospheric nitrogen to useable forms of nitrogen in a plant finds that the enzyme is neutral when the pH is 6.87.
    a. What is the significance of this pH?
    b. If the enzyme had an acid group with a $pK_a = 7.20$, what should the pH of the enzyme solution be in order to produce a molar ratio of protonated acid group to unprotonated acid group equal to 3:1?
97. Formic acid (HCOOH, $K_a = 1.77 \times 10^{-4}$) is a weak acid extracted from ants. Suppose that such an extraction resulted in 1.00 mL of volume. This 1.00 mL is then diluted to 50.0 mL with distilled water.
    a. If the resulting solution required 25.0 mL of 0.0010 M NaOH to neutralize, what would we calculate as the molarity of the formic acid solution?
    b. How many moles of formic acid were in the original 1.00-mL extract?
98. A biologist seeks to analyze a sample of lactic acid ($CH_3CH(OH)COOH$, $K_a = 1.4 \times 10^{-4}$) isolated from a tissue sample. The 20.0-mL sample required 12.5 mL of 0.086 M NaOH to neutralize. What was the molarity of the lactic acid sample? What was the initial pH, the pH midway to the equivalence point, and the pH at the equivalence point?
99. An aqueous solution starts as 0.20 M dissolved $Fe(NO_3)_2$. EDTA is added to produce a concentration of 0.10 M. Assume that the formation constant for the Fe^{2+}–$EDTA^{4-}$ complex is 2×10^{14}.
    a. Use the mass-action formation constant expression to determine the approximate concentration of Fe^{2+}.
    b. If the solution is now made 0.20 M in hydroxide, is enough Fe^{2+} still present to cause precipitation of $Fe(OH)_2$?
100. Based on our understanding of equilibrium and acid–base titrations, estimate the equilibrium constant for the titration of acetic acid with ammonia? Would this combination be suitable for titration? Justify the answer.
101. In the chapter, we claim that strong acids and bases are not buffers in the sense that they resist changes in pH upon addition of a strong acid or base but are unable to resist changes in pH with the addition of water. If we were compelled to use 400.0 mL of a 2.00 M solution of HCl to keep the pH of a solution relatively constant, how much water would we be comfortable adding before the pH is no longer "constant" enough for our own use in the procedure? In other words, how constant is constant?
102. In Sect. 16.1, we discussed buffer preparation, using the example of the calibration buffer as described in the Pharmacopeia. If we had available only KHP (potassium hydrogen phthalate), HCl, and NaOH, how would we prepare 1.00 L of a pH 6.00 buffer? ($K_{a_2} = 3.9 \times 10^{-6}$)
103. In an older method for the quantization of nitrate ion in water, the water sample is first treated with an aqueous solution of silver ions.
    a. What ion(s) is(are) typically in a sample of tap water that would react with the silver? Write the balanced net ionic reaction that would "remove" that ion from solution.
    b. Why is silver nitrate not used as the source of the silver ions? Suggest a possible compound that could be used to create a source of aqueous silver ions for this analysis.

Thinking Beyond the Calculation

104. The US Food and Drug Administration lists criteria for substances to be added to food. One criterion is for the substance to be "Generally Recognized as Safe," or GRAS. Use the Internet to determine the answer to this question: "What are the federal guidelines for the approval of such a substance?" There is a subtle distinction in the federal regulations that differentiates between the substance, itself, and its use. What is this distinction, and why is it important? Use the Internet to look up the status of several Good Buffers. Which are GRAS, and in what uses? Is there a pattern between structure and safety?
105. A researcher has produced an extract from a tropical plant that contains a monoprotic acid.
    a. The researcher titrates 100.0 mL of the plant extract with 0.100 M NaOH. The titration midpoint is reached when 25.6 mL of NaOH has been added. The pH at this point was 3.58. What is the K_a value for the acid?
    b. Graph a titration curve for this titration by determining the pH at each of the following points along the curve: 0 mL, 10.0 mL, 25.6 mL, 50.0 mL, 75.0 mL, and 100.0 mL.
    c. Which indicator would work best in this titration?
    d. By evaporating the extract, the researcher determines that there is only 0.123 g of the acid in every 100.0 mL of plant extract. What is the molecular mass of the acid?

Thermodynamics: A Look at Why Reactions Happen

Contents

Supplementary Information The online version contains supplementary material available at ▶ https://doi.org/10.1007/978-3-030-90267-4_17.

M. Mosher and P. Kelter, *An Introduction to Chemistry*,
https://doi.org/10.1007/978-3-030-90267-4_17

What We Will Learn in this Chapter

Probabilities of a particular outcome can be used to predict the outcome of a reaction (▶ Sect. 17.1).

The number of microstates and their probability of occurring determines the most likely macrostate for a system (▶ Sect. 17.2).

The entropy of the universe is always increasing (▶ Sect. 17.3).

Temperature can affect the spontaneity of a process (▶ Sect. 17.4).

The entropy of a pure, perfect crystal at 0 K is zero (▶ Sect. 17.5).

The Gibbs free energy equation allows the calculation of the spontaneity of a reaction from the enthalpy and entropy of the reaction (▶ Sect. 17.6).

The concentration of individual species in a reaction can affect the spontaneity of the overall process (▶ Sect. 17.7).

A relationship between the free energy and equilibrium constant exists (▶ Sect. 17.8).

17.1 Introduction

After a hearty breakfast, we leave for work full of energy and ready to conquer the day. However, the midmorning hours can be difficult to get through because our energy level drops a couple of hours after we eat. This is especially true if we had a big bowl of sugar-coated, sugar-injected cereal for breakfast. To make matters worse, the mid-afternoon hours are no easier—they seem to be the longest of the day. Why does this happen and what can we do to give ourselves that "burst of energy" we need when we feel so tired? Enter the snack. Whether it takes the form of a chocolate bar, a donut, or a bottle of juice, it has the effect of raising our energy level. Just like breakfast, lunch, and dinner, the snack provides our bodies with a source of glucose. How does the consumption of glucose give us energy?

Antoine Lavoisier (the scientist we met in ▶ Chap. 2, whose measurements led to the formulation of the law of conservation of mass before he was guillotined in the French Revolution) noticed that living things consume foods and transform them into the energy that maintains life. Lavoisier's views on the process seem rudimentary given our modern understanding, but they were quite revolutionary in his day. The addition of glucose to living cells is an example of this food-to-energy transformation process. To the biochemist, this is part of the broader field of **bioenergetics**, the study of the energy changes that occur within a living cell.

Bioenergetics – The study of the energy changes that occur within a living cell.

In this chapter we will discuss some of the chemical reactions in the field of bioenergetics. Although we'll soon introduce terms such as entropy, spontaneity, free energy, and equilibrium, the underlying concept of probability demands our immediate attention. Chemists benefit from understanding these topics. They use probability in many ways: to locate electrons, to determine the macroscopic properties of compounds and mixtures, and to predict the outcome of chemical reactions. We will use probabilities to discover why chemical processes occur—including the chemical transformations of the body and their relationship to our ability to live.

17.2 Probability as a Predictor of Chemical Behavior

If we entered a classroom or movie theater and noticed all the males seated on one side and all the females on the other, as shown in ◘ Fig. 17.1, would we think that some announcement, rule, or social convention had dictated that arrangement? Perhaps. But this seating pattern could also have emerged from purely random choices. In fact, there are numerous ways in which a room full of people could be sitting, and "separate seating" is just one of many possibilities.

Let's shift our focus to the room in which we are now sitting as we read this. As we know, the air in the room is a mixture of primarily oxygen and nitrogen gases. Oxygen and nitrogen molecules do not chemically interact with each other under standard conditions, so they can occupy essentially any unfurnished position in the room. Is it possible that these gases could be arranged such that all of the oxygen molecules were in one corner of the room and all of the nitrogen molecules in another, as shown in ◘ Fig. 17.2? Possible? Yes. Likely? No. Is there something that these two seemingly unrelated situations—the room of people and the room of molecules—share?

The short answer is yes, the two situations have a lot in common. To understand how this could be so, let's introduce some common terminology to help in our discussion. We refer to the **macrostate** of a system (whether

Fig. 17.1 An arrangement of students in a classroom

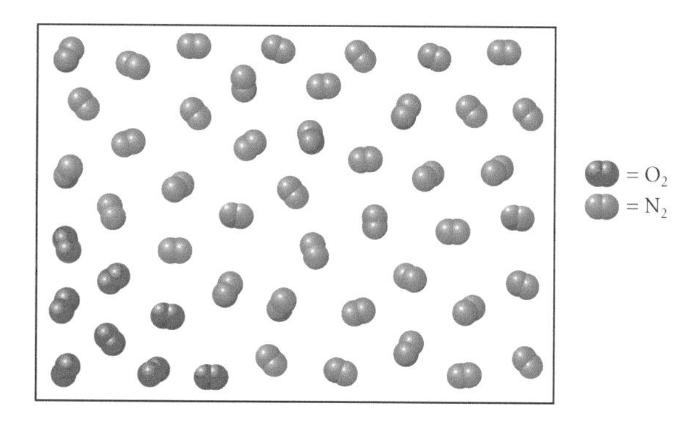

Fig. 17.2 A possible arrangement of the molecules of air within a room

seated people or molecular positions in a room) when we want to take a snapshot of the overall situation. The macrostate of the theater seating is that it is sorted by gender. The dispersed oxygen and nitrogen molecules in the room exert a certain pressure and have a certain composition; this is all part of the macrostate of the gases in the room.

Macrostate – The macroscopic state of a system that indicates the properties of the entire system.

If we wanted to describe the individuals in the theater, we might discuss which seat they occupy, which way they are facing, and whether their arms or ankles are crossed. Each of these individual descriptions is referred to as a **microstate**. For instance, each arrangement of people in the theater indicates a different microstate. How do microstates apply to molecules? The total of the microstates in a system defines, or describes, the macrostate of the system. For individual molecules, the microstate can include their translational motion (changing from one location to another), their vibrational–rotational state (this involves the atomic movements around a chemical bond), and their electron configuration (ground state, excited state, oxidation state, etc.).

Microstate – The state of the individual components within a system.

How could we quantitatively describe the macrostate of the gases in our room? We could measure their temperature and pressure. We could also assess the number of molecules of each component in the room. Given the huge number of molecules of gas in the room, we would predict an even larger number of microstates for the gas. We mentioned the hypothetical condition of all the oxygen molecules being in one corner. To bring about that unlikely macrostate of air, the oxygen molecules (being restricted to the corner) would have to assume a small number of the multitude of possible microstates in the room.

Why have we never encountered the macrostate that has all the oxygen molecules stuck in one corner of the room? One answer can be found in looking at the probability that such an event could occur. Out of all the possible macrostates, what are the chances that "all O_2 on one side, all N_2 on the other side" would occur? In short, there is a very low probability that this situation would happen. (Read on for a more detailed answer!).

17

Think back to the seating choices for a person entering a room. Which situation has more choices, the strict division by gender or "open seating"? Which situation would have the greater number of microstates that satisfy the conditions of a particular macrostate? Ludwig Boltzmann (1844–1906), an Austrian mathematician/physicist, dealt with this idea, along with its powerful implications, mathematically. We can get a sense of what he did by looking inside the tanks of oxygen that scuba divers and mountain climbers carry with them as they set off on their quests. Look closely at ◘ Fig. 17.3. This photo, taken of mountain climbers George Leigh Mallory and Andrew Irvine before their fateful June 1924 attempt to scale the summit of Mt. Everest, shows the climbers each with two tanks of oxygen strapped to their backs. Let's assume that these tanks are connected by a valve and that one of the tanks is empty and the other full. What will happen to the distribution of the gas if the valve is opened? The final volume of the two tanks, which we'll call V_f, is double the initial volume of just one tank, V_i. We can also express this by saying that the ratio $\frac{V_f}{V_i} = 2$ How many ways can the individual molecules be distributed within the two tanks? (That is, how many "microstates" are there?) To make things a lot simpler, let's assume that the tanks contain only two molecules of a gas, and we will only consider the location of the gas molecules in one tank or the other, rather than all the other factors, like vibrational states, kinetic energy, etc. If we call the molecules A and B, we can have a total of four microstates, as shown in ◘ Fig. 17.4. We will call the final number of microstates W_f (for the German word *Wahrscheinlichkeit*: "probability"), and $W_f = 4$ in this case. The ini-

Fig. 17.3 George Mallory and Andrew Irvine on Mt. Everest pack heavy cylinders of oxygen on their backs. This photo, taken in 1924 during their ascent to the summit, is the last known image of the pair. *Source* ▶ https://2.bp.blogspot.com/-H_GpGtg2ODg/WTOqsrjgGjI/AAAAAAAALkY/AwWzUU9d6DI6PYY2-b-JCbSuMj1A3LAQACLcB/s1600/_DSC0264.JPG

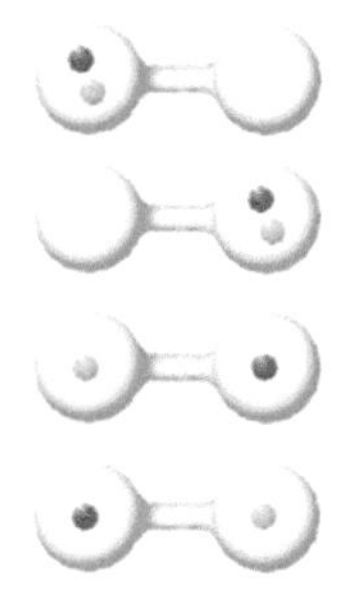

Microstate	Tank 1	Tank 2
1	AB	
2		AB
3	A	B
4	B	A

Fig. 17.4 The four possible arrangements of two molecules into two tanks

tial number of microstates, W_i, is equal to 1, representing A and B both present in one tank, before the valve was opened allowing the gases to spread apart. This means that for two molecules in two tanks, the ratio $\frac{W_f}{W_i} = \frac{4}{1}$.

How is the number of microstates, $\frac{W_f}{W_i}$, related to the volume, $\frac{V_f}{V_i}$ For N molecules,

$$\frac{W_f}{W_i} = \left(\frac{V_f}{V_i}\right)^N$$

That is the equation that Boltzmann derived. Using our numbers,

$$4 = 2^2.$$

What if our tanks contained three molecules ($N = 3$) instead of two?

$$\frac{W_f}{W_i} = \left(\frac{V_f}{V_i}\right)^N \text{ so: } \frac{W_f}{W_i} = \left(\frac{2}{1}\right)^3 = 8$$

In this case, there would be eight possible microstates. More molecules in the tanks make more microstates available—more ways in which the molecules can distribute themselves between the two tanks. This means that the probability of all of the molecules being in only one tank—a single microstate among all other possibilities—becomes smaller. This is the key point: As the number of molecules increases, the likelihood of their all being in one tank decreases sharply, and the probability of the gases being equally distributed increases. Let's look further into this point and discuss its vital implications.

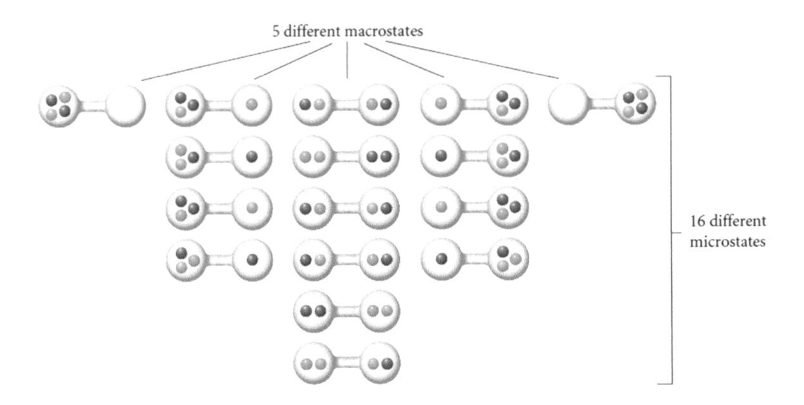

Fig. 17.5 Microstates describing the possible arrangements of four molecules between two tanks

Let's put four oxygen molecules in the full tank. Here Boltzmann's equation indicates that there should be 16 ($2^N = 2^4 = 16$) different microstates of the oxygen molecules in the two-tank system. This is exactly what we predict by drawing each of the arrangements as in Fig. 17.5. But if we look closely, we see that some of the microstates would give the same macrostate for the system. Figure 17.5 shows that only 5 unique *macro*states could arise from the 16 different *micro*states. Only 6 of these 16 microstates describe the equal distribution of the gases. The probability is therefore 6/16, or 37.5%, that the four molecules will arrange themselves into the two tanks equally. The probability of exactly equal distribution has diminished. However, there are many more microstates at or near equal distribution than microstates that are all on one side of either tank. Only 2 of the 16 microstates, or 12.5%, describe all of the molecules being on one side or the other.

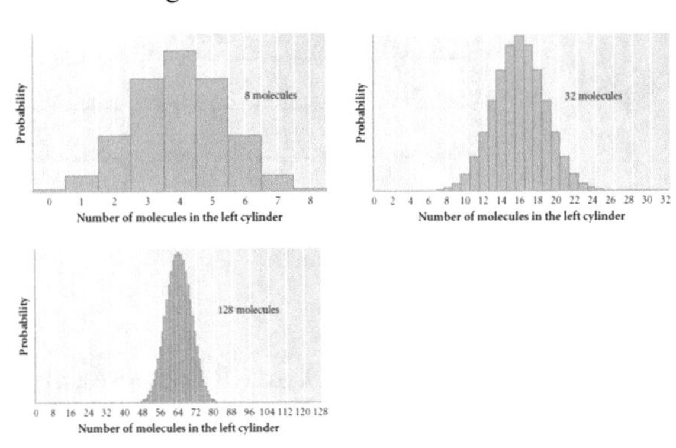

Example 17.1—Probability

A valve connects two glass jars. One of the jars contains six atoms of gaseous helium, and the other is empty. Determine the probability of the macrostate in which each jar will fill equally (that is, each jar will contain three atoms of helium) after the valve is opened. For the same system, what is the probability of finding four molecules in the left-hand jar and two in the right? What about the probability of finding all six molecules in the left-hand jar?

Asking the Right Questions

How many microstates are possible? How is this value related to the number of atoms and the number of jars? How and why is the difference between macrostate and microstate important in this example?

Solution

Because there are six molecules in this example, there are $2^6 = 64$ possible microstates. Drawing each of them and grouping them into similar macrostates leads us to conclude that there are only 7 macrostates (see Fig. 17.6). There are 20 microstates that indicate equal distribution, so the probability of equal distribution of the six molecules of gas is 20/64, or 31%. The probability of finding two molecules of gas in the right-hand jar and four in the jar on the left is 15/64, or 23%. The probability of finding all of the molecules in the left-hand jar is 1/64, or 1.6%.

Are Our Answers Reasonable?

Our drawing and counting of microstates and macrostates led to the conclusion that the probability of finding all of the molecules in one jar is much smaller than finding them evenly distributed between the jars. Our life experience matches the Boltzmann equation in asserting that the answers are reasonable.

What If…?

…we had 1000 molecules of air, the roughly 80/20 mix of nitrogen and oxygen. What is the probability that all of the air molecules will go in just our left lung?

(*Ans: 10^{-301}*)

Practice 17.1

Let's use the same valve and glass jar set-up, with one jar containing eight atoms of gaseous helium. Then the valve is opened. Determine the probability of the macrostate that shows three atoms of He in one jar and five atoms of He in the other jar.

17.3 Why Do Chemical Reactions Happen? Entropy and the Second Law of Thermodynamics

When sucrose enters the body, it is broken down into two simpler molecules, glucose and fructose. Glucose and fructose are structural isomers, both with the formula $C_6H_{12}O_6$.

$$\underset{\text{Sucrose}}{C_{12}H_{22}O_{11}(s)} + H_2O(l) \rightarrow \underset{\text{Fructose}}{C_6H_{12}O_6(aq)} + \underset{\text{Glucose}}{C_6H_{12}O_6(aq)}$$

Both molecules are used by the body to generate energy. For example, glucose in a cell undergoes **glycolysis**, the series of ten chemical transformations, shown in Fig. 17.7, that produces two molecules of pyruvate. The pyruvate is further converted, releasing energy via a different set of transformations. In addition, during glycolysis two new energy storage molecules known as ATP (adenosine triphosphate, Fig. 17.8) are produced. Glucose provides a major source of energy for living cells via the glycolysis pathway, which is the sole source of metabolic energy in some mammalian tissues and cell types. Many anaerobic (non-oxygen-consuming) microorganisms depend entirely on glycolysis for energy to carry out other biological reactions and survive.

Glycolysis – A series of biochemical reactions that convert glucose into two molecules of pyruvate. The process results in the formation of two molecules of ATP.

Metabolism – The biochemical reactions of an organism.

Glycolysis is an example of a **spontaneous process**. The rusting of iron nails, the melting of ice in a glass, and the decay of wood buried in moist soil are also spontaneous processes. That is, these processes occur without continuous outside intervention. Say a brand-new deck of cards falls off the kitchen table. The principles of probability say that there are many more ways in which the cards can land out of order than ways in which they can land sequentially. Furthermore, some of the cards will land face up and some face down. This disorder happens without our assistance—the process occurs spontaneously.

Spontaneous process – A process that occurs without continuous outside intervention.

The reverse of a spontaneous process is known as a **nonspontaneous process**. Nonspontaneous processes do not occur without continuous outside intervention. A rusty nail does not revert to a polished iron nail without con-

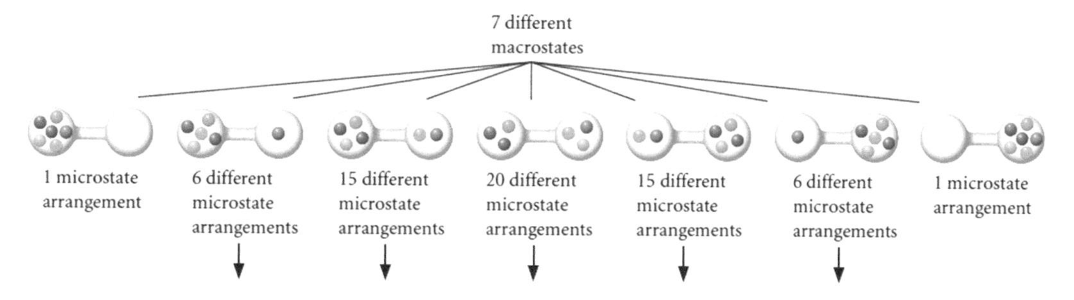

Fig. 17.6 Probability distribution for six molecules of gas in two tanks

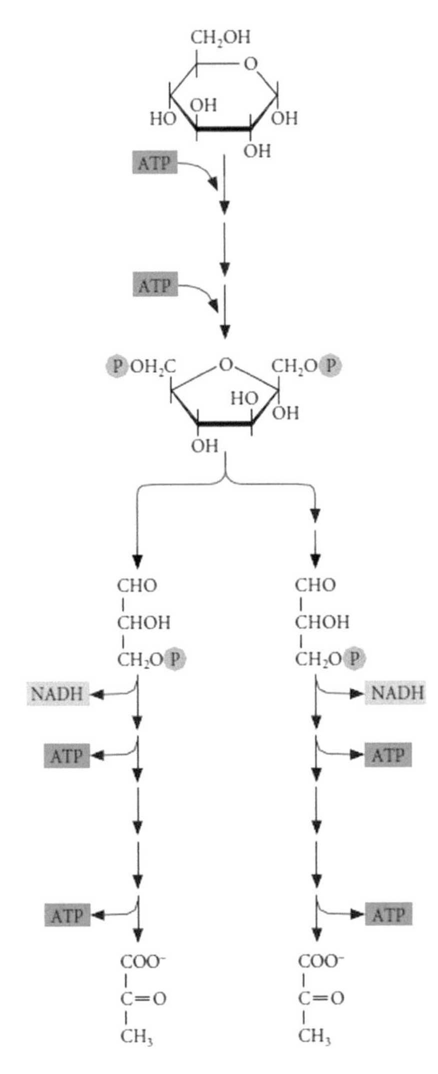

Fig. 17.7 Glucose enters the glycolysis pathway on its way to complete oxidation. The reactions along the pathway produce two molecules of pyruvate and energy

Adenosine triphosphate (ATP)

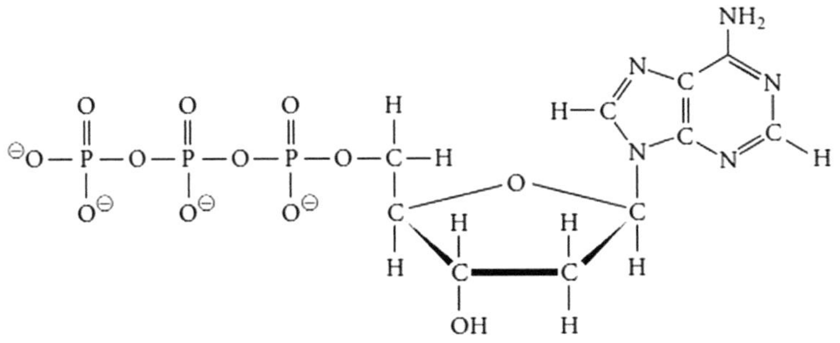

Fig. 17.8 Adenosine triphosphate (ATP) is consumed during the initial stages of glycolysis and generated during the latter stages

tinuous help. Decayed wood buried in moist soil does not re-form freshly cut pieces of wood, and a deck of cards spread out on the ground does not leap back up onto the table and into numerical and suit-based order. The reactions in chemistry can be considered either spontaneous or nonspontaneous.

Nonspontaneous process – A process that occurs only with continuous outside intervention.

▸ https://www.publicdomainpictures.net/pictures/90000/velka/cartas-13998578065ad.jpg

What happens to a copper or bronze statue when it is exposed to the environment? As illustrated by the Statue of Liberty that sits in New York Harbor (see ◻ Fig. 17.9), a green patina forms on the surface. This patina is a complex mixture of colored compounds, such as antlerite ($Cu_3SO_4(OH)_4$, blue-green), brochantite ($Cu_4SO_4(OH)_6$, pale green), chalcanthite ($CuSO_4{\cdot}5H_2O$, blue-green), cuprite (Cu_2O, dark red), and tenorite (CuO, black). All of these compounds include either Cu^+ or Cu^{+2}, so they are oxidation products of the copper metal. Although the chemistry is much more complex, we can represent this patina by showing its formation as the simple oxidation of copper:

$$2Cu(s) + O_2(g) \rightarrow 2CuO(s)$$

A chemical reaction causes the green patina to form without our assistance, so this is a spontaneous process. The reverse process,

$$2CuO(s) \rightarrow 2Cu(s) + O_2(g)$$

is nonspontaneous; it does not occur without some outside assistance. The reverse reaction of a spontaneous process is always nonspontaneous.

The spontaneous oxidation of metals due to exposure to the environment is quite common. ◻ Table 17.1 lists some of the patinas that form on other metals.

◻ **Fig. 17.9** The statue of liberty has an internal scaffold of iron and steel, but her copper exterior is coated in a green patina. *Source* ▸ https://www.publicdomainpictures.net/pictures/170000/velka/statue-of-liberty-1462950245iGm.jpg

Table 17.1 Patinas that spontaneously form on metals

Metal	Element symbol or composition	Natural color	Patina color
Aluminum	Al	Silvery white	Light gray
Brass	Cu and Zn	Gold	Dark brown to black
Bronze	Cu and Sn	Yellow to olive brown	Dark brown to black
Copper	Cu	Light red brown	Green
Iron	Fe	Lustrous silvery white	Reddish brown
Silver	Ag	White to gray	Black

Does the spontaneity of the oxidation of copper imply anything about the speed of the process? The short answer is no. It takes years for the green patina on a new copper roof to form completely. Under standard conditions, the conversion of diamond into graphite is also a spontaneous process, but luckily for people with diamond jewelry, the rate of this reaction is incredibly (almost immeasurably) slow. Rust forming on a nail and the decay of a buried log, even though they are spontaneous, are also slow processes. Biochemical reactions of glucose, on the other hand, are spontaneous and very rapid. The combustion of methane used to heat a pot of soup on the stovetop is even faster. **Thermodynamics**, the study of the changes in energy in a reaction (► Chap. 5), determines whether a process is possible. In the chapter on Chemical Kinetics (► Chap. 13), we studied the chemical kinetics of these processes to determine how fast, and by what mechanism, they occur. Both of these concepts of thermodynamics and kinetics are important for a complete understanding of why and how chemicals react with each other.

Thermodynamics – The study of the changes in energy in a reaction.

Example 17.2—Spontaneity in Common Processes

Which of these processes are spontaneous?

a. Ice melting in a hot oven.
b. Carbon dioxide and water reacting at STP to form methane and oxygen.
c. A basketball player jumping to dunk a basketball.
d. NaOH(aq) and HCl(aq) reacting when combined in a beaker.

Asking the Right Questions

What is a spontaneous process? What distinguishes it from a nonspontaneous process? Which of these processes meet our criterion for spontaneity?

Solution

The processes described in parts (a) and (d) are spontaneous, because they happen without continuous outside intervention. The combination of sodium hydroxide and hydrochloric acid is often used in chemical analysis precisely because the reaction is not only spontaneous but also rapid. The process described in part (b) is nonspontaneous, which suggests that the reverse process, the combustion of methane, is spontaneous. The basketball player in part (c) must intervene (by adding energy to oppose the force of gravity) in order to dunk the basketball. This process is nonspontaneous.

Are Our Answers Reasonable?

Our experience tells us that the answers are reasonable. Experience is an excellent teacher, and as we continue our work in chemistry, our chemistry toolbox of experiences from which we can draw will only increase!

What If…?

…we look more closely at the combustion of methane. We say that the reaction is spontaneous, yet the reaction, and related fire and heat, won't occur without a spark or other outside intervention. Why is this reaction classified as "spontaneous" if it requires outside intervention to start?

(*Ans: The reaction does not require continuous outside intervention. Rather, it just requires enough activation energy to initiate the reaction (see* ► *Sect. 13.6). Then it proceeds on its own.*)

Practice 17.2

Indicate whether each of these processes is spontaneous or nonspontaneous.

a. Potassium and water reacting.
b. Leaves falling from a tree.
c. A puddle evaporating from the sidewalk.
d. Photosynthesis.

17.3.1 Entropy

The overall **catabolism** (the biological degradation of molecules to provide smaller molecules and energy to an organism) of glucose by glycolysis is a spontaneous process similar to the combustion of glucose. When glucose is burned in air, six molecules of oxygen gas and one molecule of glucose combine to produce six molecules of water vapor and six molecules of carbon dioxide gas.

$$C_6H_{12}O_6(s) + 6O_2(g) \rightarrow 6CO_2(g) + 6H_2O(g)$$

Catabolism – The biological degradation of molecules to provide an organism with smaller molecules and energy.

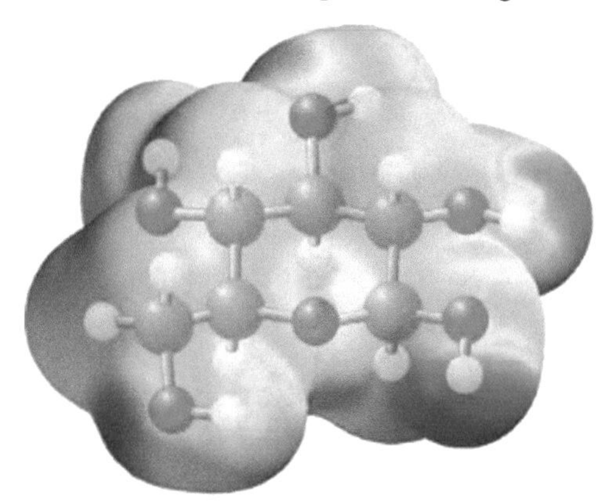

Glucose

When this reaction occurs, the number of gaseous molecules increases, which corresponds to a dramatic increase in the number of microstates that describe the system. Because of this increase, the probability that the reaction will produce products is greater than the probability that carbon dioxide and water will spontaneously form glucose and oxygen. In short, an increase in the number of microstates favors spontaneous reactions. We can describe this principle in more practical terms by introducing the concept of entropy.

Entropy (*S*) can be thought of as a measure of how the energy and matter of a system are distributed throughout the system. Investigations related to the concept of entropy began in the 1820s and 1830s with Nicolas Léonard Sadi Carnot (1796–1832) and Benoit Paul Emile Clapeyron (1799–1864). However, the concept wasn't mathematically developed until Rudolf Julius Emmanuel Clausius (1822–1888) worked on it in 1865. And although Clausius properly described entropy, its relationship to the molecular level wasn't illuminated until Boltzmann did so several decades later.

Entropy (S) – A measure of how the energy and matter of a system are distributed throughout the system.

Entropy isn't an easy concept to master, but we can gain important insight by considering the probabilities that have been the focus of our discussion. If the **multiplicity**—the number of microstates—increases, the number of ways in which energy and matter can be distributed also increases. Probability predicts that if the multiplicity of the system increases, there should be a corresponding increase in the number of ways in which energy and matter can be distributed in the system. This growth in the number of microstates increases the entropy of a system. In other words, the more probable outcome of a spontaneous process is that an increase in entropy occurs.

Multiplicity – The number of microstates within a system.

We can perform a simple experiment to help illustrate the influence of entropy. Say we have a friend place a bottle of perfume at one end of a room while we sit blindfolded in a chair on the opposite side of the room, as illustrated in ◻ Fig. 17.10. Our friend opens the bottle. Still blindfolded, can we tell if the bottle has been opened? At first, we might say it is still closed. Given a little time, though, we begin to notice the fragrance of the perfume and conclude that the bottle was opened. Why do we smell the perfume? The fragrance was released from the bottle on the opposite side of the room, so shouldn't it remain near the opened bottle? In fact, why should the perfume molecules leave the bottle at all? We know from experience that this isn't the case. What causes the perfume molecules to diffuse throughout the room and, eventually, into the noses of people at its most distant points? There is no pressure difference across the room, but still the perfume mixes spontaneously with the air. We can assume

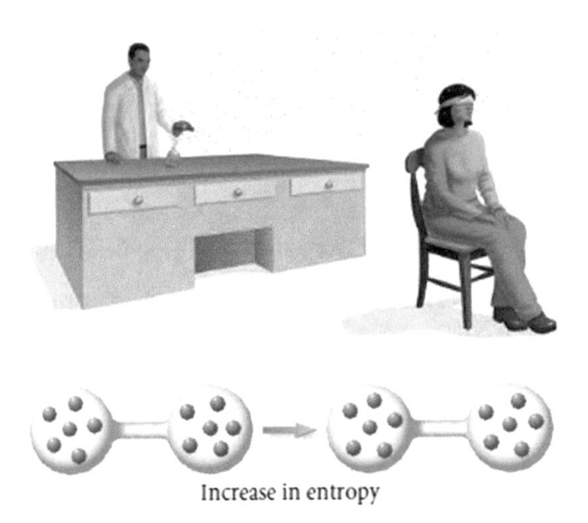

Fig. 17.10 An experiment to explain entropy. Diffusion of gases is driven by entropy

from the kinetic molecular theory that the attractive forces between the molecules of perfume and of those in the air are negligible, so there should be no significant change in the enthalpy of the process. Diffusion is neither exothermic nor endothermic for ideal gases. However, diffusion does increase the distribution of the molecules throughout the room. The entropy of the system has increased.

We need to be extremely careful when we think of entropy. In processes represented by the perfume experiment, it appears that the level of disorder of the molecules has increased. Sometimes, we incorrectly think of entropy as a measure of the disorder of a system. But even though the increase in entropy often parallels the increase in disorder, entropy is not a measure of how disorganized a system has become. Disorder is a macroscopic description of a system, whereas entropy is related to the number of microstates.

17.3.2 The Second Law

Inside the cells of our body lie the enzymes (polymers of amino acids) that release energy from glucose. As shown in Fig. 17.11, these large polymers start out as long flexible strands but, shortly after being made, fold into a small globular shape that contains pockets for binding glucose, ATP, ADP, water, and other compounds. The structure and type of amino acids around the binding pockets determine what type of reaction the enzyme will catalyze. Protein folding into the correct shape is a spontaneous process. Does this make sense? The flexible extended chain has many more motions than the folded enzyme; the number of ways to distribute energy in the system decreases as the enzyme folds. On the basis of this information alone, we might predict that protein folding should be nonspontaneous. A closer look reveals our need for a deeper understanding of entropy and spontaneity.

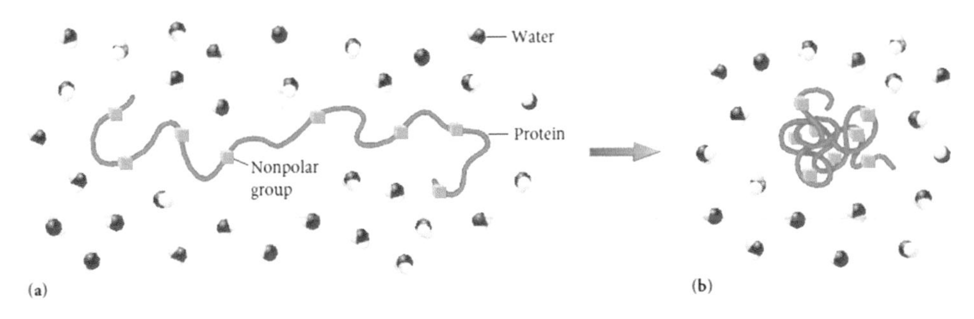

Fig. 17.11 The folding of proteins is driven by entropy. The unfolded protein (**a**) disrupts the interactions of the water molecules. The nonpolar groups are tucked inside the protein, removing them from interaction with the solvent (**b**). The folding also increases the intermolecular forces of attraction between different regions of the protein and increases the number of interactions between solvent molecules

As a rule, we say that a spontaneous process is accompanied by an increase in the entropy of the universe. This is the **second law of thermodynamics**. Mathematically, the change in the entropy of the universe is greater than zero for a spontaneous process:

$$\Delta S_{\text{universe}} > 0$$

Second law of thermodynamics – A spontaneous process is accompanied by an increase in the entropy of the universe; $\Delta S_{\text{universe}} > 0$.

However, recall the first law of thermodynamics, which says that the energy of the universe is constant ($\Delta E_{\text{universe}} = 0$). This contrasts with the second law, which implies that the entropy of the universe constantly increases. In other words, the number of possibilities for the distribution of the energy and matter of the universe constantly increases. This increase is related to the entropy changes in the system and surroundings (review the definitions of the terms universe, system, and surroundings). Because the total entropy of the universe is the sum of the change in entropy for a particular system (ΔS_{system}) and the change in entropy of the surroundings ($\Delta S_{\text{surroundings}}$), we can describe the change in entropy of the universe as follows:

$$\Delta S_{\text{universe}} = \Delta S_{\text{sytem}} + \Delta S_{\text{surrouding}}$$

- If $\Delta S_{\text{universe}} > 0$, the process is spontaneous.
- If $\Delta S_{\text{universe}} < 0$, the process is nonspontaneous.
- If $\Delta S_{\text{universe}}$ is zero, we say that the process is neither spontaneous nor nonspontaneous but is at equilibrium, a dynamic condition of energetic stability that we discussed in the Chemical Equilibrium chapter.

Because we take the sum of the change in entropy of the system and the change in entropy of the surroundings to obtain the change in entropy of the universe, ΔS_{system} could be a negative number and the overall process could still remain spontaneous. Even though an increase in the system's entropy will tend to result in a spontaneous reaction, some spontaneous reactions actually decrease the entropy of the system. This is possible only if the entropy change of the universe is positive, which means that the entropy change of the surroundings must more than make up for the negative entropy change of the system. For example, if $\Delta S_{\text{system}} = -50$ J/K·mol and $\Delta S_{\text{surroundings}} = +80$ J/mol·K, then:

$$\Delta S_{\text{system}} + \Delta S_{\text{surroundings}} = \Delta S_{\text{universe}}$$
$$-50\ \text{J/mol}\cdot\text{K} + (+80\ \text{J/mol}\cdot\text{K}) = +30\ \text{J/mol}\cdot\text{K}$$

In this case, $\Delta S_{\text{surroundings}}$ increases more than ΔS_{system} decreases, so $\Delta S_{\text{universe}}$ increases and the process is spontaneous. Table 17.2 outlines the effects of $\Delta S_{\text{universe}}$ as a function of the change in entropy of the system and surroundings. Pick some sample values for ΔS_{system} and $\Delta S_{\text{surroundings}}$, as we did just above, to help clarify the outcomes in the table.

In the case of our folding protein, we must take into account the entropy of the system and that of the surroundings if we are to properly assess the change in entropy of the universe and determine whether the process is spontaneous. As an extended chain, the enzyme's nonpolar groups must interact with the aqueous cellular environment—the surroundings—as shown in Fig. 17.12. The unfolded enzyme disrupts many of the interactions that occur among the solvent molecules (water). To reduce those disruptions within the surroundings, the folding enzyme tucks its nonpolar groups into the center of the globular structure, effectively removing their interaction with the water inside the cell. The folding allows the polar groups in the enzyme to interact with the

Table 17.2 $\Delta S_{\text{universe}}$ is the sum of ΔS_{system} and $\Delta S_{\text{surroundings}}$

$\Delta S_{\text{surroundings}}$	ΔS_{system}	$\Delta S_{\text{universe}}$	**Spontaneity**
+	+	+	Spontaneous
+	–	?	Spontaneous if $\Delta S_{\text{system}} < \Delta S_{\text{surroundings}}$
–	+	?	Spontaneous if $\Delta S_{\text{system}} > \Delta S_{\text{surroundings}}$
–	–	–	Nonspontaneous

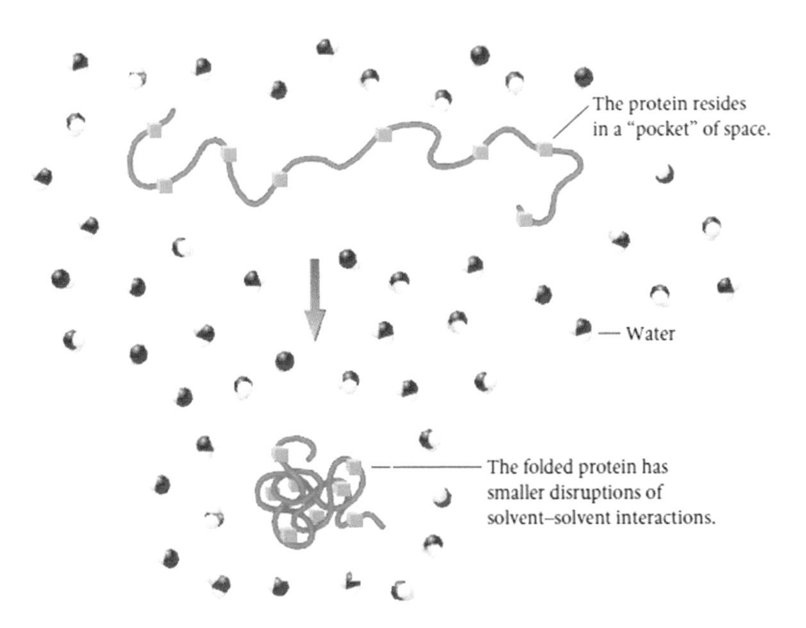

Fig. 17.12 The unfolded protein reduces the interactions of the solvent molecules. By folding, it allows those favorable interactions to take place

polar water molecules in the surroundings. The water molecules are also able to interact with each other with minimal disruption. The number of microstates for the protein decreases during the process of folding, but the number of microstates for the water and the water–protein interaction increases. The result is a $\Delta S_{\text{surroundings}}$ that is more positive than the negative ΔS_{system}. And the overall process, the folding of a protein, is spontaneous ($\Delta S_{\text{universe}} > 0$).

17

Example 17.3—Reaction Spontaneity

Determine whether these values will produce a spontaneous process.

$\Delta S_{\text{system}} = 140\,\text{J/K}\cdot\text{mol};\quad \Delta S_{\text{surroundings}} = -155\,\text{J/K}\cdot\text{mol}$

Asking the Right Questions

What entropy change defines a spontaneous process? How do we combine the entropy of the system and surroundings to determine that entropy change?

Solution

In a spontaneous process the entropy of the universe increases, so we must add the values for the system and surroundings to answer the question.

$$\begin{aligned}\Delta S_{\text{universe}} &= \Delta S_{\text{system}} + \Delta S_{\text{surroundings}}\\ &= 140\,\text{J/K}\cdot\text{mol} + (-155\,\text{J/K}\cdot\text{mol})\\ &= -15\,\text{J/K}\cdot\text{mol}\end{aligned}$$

Because our calculated value for the entropy of the universe is negative, the process is nonspontaneous.

Are Our Answers Reasonable?

Since we don't have a specific reaction to judge, we can only consider the value we calculated for the entropy of the universe. The negative value means that our answer of "nonspontaneous" is reasonable.

What If…?

…an error in the measurements was made, and the actual value for the entropy change of the surroundings is -2.06 kJ/K·mol? What is $\Delta S_{\text{universe}}$? What would ΔS_{system} need to be for the reaction to be spontaneous?

(*Ans:* $\Delta S_{universe} = -1920$ J/K mol, ΔS_{system} *would need to be greater than* +2060 J/K·mol *for the reaction to be spontaneous.*)

Practice 17.3

Determine whether the values for entropy in each of these cases will produc e a spontaneous process.

a. $\Delta S_{\text{system}} = -23$ J/K·mol; $\Delta S_{\text{surroundings}} = -55$ J/K·mol
b. $\Delta S_{\text{system}} = 38$ J/K·mol; $\Delta S_{\text{surroundings}} = 59$ J/K·mol
c. $\Delta S_{\text{system}} = -84$ J/K·mol; $\Delta S_{\text{surroundings}} = -132$ J/K·mol

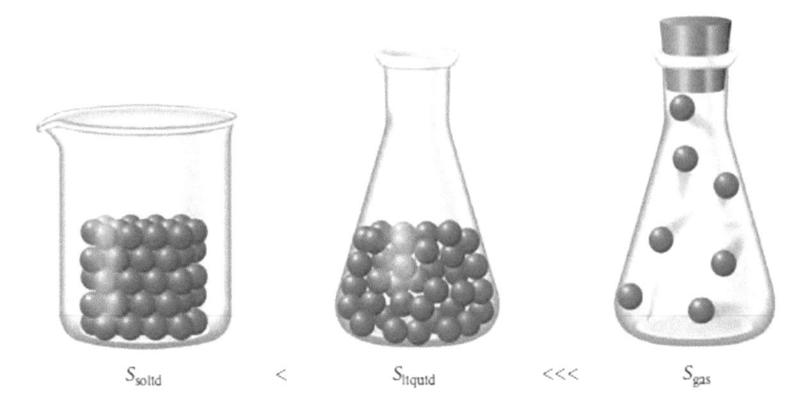

Fig. 17.13 Entropy increases as a compound changes state from solid to liquid to gas. Note that the increased molecular motions allow more microstates to exist in the compound

17.4 Temperature and Spontaneous Processes

Living organisms have been found all over the Earth, from the mouths of near-boiling geysers to the depths of the Arctic Ocean. These living organisms, just like the species that live near the tropics, require energy to survive. Many of them use glucose as a source of energy. In fact, some of the psychrotrophs (bacteria that can survive exposure to low temperatures) and psychrophiles (bacteria that thrive in low temperatures) employ a concentrated solution of glucose as an "antifreeze" because such a solution lowers their freezing point. Research by microbiologists and biochemists into the life processes of the psychrophiles indicates that these organisms, like a lot of other living things on the planet, break down glucose to produce energy through glycolysis. Does the temperature at which these organisms live affect the spontaneity of the reactions involved in glycolysis? To answer this question, we need to consider the signs on the change in entropy of the system (ΔS_{system}) and of the surroundings ($\Delta S_{surroundings}$). A positive change in the entropy of the universe is required for the process to be spontaneous.

In general, as compounds undergo changes in their physical states from solid to liquid to gas, the entropy of the *system* increases ($\Delta S_{system} > 0$), as shown in Fig. 17.13. Water molecules in an ice cube have a well-defined order. As the ice begins to melt, this well-defined order disappears and the water molecules increase their range of motions and, therefore, the number of microstates. The number of microstates that are possible in the liquid water suggests that the melting of ice corresponds to an increase in the entropy of the system. The same holds true when water is converted into steam. Conversely, the reverse of these processes (gas to liquid to solid) is typically accompanied by a loss of entropy ($\Delta S_{system} = -$).

What is the effect of energy transfer on ΔS? When steam condenses to liquid water, energy as heat flows out of the system and into the surroundings, and the kinetic energy of the particles in the surroundings increases. The motions of the atoms in the surroundings increase, and the sign of $\Delta S_{surroundings}$ is positive. On the other hand, if energy as heat flows from the surroundings to the system (liquid to vapor), we'd expect the kinetic energy of the particles in the surroundings to decrease and the motions of the atoms (and, therefore, the entropy) in the surroundings to decrease also. In short, a flow of energy as heat out of the system and into the surroundings (an exothermic process) corresponds to a positive sign for $\Delta S_{surroundings}$. An endothermic process has an opposite effect on the surroundings; endothermic processes correspond to a negative sign for $\Delta S_{surroundings}$.

NanoWorld/MacroWorld: The Big Effects of the Very Small—Industrial uses for the Extremophiles

Imagine living at the bottom of the Arctic Ocean, thinking life was grand inside glacial ice, enjoying the weather near a thermal vent, or relaxing under the crush of 1 mile of bedrock. These conditions sound fairly extreme to us, but not to a class of bacteria known as the extremophiles. Some of these microorganisms thrive near the hot bubbling mud-pots of Yellowstone National Park, others in the sulfur-laden waters near a geothermal vent at the bottom of the Atlantic Ocean. Extremophiles, some examples of which are listed in Table 17.3, survive in conditions that we humans would find extreme.

Biochemists, microbiologists, and geologists from around the world study these creatures because of the extreme conditions in which they live and to learn more

Table 17.3 Classes of extremophiles

Class	Extreme environment	Locations where they live
Acidophiles	Low pH	Sulfurous springs and acid mine drainage
Alkalinophiles	High pH	Alkaline lakes and basic soils
Anaerobes	Non-oxygen containing environments	Fermenting juices
Barophiles	High pressure	Deep sea vents and deep within the Earth
Copiotrophs	High nutrient levels	Sugar solutions
Halophiles	High ion concentration	Saline lakes and salt deposits
Hyperthermophiles	Temperatures above 70 °C	Hydrothermal vents, hot springs
Methanogens	Methane-rich environments	Deep-sea vents, oil deposits
Oligotrophs	Low nutrient levels	Desert, rocks
Psychrophiles	Low temperatures, typically below 10 °C	Glaciers, Artic Ocean, col soils
Thermophiles	Temperatures above 50 °C	Hydrothermal vents, hot springs

about the enzymes that continue to work under the equally extreme conditions inside them. For instance, millions of Americans are lactose-intolerant and have difficulty digesting the lactose in almost every dairy product, including milk and ice cream. Imagine if we could isolate beta-galactosidase (an enzyme that breaks down sugars like lactose into more easily digested compounds) from an extremophile that lived in icy environments. The extremophile's beta-galactosidase should be capable of working quite well in cold environments. By adding the isolated beta-galactosidase to milk and related products like ice cream, we could make lactose-free dairy products without having to heat them. Researchers at Pennsylvania State University have been able to show that this is possible by isolating a strain of bacteria, known as *Arthrobacter psychrolactophilus* that has a modified beta-galactosidase that works best when the temperature is 15 °C and continues to work well when the temperature is as low as 0 °C.

Thomas D. Brock isolated the first example of a true extremophile from hot springs, such as that shown in ◘ Fig. 17.14, in Yellowstone National Park in Wyoming. This bacterium, called *Thermus aquaticus*, grows most rapidly at temperatures near 70 °*C*. Other thermophiles, or heat-loving bacteria, include *Sulfolobus acidocaldarius*, which lives in sulfur-laden hot springs at temperatures as high as 85 °*C*, and *Pyrolobus fumarii*, which is isolated from deep-sea hydrothermal vents (◘ Fig. 17.15), grows only at temperatures above 90 °C, and reproduces best at 105 °C). These bacteria are of industrial interest because of their ability to grow at such high temperatures. For example, the enzyme Taq polymerase (isolated from *T. aquaticus*) is used in DNA fingerprinting because it can survive the severe temperature variations in the polymerase chain reaction used to make multiple copies of purified DNA.

The acid-tolerant extremophiles (acidophiles) are of interest because their enzymes are capable of operating in highly acidic environments. A potential application for the acidophile's enzymes is their addition to cattle feed, because they would work well in the acidic gut of an animal as an aid in digesting food. Their use would improve the usefulness of cheap food as a source of energy for the animals. The alkaliphiles (base-tolerant bacteria) thrive in basic soils such as those in the western United States and in Egypt. Proteases (enzymes that break down proteins) and lipases (enzymes that break down oils) isolated from these bacteria could find potential use in the detergent industry. Their addition to laundry detergents (which typically are basic) would improve the ability of the detergent to clean stains from clothing.

Investigators are currently searching, and finding, bacteria that live in environments we originally thought were sterile. After determining the types and properties of the enzymes that these bacteria possess, scientists are exploiting their industrial utility. The uses of these enzymes as catalysts to aid human life are endless. Will we find extremophiles that proliferate on Mars?... on Io, the volcanic innermost major moon of Jupiter?... on our own Moon? And, if so, what uses might we find for the enzymes they produce?

Fig. 17.14 The morning glory pool at Yellowstone National Park is named for the brightly colored thermophiles that flourish in this high-temperature environment. Photo by Jon Sullivan. *Source* ▶ http://pdphoto.org/PictureDetail.php?mat=pdef&pg=5277

Fig. 17.15 Hydrothermal vent at the Champagne vent of the Northwest Eifuku volcano in the Marianas Trench Marine National Monument. Extremophiles such as *Pyrolobus fumarii* and *Methanopyrus* live on the sides of these chimneys. The "smoke" flowing from the vents is actually made up of minerals from the lava under the ocean floor. *Source* ▶ https://oceanexplorer.noaa.gov/explorations/04fire/logs/hirez/champagne_vent_hirez.jpg

What is this exchange of energy to which we refer? If our process occurs under reversible conditions at a constant pressure, we can relate the energy of the process (q_{rev}) to the change in enthalpy of the process (ΔH). **Reversible conditions** occur when the process is allowed to proceed in infinitesimally small steps. At any point during the reaction, we could change the direction of the reaction with merely slight modifications. Often, reversible conditions exist during phase changes.

Reversible conditions – The conditions that occur when a process proceeds in a series of infinitesimally small steps.

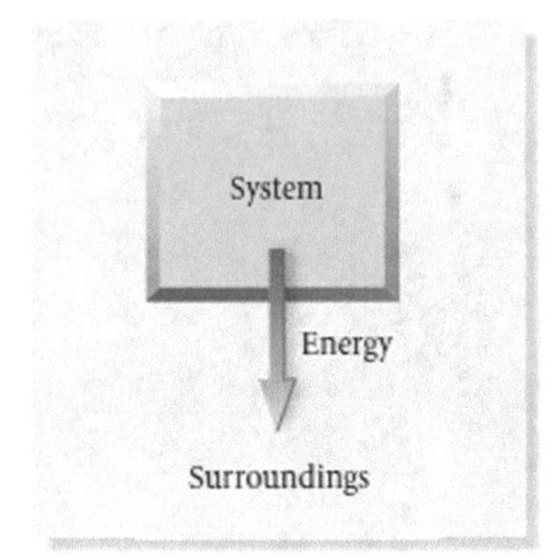

Exothermic process

Quantitatively, we can summarize our statements by saying that the change in the entropy is equal to the change in enthalpy of the phase change (a reversible process) divided by the temperature:

$$\Delta S_{\text{system}} = \frac{q_{\text{rev}}}{T} = \frac{\Delta H_{\text{system}}}{T}$$

where the temperature (T) is reported in kelvins (K) and the enthalpy (ΔH_{system}) is reported in joules per mole. Because of some assumptions we've made to arrive at this equation, its use is limited to describing heat transfers when the temperature remains constant. For example, the equation works well for describing phase transitions but poorly for describing a reaction.

Because $q_{\text{surroundings}} = -q_{\text{system}}$, and $q_{\text{system}} = \Delta H_{\text{system}}$, the value of $\Delta S_{\text{surroundings}}$ can be obtained using a similar equation.

$$\Delta S_{\text{surroundings}} = \frac{-\Delta H_{\text{system}}}{T}$$

Example 17.4—Entropy Change at a Phase Change

Instead of carrying water in their backpacks, hikers can melt ice to make drinking water. They also boil water for drinking and food preparation. What is the entropy change of the system for melting 1 mol of ice at 0.00 °C and 1 atm? What is the entropy change for melting 125 g of ice? The enthalpy of fusion at this phase change, $\Delta_{\text{fus}}H$ is 6.01 kJ/mol.

$$H_2O(s) \rightleftharpoons H_2O(l)\ \Delta_{\text{fus}}H = 6.01\,\text{kJ/mol at}\,0.00\,°\text{C}$$

Asking the Right Questions

How can we calculate the entropy change at a phase change? What quantities do we need to know? How is the second part of the example different than the first? How are they similar? The entropy change for the system or surroundings can be calculated if we know the temperature and the enthalpy of the system. Because the units of entropy are usually reported in J/K·mol, we should convert the units for the enthalpy and the temperature to match, so our calculation is simplified. The second part of the question asks us to examine the specific entropy change for a quantity of water that is not equal to 1 mol. We can do this part of the calculation by dimensional analysis.

Solution

The entropy change for the system or surroundings can be calculated if we know the temperature and the enthalpy of the system. Because the units of entropy are usually reported in J/mol·K, we can convert the units for the enthalpy and the temperature to match, so our calculation is simplified.

The change in the entropy of this reversible process can be calculated using the formula we just discussed.

$$\Delta S_{\text{system}} = \frac{\Delta H_{\text{system}}}{T} = \frac{6010\text{J/mol}}{273\text{K}} 22.0\text{J/K}\cdot\text{mol}$$

We've calculated the change in entropy (ΔS_{system} = 22.0 J/K·mol) for the phase change, so we can use this value to determine the change in entropy for melting 125 g of water.

$$\Delta S_{\text{system}} = 125\text{g} \times \frac{1\text{mol}}{18.02\text{g}} \times \frac{22.0\text{J}}{\text{mol}\cdot\text{K}} = \frac{153\text{J}}{\text{K}}$$

Are Our Answers Reasonable?
We calculated the entropy change for melting ice. What is the physical difference between liquid water and ice? Liquid water has a much greater range of motion of its particles than does ice. This means that there are many more possible microstates with the liquid, and a higher value of entropy at the same temperature. This is consistent with our answers, so they are reasonable.

What If…?
…we froze 375 g of liquid water at 0.00 °C and 1 atm? What would be the entropy change? (Should the change increase or decrease the entropy of the system?).

(Ans: −459 J/K).

Practice 17.4
Calculate ΔS_{system} for each of these processes at 25 °C. Assume that each is a reversible process.

$$I_2(g) \rightleftharpoons I_2(s) \quad \Delta_{\text{sub}}H = +62.4\,\text{kJ}.$$
$$H_2O(l) \rightleftharpoons H_2O(g) \quad \Delta_{\text{vap}}H = +40.7\,\text{kJ}.$$

▶ Here's What We Know So Far

- Spontaneous processes result in an increase in the entropy of the universe.
- The rate of a spontaneous reaction is not related to its spontaneity.
- The reverse of a spontaneous process is a nonspontaneous process.
- The change in entropy of the universe can be calculated as the sum of the change in entropy of the system and the change in entropy of the surroundings.
- The entropy of the system can be calculated for reversible processes by dividing the enthalpy for the process by the temperature of the process. ◀

17.5 Calculating Entropy Changes in Chemical Reactions

Metabolism (the biochemical reactions of an organism) releases energy. In this process, the potential energy stored in food is used by an organism, and some of the food is converted by chemical reactions into molecules that are needed for the organism to survive. All of these reactions are spontaneous in the body and vital to the living processes that occur at the cellular level. For instance, the body breaks down sucrose, perhaps contained in a sugar-laden gumdrop, into glucose and fructose. The glucose then enters a series of reactions, the glycolytic pathway being the first (◘ Fig. 17.8), as the body converts it into carbon dioxide and water. Along the way it produces ATP (◘ Fig. 17.9), a molecule used to store potential energy. Plants, on the other hand, synthesize glucose by combining carbon dioxide and water. This reaction is energetically uphill for the plant, so it uses the high-energy ATP molecule to drive the reaction to completion. Why is the breakdown of glucose spontaneous, whereas the formation of glucose is nonspontaneous? In other words, why is energy released when glucose is broken down and why is energy required to produce glucose? We can answer this question by examining the entropy changes in the combustion of glucose. Qualitatively, is the sign of ΔS_{system} positive or negative? The reactants include 1 mol of glucose molecules and 6 mol of gaseous oxygen molecules.

$$C_6H_{12}O_6(s) + 6O_2(g) \rightarrow 6CO_2(g) + 6H_2O(g)$$

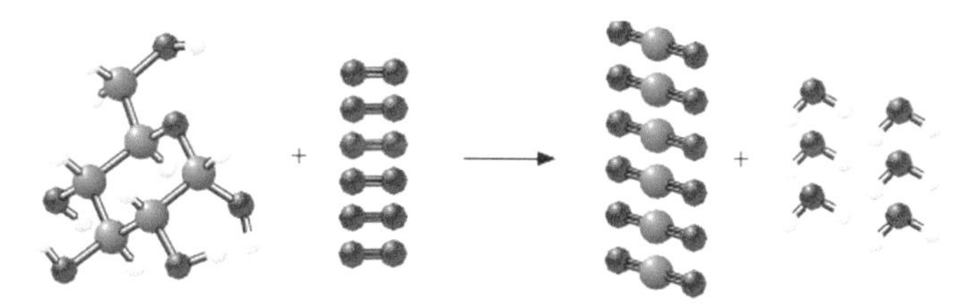

The combustion proceeds as 7 mol of reactants are converted into 12 mol of products (6 mol of gaseous carbon dioxide and 6 mol of gaseous water). Prior to the reaction, there were a large number of possible microstates because of the large number (7 mol) of reactants. Because gases occupy a larger volume in a flask than do equivalent quan-

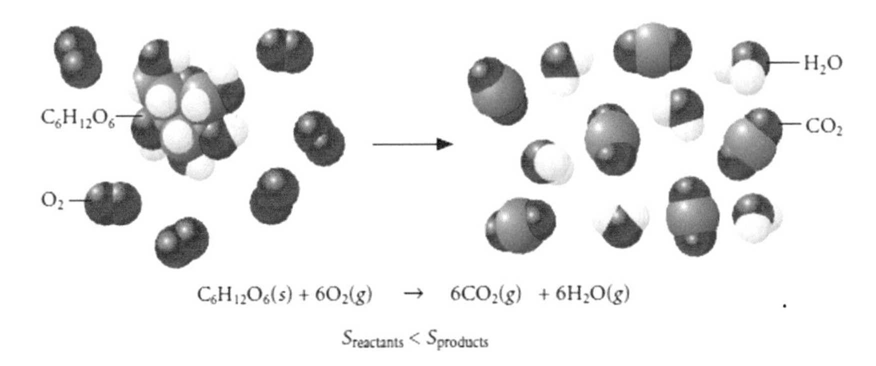

Fig. 17.16 An increase in the number of gaseous molecules increases the entropy of the system

tities of solids, the 6 mol of oxygen in our reaction occupy the majority of the locations within the flask as they rapidly travel within it. After the reaction, there are 12 mol of gas inside the flask. The number of microstates increased because of the larger number of gaseous products. The increase in the number of microstates is an increase in the entropy of the system (ΔS_{system} is positive), as shown in Fig. 17.16. If we examine the reverse of this reaction (from the viewpoint of the glucose-producing plant), the entropy of the system is lowered because we are combining 12 mol of gaseous carbon dioxide and water to make 6 mol of gaseous oxygen and 1 mol of solid sugar.

What we've discussed is a method by which one can usually predict the sign of the entropy change accurately. As a rule, the change in entropy of a reaction is positive if the number of gaseous molecules increases. Although the number of moles of solid and liquid molecules contributes to the overall number of microstates, the large volume occupied by gaseous molecules contributes much more. By simply examining a reaction, we can predict the direction of entropy change. The change is much harder to assess for reactions that do not involve gaseous molecules.

Example 17.5—Predict the Sign

1. Some camp stoves use butane as fuel. The combustion of butane is exothermic, providing heat to cook food and boil water. Predict the sign of ΔS_{system} for the combustion of butane.

$$2CH_3CH_2CH_2CH_3(g) + 13O_2(g) \rightleftharpoons 8CO_2(g) + 10H_2O(g)$$

2. The Haber process, the combination of hydrogen and nitrogen gases to form ammonia gas (NH_3) is one of the most widely used manufacturing processes because of worldwide demand for ammonia-based fertilizer. Predict the sign of ΔS_{system} for the production of ammonia.

$$3H_2(g) + N_2(g) \rightleftharpoons 2NH_3(g)$$

3. Barium hydroxide ($Ba(OH)_2 \cdot 8H_2O$) reacts with ammonium chloride (NH_4Cl) to form several products, including ammonia, water, and barium chloride. Predict the sign of ΔS_{system} for this reaction.

$$Ba(OH)_2 \cdot 8H_2O(s) + 2NH_4Cl(s) \rightleftharpoons BaCl_2(aq) + 2NH_3(g) + 10H_2O(l)$$

Asking the Right Questions

What is the relationship between the change in the number of moles of gas of reactants and products and the number of available microstates, and therefore the entropy change, of the system?

Solution

1. There are 18 mol of gaseous products and only 15 mol of gaseous reactants. This leads us to predict that the change in entropy is positive for this reaction. Based on calculations that we'll discover later, $\Delta S_{system} =$ 789 J/K·mol. This agrees with our prediction.
2. The Haber process takes 4 mol of gaseous reactants and forms 2 mol of gaseous products. The entropy change of the system is likely to be negative for this reaction. The actual ΔS_{system} value is −199 J/K·mol.
3. The reaction results in the formation of 2 mol of ammonia gas for each mole of barium hydroxide octahydrate that reacts with solid ammonium chloride. Consequently, we would predict that the change in system entropy would be positive. One factor to keep in mind is that the waters of hydration would be released as part of the process. These molecules would tend to raise the system entropy further as they join the liquid state. The actual ΔS_{system} value is 468 J/K·mol.

Are Our Answers Reasonable?

Did the sign of the entropy change of each reaction meet our expectation? In the first and third reactions, the production of more moles of gas than in the reactants led to an increase in the entropy of each system. The second reaction, the Haber Process, led to a reduction in the number of moles of gas produced, and a lower value of system entropy. These are reasonable answers, based on the relationship of entropy of the system to the change in the number of moles of gas in a reaction.

What If…?

…we have this reaction?

$$(NH_4)_2Cr_2O_7(s) \rightleftharpoons Cr_2O_3(s) + 4H_2O(l) + N_2(g)$$

What is the sign of the change in the entropy of the system?

(*Ans: positive*)

Practice 17.5

Predict the sign of the change in entropy for each of these reactions.

$$H_2O(l) \rightleftharpoons H_2O(g)$$

$$CH_3OH(l) + HCl(g) \rightleftharpoons CH_3Cl(l) + H_2O(l)$$

▶ Deepen Your Understanding Through Group Work

A member of your group runs two miles. Is that a spontaneous process?

- How do you decide?
- What are some of the chemical and thermodynamic processes the are occurring within the runner's body as she runs?
- What is happening to the environment as a result of the run? ◀

Qualitatively, our prediction of the sign of ΔS_{system} can be based on the number of gaseous molecules in the reaction. Quantitatively, the value for the change in the entropy (ΔS_{system}) is nearly as easy to determine. However, a problem arises. In order to calculate a change in a state function, we must subtract the final value from the initial value. How do we determine the initial value of entropy for a compound?

We can find the answer by returning to our Mt. Everest climbers earlier in this Chapter. Typically, such adventurers climb about 3400 m (11,400 ft) from the base camp at the foot of the mountain to reach the summit. Using this information, can we accurately say that the mountain is the tallest in the world? Mt. McKinley in Alaska, with an ascent of 5100 m (17,000 ft) appears to be taller. In order to accurately say that Mt. Everest is the tallest, we must consider that the base camp itself is 5400 m above sea level. That is, we must establish a reference point from which to compare the heights, as we've done in ◘ Fig. 17.17. In the same way, values of entropy are based on some reference point that is common to all compounds.

The **third law of thermodynamics** establishes a reference point for entropy. The law states that "the entropy of a pure perfect crystal at 0 K is zero." A "perfect" crystal, one in which all of the molecules are rigidly and uniformly aligned, has negligible kinetic energy at 0 K. In other words, if the excess kinetic energy of the crystal is zero, the crystal has zero entropy. We qualify the phrase kinetic energy with the word excess because there is still some atomic (electron) motion in a perfect crystal at 0 K.

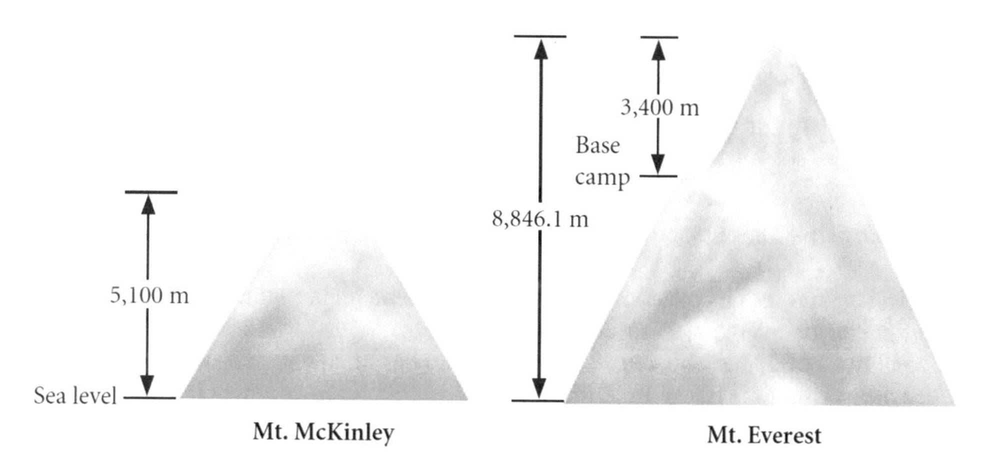

◘ **Fig. 17.17** Based on the distance traveled by the mountain climbers, Mt. McKinley appears to be taller than Mt. Everest. However, with our reference point in place, it is clear that Mt. Everest is much taller

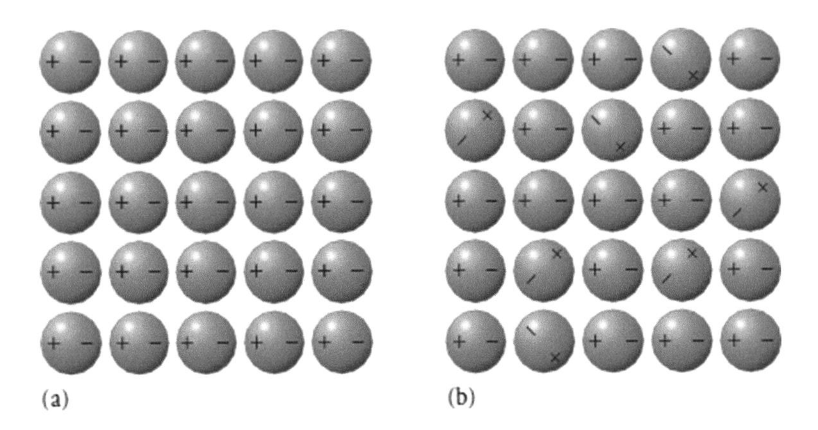

Fig. 17.18 Adding kinetic energy to a perfect crystal at 0 K increases the number of microstates. The result: The entropy of a compound is always a positive number. **a** Crystal at 0 K. **b** Crystal at some higher temperature

Table 17.4 Standard molar entropies and free energies for selected elements and compounds at 25 °C

Substance	$\Delta_f H°$ (kJ/mol)	$\Delta_f G°$ (kJ/mol)	$S°$ (J/mol·K)	Substance	$\Delta_f H°$ (kJ/mol)	$\Delta_f G°$ (kJ/mol)	$S°$(J/mol·K)
Br_2(l)	0	0	152	C_2H_5OH(l)	−278	−175	161
Br_2(g)	+31	+3	245	CH_3Cl(g)	−84	−60	234
HBr(g)	−36	−53	199	HCl(g)	−92	−95	187
CaF_2(s)	−1220	−1167	69	HCl(aq)	−168	−131	55
$CaCl_2$(s)	−796	−748	105	H_2(g)	0	0	131
$CaCO_3$(s)	−1207	−1129	93	Fe_2O_3(s)	−826	−740	90
C(graphite)	0	0	6	N_2(g)	0	0	192
C(diamond)	+2	+3	2	NO(g)	90	87	211
CO_2(g)	−394	−394	214	NO_2(g)	33	51	240
CO_2(aq)	−414	−386	118	NH_3(g)	−46	−16	192
CH_4(g)	−75	−50	186	H_2O(g)	−242	−229	189
C_2H_2(g)	+227	+209	201	H_2O(l)	−286	−237	70
C_2H_4(g)	+52	+68	220	NaOH(s)	−426	−379	64
C_2H_6(g)	−85	−33	230	H_2SO_4(aq)	−909	−745	20
CH_3OH(l)	−239	−166	127	ZnO(s)	−348	−318	44

Note that the magnitude of the standard molar entropies is larger for those molecules that are more complex: $S^{\circ}_{H_{(g)}} = 131 \text{J/mol} \cdot \text{K}$ and $S^{\circ}_{H_2(g)} = 189 \text{J/mol} \cdot \text{K}$ As more and more atoms are incorporated within a compound, the entropy of the compound increases. Similarly, note that the standard molar entropies are different for compounds in different states. For example, liquid water ($S°$ = 70 J/mol·K) has a lower entropy than gaseous water ($S°$ = 189 J/mol·K). This is understandable, because the molecules in the gaseous state have more freedom of movement and hence have a greater possible distribution of energies

Third law of thermodynamics – The entropy of a pure perfect crystal at 0 K is zero.

We convert this perfect crystal into a collection of molecules at some higher temperature by adding kinetic energy. This increases the number of microstates, resulting in an increase in entropy, as illustrated in Fig. 17.18. Under the standard conditions of 298 K and 1 atm, the compound has an associated standard molar entropy that we designate as $S°$. By measuring the change in entropy from 0 to 298 K, researchers have determined the standard molar entropies of a wide variety of compounds (see Table 17.4).

Computationally, we use these standard molar entropy values in much the same way that we used enthalpy values to calculate $\Delta H°$ for a reaction. The basic principles of Hess's law (see ▶ Chap. 5), which we used to calculate the change in enthalpy for a reaction, also work for determining the change in the entropy of a reaction. To summarize, the net standard state entropy gain (or loss) for a reaction is calculated by subtracting the total standard state entropy change for the reactants from the total standard state entropy change for the products:

$$\Delta S^{\circ} = \sum nS^{\circ} \text{ products} - \sum nS^{\circ} \text{ reactants}$$

where:

n is the stoichiometric coefficient of each of the compounds in the reaction, and,
Σ indicates that the standard state entropy ($S°$) is summed.

Just as in the determination of the change in enthalpy of a reaction, $\Delta S°$ for a reaction depends on the way the reaction is written.

Example 17.6—Calculating ΔS°

In the oxidation of glucose at 25 °C, the products are different from those we considered early in Sect. 17.4. In that discussion, glucose reacted at body temperature. Here, at 25 °C, both oxygen and carbon dioxide are gases and water is a liquid. Use the table of thermodynamic values in the appendix to calculate the change in entropy for this reaction under standard conditions.

$$C_6H_{12}O_6(s) + 6O_2(g) \rightleftharpoons 6CO_2(g) + 6H_2O(l)$$

Asking the Right Questions

What are the contributors to the change in entropy of the system? What are the changes in the number of moles of gas? What about the total number of molecules of all kinds? Does this have an impact on the system entropy? Technically, how can we calculate this entropy change from tabulated data?

Solution

The number of moles of gas is the same on both sides of the reaction arrow, so this does not seem to give us a change in entropy. We might (correctly) predict that the entropy of the system would increase slightly, because we are forming six molecules of a liquid where we had one molecule of solid reactants.

We can use the summation equation to calculate the change in entropy for the reaction.

$$\Delta S^{\circ} = \sum nS^{\circ}_{\text{products}} - \sum nS^{0}_{\text{reactants}}$$

From Table 17.4 we can obtain $S°$ values for each of the compounds in the reaction. Using the equation shown above, we can calculate the value of $\Delta S°$:

$$\begin{aligned}
6 \text{ mol } CO_2 \times 214\,\text{J/K}\cdot\text{mol} &= 1284\,\text{J/K} \\
6\,\text{mol} H_2O \times 70\,\text{J/K}\cdot\text{mol} &= \underline{420\,\text{J/K}} \\
\text{Total } S^0_{\text{products}} &= 1704\,\text{J/K} \\
1 \text{ mol } C_6H_{12}O_6 \times 212\,\text{J/K}\cdot\text{mol} &= 212\,\text{J/K} \\
6\,\text{mol} O_2 \times 205\,\text{J/K}\cdot\text{mol} &= \underline{1230\,\text{J/K}} \\
\text{Total } S^0_{\text{reactants}} &= 1442\,\text{J/K} \\
\text{Total } S^0_{\text{products}} &= 1704\,\text{J/K} \\
-\text{Total } S^0_{\text{reactants}} &= \underline{-1442\,\text{J/K}} \\
\Delta S^{\circ} &= 262\,\text{J/K}
\end{aligned}$$

Is our Answer Reasonable?

Although the number of molecules of gas is the same for the reactants and the products, we can reasonably assert that the gain in entropy is due to the formation of six molecules of liquid from every molecule of solid. The effect is not as profound as for the formation of a gas but it does contribute, as in this case. Our answer is reasonable.

What If…?

…we knew that $\Delta H°$ for this reaction is −2812 kJ/mol? Thinking ahead in the text a few pages, let's speculate: Does this large negative enthalpy favor a spontaneous reaction, or oppose it?

(*Ans: We will discuss in the next section that exothermic reactions favor spontaneous processes because they contribute energy to the surroundings, which increases the motion of the particles in the surroundings, increasing the number of microstates, and, therefore, the entropy.*)

Practice 17.6

Predict the sign of $\Delta S°$ for each of these reactions. Then calculate the values of $\Delta S°$. We will need to use the table of thermodynamic values in the appendix. Do our calculations agree with our predictions?

$$CH_3OH(l) + HCl(g) \rightleftharpoons CH_3Cl(g) + H_2O(g)$$
$$3O_2(g) \rightleftharpoons 2O_3(g)$$

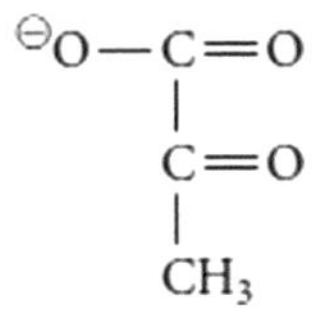

$$\begin{array}{c} {}^{\ominus}O-C=O \\ | \\ C=O \\ | \\ CH_3 \end{array}$$

Fig. 17.19 Pyruvate is a product of the glycolysis pathway

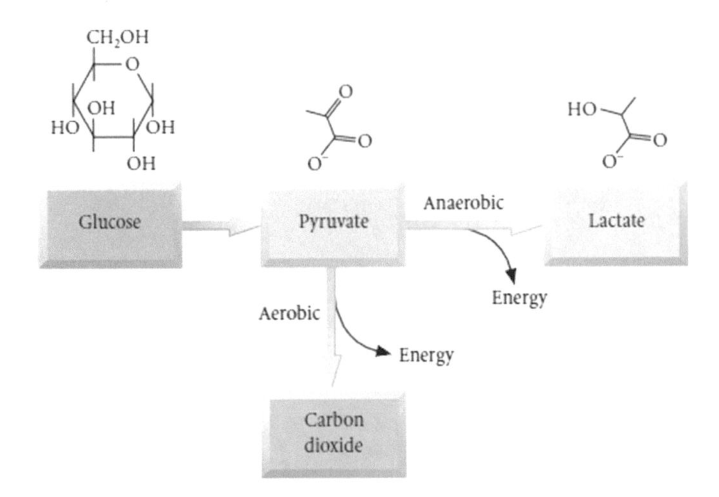

Fig. 17.20 Pyruvate cannot continue oxidation (aerobic pathway) without oxygen. Instead, energy is derived by the conversion of pyruvate to lactate (anaerobic pathway)

17.6 Free Energy

In exercising muscles, the level of oxygen often is a limiting reagent. As the muscles consume glucose to provide the energy needed to contract and relax, the limited supply of oxygen forces a buildup of pyruvate (see Fig. 17.19). This molecule can continue on the typical glycolysis pathway toward complete oxidation, but with limited oxygen this does not happen. Instead, the exercising muscle obtains energy by converting pyruvate into lactate (a reduction reaction) by the pathway shown in Fig. 17.20. The reduced form of a biological molecule known as NAD^+ (nicotinamide adenine dinucleotide) is oxidized in the process. (Remember from ▶ Chap. 5 that we must have an oxidation and a reduction for a redox reaction to occur.) Unfortunately, lactate is potentially a metabolic dead end in humans; further catabolism of lactate may not occur. Eventually, the lactate can be changed back to pyruvate by the reverse reaction and then catabolized in the presence of oxygen. While early research indicated that the body signals the presence of excess lactate (and a decrease in oxygen in the muscles) with a burning sensation, there may be some recent evidence that contradicts this claim.

The reaction of pyruvate to make lactate is energetically favorable. Some energy is harvested from the process. However, qualitatively, it is difficult to predict this by just looking at the reaction. Determining the spontaneity of the reaction by simply looking at the states of the reaction is also difficult.

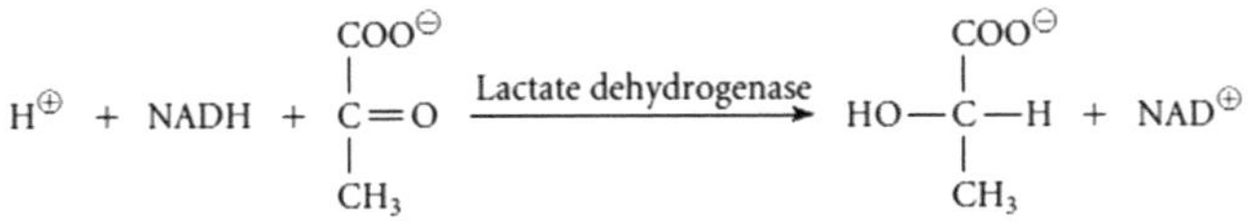

$$H^{\oplus} + NADH + \begin{array}{c} COO^{\ominus} \\ | \\ C=O \\ | \\ CH_3 \end{array} \xrightarrow{\text{Lactate dehydrogenase}} \begin{array}{c} COO^{\ominus} \\ | \\ HO-C-H \\ | \\ CH_3 \end{array} + NAD^{\oplus}$$

After some calculations, we can determine the value of $\Delta S°$ (that is, ΔS_{system}), but we also have to determine the entropy of the surroundings ($\Delta S_{surroundings}$) in order to calculate the change in entropy of the universe (system + surroundings) and, therefore, the spontaneity of the reaction. A simpler and more useful way to determine reaction spontaneity was identified by Josiah Willard Gibbs (1839–1903), professor of mathematical physics at Yale from 1871 to 1903.

Gibbs showed that calculating a property he called the **free energy (*G*)** makes possible the straightforward determination of reaction spontaneity. The equation is indicative of two ideas that we have discussed.

Free energy (*G*) – The state function that is equal to the enthalpy minus the temperature multiplied by the entropy. Used to determine the spontaneity of a process.

- Idea 1: An increase in the number of microstates available to the system favors an increase in the entropy of the universe.
- Idea 2: An exothermic reaction increases the kinetic energy of the surroundings, therefore increasing the range of motions of its particles, leading to an increase in the entropy of the surroundings and favoring an increase in the entropy of the universe.

The **Gibbs equation** is:

$$G = H - TS$$

where H is the enthalpy of the system, T is the temperature in kelvins, and S is the entropy of the system. We are interested in the *change* in the free energy of the system, so the Gibbs equation becomes:

$$\Delta G = \Delta H_{\text{system}} - \mathrm{T}\Delta S_{\text{system}}$$

at constant temperature and pressure.

Gibbs equation $\Delta G = \Delta H - T\Delta S$

At first, this equation doesn't seem to be connected to our previous discussion about entropy, but let's take a close look at where this equation comes from. We have already established that the entropy change for the universe is equal to the sum of the entropy change for the surroundings plus that for the system:

$$\Delta S_{\text{universe}} = \Delta S_{\text{surrounding}} + \Delta S_{\text{system}}$$

In addition, we know that, for a constant-temperature process, the entropy of the surroundings is equal to the negative enthalpy change for the system divided by the temperature (Sect. 17.3), allowing us to make this substitution:

$$\Delta S_{\text{surroundings}} = \frac{-\Delta H_{\text{system}}}{T}$$

So:

$$S_{\text{universe}} = \frac{-\Delta H_{\text{system}}}{T} + \Delta S_{\text{system}}$$

Now, we have conveniently defined the entropy change for the entire universe (which is very difficult to measure) in terms of changes to our system, which we can measure in the laboratory. If we multiply through by $-T$, we get this equation:

$$-T\Delta S_{\text{univers}} = \Delta H_{\text{system}} - T\Delta S_{\text{system}}$$

Notice that the right side of this equation is exactly the same as in the Gibbs equation given above. This means that $-T\Delta S_{\text{universe}}$ is the same as ΔG. Because of this relationship, a decrease in ΔG means an increase in $\Delta S_{\text{universe}}$ and vice versa. Therefore, a negative Gibbs free energy change ($\Delta G < 0$), means an increase in the entropy of the universe ($\Delta S_{\text{universe}} > 0$), which means the process will happen spontaneously. Correspondingly, a positive Gibbs free energy change ($\Delta G > 0$), means an decrease in the entropy of the universe ($\Delta S_{\text{universe}} < 0$), which means the process will not happen spontaneously. Notice also that the units of Gibbs free energy are in units of joules. Another useful property of free energy is that it tells us the maximum amount of work that can be done by a system through both enthalpy and entropy changes, and the sign of ΔG tells us whether the work will be done by the system (negative) or will need to be done on the system (positive), which is the same concept as exothermic and endothermic enthalpy changes. Table 17.5 lists the outcomes of the change in free energy based on the relationship between ΔH and ΔS.

Therefore, the Gibbs equation can be used to calculate the change in free energy for a process, because we know how to calculate ΔH and ΔS using tabulated values (see Table 17.4). All we need, then, is the temperature of the process and the application of some assumptions. First, as a requirement of the Gibbs equation, we must assume that the pressure remains constant. We also have to assume that both ΔH and ΔS are temperature independent, which is not always a good assumption, especially for ΔS.

Let's reexamine the conversion of pyruvate to lactate under anaerobic conditions. Qualitatively, we cannot use the change in the number of moles of gas from reactants to products to help us assess whether the reaction will be spontaneous, because there are no gases in the reaction.

Table 17.5 Effect of enthalpy, entropy, and temperature on the spontaneity of a process

ΔH	ΔS	Low temperature	High temperature
+	+	$\Delta G = +$; nonspontaneous	$\Delta G = -$; spontaneous
+	–	$\Delta G = +$; nonspontaneous	$\Delta G = +$; nonspontaneous
–	+	$\Delta G = -$; spontaneous	$\Delta G = -$; spontaneous
–	–	$\Delta G = -$; spontaneous	$\Delta G = +$; nonspontaneous

$$H^{\oplus} + NADH + \underset{CH_3}{\overset{COO^{\ominus}}{C{=}O}} \xrightarrow{\text{Lactate dehydrogenase}} HO{-}\underset{CH_3}{\overset{COO^{\ominus}}{C}}{-}H + NAD^{\oplus}$$

However, because the values of ΔH and ΔS have been measured for this reaction, we can determine the change in free energy. In this system, $\Delta H = -78.7$ kJ/mol and $\Delta S = -88.75$ J/mol·K. On the basis of these data alone, we identify the reaction as exothermic with a decrease in system entropy. What is the value of ΔG for the reaction at 298 K?

$$\Delta G = \Delta H - T\Delta S$$
$$\Delta G = -78{,}700\,\text{J/mol} - (298\,\text{K} \times -88.75\,\text{J/mol}\cdot\text{K})$$
$$\Delta G = -51{,}188\,\text{J/mol} = -52.3\,\text{kJ/mol}$$

Our calculations show that the free energy of this reaction decreases. One thing to keep in mind during such calculations is that in order to use the Gibbs equation, we must modify the values of ΔH and ΔS so that they have the same units. In our calculation, we expressed them in units of J/mol and J/mol·K.

We previously mentioned that the change in free energy is negatively related to the change in the entropy of the universe. That is, a negative value for the change in free energy implies a spontaneous process. For a reaction, such as the conversion of pyruvate to lactate, in which the change in free energy is a negative value ($\Delta G = -$), the entropy of the universe has a positive value ($\Delta S_{universe} = +$). The reaction is spontaneous under biological conditions. We might not have predicted this solely on the basis of our examination of $\Delta S°$ for the reaction. Similarly, a process can be considered nonspontaneous if there is an increase in the free energy of the system ($\Delta G = +$). This means that if the reaction as written, is nonspontaneous, the reverse reaction is spontaneous. For instance, the formation of lactate from pyruvate is spontaneous (a lowering of free energy), but the formation of pyruvate from lactate is nonspontaneous. Remember that "nonspontaneous" merely means that continuous work must be done to make the reaction happen. The human body does turn lactate into pyruvate but doing so has an energy cost associated with it.

Example 17.7—Spontaneity and Biochemical Reactions

Yeast cells operate under anaerobic conditions to convert glucose to pyruvate, and pyruvate to ethanol and CO_2. Humans are unable to convert pyruvate to ethanol and must make lactate instead. What is the value of the change in the free energy (ΔG) for the conversion of glucose to ethanol at 25 °C? (Assume that both the enthalpy and the entropy were also determined at 25 °C.)

$$\underset{\text{Glucose}}{C_6H_{12}O_6(s)} \rightleftharpoons \underset{\text{Ethanol}}{2C_2H_6O(l)} + 2CO_2(g)$$

$$\Delta H = -70.0\,\text{kJ/mol}$$
$$\Delta S = 534\,\text{kJ/K}\cdot\text{mol}$$

Asking the Right Questions

Based on the signs of the enthalpy and entropy values, do we expect the free energy change for the reaction to be positive or negative? In other words, do we expect the reaction to be nonspontaneous or spontaneous? Why? How do we calculate the free energy, given these data?

Solution

The entropy change for the process is positive, and the reaction is exothermic, both of which favor a spontaneous process (negative ΔG).

$$\Delta G = \Delta H - T\Delta S$$
$$\Delta G = -70{,}000\,\text{J/mol} - (298\,\text{K}) \times (534\,\text{J/K}\cdot\text{mol})$$
$$\Delta G = -229{,}132\,\text{J/mol}$$
$$\Delta G = -229\,\text{kJ/mol}$$

The reaction is spontaneous, with a rather large negative free energy change.

Is Our Answer Reasonable?

Given the signs of the changes in entropy and enthalpy, the negative value for free energy was not only chemically reasonable, but mathematically required.

What If…?

…we considered the reaction that we discussed as part of Example 17.5?

a. $4NH_3(g) + 5O_2(g) \rightleftharpoons 4NO(g) + 6H_2O(g)$.
What would be standard free energy change in this reaction? Is it spontaneous?

(*Ans: $G^o = -962$ kJ/mol. Yes, it is spontaneous.*)

Practice 17.7

Calculate ΔG at −10 °C and at 45 °C for the ice-to-water phase transition. Which do we predict to be spontaneous? Do our calculations agree with our predictions? (Assume that the enthalpy and entropy are temperature independent.)

$$H_2O(s) \rightleftharpoons H_2O(l)$$
$$\Delta H = 6.01\,\text{kJ/mol} \quad \Delta S = 22.0\,\text{J/K}\cdot\text{mol}$$

17.6.1 A Second Way to Calculate ΔG

Just like enthalpy and entropy, free energy is a state function. Remember that when we are determining the value of a state function, it doesn't matter how we calculate it because the end result is independent of the path. To illustrate this, we return to the trek up Mt. Everest. Climbers can take either the more popular south route up the Khumbu icefall or the north route along the ridge to the summit. No matter which way the climbers ascend, no matter how many steps there are in each of the two routes, and no matter how long their ascent takes the climbers, if they make it to the top they have climbed to a point 29,035 ft above sea level. As we saw before, altitude change is a state function. It doesn't matter how two climbers arrive at the summit of Mt. Everest; their change in altitude from the base camp to the top is the same.

Because free energy is a state function, we can obtain ΔG by using the same method we used to calculate ΔH. Adding a series of reactions, each with a known free energy, is often a good way to arrive at a particular calculation of the free energy, because many reactions cannot be performed directly in the lab. For example, the combustion of carbon, as occurs in our charcoal grills, can never give just carbon monoxide but rather, always results in a mixture of oxides. For another example, say we were interested in the calculation of ΔG for the formation of ice from steam at −18 °C (255 K) and 1 atm, another reaction that is difficult to do in the lab. We could sum the equations (and the free energy terms) that describe the stepwise conversion from gas to liquid and from liquid to ice to arrive at ΔG for the process. Note that the free energy terms are specific to this temperature:

$$\begin{array}{ll} H_2O(g) \rightarrow H_2O(l) & \Delta G = -13.6\,\text{kJ/mol} \\ H_2O(l) \rightarrow H_2O(s) & \Delta G = -0.39\,\text{kJ/mol} \\ \hline H_2O(g) \rightarrow H_2O(s) & \Delta G = -14.0\,\text{kJ/mol} \end{array}$$

We see that the process of converting water vapor to ice has a negative change in free energy at −18 °C. What does this value of ΔG tell us, and does it agree with what we expect for this system? Under these conditions (1 atm, −18 °C), the formation of ice from steam is spontaneous, whereas the reverse process, sublimation of solid to the vapor, is not.

The summation procedure is not limited to phase changes. For example, glycogen (a polymer of glucose used as storage for quick energy release) is degraded to glucose-1-phosphate (Glu-1-P) when the body signals that energy is needed. In order for this derivative of glucose to be used in glycolysis, it must first be converted to glucose-6-phosphate (Glu-6-P) by a two-step process. The hydrolysis reaction of each of these phosphates has been extensively studied and the free energy change noted. What is the net free energy change for the direct conversion of glucose-1-phosphate to glucose-6-phosphate? On the basis of our previous discussion, we can calculate ΔG:

$$\begin{array}{ll} \text{Glu-1-P} + H_2O \rightarrow \text{glucose} + \text{phosphate} & \Delta G = -21\,\text{kJ/mol} \\ \text{glucose} + \text{phosphate} \rightarrow \text{Glu-6-P} + H_2O & \Delta G = +14\,\text{kJ/mol} \\ \hline \text{Glu-1-P} \rightarrow \text{Glu-6-P} & \Delta G = -7\,\text{kJ/mol} \end{array}$$

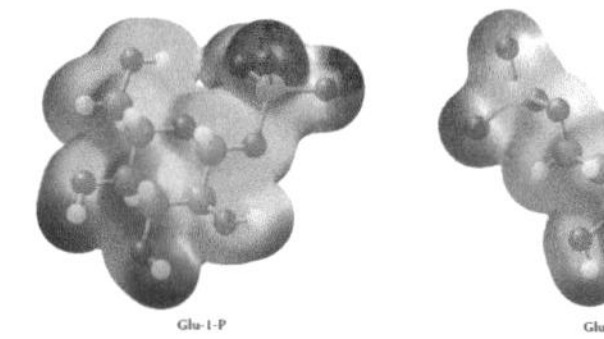

The net reaction is spontaneous. The reverse of this reaction, the conversion of glucose-6-phosphate into glucose-1-phosphate, is nonspontaneous with $\Delta G = +7$ kJ/mol. Does the result make sense? If the body needs energy, production of glucose-6-phosphate should be spontaneous.

We can use the summation method for determining the spontaneity of a reaction in cases where the free energy of each in a series of reactions is known. In some cases, we can even measure the spontaneity of supposed reactions or processes that we do not wish to perform in the laboratory. The summation of a series of reactions makes possible the quick and easy calculation of the spontaneity of a reaction.

17.6.2 Free Energy of Formation

We can also calculate the spontaneity of a reaction using standard free energies of formation ($\Delta_f G°$). The **standard molar free energy of formation ($\Delta_f G°$)** is the change in the free energy of 1 mol of a substance in its standard state as it is made from its constituent elements in their standard states. The equation relating the standard free energy of formation for any compound is analogous to the equation written for the standard enthalpy of formation. For example,

$$H_2(g) + 1/2O_2(g) \rightarrow H_2O(l)$$

The standard molar free energy of formation, $\Delta_f G°$, for all elements in their standard states is zero, just as it is for the standard molar enthalpy of formation, $\Delta_f H°$. Recall from our previous discussion on the third law of thermodynamics that the standard molar entropy of an element, $S°$, is not equal to zero.

Standard molar free energy of formation ($\Delta_f G°$) – The free energy of a compound formed from its elements in their standard states.

The tabulation of $\Delta_f G°$ values is given in Table 17.4 and in the appendix. Access to this and similar tables enables us to determine quickly the free energy change for an unknown reaction. In much the same manner that we use $\Delta S°$ and $\Delta H°$, we can find the change in the standard free energy of a reaction. The sum of the free energies of formation for the reactants is subtracted from the sum of the free energies of formation for the products:

17

$$\Delta G° = \sum n\Delta_f G°_{\text{products}} - \sum n\Delta_f G°_{\text{reactants}}$$

where ***n*** is the stoichiometric coefficient of each of the compounds in the reaction.

Example 17.8—I'm Hungry. What's for Dinner? Free Energy and Spontaneity

In a fasting organism, there is no more glucose left to make energy. Glycogen supplies become depleted. The organism must find a way to make molecules that can be catabolized for energy. One such source of energy is found in the amino acid alanine. This molecule is converted to pyruvate for use in the production of energy (remember that pyruvate is an intermediate during the complete oxidation of glucose). Given the equation for the production of pyruvic acid, the acidic form of pyruvate, from alanine, as well as the values of the free energy of formation for both, determine the spontaneity of this reaction under standard conditions.

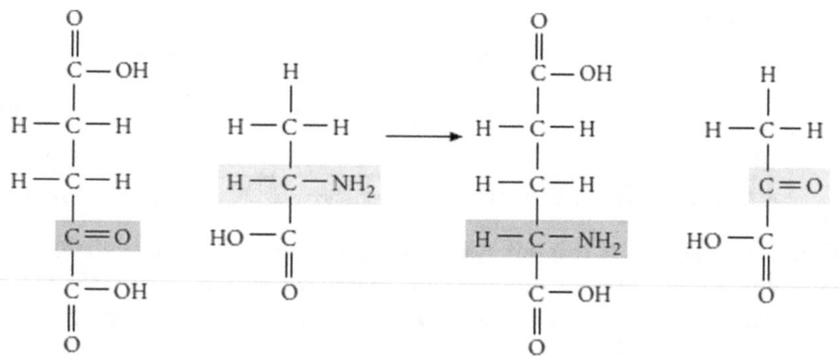

Ketoglutaric acid + Alanine → Glutamic acid + Pyruvic acid

−195.5 $\Delta_f G^\circ$ (kJ/mol) −88.5 $\Delta_f G^\circ$ (kJ/mol) −174 $\Delta_f G^\circ$ (kJ/mol) −111 $\Delta_f G^\circ$ (kJ/mol)

Asking the Right Questions

What is the assumption about the standard free energy of the products and reactants that allows us to solve for the free energy change in this exercise? How and why is this similar to our work with enthalpy and entropy?

Solution

As with enthalpy and entropy, free energy is a state function (the change is independent of the path). The existence of $\Delta_f G^\circ$ values for both the starting materials and the products means that we can use the summation equation to calculate the standard free energy change for the reaction.

$$1 \text{ mol glutamic acid} \times -174 \text{ kJ/mol} = -174 \text{ kJ.}$$
$$1 \text{ mol pyruvic acid} \times -11 \text{ kJ/mol} = \underline{-111 \text{ kJ.}}$$
$$\text{Total } \Delta_f G^\circ_{\text{products}} = -285 \text{ kJ.}$$

$$1 \text{ mol ketoglutaric acid} \times -195.5 \text{ kJ/mol} = -195.5 \text{ kJ.}$$
$$1 \text{ mol alanine} \times -8.5 \text{ kJ/mol} = \underline{88.5 \text{ kJ.}}$$
$$\text{Total } \Delta_f G^\circ_{\text{reactants}} = -284 \text{ kJ.}$$

$$\text{Total } \Delta_f G^\circ_{\text{products}} = -285 \text{ kJ.}$$
$$-\text{Total } \Delta_f G^\circ_{\text{reactants}} = \underline{-(-284 \text{ kJ}).}$$
$$\Delta G^\circ = -1 \text{ kJ.}$$

Judging by our calculations, the change in free energy is negative and the reaction is spontaneous.

Are Our Answers Reasonable?

Since the reaction does occur, a negative value for the free energy change is reasonable. Remember that the free energy change does not relate to the rate of the reaction. Rather, it only considers whether the reaction can occur without continuous outside intervention. This reaction can.

What If…?

…instead of using one mole of each substance based on the balanced equation, we used 1000 mol of each in our calculation. (Whether 1 mol or 1000 mol, the equation is still balanced!) How would that change the answer, and how would that change the meaning of the answer?

(Ans: ΔG° for 1000 mol = −1000 kJ. However, ΔG° is still 1 kJ/mol (1 kJ per mole). The answer has the same meaning; the reaction is still spontaneous to the same extent.)

Practice 17.8

Use the table of thermodynamic values in the appendix to determine whether each of these oxidations is spontaneous. We may need to balance the equations.

$$C_6H_{12}O_6(s) + 6O_2(g) \rightleftharpoons 6CO_2(g) + 6H_2O(l)$$
$$CH_3CH_2CH_3(g) + O_2(g) \rightleftharpoons CO_2(g) + H_2O(g)$$
$$Mg(s) + N_2(g) \rightleftharpoons Mg_3N_2(s)$$

▶ **Here's What We Know So Far**

- Entropy and free energy are state functions.
- The change in entropy and free energy for a reaction can be determined by subtracting the sum of these quantities for the reactants from the sum for the products.
- Negative ΔG values indicate spontaneous processes.
- The Gibbs equation ($\Delta G = \Delta H - T\Delta S$) can be used to determine the spontaneity of a process.
- Like enthalpy changes, ΔS and ΔG for a multi-step process are simply the sum of the entropy or free energy changes for each individual process. ◀

17.7 When $\Delta G = 0$; Link to Equilibrium

We've already mentioned that the complete oxidation of glucose in the body produces ATP. This molecule is a source of potential energy to force nonspontaneous reactions in the body to become spontaneous. For example, the preparation of glucose-6-phosphate from glucose and phosphate is nonspontaneous. Coupling the hydrolysis of ATP with the production of glucose-6-phosphate helps to make the overall process spontaneous. In this way, the ATP molecule is used to force a reaction to become spontaneous. Although the majority of the ATP in the body comes from the oxidation of glucose, ATP can be made in many other ways.

$$\begin{array}{lll} \text{glucose} + HPO_4^{2-} & \rightarrow \text{glucose - 6 - phosphate} + H_2O & \Delta G = +13.8\,\text{kJ} \\ ATP + H_2O & \rightarrow ADP + HPO_4^{2-} & \Delta G = -30.5\,\text{kJ} \\ \hline \text{glucose} + ATP & \rightarrow \text{glucose - 6 - phosphate} + ADP & \Delta G = -16.7\,\text{kJ} \end{array}$$

One such way is through the reaction of two molecules of ADP (adenosine diphosphate), which can occur in vigorously active muscles. When the supply of glucose runs low and the level of ATP in the muscle drops, muscle cells gather energy by converting ADP to ATP and AMP. This enables the cells to continue their activity. Let's examine more closely the preparation of one molecule each of ATP and AMP (adenosine monophosphate) from two molecules of ADP. What is the first thing that we note about the reaction as written?

$$2ADP \rightleftharpoons ATP + AMP \quad \Delta G \approx 0\text{kJ}$$

The change in the free energy of the reaction is zero. Is this reaction spontaneous? Is it nonspontaneous? We can look at a more familiar process, the phase transition of water from solid to liquid, to help sort all this out.

$$H_2O(s) \rightleftharpoons H_2O(l)$$

At temperatures below the melting point of water, the reaction is spontaneous in the direction of $H_2O(s)$; that is, water freezes. At temperatures above the melting point of water, the reaction is spontaneous in the opposite direction; ice melts. At 0 °C and 1 atm, the process is not spontaneous in either direction. At this temperature, the change in free energy for the process is zero ($\Delta G = 0$) and the reaction proceeds neither toward the products nor toward the reactants. We say that the reaction has reached **equilibrium**. We remember from the chemical equilibrium chapter that a reaction at equilibrium hasn't stopped; it is just at a point where moving toward either the reactants or toward the products is not thermodynamically favored. The reaction continues to make both products and reactants at equal rates.

Equilibrium – The state of a reaction when ΔG=0.The reaction hasn't stopped; rather, the rates of the forward and reverse reactions are equal.

Remember that ΔG is temperature-dependent ($\Delta G = \Delta H - T\Delta S$), so the value of ΔG can be changed by adjusting the temperature. That is, we can melt ice or freeze water, depending on the temperature we pick. We can think of the free energy change for a reaction as a measure of how far away it is from equilibrium. By changing the temperature, we move the system closer to, or farther away from, equilibrium. Active muscles do not change their temperature in order to drive the production of ATP from ADP, but this information can be used to calculate the temperature at which a process becomes spontaneous. Let's go back to Mt. Everest... The base camp experiences temperatures ranging from a high of about −3 °C in July to a low of about −16 °C in January. High up on the slopes of the mountain, the temperature drops even lower, to around −36 °C. For trekkers huddled in a tent, it may take a lot of effort to get a propane stove to light (especially if the temperature drops below the boiling point of propane; see Fig. 17.21). How cold must it get before the propane gas liquefies? To answer this question, we need to consider the phase transition shown below.

17

$$CH_3CH_2CH_3(l) \rightleftharpoons CH_3CH_2CH_3(g) \qquad \begin{aligned} \Delta H &= 15.1\text{kJ/mol} \\ \Delta S &= 65.4\text{J/mol}\cdot\text{K} \end{aligned}$$

When this reaction is at equilibrium, neither the forward nor the backward reaction will be spontaneous ($\Delta G = 0$). The temperature to which this corresponds is the boiling point. Because we know the equation that relates ΔG to the temperature, we can calculate the boiling point of this reaction.

$$\begin{aligned} \Delta G &= \Delta H - T(\Delta S) \\ 0 &= 15{,}100\text{J/mol} - T(65.4\text{J/mol}\cdot\text{K}) \\ T(65.4\text{J/mol}\cdot\text{K}) &= 15{,}100\text{J/mol} \\ T &= 231\text{K} \\ T &= -42^\circ\text{C} \end{aligned}$$

Fig. 17.21 A gas grill can be used to heat water on a camping trip. *Source* ► https://www.publicdomainpictures.net/pictures/290000/velka/kettle-on-gas-vintage.jpg

Our calculations suggest that if the temperature falls to −42 °C (−44 °F), the change in the free energy of the process will be zero. The reaction will be at equilibrium, and propane will be at its boiling point. If the temperature gets any lower than −42 °C, the propane will be liquid, and the portable stove will become very difficult to light.

Example 17.9—Temperature Dependence of ΔG

In one of the final reactions to produce copper metal from its ore (chalcocite), copper(I) oxide reacts with carbon to produce gaseous carbon monoxide and copper metal. Above what temperature does the reaction become spontaneous?

$$Cu_2O(s) + C(s)2Cu(s) + CO(g)$$
$$\Delta H = 59\,kJ/mol \quad \Delta S = 165\,J/K \cdot mol$$

Asking the Right Questions

What condition is necessary for a reaction to be spontaneous? How can we work with the enthalpy, entropy, and temperature data to meet that condition?

Solution

A spontaneous reaction has a negative free energy change. Since the enthalpy and entropy data are fixed (we cannot change them), the variable—which we can control—is the temperature.

$$\Delta G = \Delta H - T\Delta S$$
$$0 = 59,000\,J/mol - T(165\,J/K \cdot mol)$$
$$T(165\,J/K \cdot mol = 59,000\,J/mol$$
$$T = 358\,K$$
$$T = 85\ °C$$

At or above 85 °C, this reaction becomes spontaneous ($\Delta G < 0$). So, at room temperature, the reaction doesn't proceed without continuous outside assistance. In practice, this reaction is often run at very high temperatures to produce copper metal. The copper can be further purified by electrolytic deposition.

Is Our Answer Reasonable?

"Reasonable," in this case, means that the reaction is spontaneous above 85 °C, and is nonspontaneous below 85 °C. We can pick two temperatures to demonstrate that. How about 84 °C (357 K) and 86 °C (359 K)? Substituting,

At 84 °C (357 K),

$\Delta G = \Delta H - T\Delta S$

$= 59,000$ J/mol − 357 K(165 J/K·mol)

$= 95$ J/mol, nonspontaneous.

At 86 °C (359 K),

$\Delta G = \Delta H - T\Delta S.$

$= 59,000$ J/mol − 359 K(165 J/K·mol).

$= -235$ J/mol, spontaneous.

The answer is reasonable.

What If…?

…we have this reaction:

$$N_2(g) + 2H_2(g) \rightleftharpoons N_2H_4(l) \quad \Delta H^\circ = 104kJ$$

Under what conditions can this reaction be spontaneous?

(*Ans: None. It is endothermic and it will always be nonspontaneous.*)

Practice 17.9

What is the boiling point of methanol?... of water?

$$CH_3OH(l) \rightleftharpoons CH_3OH(g) \quad H_2O(l) \rightleftharpoons H_2O(g)$$
$$\Delta_{vap}H = 35.27kJ/mol \ \Delta_{vap}H = 40.66kJ/mol$$
$$\Delta_{vap}S = 104.6J/K \cdot mol \ \Delta_{vap}S = 109.0J/K \cdot mol$$

As the temperature changes, the value of ΔG for a reaction changes. In fact, a nonspontaneous reaction can often be made spontaneous if the temperature is adjusted enough, such that $\Delta H - T\Delta S$ becomes negative (see Table 17.5). For a muscle cell in need of a quick energy fix, changing the temperature in order to make the conversion of ADP into ATP and AMP (adenosine monophosphate) spontaneous is not an option. Thankfully, temperature isn't the only variable that can change the spontaneity of a reaction. Pressure and concentration also play a major role in determining reaction spontaneity.

17.7.1 Changes in Pressure Affect Spontaneity

The entropy change of a reaction that contains gaseous reactants and/or products depends on its pressure. This occurs because there are fewer microstates in a compressed gas than there are if the pressure of the same sample is reduced by expanding the volume of the container. Mathematically, it has been shown that the pressure causes a change in the standard free energy of a reaction in accordance with this equation:

$$\Delta G = \Delta G^{\circ} + RT \ln Q_D$$

where
- $\Delta G°$ is change in the standard state free energy,
- R is the universal gas constant, 8.3145 J/mol·K,
- T is the temperature in kelvins, and,
- Q_p is a term called the **pressure reaction quotient.**
- ΔG is the change in the nonstandard state free energy.

Pressure reaction quotient (Q_p) – The ratio of the pressures of all of the gaseous products raised to their respective stoichiometric coefficients, divided by the pressures of all of the gaseous reactants raised to their stoichiometric coefficients.

We discussed the pressure equilibrium constant (K_p) in much greater detail in chemical equilibrium chapter, where we described it as the ratio of the partial pressures of all of the gaseous products to the pressures of the gaseous reactants, raised to their respective stoichiometric coefficients. The pressure reaction quotient, Q_p, is mathematically the same expression, but is specific to the current conditions of the system rather than to the equilibrium conditions. For example, here is the equation for the Haber process for the formation of ammonia (part 2 of Example 17.5), along with the expression for its Q_p:

$$3H_2(g) + N_2(g) \rightleftharpoons 2NH_3(g)$$

$$Q_p = \frac{P^2_{NH_3}}{P_{N_2} \cdot P^3_{H_2}}$$

17

Let's consider the effect of Q_p in the equation given above. If the reaction has just started, we should have a high reactant pressure and a relatively low product pressure, because we haven't made many product molecules yet. This would give us a value of Q_p that is much smaller than 1, and the value of $RT\ln Q_p$ will be a large negative number (remember that T is *always* positive because it is in kelvins). This will tend to make the value of ΔG more negative (i.e.—more spontaneous). This makes sense because if the reaction is just starting; it should have a large driving force to make products and reach equilibrium. On the other hand, if the products have a relatively large pressure and the reactants are relatively low, then $RT\ln Q_p$ will be a large positive number and will tend to make the value of ΔG more positive, meaning that there will be less of a driving force to achieve equilibrium. For a reaction such as the fermentation of glucose, Q_p is equal to the pressure of carbon dioxide squared, because neither glucose nor ethanol is a gas.

$$\underset{\text{Glucose}}{C_6H_{12}O_6(s)} \rightleftharpoons \underset{\text{Ethanol}}{2C_2H_6O(l)} + 2CO_2(g)$$

$$Q_p = \frac{P^2_{CO_2}}{1}$$

The brewmaster at a local brewpub knows about the effect of pressure on the fermentation of glucose. If the vat is left open to the atmosphere (1 atm), the yeast carries out the fermentation at 25 °C. The reaction, as we saw earlier in this chapter, is spontaneous (−229 kJ/mol). If the brewmaster shuts the hatch on the vat and seals it, the formation of carbon dioxide begins to build up pressure as the glucose is catabolized. At some point, the pressure could reach 52.0 atm (assuming the vat is strong enough to hold that pressure). Is the reaction spontaneous at that pressure?

$$\Delta G = \Delta G^\circ + RT \ln Q_p$$
$$\Delta G = -229{,}000\ \text{J/mol} + (8.3145\ \text{J/mol·K})(298\ \text{K})\ \ln(52.0)^2$$
$$\Delta G = -229{,}000\ \text{J/mol} + 19{,}580\ \text{J/mol}$$
$$\Delta G = -209{,}420\ \text{J/mol}$$
$$\Delta G = -209\ \text{kJ/mol}$$

The reaction is still spontaneous, as we might expect for a reaction that produces energy for a living organism. However, the reaction has a lower free energy than it had at lower pressure.

17.7.2 Changes in Concentrations Affect Spontaneity

In a manner similar to that for gaseous reactions, the free energy of a reaction changes as the concentrations of the reactants and products change. Does this make sense? The equation relating the concentrations of reactants and products to the observed free energy change (ΔG) is:

$$\Delta G = \Delta G^\circ + RT \ln Q$$

Note that this equation is mathematically similar to the equation for determining the effect of changing pressure.

The only difference between this equation and the one presented in the previous section is the value of the **reaction quotient (Q)**. The reaction quotient is calculated by multiplying the current molar concentrations of all of the aqueous or gaseous products, raised to the power of their respective stoichiometric coefficients, divided by the initial molar concentrations of all of the aqueous or gaseous reactants, raised to the power of their stoichiometric coefficients. Also, recall from the chemical equilibrium chapter that solids and liquids in the reaction have a value of "1", so they don't appear in this expression. Mathematically, for the reaction.

$$r\text{A(ag)} + s\text{B(ag)} \rightleftharpoons t\text{C(aq)} + u\text{D(ag)}$$

the reaction quotient is

$$Q = \frac{[\text{C}]_0^t[\text{D}]_0^u}{[\text{A}]_0^r[\text{B}]_0^s}$$

where $[\,]_0$ = the initial molar concentration of the substance.

Reaction quotient (Q) – The ratio of the initial molar concentrations of all of the products raised to the power of their respective stoichiometric coefficients, divided by the initial molar concentrations of all of the reactants raised to the power of their stoichiometric coefficients.

What does the reaction quotient tell us about a reaction? Just as we described in the previous section, reactions with larger concentrations of reactants than products have a Q value less than 1. This means that $RT\ln Q$ is negative and that the observed free energy (ΔG) is less than the standard free energy (ΔG°). In other words, large concentrations of reactants and small concentrations of products increase the spontaneity of the forward reaction (this situation decreases the value of ΔG). When the concentration of reactants is small and the concentration of products is large ($Q > 1.0$), the value of $RT\ln Q$ is greater than zero and ΔG is more positive than ΔG°. In this case, the reaction is less spontaneous.

17.7.3 Coupled Reactions

Many reactions, such as the formation of glucose-6-phosphate from glucose, are nonspontaneous. Metabolic sequences have developed to account for this. As we've seen in this chapter, biological reactions are often coupled

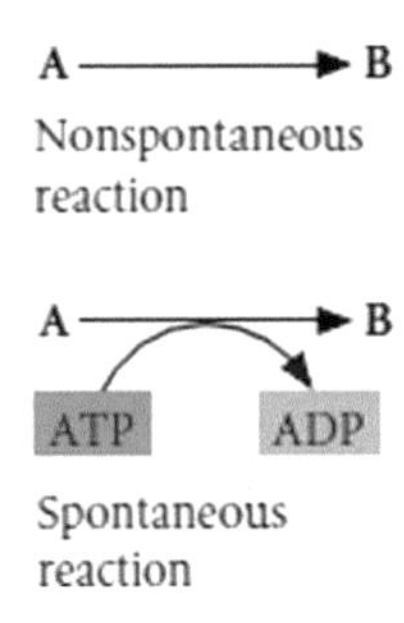

Fig. 17.22 Biochemical reactions typically involve the coupling of a spontaneous reaction with a nonspontaneous reaction. Coupling the reactions makes the nonspontaneous reaction spontaneous

with other reactions to produce a spontaneous process. Organisms typically couple a reaction that has a very large negative ΔG with reactions that have positive free energy changes, as shown in Fig. 17.22. The result is a spontaneous reaction.

In glycolysis, there are four coupled reactions. Two of these reactions use the negative free energy change of $ATP \rightleftharpoons ADP$ ($\Delta G = -30.5$ kJ/mol) to place phosphate groups on a glucose molecule before it is broken down. Then, as the resulting carbohydrate is catabolized, the organism uses the stored energy of the carbohydrate to drive the reverse of this reaction ($ADP \rightleftharpoons ATP$; $\Delta G = +30.5$ kJ/mol). In doing so, the glycolytic pathway consumes two molecules of ATP but produces four molecules of ATP using the stored energy of the carbohydrate. The net result is the accumulation of two molecules of ATP for each glucose molecule.

In this chapter, we have examined why chemical reactions proceed and have shown that it is directly related to an increase in the entropy of the universe. We routinely consider the spontaneity of a reaction as a function of the values of ΔH, ΔS, and reaction conditions such as temperature, pressure, and concentration. We've noted that the free energy change (ΔG) depends on all of these things.

The free energy change for a reaction under given conditions tells us which direction the reaction will proceed to achieve equilibrium, whether it will make more reactants or more products to get there. However, as we discussed in the kinetics chapter, the value of ΔG tells us nothing about the speed of a reaction.

Issues and Controversies—ATP—It's part of a bigger system

Adenosine triphosphate (ATP) is often referred to as the "energy currency" of most living organisms. Indeed, our body creates ATP when it has energy to store and breaks ATP down into its components when it needs to harvest some of that stored energy. Often, students only consider the creation and destruction of ATP in isolation, and this leads to the erroneous idea that energy is somehow released when the phosphate bond is broken and ATP turns into ADP plus a hydrogen phosphate group. Let's take a closer look at the simplest way to represent this reaction to see why this concept is incorrect:

$$ATP(aq) + H_2O(l) \rightarrow ADP(aq) + HPO_4^{2-}(aq)$$
$$\Delta G = -30.5 \text{kJ/mol}$$

Remember that the free energy change for a reaction is the difference between the free energy content of the products minus that of the reactants. What this equation tells us is that the products (ADP and hydrogen phosphate ion) have 30.5 kJ less free energy per mole than the ATP and water did before they reacted. The energy is released because the products are lower in energy (are more stable) than the reactants. The act of breaking the phosphate bond in the ATP molecule actually *requires* energy, because the phosphorous-oxygen bond that is being broken is quite stable—if it wasn't, it would have already fallen apart! However, the energy gained by the process of creating the aqueous phosphate ion more than makes up for the energy cost of breaking the phosphate bond. Remember also that there are lots of water molecules around and the change in entropy plays a part in the free energy change also. Notice that the reverse of the above reaction would cost 30.5 kJ of free energy per mole of ATP. So, it makes sense that when our body has energy to store, it causes a reaction to happen that requires energy input—the energy has to go somewhere! ATP is involved in many, many other reactions in our body that are much more complex to consider, but here are the basic principles that apply to all of them:

- Breaking chemical bonds ALWAYS requires energy.
- Creating chemical bonds ALWAYS releases energy.

When considering energy changes, we must always consider what is happening to the system, not just a particular molecule. Whenever a chemical reaction releases energy, it is because the products of that reaction are at a lower energy state than the reactants were.

17.8 Free Energy and the Equilibrium Constant

We've seen that changes in the pressures of, or concentrations of, reactants and products affect the value of ΔG. However, when a process is at equilibrium, like the boiling water of our intrepid explorers on Mt. Everest, or the conversion of 2 ADP molecules to ATP and AMP, the concentrations of reactants and products appear to stop changing. Although we remember from the previous ▶ Sect. 17.7 that these concentrations are changing constantly, they change at the same rate, so they appear to maintain a constant value. What is the implication of this in light of our discussion of free energy? How does free energy relate to the equilibrium constant? We remember that the free energy change for a reaction is expressed as.

$$\Delta G = \Delta G^{\circ} + RT\ln Q$$

Our thermodynamic definition says that at equilibrium, $\Delta G = 0$. Therefore, at equilibrium, the reaction quotient, Q, becomes the thermodynamic equilibrium constant, K_{eq}.

$$0 = \Delta G^{\circ} + RT\ln(K_{eq})$$

Solving for ΔG°, we obtain:

$$\Delta G^{\circ} = -RT\ln(K_{eq})$$

This equation is very useful because it enables us to calculate the equilibrium constant for a process if we know the standard free energy change for that process.

Now, having completed our discussion of thermodynamics, we are able to reveal the truth about why equilibrium constants have no units, as well as why we "ignore" solids and liquids in the expression for K. Concentration-based expressions are generally suitable representations for the equilibria that occur in chemical reactions. However, in some aqueous solutions, especially where the solute concentrations are high, chemists often make a correction using activities instead of molarities, to get the most meaningful results. For example, if we prepare a solution that contains 3.0 mol of hydrochloric acid (HCl) in a liter of water, we say that the HCl concentration is 3.0 molar. We expect this strong acid to dissociate into H^+ and Cl^- ions, and we therefore assert that in the solution, the concentration of each ion, H^+ and Cl^-, is 3.0 M. However, this is not completely true. Some of the hydrogen cations do interact with the chloride anions in solution, and the ions interact with the water solvent through ion–dipole interactions, as we discussed earlier in this text. This means that the effective concentration is likely to be somewhat different from the intended concentration, especially in relatively concentrated solutions. We've introduced this topic earlier in this text, however, it is important to revisit the meaning and implications of this concept. The effective concentration of the solute is called its **activity**. A related concept is that of **fugacity**, which is similar to activity, but is related to the effective pressure of a gas.

Activity – The effective concentration of a solute in solution.

Fugacity (*f*) – The effective pressure of a real gas

The concept of activity has two impacts that do affect the way we handle equilibrium calculations. The first is that activity coefficients have no units, which is the real reason that equilibrium constants have no units. The second is that solids and liquids are condensed matter with concentrations that depend only upon their density. For example, what is the concentration of H_2O in pure water? There are a certain number of moles in a given volume? One liter of pure water has a mass of approximately one thousand grams, which equates to approximately 55.5 mol. Since there are 55.5 mol of water in a liter of water, then pure water has a concentration of 55.5 M! Solid materials can have even higher concentrations. Since pure liquids and solids have such high concentrations, they have chemical activities approximately equal to 1.0. This is why we don't factor solids and liquids into our equilibrium constant expressions. Since their activity is equal to 1.0, they don't change the value of the equilibrium constant, so they can be ignored. On the other hand, for concentrations less than one molar, and pressures less than one atmosphere, activity is approximately equal to concentration, and fugacity is approximately equal to partial pressure. To illustrate this, let's write the equilibrium constant expression for the environmental leaching of lead into acidic waters in order to illustrate this definition, using activity (a) and fugacity (f) where appropriate.

$$PbCO_3(s) + 2H^+(aq) \rightleftharpoons Pb^{2+}(aq) + CO_2(g) + H_2O(l)$$

$$K_{eq} = \frac{a_{Pb}^{2+} \cdot f_{CO_2} \cdot a_{H_2O}}{a_{PbCO_3} \cdot a_{H^+}^2} \approx \frac{[Pb^{2+}] \cdot P_{CO_2}}{[H^+]^2}$$

because:

$$a_{Pb^{2+}} \approx \left[Pb^{2+}\right] \text{ and } a^2_{H^+} \approx \left[H^+\right]^2$$
$$f_{CO_2} \approx P_{CO_2} \text{ (the partial pressure of } CO_2 \text{) in atmospheres}$$
$$a_{PbCO_3} \approx 1.0 \text{ and } a_{H_2O} \approx 1.0$$

In this equation, $PbCO_3$ is a solid and H_2O is a liquid. As with K and K_p, we consider their activity to be 1.0 and we simply remove them from the equation.

Example 17.10—Free Energy and the Equilibrium Constant

Using the standard free energy values given, calculate the equilibrium constant at STP for the formation of SO_3

	$2SO_2(g)$	+	$O_2(g)$	⇌	$2SO_3(g)$
DG°(kJ/mol)	-300		0		-371

Asking the Right Questions

What is the value for the free energy? What is the relationship between the free energy and the equilibrium constant? If the reaction is spontaneous at standard conditions, what range of values do we expect for the equilibrium constant?

Solution

We can calculate the free energy of the reaction.

$$\Delta_{rxn}G^\circ = \Delta G^\circ_{products} - \Delta G^\circ_{reactants}$$
$$= 2(-371\,\text{kJ/mol}) - 2(-300\,\text{kJ/mol})$$
$$\Delta_{reaction}G^\circ = -142\,\text{kJ/mol}$$

In order to cancel units properly in the next equation, we must convert $\Delta_{reaction}G^\circ$ to J/mol.

$$\Delta_{reaction}G^\circ = -142,000\,\text{J/mol}$$
$$\Delta G^\circ = -RT\ln(K_{eq})$$
$$-142{,}000\,\text{J/mol} = -8.3145\,\text{J}\cdot\text{mol}^{-1}\cdot\text{K}^{-1}(298\,\text{K})\ln(K_{eq})$$

$$\ln(K_{eq}) = \frac{-142,000\,\text{J}\cdot\text{mol}^{-1}}{-8.3145\,\text{J}\cdot\text{mol}^{-1}\cdot\text{K}^{-1}(298\,\text{K})}$$
$$= 57.31$$

We can then take the inverse natural log of both sides.

$$K_{eq} = e^{57.31} = 7.76 \times 10^{24}$$

Is Our Answer Reasonable?

The value for K_{eq} is much greater than 1. This is a consequence of the negative free energy value, meaning that the reaction is spontaneous. The consistent meaning of the negative free energy and large equilibrium constant suggest that the answer is reasonable.

What If…?

… we reversed the reaction so that it is:

$$2SO_3(g) \rightleftharpoons 2SO_2(g) + O_2(g)?$$

What would be the free energy and equilibrium constant for the reaction at 25 °C?

(*Ans:* $\Delta G^\circ = +142$ *kJ/mol and* $K_{eq} = 1.29 \times 10^{-25}$)

Practice 17.10

Use the thermodynamic data in the appendix to calculate the equilibrium constant for the ionization of hydrogen sulfide.

$$H_2S(aq) \rightleftharpoons 2H^+(aq) + S^2 - (aq)$$

17

The Bottom Line

- The multiplicity of a system can be used to determine the behavior of the system. The most probable macrostate is the one that has the most contributing microstates.
- Spontaneous processes occur without outside assistance. The reverse of the spontaneous process is nonspontaneous.
- Entropy is a measure of how energy and matter can be distributed in a chemical system. Entropy is not disorder.
- The second law of thermodynamics says that the entropy of the universe continues to increase. Any process that is spontaneous must correspond to an increase in the entropy of the universe ($\Delta S_{universe} > 0$).
- The third law of thermodynamics says that the entropy of a pure perfect crystalline material is zero. This law enables us to calculate the entropy of any compound in any state.

- The free energy, ΔG, can be calculated via the Gibbs equation ($\Delta G = \Delta H - T\Delta S$) to determine the spontaneity of a process.
- When $\Delta G = 0$, the system is at equilibrium. The forward and reverse reactions still proceed, but their rates are equal.
- Coupled reactions are used by the body to make nonspontaneous processes become spontaneous. The use of ATP in this manner assists in the overall production of energy from glucose.
- The free energy change of a reaction can be determined from the equilibrium constant for that reaction and vice versa.

17.2 Probability as a Predictor of Chemical Behavior

Skill Review

1. Suppose that four pennies fall from our pocket to the floor. How many microstates for the four pennies would exist? (Consider only heads up versus tails up in the explanation.)
2. What is the probability that all four coins from Problem 1 would land heads up?
3. What is the probability that the four coins from Problem 1 would land two heads up and two tails up?
4. What is the most probable outcome for the four coins in Problem 1?
5. Contrast the meanings of the terms macrostate and microstate.
6. Define the term *multiplicity*.
7. One hundred chemistry students are about to take an exam. There are two equal 200-seat classrooms for the students. Knowing that students must have vacant seats next to them during the exam, describe at least two macrostate arrangements we could predict for the two rooms.
8. In the situation defined in Problem 7, how could the positions of the students allow us to use microstates to explain the answer?

Chemical Applications and Practice

9. Even a sample of only 1000 atoms is too small to weigh accurately, and yet we can calculate a probable distribution of their numbers between two equal containers. How many microstates are possible if 1000 atoms of He become distributed between two containers of equal size? Explain why an equal distribution becomes more probable as the number of He atoms increases.
10. A certain lecture hall is 25 m long, 12 m wide, and 15 m tall. The oxygen molecules necessary to sustain life in the room are, of course, free to circulate throughout the room. Explain, using probability and microstates, why a student sitting in the back should not be concerned that all the freely moving oxygen molecules might concentrate at the front row of the classroom.
11. Some powdered creamer is added to a cup of hot coffee. Compare the initial macrostate to the final macrostate as the creamer dissolves. Has the number of microstates for the creamer and coffee increased from the initial state to the final state? Explain the evidence we would use to explain the answer.
12. A glass of tea is cooled by placing ice cubes in it. However, after a few minutes, the ice has melted. Compare the initial macrostate to the final macrostate. Has the number of microstates for the ice and tea increased from the initial state to the final state? Explain the evidence we would use to explain the answer.
13. Which of these would we classify as increasing the number of microstates?
    a. Water in a glass evaporates.
    b. A precipitate of AgCl forms from the addition of a drop of a 0.10 M solution of $AgNO_3$ into a glass of chlorinated tap water.
    c. An icicle forms from a downspout.
14. Which of these would we classify as increasing the number of microstates?
    a. A chunk of ice melts when placed in a glass of tap water at room temperature.
    b. A glass of tap water freezes when placed in a refrigerator freezer compartment.
    b. The combustion of gasoline produces carbon dioxide and water vapor.

17.3 Why Do Chemical Reactions Happen? Entropy and the Second Law of Thermodynamics

Skill Review

15. What do we mean by the terms "system", "surroundings" and "universe"?
16. What is the relationship between the entropy change in a given system and the entropy change in the entire universe?

17. Can a process that produces lower entropy in a system occur spontaneously? If so, what must be true about the entropy of the surroundings?
18. What does a spontaneous process always increase?
19. Which of these, if any, need outside intervention to take place? Describe the necessary intervention.
    a. Combustion (combination with oxygen) of a piece of paper in the air
    b. Making a hard-boiled egg
    c. Increasing muscle mass
    d. Formation of calcium carbonate (bathtub ring) from evaporating hard water
20. Which of these, if any, need outside intervention to take place? Describe the necessary intervention.
    a. The melting of butter on a warm pancake
    b. Opening this book to this page
    c. Paddling a canoe upstream
    d. Cooking eggs for breakfast
21. What is the relationship between entropy and a nonspontaneous process?
22. What is the relationship between entropy and rate of change?
23. What is the relationship between entropy and number of microstates?
24. What is the relationship between a nonspontaneous process and the number of microstates?
25. Suppose we were studying in a residence hall. Down the hall someone is preparing popcorn. The irresistible aroma eventually reaches our room. To take our mind off this, explain how the second law of thermodynamics enables us to detect this tasty temptation from so far away.
26. A scientist on the planet Zoltan expresses the third law of thermodynamics by saying that the entropy of any compound under standard state conditions is zero. Would this change the ΔS values for the reaction of pyruvate to lactate mentioned in the chapter? Explain.
27. Which system has lower entropy: a randomly ordered pile of bricks or a house made of bricks? What is required for entropy to decrease in this particular situation?
28. Which process is spontaneous: rust becoming iron, or iron becoming rust? Rust is just another name for iron oxide, which is how most iron is found in nature. What is required to make iron oxide into iron?

Chemical Application and Practice

29. While pumping gasoline into our car, we might spill some gasoline on our hands. However, the gasoline quickly evaporates with a cooling sensation on our skin. How can this process be spontaneous if it requires the input of the energy from our skin?
30. In a previous problem we were asked about the freezing of water in a refrigerator. Water poured in a tray will spontaneously freeze in a freezer. Explain why this process arranges the water molecules in a more orderly crystal form and yet is still spontaneous.
31. Solid carbon dioxide is called dry ice. At room temperature, this material spontaneously changes from an orderly solid to a gas, bypassing the liquid phase entirely. When undergoing this change, the molecules absorb heat from the nearby surroundings. This, in turn, slows down the molecules of the surroundings. What can be said about the change in entropy for the carbon dioxide molecules relative to nearby air molecules? What can be said about the change in the entropy of the universe for this process?
32. The entropy of molecules can also depend on their relative atomic motions. The atoms in molecules can twist, rotate, and vibrate around a chemical bond. For which molecule would we be able to predict more microstates: a molecule of nitrogen or a molecule of ammonia (NH_3)? Explain the basis of the choice.
33. The following equation illustrates a gaseous reaction that is possible in the atmosphere. Which direction in the reaction shows the greater entropy for the system? Explain the choice.

34. Another reaction common in the lower atmosphere is the formation of ozone from NO_2. Which direction in the reaction shows the greater entropy for the system? Explain the choice.

$$O_2(g) + NO_2(g) \rightleftharpoons O_3(g) + NO(g)$$

17.4 Temperature and Spontaneous Processes

Skill Review

35. Classify each of these as representing either a positive change in entropy for the universe or a negative change in entropy for the universe or indicate that we would need more information to determine this.
    a. $\Delta S_{sys} > 0$; $\Delta S_{surr} > 0$
    b. $\Delta S_{sys} > 0$; $\Delta S_{surr} < 0$
    c. $\Delta S_{sys} < 0$; $\Delta S_{surr} > 0$
36. Classify each of these as representing either a positive change in entropy for the universe or a negative change in entropy for the universe or indicate that we would need more information to determine this.
    a. $\Delta S_{sys} < 0$; $\Delta S_{surr} < 0$
    b. $\Delta S_{sys} > 0$; $\Delta S_{surr} = 0$
    c. $\Delta S_{sys} < 0$; $\Delta S_{surr} = 0$
37. How does the multiplicity of a system change as the molecular motions increase in the system?
38. How does the molecular motion of the surroundings change as energy from the surroundings flows into a system?
39. How does the entropy of the surroundings change as energy flows out of a system?
40. Can we predict the change in entropy of the universe as energy flows into the system? Explain.
41. For any process that occurs under reversible conditions, what is the change in entropy for the universe?
42. What is meant by "reversible conditions"?

Chemical Applications and Practice

43. Mercury remains a liquid over a fairly wide temperature range, which is one reason why it has been used in thermometers. Classify each of these changes as a positive or a negative change in the entropy of a sample of mercury.
    a. Liquid mercury changes to a solid.
    b. Liquid mercury expands as it is heated.
    c. A small amount of mercury evaporates from the surface of a column of mercury in a sealed thermometer.
44. Chlorine gas (Cl_2) is a very toxic compound that can form as a result of the improper use of bleach and ammonia cleaning solutions. Classify each of these changes as a positive or a negative change in the entropy of a sample of this green gas.
    a. Chlorine gas condenses to a liquid.
    b. Chlorine gas is heated in a container.
    c. Chlorine gas deposits on a cold surface.
45. If we are enjoying an iced carbonated soft drink while studying, we may have noticed that the carbon dioxide used to carbonate the beverage bubbles slowly out of the beverage as it warms. Is this a positive or a negative entropy change for the CO_2? Explain.
46. In the bubbling of carbon dioxide mentioned in Problem 45, did the surroundings gain or lose heat? Is the sign for $\Delta S_{universe}$, at this temperature, positive or negative?

17.5 Calculating Entropy Changes in Chemical Reactions

Skill Review

47. Propane stoves make use of the following combustion reaction:

$$\_\_C_3H_8(g) + \_\_O_2(g) \rightarrow \_\_CO_2(g) + \_\_H_2O(g)$$

Balance the equation, report the number of moles of gas molecules on each side of the equation, and predict the change in entropy of the system.

48. Methane can be combusted by the following reaction:

$$\_\_CH_4(g) + \_\_O_2(g) \rightarrow \_\_CO_2(g) + \_\_H_2O(g)$$

Balance the equation, report the number of moles of gas molecules on each side of the equation, and predict the change in entropy of the system.

49. Repeat the prediction for Problem 47 but assume that the reaction is performed under temperatures that cause the propane and water to be liquid, yet the oxygen and carbon dioxide remain as gases.

50. Repeat the prediction for Problem 48 but assume that the reaction is performed under temperatures that cause the methane and water to be liquid, yet the oxygen and carbon dioxide remain as gases.
51. Explain why the change in the number of moles of gases, from reactant to product, is typically more influential in determining the sign on entropy changes than the change in moles of liquids and solids.
52. The third law of thermodynamics specifies a particular condition for a perfect crystal. Why must the temperature be 0 K instead of 0 °C?
53. Predict the sign on the entropy change for each of these reactions.
    a. $NH_3(g) + HCl(g) \rightarrow NH_4Cl(s)$
    b. $2HgO(s) \rightarrow 2Hg(l) + O_2(g)$
    c. $Cd(s) + O_2\ (g) \rightarrow CdO(s)$
54. Using the values for $\Delta S°$ found in the appendix, determine the actual values for the standard change in entropy for the reactions in Problem 53.
55. Predict the sign on the entropy change for each of these reactions.
    a. $SO_2(g) + O_2(g) \rightarrow 2SO_3(g)$
    b. $2NH_3(g) \rightarrow N_2(g) + 3H_2(g)$
    c. $CO(g) + 1/2\ 2H_2(g) \rightarrow CH_3OH(l)$
56. Using the values for $\Delta S°$ found in the appendix, determine the actual values for the standard change in entropy for the reactions in Problem 55.

Chemical Applications and Practices

57. Several states have implemented regulations for fuel alternatives in automobiles. One such gasoline alternative is ethanol (C_2H_5OH). Balance the following combustion reaction and determine the change in entropy for the reaction.

$$\_\_\_C_2H_5OH(l) + \_\_\_O_2(g) \rightarrow \_\_\_CO_2(g) + \_\_\_H_2O(g)$$

58. One important source of salt (NaCl) is the evaporation of ocean water. Calculate the change in the standard molar entropy as the dissolved ions precipitate the solid.

$$Na^+(aq) + Cl^-(aq) \rightarrow NaCl(s)$$

59. Graphite and diamond are both made of carbon. However, pencils (incorporating the graphite form) sell for a few cents, whereas engagement rings (displaying diamonds) often sell for thousands of dollars. Using the appendix, determine the change in entropy for the conversion of diamond to graphite.
60. Using the appendix to note the standard $\Delta S°$ value for a metal, explain why the value for the metal alone is lower than the value for any of the listed compounds of that metal.

17.6 Free Energy

Skill Review

61. Predict the relative magnitude of the temperature that would result in a spontaneous process in each of these combinations.
    a. $\Delta H > 0$; $\Delta S > 0$
    b. $\Delta H > 0$; $\Delta S < 0$
62. Predict the relative magnitude of the temperature that would result in a spontaneous process in each of these combinations.
    a. $\Delta H < 0$; $\Delta S > 0$
    b. $\Delta H < 0$; $\Delta S < 0$
63. Determine the value of ΔH in each of these cases.
    a. $\Delta G = -24.5$ kJ/mol; $\Delta S = 287$ J/mol·K; $T = 298$ K
    b. $\Delta G = 1.38$ kJ/mol; $\Delta S = 24$ J/mol·K; $T = 298$ K
    c. $\Delta G = 500.0$ kJ/mol; $\Delta S = -6439$ J/mol·K; $T = 325$ K
    d. $\Delta G = -24.5$ kJ/mol; $\Delta S = 187$ J/mol·K; $T = 39.9$ °C
64. Determine the value of ΔG in each of these cases.
    a. $\Delta H = -20.5$ kJ/mol; $\Delta S = 259$ J/mol·K; $T = 298$ K
    b. $\Delta H = 350$ kJ/mol; $\Delta S = 73$ J/mol·K; $T = 157$ K

c. $\Delta H = 299$ kJ/mol; $\Delta S = -639$ J/mol·K; $T = 325$ K
d. $\Delta H = -4505$ kJ/mol; $\Delta S = 107$ J/mol·K; $T = 19.3$ °C

65. Determine the value of ΔS in each of these cases.
   a. $\Delta G = -20.5$ kJ/mol; $\Delta H = 259$ kJ/mol; $T = 298$ K
   b. $\Delta G = 150$ kJ/mol; $\Delta H = -73$ kJ/mol; $T = 157$ K
   c. $\Delta G = 209$ kJ/mol; $\Delta H = -639$ kJ/mol; $T = 35.0$ K
   d. $\Delta G = -4505$ kJ/mol; $\Delta H = 107$ kJ/mol; $T = 135.8$ °C
66. Determine the value of T in each of these cases.
   a. $\Delta G = -20.5$ kJ/mol; $\Delta H = 259$ kJ/mol; $\Delta S = 260$ J/mol·K
   b. $\Delta G = 150$ kJ/mol; $\Delta H = -73$ kJ/mol; $\Delta S = -290$ J/mol·K
   c. $\Delta G = 209$ kJ/mol; $\Delta H = 639$ kJ/mol; $\Delta S = 560$ J/mol·K
   d. $\Delta G = -4505$ kJ/mol; $\Delta H = 107$ kJ/mol; $\Delta S = 1160$ J/mol·K
67. Determine the spontaneity of each of these reactions at 298 K.
   a. $NH_3(g)+HCl(g) \rightarrow NH_4Cl(s)$
   b. $2HgO(s) \rightarrow 2Hg(l) + O_2(g)$
   c. $Cd(s) + 1/2\, O_2(g) \rightarrow CdO(s)$
68. Determine the spontaneity of each of these reactions at 298 K.
   a. $2SO_2(g)+O_2(g) \rightarrow 2SO_3(g)$
   b. $2NH_3(g) \rightarrow N_2(g)+3H_2(g)$
   c. $CO(g) + 2H_2(g) \rightarrow CH_3OH(l)$
69. The following reaction can occur in the atmosphere:

$$NO_2(g) + N_2O(g) \rightarrow 3NO(g)$$

   a. Calculate the value of $\Delta G°$ for this reaction. Is it spontaneous at 298 K?
   b. Assuming that $\Delta H°$ and $\Delta S°$ are constant with temperature, is this reaction spontaneous at 1000 K?
   c. At what temperature does this reaction become spontaneous?
70. Hydrochloric acid can be formed by this reaction:

$$H_2(g) + Cl_2(g) \rightarrow 2HCl(g)$$

   a. Calculate the value of $\Delta G°$ for this reaction. Is it spontaneous at 298K?
   b. At what temperature does this reaction become spontaneous?

Chemical Application and Practices

71. One of the authors of this text formerly drove an automobile that could best be described as blue and rust. Consider that the rust was formed, at least partially, by the reaction illustrated by this equation:

$$3O_2(g) + 4Fe(s) \rightarrow 2Fe_2O_3(s)$$

Assuming standard conditions, what would we calculate as the $\Delta G°$ for this reaction?

72. The impressive white cliffs of Dover in England have stood without significant change from their basic structure of calcium carbonate for eons. One possible reaction for the decomposition of calcium carbonate is

$$CaCO_3(s) \rightarrow CaO(s) + CO_2(g)$$

Use the $\Delta_f G°$ values of these compounds to determine the value of $\Delta G°$ for the reaction. Comment on how thermodynamics helps explain the relative stability of this beautiful landmark.

73. When sulfur-containing coal is burned, sulfur gases pose environmental threats to air quality. Given the following two combustion reactions, determine the $\Delta G°$ value for the change from SO_2 to SO_3.

$$S(s) + {}^3\!/_2O_2(g) \rightarrow SO_3(g) \qquad \Delta G° = -371\ \text{kJ/mol}$$
$$S(s) + O_2(g) \rightarrow SO_2(g) \qquad \Delta G° = -300\ \text{kJ/mol}$$
$$SO_2(g) + {}^1\!/_2O_2(g) \rightarrow SO_3(g) \qquad \Delta G° = ?$$

74. Graphite and diamond are both composed entirely of carbon atoms. Using the following reactions at 298 K, determine the value of $\Delta G°$ for the conversion of graphite into diamond.

$$C_{diamond}(s) + O_2(g) \rightarrow CO_2(g) \qquad \Delta G^\circ = -397\ kJ/mol$$
$$C_{graphite}(s) + O_2(g) \rightarrow CO_2(g) \qquad \Delta G^\circ = -394\ kJ/mol$$
$$C_{graphite}(s) \rightarrow C_{diamond}(s) \quad \Delta G^\circ = ??$$

75. As we saw earlier in this chapter, the production of glucose requires the input of energy. This is typically shown as green plants using sunlight to help combine carbon dioxide with water to form glucose through photosynthesis. However, in some deep areas of the ocean, sunlight does not reach bacteria that may grow around thermal vents. These organisms have been shown to produce energy through oxidation of a sulfur compound, H_2S. Determine ΔH°, ΔS°, and ΔG° for this biochemically important reaction. Assume sulfur is in the standard form—rhombic allotrope.

$$H_2S(aq) + O_2(g) \rightarrow 2H_2O(l) + 2S(s)$$

76. Smog produced over a city is formed from trace amounts of automobile exhaust in a reaction with atmospheric oxygen. One of the steps in the formation of ozone is the oxidation of oxygen gas by nitrogen dioxide (a by-product of the combustion of gasoline). Determine ΔH°, ΔS°, and ΔG° for this reaction.

$$NO_2(g) + O_2(g) \rightarrow NO(g) + O_3(g)$$

17.7 When $\Delta G = 0$; a Link to Equilibrium

Skill Review

77. Determine the temperature at which the indicated system reaches equilibrium.
    a. $\Delta H^\circ = 46.4$ kJ/mol; $\Delta S^\circ = 27.6$ J/mol·K
    b. $\Delta H^\circ = 10.6$ kJ/mol; $\Delta S^\circ = 77$ J/mol·K
    c. $\Delta H^\circ = 124$ kJ/mol; $\Delta S^\circ = 295.5$ J/mol·K
78. Determine the temperature at which the indicated system reaches equilibrium.
    a. $\Delta H^\circ = 57.6$ kJ/mol; $\Delta S^\circ = 17.4$ J/mol·K
    b. $\Delta H^\circ = 94$ kJ/mol; $\Delta S^\circ = 306$ J/mol·K
    c. $\Delta H^\circ = 32.1$ kJ/mol; $\Delta S^\circ = 552$ J/mol·K
79. Write the expression for Q_p for the following systems, remembering that solids and liquids are given a value of 1:
    a. $C(s) + O_2(g) \rightarrow CO_2(g)$
    b. $2N_2O(g) \rightarrow O_2(g) + 2N_2(g)$
    c. $2C_4H_{10}(l) + 13O_2(g) \rightarrow 8CO_2(g) + 10H_2O(g)$
80. Write the expression for Q_p for the following systems, remembering that solids and liquids are given a value of 1:
    a. $C_3H_8(g)+5O_2(g) \rightarrow 3CO_2(g)+4H_2O(g)$
    b. $2LiOH(s) + CO_2(g) \rightarrow Li_2CO_3(s)+H_2O(l)$
    c. $2NH_3(g)+3N_2O(g) \rightarrow 4N_2(g)+3H_2O(l)$
81. Write the expression for Q for the following systems, remembering that solids and liquids are given a value of 1:
    a. $Al^{3+}(aq) + 3OH^-(aq) \rightarrow Al(OH)_3(s)$
    b. $H_2SO_4(aq)+2NaOH(aq) \rightarrow Na_2SO_4(aq)+2H_2O(l)$
    c. $Fe(s) + Ni(NO_3)_2(aq) \rightarrow Fe(NO_3)_2(aq) + Ni(s)$
82. Write the expression for Q for the following systems, remembering that solids and liquids are given a value of 1:
    a. $Mg(s) + CoSO_4(aq) \rightarrow MgSO_4(aq) + CO(s)$
    b. $2HC_2H_3O_2(aq)+Ba(OH)_2(aq) \rightarrow 2H_2O(l)+Ba(C_2H_3O_2)_2(aq)$
    c. $HCl(aq) + NaOH(aq) \rightarrow NaCl(aq) + H_2O(l)$
83. As oxygen is removed from the following reactions, what happens to the value of ΔG?
    a. $2KClO_3(s) \rightarrow 2KCl(s)+3O_2(g)$
    b. $C_3H_8(g)+5O_2(g) \rightarrow 3CO_2(g)+4H_2O(g)$
    c. $3CO_2(g) \rightarrow 3\ C(s)+3O_2(g)$
84. As hydrogen is added to the following reactions, what happens to the value of ΔG?
    a. $C_2H_4(g)+H_2(g) \rightarrow C_2H_6(g)$
    b. $N_2H_4(g)+H_2(g) \rightarrow 2NH_3(g)$
    c. $N_2H_4(g) \rightarrow N_2(g)+2H_2(g)$

85. At equilibrium, what is the pressure for the following process at 298 K?

$$CaCO_3(s) \rightleftharpoons CaO(s) + CO_2(g) \quad \Delta G^\circ = 131\ \text{kJ/mol}$$

86. What is the free energy change for the following reaction at 298 K, given that the pressures of NH_3 and N_2 are 1.0 atm each and the pressure of H_2 is 2.0 atm? (Hint: Use the appendix to determine the free energy change under standard conditions.)

$$2NH_3(g) \rightleftharpoons N_2(g) + 3H_2(g)$$

Chemical Applications and Practices

87. At 1 atm of pressure the boiling point of pure water is 100.0 °C. If 1000.0 g of water were brought to this point and then completely boiled away, what would we calculate as the entropy change for the water? (Note: The heat of vaporization for water is 40.7 kJ/mol.)
88. There is a historical approximation known as Trouton's rule. This approximation states that at the normal boiling point for nonpolar compounds, the standard molar entropy of vaporization is 87 J/mol·K. Using Trouton's rule, determine the approximate heat of vaporization of water. The actual molar entropy of vaporization for water is close to 41 J/mol·K. Explain why the Trouton value is so much different for polar substances such as water but is often found to be much closer to the actual value for nonpolar substances.
89. Under proper conditions, phase changes can be classified as reversible processes. The energy involved with melting or freezing of a sample is called the enthalpy of fusion. The enthalpy of fusion ($\Delta_{fus}H$) for the rare radioactive element actinium, $Z = 89$, is 10.50 kJ/mol. If the entropy of fusion is 9.6 J/mol·K, what would we calculate as the melting point for this silvery white metal?
90. One of the reasons why we are able to read this question is that the cellular functions necessary are supplied with energy from the molecule ATP (adenosine triphosphate), which couples with many biochemical reactions. The formation of ATP can be accomplished as follows:

$$ADP(aq) + H_2PO4^-(aq) \rightleftharpoons ATP(aq) + H_2O(l)$$

ΔG° for the reaction is 31 kJ/mol. Using that information, what would we estimate for the reaction quotient, Q, at 25 °C, for the reaction at equilibrium?

91. The gas commonly used in chemical laboratory burners is methane (CH_4). Shown here is the combustion reaction that takes place when we light a lab burner.

$$CH_4(g) + 2O_2(g) \rightleftharpoons CO_2(g) + 2H_2O(g)$$

   a. Use $\Delta_f G^\circ$ data to calculate the ΔG° value for the reaction.
   b. What would the value of ΔG be if the pressure of O_2 were reduced to 0.20 atm from 1.00 atm, at room conditions, with the remaining gases at standard conditions?
   c. Explain why, if ΔG has a negative value indicating a spontaneous change, we still have to bring a flame or spark to the system to get it to react in the lab.

92. Nitrogen gas and oxygen gas make up essentially 100% of our air. The following reaction has $\Delta G^\circ = 174$ kJ. Fortunately, the two do not easily combine through this reaction under the Earth's atmospheric conditions.

$$N_2(g) + O_2(g) \rightleftharpoons 2NO(g)$$

   a. What is the value of Q_p at 298 K for the reaction at equilibrium?
   b. What temperature would be needed for the reaction to be spontaneous if the pressures were returned to 1.0 atm? (Assume that enthalpy and entropy values do not change significantly over this range.)
   c. What would be the value of ΔG at 25°C if the pressure of both $N_2(g)$ and O_2 were increased to 5.0 atm each, while the pressure of NO was decreased to 0.50 atm?

17.8 Free Energy and the Equilibrium Constant

Skill Review

93. Determine the equilibrium constant, K_{eq}, associated with reactions that have these free energy changes at 25 °C.
   a. $\Delta G^\circ = -1.05$ J/mol
   b. $\Delta G^\circ = 0.230$ J/mol
   c. $\Delta G^\circ = 2.55$ kJ/mol
   d. $\Delta G^\circ = -9.80$ kJ/mol

94. Determine the free energy change, $\Delta G°$, for each of these equilibrium constants at 25 °C.
   a. $K_{eq} = 1.8 \times 10^{-5}$
   b. $K_{eq} = 6.67 \times 10^{-1}$
   c. $K_{eq} = 2.30 \times 10^{-2}$
   d. $K_{eq} = 125$
95. What is the free energy change, $\Delta G°$, in kilojoules per mole, for the formation of methanol from carbon monoxide and hydrogen gas at 25 °C?

$$CO(g) + 2H_2(g) \rightarrow CH_3OH(l) \quad K_{eq} = 13.5$$

96. Use the tables of $\Delta G°$ in the Appendix to determine the value of the equilibrium constant, K_{eq}, for the following reaction at 25 °C.

$$CaCO_3(s) \rightarrow CaO(s) + CO_2(g)$$

Chemical Applications and Practices

97. Hydrogen sulfide (H_2S), which is responsible for the odor of rotten eggs, decomposes by the reaction given below, which has an equilibrium constant at 25 °C equal to 0.020. Calculate the free energy change, in kilojoules per mole, for the decomposition reaction at 25 °C.

$$2H_2S(g) \rightarrow 2H_2(g) + 2S(s)$$

98. The ionization of HF in water has an equilibrium constant $K_{eq} = 3.5 \times 10^{-4}$ at 25 °C. Calculate the free energy change, in kilojoules per mole, for the decomposition reaction.

Comprehensive Problems

99. Programs that attempt to predict large-scale human behavior rely on statistics, including the concept of entropy, in order to produce results. Examples of such software include games, such as Sim-City, Total War, etc. While such software cannot reliably predict individual behavior, it can do reasonably well with large populations.
   a. Use the ideas of microstates and entropy to explain the Westward Expansion of the United States during the 1800's.
   a. In a purely socialist society, there is only one financial microstate, since everything is supposed to be equally distributed to all. In a capitalist society, there are many different financial microstates, from very wealthy to very poor. Based on the concepts in this chapter, which is the most probable: a capitalist society moving towards socialism or a socialist society moving towards capitalism?
100. Butane, burned in small portable lighters, combusts according to the following unbalanced equation:

$$\_\_C_4H_{10}(g) + \_\_O_2(g) \rightarrow \_\_CO_2(g) + \_\_H_2O(g)$$

First balance the equation, and then explain the answer to this question: "Does the combustion show an increase or a decrease in entropy for the system?"

17

101. Suppose a particularly noxious compound was entering a water supply at only one point. If it would stay at the entry point, it might easily be removed. However, this does not happen. Explain why environmental scientists must be aware of the second law of thermodynamics.
102. Because it was already known that a positive change in the entropy of the universe indicated a spontaneous reaction, why was it important to develop the Gibbs equation, which is also used to predict the spontaneity of a reaction? What advantage does the Gibbs equation offer?
103. In the ongoing research to find a cure or treatment for Alzheimer's disease, some investigators have focused on the appearance of plaque-like formations in brain cells. The solid masses form from protein fibers called beta-amyloids. Use Internet resources and journal sources to find out if the process is based on a positive or a negative value for the $\Delta S°$ of the plaque-forming reaction.
104. In calculations involving Gibbs energy and equilibrium, we must always be aware of the differences between ΔG and $\Delta G°$. When $\Delta G = 0$, at 25 °C, what is true about Q?
105. A certain reaction is nonspontaneous at one temperature. If raising the temperature caused the reaction to become spontaneous, what could we conclude about the entropy change of the reaction? Explain the answer.
106. In the chapter, we note that metabolism of glucose results initially in the formation of two molecules of pyruvate for every molecule of glucose.
   a. According to Fig. 17.7, what is the net production of ATP molecules for every molecule of glucose consumed?

b. How many grams of pyruvate would result from the consumption of 0.057 g of glucose? (Assuming all of the glucose is converted into pyruvate.)
c. This process, known as glycolysis, is spontaneous. How can these reactions be spontaneous if they also result in the generation of molecules with high-potential energy?

107. We know that about 76% of all chlorine atoms are chlorine-35 and about 24% are chlorine-37. Let's say that we were able to separate the two isotopes and form two samples of Cl_2, one that is pure ^{35}Cl–^{35}Cl and the other, ^{37}Cl–^{37}Cl. We then took each sample and reacted it with hydrogen gas to form HCl.

$$\mathrm{Cl_2 + H_2 \rightarrow 2HCl}$$

Based on our understanding of thermodynamics, discuss which reaction—the one with the Cl-35 isotope or the one with the Cl-37 isotope—would have the greater enthalpy of reaction, entropy, and free energy.

108. "Is there life on other worlds?" has been asked for centuries, though we have made the greatest strides in obtaining data via NASA space probes within the past 40 years. How would the biochemical reactions we've studied in this chapter be affected on Venus with its surface temperature of over 400 °C? What about the effect on Enceladus, a moon of Saturn which has a surface temperature of −330 °C?

Thinking Beyond the Calculation

109. The oxidation of glucose was discussed in detail in this chapter. A similar compound, fructose, also undergoes oxidation in biological organisms to give (eventually) CO_2 and water:

$$\_\_\mathrm{C_6H_{12}O_6(s)} + \_\_\mathrm{O_2(g)} \rightarrow \_\_\mathrm{CO_2(g)} + \_\_\ \mathrm{H_2O(l)}$$

a. Balance the equation and predict the numerical value of the change in entropy for this process.
b. Use the appendix to calculate the values of ΔH°, ΔS°, and ΔG° for this reaction. Compare the calculated value for the entropy change to the predicted value. Assume that the thermodynamic values for fructose are equivalent to those for glucose.
c. If the combustion reaction is done in the laboratory, the water is isolated as vapor. Recalculate the values of ΔH°, ΔS°, and ΔG° for the reaction where both products are gases.
d. If 5.0 g of fructose are consumed, how many liters of $CO_2(g)$ would be produced? Where does this CO_2 go in a living organism? (Assume 25 °C and 1 atm.)
e. How much heat is liberated from the combustion of 5.0 g of fructose in the laboratory?
f. In the first step of the catabolism of fructose, a phosphate is attached to the sugar unit:

$$\mathrm{C_6H_{12}O_6(aq) + ATP(aq) \rightarrow ADP(aq)}$$
$$\mathrm{+\ C_6H_{11}O_9P^{2-}(aq)} \quad \Delta G = -16.7\,\mathrm{kJ/mol}$$

We also know that ATP can be made from ADP and inorganic phosphate in the body:

$$\mathrm{ATP(aq) + H_2O(l) \rightarrow ADP(aq) + HPO_4^{2-}(aq)}\Delta G = -30.5\ \mathrm{kJ/mol}$$

What is the free energy change for the reaction of fructose with phosphate?

$$\mathrm{C_6H_{12}O_6(aq) + HPO_4^{2-}(aq) \rightarrow C_6H_{11}O_9P^{2-}(aq) + H_2O(l)} \quad \Delta G = ?$$

g. Why would an organism have a need to utilize fructose in the same manner as glucose? What is a typical source of fructose?

Electrochemistry

Contents

Supplementary Information The online version contains supplementary material available at ▶ https://doi.org/10.1007/978-3-030-90267-4_18.

M. Mosher and P. Kelter, *An Introduction to Chemistry*,
https://doi.org/10.1007/978-3-030-90267-4_18

What We Will Learn from This Chapter

Electrochemistry involves the study of reduction and oxidation reactions that are separated, yet connected so that ions and electrons can move between the two (▶ Sect. 18.2).

The oxidation state of an atom within a compound is used to determine if the atom is oxidized or reduced (▶ Sect. 18.3).

All reduction reactions must have an accompanying oxidation reaction in order to be complete (▶ Sect. 18.4).

An electrochemical cell can be used to determine the spontaneity of a redox reaction and the potential (voltage) of that reaction (▶ Sect. 18.5).

The chemical reactivity series indicates which atoms are more likely to be reduced and which are more likely to be oxidized (▶ Sect. 18.6).

The Nernst equation allows us to predict the cell potential when the conditions (concentration, temperature) are non-standard (▶ Sect. 18.7).

Electrolytic reactions occur when a potential is applied to an electrochemical cell. This results in the reaction "running backwards" (▶ Sect. 18.8).

18.1 Introduction

"I admit the deed!—tear up the planks!— here, here!—it is the beating of his hideous heart!" (Edgar Allan Poe, "The Telltale Heart"). The heart is an incredible organ. Without it, there is no life. It collects blood from the extremities and pushes it to the lungs, where oxygen and carbon dioxide are exchanged. Then it pulls the blood back and pumps it out to the various regions of the body, where the dissolved oxygen is delivered to the cells. In a normal lifetime, the four chambers of the heart rhythmically contract and relax nearly 3 billion times. The heart is always beating. What causes this fabulous muscle to contract and relax?

Muscle cells contain different concentrations of sodium and potassium cations inside (18 mM Na^+, 166 mM K^+) and outside (135 mM Na^+, 5 mM K^+) the cell. This concentration difference arises as the cells work continuously to pump Na^+ ions out and K^+ ions in. The concentration gradient (a gradually increasing difference) across the cell membrane sets up an electrical **potential**—a driving force to perform a reaction that results from a difference in electrical charge between two points. The potential—in this case, a force to restore the ion concentrations to equality—is measured in **volts** (V), just as we measure the **voltage** of a battery. The muscle cell has a very small potential (~100 mV), but it is enough to prime the cell for contraction.

Volt – The SI unit of potential.

Potential – The driving force (to perform a reaction) that results from a difference in electrical charge between two points.

Voltage – A measure of how strongly a species pulls electrons toward itself. Also known as electromotive force (emf).

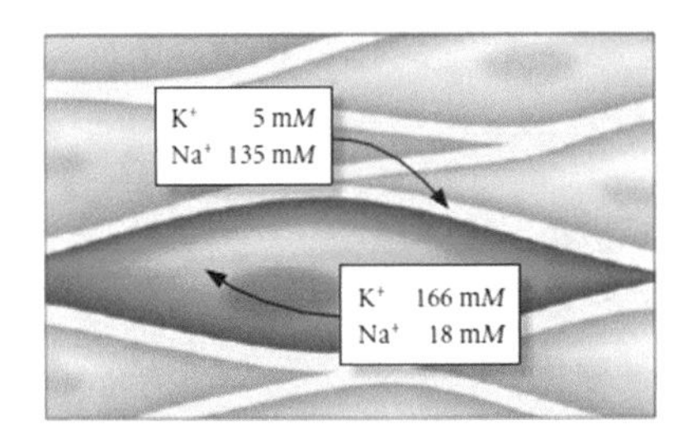

Understanding how the concentrations of sodium and potassium ions contribute to the potential inside a heart cell is of interest to the **cardiologist**, a heart specialist. More broadly, the electron exchanges that occur in chemical reactions are of interest to **electrochemists**, who create or analyze systems that allow exchanges between chemical and electrical energy. This exchange occurs via the gain of electrons (reduction) and the loss of electrons (oxidation) in reactions called **redox reactions**. Knowing how redox reactions work, how they develop a potential, and how the electrons involved can be harnessed helps scientists understand biological systems (such as muscles and nerves). It also enables them to continue to develop and improve one of the more important inventions that enhances our lives on a daily basis: the battery. The principles behind the exchange between chemical and electrical energy rely on knowing that chemical reactions can do work. In this chapter, we look at the interplay of electrical and chemical energy, along with some of its interesting and important uses.

Cardiologist – A heart specialist.

Electrochemists – Scientists who create or analyze systems that allow exchanges between chemical and electrical energy.

Redox reactions – Chemical reactions in which reduction and oxidation occur.

How Do We Know? DNA Profiling

Ionic compounds move when placed in an electric field. The Swedish chemist Arne Tiselius used this information in the 1930s to develop a separation technique. **Gel electrophoresis**, the use of electric fields to separate ions, is based on the application of an electric field across a gel-filled space. Ions, such as DNA fragments or proteins, are placed at one end of the gel, and a voltage is applied. The molecules with a net negative charge tend to migrate toward the positive pole as the positively charged molecules move toward the negative pole. The average rate of their movement is proportional to the average charge and to the average voltage applied. However, the rate of movement is inversely proportional to the size of the molecules. After a certain amount of time, the apparatus is disassembled, and the gel stained to reveal the locations of specific compounds. The molar masses of DNA fragments or proteins can be estimated by comparing their locations with those of reference samples whose molar masses are known.

Gel electrophoresis – The use of electric fields to separate ions.

Gel electrophoresis is widely used in molecular biology for separating DNA, RNA, and proteins by size. After human DNA is broken into small pieces with enzymes, it can be analyzed using gel electrophoresis to provide evidence in criminal cases, to diagnose genetic disorders, and to solve paternity cases. The sizes of the DNA pieces differ among people, so a person's "DNA fingerprint" or "DNA profile" can be constructed. DNA profiling is also used by conservation biologists to determine genetic similarity among populations or individuals. Evolutionary biologists use DNA profile information to construct hypothetical family trees indicating the relationships among species.

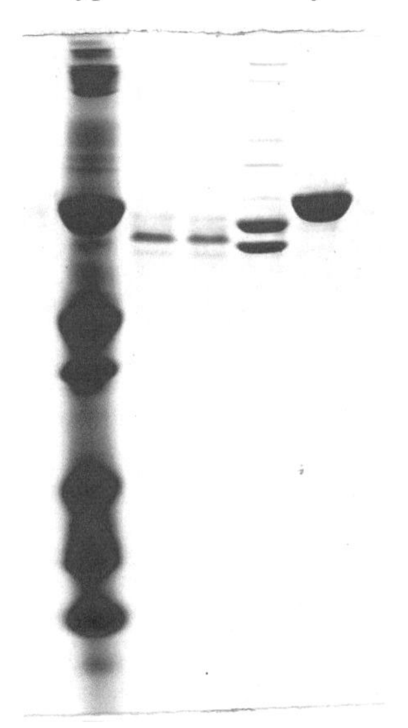

► https://upload.wikimedia.org/wikipedia/commons/5/5a/Coomassie3.jpg

18.2 What is Electrochemistry?

Worldwide, in 2018, over 4.5 billion passengers travelled on commercial airlines. There was a fatal accident about once every 5.6 million flights, resulting in 534 deaths in 2018. On those extremely rare occasions when a jetliner crashes, few may live to tell about it. However, each flight leaves a record of clues, including the "black box," for

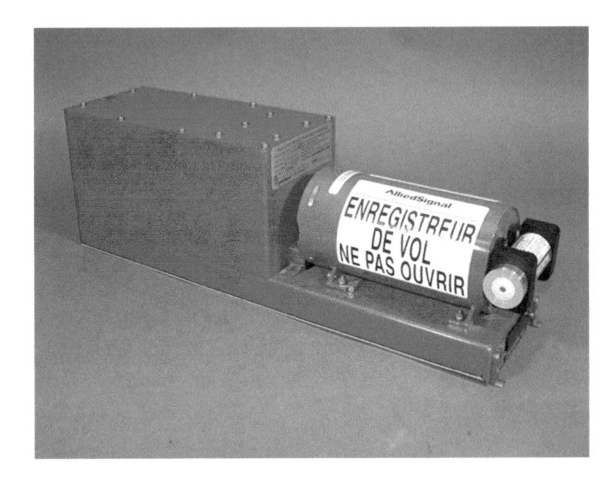

■ **Fig. 18.1** The flight data recorder. This "black box" records data about each airline flight. The information it collects can be used to help solve the mysteries of a plane crash *Source* ▸ https://upload.wikimedia.org/wikipedia/commons/6/6a/Fdr_sidefront.jpg

airline officials to decode. Known as a flight data recorder (FDR), the black box, a remarkable combination of engineering, data technology, and chemistry (■ Fig. 18.1), is made rugged enough to survive a crash. So that it too can be found after a crash, engineers must invest the same care in the design of the battery that powers the locator beacon. While constructing a battery to the exacting specifications of an FDR is challenging, the principles behind battery construction are nothing new. Like the FDR, for example, the battery in a cardiac pacemaker must also meet tough standards; it is expected to function reliably for up to ten years, to provide a steady supply of power, and not to leak chemicals into the person wearing it. The pacemaker batteries are based on the same principles of electrochemistry as the FDR.

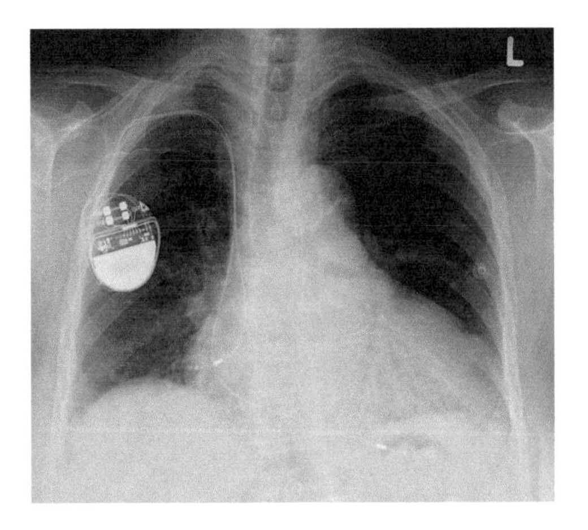

▸ https://upload.wikimedia.org/wikipedia/commons/6/66/Pacemaker.jpg

What is electrochemistry? Broadly speaking, **electrochemistry** is the study of the **reduction** and **oxidation** processes that occur at the interface between different phases of a system. Furthermore, electrochemistry typically involves reactions that take place at the surface of a solid. One important field of study within electrochemistry is called **electrodics**, the study of the interactions that occur between a solution of electrolytes and an electrical conductor, often a metal. Another important field within electrochemistry is **ionics**, the study of the behavior of ions dissolved in liquids.

Electrochemistry – The study of the reduction and oxidation processes that occur at the meeting point of different phases of a system.

Electrodics – The study of the interactions that occur between a solution of electrolytes and an electrical conductor, often a metal.

Ionics – The study of the behavior of ions dissolved in liquids.

Fig. 18.2 Count Alessandro Volta (1745–1827) invented the electrophorus, the first well-documented example of a voltaic cell, in 1775. He is also credited with the isolation of methane in 1778. *Source* ▶ https://upload.wikimedia.org/wikipedia/commons/0/0c/ETH-BIB-Volta%2C_Alessandro_%281745-1827%29-Portrait-Portr_02303.tif, ▶ http://doi.org/10.3932/ethz-a-000045652

Oxidation – The loss of electrons.

Reduction – The gain of electrons.

Electrochemical processes typically take place in an **electrochemical cell**, a device that allows the exchange between chemical and electrical energy. Two types of electrochemical cells are possible. The **voltaic cell** (also called a **galvanic cell**) is named after Alessandro Volta (Fig. 18.2). It is a type of electrochemical cell that *produces* electricity from a chemical reaction. Voltaic cells are commonly known as batteries, although technically speaking, a **battery** is two or more voltaic cells joined together in series. The second type of electrochemical cell is the **electrolytic cell**. This cell requires the addition of electrical energy to drive the chemical reaction under study. For example, the industrial production of aluminum (see ▶ Sect. 18.7) is an electrolytic process.

Electrochemical cell – A device that allows the exchange between chemical and electrical energy.

Battery – Two or more voltaic cells joined in series.

Galvanic cell – A cell that produces electricity from a chemical reaction. Also known as a voltaic cell.

Electrolytic cell – A cell that requires the addition of electrical energy to drive the chemical reaction under study.

Voltaic cell – A cell that produces electricity from a chemical reaction. Also known as a galvanic cell.

All electrochemical cells require electron exchange that can be characterized into two parts, each known as a **half-reaction**, and the sum of the half-reactions equals the complete reaction observed in the cell. One half-reaction (an **oxidation reaction**) supplies the electrons, and a second (a **reduction reaction**) uses these electrons. For this reason, the reactions that take place in electrochemical cells are also known as redox reactions. The oxidation reaction occurs at an **electrode** (typically a metal surface that acts as a collector or distributor for the electrons) known as the **anode**. The reduction reaction takes place at the **cathode**.

Half-reaction – An equation that describes the reduction or oxidation part of a redox reaction.

Oxidation reaction – In a redox reaction, the half-reaction that supplies electrons.

Anode – The electrode at which oxidation takes place.

Cathode – The electrode at which reduction takes place.

Electrode – A metal surface that acts as a collector or distributor for electrons.

Reduction reaction – In a redox reaction, the half-reaction that acquires electrons.

How does a redox reaction differ from any other kind of reaction? If we were to place the components for each of the two half-reactions into the same beaker, we wouldn't note any difference. However, because the movement of electrons is what defines the redox reaction, the half-reactions do not need intimate contact in order to produce products. As we will see later in this chapter, as long as we make sure that the half-reactions can exchange materials, the overall cell reaction will work. The driving force to complete the reaction—the cell potential—will ensure that the reaction proceeds and that we will be able to harness the power of the electrochemical cell. Our first task in understanding redox reactions and how to make use of the electron exchange is to know a redox reaction when we see it.

18.3 Oxidation States—Electronic Bookkeeping

The International Space Station and every other craft taking humans outside of the Earth's atmosphere requires electricity for lights, heaters, cameras, communications, and almost every other operation during its orbit of the Earth. In addition, the astronauts need water and oxygen to survive the trip. Electrochemists have discovered a way to provide both water for the crew and electricity for the ship. Their answer is the **fuel cell**, an electrochemical cell that uses the reaction of hydrogen with oxygen to produce electricity (◘ Fig. 18.3).

Fuel cell – An electrochemical cell that utilizes continually replaced oxidizing and reducing reagents to produce electricity.

$$2H_2(g) + O_2(g) \rightarrow 2H_2O(g)$$

Although we see that this reaction produces water, our first glance at the reaction does not reveal a process that can produce electricity. But it can. Some chemical reactions (the redox reactions), such as this one, include the shifting or transfer of electrons with one species losing electrons (oxidation) and another species gaining these electrons (reduction). Examining the **oxidation state** of the atoms helps us identify the redox reaction. All we must do is keep track of the electrons. A word of caution: There are times when the oxidation state of an atom has little physical meaning, and it can be misleading if the oxidation state is literally interpreted as an indication of where the electrons are found. For instance, when we determine the oxidation state of the oxygen atoms in formic acid (HCOOH), as we do in Example 18.1, we will find that both oxygen atoms have the same oxidation state. This should not be interpreted to mean that the same density of electrons will be found equally on both oxygen atoms. Yet even though that interpretation is wrong, oxidation states can be helpful in identifying the general distribution of the electrons in a substance.

Oxidation state – A bookkeeping tool that gives us insight into the distribution of electrons in a compound. Also known as oxidation number.

As we discussed in ▶ Chap. 4, the oxidation state of an atom in an element is zero. In monatomic anions and cations, the oxidation state is written as a superscript to the right of the atom symbol, as is done in Ca^{2+} and Br^{-}. For

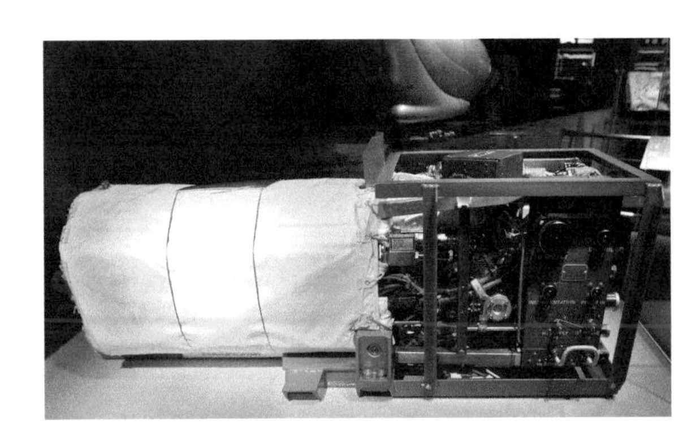

◘ Fig. 18.3 The fuel cells on the Space Shuttles (now discontinued!) were a model for future space travel in that they provide energy for use during the flight, and, as a product of the chemical reaction, also produce potable water. Photo by Steve Jurvetson *Source* ▶ https://upload.wikimedia.org/wikipedia/commons/0/08/Space_Shuttle_Fuel_Cell.jpg. *Source* ▶ https://secure.flickr.com/photos/jurvetson/8154813688/in/photostream

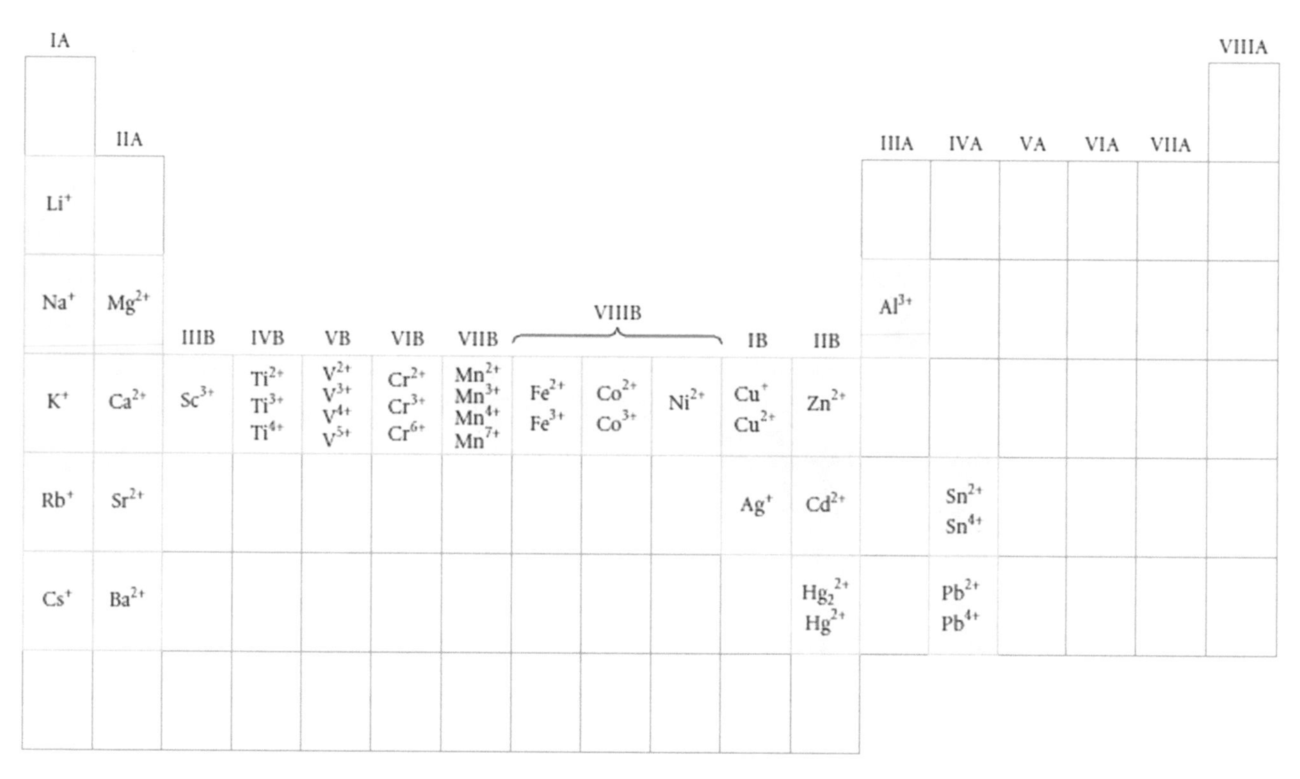

Fig. 18.4 Oxidation states of selected metals and their relationship to their location in the periodic table

compounds, oxidation states of atoms are small integers such as +2, +7, and −1 (and occasionally fractions) that indicate how an atom's electrons have shifted relative to the elemental state. In the fuel cell, both O atoms in O_2 and both H atoms in H_2 have an oxidation state of zero because these atoms are found in (neutral) elements. The oxidation state of oxygen in the product is −2, and that of each hydrogen atom is +1.

How did we know the oxidation state assignments for each atom in water? Electrochemists are familiar with chemical behavior. For example, they know that the hydrogen atom, which has a relatively low electronegativity, is assigned the oxidation state of +1 in all its compounds except metal hydrides such as NaH, in which it is −1 because the Group 1 (IA) and Group 2 (IIA) metals have even lower electronegativity values than hydrogen. With rare exceptions, all metals in compounds have positive oxidation states, as can be seen by the values in Fig. 18.4. Nonmetals, however, can have either positive or negative oxidation states, depending on the compound in which they are found. Figure 18.5 presents a set of decision rules to assist us in assigning oxidation states.

What does an oxidation state mean? We can think of it as a measure of the electron density distribution in a particular compound. A positive oxidation state indicates that electrons have shifted away from that atom.

A negative oxidation state means that electrons have shifted toward that atom. Because electrons are shared (to varying extents) between adjacent atoms, we generally expect nonzero oxidation states for each of the atoms if they are different elements. For example, in the water molecule, H is assigned as +1 and O is assigned as −2. The electrons have shifted away from hydrogen toward the oxygen, as shown in the electrostatic density map in Fig. 18.6. This is consistent with our understanding of electronegativity. The oxidation states do not mean that hydrogen has a full +1 charge and oxygen a full −2 charge. Remember that water is a covalent molecule. We can talk more comfortably about oxidation states as representations of the charge of atoms in ionic compounds. Whether in covalent or ionic compounds, we will use oxidation state as a bookkeeping tool to see where electrons shift in chemical reactions.

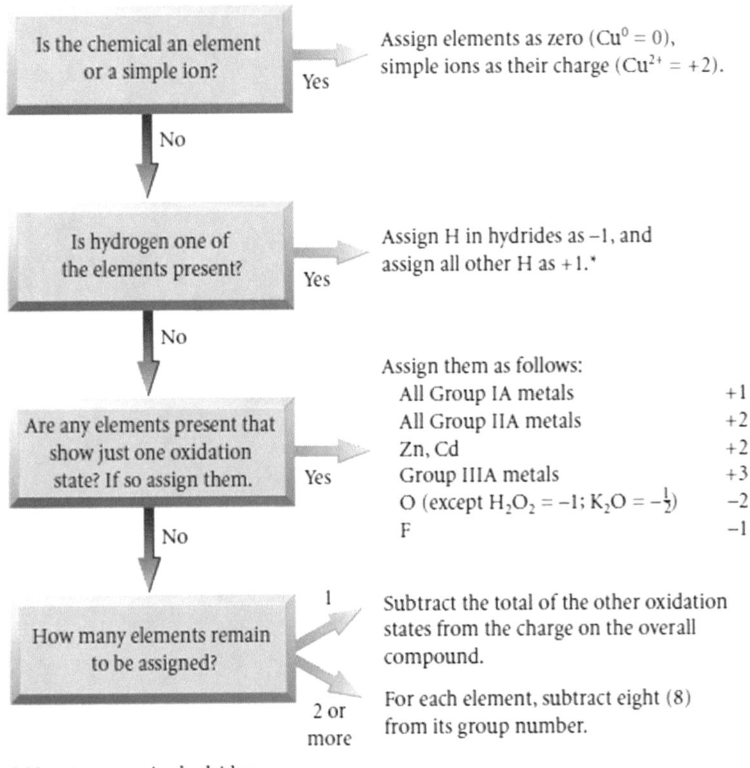

Fig. 18.5 Decision rules for assigning oxidation states

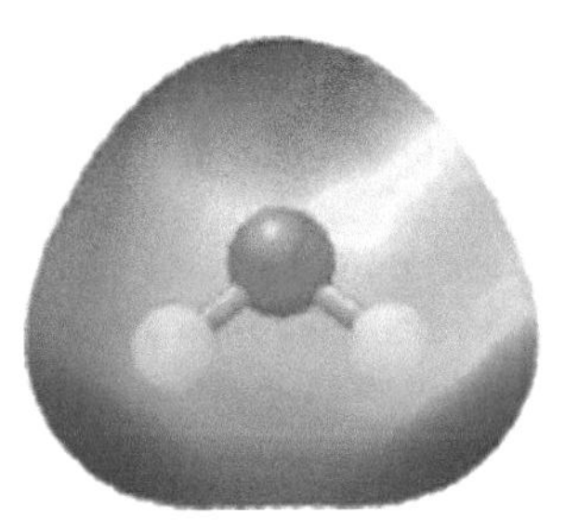

Fig. 18.6 The electrostatic potential map for water

18

Example 18.1—Oxidation States

Assign an oxidation state to each of the elements in formic acid (HCOOH). Given the Lewis structure shown below, which of the electrostatic density maps that follow represents formic acid?

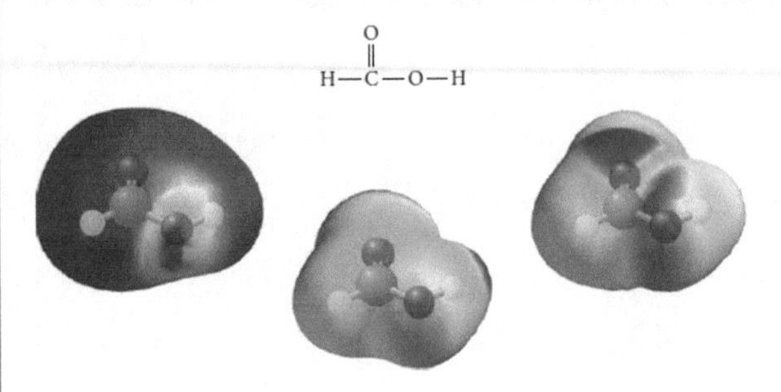

Asking the Right Questions

We have two questions here: Technically, what are the oxidation states and, conceptually, what do the oxidation states imply about the electron distribution in the molecule? In both cases, our assignment of the oxidation states and the electrostatic density map should be consistent with our understanding of electronegativity.

Solution

According to our rules for assigning oxidation states in Fig. 18.5, each oxygen is assigned an oxidation state of −2, and each hydrogen is +1. For the neutral formic acid molecule, in which the sum of all of the oxidation states must be 0, we have

$$2H = +1 \times 2 = +2$$

$$2O = -2 \times 2 = -4$$

The total is $+2+(-4)=-2$. The carbon atom must therefore have an oxidation state of +2. This means that the electron density in this molecule is focused more on the oxygen atoms than on the carbon or hydrogen atoms. Given the assignments of the oxidation states, the second electron density map is the most reasonable representation of the molecule. In the other two maps the charge density isn't in the correct location. In the first map, the electron density appears to be opposite of what we'd expect, with the red color on the map indicating regions of high electron density. The third map again shows electron density that doesn't appear to match where we have predicted the electrons to reside in the molecule.

Are our Answers Reasonable?

The assignment for oxygen's and hydrogen's oxidation states are typical; nothing unusual. The oxidation state for carbon was assigned based on those for oxygen and hydrogen. That is reasonable. Our choice of the electrostatic density map to use was based on the oxidation states of each atom in the molecule, a reasonable criterion.

What If…?

…we considered the conjugate base of formic acid, the formate anion? How would its electrostatic density map differ from that of formic acid?

(*Ans: A formal* -1 *charge around the oxygen would result in a dark red color on this atom and very little blue.*)

Practice 18.1

Determine the oxidation states of carbon in glucose ($C_6H_{12}O_6$) and carbon monoxide (CO).

We began this section discussing the reaction of hydrogen and oxygen gases in a fuel cell. By comparing the reactants and products, we see that both the H and O atoms show a change in their oxidation state. The change indicated that a transfer of electrons from one species to another has occurred. Any chemical reaction in which atoms change their oxidation states is classified as an oxidation–reduction reaction, or redox reaction for short. When O_2 and H_2 react to form water, O_2 is reduced (it gains electrons) and its oxidation state decreases from 0 to -2. Similarly, H_2 is oxidized (it loses electrons), and the oxidation state of the hydrogen atoms increase from 0 to +1:

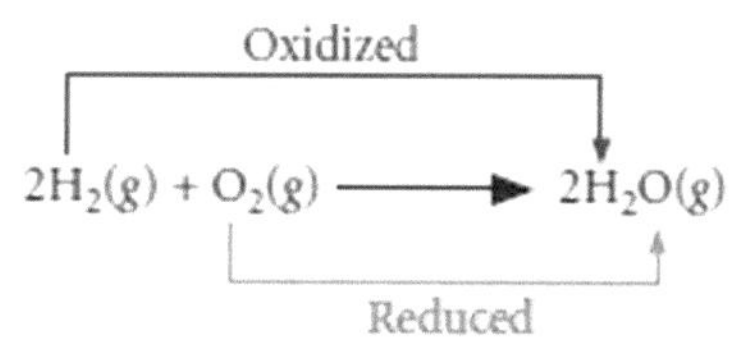

We can look at the compounds in this reaction in a different way. Oxygen (O_2) causes the oxidation of the hydrogen (H_2), so it is an **oxidizing agent**. In fact, oxygen stands as the premier oxidizing agent on planet Earth, both because of its abundance and because of its strong ability to accept electrons. We observe oxygen's effect every day our bodies metabolize glucose, or we venture out to explore a rusty, old shipwreck (rust results from the oxidation of Fe to Fe^{3+}). The hydrogen (H_2) in the fuel cell causes the reduction of the oxygen and is therefore a **reducing agent** (see Table 18.1). Looking at this another way, we can say that the oxidizing agent itself is reduced and the reducing agent itself is oxidized.

Oxidizing agent – A substance that causes the oxidation of another substance.

Reducing agent – A substance that causes the reduction of another substance.

Table 18.1 Oxidizing and reducing agents

Reactant	What happens	Examples
Oxidizing agent	Gains electrons Is reduced	Element: O_2, O_3, halogens Compound: H_2O_2 Ionic species: MnO_4^-
Reducing agent	Loses electrons Is oxidized	Element: H_2 and metals Compound: BH_3 Ionic species: NaH, $LiAlH_4$

Issues and Controversies—Destruction of the RMS Titanic

The RMS Titanic was one of three luxury liners designed by White Star Lines in an effort to outdo the competition. Everything else would be considered as a substandard luxury liner when compared to the Olympic, Titanic, and Gigantic. The Olympic (1911) and Titantic (1912) were the first two ships to be constructed, the Gigantic (1914, later dubbed the Britannic after the sinking of the Titanic) was modified with additional lifeboats and watertight doors.

▶ https://upload.wikimedia.org/wikipedia/commons/0/0e/Olympic_and_Titanic.jpg

On Sunday, April 14th, 1912, the Titanic was in the fourth day of its maiden voyage from Queenstown, Ireland (the name of the city was changed to Cobh in 1920) to New York. Despite the fact that warnings of icebergs were issued that day, the Titanic steamed ahead. At 11:40 p.m., lookouts on the Titanic spotted an iceberg directly in the path of the great ship. The First Officer, William Murdoch, ordered the engines stopped, turned hard to port, and secured the 15 watertight doors. Despite his efforts, the Titanic scraped along the side of the iceberg. Two hours and forty minutes later, the Titanic slipped below the waves taking 1502 crew and passengers with her. After years of searching, the wreck of the Titanic was finally located in 1985 in the North Atlantic at a depth of 3.9 km.

Visitors to the wreck today are amazed to see rust colored icicles of ever growing bacteria on the ship. Deemed rusticles, these dense multi-layered structures contain a high iron content (~30% Fe) consisting mostly as complex ferric oxides and hydroxides. The bacteria, known collectively as iron-related bacteria (IRB) produce these structures as they consume elemental iron from the Titanic's hull.

▶ https://upload.wikimedia.org/wikipedia/commons/3/30/Captain_Smith%27s_bathroom.jpg

An expedition to the Titanic in 1996 brought back some rusticles for further study. While there is considerable variation in the composition of rusticles, the major component is Fe^{3+} as FeO(OH) (iron oxide hydroxide, also known as goethite) and $Fe^{2+}_{3.6}Fe^{3+}_{0.9}\left(O^{2-},OH^{-},SO_4^{2-}\right)_9$ (iron oxide sulfate, also known as green rust). Where does the iron come from? Titanic's steel plates provide the iron, while the sulfur that is used to make iron oxide sulfate most likely comes from the seawater (which is greater than 880 ppm sulfur at ocean depths such as that at the site of the wreck). These elements are oxidized by IRB's to provide energy.

The chemistry of the bacterial growth on the ship is not well understood, but given the constant rate of growth, the entire ship will be completely consumed in 280–420 years. Human intervention and interference with the site could speed the rate of decay. In 1986, the United States Congress passed RMS Titanic Maritime Memorial Act in an effort to protect the Titanic's scientific, cultural, and historical significance. To further limit disturbances of the site, the United Nations Educational, Scientific, and Cultural Organization (UNESCO) has called for a law treating the Titanic and other wrecks as international maritime memorials.

Despite the actions of the United Nations and other governing bodies, salvage of the site continues. In 1986, Titanic Ventures recovered 1800 artifacts from the Titanic. RMS Titanic, Inc. (RMST), which now claims salvor-in-possession (a salvor is one who salvages) rights to the Titanic, has brought up an equal number of artifacts since 1993. RMST plans to hunt for more "high profile targets" in the future by entering the Titanic. Non-salvage commercial interests are also increasing as evidenced by visits of adventure tourists to the site.

Hydrogen peroxide is sometimes used to bleach hair or disinfect wounds. Is its decomposition into oxygen and water a redox reaction? If so, which species is the oxidizing agent and which is the reducing agent?

$$2H_2O_2(l) \rightarrow O_2(g) + 2H_2O(l)$$

Occasionally, a chemical reaction such as this one employs a single reactant as both the oxidizing and the reducing agent. Such a reaction is known as a **disproportionation**. Using our decision rules in Fig. 18.5, we can assign the following oxidation states to each atom.

Disproportionation – A reaction in which a single reactant is both the oxidizing and the reducing agent.

Oxidation

$$\overset{-1}{2H_2O_2(l)} \rightarrow \overset{0}{O_2(g)} + \overset{-2}{2H_2O(l)}$$

Reduction

$$\overset{+1\ -1}{2H_2O_2(l)} \rightarrow \overset{0}{O_2(g)} + \overset{+1\ -2}{2H_2O(l)}$$

In this reaction, the oxygen atoms in hydrogen peroxide (like any peroxide) have an oxidation state of −1. The two products of the reaction contain oxygen atoms with different oxidation states, O_2 with a higher oxidation state (zero) and H_2O with oxygen in a lower oxidation state (−2). We say that the hydrogen peroxide is both the oxidizing agent and the reducing agent.

Not all reactions require the transfer of electrons. Many of the reactions we've discussed in this text thus far are not redox reactions. For example, the neutralization of sodium hydroxide by hydrochloric acid is not a redox reaction. We can tell because the oxidation states for each of the atoms remain the same on both sides of the equation:

+1 in both compounds

$$NaOH + HCl \rightarrow NaCl + H_2O$$

−1 in both compounds

−2 in both compounds

$$\overset{+1-2+1}{NaOH} + \overset{+1\ -1}{HCl} \rightarrow \overset{+1\ -1}{NaCl} + \overset{+1-2}{H_2O}$$

Many elements show a variety of oxidation states, depending on the chemical species in which they are found. An example of this can be found in the nonmetal nitrogen, which has a range of possible oxidation states, as shown in Table 18.2. The gases found in our atmosphere, both naturally and as pollutants, have different positive oxidation states of nitrogen, since the oxygen atom in each compound is more electronegative than nitrogen and is assigned an oxidation state of −2. For the same reason, nitrates (NO_3^-) and nitrites (NO_2^-) have nitrogen with positive oxidation states. However, when combined with a less electronegative element such as hydrogen, nitrogen is assigned a negative oxidation state. Accordingly, look for a negative oxidation state of N in ammonia and in compounds containing the ammonium ion, NH_4^+.

Table 18.2 The oxidation states of nitrogen

Oxidation state	Formula	Name	Produced in nature
+5	HNO_3 NO_3^-	Nitric acid Nitrate ion	From NO_2^- by nitrifying bacteria
+4	NO_2	Nitrogen dioxide	In air by oxidation of NO
+3	HNO_2 NO_2^-	Nitrous acid Nitrite ion	From NH_3 by nitrifying bacteria
+2	NO	Nitrogen monoxide (nitric oxide)	From N_2 by lightning or volcanos
+1	N_2O	Nitrous oxide	From NO_2^- by denitrifying bacteria
0	N_2	Nitrogen	From N_2O by denitrifying bacteria
−1/3	N_3^-	Azide ion	(not found in nature)
−1	NH_2OH	Hydroxylamine	(not found in nature)
−2	N_2H_4	Hydrazine	(not found in nature)
−3	NH_3 NH_4^+	Ammonia Ammonium ion	From biological decay of proteins

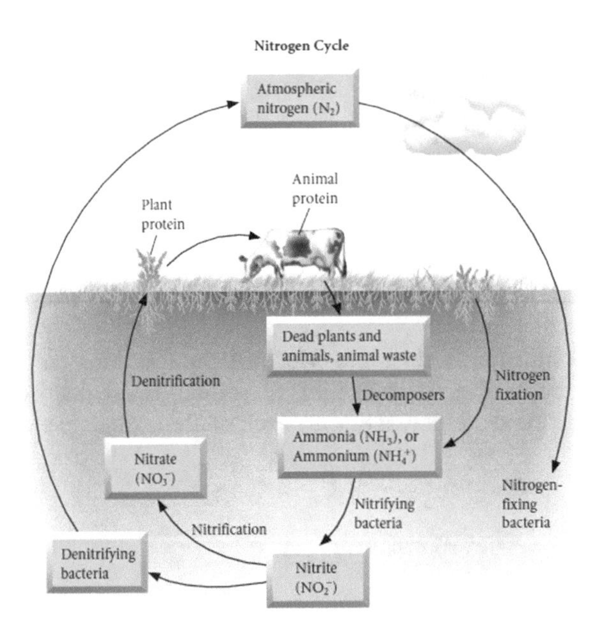

Fig. 18.7 The nitrogen cycle. The nitrogen in the environment is found in many different compounds and with many different oxidation states

18

The **nitrogen cycle** (Fig. 18.7) illustrates how nitrogen moves through its different oxidation states on our planet. We can find similar cycles where the oxidation states of sulfur and carbon change as these elements form different compounds. What is the point? Whereas oxidation states are fixed for a particular compound at a particular moment in time, chemicals in our world constantly undergo change. Oxidation states help us keep track of the changes that involve the important processes of oxidation and reduction.

Nitrogen cycle – The path that nitrogen follows through its different oxidation states on Earth.

Example 18.2—The Electrochemistry of Smog

Although oxygen and nitrogen do not react at low temperatures, they combine in a hot automobile engine to form the pollutant NO, with subsequent oxidation by atmospheric oxygen to form the poisonous brown gas NO_2, which is largely responsible for the smog found in urban areas on hot, sunny days:

$$N_2(g) + O_2(g) \rightarrow 2NO(g)$$
$$2NO(g) + O_2(g) \rightarrow 2NO_2(g)$$

Assign oxidation states to all the atoms involved in these two reactions. Which species are oxidized? Which are reduced? Do our answers make sense in terms of the relative electronegativity values of nitrogen and oxygen?

Asking the Right Questions

Which of the two atoms, nitrogen or oxygen, is more likely to be reduced (gain electrons)?

Solution

According to ◘ Fig. 18.5, we may assign all atoms in elements (such as N_2 and O_2) an oxidation state of zero. In the compounds, we assign oxygen as −2, because the compounds are not peroxides or superoxides. We assign nitrogen so that the sum of the oxidation states is zero, because NO and NO_2, as compounds, carry no charge.

Compound	Oxidation Number of N	Oxidation Number of O
N_2	0	–
O_2	–	0
NO	+2	−2
NO_2	+4	−2

The nitrogen has been successively oxidized going from N_2 to NO to NO_2, and the oxygen has been reduced as it changes from O_2 to its products.

Are our Answers Reasonable?

Oxygen is a more electronegative element than nitrogen, so we may expect it to be reduced when combined with nitrogen. Our answers confirm this, so they are reasonable.

What If…?

…our molecules had carbon, rather than nitrogen? What would be the oxidation states for carbon in CO, $CaCO_3$, CH_3COOH, $C_{12}H_{22}O_{11}$?

(*Ans: +2, +4, 0, 0*).

Practice 18.2

Rank these chemicals in order of increasing (from most negative to most positive) oxidation state of sulfur: SO_2, H_2SO_4, S_8, and Li_2S.

18.4 Redox Equations

Ira Remsen (1846–1927), one of the co-discoverers of the artificial sweetener saccharine, was particularly interested in chemistry as a boy.

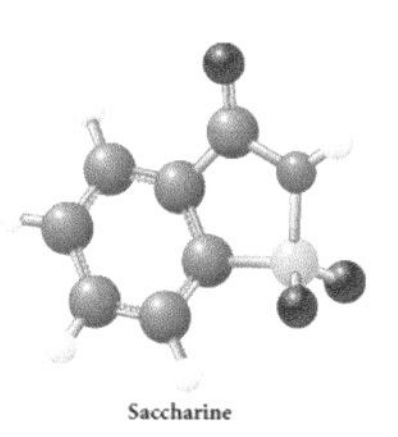

Saccharine

As an adult, he told of a childhood visit to the doctor's office where he, when left alone in the examination room, set out to discover what was meant by something he had read in a chemistry book: "Nitric acid acts upon copper." To discover this for himself, he placed a pure copper penny on the exam table and poured nitric acid from a bottle sitting on the doctor's shelf onto the penny. Remsen continues,

> But what was this wonderful thing which I beheld? The cent was already changed, and it was no small change either. A greenish blue liquid foamed and fumed over the cent and over the table. The air in the neighborhood of the performance became dark red. A great colored cloud arose. This was disagreeable and suffocating— how should I stop this? I tried to get rid of the objectionable mess by picking it up and throwing it out the window, which I had meanwhile opened. I learned another fact—nitric acid not only acts upon copper but it acts upon fingers. The pain led to another unpremeditated experiment. I drew my fingers across my trousers and another fact was discovered. Nitric acid also acts upon trousers. Taking everything into consideration, that was the most impressive experiment, and, relatively, probably the most costly experiment I have ever performed.

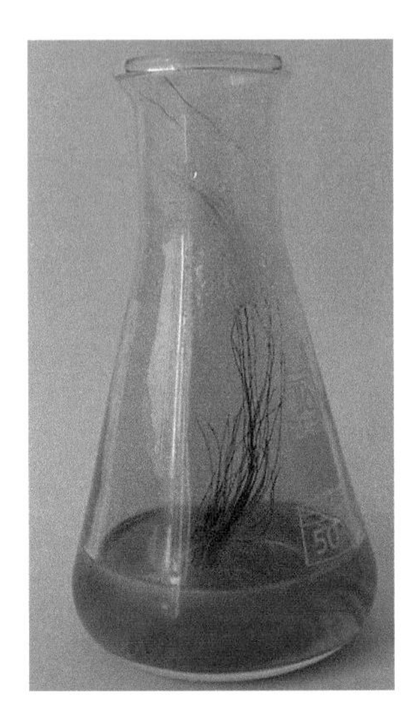

Fig. 18.8 Nitric acid acts on copper. The spontaneous reaction is evident from the generation of a blue solution and a cloud of noxious brown gas. The gas results from the reaction of NO with oxygen in the air. *Source* ▶ https://upload.wikimedia.org/wikipedia/commons/3/35/CopperReaction.JPG

The reaction that Remsen describes, which is illustrated in Fig. 18.8, is a redox reaction. The individual chemical equations that make up the action of nitric acid on copper include the formation of copper cations and NO_2. How do we know that it is a redox reaction? Examine the oxidation state of copper. Copper metal (oxidation state = 0) is being oxidized to the copper(II) ion. Simultaneously, nitrogen in nitric acid is being reduced from +5 to +2 in forming nitrogen monoxide. The NO gas released by the reaction rapidly reacts with O_2 in the air to make NO_2(g), the toxic brown fumes that so alarmed the budding chemist.

$$HNO_3(aq) + Cu(s) \longrightarrow Cu^{2+}(aq) + NO(g)$$

$$2NO(g) + O_2(g) \longrightarrow 2NO_2(g)$$

18.4.1 Half-Reactions

Ira Remsen noted that this reaction proceeded spontaneously to generate a cloud of noxious fumes. Can we predict this spontaneity by examining the reaction equation, which tells us whether the driving force (the potential) is favorable for this reaction? To assist us in answering this question, we need to extract the oxidation and reduction reactions from the overall equation. The resulting half-reactions, like so many half-reactions, are so well known that the potential for each has been measured and the results collected into a Table of Standard Reduction Potentials, such as Table 18.3. A more comprehensive table can be found in the Appendix.

Take some time and examine Table 18.3. Do we notice any patterns in the data in this table? One of the things we notice is that all of the reactions are written as reduction reactions. That is, the reactions show the consumption of electrons to make products with less positive oxidation states. Standard Reduction Potentials tables can be used to determine the potential of a reaction, be it for the silver oxide battery found in a pacemaker or for the action of nitric acid on a copper penny. Moreover, knowing the potential of the half-reactions helps us determine the spontaneity of a redox reaction, as we will see later.

Each half-reaction in the table is balanced both atomically and electrically. Half-reactions are simply what they appear to be: half of an oxidation–reduction reaction that is occurring in aqueous solution. The half-reaction listed in the table for the reduction of copper shows the reactants (copper ions and electrons), the product (copper metal), and the **standard potential (*E*°)** of the half-reaction.

$$Cu^{2+}(aq) + 2e^- \rightarrow Cu(s) \qquad E^\circ = +0.34\ V$$

Standard potential (E°) – The measure of the potential of a reaction at standard conditions.

The value of $E°$ is a measure of how strongly the reduced species on the right hand side of the reduction half-reaction pulls electrons toward itself. The standard potential is measured in volts, the SI unit of electrical potential. It is sometimes referred to as the **electromotive force (emf)** of the half-cell or, more commonly, as the voltage.

18

Table 18.3 Selected standard reduction potentials

The selected potentials shown here were obtained under standard conditions (in aqueous solution, 25 °C, all solutions 1.0 M, all gases 1.0 atm)		
Shorthand notation	**Half-cell reaction**	**Standard potential, $E°$ (V)**
$Li^+(aq) \mid Li(s)$	$Li^+(aq) + e^- \rightarrow Li(s)$	–3.04
$Na+(aq) \mid Na(s)$	$Na^+(aq) + e^- \rightarrow Na(s)$	–2.71
$Mg^{2+}(aq) \mid Mg(s)$	$Mg^{2+}(aq) + 2e^- \rightarrow Mg(s)$	–2.38
$Al^{3+}(aq) \mid Al(s)$	$Al^{3+}(aq) + 3e^- \rightarrow Alf(s)$	–1.66
$H_2O(l) \mid H_2(g)$	$2H_2O(l) + 2e^- \rightarrow H_2(g) + 2OH^-(aq)$	–0.83
$Cd(OH)_2(s) \mid Cd(s)$	$Cd(OH)_2(s) + 2e^- \rightarrow Cd(s) + 2OH^-(aq)$	–0.81
$Fe^{2+}(aq) \mid Fe(s)$	$Fe^{2+}(aq) + 2e^- \rightarrow Fe(s)$	–0.44
$H^+(aq) \mid H_2(g)$	$2H^+(aq) + 2e^- \rightarrow H_2(g)$	0
$Fe^{3+}(aq) \mid Fe(s)$	$Fe^{3+}(aq) + 3e^- \rightarrow Fe(s)$	+0.04
$Cu^{2+}(aq) \mid Cu(s)$	$Cu^{2+}(aq) + 2e^- \rightarrow Cu(s)$	+0.34
$O_2(g) \mid OH^-(aq)$	$O_2(g)_+ 2H_2O(l) + 4e^- \rightarrow 4OH^-(aq)$	+0.40
$NiO_2(s) \mid Ni(OH)_2(s)$	$NiO_2(s)_+ 2H_2O(l) + 2e^- \rightarrow Ni(OH)_2(s) + 2OH^-(aq)$	+0.49
$Ag^+(aq) \mid Ag(s)$	$Ag^+(aq) + e^- \rightarrow Ag(s)$	+0.80
$HNO_3(aq) \mid NO(g)$	$3H^+(aq) + HNO_3(aq) + 3e^- \rightarrow NO(g) + 2H_2O(l)$	+0.96
$Br_2(l) \mid Br^-(aq)$	$Br_2(g) + 2e^- \rightarrow 2Br^-(aq)$	+1.07
$O_2(g) \mid H_2O(l)$	$O_2(g) + 4H^+(aq) + 4e^- \rightarrow 4H_2O(l)$	+1.23
$Cl_2(g) \mid Cl^-(aq)$	$Cl_2(g) + 2e^- \rightarrow 2Cl^-(aq)$	+1.36
$Au^{3+} \mid Au(s)$	$Au^{3+}(aq) + 3e^- \rightarrow Au(s)$	+1.50
$F_2(g) \mid F^-(aq)$	$F_2(g) + 2e^- \rightarrow 2F^-(aq)$	+2.87

Electromotive force (emf) – A measure of how strongly a species pulls electrons toward itself in a redox process. Also known as voltage.

The values listed for $E°$ are measured under a specific set of conditions:
- Any aqueous ion is present at a concentration (technically, activity) of 1.0 M. All gases are at a pressure of 1 bar (approximately 1 atm).
- The temperature is 25 °C (298 K).

These conditions are "standard" for half-reactions and are indicated by the "°" in $E°$. If the conditions are not standard, the voltage will be different from that listed in the table (see ▶ Sect. 18.7), and the potential will be equal to E. Keep in mind that there are several "standards"! For example, standard conditions of temperature and pressure of gases (STP) refer to 0 °C (273 K) as the standard temperature.

We speak of standard reduction potential in terms of how strongly the species pulls electrons toward itself. Yet just as in a tug-of-war, we must consider what we are pulling against. In electrochemistry, this is accomplished with the use of a commonly used reference half-reaction that is compared to another half-reaction. This reference is known as the **standard hydrogen electrode reaction (SHE)**. The reaction, which is assigned a potential of zero volts, is also shown in Table 18.3 as the reduction of H^+ to H_2. To say that the reduction of Cu^{2+} to Cu^0 has a voltage of +0.34 V, as in Table 18.3, is to say that it has this voltage compared to the reduction of H^+ described by the SHE reaction. All potentials that we use in our discussions will be compared to the SHE reaction.

Standard hydrogen electrode reaction (SHE) – A reference half-reaction of the reduction of hydrogen ion to hydrogen gas, against which to compare our reduction.

The potential of a half-reaction can be used to assess the spontaneity of the half-reaction. For instance, electrons are strongly attracted to fluorine atoms. Recalling our discussion of ionization energy and electron affinity from

▶ Chap. 7, we might predict that adding electrons to F_2 should be more thermodynamically favorable than adding electrons to a Group IA metal cation such as Li^+. The half-reaction potentials for each reduction bear this out. Fluorine has a large positive reduction potential (+2.87 V), and lithium has a large negative reduction potential (−3.04 V). Michael Faraday (1791–1867), an English electrochemist, worked to illustrate how a favorable reaction could be related to the potential. Gibbs later was able to show this relationship mathematically as:

$$\Delta G^{\circ} = -nFE^{\circ}$$

where:

n – is the number of moles of electrons transferred in the reaction, and,

F – is called Faraday's constant.

A key feature of this equation is that the change in free energy, $\Delta G°$, and the cell potential, $E°$, have opposite signs (n and F are always positive). Because a negative value of free energy indicates a spontaneous process, a positive value for the cell potential must also indicate spontaneity.

Faraday's constant is a unit of electric charge equal to the magnitude of charge on a mole of electrons:

$$1 \text{ faraday} = 1\ F = 96,485 \frac{\text{coulombs}}{\text{mol}} = 9.6485 \times 10^4 \frac{\text{C}}{\text{mol}}$$

Faraday's constant – A unit of electric charge equal to the magnitude of charge on a mole of electrons.

On the basis of the relationship shown by Faraday, we can determine that 1 J equals 1 C·volt, and $1\ \text{V} = 1\ \frac{\text{joule}}{\text{coulomb}}$.

$$1\ \text{J} = 1\ \text{C} \cdot \text{V}$$
$$1\ \text{V} = \frac{\text{J}}{\text{C}}$$

Example 18.3—Spontaneity and Potential

Copper ions undergo reduction according to the following half-reaction:

$$Cu^{2+}(aq) + 2e^- \rightarrow Cu(s) \qquad E^{\circ} = +0.34\ \text{V}$$

What is the free energy change associated with this process? Is this a spontaneous half-reaction?

Asking the Right Questions

What is the relationship between $\Delta G°$ and $E°$?

Solution

The free energy change is negative; the half-reaction is spontaneous. However, this is only half of a redox reaction.

$$\Delta G^{\circ} = -nFE^{\circ}$$
$$= -\left(2\ \text{mol e}^-\right)\left(\frac{96{,}485\ \text{C}}{\text{mol e}^-}\right)\left(\frac{+0.34\ \text{J}}{\text{C}}\right)$$
$$= -65{,}609.8\ \text{J} = -66\ \text{kJ}$$

Are our Answers Reasonable?

The free energy is negative, which indicates a spontaneous process and a positive potential. That is reasonable.

What If…?

…we were given a 3-electron reduction half-reaction in which the free energy change is −155 kJ? What would be the potential for that half-reaction?

(*Ans: E°* = *+0.53 V*).

Practice 18.3

The silver cell battery used in pacemakers utilizes the following reaction with a measured potential of 1.86 V. What is $\Delta G°$ for this reaction? Is this reaction spontaneous?

$$Ag_2O(s) + Zn(s) \rightarrow 2\,Ag(s) + ZnO(s)$$

18.4.2 Balancing Redox Reactions

To balance a redox reaction such as the one describing the action of nitric acid on copper, we first determine the identity of the half-reactions. It can be hard (if not seemingly impossible!) to balance a redox equation correctly using a trial-and-error approach (▶ Chap. 3), so we often use a series of steps to accomplish the job (see ◘ Fig. 18.9). To be fair, this method is just a device to make the balancing go more quickly, rather than a rep-

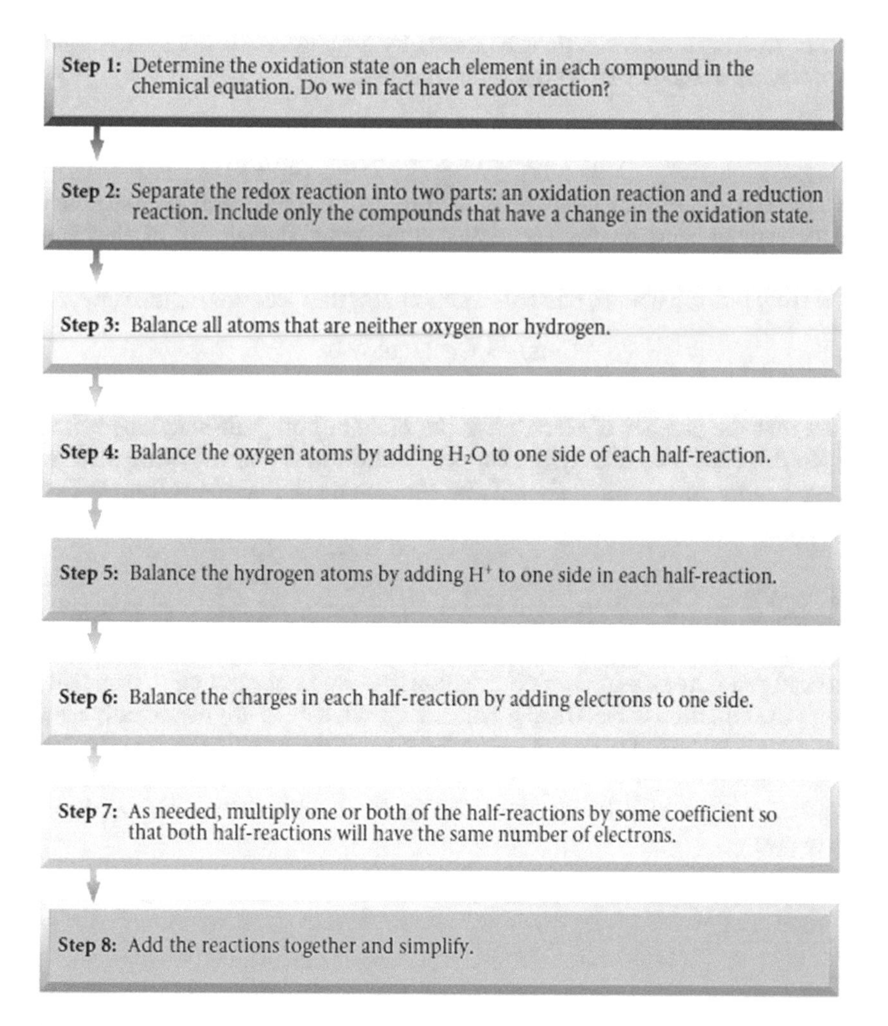

Fig. 18.9 Algorithm for balancing redox reactions in acidic solution

resentation of what actually happens at the molecular level. In the nanoworld, electron transfer processes not only occur simultaneously rather than sequentially but also occur in a fairly complex way, with charges building up at the phase changes (the so-called interfaces) in solution. Here we will focus just on the technical aspects of balancing equations.

Let's follow the algorithm in Fig. 18.9 as we balance the copper–nitric acid redox reaction:

$$HNO_3(aq) + Cu(s) \rightarrow Cu^{2+}(aq) + NO(g)$$

Step 1 in Fig. 18.9 indicates that we should determine whether we have a redox reaction. That is, does our oxidation state bookkeeping indicate that one species is undergoing oxidation and another is undergoing reduction? We determine that the equation describes a redox process by noting that the oxidation state of copper increases (from 0 to +2) while that of nitrogen decreases (from +5 to +2).

Step 2 indicates that we should separate the redox reaction into the two half- reactions. The easiest way to accomplish this is to ask: Which species are the same on both sides of the equation? These half-reactions show the oxidation (copper to copper ion) and the reduction (nitric acid to nitrogen monoxide).

$$Cu(s) \rightarrow Cu^{2+}(aq)$$
$$HNO_3(aq) \rightarrow NO(g)$$

The numbers of atoms (not including H's and O's) are then balanced on both sides of each half-reaction in Step 3. No modification of our reactions is needed for this step.

In Step 4 we balance the number of oxygen atoms by adding water molecules to the product side of the equation.

$$Cu(s) \rightarrow Cu^{2+}(aq)$$
$$HNO_3(aq) \rightarrow NO(g) + 2H_2O(l)$$

In Step 5 the number of hydrogen atoms is then balanced by adding H^+ to the reactant side. Note that in both steps 4 and 5, we did not need to modify the copper half-reaction.

$$\begin{aligned} Cu(s) &\rightarrow Cu^{2+}(aq) \\ 3H^+(aq) + HNO_3(aq) &\rightarrow NO(g) + 2H_2O(l) \end{aligned}$$

The charges are then balanced in Step 6 by adding electrons to the two half reactions. We note that the change in the oxidation number of the element being oxidized or reduced must equal the number of electrons lost or gained in the half-reaction. The copper half-reaction has electrons added as a product; the nitric acid half-reaction has electrons added as a reactant. In every redox reaction, one half-reaction gets the electrons on the right, the other on the left.

$$\begin{aligned} Cu(s) &\rightarrow Cu^{2+}(aq) + 2e^- \\ 3e^- + 3H^+(aq) + HNO_3(aq) &\rightarrow NO(g) + 2H_2O(l) \end{aligned}$$

In Step 7 we make sure that the number of electrons is the same in both half-reactions by multiplying the half-reactions by an integer. In this case, we multiply the copper reaction by 3 and the nitric acid reaction by 2. Then, in Step 8, we add the two reactions together and simplify by eliminating similar items from both sides of the equation.

$$\begin{aligned} 3\{Cu(s) &\rightarrow Cu^{2+}(aq) + 2e^-\} \\ 2\{3e^- + 3H^+(aq) + HNO_3(aq) &\rightarrow NO(g) + 2H_2O(l)\} \\ \hline 3Cu(s) + 6H^+(aq) + 2HNO_3(aq) &\rightarrow 3Cu^{2+}(aq) + 2NO(g) + 4H_2O(l) \end{aligned}$$

The result is the balanced redox reaction. One observation and one question can reasonably arise. The observation is that even though the equation is electrically balanced, each side is not electrically neutral. That is, we have the same +6 charge on each side! Now for the question: Where are the negative charges that would make each side electrically neutral? The answer comes when we remember that we have written net ionic equations, rather than complete ionic or molecular equations. In this case, the molecular form comes by adding 6 mol of nitrate ion (NO_3^-) to both sides to give

$$\begin{aligned} 3Cu(s) + 8HNO_3(aq) &\rightarrow 3Cu(NO_3)_2(aq) + 2NO(g) + 4H_2O(l) \\ 3Cu(s) + 6H^+(aq) + 2HNO_3(aq) + 6NO_3^- &\rightarrow 3Cu^{2+}(aq) + 2NO(g) + 4H_2O(l) + 6NO_3^- \\ \hline 3Cu(s) + 8HNO_3(aq) &\rightarrow 3Cu(NO_3)_2(aq) + 2NO(g) + 4H_2O(l) \end{aligned}$$

In other reactions, spectator ions such as Na^+ or Cl^- (if they are actually in the solution!) that are not listed in the net ionic equation make the system electrically neutral. The key point is that when we write equations in net ionic form, we expect them to be electrically balanced, not electrically neutral.

Example 18.4—Balancing Redox Equations in Acidic Solutions

Balance the following equation in acidic solution.

$$Cr_2O_7^{2-}(aq) + NO(g) \rightarrow Cr^{3+}(aq) + NO_3^-(aq)$$

Asking the Right Questions

Figure 18.9 can guide our thinking and technical work in balancing redox equations. Our first key question is: What is the oxidation state of each element in the compound?

Solution

Step 1: Determine the oxidation state on each element in each compound in the chemical equation. Do we in fact have a redox reaction?

Nitrogen is being oxidized from +2 to +5. Chromium is being reduced from +6 to +3. This is a redox reaction.

Step 2: Separate the redox reaction into two parts: an oxidation reaction and a reduction reaction. Include just the compounds that have a change in the oxidation state.

$$\begin{aligned} Cr_2O_7^{2-}(aq) &\rightarrow Cr^{3+}(aq) \quad \text{(reduction)} \\ NO(g) &\rightarrow NO_3^-(aq) \quad \text{(oxidation)} \end{aligned}$$

Step 3: Balance all atoms except oxygen and hydrogen.

$$\begin{aligned} Cr_2O_7^{2-}(aq) &\rightarrow 2Cr^{3+}(aq) \\ NO(g) &\rightarrow NO_3^-(aq) \end{aligned}$$

Step 4: Balance the oxygen atoms by adding H_2O to one side in each half-reaction.

$$\begin{aligned} Cr_2O_7^{2-}(aq) &\rightarrow 2Cr^{3+}(aq) + 7H_2O(l) \\ 2H_2O(l) + NO(g) &\rightarrow NO_3^-(aq) \end{aligned}$$

18

Step 5: Balance the hydrogen atoms by adding H^+ to one side in each half-reaction.

$$14H^+(aq) + Cr_2O_7^{2-}(aq) \rightarrow 2Cr^{3+}(aq) + 7H_2O(l)$$
$$2H_2O(l) + NO(g) \rightarrow NO_3^-(aq) + 4H^+(aq)$$

Step 6: Balance the charges in each half-reaction by adding electrons to one side.

$$6e^- + 14H^+(aq) + Cr_2O_7^{2-}(aq) \rightarrow 2Cr^{3+}(aq) + 7H_2O(l)$$
$$2H_2O(l) + NO(g) \rightarrow NO_3^-(aq) + 4H^+(aq) + 3e^-$$

Step 7: As needed, multiply one or both of the half-reactions by some coefficient so that the same number of electrons will appear in both half-reactions.

$$6e^- + 14H^+(aq) + Cr_2O_7^{2-}(aq) \rightarrow 2Cr^{3+}(aq) + 7H_2O(l)$$
$$2\{2H_2O(l) + NO(g) \rightarrow NO_3^-(aq) + 4H^+(aq) + 3e^-\}$$

Step 8: Add the reactions together and simplify. Note that the electrons on each side mathematically cancel, indicating the same number of electrons gained in the reduction as lost in the oxidation.

$$6e^- + 14H^+(aq) + Cr_2O_7^{2-}(aq) \rightarrow 2Cr^{3+}(aq) + 7H_2O(l)$$
$$4H_2O(l) + 2NO(g) \rightarrow 2NO_3^-(aq) + 8H^+(aq) + 6e^-$$
$$6e^- + 14H^+(aq) + Cr_2O_7^{2-}(aq) + 4H_2O(l) + 2NO(g) \rightarrow 2Cr^{3+}(aq) + 7H_2O(l) + 2NO_3^-(aq) + 8H^+(aq) + 6e^-$$

$$6H^+(aq) + Cr_2O_7^{2-}(aq) + 2NO(g) \rightarrow 2Cr^{3+}(aq) + 2NO_3^-(aq) + 3H_2O(l)$$

Is our Answer Reasonable?

The indication of "reasonable" in this case, is that the reactants and products have not been changed, and the reactants and products are balanced atomically and electronically. These criteria have been met, so the answer is reasonable.

What If…?

…instead of nitrogen monoxide, NO, we react the dichromate ion, $Cr_2O_7^{2-}$, with chloride ion, Cl^- in acidic solution? What is the balanced equation?

(*Ans:* $Cr_2O_7^{2-}(aq) + 6Cl^-(aq) + 14H^+(aq) \rightarrow 2Cr^{3+}(aq) + 3Cl_2(g) + 7H_2O(l)$)

Practice 18.4

Balance this reaction in acidic solution.

$$ClO^-(aq) + H^+(aq) + Cu(s) \rightarrow Cl^-(aq) + H_2O(l) + Cu^{2+}(aq)$$

The process we have used to balance the reaction focuses on the use of acidic solutions. Chemical reactions can also occur in basic conditions. The method we present to balance basic reactions requires a little chemical sleight-of-hand but does work effectively. Essentially, we balance the reaction as though it were in an acidic solution (by adding H^+ as necessary) and then add a quantity of hydroxide ion (OH^-) as necessary, to both sides of the equation. Mathematically, we still have our equality. Chemically, we neutralize the acid, getting water on one side and excess base on the other. Although this does not depict what goes on in the solution (we are starting with a base, not doing a titration!) we correctly end up with an excess of OH^- on one side or the other.

We will show the method by looking at the first step in a procedure to analyze methanol (CH_3OH) by reaction with the permanganate ion (MnO_4^-) in base. Before balancing, we have

$$CH_3OH(aq) + MnO_4^-(aq) \rightarrow CO_3^{2-}(aq) + MnO_4^{2-}(aq)$$

To balance the reaction in basic solution, we first balance it as though it were in acidic solution. Using our half-reaction technique, we get the following oxidation and reduction half-reactions:

$$2H_2O(l) + CH_3OH(aq) \rightarrow CO_3^{2-}(aq) + 8H^+(aq) + 6e^- \text{(oxidation)}$$
$$e^- + MnO_4^-(aq) \rightarrow MnO_4^{2-}(aq) \quad \text{(reduction)}$$

We can multiply the reduction reaction by 6 to balance electrons and add the half-reactions to get the final equation, balanced in acidic solution:

$$2H_2O(l) + CH_3OH(aq) + 6MnO_4^-(aq) \rightarrow CO_3^{2-}(aq) + 8H^+(aq) + 6MnO_4^{2-}(aq)$$

To balance in basic solution, we add an amount of OH^- to each side equal to the amount of H^+. Note that 8 $H^+(aq)$ and 8 $OH^-(aq)$ is the same as 8 $H_2O(l)$.

$$\begin{array}{rcl} 2H_2O(l) + CH_3OH(aq) + 6MnO_4^-(aq) & \rightarrow & CO_3^{2-}(aq) + 8H^+(aq) + 6MnO_4^{2-}(aq) \\ +8OH^-(aq) & & +8OH^-(aq) \\ \hline 8OH^-(aq) + 2H_2O(l) + CH_3OH(aq) + 6MnO_4^-(aq) & \rightarrow & CO_3^{2-}(aq) + 8H_2O(l) + 6MnO_4^{2-}(aq) \end{array}$$

We have 8 water molecules on the right and 2 on the left. This leaves an excess of 6 water molecules on the right, and results in our final balanced equation in basic solution:

$$8OH^-(aq) + CH_3OH(aq) + 6MnO_4^-(aq) \rightarrow CO_3^{2-}(aq) + 6H_2O(l) + 6MnO_4^{2-}(aq)$$

Example 18.5—Balancing Redox Equations in Basic Solutions

Balance the following equation in basic solution.

$$I_3^-(aq) + S_2O_3^{2-}(aq) \rightarrow I^-(aq) + SO_4^{2-}(aq)$$

Asking the Right Questions

How does this differ from balancing an equation in acid solution? What parts of the process are the same?

Solution

We can first balance the equation in acidic solution, using the multistep procedure presented in Fig. 18.9.

$$I_3^-(aq) + S_2O_3^{2-}(aq) \rightarrow I^-(aq) + SO_4^{2-}(aq)$$

The iodine changes from an oxidation state of −1/3 in I_3^- to −1 in I^-. (reduction).

The sulfur changes from an oxidation state of +2 in $S_2O_3^{2-}$ to +6 in SO_4^{2-}. (oxidation).

The balanced half-reactions in acidic solution are

$$2e^- + I_3^-(aq) \rightarrow 3I^-(aq)$$
$$5H_2O(l) + S_2O_3^{2-}(aq) \rightarrow 2SO_4^{2-}(aq) + 10H^+(aq) + 8e^-$$

We can multiply the reduction equation by 4 to equalize the electrons gained and lost in the reduction and oxidation half-reactions

$$4\{2e^- + I_3^-(aq) \rightarrow 3I^-(aq)\} =$$
$$8e^- + 4I_3^-(aq) \rightarrow 12I^-(aq)$$
$$5H_2O(l) + S_2O_3^{2-}(aq) \rightarrow 2SO_4^{2-}(aq) + 10H^+(aq) + 8e^-$$

We can add the two half-reactions to get the final, balanced equation in acidic solution.

$$5H_2O(1) + S_2O_3^{2-}(aq) + 4I_3^-(aq) \rightarrow 2SO_4^{2-}(aq) + 12I^-(aq) + 10H^+(aq)$$

To balance in base, we can add 10 OH^- to both sides.

$$5H_2O(l) + S_2O_3^{2-}(aq) + 4I_3^-(aq) + 10OH^-(aq) \rightarrow 2SO_4^{2-}(aq) + 12I^-(aq) + 10H^+(aq) + 10OH^-(aq)$$

We cancel the extra waters to give the equation that is now balanced in base.

$$10OH^-(aq) + S_2O_3^{2-}(aq) + 4I_3^-(aq) \rightarrow 2SO_4^{2-}(aq) + 12I^-(aq) + 5H_2O(l)$$

As with balancing in acid, we ask the same bottom-line question: Is the equation electrically and atomically balanced? Check to make sure that the charge is the same on both sides of the equation and that there is the number of atoms on each side. If one or both of these checks do not work, then we've made a mistake balancing the equation.

Is our Answer Reasonable?

The indication of "reasonable" is the same here as in the previous exercise 18.4, with the one difference that the equation balances using hydroxide ion rather than hydrogen ion. In this case, the reactants and products have not been changed, and the reactants and products are balanced atomically and electronically, and we used the hydroxide ion to do it. These criteria have been met, so the answer is reasonable.

What If…?

…we react perchlorate with iodide ion to form the chlorate and iodate ions in basic solution? What is the balanced equation?

(*Ans:* $3ClO_4^-(aq) + I^-(aq) \rightarrow 3ClO_3^-(aq) + IO_3^-(aq)$).

Practice 18.5

Balance the following equation in basic solution.

$$MnO_4^-(aq) + Mn^{2+}(aq) \rightarrow MnO_2(s)$$

(*Hint:* MnO_2 can be written as the product in both an oxidation and a reduction half-reaction.)

18.4.3 Manipulating Half-Cell Reactions

We have learned to balance redox reactions by recognizing that they include electron loss and electron gain. Although calculating the spontaneity of an individual half-reaction may lead to the conclusion that the half-reaction is spontaneous, this is only half of the picture. Because a redox reaction is the sum of an oxidation half-reaction and a reduction half-reaction, we determine the spontaneity of the resulting overall redox reaction only when we include both calculations in our determination of spontaneity. Remember that we are using the SHE as our reference point in these calculations.

We can then say that for two half-reactions, the sum of their change in free energy should be equal to the free energy change for the complete redox reaction:

$$\Delta_{\text{rxn}} G^{\circ} = \Delta G_1^{\circ} + \Delta G_2^{\circ}$$

If we substitute the equivalent expression ($-nFE^{\circ}$) for ΔG°, the equation becomes

$$-n_{\text{rxn}} F E^{\circ}_{\text{rxn}} = -n_1 F E_1^{\circ} + -n_2 F E_2^{\circ}$$

Because the value of F is a constant, and because the number of electrons was the same when we balanced the reaction ($n_1 = n_2 = n_{\text{rxn}}$), all terms but the cell potentials cancel:

$$E^{\circ}_{\text{rxn}} = E_1^{\circ} + E_2^{\circ}$$

This means that if the redox reactions are written so that one is a reduction and one is an oxidation, and the cell potential for the oxidation reaction relative to the SHE has the sign opposite that of its half-reaction potential when written as a reduction, then the two half-reaction potentials are additive for a balanced redox equation.

$$E^{\circ}_{\text{rxn}} = E^{\circ}_{\text{red}} + E^{\circ}_{\text{ox}}$$

From that conclusion, we can say, for example, that if copper is oxidized in a reaction, we write it as an oxidation, and just as we reverse the standard free energy value, ΔG°, when we reverse a reaction, we reverse the E°, as follows:

$$\begin{array}{ll} Cu^{2+}(aq) + 2e^- \rightarrow Cu(s) & E^{\circ} = +0.34\ \text{V (reduction)} \\ Cu(aq) \rightarrow Cu^{2+}(aq) + 2e^- & E^{\circ} = -0.34\ \text{V (oxidation)} \end{array}$$

This notion of reversing the cell potential relative to the SHE when we reverse the reaction gives a consistent picture of the thermodynamic relationship between ΔG° and E°. In addition, it is a reminder that until 1953, when IUPAC changed its protocol to list half-reaction potentials as reductions, they were formerly listed as oxidations, with opposite signs relative to the SHE.

We know that the nitric acid–copper reaction that we introduced at the beginning of this section is spontaneous, as evidenced by its effect on Ira Remsen's lungs, hands, and pants. Let's see how we can manipulate the half-reactions to confirm this. In Table 18.3, we notice that one of the reactions is the reverse of the desired reaction.

From the balancing process steps 8 and 9:

$$\begin{array}{rl} 3\{Cu(s) & \rightarrow Cu^{2+}(aq) + 2e^-\} \\ 2\{3e^- + 3H^+(aq) + HNO_3(aq) & \rightarrow NO(g) + 2H_2O(l)\} \\ \hline 3Cu(s) + 6H^+(aq) + 2HNO_3(aq) & \rightarrow 3Cu^{2+}(aq) + 2NO(g) + 4H_2O(l) \end{array}$$

From Table 18.3:

$$\begin{array}{ll} Cu^{2+}(aq) + 2e^- \rightarrow Cu(s) & E^{\circ} = +0.34\ \text{V} \\ 3H^+(aq) + HNO_3(aq) + 3e^- \rightarrow NO(g) + 2H_2O(l) & E^{\circ} = +0.96\ \text{V} \end{array}$$

To make the half-reactions from the table look like the half-reactions in the redox reaction, we need to reverse the copper reduction and write it as an oxidation. As we pointed out above, reversing the reaction also changes the sign on the potential of the reaction, or, in this case, the half-reaction.

$$\begin{array}{ll} Cu^{2+}(aq) + 2e^- \rightarrow Cu(s) & E^{\circ} = +0.34\ \text{V (reduction)} \\ Cu(aq) \rightarrow Cu^{2+}(aq) + 2e^- & E^{\circ} = -0.34\ \text{V (oxidation)} \end{array}$$

Table 18.4 Important points for potentials of redox reactions

Just as free energies can be combined for chemical reactions, an oxidation and a reduction half-cell potential can be combined to produce a chemical reaction
Just are we reverse the sign of $\Delta G°$ when we reverse a chemical reaction, we reverse the sign of $E°$ when we reverse a half-cell reaction
Although $\Delta G°$ values depend on the coefficients in the chemical equation (that is, when we double the coefficients, we double $\Delta G°$), $E°$ values do not. In a half-reaction, if the coefficients change, the number of electrons, n, will change as well, in essence cancelling the effect of the change to the coefficients l
Because of the negative sign in $\Delta G° = -nFE°$, electrochemical cell reactions with a positive voltage are spontaneous. Note that the rate is NOT related to the spontaneity of the reaction

However, no modification of the potential is necessary when we double or triple a half-reaction because the number of electrons involved in the process also doubles or triples. The free energy, however, does change when we multiply the equation, because $\Delta G° = -nFE°$, and if, as in this case, we triple n, $\Delta G°$ will triple as well. This also is consistent with our discussion of thermodynamics in ▶ Chaps. 5 and 17.

$$\begin{array}{ll} Cu(s) \rightarrow Cu^{2+}(aq) + 2e^- & E^\circ = -0.34\text{ V (oxidation)} \\ 3Cu(s) \rightarrow 3Cu^{2+}(aq) + 6e^- & E^\circ = -0.34\text{ V (oxidation)} \end{array}$$

Table 18.4 lists the key aspects of manipulating half-reactions.

For the copper–nitric acid reaction, the reaction potential is determined by adding the nitric acid reduction (+0.96 V) to the copper oxidation (−0.34 V). The resulting potential is positive (+0.62 V), identifying the reaction as a spontaneous process, as is consistent with Ira Remsen's experience.

$$\begin{array}{rcll} 3\{Cu(s) & \rightarrow & Cu^{2+}(aq) + 2e^-\} & E^\circ = -0.34\text{ V} \\ 2\{3e^- + 3H^+(aq) + HNO_3(aq) & \rightarrow & NO(g) + 2H_2O(l)\} & E^\circ = +0.96\text{ V} \\ \hline 3Cu(s) + 6H^+(aq) + 2HNO_3(aq) & \rightarrow & 3Cu^{2+}(aq) + 2NO(g) + 4H_2O(l) & E^\circ_{cell} = +0.62\text{ V} \end{array}$$

Example 18.6—Dissolving Gold

Nitric acid can be used to dissolve copper. Can nitric acid dissolve gold at standard conditions? The unbalanced reaction is shown here.

$$HNO_3(aq) + Au(s) \rightarrow Au^{3+}(aq) + NO(g)$$

Asking the Right Questions

How do we judge if a reaction can proceed? What's really being asked is, "Is the reaction shown spontaneous?"

Solution

To determine the spontaneity of the reaction, we can use Faraday's equation, which requires the potential of the reaction. We can obtain this by combining properly balanced half-reactions, which have reduction potentials given in Table 18.3. The two unbalanced half-reactions of interest are

$$\begin{array}{l} Au(s) \rightarrow Au^{3+}(aq) \\ HNO_3(aq) \rightarrow NO(g) \end{array}$$

We can balance them in acidic solution (because of the presence of nitric acid) and combine them to give

$$\begin{array}{rcll} Au(s) & \rightarrow & Au^{3+}(aq) + 3e^- & E^\circ = -1.50\text{ V} \\ 3e^- + 3H^+(aq) + HNO_3(aq) & \rightarrow & NO(g) + 2H_2O(l) & E^\circ = +0.96\text{ V} \\ \hline 3H^+(aq) + HNO_3(aq) + Au(s) & \rightarrow & NO(g) + Au^{3+}(aq) + 2H_2O(l) & E^\circ_{cell} = -0.54\text{ V} \end{array}$$

$$\begin{array}{rcll} Au(s) & \rightarrow & Au^{3+}(aq) + 3e^- & \text{(oxidation)} \\ 3e^- + 3H^+(aq) + HNO_3(aq) & \rightarrow & NO(g) + 2H_2O(l) & \text{(reduction)} \\ \hline 3H^+(aq) + HNO_3(aq) + Au(s) & \rightarrow & NO(g) + Au^{3+}(aq) + 2H_2O(l) & \end{array}$$

The negative value for the cell voltage indicates that the reaction of nitric acid and gold is not spontaneous. Nitric acid doesn't dissolve gold.

Is our Answer Reasonable?

The answer is reasonable to the extent that the conclusion is consistent with our cell voltage calculation. In fact, nitric acid does not dissolve gold.

What If…?

… we wanted to form a galvanic cell from these two reduction half-reactions? What would be the overall cell reaction and the cell voltage?

$$Ni^{2+} \rightarrow Ni(s) \quad E^{\circ} = -0.23\text{ V}$$
$$O_2(g) \rightarrow 2H_2O(l) \quad E^{\circ} = +1.23\text{ V}$$

(Ans: $O_2(g) + 4H^+(aq) + 2Ni(s) \rightarrow 2H_2O(l) + 2Ni^{2+}(aq) = +1.46\text{ V}$)

Practice 18.6

Determine the cell voltage at standard conditions for the following redox reactions:

$$CH_4(g) + H_2O(g) \rightarrow CO(g) + H_2(g)$$
$$Ag(s) + H_2O_2(aq) \rightarrow H_2O(l) + Ag^+(aq)$$

18.5 Electrochemical Cells

It may be "shocking" to learn, but it's true: The electric eel is a formidable opponent. When startled, stepped on, or hunting for food, the eel can deliver up to 1 A (1 C of charge per second) at 600 V, enough electrical energy to stun or even kill large animals. (To be accurate, the electric eel isn't an eel. It lives in fresh water and is really a fish.) How does the electric eel generate the electricity used in hunting?

The powerful shock that the eel can deliver is produced by 5000 to 10,000 specialized cells, called electroplates, in its tail (see ◘ Fig. 18.10). Using biochemical processes, the eel charges up the electroplates in much the same manner as muscle and nerve cells are charged. Then, when a nerve impulse is sent to the tail, the electroplates discharge their stored potential. The electrochemical cell seems to fit the definition. We'll explore the electrochemical cell in this section and learn how we can make use of the electrical energy of chemical reactions.

18.5.1 Electrochemical Cells in the Laboratory

Redox reactions, such as the dissolving penny, can take place inside a beaker just like other reactions. However, we can harness the energy from the electrons that are exchanged in the redox reaction if we modify the experimental setup. Take a look at what happens when we separate the copper penny from the nitric acid shown in ◘ Fig. 18.11. The brown fumes produced by the subsequent oxidation of NO to NO_2 by O_2 still waft from the beaker, and the copper ions are still generated. The redox reaction is still working. Let's examine the experimental setup closely to see why it works. We have

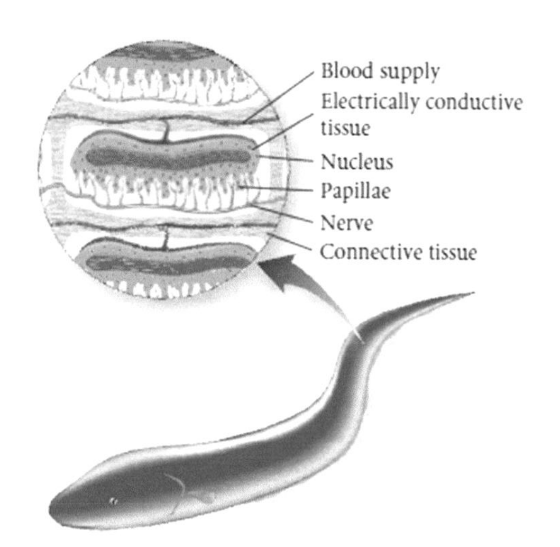

◘ **Fig. 18.10** Electroplates in the tail of the electric eel develop a potential charge that can be delivered to shock its prey. (Drawing by Rick Simonson.)

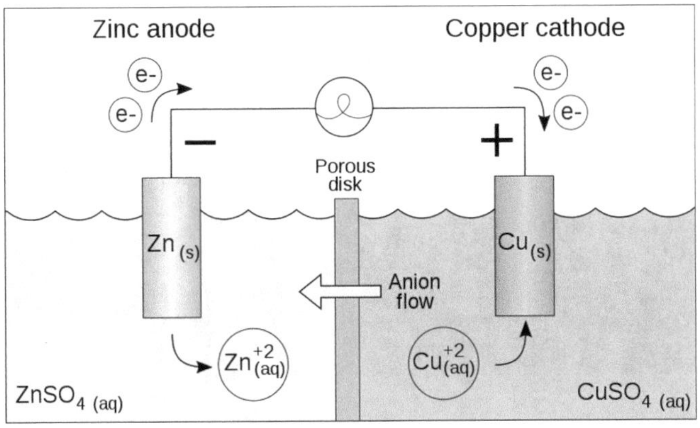

Fig. 18.11 The electrochemical cell. Ira Remsen's observations are still valid in this setup. The nitric acid still acts on the copper. Here, the porous membrane acts as the salt bridge. *Source* ▶ https://upload.wikimedia.org/wikipedia/commons/thumb/6/6d/Galvanic_cell_with_no_cation_flow.svg/800px-Galvanic_cell_with_no_cation_flow.svg.png

$$3Cu(s) + 6H^+(aq) + 2HNO_3(aq) \rightarrow 3Cu^{2+}(aq) + 2NO(g) + 4H_2O(l) \qquad E^\circ_{cell} = +0.62\ V$$

followed by the oxidation of NO to the poisonous NO_2 gas:

$$2NO(g) + O_2(g) \rightarrow 2NO_2(g) \qquad \Delta H^\circ = -112\ kJ$$

The oxidation half-reaction is in the beaker on the right, and the reduction half-reaction is on the left. The two beakers are connected with a wire that transfers the electrons from beaker to beaker. At both ends of the wire is an electrode. The electrode in the oxidation reaction is called the anode, and the electrode in the reduction reaction is called the cathode. Remember, electrons are one of the products of the copper oxidation. They need someplace to go. Providing a wire for them to travel into the nitric acid reduction reaction keeps the reaction running. If we open the circuit on the wire, the reaction stops. The wire is an essential part of this electrochemical cell.

The tube labelled **salt bridge** in the diagram contains an electrolyte such as potassium chloride or sodium nitrate and allows ions to pass from beaker to beaker. The salt bridge is needed because as electrons move from the beaker on the right (the oxidation reaction) to the beaker on the left (the reduction reaction), a strong positive charge will develop at the anode as more of the copper penny becomes Cu^{2+}. A similar thing happens in the other beaker; the H^+ is being removed to make NO and water, and the cathode becomes negatively charged. The developing charges attract the electrons as they move away from the beaker. Without the salt bridge, the electrons would stop this type of movement. We must have a salt bridge in our electrochemical cell design so that the entire system can remain electrically neutral. The finished electrochemical cell produces the same reaction that Ira Remsen observed (nitric acid acts upon copper), but the movement of the electrons through the wire can be used to light a bulb, as shown in Fig. 18.12. We have harnessed the electrons as electrical energy.

Salt bridge – A device containing a strong electrolyte that allows ions to pass from beaker to beaker.

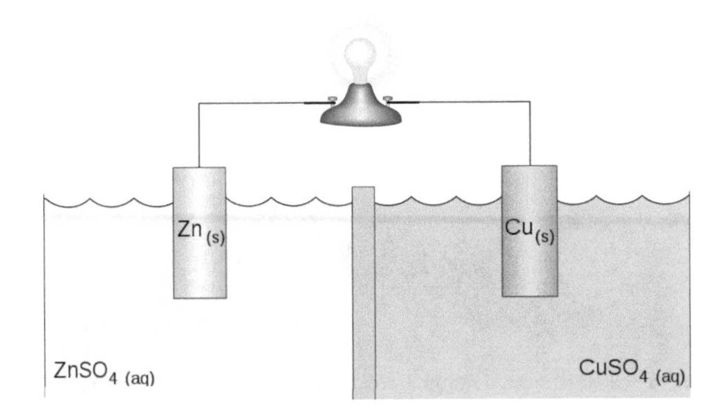

Fig. 18.12 Using the electrochemical cell to light a bulb. *Source* Fig. 18.11 and edited by Mosher

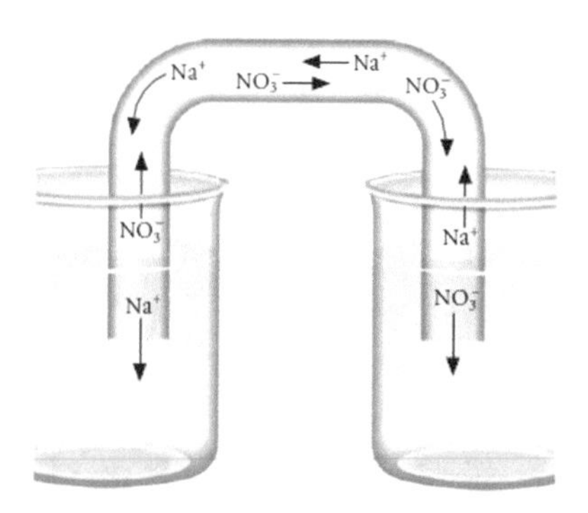

Example 18.7—Alkaline Batteries

Some electrochemical cells are constructed using an alkaline electrolyte. Here are two half-cells that can be employed:

$$Cd(OH)_2(s) \rightarrow Cd(s)$$
$$NiO_2(s) \rightarrow Ni(OH)_2(s)$$

a. Based on the information in the standard reduction potentials table (Table 18.3), write the two oxidation—reduction reactions possible from these half-reactions, and calculate their cell potentials.
b. Which could be used as the basis of an electrolytic cell?... of a galvanic cell? What conditions are required to obtain these specific cell potentials? (Recall our discussion in ▶ Sect. 18.1.)
c. When used as a voltaic cell, this set of reactions makes a useful battery that can power portable appliances and tools. Judging on the basis of the elements that make up the galvanic cell, can we identify the name of this type of battery? What significance does the reverse reaction have?

Asking the Right Questions

What information does Table 18.3 give us? How can the reactions be combined to give two overall cell reactions, and how will the cell voltages of the cell reactions differ from each other? When we ask about "conditions," we are referring to temperature and concentration. What are the temperature and concentration conditions that we use when considering standard reduction potentials?

Solution

a. The two *balanced* half-cell reactions are

$$Cd(OH)_2(s) + 2e^- \rightarrow Cd(s) + 2OH^-(aq)$$
$$E^\circ = -0.81 \text{ V}$$
$$NiO_2(s) + 2H_2O(l) + 2e^- \rightarrow Ni(OH)_2(s) + 2OH^-(aq)$$
$$E^\circ = +0.49 \text{ V}$$

If the first reaction is reversed,

$$E^\circ_{cell} = (+0.81 \text{ V}) + (+0.49 \text{ V}) = +1.30 \text{ V}$$
$$Cd(s) + NiO_2(s) + 2H_2O(l) \rightarrow Cd(OH)_2(s) + Ni(OH)_2(s)$$
$$E^\circ = +1.30 \text{ V}$$

If the second reaction is reversed,

$$E^\circ = (-0.81 \text{ V}) + (-0.49 \text{ V}) = -1.30 \text{ V}$$
$$Cd(OH)_2(s) + Ni(OH)_2(s) \rightarrow Cd(s) + NiO_2(s) + 2H_2O(l)$$
$$E^\circ = -1.30 \text{ V}$$

In both cases, the $OH^-(aq)$ and the electrons cancel when the half-cell reactions are added.

b. To obtain these voltages, the reactions must be run at 25 °C at 1 atm pressure (standard conditions). The concentration of $OH^-(aq)$ must be 1 M (solids have a constant concentration). The first oxidation–reduction reaction could form the basis of a galvanic cell, because E°_{cell} is positive. The second would be electrolytic, because E°_{cell} is negative.
c. This voltaic cell is better known as a NiCad battery. The reverse electrolytic reaction proceeds well, because once formed, the reaction products remain in contact with the electrodes. Therefore, NiCad batteries can be recharged by running the cell in reverse by applying a reverse voltage greater than the cell potential.

Are our Answers Reasonable?

The overall cell reaction is balanced, and the cell voltage is positive for the NiCad battery, so the answer is reasonable. We list this battery in Table 18.5, along with some other common batteries.

What If...?

...we multiplied each half-reaction by two? What would be the overall cell voltage for the galvanic cell?

(*Ans: It would be the same. That's why large batteries have the same voltage as smaller batteries of the same type (NiCad, in this example).*)

Practice 18.7

Draw the electrochemical cell for each of the redox reactions given in Practice 18.6. What is the potential of the galvanic cell?... of the electrolytic cell?

Table 18.5 Selected batteries

Zinc-carbon battery	Also known as a standard carbon battery. Zinc-carbon chemistry is used in all inexpensive AA, C, and D dry-cell batteries. The electrodes are zinc and carbon, with an acidic paste between them that serves as the electrolyte
Alkaline battery	Use in common batteries. The electrodes and zinc and manganese oxide, with an alkaline electrolyte
Lithium photo battery	Lithium, lithium iodide, and lead-iodide are used in cameras because of their ability to supply power surges
Lead-acid battery (rechargeable)	Used in automobiles. The electrodes are made of lead and lead oxide with a strong acidic electrolyte
Nickel–cadmium battery (rechargeable)	The electrolytes are nickel hydroxide and cadmium, with potassium hydroxide as the electrolyte
Nickel-metal hydride battery (rechargeable)	This battery is replacing nickel–cadmium batteries because it does not suffer from "voltage depression", where repeated charging after only partial discharges prevents it from fully discharging
Lithium-ion battery (rechargeable)	With a very good power-to-weight ratio, this is often found in high-end laptop computers and cell phones. Also used to power some electric vehicles
Zinc-air battery	This battery is lightweight and rechargeable
Zinc-mercury oxide battery	This is often used in hearing aids
Silver-zinc battery	This is used in aeronautical applications because the power-to-weight ratio is good
Metal-chloride battery	Used in electric vehicles
Hydrogen fuel cell	Used in electric vehicles and to power the space shuttle

18.5.2 Cell Notation

Because the balanced chemical equations require a lot of time to write, shorthand notation (called **cell notation**) is often used when describing electrochemical cells. It may look hard to do, but cell notation is as easy as knowing our ABCs—that is, anode, bridge, cathode. Consider the reaction that takes place in the silver oxide cell found in pacemaker batteries:

$$Ag_2O(s) + Zn(s) \rightarrow 2Ag(s) + ZnO(s) \qquad E^{\circ} = +1.86\ V$$

Cell notation – The chemist's shorthand used to describe electrochemical cells.

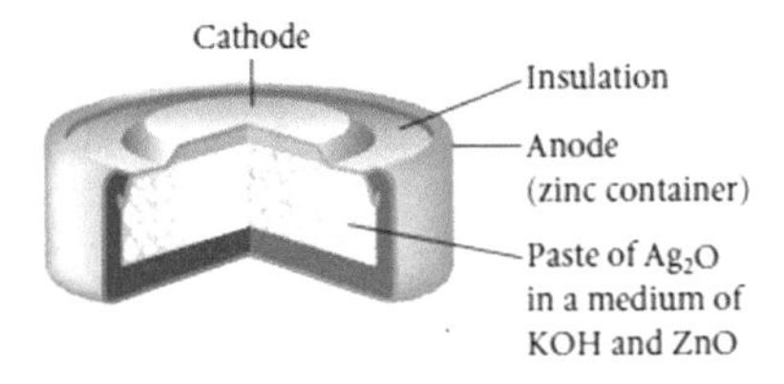

A closer look at the half-reactions tells us that silver is being reduced and zinc is being oxidized. The silver metal must be the cathode, and the zinc metal must be the anode. We write each half-reaction, and then, without balancing them, we construct the overall shorthand notation. Note that the different sides of each half-reaction are separated by a vertical line and that the bridge is represented by a double vertical line.

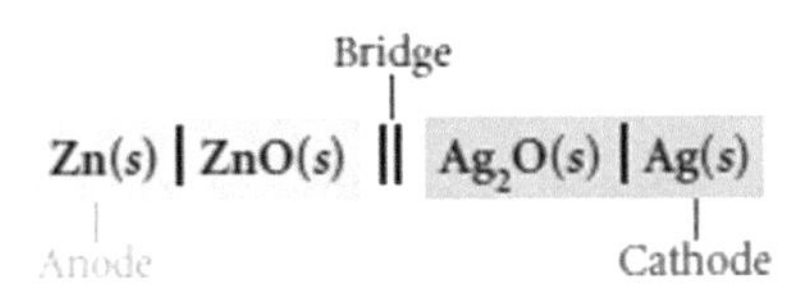

$$Zn(s) \rightarrow ZnO(s) \quad \text{(oxidation, anode)} \quad Zn(s)|ZnO(s)$$
$$Ag_2O(s) \rightarrow Ag(s) \quad \text{(reduction, cathode)} \quad Ag_2O(s)|Ag(s)$$

18

Example 18.8—Shorthand Notation

Write the shorthand cell notation for the following voltaic cell.

$$Cu^{2+}(aq) + Mg(s) \rightarrow Cu(s) + Mg^{2+}(aq)$$

Asking the Right Questions

What goes in each section of the cell notation, from left to right? How do we know which metal is the cathode and which is the anode?

Solution

Our goal is to write the ABCs—anode, bridge, and cathode. In order to do this, we need to know which half-reaction is the oxidation and which is the reduction. The copper ion gains two electrons, so it is reduced. The copper is the cathode. The magnesium loses two electrons, so it is oxidized. It is the anode. We now have enough information to write the cell notation:

Bridge

$Mg(s)\,|\,Mg^{2+}(aq)\,||\,Cu^{2+}(aq)\,|\,Cu(s)$

Anode   Cathode

Is our Answer Reasonable?

"Reasonable," in this context, means that we have used the proper structure to represent the system, and we have properly identified the system itself – that is, the correct half-reactions for the anode and cathode. Given the correct processes, the answer is reasonable.

What If...?

...we have these half-reactions?

$$Co^{2+}(aq) \rightarrow Co(s) \quad E^\circ = -0.28\ V$$
$$Cu^{2+}(aq) \rightarrow Cu(s) \quad E^\circ = +0.34\ V$$

What would be the shorthand cell notation for a *galvanic* cell?

(*Ans: $Co(s)|Co^{2+}(aq)||Cu^{2+}(aq)|Cu(s)$*).

Practice 18.8

Create a galvanic cell using these half-reactions and write the cell notation. Check to make sure our reaction is written as a spontaneous redox reaction.

$$Fe^{2+}(aq) + 2e^- \rightarrow Fe(s)$$
$$Al^{3+}(aq) + 3e^- \rightarrow Al(s)$$

18.5.3 Batteries

Commercial batteries come in many shapes and sizes and are based on a wide variety of chemical processes (Table 18.5). The best battery for any application is usually chosen by considering power output, cost, convenience, size, and whether the battery will be rechargeable. The chemistry of commercial batteries is often rather complex compared to the simple electrochemical cells used in the teaching lab to convey the basic principles.

18.5.4 The Chemistry of Some Common Batteries

Nickel metal hydride (NiMH) rechargeable batteries are used in many cellular phones (Fig. 18.13). During the charging phase, an external source of electricity causes water in the electrolyte (often aqueous potassium hydroxide) to react with a rare earth– or zirconium metal–based alloy at what will be the negative electrode of the battery when it is in operation. This generates hydrogen atoms that are absorbed into the alloy, and releases hydroxide ions:

$$Alloy + H_2O(l) + e^- \rightarrow Alloy{-}H(s) + OH^-(aq) \qquad \text{(reduction)}$$

At the other electrode, which will be the positive electrode when the battery is powering the phone, nickel hydroxide reacts with hydroxide ions to form nickel oxyhydroxide, which has nickel in what for it is an unusual +3 oxidation state:

$$Ni(OH)_2(s) + OH^-(aq) \rightarrow NiOOH + H_2O + e^- \qquad \text{(oxidation)}$$

When the battery is in use, the hydrogen atoms that were absorbed into the alloy at the negative electrode are released, combining with hydroxide ions to form water and supply the electrons that flow through a circuit to power the phone.

$$Alloy{-}H(s) + OH^-(aq) \rightarrow Alloy + H_2O(l) + e^- \qquad \text{(oxidation)}$$

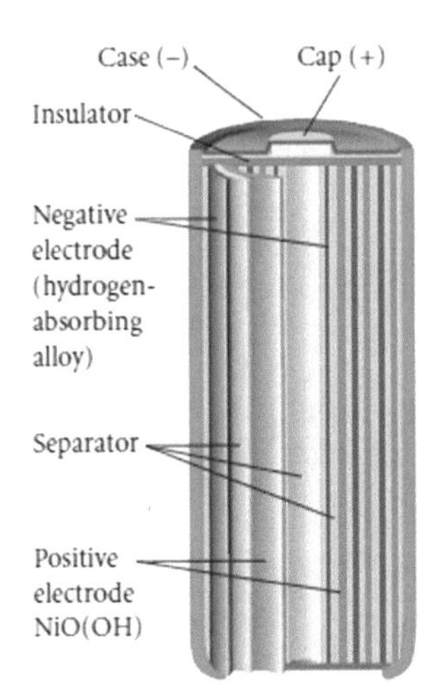

Fig. 18.13 Cutaway view of a nickel-metal hydride battery

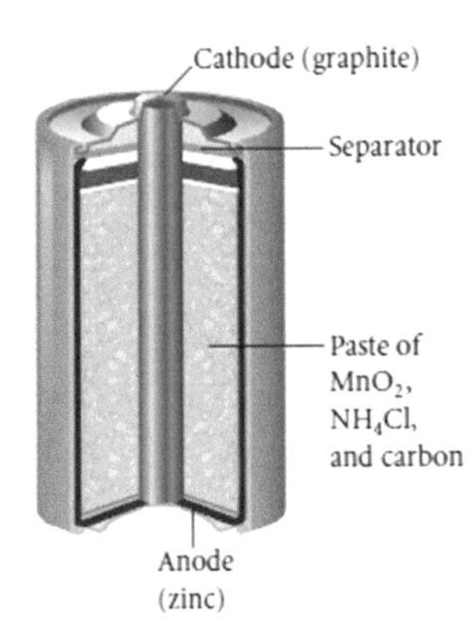

Fig. 18.14 Cutaway of a common dry cell battery

At the positive electrode, nickel oxyhydroxide is reduced back to nickel hydroxide by the electrons that arrive through the circuit, having done their work for us:

$$\mathrm{NiOOH(s) + H_2O(l) + e^- \rightarrow Ni(OH)_2(s) + OH^-(aq)} \text{ (reduction)}$$

The cycle of charge and discharge can be repeated many times, to power all the talking and text messaging on the move that is such a pervasive part of modern life.

A typical nonrechargeable "alkaline" battery for a flashlight (Fig. 18.14) uses the oxidation of zinc metal into zinc ions to generate the electrons for the electric current:

$$\mathrm{Zn(s) \rightarrow Zn^{2+}(aq) + 2e^-} \qquad \text{(oxidation)}$$

When electrons flow back into the battery at the other electrode, they combine with manganese dioxide:

$$\mathrm{2MnO_2(s) + 2H_2O(l) + 2e^- \rightarrow 2MnO(OH)(s) + 2OH^-(aq)} \text{ (reduction)}$$

Therefore, the indirect reaction of zinc with manganese dioxide is the source of the energy that lights the bulb.

Our laptop computer and cell phone may be powered by a "lithium-ion battery" (Fig. 18.15). These batteries use lithium oxide mixed with other metal oxides as the positive electrode, and crystalline graphite with lithium ions intercalated within it as the negative electrode. Unlike most conventional batteries, however, the lithium-ion battery is not powered by a redox reaction. Instead, lithium ions move back and forth within the battery during the cycle of charging and recharging, accompanied by electrons moving in the external circuit. During charging, the lithium ions are driven into the graphite cathode by application of the external electric current. When the battery is used as a source of power, the ions drift back to the lithium oxide anode. As the ions leave the cathode, electrons must travel around the external circuit, ensuring that there is no overall transfer of electric charge from cathode to anode as the battery is discharged.

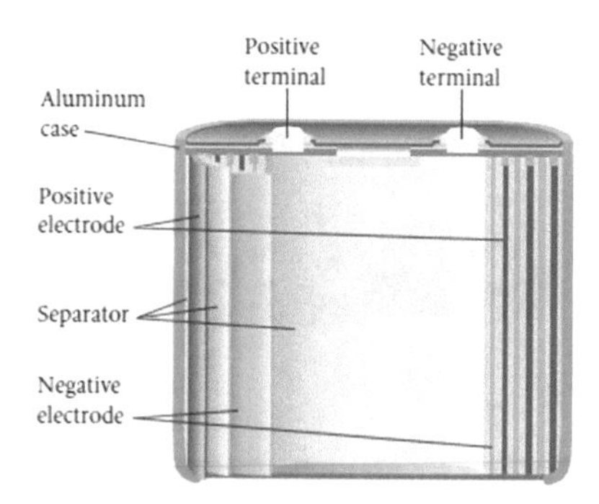

■ **Fig. 18.15** Cutaway of a lithium ion battery

18.6 Chemical Reactivity Series

Plumbers often use metal pipes to deliver water from the main water line to our sink faucet. In fact, plumbers used to use lead pipes, but because the lead in the pipe can leach into the water and cause heavy metal poisoning, that practice isn't followed anymore. Even though plastic pipes made from polyvinyl chloride (PVC) are much cheaper than metal pipes, there is still a demand for copper pipes. Why do plumbers use copper for pipes? The demand is mostly due to the durability of copper compared to plastic, but why aren't iron, lithium, calcium, or magnesium pipes used for plumbing?

The Table of Standard Reduction Potentials (also known as Standard Electrode Potentials) (■ Table 18.3) provides not only cell potentials but also a ranking of reducing agents and oxidizing agents. This ranking is called a **reactivity series.** ■ Table 18.6 shows a relative listing of some of the more common metals and includes hydrogen as a reference point. The strongest reducing agents (those elements that are most easily oxidized) are the most reactive metals and can be found at the top of the table: Li, Na, and Mg.

Reactivity series – A ranking of the electrochemical reactivity of some elements.

■ **Table 18.6** Reactivity series of the metals

	Ion	Atom	
Ions difficult to displace	Li^+	Li	Metals that react with water
	K^+	K	↓
	Ca^{2+}	Ca	
	Na^+	Na	
	Mg^{2+}	Mg	
	Al^{3+}	Al	
	Zn^{2+}	Zn	Metals that react with acid
	Fe^{2+}	Fe	↓
	Ni^{2+}	Ni	
	Pb^{2+}	Pb	
	H^+	H_2	
	Cu^{2+}	Cu	
	Ag^+	Ag	Metals that are highly unreactive
	Au^{3+}	Au	↓

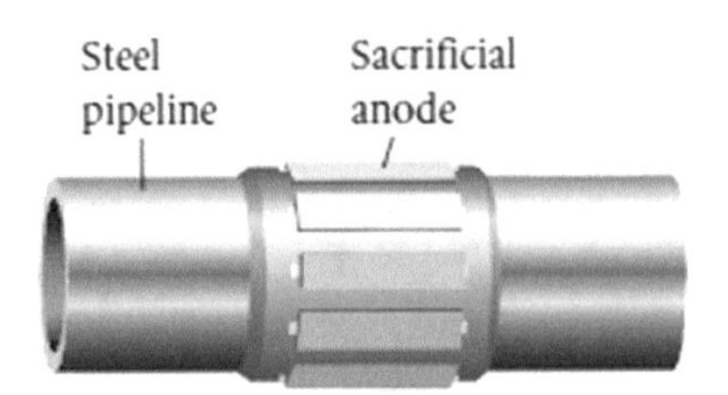

Fig. 18.16 A sacrificial anode on a steel pipeline

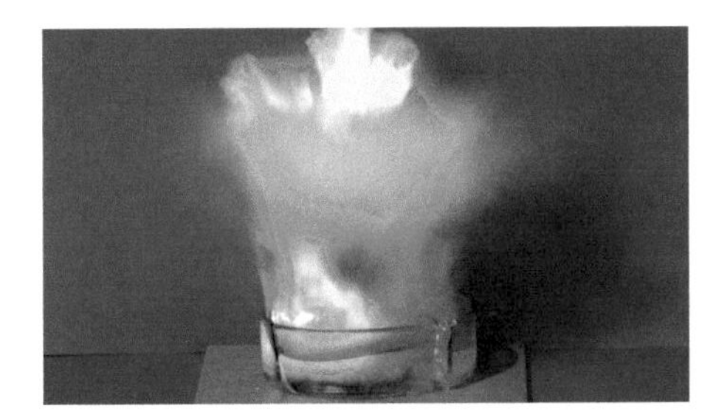

▶ https://upload.wikimedia.org/wikipedia/commons/d/d6/Naatriumi_reaktsioon_veega_purustab_klaasist_anuma.jpg

Because these metals are such good reducing agents, we would not want an earring made out of them; a better choice would be a relatively unreactive metal (and a weak reducing agent) such as gold, copper, or silver, found at the bottom of the table. "Fourteen-carat" gold (which means that 14/24 of the sample is gold) is an alloy of gold, copper, and silver. "Twenty-four-carat" gold is pure gold. The reactivity series has social importance beyond its significance to earrings. Underwater steel pipelines, which contain substantial amounts of iron, electrochemically corrode (the metal deteriorates via oxidation) as a consequence of the interaction of the iron with water, salt, and oxygen dissolved in the water. This **corrosion** can be minimized by putting, for example, magnesium strips in direct contact with the pipeline, shown in Fig. 18.16. The magnesium, which is more chemically active than iron, will preferentially oxidize, ideally leaving the iron in its elemental form. The magnesium, in effect, is sacrificed for the good of the pipeline and therefore is known as a **sacrificial anode**. Other sacrificial anodes include aluminum wrapped around steel in hot water heaters and zinc coating the propellers and rudders of ships.

Corrosion – The deterioration of a metal as a consequence of oxidation.

Sacrificial anode – A material that will oxidize more easily than the one we seek to protect from oxidation.

Example 18.9—Which Is More Reactive?

Iron, especially in the form of stainless steel, can be used in jewelry, such as in the post of an earring. Where does iron fall on the reactivity series? Based on the potential for the oxidation of iron to iron(III) ion, why is iron suitable (or not suitable) for jewellery?

Asking the Right Questions

What is the relationship between standard reduction potential and reactivity?

Solution

From Table 18.3 we can see that the oxidation of metallic iron, Fe(s), has a half-reaction potential of +0.04 V. Metallic copper (−0.34 V), silver (−0.80 V), and gold (−1.50 V) have negative potentials. From this information we can conclude that iron is more reactive than copper, silver, or gold. In this sense *pure* iron would not be suitable, and we know that iron spontaneously reacts to make rust when in contact with moisture and air. Stainless steel, however, is an alloy formulated to inhibit corrosion. Some formulas have as much as 18% chromium and 8% nickel added to the iron. The half-reaction potential for stainless steel is different, and it is difficult to predict reactivity from the $E°$ values for its constituent elements.

Is our Answer Reasonable?

We can judge from the standard reduction potentials, which indicate that our answer is reasonable, and our experience, in which iron rusts.

What If...?

...we wanted to use magnesium to make the post for an earring? Would it be a reasonable choice? Explain.

(*Ans: Magnesium has a very low standard reduction potential and will therefore tend to oxidize much more readily than most other common metals.*)

Practice 18.9

Arrange the following in decreasing order of reactivity:

Na Al Ca Cu

18

18.7 Not-So-Standard Conditions: The Nernst Equation

Native copper and other copper objects can be cleaned with relatively dilute solutions of nitric acid. Concentrated nitric acid is too strong for the job, as Ira Remsen noted, so ancient coins should not be cleaned with concentrated nitric acid. A very dilute solution in the hands of a professional, however, can transform a 1600-year-old coin into a masterpiece that looks as new as the day it was minted. How does lowering the concentration of nitric acid change the reactivity? It does reduce the rate of the reaction as we uncovered in our discussion about kinetics, but is there an effect on the potential of the reaction as well? What about the temperature of the reaction? We know that in general, the rate of a reaction increases as the temperature is raised. Does cold, dilute nitric acid still react with copper? We can understand these relationships, which are at nonstandard conditions, by considering a simpler system in which the copper ion reacts with zinc metal.

If we put a zinc strip into a solution of copper(II) chloride, we note that a dark film immediately begins to form on the zinc strip (◘ Fig. 18.17a). Within an hour, the entire zinc strip has oxidized into the solution, and we are left with a brown mass of copper metal (◘ Fig. 18.17b). We also note the disappearance of the blue color that is characteristic of Cu^{2+} in water. This is consistent with the following half-reactions and overall cell reaction:

$$\begin{array}{rcll} Cu^{2+}(aq) + 2e^- & \rightarrow & Cu(s) & E^\circ = +0.34\ V \\ Zn(s) & \rightarrow & Zn^{2+}(aq) + 2e^- & E^\circ = +0.76\ V \\ \hline Cu^{2+}(aq) + Zn(s) & \rightarrow & Zn^{2+}(aq) + Cu(s) & E^\circ_{cell} = +1.10\ V \end{array}$$

We know from ▶ Sect. 18.3 that we can relate the free energy of a system to the cell potential:

$$\Delta G = -nFE$$

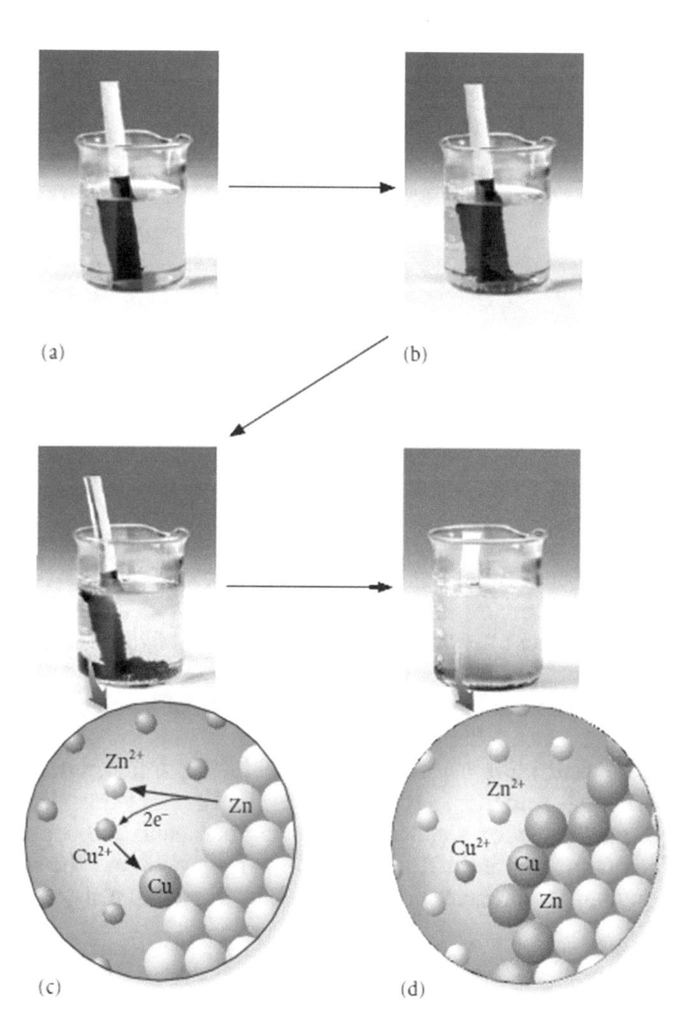

◘ **Fig. 18.17** Zinc metal reacting with copper(II) chloride solution

We also discussed in ▶ Sect. 17.6 the relationship between free energy change and the standard free energy change:

$$\Delta G = \Delta G^{\circ} + RT \ln Q$$

in which Q = the reaction quotient, the ratio of the concentrations (more properly, the activities) of the products over the reactants; R is the universal gas constant; and T is the temperature in kelvins. For our current example,

$$\Delta G = \Delta G^{\circ} + RT \ln \frac{[Zn^{2+}]_0}{[Cu^{2+}]_0}$$

We can substitute for ΔG the expression for the cell potential, $\Delta G = -nFE$:

$$-nFE = -nFE^{\circ} + RT \ln \frac{[Zn^{2+}]_0}{[Cu^{2+}]_0}$$

Dividing each term by $-nF$ enables us to solve for the cell potential at nonstandard conditions:

$$E = E^{\circ} - \frac{RT}{nF} \ln \frac{[Zn^{2+}]_0}{[Cu^{2+}]_0}$$

This is one form of the **Nernst equation**, named after Walther Hermann Nernst (1864–1941), a German chemist who studied the effect of concentration on the potential of an electrochemical cell. We can write a more general form of the equation, taking into account the reaction quotient for any process:

$$E = E^{\circ} - \frac{RT}{nF} \ln Q$$

Nernst equation – The equation used to determine the cell potential at nonstandard conditions.

Because most reactions are conducted at standard temperature (25 °C, 298 K), we can calculate the term $\frac{RT}{F}$.

$$\frac{RT}{F} = \frac{8.3145 \text{ J} \cdot \text{mol}^{-1} \cdot \text{K}^{-1}(298 \text{ K})}{96485 \text{ C} \cdot \text{mol}^{-1}} = 0.0257$$

which we can substitute into the Nernst equation in this form:

$$E = E^{\circ} - \frac{0.0257}{n} \ln Q$$

Although our calculators can easily handle the natural logarithm ("ln") in the equation, we typically convert the equation to the more familiar base 10 "log" by multiplying by 2.3026 (that is, log(10) = 1, and ln 10 = 2.3026, so log = ln/2.3026). To account for that, we multiply the coefficient, $\frac{0.0257}{n}$, by 2.3026, which gives us a common form of the Nernst equation:

$$E = E^{\circ} - \frac{0.0592}{n} \log Q$$

While this form of the Nernst equation is often used in calculations, we should remember that it assumes the temperature is 298 K. We must use the full version of the Nernst equation for all other temperatures. In any case, with this equation, we can determine the effect of lowering the concentration of nitric acid on the potential of the copper oxidation.

The Nernst equation can also be used to measure the concentration of a solution if the standard cell potential and the actual potential are known. Electrochemists take advantage of this use of the Nernst equation via ion-selective electrodes, in which the concentration of an ion such as chloride, ammonium, cadmium, nitrate, or hydrogen (in a pH electrode) is determined. And there's something else that's interesting about the Nernst equation. For example, let's consider the hypothetical redox reaction between copper metal and copper ions in solution. The potential of the reduction half-reaction and that of the oxidation half-reaction are the same at standard conditions, and, as we'd expect, no net potential should be noticed for an electrochemical cell containing these reactions.

$$\begin{array}{rcll} Cu^{2+}(aq) + 2e^- & \rightarrow & Cu(s) & E^{\circ}_{red} = +0.34 \text{ V} \\ Cu & \rightarrow & Cu^{2+}(aq) + 2e^- & E^{\circ}_{ox} = -0.34 \text{ V} \\ \hline Cu(s) + Cu^{2+}(aq) & \rightarrow & Cu^{2+}(aq) + Cu(s) & E^{\circ}_{cell} = 0.00 \text{ V} \end{array}$$

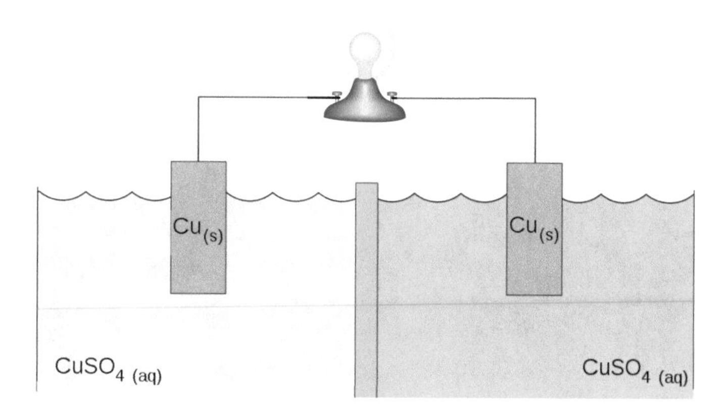

Fig. 18.18 The concentration cell. The voltage observed in the copper concentration cell is due to the differences in concentration of Cu^{2+} at the anode and cathode 18.12 *Source* Mosher

However, what would happen if we increased the concentration of the reactant copper ions to 2.0 M instead of the standard 1.0 M? Doing so changes the distribution of the species in the reaction. Using the Nernst equation, we can calculate the result of our modification.

$$E = E^{\circ} - \frac{0.0592}{n} \log\left(\frac{1.0}{2.0}\right)$$
$$E = 0.00 - \frac{0.0592}{2} \log(0.5)$$
$$E = 0.00 - (-0.0089)$$
$$E = +0.0089 \text{ V}$$

The electrochemical potential of the reaction is a nonzero value. By adjusting the concentrations of the products and reactants, we have created an electrochemical cell. This type of cell is called a **concentration cell** because the concentrations are driving the potential of the cell (Fig. 18.18). This is the same type of potential that develops across the membrane of a muscle cell or nerve cell.

Concentration cell – A cell in which different concentrations of identical ions on both sides of the cell provide the driving force for the reaction.

How does our cell notation change to reflect the fact that the conditions are not-so-standard? By indicating the concentrations in parentheses immediately after the species, we can immediately show how the reaction should be written. For example,

$$\text{Cu(s)}|\text{Cu}^{2+}\text{(aq)(1.0 M)} \parallel \text{Cu}^{2+}\text{(aq)(2.0 M)}|\text{Cu(s)}$$

Example 18.10—Heart Cell Potential

At the beginning of this chapter, we mentioned the electrical signals in the heart muscle. Given the differing concentrations of potassium ions inside and outside the heart cells, what is the electrochemical potential that corresponds to this concentration gradient? Assume that the electrochemical cell in the body can be represented by the following cell notation. (In truth, no elemental potassium exists in the human body. We use this concentration cell as a model merely to estimate the potential that is obtained in a heart muscle cell.)

$$\text{K(s)} + \text{K}^{+}\text{(aq)} \rightarrow \text{K(s)} + \text{K}^{+}\text{(aq)} \quad E^{\circ}_{\text{cell}} = 0.00 \text{ V}$$

Asking the Right Questions

What is the reaction that is the basis of this concentration cell? How do we use the Nernst equation, along with the reactant and product concentrations, to determine the cell voltage?

Solution

This concentration cell is based on the potassium half-reaction. The complete reaction is

$$\text{K(s)}|\text{K}^{+}\text{(aq)(0.005 M)} \parallel \text{K}^{+}\text{(aq)(0.166 M)}|\text{K(s)}$$

There is one electron involved in the reaction. Using the information from the Nernst equation, we get a potential of +0.090 V.

$$E_{\text{cell}} = E^{\circ}_{\text{cell}} - \frac{0.0592}{n}\log\left(\frac{[K^{+}_{\text{product}}]_0}{[K^{+}_{\text{reactant}}]_0}\right)$$

$$E_{\text{cell}} = 0.00 - \frac{0.0592}{1}\log\left(\frac{0.005}{0.166}\right)$$

$$E_{\text{cell}} = 0.00 - 0.0592\log(0.03012)$$

$$E_{\text{cell}} = 0.00 - 0.0592(-1.521)$$

$$E_{\text{cell}} = 0.00 - (-0.09005)$$

$$E_{\text{cell}} = +0.090\ \text{V}$$

In the heart cell, the sodium gradient provides a potential of −0.052 V. The net potential across the membrane of the heart cell is +0.038 V.

Is our Answer Reasonable?

Aside from checking the calculations, the best way to judge if our answer is reasonable is to note the positive cell potential, which means that the process proceeds. That is a reasonable answer.

What If...?

...we looked at fate of the reaction over time? Which side of the reaction would increase in concentration as the potential lowers and the reaction eventually stops?

(*Ans: The reaction will stop when the concentrations of potassium ion in the two half-reaction cells are equal. This will happen when enough potassium ion in the cathode side is reduced to lower its concentration to equal that of the oxidizing potassium in the anode side.*)

Practice 18.10

What is the potential of this electrochemical cell written in shorthand cell notation? What is the half-reaction listed on the right-hand side? Is this a voltaic cell? (Assume that the pressure of hydrogen gas is 1.0 atm in each half-cell.)

$$H_2(g)|H^+(aq)(1.0\,M)\ ||\ H^+(aq)(0.10\,M)|H_2(g)$$

When we listen to a battery-powered radio, the sound tends to get softer as the radio is used. What is happening inside the battery that causes this power loss? During the progress of the reaction inside the battery, the reactants are being used up as the products are being formed. According to the Nernst equation, as the value of Q gets larger, the modification to the standard cell potential ($E°$) gets more and more negative and closer and closer to zero. What is the result? As the reaction proceeds, the potential of the battery decreases as the system inches ever closer to equilibrium, and the music gets softer. At some point, the battery doesn't have enough voltage to run the radio. It has not yet reached equilibrium, but it is below the threshold that will allow the radio to operate. We say that the batteries are dead. As we have pointed out already, some batteries can be recharged, because their discharge reactions can be run in reverse. Rechargeable batteries can be charged only a finite number of times, typically in the range of 500–1000 times for household batteries, because the surfaces of the recharged electrodes do not form as cleanly as the original surface, and they eventually become too worn to be useful.

18.7.1 The Nernst Equation and the Equilibrium Constant

Measuring the standard cell potential is a very powerful way to solve for the equilibrium constant of a reaction. Here's how. When a redox reaction proceeds without any intervention, it will inevitably reach equilibrium. At that point the value of E_{cell} must become zero, and the free energy change for the process also becomes zero—our thermodynamic definition of equilibrium, introduced in ▶ Chap. 14. When this occurs, the Q value in the Nernst equation is equal to the equilibrium constant K. We can show this reasoning by using the following equations:

$$\Delta G = \Delta G^{\circ} + RT\ln Q$$

At equilibrium, $\Delta G = 0$ and $Q = K$.

so

$$0 = \Delta G^{\circ} + RT\ln K$$

$$\Delta G^{\circ} = -RT\ln K$$

We know that $\Delta G° = -nFE°$. Substituting $-nFE°$ into the previous equation yields

$$-RT\ln K = -nFE^{\circ}$$

Solving for K, we get $\ln K = \frac{nFE^\circ}{RT}$

Calculating $\frac{F}{RT}$ at standard conditions (as we did for the Nernst equation), we find that

$$\ln K = 38.92\ nE^\circ$$

Converting from natural ln to base 10 log (by dividing by 2.3026, as we did in the Nernst equation, yields

$$\log K = 16.9\ nE^\circ$$

To clearly show the connection to the Nernst equation, we will invert 16.9 and put the result in the denominator.

$$\log K = \frac{nE^\circ}{0.0592}$$

We can use the equation in this way. Alternatively, we can take the antilog of both sides and use it in this way:

$$K = 10^{\frac{nE^\circ}{0.0592}}$$

Which form of the equation we use depends mostly on our comfort level. We can solve for the equilibrium constant either way. The key point is that our understanding of the thermodynamic meaning of equilibrium enables us to relate cell potential to the equilibrium constant. The rest is just manipulating equations to get where we want to go. Let's see how we use this relationship to determine the equilibrium constant for the copper–nitric acid reaction performed by Ira Remsen. Recall that the reaction is

$$3Cu(s)+6H^+(aq)+2HNO_3(aq) \rightarrow 3Cu^{2+}(aq)+2NO(g)+4H_2O(l) \quad E^\circ_{cell} = +0.62\ V$$

$$K = 10^{\frac{nE^\circ}{0.0592}}$$

$$K = 10^{\frac{6(+0.62)}{0.0592}} = 10^{62.8} \approx 10^{63}$$

Alternatively,

$$\log K = \frac{nE^\circ}{0.0592} = \frac{6(+0.62)}{0.0592} = 62.8$$

Raising both sides to the power of 10 yields

$$K \approx 10^{63}$$

This is yet another confirmation that the reaction does, in fact, proceed toward products. As we have this discussion, however, please keep in mind the difference between thermodynamics and kinetics. Thermodynamics answers the question "Can a process occur spontaneously?" It says absolutely nothing about speed. Kinetics addresses the issues of rates and mechanisms. All that our calculations tell us is that nitric acid can react spontaneously with the copper penny. They don't say how fast. That's a question of kinetics.

Example 18.11—Equilibrium Constants and Cell Potential

Vanadium(V) ion can be reduced stepwise (that is, to V^{4+}, V^{3+} and, finally, V^{2+}) by reaction with a "Jones Reductor," a zinc–mercury amalgam. The reaction for the reduction of V^{5+} to V^{4+} ion includes the following two half-reactions:

$$VO_2^+(aq) + H^+(aq) \rightarrow VO^{2+}(aq) + H_2O(l)\ E^\circ = +1.00\ V$$
$$Zn^{2+}(aq) \rightarrow Zn(s)\ E^\circ = -0.76\ V$$

Calculate the equilibrium constant for this reaction.

Asking the Right Questions

Where do we want to be, and how do we get there? There are a number of steps to solving this problem. How do we determine the value of K and E°? We should start with a balanced redox equation for the reduction of VO_2^+ to VO^{2+}, in which V^{5+} is reduced to V^{4+}, and Zn^0 is oxidized to Zn^{2+}.

Solution

Our order of operations is

1. Balance the redox reaction.
2. Calculate the E° value for the reaction.
3. Calculate the equilibrium constant, knowing the number of electrons exchanged in the reaction, along with the value for E°.

The balanced half-reactions and overall cell reaction are

$$
\begin{array}{llr}
2VO_2^+(aq) + 4H^+(aq) + 2e^- \rightarrow 2VO^{2+}(aq) + 2H_2O(l) & & E^\circ = +1.00\ V \\
Zn(s) \rightarrow Zn^{2+}(aq) + 2e^- & & E^\circ = +0.76\ V \\
\hline
2VO_2^+(aq) + 4H^+(aq) + Zn(s) \rightarrow 2VO^{2+}(aq) + Zn^{2+}(aq) + 2H_2O(l) & & E^\circ = +1.76\ V
\end{array}
$$

$$K = 10^{\frac{nE^\circ}{0.0592}}$$

$$K = 10^{\frac{2(+1.76)}{0.0592}} = 10^{59.5} \approx 10^{60}$$

Alternatively,

$$\log K = \frac{nE^\circ}{0.0592} = \frac{2(+1.76)}{0.0592} = 59.5$$

$$K \approx 10^{60}$$

Is our Answer Reasonable?

At this point, because we know that the reduction of vanadium ion does occur, the equilibrium constant should be greater than 1. How much greater will depend on the cell voltage and on the number of electrons transferred in the process. In any case, our answer is reasonable.

What if…?

…we are given the value for the equilibrium constant of a 3-electron transfer as being 4.6×10^{-18}? What is the cell voltage of the reaction?

(*Ans:* $E^\circ = -0.34$ V).

Practice 18.11

What is the equilibrium constant for the reaction of copper metal with zinc ion, discussed at the beginning of this section?

18.8 Electrolytic Reactions

Some metals, including copper, gold, and silver, are found in their pure elemental state in the environment. On the other hand, aluminum metal, based on its reactivity is found only chemically combined in ores such as bauxite (hydrated aluminum oxide; $Al_2O_3{\cdot}H_2O$ or $Al_2O_3{\cdot}3H_2O$). In fact, aluminum, largely in the form of the aluminum oxides and silicates, makes up 8.1% of the Earth's crust. However, we know that pure aluminum can be produced, because it is a major component in so many common products: the can in which we store our soda, the wrap in which we put our fish for freezing, and the lightweight bicycle we ride down the street. How is aluminum metal made from bauxite?

The basic process, called electrolysis, entails passing a current through a solution of metal ions in an electrochemical cell in the direction opposite to the spontaneous reaction. Doing so forces the nonspontaneous reaction to occur. This process, also known as **electrowinning**, is responsible for the manufacture and purification not only of aluminum but also of many other metals.

$$Al^{3+} + 3e^- \rightarrow Al$$

$$Cu^{2+} + 2e^- \rightarrow Cu$$

$$Ag^+ + e^- \rightarrow Ag \qquad etc\ldots.$$

18

Electrowinning – The isolation of pure metals from a solution of metal ions.

Electrowinning is the most inexpensive method for making aluminum and magnesium metals. Electrowinning of metals such as aluminum and magnesium is the isolation of pure metals from a solution of metal ions, and it has been known for quite some time. Humphrey Davey, an English chemist, used this process to isolate metallic sodium from NaCl in 1807. However, the electrowinning of aluminum wasn't invented until 1886. Working independently, Charles M. Hall (an American chemist) and Paul Heroult (a French chemist) discovered that if bauxite ore was purified to alumina (Al_2O_3) and dissolved in molten cryolite (Na_3AlF_6), metallic aluminum could be made. In the **Hall–Heroult process**, an electrical current is passed into a molten mixture of alumina and cryolite to make molten aluminum (Fig. 18.19). The overall reaction used in the Hall–Heroult process, shown below, is much simpler than the reactions that take place at the electrodes. Why can't an electric current be applied directly to an aqueous solution of Al^{3+} in order to produce aluminum metal? Because the potential is so large, instead of obtaining aluminum metal from the aqueous solution, the water undergoes electrolysis to form hydrogen and oxygen gases.

$$2Al_2O_3(l) + 3C(s) \rightarrow 4Al(l) + 3CO_2(g) \qquad E^\circ \approx -2.1\ V$$

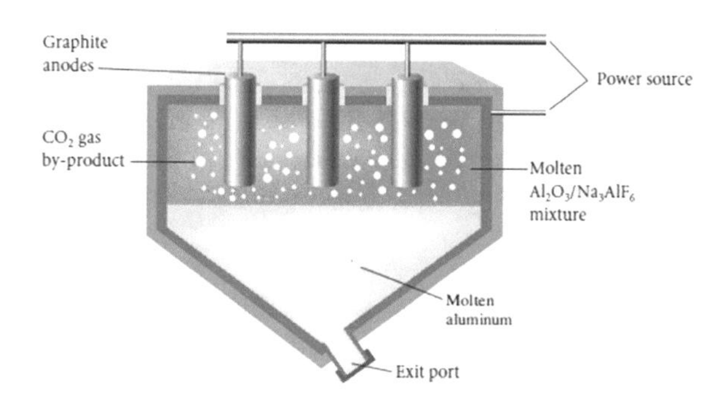

Fig. 18.19 The Hall–Heroult process. This process is still used today to satisfy the world's demand for aluminum products. The product, molten aluminum, is relatively dense and settles to the bottom of the electrochemical cell, where it is drawn off and poured into castings

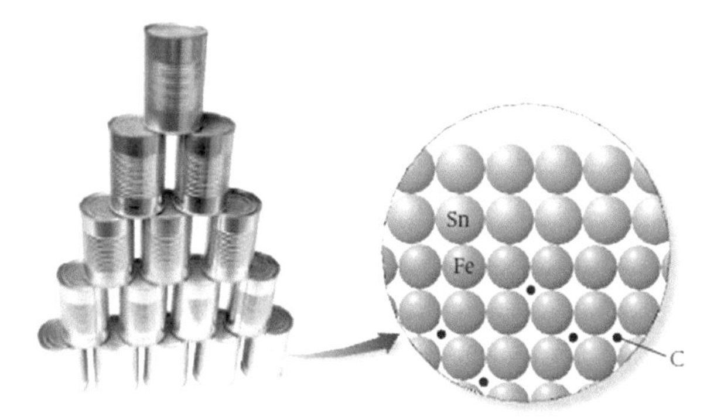

Fig. 18.20 Tin cans are actually steel cans electroplated with a thin coat of tin. The tin resists corrosion and helps keep the contents fresh

Hall–Heroult process – The most widely used process for the preparation of aluminum from bauxite.

The negative potential tells us that this redox reaction is not spontaneous. In fact, the reverse reaction—that is, $4Al(l) + 3CO_2(g) \rightarrow 2Al_2O_3(l) + 3C(s)$—is quite spontaneous ($E° \approx +2.1$ V). Modifying the concentrations and temperatures helps a little to make the potential of the overall reaction less negative, via the Nernst equation, but not enough to make the reaction spontaneous. If we apply a positive potential to the reaction that is larger than the negative potential expressed by the electrochemical cell, we can force the reaction to go forward.

Examination of a simpler redox reaction can be helpful here. For example, in concept, if we wish to make the copper–iron redox reaction spontaneous, we must supply a potential of just over +0.78 V to compensate for the $E°$ of −0.78 V.

$$Cu(s) + Fe^{2+}(aq) \rightarrow Fe(s) + Cu^{2+}(aq) \quad E^{\circ} = -0.78 \text{ V}$$

When we do this, however, we note experimentally that the reaction still doesn't proceed toward products. If we supply a still greater positive potential, the reaction does proceed. The extra voltage required is called the **overpotential** of the reaction. Overpotentials can be fairly high, especially when the products of the reaction are gases. Unfortunately, we can't easily predict what the overpotential for a particular reaction will be. Instead, we measure the overpotential experimentally.

Overpotential – The extra potential needed, above that which is calculated, in order to make an electrochemical process proceed.

18.8.1 The Applications of Electrolysis

The average American uses over 100 tin cans each year. From what are tin cans made? That might sound like a silly question, but "tin" cans are actually made of steel coated with a very thin layer of tin (Fig. 18.20). Approximately 0.25% of the mass of a tin can is actually tin, and chromium is becoming more common as a coating on the steel. Without the coating, the steel would rust and the contents would spoil. The tin coating on a steel can isn't applied like paint on a house. How is the tin applied to the steel?

The most common use for electrolysis is **electroplating**, or depositing metals onto a conducting surface. The result is a coating that is very tightly integrated into the surface of the metal. Because of this tight integration into the metal surface, the coating resists flaking and peeling. This coating makes the item more attractive and imparts some corrosion resistance or chemical resistance to the surface. Electroplating is used in the manufacture of inexpensive jewellery and chrome bumpers for our car, but most common electroplating today occurs in the manufacture of tin cans.

Electroplating – The process of depositing metals onto a conducting surface.

18.8.2 Calculations Involving Electrolysis

The process used to coat a steel can with tin is an example of electrolysis. How is it done? The steel can is hooked up to a power supply and dipped into a solution of tin ions (Sn^{2+}). A block of tin metal is also placed in the solution and connected to the power supply. Then a current is applied to the can (◘ Fig. 18.21). In the terminology of the electrolytic cell, the block of tin metal becomes the anode and the can becomes the site of reduction (the cathode) for tin ions.

Michael Faraday (remember him from our discussion on potentials and spontaneity) noted that the amount of current applied to a cell is directly proportional to the amount of metal that can be deposited in an electrolytic reaction. We can represent this mathematically as

$$g = \frac{A \cdot s \cdot (\underline{M})}{F} \left(\frac{\text{mol metal}}{\text{mol e}^-} \right)$$

where

A is the number of amperes applied to the can (1 ampere = 1 coulomb of charge per second)
s is the number of seconds that the current is applied.
$\underline{M}$ is the molar mass of the metal.
F is Faraday's constant (96,485 coulombs/mol electrons).

the ratio (mol metal/mol e−) is the mole ratio of the reduction half-reaction.

It is helpful to look at this calculation as an extended unit conversion problem. Exercise 18.12 shows how this is done. Note that the unit conversion problem is the same as the equation shown above.

Example 18.12—Golden Forks

To save money and still have a beautiful set of dinnerware, a chemist decides to electroplate the metal dinnerware with gold. How many grams of gold will be electroplated on a fork if 2.5 A is applied to the fork for 20 s?

Asking the Right Questions

What is the relationship between current and the mass of gold electroplated on to the fork?

Solution

The number of grams can be calculated by starting with the current and performing a unit conversion. Note that the unit amperes (amps, A) can be written as coulombs/second (C/s). We also need to examine the reduction half-reaction from ◘ Table 18.3 to determine the number of electrons involved in the process.

$$Au^{3+}(aq) + 3e^- \rightarrow Au(s)$$

$$\text{Amperes} \times \text{time} \times \frac{1}{\text{faraday}} \times \text{mole ratio} \times \underline{M} = \text{grams}$$

$$\frac{2.5\ \text{C}}{\text{s}} \times 20\ \text{s} \times \frac{1\ \text{mol e}^-}{96,485\ \text{C}} \times \frac{1\ \text{mol Au}}{3\ \text{mol e}^-} \times \frac{196.97\ \text{g Au}}{1\ \text{mol Au}} = 0.034\ \text{g Au}$$

This means that our chemist would need to make sure to buy at least 0.034 g of gold per fork.

Is our Answer Reasonable?

On a dimensional analysis level, the units work. Also, the quantity of gold is relatively small. On these two bases, the answer seems reasonable.

18

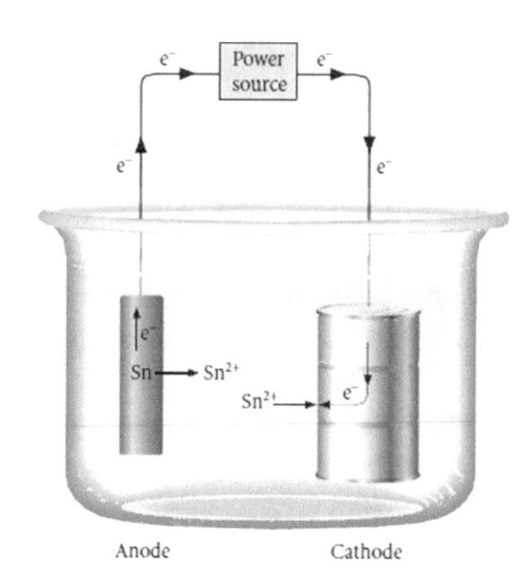

Fig. 18.21 Electroplating a tin can. The tin can acts as the cathode in the electrolysis experiment

What If…?

…we were electroplating chromium on to a trophy from a solution of Cr(III) ion. We require 0.624 g of chromium, and we have 900.0 min to complete the process. Only 20.0% of the energy that is input to the reaction goes toward the reduction of the chromium. The rest goes toward the concurrent reduction of hydrogen gas. What is the minimum current that we must use to meet these conditions?

(*Ans:* 3.22 amps).

Practice 18.12

How many grams of tin will be deposited from a solution of tin(II) nitrate on a steel can if 0.45 A are applied to the can for 1.5 h?

The calculations can also be done in reverse. If we want to coat our steel can with a certain number of grams of tin, we can use either the equation or the unit conversion method to calculate the number of amps that need to be applied.

As we've seen in this chapter, electrochemistry is a very useful topic, especially in today's society. Learning how muscles in our body begin a contraction can help us understand how the heart works and why it is so important to maintain sufficient levels of electrolytes during physical exercise. And as we become increasingly mobile, the demand for longer-lasting batteries will only increase. Batteries power our cell phones, our smart watches, and even our cars. Electroplating the surfaces of many everyday items makes them appear expensive and confers resistance to corrosion. Everywhere we look, there's electrochemistry!

The Bottom Line

- Redox reactions involve both a reduction and an oxidation half-reaction.
- Redox reactions can be identified by determining the oxidation state of the atoms involved in a reaction.
- Redox reactions can be balanced by summation of balanced half-reactions.
- Positive cell potentials indicate a spontaneous reaction and are related to the free energy change by $\Delta G° = -nFE°$.
- Electrochemical cells require both a path for the electrons and a path for other ions.
- The oxidation reaction takes place at the anode. The reduction reaction takes place at the cathode.
- Half-reaction potentials can be used to determine the relative reactivity of metals. Organization of the metals in this fashion is known as a chemical reactivity series.
- The Nernst equation relates the actual potential of a redox reaction to conditions other than the standard conditions.
- Cell potentials enable us to calculate equilibrium constants.
- Electrowinning and electroplating are examples of electrolysis reactions. In electrolysis, a positive potential that includes the overpotential is applied to force the reaction to run in reverse.

Section 18.2 What is Electrochemistry?

Skill Review

1. Explain why the descriptive term battery is often used, but is technically not correct, to describe these most commonly purchased sources of electricity.
2. Provide two other names for an electrochemical cell.
3. What is the difference between electrodics and ionics when discussing the definition of electrochemistry? The introduction to the chapter discusses potassium and sodium ion differences in cells in the heart. Is this an example of electrodics or ionics?
4. We study the electrochemical changes that occur when we add hydrochloric acid to water. Is this an example of an electrodic or ionic study? Explain the answer.

Chemical Applications and Practices

5. Every electrochemical cell is developed around two types of chemical reactions. Name and describe both of these connected reactions.
6. List three properties of electrochemical cells that are considered when designing an appropriate power source.

Section 18.3 Oxidation States—Electronic Bookkeeping

Skill Review

7. Determine the oxidation number of each atom in the structure of dimethylsulfoxide (DMSO).

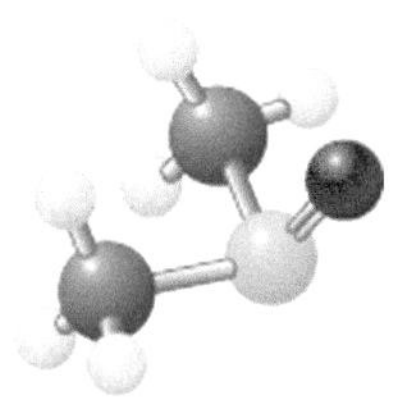

8. Determine the oxidation number of each atom in the structure of periodic acid.

9. In which of the following compounds would the chlorine atom have the most positive oxidation number? In which would chlorine have the most negative oxidation number?

   Cl_2 ClO_2 $NaClO_4$ HCl

18

10. Which species in the following list shows the nitrogen atom in its most reduced form? Which depicts nitrogen in its most oxidized form?

    N_2 HNO_3 NH_3

11. Put the following compounds in order from the lowest to the highest oxidation number for nitrogen.

    NO N_2O NO_2 N_2H_4 NH_3

12. Put the following compounds in order from the lowest to the highest oxidation number for carbon.

    $C_6H_{12}O_6$ CO_2 CH_3OH CH_4 C_6H_6

13. Determine the oxidation state for each atom in the following compounds:

    a. $KMnO_4$ b. $LiMnO_2$ c. NH_4ClO_4

14. Determine the oxidation state for each atom in the following compounds:
    a. K_2MnCl_4 b. Cr_2O_3 c. $C_{12}H_{22}O_{11}$
15. Use the following four terms or expressions to identify each of the chemical situations indicated: (*is oxidized; is reduced; is an oxidizing agent; is a reducing agent*). Use as many terms as apply.
    a. An atom has gained an electron.
    b. An atom increases its oxidation number.
    c. The oxidation number of an atom changes from −2 to −3.
16. Use the following four terms or expressions to identify each of the chemical situations indicated: (is oxidized; is reduced; is an oxidizing agent; is a reducing agent). Use as many terms as apply.
    a. An atom decreases its oxidation number.
    b. An atom loses two electrons.
    c. The oxidation number of an atom changes from +3 to +5.
17. Butane, C_4H_{10}, burns in oxygen to form carbon dioxide and water.
    a. Is this an oxidation-reduction reaction? Prove the answer by assigning oxidation states to each element in each compound.
    b. Is this electrochemistry? Explain the answer.
18. Supply the oxidation number for each element in the combustion of hexane, C_6H_{14}, with oxygen.
19. Supply the oxidation number of each atom, on both sides of the reaction arrow, in the equation

$$3Mg(s) + 2H_3PO_4(aq) \rightarrow Mg_3(PO_4)_2(aq) + 3H_2(g)$$

20. Supply the oxidation number of each atom, on both sides of the reaction arrow, in the equation

$$2AgNO_3(aq) + Cu(s) \rightarrow Cu(NO_3)_2(aq) + 2Ag(s)$$

Chemical Applications and Practices

21. The main active ingredient in commercial household bleach is the hypochlorite ion, ClO^-.
    a. What is the oxidation number of the Cl atom in the ion?
    b. Hypochlorite is known as a good oxidizing agent. Would that property indicate that the Cl tends to gain or lose electrons as it reacts? Explain the logic of the response.
22. The combustion of propane in a portable burner is a common redox reaction. Examine each component of the equation. Assign an oxidation number to each atom, and determine which component is acting as the reducing agent.

$$C_3H_8(g) + O_2(g) \rightarrow CO_2(g) + H_2O(g)$$

23. The brilliant red color of many fireworks is due to the presence of strontium. However, strontium metal is not typically found in its pure state. The following redox reaction depicts the isolation of strontium from molten strontium chloride.

$$SrCl_2(l) \rightarrow Sr(s) + Cl_2(g)$$

    a. Assign an oxidation number to each of the atoms in the reaction.
    b. What has been reduced in the reaction?
24. Sodium thiosulfate is familiar to photographers as "hypo." It helps dissolve some of the silver salts used in developing photographs. Aqueous solutions of sodium thiosulfate can undergo disproportionation reactions, as shown here in this unbalanced reaction:

$$S_2O_3^{2-}(aq) + H^+(aq) \rightarrow H_2O(l) + S(s) + SO_2(g)$$

    a. What is the change in the oxidation number of S in $S_2O_3^{2-}$ in the oxidation portion of the reaction?
    b. What is the change in the oxidation number of S in $S_2O_3^{2-}$ in the reduction portion of the reaction?
25. Four equations can be written to describe the rusting of iron:

$$O_2(aq)_{n+}2H_2O(l) + 4e^- \rightarrow 4OH^-(aq)$$
$$Fe(s) \rightarrow Fe^{2+}(aq) + 2e^-$$

$$Fe^{2+}(aq) + 2OH^-(aq) \rightarrow Fe(OH)_2(s)$$
$$Fe(OH)_2(s) + OH^-(aq) \rightarrow FeO(OH)(s) + H_2O(l) + e^-$$

a. Which of these are reduction or oxidation half-reactions?
b. What other kind of equation is present here?
c. Combine the equations to give the overall equation that describes the formation of rust, FeO(OH)(s), from iron, Fe(s).

26. In the chapter, we discussed the reaction of hydrogen and oxygen to give water. How is it possible for this reaction to proceed either as an explosion (as in the space shuttle main engines) or as a gentle, readily managed source of electricity?

Section 18.4 Redox Equations

Skill Review

27. The superscript on the symbol E° refers to standard conditions for electrochemical reactions. What specifically does that indicate for the concentrations and pressure of the reacting system?
28. The voltage values (emf) given on a standard reduction table are referenced to a standard called SHE.
    a. To what do the letters refer?
    b. What is the voltage of the reference?
29. Complete each of these statements using the word *positive, negative, spontaneous, or nonspontaneous.*
    a. When E° is negative, the electrochemical reaction is ________.
    b. When E° is positive, the value for ΔG° is ________.
    c. When a reaction is spontaneous, the values for ΔG° will be ________, and the values for E° will be ________.
    d. Nonspontaneous redox reactions have a ________ value for E°.
30. The typical battery used in a standard flashlight produces approximately 1.5 V. If the value for ΔG° of the reaction were – 289,500 J, what would we calculate as the moles of electrons exchanged in the balanced redox reaction?
31. Combine these half-reactions in such a way that a galvanic cell results and calculate the cell potential.

$$Fe^{3+} + e^- \rightarrow Fe^{2+} \qquad E^\circ = 0.77\ V$$
$$Fe^{2+} + 2e^- \rightarrow Fe \qquad E^\circ = -0.44\ V$$

32. Combine these half-reactions in such a way that a galvanic cell results and calculate the cell potential.

$$Sn^{2+} + 2e^- \rightarrow Sn \qquad E^\circ = -0.14\ V$$
$$Sn^{4+} + 2e^- \rightarrow Sn^{2+} \qquad E^\circ = +0.15\ V$$

33. Calculate the free energy change for the cell in Problem 31.
34. Calculate the free energy change for the cell in Problem 32.
35. Combine these reactions in such a way that a galvanic cell results, and calculate the cell potential:

$$\left[PtCl_4^{2-}\right] + 2e^- \rightleftharpoons Pt + 4Cl^- \qquad E^\circ = 0.755\ V$$
$$Ce^{4+} + e^- \rightleftharpoons Ce^{3+} \qquad E^\circ = 1.61\ V$$

18

36. Combine these reactions in such a way that a galvanic cell results, and calculate the cell potential:

$$AuBr4^- + 3e^- \rightleftharpoons Au + 4Br^- \qquad E^\circ = 0.86\ V$$
$$Fe^{3+} + e^- \rightleftharpoons Fe^{3+} \qquad E^\circ = 0.77\ V$$

37. Calculate the free energy change for the cell in problem 35.
38. Calculate the free energy change for the cell in problem 36.
39. What is the free energy change for the following reaction at standard conditions?

$$PbO_2 + 4H^+ + 2Hg + 2Cl^- \rightarrow Pb^{2+} + 2H_2O + Hg_2Cl_2 \qquad E^\circ_{cell} = 1.12\ V$$

40. What is the free energy change for the following reaction at standard conditions?

$$O_2 + 4H^+ + 2Ni \rightarrow 2H_2O + 2Ni^{2+} \qquad E^\circ_{cell} = 1.46\ V$$

41. What are three considerations or aspects that must be "balanced" in a balanced redox reaction?

42. In this balanced redox reaction, chlorine is shown to replace bromide ions from a solution.

$$Cl_2(aq) + 2Br^-(aq) \rightarrow 2Cl^-(aq) + Br_2(aq)$$

What is the value of n in the overall reaction?

43. Balance these half-reactions in acidic solution. Is each a reduction or an oxidation? Which substance is oxidized and which is reduced?
   a. $CO_2 \rightarrow H_2C_2O_4$
   b. $Np^{4+} \rightarrow NpO_2^+$
44. Balance these half-reactions in acidic solution. Is each a reduction or an oxidation? Which substance is oxidized and which is reduced?
   a. $I_2 \rightarrow IO_3^-$
   b. $NO_3^- \rightarrow NO$
45. Balance these redox equations in acidic solution:
   a. $Sn^{2+} + Cu^{2+} \rightarrow Sn^{4+} + Cu^+$
   b. $S_2O_3^{2-} + I_3^- \rightarrow S_4O_6^{2-} + 3I^-$
   c. $SO_3^- + Fe^{3+} \rightarrow SO_4^{2-} + Fe^{2+}$
46. Balance these redox equations in acidic solution:
   a. $Al + Cu^{2+} \rightarrow Al^{3+} + Cu$
   b. $UO_2^{2+} + Ag + Cl^- \rightarrow U^{4+} + AgCl$
   c. $H_2SO_4 + HBr \rightarrow SO_{2+} Br_2$
47. Balance this reaction in basic solution and calculate $E°$ for the process:

$$Mn^{2+}(aq) + NaBiO_3(s) \rightleftharpoons Bi^{3+}(aq) + MnO_4^-(aq)$$

Note: $NaBiO_3(s) + 6H^+(aq) + 2\,e^- \rightleftharpoons Bi^{3+}(aq) + Na^+(aq) + 3\,H_2O$ $E° = +1.6$ V

48. We discussed corrosion in this chapter as well as in Sect. 8.9, in which we introduced the chemical reactivity series. Balance this reaction describing one process for the corrosion of iron in the presence of oxygen and water and calculate $E°$:

$$Fe(s) + O_2(g) + H^+(aq) \rightleftharpoons Fe^{2+}(aq) + H_2O(l)$$

49. Balance the redox equation that illustrates the reaction of solid copper and dichromate, first in acidic and then in basic solution.

$$Cu(s) + Cr_2O_7^{2-}(aq) \rightarrow Cu^{2+}(aq) + Cr^{3+}(aq)$$

50. Balance the redox equation that illustrates the reaction of permanganate and methanol, first in acidic and then in basic solution.

$$MnO_4^- + CH_3OH \rightarrow CO_3^{2-} + MnO_4^{2-}$$

51. Balance this redox reaction, first in acidic and then in basic solution.

$$ClO_4^- + I^- \rightarrow ClO^- + IO_3^-$$

52. Balance this redox reaction, first in acidic and then in basic solution.

$$Zn + NO_3^- \rightarrow Zn^{2+} + NH_3$$

Chemical Applications and Practices

53. Using the standard reduction potentials found in the appendix, locate the half-cell reaction for zinc. Zinc is often used in the production of dry cell batteries. It is also used to protect other metals from oxidation.
   a. What is the $E°$ value for this reaction?
   b. What would be the value of $\Delta G°$ for the half-reaction?
   c. What would be the $E°$ value if the reaction stoichiometry were doubled?
54. Splitting water into hydrogen gas and oxygen gas is one technique being investigated as a way to produce hydrogen gas for use in fuel cells. The reaction is shown here as

$$2H_2O \rightarrow 2H_2 + O_2$$

If the $E°$ value for this nonspontaneous reaction were approximately −2.00 V, what would we calculate as the value for $\Delta G°$?

55. Use the Standard Table of Reduction Potentials to answer the following questions:
   a. Which of the three alkali metal ions Na, K, and Li has the least potential to attract electrons?
   b. Of the three halogens which would be the strongest oxidizing agent, F_2, Cl_2, or Br_2?
   c. If we prepared a battery by connecting two of the following three half-cells, which combination could produce the highest potential? Zn, Cu, Ni
56. Suppose we attempted to build a battery using lead (and Pb^{2+}) with chromium (and Cr^{3+}).
   a. Write out the half-cell reduction reactions and their $E°$ values for each.
   b. Which of the two would provide the reduction reaction, and which would provide the oxidation reaction?
   c. What would be the total $E°$ of the overall redox reaction?
   d. How many moles of electrons would be exchanged in the overall reaction?
57. The "Zebra" cell was invented in 1985 in South Africa as part of the "ZEolite Battery Research Africa" project and is used in all-electric municipal buses and other vehicles. It uses nickel and molten sodium as its electrodes. Its overall cell reaction is

$$Ni + NaCl \rightleftharpoons NiCl_2 + Na$$

What are the $\Delta G°$ and $E°$ for the Zebra cell?

58. We listed the zinc-air battery in Table 18.5 as being lightweight and rechargeable. As such, it is used hearing aids, as well as in the military as a portable power source for "Micro Unmanned Aerial Vehicles" (MAVs) and "Unmanned Aerial Vehicles" (UAV's). At the anode, zinc is converted to zinc oxide, and at the cathode, oxygen is reduced to hydroxide ion.
   What are the values of $\Delta G°$ and $E°$ for the Zebra cell? The reaction "in the electrolyte" forms the zinc oxide but does not involve electron exchange. We may want to start by determining the overall equation.

Anode :	$Zn(OH)_4^{2-} + 2e^- \rightleftharpoons Zn + 4OH^-$		$E° = 1.25$ V
In the electrolyte:	$Zn(OH)_4^{2-} \rightleftharpoons ZnO + H_2O + 2OH^-$		
Cathode :	$O_2 + 2H_2O + 4e^- \rightleftharpoons 4OH^-$		$E° = 0.4$ V
Overall :	$\mathbf{2Zn + O_2 \rightleftharpoons 2ZnO}$		

59. The oxidation of copper metal gives rise to a beautiful green patina. An equation that illustrates the oxidation of copper is shown here. Balance this redox reaction.

$$Cu(s) + CO_2(g) \rightarrow CuO(s) + C_2O_4^{2-}(aq)$$

60. One technique used for the detection of ethanol involves the following redox reaction with dichromate ions. Balance this redox reaction.

$$C_2H_5OH(aq) + Cr_2O_7^{2-}(aq) \rightarrow CH_3CO_2H(aq) + Cr^{3+}(aq)$$

61. As with acid–base reactions, careful addition of a solution containing an oxidizing agent to a solution containing a reducing agent can be used for titration analysis. Standardizing a solution of potassium permanganate to be used later in a redox titration often involves reaction with a known amount of sodium oxalate. Balance the redox reaction used in the standardization process.

$$C_2O_4^{2-}(aq) + MnO_4^-(aq) \rightarrow CO_2(g) + Mn^{2+}(aq) + H_2O(l)$$

62. Once a standard solution of potassium permanganate is prepared, it can be used to determine the concentration of iron in an unknown sample. For example, the iron content of a small steel sample could be obtained through a titration reaction with a standard permanganate solution. The redox reaction that would take place in the analysis is shown here (unbalanced, and after the iron has been prepared as +2 ion). Balance the redox titration and identify the reducing agent.

$$MnO_4^-(aq) + Fe^{2+}(aq) \rightarrow Mn^{2+}(aq) + Fe^{3+}(aq) + H_2O(l)$$

63. One step in a common method for the analysis of vitamin C in juice drinks is to oxidize the vitamin C (ascorbic acid) with aqueous I_2. Balance this redox reaction.

$$C_6H_8O_6(aq) + I_2(aq) \rightarrow I^-(aq) + C_6H_6O_6(aq)$$

64. There are many half-cell reaction combinations that could be used to make batteries. One such reaction would be to use system consisting of silver(II) oxide and zinc. The reaction shown below would have to occur in a basic environment. Balance this redox reaction.

$$Zn(s) + AgO(s) \rightarrow Zn(OH)_2(s) + Ag(s)$$

Section 18.5 Electrochemical Cells

Skill Review

65. Define the following terms: cell, half-reaction, galvanic cell, voltaic cell, electromotive force.
66. Compare and contrast an electrolytic cell with a galvanic cell on the basis of sign on the $E°$ value, ability to do work, chemical process taking place at the anode, and spontaneity.
67. The following notation refers to a specific voltaic cell. Identify the species that would serve as the anode and the species that is the oxidizing agent.

$$Zn(s)|Zn^{2+}(aq) \parallel Cu^{2+}(aq)|Cu(s)$$

68. The following notation refers to a specific voltaic cell. Identify the species that would serve as the anode and the species that is the oxidizing agent.

$$Mg(s)|Mg^{2+}(aq) \parallel Co^{2+}(aq)|Co(s)$$

69. The following schematic diagram represents two metals and their cations in separate beakers joined by a $NaNO_3$ salt bridge. If the metals were lead and silver (with their respective cations Pb^{2+} and Ag^+), which beaker would require the silver and which the lead if the nitrate ions were moving from left to right through the salt bridge? Justify the answer.

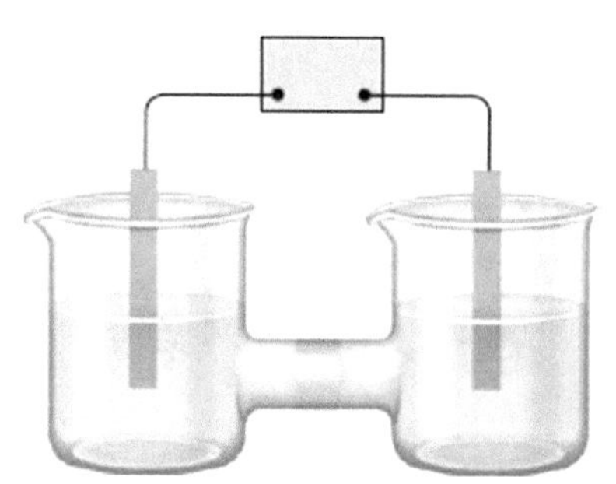

70. The following schematic diagram represents two metals and their cations in separate beakers joined by a $NaNO_3$ salt bridge. If the metals were copper and iron (with their respective cations Cu^{2+} and Fe^{3+}), which beaker would require the copper and which the iron if the nitrate ions were moving from left to right through the salt bridge? Justify the answer.

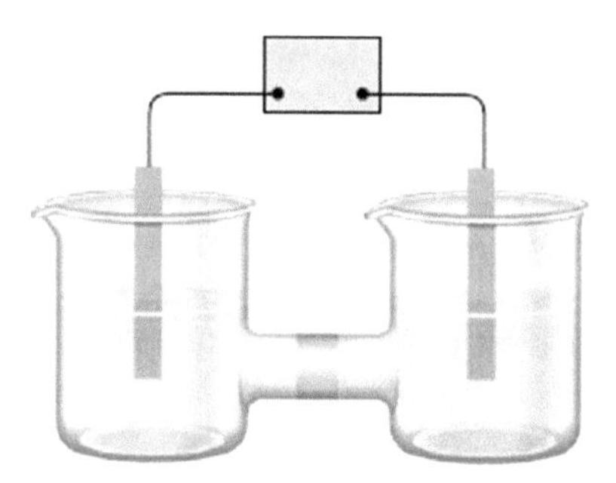

Chemical Applications and Practices

71. Diagram a battery consisting of two beakers and a salt bridge that makes use of the two half-reactions silver and gold.
    a. Predict the $E°$ value for the battery.
    b. What is the value of n for the balanced equation?
    c. Label the cathode.

d. Exactly what species is acting as the oxidizing agent?
e. Represent the battery in the ABC notation.

72. Search Internet references using the keywords "hydrogen fuel cell" and report on the basic operation of such a cell. Give particular emphasis to the electrolyte used in such a cell. Be sure to report the necessary URL for the references.
73. Small button-sized batteries can be made using mercury and zinc. If the two half-reactions are as follows, what would we report as the ABC notation for the battery? (Mercury in the batteries is considered a toxic substance and should be carefully recycled.)

$$Zn(s) \rightarrow ZnO(s)$$
$$HgO(s) \rightarrow Hg(l)$$

Note that the reactions are not balanced. The reaction takes place in a basic medium.

74. Write out the standard reduction half-cell reactions and $E°$ values for Zn, Ni, Pb, and $H_2(g)$.
   a. Select the two half-reactions that, when combined, would produce the battery with the greatest theoretical overall $E°$ value, and report that value.
   b. Which, if any, of those represented would have electrons move along an external wire when connected with the hydrogen half-cell, toward the hydrogen electrode?

Section 18.6 Chemical Reactivity Series

Skill Review

75. Use the Standard Table of Reduction Potentials to answer the following questions.
   a. Which is the better reducing agent, Ba or Ca?
   b. Which is the better oxidizing agent, Pb^{2+} or Ni^{2+}?
   c. In which direction, to the left or to the right, would the following reaction take place spontaneously? (Assume standard conditions.)

$$Ag + Fe^{3+} \rightarrow Fe^{2+} + Ag^{+}$$

76. Under standard conditions one of the following two reactions will not occur spontaneously. Use the Standard Table of Reduction Potentials to explain how we could correctly predict which one will not take place.

$$Fe + Sn^{2+} \rightarrow Fe^{3+} + Sn$$
$$Cu^{2+} + Fe \rightarrow Fe^{3+} + Cu$$

Chemical Applications and Practices

77. Concrete used in road construction is reinforced using steel bars in a grid, known collectively as "rebar." After many years, rebar will begin to corrode, and the concrete will weaken. Studies are ongoing regarding the best metal to be a sacrificial anode to prevent corrosion of the rebar. Which would work better for the purpose, magnesium or zinc? Although different steels vary in their metallic and carbon composition, they contain iron as the primary metal. Assume the steel is made up of 100% iron.
78. The reinforcing steel rods, or rebar, in concrete are especially susceptible to corrosion when exposed to seawater, because of the interaction of the chloride ions in seawater with the iron in the rebar steel. Propose reduction and oxidation half-reactions and an overall cell reaction for the process.
79. Suppose we have a shiny piece of metal that is unlabeled. It is known to be either aluminum or tin. We also have a solution of 1.0 M nickel(II) nitrate. Suggest a chemical test that we could use to determine the identity of the metal using the solution. Explain the expected results and the basis of the conclusion.
80. One technique to protect against oxidation of structures such as ship hulls and underground iron pipes is to place the structure in contact with a metal that will oxidize more easily than the iron in the structure. The more active metal is then referred to as a sacrificial anode. Neglecting such issues as cost and availability, would copper or zinc make the better sacrificial anode to protect an iron-based structure? Explain the basis of the choice.

Section 18.7 Not-So-Standard Conditions: The Nernst Equation

Skill Review

81. Calculate the E_{cell} for the oxidation of Zn by Pb^{+2} if the equilibrium concentrations of the ions are $[Pb^{2+}]=3.0\times10^{-3}$ M and $[Zn^{2+}]=2.0\times10^{-3}$ M.
82. Calculate the E_{cell} for the galvanic cell based on these half-reactions at 25 °C, in which

$$\left[FeO_4^{2-}\right] = 1.0\times10^{-3}M \quad [O_2] = 1.0\times10^{-5}M$$
$$\left[Fe^{3+}\right] = 5.0\times10^{-4}M \quad pH = 5.1$$

$$FeO_4^{2-}(aq) + H^+(aq) \rightleftharpoons Fe^{3+}(aq) + H_2O(l) \quad E^\circ = +2.20\ V$$
$$O_2(g) + H^+(aq) \rightleftharpoons H_2O(l) \quad E^\circ = +1.23\ V$$

83. The constant used in the Nernst equation, 0.0257, is derived from the combination of three other constants. Obtain the three values, assuming standard temperature, and derive the constant used in the equation. Some texts refer to the Nernst equation using the base 10 logarithmic scale. What would be the value of the constant in that scale?
84. The simplified equation presented here represents the redox reaction taking place inside the typical, non-alkaline flashlight battery. Use the principles described in the Nernst equation to answer the questions that follow.

$$Zn(s) + 2MnO_2(s) + 2NH_4^+(aq) \rightarrow Zn^{2+}(aq) + Mn_2O_3(s) + 2NH_3(g) + H_2O(l)$$

  a. What would be the effect on the spontaneity of the reaction if the concentration of $NH_4^+(aq)$ were decreased?
  b. What would be the effect on the voltage of the cell if the concentration of $NH_4^+(aq)$ were decreased?
  c. What is the value of n in the equation?
  d. At equilibrium what is the value of E?
85. Calculate the value of E_{cell} for the reaction of iron and copper(II), given the specific concentrations listed.
  a. $Fe(s) \mid Fe^{2+}(0.10\ M) \parallel Cu^{2+}(0.10\ M) \mid Cu(s)$
  b. $Fe(s) \mid Fe^{2+}(1.5\ M) \parallel Cu^{2+}(0.10\ M) \mid Cu(s)$
  c. $Fe(s) \mid Fe^{2+}(0.10\ M) \parallel Cu^{2+}(1.5\ M) \mid Cu(s)$
86. Calculate the value of E_{cell} for the reaction of cobalt(II) chloride and zinc metal, given the specific concentrations listed.
  a. $Zn(s) \mid Zn^{2+}(0.10\ M) \parallel Co^{2+}(0.10\ M) \mid Co(s)$
  b. $Zn(s) \mid Zn^{2+}(2.5\ M) \parallel Co^{2+}(0.050\ M) \mid Co(s)$
  c. $Zn(s) \mid Zn^{2+}(0.010\ M) \parallel Co^{2+}(0.10\ M) \mid Co(s)$

Chemical Applications and Practices

87. Suppose that at night on a campout our flashlight dims and finally stops working. Our friend remarks, "I guess the battery is dead." In what three other chemical ways, using Gibbs free energy, equilibrium, and potential, could we state the same conclusion as our friend?
88. A quick source of hydrogen gas in the lab is to carefully place some solid magnesium in hydrochloric acid.

$$Mg(s) + 2HCl(aq) \rightarrow MgCl_2(aq) + H_2(g)$$

If the Mg^{2+} concentration were 1.0 M and the HCl concentration were 0.10 M (assume completely dissociated HCl), what would we calculate as the E value for the reaction at 25°C?

$$Mg(s)|Mg^{2+}(aq) \parallel H^+(0.1M)|H_2(1\ atm)$$

Section 18.8 Electrolytic Reactions

Skill Review

89. Calculate the number of grams of gold electroplated onto a surface given these conditions. (Assume that the solution of Au^{3+} ions is concentrated enough to complete the electrolysis.)
    a. 1.25 A for 60 s
    b. 2.11 A for 2.33 h
    c. 0.75 A for 1 d
90. Calculate the time required to electroplate 1.0 g of tin metal onto a surface given these conditions. (Assume that the solution of Sn^{2+} ions is concentrated enough to complete the electrolysis.)
    a. 2.25 A
    b. 0.11 A
    c. 1.38 A
91. Electroplating chromium on an industrial scale requires a solution of "chromic acid" (more correctly called chromium(VI) oxide), CrO_3, and sulfuric acid. The typical current is on the order of kiloamperes per square meter. If we pare this down a bit, how much chromium can be plated out of a Cr^{6+} solution with a current of 3.00 amps after 25.0 min?
92. How many grams of copper can be reduced by applying a 4.00 amp current for 53.0 min to a solution containing Cu^{2+} ions?
93. Alessandro Volta is given historical precedence in the discovery of the first battery in the sense that we think of batteries today. The "Voltaic pile" consisted of dissimilar metals joined by salt-moistened paper strips. Davy and his assistant Michael Faraday later refined this. At that time, early electroplating businesses began developing in England. If such a business silver-plated a teaspoon, would the teaspoon be the cathode or the anode in such a process? If the spoon received 0.33 g of silver after being plated for 1.0 h, what amperage was used?
94. In addition to isolating sodium, Sir Humphrey Davy isolated potassium, calcium, and magnesium from impure natural ore samples.
    a. If he used a current of 1.00 A for 1.00 h, how many grams of magnesium metal could he obtain from molten $MgCl_2$?
    b. Using the same electrical set up, how many grams of potassium metal could Sir Humphrey isolate from molten KCl?

Chemical Applications and Practices

95. In the typical lead storage battery found in most automobiles, lead is oxidized to Pb^{2+} (in the form of $PbSO_4$). In recharging, the reaction is reversed. If a battery were recharged for 30.0 min at a current of 8.00 A, how many grams of lead would be reduced from $PbSO_4$ during the process?
96. Germanium has become a valuable metal in semiconductor fields. If a 1.00 A current were used for 1.00 h to plate out 0.677 g of germanium, what would we calculate as the oxidation number of the germanium ions in the plating solution?

Comprehensive Problems

97. Use the illustration below as the starting point to draw the electrolytic cell that would be used to plate copper metal onto a steel saucepan. Be sure to indicate the location of the cathode and the anode, the location of the copper electrode, and the steel saucepan.

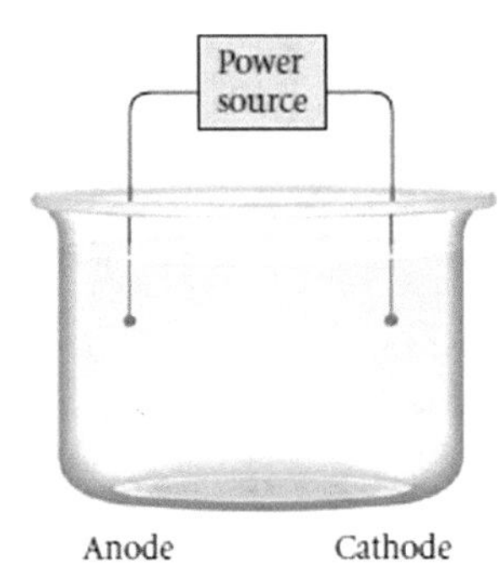

98. The Nernst equation can be used just as well with half-cell reactions as with balanced redox reactions. An application of this is the use of potential differences in the acid level of solutions compared to a standard when producing the probes for pH meters. Using the half-reaction $2H^+(aq) + 2e^- \rightarrow H_2(g)$ at 1 atm, what would we calculate, using the Nernst equation, as the value for E in these situations?
    a. Pure water, pH = 7.00
    b. An acid solution with pH = 2.00
    c. A 0.10 M solution of nitric acid
    d. A 0.10 M solution of acetic acid, $K_a = 1.8 \times 10^{-5}$
99. Potassium ferrate (K_2FeO_4) is a powerful oxidizing agent ($E° = 2.20$ V relative to SHE in acidic solution) in which the iron(VI) ion is reduced to the iron(III) ion. It oxidizes water to oxygen gas.
    a. Determine the cell potential and equilibrium constant for the reaction at standard conditions for the oxidation of water by ferrate ion in acidic solution.
    b. Balance the reaction in basic solution.
    c. Calculate the cell potential, given the following initial concentrations:

$$\left[FeO_4^{2-}\right] = 1.5 \times 10^{-3}\text{M}; \left[Fe^{3+}\right] = 1.1 \times 10^{-3}\text{M}; P_{O_2} = 8.3 \times 10^{-5} \text{ atm; pH} = 2.8$$

    d. The reaction results in the production of a yellow solid that is especially apparent at high pH. Can we suggest what this might be? Further, can we suggest how this solid might help if the ferrate ion were used to treat wastewater contamination?
100. Potassium permanganate ($KMnO_4$) is a useful analytical reagent for determining the percentage of iron in an iron ore. The procedure includes dissolving the iron with HCl and then converting all of the iron in the ore to Fe^{2+} using several reagents. The titration of the resulting Fe^{2+} with MnO_4^- is

$$Fe^{2+} + MnO_4^- \rightarrow Fe^{3+} + Mn^{2+}$$

A sample of the original iron ore weighing 3.852 g was processed for titration in a total volume of 300.0 mL of solution. A 100.0 mL aliquot (accurately measured portion) was removed and titrated with a 0.1025 M $KMnO_4$ solution. A total of 23.14 mL was required to the light pink endpoint. What is the percent of iron in the ore sample?
101. We noted in Chapter 18, Thermodynamics, that we can have "standard conditions" at a given pressure, but the temperature can vary and must be stated as part of specifying our standard values. If we define our standard values at 1 bar pressure and 3000 °C, how would that affect the values for our "new" standard reduction potentials compared to the ones we commonly use at 25 °C?
102. We know from our own experience that rechargeable batteries can be recharged hundreds of times but eventually wear down and must be replaced. Why can't they be recharged an infinite number of times? In particular, investigate the way in which metals plate back onto the cathode when the recharging takes place.
103. Chrome plating is the process by which Cr^{6+} is electroplated on automobile surfaces to give them a nice luster. The chromium layer can be as little as 10.0 μm thick. How many grams of CrO_3 would be necessary to plate the back surface of a car mirror that has a coverable area of 150 cm^2? (d = 7.2 g/cm^3.)
104. Calculate the $E°$, $\Delta G°$ and K at 298 K for the hydrogen/oxygen fuel cell described in ▶ Sect. 18.3. Once we have done so, can we propose a fuel cell that might be more effective for use in automobiles? What are the criteria that we use to make that judgment?

Thinking Beyond the Calculation

105. In our discussion of acid–base chemistry, we learned that water undergoes autoprotolysis to form the hydrogen and hydroxide ions.
    a. What is the value for $E°$?
    b. What is the value for K at 25 °C?
    c. How does the value for K relate to K_w at 25 °C?
    d. What is the value for K at 50 °C?
    e. How and why is the value for K different at the higher temperature, and how does this effect the pH of a neutral solution?

106. A battery designer wishes to prepare a solution-based battery for use in a new automobile. The designer chooses iron and zinc as the two metals for study.
  a. Draw the setup (using beakers and a salt bridge) that indicates the location of the iron and zinc electrodes, the iron(II) and zinc(II) solutions, and the flow of electrons.
  b. Which half-reaction is the oxidation reaction?
  c. What is the cell potential for the reaction if the initial concentrations of iron(II) and zinc(II) are 0.25 M at 25 °C?
  d. What would be the cell potential if the concentrations of both solutions in part c were increased to 1.0 M?... to 2.0 M?
  e. If the battery is cooled to 10 °C, is there a change in the cell potential? If so, what is the new potential?
  f. Describe at least one advantage to using a battery made from these metals.
  g. Describe at least one disadvantage to the use of iron in a battery.

Coordination Complexes

Contents

Supplementary Information The online version contains supplementary material available at (▶ https://doi.org/10.1007/978-3-030-90267-4_19).

M. Mosher and P. Kelter, *An Introduction to Chemistry*,
https://doi.org/10.1007/978-3-030-90267-4_19

What We Will Learn From This Chapter

- Coordination complexes result from the association of ligands with a central metal (▶ Sect. 19.2).
- Ligands can bind to the central metal via one, two or more coordinate covalent bonds (▶ Sect. 19.3).
- The number of ligands around a metal gives rise to the coordination number (▶ Sect. 19.4).
- Common geometries of coordination complexes include square planar, tetrahedral, and octahedral (▶ Sect. 19.5).
- The arrangement of the ligands and the charge on the metal give rise to a variety of isomers (▶ Sect. 19.6).
- A set of nomenclature rules have been developed that allow the name of a coordination complex to be determined (▶ Sect. 19.7).
- Crystal field theory explains why some transition metal complexes have a different color based on the ligands in the complex (▶ Sect. 19.8).
- Coordination complexes can undergo redox reactions and ligand exchange reactions (▶ Sect. 19.9).

19.1 Introduction

Many of the nutritional supplements we consume on a daily basis contain iron. Why? Iron is an essential nutrient with the U.S. recommended daily allowance (RDA) for men 19 to 50 years old of 8 mg and women of the same age of 18 mg, though when pregnant, women require more, with an RDA of 28 mg per day. Iron deficiency is a long-standing problem in the U.S., with rates ranging from 12 to about 30%, depending upon race and socioeconomic status. Iron isn't a vitamin; rather, it's called a mineral—the term used for any inorganic element important for health. It turns out that few items claiming to be rich in iron actually contain the neutral metal. Most items typically contain iron ions in a salt, often iron sulfate. Why is iron so important to good health? Most of the iron in the body is contained in a protein known as hemoglobin within the red blood cells (▶ Chap. 22). Specifically, iron ions are chemically bound in hemoglobin in a form called a coordination complex.

Within hemoglobin, the iron ions are bonded to other atoms in an elegant and complex structure known as a heme (see ◘ Fig. 19.1). The arrangement of groups attached to the iron enables it to collect an oxygen molecule and later release it at an appropriate time. Close examination of the heme indicates that the conformational changes illustrated in ◘ Fig. 19.2 occur during oxygen binding. We visited a simpler form of this complex in ▶ Chap. 14 when we discussed the equilibrium involving oxygen and myoglobin. Without this specific complex structure, whether in hemoglobin or myoglobin, iron would be incapable of playing the role of oxygen carrier.

In a typical **coordination complex** (or simply complex), a central metal atom or ion is chemically bonded to several other components. For example, natural water often contains iron in the form of a chemical combination of an Fe^{3+} ion and six water molecules, $[Fe(H_2O)_6]^{3+}$, as shown in ◘ Fig. 19.3. The bonding, the structure, and the many biological, medical, and industrial applications of coordination complexes are the subject of this chapter.

Coordination complex – A metal bonded to two or more ligands via coordinate covalent bonds.

19

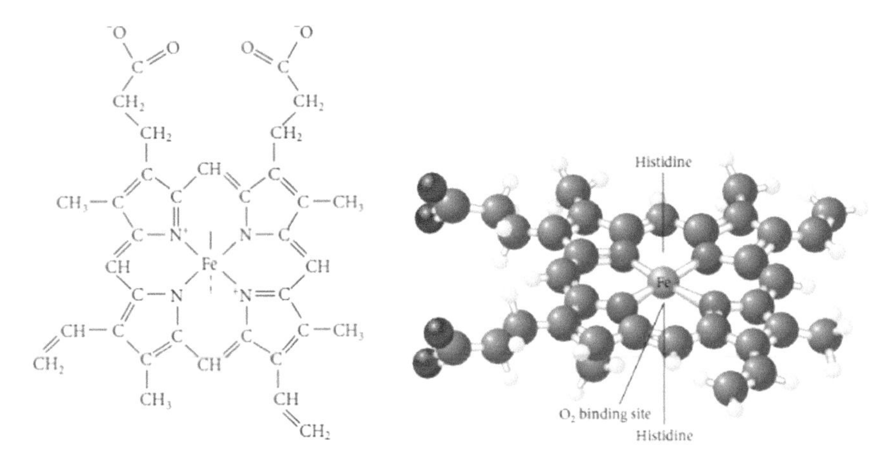

◘ **Fig. 19.1** The iron is bonded to other atoms in an elegant and complex structure known as a heme. The hemoglobin protein uses two amino acid residues to hold the iron within the protein's structure. One of the amino acid residues can move out of the way to allow an oxygen molecule to bind to the iron

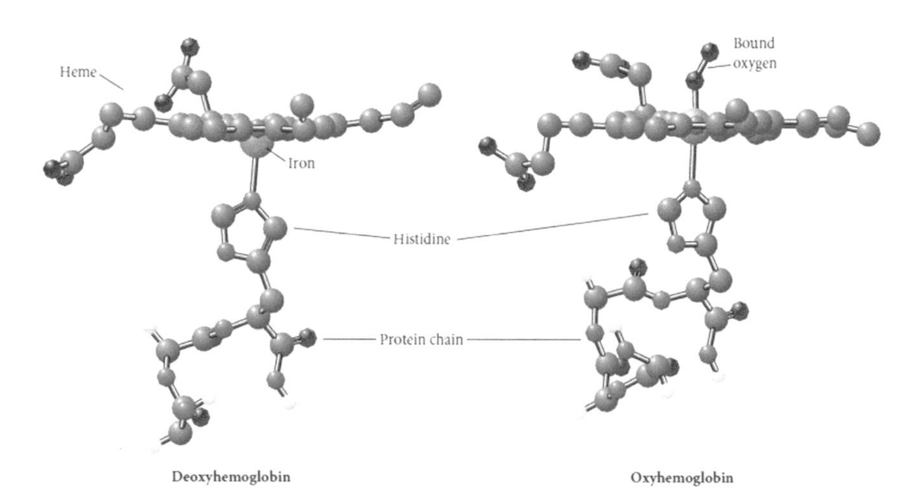

Fig. 19.2 Conformational changes take place during the binding of oxygen to the iron ion in hemoglobin. These changes act like a switch in the hemoglobin protein, activating it and increasing its ability to bind to three other molecules of oxygen. In the structure shown here, the oxygen molecule (red) is bound in oxyhemoglobin. When oxygen is not bound to hemoglobin, the protein chains pull the iron so that the heme sits slightly above the center of the iron ion

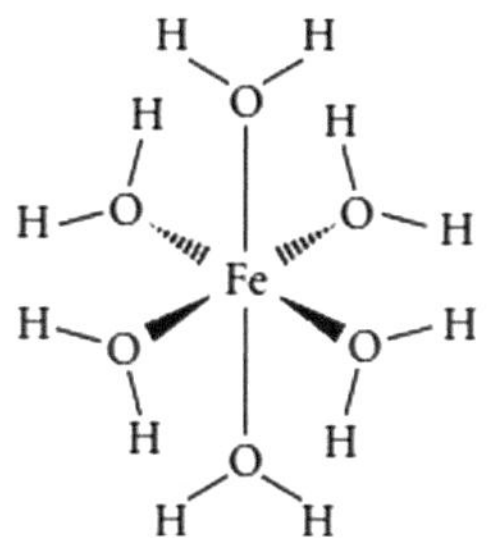

Fig. 19.3 Natural water often contains iron in the form of a chemical combination of an Fe^{3+} ion and six water molecules, $[Fe(H_2O)_6]^{3+}$

19.2 Bonding in Coordination Complexes

A common feature of bonding within metal complexes is the **coordinate covalent bond**, which we discussed in ▶ Chap. 8. Recall that the coordinate covalent bond forms when both bonding electrons originate from one atom. This means that a molecule with a lone electron pair (a Lewis base) could potentially be a component of a coordination complex. Lewis bases that form a coordinate covalent bond with a metal or metal ion are known as **ligands**. In the aqueous iron complex found in natural water that contains iron, each water molecule donates a pair of electrons to the iron(III) ion and is therefore the Lewis base. The iron(III) ion behaves as a Lewis acid, an electron pair acceptor. The resulting bond between the oxygen of water and the iron(III) ion is a coordinate covalent bond. With six equivalent water molecules donating electron pairs to create six bonds to the central Fe^{3+} ion, the complete complex ion is $[Fe(H_2O)_6]^{3+}$.

Coordinate covalent bond – A covalent bond that results when one atom donates both electrons to the bond.

Ligand – A Lewis base that donates a lone pair of electrons to a metal center to form a coordinate covalent bond.

$$Fe^{3+} + H_2O \longrightarrow [Fe{-}OH_2]^{3+}$$

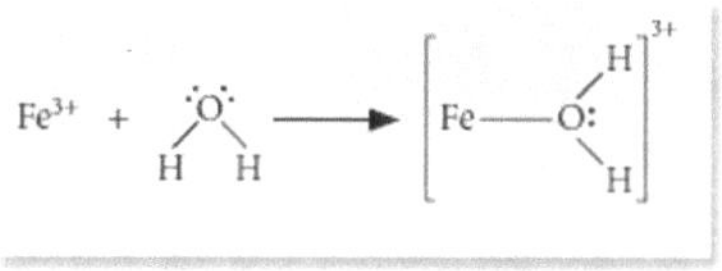

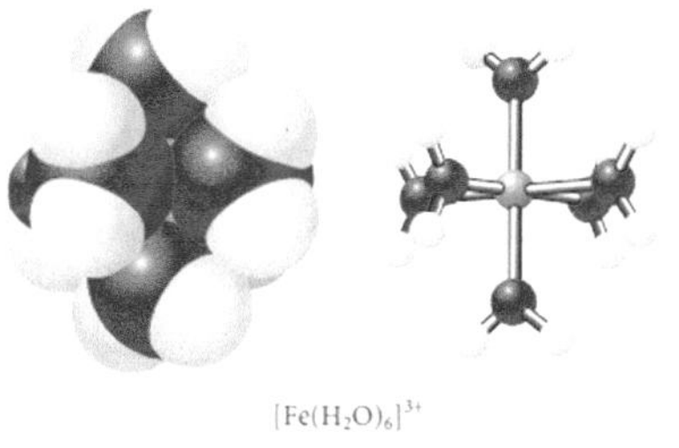

$[Fe(H_2O)_6]^{3+}$

In general, we can use the following equation to illustrate the formation of a coordination complex via donation of a lone pair of electrons from a ligand, L, to a metal center, M,

$$M + :L \rightarrow M - L$$

The chemistry of hemoglobin is an example of the formation of such a coordination complex. The iron ion in the hemoglobin molecule forms a bond with an oxygen molecule and ultimately transports the oxygen molecule from the lungs to cells throughout the body. Using our shorthand, we can represent the bonding with oxygen by the following equation:

$$\text{HbFe}\ \ddot{O}=\ddot{O} \rightleftharpoons \text{HbFe}-\ddot{O}-\ddot{O}$$

where Hb represents the complex structure of the hemoglobin molecule. Where does the oxygen come from? In the lungs, the oxygen **coordinates** to the hemoglobin in a red blood cell—that is, forms a coordinate covalent bond to it. Under slightly different conditions in the cell, the oxygen is released in the reverse process and becomes available for respiration.

Coordinates – Forms a coordinate covalent bond.

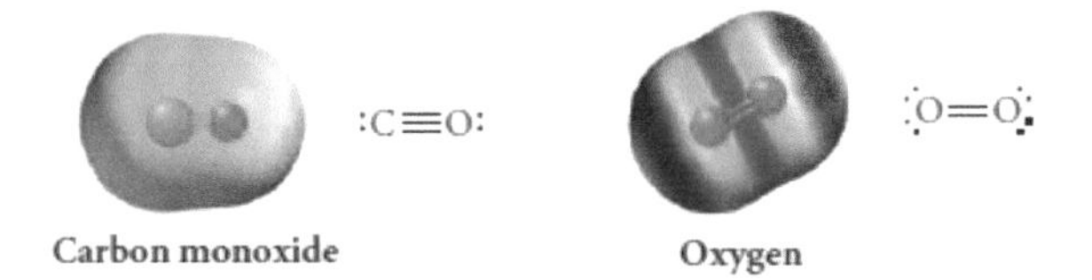

Carbon monoxide can also coordinate to hemoglobin. In fact, it does so more strongly than oxygen. A close look at the structure that results indicates that the lone pair of electrons on the carbon forms the coordinate covalent bond with the iron in hemoglobin. Equilibrium principles tell us that because the equilibrium constant for the reaction with CO is much greater (about 200 times greater) than that for the reaction with oxygen, the presence of small amounts of CO in the body limits the amount of O_2 that can be carried by hemoglobin. Excessive amounts of carbon monoxide in the body can result in suffocation. The National Institute of Occupational Safety and Health (NIOSH) cites an immediate danger to life with an exposure of 1500 parts per million of CO in air and specifies an 8-h exposure limit of 35 ppm in air.

$$\text{HbFe}\ :C\equiv O: \longrightarrow \text{HbFe}-C\equiv O:$$

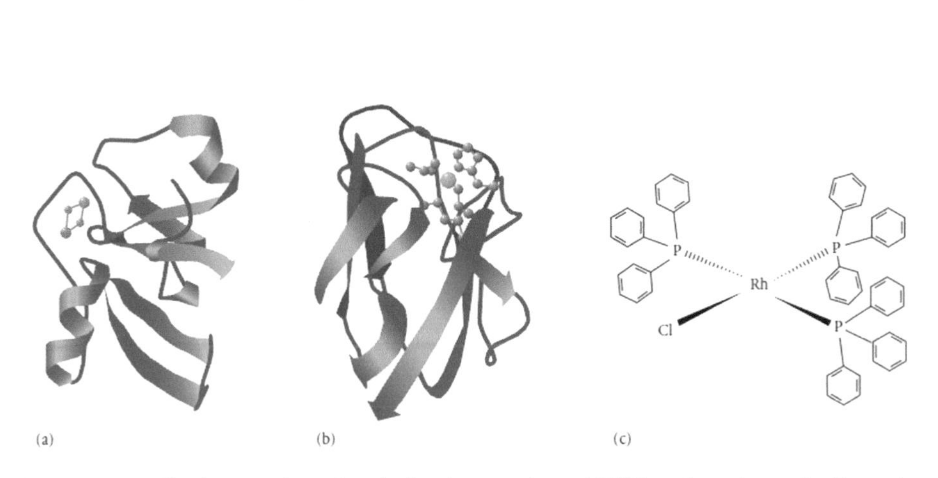

Fig. 19.4 Three important coordination complexes: ferrodoxin, plastocyanin, and Wilkinson's catalyst. **a** An illustration of the protein ferrodoxin indicates the location of an Fe_2S_2 cluster—shown in orange. **b** A drawing of the protein plastocyanin contains a copper ion—green—held in place by the amino acids in the structure of the protein. **c** Wilkinson's catalyst contains a rhodium ion surrounded by triphenylphosphine ligands

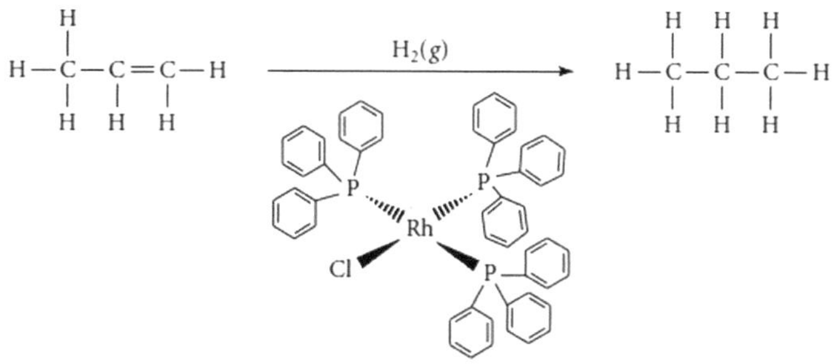

■ Fig. 19.5 The reaction of hydrogen with an alkene in the presence of Wilkinson's catalyst. This reaction can be used to convert low-melting-point vegetable oils into semisolid fats for use in making margarine

The coordinate covalent bond is very common in compounds involving metals and nonmetals. These types of compounds, as shown in ■ Fig. 19.4, are often found at critical reaction centers in biological systems and in many industrial processes. For example, cytochromes and ferredoxins are iron-containing coordination complexes in nonanimal biological systems, where they assist in reduction and oxidation reactions (see ▶ Chap. 18) during photosynthesis. In some plants, nitrogen is converted into a usable form by coordination of an N_2 molecule to a molybdenum ion in the enzyme nitrogenase. In industrial processes, metal ions can be used to form coordinate covalent bonds with small molecules in order to make the small molecule react in certain ways.

Wilkinson's catalyst, shown in ■ Fig. 19.5, is a coordination complex of rhodium used to promote reaction of H_2 with alkenes (in polyunsaturated fats) to make margarine. In another industrial process used to make the class of molecules known as aldehydes (which are used in essential oils—volatile compounds that have odors characteristic of plants, polymer formation, and food additives), carbon monoxide forms a coordination complex with a cobalt atom as it is transformed into the C=O fragment of an aldehyde.

The coordination complexes of transition metals have some particularly interesting properties. The color of the complex, such as that found in rubies, depends on the nature and number of ligands surrounding the metal complex. The magnetism of coordination complexes, such as the magnetism of the iron oxide used in videotape, depends on the nature and number of ligands surrounding the metal. The biochemical reactions that are possible with coordination complexes found in the body are also influenced by the ligands.

19.3 Ligands

A coordination complex such as $[Fe(H_2O)_6]^{3+}$ consists of a central Fe^{3+} metal ion and coordinated water molecules. The species, such as the water molecules, that coordinate to the metal are called ligands. ■ Table 19.1 lists some common ligands. Remember that ligands *donate* a lone pair of electrons to the metal or metal ion to form the coordinate covalent bond. A metal ion will tend to form coordinate covalent bonds with a specific number of ligands, often either four or six.

Example 19.1—Identifying Possible Ligands

Which of the following could act as ligands in forming coordination complexes?

a. CH_4
b. N_3^-
c. BH_3
d. CS_2
e. Li^+

Asking the Right Questions
What is the criterion for a species to be a ligand? Which of these species meets that criterion?

Solution
A ligand donated a lone, or nonbonded, pair of electrons to the central atom. Which of these species has a nonbonded pair of electrons to donate? Species (b) and (d) each has a nonbonded pair of electrons, and, therefore, could be ligands.

Are the Answers Reasonable?
Methane, borane, and the lithium ion do not have a lone pair of electrons, and are not found as ligands, while the other compounds are ligands, so the answers make sense.

Table 19.1 Selected common ligands and their names

The atoms **shown in bold** *donate a lone pair of electrons to form a coordinate covalent bond*					
Monodentate ligands					
$\mathbf{Cl}^-$	Chloro	$\mathbf{Br}^-$	bromo	$\mathbf{I}^-$	iodo
$\mathbf{C}N^-$	Cyano	$\mathbf{N}O_2^-$	nitro	$N\mathbf{O}_3^-$	nitrato[a]
$\mathbf{S}CN^-$	Thiocyanato	$N\mathbf{O}_2^-$	nitrito	$\mathbf{O}SO_3^{2-}$	sulfato
$\mathbf{S}SO_3^{2-}$	Thiosulfato	$\mathbf{O}^{2-}$	oxido	$\mathbf{F}^-$	fluoro
$\mathbf{N}H_3$	Ammine	$H_2\mathbf{O}$	aqua	$\mathbf{N}O$	nitrosyl
$\mathbf{C}O$	Carbonyl	$\mathbf{O}H^-$	hydroxo		
Polydentate ligands					
$H_2\mathbf{N}CH_2CH_2\mathbf{N}H_2$		ethylenediamine (en)			
$(^-\mathbf{O}OCCH_2)_2\mathbf{N}CH_2CH_2\mathbf{N}(CH_2CO\mathbf{O}^-)_2$		ethylenediaminetetracetato (EDTA)			
$^-\mathbf{O}OCCO\mathbf{O}^-$		oxalate (ox)			

[a] can be monodentate or bidentate

What If…?

…a friend suggested the compound shown below? Given our discussion to this point, could this coordination complex exist? Explain.

$Co(CN)_5^{3-}$

(Ans: Yes, because each cyanide ion has a lone pair on the nitrogen atom to donate to the central cobalt atom.)

Practice 19.1

Indicate the formula of a coordination complex that might result from the combination of these metals and ligands.

a. Fe^{2+} and $6Cl^-$
b. Ni^{2+} and $4NH_3$
c. Zn^{2+} and $6H_2O$

Some of the molecules and ions that are commonly observed to act as ligands are shown in Fig. 19.6. Oxygen, nitrogen, sulfur, and phosphorus atoms in molecules are common donor atoms in ligands because they possess a lone pair of electrons that can be donated to the metal center. Some ligands have nonbonded pairs of electrons in different locations throughout their structure. Often, any of these pairs of electrons can be used to create a coordinate covalent bond. In some cases, only one specific pair of electrons is used to create the bond. For instance, cyanide (CN^-) usually binds via the pair of electrons on the carbon atom. It does so in the "Prussian blue" dye used in dyes and paints, where the complex $[Fe(CN)_6]^{4-}$ is formed.

Ligands can also use more than one nonbonded pair of electrons to bond to a metal center. If more than one atom has a nonbonded pair of electrons, it may use each of those pairs to form independent coordinate covalent bonds to the metal. In such cases, if the ligand forms two bonds to the metal, we say it is **bidentate** ("two-toothed"); we call it **tridentate** ("three-toothed") if it forms three bonds, and so on. For example, the bidentate oxalate ion C_2O_4 can bind to the iron(III) ion in this fashion, as shown in Fig. 19.7. Bathroom stain removers often contain oxalate salts, because the oxalate ion will coordinate to iron ions in rust and help wash the stain away. Another common bidentate ligand that donates more than one pair of electrons is ethylenediamine $H_2NCH_2CH_2NH_2$ (often abbreviated en) in which each nitrogen has a lone pair of electrons that can be donated to the same metal ion.

Bidentate – Capable of forming two coordinate covalent bonds to the same metal center.

Tridentate – Capable of forming three coordinate covalent bonds to the same metal center.

Fig. 19.6 Examples of ligands. The lone pairs on these ligands can be used to create a coordinate covalent bond to a metal

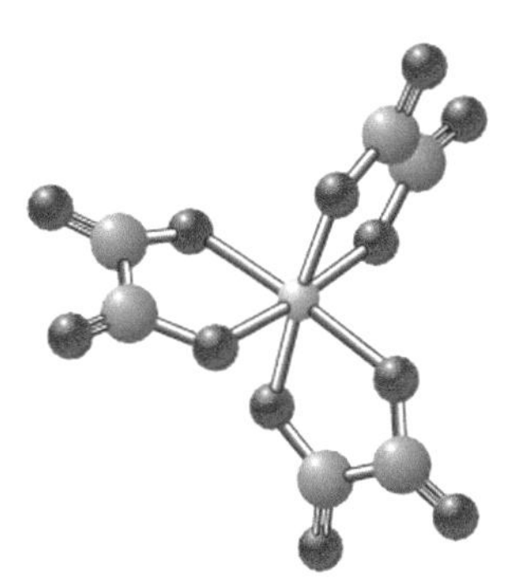

Fig. 19.7 The oxalate ion is a bidentate ion. Each oxalate is capable of donating lone pairs of electrons from two different atoms to form two coordinate covalent bonds. Three oxalates can bind to one iron(III) ion

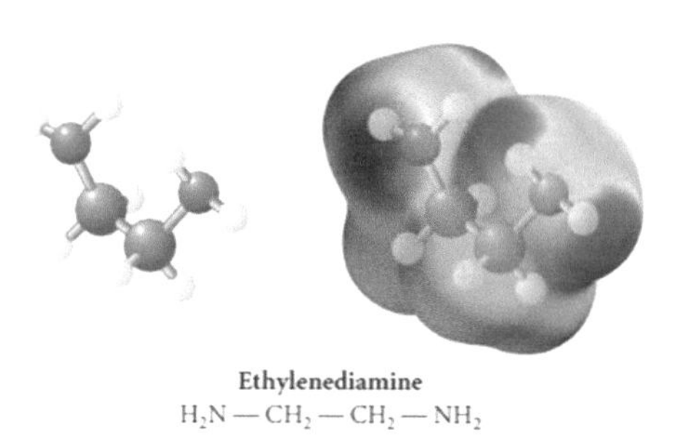

The **polydentate** ligands (those that form more than one coordinate covalent bond to the metal center) are known as **chelates** (from the Greek *chele* for "claw") because of the way they clamp onto the metal center and bond tightly. For example, the porphyrin ligand is a planar, **tetradentate** ligand found in hemoglobin, vitamin B12, and chlorophyll, as shown in Fig. 19.8. The porphyrin chelates to iron in hemoglobin, cobalt in vitamin B12, and magnesium in chlorophyll. The chelates are so good at holding the metal firmly within their grasp that it is difficult to remove the iron from hemoglobin without destroying the hemoglobin molecule itself. Note that the porphyrin leaves two sites on the metal open so that it may bind to other ligands. It is the open locations on the metal within the porphryin chelate that make chemical reactions possible.

Polydentate – A ligand that contains two or more lone pairs on different atoms, each forming a coordinate covalent bond to a metal center.

Chelate – A polydentate ligand that forms strong metal–ligand bonds.

Tetradentate – Capable of forming four coordinate covalent bonds to the same metal center.

The most important industrial ligand, discussed in ▸ Chaps. 15 and 16, is ethylenediaminetetraacetate, which is abbreviated $EDTA^{4-}$. The structure of this tetraanion, shown in Fig. 19.9, illustrates that both nitrogen atoms, and one oxygen atom from each end of the compound, have lone pairs of electrons. It acts as a chelate by forming up to six coordinate covalent bonds to a metal center, depending on the size of the metal ion. For example, $EDTA^{4-}$ forms six bonds to calcium ions, as shown in Fig. 19.10. The exceptionally large formation constant for the reaction of $EDTA^{4-}$ with many metal ions illustrates that the formation of these bonds is quite favorable.

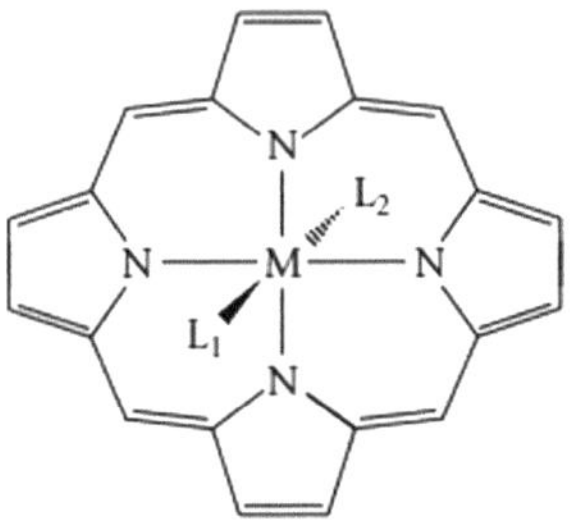

■ Fig. 19.8 The basic porphyrin structure, showing a metal at the center coordinated in six different positions. Compare this structure to that found in the hemoglobin protein from ■ Fig. 19.1

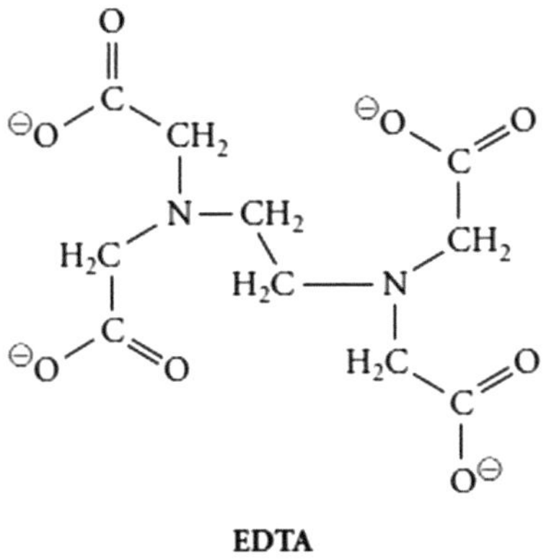

■ Fig. 19.9 $EDTA^{4-}$. The four negatively charged oxygens and the two nitrogens can donate their lone pairs to a metal. This ligand is hexadentate

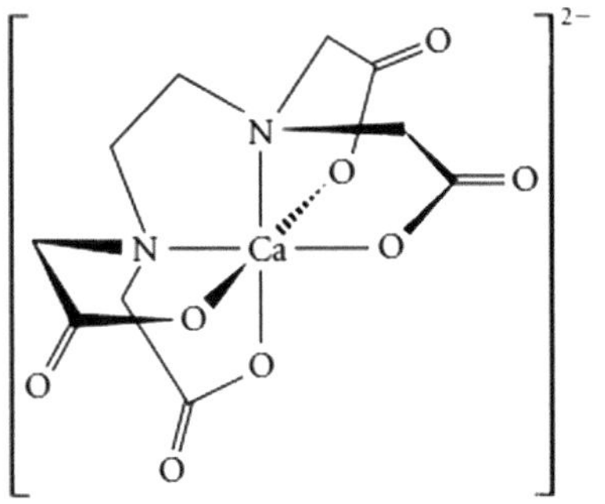

■ Fig. 19.10 $EDTA^{4-}$ coordinated to calcium ion

19.4 Coordination Number

The number of donor atoms to which a given metal ion bonds is known as its **coordination number**. Coordination numbers of 4 and 6 are most common, but coordination numbers as low as 2 and as large as 8 are not unusual. For example, in the film development process, excess silver ion is removed from photographic film by using the thiosulfate ion ($S_2O_3^{2-}$), which forms a coordination complex with a coordination number of 2 (see ■ Fig. 19.11). In another example, cisplatin, a prominent anticancer drug, has at its center a platinum(II) ion with a coordination number of 4, as shown in ■ Fig. 19.12. As we discussed earlier, when an Fe^{3+} ion is present in water, it is surrounded by 6 water molecules in a complex, $[Fe(H_2O)_6]^{3+}$. And iron's coordination number is 6. If iron forms a complex with the oxalate ligand, $[Fe(C_2O_4)_3]^{3-}$, its coordination number is still 6, because each oxalate ligand donates two electron pairs.

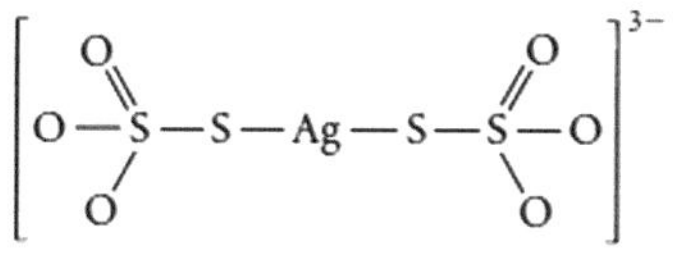

$$\left[O{-}\overset{O}{\underset{O}{S}}{-}S{-}Ag{-}S{-}\overset{O}{\underset{O}{S}}{-}O\right]^{3-}$$

Fig. 19.11 A two-coordinate complex of silver

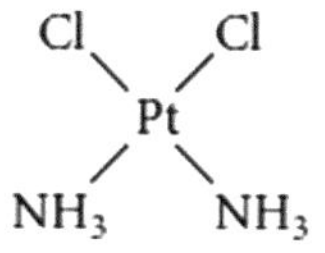

$$\begin{matrix} Cl & & Cl \\ & Pt & \\ NH_3 & & NH_3 \end{matrix}$$

Fig. 19.12 Four-coordinate cisplatin

Coordination number – The number of coordinate covalent bonds that form in a complex.

What makes a metal ion have a certain coordination number? It is primarily a function of the nature of metal ions. Large metal ions have space around them within which more ligands can fit. Not as many ligands can fit around a small metal ion. Other factors, such as the charges on the ligands and metal and the electron configuration of the metal ion, determine the most likely coordination number for a metal ion.

How Do We Know?—What is the Nature of the Structure, Bonding, and Reactivity in Cisplatin?

Cancer is often a devastating disease. Research has made significant strides in understanding the disease in its many forms, including advances in detection, treatment, and care that have improved the quality of life for its victims. One prominent drug used in cancer treatment, especially testicular and ovarian cancers, is cisplatin, a coordination complex of platinum(II) that is shown in Fig. 19.12. The anticancer activity of cisplatin was discovered as a result of careful observation and basic interpretation of an experiment that had nothing to do with fighting cancer. But good science nonetheless led to the discovery.

Dr. Barnett Rosenberg at Michigan State University investigated the effect of an electric current on a culture of *Escherichia coli* bacteria. This bacterium, commonly found in the gastrointestinal tracts of many living creatures, is often used in initial biochemical studies because of its ready availability. Rosenberg observed that when an electric current was applied to solutions of the bacteria, cell division in the vicinity of a platinum electrode was inhibited. Studying this interesting result further, he recognized that it was not due to the electric current, which was the focus of the investigation. Noting that a compound known as *cis*-diamminedichloroplatinum (cisplatin) was being produced in the vicinity of the platinum electrode used for his experiment, Rosenberg reasoned that the cisplatin must be responsible for inhibiting the cell division. It was later determined that this compound, when given to cancer patients, can significantly reduce the size of their tumors and can even cause the disease to go into remission. Since 1970, the survival rate for testicular cancer patients has increased from 10% to over 95%.

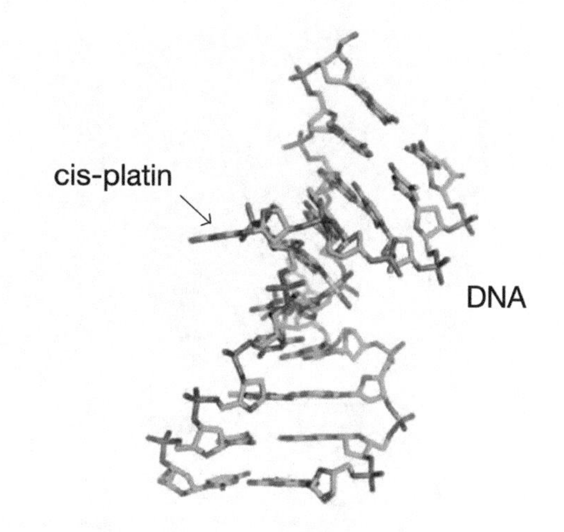

How does cisplatin exhibit this remarkable biological activity? Cisplatin is a four-coordinate, square planar complex, as is commonly observed for many metal complexes with eight *d* orbital electrons. The platinum(II) metal center is fairly unreactive, which allows the neutral cisplatin complex to remain largely intact through injection, circulation, and penetration into the nucleus of a cancerous cell. The chloro ligands are eventually replaced by water molecules. This provides an opportunity for the platinum center to coordinate to DNA molecules. Ultimately, the platinum center binds to a nitrogen atom from each of two guanine units in a single DNA strand (see ▶ Chap. 22). The geometry of the ligands around the platinum center is vital to this biological activity. In fact, the chloro ligands must be *cis* to each other—on the same side of the complex. Once coordinated to the guanine nitrogens, the inert platinum center remains bound, tying two points on the DNA strand together. Such binding inhibits reproduction of DNA during cell division and restricts use of the DNA for normal cellular functions. Ultimately these effects hinder the growth of the cancer cells.

19.5 Structure

Mercury in the environment poses a particular threat to life, both to aquatic species and to those species that consume them. The most insidious form of mercury is dimethyl mercury, $[Hg(CH_3)_2]$. Transformed from mercury metal in the environment, dimethyl mercury is a very toxic product. The mercury(II) ion in this complex coordinates with the methyl groups, $:CH_3^-$, to form the complex $[H_3C—Hg—CH_3]$. Given the VSEPR rules, the complex is *linear*, with a C—Hg—C bond angle of 180°. This structure is typical of metals with a coordination number of 2.

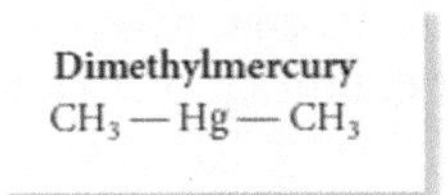

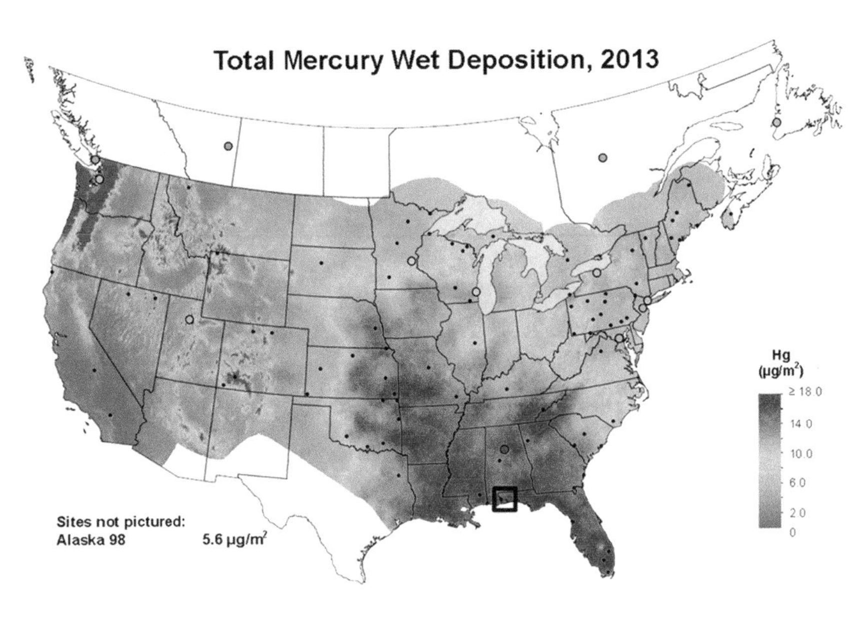

▶ http://nadp.sws.uiuc.edu/maplib/pdf/mdn/hg_Conc_2013.pdf

Is there only one geometry for a given coordination number? We observe that two geometries are common for metals that have a coordination number of 4. Those geometries, **tetrahedral** and **square planar**, are shown in ◘ Fig. 19.13. The tetrahedral geometry is the one predicted by VSEPR (see ▶ Sect. 9.4) when there are four electron sets around the central atom. In this geometry, the bond angles are close to 109°. The square planar geometry arises as a consequence of the electron configuration of the metal ion. Square planar geometry, with bond angles of 90°, is often observed for metal ions such as $Ni^{2+}$and Pt^{2+} that have an outer nd^8 electron configuration. That is, they have eight electrons in their outermost *d* orbitals. An example of the square planar geometry is found in the anticancer drug cisplatin, which we introduced in the previous section. The square planar geometry is crucial in the interaction of this drug with DNA.

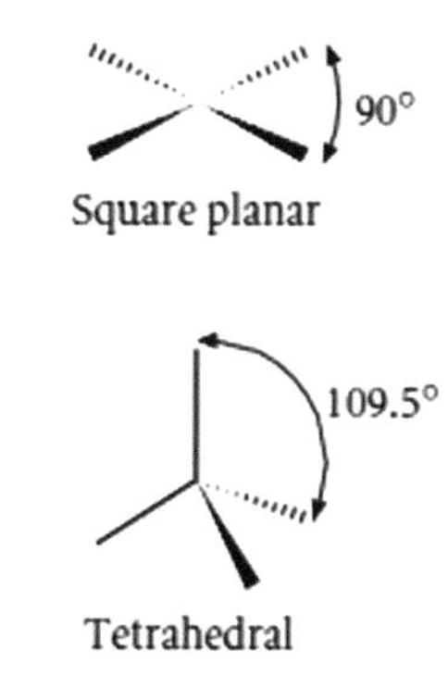

Fig. 19.13 Common geometries with a coordination number of 4

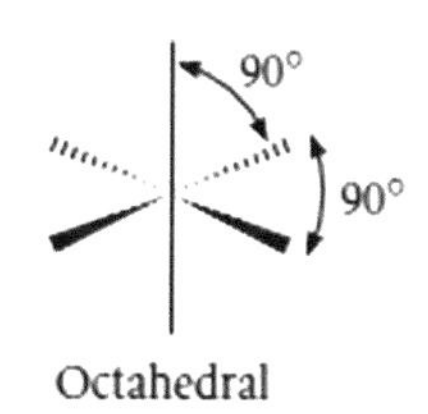

Fig. 19.14 The octahedral geometry; all six positions in this geometry are identical

Square planar – The geometry indicated by four electron groups (lone pairs and/or bonds) positioned symmetrically in a plane around a central atom such that the bond angles are 90°.

Tetrahedral – The geometry indicated by four electron groups (lone pairs and/or bonds) positioned symmetrically around a central atom such that the bond angles are 109°.

The final geometry common to the coordination compounds is the **octahedral** complex, an example of which is shown in Fig. 19.14. In this geometry, the six ligands occupy equivalent positions around the metal center. For example, hexaaquairon(III), $[Fe(H_2O)_6]^{3+}$, includes six water ligands symmetrically surrounding the central Fe^{3+} ion. We can think of the ligands as sitting at positions on each end of the coordinate x, y, and z axes with bond angles of 90°. The iron in hemoglobin, shown in Fig. 19.1, occupies a nearly octahedral coordination environment with four sites occupied by nitrogen atoms from a porphyrin ligand. One site is occupied by a histidine (an amino acid that makes up one part of the hemoglobin molecule), and the final site is occupied by a molecule of oxygen.

Octahedral – The geometry indicated by six electron groups (lone pairs and/or bonds) positioned symmetrically around a central atom such that the bond angles are 90°.

Example 19.2—Coordination Number and Geometry

Give the coordination number of the central metal and predict the geometry of the following compounds.

a. $[Al(H_2O)_6]^{3+}$
b. $[Ag(CN)_2]^-$
c. $Na_2[PdCl_4]$
d. $[Co(ox)_3]^{3-}$

Asking the Right Questions?

What is the ligand, and what is the central metal to which it is attached? What is the importance of the coordination complex compared to the coordination compound?

Solution

In each species, the central atom is a metal, and the ligands are attached to it. The species in brackets represents the coordination complex. It is only the coordination complex that we should consider when answering a question about the coordination number. Remember that oxalate (ox) is a bidentate ligand.

Once we know the coordination number of the complex, we can determine the geometry of the complex. For coordination number 2 or 6, the geometry is linear or octahedral, respectively. For four-coordinate complexes, either tetrahedral or square planar geometry is expected. Remember that metals with eight *d* orbital electrons typically form square planar complexes.

	a	b	c	d
Coordination Number	6	2	4	6
Geometry	octahedral	linear	square planar	octahedral

Are the Answers Reasonable?

We have accounted for both the coordination number and the resultant geometry of each complex, noting especially the effect of the electron configuration of the palladium ion on the geometry of its complex. The answers are reasonable on the basis of the correlation between the coordination number and the geometry.

What If…?

…we have $[CuCl_4]^{2-}$? What is the geometry of the complex?

(*Ans: Tetrahedral*)

Practice 19.2

Determine the coordination number and geometry for each of the following compounds.

a. $[Cr(CO)_6]$
b. $[Fe(H_2O)_6]^{2+}$
c. $K_2[TiCl_4]$
d. $[Cu(NH_3)_4]^+$

19.6 Isomers

Does the formula of a coordination compound relate the exact three-dimensional structure of the compound? As we'll discuss in ▶ Chap. 21, the formula tells us very little, if anything, about the structure, identity, or properties of a compound. If the atoms can be arranged in different ways to create compounds with a different structure (while maintaining the same chemical formula), we say that the compounds are **isomers** of each other. Each arrangement of atoms would exist as a different compound, with different properties, but have the same chemical makeup. In coordination compounds, there are many different types of isomers that are possible. They are divided into two main categories on the basis of whether there is a change in the structure or in the geometry of the complex. Structural differences are observed in the linkage isomers, ionization isomers, and coordination sphere isomers. Differences in geometric isomers arise from the nonspecific nature of the formation of a coordinate covalent bond between a ligand and a metal center.

Isomers – Compounds that have the same chemical formula but different structure or geometry.

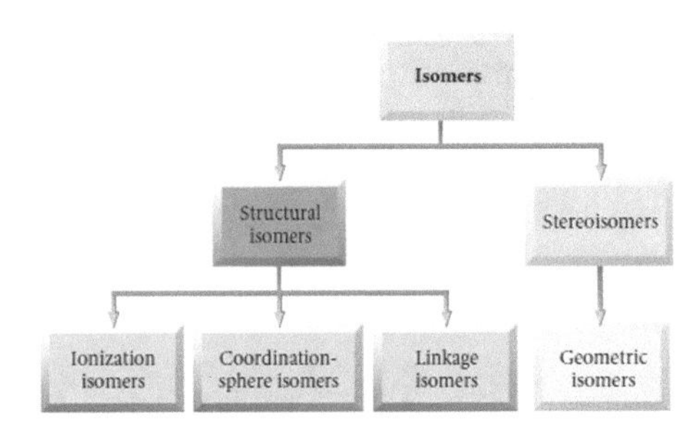

Some ligands coordinate to a metal through one of several donor atoms. For example, the thiocyanate ion (SCN^-) can coordinate to a chromium(III) ion either through the nonbonded pair of electrons on the nitrogen or through the nonbonded pair of the electrons on the sulfur atom:

$[Cr(H_2O)_5 — SCN]^{2+}$ $[Cr(H_2O)_5 — NCS]^{2+}$

These two different compounds are called **linkage isomers**. They differ only in the atom that participates in the coordinate covalent bond with the metal.

Linkage isomers – Isomers that differ in the point of attachment of a ligand to a metal.

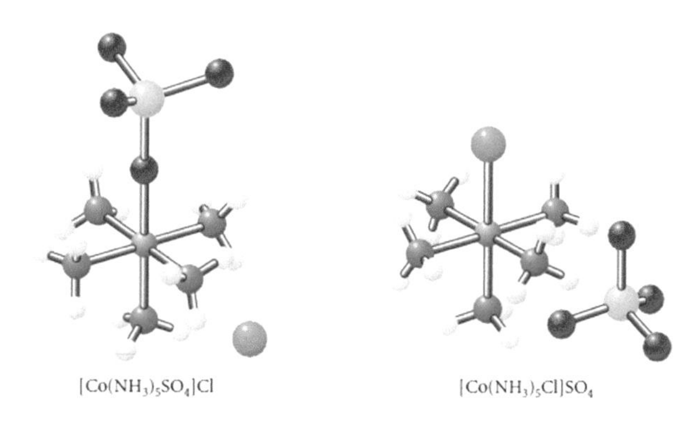

Fig. 19.15 Ionization isomers

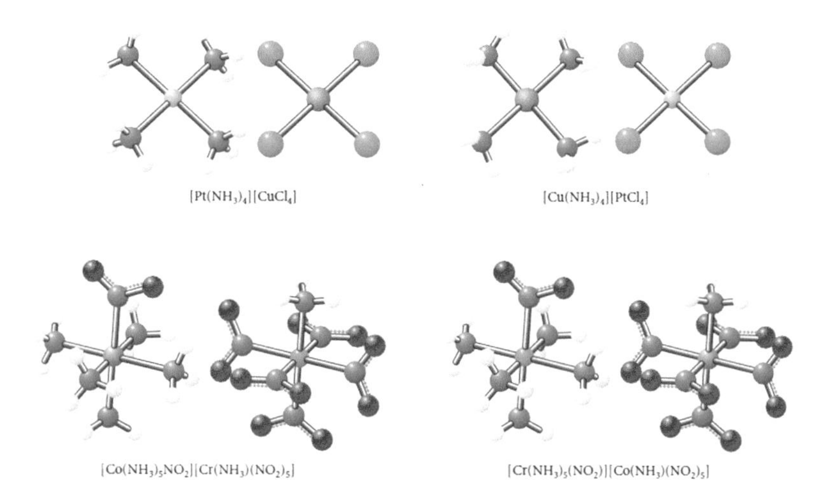

Fig. 19.16 Examples of coordination sphere isomers

Another isomer that is common among the coordination compounds is known as the **ionization isomer**. These are compounds in which a ligand and an ion (of the same charge as the ligand) exchange roles. For example, examine the cobalt complexes shown in Fig. 19.15. Note that the counter ion and a ligand have changed positions in the two compounds. It is that change that makes these two compounds ionization isomers.

Ionization isomers – Isomers that differ in the placement of counter ions and ligands.

Coordination sphere isomers contain different ligands in the coordination spheres of cations and anions. In the coordination sphere isomers, both the cation and anion are coordination complexes, but the metal centers have changed positions. Examples are shown in Fig. 19.16. Note that each of these complexes includes a cation that is a coordination complex and an anion that is a coordination complex. The metal ions in each case have switched location from the cation to the anion.

Coordination sphere isomers – Substances that contain different ligands in the coordination spheres of complex cations and complex anions.

Geometric isomers, the other main class of isomers, are substances in which all of the atoms are attached with the same connectivity, or bonds, but the geometric orientation differs. Square planar four-coordinate complexes of the general composition MA_2B_2 and octahedral complexes of the composition MA_2B_4, where M is the metal and A and B are different ligands, can exist as geometric isomers labeled either *cis* or *trans*. As shown in Fig. 19.17 for the platinum complex $[Pt(NH_3)_2Cl_2]$, the ***cis* isomer** has two identical ligands next to each other. The ***trans* isomer** has two identical ligands on opposite sides of the metal. Octahedral complexes can exist as *cis* or *trans* isomers in complexes such as $[Co(NH_3)_4Cl_2]^+$ (Fig. 19.18). In fact, in the days when Alfred Werner (1866–1919) first defined the nature of coordination complexes (for which he won the 1913 Nobel Prize in chemistry), he noted

cis-Diamminedichloroplatinum(II) *trans*-Diamminedichloroplatinum(II)

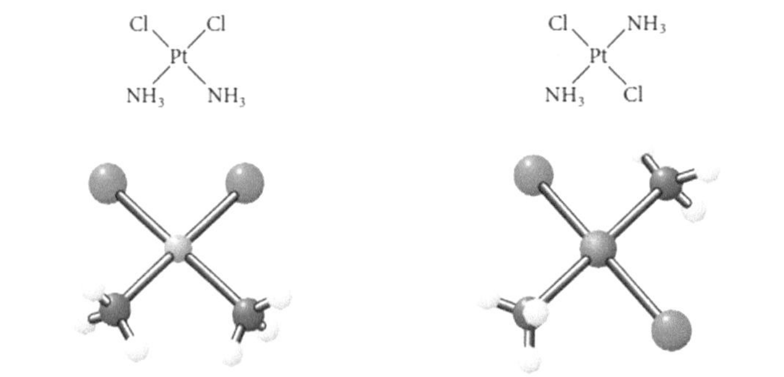

Fig. 19.17 *Cis* and *trans* isomers are geometric isomers. In the *cis* isomer shown here, the two chloro ligands are on the same side of the complex. The chloro ligands are on opposite sides in the *trans* complex

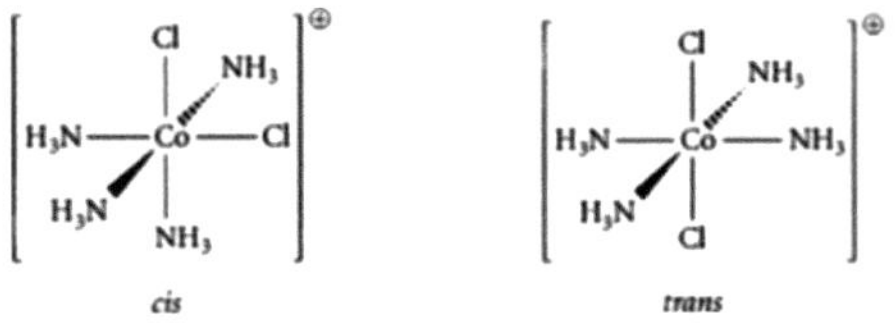

Fig. 19.18 *Cis* and *trans* isomers are also possible in octahedral complexes

Table 19.2 Isomers of coordination complexes

Isomer	Example	Explanation
Ligand	$[Co(NH_3)_5—ONO]^+$ $[Co(NH_3)_5—NO_2]^+$	Ligand bonds through different donor atoms
Ionization	$[Pt(NH_3)_3Cl]Br$ $[Pt(NH_3)_3Br]Cl$	Anion and ligand interchanged
Coordination sphere	$[Co(en)_3][Cr(ox)_3]$ $[Cr(en)_3][Co(ox)_3]$	Distribution of coordinating ligands differs
Geometric	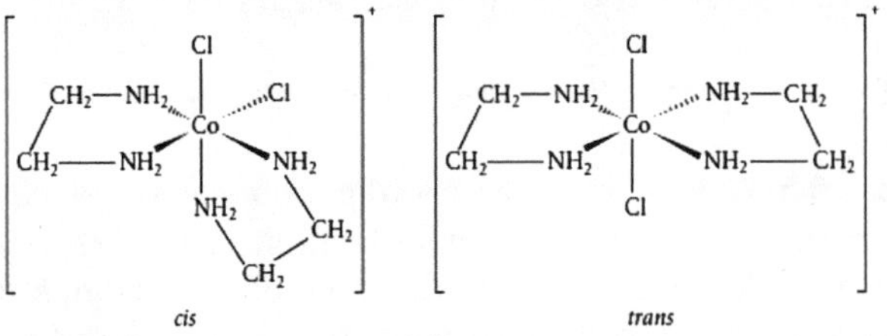 *cis* *trans*	Orientation of ligands around the metal center differs

the two isomers of $[Co(NH_3)_4Cl_2]Cl$. He called one the Praseo complex and the other the Violeo complex. These two isomers were identified easily and named Praseo and Violeo due to the dramatic difference in their color. To be sure, the color of a coordination complex is markedly influenced by the geometric arrangement of the ligands around the metal center. Table 19.2 gives a summary of the various isomers common among coordination compounds.

Geometric isomers – Substances in which all of the atoms are attached with the same connectivity, or bonds, but the geometric orientation differs.

cis Isomer – An isomer containing two similar groups on the same side of the compound.

trans Isomer – An isomer containing two similar groups on opposite sides of the compound.

▶ Here's What We Know so Far

- Coordination complexes are present in simple metal ions in solution and as the reaction center in many biological molecules.
- A Lewis base that donates a lone pair of electrons to a metal to form a coordinate covalent bond forms a coordination complex.
- The coordination numbers of various metal centers are commonly observed to be 2, 4, or 6.
- Common coordination geometries are linear, tetrahedral, square planar, and octahedral.
- A variety of isomer classes are observed for metal complexes. ◀

19.7 Formulas and Names

Coordination compounds are quite varied in their structure, bonding, ability to form isomers, and other features. For this reason, assigning specific names to these compounds must be done with care. Just as in naming binary compounds (▶ Chap. 2) and organic compounds (▶ Chap. 22), IUPAC rules will guide us in constructing the proper name.

19.7.1 Formulas

Coordination complexes (the metal center and the ligands) can be neutral, anionic, or cationic in nature. To ensure continuity from one structure to another, we often write the formula of a coordination compound in accordance with a set of rules. Those rules are outlined in ◘ Table 19.3.

19.7.2 Nomenclature

Systematic names for coordination compounds are constructed in accordance with a set of rules. Those rules are outlined in ◘ Table 19.4. Let's use these rules to name a few examples. Consider that we are interested in naming the neutral complex $[PtCl_2(NH_3)_2]$. Step 1 from ◘ Table 19.4 indicates that we should name the compound using the rules in step 3.

The ligands are named alphabetically before we name the metal. Thus the ammine ligand is named before the chloro ligand. Moreover, there are two of each of these ligands, so we should write down

diamminedichloro

without any spaces. Then, in step 3d, we write the name of the metal and its charge in parentheses immediately afterward. Because the charge on the entire complex is zero, and the complex is neutral, we do not add a suffix to the name:

diamminedichloroplatinum(II)

The name we have created for the formula indicates the number and type of each ligand, and the metal and its oxidation state. If we knew the three-dimensional arrangement of the atoms in this complex, we could also include that information in the name. For example, if we knew that the structure indicated a *cis* arrangement of the atoms, we could designate that by writing *cis*-diamminedichloroplatinum(II). Without that knowledge, though, we can provide only the name of the formula.

◘ **Table 19.3** Rules for writing formulas for coordination compounds

1. The metal is written first, followed by the ligands
2. Anionic ligands are written before the neutral ligands, each in alphabetical order
3. The complex is enclosed in brackets
4. Polyatomic ligands, such as NH_3 or NO_2^- are enclosed in parentheses
5. Ionic compounds containing coordination complexes are written in the traditional way: cation on the left, anion on the right

Table 19.4 Rules for naming coordination complexes

1. Name the cation, and then the anion. If the complex is neutral, name it using step 3
2. Is the cation a complex ion?
3. Yes—name it using these rules, and then skip to step 5. No—skip to step 4
   a. Name the ligands first using Table 19.1
   b. Name the ligands in alphabetical order. Anionic ligands often end in "o," as in bromo, hydroxo, and sulfato
   c. If more than one ligand of the same type is present, a prefix indicates the number of units. For simple ligands, the prefixes di-, tri-, tetra-, penta-, and hexa- are used. For complex ligands, the prefixes bis-, tris-, tetrakis-, pentakis-, and hexakis- are used
   d. Name the metal last, with its oxidation number in parentheses in Roman numerals
      (1) If the complex is negatively charged, the name of the metal ends in -ate
      (2) If the complex is positively charged or neutral, no suffix is added to the metal name
4. Name the cation using the conventions described in ▶ Chap. 3
5. Is the anion a complex ion?
6. YES—Name it using the rules in step 3, and then stop. NO—Go to step 7
7. Name the anion using the conventions described in ▶ Chap. 2, and then stop

If we wish to name $[Co(NH_3)_6]Cl_3$, we can do so by following the rules. In step 1, we note that the compound is made up of a cationic complex and some anionic counter ions. Step 2 in Table 19.4 directs us to write the name of the complex (the cation). The ligands are named first, like this:

hexaammine

Then in step 3d we name the metal and its charge. Again, we do not add a suffix, because the complex is a cation:

hexaamminecobalt(III)

Finally, steps 6 and 7 tell us to add the counter ions by naming them as we did in ▶ Chap. 2. Note that we don't add a prefix to their name.

hexaamminecobalt(III)chloride

An anionic complex is not much different. Suppose we wished to name $K_2[NiCl_4]$. In step 1, we note that the anion is a complex and the cation is a counter ion. Accordingly, we name the cation first.

potassium

The complex anion is named using step 3. The ligands go first (note that there are four of them):

potassium tetrachloro

And the metal is part of an anion, so it gets the suffix *-ate*:

potassium tetrachloronickelate(II)

$K_2[NiCl_4]$ $[Co(NH_3)_6]Cl_3$ $[Co(NO_2)_2(NH_3)_4]_2SO_4$

Example 19.3—Naming Compounds

Give the name for each of these compounds.

a. $[Cr(NH_3)_2(en)_2]SO_4$
b. $(NH_4)_3[Fe(CN)_6]$

Asking the Right Questions

How do you approach naming coordination compounds? What are the important things to know about the ligands and the metal center?

Solution

Naming coordination compounds requires us to identify the oxidation states of the metal center in addition to understanding and knowing the charges on the individual ligands and ions. To answer this question accurately, we separate the compounds into their two halves (cation and anion). For the half containing the metal ion, determine its oxidation state, and then use the rules in Table 19.4 to name it.

a. diamminebis(ethylenediamine)chromium(II) sulfate
b. ammonium hexacyanoferrate(III)

Are the Answers Reasonable?

Do you have the cation name first? Are the ligands properly named? does the metal center have the proper name and oxidation state? These are the things that combine to let us know that our answers are reasonable.

What If…?

…we have hexaamine cobalt(III) chloride? What is the formula for this coordination complex? What about tris(ethylenediamine)cobalt(III)chloride?

(Ans: $[Co(NH_3)_6]Cl_3$, $[Co(H_2NCH_2CH_2NH_2)_3Cl_3]$)

Practice 19.3

Name each of these compounds.

a. $[CrCl_2(NH_3)_4]_2SO_4$
b. $Na_2[Ni(CN)_4]$

Draw the formula for each of the these coordination compounds.

c. calcium hexafluoroferrate(II)
d. tetraamminedicarbonylmanganese(II) sulfate

19.8 Color and Coordination Compounds

Many painters create images that both catch our eye and induce a feeling in our mind. The goal of the painter is often to portray more than just a staged scene. Van Gogh's painting Starry Night (see Fig. 19.19) evokes a certain sense of wonder. The effect is generated by the brush strokes and the colors applied to the canvas. Are chemical compounds responsible for the color of paint? The answer is a resounding yes. Colors used in paintings are often constructed from ancient formulations that include compounds of transition metals. One of the general characteristics of transition metals is the color of their compounds. Compare table salt (NaCl), baking soda ($NaHCO_3$), chalk ($CaCO_3$), and Epsom salts ($MgSO_4$) to ruby (with Cr_2O_3), emerald (with Cr_2O_3), and rust ($Fe_2O_3{\cdot}nH_2O$, or hematite). Chromium was so named because of the variety and brilliance of the colors of its compounds. What causes compounds that contain transition metals to have such striking colors?

19.8.1 Transition Metals and Color

Because color is a common feature of transition metal compounds, it seems reasonable that there must be some common characteristics that give rise to their colors. The presence of color in most transition metal compounds can be attributed to the presence of partially filled *d* orbitals in the compounds and to the influence of the coordi-

Fig. 19.19 Van Gogh's Starry Night uses the colors of transition metals to evoke the feeling of magic in the night sky. *Source* ▶ https://upload.wikimedia.org/wikipedia/commons/thumb/e/ea/Van_Gogh_-_Starry_Night_-_Google_Art_Project.jpg/970px-Van_Gogh_-_Starry_Night_-_Google_Art_Project.jpg

nation environment on the energies of the *d* orbitals. Let's see how the color of transition metal compounds is related to fundamental atomic structure principles.

We perceive color when our eye detects light rays that differ from the ordinary distribution of those present in white light. For example, a sweater appears blue if white light strikes it, colors complementary to blue are absorbed, and the remaining light is reflected. A glass of fruit punch may appear red if white light strikes it and the colors complementary to red are absorbed. The remaining light, which appears red to our eye, is transmitted through the liquid and gives the liquid its red color. Color, then, is the array of light rays that our eyes observe being reflected from or transmitted through an object.

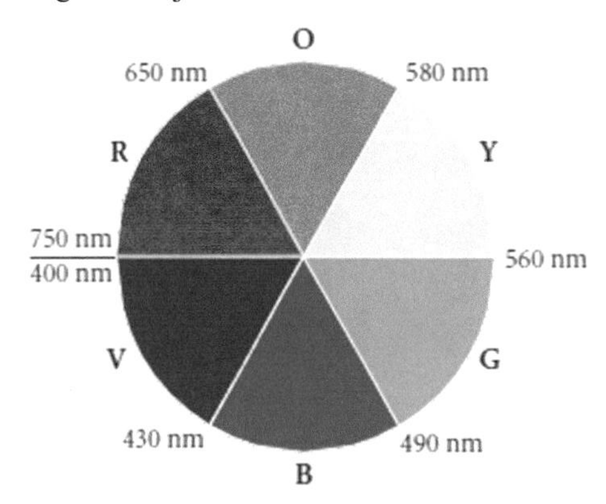

There are many objects that can absorb or transmit certain wavelengths of light but not be "colored." This occurs when the light being absorbed or transmitted is outside of the visible region (400–700 nm) of the electromagnetic spectrum. Ultraviolet (UV) light has shorter wavelengths, typically defined as light in the wavelength range of 200–400 nm. A photon of UV light carries a lot of energy and may be damaging to substances it strikes. Infrared light has longer wavelengths than visible light.

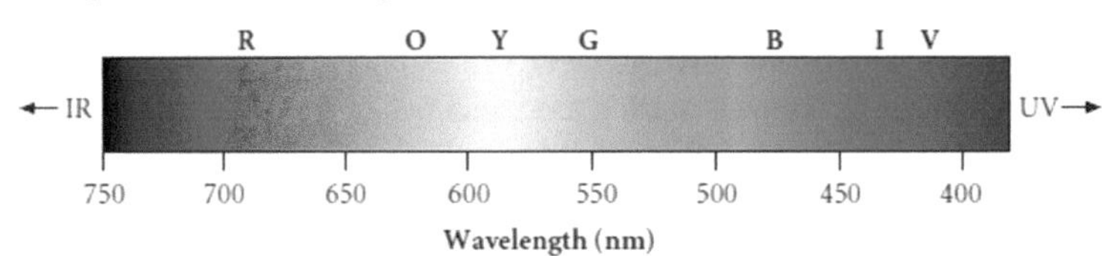

Given our understanding of light, we can begin to ask questions about how a compound can appear to be colored. For example, what characteristic change occurs in a substance when light is absorbed? Answering this question will enable us to understand why lime (CaO) is colorless but rust (Fe_2O_3) is colored. Several core principles are involved.

Absorption of a photon of visible light by a compound results in the excitation of an electron from a low energy orbital to a higher energy orbital within the substance. In order for the photon to be absorbed, the difference in orbital energy levels must match the energy of the photon absorbed. Finally, the excited electron must be able to be excited to the higher energy orbital—that is, the higher energy orbital must not be full. For example, when visible light strikes rust, an electron in a lower energy orbital absorbs a photon of blue light. The electron is excited to a higher energy orbital. (Usually, the high-energy electron dissipates its energy as heat and relaxes back to its original energy level, ready to absorb again!) Orange light is reflected, and that is what our eyes detect. Rust is orange-colored.

19

In many transition metal compounds, of which rust is one, the same process occurs. What electronic transitions take place to allow the absorption of visible light? The *d* orbitals of the transition metal are often arranged such that they have a slight difference in energy. One example is shown in Fig. 19.20. This difference in energy is

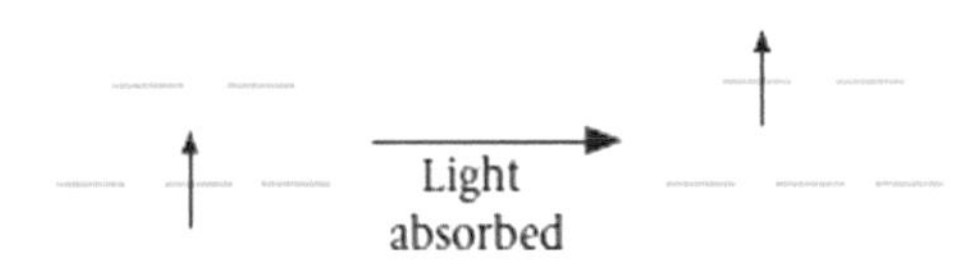

Fig. 19.20 A slight difference in the energy of the five d orbitals on a transition metal in a complex allows visible light to be absorbed. This causes an electron to be excited to the higher-energy d orbitals

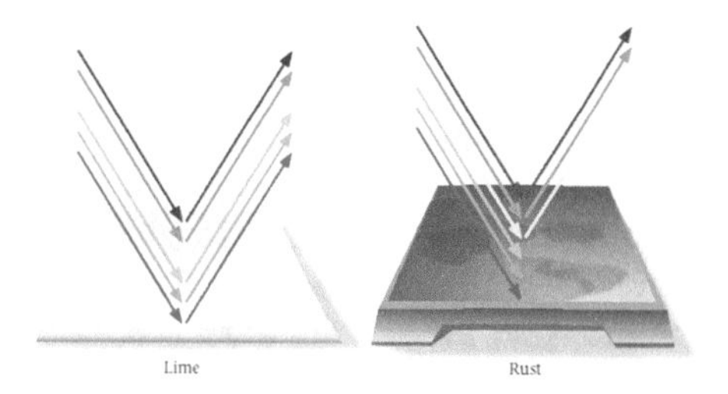

Fig. 19.21 The color of lime and that of rust are a result of the absorption of specific wavelengths of light in the visible region of the electromagnetic spectrum. Calcium ions in lime do not possess electrons in d orbitals and cannot absorb light in the visible region. Iron ions in rust have d orbital electrons and do absorb visible light

roughly that of a photon in the visible region of the electromagnetic spectrum. If the *d* orbitals are partially filled, as they are in the transition metal ions, a photon can cause an electron in the lower energy orbitals to jump to an empty (or partially filled) higher energy *d* orbital.

Let's use this information to compare the colors of rust (Fe_2O_3) and chalk ($CaCO_3$). The iron(III) ion in rust has the electron configuration [Ar]$3d^5$. It contains partially filled d orbitals as its valence orbitals. These *d* orbitals on the metal, surrounded by a field of oxide ions, are split into different energy levels in ways that we will discuss in the next section. When visible light strikes rust, some wavelengths of light are absorbed, and an electron is excited to a higher empty energy level. The calcium ion in lime is isoelectronic with argon. It does not have partially filled *d* orbitals, cannot absorb visible light energy, and reflects all of the visible light to our eyes. It appears to be white, as shown in Fig. 19.21.

19.8.2 Crystal Field Theory

We can understand the nature of the *d* orbital splitting from careful analysis of geometry and *d* orbital shape. In a free, gaseous metal atom or ion, there is no difference in energy among the orbitals of the same sublevel. We say that these orbitals, such as the 3*d* orbitals on gaseous iron, are degenerate. But something happens when we bring other atoms (such as negatively charged anions) up close to the metal. They cause distortions in the orbitals of the metal. In a six- coordinate octahedral complex, for example, ions approach the central metal from each end of the three coordinate axes. Their orientation causes the orbitals on the metal to lose their degenerate nature.

How do ligands cause the energy levels in the d-orbitals to split? Let's look just at the xy plane in an octahedral complex. The octahedral ligands are oriented along the *x* axis and the *y* axis. The *d* orbitals on a metal at the center of our plane are oriented so that one orbital (the d_{xy}) is between the *x* and *y* axes, and the other is lined up with the *x* and *y* axes (see Fig. 19.22).

Which electron would have lower energy, an electron in the $d_{x^2-y^2}$ orbital pointed directly at the negatively charged ligand, or an electron in a d_{xy} orbital, which is pointed between the negatively charged ligands? Because like charges repel, an electron in the $d_{x^2-y^2}$ orbital would be higher in energy. We would represent this on an energy-level diagram as follows:

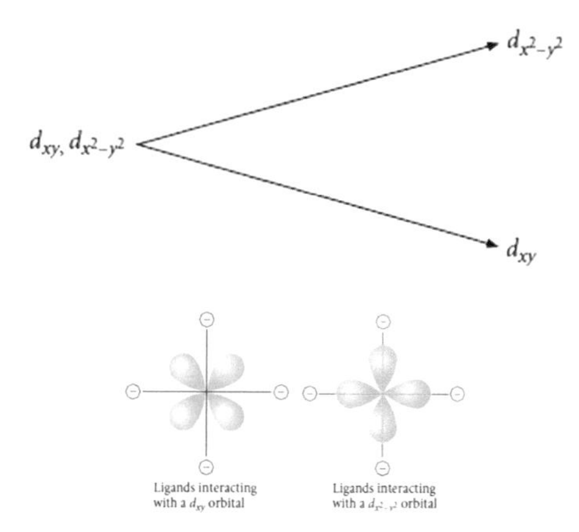

Fig. 19.22 As ligands approach the d orbitals, they affect the relative energy levels of the 5*d* orbitals. Some, like the $d_{x^2-y^2}$ orbital, have an increased energy due to the repulsion of similar charges

19.8.3 Octahedral Crystal Field Splitting

For a metal center in an octahedral crystal field, an electron in the $d_{x^2-y^2}$ orbital would be at higher energy than an electron in the d_{xy} orbital. We can extend this analysis to three dimensions and include all *d* orbitals as shown in ◘ Fig. 19.23. Note that the $d_{x^2-y^2}$ and d_{z^2} orbitals are pointed right at the negatively charged ligands, but the d_{xy}, d_{xz}, and d_{yz} orbitals are pointed between the axes. The d_{xy}, d_{xz}, and d_{yz} orbitals will be at lower energy than the $d_{x^2-y^2}$ and d_{z^2} orbitals. Based on the symmetry of the system, the d_{xz}, d_{yz}, and d_{xy} orbitals are also at the same energy level. Although it is less obvious, the $d_{x^2-y^2}$ and d_{z^2} orbitals are at the same energy level, which is higher than that of the d_{xy}, d_{xz}, and d_{yz} orbitals.

In an octahedral array of anions or ligands, the *d* orbital energies of the central metal are split into a lower energy set (d_{xz}, d_{xy}, d_{yz}) that we label the t_{2g} set and an upper energy set ($d_{x^2-y^2}$, d_{z^2}) that we label the e_g set, as shown in the top half of ◘ Fig. 19.24. The energy difference between the t_{2g} orbitals and the e_g orbitals in the octahedral field, the **crystal field splitting energy**, is given the symbol Δ_o, where the subscript "o" indicates the splitting energy for an octahedral complex. The magnitude of Δ_o, the difference in energy between the t_{2g} and e_g orbitals, depends on the nature of the central metal and the nature of the ligands.

Crystal field splitting energy – The difference in energy between the *d* orbital sets that arises because of the presence of ligands around a metal; symbolized by Δ.

19.8.4 Tetrahedral Crystal Field Splitting

A tetrahedral ligand field can be viewed as one with the metal ion at the center of a cube and ions at alternate corners of a cube, with the coordinate axes going through the centers of the faces of the cube as shown in ◘ Fig. 19.25. In this situation, none of the *d* orbitals of a central atom are pointed at the ions. The d_{xy}, d_{xz}, and d_{yz} lobes are closer to the ions than the $d_{x^2-y^2}$ and d_{z^2} lobes. Thus, the d_{xy}, d_{xz}, and d_{yz} orbitals are at a higher energy than the $d_{x^2-y^2}$ and d_{z^2} orbitals. As shown in ◘ Fig. 19.25, the tetrahedral crystal field splitting is qualitatively the inverse of the octahedral field splitting because the orbitals that were between the ligands in the octahedral ar-

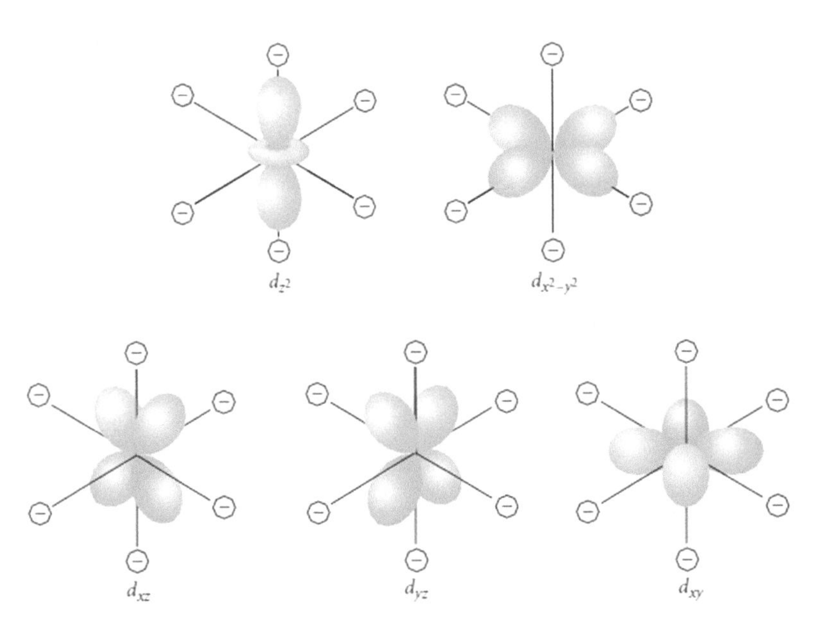

◘ **Fig. 19.23** *d* orbitals in an octahedral crystal field

e_g / t_{2g} — Δ_o — Octahedral crystal field splitting

t / e — Δ_t — Tetrahedral crystal field splitting

◘ **Fig. 19.24** Crystal field splitting of metal *d* orbitals for octahedral and tetrahedral geometries

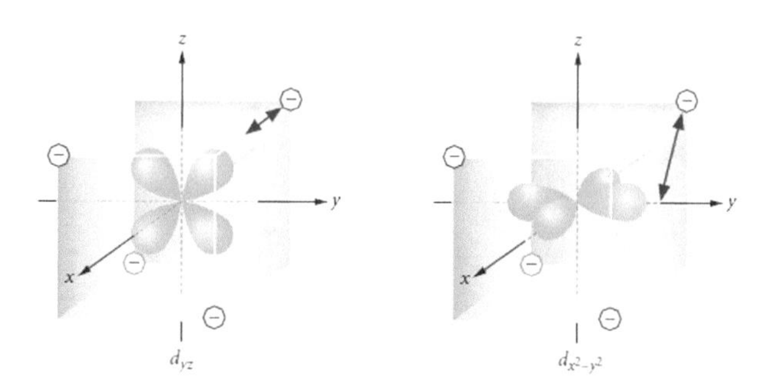

Fig. 19.25 Tetrahedral crystal field splitting (arrows show the distance from orbital lobe to ions)

Fig. 19.26 The value of Δ_t is 4/9 that of Δ_o, the distance in energy from the $d_{x^2-y^2}$ to the d_{xy} orbital

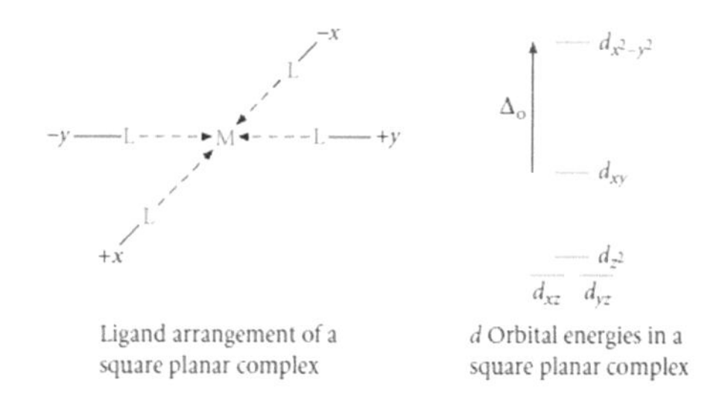

Fig. 19.27 Square planar coordination complex

rangement are directed towards the ligands in the tetrahedral arrangement, and vice-versa. We label the lower orbital set "e" and the upper orbital set "t". As the comparison in Fig. 19.26 shows, the magnitude of the splitting, Δ_t, is estimated to be 4/9 the size of Δ_o, so the magnitude of tetrahedral crystal field splitting is smaller than that of octahedral field splitting.

19.8.5 Square Planar Crystal Field Splitting

We can define the square planar geometry as one with the four ligands in the *xy* plane as in an octahedral complex, but with the two opposing ligands on the *z* axis completely removed. The crystal field effect in the *xy* plane remains the same, but the effect in the *z* direction is removed. This results in stabilization of the *d* orbitals with *z* axis components. The resulting crystal field splitting diagram is shown in Fig. 19.27. The difference in energy between d_{xy} and $d_{x^2-y^2}$ is still Δ_o. The other energy differences are considerably smaller.

19.8.6 Orbital Occupancy

The crystal field energy-level diagrams are the basis for understanding both physical and chemical properties of transition metal coordination complexes. However, as with all other chemicals, these properties are related to the electron configuration of the species. This configuration still follows the Aufbau rule, Pauli's exclusion principle, and Hund's rule of maximum multiplicity. However, the splitting between the energy levels of the *d* orbitals becomes similar to the electrostatic repulsion that occurs when two electrons occupy the same orbital. This repulsion is called the **pairing energy (*P*)**. Depending on the values of Δ and *P*, it may be favorable to place an electron in a higher energy orbital rather than pair it with another electron in one orbital where the negatively charged

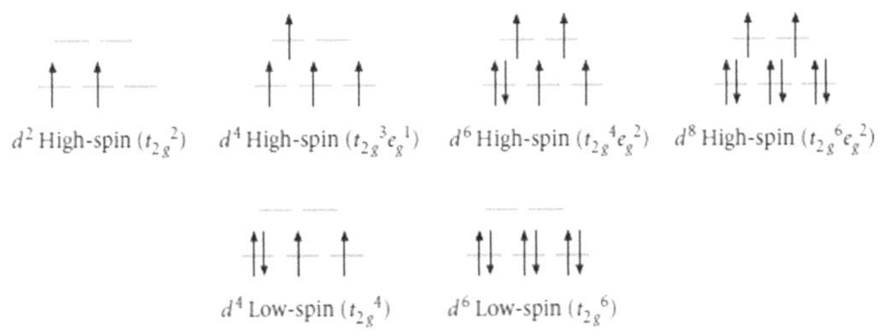

d^2 High-spin (t_{2g}^2) d^4 High-spin ($t_{2g}^3e_g^1$) d^6 High-spin ($t_{2g}^4e_g^2$) d^8 High-spin ($t_{2g}^6e_g^2$)

d^4 Low-spin (t_{2g}^4) d^6 Low-spin (t_{2g}^6)

Fig. 19.28 Electron distribution in d^2, d^4, d^6, d^8 octahedral coordination complexes

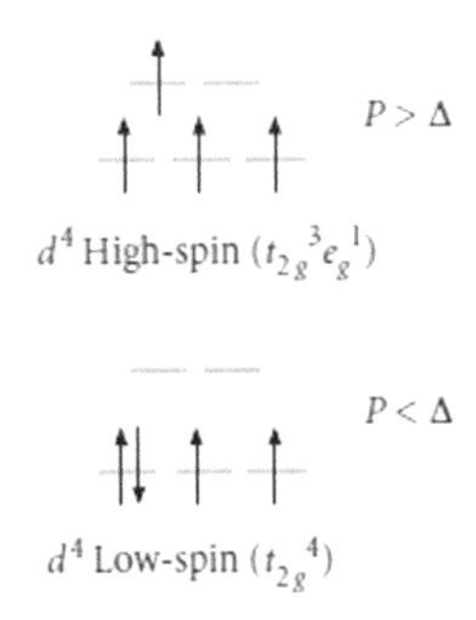

Fig. 19.29 The relative magnitude of the pairing energy and the crystal field energy determine the electron configuration

electrons will repel each other. It is also important to note that the orbital splitting (Δ) in each kind of crystal (octahedral, tetrahedral and square planar) will vary based on the nature of the ligand. In other words, the ability of the ligand to donate electrons to the metal's *d* orbitals will influence the size of the splitting.

Pairing energy (*P*) – The energy required to spin-pair two electrons within a given orbital.

19.8.7 Octahedral Complexes

An octahedral coordination complex containing a central metal atom with one *d* electron, such as Ti^{3+} or V^{4+}, will have one electron occupying one of the t_{2g} orbitals. For a central metal with a d^2 configuration, such as V^{3+}, one electron will occupy each of two t_{2g} orbitals, as shown in Fig. 19.28. This configuration can be symbolized t_{2g}^2, where the superscript represents the number of electrons in the orbital set labeled t_{2g}. An octahedral complex with three *d* electrons would give rise to a t_{2g}^3 orbital occupancy. However, if another electron is added, it may either pair up with one of the electrons already in a t_{2g} orbital at an energy cost of $+P$, or occupy one of the e_g orbitals at an energy cost of $_+\Delta_o$. The electron will tend to attain the lowest energy situation and hence will occupy the e_g orbital if Δ_o is less than the pairing energy, *P*. Therefore, depending on the values of Δ_o and *P*, the configuration could either be t_{2g}^4 with two unpaired electrons or $t_{2g}^3e_g^1$ with four unpaired electrons, as shown in Fig. 19.29.

These two configurations of electrons in the *d* orbitals are termed **low-spin**, (containing the minimum number of unpaired electrons, in this case two) or **high-spin** (with the maximum number of unpaired electrons, in this case four), respectively. For example, chromium complexes containing the chromium(II) ion (four *d* electrons) may be either high-spin with four unpaired electrons, as in $[Cr(H_2O)_6]^{2+}$, or low-spin with two unpaired electrons, as in $[Cr(CN)_6]^{4-}$.

19

Low-spin – A coordination complex with the minimum number of unpaired electrons.

High-spin – A coordination complex with the maximum number of unpaired electrons.

Continuing with our orbital occupancy evaluation, a d^5 configuration could be either t_{2g}^5 (low-spin) or $t_{2g}^3e_g^2$ (high-spin). At the d^6 configuration, electron pairing must occur, but either low-spin (t_{2g}^6) or high-spin ($t_{2g}^4e_g^2$) may exist. A d^7 configuration could also be either low-spin ($t_{2g}^6e_g^1$) or high-spin ($t_{2g}^5e_g^2$). Like the electron configurations with one, two, or three electrons, only one option is available for occupancy of the t_{2g} and e_g orbital sets for the d^8 ($t_{2g}^6e_g^2$), d^9 ($t_{2g}^6e_g^3$), and d^{10} ($t_{2g}^6e_g^4$) configurations.

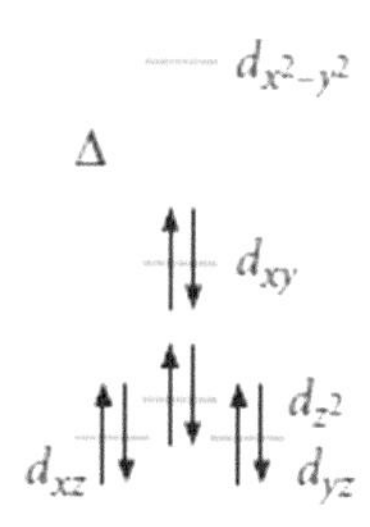

Fig. 19.30 Electron distribution in square planar *cis*-$[PtCl_2(NH_3)_2]$

19.8.8 Tetrahedral Complexes

Although it would seem that similar high-spin and low-spin electron configurations would be possible for the d^3 – d^6 metals in tetrahedral complexes, such situations are not observed. The magnitude of Δ_t is usually less than the pairing energy P. Recall that Δ_t for a tetrahedral complex is only about 4⁄9 that of Δ_o for an octahedral complex (comparing the same ligands in each case). Because P is always greater than Δ_t, high-spin complexes are nearly always the only possibility for tetrahedral complexes.

19.8.9 Square Planar Complexes

Similar analyses could be done with square planar complexes. However, it turns out that most square planar complexes occur for metal centers and ligands that generate a low-spin configuration. For example, cisplatin (*cis*-$[PtCl_2(NH_3)_2]$) has the orbital occupancy shown in Fig. 19.30.

Example 19.4—Drawing Crystal Field Diagrams

Draw crystal field splitting diagrams with electron occupancy for $[Mn(H_2O)_6]^{2+}$ (high-spin).

Asking the Right Questions

What is the relationship between the number of unpaired electrons the spin configuration, and the splitting of the d orbitals of the metal?

Solution

Manganese(II) has a d^5 configuration. With a coordination number of 6, an octahedral crystal field is expected. A high-spin d^5 ($t_{2g}^3e_g^2$) configuration places the electrons as shown here.

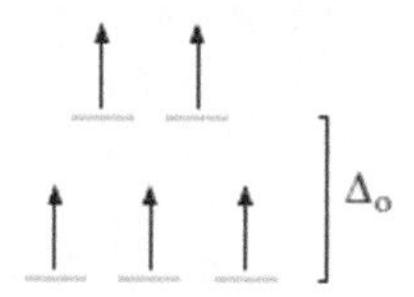

Is Our Answer Reasonable?

In this example, "reasonable" means that the crystal field splitting is consistent with the number of unpaired electrons, which is consistent with the oxidation number of the magnesium ion. The conclusions from each piece of information lead to the next piece, so your answer is reasonable.

What If…?

…we have the $[CoF_6]^{3-}$ ion? Describe the crystal field splitting in the cobalt ion?

(*Ans: This is a high-spin configuration, with* Co^{3+} *having 4 unpaired electrons.*)

Practice 19.4

How many unpaired electrons would you predict for the low-spin complex $[Fe(CN)_6]^{4-}$? Show the basis of your prediction using the crystal field splitting diagram for the iron metal ion.

19.8.10 The Result of *d* Orbital Splitting

A really dramatic example of color differences occurs in the two complexes of Co^{2+} shown in Fig. 19.31. When water is coordinated to Co^{2+} in the $[Co(H_2O)_6]^{2+}$ ion, the color is red. When chloride ligands coordinate to Co^{2+} in $[CoCl_4]^{2-}$, the color is blue. How does this fit into our understanding of color? The valence electron configu-

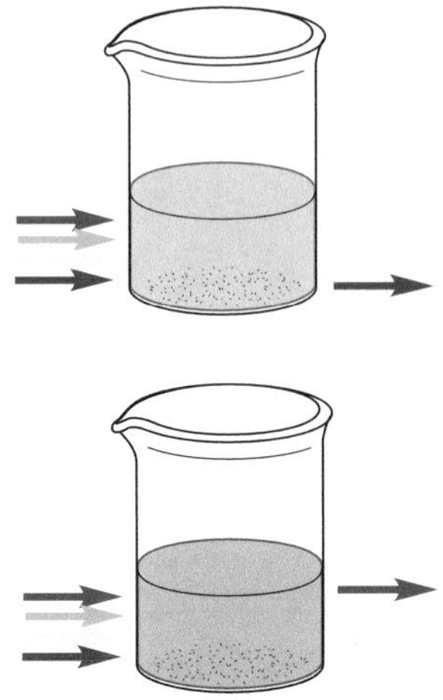

Fig. 19.31 Absorption of light (arrow) by cobalt complexes. The red solution is a solution of $[Co(H_2O_6)^{2+}]$; the blue solution is the coordination of chloride with cobalt(II) in $[CoCl_4]^{2-}$. Note that different wavelengths of light are absorbed by these complexes

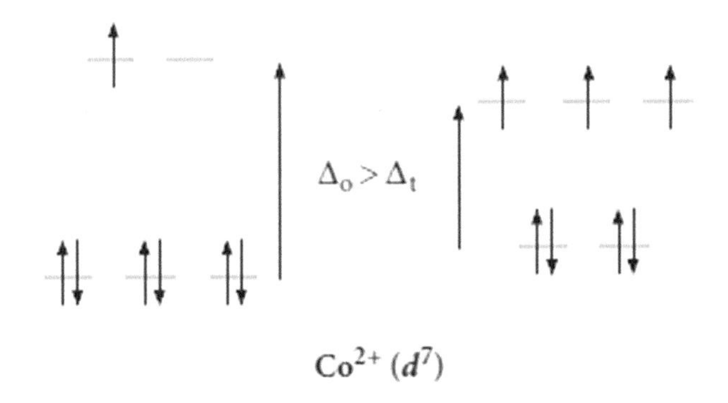

Fig. 19.32 The electron configurations of two cobalt complexes, $[Co(H_2O)_6]^{2+}$ and $[CoCl_4]^{2-}$. The difference in the magnitude of the crystal field energy determines what wavelengths of light they absorb

rations of all of these species are shown in Fig. 19.32. For $[Co(H_2O)_6]^{2+}$ with an octahedral ligand field, the red color arises when an electron is excited from a t_{2g} orbital to an e_g orbital by absorption of a photon of energy Δ_o. For $[CoCl_4]^{2-}$ with a tetrahedral ligand field, the blue color arises when an electron is excited from an e orbital to a t orbital by absorption of a photon of energy Δ_t. Because, in general, Δ_t is smaller than Δ_o, the photon absorbed by $[CoCl_4]^{2-}$ must be lower in energy than the photon absorbed by $[Co(H_2O)_6]^{2+}$. In this case, $[Co(H_2O)_6]^{2+}$ absorbs higher-energy blue light and appears red, and $[CoCl_4]^{2-}$ absorbs lower-energy red light and appears blue.

The absorption spectra in Fig. 19.33, which show how much light of various wavelengths is absorbed, exhibit a maximum for $[CoCl_4]^{2-}$ at 660 nm (equivalent to 181 kJ/mol) and a maximum for $[Co(H_2O)_6]^{2+}$ at 510 nm (equivalent to 221 kJ/mol). Recall that longer wavelength corresponds to lower energy, $E = \frac{hc}{\lambda}$, so the $[CoCl_4]$ complex absorbs lower-energy photons. The key point is that based on the model and with experimental evidence to support it, the geometry of the ligand field around a metal center influences the size of Δ.

The nature of the ligands around the metal center also influences the magnitude of Δ. Figure 19.34 shows solutions of three cobalt(III) complexes with different ligands around the cobalt center. The wavelength of light that is absorbed decreases, as shown in Fig. 19.35, as the unique ligand varies from Cl^- to H_2O to NH_3. As the ligand varies, the value of Δ_o increases and the wavelength of maximum absorbance decreases. Using this type of information from a wide variety of complexes, a **spectrochemical series** was developed that illustrates the effect of a particular ligand on the value of Δ. The series is

$$Cl^- < F^- < OH^- < H_2O < NH_3 < NO_2^- < CN^- < CO$$

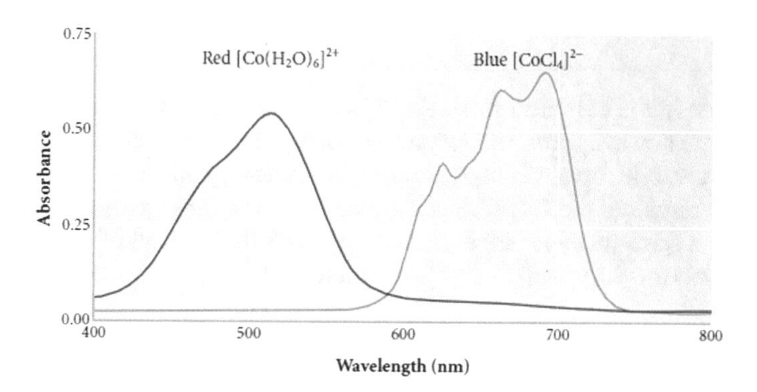

Fig. 19.33 Absorption spectra of $[Co(H_2O)_6]^{2+}$ and $[CoCl_4]^{2-}$

Fig. 19.34 The influence of coordinated ligands on color of coordination complexes. Each complex of cobalt(III) contains five NH_3 ligands and one other ligand. From left to right: $[Co(NH_3)_6]^{3+}$, $[Co(NH_3)_5(H_2O)]^{3+}$, and $[Co(NH_3)_5Cl]^{2+}$

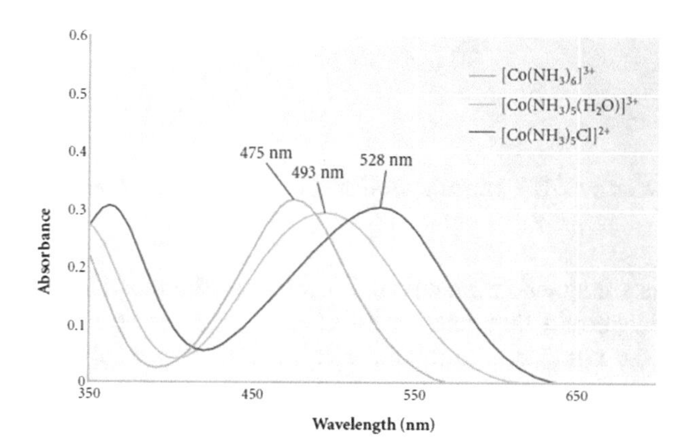

Fig. 19.35 Absorption spectra of $[Co(NH_3)_6]^{3+}$, $[Co(NH_3)_5(H_2O)]^{3+}$, and $[Co(NH_3)_5Cl]^{2+}$, showing higher-energy, shorter wavelength absorption in the order $[Co(NH_3)_6]^{3+} < [Co(NH_3)_5(H_2O)]^{3+} < [Co(NH_3)_5Cl]^{2+}$. (Spectra provided by Jerry Walsh.)

Spectrochemical series – The series of ligands organized with respect to their effect on the value of the crystal field splitting energy.

Conveniently, this series is generally the same for all metals. Knowing this information, we can predict that the wavelength of light absorbed by $[FeF_6]^{3-}$ would be longer (lower in energy) than the wavelength of light absorbed by $[Fe(H_2O)_6]^{3+}$ because a water ligand produces a larger value of Δ than the fluoride ligand.

Let's use this information to answer a practical question: Why is the blood in your veins dark red, whereas the blood in arteries is bright red? When blood in a vein is exposed to air, its color changes to bright red. Although the exact origin of this color change is fairly complex, it turns out that the crystal field splitting increases when oxygen binds to the iron center. When oxygen binds to the Fe^{2+} center, higher-energy light is absorbed, longer-wavelength light is transmitted by the hemoglobin, and the blood changes color. Oxyhemoglobin (with a larger crystal field splitting) absorbs higher-energy blue light and is red in the arteries; deoxyhemoglobin (with a smaller crystal field splitting) absorbs lower energies and appears bluish-red in the veins.

19.8.11 Magnetism

When magnetism is mentioned, it is common to think of bar magnets and their attraction for metal objects. This common form of magnetism shown by iron, nickel, and cobalt is known as **ferromagnetism**. However, there is a more common but also more subtle form of magnetism called **paramagnetism** (▶ Chap. 9). This property exists in any substance that contains unpaired electrons. A substance that is paramagnetic is attracted to a magnetic field, though less strongly than in a ferromagnetic substance. A material that has all of its electrons paired exhibits **diamagnetism**, and these materials are very weakly repelled by a magnetic field.

Ferromagnetism – Occurs when paramagnetic atoms are close enough to each other (such as in iron) that they reinforce their attraction to the magnetic field, so the whole is, in effect, greater than the sum of its parts.

Paramagnetism – The ability of a substance to be attracted into a magnetic field. This attraction arises because of the presence of unpaired electrons within the molecule.

Diamagnetism – The ability of a substance to be repelled from a magnetic field. This property arises because all of the electrons in the molecule are paired.

The origin of paramagnetism is in the spin of the unpaired electrons in a substance. In the absence of a magnetic field, the spins of the unpaired electrons are randomly oriented. When these unpaired electrons are placed in a magnetic field, their spins align with the field and result in a net attraction. The experimental characterization of transition metal coordination complexes was greatly aided by measurement of the paramagnetism of various species. Measurement of the strength of the paramagnetism provides an experimental quantity called the **magnetic moment** (μ). In many cases, μ is related to the number of unpaired electrons, n, and is nearly equal to or slightly greater than the theoretical value given by the formula:

$$\mu = \sqrt{n\,(n + 2)}$$

where $\boldsymbol{n}$ = number of unpaired electrons.

Magnetic moment (μ) – The strength of the paramagnetism of a compound. Can be used to determine the number of unpaired electrons.

The complex $[Mn(H_2O)_6]^{2+}$ has a magnetic moment of 5.9, whereas the magnetic moment of $[Mn(CN)_6]^{4-}$ is 2.2. Both have five d electrons. Why should they have such different magnetic moments? Using the relationship between magnetic moment and the number of unpaired electrons, $\mu = [n(n+2)]^{1/2}$, we find that if $n = 5$, then μ is 5.92 that agrees with the observed magnetic moment for the aqua complex. Thus the $[Mn(H_2O)_6]^{2+}$ complex is a high-spin d^5 complex, as shown in ◘ Fig. 19.36. Using the same relationship, if $n = 1$ (as it would be in the low-spin case), we expect μ to be 1.73, which is close to the value observed in the cyano complex. Therefore, the $[Mn(CN)_6]^{4-}$ complex must be a low-spin complex. This makes sense when we note that CN^- imparts a strong ligand field (high in the spectrochemical series) and H_2O imparts a weak ligand field. For CN^-, the magnitude of Δ is large enough that the electrons will be paired up in the t_{2g} orbitals rather than unpaired and occupying the higher energy e_g orbitals (◘ Fig. 19.37).

A fascinating change related to the paramagnetism of the iron ion in hemoglobin occurs upon binding of oxygen. Deoxyhemoglobin contains a high-spin, paramagnetic Fe^{2+} ion with four unpaired electrons. Upon binding with oxygen, the increased ligand field causes an increase in Δ, and the Fe^{2+} becomes low-spin and diamagnetic. This change in spin state, which can be followed by measurement of its magnetism, is critical to the ability of the iron to bind oxygen and release it under physiological conditions, as we will discuss later.

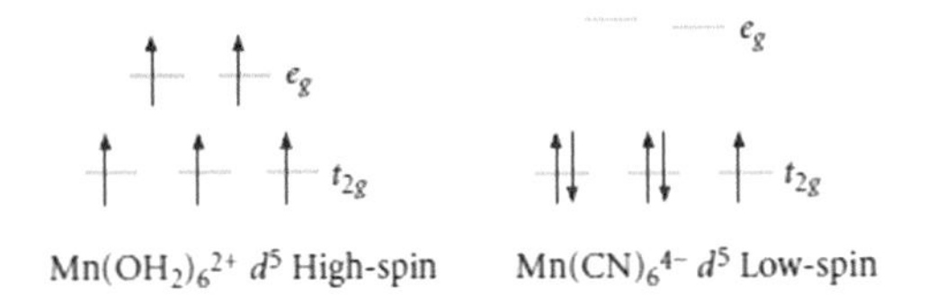

◘ **Fig. 19.36** High-spin and low-spin manganese complexes. In the high-spin case, the crystal field energy is less than the pairing energy. In the low-spin case, the magnitude of is greater than the pairing energy

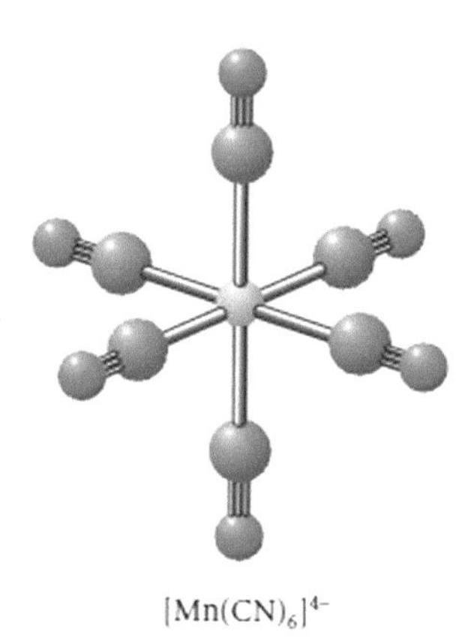

Fig. 19.37 Manganese with six ligands forms an octahedral complex

19.9 Chemical Reactions

Aside from possessing interesting color and magnetism properties, transition metal coordination complexes also undergo useful chemical reactions. Of particular biological importance is a class of reactions known as ligand exchange reactions. This class of reactions includes the coordination of oxygen and its release from an iron ion in hemoglobin. Another class of reactions, known as electron transfer reactions, are also important in biological processes. This class of reactions is quite common in photosynthesis and respiration.

19.9.1 Ligand Exchange Reactions

Why do coordination complexes form? All chemical reactions proceed when the total free energy of the system, ΔG, decreases. Because the free energy change depends on two factors, entropy and enthalpy, these must be considered in a reaction involving the association of a ligand with a metal. Consider the formation of hexaamminenickel(II) from the aqua complex:

$$\left[Ni(H_2O)_6\right]^{2+} + 6NH_3 \rightarrow \left[Ni(NH_3)_6\right]^{2+} + 6H_2O \quad K = 4 \times 10^8$$

Because the two different ligands, NH_3 and H_2O, are similar in size and the same number of each are involved in the reaction, the change in entropy for the reaction is small. The driving force for this reaction must then rely on the stability of the resulting coordinate covalent bonds. In this example, the Ni^{2+}—N bond is stronger than the Ni^{2+}—O bond, so ΔG is relatively large and negative because of the favorable enthalpy change in nickel–nitrogen bond formation. This translates into a large equilibrium constant for the formation of $[Ni(NH_3)_6]^{2+}$ via the equation $\Delta G = -RT \ln K$.

Consider the following reaction of our nickel complex. Each mole of this complex can react with three moles of ethylenediamine to form $[Ni(en)_3]^{3+}$. Both complexes contain Ni^{2+}—N bonds. Does it make sense that this reaction should also have a very large equilibrium constant?

$$\left[Ni(NH_3)_6\right]^{2+} + 3en \rightarrow [Ni(en)_3]^{2+} + 6NH_3 \quad K = 5 \times 10^9$$

There doesn't appear to be much change in the enthalpy of this reaction (the change in enthalpy is expected to be small because the bonds created in the product are similar to the bonds broken in the reactant), so we must focus on changes in entropy for this reaction. What predictions can we make about the entropy change of this reaction? A very favorable entropy change is observed; the reaction produces three more moles of product than moles of reactant. This exceptionally high favorability for forming complexes with polydentate ligands is known as the **chelate effect**.

Chelate effect – An unusually large formation constant due to a favorable entropy change for the formation of a complex between a metal center and a polydentate ligand.

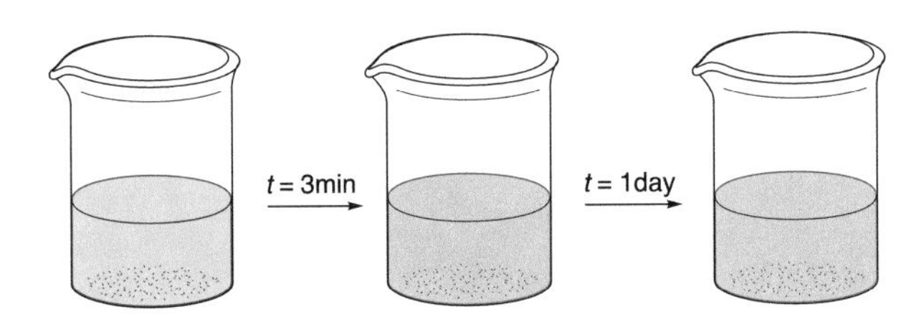

Fig. 19.38 Rates of ligand exchange. $[CoCl_4]^{2-}$ exchanges chloro ligands immediately on mixing

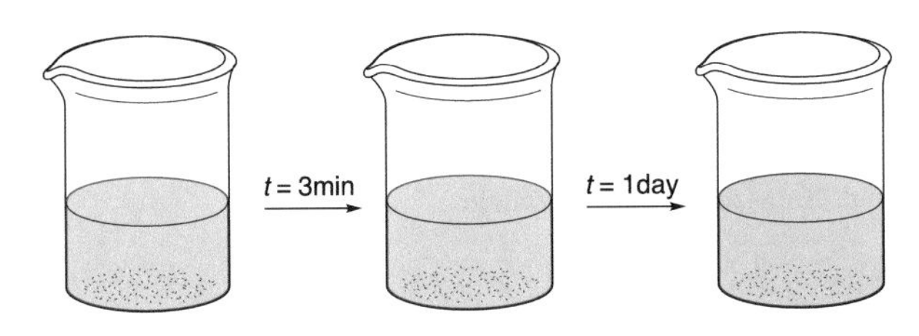

Fig. 19.39 $[CrCl_2(H_2O)_4]^+$ exchanges ligands slowly, taking a day or more to fully exchange its ligands

Although we can predict the direction of a reaction by using thermodynamics, only an examination of the kinetics of the reaction will determine the rate of the reaction. Some coordination complexes exchange their ligands very rapidly and are referred to as **labile**; others do so more slowly and are referred to as **inert**. For example, the blue $[CoCl_4]^-$ complex rapidly exchanges chloro ligands with water to produce the red $[Co(H_2O)_6]^{2+}$ complex, both of which are shown in Fig. 19.38. In contrast, the chromium complexes shown in Fig. 19.39, green $[CrCl_2(H_2O)_4]^+$ and purple $[Cr(H_2O)_6]^{3+}$, take at least a day to exchange ligands. In the interaction between cisplatin and DNA molecules, it is important for the platinum complex to bond to the DNA and remain bonded long enough for the complex to be toxic to the system. The platinum–DNA complex is inert—and must be inert to exhibit the kind of anticancer activity that it does.

Labile – The opposite of inert; said of a compound that exchanges ligands rapidly.

Inert – The opposite of labile; said of a compound that is very slow to exchange ligands.

The kinetic and thermodynamic factors exhibited in the reactivity of coordination complexes are intriguingly complementary in hemoglobin. The Fe^{2+} metal center is kinetically labile, which is important for the rapid exchange of oxygen ligands:

$$\mathrm{HbFe^{2+} + O_2 \rightleftharpoons HbFeO_2^{2+}} \quad \text{rapid}$$

However, in spite of the fact that Fe^{2+} can exchange its ligands rapidly, the chelate effect of the tetradentate porphyrin ligand maintains stability of the iron–porphyrin portion of the complex.

$$\mathrm{HbFe^{2+} + 6H_2O \rightleftharpoons Hb + \left[Fe(H_2O)_6\right]^{2+}} \quad K \ll 1$$

19

The balance of chemical characteristics for molecules in living systems is exquisite. This is why the scientific challenge of generating a chemical substitute for hemoglobin to transport oxygen in the bloodstream has been formidable.

19.9.2 Electron Transfer Reactions

Transition metals typically exhibit several stable oxidation states, and the + 2 and + 3 states are fairly common. Because many transition metals exhibit stability in two or more oxidation states, transition metal complexes can play important roles in electron transfer processes. For example, cytochromes are electron transfer agents in biological systems. Within a cytochrome protein, an iron ion coordinates to a porphyrin ring. The other sites on the octahe-

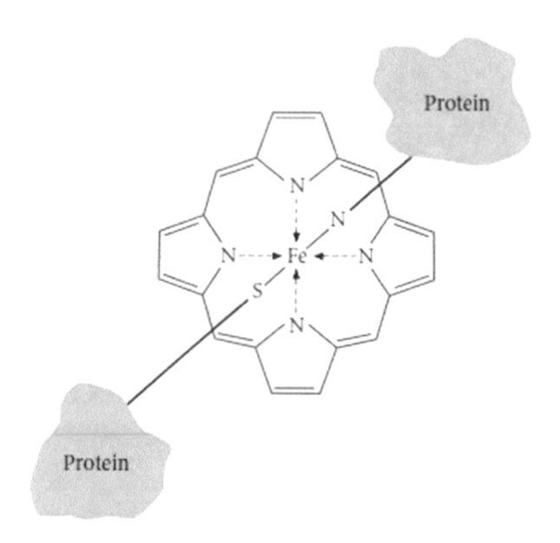

Fig. 19.40 Iron coordination in cytochrome electron transfer protein

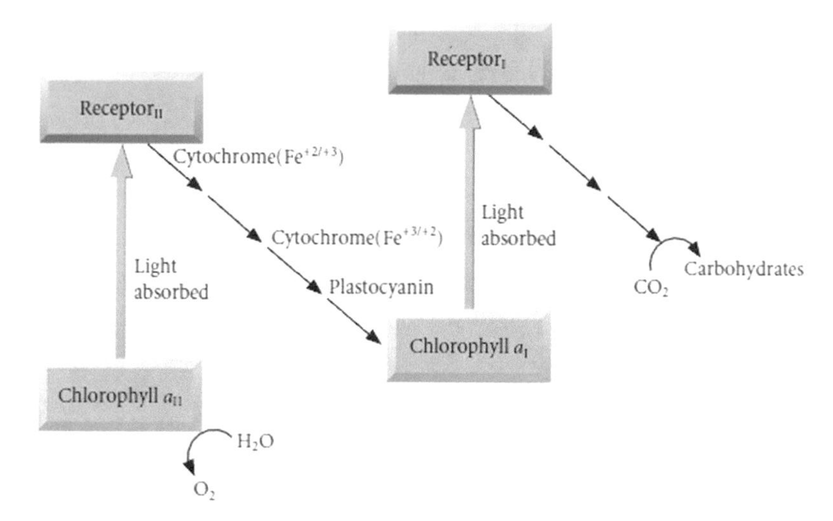

Fig. 19.41 The role of iron (cytochrome) and copper (plastocyanin) oxidation states in electron transfer reactions in photosynthesi

dral complex are occupied by ligands that are part of the protein structure, as shown in Fig. 19.40. During respiration and photosynthesis, the iron changes oxidation state from Fe^{2+} to Fe^{3+} to Fe^{2+} as the cytochrome shuttles electrons between two biological reaction sites (Fig. 19.41).

The Bottom Line

- Coordination complexes are present in simple metal ions in solution and as the reaction center in many biological molecules.
- A Lewis base that donates a lone pair of electrons to a metal to form a coordinate covalent bond acts as a ligand in producing a coordination complex.
- The coordination numbers of various metal centers are commonly observed to be 2, 4, or 6.
- Common coordination geometries are linear, tetrahedral, square planar, and octahedral.
- A variety of isomer classes are observed for metal complexes.
- The proper names and formulas for coordination complexes are specified by IUPAC rules.
- In transition metal complexes, the *d* orbitals are no longer degenerate but split into two or more energy levels, depending on coordination geometry.
- The electron configuration for octahedral complexes gives rise to high-spin and low-spin complexes for d^4 to d^7 metal centers.
- The color of many transition metal compounds arises when a photon of visible light is absorbed and an electron is excited to a higher energy *d* orbital.

- The order of ligands, in terms of their influence on the magnitude of the *d* orbital splitting and the energy of the photon of light absorbed, is defined as the spectrochemical series.
- The number of unpaired electrons determines the magnetic moment of the complex.
- Metal centers that exchange ligands rapidly are called labile; those that exchange ligands slowly are called inert.
- Chelate complexes have high formation constants because of an entropy effect.
- Because transition metal complexes often exhibit several oxidation states, transition metal complexes are good electron transfer (redox) agents.

19.2 Bonding in Coordination Complexes

Skill Review

1. Define each of these terms in your own words:
   a. ligand b. coordinate covalent bond c. Lewis acid
2. Define each of these terms in your own words:
   a. Lewis base b. complex c. metal center
3. For each of these coordination complexes, state the number of ligands, the oxidation number of the coordinated metal, and the number of coordinate covalent bonds that are formed within the complex.
   a. $[Fe(H_2O)_6]^{2+}$
   b. $[Co(NH_3)_6]^{2+}$
   c. $[Zn(NH_3)_4]^{2+}$
   d. $[Pt(CO)_4]$
4. For each of these coordination complexes, state the number of ligands, the oxidation number of the coordinated metal, and the number of coordinate covalent bonds that are formed within the complex.
   a. $[Fe(CN)_6]^{3-}$
   b. $[CuCl(NH_3)_3]^{+}$
   c. $[Ni(H_2O)_6]^{2+}$
   d. $[Au(NH_3)_2]^{2+}$
5. Diagram a structure that depicts a metal ion (M) surrounded by four symmetrically arranged ligands (L).
6. Diagram a structure that depicts a metal ion (M) surrounded by six symmetrically arranged ligands (L). Why are coordination complexes rarely found with more than six ligands?

19.3 Ligands

Skill Review

7. Diagram the Lewis dot structure for ammonia (NH_3). Explain why this molecule is classified as a Lewis base.
8. Diagram the Lewis dot structure for methylamine (CH_3NH_2). Is this molecule a Lewis base?
9. Would you predict the following molecule to be a ligand? If so, indicate whether it would be monodentate, bidentate, tridentate, or tetradentate.

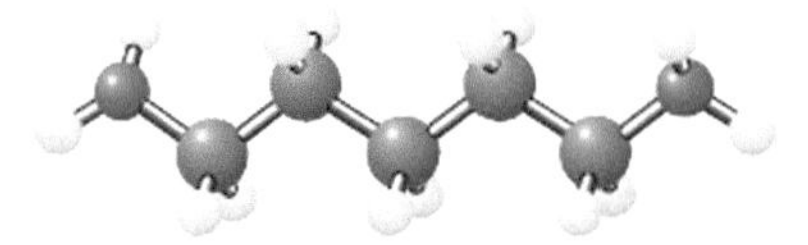

10. Would you predict the following molecule to be a ligand? If so, indicate whether it would be monodentate, bidentate, tridentate, or tetradentate.

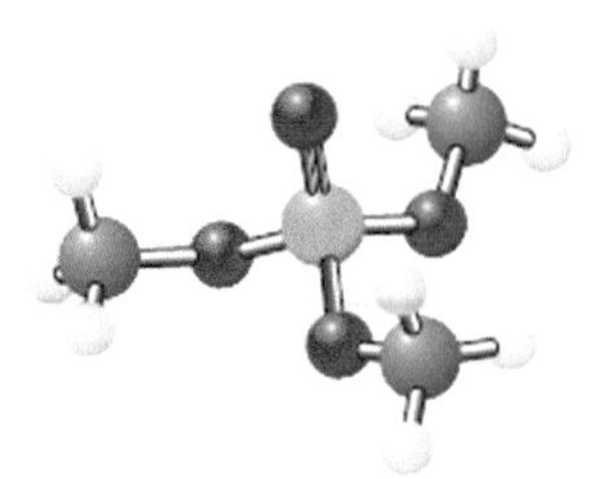

11. Which of these are not likely act as ligands in forming coordination complexes?
    a. H_2O b. CN^- c. Ca^{2+} d. O_2 e. C_2H_6
12. Which of these can act as a ligand in forming coordination complexes?
    a. CH_4 b. Br_2 c. O^{2-} d. H_2O_2 e. Li^+

19

Chemical Applications and Practices

13. Explain why the structure of ethylenediammine (en) enables it to act as a bidentate ligand, whereas nitrogen gas, which also contains two nitrogen atoms, cannot act as a bidentate ligand.
14. $EDTA^{4-}$ can coordinate with metal ions at six different sites. $EDTA^{4-}$ is often used as a food preservative in such food products as mayonnaise. Explain how $EDTA^{4-}$ functions in such an effective manner.
15. The formula $Co(A_2B_2)$ represents a coordination complex with two ligands, A and B. The Co ion is coordinated through six sites. If two A ligands occupy two of those sites, what must be true about the bonding for ligand B?
16. Amino acids can act as bidentate ligands. This is due to the presence of both oxygen and nitrogen. For example, alanine (Ala) could form the metal complex $[Fe(ala)_3]^{3+}$. Use the alanine structure to draw a structure of the metal complex.

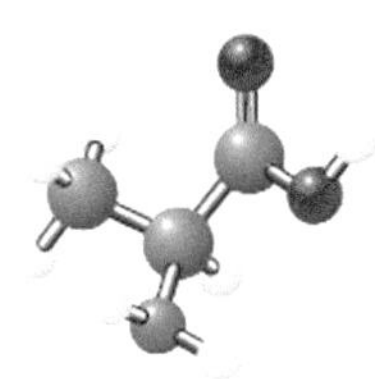

19.4 Coordination Number

Skill Review

17. Determine the oxidation state and the coordination number of the central metal in each of these coordination complexes:
    a. $[Cr(NH_3)_5Br]^{3+}$
    b. $[Mn(NH_2CH_2CH_2NH_2)_3]^{2+}$
    c. $[Cd(NH_2CH_3)_4]^{2+}$
18. Determine the oxidation state and the coordination number of the central metal in each of these coordination complexes:
    a. $[Co(NH_3)_6]^{3+}$ b. $[Pd(en)Cl_2]$ c. $[Mo(ox)_3]^{3-}$
19. Explain how it is possible that the central metals in these two complexes can have the same coordination number even though the number of ligands differs in the complexes.

$$[Fe(ox)_3]^{3-} \quad [Co(SCN)_2(H_2O)_4]^{+}$$

20. Explain how it is possible that the central metal in these two complexes can have the same oxidation state even though the coordination number differs in the complexes.

$$[Fe(en)_3]^{3+} \quad [ScI_4]^{-}$$

Chemical Applications and Practices

21. Many heavy-metal-containing salts are not very water soluble. One way to increase the solubility is to form a complex. Examine the structure of the bidentate oxinate ion shown here. If two of these complexed with a lead ion, what would be the coordination number of the lead?

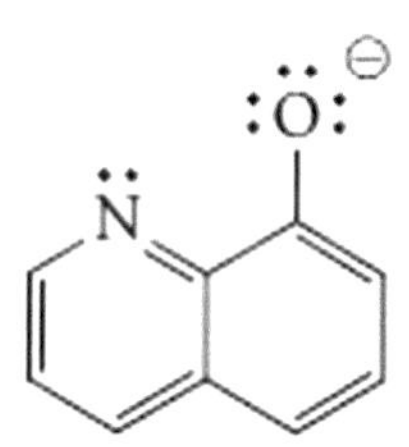

Oxinate ligand

22. One method used to determine the "hardness" of water samples is to titrate the sample with disodium EDTA. This reaction forms a complex with the calcium ions in the water. Using the structure of $EDTA^{4-}$ depicted in Fig. 19.9, determine what the coordination number of the calcium ion would be if it combined with one disodium EDTA ion. Diagram the structure that justifies your answer.

19.5 Structure

Skill Review

23. Diagram the structures of the complexes listed in Problem 3 and identify the geometry (bond angles and overall shape) of each structure.
24. Diagram the structures of the complexes listed in Problem 4 and identify the geometry (bond angles and overall shape) of each structure.
25. Indicate the oxidation number, coordination number, and geometry for the cobalt ion in the following compound.

$$[CoCl(NO_2)(en)_2]Cl$$

26. Indicate the oxidation number, coordination number, and geometry for the silver ion in the following compound.

$$[Ag(NH_3)_4]Br_2$$

Chemical Applications and Practices

27. When nickel ore is refined, it must be removed from other metals. This can be done by forming a coordination complex between nickel and carbon monoxide. The highly poisonous nickel–carbon monoxide complex $Ni(CO)_4$ evaporates easily and can therefore be used to separate nickel from its impurities. What is the coordination number for nickel, and what two possible geometries could you predict for this structure?
28. The readily available electron pairs found in the oxygen, nitrogen, and some sulfur atoms of amino acids provide bonding sites for metals. The metal–amino acid combinations serve as the basis for many important biochemical processes. For example, a certain copper-containing enzyme utilizes an octahedral structure around a Cu^{2+} ion to assist in the transport of electrons within cells. Draw the structure of a copper(II) complex that would form if three glycine amino acids (shown below) formed coordinate covalent bonds with the copper. (Assume that each glycine is bidentate.)

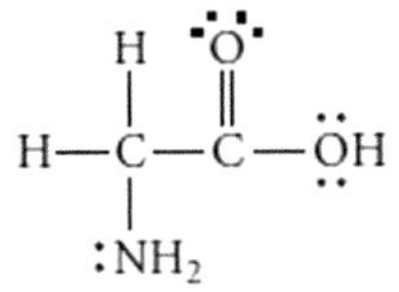

H O
H—C—C—OH
:NH₂

Glycine

19.6 Isomers

Skill Review

29. Define the type of isomer present, if isomerization exists, in each of these pairs of complexes.
    a. *trans*-$[Pt(NH_3)_2Cl_2]$ and *cis*-$[Pt(NH_3)_2Cl_2]$
    b. $[Pt(CN)_2(NH_3)_4]Cl_2$ and $[PtCl_2(NH_3)_4](CN)_2$
    c. $[Fe(H_2O)_6][CuBr_4]$ and $[Cu(H_2O)_6][FeBr_4]$
30. Define the type of isomer present, if isomerization exists, in each of these pairs of complexes.
    a. $[Cu(NH_3)_3(ONO)]$ and $[Cu(NH_3)_3(NO_2)]$
    b. $[Mn(H_2O)_4Cl_2]Br_2$ and $[Mn(H_2O)_4Br_2]Cl_2$
    c. *cis*-$[PdCl_2(NH_3)_2]$ and *trans*-$[PdCl_2(NH_3)_2]$
31. Diagram two square planar geometric isomers with the formula $[PtI_2(NH_3)_2]$. Label the *cis* isomer. It is not possible to diagram two tetrahedral geometric isomers of $[Pt(NH_3)_2I_2]$. Explain why.
32. Diagram all of the possible isomers of $[Co(NH_3)_2(SCN)_2]$.
33. How many different isomers of $[NiCl_3F_3]^{4-}$ can be drawn? Show the structure of each.
34. Show the structures of the coordination sphere isomers for $[Co(NH_3)_6][Cr(NO_2)_6]$.

Chemical Applications and Practices

35. Ethylenediammine (en) is a bidentate ligand, so it forms two attachments to metal ions in coordination complexes. However, the square planar complex $[PtI_2(en)]$ exhibits only one type of geometric isomer. Draw possible structures for $[PtI_2(en)]$ and for a complex between iron(III) and en.
36. Diagram the Lewis dot structure for the cyanate ion (OCN^-). Show how this ion would make it possible to have two different forms of the following complex: $[Co(OCN)(NH_3)_5]^{2+}$. What type of isomerism does this example illustrate?

19.7 Formulas and Names

Skill Review

37. Provide names for these complex ions:
    a. $[Co(CN)(en)_2(NH_3)]^{2+}$
    b. $[Cr(C_2O_4)_2(NH_3)_2]^-$
    c. $[Fe(NO_2)_6]^{3-}$
    d. $[CoCl_3(H_2O)_3]$
38. Provide names for these complex ions:
    a. $[Mn(en)_3]^{2+}$
    b. $[Ni(H_2O)_4(NH_3)_2]^{2+}$
    c. $[Cr(NO_2)_6]^{3-}$
    d. $[V(SCN)_2(H_2O)_4]$
39. Write the chemical formula for each of these compounds and complex ions:
    a. tetraammineaquachlorocobalt(III)
    b. *trans*-diaquabis(ethylenediamine)copper(II) chloride
    c. sodium tetrachlorocobaltate(II)
    d. pentacarbonylchloromanganese(I)
40. Write the chemical formula for each of these compounds and complex ions:
    a. tetraaquadichlorocopper(II)
    b. potassium *cis*-dibromooxalatoplatinate(II)
    c. tetraamminenickel(II) sulfate
    d. tetraaquathiosulfatoiron(III) nitrite
41. Which of these species would produce the greater number of ions per mole when dissolved in water?

$$K_2\left[Cr(C_2O_4)_2(H_2O)_2\right] \text{ or tetraamminediaquachromium(III)nitrate}$$

42. Which of these species possesses the larger positive charge on the complex ion?

tetraaquacopper(II) nitrate or dichlorobis(ethylenediamine)iron(III) bromide

19.8 Color and Coordination Compounds

Skill Review

43. What is the electron configuration for each of these transition metal ions?
    a. Fe^{2+} b. Cr^{2+} c. Zn^{2+}
44. What is the electron configuration for each of these transition metal ions?
    a. Pd^{4+} b. Ag^+ c. Mn^{2+}
45. Consider the following two transition metal ions as free gaseous ions. Which would have the greater number of unpaired *d* electrons, Fe^{3+} or Cu^{2+}?
46. Which free ion has the greater number of unpaired *d* electrons, Ti^{2+} or Co^{2+}?
47. Draw the orbital diagram for the *d* orbitals in an octahedral complex containing each of these metal centers. (Assume that $P < \Delta_o$.)
    a. Fe^{3+} b. Co^{2+} c. Ni^{2+}

48. Draw the orbital diagram for the d orbitals in an octahedral complex containing each of these metal centers. (Assume that $P > \Delta_o$.)
    a. Mn^{2+} b. Fe^{2+} c. Cr^{2+}
49. Repeat Problem 47, but assume that the metal centers are involved in tetrahedral complexes. Although tetrahedral complexes typically have $\Delta_t > P$, what would you draw if the tetrahedral complex existed with $P > \Delta_t$?
50. Repeat Problem 48, but assume that the metal centers are involved in square planar complexes. Assume that $P > \Delta$ in this problem.
51. Draw the orbital diagram for the metal center in each of these complexes. Use the information in the spectrochemical series to assist you in placing the orbitals.
    a. $[FeCl_4]^-$ b. $[Co(CN)_6]^{3-}$ c. $[Mn(CO)_6]^+$
52. Draw the orbital diagram for the metal center in each of these complexes. Use the information in the spectrochemical series to assist you in placing the orbitals.
    a. $[CuF_6]^{4-}$ b. $[Ni(OH)_6]^{4-}$ c. $[Cr(NO_2)_6]^{4-}$
53. Which of the complexes in Problems 51 and 52 is(are) paramagnetic?
54. Calculate the magnetic moment for each of the complexes in Problem 52.

Chemical Applications and Practices

55. Which one of these complexes would you predict to absorb blue light: $[M(CN)_6{}^{2-}]$, $[M(H_2O)_6{}^{4+}]$, $[MCl_6{}^{2-}]$, or $[M(NH_3)_6{}^{4+}]$?
56. Which of these complexes would you predict to absorb the longest wavelength of visible light: $[M(CN)_6{}^{2-}]$, $[M(H_2O)_6{}^{4+}]$, $[MCl_6{}^{2-}]$, or $[M(NH_3)_6{}^{4+}]$?
57. The colors of common gemstones are due to the presence of transition metal ions. The color is produced when the metal ion absorbs visible light. Would you predict the common gemstones to have different "colors" under infrared light?
58. Coordination compounds with Zn^{2+} ions typically are white or colorless. Explain why this particular metal does not form brightly colored compounds the way many other transition metals do.
59. The crystal field theory provides an explanation of color in various coordination complexes. For example, $[Cr(H_2O)_6]^{3+}$ can be detected as a violet color when dissolved.
    a. What colors would the complex be absorbing?
    b. How many unpaired electrons does the central chromium ion have?
    c. The compound $[Cr(NH_3)_6]^{3+}$, when dissolved, appears yellow. Would you expect it to absorb light at a higher or lower frequency than $Cr(H_2O)_6]^{3+}$? Explain.
    d. Which ligand, NH_3 or H_2O, is causing the greater value of Δ_o?
60. Compare the two iron complexes $[Fe(H_2O)_6]^{2+}$ and $[Fe(CN)_6]^{4-}$.
    a. Which is more likely to be paramagnetic?
    b. Which is more likely to absorb light of greater energy?
    c. Which is more likely to be "high-spin"?

19.9 Chemical Reactions

Skill Review

61. Explain why the chelate effect typically provides for a very favorable entropy change when a ligand exchange reaction involves a complex going from a nonchelated complex to a chelate complex.
62. If a ligand exchange reaction produced a large positive ΔG value, would you expect the reaction to have a large or a small equilibrium constant? Justify your choice.
63. A common chemical demonstration is to change a light blue solution containing $[Cu(H_2O)_4]^{2+}$ quickly to a deep purple solution by changing $[Cu(H_2O)_4]^{2+}$ into $[Cu(NH_3)_4]^{2+}$ with the addition of ammonia to the solution. Would you consider the first compound labile or inert? Is the exchange of oxygen in hemoglobin considered to be representative of a labile or an inert complex?
64. Except through loss of blood, the level of iron in humans is fairly constant. One way that iron is moved throughout the body, particularly from the liver, is within a molecule known as ferritin. The iron(III) is held in a six-coordinate system through bonds to oxygen and nitrogen that are part of several amino acids. Would it be more logical for this molecule's iron site to be labile or inert with respect to other metals? Explain. (Remember, the terms *inert* and *labile* refer to kinetic considerations, not to equilibrium predictions.)

Comprehensive Problems

65. In addition to the coordination complexes of rhodium, cobalt, and molybdenum described at the start of the chapter, use other resources (the Internet or journals) to select another transition metal that has a catalytic role in a chemical reaction.
66. Visit the pharmacy section of a grocery store and list the metals found in a mineral supplement. Use a reference (the Internet or journals) to determine the primary biochemical function of two of the metals from your list.
67. If a complex were assembled from a Co^{3+} ion, four NH_3, and two Cl^-, would you expect to see a neutral compound, an anion, or a cation? Show a formula that justifies your answer.
68. The following complex contains an iron ion in the +3 state. However, the resulting charge has been omitted from the complex. Assign the charge for the complex and indicate how many counter ions (either Na^+ or Cl^-) would have to be ionically bonded to the complex to form a neutral compound.

$$[FeBr_4(H_2O)_2]$$

69. The appearance of a visible color from a compound is associated with three fundamental phenomena. Describe a color-producing compound's properties associated with:
    a. electron excitation
    b. the energy of the photon of light being absorbed
    c. the relative occupancy of lower and higher level orbitals
70. If 35.7 g of the complex $Ca_3[Fe(C_2O_4)_3]_2$ formed as a result of using an oxalate-containing rust remover, how many grams of iron would be removed?
71. Polydentate ligands have been very effective in applications of soil chemistry. EDTA has been added to soil near citrus trees to concentrate iron. Some polydentates have been used to extract heavy metals from soil samples for further analysis. Some internal digestive functions use polydentate ligands to extract metals from foods we eat. Does this indicate that the relative equilibrium constant for these reactions is larger or smaller than one?

Thinking Beyond the Calculation

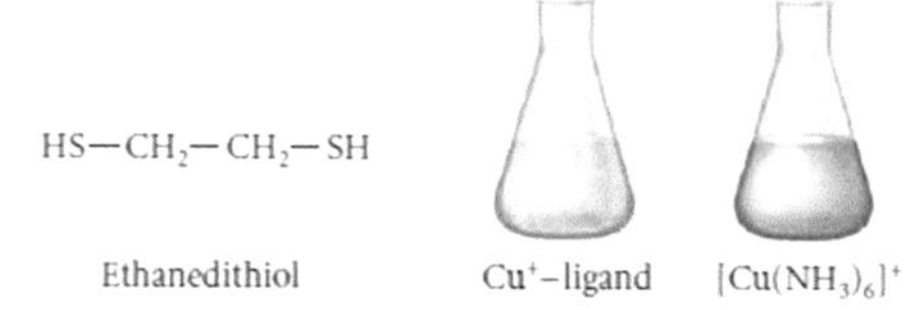

72. A new ligand is developed for use in a study to mimic the electron transport reactions of copper metal.
    a. Draw the octahedral complex that would result from the use of this ligand and copper(II) ions.
    b. Draw the *d* orbital diagram for an octahedral complex of this ligand.
    c. Compare the color of this octahedral complex to that of $[Cu(NH_3)_6]^{2+}$. Where does this ligand most likely fall in the spectrochemical series?
    d. Under certain conditions, 1.00 g of copper(II) produces 3.96 g of the copper–ligand complex. What is the coordination number of the complex under these conditions?
    e. The equilibrium constant for the ligand exchange of this new ligand and chloride ions is 1.45×10^{-4}. What does this indicate about the new ligand?

Nuclear Chemistry

Contents

Supplementary Information The online version contains supplementary material available at ▶ https://doi.org/10.1007/978-3-030-90267-4_20.

M. Mosher and P. Kelter, *An Introduction to Chemistry*,
https://doi.org/10.1007/978-3-030-90267-4_20

What We Will Learn in This Chapter

- Nuclides differ in the number of neutrons in the nucleus of a particular isotope (▶ Sect. 20.2).
- Alpha, beta, and gamma are three types of radiation that are emitted from an unstable nucleus (▶ Sect. 20.3).
- A decay series is a stepwise progression of a radioactive nuclide toward stability (▶ Sect. 20.3).
- Ionizing radiation is harmful to living things (▶ Sect. 20.4).
- Radioactive nuclei decay by first order kinetics (▶ Sect. 20.5).
- The difference in the mass of the particles that make up a nucleus and the mass of the nucleus is known as the binding energy that holds the nucleus together (▶ Sect. 20.6).
- Nuclear fission can be used to generate electricity (▶ Sect. 20.8).
- Radioactive nuclei can be used in medicine to identify and treat disease (▶ Sect. 20.9).

20.1 Introduction

The odds are that you know somebody who has fought cancer. If you have talked with a friend or relative who is a cancer patient, you may have heard that radioactive substances are used in the process of producing images of internal organs. You may know that radiation can shrink a tumor or kill cancer cells. You may even be aware of the dramatic procedure using full-body radiation that is given before a bone marrow transplant. These applications illustrate how a health care team can use **nuclear radiation** for the benefit of a cancer patient.

Nuclear radiation – The particles and/or energy emitted during radioactive decay.

Although radiation can save lives, it can also damage and kill. Back in the early 1900s, early radiologists held film plates in an X-ray beam to get pictures of their patients. These radiologists often developed sores on their hands that would not heal, and some eventually lost parts of their fingers. In August 1945, people in the Japanese cities of Hiroshima and Nagasaki were exposed to huge bursts of radiation from the only use of atomic bombs in warfare. Tens of thousands died immediately, and several hundred thousand succumbed in the months and years thereafter from a then-unknown disease that we now call radiation sickness. In the United States, those who worked in the nuclear weapons industry are now disproportionately contracting lung, stomach, lymphatic, and other cancers.

How can nuclear radiation both cure and cause cancer? The answer to this seeming paradox rests on our understanding of **radioactivity** (the release of particles and energy accompanying a nuclear change) and how it is produced from nuclear processes. In this chapter we will examine types of radioactive decay, touching on the mathematics of half-lives, the relationship between mass and energy, and the interactions of ionizing radiation with matter. We also will examine nuclear fission, a process that helps produce a whole host of substances useful in nuclear medicine. The story of radioactivity begins, however, at the tiny center of the atom—its nucleus.

Radioactivity – The emission of radioactive particles and/or energy.

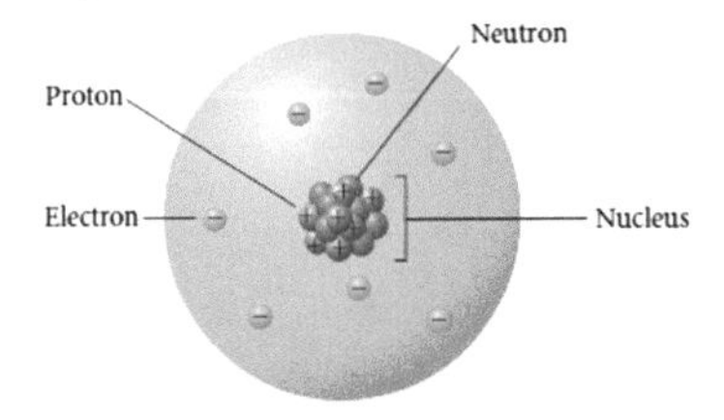

20.2 Isotopes and More Isotopes

Nuclei that occur naturally on Earth can be as small as the nucleus of a hydrogen atom (a single proton) or as large as uranium (92 protons plus 146 neutrons). All of these naturally occurring atoms, and all those that are human-made, are represented by writing their element symbol in the manner described in ▶ Chap. 2. The symbol is placed next to a superscripted mass number and a subscripted atomic number. Some of these nuclei are stable—that is, they do not spontaneously decompose—and others are radioactive, decomposing to other nuclei. As we will soon see, the size and makeup of the nucleus determine whether it is stable or radioactive.

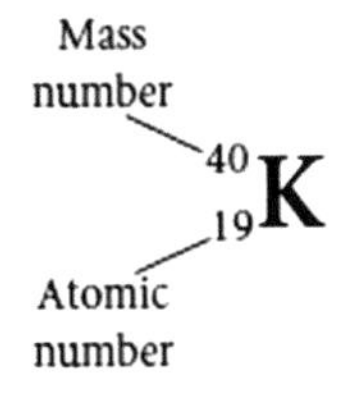

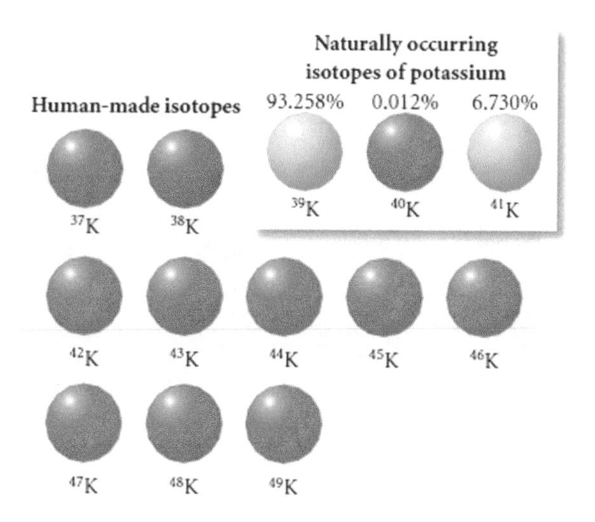

Fig. 20.1 Potassium has many isotopes, most of which are listed here. Those colored orange are radioactive. Natural abundances are provided for those isotopes that occur in nature. Note that 99.988% of the potassium atoms on Earth are not radioactive

Recall from ▸ Chap. 2 that we can use nuclide notation, in which we list the symbol for the element, accompanied by its atomic number and the mass number, to indicate the isotope of the element that we wish to describe. For example, consider the element potassium, whose atomic number (Z) is 19. As we know from the periodic table, this element has 19 protons. If the nucleus of a potassium atom has 21 neutrons, we represent this in nuclide notation as $^{40}_{19}K$. The number 40, the mass number, is the sum of 19 protons and 21 neutrons. Because "19 protons" is always potassium, we can more simply represent $^{40}_{19}K$ as ^{40}K, K-40 or potassium-40. In nuclear chemistry, we are often not concerned with the number of electrons around a particular nucleus. We are interested only in the nucleus of the atom, so we often omit charges (even though they may exist). This means that in our chemist's shorthand for the discussion of changes in the nucleus, we'll consider $^{40}K^+$ as ^{40}K. The nucleus is our only focus here.

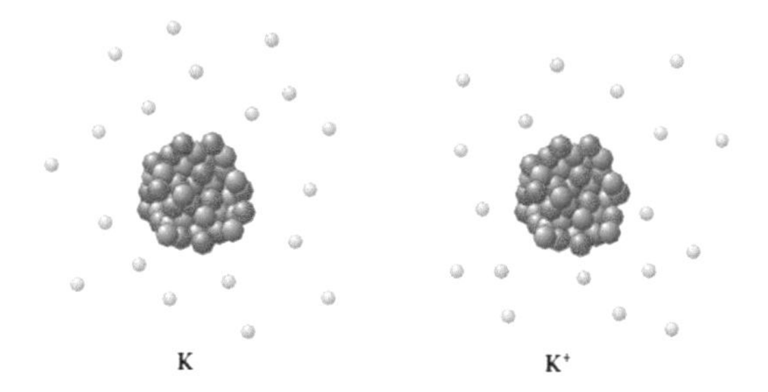

Potassium atoms come in several varieties, shown in ▫ Fig. 20.1, some with fewer than 21 neutrons and some with more. These varieties are known as isotopes, and each isotope is known as a nuclide of that element, as we learned in ▸ Chap. 2.

Example 20.1—Decoding Isotopes

Describe the differences and similarities in nuclear structure among the members of each of these sets:

a. $^{12}_{6}C$, $^{13}_{6}C$, and $^{14}_{6}C$

b. ^{40}Ar, ^{40}K, and ^{40}Ca

c. $^{40}Ca^{2+}$, and ^{40}Ca

Asking the Right Questions

What do the numbers at the bottom of each symbol represent? What about those at the top? How do we find the number of neutrons using nuclide notation? What about the number of electrons?

Solution

The numbers at the bottom of each representation are a restatement of the element symbol. The mass number is the sum of the number of neutrons and protons. To find the number of neutrons in a particular nuclide, we subtract the atomic number from the mass number.

a. $^{12}_{6}C$, $^{13}_{6}C$, and $^{14}_{6}C$ are isotopes of carbon, each containing six protons. But each has a different mass number and hence a different number of neutrons: 6, 7, and 8, respectively.

b. ^{40}Ar, ^{40}K, and ^{40}Ca have the same mass number but different numbers of protons. Ar has 18p and 22n, K has 19p and 21n, and Ca has 20n and 20p.
c. $^{40}Ca^{2+}$ and ^{40}Ca differ only in the number of electrons, so there is no difference in *nuclear* structure. The ^{40}Ca atom loses two electrons to form the $^{40}Ca^{2+}$ ion.

Are the Answers Reasonable?
We used nuclide notation to show differences in the number of protons, neutrons and electrons in the various parts of the example. One way of assessing that our answers are reasonable is to connect this with our work with carbon, argon, and calcium in previous chapters. The data for the elements is consistent with that from our prior work – it is reasonable on that basis.

What If…?
…we compared $^{56}Fe^{2+}$ with $^{55}Mn^{4+}$? Which has more electrons? Protons? Neutrons?
(Ans: $^{56}Fe^{2+}$ has 26 protons, 30 neutrons and 24 electrons. $^{55}Mn^{4+}$ has 25 protons, 30 neutrons, and 21 electrons.)

Practice 20.1
Indicate the number of protons and neutrons in each of the following nuclides.
a. ^{32}S
b. $^{23}_{11}Na$
c. radon-222
d. Tc-98.

Potassium is considered an essential nutrient. Deficiencies in potassium can result in muscle pain, angina, and even heart problems. Eating bananas is one way in which we can ensure a healthy supply of potassium in our bodies. From Fig. 20.1 we can see that three isotopes of potassium occur naturally: $^{30}_{19}K$, $^{40}_{19}K$, and $^{41}_{19}K$. However, whereas potassium-39 and potassium-41 possess stable nuclei, $^{40}_{19}K$ is radioactive. This means that when we consume a banana, we get a measurable amount of radioactive potassium-40. How much radioactive potassium-40 do we consume? The natural abundance of potassium-40 is only 0.012%, or approximately 1 atom in 10,000. A typical banana has approximately 450 mg of potassium. Therefore, with each banana we eat, we ingest approximately 0.054 mg of radioactive potassium-40.

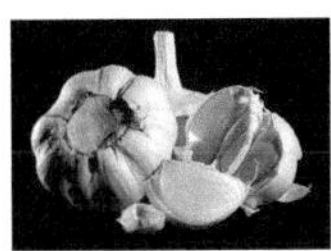

▶ https://upload.wikimedia.org/wikipedia/commons/thumb/9/9c/Garlic_Bulbs_2.jpg/320px-Garlic_Bulbs_2.jpg; ▶ https://upload.wikimedia.org/wikipedia/commons/thumb/d/d3/Green_Pepper.jpg/320px-Green_Pepper.jpg; ▶ https://upload.wikimedia.org/wikipedia/commons/9/9b/Cavendish_Banana_DS.jpg

Although potassium has 18 known isotopes, most do not occur in nature and must be produced in a laboratory. Each of these isotopes behaves essentially the same in chemical reactions, which involve the interaction of orbiting electrons with other atoms. For example, on exposure to oxygen or moisture, potassium metal ionizes to form K^+. All isotopes of potassium behave in this manner. Because our planet is wet and blanketed in oxygen, *all* naturally occurring potassium isotopes are found in nature as K^+. Remember, however, that we often omit the charge when writing nuclides, and you are likely to see ^{40}K rather than $^{40}K^+$. The key idea here is that we differentiate between chemical reactions (electron interactions between atoms) and nuclear processes (changes within the nucleus of an individual atom). Potassium-39 is a stable nuclide, but potassium metal certainly is not a stable chemical when in the presence of water. On the other hand, argon is chemically inert, but it has isotopes that are radioactive.

Natural isotopic abundances for potassium or for any element can be found in many of the chemistry handbooks in the library or on the internet. Table 20.1 offers a brief look at what you'll find there. As we just saw, the nuclei of elements present in nature are not necessarily stable. In fact, some elements, such as uranium and radon, exist naturally only in radioactive forms. Other elements, such as potassium and carbon, have both stable and radioactive isotopes. Still others, such as aluminum, have only one stable naturally occurring form.

Table 20.1 Natural isotopic abundances for selected elements

Element	Isotope	Abundance (%)
Carbon	^{12}C	98.90
	^{13}C	1.10
Oxygen	^{16}O	99.762
	^{17}O	0.038
	^{18}O	0.200
Magnesium	^{24}Mg	78.99
	^{25}Mg	10.00
	^{26}Mg	11.01
Aluminum	^{27}Al	100
Potassium	^{39}K	93.2581
	^{40}K	0.0117
	^{41}K	6.7302
Iron	^{54}Fe	5.8
	^{56}Fe	91.72
	^{57}Fe	2.2
	^{58}Fe	0.28
Silver	^{107}Ag	51.84
	^{109}Ag	48.16

20.3 Types of Radioactive Decay

There is no way for you hold a potassium atom in your hand and peer into its nucleus. But if you could keep an eye on a few ^{40}K atoms for a period of time (perhaps over a billion years), you would observe that some of the potassium atoms had been replaced by atoms of calcium. How do we account for this nuclear sleight-of-hand?

Let's revisit the composition of the nucleus. Recall that, in Chap 2, we described the nucleus as a combination of positively charged protons and neutral neutrons. It turns out to be quite a bit more complicated than that. In the years since Thomson, Rutherford, Chadwick, and others first discovered that atoms were composed of subatomic particles, physicists and chemists have continued to refine the model of the subatomic world. We now know that the protons and neutrons, which we once thought were fundamental particles, are actually made of even smaller bits of matter called "quarks". Physicists Murray Gell-Mann and George Zweig proposed this discovery independently in 1964. This discovery led to the notion that protons, neutrons, and even electrons (which are also made of quarks) can be interconverted in certain circumstances. For instance, a proton can change into a neutron when it combines with an electron.

This brings us to the question of nuclear decay—what kinds of changes do nuclei commonly undergo? As we discussed in ▶ Chap. 2, we have known for nearly a century that nuclei can spontaneously fall apart to produce alpha particles, beta particles and gamma radiation. **Alpha particles** are essentially helium nuclei, composed of two protons and two neutrons. Beta particles are fast-moving electrons that are ejected from inside the nucleus. In this case, it is important to note that we are not talking about the electrons that normally orbit the nucleus. Beta particles are electrons that have been produced by an inter-nuclear transformation. Gamma rays, unlike alpha and beta particles, are high-energy electromagnetic radiation. They are produced when protons, neutrons and electrons interconvert inside the nucleus. In fact, there are other combinations of quarks inside a nucleus that can result in the formation of other subatomic particles, which we will discuss a little later. Now, let's take a look at what happens to nuclei when they either emit or combine with subatomic particles and energy.

Alpha particles (α particles) – Particles emitted from the nucleus of a radioactive element during the process of alpha decay. They are helium nuclei (2 protons, 2 neutrons), with a +2 charge.

20.3.1 Beta-Particle Emission

In the case of potassium transforming into calcium, the process that is happening is **beta-particle emission**, a type of radioactive decay. In beta emission, the nucleus of $^{40}_{19}K$ ejects a **beta particle**, $^{0}_{-1}\beta$, that travels at 90% the speed of light. A new nucleus, $^{40}_{20}Ca$, is formed that has one more proton. Because calcium-40 has an energetically stable nucleus (you can verify this in a table of radioisotopes), no further *nuclear* reactions take place. For now, let's write the **nuclear equation** for beta-minus emission, or, more commonly, "beta emission" as

$$^{40}_{19}K \rightarrow {}^{40}_{20}Ca + {}^{0}_{-1}\beta$$

Beta-particle emission – A naturally occurring type of radioactive decay wherein an electron is ejected at high speed from the nucleus, typical of nuclei that have too many neutrons to be energetically stable. An antineutrino accompanies this emission, and sometimes one or more gamma rays as well. Also known as beta emission.

Beta particles (β particles) – Particles emitted from the nucleus of a radioactive atom during the process of beta decay. These particles are high-speed electrons.

Nuclear equation – An equation showing a nuclear transformation, where the atomic and mass numbers are provided.

We note that the sums of the atomic masses, as well as the sums of the atomic numbers, are the same on both sides of our equation. The nuclide that results from the beta emission, $^{40}_{20}Ca$, is comparable in size to potassium. In addition, a tiny product is ejected with mass number zero and a negative charge—an electron. In the context of nuclear decay, this nuclear electron is called a beta emission particle. The subscript –1 in $^{0}_{-1}\beta$ may look strange as an atomic number. Interpret it as a "negative one charge" rather than a "minus one proton."

Beta emission, like many other of the nuclear reactions we will study, has the following features:

- A nuclide of one element, through the process of **radioactive decay**, is transformed into a nuclide of another. The resulting **daughter nuclide** may either be stable or radioactive.
- The sum of the atomic numbers on one side of the nuclear equation is equal to the sum on the other. For the beta emission of ^{40}K, the sum of the mass numbers on each side of the equation is 40, and the sum of the atomic numbers on each side of the nuclear equation is 19.
- Energy is released in radioactive decay reactions. **Gamma rays** of varying energy nearly always accompany the nuclear reaction.

Radioactive decay – The process by which an unstable nucleus becomes more stable via the emission or absorption of particles and energy.

Daughter nuclide – An isotope that is the product of a nuclear reaction.

Gamma rays (γ rays) – A high-energy form of electromagnetic radiation that is emitted from the nucleus. Gamma rays sometimes accompany alpha and beta decays.

How can an electron be emitted from a nucleus when there are no electrons in the nucleus? To see how this could happen, let's venture down from the macroworld into the nanoworld to look at the beta emission of a single neutron. Neutrons aren't something you can keep in a bottle in your chemistry lab. But if you had the proper specialized equipment, you could demonstrate that a neutron decays, through rearrangement of quarks, to form a proton, an electron, and an **antineutrino**, $^{0}_{0}\nu$

$^{1}_{0}n$ ⟶ $^{1}_{1}p$ + $^{0}_{-1}\beta$ + $^{0}_{0}\nu$

In nucleus — In nucleus — β expelled — Expelled

$$^{1}_{0}n \rightarrow {}^{1}_{1}p + {}^{0}_{-1}\beta + {}^{0}_{0}\nu$$

The net effect of beta decay is the transformation of a neutron into a proton with the release of an electron, an antineutrino, and energy. This is consistent with the beta decay we wrote for $^{40}_{19}K$, where the product nuclide has one more proton.

Antineutrino – A subatomic particle produced in beta-minus decay that has no charge, has essentially no mass, and interacts only rarely with matter.

The antineutrino, ${}_{0}^{0}\nu$, is an example of **antimatter**. Each antimatter particle has a mate in our world of "real" matter. For the antineutrino, this mate is the **neutrino**, or "little neutron"—a particle similarly hidden within the neutron. With no charge and probably very little mass, the neutrino required a sophisticated piece of scientific detective work to prove its existence. However, because this ghostly particle and its antimatter mate interact little if at all with matter, neutrinos typically are omitted from nuclear equations.

Antimatter – Particles that have the same mass as, but charges opposite to, corresponding matter. Antimatter particles such as the positron and antineutrino are similar to the electron and neutrino, respectively, but have opposite characteristics.

Neutrino – A subatomic particle that is produced in beta-plus decay and that has no charge, has essentially no mass, and interacts only rarely with matter.

Example 20.2—Beta Emissions In and Around Us

A look at the world around us reveals a lot of naturally occurring radiation. Elements responsible for this radiation include carbon-14, potassium-40, and hydrogen-3 (tritium).

a. Write the nuclear equation for a beta emission by carbon-14.
b. If ${}_{2}^{3}He$ is formed via a beta emission, from which radioisotope was it produced?

Asking the Right Questions

What us a beta particle? What is the effect of a beta emission on the number of protons (Z) and the mass number (A) of the nuclide from which it is ejected?

Solution

A beta emission, the emission of an electron from a nucleus that has become more stable by the conversion of a neutron to a proton (plus a beta particle!), follows a set pattern: an increase of + 1 in Z and no change in A.

a. ${}_{6}^{14}C \rightarrow {}_{-1}^{0}\beta + {}_{7}^{14}N$
b. ${}_{1}^{3}H \rightarrow {}_{-1}^{0}\beta + {}_{2}^{3}He$

Are Our Answers Reasonable?

The masses and charges in the nucleus must balance if the answer is reasonable. That is the case with our answers.

What If...?

...the result of an emission of a beta particle from a nucleus results in nucleus of ${}_{56}^{134}Ba$? What was the original nuclide that emitted the beta particle?

(*Ans:* ${}_{55}^{134}Cs$)

Practice 20.2

Write the nuclear equation for the beta emission of ${}^{60}Co$.

20.3.2 Alpha-Particle Emission

There are two other common forms of radioactivity: alpha-particle emission and gamma-ray emission. To appreciate how these differ, let's revisit potassium-40. Imagine again that you were watching several individual atoms of the nuclide. No matter how long you waited, you would not observe an **alpha decay**, the emission of a helium nucleus (${}_{2}^{4}He$) from a larger nucleus. Why does potassium-40, like many radioisotopes, not exhibit this form of radioactivity? The simple answer is that the nucleus is made more energetically stable by beta emission than by alpha decay. Why? The answer to this is a little more complex and will be the focus of Sects. 20.5 and 20.6.

Alpha decay – A type of radioactive decay where an alpha particle is emitted from the nucleus of a radioactive nuclide. Common for elements whose nuclei are larger than bismuth, alpha decay is often accompanied by the release of a gamma ray.

Which elements decay by alpha emission? This information is often accessed using the internet or a chemistry handbook. However, there are many examples of this type of decay. Radon is one example. The nuclear reaction for the alpha decay of radon-222 is

$$^{222}_{86}\text{Rn} \rightarrow ^{218}_{84}\text{Po} + ^{4}_{2}\text{He}$$

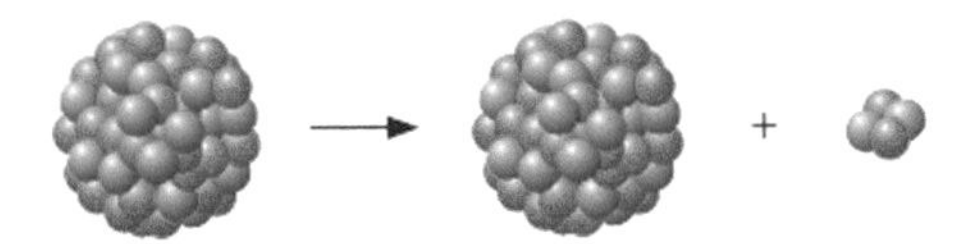

Energy is given off, along with an **alpha particle** ($^{4}_{2}$He), which is energetic but travels more slowly than a beta particle, at only about 5–10% of the speed of light. You also may see $^{4}_{2}\text{He}^{2+}$, $^{4}_{2}$He or simply α used to denote the alpha particle. Is alpha emission possible for a hydrogen or helium atom? No, because these atoms are too small to emit an alpha particle and still have any protons left for the remaining nucleus. In fact, alpha emission typically does not occur for small nuclei. It is far more common for elements above bismuth ($Z=83$).

20.3.3 Gamma-Ray Emission

For both alpha and beta emissions, the nuclear equations we have written so far are not complete, because a gamma ray is usually produced as well. When we include this gamma ray ($^{0}_{0}\gamma$) in the alpha decay of radon, the nuclear equation becomes

$$^{222}_{86}\text{Rn} \rightarrow ^{218}_{84}\text{Po} + ^{4}_{2}\text{He} + ^{0}_{0}\gamma$$

A gamma ray, as indicated in Fig. 20.2, is a high-energy photon—a form of electromagnetic radiation with a short wavelength (typically less than a picometer) traveling at the speed of light. Loss of energy from the system, in the form of a gamma ray, is favorable because the products of the reaction would then have less energy than the starting material. In essence, the free energy of the system is lowered.

Gamma rays carry off excess energy in an amount depending on the particular nuclide. In some cases, lower-energy gamma rays are identical to X-rays; however, we mentally distinguish between the two by noting that the former originated *inside* the nucleus and the latter *outside* of it. For medical purposes, as we will see in the final section of this chapter, both X-rays and gamma rays can accomplish similar diagnostic and therapeutic tasks.

Few nuclides are pure or almost pure gamma emitters. A notable exception is technetium-99 m, where the m indicates a **metastable state**. It represents a misarrangement of protons and neutrons in a nucleus after a neutron has become a proton. In beta emitters, this arrangement occurs so rapidly that it appears to be simultaneous with the original beta emission. In some cases, however, it is slow enough to be obvious:

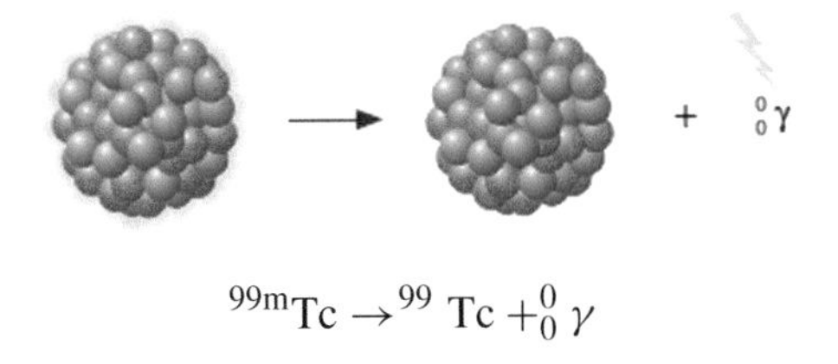

$$^{99m}\text{Tc} \rightarrow ^{99}\text{Tc} + ^{0}_{0}\gamma$$

Because it is relatively short-lived, technetium-99 m is generated on demand in nuclear medicine imaging.

Metastable state – An energetically unstable arrangement of protons and neutrons in a nucleus after a neutron has become a proton.

20

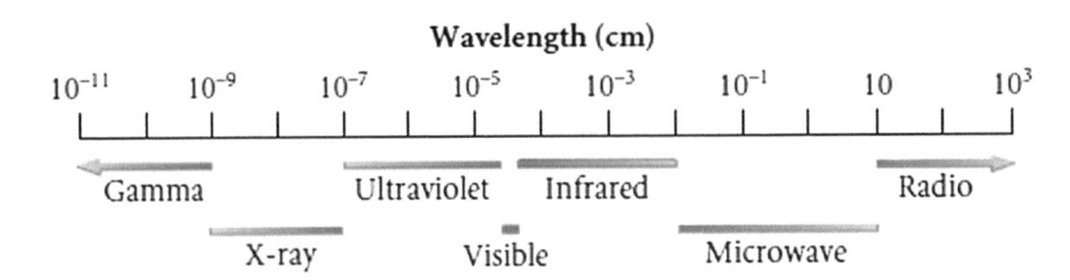

Fig. 20.2 Gamma rays are at the high-energy end of the electromagnetic spectrum. They have very short wavelengths on the order of ten-trillionths of a meter or less. The energy of cosmic rays is comparable to that of gamma rays, but cosmic rays are particles and are not part of the electromagnetic spectrum

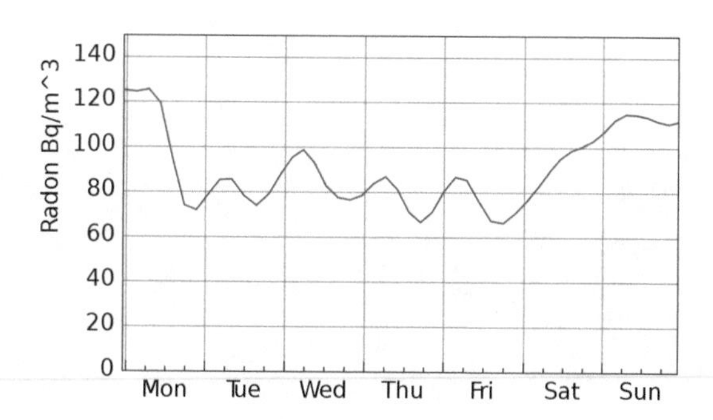

Fig. 20.3 Radon test kits can be purchased and used to measure the concentration of radon in a basement. In a radon-troubled basement, the level of radon gas fluctuates over the course of a week. *Source* ▶ https://upload.wikimedia.org/wikipedia/commons/5/5c/Radon-indoor-one-week.svg

Example 20.3—Alpha Emissions In and Around Us

Radon, discussed earlier, is present in the atmosphere at a concentration of about 1 part in 10^{21}. In some caves and basements, however, test kits can determine the exact concentration over time (Fig. 20.3). Write nuclear equations for the alpha decay of radon-220, radon-222, and radon-219. A gamma ray accompanies each of these processes.

Asking the Right Questions

What is an alpha particle? What is alpha decay? How do you symbolize the process?

Solution

Alpha decay is the emission of a helium nucleus, ${}^{4}_{2}\text{He}$, from the larger nucleus. As is common with many nuclear emissions, a gamma ray—a high-energy photon—is also released.

$$ {}^{222}_{86}\text{Rn} \rightarrow {}^{218}_{84}\text{Po} + {}^{4}_{2}\text{He} + {}^{0}_{0}\gamma $$

$$ {}^{220}_{86}\text{Rn} \rightarrow {}^{216}_{84}\text{Po} + {}^{4}_{2}\text{He} + {}^{0}_{0}\gamma $$

$$ {}^{219}_{86}\text{Rn} \rightarrow {}^{215}_{84}\text{Po} + {}^{4}_{2}\text{He} + {}^{0}_{0}\gamma $$

Are Our Answers Reasonable?

Again, the masses and charges in the nucleus must balance if the answer is reasonable. That is the case with our answers.

What If…?

…thorium-224 undergoes alpha decay, and the resultant nuclide itself undergoes alpha decay? What nuclide remains? (Note: this nuclide also undergoes alpha decay)

(*Ans:* ${}^{216}_{86}Rn$)

Practice 20.3

Write nuclear equations for the alpha decay of ^{218}Po and ^{230}Th.

20.3.4 Other Types of Radioactive Decay

The three types of radioactive decay that we have discussed are the most common, but two other modes of nuclear decay are also important to a comprehensive picture of radioactive processes. **Electron capture (EC)** is the combination of an inner-orbital electron and a proton from the nucleus to form a neutron. The mass of the nuclide doesn't change during the process because a proton and a neutron are similar in mass, but the atomic number decreases by one as the proton is changed to a neutron. Typically, this process is accompanied by the emission of X-rays from the nuclide. The radioactive decay of iodine-125, which is used to diagnose problems with the pancreas and intestines, occurs by the process of electron capture:

$$ {}^{125}_{53}\text{I} + {}^{0}_{-1}\beta \rightarrow {}^{125}_{52}\text{Te} $$

Electron capture (EC) – A type of radioactive decay that occurs when an inner-core electron is captured by a proton from the nucleus to form a neutron. The process is usually accompanied by the emission of X-rays.

In **positron emission** (${}^{0}_{+1}\beta$), a proton decays into a neutron and a positron. A neutrino accompanies this emission, and usually one or more gamma rays as well. A **positron**, or positive electron, is a particle that has the same mass

Table 20.2 Radioactive decay processes

Type of decay	Emission	Change in atomic number	Change in mass number
Alpha-particle emission	${}^{4}_{2}He$	–2	–4
Beta-particle emission	${}^{0}_{-1}\beta$	+1	0
Gamma-ray emission	${}^{0}_{0}\gamma$	0	0
Positron emission	${}^{0}_{+1}\beta$	–1	0
Electron capture	X-ray	–1	0

as an electron but carries a charge of + 1. Interestingly, the positron typically doesn't have a very long life, because when it comes into contact with an electron, the two particles combine to form two gamma rays. This type of radioactive decay is important in the lighter elements such as aluminum-26.

$$ {}^{26}_{13}Al \rightarrow {}^{0}_{1}\beta + {}^{26}_{12}Mg $$

Positron – The antimatter equivalent of an electron. Positrons have a positive charge and the same mass as an electron.

Positron emission – See beta-plus emission.

Table 20.2 lists five types of radioactive decay that are important in understanding nuclear decay processes.

20.3.5 Decay Series

Radon, discussed in Exercise 20.3, is one example in which the products of the nuclear reaction are still radioactive. All isotopes of the elements past bismuth (Z = 83) are radioactive. So far we have described nuclear decay as though it were a one-step process. However, a nuclide may decay to form a second radioactive nuclide, which in turn may decay more than a dozen times in a stepwise progression toward stability that is called a **decay series**.

Decay series – A series of nuclear reactions that a large nuclide undergoes as it changes from an unstable and radioactive nucleus to a stable nucleus.

► https://upload.wikimedia.org/wikipedia/commons/7/7a/Yellowcake_in_a_test_tube.jpg

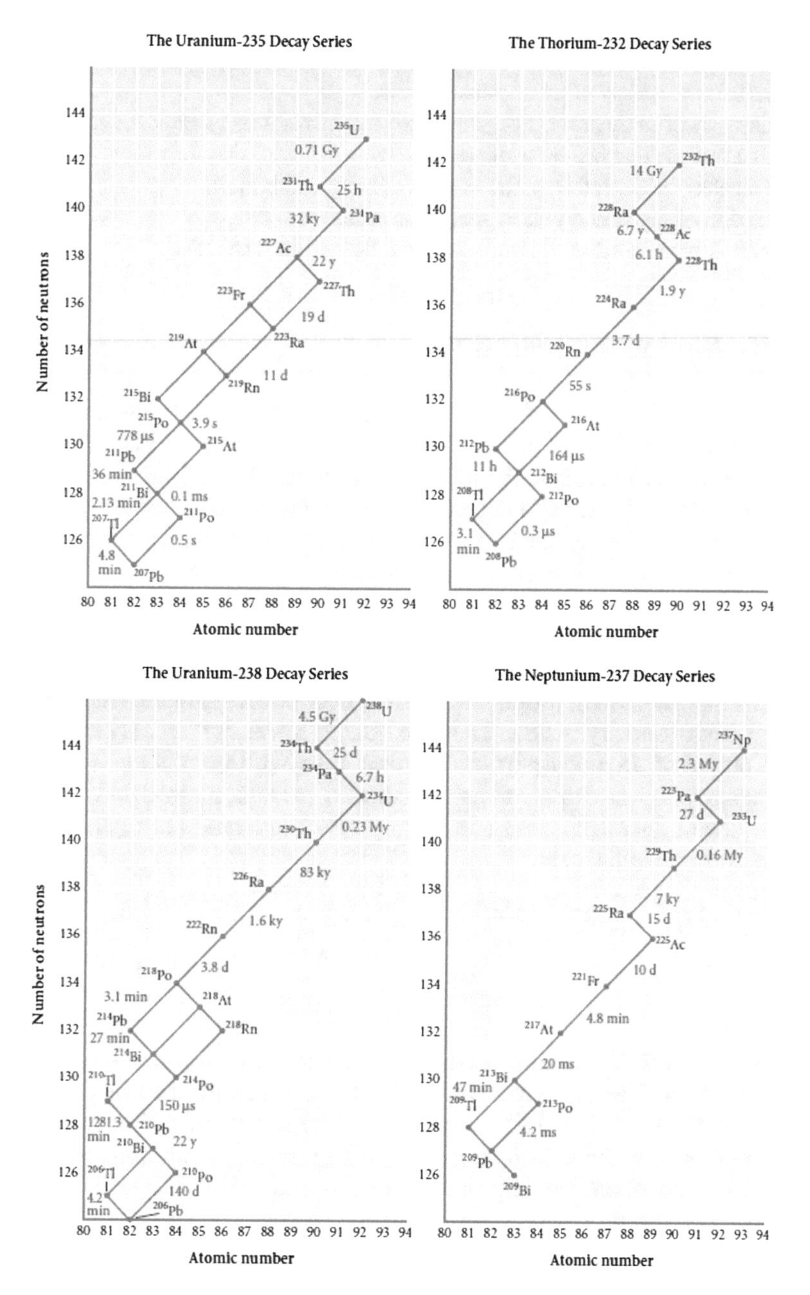

Fig. 20.4 The four natural decay series. Each decay series begins at the nuclide listed at the top right-hand side of the series and proceeds to the lower left-hand side by nuclear decay. The half-life of many of these transitions are indicated in red numbers. Gamma emissions accompany many of these decays

Consider the element uranium. Two natural isotopes exist for this element, ^{238}U and ^{235}U, with natural abundances of 99.28% and 0.72%, respectively. These isotopes exhibit a fairly extensive decay series. Both series consist of alpha and beta emissions, as shown in Fig. 20.4. The accompanying gamma rays are not shown in these series. Why do the lines in the decay series zig-zag? Alpha decays decrease Z by 2, thus moving the line to the left. But a beta-minus emission increases Z by 1 unit. If beta emission follows an alpha emission, there is a zag back to the right. The series that begins with ^{235}U ends with the stable lead-207 isotope, and the ^{238}U series forms ^{206}Pb. There are two other naturally occurring decay series, one that begins with ^{237}Np to form ^{209}Bi, and one in which ^{232}Th decays to ultimately form ^{208}Pb. Along the way, these decay chains provide dozens of radioisotopes that we typically find in our biosphere. This is one of the sources of naturally occurring radioactive isotopes. Of particular concern is radon, a radioactive gas that can collect in basements dug into soil that is rich in uranium ores. Is radon harmful to humans? Understanding the relationship between radioactivity and human health will help us answer this question.

▶ Here's What We Know So Far

- Radioactivity results from the decay of an unstable nucleus.
- There are three main types of radioactive decay and three common forms of radioactivity.
- The alpha particle is a fast-moving helium nucleus.
- The beta particle is an electron ejected from the nucleus of an unstable atom.
- The gamma ray is a burst of high-energy electromagnetic radiation.
- A decay series is a stepwise progression of a radioactive nuclide toward stability.

20.4 Interaction of Radiation with Matter

Taking a walk on a crisp, sunny day is one of the pleasures of autumn. We notice any cloud that blocks the Sun. Not only does the shade reduce the amount of light hitting your eyes, but your skin also registers the change. The energy exchanges, from the infrared through the ultraviolet regions of the electromagnetic spectrum, are apparent and profound. However, when it comes to detecting the small amounts of alpha particles, beta particles, and gamma rays that bombard you on a daily basis, your senses don't help. For instance, you cannot detect the alpha particles that radioactive radon, an odorless, colorless, and tasteless gas, emits.

Though seemingly invisible, the different types of radiation form quite a nuclear arsenal, as summarized in ◻ Fig. 20.5. We can envision alpha particles as the cannon balls of the group. With their greater size and +2 charge, they do not travel very far before they smash into other atoms. The typical collision results in the capture of two electrons to form a neutral helium atom. In contrast, beta particles, being smaller and traveling more rapidly, are the equivalent of high- velocity bullets. They are able to travel significantly longer distances before a collision, but they lack the punch of an alpha particle. Gamma rays, with no mass and no charge, are akin to laser weapons. The high-frequency photons can travel great distances through matter, now and then searing something in their path. What about the antineutrinos that accompany beta emissions? Having neither mass nor charge, they are not absorbed and are considered harmless.

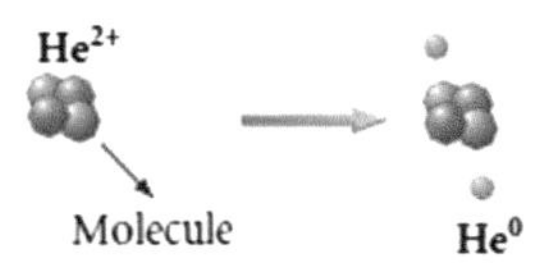

The type and the energy of the radiation dictate what must be used to shield us to the greatest extent possible (see ◻ Table 20.3). Alpha particles penetrate matter the least, being stopped by just a few centimeters of air, by the outer layer of your skin (which is mostly dead cells), or by a piece of paper. Beta particles penetrate more deeply and can pass through several pieces of paper, through a thin sheet of aluminum, or about a centimeter into your skin. In contrast, gamma rays can pass right through you. Shielding your body from them requires several inches of aluminum or lead, and even these may not do the job.

A visit to the dentist reveals an interesting feature of stopping damage from nuclear radiation. Before an X-ray of a patient's teeth is obtained, the patient is typically draped with a sheet of lead. Why is lead one of the materials employed to shield us from high-energy electromagentic radiation, such as X-rays or nuclear radiation? The high density of lead, 11.3 g/cm^3, means that only two inches of lead will easily shield you from alpha and beta radiation, as well as from most gamma radiation. However, lead is not unique in this ability. Gold $\left(d = 19.3\,\text{g/cm}^3\right)$ is more dense than lead and would work even better. Bricks or blocks of a moderately dense material such as concrete also would do the trick. However, no dentist is likely to place an apron of gold or a foot of concrete over your abdomen before taking dental X-rays. Given its density and price, lead is often the shielding medium of choice.

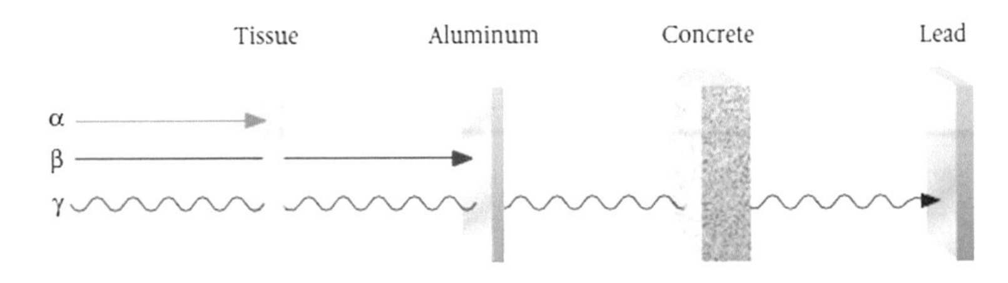

◻ **Fig. 20.5** Alpha, beta, and gamma radiation differ in their penetration. Gamma rays are the most highly penetrating, and alpha particles the least

Table 20.3 Differences in penetration and shielding for different types of radiation

Type	Examples	Penetration in dry air	Penetration in skin or tissue	Shielded by…	Q[a]
Alpha	Uranium, plutonium, americium	2–4 cm	0.05 mm	Paper, air, clothing	20
Beta	Potassium-40, cesium-137	200–300 cm	4–5 mm	Heavy clothing	~1
Gamma	Technetium, cobalt-60	500 m	50 cm	Lead, concrete	~1
Fast neutrons	Accelerators	Several hundred feet	High	Water, plastic, concrete	20

[a]Q is the relative biological effectiveness, a factor that indicates relative amounts of damage to living tissue

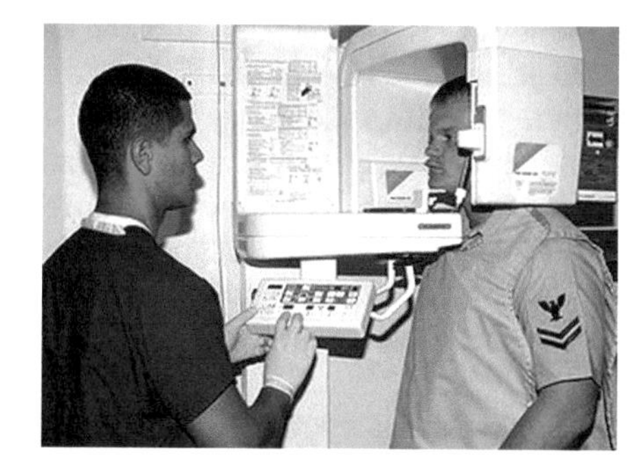

▶ https://upload.wikimedia.org/wikipedia/commons/thumb/6/6b/Dental_X-Ray.jpg/320px-Dental_X-Ray.jpg

Although alpha, beta, and gamma radiation differ in penetrating ability, they are alike in the effects they produce on the molecular level. All three are known as **ionizing radiation**; that is, they are capable of forming ions by knocking electrons out of atoms and molecules. The damage they cause is the reason they can be detected with, for example, a Geiger counter or film badge.

Ionizing radiation – Radiation such as alpha and beta particles, gamma rays, or X-rays that is capable of removing an electron from an atom or a bond when it interacts with matter.

The consequences of ionizing radiation can be negligible or severe, depending on how many molecules are damaged inside the body. Although small amounts of radiation typically lead to only negligible damage that can be repaired by the body, large doses of radiation can be life-threatening. How does ionizing radiation cause damage? Between about 50% and 70% of your body is water, so a scenario of particular interest occurs when ionizing radiation strikes a water molecule. The blow can knock off an electron to form a highly reactive species with an unpaired electron - a cation radical:

$$H_2O \xrightarrow{\text{ionizing radiation}} [H_2O]^{\oplus} + e^{\ominus}$$

Water (neutral) **Water (cation radical)**

The resulting radical cation undergoes autoprotolysis to form hydroxyl free radicals:

$$H_2O^{\oplus} + H_2O \rightleftharpoons H_3O^{\oplus} + OH$$

The fate of the hydroxyl free radicals is particularly damaging to the cells within your body. If they encounter a molecule of DNA in a dividing cell, they may damage a section of the genetic code (Fig. 20.6). The damage

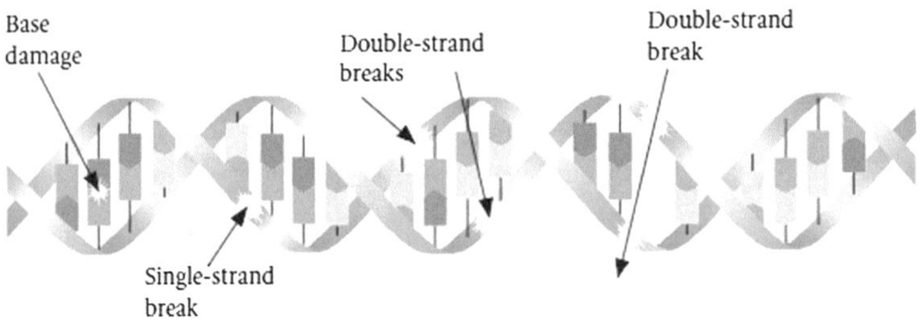

Fig. 20.6 Examples of chromosomal damage from radiation

could result in death of the cell, triggering the formation of mutant proteins (proteins with a non-natural primary structure; see ▶ Chap. 22) or triggering an abnormal function of the cell leading to cancer. In short, nuclear radiation can be carcinogenic. We know the following things happen when exposed to damaging radiation: (1) The more radiation a person is exposed to, the greater the likelihood of that person's developing cancer. (2) These cancers may not show up until decades after the time of exposure.

Is there a threshold below which radiation is safe? Recent studies have convincingly demonstrated that the damage inflicted by low-level radiation upon workers in the nuclear industry and upon World War II nuclear bomb survivors have been greatly underestimated. Similarly, the link between fetal X-rays (an abandoned medical practice) and childhood cancer has been established. Fortunately, because cancer cells grow and divide, they also are susceptible to radiation. Carefully measured doses of radiation directed at cancerous cells can result in their death.

Overall, the biological effects of nuclear radiation depend on the quantity of energy transferred to the cells and tissues. In the United States, the **rem** or "roentgen equivalent in man" is the unit for estimating the damage. Other units that measure radiation include the **becquerel**, **curie**, **roentgen**, and **rad**. Of these, only the becquerel (Bq) is an SI unit. Scientists in most parts of the world employ two other SI units, the **sievert (Sv)** and the **gray (Gy)**, which are related to the rem and the rad, respectively, as shown in Table 20.4.

Rem – A unit that measures the "equivalent dose" of radiation; that is, it takes into account the interaction of radiation with human tissue. The word stands for "roentgen equivalent in man." The rem is not an SI unit but is related to the SI unit, the sievert.

Roentgen – A unit used to measure exposure to radiation. One roentgen is equal to 2.58×10^{-4} C/kg of dry air at STP.

Becquerel (Bq) – An SI unit of activity equivalent to one nuclear disintegration per second.

Curie (Ci) – A larger unit of activity than the becquerel, equivalent to 3.7×10^{10} Bq.

Sievert (Sv) – A unit that measures the "equivalent dose" of radiation; that is, it takes into account the interaction of radiation with human tissue. One sievert equals 100 rem.

Gray (Gy) – A measure of absorbed radiation equal to 100 rad.

Rad – A unit of energy absorbed by irradiated material equal to 0.01 J/kg of exposed material.

Table 20.4 Units for measuring radiation

Measure of	Name	Abbreviation	Definition
Activity	Becquerel	Bq	1 disintegration per second
Activity	Curie	Ci	1 curie $= 3.7 \times 10^{10}$ Becquerel
Exposure	Roentgen	R	1 roentgen $= 2.58 \times 10^{-4}$ coulombs of charge per kg of tissue
Absorbed dose	Radiation absorbed dose	Rad	1rad $= 1 \times 10 - 2$ J of energy deposited per kg of tissue
Absorbed dose	Gray	Gv	1 gray = 100 rad
Dose equivalent	roentgen equivalent in man	Rem	Q $\times$ absorbed dose
Dose equivalent	Sievert	Sv	1 sievert = 100rem

Table 20.5 Health effects of acute radiation exposure

Exposure (rem)	Health effect	Time to onset of symptoms
10	Burns, changes in blood chemistry	
50	Nausea	Hours
75	Vomiting, hair loss	2–3 weeks
100	Hemorrhage	
400	Death	Within 2 months
1000	Internal bleeding, death	Within 1–2 weeks
2000	Death	Within hours

The quantity of energy absorbed by tissues is directly related to the time of exposure to ionizing radiation. In fact, time is an important part of the decision-making process in medical diagnosis and treatment. For example, prostate cancer in elderly men is often treated by the implantation of metal "seeds" coated with ^{125}I, ^{103}Pd or, more recently, ^{131}Cs. Why use these nuclides? They are sufficiently radioactive for only the time needed to control the cancer without creating new cancers. How do we know how long they will be radioactive? The most important measure that we use to judge the length of time a substance is radioactive is its half-life.

The U.S. Environmental Protection Agency suggests that the average person receives an annual dose of 0.3 rem of radiation from natural sources. Over the course of a lifetime, this is predicted to result in 5 or 6 deaths due to cancer per 10,000 people. This sounds shocking until we consider that the rate of deaths due to cancer from nonradioactive sources is predicted to be about 2000 people per 10,000. Larger doses received in one exposure have a much more deleterious effect on human health, as shown in Table 20.5. Acute exposures, such as those that result from accidents in nuclear power plants and those that resulted from the U.S. bombing of Japan in World War II, cause severe damage to the human body, often resulting in lifelong health problems and even death.

20.5 The Kinetics of Radioactive Decay

The half-life, $t_{1/2}$, of a radioactive isotope is the period of time it takes for exactly half of the original nuclei in a radioactive sample to decay. Table 20.6 illustrates that half-lives can vary widely among the radioactive isotopes. They can be as short as a few microseconds and as long as a few billion years.

How can the half-life tell us how much radioactivity remains after a given time? Recall that after one half-life, one-half of the sample has reacted and only one-half of the sample remains. After a second half-life,

Table 20.6 Half-lives of some radioactive elements

Element	Nuclide	Half-life
Nobelium	^{250}No	250 μs
Technetium	^{99m}Tc	6.0 h
Thallium	^{201}Tl	21.5 h
Radon	^{222}Rn	3.8 d
Iodine	^{131}I	8.040 d
Palladium	^{103}Pd	16.97 d
Cobalt	^{60}Co	5.271 y
Hydrogen	^{3}H	12.3 y
Carbon	^{14}C	5730 y
Radium	^{226}Ra	1.6×10^3y
Uranium	^{238}U	4.5×10^9y

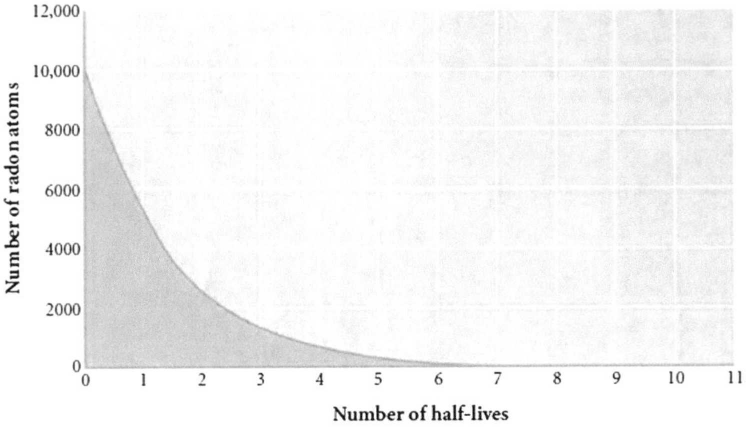

Fig. 20.7 The kinetics of radioactive decay. The half-life of a radioisotope, such as radon, is the time required for half of the nuclei in the radioactive sample to decompose

$1/2 \times 1/2 = 1/4$ of the sample is left. After 3 half-lives, 1/2 of 1/2 of 1/2, or $(1/2)^3 = 1/8$, of the sample remains. More generally, then, for n half- lives, the fraction of the original sample remaining is $(1/2)^n$. This trend is shown in Fig. 20.7 and is valid for all radioactive decay processes.

Palladium-103, used to coat the seeds implanted in the prostate, decays by electron capture, in which an inner-orbital electron is captured by a proton in the nucleus to form a neutron. The half-life of Pd-103 is 16.97 days. How long will it take for the radiation to be diminished to 1.00% of its original value so it is considered safe for radiation workers, the prostate cancer patient, and his family? A look at Fig. 20.7 indicates that this will take between 6 and 7 half-lives, or between 102 and 119 days, for Pd-103.

Such approximations are often sufficient. However, when necessary, we can also solve explicitly for n.

$$(1/2)^n = 1/100 = 0.0100$$

Taking the natural logarithm (ln) of both sides yields

$$\begin{aligned} n[\ln(1/2)] &= \ln(0.0100) \\ n(-0.693) &= -4.605 \\ n &= 6.645 \text{ half-lives} \end{aligned}$$

Solving for the time needed, we find that

$$\begin{aligned} t &= 6.645 \text{ half-lives} \times 16.97 \text{ days/half-life} \\ &= 113 \text{ days} \end{aligned}$$

There is another approach based on the understanding that radioactive isotopes decay via first-order kinetics. Recall that the relationship among concentration, time, and half-life for any first-order process is shown by

$$\ln \frac{[A_t]}{[A_i]} = -kt$$

where A_t = the amount of substance remaining.
A_i = the initial amount of substance.
k = the first-order rate constant for the reaction (in this case, the decay).
t = time.
We also know that the rate constant for a first-order reaction can be determined by

$$k = \frac{0.693}{t_{1/2}}$$

This means that we can solve explicitly for the rate constant if we are given the half-life, $t_{1/2}$. In our example,

$$k = \frac{0.693}{t_{1/2}} = \frac{0.693}{16.97 \text{ day}} = 0.04084 \text{ day}^{-1}$$

Assuming that we start with $A_i = 1.000$, we can determine that if we only have 1% remaining, $A_t = 0.0100$. Then

$$\ln\frac{[0.0100]}{[1.00]} = -0.04084t$$
$$-4.605 = -0.04084t$$
$$113 = t$$

so $t = 113$ days, and we obtain the same answer (113 days) by either method.

When would the entire Pd-103 sample be gone? This is a question we cannot precisely answer, although we can come very close. Why are we unable to tell when all of the Pd-103 is gone? After each half-life, half of the number of radioactive atoms remaining from a previous half-life are still remaining. Sooner or later, after a very large number (roughly 80) of half-lives have passed, only two radioactive atoms remain for every mole we started with. Half-life is a statistical measure. Probabilities, which don't apply with a sample size of two, will not accurately describe the rate of the reaction.

Example 20.4—Half-life Calculations: Here Today, Gone Tomorrow?

After exercising on a treadmill, a patient was given thallium-201 for a diagnostic scan of his heart (Fig. 20.8). How long will it take for 95.0% of the thallium to have decayed? The half-life of thallium-201 is 73.1 h.

Asking the Right Questions

Can you suggest two ways to solve the example? One relates to finding the number of half-lives, the other to finding the rate constant.

Solution

Using the first method, we find the number of half-lives that pass until 5.0%, or a fraction of 0.050, of the Tl-201 remains.

$$(1/2)^n = 0.050$$
$$n[\ln(1/2)] = \ln(0.050)$$
$$n(-0.693) = -3.00$$
$$n = 4.32 \text{ half-lives}$$
$$t = 4.32 \text{ half-lives} \times 73.1 \text{ h/half-life} = 316 \text{ h}$$

We can also proceed using the second method, via the rate constant and the first-order rate law:

$$k = \frac{0.693}{t_{1/2}} = \frac{0.693}{73.1\text{ h}} = 0.00948\text{ h}^{-1}$$
$$\ln\frac{[0.050]}{[1.00]} = -0.00948t$$
$$-3.00 = -00948t$$
$$316\text{ h} = t$$

Are Our Answers Reasonable?

On the surface, both methods resulting in the same answer should give us confidence in the answer. On a deeper level, would we expect an answer of around 316 h for 95% of an isotope to decay in which the half-life is 73.1 h? Knowing the meaning of a half-life, in which 50% of the remaining isotope decays, one half-life = 50%, 2 half-lives = 75%, 3 half-lives = 87.5%, and 4 half-lives = 93.75%, having 5.0% remaining (95.0% decayed) means a little more than 4 half-lives. Our answer is therefore reasonable.

What If…?

…we have a 75.00-g sample of sample of ^{232}Th, which has a half-live of 1.40×10^{10} years? How much of the sample will decay in 1.00 billion years?

(*Ans*: 3.62 g).

Practice 20.4

The half-life of ^{198}Au is 2.69 days. How long would it take for 99% of a gold-198 sample to decay?

The previous problem involved percentages. But you also can work half-life problems given a starting mass or a starting number of atoms. Both are measures of the radioactivity present. Similarly, you can work half-life problems given a unit of activity such as the number of disintegrations per second (becquerels) or curies, because these also measure the amount of radioactivity and are proportional both to the mass and to the number of atoms. Such variations in units reflect the differing needs of real-world situations where you are likely to encounter radioisotopes.

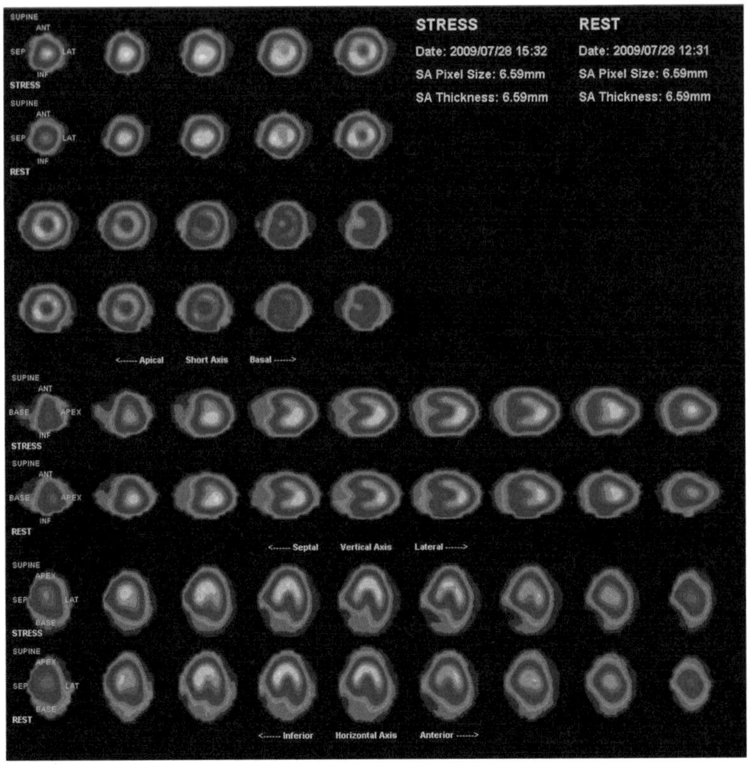

Fig. 20.8 A cardiac stress test on a patient requires analysis of the heart at rest and after vigorous exercise, such as one gets from exercising on a treadmill. *Source* ▶ https://upload.wikimedia.org/wikipedia/en/0/02/NI_mpi2.jpg

The half-life of a radioactive isotope can also help you to determine how long the radioisotope will be useful or, perhaps, to weigh the hazards associated with that particular isotope. For example, would you rather be around a radioisotope that decayed to 1% of its activity in 10 s or one that did so in 10 centuries? For the latter, the rate of decay is much lower, and you would be bombarded with far fewer alpha particles, beta particles, or gamma rays in a given time period. This principle is important in the use of technetium in medical imaging. We have noted that technetium-99 m is a gamma emitter. It has a half-life of 6.01 h, which is long enough for a medical procedure, but short enough that the substance doesn't persist very long.

$$^{99\,\mathrm{m}}_{43}\mathrm{Tc} \rightarrow ^{99}_{43}\mathrm{Tc} + ^{0}_{0}\gamma \quad t_{1/2} = 6.01\,\mathrm{h}$$

The product nuclide, $_{43}{}^{99}$Tc, is still radioactive. However, with its half-life of 213,000 years, the activity of Tc-99 as a beta emitter is low. In the two and a half weeks it takes for the majority of Tc-99 to be completely eliminated from the body through biological processes, it does little damage.

20.6 Mass and Binding Energy

By now you might be a bit suspicious. Many nuclei are unstable and spontaneously decay. Significant amounts of energy are released during alpha, beta, and gamma emission—enough to ionize molecules or to kill cancer cells—but we haven't talked about the source of the energy. Where does it come from, and how does this help explain the decay processes that nuclei undergo?

The place to start to find an answer is with Einstein's famous equation,

$$E = mc^2$$

in which the constant, c, is equal to the speed of light. This relationship illustrates that a particular mass, when completely converted, is equivalent to a surprisingly large quantity of energy. Any chemical reaction that is accompanied by a loss in energy, such as in an exothermic reaction, also has a corresponding loss in mass. However, such mass losses are so miniscule that we cannot detect them using conventional instruments. In contrast, the changes is mass due to a nuclear reaction, though tiny, are quite measurable.

Let's explore this connection by examining the formation of a nitrogen-14 nucleus from its individual nuclear particles:

$$7p + 7n \rightarrow {}^{14}\text{N nucleus}$$

A mole of nitrogen-14 nuclei (without any electrons) weighs 13.99540 g. What is the mass of 7 separate moles of protons (1.00727 g/mol) and 7 mol of neutrons (1.008665 g/mol)?

$$\begin{aligned}
&7\text{ mol of protons} \times 1.00727\text{ g/mol} = 7.05089\text{ g}\\
&7\text{ mol of neutrons} \times 1.008665\text{ g/mol} = 7.060655\text{ g}\\
&\text{Total mass} = 7.05089\text{ g} + 7.060655\text{ g} = 14.11154\text{ g}
\end{aligned}$$

This exceeds the mass of a mole of nitrogen-14 nuclei by (14.11154 g – 13.99540 g) = 0.11614 g. This **mass defect**, the mass difference between the individual protons and neutrons and the composite nucleus, was used in binding the protons and neutrons together within the nucleus. The **binding energy**, expressed as a positive number, is the energy required to dismantle the nucleus into its individual protons and neutrons. Very often, this is expressed in terms of binding energy per **nucleon** by dividing the energy by the number of protons and neutrons in the nucleus.

Mass defect – The loss in mass that occurs when a nucleus is formed from its protons and neutrons.

Binding energy – The energy released when a nucleus is formed from protons and neutrons. Binding energies are expressed as a positive number.

Nucleon – The name given to a particle (proton or neutron) that is part of a nucleus. For example, ^{13}C contains 13 nucleons, 6 protons, and 7 neutrons.

To ensure that our calculations give us positive numbers for binding energy, we typically use Einstein's equation in a different form:

$$\Delta E = |\Delta m|c^2$$

where ΔE is the binding energy, and,

$|\Delta m|$ is the absolute value of the change in mass in kilograms.

We use kilograms because our SI unit of energy, the joule, is defined as kg m/s^2. For nitrogen-14, using our mass defect of 0.11614 g(= 1.1614×10^{-4} kg), the binding energy can be calculated as

$$\Delta E = |\Delta m|c^2 = \left|-1.1614 \times 10^{-4}\,\text{kg}\right| \times \left(2.9979 \times 10^8\,\text{m/s}\right)^2 = 1.04 \times 10^{13}\,\text{J}$$

Example 20.5—The Energy Advantage of Nuclear Reactions

When bombarded with neutrons, uranium-235 can split to form bromine-87, lanthanum-146, and three neutrons. The masses reported here are those of the bare nuclei.

$$\begin{array}{lccccccccc}
 & {}^{235}_{92}\text{U} & + & {}^{1}_{0}\text{n} & \rightarrow & {}^{87}_{35}\text{Br} & + & {}^{146}_{57}\text{La} & + & 3{}^{1}_{0}\text{n}\\
\text{g/mole:} & 234.993455 & & 1.008665 & & 86.901568 & & 145.89440 & & 3 \times (1.008665)
\end{array}$$

Calculate the mass defect and the energy released in kJ/mol.

Asking the Right Questions

What is the mass defect? Would you expect the energy equivalent to be relatively small or large compared to the energy released in a conventional chemical reaction, such as combustion?

Solution

The mass defect is the difference in mass between the ending and starting materials.

$$\begin{aligned}
\Delta m &= (86.901568 + 145.89440 + 3.0260)\\
&\quad - (234.993455 + 1.008665)\\
&= -0.18015\text{ g/mol} = -1.8015 \times 10^{-4}\text{ kg/mol}
\end{aligned}$$

The energy equivalent to this mass is $\Delta E = |\Delta m|c^2$.

$$\Delta E = |\Delta m|c^2 = \left|-1.8015 \times 10^{-4}\ \text{kg/mol}\right| \times \left(2.9979 \times 10^8\ \text{m/s}\right)^2$$
$$= 1.621 \times 10^{13}\ \text{J/mol}$$
$$= 1.621 \times 10^{10}\ \text{kJ/mol}$$

This is many orders of magnitude greater than the energy released by an equivalent mass of materials in a combustion reaction.

Is Our Answer Reasonable?

The key lies in the quantity of energy released. Nuclear processes release much greater quantities of energy than do chemical processes such as combustion. Our answer of more than a billion kJ/mol, which is about 7 orders of magnitude greater than for a combustion reaction, is therefore reasonable.

What if…?

…we have 0.30 mol of $^{75}_{33}\text{As}$. What is the mass defect and total binding energy in kJ/mol? The atomic mass of $^{75}_{33}\text{As}$ is 74.921597 g/mol.

(*Ans*: mass defect = −0.682243 g/mol = -6.82243×10^{-4} kg/mol, binding energy for 0.30 mole = 1.840×10^{10} kJ/mol)

Practice 20.5

Calculate the energy released in this process.

$$^{224}_{88}\text{Ra} \rightarrow ^{220}_{86}\text{Rn} + ^{4}_{2}\text{He}$$

Some atoms are more thermodynamically stable than others. A table of binding energies shows only that the values generally increase as the atoms get heavier. However, if you recalculate the binding energy for each atom and report the values per nucleon (proton or neutron), the stabilities pop right out at you. Nuclei with greater binding energies per nucleon are more stable. Note in Fig. 20.9 that the lightest elements, those with mass numbers of 20 or less, have the lowest binding energies per nucleon. In comparison with iron, helium simply does not have enough nucleons to be as strongly glued together. The process that liberates energy on the sun, **fusion** (or nuclear fusion), is energetically favorable because lighter elements such as helium are joined to form heavier ones that have a more favorable binding energy per nucleon.

Fusion (or nuclear fusion) – A type of nuclear reaction wherein small nuclei are joined to form a larger nucleus with the release of energy. Nuclear fusion powers the stars.

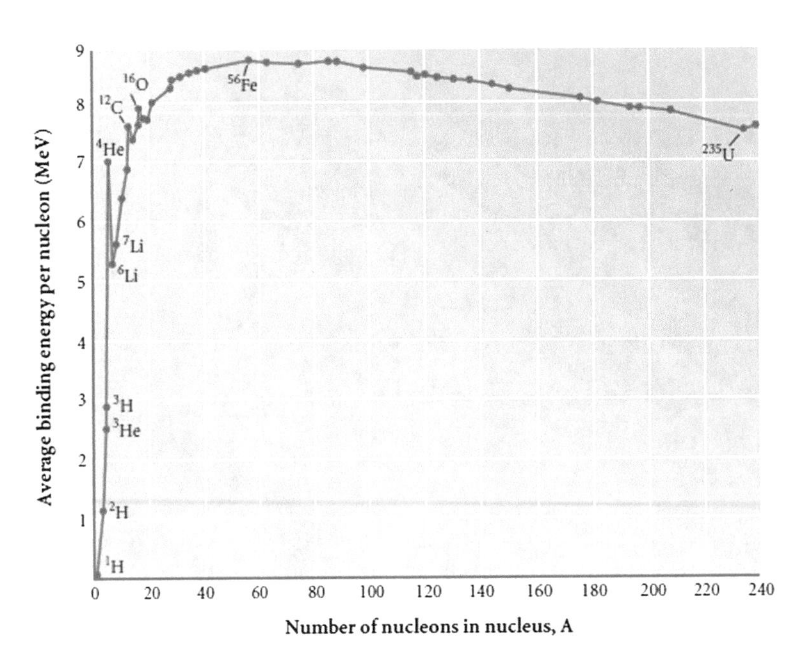

Fig. 20.9 The binding energy per nucleon reaches a maximum near iron, atomic number 26

20

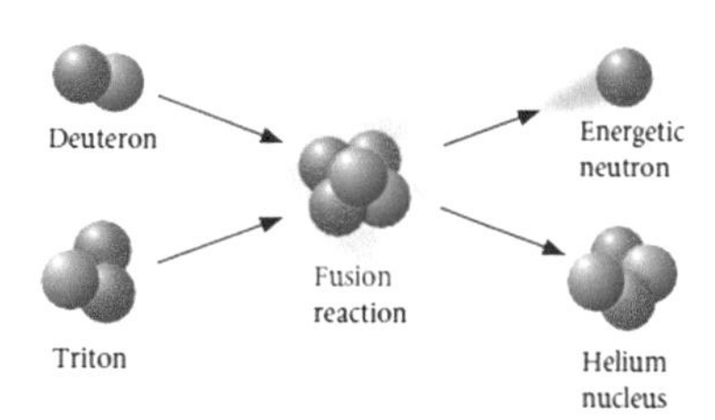

The heaviest elements also have somewhat lower stability. Uranium, with so many protons and neutrons packed into its nucleus, has a lower binding energy per nucleon than iron or cobalt. As we will see in ▶ Sect. 20.7, it is energetically favorable to split heavier nuclei into smaller ones via **fission** (or nuclear fission) to form nuclei with a more favorable binding energy per nucleon.

Fission (or nuclear fission) – A type of nuclear reaction wherein a large nucleus splits into two or three smaller nuclei with the release of energy.

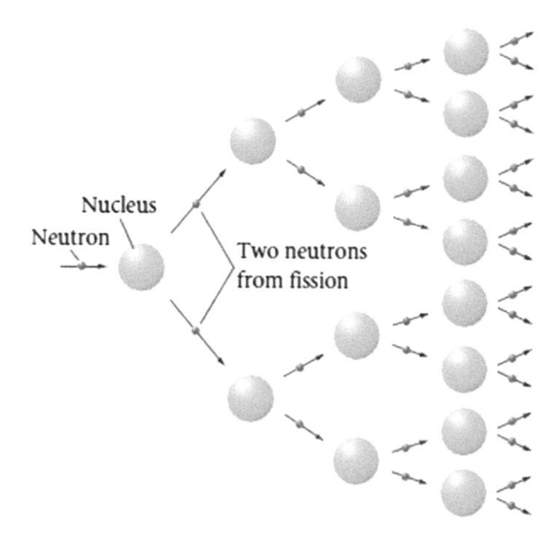

Finally, look at the maximum of the curve and you will find elements with the most stable nuclei—those with mass numbers around 60, such as iron and nickel. Other elements also have surprisingly high binding energies given their mass number, such as ^{4}He (an alpha particle), ^{12}C, and ^{16}O. Alternatively, one might argue that the values for ^{6}Li and ^{14}N are surprisingly low.

Example 20.6—Working With Binding Energy

What is the binding energy, in J/nucleus and J/nucleon, for ${}_{51}{}^{124}$Sb (atomic mass = 123.905938 g/mol)?

Asking the Right Questions

How do you calculate the binding energy? How does this relate to the individual protons and neutrons?

Solution

The binding energy is the energy required to separate the nucleus into its individual protons and neutrons.

Mass of the nucleons in a mole of antimony

$$= 51 \text{ protons } \times 1.007276\,\text{g/mol protons} = 51.37108\,\text{g}$$

$$73 \text{ neutrons } \times 1.008665\,\text{g/mol neutrons} = 73.63255\,\text{g}$$

$$\text{Total mass} = 125.00363\,\text{g}$$

$$\Delta m = 125.00363 - 123.905938$$

$$= 1.097687\,\text{g/mol} = 1.097687 \times 10^{-3}\,\text{kg/mol}$$

The energy equivalent to this mass is $\Delta E = |\Delta m|c^2$.

$$\Delta E = |\Delta m|c^2$$

$$= \left|1.097687 \times 10^{-3}\,\text{kg/mol}\right| \times \left(2.9979 \times 10^{8}\,\text{m/s}\right)^2$$

$$= 9.86536 \times 10^{13}\,\text{J/mol}$$

This is the binding energy for a *mole* of nuclei. For one nucleus,

$$\frac{9.86536 \times 10^{13}\,\text{J}}{\text{mol}} \times \frac{1\,\text{mol}}{6.022 \times 10^{23}\,\text{nuclei}}$$

$$= 1.6382 \times 10^{-10}/\,\text{nucleus}$$

There are 124 nucleons (protons and neutrons) per nucleus, so

$$\frac{1.6382 \times 10^{-10}\,\text{J}}{\text{nucleus}} \times \frac{1\,\text{nucleus}}{124\,\text{nucleons}} = 1.3211 \times 10^{-12}\,\text{J/ nucleon}$$

Are our Answers Reasonable?

As in the previous example, the energy per mol for the nuclear process is much larger than for any chemical process, so the answer is reasonable. In addition, the energy *per nucleon* is very small, which is also reasonable.

What If…?

…we have a sample of ^{106} Cd? What is the binding energy per nucleon? The atomic mass of (^{106}Cd = 105.90646 g/mol)

(*Ans:* 1.331×10^{-12} *J/ nucleon*)

Practice 20.6

What is the binding energy in kJ/mol and J/nucleon for ^{251}Fm (atomic mass = 251.08157)?

20.7 Nuclear Stability and Human-Made Radioactive Nuclides

"Here today, gone tomorrow" doesn't apply to most of the atoms that make up our world. That is fortunate for us. In fact, the majority of the atoms on our planet that are here today can be expected to be here tomorrow. Although it may be hard to locate a specific atom from one day to the next, you can be reasonably sure that it is still here. Why? Most atoms on Earth are not radioactive.

What factors seem to affect nuclear stability? Measurements indicate that nature favors even numbers of protons. Elements such as helium, oxygen, iron, and lead that have even atomic numbers tend to be more abundant than their odd neighbors. For example, of the eight elements that make up over 99% of Earth's total mass, only one (aluminum) has an odd atomic number. Still more favored are nuclei that have even numbers of both protons and neutrons. Perhaps the most dramatic case is $^{4}_{2}$He, the alpha particle. Given this stability, it is not surprising that helium is the second most abundant element in the universe.

Experimental data confirm that certain numbers of either protons or neutrons (called "magic numbers") are favored: 2, 8, 20, 28, 50, 82, and 114. The elements helium ($Z = 2$), oxygen ($Z = 88$), calcium ($Z = 20$), and nickel ($Z = 28$) have more stable nuclei than their neighbors. Check the graph of binding energy per **nucleon** (proton or neutron in the nucleus) versus the mass number shown in ◘ Fig. 20.9 to see that these elements sit at local maxima. The nuclides $^{16}_{8}$O and $^{40}_{20}$Ca are "doubly magic"; they have magic numbers for both protons and neutrons. In 2000, a French research team created another doubly magic nucleus, Ni-48, which contains 28 protons and 20 neutrons.

When you examine a larger set of nuclides, other trends appear. According to the US Department of Energy, there are 254 stable isotopes, a few naturally occurring radioactive elements, and hundreds of synthetic isotopes. Many of these find application in nuclear medicine. ◘ Fig. 20.10 sketches the band of stable isotopes for each element with at least one stable isotope. Note the following points:

- At higher atomic numbers, stable nuclei have increasingly more neutrons than protons.
- Some radioactive elements have too many neutrons relative to the stable isotopes. These tend to decay by beta emission, where a neutron changes into a proton and an electron.
- Some radioactive elements have too few neutrons relative to the stable isotopes. These tend to decay by positron emission, where a proton changes into a neutron and a positron.

The plot ends at bismuth, $Z = 83$, for no elements beyond this have stable isotopes. Alpha emission is typical for heavier elements that are unstable, simply because too many nucleons are present, be they protons or neutrons. ◘ Table 20.7 lists some guidelines that are helpful in determining the type of decay for a particular nucleus.

The radioactive decay can occur by the process of positron emission, or **beta-plus emission**, $^{0}_{+1}\beta$. This process does not occur naturally on Earth and was not discovered until scientists began creating new isotopes in the laboratory. This is done by slamming high-energy particles, ranging from protons and neutrons to atomic nuclei, into a target nucleus. This can result in changing the nucleus's mass number and giving it additional energy. In general, the laboratory process is.

$$\text{Nucleus} + \text{small particle} \rightarrow \text{bigger nucleus}$$

The product nucleus that is formed may undergo radioactive decay.

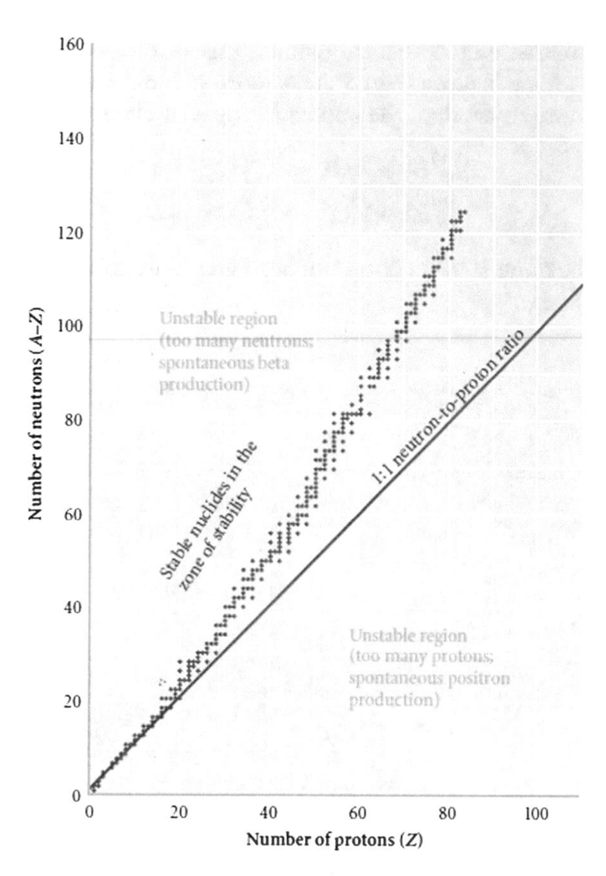

Fig. 20.10 The band of known isotopes for each element with at least one stable isotope (represented by red dots). n/p is the neutron-to-proton ratio. Note that the n/p ratio is greater than 1 for most stable isotopes

Table 20.7 Predicting nuclear decay

Type of decay	Reason for instability	Change in n/p ratio
Alpha emission	Nucleus too heavy	Increase (small for heavy nuclides)
Beta emission	n/p too high	Decrease
Positron emission	n/p too low	Increase
Gamma	Too much energy Nucleus energetically excited	None
Electron capture	n/p too low	Increase

Beta-plus emission – A type of radioactive decay wherein a positron is ejected at high speed from the nucleus, typical of nuclides that have too few neutrons. A neutrino accompanies this emission, and usually one or more gamma rays as well. Also known as positron emission.

For example, in 1930 phosphorus-30 was synthesized in the laboratory by the bombardment of an aluminum target with alpha particles, ${}^{4}_{2}\mathrm{He}$:

$$ {}^{27}_{13}\mathrm{Al} + {}^{4}_{2}\mathrm{He} \rightarrow {}^{30}_{15}\mathrm{P} + {}^{1}_{0}\mathrm{n} $$

More modern work includes the formation of tiny amounts of superheavy elements, which we will define as those beyond atomic number 106. Reactions include:

$$ {}^{208}_{82}\mathrm{Pb} + {}^{86}_{36}\mathrm{Kr} \rightarrow {}^{293}_{118}\mathrm{Og} + {}^{1}_{0}\mathrm{n} \qquad (t_{1/2} < 1\ \mathrm{ms}) $$
$$ {}^{249}_{97}\mathrm{Bk} + {}^{22}_{10}\mathrm{Ne} \rightarrow {}^{267}_{107}\mathrm{Bh} + 4{}^{1}_{0}\mathrm{n} \qquad (t_{1/2} = 17\ \mathrm{s}) $$

In 2010, a joint team of U.S. and Russian scientists in Dubna, Russia, created two isotopes of element 117 known as tennessine (Ts). The heavier isotope is more than 5 times more stable than the lighter isotope, and with its half-life of 78 ms, it is dramatically more stable than the known isotopes of element 118, known as oganesson (Og).

$$^{249}_{97}Bk + ^{48}_{20}Ca \rightarrow ^{294}_{117}Ts + 3^{1}_{0}n$$
$$^{249}_{97}Bk + ^{48}_{20}Ca \rightarrow ^{293}_{117}Ts + 4^{1}_{0}n$$

Some nuclides are expected to be quite stable, others not so. The nature of the nucleus becomes clearer when we study the half-lives of these nuclides.

Example 20.7—Ten Tin Isotopes

Tin has more stable isotopes than any other element. Explain why you might expect this to be the case. Then examine the radioisotopes of tin and discuss their decay modes.

Stable isotopes ^{112}Sn, ^{114}Sn, ^{115}Sn, ^{116}Sn, ^{117}Sn, ^{118}Sn, ^{119}Sn, ^{122}Sn, ^{124}Sn.

(^{120}Sn, ^{118}Sn, and ^{116}Sn are the most abundant.)

Radioisotopes ^{121}Sn, ^{123}Sn, ^{125}Sn, ^{126}Sn, ^{127}Sn.

that decay by β^- (plus others with higher mass and short half-life).

Radioisotopes that ^{110}Sn, ^{111}Sn, ^{113}Sn.

decay by β^+or by (plus others with lower mass and short half-life).

electron capture.

Asking the Right Questions

Data such as these exist for every element; in general, they are most conveniently accessed on the Web or in a chemistry handbook. We cannot reason out which isotopes actually exist. Although tin's location provides some guidance, we must still look them up.

Solution

The reasons for the stability of so many isotopes of tin include: tin is in the middle of the periodic table where nuclei are more stable and where a few extra neutrons do little to upset the balance of nuclear forces; it has an atomic number of 50, one of the "magic numbers"; and, 8 of the 10 isotopes have even numbers of protons and neutrons, and the even isotopes are the most abundant—another indication of their stability.

Are our Answers Reasonable?

In terms of its location on the periodic table, its number of protons and neutrons, and the particular atomic number of 50, yes, the answers are reasonable. You may want to compare the nuclear stability to that of a transuranium element to see the impact of its much more massive nucleus on nuclear stability.

What If…?

…we asked the same question about plutonium? Explain its nuclear stability, or lack thereof.

(*Ans: There are too many nucleons—both protons and neutrons—for any plutonium nuclei to be stable. The nuclei typically decay via alpha particle emission, electron capture, or spontaneous emission.*)

Practice 20.7

Predict whether each of the following isotopes might be stable or radioactive.

a. ^{79}Br
b. ^{101}Ru
c. ^{136}Ba
d. ^{180}Ta

20.8 Splitting the Atom: Nuclear Fission

20

With the discovery of fission came the birth of new radioisotopes and of a new consciousness that the nuclear age was upon us. Nuclear fission was discovered in the 1930s through the work of scientists such as Enrico Fermi, Fritz Strassman, Otto Hahn, and Lise Meitner. Work on fission continued in the early 1940s in both Germany and the United States, culminating in the deployment by the United States of the first atomic bomb used in war, fueled by uranium-235, on the town of Hiroshima, Japan, on August 6, 1945, and of a second combat-based atomic bomb, fueled by plutonium-239, on Nagasaki, Japan, on August 9, 1945. The war ended shortly after the second bomb was dropped, but the nuclear age had just begun.

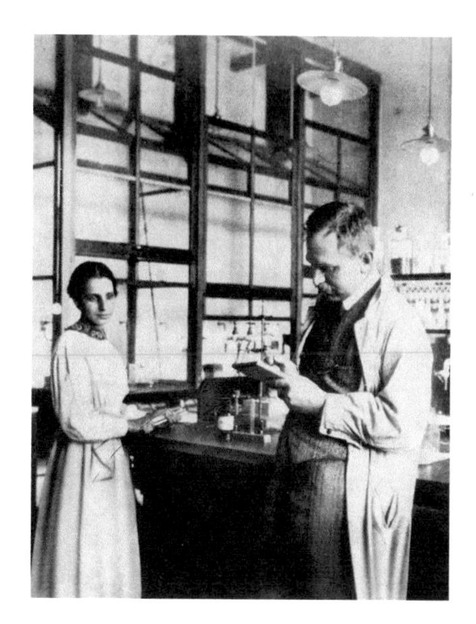

▸ https://upload.wikimedia.org/wikipedia/commons/thumb/2/2d/Otto_Hahn_und_Lise_Meitner.jpg/586px-Otto_Hahn_und_Lise_Meitner.jpg

Issues and Controversies—The Manhattan Project

In the early 1930s, it was discovered by scientists that a nuclear reaction could potentially undergo a chain reaction. Publication of this information in the scientific community led to the fears that Nazi Germany could develop this discovery into a nuclear weapon. By the late 1930s, the feasibility of an atomic bomb was becoming real. Albert Einstein, and other scientists who feared what would happen if the Nazis obtained an atomic bomb, strongly urged the US government to develop this technology as quickly as possible. While initial studies began in 1939 with members from the US, UK, and Canada, the Manhattan Project, as it became known, was not officially started until after the US had entered World War II.

The project included research at over 30 different secret locations throughout the United States. At the height of the project, more than 130,000 people were employed. Some of the noteworthy locations included the Hanford Site in central Washington State for plutonium production, the Oak-Ridge site in Tennessee for the enrichment of uranium, and the Los Alamos National Laboratory in New Mexico for the location for weapons development and design. By late 1944, scientists were sure that they could produce the world's first nuclear explosion.

▸ https://upload.wikimedia.org/wikipedia/commons/2/2d/The_gadget_in_the_Trinity_Test_Site_tower_%281945%29.jpg

The Manhattan Project realized its goal on July 16, 1945 with the detonation of the Trinity "gadget". (Nazi Germany had surrendered to the Allies on May 7–8, 1945, just two months before the test.) The explosion, detonated at a site 210 miles south of Los Alamos, was equivalent to 20 kilotons of TNT and was felt and heard as far away as 200 miles. Robert Oppenheimer, the director of the Los Alamos portion of the project, was so relieved that the gadget detonated that all he could say was "It worked." However, clearly the magnitude of the explosion was greater than any of them had imagined.

▶ https://upload.wikimedia.org/wikipedia/commons/f/fc/Trinity_Detonation_T%26B.jpg

The Manhattan Project also produced two other atomic bombs. A uranium-235 device known as Little Boy and a plutonium device code-named "Fat Man". These bombs were dropped on industrial cities in Japan and resulted in the surrender of that nation and the end of World War II. The blasts killed over 140,000 people immediately, and due to radiation sickness, that death toll had risen to over 340,000 just five years later. The decision to use these weapons was initially, and has continued to remain, a controversial topic, with supporters and valid reasoning on both sides of the issue.

Why do ^{235}U and ^{239}Pu split and release energy? The answer lies in part in the thermodynamics of nuclear stability. Refer to ◘ Fig. 20.9. Uranium nuclei are not so thermodynamically stable as iron, bromine, and other elements in the region of greatest stability near the top of the curve. The answer also lies in considering the precarious balance that exists in large nuclei such as uranium and plutonium. These nuclei are very heavy and are held together by the strong force between nucleons. However, opposing the strong force are the many proton–proton repulsions in an atom of this size. For some atoms, the injection of extra mass into the nucleus can tip the balance in favor of the proton repulsions and send the nucleus flying apart into two or more pieces. This is what happens with ^{235}U and ^{239}Pu. A neutron, with no charge, is an ideal particle to shoot into a nucleus. Once it slips into the nucleus, a heavier nuclide is formed that fragments in a matter of nanoseconds. ◘ Fig. 20.11 illustrates the process for the fission of uranium-235.

$$^{239}_{94}Pu + ^{1}_{0}n \rightarrow \left[^{240}Pu\right] \rightarrow ^{70}_{30}Zn + ^{167}_{64}Gd + 3^{1}_{0}n + \text{energy}$$

$$^{235}_{92}U + ^{1}_{0}n \rightarrow \left[^{236}U\right] \rightarrow ^{139}_{56}Ba + ^{94}_{36}Kr + 3^{1}_{0}n + \text{energy}$$

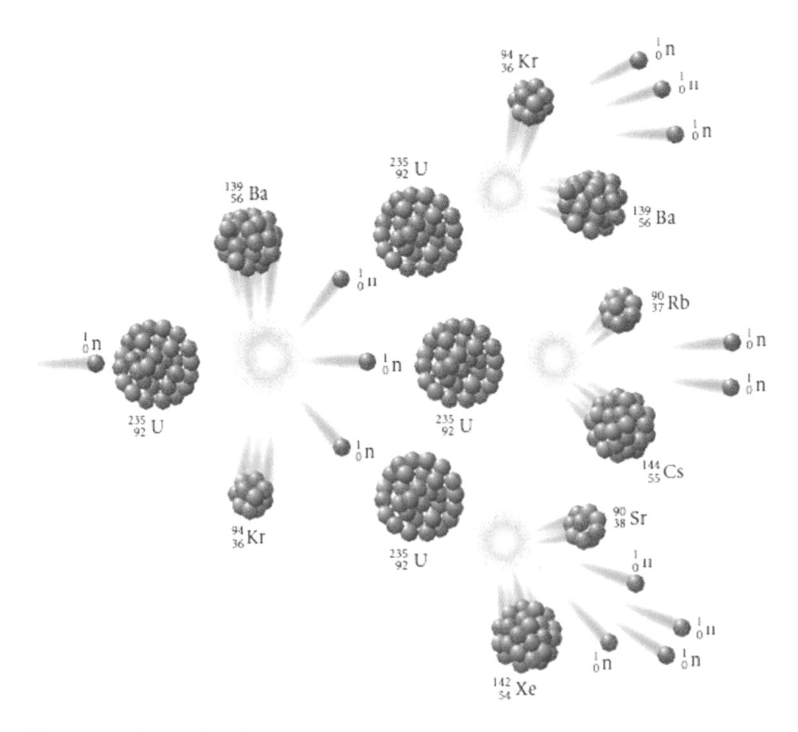

◘ **Fig. 20.11** A nucleus of ^{235}U undergoing nuclear fission

Nuclear reactions such as these were initially of interest because the tremendous energy they released could be unleashed in a weapon. Since the violent birth of fission in 1945, these reactions have found a variety of other, more humanitarian uses. Today, using non-weapons-grade fissionable fuel, they provide the energy in nuclear power plants worldwide, they power spacecraft, ships and submarines, and they are the source of many isotopes used in nuclear medicine.

20.8.1 Nuclear Reactors as a Vital Source of Electricity

Roughly 440 nuclear reactors are alternatives to the pollution and greenhouse gas emissions caused by coal-burning and oil-burning power generation. The world's first nuclear power plant went online in 1954 in the Russian city of Obninsk. This was immediately followed by the construction of a nuclear facility in Sellafield, England. It wasn't until 1957 that the first full-scale power plant in the United States began operation in Shippingport, Pennsylvania. Municipal power generation by nuclear fission is not new. However, the use of nuclear fission has been controversial because of safety concerns, the two most important being the possible accidental release of radiation, and the disposal and long-term (thousands, and perhaps millions, of years!) safeguarding of radioactive wastes. Accidental releases of radiation occurred on a relatively small scale at the Three Mile Island nuclear facility in Pennsylvania in 1979 and on a much larger scale just outside the Ukrainian town of Chernobyl, in 1986. Still, much of the world uses nuclear power to meet just over 10% its energy needs, as shown in ◘ Fig. 20.12.

The goal of any large power plant is to generate electricity by turning a turbine, which converts the mechanical energy into electricity. Steam, resulting from the heating of water, supplies the energy to turn the turbine. The essential difference among the different types of power plants is the fuel source that creates the steam from water. In conventional nuclear reactors, this energy is supplied by the controlled fission of ^{235}U (we say that the fission "goes critical").

Conventional nuclear reactors have three essential parts, as shown in ◘ Fig. 20.13. The nuclear reactor part comprises 100 to 200 fuel rods that contain the fissionable uranium. A series of movable control rods, typically made of boron or cadmium, absorb neutrons as a way of controlling the rate of fission. The rods are located at the bottom of a pool of water, which acts as a moderator to slow these neutrons so that they can be captured by the uranium in the fuel rods. The other two parts of the system, the steam generator and the turbine and condenser, are common to many types of conventional power plants. Other types of nuclear reactors also exist.

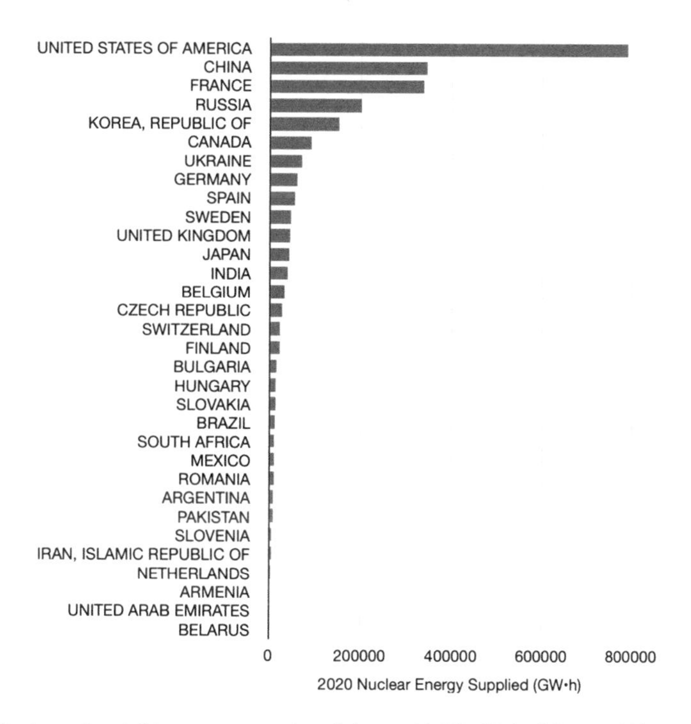

◘ **Fig. 20.12** Nuclear power plants are located in many countries of the world. The United States, China, France, and the Russian Federation possess the majority of the plants

Fig. 20.13 The essential parts of a nuclear power plant. The nuclear reactor, submersed in a pool of water, contains the uranium-based fuel rods and a series of control rods to slow the fission process. The high-pressure water, heated by the fission, travels to a steam generator where the heat vaporizes water and creates high-pressure steam. The steam is used to turn a turbine and generate electricity. The steam is then condensed by passing large amounts of water in a cooling pond through the condenser

Breeder reactors, for example, are so named because they produce more fissionable fuel than they consume. In these reactors, relatively abundant ^{238}U is bombarded with neutrons to "breed" ^{239}Pu, along with the emission of β-particles.

$$^{238}_{92}\text{U} + ^{1}_{0}\text{n} \rightarrow ^{239}_{92}\text{U} \xrightarrow{\beta} ^{239}_{93}\text{Np} \xrightarrow{\beta} ^{239}_{94}\text{Pu}$$

Prototype large breeder reactors either have been built or are being built in China, France, Scotland, the United States, India, Japan, and the former Soviet Union. Breeder reactors are not currently used in the commercial production of energy because of the exceptionally long (24,400 years) half-life of ^{239}Pu. In these reactors, sodium metal is used as a coolant because it can transfer heat away from the reactor core much better than water and has a much higher boiling point, allowing it to remain liquid without being pressurized. Because of the high operating temperature of the liquid sodium, along with sodium's capacity to absorb neutrons (becoming radioactive after it travels by the core), sodium is viewed as particularly hazardous. The future of breeder reactors is, at the very least, uncertain.

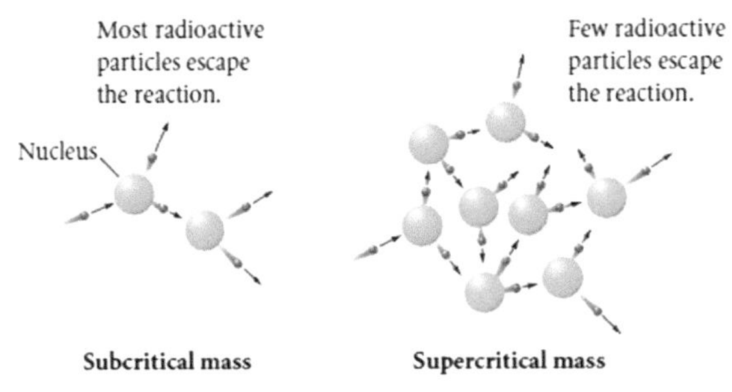

Here are the details of fission as we understand it today:

- Fission reactions release energy. The masses of the product nuclei are less than those of the starting nuclei, and the source of the energy is this mass difference. The energy released is orders of magnitude higher than that of "ordinary" chemical reactions.
- With rare exception, fission is not a naturally occurring process. We initiate it in some nuclei by brute force—that is, by smashing them with a high-energy neutron. The impact simply drives the nucleus apart. A few nuclei can be induced to undergo fission after they capture lower-energy neutrons. Furthermore, the only *naturally occurring* fissionable nucleus is ^{235}U, which is present in only 0.72% of all uranium atoms. The low availability of fissionable fuels slowed the development of fission. It also spurred the breeding of human-made fissionable fuels such as ^{233}U and ^{239}Pu.
- Once induced, fission releases more neutrons, typically 2 to 3 neutrons, per event. These daughter neutrons usually are traveling fast and may escape without further interaction with a fissionable nucleus. In this in-

stance, we have a condition known as a **subcritical mass**. But if enough fissionable nuclei are nearby (a **critical mass**) or if the neutrons are slowed, these neutrons can continue the fission process in the absence of a neutron source. A self-sustaining **chain reaction** is possible. If too much fissionable material (a **supercritical mass**) is present, the reaction goes out of control.
- Nuclei can split in more than one way, forming a whole host of fission products. The split is usually into two or sometimes three fragments. This means that fission is "messy" because of the many products.
- Fission is also messy because the products are usually radioactive. They tend to be neutron-rich and beta emitters. In the case of nuclear reactors, the radioactive products end up in high-level and low-level nuclear waste—in other words, storage. In the case of weapons testing above ground, they result in nuclear fallout.

Subcritical mass – A mass of a radioactive isotope that is too small to sustain a chain reaction.

Chain reaction – In nuclear chemistry, a reaction that is self-sustaining as one event in turn causes more events.

Critical mass – The amount of fissionable fuel needed to sustain a chain reaction.

Supercritical mass – A mass of a radioactive isotope that not only supports a chain reaction but causes the majority of the nuclei to undergo unfettered radioactive decay within a very short period of time, releasing huge amounts of energy.

Example 20.8—Fission: A Chain Reaction

Draw a sketch to show why the fission of ^{235}U can be called a chain reaction. What factors do you think will influence whether the fission chain reaction will merely sustain itself or will be explosive?

Asking the Right Questions

What initiates fission? What is necessary for a chain reaction to occur?

Solution

Fission of ^{235}U is initiated by neutrons and produces neutrons. This can result in a chain reaction if enough of the neutrons released are able to interact with more ^{235}U.

As shown here, the fission events quickly multiply, and the reaction becomes explosive. But if each fission event could be controlled to produce only one more fission event (instead of three), the chain reaction would simply sustain itself.

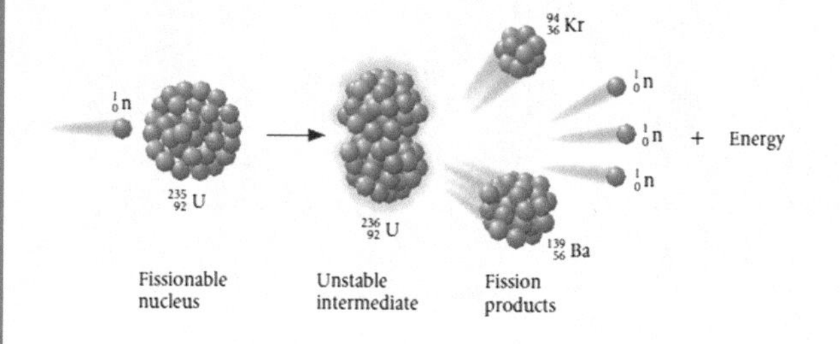

Are Our Answers Reasonable?

The key for us is to note the number of neutrons that are produced in the reaction. If more than one is produced, the result could make the chain reaction a reasonable outcome.

What If…?

…a block of the fissionable material were sent out into space, far past the Earth's gravitational field. Speculate about the ability of the ^{235}U nucleus to undergo fission in deep space.

(*Ans: There should be little difference in the fission on Earth as in deep space.*)

Practice 20.8

Draw a sketch that would illustrate the fission reaction of ^{235}U that is just sustainable. (*Hint:* A sustainable reaction is not a supercritical chain reaction.)

20.9 Medical Uses of Radioisotopes

Now used almost routinely for diagnosis, radioisotopes enable us to image internal organs, bone, and tissue structures. We can watch biological processes such as oxygen uptake and brain activity. Radioisotopes also make possible the treatment of tumors without anesthesia or invasive surgery. There are more than 40 million nuclear medicine procedures carried out worldwide each year.

20.9.1 Tracer Isotopes for Diagnosis

We have already mentioned that you can image certain internal structures such as bones (or bullets or swallowed coins) using X-rays. The bones, by absorbing the incoming radiation, show up as shadows on the X-ray films or detectors. However, the heart, liver, and thyroid gland do not show up very well on X-ray films because they absorb so little of the X-rays.

Imaging these organs is the job of radioactive tracers, more technically called diagnostic **radiopharmaceuticals**. Radiopharmaceuticals are used in small amounts, so they release only low amounts of radiation into the body. They are considered safe to use for diagnostic purposes. Through introduction into the body by mouth, inhalation, or injection, the radiopharmaceutical can be used to outline the organs of interest. This imaging occurs by measuring the emitted radiation outside the patient. Either a "hot spot" (a tumor that preferentially absorbs the radioisotope) or a "cold spot" (a place where the surrounding tissues preferentially absorb the radioisotope) will produce the desired result. Using an alpha or beta emitter as a radioactive source would not work, because these particles would be absorbed by the tissues inside the body before they could be detected outside the body.

Radiopharmaceuticals – Compounds containing radioactive nuclides that are used for imaging studies in nuclear medicine.

To understand imaging, we no longer can ignore the chemical form of the nuclide, as we have done throughout most of this chapter. The art of getting a good image lies in understanding the chemical behavior of the nuclide in the body. For example, although you can make elemental iodine, $^{131}I_2$, from radioactive ^{131}I, you cannot feed this chemical to a patient to image the thyroid gland, because iodine is chemically reactive and would damage the mouth and stomach. Similarly, an organic fat-soluble compound containing iodine would tend to concentrate in the lymph system rather than travel via the bloodstream to the thyroid. The chemical form of choice for thyroid studies is the iodide ion, $^{131}I^-$, in the water-soluble form of sodium iodide, $Na^{131}I$, which can be swallowed in a salty "cocktail." Each radioactive substance must be prepared in a carefully tailored chemical form, because each organ in the body has a different chemical profile.

Example 20.9—^{123}I and ^{131}I: Cousins But Not Twins

Radioactive ^{131}I is used for both treatment and diagnostic scans, whereas radioactive ^{123}I is used only for diagnosis. Account for this difference after looking up the decay mode for each nuclide.

Asking the Right Questions

What might make ^{123}I more useful than ^{131}I for long-term use? What particle emissions are more effective in destroying tumors than others?

Solution

Both isotopes of iodine behave the same chemically in the body (they both are taken up by a normal thyroid gland), so the difference must lie in their nuclear properties. A table of radioactive decay shows that both ^{131}I and ^{123}I release gamma rays. Because gamma rays easily penetrate matter, they can be detected outside the body. Therefore, both can be used for diagnosis. For destruction of a tumor, you need alpha or beta particles that damage tissue at short distances. ^{131}I is a beta emitter; ^{123}I is not, decaying instead by electron capture.

Is the Answer Reasonable?

"Reasonable," in this discussion means the claim that beta emission is more effective than electron capture at destroying tumors is valid. This is the case, although new treatments are being introduced, including microwave-based, as well as proton beam-based therapies.

What if…?

…we wanted to confirm the products of the decay of each of the iodine isotopes? What nuclides would we expect to find in each decay process?

(*Ans:* ^{131}Xe *and* ^{123}Te,)

Practice 20.9

Understanding the biological action of salicylic acid can be enhanced by imaging a patient fed with a radiopharmaceutical. Which of the following salicylic acid derivatives would *not* be a suitable choice for such a study?

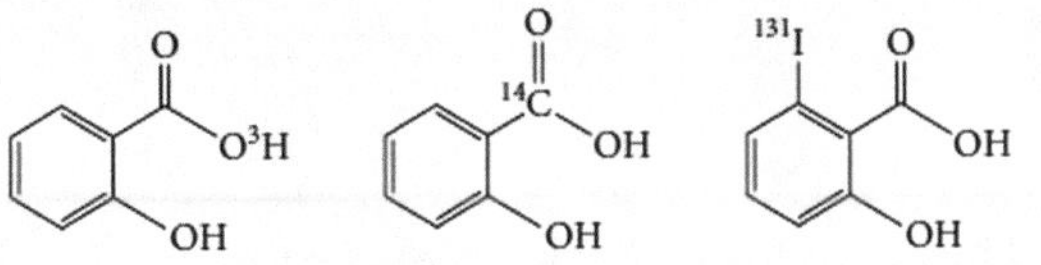

O, O³H, OH   O, ^{14}C, OH, OH ^{131}I, O, OH, OH

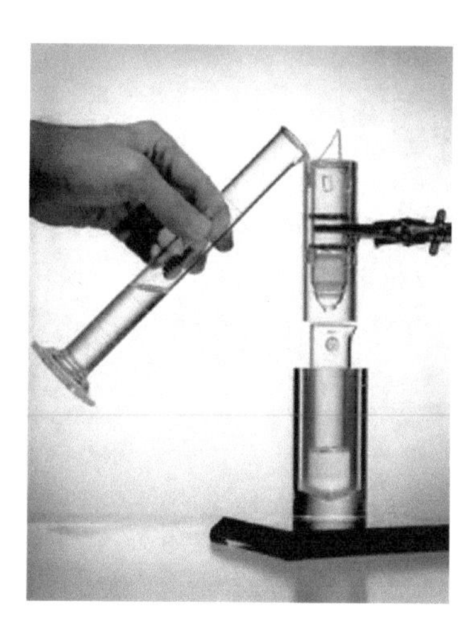

Fig. 20.14 This photograph shows the early column chromatography apparatus used to separate Tc-99 m from the parent isotope (Mo-99). The molybdenum remains on the column, and the technetium passes through. The early work on technetium-99 m was done at Brookhaven National Lab. *Source* ▶ https://upload.wikimedia.org/wikipedia/commons/2/25/First_technetium-99m_generator_-_1958.jpg

In many cases, technetium-99 m is the nuclide of choice. The existence of technetium was predicted by Mendeleev as "eka-manganese," and it is the only non-rare-Earth element he foresaw that was not discovered in his lifetime. Although it was first produced in 1939 by Glen Seaborg and Emilio Segrè, it was not used in nuclear medicine until the 1960s. Technetium has a versatile chemistry and is dispensed in dozens of chemical compounds for imaging different parts of the body. Furthermore, technetium-99 m emits only gamma rays, which are little absorbed by the body and therefore expose the patient to only a small radiation dose. Because it has such a short half-life, Tc-99 m cannot be stored in a flask on a shelf. Rather, it usually is generated in the hospital from a source of molybdenum called a "molybdenum (or sometimes technetium) cow":

$$^{99}\text{Mo} \rightarrow {}^{99\,\text{m}}\text{Tc} + {}^{0}_{-1}\beta + {}^{0}_{0}\gamma$$
$$^{99\,\text{m}}\text{Tc} \rightarrow {}^{99}\text{Tc} + {}^{0}_{0}\gamma$$

The molybdenum-99 isotope used in the cow is not naturally occurring on Earth. It is generated as a fission product of uranium in nuclear reactors and shipped to hospitals worldwide. The $^{99\text{m}}$Tc nuclide is separated from the molybdenum as it is needed in a process called "milking the cow." Fig. 20.14 shows the earliest chromatographic column used to do a hands-on separation of $^{99\text{m}}$Tc from ^{99}Mo. Today such processes are automated. The half-life of ^{99}Mo is a brief 67 h, so it is shipped to medical suppliers for immediate distribution to hospitals.

How Do We Know?—Imaging with Positron Emission Tomography (PET)

Why does nuclear medicine employ such exotic nuclei as technetium and palladium yet seemingly ignore the elements that make up most of our body, including hydrogen, oxygen, carbon, nitrogen, sulfur, and potassium? All of these elements play biochemical roles in the development of cells and tissues, yet none has been mentioned as a diagnostic or therapeutic tool. Why the omission?

Think back on what is needed to do imaging. First is a source of radiation that penetrates well enough to be detected outside the body. Gamma emitters usually are the nuclides of choice. Second, you need availability of the nuclide in sufficient quantity to do a study. Third, you need a half-life that is reasonably short, and if it is very short, the nuclide must be generated on site. Finally, you need to create a chemical form of the nuclide that will give either a "hot spot" or a "cold spot" in the area of medical interest.

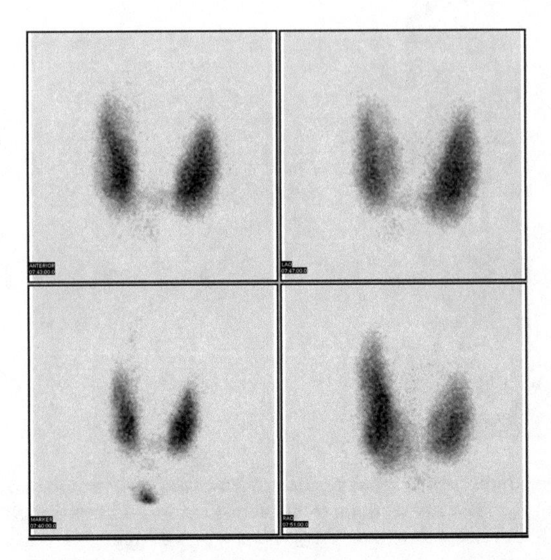

▶ https://upload.wikimedia.org/wikipedia/commons/e/e0/Thyroid_scan.jpg

Carbon-14, though biologically active, has too long a half-life and is a pure beta-minus emitter (see Fig. 20.15). However, ^{11}C, with a n/p that is lower than those of the stable isotopes, is a positron (beta-plus) emitter. Other positron emitters include ^{15}O, ^{13}N, and ^{30}S. Positrons themselves do not penetrate very far. But when a positron encounters an electron, which happens almost immediately, an annihilation occurs whereby the particle (positron) and antiparticle (electron) are converted into energy:

$$^{0}_{+1}\beta + e^{-} \rightarrow 2^{0}_{0}\gamma$$

The photons from the two gamma rays are emitted in exactly opposite directions. When a gamma detector is positioned both above and below the patient, if each one simultaneously records an event, then a positron was annihilated. By feeding the data from the detectors into a computer, it is possible to reconstruct an image of where the positron emission took place.

Positron emission imaging is better known as a PET scan, short for **positron emission tomography**. It is a much trickier procedure than other types of nuclear imaging, largely because positron emitters tend to have half-lives on the order of minutes. To have enough radioactive material present, they must be produced nearby or on site at a hospital. In either case, technicians are needed to maintain the production equipment. PET scans also require the injection of radioactive material, which is not the case for MRI or CT scans.

Positron emission tomography – A medical imaging technique that images metabolic processes within the body. Also known as a PET scan.

The payoff with PET, though, is impressive. It produces "functional imaging"—that is, images of chemical processes in action. For example, it can record the brain in action during a seizure or when the patient is hearing music, thus decoding the neural pathways. Using glucose labeled with ^{11}C or with ^{18}F (see Fig. 20.16), PET can reveal brain metabolism. Similarly, ^{18}F-labeled estrogen can be used within a patient to show how tumors grow. The color-enhanced real-time images are far more dramatic than the black-and-white slices produced by other means.

	^{9}C	^{10}C	^{11}C	^{12}C	^{13}C	^{14}C	^{15}C	^{16}C	^{17}C
Type of decay	EC	EC	+β or EC	Stable	Stable	−β	−β	−β	−β
Half-life	0.127 s	19.3 s	20.3 min			5715 years	2.45 s	0.75 s	0.19 s
	n/p ratio too low					n/p ratio too high			

Fig. 20.15 The isotopes of carbon

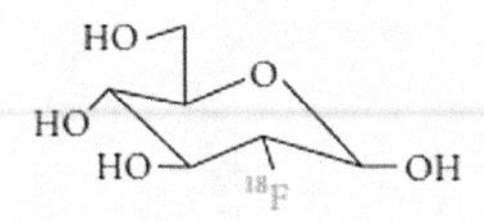

Fig. 20.16 ^{18}F-labeled 2-deoxyglucose, or FDG, is used to study glucose metabolism in the body. It can be used to differentiate benign tumors from cancerous ones, because the latter use glucose at a higher rate. After a patient consumes this radiopharmaceutical, tumors within her or his body show up as white spots where glucose metabolism is higher

One of the more widely used technetium compounds is sodium pertechnetate, $NaTcO_4$. The pertechnetate ion, TcO_4^-, has properties similar to those of the chloride ion, Cl^-, and concentrates in brain tumors, in the thyroid and salivary glands, and in areas of the body where blood is pooling (as happens in internal bleeding). Similarly, technetium pyrophosphate, TcP_2O_7, can be used to image the heart to see the extent of damage to heart muscle after a heart attack.

Although it is well developed, nuclear medicine is still a relatively young field; radioisotope tracers were developed in the 1930s and put into widespread use in the 1950s and 1960s. In the 1980s, the ready availability of computers to help process images led to explosive growth in the field that continues today. There are announcements of new techniques, new types of images, or methods that require lower amounts of radiation almost daily. (There are also cautions that costly scans are being used too routinely to warrant either the risks or the costs involved.) The odds are that you know somebody who has waged battle with cancer. Nuclear medicine is undoubtedly a part of that person's health history.

The Bottom Line

- Each element is composed of atoms containing the same number of protons. These may contain isotopes with differing numbers of neutrons.
- Some nuclear configurations are unstable. They decay in a stepwise progression toward stable nuclei.
- There are three main types of radioactive decay: alpha-particle emission, beta-particle emission, and gamma-ray emission.
- Ionizing radiation can interact with living tissue and cause damage to the DNA of a cell. This damage may be repaired and cause no harm or, in some cases, may lead to cancer.
- Radioactive decay occurs via first-order kinetics.
- Energy is released in radioactive decay processes as a consequence of the mass defect in nuclei.
- Nuclei with a "magic" number of protons and/or neutrons (2, 8, 20, 28, 50, or 82) are stable. Nuclei with even numbers of protons and/or neutrons are also more likely to be stable.
- Nuclear fission is the splitting of heavier nuclei into lighter ones. Nuclear fusion results when smaller nuclei combine into heavier nuclei.
- Radioisotopes can be used in medicine for imaging the body and for treating and eliminating cancerous tissues.

Section 20.2 Isotopes and More Isotopes

Skill Review

1. What is the atom with smallest atomic number? The smallest mass number?
2. Can different elements both have the same number of protons? The same number of neutrons? Explain.
3. Can an atom have no neutrons? Explain.
4. Can an atom have no protons? No electrons? Explain.
5. Can a helium nuclide have a smaller mass number than a hydrogen nuclide? Explain.
6. Can a carbon nuclide have a smaller mass number than a nitrogen nuclide? Explain.
7. Identify the number of protons, neutrons, and electrons in each of the following isotopes.
   a. ^{12}N b. ^{124}Sb c. ^{152}Eu d. ^{9}Be
8. Identify the number of protons, neutrons, and electrons in each of the following isotopes.
   a. ^{7}Li b. ^{122}Cs c. ^{17}O d. ^{18}F

Chemical Applications and Practices

9. A person who weighs 60 kg (132 lb) has about 120 g of potassium in his or her body. How many grams of K-40 does this include? K-40 has a natural abundance of 0.0118%.
10. No isotopes of potassium are chemically stable in the presence of water and/or oxygen. Potassium-39 and potassium-41 are *stable* isotopes. Explain these different meanings of the word stable.
11. Fallout from a nuclear weapon includes the radioactive nuclide Sr-90.
    a. In the biosphere, which chemical form for Sr-90 is more likely, Sr^{2+} or Sr?
    b. Once Sr-90 lands downwind, it is extremely difficult to separate from the plants and soils. Suggest reasons why.
12. Strontium-90 from fallout gets into the food chain and eventually can end up in cows' milk. Can you remove the radioactivity from cows' milk by boiling it? Why or why not?

Section 20.3 Types of Radioactive Decay

Skill Review

13. Both gamma rays and infrared radiation are forms of electromagnetic radiation. How do they differ?
14. Beta decay involves the emission of a high-speed electron from an atom, yet the overall charge on the atom does not become less negative. Explain why.
15. Write nuclear equations for the following processes:
    a. An alpha particle (along with a gamma ray) is emitted from plutonium-239.
    b. Carbon-14 undergoes beta decay.
    c. Cesium-137 emits a beta particle with an accompanying gamma ray.
16. Write nuclear equations for the following processes:
    a. Plutonium-238 emits an alpha particle with an accompanying gamma ray.
    b. Radon-222 is produced from the decay of a radium isotope with the emission of a gamma ray.
    c. Radium-225 emits a gamma ray followed by an alpha particle.
17. Write nuclear equations for the following processes:
    a. Polonium-215 (with a gamma ray) is produced by an alpha emission.
    b. Strontium-90 decays by beta emission. Little or no gamma radiation is released.
    c. Tc-99 decays by beta-minus emission.
18. Write nuclear equations for the following processes:
    a. Nitrogen-14 is formed from a radioisotope of carbon.
    b. Cadmium-110 is formed from a radioisotope of silver.
    c. Technetium-99 is formed from Technetium-99 m.

Chemical Applications and Practices

19. Samarium-146 is the lightest element found naturally on our planet to undergo alpha emission. Write the equation for this alpha decay. There is no accompanying gamma ray.
20. Iodine-131, used in medical imaging, undergoes beta decay. Write the nuclear equation for this reaction.
21. Darlene Hoffman, an award-winning nuclear chemist, postulated the existence of Pu-244 before it was discovered. Into which of the four natural decay series does it fit?
22. A chemistry source states that radon-219 is produced in the actinium-227 radioactive decay series. Into which of the four decay series mentioned in Section 20.2 does actinium-227 fit?

Section 20.4 Interaction of Radiation with Matter

Skill Review

23. Classify the following as ionizing or nonionizing radiation: cosmic rays, infrared radiation, gamma rays, visible light, microwaves, X-rays.
24. You can cook food using microwaves, and you can sterilize food using gamma rays. Why do these two types of radiation produce such different results?
25. Suppose that you had administered a gamma emitter to a patient in order to diagnose how well his or her heart was functioning. Name three things you could do to minimize your exposure to the radiation.
26. Which cells in your body are most susceptible to radiation? Why?
27. Explain the similarities and differences between:
    a. a curie and a becquerel
    b. a rad and a rem
28. Explain the similarities and differences between:
    a. a rem and a sievert
    b. a curie and a rem
29. Smoke detectors use only a small quantity of americium, less than 35 kBq. How many disintegrations per second is this?
30. Using the information in the previous problem, show that the result is comparable to 1 microcurie.

Chemical Applications and Practices

31. In the mid-1990s, a watch was advertised that glowed in the dark. The source of the glow was the radioisotope tritium interacting with a luminous paint. The annual dose for a per- son wearing the watch as estimated at 4.0 microsieverts.
    a. How many rem is this?
    b. Do you think this amount of radiation warrants concern?
32. In the previous problem, we noted that tritium, a radioisotope of hydrogen, was used in some watches.
    a. What mode of decay would you predict for tritium?
    b. Given that radiation escapes from the watch case, does this evidence support your prediction?
33. Three metals—aluminum, iron, and cadmium—are proposed as materials that could be used to shield nuclear radiation.
    a. What property of these materials would you need to look up to determine which one would have to be used in the greatest thickness?
    b. What else might you need to know about a substance before you use it in shielding?
34. If radon-222 gas decays in your lungs to produce solid polonium, which is then trapped there, how many decays does the polonium progress through before reaching the stable isotope of lead-206? How many alpha and beta particles are emitted in the process?

Section 20.5 The Kinetics of Radioactive Decay

Skill Review

35. Here are the decay plots for two different hypothetical radioactive nuclei. The plot denoted by the red line is A; that denoted by the blue line is B. From these graphs, determine:
    a. Which nuclide has the longer half-life?
    b. What is the half-life of the nuclide indicated by the red line?
    c. Which one has the higher activity?
    d. Which one would be more dangerous if swallowed?

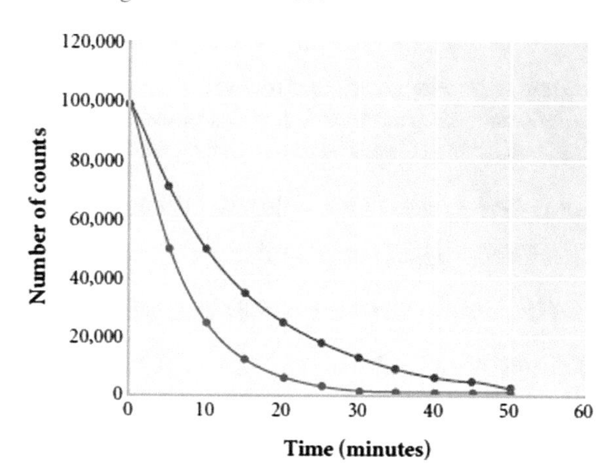

36. Using the plot in the previous problem, estimate the half-life of each nucleus. What does a shorter half-life imply?
37. If 75% of a radioactive sample is gone after 30 days, what is the half-life of the nuclide?
38. If 50% of a radioactive sample remains after 30 days, what is the half-life of the nuclide?
39. Tritium has a half-life of 12.3 years. What fraction of a sample of tritium will remain after 12.3 years? After 24.6 years?
40. What percent of the original sample of strontium-90 (half-life = 28.9 years) will remain
    a. after 14 years?
    b. after 49.6 years?
    c. after 1000 years?
41. The half-life of strontium-90 is 28.9 years. How long would it take for the activity of a Sr-90 sample to diminish by 87.5%?
42. The half-life of strontium-85 (about 64 days) is considerably shorter than that of strontium-90. How long would it take for the activity of a sample of Sr-85 to diminish by 93.75%? Why do you think that Sr-85 is used diagnostically for bone scans, but Sr-90 is not?

Chemical Applications and Practices

43. A 10-mCi sample of a tracer isotope is used to diagnose blood flow from the heart. If it is desirable that nearly all of the radioactivity (99%) be gone after 3 days, approximately what half-life should the isotope have?
44. It is estimated that the nuclear accident at Chernobyl released 1.85×10^{18} becquerels, a large amount of radiation. How many curies is this? Why is it not easy to translate either of these values into the number of radioactive atoms present?
45. A 0.1-microcurie sample of polonium-210 is used to demonstrate alpha decay in the lecture hall, because there is little accompanying gamma radiation. The half-life of this isotope is 138 days. About how often does an instructor need to buy a new source? State any assumptions you made in arriving at your answer.
46. The world uranium reserves are currently estimated at 3.4 million tonnes, where a tonne is a metric ton, or 1000 kg. Current knowledge places Earth at 4.5 billion years old. How much uranium was present at the time our world formed?
47. The half-life of plutonium-239 is about 24,000 years. After approximately how much time will over 99% of a sample of plutonium-239 have decayed? How can this element exist on our planet if its half-life is so short?
48. Plutonium oxide, used to power the Cassini space-probe to Saturn, was stored in corrosion-resistant materials designed to contain the fuel for 10 half-lives, or 870 years.
    a. What is the half-life of Pu-238?
    b. Why do you think the Fig. 10 half-lives was selected?
    c. The Cassini batteries contained 72 lb (33 kg) of Pu-238 in the form of plutonium dioxide. How much Pu-238 (in kilograms and in grams) will remain after 870 years?

Section 20.6 Mass and Binding Energy

Skill Review

49. a. Why don't the masses of the neutrons and protons that make up an oxygen-16 atom add up to the mass of the oxygen nucleus?
    b. Is this true for all isotopes of oxygen?
50. a. Is an atom of carbon likely ever to fall apart into its component protons, electrons, and neutrons?
    b. Explain why or why not.
51. Explain how the mass defect and the binding energy are related.
52. Is mass conserved in nuclear reactions, such as alpha decay, that proceed spontaneously? Why or why not?
53. Calculate the mass defect and the resulting energy for the following nuclear reaction:

$$({}^{1}_{0}\text{n} = 1.008665\ \text{g/mol};\ {}^{4}\text{He} = 4.002603\ \text{g/mol})$$
$$^{170}\text{Ir}(169.974970\ \text{g/mol})\ \text{decays to}\ ^{166}\text{Re}\ (165.965740\ \text{g/mol})$$

54. Calculate the mass defect and the resulting energy for the following nuclear reaction:

$${}^{40}_{19}\text{K}(39.963999\ \text{g/mol}) \rightarrow {}^{40}_{20}\text{Ca}(39.962591\ \text{g/mol}) + {}^{0}_{-1}\beta$$

Chemical Applications and Practices

55. The masses of the neutral atoms involved for an alpha emission from U-238 are shown below. These values include the masses of the electrons.
    a. Would you expect the U-238 or its decay products to have more mass?
    b. What is the source/result of any difference in these masses?

${}^{4}_{2}\text{He}$	4.002603 u
${}^{238}_{92}\text{U}$	238.050784 u
${}^{234}_{90}\text{Th}$	234.040945 u
e^-	0.005485799 u

56. Use the data in the previous problem to calculate the mass defect for the process. What is the source/result of any difference in these masses?
57. Describe the similarities and differences between beta-minus and beta-plus emission.
58. The stable isotopes of carbon are ^{12}C and ^{13}C. What decay mode would you predict for ^{14}C?

Section 20.7 Nuclear Stability and Human-made Radioactive Nuclides

Skill Review

59. a. What does "doubly magic" mean, in reference to nuclides?
    b. Give two examples of nuclei that are doubly magic and two examples of nuclides that have no magic numbers at all. How would you expect these nuclei to differ?
60. Suggest a reason why the heavier elements have proportionately more neutrons than the lighter ones.

Chemical Applications and Practices

61. Element 114 was recently discovered. Why was this element sought, but not the neighboring elements 115 and 113?
62. Some claims to the discovery of element 118 have been made. Give reasons why this element may be considered to have both nuclear and chemical stability.

Section 20.8 Splitting the Atom: Nuclear Fission

Skill Review

63. Why is alpha emission a better prediction for the mode of radioactive decay for uranium or plutonium than it is for iron, carbon, or hydrogen?
64. Answer the following questions for the fission of plutonium:

$$^{239}_{94}Pu + ^{1}_{0}n \rightarrow \left[^{240}_{94}Pu\right] \rightarrow ^{136}_{51}Sb + ^{100}_{43}Tc + 4^{1}_{0}n + \text{energy}$$

    a. What is the significance of the fact that neutrons are produced?
    b. What is the source of the energy in this equation?
65. Iron and cobalt are not expected to undergo nuclear fission. Why?
66. Would you expect helium to undergo nuclear fission? Will hydrogen-1 undergo fission?
67. Write nuclear reactions for the fission of ^{235}U to form:
    a. ^{94}Kr and ^{139}Ba and neutrons
    b. ^{80}Sr and ^{153}Xe and neutrons
68. An isotope of the element technetium can be produced by bombarding molybdenum-97 with deuterium nuclei. Two neutrons are also formed. Write the nuclear equation.

Chemical Applications and Practices

69. On our planet, both U-235 and U-238 occur naturally.
    a. How do these isotopes differ?
    b. What is the natural abundance of each?
    c. Propose a reason why it is very difficult to separate these two isotopes.
70. Using "conventional" explosives such as TNT, you can make tiny explosive devices as well as huge ones. Is it possible to make a similarly tiny nuclear bomb? Why or why not?
71. When fission of U-235 or Pu-239 occurs, elements such as americium, californium, and berkelium are not found in the fallout. Explain why.
72. One particularly nasty component of nuclear fallout is strontium-90. Explore the reactivity of this particular isotope and explain why it may be harmful to living creatures.

Section 20.9 Medical Uses of Radioisotopes

Skill Review

73. What questions should you ask about a radionuclide to be injected for diagnostic purposes?
74. Why aren't alpha emitters useful for diagnostic scans, such as a scan of the heart or of the thyroid gland?
75. Technetium-99 m samples should not be stored overnight but, rather, should be freshly prepared each day for diagnostic scans. Why?
76. Why is molybdenum-99 not used directly as a component of a radiopharmaceutical?

Chemical Applications and Practices

77. A patient was injected with 10 mg of fluorine-18–labeled glucose for a PET scan. Fluorine-18 has a half-life of 110 min and disintegrates by positron emission. Write the nuclear equation for the decay.
78. Using the information in the previous problem, determine the amount of time needed to reduce the radioactivity of fluorine-18 to 1/16 of its original activity.
79. In the chapter, it was mentioned that it takes about two and a half weeks for Tc-99 to be eliminated from the body. The Department of Energy reports that it takes approximately 60 h for the body to eliminate half of the technetium. Are these two figures consistent with each other?
80. Gallium-67 citrate is used as a radiopharmaceutical for diagnosing tumors and infections.
    a. What type of radioactive decay would you predict for this nuclide?
    b. A typical activity of a radionuclide used in the treatment of an adult lymphoma is on the order of 10 mCi. After how many days would radiation levels drop to less than a millicurie?

Comprehensive Problems

81. Use the Internet to research the connection between smoking and exposure to the nuclides polonium-210 and lead-210.
82. For the same dose of radiation, which has a higher dose equivalent, strontium-90 or radon-222?
83. How do you know whether or not a gamma ray accompanies an alpha or a beta emission?
84. Write nuclear equations for the following processes:
    a. A positron is emitted by oxygen-15.
    b. Boron-11 is formed by positron emission.
    c. A positron is emitted by chlorine-35.
    d. Oxygen-18 is formed by positron emission.
85. One of the radioactive decay series is shown below. Identify the mode of radioactive decay at each step.

$$^{232}_{90}\mathrm{Th} \rightarrow {}^{228}_{88}\mathrm{Ra} \rightarrow {}^{228}_{89}\mathrm{Ac} \rightarrow {}^{228}_{90}\mathrm{Th} \rightarrow {}^{224}_{88}\mathrm{Ra} \rightarrow {}^{220}_{86}$$
$$\mathrm{Rn} \rightarrow {}^{216}_{84}\mathrm{Po} \rightarrow {}^{212}_{82}\mathrm{Pb} \rightarrow {}^{212}_{83}\mathrm{Bi} \rightarrow$$
$$^{212}_{84}\mathrm{Po} \rightarrow {}^{208}_{82}\mathrm{Pb}$$

86. In the radioactive decay series in the previous problem, lead-212 was formed. Why didn't the decay series stop at lead-212?
87. In the radioactive decay series given in Problem 85, which elements are represented by the symbols Rn and Ra? Which one is a gas? Which one is a metal? Which one is chemically inert?
88. Radon is produced in three of the naturally occurring decay series. Which three? Which isotopes of radon are formed? How would you expect these isotopes to differ? How would you expect them to be the same?
89. Of the three radon isotopes mentioned in the previous problem, only radon-222 is a health hazard. Propose a reason why, and then research your answer to see whether you are correct.
90. Many tropical islands are volcanic in origin and contain uranium in the rocks and minerals beneath the soils. However, radon is less likely to be a problem in homes built in the tropics. Propose two reasons why (and more if you can).
91. How does the nucleus of a carbon atom compare in density with that of elemental lead ($d = 11.3\ \mathrm{g/cm^3}$)? To answer this question, calculate the volume of a ^{12}C nucleus, assuming that the nucleus is spherical and that the radius is 1.2×10^{-13} cm. The mass of the nucleus in ^{12}C is 11.96709 u, or 1.98718×10^{-23} g.
92. The transformation of elements into other elements also takes place in stars. Write the nuclear equation for the formation of oxygen-16 when carbon-12 is hit with an alpha particle. A gamma ray is also released in this reaction.

93. Does food irradiation make the food radioactive? Find an answer to this question using the resources of the World Wide Web. Cite your sources.
94. Look up on the Internet the current maximum allowed exposure of workers in the nuclear industry. The Department of Energy (DOE) sets this standard. How does this standard vary for some individuals?
95. Irene Curie and her mother Marie are not the only scientists to have won the Nobel Prize for their pioneering work in nuclear chemistry and physics. Others include Ernest Rutherford, Ernest O. Lawrence, and Emilio G. Segrè. Use the Internet to find out why these and/or other prizes for nuclear work were awarded.
96. We pointed out in the text that palladium-103 (half-life = 16.97 days) is used in the treatment of prostate cancer, replacing iodine-125 (half-life = 59.4 days), though both are widely used. Recently, Cs-131 (half-life = 9.7 days) has been used. Why are these different products being used? That is, what are the advantages and disadvantages of each?
97. Palladium-103 has a half-life of 16.97 days.
    a. What is the half-life in years?
    b. How many protons, neutrons, and electrons are in an atom of palladium-103 in $PdCl_2$?
    c. A researcher develops cisplatin ($Pd(NH_3)_2Cl_2$) using the palladium-103 isotope. Assume that the process involves the direct conversion of palladium(II) chloride to cisplatin in one step, but the process itself requires 24 h to complete. How many grams of radioactive cisplatin would remain if the researcher started with 100.0 g of pure palladium-130(II) chloride?

Thinking Beyond the Calculation

98. Americium oxide is typically used in household smoke detectors. A document reports that a gram of americium oxide provides enough active material for "more than 5000 household smoke detectors." The particular isotope used in this application is americium-241.
    a. How much americium is present in a typical smoke detector?
    b. Give two reasons why Am-240 and Am-242 would not be appropriate isotopes to use.
    c. Americium-241 decays by alpha emission. Write the nuclear reaction for this process.
    d. The half-life of americium-241 is 432.2 years. How long will it take the reactivity of a sample of this nuclide to drop to 1% of its original activity?
    e. Beta-particle emission by americium-242 is found in 83% of the sample. The rest of an americium-242 sample decays by electron capture. Write nuclear reactions for these processes.

Carbon

Contents

Supplementary Information The online version contains supplementary material available at ▶ https://doi.org/10.1007/978-3-030-90267-4_21.

M. Mosher and P. Kelter, *An Introduction to Chemistry*,
https://doi.org/10.1007/978-3-030-90267-4_21

What We Will Learn in This Chapter

- Organic compounds are composed of non-metals and especially include carbon (▶ Sect. 21.1).
- Diamond, graphite, and fullerene are allotropes of carbon (▶ Sect. 21.2).
- Oil is a major source of organic compounds (▶ Sect. 21.3).
- Organic compounds are named from a set of rules based on their structure (▶ Sect. 21.4).
- Fractional distillation can be used to separate the components in crude oil (▶ Sect. 21.5).
- The arrangement of atoms within an organic molecules imparts specific properties to the molecule (▶ Sect. 21.8).
- Alkenes are precursor molecules to the formation of polymers (▶ Sect. 21.9).
- Each functional group has a particular reactivity that can be used to create new organic molecules (Sects. 21.11 and 21.12).
- Organic chemistry and the reactions of organic compounds can be used to discover new drugs to treat disease (▶ Sect. 21.15).

21.1 Introduction

Deep beneath the surface of the Earth, a spinning drill bit from an oil exploration platform breaks through the hard rock and hits pay dirt—oil. The dark liquid escapes from its high-pressure vault to emerge at the surface as a gushing fountain of chemical possibilities. The famous "gushers" of the past (see ◘ Fig. 21.1) may be rare now, but oil remains the chemical foundation of a huge industry sustaining our modern way of life. The gas and diesel fuels that power our vehicles, along with plastics, paints, modern textiles, and a vast range of medicines, are derived from the petrochemical industry. The raw material that flows out of the Earth and into that industry is crude oil, petroleum. The element at the heart of the chemistry of oil is carbon. This carbon is not pure but is bonded together with hydrogen and other elements to form a rich mixture of molecules that can be separated, modified, and exploited to make so many of the products that sustain our modern way of life.

Coal, another carbon-based substance, is mined from natural outcroppings both above and below ground, such as the one shown in ◘ Fig. 21.2. The use of this resource helped fuel the industrial revolution of eighteenth and nineteenth centuries. Since then, coal has been employed to make a wide range of useful carbon-based substances, including gasoline, diesel and jet fuels, and some chemicals. Although its use is diminishing rapidly, coal also remains one of the most important fuels for generating electricity (see ◘ Table 21.1). It is worthwhile to note that as vital as crude oil and coal have been, and still are, to the worldwide generation of electricity, their supply is limited and alternative energy sources are becoming cheaper and more readily useable.

Our planet is made most interesting by the chemistry of carbon. In addition to its vital role as the primary component of fossil fuels, the chemistry of carbon is also important to perhaps the most precious treasure of all, life. All of the key molecules of life are built around a framework of bonded carbon atoms. These carbon-based

◘ **Fig. 21.1** Black gold. An oil derrick becomes engulfed in a spray of crude oil in 1909. *Source* ▶ https://upload.wikimedia.org/wikipedia/en/0/08/Well2-6-Kern%2B1909.jpg

Fig. 21.2 Coal is obtained from huge open-pit mines and from deep underground in large pockets. This natural resource is useful in heating homes, in fueling electric plants, and in making steel. Image provided by Amcyrus2012. *Source* ▶ https://upload.wikimedia.org/wikipedia/commons/a/a4/Bituminous_Coal.JPG

Table 21.1 2019 Fuel sources of energy in the United States

Source	Percent of total BTU* (%)
Coal	11.3
Natural gas	32.1
Petroleum	36.8
Hydroelectric	2.6
Nuclear	8.5
Solar	1.0
Wind	2.6
Biomass	4.9
Total production	100.3 quadrillion BTU (quadrillion = 10^{15})

*1 BTU = 1055 J

molecules were initially known as **organic compounds** because this class of compounds was found in living organisms. It was later determined that organic compounds do not appear exclusively in living organisms, but the name stuck. In fact, the oil and coal we use as a source of carbon-based compounds were themselves formed from the carbon-based chemicals in ancient animals and plants. For this reason, the study of carbon compounds, known as **organic chemistry**, focuses on the reactions and properties of a vast number of organic compounds.

Organic compound – Carbon-based compounds.

Organic chemistry – The chemistry of carbon-containing compounds, especially those derived from living things or fossil fuels.

What is the nature of these organic chemicals? How do we make them? How do we chemically manipulate them to produce the stunning range of goods based on the crude oil that generates over a quarter of a trillion dollars in sales for oil and gasoline companies per year?

21.2 Elemental Carbon

Carbon is a very versatile element that can be used to prepare a nearly endless array of carbon-based compounds. It also exists in three different and distinct forms as the element. Different forms of the same element, known as **allotropes**, are not exclusive to carbon. Other elements, such as phosphorus, oxygen (as O_2 and O_3), and sulfur, also can be found in a variety of allotropes. The allotropes of carbon are diamond, graphite, and fullerenes.

Allotropes – Different forms of the same element.

Fig. 21.3 A raw uncut diamond. Diamond is an allotrope of carbon that humans find particularly valuable when cut and polished. Photo taken by Tõnu Pani. *Source* ▶ https://upload.wikimedia.org/wikipedia/commons/d/d8/Diamond_crystal%2C_0%2C7_ct.jpg

21.2.1 Diamond

Figure 21.3 illustrates the first of the three allotropes of carbon, diamond. The carbon within a diamond is bonded to four neighboring carbons in a giant covalent network, as shown in Fig. 21.4. The bonds in a diamond are the result of the overlap of identical sp^3 hybridized orbitals that are arranged tetrahedrally around each atom. In fact, this makes diamond one of the hardest substances known, because the network of many strong covalent bonds must be disrupted to break or distort a piece of diamond. The strong bonding in diamond explains why it forms such an enduring gemstone and why it can be used to cut other materials.

21.2.2 Graphite

Compare the structure of diamond to that of graphite, another of the allotropes of carbon shown in Fig. 21.5. The sp^2 hybridized carbon atoms in graphite share covalent bonds to three neighboring atoms, creating a repeating hexagonal network of carbon atoms in extended layers. Each atom in this structure contributes one unbonded electron to a system of delocalized π electrons above and below each hexagonal layer. Unlike the sp^3 hybridized carbons in diamond, the delocalized and mobile electrons in graphite allow it to conduct electricity. This means that graphite can be used as the electrodes in industrial processes such as the Hall–Heroult process for producing aluminum. The layered structure of graphite also makes it soft and slippery. Each layer is only weakly attracted to the layers above and below. This enables us to use graphite as pencil "lead"; portions of the layers will slide off the tip of the pencil and onto the paper. Graphite's slipperiness also makes it useful as a solid lubricant in special-

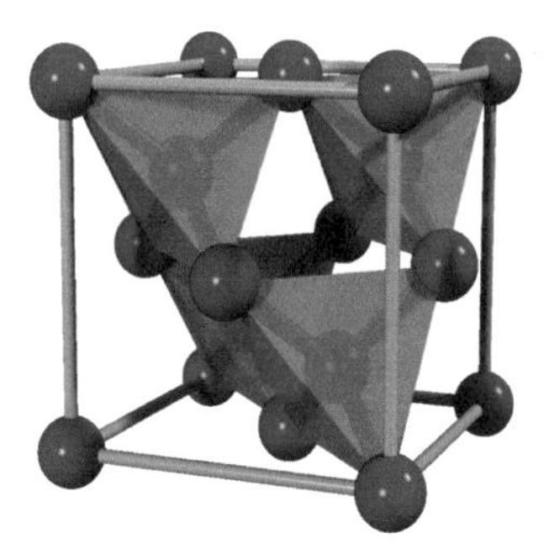

Fig. 21.4 The unit cell of diamond. The tetrahedral pattern of bonding is shown by the orange outlines. Image constructed by Pieter Kuiper. *Source* ▶ https://upload.wikimedia.org/wikipedia/commons/0/0a/Diamond_structure.gif

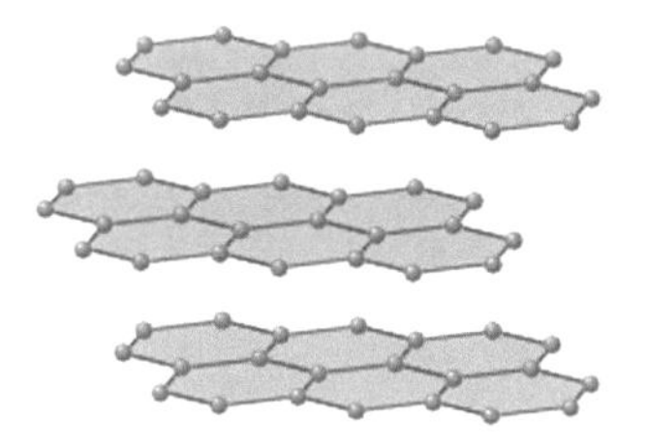

Fig. 21.5 The structure of graphite

ized machinery. Diamond, by contrast, in which the entire structure is a network of covalent bonds, is a terrible lubricant but an excellent grinding agent.

21.2.3 Fullerenes

A third allotrope of carbon, shown in Fig. 21.6, was discovered in 1985. This allotrope consists of sp^2 hybridized carbon atoms folded into structures that resemble balls and tubes. The core member of this class of molecules is called Buckminsterfullerene (C_{60}) because of its similarity to the geodesic domes designed by the architect Buckminster Fuller. Buckminsterfullerene (also known as "buckyball") is prepared when an electric discharge arcs between two graphite rods surrounded by helium gas at high pressure. Although C_{60} and graphite are similar in the presence of sp^2 hybridized carbon atoms, the carbon atoms in buckminsterfullerene are held in both six-membered and five-membered rings. The five-membered rings cause a curvature in the overall structure of the compound.

Buckminsterfullerene is a member of an increasingly varied group of related structures known as **fullerenes**. These include cage-like molecules both smaller and larger than the C_{60} cage, as well as tubes of bonded atoms that can occur either alone or in concentric nested patterns, as shown in Fig. 21.6. Many research teams are busy trying to find medical and technological applications for these structures.

Fullerenes – Forms of carbon based on the structure of buckminsterfullerene, C_{60}.

► https://upload.wikimedia.org/wikipedia/commons/1/18/Buckminster_Fuller_dome_in_Carbondale.jpg

21.3 Crude Oil—The Basic Resource

Crude oil, also known as petroleum, is not an allotrope of carbon. Instead, it is a very complex mixture of hundreds of compounds containing carbon and other atoms. In general terms, crude oil contains many different kinds of **hydrocarbon** molecules (molecules composed of only carbon and hydrogen atoms), together with smaller amounts of molecules derived from hydrocarbons, often including atoms of sulfur, nitrogen, oxygen, and various metals. Table 21.2 lists the typical classes of compounds that can be found in crude oil.

Hydrocarbons – Compounds composed of only hydrogen atoms and carbon atoms.

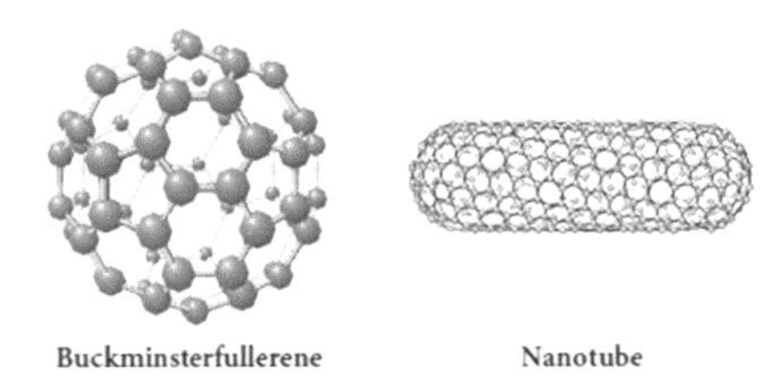

Fig. 21.6 Fullerenes, one of the allotropes of carbon, includes ball-shaped structures and hollow tube structures. Each of the carbon atoms in these structures is sp^2 hybridized

Table 21.2 Composition of crude oil

Class of compound	Percent of oil (%)	Typical example
Aliphatics (hydrocarbons containing no double or triple bonds)	25	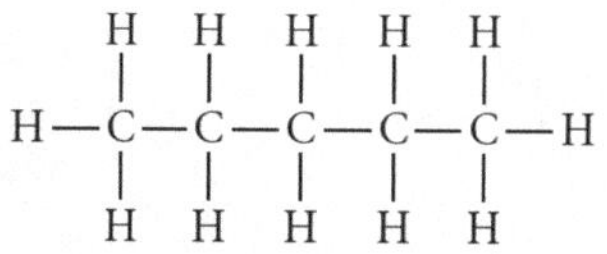 $CH_3-CH_2-CH_2-CH_2-CH_3$ (structural formula)
Aromatics (containing aromatic groups)	17	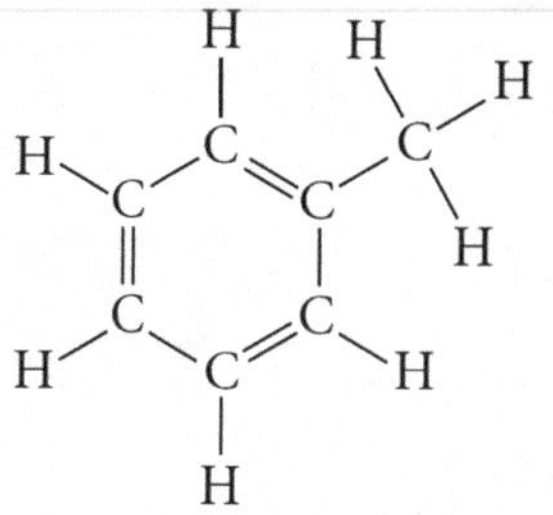 (structural formula)
Naphthenes (cyclic hydrocarbons with all single-bonded carbons)	47	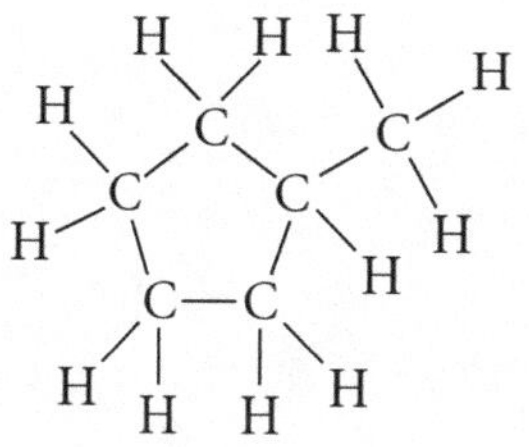 (structural formula)
Sulfurous	<8	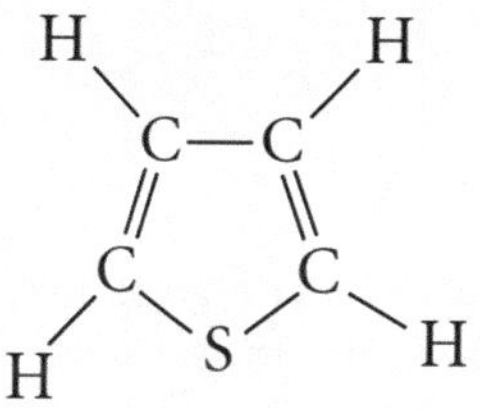 (structural formula)
Nitrogenous	<1	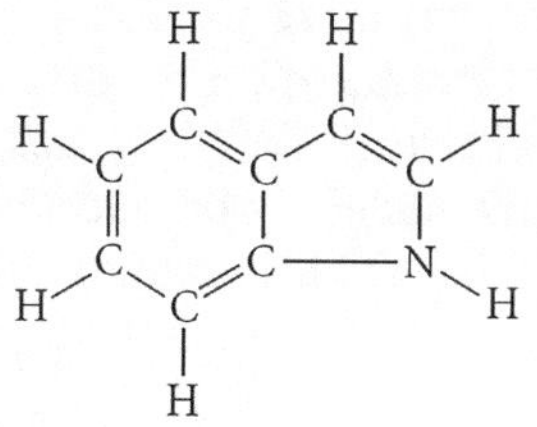 (structural formula)
Oxygenated	< 3	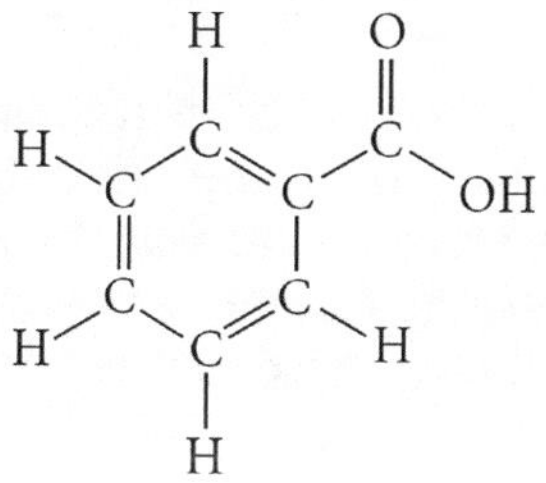 (structural formula)
Metals	<<1	Fe, Mn, Zn, V

Crude oil is generally found deep beneath the surface of the Earth, trapped between layers of rock in pockets like that shown in Fig. 21.7. It is found nearly everywhere in the world, under both land and sea. Unfortunately, removing the oil from these pockets is an involved and expensive process, but it is done because of the incredible

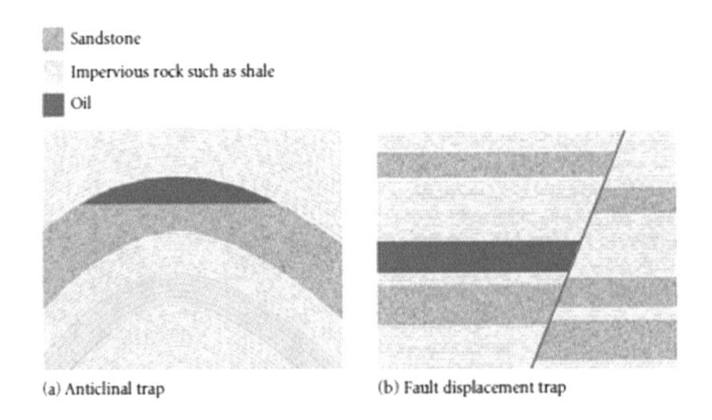

Fig. 21.7 Geology of a crude oil reservoir

Table 21.3 Top 10 oil producers in 2019

Country	Million barrels per day
United States	19.47
Saudi Arabia	11.62
Russia	11.49
Canada	5.50
China	4.89
Iraq	4.80
United Arab Emirates	4.13
Brazil	3.67
Iran	3.20
Kuwait	2.95

usefulness of the compounds found in crude oil. Table 21.3 lists the countries that produce the most crude oil and indicates how much oil they ship each day. Note how many of these countries are global "hot spots" of regional conflict.

Why do we so passionately seek out oil? Why are nations willing to go to war over it? Why are national economies dependent on the prevailing price of oil? The answers include our heavy use of oil to produce the gasoline and other fuels for cars, trucks, trains, boats, and jet airplanes. But the uses of oil go well beyond these things. We are surrounded by products made from oil, including plastics, clothing, furniture, carpeting, eyeglasses, and compact disks; the list is almost endless.

21.4 Hydrocarbons

The carbon atoms of hydrocarbons are bonded into straight chains, branched chains, rings, or more complex combinations of these three basic structures (Fig. 21.8). The hydrogen atoms are bonded to the carbon atoms so that each carbon has a total of four covalent bonds. Some of the carbon atoms in hydrocarbons can be bonded together by carbon–carbon double bonds or carbon– carbon triple bonds, reducing the number of hydrogen atoms that can bind to the carbon atoms.

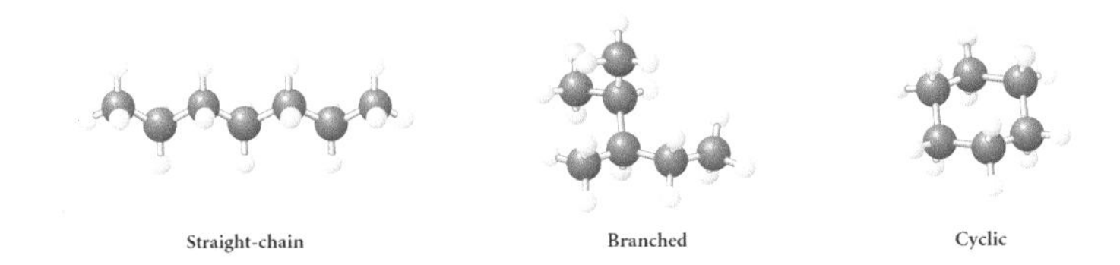

Fig. 21.8 The three classes of hydrocarbons include the straight-chain, branched, and cyclic structures

Hydrocarbons that contain no double or triple bonds are known as **saturated hydrocarbons**, or **aliphatic compounds**. They are "saturated" with hydrogen, which means that the carbon atoms are bonded to the maximum possible number of hydrogen atoms, because none of the electrons are involved in double or triple bonds. This is the origin of the term *saturated fats*, which might be familiar to you from food labels. All fats include hydrocarbon chains as part of their structure, and the saturated fats have no C=C double bonds.

Aliphatic compounds – Compounds in which all of the carbon atoms are sp^3 hybridized and contain only single bonds.

Saturated hydrocarbons – Hydrocarbons that do not contain double or triple bonds.

21.4.1 Alkanes

Saturated hydrocarbons are also known as **alkanes**, and they make up the largest fraction of any class of crude oil compounds (see ◘ Table 21.2). The simplest possible alkanes are the **normal alkanes**, which have no branched chains or rings of carbon atoms. The simplest normal alkane—and also the simplest hydrocarbon— contains a single carbon atom bonded to four hydrogen atoms. This molecule is known as methane.

Alkanes – Saturated hydrocarbons.

Normal alkanes – Alkanes that have no branched chains or rings of carbon atoms.

Methane is the principal component of "natural gas," which is commonly found trapped above petroleum deposits and is piped into homes, offices, and factories as a fuel for heating and cooking.

Methane

Ethane

The next-simplest hydrocarbon has two carbon atoms bonded together, with each carbon atom bonded to three hydrogen atoms. This is ethane. As you can see from ◘ Table 21.4, the normal alkanes form a regular series in which each member has one more $-CH_2-$ group, a **methylene group**, than the preceding member of the series. These alkanes are members of a **homologous series** of compounds; this means they all share the same general formula, $C_nH_{(2n+2)}$ for the normal alkanes. They have similar chemical and physical properties, which change in a gradual manner as we move through the series.

Methylene group – The $-CH_2-$ group.

Homologous series – A series of compounds sharing the same general formula.

◘ **Table 21.4** The first ten normal alkanes

Name	Number of carbons	Molecular formula	Structural formula	Boiling point (°C)
methane	1	CH_4	CH_4	–161.5
Ethane	2	C_2H_6	CH_3CH_3	−88.6
Propane	3	C_3H_8	$CH_3CH_2CH_3$	–42.1
Butane	4	C_4H_{10}	$CH_3CH_2CH_2CH_3$	−0.5
Pentane	5	C_5H_{12}	$CH_3(CH_2)_3CH_3$	36.0
Hexane	6	C_6H_{14}	$CH_3(CH_2)_4CH_3$	68.7
Heptane	7	C_7H_{16}	$CH_3(CH_2)_5CH_3$	98.5
Octane	8	C_8H_{18}	$CH_3(CH_2)_6CH_3$	125.6
Nonane	9	C_9H_{20}	$CH_3(CH_2)_7CH_3$	150.8
Decane	10	$C_{10}H_{22}$	$CH_3(CH_2)_8CH_3$	174.1

Although we call the normal alkanes **straight-chain alkanes**, in reality the carbon chain zig-zags, as shown in ◘ Fig. 21.8. This is because the four bonds of each carbon atom are formed when four *sp*^{3} hybridized orbitals of the atom overlap with the orbitals of neighboring carbon atoms and hydrogen atoms. The bonds around each carbon atom adopt a tetrahedral arrangement.

Straight-chain alkanes – Molecules that contain only hydrogen atoms and *sp*^{3} hybridized carbon atoms arranged in linear fashion.

The major direct use we make of the alkanes in petroleum is as fuels. Hydrocarbons burn in oxygen (or in air, which is one-fifth oxygen) to generate mostly carbon dioxide and water and release considerable amounts of heat. The combustion of hydrocarbons helps us heat our homes, cook our food, and power our cars. Like most of the other hydrocarbons in crude oil, the alkanes also serve as valuable and versatile "feedstock" molecules—molecules that can be modified by the chemical industry to generate many useful products.

21.4.2 Branched-Chain Alkanes: Isomers of the Normal Alkanes

Crude oil also contains a significant proportion of alkanes with "branched" chains. Each branched-chain alkane has the same formula as a corresponding straight-chain (normal) alkane, so each one can be regarded as a **structural isomer** of a normal alkane. Structural isomers always have the same formula but differ in the way the atoms are attached. We say that they share the same molecular formula but have different molecular structures. Structural isomers do not have the same properties. For instance, the structural isomers pentane and 2,2-dimethylpropane have the same formula (C_5H_{12}). Their boiling points, however, are much different: 36 °C for pentane and 10 °C for 2,2-dimethylpropane.

Structural isomers – Molecules with the same formula but different structures.

2,2,4-Trimethylpentane (isooctane)	CH_3–C(CH_3)$_2$–CH_2–CH(CH_3)–CH_3
2-Methylbutane (isopentane)	CH_3–CH_2–CH(CH_3)–CH_3
2-Methylpentane (isohexane)	CH_3–CH_2–CH_2–CH(CH_3)–CH_3
2,4-Dimethylpentane	CH_3–CH(CH_3)–CH_2–CH(CH_3)–CH_3

Example 21.1—Finding the Isomers

How many structural isomers of pentane (C_5H_{12}) exist? How many structural isomers are there for hexane (C_6H_{14})?

Asking the Right Questions
What is the definition of "structural isomers"? Is there a relationship between the number of structural isomers and the formula for an organic molecule?

Solution
There isn't an easy relationship between the number of structural isomers and the formula; however, we can find all the isomers by drawing the possible different carbon skeletons then adding the hydrogen atoms needed to ensure that each carbon atom has four bonds.

Isomers of pentane:

CH_3—CH_2—CH_2—CH_2—CH_3

CH_3—CH(CH_3)—CH_2—CH_3

CH_3—C(CH_3)$_2$—CH_3

Isomers of hexane:

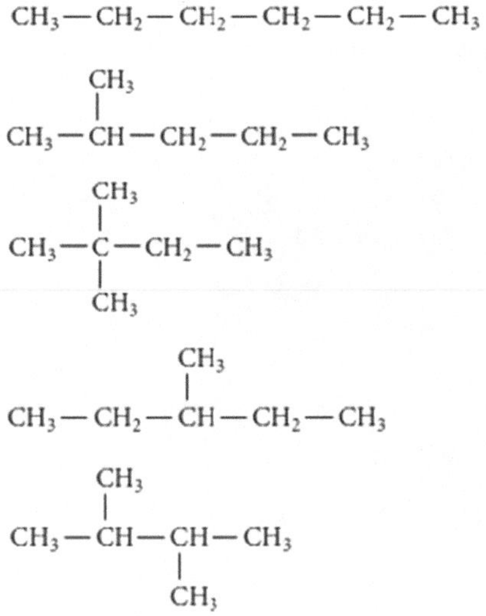

$$CH_3-CH_2-CH_2-CH_2-CH_2-CH_3$$

$$\begin{array}{ccccccccc} & & CH_3 & & & & & & \\ & & | & & & & & & \\ CH_3 & - & CH & - & CH_2 & - & CH_2 & - & CH_3 \end{array}$$

$$\begin{array}{ccccccc} & & CH_3 & & & & \\ & & | & & & & \\ CH_3 & - & C & - & CH_2 & - & CH_3 \\ & & | & & & & \\ & & CH_3 & & & & \end{array}$$

$$\begin{array}{ccccccccc} & & & & CH_3 & & & & \\ & & & & | & & & & \\ CH_3 & - & CH_2 & - & CH & - & CH_2 & - & CH_3 \end{array}$$

$$\begin{array}{ccccccc} & & CH_3 & & & & \\ & & | & & & & \\ CH_3 & - & CH & - & CH & - & CH_3 \\ & & & & | & & \\ & & & & CH_3 & & \end{array}$$

Are Our Answers Reasonable?

Examining each of the structures illustrates that we have *systematically* determined each possibility. Therefore our answer is reasonable.

What If…?

…we wanted to know about C_3H_8O? How many isomers would you predict?

(*Ans*: 3)

Practice 21.1

Draw all of the isomers of heptane (C_7H_{16}).

21.4.3 Naming the Alkanes

With well over 16 million known organic chemicals, it is useful to have an internationally agreed-upon set of nomenclature rules. We hinted at such a system with the first 10 normal alkanes in ◘ Table 21.4. These names form the basis of the more complex names given to branched-chain alkanes. This molecule is called 3-methylheptane. Why?

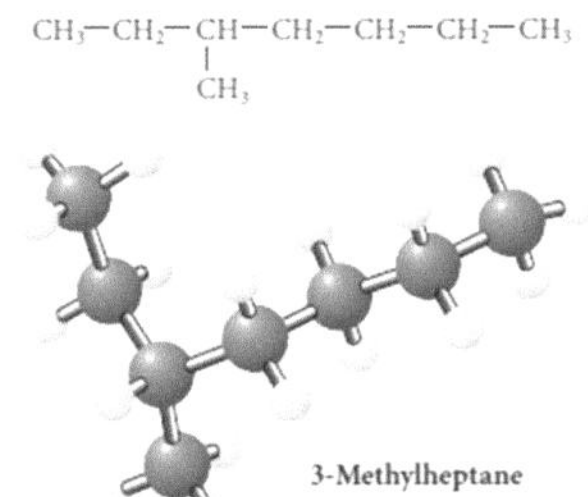

1. We identify and name the longest straight chain in the molecule, which in this case is derived from heptane.
2. We name the branch, using the name of the alkane with the same number of carbon atoms as are in the branch (methane, in this case). However, we replace the *-ane* ending with -yl to indicate that the group is a branch. The word methyl then is added to the front of our molecule's name, which becomes methylheptane.
3. We identify the location of the branch by associating it with the numbered carbon atom of the main chain to which it is attached. We number the main chain from the end that gives the branch point the lowest number. The final name is 3-methylheptane.

Various other rules guide us in naming more complex alkanes. These rules, collectively, are known as IUPAC nomenclature rules after the International Union of Pure and Applied Chemistry, whose members developed the naming system. All compounds in the world can be named using IUPAC nomenclature, but many compounds have less structured names. For instance, as we learned in ▶ Chap. 2 dihydrogen monoxide is commonly known as water. The common names, unfortunately, have few rules—and many exceptions. It is best to learn the common names as we are exposed to the compound, though with all but the most common compounds, we will use the IUPAC nomenclature.

21.4.4 Cyclic Alkanes

Carbon atoms can be bonded into rings, to form **cyclic alkanes**. The four simplest cyclic alkanes are shown below. Cyclohexane is used in the preparation of nylon fiber and is a nonpolar solvent. Cyclopentane is increasingly employed as a foaming agent—a substance that is needed to create foam—in the preparation of insulation for use in refrigerators. It replaces some of the chlorofluorocarbons implicated in ozone depletion.

Cyclic alkanes – Alkanes that possess a group of carbon atoms joined in a ring.

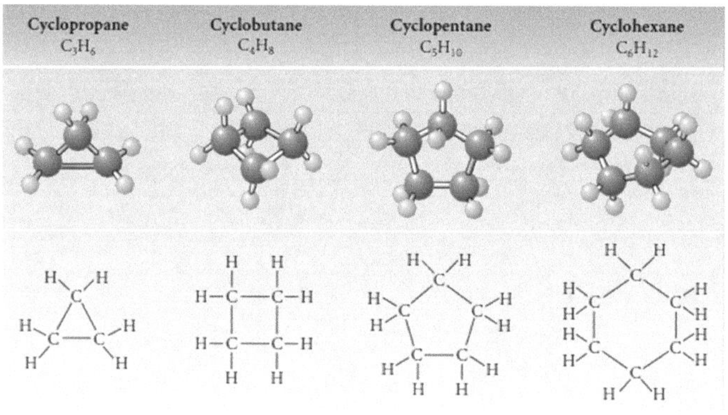

Cyclopropane C_3H_6 | Cyclobutane C_4H_8 | Cyclopentane C_5H_{10} | Cyclohexane C_6H_{12}

21.4.5 Alkenes

Alkenes are hydrocarbons that contain at least one carbon-to-carbon double bond (C=C). The members of this class of compounds are important to the chemical industry as starting materials in the manufacture of a host of organic compounds. Only tiny amounts of alkenes are found in crude oil, but certain alkenes are made from crude oil in huge quantities, in processes that we will consider shortly. The simplest three alkenes are shown below.

Alkenes – Hydrocarbons containing at least one C=C double bond.

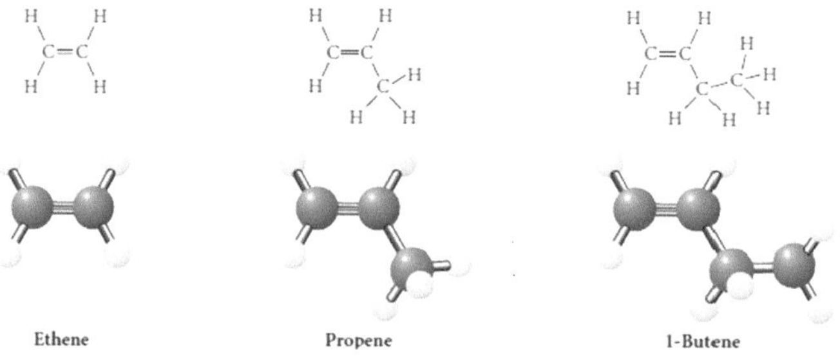

Ethene Propene 1-Butene

The names of the alkenes are derived from the names of the corresponding alkanes, but with *-ene* at the end instead of –ane. The simple alkenes, those with only one double bond, form a homologous series with the general formula C_nH_{2n}. This general formula is also the same as that of the cyclic alkanes.

Alkenes are known as **unsaturated hydrocarbons** because, unlike the alkanes, they are not "saturated" with hydrogen. Each C=C bond in an alkene can react with one hydrogen molecule, under suitable conditions, to generate the corresponding alkane. For example, the **hydrogenation** (addition of a molecule of hydrogen) of ethene gives ethane. This is a general reaction of the alkenes. Platinum metal is a catalyst in this reaction and therefore is indicated over the arrow in the chemical equation that represents the reaction.

Hydrogenation – The addition of a molecule of hydrogen to a compound.

Unsaturated hydrocarbons – Hydrocarbons that contain C=C or C≡C bonds.

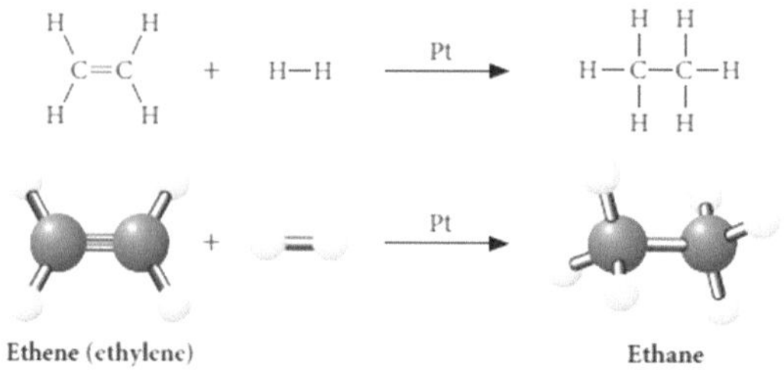

$$H_2C{=}CH_2 + H{-}H \xrightarrow{Pt} H_3C{-}CH_3$$

Ethene (ethylene) Ethane

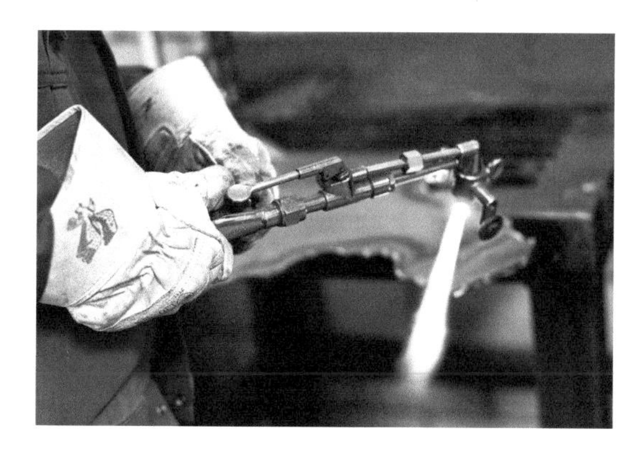

Fig. 21.9 Photograph of oxy-acetylene welding. High temperatures are obtained during the combustion of acetylene. This melts the metals, allowing a strong weld to form. Photo by Jonas Boni. *Source* ▶ https://upload.wikimedia.org/wikipedia/commons/a/a5/Oxy-fuel_cutting1.jpg

21.4.6 Alkynes

Hydrocarbons with a carbon-to-carbon triple bond are known as **alkynes**. Those with only one triple bond form a homologous series with the general formula $C_nH_{(2n-2)}$. The first three members of this series are shown below. Ethyne (acetylene) is burned with oxygen to release the energy that generates very high temperatures in oxy-acetylene welding, shown in ■ Fig. 21.9.

Alkynes – Hydrocarbons containing a C≡C bond.

H—C≡C—H
Ethyne (acetylene)

H—C≡C—CH₃ (drawn with H atoms on the terminal carbon)
Propyne

H—C≡C—CH₂—CH₃ (drawn with H atoms on each carbon)
1-Butyne

Example 21.2—Care with General Formulas

We have seen that the general formula for the alkenes is C_nH_{2n}. Does this mean that every hydrocarbon with that general formula must be an alkene?

Asking the Right Questions

Based on what we know about organic compounds, what are the general formulas for an alkane, cyclic alkane, alkene, and alkyne?

Solution

Can we draw any hydrocarbon structure that has twice as many hydrogen atoms as carbon atoms, but no double bonds? It cannot be done using straight chains or branched chains of carbon atoms, but it *can* be done if the carbon chain bonds back on itself, to form a cyclic hydrocarbon. As we see from the examples below, cyclic alkanes share the general formula C_nH_{2n} with alkenes.

Cyclopropane
C_3H_6

Cyclobutane
C_4H_8

Are Our Answers Reasonable?

The number of bonds to carbon and hydrogen has not been violated in the cyclic structures, so our answers are reasonable.

What If…?

…we considered the formula of an alkyne? Does the general formula of an alkyne always indicate that we have an alkyne functional group?

(*Ans: No. The stucture could have two rings, two alkenes, or any combination and still have the formula of an alkyne.*)

Practice 21.2

Provide a structural drawing for a cyclic hydrocarbon that would be an isomer of hexene.

The general formula C_nH_{2n} is not unique to alkenes. We saw in Exercise 21.2 that each alkene with three or more carbon atoms will have a cyclic alkane as one of its isomers. For example, cyclobutane is an isomer of butene:

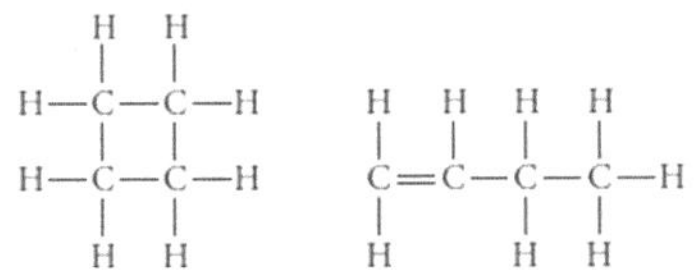

Cyclobutane
C_4H_8

1-Butene
C_4H_8

Isomers do not always have to belong to the same homologous series; they just have to have the same atoms bonded in different ways.

The naming of alkenes and alkynes is complicated by the need to distinguish exactly where the double or triple bond lies within the molecule. This is handled by attaching a number in front of the name of the molecule. The number corresponds to the location of the first carbon in the double or triple bond. Just as in naming branched alkanes, the number we use must be the lowest number that can be generated by numbering the longest carbon chain. For example, $CH_3CH{=}CHCH_2CH_3$ is 2-pentene, not 3-pentene. In this way, we see that there are only two alkenes with the formula C_4H_8: 1-butene ($CH_2{=}CHCH_2CH_3$) and 2-butene ($CH_3CH{=}CHCH_3$).

21.4.7 Geometric Isomers

Geometric isomers are common in the alkenes. What is a **geometric isomer**? Looking at the boiling points of two of the isomers of C_4H_8 shown below will help us arrive at an answer to this question.

Geometric isomers – Molecules that differ in their geometry but have identical chemical formula and points of attachment. See also *cis* and *trans*.

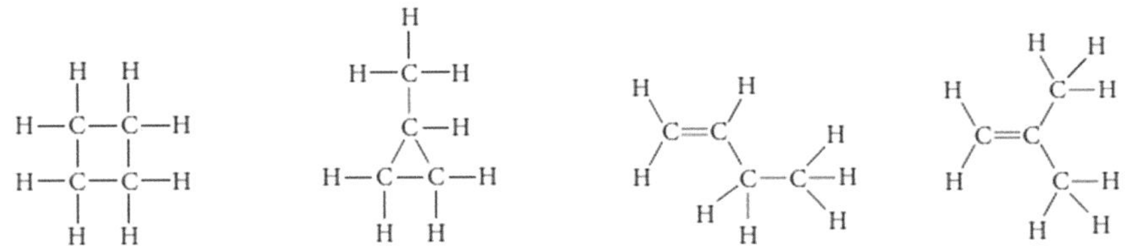

A B C D

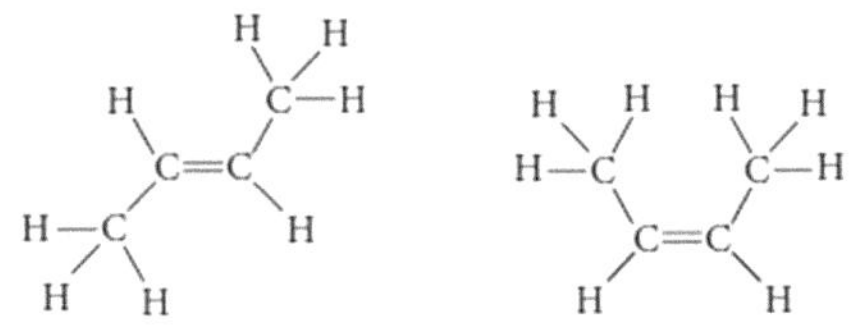

E F

We see that two cyclic molecules, cyclobutane (A) and methylcyclopropane (B), are the only alkanes we can draw. The remainder of the molecules with the formula C_4H_8 are alkenes, including 1-butene (C), 2-methylpropene (D), and two different 2-butenes (E and F). Are the two 2-butenes really different? We could imagine that if we rotated the C=C bond around, we would have the same structures. However, the C=C bond doesn't allow rotation to occur, so these two molecules cannot be redrawn to show the same structure. In fact, their physical properties indicate to us that they are completely different. This is where the boiling points prove important. Molecule (F) has a boiling point of 3.7 °C, whereas molecule (E) has a boiling point of 1 °C, giving evidence for two different molecules.

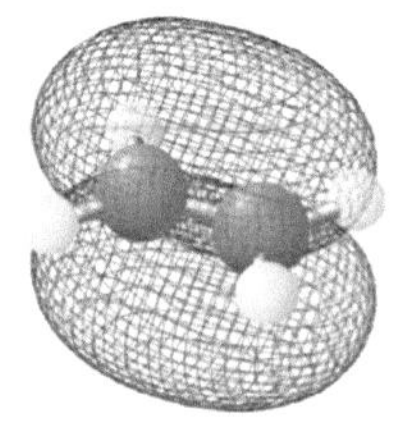

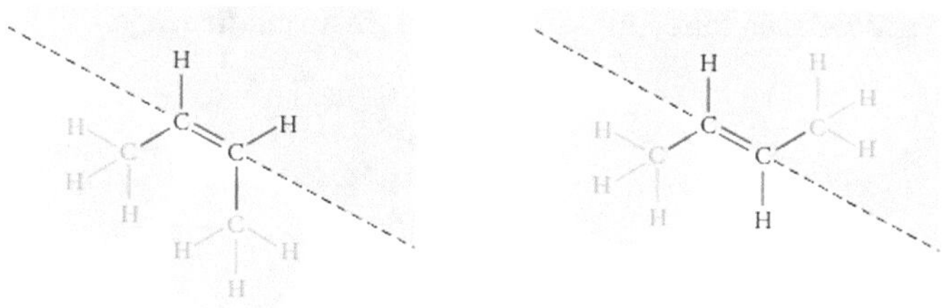

cis *trans*

Fig. 21.10 Cis and trans isomers of the alkenes. The cis isomer contains groups on the same side of the alkene. The trans isomer contains groups on opposite sides of the alkene

How can we distinguish between these two different molecules (E and F)? We use the designations *cis* and *trans* to describe their structures, as illustrated in Fig. 21.10. These prefixes are always given in italics. In the ***cis*** form, the groups on either side of the C=C are on the same side of the molecule (Fig. 21.10). (As a remembering device, think of *cis* as "sisters.") The molecule with the 3.7 °C boiling point is *cis*-2-butene. In the ***trans*** form, the groups on either side of the C=C are on opposite sides of the molecule (Fig. 21.10). (Think about the groups having to transfer from one side to the other.) This molecule is *trans*-2-butene.

Cis – A molecule containing groups that are fixed on the same side of a bond that cannot rotate.

Trans – A molecule containing groups that are fixed on opposite sides of a bond that cannot rotate.

In general, geometric isomers are isomers that differ in the location of the atoms only because of the geometry of the carbon to which they are attached. *Cis* and *trans* isomers are geometric isomers because the carbons of the alkene group have a fixed geometry.

Example 21.3—*Cis* and *Trans*

Consuming *trans* fatty acids is considered bad for your health. They arise when vegetable oil is hydrogenated. The resulting partially hydrogenated vegetable oils contain a small amount of *trans* fats. Is the compound below a *cis* or a *trans* fatty acid?

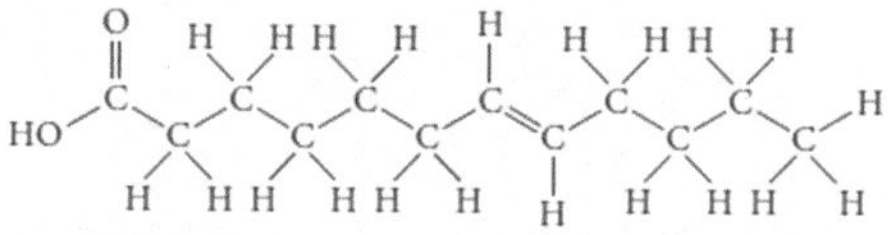

Asking the Right Questions

What functional group requires us to consider *cis* or *trans* isomerism? Does that group exist in this molecule? If so, does it possess at different groups on each carbon?

Solution

This compound is a *trans* fatty acid. The groups on either end of the C=C are on opposite sides of the molecule.

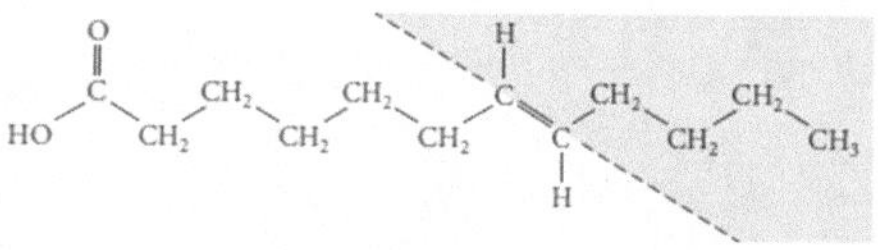

Are Our Answers Reasonable?

One of the large groups is in the shaded area and one is not. They are opposite each other, thus our answer of *trans* is reasonable.

What If…?

…we were concerned with 2-methyl-2-butene? How many geometric isomers does this compound possess?

(*Ans: None, because the arrangement on one end of the alkene functional group does not matter.*)

Practice 21.3

Draw all of the alkenes with the formula C_5H_{10} and indicate whether each is *cis* or *trans*.

21.4.8 Aromatic Hydrocarbons

Aromatic hydrocarbons contain one or more aromatic rings, which were discussed in ▶ Chap. 9. A common aromatic compound is benzene. This compound and its derivatives are used in the production of a vast array of consumer products such as polystyrene and other plastics, nylon, detergents, and pharmaceuticals. The molecule consists of three alternating carbon–carbon double bonds in a six-carbon ring. Benzene and naphthalene are common examples.

Aromatic hydrocarbons – Hydrocarbons that contain one or more aromatic rings.

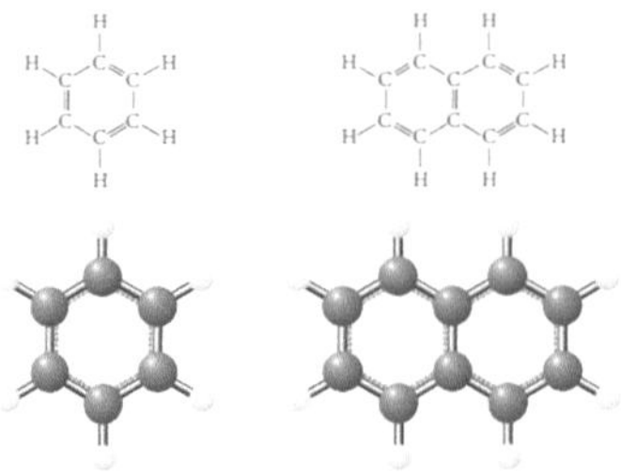

Benzene

Naphthalene

21.4.9 Alkyl Groups

Many hydrocarbons can be regarded as being composed of a main chain of carbon atoms with various hydrocarbon branches attached. These branches are known as **alkyl groups**, and many have a name derived from the name of the alkane on which they are based. The names of alkyl groups are placed in front of the parent name in alphabetical order.

Alkyl group – A hydrocarbon group, such as $-CH_3$ or $-C_2H_5$, within the structure of an organic compound.

$-CH_3$ is a methyl group.
$-C_2H_5$ is an ethyl group.
$-C_3H_7$ is a propyl group.
$-C_4H_9$ is a butyl group.
$-C_5H_{11}$ is a pentyl group.
$-C_6H_{13}$ is a hexyl group.
$-C_7H_{15}$ is a heptyl group.
$-C_8H_{17}$ is an octyl group.
–Ph (C_6H_5) is a phenyl group.

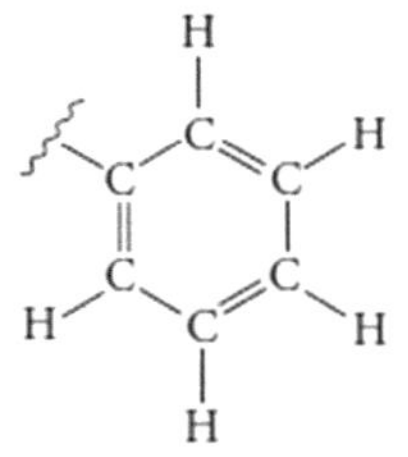

Phenyl group

Example 21.4—Naming with Alkyl Groups

Fill in the blanks, corresponding to the type of alkyl group, to complete the names of the hydrocarbons shown.

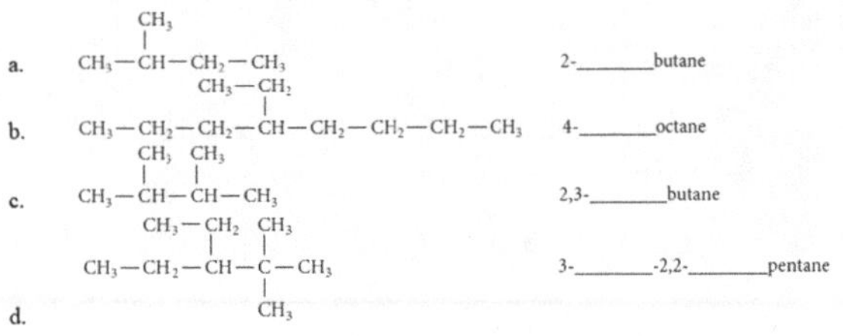

a. $CH_3-CH(CH_3)-CH_2-CH_3$ 2-______butane

b. $CH_3-CH_2-CH_2-CH(CH_2-CH_3)-CH_2-CH_2-CH_2-CH_3$ 4-______octane

c. $CH_3-CH(CH_3)-CH(CH_3)-CH_3$ 2,3-______butane

d. $CH_3-CH_2-CH(CH_2-CH_3)-C(CH_3)_2-CH_3$ 3-______-2,2-______pentane

Asking the Right Questions

What are the names of the alkyl groups? We must first identify the longest continuous carbon chain, then identify the alkyl groups on that chain.

Solution

a. 2-**methyl**butane
b. 4-**ethyl**octane
c. 2,3-**dimethyl**butane
d. 3-**ethyl**-2,2-**dimethyl**pentane

Are Our Answers Reasonable?

We have followed the rules for naming the groups, so our answers are reasonable.

What If...?

...we were asked to determine the number of carbons in a molecule? How many carbon atoms are in 2,2,4-trimethylpentane?
(*Ans*: 8)

Practice 21.4

Draw the structure of 2,2-dimethyl-4-propyloctane.

21

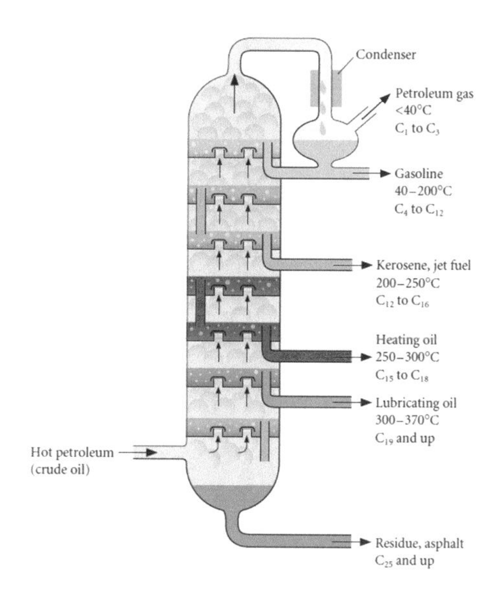

Fig. 21.11 A diagram of the fractional distillation of oil. Note each of the different types of fractions that can be isolated

21.5 Separating the Hydrocarbons by Fractional Distillation

Crude oil contains a wide variety of carbon-based compounds. Separating those compounds so that they can be used to make fabrics, medicines, and other compounds we need everyday is an important procedure. The first step in this process is to separate the oil into several "fractions" using fractional distillation, as shown in Fig. 21.11. Each of the fractions contains a selection of hydrocarbons with generally similar physical properties. Because a hydrocarbon's boiling point is broadly related to its molecular size, each fraction contains molecules that are similar in molecular size. The oil is heated to around 400 °C, causing much of it to vaporize. The tarry residue that remains is a mixture of very large hydrocarbon molecules. This residue is collected as one of the basic fractions produced by the fractional distillation process.

The vaporized portion of the oil rises up the fractionating column (Fig. 21.11), which can be around 60 m in height. The vapor cools as it rises, causing the various fractions of the oil to condense back into liquid form at different heights up the column, depending on their boiling points. At selected levels, the hydrocarbons that are in liquid form gather on collecting trays and are piped away. The smallest (and lowest-boiling) hydrocarbons do not condense into liquid at all but emerge at the top of the column as "refinery gas."

How Do We Know?—Which Hydrocarbons Are in Crude Oil?

On March 24, 1989, the oil tanker *Exxon Valdez* ran aground in Prince William Sound, Alaska, spilling over 40 million liters of crude oil into the sea. Despite the massive clean-up effort, small amounts of crude oil are still found along the shores near the accident. Toward the end of 1990, samples of these oil residues were collected as part of research to monitor the fate of the oil as the environment slowly recovered. It turned out, however, that some of the samples could not have come from the *Exxon Valdez*, because they contained the wrong mixture of hydrocarbons. The debris from older oil spills was complicating the picture. How was this determined? It was done using a technique called **gas chromatography (GC)**, one of the three most common methods for analyzing the hydrocarbon content of crude oil.

gas chromatography (gc) – Analytical technique in which components of a mixture are separated by their different flow rates through a chromatography column, driven by a carrier gas.

In gas chromatography, a sample such as crude oil that is to be analyzed is converted into vapor by heating and then carried by a flow of inert gas (such as He) through a heated column packed with solid powder such as silica, as shown in Fig. 21.12. In some cases, the particles of solid in the column may be coated with a nonvolatile liquid, a technique known as **gas–liquid chromatography (GLC)**. As the sample flows through the column, the components of the mixture travel at different speeds because of the differing extents to which they are adsorbed onto the solid phase or are soluble in the liquid phase (in GLC). The components emerge from the column separately and are detected, often by their effect on a detecting flame. The apparatus will have previously been calibrated using a range of known hydrocarbons,

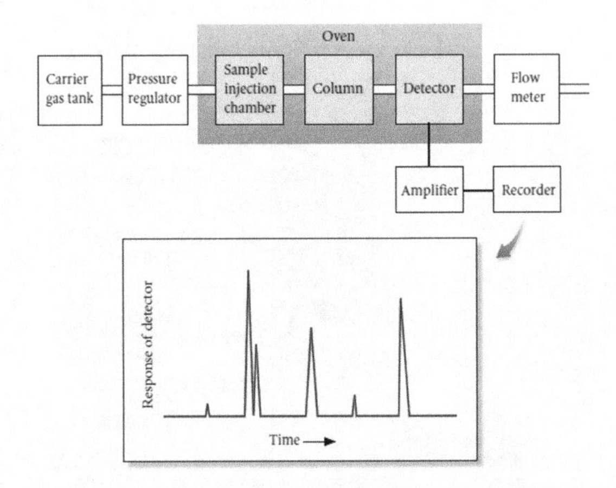

Fig. 21.12 The operating principles of gas chromatography and a sample readout (a gas chromatogram)

so specific hydrocarbons in the sample can be identified by comparison with the results of the calibration.

gas–liquid chromatography (glc) – A form of gas chromatography in which the solid phase is covered with a thin coating of liquid.

An example of the results of an analysis of one type of crude oil is shown in Fig. 21.13. The technique is very powerful, particularly in providing a GC "fingerprint" of different samples from different origins, but it has significant limitations. In particular, it cannot readily distinguish between some of the different structural isomers of hydrocarbons. For this reason, more thorough analysis of the hydrocarbons in crude oil also makes use of the techniques of mass spectrometry and nuclear magnetic resonance spectroscopy.

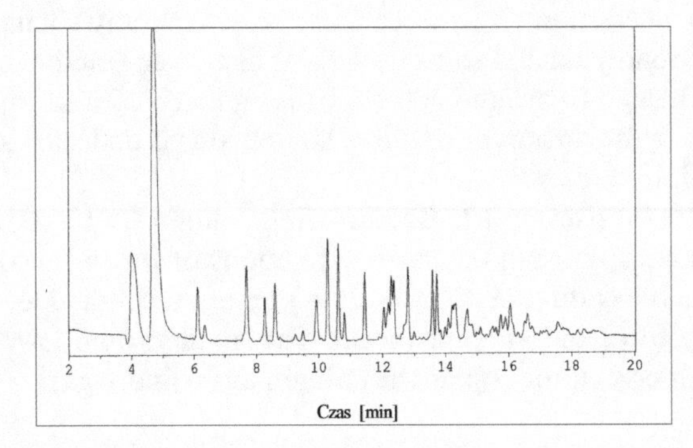

Fig. 21.13 A gas chromatogram showing the individual components in a mixture. Image by Joanna Kośmider. *Source* ▶ https://upload.wikimedia.org/wikipedia/commons/9/91/Chromatogram_KF.png

This primary fractional distillation process is typically used to separate the crude oil into six major fractions, summarized in Table 21.5. These fractions can be subjected to secondary fractional distillation or can be fed into further processing steps to convert them into useful products. Table 21.5 also lists some of the major uses made

Table 21.5 Products of the fractional distillation of crude oil and their uses

Fraction	Formulas	Uses	Boiling point (°C)
Natural gas	CH_4 to C_4H_{10}	Fuel, cooking gas	<0
Petroleum ether	C_5H_{12} to C_6H_{14}	Solvent for organic reactions	30–70
Gasoline	C_6H_{14} to $C_{12}H_{26}$	Fuel, solvent	70–200
Kerosene	$C_{12}H_{26}$ to $C_{16}H_{34}$	Rocket and jet engine fuel, domestic heating	200–300
Heating oil	$C_{15}H_{32}$ to $C_{18}H_{38}$	Industrial heating, fuel for producing electricity	300–370
Lubricating oil	$C_{16}H_{34}$ to $C_{24}H_{50}$	Lubricants for automobiles and machines	>350
Residue	$C_{20}H_{42}$ and up	Asphalt, paraffin	Solid

of each fraction. We will explore some of these uses in more detail, and in doing so we will investigate the chemical properties and reactivities of the hydrocarbons in each fraction.

21.6 Processing Hydrocarbons

The hydrocarbons within the fractions obtained by distillation are normally subjected to further processing before being used as fuels or as feedstocks (raw material) for the chemical industry. The two major types of processing are known as cracking and reforming.

21.6.1 Cracking

Many of the most useful hydrocarbons are the smaller ones, with fewer than 10 carbon atoms. At the refinery, long-chain hydrocarbons are isolated from the crude oil by fractional distillation. Then they are broken into a mixture of shorter-chain hydrocarbons, including some branched and unsaturated hydrocarbons, by means of a process called **cracking**. This process uses heat to break some of the carbon–carbon bonds in the long-chain hydrocarbons. The products of cracking often contain short saturated and unsaturated hydrocarbons. There are various types of cracking:

- **Thermal cracking** uses heat alone.
- **Catalytic cracking** uses heat plus a catalyst.
- **Hydrocracking** uses heat plus a catalyst in the presence of hydrogen gas, encouraging formation of saturated rather than unsaturated products.
- **Steam cracking** uses heat plus steam and a catalyst.

Cracking – The breaking up of hydrocarbons into smaller hydrocarbons, using heat and in some cases steam and/or catalysts.

Catalytic cracking – A type of cracking wherein heat and a catalyst are used.

Hydrocracking – A type of cracking that uses heat plus a catalyst in the presence of hydrogen gas.

Steam cracking – A form of cracking in which heat, steam, and a catalyst are used.

Thermal cracking – A form of cracking that uses heat alone.

21.6.2 Reforming

Another crucial process at the refinery is called **reforming**. In this process hydrocarbons distilled from petroleum are passed through a catalyst bed and converted into more useful forms. When a mixture is reformed, the predominantly straight-chain hydrocarbons are made into a mixture containing more aromatic and branched-chain hydrocarbons.

Reforming – A process that changes a mixture of predominantly straight-chain hydrocarbons into a mixture containing more aromatic and branched-chain hydrocarbons.

Reforming is a key step in making the gasoline we use as automobile fuel. Gasoline must contain a suitable proportion of branched and aromatic hydrocarbons to prevent the problem known as **knocking**. This is the rapid explosion of the gas–air mixture as it is compressed in the automobile cylinder before the spark plug creates the spark that is intended to cause the mixture to explode. In some cases, knocking damages a car's engine, but most often it results only in poor engine performance and reduced gas mileage. Addition of reformed distillates to gasoline reduces knocking.

Knocking – The rapid explosion of a gas–air mixture as it is compressed in an automobile cylinder before the spark plug creates the spark that is intended to cause the mixture to explode.

21.6.3 Gasoline

The gasoline we use to power automobiles is a complex mixture of well over a hundred different hydrocarbons, which can include molecules such as pentane, benzene, ethylbenzene, toluene (methylbenzene), and xylene (dimethylbenzene). The precise composition of the mixture varies with the brand of gasoline, the location, and the time of year, but the hydrocarbon molecules in automobile gas generally contain between 5 and 17 carbon atoms, many being in the C_5-to-C_9 range. One factor influencing the optimum mixture of hydrocarbons is the need to produce a fuel with a vapor pressure that allows it to evaporate to an appropriate degree under the conditions in which it will be used.

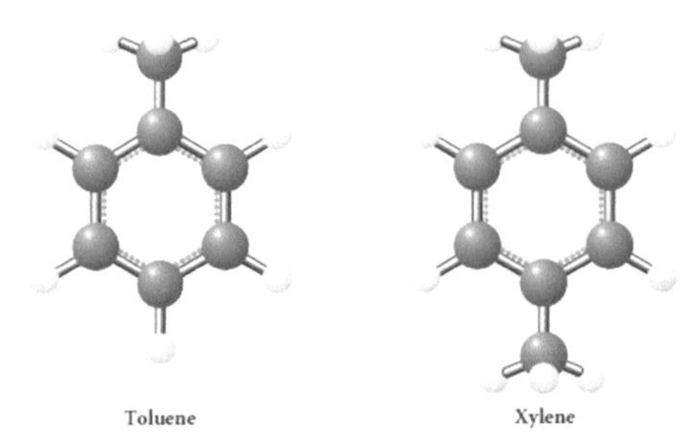

The extent to which a particular type of gasoline will burn smoothly in an engine (without knocking) is indicated by using the **octane rating**. This is based on the early twentieth-century discovery that pure isooctane produced minimal knocking in an engine, whereas pure heptane produced a very high level of knocking. On the octane scale, 2,2,4-trimethylpentane (also known as isooctane) is given the octane rating of 100, and heptane is given the octane rating of 0. By comparing mixtures of isooctane and heptane to different formulations of gasoline, an octane rating for the gasoline can be assigned. For example, a high-quality gasoline, which produces a level of knocking equivalent to a mixture of 90% isooctane and 10% heptane, is assigned an octane rating of 90. Low-quality gasoline can be improved by adding other compounds with octane ratings that are higher than 100, such as ethanol (octane rating 112) or toluene (octane rating 118). Brands of automobile gasoline are typically available with octane ratings in the range of 83–98.

Octane rating – A number assigned to motor fuel (gas) on the basis of its "anti-knock" properties.

$$CH_3-C(CH_3)_2-CH_2-CH(CH_3)-CH_3$$

2,2,4-Trimethylpentane (isooctane)

$$CH_3-CH_2-CH_2-CH_2-CH_2-CH_2-CH_3$$

Heptane

21.7 Typical Reactions of the Alkanes

Globally, the most significant reaction of the alkanes is their combustion in air. As we have already seen, the complete combustion of alkanes generates carbon dioxide and water vapor and releases considerable amounts of energy. The energy can be used to provide heat, both for warmth and for cooking, to generate electricity, or to power vehicles.

The chemical industry also makes substantial use of alkanes in **substitution reactions**, in which one or more hydrogen atoms are replaced by other types of atoms, most commonly halogens such as bromine and chlorine. These substitution reactions of the alkanes are promoted by ultraviolet light, which causes the halogen molecules to split into two atoms:

$$Cl_2 \xrightarrow{h\nu} Cl\cdot + Cl\cdot$$

Substitution reactions – Reactions in which one or more atoms (often hydrogen atoms) are replaced by other types of atoms (often halogens or a hydroxyl group).

The halogen atoms generated in this process are known as **free radicals**, each having an unpaired electron. The presence of the unpaired electron makes them highly reactive. In the presence of an alkane, the free radicals can readily attack the C–H bonds. The result is a halogenated alkane and a molecule of the hydrogen halide.

Free radicals – Chemical species that carry an unpaired electron and so are highly reactive.

For example, the simplest hydrocarbon, methane, can be converted into four different organic products and HCl by successive substitution reactions with chlorine:

$$CH_4 + Cl_2 \xrightarrow{hv} CH_3Cl \text{ (Chloromethane)} + HCl$$

$$CH_3Cl + Cl_2 \xrightarrow{hv} CH_2Cl_2 \text{ (Dichloromethane)} + HCl$$

$$CH_2Cl_2 + Cl_2 \xrightarrow{hv} CHCl_3 \text{ (Trichloromethane (chloroform))} + HCl$$

$$CHCl_3 + Cl_2 \xrightarrow{hv} CCl_4 \text{ (Tetrachloromethane (carbon tetrachloride))} + HCl$$

These four products can be put to a variety of uses. Chloromethane is used in the manufacture of silicone-based polymers and as a solvent in the production of certain types of rubber. Dichloromethane is widely used as a solvent in the chemical industry and within products such as paint strippers.

Trichloromethane, also known as chloroform, was one of the first anesthetics, pioneered as such in 1847 by the Scottish physician Sir James Simpson. Because of the sweetness of this molecule, it was also used as the solvent in early cough medicines. However, this compound was found to be both toxic and carcinogenic, and ethanol has taken its place in medicines. Chloroform is still used as an industrial solvent. Tetrachloromethane, also known as carbon tetrachloride, once was the solvent most commonly used for the dry cleaning of clothes and other fabrics. It was also used as a fire extinguisher agent because it is not flammable. Unfortunately, it too is toxic and carcinogenic. It has now largely been abandoned for dry cleaning and fire extinguishing in favor of less toxic compounds such as liquid carbon dioxide. However, it is still used as a solvent in various industrial processes.

Another important reaction of the alkanes is their **dehydrogenation** to produce the corresponding alkenes. In a dehydrogenation reaction, a molecule of hydrogen is removed from the starting material. The result is an unsaturated hydrocarbon, an alkene. For example, dehydrogenation of ethane produces ethene. Industrially, the dehydrogenation of ethylbenzene results in styrene, a monomer used to make Styrofoam. The dehydrogenation of propane produces propene, and so on:

Dehydrogenation – The removal of hydrogen.

$$\underset{\text{Ethane}}{CH_3{-}CH_3} \xrightarrow[500^\circ C]{Pt} \underset{\text{Ethene}}{H_2C{=}CH_2} + H_2$$

$$\underset{\text{Propane}}{CH_3{-}CH_2{-}CH_3} \xrightarrow[500^\circ C]{Pt} \underset{\text{Propene}}{CH_3{-}CH{=}CH_2} + H_2$$

21.8 The Functional Group Concept

The processes of adding chlorine atoms or C=C double bonds to the basic structure of alkanes are examples of adding **functional groups** to a hydrocarbon framework. A functional group is an atom or group of atoms with a characteristic set of chemical properties. Chemists often make sense of the seemingly infinite variety of chemical reactions in organic chemistry by treating each molecule as a hydrocarbon framework with a particular set of functional groups incorporated within that framework.

Functional groups – Groups of atoms, or arrangements of bonds, that bestow a specific set of chemical and physical properties on any compound that contains them.

A chlorine atom bonded to a hydrocarbon gives rise to a molecule that contains the **alkyl halide** functional group, a term that covers any of the *halogen* atoms bonded to a hydrocarbon framework. These and many other examples of functional groups are listed in ◘ Table 21.6. For example, the C=C double bond (the alkene functional group) is a functional group whose properties depend on the type of bonding between the atoms involved. The C≡C triple bond (the alkyne functional group) is another example.

Alkyl halide – A functional group in which a halogen atom is bonded to an alkyl group.

Much of the chemistry done with hydrocarbons from oil involves the addition of specific functional groups. Chemists can modify the structure of the basic hydrocarbon chain, adjusting its length and degree of branching, and then add whatever selection of functional groups will create the molecules they are seeking.

Example 21.5—Functional Groups and You!

Circle and name each of the functional groups in cyclexanone, a pharmaceutical agent used to prevent coughing (an antitussive agent).

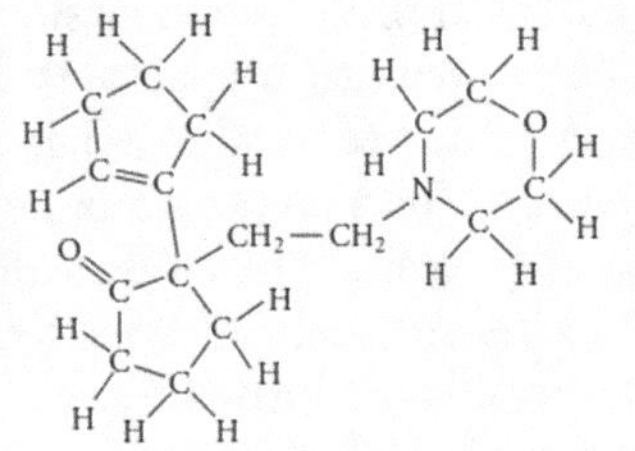

Asking the Right Questions

We will use ◘ Table 21.6 to identify the combinations of atoms that make up a functional group.

Solution

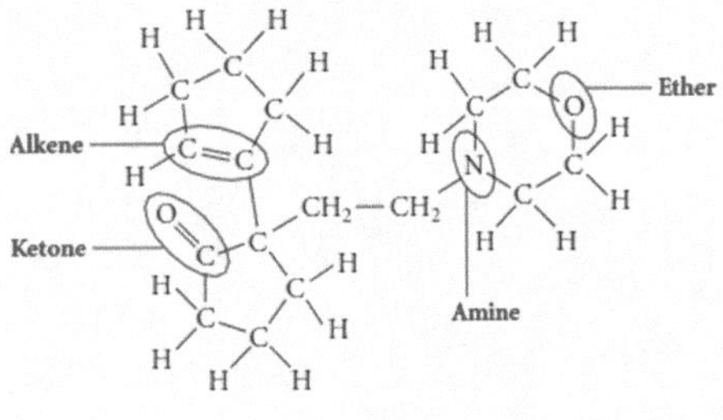

Alkene

Ketone

Ether

Amine

Are Our Answers Reasonable?

Based on the information in ◘ Table 21.6, our answers seem to be a reasonable assessment of the functional groups in cyclexanone.

What If…?

…we were asked to identify the geometric isomerism in cyclexanone? What is the geometry of the alkene?
(*Ans: Trans*)

Practice 21.5

Circle and name each of the functional groups in tyrosine, one of the building blocks used to make proteins.

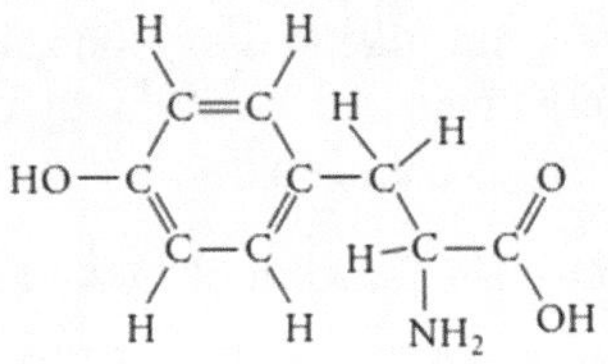

21.9 Ethene, the C=C Bond, and Polymers

One of the most versatile of the hydrocarbons produced by catalytic cracking is ethene (ethylene). It serves as the central feedstock molecule that is converted into a vast range of organic chemicals used in modern life. The chemistry of ethene is dominated by the reactivity of its C=C double bond. Especially important is the **addition reaction**, in which two parts of another chemical species become added to the atoms at either end of the double bond. The C=C of ethene is converted into a single bond in the process. The pi electrons of the double bond end up participating in the new bonds holding the groups that have become added "across" the double bond. We've already examined one of the addition reactions with ethene, hydrogenation.

Addition reaction – A type of reaction in which atoms are added to a double or triple bond.

Table 21.6 Selected functional groups

Structure	Group	Example	3D structure	Name	Selected Uses
$>C=C<$	Alkene	$CH_2{=}CH_2$		Ethene	Refrigerant, production of polyethylene
$-C\equiv C-$	Alkyne	$HC\equiv CH$		Acetylene	Welding, cutting, brazing
$-C-O-H$	Alcohol	CH_3CH_2OH		Ethanol	Liquors, industrial solvent
$C-O-C$	Ether	$CH_3CH_2OCH_2CH_3$		Diethyl ether	Industrial solvent
$C-C(=O)-H$	Aldehyde	$CH_2{=}O$		Formaldehyde	Bactericide, fungicide, chemicals production
$C-C(=O)-C$	Ketone	$CH_3-C(=O)-CH_2-CH_3$		Methyl ethyl ketone	Solvent in rubber industry
$C-C(=O)-OH$	Carboxylic acid	$H-C(=O)-OH$		Formic acid	Manufacture of textiles, pesticides, electroplating

(continued)

Table 21.6 (continued)

Structure	Group	Example	3D structure	Name	Selected Uses
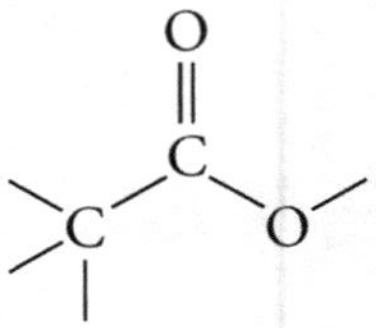 $-C-C(=O)-O-$	Ester	$CH_3-C(=O)-O-CH_3$	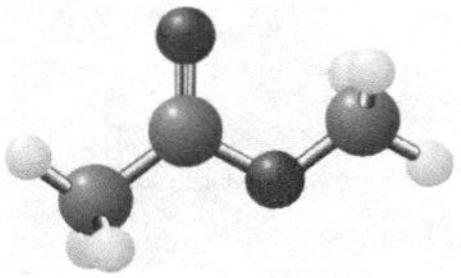	Methyl acetate	Paint remover, pharmaceutical manufacture
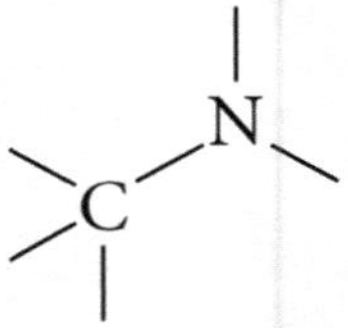 $-C-N<$	Amine	$CH_3(CH_2)_3NH_2$	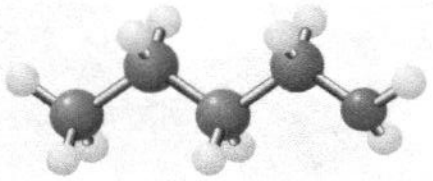	Butylamine	Manufacture of rubber, insecticides
$-C\equiv N$	Nitrile	$CH_3-C\equiv N$	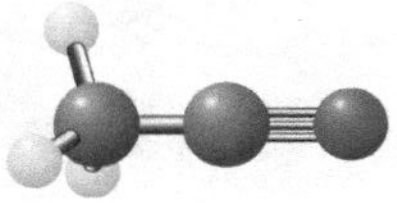	Acetonitrile	Industrial solvent
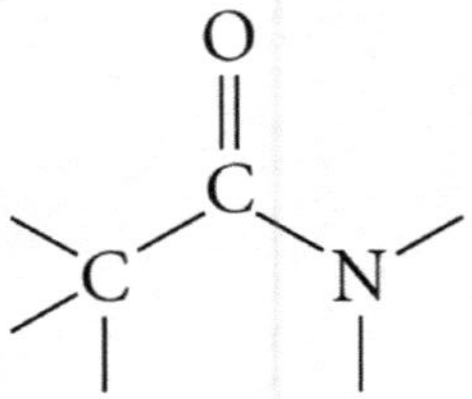 $-C-C(=O)-N<$	Amide	$H-C(=O)-NH_2$	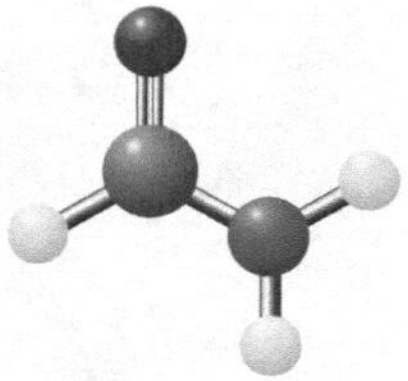	Formamide	Manufacture of paper, glue
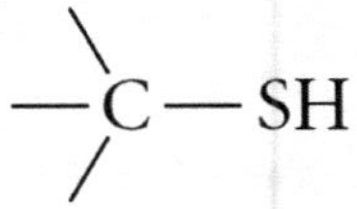 $-C-SH$	Thiol	$CH_3CH_2CH_2SH$	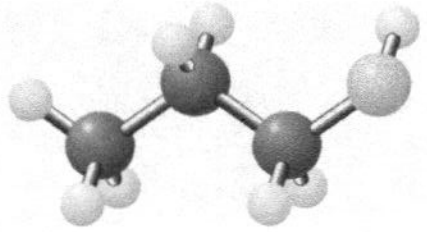	Propanethiol	Herbicide, flavoring agent

As another example, ethene can be converted into ethanol (CH_3CH_2OH) by reaction with water in the presence of phosphoric acid (H_3PO_4) catalyst at 330 °C. Industrially, ethanol can be made in large quantities using this process. However, because of the value of ethene as a feedstock for other organic molecules, much of the ethanol that is made in the United States today comes from the fermentation of sugars found in corn. Ethanol is an example of an **alcohol**, a compound carrying the **hydroxyl group** (–OH group).

Alcohol – A chemical containing the hydroxyl (–OH) group.

Hydroxyl group – The –OH functional group.

$$H_2C{=}CH_2 \xrightarrow[H_3PO_4,\ 330^\circ C]{H_2O} H{-}CH_2{-}CH_2{-}OH$$

Another addition reaction converts ethene into dichloroethane (ethylene dichloride), an intermediate in the production of the seemingly ubiquitous material polyvinyl chloride (PVC), used for materials in healthcare, building, and electronics, among other applications.

$$H_2C{=}CH_2 \xrightarrow{Cl_2} H{-}CHCl{-}CHCl{-}H$$

Ethene is the starting material for a great many of the plastics that are such a common feature of modern life. The simplest of these plastics is **polyethene**, which is made when huge numbers of ethene molecules participate in addition reactions with themselves to generate long chain-like polyethene molecules. In the reaction, one molecule of ethene adds to an adjacent molecule of ethene, which adds to another molecule of ethene, and so on. The result is a very long chain of carbon atoms.

Polyethenes – Organic polymers made when huge numbers of ethene molecules participate in addition reactions with themselves to generate long chain-like molecules. Also known as polyethylenes.

$$H_2C{=}CH_2 \xrightarrow{\text{Polymerize}} \sim CH_2{-}[CH_2{-}CH_2]_n{-}CH_2 \sim$$

Polyethene (polyethylene)

Polyethene, which is more commonly known as **polyethylene**, is an example of a **polymer** (literally, "many units"), a compound composed of large molecules made by the repeated bonding together of smaller **monomer** (literally, "one unit") molecules, ethene in this case. Because the reaction that bonds the ethene monomers together is an addition reaction, polyethylene is known as an **addition polymer**, formed by the process of **addition polymerization**.

Addition polymer – A polymer formed by the process of addition polymerization.

Addition polymerization – A type of polymerization involving addition reactions among the monomers involved.

Monomer – One of the small chemical units that combine with other monomers to form polymers.

Polyethylenes – See polyethenes.

Polymer – A chemical composed of large molecules made by the repeated bonding together of smaller monomer molecules.

Different types of polyethylene can be made by varying the temperatures and pressures of the reaction, as well as changing the type of catalysts and initiating substances used. These different forms are classified into two main types: **high-density polyethylene (HDPE)**, in which the molecules are long and linear with no branch points, and **low-density polyethylene (LDPE)**, with shorter, branched-chain molecules, as shown in Fig. 21.14. The length of the polyethylene chain imparts different properties to the resulting molecule. For instance, because we know that shorter alkanes have lower melting points, you might expect a shorter polyethylene chain to have a lower melting point than a longer polymer.

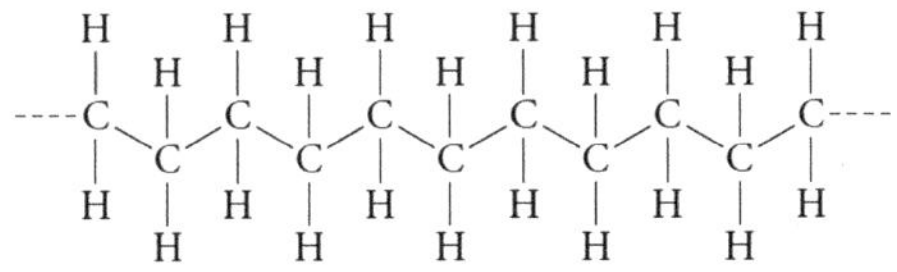

■ **Fig. 21.14** HDPE and LDPE. The molecular structures of HDPE and LDPE are identical, but their processing gives them unique properties for use in different materials

High-density polyethylene (HDPE) – A form of polyethylene in which the molecules are linear, with no branch points.

Low-density polyethylene (LDPE) – A form of polyethylene with branched-chain molecules that are shorter than the unbranched chains found in HDPE.

The high-density part of HDPE is a result of the close packing of the linear molecules. This makes HDPE a strong plastic, used to make such things as blow-molded bottles and toys. LDPE has a more flexible, less crystalline structure; it is used to make such things as plastic trash and grocery bags, packaging films, insulation sleeves around electrical cables, and squeeze bottles. The properties of polyethylene can be modified by incorporating some other atoms, such as chlorine or sulfur, into its structure as it is formed. This can make the resulting modified polyethylene more resistant to oxidation, for example, or more flexible. Many of the polyethylene-based materials around us are actually composed of such modified polyethylenes.

We can prepare polymers that have properties different from those of polyethylene, and yet are closely related to it in chemical structure, by using monomers in which one or more of the hydrogen atoms of ethene are replaced with other atoms or larger chemical groups. ■ Table 21.7 lists some of these.

■ **Table 21.7** Polymerization of some common monomers

Monomer 3D structure	Polymer	Uses
Vinyl chloride	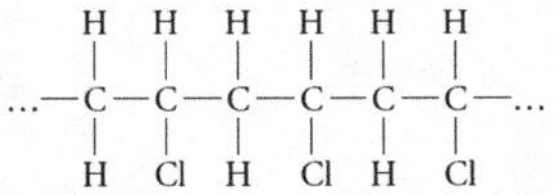 Polyvinylchloride (PVC)	Indoor plumbing, toys, plastic wrap, vinyl siding
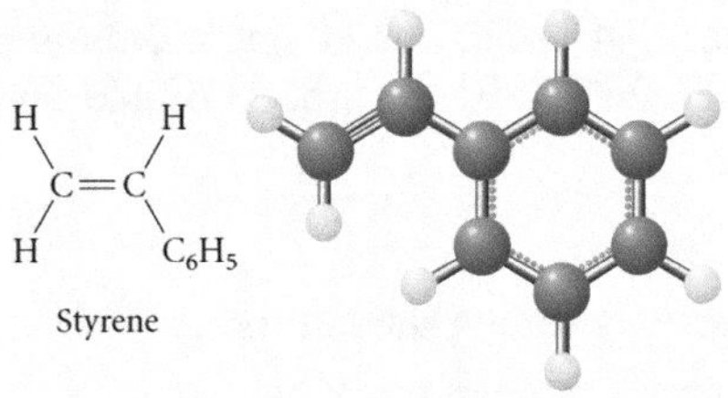 Styrene	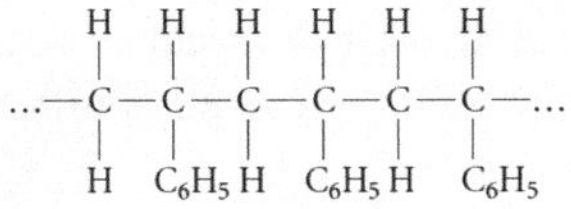 Polystyrene	Outdoor furniture, insulation, packing "peanuts"
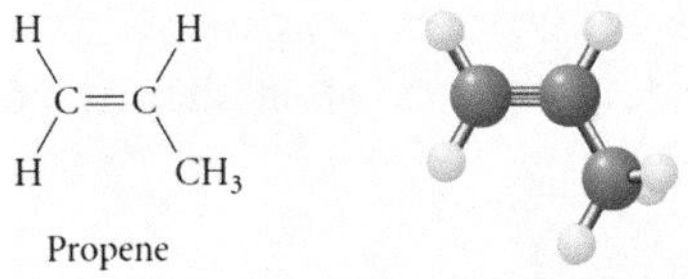 Propene	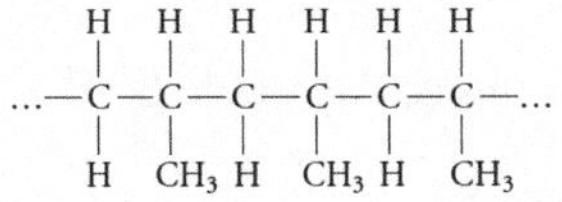 Polypropylene	Contained in indoor–outdoor carpeting
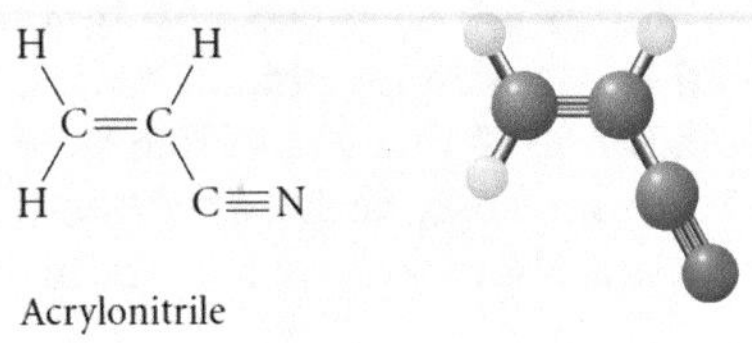 Acrylonitrile	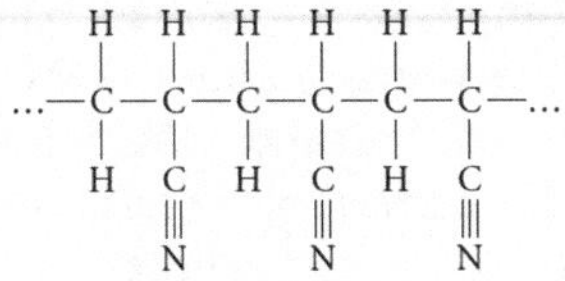 Polyacrylonitrile	Orlon, Acrilon clothing, yarns, wigs

21.10 Alcohols

As the alcohol found in alcoholic beverages, ethanol is the best known of the alcohols. Ethanol can be made by hydrating ethene. However, because of the value of ethene as a feedstock for other organic molecules, and because of the toxicity of some of the by-products of this process, much of the ethanol that is typically used for human consumption is made by the biological activity of yeast cells. In this reaction, the yeast consumes glucose in a process known as **fermentation**. The waste products of the process are ethanol and carbon dioxide. These waste products are toxic to the yeast and are excreted from the cell. Fortunately, the industrial chemist can easily harvest the ethanol by distilling it from the solution. The chemist also can capture the CO_2 gas given off in the reaction in order to make other products.

Fermentation – The biological process that converts glucose into ethanol and carbon dioxide.

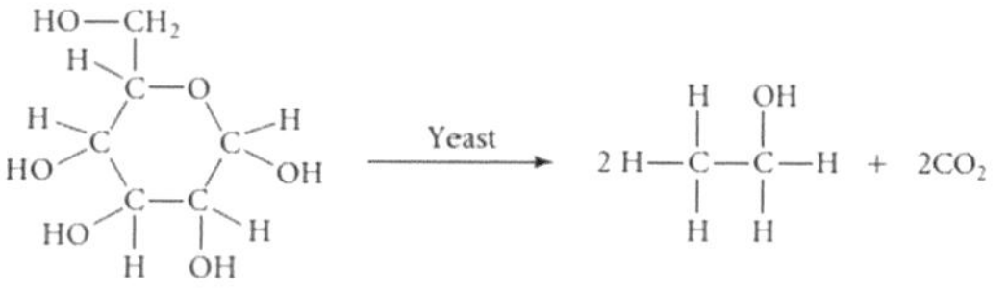

The ethanol that is distilled from the fermentation mixture unavoidably contains a small amount (roughly 5%, by volume) of water. For beverages, this water is not a problem. However, if the ethanol is to be used as an additive to fuel used in automobiles, the water must be removed. The process used to make 100% ethanol by fermentation often leaves behind traces of toxic compounds, which render the pure ethanol undrinkable.

Other alcohols of note include methanol, 2-propanol, the "dihydroxy alcohol" 1,2-ethanediol, and the "trihydroxy alcohol" 1,2,3-propanetriol, whose structures and uses can be found in ▫ Table 21.8. Note the way in which numbers are used to indicate which carbon atom carries the –OH (hydroxy) group or groups in alcohol structures. We can also use the structures themselves to explain the way in which alcohols are classified into categories known as primary, secondary, or tertiary alcohols, a type of classification that can be applied to other functional groups as well.

Primary alcohols have only one carbon atom bonded to the carbon atom that carries the hydroxyl group. This means that ethanol is a primary alcohol. **Secondary alcohols** have two carbon atoms bonded to the carbon atom that carries the hydroxyl group, so 2-propanol is a secondary alcohol. **Tertiary alcohols** have three carbon atoms bonded to the carbon atom that carries the hydroxyl group, so 2-methyl-2-propanol is a tertiary alcohol. This classification system is well correlated with the reactivity of the alcohols. For instance, primary alcohols undergo many reactions more rapidly than do secondary or tertiary alcohols. For example, primary alcohols react fastest, and tertiary alcohols slowest, in the formation of esters.

Primary alcohols – Alcohols that have no more than one carbon atom bonded to the carbon atom that carries the hydroxyl group.

Secondary alcohols – Alcohols that have two carbon atoms bonded to the carbon atom that carries the hydroxyl group.

Tertiary alcohols – Alcohols that have three carbon atoms bonded to the carbon atom that carries the hydroxyl group.

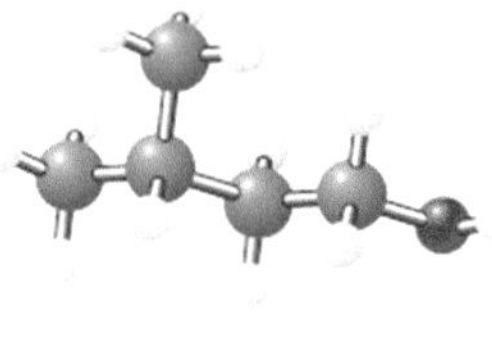
Primary alcohol

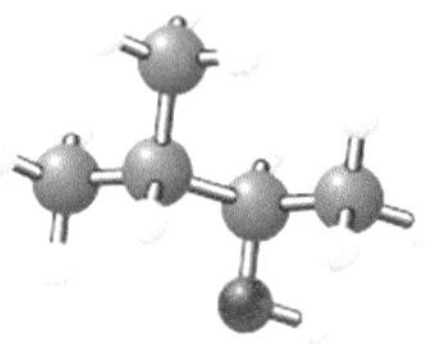
Secondary alcohol

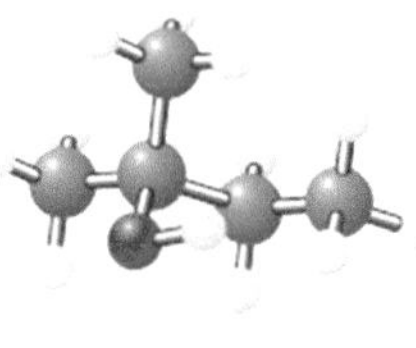
Tertiary alcohol

Table 21.8 Structures and uses of selected alcohols

Common name	Lewis structure	3D structure	Selected uses
Methanol	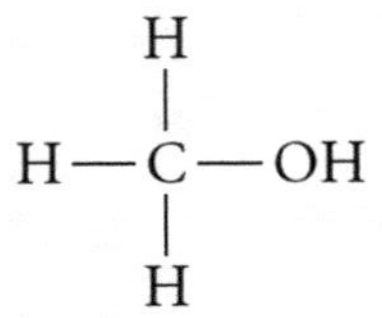	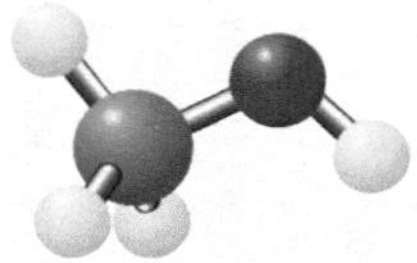	Fuel, fuel additive
2-propanol (isopropyl alcohol)	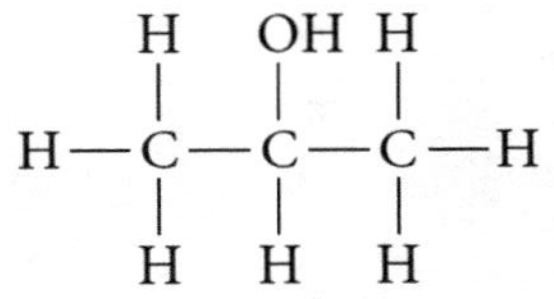	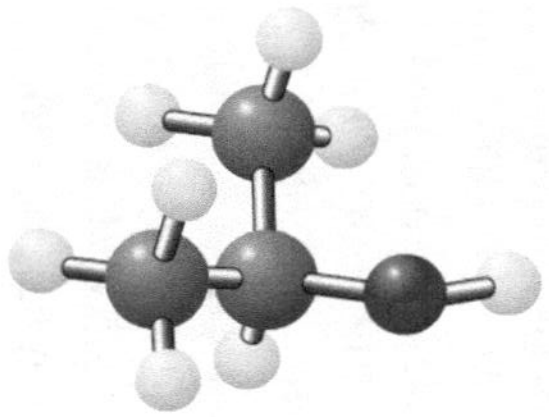	Rubbing alcohol
2-methyl-2-propanol (tertiary butyl alcohol)	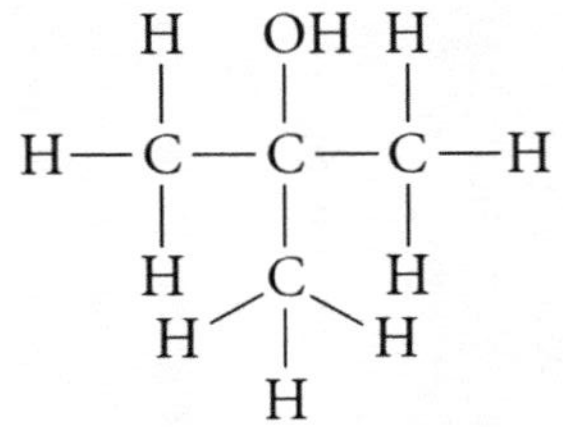	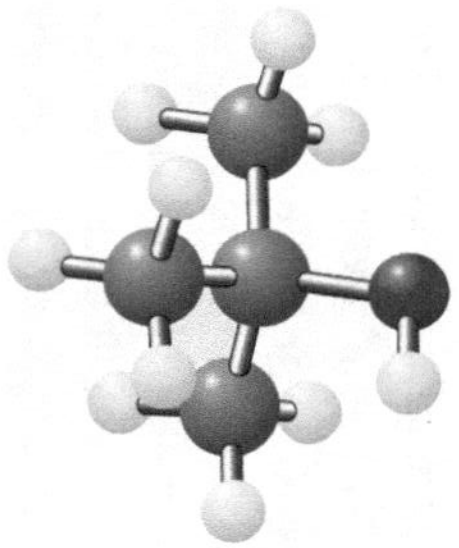	Solvent in chemical reactions
1,2-ethanediol (ethylene glycol)	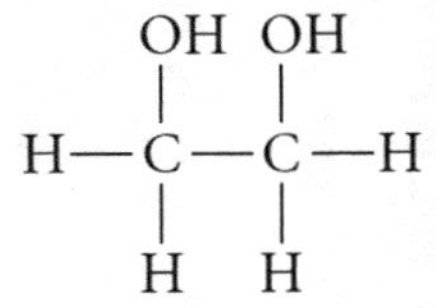	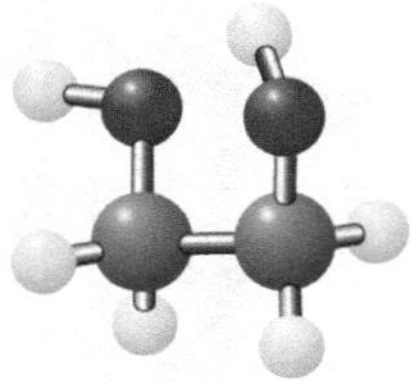	Antifreeze, de-icing agent

(continued)

Table 21.8 (continued)

Common name	Lewis structure	3D structure	Selected uses
1,2,3-propanetriol (glycerol)	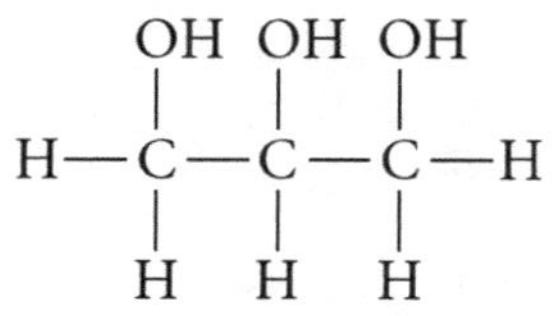	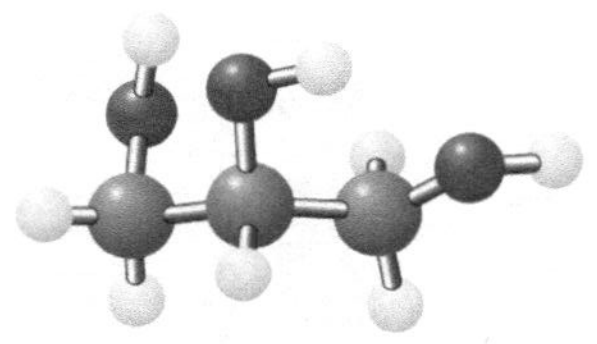	Found in cosmetics and foods
phenol	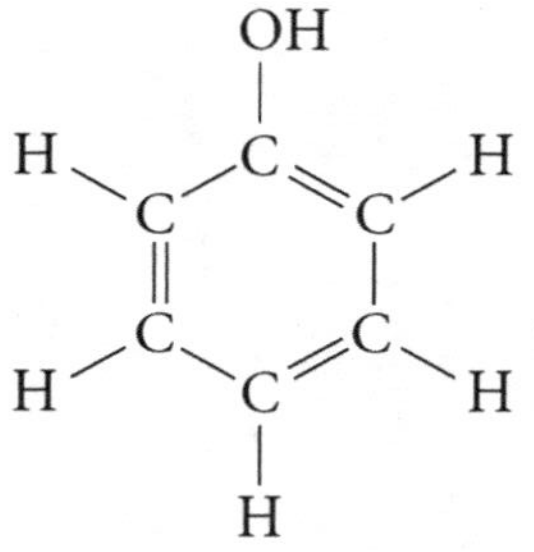	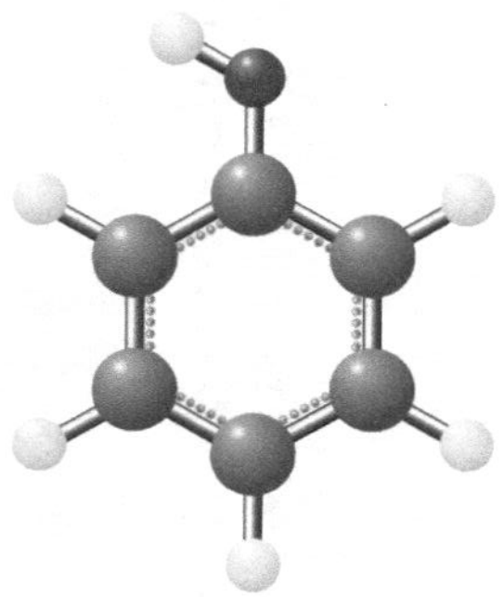	Making resins for plywood manufacture

21.11 From Alcohols to Aldehydes, Ketones, and Carboxylic Acids

In addition to determining the rate of reaction, the distinction between primary, secondary, and tertiary alcohols is important in determining what types of molecules can be made from the alcohol. For example, primary alcohols can undergo oxidation to form **aldehydes**, which carry a **carbonyl functional group** (C=O) with at least one H atom bonded to the carbonyl carbon atom. The oxidation is usually performed with a highly oxidized metal such as chromium(VI). In the lab, however, special reactions must be performed to prevent the overoxidation of the aldehyde. Methanol, for example, can be oxidized to the aldehyde methanal ($H_2C{=}O$), also known as formaldehyde. The solution formed when formaldehyde is dissolved in water is most commonly known as formalin, used to preserve biological specimens.

Aldehydes – Compounds that carry a carbonyl functional group (C=O) with at least one H atom bonded to the carbonyl carbon atom.

Carbonyl functional group – A functional group containing the group C=O attached to either hydrogen atoms (as in the aldehydes) or carbon atoms (as in the ketones).

Methanol (wood alcohol) —Oxidation→ Methanal (formaldehyde)

Note that this reaction is similar to the dehydrogenation reactions of the alkanes. As another example, ethanol can be oxidized to ethanal. Ethanal, also known as acetaldehyde, is used primarily in the production of polymers (► Sect. 21.12) and medicines.

Ethanol (ethyl alcohol; grain alcohol) —Oxidation→ Ethanal (acetaldehyde)

An aldehyde is named by taking the name of the alcohol from which it is derived and replacing the ending -ol with -al. No number is needed as part of the name to indicate where the aldehyde group occurs in the compound, because it always occurs at the end of a carbon chain.

The hydroxyl group of secondary alcohols can be oxidized to a carbonyl group. In this case, compounds called **ketones** are formed. Ketones differ from aldehydes in that they have two carbon atoms bonded to the carbonyl carbon atom. For example, 2-propanol can be oxidized to propanone (acetone). Propanone is an excellent solvent because of its ability to dissolve both polar and nonpolar molecules. It has a low boiling point and a low toxicity, finding use as a solvent for paints and as a fingernail polish remover. For these reasons, around 10 billion kilograms of propanone are used each year in industry.

Ketones – Compounds that have two carbon atoms bonded to a carbonyl carbon atom.

2-Propanol (isopropanol) —Oxidation→ Propanone (acetone)

Tertiary alcohols do not react with oxidizers to make carbonyl compounds. For instance, there is no reaction when we attempt to oxidize 2-methyl-2-propanol under the conditions typical for primary and secondary alcohols. This lack of reactivity occurs because the carbon bearing the –OH group does not have a hydrogen that can be removed to make the carbonyl (C=O).

2-Methyl-2-propanol —Oxidation→ No reaction

Aldehydes have a characteristic biting and often unpleasant aroma. Ketones, on the other hand, have an aroma that is not so unpleasant, but more medicinal. Chemically, though, the important difference between aldehydes and ketones is that the aldehydes can be oxidized further to form **carboxylic acids**. The carboxylic acids carry the **carboxyl functional group**, the combination of a carbonyl and a hydroxyl group. Ethanal, for example, can be easily oxidized to ethanoic acid. All carboxylic acids are named by naming the molecule as if it didn't contain a carboxylic acid functional group. Then, the ending -e is dropped and replaced with -oic acid.

Carboxylic acids organic – acids that carry the carboxyl functional group (–COOH).

Carboxyl functional group – The –COOH functional group.

Ethanal —Oxidation→ Ethanoic acid

Ethanoic acid (also known as acetic acid) is the organic acid that we use as vinegar when it is in the form of a weak solution (typically about 4 to 5% by mass) in water. Some bacteria can perform the conversion of ethanol to ethanoic acid, which can cause wines to spoil when they "turn to vinegar" due to bacterial contamination. However, this process can also be exploited purposefully to produce the wine vinegars that are widely used in cooking. Industrially, ethanoic acid is used in the manufacture of plastics, pharmaceuticals, dyes, and insecticides.

The carboxyl functional group is weakly acidic due to the dissociation of a hydrogen ion from the oxygen. The result is the formation of a **carboxylate anion** and a hydrogen ion (i.e., a proton). Although it would appear that any hydrogen on an oxygen could dissociate in a similar manner, the carboxylic acids have a special feature that alcohols do not have (hence the latter are not acidic). Let's explore this property further. Why does the carboxyl functional group dissociate?

Carboxylate anion – An anion resulting from the removal of a hydrogen ion from the oxygen in a carboxylic acid functional group.

Ethanoic acid ⇌ Ethanoate + $H^{\oplus}$

When we draw Lewis dot structures of compounds, we must always consider the possibility that the electrons could be placed in different locations (see ▶ Chap. 9). This would generate a resonance structure of the compound. The carboxylate anion possesses this ability. In fact, we can draw two good resonance structures for this ion. The resonance spreads the large negative charge across the functional group and helps to stabilize the carboxylate by lowering the energy of the anion.

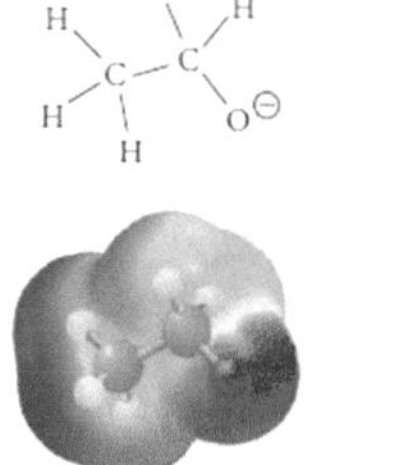

Ethoxide. The red color in the electron density map is localized on the oxygen. The electron density isn't dispersed and ethoxide is not as stable as ethanoate.

Ethanoate. The red color (high electron density) is dispersed across more atoms in this structure. Thus, this is a more stable anion than ethoxide.

Why doesn't the alcohol functional group also dissociate? If a hydrogen ion were to leave the molecule, a negative charge would reside on the oxygen as shown in both the Lewis dot structure and the electron density map. Unfortunately, there aren't any other resonance structures that we can draw for this **alkoxy anion**. Therefore, the electrons cannot be spread to other atoms by resonance, and the resulting alkoxyl anion is not energetically stabilized.

Alkoxy anion – An anion that contains an alkyl group attached to an oxygen anion, such as CH_3O^- or $CH_3CH_2O^-$.

21.12 From Alcohols and Carboxylic Acids to Esters

An extremely important reaction, widespread throughout both natural and industrial chemistry, takes place between the alcohol and carboxylic acid functional groups. These two groups can react, often with an acid catalyst such as sulfuric acid, to eliminate a molecule of water in a reaction known as a **condensation reaction**. In the reaction, the carbonyl carbon from the carboxyl group becomes bonded to the oxygen atom from the alcohol group in what is known as an ester linkage or ester bond. Compounds containing this arrangement of atoms (carbonyl–oxygen–carbon) are known as **esters**. The simplest possible ester is methyl methanoate (also known as methyl formate), formed when methanol reacts with methanoic acid. Note that the first part of the name of an ester is typically an alkyl group name, and the second part of the name is derived from the carboxylic acid component by changing the ending of the carboxylic acid name. The *-ic acid* is changed to *-ate*.

Condensation reaction – A reaction in which two molecules become joined in a process accompanied by the elimination of a small molecule such as water or HCl.

Esters – Compounds formed when an alcohol and a carboxylic acid become bonded as a result of a condensation reaction.

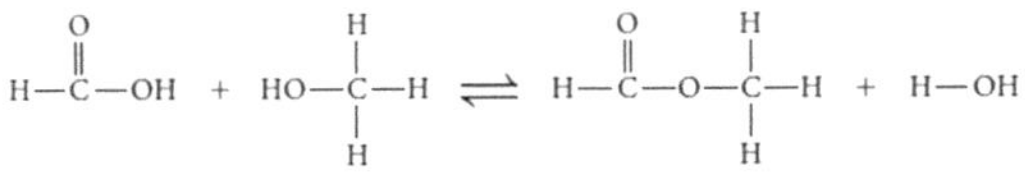

Methanoic acid + Methanol ⇌ Methyl methanoate (methyl formate) + H—OH

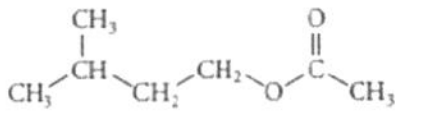

Isopentyl ethanoate

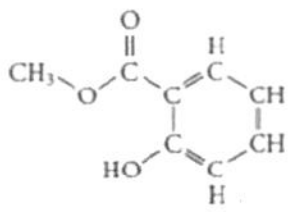

Methyl salicylate

Compounds that contain the ester functional group often have a very fruity aroma and in nature are found in plant oils. For example, isopentyl ethanoate has the odor of ripe bananas, and methyl salicylate smells like wintergreen. For this reason, esters are often used in industry to lend flavors and odors to products that we buy. Esters are also used in industry as solvents, due to their ability to dissolve both polar and nonpolar solutes. For example, ethyl ethanoate (commonly known as ethyl acetate) is found in cleaners and glues.

Example 21.6—Nomenclature of Esters

Provide names for each of these esters.

a.

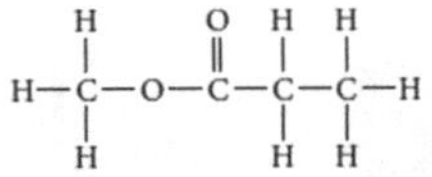

b.

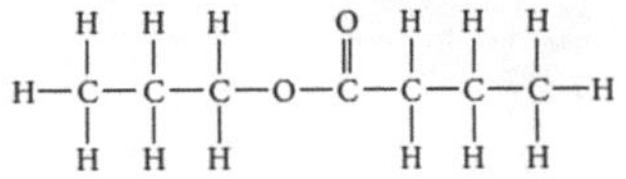

c.

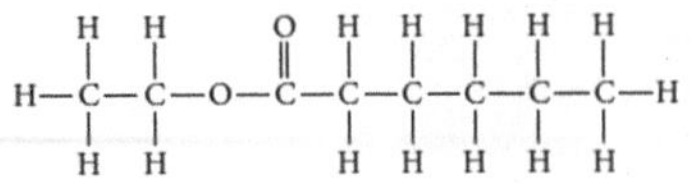

d.

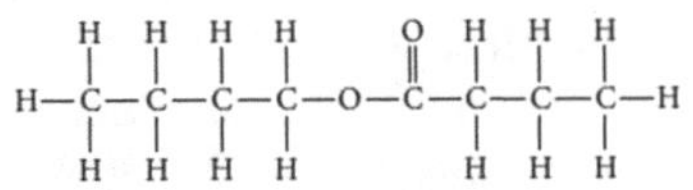

Asking the Right Questions

What are the rules for naming an ester? Esters come in two parts. What part of each ester is the alcohol portion (the portion of the name that comes first)?

Solution

a. methyl propanoate
b. propyl butanoate
c. ethyl hexanoate
d. butyl butanoate.

Are the Answers Reasonable?

We have identified the two portions of each name, so our answers are reasonable.

What If…?

…we were asked to name the ester formed from phenol and ethanoic acid?

(*Ans: phenyl ethanoate*)

Practice 21.6

Shown below is the structure of benzoic acid. Can you draw the structure of ethyl benzoate?… of benzyl ethanoate?

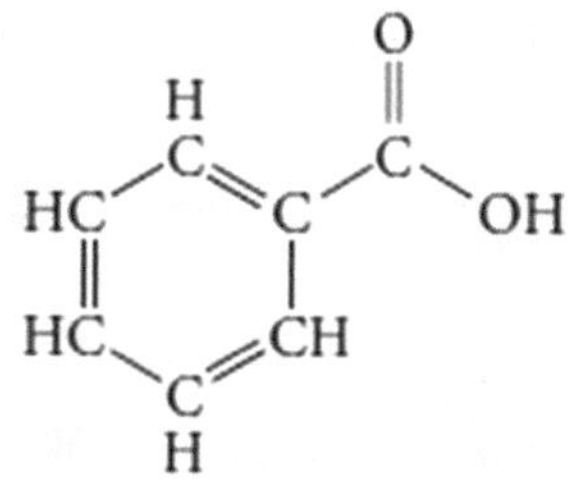

21.13 Condensation Polymers

Industrial preparation of polyethylene, polyvinylchloride, and polystyrene occurs by addition polymerization. But that is not the only way to make a polymer. A major class of polymers that are very useful in our lives are constructed by a condensation reaction. Condensation reactions entail the removal of a small molecule as two larger molecules are connected. If a polymer is the product of the reaction, we call the process **condensation polymerization**.

Condensation polymerization – Formation of a polymer by condensation reactions.

Esters can be made by condensation reactions. The process, **esterification**, can be used to link a variety of monomers into long polymer chains. The products of the reaction are **polyesters**, which may be familiar to you as parts of clothes, bedding, fabrics, upholstery, ropes, belts, and so on. They are called polyesters because of the many ester linkages that hold the monomer units within the structure of the polymer. One of the most common polyesters is polyethylene terephthalate (PETE), which as used for such things as soft drink and mouthwash bottles.

Esterification – The process by which an ester is formed when the –OH group of an alcohol participates in a condensation reaction with the –COOH group of a carboxylic acid.

Polyesters – Polymers in which the monomers are held together by repeated ester linkages.

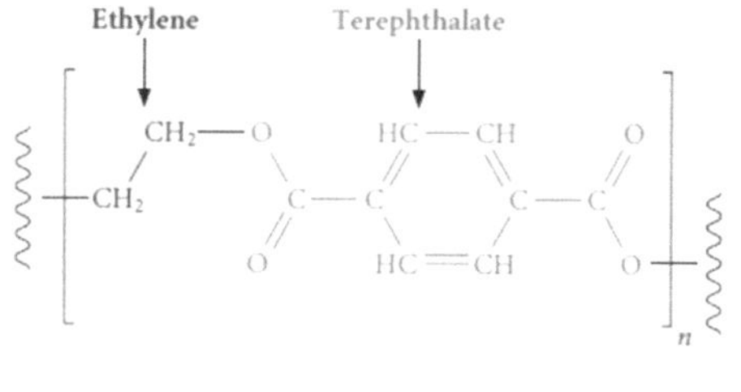

Ethylene

Terephthalate

Another functional group that is a part in many important condensation polymerization reactions is the **amino functional group**, $-NH_2$. Many **amines**, unfortunately, have a particularly foul odor. For instance, the common names of some of the amines indicate the types of environments from which they were first isolated; putrescine (1,4-butanediamine) and cadaverine (1,5-pentanediamine) are examples. When an amine group participates in a condensation reaction with a carboxylic acid group, an **amide bond** is formed.

Amine – A compound containing the $-NH_2$ functional group.

Amino functional group – A functional group containing an $-NH_2$ attached to an sp^3 hybridized carbon atom.

Amide bond – A bond formed by a condensation reaction between an amino group ($-NH_2$) and a carboxylic acid group (–COOH).

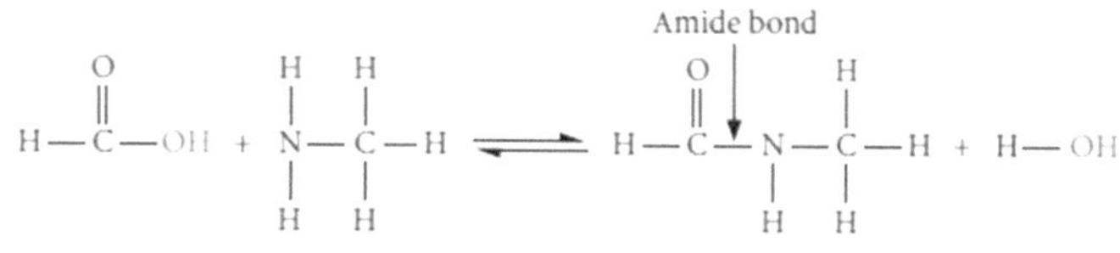

Amide bond

Methanoic acid (formic acid)

Methanamine

N-Methylmethanamide (N-methylformamide)

The amide bond holds together the monomer units of polymers known as the **polyamides**. The fabrics we know as nylons, for example, are polyamides. Nylons are named with numbers to help indicate the structure of the polymer. For example, Nylon [6,6] is made of monomers that each contain six carbon atoms.

Polyamides – Polymers whose monomers are bonded together by amide bonds.

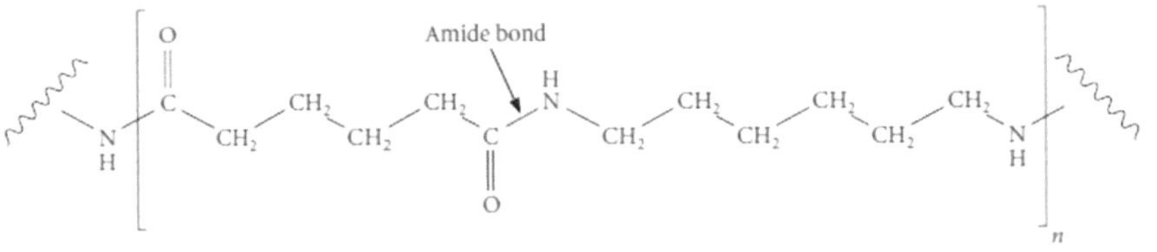

Amide bond

$$\left[-NH-C(=O)-CH_2-CH_2-CH_2-CH_2-C(=O)-NH-CH_2-CH_2-CH_2-CH_2-CH_2-CH_2- \right]_n$$

Nylon 6-6

21.14 Polyethers

Another important category of polymers contains the **ether** functional group at regular intervals along the polymer chain. The ether group is –C–O–C–, an oxygen atom linking two carbon atoms. Three simple examples of ethers are ethyl methyl ether, diethyl ether, and methyl t-butyl ether (MTBE). Note that the name is created by indicating the two alkyl groups attached on either side of the oxygen, and then adding the word “ether”. Diethyl ether is the “ether” that, like the chloroform mentioned earlier, was used as an anesthetic in the early days of medicine. It is often still used in biology classes to anesthetize and determine the characteristics of *Drosophila* flies. MTBE is the ether used as a fuel additive. Its octane rating is 116, and for this reason it can help boost the rating on otherwise poor-quality gasoline.

Ether – Any compound containing the ether functional group (–C–O–C–).

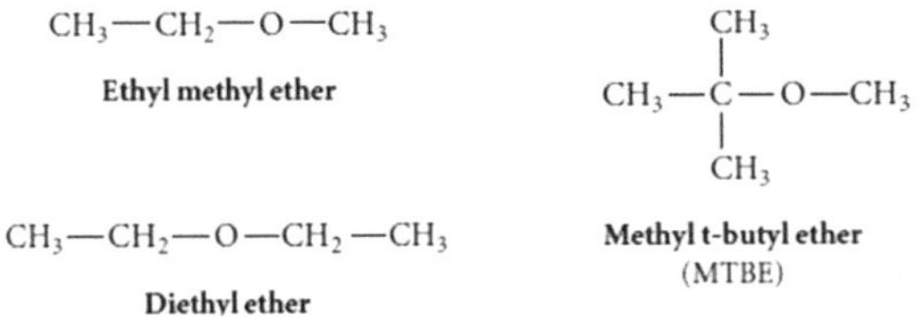

$$CH_3-CH_2-O-CH_3$$

Ethyl methyl ether

$$CH_3-CH_2-O-CH_2-CH_3$$

Diethyl ether

$$CH_3-C(CH_3)_2-O-CH_3$$

Methyl t-butyl ether
(MTBE)

The properties of ethers make them ideal as solvents. Because they lack the ability to form hydrogen bonds with themselves, they have boiling points similar to those of the alkanes. However, the presence of an oxygen atom within the structure indicates that they can participate in hydrogen bonding as acceptors with other molecules that can donate a hydrogen atom. This means that they make good solvents for chemical reactions.

Polyethers, which incorporate repeating ether linkages within very long chain-like molecules, are widely used in adhesive resins. An example of a polyether containing the ether functional group is found in the **epoxy resins**.

Epoxy resins – Polymer resins containing the epoxide group, in which two carbons atoms and an oxygen atom form a cyclic ether ring.

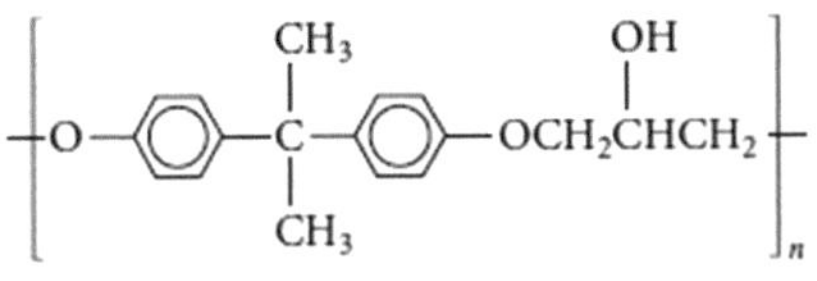

$$\left[-O-C_6H_4-C(CH_3)_2-C_6H_4-OCH_2CH(OH)CH_2- \right]_n$$

These rather viscous polymers, used mostly as construction and household adhesives, contain an ether linkage formed between one carbon atom that is part of a phenyl ring, and another that is part of a methylene (CH_2) group. The monomers that make up an epoxy resin contain the phenol functional group and a triangular ether structure known as an **epoxide**.

Epoxide – A compound containing a triangular ring of two carbon atoms linked by an oxygen atom, known as the epoxide group.

CH_3 | HO–C_6H_4–C–C_6H_4–OH | CH_3

A phenol

HO—CH_2—CH—CH_2 (with O bridging CH and CH_2)

An epoxide

▶ **Here's What We Know So Far**

- The major components of crude oil include hydrocarbons.
- Crude oil has many uses, including serving as fuel and supplying feedstock molecules for industrial processes.
- Manipulation of the hydrocarbons can yield a variety of molecules containing different functional groups.
- Functional groups are common specific arrangements of atoms within a molecule that impart similar physical and chemical properties to the molecule.
- Organic molecules are named by applying a set of rules established by the International Union of Pure and Applied Chemistry (IUPAC). ◀

21.15 Handedness in Molecules

Symmetry abounds in nature. Take, for instance, the symmetry of a person. The left-hand side of your friend appears to be a mirror image of the right-hand side. Proof of this can be found in your own hands. Hold your hand to a mirror such as that shown in ◘ Fig. 21.15. What do you see? Your other hand is the mirror image. What's more interesting is that your left and right hands (despite being mirror images) are not superimposable. That is to say, your left hand and your right hand cannot be lined up perfectly. Try placing a left-handed glove on your right hand. It just doesn't fit.

Molecules can also exhibit this handedness. In chemistry, we say that "handed" molecules are **chiral**, a term derived from the Greek *cheir* ("hand"). A chiral object has a nonsuperimposable mirror image. Those molecules that lack **chirality** are said to be **achiral**, and they have a superimposable mirror image. Every chiral molecule possesses a twin that has the same chemical formula, the same structure, but a different arrangement of atoms in three-dimensional space.

Chiral – A term used to denote molecules that have a nonsuperimposable mirror image.

Achiral – A term used to describe molecules that have a superimposable mirror image.

Chirality – The quality of being chiral—that is, of having a nonsuperimposable mirror image.

What makes a molecule chiral? Carbon-based molecules are chiral when they contain a carbon atom that is attached to four different groups. These **chiral centers** exist in many molecules in nature. Let's examine a derivative of methane to explain this feature. Bromochlorofluoromethane (CHBrClF) contains a carbon atom attached to four different groups. A mirror image of this molecule (see ◘ Fig. 21.16) shows this molecule to have handedness. These two molecules are **stereoisomers**. They are a special kind of structural isomers that differ in their 3-D arrangements of atoms, rather than in the order in which the atoms are bonded.

Chiral center – A carbon attached to four different groups.

Stereoisomers – Structural isomers that differ in their three-dimensional arrangements of atoms, rather than in the order in which the atoms are bonded.

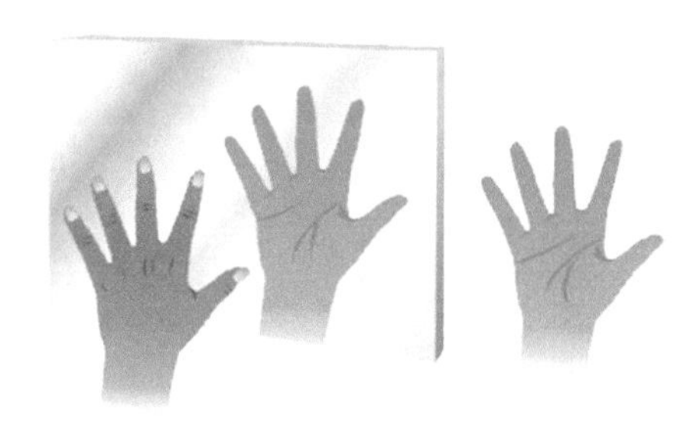

◘ **Fig. 21.15** Chiral objects have a nonsuperimposable mirror image. The mirror image of an achiral object is superimposable

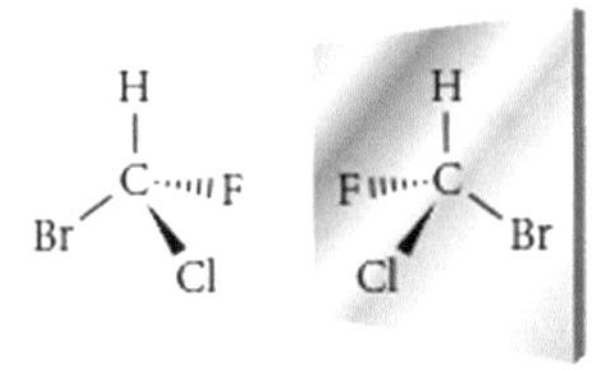

Fig. 21.16 Some organic molecules are also chiral. The image and the original are nonsuperimposable

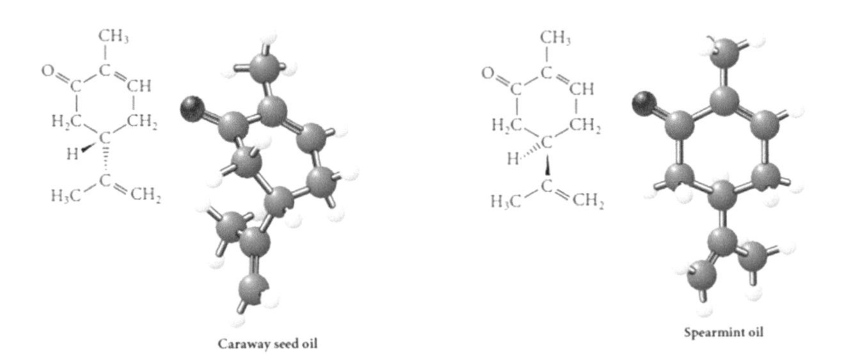

Fig. 21.17 The isomers of carvone

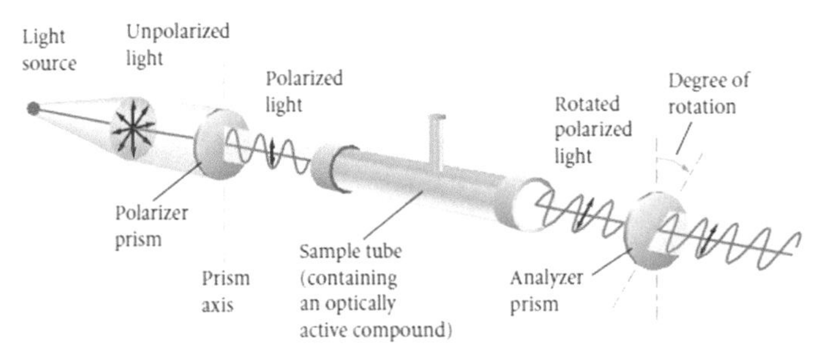

Fig. 21.18 Plane polarized light is rotated as it interacts with chiral molecules

The right-handed molecule and the left-handed molecule have almost identical physical properties because they have the same chemical structure. They are different, however. For instance, the receptors in your nose perceive different aromas based on their interaction with molecules. Carvone, shown in Fig. 21.17, a molecule your nose might encounter in the kitchen, possesses a chiral carbon and therefore exists as two stereoisomers. One of the isomers smells like caraway seeds; the other smells like spearmint.

Chiral molecules also exhibit the ability to rotate the plane of polarized light. Light, passed through a polarizing filter (such as the one in most sunglasses), is composed of electromagnetic waves that are aligned in one direction (such as the up-and-down direction). This polarized light interacts with the two stereoisomers of a chiral molecule to the same degree, but one of the isomers rotates the light to the left and the other to the right, as shown in Fig. 21.18.

Example 21.7—Chiral Drugs

Thalidomide was used as a drug that relieved morning sickness in pregnant women in the 1950s and 1960s. Although the FDA was hesitant (because of some neurological side effects) to approve the drug for use in the United States, thalidomide found widespread use in Europe and Canada. On the basis of the structure given here, identify the chiral center in the molecule.

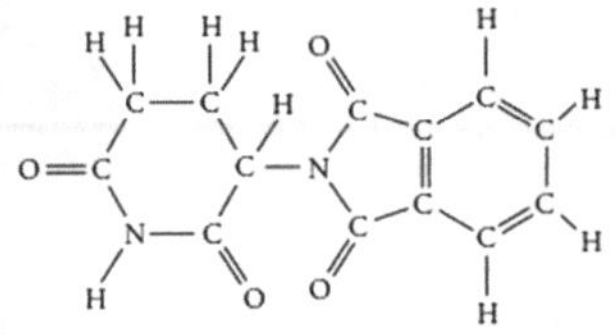

21

Asking the Right Questions
What is a chiral center? We'll examine the molecule, looking for carbons that are bonded to four different groups. A helpful starting point is to ignore all carbons that contain double bonds or more than one hydrogen atom.

Solution

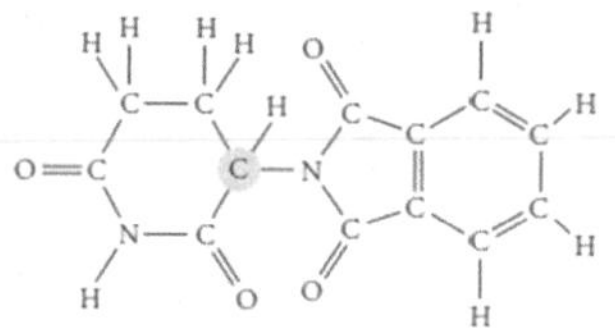

Are Our Answers Reasonable?
We can confirm that there are four different things attached to the carbon we've chosen, so the answer is reasonable.

What If…?
…we were asked to identify the number of chiral centers in carvone (see Fig. 21.17). How many chiral centers are there?
(*Ans: 1*)

Practice 21.7
Identify the chiral centers in each of the molecules below.

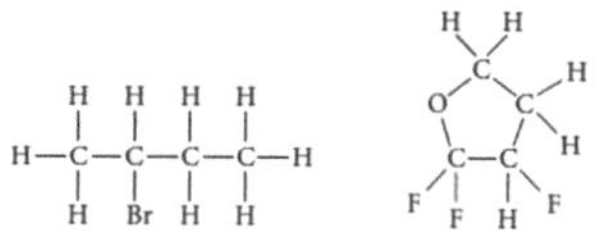

21.16 Organic Chemistry and Modern Drug Discovery

Perhaps one of the most profound impacts of organic chemistry on modern society has been its key role in pharmaceutical science. Historically, chemists and medical researchers have analyzed natural product extracts for effectiveness in curing diseases. In many cases, these plant extracts are the same ones identified in folk medicine as curatives for common maladies. Ultimately, there must be some component or collection of compounds present in the extract mixture that is responsible for their effectiveness. Organic chemists can purify the active constituents of complex mixtures originating from such natural sources, and very often, this active constituent turns out to be a single compound. These compounds are nearly always based on a carbon framework. Organic chemists can determine the structure of these compounds and can even synthesize them using simple starting materials. They can also make clever improvements on these natural products by rationally changing structural features (and, therefore, specific properties) of the molecule.

How do drugs work? Cellular processes, such as intercellular communication, metabolism, and construction of the cell structure can be disrupted by the binding of drug molecules to specific proteins and other biomolecules essential to those functions. The organic chemical Taxol™ is arguably one of the most important anticancer compounds to have been found in a plant extract over the last 50 years. Let's briefly trace the discovery of Taxol to show how drug development occurs in industry.

21.16.1 Taxol

In 1962, National Cancer Institute scientists found that extracts from the bark of the Pacific yew tree (*Taxus brevifolia*) had the ability to kill some types of cancer. Nine years later, the compound responsible for this promising activity, named paclitaxel or Taxol, was isolated from the yew extract, and its structure was determined to be the one shown in Fig. 21.19. Taxol acts by tight association with cellular microtubules, which are protein as-

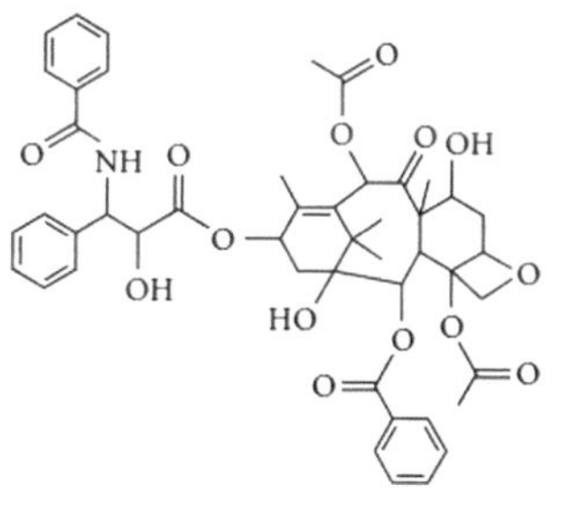

Fig. 21.19 The chemical structure of Taxol. Can you identify each of the 11 chiral centers?

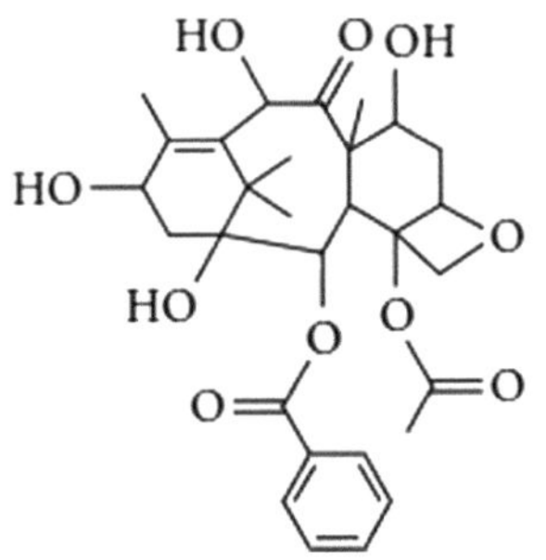

■ Fig. 21.20 The structure of 10-deacetylbaccatin III

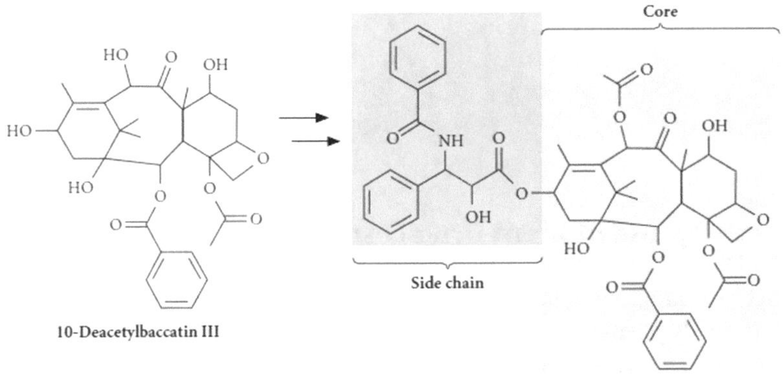

■ Fig. 21.21 The semisynthesis of Taxol

semblies essential to cell division. The idea behind using Taxol for anticancer chemotherapy is to stop the uncontrolled cell division responsible for tumor growth.

Given Taxol's importance, scientists needed large quantities of this compound for experimental treatment of cancer patients and also for basic research. When large amounts of a natural product are required, chemists can either purify it in bulk from its natural source or synthesize it from readily available materials. In the case of Taxol, neither of these two avenues alone was feasible.

At the time of its discovery, the only known source of Taxol was the bark of mature Pacific yew trees. Unfortunately, the yield of Taxol obtained from the Pacific yew is 100 mg per kilogram of bark—a very small amount. Even worse, bark removal results in the death of the tree. The Pacific yew tree is mainly found in the Pacific Northwest, and large-scale harvesting of these trees for Taxol would quickly result in extinction of the tree species. Another way to produce Taxol was desperately needed.

A potential solution to this problem might have been to synthesize Taxol from scratch (an approach known as **total synthesis**), therefore eliminating the threat to the Pacific yew tree. The problem is that the structure of Taxol is enormously complex (compare the structure in ■ Fig. 21.19 to the structures of the heptane isomers from Practice 21.1). What was needed was the development of an *efficient* total synthesis of Taxol, and synthetic organic chemists accepted the challenge. After several years of effort, total syntheses of Taxol were accomplished. Yet even though this was a great feat for modern organic chemistry, these total syntheses did little to address the supply problem. All of these total syntheses were long, technically complex, very costly, and consequently incapable of cheaply producing large quantities of Taxol. In summary, neither isolation from the Pacific yew tree nor total synthesis was an environmentally or economically viable method for the mass production of Taxol.

Total synthesis – The preparation of a molecule from readily available starting materials. Typically the total synthesis involves many steps.

During the course of Taxol research, chemists found that a biosynthetic intermediate of Taxol (10-deacetylbaccatin III, shown in ■ Fig. 21.20) could be isolated from the needles of a relative of the Pacific yew tree, in yields of up to 1 g per kilogram of needles. The tree first synthesizes the biosynthetic intermediate 10-deacetylbaccatin III and then converts it to Taxol. And crucially, harvesting these needles for 10-deacetylbaccatin III is not fatal to the tree. Chemists could then take advantage of this renewable resource for 10-deacetylbaccatin III by finding a way to convert it into Taxol. This is one reason why the efforts to achieve total synthesis were so important to Taxol production. All of the previous total syntheses combined the "core" with the "side chain" to give the full structure of Taxol. Pharmaceutical scientists adopted this same strategy by converting 10-deacetylbaccatin III into something synthetically useful and then attaching the "side chain," as shown in ◘ Fig. 21.21. This method is called a semisynthetic method: A portion of the structure of the desired natural product is isolated from a natural source and then converted into the natural product via synthetic chemistry.

In 1992, Taxol was approved for use in the United States to treat ovarian cancer, breast cancer, and certain types of lung cancer. Along with the drug cisplatin, Taxol is often the first drug used in chemotherapy against cancer. Today, the pharmaceutical industry uses a version of the semisynthetic method to produce Taxol.

Example 21.8—Taxol as a Drug

One problem often associated with drugs is side effects. This problem can come from a lack of complete specificity for the target. In treating cancer, we would like to kill the cancer cells but leave the patients' normal cells alone. Cancer tissue grows very rapidly, but other than that, these cells don't have many other unusual features compared to normal cells. This is one of the reasons why cancer is such a difficult disease to treat. Why might Taxol have side effects?

Asking the Right Questions

Cancer cells have acquired mutations in their genetic code that enable them to grow uncontrollably. Despite this, all of the normal cellular machinery is still present in cancer cells. First consider the mode of action of Taxol and then think about how this might affect normal cells.

Solution

Taxol acts by binding to microtubules, which are proteins directly involved in cell division. When Taxol binds to the microtubules, the cancer cells cannot divide to give two daughter cells. This can stop the growth of tumors, which are masses of cells that divide uncontrollably. Because it is nearly impossible to administer Taxol so that only cancer cells in the patient's body receive it, the drug is likely to interact with normal cells as well and bind to their microtubules. As a result, these normal cells cannot divide, which could result in the disruption of otherwise healthy tissue.

Are Our Answers Reasonable?

Since cancer cells and normal cells have so many similarities, it seems reasonable that an anti-cancer drug might cause side effect by reacting with those normal cells.

What If…?

…we were asked to develop a drug that acts only on cancerous cells? What basic difference exists that we could exploit in our drug design?

(*Ans: The difference in growth rate is a good target for an anticancer drug.*)

Practice 21.8

m-Amsacrine is another anticancer drug that exhibits biological activity by binding to DNA inside cancer cells. Once bound, it causes proteins to break the DNA molecule into fragments. The result is the death of the cancer cell. Would you predict *m*-amsacrine to show side effects?

Example 21.9—Nomenclature of Esters

Taxol is a very complicated molecule, and organic chemists typically cope with structural complexity by first looking at its most important features: the functional groups. Look at the structure of Taxol in ◘ Fig. 21.19 and identify each of the functional groups we have learned about in this chapter.

Asking the Right Questions

Identifying functional groups indicates that we must know the functional groups and how the atoms are placed in each.

Solution

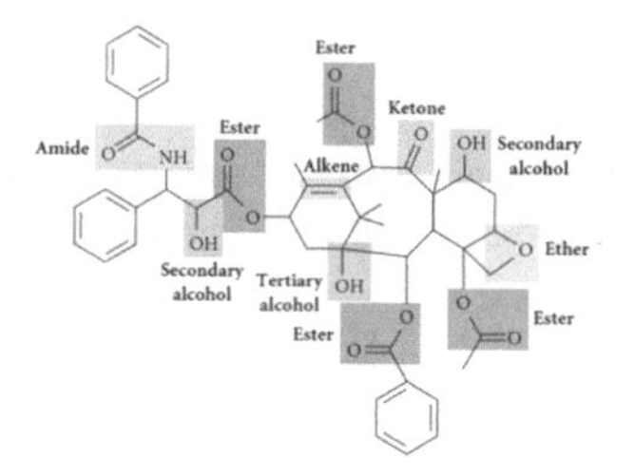

Are Our Answers Reasonable?

We have identified the two portions of each name, so our answers are reasonable.

What If…?

…we were asked to identify chiral centers in Taxol? How many exist in this molecule?

(*Ans: eleven*)

Practice 21.9

Two other functional groups are related to the ester. What are they and why are they related?

The Bottom Line

- Organic chemistry is the study of carbon-containing compounds, particularly those derived from living things or fossil fuels.
- The element carbon occurs in the form of three allotropes: the diamond, graphite, and fullerene forms.
- Crude oil is our major source of carbon-containing compounds, which are processed by the chemical industry into a wide range of products.
- Hydrocarbons are compounds composed of carbon and hydrogen only. They include the alkanes, alkenes, alkynes, and aromatic hydrocarbons.
- Organic compounds can exist as structural and geometric isomers.
- Hydrocarbons are separated into fractions by fractional distillation of crude oil.
- Cracking and reforming are two important methods of preparing specific organic molecules.
- Alkanes can undergo substitution and dehydrogenation reactions.
- All organic compounds can be regarded as carrying zero, one, or more distinctive functional groups on a hydrocarbon frame.
- Each functional group is associated with a distinctive set of chemical characteristics.
- Many modern materials are polymers composed of thousands of organic monomer units linked together.
- Molecules may exhibit chirality, which may lead to different chemical properties in two stereoisomers.
- Organic chemists can isolate and purify the active constituents of complex mixtures in order to make products such as pharmaceuticals.

21.2 Elemental Carbon

Skill Review

1. List the three allotropes for carbon and indicate the hybridization of the carbon atoms in each.
2. Explain the differences in the properties of amorphous (lacking a defined crystalline shape, disordered) carbon and graphite.

Chemical Applications and Practices

3. Carbon in the diamond form and carbon in the graphite form have widely differing uses. We pay hundreds, even thousands, of dollars for diamond jewelry, but only pennies for carbon in pencils. Explain, using the differences in bonding, why diamonds are held together more strongly than the carbon atoms in graphite.
4. Using hybridization, explain how carbon forms both sigma and pi bonds in graphite, but only sigma bonds in the diamond form.
5. Diamonds are typically measured in carats (1 carat is 200 mg). If an industrial diamond contained only carbon, how many atoms of carbon would be found in a 2.00-carat diamond?
6. Graphite can be converted to diamond with great pressure. The type of diamond used in polishing powder can be produced with pressures on the order of 1000 MPa.
   a. What is this pressure in terms of atmospheres?
   b. Why is pressure a factor in conversion of the graphite form of carbon to the diamond form?

21.3 Crude Oil—The Basic Resource

Skill Review

7. Judging on the basis of Table 21.2, what is the main difference between the aliphatic and the aromatic compounds found in crude oil?
8. Judging on the basis of Table 21.2, what is the difference between the structures and formulas for compounds in the aliphatic and naphthene classes?

Chemical Applications and Practices

9. Sulfurous compounds occur to a minor extent in crude oil, and their presence is indicative of the source of the crude. Explain how the decomposition of biological material could give rise to sulfur-containing compounds.
10. Why do you think that there is such a wide diversity of compounds in crude oil?

21.4 Hydrocarbons

Skill Review

11. Provide the correct name for each of these:
    a. $CH_3(CH_2)_5CH(CH_3)_2$
    b. $CH_3CH_2CH_2CH(CH_3)_2$
    c. $CH_3CH_2CH(CH_2CH_3)_2$
    d. $CH_3CH_2CH_2CH_2CH_3$
12. Provide the correct name for each of these:
    a. $CH_3CH(CH_3)_2$
    b. $CH_3CH_2CH_3$
    c. $CH_3CH_2CH_2C(CH_3)_3$
    d. $(CH_3)_2CHCH_2CH(CH_3)_2$
13. Diagram the structure of the compounds listed in Problem 11.
14. Diagram the structure of the compounds listed in Problem 12.
15. Name each of the following hydrocarbons.

```
        CH3
        |
a.   H2C    CH3          b.    H  H  H               c.  H3C   H H  H
        \  /                    \ |  |                      \ H   \ /
         C                      C-C-C-CH3                  C-    C
        /  \                   //  |  |                  H/   \   \H
     H2C    CH3            H-C    H  H                         C=C
      \                       \                               /   \
      CH3                      H                             H     H
```

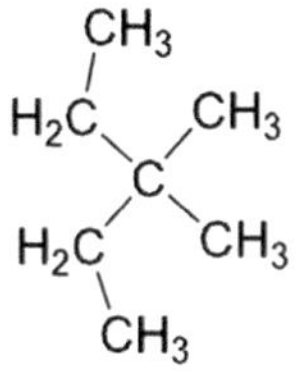

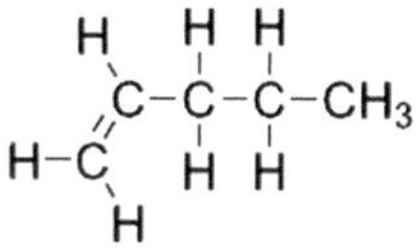

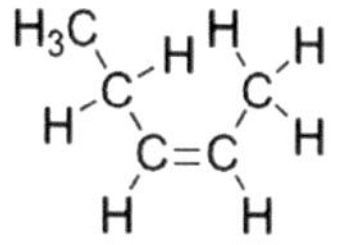

16. Name each of the following hydrocarbons.

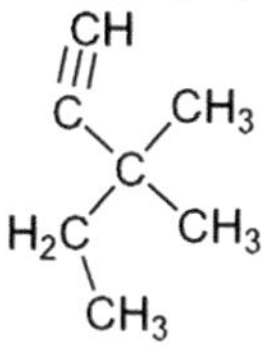

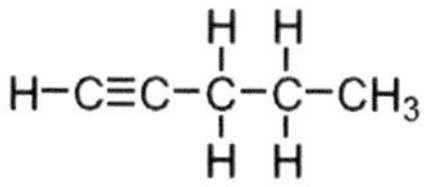

```
      CH
      |||
a.    C   CH3        b.      H  H              c.        H          H
       \ /                   |  |                        |          |
        C            H-C≡C-C-C-CH3                H3C-C-C≡C-C-H
       / \                   |  |                        |          |
    H2C   CH3                H  H                        H          H
     \
     CH3
```

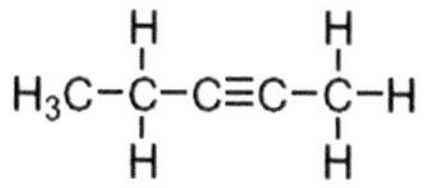

17. Complete this table:

	Alkane	Alkene	Alkyne
Type of C to C bond			
Type of hybridization			

18. Complete this table:

	Alkane	Alkene	Alkyne
Angles between bonds to other atoms			
If made of only two carbons, will have ___ total hydrogen atoms			
Saturated or unsaturated			

19. a. Diagram the structure of each of these:
    pentane
    2-methylbutane
    2,2-dimethylpropane
    Also give the empirical formula of each.
    b. Remembering that intermolecular forces affect boiling point, match each structure with its corresponding boiling point (all are in degrees Celsius): 9.5, 36, 28.
20. a. Diagram the structure of each of these:
    3-ethyl-2-pentene
    cyclooctene
    3-hexyne
    Also give the molecular formula of each
    b. Remembering that intermolecular forces affect boiling point, match each structure with its corresponding boiling point (all are in degrees Celsius): 82, 94, 148.
21. One of the high-molecular-mass hydrocarbons that helps form the protective waxy skin of an apple contains 28 carbon atoms. What is the formula of this noncyclic alkane?

22. What is the formula of a saturated noncyclic hydrocarbon containing 8 carbons? Assume that a different hydrocarbon contains two C=C groups and has a total of 10 carbon atoms. What is its formula?
23. Diagram five isomers of octane that all have a six-member continuous carbon chain.
24. Diagram three isomers of octane that all have a five-member continuous carbon chain.
25. Diagram all of the alkenes that could have the formula C_5H_{10}.
26. Diagram all of the alkynes that could have the formula C_5H_8.
27. How many different mono-substituted methylnaphthalenes exist?
28. How many different mono-substituted methylanthracenes exist?
29. Which, if any, of these hydrocarbons could be cycloalkanes?
    a. C_2H_6
    b. C_6H_{12}
    c. $C_{12}H_{26}$
    d. C_7H_{14}
30. How many hydrogen atoms are found in one molecule of cyclodecane?
31. Identify each of these molecules as a cis or a trans isomer. 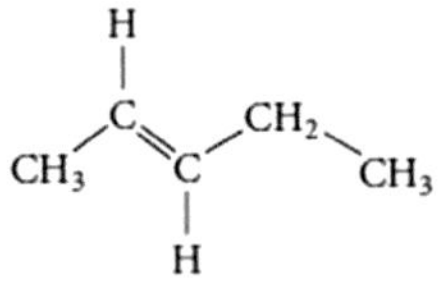 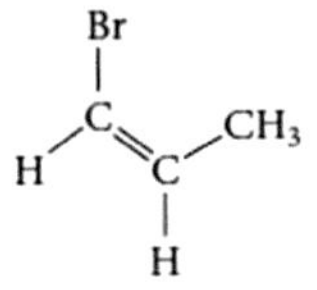
32. Identify each of these molecules as a cis or a trans isomer.

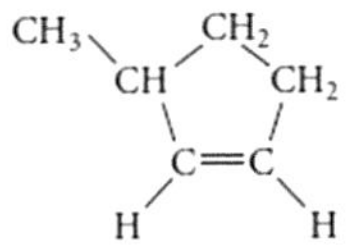

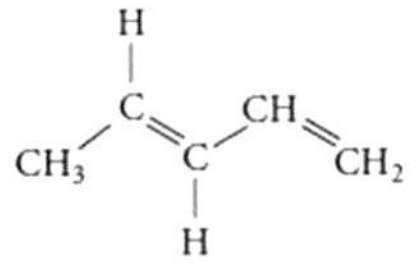

Chemical Applications and Practices

33. Ethyne (also known as acetylene) can be produced by the reaction between calcium carbide (CaC_2) and water. In addition to ethyne, calcium hydroxide is a product. Write and balance the equation for the production of this important alkyne.
34. The combustion of methane is an important reaction that heats our homes. Incomplete combustion, however, involves the reaction of methane with oxygen to produce carbon monoxide and water. Write and balance the equation for the incomplete combustion of methane.
35. Cyclohexane and benzene both have a ring structure containing six carbon atoms. Compare the two structures in terms of:
    a. The number of hydrogen atoms in each
    b. Carbon orbital hybridization
    c. The angles between carbon atoms
    d. The number of double bonds
    e. Resonance structures
36. At least three alkene isomers that contain two C=C can be drawn for C_5H_8. Draw these isomers and identify the one that is stabilized most by resonance structures.

21.5 Separating the Hydrocarbons by Fractional Distillation

Skill Review

37. Mixtures of hydrocarbons, and of other types of compounds, can be separated into their individual components by gas chromatography. Briefly explain the main steps by which this technique makes possible the identification of compounds in a mixture.
38. For specific uses, hydrocarbon fractions obtained from crude oil may be further modified. The two major industrial processes that provide useful hydrocarbon products are cracking and reforming. Compare and contrast these two important techniques.
39. What property of hydrocarbons serves as the basis for a separation, during fractional distillation, of the components in crude oil? What is the relationship between this property and the structure of the components?
40. Although gas chromatography is often used to resolve mixtures of hydrocarbons, it cannot be relied on in every case. What is a chief limitation of GC when applied to hydrocarbon separations?

21.6 Processing Hydrocarbons

Skill Review

41. The octane rating system has been used as a standard to rate the burning efficiency of other fuels. What is the formula of octane? Use the formula to write and balance the reaction for the combustion of octane.
42. The actual isomer of octane used in the rating system is "isooctane." The descriptive name for the compound is 2,2,4-trimethypentane. Diagram the structure of this compound and write a balanced reaction for the combustion of isooctane.

Chemical Applications and Practices

43. Another compound, besides isooctane, used in fuel studies was found to have the following percentage composition: 84.0% carbon and 16.0% hydrogen. The molar mass of the alkane is 100.0. From these data, what would you calculate as the molecular formula of the compound?
44. Calculate the mass percent carbon and the mass percent hydrogen of isooctane.
45. Currently, many automobiles run on fuel with a minimum octane rating of 87. To what mixture of isooctane and heptane does this fuel correspond?
46. Some additives to gasoline have a much higher octane rating. For instance, a particular alcohol has an octane rating of 116. What quantity of this alcohol would need to be added to crude gasoline (octane rating 55) to give a product with an octane rating of 92?

21.7 Typical Reactions of the Alkanes

Skill Review

47. Using methane and bromine, show the balanced equation to form dibromomethane.
48. Using butane as a reactant, show the balanced equation that produces 2-butene.
49. Provide the structural diagram and the name for all of the products that would result from the substitution reaction between Cl_2 and ethane.
50. Provide the structural diagram and the name for all of the products that would result from the dehydrogenation of butane.
51. What are the typical products that result from the combustion of an alkane?
52. A reaction produces a molecule with a C=C bond. If the product was made from an alkane, what would be the classification of this reaction?

Chemical Applications and Practices

53. The combustion of propane to form carbon dioxide and water can produce approximately 2200 kJ per mole. If a burning propane torch used 42.0 g of propane, how many kilojoules would be released?
54. Why do arctic hikers use propane as fuel for their cook stoves instead of pentane?
55. Indicate the product(s) for each of the following reactions:

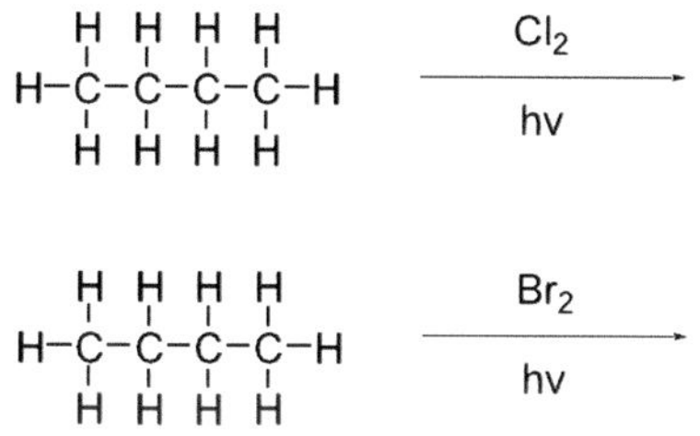

```
  H H H H        Cl2
  | | | |     --------->
H-C-C-C-C-H
  | | | |        hv
  H H H H

  H H H H        Br2
  | | | |     --------->
H-C-C-C-C-H
  | | | |        hv
  H H H H
```

56. Indicate the product(s) for each of the following reactions:

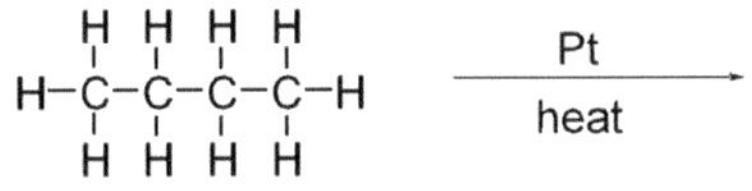

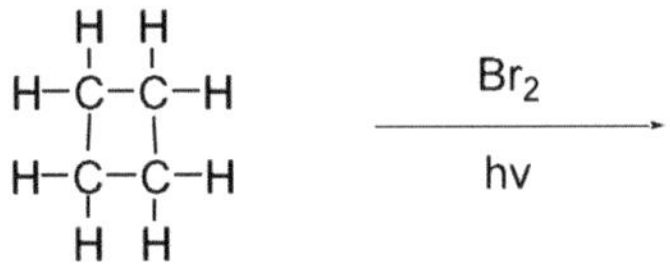

$$CH_3CH_2CH_2CH_3 \xrightarrow[\text{heat}]{\text{Pt}}$$

$$\text{cyclobutane} \xrightarrow[h\nu]{Br_2}$$

21.8 The Functional Group Concept

Skill Review

57. Provide a suitable chemical formula and an example structure of a molecule for each of these functional groups.
    a. alcohol
    b. aldehyde
    c. alkene
    d. ketone
58. Provide a suitable chemical formula and an example structure of a molecule for each of these functional groups.
    a. carboxylic acid
    b. alkyne
    c. ether
    d. cyclic alkane
59. Name the functional group in each molecule.
    a. CH_3CH_2CN
    b. $(CH_3)_2CHOH$
    c. $CH_3CH{=}CHCH_3$
    d. $CH_3CH_2OCH_3$
60. Name the functional group in each molecule.
    a. $HC{\equiv}CCH_2CH_3$
    b. $HSCH_2C(CH_3)_3$
    c. CH_3CH_2COOH
    d. $CH_3CH_2CH_2CHO$
61. Circle and identify each of the functional groups.

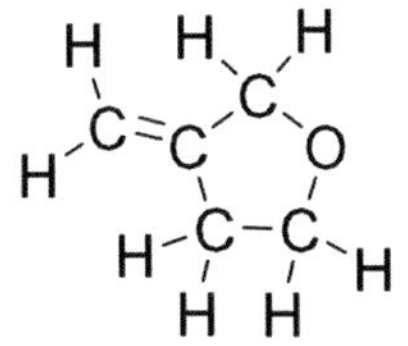

62. Circle and identify each of the functional groups.

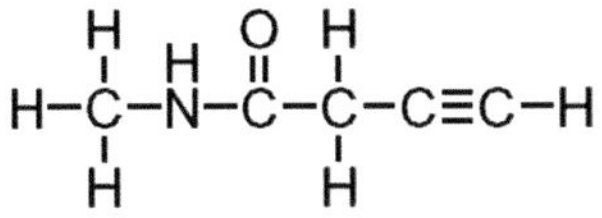

Chemical Applications and Practices

63. The structure of the common analgesic acetaminophen follows. Identify any and all functional groups shown in the structure.

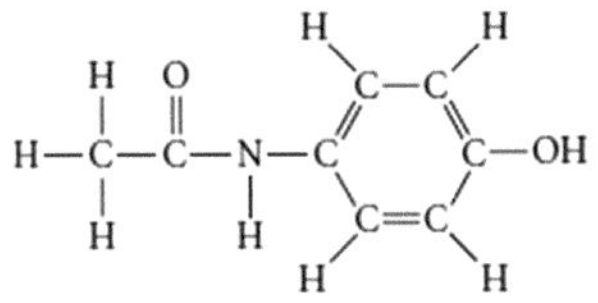

64. The structure for tetracycline follows. Identify any and all functional groups shown in the structure.

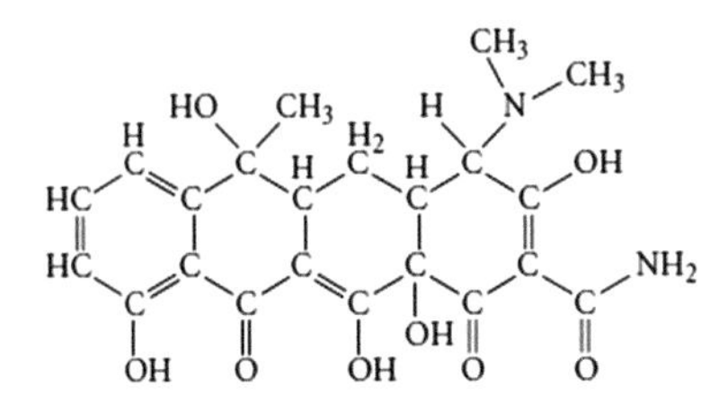

65. The partially complete structure (showing all atoms except hydrogen) for the pesticide malathion follows. Reproduce the structure and add the missing hydrogen atoms to the molecule.

```
                      O
                      ‖
            S         C—O—C—C
    C—O     ‖         |
        \P—S—C   O
    C—O/          |   ‖
                  C—C—O—C—C
```

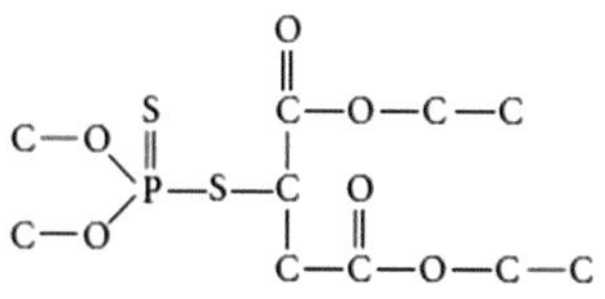

66. The partially complete structure (showing only carbons and oxygens) for carvone follows. Reproduce the structure and add the missing hydrogen atoms to the molecule.

```
                         O
                        //
    C          C—C
     \        /    \
      C—C          C—C
     //       \    //
    C          C—C
```

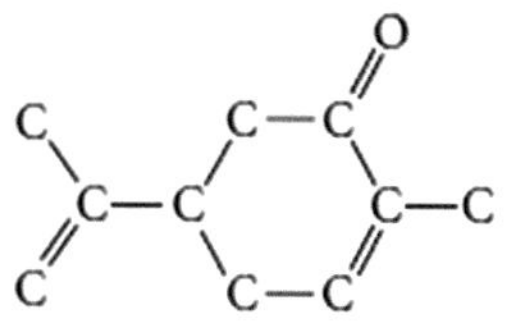

21.9 Ethene, the CPC Bond, and Polymers

Skill Review

67. How many carbon–carbon π bonds are present in each of these noncyclic molecules?
    a. C_3H_8 b. C_6H_8 c. $C_{10}H_8$
68. How many carbon–carbon π bonds are present in each of these noncyclic molecules?
    a. C_4H_8 b. C_5H_8 c. C_4H_6
69. Diagram the structure of each of these compounds:
    a. 3-Methyl-2-pentene
    b. 2-Methyl-4-propyl-3-heptene
70. Name these compounds:

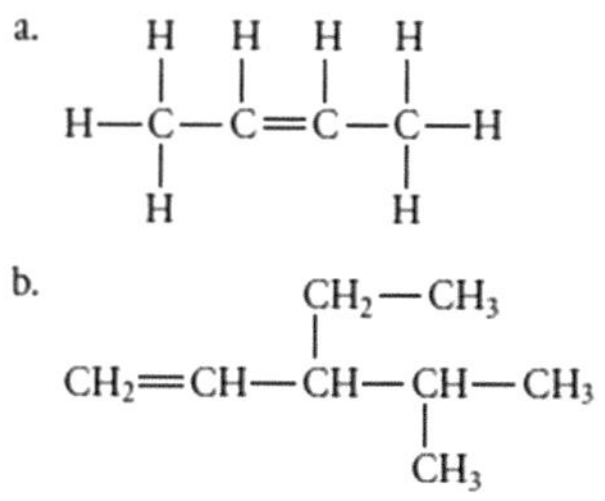

```
a.     H   H   H   H
       |   |   |   |
    H—C—C=C—C—H
       |           |
       H           H

b.                CH2—CH3
                  |
    CH2=CH—CH—CH—CH3
                       |
                       CH3
```

71. Identify the major product(s) of the following reaction.

```
       H  H  H  H
       |  |  |  |           H2O
    H—C—C=C—C—H      ———————————→
       |        |          H3PO4
       H        H           heat
```

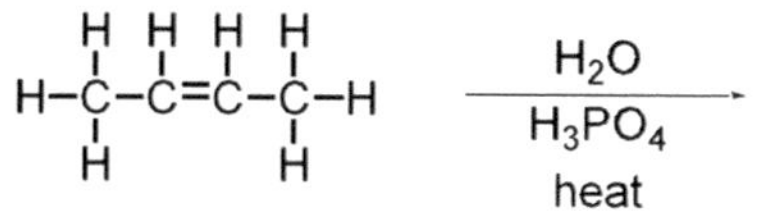

72. Identify the major product(s) of the following reaction.

```
       H  H  H  H
       |  |  |  |           HCl
    H—C—C=C—C—H      ———————————→
       |        |
       H        H
```

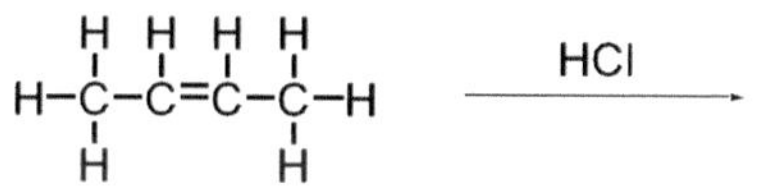

Chemical Applications and Practices

73. One gram of an important hydrocarbon, when completely combusted, produced 2.93 g of carbon dioxide and 1.80 g of water. The molar mass of the compound is 30.0. Is this compound an alkane or an alkene? What is the formula of the compound?
74. Chemists often refer to the addition reaction of alkenes as "adding across the double bond." Explain what aspect of the double bond between carbon atoms favors this process.

75. a. Diagram the structure of propene.
    b. Show the balanced equation that represents the conversion of propene into the alcohol 1-propanol.
76. One way to distinguish an alkane from an alkene is to halogenate the double bond. If propene reacted with Br_2, what compound would be formed?
77. Polypropylene is an addition polymer with many uses, one of which is in indoor–outdoor carpeting.
    a. What is the structure of the monomer used in this polymer?
    b. Diagram a section of polypropylene using four joined monomers.
78. Clothing materials made from Orlon® contain the addition polymer polyacrylonitrile shown here.

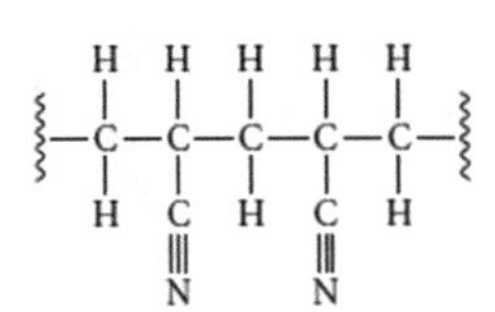

After examining the structure, diagram the structure of the repeating monomer.

79. HDPE and LDPE are both plastics that can be recycled. HDPE is coded as 2, and LDPE is coded as 4. What contrasting structural factor in the molecules of these two polymers causes their differing properties?
80. Indicate which of these statements is(are) true? For those that are false, explain why.
    a. Addition reactions always involve compounds with multiple bonds.
    b. All addition reactions produce polymers, but not all polymers are addition polymers.
    c. All addition polymers are based on reactions between molecules with multiple bonds.
    d. Addition polymers must contain double bonds.

21.10 Alcohols

Skill Review

81. After looking at the structure of a compound, a student incorrectly named it 2,5-pentanediol. Provide the correct name for the compound.
82. Another compound was mislabeled as 1,4-propanediol. Explain the mistake in this name.
83. Diagram the structure of 1,3-pentanediol. Classify each –OH group as primary, secondary, or tertiary.
84. Diagram the structure of 2-butanol. Would this alcohol be classified as a primary, secondary, or tertiary alcohol?
85. Which of the following alcohols would you expect to react fastest in a reaction?

```
  H H H H            H H H OH
H-C-C-C-C-H        H-C-C-C-C-H
  H H | H            H H H H
      OH
```

86. Which of the following alcohols would you expect to react fastest in a reaction?

```
  H H                  H H H H
H-C-C-H              H-C-C-C-C-H
H-C-C-OH               H H | H
  H CH3                    OH
```

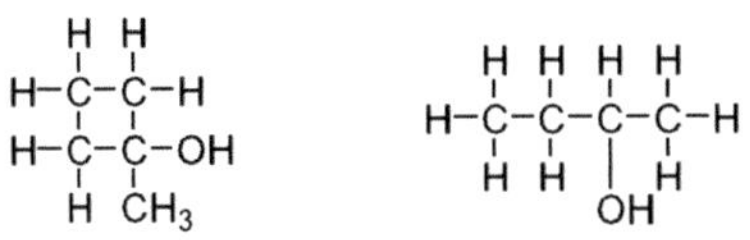

Chemical Applications and Practices

87. Phenol is an example of an aromatic alcohol containing six carbons, six hydrogens, and one oxygen. It is used in producing disinfectants and some polymers. Diagram the structure of this important compound.
88. Alcohols can undergo a reaction known as dehydration. In this reaction, the alcohol loses a molecule of water to form an alkene. Provide the name of the alkene formed from the dehydration of each of these alcohols:
    a. Ethanol
    b. 1-Propanol
    c. 2-Propanol
    d. 1-Butanol

21.11 From Alcohols to Aldehydes, Ketones, and Carboxylic Acids

Skill Review

89. Draw the structure of 1-propanol, and identify the product that would result from oxidation of this compound.
90. Diagram the structure of butanal. What alcohol can be oxidized to produce this compound?
91. Propanoic acid is used to make a type of mold inhibitor. Diagram the structure of propanoic acid. What aldehyde could be oxidized to form this acid?
92. Isopropyl alcohol (2-propanol) is the main ingredient in rubbing alcohol. This compound can be oxidized to a ketone, but not to a carboxylic acid. Draw the structure of the alcohol and the ketone and explain why it can't be used to produce the corresponding three-carbon carboxylic acid.

Chemical Applications and Practices

93. Methyl ethyl ketone is often used as a solvent for organic compounds. Diagram the structure of this important organic compound. What secondary alcohol could be oxidized to form this compound?
94. An unknown compound was oxidized to form pentanoic acid. The unknown compound was either 2-pentanone or pentanal. Identify the starting compound.
95. Carboxylic acids are generally considered only weakly acidic. Show the acid dissociation reaction of propanoic acid in water.
96. An unknown compound was found to produce an acidic solution and contained four carbon atoms. The compound could be prepared from a straight-chain alcohol. Identify the unknown compound and the starting alcohol.

21.12 From Alcohols and Carboxylic Acids to Esters

Skill Review

97. The ester that produces a banana aroma is called 3-methylbutyl ethanoate. Diagram the structures of the starting materials that produce the ester, and show the balanced equation for the production of isopentyl acetate.
98. An apple-like aroma can be produced from the combination of methanol and butanoic acid. Diagram the structure and name this ester.
99. An apricot aroma can be produced from the ester pentyl butanoate. Using structures, show the balanced equation that produces this pleasant fruity aroma.
100. Tristearin is the animal fat associated with beef. The formula of stearic acid is $CH_3(CH_2)_{16}COOH$. Show the reaction between stearic acid and 1,2,3-propanetriol to form the triple ester known as tristearin.

Chemical Applications and Practices

101. Linoleic acid is used to form the major oil found in corn. This unsaturated acid has two double bonds per molecule. The double bonds are found between carbon atoms number 9 and 10 and between carbon atoms number 12 and 13. The formula is $C_{18}H_{32}O_2$. Draw the diagram of this linear unsaturated carboxylic acid.
102. Oleic acid (a major component of animal fats) contains a *cis* alkene and 18 carbon atoms and has the formula $C_{18}H_{34}O_2$. The alkene is found between carbons number 9 and 10. Draw the diagram of this linear carboxylic acid.

21.13 Condensation Polymers

Skill Review

103. The simplest amino acid is glycine. Use the structure shown below to diagram a condensation polymer made from three glycine monomers. What molecule has been split out as the polymer forms?

$$H_2N-\underset{\displaystyle H}{\overset{\displaystyle H}{C}}-\overset{\displaystyle O}{\overset{\|}{C}}-OH$$

104. Why is the formation of a condensation polymer impossible in these cases?
    a. Ethanoic acid + $NH_2CH_2CH_2CH_2CH_2NH_2$
    b. Ethanoic acid + $CH_3CH_2NH_2$
    c. $HOOC–CH_2–COOH_+CH_3CH_2NH_2$

Chemical Applications and Practices

105. Thiokol, dating back to the 1920s, was the first synthetic rubber commercially produced in the United States. It can be formed as a condensation polymer from the monomer $ClCH_2CH_2OCH_2CH_2Cl$. The reaction takes place when sodium polysulfide (Na_2S_2) reacts with the monomer. Assuming that Na_2S_2 is a source of –S–S–, draw the structure of the polymer. What smaller compound is also produced from the reaction?
106. In addition to water, a by-product from the condensation polymerization of 5-aminopentanoic acid is common. What is the structure of this by-product?
    $NH_2CH_2CH_2CH_2CH_2COOH$ (5-aminopentanoic acid)

21.14 Polyethers

Skill Review

107. Using the information and examples in the chapter, devise the name of these ethers.
    $CH_3{-}O{-}CH_3$ $\quad CH_3{-}O{-}CH_2{-}CH_2{-}CH_2{-}CH_3$
108. Using the information and examples in the chapter, draw the structure for these ethers.
    a. ethyl methyl ether
    b. propyl methyl ether
    c. butyl ethyl ether

21.15 Handedness in Molecules

Skill Review

109. Indicate whether chirality exists in each of these objects.
    a. a pencil
    b. your feet
    c. your textbook
    d. a fork
110. Indicate whether chirality exists in each of these objects.
    a. a CD
    b. a butterfly
    c. a bottle
    d. a sportscar
111. Identify the chiral center in each of these molecules. 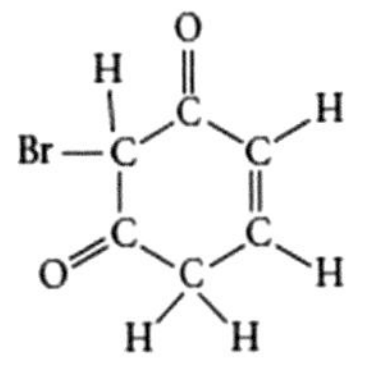 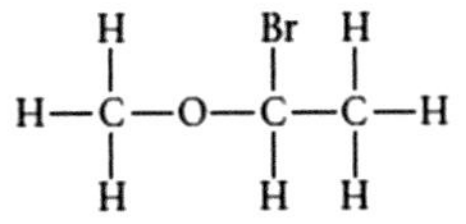
112. Identify the chiral center in each of these molecules.

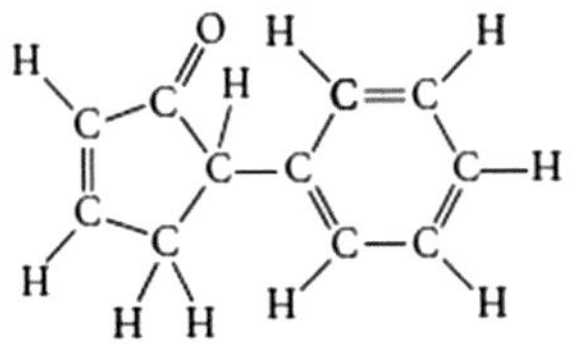

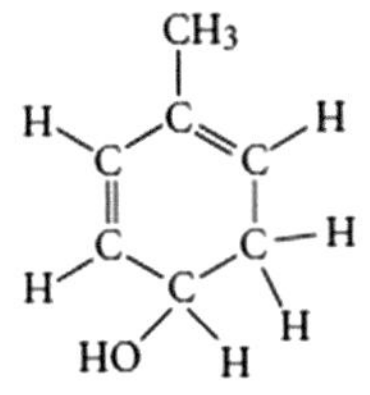

21.16 Organic Chemistry and Modern Drug Discovery

Chemical Applications and Practices

113. The total synthesis of a potential drug requires nine steps, each with a 95% yield. What is the overall yield of the final drug made from this synthesis?
114. Acetylsalicylic acid, also known as aspirin, is a derivative of a natural compound found in the bark and leaves of a willow tree. Explain how a researcher might have originally surmised that an active substance was to be found in the leaves and bark of the willow tree.

Comprehensive Problems

115. Why do gasoline manufacturers adjust the gasoline mixture for automobiles in accordance with the season of the year?

116. Using information from this chapter, describe how the acidity of salicylic acid, shown below, can be reduced by modification of the compound via a chemical reaction.

Salicylic acid

117. Gasoline is delivered to a farmer for use in the farm's combines. A cylindrical storage tank with a circumference of 6.0 ft is used to store the gasoline until it is used. The farmer uses a yardstick to measure the depth of the gasoline in the storage tank.
    a. How many gallons of gasoline are in the tank if the yardstick measures the gasoline to a depth of 22 inches?
    b. 2,3-dimethylhexane is one of the components in the gasoline. Draw the structure of this compound.
118. The compound responsible for the odor of wintergreen was discussed in the chapter.
    a. What two compounds (an alcohol and a carboxylic acid) could be condensed to prepare methyl salicylate?
    b. If a company that manufactures a muscle pain relief cream requires 352 kg of methyl salicylate, how many liters of the alcohol would be needed? (Assume the density of the alcohol is 0.789 g/mL.)
119. In Table 21.5, we mention that one of the fractions distilled from crude oil is known as petroleum ether. This fraction is used in the organic research lab to recrystallize reaction products.
    a. Draw and name at least two compounds that could be found in this distillation fraction.
    b. Why does the name of this fraction appear to be a misnomer?
120. The addition of a halogen to an alkene is a good way to produce an alkyl halide in the laboratory. But often, a researcher is interested in adding only one halogen atom to the compound. In such a case, a hydrogen halide is used for the addition reaction.
    a. If a research student fills a 2.5 L flask with 0.25 g gaseous hydrogen chloride and 0.48 g of gaseous trans-2-butene, what mass of the addition product can be made?
    b. What is the name of that product?
    c. What is the concentration, in molarity, of the product in the flask?
121. The organic chemistry industry of the late part of the 1800s was interested in creating new dyes for fabrics.
    a. What features of an organic compound would make it suitable for use as a dye?
    b. If a particular organic dye was yellow in color, what range of wavelengths would be reflected by the compound?
    c. What range of energies, in kJ/mol, is associated with these wavelengths?
122. Based on your understanding of carbon's allotropes, which do you think is thermodynamically more stable (graphite or diamond)?
123. Can an alkyne exhibit *cis/trans* isomerism?
124. A researcher treats a solution of *cis*-2-butene with aqueous phosphoric acid and obtains the product. After careful examination, she realizes that there are actually two products formed. Explain this result.

Thinking Beyond the Calculation

125. An unknown organic compound is identified. Reaction of this unknown with two molecules of ethanol produces a diester. The unknown organic compound can be prepared by oxidizing 1,2-ethanediol.
    a. What is the structure of the unknown?
    b. What properties (chemical and physical) would you predict this unknown to possess?
    c. If 10 g of the unknown produce 10 g of the diester, what is the percent yield of the reaction?
    d. When the diester is placed in aqueous solution containing a small amount of catalyst, the solution becomes acidic. Provide a balanced reaction that explains this result.
    e. A polymer can be made when the unknown organic compound is condensed with ethylene glycol ($HOCH_2CH_2OH$). Draw the structure of this polymer.

The Chemistry of Life

Contents

Supplementary Information The online version contains supplementary material available at ▶ https://doi.org/10.1007/978-3-030-90267-4_22.

M. Mosher and P. Kelter, *An Introduction to Chemistry*,
https://doi.org/10.1007/978-3-030-90267-4_22

What We Will Learn in This Chapter

- The genetic code for life is written in the types of molecule residues that make up DNA (Sects. 22.1, 22.2).
- Proteins are polymers of amino acids that have many different functions. (▶ Sect. 22.3).
- The order of the nitrogenous bases in DNA determines the order of amino acids in a protein (▶ Sect. 22.4).
- Proteins have many different functions in the body, from structure to signaling (Sects. 22.5, 22.6).
- There are many other organic molecules that have function within the body, including the carbohydrates and lipids (Sects. 22.7–22.8).
- Biochemical reactions within the body contribute to the metabolism. (▶ Sect. 22.9).
- Chirality is very important in biological systems. Only the correct handedness of a molecule will interact with the appropriate biological molecule (▶ Sect. 22.10).

22.1 Introduction

What is life? That is surely one of the most significant and most challenging of the many questions we ask throughout this book. From a chemical point of view, the answer must be focused on the self-sustaining network of chemical changes within living cells. Life is an adventure of changes. From the moment of conception, through birth, growth, adolescence, maturity, aging and death, we are sustained by an endless process of change. The chemistry of life is the chemistry of these changes and, as you might expect, it is exceptionally complex. The human body is one of the most intricate chemical systems we know, and yet there is a wonderful simplicity at the root of it all. Our introduction to chemistry has given us the basic tools with which to examine the chemistry of life in some detail. At the root of this detail is a chemical we call DNA. This substance serves as the set of instructions that control the chemical processes occurring in living things. To understand the chemical reactions that take place in the human body, we must understand the nature of DNA because the chemistry of life is based on DNA and the chemical species it produces.

22.2 DNA—The Basic Structure

In April 2003, the International Human Genome Sequencing Consortium, announced the successful completion of the Human Genome Project. They had deciphered the sequence in which simple chemical components can be linked up into the genetic material of a human. All of the genes that determine so much about what we are had at last become subject to direct chemical and biological scrutiny. Throughout the world, researchers are now investigating the different versions of genes that underlie many of the differences among us. The information they are discovering is being used to develop new medicines and biotechnologies and to understand more about how life works. What exactly are genes and why are they the focus of such intense scrutiny?

To best uncover the meaning and significance of the gene, we will begin with DNA. **Deoxyribonucleic acid (DNA)** is the name for a series of polymers composed of chemicals called **nucleotides**. In DNA, each nucleotide is itself composed of a phosphate group bonded to a **deoxyribose** sugar group, bonded to one of four **nitrogenous organic bases** (schematically shown in ◘ Fig. 22.1). The four nitrogenous organic bases, shown in ◘ Fig. 22.2, are adenine, guanine, thymine, and cytosine.

Deoxyribonucleic acid (DNA) – A huge nucleotide polymer that has a double-helical structure. Each strand of the DNA polymer complements the other by forming base pairs. DNA provides instructions for the synthesis of proteins and enzymes that carry out the biological activities of the cell.

Deoxyribose – A dehydrated carbohydrate used in the construction of DNA.

Nitrogenous organic base – The portion of a nucleotide where hydrogen bonding holds the strands of DNA together in a duplex.

Nucleotides – Monomers of the nucleic acids. A nucleotide contains a five-carbon sugar, a phosphate, and a nitrogenous organic base.

DNA is a polymer of the four nucleotides bonded together in the manner shown in ◘ Fig. 22.3. The phosphate groups on the nucleotides participate in condensation reactions to form phosphodiester bonds that bind the individual nucleotides into the DNA chain. Recall that condensation reactions result in the coupling of two molecules

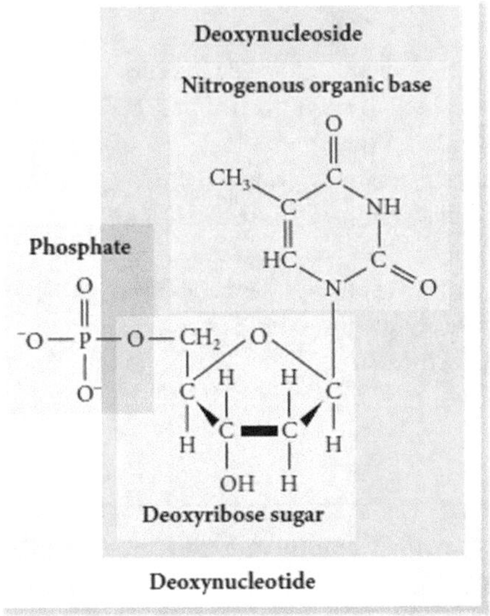

Deoxynucleoside

Nitrogenous organic base

Phosphate

Deoxyribose sugar

Deoxynucleotide

Fig. 22.1 A nucleotide contains a nitrogenous organic base, a deoxyribose sugar, and a phosphate

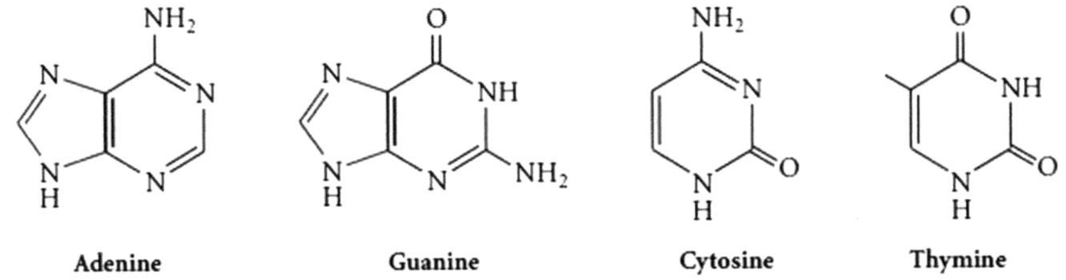

Adenine Guanine Cytosine Thymine

Fig. 22.2 The four nitrogenous organic bases found in DNA

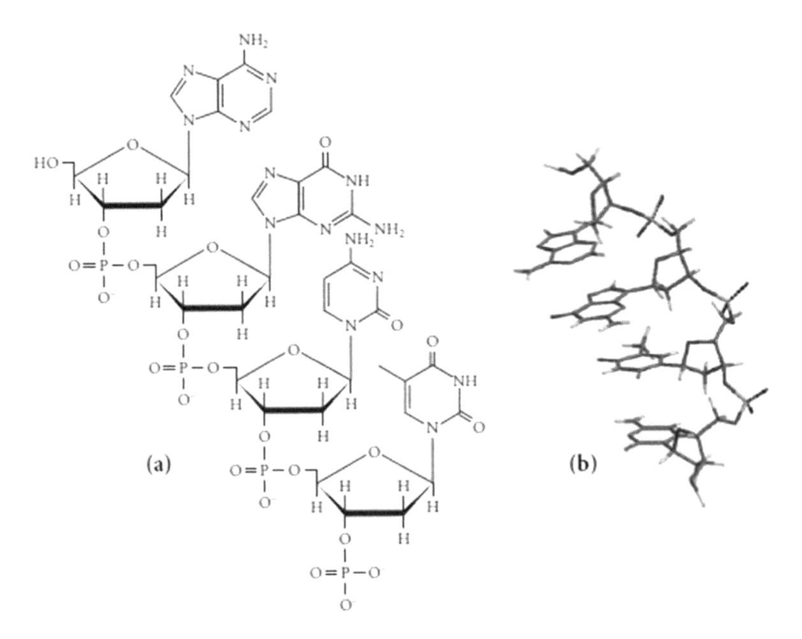

Fig. 22.3 The structure of DNA. **a** A line-drawing of a strand of DNA showing the four nitrogenous bases, **b** a computer drawn image of the same structure showing how a strand of DNA naturally twists when the nitrogenous bases are on the same side of the molecule

with the elimination of a molecule of water (▶ Chap. 21). In DNA, the phosphate's —OH group is lost as the sugar's oxygen adds to the phosphorus atom (Fig. 22.4).

The resulting condensation product is known as single-stranded DNA. However, the DNA that makes up genes is more complex in a very beautiful and significant way. This occurs when a second strand of DNA (up-

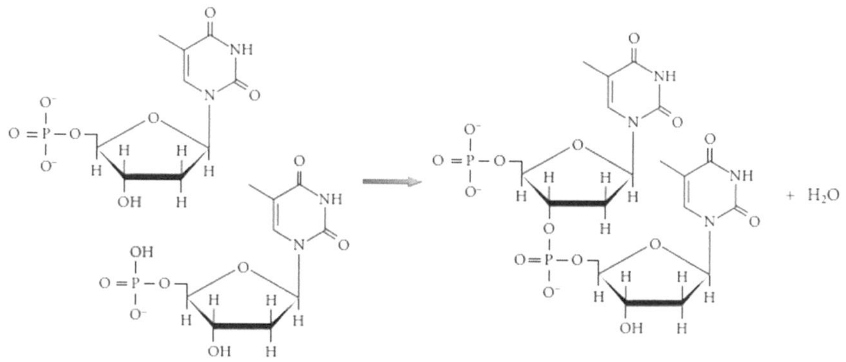

■ **Fig. 22.4** The condensation of two nucleotides and the elimination of water form the backbone of the DNA molecule

■ **Fig. 22.5** Base pairs in DNA, looking down the axis of the DNA molecule. Note the proximity of the oxygen atom to the amine group on the opposite base, and that of the nitrogen to the opposite amine

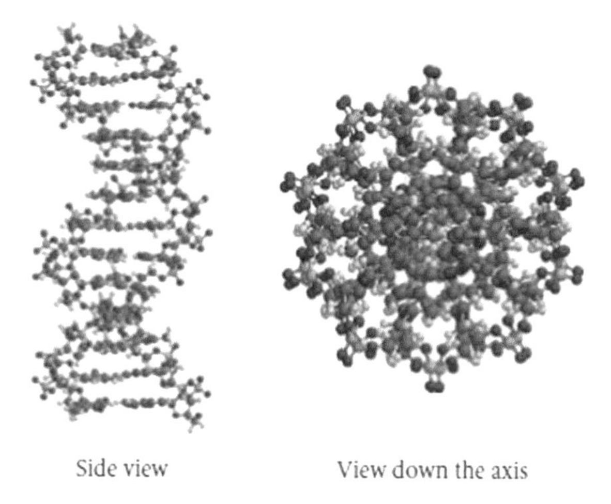

■ **Fig. 22.6** The double helix of DNA

side down in relation to the first strand) interacts with the first strand. Why does DNA occur as pairs of strands? Double-stranded DNA is much more stable than a single strand, because the nitrogenous bases are able to participate in hydrogen bonds with each other. The hydrogen bonding results in a **base pair**. As a consequence of their relative size, the amount of space available between the two DNA strands, and the types of hydrogen bonds in the bases, adenine always base-pairs with thymine, and cytosine always base-pairs with guanine (■ Fig. 22.5). We say that these pairs of bases are complementary; that is, they pair in a reciprocal fashion with one another.

Base pair – A pair of nucleic acids that are intermolecularly bound to each other through noncovalent forces of attraction, for example, the combination of A and T or of G and C in a DNA duplex.

To improve the hydrogen bonding, the two strands of DNA wind around each other to form a base-paired double helix, as shown in ■ Fig. 22.6. This structure had eluded scientists for quite some time until James Watson and Francis Crick obtained the 1953 X-ray diffraction photograph shown in ■ Fig. 22.7. The image of this diffrac-

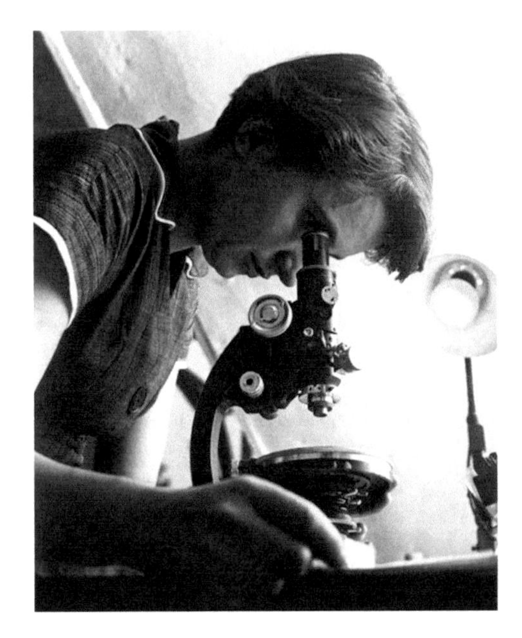

Fig. 22.7 Rosalind Franklin (1920–1958) helped solve the double-helical structure of DNA. Her contributions to this discovery were never acknowledged during her life. She died of cancer at a very early age. *Source* ► https://upload.wikimedia.org/wikipedia/commons/f/fd/Rosalind_Franklin_%28retouched%29.jpg

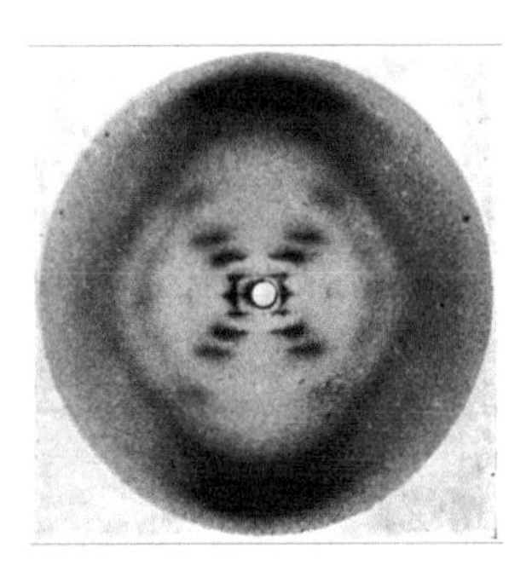

Fig. 22.8 An X-ray diffraction pattern of DNA. *Source* ► https://upload.wikimedia.org/wikipedia/commons/d/db/Fig-1-X-ray-chrystallography-of-DNA.gif

tion pattern was made by the chemist Rosalind Franklin and given to Watson and Crick without her knowledge. It turned out to be the key piece of data used to determine the structure of the double helix. In fact, the structure of the DNA double helix has become an icon of science (Fig. 22.8). Why is this structure so important? Within the double helix, we find the secret of how life is able to reproduce—and also the secret of how mere chemicals can contain the coded information that controls the structure and activities of all living things.

Example 22.1—What Exactly is DNA?

We often read and hear the term molecules of DNA. Scientists use the phrase routinely, and "DNA molecules" have become such a hot news story that they are referred to regularly in magazines, newspapers, on-line, and on TV. But when we closely examine the structure of DNA as we will do in this chapter, we will see that the phrase a molecule of DNA is not strictly accurate. Why not?

Asking the Right Questions
What is the definition of a molecule, and what other types of particles participate in chemical processes, other than molecules?

Solution
Several factors introduce complications here. First, the negative charges on the phosphate groups mean that DNA, in the form in which it exists in living things, is a multi-charged ion, not a molecule at all. If hydrogen ions combined with all these negative charges, it would become a molecule, but this is not the structure we observe in living things. Second, when people speak of "DNA molecules" they are often referring to double-helical DNA, rather than to the single-stranded form. The DNA double helix is held together by largely noncovalent forces from

hydrogen bonds, so it is really a complex formed from two intertwined molecules (or multi-charged ions). It cannot properly be described as a molecule at all.

Is Our Answer Reasonable?

Although we examined the definition of a molecule in this example, there is more to it. Many people, including some scientists, use the term "molecule" as a general term which actually includes atoms and ions. For example, they may talk of chemistry "at the molecular level" but they are not excluding atoms and ions from their discussions at that level. We need to appreciate that scientific terms are sometimes used loosely to describe something somewhat different from their formal definition when used strictly.

What If…?

…we had been asked to explain why deoxyribonucleic acid (DNA) has "acid" in its name?

(*Ans*: *The phosphate groups of the DNA backbone have acidic OH groups, although in DNA within the cell these have generally lost their hydrogen ions to form* O^- *groups so the DNA in its physiological state could be more rationally described as the conjugate base of an acid.*)

Practice 22.1

Suppose that a new base, shown in the margin, was discovered by scientists. To which of the four nitrogenous organic bases would this new base be most likely to form a stable base pair?

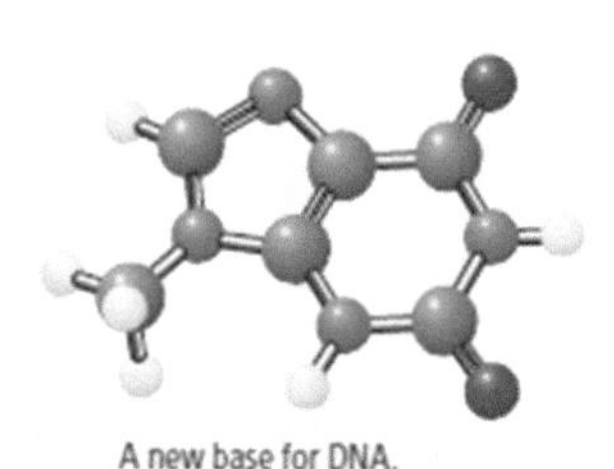

A new base for DNA.

22.2.1 DNA Replication—The Secret of Reproduction

One of the most important characteristics of living creatures (animal or plant) is that they make more of themselves. In other words, they "reproduce." Reproduction in this sense depends on the ability of double-helical DNA to copy itself or "replicate."

Because the bases form complementary pairs in a double strand of DNA, the sequence of nucleotide bases in one strand can be readily determined by examining the other strand of DNA. In other words, the A on one strand is paired with a T on the other, each T with an A, each G with a C, and each C with a G. This means that within a reproducing cell, the double helix can unravel and separate into individual strands. Then, each strand can serve as the template on which a new complementary strand assembles, as shown in ◘ Fig. 22.9. The chemical reactions involved in this DNA **replication** are catalyzed by molecules within the cell known as enzymes. However, the structure of the existing DNA strands determines the sequence of the new DNA strands. Nucleotides carrying the appropriate bases are the only ones that can bind to the existing strand in a manner that allows the enzymes to link them up into a new DNA strand. The result of this process is two identical strands of DNA, one for the original cell and one for the new cell when it is formed.

Replication – The process by which DNA is duplicated prior to cell division.

Example 22.2—Using the Template

If the DNA strand shown below serves as the template strand during DNA replication, what will be the sequence of the new double-helical DNA that results?

AATTGCGGGTCCGACC

Asking the Right Questions

What bases of DNA can pair together? These are the rules that govern the replication of DNA.

Solution

Remember the rules of base-pairing. Only two types of base pairs are allowed: AT and GC (either way round), so the double-helical DNA will have the following sequence:

AATTGCGGGTCCGACC
TTAACGCCCAGGCTGG

Is Our Answer Reasonable?

We followed the rules of base pairing, so our answer is reasonable.

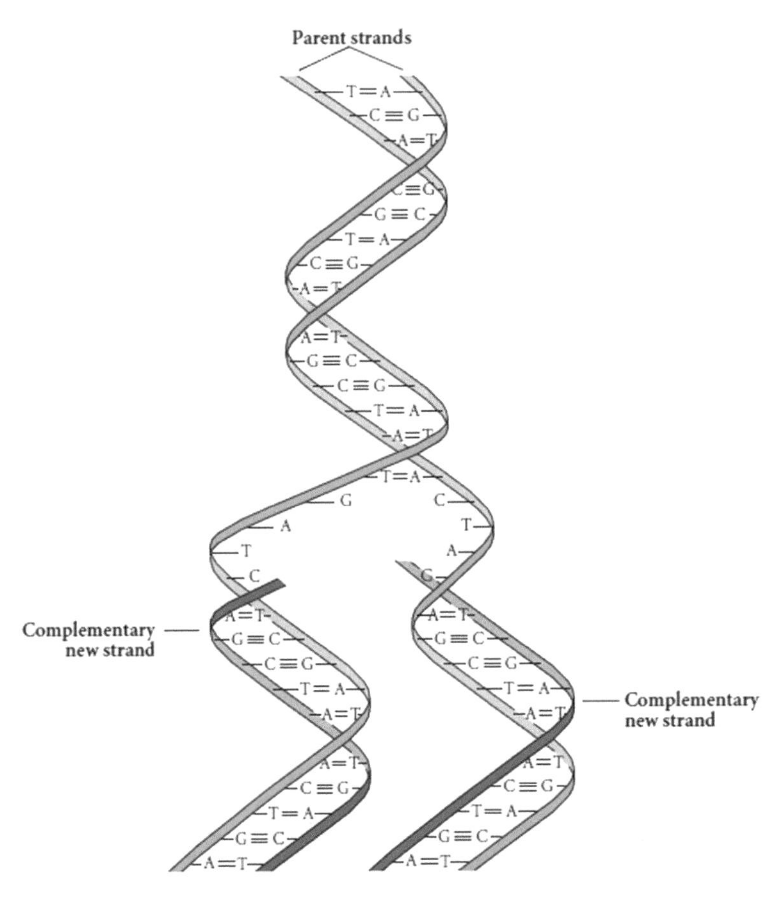

Fig. 22.9 Single-stranded DNA can serve as a template for the creation of a new strand

What If…?

…the bonds between DNA strands had been covalent bonds? What do you think may have been the implications for the suitability of DNA as the genetic material of life?

(*Ans: covalent bonds would link the strands together so that they could not be replicated in the normal fashion. Because each strand can "recognize" its partner, this allows the system to identify if any damage has occurred to the DNA.*)

Practice 22.2

A small piece of one of the strands of DNA is added to a single strand of DNA. Indicate the correct alignment of the small piece with the long strand.

...AAAATGCTGGCATAGCGTTCCAGATACGGACTGACTGC...

CTATGC

DNA is located inside cells within a structure known as the nucleus. It is a beautiful chemical structure with the ability to act as a template on which new copies of itself are formed. But what makes DNA so important to life? The answer is that it carries coded information within its base sequence—information that can be decoded to specify which molecules are made in a cell. These molecules, known as **proteins**, then perform most of the chemical activities that actually sustain life.

Protein – A large polymer of amino acids. May contain one or more polypeptides and often involves complex and intricate folds of the chain.

22.3 Proteins

Proteins are polymers that are formed when molecules called amino acids undergo condensation reactions. The amine on one amino acid forms an **amide bond** with the carboxylic acid on another amino acid. This process is repeated many times as the protein is formed. Some of the important amino acids found in living cells are shown in

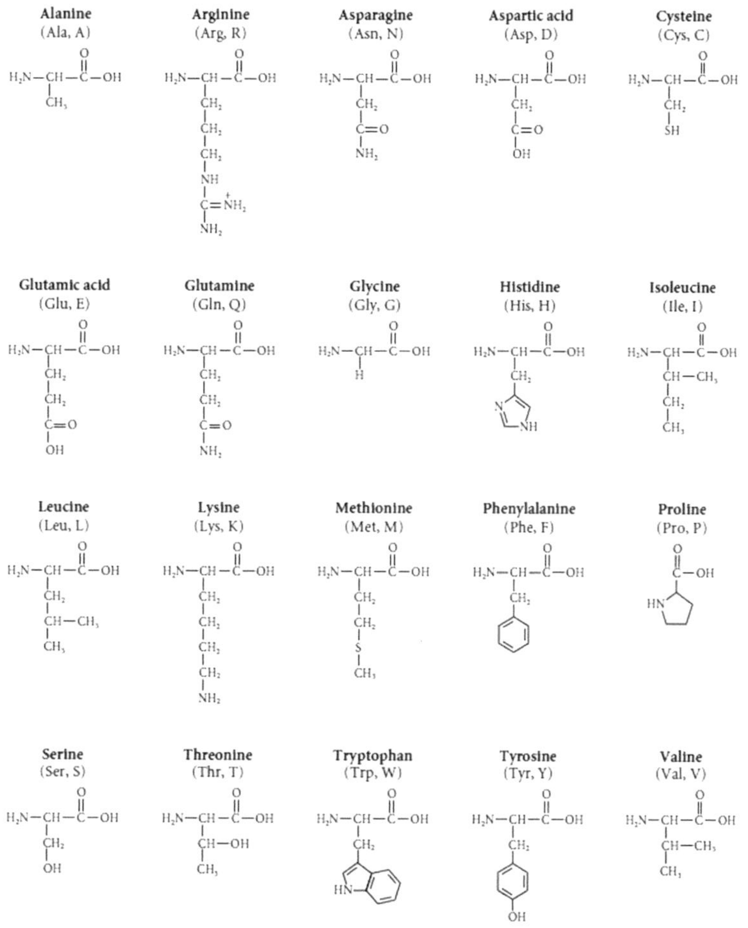

Alanine (Ala, A) | Arginine (Arg, R) | Asparagine (Asn, N) | Aspartic acid (Asp, D) | Cysteine (Cys, C)

Glutamic acid (Glu, E) | Glutamine (Gln, Q) | Glycine (Gly, G) | Histidine (His, H) | Isoleucine (Ile, I)

Leucine (Leu, L) | Lysine (Lys, K) | Methionine (Met, M) | Phenylalanine (Phe, F) | Proline (Pro, P)

Serine (Ser, S) | Threonine (Thr, T) | Tryptophan (Trp, W) | Tyrosine (Tyr, Y) | Valine (Val, V)

Fig. 22.10 Twenty common amino acids. The names of the amino acids are also shown with their three-letter abbreviation and one-letter code

Fig. 22.10. Note that all these amino acids contain very similar arrangements of the carboxylic acid and amine functional groups.

Amide bond – A bond constructed from the condensation of an amine and a carboxylic acid.

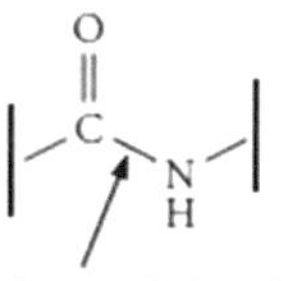

The amide bond

The difference between them can be found in the group of atoms attached to the same carbon as the amine. These groups are often referred to as **residues** because they are what remains after the protein is broken up into its individual amino acids. Over 20 different residues are common within your body, and each has been given a specific name.

Residue – The portion of an amino acid that differs from other amino acids.

Proteins are made in living cells by the sequential addition of amino acids to a lengthening **polypeptide** chain. This sequence of the protein constitutes the first level of structure of the protein. As indicated in Fig. 22.11, it is often called the **primary structure** of the protein. Just like every individual person, each type of protein within a living cell is unique thanks to its amino acid sequence. The sequence of the amino acids determines the resulting shape of the protein and the resulting function of the protein.

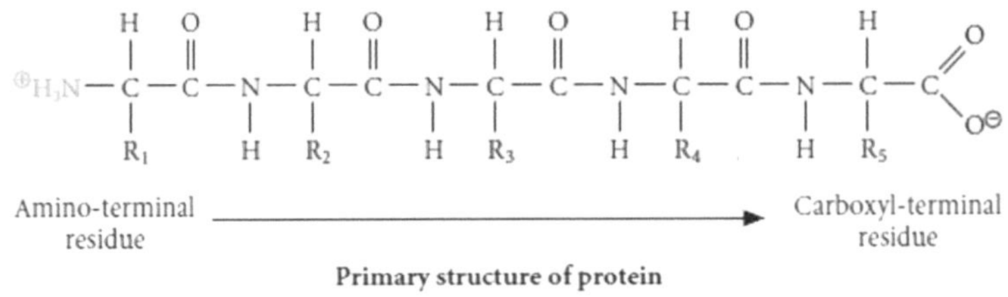

$^{\oplus}H_3N-C(H)(R_1)-C(=O)-N(H)-C(H)(R_2)-C(=O)-N(H)-C(H)(R_3)-C(=O)-N(H)-C(H)(R_4)-C(=O)-N(H)-C(H)(R_5)-C(=O)O^{\ominus}$

Amino-terminal residue → Carboxyl-terminal residue

Primary structure of protein

Fig. 22.11 Primary structure of a protein. The primary structure is the order of the amino acid monomers in the strand of the protein

Polypeptide – A polymer of amino acids linked by amide bonds.

Primary structure – The sequence of amino acids within a protein.

We can now appreciate that both DNA and proteins are chemicals whose structure is determined by the sequence in which monomer units are bonded into polymers. Each DNA polymer has a unique base sequence (also known as its nucleotide sequence). Each protein has a unique amino acid sequence. The key to understanding how DNA can specify which proteins a living thing contains is to understand how the base sequence of DNA can specify the amino acid sequence of a protein. This is achieved by the operation of a simple chemical code.

22.4 How Genes Code for Proteins

A **gene** is a section of DNA that encodes a specific protein molecule. This means that the base sequence of the gene determines the amino acid sequence of the resulting protein. The whole process of converting the genetic code into a particular protein is known as gene expression, which occurs in two distinct phases: **transcription**, in which an RNA copy of the gene is made, and **translation**, in which the RNA copy directs production of the protein.

Gene – A specific region of the DNA polymer that codes (carries instructions) for a single protein.

Transcription – The process by which mRNA is synthesized from a gene.

Translation – The process by which a protein is synthesized from mRNA.

What is RNA? **Ribonucleic acid (RNA)** is a substance that bears a close resemblance to DNA. In fact, there are only two main differences between DNA and RNA (see Fig. 22.12). The sugar in RNA is ribose, which carries one more oxygen atom than the deoxyribose found in DNA. Also, in RNA the nitrogenous base uracil, shown in Fig. 22.13, is found in place of the thymine of DNA. Uracil can participate in an A–U base pair as shown in Fig. 22.14, in the same way that thymine can participate in an A–T base pair. The result is that RNA can form base pairs with a complementary strand of DNA.

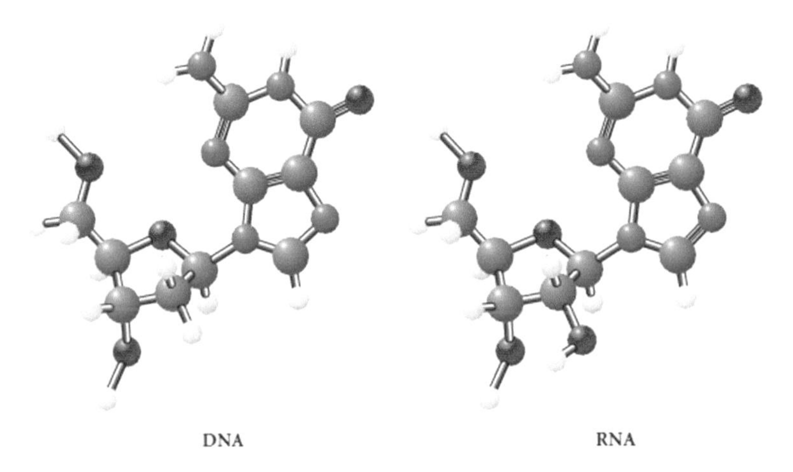

Fig. 22.12 The difference between DNA and RNA lies in the ribose sugar

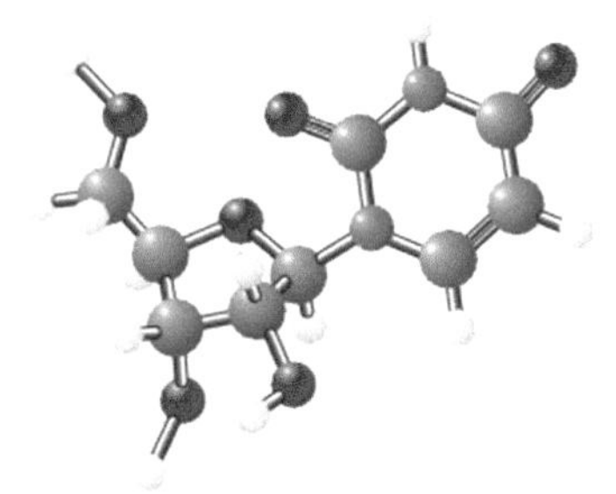

Fig. 22.13 The different monomer used in RNA, shown here without the attached phosphate group, is similar in structure to thymidine

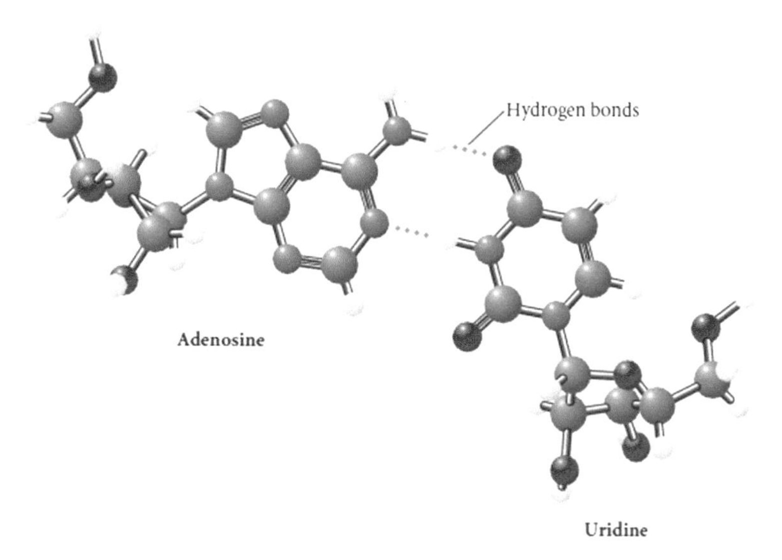

Fig. 22.14 An A–U base pair. Note that the hydrogen bonding is similar to that found in an A–T base pair. Uracil and adenine attached to a ribose sugar unit (without a phosphate) are known as uridine and adenosine, respectively

Ribonucleic acid (RNA) – A nucleotide polymer that contains the information from a single gene and either transfers that information to the ribosomes (mRNA) or recognizes and constructs the protein product (tRNA).

Nucleoside
Nitrogenous organic base
Phosphate
Ribose sugar
Nucleotide

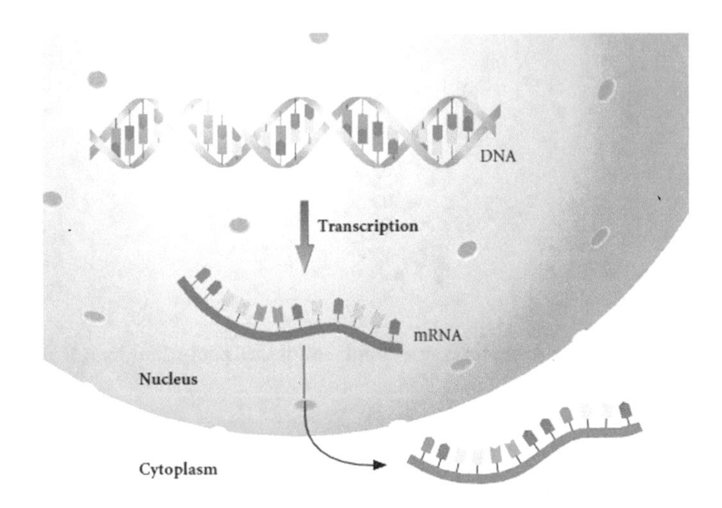

■ **Fig. 22.15** mRNA is made in the nucleus and then transferred to the cytoplasm for protein synthesis

22.4.1 Transcription—Making the Message

The process of gene transcription occurs in a very similar way to DNA replication. Inside the nucleus of the cell, the DNA unwinds into the individual strands of DNA, and a new, complementary strand of RNA is made. The complementary strand is manufactured using enzymes via condensation reactions in much the same manner as in the manufacture of DNA. The single-stranded RNA copy of a gene that is produced in transcription is known as **messenger RNA (mRNA)**. This copy carries the genetic "message" from the nucleus to the ribosomes (see below) in the cytoplasm, where the message is used (decoded) to make protein. mRNA gets its name from the message held within the genes to the site of protein synthesis, as illustrated in ■ Fig. 22.15.

Messenger RNA (mRNA) – A short-lived single strand of ribonucleic acid that transfers the information from a gene to the ribosomes where a protein is constructed.

22.4.2 Translation—Making Proteins

Once at the site of protein synthesis, the mRNA is read by another class of RNA molecules known as **transfer RNA (tRNA)**. Each tRNA molecule includes a set of three nitrogenous bases that can form base pairs with three sequential bases of mRNA. Each set of three sequential bases within the mRNA is known as a **codon**. The codon binds to the complementary **anticodon** on the tRNA. What is the purpose of the tRNA? Each tRNA also carries a specific amino acid at its amino acyl binding site. When the tRNA anticodon binds to the codon on the mRNA, the amino acid on that tRNA becomes incorporated into the growing protein chain. If the anticodon and codon are not complementary, the tRNA leaves without donating its amino acid to the protein. The code, indicating the relationship between each specific codon and amino acid is called the genetic code and is summarized in ■ Table 22.1. The entire process of translation is outlined in ■ Fig. 22.16.

Codon – A three-nucleic-acid sequence on an mRNA that is translated into a particular amino acid during protein synthesis.

Anticodon – The three-nucleic-acid-sequence on a tRNA polymer that recognizes and binds to the codon on an mRNA polymer.

Transfer RNA (tRNA) – A nucleic acid polymer that recognizes the codon on mRNA and adds the corresponding amino acid to the growing protein.

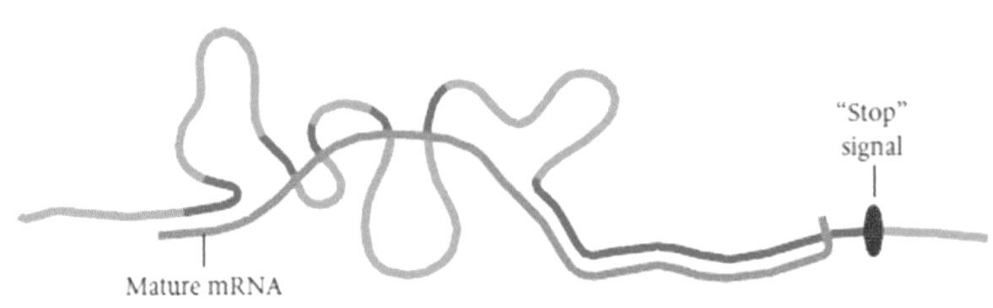

Table 22.1 The Genetic Code, The three-letter codes listed are mRNA codons. For example, AAA is the mRNA codon for the amino acid lysine

	U		C		A		G		
U	UUU UUC UUA UUG	phenyl-alanine Leucine	UCU UCC UCA UCG	Serine	UAU UAC UAAUCG	Tyrosine stop codon	UGU UGC UGA UGG	Cysteine Stop codon	U C A G
C	CUU CUC CUA CUG	Leucine	CCU CCC CCA CCG	Proline	CAU CAC CAA CAG	Histidine Glutamine	CGU CGC CGA CGG	Arginine	U C G A
A	AUU AUC AUA AUG	isoleucine methionine, initiation codon	ACU ACC ACA ACG	Threonine	AAU AAC AAA AAG	Asparagine Lysine	AGU AGC AGA AGG	Serine Arginine	U C A G
G	GUU GUC GUA GUG	valine	GCU GCC GCA GCG	Alanine	GAU GAC GAA GAG	Aspartic acid Glutamic acid	GGU GGC GGA GGG	Glycine	U C A G

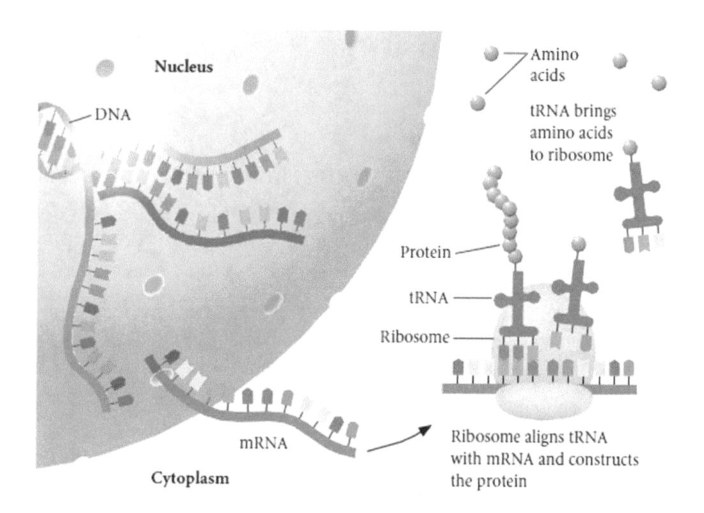

Fig. 22.16 The overall process of translation

The mRNA contains a special "start" sequence that identifies the beginning of a protein. In fact, every protein that is made begins with the amino acid methionine. From that point onward, the amino acids are identified by each codon until the final "stop" codon is reached. Sometimes all of the mRNA is translated into a protein. Some genes make an mRNA that contains a lot of extraneous nucleotides.

To hold the molecules close together and to assist in catalyzing the process of adding amino acids into the protein, the process of translation takes place on giant assemblies of protein and RNA known as **ribosomes** (see Fig. 22.17). As the ribosome moves along an mRNA molecule, the appropriate tRNAs are able to bind to special sites on the ribosome, allowing enzymes to link the amino acids they carry into a new protein chain. The end result is the translation of the base sequence of the gene (via the mRNA) into the amino acid sequence of the protein.

Ribosomes – Large molecular structures within the cytoplasm of a cell that aid the construction of a protein from mRNA, tRNA, and the amino acid building blocks.

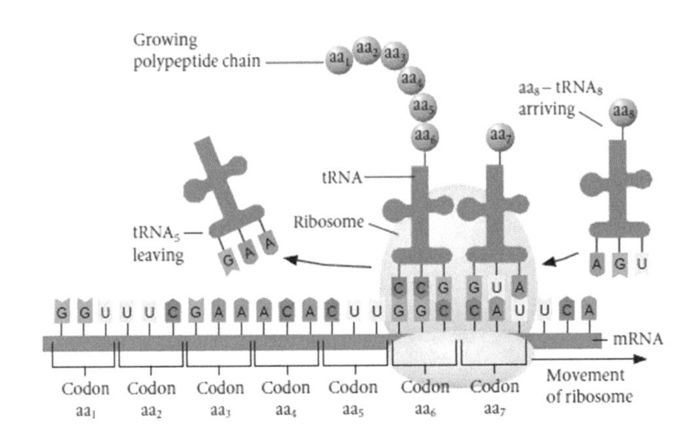

Fig. 22.17 The ribosome slides along the mRNA. tRNAs then bring specific amino acids to the ribosome and allow the protein chain to be constructed. In this figure, the amino acids (aa) and the tRNAs are numbered to indicate their position

Example 22.3—Examining the Genetic Code

Examine the genetic code shown in Table 22.1. Do all codons encode amino acids? Does each codon specify a unique amino acid?

Asking the Right Questions

Because there are four different bases and three positions, we can figure out the possible number of arrangements of these bases into the three positions.

Solution

There are 64 possible codons—different ways to arrange the four bases into sets of three. There are only about 20 amino acids found in proteins, so there must be more codons available than amino acids. Examining the genetic code reveals that each amino acid can be encoded by several alternative codons. It also reveals that three codons act as "stop" signals, indicating the point at which the synthesis of a protein chain should end.

Are Our Answers Reasonable?

There must be at least 20 codons available to encode 20 amino acids, but with 64 possible codons it makes perfect sense that each amino acid should be encoded by more than one codon.

What If…?

…codons were only two bases long, how many different amino acids could be incorporated in proteins, if at least one codon had to be a stop codon.

(*Ans: If you draw up a code table with only first and second letters required you can see that there would 16 possible combinations, which would be enough to encode only 15 amino acids plus one stop codon.*)

Practice 22.3

What is the primary structure of a protein made using the following mRNA sequence?

AUGUGGCCAAAAUUGGACAUGUUCGACUAG

The human genome contains between about 20,000 and 30,000 genes. These genes are able to encode an even greater number of proteins, because the RNAs that are originally made from the genes can be edited to make many different proteins in a kind of enzymatic "cut-and-paste" process. The way in which genes encode proteins is an astonishing demonstration of the power of chemistry to sustain the complex processes that underpin all life. To understand more fully just how powerfully genes influence the chemistry of life, we need to look at the proteins encoded by our genes and investigate the things these proteins can do.

22.4.3 Protein Folding

As soon as a protein chain begins to be formed, it starts to fold into a specific three-dimensional conformation. The folding process is governed principally by noncovalent interactions among the amino acids themselves, and also among the amino acids and the water molecules surrounding the protein. Localized regions within a polypeptide chain that fold in a particular way are examples of a protein's **secondary structure**. The most significant secondary structures are the alpha helix (α helix) and the beta pleated sheet (β sheet), shown in Fig. 22.18. The α helix forms as a consequence of hydrogen bonds between the N–H and C=O groups of the polypeptide chain, holding the chain in the form of a helix. The β sheet is also held together by hydrogen bonds between N–H and C=O groups, but with the hydrogen bonding occurring between neighboring portions of a polypeptide chain.

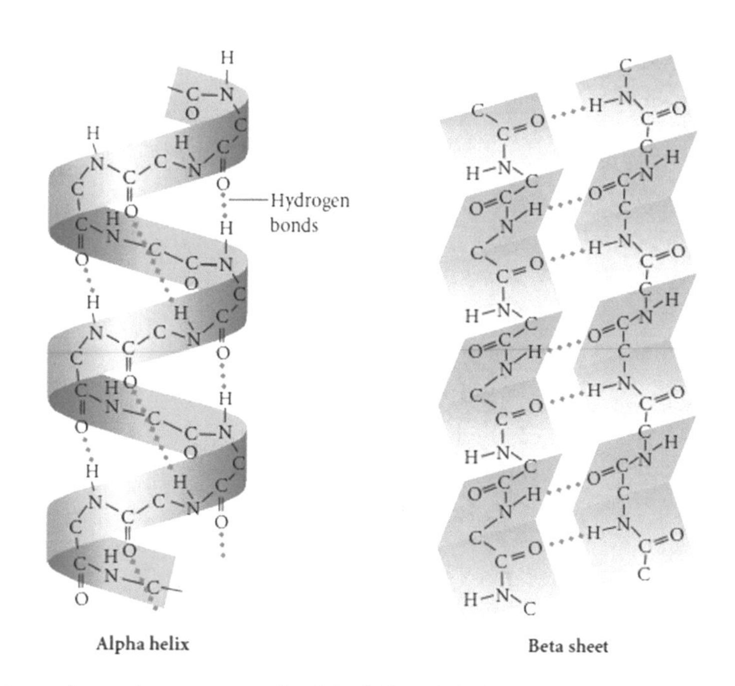

Fig. 22.18 Two common types of secondary structures: the alpha helix and the beta sheet. These structures are held together by hydrogen bonding (shown as dotted lines)

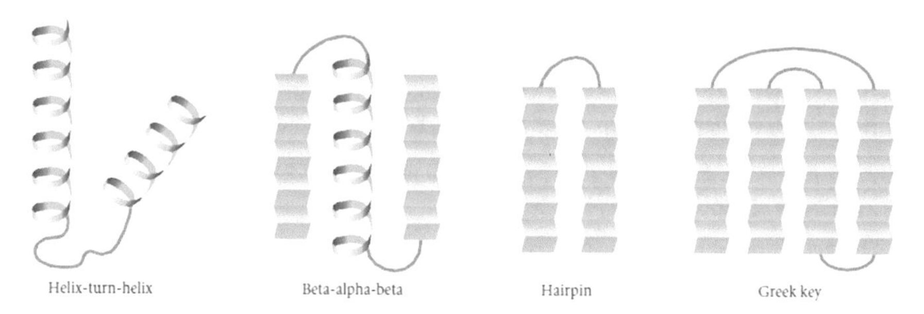

Fig. 22.19 Some common tertiary structures. These structures are held in place by intermolecular forces of attraction such as London forces, dipole–dipole interactions, and hydrogen bonding

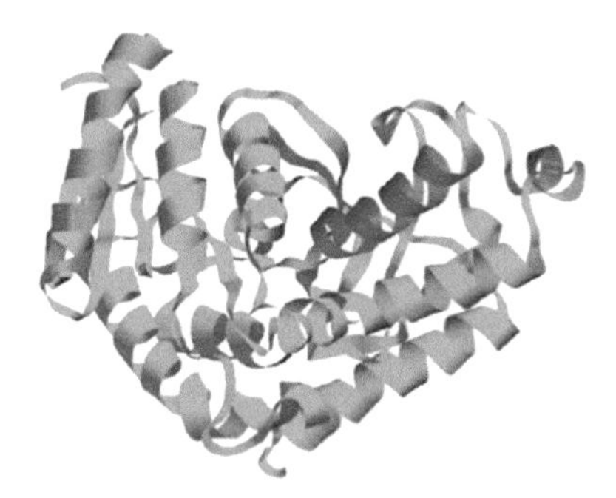

Fig. 22.20 Lactate dehydrogenase is a globular enzyme involved in the biochemical process of harvesting energy from sugars

Secondary structure – The specific folds within a polypeptide chain. Common secondary structures include the alpha helix and the beta pleated sheet.

Regions of specific secondary structure are linked by turns in the polypeptide chain, and by less ordered structures, to form the overall three-dimensional conformation of a protein. The folding of the secondary structures into a three-dimensional conformation is known as the protein's **tertiary structure**. Just as is true of α helices and β sheets, there are many common tertiary structures. Some of these are shown in Fig. 22.19. However, many proteins have what can loosely be described as a "globular" tertiary structure; see the example shown in Fig. 22.20.

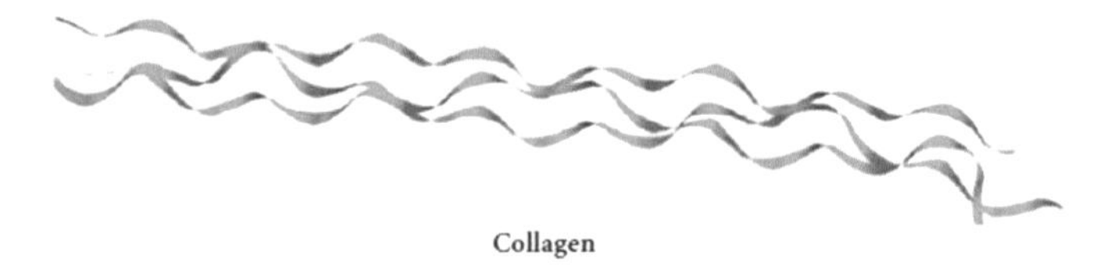

Fig. 22.21 Collagen is a fibrous protein

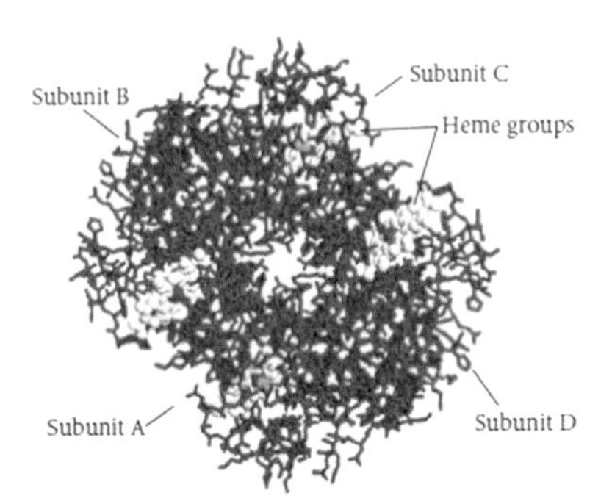

Fig. 22.22 Hemoglobin. There are four subunits, each with its own iron-containing heme, that come together to form hemoglobin. The subunits are held together to form the quaternary structure of the protein

This is the general three-dimensional shape found for most of the proteins that act as enzymes. Other proteins have linear tertiary structures, largely composed of one or more extended helices or sheets. Many such proteins form strong fibers that contain several protein molecules intertwined. For example, the protein **collagen**, shown in Fig. 22.21, gives strength to connective tissues such as tendons, ligaments, and bones. The structural strength of collagen is a result of three helical proteins wound around one another to form a triple helix.

Collagen – A structural protein that exhibits a linear tertiary structure formed from intertwined alpha helixes. It is found in connective tissues in the body.

Tertiary structure – The way in which secondary structures fold together within a polypeptide chain.

When several polypeptide chains are held together by forces of attraction, as in collagen, we say that a protein with **quaternary structure** has been formed. Many globular proteins also possess quaternary structure. For example, hemoglobin, the protein that carries oxygen to your muscles, is made of four polypeptide chains that have folded around each other as shown in Fig. 22.22. Not all proteins have quaternary structure. For example, myoglobin, the oxygen storage protein that resides in your muscles, consists of only a single polypeptide chain. The precisely folded structure of proteins can be disrupted, or **denatured**, by heat or by variations in the chemical surroundings, including changes in pH and in the concentration of salt ions. Denaturation reduces or destroys a protein's biochemical activity, depending on how severe it is. This explains, for example, why the protein albumin in an egg comes out of solution and turns white upon heating.

Denature – To destroy the biochemical activity of a protein.

Quaternary structure – The way in which two or more polypeptide chains have folded to make a protein.

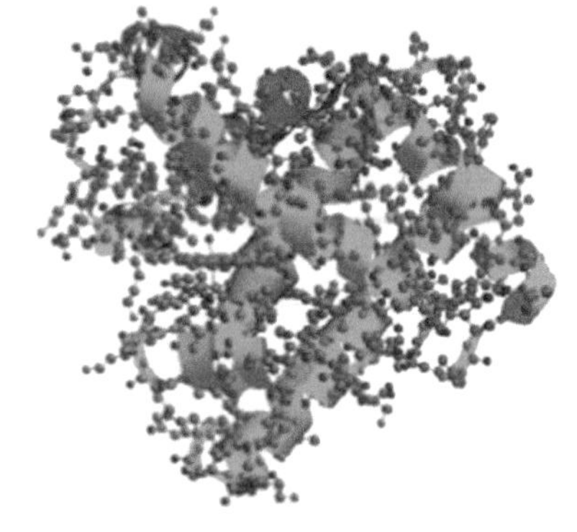

▶ **Here's What We Know So Far**

- DNA is a polymer of nucleotides made from two complementary single strands wrapped around each other to form a double helix.
- Proteins are polymers of amino acids linked through an amide bond. They contain primary, secondary, tertiary, and sometimes quaternary structure.
- Proteins are made by translating the genetic code from mRNA. tRNA supplies the specific amino acids to the growing polypeptide chain. The construction occurs at the ribosomes within the cytoplasm of a cell.
- Primary structure is the sequence of amino acids that make up the protein.
- Secondary structures are the specific regions of a polypeptide chain that fold into an α helix or a β sheet.
- Tertiary structures result from folding of the secondary structures within a polypeptide chain into a three-dimensional shape. They can be globular or linear in arrangement.
- Quaternary structure results when two or more polypeptide chains are folded into a specific shape. Not all proteins have quaternary structure. ◀

22.5 Enzymes

The covalent bond that links glucose and fructose together in a molecule of sucrose can be hydrolyzed by water, although the reaction is quite slow. In the laboratory, we can speed the reaction by adding a little acid to an aqueous solution of sucrose. The reaction can be monitored, and the rate of the catalyzed reaction measured. Why is this reaction important? In the body, much of the "fuel" that we ingest is sucrose. This sugar is broken down into its components, glucose and fructose. As we noted in ▶ Chap. 17, both of the sugars are metabolized to provide energy to operate other biological processes and sustain life. Because these molecules are vital to our existence, any reaction that makes them, or metabolizes them, needs to be very rapid. Therefore, almost every chemical reaction within our cells is quite fast; some are so fast that they must pause and wait for molecules to arrive at the reaction site. What type of compound makes these reactions fast? Specific types of proteins in the body, known as enzymes, catalyze these reactions.

A few key principles govern enzyme catalysis. This process is summarized in ◻ Fig. 22.23.

- Enzymes contain specific sites to which certain chemicals can bind.
- The structure of the sites enhances the binding of specific molecules and ions through the same type of intermolecular forces of attraction that we discussed earlier in this text.
- One of the binding sites on an enzyme includes the **active site** where the **substrate** binds.
- The enzyme next catalyzes the reaction of the substrate and produces the product, which is then released from the enzyme.

Active site – The location on an enzyme where the reaction of the substrate is catalyzed.

Substrate – A reactant molecule that binds to an enzyme.

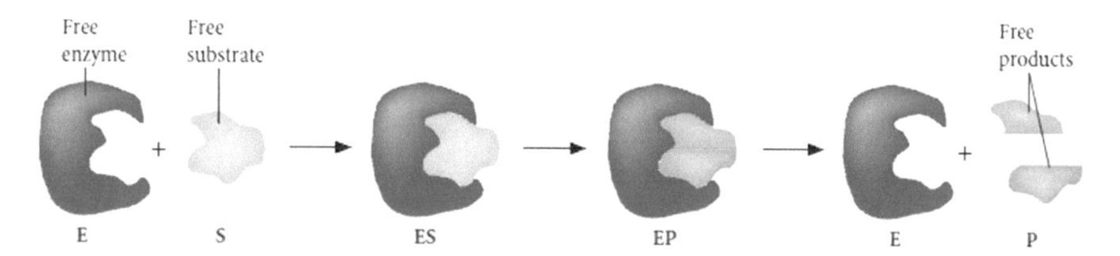

◻ **Fig. 22.23** Substrates bind to the active site of an enzyme. The enzyme then catalyzes the reaction and releases the products. The enzyme is then primed for the next reaction

22.5.1 Cofactors, Coenzymes, and Protein Modifications

Regulation and control of the activity of an enzyme is often necessary. Imagine, for instance, what would happen if all of the sucrose you just consumed at lunch were immediately converted into energy. That energy would be used up quite quickly in the biological processes that keep you alive. After a very short time, your body would be out of energy, and the basic reactions that keep you functioning would stop. This suggests that the reactions within a cell must be regulated—turned on and off—so that they are performed only when needed. This regulation often involves chemical modification of the enzyme. The major modifications are as follows:

Regulation – The control of the function of a gene or enzyme.

- Incorporation of specific metal ions at appropriate sites, to form metalloproteins, in which metal ions are tightly bound within the protein structure, or metal-activated proteins, which are activated by loosely bound metal ions picked up from the surrounding solution
- Incorporation of loosely bound small chemical groups called coenzymes at specific sites within the protein
- Incorporation of tightly bound prosthetic groups, often held to the protein by covalent bonds
- Covalent linking of phosphates to specific sites to form phosphoproteins
- Covalent linking of carbohydrates at specific sites to form glycoproteins
- Covalent bonding of lipids at specific sites to form lipoproteins

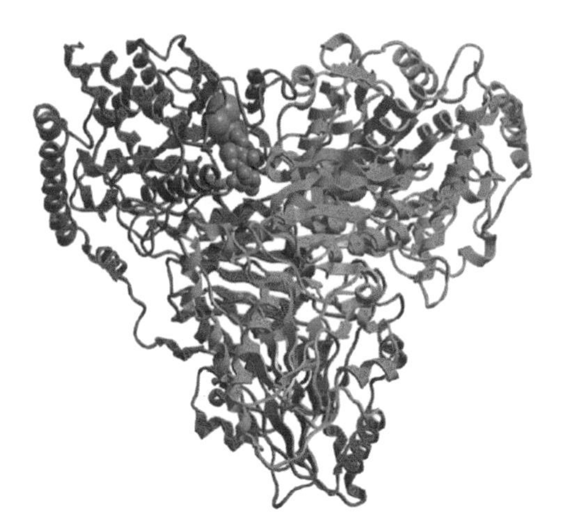

The need for many enzymes and other proteins to be modified in these ways explains many of our nutritional needs for specific minerals and vitamins. For instance, much of the iron that we must consume in our diet is needed to provide iron ions for incorporation into metalloproteins such as hemoglobin. Many of the vitamins we must consume in tiny amounts are needed to form the coenzymes that combine with specific enzymes and allow them to function.

22.5.2 The General Classes of Enzymatic Reactions

Enzymes catalyze many thousands of different chemical reactions. Fortunately, these reactions can be classified into just six categories on the basis of the general type of reaction that is being catalyzed.

- Oxidoreductases are enzymes that catalyze redox reactions.
- Transferases are enzymes that catalyze the transfer of groups from one molecule to another.
- Hydrolases are enzymes that catalyze hydrolysis reactions.
- Lyases are enzymes that catalyze elimination reactions that remove hydrogen atoms or functional groups and form alkenes (C=C) in the substrate.
- Isomerases are enzymes that catalyze the interconversion between isomers.
- Ligases are enzymes that catalyze the formation of new bonds linking substrates together.

Example 22.4—What is It Doing?

Examine the following reaction and classify the enzyme's activity into one of the six types of reactions common among enzymes.

Benzyl alcohol $\xrightarrow{\text{Enzyme}}$ Benzoic acid

Asking the Right Questions

What are the key features of each enzyme reaction type? You need to remind yourself of this first.

Solution

At first glance, it appears that the substrate (benzyl alcohol) might be converted into the product (benzoic acid) by a lyase, because there is a double bond formed in the product. However, this double bond is not an alkene. Further examination of the oxidation states of the carbon atoms in the molecule reveals that one of them is undergoing oxidation. Therefore, this reaction is catalyzed by an oxidoreductase.

Is Our Answer Reasonable?

Once the reactions has been identified as an oxidation then it follows that that the enzyme must be an oxidoreductase, so the answer is reasonable.

What If…?

…benzyl alcohol was acted on by another oxidoreductase enzyme, what other product do you think could be produced?

(*Ans: an oxidoreductase must catalyse an oxidation or a reduction. Oxidation of the alcohol group could produce an aldehyde if it did not proceed all the way to a carboxylic acid. You might also suggest that reduction of the* CH_2OH *group could, in principle, convert it to a methyl group.*)

Practice 22.4

Indicate what type of enzyme is responsible for each of these transformations.

Cyclohexane $\xrightarrow{\text{Enzyme}}$ Benzene

Benzene $\xrightarrow[\text{Enzyme}]{\text{Acetal CoA}}$ $C_6H_5C(=O)CH_3$

$H_2N{-}CH(CH_3){-}C(=O){-}OH + H_2N{-}CH(CH_2SH){-}C(=O){-}OH \xrightarrow{\text{Enzyme}} H_2N{-}CH(CH_3){-}C(=O){-}N(H){-}CH(CH_2SH){-}C(=O){-}OH$

NanoWorld/MacroWorld—The Big Effects of the Very Small: Vitamins and Disease

Vitamins – are small molecules that we need for proper health; most either are not normally synthesized in our bodies or are made in insufficient amounts. We humans obtain most of our needed vitamins from food. The exception to this is that our bodies can make vitamin D and vitamin K. In all, we need 13 vitamins to sustain our life: vitamins A, C, D, E, and K and the B vitamins (thiamine, riboflavin, niacin, pantothenic acid and biotin, vitamin B6, vitamin B12, and folate). Each of these vitamins plays an interesting role in the biochemistry of life. Details of these vitamins are shown in ◘ Table 22.2.

Vitamins – Small molecules that we need for proper health. Normally, vitamins either are not synthesized in our bodies or are made in insufficient amounts.

Vitamin D

Vitamin D (calciferol) promotes retention and absorption of calcium and phosphorus, primarily in the bones. Too much vitamin D in the body may have the opposite effect of taking calcium from the bones and depositing it in the heart or lungs, making them function less efficiently. Because vitamin D is essential for the body's utilization of calcium, a deficiency may result in severe loss of calcium and, consequently, a softening and weakening of bones (osteomalacia). Extreme vitamin deficiency gives rise to a disease known as rickets.

Like most vitamins, vitamin D may be obtained in the recommended amount with a well-balanced diet, includ-

Table 22.2 Selected vitamins and their sources in foods

Vitamin	Deficiency disease	Source in food
A 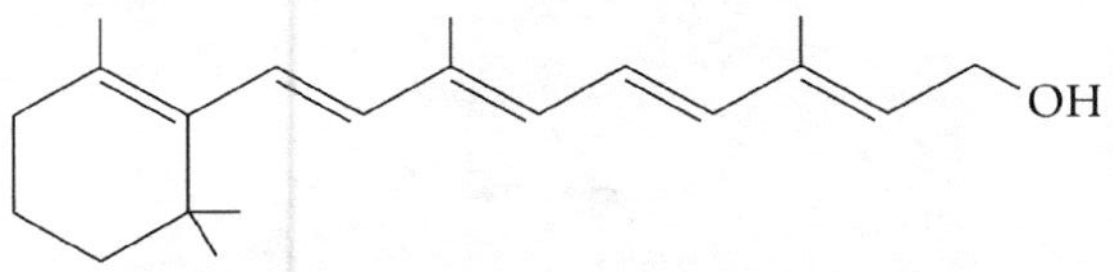	Night blindness, xerophthalmia	Eggs, whole milk, cheese, liver, green and yellow vegetables (carrots, squash, spinach)
B1 (thiamine) 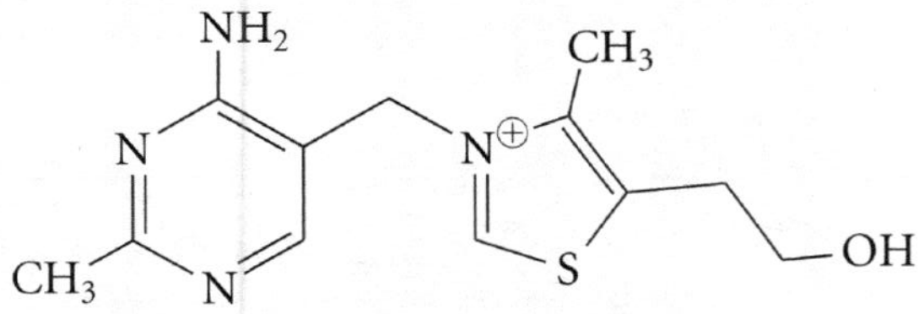	Beriberi	Ham, pork, milk, fortified cereals, peanuts, liver, yeast
B2 (riboflavin) 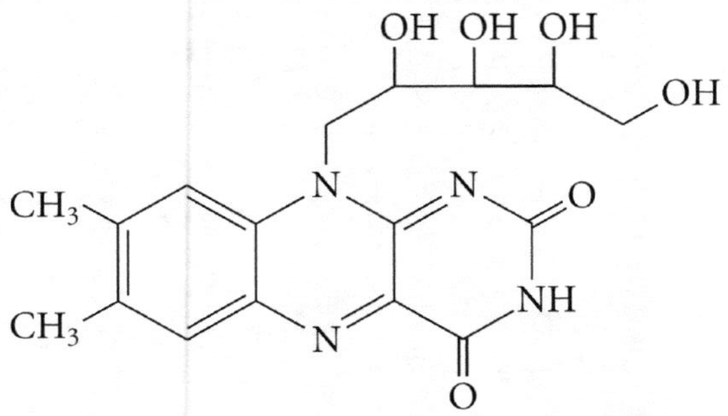	Ariboflavinosis	Liver and other organ meats, milk, green vegetables, fortified cereals, yeast
Niacin 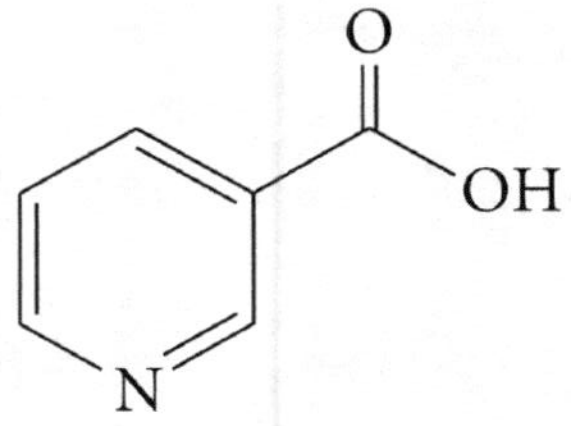	Pellagra	Peanuts, lean meats, fish, poultry, bran, yeast, liver

(continued)

Table 22.2 (continued)

Vitamin	Deficiency disease	Source in food
B6 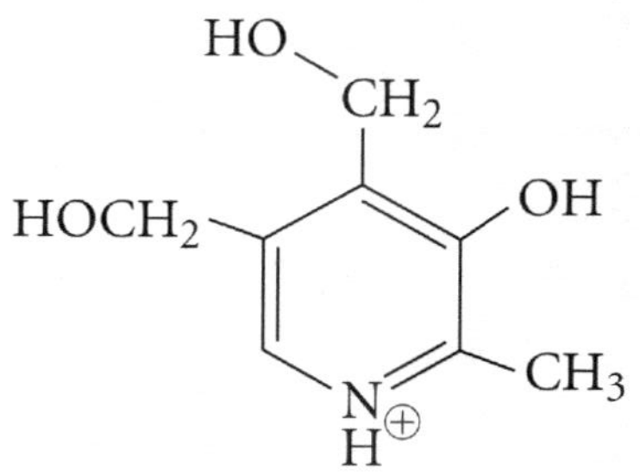	Vitamin B6 deficiency	Whole-grain cereal, fish, legumes, liver and other organ meats, yeast
B12 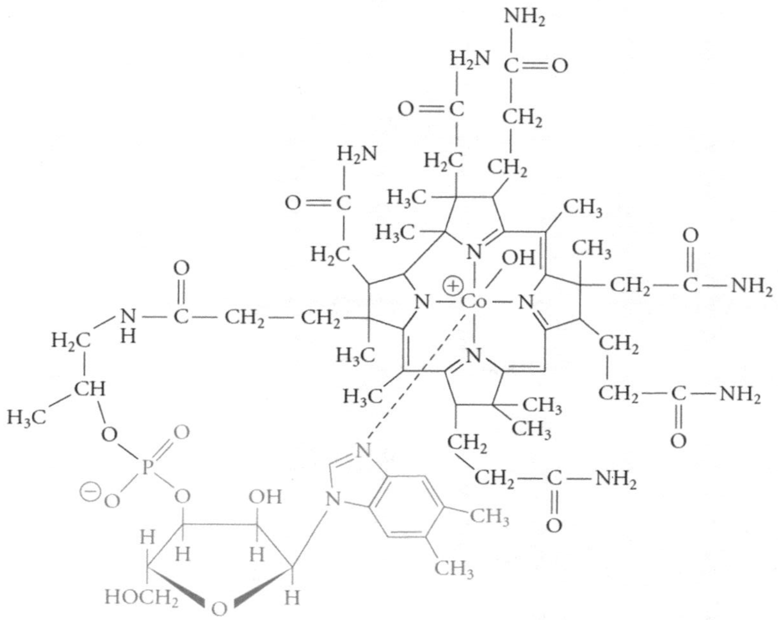	Pernicious anemia	Eggs, milk, liver

(continued)

Table 22.2 (continued)

Vitamin	Deficiency disease	Source in food
C 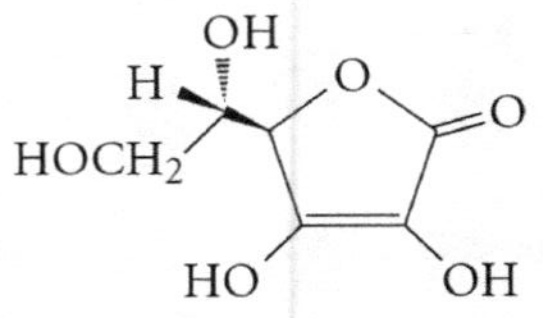 OH, H, O, O, $HOCH_2$, HO, OH	Scurvy	Fresh fruits and vegetables (oranges and other citrus Fruits, Brussels sprouts, cabbage, etc.)
D 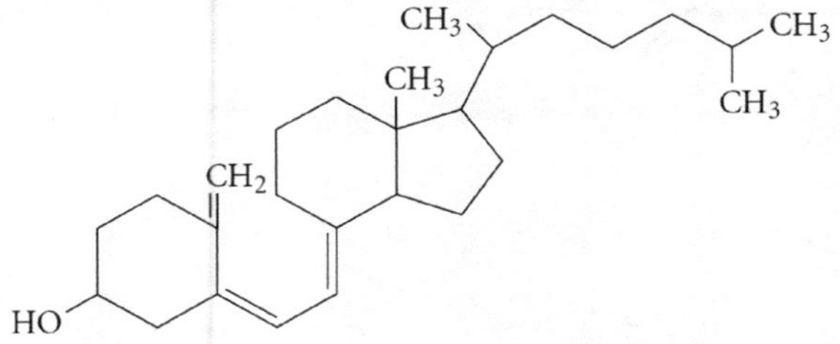 CH_3, CH_3, CH_3, CH_3, CH_2, HO	Rickets	Fortified milk, fish liver oil
E 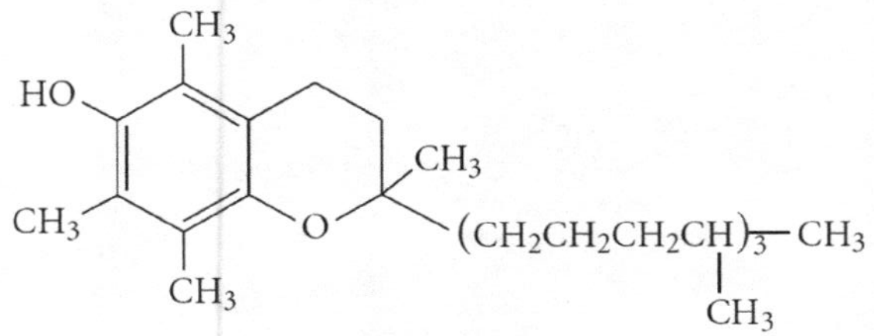 CH_3, HO, CH_3, CH_3, O, CH_3, $(CH_2CH_2CH_2CH)_3$—CH_3, CH_3	Vitamin E deficiency	Vegetable seed oil, egg yolk, cereals, beef liver
K 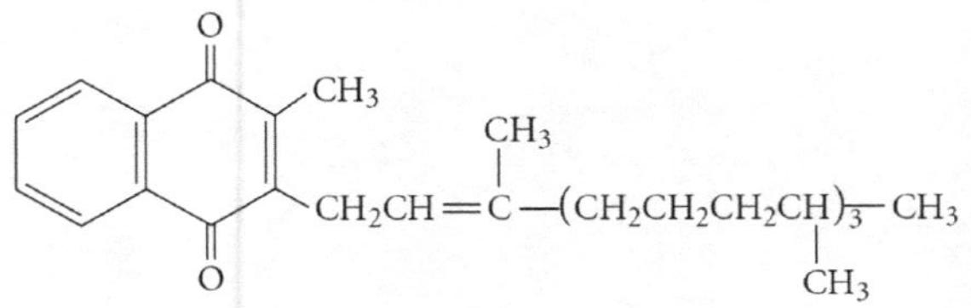 O, O, CH_3, CH_3, $CH_2CH{=}C$—$(CH_2CH_2CH_2CH)_3$—CH_3, CH_3	Vitamin K deficiency	Leafy green vegetables, liver

ing some enriched or fortified foods such as milk. In addition, the liver and kidneys manufacture vitamin D when the skin is exposed to sunshine. Deficiencies in this vitamin arise primarily from insufficient exposure to sunlight. That is why it is recommended that we get at least 10 to 15 min of sunshine three times a week.

Vitamin A

Vitamin A (retinol) is supplied by many foods of both animal and vegetable origin. Vegetables sources, such as carrots, pumpkin, and brocolli, actually contain a precursor of retinol called beta-carotene, which is converted into retinol by the body. In the body, vitamin A is used in the rods and cones within your eyes to absorb light and enable you to see. The vitamin also plays an important role in bone growth, reproduction, and cell division. Vitamin A has been implicated in the regulation of your immune system by facilitating the action of lymphocytes, a type of white blood cell that fights infection.

Deficiencies in vitamin A rarely occur in the United States but are very common in poor and developing countries. Night blindness and xerophthalmia (damage to the cornea of the eye) are the results of this deficiency. These conditions lead to as many as a half-million cases of blindness each year and, according to the World Health Organization, account for 70% of all childhood blindness in developing countries. Recent advances in treating this preventable disease have included the development of so-called golden rice, a genetic hybrid of rice that produces beta-carotene. Its use in countries where the staple food is rice has been quite successful in reducing the number of cases of blindness.

Vitamin E

Vitamin E (alpha-tocopherol) is found in corn, nuts, seeds, olives, spinach, asparagus, and other green leafy vegetables and in products made from them, such as margarine. This vitamin is often touted as "the body's antioxidant." Considerable evidence suggests that it prevents the natural oxidation of lipoproteins, which play an important role in the development of atherosclerosis, the disease process that leads to heart attacks and strokes.

Vitamin C

Vitamin C (ascorbic acid) has many functions in the body, including assisting in the absorption of iron. This vitamin is found in citrus fruits, such as lemons and limes, and in most other vegetables. It is a component of enzymes involved in the synthesis of collagen. Without vitamin C, the enzymatic preparation of collagen still proceeds, but the resulting collagen protein doesn't include the modified amino acid known as hydroxyproline. Collagen with hydroxyproline forms strong intermolecular forces of attraction between strands of the collagen protein. Without hydroxyproline, the collagen cannot form these strong attractive forces. The result is nothing short of the breakdown of the human body. Deficiency in vitamin C, known as scurvy, is evidenced by sore joints, loss of teeth, and aching muscles. Severe deficiencies and untreated deficiencies result in respiratory failure.

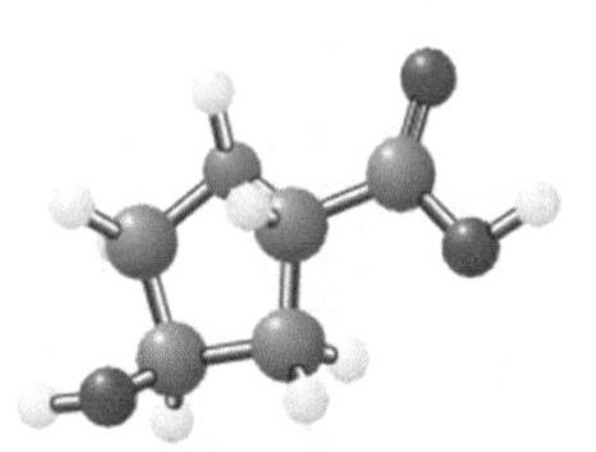

4-Hydroxyproline

Abnormally large doses of vitamin C appear to be harmless at the least, and they may be helpful. Because vitamin C is water soluble, excess vitamin C can be easily excreted from the body. Linus Pauling (remember him from electronegativity) suggested that vitamin C may play a crucial role in the maintenance of a healthy immune system. He proposed that massive doses of vitamin C (as high as 1000 mg/day) may even reduce the frequency and severity of the common cold. Recent research suggests that this may, in fact, be true.

Vitamin B1

Vitamin B1 (thiamine) is found in cereals, breads, and pasta. This vitamin, like vitamin C, is one of the water-soluble vitamins. Deficiency in vitamin B1 leads to fatigue, psychosis, and even nerve damage. Extremely severe deficiencies result in beriberi, a condition that is characterized by nerve degeneration and muscle disease and that affects the function of the heart. This disease is prevalent in developing countries such as those in eastern and southern Asia. In developed countries, it is rarely found.

22.6 The Diversity of Protein Functions

One way of summarizing the relationship between genes and proteins—the two central categories of chemicals that sustain life—is to say that "genes hold the instructions while proteins do the work." In a human, our 20,000 to 30,000 genes encode the structure proteins, and each protein has its own specialized task, its own little bit of "chemical work" to do. There is often simplicity at the heart of complexity, however, and proteins are no exception. Almost everything they achieve can be explained in terms of three fundamental activities. Selective binding, catalysis, and conformational change are the three keys to understanding the activities of proteins. Here is a brief list of the main tasks these three operational principles enable proteins to achieve.

22.6.1 Proteins as Enzymes

Some proteins contain active sites that bind substrates and catalyze reactions to make products, the main subject of ► Sect. 22.5.

22.6.2 Proteins as Transporters

Many proteins bind to specific chemicals at particular sites and then release their cargo at other sites. The protein hemoglobin, for example, binds to oxygen molecules in the lungs, transports the oxygen through the blood, and then releases it in the tissues of the body that need oxygen to survive.

22.6.3 Proteins as "Movers and Shakers"

What in your body makes you move? You move because you have muscles. Muscles are composed of filamentous contractile proteins. These are proteins that respond to stimuli from nerves to undergo conformational changes that result in their being ratcheted past one another and causing the whole structure of a muscle to contract. Reversal of the process enables the muscles to relax.

22.6.4 Proteins as Scaffolding and Structure

Much of the shape and structure of cells, tissues, organs, and the body as a whole is maintained by strong fibrous structural proteins that form fibers and pillars and sheets of material. Hair, skin, connective tissue, and an intricate skeleton of tubules within all cells are composed of structural proteins.

22.6.5 Proteins as Messengers

Proteins that are made at one place in the body can be transported through the blood to other parts of the body, where they bind to other chemicals to initiate specific biochemical responses. Some such proteins are known as **hormones**. The protein insulin, for example, is a hormone that is released from the pancreas and then binds to cells and assists in the uptake of glucose into these cells. A deficiency in this hormone or in its production causes diabetes.

Hormones – Small molecules, such as steroids, used by living organisms for intercellular communication.

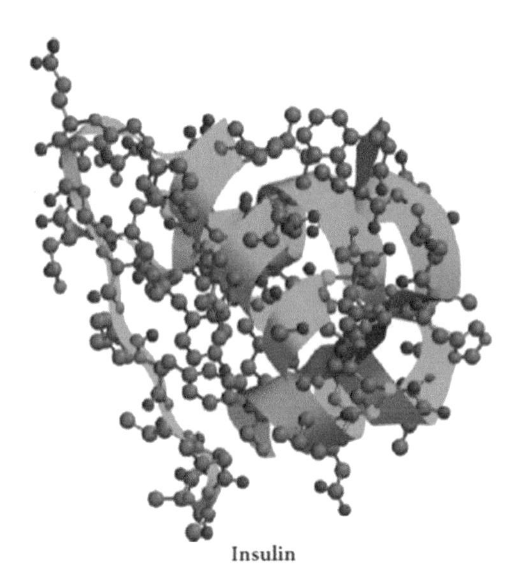

Insulin

22.6.6 Proteins as Receptors

The outermost portion of the cell, known as the cell membrane, is studded with proteins that act as receptor molecules. These receptors bind to specific messenger molecules, some of which are proteins like insulin, and then participate in acts of catalysis or conformational change that initiate the effect that the messenger elicits. Protein receptors also exist inside of cells. The interactions between receptors and messengers are vital in controlling cell growth. This makes disorders in these systems very significant in the onset of serious diseases such as cancer.

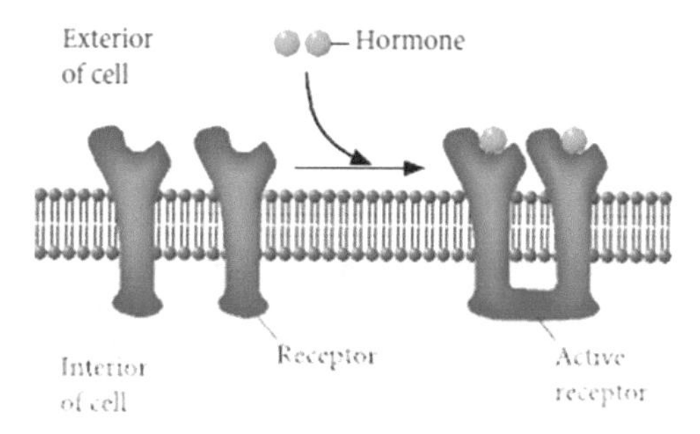

22.6.7 Proteins as Gates and Pumps

Cell membranes act as very selective barriers, through which only specific chemicals can pass, and only at appropriate rates and at appropriate times. Much of this selective movement of chemicals into and out of cells is maintained by proteins spanning the cell membrane. Some of these proteins can simply open or close a hole in the membrane, allowing selected chemicals to pass. Others act more like pumps; that is, they transport a chemical across a membrane in such a way that a significant difference can build up between the concentration of the chemical on one side of the membrane and its concentration on the other side.

22.6.8 Proteins as Controllers

A whole battery of proteins known as regulatory proteins serve the function of controlling other proteins, and controlling genes, by binding to them and switching them on or off. In most cells of the body, for example, only a small proportion of the genes in the cell nucleus must be active at any one time. Switching the right genes on and off at the correct times is largely achieved by the activities of regulatory proteins.

22.6.9 Proteins as Defenders

The immune system, which defends us against disease, is a network of interacting cells and chemicals, and many of its most significant chemicals are proteins. Most famous of all are the **antibodies**—proteins that selectively bind to foreign chemicals and assist in their elimination from the body. Allergies such as hay fever result when this system goes wrong by eliciting too strong a reaction against relatively harmless foreign chemicals. When it goes wrong by attacking the body's own tissues, serious autoimmune diseases such as rheumatoid arthritis can result.

Antibodies – Large biological molecules that are produced during an immune response, travel to the site of infection, bind to foreign material, and signal the destruction of the foreign material.

22.6.10 Glorious Complexity

The actions of genes and proteins are part of a very complex picture of the biochemistry of life. All of that complexity is, at heart, simply chemistry, driven by the interactions of energy, charge, and force. However, genes and proteins aren't the only players. There are many other compounds involved in the smooth operation of a living being. Carbohydrates and lipids are two important classes of the compounds of life.

22.7 Carbohydrates

Most people love carbohydrates. Runners eat copious amounts before a very long race. Snackers munch on carbohydrates in front of the TV or laptop. Toddlers often will drop everything else to get a small taste of a **carbohydrate**. What are carbohydrates and why do our bodies require them? This class of compounds shares the general formula $C_x(H_2O)_y$—historically, they were known as hydrates of carbon, hence their name. The simplest carbohydrates are **sugars**, which are known as triose sugars if they contain three carbon atoms per molecule, tetrose sugars with four carbons, pentose sugars with five carbon atoms per molecule, hexose sugars with six carbon atoms per molecule, and so on. In addition, the carbohydrates can be classified on the basis of the type of functional group (see ▶ Chap. 21). For example, glucose is an aldose because it contains an aldehyde functional group. Fructose is a ketose because it contains a ketone functional group. The glucose used in medical drip-feeding is an aldohexose (from aldose and hexose) and is the main form in which carbohydrate circulates in our blood.

Carbohydrates – Compounds that have the general formula $C_x(H_2O)_y$. These compounds serve living organisms as structural molecules (cellulose), as cellular recognition sites, and for energy storage (amylose and starch).

Sugars – The simplest carbohydrates.

a.

Triose	Tetrose	Pentose	Hexose
			CHO
		CHO	H—C—OH
	CHO	H—C—OH	HO—C—H
CHO	H—C—OH	H—C—OH	H—C—OH
H—C—OH	H—C—OH	H—C—OH	H—C—OH
CH_2OH	CH_2OH	CH_2OH	CH_2OH
Glyceraldehyde	Erythrose	Ribose	Glucose

b.

Ketose		Aldose
	CH_2OH	CHO
	C=O	H—C—OH
CH_2OH	HO—C—H	HO—C—H
C=O	H—C—OH	H—C—OH
H—C—OH	H—C—OH	H—C—OH
CH_2OH	CH_2OH	CH_2OH
Erythrulose	Fructose	Glucose

The familiar table sugar we may add to coffee is sucrose and is formed when two simpler hexose sugars, glucose and fructose, become linked together in a condensation reaction, as shown in ◘ Fig. 22.24. You'll immediately notice a difference between the sugars in ◘ Fig. 22.24 and those from our previous discussion. Why? Most carbohydrates undergo a condensation reaction within their own structures when they are placed in aqueous solution. The result is a group of cyclic structures that are more stable than the straight-chain carbohydrates.

Sugars such as glucose and fructose are also known as **monosaccharides** (for "single sugars"), whereas those like sucrose, made when two monosaccharides combine, are **disaccharides**. Note that two cyclic sugars, one glucose and one fructose, have come together to form the disaccharide, as shown in ◘ Fig. 22.24.

Disaccharide – A carbohydrate made from the condensation of two smaller sugars.

Monosaccharide – A small carbohydrate monomer. Typically contains between three and nine carbon atoms.

One major role for carbohydrates in living things is to serve as energy storage compounds. For long-term storage, monosaccharides are linked up into giant polymers known as **polysaccharides**. The major storage polysaccharide in plants is starch, composed of many glucose monomers linked in the manner shown in ◘ Fig. 22.25. Foods such as rice or potatoes supply us with energy largely thanks to the deposits of starch within them, whose role in the plant is to provide the energy needed to sustain its growth. Animals, including humans, contain a slightly different polymer of glucose called glycogen in which the glucose monomers are linked together in a slightly different fash-

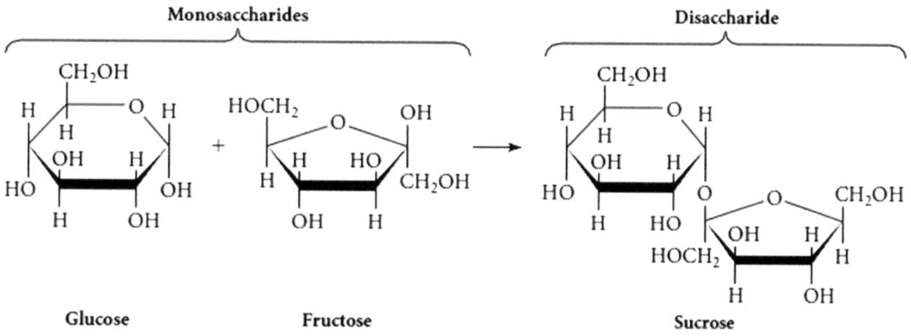

Fig. 22.24 The condensation of glucose with fructose gives sucrose. Monosaccharides include glucose and fructose. Sucrose is a disaccharide

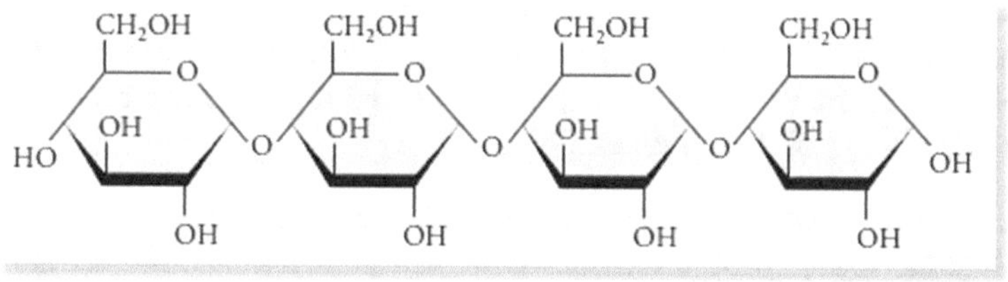

Fig. 22.25 Starch is a glucose polymer

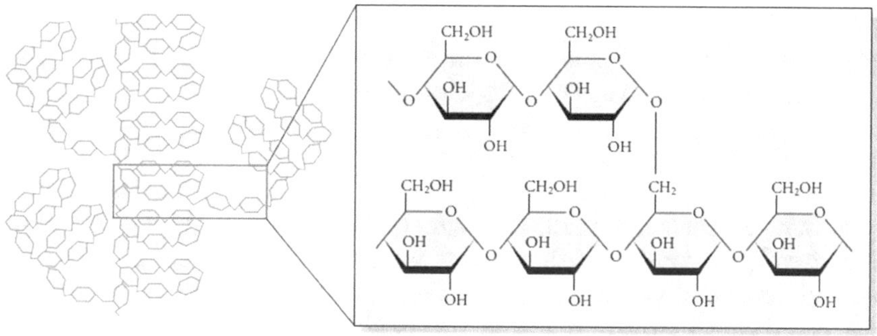

Fig. 22.26 Glycogen is also a polymer of glucose, with occasional branching of new glucose strands

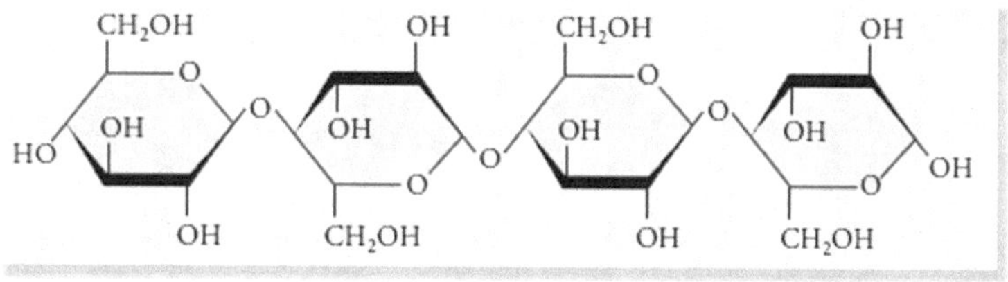

Fig. 22.27 Cellulose is a polymer of glucose units linked together as shown. Note the position of the OH on the glucose monomer at the right. It is pointed in a direction different from the direction it would point in starch

ion, as shown in Fig. 22.26. Another important polysaccharide is cellulose, which is a polymer of glucose that forms much of the framework of plant cell walls. The glucose monomers are linked together in a completely different manner, as shown in Fig. 22.27. Humans lack the enzyme capable of hydrolyzing cellulose into glucose, so cellulose contains no nutritional value to us.

Polysaccharide – A polymer of carbohydrate monomers.

22.8 Lipids

Lipids are a very broad class of compounds that differ from most other classes of compounds. For example, the molecules known as carbohydrates exhibit similarity in structure and function. Lipids do not have similarity in structure or function. A compound is characterized as a **lipid** solely on the basis of its solubility in nonpolar solvents. Lipids include fats, oils, some hormones, and some vitamins. Lipids function as important energy storage molecules, act as intercellular signaling molecules, and also play many structural roles in living things. For example, the thin membranes around every living cell are composed of lipids.

Lipids – Biological compounds characterized by their ability to dissolve in nonpolar solvents. They include fatty acids, some vitamins, and the steroid hormones.

Some of the simplest lipids are the **fatty acids**, which are long-chain carboxylic acids that may contain one or more C=C double bonds. ◘ Table 22.3 lists some of the common fatty acids. When three fatty acids combine with the tri-alcohol glycerol in an esterification reaction, a **triglyceride (triacylglycerol)** is formed. ◘ Figure 22.28 shows an example of this reaction. Triglycerides that are solid at room temperature are called **fats**, whereas those that are liquid at room temperature are **oils**. We explored this distinction earlier when examining the effects of C=C double bonds on the melting points of fats and oils.

Fatty acids – Carboxylic acids that contain long carbon chains. Commonly serve as major building blocks for the production of cell membranes.

Fat – A triglyceride that is made from carboxylic acids with long carbon chains and has a relatively high melting point. Typically derived from animal sources.

Oil – A triglyceride made from carboxylic acids with long carbon chains and a relatively low melting point. Typically derived from plant sources.

Triglyceride (triacylglycerol) – A molecule made from three fatty acids and a molecule of glycerol. Functions as structural material in cell membranes and as an energy storage molecule.

Triglycerides can be modified in various ways to form compound lipids, such as the **phospholipids** that are important components of cell membranes (see ◘ Fig. 22.29). Living things also contain a great many derived lipids, which are fatty substances that can be made from simpler lipids but have their own unique structures. One of the major classes of derived lipids consists of the **steroids**, which share a characteristic four-ring structure and include cholesterol and various steroid hormones such as estrogen. Cholesterol has a bad reputation because of the link between excessive blood cholesterol levels and cardiovascular disease, but it is an essential component of cell membranes. Like almost every chemical associated with life, cholesterol is "good" in the right amounts and right places but "bad" in the wrong amounts and wrong places.

◘ **Table 22.3** Some common fatty acids

Common Name	Formula	Melting point (°C)
Lauric acid	$CH_3(CH_2)_{10}COOH$	46
Myristic acid	$CH_3(CH_2)_{12}COOH$	55
Palmitic acid	$CH_3(CH_2)_{14}COOH$	64
Stearic acid	$CH_3(CH_2)_{16}COOH$	69
Oleic acid	$CH_3(CH_2)_7CH{=}CH(CH_2)_7COOH$	13
Linoleic acid	$CH_3(CH_2)_4CH{=}CHCH_2(CH_2)_7COOH$	–5

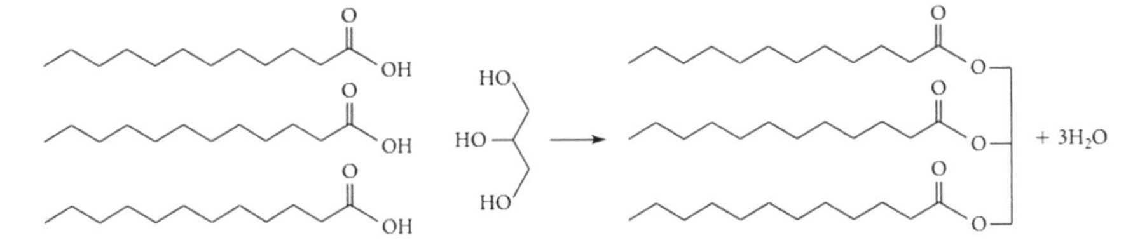

◘ **Fig. 22.28** Triacylglycerides result from the condensation of three fatty acids and glycerol

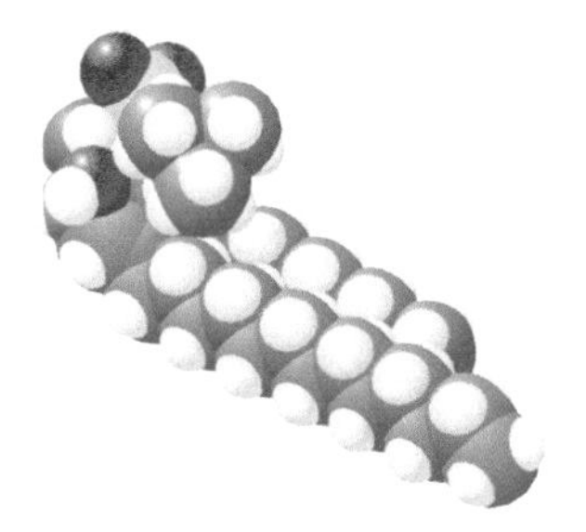

Fig. 22.29 Sphingomyelin is one of the components of cellular membranes

Phospholipid – A phosphate-modified lipid.

Steroids – Molecules composed of three 6-member carbon rings and one 5-member carbon ring fused together. These compounds are part of the lipid fraction of a cell and function as structural materials and intercellular communicators.

22.9 The Maelstrom of Metabolism

Chemists have a rather simple chart to look at on their walls; we call it the periodic table. Biochemists have a much more complex chart to admire. The biochemical pathways chart is so complex that, even devoting a full wall of our bedroom to it, we struggle to make out the details. The most commonly used version is more than 1 square meter in size, and even at that scale it still looks dense and complex. A much simpler version of a portion of the chart is shown in Fig. 22.30. The biochemical pathways chart summarizes the major sequences of chemical reactions of life, each particular sequence being called a **biochemical pathway**. The individual pathways are interconnected at various points, offering yet another example of glorious complexity: The chemical reactions enable the atoms needed for life to flow through all the different molecules and ions found in living things. The entire network of chemical reactions involved in life is called **metabolism**, and the individual chemicals are called **metabolites**.

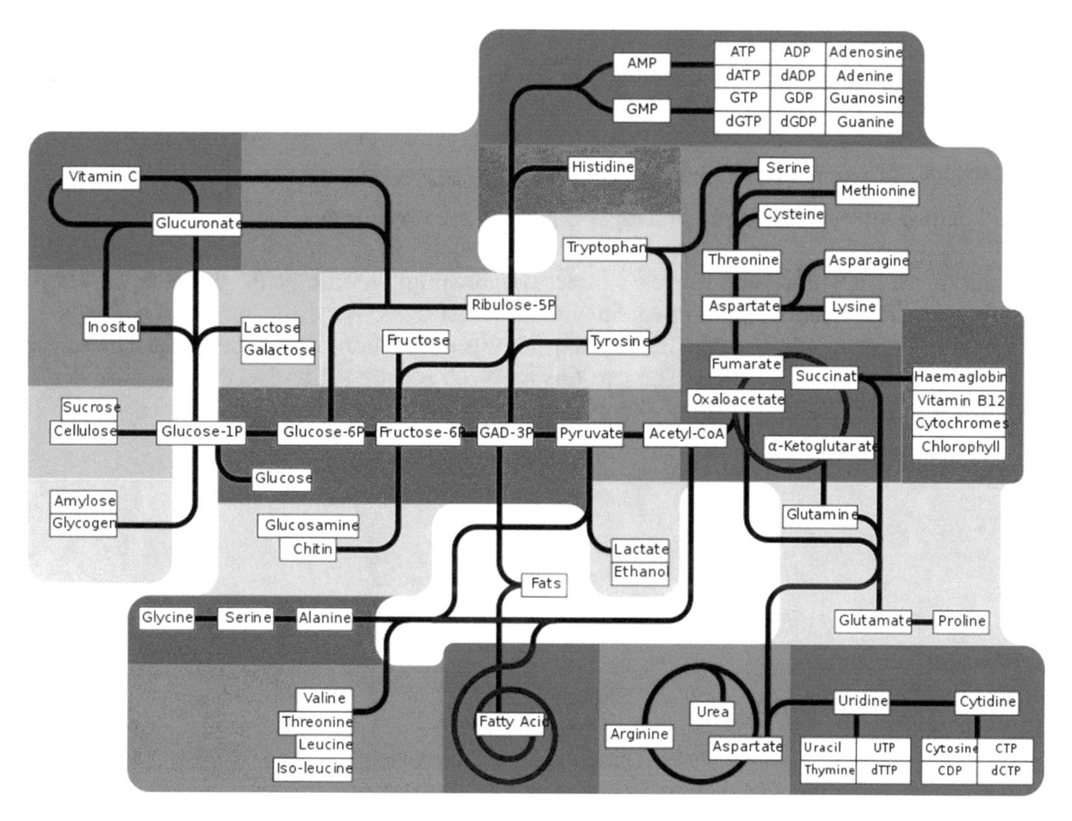

Fig. 22.30 A simplified biochemical pathways chart by Fred the Oyster. CC BY-SA 4.0. *Source* ▶ https://upload.wikimedia.org/wikipedia/commons/5/5d/Metabolism_pathways_%28partly_labeled%29.svg

Biochemical pathway – A sequence of biochemical reactions that lead to the destruction or production of a particular metabolite.

Metabolism – The biochemical reactions inside living cells.

Metabolite – A product of each step in a biochemical pathway.

Each arrow on the biochemical pathways chart represents a chemical reaction catalyzed by a specific enzyme (the arrows in the simplified version in ◘ Fig. 22.30 may represent multiple reactions). Looking at the chart, we can really begin to appreciate the true complexity of life, the reasons why we need around 20,000 to 30,000 genes (always remembering that many genes code for proteins other than enzymes), and the breathtaking achievement of the chemistry within each living thing. We can also appreciate the other breathtaking achievement: that in a little over 70 years, scientists have been able to sort this all out!

22.9.1 Breaking Things Down and Building Them up Again

The food we eat is a mixture of water, carbohydrate, protein, lipids, vitamins, and minerals. It provides us with chemical raw materials needed to build the chemicals of our bodies, and it also acts a source of energy. To release the energy, and also to use much of our food as raw materials, our bodies must break the chemicals in the food down into simpler forms. This aspect of metabolism is called **catabolism**. Building the raw materials back up into our own carbohydrates, proteins, lipids, and other essential chemicals is the aspect of metabolism called **anabolism**.

Anabolism – The metabolic process of constructing molecules from smaller molecules using high-energy molecules that have been made by catabolism.

Catabolism – The metabolic process of breaking down molecules to form smaller molecules and produce high-energy molecules.

22.10 Biochemistry and Chirality

There is a crucial aspect of the chemistry of life that we have not even mentioned yet, and it is also an important aspect of the structure of many chemicals not necessarily involved in living systems. Many of the molecules in our bodies are chiral, a term we introduced in ▸ Chap. 21. For instance, the mirror image forms of the general structure of an amino acid are shown in ◘ Fig. 22.31. These forms will be nonsuperimposable, provided that the R group is not a hydrogen atom. All amino acids apart from glycine are chiral, and the nonsuperimposable forms of molecules such as this are known as **enantiomers**.

Enantiomers – A pair of molecules that are nonsuperimposable mirror images.

Chirality is found in the body in a wide range of molecules, including amino acids, proteins, carbohydrates, and even DNA and RNA. The amino acids in proteins have, almost exclusively, the same handedness. That is, our bodies have evolved to utilize only one of the two enantiomers possible in the amino acids. The same is true for most of the carbohydrates in living things. The carbohydrates of life are similarly almost exclusive in their same handedness. Why is this so? This is another of the wonderful mysteries of life.

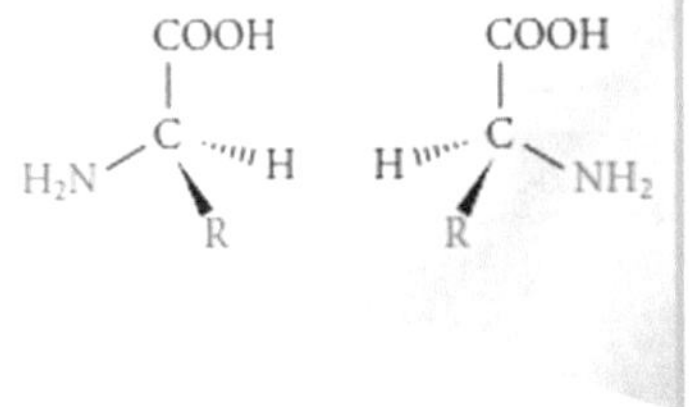

◘ Fig. 22.31 Mirror images of a generic amino acid. The "R" in the structure refers to the remaining portion of the amino acid

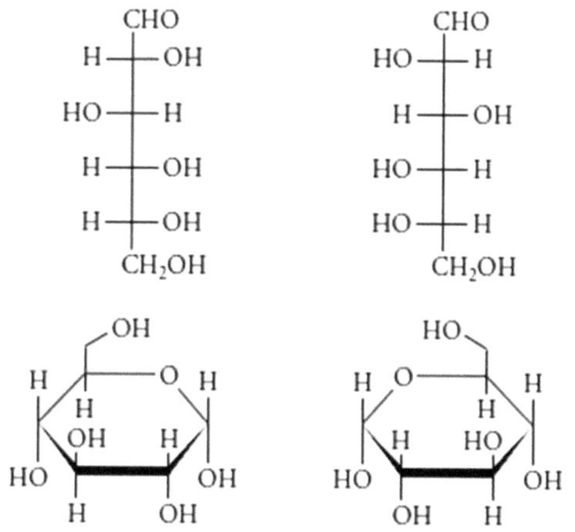

D-Glucose L-Glucose

Why is chirality in chemicals important? Because it controls the way in which molecules interact. For example, the binding sites of enzymes are constructed from the chiral amino acids within protein molecules. Therefore, enzymes themselves are chiral. They will interact differently with the different enantiomers of the substrates to which they bind. In many cases, one enantiomer of the substrate will be able to bind to the protein, whereas the other will not fit into the active site. These distinctions can be highly significant.

From 1958 to 1962, large numbers of pregnant women in many countries began to take a drug called thalidomide, shown in Fig. 22.32. This drug was prescribed in an effort to control the "morning sickness" that often accompanies the early stages of pregnancy. The large batches of thalidomide prepared on an industrial scale for distribution around the world contained both enantiomers. One of these was very effective in controlling morning sickness. Unfortunately, the other enantiomer caused a range of profound abnormalities in the developing fetus. Babies born to mothers who had taken thalidomide were badly deformed. These dangers became evident before thalidomide received approval for use in the United States, which was therefore fortunate enough never to experience the tragic wave of "thalidomide babies" that occurred in other countries.

Thalidomide has since been found to be useful in treating some aspects of various diseases, including leprosy, rheumatoid arthritis, AIDS, and some cancers, especially those related to blood marrow. It remains in use in carefully controlled circumstances, but never for the treatment of women who are, or ever may become, pregnant. Recently, a compound with a structure that is very similar to that of thalidomide has been identified as a potential treatment for myelodysplastic syndrome, a malignant disorder that affects blood cell production. The company manufacturing the drug is seeking approval from the U.S. Food and Drug Administration to market the compound under the name Revlimid, which was approved in 2006. Initial studies have indicated that Revlimid has fewer of the harmful side effects caused by thalidomide, and its use could be beneficial to patients afflicted with the malignant blood disorder.

Many scientists were perplexed at why the safe enantiomer of thalidomide was not prepared and used in its pure form. Preparations of the safe enantiomer can indeed be made, but over time, the pure enantiomers spontaneously convert into a mixture of the two enantiomers, until an interconverting 50/50 mixture of both forms, known as a **racemic mixture**, is obtained. Unfortunately, the process of **racemization** occurs with many chiral molecules.

Racemic mixture – A 1:1 mixture of a molecule and its enantiomer.

Racemization – The reaction that describes the preparation of a racemic mixture from a single enantiomer.

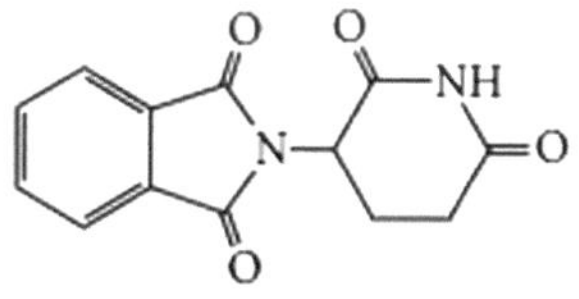

Fig. 22.32 Thalidomide is a chiral molecule

22.11 A Look to the Future

Talking about thalidomide reminds us that one major reason behind our interest in the chemistry of life is the desire to prevent and treat the diseases and discomforts that can afflict us. Why do we get ill and what can chemistry do about it? The major causes of illness and disease are

- *Infection*—occurs when our bodies become home to harmful microorganisms
- *Abnormal growth of tissues*—such as cancerous tumors and less damaging "benign" tumors
- *Abnormal production of important biochemicals*—such as hormones and the neurotransmitters that make it possible for signals to pass between nerves
- *Genetic diseases*—associated with one or more abnormal genes or larger abnormalities in genetic material
- *Aging*—the degeneration of maintenance and repair functions within the body

Every doctor's shelf carries a large volume called a pharmacopeia (see ▶ Sect. 15.1), a book that lists the drugs available to fight all manner of diseases and includes instructions on dosage, uses, and possible adverse reactions. However, many of these drugs have limited use, exhibit lessening effectiveness, or are simply ineffective at treating or curing diseases and maladies that affect humans. The future of useful therapeutic agents relies on the preparation and implementation of alternatives to these existing drugs. Some of the important current advances in this area are noted here.

22.11.1 Antibiotics

Most of us are treated with **antibiotics** several times in our lives. These are chemicals that can inhibit the growth of microorganisms or even destroy them. The first antibiotic that was isolated and used on patients was penicillin (◘ Fig. 22.33). This compound inhibits the synthesis of bacterial cell walls by selectively binding to enzymes needed for such synthesis. Nowadays we have a vast armory of antibiotics that work in different ways, but very often by binding to and inhibiting specific enzymes. For example, the tetracycline antibiotics are all derived from the basic structure of tetracycline (◘ Fig. 22.34). They work by binding to and inhibiting enzymes that catalyze protein synthesis in bacteria.

Antibiotics – Substances that kill, harm, or retard the growth of microorganisms.

22.11.2 Anticancer Agents

One major strategy for treating cancer is to prevent the replication of DNA within cancerous cells. If DNA replication is prevented, the cells cannot multiply by cell division. Various drugs have been developed that achieve this effect by forming crosslinks between the two strands of DNA or by adding bulky groups to one strand, thus preventing the double helix from unwinding in the manner required for replication. One of the oldest antican-

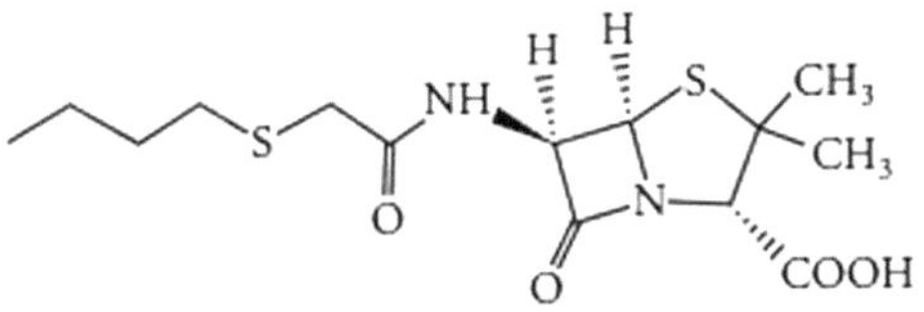

H H NH S CH_3 CH_3 S O N O COOH

◘ Fig. 22.33 Penicillin was first identified on moldy slices of bread

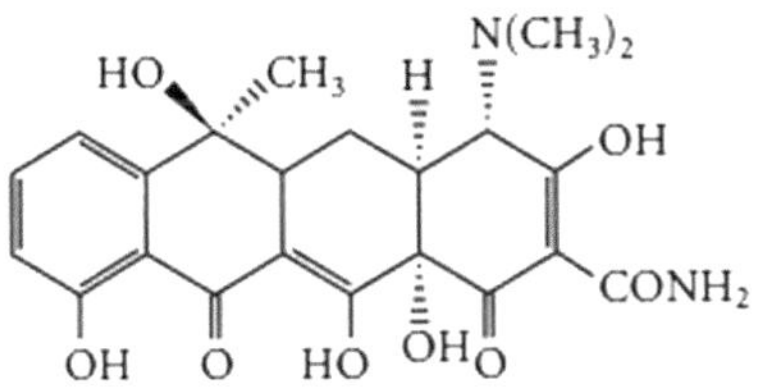

$N(CH_3)_2$ HO CH_3 H OH $CONH_2$ OH O HO OH O

◘ Fig. 22.34 Tetracycline is an antibiotic useful for treating many human infections

22

cer drugs is mechlorethamine, which forms crosslinks between the DNA strands. A more recent anticancer agent called m-amsacrine inserts into the DNA helix and causes the cell's natural maintenance enzymes to break the DNA strand into little pieces. m-Amsacrine is very effective in treating some forms of childhood leukemia.

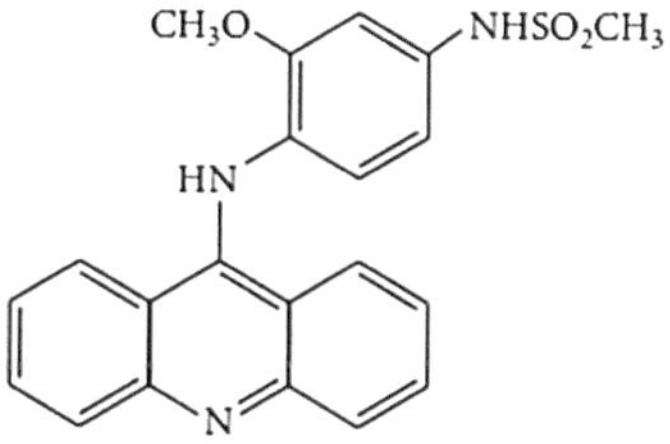

m-Amsacrine
(m-AMSA)

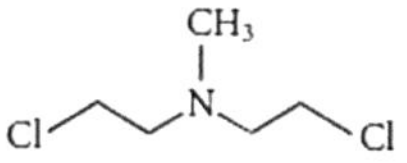

Mechlorethamine
(also known as nitrogen mustard)

22.11.3 Hormones and Their Mimics

Insulin, injected under the skin, can be used to treat diabetic patients. Many other nonprotein hormones are also used to treat disease. One major category of such hormones consists of the steroid hormones, such as estrogen and testosterone. Natural and synthetic steroids are used to treat such conditions as cancer, rheumatoid arthritis, allergies, and a wide range of hormone-deficiency diseases. Synthetic steroids are also the active ingredients in contraceptives and in the hormone replacement therapy sometimes used to combat the effects of menopause in women.

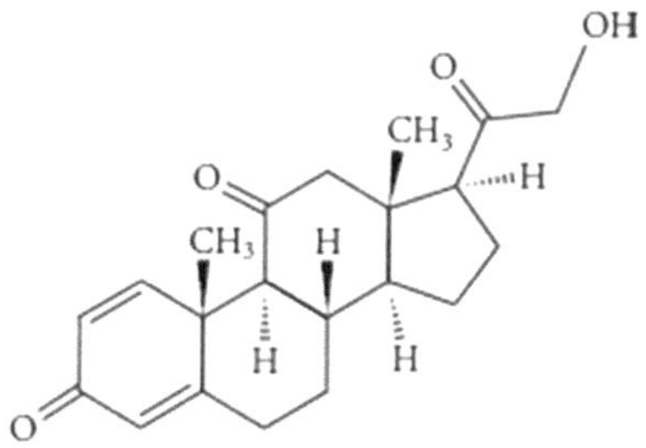

22.11.4 Neurotransmitters

The nervous system is composed of billions of nerve cells that communicate with one another via the release of small organic compounds known as neurotransmitters. These chemicals bind to specific receptors on the surface of neighboring nerve cells and either activate or inhibit the transmission of nerve impulses. Many degenerative and psychiatric conditions are associated with abnormalities in neurotransmitter function. The pronounced tremors of Parkinson's disease, for example, are due to the degeneration of nerve cells that use a molecule known as dopamine as their neurotransmitter. The condition can be dramatically alleviated by administering a precursor of dopamine called L-Dopa, which is converted to dopamine by enzymes in the body.

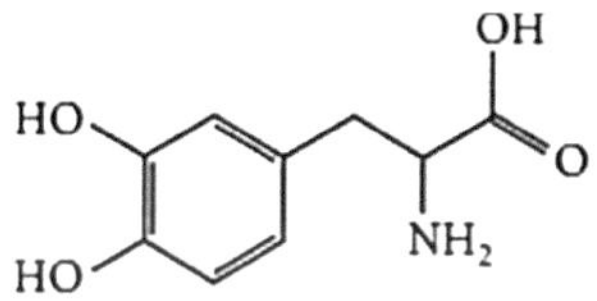

L-Dopa

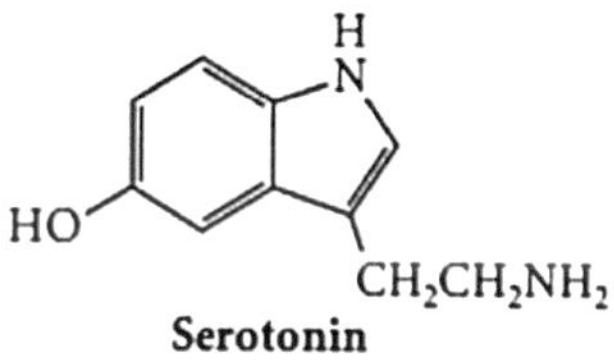

Serotonin

One of the most significant neurotransmitters is serotonin (5-hydroxytryptamine). A large new range of drugs that appeared in the 1990s are useful because of their ability to modulate the effects of this neurotransmitter in various cells and tissues.

Another set of significant neurotransmitters are the endorphins, a class of short polypeptides. Long-distance runners know the calming effects of endorphins. They give rise to the (nonadditive) "runner's high" that relieves stress for hours after a long run. Many of the endorphins have also been implicated in other significant biological effects. For instance, "Substance P" shows remarkable activity in ocular wound repair.

22.11.5 Genetic Disease

Genetic diseases are those that are clearly associated with one or more abnormal genes or larger abnormalities in genetic material. Genetic abnormalities give rise to sickle-cell anemia, retinitis pigmentosa, obesity, and a host of other diseases. How does damage in the gene cause disease? For each defective gene, there will be a correspondingly defective protein where at least one amino acid has been replaced with an incorrect amino acid. Unfortunately, it is also possible to have abnormalities in the **chromosomes** in which all of our DNA is packaged. Humans contain 23 pairs of chromosomes, one copy of each pair inherited from each parent. Having damaged, extra, or missing chromosomes can cause serious diseases.

Chromosomes – The packaged functional units of DNA.

One of the main driving forces behind the Human Genome Project that we mentioned at the beginning of this chapter is the desire to identify all the genes that are involved in genetic diseases. In doing so, scientists can learn exactly what has gone wrong with them and, in many cases, discover how to compensate for the defective proteins encoded by the genes. Eventually, we may even be able to develop ways to repair the genes and restore their normal state. The idea of actually repairing genetic damage is known as **gene therapy**. This advancement in science is currently in the early stages of development and trials. Someday, in the not-too-distant future, gene therapy may be performed on unborn babies to treat their diseases before we welcome them into this oh, so interesting world.

Gene therapy – A relatively new area of biochemistry wherein attempts are made to cure or treat genetic diseases by modification of a patient's damaged genes.

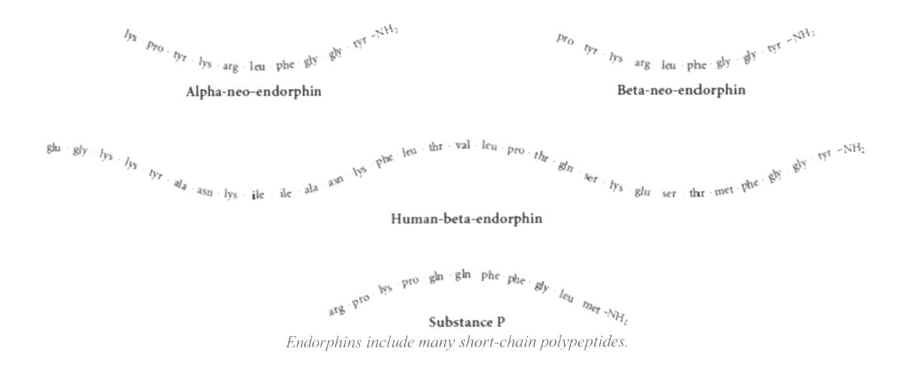

Endorphins include many short-chain polypeptides.

The Bottom Line

- DNA is a polymer of nucleotides made from two complementary single strands wrapped around each other to form a double helix.
- Proteins are polymers of amino acids linked through an amide bond. They contain primary, secondary, tertiary, and sometimes quaternary structure.
- Proteins are made by translating the genetic code from mRNA. tRNA supplies the specific amino acids to the growing polypeptide chain. The construction occurs at the ribosomes within the cytoplasm of a cell.
- Enzymes are proteins that contain an active site and catalyze biochemical reactions.
- Carbohydrates function as structural features such as cell walls in plants and as energy storage molecules in all living organisms.
- Lipids are a diverse class of molecules with a variety of functions.
- Biological systems exploit the use of enantiomers. Most amino acids and carbohydrates in living organisms are only one of two possible enantiomers.
- The promising future activities in the chemistry of medicines include development of new antibiotics, new drugs to treat neurological disorders and cancer, and gene therapy.

Section 22.2 DNA—The Basic Structure

Skill Review

1. Define the term DNA.
2. Define the term nucleic acid.
3. Indicate the complement for the following DNA sequence:

ATTAAAAAGGGACTA

4. Indicate the complement for the following DNA sequence:

GGCGAATTAGCCCA

Chemical Applications and Practices

5. A researcher has determined that a particular sequence in double-stranded DNA can be cleaved by a laboratory method. The method is specific for breaking DNA in the middle of an ATTA sequence. How many breaks would occur if the following single strand of DNA were paired with its complement?

ATTAAAGCCTAATTACCATAAT

6. A short segment of DNA is prepared as a way to locate a particular sequence in double-stranded DNA. To how many locations would the following segment of DNA bind in the double-stranded DNA shown?
Segment: ATGCA
DNA: TCGATTACGTATGCATTACGT

Section 22.3 Proteins

Skill Review

7. Cite two amino acids that differ by one carbon in their residues.
8. Cite two amino acids that have acidic residues.
9. Show the two products that could result from the condensation of valine and phenylalanine. Identify the amide bonds in each peptide.
10. Draw the three peptides that could be made from the condensation of two glycines and one proline. Identify the amide bonds in each peptide.

Section 22.4 How Genes Code for Proteins

Skill Review

11. Define the term codon in your own words.
12. Describe the process of transforming the information in a gene into a protein.
13. Explain how a mistake in the preparation of an mRNA could result in a protein that contains the wrong amino acid.
14. Is it possible to have a mistake in a gene without a mistake in the protein made by that gene? Explain.

Chemical Applications and Practices

15. A researcher isolates a fragment of a particular mRNA. Is this fragment at the beginning, end, or somewhere in the middle of the part that is translated into a protein?

UUUCGAAGUAUGGGUGGAGAUUCUCCCGCG

16. A particular genetic disorder causes an extra nucleotide to be inserted into the mRNA of a gene. Assuming that the entire mRNA is shown below and that the site of insertion is indicated, what is the primary structure of a protein that would result from the insertion of each of the four nucleotides?

↓
AUG–AAU–UAU–CGG–AUU–UAA

17. The skeleton inside a cell is made up of microfilaments—polymers of the protein known as actin. Given the function of the protein, predict whether actin is a globular or a fibrous protein.

18. Keratin is a protein used in defining and holding the structure of skin, hair, and claws. Given the function of the protein, predict whether actin is a globular or a fibrous protein.

Section 22.5 Enzymes

Skill Review

19. Identify the class of enzyme based on the following reaction:

$$C_6H_5{-}CH_2CH_2OH \xrightarrow{\text{Enzyme}} C_6H_5{-}CH_2{-}C(=O){-}H$$

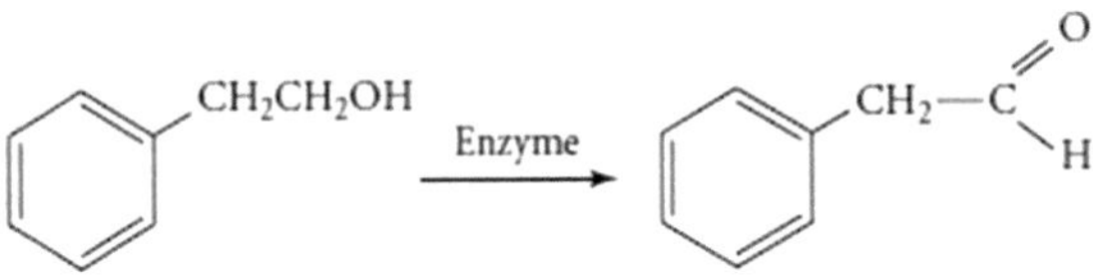

20. Identify the class of enzyme based on the following reaction:

$$H_2N{-}CH(CH_3){-}C(=O){-}N(H){-}CH(H){-}C(=O){-}N(H){-}CH(H){-}C(=O){-}OH$$

$$\xrightarrow{\text{Enzyme}} H_2N{-}CH(H){-}C(=O){-}OH \quad H_2N{-}CH(H){-}C(=O){-}OH \quad H_2N{-}CH(H){-}C(=O){-}OH$$

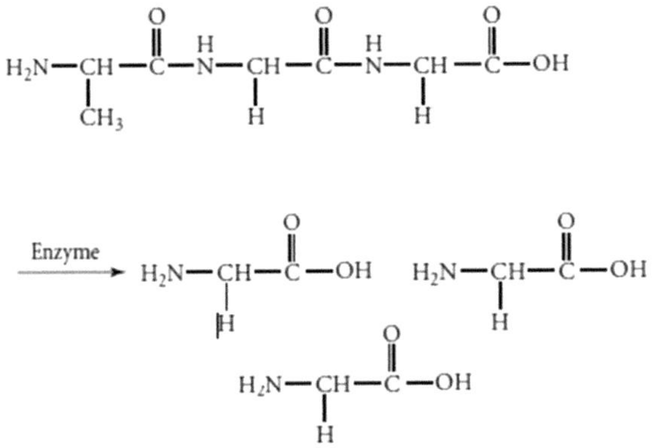

21. In some enzymatic reactions, the product is able to bind to the active enzyme and inhibit its activity. Why would this be a useful feature for an enzyme? Explain.
22. In some metabolic pathways (biochemical reactions involving multiple reactions), large concentrations of the final product can turn off the activity of the first enzyme in the pathway. What advantage would this type of regulation have over that described in the previous problem? Explain.

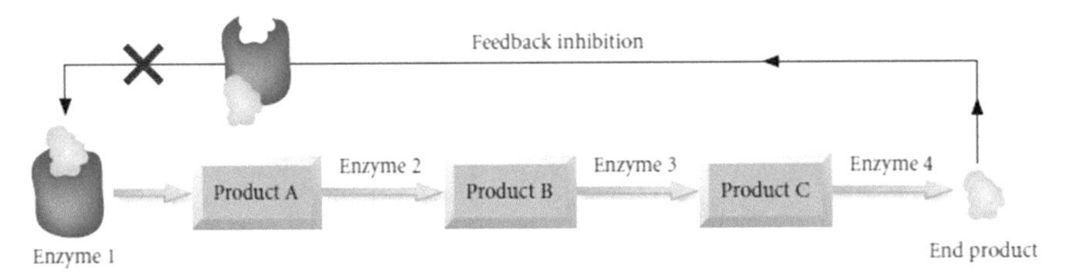

Section 22.7 Carbohydrates

Skill Review

23. Identify of the following carbohydrate as either a triose, a tetrose, a pentose, or a hexose.

$$HOCH_2{-}C(H)(OH){-}C(OH)(H){-}CHO$$

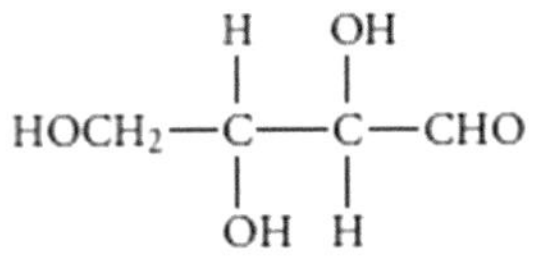

24. Identify the following carbohydrate as either a triose, a tetrose, a pentose, or a hexose.

$$HOCH_2{-}C(H)(OH){-}C(=O){-}C(H)(OH){-}C(OH)(H){-}CH_2OH$$

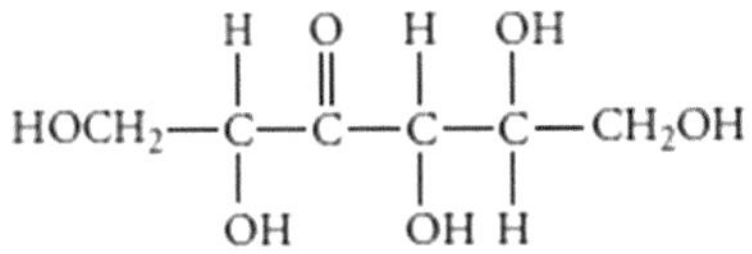

25. Identify the carbohydrate in Problem 23 as either an aldose or a ketose.
26. Identify the carbohydrate in Problem 24 as either an aldose or a ketose.

27. Which simple carbohydrate do starch, glycogen, and cellulose have in common?
28. A very serious malady exists in humans who lack the enzyme responsible for hydrolyzing glycogen. Explain why this deficit would be harmful to health.

Chemical Applications and Practices

29. High-fructose corn syrup is a common ingredient in some fruit drinks. What is the most likely source of this fructose? Identify fructose as either a triose, a tetrose, a pentose, or a hexose. Identify fructose as an aldose or a ketose.
30. Maltose, shown below, is a disaccharide. Which two simple carbohydrates condense to make maltose?

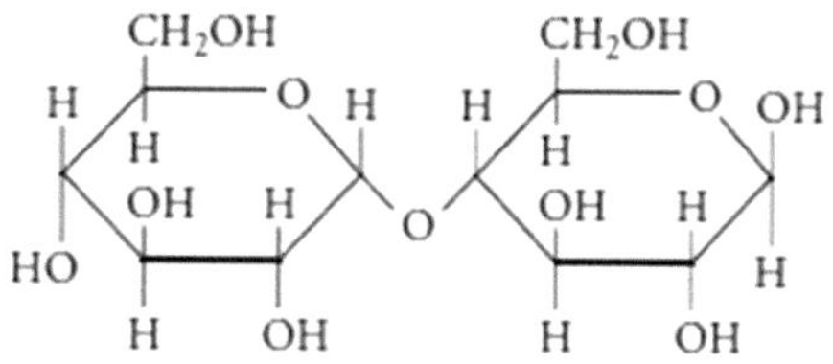

CH_2OH CH_2OH
H O H H O OH
H H
OH H O OH H
HO H
H OH H OH

Section 22.8 Lipids

Skill Review

31. Explain the distinction between fats and oils.
32. A company wishes to manufacture margarine using vegetable oil. Explain how this is done.
33. Would you expect the triacylglycerol shown below to be a component of a fat or of an oil?

O O O O O O

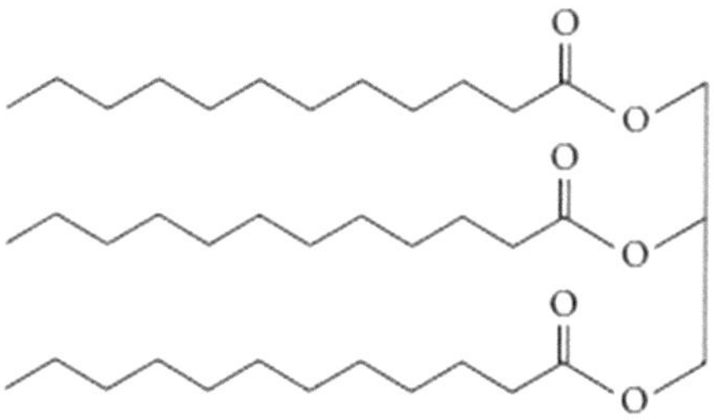

34. Would you expect the triacylglycerol shown below to be a component of a fat or of an oil?

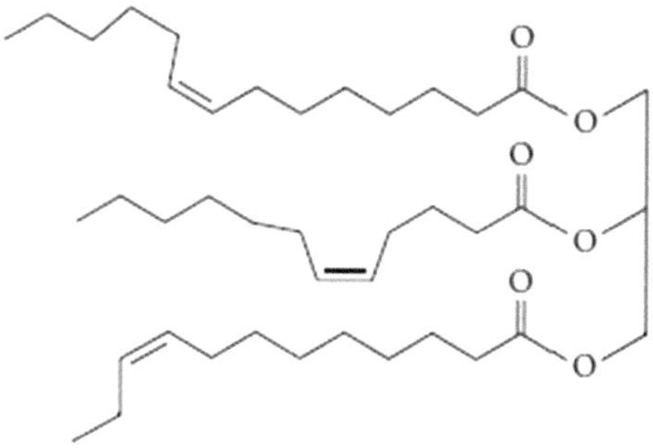

O O O O O O

Chemical Applications and Practices

35. Triacylglycerols and phospholipids make up the cell membranes in living organisms. Which end of these molecules would you predict to associate closely with the outside of the cell (in contact with water)?

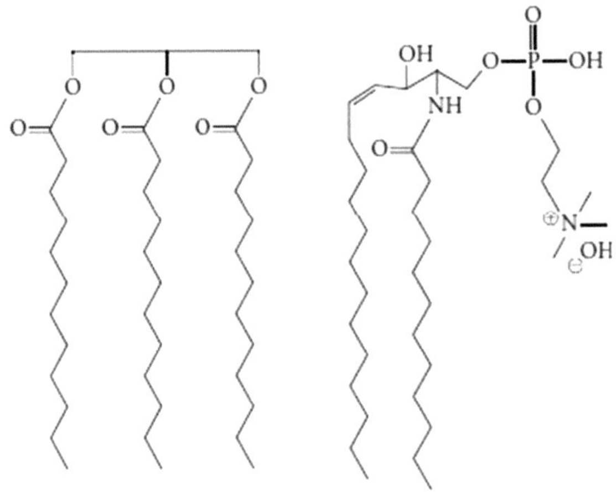

36. The interior of a cell is known as the cytoplasm. It contains many different types of molecules, but the bulk of the molecules are water. Explain how a cell membrane could be made from phospholipids if the exterior and interior of a cell are both made up of water molecules.

Section 22.10 Biochemistry and Chirality

Skill Review

37. Draw both enantiomers of serine and of phenylalanine.
38. Draw both enantiomers of glucose as a straight-chained molecule and as a cyclic monosaccharide.
39. What are the advantages of an enzyme's possessing a chiral active site?
40. If every enzyme possesses a chiral active site, why do both enantiomers of thalidomide exhibit biological activity?

Chemical Applications and Practices

41. Serotonin is a neurotransmitter involved in sleep, sensory perception, and the regulation of body temperature. From which amino acid is it probably likely manufactured in the body? Does serotonin have an enantiomer?

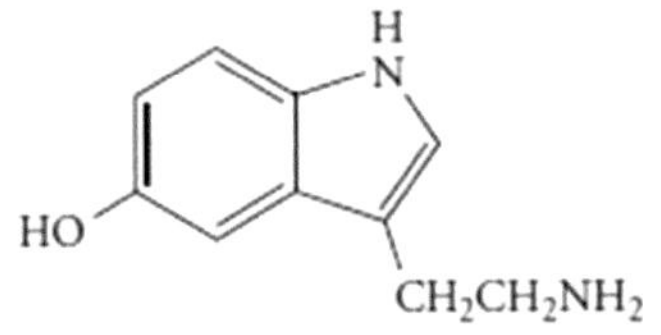

42. Nicotine is a common stimulant. In doses greater than 50 mg, it can be lethal to humans. Nicotine is excreted via urine by protonation, at which point the resulting cation is very soluble in water. Eating a meal causes the pH of urine to drop.
    a. Identify the chrial center in nicotine.
    b. Why would you predict that a smoker would crave a cigarette immediately after a meal?

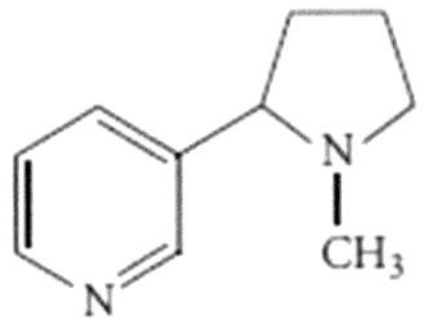

22

Comprehensive Problems

43. Indicate the mRNA and protein that would result from the transcription and translation of this DNA sequence.

TACAGCGCTTAAATTCCGACGAATAA

44. What sequence of the gene would produce this polypeptide?

gly-lys-glu-arg-lys-his-his

45. Explain why the melting points (shown in Table 22.3) increase from lauric acid, to myristic acid, to palmitic acid, to stearic acid.
46. Explain why the melting point of oleic acid is less than the melting point of stearic acid.
47. Would you predict the opposite enantiomer of the compound we call table sugar to be sweet? Explain
48. Termites consume wood as food. What polysaccharide is present in wood? Knowing this, and knowing that termites lack the enzyme that can digest that particular polysaccharide, explain what must be present within the termite's stomach in order for it to derive energy from its food.

Thinking Beyond the Calculation

49. Chocolate-covered cherries are made by coating a cherry with a paste of sucrose and the enzyme invertase. The cherries are stored after they are manufactured until the enzyme has had a chance to partially hydrolyze the sucrose. The resulting syrup is called invert sugar.
    a. Draw the reaction of sucrose and invertase.
    b. On the basis of your understanding of enzymes, would you predict that maltose would undergo a similar reaction with invertase.
    c. Genetic modification of invertase places an extra lysine in the protein. What codon and anticodon could produce this modification?
    d A researcher wishes to manufacture 10.0 kg of invert sugar from sucrose. How many grams of sucrose would be required to accomplish this feat?
    e. If the same researcher were able to separate all of the glucose from the invert sugar, how many grams of fructose could be made from 15.5 kg of sucrose?
    f. The same researcher is interested in finding a hydrolase that will accept cellulose as the substrate. What benefit would this have to humans?

Printed in the United States
by Baker & Taylor Publisher Services